ELSEVIER'S DICTIONARY OF TECHNOLOGY

SPANISH – ENGLISH

ELSEVIER
DICCIONARIO DE TECNOLOGÍA
ESPAÑOL – INGLÉS

Part 2
Macho – Z

ELSEVIER'S DICTIONARY OF TECHNOLOGY

SPANISH – ENGLISH

ELSEVIER
DICCIONARIO TECNOLOGÍA
ESPAÑOL – INGLÉS

Part 2
Macho – Z

compiled by

Arthur E. Thomann
Friendswood, Texas, U.S.A.

ELSEVIER
Amsterdam – London – New York – Tokyo   1993

ELSEVIER SCIENCE PUBLISHERS B.V.
Sara Burgerhartstraat 25
P.O. Box 211, 1000 AE Amsterdam, The Netherlands

ISBN: 0-444-88070-4

In the following list:

Unindented abbreviations indicate a field or specialty.

Abbreviations indented two spaces indicate a country or geographical area, and are normally followed by a period.

Abbreviations indented four spaces correspond to a part of speech or other grammatical, etc. indication.

En la enumeración que sigue:

Las abreviaturas que aparecen sin sangrar indican una rama o especialidad.

Las abreviaturas que aparecen sangradas dos espacios indican un país o región geográfica, y van seguidas generalmente por un punto.

Las abreviaturas que aparecen sangradas cuatro espacios corresponden a partes de la oración u otra indicación de tipo gramatical, etc.

---

a.acond	(air conditioning)	aire acondicionado	dren	(drainage)	drenaje
accntg	accounting	(contabilidad)	drwng	drawing	(dibujo)
accoust	accoustics	(acústica)	EELU.	(United States)	Estados Unidos
acust	(accoustics)	acústica	econ	economics	economía
a	adjective	adjetivo	electr	electricity	electricidad
adv	adverb	adverbio	electron	electronics	electrónica
admin	(management)	administración	engr	engineering	(ingeniería)
aeron	aeronautics	aeronáutica	entom	entomology	entomología
agric	agricultura	agriculture	environm	environmental	(ambiente,tal)
air.cond	air conditioning	(aire acondicionado)	equip	equipment	equipo
aislac	(insulation)	aislación	Esp.	(Spain)	España
alambr	(wire)	alambre	estruct	(structural)	estructural
ambient	environment(al)	ambiente; ambiental	explos	explosives	explosivos
anatom	anatomy	anatomía	extrus	extrusion	extrusión
Ant.	Antilles	Antillas	f	feminine	femenino,na
apicult	apiculture	apicultura	f.c.	(railways)	ferrocarriles
archit	architecture	(arquitectura)	fabr(ic)	fabrication	fabricación
Arg.	Argentina	Argentina	fam	familiar (colloquial)	familiar
arquit	(architecture)	arquitectura	fans	fans	(ventiladores)
art	article	artículo	fasten	fasteners	(sujetadores)
astron	astronomy	astronomía	fig	figurative	figurativo,va
autom	automobiles	automóviles	filol	(philology)	filología
avicult	aviculture	avicultura	fin	finance	finanzas
aviac	(aviation)	aviación	fis	(physics)	física
aviat	aviation	(aviación)	fisc	fiscal (taxes)	fiscal (impuestos)
boil	boilers	(calderas)	for	forensic	forense
Bol.	Bolivia	Bolivia	fotogr	(photography)	fotografía
botan	botany	botánica	fund	(melting)	fundición
bridg	bridges	(puentes)	ganad	(cattle)	ganadería
byprod	byproducts	(subproductos)	G.B.	Great Britain	Gran Bretaña
CA.	Central America	Centroamérica	geogr	geography	geografía
cald	(boilers)	calderas	geol	geology	geología
calef	(heating)	calefacción	geom	geometry	geometría
Car.	Caribbean	Caribe	gob	(government)	gobierno
carb	carburetors	carburadores	gov	government	(gobierno)
carp	carpentry	carpintería	gram	grammar	gramática
cattle	cattle	(ganadería)	grind.med	grinding media	cuerpos moledores
chem	chemistry	(química)	gruas	(cranes)	grúas
Chi.	Chile	Chile	heat	heating	(calefacción)
chronol	chronology	cronología	herram	(tools)	herramientas
civ.engr	civil engineering	ingeniería civil	hidr	(hydraulics)	hidráulica
clav	(nails)	clavos	hormig	(concrete)	hormigón
clim	climate	clima	hydr	hydraulics	(hidráulica)
cojin	(bearings)	cojinetes	il(l)um	illumination	iluminación
coke	coke	(coque)	imprent	(printing)	imprenta
col.cont	(continuous casting)	colada continua	ind	industry	industria
Col.	Colombia	Colombia	ind.engr	industrial engineering	(ingeniería industrial)
com	commerce	comercio	ing	(engineering)	ingeniería
comb.int	(internal combustion)	combustión interna	ing.civil	(civil engineering)	ingeniería civil
combust	combustion	combustión	ing.sanit	(sanitary engineering)	ingeniería sanitaria
comput	computers	computadoras	instal	installation	instalación
comun	communications	comunicaciones	instrum	instruments	instrumentos
concr	concrete	(hormigón)	insul	insulation	(aislación)
con.cast	continuous casting	(colada continua)	int.comb	internal combustion	(combustión interna)
cond	conductors	conductores	journ	journalism	(periodismo)
constr	construction	construcción	L.A.	Latin America	Latinoamérica
contab	(accounting)	contabilidad	labor	labor	laboral
coque	(coke)	coque	lam	(rolling)	laminación
cranes	cranes	(grúas)	legal	legal	legal
criog	(cryogenics)	criogénica	light	lighting	(iluminación-alumbrado)
cronol	(chronolgy)	cronología	lumb	lumber	(maderas)
cryog	cryogenics	criogénica	m	masculine	masculino,na
cuerp.mol	(grinding media)	cuerpos moledores	mf	masculine-feminine	masculino-femenino
Cub.	Cuba	Cuba	mader	lumber	(maderas)
culin	culinary	culinaria	magnet	magnetism	magnetización
deport	(sports)	deportes	manuf	manufacturing	manufactura (fabricación)
dib	(drawing)	dibujo	marit	maritime	marítimo,ma
drain	drainage	(drenaje)	mat	(mathmatics)	matemática

math	mathmatics	(matemática)
mec(h)	mechanics	mecánica
med(ic)	medical	médico,ca
melt	melting	(fundición)
metal	metal(lurgy)	metal(urgia)
meteorol	meteorology	meteorología
Mex.	Mexico	México
milit	military	militar
min	mining	(minería)
miner	minerals	minería, minerales
mngm	management	(administración)
mot	motors	motores
mus	music	música
n	noun	(substantivo)
nails	nails	(clavos)
neut	neuter	neutro
NLA,	Northern Latin America	(Latinoamérica Norte)
neumat	(pneumatics-tires)	neumática,cos
nucl	nuclear	nuclear
numism	numismatics	numismática
optic	optics	óptica
ornit(h)	ornithology	ornitología
paint	paint	(pintura)
papel	(paper)	papel
paper	paper	(papel)
period	(journalism)	periodismo
PR.	Puerto Rico	Puerto Rico
Per.	Peru	Perú
pers	personnel	personal
petrol	petroleum, oil	petróleo
philol	philology	(filología)
photogr	photography	(fotografía)
phys	physics	(física)
pil(es)	piles	pilotes
pl	plural	plural
pint	(paint)	pintura
plast	plastics	(materiales) plásticos
pneumat	pneumatics	(neumática)
pol	politics	política
prep	preposition	preposición
print	printing	(imprenta)
prod	production	producción
psic	(psychology)	psicología
psych	psychology	(psicología)
publ	publishing	publicación(es)
puent	(bridges)	puentes
quim	(chemistry)	química
RPl.	River Plate	rioplatense
rail	railways	(ferrocarriles)
recr	recreation	recreación
refract	refractories	refractarios
refrig	refrigeration	refrigeración
relig	religion	religión
roads	roads	(vialidad)
roll	rolling	(laminación)
s	(noun)	substantivo
SA.	South America	Sud América
SLA	Southern Latin America	(Latinoamerica Sur)
safety	safety	(seguridad)
sanit	sanitary, sanitation	sanitario(s)
san.engr	sanitary engineering	(ingeniería sanitaria)
segurid	(safety)	seguridad
seguros	(insurance)	seguros
sing	singular	singular
sold	(welding)	soldadura
Spa.	Spain	(España)
sports	sports	deportes
steam	steam	(vapor-calderas)
steel	steel	(acero)
subprod	(by-products)	subproductos
sujet	(fasteners)	sujetadores
telecom	telecommunications	telecomunicaciones
termol	(thermology)	termología
textil	textiles	textiles
thermol	thermology	(termología)
tires	tires	neumáticos
tocad	(toiletries)	tocador
toilet	toiletries	(tocador)
tools	tools	herramientas
topogr	topography	topografía
transf	transformers	transformadores
transp	transportation	transportes
trat	(treatment)	tratamiento
treat	treatment	(tratamiento)
trefil	(wiredrawing)	trefilería
tub	tubing (pipes)	tuberás
turb	turbines	turbinas
USA.	United States	Estados Unidos
Uru.	Uruguay	Uruguay
v	verb	verbo
Ven.	Venezuela	Venezuela
ventil	(fans)	ventiladores
vest	vestments (clothing)	vestimenta (ropa)
vet(er)	veterinary	veterinaria
vial	(roads)	vialidad (caminos)
weld	welding	(soldadura)
wire	wire	(alambre)
wiredrwng	wiredrawing	(trefilería)

macho-hembra a - [tub] male-female
— m intermedio m - [mec] middle tap
—macho a - [tub] male-male
— m manual - [mec] hand tap
— m mecánico - [mec] machine tap
— m — abrir v tuerca(s) f - [herram] screw tap
— m — terminación f - [mec] finishing tap
— m — terraja f - [mec] tap
— m — f desarmable - [herram] collapsible
   tap
— m — f plegadizo—[herram] collapsible tap
— m pescador - [petról] fishing tap; screw grab
— m sacasonda(s) m - [petról] tapered tap
machón m - . . . • [metal-prod] buck stay •
   [arquit] butress; pillar • [constr] pier
machuelo n - [mec] véase macho m
madera f corriente - [madera] plain wood
— f de acarreo m - [hidr] driftwood
— f — álamo m - [mader] poplar wood
— f — balsa f - [mader] balsa wood
— f — f pulverizada - [mader] powdered
   balsa wood
— f — cinco por diez (centímetros m) - [mader]
   two by four (inch), wood, o timber
— f — contrahilo - [mader] end grain wood
— f — — m para piso(s) m - [constr] end grain
   wood flooring
— f — pino m - [mader] pine, wood, o lumber
— f desgastada - [mader] worn wood
— f dura - [mader] hardwood (timber)
— f fijada - [mec] fastened wood
— f gastada - [mader] worn wood
— f laminada - [mader] laminated wood • plywood
— f machihembrada - [mader] matchboard; match
   board; dovetailed lumber
— f para piso(s) m - [constr] wood flooring
— f — — m para uso m interior - [constr] interior, o inside, (use) wood floor(ing)
— f sin tratar - [mader] untreated, lumber, o
   timber, o wood
— f sólida - [constr] solid timber
— f sujetada - [mec] fastened, wood, o lumber
— f terciada - [mader] plywood
maderamen - [constr] timber(ing) work
maderas f - [transp] crating
madero m - [constr] . . .; large wood member
— m para anclaje m - [constr] crib brace
— m grande - [constr] large, timber, o wood
   member
— m para guía f - [constr] guide timber
— m — guiar v - [constr] guide timber
— m pesado m - [mader] heavy, timber, o lumber
madrastra f - . . . • [metal-prod] mantle ring;
   bustle pipe
madre f - . . . • [hidr] . . .; course
— f natura - mother nature
madriguera f - . . . • nest
madrugada f - [cronol] wee hour(s)
maduración f - . . .; maturing,ration; ripening
madurado,da a - ripened; matured
madurar v - . . .; to season
maduro,ra a - . . .; seasoned
maestría f - [ind] craftsmanship
maestro m - . . . • [educ] teacher • [ind] journeyman • foreman • artisan
— m destacado - [educ] outstanding teacher
— m ciencia(s) - [educ] Master of Science; M S
— m mecánico - [mec] master mechanic
— m para colada f - [metal-prod] tapping, i
   casting, foreman, o first helper
— m — combustión f - [combust] combustion, engineer, o jouneyman
— m — — f para horno m con fosa f - [metal-
   -lam] soaking pit combustion engineer
— m — forja f - [metal-fabr] forger
— m — horno m - [metal-prod] (s)melter
— m — — m alto - [metal-prod] blast furnace,
   foreman, o superintendent
— m — laminación f - [metal-lam] roller
— m — — f en caliente - [metal-lam] hot mill
   roller
— m — — f — frío - [metal-lam] cold mill
   roller
— m — mantenimiento m - [ind] maintenance
   journeyman
maestro m para mantenimiento m, asignado, o destacado—[ind] maintenance assigned journeyman
— m para obra f - [constr] project director
maestro,tra a - . . . • véase también principal
máfico,ca m - [miner] maphic
 magistral a - . . . • [fig] beautiful
magmatismo m - [geol] magmatism
Magna-Die* m - [mec] Magna-Die
magnesio m en polvo - [quím] powdered magnesium
— m pulverizado - [quím] powdered magnesium
magnesita f - [miner] . . .; magnesite
— f calcinada - [metal-prod] burned magnesite
— f — totalmente - [metal] dead burned magnesite
— f con calidad f alta - [miner] high quality
   magnesite|
— f — — f baja—[miner] low quality magnesite
— f granulada - [metal-prod] granulated magnesite
— f quemada - [metal-prod] burned magnesite
magnético,ca con recocido m - [metal-trat] annealed magnetic
— a extra dulce - [metal-prod] extra mild magnetic
— a — con permeabilidad f alta - [metal-
   -prod] extra mild high permeability magnetic
— a recocido,da - [metal-trat] magnetic annealed
magnetímetro m - véase magnetómetro m
magnetismo m inicial - [geol] initial magnetism
— m personal - charisma | a - [con] charismatic
— m remanente - [magnet] residual magnetism
— m residual - [magnet] residual magnetism
magnetita f - [miner] . . .; magnetic iron ore;
   ferrous-ferric oxide
— f compacta - [geol] compact magnetite
— f diseminada - [miner] diseminated magnetite
magnetítico,ca a - [miner] magentitic
magnetización f - [magnet] . . .; magnetizing
— f circular - [magnet] circular magnetization
— f primaria - [magnet] primary magnetization
— f virgen - [magnet] virgin magnetization
magnetizado,da a - [magnet] magnetized
magnetizante a - magnetizing
magneto m para tensión f alta - [comb.int] high
   tension magneto
— m — f en volante m - [comb.int] high
   tension flywheel magneto
— m — voltaje m alto - [Comb.int] high tension
   magneto
— m — volante m - [comb.int] flywheel magneto
magnetohidrodinámica f - [electr-prod] magneto-
   hydrodynamic
magnetométrico,ca a - [miner] magnetometric
magnetomotriz a - [electr] magnetomotive
magnetoscopía f - . . . •
magnetoscópico,ca a - magnetoscopic
magnificación f - . . . • compounding
magnificado,da a - magnified • compounded
magnificar f - . . . • to compound
magnífico,ca a - . . . •; wonderful; imposing;
   awesome
magnitud f - . . . • value • size
— f corriente - current, o common, o ordinary,
   magnitude
— f de carga f - [constr] load magnitude
— f — — f compresiva - [Constr] compressive
   load magnitude
— f — deformación f - deflection magnitude
— f — movimiento m - movement magnitude
— f — obra f - [constr] project size
— f — presión f - pressure magnitude
— f — — f lateral - [Constr] lateral pressure
   magnitude
— f mecánica - [mec] mechanical magnitude
— f práctica - practical magnitude
— f corriente - [ind] current practical magnitude
magullado,da a - [segurid] bruised
magulladora f - [segurid] . . .; bruising
maíz f - [botán] . . .; corn
— m cafre - [botán] Kafir, corn, o maize
— m cosechado - [agric] harvested, corn, o maize

majado,da a - crushed
majadura f - . . .; crush(ing)
majar v - . . .; to crush
majestuoso,sa - . . .; imposing
mal m - . . .; woe | adv - poorly
— ajustado,da a - misadjusted; poorly, o badly, adjusted
— alineado,da - - misaligned; poorly aligned; out of alignment
— aplicado,da a - misapplied; poorly applied
— asentado,da a - poorly seated
— centrado,da a - offcenter; poorly centered
— colocado,da a - misplaced; poorly placed
— cortado,da a - poorly, o improperly, cut
— de muchos consuelo de tontos - fool's, o cold, comfort
— documentado,da a - poorly documented
— — sangrado,da a - [metal-prod] poorly tapped; incomplete cast(ing)
— seleccionado,da - poorly chosen
— tapado,da a - poorly covered • [metal-prod] poorly plugged
— ventilado,da a - poorly ventilated
— m emotivo - [medic] emotional malady
— m mecánico - [mec] mechanical, ill, o ailment • woe
— m mental - [medic] mental malady
malabarismo m - . . . . stunt • [deport] fancy footwork
malacate m - [grúas] . . . • [mec] capstan • hoist; windlass; winch • [petról] gin pole; drawworks • [agric] sweep
— m auxiliar - [grúas] auxiliary winch
— m — con caída f libre - [grúas] free fall auxiliary winch
— m — una (sóla) velocidad f - [grúas] single speed auxiliary winch
— m — — . . . velocidad(es) f - [grúas] . . . speed auxiliary winch
— m — — f única - [grúas] single speed auxiliary winch
— m — sin caída f libre - [grúas] non free fall auxiliary winch
— m Braden - [grúas] Braden winch
— m con caída f libre - [grúas] free fall winch
— m costa adentro - [petról] on land drawworks
— m — afuera - [petról] offshore drawworks
— m eléctrico - [grúas] electrical winch • [petról] electrical drawworks
— m hidráulico - [grúas] hydraulic winch
— m — montado - [grúas] mounted hydraulic winch
— m — — sobre bastidor m - [grúas] frame mounted hydraulic winch
— m lubricado - [petról] lubricated drawworks
— m mecánico - [grúas] mechanical hoist
— m montado - [grúas] mounted winch • [petról] mounted drawworks
— m — sobre bastidor m - [grúas] frame mounted winch
— m — — subestructura - [petról] substructure mounted drawworks
— m para cable m - [cabl] calf wheel
— — m para tubería f - [petról] (pipe) calf wheel
— m — cantera f - [miner] quarry, hoist, o winch
— m — cuchara f - [petról] sand reel
— f — emparvadora f - [agric-equip] press, o baler, sweep, o winch
— m — f para heno m - [agric-equip] hay, press, o baler, sweep, o winch
— m — — f — pasto m - [agric-equip] hay, press, o baler, sweep, o winch
— f — enfardadora f - [agric-equip] hay press sweep
— m — equipo m rotativo - [petról] rotary drawworks hoist
— m — grúa f - [grúas] crane hoist
— m — herramienta(s) f - [petról] bull wheel
— m — perforación f - [petról] drilling hoist
— m — tubería f - [petról] calf wheel
— m paralizado - [petról] shut down drawworks

malacate m poderoso - [petról] powerful drawworks
— m principal - [grúas] main winch
— m — con caída f libre - [grúas] free fall main winch
— m — una (sóla) velocidad f - [grúas] single speed main winch
— m — — velocidad f única - [grúas] single speed main winch
— m — — . . . velocidad(es) f - [grúas] . . . speed main winch
— m — Gearmatic - [grúas] Gearmatic main winch
— m — sin caída f libre - [grúas] non free fall main winch
— m rotativo - [petról] rotary drawworks
— m sin caída f libre - [grúas] non free fall winch
— m sobre tierra f - [petról] on land drawworks
— m vibratorio - [petról] vibrating drawworks
malcomprender* v - to misunderstand
malcomprendido,da* a - misunderstood
maleabilidad f - . . .; ductility
maleabilización f - [metal-prod] malleabilization
maleabilizador,ra a—[metal-prod] malleabilizing
maleable a - [metal-prod] . . .; ductile • soft
malecón m - [hidr] . . .; pier; jetty; breakwater; sea wall; dam; mound
malestar m - [medic] . . .; illness
— m cualquiera - [medic] illness of any kind
maleza f - [botán] . . .; undergrowth
— f acuática - [sanit] aquatic weed (nuisance)
malfuncionado,da a - malfunctioned • gone awry
malfuncionamiento m - . . .; malfunctioning • going awry
malgastar v tiempo m - [labor] to waste @ time
malo,la a - . . . • faulty
malogrado,da a - . . .; wasted; lost; missed; marred
malograr v - . . . • to mar
malogro m - . . .; marring
maltratado,da a - mistreated; maltreated • manhandled
maltratamiento m - . . .; mistreating,tment; ill treatment • manhandling • abuse
maltratar v - . . .; to mistreat • to manhandle
maltrato m - . . . • abuse
malversación f de fondo(s) m - [admin] fund(s) malveresation; graft
malla f - [mec] . . .; sieve • mat • [alambre] net(ting); fabric • [textil] netting; network
— f aluminizada - [alambre] aluminized (wire) mesh
— f con unión(es) f - [metal-fabr] fastening netting
— f — — m tipo m doble - [metal-fabr] double type fastening netting
— f — — m — m — torcido - [metal-fabr] double wound type fastening netting
— f de alambre m - [metal-fabr] wire fabric • netting • fence • screen wire • [constr] wire mesh • screen cloth • welded mesh
— f — — m aluminizada - [alambre] aluminized wire, mesh, o fabric
— f — — m con unión(es) f - [alambre] fastening netting
— f — — m — — f tipo m doble torcido - [alambre] double wound type fastening netting
— f — — m de acero m - [alambre] steel wire fabric
— f — — m galvanizado - [alambre] galvanized fence,cing
— f — — m — hexagonal - [alambre] galvanized hexagon netting
— f — — m hexagonal—[alambre] hexagon netting
— f — — m metálico - [alambre] metallic wire mesh
— f — — m para gallinero m - [avicult] poultry fence
— f — — m soldada - [alambre] welded wire fabric
— f — — m — de acero m - [alambre] welded steel wire fabric

malla f̲ de plástico m̲ - [plást] plastic, screen,
o mesh
— f̲ de tipo m̲ torcido - [alambre] wound type
netting
— f —— m— triple - [alambre] triple wound
type netting
— f elástica - [alambre] wire mesh net(ting)
— f̲ — de alambre m̲ - [alambre] wire mesh
net(ting)
— f estándar - [mec] standard, sieve, o mesh
— f̲ según Sociedad f̲ Estadounidense para Ensayos m̲ y Materiales m̲ - [mec] American Society for Testing and Materials standard mesh
— f fina - [alambre] fine mesh
— f̲ mecanizada - [sanit] bar screen
— f̲ — vertical - [sanit] vertical bar screen
— f̲ para cerca f̲ - [alambre] woven wire fencing
— f̲ — criba(s) f̲ - [alambre] screen, cloth, o
mat
— f —— f Gyrex - [metal-prod] (Gyrex) ore
screen mat
— f —— f para coque m̲ - [coque] coke screen
mat
— f —— f —mineral m̲ - [miner] ore screen
mat
— f̲ — filtro m̲ - [mec] filter screen
— f̲ —— m̲ para agua m̲ - water filter screen
— f̲ — gunitado* m̲ - [metal-prod] guniting mesh
— f̲ — pasarela(s) f̲ - [constr] walk(way) mesh
— f̲ —— f elevada(s) - [constr] skywalk(way)
mesh
— f — pavimento(s) m̲ - [constr] pavement mesh
— f̲ — portón(es) m̲ - [constr] gate, fabric, o
mesh
— f — zaranda(s) f̲ - [mec] screen mesh
— f̲ soldada - [alambre] welded, fabric, o mesh
— f̲ — de acero m—[alambr] welded steel fabric
— f̲ ——— m̲ para armadura f̲ - [constr] welded steel fabric reinforcing
— f —— m —— m̲ para hormigón m̲ -
[constr] welded steel wire fabric for concrete reinforcing
— f —— alambre m̲ - [metal-fabr] welded wire
fabric
— F ——— m̲ aluminizado - [metal-fabr] aluminized welded wire fabric
— f — para armadura f̲ - [alambre] welded wire
reinforcing fabric
— f —— para hormigón m̲ - [constr] welded wire fabric for concrete reinforcing
— f torcida - [alambre] wound netting
— f̲ — triple - [alambre] triple wound netting
— f̲ vertical - [alambre] vertical screen
mallete m̲ - [herram] mallet • [cadenas] stud
mamelón m̲ - nipple
mampara f̲ - [constr] . . . • bulkhead • curtain
wall • screening
— f no inflamable - nonflammable screen(ing)
— f̲ para seguridad f̲ - [constr] safety wall
— f̲ provisoria f̲ - [constr] temporary bulkhead
mamparo m̲ - . . . • [constr] fire wall
— m̲ longitudinal - [mec] longitudinal partition
— m̲ transversal - [mec] crosswise partition
mampostería f̲ - [constr] . . .; brickwork
— f con roca - [constr] stone masonry
— f̲ de bloque(s) m̲ - [constr] block wall
— f̲ — ladrillo(s) m̲ - [constr] brick masonry
— f̲ en seco - [constr] dry (stone) masonry
— f̲ exterior - [constr] outside, o exterior,
masonry, o brickwork
— f — para horno m̲ - [combust] furnace, outside, o exterior, brickwork, o masonry
— f interior - [constr] inside, o interior, masonry, o brickwork
— f — para horno m—[combust] furnace, inside,
o interior, masonry, o brickwork
mampostero m̲ - [constr] . . .; stone mason
— m̲ de localidad f̲ - [constr] local mason
manado,da a̲ - flowed; oozed
manantial m̲ - . . . . • [petról] seepage
— m̲ artesiano - [hidr] artesian spring
— m̲ rugiente - [hidr] roaring spring
— m̲ subterráneo - [geol] underground spring

manantío m̲ - . . .; oozing
mancomunación f̲ - pooling
— f de interés(es) - [fin] interest(s) pooling
mancha f̲ - . . .; speck • [sold] smudge
— f circular - [sold] circular, o round, spot,
o indication
— f̲ — aceite m̲ - oil, spot, o stain
— f̲ — cobre m̲ - copper spot
— f̲ — recocido m̲ - [metal-trat] annealing spot
— f̲ debida a porosidad f̲ - porosity spot
— f̲ en, banda f̲, o cinta f̲, o chapa f̲, o fleje
m̲ - [metal-laml] strip spot
manchado,da a̲ - . . .; stained
manchita f̲ - . . .; speck
mandadero,ra m/f̲ - [comunic] carrier
mandado,da a̲ - sent • [mec] driven • [transp]
shipped • [comput] commanded; prefiérase ordenado,da
— a a distancia f̲ - [mec] remotely controlled
— a con correa f̲ - [mec] belt driven
— a — pedal m̲ - [mec] foot, o pedal, operated
— a por engranaje m̲ - [mec] gear driven
— a remotamente - [mec] remotely controlled
mandar v̲ - . . . . [mec] to, drive, o operate •
[transp] to ship • [comput] to command
— v̲ con engranaje(s) m̲ - [mec] to gear drive
— v̲ — pedal m̲ - [mec] to pedal drive
— v̲ desde distancia f̲ - [mec] to control remotely
— v̲ embrague m̲ - [mec] to actuate @ clutch
— f̲ — m̲ con pedal m̲ - [mec] to pedal actuate
@ clutch
— v̲ remotamente - [mec] to control remotely
— v̲ válvula f̲ - [mec] to, actuate, o operate, @
valve
mandarria f̲ - [herram] . . .; hand maul; about
sledge
mandatario m̲ - [legal] . . .; proxy
mandato m̲ - . . . . • [legal] mandate; commission;
term • [comput] command; prefiérase orden f̲
— m̲ colacionado—[comput] reported back command
— m̲ cumplido - [comput] performed command
— m̲ recuperado - [comput] recovered, o reported
back, command
— m̲ retornado - [comput] reported back command
mandíbula f̲ - [anatom] . . . • [mec] jaw, alligator jaw; tongs • [herram] leg • [metal-
-prod] pincer jaw
— f abierta f̲ - [mec] open(ed) jaw
— f̲ accionada con resorte m̲ - [mec] spring-released jaw
— f ——— m̲ con mando m̲ con pedal m̲ - [mec]
spring-released pedal-actuated jaw
— f aislada - [herram] insulated jaw
— f̲ ajustable - [mec] adjustable, o swivel, jaw
— f̲ con mando m̲ con pedal m̲ - [mec] pedal actuated jaw
— f — rosca f̲ - [mec] threaded jaw
— f̲ conectada f̲ - [mec] connected jaw
— f̲ de cobre m̲ - [mec] copper jaw
— f̲ — placa f̲ - [mec] (face)plate jaw
— f̲ portaelectrodo(s) m̲ - [sold] (electrode)
holder jaw
— f — pulverizadora f̲ - [mec] pulverizer jaw
— f̲ — tornillo m̲ - [herram] vise jaw
— f —— m̲ para banco m̲ - [herram] vise jaw
— f̲ fija - [mec] fixed, o stationary, jaw
— f̲ giratoria - [mec] swivel(ling), o rotating,
jaw
— f hembra - [mec] female jaw
— f̲ inferior - low(er) jaw
— f̲ macho - [mec] male jaw
— f̲ movible - [mec] movable jaw
— f̲ oscilante - [mec] oscillating, o swinging,
jaw
— f para arranque m̲ - [comb.int] starting jaw
— f̲ —— m̲ con manija f̲ - [comb] crank starting jaw
— f ranurada - [mec] slotted jaw
— f̲ recia - [mec] rugged jaw
— f̲ — para contacto m̲ - [sold] rugged contact
jaw
— f reemplazable - [mec] replaceable jaw

mandíbula f removible - [mec] removable jaw
— f **renovable** - [mec] replaceable, o renewable, jaw
— f **rotativa** - [mec] rotating, o swivel, jaw
— f **sujetada** - [mec] clamped, o held, jaw
— f **superior** - [mec] high(er), o upper, jaw
— f **vencida** - [mec] spread jaw
— f **y ojo** m - [mec] jaw and eye
mandil m expulsor m = [agric-equip] ejector
— m — **trasero** - [agric-equip] ejector tailgate • back apron
— m **frontal**—(front) apron
— m **trasero** - tailgate
— m — **expulsor** - [agric-equip] ejector tailgate
mando m - . . . . • [labor] authority; supervision,sor; supervisory personnel • [milit] command • [mec] drive; control • operator • [comput] command; **prefiérase orden** f
— m **automático** - [mec] automatic control
— m **auxiliar** - [mec½ auxiliary drive
— m **Bendix** - [mec] Bendix drive
— m **centralizado** - [ind] centralized control
— m **con árbol** m - [mec] shaft drive,ving
— m — — m **con leva(s)** f - [mec] camshaft drive,ving
— m **con bomba** f - [bombas] pump drive,ving
— m — **cadena** f - [mec] chain drive,ving
— m — **correa** f - [mec] belt drive,ving
— m — — f **en V** - [mec] V-belt drive,ving
— m — **engranaje(s)** m - [mec] gear drive,ving
— m — — m **cónico(s)** - [mec] bevel(ed) gear drive,ving
— m — **palanca** f - [mec] lever operating,tion
— m — — f **para regulador** m—[válv] control lever operating,tion
— m — **pedal** m - [mec] pedal operating,tion | a - (con) pedal operated
— m — **pulsador(es)** m - [electrón] push button operating,tion | a - [electrón] (con) push button operated
— f — **rueda** f **dentada** - [mec] sprocket drive,ving
— m — — f — **y cadena** f - [mec] sprocket and chain drive,ving
— m — **sin fin** m - [mec] worm drive,ving
— m — — m **y rueda** f **dentada** - [mec] worm gear drive,ving
— m — **solenoide** - [mec] solenoid drive,ving | a - [electr] (con) solenoid operated
— m — **tornillo** m - [mec] screw drive,ving
— m — — m **sin fin** - [mec] worm drive,ving
— m — **transmisión** f - [mec] transmission drive,ving
— m — **turbina** f - [mec] turbine drive,ving
— m **cónico** helicoidal - [mec] spiral bevel drive,ving
— m **cumplido** - [comput] performed command
— m **de tipo** m **planetario** - [mec] planetary type drive,ving
— m **desde distancia** f - [mec] remote control | a - [mec] (con) remotely controlled
— m **directo** - [mec] direct drive,ving • [admin] authority line
— m **doble** - [mec] double, drive,ving, o control,ling
— m **en cuatro ruedas** f - [autom-mec] four wheel drive,ving
— m — . . . **rueda(s)** f - [autom-mec] . . . wheel drive,ving
— m **fijo** - [grúas] seated control
— m **final** - [mec] final drive,ving
— m **hidráulico** - [mec] hydraulic, o fluid, drive,ving
— m **Hotchkiss**—[autom-mec] Hotchkiss drive,ving
— m **mecánico** - [mec] mechanical, drive,ving, o operating,tion • motorized operator
— m — **con correa** f - [mec] belt motorized opoerating,tion
— m — — — f **articulada** - [mec] link belt motorized operating,tion
— m **para aguilón** m - [grúas] boom control(ling)
— m — **avión** m - [aeron] aircraft control(ling)

mando m **para bomba** f - [bombas] pump drive,ving; pump control(ling)
— m — — f **para aceite** m - [bombas] oil pump, drive,ving, o control(ling)
— m — — f — — m **lubricante** - [bombas] lubricating oil pump, drive,ving, o control
— m — — f **principal** - [bombas] main pump control(ling)
— m — — — f — **para aceite** m - [bombas] main oil pump control(ling)
— m — — f — — m **lubricante** - [bombas] main lubricating oil pump control(ling)
— m — **cable** m - [cabl] cable control(ling)
— m — — m **a gancho** m - [grúas] whip control
— m — **cepilladora** f - [mec] planer drive,ving
— m — **desplazamiento** m - [grúas] travel drive
— m — — m **lateral** - [grúas] cross travel drive,ving
— m — — — m **para carrillo** m - [grúas] trolley cross travel drive,ving
— m — — m **transversal**—cross travel drive
— m — — — m **para carrillo** m - [grúas] trolley cross travel drive,ving
— m — **eje** m - [mec] shaft drive,ving
— m — **elevador** m - [grúas] hoist drive,ving • elevator drive,ving
— m — **embrague** m - [mec] clutch, drive,ving, o actuating,tion, o operating,tion
— m — — m **con pedal** m - [mec] clutch pedal, actuating,tion, o operating,tion
— m — **empujadora** f - [mec] pusher drive,ving
— m — — m **con diente(s)** m **recto(s)** - [válv] spur gear operating,tion
— m — **estabilizador** m - [grúas] stabilizer control(ling)
— m — **fuente** f - source control(ling)
— m — — f **para energía** f - [sold] power source control(ling)
— m — **grúa** f - [grúas] crane control(ling)
— m — **implemento** m - [agric] implement drive,ving
— m — — m **montado** - [agric-equip] mounted implement drive,ving
— m — **impulsión** f - [grúas] travel drive,ving
— m — **interruptor** m - [electr-oper] switchgear operating,tion
— m — **motor** m - [electr-mot] motor control(ling) • [comb.int] engine control(ling)
— m — **movimiento** m - [grúas] movement, o travel, drive,ving
— m — — m **lateral** - [grúas] cross travel drive,ving
— m — — — m **para carrillo** m - [grúas] trolley cross travel drive,ving
— m — — m **transversal** - [grúas] cross travel drive,ving
— m — — — m **de colector** m - [grúas] trolley cross travel drive,ving
— m —   **rotación** f - [grúas] swing, drive,ving, o control • [válv] valve control(ling)
— m — **válvula** f - [válv] valve control(ling) • valve wheel
— m — **vehículo** m - [autom] carrier control
— m — — m **transportador** - [autom] carrier control(ling)
— m — **velocidad** f - [comb.int] speed control
— m — **ventilador** m - [mec] fan drive,ving
— m **planetario** - [mec] planetary drive,ving
— m **principal** - [mec] main drive,ving • main control(ling)
— m **remoto** - [mec] remote control | a - [mec] (con) remotely controlled
— m **simple** m - single control(ling) • : :
— m **solidario** - [mec] solidary drive,ving
— m **tipo** m **Hotchkiss** - [autom] Hotchkiss drive
— m **reversible** - [mec] revesible drive,ving
mandril m - [herram] . . .; collet; treblet; Jacob's chuck • [constr-pil] mandrel • [sold] chuck • [mec] plunger
— m **acanalado** - [tub] fluted swage
— m **amortiguador** - [mec] buffing mandrel
— m — **para pulidora** f - [herram] buffing mandrel

mandril m con dos segmentos m - [metal-lam] two-
-segment mandrel
— m — extensión f - [mec] elastic clutch
— m — rodillo(s) m - [metal-lam] rolling mandrel • [herram] roll carrier
— m contraído - [mec] contracted mandrel
— m de acero m - [constro-pil] steel mandrel
— m — latón m - [mec] brass drift
— m desabollador - [tub] pipe swage
— m doble - [mec] double mandrel
— m electromagnético - [mec] electromagnetic
chuck
— m según norma f - [herram] standard electromagnetic chuck
— m elíptico - [herram] elliptic(al) chuck
— m enroscado - [mec] screw chuck
— m equipado con rodillo(s) m - [mec] roller
equipped mandrel
— m excéntrico - [herram] eccentric chuck
— m expansivo - [mec] expander
— m expansible - [mec] expanding mandrel
— m — de acero m - [constr-pil] expanding
steel mandrel
— m — doble - [mec] double expanding mandrel
— m golpeado - [mec] struck driver
— m independiente m - [mec] independent chuck
— m intercambiable - [mec] interchangeable
mandrel
— m interior - [constr-pil] internal mandrel
— m — expansible - [constr-pil] expanding internal mandrel
— m magnético - [mec] magnetic chuck
— m neumático—[herram] air, o pneumatic, chuck
— m para barreno m - [herram] boring chuck
— m — bobinadora f—[metal-lam] coiler mandrel
— m — cambio m rápido - [metal-lam] quick-
-change mandrel
— m — m de desbobinadora f - [metal-lam]
quick-change decoiler mandrel
— m — casquillo m - [herram] socket mandrel
— m — m giratorio - [herram] swivel socket
mandrel
— m — m para cable m - [herram] swivel
rope socket mandrel
— m — m para cable m - [herram] rope socket
mandrel
— m — comprobación f - [cojin] gaging plug
— m — conformación f - [herram] forming mandrel
— m — desbobinadora f - [metal-lam] decoiler
mandrel
— m — f de tipo m para cambio m rápido -
[metal-lam] quick-change type decoiler mandrel
— m — electrodo m - [sold] electrode chuck
— m — — m de tungsteno m - [sold] tungsten
electrode chuck
— m — expansión f - [mec] expanding mandrel •
[tub] expander
— m — f de tubería f - [tub] pipe expander
— m — f tubo(s) m - [tub] pipe expander
— f — máquina f para barrenar - [herram] boring chuck
— m — plantilla f - [mec] jig plunger
— m — rodillo(s) m - [herram] roller swage
— m — soldadura f - [sold] welding chuck
— m — soldar - [sold] welding chuck
— m — taladro m - [herram] drill, o boring,
chuck
— m — torno m - [herram] lathe chuck
— m — tornillo m - [mec] screw chuck
— m — tubería f - [tub] pipe mandrel
— m — f para entubación f - [petról] casing, mandrel, o swage
— m provisto con rodillo(s) m - [mec] roller
equipped mandrel
— m retráctil - [constr-pil] retractable mandrel
— m según norma f - [herram] standard chuck
— m sin llave f - [mec] keyless chuck
— m universal - [mec] universal chuck
— m usado - [mec] used, mandrel, o drift
mandrilado m - [mec] drifting • gear bore

mandrilado m con rodillo(s) m - [metal-fabr]
(mandrel) rolling
— m — m con fuerza f axial - [tub] axial
force (mandrel) rolling
— m — m — f radial - [tub] radial
force (mandrel) rolling
— m — m — f radial y axial - [tub]
radial and axial force (mandrel) rolling
mandrilado,da a - [mec] drifted
mandrilador m - [mec-tornos] broaching machine
mandrilar v - [mec] to drift
mandrinadora* f - [herram] véase taladro m
mandrinar* v - [mec] véase taladrar
maneado,da a - hobbled
manecilla f - [instrum] . . .; pointer; index
hand
— f de reloj m - [instrum] clock hand
manejado,da apropiadamente - handled properly
— a con seguridad f - [segurid] handled safely
— a debidamente - handled properly
— a fácilmente - handled easily
— a manualmente - [mec] handed manually
manejar v - . . .; to operate
— v agua m - to handle @ water
— v apropiadamente - to handle properly
— v combustible m - [combust] to handle @
fuel properly
— v automóvil m - [autom] to, drive, o operate,
@, car, o automobile
— v balde m para achique m - [nav] to operate @
bailer
— v cereal(es) m - [agric] to handle @ grain
— v combustible m - [combust] to handle @ fuel
— v con seguridad f—[segurid] to handle safely
— v dato(s) m - to handle, data, o information
— v debidamente - to handle properly
— v — combustible m - [combust] to handle @
fuel properly
— v equipo m - [ind] to handle @ equipment
— v grano(s) m - [agric] to handle @ grain
— v información f - to handle @ information
— v — f técnica - to handle @ technical, information, o data
— v lingotera(s) f - [metal-prod] to handle @
ingot mold(s)
— v material(es) m - [ind] to handle @ mate-
rial(s)
— v — m sobre piso m para carga f - [metal-
-prod] to handle @ charging floor material(s)
— v vapor m - [cald] to handle @ steam
— v vehículo m - [autom] to handle @ vehicle
manejo m - . . . .; manipulating,tion; operation •
use • treatment
— m apropiado - proper handling
— m — de combustible m - proper fuel handling
— m bueno - good handling
— m correcto - proper handling
— m — agregado(s) m - [constr] aggregate(s)
handling
— m de agua(s) m - [hidr] water handling
— m — automóvil m - [autom] automobile, driving, o operating,tion, o handling
— m — bidón(es) m - [ind] (steel) drum(s)
handling
— m — bobinadora f - [metal-lam] coiler operation
— m — cable m - [cabl] cable handling
— m — m para pistola f - [sold] gun cable
handling
— m — ceniza(s) f - [combust] ash(es) handling
— m — cereal(es) m - [agric] grain(s) handling
— m — colada f - [metal-prod] tapping handling
— m — combustible m - [combust] fuel handling
— m — chatarra f - [metal-prod] scrap handling
— m — desecho(s) m - [nucl] waste(s) (materials) handling
— m — electrodo(s) m - [sold] electrode(s),
handling, o manipulation
— m — m por soldador m - [sold] operator
electrode(s) handling
— m — equipo m - [ind] equipment, handling, o
operation
— m — flujo m laminar - [metal-lam] strip flow,
handling, o operation

manejo m de fundente m - [sold] flux handling
— m — grano(s) m - [agric] grain handling
— m — información f - information, o data, handling
— m — f técnica - technical, information, o data, handling
— m — lingotera(s) f - [metal-prod] ingot mold(s) handling
— m — material(es) m - [ind] material(s) handling
— m — m de desecho m - [ind] waste material(s) handling
— m — m — m radiactivo(s) m - [nucl] radioactive waste material(s) handling
— m — m — residuo m - [ind] waste material(s) handling
— m — m — m radiactivo(s) m - [nucl] radioactive waste material(s) handling
— m — m sobre piso m - [ind] floor material(s) handling
— m — m — m para carga f - [metal-prod] charging floor material(s) handling
— m — molde(s) m - [ind] mold(s) handling
— m — pistola f - [sold] gun handling
— m — regulación f - [ind] control operation
— m — f automática - [ind] automatic control operation
— m — residuo(s) m - [ind] waste(s) (materials) handling
— n — sistema m - [ind] system operation
— m — m para refrigeración f - [ind] cooling system operation
— m — tubería f - [tub] pipe handling
— m — vapor m - [cald] steam handling
— m — vehículo m - [autom] vehicle handling
— m debido - proper handling
— m — de combustible m - [combust] proper fuel handling
— m diario - day-by-day handling
— m excelente - excellent handling
— m fácil - easy handling
— m — de electrodo(s) m - [sold] easy electrode(s) handling
— m malo - bad handling
— m por soldador m - [sold] operator, handling, o manipulation
— m seguro m - [segurid] safe handling
manera f - . . .; fashion
— f básica - basic way
— f categórica - categorical way | adv - (en) véase categóricamente
— f coherente - coherent manner | adv - (en) coherently
— f — estructuralmente - structurally coherent manner
— f de ser - [fam] mood • way of being
— f fácil - easy way
— f ideal - ideal way
— f idéntica - identical manner
— f más fácil - easier way
— f significativa - significant, manner, o way
— f terminante - positive, o final, way | adv - (en) véase terminantemente
— f tradicional - traditional, manner, o way
— f uniforme - uniform, manner, o way | adv - (en) uniformly; véase uniformemente
manga f - . . . • [metal-prod] bustle; tuyere
— f aisladora - [electr-instal] insulation boot
— f aplastada - [mec] collapsed hose
— f de acero m - [mec] steel sleeve
— f — m prensado - [tub] pressed steel sleeve
— f de cuero m - [segurid] leather sleeve
— f — m con cromo m - [segurid] chrome leather sleeve
— f — lona f - [ind] canvass hose
— f — plomo m - [electr-instal] lead sleeve
— f deformada - [ind] deformed hose
— f doble - [ind] double, o Dualine, hose
— f exterior - [mec] external, o outer, sleeve
— f — de plomo m - [electr-instal] external, o outer, lead sleeve
— f metálica - [constr] metal sleeve

manga f metálica articulada - [constr-hormig] elephant trunk chute
— f neumática - [ind] air hose
— f para agua m - [hidr] water hose
— f — aire m - [mec] air hose
— f — m para entrada f - [neumática] input air hose
— f — alambre m - [sold] wire hose
— f — alimentación f - [sold] feed(ing) hose
— f — — f continua - [mec] continuous feed(ing) hose
— f — — f de alambre m - [sold] continuous wire feed(ing) hose
— f — — f de alambre m - [sold] wire feed(ing) hose
— f — barra f - [mec] bar, sleeve, o hose
— f — — f taponadora - [metal-prod] véase camisa f para barra f taponadora
— f —, entrada f, o admisión f - [mec] input hose
— f — — f para aire m - [neumática] input air hose
— f — fundente m - [sold] flux hose
— f — — m larga - [sold] long flux hose
— f — inflar neumático(s) m - [autom-neumát] tire inflating hose
— f — oxígeno m - [ind] oxygen hose
— f — — m con conexión(es) f para oxígeno m y agua m - [sold] oxygen hose with (proper) oxygen and water fittings
— f — posicionador* m - [mec] placer sleeve
— f — radiador m - [autom] radiator hose
— f — recuperación f - [ind] recovery hose
— f — succión f - [ind] suction hose
— f partida - split sleeve
manganésico,ca a - [quím] manganese,sic
mangánico,ca a - [quím] . . .; manganesian
— a con carbono m - [quím] carbon-manganese,sic
manganeso m bajo,- [metal- low manganese | a - [metal] (con) low manganese
— m calcinado - [miner] calcined, o burned, manganese
— m — molido - [miner] ground, calcined, o burned, manganese
— m con carbono m - [quím] carbon manganese
— m mediano - [metal] medium manganese | a - [metal] (con) medium manganese
— m molido - [miner] ground manganese
— m reducido - [metal] low manganese | a - (con) low manganese
mangánico,ca a - véase manganésico,ca
mango m - [herram] . . .; grip; tang; handle bar • pad; ear
— m de pistola f - pistol grip
— m en T - [herram] T, o tee, handle
— m — — largo - [mec] long, T, o tee, handle
— m para azuela f - [herram] adz handle
— f — broca f - [petról] auger handle
mangón m - . . . • [mec] universal coupling; véase también mango m para acoplamiento m
manguera f - . . .; véase también manga f
— f acoplada - [mec] coupled hose
— f armada - [mec] assembled hose
— f colgante - [mec] hanging hose
— f conectada - [mec] connected hose
— f conectiva - [mec] connecting hose
— f de acero m - [tub] steel hose
— f — m con resistencia f alta (a tensión f) - [tub] high tensile steel hose
— f — caucho m - [plást] rubber hose
— f — — m moldeado - molded rubber hose
— f — lona f - canvass hose
— f desconectada - [mec] disconnected hose
— f deteriorada - bad hose
— f flexible - [mec] flexible hose
— f hidráulica—[hidr] hydraulic, o water, hose
— f inferior - [mec] lower hose
— f — para radiador m - [comb.int] radiator lower, o lower radiator, hose
— f moldeada - [mec] molded hose
— f — de caucho m - molded rubber hose
— f neumática - [mec] pneumatic, o air, hose
— f para agua m - [hidr] water, o hydraulic,

hose

manguera f para agua m para venturi - [metal-
-prod] Venturi water hose
— f — — m para toberón m - [metal-prod]
cooling plate water hose
— f — aire m - [mec] air hose
— f — m comprimido - [ind] (compressed) air
hose
— f — balancín m - [metal-prod] rocker hose
— f — m para distribuidor m - [metal-prod]
distributor rocker hose
— f - bomba f - [bombas] pump hose
— f — caja f - [ind] housing hose
— f — f para regulación f - [mec] control
housing hose
— f — f principal - [mec] main housing hose
— f — f — para regulación f - [ind] main
control housing hose
— f — calentador m - [comb.int] heater hose
— f — cementación f - cementing hose
— f — circuito m - circuit hose
— m — m para freno m - [mec] brake circuit
hose
— f — m — m por aire m - [mec] air
brake circuit hose
— f — combustible m - [comb.int] fuel hose
— f — compuerta f - [constr] gate hose
— f — conexión f - [mec] connecting hose
— f — depósito m - [mec] tank hose
— f — derivación f - [comb.int] by-pass hose
— f — f a termostato m - [comb.int] thermostat by-pass hose
— f — drenar n - [comb.int] drain(ing) hose
— f — cárter m - [comb.int] crankcase drain
hose
— f — entrada f - [comb.int] inlet, o intake,
hose
— f — f para combustible m - [comb.int]
fuel inlet hose
— f — f — m para bomba f - [comb.int]
pump fuel inlet hose
— f — f — m — f para inyección f -
[comb.int] injection pump fuel inlet hose
— f — enfriador m - [comb.int] cooler hose
— f — m para aceite - [comb.int] oil cooler
hose
— f — equipo m - [ind] equipment hose
— f — m rotativo m - [petról] rotary (system drilling) hose
— f — filtro m - [comb.int] filter hose
— f — m para carburador m - [comb.int] filter to carburetor hose
— f — freno m - [mec] brake hose
— f — m para aire m - [mec] air brake hose
— f — inyección f - [petról] rotary hose
— f — jardín m - [domést] garden hose
— f - motor m - [electr-mot] motor hose •
[comb.int] engine hose
— f — m auxiliar - [electr-mot] auxiliary
motor hose
— f — — m auxiliar con velocidad f variable -
[electr-mot] variable speed auxiliary motor
hose
— f — — m con velocidad f variable - [electr-
-mot] variable speed motor hose
— f — — m hidráulico - [hidr] hydraulic motor
hose
— f — — m — con velocidad f variable—[hidr]
variable speed hydraulic motor hose
— f — perforación f - [petról] dril(ling) hose
— f — presión f - [mec] pressure hose
— f — f alta - [mec] high pressure hose
— f — f baja - [mec] low pressure hose
— f — f intermedia - [mec] medium pressure
hose
— f — propano m - [combust] propane hose
— f — provisión f - [combust] supply hose
— f — f de gas m—[combust] gas supply hose
— f — radiador m - [comb.int] radiator hose
— f — recuperación f recovery hose
— f — refrigeración f cooling hose
— f — f para vástago m - [ind] rod cooling
hose

manguera f para refrigeración f para vástago m
para válvula f - [metal-prod] valve rod
cooling hose
— f — — m — f para viento m caliente para estufa f - [metal-prod] stove hot
blast valve rod cooling hose
— f — f — lanza f - [metal-prod] lance
cooling hose
— f — f — f para oxígeno m - [metal-
-prod] oxygen lance cooling hose
— f — tobera f - [metal-prod] tuyere hose
— f — vapor m - vapor hose • [cald] steam hose
— f — m de propano m - [combust] propane
vapor hose
— f — Venturi m - [metal-prod] Venturi, o
scrubber, hose
— f purgada - [ind] purged hose
— f sucia - [ind] dirty hose
— f reforzada - [mec] reinforced hose
— f — para equipo m rotativo - [petról] rotary
system drilling hose
— f rota - [ind] broken hose
— f superior - [mec] upper, o top, hose
— f — para radiador m - [comb.int] radiator
upper, o upper radiator, hose
— f suelta - loose hose

manguito m - . . . . • [mec] . . .; bushing; boot;
sleeve; sleevelet; sleeving; pad; thimble •
[tub] coupling • [metal-prod] maneuver ferrule • [hormig] pipe sleeve • [combust]
(boat) puller • [vest] . . .
— m ahusado - [mec] tapered spindle
— m aislador - [electr-instal] (insulating)
sleeve,ving
— m aislante - [electr-instal] (insulating)
sleeve,ving
— m apropiado - [mec] appropriate, o suitable,
sleeve,ving
— m — usado - [mec] used appropriate sleeve
— m con calidad f especial - [petról] premium
sleeve
— m — rosca f - [tub] tap(ped) coupling
— m — f ahusada - [tub] taper tap(ped)
coupling
— m — f cónica—[tub] taper(ed) tapped
coupling
— m corrido - [mec] slid, o moved, sleeve
— m cortador - [mec] cutter quill
— m de acronitrilo-butadieno-estireno - [tub]
acronitrile-butadiene-styrene sleeve
— m — neopreno m - [mec] neoprene sleeve
— m — silicono* m - [electr-equip] silicone
sleeve,ving
— m desengranado - [mec] disengaged sleeve
— m engranado - [mec] engaged sleeve
— m extractor m - [combust] boat puller
— m — para navecilla f - [combust] boat puller
— m movido - [mec] slid, o moved, sleeve
— m para acoplamiento m - [mec] coupling sleeve
• [tub] sleeve coupling
— m — — m corrido - [mec] slid, o moved,
coupling sleeve
— m — — m desengranado - [mec] disengaged
coupling sleeve
— m — — m engranado - [mec] engaged coupling
sleeve
— m — — m movido—[mec] moved coupling sleeve
— m — árbol m - [mec] shaft sleeve
— m — m para impulsión f - [bombas] drive
shaft sleeve
— m — m — f para bomba f - [bombas]
pump drive shaft sleeve
— m — m — — f para aceite m -
[bombas] oil pump drive shaft sleeve
— m — cabeza f para inyección f - [petról]
swivel sleeve
— m — cable m - [cabl] cable sleeve
— m — — m aéreo - [telecom-cond] overhead
cable sleeve
— m — cambio m - [metal-lam] changing sleeve
— m — m de rodillo(s) m - [metal-lam] roll
changing sleeve
— m — camisa f - [petról] liner sleeve

manguito m para central f - [telecom-equip] exchange sleeve
— m — cierre m - [mec] seal(ing), o closing, sleeve
— m — m para grasa f - [mec] grease seal sleeve
— m — m — — m para cojinete m - [mec] bearing grease seal sleeve
— m — cilindro m - [mec] cylinder sleeve
— m — cojinete m - [mec] bearing sleeve
— m — m lubricado - [mec] lubricated bearing sleeve
— m — corte m - [mec] cutter quill
— m — conducto m - [electr-equip] conduit sleeve
— m — — m para cable m - [electr-equip] cable conduit sleeve
— m — conexión f - [mec] connecting sleeve
— m — f hidráulica f - [mec] water connecting,tion sleeve
— m — f neumática - [mec] air connecting,tion sleeve
— f — f e hidráulica - [mec] air and water connecting,tion sleeve
— m — f — — para cojinete m - [mec] bearing air and water connecting,tion sleeve
— m — f para aire m y agua m - [mec] air and water connecting,tion sleeve
— m — f — — m — — m para cojinete m - [mec] bearing air and water connecting,tion sleeve
— m — desgaste m - [mec] wear, bushing, o sleeve
— m — — m para muñón m - [metal-prod] trunnion wear bushing
— m — — m — — m para cuchara f - [metal-prod] ladle trunnion wear bushing
— m — — m — — m — — f para colada f - [metal-prod] (pouring) ladle trunnion wear bushing
— m — desplazador m - [mec] shifter sleeve
— m — dilatación f - [mec] expansion sleeve
— m — distribución f - [mec] distribution sleeve
— m — eje m - [mec] axle sleeve
— m — embrague m - [mec] clutch sleeve
— m — — m, corredizo, o deslizable - [mec] sliding clutch sleeve
— m — empalme m - [electr-instal] connecting,tion sleeve
— m — empaquetadura f—[petról] packing sleeve
— m — — f para camisa f - [petról] liner packing sleeve
— m — empuje m - [mec] thrust sleeve
— m — — m para regulador m con bola(s) f - [comb.int] fly-weight thrust sleeve
— m — encaje m - [mec] fit(ting) sleeve
— m — — m ajustado - [mec] tight(ly, o snugly, fitting sleeve
— m — entrerrosca f - [mec] nipple sleeve
— m — — f para conexión f - [mec] connecting nipple sleeve
— m — fijación f - [mec] fastening, o holding, sleeve
— m — junta f - [mec] joint, o union, sleeve
— m — f para manguera f—[mec] hose, joint, o union, sleeve
— m — manguera f - [mec] hose sleeve
— m — muñón m - [meta-prod] trunnion bushing
— m — navecilla f - [mec] boat puller
— m — — f para horno m - [combust] furnace boat puller
— m — f — — — m para combustión f - [combust] combustion furnace boat puller
— m — protección f - [mec] protective sleeve
— m — regulador m - [comb.int] throttle, o governor, sleeve
— m — retén m - [mec] seal sleeve
— m — — m para grasa f - [mec] grease seal sleeve
— m — separación f - [mec] spacer (nipple)
— m — tambor m - [mec] drum sleeve
— m — — m para equilibrio m - [mec] balancing drum sleeve
— m — tubo,bería f - [tub] pipe,ping nipple
— m — unión f - [mec; sleeve coupling; nipple; pipe coupling
— m portabroca(s) m - [mec] chuck; chock
— m protector m - [mec] protective sleeve
— m roscado - [tub] nipple
— m para tubo,bería f - [tub] pipe nipple
— m separador m - [tub] spacing nipple; spacer
— m para carrete m - [mec] reel spacer
— m usado - [mec] used sleeve
mango m para herramienta f - [herram] tool helve
— m hueco - [herram] hollow, helve, o handle
— m para caja f - [mec] box handle
— m — — f sujetadora—[trefil] grip box handle
— m — cucharón m - [constr] dipper, stick, o handle
— m — trinquete m - [herram] ratchet handle
— m — llave f - [herram] wrench handle
— m — soltador m - [mec] release handle
— m — soltar - [mec] release handle
— m — — v sujetador m - [mec] grip release handle
— m soltador - [mec] release handle
maní m - [botán] peanut; groundnut
manía m - [medic] . . .; complex
— f de inferioridad f - [medic] inferiority complex
— f — persecución f - [medic] persecution complex
— f — superioridad f - [medic] superiority complex
manifestación f - . . .; expression; expressing; manifesting; stating; saying • purpose; intent
manifestado,da a - manifested; represented; stated; said
manifestar v - . . .; to express; to state; to represent; to say
— v conformidad f - to, manifest, o express, @ agreement
manifiesto m - [transp] manifest
manigueta f - [mec] . . .; knob
manija f - [mec] . . .; knob
— f a conmutador (de) - handle to switch
— f colocada - [mec] installed handle
— f cromada - [mec] chromium (plated) handle
— f girada - [mec] turned handle
— f para válvula f - [mec] turned valve handle
— f — — — f para regulación f - [mec] turned control valve handle
— f — totalmente - [mec] handle turned all @ way; fully turned handle
— f — — hasta tope m - [mec] handle turned, all @ way, o fully, to @ stop
— f instalada - [mec] installed handle
— f levantada - [mec] raised handle
— f — totalmente - [mec] handle raised, fully, o all @ way
— f operada ,- [mec] operated handle
— f para bomba f - [bombas] pump handle
— f — — f operada—[mec] operated pump handle
— f — capó m - [autom] hood handle
— f — conmutador m - [electr-equip] switch handle
— f — interruptor m - [electr-equip] breaker, o switch, handle
— f — puerta f - [constr] door handle
— f — válvula f - [válv] valve handle
— f — — f girada - [mec] turned valve handle
— f — — f para regulación f - [mec] control valve turned handle
— f soltada - [mec] released handle
manilla f - [penal] shackle • [mec] handle
— f para interruptor m - [electr-instal] switch handle
maniobra f - . . . . • handling; manipulation
— f brusca - sudden, o severe, maneuver
— f de vehículo m - [transp] vehicle maneuver
— f — viga f - [constr] beam handling
— f — volante m - [autom] steering wheel ma-

neuver(ing)
maniobra f evasiva - [autom] evasive maneuver
— f para cambio m - change maneuver(ing)
— f — — m brusco - [autom] abrupt change maneuver(ing)
— f — — m — de vía f - [autom] abrupt lane change maneuver(ing)
— f — — de vía f - [autom] lane change maneuver(ing)
— f — conducción f - [autom] handling (maneuver)
— f — parada f - [autom] stop(ping) maneuver • [ind] shut-down, maneuver, o operation
— f severa - [autom] severe maneuver(ing)
— f transitoria - temporary, o transient, maneuver(ing)
— f rápida - quick transient maneuver(ing)
maniobrabilidad f - . . .; manageability • [autom] handling; maneuvering
maniobrable a - . . .; manageable
maniobrado,da a - maneuvered; handled • managed
maniobrar v - . . .; to operate; to handle
— v vehículo m - [autom] to, maneuver, o handle, @ vehicle
— v ventaja f - [deport] to squeeze out @ lead
— v viga f - [constr] to handle @ beam
— v volante m - [autom] to handle (@ steering wheel)
manipulación f - . . . • handling; work(ing) • [telecom] keying
— f acertada - skillful, handling, o managing
— f de arrastre(s) m - [hidr] debris handling
— f — bidón(es) m—[ind] (steel) drum handling
— f — bobina(s) f - [metal-lam] coil handling
— f — carbón m - [miner] coal handling
— f — carga f - [grúas] load handling
— f — cereal(es) m - [agric] grain handling
— f — electrodo m - [sold] electrode, manipulation, o handling
— f — grano(s) m - [agric] grain handling
— f — hormigón m—[constr] concrete handling
— f — material(es) m—[ind] materials handling
— f — — m de desecho m - [nucl] waste material(s) handling
— f — — m — — radiactivo(s) m - [nucl] radioactive waste materials handling
— f — — m — residuo m - [nucl] waste material(s) handling
— f — — — m radiactivo(s) - [nucl] radioactive waste materials handling
— f — — m entre sección(es) f de planta f - [ind] intra-plant material(s) handling
— f — tambor(es) m - [ind] drum handling
— f — tono m - [electrón] tone keying
— f — tubería f - [tub] pipe,ping handling
— f — — f de acero m - [tub] steel pipe,ping handling
— f — f — — m corrugado - [tub] corrugated steel pipe,ping handling
— f — — f para entubación f - [petról] casing handling
— f — variable(s) f - variable(s) manipulating,tion
— f dolosa - tampering; deceitful handling
— f eficiente - efficient handling
— f en nivel m alto - [electrón] high level keying
— f — — bajo - [electrón] low level keying
— f — — m lógico - [electrón] logic level keying
— f fácil - easy, manipulation, o handling
— f impropia - improper, manipulation, o handling
— f — de electrodo - [sold] improper electrode handling
— f inapropiada - improper, manipulation, o handling • tampering
— f — de electrodo m - [sold] improper electrode, manipulting,tion, o handling
— f — por personal m no autorizado - tampering
— f más fácil - easier handling
— f mecánica - [mec] mechanical handling
— f mecanizada - mechanical handling

manipulación f neutral - [electrón] neutral keying
— f normal - [ind] normal handling
— f — apilamiento m estratificado - [ind] bedding manipulation
— f — desplazamiento m - [electrón] shift keying
— f — — m de audiofrecuencia f - [electrón] audio frequency shift keying
— f — entrada f - [electrón] input keying
— f — nivel m - [electrón] level keying
— f — salida f - [electrón] output keying
— f ociosa - tampering
— f posterior - [ind] later handling
— f práctica - practical handling
— f segura f - [segurid] safe handling
manipulado,da a - handled • [electrón] keyed
— a acertadamente - handled skillfully
— a con eslinga f - [cadenas] sling handled
— a — seguridad f - [segurid] handled safely
— a cuidadosamente - handled carefully
— a dolosamente - tampered
— a eficientemente - handled efficiently
— a en obra f - [constr] field handled
— a fácilmente - handled easily
— a inapropiadamente - improperly handled; tampered
— a normalmente - handled normally
— a ociosamente - handled carelessly; tampered
— a para desplazamiento m de audiofrecuencia f - [electrón] audio frequency shift keyed
— a posteriormente - handled latter
— a prontamente - handled promptly
— a rápidamente - handled, quickly, o promptly
— a seguramente - [segurid] handled safely
manipulador m - manipulator - [sold] jig • [instrum] key • [labor] handler • [electrón] keyer
— m conectado - [electrón] connected keyer
— m corredizo - [mec] traveling manipulator
— m de mesa f - [mec] table manipulator
— m — — f basculante - [mec] tilting table manipulator
— m — — f —, corrediza, o deslizante, o movible - [mec] traveling tilting table manipulator
— m — — f corrediza - [mec] traveling table manipulator
— m — — f, deslizante, o movible - [mec] traveling table manipulator
— m — tipo m de, guardas, o guías, laterales - [mec] side guard type manipulator
— m — tren - [metal-lam] mill manipulator
— m — — m para planchúela(s) f - [metal-lam] slabbing mill manipulator
— m deslizante - [mec] traveling manipulator
— m lógico - [electrón] logic(al) keyer
— m — conectado - [electrón] connected logic keyer
— m movible - [med] traveling manipulator
— m para corrimiento m - [electrón] shift keyer
— m — — m de frecuencia f - [electrón] frequency shift keyer; FSK
— m — desplazamiento m - [electrón] shift keyer
— m — — m de frecuencia f - [electrón] frequency shift keyer; FSK
— m — entrada f—[metal-lam] entry manipulator
— m —, guarda(s) f, o guía(s) f, lateral(es) - [mec] side guard(s) manipulator
— m — lingote(s) m - [metal-lam] ingot manipulator
— m — material(es) m - [ind] material(s) handler
— m — nivel m - [electrón] level keyer
— m — — m lógico - [electrón] logic(al) level keyer
— m — — m — conectado - [electrón] connected logic(al) level keyer
— m — pieza(s) f - [ind] part(s) manipulator
— m — punta(s) f - [metal-lam] crop(s) handler
— m — recorte(s) m - [metal-lam] crop(pings) handler

manipulador m para salida f - [metal-lam] delivery manipulator
— m — tambor(es) m - [ind] drum handler
— m — tono(s) m - [electrón] tone keyer
— m — m conectado - [electrón] connected tone keyer
— m — — m inoperante - [electrón] inoperative tone keyer
— m — — m operado - [electrón] operated tone keyer
— m — — m proyectado - [electrón] designed tone keyer
— m — — m substituido - [electrón] substituted tone keyer
— m — traviesa(s) f - [f.c.] tie handler
— m sin dedo(s) m fingerless manipulator
manipular v - . . .; to operate • [electrón] to key
— v acertadamente - to handle skillfully
— v arrastre(s) m - [hidr] to handle @ debris
— v carga f - [grúas] to handle @ load
— v cereal(es) v - [agric] to handle @ grain
— v con eslinga f - [cadenas] to sling handle
— v — seguridad f-[segurid] to handle safely
— v cuidadosamente - to handle carefully
— v chatarra f - [metal-prod] to handle @ scrap
— v dolosamente - to tamper (with)
— v eficientemente - to handle efficiently
— v en obra f - [constr] to field handle
— v escombro(s) m - [hidr] to handle @ debris
— v fácilmente - to handle easily
— v fluido - to handle @ fluid
— v grano(s) m - [agric] to handle @ grain
— v hormigón m - [constr] to handle @ concrete
• to work @ concrete
— v inapropiadamente - to handle improperly •
to tamper (with)
— v material(es) m - [ind] to handle @ material(s)
— v — m sobre piso m - [ind] to handle @ floor material(s)
— v — m — — m para carga f - [ind] to handle @ charging floor material(s)
— v materias f primas - [ind] to handle @ raw materials
— v normalmente - to handle normally
— v ociosamente - to handle carelessly; to tamper (with)
— m para desplazamiento m - [electrón] to shift key
— f — — m de audiofrecuencia f - [electrón] to audio frequency shift key
— v plancha f - [mec] to handle @ plate
— f posteriormente - to handle later
— v prontamente - to handle promptly
— v rápidamente - to handle, promptly, o quickly
— v retén m para cierre m - [mec] to handle @ seal retainer
— v seguramente - [segurid] to handle safely
— v salida f - [electrón] to key @ output
— m sólido(s) m - [ind] to handle @ solid(s)
— v tambor(es) m - [ind] to handle @ drum(s)
— v tono(s) m - [electrón] to key @ tone(s)
— v variable(s) f—to manipulate @ variable(s)
— v viga f - [constr] to handle @ beam
manipulear* v - véase manipular v
manipuleo m - . . .; handling • work(ing)
— m con eslinga(s) f - [cadenas] sling handling
— m — seguridad f - [segurid] safe handling
— m de bidón(es) m - [ind] drum handling
— m — carbón m - [miner] coal handling
— m — cereal(es) m - [agric] grain handling
— m — chatarra f - [metal-prod] scrap handling
— m — desecho(s) m - waste(s) (materials) handling
— m — fiador m - [mec] retainer handling
— m — — m para cierre m - [mec] seal retainer handling
— m — — m — sello m - [mec] seal retainer handling
— m de fluido m - fluid handling
— m — grano(s) m - [agric] grain(s) handling

manipuleo m de hierro m - [metal] iron handling
— m — — m prerreducido - [metal-prod] prereduced iron handling
— m — hormigón m - [constr] concrete handling • concrete working
— m — materia(s) f prima(s) - [ind] raw material(s) handling
— m — material(es) m - [ind] material(s) handling
— m — — m sobre piso m - [metal-prod] floor material(s) handling
— m — — m — — m para carga f - [metal-prod] charging floor material(s) handling
— m — mineral m - [metal-prod] ore handling
— m — plancha(s) f - [mec] plate(s) handling
— m — residuo(s) m - [ind] waste material(s) handling
— m — retén m - [mec] retainer handling
— m — — m para, cierre m, o sello m - [mec] seal retainer handling
— m — sólido(s) m - solid(s) handling
— m — tambor(es) m - [mec] drum(s) handling
— m — tocho(s) m - [metal-prod] bloom(s) handling
— m — viga(s) f - [constr] beam(s) handling
— m descuidado - careless, o rough, handling
— f eficiente - efficient handling
— m — obra f - [constr] field handling
— m fácil - easy, o convenient, handling
— m mínimo - minimum (of) handling
— m normal - normal handling
— m pronto - prompt handling
— m rápido - quick, o fast, o prompt, handling
— m rudo - rough handling
— m seguro - [segurid] safe handling
manivela f - [mec] . . .; hand crank; lever; handle; crank, arm, o handle; véase también perilla f
— f accionada - [mec] actuated, o activated, crank
— f actuada - [mec] actuated crank
— f con cabilla f - [mec] pin handle
— f girada - [mec] turned handle
— f manual - [mec] hand crank
— f para, activación f, o actuación f - [mec] activating, o actuating, crank
— f — amperaje m - [sold] current handle
— f — árbol m - [mec] shaft crank
— f — — m para freno m - [mec] brake shaft handle
— f — arranque m - [comb.int] starting crank
— f — campana f - [mec] bell crank
— f — — f para árbol m - [mec] shaft bell crank
— f — f para desplazador m - [mec] shifter bell crank
— f — f — — m mecánico - [mec] mechanical shifter bell crank
— f — f — dispositivo m para cambio(s) m - [mec] shifter bell crank
— f — f — — m mecánico para cambio(s) m - [mec] mechanical shifter bell crank
— f — cojinete m - [mec] bearing crank
— f — conmutador m - [electr-equip] switch handle
— f — — m centrado - [electr-instal] centered switch handle
— f — contrapeso m—[mec] counterbalance crank
— f — desplazador m - [mec] shifter crank
— f — — m mecánico - [mec] mechanical shifter crank
— f — dispositivo m - [mec] device('s) crank
— f — — m para cambio(s) m - [mec] shifter, o shifting (device), crank
— f — embrague m - [mec] clutch crank
— f — freno m - [mec] brake crank
— f — mando m - [mec] actuating, o driving, crank
— f — llave f - [electr-equip] switch handle
— f — — f selectora - [electr-equip] selector switch handle
— f — operación f - [mec] operating crank
— f — regulación f - [sold] control handle

manivela f para regulación f para amperaje m -
[sold] current control handle
— f - . . . - f permanente - [electr-oper] permanent, o continuous, control crank
— f — — f de amperaje m - [electr] continuous amperage control crank • [sold] continuous current control crank
— f regulador m - control handle
mano f - . . . • [pint] . . .; coating • [vial]
lane • [Arg.] right of way
— f aislada f - [segurid] insulated hand
— f aplicada - [pint] applied coat(ing)
— f — apropiadamente - [pint] appropriately
applied coat
— f apropiada - [pint] appropriate coat(ing)
— f contraria f - [mec] opposite hand • [vial]
opposite direction
— f de epoxia f - [pint] epoxy coat(ing)
— f — fondo m - [pint] primer (coating); un-
dercoat(ing)
— f — — m con esmalte m - [pint] enamel un-
dercoat(ing)
— f — fosfato(s) m - phosphate coat(ing)
— f — obra f - [labor] labor • force; manpower
• hand • workmanship
— f — f costosa - [labor] expensive, o high
cost, labor, o manpower
— f — f de calidad f - [ind] quality workmanship
— f — f — personal - [ind] personnel labor
— f — f — — m operativo - [ind] operating
personnel labor
— f — f — primera - [labor] first quality
workmanship
— f — f — f calidad f - first quality
workmanship
— f — — f deficiente - [ind] defective labor
— f — — f diestra - [ind] skilled labor
— f — — f directa - [labor] direct labor
— f — — f escasa - [labor] scarce labor
— f — — f especializada - [ind] skilled labor
— f — — f esmerada - craftsmanship
— f — — f excesiva - [ind] excessive labor
— f — — f exigida - required labor
— f — — f externa - [labor] outside labor
— f — — f indirecta - [labor] indirect labor
— f — — f intensiva - [labor] intensive labor
— f — — f local - [labor] local labor
— f — — f mayor - [labor] more manpower
— f — — f menor - [labor] less manpower
— f — — f operativa - [ind] operating labor
— f — — f para amoladura f - [mec] grinding
manpower
— f — — f para amoladura f de rodillo(s) m -
[metal-lam] roll grinding manpower
— f — — f para conservación f - [ind] maintenance labor
— f — — f — — f por tonelada f - [ind]
maintenance labor per ton
— f — — f — f — f producida - [ind]
maintenance labor per ton produced
— f — — f — granallado - [mec] shot blasting manpower
— f — — — m de rodillo(s) m - [metal-
-lam] roll(er) shot blasting manpower
— f — — f para horno m - [metal-prod] furnace
hand • furnace manpower
— f — — f — mantenimiento m - [ind] maintenance labor
— f — — f — — m por tonelada f - [ind]
maintenance labor per ton
— f — — f — m — — f producida - [ind]
maintenance labor per ton produced
— f — — f neumático(s) m - [autom-neumát]
tire workmanship
— f — — f para rectificación f - [mec] grinding manpower
— f — — f — f de rodillo(s) m - [metal-
-lam] roll(er) grinding manpower
— f — — f torneado m - [mec] turning manpower
— f — — f — m de rodillo(s) m - [metal-
-lam] roll(er) turning manpower

mano f de obra f pobre - [labor] poor workmanship
— f — f requerida - [labor] required labor
— f — operador m - [labor] operator('s) hand
— f — pintura f - [pint] paint coat(ing)
— f — reloj m - [instrum] clock, o watch, hand
— f — soldador m - [sold] welder('s), o op-
erator('s), hand
— f derecha - right hand
— f desnuda - bare hand
— f diestra f - right hand
— f en taller m - [pint] shop coat
— f enguantada - gloved hand
— f exterior - [pint] top coat
— f — acrílica - [pint] acrylic top coat
— f final - [pint] top, o finishing, coat
— f — acrílica - [pint] acrylic top coat
— f — aplicada - [pint] applied top coat
— f — apropiada - [pint] appropriate top coat
— f — de esmalte m - [pint] enamel finish(ing)
coat
— f — — m de vinilo m - [pint] vinyl enamel finish(ing) coat
— f — — m vinílico - [pint] vinyl enamel
finish(ing) coat
— f — pintura f - [pint] final, o finishing, paint coat
— f — — f aplicada - [pint] applied final
point coat
— f — — f omitida - [pint] omitted, final,
o finish(ing), paint coat
— f interior - [pint] undercoat
— f — con base f de cinc m - [pint] zinc based
paint unercoat
— f izquierda - left hand
— f — de operador m - [ind] operator('s) left
hand
— f lavada - washed hand
— f libre - free hand
— f opuesta - opposite hand
— f para imprimación f - [pint] primer (coat)
— f — f aplicada f - [pint] applied primer
coat
— f — f — en obra f - [pint] field applied
coat
— f — — f — taller m - [pint] shop (applied) primer coat
— f — f en taller m - [pint] shop primer, o
primer shop, coat
— f — f metálica - [pint] shop metallic
primer coat
— f — f cocida en horno m - [pint]
baked on metallic prime coat
— f — — — m aplicada en taller
m - [pint] shop baked on metallic primer coat
— f — — f — — m en taller m—[pint]
shop baked on metallic primer oat
— f — terminación f - [pint] finish(ing) coat
— f primera - first hand • [pint] first coat;
undercoat
— f — de esmalte m - [pint] enamel undercoat
— f segunda - second hand • [pint] second coat
— f — epoxia f - [pint] epoxy second coat
— f siniestra - left hand
manómetro m - [instrum] . . .; gauge
— m acoplado - [instrum] coupled, o attached,
pressure gage
— m conectado - [instrum] connected, o attached, pressure gage
— m de Marsh - [mec] Marsh gage
— m indicador - [instrum] dial indicator; indicating gage
— m — con esfera f - [instrum] dial indicator
(gage)
— m — para aceite m - [instrum] oil indicating
gage
— m — para presión f - [instrum] pressure indicating gage
— m — — — f de aceite m - [comb.int] oil
pressure indicating gage
— m — — — f — — m en caja f para cambio(s)
m - [mec] gear box oil pressure indicating
gage

manómetro m̲ indicador para presión f̲ de aceite m̲
en caja f̲ para velocidades f̲ - [instrum] gear
box oil pressure indicating gage
— m̲ —— f̲ —— m̲ en embrague m̲ - [mec]
clutch oil pressure indicating gage
— m̲ —— f̲ —— m̲ — motor m̲ - [comb.int]
motor oil pressure indicating gage
— m̲ instalado - [instrum] mounted gage
— m̲ montado - [instrum] mounted gage
— m̲ — en obra f̲ - [instrum] field installed
gage
— m̲ para abastecimiento m̲ - [mec] supply gage
— m̲ —— m̲ de aire m̲ - [instrum] air supply
gage
— m̲ — aceite m̲ - [mec] oil gage
— m̲ — agua m̲ - [instrum] water gage
— m̲ — aire m̲ - [mec] air (pressure) gage
— m̲ — alarma f̲ - [instrum] alarm (pressure)
gage
— m̲ — caldera f̲ - [cald] boiler gage
— m̲ — combustible m̲ - [instrum] fuel (pressure) gage
— m̲ — freno m̲ - [mec] brake (pressure) gage
— m̲ — gas m̲ - [instrum] gas (pressure) gage
— m̲ — inflación f̲ - [autom-neumát] (tire) inflation (pressure) gage
— m̲ para motor m̲ - [comb.int] engine gage •
[electr-mot] motor gage
— m̲ —— m̲ verificado - [comb.int] checked engine gage • [electr-mot] checked motor gage
— m̲ — nivel m̲ - [instrum] level gage
— m̲ —— m̲ de agua m̲ - [instrum] water (level)
gage
— m̲ —— m̲ — aceite m̲ - [comb.int] oil level
gage
— m̲ —— m̲ — líquido m̲ - [mec] liquid level
gage
— m̲ — presión f̲ - [instrum] pressure gage
— m̲ —— f̲ acoplado - [instrum] coupled, o attached, pressure gage
— m̲ —— f̲ conectado - [instrum] connected
pressure gage
— m̲ —— f̲ de aceite m̲ - [mec] oil pressure
gage
— m̲ —— f̲ —— m̲ en caja f̲ - [instrum] box,
o case, oil pressure gage
— m̲ —— f̲ —— m̲ —— f̲ para cambio(s) m̲ -
[instrum] gear box oil pressure gage
— m̲ —— f̲ —— m̲ — caja f̲ — velocidad(es)
f̲ - [mec] gear box oil pressure gage
— m̲ —— f̲ —— m̲ — embrague m̲ - [mec]
clutch oil pressure gage
— m̲ —— f̲ —— m̲ — motor m̲ - [comb.int]
engine oil pressure gage • [electr-mot] motor
oil pressure gage
— m̲ —— f̲ — combustible m̲ - [instrum] fuel
pressure gage
— m̲ —— f̲ — tipo m̲ con, cuadrante m̲, o esfera f̲ - [instrum] dial type pressure gage
— m̲ —— f̲ en tubería f̲ - pipe pressure gage
— f̲ —— f̲, instalado, o montado - [ind] installed, o mounted, pressure gage
— m̲ —— f̲, —, o —, en obra f̲—[ind] field,
installed, o mounted, pressure gage
— m̲ —— f̲ para abastecimiento m̲ - [instrum]
supply pressure gage
— m̲ —— f̲ —— m̲ de agua m̲—[instrum] water
supply pressure gage
— m̲ —— f̲ —— m̲ — aire m̲ - [instrum] air
supply pressure gage
— m̲ —— f̲ — agua m̲ para freno m̲ -[mec] brake
water pressure gage
— m̲ —— f̲ —— m̲ —— llanta f̲ - (mec] rim
water pressure gage
— m̲ —— f̲ —— m̲ —— f̲ para freno m̲—[mec]
brake rim water pressure gage
— m̲ —— f̲ — aire m̲ - [instrum] air pressure
gage
— m̲ —— f̲ — bomba f̲ - [bombas] pump pressure
gage
— m̲ —— f̲ —— f̲ para inyección f̲ - [petról]
mud pump pressure gage
— m̲ —— f̲ —— f̲ para lodo m̲ - [petról] mud
pump pressure gage

manómetro m̲ para presión f̲ para evaporador m̲ -
[ind] evaporator pressure gage
— m̲ —— f̲ —— m̲ verificado m̲ - [ind]
checked evaporator pressure gage
— m̲ —— f̲ para tubería f̲ - [tub] piping
pressure gage
— m̲ —— f̲ para perforación f̲ -
[petról] drill(ing) pipe pressure gage
— m̲ —— f̲ —— entubación f̲ - [petról]
casing (pipe) pressure gage
— m̲ —— f̲ — vapor m̲ - [cald] steam pressure
gage
— m̲ — provisión f̲ - [instrum] supply gage
— m̲ —— f̲ de agua m̲ - [instrum] water supply
gage— m̲ —— f̲ de aire m̲ - [instrum] air
— m̲ —— f̲ — aire m̲ - air supply gage
— m̲ — prueba(s) f̲ - [instrum] test(s) gage
— m̲ — regulador m̲ - [mec] control(ler) gage
— m̲ —— m̲ para presión f̲ - [mec] pressure
control(ler) gage
— m̲ — suministro m̲ - [instrum] supply gage
— m̲ —— m̲ de agua m̲ - [instrum] water supply
gage
— m̲ —— m̲ de aire m̲ - [instrum] air supply
gage
— m̲ — temperatura f̲ - [instrum] temperature
gage
— m̲ —— f̲ de aceite m̲ - [instrum] oil temperature gage
— m̲ —— f̲ —— m̲ hidráulico - [instrum] hydraulic oil temperature gage
— m̲ —— f̲ — agua m̲ - [instrum] water temperature gage
— m̲ —— f̲ —— m̲ en motor m̲ - [comb.int]
engine water temperature gage
— m̲ —— f̲ en culata f̲ - [comb.int] head temperature gage
— m̲ —— f̲ —— f̲ de cilindro m̲ - [comb.int]
cylinder head temperature gage
— m̲ —— f̲ de tipo m̲ con, cuadrante m̲, o esfera f̲ - [instrum] dial type temperature gage
— m̲ —— f̲, instalado, o montado - [instrum]
installed, o mounted, temperature gage
— f̲ —— f̲, —, o —, en obra f̲ - [instrum]
field mounted temperature gage
— m̲ — tensión f̲ - [instrum] tension gage
— m̲ — termostato m̲ - [instrum] thermostat gage
— m̲ —— m̲ para gas m̲ - [instrum] gas thermostat gage
— m̲ — tubería f̲ - [tub] line, o pipe, gage
— m̲ —— f̲ para aceite m̲ - [mec] oil line gage
— m̲ — vacío m̲ - [instrum] vacuum gage
— m̲ — vapor m̲ - [cald] steam gage
— m̲ protegido - [instrum] protected gage
— m̲ registrador m̲ - [instrum] recording gage
— m̲ — para presión f̲ - [instrum] recording
pressure gage
— m̲ —— presión f̲ - [instrum] pressure recording gage
— m̲ —— —— m̲ en ensanchador m̲ - [instrum] expander recording pressure gage
— m̲ —— —— f̲ — escariador m̲ - [instrum] expander recording pressure gage
— m̲ regulador - [instrum] regulator,ting gage
— m̲ — para propano m̲ - [combust] propane regulator,ting gage
— m̲ verificado - [instrum] checked gaged
— m̲ —— operación f̲ - [instrum] operation
checked gage
— m̲ —— —— f̲ apropiada - [instrum] proper
operation checked gage
manopla f̲ - . . . . • [segurid] mitten; glove
— f̲ para trabajo m̲ - [segurid] work(ing) glove
manorreductor m̲ - [metal-prod] oxygen lance
(portable) pressure reducer
manoseado,da a - handled excessively; dog eared
manosear v - to handle excessively; to dog ear
manoseo m̲ - excessive handling; dog earing
manostato m̲ - [instrum] pressurestat; pressure,
control(ler), o regulator
mantención* f̲ - véase mantenimiento m̲; manutención f̲
— f̲ de equipo m̲ - [ind] equipment, maintaining,
o maintenance, o servicing

mantener v - . . . . • to hold (up); to uphold; to keep (up) • to support • to insist • [mec] to clinch • to bear • to lock in • [ind] to service • [comput] to strap
— v **a día** m - to, update, o keep up to date
— v — **mano** f - to keep on hand
— v — **mínimo** m - to hold to @ minimum
— v — **par** - to keep up with
— v — **tanto** - to keep, up, o in touch, (with)
— v — **vista** f - to keep in, sight, o view
— v **abierto,ta** - to, keep, o hold, open
— v **acondicionador** m **(para aire** m**)** - [ambient] to maintain @ (air) conditioner
— v **actualizado,da** - to keep updated
— v **adecuadamente** - to maintain adequately
— v **ajustado,da** a - to keep tight(ened)
— v **alejado,da** - to keep, away, o off
— v **alerta** - to, be, o keep, alert
— v **alineación** f - to, maintain, o hold, @, alignment, o line
— v **alineado,da** - [mec] to keep, aligned, o in alignment
— v **almacenado,da**—to keep in, stock, o storage
— v **alto,ta** - to, maintain, o keep, high
— v **amperaje** m - [sold] to hold @, amperes, o heat
— v — m **en forma** f **continuada** - [sold] to hold @, amperage, o heat, constantly
— v **angosto,ta** a - to keep narrow
— v **ángulo** m - to, keep, o hold, @ angle
— v — m **de electrodo** m - [sold] to, keep, o hold, @ electrode angle
— v **apagado,da** - to keep, out, o unlit
— v **aparato** m - [ind] to maintain @, device, o apparatus. o unit
— v **apretado,da** - to keep tight
— v **apropiadamente** - [ind] to maintain properly
— v **arco** m - [sold] to hold @ arc
— v — m **corto** - [sold] to hold @ short arc • to keep @ arc short
— v — m **largo** - [sold] to hold @ long arc • to keep @ arc long
— v — m **muy corto** - [sold] to hold @ very short arc
— v — n — **largo** - [sold] to hold @ very long arc
— v **bajo,ja** - to, keep, o hold, low
— v **caudal** m - to maintain @ flow
— v **cubierta** f - to keep under @ cover • [nav] to keep under @ deck
— v — **gobierno** m - to, keep, o hold, under @ control
— v **bobina** f **sin conectar** - [electr-oper] to hold @ coil open
— v **calidad** f - to maintain @ quality
— v — f **constante** - [ind] to maintain @ constant quality
— v **calidad** f **de soldadura** f - [sold] to maintain @ weld(ing) quality
— v **caliente** - to keep, hot, o warm
— v **calor** m - to keep @ heat
— v **campo** m **con corriente** f - [sold] to keep @ field warm
— v — m — **temperatura** f - [sold] to keep @ field warm
— v — m **en circuito** m - [sold] to keep @ field warm
— v **cimiento** m - [constr] to maintain @ foundation
— v — m **para estructura** f - [constr] to maintain @ structure foundation
— v **cómodamente** - to maintain comfortably
— v **con presión** f - [mec] to hold with pressure
— v — f **manual** - [mec] to hold with @ hand pressure
— **confidencial(mente)**—to keep confidential(ly)
— v **conformación** f - [mec] to retain, o maintain, @ shape
— v — f **de bóveda** f - [constr] to, keep, o maintain, @ arch shape
— f — **tubería** f - [constr] to maintain @ pipe shape
— v — f — — f **abovedada** - [constr] to maintain @ pipe arch shape

mantener v **constante** - to, hold, o keep, constant, o steady
— v — **voltaje** m - [electr-oper] to, hold, o maintain, @ voltage steady
— v — — m **para soldadura** f - [sold] to, hold, o maintain, @ welding voltage steady
— v **constantemente** - to maintain constantly
— v **correctamente** - [ind] to maintain correctly
— v **corto arco** m - [sold] to keep @ short arc • to keep @ arc short
— v **cota** f - [constr] to maintain @ elevation
— v **demasiado cerca** - to hold too close
— v **deprimido** - to hold (down)
— v — **botón** m - [mec] to hold (down) @ button
— v — — m **para puesta** f **en marcha** - [comb.-int] to hold (down) @ starter button
— v — **gatillo** m - [mec] to hold (down) @ trigger; to press (down) @ trigger
— v **desde** - to stretch back (from)
— v **disponible** - to keep on hand
— v **efectividad** f - to maintain @ effectiveness
— v **electrodo** m - [sold] to, keep, o space, o position, @ electrode
— v — m **en posición** f **perpendicular** - [sold] to hold @ electrode perpendicular(ly)
— v **elevación** f - [constr] to maintain @ elevation
— v **en caución** f - [legal] to hold in escrow
— v — **contacto** m - to keep in, contact, o touch
— v — **delantera** f - [deport] to keep in @ lead
— v — **existencia** f - [com] to (keep in) stock
— v — **forma** f **continuada** - to hold constantly
— v — **garantía** f - [legal] to hold in escrow
— v — **nivel** m **requerido** - to maintain at @ required level
— v — **operación** v - [ind] to keep in operation
— v — — f **segura** - to maintain in safe operation
— v — **poder** m - to hold in @ possession
— v — **posición** f - to hold in @ position
— v — — f **correcta** - [sold] to hold in @ correct position
— v — **servicio** m - [ind] to remain in service
— v — **sitio** m - to, keep, o hold, in @, place, o position
— f — **vanguardia** f - to, hold, o keep, @ lead
— v **equilibrio** m - to maintain @ equilibrium
— v **equipo** m - [ind] to maintain @ equipment
— v **escalón** m - [sold] to maintain @ shelf
— v **especificación** f - to maintain @ specification
— v **espectrómetro** m - [instrum] to maintain @ spectrometer
— v **estanqueidad** f - to retain @ watertightness
— v **estructura** f - to maintain @ structure
— v — f **para drenaje** m - [constr] to maintain @ drainage structure
— v **excluido,da** - to lock (with)out
— v **existencia(s)** f - [com] to maintain, @ inventory, o in stock
— v — — f **pieza(s)** f **para reemplazo** m - [ind] to stock @ replacement part(s)
— v — — f — f — **repuesto** - [ind] to stock @ replacement part(s)
— m **fijo** - to remain, put, o set
— v **firme** - to (hold) steady
— v **firmemente** - to hold, firmly, o securely
— v **fondo(s)** m - [fin] to hold @ fund(s)
— v **forma** f - to hold @ shape
— v **fresco** - to keep, cool, o fresh • to remain, cool, o fresh
— v **frío,ría** - to keep cold
— v **fundido,da** - [sold] to keep molten
— v **fusión** f - to maintain @ fusion
— v — f **buena** - [sold] to maintain good fusion
— v **horizontal** - to hold horizontal(ly)
— v — **electrodo** m - [sold] to hold @ electrode horizontal(ly)
— v **hermeticidad** f - to retain @ watertightness
— v **horno** m - [combust] to maintain @ furnace
— v — m **apagado** - [combut] to keep @ furnace unlit
— v — f **con fosa** f **apagado** - [metal-lam] to keep @ soaking pit unlit

mantener v inclinación f - [constr] to hold @ grade
— v incolume - to, hold, o keep, intact
— v inconspicuidad* f - to maintain @ low profile
— v intacto,ta - to, hold, o keep, intact
— v integridad f - to maintain @ integrity
— v — f estructural - to maintain @ structural integrity
— v largo - to maintain @ length
— v — m correcto—to maintain @ correct length
— v — — de arco m - [sold] to maintain @ correct arc length
— v — m de arco m - [sold] to maintain @ arc length
— v limpio,pia - to keep clean
— v máquina f - [ind] to maintain @ machine
— v maquinaria f—[ind] to maintain @ machinery
— v más cerca - to hold closer
— v máximo m - to maintain @ maximum
— v mínimo m - to maintain @ minimum
— v motor m - [comb.int] to maintain @ engine • [electr-mot] to maintain @ motor
— v — m con marcha f lenta - [comb.int] to (hold) idle @ engine • [electr-mot] to hold (idle) @ motor
— v nivel m - [constr] to maintain @ grade
— v onda f - [electrón] to maintain @ wave
— v — f portadora - [electrón] to maintain @ carrier wave
— v operación f - [ind] to maintain @ operation
— v — f de alternador m - [electr-oper] to maintain @@ alternator (in) operation
— v — f segura - to maintain @ safe operation
— v oprimido,da - [mec] to hold, down, o in
— v — gatillo m - [sold] to hold @ trigger, in, o down
— v palanca f - [mec] to hold @ lever
— v — f para regulación f - [mec] to hold @ control lever
— v — f — — f hidrostática - [mec] to hold @ hydrostatic lever
— v paridad f - to maintain @ parity
— v paso m - to keep @, pace, o step
— v pavimento m - [constr] to maintain @ pavement
— v — m asfáltico - [constr] to maintain @ asphalt pavement
— v pendiente f - [constr] to maintain @ grade
— v peso m - to maintain @ weight
— v — m de bola f - [cuerp.moled] to maintain @ ball weight
— v póliza f - [seguros] to maintain @ policy
— v posición f - to hold (a position)
— v — f inconspicua—to maintain @ low profile
— v presión f - [mec] to, maintain, o hold, @ pressure • [autom-neumát] to hold @, pressure, o air
— v — f adecuada - [mec] to maintain @ adequate pressure|
— v — f constante - to maintain @ constant pressure
— v — f de inflación f - [autom-neumát] to maintain @ inflation pressure
— v — en red f - [mec] to maintain @ system's pressure
— v — f — sistema m - [mec] to maintain @ system's pressure
— v — f negativa - to maintain @ negative pressure
— v — f positiva - to maintain @ positive pressure
— v quemador m - [combust] to hold @ burner
— v — m con flujo m mínimo - [combust] to hold @ burner at @ minimum flow
— f red f - [ind] to maintain @ system
— v — f para combustible m - [comb.int] to maintain @ fuel system
— v reducido - to, hold, o keep, down
— v relleno m - [constr] to keep @ fill
— v rígidamente - to hold rigidly
— m ritmo m - [deport] to keep up @ pace
— v separación f - [mec] to maintain @ gap

mantener v separado(s) - to, hold, o keep, separate, o apart
— v sin conectar v - [electr-oper] to hold open
— v sin rotar - [mec] to keep from, turning, o rotating
— v sistema m - [ind] to maintain @ system
— v — m para estrangulador m - [petról] to maintain @ choke system
— v — — — para perforación f - [petról] to maintain @ drilling choke system
— v — m — perforación f - [petról] to maintain @ drilling system
— v soldadora f - [sold] to maintain @ welder
— v soldadura f - [sold] to maintain @ weld
— v subrasante f - [constr] to maintain @ subgrade
— v suelo m—[constr] to hold @, soil, o ground
— v sujetador m - [mec] to hold @ cam
— v temperatura f - to maintain @ temperature
— v — f máxima - to maintain @ maximum temperature
— v — f mínima - to maintain @ minimum temperature3
— v — f para forja f - [metal-prod] to maintain @ forging temperature
— v tendencia f - to maintain @ tendency
— v tibio,bia - to keep warm
— v turboalimentador m - [comb.int] to maintain @ turbocharger
— v unido(s) - to, hold, o keep, o secure, together
— v — conjunto m - to, keep, o hold, together @ assembly, o @ assembly together
— v — — m de cuña f - [mec] to, keep, o hold, together, @ wedge assembly
— v firmemente - to hold together tightly
— f validez f - to remain valid
— f velocidad f - to maintain @, speed, o velocity • to lock in on @ speed
— v — f para traslación f - [autom] to maintain @ ground speed
— v — f — — precisa - [autom] to maintain @ accurate ground speed
— v ventaja f - to maintain @ advantage
— v vertical - to,hold, o keep, vertical(ly)
— v — electrodo m - [sold] to hold @ electrode vertical(ly)
— v vigente - to remain, valid, o in force
— v vista f - to, hold, o keep, @ eye (on)
— v volante m - [autom] to, hold, o keep, @ steering wheel
— v — m para dirección f - [autom] to, hold, o keep, @ steering wheel
— v voltaje m - [electr-oper] to, maintain, o hold, @ voltage
— v — m para soldadura f - [sold] to, maintain, o hold, @ welding voltage
mantenerse v alerta - to, be, o remain, alert
— v apareado - to keep pace (with)
— v apartado,da - to, remain, o stand, clear, o separate, o apart
— v ausente - to remain absent • to sit out
— v delante - to, remain, o stay, ahead
— fijo - [mec] to lock tight (onto)
— v ocupado,da - to, be, o keep, busy
mantenible a - maintainable • [electrón] strappable
mantenido,da a - maintained; sustained • kept up; serviced • upheld; held up; insisted • locked in • [comput] strapped
— a a día - kept up to date; updated
— a a mano - kept, handy, o at hand
— a a par - kept up with
— a abierto,ta - held open
— a adecuadamente - maintained adequately
— a alejado,da - kept, away, o off
— a almacenado,da - kept, stored, o in storage
— a alto,ta - kept high
— a apropiadamente - [ind] maintained properly
— a bajo.ja - kept low
— a constantemente - maintained constantly
— a bajo cubierta f - kept under @ cover • [transp] kept under @ deck

mantenido,da a̲ bajo gobierno m̲ - kept under @
control
— a̲ cómodamente - maintained comfortably
— a̲ con presión f̲ manual - [mec] held with @
hand pressure
— a̲ correctamente - [ind] maintained correctly
— a̲ demasiado cerca - kept, o̲ hold, too close
— a̲ deprimido,da - held down
— a̲ disponible - kept, available, o̲ on hand
— a̲ en caución f̲ - [legal] kept in escrow
— a̲ — contacto - kept in, contact, o̲ touch
— a̲ — existencia - kept, in stock, o̲ on hand;
stocked
— a̲ — forma f̲ continuada - held continuously
— a̲ — garantía f̲ - [legal] kept in escrow
— a̲ — memoria f̲ - kept, o̲ held, in @ memory
— a̲ — movimiento m̲ - kept in @ motion
— a̲ — sitio m̲ - kept in, place, o̲ position
— a̲ — poder - held in @ possession • [pol]
kept in power
— a̲ — posición f̲ - held, o̲ kept, in @ position
— a̲ — f̲ correcta - held, o̲ kept, in @ (correct) position
— a̲ firmemente - held, o̲ kept, securely
— a̲ — unido,da(s) - held together tightly
— a̲ incolume - kept intact
— a̲ intacto,ta - kept intact
— a̲ limpio,pia - kept clean
— a̲ más cerca - held closer
— a̲ reducido,da - held, o̲ kept, down
— a̲ rígidamente - held rigidly
— a̲ separado,da - held, separate, o̲ apart
— a̲ sin conectar - [electr-oper] held open
— a̲ sin rotar - [mec] kept from, turning, o̲ rotating
— a̲ unido,da(s) - kept, o̲ held, together
mantenimiento n - . . . .; maintaining; holding
(up) upholding; locking in; maintenance operation - [com] validity • [comput] strapping
— m̲ a día - keeping updated; updating
— n̲ a mano - [ind] keeping on hand
— m̲ a par - keeping up with
— m̲ abierto - keeping, o̲ holding, open - keeping clean
— m̲ adecuado - adequate, o̲ proper, maintenance
— m̲ adscripto - [ind] assigned maintenance
— m̲ alejado,da - keeping, away, o̲ off (of)
— m̲ almacenado,da - storing; keeping in storage
— m̲ alto - high, keeping, o̲ holding
— m̲ apropiado,da - proper, o̲ adequate, maintenance
— m̲ — para red f̲ - good system maintenance
— m̲ — — f̲ para combustible m̲ - [comb.int]
good fuel system maintenance
— m̲ — — sistema - good system maintenance
— m̲ asignado - [ind] assigned maintenance
— m̲ — eléctrico - [ind] electrical assigned, o̲
assigned electrical, maintenance
— m̲ — electro-mecánico - [ind] assigned electro-mechanical, o̲ electro-mechanical assigned, maintenance
— m̲ — mecánico - [ind] assigned mechanical, o̲
mechanical assigned, maintenance
— m̲ — modificado - [ind] altered, o̲ modified,
assigned maintenance
— m̲ — para horno(s) m̲ - [ind] furnace assigned
maintenance
— m̲ — — m̲ alto(s) - [metal-prod] blast
furnace assigned maintenance
— m̲ — — m̲ para acero m̲ - [metal-prod]
steel furnace(s) assigned maintenance
— m̲ — — m̲ Siemens-Martin - [metal-prod] open
hearth (furnaces) assigned maintenance
— m̲ — — laminadora f̲ - [metal-lam] (rolling)
mill assigned maintenance
— m̲ — — — f̲ para temple m̲ - [metal-lam] temper mill assigned maintenance
— m̲ bajo m̲ - low, keeping, o̲ holding
— m̲ bajo cubierta f̲ - [ind] under cover, keeping, o̲ storage,ring • [transp] under deck,
keeping, o̲ storing
— m̲ — gobierno m̲ - keeping under @ control
— m̲ bueno - [ind] good maintenance

mantenimiento m̲ bueno para, red f̲, o̲ sistema m̲ -
good system maintenance
— m̲ central - [ind] central maintenance
— m̲ cómodo - comfortable maintenance
— m̲ con presión f̲ manual - [mec] hand pressure,
holding, o̲ maintaining
— m̲ constante - constant maintenance,taining
— m̲ continuado - continuous maintenance • continuous holding
— m̲ — de amperaje m̲ - [sold] constant heat
holding
— m̲ correctivo - corrective maintenance
— m̲ — intentado - attempted corrective maintenance
— m̲ correcto - [ind] correct maintenance
— m̲ costoso - costly, o̲ expensive, maintenance
— m̲ — para calzada f̲ - [vial] expensive roadway maintenance
— m̲ — para carretera f̲ - [vial] expensive
roadway maintenance
— m̲ de actividad(es) f̲ - [labor] activity,ties
maintenance
— m̲ — — f̲ en relación(es) f̲ laboral(es) -
[labor] labor relations activity,ties maintenance
— m̲ — alineación f̲ - [constr] alignment, o̲
line, maintaining
— m̲ — amperaje m̲ - [sold] heat holding
— m̲ — arco m̲ - [sold] arc, holding, o̲ drawing
— m̲ — calidad f̲ - [ind] quality maintenance
— m̲ — — f̲ constante - [ind] constant quality
maintaining,nance
— m̲ — calor m̲ - heat, keeping, o̲ maintaining
— m̲ — centro(s) m̲ - center(s) maintenance
— m̲ — — m̲ directo(s) - [electrón] direct
center(s) maintenance
— m̲ — circuito m̲ - circuit maintenance
— m̲ — — m̲ lógico - [electrón] logical circuit
maintenance
— m̲ — — m̲ (en) general - [electrón] general logical circuit maintenance
— m̲ — conformación f̲ - shape maintaining
— m̲ — cota - [constr] elevation maintaining
— m̲ — efectividad f̲ - effectiveness, maintaining, o̲ maintenance
— m̲ — electrodo m̲ - [sold] electrode, holding,
o̲ keeping
— m̲ — elevación f̲ - [constr] elevation maintaining
— m̲ — emergencia f̲ - [ind] emergency maintenance
— m̲ — equilibrio m̲ - [mec] equilibrium, o̲
balance, maintaining
— m̲ — especificación f̲ - specification maintaining
— m̲ — espectrómetro m̲ - [instrum] spectrometer, maintenance, o̲ maintaining
— m̲ — estanqueidad f̲ - watertightness maintaining
— m̲ — existencia(s) f̲ - inventory maintaining
— m̲ — fondo(s) - [fin] fund(s) holding
— m̲ — forma f̲ - shape holding
— m̲ — hermeticidad f̲ - watertightness maintaining • airtightness maintaining
— m̲ — integridad f̲ - [constr] integrity, retaining, o̲ maintaining
— m̲ — largo(r) m̲ - length maintenance,taining
— m̲ — — m̲ correcto - correct length maintaining
— m̲ — — m̲ para arco m̲ - [sold] correct arc
length maintaining
— m̲ — — m̲ para arco m̲ - [sold] arc length
maintaining
— m̲ — máximo m̲ - maximum maintaining
— m̲ — mínimo m̲ - minimum maintaining
— m̲ — oferta(s) f̲ - [com] bid(s) validity
— m̲ — onda f̲ - [electrón] wave maintaining
— f̲ — — f̲ portadora - [electrón] carrier
(wave) maintaining
— m̲ — operación f—[ind] operation maintaining
— m̲ — — f̲ de alternador m̲ - [electr-oper]
alternator operation maintaining
— m̲ — palanca f̲ - [mec] lever holding

mantenimiento m̲ de palanca f̲ hidrostática—[mec] hydrostatic lever holding
— m̲ —— f̲ **para regulación** f̲ - [mec] control lever holding
— m̲ —— f̲ **hidrostática** - [mec] hydrostatic control lever holding
— m̲ — **paridad** f̲ - parity, maintainging,tenance
— m̲ — **paso** m̲ - pace, o̲ step, keeping
— m̲ — **pavimento** m̲ - [constr] pavement maintenance
— m̲ —— m̲ **asfáltico** - [constr] asphalt pavement maintenance
— m̲ — **pendiente** f̲ - [constr] grade maintaining
— m̲ — **peso** m̲ - weight maintaining
— m̲ —— m̲ **para bola** f̲ - [cuerp.moled] ball weight maintaining
— m̲ — **póliza** f̲ - [seguros] policy maintaining
— m̲ — **presión** f̲ - [mec] pressure maintaining • [autom-neumát] pressure, o̲ air, holding
— m̲ —— f̲ **adecuada** - [mec] adequate pressure maintaining
— m̲ —— f̲ **en red** f̲ - [mec] system pressure maintaining
— m̲ —— f̲ **en sistema** m̲ - [mec] system pressure maintaining
— m̲ —— f̲ **negativa** - negative pressure maintaining
— m̲ —— f̲ **positiva** - positive pressure maintaining
— f̲ — **relación(es)** f̲ - relations(ships) maintaining
— m̲ —— f̲ **sindical(es)** f̲ - [labor] union relations(hips) maintaining
— m̲ — **retribución(es)** f̲ - [labor] rate(s) maintaining
— m̲ —— f̲ **básica(s)**—[labor] base,sic rate(s) maintaining
— m̲ —— f̲ **para estímulo** m̲ - [labor] incentive rate(s) maintaining
— m̲ — **ritmo** m̲ - [deport] pace keeping (up)
— m̲ — **rutina** f̲ - routine maintenance,taining
— m̲ — **servicio(s)** m̲ - [ind] service(s), o̲ utility,ties, maintaining
— m̲ — **sistema** m̲ - [ind] system maintaining
— m̲ —— m̲ **para estrangulador** m̲ - [mec] choke system, maintaining, o̲ maintenance
— m̲ —— m̲ **para perforación** f̲ - [petról] drilling choke system maintaining
— m̲ —— m̲ **para perforación** f̲ - [petról] drilling system, maintenance, o̲ maintaining
— m̲ —— m̲ **refrigeración** f̲ - [comb.int] cooling system maintenance,taining
— m̲ — **subrasante** f̲ - [constr] subgrade, maintenance, o̲ maintaining
— m̲ — **suelo** m̲ - [constr] ground holding
— m̲ — **temperatura** f̲ - temperature maintaining
— m̲ —— f̲ **máxima** - maximum temperature maintaining
— m̲ —— f̲ **mínima** - minimum temperature maintaining
— m̲ — **tendencia** f̲ - tendency maintaining
— m̲ — **velocidad** f̲ - velocity, o̲ speed, maintaining • locking in on @ speed
— m̲ —— f̲ **para traslación** f̲ - [autom] ground speed maintaining
— m̲ —— f̲ —— f̲ **precisa** - [autom] accurate ground speed maintaining
— m̲ — **ventaja** f̲ - advantage maintaining
— m̲ — **volante** m̲ - [autom] steering wheel holding
— m̲ —— m̲ **para dirección** f̲ - [autom] steering wheel holding
— m̲ **debido** - proper, o̲ good, maintenance
— m̲ — **para red** f̲ - good system maintenance
— m̲ —— f̲ **para combustible** n̲ - [comb.int] good fuel system maintenance
— m̲ —— **sistema** m̲ - good system maintenance
— m̲ **demasiado cerca** - holding too close
— m̲ **deprimido** - holding down
— m̲ — **de botón** m̲ [mec] (push)button holding (down)
— m̲ —— m̲ **para puesta** f̲ **en marcha** f̲ - [comb.int] starter button holding (down)

mantenimiento m̲ destacado - [ind] assigned maintenance
— m̲ **diario** - [ind] daily maintenance (procedure)
— m̲ —— **mínimo** - [ind] minimum daily maintenance
— m̲ **disponible** - [ind] keeping on hand
— m̲ **eléctrico** - [ind] electrical maintenance
— m̲ — **para usina** f̲ - [electr-prod] power plant electrical maintenance
— m̲ —— f̲ **para planta(s)** f̲ **siderúrgica(s)** - [electr-prod] steel plant power plant electrical maintenance
— m̲ **en caución** f̲ - [legal] holding in escrow
— m̲ — **contacto** m̲—keeping in, touch, o̲ contact
— m̲ — **existencia** f̲ - [ind] stocking; keeping in stock
— m̲ —— f̲ **de pieza(s)** f̲ - [ind] part(s), stocking, o̲ holding in stock
— m̲ —— f̲ —— f̲ **para reemplazo** m̲ - [ind] replacement part(s),stocking, o̲ holding in stock
— m̲ —— f̲ —— f̲ **para repuesto** m̲ - [ind] replacement part(s), stocking, o̲ holding in stock
— m̲ — **garantía** f̲ - [fin] holding in escrew
— m̲ — **forma** f̲ **continuada** - continuous holding
— m̲ — **memoria** f̲ - [electrón] holding in @ memory
— m̲ — **movimiento** m̲ - keeping in @ motion
— m̲ — **posesión** f̲ - holding in @ possession
— m̲ — **posición** f̲ - holding in @ position
— m̲ — **sitio** m̲ - holding in @, place, o̲ position • keeping in @, place, o̲ position
— m̲ **entre estación(es)** f̲ - [ind] (inter)seasonal maintenance
— m̲ **estacional** - [ind] seasonal maintenance
— m̲ **excelente** - excellent maintenance
— m̲ **exigido** - [ind] required maintenance
— m̲ **explicado** - [ind] explained maintenance
— m̲ **exterior** - [ind] outside, o̲ out-of-doors, maintenance
— m̲ **fácil** - [ind] easy, maintenance, o̲ service
— m̲ **firme** - secure holding
— m̲ **garantizado** - guaranteed maintenance
— m̲ — **para oferta** f̲ - bid('s) guaranteed maintenance
— m̲ — **propuesta** f̲ - proposal('s) guaranteed maintenance
— m̲ **general** - [ind] general maintenance
— m̲ — **en planta** f̲ - [ind] general plant maintenance; plant general maintenance
— m̲ **impropio** - [ind] improper maintenance
— m̲ **inadecuado** - [ind] inadequate maintenance
— m̲ **inapropiado** - [ind] inapropriate, o̲ improper, maintenance
— m̲ **incólume** - intact keeping
— m̲ **industrial** - [ind] industrial, o̲ factory, o̲ plant, maintenance
— m̲ — **exterior** - [ind] outside, o̲ out-of-doors, industrial maintenance
— m̲ — **interior** - [ind] inside, o̲ indoors, factory, o̲ industrial, maintenance
— m̲ — **puertas** f̲ **afuera** - [ind] outside, o̲ out-of-doors, industrial, o̲ factory, maintenance
— m̲ **intacto** - intact, keeping, o̲ holding
— m̲ **integral** - [ind] integral maintenance
— m̲ **intentado** - attempted maintenance
— m̲ **interior** - [ind] inside, o̲ indoors, maintenance
— m̲ **más cerca** - holding, o̲ keeping, closer
— m̲ **más rápido** - [ind] faster, o̲ quicker, maintenance
— m̲ **máximo** - [ind] maximum maintenance
— m̲ **mayor** - [ind] major maintenance
— m̲ **mecánico** - [ind] mechanical maintenance
— m̲ — **para usina** f̲ - [electr-prod] power plant mechanical maintenance
— m̲ —— f̲ **para planta** f̲ **siderúrgica** - [electr-prod] steel plant power plant mechanical maintenance
— m̲ **mensual** - [ind] monthly maintenance
— m̲ **metalúrgico** - [metal] metallurgical maintenance

mantenimiento m mínimo - minimum maintenance
— m normal - [ind] normal maintenance
— m para acondicionador m para aire m - [ambient] air conditioner maintenance
— m — alcantarilla f - [constr] culvert maintenance
— m — aparato m - [ind] apparatus maintenance
— m — bomba f - [bombas] pump maintenance
— m — cadena f - [cadenas] chain maintenance
— m — calzada f - [constr] roadway maintenance
— m — cambiador m - [mec] changer maintenance
— m para toma(s) f - [electr-equip] tap changer maintenance
— m — camino(s) m - [constr] road maintenance
— m — carretera(s) f - [constr] highway maintenance
— m — cierre(s) f - [petról] seal maintenance
— m — cilindro m - [mec] cylinder maintenance
— m — m para eter m - [comb.int] ether cylinder maintenance
— m — cimiento(s) m - [constr] foundation(s) maintenance
— m — — m para estructura f - [constr] structure foundation(s) maintenance,taining
— m — combustible m - [comb.int] fuel maintenance
— m — computadora f - [comput] computer maintenance
— m — conformación f - shape maintaining
— m — — f de bóveda f - [constr] arch shape maintaining
— m — — f — tubería f - [constr] pipe shape maintaining
— m — — f — — f abovedada - [constr] pipe--arch shape maintaining
— m — edificio(s) m - [constr] building maintenance
— m — eje f - [mec] axle maintenance
— m — — m motor - [autom-mec] drive,ving axle maintenance
— m — — m en tándem - [autom-mec] tandem drive,ving axle maintenance
— m — — m para impulsión f - [autom-mec] drive,ving axle maintenance
— m — equipo m - [ind] equipment maintenance
— m — — m para escritorio m - [com] office equipment maintenance
— m — — m — oficina f - [com] office equipment maintenance
— m — — m — producción f - [ind] production equipment maintenance
— m — — m — refrigeración f - [ind] cooling equipment maintenance
— m — — m — trabajo m - [ind] operating, o work, equipment maintenance
— m — — m — combustion f - [combust] combustion equipment maintenance
— m — — m — medición f - [ind] metering equipment maintenance
— m — — — m — f para combustible m - [combust] fuel metering equipment maintenance
— m — — m — regulación f - [ind] control, o regulating, equipment maintenance
— m — — — f para combustible m - [combust] fuel regulating equipment maintenance
— m — freno m - [mec] brake maintenance
— m — horno(s) m - [ind] furnace maintenance
— m — instrumento(s) m - [instrum] instrument maintenance
— m — — m para combustión f - [combust] combustion instrument(s) maintenance
— m — laminación f - [metal-prod] (rolling) mill maintenance
— m — laminador m - [metal-lam] (rolling) mill maintenance
— m — — m para temple m - [metal-lam] temper mill maintenance
— m — locomotora(s) f - [f.c.] locomotive, o engine, maintenance
— m — maquina(s) f - [ind] machine(s) maintenance
— m — maquinaria f - [ind] machinery maintenance

mantenimiento m para motor m - [comb.int] engine maintenance • [electr-mot] motor maintenance
— m — — m con marcha f lenta - [comb.int] idling engine maintenance • [electr-mot] idling motor maintenance
— m — ordenadora f - [comput] computer maintenance
— m — pieza(s) f - [mec] part(s) maintenance
— m — planta f - [ind] plant maintenance
— m — playa(s) f - [ind] yard maintenance
— m — f y camino(s) m - [ind] yard(s) and road(s) maintenance
— m — pozo m - [petról] well servicing
— m — propiedad(es) f - property maintenance
— m — red f - [ind] system maintenance
— m — f para combustible(s) m - [comb.int] fuel system maintenance
— m — f — refrigeración f - [comb.int] cooling system maintenance
— m — refractario(s) m - [metal-prod] refractory,ries maintenance
— m — refrigeración f - [ind] cooling equipment maintenance
— m — relleno m - [constr] fill, maintenance, o keeping
— m — reparación f - [ind] repair maintenance
— m — — f para equipo m - [ind] equipment repair maintenance
— m — — f — — m para combustión f - [combust] combustion equipment repair maintenance
— m — — f — instrumento(s) m - [instrum] instrument(s) repair maintenance
— m — — f — — m para combustión f - [combust] combustion instrument repair maintenance
— m — soldadora f - [sold] welder maintenance
— m — — f con motor m con combustión f interna - [sold] engine welder maintenance
— n — sujetador m - [mec] cam, o holder, maintenance
— m — transformador m - [electr-equip] transformer maintenance
— m — transporte(s) m - [transp] transportation maintenance
— m — tren m (laminador) - [metal-lam] (rolling) mill maintenance
— m — — m semicontinuo - [metal-lam] semicontinuous mill maintenance
— m — turboalimentador m - [comb.int] turbocharger maintenance
— m — usina f - [electr-prod] power plant maintenance
— m — — f para planta(s) f siderúrgica(s) - [electr-prod] steel plant power plant maintenance
— m — válvula f - [válv] valve maintenance
— m — — f para escape m - [válv] release valve maintenance
— m — — — m para aire m - [válv] air release valve maintenance
— m — — f — regulación f - [válv] control valve maintenance
— m — — — f para aire m - [válv] air control valve maintenance
— m — — f — salida f - [válv] release valve maintenance
— m — — — f para aire m - [válv] air release valve maintenance
— m — vehículo(s) m - [autom] vehicle(s) maintenance
— m — vía(s) f férrea(s) - [ind] track maintenance
— m perpetuo - [ind] perpetual maintenance
— m pobre - [ind] poor maintenance
— m por tonelada f - [ind] maintenance per ton
— m — f producida - [ind] maintenance per ton produced
— m preventivo - [ind] preventive maintenance
— m — eléctrico - [ind] electric preventive maintenance
— m — excelente - [ind] excellent preventive maintenance
— m — fácil - easy preventive maintenance

mantenimiento m preventivo mayor - [ind] major
preventive maintenance
— m — mecánico - [ind] mechanical preventive
maintenance
— m — para cambiador m - [electr-equip]
changer preventive maintenance
— m — — — m para toma(s) f - [electr-equip]
tap changer preventive maintenance
— m — — equipo m - [ind] equipment preventive
mainenance
— m — — — m para combustión f - [combust]
combustion equipment preventive maintenance
— m — — fuente f para energía f - [electr-
-prod] power source preventive maintenance
— m — — instrumento(s) m - [instrum] instru-
ment(s) preventive maintenance
— m — — — m para combustión f - [combust]
combustion instrument(s) preventive maintenance
— m — — transformador m - [electr-equip]
transformer preventive maintenance
— m — recomendado - [ind] recommended preventive maintenance
— m programado - [ind] scheduled maintenance
— m puertas f adentro - [ind] inside, o indoors, maintenance
— m — f afuera - [ind] outside, o outdoors,
maintenance
— m rápido - [ind] fast, o quick, maintenance
— m razonable - [ind] reasonable maintenance
— m recomendado - [ind] reocmmended maintenance
— m reducido - [ind] reduced, o little, maintenance
— m refractario - [metal-prod] refractory,ries
maintenance
— m regular - [ind] regular maintenance
— m requerido - [ind] required maintenance
— m rígido - [mec] rigid holding
— m semanal - [ind] weekly maintenance
— m separado - holding, separate, o apart
— m sin conectar - [electr-oper] holding open
— m sistemático - [ind] systematic maintenance
— m unido - holding, o keeping, together, o
united(ly)
— m — de conjunto m - [mec] assembly, holding,
o keeping, together, o united
— m — — — m para cuña f - [mec] wedge assembly, holding, o keeping, together, o united
mantilla f - . . . • [textil] mantle
— f litográfica - [textil] lithographic mantle
mantillo m - . . . • [suelos] . . .; loam; top
soil
— n conservado - [suelos] saved top soil
— m valioso - [suelos] valuable top soil
manto m - . . . • [geol] layer; stratum,ta;
sheet • [hydr-diques] blanket • [miner] seam
— m anterior - previous layer
— m completo - [cerám-prod] complete blanket
— m — para hornada f - [cerám-prod] complete
batch blanket
— m de escoria f - [sold] slag blanket
— m — grava f - [hidr] gravel blanket
— m — — f para filtro m - [hidr] gravel filter, o filter gravel, blanket
— m — material m impermeable - [geol] impervious material, blanket, o seam
— m — — — m permeable - [geol] pervious material, blanket, o seam
— m — piedra f - [hidr] stone blanket
— m — — f triturada - [hidr] crushed stone
blanket
— m — f — para filtro m - [hidr] cushed
stone filter blanket
— m — ripio m - [hidr] gravel blanket
— m — — — m para filtro m - [hidr] gravel filter, o filter gravel, blanket
— m delgado - [domést] thin blanket • [miner]
thin seam
— m — de material m—[geol] material thin seam
— m — — impermeable - [geol] impervious
material thin seam
— m — — — m permeable - [beol] previous material thin seam

manto m delgado impermeable - [geol] impervious
thin, o thin impervious, seam
— m — permeable - [geol] pervious thin, o
thin pervious, seam
— m grueso - [geol] thick seam
— m — de material m - [geol] material thick
seam
— m — — — m impermeable - [geol] impervious
material thick seam
— m — — — m permeable - [geol] pervious
material thick seam
— m — impermeable - [geol] impervious thick, o
thick impervious, seam
— m — permeable - [geol] pervious thick, o
thick pervious, seam
— m impermeable - [geol] impervious seam
— m — para evitar socavación f de dique m -
[hidr] dam (impervious(blanket
— m inferior - lower layer
— m intermedio - intermediate layer
— m limo-loesoide - [geol] silt-loessial layer
— m para estanque m - [petról] tank mantle
— m — filtro m - [hidr] filter blanket
— m — hornada f - [cerám] batch blanket
— m impermeable - [geol] impervious seam
— m permeable - [geol] pervious seam
— m superior - upper layer • [geol] upper seam
— m uniforme - [constr] uniform blanket
— m útil - useful layer
manuable a - . . .; maneuverable • portable
manual m - . . .; guide; reference, o little
gray, book | a - . . .; handmade • manually
controlled • hand held
— m con instrucción(es) f - [ind] instruction,
o operating, manual, o book, o guide • training guide
— m — — f de fabricante m - [ind] manufactur-
er('s) operating manual
— m — — f — m de motor m - [comb.int]
engine manufacturer('s) operating manual •
[electr-mot] motor manufacturer('s) operating
manual
— m — — f — — m — m diesel - [comb.int]
diesel engine manufacturer('s) operating
manual
— m — — f para alimentadora f - [sold] feeder
instruction manual
— m — — f — — f para alambre m - [sold]
wire feeder, instruction, o operation, manual
— m — — — conservación f - [ind] maintenance, o service, manual
— m — — f para entrevista(s) f - [labor] contact training guide
— m — — f — — f sobre seguridad f -
[segurid] safety contact training guide
— m — — f para fuente f para energía f -
[electr-prod] power source instruction manual
— m — — f — manipulador m (para tonos m) -
[electrón] (tone) keyer instruction manual
— m — — f — motor m - [comb.int] engine, instruction, o operating, manual • [electr-mot]
motor, operating, o instruction, manual
— m — — f — — diesel - [comb.int] diesel
engine, instructions, o operating, manual
— m — práctica(s) - [ind] practice manual
— m — — f usual(es) - [constr] standard practice manual
— m — procedimiento(s) m - [ind] procedure(s)
manual
— m — — m para ajuste m - [ind] adjustment(s)
procedure(s) manual
— m — — m para diseño m - [ind] design(ing)
procedure(s), manual, o handbook
— m — — — m para soldadora f - [sold]
welder design(ing) procedure(s) handbook
— m — — — m — — f con arco m - [sold]
arc welder design(ing) procedure(s) handbook
— m — — m — — — soldadura f - [sold]
welding design procedure(s) handbook
— m — — m — — — f con arco m - [sold]
arc welding design procedure(s) handbook
— m — — m — m y Práctica f de Soldadura f
por Arco m - [sold] Procedure Manual of Arc

Welding Design and Practice
**manual m con procedimientos m para práctica f de
soldadura f con arco m** - [sold] arc welding
practice procedure handbook
— m — m **soldadura f con arco m** - [sold]
arc welding procedure(s) handbook
— n — — m **usuales** - [ind] standard practice
manual
— m — **repuesto(s) m** - [ind] part(s) manual
— m **controlado** - controlled manual
— m **de fabricante m** - manufacturer('s) manual
— m — **de motor m** - [comb.int] engine
manufacturer('s) manual • [electr-mot] motor
manufacturer('s) manual
— m — — m **diesel** - [comb.int] diesel
engine manufacturer('s) manual
— m — m **para operación f** - [ind] manufactu-
rer('s) operating manual
— m — — f **de motor m** - [comb.int] engine manufacturer('s) operating manual •
[electr-mot] motor manufacturer('s) operating
manual
— m — **Instituto m Estadounidense para cons-
trucciones con Acero m** - [constr] American
Institute of Steel Construction manual
— m — **práctica(s) f** - practice(s) manual
— f — — f **recomendado** - [ind] recommended
practice manual
— m — **proveedor m** - [ind] supplier('s) manual
— m — — m **sobre garantía f (de calidad f)** -
[ind] supplier('s) quality guaranty manual
— m **entregado** - delivered manual
— m **Estadounidense para Proyección f para conservación f de Suelo(s) m** - [suelos] United
States Soil Conservation Design Manual
— m **para adquisición(es) f** - acquisitions, o
procurement, manual
— m — **administración f** - management manual
— m — — f **para proyecto(s) m** - project management manual
— m — **alimentadora f** - [ind] feeder manual
— m — — f **para alambre m** - [sold] wire feeder
manual
— m — — f — **electrodo(s) m** - [sold] wire, o
electrode(s) feeder manual
— m — **Calefacción, Ventilación f y Acondicionamiento m de Aire m** - [ambient] Heating,
Ventilating (and) Air Conditioning Guide
— m — **Código m Eléctrico Estadounidense** -
[electr-instal] National Electrical Code
Handbook
— m — **cojinete(s) m** - bearing(s) manual
— m — **conservación f** - [ind] maintenance, o
service, manual
— m — — f **general** - [ind] general, maintenance, o service, manual
— m — **construcción f con acero m** - [constr]
steel construction, handbook, o manual
— m — **contabilidad f** - [contab] accounting
manual
— m — **control m de calidad f** - [ind] quality
control manual
— m — **Drenaje m y Productos m para Construcción f** - [constr] Handbook of Drainage and
Construction Products
— m — **ensayo(s) m** - [ind] tests,ting manual
— m — **física f** - [mec] physics manual
— m — **fuente f para energía f** - power source
— m — **ingeniería f** - [ing] engineering, handbook, o practice(s) manual
— m — **garantía f** - guaranty manual
— m — — f **de calidad f** - quality guaranty
manual
— m — **Ingeniero(s) m** - Engineer's Handbook
— m — — m **Civil(es) m** - [constr] Civil Engineers' Handbook
— m — **Mecánicos m** - [mec] Mechanical
Engineer's Handbook
— m — **instrucción f** - instruction manual
— m — — f **para alimentador m** - [mec] feeder
instruction manual
— f — — f — m **para alambre m** - [mec] wire
feeder instruction manual

**manual m para instrucción f para alimentador m
para electrodo(s) m** - [sold] electrode(s), o
wire, feeder instruction manual
— m — **manipulador m** - [electrón] keyer manual
— m — — m **para tono(s) m** - [electrón] Tone
Keyer manual
— m — **mantenimiento m** - [ind] maintenance
manual
— m — **maquinaria(s) f** - [ind] machinery manual
— m — **motor m** - [comb.int] engine manual •
[electr-mot] motor manual
— m — — m **diesel** - [comb.int] diesel engine
manual
— m — **operación f** - [ind] operating,tion
manual
— m — **operador m** - [ind] operator('s) manual
— m — — m **para unidad f motriz** - [ind] power
unit operator('s) manual
— f — **orientación f** - orientation manual
— m — — f **para cargo m** - [labor] position
orientation guide
— m — **pieza(s) f para servicio m** - [ind] service part(s) manual
— m — **práctica(s) f** - [ind] practice(s) manual
— m — — f **para inginiería f** - engineering
practice(s) manual
— m — **producto(s) m** - [ind] product(s), manual, o handbook
— m — — m **para drenaje m** - [hidr] drainage
product(s), manual, o handbook
— m — — m — **construcción f** - [constr] construction product(s), manual, o handbook
— m — **propietario m** - owner('s) manual
— m — — m **de automóvil** - [autom] car owner's
manual
— m — **proyección f** - [constr] design(er's)
manual, o handbook
— m — — f **de cauce(s) m** - [hidr] channel(s)
design handbook
— m — — f — **cloaca(s) f** - [sanit] sewer(s)
design manual
— f — — f **para conservación f** - [ind] maintenance design manual
— m — — f — — f **para suelo(s) m** - [suelos]
soil conservation design manual
— m — **puesta f en marcha** - [ind] start-up,
manual, o handbook
— m — **reparación(es) f** - repair(s) manual
— m — **rodamiento(s) m** - [cojin] bearings, manual, o handbook
— m — **ruedas f** - [f.c.-ruedas] wheel manual
— m — — f **y ejes m** - [f.c.-ruedas] (A. A. R.)
Wheel(s) and Axle(s) Manual
— m — — f — m **de Asociación f Estadounidense de Ferrocarriles m** - [f.c.-ruedas] Association of American Railroads Wheel and
Axle Manual
— m — **seguridad f** - [segurid] safety manual
— m — — f **física radiológica** - [nuecl] radiological physics safety manual
— m — — f **para física f** - [segurid] physics
safety manual
— m — — f **radiológica** - [nucl] radiological
safety manual
— m — — f — **para salud f** - [nucl] radiological health safety manual
— m — **servicio m** - [ind] service manual
— m — **soldadura f** - [sold] welding manual
— m — — f **con arco m** - [sold] arc welding,
manual, o handbook
— m — **taller m** - [ind] shop manual
— m — — m **para cojinete(s) m** - [cojin] bear-
ing(s) shop manual
— m — — m — m **con rodillo(s) m** - [cojin]
roller bearing(s) shop manual
— m — — m — **rodamiento(s) m** - [cojin] bear-
ing(s) shop manual
— m — **tubería f** - [tub] pipe,ping manual
— m — — f **Multi-Plate** - [tub] Multi-Plate
(pipe) manual
— m — **tungsteno m** - [metal] tungsten manual
— m — **volframio m** - [metal] tungsten manual
— m **proyectado** - designed manual

manual m recibido - [ind] received manual
— m sobre neumático(s) m - [autom-neumát] tire, manual, o book
manualmente adv - . . . • [mec] finger tight
manubrio m - [mec] . . .; handwheel; turn crank; crank handle • [autom-mec] steering wheel
— m para ajuste m - [mec] adjustment (turn) crank handle
— m — giro m - [mec] turn (crank) handle
— m — regulación f - [mec] control handle
— m — válvula f - [válv] valve wheel
manufactura f - [ind] manufacturing; fabrication • poroduction • factory
manufacturado* m - véase producto m
manufacturado,da a - . . . • man made
manufacturero m - [ind] manufacturer
manutención f - . . .; véase también mantenimiento m
— f corriente - current, o routine, maintenance
— f rutinaria - [ind] routine maintenance
mañana f apacible - [meteorol] peaceful morning
— f fresca - [meteorol] cool morning
mapa m de durabilidad f - [constr] durability map
— m — f para proyección f - [constr] durability design map
— m de perfil m - [dib] profile map
— m — recorrido m - [f.c.] route map
— m detallado - detailed map
— m geológico - [geol] geological map
— m para guía - guide map
— m — f para durabilidad f para proyección f - [constr] durability design guide map
— m — f — proyección f - [constr] design guide map
— m parcial - [geogr] partial map
— m topográfico - [topogr] topographic(al) map
— m — de sitio m - [ind] site topograhic(al) map
— m — detallado - [topogr] detailed topographic(al) map
maqueta f - . . . • (physical) model
— f en escala f reducida - reduced scale (physical) model
— f topográfica - [topogr] topographic model
maquila f - [agric] . . .; multure; miller's, fee, o share - [ind] véase fabricación f; elaboración f; manufactura f
máquina f - [ind] . . .; machinery piece; engine • fixture • unit - [f.c.] engine • [vapor] engine
— f actual - [ind] present machine
— f afiladora - [mec] dresser; dressing machine
— f — para barrenos,nas - [petról] bit dresser
— f — para trépano(s) m - [petról] bit dresser
— f agrícola - [agric] agricultural, o farm, machine
— f ajustada - [mec] adjusted machine
— f alzapuerta(s)—[coque] door lifting machine
— f amoladora - [herram] grinding machine
— f asociada - [ind] related equipment
— f ataludadora - [constr] sloping machine; sloper
— f automática - [ind] automatic machine
— f — para amarrar paca(s) f - [agric-equip] automatic bale tying machine
— f — — bola(s) f - [cuerp.moled] automatic ball machine
— m — — — f para molienda f - [cuerp.moled] automatic grinding ball machine
— f — — fabricación f - [ind] automatic fabricating machine
— f — — f de tornillo(s) m - [ind] automatic screw (making) machine
— f auxiliar - [mec] auxiliary machine
— f avanzada gradualmente—[mec] jogged machine
— f barrenadora - [herram] drilling, o boring, machine; drill
— f básica - [mec] basic machine
— f calcinadora - [miner] calcining machine
— f — con chimenea f circular - [miner] annular stack calcining machine
— f calibradora - [ind] sizing machine

máquina f cansada - [mec] tired machine • [deport] tired car
— f cargadora - [ind] loading machine; loader • [metal-prod] charging, machine, o car
— f — a nivel m de piso m - [metal-prod] floor type charging machine
— f cepilladora - [herram] planing, o brush, machine
— f combadora - [metal-lam] cambering machine
— f — para riel(es) m - [metal-lam] rail cambering machine
— f comercial m - [com] business machine
— f compacta - [mec] compact machine
— f compaginadora f - [imprent] collating machine
— f completa - [mec] complete machine
— f con banda f - [herram] band machine
— f — — f con regulación f manual - [mec] manually controlled band machine
— f — — f para servicio m según norma f - [herram] standard duty band machine
— f — condensador(es) m - condensor(s) equipped machine
— m — costo m bajo - [ind] low cost machine
— m — m — con motor m con combustión f interna - [electr] low cost engine driven machine
— f — dos líneas f - [metal-col.cont] double, strand, o line, unit
— f — línea f doble - [metal-col.cont] double, strand, o línea, unit
— f — . . . línea(s) f - [metal-col.cont] . . . strand, unit, o machine
— f — movimiento m, giratorio, o rotativo - [mec] rotating machine
— f — tamaño m natural - [ind] full size machine
— f — — m — para probar rueda(s) f - [mec-ruedas] full size wheel testing machine
— f — — m natural para prueba(s) f - [ind] full size testing machine
— f conformadora - [herram] forming, o shaping, machine
— f construida - [mec] built machine
— f corchadora - [cabl] layer; laying machine
— f corrugadora f - [ind] corrugating machine
— f corrugadora para lámina(s) f - [metal-fabr] sheet corrugating machine
— f cortadora f - [mec] cutting machine • [ind] cut-off machine
— f — para alambre m - [mec] wire cutting machine
— f — para carbón m - [coque] coal, cutter, o cutting machine
— f — — galletita(s) f - [culin] cookie cutting machine
— f — — tubería f - [tub] pipe cutting machine
— f — sobre banda f - [culin] over-the-band cutting machine
— f cortatubos m - [tub] pipe cutting machine
— f curvadora - [ind] cambering, o curving, machine
— f — para riel(es) m - [metal-lam] rail cambering machine
— f de tipo m reciente - [ind] late model machine
— f — — m para emparedados m - [culin] late model sandwich machine
— f — — — sandwiches m - [culin] late model sandwich machine
— f — — m vertical - [ind] vertical type, machine, o unit
— v — vapor m - [cald] steam engine
— f desalineada - [mec] misaligned, o out-of-alignment, machine
— f desaplomada - [mec] out of plumb machine
— f deshojadora - [metal-lam] pack opening machine
— f deshornadora - [coque] pusher
— f detenida - [ind] stopped, o halted, machine
— f diseñada - [ind] designed machine
— f dispensadora f - [mec] issuing, o spitting,

o dispensing, machine

máquina f dispensadora automática - automatic dispensing machine
— f —— **para boleta(s) f** - automatic ticket, dispensing machine, o spitter
— f **disponible** - [mec] avilable machine
— f **dobladora** - [mec] bending, o folding, machine
— f **duradera** - [mec] durable machine
— f **elaboradora** - [ind] processing, machine, o equipment
— f **eléctrica** - [mec] electrical machine
— f — **para corte m** - [ind] electric cutting machine
— f —— **soldadura f** - [sold] electric welding machine
— f —— **rotativa** - [electr-equip] rotary electric machine
— f **elevadora** - [mec] hoisting, machine, o engine
— f **embridadora** - [ind] flanging machine
— f **en marcha** - [ind] operating, o running, machine
— f — **operación f** - [mec] operating machine
— f **encabezadora** - [trefil] heading machine
— f **encajonadora** - [ind] boxing machine
— f **enderezadora** - [mec] straighening machine; straightener
— f — **continua** - [metal-fabr] continuous straightening machine
— f — **para alambre m** - [mec] wire straightening machine
— f —— **barra(s) f** - [mec] bar straightening machine
— f —— **perfil(es) m estructural(es)**—[metal-lam] structural(s) straightening machine
— f —— **riel(es) m** - [metal-lam] rail(s) strightening machine
— f —— **tubo(s) m** - [tub] pipe, straightener, o straightening machine
— f **engrasada** - [lubric] greased machine
— f **enorme** - [ind] huge machine
— f **ensacadora** - [ind] bagging machine
— f **ensambladora** - [mec] jointer
— f **escarpadora** - [metal-lam] scarfing machine
— f **escopleadora** - [mec] notching, o slotting, machine
— f **especial** - [ind] special machine
— f **estadunidense** - [ind] American, o United States, made machine
— f **estampadora f** - [ind] stamping machine
— f — **para riel(es) m** - [metal-lam] rail stamping machine
— f **estándar** - [ind] standard machine
— f — **para ensayo(s) m** - [ind] standard testing machine
— f **estañadora** - [metal-trat] tinning machine
— f **estiradora para alambre m** - [trefil] wire-drawing machine
— f **excavadora** - [constr] excavating machine • [miner] mining machine
— f **forjadora** - [metal-prod] forging machine
— f — **automática** - [metal-prod] automatic forging machine
— f —— **con golpe m único** - [metal-prod] automatic single blow forging machine
— f —— **troquel m sólido** - [metal-prod] automatic single die forging machine
— f ———— **m con golpe m único**—[metal-prod] automatic single blow single die forging machine
— f — **con golpe m único** - [metal-prod] single blow forging machine
— f — **troquel m sólido** - [metal-prod] solid die forging machine
— f ———— **m con golpe m único** - [metal-prod] single blow solid die forging machine
— f — **corriente** - [metal-prod] standard forging machine
— f **formadora f** - [mec] forming, o shaping, machine, o unit
— f **fotográfica** - [fotogr] camera
— f **fourdrinier** - [papel] fourdrinier machine

máquina f fourdrinier con cinta f sin fin de malla f de alambre m (para fabricación f de papel m) - [papel] fourdrinier paper machine
— f — **para fabricación f de papel m** - [papel] fourdrinier paper machine
— f **fresadora** - [herram] milling machine
— f — **para extremo(s) m** - [metal-lam] end milling machine
— f ——— **m de riel(es) m** - [metal-lam] rail end milling machine
— f **giradora** - [mec] turning machine
— f — **para lingote(s) m** - [metal-lam] ingot turning machine
— f **gobernada** - [autom] controlled car
— f **granalladora** - [ind] Pangborn machine
— f **grande** - [ind] large, o big, machine
— f **helicoidal** - [ind] helix machine
— f **múltiple** - [ind] multiple helix machine
— f **herramienta** - [ind] machine tool
— f — **con potencia f elevada** - [ind] high powered machine tool
— f — **ultramoderna** - [ind] ultra-modern machine tool
— f **hidráulica** - [mec] hydraulic machine
— f — **para sujeción f** - [mec] hydraulic clamping machine
— f **horadadora f** - [ind] drilling, o piercing, machine • [miner] (rock-)boring machine
— f **horizontal** - [ind] horizontal machine
— f **impulsada** - [mec] driven machine
— f **inclinada** - [mec] slanting, o inclined, machine, o engine
— f **inspeccionada** - [mec] inspected machine
— f **integrada,gral** - [mec] self contained machine
— f **inversora** - [mec] reversing machine
— f **lavadora** - [mec] washing machine • scrubber
— f **limpia** - [mec] clean machine
— f **limpiada** - [mac] cleaned machine
— f **limpiadora** - [mec] cleaning machine
— f — **para tubería(s) f** - [tub] pipe cleaning machine • [petról] (pipe) cleaning machine
— f **lingoteadora** - [metal-prod] (pig) casting machine
— f **lingotera** - [metal-prod] pigcasting machine
— f **llenadora** - [ind] filling machine
— f **mantenida** - [ind] maintained machine
— f **más veloz** - faster machine • fastest machine • [deport] faster car; fastest car
— f **mayor** - [ind] larger machine
— f **mejor** - [mec] better machine • best machine • [autom] best, machine, o car
— f **metalizadora** - [metal-fabr] metallizing machine
— f **minadora** - [miner] mining machine
— f **moldeadora** - [metal-fabr] molding machine
— f — **estadounidense** - [plást] American made molding machine
— f —— **para inyección f** - [plást] American made injection molding machine
— f — **horizontal** - [plást] horizontal molding machine
— f —— **para inyección f** - [plást] horizontal injection molding machine
— f — **para inyección f** - [plást] injection molding machine
— f ——— **f para . . . color(es) m** - [plást] . . .-color injection molding machine
— f — **vertical** - [plást] vertical molding machine
— f —— **para inyección f** - [plást] vertical injection molding machine
— f **molduradora** - [mec] mortizing machine
— f **movida intermitentemente** - [mec] jogged machine
— f **múltiple** - [ind] multiple machine
— f **normal** - [ind] normal, o standard, machine
— f **nueva** - [ind] new machine
— f **operada** - [ind] operated, o run, machine
— f **oscilante** - [mec] oscillating machine
— f — **para ataludar** - [constr] oscillating, shaper, o shaping machine
— f **papelera f** - [papel] paper [making] machine

máquina f papelera completa - [papel] complete
paper (making) machine
— f **para acanalar** - [mec] corrugating machine
— f — **aceitar** - [ind] oiling machine
— f — v **con emulsión** f - [ind] emulsion
oiling machine
— f — **achaflanar** - [mec] chamfering, o beveling, machine
— f — **apuestas** f - [juego] slot machine
— f — **ataludar** v - [constr] sloping machine;
sloper
— f — **bandeja(s)** f - [culin] pan machine; machine for @ pan(s)
— f — **bañar** v - [culin] icing machine
— f — **baño** m - [culin] icing machine
— f — m **superior**—[culin] top icing machine
— f — **barrenar** v - [metal-prod] bott mechanism
— f — **bizcocho(s)** m - [culin] biscuit machine
— f — **bola(s)** f - [cuerp.moled] ball machine
— f — f **actual** - [cuerp.moled] present ball
machine
— f — f **nueva** - [cuerp.moled] new ball machine
— f — — f **para molienda** f - [cuerp.moled]
grinding ball machine
— f — — f **vieja** - [cuerp.moled] old ball machine
— f — **cajón(es)** n - [ind] box machine
— f — **calcinación** f—[miner] calcining machine
— f — — f **con chimenea** f **anular** - [miner] annular stack calcining machine
— f — **calcular** - [com] calculator
— f — **cantear** v - [mec] trimming machine
— f — **carga(r)** - [ind] charging machine
— m — **cierre** m - [ind] closing, o capping, machine
— f — — m **de saco(s)** m **[de papel]** - [ind]
(paper) bag capping machine
— f — **cocción** f - [culin] baking machine •
cooking machine
— f — — f **de galletita(s)** f - [culin] cookie
baking machine
— f — **colada** f - [metal-prod] casting machine
— f — — **con** . . . **línea** f **doble** - [col.cont]
double line casting, machine, o unit
— f — — f **con** . . . **línea(s)** f - [col.cont]
. . . line casting, machine, o unit
— f — — f **continua** - [col.cont] continuous,
caster, o casting machine
— f — — f — **con** . . . **línea(s)** f - [col.-
cont] . . . strand continuous casting machine
— f — — f — **de tipo** m **vertical** - [col.cont]
vertical (type) continuous casting machine
— f — — — m — **con** . . .
línea(s) f - [col.cont] . . .-strand vertical
type continuous casting, machine, o unit
— f — — f — **vertical** - [col.cont] vertical
(type) continuous casting machine
— f — — f — **con** . . . **líneas** f - [col.-
-cont] . . . strand vertical continuous casting machine
— f — — f **vertical** - [col.cont] vertical
casting machine
— f — **colar** v - [metal-prod] (pig) casting
machine
— f — — **arrabio** m - [metal-prod] pig casting
machine
— f — — v **lingote(s)** m - [metal-prod] ingot,
o pig, casting machine
— f — . . . **color(es)** m - [mec] . . .-color
machine
— f — **comercio** m - [com] business machine
— f — **comprobación** f - [mec] testing machine
— f — **comprobar presión(es)** f - [ind] pressure
testing machine
— f — **corrugar** v - [ind] corrugating machine
— f — — **chapa(s)** f - [ind] (sheet) corrugating machine
— f — — **lámina(s)** f - [metal-fabr] sheet cor-
corrugating machine
— f — **cortar alambre** m - [ind] wire cutting
machine
— f — — **bizcocho(s)** m - [culin] cookie cutting machine

máquina f para cortar bizcocho(s) m con alambre
m - [culin] wire cutting cookie machine
— f — **corte** m - [mec] cutting, o cut-off, machine
— m — — m **antes de banda** f - [culin] before @
band cutting machine
— m — — m **con alambre** m - [mec] wire cutting
machine
— f — — v — **llama** f - [sold] flame cutting,
o burn(ing) off, machine
— f — **corte** m - [sold] cut(ting), o burn-off,
machine
— f — — m **de galletita(s)** f - [culin] cookie
cutting machine
— f — — m — — f **con alambre** m - [culin]
wire cutting cookie machine
— f — — m **sobre banda** f - [culin] over @ band
cutting machine
— m — — f — — f **con alambre** m - [culin]
over-@-band wire cutting machine
— f — **dolomí(t)a** - [metal-prod] dolomite machine
— f — **elaborar metal(es)** m - [metal-fabr] metalworking machine
— f — **emparedados** m - [culin] sandwich machine
— f — **empuje** m - [coque] véase **deshornadora** f
— f — **encajar tubo(s)** m - [mec] pipe fitting
machine
— f — — — m **sin aterrajar** - [petról] bucking
on machine
— m — **encajonar** v - [ind] boxing machine
— f — **encarrujar** v - [mec] corrugating machine
— f — — v **chapa(s)** f - [metal-fabr] sheet
corrugating machine
— f — **enderezar** v - [mec] straightening machine
— f — — v **barra(s)** f - [mec] bar straightening machine
— f — **engrasar** v - [mec] greasing machine
— f — **enroscar** - [mec] threading machine •
[petról] bucking on machine
— f — **ensayo(s)** m - [mec] testing machine
— f — — m **para rueda(s)** f - [mec-ruedas]
wheel testing machine
— f — — m — — f **con pestaña** f—[mec-ruedas]
flange(d) wheel testing machine
— f — — m — — f — — f **doble**—[mec-ruedas]
double flange(d) wheel testing machine
— f — **escoplear** - [herram] mortising machine;
mortiser
— f — — v **ranura(s)** f - [mec] groove slotting
machine
— f — — v — f **para chaveta(s)** f - [herram]
key groove slotting machine
— f — **escribir** v - [com] typewriter
— f — **escritorio** m - [com] office, o business,
machine
— v — **estampar** v - [mec] stamping machine
— f — — v **bizcocho(s)** m - [culin] biscuit
stamping machine
— f — **expansión** f - [vapor] expansion engine
— f — **extracción** f - [mec] extracting machine
• [miner] winding engine
— f — **fabricación** f - [ind] manufacturing
machine
— f — — f **de cartón** m - [papel] cardboard
manufacturing machine
— f — — f — **papel** m - [papel] paper, manufacturing, o making, machine
— f — — f — **tornillo(s)** m - [mec] screw
(making) machine
— f — **fabricar** v - [ind] manufacturing machine
— f — — v **cartón** m - [papel] cardboard, manufacturing, o making, machine
— f — — v **papel** m - [papel] paper, manufacturing, o making, machine
— f — — v **tornillo(s)** m—[mec] screw (making)
machine
— f — **faceta(s)** f - [metal-prod] fluting machine
— f — **fin(es)** f **especial(es)** - [ind] special
purpose(s) machine

máquina f para fin(es) m múltiple(s) - [mec]
multi-purpose machine
— f — forja(r) - [metal-prod] forging machine
— f̄ — formación f - [mec] forming, o shaping,
machine
— f — fresar v - [herram] milling machine
— f̄ — — v cajera(s) f - [mec] key seating machine
— f — — v extremo(s) n - [mec] end(s) milling
machine
— f — — v — m de riel(es) m - [metal-lam]
rail end milling machine
— f — — v ranura(s) f - [mec] key seating
machine
— f — galvanización f - [metal-trat] galvanizing rig
— f — galletita(s) f - [culin] cookie machine
— f̄ — — f para corte m antes de banda f -
[culin] before @ band cookie cutting machine
— f — granallar v - [metal-lam] shot blasting,
abrader, o machine
— f — — v cilindro(s) m - [metal-lam] cylinder shot blast, abrader, o machine
— f — granulación,lar - [metal-prod] granulating machine
— f̄ — — f de escoria f - [metal-prod] slag
granulating machine
— f — hielo m - ice machine
— f̄ — horadación f - [constr] tunneling machine
— f — horadar v - [constr] tunneling machine
— f̄ — inversión f - [mec] reversing machine
— f̄ — inyección f̄ - [mec] injecting,tion machine
— f — lanzar,zamiento - [ind] throwing machine
— f̄ — — v dolomí(t)a f - [metal-prod] dolomite throwing machine
— f — lavar v - [domést] washing machine
— f̄ para levantar v puerta(s) f - [coque] door
lifting machine
— f — limpieza f - cleaning machine
— f̄ — — f con chorro m de aire m - [ind]
(air) blast cleaning machine
— f — lingote(s) m - [metal-prod] pig casting
machine pig caster
— f — lingotillo(s) m - [metal-prod] pig casting machine; pig caster
— f — metalistería f - [metal-fabr] metalworking machine
— f — moldear v - [metal-prod] molding machine
— f̄ — moldeo m - [metal-prod] pig casting machine • molding machine
— f — — m por inyección f - [plást] injection
molding machine
— f — moldurar f - [mec] molding machine
— f̄ — ondular - [mec] corrugating machine
— f̄ — ordeñar,ñe - [ganad] milking machine
— f̄ — papel m - [papel] paper (making) machine
— f̄ — perforación f - [mec] drilling machine
— f̄ — plástico(s) m - [plást] plastic(s) machine
— f — probar rueda(s) f - [mec-ruedas] wheel
testing machine
— f — — v — f para riel(es) m - [mec-ruedas]
track wheel testing machine
— f — procesamiento m - [ind] processing machine
— f — — m de plástico(s) m - [plást] plas-
tic(s) processing machine
— f — producción f - [ind] production machine
— f̄ — — f de bola(s) f - [metal-prod] ball
(making) machine
— f — prueba(s) f - [ind] testing machine
— f̄ — recalcar v - [metal-fabr] swaging machine; swager
— f — recortar v - [mec] trimming machine
— m — recubrir,rimiento - [culin] topping machine
— f — refrentar,tación - [mec] facing machine
— f̄ — — v extremo(s) m - [mec] end(s) facing
machine
— f — — m — m de tubería f - [petról] tube
ende(s) refacing machine; shoulder dressing

tool(s)
máquina f para refrentar extremo(s) m de tubería
f - [petról] tool joint, refacing machine, o
shoulder dressing tool
— f — refrigeración f - cooling, o ice, machine
— f — rotar v barra(s) f - [metal-lam] bar
rotating machine
— f — sandwich(es) m - [culin] sandwich machine
— f — — m con . . . cabeza(s) f - [culin]
. . ., head sandwich machine
— f — servicio m - [ind] uitility, machine, o
unit
— f — sinterización f - [metal-prod] sintering
machine
— f — sinterizar v - [metal-prod] sintering
machine
— f — skip* m - [metal-prod] skip hoist
— f̄ — soldadura f - [sold] welding machine
— f̄ — — f eléctrica - [sold] electric welding
machine
— f — — f con arco m - [sold] arc welding machine
— f — soldar v - [sold] welding machine •
pl - welding equipment
— f̄ — sopleteado m—[sold] burning off machine
— f̄ — sujeción f - [mec] clamp(ing) machine
— f̄ — taladrar v - [mec] drill(ing) machine
— f̄ — — v piquera f - [metal-prod] taphole
drill(ing machine)
— f — tapar v - [metal-prod] clay, o mud, gun
— f̄ — — v piquera f - [metal-prod] (taphole),
clay, o mud, gun
— f — tarjeta(s) f - [com] card machine
— f̄ — — f de crédito m - [com] credit card
machine
— f — tensión* f alta - [electr-equip] high
voltage machine
— f — tornillo(s) m [ind] screw machine
— f̄ — trefilar v - [trefil] drawing machine
— f̄ — barra(s) f - [trefil] wire drawing
machine
— f — varilla(s) f - [metal-lam] rod machine
— f̄ — — f de acero m - [metal-lam] steel rod
machine
— f — — f — — m con carbono m - [metal-lam]
carbon steel rod machine
— f — — f — hierro m - [metal-lam] iron rod
machine
— f — — f — — m con carbono m - [metal-lam]
carbon iron rod machine
— f — volcar v - [metal-prod] casting machine
• tilt caster
— f — — v cuchara(s) f - [metal-prod] ladle
tilt(ing), machine, o engine
— f — voltaje m alto - [electr-equip] high
voltage machine
— f — — m constante - [sold] fixed, o constant, voltage machine
— f — — m graduable - [sold] adjustable voltage, machine, o welder
— f — zunchar v - [transp] banding machine
— f̄ parada - [ind] stopped, o halted, o idle,
machine
— f pequeña - [ind] small (size) machine
— f̄ perforadora - [herram] perforating, o
drilling, machine
— f — para butano m - [petról] butane drilling machine
— f pesada - [mec] heavy, o weighty, machine
— f̄ — transportada - [transp] transported, o
hauled, heavy machine
— f portátil - [ind] portable, machine, o unit
— f̄ preparada - [ind] prepared machine • re-
tooled machine • [deport] prepared racer
— f preparadora - [ind] preparing machine
— f̄ — para pasta f - [papel] pulper
— f primera - [ind] first, o early, machine
— f̄ proyectada - [ind] designed machine
— f̄ que opera v sobre banda f - [culin] over @
band operating machine
— f reacondicionada - reconditioned machine

máquina f rebabadora

máquina f rebabadora - [metal-prod] scarfing machine
— f recalcadora - [metal-prod] swaging machine
— f recia - [ind] rugged, o sturdy, machine
— f recogedora - [metal-prod] reclaimer
— f recortadora - [ind] trimming machine
— f rectificadora - [electr-equip] rectifier (machine)
— f refrentadora - [herram] facing machine
— f — para extremo(s) m - [mec] end facing machine
— f regulada - [mec] controlled machine
— f remolcada - [ind] towed machine
— f reparadora f - [mec] repair(ing) machine
— f — para barreno(s) m - [petról] bit dresser
— f — — trépano(s) m - [petról] bit dresser
— f roscadora - [tub] threading machine
— f — para tubo(s) m - [tub] pipe threading machine
— f rotativa - [mec] rotating machine • [petról] rotary machine
— f — para corte m - [culin] rotary cutting machine
— f — — — m de galletita(s) f - [culin] rotary cookie (cutting) machine
— f — — — f antes de banda f - [culin] before @ band rotary cookie (cutting) machine
— f — — m — — f sobre banda f - [culin] over-@-band rotary cookie (cutting) machine
— f — — enderezar v riel(es) m - [metal-lam] rotary rail strightening machine
— f — para galletita(s) f - [culin] rotary cookie machine
— f sacapuerta(s) f - [coque] door (removing) machine
— f — convencional - [coque] conventional, o multi-spot, door (removing) machine
— f — en carro m guía (para coque m) - [coque] coke guide door (removing) machine
— f — single-spot - [coque] single-spot door (removing) machine
— f según norma f - [ind] standard machine
— f segura - [ind] safe machine
— f selladora f - [ind] sealing machine
— f semiautomática - [ind] semiautomatic machine
— f sencilla f - [mec] simple machine
— f sin arrancar - [sold] unstarted, o not started, machine
— f soldadora - [sold] welding, machine, o unit • welder
— f — eléctrica - [sold] electrical, welder, o welding machine
— f sujetadora - [mec] clamp(ing) machine
— f taladradora - [herram] drilling machine
— f — para riel(es) m - [herram] rail drilling machine
— f taqueteadora - [electrodos] slug press
— f terminadora - [herram] finishing machine • ending machine
— f texturadora* - [herram] texturing machine
— f torcedora - [mec] twisting machine • [cabl] strander; stranding machine
— f transformadora - [electr-equip] transformer machine
— f — rectificadora - [electr-equip] transformer rectifier machine
— f trazadora - [instrum] plotter
— f trefiladora - [trefil] (wire) drawing machine; wire mill
— f — en continuo - [trefil] continuous drawing machine
— f — para velocidad f alta - [trefil] high speed drawing machine
— f trenzadora - [cabl] stranding machine; strander • layer
— f trilladora - [agric-equip] threshing, machine, o rig; thresher
— f tumbadora - [mec] turning, o dumping, machine
— f ultramoderna - [ind] ultra-modern machine
— f vertical - [mec] vertical machine

máquina f vieja - [ind] old machine
— f volteadora - [ind] turning machine
— f volvedora - [ind] turning machine
— f — para lingote(s) m - [metal-lam] ingot turning machine
— f — — riel(es) m - [metal-lam] rail turning machine
maquinación f - . . . • [ind] véase también mecanización f
maquinar v - . . . • véase también mecanizar v
maquinaria f - [ind] . . .; equipment • [petról] drawworks
— f adquirida - [ind] acquired machinery
— f agrícola - [agric-equip] agricultural, o farm, machinery, o equipment
— f comprada - [ind] purchased machinery
— f conservada - [ind] maintained machinery
— f costosa - [ind] expensive machinery
— f de fundición f - [ind] cast iron, machinery, o equipment
— f — — f gris - [ind] gray cast iron, machinery, o equipment
— f desalineada - [ind] out of alignment machinery
— f desaplomada - [mec] out of plumb machinery
— f disponible - [ind] available machinery
— f empujadora - [ind] pushing machinery
— f en tajo m - [miner] machinery in @ cut
— f encerrada - [mec] enclosed machinery
— f generada - [ind] generated machinery
— f industrial - [ind] industrial, machinery, o equipment
— f instalada - [ind] installed machinery
— f mantenida - [ind] maintained machinery
— f masiva - [ind] massive machinery
— f minera - [miner] mining machinery
— f moderna - [ind] modern machinery
— f — para horno m - [combust] furnace, o oven, modern machinery
— f — — m para coque m - [coque] modern coke oven machinery
— f nacional - [ind] national machinery
— f nueva - [ind] mew machinery
— f operada - [ind] operated machinery
— f para anclaje m - [ind] anchoring machinery
— f — construcción f - [constr] construction machinery
— f — — f de ferrocarril(es) m - [f.c.] railway construction machinery
— f — horno m - [combust] oven, o furnace, machinery
— f — — m para coque m - [coque] coke oven machinery
— f — manejo m - [ind] handling machinery
— f — — m de ceniza(s) f - [ind] ash handling machinery
— f — perforación f - [petról] drilling machinery
— f — plástico(s) m - [plást] plasctic(s) machinery
— f — plataforma f - [petról] rig machinery
— f — — f para perforación f - [petról] drilling rig machinery
— f — procesamiento m - [ind] processing machinery
— f — — m de plástico(s) m - [plást] plastic(s) porocessing machinery
— f — producción f - production machinery
— f — — f de nieve f - [deport] snowmaking, machinery, o equipment
— f — soldar v - [sold] welding, machinery, o equipment
— f — trabajo m - [ind] work(ing) machinery
— f — yacimiento m - [miner] field machinery
— f — — m petrolífero m - [petról] oil field machinery
— f perdida f - [seguros] lost machinery
— f protegida - [ind] protected machinery
— f reacondicionada - [ind] reconditioned machinery
— f segura - [ind] safe machinery
— f superior - [mec] upper machinery • superior machinery

maquinarias f, equipo(s) m, e instalación(es) f -
[ind] machinery, equipment, and installations
maquinista m - . . . . • operator • [f.c.] engine
driver; engineer • véase también operador m;
conductor m
— m para locomotora f - [f.c.] locomotive, o
railway, engineer
— m — motoestibadora f - [ind] fork lift truck
operator
mar adentro a - offshore | adv - out at sea •
into @ ocean
— m Caribe - [geogr] Caribbean Sea
— f de gente - ocean of people
— f en coche adv - ad infinitum
— m subterráneo - underground, sea, o ocean
— m tempestuoso - stormy, o rough, sea
maraña f - . . . ; maze • snag
— f de raíz,ces f - [constr] matted root(s)
maratón f de venta(s) f - [com] sale-a-thon
marbete m - . . . ; nameplate • marker • identification, sticker, o tag • paste-in
— m identificador - identification sticker
— m — para fusible(s) m - [electr-instal] fuse
identification, marker, o sticker
— m negativo - [electr-instal] negative, marker, o sticker
— m para acumulador—[electr-bat] battery label
— m — conmutador m - [electr-instal] switch,
nameplate, o marker
— m — — m para régimen m - [electr-instal]
rate switch nameplate
— m — — m — — m para carga f - [electr-
-instal] charging rate switch nameplate
— m — disyuntor m - [electr-instal] breaker,
nameplate, o marker
— m — equipo m - [ind] equipment, nameplate, o
marker
— m — — m automático - [sold] automatic
equipment, nameplate, o marker
— m — fuente f para energía f - [sold] power
source, nameplate, o marker
— m — fusible m - [electr-instal] fuse, nameplate, o identification, marker, o sticker
— m — identificación f - identification, nameplate, o marker, o sticker, o tag
— m — — f para bulto m - [transp] pack(age)
identification, nameplate, o marker, o tag
— m — interruptor m - [electr-instal] breaker,
nameplate. o marker
— m — — m para arranque m - [comb.int]
start(ing) switch, nameplate, o marker
— m — producto m - [ind] product label
— m positivo - [electr-instal] positive, marker, o sticker
marca f - . . . ; marker; marking; check • streak
• notation • [mec] furrow • [comb.int] bead •
[deport] record • véase también marcación f
— f a tierra f - [electr-instal] to ground
mark(ing)
— f absoluta - [deport] absolute record
— f alta - high mark
— f — en varilla f - [mec] rod, o stick, o
gage, high mark
— f — en varilla f medidora - [mec] dipstick,
o gage, high mark
— f aprobada - [com] approved mark
— f baja - low mark
— f baja en varilla f - [mec] rod, o stick, o
gage, low mark
— f — — f medidora - [mec] dipstick, o
gage, low mark
— f bien conocida - [com] well known brand
— f clara - clear mark(ing)
— f con clarioncillo m - [dib] grease pencil
mark
— f — punzón m - [mec] punch mark
— f — — m alineada - [mec] aligned punch mark
— f — símbolo m - symbol mark(ing)
— f conocida - [com] known brand
— f — nacionalmente - [com] nationally known
brand
— f — agua m - [papel] watermark
— f de fábrica f - [legal] trademark; trade

name • [ind] manufacturer(s) brand
marca f de calidad f - grade marking; hallmark
— f — camión m - [autom] truck, make, o brand
— f — cuchara f - [hormig] trowel mark
— f — ensambladura f - jointing mark
— f — fábrica f - [ind] brand (name)
— f — — f comprada - [legal] purchased trade
mark
— f — — con letra(s) f blanca(s) - [ind]
white trademark
— f — — f — — f hundida(s) - [ind] inset
white trademark
— f — — f — — f hundida(s) - [ind] inset
trademark
— f — — f — — f negra(s) - [ind] black
trademark
— f — — f — — f hundida(s) - [ind] inset
black trademark
— f — — f de fabricante m - [ind] manufactu-
rer('s) trademark
— f — — f de material m - [ind] material
trade
— f — fabricación f - [ind] trademark
— f — — f comprada - [legal] purchased trademark
— f — — f creada - [legal| developed trade-
mark
— f — — f poseida - [legal] owned trademark
— f — — f registrada - [legal] registered
trademark
— f — — f retenida - [legal] held trademark
— f — — f usada - [legal] used trademark
— f — fabricante m - [ind] manufacturer('s),
trademark, o brand
— f — freno m - [mec] brake, make, o brand
— f — garantía f - [ind] hallmark
— f — infinidad f - [instrum] infinity reading
— f — ítem m - item check(ing)
— f — ladrillo m - [metal-lam] brick mark
— f — límite m - limit mark(ing)
— f — lleno - full mark
— f — — en medidor m - [instrum] gage full
mark
— f — — — — m para aceite m - [instrum]
oil gage full mark
— f — — — varilla f - [comb.int] gage full
mark
— f — — — — f medidora - [comb.int] gage, o
dipstick, full mark
— f — material m - [ind] material trademark
— f — medida f - measure mark(ing)
— f — mensura f - measure mark(ing)
— f — mordaza f - [mec] clamp (bite) mark
— f — motor m - [comb.int] engine, mark, o
make
— f — neumático m - [autom-neumát] tire brand
— f — paleta f - [hormig] trowel mark
— f — pureza f - hallmark
— f — refractario(s) m - [cerám] refractory,ries, brand, o make
— f — tipo m - [ind] type mark(ing)
— f — vacío m - empty mark
— f — velocidad f - speed record
— f — vibración f - [metal-trat] chatter mark
— f destruida - destoyed mark
— f distintiva - distinctive mark • brand
— f en alambre m - [mec] wire mark
— f — borde m de volante m - [comb.int] flywheel rim mark(ing)
— f — — m — — m para avance m para regulación f de encendido m - [Comb.int] flywheel
rim advance timing mark
— f — borne m - [sold] stud mark(ing)
— f — cárter m - [comb.int] crankcase marking
— f — — m para regulación f para encendido m
- [comb.int] crankcase timing mark
— f — cilindro m - [metal-lam] rolled-in mark
— f — collar m - [mec] collar mark
— f — diente m - [mec] tooth mark(ing)
— f — medidor m - [mec] gage mark
— f — — m para aceite m - [instrum] oil gage
mark
— f — pulgada(s) f - inch mark

marca f en tinta f - ink mark(ing)
— f en tubo m - [tub] pipe mark
— f en varilla f - [mec] rod mark(ing) •
[comb.int] dipstick mark(ing)
— f — — f medidora - [comb.int] dipstick, o
(bayonet) gage, mark(ing)
— f — volante m—[comb.int] flywheel mark(ing)
— f — — m para avance m para regulación f para encendido m - [comb.int] flywheel advance
timing mark
— f eclipsada - eclipsed mark
—/espacio f - [electrón] mark/space
—/— f activada - [electrón] gated mark/space
— f europea - [ind] European brand
— f indeleble - indelible mark(ing)
— f industrial - [legal] trade mark
— f — adquirida - [legal] acquired, o purchased, trade mark
— f — comprada - [legal] purchased trade mark
— f — creada - [legal] developed trade mark
— f — gestionada - [legal] applied for trade
mark
— f — obtenida - [legal] obtained trade mark
— f — poseída - [legal] owned trade mark
— f — reconocida - [ind] recognized trade mark
— f — registrada—[legal] registered trademark
— f — retenida - [legal] held trade mark
— f — solicitada - [legal] applied for trade
mark
— f — usada - [legal] used trade mark
— f — vendida - [legal] sold trade mark
— f inferior - [instrum] lower reading
— f legible - legible mark(ing)
— f mejor - better mark(ing) • [com] better
brand • [deport] better record
— f mejorada - [deport] bettered, o broken,
record
— f para alineación f - [mec] boss; alignment
mark
— f — apareamiento m - [mec] match mark(ing)
— f — automóvil(es) m de serie - [deport]
stock car record
— f — — m - stock - [deport] stock car record
— f — avance m - [comb.int] advance mark(ing)
— f — — m para regulación f de encendido m -
[comb.int] advance timing mark
— f — categoría f - [deport] class record
— f — contraste - hallmark
— f — ensayo(s) m - test mark(ing)
— f — guía - [mec] guide mark(ing)
— f — identificación f - identification mark-
-(ing) • trade mark
— f — — f de material m - [ind] material
identifying trade mark
— f — referencia f - [mec] reference, o bench,
mark
— f — regulación f para encendido m - [comb.-
-int] timing mark(ing)
— f — — f — — f sobre engranaje m de cigüeñal m - [comb.int] crankshaft gear timing
mark
— f — seguridad f - [segurid] safety record
— f — tierra f - [electr-instal] to ground
marking
— f puesta - check; placed mark
— f punzada - punch mark
— f reconocida - recognized mark • [ind] reputable brand
— f registrada - [legal] (registered) trade
mark • brand name
— f — adquirida - [legal] acquired, o purchased, trade mark
— f — comprada - [legal] purchased trade mark
— f — conocida - known, trade mark, o brand
— f — — nacionalmente - [com] nationally
known, trade mark, o brand
— f — creada - [legal] developed trade mark
— f — gestionada - [legal] applied for trade
mark
— f — obtenida - [legal] obtained trade mark
— f — por fabricante m - [ind] manufacturer's
trade mark

marca f registrada poseída - [legal] owned trade
mark
— f — retenida - [legal] held trade mark
— f — solicitada - [legal] applied for trade
mark
— f — usada - [legal] used trade mark
— f — vendida - [legal] sold trade mark
— m rumbo(s) - pacesetter
— m — en tenología f - [ind] technology pacesetter
— m — — f alta - [ind] high technology
pacesetter
— f y modelo m - [ind] make and model
marcación f - . . . . • marking • registering •
checking; posting • [ind] label(l)ing; branding • [metal-trat] (sk)etching
— f clara - clear mark(ing)
— f con fuego m - branding
— f — punzón m - [mec] punch marking
— f — símbolo m - symbol marking; marking
with @ symbol
— f continua - continuous marking
— f de artículo m - article marking • item
marking
— f — bola f - [cuerp.moled] ball marking
— f — calidad f - [ind] quality marking; hallmarking
— f — carril(es) m - [metal-lam] rail marking
— f — contraste m - [ind] hallmarking
— f — curso m - course marking
— f — desbaste(s) m—[metal-lam] bloom marking
— f — dígito(s) m - [comput] digit(s) marking
— f — equipo m - [ind] equipment marking
— f — garantía f - guaranty marking • hallmarking
— f — huella(s) f - [mec] groove marking
— f — ítem(es) f - item marking • item checking
— f — lata(s) f - [ind] can marking
— f — límite(s) m - limit(s) marking
— f — material(es) m - [ind] material(s) marking
— f — medida(s) f - measure(ment)s marking
— f — mensura(s) f - measurement(s) marking
— f — número(s) m - number(s) marking •
[comput] number entering
— f — perfil(es) m - [metal-lam] shape(s)
marking
— f — — m estructural(es) - [metal-lam]
structural(s) (shape(s) marking
— f — presión f - [mec] - pressure marking •
pressure dialing
— f — pureza f - hallmarking
— f — neumático(s) m - [autom-neumát] tire(s),
marking, o branding
— f — riel(es) m - [metal-lam] rail(s) marking
— f — — m en relieve m - [metal-lam] rail(s)
(relief) marking
— f — tipo m - [ind] type mark(ing)
— f — vía(s) f—[vial] lane marking; stripping
— f — — f para circulación f - [vial] lane
marking; stripping
— f destruida - destroyed marking
— f en cabeza f de perno m - [mec] bolt head
marking
— f — pared f - [autom-neumát] wall marking
— f — — f lateral - [autom-neumát] sidewall
marking
— f — superficie f - [metal-lam] surface, o
ragging, mark
— f — tubo(s) m - [tub] tube, o pipe, marking
— f entre operación(es) f - [ind] marking between operation(s)
— f existente - existing marking
— f indeleble - indelible marking
— f ininterrumpida - uninterrupted, o continuous, marking
— f legible - legible marking
— f numerada - numbered marking
— f — en secuencia - consecutively, o se-
quentially, numbered mark(ing)s
— f para apareamiento m - match(ing) mark(ing)s
— f — categoría f - category marking
— f — corte m - cut marking; sketching

marcación f para identificación f - [ind] identification marking
— f — regulación f - [mec] control marking
— f — para encendido m - [comb.int] timing
marking
— f — velocidad(es) f - [autom-neumát] speed
marking
— f por categoría f para velocidad(es) f -
[autom-neumát] speed category marking
— f punzada - [mec] punched mark(ing)
marco m recio - [mec] rugged frame
— m — de viga(s) f acanalada(s) conformada(s)
- [mec] rugged formed channel frame
— f verificada - [ind] checked marking
marcadamente adv - . . .; sharply • dramatically
marcado* m - véase marcación f
marcado,da a - . . . . • distinct; pronounced •
registered • checked; posted • label(l)ed;
branded; tagged; stamped • dramatic; outstanding
— a a tierra f - [electr-instal] marked to @
ground
— a apareadamente - [mec] match marked
— a claramente - marked clearly
— a con fuego m - branded
— a — punzón m - [mec] punch marked
— a — símbolo m - symbol marked; marked with @
symbol
— a en forma legible - legibly marked
— a — pared f - wall marked
— a — — f lateral - [autom-neumátt] sidewall,
marked, o branded
— a indeleblemente - marked indelibly
— a para apareamiento m - [mec] match marked
— a para tierra f - [electr-instal] marked to @
ground
marcador m - . . .; checker; flag; véase también
marcación f • [deport] scoreboard
— m con disco(s) m - [mec] disk marker
— m en caliente - [metal-lam] hot marker
— m — — en laminación f - [metal-lam] rolling
hot marker
— m — laminación f—[metal-lam] rolling marker
— f — mesa f - [mec] table marker
— m — — f para inspección f - [ind] inspection table marker
— m paralelo - parallel marker • [instrum] surface gage
marcador,ra a - marking • marring
marcar v - . . . . • to point (out); to pinpoint •
to register • to check; to post • [telecom]
to dial
— v a tierra f - [electr-instal] to mark to @
ground
— v amperaje m - [sold] to dial @, amperage, o
current
— v — m deseado - [sold] to dial @ desired,
amperage, o current
— v apareadamente - [mec] to match mark
— v artículo m - to mark @, article, o item
— v bola f - [cuerp.moled] to mark @ ball
— v borne m - [sold] to mark @ stud
— v cabeza f - [mec] to mark @ head
— v — f de perno m - [mec] to mark @ bolt head
— v calidad f - [ind] to mark @, quality, o
grade
— v claramente - to mark clearly
— v con fuego - to brand
— v — punzón m - to punch @ mark
— v — símbolo m - tp symbol mark; to mark with
@ symbol
— v curso - to, mark, o map, @ course
— v diente m - [mec] to mark @ tooth
— v dígito m - [comput] to enter @ digit
— v en pared f - to mark on @ wall
— v — — f lateral - [autom-neumát] to sidewall brand
— v — relieve - to relief mark • to brand
— v equipo m - [ind] to mark @ equipment
— v indeleblemente - to mark indelibly
— v ítem m - to, mark, o check, @ item
— v límite m - to mark @ limit
— v material - [ind] to mark @ material

marcar v medida f - to measure mark; to mark @
measurement
— v mensura f - to measure mark
— v neumático m - [autom-neumát] to, mark, o
brand, @ tire
— v número m - [telecom] to dial @ number •
[comput] to enter @ number
— v — en secuencia f - to consecutively mark
@ number
— v para apareamiento m - [mec] to match mark
— f cabeza f - [mec] to mark @ head
— v — f de perno m - [mec] to mark @ bolt head
— v — tierra f - [electr-instal] to mark to
ground
— v paso - to keep step • to pace
— v presión f - [mec] to dial @ pressure
— f — f deseada - [mec] to, mark, o dial, @
desired pressure
— v regulación f para encendido m - [comb.int]
to mark @ timing
— v rumbo(s) m - to show @ way • to pioneer •
to (set @) pace • to set @ standard
— v tipo m - [ind] to mark @ type
— v vías f - [vial] to mark @ lane(s); to
stripe
— v — f para circulación f - [vial] to mark @
lane(s); to stripe
marcarrumbos* m - pacesetter
marco n - . . . .; framework • setting • border •
[constr] . . .; buck; mounting; cradle
— m con celosía f - [constr] lattice frame
— m de acero m - [mec] steel frame
— m — — m estructural - [constr] structural
steel frame
— m — — m prensado -- [constr] pressed steel
frame
— m — alambre m - [mec] wire frame
— m — madera f - [carp] wood buck
— m — metal m - [constr] metal frame
— m — — m prensado - [constr] pressed metal
frame
— m — Mercado m Andino - [pol] Andean Pact
framework
— m — viga(s) f - [mec] beam frame
— m — — f acanalada(s) - [mec] channel (beam)
frame
— m — — f — conformada(s) - [mec] formed, o
shaped, channel (beam) frame
— m económico - [econ] economic frame(work)
— m empotrado - [constr] wall thimble - [hidr]
thimble
— m — de hierro m - [hidr] iron wall thimble
— m — — m fundido - [hidr] cast iron wall
thimble
— m — — m — con níquel m - [hidr] Ni-Resist
cast iron wall thimble
— m — en pared f - [hidr] wall thimble
— m enrejado - [constr] lattice(d) frame
— m favorable - favorable setting
— m guía - [mec] guide frame
— m laminar - [electr-equip] lamination
— m lateral - [estruct] side frame
— m metálico - [mec] metal(lic) frame
— m para compuerta f - [hidr] gate frame
— m — excitador m - [electr-mot] exciter frame
— m para fijar armazón m para compuerta f -
[hidr] gate thimble
— m — petaca f - [metal-prod] cooling plate
frame
— m — piquera f - [metal-prod] iron notch, o
taphole, frame
— m — clasificación f - classification framework
— m — política f - [econ] policy framework
— m — — f económica - [econ] economic policy
framework
— m — puerta f - [constr] door, frame, o buck
— m — resorte(s) m - [mec] spring frame
— m — tapicería f - [mec] upholstering frame
— m — tejido m - [constr] (wire) screen frame
— m — — m de alambre m - [constr] wire screen
frame
— m — tobera f - [metal-prod] tuyere frame

marco m para toberón m—[metal-prod] (tuyere) cooler frame
— m que abarca (toda) empresa f (íntegra) - [admin| organization wide framework
— m protector - [mec] protective frame
— f para caja f para refrigeración f - [metal-prod] cooling plate protective frame
— m recio - [mec] rugged frame
— m reforzado - [mec] reinforced, o heavy duty, frame, o thimble
— m empotrado - [constr] heavy duty wall thimble
— m residencial - residential setting
— m rígido - [mec] rigid frame
— m tranquilo - tranquil, o restful, setting
— m útil - useful framework
— m vaciado - [constr] (poured) thimble
— m vaciado para armzón m para compuerta f - [hidr] gate thimble
marcha f - . . . . • operation; process • [mec] gear; véase también velocidad f • [autom] ride • [deport] pace
— f abierta - open speed
— f acelerada - accelerated speed
— f — en vacío - [mec] high idle speed
— f — sin carga f - [mec] high idle speed
— f acortada - [metal-prod] reduced blast
— f alta - [mec] high, speed, o setting
— f analítica - [quím] analytic progression
— f analítica actualizada - [quím| updated analytic progression
— f atrás - [autom-mec] reverse (speed)
— f aminorada - reduced, o slowed (down) speed
— f atrás - [mec] reverse (speed, o gear)
— f — lograda eléctricamente - [mec] electrically accomplished reverse (speed)
— f cambiada - [mec] shifted, gear, o speed
— f con velocidad f alta - [autom] high speed travel
— f — — f baja - [autom] low speed travel
— f — — f lenta f - [autom- slow speed travel
— f — — f — sin carga f - [comb.int] low idle speed, running, o operation
— f — — f plena - full speed running
— f — — f reducida - [autom] low speed, travel, o operation
— f de análisis m - [quím] analysis,ses progress
— f — — f por vía f húmeda - [quím] wet process analysis progress
— f — — m — f química - [quím] chemical process, analysis progress, o progress analysis
— f — — m — — f química de nitrógeno m - [quím] nitrogen chemical process, analysis progress, o progress analysis
— f — — m — — f — nitruro m - [quím] nitride chemical process, analysis progress, o progress analysis
— f — espera f - [mec] véase marcha, lenta, o sin carga
— f — horno m - [ind] furnace operation
— f — motor m - [comb.int] engine, operation, o running • [electr-mot] motor, operation, o running
— f — taller m - [ind] shop operation
— f — vía f química - [quím] chemical process progress
— f elevada - [metal-prod] high operating rate
— f en seco—dry, operation, o running
— f — vacío - [mec] idling
— f — — m desigual - [mec] uneven, o rough, idling
— f — — m — de motor m - [comb.int] uneven, o rough. engine idling • [electr-mot] uneven, o rough, motor idling
— f fácil - easy, operation, o pace, o speed
— f gradual - [mec] jog(ging)
— f irregular - irregular, operation, o

speed • [comb.int] surging
marcha f lenta - idle,dling (speed)
— f — de motor m - [comb.int] engine idling • [electr-mot] motor idling
— f — desigual - [mec] uneven, o rough, idling
— f — — de motor m - [comb.int] rough, o uneven, engine idling • [electr-mot] rough, o uneven, motor idling
— f — prolongada - [mec] prolonged idling
— f — sin carga f - [mec] low idle speed
— f — suave - [mec] smooth idling
— f — — de motor m - [comb.int] smooth engine idling; [electr-mot] smooth motor idling
— f máxima - [mec] high setting
— f normal - [mec] normal operation
— f pareja - [deport] steady pace
— f plena - full speed • full throttle • full swing
— f por gravedad - [mec] coasting
— f — impulso m propio - [mec] coasting
— f — inercia f - [mec] coasting
— f — —,a velocidad f menos que media, o a menos que media velocidad - [mec] coasting below half speed
— f — vía f húmeda - [quím] wet process progression
— f — — f química - [quím] chemical process progression
— f silenciosa - [mec] quiet operation
— f sin carga f - [mec] idling
— f — — f de bomba f para aceite m - [comb-int] oil pump idling
— f — — motor m - [comb.int] engine idling • [electr-mot] motor idling
— f — — f lenta - [mec] slow idling
— f — — f rápida - [mec] fast idling
— f — — f suave - [mec] smooth idling
— f — — f — de motor m - [comb.int] smooth engine idling • [electr-mot] smooth motor idling
— f suave f - [mec] smooth operation • [autom] smooth ride,ding
marchado,da a - [mec] run; operated
— a con velocidad f lenta - run at @ idle speed
— a — — f — sin carga f - [mec] run at @ low idle speed
— a en seco - [mec] run, o operated, dry
— a por inercia - [mec] coasted
marchando adv - [mec] operating; running
marchar v - . . . .; to operate • to come
— v como reloj m - to go like clockwork
— v con marcha f lenta - [mec] to, operate, o run, at @ (s)low speed
— v — velocidad f lenta - to run at @ (s)low speed
— v — — f — sin carga - [mec] to run at @ (s)low idle speed
— v — — f plena - [mec] to, run, o operate, at full, speed, o throttle
— v en seco - [mec] to run dry
— v gradualmente - [mec] to, jog, o inch, (along)
— v hacia adelante - [mec] to drive forward
— v irregularmente - [comb.int] to surge
— v ladrillo(s) m de boca - [metal-prod] to lose @ tap hole brick(s)
— v motor m - [comb.int] to, run, o operate, @ engine • [electr-mot] to, run, o operate, @ motor
— v — m con marcha f lenta - [comb.int] to. run, o operate, @ engine at low speed • [electr-mot] to, run, o operate, @ motor, at low speed
— v — — — f plena - [comb.int] to, run, o operate, @ engine at full speed • [electr-mot] to, run, o operate, @ motor at full speed
— v por impulso m propio - [mec] to coast
— v — inercia f - [mec] to coast
— v — f con velocidad f menos que media - [mec] to coast below half speed
— v — gravedad f - [mec] to coast
— v sin presión f - to operate without pressure

marchar v sin presión f alta - [mec] to operate
without @ high pressure
— v — — f — en tragante m - [metal-prod] to
operate without throat (high) pressure
— v velozmente - to go quickly • to buzz along
— v ventilador m - to operate @ fan
marea f alta - [mar] . . .; flow
marejada f - . . . • surge
marga f - [suelos] . . . • [geol] chalk
— f arcillosa - [suelos] cleay(ey) loam
margen m - . . .; rim • ratio • [hidr] bank;
shore • [deport] edge
— m alto de rentabilidad f - [econ] high pro-
fit(ability) ratio
— m amplio - wide, o large, o generous, margin
— m — para seguridad f - [segurid] wide, o
large, o gencrous, safety margin
— m — para sobrecarga(s) f - generous overload
capacity
— m bajo - low margin • low ratio
— m — de rentabilidad f - [econ] low pro-
fit(ability) ratio
— m de cauce m - [hidr] channel bank
— m — m para arrimo m - [hidr] approach
channel bank
— m — m protegido - [hidr] protected channel slope
— m — dumping - [econ] dumping margin
— m — preferencia f - preference margin
— m — rentabilidad f - [econ] profit(ability)
ratio
— m — victoria f - victory margin
— f derecha - [hidr] right bank
— m excepcional - exceptional margin
— m excepcional para seguridad f - exceptio-
nal(ly large) safety margin
— m exiguo - [econ] low ratio
— m — de rentabilidad f - [econ] low profitability ratio
— m grande - wide margin
— m — para seguridad f - wide safety margin
— m izquierdo - [hidr] left bank
— m para seguridad f - safety, margin, o factor
• freeboard
— m — — amplio - wide, o large, safety margin
— m — — f excepcionalmente amplio - exceptionally, wide, o large, safety margin
— m — — f liberal - liberal safety margin
— m — sobrecarga(s) f - overload capacity
— m protegido - [hidr] protected bank
marginal a - . . .; fringe; border(line)
marina f - . . . • [milit] navy
— f de guerra f - [milit] (war) navy
— f mercante - [nav] merchant, navy, o marine
marino,na a - . . . . • off shore; véase también
costa afuera
mariposa f - . . . • [mec] butterfly (valve) •
[comb.int] throttle body
— f normal - [valv] normal butterfly (valve)
— f — para venturi m - [metal-prod] normal
venturi butterfly (valve)
— f para acelerador m - [comb.int] throttle
shaft
— f — Askania f - [metal-prod] Askania butterfly (valve)
— f — — f para emergencia f - [metal-prod]
[metal-prod] Askania emergency buttterfly
(valve
— f — — f — — f para Venturi m - [metal-
-prod] Venturi Askania emergency butterfly
(valve]
— f — carburador m - [comb.int] carburetor
throttle
— f — — emergencia f - [metal-prod] emergency
(line) butterfly (valve)
— f — lavadora f - [metal-prod] washer, o
scrubber, butterfly (valve)
— f — nivel m - [metal-prod] level butterfly
(valve)
— f — — m para agua m - [metal-prod] water
level butterfly (valve)
— f — m — — m en Venturi m - [metal-prod]
Venturi water level butterfly (valve)

mariposa f para Venturi - [metal-prod] Venturi
(washer) butterfly (valve)
marisma f - [hidr] . . .; tidal marsh
marítimo,ma a - . . .; marine; ocean
marlo m - [botán] cob
marmita f para alquitrán m - [constr] tar kettle
— f — succión f - [petról] suction pot
— f — válvula f - véase taza f para válvula f
marquesina f - [constr] . . .; (free standing)
canopy
— f autosoportante - [constr] free-standing
canopy
marrón a gris - gray-brown
martelina f - [herram] . . .; bush hammer
martellinado m - bush hammering
martensita* f - [metal] martensite
martensitación* f - [metal-prod] martempering
martensítico,ca* a - [metal] martensitic
martillado m - [mec] (hammer) driving
— m con perdigón(es) m - [metal-trat] shot
peening
— m de separador m - separator, o divider, hammering, o peening • set driving
martillado,da a - [mec] hammered; pounded; hammer driven; driven; peened; tapped • planished*
— a con perdigón(es) m - [metal-trat] shot
peened
— a levemente - [mec] tapped (lightly)
— a moderadamente - [mec] hammered moderately
martillar v - [mec] . . .; to peen; to pound; to
hammer drive • to tap • to planish*
— v acabadamente - [sold] to peen thoroughly
— v aleta f - [mec] to hammer @ lug
— v con perdigón(es) m - [metal-trat] to shot
peen
— v extremo m - [mec] to peen @ end
— v — m de perno m - [mec] to peen @ bolt end
— v levemente - [mec] to tap (lightly)
— v moderadamente - [mec] to hammer moderately
— v punto m de soldadura f - [sold] to peen @
tack
— v separador m - [mec] to drive @, divider, o
set
martilleo m - [mec] . . .; pounding; peening;
tap(ping) • planishing*
— m de aleta f - [mec] lug hammering
— m — extremo m - [mec] end, hammering, o
peening
— m — m de perno m - [mec] bolt end, hammering, o peening
— m leve - [mec] (s(light) tapping
— m moderado - [mec] moderate hammering
martillo m - [herram] . . .; mallet; sledge
— m blando - [herram] soft hammer
— m cincel(ador) m - [herram] chipping hammer
— m — neumático - [herram] pneumatic chipping
hammer
— m con cabeza f blanda - [herram] soft-faced,
o soft-headed, hammer
— m — — f de metal m antifricción - [herram]
Babbitt hammer
— m — — f doble - [herram] peen(ing) hammer
— m — cotillo m de metal m blanco - [herram]
Babbit (striking face) hammer
— m corredizo - [herram] slide hammer
— m — extractor - [herram] slide hammer puller
— m de cuero m sin curtir - [herram] rawhide
hammer
— m — metal m blanco - [herram] Babbit hammer
— f — vapor m - [mec] steam hammer
— m estampador - [herram] stamping hammer
— m giratorio - [herram] swinging hammer
— m grande - [herram] large hammer
— m — de madera f - [jerram] maul
— m mecánico - [herram] monkey hammer
— m neumático - [herram] pneumatic, o air, hammer • jack hammer
— m — liviano - [herram] light, pneumatic, o
air, hammer
— m — pesado - [herram] heavy, pneumatic, o
air, hammer
— m para acanalar v - [herram] creasing hammer

martillo m̲ para caída f̲ libre—[mec] drop hammer
— m̲ para calafatear,teo - [herram] caulking hammer
— m̲ — cincelar,ladura - [herram] chipping hammer
— m̲ — desbaste,tadura - [herram] cogging hammer
— m̲ — estampar,padura f̲ - [herram] stamping hammer
— m̲ para estriar - [herram] creasing hammer
— m̲ — forja,jar,jadura - [herram] forging hammer
— m̲ — mecánico m̲ - [herram] machinist's hammer
— m̲ — picar - [sold] (chipping) hammer
— m̲ — realce,lzar - [herram] embossing hammer
— m̲ — reborde(s) m̲ - [herram] seam hammer • creasing hammer
— n̲ — retacar - [herram] caulking hammer
— m̲ — trituradora f̲—[constr] (crusher) hammer
— m̲ perforador - [constr] jack hammer
— m̲ pesado - [herram] heavy hammer
— m̲ picador - [herram] jack hammer • pneumatic, o̲ air, hammer
— m̲ pilón - [mec] forging press
— m̲ — por vapor m̲ - [constr] steam hammer
— m̲ portabroca(s) - [herram] drill chuck
— m̲ portapeine(s) - [herram] chaser holder hammer
— m̲ suave - [herram] soft hammer
— m̲ triturador - [herram] breaking, o̲ crushing, hammer
Martin m̲ - [metal-prod] véase Siemens-Martin
martinete m̲ - [mec] . . .; ram - [constr-pil] (pile) driving rig; pile driver; rig; pile hammer; driver; pile driving hammer; driver • [ind] drop press
— m̲ con efecto m̲ doble - [Constr] double acting hammer
— m̲ corriente - [constr] standard rig (and driving hammer)
— m̲ — para hincadura f̲ - [constr-pil] standard (pile) driving hammer
— m̲ — — f̲ para pilote(s) m̲ - [constr-pil] standard pile driving hammer
— n̲ diesel m̲ - [constr-pil] diesel (pile driving) hammer
— m̲ estándar (para hincadura f̲) - [constr-pil] standard pile driving hammer
— m̲ — — — f̲ para pilote(s) m̲ - [constr-pil] standard pile driving hammer
— m̲ forjador - [herram] drop press
— f̲ — con efecto m̲ doble - [constr-equip] double acting pneumatic hammer
— m̲ — para hincadura,car - [constr-pil] pneumatic (pile) driver
— m̲ — — para pilote(s) m̲ - [constr-pil] pneumatic pile driver
— m̲ — forja,jar,jadura - [ind] forge hammer
— m̲ para hincar,cadura - [constr-pil] (pile) driving hammer
— m̲ — — f̲ para pilote(s) m̲ - [pil] pile driving, hammer, o̲ rig
— m̲ para vapor - [constr] steam, hammer, o̲ (pile) driver
— m̲ — — m̲ para hincadura f̲ - [constr] steam pile driver
— m̲ por caída - [constr-equip] drop hammer
— m̲ — gravedad f̲ - [constr-equip] drop hammer
— m̲ tipo m̲ báscula - [mec] helve hammer
más adv - . . .; further
— adv a menudo - most often • oftener
— adv abajo - below • later • hereinafter
— adv acarreo m̲ - [transp] plus, cartage, o̲ hauling, o̲ drayage
— adv — m̲ de container* m̲ - [transp] plus container drayage
— adv acercado,da (entre sí) - closer together
— adv adelantado,da - foremost
— adv adelante - later • below • later on • further front
— adv alejado,da - farthest (away)
— adv almacenamiento m̲ - [ind] plus storage •

more storage
más adv alto,ta - higher • highest
— adv — que arco m̲ - [sold] higher than @ arc
— adv — — indicado,da - higher than, given, o̲ listed
— adv allá - past • beyond
— adv — que sello m̲ - [mec] beyond, o̲ past, @ seal
— adv allegado,da (entre sí) - closer together
— adv amplio,plia - broader • more detailed
— adv apartado,da - farther apart
— adv aproximado,da - nearest; closest
— adv áspero,ra - rougher • harsher
— adv atrás - farther back • farthest back
— adv bajo,ja - lower • lowest
— adv — que indicado,da - lower than listed
— adv — — soldadura f̲ - [sold] lower than @, weld, o̲ work
— adv barato,ta - cheaper; less, costly, o̲ expensive
— adv bien - rather • instead
— adv breve - briefer; shorter • briefest; shortest
— adv caja f̲ para exportación f̲ - [transp] plus export packing case
— adv calor m̲ - more heat
— adv camión m̲ - [autom] more truck • [transp] plus truck(ing)
— adv — m̲ terrestre - [transp] plus inland truck(ing)
— adv cargo m̲ máximo por acarreo m̲ - [transp] plus maximum, cartage, o̲ haulage, charge
— adv — m̲ mínimo—[transp] plus minimum charge
— adv — — — por acarreo m̲ [transp] plus minimum, cartage, o̲ haulage, charge
— adv — m̲ por almacenaje m̲ - [ind] plus storage charge
— adv — m̲ — empaque m̲ - [transp] plus packing charge
— adv — m̲ — — m̲ para exportación f̲—[transp] plus export packing charge
— adv — m̲ — empleo m̲ de containers*—[transp] plus @ containerization charge
— adv — m̲ — entrega f̲ - [ind] plus delivery charge
— adv — m̲ — — f̲ en muelle m̲ - [transp] plus pier delivery charge
— adv — m̲ — estiba f̲ - [ind] plus stowing charge
— adv — m̲ — — f̲ en dos pisos m̲ - [ind] plus double stowing charge
— adv — m̲ — expedición f̲ - [transp] plus @ shipping charge
— adv — m̲ — — f̲ aérea - [transp] plus air (shipment) charge
— adv — m̲ expreso m̲ aéreo - [transp] plus air express charge(s)
— adv — m̲ — flete m̲ - [transp] plus freight charge(s)
— adv — m̲ — — m̲ por camión m̲ (terrestre) - [transp] plus inland truck freight charge
— adv — m̲ — — m̲ — expedición f̲ terrestre - [transp] plus inland freight forwarding charge(s)
— adv — m̲ — flete m̲ - [transp] plus freight charge(s)
— adv — m̲ — — m̲ terrestre - [transp] plus inland freight charge(s)
— adv — m̲ — paquete m̲ postal - [transp] plus parcel post (shipping) charge(s)
— adv — m̲ — recolección f̲ - [transp] plus pick-up, fee, o̲ charge(s)
— adv — m̲ — — f̲ con camión m̲ - [transp] plus truck pick-up fee
— adv — m̲ — uso m̲ de container* m̲ - [transp] plus @ container(ization), charge, o̲ fee
— adv — m̲ — vía f̲ aérea - [transp] plus @ air mail, charge, o̲ fee
— adv carrete m̲ - [ind] plus reel
— adv cerca (de) - closer (to)
— adv — de pieza f̲ - [mec] nearer @ part
— adv — — f̲ para soldar(se) v̲ - [sold] closer to @, job, o̲ work

más <u>adv</u> **cerca de trabajo** <u>m</u> - closer to @ work
— <u>adv</u> — **entre sí** - closer together
— <u>adv</u> **común** - more common • most common
— <u>adv</u> **considerado,da** - more considerate; kinder • most considerate; kindest
— <u>adv</u> **correo** <u>m</u> **aéreo** - [Trans] plus air mail
— <u>adv</u> **corto,ta** - shorter • shortest
— <u>adv</u> **costoso,sa** - more, expensive, <u>o</u> costly • most, expensive, <u>o</u> costly
— <u>adv</u> **crítico,ca**—more critical • most critical
— <u>adv</u> **chistoso,sa** - funnier; zanier
— <u>adv</u> **de** - more than
— <u>adv</u> **débil** - weaker • thinner; skinnier • weakest • thinnest; skinniest • decreased
— <u>adv</u> **demoroso,sa** - slower; more time consuming
— <u>adv</u> **derecho(s)** <u>m</u> - [com] plus @ fee(s)
— <u>adv</u> — <u>m</u> **consular(es)** - [fisc] plus consular fees
— <u>adv</u> **desperdicio(s)** - [ind] more scrap
— <u>adv</u> **destacado,da** - outstanding; leading
— <u>adv</u> **detallado,da** - more detailed
— <u>adv</u> **difícil** - harder; more difficult
— <u>adv</u> **difundido,da** - better known • best known
— <u>adv</u> **distante** - farthest (away)
— <u>adv</u> **dramático,ca** - more dramatic • most dramatic
— <u>adv</u> **duro,ra** - harder • hardest
— <u>adv</u> **económico,ca** - more economical; cheapest; less expensive; lowest priced
— <u>adv</u> **embalaje** <u>m</u> - [transp] plus packing
— <u>adv</u> — **para exportación** <u>f</u> - [transp] plus export packing
— <u>adv</u> **empaque** <u>m</u> - [transp] plus, packing, <u>o</u> crating, <u>o</u> boxing
— <u>adv</u> **encajonamiento** <u>m</u> - [transp] plus, boxing, <u>o</u> crating
— <u>adv</u> **entrega** <u>f</u> - [transp] plus delivery charge(s)
— <u>adv</u> — <u>f</u> **en aeropuerto** <u>m</u> - [transp] plus, airport delivery, <u>o</u> delivery at @ airport
— <u>adv</u> — — **muelle** <u>m</u> - [transp] plus pier delivery
— <u>adv</u> **estable** - more stable • most stable
— <u>adv</u> **exceso** <u>m</u> - plus excess
— <u>adv</u> — <u>m</u> **de flete** <u>m</u> - [transp] plus excess freight
— <u>adv</u> **experimentado,da** - more experienced • [labor] senior
— <u>adv</u> **expreso** - [transp] plus express (charges)
— <u>adv</u> — <u>m</u> **aereo** - [transp] plus air express
— <u>adv</u> **facil** - (much) easier
— <u>adv</u> — **uso** <u>m</u> - easier to use
— <u>adv</u> **fácilmente** - easier • easiest
— <u>adv</u> **flete** <u>m</u> - [transp] plus freight
— <u>adv</u> — <u>m</u> **aéreo** - [transp] plus air freight
— <u>adv</u> — <u>m</u> **como partida** <u>f</u> **separada** - [transp] plus freight as @ separate item
— <u>adv</u> **flete** <u>m</u> - [transp] plus freight
— <u>adv</u> — <u>m</u> **interior** - [transp] plus inland freight
— <u>adv</u> <u>m</u> — **pagado por, adelantado, <u>o</u> anticipado** - [transp] plus prepaid inland freight
— <u>adv</u> — <u>m</u> **marítimo** - [transp] plus ocean freight
— <u>adv</u> — <u>m</u> **pagado por, adelantado, <u>o</u> anticipado** - [transp] plus prepaid freight
— <u>adv</u> — <u>m</u> **terrestre** - [transp] plus inland freight
— <u>adv</u> — <u>m</u> — **pagado por, adelantado, <u>o</u> anticipado** - [transp] plus prepaid inland freight
— <u>adv</u> — <u>m</u> **y seguro** <u>m</u> - [transp] plus freight and insurance
— <u>adv</u> **fluido,da** - more fluid • most fluid
— <u>adv</u> **franqueo** <u>m</u> - [transp] plus postage
— <u>adv</u> — <u>m</u> **por paquete** <u>m</u> **postal** - [Transp] plus postage per parcel post
— <u>adv</u> — — <u>m</u> — **aéreo** - [transp] plus postage per air parcel post
— <u>adv</u> — **vía** <u>f</u> **aérea** - [transp] plus air mail postage
— <u>adv</u> **frecuente** - more frequent • most frequent • more commonly; most commonly
— <u>adv</u> **fuerte** - stronger • increased

más <u>adv</u> **gasto(s)** <u>m</u> **para expedición** <u>f</u> - [transp] plus shipping charge(s)
— <u>adv</u> — <u>m</u> — **reexpedición** <u>f</u> - [transp] plus forwarding charge(s)
— <u>adv</u> **giro** - further, turning, <u>o</u> rotating
— <u>adv</u> **grande** - bigger; larger size
— <u>adv</u> **grave** - more grave; evítese **graver***
— <u>adv</u> **grúa** <u>f</u> - [grúas] more crane
— <u>adv</u> **grueso,sa** - thicker
— <u>adv</u> **imperativo,va** - more imperative; especially true
— <u>adv</u> **imprevisto(s)** <u>m</u> - plus contingency,cies
— <u>adv</u> **instalación(es)** <u>f</u> - plus installation(s)
— <u>adv</u> **largo,ga** - longer
— <u>adv</u> **lejos** - farther, apart, <u>o</u> away
— <u>adv</u> **lentamente** - slower; more slowly
— <u>adv</u> — **que máximo** <u>m</u> - slower than maximum
— <u>adv</u> **liviano,na** - lighter (in weight)
— <u>adv</u> — **en peso** <u>m</u> - lighter in weight
— <u>adv</u> **mano** <u>f</u> **de obra** - [labor] more manpower
— <u>adv</u> **material(es)** <u>m</u> - [ind] plus material(s)
— <u>adv</u> — <u>m</u> **para embalaje** <u>m</u> - [ind] plus packaging material(s)
— <u>adv</u> — <u>m</u> — — <u>m</u> **para exportación** <u>f</u>—[transp] plus export packaging material(s)
— <u>adv</u> **moderno,na** - more modern • most modern
— <u>adv</u> **necesitado,da** - most needy,dful • more needy,dful
— <u>adv</u> **o menos** - more or less; or so • [matem] plus or minus • [cronol] circa
— <u>adv</u> **paquete** <u>m</u> **aéreo**—[transp] plus air parcel
— <u>adv</u> — <u>m</u> — **asegurado** - [transp] plus insured air parcel
— <u>adv</u> — <u>m</u> **asegurado** - [transp] plus insured parcel
— <u>adv</u> — <u>m</u> **postal aéreo** - [transp] plus air parcel post
— <u>adv</u> — <u>m</u> — — **asegurado** - [transp] plus insured air parcel post
— <u>adv</u> — <u>m</u> — **asegurado** - [transp] plus insured parcel post
— <u>adv</u> **pequeño,ña** - smaller • smallest
— <u>adv</u> **personal** - [labor] more, personnel, <u>o</u> manpower
— <u>adv</u> — <u>m</u> **especializado** - [ind] more skilled, manpower, <u>o</u> labor
— <u>adv</u> **plano,na** - flatter
— <u>adv</u> **porte** <u>m</u> - [transp] plus postage
— <u>adv</u> — <u>m</u> **por encomienda** <u>f</u> **postal** - [transp] plus parcel post (postage)
— <u>adv</u> — <u>m</u> — — <u>f</u> — **aérea** - [transp] plus air parcel post (postage)
— <u>adv</u> **porte** <u>m</u> - [transp] plus postage
— <u>adv</u> — <u>m</u> **por paquete** <u>m</u> **postal** - [transp] plus parcel post (postage)
— <u>adv</u> — <u>m</u> — — — **aéreo** - [transp] plus air parcel post postage
— <u>adv</u> — <u>m</u> — **vía** <u>f</u> **aérea** - [transp] plus air mail postage
— <u>adv</u> **potente** - more powerful
— <u>adv</u> **prima** <u>f</u> - [seguros] plus premium
— <u>adv</u> — <u>f</u> **por seguro** <u>m</u> - [seguros] plus insurance premium
— <u>adv</u> **probable** - more probable • most probable
— <u>adv</u> **profundo,da** - deeper • deepest
— <u>adv</u> **pronto** - sooner
— <u>adv</u> **próspero,ra** - more prosperous • most prosperous
— <u>adv</u> **próximo,ma** <u>adv</u> - nearer; closer • nearest; closest • immediately off
— <u>adv</u> — **a estructura** <u>f</u> [constr] closest to, <u>o</u> <u>o</u> immediately off @, structure
— <u>adv</u> — **a núcleo** <u>m</u> **laminar** - nearest @ lamination
— <u>adv</u> . . . **pulgada(s)** <u>f</u> - plus . . . inches
— <u>adv</u> **que** - more than
— <u>adv</u> — **casual** - more than passing
— <u>adv</u> — **de costumbre** - more than, usual, <u>o</u> normal
— <u>adv</u> — **requerido,da por arco** <u>m</u> - [sold] more than required by @ arc; above arc voltage
— <u>adv</u> **rápidamente** - faster; quicker • fastest; quickest

más adv rápido,da - faster; quicker • fastest; quickest
— adv — **que máximo** m - faster than @ maximum
— adv — **mínimo** m - faster than @ minimum
— adv **reciente** - more recent • later; newer • most recent; latest; newest
— adv **recio,cia** - toughest; sturdiest • tougher • sturdier
— adv **recolección** f - [transp] plus pick-up
— adv — **f con camión** m - [transp] plus truck pick-up
— adv **remoto,ta** - remoter • remotest
— adv **reñido,da** - most fought • [deport] most (hotly) contested
— adv **rigurosamente** - more rigorously • most rigrorously
— adv **rudo,da** - tougher • toughest
— adv **ruidoso,sa** - noisier; more noisy • noisiest; most noisy
— adv **satisfactorio,ria** - more satisfactory • most satisfactory
— adv **seguro(s)** m - [seguros] plus insurance
— adv — m **pagado** - [seguros] plus, insurance paid, o paid insurance
— adv — m — **por, adelantado,** o **anticipado** - [seguros] plus prepaid insurance
— adv — m **sobre** . . . - [seguros] plus insurance on . . .
— adv **separado,da** - farther, apart, o separate
— adv **serio,ria** - more serious • most serious
— adv **severo,ra** - more severe; tougher • most severe; toughest
— adv **sobrecarga** f - plus overload
— adv **tardío,día** - later • latest
— adv **tiempo** m - more, o longer, time
— adv **transporte** m - [transp] plus transportation
— adv — m **a aeropuerto** m - [transp] plus, transportation to, o delivery at, @ airport
— adv **tres pulgadas** f - [autom-neumát] plus three (inches)
— adv **una pulgada** f - [autom-neumát] plus one (inch)
— adv **veloz** - speedier; quicker • speediest; quickest
— adv **vil** - meaner • meanest
— adv **voltio(s)** m - [electr-oper] more volts
— adv — **que requerido(s) por arco** m - [sold] more volt(s) than required by the arc; above arc voltage
masa f - . . . ; bulk • pat • [miner] pocket • [electr-instal] **véase tierra** f
— f **aislante** - [electr-instal] insulating mass • Chatterton's compound
— f — **para relleno** m - insulating filling mass
— f — **no drenante*** - [electr-instal] non--draining (type) insulation
— f **caliente** - hot mass
— f **circundante** - [constr] surrounding, o abutting, mass
— f — **de tierra** f - [constr] abutting, o surrounding, soil mass
— f **compuesta** - compound mass
— f **conductora** f - [electr-instal] conductive mass
— f **confinada** - confined mass
— f **contigua** - [constr] contiguous, o abutting, o surrounding, mass
— f — **de tierra** f - [constr] surrounding, o abutting, soil mass
— f — **composición** f - [quím] composition, o compound, mass
— f — **compuesto** m - compound mass
— f — — m **viscoso** - viscous compound mass
— f — — **explosivo(s)** m - [explos] explosive('s) mass
— f — **fibra(s)** f - fiber(s) mass
— f — **hierro** m - [metal-prod] iron mass
— f — **material** m **compuesto** - [quím] composition mass
— f — **mineral** m - [miner] ore mass
— f — **motor** m - [comb.int] engine('s) mass
— f — — m **rotativo** - [comb.int] rotary engine mass

masa f **de piedra** f - [constr] stone mass
— f — — f **pómez** - [miner] pumice, material, o stone, pocket
— f — — f **pulverizada** - [miner] powdered pumice, material, o stone, pocket
— f — **relleno** m - filling mass
— f — **tierra** f - [constr] earth, o soil, mass
— f — — f **confinada** - [constr] confined earth mass
— f **descompuesta** - decomposed mass
— f **elástica** - resilient mass
— f **en movimiento** m - moving mass
— f — — **en motor** m - [comb.int] engine's moving mass • [electr-mot] motor's moving mass
— f — — m — m **rotativo** - [comb.int] rotary engine's moving mass
— f **especificada** - specified mass
— f **explosiva** - [explos] explosive mass
— f **fría** - [sold] cold mass
— f **informe** - jumble
— f **íntegra** - entire, mass, o bulk
— f **máxima** - maximum mass
— f **mínima** - minimum mass
— f **para sellado** m - [telecom-instal] sealing mass
— f **porosa** - porous mass
— f **similar a humus** m - [sanit] earthlike humus
— f **viscosa** - viscous mass
— f **volumétrica** - [miner] bulk density
máscara f - . . . . • [sold] shield; headshield; face shield
— f **buena** - [sold] good shield
— f **fijada a casco** m **para soldadura** f **por arco** m - [sold] arc welding headshield
— f **limpia** - [segurid] clean mask
— f **facial** - [sold] face shield
— f **limpiada** - [segurid] cleaned mask
— f **moldeada** - [sold] molded shield
— f **para soldador** m - [sold] welding shield
— f — **soldadura** f - [sold] welding headshield
— f — — f **por arco** m - [sold] arc welding headshield
— f — **soldar** v - [sold] (welding) head shield
— f **portátil** - [sold] hand (held) shield
— f — **para soldadura** f - [sold] (portable, o welding) hand shield
— f **protectora** - [sold] (protective) shield
— f **que cubre rostro** m **entero** - [segurid] full--face shield
— f **sucia** - [segurid] dirty mask
masera f - . . . . • [constr] mortar box
maserota* f - [metal-prod] **véase mazarota** f
masilla f **asfáltica** - [constr] asphalt mastic
masillar v - [constr] to putty
masivo,va a - massive
mástil m - . . . . • [náut] . . .; spar
— m **armado** - [petról] frame mast
— m — **tipo A** - [petról] A-frame mast
— m **con extensión** f - [petról] telescoping mast
— m — — f **enchufable** - [petról] telescoping mast
— m — — f **enchufada** - [petról] telescoping mast
— m **en forma** f **de A** - [petról] A-frame mast
— m **extensible** - [petról] telescoping mast
— m **para carga** f - [grúas] gin pole
— m **tipo A** - [petról] A-frame mast
mastique m **aprobado** - [tub] approved mastic
— m — **para junta(s)** f - [tub] approved jointing compound
— m **asfáltico** - [petról] asphalt(ic) mastic
— m — **fibroso** - fibrated asphalt mastic
— m **de asfalto** - [petról] asphalt mastic
— m — — m **fibroso** - [constr] fibrated asphalt mastic
— m **para junta(s)** f - [tub] jointing compound
mata f - . . . . • [miner] matte
matafuego(s) m **con bióxido** m **de carbono** m - [segurid] carbon dioxide (type) fire extinguisher
— m — **espuma** f - [segurid] foam (type) (fire) extinguisher
mate m - . . . . | a - [fotogr] flat • [pint] . . . ; matte; dead

materia f carbonosa—[miner] carbonaceous matter
— f contaminadora - [sold] contaminant
— f de convenio m - (matter) hereby covered
— f — partida - [ind] basic raw material
— f — seguridad f - [segurid] safety matter(s)
— f en descomposición f - decaying matter
— f — media agua - [hidr] suspended matter
— f — suspensión f - [hidr] suspended matter
— f extraña - extraneous, o foreign, matter, o
material
— f fecal - [sanit] fecal matter; feces
— f financiera - [fin] financial matter
— flotante - [hidr] floating matter; flotsam
— f foránea - foreign matter
— f líquida - liquid matter • [metal-prod]
salamander
— f mineral - [miner] mineral matter
— f inorgánica - [hidr] inorganic matter; in-
organics
— f — contaminadora - [sold] inorganic contaminant
— f prima - [ind] raw material
— f — autofundente - [miner] self-fluxing raw
material
— f — complementaria - [ind] complementary raw
material
— f — con azufre m alto - [miner] high sulfur
raw material
— f — — — m bajo - [miner] low sulfur raw
material
— f — económica - [ind] economical, o low
cost, raw material
— f — elaborada - [ind] processed, o manufactured, raw material
— f — embarcada - [ind] shipped raw material
— f — en bruto - [ind] crude raw material
— f — en estado m bruto - [ind] crude state
raw material
— f — endurecida - [ind] hardened raw material
— f — manipul(e)ada - [ind] handled raw material
— f — molida - [ind] ground raw material
— f — — finamente - [ind] finely ground raw
material
— f — para horno m—[ind] furnace raw material
— f — — m eléctrico - [metal-prod] electric furnace raw material
— f — — planta f - [ind] plant raw material
— f — — f para sinter(ización) - [metal-
-prod] sinter(ing) plant raw material
— f — recibida - [ind] received, o accepted, o
incoming, raw material
— f — semiprocesada - [ind] semi-processed raw
material
— f primaria - [ind] véase materia f prima
— f primera - [ind] véase materia f prima;
crude material
— f sintética - [ind] synthetic matter
— f — orgánica - [ind] organic synthetic (material
— f suspendida - [hidr] suspended matter; flotsam
— f vegetal - vegetable matter
— f volátil - [miner] volatile matter
material m - . . . .; matter
— m a granel - [ind] bulk material
— m abrasivo - [sold] abrasive material
— m absorbente - absorbent material • dope
— m — para humedad f - moisture absorbing material
— m acabado - finished material
— m — con nitrógeno m - [metal-trat] nitrogen
finished material
— m — de trefilar - [trefil] just, o freshly,
drawn material
— m aceptado - [ind] accepted material
— m — para construcción f - [constr] accepted
building material
— m ácido - acid material
— m acopiado - [ind] gathered material
— m adecuado - adequate, o suitable, material
— m — para relleno m - adequate, o proper, o
appropriate, backfill(ing) material

material m adhesivo - adhesive (material)
— m — incombustible - non-combustible adhesive
material
— m adquirido - [ind] acquired, o purchased,
material
— m aeronáutico - [aeron] aeronautical material
— m aglomerado - [vial] agglomerated material
— m agrietado - [constr] cracked material
— m ahorrado - saved material
— m aislador* - véase material m aislante
— m aislante - insulating,tion material • insulator
— m — m aprobado - [electr-instal] approved
insulating material
— m altamente abrasivo - highly abrasive material
— m — orgánico - [geol] highly organic material
— m aluvial - [geol] alluvial material
— m angular - [constr] angular material
— m aguloso - [constr] angular material
— m anisotrópico - anisotropic material
— m apartado - [hidr] separated, o diverted,
material
— m apisonado - [constr] tamped material
— m — para filtración f tamped filtering material
— m aplicado - applied material
— m — en caliente - hot-applied material
— m — — frío - cold-applied material
— m aportado - [sold] borrow, o added, material
— m apropiado - [constr] appropriate, o suitable, material
— m — para relleno m - appropriate, o proper,
o adequate, backfill(ing) material
— m arrastrado - [hidr] stream debris
— m — por curso m de agua m - [hidr] stream
debris
— m arrollado - [metal-lam] coiled material
— m asegurado - [seguros] insured material
— m asfáltico - [constr] asphalt(ic) material
— m aterronado - caked material
— m autofundente - [miner] self-fluxing material
— m auxiliar - auxiliary material
— m avanzado - [ind] advanced material
— m básico - base,sic material
— m bien apisonado - [constr] well, tamped, o
compacted, material
— m — clasificado - [constr] well graded material
— m — compactado - [constr] well, compacted, o
tamped, material
— m bitumástico* - bitumastic material
— m bituminoso - [constr] bituminous material
— m — líquido - [petról] liquid bituminous
material
— m — semisólido - [petról] semisolid bituminous material
— m — sólido - [petról] solid bituminous material
— m blando - soft material
— m calcáreo - [geol] calcareous material
— m calentado - [constr] heated material
— m calibrado - [metal-lam] gage(d) material
— m caliente - hot material
— m — para soldadura f - [sold] hot welding
material
— m capaz de deteriorar(se) - material subject
to deterioration • degradable material
— m carbonoso - [miner] carbonaceous material
— m cementado - [suelos] cemented material
— m certificado - certified material
— m clasificado - [ind] classified material •
[constr] graded material
— m — uniformemente - [constr] uniformly graded material • uniformly classified material
— m cohesivo - cohesive material
— m colocado - placed material
— m coloidal - [hidr] colloidal material
— m colorante - color(ing), material, o mixture
— m combinado - combined material(s)
— m combustible - combustible material

material m compactado - [constr] compacted, o tamped, material
— m — cuidadosamente - [constr] carefully, compacted, o tamped, material
— m —, levemente, o ligeramente - [constr] lightly, compacted, o tamped, material
— m comparable - comparable material
— m competidor,titivo - competing,titive material
— m complementario - complementary material
— m componente - component material
— m completado - completed material
— m comprado - purchased, o bought, material
— m comprobatorio - proving, o supporting, material
— m compuesto - [constr] composite material
— m con acabado m con nitrógeno - [metal-prod] nitrogen finished material
— m — base f arenisca - [geol] sandy base material
— m — — f de cromo m - [metal-prod] chromium base(d) material
— m — — f — — m magnesita - [quím] chromium magnesite base material
— m de base - base material
— m — calidad f - quality material
— f — — f excepcional - premium material
— m con capacidad f portante - [constr] bearing capacity material
— m — — f — buena - [constr] good bearing capacity material
— m — — f — no uniforme - [constr] non-uniform bearing capacity material
— f — — f — pobre - [constr] poor bearing capacity material
— m — — f — uniforme - [constr] uniform bearing capacity material
— m — carbono m - [metal-prod] carbon material
— m — — m alto - [metal-prod] high carbon material
— m — — m bajo - [metal-prod] low carbon material
— m — costo m reducido - low cost material
— m — densidad f alta - high density material
— m — — f baja - low density material
— m — dureza f - [metal-lam] hardness material
— m — — f total - [metal-lam] total hardness material
— m — espesor m especificado - specified thickness material
— m — grano(s) m anguloso(s) - gritty material
— m — hidrógeno m - [sold] hydrogen material
— m — — m alto - [sold] high hydrogen material
— m — — m bajo - [sold] low hydrogen material
— m — nitrógeno m - [metal-prod] nitrogen material
— m — origen m mineral - [miner] mineral origin material
— m — — m nacional - [com] national, o local, origin material
— m — — m norteamericano—[com] United States origin material
— m — peso m reducido - light weight material
— m — proporción f alta de hidrógeno m - [metal] high hydrogen material
— m — — f baja de hidrógeno m - [metal] low hydrogen material
— m — resistencia f alta (a impacto(s) m - [metal-prod] high impact material
— m — — f baja (a impacto(s) m - [metal-prod] low impact material
— m — — f dieléctrica - [metal] dielectric strength material
— m — — f — alta - [metal] high dielectric strength material
— m — — f — baja - [metal] low dielectric strength material
— m — tamaño m grande - large size material
— m — — m mayor - larger size material
— m — — m menor - smaller size material
— m — — m pequeño - small size material
— m — — m único—one, o single, size material
— m — tenor alto de carbono m - [metal-prod] high carbon material

material m con tenor m bajo de carbono m - [metal-prod] low carbon material
— m — terminación f con nitrógeno m - [metal-prod] nitrogen finish(ed) material
— m — tolerancia f mínima - close tolerance material
— m congelado - frozen material
— m consistente - consistent material
— m consolidado - consolidated materail
— m consumible - consumable (material)
— m contaminador - contaminant
— m entrado - entered contaminant
— m copolímero - copolymer material
— m coquizado - [metal] coked material
— m cribado - [constr] graded material • screened material
— m cubierto - covered material
— m cubridor - [miner] covering material
— m dado - given material
— m dañado - damaged material
— m — en tránsito m - [transp] material damaged in transit
— m — — transporte m - [transp] material damaged in transport(ation)
— m — para retención f de agua m - [constr] damaged water stop(ping) material
— m de acero m - [metal-lam] steel plate stock
— m — — m dulce - [metal-lam] mild steel plate stock
— m — asfalto m - [constr] asphalt material
— m — — m y membrana f - [constr] asphalt and membrane material
— m — — m — — f a prueba f de agua m - [constr] asphalt and membrane waterproofing material
— m — — m — — para impermeabilización f - [constr] asphalt and membrante waterproofing material
— m — base f - [ind] base,sic material • [constr] bedding material
— m — — f para macadam m - [constr] macadam base, material, o course
— m — calidad f - quality material
— f — — f primera - first, quality, o rate, material
— m — categoría f - quality material • [metal-lam] prime material
— m — cromo m - [metal] chromium material
— m — — m magnesita - [metal] chromium magnesite material
— m — fibra f - [plást] fiber material
— m — — f de vidrio m - [plást] fiberglass material
— m — — f — — m y resina f (poliestérica) - [plást] fiberglass and (polyester) resin material
— m — fondo m - background material
— m — importación f - [ind] imported material
— m — primera calidad f - prime quality material • [metal-lam] prime material
— m — rechazo m - [miner] rejected material • [ind] reject(s)
— m — resina f (poliestérica) - [plást] (polyester) resin material
— m — competencia f - competitive material
— m — desecho(s) m - [ind] waste material(s)
— m — — m radioactivo(s) - [nucl] radioactive waste material(s)
— m — desperdicio m - [miner] waste material
— m — hierro m - [miner] iron material
— m — — m magnético - [miner] magnetic iron material
— m — polietileno m - polyethylene material
— m — recirculación f - [ind] véase material m reprocesado
— m — rayón - [textil] rayon material
— m — — m resistente - [autom-neumát] tough rayon material
— m — — m para banda f (para rodamiento m) - [autom-neumát] tough rayon belt material
— m — residuo(s) - [ind] waste material(s)
— m — — m radiactivo(s) - [nucl] radioactive

waste material(s)
material m̲ de subcontratista m̲ - [constr] subcontractor's material
— m — **tipo m̲ cohesivo** - cohesive type material
— m̲ **defectuoso** - [ind] defective material
— m̲ — **embarcado m̲** - [ind] shipped defective
material
— m — — **anteriormente** - [ind] previously
shipped defective material
— m̲ **deficiente** - [ind] deficient material
— m̲ **degradado** - [ind] degraded, o̲ downgraded,
material
— m — **procesado** - [ind] processed downgraded
material
— m̲ **delgado** - thin material
— m̲ **depurado** - [sanit] treated material
— m̲ **descartable** - [ind] expendable material
— m̲ **descomponible** - [constr] decomposable material
— m̲ **descriptivo** - descriptive material
— m̲ **descri(p)to** - described material
— m̲ **desecado** - dried, o̲ dewatered, material
— m̲ **desplazable** - displaceable material • removable material
— m **desulfurante** - [metal-prod] desulfurizing
material
— m̲ **determinado** - specific, o̲ particular, material
— m̲ **detrítico** - [hidr] detrital material; de-
tritus
— m̲ **dieléctrico** - dielectric material
— m — **higroscópico** - hygroscopic dielectric
material
— n — **no higroscópico** - non-hygroscopic dielectric material
— m̲ **diferente** - different, o̲ other, material
— m̲ **disponible** - available material
— m̲ **distinto** - different, o̲ other, material •
varios material(s)
— m — **para relleno m̲** - [constr] different
backfill(ing) material
— m̲ **distribuido** - distributed material
— m̲ **diverso** - sundry, o̲ various, material(s)
— m̲ **duro** - hard material
— m̲ — **en circuito m̲ magnético**—[electr-instal]
magnetic circuit hard material
— m — **inflexible**—hard unyielding material
— m — **magneticamente** - [electr-equip] magnetically hard material
— m̲ **ebonítico** - [plást] ebonitic material
— m̲ **económico m̲** - economical, o̲ low-cost, o inexpensive, material
— m **efervescente** - boling material • [metal-
-prod] rimming material
— m̲ **electrohipobárico** - electrohypobaric material
— m̲ **embalado** - [ind] packed material
— m̲ **embarcado** - [ind] shipped material
— m̲ — **anteriormente** - previously shipped material
— m̲ **empleado** - material used
— m̲ — **para reparación f̲** - [ind] repair material used; material used for repair
— m̲ **en caliente** - [metal-lam] hot (rolled)
material
— m̲ **en electrodo m̲** - [sold] electrode material
— m̲ — **estado m̲ bruto** - [ind] crude state material
— m — — m **natural** - natural (state) material
— m̲ — **exhibición f̲** - [com] display material
— m̲ — **existencia** - [com] stock
— m̲ — **forma f̲ de laja(s) f̲** - [geol] spell
shaped material
— m̲ — **hierro m̲** - [metal-prod] iron material
— m̲ — **piso m̲** - [constr] floor material
— m̲ — — m̲ **para carga f̲** - [metal-prod] charging floor material(s)
— m̲ — — m — — f̲, **manejado**, o̲ **manipulado** -
[metal-prod] handled charging floor material
— m̲ — **proceso m̲** - [ind] material in process •
good(s) in process inventory
— m̲ — **suspensión f̲** - [hidr] suspended material
— m̲ — **tránsito m̲** - material in transit

material m̲ encorvado - [ind] buckled material
— m̲ **energético** - energy,getic material(s)
— m̲ **enfriado** - cooled material
— m̲ **enrollado** - [metal-lam] coiled material
— m̲ **ensayado** - [ind] tested material
— m̲ **entrado** - [ind] incoming material
— m̲ **entre sección(es) f̲ (de planta f̲)** - [ind]
intra-polant material
— m̲ **equivalente** - equivalent, o̲ comparable, material
— m̲ **erosionado** - eroded material
— m̲ **escogido** - chosen, o̲ selected, material
— m̲ **especial** - special(ty) material
— m̲ — **para polea f̲** - [grúas] special sheave
material
— m̲ **especificado** - specified material
— m̲ — **para relleno m̲** - [constr] specified
backfill(ing material)
— m̲ **específico** - specific, o̲ particular, material
— m — **seleccionado** - [ind] selected, o̲ chosen,
material
— m̲ **esponjoso** - spongy material
— m̲ **estable** - stable material
— m̲ **estándar** - standard material
— m̲ — **para relleno m̲** - [constr] standard
backfill(ing material)
— m̲ **estructural** - [estruct] structural material
— m̲ **evaluado** - evaluated material
— m̲ — **para relleno m̲** - [constr] evaluated
backfill(ing) material
— m̲ **excavado** - [constr] excavated material;
excavation
— m̲ — **en exceso m̲** - [constr] excess(ive) excavated material
— m̲ — **granular** - [constr] granular excavated
material
— m̲ — **sin granular** - [constr] ungramulated excavated material
— m̲ **excedente** - excess material
— m̲ **exotérmico** - [metal-prod] exothermic material
— m̲ **expuesto** - exposed material
— m̲ **extranjero**—[ind] foreign (origin) material
— m̲ **extraño** - extraneous, o̲ foreign, material
— m̲ —, **quitado**, o̲ **removido**, o̲ **sacado** - [ind]
removed foreign material
— m̲ **fabricado** - [ind] manufactured material •
fabricated material
— m̲ **facturado** - [com] invoiced, o̲ billed, material
— m̲ — **(pero) no despachado** - [com] not shipped
billed material
— m̲ **fenólico** - [quím] phenol(ic) material
— m̲ — **laminado** - [plást] rolled, o̲ laminated,
phenol(ic) material
— m̲ — — **con calidad f̲** - [plást] quality laminated phenol(ic) material
— m̲ — — — f̲ **alta** - [plást] high (quality)
grade laminated phenol(ic) material
— m̲ **ferromagnético** - [metal] ferromagnetic material
— m̲ **ferroso** - [metal] ferrous material
— m̲ — **amolado** - [mec] ground ferrous material
— m̲ **ferroviario** - [f.c.] railway materials •
[metal-lam] rails and fittings
— m̲ **fijado** - [mec] fixed, o̲ fastened, material
— m̲ **filtrado** - [ind] filtered material
— m̲ **filtrador** - [hidr] filter(ing) material
— m̲ **filtrante** - [hidr] filter(ing), material, o̲
media
— m̲ — **apisonado** - [constr] tamped filter(ing)
material
— m̲ — **fino** - [hidr] fine filter(ing) material
— m̲ — **grueso**—[hidr] coarse filtering material
— m̲ — **para subdrenaje m̲** - [hidr] subdrain fil-
teer(ing) material
— m̲ — **impermeable** - [hidr] non-pervious fil-
ter(ing) material
— m̲ — — **clasificado** - ungraded pervious fil-
ter(ing) material
— m̲ — **permeable** - [hidr] pervious filter(ing)
(material)

material m filtrante permeable clasificado -
  graded pervioius filter(ing) material
— m final - final material
— m fino - fine material
— m firme - firm, o solid, material
— m fisionable* - [nucl] fissionable material
— m flotado - [miner] floated material
— m fluorescente - fluorescent material
— m fotográfico—[fotogr] photographic material
  • photographic supply,lies
— m frágil - brittle material
— m frío - cold material
— m galvanizado - [metal-fabr] galvanized material
— m gaseoso - [sold] gaseous material
— m generador - generating, o forming, material
— m — de gas(es) m - gas forming material
— m — vapor(es) m - vapor forming material
— m gráfico - [imprent] graphics; graphic material
— m granular - [constr] granular material
— m — apisonado - [constr] tamped granular material
— m — arrastrado - [hidr] washed out, o carried, granular material
— m — compactado - [constr] compacted granular
  material
— m — — cuidadosamente - [constr] carefully
  compacted granular material
— m — fino - [constr] fine granular,lated
  material
— m — grueso - [constr] coarse granular,lated
  material
— m — obtenido en sitio m - [constr] granular
  site material
— m — para lecho m - [constr] granular bedding
  material
— m — — m para tubería f - [constr] granular pipe bedding material
— m — — relleno m - [constr] granular back-
  fill(ing) material
— m — rodado - [constr] rounded granular material
— m grueso - [constr] coarse material • [metal-
  -lam] heavy material
— m helado - frozen material
— m hidráulico - [constr] hydraulic material
— m — para relleno m - [constr] hydraulic
  fill(ing) material
— m higroscópico - hygroscopic material
— m homologado - [ind] homologated material
— m humeante - [combust] smoking material
— m hundido - [miner] sunk material
— m ideal - ideal material
— m idéntico - identical material
— m identificado - [ind] identified material
— m igual - same, o identical, material
— m impermeable - impervious material
— m importado - imported material
— m inaceptable - unacceptable material
— m inapropiado - [constr] unsuitable, o inappropriate, material
— m incluida instalación f - [constr] furnished
  and installed material
— m incombustible - noncombustible, o noninflammable, material
— m indeseable - undesirable material
— m — para, cimentación f, o cimiento(s), o
  fundación, o fundamento m - [constr] undesirable foundation material
— m inducido - induced material
— m inerte—inert material • [miner] overburden
  • ground material
— m inestable - [constr] unstable material
— m inferior—inferior, o substandard, material
— m — para cimiento(s) m - [constr] inferior
  foundation material
— m — — fundación f - [constr] inferior foundation material
— m inflamable - [segurid] flammable material •
  fire hazard
— m inflexible - unyielding material
— m — para fundación f - [constr] unyielding

  foundation material
material m inorgánico - [geol] inorganic material
— m inspeccionado - inspected material
— m interior - inside, o inner, material •
  [petról] bore material
— m — con aleación f de hierro m - [petról]
  iron alloy bore material
— m — — f — — m con carbono m alto -
  [petról] high carbon iron alloy bore material
— m — — f — — m — cromo m alto -
  [petról] high chromium iron alloy bore material
— m — — f — m — — m y carbono m alto
  - [petról] high-chromium high-carbon iron alloy bore material
— m inundante - stuffing material
— m inútil - useless material • [miner] ground
  material • clutter
— m laminado - [metal-lam] rolled material
— m — con espesor m mayor - [metal-lam] thicker stock
— m — — — m menor - [metal-lam] thinner
  stock
— m — en caliente - [metal-lam] hot rolled
  material
— m — — frío - [metal-lam] cold rolled material
— m — grueso - [metal-lam] thick stock
— m — plano - [metal-lam] flat sheet material
— m levemente frío m - slightly cold material
— m ligante* - binding material; binder
— m — químico - chemical, binding material, o
  binder
— m ligeramente frío - slightly cold material
— m limpiado - cleaned (up) material
— m líquido - liquid material • [metal-prod]
  salamander
— m liviano - [metal-lam] light (gage, o
  weight) material
— m lixiviado - [miner] lixiviated material
— m local - local material
— m — disponible - available local material
— m — obtenible - available local material
— m llevado - [ind] carried material
— m magnéticamente duro - [electr] magnetically
  hard material
— m — — en circuito m - [electr] circuit magnetically hard material
— m — — — m magnético - [electr] magnetic
  circuit magnetically hard material
— m magnético - magnetic material
— m — en circuito m - [electr] circuit magnetic material
— m — — — m magnético - [electr] magnetic
  circuit magnetic material
— m — duro - [electr] hard magnetic material
— m marcado - [ind] marked material
— m más liviano - lighter material
— m — pesado - heavier material
— m médico - [medic] medical, material, o supply,lies
— m mejor - better material • best material
— m metálico - [metal-prod] metallic material
— m mezclado - [ind] mixed material
— m mineral - [miner] mineral material
— m moderadamente abrasivo - moderately abrasive material
— m molido - [ind] ground, o crushed, material
— m — finamente - [ind] finely ground material
— m móvil - [f.c.] rolling stock; equipment
— m — actual - [f.c.] present, o existing, equipment, o rolling stock
— m — nuevo - [f.c.] new, equipment, o rolling
  stock
— m muy abrasivo - very, o highly, abrasive
  material
— m nacional - [ind] domestic, o local, o national, material
— m natural - natural material
— m necesario - needed, o necessary, material
— m no abrasivo - nonabrasive material
— m no aprobado - [ind] unapproved material

material m no cedente - [constr] unyielding material
— m — conductor - [electr] non-conducting material
— m — conforme - [ind] non-approved material
— m — corriente - [ind] non-current, o non--standard, material, o item
— m — despachado - [com] unshipped, material, o item
— m — higroscópico - non-hygroscopic material
— m — inflamable - nonflammable material
— m — magnético - non-magnetic material
— m — metálico - [miner] nonmetallic material
— m — sensible - nonsensitive material
— m normal - normal, o standard, material
— m nuclear - [nucl] nuclear material
— m nuevo - [ind] new material
— m obtenible - available material
— m — en plaza f - locally available material
— m — — sitio m - [constr] site material
— m opaco - opaque material
— m orgánico - [geol] organic material • [quím] organics
— m original - original, o parent, material
— m originario - material originating
— m — de Estados Unidos m de América f de Norte - United States origin(ated) material
— m para absorción f - absorbing material
— m — — f alta - [electr] high absorbing material
— m — — f — de humedad f - [electr] high moisture absorbing material
— m — — f baja - low absorbing material
— m — — f — de humedad f - low moisture absorbing material
— m — acabado m - finishing material
— m — accesorio m - [tub] fitting material
— m — afino m - refining material
— m — m, empleado, o usado, o utilizado - [metal-prod] refining material used
— m — aislación f - [electr] insulating material
— m — — f aprobado - [electr] approved insulating material
— m — albañilería f - [constr] masonry material • masonry unit
— m — alcantarilla(s) f - [constr] culver material
— m — aleación f - [metal-prod] alloying material
— m — almacenamiento m - [ind] storing,rage material
— m — aportación f - [sold] borrow material
— m — aportar - [sold] borrow material
— m — apoyo m - [constr] bedding material
— m — árbol m - [mec] shaft, o axle, material
— m — armadura f - [constr] reinforcement,cing material
— f — — f de acero m - [constr] steel reinforcing material
— m — asiento m - [constr] bedding material
— m — — m para tubería f - [tub] pipe bedding material
— m — auxilio(s) m primero(s) - [medic] first aid, materials, o supplies
— m — cable m - [cabl] cable material
— m — — m para arrastre m - [constr] dragline material(s)
— m — — m — — m para cucharón m con almeja f - [constr-equip] clamshell dragline material
— m — — m — — m — grúa f - [constr-equip] crane dragline material
— m — calce m - [mec] shim, material, o stock
— m — canal m - [metal-prod] runner material
— m — — m para colada f - [metal-prod] runner lining material
— m — carcasa f - [ind] housing material
— m — carga f - [ind] charge,ging material
— m —, cimentación f, o cimiento(s) m - [constr] foundation material
— m — circuito m - [electr] circuit material
— m — — m magnético - [electr] magnetic circuit material

material m para cloración f - [quím] chlorinating material
— m — comercialización f - [com] merchandising material
— m — comunicación(es) m - [telecom] communication(s) material(s)
— m — conducto m - [constr] conduit material
— m — construcción m - [constr] construction, o building, material, o product
— m — — f compuesto - [constr] composite construction material
— m — consumo m - [contab] consumable
— m — conversión f - [mec] conversion material • converting material
— m — cortarse v - [trefil] cutoff blank
— m — cucharón m - [metal-prod] ladle material
— m — m con almeja f - [constr-equip] crane clamshell material
— m — m — — f para grúa f - [constr-equip] crane clamshell material
— m — chapa(s) f - [metal-prod] sheet, o strip, material
— m — — f para alcantarilla f - [tub] culvert sheet material
— m — curación f - [hormig] curing material
— m — chaqueta f - jacket material
— m — — f para cable m - [telecom-cond] cable jacket material
— m — desarrollo m - development material
— m — descarte m - [metal-prod] scrap material
— m — diseño m - [dib] design material(s)
— m — dispositivo m - [mec] device('s) material
— m — — m para sujeción f - [sold] (holding) fixture material(s)
— m — drenaje m - drainage material
— m — eje m - [mec] axle material
— m — elaboración f - [mec] stock
— m — electrodo m - [sold] electrode material
— m — embalaje m - packing,ckaging material
— m — embutición f - [metal-prod] drawing material
— m — — f profunda - [metal-fabr] deep drawing material
— m — empalme(s) m - [electr-instal] jointing material
— m — encofrado m - [constr-hormig] form material(s)
— m — enlechado m - [constr] grouting material
— m — enseñanza f - teaching, o training, material
— m — entibación f - [constr] shoring, o lining, material
— m — escritorio m - [com] office supply,lies
— m — estampación,pado - [metal-fabr] drawing material
— m — — extra profundo - [metal-fabr] extra deep drawing material
— m — — profundo,da - [metal-fabr] deep drawing material
— m — exhibición f - [com] display material(s)
— m — explotación f - [petról] operating supply,plies
— m — fabricación f - [mec] fabricating supply,plies
— m — — f de tornillo(s) m - [metal-fabr] screw stock
— m — filtración f - filtering material
— m — filtro m - [mec] filter material
— m — fundación,damento - [constr] foundation material
— m — fusión f - [metal-prod] melting stock
— m — generador m - [electr-prod] generator material(s)
— m — — m atómico - [electr-prod] atomic generator material(s)
— m — grúa f - [grúas] crane material(s)
— m — — f para cable m para arrastre m - [constr-equip] dragline crane material(s)
— m — — f sobre torre f - [grúas] tower crane material(s)
— m — hierro m - [metal-prod] iron material(s)

material m̲ para hogar m̲ - [domést] home material
• [combust] firebox material
— m̲ — **horno m̲ alto** - [metal-prod] blast furnace material • blast furnace feed(ing)
— m̲ — **ingeniería f** engineering material
— m̲ — **instalación f** - [constr] installation
material
— m̲ — — f **de buceta(s) f** - [metal-prod] nozzle installation material
— m̲ — — f — **encofrado(s) m̲** - [constr] form
material(s)
— m̲ — — f — — m **para, frente(s) m̲, o su-**
perficie(s) f - [constr] face form materials
— m̲ — **instrucción f** - [educ] instruction, o
training, material
— m̲ — — f **necesario**—needed training material
— m̲ — **junta(s) f** - joint material
— m̲ — — f **para dilatación f** - [constr] expansion joint(s) material
— m̲ — **labrar f** - [trefil] stock
— m̲ — — **trefilado** - [trefil] drawn stock
— m̲ — **laminación f** - [metal-lam] material to
be rolled; rolling material
— m̲ — **lectura f** - reading material • [educ]
reading assignment
— m̲ — **lechada f** - [constr] grout(ing) material
— m̲ — **lecho m̲** - [constr] bedding material
— m̲ — — m **para tubería f** - [tub] pipe bedding
material
— m̲ — **malla f** - [mec] screen material
— m̲ — **mantenimiento m̲** - [ind] maintenance, material, o supply
— m̲ — **máquina f para fabricar tornillos m̲** -
[ind] screw machine stock
— m̲ — **mercadeo m**—[com] merchandising material
— m̲ — **mezcla f** - [constr] aggregate
— m̲ — **montaje m** - [ind] mounting, o erection,
o erecting, material(s)
— m̲ — — **de aguilón m̲** - [grúas] boom mounting material(s)
— m̲ — — m — **cucharón m̲** - [grúas] scoop
mounting material(s)
— m̲ — — m — — m **con almeja f** - [grúas]
clamshell mounting material(s)
— m̲ — — m — — f **sobre aquilón m** -
[grúas] clamshell on @ boom mounting material
— m̲ — **neumático(s) m̲** - [autom-neumát] tire ma-
terial(s)
— m̲ — **núcleo(s) m̲** - core material(s)
— m̲ — **oficina f** - [com] office supply,lies
— m̲ — **pavimento m** - [constr] paving,vment, o
road surfacing, material(s)
— m̲ — — m **asfáltico** - [constr] asphalt pavement material(s)
— m̲ — **perno(s) m̲** - [mec] bolt material
— m̲ — — m **especial(es)** - [mec] special bolt
material
— m̲ — **piso(s) m̲** - [constr] flooring; floor ma-
terial(s)
— m̲ — — m **para uso m interior** - [constr] interior use flooring
— m̲ — **placa(s)** - [metal-lam] plate material
— m̲ — **plancha(s) f**—[metal-lam] plate material
— m̲ — **planta f** - [ind] plant material(s) •
mill supply,lies
— m̲ — — f **industrial** - [ind] industrial mill
supply,lies
— m̲ — **plegado m̲** - folding material
— m̲ — — m **de aguilón m̲** - [grúas] boom folding
material
— m̲ — **plegar v** - folding material
— m̲ — **polea f** - [grúas] pulley, o sheave, material
— m̲ — **presentación f** - [com] display material
— m̲ — **producción f** - [ind] production supply
— m̲ — **propaganda f**—[com] advertising material
— m̲ — **protección f** - [segurid] protection, o
safety, equipment, o material(s)
— m̲ — — f **para arco m** - [sold] arc-shielding
material
— m̲ — **proyección f** - [ind] design material(s)
• [fotogr] projection, material, o supplies
— m̲ — **recalcado m̲** - [metal-fabr] swaging ma-

terial
— m̲ — **recubrimiento m̲** - [constr] facing, o
coating, material
— m̲ — **reducción f directa** - [miner] direct reduction material
— m̲ — **reemplazo m̲** - [ind] replacement,cing material
— m̲ — — m **para conservación f** - [constr]
maintenance replacement material
— m̲ — **regulación f** - regulating material
— m̲ — **relleno m̲** - backfill(ing), o soil, o
fill(ing),material; (back)fill; bedding material
— m̲ — — m **bien drenado** - [constr] well
drained backfill(ing material)
— m̲ — — m **colocado** - [constr] placed backfill (material)
— m̲ — — **compactado** - [constr] compacted, o
tamped, backfill (material)
— m̲ — — m **con calidad f** - [constr] quality
backfill(ing material)
— m̲ — — f **superior** - [constr] high, o
top, quality, backfill(ing material)
— m̲ — — **drenado** - [constr] drained back-
fill(ing material)
— m̲ — — m **especificado** - [Constr] specified
backfill(ing material)
— m̲ — — m **estándar** - [constr] standard back-
fill(ing material)
— m̲ — — m **evaluado** - [constr] evaluated
backfill(ing material)
— m̲ — — m **lubricado** - [constr] lubricated
(back)fill(ing material)
— m̲ — — m — **por lluvia f** - [constr] rain
lubricated (back)fill(ing material)
— m̲ — — m **para estructura f** - [constr] structural backfill(ing material)
— m̲ — — **tubería f** - [constr] pipe back-
fill(ing material)
— m̲ — — m — **uso m corriente** - [constr] commonly used backfill(ing material)
— m̲ — — m **proyectado** - [constr] designed
backfill(ing material)
— m̲ — — m **regulado** - [constr] controlled
backfill(ing material)
— m̲ — — m **seleccionado** - [constr] selected, o
chosen, backfill(ing material)
— m̲ — **remoción f** - removal,ving material
— m̲ — — f **automática** - automatic, o self, removal, material
— m̲ — — f — **de contrapeso m̲** - [grúas] counter weight self removal material
— m̲ — — f **de contrapeso m̲** - [grúas] counterweight removal,ving material
— m̲ — **reparación f** - [ind] repair material(s)
— m̲ — — f **empleado(s)** - [ind] repair mate-
rial(s) used
— m̲ — — f, **usado(s), o utilizado(s)** - [ind]
repair material(s) used
— m̲ — **respaldo m̲** - [constr] backing materials
• [sold] véase planchuela † para respaldu ⅲ
— m̲ — **retén m̲** - [constr] stop, o retainer,
material
— m̲ — — m **para agua m̲** - [constr] waterstop
material
— m̲ — **revestimiento m̲** - lining material •
[constr] façade, o face,cing, material
— m̲ — **rodillo m̲** - [metal-lam] roll material
— m̲ — **servicio m̲ pesado** - heavy duty material
— m̲ — **sobrecarga f** - [constr] surcharge fill
— m̲ — **soldadura f** - [sold] weld(ing), mate-
rial(s), o supply,lies
— m̲ — — f **con proporción f alta de hidrógeno**
• [sold] high hydrogen welding material
— m̲ — — f — — f **baja de hidrógeno m̲**—[sold]
low hydrogen welding material
— m̲ — — **con arco m̲** - [sold] arc welding supply,lies
— m̲ — **soldarse v** - [sold] material to be,
welded, o soldered • weld material
— m̲ — **subdrenaje m̲** - [hidr] subdrain(age) material
— m̲ — **superficie f** - surface,cing material

material m **para tambor(es)** m [mec] drum material
— m — — **para excavación** f - [constr-equip] digging, o excavating, drum material(s)
— m — **techado(s)** m - [constr] roofing material
— m — **techar** v - [constr] roofing material(s)
— m — **terminación** f - finishing material(s)
— m — **terraplén** m - [constr] embankment, o fill, material(s)
— m — m **colocado** - [constr] placed, embankment, o fill, material(s)
— m — **tubería** f - [tub] tube, o pipe, material
— m — f **para conducción** f - [petról] pipeline material(s)
— m — **tuerca** f - [mec] nut material
— m — f **especial(es)** - [mec] soecial nut material
— m —, **usarse**, o **utilizarse** - [ind] material to be used
— m — **uso** m **corriente** - commonly used material
— m — **vía** f - [f.c.] track material(s)
— m — **zunchar** v - [transp] strapping material
— m **pasado por, cedazo, o tamiz** m - [constr] passing material
— m **peor** - worse material • worst material
— m **perforado** - [mec] perforated material
— m **permeable** - [constr] pervious material
— m — **para relleno** m - [constr] permeable backfill(ing material)
— m **pesado** - heavy material
— m **piezoeléctrico** - [electrón] piezoelectric material
— m **pintado** - [ind] painted material
— m **pizarroso** - [miner] shal(e)y material
— m **plano** - [metal-lam] flat, material, o stock
— m **plástico** - [plást] plastic (material)
— m — **especial** - [plást] special plastic (material)
— m — **laminado** - [plást] laminated plastic (material)
— m — **negro** - [plást] black plastic (material)
— m — **para uso** m **múltiple** - [plást] multiuse plastic (material)
— m — **reforzado** - [plást] reinforced plastic (material)
— m — — **con fibra(s)** f **de vidrio** m - [plást] fiber glass reinforced plastic (material)
— m **por labrar** - [mec] stock
— m — — **avanzado gradualmente** m - [mec] inched stock|
— m — — v **con tolerancia** f **mínima** - [ind] accurate size stock
— m — — v **plano** - [ind] flat stock
— m — **trefilar** v - [trefil] stock
— m — — v **plano** - [trefil] flat stock
— m **precisado** - [ind] required material
— m **preciso** - [ind] required material • particular material
— m **prefabricado** - [ind] prefabricated material
— m — **para entibación** f - [constr] prefabricated lining material
— m — — f **de túnel(es)** m - [constr] prefabricated tunnel lining material|
— m **preferido** - preferred material
— m **premezclado** - [ind] premixed material
— m **premoldeado** - [constr] preformed, o premolded, material
— m — **para junta(s)** f - [mec] preformed joint material
— m — — — f **para dilatación** f - [constr] preformed expansion joint material
— m **prerreducido** - [metal-prod] prereduced material
— m **primero** - [ind] raw material
— m **principal** - [ind] main material
— m **probado** - tested, o proven, material
— m **procesado** - [ind] processed material
— m **protector** - [ind] protective material
— m **protegido** - [ind] protected material
— m **proviniente de desgaste** m - [mec] wear material
— m **provisto** - provided, o furnished, material
— m — e **instalado** - [ind] furnished and installed, material • furnished and installed basis

material m **proyectado** - designed material
— m **publicitario** - advertising material
— m **que produce hidrógeno** m **alto** - [sold] high hydrogen (producing) material
— m — — m **bajo** - [sold] low hydrogen (producing) material
— m — **protege** v - protecting,tive material • [sold] shielding material
— m — — v **arco** m - [sold] arc shielding material
— m — **sirve para formación** f **de escoria** f - [sold] slagging material
— m **quebradizo** - brittle material
— m **quirúrgico** - [medic] surgical, material, o supply,lies
— m **quitado** - removed material
— m **radiactivo** - [nucl] radioactive material
— m — **inducido** - [nucl] induced radioactive material
— m — **natural** - [nucl] natural radioactive material
— m **recarburante** - recarburizing material
— m **recibido** - received material
— m **recirculado** - [ind] recycled material
— m **recto** - [trefil] straight material
— m **recubridor** - [constr] facing material
— m **recuperable** - recoverable, o salvageable, material
— m **recuperado** - [ind] recovered, o salvaged, material • recycled material
— m **rechazado** - rejected material
— m **reducido** - [miner] reduced material
— m — **directamente** - [miner] directly reduced material
— m — **parcialmente** - [miner] partially reduced material; material partially reduced
— m — — **recuperado** - [metal-prod] recovered partially reduced material •
— m — **recuperado** - [metal-prod] recovered reduced material
— m **reemplazado** - replaced material
— m **refractario** - [refract] refractory material
— m — **ácido** - [refract] acid, o siliceous, refractory material
— m — **silicoso** - [refract] siliceous, o acid. refractory material
— m **removido** - removed material
— m **reprocesado** - [ind] reprocessed, o recirculated, o recycled, material
— m **requerido** - required material • [ind] required equipment
— m **resistente** - resistant material
— m — **a abrasión** f - [petról] abrasion resistant material
— m — — **corrosión** f - [petról] corrosion resistant material
— m — **de rayón** m - [textil] tough rayon material
— m — — — m **para banda** f **para rodamiento** m - [autom-neumát] tough rayon belt material
— m **restante** - remaining, o other, material
— m **retenido** - [suelos] retained, o withheld, material
— m **rival** - rival, o competing,titive, material
— m **rodado** - [constr] rounded material
— m **rodante** - [f.c.] rolling stock; equipment
— m — **actual** - [f.c.] existing, rolling stock, o equipment
— m — **ferroviario** - [f.c.] railway rolling stock
— m — **nuevo** - [f.c.] new, rolling stock, o equipment
— m **sacado** - removed material
— m — **de tubería** f - [ambient] material removed from @ duct
— m **salido** - [contab] outgoing material
— m **sedimentable,tario** - [hidr] sedimentary material
— m **según norma** f - standard material
— m **seleccionado** - selected, o chosen, material
— m **sellador** - sealing material; sealant
— m — y **aislador** - [aislac] sealing-insulating material

material m sellante* - [mec] sealing material
— m — **en forma** f **de cinta** f - [constr] tape sealant
— m **semiprocesado** - semi-processed material
— m **semisólido** - semisolid material
— m **sensible** - sensitive material
— m — **a mecanización** f - [mec] machining sensitive material
— m — — — f **por abrasión** f - [mec] abrasion machining sensitive material
— m **separado** - separated material • [constr] screened out material
— m **similar** - similar material • competitive material
— m **sin analizar** - [ind] unanalyzed material
— m — **clasificar** - [constr] unclassified, o bank run, material
— m — **cribar** - [constr] bank run material
— m — **decapar**—[metal-trat] unpickled material
— m — **quemar**—[combust] unburned material
— m **sintético** - [plást] synthetic material
— m **sobrante** - surplus material • [f.c.] extra stock
— m **sólido** - solid material
— m **substitutivo** - alternate material
— m **suelto** - loose material
— m — **transportado** - [miner] transported, o hauled, loose material
— m **suficiente** - sufficient material • [ind] sufficient stock
— m **sujetado** - [mec] fastened material
— m **suministrado** - supplied material
— m **suspendido** - [hidr] suspended material
— m **tal** - such material
— m **tamizado** - [constr] screened, o passed, material
— m **templado** - [metal-trat] tempered, o hardened, material
— m **tenaz** - tough material
— m **terminado** - finished, o completed, material
— m — **con nitrógeno** m - [metal-trat] nitrogen finished material
— m **térreo** - [miner] earthy material
— m **típico** - typical material
— m **trabajable*** - workable material
— m **trabajado** - [ind] worked material
— m **transportado** - [ind] transported, o carried, material
— m — **desde fuera de obra** f—[constr] brought, o imported, material
— m — **desde punto(s)** m **distinto(s)** - [constr] overhaul
— m — — **sitio(s)** m **distinto(s)** - [constr] imported material
— m **trefilado** - [trefil] drawn material
— m **tubular** - [tub] pipe(-type) material
— m — **montado** - [tub] mounted pipe-type material
— m **usado** - material used
— m — **para reparación** f - [ind] repair material used
— m **utilizado** - used material(s) • work material(s)
— m — **para reparación** f - [ind] repair material(s) used
— m **vario(s)** - sundry, o miscellàneous, material(s)
— m **volátil** - [coque] volatile material(s)
— m — **captado** - [coque] captured volatile(s)
— m — **condensado(s)** - [coque] condensed volatile(s) (materials)
— m — **en carbón** m - [combust] coal volatile material
— m — — — m **seco** - [combust] dry coal volatile material
— m — — m **vegetal** - [combust] charcoal volatile material
— m — — — m **seco** - [combust] dry charcoal volatile material
materiales m - materials; matter • supplies • [com] commodity,ties • [ind] stock(s) • resource(s) • [mec] véase también equipo(s)
materialización f - . . . . • implementation

materialización f **de proyecto** m - project implementation
matorral m - . . .; brush
matraz f - [ind] . . .; flask
— m **para destilación** f - [ind] distilling flask
matricería f - [mec] die making • die shop • dies
matricero m - [mec] die maker
— m **diestro** - [mec] skilled die maker
matrícula f - . . .; enrollment; registration • register • [autom] license (plate) • [nav] registry
matriculado,da a - matriculated; enrolled; registered • [autom] licensed
matricular v - . . . • [autom] to register
matriz f - . . .; jig • [imprent] stencil; reproducible • [legal] original, deed, o record | a - [legal] parent
— f **actual** - [mec] present insert • present die
— f **ajustada** - [mec] set die
— f **almacenada** - [mec] stored die
— f **armada** - [mec] assembled die
— f **cambiable** - [mec] changeable die
— f **cantonera** - [mec] corner die
— f **cerrada** - [mec] closed die
— f **con orificio** m - [mec] cored die
— f — — m **en basto** - [mec] rough cored die
— f — **punto(s)** m - [electrón] dot matrix
— f **confeccionada** - [mec] made, o built, die
— f **convencional** - [mec] conventional die
— f **correspondiente** - [metal-fabr] respective die
— f **corriente** - [mec] common, o conventional, die
— f **de conexión(es)** f - [dib] wiring reproducible
— f — **diagrama** m - [dib] diagram reproducible
— f — — m **para conexión(es)** f - [dib] wiring, diagram, o drawing, reproducible
— f — **tipo oscilante** - [mec] rocker type die
— f **esquinera** - [mec] corner die
— f **fabricada** - [mec] made, o manufactured, die
— f **hecha** - [mec] made, o built, die
— f **inferior** - [mec] lower die • inferior die
— f **insumo-producto** - input-product pattern
— f **individual** - [mec] individual die • die component
— f — **de tipo** m **oscilante** - [mec] rocker type individual die
— f **intercambiable** - [mec] interchangeable die
— f **maestra** - [mec] master die
— f **Magna-Die** - [mec] Magna-Die die
— f — — **de Whistler**—[mec] Whistler Magna-Die
— f **magnética** - [mec] magnetic die
— f **modular*** - [mec] changeable die
— f **oscilante** - [mec] rocker die
— f — **universal** - [mec] universal rocker die
— f **para bizcocho(s)** m - [culin] biscuit die
— f — **buje(s)** m - [mec] bushing die
— f — **conjunto** m - [mec] set, o assembly, die
— f — — m **maestro** - [mec] master set die
— f — **corrugación(es)** f—[mec] corrugating die
— f — **corrugar** v - [mec] corrugating die
— f — **cortar sobrante(s)** m - [metal-fabr] trimming die
— f — **corte** m - [mec] cutting die • notching die • [tub] cutoff die
— f — **curvar** v - [mec] curving die
— f — **doblamiento** m - [mec] bending die
— f — **doblar** - [mec] bending die
— f — — v **pestaña** f—[mec] flange bending die
— f — — v **pestaña** f **para fregadero** m - [mec] sink flange bending die
— f — v — f — **pileta** f - [mec] sink flange bending die
— f — — v — f **esquinera** - [mec] corner flange bending die
— f — **embridar** v - [mec] flanging die
— f — **embutir** v - [mec] drawing die
— f — — v **forma(s)** f - [mec] form drawing die
— f — — v **pestaña(s)** f - [mec] flange drawing die
— f — — v — f **para, fregadero** m, o **pileta** f

[mec] sink flange drawing die
matriz f para entalladura f - [mec] notch(ing) die
— f — estampar v - [mec] drawing die
— f — v forma(s) f - [mec] form drawing die
— f — freno m - [mec] brake die
— f — hacer f rebaje(s) m - [mec] cutting die
— f — muesca(s) f - [mec] notch(ing) die
— f — perforación(es) f - [mec] punching die
— f — pestañar v - [mec] flanging die
— f — pestañas f - [mec] flange die
— f — f, cantonera(s), o esquinera(s) - [mec] corner bending die
— f — f para, fregadero m. o pileta f - sink flange drawing die
— f — prensa f en 0 - [mec] 0-press die
— f — prensadura f - [mec] pressing die
— f — punzadura f - [mec] punch(ing) die
— f — f para desagüe m - [mec] drain hole punching die
— f — punzar v - [mec] punch(ing) die
— f — v para desagüe m - [mec] drain hole punching die
— f — recortar v - [metal-fabr] trimming die
— f — recubrimiento m - [metal-fabr] coating die
— f — f m con orificio m en basto - [mec] rough cored coating die
— f — sobrante(s) m - [mec] trimming die
— f — terraja f - [herram] threading (machine) die
— f — trefilado m - [alambre] wire drawing die
— f perforadora - [mec] perforating, o drilling, die
— f magnética - [mec] magnetic perforating die
— m proyectado - [mec] designed die
— f puesta a punto - [mec] set die
— f rígida - [mec] rigid die • permanent tool
— f según medida f - [mec] custom (made) die • custom tooling
— f soldadora - [sold] welding die
— f superior - [mec] upper die
— f — de tipo m oscilante - [mec] rocker type upper die
— f tenaz - [mec] tough matrix
— f vítrea - [geol] vitreous matrix
matrizado* m - [mec] forming; drawing
— m de pieza f - [metal-fabr] part forming
matrizado,da a - [metal-fabr] drawn; formed
matrizar v - [metal-fabr] to draw; to form
máxima y mínima a - high-low; hi-lo
maximización* f - maximization,zing
— f de solidez f - ruggedness maximization
maximizar* v - to maximize
— v beneficio m - to maximize @ benefit
— v solidez f - to maximize @ ruggedness
máximo m - . . . . • most • peak
— m absoluto - absolute maximum
— m admisible - maximum allowable
— m admitido - allowed maximum
— m de azufre - [metal-prod] maximum sulfur
— m — m tolerado - [metal-prod] maximum allowable sulfur
— m — carbono m - [metal-prod] maximum carbon
— m — comodidad - maximum, o utmost, comfort
— m — durabilidad f - maximum, o utmost, durability
— m — frecuencia f - [electrón] maximum frequency
— m — f desplazado - [electrón] shifted maximum frequency
— m — protección f - maximum, o utmost, protection
— m — resistencia f - maximum resistance • maximum strength
— m — salida f - maximum outflow
— m — f de caudal m de aire m - maximum air outflow
— m definitivo - definite maximum
— m en variación f - maximum variation
— m especificado - specified maximum

máximo m estándar - standard maximum
— m exigible - maximum, requirable, o demandable
— m fuera de brazo(s) m - [mec] maximum outside of @ fork(s)
— m mantenido - maintained maximum
— m nominal - [electr] fully rated
— m para artefacto(s) m - [electr-instal] maximum for @ accessory,ries
máximo,ma a - maximum • allowable; ultimate; peak; utmost • more • heavy
mayor m - . . . . [milit] major | a - . . . . • most; higher; better; faster; greater • senior • largest; larger (of two)
— a a menor a - tapered • decreasing
— a cuantía f - larger, o major, quantity, o amount
— a de dos catetos m - [geom] larger of two legs • [sold] larger of two legs
— m de edad f - [legal] major; of age
— a disponible - largest available
— m fabricante - [ind] largest, manufacturer, o producer
— m — en mundo m - [ind] world's largest, manufacturer, o producer
— a posible - largest possible
— adv que . . . - greater than . . .
— a — óptimo - higher than optimum
— a resistencia f - greater resistance • greatest resistance
— a — f a abrasión f - [sold] greatest, o best, abrasion resistance
— a valor m - [fisc] value-added
mayordomo m - . . . . • [labor] foreman; véase también capataz m
mayoría f - . . . | [adv] most; almost all
— f de aplicación(es) f - most applications
— f — caso(s) - most, cases, o conditions
— f — condición(es) f - most condition(s)
— f — caso(s) m - most, conditions, o situations
— f — distribuidor(es) m - [com] most dealers • dealer mainstream
— f — edad - [legal] majority; of age
— f — trabajo(s) m - most job(s)
— f — voto(s) m - majority of votes
— f grande - large majority • overwhelming majority
mayorista m - [com] wholesaler
mayormente adv - largely; mainly; mostly; basically; primarily • extensively; materially
mayúscula f - [gram] . . . . • [imprent] upper case
maza f - [herram] . . .; mall(et) • [constr] driving, head, o hammer • [constr-equip] monkey • [ruedas] . . .; boss; véase también cubo m
— f de engranaje m - [mec] gear, hub, o nave
— f — martinete m - [mec] monkey • [constr-pil] driving, head, o hammer
— f — m operada con aire m comprimido - [constr-pil] air operated driving head
— f — rueda f dentada - [mec] sprocket, hub, o nave • gear boss
— f para hincadora f con martinete m - [constr-pil] (pile driver) driving, head, o hammer
— f y rayos m (de rueda f) - [mec-ruedas] wheel spider
mazarota f - [metal-prod] . . . . • hot top; sink; head; riser • upper part of @ ingot • fin • [refract] gate(s)
— f con revestimiento m - [metal-prod] lined hot top
— f — m de ladrillo(s) m - [metal-prod] brick lined hot top
mazo m - [herram] . . .; hand maul; mall; sledge • block • bettle • [electr-mot] harness • [mec-ruedas] véase también cubo m
— m a retardador m - [sold] timer harness
— m — m para frecuencia f alta - [sold] pre-flow timer harness
— m anterior - [electr-instal] forward harness

mazo m̲ con brazo(s) m̲ radial(es) - [mec] spider hub
— m̲ — . . . conductor(es) m̲ - [electr-instal] . . . wire harness
— m̲ de cable(s) m̲ - [electr-instal] (wire, o̲ wiring) harness
— m̲ — m̲ delantero - [electr-instal] forward wiring harness
— m̲ — m̲ fijado - [electr-instal] attached harness
— m̲ — m̲ instalado - [electr-instal] installed harness
— m̲ — m̲ para avance m̲ - [electr-instal] advance, o̲ forward, wiring harness
— m̲ — m̲ — enchufe m̲ - [electr-instal] plug lead harness
— m̲ — m̲ — m̲ para conversión f̲—[electr--instal] conversion plug lead harness
— m̲ — m̲ para entrada f̲ - [electr-instal] plug lead harness
— m̲ — m̲ — m̲ — f̲ a enchufe m̲ - [electr--instal] plug lead harness
— m̲ — m̲ — f̲ — — m̲ para conversión f̲ - [electr-instal] conversion plug lead harness
— m̲ — m̲ — m̲ marcha f̲ atrás - [electr-instal] reverse, o̲ rear, wiring harness
— m̲ — m̲ — m̲ retroceso m̲ - [electr-instal] reverse, o̲ rear, wiring harness
— m̲ — m̲ — m̲ posterior - [electr-instal] rear wiring harness
— m̲ — m̲ — m̲ principal - [electr-instal] main harness
— m̲ — m̲ — m̲ trasero - [electr-instal] rear wiring harness
— m̲ — conductor(es) m̲ - [electr-instal] (wire, o̲ wiring) harness; wire assembly; lead
— m̲ — — m̲ a retardador m̲ - [sold] timer harness
— m̲ — — m̲ — m̲ para frecuencia f̲ alta - [sold] pre-flow timer (lead) harness
— m̲ — — m̲ conectado a tierra f̲ - [electr--instal] grounded wiring harness
— m̲ — — m̲ delantero - [electr-instal] front, o̲ forward, wiring harness
— m̲ — — m̲ para chasis m̲ - [autom-electr] chassis wiring harness
— m̲ — — m̲ — entrada f̲ - [electr-instal] lead harness
— m̲ — — m̲ — fuerza f̲ motriz - [agric-equip] power unit wiring harness
— m̲ — — m̲ — interruptor m̲ - [comb.int] switch, o̲ breaker, wire, harness, o̲ assembly
— m̲ — — — m̲ para detención f̲ - [comb.--int] stop(ping), switch, o̲ breaker, wire assembly
— m̲ — — — remolque m̲ - [autom-electr] trailer wire,ring harness
— m̲ — — — suministro m̲ posterior (de gas m̲) - [sold] after flow harness lead
m̲ — — m̲ — vehículo m̲ - [autom-electr] vehicle wiring harness
— m̲ — — m̲, posterior, o̲ trasero m̲ - [electr--instal] rear wiring harness
— m̲ — eje m̲ - [mec] axle, o̲ shaft, hub
— m̲ — émbolo m̲ - [mec] piston hub
— m̲ — hierro m̲ - [herram] (iron) sledge (hammer)
— m̲ — madera f̲ - [herram] (wooden) mallet
— m̲ — supercarreteras f̲ interestatales—[vial] interstate (highways) hub
— m̲ — trituradora f̲ - [constr-equip] crusher head
— m̲ delantero m̲ - [electr-instal] forward, o̲ front, (wiring) harness
— m̲ fijado - [electr-instal] attached harness
— m̲ girado - [mec] rotated hub
— m̲ instalado - [mec] installed hub • [electr--instal] installed harness
— m̲ moldeador* - véase recalcador m̲
— m̲ — para barrenos,nas - véase recalcador m̲ para barrenos,nas
— m̲ — encablado* m̲ - [electr-instal] wiring harness

mazo m̲ para suministro m̲ posterior (de gas m̲) - [sold] after-flow harness
— m̲ posterior - [electr-instal] rear (wire) harness
— m̲ principal - [electr-instal] main harness
— m̲ trasero - [electr-instal] rear harness
— m̲ usado - [mec] used mallet
mecánica f̲ de fluido(s) m̲ - fluid meachanics
— f̲ de proceso(s) m̲ - [ind] process mechanics
— f̲ de suelo m̲ - [suelos] soil(s) mechanics
— f̲ general - general mechanics
— f̲ — de suelo m̲ - [suelos] general soil(s) mechanics
— f̲ pura - pure mechanics
— f̲ — de fluido(s) m̲ - pure fluid mechanics
mecánicamente adv - . . . • power driven
mecánico m̲ - . . .; repairman; serviceman
— m̲ aficionado - amateur, o̲ home, mechanic
— m̲ ayudante - [mec] assistant mechanic
— m̲ calificado - [mec] qualified mechanic
— m̲ — para refrigeración f̲ - [ind] qualified refrigeration mechanic
— m̲ capacitado - [mec] qualified mechanic
— m̲ general—[mec] overall, o̲ general, mechanic
— m̲ jefe m̲ - [mec] head mechanic
— m̲ llamado - [ind] called, mechanic, o̲ serviceman, o̲ repairman
— m̲ maestro - [mec] master mechanic
— m̲ para conservación f̲ - maintenance mechanic
— m̲ — f̲ de laminador m̲ - [metal-lam] millwright
— m̲ — mantenimiento m̲ - [ind] maintenance mechanic
— m̲ — refrigeración f̲ - [ind] refrigeration, mechanic, o̲ serviceman, o̲ repairman
— m̲ principal - [mec] master mechanic
mecánico,ca a̲ - . . .; motorized
mecanismo m̲ - . . .; fixture • gear • unit • train
— m̲ accionado con cuña(s) f̲ - [mec] wedge operating mechanism
— m̲ — resorte(s) m̲ - [mec] springloaded mechanism
— m̲ accionador - [mec] driving mechanism
— m̲ — para rodillo m̲ - [metal-lam] roll driving mechanism
— m̲ — — m̲ extractor - [col.cont] withdrawal roll driving mechanism
— m̲ activador m̲ - activating, o̲ driving, mechanism
— m̲ alimentador m̲ - [mec] feed(ing) mechanism
— m̲ — para alambre m̲ - [sold] wire feed(ing), mechanism, o̲ unit, o̲ device
— m̲ alzador - [mec] lifting mechanism
— m̲ automático - automatic, mechanism, o̲ device
— m̲ — para marcha f̲ en vacío m̲ - [mec] automatic idling device; automatic idler
— m̲ — — — f̲ lenta - [wold] automatid idler
— m̲ — — — f̲ sin carga f̲ - [mec] automatic, idler, o̲ idling device
— m̲ auxiliar -auxiliary, mechanism, o̲ device
— m̲ — para elevación f̲ - [grúas] auxiliary hoist(ing device)
— m̲ — — impulsión f̲ - [grúas] auxiliary drive
— m̲ — — — f̲ para elevación f̲ - [grúas] auxiliaray hoist drive
— m̲ basculante - [mec] rocking, o̲ tilting, mechanism, o̲ drive
— m̲ blindado - [mec] enclosed, o̲ shielded, mechanism, o̲ device
— m̲ — para avance m̲ - [mec] enclosed travel mechanism
— m̲ cambiador—[mec] changing, o̲ shifting, unit
— m̲ — eléctrico - [mec] electric, changing, o̲ shiftring, unit
— m̲ — — nuevo - [autom-mec] new electric shift(ing) unit
— m̲ — nuevo - [mec] new shift(ing) unit
— m̲ — para . . . velocidad(es) f̲ - [autom-mec] . . . speed shifting unit
— m̲ con engranaje(s) m̲ para reducción f̲ - [mec] gear reduction, box, o̲ unit
— m̲ — movimiento m̲ - moving, o̲ operating, me-

chanism
mecanismo m con resorte m - [mec] springloaded
mechanism
— m **contra traba(s) f** - [mec] lock mechanism
— m **corriente** - [mec] standard, mechanism, o
device
— m — **para impulsión f** - [mec] standard drive
— m **de engranaje(s) m** - [mec] gear, unit, o device; gearing
— m — **leva f** - [mec] cam mechanism
— m — **puesto m para cambio(s) m** - [f.c.] (reversing) switchstand mechanism
— m — **relojería f** - [mec] clockwork
— m — **resorte m** - [mec] spring mechanism
— m **desacoplador** - [mec] lockout (unit, o mechanism)
— m **desplazador** - [mec] shift(ing) mechanism;
mechanical shift(er)
— m **divisor** - dividing mechanism • indexing mechanism
— m **eléctrico** - [mec] electric(al), mechanism,
o unit, o device
— m — **para cambio(s) m** - [mec] electric
shift(ing), device, o unit, o mechanism
— m — — **m reparado** - [mec] repaired electric shift(ing), unit, o mechanism
— m — — **impulsión f** - [hydr-compuertas] electric wrench
— m **elevador** - [mec] lifting, o hoisting, device, o mechanism; hoist
— m — **para aguilón m** - [grúas] hoist worm boom
— m — **bobina(s) f** - [metal-lam] coil lifting mechanism
— m — — **f de hojalata f** - [metal-trat]
tin plate coil lifting, device, o mechanism
— m **entre eje(s) m** - [mec] inter-axle mechanism
— m — — **m para transmisión f** - [autom-mec]
inter-axle driveline
— m **esparcellama(s) f** - [metal-prod] flame
spreading, device, o mechanism
— m **extractor** - [mec] extractor,ting unit
— m — **para puerta(s) f** - [coque] door extractor, device, o unit
— m **impulsor m** - [mec] drive,ving, device, o
unit, o mechanism
— m — **para alambre m** - [sold] wire drive,ving,
mechanism, o unit
— m — **banda f** - [mec] belt, o band, driving
mechanism
— m — — **f transportadora** - [mec] conveyor
belt driving mechanism
— m — — **correa f** - [mec] belt driving mechanism
— m — — — **f transportadora** - [mec] conveyor
belt driving mechanism
— m — — **elevador m** - [ind] hoist driving
mechanism
— m — — — **m con cangilón(es) m** - [constr]
bucket conveyor driving mechanism
— m **planetario** - [mec] planetary drive
— m **interruptor** - [electr-equip] breaker, device, o mechanism
— m — **por rotura f** - [electr-equip] breaker
block
— m **propio** - [mec] self drive
— m **mecanizado** - [mec] motorized device
— m **motor** - [mec] power unit
— m **motorizado** - [ind] motor driven mechanism
— m — **con desplazamiento m en dos sentidos m** -
[mec] two axes motorized, slide, o mechanism
— m — — **en sentido m único** — [mec] single
axis motorized, slide, o mechanism
— m — — — **m lateral** - [mec] motorized cross
slide
— m — — — **m único** - [mec] single axis motorized slide
— m **movido por cuña(s) f** - [mec] wedge operating mechanism
— m **normalizado** - [mec] standardized drive
— m — **para impulsión f** - [grúas] standardized
drive
— m — — — **f de puente m** - [grúas] standardized bridge drive

mecanismo m nuevo - new, mechanism, o device
— m — **para cambio(s) m** - [mec] new shift(ing),
device, o unit
— m **operado** - [mec] operated, mechanism, o device
— m **oscilador** - [mec] oscillating, mechanism, o
device, o assembly
— m **para acción f retardada** - [comb.int] delayed action, device, o fixture
— m — **accionamiento m** - operating mechanism
— m — — **de copiador m para elevación f para
portacarga(s) m** - [metal-prod] camp type skip
hoist operating mechanism limit(ing) switch
— m — — **m para máquina f** - [mec] machine
drive,ving mechanism
— m — — — **f sinterizadora** — [metal-prod]
sintering machine drive,ving mechanism
— m — **accionar v rodillo m** - [col.cont] roll
driving mechanism
— m — — **v m extractor** - [col.cont] withdrawal roll drive,ving mechanism
— m — **alimentación f** - [mech] feed(ing) mechanism
— m — — **f gradual** - [Trefil] inch feed(ing)
mechanism
— m — — **f para alambre m** - [sold] wire
feed(ing) mechanism
— m — **arranque m** - [mec] starter,ting, mechanism, o gear
— m — **avance m** - [mec] travel, gear, o mechanism • [constr-equip] travel mechanism;
traction drive • [sold] travel
— m — — **gradual** - [mec] inching mechanism
— m — — **bajada f** - [mec] lowering, mechanism, o
device
— m — — **f para pescante m** - [grúas] jib lowering, mechanism, o device
— m — — **f planetaria f** - [grúas] planetary
lowering, mechanism, o device
— m — — **f — para pescante m** - [grúas] jib
planetary lowering (device)
— m — **barra f** - [mec] bar, o rod, device
— m — — **f taponadora** - [metal-prod] stopper
rod, mechanism, o rigging, o device
— m — **biela f** - [mec] crankshaft, gear, o
mechanism
— m — — **f y patín m** - [mec] crank(shaft) gear
— m — **bombeo m** - [petról] pumping jack
— m — **cambiador m** - [mec] changer mechanism
— m — — **m para toma(s) f** - [electr-equip]
tap changer mechanism
— m — **cambio(s) m** - [mech] shifter,ting, mechanism, o unit, o system, o device
— m — — **m para marcha f** - [mec] shifting, mechanism, o device; gear shifting device
— m — — **m — eje m** - [autom-mec] axle shifting, unit, o device
— m — — **m para embrague m** - [mec] clutch,
shifter, o shifting device
— m — — **m reemplazado** - [mec] replaced,
shifter, o shifting device
— m — **carrete m** - [mec] reel, unit, o device
— m — — **m para alambre m** - [sold] wire reel,
unit, o device
— m — **cierre m** - [mec] closing device
— m — — **m para solenoide m** - [electr-equip]
closing solenoid
— m — **colada(s) f** - [metal-prod] tapping, o
casting, mechanism
— m — **comercialización f** - [com] marketing, o
merchandising, mechanism
— m — **corte m** - [metal-lam] cutting, o shearing, mechanism
— m — **descenso m** - [grúas] lowering device
— m — — **m para pescante m** - [grúas] jib lowering device
— m — — **m planetario** - [grúas] jib planetary
lowering (device)
— m — **desplazador m** - [mec] shifter,ting, mechanism, o device
— m — **detonación f** - [explos] detonation,ting,
mechanism, o device
— m — **dirección f** - [autom-mec] steering (con-

trol, o gear)
mecanismo m **para dirección** f **irreversible** -
[autom-mec] irreversible steering gear
— m — **distribución** f - [electr-equip] switching gear
— m — **elevación** f - [grúas] hoisting, o lifting, mechanism; hoist • [hidr-compuertas]
lift • [grúas] hoist(ing) drive
— m — — f **con volante** m **manual** - [hidr-
-compuertas] hand wheel lift(er)
— m — — f — — m **sobre pedestal** m -
[hidr-compuertas] hand wheel pedestal mounted
lift(ing device)
— m — — t — m — — — m **con cojinete** m -
[hidr-compuertas] hand wheel pedestal mounted
lift with @ bearing(s)
— m — — f **para compuerta** f - [hidr-compuert]
gate lift(ing device)
— m — — f — — f **rodante** - [hidr-compuertas]
roller gate lift
— m — — f **sobre pedestal** m - [hidr-compuert]
pedestal mounted lift
— m — **formación** f - [mec] forming mechanism
— m — — f **para sínter** m - [metal-prod] (sinter) formation mechanism
— m — **freno** m - [mec] brake mechanism
— m — **giro** m - [grúas] rotation, o revolving,
mechanism, o system
— m — **(hacer) avanzar** v **planilla** f - [instrum]
chart drive
— m — **igualación** f—[mec] equalizing mechanism
— m — — f **de carga** f - [mec] load equalizing
mechanism
— m — **impulsión** f - [mec] drive,ving, mechanism, o unit, o train
— m — — f **con engranaje(s)** m - [mec] gear
drive,ving (mechanism)
— m — — f — — m **planetario(s)** - [mec] planetary gear drive,ving (mechanism)
— m — — f **para alambre** m - [sold] wire, feed
drive unit, o drive,ving mechanism
— m — — f **desplazamiento** m - [mec] travel
drive,ving mechanism
— m — — f — — m **transversal** - [mec] cross
travel drive,ving (mechanism)
— m — — f — — m **de carrillo** m - [grúas]
trolley cross travel drive,ving (mechanism)
— m — — f **planilla** f - [instrum] chart,
drive, f driving mechanism
— m — — f — — f **para . . . hora(s)** f -
[instrum] . . . hour chart drive (mechanism)
— m — — f **para puente** m - [grúas] bridge,
drive, o driving mechanism
— m — — f — — m **para grúa** f - [grúas] crane
bridge, drive, o driving mechanism
— m — **inversión** f - [mec] reversing mechanism
— m — **línea** f **motriz** - [autom-mec] driveline
— m — — f **conectado** - [autom-mec] connected driveline
— m — **mando** m - operating mechanism; drive
— m — **maniobra(s)** f - operating mechanism
— m — **mesa** f - [mec] table, mechanism, o gear
— m — — f **giratoria** - [mec] turntable gear
— m — **movimiento** m - drive,ving mechanism
— m — — m **de émbolo** m - [mec] piston driving
mechanism - [metal-prod] (mud, o clay, gun)
piston driving mechanism
— m — — m **lateral** - [mec] cross slide,ding
mechanism
— m — — m **transversal**—[mec] cross slide,ding
mechanism
— m — **oscilación** f - [constr] swing gear
— m — **propulsión** f - [mec] drive
— m — **plataforma** f **giratoria** - [mec] turntable
gear
— m — **reacción** f - reaction mechanism
— m — — f **para desoxidación** f - [metal-prod]
deoxidation reaction mechanism
— m — **reducción** f - [mec] reduction mechanism;
gear unit
— m — — f **para rotación** f - [grúas] swing reduction (mechanism)
— m — **regulación** f - control mechanism •
[electr-equip] switchgear

mecanismo m **para regulación** f **para entrada** f
para aire m - [mec] air intake control mechanism
— m — — f — — f — — m **tipo** m **celosía** f -
[mec] louver type air intake control mechanism
— m — — f **recorrido** m **de cubeta** f—[metal-
-prod] skip travel control mechanism
— m — — f — — m **montacargas** m - [metal-
-prod] skip travel control mechanism
— m — — f — — m **skip** m - [metal-prod]
skip travel control mechanism
— m — **reloj** m - [mec] clockwork
— m — **rodillo** m - [mec] roll mechanism
— m — — m **para arrastre** m - [mec] pinch
roll mechanism
— m — — m **para transferencia** f -
[cuerp-moled] transfer pinch roll mechanism
— m — — m **para transferencia** f - [cuerp.-
moled] transfer roll mechanism
— m — **rotación** f - [grúas] swing (mechanism) •
rotation mechanism
— m — **rotadora** f - [mec] rotator mechanism
— m — —/**alimentadora** f - [cuerp.moled] rota-
tor/feeder mechanism
— m — **seguridad** f - [segurid] safety control
— m — **seguro** m **contra rotación** f - [mec] rotation lock mechanism
— m — **soplador** m - [mec] blower mechanism
— m — — m **auxiliar** - [mec] auxiliary blower
mechanism
— m — **tablón(es)** m - log mechanism
— m — — m **para cierre** m - stop log mechanism
— m — **taponadora** f - [metal-prod] stopper rod,
mechanism, o rigging
— m — **torsión** f - [mec] torque mechanism
— m — **traba** f - [mec] locking mechanism
— m — — f **accionado con resorte** m - [mec]
springloaded locking mechanism
— m — — f **con resorte** m - [mec] springloaded
locking mechanism
— m — — f **contra rotación** f - [mec] rotation
lock mechanism
— m — **tracción** f - [grúas] running gear
— m — **transmisión** f - [autom-mec] drive(line)
— m — — f **conectado** - [autom-mec] connected
drive(line)
— m — — f **entre eje(s)** m - [autom-mec] inter-
-axle driveline
— m — **para vehículo** m - [autom-mec] vehicle driveline
— m — **traslación** f - [constr] travel mechanism
— m — **trinquete** m - [mec] pawl mechanism
— m — — m **para traba** f - [mec] lock pawl
mechanism
— m — **vinculación** f **sincrónica** - [metal-lam]
synchrotic mechanism
— m — **vuelco** m - [mec] tilting mechanism
— m **principal** - [mec] main mechanism
— m **para elevación** f - [grúas] main hoisting
mechanism
— m — — **impulsión** f - [grúas] main drive,ving
mechanism
— m — — f **para elevación** f - [grúas] main
hoist(ing) drive, o driving mechanism
— m **reductor** - [mec] reducer,cing,ction mechanism
— m **reforzado** - reinforced, o heavy duty, mechanism
— m **para impulsión** f - [mec] reinforced, o
heavy duty, drive, o driving train
— m — — f **de alambre** m - [sold] heavy duty
wire, drive, o driving mechanism
— m **removedor** - [coque] remover, o extractor,
unit
— m **para puerta(s)** f - [coque] door, remover, o extractor, unit
— m **reversible** - [f.c.] reverisble mechanism
— m **en puesto** m **para cambio(s)** m - [f.c.]
reversing switchstand mechanism
— m **sincrótico*** - [metal-lam] synchrotic mechanism
— m **totalmente blindado** - [mec] fully enclosed
mechanism

mecanismo m totalmente blindado para avance m -
[mec] fully enclosed travel mechanism
— m ——— traslación f - [mec] fully enclosed travel mechanism
mecanizibilidad* f - [mec] machinability
mecanizable adv - [mec] machinable; workable
mecanización f - [mec] . . .; machining; milling
• mechanical working • machine work • workability
— f anterior - [mec] previous, o prior, machining
— f anticipada - [mec] premachining
— f completa - complete, mechanization, o machining
— f con banda f - [mec] band machining
— f —— f abrasiva - [mec] abrasive band machining
— f —— precisión f - [mec] precision machining
— f de árbol m - [mec] shaft machining
— f —— borde m - [mec] edge machining
— f —— cavidad f - [mec] pocket machining
— f depresión f - [mec] pocket machining
— f —— eje m - [mec] axle machining • shaft
machining
— f —— muñón m - [mec] journal machining
— f —— paleta f - [mec-ventil] blade machining
— f —— ranura f - [mec] groove machining
— f —— rodillo(s) m—[metal-lam] roll machining
— f —— soldadura f - [sold] weld(ing), machining, o mechanization
— f —— superficie f - [mec] surface machining
— f —— f aplanada - [mec| flat surface ma-
chining
— f —— zona f - [mec] area machining
— f —— f por abrasión f—[mec] area machine,
abrading, o abrasion
— f durante período m más extendido - [mec]
extended workability
— f en basto m - [mec] scalping
— f —— desbaste m - [f.c.-ruedas] rough
— f fácil - [mec] easy workability • free machining
— f final - [mec] final, o finish, machining
— f —— con abrasión f - [mec] final abrasion
machining
— f libre - [mec] free machining
— f para acabado - [mec] finish(ing) machining
— f —— m por abrasión f - [mec] abrasion
finish(ing) machining
— f —— trabajo(s) con volumen m reducido -
[sold] low volume, mechanization, o machining
— f por abrasión f - [mec] abrasion machining
— f —— f introducida - [mec] introduced
abrasion machining
— f —— f para acabado m - [mec] abrasion
finish(ing) machining
— f preliminar - [metal-prod] scalping
— f superior - [mec] superior, machinability, o
workability
— f tenaz - [mec] rough machining
— f total - [mec] total machining • full power
mecanizadamente adv - [mec] under power
mecanizado* m - véase mecanización f
mecanizado,da a - [mec] mechanized • machined;
worked • power controlled • [sold] carried
mechanically
— a con precisión f - [mec] precision machined
— a en basto - [mec] rough machined • scalped
— a por abrasión f - [mec] abrasion machined
— a totalmente - [mec] full(y) machined • full
power
mecanizar v - . . .; to machine; to work • to
mechanize
— v árbol m - [mec] to machine @ shaft
— v borde m - [mec] to machine @ edge
— v con precisión f—[mec] to precision machine
— v chaflán m - [mec] to machine @ bevel
— v depresión f - [mec] to machine @ pocket
— v eje m - [mec] to machine @, axle, o shaft
— v en basto - [mec] to rough machine; to scalp
— v metal m - [mec] to machine @ metal
— v metal ferroso - [mec] to machine @ ferrous
metal

mecanizar v metal m no ferroso - [mec] to machine @ non-ferrous metal
— v muñón m - [papel] to machine @ journal
mecanizar v - [mec] . . .; to work
— v paleta f - [mec-ventil] to machine @ blade
— v por abrasión f - [mec] to abrasion machine
— v ranura f - [mec] to machine @ groove
— v superficie f - [mec] to machine @ surface
— v —— f aplanada - [mec] to machine @ flat
surface
— v zona f - [mec] to machine @ area
mecanografía f - [com] . . .; typing
mecedora f - [domést] rocking chair; rocker
mecer v engranaje m - [mec] to rock @ gear
— v suavemente - to rock gently
mecido,da a - rocked
— a suavemente - rocked gently
mecimiento* m - rocking; prefiérase mecedura f
— m —— engranaje m - [mec] gear rocking
— m suave - [mec] gentle rocking
mecha f - . . . • [herram] bit; drill; broach;
véase también broca f
— f con combustión f lenta - [explos] safety,
o slow burning, fuse
— f —— barrena f para roca f - [herram] rock
drill bit
— f para, encendido m, o ignición f - [combust]
lighting wick
— f —— lámpara f - lamp wick
— f —— perforadora f - [herram] drill(ing) bit
— f —— f para roca f - [herram] rock drill
bit
— f —— sacar núcleo(s) m - [herram] core drill
— f —— seguridad f - [explos] safety fuse
— f —— taladro m - [herram] drill bit
— f —— m para roca f - [herram] rock drill
bit
— f sacanúcleo(s) - [herram] core drill
— f sacatestigo(s) m - [herram] core drill
mechero m - [combust] . . .; burner
— m para gas m - [combust] gas burner
— m —— laboratorio m - [combust] Bunsen burner
— m —— petróleo m - [combust] oil burner
— m principal - [combust] main burner
media f anual - annual, o yearly, average
— f aritmética - [arit] arithmetical, average,
o mean
— a bobina f - [electr-mot] half, coil, o winding
— a —— f para excitatriz f - [electr-gen] exciter half, coil, o winding
— a —— f —— generador m - [electr-prod] generator half, coil, o winding
— a —— f —— motor - [electr-mot] motor half,
coil, o winding
— f caña f - [geom] half, round, o oval • [mec]
dowel • chamfer; groove
— a cupla* f - [mec] half coupling
— a embutición f - [mec] half mortise
— f geométrica - [geom] geometric mean
— a hora f - [chron] half hour
— a luna f - [astron] half moon • crescent
— f onda f - [electrón] half wave
— f ponderada - [mat] pondered, average, o mean
— a vuelta f - half turn
mediacaña f escariadora - [herram] side rasp
medicación f mercantil - [com] mercantile media-
tion
mediado(s) m de carrera f - [deport] mid-race;
race midpoint
— m —— f rally - [deport] rally midpoint
mediador m mercantil - [com] mercantile mediator
— m exclusivo - [com] exclusive mercantile
mediator
— m no exclusivo - [com] nonexclusive mercantile mediator
— m no exclusivo - [com] nonexclusive mediator
medianamente adv - . . .; moderately
medianera f - [constr] sidewall
mediano,na a - . . .; median; average; fair
— a en carbono m - [metal] medium carbon
mediante adv - . . .; through • whereby | [prep]
by

mediante adv carta † de crédito m - [fin] by @
letter of credit
— adv corte(s) - [electr-oper] by @ cut-off(s)
— adv — m en circuito m - [sold] by shortcir-
cuiting (cut-off)
— adv empleo m - using; by @ use of
— adv extrusión f - [ind] by extrusion
— adv pala(s) f - [mec] by spading
— adv termostato m - with @ thermostat; thermostatically
medición f - . . .; metering; gaging • sizing
— f automática - austomatic measuring • auto-
metrics*
— f con ayuda † de integrador m - [instrum] integrator assisted measuring
— f — — — m electrónico - [instrum]
electronic integrator assisted measuring
— f — indicador m con esfera f - [mec] dial
indicator measuring
— f — integrador m - [instrum] integrator
measuring
— f — — m electrónico - [instrum] electronic
integrator measuring
— f — paso(s) m - pacing (off)
— f continua - continuous measuring
— f de ácido m - [ind] acid measuring
— f — actuación f - [admin] performance measuring
— f — altura f - height measuring
— f — ancho(r) m - [mec] width measuring
— f — calidad f - quality measuring
— f — campo m magnético - [electr] magnetic
field measuring
— f — capacidad f - capacity, measuring, o
sizing
— f — carga f - [constr] load measuring
— f — — f muerta - [constr] dead load measuring
— f — — f viva - [constr] live load measuring
— f — cloaca f - [sanit] sewer, measuring, o
gaging
— f — combustible m - [combust] fuel measuring
— f — comportamiento m - [admin] performance
measuring
— f — conductibilidad - [electr-cond] conductivity measuring
— f — contragolpe m - [mec] backlash measuring
— f — cordón m - [sold] bead measuring
— f — — m plano m flat bead measuring
— f — — m — con cateto(s) m igual(es) -
[sold] equal, bead, o weld, measuring
— f — m — — — m — en ángulo m interior -
[sold] flat fillet with equal legs measuring
— f — — m — en ángulo m interior - [sold]
flat fillet measuring
— f — deformación f - [mec] deformation measuring
— f — densidad f - density measuring
— f — desempeño m - [admin] performance neas-
uring
— f — distancia f - distance measuring
— f — eficiencia f - [admin] efficiency measuring
— f — entrada f - entrance measuring • [hidr]
inlet, measuring, o gaging
— f — entrehierro m - [electr-mot] (air) gap
measuring
— f — espesor m - [mec] thickness measuring •
[metal-lam] gage measuring
— f — — m de pared f - wall thickness measuring
— f — fenómeno m - phenomenon measuring
— f — flujo m - [hidr] flow, measuring, o
gaging
— f — — m magnético - [electr] magnetic flow
measuring
— f — frecuencia f alta - [electrón] high frequency measuring
— f — — f elavada - [electrón] high frequency
measuring
— f — fuerza f - force, o strength, measuring
— f — — f coercitiva - coercive force measuring

medición f de grosor m - thickness measuring
— f — m de banda f - [mec] band thickness
measuring
— f — — m — — f para freno m - [mec] brake
band thickness measuring
— f — — m de cinta f - [mec] band thickness
measuring
— f — — m — f para embrague m - [mec]
clutch lining thickness measuring
— f — holgura f - [mec] clearance measuring
— f — — f de cubierta f - [mec] cover clearance measuring
— f — — f — f para cojinete m - [mec]
bearing cover clearance measuring
— f — humo(s) m - [ind] fume(s) measuring
— f — inducción f - [electr-oper] induction
measuring
— f — — f magnética - [electr-oper] magnetic
induction measuring
— f — intensidad f - intensity measuring
— f — — f de campo m - [electr-oper] field,
intensity, o strength, measuring
— f — — f — — m magnético - [electr-oper]
magnetic field intensity measuring
— f — — f magnética - [electr-oper] magnetic
intensity measuring
— f — juego m - [mec] play measuring
— f — — m longitudinal - [mec] end play measuring
— f — — m — de árbol m - [mec] shaft end
play measuring
— f — — — — m para entrada f (de
fuerza f) f - [mec] input shaft end play
measuring
— f — largo(r) m - [mec] length measuring
— f — líquido(s) m liquid(s) measuring
— f — — m cloacal(es) m - [sanit] sewage
measuring
— f — — m — doméstico(s) - [sanit] domestic
sewage measuring
— f — movimiento m - movement measuring
— f — — m de entrerrosca f - [mec] nipple
movement measuring
— f — — m — — f para tubo(s) m - [mec]
pipe nipple movement measuring
— f — nivel m - level measuring
— f — pieza f - [mec] part, o component,
measuring
— f — precarga f - [mec] preload measuring
— f — — f para cojinete m - [mec] bearing
preload measuring
— f — — f — — m para piñón m - [mec] pinion
bearing preload measuring
— f — presión f - pressure measuring
— f — — f para entrada f - intake pressure
measuring
— f — — f — f de agua m - water intake
pressure measuring
— f — — f — salida f - outlet pressure measuring
— f — — f — — f de agua m - water outlet
pressure measuring
— f — profundidad f - depth measuring
— f — propiedad(es) f—property,ties measuring
— f — — f magnética(s) - [electr-oper] magnetic property,ties measuring
— f — — f — de imán - [electr-oper] magnet's
magnetic property,ties measuring
— f — — f — — — m permanente - [electr-
-oper] permanent magnet magnetic property,ties measuring
— f — — f — — producto(s) m - [metal-prod]
product(s) magnetic property,ties measuring
— f — — f — — — m masivo(s) - [metal-prod]
massive product(s) magnetic property,ties
measuring
— f — — f de imán m (permanente) - [metal-
-prod] (permanent) magnet('s) property,ties
measuring
— f — rendimiento m—performance('s) measuring
— f — resistencia f - resistance('s) measuring
— f — — f de aislación f - [electr-equip] in-

sulation resistance measuring

medición f de **resistencia** f **eléctrica** - [electr] electrical resistance measuring

— f — f **efectiva** - [quím] resistivity measuring

— f — f — — **de agua** m - [hidr] water resistivity measuring

— f — f — — — **muestra** f - sample('s) resistivity measuring

— f — f — — — — f **de suelo** m - [suelos] soil sample('s) resistivity measuring

— f — f — — **infinita** f - infinite resistivity measuring

— f — **rigidez** f - rigidity measuring

— f — **soldadura** f - [sold] weld measuring

— f — — f en **ángulo** m **interior** - [sold] fillet (size) measuring

— f — **tamaño** m - size('s) measuring

— f — — m de **soldadura** f - [sold] weld size measuring

— f — — m — — f en **ángulo** m **interior** - [sold] fillet size measuring

— f — **temperatura** f - temperature measuring

— f — — f de **baño** m - bath temperature measuring

— f — **tiempo** m - time measuring

— f — **tierra(s)** f - [topogr] survey(ing)

— f — **trabajo** m - work measuring

— f — **tubería** f - [tub] tube,bing, measuring, o sizing

— f — **uniformidad** f - uniformity measuring

— f — — f en **inducción** f - [electr-oper] induction uniformity measuring

— f — — f — — f **magnética** - [electr-oper] magnetic induction uniformity measuring

— f — **variación** f - variation measuring

— f — — f en **flujo** m - flow variation measuring

— f — — f — — m **magnético** - [electr-oper] magnetic flow variation measuring

— f — **viaje** m - trip measuring

— f — — m **redondo** - round trip measuring

— f — **voltaje** m - [electr-oper] voltage measuring

— f **efectuada** - made, o performed, measuring

— f **en escala** f **cabal** - full scale measuring

— f — **obra** f - field measuring

— f **entre borne(s)** m - [electrón] across @ stud(s) measuring

— f **exacta** - exact, o accurate, measuring

— f **fácil** - easy measuring

— f **gráfica** - graphic measuring

— f **magnética** - [electr] magnetic measuring

— f **más fácil** - easier measuring

— f — **rápida** - quicker, o faster, measuring

— f **parcial** - partial measuring

— f **precisa** - exact, o accurate, o precise, measuring • fine metering

— f — **de voltaje** m - [electr-oper] accurate, o exact, o precise, voltage measuring

— f **rápida** - rapid, o quick, o fast, measuring

— f **real** - actual, measuring, o gaging

— f — **de cloaca(s)** f - [hidr] actual sewer(s), measuring, o gaging

— f — — **entrada** f - [hidr] intake, o inlet, actual, measuring, o gaging

— f **ultrasónica** - [electrón] ultrasound,sonic measuring

médico m **individual** - [medic] individual physician

— m **para planta** f - [ind] plant physician

medida f - . . . .; dimension • length • extent; degree • value • rate • gaging; sizing • measure; precaution; action • [instrum] reading | adv - (con) measurably; reservedly

— f **absoluta** - absolute measurement

— f **aceptable** - acceptable, measurement, o size

— f — **para neumático(s)** m - [autom-neumát] acceptable tire, measurement, o size

— f — — m **P-métrico(s)** - [autom-neumát] acceptable P-metric tire, measurement, o size

— f — — — m — **substitutivo(s)**—[autom-neum] acceptable P-metric substitute tire size(s)

medida f **aceptable para neumático(s)** m **substitutivo(s)** - [autom-neumát] acceptable substitute tire size

— f **acostumbrada** - [segurid] common precaution

— f — **para seguridad** f - [segurid] common safety precaution

— f **adicional** - additional, size, o measurement • [segurid] additional precaution

— f **alfa-numérica** f - [autom-neumát] alpha-numeric size

— f — **para neumático(s)** m - [autom-neumát] alpha-numeric tire, size, o measurement

— f **anotada** - [instrum] noted, o recorded, reading

— f **anterior** - previous size • older size

— f — **para neumático** m - [autom-neumát] older tire size

— f **antidumping** - [econ] antidumping measure

— f **apropiada**—appropriate, measurement, o size

— f **aumentada** - increased measurement • spread dimension

— f **automática** - automatic measurement

— f **básica** - basic, measurement, o dimension

— f **combinada** - combined, measurement, o size

— f **comercial** - commercial, o trade, size

— f **comparable**—comparable, measurement, o size

— f **comparada** - compared, measurement, o size

— f **con gancho m recogido** - [cadenas] drawn up (hook) dimension

— f — **indicador** m **con esfera** f - [mec] dial indicator measurement

— f **considerable** - substantial degree | adv - (en) substantially

— f **continua** - continuous measurement

— f **contra dumping** m - [econ] antidumping measure

— f **correcta para neumático** m - [autom-neumát] correct tire size

— f **correctiva** - corrective, o remedial, measure, o action, o procedure

— f — **posible** - [segurid] possible corrective action

— f **de abertura** f - [constr] opening, measurement, o size

— f — **accesorio** m - [mec] fitting, dimension, o size

— f — **ácido** m - [ind] acid measure,rement

— f — **actuación** f - [admin] performance measurement

— f — **alcantarilla** f - [constr] culvert, measurement, o size

— f — **altura** f - [mec] height measurement

— f — **ancho(r)** m - [mec] width measurement

— f — **árbol** m - [mec] shaft, measurement, o size

— f — **bancada** f - [tornos] bench dimension

— f — — f **para torno** m - [tornos] lathe bench dimension

— f — **bobina** f - coil, measurement, o size

— f — **cadena** f - [cadenas] chain, measurement, o size

— f — **calidad** f - quaility measurement

— f — **cámara** f - chamber, measurement, o size

— f — — f **para inspección** f - [constr] manhole size

— f — **campo** m - field measurement

— f — — m **magnético** - [electr] magnetic field measurement

— f — **cateto** m - [sold] leg, size, o length

— f — — m **de soldadura** f - [sold] weld, measurement, o size

— f — — m — — f en **ángulo** m **interior** - [sold] fillet size

— f — **caudal** - [hidr] flow measurement

— f — **cloaca** f - [sanit] sewer, measurement, o size

— f — **conductibilidad** f - [electr-cond] conductivity measurement

— f — **conducto** m - [constr] conduit, measurement, o size • [hidr] waterway area

— f — — m **para agua** m - [hidr] waterway area

— f — **conjunto** m - assembly, o overall, size

— f — — m **de calce** m - [mec] shim pack size

medida f de contragolpe m - [mec] backlash measurement
— f — cordón m - [sold] bead, o weld, measurement
— f — — m cóncavo - [sold] concave, bead, o weld, measurement
— f — — m — en ángulo m interior - [sold] concave fillet measurement
— f — — m convexo - [sold] convex, bead, o weld, measurement
— f — — m — en ángulo m interior - [sold] convex fillet measurement
— f — — m plano—[sold] flat bead measurement
— f — — m — con cateto(s) m igual(es) - [sold] flat equal leg bead measurement
— f — — — m — en ángulo m interior - [sold] equal leg flat fillet bead measurement
— f — — m — en ángulo m interior - [sold] flat fillet (weld) measurement
— f — corrosión f - corrosion rate
— f — corrugación f [mec] corrugation, measurement, o size, o dimension
— f — deformación f -[mec] deformation measurement
— f — densidad f - density measurement
— f — — f de material m - material density measure(ment)
— f — — — m para relleno m - [constr] backfill material density measure(ment)
— f — — f — relleno m - [constr] backfill material density measurement
— f — despunte(s) m - [metal-lam] crop(ping) measurement
— f — distancia f - distance measurement
— f — edificio m - [constr] building, size, o measurement
— f — efectividad f - efectiveness, o performance, measure(ment)
— f — — f de entrevista f - [labor] contact performance measure(ment)
— f — — f — — f sobre seguridad f - [segurid] safety contact performance measure
— f — eficiencia f - [ind] efficiency measurement
— f — — f productiva - [ind] production efficiency measurement
— f — eje m - [mec] axle, o shaft, measurement
— f — entrada f - [constr] inlet, measurement, o geometry • [hidr] inlet measurement
— f — entrehierro m - [electr-mot] air gap measurement
— f — espesor m - thickness measurement
— f — — m de pared f - wall thickness measurement
— f — — m útil de cordón m - [sold] throat, measurement, o dimension
— f — — m — — m para raíz f - [sold] root pass throat, measurement, o dimension
— f — — — m — — m primero - [sold] first pass throat, measurement, o dimension
— f — estructura f - [constr] structure, measurement, o dimension
— f — excavación f - [constr] excavation, measurement, o dimension
— f — fenómeno m - phenomenon measurement
— f — flujo m - flow measurement
— f — — m magnético - [electr-oper] magnetic flow measurement
— f — frecuencia f - [electrón] frequency measurement
— f — — f, alta, o elevada - [electrón] high frequency measurement
— f — fuerza f—force, o strength, measurement
— f — — f coercitiva - coercive force measurement
— f — garganta f - [mec] throat, measurement, width, o size
— f — grano(s) m - véase granulometría f
— f — grieta f - [ind] crack measurement
— f — grosor m - thickness measurement
— f — — m de banda f - [mec] band thickness measurement • [metal-lam] band, o strip,

thickness measurement
medida f de grosor m de cinta f - [mec] lining thickness measurement
— f — — m — — f para embrague m - [mec] clutch lining thickness measurement
— f — — m de cinta f - [mec] lining thickness measurement
— f — holgura f - [mec] clearance measurement
— f — — f de cubierta f - [mec] cover clearance measurement
— f — — f — — f para cojinete m - [mec] bearing cover clearance measurement
— f — horno m alto - [metal-prod] blast furnace, measurement, o size
— f — horquilla f - [mec] fork, o yoke, measurement
— f — humo(s) m - [ind] smoke, o fume(s), measurement
— f — imán m - magnet, measurement, o size
— f — inducción f - [electr] induction measurement
— f — — f de flujo m - [electr] flow induction measurement
— f — — f — — m magnético - [electr] magnetic flow, induction, o density, measurement
— f — inducción f magnética - [electr] magnetic, induction, o flow density, measurement
— f — intensidad f - intensity measurement
— f — — f de campo m - [electr] field intensity measurement
— f — — f — — m magnético - [electr] magnetic field intensity measurement
— f — — f magnética - [electr] magnetic intensity measurement
— f — juego - [mec] play measurement
— f — — m longitudinal - [mec] end play measurement
— f — — m — de árbol m - [mec] shaft end play measurement
— f — — m — — — m para entrada f - [mec] input shaft end play measurement
— f — — m — — m para entrada f de fuerza - [autom-mec] input shaft end play measurement
— f — junta f - [sold] joint, measurement, o dimension, o size
— f — lado m - side measurement
— f — — m de cordón m - [sold] bead side measurement • leg size
— f — laminación f - [metal-lam] mill length
— f — largo(r) m - [mec] length measurement
— f — letrero m - [vial] sign, size, o measurement
— f — lingotera f - [metal-prod] ingot mold, size, o measurement, o dimension
— f — luz f - [ilumin] light measurement • [mec] clearance, measurement, o dimension
— f — llanta f - [autom-neumát] rim size
— f — — f nueva - [autom-neumát] new rim size
— f — molde m - mold size
— f — movimiento m - movement measurement
— f — — m de entrerrosca f - [tub] nipple movement measurement
— f — — m — — f para tubo m - [tub] pipe nipple movement measurement
— f — muestra f - sample, measurement, o size, o dimension
— f — muro m - [constr] wall, measurement, o dimension
— f — neumático m - [autom-neumát] tire size
— f — — m inflado - [autom-neumát] inflated tire dimension(s)
— f — — m nuevo—[autom-neumát] new tire size
— f — — m P-métrico - [autom-neumát] P-metric tire size
— f — — m — substitutivo - [autom-neumát] P-metric substitute tire size
— f — — m para automóvil m - [autom-neumát] automobile, o car, tire size
— f — — m — — m para pasajero(s) m—[autom-neumát] passanger, car, o automobile, tire size
— f — — m — reemplazo m - [autom-neumát]

substitute tire size
medida f de nivel m - level measurement
— f — **obra f** - [constr] job, o building, o
structure, measurement, o dimension
— f — **orificio m** - [mec] opening size
— f — **palanca f** - [mec] lever, measurement, o
size
— f — **palanquilla f** - [metal-lam] billet measurement
— f — — f **de colada f continua** - [col.cont]
continuous cast(ing) billet measurement
— f — **pared f** - [constr] wall, measurement, o
dimension
— f — **paso m** - [mec] pitch, measurement, o
dimension(s)
— f — — m **en relación f con profundidad f** -
[mec] pitch, versus, o to, depth dimension(s)
— f — **pérdida(s)** - less(es) measurement
— f — **permeabilidad f** - [electr] permeability
measurement
— f — — f **en corriente f alterna** - [electr]
alternate current permeability measurement
— f — — f — f **continua** - [electr] direct
current permeability measurement
— f — **pieza f** - [mec] part, o component, measurement, o dimension
— f — **placa f** - [mec] plate, measurement, o
dimension
— f — — f **para lingotera f** - [metal-prod] ingot mold stool, measurement, o dimension
— f — **polo m** - [electr] pole measurement
— f — **poste m** - [constr] post, measurement, o
size
— f — **presión f** - [instrum] pressure measurement
— f — — f **para salida f** - outlet pressure
measurement
— f — — f — — f **para agua m** - water outlet
pressure measurement
— f — **profundidad f** - depth measurement
— f — **propiedad(es) f** - property,ties measurement
— f — — f **magnética(s)** - [electr] magnetic
property,ties measurement
— f — — f — **de imán m** - [magnet] magnet('s)
magnetic property,ties measurement
— f — — — m **permanente** - [electr]
permanent magnet('s) magnetic property,ties
measurement
— f — — f — **de producto m** - [metal] product
magnetic property,ties measurement
— f — — f — — m **masivo** - [metal] massive
product magnetic property,ties measurement
— f — — f **de imán m** - [metal-prod] magnet('s)
property,ties measurement
— f — — f — — f **permanente** - [metal] permanent magnet('s) property,ties measurement
— f — **rendimiento m** - performance measurement
— f — **resistencia f** - resistance measurement
— f — — f **de aislación f** - [electr] insulation resistance measurement
— f — — f **eléctrica** - [electr] electric resistance measurement
— f — — f — f **efectiva** - [quím] resistivity
measurement
— f — — f — — de agua m - [hidr] water resistivity measurement
— f — — f — — — muestra f - sample resistivity measurement
— f — — f — — — — f de suelo m - suelos]
soil('s) sample resistivity measurement
— f — — f — — infinita f - infinite resistivity measurement
— f — **riel m** - [metal-lam] rail, measurement, o
size
— f — **rigidez f** - rigidity measurement
— f — **rueda f** - [mec-ruedas] wheel size
— f — — f **en existencia f** - [mec-ruedas]
stock wheel(s), measurement, o size
— f — **salida f** - [constr] outlet, o exit, measurement, o geometry
— f — **separación f** - spacing dimension
— f — — f **para raíz f** - [sold] root spacing,
measurement, o dimension

medida f de separación f en relación f con profundidad f - [mec] pitch, versus, o to, depth
dimension
— f — **soldadura f** - [sold] weld, size, o geometry
— f — — f **en ángulo m interior** - [sold] fillet (weld) size measurement
— f — **sujetador m** - [mec] fastener, measurement, o size
— f — **superficie f** - surface('s) measurement
— f — — f **de bobina f** - [instrum] coil surface measurement
— f — — f — — f **para medición f** - [instrum]
measuring coil surface measurement
— f — — f **de fusión f** - [sold] leg size
— f — — f — f **útil** - [instrum] measuring coil
useful surface
— f — — f **de bobina f para medición f** -
[instrum] measuring coil useful surface measurement
— f — **tafilete m** - [vest] headband. size
— f — **temperatura f** - temperatue measurement
— f — — f **de baño m** - bath temperature measurement
— f — **tiempo m** - time measurement
— f — **tierra(s) m** - [topogr] (land) survey
— f — **tolerancia f** - tolerance, measurement, o
dimension
— f — **tornillo m** - [mec] screw, size, o measurement
— f — — m **con casquete m** - [mec] cap screw,
size, o measurement
— f — **trabajo m** - work measurement
— f — **tubo,bería** - [tub] tube,bing, o pipe,
measurement
— f — **variación f** - variation measurement
— f — — f **en flujo m** - flow variation measurement
— f — — f — — m **magnético** - [electr] magnetic flow variation measurement
— f — **viaje m** - [comput] trip measurement
— f — — m **redondo** - [comput] round trip measurement
— f — **voltaje m** - [electr-oper] voltage measurement
— f **en entrada f** - [comput] inbound, o input,
measurement
— f — **escala f cabal** - full scale measurement
— f — **salida f** - [comput] outbound measurement
— f **entre borne(s) m** - [electrón] measurement,
across, o between stud(s)
— f **exacta** - exact measurement
— f **extensiva** - extensive measure(ment)
— f **exterior** - outside, measurement, o dimension • pl - overall dimensions
— f — **máxima** - maximum, o largest, outside,
measurement, o dimension
— f — **mayor** - larger autside dimension •
largest outside dimension
— f — **menor a** - smaller outside dimension •
smallest outside dimension
— f — **mínima** - smallest, o minimum, outside dimension
— f **fácil** - easy measurement
— f **final** - final, dimension, o measurement
— f — **de separación f** - [mec] final gap dimension
— f **general** - general, o overall, dimension, o
measurement • clearance diagram
— f — **para seguridad f** - [segurid] general
safety precaution
— f **horizontal** - horizontal, dimension, o measurement
— f **igual** - equal, dimension, o measurement
— f **indicada** - [instrum] (noted) reading
— f **notada** - [instrum] noted reading
— f — **en ohmímetro m** - [instrum] ohmmeter
reading
— f **indirecta** - indirect measurement
— f **individual** - individual measurement
— f **infinita** - infinite measurement
— f — **de resistencia f** - resistance infinite
measurement

medida f infinita de resistencia f eléctrica -
[electr-oper] infinite resistance measurement
— f — — — f efectiva - [electr-oper] infinite resistivity measurement
— f inglesa - [metric] English measurement(s)
— f inicial - initial, measurement, o dimension
— f — de separación f - [mec] initial gap, dimension, o measurement
— f interior - inside, measurement, o dimension
• [cadenas] inside length
— f — menor - smaller inside, measurement, o
dimension • smallest inside, dimension, o
measurement
— f — máxima - maximum, o largest, inside,
measurement, o dimension
— f — mayor - larger inside, measurement, o
dimension • largest inside, measurement, o
dimension
— f — mínima - minimum, o smallest, inside,
measurement, o dimension
— f lateral - side, measurement, o dimension
— f legal - [metric] legal, measurement, o size
— f lineal - [metric] linear measure(ment)
— f magnética - [electr] magnetic measurement
— f más extensiva - more extensive measurement
— f — fácil - easier measurement
— f — rápida - faster, o quicker, measurement
— f máxima - maximum measurement
— f — para horquilla f - [mec] fork maximum, o
maximum fork, measurement
— f — — neumático - [autom-neumát] maximum
tire, o tire maximumn, measurement, o size
— f mayor - larger, measurement, o dimension •
largest, measurement, o dimension • more extensive measure
— f menor - smaller, measurement, o dimension •
smallest, measurement, o dimension
— f métrica - [metric] metric, measurement, o
size
— f mezclada - mixed, measurement, o size
— f mínima - smaller, measurement, o size
• minimum, o smallest, mesasurement, o dimension, o size
— f — para neumático m - [autom-neumát] minimum tire size
— f necesaria - needed measure(ment) | adv -
(en) as needed
— f neta - struck measure
— f nominal - nominal, measurement, o dimension
• [cadenas] trade size
— f normal - normal, measurement, o dimension •
standard size
— f — para neumático m - [autom-neumát] tire
standard, o standard tire, size
— f normalizada - standarized, measurement, o
dimension, o size
— f — para fabricación f - [ind] standardized,
fabrication, o manufacturing, measurement, o
dimension, o size
— f nueva - new size
— f observada - observed size • observed rate
— f obtenida - [instrum] reading
— f optativa - optional size
— f — para neumático m - [autom-neumát] tire
optional, o optional tire, size
— f original - original, measurement, o size
— f P-métrica - [autom-neumát] P-metric size
— f para canal m - [ind] channel, measurement,
o size
— f — — m para colada f - [metal-prod] runner
measurement
— f — esmerilador m - [mec] grind(ing) size
— f — fabricación f - fabrication,ting, dimension, o size
— f — flotación f - [autom-neumát] flotation
size
— f — horquilla f - [mec] fork, measurement,
o size
— f — — neumático m - [autom-neumát] tire size
— f — — m estándar - [autom-neumát] standard
tire size
— f — — m inflado - [autom-neumát] inflated
tire, dimension, o size

medida f para neumático m original - [autom-
-neumát] original tire size
— f — — m para camión f - [autom-neumát]
truck tire size
— f — — — m liviano - [autom-neumát]
light truck tire size
— f — plazo m corto - short term measure
— f — — m — tomada - taken short term measure
— f — — m largo - long term measure
— f — — m — tomada - taken long term measure
— f — precaución f - precaution(ary) measure
— f — — f seguida - followed precautionary
measure
— f — reemplazo m - replacement size
— f — regulación f - control, o setting, measure
— f — — f de flotador m - [comb.int] float,
control, o setting dimension
— f — seguridad f - [seguird] safety, precaution, o measure
— f — — t para descongelación f - [sold]
thawing safety, precaution, o measure
— f — — f específica - [seguird] specific
safety, precaution, o measure
— f — — f para soldadora f - [sold] welder
safety precaution
— f — — — f transformadora - [sold]
transformer welder (safety) precaution
— f pequeña - small, measurement, o dimension
— f plana - flat dimension
— f — grande - large flat dimension
— f — máxima - maximum, o largest, flat dimension
— f — mayor - larger flat dimension • largest
flat dimension
— f — menor - smaller, o smallest, flat dimension
— f — mínima - minimum, o smallest, flat dimension
— f — pequeña - small flat dimension
— f popular - popular, measurement, o size
— f práctica - practical measure
— f — de efectividad f - practical, efective,
o performance, measure
— f — — — f en entrevista(s) f - [admin]
practical contact performance measure
— f — — — f — — f sobre seguridad f -
[seguird] practical safety contact performance measure
— f — — — f sobre seguridad f - [seguird]
practical safety contact measure
— f posible - possible, measure, o action
— f precaucionaria - precautionary measure
— f precisa - precise, o accurate, measurement
— f — de voltaje m - [electr-oper] precise, o
accurate, voltage measurement
— f preventiva - preventive (measure) conservative measure
— f principal - main, measurement, o dimension
— f rasa - struck measure
— f razonable - reasonable, measure, o step
— f real - actual, o true, measurement, o size
— f — de cordón m - [sold] true bead size
— f — — — m en ángulo m (interior) - [sold]
true fillet (bead) size
— f recogida - [cadenas] drawn-up dimension
— f recomendada - recommended, measurement, o
dimension, o size
— f — para neumático m - [autom-neumát] recommended tire size
— f — para rueda f - [autom] recommended wheel
size
— f saliente - [comput] outbound measurement
— f según norma f—standard, measurement, o size
— f significante - significant measure(ment)
— f significativa - significant measure(ment)
— f substitutiva—[autom-neumát] substitute size
— f — aceptable - [autom-neumát] acceptable
substitute size
— f — P-métrica - [autom-neumát] P-metric substitute size

medida f **sugerida** - suggested, measurement, o
dimension, o size
— f **para árbol** m - [mec] suggested shaft,
measurement, o size
— f —— **eje** m - [mec] suggested, axle, o
shaft, size
— f **para rueda** f - suggested wheel size
— f **tal como laminado,da** a - [metal-lam] mill
length
— f **temporaria** - temporary, o interim, measure
— f **tomada** - taken measurement • taken action
— f — **en obra** f - field measurement
— f **total** - overall, measurement, o dimension
— f **transversal** - cross sectional, o across,
measurement, o dimension
— f **vertical** - vertical, measurement, o dimension, o size
medidas f **varias** - sundry, o various, sizes
medido,da a - measured, o sized; gaged
— a con **indicador** m con esfera f - [mec] measured with @ dial indicator
— a con **paso(s)** m - stepped off; paced
— a **continuamente** - measured continuously
— a **entre borne(s)** m - [electrón] measured, between studs, o across
— a **más fácilmente** - measured easier
— a **rápidamente**—measured, faster, o quicker
— a **parcialmente** - measured partially
— a con **precisión** f - measured accurately
medidor n - [instrum] . . .; gage(r)
— m con **expendio** m de boleta(s) f - [autom]
ticket issuing meter
— m — **núcleo** m **móvil** - [instrum] moving,
core, o iron, meter
— m de **empresa** f—[electr-instal] company meter
— n — f **para energía** f - [electr-instal]
power company meter
— m **diferencial** - [instrum] differential, meter, o instrument
— m **eléctrico** - [instrum] electric meter
— m **explosivo** - [instrum] explosive meter
— m **indicador** - [instrum] indicating meter
— m **optativo** - [instrum] optional meter
— m **para aceite** m - [mec] oil gage
— m — **ácido** m - [instrum] acid, meter, o gage
— m — **agua** m - [hidr] water, meter, o gage
— m — **aparcamiento** m - [autom] parking meter
— m — **carbono** m - [instrum] carbometer
— m — **caudal(es)** m - flow meter • flux meter
— m —— m **para aire** m - [instrum] air flow
meter
— m —— m — **gas** m - [combust] gas flow meter
— m —— m — **vapor** m - [instrum] steam flow
meter
— m **combustible** m - [comb.int] fuel gage
— m **combustión** f - [combust] combustion
meter
— m **concentración** f - [ind] concentration
meter
— m —— f de **ión(es)** m - [hidr] ion concentration meter
— m —— m —— m **de hidrógeno** m - [hidr]
hydrogen ion concentration meter
— m **conductibilidad** f - [instrum] conductivity meter
— f **corriente** f - [instrum] flow meter •
[electr-equip] current, o electric, o watt-
-hour, meter
— m —— f **para aire** m - [instrum] air flow
meter
— m — **densidad** f - [instrum] density meter
— m — **desviación** f - [petról] drift meter
— m — **energía** f - [electr-instal] power meter
— m — **esfuerzo** m - [instrum] strain gage
— m — **espesor** m - [instrum] thickness meter;
gage
— m — **con rayo(s)-X** - [instrum] X-ray gage
— m — **fase(s)** f - [electr-equip] phase meter
— m — **flujo** m - [instrum] flow, o flux, meter
— m —— m **para aires** m - [instrum] air flow
meter
— m —— m **para combustible** m - [instrum] fuel
flow meter

medidor m **para flujo** m de gas m - [instrum]
gas flow meter
— m —— m —— m **natural** - [instrum] natural
gas flow meter
— m —— m — **vapor** m - [instrum] steam flow
meter
— m — **frecuencia** f - [instrum] frequency meter
— m — **gas** m - [instrum] gas meter
— m —— m **natural** - [instrum] natural gas
meter
— m — **gasto** m - [instrum] flow, o flux, meter
— m —— m **para combustible** m - [instrum] fuel
flow meter
— m —— m — **fuel oil** - [instrum] fuel oil
flow meter
— m —— m — **gas** m - [instrum] gas flow meter
— m —— m —— m **natural** - [instrum] natural
gas flow meter
— m —— m —— m **para estufa** f—[metal-prod]
gas to stove flow meter
— m —— m **para gas** m **y aire** m - [instrum]
gas and air flow meter
— m — **grieta(s)** f - [instrum] crack gage
— m — **humo(s)** m - [instrum] fumes, o smoke,
meter
— m — **inducción** f - [instrum] induction meter
— m — **integración** f - integrating,tion meter
— m — **inyección** f - [instrum] injection meter
— m — f **de vapor** m - [instrum] steam injection (flow) meter
— n —— f —— m **en viento** m - [metal-prod]
blast steam injection flow meter
— m — **largo(r)** m - [instrum] length gage
— m — **metraje** m - [instrum] measuring meter
— f —— m **de pozo** m - [petról] well measuring
meter
— m — **motor** m - [instrum] motor meter
— m — **neumático(s)** n - [autom-neumát] tire
gage
— m — **nivel** m - [instrum] level, meter, o gage
— m — **orificio** m - [instrum] orifice, o opening, meter
— m — **permeabilidad** f - [instrum] permeability
meter, o gage; Magna-Gage; ferrite gage
— m — **presión** f - [instrum] pressure, gage, o
meter
— m —— f **para aceite** m - [mec] oil pressure
gage
— m —— f **para rodillo** m - [metal-lam] roll
pressure, gage, o meter
— m —— f — **suelo** m - [suelos] soil pressure
gage, o meter
— m —— f — **tragante** m - [metal-prod] throat
pressure, gage, o meter
— m —— f —— m **para horno** m - [metal-prod]
furnace throat pressure, gage, o meter
— m — **producción** f - [ind] productimeter
— m —— f **(de) Selsyn** - [instrum] Selsyn productimeter
— m — **rayos-X** - [instrum] X-ray meter
— m — **resistencia** f - [instrum] resistance
meter
— m —— f **eléctrica** - [instrum] electric resistance meter
— m —— f — **efectiva** - [instrum] resistivity
meter
— m —— f —— **para agua** m - [instrum] water
resistivity meter
— m —— f ——— **suelo** m - [suelos] soil(s)
resistivity meter
— m — **temperatura** f - [instrum] temperature,
meter, o gage; thermometer
— m —— f **para agua** m - [instrum] water temperature gage
— m —— f **para culata** f - [int.comb] head
temperature gage
— m —— f —— f **para cilindro** m - [comb.-
-int] cylinder head temperature gage
— m **portátil**—[instrum] portable, meter, o gage
— m — **para resistencia** f - [instrum] portable
reistance meter
— m —— f **eléctrica** - [instrum] portable
electric resistance meter

medidor m **portátil para resistencia** f **eléctrica efectiva** - [instrum] portable resistivity meter
— m ———— f ——— **de agua** m - [instrum] portable water resistivity meter
— m ———— f ——— **suelo** m - [instrum] portable, earth, o soil, resistivity meter
— m **registrador** - [instrum] recording, meter, o gage
m **tipo bayoneta** f (**para aceite**) m [comb.int] bayonet (oil) gage
— m **usado** - [instrum] used meter
— m **volumétrico** - [instrum] volumetric, meter, o gage
— m — **con desplazamiento** m - [instrum] displacement volumetric, meter, o gage
— m ——— m **positivo** - [instrum] positive volumetric displacement, meter, o gage
— m — m — **tipo Roots** - [instrum] Roots (type) positive displacement volumetric meter
— m — **con diafragma** m - [instrum] diaphragm volumetric, meter, o gage
— m ——— m **para presión** f **alta** - [instrum] high pressure diaphragm volumetric meter
medidor,ra a - measuring; metering
— a **para ácido(s)...** • - [ind] acid measuring
medio m - . . . • facility; medium; tool • [fig] sphere; world; circle; set
— a **abierto,ta** - half open
— m **acuoso** - aqueous medium
— m **altamente corrosivo** m - [metal] highly corrosive medium
— m **ambiente** - [amtient] (ambient) environment
— m — **atmosférico** - atmospheric environment
— m **avanzado** - advanced mean(s)
— m **audiovisual** - audiovisual, o multi-media, means
— a **casquillo** m - [mec] half bushing
— m **conveniente** - convenient means
— m — **para acceso** m - convenient access means
— a **cojinete** m - [mec] half bearing
— m **confiable** - dependable means
— m **corrosivo** - corrosive means
— m **de escala** f - [instrum] range, center, o middle
— m — **sección** f - [instrum] range center
— m — **temporada** f - [meteorol] season middle
— m **denso** - [miner] dense medium,dia
— a **diámetro** m - half diameter
— m — **de alcantarilla** f - [tub] culvet (one) half diameter
— m **directo** - direct, means, o medium,dia
— m — **para acceso** m - direct access means
— m **distinto** - different, o alternate, means
— m **económico** - economic(al) means
— m — **avanzado** [econ] advanced economic means
— m **efectivo** - effective means
— a **embutido,da** - [mec] half mortised
— m **excelente** - excellent means
— m — **para precalentamiento** m - [sold] excellent preheating means
— m **expeditivo** - expeditive means • short cut
— m **físico** - physical, means, o agent, o surrounding(s)
— m **ilustrative** - illustrative means; prop
— m — **singular** - unique, illustrative means, o prop
— m **informativo** - [public] media
— m **líquido** - liquid medium
— a **lleno,na** - half full
— m **magnético** - [miner] magnetic means
— a **manchón** m - [tub] half coupling
— m **mecánico** - [mec] mechanical mean(s) • mechanical agent • mechanical surrounding(s)
— m **normal** - [metal-lam] normal (roll) center
— a **oeste** m (**estadounidense**) - [geogr] Midwest
— m **oficial** - [pol] official means | a [ind] semiskilled workman
— a **Oriente** m - [geogr] Middle East
— a **óvalo** m - half oval
— m **para acceso** m - access means
— m ——— m **conveniente** - convenient access means

medio m **para acceso** m **directo** - direct access means; access direct means
— m ——— m **económico** - economical access means
— m — **comprobación** f - proving, o testing, o checking, means
— m ——— f **de rodillo(s)** m - [metal-lam] roll checking means
— m — **comunicación** f - [telecom] communication means
— m — **difusión** f - [comunic] medium,dia
— m — **embarque** m - [transp] shipping means
— m — **fijación** f - fixing,xation means
— m — **pago** m - [fin] money supply; payment means
— m — **precalentamiento** m - [sold] preheating means
— m — **transporte** m - [transp] transportation, means, o mode
— m ——— m **disponible** - [transp] available transportation means
— m ——— m **satisfactorio** - [transp] satisfactory transportation mean(s)
— m **publicitario** - [comunic] media
— a **punto** - semicircular
— m **químico** m - chemical means
— m **real(es)** - [econ] real means; actual means
— m — **para pago** m - [econ] actual money supply
— a **siglo** m - half century
— a **tambor** m - [mec] half drum
— m **tramo** - [constr] half section
— a **troquel** m - [mec] half die
— m **ultrasónico** - [electrón] ultrasonic,sound mean(s)
medio,dia - half • mean; average • middle • all day
medios m - means
medir v - . . .; to size • to gage; to meter • to read
— v **calidad** f - to measure quality
— f **capacidad** f - to, measure, o size, capacity
— v **carga** f - [constr] to measure load
— v — f **muerta** - [constr] to measure dead load
— v — f **viva** - [constr] to measure live load
— v **con indicador** m - [instrum] to measure with indicator
— v ——— m **con esfera** f - [instrum] to measure with dial indicator
— v **con paso(s)** m - to pace; to step off
— v — **precisión** f - to measure, precisely, o accurately
— v **continuamente** - to measure continuously
— v **contragolpe** m - [mec] to measure backlash
— v **desempeño** m - [labor] to measure performance
— v **distancia** f - to measure distance
— v **eficiencia** f — to measure efficiency
— f — f **productiva** - [ind] to measure production efficiency
— v **entre borne(s)** m - [electrón] to measure, across, o between studs
— v **espesor** m - to measure thickness
— v — m **de pared** f - to measure wall thickness
— v **fenómeno** m - to measure phenomenon
— v **grosor** m - to measure thickness
— v — m **de banda** f - [mec-frenos] to measure band thickness
— v ——— **cinta** f - [mec] to measure lining thickness
— v ——— f **para embrague** m - [mec] to measure clutch lining thickness
— v **holgura** f - [mec] to measure clearance
— f — f **para cubierta** f - [mec] to measure cover clearance
— v — f ——— f **para cojinete** m - [mec] to measure bearing cover clearance
— m **juego** m - [mec] to measure play
— v — m **longitudinal** - [mec] to measure end play
— v — m — **de árbol** m **para entrada** f **de fuerza** f - [autom-mec] to measure input shaft end

play

medir v líquido(s) - to measure @ liquid(s)
— v — m cloacal(es) - [sanit] to measure @ sewage
— v — m — doméstico(s) - [sanit] to measure @ domestic sewage
— v más fácilmente - to measure more easily
— v — rápidamente - to measure, faster, o quicker
— v movimiento m - to measure @ movement
— n — de entrerrosca f - [mec] to measure @ nipple movement
— v — m — f para tubo m - [mec] to measure @ pipe nipple movement
— v nivel m para aceite m - [mec] to, measure, o gage, o oil level
— v parcialmente - to measure partially
— v precarga* f - [mec] to measure @ preload
— v producción f—[ind] to measure @ production
— v rápidamente - to measure quickly
— v resistencia f - to measure @ resistance
— f — f eléctrica - to measure @ electric resistance
— v — — f efectiva - [quím] to measure @ resistivity
— f — — f — de agua m - [hidr] to measure @ water resistivity
— v — f — — muestra f - [hidr] to measure @ sample resistivity
v — f — — — — f de agua m - [quím] to measure @ water sample resistivity
— v — f — — infinita - [quím] to measure @ infinite resistivity
— v rigidez f - to measure @ rigidity
— v tiempo m - to measure @ time
— v trabajo m - to measure @ work
— v viaje m - [comput] to measure @ trip
— v — m redondo - [comput] to measure @ round trip
— v voltaje m - [electr-oper] to measure @ voltage
— v — m con precisión f - [electr-oper] to measure @ voltage, precisely, o accurately
medirse v - to measure oneself • to compete
mediterráneo,nea a - [geogr] . . .; midcontinent
medra f - . . . . • thriving; prospering
medrado,da a - thrived; prospered
medrante a - thriving; prospering
megavatio m - [electr] megawatt
megohmmetro m - [instrum] megohmmeter
— m para ensayo(s) m - [instrum] megohmmeter tester
— m — — m a tierra f - [instrum] megohmmeter ground tester
mejicanización f - Mexicanization
mejicanizado,da - Mexicanized
mejicanizar v - to Mexicanize
Méjico m - [geogr] Mexico
mejor a - . . . .; finest • less
— a de caso(s) m - best case
— a equivalencia m - best, match, o equivalence
— a — de color(es) m - [pint] best color match
— a postor - [com] low, o best, bidder
— a postura f - [com] lowest, o best, o successful, bid
— a resistencia f - best resistance
— a — a abrasión f - [sold] best abrasion resistance
mejora f - . . . .; improving; betterment; bettering; advance • refinement • [deport] breaking • véase también mejoramiento m
— f asombrosa - startling improvement
— f básica - basic improvement
— f considerable - substantial improvement
— f contundente - quantum, improvement, o leap
— f de cristal m - crystal improvement,ving
— f — diamante n - diamond imorovement.ving
— f — marca f - [deport] record, improvement, o breaking, o bettering
— f — mordaza f - [mec] clasp improvement,ving
— f — producto m - product, o breed, improvement,ving
— f — recurso(s) m - [econ] resource(s) improvement,ving

mejora f de tipo m de diamante - diamond type improvement,ving
— f dramática - dramatic improvement
— f efectuada - made improvement
— f en actitud - [admin] attitude improvement
— f — alineación f - alignment improvement
— f — aptitud f - aptitude improvement,ving
— f — calidad f - quality improvement,ving
— f — — f de producto m - [ind] product quality improvement
— f — cantidad f - [ind] quantity improvement
— f — — f de producto m - [ind] product quantity improvement
— f — capacitación f - [labor] qualification, o training, improvement
— f — carretera f - [constr] highway improvement
— f — cauce n - [constr] channel improvement
— f — comportamiento m - [labor] performance improvement
— f — conducción f - [autom] driving improvement
— f — conformación f - contour improvement
— f — — f de cordón m - [sold] bead contour improvement
— f — conocimiento(s) m - knowledge improvement
— f — construcción f - [constr] construction, o building, improvement
— f — — f con lengüeta f - [mec] tongue construction improvement
— f — costo(s) m - [ind] cost(s) improvement
— f — desempeño m - [admin] performance improvement
— f — destreza f - [labor] skill improvement
— f — diseño m - [ind] design improvement
— f — eficiencia f - efficiency improvement
— f — equipo m - [ind] equipment improvement
— f — factor m - factor improvement
— f — — m para potencia f - [sold] power factor imporovement
— f — — — f en fuente f para energía f - [sold] input source power factor improvement
— f — habilidad f - [labpr] skill improvement
— f — horno m - [ind] furnace improvement
— f — — m eléctrico - [metal-prod] electric furnace improvement
— f — ingeniería f - [ind] engineering improvement
— f — invariabilidad f - consistency improvement
— f — kilometraje m - mileage improvement
— f — navegación f - [nav] navigation improvement
— f — operación f - operation improvement,ging
— f — peso m - weight improvement
— f — — m específico - specific weight improvement
— f — — m — de petróleo m - [petról] oil gravity improvement
— f — precio(s) f - price improvement
— f — preparación f - preparation improvement
— f — — f de carga(s) f - [metal-prod] charge preparation improvement
— f — procedimiento(s) m - [sold] procedure, o process, improvement, o development
— f — productividad f - [ind] productivity improvement
— f — producto m - [ind] product improvement
— f — proyección f - design, o engineering, improvement
— f — reacción f - reaction improvement
— f — reacción f a manioba(s) f - [autom-mec] steering response improvement,ving
— f — rendimiento m - performance, o yield improvement,ving
— f — — m de horno m - [ind] furnace performance imporovement,ving
— f — — m — m eléctrico - [ind] electric furnace performance improvement,ving
— f — seguridad f - [segurid] safety improve-

ment
mejora f en toma f de curva f - [autom] cornering improvement,ving
— f — **trabajo m** - [labor] work improvement
— f —— m **ejecutado** - [labor] work being done
improvement
— f — **tracción f** - [mec] traction improvement
— f — **vecindario m** - [pol] neighborhood improvement
— f — **viraje(s) m** - [autom] cornering improvement
— f — **visibilidad f** - visibility improvement
— f **futura** - future, improvement, o development
— f — **en cauce m** - [constr] future channel improvement
— f —— **procedimiento(s) m** - [ind] future,
process, o procedure, improvement
— f — **para carretera f** - [constr] future highway improvement
— f **hecha** - made improvement
— f **más reciente** - latest improvement
— f **municipal** - [pol] municipal, improvement, o
betterment
— f **para mejorar v tarea(s) f para conservación
f** - maintenance-minded advance
— f **reciente** - recent, o late, improvement
— f **señalada** - marked, o major, improvement
— f **significativa** - significant improvement
— f **singular,**- unique improvement
— f **vecinal,**- [pol] neighborhood improvement
mejorado,da a - improved • enhanced • [deport]
bettered; broken
— a **para arado m** - [metal-prod] improved plow
— a **y galvanizado,da a** - [metal-prod] improved
galvanized
— a — **pulido,da a** - [metal-prod] improved polished
mejorador,ra a - improver,ving
— a **para eficiencia f** - efficiency improving
mejoramiento m - . . .; improving; betterment;
véase también **mejora f** - [deport] braking •
[hidr] reclamation
— m **básico** - basic improvement
— m **continuo** - continuous,nuing, improvement
— m **de acceso m** - access improvement • [hidr]
inlet improvement
— m — **activo m fijo** - [econ] fixed asset(s)
improvement • capital improvement
— m — **agua m** - water improvement
— m — **aptitud** - aptitude improvement
— m — **costa,te,to** - [ind] cost improvement
— m — **entrada f** - [hidr] inlet improvement
— m — **invariabilidad f** - invariability, o consistency, improvement
— m — **marca f** - [deport] record breaking
— m — **recurso(s) m** - [econ] resource(s) improvement
— m — **salida f** - [hidr] outlet improvement
— m **en apariencia f** - appearance improvement
— m — **apoyo m** - [constr] bearing improvement
— m ——— m **para ariete m** - [constr] ram bearing improvement
— m ——— m —— m **hidráulico** - [constr] hydraulic ram bearing improvement
— m — **aspecto m** - appearance improvement
— m — **calidad f** - [ind] quality, improvement,
o gain
— m ——— f **de producto m** - [ind] product quality improvement
— m — **cantidad f** - [ind] quantity improvement
— m ——— f **de producto m** - [ind] product quantity improvement
— m — **capacitación f** - [labor] training, o
qualification, improvement
— m — **carga f** - [constr] bearing improvement •
[transp] loading improvement
— m — **comportamiento m** - performance improvement
— m — **declive m** - [constr] grade improvement
— m — **desagüe(s) m** - [hidr] drainage improvement
— m — **destreza f** - skill, improvement, o
sharpening

mejoramiento m — **eficiencia f** - efficiency improvement
— m — **kilometraje m** - mileage improvement
— m — **navegación f** - [nav] navigation improvement
— m — **operación f** - operation improvement
— m — **permeabilidad f** - [metal-prod] permeability improvement
— m —— f **de columna f** - [metal-prod] column
permeability improvement
— m —— f —— f **de carga f** - [metal-prod]
burden (column) permeability improvement
— m — **posición f** - position improvement
— m — **preparación f** - preparation improvement
— m —— f **de carga f** - [metal-prod] charge
preparation improvement
— m — **rendimiento m** - yield improvement
— m — **productividad f** - [ind] productivity
improvement
— m — **reacción f** - response improvement
— m —— f **a maniobra f** - [autom-neumát]
steering response improvement
— m — **rendimiento m** - performance improvement
— m — **resistencia f** - resistance improvement
— m —— f **mecánica** - [metal-prod] mechanical
resistance improvement
— f —— f **mecánica de sínter m** - [metal-prod]
sinter mechanical resistance improvement
— m — **seguridad f** - [segurid] safety improvement
— m **profesional** - professional improvement
mejorar v - . . .; to upgrade • [deport] to exceed; to break
— v **actitud f** - to improve @ attitude
— f **activo m fijo**—to improve @ fixed asset(s)
— v **alineación f** - to improve @ alignment
— v **apariencia f** - to improve @ appearance
— v **apoyo m** - [constr] to improve @ bearing
— v — m **para ariete m** - [constr] to improve @
ram bearing
— v — m —— m **hidráulico** - [constr] to im-
prove @ hydraulic ram bearing|
— v **aspecto m** - to improve @ appearance
— v **calidad f** - to improve @ quality
— v — f **de producto m** - to improve @ product
quality
— v **cantidad f** - to improve @ quanity
— v — f **de producto m** - to improve @ product
quantity
— v **capitalización f** - [fin] to improve @ capitalization
— v **carga f** - [constr] to improve @ bearing •
[transp] to improve @ load(ing)
— v **comportamiento m** - to improve @ performance
— v **conducción f** - [autom] to improve @ driving
— v **conformación f** - to improve @ contour
— v — f **de cordón m** - [sold] to improve @ bead
contour
— v **conocimiento(s) m** - to, improve, o broaden,
@ knowledge
— v **considerablemente**—to improve significantly
— v **construcción f** - [constr] to improve @,
construction, o building
— v — f **de lengüeta f** - [mec] to improve @
tongue construction
— v **costa,te,to** - to improve @ cost
— m **de producto m** - to improve @ product
cost
— v **cristal m** - [cerám] to improve @ crystal
— v **declive m** - [constr] to improve @ grade
— v **desempeño m** - [admin] to improve @ performance
— v **destreza f**—to, improve, o sharpen, @ skill
— v **diamante m** - to improve @ diamond
— v **diseño m** - to improve @ design
— v **eficiencia f** - to improve @ efficiency
— v **equipo m** - [ind] to improve @ equipment
— v **factor m** - to improve @ factor
— v — m **para potencia f** - [sold] to improve @
power factor
— v —— f **para fuente f** - [sold] to improve @ source power factor
— v — m —— f —— f **para energía f**—[sold]

to improve @ input source power factor
mejorar v **habilidad** f - to improve @ skill
— v **horno** m - [ind] to improve @ furnace
— v — m e**léctrico** - [metal-prod] to imrove @
electric furnace
— v **invariabilidad** f - to improve @, invariability, o consistency
— v **kilometraje** m - to improve @ mileage
— v **labrabilidad** f - [mec] to improve @ machinability
— v **marca** f - [deport] to, break, o improve, o
better, @ record
— v **mensurablemente** - to improve measurably
— v **mordaza** f - [mec] to improve @ clamp
— v **operación** f - to improve @ operation
— v **percepción** f - to sharpen @ skill
— v **peso** m - to improve @ weight
— v — m e**specífico** - to improve @ specific
weight
— v — m — de **petróleo** m - [petról] to improve
@ oil gravity
— v **posición** f - to improve @ position
— v **productividad** f - [ind] to improve @ productivity
— v **producto** m - [ind] to improve @, product, o
breed
— v **reacción** f - to improve @ response
— v — f a **maniobra(s)** f [autom-neumát] to imporove @ steering response
— v **recurso(s)** m - [econ] to improve @ resource
— v **rendimiento** m - [mec] to improve @ performance
— v — m de **horno** m - [ind] to improve @ furnace performance
— v — m — — m e**léctrico** - [metal-prod] to
imporove @ electric furnace performance
— v **seguridad** f - [segurid] to improve @ safety
— v **tipo** m - to improve @ type
— v — m de **diamante** m - to improve @ diamond
type
— v **toma** f de **curva** f - [autom] to improve @
cornering
— v **trabajo** m - to improve @, work, o job
— v **tracción** f - [mec] to improve @ traction
— v **viraje(s)** m—[autom] to improve @ cornering
— v **visibilidad** f - to improve @ visibility
mejoría f - . . . . ; [medic] . . . ; recovery
— f en **aspecto** m - appearance improvement
melaza f **diluida** - diluted molasses
mella f - . . . ; gab • crinkle • inroad
— f **creada** - [mec] created, notch, o nick
— f **posible** - [mec] possible, notch, o nick
— f — **creada** - [mec] checked possible nick
— f **verificada** - [mec] checked, notch, o nick
mellado,da a - notched; nicked; dented
melladura f - véase **mella** f
mellar v - . . . ; to nick; to dent
membrana f - . . . . • [mec] diaphragm
— f de **papel** m - [aislac] paper membrane
— f — — m **con resina** f - [aislac] resin paper
membrane
— f **exterior** - outside, o outer, membrane
— f **interior** - inside, o inner, membrane
— f **para regulador** m - control membrane
— f — — m **para caudal** m - flow control membrane
— f — **tapajunta(s)** m - [constr] flashing membrane
— f **resinosa** - [hormig] resinous membrane
memorable a - . . . ; eventful
memorándum m **interno** - [com] internal, o inter-
office, memorandum
memoria f - . . . . • [legal] annual, o public,
report; statement; proceeding(s); report
— f **con velocidad** f **alta** - [comput] high speed
memory
— f de **cálculo(s)** m - calculation, o computation, sheet, o statement
— f — **cómputo(s)** m - computation sheet
— f de **microplaqueta** f - [comput] chip memory
— f **descriptiva** - [legal] descriptive statement
• memorial
— f **para entrada** f - [comput] input memory

memoria f **para salida** f - [comput] output memory
— f **perdida** - lost memory
— f **programada** - [comput] programmed,mable
memory
— f — **para lectura** f **únicamente** - [comput]
programmable read only memory; P R O M
— f — **únicamente para lectura** f - [comput]
programmable reading only memory; P R O M
memorización f - memorization,zing
— f de **código** m - code, memorization,zing
— f — — m **numérico** - numerical code memorization,zing
memorizar v - to memorize
— v **código** m - to memorize @ code
— v — m **numérico**—to memorize @ numerical code
mena f - [miner] . . . ; iron ore
— f **básica** - [miner] basic ore
— f **calcopirítica** - [miner] chalcopyritic ore
— f **compacta** - [miner] compact ore
— f **cruda** - [miner] raw, o crude, ore
— f **con galena** f - [miner] galena ore
— f — **grano(s)** m **grueso(s)** - [miner] coarse
grain ore
— f — **hierro** m - [miner] iron ore
— f **diseminada** - [miner] dispersed ore
— f **esfalerítica** f - [miner] sphalerite ore
— f **magnética** - [miner] magnetic ore
— f **piritosa** - [miner] pyritic ore
— f — **compacta** - [miner] compact pyritic ore
— f — **diseminada** - [miner] dispersed pyritic
ore
mención f - . . . . • discussing,sion • featuring
— f **especial** - special honor
— f **merecida** - merited, o deserved, mention
— f **pública** - [admin] public, mention(ing), o
recognition
mencionado,da a - mentioned • discussed • featured • [legal] aforesaid; said
mencionar v - . . . . • to discuss; to feature, to
cover
meneado,da a - wagged; flicked
— a **furiosamente** - [autom-limpiaparabrisa]
flicked furiosamente
menear v - . . . ; to flick
— v **cabeza** f - to shake @ head
— v **furiosamente** - [autom-limpiaparabrisas] to
flick furiosly
meneo m - . . . ; flicking
— m **furioso** - [autom-limpiaparabrisas] furious
flicking
mengua f - . . . ; decline; impairment,ring
— f en **capacidad** f - capacity impairment,ring
— f — f **para carga** f - load capacity impairment,ring
— f — **recurso(s)** m - resource(s) decline
— f — m **natural(es)** - natural resource(s)
decline
menguar v - . . . ; to reduce; to impair
— v **capacidad** f - to impair @ capacity
— v — f **para carga** f - to impair @ load(ing)
capacity
menisco m **cóncavo** ⊢ [fís] concave meniscus
— m **convexo** - [fís] convex meniscus
menor m - minor | a - smallest; smaller; poorer;
decreased; lower • worse; worst • light
— a **cuantía** f - [legal] lesser, o minor, quan-
tity, o amount | a - [legal] small claims
— de adv - less(er) than
— adv **posible** - least, o lowest, possible
— adv **que** - less than
— adv — **máximo,ma** a - less than, maximum, o
full
menos adv **almacenamiento** m - less storage
— adv **áspero,ra** a - less, rough, o harsh
— adv **calor** - less heat
— adv **camión** m - [transp] less truck
— adv **comisión** f - [com] less commission
— adv **costoso,sa** a - less, costly, o expensive
• least. costly, o expensive
— adv **crítico,ca** - less critical
— adv **daño** m - less damage
— adv **definido,da** - less (sharply) defined
— adv **demoroso,sa** - less time consuming

menos adv desarrollado,da a - less developed •
[econ] underdeveloped
— adv descuento m - less discount
— adv — m de . . . sobre valor m de factura f
- [com] less . . . discount on invoice value
— adv — de . . . porciento m—[com] less . . .
per cent discount
— adv — m por pago m a contado - [com] less
cash discount
— adv desperdicio(s) m - [ind] less scrap
— adv dramático,ca a - less dramatic
— adv económico,ca a - less economical
— adv efectivo,va - less effective
— adv fluido,da - less fluid
— adv frecuente a - less frequent
— adv gasto(s) m - [fin] less, o net of, expen-
se(s)
— adv grave a - less grave
— adv grúa f - [grúas] less crane
— adv inclinado,da a - less slanting; flatter
— adv mano f de obra - [labor] less, labor, o
manpower
— adv necesitado,da—less needed • least needed
— adv oneroso,sa a - easier
— adv — a en tiempo m - less time consuming
— adv personal m - [labor] less, personnel, o
manpower
— adv — especializado - [ind] less skilled,
labor, o manpower
— adv pieza(s) f usadas para (reparar) otras
máquina(s) f - [mec] cannibalized part(s)
— adv . . . por ciento - [com] less . . . per
cent
— adv profundo,da a - shallower
— adv que - less than
— adv — circular - less than circular
— adv — semicircular - less than semicircular
— adv — semicircunferencial - less than semicircular
— adv próspero,ra a - less prosperous
— adv serio,ria - less serious
— adv severo,ra - less severe • least severe
— adv tendencia f - less tendency
— adv tiempo m - less time • least time
menoscabado,da a - impaired; degraded • [fam]
deflated
menoscabador,ra a - . . .; degrader
menoscabar v - . . .; to degrade
menoscabo m - impairment,ring; degradation,ding
— m — inversión f—[fin] investment impairment
— m — propiedad f - property, impairment, o
degradation
mensaje m a comprador m - message to @ purchaser
— m almacenado - [comput] stored message
— m anterior - earlier, o former, o previous,
message
— m codificado - [telecom] coded message
— m en código m - [telecom] code message
— m nuevo - new message
— m sobre alarma f - [electrón] alarm message
— m — estado m - [electrón] status message
mensajero,ra a - messenger; carrier
mensualmente adv - . . .; each month; on @
monthly basis|
ménsula f - [mec] . . .; gib • [arquit] haunch;
corbel • [constr] shelf
— f achaflanada - [mec] beveled bracket
— f — para crear voladizo m - [constr] beveled
block-out bracket
— f colgante - [mec] hanger,ging bracket
— f — con resorte m - [mec] spring hanger,ging
bracket
— f compensadora - [mec] equalizer,zing bracket
— f — para freno m—[mec] brake equalizer,zing
bracket
— f con resorte m - [mec] spring (loaded)
bracket
— f de acero m - [mec] steel, bracket, o gib
— f — m fundido - [mec] cast steel, bracket, o gib
— f delantera - [mec] front bracket
— f en forma de L - [mec] L-shaped bracket
— f enderezadora - [mec] straightening bracket

ménsula f giratoria - [mec] swivel bracket
— f — delantera - [mec] front swivel bracket
— f — trasera - [mec] rear swivel bracket
— f inferior - [mec] lower, o bottom, bracket, o gib
— f para coraza f - [ind] bottom shell
bracket
— f para adaptación f - [mec] adapter bracket
— f — ajuste m—[mec] adjusting,tment bracket
— f — alimentación f - [mec] feed(er) bracket
— f — f desde frente m - [mec] front
feed(er) bracket
— f — f — atrás - [mec] rear feed(er)
bracket
— f — anclaje m - [mec] anchor bracket
— f — árbol m - [mec] shaft bracket
— f — m para freno m - [mec] brake shaft
bracket
— f — m — desplazador m - [mec] shift(er)
shaft bracket
— f — m — mecanismo m para cambio(s) m -
[mec] shift(er) shaft bracket
— f — banda f - [mec] strap bracket • strip
bracket
— f — bomba f - [bombas] pump bracket
— f — f rociadora—[mec] spray pump bracket
— f — f — para camisa f - [mec] liner
spray pump bracket
— f — f — — émbolo m - [mec] piston spray
pump bracket
— f — f — — m y camisa f - [mec] piston (and) liner spray pump bracket
— f — cable(s) m - [electr-instal] bracket
— f — cojinete m - [mec] bearing bracket
— f — m para embrague m - [mec] clutch
bearing bracket
— f — colgador m - [mec] hanger bracket
— f — colgar v - [mec] hanger,ging bracket
— f — conmutador m - [electr-instal] breaker,
o switch, bracket
— f — coraza f - [ind] shell bracket
— f — crear voladizo m - [constr] block-out
bracket
— f — defensa f - [mec] guard bracket
— f — f para cable m - [mec] rope guard
bracket
— f — f — — m para desenroscadora f -
[petreól] breakout rope guard bracket
— f — f — — m — enroscadura f - [petról]
make-up rope bracket
— f — depósito m - tank bracket
— f — m para combustible m - [comb.int]
fuel tank bracket
— f — desplazador m - [mec] shifter bracket
— f — electrodo m - [mec] electrode bracket
— f — elevación f - [mec] lifter.ting bracket
— f — enganche m - [mec] hitch(ing) bracket
— f — freno m - [mec] brake bracket
— f — guía(dera) f - [mec] guide(ing) bracket
— f — horquilla f - [mec] yoke bracket
— f — f para desplazador m - [mec] shifter
yoke bracket
— f — interruptor m - [electr-instal] switch,
o breaker, bracket
— f — lanza f - [mec] lance bracket • [transp]
tongue bracket
— f — montaje m - [mec] mounting bracket
— f — m para carrete m - [sold] reel mounting bracket
— f — m — motor m - [mec] motor mount(ing)
bracket
— m — m — panel m - [mec] panel mounting
bracket
— f — m — — m para selector m para volta-
je(s) m - [sold] voltage selector panel
mounting bracket
— f — m — platino(s) m - [comb.int] points
mounting bracket
— f — m — quemador m - [combust] burner
mounting panel
— m — m — radiador m - [comb.int] radiator
mounting bracket
— f — m — resistencia f - [electr-instal]

resistance,stor mounting bracket
ménsula f **para montaje** m **para tablero** m - [mec] panel mounting bracket
— f —— m —— m **para selector** m - [sold] selector panel mounting bracket
— f —— m —— m **para voltaje(s)** m - [sold] voltage selector panel mounting bracket
— f —— m **ventilador** m - [mec] fan mounting bracket
— f **muñonera** f - [mec] trunnion bracket
— f **palanca** f - [mec] lever bracket
— f —— f **desplazadora** - [mec] shifter lever, support, o bracket
— f —— t **para freno** m - [mec] brake lever bracket
— f —— f —— m **para carrete** m - [mec] reel brake lever bracket
— f —— f —— m **para núcleo** m - [petról] core reel brake lever bracket
— f —— f — **regulación** f - [mec] throttle, o control, lever bracket
— f —— f **soltadora** - [mec] release lever bracket
— f **pared** f - [constr] wall bracket
— f **pesa** f - [agric-equip] weight bracket
— f **placa** f - [mec] plate bracket
— f —— f **con ojo,jete** m - [mec] eye plate bracket
— f —— f —— m **con resorte** m—[mec] spring eye plate bracket
— f —— f **para acelerador** m - [comb.int] throttle plate bracket
— f **platino(s)** m - [comb.int] points bracket
— f **polín** m - [mec] roller bracket
— f —— m **levantador** - [mec] (lifting) roller bracket
— f —— m **para leva** f - [mec] cam roller bracket
— f **posición** f - [mec] position, o placer, bracket
— f **poste** m - [constr] post bracket
— f **regulador** m - [mec] throttle bracket
— f **riel** m - [constr] track bracket
— f **rodillo** m - [mec] roll(er) bracket
— f —— m **enderezador** - [mec] straightening roll(er) bracket
— f —— m **para alimentación** f - [mec] feeding roll(er) bracket
— f —— m **leva** f—[mec] cam roller bracket
— f **roldana** f - [mec] roller bracket
— f —— f **para cuerda** f - [mec] rope roller bracket
— f **selector** m - [mec] selector, bracket, o stud
— f **soltador** m - [mec] release bracket
— f **soporte** m - [mec] support(ing), o hanger, bracket, o mounting
— f —— m **colgante** - [tub] hanger bracket
— f —— m **árbol** m - [mec] shaft support bracket
— f —— m —— m **para desplazador** m - [mec] shifter shaft support bracket
— f —— m —— m **mecanismo** m **para cambios** - [mec] shifter shaft support bracket
— f —— m **cable** m - [mec] wire support bracket
— f —— m **elevador** m - [mec] hoist support bracket
— f —— m **palanca** f - [mec] lever support bracket
— f —— m —— f **desplazadora** - [mec] shifter shaft support bracket
— f —— m **tubería** f - [mec] pipe,ping support bracket
— f **sostén** m - [mec] support bracket
— f **tablero** m - [mec] panel bracket
— f —— m **para selección** f - [ind] selector panel bracket
— f —— m —— f **para voltaje** m - [sold] voltage selector panel bracket
— f **válvula** f - [mec] valve bracket
— f —— f **para gas** m - [combust] gas valve

bracket
ménsula f **para ventilador** m - [mec] fan bracket
— f **vertedero** m - [ind] chute bracket
— f —— m **para descarga** f - [mec] discharge chute bracket
— f **portacable(s)** m - [electr-instal] cable support(ing) bracket
— f **rígida** - [mec] rigid bracket
— f **superior** - [mec] top, o upper, bracket, o gib
— f —— **coraza** f - [ind] top shell bracket
— f **trasera** - [mec] rear bracket
— f **y roldana** f - [mec] roller and bracket
— f —— f **para cuerda** f - [mec] rope roller and bracket
— f —— f **soga** f - [mec] rope roller and bracket
mensurable a - . . . .; measurable
mentonera f - [segurid] chin starp; hat band
menú m **auténtico** - [culin] authentic menu
— m **gastronómico** - [culin] gourmet menu
— m **regional** - [culin] regional, o authentic, menu
menudeado,da a - repeated • [fig] dotted
menudar v - . . . • to repeat • to dot
— v **en pista** f - [deport] to dot (on) @ course
menudencia f **de grano** m - [miner] grain minuteness
— f —— m **de hierro** m - [miner] iron grain minuteness
menudo m - . . . . • [miner] breeze; fines
— m **de carbón** m - [miner] coal, breeze, o fines
— m **coque** m - [coke] coke fines • buckwheat, o small, coke
— m **criba** f - [miner] screening(s)
— m **fundente** m - [ind] flux fines
— m **piedra** f - [constr] stone screening(s)
— f —— f **triturada** - [constr] crushed stone screening(s)
mercadear v - [com] . . .; to market
— v **elemento(s)** m - [com] to market @ supply,plies
— v **equipo(s)** m - [com] to market @ equipment
— v **provisión(es)** f - [com] to market @ supply,lies
mercadeo* m - [com] marketing; merchandising
— n **de elemento(s)** m - [com] supply,lies marketing
— m —— m **para soldadura** f - [sold] welding supplies marketing
— f **equipo** m - [com] equipment marketing
— m —— m **para soldadura** f - [sold] welding equipment marketing
— m **provisión(es)** f - [com] supplies marketing
— m —— f **para soldadura** f - [sold] welding supplies marketing
— m **doméstico** - [com] domestic marketing
— m **exterior** f - [com] overseas marketing
— m —— m **de, elemento(s)** m, o **provisión(es)** f **para soldadura** f - [sold] welding supplies overseas marketing
— m —— m **equipo(s)** m - [com] equipment overseas marketing
— m —— m —— m **para soldadura** f - [sold] welding equipment overseas marketing
— m **internacional** - [com] international marketing
— m **interno** - [com] internal, o domestic, marketing
— m **nacional** - [¢om] domestic marketing
mercader m - [com] . . .; marketer
mercadería f - [com] . . .; material(s)
— f **adquirida** - [com] acquired, o procured, o purchased, goods
— f **cargada** - [transp] loaded, merchandise, o goods
— f **de calidad** f **(alta)** - [com] (high) quality merchandise
— f —— f **baja** - [com] low quality merchandise
— f —— f **excepcional** - [com] exceptional, o premium, quality merchandise

mercadería f de calidad f primera,rísima -
[com] first, o premium, quality merchandise
— f —— f segunda - [com] second quality
goods; seconds
— f — importación f - imported, merchandise, o
goods
— f devuelta - [com] returned, mechandise, o
goods
— f embalada - [com] packed, merchandise, o
goods
— f en tránsito m - [transp] in transit, merchandise, o goods
— f importada - [com] imported, merchandise, o
goods
— f para reventa f - [com] merchandise, o
goods, for resale
— f repuesta - [com] replaced, mechandise, o
goods
— f trincada - [transp] lashed, merchandise, o
goods
mercado m abierto - [fin] open, o public, market
• aftermarket
— m aislado - [com] isolated market
— m alejado - [com] distant market
— m ampliado - [com] expanded market
— m amplio - [com] broad market
— m analizado - [com] analyzed market
— m andino m - [com] Andean market
— m automotor - [autom] automotive market
— m cambiario - [fin] exchange market
— m comercial - [com] commercial market
— m — para cambio(s) m - [fin] commercial exchange market
— m común - [econ] common market
— m — de América f Central - [econ] Central
American common market
— m comunal - [econ] Community market
— m con rendimiento m alto - [com] high performance market
— m concentrado - [com] concentrated market
— m conquistado - [econ] conquered market
— m considerable - [com] significant market
— m consumidor - [com] consumer market
— m contiguo - [com] contiguous, o adjacent,
market
— m creado - [com] carved, o developed, market
— m de accesorio(s) m - [autom] aftermarket
— m —— m para automóviles m - [autom] automotive aftermarket
— m — cambio(s) m - [fin] (foreign) exchange
market
— m — capital(es) m - [fin] capital(s) market
— m — Comunidad f - [econ] Community market
— m — cristal(es) m—[óptica] lens market
— m — chatarra f - [metal-prod] scrap market
— m —— f desequilibrado - unbalanced scrap
market
— m — detalle m - [com] retail market • aftermarket
— m — equipo m original - [autom] original
equipment market
— m — este - [com] Eastern market
— m — exportación f - [com] export market
— m — lente(s) m - óptica] lens market
— m — neumáticos m—[autom-neumát] tire market
— m —— m con rendimiento m alto - [autom-
-neumát] (high) performance (tire) market
— m — norte m - [com] Northern market
— m — oeste m - [com] Western market
— m — perfil(es) m - [metal-lam] shapes market
— m — recurso(s) m - resource(s) market
— m —— m humano(s) m - [labor] human re-
source(s) market
— m — sector m privado - [com] private sector
market
— m —— m público—[com] public sector market
— m — sud - [com] Southern market
— m — título(s) m - [fin] securities market
— m — valor(es) m - [fin] stock, o securities,
market, o exchange
— m desequilibrado - unbalanced market
— m — para chatarra f - metal-prod] unbalanced
scrap market

mercado m distante - [com] distant market
— m específico - [com] specific market
— m estadounidense - [econ] United States market
— m estudiado - [com] studied market
— m —, anticipadamente, o de antemano m - [com]
prestudied market
— m exterior - [com] foreign market
— m externo - [Econ] external, o foreign, market
— m — estudiado, anticipadamente, o de antemano
- [com] prestudied foreign market
— m financiero - [fin] financial market
— m — ampliado - [fin] expanded financial market
— m — amplio - [fin] broad, o expanded, financial market
— m — para cambio(s) m - [fin] financial exchange market
— m general - [com] aftermarket
— m industrial - [com] industrial market
— m — y comercial - [com] industrial and commercial market
— m — y consumidor - [com] industrial and consumer market
— m inicial - [com] initial market
— m integrado - [econ] integrated market
— m interior - [econ] domestic market
— m internacional - [com] international, o
foreign, market
— m interno - [com] internal, o domestic, market • local market
— m libre - [fin] free market
— m local - [com] local market
— m — estudiado, anticipadamente, o de antemano - [com] prestudied local market
— m marino - marine market
— m meridional - [com] Southern market
— m mundial - [com] world(wide) market
— m — para acero m - [metal-prod] world)wide)
steel market
— m nacional - [com] domestic, o internal, o
local, market
— m natural - [econ] natural market
— m — significativo - [com] significant natural market
— m negro m - [com] black market • [fin] black,
o parallel, market
— m norte - [com] Northern market
— m nuevo - [com] new market
— m occidental - [com] Western market
— m oficial - [fin] official market
— m — para cambio(s) m - [fin] official exchange market
— m óptico - [óptica] lens market
— m para acero m - [metal-prod] steel market
— m — construcción f - [constr] construction
market
— m — diamante(s) m - [com] diamond market
— m —— m industrial(es) - [com] industrial
diamond(s) market
— m — horno(s) m - [ind] furnace(s) market
— m —— m eléctrico(s) m - [metal-prod] electric furnace(s) maraket
— m — neumático(s) m - [autom-neumát] tire
market
— m —— m radial(es) - [autom-neumát] radial
(tire) market
— m — recurso(s) m humano(s) - [labor] human
resoures market
— m — reemplazo m - [com] replacement market
— m —— m de neumático(s) m - [autom-neumát]
tire replacement market
— m — superabrasivo(s)* m - [quím] superabra-
sie(s) market
— m paralelo - [fin] parallel, o unofficial, o
free, market
— m penetrado - [com] penetrated market
— m principal - [com] main market
— m propio - [com] own market
— m próximo - [com] close, o near, o adjacent,
market
— m real - [com] actual, o true, market
— f significativo - [com] significant market

mercado m total - [com] total market
— m venezolano - [com] Venezuelan market
mercadotecnía* f - [com] véase comercialización
  f; [Mex.] marketing*
mercancía f - . . .; véase también mercadería f
— f adquirida - [com] acquired, o procured,
  merchandise, o goods
— f en tránsito m - ⌊com⌋ goods in transit
merceología* f - véase tecnología f industrial
mercúrico,ca* a - [quím] mercuric
merecedor m de crédito m - [com] credit worthy
merecer v - . . . • to earn • to be credited
  with • [deport] to, win, o capture. o claim;
  to nail down
— v atención f—to, merit, o deserve, attention
— v clasificación f - to, merit, o deserve, o
  carry, @ rating
— v — f buena - to merit @ good rating; to do
  well
— v distinción f - to, merit, o deserve, o
  earn, @, distinction, o status
— v mención f - to, merit, o deserve, @ mention
— v respeto m - to, merit, o deserve, respect,
  o attention
merecido,da a - . . .; merited; earned •
  [deport] won; claimed; captured; nailed down
merecimiento m - . . .; meriting; deserving •
  earning; worthiness • [deport] winning; capture,ring; claiming; nailing down
— m crediticio - [fin] credit worthiness
— m de comprador m - [fin] buyer('s) credit
  worthiness
— m de clasificación f - rating, meriting, o
  deserving, o carrying
— m comprador m - [fin] buyer's worthiness
— m distinción f - distinction, o status,
  earning
merendero m - [culin] . . .; buffet; refreshment
  patio
merga* f - véase fusión f; consolidación f
merienda f - [culin] . . .; quick lunch
merma f - . . .; shortage; loss
— f en potencia f - [mec] power, decrease, o
  loss
mermado,da a - decreased • lost
mermar v - . . .; to lose
— v potencia f - [mec] to, decrease, o lose, @
  power
mes m de funcionamiento m - month of operation
— m — invierno m - [meteorol] winter month
— m — otoño m - [meteorol] autumn, o fall,
  month
— m — primavera f - [meteorol] spring month
— m — verano m - [meteorol] summer month
— m estival - [meteorol] summer month
— m invernal - [meteorol] winter month
— m otoñal - [meteorol] autumn, o fall, month
— m pasado - [cronol] last month
— m primaveral - [meteorol] spring month
— m reciente - recent month
— m venidero - [cronol] coming month
— m veraniego - [meteorol] summer month
— m vernal - [meteorol] spring month
mesa f - . . . • [mec] plate; platen
— f adicional • additional table
— f alimentadora f - [mec] feeder,ding table
— f — para planta f - [ind] plant feeder table
— f — — — f para sínter m - [metal-prod]
  sinter(ing) plant feeder table
— f almacenada - stored table • [petról] stored
  (rotary) table
— f almacenadora f - storage,ring table
— f anterior - previous, o front, table
— f — de laminador m - [metal-lam] front mill
  table
— f — y posterior m - [metal-lam] front and
  back table(s)
— f — — — de laminador m - [metal-lam] mill
  front and back, o front and back mill, tables
— f apiladora f - [ind] piler,ling table
— f automática f - [ind] automatic table
— f — para alimentación f - [mec] automatic, o
  traveling, feed(ing) table

mesa f auxiliar - [mec] auxiliary, o trailer, o
  trailing, table
— f —, corrediza, o deslizante, o movible, o
  viajera - [ind] traveling trailer table
— f bascula f—scales, o weighing, table
— f basculante - [mec] tilting table
— f —, corrediza, o deslizante, o movible, o
  viajera - [ind] traveling tilting table
— f con accionamiento m eléctrico - [petról]
  electric drive(n) table
— f — — m independiente - [petról] independent electric drive(n) table
— f — — m independiente - [petról] independent drive(n) table
— f — disco(s) m - [ind] disk table
— f — engranaje(s) m - [mec] gear table
— f — — m inmovilizada - [ind] locked gear
  table
— f — m trabada - [mec] locked gear table
— f — impulsión f - [mec] drive(n) table
— f — — f con árbol m para transmisión f -
  [ind] line shaft driven table
— f — patín(es) m - [ind] skid bed
— f — rodaja(s) f - [mec] caster, bed, o table
— f — rodillo(s) m - [ind] roller table •
  transfer table
— f — — m basculante - [ind] tilting roller
  table
— f — — m —, corrediza, o deslizante, o movible - [ind] traveling tilting roller table
— f — — m corrediza - [ind] traveling roller
  table
— f — — m — para entrada f para laminador m
  - [metal-lam] mill traveling approach roller
  table
— f — — m — — laminador m - [metal-lam]
  mill traveling roller table
— f — — m deslizante - [ind] traveling roller
  table
— f — — m — para entrada f para laminador m
  - [metal-lam] mill traveling approach roller
  table
— f — — m — para laminador m - [metal-lam]
  mill traveling roller table
— f — — m doble - [ind] double roller table
— f — — m frontal - [ind] front roller table
— f — — m movible - [ind] traveling roller
  table
— f — — m — para entrada f - [ind] traveling
  approach rolling table
— f — — m — — f para laminador m -
  [metal-lam] mill traveling approach roller
  table
— f — — m — — laminador m - [metal-lam]
  mill traveling roller table
— f — — m para apiladora f - [ind] piler
  rolling table
— f — — m — entrada f - [metal-lam] approach
  roller table
— f — — m — — f, corrediza, o deslizante -
  [ind] traveling approach roller table
— f — — m — entrada f - [ind] approach roller table
— f — — m — — f para laminador m - [metal-
  lam] mill approach roller table
— f — — m — — f — tren m desbastador -
  [metal-lam] roughing mill approach roller
  table
— f — — m inspección f - [ind] inspection
  roller table
— f — — m sierra f - [metal-lam] saw roller table
— f — — m — — f en caliente - [metal-lam]
  hot saw roller table
— f — — m — — f para corte m en caliente -
  [metal-lam] hot saw roller table
— f — — m portátil - [mec] portable roll(er)
  table
— f — — m posterior - [mec] rear roller table
— f — — m separador(es) - [metal-lam] roller
  spacer table
— f — — m suplementaria - [mec] tail roller
  table

mesa f con rodillos m tipo carrete m - [mec]
spool (type) roller table
— f — — m trasera - [ind] back roller table
— f — roladana(s) f - [ind] caster bed
— f — ruedas f - [ind] caster bed
— f — . . . sección(es) f - [mec] . . . section table
— f concentradora - [miner] concentrating table
— f conectada con tierra f - [sold] grounded
table
— f corrediza - [mec] traveling table
— f — para laminador m - [metal-lam] mill
traveling table
— f cruzada - [mec] cross(ed) table
— f cubímetro - [metal-prod-sinter] measuring
table
— f cuna - [mec] cradle table
— f — para tocho(s) m - [metal-lam] bloom
cradle table
— f de madera f - [domést] wood(en) table
— f — plástico m - [domést] plastic table
— f depresora - [mec] depressing table
— f — de cizalla f - [mec] shear depressing
table
— f delizante - [mec] sliding table • [ind]
traveling table
— f — con rodillo(s) m - [metal-lam] approach
roller table
— f — — — m para acceso m - [ind] traveling
approach roller table
— f — — m — — m a laminador m - [metal-
-lam] traveling mill approach roller table
— f deslizante - [mec| sliding table • [ind]
traveling table
— f — para laminador m - [metal-lam] mill
traveling table
— f doble - [mec] double table
— f — con rodillo(s) m - [mec] double roller
table
— f elevadora - [ind] lifting table • cradle
car; drop pit table
— f — e inclinante - [ind] lifting and tilting
table
— f — con vuelco m - [ind] lifting and tilting
table
— d — y basculante - [ind] lifting and tilting
table
— f — — tumbadora - [ind] lifting and tilting
table
— f — aire m libre (para excursionistas m) -
picnic table
— f en forma f de artesa - [domést] dough box
(table) • [mec] trough table
— f — . . . sección(es) f - [mec] . . .-sec-
tion table
— f enfriadora - [ind] cooling table
— f espaciadora - [mec] spacer table
— f esquinera - [domést] corner table
— f extensible - [mec] run-out table
— f — para entrega f - [mec] run out (delivery) table
— f — — — f para cepilladora f - [mec]
planer run-out (delivery) table
— f filtro - filter table • table filter
— f frontal - [mec] front table
— f — con rodillo(s) m - [ind] front roller
table
— f general - [legal] receiving office
— f giratoria - [mec] revolving table; turntable • [petról] rotary table • [ind] rotating table
— f — inferior - [mec] lower turntable
— f — para laminador m - [metal-lam] mill
turntable
— f — — — m para plancha(s) f - [metal-lam]
plate line turntable
— f — línea f para plancha(s) f - [metal-
-lam] plate line turntable
— f — — lingote(s) m - [metal-lam] ingot
turntable
— f — — selladura f - [mec| rotating sealing
table
— f — superior - [mec] upper turntable

mesa f inclinada - [mec] skew table
— f inclinante - [mec] tilting table
— f indicadora - indicating table
— f — para estado m - [instrum] status (indicating) table
— f — — — m de luz,ces f - [instrum] light
status (indicating) table
— f inmovilizada - [mec] immobilized, o locked,
table
— f — en sitio m - [mec] locked-in-position
table
— f inversora f - [mec] reversing table
— f medidora - [ind] measuring table
— f movible - movable, o traveling, table •
portable table
— f — para rodillo m - [metal-lam] mill traveling table • [mec] portable roll table
— f oblicua - [mec] skew table
— f para acceso m - [ind] approach table; véase
también mesa f para entrada f
— f — alimentación f - [mec] feeder, o run-in,
o in loading, table; apron feeder; table •
approach table
— f — — f para horno m - [ind] furnace, feeding, o charging, table, o equipment
— m — — f para laminador m - [metal-lam] mill
approach table
— f — almacenamiento m - storage,ring table
— f — aproximación f - [metal-lam] approach
table
— f — — f para, descabezadora f, o despuntadora f - [metal-fabr] crop shear approach
table
— f — — f — escarpadora f - [metal-lam]
scarfer approach table
— f — — — horno m - [metal-lam] furnace
approach table
— f — — f — rebabadora f - [metal-lam] scarfer approach table
— f — atadora f - [agric-equip] binder deck
— f — báscula f - [ind] scale table
— f — café m - [comést] coffee table
— f — calibradora f - [mec] gage table
— f — — f para cizalla f - [mec] shear gage
table
— f — — — tijera f - [mec] shear gage
table
— f — calibre—[mec] gage table
— f — — m para, cizalla f, o tijera f - [mec]
shear gage table
— f — carga f - [mec] load(ing), o charging, o
feeder, table
— f — para horno m - [metal-prod] furnace, charging, o feeding, table
— f — cargadora f - [agric-equip] loader deck
— f — carta(s) f - [domést] card table
— f — cepilladora f - [mec] planer table
— f — cocina f - [domést] kitchen table
— f — comedor m - [domést] dining (room) table
— f — compensación f—[metal-lam] looper table
— f — demora(r) - [metal-lam] delay table
— f — descarga f - [mec] unloading, o discharge,ging, table
— f — — f para horno m - [metal-lam] furnace
discharge,ging table
— f — despuntadora f - [metal-lam] cropping, o
crop shear, table
— f — dibujo m - [dib] drawing, o drafting,
table, o board
— f — diseño m - [dib] drawing, o drafting,
table, o board
— f — empaque m - [ind] packing table
— f — enfriamiento m - [ind] cooling table
— f — enhebrado* m - [ind] threading table
— f — entornadura* f de borde(s) m - [mec]
edge turn roll(ing) table
— f — entrada f - [mec] approach, o run-in, o
feeding, o entrance, o entry, o feeder, table
• receiving table • [pol] receiving and routing office
— f — — f corrediza - [mec] traveling approach table
— f — — f — para laminador m - [metal-lam]

mill traveling approach table
mesa f̲ **para entrada** f̲ **deslizante** - [metal-lam] traveling approach table
— f̲ — — f̲ — **para laminador** m̲ - [metal-laml mill traveling approach table
— f̲ — — f̲ — **en forma de artesa** f̲ - [mec] trough, entry, o̲ approach, table
— f̲ — — f̲ **movible** - [mec] traveling approach table
— f̲ — — f̲ — **para laminador** m̲ - [metal-lam] mill traveling, approach, o̲ run-out, table
— f̲ — — f̲ — **enderezadora** f̲ - [mec] straightener approach table
— f̲ — — f̲ **entornadora** f̲ **(para bordes** m̲**)** - [mec] edge turn roll(ing) approach table
— f̲ — — f̲ **para escarpadora** f̲ - [metal-lam] scarfer, approach, o̲ run-in, table
— f̲ — — f̲ — — f̲ **para palanquilla** f̲—[metal--lam] billet scarfer approach table
— f̲ — — f̲ — **horno** m̲ - [metal-lam] furnace entry (approach) table
— f̲ — **laminadora** f̲ - [metal-lam] mill (entry) approach table
— f̲ — — f̲ — — f̲ **para palanquilla** f̲—[metal--lam] billet mill approach table
— f̲ — **lecho** m̲ - [metal-lam] bed, approach, o̲ run-in, table
— f̲ — — f̲ — — m̲ **para enfriamiento** m̲ - [metal-lam] cooling bed, approach, o̲ run-in, table
— f̲ — — f̲ — **lingote(s)** m̲ - [metal-lam] ingot run-in table
— f̲ — — f̲ — **niveladora** f̲ - [mec] leveler approach table
— f̲ — — f̲ — **máquina** f̲ **niveladora** - [metal--lam] straightening machine, approach, o̲ run--in, table
— f̲ — — f̲ — — f̲ — **para riel(es)** m̲—[metal--lam] rail straightening machine, approach, o̲ run-in, table
— f̲ — — f̲ **rebabadora** f̲ - [metal-lam] scarfer, approach, o̲ run-in, table
— f̲ — — f̲ — — f̲ **para palanquilla** f̲—[metal--lam] billet scarfer approach table
— f̲ — — f̲ — **rodillo(s)** n̲ - [metal-lam] roller(s), approach, o̲ run-in, table
— f̲ — — f̲ — m̲ **para sierra** f̲ - [metal-lam] saw, approach, o̲ run-in, table
— f̲ — — f̲ — — f̲ — — f̲ **para corte** m̲ - [metal-lam] cutting saw, approach, o̲ run-in, table
— f̲ — — f̲ — — m̲ — — f̲ — — m̲ **en caliente** - [metal-lam] hot saw roller, approach, o̲ run-in, table
— f̲ — — f̲ — — m̲ — — f̲ **en caliente** - [metal-lam] hot saw approach roller table
— f̲ — — f̲ — **sierra** f̲—[mec] saw run-in table
— f̲ — — f̲ — — f̲ **en caliente** - [meal-lam] hot saw approach table
— f̲ — — f̲ — — f̲ **oara corte** m̲ - [metal-lam] saw approach table
— f̲ — — f̲ — — f̲ — — m̲ **en caliente** - [metal-lam] hot saw approach table
— f̲ — — f̲ — **tijera** f̲ - [metal-lam] shear, approach, o̲ run-in, table
— f̲ — — f̲ **para transferidor** m̲ - [metal-lam] transfer, approach, o̲ run-in, table
— f̲ — — f̲ — **transportador** m̲ - [mec] transfer run-in table
— f̲ — — f̲ — — m̲ **para riel(es)** m̲ - [metal--lam] rail transfer run-in table
— f̲ — — f̲ **por gravedad** i̲ - [mec] gravity, entry, o̲ run-in, o̲ approach, table
— f̲ — — f̲ **en forma** f̲ **de artesa** f̲ - [mec] trough gravity, entry, o̲ run-in, table
— f̲ — — f̲ **y salida** f̲ - [mec] approach and delivery table • [legal] receiving and distribution, office, o̲ table
— f̲ — **entrega** f̲ - [metal-lam] delivery, o̲ run--out table
— f̲ — — f̲ **para niveladora** f̲ - [mec] leveler run-out table
— f̲ — **escuadrar** - [mec] edging, o̲ resquaring,

mesa f̲ **para estado** m̲ - [instrum] status table
— f̲ — — m̲ **de luz,ces** f̲ - [instrum] light(s) status table
— f̲ — **extrusión*** f̲ - extrusion table
— f̲ — — f̲ **de electrodo(s)** m̲ - electrode(s) extrusion table
— f̲ — **horno** m̲ - [metal-lam] furnace table
— f̲ — **inspección** f̲ - [ind] inspection table
— f̲ — **inversión** f̲ - [mec] reversing table
— f̲ — **juego** m̲ - [juego] gaming, o̲ gambling, table
— f̲ — — f̲ **con carta(s)** f̲—[domést] card table
— f̲ — **laminador** m̲ - [metal-lam] mill table
— f̲ — **lazo** m̲ - [metal-lam] loop table
— f̲ — **luz,ces** f̲ - [instrum] light(s) table • [domést] night stand
— f̲ — **macho(s)** m̲ - [mec] core bench
— f̲ — **medida(s)** f̲ - [mec] gage table; véase **mesa** f̲ **para calibre(s)** m̲
— f̲ — **naipe(s)** m̲ - [domést] card table
— f̲ — **negociación(es)** f̲ - [labor] bargaining table
— f̲ — **núcleo(s)** m̲ - [mec] core bench
— f̲ — **operador** m̲ - [ind] operator('s), table, o̲ desk, o̲ pulpit
— f̲ — **pasillo** m̲ - [domést] hall table
— f̲ — **patio** m̲ - [domést] patio table
— f̲ — **prensa** f̲ - [mec] press table
— f̲ — **prueba(s)** f̲ - [ind] test(ing), table, o̲ rack
— f̲ — **recepción** f̲ - [ind] receiving table
— f̲ — — f̲ **para lingote(s)** m̲ - [metal-lam] ingot receiving table
— f̲ — **reparación(es)** f̲ - [mec] repair, table, o̲ stand
— f̲ — **retardo** m̲ - [metal-lam] delay table
— f̲ — **retroceso** m̲ - [mec] back-up, o̲ pull--back, table
— f̲ — — m̲ **para cizalla** f̲ - [mec] shear pull--back table
— f̲ — — m̲ — **tijera** f̲ - [mec] shear pull-back table
— f̲ — **rodillo(s)** m̲ - [mec] roll(er) table
— f̲ — **sala** f̲ - [domést] parlor table • end table
— f̲ — **salida** f̲ - [mec] delivery, o̲ exit, o̲ run-out, o̲ discharge, o̲ unloading, table
— f̲ — — f̲ **cepilladora** f̲ - [mec] planer, run-out, o̲ delivery, table
— f̲ — — f̲ — — f̲ **y para entrada** f̲ **para ciza-lla** f̲ - [ind] planer run-out and shear, approach, o̲ run-in, table
— f̲ — — f̲ **para cizalla** f̲ - [mec] shear, delivery, o̲ run-out, table
— f̲ — — f̲ **enderezadora** f̲ - [metal-lam] straightener, run-out, o̲ delivery, table
— f̲ — — f̲ — **escarpadora** f̲ - [metal-trat] scarfer, run-out, o̲ delivery, table
— f̲ — — f̲ — — f̲ **para palanquilla** f̲—[metal--trat] billet scarfer, delivery, o̲ run-out, table
— f̲ — — f̲ — **horno** m̲ - [metal-lam] furnace, delivery, o̲ run-out, o̲ discharging, table
— f̲ — — f̲ — **lecho** m̲ - [metal-lam] bed run--out table
— f̲ — — f̲ — — m̲ **para enfriamiento** m̲ - [metal-lam] cooling bed run-out table
— f̲ — — f̲ — **máquina** f̲ - [mec] machine run--out table
— f̲ — — f̲ — — f̲ **enderezadora** - [metal-lam] straightening machine, run-out, o̲ delivery, table
— f̲ — — f̲ — — f̲ — **para riel(es)** m̲—[metal--lam] rail straightening machine, run-out, o̲ delivery, table
— f̲ — — f̲ — **niveladora** f̲ - [mec] leveler run-out table
— f̲ — — f̲ — **rebabadora** f̲ - [metal-trat] scarfer, run-out, o̲ delivery, table
— f̲ — — f̲ — — f̲ **para palanquilla** f̲—[metal--lam] billet scarfer delivery table
— f̲ — — f̲ — **sierra** f̲ - [metal-lam] saw run-

-out table
mesa f̲ para salida f̲ para tijera f̲ - [mec] shear
run-out table
— f̲ —— f̲ — **transbordador** m̲ - [mec] transfer
run-out table
— f̲ —— f̲ — **transferidor** m̲ - [mec] transfer
run-out table
— f̲ —— f̲ — **transportador** m̲ - [mec] conveyor
run-out table
— f̲ —— f̲ —— m̲ **para riel(es)** m̲ - [metal-
-lam] rail transfer run-out table
— f̲ — **sierra** f̲ - [mec] saw table
— f̲ —— f̲ **basculante**—[herram] drop saw table
— f̲ —— f̲ **en caliente** - [metal-lam] hot saw
table
— f̲ —— f̲ **levadiza** f̲ - [herram] drop saw
table
— f̲ —— f̲ **para corte** m̲ - [metal-lam] cutting
saw table
— f̲ —— f̲ —— **en caliente** - [metal-lam] hot
(cutting) saw table
— f̲ — **soldadora** f̲ - [sold] welder table
— f̲ — **soldadura** f̲ - [sold] welding table
— f̲ — **trabajo** m̲ - work(ing) table
— f̲ — **transferencia** f - [mec] transfer (table)
— f̲ — **trazado** m̲ - [dib] layout table • [mec]
marking-off table
— f̲ **plegabie** - [domést] folding table
— f̲ — **para cizalla** f̲ - [mec] shear, folding, o
depressing, table
— f̲ **por gravedad** f̲ - [mec] gravity table
— f̲ **portátil** - [domést] portable table
— f̲ — **para entrada** f̲ - [mec] portable, run-in,
o approach. table
— f̲ —— f̲ **y salida** f̲ - [mec] portable, approach and delivery, o run-in and run-out,
table
— f̲ —— **rodillo(s)** m̲ - [mec] portable roll(s)
table
— f̲ —— **salida** f̲ - [mec] portable, delivery,
o run-out, table
— f̲ **posterior** - [mec] back, o rear, table
— f̲ — **para laminador** m̲ - [metal-lam] mill,
back, o rear, table
— f̲ —— **rodillo(s)** m̲ - [mec] back, o rear,
roll(er)s table
— f̲ **puesta a tierra** f̲ - [sold] grounded table
— f̲ **ratona** - [domést] coffee table
— f̲ **receptora** - [mec] receiving table
— f̲ — **para lingote(s)** m̲ - [metal-lam] ingot(s)
receiving table
— f̲ **redonda** - [domést] round table • [legal]
panel discussion
— f̲ — **minera** - [miner] mining round table
— f̲ — **sobre minería** f̲ - [miner] mining round
table
— f̲ **rotativa**—[petról] rotary, table, o machine
— f̲ **rotatoria** - rotary table
— f̲ — **almacenada** - [petról] stored rotary
table
— f̲ — **con accionamiento** m̲ **directo** - [petról]
direct drive(n) rotary (table)
— f̲ — m̲ **eléctrico** - [petról] electric
drive(n) rotary (table)
— f̲ —— m̲ — **independiente** - [petról] independent electric drive(n) rotary (table)
— f̲ — m̲ **independiente** - [petról] inde-
poendent drive(n) rotary (table)
— f̲ — **impulsión** f̲ **directa** - [petról] direct
drive(n) rotary (table)
— f̲ — **propulsión** f̲ **directa** - [petról] direct drive(n) rotary (table)
— f̲ — **(de) National** - [petról] National rotary
(table)
— f̲ — **independiente** - [petról] independent
rotary (table)
— f̲ — **inspeccionada** - [petról] inspected rotary (table)
— f̲ — **para malacate** m̲ - [petról] drawworks
rotary (table)
— f̲ **separadora** - [mec] separator,ting table •
spacer table
— f̲ **suplementaria** f̲ - trailer, o tail, table

mesa f̲ suplementaria con rodillo(s) m̲ - [mec]
tail roller table
— f̲ — **corrediza** - [mec] traveling trailer
table
— f̲ — **deslizante** - [mec] traveling trailer
table
— f̲ — **movible,**- [mec] traveling trailer table
— f̲ **tipo** m̲ **Fourdrinier** - [papel] Fourdrinier
type table
— f̲ **transbordadora** - [mec] transfer table
— f̲ **transportadora** f̲ - [mec] conveyor, o transfer, table • [metal-lam] run-out table
— f̲ **transversal** - [mec] cross(ed) table
— f̲ **trasera** - back, o rear, table
— f̲ — **con rodillo(s)** m̲ - [mec] back, o rear,
roller table
— f̲ **tumbadora** - [mec] tilting table
— f̲ **vibratoria** - [mec] vibrating table
meseta f̲ - [topogr] . . . • [constr] . . .;
resting platform
— f̲ **desértica** - [geogr] high desert
mesita f̲ **de madera** f̲ - [domést] small wood(en)
table
mesurado,da a̲ - measured
meta f̲ - [deport] . . .; home; end; final checkpoint
— f̲ **alcanzada** - reached, o accomplished, goal
— f̲ **lograda** - reached, o accomplished, goal
metaestable* a̲ - metastable
metal m̲ - . . . • [metal] brass
— m̲ **acondicionado** - [metal] conditioned metal
— m̲ **adyacente** - adjacent metal
— m̲ **afinado** - [metal-prod] refined metal
— m̲ **agrietado** - [metal] cracked metal
— m̲ **aislado** - [mec] insulated metal
— m̲ **aleado** - [metal] (metal) alloy • alloyed
metal
— m̲ — **débilmente** - [metal] low alloy(ed) metal
— m̲ **altamente resistente** - [metal-prod] high
resisting metal
— m̲ —— **a corrosión** f̲ - [metal-prod] high
corrosion resisting metal
— m̲ **amarillo** - [metal] brass
— m̲ **antifricción** f̲ - [metal-prod] antifriction
metal; Babbit (metal)
— m̲ **aportado** - [sold] (weld) deposit; weld, o
deposited, meta; added, o filler, metal
— m̲ — **por pasada** f̲ - [sold] metal deposited
oper pass
— m̲ — **suficiente**—[Sold] sufficient, o enough,
filler metal
— m̲ **arrollado** - wound metal
— m̲ — **en espiral** - spiral wound metal
— m̲ **Babbit** - [metal] Babbit metal
— m̲ **base** - [sold] base metal • parent metal
— m̲ **básico** - base, o basic, metal
— m̲ **bien aislado** - [metal] well insulated metal
— m̲ **blanco** - [metal] white metal • [mec] Babbit, o antifriction, o bearing, metal •
Babbit
— m̲ **blando** - [metal- soft metal
— m̲ **brillante** - [metal] bright, o shiny, metal
— m̲ **caliente** - [metal-prod] hot metal • hot pig
iron
— m̲ **carcomido** - [metal] eaten (out) metal
— m̲ **cayente** - [metal] falling metal
— m̲ **cementado** - [metal-prod] precipitated metal
— m̲ **colado** - [metal-prod] poured, o cast, metal
— m̲ **común** - [metal] common metal
— m̲ **con aleación** f̲ - [metal-prod] alloy metal
— m̲ —— f̲ **alta** - [metal-prod] high, alloy, o
analysis, metal
— m̲ —— f̲ **baja** - [metal-prod] low, alloy, o
analysis, metal
— m̲ — **calibre** m̲ **superior en . . . números(s)** m̲
- [metal-lam] metal . . . gage(s) heavier
— m̲ — **contenido** m̲ **alto** - [metal] high content
metal
— m̲ —— m̲ — **de aleación** f̲ - [metal-prod]
high alloy (content) metal
— m̲ —— m̲ **bajo** - [metal-prod] low content
metal
— m̲ —— m̲ — **de aleación** f̲ - [metal] low alloy (content) metal

metal m **con espesor** m **mayor** - [metal-lam] heavier (gage) metal
— m —— m **menor** - [metal-lam] lighter (gage) metal
— m **con liga** f - [metal-prod] alloyed metal
— m —— **metal** a - metal to metal
— m —— **peso** m **reducido** - [metal- lightweight metal
— m —— **tenor** m **alto de aleación** f—[metal-prod] high alloy metal
— m —— **bajo de aleación** f - [metal-prod] low alloy metal
— m **contra metal** m - [mec] metal to metal
— m **corroído** - [quím] corroded, o rusted. metal
— m **corrugado** - [metal-fabr] corrugated metal
— n **de litio** m - [metal] lithium metal
— m —— **aportación** f - [sold] weld metal
— f —— f **agregado**—[sold] added filler metal
— f —— f **añadido** - [sold] added filler metal
— m —— **aporte** m - [sold] weld deposit
— m —— **edad** f **espacial** - [metal] space,cial age metal
— m —— **electrodo** m - [sold] electrode, o wire, metal
— m —— **paleta** f - [mec] blade metal • [ventil] blade iron
— f —— **tubería** f - [tub] pipe,ping metal
— m —— **calentado con resistencia** f **eléctrica** - [sold] electrical resistance heated pipe metal
— m **delgado** - [metal-lam] thin metal
— m **depositado**—[sold] deposited metal; deposit
— m —— **por pasada** f - [sold] metal deposited per pass
— m **derretido** - [sold] molten, metal, o pool
— m **desperdiciado** - waste metal
— m **desplegado** - [metal-fabr] expanded metal; metal lath • [constr] (metal), lath, o mesh
— m —— **rómbico** - [constr] diamond mesh lath
— m **diferente** - [metal] different, o other, metal
— m **disinto** - [metal] different, o other, metal
— m **diverso** - [metal] different, o other, metal
— m **doblado** - [mec] bent, o folded, metal
— m **duro** - [metal] hard metal
— m **eliminado** - [mec] eliminated, o removed, metal
— m **emplomado** - [metal-prod] terne metal
— m **en exceso** - [sold] excess(ive) metal
— m —— **fusión** f - [sold] molten, metal, o weld
— m **enderezado** - [mec] straightened metal
— m **estirado** - [metal-fabr] expanded metal • stretched metal
— m **excedente** - [sold] excess(ive) metal
— m **excelente** - excellent metal
— m **excesivo** - excess(ive), o too much, metal
— m —— **aportado** - [sold] excess(ive), o too much, deposited, metal
— m **expandido** - [metal-fabr] expanded, metal, o (mesh) lath
— m **expuesto** - exposed metal
— m **fuertemente aleado** - [metal-prod] high alloy metal
— m **fundido** - [metal-prod] (hot) pig iron • [sold] molten, o fused, metal, o pool, o puddle • [segurid] hot metal
— m —— **alimentado** - [metal-prod] fed molten metal
— m —— **cayente** - [sold] falling molten metal
— m —— **con ley** f **alta** - [metal-prod] high grade molten metal
— m —— **de varilla** f - [sold] filler rod
— m —— f **para aportación** f - [sold] molten filler rod
— m —— **debajo de arco** m - [sold] molten metal beneath @ arc
— m **fundido en horno** m - [metal-prod] furnace molten metal
— m —— —— m **para reverbero** m - [metal-prod] reverberatory furnace molten metal
— m —— **matriz** f - [metal-prod] die cast metal
— m —— **excesivo** - [sold] excess(ive) molten metal

metal m **fundido procesado** - [metal-prod] processed molten metal
— m —— **solidificado** - [sold] solidified molten metal
— m **grueso** - [metal-lam] thick metal
— m **herrumbrado** - [metal] rusted (out) metal
— m —— **totalmente** - [metal] (totally) rusted (out) metal
— m **hiperaleado** - [metal-prod] high metal alloy • high alloy metal
— m **hipoaleado** - [metal-prod] low metal alloy • low alloy metal
— m **inferior** - [metal-prod] inferior, o low grade, o bottom, metal
— m **interior** - [metal-prod] inner metal
— m **laminado** - [metal-lam] rolled, o sheet, metal
— m **laminar** - [metal-lam] laminar metal; ply-metal
— m **ligero** - [metal-prod] light (weight) metal
— m **líquido** - [metal-prod] liquid metal • hot pig iron
— m **liso** - [metal-lam] smooth metal
— m **liviano** - [metal-prod] light (weight) metal
— m **monel** - [metal-prod] monel (metal
— m **Muntz** - [metal-prod] Muntz metal
— m **nivelado** - [mec] leveled metal
— m **no ferroso** - [metal-prod] non-ferrous metal
— m —— **mecanizado** - [mec] machined non-ferrous metal
— m **original** - [sold] original, o parent, metal • véase también **metal** m **de base** f
— m **oxidado** - [metal] rusted (out) metal
— m —— **totalmente** - [metal] (totally) rusted (out) metal
— m **para aleación** f - [metal-prod] alloy(ing) metal; alloy content • [sold] alloy
— f —— **aportación** f - [sold] deposit; weld metal; weld deposit; filler metal
— m —— f **con carbono** m - [sold] carbon weld deposit
— m —— f —— **alto** - [sold] high carbon weld deposit
— m —— f —— **bajo** [sold] low carbon weld deposit
— m —— f —— m **excepcionalmente bajo** - [sold] unusually low carbon weld deposit
— m —— f **excesivo** - [sold] excess(ive) weld metal
— m —— f **fundido**—[sold] molten filler metal
— f —— f **para proceso** m **para solidificación** f - [sold] freezing (process) weld metal
— m —— f **requerido** - [sold] required weld metal
— m —— f **sin diluir** - [sold] undiluted weld metal; all weld metal
— m —— f —— **típico** - [sold] typical undiluted weld deposit
— m —— f **sólido** - [sold] solid weld metal
— m —— f **solidificado** - [sold] solid(ified) weld metal
— m —— f **típico** - [sold] typical weld metal
— f —— f —— **sin diluir** - [sold] typical undilued weld metal
— m —— **total** - [sold] total weld metal
— m —— **aporte** m - véase **metal** m **para aportación** • deposit
— m —— **base** f - [sold] base, o parent, metal
— m —— f **con espesor, diferente, o distinto, o diverso** - [sold] different thickness base metal
— m —— f **frío** - [sold] cold base metal
— m —— f **fundido** - [sold] molten base metal
— m —— f **próximo**—[sold] adjacent base metal
— f —— f **subyacente** - [sold] underlying base metal
— m —— **brida** f - [metal-fabr] flange sheet
— m —— **calentar(se)** - metal to be heated
— m —— **elaborar** - [metal-fabr] metal to be worked; stock
— m —— **labrar** - [metal-fabr] metal to be worked • stock

metal m para liga f—[metal-prod] alloying metal
— m — paleta(s) f - [mec] blade metal
— m — relleno m - [sold] filler metal
— m — soldadura f - [sold] weld, metal, o deposit
— m — — f aportado - [sold] deposited weld metal; weld metal deposited
— m — — f — por pasada f - [sold] weld metal deposited per pass
— m — — f caliente - [sold] hot weld metal
— f — — f — aportado - [sold] deposited hot weld metal
— m — — f — depositado - [sold] deposited hot weld metal
— m — — f depositado por pasada f - [sold] weld metal deposited per pass
— m — — f desperdiciado - [sold] wasted weld metal
— m — — f en proceso m de solidificación f - [sold] freezing weld metal
— m — — f excesivo m - [sold] excess(ive[weld metal
— m — — f fundido - [sold] molten weld metal
— m — — f sano - [sold] sound weld metal
— f — — f solidificado - [sold] solid(ified) weld metal
— m — — f sólido - [sold] solid weld metal
— m perforado - [metal-fabr] perforated metal
— m — por corrosión f - [metal- rusted out metal
— m — totalmente - [metal] rusted out metal
— m — — por corrosión f - [metal (totally) rusted out metal
— m pesado - [metal] heavy metal
— m precipitado - [metal-prod] precipitated metal
— m prensado - [metal-prod] pressed metal
— m pulverizado - powdered metal
— m ranurado - [mec] grooved metal
— m raro - [miner] rare metal
— m raspado - [mec] scratched metal
— m reaccionante - [quím] reacting metal
— m reducido - [metal-prod] reduced metal
— m — directamente - [metal-prod] directly reduced, metal, o ore
— m refinado - [metal-prod] refined metal
— m relativamente liso m - [metal-lam] relatively smooth metal
— m remendado - [mec] mended metal
— m resistente - [metal-prod] resistant metal
— m — a calor m - [metal-prod] heat resisting metal
— m sin diluir - undiluted metal
— m soldador - véase metal m para soldadura f
— m solidificado - [sold] solidified metal
— m suave - [metal-prod] soft metal
— m subyacente - [mec] underlying metal
— m suficiente(mente) caliente - [sold] suficiently hot, o hot enough, metal
— m — frío - [sold] sufficiently cold, o cold enough, metal
— m superficial - [metal-prod] surface, o top, metal
— m superior - [metal-prod] superior metal • [mec] top, o upper, o surface, metal
— m terne - [metal-trat] terne metal
— m típico - [metal-prod] typical metal
— m — para soldadura f - [sold] typical weld metal
— m tocado - [sold] touched metal
— m total - [metal] total metal
— m unido - [metal] joined metal
— m vertical - vertical metal
— m vertido—[metal-prod] poured, o cast, metal
— m virgen - [metal-prod] virgin metal
Metaldom f - véase Complejo m Metalúrgico Dominicano
metálico,ca a - metallic; metal
— a sinterizado,da a - sintered metallic
metalistería f - [metal] metal work(ing)
metalización f - [metal-prod] . . .; metallization,zing
— f alta - [metal-prod] high metallization,zing

metalización f baja - [metal]prod] low metallization,zing
— f bajo control - [metal-prod] under control, o controlled, metalization,zing
— f — fiscalización f - [metal-prod] under control, o controlled, metallization,zing
— f de alambre m - [metal-alambre] wire metallization,zing
— f — hierro m - [metal-prod] iron metallization,zing
— f — pella(s) f - [miner] pellet(s) metallization,zing
— f — — f de óxido m - [miner] oxide pellets metallization,zing
— t por proyección f - [sold] projection metallization,zing
— f regulada - [metal-prod] controlled metallization,zing
— f uniforme - [metal-prod] uniform metallization,zing
metalizado,da a - [metal-prod] metallized
metalizar v - [metal-prod] . . .; to metallize
— v con aluminio m - [metal] to metallize with aluminum
— v hierro m - [metal-prod] to metallize @ iron
metalmecánico,ca a - metal-mechanical; metallurgical and mechanical
metalográfico,ca a - [metal] metallographic
metalurgia f - . . .; smelting
— f buena - [metal] good metallurgy
— f de boro m - [metal-prod] boron metallurgy
— f — hierro m - [metal] iron, o ferrous, metallurgy
— f — polvo m - [metal] powder metallurgy
— f — proceso m - [metal-prod] process(es) metallurgy
— f extractiva - [miner] extraction,tive metallurgy
— f — de importancia f - [miner] large scale, o important, extraction,tive metallurgy (project)
— f mala - [metal] bad, o poor, metallurgy
— f pobre - [metal-prod] poor metallurgy
— f verificada - [metal] verified metallurgy
metalurgista m jefe - [metal-prod] chief, o head, metallurgist
— m supervisor - [metal-prod] supervising metallurgist
metamérico,ca a - [quím] metameric
metamorfismo m dinámico - [geol] dynamic metamorphism
metanol m - [quím] methanol
metasilicato m - [quím] metasilicate
meteorización f - . . .; [geol] meteorization
meteorológico,ca a - [meteorol] . . .; weather • nature
meter v - . . . • to stuff; to place • to move in(to)
— v agua n - to leak water (into)
— v — en horno m - [ind] to leak water into @ furnace
— v entre paréntesis m - [gram] to enclose, o place, between poarenthesis
— v sínter m - [metal-prod] to charge @ sinter
— v y sacar v - to move in and out
meterse v detrás de volante m - [autom] to slide behind @ wheel
metido,da a - . . . . • stuffed; moved in • placed
metiletilcetona f - [quím] methyl-ethyl-ketone
metiletilcetónico,ca a - [quím] methyl-ethyl--ketonic
metilcetona f - [quím] methyl-ketone
metilcetónico,ca a - [quím] methyl-ketonic
metiletílico,ca a - [quím] methyl-ethylic
metiletilo m - [quím] methyl-ethyl
metimiento m - . . .; moving in • stuffing
método m - . . .; practice; procedure; process; approach • system
— m aceptable - acceptable method
— m actual - corrent, o present, o existing, method
— m — para remoción f - [quím] present, o existing, removal method

método m **alternativo** - alternate method; alternative
— m **para selección** f - alternate selection method
— m **anterior** - previous, o former, method
— m **aplicable** - applicable method
— m **aplicado** - applied method
— m **erróneamente** - misapplied method
— m **apropiado** - appropriate, o proper, o suitable, method
— m — **para conexión** f **con tierra** f - [sold] appropriate, o proper, grounding method
— m — **puesta** f **a tierra** f - [sold] proper grounding method
— m **automático** - [ind] automatic method
— m — **para esmerilado** m - [mec] automatic grinding method
— m — — m **de bloque** m - [trefil] automatic block grinding method
— m — — m — — m **cuneiforme** - [trefil] automatic wedge block grinding method
— m — — m — **para sujeción** f - [trefil] automatic wedge block grip block grinding method
— m **básico** - basic method
— m — **para horadación** f - [constr] basic boring method
— m — **perforación** f - [constr] basic drilling method
— m **científico** - scientific method
— m **común** - common, o standard, method
— m **con anillo** m - ring method
— m — — m **para compresión** f - ring compression method
— m — **barra(s)** f **de aguja** f - needle beam method
— m — **burbuja(s)** f **con gas** m - soap bubble method
— m — **cable** m - [petról] cable method
— m — — m **con herramienta(s)** f - [petról] cable tool method
— m — **celda** f **para carga** f - [metal-lam] roll assembly method
— m — **cera** f - wax method
— m — — f **perdida** - lost wax method
— f **con clasificadora** f - [miner] clasified method
— m — — f **hidráulica** - [miner] jig method
— m — **convertidor** m - [metal-prod] converter method
— m — **costo** m **relativamente reducido** - relatively low-cost method
— m — **desmontaje** m - [mec] dismantling method
— m — — m **de rodillo(s)** m - [metal-lam] roll dismantling method
— m — **electrodos** m **múltiples** — [sold] multiple, electrode(s), o wire(s), method
— m — **ensayo** m - testing method
— m — — m **destructivo** - destructive testing method
— m — — m **para soldadura(s)** f - [sold] destructive weld testing method
— m — — m **fluorescente** - fluorescent testing method
— m — — m **no destructivo** - non-destructive testing method
— m — — — **para soldadura(s)** f - [sold] nondestructive weld testing method
— m — **ocular** - visual test(ing) method
— m — — m **para soldadura(s)** f - [sold] weld test(ing) method
— m — — m **con partícula(s)** f **magnética(s)** - [metal] magnetic paraticle(s) testing method
— m — **por medio de partícula(s)** f **magnética(s)** - [metal] magnetic particle(s) testing method
— m — — m **radiográfico** - [electrón] radiographic test(ing) method
— m — — m **ultrasónico** - [electrón] ultrasound, sonic test(ing) method
— m — **enterramiento** m **directo** - [electr-instal] direct burial method
— m — **escudo** m - [constr] shield method

método m **con escudo** m **para avance** m - [constr] shield method (tunneling)
— m **con estímulo(s)** m - [labor] incentive(s) method
— m **con examen** m - [metal] examination, o inspection, method
— m — m **con partícula(s)** f **magnética(s)** - [metal] magnetic particle(s), examination, o inspection, method
— m — — m **por medio de partícula(s)** f **magnética(s)** - [metal- magnetic particle(s), examination, o inspection, method
— n — **extremo** m **libre** - [constr-pil] free-end method
— m — **flotación** f - [miner] flotation method
— m — — f **selectiva** - [miner] selective flotation method
— m — **granallado** m - [mec] shot, blasting, o peening, method
— m — **horadación** f - [constr] tunneling, o piercing, method
— m — **impulso-eco** m - [electrón] impulse-echo method
— m — **incentivo(s)** m — [labor] incentive method
— m — — m **habitual(es)** m - [labor] habitual incentive(s) method
— m — **infiltración** f - [hidr] infiltration method
— m — **inserción** f - inserting, tion method
— m — — f **con gato(s)** m - [constr] jacking method
— m — **inspección** f - inspection method
— m — — f **metalográfica** - [metal] metallographic inspection method
— m — — f **destructiva** - [metal] destructive metallographic inspection method
— f — — f — **no destructiva** - [metal] non-destructive metallographic inspection method
— m — — f — **totalmente no destructivo** - [metal] totally, o completely, nondestructive metallographic inspection method
— m — — f **no destructiva** - [ind] nondestructive inspection method
— m — — f — — **de soldadura(s)** f **sobre acero** m - [sold] steel weld(ment) nondestructive inspection method
— m — — **por medio de partícula(s)** f **magnética(s)** - [metal] magnetic particle(s) inspection method
— m — — f **química** - [quím] chemical inspection method
— m — — f — **destructiva** - [quím] destructive chemical inspoection method
— m — — f — **no destructiva** - [metal] non-destructive chemical inspection method
— m — — f **totalmente no destructiva** - [quím] totally, o completely, non-destructive chemical inspection method
— m — — f **radiográfica** - [electrón] radiographic inspection method
— m — — f **recomendado** - recommended inspection method
— m — — f **totalmente no destructiva** - [quím] totally, o completely, nondestructive inspection method
— m — — f **ultrasónica** - [electrón] ultrasound, sonic inspection method
— m — — **visual** - visual inspection method
— m — **línea** f **recta** - [contab] straight line method
— m — **malla** f - [ind] mesh method
— m — — f **elástica (de alambre** m**)** - [constr] wire mesh net method
— m — **mesa** f **concentradora** - [miner] concentrating table method
— m — **movimiento** m **en línea** f **recta** - [metal-fabr] straight line movement method
— m — **polvo** m - [ind] powder method
— m — — m **húmedo** - [ind] wet powder method
— m — — m **seco** - [ind] dry powder method
— m — **prensa** f - [mec] press method
— m — **prisionero(s)** m **(roscados)** - [sold] studding method

— m con red f - net method
— m — — f de malla f - mesh net method
— m — retroceso m - [sold] back step(ping) method
— m — soldadura f - [sold] welded method
— m — — f automática - [sold] automatic welding method
— m — sólido(s) m - solids method
— m — — m fluido(s) - fluid solid(s) method
— m — templado m - [metal-trat] tempering method
— m — — m húmedo - [metal-trat] aquatemp tempering method
— m — tiempo m - time method
— m — tornillo m - [mec] vise method • screw method
— m — — m para banco m - [mec] vise method
— m — torsión f - [mec] torque,quing method
— m — — f cruzada - [mec] cross torque,quing method
— m — valor(es) m - value(s) method
— m — — extremo(s)—extreme value(s) method
— m — viga(s) f - [constr] beam method
— m — — f equivalente(s) - [constr-pil] equivalent beam(s) method
— m — zanja f - [constr] trench method
— m — — f abierta - [constr] open trench method
— m conocido - known method
— m contable - [contab] accounting method
— m — conformado - [contab] conformed accounting method
— m convencional - conventional method
— m — para drenaje m - [hidr] conventional drainage method
— m — — — m de franja f divisoria - [constr] conventional median drainage method
— m — — — m — — f medianera - [constr½] conventional median drainage method
— m conveniente - convenient method
— m corriente - current, o conventional, o standard, method
— m — para propaganda f - [com] current advertising method
— m — área m - area method
— m — Burton - [petról] Burton('s) method
— m — Coulomb - [suelos] Coulomb('s) method
— m — Frasch - [petról] Frasch('s) method
— m — Perkins - [petról] Perkins(') method
— m — Ranikne - [suelos] Rankine('s) method
— m — toro - [matem] torus method
— m destructivo - destructive method
— m — para ensayo(s) m - destructive testing method
— m diferente - different method
— m — para soldadura f - [sold] different welding method
— m disponible - available method
— m distinto—different method; alternate means
— m diverso - different method
— m — para soldadura f - [sold] different welding method
— m doble - double method • duplex method
— m económico - economical method
— m — para instalación f - economical installation method
— m educativo - [educ] educational method
— m eficiente - efficient method
— m empírico - empirical method
— m empleado - employed, o used, method
— m — anteriormente - previously used method
— m — en obra f - [constr] field method
— m — previamente - previously used method
— m — en obra f - [constr] field method
— m estándar - standard method
— m — para ensayo(s) m - standard test(ing) method
— m exacto - exact, o accurate, method, o approach
— m excelente - excellent method
— m exclusivo - exclusive method
— m expeditivo - expeditious method; short cut
— m Flo-Through - [contab] flow-thru method

método Flo-Through para contabilización f - [contab] flow-through accounting method
— m fluorescente - fluorescent method
— m funcional - [ind] functional, o operating, method
— m habitual - habitual method
— m — para establecimiento m de incentivo(s) m - [labor] habitual incentive(s) establishing method
— m húmedo m - wet method
— m — en temple(s) m - [metal-trat] aquatemp method
— m igual - same, o equal, method
— m lógico - logical method
— m mal aplicado - misapplied method
— m Manning-Kutter - [hidr] Manning-Kutter method
— m manual - [sold] manual, o hand, method
— m mejor - better, o improved, method
— m — para soldadura f - [sold] better, o improved, welding method • best welding method
— m mejorado - improved method
— m matalográfico - metallographic method
— m no destructivo - non-destructive method
— m — para ensayo(s) m - non-destructive testing method
— m nomográfico - nomographic method
— m normal - normal, o standard, method
— f — empleado - used normal method
— m — — en obra f - [constr] normal field method
— m nuevo - new method
— m operativo - operative,tional method
— m — para producción f - [ind] production operative,tional method
— m para acceso m - access method
— m — — — m con disco m - [electrón] dial-up method
— m — adición f - [ind] addition method
— m — adjudicación f - awarding method
— m — f de punto(s) m - [deport] (points), awarding, o scoring, method, o structure
— m — amoladura f - [mec] grinding method
— m — — f de cuchilla f - [mec] blade grinding method
— m — — f — f para cizalla f - [mec] shear blade grinding method
— m — — f — hoja f - [mec] blade grinding method
— m — — f — f para cizalla f - [mec] shear blade grinding method
— m — — f — rodillo(s) m - [metal-lam] roll grinding method
— m — análisis m - analysis method
— m — — f para tubería f - [tub] pipe analysis method
— m — — — f delgada - [tub] thin pipe analysis method
— m — aplastamiento m - [mec] mashing method
— m — — m con golpe(s) m - [mcc] cold flowing method
— m — aplicación f - application method
— m — — f para recubrimiento m - [meal-trat] lining application method
— m — — f — — m para acero m - [metal-trat] steel lining application method
— m — — f — — placa f - [metal-trat] stool lining application method
— m — — f — — m — — f para lingotera f - [metal-prod] ingot mold stool lining application method
— m — — f por vía f húmeda - [ind] wet application method
— m — — f — — f seca - [ind] dry application method
— m — aprendizaje m - [ind] apprenticeship, o training, o learning, method
— m — arrendamiento m - lease,sing method
— m — asentamiento m - settling method • [constr] bedding method
— m — asiento m - [constr] bedding method
— m — avance m - advance,cing, o progress, method

método m para avance m gradual - [mec] inching
method
— m — **cálculo** m - calculation, o computation,
method
— m — — m **para carga** f - [constr] load computing,tutation method
— m — — m — — f **para zanja** f - [constr]
trench load computing,tation method
— m — — **calentamiento** m - [ind] heating method
— m — **calibración** f - [instrum] calibrating, o
gaging, method
— m — — f **para integrador** m - [instrum] integrator, calibrating, o gaging, method
— m — — f — — m **para caudal** m - [instrum]
flow, integrator, calibrating, o gaging,
method
— m — — f — — m — **flujo** m - [instrum] flow
integrator, calibrating, o gaging, method
— m — **calificación** f - qualifying method
— m — — f **para propuesta(s)** f - proposal(s),
o bid(s), qualifying method
— m — **cambio** m - [mec] change,ging method -
replacing,cement method
— m — — m **para elemento** m - [ind] element, o
part, replacement method
— m — — m **refrigerado** - [ind] cooled
part replacing,cement method
— m — **capacitación** f - [labor] training method
— m — **carga** f - [transp] loading method •
[electr-oper] charging method
— m — — f **para horno** m - [ind] furnace charging method
— m — — f **para mineral** - [miner] ore loading
method • [metal-prod] ore charging method
— m — **colocación** f - placement,cing method •
[constr] bedding method • laying method
— m — — f **de material(es)** m - [ind] material
placement,cing method
— m — **combustión** f - [combust] combustion
method
— m — **comparación** f - comparison method
— m — — f —, **rendimiento** m, o actuación t, o
desempeño m - [ind] performance comparison
method
— m — — f — **voltaje** m - [instrum] voltage
comparison method
— m — — f — — m **en bobina** f - [instrum]
coil voltage comparison method
— m — — f — — m — — f **para medición** f -
[instrum] measuring coil voltage comparison
method
— m — — f — — m **para medición** f - [instrum]
measuring voltage comparison method
— m — **comprobación** f - proving, o testing,
method
— m — — f **de ferrita** f - [miner] ferrite
testing method
— m — — f — **sinter(ización)** - [miner] sin-
ter(ing) test(ing) method
— m — **cómputo** m - computing,tation method
— m — — m **para carga** f - [constr] load(ing
computing,tation method • [ind] charge,ging
computing,tation method
— m — — m — — f **para zanja** f - [constr]
trench load computing,tation method
— m — **concentración** f - [miner] concentrating,tion method
— m — **conexión** f - connecting,tion method
— m — — f **con tierra** f - [sold] grounding
method
— m — **conformación** f - [mec] forming, o shaping, method • [cojín] winging method
— m — — f **múltiple** - [cojin] multiple winging
method
— m — — f **con rueda** f, **amoladora**, o **esmerila-**
dora - [mec] grinding wheel, forming, o shaping, method
— m — **conservación** f - [ind] maintenance,
method, o procedure
— m — **construcción** f - [constr] construction
method
— m — — f **con horadación** f - [constr] tunneling consruction method

método m para construcción f para puente(s) m -
[constr] bridging method
— m — **control** m - control(ling) method
— m — **conversión** f - [fin] conversion, o
translation, method
— m — — f **de transacción(es)** - [fin] transac-
tion(s), conversion, o translation, method
— m — — f — — f **en moneda** f **extranjera** -
[fin] foreign currency transaction(s), conversion, o translation, method
— m — **depreciación** f - [contab] depreciation
method
— m — — f **proporcional** - [contab] proportional, o straight line, depreciation method
— m — **detección** f - detecting,tion method
— m — — f **supersensible** - supersensitive detecting,tion method
— m — **determinación** f - determination,ning
method
— m — — f **de humedad** f - moisture determination,ning method
— m — — f — — f **en mezcla** f - blend moisture determination,ning method
— m — — f **de módulo** m - modulos determination
method
— m — — f **en obra** f - [constr] field determination,ning method
— m — **diseño** m - design(ing) method
— m — — m **con anillo** m - ring design(ing)
method
— m — — m — — m **para compresión** f - ring
compression design(ing) method
— m — **distribución** f—distributing,tion method
— m — **drenaje** m - [hidr] drainage,ning method
— m — — m **convencional** - [hidr] conventional
drainage,ning method
— m — — m — **para franja(s), divisoria(s)**, o
medianera(s) - [vial] conventional median
drainage,ning method
— m — **educación** f—[educ] education(al) method
— m — **elaboración** f - [ind] manufacture,ring
method
— m — **eliminación** f - [sanit] disposal method
— m — — f **con basurero** m - [sanit] landfill
disposal method
— m — — f **de tipo** m **con basurero** m - [sanit]
landfill type disposal method
— m — **embarque** m - [transp] shipping method
— m — **embridado** m - [metal-fabr] flanging
method
— m — **enfriamiento** m - [ind] cooling method
— m — — m **de lingote(s)** m - [metal-prod] ingot cooling method
— m — **ensayo(s)** m - test(ing) method
— m — — m **con líquido** m - [ind] liquid testing method
— m — — m — — m **penetrante** - [sold] liquid
penetrant, o penetrating liquid, testing
method
— m — — m — **partícula(s)** f **magnética(s)** -
magentic particle(s) testing method
— m — — m **con penetrante** m **líquido** - liquid
penetrant(s) testing method
— m — — m — **absorción** f **de corriente** f **con-**
tinua - [electr-oper] direct current absorption test(ing) method
— m — — m **de ferrita** f - [miner] ferrite
test(ing) method
— m — — m **para carretera** f - [constr] highway
test(ing) method
— m — — m **satisfactorio** - satisfactory test-
-ing) method
— m — — m **verificado** - tested, o checked,
test(ing) method
— m — **escorificación*** f - [metal-prod] slagging method
— m — **esmerilado** m - [mec] grinding method
— m — **esmerilar** v - [mec] grinding method
— m — — m **automático** - [mec] automatic grinding method
— m — — m — **de bloque** m - [trefil] automatic
block grinding method
— m — — m — — m **cuneiforme** -[trefil]

automatic wedge block grinding method
**método m para esmerilado m automático de bloque
m cuneiforme para sujeción** f - [trefil] au-
tomatc wedge grip block grinding method
— m — m **de ranura** f - [trefil] groove
grinding method
— m — **esmerilar** v - [mec] grinding method
— m — v **bloque** m - [mec] block grinding
method
— m — v — m **sujetador** - [mec] grip block
grinding method
— m — **especificación** f - specification,fying
method
— m   **cstablecimiento** m - establishing method
— m — m **de estímulo(s)** m - [labor] incentive establishing method
— m — **estimación** f - estimating method
— m — f **de tiempo** m—time estimating method
— m — **estudio** m - study(ing) method
— m — m **de tiempo(s)** m - [labor] time study
method
— m — **evacuación** f - [sanit] disposal method
— m — **evaluación** f - evaluating,tion method
— m — f **de comportamiento** m - performance
evaluating,tion method
— m — **exfiltración** f—[hidr] exfiltrating,tion
method
— m — **extracción** f - extracting,tion method
— m — **en tajo** m **abierto** - [miner] open pit
extracting,tion method
— m — **fabricación** f - [ind] manufacture,ring,
o fabricating,tion, method, o procedure
— m — **fijación** f - fixing, o fastening, o attaching, method
— m — f **con aplastamiento** m **con golpe(s)** m
- [mec] cold flowing method attaching,chment
— m — **granallado** m—[mec] shot blasting method
— m — m **para rodillo(s)** m - [metal-lam]
roll(s) shot blasting method
— m — **hincadura** f—[constr-pil] driving method
— m — **horadación** f - [constr] tunneling, o
boring, method
— m — f **e inserción** f **(con gatos** m**)** -
[constr] boring and jacking method
— m — **implantación** f—implanting,tation method
— m — **impulsión** f - [mec] driving method
— m — **inserción** f **(con gatos** m**)** - [constr]
jacking (installation) method
— m — **inspección** f - inspection method
— n — f **aplicable** - [ind] applicable inspection method
— m — f **con partícula(s)** f **magnética(s)** -
[metal] magnetic particles inspection method
— m — f — f — **con polvo** m **húmedo** -
wet powder magnetic particle(s) inspection
(method)
— m — f — f — — m **seco** - dry powder magnetic particle(s) inspection (method)
— m — f **destructiva** - [sold] destructive
inspection method
— m — f — **para soldadura(s)** f - [sold]
weld(ment) destructive inspection method
— m — f — — f **sobre acero** m - [sold]
steel weldment destructive inspection method
— m — **instalación** f—[mec] installation method
— m — f **de revestimiento** m - [constr] liner
installation method
— m — f — m **para cloaca(s)** f - [sanit]
sewer liner installation method
— m — **instrucción** f - [educ] instruction, o
training, method
— m — f **sobre seguridad** f - [segurid] safety training method
— m — f — **en trabajo** m - [segurid] job
safety training method
— m — **investigación** f - investigation, o survey, method
— m — f **de resistencia** f - [mec] strength,
o resistance, investigation method
— m — f — f **eléctrica** f - [electr-oper]
electric resistance investigation method
— m — f — f — **efectiva** - [hidr] resistivity, investigaction, o survey, method

método m para investigación f **en obra** f—[constr]
field investigation method
— m — — f — — f **de resistencia** f - [constr]
field resistance, investigation, o survey,
method
— m — — f — — f — f **eléctrica** - [electr]
field electric resistance method
— f — — f — — f — f **efectiva**—[hidr]
field resistivity survey method
— m — **laboratorio** m - [ind] laboratory method
— m — **lecho** m - [constr] bedding method
— m — **locación** f - locating method
— m — f **supersensible** - supersensitive locating method
— m — **mancomunación** f **de interés(es)** m - [fin]
interest(s) pooling method
— m — **manipuleo** m - handling method
— m — **mantenimiento** m—[ind] maintenance method
— m — **manufactura** f - [ind] manufacture,ring
method
— m — **medición** f - measuring method
— m — f **de fuerza** f - force measuring method
— m — f — f **coercitiva** - coercive force
measuring method
— m — f **de propiedad(es)** f - property,ties
measuring method
— m — f — f **magnética(s)** f - magnetic
property,ties measuring method
— m — f — f — **de imán** m - [instrum]
magnet magnetic property,ties measuring method
— m — f — f — — f **permanente** -
[instrum] permanent magnet magnetic property,ties measuring method
— m — f — f — **producto(s)** m **masi-
vo(s)** - [magnet] massive product(s) magnetic
property,ties measuring method
— m — f **supersensible** - supersensitive
measuring method
— m — **montaje** m - [mec] mounting method
— m — m **para prolongación** f - [sold] extension mounting method
— m — m — **rodillo** m - [metal-lam] roll assembly,ling method
— m — **muestreo** m - [ind] sampling method
— m — **normalización** f - standardization method
— m — **operación** f - operating,tion method •
practice
— m — f **de horno** m - [metal-prod] furnace,
operating,tion method, o practice
— m — **pago** m - [com] payment method • payments
promptness
— m — **panetración** f - penetration method
— m — **perforación** f - [constr] boring method
— m — f **múltiple** - [mec] multiple perforating,tion method
— m — **preparación** f - preparation method
— m — f **de base** f - [constr] bedding method
— m — f **de lecho** m - [constr] bedding method
— m — f — m **para estructura** f - [constr]
structure bedding (preparation) method
— m **preparar** v **extremo(s)** m - [mec] end(s) preparation method • [electr-instal] terminating
method
— m — **procesamiento** m - [ind] processing method
— m — **producción** f - production, o fabrication,
method
— m — **programación** f - [electrón] programming
method
— m — f **de código** m - [electrón] code programming method
— m — **propaganda** f - [com] advertising method
— m — **proyección** f - design(ing) method
— m — f **contra corrosión** f - [constr] corrosion design method
— m — **punz(on)ado** m - [metal-fabr] punch(ing)
method
— m — m **y troquelado** m - [metal-fabr] punch
and die method
— m — **rectificación** f - [mec] grinding method •
dressing method
— m — f **de rodillo(s)** m - [metal-lam] roll,
grinding, o dressing, method
— m — **reducción** f **de costo(s)** m - [ind] cost(s)

control, method, o skill
método m para reemplazo m - replacement method
— m — — m **de elemento m refrigerado** - [ind] cooled part replacement method
— m — **refrigeración f** [ind] cooling method
— m — **regulación f** - control method
— m — **f de ambiente m** - [ambient] environmental control method
— m — **relación(es) f humana(s)** - [labor] human relations, method, o approach
— m — — **f para resolver problema(s) m** - [labor] human relations approach
— m — **relleno m**—[constr] backfill(ing) method
— m — **remoción f** - removal method
— m — — **f de anhidrido m sulfuroso** - [quím] hydrogen sulphide removal method
— m — **reparación f** - [ind] repair(ing) method
— m — — **f de cojinete m** - [cojin] bearing repair(ing) method
— m — **revestimiento m** - relining method
— m — — m **nuevo** - [ind] relining method
— m — **secado,camiento m** - drying method
— m — **selección f** - selection method
— m — **simplificación f** - simplification method
— m — — **f de trabajo m** - work simplification method
— m — **sinterización f** - [miner] sintering method
— m — **soldadura f** - [sold] welding method
— m — — **f con arco m sumergido** - [sold] submerged arc welding method
— m — — **f — retroceso m** - [sold] back step welding method
— m — — **en segmento(s) de 13 milímetros m (media pulgada f)** - [sold] half inch at @ time method
— m — **temple m** - [metal-trat] tempering, o hardening, method
— m — **toma f de muestra(s) f** - [ind] sampling method
— m — **torneadura f** - [mec] turning method
— m — — m **de rodillo(s) m** - [metal-lam] roll turning method
— m — **trabajo m** - work(ing), o operating, method
— m — — m **según norma f** - [ind] standard, work(ing), o operating, method
— m — **transporte m** - [transp] transportation, o hauling, method
— m — — m **para mineral m** - [miner] ore transportation method
— m — — m — — m **de hierro m** - [miner] iron ore transportation method
— m — **tratamiento m** - [ind] treatment method
— m — — m **térmico** - [metal-trat] heat, o thermal,mic, treating,tment method
— m — **troquelado m** - [metal-fabr] die method
— m — **verificación f** - [ind] check(ing) method
— m — **vigilancia f** - [ind] surveillance method
— m — — **f de ambiente m** - [ambient] environmental surveillance method
— m **paso a paso** - step-by-step method
— m **por proporción f** - proportion method
— m — — **f de diámetro m** - diameter proportion method
— m — **corte m abierto**—[miner] open cut method
— m — **diseño m de fatiga f** - [constr] working stress, method, o design
— m — — **tensión f** - [constr] working stress, method, o design
— m — — **f — trabajo m** - [constr] working stress, method, o design
— m **práctico** - practical method
— m — **para instalación f** practical installation method
— m **preciso** - precise, o accurate, method
— m — **para análisis m**—precise analysis method
— m **previo** - previous, o former, method
— m **primario** - primary method
— m — **para inspección f** - primary inspection method
— m **principal** - main, o major, method
— m **Próctor** - [constr] Proctor method

método m proporcional - proportional method • [contab] straight line method
— m **químico** - [quím] chemical method
— m **racional** - rational method
— m — **de Armco** - [tub] Armco rational method
— m — **para proyección f** - [constr] rational design method
— m **radiográfico** - [electrón] radiographic method
— m **rápido** - rapid, o quick, method
— m — **para diseño m** - quick design method
— m **recomendado** - recommended method
— m **satisfactorio** - satisfactory method
— m — **para ensayo(s) m** - satisfactory test-(ing) method
— m **seco** - dry method
— m **según norma f** - standard method
— m — — **f para ensayo(s) m** - standard test-(ing) method
— m **seguro** - [segurid] safe method • foolproof approach
— m **selectivo** - selective method
— m **separado** - separate method
— m — **para tendido m de tubería f** - [tub] stove pipe method
— m **Siemens-Martin** - [metal-prod] open hearth method
— m **simplificado** - simplified method; short cut
— m — **con extremo m libre** - [constr-pill] free--end simplified method
— m **substitutivo** - alternative method
— m **supersensible** - supersensitive method
— m **temporario** - temporal,rary method
— m — **para conversión f** - [fin] temporary conversion method
— m — — **f de transacción(es) f** - [fin] transaction, conversion, o translation, temporary method
— m — — **f en moneda f extranjera** - [fin] foreign currency transaction, conversion, o translation, temporary method
— m **totalmente no destructivo** - completely, o totally, nondestructive method
— m **transitorio** - temporary, o transitory method
— m **ultrasónico** - [electrón] ultrasonic,sound method
— m **único** - single, o only, method
— m — **para calibración f** - [instrum] single gaging method
— m **usado** - method used
— m — **anteriormente** - previously used method
— m — **previamente** - prevvioulsy used method
— m **variante** - alternate,tive method
— m **visual** - visual method
métodos m varios - various, o sundry, methods
— m **y desarrollo m** - methods and development
metodología f para estudio m - study methodology • survey(ing) method
— **f — calificación f** - qualifying method
metro m - [metric] . . . • [transp] servicio m urbano para transporte m
— m **cúbico** - [metric] cubic meter
— m **—/kilómetro m** - [transp] cubic meter per kilometer • úsese cubi yard/highway mile; CYHM
— m **de soldadura f** - [sold] meters of weld
— m — — **f por hora f** - [sold] meters of weld per hour • úsese feet of weld per hour
— m **lineal** - [metric] linear meter • úsese linear feet
— m — **laminado** - [metal-lam] rolled linear meters; úsese linear feet rolled
metros m por segundo m - meter(s) per second
mexicano,na a - [pol] Mexican
Mexico m - [pol] Mexico; véase también **Méjico m**
mezcla f - . . .; mix • blend(ing) • [sold] admixture • [ind] compounding • [quím] formulating,tion
— **f acabada** - thorough, mix, o mixing,xture
— **f aceptable** - acceptable mix(ture)
— **f agua-aceite m** - water-oil blend
— **f — detergente** - [metal-lam] water-oil

detergent blend
mezcla f aire-gas - [combust] air-gas blend
— f **aluminotérmica** - [sold] thermit mixture
— f **apropiada**—appropriate, o proper, mix(ture)
— f **artificial** - [miner] artificial blend
— f **asfáltica** - [constr] asphalt(ic) mix
— f **benzol-toluol-xilol** - [quím] benzol-toluol-
-xylol blend
— f **binaria** - [miner] binary mixture
— f **bituminosa** - [constr] bituminous mixture
— f **caliente** - hot mix(ture)
— f **combustible** - [combust] burning mixture
— f **con morrillo m** - [constr] gravel mix
— f — — m **menudo** -[constr] pea gravel mix
— f **continua** - continuous blend(ing)
— f **cribada** - [constr] graded, mix, o blend
— f **de aceite m** y compuesto m - [trefil] oil
and compound mix(ture)
— f — **ácido m** - [quím] acid, blend, o mixture
— f — — m **sulfúrico** - [quím| sulfuric acid
blend
— f — — — **fluorhídrico** - [quím] sulfuric
hydrofluoric (acid) blend
— f — **agua m** y aceite m - water-oil blend
— f — **aire-m** y **combustible m** - [comb.int] air-
-fuel mixture
— f — **arcilla f** - [cerám] clay(s) mix(ing)
— f — — f y **paja f** - [constr] cob
— f — **arena f** y **arcilla f** - [geol] sand-clay,
mix(ture), o composition
— f — — f (**seca**) y **cemento m** - [constr] dry
sand-cement mixture
— f — — — f y **cemento m** - [constr] sand-cement
mixture
— f — **asfalto m** - [constr] asphalt mix(ture)
— f — — m **fibroso y carbón m** - [constr] fi-
brous asphalt-coal mix(ture)
— f — **carbón(es) m** - [metal-prod] coals blend
• coal blending
— f — — m **para coquizar v** - [coque] coking
coal blend
— f — **caudal(es) m** - [hidr] flow, o stream,
mixing
— f — **cemento m diluida** - véase **lechada f**
— f — — m y **arena f** - [constr] cement-sand
mix(ture)
— f — **combustible(s) m** - [comb.int] fuel(s)
mixture
— f — — m **apropiada** - [comb.int] appropriate,
o proper, fuel(s) mixture
— f — — m **pobre(s)** - [comb.int] poor fuel
mix(ture)
— f — — m **rica** - [comb.int] rich fuel mixture
— f — — m y **aire m** - [comb.int] fuel-air
mixture
— f — — m y **aire m pobre** - [comb.int] poor
fuel-air mixture
— f — — — m **rica** - [comb.int] rich
fuel-air mixture
— f — **compuesto m** - [quím] compound mix(ing)
— f — **fécula f** (**de maíz m**) - [culin] corn
starch mixture
— f — — — m y **goma f guar** - [quím]
corn starch and guar gum blend
— m — **flujo(s) m** - [hidr] flow(s), o stream(s)
mixing
— f — **fundente(s) m** - [sold] flux(es) blend
— f — — m **neutral(es)**—[sold] neutral fluxes
blend
— f — **gas(es) m** - [combust] mixed gas(es) •
gas(es) mix(ture)
— f — — m **de hidrógeno m** - [quím] hydrogen
gas mixture
— f — — m **inerte(s)** - [quím] inert gas(es)
blend
— f — — m **natural y aire m** - [combust] natural gas/air blend
— f — — m y **aire m** - [combust] gas-air, mixture, o blend
— f — **grafito m** y **escoria f** - [metal-prod]
kish (in hot metal car)
— f — **hidrógeno m** y **nitrógeno m** - [quím] hydrogen and nitrogen blend

mezcla f de hormigón m - [constr] concrete mix
— f — **ingrediente(s) m** - ingredient(s) blend-
-(ing)
— f — **insecticida(s) m** - [quím| insecticide(s)
mix(ing)
— f **para marcha f sin carga f** - [comb.int]
idle,ling mixture
— f — **materia(s) f prima(s)** - [ind] raw mate-
rial(s) blending
— f — — f — **molida(s)** - [ind] ground raw
materials blending
— f — **material(es) m** - [ind] material(s),
mix(ing), o blend(ing)
— f — **metal n** - [metal] metal(s) blending
— m — — m **para aportación f con (metal m) de
base f** - [sold] admixture • puddling
— f — — m **de base f con (metal m) aportado** -
[sold] (base metal into weld metal) admixture
— f — **en superficie f final** - [sold] final
surface admixture
— f — **morrillo m** - [constr] gravel mix
— f — — m **menudo** - [constr] pea gravel mix
— f — — m **neumático(s) m** - [autom-neumát] tire(s)
mix(ing)
— f — **pasto(s) m** - [botán] grass(es) mixing
— f — **producto(s) m** - product(s), blend(ing),
o mix(ing), o mixture
— f — — m **de fisión f** - [nucl] fusion pro-
duct(s), mix(ture), o mixing
— f — **reductor(es) m** - [quím] reducer(s) blend
— f — **sólido(s) m** - solid(s), blend, o mix
— f — — m **reductor(es) m** - [quím] solid re-
ducer(s) blend(ing)
— f — **talco(s) m** - [miner] talc(s) blend(ing)
— f — **termita f** - [sold] thermit mixture
— f — **trozo(s) m** - [constr] chunk mix(ing)
— f — — m **de asfalto m** - [constr] asphalt
chunk(s) mix(ing)
— f **débil** - [comb.int] lean mixture
— f **delgada** - véase **mezcla f pobre**
— f **detergente** - detergent blend
— f — **de agua m** y **aceite m** - water-oil detergent blend(ing)
— f **empleada** - [constr] mortar used
— f **en parte(s) f igual(es)** - half and half
mixture
— f **encendida** - [comb.int] ignited mixture
— f **equilibrada** - balanced mix(ture)
— f **especial** - special, blend, o mix(ture)
— f — **de fundente** - [sold] special(ly)
blend(ed) flux
— f **estoquiométrica** - [metal] stoichimetrical
blend(ing)
— f — **de vapor m** y **metano m** - [metal] steam-
-methane stoichiometrical blend
— f **excedente** - excess mix(ture)
— f **excesiva** - [sold] excessive admixture
— f **explosiva** - [quím] explosive mixture •
[comb.int] mixture
— f — **de gas m** y **aire m** - [combust] explosive
gas-air mixture
— f **fría** - cold mix(ture)
— f **fresca** - [constr] fresh cold mix
— f **fresca** - fresh mix
— f **gaseosa** - [quím] gas blend
— f **hasta homogeneidad f** - [quím] mixing to
homogeneity
— f **homogénea** - homogenous, blend, o mixture;
blended mixture
— f — **de concentrado(s) m** - [miner] blended
concentrate(s) mixture
— f — — **materia(s) f prima(s)** - [ind] raw
material(s) even blending
— f — — — f **molida(s)** - [ind] ground raw
material(s) even blending
— f — — **mineral(es) m** - [miner] blended ore
mixture
— f — — — m **concentrado(s)** - [miner] blended
concentrated ore(s) mixture
— f — — — m y **concentrado(s) m** - [miner]
ore(s) and concentrate(s) blended mixture
— f — **uniforme** - [constr] homogenous uniform
mixture

mezcla f inapropiada - inappropriate, o improper, mix(ture)
— f inerte - [quím] inert blend • [comb.int]
inert (gas) blend
— f — de gas(es) m - [quím] inert gas blend
— f íntima - intimate blend(ing)
— f labrable - [hormig] workable mix
— f mínima - [sold] minimum admixture
— f — de metal m para aportación f con (metal
m de) base f - [sold] minimum admixture
— f neutral - neutral blend(ing)
— f para coquizar v - [coque] coking blend •
coal blend
— f — hormigón m - [constr] ready-mix
— f — horno m - [ind] furnace blend
— f — — m alto - [metal-prod] blast furnace
blend
— f — sinter(izar) - [ind] sinter(ing) blend
— m — tiempo m frío - [lubric] cold weather
mix(ture)
— f — torta(s) f - [culin] cake mix
— f pobre - [comb.int] poor, o lean, mix
— f — de combustible m y aire m - [comb.int]
poor fuel-air, blend, o mixture
— f reducida - [sold] low admixture
— f rejuntante - [constr] pointing mixture
— f rica - [comb.int] rich, blend, o mixture
— f — de combustible m y aire m - [comb.int]
rich fuel-air mixture
— f saturante - saturating blend; saturant
— f — bituminosa - bituminous saturant
— f seca - [constr] dry, mix(ture), o mixing;
aggregate
— f — para hormigón m - [constr] dry concrete
mix
— f — — m en saco(s) m - [constr] (dry
sacked concrete
— f — — recubrimiento m - [sold] coating dry
mix
— f suave - [hormig] smooth mix(ture)
— f sulfonítrica - [quím] sulfonitric mixture
— f usada - [constr] used mortar
— f utilizada - [constr] used mortar
mezclable a - miscible
— a con agua m - water miscible
— a en agua m - water miscible
mezclado m - véase mezcladura f
mezclado,da a - . . .; blended • interspersed •
[quím] formulated
— a acabadamente - mixed thoroughly
— a aceptablemente - acceptably mixed
— a apropiadamente - properly mixed
— a en seco - dry mix(ed)
— a hasta homogeneidad f - [quím] mixed to homogeneity
— a inapropiadamente - inappropriately, o improperly, mixed
— a íntimamente - intimately, mixed, o blended
— a según pedido m - specially blended
mezclador m - . . . • [miner] mulling
mezclador,ra a - . . . • [miner] mulling
mezcladora f - [sinter] fluffer mill; muller
— f asfáltica - [constr] asphalt mixer
— f — portátil - [constr] portable asphalt
mixer
— f cilíndrica - [metal-prod] cylindrical mixer
• [quím] cylindrical blender
— f — fija - [metal-prod] stationary cylindrical mixer
— f con . . . brazo(s) m para masa f - [culin]
. . . arm dough mixer
— f — . . . — pasta f - [culin] . . .-arm
dough mixer
— f — — m único - [culin] single arm mixer
— f — — m — para baño m maría - [culin]
jacketed single arm mixer
— f — — m — — masa f - [culin] single arm
dough mixer
— f — — m — — pasta f - [culin] single arm
dough mixer
— f — husillo(s) m - [culin] spindle mixer
— f — inclinación f manual - [culin] hand
tilt(ing) mixer

mezcladora f con regulación f, común, o corriente - [culin] regular control mixer
— f — — f doble - [culin] dual control mixer
— m — — f para temperatura f - [culin] temperature controlling mixer; votator*
— f — — f, simple, o única - [culin] single
control mixer
— f — velocidad f alta - [culin] high speed
mixer
— f — — f baja - [culin] (s)low speed mixer
— f — — f — inclinable - [culin] tilt(ing)
(s)low speed mixer
— f — — f — — manualmente - [culin] hand, o
manually, tilt(ing) (s)low speed mixer
— f — — f reducida - [culin] (s)low speed
mixer
— f continua - [culin] continuous mixer
— f — para torta(s) f - [culin] continuous
cake mixer
— f de acero m - [culin] steel mixer
— f — — m inoxidable - [culin] stainless
steel mixer
— f — — m — con baño m maría f - [culin]
jacketed stainless steel mixer
— f — — m — — m — con brazo m único -
[culin] single arm jacketed stainless steel
mixer
— f estacionaria—[metal-prod] stationary mixer
— f fija - [metal-prod] stationary mixer
— f inactiva - inactive mixer
— f inclinable - [culin] tilting mixer
— f — manualmente - [culin] hand, o manually,
tilt(ing) mixer
— f instantánea - [hidr] flash mixer
— f intensiva - [sold] instensive mixer
— f manual - [domést] hand mixer
— f para argamasa f - [constr] mortar mixer
— f — arrabio m - [metal-prod] metal, o pig
iron, mixer
— f — baño m maría - [culin] jacketed mixer
— f — batido(s) m - [culin] batter mixer
— f — — m para oblea(s) f - [culin] wafer
batter mixer
— f — carbonato m de sodio m - [ind] soda ash,
mixer, o mixing tank
— f — — m sódico (anhidro) - [ind] soda ash,
mixer, o mixing tank
— f — gas m - gas, mixer, o blender
— f — hormigón m - [constr] concrete mixer
— f — ingrediente(s) m - [culin] ingredient(s)
mixer
— f — inyección f - [petról] mud mixer
— f — lodos m - [petról] mud mixer
— f — metal(es) - [metal-prod] metal(s) mixer
— f — — m caliente(s) - [metal-prod] hot
metal mixer
— f — — m fundido - [metal-prod] hot metal
mixer
— f — — m líquido - [metal-prod] hot metal
mixer
— m — mortero m - [constr] mortar mixer
— f — oblea(s) f - [culin] wafer mixer
— f — pasta f - [culin] paste, o dough, mixer
— f — torta(s) f - [culin] cake mixer
— f — velocidad f alta - [culin] high, o fast,
speed mixer
— f — — f — para ingrediente(s) m - [culin]
high, o fast, speed ingredient(s) mixer
— f — — f baja - [culin] (s)low speed mixer
— f — viento m - [metal-prod] (blast) mixer
— f pavimentadora - [vial] paving mixer • retread paver
— f por balde(s) f - [constr] batch mixer
— f — — m rotativa - [constr] rotary batch
mixer
— f rotativa - [constr] rotary mixer
mezcladura f - . . .; blending • [miner] mulling
— f de lodo(s) m - [sanit] sludge mixing
— f — muestra(s) f - sample(s) mixing
— f manual - [hormig] hand, o manual, mixing
— f original - [constr] original mixing
mezclar v - . . . . • to intersperse • [metal] to
alloy • [quím] to formulate

mezclar v acabadamente - to mix throughly
— v aceptablemente - to mix acceptably
— v apropiadamente - to mix, appropriately, o
properly
— v arcilla(s) f - to mix @ clay(s)
— v cabalmente - to mix thoroughly
— v caudal(es) m - to mix @, stream, o flow
— v combustible(s) m - [combust] to, mix, o
blend, @ fuel(s)
— v compuesto m - [quím] to mix @ compound
— v flujo m - to mix @, flow, o stream
— v hasta homogeneidad f - [quím] to mix to @
homogeneity
— v inapropiadamente - to mix, inappropriately,
o improperly
— v ingrediente(s) m - to blend @ ingredient(s)
— v insecticida m - [quím] to, mix, o blend, @
insecticide
— v intimamente - to blend intimately
— v material(es) m - [ind] to mix @ material(s)
— v medida(s) f to mix @ size(s)
— v — f de neumático(s) m - [autom-neumát] to
mix @ tire size(s)
— v neumático(s) m - [autom-neumát] to mix @
tire(s)
— v producto(s) m - [ind] to, mix, o blend, @
product(s)
— v tipo(s) m para construcción f de neumáticos
[autom-neumát] to mix @ tire construction
(type)
— v torta(s) f - [culin] to mix @ cake(s)
mica f blanca - [miner] white mica
micarta f - [electr-instal] micarta
micro m - véase análisis m micrográfico
microagrietamiento m - [sold] micro-cracking
microalambre m - [alambre] microwire; (very)
fine wire
microaleación f - [metal] microalloy(ing)
microamperímetro m - [instrum] microammeter
microanálisis m - microanalysis
microbus m - [transp] minibus
microcomputador m - [electrón] microcomputer;
véase también microordenador m
microconmutador m - [electr-equip] microswitch
microestructura f austenítica - [metal] austenitic microstructure
— f ferrítica - [metal] ferritic microstructure
— f magnética - [metal] magnetic microstructure
— f martensítica - [metal] martensitic microstructure
microficha f - . . . ; microfiche
microfilmado,da a - [fotogr] microfilmed
microfisura f - microfissure
microfosil m - [geol] microfossil
microfusible m - [electr-equip] microfuse
— m para excitador,tatriz - [electr-prod] exciter microfuse
micrografía f de revestimiento m - [metal-trat]
coating micrography
micrométrico,ca a - . . . ; vernier
micrómetro m exterior - outside micrometer
— m interior - [instrum] inside micrometer
— m para carátula f - [extrusión f de
electrodos] excentricity gage
— m — medida(s) f interior(es) - [instrum]
inside micrometer
— m — profundidad(es) f - [instrum] depth micrometer
— m volante - [instrum] flying micrometer
micronivelador,ra a - microleveling
microordenador m - [comput] microcomputer
microplaqueta f - [electrón] microchip
— f para memoria f - [comput] memory chip
— f — f programada - [electrón] programmed,mable memory chip
— f — f para lectura f - [electron] programmable read(ing) memory, o P R M, chip
— f — f — f únicamente - [electrón]
programmable read(ing) only memory, o
P R O M, chip
— f programada - [electrón] programmed chip
— f — en fábrica f - [electrón] factory programmed chip

microprocesador m - [electrón] microprocessor
microrretardo - [electrón] micro delay
microscopio m petrográfico - [instrum] petro-
graphic microscope
— m por reflexión f - [instrum] reflection microscope
microseparación f - [metal] microseparation
micum m - [metal-prod] micum
miembro m actual - present, o current, member
— m de organización f - [admin] organization
member
— m asistente - attending member
— m concurrente - [legal] attending member
— m de Banco m Interamericano para Desarrollo m
- [pol] Interamerican Development Bank member
— m — club m automovilístico - [autom] car, o
automobile, club member
— m — cuadrilla f - [ind] crew, o gang, member
— m — equipo m - [deport] team member
— m — Fondo m Monetario Internacional - [pol]
International Monetary Fund member
— m — junta f directiva - [legal] board of
directors, o managing board, member
— m — prensa f - [public] press member
— m — grupo m - [labor] team, o group, member
— m — parlamento m - [pol] parliament member;
member of parliament; M P
— m para tensión f - [mec] tension member
— m presente - present, o attending, member
mientras adv en - while at
— adv no se está soldando - [sold] while not
welding
— adv — suelda - [sold] when, o while, not
welding
— adv que - . . . . • whereas
— adv opera(n) - in operation
— adv se coloca relleno m - [constr] while, o
during, backfilling
— adv — está soldando - [sold] when, o
while, welding
— adv — suelda - [sold] when, o while, welding
mientras adv tanto - . . . . • at the same time
migajón n - . . . • [suelos] loam
migración f de hidrógeno m - [metal] hydrogen
migration
— f — m liberado - [metal] liberated hydrogen migration
mil m - . . . . • [fin] grand
miles m — libras f por pie m cuadrado - kip(s)
per square foot
milésimo n - [metric] mill
— m de dólar m - [fin] mill
miliamperímetro m - [electr] . . .; milliammeter
miligramo(s) m por litro m - milligram(s) per
liter; part(s) per million
milímetro(s) m a . . . potencia f por milímetro
• millimeter(s) to @ . . . power per millimeter
— m cuadrado - [metric] square millimeter
— m cúbico - [metric] cubic millimeter
milipulgada f - [metric] mil
— f circular - [alambre] circular mil
milisegundo m - [cron] millisecond
militar m - . . . . | a - [milit] defense related
| v - to militate
milivoltio m - [electr-oper] millivolt; mV
milla f de etapa f para velocidad f alta -
[deport] high speed stage mile
mimbre m - . . . .; cane
mimeografiado,da a - [impr] mimeographed
mina f - [miner] . . . • [lápiz] lead
— f abandonada - [miner] abandoned mine
— f activa - [miner] active mine
— f ampliada - [miner] expanded, o enlarged,
mine
— f bajo cielo m abierto - [miner] open pit,
mine, o mining; quarry
— f con ley f baja - [miner] low grade mine
— f chica - [miner] small mine
— f de carbón m - [miner] coal, mine, o pit
— f — m bajo cielo m abierto - [miner] open
pit coal mine
— f — m en tajo m abierto - [miner] open

pit coal mine
mina f desagotada - [miner] drained mine
— f drenada - [miner] drained mine
— f en actividad - [miner] active mine
— f — tajo m abierto - [miner] open pit mine; quarry
— f entrada - [miner] entered mine
— f existente - [miner] existing mine
— f explotada - [miner] exploited mine
— f — bajo cielo m abierto - [miner] open pit mine; quarry
— f inactiva - [miner] inactive mine
— f para lápiz f - pencil lead
— f profunda - [miner] deep (level) mine
— f subterránea - [miner] (underground) mine
minado,da a - [miner] mined
— a bajo cielo m abierto - [miner] open pit mined; quarried
— a con costo m reducido - [miner] low cost mined
— a subterráneamente—[miner] underground mined
minadora f - [miner] mining machine
minar v bajo cielo m abierto - [miner] to open pit mine; to quarry
— v carbón m - [miner] to mine @ coal
— v con costo m reducido - [miner] to low cost mine
— v diamante(s) m—[miner] to mine @ diamond(s)
— v hulla f - [miner] to mine @ coal
— v piedra f - to, mine, o quarry, stone
— v — en cantera f - [miner] to quarry @ stone
— v subterráneamente - [miner] to underground mine
minera f - [miner] mining, company, o concern
mineración* f - [miner] mining; véase también minería f
mineral m ácido - [miner] acid ore
— m adjudicado—[miner]awarded, o assigned, ore
— m aglomerado - [miner] agglomerated ore
— m aglomerado en frío - [miner] balled ore
— m alto en azufre - [miner] high sulfur ore
— m — — fósforo - [miner] high phos(phorus) ore
— m analizado—[miner] analyzed, o assayed, ore
— m arcilloso - [miner] argillaceous, o clayey, ore
— m arsenioso - [miner] arsenical ore
— m autofundente - [miner] self-fluxing ore
— m base,sico - [miner] base,sic ore
— m beneficiado - [miner] beneficiated ore
— m beneficiador - [miner] beneficiating ore
— m brasileño - [miner] Brazilian ore
— m bruto - [miner] raw, o crude, ore
— m calcáreo - [miner] calcareous, o calcium bearing, ore
— m calcífero - [miner] claciferous ore
— m calcinado - [miner] clacined, o roasted, o burned, ore
— m calcopirítico - [geol] chalcopyritic ore
— m calculado - [miner] calculated ore
— m calizo - [miner] calcareous ore • calciferous ore
— m clasificado - [miner] classified, o graded, o sized, ore
— m — por zarandeo m - [miner] screen classified ore
— m cloritizado - [miner] chloritized, ore, o mineral
— m compacto - [miner] compact ore
— m componente - [geol] component mineral
— m con banda(s) f - [miner] banded ore
— m con calidad f - [miner] quality ore
— m — — f alta - [miner] high grade ore
— m — — f baja - [miner] low grade ore
— m — concentración f - [miner] concentration ore
— m — — f alta - [miner] high concentration ore
— m — — f baja - [miner] low concentration ore
— m — — f elevada - [miner] high concentration ore
— m — contenido m alto - [miner] high content

ore
mineral m con contenido m alto de hierro m - [miner] high iron content ore
— m — — m — — impureza(s) f - [miner] high impurity,ties content ore
— m — — m bajo - [miner] low content ore
— m — — m — de hierro m - [miner] low iron content ore
— m — — m — — impureza(s) f - [miner] low impurity,ties content ore
— m — azufre m alto - [miner] high sulfur ore
— m — m bajo - [miner] low sulfur ore
— m — fósforo m - [miner] phos(phorus) ore
— m — — m alto - [miner] high phos(phorus) ore
— m — — m bajo - [miner] low phos(phorus) ore
— m — granulometría f regular - [miner] constant grain size ore
— m — ley f alta - [miner] high, grade, o content, o concentration, ore
— m — — f baja - [miner] low, grade, o content, o concentration, ore
— m — proporción f, alta, o elavada, de ganga f - [miner] high gangue ore
— m — — f, baja, o reducida, de ganga f - [miner] low gangue ore
— m concentrado - [miner] concentrated ore
— m cribado - [miner] screened ore
— m cristalino - [miner] crystal(line) ore
— m — frágil - [miner] brittle cystalline ore
— m cubicado - [miner] measured, o computed, ore
— m de Breccia f - [miner] Breccia ore
— m — cromo m - [miner] chrome,mium ore
— m — cuarzo m - [miner] quartz ore
— m — — m con banda(s) f - [miner] banded quartz ore
— m — — m veteado - [miner] banded quartz ore
— m — galena f - [geol] galena ore
— m — ganga f - [geol] gangue, mineral, o ore
— m — hierro m - [miner] iron ore; ironstone
— m — — m arcilloso - [miner] ironstone clay
— m — — m arsenioso - [miner] arsenical iron ore
— m — — m calibrado - [miner] sized iron ore
— m — — m con ley f alta - [miner] high, grade, o content, iron ore
— m — — m — — f baja f - [miner] low, grade, o content, iron ore
— m — — m exportado - [miner] exporte iron ore
— m — — m grueso - [miner] coarse iron ore
— m — — m importado - [miner] imported iron ore
— m — — m para acería f - [miner] steel plant, o open hearth, iron ore
— m — — m — horno m alto - [miner] blast furnace iron ore
— m — — m — — m Siemens-Martin - [miner] open hearth iron ore
— m — — m pardo - [miner] brown iron ore
— m — — m prerreducido - [miner] prereduced iron ore
— m — — m puro (químicamente) - [miner] ferrite
— m — — m reducido - [metal-prod] reduced iron ore
— m — — m — directamente - [metal-prod] directly reduced, o reduced directly, iron ore
— m — manganeso - [miner] manganese ore
— m — — m molido - [miner] ground manganese ore
— m — reserva - [metal-prod] reserve ore
— m diseminado - [miner] dispersed ore
— m dispersado - [miner] dispersed ore
— m disperso - [miner] scattered ore
— m doméstico - [miner] domestic ore
— m eluvial* m - [miner] eluvial ore
— m — en terrón(es) m - [miner] lump eluvial ore
— m en pella(s) f - [miner] pelletized ore
— m ensayado - [miner] assayed ore • tested ore
— m esfalerítico - [geol] sphalerite ore

mineral m especial - [miner] special ore
— m especialmente clasificado - [miner] specially graded ore
— m estéril - [miner] sterile ore
— m — cargado - [miner] loaded sterile ore •
[metal-prod] charged sterile ore
— m exportado - [miner] exported ore
— m fenocristal - [miner] phenocrystal, mineral, o ore
— m ferrífero - [miner] iron bearing ore
— m fino - [miner] fine ore
— m — concentrado - [miner] concentrated fine
ore
— m fosforoso - [miner] phosphorous ore
— m frágil - [miner] brittle ore
— m grueso - [miner] coarse, o lump, ore • ore
lump(s)
— m — de pila f - [metal-prod] coarse stockpile ore
— m hematítico - [miner] hematite,tic ore
— m — pardo - [miner] brown hematite,tic ore
— m heterogéneo - [miner] heterogenous ore
— m hidrotermal - [geol] hydrothermal, mineral,
o ore
— m homogeneizado - [miner] homogenized ore •
blended ore
— m importado - [miner] imported ore
— m indicado - [miner] indicated orc; known ore
— m inerte m - [miner] inert, mineral, o ore
— m liviano - [miner] light, mineral, o ore
— m máfico - [miner] maphic, mineral, o ore
— m — cloritizado - [miner] chloritized maphic, mineral, o ore
— m magnético - [miner] magnetic ore
— m medido - [miner] measured ore
— m metálico - [miner] metallic, mineral, o ore
— m metalizado - [miner] metallized ore
— m micáceo - [miner] micaceous ore
— m molido m - [miner] ground, o crushed, ore
— m muy concentrado - [miner] high concentration, o highly concentrated, ore
— m muy metalizado - [miner] highly metallized
ore
— m normal—[miner] normal, o conventional, ore
— m oolítico* - [miner] oölitic ore
— m — ácido - [miner] acid oölitic ore
— m opaco - [miner] opaque, mineral, o ore
— m — de hierro m - [miner] opaque iron ore
— m oxidado - [miner] oxidized ore
— m — con calidad f alta - [miner] high grade
oxidized ore
— m — — — f baja - [miner] low grade oxidized ore
— m natural - [miner] natural ore
— m — para embarque m directo - [miner] direct, loading, o shipping, natural ore
— m para acería f - [metal-prod] steel plant, o
open hearth, ore
— m — afino m - [metal-prod] refining ore
— m — — m de baño m - [metal-prod] feed ore
— m — embarque m directo - [miner] direct,
shipping, o loading, ore
— m — horno m - [metal-prod] furnace ore
— m — — m alto - [miner] blast furnace ore
— m pardo - [miner] brown ore
— m pendiente de liquidación f - [miner] settlement pending ore
— m — — — f final - [miner] final settlement
pending ore
— m pesado - [miner] heavy, mineral, o ore
— m piritoso - [miner] pyrite,tic ore
— m — compacto—[miner] compact pyrite,tic ore
— m — diseminado - [geol] dispersed pyrite,tic
ore
— m portador - [miner] bearing ore
— m — de fósforo m - [miner] phos(phorus)
bearing ore
— m prerreducido - [metal-prod] prereduced ore
— m primario - [miner] prime,mary, mineral, o
ore
— m pulverizado - [miner] powdered, o pulverized, mineral, o ore
— m reducido - [metal-prod] reduced ore

mineral m reducido directamente - [miner] directly reduced ore
— m — sin aglomerar - [miner] unpelletized reduced ore
— m residual - [miner] residual, mineral, o ore
— m rodado - [miner] scattered, o boulder, ore
— m secundario - [miner] secondary, mineral,o
ore
— m semielaborado - [miner] semiprocessed ore
— m sin beneficiar - [miner] unbeneficiated ore
— m — cribar - [miner] unscreened ore
— m — peletizar - [miner] unpelletized ore
— m sinterizado - [metal-prod] sinter(ed ore)
— m sueco - [miner] Swedish ore
— m suelto - [miner] loose ore
— m — cargado - [miner] loaded loose ore •
[metal-prod] charged loose ore
— m sulfuroso - [geol] sulfurous, mineral, o
ore
— m transportado - [miner] transported ore
— m triturado - [miner] crushed ore
— m veteado - [miner] banded ore
— m zarandeado - [miner] screened ore
mineralizado,da a - mineralized
mineralógico,ca a pesado,da - [geol] heavy min-
eral(ogical)
minería f - [miner] . . .; mining industry •
extracting,tion • mining project
— f bajo cielo m abierto - [miner] open pit
mining
— f creativa - [miner] creative mining
— f de carbón m - [miner] coal mining
— f — — m bajo cielo m abierto - [miner]
open pit coal mining
— f — — m en tajo m abierto - [miner] open
pit coal mining
— f — diamante(s) m - [miner] diamond mining
— f — hierro m - [miner] iron (ore) mining
— f — hulla f - [miner] coal mining
— f — importancia f - [miner] important, o
large scale, mining (project) • developed
mining industry
— f — mineral m de hierro m - [miner] iron ore
mining
— f desarrollada - [miner] developed mining
(industry)
— f en escala f grande - [miner] large scale
mining
— f — — f pequeña - [miner] small scale mining
— f — país m - [miner] country('s) mining
— f — tajo m abierto - [miner] open pit mining
— f grande - [miner] large scale mining
— f — por contrato m - [miner] large scale
contrtact, o contract large scale, mining
— f hidráulica - [miner] hydraulic mining
— f industrial - [miner] industrial mining
— f mediana - [miner] medium scale mining
— f nacional - [miner] national mining industry
— f por contrato m - [miner] contract mining
— f profunda - [miner] deep (mine) mining
— f reducida - [miner] small scale mining
— f subterránea - [miner] underground mining
minero-metalúrgico a - mining-metallurgical
mingitorio m, recto, o vertical - [sanit] stall
urinal
miniacería f - [metal-prod] mini steel plant
minicamión* m - [autom] minitruck
minicamioneta* f - [autom] minipickup
— f de serie f - [autom] stock minipickup
— f sin modificación(es) f - [autom] showroom
stock minipickup
minicamionetita* f - [autom] (small) minipickup
minigimansio* m - [Deport] mini-gymnasium
minimización* f - minimization; véase también
reducción f hasta mínimo m
— f de comunicación f verbal - [electrón]
voice traffic minimization,zing
— f — — f vocal - [electrón] voice traffic
minimization,zing
— f — corrosión f - [mec] corrosion minimization,zing
— f de daño m - damage minimization,zing

minimización f de deslizamiento - [autom] sliding minimization,zing
— f — — m sobre agua m - [cutom] hydroplaning
minimiization,zing
— f — distorsión f - [mec] distortion minimization,zing
— f — efecto(s) m—effect(s) minimization,zing
— f — — m de corrosión f - [metal] corrosion
effect(s) minimization,zing
— f — erosión f - erosion minimization,zing
— f — escombro(s) m - [constr] debris, o rubble, minimization,zing
— f — existencia(s) f - [com] stock(s), o inventory, minimization,zing
— f — inversión f - [fin] investment minimization,zing
— f — — f en material(es) m - [ind] inventory
investment minimization,zing
— f — material(es) m - [ind] materials, o inventory, minimization,zing
— m — — m sobrante(s) - [ind] surplus mate-
rial(s) minimization,zing
— f — peligro m - [segurid] danger minimization,zing
— f — peso m - weight minimization,zing
— f — problema m - problem minimization,zing
— f — — m de distorsión f - [mec] distortion
problem minimization,zing
— f — resistencia f - resistance minimization,zing
— f — riesgo m - [segurid] risk, o hazard, minimization,zing
— f — tiempo m - [ind] time minimization,zing
— f — — m de detención f - [ind] downtime
minimization,zing
— f — — m — — f de torre f para perforación
f - [petról] rig downtime minimization,zing
— f — — m inactivo - [ind] downtime minimization,zing
— f — tráfico m - [electrón] traffic minimization,zing
— f — — m, verbal, o vocal - [electrón] voice
traffic minimization,zing
minimizado,da* a - minimized
minimizar* v - to minimize
— v agrietamiento m - to minimize @ cracking
— v — m de calzada f - [constr] to minimize @
pavement cracking
— v — m — pavimento m - [constr] to minimize
@ pavement cracking
— v cantidad f - to minimize @, quantity, o
amount
— v — f de humo m - to minimize @ smoke amount
— v — f — polvo m - to minimize @, dust, o
dirt, amount
— v — f — tierra f—to minimize @ dirt amount
— v ciclo m - to minimize @ cycle,cling
— v — m para marcha/detención f - [mec] to
minimize @ on/off cycling
— v — m — —/— para compresor m - [mec] to
minimize @ compressor on/off cycling
— v comunicación f, verbal, o vocal—[electrón]
to minimize @ voice traffic
— v conservación f - [ind] to minimize @ maintenance
— v corrosión f - [metal] to minimize @ corrosion
— v costo m - to minimize @ cost
— v — m para conservación f - [ind] to minimize @ maintenance cost(s)
— v — m — mantenimiento m - [ind] to minimize
@ maintenance cost(s)
— v — m para tiempo m inactivo - [ind] to minimize @ downtime cost(s)
— v daño(s) - to minimize @ damage(s)
— v desgaste m - [mec] to minimize @ wear
— v deslizamiento m - to minimize @ slide,ding
— v — m sobre agua m - [autom] to minimize @
hydroplaning
— v distorsión f - to minimize @ distortion
— v efecto m - to minimize @ effect
— v — m de corrosión f - [mec] to minimize @
corrosion efecto(s)

minimizar v erosión f - to minimize @ erosion
— v escombro(s) m - [constr] to minimize @,
debris, o rubble
— v formación f de arco m - [sold] to minimize
@ arcing
— v fuga f - to minimize @ leak(age)
— v inversión f - [fin] to minimize @ investment
— v — f en material(es) m - [ind] to minimize
@ material(s) investment
— v grieta f - to minimize @ crack
— f — f en calzada - [constr] to minimize @
pavement crack
— v — f — pavimento m - [constr] to minimize
@ pavement crack
— v humo m - to minimize @ smoke
— v mantenimiento m - [ind] to minimize @ maintenance
— v parada f - [ind] to minimize @ downtime
— v peligro m - [segurid] to minimize @ danger
— v peso m - to minimize @ weight
— v polvo m - to minimize @, dust, o dirt
— v resistencia f - to minimize @ resistance
— v riesgo(s) m - [segurid] to minimize @,
risk(s), o hazard(s)
— v salpicadura f—[sold] to minimize @ spatter
— v — f fina - [sold] to minimize @ fine spatter
— v suciedad f - to minimize @ dirt
— v tendencia f - to minimize @, tendency, o
trend
— v — f a agrietamiento m - [sold] to minimize
@ cracking tendency
— v tiempo m - to minimize @ time
— v — m de detención f - [ind] to minimize @
downtime
— v — m — — f de torre f para perforación f
- [petról] to minimize @ rig downtime
— v — m de inactividad f - [ind] to minimize @
downtime
— v — m — parada f - [ind] to minimize @
downtime
— v — m inactivo—[ind] to minimize @ downtime
— v tierra f - to minimize @ dirt
— v tráfico m, verbal, o vocal - [electrón] to
minimize @ voice traffic
mínimo m absoluto - absolute minimum
— m admisible - allowable minimum; minimum allowable
— m alcanzado - reached, o attained, minimum
— m de agua m - water minimum
— m — carbono m - carbon minimum
— m — conservación f - [ind] minimum maintenance • maintenance minimum
— m — mantenimiento m - [ind] minimum maintenance • maintenance minimum
— m — excavación f - [constr] minimum excavation
— m — frecuencia f - [electrón] minimum frequency • frequency minimum
— m — — f desplazado - [electrón] shifted minimum frequency
— m — instalación(es) m - [sold] minimum fixturing
— m — manipuleo m - [ind] minimum handling
— m — mantenimiento m - [ind] minimum maintenance
— m — picadura(s) f - [metal] minimum, pitting, o pinholing
— m — recubrimiento m - [constr] minimum cover
height
— m — resistencia f - minimum resistance •
minimum strength
— m definitive - definite minimum
— m en variación f - variation minimum • minimum variation
— m especificado - specified minimum
— m estándar - standard minimum
— m excavado - [constr] excavated minimum
— m exigible - minimum, required, o requirable
— m mantenido - maintained minimum
— m muy reducido - extreme, o very low, minimum
— m satisfactorio - satisfactory requirement

mínimo m según norma f - standard minimum
— m — Sociedad f Estadounidense para Soldadura
f - [sold] American Welding Society minimum
minio m - [pint] . . .; red oxide
miniplanta* f - [ind] miniplant
— f automatizada - [ind] automated miniplant
— f — para acero m - [metal-prod] automated
steel miniplant
— f — altamente - [ind] highly automated miniplant
— f económica - [ind] economic(al) miniplant
— f integrada - [ind] integrated miniplant
— f para acero m - [metal-prod] steel miniplant
— f — — m automatizada - [metal-prod] automated steel miniplant
— f — — m — altamente - [metal-prod] highly
automated steel miniplant
— f — f de chatarra f - [metal-prod] scrap
based miniplant
— f primitiva - [metal-prod] early, o primitive, miniplant
— f semi-integrada - [metal-prod] semi-integrated miniplant
— f siderúrgica - [metal-prod] steel miniplant
ministerio m - . . . . [pol] . . .; véase también departamento m; secretaría f
— m de Agricultura f - [pol] Agriculture Deparment; Department of Agriculture
— m — Comercio m - [pol] Department of Commerce
— m — Defensa f - [pol] Department of Defense
— m — Nacional - [pol] National Defense Department; National Department of Defense
— m — Economía f - [pol] Department of Economy
— m — Energía f - [pol] Department of Energy
— m — — f y Minas f - [pol] Department of
Energy and Mines
— m — Estado m - [pol] Department of State;
véase también Ministerio de Relaciones Exteriores
— m — Finanzas f - [pol] Department of Finance
— m — Fomento m - [pol] Department of, Promotion, o Development
— m — — m para Vivienda f y Urbanización f -
[pol] Department of Housing and Urban Devel-
pment
— m — Gobierno m - [pol] Government Department
• Department of State; Department of Interior
— m — Hacienda - [pol] Department of @ Treasury
— m — — f y Crédito m Público - [pol] Department of @ Treasury and Public Credit
— m — Industria f - [pol] Department of Industry
— m — — f y Comercio m - [pol] Department of
Industry and Commerce
— m — — f y Energía f [pol] Department of
Industry and Energy
— m — Interior—[pol] Department of @ Interior
— m — Justicia f - [pol] Department of Justice
— m —, Minas, o Minería f - [pol] Department
of Mines
— m — Obras f Públicas - [pol] Department of
Public Works
— n — — f y Servicio(s) m Público(s) - [pol]
Department of Public Works and Services
— n — — f — y Transportes m - [pol] Depàrt-
ment of Public Works and Transportation
— m — Producción f - [pol] Department of Production
— m — Recursos m Naturales - [pol] Department
of Natural Resources
— m — Relaciones f Exteriores - [pol] Department of Foreign Affairs; Department of State
— m — Salud f Pública - [pol] Department of
(Public) Health
— m Estadounidense - [pol] United States Department
— m estatal - [pol] state department
— m para Conservación f de Ambiente m - [pol]
Department of @ Environment
— n — Planeamiento m y Coordinación f - [pol]
Planning and Coordination Department

ministro m - . . . . • [pol] minister; secretary •
[rel] minister
— m de departamento m - [pol] department secretary; cabinet minister
— m — estado m - [pol] minister; secretary
— m — gobierno m - [pol] secretary of state
— m — hacienda f - [pol] secretary of @ treasury
— m — obras f públicas - [pol] public works
secretary
— m — — f — y transporte(s) m - [pol] public
works and transportation secretary
— m — relacion(es) f exterior(es) - [pol] secretary of foreign affairs; secretary of state
— m — transportes m - [pol] secretary of
transportation
minoración f - . . .; reducing,ction
— f de marcha f - slowing down
— f — régimen m de marcha f - slowing down
— f — velocidad f - slowing down
minorado,da a - lessened; reduced
minorar v - . . .; to reduce • to understate •
[metal-prod] to turn down
— v petaca f - [metal-prod] to turn down @
cooling plate
minorista m - [com] retailer; dealer | a - [com]
retail
minoritario,ria a - minority
minucioso,sa a - . . .; minute
minuto m concedido - [labor] allowed minute
— m de producción f - [labor] production minute
— m deficitario - [labor] deficit minute
— m empleado - used minute; minute used
— m estándar - [labor] standard minute
— m — de producción f - [labor] standard production minute
— m óptimo - [labor] optimum minute
— m — según norma f - [labor] optimum standard minute
— m para colada f - [metal-prod] minute(s) to
tapping (time)
— m — sangría f - [metal-prod] minute(s) to
tapping time
— m según norma f - [labor] standard minute
— m — — f para producción f - [labor] standard production minute
mira f - . . . . • scope • [instrum] leveling,
rod, o pole, o staff
— f amplia - broad, view, o scope
— f hacia frente m - forward look
— f más amplia - broader, view, o scope
mirada f escrutadora - close, o careful, look
— f escudriñante - close, o scrutinizing, look
— f hacia adelante - forward look
— f retrospectiva - look(ing) back; retrospec-
tive look
mirador m - . . . . • [ind] peep sight
— m en puerta f - [ind] door (peep) sight
— m — — f para carga f - [ind] charge,ging
door (peep) sight
— m — — f — — f de horno m Siemens-Martin -
[metal-prod] open hearth (furnace) wicket
(hole)
mirando v desde arriba - looking down
— v — atrás - looking from behind | [adv] fac-
ing @ rear
— v hacia - looking towards
— v — afuera - looking out • facing away from
— v — — desde - looking out of • facing away
from
— v — frente - looking, forwards, o toward @
front
— v — respaldo - looking towards @ rear
— v hacia tablero m - looking towards @ panel
— v hacia tablero m para regulación f - looking
towards # control panel
— v — — m — — f desde frente m - facing @
control panel
mirar v a otro lado m - to look elsewhere
— v arco m - [sold] to watch @ arc
— v hacia adelante - to look forward
— f — atrás - to look backward • to look over
@ shoulder

mirar v **hacia afuera** - to, look, o face, away
— v — — **desde** - to, look o face, away from
— v **por encima de hombro** m - to look over @ shoulder
— v **retrospectivamente** - to look back
mirilla f - . . . . • [ind] (inspection) port; sight glass; peephole assembly; peep sight
— f **en compresor** m - [ind] compressor peep sight
— f — — m **verificada** - [ind] checked compressor peep sight
— f — **puerta** f - door peep sight
— f — — f **para carga** f - [ind] charg(ing) door peep sight
— f — — — f **para horno** m (Siemens-Martin) - [metal-prod] (open hearth furnace) wicket (hole)
— f — **observación** f - [ind] peep sight (glass)
— f — f **de nivel** m **de aceite** m - [mec] oil level, peep, o sight, glass
mirón m - . . . . • [constr] sidewalk superintendent
miscible a **con agua** m - water miscible
— a **en agua** m - water miscible
misión f - . . . . • job; responsibility
— f **extranjera** - foreign mission
misma adv **calidad** f - same quality
— f **carga** f - same load(ing)
— f **composición** f - same make-up • [quím] same analysis
— adv **forma** - same shape • same way
— adv **frecuencia** f - [electrón] same frequency
— adv **industria** f - same industry
— adv **información** f - same information
— adv **manera** f - same manner
— adv **norma** f - same standard
— adv —, **alta**, o **elevada** - same high standard
— adv **pieza** f - [ind] same part
— adv **ubicación** f - same location
mismo adv **bulto** - same parcel
— adv **costado** m - same side
— adv **curso** m **de agua** - [hidr] same watercourse
— adv **diámetro** m - same diameter
— adv **efecto** - same effect • single purpose
— adv **eje** m - [mec] same axle
— adv **electrodo** m - [sold] same electrode
— adv **empaque** - same packing
— adv **espesor** - same thickness
— adv **fondo** - same bottom • very bottom
— adv **lado** - same side
— adv **material** - same material
— adv **método** m - same method
— adv **nivel** m - same level | (con) even; flush
— adv **orden de magnitud** f - similar size
— adv **plano** - (en) flush
— adv **tipo** m - same, type, o kind
— adv **período** m - same period
— adv — m **de tiempo** m - same time period
— adv **que** - same, o just, like, o as
— adv **tenor** m - [legal] same, o identical, content; reading the same
— adv **tiempo** m - same time
— adv **trabajo** m - same, work, o task, o job
— adv **voltaje** m - [electr-oper] same voltage
misterio m **mecánico** - [mec] mechanical mystery
— m **para fabricación** f - [ind] manufacture, ring mystery
— m — — f **de neumático** - [autom-neumát] tire, construction, o manufacture, mystery
mitad f - . . . . • side | adv - halfway
— f **aguas** f **abajo** - downstream half
— f — f — **de tubería** f - [tub] pipe downstream half
— f **aguas** f **arriba** - upstream half
— f — f — **de tubería** f - [tub] pipe upstream half
— f **anterior** - front half • previous half
— f — **de lado** m - side front half
— f — — — m **derecho** - right side front half
— f — — — m **izquierdo** - left side front half
— f **(de) camino** m - halfway
— f **con corriente** f - [electr-oper] hot side
— f **de acoplamiento** m - [mec] coupling half

mitad f **de altura** f - half @ height • [geom] centerline • [tub] centerline • mid diameter
— f — — f **de costado** - halfway up @ side
— f — — f — **tubería** f - [tub] half way up @ pipe side; pipe centerline
— f — **ancho(r)** m - half width
— f — **anillo** m - [mec] ring half
— f — **aro** m - [mec] ring half
— f — **banda** f - [tub] band half
— f — — f **para acoplamiento** m - [tub] coupling band half
— f — **caja** f - [mec] case half
— f — — f **para diferencial** - [autom-mec] differential case half
— f — — — m **instalada** - [autom-mec] installed differential case half
— f — — f **instalada** - [autom-mec] installed case half
— f — — f **separada** - [mec] separated case half
— f — **camino** m - midway; half way
— f — **carrera** f - [mec] travel midpoint • [deport] race midpoint
— f — **cojinete** m - [mec] bearing half
— f — **concavidad** f - concavity, half, o middle part
— f — **costura** f - [mec] seam center
— f — **cuerda** f - span, half, o center
— f — **distancia** f - half way
— f — **espesor** m - half thickness
— f — — m **de banda** f **para rodamiento** m - [autom-neumát] (one) half of tread depth
— f — **línea** f - line half • [electr-instal] line side
— f — — f **de conductor(es)** m - [electr-cond] line side
— f — f — — m **eléctrico(s)** - [electr-cond] power line, half, o side
— f — — f **para energía** f - [electr-cond] power line, half, o side
— f — **luz** - span half • half span
— f — **material** m **plástico** - plastic half
— f — **noche** - night, middle, o half
— f — **plástico** m - plastic half
— f — **precipitador** m - [metal-prod] precipitator, half, o middle
— f — **profundidad** - half depth
— f — — f **de banda** f **para rodamiento** m - [autom -neumát] (one) half of tread depth
— f — **radio** m - [geom] half @ radius
— f — **recorrido** m - half way; midway
— f — **resistencia** f - [mec] half resistance • half strength
— f — — f **máxima** - [mec] half @, maximum, o ultimate, resistance, o strength
— f — — f — **admisible** - [mec] half @, ultimate, o maximum, allowable strength
— f — **temporada** - [deport] mid season
— f — **tomacorriente** m - [electr-instal] receptacle half
— f — — m **doble** - [electr-instal] duplex, o double, o dual, receptacle half
— f — **tramo** m - span half • mid span
— f — **transformador** m - [electr-transf] transformer half
— f — **tubería** f - [tub] pipe, ping half
— f — **velocidad** f - half @ speed
— f — **viento** m - half of @ wind
— f **derecha** - right half
— f — **de caja** f [sold] right mount (half)
— f — — f **para pistola** f - [sold] right (half) of @ gun mount
— f **embridada** - [mec] flanged half
— f **inferior** - lower, o bottom, half
— f — **de banda** f - [mec] band, lower, o bottom, half
— f — **de carcasa** f - [electr-mot] frame, bottom, o lower, half
— f — — **diafragma** m - [constr] diaphragm bottom half
— f — — **tubería** f - [tub] pipe, bottom, o lower, half
— f **izquierda** - left half

mitad f izquierda de caja f - [mec] case left
half
— f — — — f para pistola f - [sold] left
(half of @) gun mount
— f lisa - [mec] smooth, o plane, half
— f plana f - plane half
— f posterior - rear, o back, half
— f — de lado m derecho - left side rear half
— f primera - first half
— f rebordeada - [mec] flanged half
— f rígida - [mec] rigid half
— f río m abajo - [hidr] downstream half
— f — m arriba - [hidr] upstream half
— f segunda - second half
— f separada - [mec] separated half
— f superior - top, o upper, half
— f — de banda f - [mec] band, top, o upper,
half
— f — — — f para acoplamiento m - [tub] coupling band, top, o upper, half
— f — — carcasa f - [electr-mot] frame, top,
o upper, half
— f — — diafragma - [constr] diaphragm, top,
o upper, half
— f — — pared f - [constr] wall, top, o upper, half
— f — — — f lateral - [constr] sidewall,
top, o upper, half
— f — — tubería f - [tub] pipe,ping, top, o
upper, half
— f última - last half
mitigar v - . . .; to alleviate; to reduce
mitón m - [segurid] . . .; mitten
mitrado,da a - [mec] véase ingleteado,da; con
inglete m
mixto,ta a - . . . • [fin] binational
mobiliario m exigido - [domést] required furniture; furniture required
— m requerido - [domést] required furniture;
furniture required
— m y vehículo(s) m - [contab] furniture and
vehicles
mocetón m - . . .; kid
moción f de renuncia f - [legal] resignation motion
— f — — f de representación f - [legal] representation resignation motion
— f — — f — — f profesional - [legal] professional representation resignation motion
mocho,cha a - . . .; blunt; dull
modalidad f - . . .; modality; condition • basis
• pattern; guideline(s) • [comput] mode
— f autoiniciada - [comput] self-initiated mode
— f automática - [electrón] automatic mode
— f cíclica—[econ] cyclic, modality, o pattern
— f — de negocio(s) m - [econ] cyclic business
modality
— f con retorno m - [comput] loopback mode
— f con tono(s) m - [comput] tone mode
— f — — m para ensayo(s) m - [comput] test
tone mode
— f de contrato m - contract type
— f — corriente f - [hidr] flow pattern
— f — operación f - operating, mode, o pattern
— f — — f normal - [comput] normal operating,
mode, o pattern
— f — tanto m alzado - [fin] lump sum basis
— f dirigible - [comput] addressable mode
— f de viraje(s) m - [autom] cornering, mode, o
pattern
— f encaminable - [comput] addressable mode
(select)
— f local - [comput] local mode
— f normal - [comput] normal mode
— f para actualización f - [comput] update,ting
mode
— f — — f automática - [comput] automatic
update,ting mode
— f — almacenamiento m - [comput] storage,ring
mode
— f — autoiniciación f - [comput] self-initiated,ting mode
— f para compra-venta f - [com] purchase-resale

pattern
modalidad f para contrato m - contract modality
— f para corriente f alterna - [electrón] alternating current mode
— f — — f continua - [electrón] direct current mode
— f — eligibilidad f - eligibility guideline
— f — ensayo(s) m - [electrón] test(ing) mode
— f — — seleccionado - [electrón] selected
test(ing) mode
— f — instalación f - installation, mode, o
pattern, o practicality
— f — interrogación f - [comput] interrogate
mode
— f — operación f - [comput] operating mode
— f — — f con retorno m - [comput] loopback
operating mode
— f — — f — terminación f silenciosa -
[comput] quiet termination operating mode
— f — — — f sin ruido(s) m - [comput]
quiet, o noiseless, termination operating
mode
— f — — f con tono m para prueba(s) f -
[comput] test(ing) tone operating mode
— f — prueba(s) f - [comput] test(ing) mode
— f — — f seleccionada - [comput] selected
test(ing) mode
— f — puesta f en marcha f - [sold] starting
mode
— f — reajuste m - [electrón] reset(ting) mode
— f — señalización f—[comput] signalling mode
— f — — f automática - [comput] automatic
signalling mode
— f — — f para, ensayo(s) m, o prueba(s) f -
[comput] test(ing) signalling mode
— f — terminación f—[comput] termination mode
— f — — f silenciosa - [comput] quiet termination mode
— f — — f sin ruido(s) m - [comput] quiet
termination mode
— f seleccionada - [comput] selected mode
— f silenciosa - [comput] quiet, o silent, mode
— f sin ruido(s) m - [comput] quiet mode
modelación f - modeling
— f de circuito m - [comput] circuit modeling
— f — m para molienda f - [cuerp.moled]
grinding circuit modeling
— f en sitio m - site, modeling, o forming
modelado,da a - modeled; formed; shaped
— a en sitio m - site modeled
modelador m - . . .; molder
modelar v - . . .; to shape; to form
— v circuito m - [comput] to model @ circuit
— v — m para molienda f - [cuerp.moled] to
model @ grinding circuit
— v en sitio m - to site, model, o form
modelista m - [inf] modeler; molder
— m mecánico - [ind] pattern, modeler, o molder
modelo m - . . . . • mock-up • [mec] mold • size;
version • [ind] product • [legal] draft
— m anterior - previous, o prior, model •
[sold] earlier, model, o machine
— m anticuado - [sold] older welder; antique
— m antiguo - obselete model; old style
— m autónomo - autonomous model
— m básico - basic model
— m de eje m - [autom-mec] basic axle model
— m combinado - combination model
— m comparable - [ind] comparable model
— m competidor - [ind] competing,titive model
— m con alimentación f de corriente f alterna -
[sold] alternating current input model
— m — — f — — f continua - [sold] direct
current, o D C, input model
— m — amperaje(s) m muy bajo(s) - [sold] low
amperage, o aircraft, model
— m — bomba f - [bombas] pump model
— m — — f únicamente - [bombas] pump model
only
— m — capa(s) f - [sold] layered mock-up
— m — — f ejecutada(s) cuidadosamente—[sold]
carefully layered mock-up
— m — contactador m - [sold] contactor model

modelo m con contactador m en salida f - [sold]
output contactor model
— m — correa f - [mec] belt(ed) model
— m — energía f auxiliar - [sold] auxiliary
power unit
— m — — f con . . . voltio(s) m - [sold]
. . . volt(s) auxiliary power unit
— m — impulsión f - [mec] driven model
— m — — f con correa f - [mec] belted model
— m — interruptor m (automático) - [sold] contactor model
— m — — m en salida f - [sold] output contactor model
— m — mandíbula f - [mec] jaw model
— m — — f ajustable - [mec] adjustable, o
swivel, jaw model
— m — — f fija f - [mec] fixed jaw model
— m — motor m con combustión f interna - [sold]
engine driven model
— m — — m con explosión f - [sold] engine
driven model
— m — — m eléctrico - [sold] (electric) motor
driven, model, o welder, o machine
— m — — m para gasolina f - gasoline engine
model
— m — propósito m doble - [sold] dual purpose,
o combination, model
— m — reducción f - [mec] reducing,ction model
— m — — f simple - [autom-mec] single reduction model
— m — relé m - [sold] relay model
— m — — inversor (de campo m) - [sold]
(field) reversing relay model
— m — tamaño m estándar - [mec] standard, o
full size, model
— m correspondiente m - corresponding, model, o
pattern
— m de alimentador m - [mec] feeder model
— m — — m para alambre m - [sold] wire feeder
model
— m — bomba f - [bombas] pump model
— m — caja f - [mec] housing model
— m — cambiador m - [electr-equip] changer
model
— m — — m para toma(s) f - [electr-equip] tap
changer model
— m — camión m - [autom] truck model
— m — camioneta f - [autom] pickup model
— m — carcasa f - [mec] housing model
— n — circuito m - circuit model
— m — — m para computador m - [comput] computer circuit model
— m — combinación f - combined,nation model
— m — competencia f - [ind] competing,titive
model
— m — computador m - [comput] computer model
— m — eje m - [mec] axle model
— m — equipo m—[ind] equipment, model, o type
— m — extrusor* m - [plást] extruder model
— m — freno m - [mec] brake model
— m — grúa f - [grúas] crane model
— m — manipulador m - [electrón] keyer model
— f — — m para tono(s) m - electrón] tone
keyer model
— m — máquina f - [ind] machine model
— m — motor m - [comb.int] engine model •
[electr-mot] motor model
— m — mordaza f - [mec] clamp model
— m — pie m - [domést] console (model)
— m — soldadora f - [sold] welder model
— m — — f para amperaje(s) m (muy) bajo(s) -
[sold] low amperage, o aircraft, model
— m — — f para trabajo(s) m de aviación f -
[sold] aircraft model (welder)
— m — utilidad f - [legal] registered model
— m diesel - [comb.int] diesel model
— m distinto - different, o separate, model
— m doble - double, model, o pattern
— m elaborado - prepared model
— m en existencia f - avilable model
— m — tamaño m natural - (full scale) mock-up
— m — — m real - (full scale) mock-up
— m — tándem - tandem model

modelo m en tándem con reducción f - [autom-mec]
reduction tandem model
— m — — — f simple - [autom-mec] single
reduction tandem model
— m especializado - [econ] specialized model
— m específico - specific, o particular, model
— m estándar - [autom] standard, o full-size,
model
— m ...fásico - [electr-equip] . . . phase
model
— m fuerte - strong model
— m identificado - identified model
— m impulsado - [mec] driven model
— m — con correa f - [mec] belt driven model
— m industrial - [legal] industrial model
— m más antiguo - older model • oldest model
— m — grande - larger model
— m — nuevo - newer model
— m matemático - [matem] mathmatic(al), model,
o pattern
— m — apropiado - [comput] appropriate mathmatical model
— m — elaborado - prepared mathmatical model
— m — preparado - prepared mathmatical model
— m mayor - larger model
— m motogenerador - [sold] motor-generator
model
— m mundial - world(-wide) model
— m no desarmable - [mec] no-take-apart, model,
o version
— m nuevo - new model • new style
— m para ambas corrientes f - [electr-equip]
alternating (current)/direct current model;
AC/DC model; combination (alternating/direct
current) model • [sold] combination model
— m para . . . amperio(s) m - [electr-equip]
. . . ampere, model, o size
— m — aplicación f radiotelefónica—[electrón]
radio(telephone) application model
— m — aviación f - [sold] aircraft model
— m — carta f - letter, model, o draft
— m — — f de, intento, o intención f - [com]
intent letter draft
— m — — f para compromiso m - commitment letter draft
— m — circuito m - [comput] circuit model
— m — circulación f - [vial] traffic pattern
— m — correlación f - correlation model
— m — — f múltiple - multiple correlation
model
— m — — f, sencilla, o única - single correlation model
— m — corriente f alterna - [sold] alternating
current, o A C, model
— m — — f con motor m eléctrico - [sold]
alternating current motor driven model
— m — — f solamente - [sold] straight alternating current model
— m — — f continua - [sold] direct current, o
D C, model
— m — escritorio m - [comput] desk-top, model,
o unit
— m — fin m doble - [sold] double purpose, o
combination, model
— m — fundición f - [metal-prod] foundry,
model, o pattern
— m — montar sobre anaquel m - [comput] rack
model
— m — propósito m doble - [sold] combination
model; double purpose model
— m — radio(telefonía) f - [comput] radio
(telephone) model
— m — taller m - [ind] shop model
— m — . . . tonelada(s) f - . . .-ton model
— m — trabajo(s) m para aviación f - [sold]
aircraft model
— m — tráfico m - [vial] traffic pattern
— m — . . . velocidad(es) f—[mec] . . . speed
model
— m — . . . voltio(s) m - [electr-oper] . . .
volt model
— m patrón m - [sold] standard, model, o unit
— m periférico - periferal model

modelo m popular - popular model
— m **posterior** - later model • [sold] later welder
— m **preparado** - prepared model
— m **primero** - first, o early, model
— m **primitivo** - early, o primitive, model
— m **provisto** - provided, o supplied, model
— m **reciente** - recent, o late, model
— m — **de bomba** f - [bombas] pump late model
— m **remolcable** - road, o towable, model
— m **siguiente** - following model
— m **sin depósito** m - [mec] tankless model
— m — — m **para fundente** m - [sold] model without @ flux tank
— m **temprano** - early model
— m **trifásico**—[electr-equip] three-phase model
— m **último** m - late model • last model
— m **viejo** - [ind] old model
moderadamente adv **abrasivo,va** - moderately abrasive
— adv **corrosivo,va** a - moderately corrosive
— adv **permeable** a - moderately pervious
— adv **pesado,da** a - moderately heavy
— adv **progresivo,va** - moderately progressive
moderado,da - . . . . • mild
moderador m **para velocidad** f - [f.c.] retarder
— m — — f **por gravedad** f - [f.c.] gravity retarder
moderar v **marcha** f - to slow down
— v **velocidad** f - to slow down @ speed
modernización f - . . .; modernizing • improvement,ving; upgrade,ding
— f **de código** - [ind] code medernization,zing
— f — — m **para construcción** f - [constr] building, o construction, code modernization
— f — **planta** f - [ind] plant modernization
— f — — f **para alquilación** f - [petról] alkylization,zing plant modernization,zing
— f — **sistema** - system, modernization,zing, o upgrade,ding
— f **ferroviaria** - [f.c.] railway, modernizing, o upgrading
— f **futura** - future, modernization,zing, o upgrading
— f — **de sistema** m - future system, o system future, modernization,zing, o upgrade,ding
— f **urbana,nística** - urban modernization
modernizado,da - modernized; upgraded
modernizar v - . . .; to upgrade
— v **sistema** m - to, modernize, o upgrade, @ system
moderno,na a - . . .; upgraded; advanced
módico,ca a - . . .; modest
modificación f - . . .; modifying; alteration; altering; change,ging (over); deviation; variant • [legal] amendment; revision
— f **constante** - constant change,ging
— f **cortada manualmente** - hand cut modification
— f — — **en neumático** m - [autom-neumát] tread hand-cut, modification, o change
— f **de ajuste** m - adjustment change,ging • [mec] set(ting) change,ging
— f — — m **en amperaje** m - [electr-oper] amperage setting change,ging • [sold] current setting change,ging
— f — **amperaje** m - [electr-oper] amperage change,ging • [sold] current, variation, o change,ging
— f — **ángulo** m **para ataque** m - [mec] attack angle change,ging
— f — — m **para boquilla** f - [mec] nozzle attack angle change,ging
— f **automóvil** - [autom] car change,ging
— f — — m **con tracción** f **en . . . rueda(s)** f - [autom] . . .-wheel-drive (car) change,ging
— f — **banda** f **para rodamiento** m - [autom-neum] tread modification
— f **de carta** f **para crédito** m - [fin] credit letter, modification, o amendment, o change
— f — **cimiento(s)** m - [constr] foundation change
— f — **conexión** f - connection change
— f — **detalle(s)** m - detail(s), modification, o change

modificación f **de distancia** f - distance change,ging
— f — **documento** m - [legal] document change,ging
— f **de eje** m - [mec] axle change,ging
— f — — m **para transmisión** f - [autom-mec] transaxle, change,ging, o modifying,fication
— f — **entrada** f - entrance change,ging • [electrón] input change,ging
— f — **equipo** m - [ind] equipment, change,ging, o modifying,fication
— f — — m **para laminador** m - [metal-lam] mill equipment, change,qing, o modifying,fication
— f — — — m — — m **en caliente** - [metal-lam] hot reduction mill equipment change,ging
— f — — — m — — m **frío** - [metal-lam] cold reduction mill equipment change,ging
— f — — — m **para palanquilla** f—[metal-lam] billet mill equipment change,ging
— f — — — m — — **temple** m - [metal-lam] temper mill equipment change,ging
— f — — — m — — — m **doble** - [metal-lam] double temper mill equipment change,ging
— f — — — m — — — m **simple**—[metal-lam] single temper mill equipment change,ging
— f — — m **eléctrico** - [electr-equip] electrical equipment change,ging
— f — — m — **para laminador** m - [metal-lam] mill electrical equipment change,ging
— f — — — m **para palanquilla** f - [metal-lam] billet mill electrical equipment change,ging
— f — — — m — — **temple** m - [metal-lam] temper mill electrical equipment change
— f — — m — — m — — m **doble** - [metal-lam] double temper mill electrical equipment change,ging
— f — — m — — m — — m **simple** - [metal-lam] single temper mill electrical equipment change,ging
— f — — m **estándar** - [ind] standard equipment, change,ging, o modifying,fication
— f — — m **existente** - [ind] existing equipment, change,ging, o modifying,fication
— f — — m **mecánico** - [ind] mechanical equipment, change,ging, o modifying,fication
— f — — m — — **laminador** m - [metal-lam] mill mechanical equipment, change,ging, o modifying,fication
— f — — m — — m **en caliente**—[metal-lam] hot (reduction) mill mechanical equipment, change,ging, o modifying,fication
— f — — m — — m — **frío** - [metal-lam] cold (reduction) mill mechanical equipment, change,ging, o modifying,fication
— f — — m — — m **para palanquilla** f - [metal-lam] billet mill mechanical equipment, change,ging, o modifying,fication
— f — — m — — m — — **temple** m—[metal-lam] temper mill mechanical equipment change,ging
— f — — m — — m — — m **doble** - [metal-lam] double temper mill mechanical equipment, change,ging, o modifying,fication
— f — — m — — m — — m **simple** - [metal-lam] single temper mill mechanical equipment, change,ging, o modifying,fication
— f — **estructura** f - [constr] structure modifying,fication
— f — **fórmula** f - [quím] formula, change,ging, o modifying,fication
— f — **herramienta(s)** f - [herram] tool(s), change,ging, o modifying,fication
— f — **hoja** f - [metal-lam] sheet change,ging
— f — **horno** m - [ind] furnace change,ging
— f — **inmueble(s)** m - [constr] real estate, change,ging, o altering,ration
— f — **instalación(es)** f - [ind] installation, o facility,ties, change,ging, o modifying,fication
— f — — f **accesoria(s)** f - [ind] appurtenance modifying,fication
— f — **laminador** m - [metal-lam] mill,

change,ging, o modifying,fication
modificación f **de laminador** m **con puente** m **doble**
- [metal-lam] double bridge mill change,ging
— f — — m **en caliente** - [metal-lam] hot,
(strip, o reduction) mill, change,ging, o modifying,fication
— f — — m **en frío** - [metal-lam] cold, (strip,
o reduction) mill, change,ging, o modifying,fication
— f — — m **en tándem** - [metal-lam] tandem
mill, change,ging, o modifying,fication
— f — — m **en caliente** - [metal-lam]
tandem hot mill, change,ging, o modification
— f — — m — — — **frío** - [metal-lam] tandem
cold mill, change,ging, o modifying,fication
— f — — m **para banda(s)** f - [metal-lam] strip
mill, change,ging, o modifying,fication
— f — — m — — f **en caliente** m - [metal-lam]
hot strip mill, change,ging, o modifying
— f — — m — — f **en frío** - [metal-lam] cold
strip mill, change,ging, o modifying,fication
— f — — m — **cinta(s)** f - [metal-lam] strip
mill, change,ging, o modifying,fication
— f — — m — — f **en caliente** - [metal-lam]
hot strip mill, change,ging, o modifying
— f — — m — — f **en frío** - [metal-lam] cold
strip mill, change,ging, o modifying,fication
— f — — m — **chapa** f—[metal-lam] strip mill,
change,ging, o modifying,fication
— f — — m — — f **en caliente** - [metal-lam]
hot strip mill, change,ging, o modifying
— f — — m — — f **en frío** - [metal-lam] cold
strip mill, change,ging, o modifying,fication
— f — — m — **fleje** m—[metal-lam] strip mill,
change,ging, o modifying,fication
— f — — m — — m **en caliente** - [metal-lam]
hot strip mill, change,ging, o modifying
— f — — m — — m — **frío** - [metal-lam] cold
strip mill, change,ging, o modifying,fication
— f — — m **palanquilla** f - [metal-lam] billet mill, change,ging, o modifying,fication
— f — — m **planchón(es)** m - [metal-lam]
slabbing mill, change,ging, o modifying
— f — — m **temple** m - [metal-lam] temper
mill, change,ging, o modifying,fication
— f — — m **doble** - [metal-lam] double
temper mill, change,ging, o modifying
— f — — m — — m **simple** - [metal-lam] single
temper mill, change,ging, o modifying
— f — — m **para tocho(s)** m—[metal-lam] blooming mill, change,ging, o modifying,fication
— f — **mantenimiento** m - [ind] maintenance,
change,ging, o altering,ration
— f — — m **asignado** - [ind] assigned maintenance, change,ging, o altering,ration
— f — — m **destacado** - [ind] assigned maintenance, change,ging, o altering,ration
— f — **norma** f - [admin] standard, o policy,
change,ging, o altering,ration, o revision
— f — **número** m - [ind] number change,ging
— f — — m **de serie** - [ind] serial number
change,ging
— f — **plancha** f - [mec] plate change,ging
— f — — f **para asiento** m - [mec] seat plate
change,ging
— f — — f — — m **para compresor** m - [ind]
compressor seat plate change,ging
— f — — f — — m — **ventilador** n - [ind] fan
seat plate change,ging
— f — — f **cubierta** f - [ind] covering
plate change,ging
— f — **producto** m - [ind] product, change,ging,
o modifying,fication
— f — **quemador** m - [combust] burner, modifying,fication, o change,ging, o replacement,cing
— f — **receta** f - [culin] recipe, change,ging,
o modifying,fication • [quím] prescription,
change,ging, o modifying,fication
— f — **ajuste** m - [mec] adjustment, o setting,
change,ging, o modifying,fication
— f **en** - véase también **modificación** f **de**
— f — **cadena** f - [cadenas] chain change,ging

modificación f **en cadena** f **para transportador** m
- [ind] conveyor chain change,ging
— f — — f — — m **para arrastre** m - [ind]
drag conveyor chain change,ging
— f — **cambiador** m [ind] changer change,ging
— f — — m **para toma** f **de voltaje** m - [electr-
-distrib] voltage intake changer change,ging
— f — — f — — — f — m **transformador** m -
[electr-disrib] voltage intake-transformer
changer change,ging
— f — **carrocería** f - [autom] body change,ging
— f — **construcción** f - [constr] construction
change
— f — — f **de horno(s)** m - [combust] furnace
construction change
— f — **contabilización** f - [contab] accounting
change
— f — **contrato** m **social** - [legal] social, o
incorporation, contract change,ging
— f — **diseño** m - design change,ging
— f — — m **de horno(s)** m - [combust] furnace
design change,ging
— f — — m **para construcción** f - [constr] construction design change,ging
— f — — m — — f **de horno(s)** m - [combust]
furnace construction design change,ging
— f — **documento** m - document change,ging
— f — **enfriador** m - [ind] cooler change,ging
— f — **equipo** m - [ind] equipment change,ging
— f — **escritura** f **social** - [legal] social, o
incorporation, contract change,ging
— f — **especificación(es)** f - specification(s)
change,ging
— f — — f **para refractario(s)** m - [combust]
refractory,ries specification change,ging
— f — — f — — m **para precalentador** m -
[combust] preheater refractory,ries specification change,ging
— f — **estructura** f - structure, change,ging, o
modifying,fication
— f — — f **de motor** m - [ind] motor structure
change,ging
— f — — f — — m **para enfriador** m - [ind]
cooler motor structure change,ging
— f — **generador** m - [electr-prod] generator
change,ging
— f — — m **para emergencia** f - [electr-prod]
emergency generator change,ging
— f — **hoja** f - sheet change,ging
— f — **inducción** f - [electr-oper] induction
change,ging
— f — — f **magnética** - [electr-oper] magnetic
induction change,ging
— f — — f — **medida** - [electr-oper] measured
magnetic induction change,ging
— f — — f **medida** - [electr-oper] measured induction change,ging
— f — **instalación** f - [ind] installation
change,ging
— f — **laminador** m - [metal-lam] mill change
— f — — m **para banda(s)** f - [metal-lam] strip
mill change,ging
— f — — m — **cinta(s)** f - [metal-lam] strip
mill change,ging
— f — — m — **chapa(s)** f - [metal-lam] strip
mill change,ging
— f — — m — — f **en caliente** - [metal-lam]
hot strip mill change,ging
— f — — m — — f — **frío** - [metal]lam] cold
strip mill change,ging
— f — — m — **fleje** m - [metal-lam] strip mill
change,ging
— f — **licencia** f - [com] license, o permit,
change,ging
— f — — f **para exportación** f - [com] export
permit license
— f — — f — **importación** f - [com] import
permit change,ging
— f **en locomotora** f - [f.c.] locomotive, modification, o change
— f — — f **para apagamiento** m - [coque]
quench(ing) locomotive, change,ging, o modifying,fication

modificación f **en molino** m **para cemento** m -
[cement] cement mill change,ging
— f **en neumático** m - [autom-neumát] tire, modifying,fication, o change,ging
— f **en permiso** m **para exportación** f - [com] export permit change,ging
— f —— m — **importación** f - [com] import
permit change,ging
— f — **perno** m - [mec] bolt change,ging
— f —— m **para anclaje** m - [mec] anchor bolt
change,ging
— f —— m —— m **para precalentador** m -
[mec] preheater anchor bolt change,ging
— f —— m —— m — **separador** m - [mec] separator anchor bolt change,ging
— f — **política** f - [admin] policy, revision, o
change,ging
— f — **posición** f - position change,ging
— f — f **de electrodo** m - [electr-instal]
electrode repositioning
— f — **precio** m - [com] price change,ging
— f — **propuesta** f - [com] proposal, o bid,
change,ging
— f — **proyección** f - [ind] design, change,ging
— f — **proyecto** m - project change,ging
— f — **regulación** f - [mec] setting change,ging
— f —— f **de amperaje** m - [sold] current, o
amperage, setting, o adjustment, change,ging
— f — **revestimiento** m **para molino** m - [mec]
mill lining change,ging
— f — **sala** f - [ind] room change,ging
— f —— f **para motor** m - [ind] engine, o motor, room change,ging
— f — f —— m **diesel** - [ind] diesel engine room change,ging
— f — **subestación** f - [electr-prod] substation
change,ging
— f — **suspensión** f - [autom-mec] suspension,
modifying,fication, o change,ging
— f — **tabla** f - table change,ging
— f —— f **de materia(s)** - [public] content(s)
table change,ging
— f — **tipo** m - [imprent] type change,ging
— f —— m **de letra** f - [imprent] type, face,
o style, change,ging
— f — **toma** f - [ind] intake change,ging
— f —— f **de voltaje** m - [electr-oper] intake-transformer change,ging
— f —— f ——**transformador** m - [ind] intake-transformer change,ging
— f —— **para agua** m - [hidr] water intake
change,ging
— f — f —— m **de río** m - [hidr] river
water intake change,ging
— f — **trazado** m - [vial] realignment
— f — **valor** m - value change,ging
— f —— m **de contrato** m - [legal] contract
value change,ging
— f — **volante** m - [mec] flywheel change,ging
— f —— m **para separador** m - [mec] separator
flywheel change,ging
— f **estructural** - structure,ral, change,ging, o
modifying,fication
— f **fundamental** - fundamental, o basic, change
— f **hecha** - made, change, o modification
— f **lenta** - slow change,ging
— f **más reciente** - more recent, modificatión,
o change,ging • latest, modification,fying, o
change,ging
— f **mayor** - major, modification,fying, o change
— f **medida** - measure, modification, o change
— f **menor** - minor, modification,fying, o change
— f **necesaria** - necessary, o needed, modification,fying, o change
— f **procedente** - proper change,ging
— f **sugerida** - suggested change,ging
modificado,da a - modified; changed; altered •
[legal] revised
— a **constantemente** - changed constantly
— a **lentamente** - changed slowly
modificador,ra a **de estructura** f - structure
modifying
modificar v - . . . .; to, alter, o change • to,

revise, o amend • [gram] . . .; to qualify
modificar v **ajuste** m - to change @, setting, o
adjustment
— v — m **de amperaje** m - [sold] to, adjust, o
change, @, current, o amperage, setting
— v **automóvil** - [autom] to modify @, automobile,
o car
— v — m **con tracción** f **en cuatro rueda(s)** f
[autom] to modify @ four-wheel-drive car
— v **cambiador** m - [mec] to modify @ changer
— v — m **para toma** f - [electr-oper] to, modify,
o change, @ voltage intake changer
— v — m — f — **voltaje-transformador** -
[electr-oper] to change @ voltage intake-
-transformer changer
— v **carrocería** f - [autom] to modify @ body
— v **conexión** f - [electr-instal] to change @
connection
— v **constantemente** - to change constantly
— v **continuamente** - to change continually
— v **distancia** f - to change @ distance
— v **documento** m - [legal] to revise @ document
— v **eje** m - [mec] to, modify, o change, @ axle
— v — m **para transmisión** f - [autom-mec] to,
modify, o change, @ transaxle
— v **enfriador** m - [ind] to, modify, o change, @
cooler
— v **entrada** f - [mec] to change @ entrance •
[electrón] to change @ input
— v **entrehierro** m - [electr-mot] to change @
(air) gap
— v **equipo** m - [ind] to, modify, o change, @
equipment
— v — m **según norma** f - [ind] to modify @
standard equipment
— v — m **estándar** - [ind] to modify @ standard
equipment
— v **estructura** f - to modify @ structure
— v — f **de motor** m - [ind] to change @, motor,
o engine, structure
— v — f —— f **para enfriador** m - [ind] to,
modify, o change, @ cooler motor structure
— v **generador** m - [electr-prod] to, modify, o
change, @ generator
— v — m **para emergencia** f - [electr-prod] to,
modify, o change, @ emergency generator
— v **herramienta** f - [herram] to modify @ tool
— v **hoja** f - to change @ sheet
— v **inmueble** m - [constr] to alter @ real estate
— v **instalación** f - [ind] to, modify, o change,
o alter, @, installation, o facility
— v **largo(r)** m - to, modify, o change, @ length
— v — m **sobresaliente** - [sold] to change @
stickout
— v — m — **de electrodo** m - [sold] to change @,
electrode, o wire, stickout
— v **lentamente** - to, modify, o change, slowly
— v **molino** m—[ind] to, modify, o change, @ mill
— v — m **para cemento** m - [cement] to change @
cement mill
— v **norma** f - [admin] to, revise, o modify, @,
standard, o policy
— v **número(s)** m - to, modify, o change, @
number(s)
— v — m **de serie** - [ind] to change @ serial
number
— v **perno** m - [mec] to, modify, o alter, @ bolt
— v — m **para anclaje** m - [mec] to, modify, o
alter, @ anchor bolt
— m — m —— m **para precalentador** m - [mec]
to, modify, o change, @ preheater anchor bolt
— v — m —— m — **separador** m - to. change, o
modify, o alter, @ separator anchor bolt
— v **plancha** f - [mec] to, change, o alter, @
plate
— v — f **para asiento** m - [ind] to, change, o
alter, @ seat plate
— v — f —— m **para ventilador** m - [mec] to,
change, o alter, @ fan seat plate
— v — f — **cubierta** f - [mec] to, change, o al-
yrt, @ cover(ing) plate
— v **política** f - [admin] to, change, o alter, o
revise, @ policy

modificar <u>v</u> posición <u>f</u> - to change @ position
— <u>v</u> — <u>f</u> de electrodo <u>m</u> - [sold] to reposition @ electrode
— <u>v</u> producto <u>m</u> - [ind] to, modify, <u>o</u> change, <u>o</u> alter, @ product
— <u>v</u> proyección <u>f</u> - to modify @ design
— <u>v</u> proyecto <u>m</u>—to, modify, <u>o</u> change, @ project
— <u>v</u> receta <u>f</u> - [culin] to, modify, <u>o</u> change, @ recipe • [quím] to, modify, <u>o</u> change, @ pre-scription
— <u>v</u> regulación <u>f</u> - to change @ setting
— <u>v</u> — <u>f</u> de amperaje <u>m</u> - [sold] to change @ current setting
— <u>v</u> sala <u>f</u> - [ind] to change @ room
— — <u>f</u> para motor - [ind] to change @, motor, <u>o</u> engine, room
— <u>v</u> — <u>f</u> — — <u>m</u> diesel - [ind] to change @ diesel engine room
— <u>v</u> suspensión <u>f</u> - [autom-mec] to, modify, <u>o</u> alter, @ suspension
— <u>v</u> tabla <u>f</u> - to change, <u>o</u> alter, @ table
— <u>v</u> — <u>f</u> de materia(s) <u>f</u> - to, change, <u>o</u> alter, @ content(s) table
— <u>v</u> tipo <u>m</u> - to, modify, <u>o</u> change, @ type
— <u>v</u> — <u>m</u> de letra <u>f</u> - [imprent] to, modify, <u>o</u> alter, @ type face
— <u>v</u> toma <u>f</u> - [ind] to, change, <u>o</u> alter, @ in-take
— <u>v</u> — <u>f</u> de voltaje <u>m</u> - [electr-oper] to, mod-ify, <u>o</u> change, <u>o</u> alter, @ voltage intake
— <u>v</u> — <u>f</u> — — <u>m</u> transformador - [ind] to, mod-ify, <u>o</u> change, @ intake-transformer
— <u>v</u> trazado <u>m</u> - [vial] to realign • to relocate
— <u>v</u> volante <u>m</u> - [mec] to change @ flywheel
— <u>v</u> — <u>m</u> para separador <u>m</u> - [mec] to change @ separator flywheel
— <u>v</u> valor <u>m</u> - to, modify, <u>o</u> change, @ value
— <u>v</u> — <u>m</u> de contrato <u>m</u> - [legal] to, modify, <u>o</u> change, <u>o</u> alter, @ contract value
modillón <u>m</u> - [arquit] . . .; offset
modo <u>m</u> - . . .; fashion
— <u>m</u> de control <u>m</u> - [comput] control mode
— <u>m</u> — coordenado - [comput] coordinated control mode
— <u>m</u> distinto - different way | <u>adv</u> - differ-ently
— <u>m</u> general - general manner
— <u>m</u> uniforme - uniform manner
modulación <u>f</u> de amplitud <u>f</u> - [electrón] ampli-tude modulation
— <u>f</u> — frecuencia <u>f</u> - [electrón] frequency mod-ulation
modulador,ra <u>a</u> - . . .; modulatory
modular <u>a</u> - . . .; module basis • [mec] reusable
módulo <u>n</u> - . . . → mode
——almacén <u>m</u> - [nucl] storage module
— <u>m</u> alto - high module,lus
— <u>m</u> — para ruptura <u>f</u> - high breaking modulus
— <u>m</u> con altura de . . . • [mec] . . . net height module,lus
— <u>m</u> — prisionero(s) <u>m</u> - [mec] studded module
— <u>m</u> — Young - Young's modulus
— <u>m</u> determinado - determined modulus
— <u>m</u> elástico - véase módulo <u>m</u> de elasticidad <u>f</u>
— <u>m</u> individual - individual module,lus
— <u>m</u> — para cilindro <u>m</u> - [petról] individual cylinder module
— <u>m</u> manual - [comput] manual, mode, <u>o</u> modulus
— <u>m</u> para alivio <u>m</u> - relief module,ulus
— <u>m</u> — para traspaso <u>m</u> - cross-over relief module,ulus
— <u>m</u> — almacenamiento <u>m</u> - [nucl] storage module
— <u>m</u> — aspiración <u>f</u> - [mec] suction module
— <u>m</u> — <u>f</u> con perno(s) <u>m</u> prisionero(s)—[mec] studded section module
— <u>m</u> — — <u>f</u> con prisionero(s) <u>m</u> - [mec] studded suction module
— <u>m</u> — cilindro <u>m</u> - [petról] cylinder module
— <u>m</u> — — <u>m</u> individual - [petról] individual cylinder module
— <u>m</u> — compresión <u>f</u> - compression modulus
— <u>m</u> — — <u>f</u> contenida - [constr] confined com-pression modulus

módulo <u>m</u> para compresión <u>f</u> encerrada - [constr] enclosed, <u>o</u> confined, compression modulus
— <u>m</u> — — <u>f</u> restringida - [constr] confined, <u>o</u> restricted, compression modulus
— <u>m</u> — corte <u>m</u> transversal - (cross) section modulus
— <u>m</u> — descarga <u>f</u> - [mec] discharge module,ulus
— <u>m</u> — <u>f</u> con perno(s) <u>m</u> prisionero(s) <u>m</u> - [mec] studded discharge module,lus
— <u>m</u> — — <u>f</u> con prisionero(s) <u>m</u> - [mec] studded discharge module,lus
— <u>m</u> — elasticidad <u>f</u> - elasticity modulus
— <u>m</u> — — <u>f</u> de suelo <u>m</u> - [suelos] soil elasti-city modulus
— <u>m</u> — — <u>f</u> para aluminio <u>m</u> - alumina elasticity modulus
— <u>m</u> — — <u>f</u> — pared <u>f</u> - [tub] wall elasticity modulus
— <u>m</u> — — <u>f</u> — — <u>f</u> de tubería <u>f</u> - [tub] pipe wall elasticity modulus
— <u>m</u> — <u>f</u> para acero <u>m</u> - steel elasticity mod-ulus
— <u>m</u> — gobierno <u>m</u> - control module,lus
— <u>m</u> — reacción <u>f</u> - reaction modulus
— <u>m</u> — — <u>f</u> de suelo <u>m</u> - [suelos] soil reaction modulus
— <u>m</u> — regulación <u>f</u> - control module,lus
— <u>m</u> — regulador <u>m</u> - control(ler) module,lus
— <u>m</u> — — <u>m</u> para función <u>f</u> única - [comput] sin-gle function control(ler) module,lus
— <u>m</u> — ruptura <u>f</u> - breaking modulus
— <u>m</u> — sección <u>f</u> - section modulus
— <u>m</u> — secante - [geom] secant modulus
— <u>m</u> — suelo <u>m</u> - [suelos] soil modulus
— <u>m</u> — — <u>m</u> determinado - [suelos] determined soil modulus
— <u>m</u> — traspaso <u>m</u> - cross-over module,lus
modus <u>m</u> operandi - modus operandi
— vivendi - modus vivendi; way of living
moho <u>m</u> - . . .; mold • [metal] rust
— <u>m</u> blanco - white mold
mojado,da <u>a</u> - wet; damp
mojadura <u>f</u> de ropa <u>f</u> - clothing, wetting, <u>o</u> soak-ing
mojar <u>v</u> - . . .; to soak
— <u>v</u> ropa <u>f</u> - to soak @ clothing
moldaje <u>m</u> - [metal-prod] véase pieza <u>f</u> fundida
molde <u>m</u> - . . .; form • [mec] template; model; matrix; frame • [metal-prod] mold; cast • [domést] pan; tin
— <u>m</u> alineado - [constr] aligned form
— <u>m</u> aluminotérmico - [sold] thermit mold
— <u>m</u> asfáltico - [nucl?] asphalt mold
— <u>m</u> (de) botella <u>f</u> - [metal-prod] bottle top (ingot) mold
— <u>m</u> con extremo <u>m</u> estampado - [puentes] swaged end form
— <u>m</u> — — <u>m</u> — para cubierta <u>f</u> - [puentes] swaged end deck form
— <u>m</u> — — — <u>f</u> para puente <u>m</u> - [puentes] swaged end bridge deck form
— <u>m</u> — mazarota <u>f</u> - [metal-prod] hot top (ingot) mold
— <u>m</u> con tope <u>m</u> abierto - [metal-prod] open top (ingot) mold
— <u>m</u> conformado - [constr] fitted, <u>o</u> shaped, form
— <u>m</u> corrugado - [metal-prod] corrugated mold
— <u>m</u> para lingote(s) <u>m</u> - [metal-prod] corru-gataed (ingot) mold
— <u>m</u> de acero <u>m</u> - [constr] steel, mold, <u>o</u> form
— <u>m</u> — — <u>m</u> para piso <u>m</u> - steel, floor, <u>o</u> deck, form
— <u>m</u> — — — <u>m</u> para puente <u>m</u> - [constr] steel bridge, floor, <u>o</u> deck, form
— <u>m</u> — administración <u>f</u> - [admin] management pattern
— — <u>f</u> para trabajo <u>m</u> - [admin] work mana-gement pattern
— <u>m</u> — arena <u>f</u> - sand mold
— <u>m</u> — — <u>f</u> mojada - [metal-prod] wet sand mold
— <u>m</u> — — <u>f</u> seca - [metal-prod] dry sand mold
— <u>m</u> — asfalto <u>m</u> - [mec] asphalt mold
— <u>m</u> — cobre <u>m</u> - copper mold

molde m de cobre m de tipo m ajustable - adjustable type copper mold
— m — tipo m ajustable - [metal-prod] adjustable type mold
— m — m oscilante - [col.cont] oscillating
type mold
— m desarrollado - developed pattern
— m deslizable - [constr] sliding form
— m fijado - [constr] fastened form
— m impermeable - [constr] tight form
— m — contra pérdida(s) f de enlechado m -
[constr] grout tight form
— m liso - [metal-prod] smooth, o plane, mold
— m metálico - [metal-prod] metal, mold, o form
— m — procesado - [metal-prod] processed metal
form
— m para asfalto m - [mec] asphalt mold
— m — crepes m suzettes - [culin] crepes pan
— m — hormigón m - [constr] concrete form
— m — limpieza f - [metal-prod] cleaning mold
— m — f para lingotera(s) f - [metal-prod]
ingot mold cleaning pulpit
— m — lingote(s) m - [metal-prod] ingot mold
— m — lingoteadora f - [metal-prod] pig casting machine mold
— m — máquina f lingoteadora - [metal-prod]
pig casting machine, mold, o chill
— m para matrizar - [mec] stamping die
— m — neumático(s) m—[autom-neumát] tire mold
— m — piso m - [constr] deck form
— m — m para puente(s) m - [constr] bridge
deck form
— m — producción f - [ind] production mold
— m — puente(s) m - [constr] bridge form
— m permanente - [constr] permanent, o stay-in-
-place, mold, o form
— m — para piso m - [constr] permanent deck
form
— m — — m para puente m - [constr] permanent bridge deck form
— m — de acero m - [constr] permanent steel
form
— m — m para calzada f - [constr] permanent steel deck form
— m — m — f para puente(s) m -
[constr] oermanent steel bridge deck form
— m — m — piso m - [constr] permanent
steel deck form
— m — m — m para puente(s) m -
[constr] permanent steel bridge deck form
— m recto - [col.cont] straight mold
— m — con nivel m alto - [col.cont] high level
straight mold
— m — — m bajo - [col.cont] low level
straight mold
— m reforzado - [ind] reinforced, o heavy duty,
mold
— m, removido, o sacado - [constr] removed form
— m seco - dry mold • [constr] dry form
— m según norma f - standard mold
— m termita* - [sold] thermit mold
moldeado,da a - molded • [metal-prod] cast
— a en sitio m - [constr-pil] cast-in-place
— a por inyección f - [metal-fabr] extruding;
extrusion
moldeador m - molder
moldeadora f—[metal-prod] (pig) casting machine
— f automática - [metal-prod] automatic casting
machine
— f — de lingote(s) m - [metal-prod] pig casting machine
moldeamiento* m - véase moldeado m
moldear v nervadura f - [metal-lam] to cast @
rib
— v plástico m - [plást] to mold @ plastic
— v por inyección f - [metal-fabr] to extrude
moldeo m - molding • branding
— m por inyección f - [mec] extruding; extrusion; injection molding
— m — f en . . . color(es) m - [plást]
. . .-color injection molding

moldeo m de plástico(s) m - [plást] plastic('s)
molding
moldería f - [metal-prod] foundry • molding(s);
casting(s) • foundry pig iron • pouring
moldero* m - [metal-prod] molder
moldura f - [metal-prod] . . .; moulding; stop •
[constr] spline • [estruct] cove • [hormig]
reglet • [arquit] bead • [tub] shoulder •
[carp] molding • [autom] trim
— f de caucho m - [constr] rubber base
— f — — m para zócalo - [constr] rubber cove
base
— f — tipo m con presión f - [carp] snap-on
molding
— t integral - [constr] integral stop
— f para vidrio(s) m - [carp] glazing molding
— f — — m de tipo m con presión f - [carp]
snap-on glazing molding
— f — zócalo m - [constr] cove • wall base
— f saliente - [mec] ledge
moldurar v - . . .; to mold
molécula f de carbono m - [quím] carbon molecule
— f — hidrocoloide* m - [quím] hydrocolloid
molecule
— f — hidrógeno m - [quím] hydrogen molecule
— f — óxido m - [quím] oxide molecule
— f — m de carbono m - [quím] carbon oxide
molecule
moledora f - [ind] mill
— f con barra(s) f - [miner] bar mill
— f — bola(s) f - [miner] ball mill
moledura f - [mec] . . .; milling
moler v - . . . • to crush
— v cemento m - [cement] to grind @ cement
— v crudo m - [cement] to grinde @ raw material
— v en seco - [ind] to dry grind
— v fibra f - [mec] to grind @ fiber
— v — f de vidrio m - [cerám] to, grind, o
mill, @ fiberlass
— v finamente - to grind finely
— v gruesamente - to grind coarsely
— v materia f prima - [ind] to grind @ raw material
molestado,da a - bothered; disturbed • annoyed;
irritated; nagged
molestar v - . . .; to molest • to, irritate, o
nag • to interfere
molestarse v - to be(come) annoyed
molestia f - . . .; bothering; nuisance; annoyance; irritation; nagging
— f para motorista m - [vial] motorist incon-
veience
molesto,ta a - . . .; annoyed • nagging; irritating • pesky
moleteado,da a - [mec] knurled
moleteadura f - [mec] knurling
moletear v - [mec] to knurl
molibisulfuro* m - [quím[molydisulphide
molido,da a - [mec] ground; milled; crushed
— a en seco - dry, ground, o milled, o crushed
— a fino,na - ground fine(ly)
— a grueso,sa - ground coarse(ly)
molienda f - [mec] . . .; crushing
— f autógena - [miner] autogenous grinding
— f de carbón m [miner] coal crushing
— f — cemento m - [cement] cement grinding
— f — crudo m - [cement] raw (material) grinding
— f — fibra f - [mec] fiber(s) grinding
— f — — f de vidrio m - [cerám] fiberglass,
grinding, o milling
— f — materia f prima - [ind] raw material(s)
grinding
— f — mineral - [miner] ore grinding
— f — pigmento(s) m - [pint] pigment(s) grinding
— f — — m para pintura f - [pint] paint pigment grinding
— f en seco - dry, milling, o grinding
— f — — de cemento m - [cement] cement dry
grinding
— f fina - fine, milling, o grinding
— f gruesa - coarse, milling, o grinding

molienda f húmeda - [ind] wet grinding
molinete m - . . . . • [mec] gin • crab • reel •
pinwheel `winch • vane; whirl
— m eléctrico - [ind] electric, vane, o whirl
molino m - [mec] . . .; grinding equipment •
[metal-lam] véase laminador m - [metal-prod]
(mud gun) mixer
— m Aerofall - [miner] Aerofall mill
— m autógeno - [miner] autogenous mill
— m con barra(s) f - [miner] rod mill
— m — bola(s) f - [miner] ball mill
— m con guijarro(s) m - [miner] pebble mill
— m con martillo(s) m - [coque] hammer mill
— m — — m giratorio(s) - [miner] swinging, o
rotating, hammer mill
— m con mazo(s) m - [miner] stamp(ing) mill
— m — piedra(s) f - [ind] stone mill • [miner]
pebble mill
— m — — f para harina f - [ind] stone burr
mill
— m — pisón(es) m - [miner] stamp(ing) mill
— m — viento m - [ind] windmill
— m en seco - [ind] dry pan, mill, o grinder
— m harinero - [ind] flour, o burr, mill
— m húmedo - [miner] wet, mill, o grinding
equipment
— m nuevo - [ind] new mill
— m para arcilla f - [ind] pug, o clay, mill
— m — carbón m - [coke] coal (grinding) mill
— m — cemento m - [cerám] cement mill
— m — — m modificado - [cement] modified, o
changed, cement mill
— m — crudo m - [cerám] raw (materials) mill
— m — forraje(s) m - [agric-equip] roughage, o
feed, mill
— m — harina f - [ind] flour, o burr, mill
— m — mineral - [miner] ore mill • stamp(ing)
mill
— m — recuperación f - [ind] recuperating,tion
mill; recovery,ring mill
— m — trituración f - [mec] grinding mill
— m — viento m - [ind] windmill
— m proyectado - [mec] designed mill
— m triturador - [ind] grinding mill
molturación f - [ind] grinding
momentáneo,nea a - . . . .; transient
momento m - . . . .; time being • [mec] momentum;
impetus
— m ansioso - anxious moment
— m anulador - cancelling moment
— m cero - [fís] zero moment
— m dado - [cronol] given, moment, o time point
— m de apuro m - straight; tough situation
— m — compra m - [com] purchase, time, o date
— m — flexión f - [metal] flexing, o bending,
moment(um)
— m — — f resistido - resisted bending moment
— m — giro m - [mec] turning moment
— m — inercia f - [mec] inertia moment(Um)
— m — — f en corte m transversal - [mec]
section modulus
— m — inercia f - [mec] inertia moment(um)
— m — — f de pared f - [tub] wall inertia
moment(um)
— m — — f — — f de tubo,bería - pipe wall
inertia moment(um)
— m — pared f - [tub] wall moment(um)
— m — recibir(se) - time of receipt | adv -
(en) upon receipt
— m — resistencia f - resistance,ting momentum
— m — siniestro m - [seguros] time of, event,
o damage
— m — torsión f - [mec] torque (momentum)
— m — — f con carga f - [mec] (nominal) torque (momentum) with @ load
— m — — m de motor m - [mec] motor, o engine,
torque
— m — — f — — m en operación f - [mec]
motor, operating, o running, torque
— m — — f en operación f - [mec] running
torque
— m — — f inicial - [mec] starting torque
— m — — m nominal - [mec] rated, o nominal,

torque
momento m de torsión f nominal de motor m -
[mec] motor, rated, o nominal, torque
— m — f — — m en operación f - [mec]
running, o operating, motor, nominal, o
rated, torque
— m — — f — en operación f - [mec] rated
running, o running rated, torque
— m — — f variable - [mec] variable torque
— m — volante m - [mec] flywheel momentum
— m — — m para turbogenerador m - [electr-
-prod] turbogenerator flywheel momentum
— m — — m generador m - [electr-prod]
generator flywheel momentum
— m debido a excentricidad f - [mec] eccen-
tricity momentum
— m determinado - [cronol] given, moment, o
time point
— m económico - [econ] economic situation
— n en que - at which point
— m eufórico - high
— m flexor m - [mec] bending momentum
— m impuesto m - [mec] imposed momentum
— m indebido - wrong time
— m inolvidable - unforgettable moment • bright
spot
— m magnético - [electr] magnetic moment(um)
— m máximo - maximum moment(um)
— m — de carga f de viento m - [constr] maximum wind load moment(um)
— m — para torsión f - [mec] breakdown torque
— m — para vuelco m - maximum tipping moment
— m — — — m sobre suelo m - [electr-instal]
maximum to ground tipping momentum
— m mínimo - minimum moment • minimum momentum
— m — de inercia f - [constr] minimum inertia
moment(um)
— m oportuno - opportune time • right, time, o
instant
— m para carga f - load(ing), moment, o time
— m — — f de viento m - [constr] wind load
moment(um)
— m — colar v - [metal-prod] tapping time
— m — inspección f - inspection time
— m — puesta f en, funcionamiento m, o marcha
f - [ind] start(ing) up time
— m — recepción f - acceptance moment
— m — registro,tración - time of registration
— m — substitución f - substition time
— m político - [pol] political oportunity
— m positivo - [fis] positive moment(um)
— m — de flexión f - positive bending momentum
— m — máximo - maximum positive moment(um)
— m resistente - [constr] zero moment strength
— m — cero de pared f - [contr] zero wall moment strength
— m — — — — f de tubería f - [constr] zero
pipe wall moment strength
— m — de pared f - [constr] wall moment(um)
strength
— m — — — f de tubería f - [constr] pipe
wall moment strength
— m — suficiente - [constr] sufficient, o adequate, moment strength
— m torsional - [mec] (tightening) torque;
véase también par m motor
— m — alto - [mec] high torque
— m — autoalineador - [autom] self-aligning
torque
— m — — negativo - [autom-neumát] negative
self-aligning torque
— m — — positivo - [autom-neumát] positive
self-aligning torque
— m — bajo - [mec] low torque
— m — distribuido - [mec] distributed torque
— m — hidráulico - [mec] hydraulic torque
— m — para sujetador m-[mec] fastener torque
— m — transmitido - [mec] transmitted torque
— m torsor m - [mec] torque; véase también momento torsional
moneda f con convertibilidad f libre - [fin]
free convertibility currency
— f convertible libremente - [fin] freely con-

vertible currency
moneda f de curso m legal - [fin] legal tender
— f — **Estados m Unidos** - [fin] United States'
currency
— f — **origen** - [fin] original currency
— f — **pago m** - [fin] payment currency
— f — **país m** - [fin] country's currency
— f — **m de origen** - [fin] country of origin
currency
— f — **República f Dominicana** - [fin] Dominican
(Republic) currency
— f — **póliza f** - [seguros] policy currency
— f **desvalorizada f** - [fin] devalued currency
— f **devaluada*** - [fin] devalued currency
— f **dominicana f** - [fin] Dominican (Republic)
currency
— f **estadounidense** - [fin] United States currency
— f **extranjera** - [fin] foreign, currency, o exchange
— f — **convertida f** - [fin] converted, o translated, foreign currency
— f **fijada** - [fin] established currency
— f — **por póliza f** - [seguros] policy (established) currency
— f **inconvertible** - [fin] inconvertible, o non-
-convertible, currency
— f **local** - [fin] local currency
— f —, **devaluada, o desvalorizada** - [fin] devalued local currency
— f — **legal** - [fin] legal local currency
— f **mejicana** f - [fin] Mexican currency
— f **mexicana f** - Mexican currency
— f **nacional** - [fin] national, o local, currency
— f **original** - [fin] original currency
— f **para cambio m** - [fin] exchange currency
— f — **pago m** - [fin] payment currency
— f **propia f** - [fin] own currency
— f **según póliza f** - [seguros] policy currency
— f, **usada, o utilizada** - [fin] used currency
monetario,ria a - . . . . • dollar
— a **mundial** - [econ] world currency
monitor m con circuito m cerrado - [electrón]
closed circuit monitor
monitor m para corriente f - [electron] current,
monitor, o sensor
— m — **falla(s) f** - [electrón] failure monitor
— m — **granulometría f** - [miner] particle size
monitor
— m — **partícula(s) f** - [electrón] particle(s)
monitor
— m — **segadora f trilladora** - [agric-equip]
combine monitor
— m — **seguridad f** - safety monitor • [combust]
flame monitor
— m — **tamaño m** - [electrón] size monitor
— m — m **para partícula(s) f** - [electrón]
particle(s) size monitor; P S M
— m — **televisión f** - [electrón] television
monitor
mono m - . . . . • [vest] coveralls
— m **enterizo** - [vest] coveralls
monocarril m - [mec] monorail • [metal-prod]
monorail (hoist)
— m **aéreo** - [ind] air, o overhead, monorail
— m **para horno m** - [ind] furnace monorail
(hoist)
monocilíndrico,ca a - [comb.int] single cylinder
monoclinal a - [geol] . . . .; monoclinal,ne
monocrotofos* m - [quím] monocrotophos
monodisco a - [mec] single, disk, o plate
monoetanolamina* f - [quím] monoethanolamine
monoetilamina* f - [quím] monoethylamine; amine
monoethyl
monoetilo m - [quím] monoethyl
monofásico,ca a - [electr-prod] single phase(d)
monograma m individual - [ind] individual monogram
— m **particular** - [ind] individual, o specific,
monogram
monomérico,ca a - [quím] monomeric
monómero m - [quím] monomer

mononitrotolueno m - [quím] mononitrotoluene
monopila a - [metal-trat] single stack
monopolio m - [econ] . . . • cartel
monopolización f - monopolizing,zation; dominating,tion
— f **de conversación f** - conversation, monopo-
lizating,tion, o dominating,tion
monopolizar v conversación f - to, monopolize,
o dominate, @ conversation
monorriel m aéreo - [ind] air, o overhead, monorail
monstruosidad f - . . . . • freak
montacarga(s) m - [ind] . . .; freight elevator;
hoisting engine • [metal-prod] skip (bridge)
• [constr] lift
— m **con cadena f** - [ind] chain hoist
— m — **cajón n** - [ind] (box) hoist
— m — **engranaje m** - [ind] gear elevator
— m — m **helicoidal** - [ind] helical gear
elevator
— m — m **sin fin** - [ind] helical gear elevator
— m — **aguilón m** - [grúas] boom hoist
— m — **horno m** - [ind] furnace skip
— m — m **alto** - [metal-prod] blast furnace
skip
— m — **obra f** - [constr] construction, lift, o
hoist
— m — **suministro m** - [constr] materials hoist
montaceniza(s) f - [ind] ash(es) hoist
montado,da a - mounted; attached (to); built in;
assembled; in place • [mec] erected; set up;
installed; put together • inserted • popped
in • based • [petról] rigged up
— a **a horcajada(s) f** straddle mounted; strad-
dled
— a **apropiadamente** - properly mounted
— a **con pasador m** - [mec] pin connected
— a **correctamente** - mounted, correctly, o
appropriately, o properly
— a **dentro de bomba f** - [bombas] mounted in(to)
@ pump
— a — **volante m** - [mec] mounted, in(to), o
within, @ flywheel
— a **detrás** - rear mounted; mounted behind
— a **directamente** - [mec] mounted directly
— a — **sobre base f** - [ind] mounted directly
to @ base
— a — — **f de soldadora f** - [sold] mounted
directly to @ welder base
— a — — **soldadora f** - [sold] mounted directly
to @ welder
— a **en bronce** - [válv] bronze mounted
— a — **cabina f** - [ind] cab mounted
— a **contraposición f** - mounted in opposition
— a — **costado m** - side mounted
— a — **emplazamiento m** - [ind] project site,
mounted, o erected
— a **extremo m** - end mounted
— a — **forma f ajustable** - [mec] adjustably
mounted
— a — — **f regulable** — [mec] adjustably mounted
— a — **obra f** - [constr] field mounted • locally mouned • assembled at @ (project) site
— a — **par(es) m** - [mec] mounted in pair(s)
— a — **posición f** in position mounted
— a — — **f horizontal** - mounted horizontally
— a — — **vertical** - mounted vertically
— a — **serie f** - series mounted
— a — **sitio m** - [constr] on site mounted;
assembled at @ site
— a — **tope m** - [mec] top mounted
— a — **tornillo m para banco m** - [mec] vise
mounted
— a — **voladizo** - [mec] cantilevered; offset
mounted; blocked out
— a **horizontalmente** - mounted horizontally
— a **incorrectamente** - incorrectly, o improperly, mounted
— a **nuevamente** - reassembled
— a — **en plantilla f** - [mec] reassembled on @
templet
— a — **sobre plantilla f** - [mec] reassembled on
@ templet

@ templet

montado,da para telemando m - [mec] remote
mounted
— a **paralelamente** - mounted parallely
— a **permanentemente** - [mec] mounted permanently
— a **rígidamente** - [mec] mounted rigidly
— a **según medida** - custom mounted
— a **separadamente** - [ind] separately mounted
— a **simplemente** - [mec] simply mounted
— a **sobre** - mounted on • assembled, o affixed,
to
— a — **almohadilla** f - [mec] cushion, o pad,
mounted
— a — — f **de caucho** m - [mec] rubber, cushion, o pad, mounted
— a — **anaquel** m - [electrón] rack mounted
— a — **bandeja(s)** - [mec] pelletized
— a — **base** f - [mec] pad mounted
— a — — f **rodante** - [mec] undercarriage
mounted • [transp] trailer mounted
— a — **bastidor** m - [mec] frame, o pad. o rack,
mounted
— a — **bisagra(s)** f - [mec] hinged; hinge
mounted
— a — **caja** f - [mec] housing mounted
— a — — f **para eje** m - [autom-mec] mounted to
@ axle housing
— a — — f — **portadiferencial** m - [autom-mec]
assembled to @ differential carrier case
— a — **camión** m (automóvil) - [transp] truck
mounted
— a — **camisa** f - [mec] liner mounted
— a — **carrito** m - [mec] carriage mounted
— a — **cárter** m - [comb.int] crankcase mounted
— a — **cojinete(s)** m - [mex] bearing mounted
— a — — m **con rodillo(s)** m - [mec] roller
bearing mounted
— a — — m — — m **refrentados con estelita** f -
[mec] stellite faced roller bearing mounted
— a — **conmutador** m - [mec] switch mounted
— s — — m **para selección** f - [electr-instal]
selector switch mounted
— a — **cubierta** f - [mec] cover mounted
— a — — f **para distribuidor (para fuerza** f) -
[autom-mec] mounted to @ (power) divider
— a — **eje** m - [mec] axle mounted
— a — **ficha** f - [com] card mounted; carded
— a — — f **de cartón** m - [com] carded
— a — **fondo** m - [mec] bottom mounted
— a — **frente** f - [mec] front mounted
— a — **mano** f **derecha** - [mec] right hand mounted
— a — — f **izquierda** - [mec] left hand mounted
— a — **muelle(s)** f - [mec] spring mounted •
spring loaded
— a — **oruga(s)** f - [mec] crawler mounted
— a — **pared** f - [mec] wall mounted
— a — **parte anterior** - [mec] front mounted
— a — — f **de adelante** - [mec] front
(end) mounted
— a — — f **atrás** - [mec] rear (end) mounted
— a — — f **delantera** - [mec] front end mounted
— a — — f **inferior** - [mec] bottom mounted
— a — — f **posterior** - [mec] rear mounted
— a — — f **superior** - [mec] top mounted
— a — **patín(es)** m - [mec] skid, o pallet,
mounted
— a — **piso** m - [mec] floor mounted
— a — **placa** f - [mec] plate mounted
— a — — f **para asiento** m - [mec] bedplate
mounted
— a — **plantilla** f - [mec] templet mounted
— a — **portadiferencial** m - [autom-mec] differential carrier mounted
— a — **respaldo** m - [mec] rear, o back, mounted
— a — **riel(es)** m - [mec] rail mounted
— a — **rueda(s)** f - [mec] wheel mounted • [ind]
dolly mounted
— a — **soporte** m - [mec] support, o holder,
mounted
— a — **subestructura** f - [mec] substructure
mounted
— a — **suelo** m - [mec] floor mounted
— a — **tablero** m - [mec] panel mounted

montado,da sobre tableto m **para regulación** f -
[ind] control panel mounted
— a — **tapa** f - [mec] cover mounted
— a **temporariamente** - [mec] mounted temporarily
— a **verticalmente** - mounted vertically
montador m - . . .; fitter; erectpr; rigger; as-
sembler; set-up man • [ind] millwright
[constr] steel worker
— m **de fabricante** m - manufacturer's erector
— m — — m **en obra** f - manufacturer's field
erector
— m **en obra** f - [ind] field erector
— m **mecánico** - [mec] millwright
— m **para tubería(s)** f - [tub] pipefitter
— m **soldador** - [tub] pipefitter weldor
— m — **para tubería(s)** f - [tub] pipefitting
weldor
montaje m - . . .; erecting,tion; set(ting) up;
rigging; fitting; installing,lation • staging • insertion • popping on • attachment •
pasting • [trefil] casing • [fotogr] montage
• [mec] véase también **soporte** m
— m **a horcajadas** - straddle mounting
— m **apareado** - [mec] true mounting
— m — **sobre derecha** f **e izquierda** f - [mec]
true left and right mounting
— m **apropiado** - [mec] proper, mounting, o assembling,ly
— m **cambiado** - [mec] changed mounting
— m **certificado** - certified, mounting, o erecting,tion
— m **completado** - [ind] completed erection
— m **con pasador** m - [mec] pin connection
— m — **perno(s)** m - [mec] bolted assembly,ling
— m — — m **con resistencia** f **alta** - [constr]
assembly with high-strength bolts
— m — — m **sobre brida** f - [mec] bolt(ed)
flange(d) mounting
— m **contra vibración(es)** f - [mec] vibration
mount(ing)
— m **correcto** - [constr] correct, o proper, assembly,ling, o mounting
— m **de acero** m **estructural** - [constr] structural steel mounting
— m — **adaptador** m - [mec] adapter mounting
— m — **aguilón** m - [grúas] boom, mounting, o
coonecting,tion
— m — **alternador** m - [electr-instal] alternator mounting
— m — **apoyo** m - [grúas] bearing mounting
— m — — m **para rotación** f - [grúas] swing
bearing mounting
— m — **árbol** m - [mec] shaft mounting
— m — — m **para salida** f **(de fuerza** f) -
[autom-mec] output shaft mounting
— m — **aro** m - [mec] ring, o circle, mounting
— m — — m **con rodillo(s)** m - [grúas] roller
circle mounting
— m — — m **con rodillo(s)** m **activo(s)** - [grúas]
live roller circle mounting
— m — — m **para rodillo** m **motor** - [grúas] live
roller circle mounting
— m — **arrancador** m - [comb.int] starter mounting
— m — — m **con arrollamiento** m **automático** -
[comb.int] recoil starter mounting
— m — **ataguía** f - [hidr] cofferdam erection
— m — **barra** f - [mec] rod assembly
— m — — f **taponadora** f - [metal-prod] stopper
rod, mounting, o assembling,bly
— m — **base** f - [mec] base erection
— m — **bloque** m - [mec] block mounting
— m — — m **para contacto** m - [electr-instal]
contact block mounting
— m — — m — **enclavamiento** m - [electr-instal]
interlock(ing) block mounting
— m — **bobina** f - [electr-mot] coil mounting
— m — **bóveda** f - [constr] arch mounting
— m — **cabeza** f - [mec] head mounting
— m — **caja** f - [mec] case, o housing, mounting
— m — — f **para engranaje(s)** m - [mec] gear
box mounting
— m — **canal** m **para sangría** f - [metal-prod]

(tapping) spout installation
montaje m de carburador m - [comb.int] carburetor mounting
— m — carrillo m - [ind] trolley, mounting,
o attaching,chment
— m — casco m - [mec] shell mounting
— m — casquete m - [mec] cap, o cover,
mounting
— m — m para extremo m - [mec] end cover
mounting
— m — cauce m - [hidr] channel erecting,tion
— m — cojinete m - [mec] bearing mounting
— m — m para rotación f - [grás] swing
bearing mounting
— m — compartimiento m - [mec] compartment
mounting
— m — m de caja f - [mec] case compart-
tment mount(ing)
— m — m — m para soldadora f - [sold]
welder case compartment mount(ing)
— m — compresor m—[ind] compressor mounting
— m — compuerta f - [hidr] (sluice) gate
erection
— m — m f para toma f - [hidr] intake
(sluice) gate erection
— m — m f para vertedero m - [hidr] spillway gate erection
— m — condensador m - [comb.int] condenser
mounting
— m — conducto m - [electr-instal] conduit
mounting
— m — m para cable(s) m - [electr-instal]
(cables) conduit mounting
— m — conjunto m - [mec] assembly mounting
— m — conmutador m - [electr-instal] switch
mounting
— m — m para selección f - [electr-equip]
selector switch mounting
— m — contacto m - [electr-instal] contact
mounting
— m — coraza f - [ind] shell erection
— m — m f en obra f - [ind] shell field
erection
— m — m f — taller m -[ind] shell shop
erection
— m — cubierta f - [mec] roof, o cover, o
shroud, mounting
— m — cubresoporte m - [mec] bracket cover
mount(ing)
— m — chasis m - [mec] chassis mounting
— m — deflector m - [mec] baffle mounting
— m — depósito m - [comb.int] tank mounting
— m — m para gasolina f - [comb.int] gasoline tank mounting
— m — m sobre ménsula f - [comb.int] tank
to bracket mounting
— m — depurador m (para aire m) - [comb.int]
(air) cleaner mounting
— m — dispositivo m - [mec] device mounting
• [sold] fixture mounting
— m — disyuntor m - [electr-instal] breaker
mounting
— m — eje m - [mec] axle mounting
— m — enclavamiento m - [electr-instal] in-
terlock(ing) mounting
— m — equipo m - [ind] equipment, mounting,
o erecting,tion
— m — m eléctrico - [electr-instal] electrical equipment, mounting, o erecting,tion
— m — m — para laminador m - [metal-lam]
mill electrical equipment mounting
— m — m — m — m para palanquilla f -
[metal-lam] billet mill electrical equipment, mounting, o erecting,tion
— m — m para perforación f - [petról]
rig(ging) up
— m — estabilizador m - [electr-instal] stabilizer mounting
— m — fiador m - [mec] catch, o retainer,
mounting
— m — freno m - [autom-mec] brake mounting
— m — gatillo m - [sold] trigger mounting

montaje m de herramienta f - [herram] tool
mounting - tool insertion
— m — implemento m - [agric-equip] implement
mounting
— m — indicador m - [vial] sign mounting
— m — interruptor m - [electr-instal] switch
mounting
— m — m magnético - [electr-instal] magnetic switch mounting
— m — línea f - line, erection, o mounting
— m — f de cizalla(s) f - [metal-lam] shear
line erection
— m — f de tijeras f - [metal-lam] shear
line erection
— m — magneto m - [comb.int] magneto mounting
— m — malacate m - [mec] winch mount(ing) •
[petról] drawworks mounting
— m — m hidráulico - [grúas] hydraulic,
hoist, o crane, mounting
— m — m — m sobre bastidor m - [grúas] hydraulic, hoist, o winch, frame mounting
— m — m — m sobre bastidor m - [grúas] winch, o
hoist, frame mounting
— m — manivela f - [mec] handle mounting
— m — manómetro m - [instrum] gage mounting
— m — m para aire m - [sold] air pressure
gage assembly
— m — material m - [ind] material(s) mounting
— m — m tubular - [tub] tubular, o pipe-
-type material(s) mounting
— m — mazarota f - [metal-prod] hot topping
— m — ménsula f - [mec] bracket mounting
— m — f sobre culata f - [comb.int] bracket
top head mounting
— m — motor m - [electr-equip] motor mounting
• [comb.int] engine mounting
— m — m para arranque m - [electr-mot]
starting motor mounting
— m — nave f espacial - [astronáut] (space)
vehicle, mounting, o assembly
— m — neumático m - [autom-neumát] tire
mounting
— m — núcleo - [mec] core mounting
— m — m laminar - [electr-equip] lamination
mounting
— m — panel m - [mec] panel mounting
— m — m lateral - [mec] side panel mounting
— m — parte f superior - [mec] top part, o
cap, mounting
— m — f — de cojinete m - [mec] bearing
cap mounting
— m — f — — m central - [mec] center
bearing cap mounting
— m — f — — m — principal - [comb.-
-int] main center, o center main, bearing cap
mounting
— m — f — — m frontal - [mec] front
bearing cap mounting
— m — f — — — m — principal - [mec]
main front, o front main, bearing cap mounting
— m — f — — — m posterior - [mec] main
rear, o rear main, bearing cap mounting
— m — f — — — m — principal - [mec]
main rear, o rear main, bearing cap mounting
— m — f — — — m principal - [mec] main
bearing cap mounting
— m — pasarela f - [mec] walkway, mounting, o
assembly,bling, o erecting,tion
— m — pescante m - [grúas] jib mounting
— m — pieza f - [mec] part assembling,bly
— m — f constitutiva - [mec] component, assembling,bly, o mounting
— m — f para pote m - [válv] pot part(s)
assembling,bly
— m — f — — m para válvula f - [válv]
valve pot part(s) assembling,bly
— m — f — válvula f - [válv] valve part(s)
assembling,bly
— m — f polar - [electr-mot] pole piece
mounting
— m — placa f - [mec] plate mounting

montaje m de placa f para escobilla f - [electr-mot] brush plate mounting
— m —— f para identificación f [ind] name-plate mounting
— m — planta f - [ind] plant erecting,tion
— m — plataforma f - [mec] platform erection
— m — platino(s) m - [comb.int] point(s) mounting
— m —— m y condensador m - [comb.int] point(s) and condenser mounting
— n — polea f - [mec] pulley mounting
— m —— f para arranque m - [comb.int] starting pulley mounting
— m — portaescobilla(s) f - [electr-mot] brushholder mounting
— m — pote m - [válv] pot assembling,bly
— m —— m para válvula f - [válv] valve pot assembling,bly
— m — prolongación f - [sold] extension mounting
— m — quemador m - [combust] burner mounting
— m — rectificador m - [electr-instal] rectifier mounting
— m — regulador m - control, o throttle, o governor, mounting
— m —— m para gobierno m - [comb.int] governor control mounting
— m —— m para marcha f sin carga f - [mec] idler mounting
— m — rejilla f - [mec] screen, o grid, mounting
— m —— f para volante m - [comb.int] flywheel screen mounting
— m — resóstato m - [electr-instal] rheostat mounting
— m — resistencia f - [electr-instal] resistor mounting
— m — respaldo m - [mec] back mounting
— m — retén m - [mec] stop, o retainer, o hook, mounting
— m — rodillo m - [mec] roller mounting • [metal-lam] roll assembling,bly
— m —— m vivo - [mec] live roller mounting
— m —— m motor - [mec] live roller mounting
— m — segadora f - [agric-equip] reaper mounting
— m —— f trilladora - [agric-equip] combine mounting
— m — silenciador m - [comb.int] muffler mounting
— m — soldadora f - [sold] welder mount(ing)
— m — soporte m - [mec] bracket, o holder, o support, mounting
— m — tablero m - [instrum] board mounting
— m — talón m - [mec] lug mounting
— m — tapa f - [mec] cover, o cap. mounting
— m —— f guardapolvo(s) - [electr-mot] dust cap mounting
— m —— f para engranaje(s) m - [comb.int] gear cover mounting
— m —— f — platino(s) m - [comb.int] points cover mounting
— m —— f — relé m - [sold] relay cover mounting
— m — termostato m - [instrum] thermostat mounting
— m — tomacorriente m - [electr-instal] receptacle mounting
— m — tope m - [mec] stop mounting
— m —— m para brazo m - [mec] arm stop mounting
— m — torre m - [petról] rig mounting
— m —— f para perforación f - [petról] rigging up
— m — tragante m - [metal-prod] shaft assembling,bly
— m — tubería f - [hidr] pipe erecting,tion
— m —— f para henchimiento n - [hidr] filling pipe, assembly, o erection
— m — unión f - [mec] joint(s) assembling,bly
— m —— m con perno(s) m - [mec] bolt(ed)

montaje m de unión(es) f con perno(s) m con resistencia f alta - [mec] high strength bolt(ed) joint(s) assembly
— m — válvula(s) f - [válv] valve mounting • valve assembling,bly
— m —— f mariposa - [válv] butterfly valve, assembling,bly, o erecting,tion
— m —— f — para admisión f - [válv] intake butterfly valve erecting,tion
— m —— f piloto - [válv] pilot valve mounting
— m — ventilador m - [mec] fan mounting
— m — viga f - [constr] beam mounting
— m — volante m - [mec] flywheel mounting
— m dentro de bomba f —assembling,bly in @ pump
— m —— volante m - [comb.int] mouting within @ flywheel
— m detrás - [mec] rear mounting
— m directo - [mec] direct mounting
— m — sobre soldadora f - [sold] direct mounting on @ welder
— m electro mecánico - [ind] electro-mechanical erecting,tion
— m en caliente - [f.c.-ruedas] hot fitting
— m — círculo m - [mec½] circle mounting
— m —— m de rodillo m - [grúas] roller circle mounting
— m — emplazamiento m - [ind] (project) site, mounting, o erecting
— m — obra f - [ind] project, o on site, o field, o in place, erecting,tion, o mounting, o assembling,bly
— m —— de coraza f - [metal-prod] shell field erection
— m — posición f - in @ position mounting
— m —— f horizontal - horizontal mounting
— m —— f sobrecabeza - [sold] overhead, fabricating,tion, o mounting
— m — vertical - vertical mounting
— m — sitio m - [constr] in place, assembling,bly, o mounting
— m — taller m - [mec] shop assembling,bly
— m —— m de coraza f - [metal-prod] shell shop erection
— m — tándem—[sold] tandem, setup, o mounting
— m — tornillo m (para banco m) - [mec] vise mounting
— m — voladizo m - [mec] offset mounting • blocking out
— m encerrado - [sold] enclosed mounting
— m estructural - [constr] structural mounting
— m — en intemperie f—[constr] outdoor structural, erection, o mounting
— m fácil - easy mounting
— m frontal - [mec] (on) front mounting
— m — de motor m - [comb.int] engine front mounting • [electr-mot] motor front mounting
— m horizontal - horizontal mounting
— m inapropiado - improper mounting
— m incorrecto—incorrect, o improper, mounting
— m inseguro - [mec] insecure mounting
— m liviano - [mec] lightweight mounting
— m para rollo m - [mec] lightweight roll mounting
— m —— m de alambre m - [mec] lightweight wire roll mounting
— m nuevo - [mec] remounting • reassemblng,bly
— m — sobre plantilla f - [mec] reassembling, o reassembly, on @ templet
— m para aislación f - [electr-instal] insulation mounting
— m — amortiguación,guamiento - [mec] shock mounting
— m — amplificador m—[mec] amplifier mounting
— m —— m magnético - magnetic amplifier mount(ing)
— m — bastidor m - [mec] frame mount(ing)
— m —— m inferior - [mec] lower, o sub-frame, mount(ing)
— m —— m — unificado - [mec] unitized sub-frame mount(ing)

montaje m para cabeza f - [sold] head mounting
— m — caja f - [mec] housing mounting
— m — ensayo(s) m - [mec] trial(s) assembly
— m — estabilizador m - [mec] stabilizer mounting
— m — mando m - [mec] control mounting
— m — motor m - [comb.int] engine mounting • [electr-mot] motor mounting
— m — núcleo m - core mounting
— m — — m laminar - [electr] lamination mounting
— m — pestillo m - [mec] catch, o latch, mounting
— m — portafusible(s) m - [electr-instal] fuse holder mounting
— m — portarresistor* m - [electr-instal] resistor holder mounting
— m — producción f - [ind] production assembly
— m — prueba(s) f - [mec] trial, o test, assembly
— m — resistor m - [electr-instal] resistor mounting
— m — suspensión f - [mec] suspension mounting
— m — tambor m - [mec] drum mounting
— m — — m Speed Feed - [sold] Speed Feed drum mounting
— m — telemando m - [mec] remote mounting
— m — transformador m - [electr-instal] transformer mounting
— m — unidad* f para cambio(s) m - [mec] shifting unit mounting
— m paralelo - parallel mounting
— m permanente - [mec] permanent mounting
— m posterior - [mec] rear mounting • later mounting
— m — de motor m - [comb.int] rear engine mounting • later engine mounting • [electr-mot] rear motor mounting • later motor mounting
— m provisorio - [ind] temporary mounting • [constr] false work
— m quitado - [mec] removed mounting
— m reemplazado - [mec] replaced mounting
— m removido - [mec] removed mounting
— m rígido - [mec] rigid mounting
— m sacado - [mec] removed mounting
— m según medida - [mec] custom mounting
— m sencillo - [mec] simple mounting
— m — para cabeza f - [sold] simple head mounting
— m simulado - [sold] simulated mounting • simulated setup
— m sobre - assembling, o affixing, to, o on
— m — aguilón m - [grúas] boom mounting
— m — almohadilla f - [mec] cushion mounting
— m — — f de caucho m - [mec] rubber cushion mounting
— m — anaquel m - [electrón] rack mounting
— m — astillero m - [mec] rack mounting
— m — base f rodante - [mec] trailer, o undercarriage mounting
— m — bastidor m - [mec] frame mounting • [sold] bed-plate mounting • rack mounting
— m — brida f - [mec] flange mounting
— m — caja f - [mec] housing mounting
— m — — f para eje m - [autom-mec] assembling to @ axle mounting
— m — — f — portadiferencial m - [autom-mec] assembling,bly on @ differential carrier case
— m — cárter m - [comb.int] crankcase mounting
— m — costado m - [mec] side mounting
— m — cubierta f - [mec] assembling to @ cover
— m — — f para distribuidor m - [autom-mec] assembling to @ divider cover
— m — — f — m para fuerza f - [cutom-mec] assembling to @ power divider cover
— m — derecha f - [mec] mounting on @ right
— m — — f e izquierda f - [mec] mounting on @ right and @ left
— m — — f — — f en forma f apareada - [mec] true left and right mounting
— m — eje m - [mec] axle mounting
— m — ficha f (de cartón m) - [com] carding

montaje m sobre izquierda f - [mec] mounting on @ left
— m — mano f derecho - [mec] right hand mounting
— m — — f izquierda - [mec] left hand mounting
— m — pared f - wall mounting
— m — parte f anterior - [mec] front mounting
— m — — f posterior - [mec] rear mounting
— m — pata(s) f - [mec] leg, o foot, mounting
— m — patín(es) m - [mec] skid mounting
— m — pescante m - [grúas] jib mounting
— m — pie m - [mec] foot mounting
— m — plantilla f - [mec] assembly on @ templet
— m — portadiferencial m - [autom-mec] assembling,bly on @ differential carrier
— m — respaldo m - [mec] back mounting
— m — soporte m - [mec] support, o holder, mounting
— m — subestructura f - [petról] substructure mounting
— m superior - [mec] top, o upper, mounting
— m — de magneto m - [comb.int] upper magneto mounting
— m — — regulador m - [mec] upper governor mounting
— m — — — m y magneto m - [mec] upper governor and magneto mounting
— m supervisado - [mec] supervised erection
— m temporario - [mec] temporary mounting
— m terminado - [ind] completed erection
— m trasero - [mec] rear mounting
— m — de motor m - [electr-mot] motor rear mounting • [comb.int] engine rear mounting
— m vertical - [mec] vertical, mounting, o assembling,bly
— m y empernado - [mec] assembling and bolting
montajes m - [ind] riggers
montante m - [mec] . . .; vertical member; stud; stanchion; derrick; king post; buck stay; cramp; mullion; muntin • derrick • [grúas] mast • leg • [autom] side piece
— m para aguilón m - [grúas] boom mast
— m — batería f - [coque] battery buck stay
— m — caballete m - [grúas] gantry leg
— m — — m verificado - [grúas] checked gantry leg
— m — contención f - [constr] buck stay
— m — — f para batería f - [coque] battery buck stay
— m — elevador m - [grúas] hoist mast
— m — grúa f - [grúas] crane post; leg pin
— m — — f para torre f - [grúas] derrick crane post
— m — — f — — f para perforación f - [petról] derrick crane post
montaña f - [geogr] . . .; mount
— f abrupta—[topogr] abrupt, o steep, mountain
— f, intransponible, o imposible de trasponer - [topogr] impassible mountain
Montañas f Rocosas - [geogr] Rocky Mountains; Rockies
montañoso,sa - , , , ,; hilly
montar v - . . . . • [mec] to assemble; to erect • to paste • to, fit, o set, up • to insert; to install • to put together • to adjust
— v a horcajada(s) f - to straddle (mount)
— v acero m - [estruct] to erect @ steel
— v — m estructural - [estruct] to erect @ structural steel
— v aguilón m - [grúas] to, erect, o connect, @ boom
— v apropiadamente - [ind] to mount properly
— v árbol m - [mec] to mount @ shaft
— v — m para salida f (de fuerza f) - [autom-mec] to, mount, o install, @ output shaft
— v ataguía f - [hidr] to erect @ cofferdam
— v base f - [mec] to, mount, o erect, @ base
— v bobina f - [electr-mot] to mount @ coil
— v busa f - [metal-prod] to install @ blowpipe
— v cabezal m - [mec] to mount @ header
— v cabina f - [autom] to mount @ cab

montar v cable m - [constr] to mount @ cable
— v caja f - [mec] to mount @ housing
— v — f para soldadora f - [sold] to mount @ welder case
— v carrillo m - [ind] to, mount, o attach, @ trolley
— v casquete m - [mec] to mount @, cap, o cover
— v — m para extremo m - [mec] to mount @ end, cap, o cover
— v compartimiento m - [mec] to mount @ compartment
— v — m para caja f - [mec] to mount @ case compartment
— v — m — — f para soldadora f - [sold] to mount @ welder case compartment
— v compuerta f - [hidr] to erect @ (sluice) gate
— v — f para toma f - [hidr] to, erect, o install, @ intake (sluice) gate
— v — f — vertedero m - [hidr] to erect @ spillway (sluice) gate
— v con lanza f hacia atrás - [sold] to mount with @ handle towards @ rear
— v conducto m - [electr-instal] to mount @ conduit
— v — m para cable(s) m - [electr-instal] to, mount, o install, @ (cable) conduit
— m conjunto m - [mec] to mount @ assembky
— v conmutador m - [electr-instal] to mount @ switch
— v — m para selección f - [electr-instal] to mount @ selector switch
— v correctamente - to mount, correctly, o properly
— v cubresoporte m - [mec] to mount @ bracket cover
— v deflector m - [mec] to mount @ baffle
— v dentro de bomba f - [bombas] to assemble in(to) a pump
— v — — volante m - [comb.int] to mount with(in) @ flywheel
— v detrás - to rear mount • to mount behind
— v directamente - [mec] to mount directly
— v — sobre base f - [mec] to mount directly to @ base
— v — — f de soldadora f - [sold] to mount directly to @ welder base
— v — — soldadora f - [sold] to mount directly to @ welder
— v dispositivo m - [mec] to mount @ device • to fit @ attachment
— v en emplazamiento m - [ind] to project site, mount, o erect
— v — obra f [constr] to assemble at @ site
— v — posición f - to mount in(to) @ position
— v — — i horizontal - to mount horizontally
— v — — f vertical - to mount vertically
— v — sitio m - [constr] to assemble at @ site
— v — tornillo m (para banco f) - [mec] to mount in @ vise
— v — voladizo - [mec] to offset mount; to block out • [constr] to cantilever
— v equipo m - [ind] to mount @ equipment
— v — m para perforación f—[petról] to rig up
— v herramienta f - [herram] to, mount, o insert, @ tool
— v horizontalmente - to mount horizontally
— v implemento m - [agric-equip] to mount @ implement
— v inapropiadamente - to mount improperly
— v incorrectamente - to mount, incorrectly, o improperly
— v indicador m - [constr-vial] to mount @ sign
— v malacate m - [mec] to mount @ winch • [petról] to mount @ drawworks
— v — m hidráulico - [grúas] to mount @ hydraulic winch
— v — m — sobre bastidor m - [grúas] to frame mount @ hydraulic winch
— v — sobre bastidor m - [grúas] to frame mount @ winch
— v — m — estructura f - [petról] to substructure mount @ drawworks

montar v manómetro m - [instrum] to mount @ gage
— v material m - [mec] to mount @ material
— v — m tubular - [tub] to mount a, tubular, o pipe type, material
— v motor m - [electr-mot] to, mount, o erect, @ motor • [comb.int] to, mount @ engine
— v neumático m - [autom-neumát] to mount @ tire
— v muevamente - [mec] to, reassemble, o reinstall
— v — sobre plantilla f - [mec] to reassemble on @ templet
— v para telemando m - [mec] to mount for @ remote control
— v paralelamente - [mec] to mount parallely
— v pasarela f - [mec] to assemble @ walkway
— v permanentemente - [mec] to mount permanently
— v pieza f - [mec] to, mount, o assemble, @ part
— v — f para pote m (para válvula f) - [válv] to, assemble, o mount, @ (valve) pot part
— v — f — válvula f - [válv] to assemble @ valve part
— v planta f - [ind] to erect @ plant
— v plataforma f - [mec] to erect @ platform
— v pote m - to, assemble, o install, @ pot
— m — para válvula f - [válv] to assemble @ valve pot
— v retén m - [mec] to mount @ retainer
— v rígidamente - [mec] to mount rigidly
— v segadora f - [agric-equip] to mount @ harvester
— v — f trilladora - [agric-equip] to mount @ combine
— v según medida f - [mec] to custom mount
— v sobre - to, mount, o assemble, on; to affix to
— v — almohadilla(s) f - [mec] to cushion mount
— v — f de caucho m - [mec] to rubber cushion mount
— v — anaquel m - [electrón] to rack mount
— v — bandeja(s) f - [transp] to palletize
— v — base f - [mec] to pad mount
— v — f rodante - [ind] to, undercarriage, o trailer, mount
— v — bastidor m - [mec] to frame mount
— v — caja f - [mec] to mount on @ housing
— v — f para eje m - [autom-mec] to assemble to @ axle housing
— v — — f — portadiferencial m - [autom-mec] to assemble on @ differential carrier case
— v — camisa f - [mec] to fit on @ liner
— v — cárter m - [comb.int] to mount on @ crankcase
— v — costado m - [mec] to side mount
— f — cubierta f - [mec] to assemble on @ cover
— v — — f para distribuidor m - [mec] to, assemble, o mount, to @ divider cover
— v — — f — m para fuerza f - [auto-mec] to assemble to a power divider cover
— v — eje m - [mec] to axle mount
— v — ficha f (de cartón m) - [com] to card
— v — mano f derecha - [mec] to right hand mount
— v — — f izquierda - [mec] to left hand mount
— v — pared f - [constr] to wall mount
— v — parte, anterior, o delantera - [mec] to front (end) mount
— v — — f, posterior, o trasera - [mec] to rear (end) mount
— v — patín(es) m - [mec] to skid(s) mount
— v — plantilla f - [mec] to assemble on @ templet
— v — portadiferencial m - [autom-mec] to assemble on @ differential carrier
— v — subestructura f - [petról] to mount on @ substructure
— v soldadora f - [sold] to mount @ welder
— v sonda f - [metal-prod] to install @ stock

rod
montar v temporariamente - [mec] to mount temporarily
— v **tomacorriente** m - [electr-instal] to mount
@ receptacle
— v **torre** f - [constr] to erect @ tower
— v — f **para perforación** f - [petról] to rig
up
— v **válvula** f - [mec] to mount @ valve • to assemble @ valve
— v — f **mariposa** - [válv] to, mount, o erect,
o install, @ butterfly valve
— v — f **para toma** f - [válv] to, erect, o
install, @ intake butterfly valve
— v — f **piloto** - [mec] to mount, o install, @
pilot valve
— v **ventilador** m - [mec] to mount @ fan
— v **viga** f - [constr] to, mount, o install, @
beam
— v **verticalmente** - to mount vertically
monte m - . . . • brush
— m **bajo** - [topogr] (low) scrub country
montículo m **de fundente** m - [sold] flux mound
monto m - . . . • [com] amount; value; total;
quantity • size • cost
— f **afectado** - affected amount
— m **cubierto** - [seguros] covered amount
— m **de amortización** f - amortization, o depreciation, amount
— m — **asignación** f - allowance amount
— m — — f **para propaganda** f - [com] advertising allowance amount
— m — — f — — f **con participación** f - [com]
co-op (advertising) allowance amount
— m — **capital** m - [fin] capital, sum, o amount
— m — **crédito** m - [fin] credit, value, o amount
— m — m **solicitado** - [fin] requested credit
value
— m — **depreciación** f - depreciation, value, o
amount
— m — **equipo** m - [ind] equipment value
— m — m **adquirido** - purchased equipment
value
— m — — m **importado** - [ind] imported equipment value
— m — — m **adquirido** - purchased imported
equipment value
— m — **importación(es)** f - [com] import(s) value
• total import(s)
— m — **incremento** m - increase value
— m — **inversión** f - [fin] investment, value, o
amount
— m — — f **total** - [fin] total investment value
— m — **mano** f **de obra** f - [labor] labor cost(s)
— m — **mora** f - [fin] default amount
— m — **pérdida** f - [seguros] loss amount
— m — — f **cubierta** - [seguros] covered loss
amount
— m — **producción** f - production value
— m — — f **de baritina** f - [miner] barytine
production value
— m — — f — **cobre** m - [miner] copper production value
— m — **respuesto(s)** m - [ind] spare(s) (parts)
value
— m — — m **adquirido(s)** - [ind] purchased
spare(s) (parts) value
— m — — m **importado(s)** - [ind] imported
spare(s) (parts) value
— m — — m **adquirido(s)** m - [ind] purchased imported spare(s) (parts) value
— m **disponible** - available, amount, o value
— m **equivalente** - equivalent, value, o amount
— m **estimado** - estimated, value, o amount
— m **invertido** - [fin] invested, value, o amount
— m **máximo** - [fin] maximum, value, o amount
— m **mínimo** - [fin] minimum, value, o amount
— m **neto** - net, value, o amount
— m — **de indemnización** f - net indemnity amount
— m **para apoyo** m - [fin] support, value, o
amount
— m **por financiar** - [fin] amount to be financed
— m **provisto** - provided amount

monto m **subscripto** - [fin] subscribed amount
— m **total** - total, value, o amount
— m — **de inversión** f - [fin] total investment, value, o amount
— m — **estimado** - total estimated, value, o
amount
— m **único** - only, o single, value, o amount
montomorillonita f - [geol] montmorillonite
montón m - . . . • dump • lump • jumble • cob •
[fam] bundle
— m **ahorrado** - [fam] saved bundle
— m **considerable** - large amount • quite @ pile
— m **de fundente** m - [sold] flux mound
— m — **material(es)** m - [ind] stockpile
montoncillo m - [agric] (small) hill
montura f - saddle • [mec] mount(ing); fit(ting)
• [petról] rig
— f **de cable** m - [constr] cable mounting
— f — **viga** f - [constr] beam mounting
— f **en voladizo** - [mec] blocking out; offset
mounting; cantilevering
— f **para cabezal** m - [mec] header mounting
— m — **depósito** m - [mec] tank saddle
— f — **motor** m - [comb.int] engine mount(ing)
— f **sobre camisa** f - [mec] fitting on @ liner
monumento m **a caído(s)** m **en guerra(s)** f - [pol]
war memorial
monzonita* f - [miner] monzonite
mora f - . . . . • [fin] default • arrears
— f **prolongada** - [fin] protracted default
morada f - . . . . • habitat
moral f **bajada** - [admin] lowered morale
— f **decaída** - [admin] dropped morale
— f **más baja** - [admin] lower morale
— f **socavada** - lowered, o undercut, morale
moralidad f **mercantil** - [com] concept
morcilla f - . . . . • [metal-prod] bustle pipe
mordaza f - [mec] . . .; clip; grip; jaw (plate,
o mandrel) • [sold] contact jaw • [tub] jaw;
pipe vise • [petról] slip • [herram] work jaw
— f **abierta** - [mec] open, jaw, o clamp
— f **adaptable** - [mec] versatile clamp
— f **aisladora** - [electr-equip] cleat insulator
— f **ajustable** - [mec] adjustable, o maneuverable, clamp
— f — **Adjusta-Clamp** - [mec] adjustable clamp;
(Adjusta-Clamp) clamp
— f — **con tornillo** m - [mec] screw adjusted
clamp
— f — — — m **para sujetador** m - [mec] screw
adjusted cam clamp
— f **ajustada** - [mec] adjusted, clamp, o grip
— f **asegurada** - [mec] attached clamp
— f **automática** - [mec] automatic, clamp, o grip
— f — **para alambre** m - [trefil] automatic wire
grip
— f — — — m **trefilado** - [trefil] drawn wire
automatic grip
— f **compacta** - [mec] compact clamp
— f **con abertura** f **grande** - [mec] large opening
clamp
— f — — **ajuste** m - [mec] adjustable,ted clamp
— f — — m **con tornillo** m - [mec] screw adjusted clamp
— f — — m — m **para sujetador** m - [mec]
screw adjusted cam clamp
— f — **aleación** f - [mec] (steel) alloy clamp
— f — **brazo(s)** m **afinado(s)** - [mec] sharp leg
clamp
— f — — m **corto(s)** - [mec] short leg clamp
— f — — m **para pieza(s)** f **estructural(es)**
- [mec] short leg structural clamp
— f — **conectador(es)** m - [mec] connector clamp
— f — — m **para cadena** f - [mec] chain connector clamp
— f — **costo reducido** - [mec] inexpensive, o
low cost, clamp
— f — **hocico** m **corto** - [mec] short nose clamp
— f — **mandíbula ajustable** - [mec] swivel jaw
clamp
— f — — f **fija** - [mec] fixed jaw clamp
— f — **peso** m **reducido** - [mec] lightweight clamp
— f — **resorte** m - [mec] spring clip

mordaza f con tratamiento m térmico - [mec] heat treated clamp
— f — — m — para elevación f - [ind] heat treated, lifting, o hoisting, clamp
— f — — — — f de plancha(s) f—[metal- -lam] heat treated plate lifting clamp
— f contra (re)torcedura(s) f—[mec] twist grip
— f — — f en palanca f para rotación f - [grúas] swing lever twist grip
— f cuneiforme - [mec] wedge (shaped) grip
— f — para avance m gradual - [trefil] inching wedge grip
— f de acero m - [mec] steel, clamp, o grip
— f — — m con aleación f - [mec] alloy clamp
— f — cobre m - [mec] copper, clamp, o jaw
— f — modelo m estándar - [mec] standard model clamp
— f — plomo m - [electr-instal] lead clip
— f — tipo m estructural - [mec] structural, type, o style, clamp
— f económica - [mec] inexpensive, o low cost, clamp
— f elevada - [mec] raised, o lifted, clamp
— f empaquetada - [mec] packed, grip, o clamp
— f engranada - [mec] engaged clamp
— f estándar - [mec] standard clamp
— f estibadora f - [ind] stacking clamp
— f — para alambre m — [ind] comealong
— f extrarresistente—[mec] high-strength clamp
— f fijada - [mec] attached. o fixed, clamp
— f forjada - [mec] forged clamp
— f — con martinete m—[mec] drop forged clamp
— f — — m para elevación f - [mec] drop forged, lifting, o hoisting, clamp
— f — — — f de plancha(s) f—[metal- -lam] drop forged plate lifting clamp
— f — — m con tratamiento m térmico—[mec] drop forged heat treated lifting clamp
— f — — m — m para elevación f - [ind] drop forge heat treated, lifting, o hoisting, clamp
— f — — — m — m — — f de plancha(s) f - [metal-lam] drop forged heat treated plate lifting clamp
— f — para elevación f - [ind] forged, hoist- ing, o lifting, clamp
— f girada - [mec½ turned clamp
— f guiada - [mec] guided clamp
— f horizontal - [mec] horizontal clamp
— f — para plancha(s) f - [metal-lam] horizon- tal plate clamp
— f instalada - [mec] installed clamp
— f íntegra - [mec] entire, o whole, clamp
— f liviana - [mec] light clamp
— f manual - [mec] hand, grip, o clamp
— f mejorada - [mec] improved clamp
— f negra - [electr-instal] black clip
— f no estropeadora f - [mec] non-marring clamp
— f — trabadora - [mec] non-locking clamp
— f para alambre m - [mec] wire, grip, o clamp
— f — — m trefilado - [trefil] drawn wire, grip, o clamp
— f — arrastre m - [mec] drag(ging) clamp
— f — avance m gradual - [mec] inch(ing) grip
— f — barra f conect(ad)ora - [mec] tie rod clamp
— f — — f para acoplamiento m - [mec] tie rod clamp
— f — — f — — m aflojada - [mec] loosened tie rod clamp
— f — — f — — m ajustada - [mec] tightened tie rod clamp
— f — — f — — m apretada - [mec] tightened tie rod clamp
— f — brazo m - [mec] arm clamp
— f — — m para ajustador m para terraja f - [mec] gage,ging arm clamp
— f — — m — — m para terraja f - [mec] stock gage,ging arm clamp
— f — brida f - [mec] flange clamp
— f — — f abierta - [mec] open(ed) flange clamp
— f — cubierta f - [mec] cover spring

mordaza f para cubierta f para cursor m - [mec] slide cover clamp
— f — — m para hilera f - [trefil] die slide clamp
— f — elevación f - [grúas] lifting, o hoist- ing, clamp
— f — — f para plancha(s) f - [metal-lam] plate lifting clamp
— f — extremo m - [sold] end clip
— f — — m de conductor m - [sold] lead end clip
— f — fines m múltiples - [mec] multi-purpose clamp
— f — mandril m - [mec] chuck jaw
— f — manipuleo m - [mec] handling clamp
— f — — m de material(es) m - [ind] materials handling clamp
— f — morsa f - [mec] vise grip
— f — plancha(s) f - [metal-lam] plate clamp
— f — — f con ajuste m - [mec] adjustable plate clamp
— f — — f — — m con tornillo m - [mec] screw adjusted plate clamp
— f — — — — m para sujetador m - [mec] screw adjusted cam plate clamp
— f — — f horizontal - [mec] horizontal plate clamp
— f — plato m - [mec] chuck jaw
— f — soldadura f - [sold] welding clamp
— f — sujeción f - [mec] holding clamp
— f — — para suela f - [mec] base plate holding clamp
— f — . . . tonelada(s) f - [mec] . . . ton clamp
— f — tornillo m (para banco m) - [mec] vise, gripo, o jaw
— f — tubo(s) m - [tub] pipe grip
— f — viga(s) f - [mec] beam, o girder, clamp
— f — — f maestra f - [mec] girder clamp
— f perimétrica - [sold] girth clamp
— f — para soldadura f - [sold] girth welding clamp
— f por contacto m - [sold] contact jaw
— f proyectada - [mec] designed clamp
— f, quitada, o removida - [mec] removed clamp
— f roja - [electr-instal] red clip
— f sacada - [mec] removed clamp
— f soltada - [mec] released, o loosened, clamp
— f trabada - [mec] locked clamp
— f — en posición f cerrada - [mec] closed locked clamp
— f trabadora - [mec] locking clamp
— f útil - [mec] useful clamp
mordedor,da a - biting • [papel] nipping
mordedura f - . . .; biting; nip(ping) • [mec] grip(ping); pinch(ing) • gripping force • [sold] overlap(ping); overhang(ing)
— f a fondo m - [mec] full engaging,gement
— f abierta - [papel] open(ed) nip(ping)
— f ajustable - [mec] adjustable grip
— f apropiada - [mec] appropriate grip(ping)
— f — de banda f (para rodadura f) - [autom- -neumát] proper tread head
— f automática - [mec] automatic clamping
— f cero - [mec] zero grip
— f completa - [mec] full engaging,gement
— f de acanaladura(s) f—[cabl] pinching groove
— f — alambre m - [Trefil| wire grip(ping)
— f — banda f para rodadura f - [autom-neumát] tread lead
— f — conductor m - [electr-instal] wire pinching
— f — sujetador m - [mec] cam grip(ping)
— f definida - [mec] definite pinch(ing)
— f fuerte - [mec] tight grip(ping)
— f máxima - [mec] maximum grip • grip(ping) range
— f mayor - [mec] greater, o large, o wider, grip, o bite
— f nula - [mec] zero grip
— f perdida - [mec] lost, pinch, o grip
— f segura - [mec] secure, bite, o grip
— f total - [mec] total, bite,ting, o gripping

morder v - . . . • to grip; to clip • [cabl] to pinch
— v a fondo - [mec] to engage fully
— v acanaladura f - [cabl] to pinch @ groove
— v alambre m - [trefil] to grip @ wire
— v automáticamente - [mec] to clamp automatically
— v completamente - [mec] to engage fully
— v con mordaza f - [mec] to engage @ clamp
— v conductor m - [electr-instal] to pinch @ wire
— v fuertemente - [mec] to grip tightly
— v plenamente - [mec] to engage fully
— v totalmente - [mec] to engage, fully, o completely
mordido,da a - bit; nipped • [mec] gripped
— a a fondo - [mec] engaged fully
— a automáticamente - [mec] clamped automatically
— a completamente - [mec] engaged fully
— a fuertemente - [mec] gripped tightly
— a plenamente - [mec] engaged fully
— a totalmente - [mec] engaged fully
mordiente a - . . . • [mec] fixing • nipping
morir v instantáneamente - [medic] to die instantly
moroso,sa a - . . .; time consuming
morrillo m - [geol] . . .; boulder • gravel
— m menudo - [geol] pea gravel
morro m - . . . • [metal-prod] (tuyere) nose
— m inclinado - [metal-prod] angle, o bull, nose tuyere; deflector (type) tuyere
morsa* f - [mec] vise; véase **tornillo** m **para banco** m
mortaja f - . . . . [mec] notch; groove
mortajado* m - [carp] mortising
mortajado,da a - . . . . véase **amortajado,da** a
— a con exactitud - [carp] mortised accurately
mortajadora* f - [herram] mortiser; key seater; slotter
mortajar* v - [carp] to mortise
mortero m - [milit] . . . • [constr] . . .; mix
— m aluminoso - [constr] alumina mortar
— m — granulado - [constr] granular alumina mortar
— m blanco - [constr] white mortar
— m — para tomar - [constr] white pointing mortar
— m blando m - [constr] slush • grout
— m, de, o con, cemento m - [constr] cement mortar
— m, —, o —, — m y arena f - [constr] sand and cement mortar • cement and sand mortar
— m — hormigón m - [constr] concrete mortar
— m endurecido - [constr] hardened, o stiffened, mortar
— m excedente - [constr] excess mortar
— m empleado - [constr] mortar used
— m para barro m - [miner] pug mill
— m — rejuntar - [constr] pointing mortar
— m — revestimiento m - [tub] lining mortar
— m — — m interior - [tub] (inner) lining mortar
— m — toma f (para juntas f) - [constr] pointing mortar
— m refractario - [refract] refractory mortar
— m usado - [constr] used mortar • mortar used
— m utilizado - [constr] mortar used
mosaico m - [constr] . . . • tile
mosquetón m - [mec] snap (hook)
— m con pasador m - [mec] bolt snap
— m — pestillo m - [mec] bolt snap
mosquitero m - [constr] . . . • screen
mosquito m eliminado - [hidr] eliminated mosquito
mostrado,da a - . . .; shown; showed • developed
— a claramente - shown clearly
— a compatible - shown, o proven, compatible
— a en dibujo m - [dib] shown on @ drawing
— a — ilustración f - shown in @ illustration
— a más abajo - shown below
— a — arriba - shown above
mostrador m para horno m - [metal-prod] furnace (building) charging side

mostrar v - . . . . • to develop
— v aceite m - to show @ oil
— f cabeza f para inyección v - [petról] to show @ swivel
— claramente v - to show clearly
— v desde cerca - to show close-up
— v — — cabeza f para inyección f - [mec] to show @ swivel close-up
— v — — calibrador m - [mec] to show @ gage close-up
— v — — placa f - [mec] to show @ plate close-up
— v — — — f para identificación f - [mec] to show @ identification plate close-up
— v — — zona + - to show @ area close-up
— v en dibujo m - [dib] to show on @ drawing
— f — iludstración f - to show in @ illustration
— v información f - tp show @, information, o data
— v más abajo - to show below
— v — arriba - to show above
— v para vender v - to show to sell
— v placa f para identificación f t- to show @ identification plate
— v presión v - to show @ pressure
— v — f de aceite m - [mec] to show @ oil pressure
— v — f — aire m - [mec] to show @ air pressure
— v — f en acumulador m - [mec] to show @ accumulator pressure • [electr] to show @ battery pressure
— v rojo - [instrum] to show @ red
— v saca f - to show @ removal
— v — f de tapón m - [mec] to show @ plug removal
— v — f — — m para filtro m - [petról] to show @ filter plug removal
— v sed f - to, show, o develop, @ thirst
— v remoción f - [mec] to show @ removal
— v — f de tapón n- [mec] to show @ plug removal
— v — f — — m para filtro m - [petról] to show @ filter plug removal
— v ubicación f - to show @ location
— v vista f desde cerca - to show @ close-up
— v — f — — de cabeza f - [mec] to show @ head close-up
— v — f — — — f para inyección f - [petról] to show @ swivel close-up
— v — f — — calibrador m - [mec] to show @ gage close-up
— v — f — — placa f - [mec] to show @ plate close-up
— v — f — — — f para identificación f - [mec] to show @ identification plate close-up
— v — f — — — zona f - to show @ area close-up
— v y vender v - [com] to show and sell
mostrarse v - to show, one, o it, self • to take @ approach
— v afirmativo,va - to assert (one, o it, self)
— v arrogante - smart aleck approach taking
— v asertivo,va - to be assertive
— v capaz - to prove to be qualified
— v — para tarea f - to rise to @ task
— v compatible - to prove compatible
— v resentido,da - to feel resentful; to run out of @ town on @ rail
moteado,da a - . . .; spotty; véase también atruchado,da
motilla f - [arquit-arco] corbel
motín m - [pol] . . .; uprising; upheaval
motivación f - motivating,tion
— f aumentada - increased motivation
— f de personal m - [admin] personnal, o employee, o people, motivating,tion
— f encontrada - [labor] conflicting motivation
— f humana - [labor] human motivation
— f inadecuada - [labor] inadequate motivation
— f para inversión f - [fin] investment motiva-

tion
motivación f total - [labor] over-all, o total,
  motivation
motivado,da a - motivated
motivador,ra a - motivating,tive; motive
motivar v - to motivate
— v inversión f—[fin] to motivate @ investment
— v personal m - [labor] to motivate @, personell, o employee(s), o people
motivo m - . . .; reason
— m de excepción f - reason for @ exception
motivo,va a - . . .; motivational,tive
motobarredora f - [ind] motor sweeper
motobomba f - [bombas] motor pump; power pump
— f para presión f de aceite m - [grúas] oil
  lift pump motor
motódromo m - [autom] . . .; speedway
— m para motocicleta(s) f - [deport] motorcycle
  speedway
motoestibadora f - [ind] payloader; fork lift
  truck; fork floor tractor; lift truck; tow-
  motor; motor stacker
— f transformada - [ind] transformed fork lift
  truck
motogenerador m Lincolnweld—[sold] Lincolnweld
  motor-generator
— m para corriente f alterna - [sold] alternating current motor-generator
— — f — para servicio m pesado - [sold]
  heavy duty alternating current motor-generator
— m — f continua - [sold] direct current
  motor-generator
— m — f — para servicio m pesado - [sold]
  heavy duty direct current motor-generator
— m — excitación f - [electr-prod] motor-generator exciter
— m — f para . . . voltio(s) m - [electr-
  -prod] . . .-volt motor-generator exciter
— m - . . . hercio(s) m - [sold] . . .-Hertz
  machine
— m — regulador m, giratorio, o rotativo -
  [metal-lam] Rototrol motor generator
— n — m, —, o —, para laminador m -
  [metal-lam] mill Rototrol motor-generator
— m — rodillo m - [metal-lam] roll motor-
  -generator
— m — Rototrol m - [metal-lam] Rototrol motor-
  -generator
— m — m para laminador m - [metal-lam] mill
  Rototrol motor-generator
— m — servicio m pesado - [sold] heavy duty
  motor-generator
motogeneradora* f - [sold] véase motogenerador m
motón m - [mec] . . .; sheave
— m bajado - [grúas] lowered, o dropped, block
— m caído - [grúas] dropped, o fallen, block
— m colgante - [mec] hanging, o dangling,
  block, o sheave
— m con gancho m - [cabl] hook block
— m - . . . polea(s) f para gancho m - [grúas]
  . . .-sheave hook block
— m - . . . — f — motón m para gancho m para
  grúa f - [grúas] . . .-sheave crane hook block
— m — polea f única - [grúas] single sheave
  block
— m — f única para cable m en . . . sec-
  ción(es) f para elevación f para pescante m -
  [grúas] . . .-part jib hoist(ing) line single
  sheave block
— m — — f — para gancho m - [grúas] single
  sheave hook block
— m — roldana f - [cabl] sheave block
— f — una polea f para gancho m - [grúas] single sheave hook block
— m — varias poleas f - [grúas] multi-line
  block
— m de acero m - [mec] steel block
— m — m para cajera f - [mec] steel shell
  block
— m — madera f - [mec] wood(en) (tackle) block
— m — f para cajera f - [mec] wood(en) shell
  block

motón m de metal m - [mec] metal (tackle) block
— m — nilón m - [mec] nylon block
— m — — m para cajera f - [mec] nylon shell
  block
— m elevado - [grúas] raised, o overhead, block
— m en cajera f - [mec] shell block
— m giratorio - [mec] swivel, block, o sheave
— m inspeccionado - [mec] inspected, block, o
  sheave
— m movible - [cabl] travelling block
— m para aparejo m - [grúas] tackle block
— m — m para entubación f - [petról] casing block
— m — cable m - [grúas] cable block
— m — — m de manila f - [grúas] manila rope
  block
— m — principal - [grúas] main load block
— m — — m — para carga f - [grúas] main load
  block
— m — cabo m - [cabl] rope block
— m — cajera f - [grúas] shell block
— m — — f de acero m—[mec] steel shell block
— m — — f — madera f - [mec] wood(en) shell
  block
— m — f — nilón m—[mec] nylon shell block
— m — carga f - [grúas] load block
— m — corona f (para torre f) - [petról]
  crown block
— m — — f para achicador m - [petról]
  bailing crown block
— m — cubierta f - [grúas] cover block
— m — f desmontable - [grúas] snatch block
— m — elevación f - [grúas] hoist(ing) block
— m — entubación f - [petról] casing block
— m — gancho m - [grúas] hook block
— m — m con . . . polea(s) f - [grúas] . . .
  sheave hook block
— m — — m de National f - [grúas] National
  hook block
— m — — m auxiliar - [grúas] auxiliary hook
  block
— m — — m inspeccionado - [grúas] inspected
  hook block
— m — — m para grúa f - [grúas] crane hook
  block
— m — — m principal - [grúas] main hook block
— m — garrucha f - [petról] pulley block
— m — — f para polea f - [petról] shave pulley block
— m — — f — — f para cuchara f - [petról]
  sand sheave pulley block
— m — maniobra(s) f - [grúas] snatch block
— m — punta f - [grúas] point sheave
— m — — f de aguilón m - [grúas] boom point
  sheave
— m — tubería f - [petról] tubing block
— m — — f para bombeo m - [petról] tubing
  block
— m — — f para entubación f - [petról] casing, block, o pulley
— m principal - [grúas] main block
— m — elevado - [grúas] hoisted main block
— m — para carga f - [grúas] main load,
  sheave, load, o block
— m subido - [grúas] raised block
— m verificado - [grúas] checked block
— m viajero - [grúas] traveling block
— m — verificado - [grúas] checked traveling
  block
— m y aparejo m - [cabl] véase aparejo m
motoneta f - [transp] scooter
motoniveladora f - [constr] self-powered, o
  tractor, o motor, grader; bullgrader; scraper
— f articulada - [constr-equip] articulated
  grader
motopala f - [constr-equip] (power) scraper;
  earth mover
motor m - . . . . • [comb.int] engine
— m a marcha f plena - [comb.int] engine (running) at full speed
— m — prueba f de explosión(es) f - [electr-
  -mot] explosion proof motor • [comb.int] explosion proof engine

motor m̲ a prueba f̲ de goteo m̲ - [electr-mot]
drip proof motor
— m̲ acelerado - [comb.ind] accelerated, o̲
speeded, o̲ revved (up), engine • [electr-mot]
accelerated, o̲ speeded, o̲ revved (up), motor
— m̲ — rápidamente - [electr-mot] rapidly accelerated motor • [comb.int] rapidly accelerated engine
— m̲ acoplado - [mec] coupled motor • [comb.int]
coupled engine
— m̲ — directamente - [electr-mot] directly
coupled motor • [comb.int] directly coupled
engine
— m̲ activado - [electr-mot] energized motor •
[comb.int] energized engine
— m̲ ahogado - [comb.int] flooded engine
— m̲ ajustado - [electr-mot] adjusted motor •
[comb.int] adjusted engine
— m̲ alineado - [electr-mot] aligned motor •
[comb.int] aligned engine
— m̲ alternativo m̲ - [comb.int] reciprocating
engine
— m̲ anegado - [comb.int] flooded engine
— m̲ apagado - [comb.int] turned off engine
— m̲ arrancado m̲ - [comb.int] started engine
— m̲ asentado - [electr-mot] broken in motor •
[comb.int] broken in engine
— m̲ atendido - [electr-mot] serviced motor •
[comb.int] serviced engine
— m̲ automotor - [autom] automotive engine
— m̲ auxiliar - [electr-mot] auxiliary motor •
[comb.int] auxiliary engine
— m̲ — con velocidad f̲ variable - [electr-mot]
variable speed auxiliary motor • [comb.int]
variable speed auxiliary engine
— m̲ — para laminador m̲ - [metal-lam] auxiliary
mill motor
— m̲ bien construido - [electr-mot] well, made,
o̲ built, motor • [comb.int] well built engine
— m̲ blindado - [electr-mot] (totally) enclosed
motor
— m̲ — sin ventilación f̲ - [electr-mot] enclosed non-ventilated motor
— m̲ — totalmente - [electr-mot] totally enclosed motor
— m̲ — — sin ventilación f̲ - [electr-mot] totally enclosed fan cooled motor
— m̲ caldeado - [electr-mot] heated motor
— m̲ calentado - [comb.int] heated engine •
warmed (up) engine
— m̲ caliente - [electr-mot] hot motor • [comb.-
-int] warm engine • hot, o̲ heated, engine
— m̲ — detenido - [electr-mot] stopped hot motor • [comb.int] stopped, hot, o̲ warm, engine
— m̲ calzado - [electr-mot] shimmed motor •
[comb.int] shimmed engine
— m̲ Caterpillar - [comb.int] Cat(erpillar) engine
— m̲ cebado - [comb.int] choked engine
— m̲ colocado - [electr-mot] installed motor •
[electr-mot] installed motor
— m̲ completo - [electr-mot] complete, o̲ entire,
motor • [comb.int] complete, o̲ entire, engine
— m̲ compuesto m̲ - [electr-mot] compound (wound)
motor
— m̲ — anillo(s) m̲ - [mec] ring motor
— m̲ — — m̲ deslizante(s) - [mec] slip ring
motor
— m̲ — arrancador m̲ (eléctrico) - [comb.int]
electric(al) starter,ting engine
— m̲ — bobinado m̲ - [electr-mot] wound motor
— m̲ — en derivación f̲ - [electr-mot] shunt
wound motor
— m̲ con , , , caballo(s) m̲ (de fuerza) f̲ -
[electr-mot] . . . horse power motor • [comb.
-int] . . . horse power engine
— m̲ — carrera f̲ corta - [comb.int] short
stroke engine
— m̲ — — f̲ larga - [Comb.int] long stroke engine
— m̲ . . . cilindro(s) m̲ - [comb.int] . . .
cylinder engine
— m̲ . . . — m̲ con enfriamiento m̲ con aire -

[comb.int] . . . cylinder air cooled engine
motor m̲ con . . . cilindro(s) m̲ en línea f̲ -
[comb.int] straight . . .-cylinder engine
— m̲ . . . — en V - [comb.int] V-... cylinder engine
— m̲ . . . — enfriado con aire - [comb.int]
...-cylinder air cooled engine
— m̲ — corriente f̲ cortada - [electr-mot] de-
-energized motor
— m̲ — cuatro cilindros m̲ - [comb.int] four-
-cylinder engine
— m̲ — tiempos m̲ - [comb.int] four stroke
engine
— m̲ — culata f̲ en L - [comb.int] L-head engine
— m̲ — chispa f̲ cortada - [comb.int] turned off
engine
— m̲ — desplazamiento m̲ fijo - [comb.int] set,
o̲ fixed, displacement engine
— m̲ — devanado m̲ - [electr-mot] wound motor
— m̲ — m̲ en derivación f̲ - [electr-mot]
shunt wound motor
— m̲ — dos cilindros m̲ - [comb.int] two-cylin-
der engine
— m̲ — — rotores m̲ - [comb.int] two-rotor
engine
— m̲ — émbolo(s) m̲ - [comb.int] piston engine
— m̲ — m̲ axial(es) - [comb.int] axial piston
engine
— m̲ — encendido m̲ regulado - [comb.int] timed
engine
— m̲ — — m̲ — nuevamente - [comb.int] retimed
engine
— m̲ — enfriamiento m̲ - [electr-mot] cooled
motor • [comb.int] cooled engine
— m̲ — — m̲ con agua m̲ - [elecrr-mot] water
cooled motor • [comb.int] water cooled engine
— m̲ — — m̲ — aire m̲ - [electr-mot] air cooled
motor • [acomb.int] air cooled engine
— m̲ — — m̲ — ventilador m̲ - [electr-mot] air
cooled motor • [comb.int] air cooled engine
— m̲ con engranaje(s) m̲ - [mec] gear motor
— m̲ — — m̲ reductor(es) - [mec] reducing gears
motor
— m̲ — — m̲ sincrónico(s) - [electr-mot] synchronous gear, o̲ Synchrogear, motor
— m̲ — enrollamiento m̲ - [electr-mot] wound
motor
— m̲ — — m̲ en derivación f̲ - [electr-mot]
shunt wound motor
— m̲ con escobilla(s) f̲ - [electr-mot] brush
motor
— m̲ con excitación, compound, o̲ compuesta -
[electr-mot] compound wound motor
— m̲ — expansión f̲ [comb.int] expansion engine
— m̲ — — f̲ múltiple - [comb.int] multiple expansion engine
— m̲ . . . H P - [electr-mot] . . . horse
power motor • [comb.int] . . . horse power
engine
— m̲ — inducción f̲ - [electr-mot] induction
motor
— m̲ — — f̲ con jaula f̲ de ardilla f̲ - [electr-
-mot] squirrel cage induction motor
— m̲ — — f̲ eléctrica - [electr-mot] electric
induction motor
— m̲ — — f̲ . . .fásico con jaula f̲ de ardilla
- [electr-mot] . . . phase squirrel cage induction motor
— m̲ — — f̲ trifásico con jaula f̲ de ardilla f̲
- [electr-mot] three phase squirrel cage induction motor
— m̲ — lubricación f̲ - [mec] lubricated motor •
[comb.int] lubricated engine
— m̲ — — f̲ permanente - [mec] permanently lubricated motor • [comb.int] permanently lubricated engine
— m̲ — movimiento m̲ alternativo - [comb.int] reciprocating engine
— m̲ — ocho cilindro(s) m̲ - [comb.int] eight-
-cylinder engine
— m̲ — — en línea - [comb.int] straight
eight-cylinder engine
— m̲ — — — m̲ en V - [comb.int] V-eight cylin-

der engine

motor m con potencia f menor que un H P -
[electr-mot] fractional (horsepower) motor •
[comb.int] fractional (horsepower) engine
— m — **regulación f** - [mec] controlled motor
— m — **f independiente** - [mec] independently
controlled motor
— m — **rendimiento m alto** - [comb.int] high
output engine • [electr-mot] high output motor
— m — **m elevado** - [comb.int] high output
engine • [electr-mot] high output motor
— m — **. . . revoluciones f por minuto m** -
[electr-mot] . . ., revolutions per minute, o
r p m, motor • [comb.int] . . . revolutions
per minute engine
— m — **dos rotores m** - [comb.int] two rotor
engine
— m — **seis cilindros m** - [comb.int] six-cylin-
der engine
— m — **. . . tiempos m** - [comb.int] . . . cycle
engine
— m — **un cilindro m** - [Comb.int] one cylinder
engine
— m — **rotor m** - [comb.int] one, o single,
rotor engine
— m — **válvula(s) f** - [comb.int] valve engine
— m — **f en corredera(s) f** - [comb.int]
sleeve valve engine
— m — **f culata f** - [comb.int] overhead
valve engine
— m — **varios cilindros m** - [comb.int] multi-
cylinder engine
— m — **. . . velocidad(es) f** - [electr-mot]
. . . speed motor
— m — **velocidad f variable** - [electr-mot]
variable speed motor
— m — **voltaje m alto** - [electr-mot] high voltage motor
— m — **m bajo** - [electr-mot] low voltage
motor
— m — **m, intermedio, o mediano** - [electr-
-mot] intermediate, o medium, voltage motor
— m **conectado** - [electr-instal] connected, o
wired, motor • [comb.int] connected engine
— m **confiable** - [electr-mot] reliable motor •
[comb.int] reliable engine
— m **dañado** - [electr-mot] damaged motor •
[comb.int] damaged engine
— m **de butano m** - [petról] butane engine
— m — **m para perforación f** - [petról] butane drilling engine
— m — **derecha f** - [electr-mot] right (hand)
motor • [comb.int] right (hand) engine
— m — **combustión f interna** - [comb.int] internal combustion engine
— m — **explosión f** - [comb.int] explosion, o
internal combustion, engine
— m — **gas m** - [comb.int] gas (fuel) engine
— m — **gasolina f** - [comb.int] gasoline engine
— m — **f ajustado** - [comb.int] adjusted gasoline engine • timed gasoline engine
— m — **f con . . . cilindro(s) m** - [comb.int]
. . . cylinder gasoline engine
— m — **f . . . — m enfriado con aire m** -
[comb.int] air cooled . . . cylinder gasoline
engine
— m — **f enfriado con aire m** - [comb.int]
air cooled gasoline engine
— m — **f fijo** - [comb.int] stationary gasoline engine
— m — **f para recambio m** - [comb.int] replacement gasoline engine
— m — **f para reposición f** - [comb.int] replacement gasoline engine
— m — **nafta f** - [comb.int] véase **motor m de
gasolina f**
— m — **petróleo m** - [comb.int] oil engine
— m — **m Hesselman** - [comb.int] Hesselman
oil engine
— m — **queroseno m** - [comb.int] kerosene engine
— m — **reacción f** - [comb.int] jet engine
— m — **tipo m blindado m** - [electr-mot] enclosed (type) motor

motor m de tipo m blindado enfriado - [electr-
-mot] enclosed cooled type motor
— m — — **m con ventilador m** - [electr-
-mot] enclosed fan cooled type motor
— m — — **m blindado sin ventilación f** -
[electr-mot] enclosed unventilated type motor
— m — — **m totalmente** - [electr-mot] totally enclosed type motor
— m — — **m sin ventilación f** - [electr-
-mot] totally enclosed unventilated type
motor
— m — — **m enfriado** - [electr-mot] totally enclosed cooled type motor
— m — — — **m con ventilador m** - [electr-
-mot] totally enclosed fan cooled type motor
— m — — **m con culata f en L** - [comb.int]
L-head type engine
— m — — **m jaula f de ardilla f** - [electr-
mot] squirrel cage type motor
— m — — **m con . . . tiempo(s) m** - [cobm.int]
. . . cycle type engine
— m — — **m horizontal** - [electr-mot] horizontal type motor - [combust.int] horizontal
type engine
— m — — **m para laminación f** - [metal-lam]
mill type motor
— m — — **m sin ventilación f** - [electr-mot]
unventilated type motor
— m — — **m vertical** - [electr-mot] vertical
type motor • [comb.int] vertical type engine
— m — **vapor m** - [vapor] steam engine
— m **defectuoso** - [electr-mot] defective, o
faulty, motor • [comb.int] defective, o
faulty, engine
— m **derecho m** - [electr-mot] right motor •
[comb.int] right engine
— m **desactivado** - [electr-mot] de-energized
motor
— m **desarmado** - [electr-mot] unassembled motor •
disassembled motor • [comb.int] unassembled
engine • disassembled engine
— m **descalabrado** - [electr-mot] destroyed motor
• [comb.int] destroyed engine
— m **desconectado** - [electr-mot] disconnected, o,
turned, o shut, off, motor • [comb.int] disconnected, o turned off, engine
— m **detenido** - [electr-mot] stopped motor •
[comb.int] stopped, o turned off, engine •
stalled engine
— m **detrás de conductor m** - [autom] rear mounted engine
— m **devanado** - [electr-mot] wound motor
— m — **en derivación f** - [electr-mot] shunt
wound motor
— m — **paralelo** - [electr-mot] parallel wound
motor
— m — **serie** - [electr-mot] series wound
motor
— m **diesel** - [comb.int] diesel engine
— m — **caterpillar** - [comb.int] caterpillar
diesel engine
— m — **con arrancador m (eléctrico)** - [comb.int]
(electric) starting diesel engine
— m — **. . . cilindro(s) m** - [combust.int]
. . . cylinder diesel engine
— m — **convertidor m** - [comb.int] converter
diesel engine
— m — — **m para, par motor, o torsión f** -
[comb.int] torque converter diesel engine
— m — **transmisión f** - [comb.int] transmission diesel engine
— m — **f para . . . velocidad(es) f** -
[comb.int] . . .speed transmission diesel
engine
— m — **. . . tiempos m** - [comb.int] . . .
cycle diesel engine
— m — **confiable** - [comb.int] reliable diesel
engine
— m — **fabricado** - [comb.int] manufactured
diesel engine
— m — **fijo** - [comb.int] stationary diesel engine
— m — **para equipo m** - basic diesel engine

motor m diesel para equipo m básico - [grúas] basic, crane, o machine, diesel engine
— m ——, recambio m, o reposición f, o repuesto m - [comb.int] replacement diesel engine
— m — Perkins - [comb.int] Perkins diesel engine
— m — seguro - [comb.int] reliable diesel engine
— m diesel turboalimentado* - [comb.int] turbocharged diesel engine
— m — con . . . cilindro(s) - [comb.int] . . . cylinder turbocharged diesel engine
— m dotado a con - [electr-mot] motor fitted with • [comb.int] engine fitted with
— m eléctrico - [electr-mot] electric motor
— m — completo - [electr-mot] complete electric motor
— m — — para unidad f - [electr-mot] unit('s) complete electric motor
— m — — f para bombeo m - [bombas] pumping unit complete electric motor
— m — con corriente f cortada - [electr-mot] de-energized electric motor
— m — inducción f - [electr-mot] induction electric motor
— m — con velocidad f variable - [electr-mot] variable speed electric motor
— m — desactivado - [electr-mot] de-energized electric motor
— m — desconectado - [electr-mot] de-energized electric motor
— m — montado - [electr-mot] mounted electric motor
— m — — en forma f corriente - [electr-mot] conventionally, o conventional manner, mounted electric motor
— m — — sobre extremo m - [electr-mot] end mounted electric motor
— m — — — fondo m - [electr-mot] bottom mounted electric motor
— m — — — parte f posterior - [electr-mot] rear mounted electric motor
— m — — — — f superior - [electr-mot] top mounted electric motor
— m — — tope m - [electr-mot] top mounted electric motor
— m — para arranque m - [comb.int] electric starting motor
— m — — corriente f alterna - [electr-mot] alternating current electric motor
— m — — f continua - [electr-mot] direct current electric motor
— m — — impulsión f - [electr-mot] electric drive,ving motor • [hidr-compuertas] electric wrench
— m — —, recambio m, o reposición f, o repuesto m - [electr-mot] replacement electric motor
— m — — unidad f - [electr-mot] unit('s) electric motor
— m — — f para bombeo m - [electr-mot] pumping unit electric motor
— m — reforzado m - [electr-mot] heavy duty electric motor
— m — para impulsión f - [electr-mot] heavy duty drive,ving motor • [hidr-compuertas] heavy duty electric wrench
— m — — — f de compuerta(s) f - [electr-mot] heavy duty electric gate wrench
— m — — — f de varilla f - [hidr-compuer] heavy duty electric stem wrench
— m — — — f para compuerta f - [hidr-compuer] heavy duty electric gate stem wrench
— m — reversible - [electr-mot] reversible electric motor
— m electrónico - [electrón] electronic motor
— m en extremo m - [electr-mot] end motor
— m — — hacia mesa f (rotatoria) f - [petról] rotary end motor
— m — — — perforador m - [petról] driller's end motor

motor m en funcionamiento m - [electr-mot] running motor • [comb.int] running engine
— m — lado m hacia perforador m - [petról] driller('s) side motor
— m — marcha f - [electr-mot] running motor • [comb.int] running engine
— m — operación f - [electr-mot] running motor • [comb.int] running engine
— m — serie y paralelo - [electr-mot] compound wound motor
— m — Y - [electr-mot] Y motor
— m encerrado - [electr-mot] enclosed motor
— m — para ventilador m - [electr-mot] enclosed fan motor
— m enfriado - [electr-mot] cooled motor • [comb.int] cooled (down) engine
— m — con agua m - [comb.int] water cooled engine
— m — — aire m - [electr-mot] air cooled motor • [comb.int] air cooled engine
— m — — tolva f - [comb.int] hopper cooled engine
— m — — ventilador m - [electr-mot] fan cooled motor • [comb.int] fan cooled engine
— m engargantado - [autom] geared engine
— m — inapropiadamente - [autom] lugged engine
— m entero m - [electr-mot] entire motor • [comb.int] entire engine
— m estallado - [comb.int] blown engine
— m estrangulado - [electr-mot] starved motor • [comb.int] choked engine
— m fabricado - [electr-mot] manufactured motor • [comb.int] manufactured engine
— m fallado - [electr-oper] failed motor • [comb.int] failed, o faulty, engine
— m . . . fásico - [electr-mot] . . . phase motor
— m . . . — con jaula f de ardilla f - [electr-mot] . . . phase squirrel cage motor
— m . . . fásico de tipo m con jaula f de ardilla f - [electr-mot] . . . phase squirrel cage type motor
— m . . . fásico para ventilador m - [electr-mot] . . . phase fan motor
— m fijo - [electr-mot] stationary motor • [comb.int] stationary engine
— m fiscalizado - [electr-mot] monitored motor • [comb.int] monitored engine
— m fornido - [electr-mot] husky, o rugged, motor • [comb.int] husky, o rugged, engine
— m frío - [electr-mot] cold motor • [comb.int] cold engine
— m fuera de borda - [nav] outboard motor
— m fundido - [comb.int] blown engine
— m generador - [electr-mot] motor generator
— m — para corriente f continua - [electr-prod] direct current motor-generator
— m — para excitación f - [electr-mot] motor-generator exciter
— m — — grúa f—[grúas] crane motor-generator
— m girado m - [electr-mot] turned, o rotated, motor • [comb.int] turned, o rotated, engine
— m hecho, girar, o rotar—[electr-mot] turned, o rotated, motor • [comb.int] turned, o rotated, engine; cranked engine
— m hermético - [electr-mot] sealed motor
— m hidráulico - [hidr] hydraulic, motor, o engine
— m — con desplazamiento m fijo - [mec] fixed displacement hydraulic motor
— m — — m — con émbolo m (axial) - [mec] fixed displacement (axial piston) hydraulic motor
— m — — émbolo m axial - [mec] axial piston hydraulic motor
— m — — m — con desplazamiento m fijo - [mec] fixed displacement axial piston hydraulic motor
— m — — regulación f (independiente) - [mec] (independently) controlled hydraulic motor
— m — — velocidad f variable - [hidr] variable speed hydraulic motor
— m — operado - run, o operated, hydraulic,

motor
motor m **hidráulico regulado** - [mec] controlled
hydraulic motor
— m — — **independientemente** - [mec] independently controlled hydraulic motor
— m **hidrostático** - hydrostatic motor
— **horizontal** - [electr-mot] horizontal motor •
[comb.int] horizontal engine
— m **impulsado** - [electr-mot] driven motor
— m **impulsor** - [mec] drive,ver,ving motor
— m **inclinado** - [comb.int] slanting, o inclined, engine
— m **independiente** - [electr-mot] independent,
o separate, motor • [comb.int] independent,
o separate, engine
— m **individual** - [electr-mot] individual motor
• [comb.int] individual engine
— m — **reversible** - [electr-mot] individual reversible motor
— m **inferior** - [electr-mot] lower, o bottom,
motor • [comb.int] lower, o bottom, engine
— n — **optativo** - [electr-mot] lower alternate
motor • [comb.int] lower alternate engine
— m — **para laminador** m - [metal-lam] lower
mill motor
— m **instalado** - [electr-mot] installed motor •
[comb.int] installed engine
— m **izquierdo** - [electr-mot] left motor •
[comb.int] left engine
— m **lento** - [electr-mot] slow motor—[comb.int]
slow (speed) engine
— m **limpiado** - [electr-mot] cleaned motor •
[comb.int] cleaned engine
— m — **con vapor** m - [electr-mot] steam cleaned
motor • [comb.int] steam cleaned engine
— m **limpio** - [electr-mot] clean motor • [comb.
-int] clean engine
— m **liviano** - [electr-mot] light motor • (comb.
-int] light engine
— m **lubricado** - [electr-mot] lubricated motor •
[comb.int] lubricated engine
— m — **permanentemente** - [electr-mot] permanently lubricated motor
— m **mal construido** - [electr-mot] poorly, made,
o built, motor • [comb.int] poorly, made, o
built, engine
— m **mantenido con marcha** f **lenta** - [comb.int]
idled engine
— m **marino** - [electr-mot] marine motor • [comb.
-int] marine engine
— m **moderno** - [electr-mot] modern motor •
[comb.int] modern engine
— m **monofásico** - [electr-mot] single phase
motor
— m **montado** - [electr-mot] mounted motor •
[comb.int] mounted engine
— m — **forma** f **corriente** - [electr-mot] conventionally, o conventional manner, mounted
motor • [comb.int] conventionally, o conventional manner, mounted engine
— m — **sobre parte** f **superior** - [electr-mot]
front mounted motor • [comb.int] front mounted engine
— m — — **extremo** m - [electr-mot] end mounted
motor • [comb.int] end mounted engine
— m — — **fondo** m - [electr-mot] bottom mounted
motor • [comb.int] bottom mounted engine
— m — — **frente** m - [electr-mot] front mounted
motor • [comb.int]| front mounted engine
— m — **parte** f **posterior** - [electr-mot] rear
mounted motor • [comb.iunt] rear mounted engine
— m — — — f **superior** - [electr-mot] top
mounted motor • [comb.int] top mounted engine
— m — — **patín(es)** m - [comb.int] skid mounted
engine
— m — — **tope** m - [electr-mot] top mounted
motor • [comb.int] top mounted engine
— m — — **trineo** m - [comb.int] skid mounted
engine
— m **movido** - [electr-mot] moved motor • [comb.-
int] moved engine
— m **mudado** - [electr-mot] moved motor • [comb.-

int[moved engine
motor m **multicilíndrico** - [comb.int] multi-cyl-
inder engine
— m **neumático** - [neumát] pneumatic motor
— m **no rectificado** - [comb.int] unground engine
• standard bore (engine)
— m **nuevo** - [electr-mot] new motor • [comb.int]
new engine
— m **operado** - [electr-mot] operated, o run,
motor • [comb.int] operated, o run, engine
— m — **con marcha** f **lenta** - [comb.int] idled
engine; engine run at @ low speed
— m — — f **plena** - [electr-mot] engine operated at @ full speed • [comb.int] engine
operated at @ full, speed, o throttle
— m **optativo** - [electr-mot] optional, o alternate, motor • [comb.int] optional, o alternate, engine
— m **Orbitrol*** - [electr-mot] Orbitrol* motor
— m **original** - [electr-mot] original motor •
[comb.int] original engine
— m **para accionamiento** m [electr-mot] driving,
o operating, motor • [comb.int] driving, o
operating, engine
— m — **activación** f - activating motor
— m — — m **de dispositivo** - [mec] device activating motor
— m — — m — — m **para ajuste** m - [mec] adjusting device activating motor
— m — — f — — m — — m **para voltaje** m -
voltage adjusting device activating motor
— m — — f **de elemento(s)** m - [electr-mot]
element(s) activating motor
— m — — f — — m **para regulación** f -
[electr-mot] control(ling) element(s) activating motor
— m — **alimentación** f - [ind] feed(ing) motor
— m — — f **acelerado (rápidamente)** - [mec]
(rapidly) accelerated feed(ing) motor
— m — — f **de alambre** m—[sold] wire feed(ing)
motor
— m — — f — **electrodo** m - [sold] electrode,
o rod, feed(ing) motor
— m — **alimentadora** f - [avicult] feeder motor
— m — — f **para alambre** m - [sold] wire feed-
(ing) motor
— m — **amperaje** m **constante** - [electr-mot]
constant, amperage, o current, motor
— m — **apartador(a)** - [mec] pull-off motor
— m — **arrancador** m - [comb.int] starter motor
— m — — m **engranado** - [comb.int] engaged
starter motor
— m — — m **reengranado** - [comb.int] re-engaged
starter motor
— m — **arranque** m - [comb.int] starting motor •
cranking, o engine (starting), motor
— m — — m **de motor** m - [comb.int] engine
starter,ting motor
— m — — m — **tipo** m **con acoplamiento** m **con**
pedal - [comb.int] manual shift type starting
motor
— m — — m **eléctrico** - [comb.int] electric
starter,ting motor
— m — — m **engranado** - [comb.int] engaged
starter motor
— m — — m **fácil** - [comb.int] easy starting
engine
— m — — m **sincrónico** - [electr-mot] synchronous starter,ting motor
— m — **avance** m - [sold] travel motor
— m — **avión** m - [aeron] aircraft engine
— m — **bomba** f - [bombas] pump motor
— m — — f **para agua** m - [hidr] water pump
motor
— m — — f — — m **para venturi** m - [metal-
-prod] scrubber water pump motor
— m — — f — **vacío** m - [mec] vacuum pump
motor
— m — — f — **venturi** m - [metal-prod] scrubber pump motor
— m — **cargadora** f - [metal-prod] skip hoist
motor

motor m para carrillo m - [grúas] trolley motor
— m — **carrera(s)** f - [autom] race,cing engine
— m — **cinta** f - [ind] belt motor
— m —— f **transportadora** - [ind] conveyor (belt) motor
— m —— f **para entrada** f - [ind] entry conveyor (belt) motor
— m —— f —— f **para prensa** f - [ind] press entry conveyor (belt) motor
— m —— f **para prensa** f - [ind] press slide conveyor (belt) motor
— m —— f —— **salida** f - [ind] exit conveyor (belt) motor
— m —— f —— f **para prensa** f - [ind] press exit conveyor (belt) motor
— m — **comedero** m - [avicult] feeder motor
— m — **corriente** f **alterna** - [electr-mot] alternating current motor
— m —— f **con voltaje** m, **alto, o elevado** - [electr-mot] high voltage alternating current motor
— m —— f —— m, **bajo, o reducido** - [electr-mot] low voltage alternating current motor
— m —— f **constante** - [electr-mot] constant current motor
— m —— f **continua** - [electr-mot] direct current motor
— m —— f **con engranaje(s)** m **reductor(es)** - [electr-mot] direct current (reducing) gear(s) motor
— m —— f —— **voltaje** m, **alto, o elevado** - [electr-mot] high voltage direct current motor
— m —— f —— m, **bajo, o reducido** - [electr-mot] low voltage direct current motor
— m — **desempañador** m - [autom] wiper motor
— m —— m **frontal** - [autom] front wiper motor
— m — **destilado*** m - [comb.int] distillate engine
— m — **dirección** f - [autom] steering motor
— m —— f **mecanizada** - [autom] power steering motor
— m — **dispositivo** m - [mec] device('s) motor
— m —— m **para avance** m - [sold] travel (device) motor
— m —— m — **cambio(s)** m - [autom-mec] shift unit motor
— m — **efecto** m **único** - [comb.int] single acting engine
— m — **elevación** f - [mec] hoisting motor
— m —— m **para aguilón** m - [grúas] boom hoist(ing) motor
— m —— m — **cable** m - [grúas] line hoisting motor
— m —— m —— m **fijo** - [grúas] fast line hoist(ing) motor
— m — **elevador** m - [metal-prod] hoist motor
— m —— m **para montacarga(s)** m - [metal-prod] skip hoist(ing) motor
— m —— m — **motón** m—[grúas] main hoist(ing) motor
— m —— m **para skip*** m - [metal-prod] skip hoist(ing) motor
— m —— m **principal** - [grúas] main hoist(ing) motor
— m — **émbolo** m - [petról] plunger motor
— m — **enfriador** m - [ind] cooler, motor, o engine
— m — **estufa** f - [metal-prod] stove motor
— m — **gancho** m - [grúas] (crane), hook, o hoist, motor
— m —— m **para grúa** f - [grúas] crane hook block
— m —— m —— f **para nave** f - [ind] bay crane hook motor
— m —— m —— f —— f **para colada** f - [metal-prod] cast house crane hook motor
— m — **gasolina** f - [comb.int] gasoline engine
— m — **giro** m - [grúas] swing motor
— m — **grúa** f - [grúas] crane motor
— m —— f **sobre torre** f - [grúas] tower crane block

motor m para . . . hercios m - [electr-mot] . . . Hertz motor
— m — **impulsión** f - [mec] drive,ver,ving motor • [petról] prime mover • [hidr-compuertas] wrench
— m —— f **para alambre** m - [sold] wire drive motor
— m —— f —— m **y caja** f **para engranaje(s)** m - [sold] wire drive motor and gear box
— m —— f — **alimentadora** f - [avicult] feeder drive,ving motor
— m — **impulsión** f **para válvula** f - [válv] valve drive,ving motor
— f —— f —— f **para regulación** f - [válv] control valve drive,ving motor
— m —— f —— f —— f **para suministro** m - [válv] supply control valve drive,ving motor
— m —— f —— f —— m **de gas** m - [combust] gas supply control valve drive,ving motor
— m —— f **y caja** f **para engranaje(s)** m—[mec] drive,ving motor and gear box
— m — **impulso** m - [petról] power package
— m — **inclinación** f - [ind] tilting motor
— m — **jeep** m - [autom-mec] jeep engine
— m — **laminación** f - [metal-lam] mill motor
— m — **laminador** m - [metal-lam] mill motor
— m — **limpiaparabrisa(s)** m—[autom] windshield wiper motor
— m —— m **frontal** - [autom] front wiper motor
— m —— m **trasero** - [autom] rear wiper motor
— m — **máquina** f **para skip** m - [metal-prod] skip hoist motor
— m — **medidor** m - [instrum] meter motor
— m —— m **registrador** - [instrum] recording meter motor
— m — **molino** m - [mec] mill motor
— m — **montacarga(s)** f - [metal-prod] skip hoist motor
— m — **nafta** f - [comb.int] gasoline engine
— m — **oscilación** f - [grúas] swing motor
— m — **perforación** f - [petról] drilling engine
— m — **planta** f - [ind] plant, o mill, motor
— m —— f **industrial** - [ind] industrial, plant, o mill, motor
— m — **prensa** f - [mec] press motor
— m — **puente** m - [grúas] bridge motor
— m — **recambio** m - [electr-mot] replacement motor • [comb.int] replacement engine
— m — **recolección** f - gathering motor • close motor
— m — **reemplazo** m - [electr-mot] replacement motor; alternate motor • [comb.int] replacement, o alternate, engine
— m — **regulación** f - [mec] control(ling) motor
— m —— f **para dirección** f - [autom-mec] steering control(ling) motor
— m —— f —— f **mecanizada** - [autom-mec] power steering control(ling) motor
— m —— f **para gas** m - [combust] gas control motor
— m —— f **para servodirección** f - [autom-mec] power steering control(ling) motor
— m — **reposición** f - [electr-mot] replacement motor • [comb.int] replacement engine
— m — **rodillo** m - [metal-lam] roll, o mill, motor
— m —— m **inferior** - [metal-lam] lower, o bottom, roll, o mill, motor
— m —— m **para laminador** m - [metal-lam] lower, mill, o roll, motor
— m —— m **superior** - [metal-lam] upper, o top, roll, o mill, motor
— m —— m — **para laminador** m - [metal-lam] upper, mill, o roll, motor
— m — **rotación** f - [grúas] swing motor
— m — **servodirección** f - [autom-mec] power steering motor
— m — **soplador** m - [ind] blower motor
— m —— m **para combustión** f - [combust] combustion blower motor
— m — **todo uso** m - [electr-mot] general purpose motor

motor m **para tractor** m - [comb.int] tractor engine
— m — **transportador** m - [ind] conveyor motor
— m — — m **para entrada** f - [ind] entry conveyor motor • transfer motor
— m — — m — **riel(es)** m - [metal-lam] rail(s) transfer motor
— m — — **salida** f - [mec] exit, o delivery, conveyor motor
— m — **unidad** f - [electr-mot] unit motor • [comb.int] unit engine
— m — — f **motriz** - [agric-equip] power unit engine
— m — **uso** m **general** - [electr-mot] general purpose motor • [comb.int] general purpose engine
— m — **válvula** f - [válv] valve motor
— m — — f **para regulación** f - [válv] control(ling) valve motor
— m — — f — — f **para suministro** m - (válv) supply control(ling) valve motor
— m — — f — — f — m **de gas** m—[combust] gas supply control(ling) valve motor
— m — — f **para viento** m **(caliente)** - [metal-prod] (hot) blast valve motor
— m — — f **reguladora** f - [válv] control(ling) valve motor
— m — — f **séptum** - [metal-prod] septum valve motor
— m — **vehículo** m - [autom-mec] vehicle engine
— m — — m **transportador** - [grúas] carrier engine
— m — **velocidad** f, **alta**, o **elevada** - [electr-mot] high speed motor • [comb.int] high speed engine
— m — — f, **baja**, o **reducida** - [electr-mot] low speed motor • [comb.int] low speed engine
— m — — f **constante** - [elecrr-mot] constant speed motor • [comb.int] constant speed engine
— m — — f **variable** - [electr-mot] variable speed motor • [comb.int] variable speed engine
— m — **ventilador** m - [mec] fan, o blower, motor
— m — — m **activado** - [electr-mot] energized fan motor
— m — — m **engrasado** - [mec] grased fan motor
— m — — m **para estufa** f - [metal-prod] stove fan motor
— m — — m **principal** - [ind] main fan motor
— m — — m — **para combustión** f - [comb] main combustion fan motor
— m — — m **reemplazado** - [mec] replaced fan motor
— m **parado** - [electr-mot] stopped motor • [comb.int] stopped engine • stalled engine
— m **Perkins** - [comb.int] Perkins engine
— m **pesado** - [electr-mot] heavy motor • [comb.-int] heavy engine
— m **portátil** - [electr-mot] portable motor • [comb.int] portable engine
— m **preparado** - [electr-mot] prepared motor • [comb.int] prepared engine
— m — **en fábrica** f - [electr-mot] factory prepared motor • [comb.int] factory prepared engine
— m **primario** - [electr-mot] primary motor • [comb.int] primary engine • [petról] prime mover
— m **principal** - [electr-mot] main motor • [comb.int] main engine
— m — **para accionamiento** m - [electr-mot] main drive,ving motor • [comb.int] main drive,ving engine
— m — **para elevación** f - [grúas] main hoist-(ing) motor
— m — — **impulsión** f - [electr-mot] main drive,ving motor • [comb.int] main drive,ving engine
— m — — **laminador** m - [metal-lam] main mill motor
— m **protegido** - [electr-mot] protected, o shielded, motor • [comb.int] protected engine

motor m **proyectado** - [electr-mot] designed motor • [comb.int] designed engine
— m **puesto en marcha** f - [electr-mot] started motor • [comb.int] started, o turned on, engine
— m **quemado** - [electr-mot] burned out motor • [comb.int] burned out engine
— m **rápido** - [electr-mot] fast, o high speed, motor • [comb.int] fast, o high speed, engine
— m **rearmado** - [electr-mot] reassembled motor • [comb.int] reassembled engine
— m **recalentado** - [electr-mot] overheated motor • [comb.int] overheated engine
— m **recio** m - [electr-mot] tough motor • [comb.-int] tough engine
— m **reconstruido** - [eletr-mot] rebuilt motor • [comb.int] rebuilt engine
— m — **ensayado** - [electr-mot] tested rebuilt motor • [comb.int] tested rebuilt engine
— m **rectificado** - [comb.int] rebored engine
— m **reductor** - [electr-mot] reducer,cing motor
— m **reemplazado** - [electr-mot] replaced motor • [comb.int] replaced, o transplanted, engine
— m **regulado** - [electr-mot] controlled motor • [comb.int] controlled engine
— m — **independientemente** - [electr-mot] independently controlled motor • [comb.int] independently controlled engine
— m **reparado** - [electr-mot] repaired motor • [comb.int] repaired engine
— m **reversible** - [electr-mot] reversible motor
— m **revolucionario** - [electr-mot] revolutionary motor • [comb.int] revolutionary engine
— m **rotado** - [electr-mot] rotated, o turned, motor • [comb.int] rotated, o turned, engine • cranked engine
— m **rotativo** - [comb.int] rotary engine
— m — **con . . . rotor(es)** m - [comb.int] . . .-rotor rotary engine
— m **rudo** - [electr-mot] rugged, o husky, motor • [comb.int] rugged, o husky, engine
— m **seguro** - [electr-mot] safe, o reliable, motor • [comb.int] safe, o reliable, engine
— m **Selsyn** - [electr-mot] synchronous, o Selsyn, motor
— m **semi-diesel** m—[comb.int] semidiesel engine
— m **separado** - [electr-mot] separate motor • [comb.int] separate engine
— m **sin abastecer** - [comb.int] dry, o unserviced, engine
— m — **carga** f - [electr-mot] unloaded motor • [comb.int] unloaded, o no load, engine
— m — — f **para corriente** f **alterna** - [electr-mot] unloaded alternating current motor
— m — — f **continua** - [electr-mot] unloaded direct current motor
— m — **ventilación** f - [electr-mot] unventilated, o nonventilated, motor
— m **sincrónico** - [electr-mot] synchronous, o Selsyn, motor
— m — **para arranque** m **sin carga** f - [electr-mot] unloaded start(ing) synchronous motor
— m **sobrecalentado** m - [electr-mot] overheated motor • [comb.int] overheated engine
— m **sobrecargado** - [electr-mot] overloaded motor • [comb.int] overloaded engine
— m **sólido** m - [electr-mot] rugged motor • [comb.int] rugged engine
— m **superior** - [electr-mot] upper, o top, motor [comb.int] upper, o top, engine
— m — **optativo** - [electr-mot] upper, o top, optional, o alternate, motor • [comb.int] upper, o top, optional, o alternate, engine
— m — **para laminador** m - [metal-lam] upper, o top, mill, o roll, motor
— m **Synchrogear**—[electr-mot] Synchrogear motor
— m **térmico** - heat engine
— m **totalmente blindado** - [electr-mot] totally enclosed motor
— m — — **para ventilador** m - [electr-mot] totally enclosed fan motor
— m — **encerrado** - [electr-mot] totally en-

closed motor
motor m **totalmente encerrado para ventilador** m -
[electr-mot] totally enclosed fan motor
— m **trifásico** - [electr-mot] three-phase motor;
Y motor
— m — **con jaula** f **de ardilla** f - [electr-mot]
three-phase squirrel cage motor
— m — **de tipo** m **con jaula** f **de ardilla** f -
[electr-mot] three-phase squirrel cage type
motor
— m — **en Y** - [electr-mot] three-phase Y motor
— m — — **para ventilador** m - [electr-mot] three-
-phase fan motor
— m **turboalimentado*** - [comb.int] turbocharged
engine
— m **univeral** - [electr-mot] universal motor
— m **variable** - [electr-mot] variable motor
— m **verificado** - [electr-mot] checked motor •
[comb.int] checked engine
— m **vertical** - [electr-mot] vertical motor •
[comb.int] vertical engine
— m **viejo** - [electr-mot] old motor • [comb.int]
old engine
motorización f **completa** - [ind] complete mechanization
motorizado,da a - motorized; power; motor operated; mechanized
motorreductor* m - [mec] motoreducer*
mototopadora* f - [constr-equip] bulldozer
mototrailla* f - [constr-equip] self-powered, o
tractor, scraper
motovariador m - [mec] motovariator
motovariador,ra* a - motovariating,tor
motriz a - . . .; power; drive
mover v - . . . • to activate; to actuate; to
wiggle • [mec] to shift • to jog • [instrum]
to set • [electr-oper] to throw (@ switch)
— v **acelerador** m—[comb.int] to move @ throttle
— v **acoplamiento** m - [mec] to move @ coupling
— v **antorcha** f - [sold] to move @ (arc) torch
— v — **hasta comienzo** m **de junta** f - [sold] to
move @ (arc) torch to @ joint beginning
— v — f **con dos electrodos** m **de carbono** m -
[sold] to move @ (arc) torch
— v — f — — m — — **hasta comienzo** m **de**
junta f - [sold] to move @ (arc) torch to @
joint beginning
— v **árbol** m - [botán] to move @ tree • [mec] to
move @ shaft
— v — m **para entrada** f **(de fuerza** f**)** - [mec]
to move @ input shaft
— v **arco** m - [constr] to move @ arch • [sold]
to move, o to play, @ arc
— v — m **en ambos sentidos** - [sold] to, move, o
play, @ arc back and forth
— v — m — — — m **a (lo) largo de costura** f -
[sold] to, move, o play, @ arc back and forth
along @ seam
— v **armadura** f - [electr-mot] to, move, o rock,
@ armature
— v **biela** f - [comb.int] to move @ connecting
rod
— v **brazo** m - [mec] to move @ arm
— v — m **hacia adentro y hacia afuera** - [mec]
to move @ arm in and out
— v **caja** f - [mec] to move @, box, o case
— v — f **portahilera** - [trefil] to move @ die
box
— v **campana** f - to move @ bell • [mec] to move
@ hood
— v **con gato** m - [mec] to jack
— v **conductor** m—[electr-instal] to move @ lead
— v — m **para osciloscopio** m - [electrón] to
move @ oscilloscope lead
— v **conmutador** m - [electr-oper] to, move, o
throw, @ switch
— v — m **con palanca** f - [electr-oper] to,
move, o throw, @ toggle switch
— v — m **en ambos sentidos** m - [electr-oper] to
move @ switch back and forth
— v **corona** f **dentada**—[mec] to move @ ring gear
— v **de lado a lado** - [mec] to move, back and
forth, o from side to side

mover v **demasiado** - to move too much
— v — **rápidamente** - to move too quickly
— v **desde** - to move (away) from
— v — **posición** f **neutral**—[mec] to move (away)
from @ neutral position
— v **dirección** f - [autom] to steer; to move @
steering wheel
— v **eje** m - [mec] to move @, axle, o shaft
— v — m **para entrada** f **(de fuerza** f**)** - [autom-
-mec] to move @ input shaft
— v **electrodo** m - [sold] to move @ electrode
— v **émbolo** m - [mec] to move @ piston
— v **embrague** m - [mec] to move @ clutch
— f — m **deslizante** - [mec] to move @ sliding
clutch
— v **en ambos sentidos** - to move back and forth
— m — — — **con (movimiento** m **de) vaivén** m -
[mec] to work back and forth
— v — **cualquier sentido** m - [mec] to move in
any direction
— v — **sentido** m **axial** - to move axially
— v — **árbol** m **para entrada** f **(de**
fuerza f**)** - [autom-mec] to move @ input
shaft axially
— v — **suelo** m - [suelos] to move in @ soil
— v **encabezadora** f - [trefil] to move @ header
— v **entrerrosca** f - [tub] to move @ nipple
— v — f **para tubería** f - [mec] to move @
pipe nipple
— v **excesivamente** - to move excessively
— v **hacia adelante** - to move, forward, o to @
front
— v **fácilmente** - to move easily
— v **furiosamente** - [autom-limpiaparabrisas] to
flick furiously
— v **hacia** - to move toward(s)
— v — **abajo** - to move down(wards)
— v — **adelante** - to move forward(s)
— v — **adentro** - to move in(wards)
— v — — **y (hacia) afuera** - to move in and out
— v — — — **de arco** m - [sold] to move in
and out of @ arc
— v — **afuera** - to move, out(wards), o in @
outward direction
— v — **arriba** - to move up(wards)
— v — — **y (hacia) abajo** - to move up and down
— v — — **(hacia) otro lado** m - to move up
and across
— v — — — — — — m **de soldadura** f - [sold]
to move up (and) across @ weld
— v — **atrás** - to move, back(wards), o towards
@ rear
— v — **costado** m - to move sideways
— v — **dentro y fuera** - to move in and out
— v — **izquierda** f - to move (towards) @ right
— v — **izquierda** f - to move (towards) @ left
— v — **lado** m **contrario** - to move to @ opposite
side
— v **otro lado** - to move across
— v — — — — m **de soldadura** f - [sold] to move
across @ weld
— v **hasta** - to move (in)to
— v — **posición** f - to move (in)to @ position
— v **horquilla** f - [mec] to move @ fork
— v — f **para cambio(s)** m - [mec] to move @
shift(ing) fork
— v **intermitentemente** - to jog (intermittently)
— v — **encabezadora** f—[trefil] to jog @ header
— v — **máquina** f - [mec] to jog @ machine
— v **interruptor** m - [electr-oper] to throw @
switch
— v **lentamente** - to move slowly; to creep • to
ease
— v — **antorcha** f **(con dos electrodos** m **de car-**
bono m**)** - [sold] to move @ torch slowly
— v **leva** f - [mec] to move @ cam
— v **libremente** - to move freely
— v — **de lado a lado** - [mec] to move freely
from side to side
— v **llave** - [electr-oper] to, move, o throw, @
switch
— m **manguito** m - [mec] to, move, o slide, @
sleeve

mover v̲ manguito m̲ para acoplamiento m̲ - [mec]
to, move, o̲ slide, @ coupling sleeve
— v̲ manivela f̲ - [mec] to crank
— v̲ manualmente - to move, manually, o̲ by hand
— v̲ máquina f̲—[mec] to, move, o̲ jog, @ machine
— v̲ material(es) v̲ - [constr] to move @ mate-
rial(s)
— v̲ motor v̲ - [electr-mot] to move @ motor •
[comb.int] to move @ engine
— v̲ palanca f̲ - to move @ lever • [electr-oper]
to, move, o̲ throw, @ switch • [autom] to move
@ stick
— v̲ — para regulador m̲ - [comb.int] to move @
throttle lever
— v̲ — f̲ hacia adelante - [mec] to move @
lever forward(s)
— v̲ — f̲ — atrás - [mec] to move @ lever back-
(wards)
— v̲ — para cambio m̲ de marcha f̲ - [autom]
to move @, shift, o̲ stick
— v̲ patrón m̲ para contacto m̲ - [mec] to move @
contact pattern
— v̲ piñón m̲ - [mec] to move @ pinion
— v̲ — m̲ impulsor - [mec] to move @ drive,ving
pinion
— v̲ por curso m̲ - [mec] to move through @ range
— v̲ — todo curso m̲ - [mec] to move through @
(whole) range
— v̲ producto m̲ - to move @ product
— v̲ rápidamente - to move, rapidly, o̲ fast • to
flick
— v̲ regulador m̲ - [comb.int] to move @ throttle
— v̲ — m̲ para amperaje m̲ - [sold] to adjust @
current control
— v̲ — m̲ en ambos sentidos m̲ - [electr-oper] to
move @ switch back and forth
— v̲ sujetador m̲ - [mec] to move @ cam
— v̲ tablestaca f̲ - [constr] to move @ sheet
— v̲ tablestacado - [constr] to move @ sheeting
— v̲ tierra f̲ - [constr] to move @ earth
— v̲ torre f̲ - to move @ tower
— v̲ tuerca f̲ - [mec] to move @ nut
— v̲ uniformemente - to move, evenly, o̲ uniformly
— v̲ volante m̲ - [mec] to move @ flywheel •
[autom-mec] to move @ steering wheel
— v̲ — m̲ para dirección f̲ - [autom-oper] to
move @ steering wheel
movible a̲ - . . .; moving; traveling • portable
• [mec] working • [com] subject to escalation
movido,da a̲ - moved; wiggled • driven • shifted;
jogged • powered
— a̲ de lado a lado - moved across
— a̲ desde - moved (away) from
— a̲ desde posición f̲ neutral - [mec] moved from
@ neutral position
— a̲ en ambos sentidos m̲ - moved back and forth
— a̲ — sentido m̲ axial - [mec] moved axially
— a̲ excesivamente - moved excessively
— a̲ fácilmente - moved easily
— a̲ furiosamente - [autom-limpiaparabrisas]
flicked furiously
— a̲ hacia - moved to(wards)
— a̲ — abajo - moved down(wards)
— a̲ — adelante - moved, forwards, o̲ to(wards)
@ front
— a̲ — adentro - moved in(wards)
— a̲ — y afuera - moved in and out
— a̲ — — — de arco m̲ - [sold] moved in and
out of @ arc
— a̲ — afuera - moved, out(wards, o̲ in @ outward direction
— a̲ — arriba - moved up(wards)
— a̲ — atrás - moved, back(wards), o̲ to @ rear
— a̲ — costado - moved, sideways, o̲ to @ side
— a̲ hasta - moved, up to, o̲ into
— a̲ — posición f̲ - moved (in)to @ position
— a̲ intermitentemente - moved intermittently •
[mec] jogged
— a̲ lentamente - moved slowly • eased
— a̲ — a largo m̲ de junta f̲ - [sold] moved
slowly along @, joint, o̲ seam
— a̲ libremente - moved freely

movido,da - libremente de lado a lado - [mec-
moved freely from side to side
— a̲ manualmente - moved, manually, o̲ by hand
— a̲ por curso m̲ - [mec] moved through @ range
— a̲ — eje m̲ para transmisión f̲ - [mec] line
shaft driven
— a̲ — todo curso m̲ - [mec] moved, across, o̲
through, @ (full) range
— a̲ rápidamente - moved rapidly • flicked
— a̲ uniformemente - moved, evenly, o̲ uniformly
moviéndose a̲ furiosamente—[mec-limpiaparabrisa]
flicking furiously
móvil m̲ - . . .; motivation; appeal | a̲ - . . .;
moving • portable • [mec] follower
— m̲ para compra(s) f̲ - [com] buying, o̲ purchasing, motive,vation
movilidad f̲ - . . . • portability
— f̲ máxima - maximum, movability, o̲ mobility •
maximum portability
movilización f̲ - . . . • handling
movimiento m̲ - . . . . • action • flow • twist •
[mec] travel • jogging • shifting • wiggling
• swivel,ling • [sold] motion • [f.c.] traffic (department)
— m̲ acelerado - [fís] accelerated motion
— m̲ — uniformemente - [fís] uniformly accelerated, motion, o̲ movement
— m̲ acometedor - jabbing action
— m̲ — rápido - fast jabbing action
— m̲ alternativo - [mec] reciprocating movement
— m̲ amplio - wide, o̲ sweeping, action
— m̲ — de tejido m̲ - [sold] wide weave,ving
— m̲ ascendente - upward, motion, o̲ movement
— m̲ — de arco m̲ - [sold] arc upward motion
— m̲ — leve - [sold] slight upward motion
— m̲ — ligero - [sold] slight upward motion
— m̲ — y descendente - [sold] up(ward) and
down(ward) motion
— m̲ axial - [mec] axial movement
— m̲ bruto - [fin] gross flow
— m̲ bruto de, caja, o̲ efectivo - [fin] gross
cash flow
— m̲ — de fondo(s) m̲ - [fin] gross cash flow
— m̲ centrípeto - [mec] in(ward) movement
— m̲ circular - swirl(ing)• [sold] circular, o̲
rotary, motion; circulating motion
— m̲ —, leve, o̲ ligero - [sold] slight circular,culating motion
— m̲ — para junta(s) f̲ - [sold] joint(s) circular, motion, o̲ movement
— m̲ — — — f̲ con chaflán m̲ - [sold] butt
joints circular, motion, o̲ movement
— m̲ — — — en V - [sold] V-butt joint
circular, motion, o̲ movement
— m̲ circulatorio - circulating, o̲ rotating,
movement, o̲ motion
— m̲ como con serrucho m̲ - [mec] sawing motion
— m̲ con camión(es) m̲ - [transp] truck transportation
— m̲ — mano f̲ - hand, o̲ manual, moving,vement
— m̲ — reloj m̲ - [mec] clockwise movement
— m̲ continuo - continuous, motion, o̲ movement
— m̲ contra reloj m̲ - [mec] counterclockwise
movement
— m̲ cuidadoso - careful, o̲ deliberate, movement
— m̲ dado - [autom] given maneuver
— m̲ de árbol m̲ - [mec] shaft movement
— m̲ — — m̲ para entrada f̲ (de fuerza f̲) -
[autom-mec] input shaft, moving, o̲ movement
— m̲ — acelerador m̲ - [comb.int] throttle, moving, o̲ movement
— m̲ — acoplamiento m̲ - [mec] coupling moving
— m̲ — agua m̲ - [hidr] water movement,ving
— m̲ — arco m̲ - [sold] arc, weaving, o̲ playing,
o̲ moving,vement
— m̲ — armadura f̲ - [electr-mot] armature,
moving, o̲ rocking
— m̲ — biela f̲ - [comb.int] connecting rod
movement
— m̲ — boquilla f̲ - [mec] nozzle movement,ving
— m̲ — caja f̲ - [ind] box, o̲ case, movement •
[com] cash, movement, o̲ handling, o̲ flow, o̲
operation(s), o̲ transaction(s)

movimiento m de caja f portahilera(s) - [trefil] die box movement,ving
— m — **cambio m para velocidad f** - [autom-mec] gear change movement • catch motion
— m — **campana f** - bell movement,ving • [ind] hood movement,ving
— m — **carga(s) f** - [constr] load movement,ving • [transp] freight movement,ving
— m — **compresión f axial** - [mec] axial compression movement,ving
— m — **conductor(es) m** - [electr-instal] leads movement,ving
— m — — **m para osciloscopio m** - [electrón] oscilloscope lead movement,ving
— m — **conmutador m** - [electr-oper] switch flick(ing)
— m — **contacto m** - [electr-oper] contact movement,ving
— m — **contracción f** - contracting,tion movement • [sold] shrinkage,king movement
— m — — **f libre** - free contraction movement • [sold] free shrinkage,king movement
— m — **cordón(es) m** - [cabl] strand(s) movement
— m — **corona f** - [mec] ring gear movement
— m — — **f dentada** - [mec] ring gear movement
— m — **chicoteo m** - [sold] whipping action
— m — — **m hacia arriba y (hacia) abajo** - [sold] up(ward) and down(ward) whipping motion
— m — — **m sencillo** - [sold] straight whipping motion
— m — — **m simple** - [sold] straight whipping motion
— m — **chorro m** - [mec] jet movement
— m — **circulación f** - [vial] traffic movement
— m — **dirección f** - [autom-mec] steering (action)
— m — **divisa(s) f** - [fin] (foreign) currency, transfer, o movement
— m — **efectivo** - [fin] cash flow
— m — **eje m** - [mec] axle movement • shaft movement
— m — **émbolo m** - [mec] piston movement
— m — **embrague m** - [mec] clutch movement
— m — — **m deslizante** - [mec] sliding clutch movement
— m — **encabezadora f** - [trefil] header jogging
— m — **entrerrosca f** - [tub] nipple movement
— m — — **f para tubo(s) m** - [tub] pipe nipple movement
— m — **equipo m** - [ind] equipment moving
— m — — **m móvil** - [ind] mobile equipment traffic
— m — **esquina f** - corner movement
— m — **fondo(s) m** - [fin] fund(s) movement; cash flow
— m — — **m procedente(s) de operación(es) f** - [fin] cash flow from @ operation(s)
— m — **freno m** - [mec] brake, movement, o travel
— m — **ganado m** - [ganad] livestock movement
— m — **grúa f** - [grúas] crane movement • crane function
— m — **hogar m** - [metal-prod] hearth movement
— m — **horquilla f** - [mec] fork movement,ving
— m — — **f para cambio(s) m** - [mec] shift(ing) fork movement,ving
— m — **inclinación f** - [metal-prod] tilting, movement,ving, o mechanism
— m — — **f mecánica** - [ind] motor tilting
— m — **lado a lado** - [mec] side to side movement • swaying
— m — **leva f** - [mec] cam movement,ving
— m — **manivela f** - [mec] cranking
— m — **máquina f** - [mec] machine movement,ving • machine jogging
— m — **material(es)** - [ind] materials handling
— m — **materias f primas** - [ind] raw materials handling
— m — **motor m** - [electr-mot] motor move,ving • [comb.int] engine move,ving
— m — **muñeca f** - wrist, movement, o motion
— m — **palanca f** - lever movement,ving • [autom] stick movement,ving

movimiento m de palanca f para conmutador m - [electr-oper] switch flick(ing)
— m — — **f reguladador m** - [comb.int] throttle lever move,ving
— m — — **f hacia adelante** - [mec] lever forward movement,ving
— m — — **f cambio m de marcha f** - [autom-mec] stick movement,ving
— m — **pasador m** - [mec] pin movement,ving
— m — — **excéntrico** - [mec] eccentric pin movement,ving
— m — **patrón m** - [mec] pattern movement,ving
— m — — **m para contacto m** - [mec] contact pattern movement,ving
— m — **peatón(es) m** - [vial] pedestrian(s), o people, movement,ving
— m — **piñón m** - [mec] pinion movement,ving
— m — — **m impulsor** - [mec] drive,ving pinion movement,ving
— m — **población f** - [pol] population, movement,ving, o shift(ing)
— m — **producto m** - [com] product movement,ving
— m — **regulador m** - [mec] control movement • [comb.int] throttle movement,ving
— m — **retorno m** - [mec] return movement
— m — **rotación f** - [mec] rotation(al) movement
— m — — **f necesario** - [mec] necessary, o required, rotation movement
— m — **suelo m** - [suelos] soil movement
— m — **sujetador m** - [mec] fastener movement • cam movement,ving
— m — **suspensión f** - [autom-mec] suspension movement
— m — **tablestaca f** - [constr] sheet(ing) movement,ving
— m — **tablestacado m** - [constr] sheeting movement,ving
— m — **tapón m** - [mec] plug movement,ving
— m — — **m para estrangulador m** - [mec] choke plug movement,ving
— m — **tejido m** - [sold] weaving motion; side to side weave,ving; weave,ving
— m — — **m angosto** - [sold] narrow (side-to-side) weave,ving
— m — — **m triangular** - [sold] triangular weave,ving
— m — — **m — angosto** - [sold] narrow, o small, triangular weave,ving
— m — **tierra f** - [constr] earth movement,ving; grading; earthwork(ing)
— m — **torre f** - tower movement,ving
— m — **torsión f** - twist(ing motion)
— m — — **f de muñeca f** - wrist movement,ving
— m — — **f en sentidos m opuestos** - back and forth twisting motion
— m — — **f leve de muñeca f** - slight wrist, motion, o movement
— m — **tránsito m** - [transp] traffic, flow, o movement
— m — **traslación f** - [med] translation movement
— m — — **f lenta** - [mec] slow translation movement
— m — — **f necesario** - [mec] necessary, o required, translation movement
— m — **tubería(s) f** - [tub] piping movement
— m — — **f después de colocación f** - [tub] pipe after laying movement
— m — — **f permitido** - [tub] allowed pipe movement
— m — **tuerca f** - [mec] nut movement,ving
— m — **vagón(es) m** - [f.c.] car, movement,ving, o operation
— n — **vaivén m** - back and forth, o oscillating, motion • [sold] weave,ving
— m — **varilla f** - [mec] rod movement,ving
— m — — **f hacia adentro y (hacia) afuera** - [sold] in and out rod movement
— m — **vehículo(s) m** - [transp] vehicle(s) movement,ving
— m — **volante m** - [mec] flywheel movement,ving • [autom] steering wheel movement,ving; steering

movimiento m de volante m **para dirección** f -
[autom] steering wheel movement,ving
— m — **zigzaqueo** m - zigzaging movement,ving •
[sold] whipping movement,ving
— m — — m **sencillo** - [sold] straight whipping
movement,ving
— m — — m **simple** - [sold] straight whipping
movement,ving
— m **debajo de babeta** f - [constr] under @
flashing movement,ving
— m — — **tapajunta(s)** m - [constr] under @
flashing movement,ving
— m **decreciente**—[mec] decreasing movement,ving
— m **definido** - [sold] definite, o deliberate,
movement,ving, o motion
— m **descendente** - downward, movement,ving, o
motion
— m — **de arco** m - [sold] arc downward(s), movement,ving, o motion
— m **descontado** - [fin] discounted flow
— m — —, **caja** f, o **efectivo** m - [fin] discounted cash flow
— m — — **fondo(s)** m - [fin] discounted cash
flow
— m **desde** - movement,ving away (from)
— m — **posición** f **neutral**—[mec] movement,ving
(away) from @ neutral position
— m **dextrogiro** - clockwise, movement, o motion
— m **dextrorso** - clockwise, movement, o motion
— m **en ambos sentidos** m - back and forth movement,ving
— m — **chicote(o)** m - [sold] whipping motion
— m — **línea** f **recta** - [mec] straight line,
movement, o motion
— m — **playa(s)** f - [f.c.] yard switching •
yard towing
— m — **sentido** m **axial** - [mec] axial movement
— m — — m **axial de eje** m - [autom-mec] axial
shaft movement,ving
— m — — m — — — m **para entrada** f (**de**
fuerza f) - [autom-mec] axial input shaft
movement,ving
— m — — — — **árbol** m **para entrada** f (**de**
fuerza f) - [autom-mec] axial input shaft
movement,ving
— m — **sentido** m **contrario a agujas** f **de reloj**
m - counterclockwise, movement, o motion
— m — — m **de agujas** f **de reloj** m - [mec]
clockwise, movement, o motion
— m — **taller** m - [transp] (in-)shop towing
— m — **vaivén** m - side-to-side movement,ving
— m — **zigzag** - [sold] whipping motion; weave
— m **entre engranaje** m **y desengranaje** m - [mec]
movement between engagment and disengagement
— m **evitado** - avoided, o prevented, movement
— m **excesivo** - excessive movement
— m **fácil** - easy movement,ving; mobility
— m — **en obra** f - [ind] on, site, o job, mobility
— m **favorable** - favorable, movement, o motion
— m **ferroviario** - [ind] railway transportation
— m **giratorio** - [grúas] swing movement
— m **hacia** - movement,ving toward(s)
— m — **abajo** - downward movement,ving
— m — **adelante** - forward movement,ving; moving
to(wares) @ front
— m — **adentro** - inward movement; moving in
— m — — **y (hacia) afuera** - in and out movement,ving
— m — — — (—) — **de arco** m - [sold] in-and-
-out of @ arc movement
— m — — — (—) — — **varilla** f (**para apor-**
tación f) - [sold] in-and-out (filler) rod
movement
— m — **afuera** - out(ward) movement,ving
— m — **arriba** - up(ward) movement,ving
— m — — **y (hacia) abajo** - up(wards) and
down(wards) movement
— m — **atrás** - backward, o to(wards) @ rear,
movement,ving
— m — **derecha** f - (towards @) right movement •
clockwise motion
— m — **izquierda** f - (towards @) left movement

• counterclockwise, movement, o motion
movimiento m **horizontal** - horizontal movement
— m **improductivo** - lost motion
— m **independiente** - [grúas] independent movement
— m — **horizontal** - [grúas] independent horizontal movement
— m — **vertical** - [grúas] independent vertical
movement
— m **individual** - independent movement • [mec]
independent drive,ving
— m **ininterrumpido** - uninterrupted movement
— m **intermitente** - intermittent movement •
[mec] jogging
— m — **de encabezadora** f - [trefil] header
jogging
— m — — **máquina** f - [mec] machine jogging
— m **interrumpido** - interrupted movement
— m **lateral** - lateral, o sideways, movement
— m — **de crisol** m - [metal-prod] hearth, lateral, o sideways, movement
— m — — **hogar** m - [metal-prod] hearth lateral
movement
— m — — **suelo** m - [suelos] soil, lateral, o
sideways, movement
— m — **mecánico** - [mec] motor transversing
— m **lento** - slow, movement,ving, o motion;
easing
— m — **a largo** m **de junt(ur)a** f - [sold] slow
movement along @ joint
— m — — — n — **soldadura** f - [sold] slow
movement along @, weld, o joint
— m **leve** - slight, movement, o motion
— m — **de muñeca** f - slight wrist movement
— m — **tejido** m - [sold] slight weave,ving
— m — **vaivén** m - slight back and forth,
movement, o motion
— m **libre** - free movement,ving
— m — **de contacto** m - [electr-oper] contact
free movement,ving
— m — — **lado** m **a lado** m - [mec] free movement,ving from side to side
— m **ligero** - [mec] slight, movement,ving, o
motion
— m **ligero de raspado** m - [sold] (s)light moving scratch
— m — — **tejido** m - [sold] slight weave,ving
— m **lineal** - linear, movement,ving, o motion
— m — **de bloque** m **deslizable** - [mec] slide(r)
block linear, movement,ving, o motion
— m **medido** - measured movement,ving, o motion
— m — — **entrerrosca** f - [tub] measured nipple, movement,ving, o motion
— m — — — f **para tubo(s)** m - [tub] measured
pipe nipple, movement,ving, o motion
— m **mínimo** - minimum, movement,ving, o motion
— m **muy leve** - very light movement,ving
— m **necesario** - necessary movement,ving
— m **neto** - net movement,ving • [fin] net flow
— m — —, **caja** f, o **efectivo** m - [fin] net
cash flow
— m — — **fondo(s)** m - [fin] net cash flow
— m **oblicuo** m - [mec] oblique, o skew, movement
— m — **hacia derecha** f - [mec] skew right movement
— m — — **izquierda** f—[mec] skew left movement
— m **ondulatorio** - waving,vy, movement,ving, o
motion
— m **oscilante** - oscillating, o pendular, movement,ving, o motion
— m **oscilatorio** - oscillating, movement,ving, o
motion
— m — **longitudinal**—[sold] straight whip(ping)
— m **para avance** m—[mec] forward, movement,ving
• feed(ing) motion
— m — **raspado** m - [sold] scratch(ing), movement,ving, o motion
— m — — m **cruzando costura** f - [sold] across
@ seam scratch(ing), movement,ving, o motion
— m **pendular** - pendular movement; swaying
— m **permitido** - allowed,wable movement,ving
— m **por carretera** f - [transp] highway towing
— m — **curso** m - [mec] through, o over, @ range

movement,ving
movimiento m **por todo curso** m - [mec] through, o over, @ range, movement,ving, o motion
— m **radial** - [mec] radial movement,ving
— m **rápido** - rapid, o fast, o quick, movement, o motion • flicking
— m **restringido** - restricted movement,ving
— m — **de cordón** m - [cabl] strand restricted movement,ving; strand movement restriction
— m **retardado** - delayed movement • lagging
— m **relativo**—relative, movement,ving, o motion
— m — **de esquina** f - [constr] corner relative movement,ving
— m — **desfavorable** - unfavorable relative, movement,ving, o motion
— m — **favorable** - favorable relative, movement,ving, o motion
— m **retardado** - delayed, movement,ving, o motion • lag(ging) • decreased movement,ving
— m **rotativo** - [mec] rotary, movement,ving, o motion
— m **seguro** - [segurid] safe movement,ving
— m **siniestrogiro** - counterclockwise, movement,ving, o motion
— m **siniestrorso** - counterclockwise, movement,ving, o motion
— m **total** - total movement,ving
— m **transversal** - transverse movement,ving • [mec] cross slide,ding
— m — **de suelo** m - [suelos] soil transverse movement,ving
— m **traslatorio** - [agric] ground drive
— m **uniforme** - uniform, o even, movement,ving, o motion
— m **variable** - [fís] variable motion
— m **vertical** - vertical movement,ving
mozo m - . . .; bellman
— m **para limpieza** f - [ind] clean-up man
mucha adv **intensidad** f - much intensity
— f **protección** f - much, o over-, protection
muchacho m - . . .; kid
muchas adv **aplicaciónes** f - many applications
— adv — f **distintas** - many (different) applications
— f **ranuras** f - many grooves | adv - (con) severely grooved
mucho,cha a - . . . | adv - widely; heavily
— adv **espacio** m - much space
— adv **más** - much, o @ great deal, o @ lot, more
— adv **menor** - much, smaller, o less
— adv **metal** m **aportado** - [sold] much deposited, weld, o metal
— adv **tiempo** m - much, o lots of, time
— adv **trabajo** m - much, work, o labor
— adv — m **manual** - much, o lots of, manual, o hand, work, o labor
— m **tráfico** - [vial] much, o lots of, traffic
— adv **uso** m - much, o extensive, use
— adv **vacío** m - [neumát] much vacuum
muchos adv **aumentos** m - high magnification
— adv **sentidos** m - many ways
muchos,chas adv - many
mudado,da a - moved
— a **fácilmente** - moved easily
mudanza f - . . .; move
— f **de motor** m - [electr-mot] motor move,ving • [comb.int] engine move,ving
— f — **válvula** f - [válv] valve move,ving
— f **fácil** - easy move,ving
mudar v **fácilmente** - to move easily
— v **motor** - [electr-mot] to move @ motor • [comb.int] to move @ engine
— v **válvula** f - [válv] to move @ valve
mueble m **para archivo** m - [com] file; filing cabinet
— m — **escritorio** m - [com] office furniture
— m — **patio** m - [domést] patio furniture
— m — **venta** f - [com] furniture for sale • selling station
— m — — f **de neumático(s)** m - [autom-neumát] tire selling station
— m **usado** - [domést] used furniture
mueblero,ra a - furniture • [Méx.] upholstering

muebles m, **útiles** m **y enseres** m - [contab] furniture, tools and fixtures
— m **y enseres** m - [contab] furniture and fixtures
— m — — m **para hogar** m - [domést] home furniture and fixtures
— m — **útiles** m - [contab] furniture and fixtures
muela f - [mec] . . .; grinding stone • wheel
— f **abrasiva** - [mec] abrasive (grinding) wheel
— f — **convencional** - [mec] conventional abrasive (grinding) wheel
— f **convencional** - [mec] conventional (grinding) wheel
— f **de carburo** m **con silicio** m - [mec] silicon carbide grinding wheel
— f — **esmeril** m - [mec] emery (grinding) wheel • emery grinder
— f — **óxido** m **de aluminio** m - [mec] aluminum oxide grinding wheel
— f **para afilar herramienta(s)** f - [mec] tool grinding stone
— f **plana** f - [mec] flat grinding stone
— f — **para afilar herramienta(s)** f - [mec] flat tool grinding, stone, o wheel
— f **preparada** - [mec] dressed grinding wheel
— f **rectificada** - [mec] dressed grinding wheel
— f **resinosa** - [mec] resinous grinding wheel
— f **vitrificada**—[mec] vitrified grinding wheel
muellaje m - [transp] . . .; dock charge(s)
muelle m - [transp] . . .; dock • [nav] wharf • | a - . . .; plush • [mec] **véase resorte** m
— m **accionador** - [mec] activating spring
— m **accionador para anillo** m **portarrodillos** m - [mec] roller cage activating spring
— m **aislador** m - [mec] insulating spring
— m **cilíndrico** - [mec] coil spring
— m **compresible** - [mec] compressible spring
— m **comprimible** - [mec] compression spring
— m **comprimido** - [mec] compressed spring
— m **cónico** - [mec] conical spring
— m **corto** - [mec] short spring • [náut] short, wharf, o dock
— m **chato** - [mec] flat spring
— m **de acero** m - [mec] steel spring • [transp] steel, dock, o wharf
— m — — m **en espiral** - [mec] wound steel spring
— m — — — **apretado** - [mec] tightly wound steel spring
— m **delantero** m - [autom-mec] front spring
— m **eléctrico** - electrical spring
— m **en espiral** - [mec] coil, o helical, o wound, spring
— m — — **apretado** - [mec] tightly wound spring
— m **flotante** - [náut] floating dock
— m **helicoidal** - [mec] helical spring
— m **igualador** - [mec] equalizing spring
— m **inferior** - [mec] lower, o bottom, spring
— m **motor** - [mec] power spring
— m — **anillo** m **portarrodillos** m - [mec] roller cage spring
— m — **cabotaje** m - [náut] coastal, o coastwise, (shipping) wharf
— m — **carbón** m - [transp] coal, wharf, o dock
— m — — m **y mineral(es)** - [transp] coal and ore dock
— m — **carga** f - [constr] (loading) dock
— m — — f **general** - [ind] general purpose, wharf, o dock
— m — — f **para (auto)camión(es)** m - [constr] truck, dock, o wharf
— m — **cargar** - [transp] loading dock
— m — — v **alquitrán** m - [transp] tar loading dock
— m **para compresión** f—[mec] compression spring
— m — **contactador** m - [mec] contactor spring
— m — — m **principal** - [electr-equip] main contactor spring
— m — **contacto** m - [electr-equip] contact spring
— m — — m **principal** - [electr-equip] main contact spring

muelle m para contrafuerte m - [constr] abutment pier
— m — coque m - [ind] coke, wharf, o pier
— m — descarga f - [ind] unloading, o receiving, dock, o wharf
— m — embarque m - [transp] (shipping) dock
— m — enclavamiento m - [electr-equip] interlock(ing) spring
— m — gatillo m - [mec] trigger spring
— m — manipuleo m - [ind] handling dock
— m — — m de material(es) m - [ind] materials (handling) dock
— m — material(es) m - [ind] materials handling dock
— m — mineral(es) m - [ind] ore dock
— m — portaescobilla(s) m - [electr-mot] brushholder spring
— m — puerta f - [constr] door spring
— m — f para tablero m - [mec] panel door spring
— m — f — m para selección f - [sold] selector panel door spring
— m — f — m — f de voltaje(s) m - [sold] voltage selector panel door spring
— m — reactor m - [electr] reactor spring
— m — retención f - [mec] retaining spring
— m — f para puerta f - [constr] door retaining spring
— m — f — f para panel m - [mec] panel door retaining spring
— m — f — f — tablero m - [sold] panel door retaining spring
— m — f — f — m para selección f - [sold] selector panel door retaining spring
— m — f — f — m — f de voltaje m - [sold] voltage selector panel door retaining spring
— m — retorno m - [mec] return spring
— m — — m para regulador m - [mec] control return spring
— m — m — m para marcha f sin carga f - [mec] idler return spring
— m — rodillo m - [mec] roll(er) spring
— m — m loco - [mec] idle roll(er) spring
— m — tensión f - [mec] tension spring
— m — válvula f - [válv] valve spring
— m — f con aguja f - [mec] needle valve spring
— m — f con aguja f para marcha f lenta - [mec] idler needle valve spring
— m — varilla f - [mec] rod spring
— m — f para acelerador m - [mec] throttle rod spring
— m — vástago m - [mec] rod spring
— m principal m - [mec] main spring
— m semielíptico - [mec] semielliptic spring
— m superior - [mec] upper, o top, spring
— m trasero - [mec] back, o rear, spring
— m tres cuartos elíptico - [mec] three quarters elliptic spring
muerdo m - [mec] bit(ting); grip(ping)
— m fuerte - [mec] tight, bit(ting), o gripping
— m seguro - [mec] secure, bit(ing), o gripping
muerte f - . . .; dying
— f instantánea - [medic] instant, o sudden, death, o dying
— f en vía f pública - [segurid] highway death
— m por accidente - [segurid] accidental death
— f — causa(s) f natural(es) - [segurid] natural cause(s) death
— m posible - [segurid] possible death
— f repentina - [segurid] sudden death
— f súbita - [segurid] sudden death
muerto m - dead person • [nav] deadman
muerto,ta a - . . . . • [electr] no load
muesca f - [mec] . . .; nick; break • [mec] . . . • score; gab; chap; furrow; gain - housing
— f en carrete m - [mec] reel, o spool, notch
— f — engranaje m - [mec] gear notch • engaging scarf
— f — ménsula f - [mec] bracket notch
— f — — f para árbol m - [mec] shaft bracket notch

muesca f en ménsula f para árbol m para desplazador m - [mec] shifter shaft bracket notch
— f — f — desplazador m - [mec] shifter bracket notch
— f — muelle m motor - [mec] power spring notch
— f — polea f - [mec] sheave, o pulley, notch
— f — f para carrete m - [mec] reel pulley notch
— f doble - [mec] double notch
— f en resorte m - [mec] spring notch
— f — motor - [mec] power spring notch
— f — sector m - [mec] sector notch
— f — V - [sold] V notch
— f engranada - [mec] engaged notch
— f esquinera - [mec] corner notch
— f para chaveta f - [mec] keyway
— f — regulación f - [mec] control notch
— f trinagular - [sold] V notch
muescado,da a - notched
muestra f - . . . • show(ing) • [petról] core
— f a azar m - [ind] random sample
— f analizada - analyzed sample
— f bajo prueba - sample being tested
— f clara - clear showing
— f colada - [metal-prod] cast sample
— f con dureza f Brinnel - [metal-prod] Brinnel (hardness) sample
— f — naturaleza f inalterada - unaltered (nature, o type) sample
— f — f alterada - altered (nature, o type) sample
— f cortada - cut, o clipped, sample
— f cuadrada - square sample
— f de aceite m - oil sample • oil showing
— f — acero m - [metal-prod] steel sample
— f — m aluminizado - [metal-prod] aluminized steel sample
— f — aire m - air sample
— f — ajuste m - [ind] setting, o adjustment, sample
— f — área m - area sample
— f — corregida - corrected area sample
— f — arena f - [petról] (sand) cutting
— f — arrabio m - [metal-prod] pig iron sample
— f — ataque m - [quím] etching sample
— f — m profundo - [metal] deep etching sample • deep, o destructive, attack sign
— f — barco m - [transp] hold sample
— f — cabeza f para inyección f - [petról] swivel, sample, o showing
— f — colada f - [metal-prod] cast(ing) sample
— f — compatibilidad f - compatibility proving
— f — chapa f - [metal-lam] strip sample
— f — f aluminizada - [metal-trat] aluminized, sheet, o strip, sample
— f — f de acero - [metal-lam] steel, sheet, o strip, sample
— f — f — m aluminizado - [metal-trat] aluminized steel, sheet, o strip, sample
— f — dato(s) m - data showing
— f — ensayo m - [ind] production specimen
— f — escoria f - [metal-prod] slag sample
— f — formación f - [petról] (formation) cutting
— f — formulario - [com] form sample
— f — fundición f - [metal-prod] cast(ing) sample
— f — información f - information, o datum,ta, showing, o sample
— f — lámina f - [metal-lam] sheet sample
— f — f aluminizada - [metal-trat] aluminized sheet sample
— f — f de acero m - [metal-lam] steel sheet sample
— f — f — m aluminizado - [metal-trat] aluminized steel sheet sample
— f — líquido m - liquid sample
— f — m para ensayo(s) m - liquid test(ing) sample; testing liquid sample
— f — material(es) - [ind] material(s) sample

muestra f de material m polietilén(ic)o m -
[electr-cond] polyethylene material sample
— f — núcleo m - [petról] core sample
— f — óxido m - oxide sample
— f — placa(s) f - plate(s) sample
— f — — f para identificación f - identifica-
rion plate(s) sample
— f — presión f - [mec] pressure showing
— f — — f de aceite m - [mec] oil pressure
showing
— f — — f — aire m - [mec] air pressure
showing
— f — — f en acumulador m - [mec] accumulator
pressure showing
— f — producción f - [ind] production sample
— f — — f actual - [ind] current production
sample
— f — puesta f a punto m—[ind] setting sample
— f — remoción f - removal show(ing)
— f — — f de tapón n - [mec] plug removal
show(ing)
— f — — — m para filtro m - [petról]
filter plug removal show(ing)
— f — resistencia f - resistance show(ing)
— f — — f eléctrica - [hidr] resistivity,
showing, o measuring
— f — — f efectiva f de muestra f - [hidr]
sample resistivity, sample, o measuring
— f — — f — — — f de agua m -
[hidr] water sample resistivity, sample, o
measuring
— f — saca(da) f - removal show(ing)
— f — — f de tapón m - [mec] plug removal
show(ing)
— f — — f — — m para filtro m - [petról]
filter plug removal show(ing)
— f — sedimento(s) m - [hidr] sediment sample
— f — sinter m - [metal-prod] sinter sample
— f — sondeo m - sounding sample
— f — suelo m - [suelos] soil sample
— f — — m seleccionada - [suelos] selected
soil sample
— f — tolva(s) f - [ind] bin(s) sample
— f — tubería f—[tub] tube,bing, o pipe,ping,
— sample
— f — tubo m - [tub] tube, o pipe, sample
— f — ubicación f - location showing
— f — varilla(s) f - [sold] rod(s) sample
— f — vista f desde cerca - close-up show(ing)
— f — — f — de cabeza f para inyección f -
- [petról] swivel close-up show(ing)
— f — — f — — calibrador m - Lmec] gage
close-up show(ing)
— f — — — placa f - [mec] plate close-up
show(ing)
— f — — — f para identificación f - [mec]
identification plate close-up show(ing)
— f — — — zona f - [ind] area close-up
show(ing)
— f devuelta - [com] returned sample
— f doble — [ind] double sample • [petról]
double core
— f embebida - [constr] confined sample
— f en ilustración f - show(ing) in @ illustration
— f — tamaño m final - [ind] production size
specimen
— f ensayada - [ind] tested sample
— f espectrográfica - spectrographic sample
— f fundida - [metal-prod] cast sample
— f global - global, o bulk, sample
— f gratis - [com] free, o no charge, sample
— f inalterada - unaltered, o unchanged, o undisturbed, sample
— f longitudinal - longitudinal sample
— f metalográfica - [metal- metallographic sample
— f mineralógica f—[miner] mineralogical sample
— f múltiple - multiple sample
— f navicular - boat shaped sample
— f para corte m - [suelos] cut sample
— f — ensayo(s) m - test(ing), piece, o sample
— f — — m de tracción f—traction test sample

muestra f patrón m - standard sample
— f — de ferro-nobio - [metal] standard ferro-
-columbium sample
— f — — — — manganeso - [metal] standard
ferro-columbium-manganese sample
— f perforada - [mec] drilled, sample, o specimen
— f periódica - periodic sample
— f porosa - porous sample
— f preliminar - preliminary sample
— f probada - tested sample
— f recortada - cut, o clipped, sample
— f recuperada - recovered sample
— f representativa - representative sample
— f seca - dry sample
— f secada - dried sample
— f seleccionada - selected sample
— f simulada - simulated sample
— f — de producción f - [ind] simulated production, sample, o specimen
— f sin valor - [com] no value sample; sample
without value
— f — — m comercial - [com] sample, without,
o with no, commercial value
— f suspendida - suspended, o hung, sample
— f tomada en obra f - field sample
— f total - total sample
— f transversal - transverse sample
— f única - single, sample, o specimen
muestrear v - to (take @) sample
muestrario m - . . .; sample(s) kit
muestrereado,da* a - sampled
— a periódicamente - [ind] sampled periodically
muestrereador* m - sampler • [petról] véase
sacamuestras m
— m automático - [miner] automatic sampler
muestrear* v - [ind] to (take @) sample; véase
— sacar v muestra(s)
— v aire m - to sample @ air
— v periódicamente - to sample periodically
muestreo m automático - [ind] automatic sampling
— m — en planta f - [ind] plant automatic sampling
— m — — — f para sinterización f - [metal-
-prod] sintering plant automatic sampling
— m continuo - [ind] continuous sampling
— m — de aire m - continuous air sampling
— m de aire m - air sampling
— m — aislación f - insulation sampling
— m — atmósfera f - atmosphere sampling
— n — — f en horno m - [combust] furnace atmosphere sampling
— m — gas m - [combust] gas sampling
— m — — m en tragante m - [metal-prod] throat
gas sampling
— m — mineral(es) m - [miner] ore(s) sampling
— m en acería f - [metal-prod] steel plant
sampling
— m — planta f - [ind] plant sampling
— m — — f para sinterización f - [metal-prod]
sintering plant sampling
— m espectrográfico - spectrographic sampling
— m periódico - [ind] periodic sampling
— m simple - [ind] simple sampling
mufa* f - [telecom-cond] muff; véase mufla f
mufla f - [combust] . . .; muffle
mugriento,ta a - . . . • black
mujer f - . . . • [fam] gal; female
Mujeres f de Negocios y Profesionales - Business
and Professional Women
multa f - . . .; penalty
— f eventual - [fisc] eventual, fine, o penalty
Multi-Placa* a - Multi-Plate
multicompañía f - véase compañía f múltiple;
compañías f varias
multidireccional a - multi-directional
multifásico,ca a - [electr] multiphase
multifilar* a - [alambre] multiwire; multiple
wire
multifrecuencia f - [electrón] multifrequency
— f con tono m doble - [electrón] dual tone
multifrequency
— f — — m doble con . . . dígito(s) m -
[electrón] . . . digit dual tone multifrequency

[comput] . . . digit dual tone multifrequency
multifrecuencia f **estándar** - [electrón] standard multifrequency
— f **para tono** m **doble** - [electrón] standard dual tone multifrequency
multilingüe a - multilingual
multiple a - . . .; multi. . . . • gang • [comb.-int] manifold | m - [comb.int] manifold
— m **a bloque** a - [comb.int] manifold to block
— m **colector** - [mec] collector manifold
— m — **para retorno** m (**de aceite** m) - [mec] (oil) return collector manifold
— m **doble** - [comb.int] dual manifold
— m **enfriado** - [mec] cooled manifold
— m — **con agua** m - [mec] water cooled manifold
— m **Oilmaster** - [petról] Oilmaster manifold
— m — **para regulación** f - [petról] Oilmaster control manifold
— m — — f **motriz** - [petról] Oilmaster power control
— m **para aceite** m - [mec] oil manifold
— m — **admisión** f - [comb.int] intake, o inlet, manifold
— m — — f **doble** - [comb.int] dual, intake, o inlet, manifold
— m — **aspiración** f - [mec] suction manifold
— m — **camisa** f - [mec] liner manifold • spray manifold
— m — — f **y émbolo** m - [mec] piston-liner manifold
— m — **distribución** f - [mec] (distribution) manifold
— m — — f **enfriado con agua** m - [mec] water cooled (distribution) manifold
— m — **escape** m - [comb.int] exhaust manifold
— m — **limpieza** f - [ind] cleaning manifold
— m — **motor** m - [comb.int] engine manifold
— m — **regulación** f - [petról] control manifold
— m — — f **para potencia** f - [petról] power control manifold
— m — **retorno** m - [mec] return manifold
— m — — m **para aceite** m - [mec] oil return manifold
— m — **rociador** m - [mec] spray manifold
— m — — m **para camisa** f - [mec] liner spray manifold
— m — — — f **y émbolo** m - [mec] piston--liner spray manifold
— m — **rociar** v - [mec] spray(ing) manifold
— m — — v **camisa** f - [mec] liner spray manifold
— m — **succión** f - [mec] suction manifold
— m — **tubería** f - [mec] pipe manifold
— m — — f **para limpieza** f - [ind cleaning (pipe) manifold
— m **rociador** - [mec] spray, manifold, o header
— m — **para camisa** f - [mec] liner spray, manifold, o header
— m — — m — — f **para émbolo** m - [mec] piston-liner spray manifold
— m — **camisa émbolo** m - [mec] piston spray manifold
multiplicación f - . . . • [mec] . . .; ratio
— f **alta** - high ratio • [mec] high gear
— f **baja** - low ratio • [mec] low gear
— f **de torsión** f - [mec] torque multiplication
— f **y desmultiplicación** f (**de engranajes** f) - [mec] gearing
multiplicado,da a - multiplied
— **a por** - [mat] multiplied by; times; by
multiplicador m - . . . • [mec] booster
— m **para fuerza** f - [mec] (power) booster
multipunto* m - multipoint
multitud f - . . .; large crowd
mullido,da a - [agric] mulched
mullir v - . . . • [agric] to mulch
mundial a - . . .; world-wide; earth-wide
mundialmente adv - world wide
mundo m **civilizado** - civilized world
— m **de carreras** f - [deport] racing world
— m **de zaranda** f - [miner] scrrening(s)
munición f - . . . • [mec] ball
— f **de acero** m - [cojin] steel ball

municiones f - [milit] munitions • [mec] shot
municipal m - . . . | a - . . .; local; borough
municipalidad f - [pol] . . .; township; city
— f **con recurso(s)** m **limitado(s)** - [pol] limited resources, o budget conscious, city
municipio m - [pol] . . .; township; borough
— m **con concejo** m (**municipal**) **y administrador** m - [pol] council-manager form of government
muñeca f - [anatom] . . . • [domést] . . . • [mec-tornos] stock
— f **corrediza** - [mec-tornos] véase contrapunta
— f **movible** - [mec-tornos] tail stock
muñequera f - wristlet
muñequilla f - [mec] journal
muñón m - [mec] gudgeon; lug; peg; stud; socket • bearing • journal; trunnion; spindle; neck [medic] . . .; stub
— m **ahusado** - [mec] tapered shaft
— m **central** - [mec] central gudgeon
— m **con brida** f - [mec] flanged gudgeon
— m **cónico** - [mec] tapered, shaft, o gudgeon
— m **de bancada** f - [mec] véase cojinete m de cigüeñal m
— m — **cigüeñal** f - [comb.int] crankshaft journal; crankpin
— m — **cuchara,rón para colada** f - [metal-prod] trunnion pin • ladle trunnion
— m — **eje** m - [mec] shaft journal • [cojin] hub
— m — **rodillo** m - [metal-lam] roll neck
— m — **rueda** f - [autom] wheel spindle
— m **embridado** - [mec] flanged gudgeon
— m **encurecido** - [mec] hardened trunnion
— m **esférico** - [mec] ball gudgeon
— m **mecanizado** - [mec] machined journal
— m **metálico** - [mec] metal stub
— m **para malacate** m - [petról] wheel gudgeon
— m — m **para herramienta** f - [petról] bull wheel gudgeon
— m — — **tubería** f - [petról] calf wheel gudgeon
— m — **tambor** m **para herramienta(s)** f—[petról] bull wheel gudgeon
— m **templado** - [mec] hardened trunnion
muñonera f - [mec] . . .; trunnion hole • gab; socket
murete m - [vial] . . .; dike
— m **en franja** f **divisoria** - [vial] median strip dike
— m — — f **medianera**—[vial] median strip dike
— m **interceptor** - [hidr] cut-off wall
— m **longitudinal** - [vial] (median) dike
— m — **en divisoria** f - [vial] median dike
— m **separador** - [constr] barrier wall
— m **y sumidero** m - [hidr] dike and catch basin
muro m **alero** - [constr] wing wall; wingwall
— m — **de hormigón** m - [constr] concrete wing wall
— m — — m **armado** - [constr] reinforced concrete wing wall
— m **alto** - [constr] high wall
— m **anterior** - [constr] front wall
— m **armado** - [constr] assembled wall
— m **bajo** - [constr] low wall
— m **Bin-Wall** - [metal-fabr] Bin-Wall
— m **celular** - [metal-fabr] bin-wall • [constr] cellular wall
— m **combado** - [constr] crooked, o warped, wall
— m **con altura** f **diversa** - [constr] different, o various, height(s) wall(s)
— m **con contrafuerte** - [constr] butress wall
— m — **encostillado** m - [constr] lagging wall
— m — — m **de acero** m - [constr] steel lagging wall
— m — **inclinación** f **de** - [constr] wall on . . . batter
— m **conformado**—[constr] formed, o shaped, wall
— m **contra explosión(es)** f - [constr] explosion wall
— m **curvado** - [constr] curved wall
— m **curvo** - [constr] curved wall
— m **de acero** m - [constr] steel wall
— m — — m **para contención** f - [constr] steel

retaining wall
muro m de acero m para contención f - [constr]
steel retaining wall
— m — **ala** - [constr] wingwall; wing wall
— m — — m **con borde(s) m rectangulares** -
[constr] square edge wing wall
— m — **aleta f** - [constr] wing wall; wingwall
— m — — f **de hormigón m** - [constr] concrete
wing wall
— m — **arcón(es) m** - [constr] bin wall
— m — — m **de acero m** - [constr] steel bin
wall
— m — **caja f** - [constr] bin, o crib, wall
— m — — f **convencional** - [constr] conventional, crib, o bin, wall
— m — **cajas f de acero m** - [constr] steel bin
wall
— m — **cajón(es) m** - [constr] bin wall
— m — **cortina f** - [constr] curtain wall
— m — **pie m** - [constr] toe wall
— m — **hormigón m** - [constr] concrete wall
— m — — m **armado** - [constr] reinforced concrete wall
— m — — m **conformado** - [constr] formed concrete wall
— m — — m **vaciado** - [constr] poured concrete
wall
— m — **pie m** - [constr] toe wall
— m — **piedra f** - [constr] stone wall
— m — — f **para refuerzo m** - [constr] stone
reinforcing wall
— m — **ribera f** - [constr] bulkhead
— m — — f **intermedio,diario** - [constr] intermediate bulkhead
— m — **roca f** - [constr] stone wall
— m — — f **para refuerzo m** - [constr] stone
reinforcing wall
— m — — f **reconstruido** - [constr] reconstructed stone wall
— m — — f — **para cabecera f** - [constr] reconstructed stone headwall
— m — **tipo m para defensa f** - [constr] bulkhead type wall
— m — — m **por gravedad f** - [constr] gravity
type wall
— m **drenado** - [constr] drained wall
— m **efectivo** - [constr] effective wall
— m **en voladizo m** - [constr] cantilever wall
— m **encofrado** - [constr] cribbing
— m **estable** - [constr] stable wall
— m **frontal** - [constr] front wall
— m **inclinado** - [constr] inclined, o slanting, front wall
— m **inclinado** - [constr] inclined, o slanting,
o batter(ed), wall
— m **instalado** - [constr] installed wall
— m **interceptador m** - [hidr] cutoff wall
— m **interceptador contra filtración f** - [hidr]
seepage cutoff wall
— m — **alto** - [constr] high cutoff wall
— m — **bajo** - [constr] low cutoff wall
— m — **de acero m** - [hidr] steel cutoff wall
— m — **permanente** - [constr] permanent cutoff
wall
— m **interior** - [constr] inside wall
— m — **para impermeabilización f** - [constr]
core wall
— m **lateral** - [constr] side wall; sidewall
— m — **de hormigón m** - [constr] concrete sidewall
— m **medianero** - [constr] median wall
— m — **para retención f** - [constr] median retaining wall
— m **para apoyo m** - [constr] abutment, o supporting wall
— m — **cabecera f** - [constr] headwall
— m — — f **bajo** - [constr] low headwall
— m — — f **con borde(s) m rectangular(es)** -
[constr] square edge headwall
— m — — f **complicado** - [constr] elaborate, o
complicated, headwall
— m — — f **con altura f media(na)** - [constr]
half-, o half height, headwall

muro m para cabecera f con borde m **redondeado** -
[constr] rounded edge headwall
— m — — f **construido** - [constr] constructed,
o built, headwall
— m — — f **de fábrica f** - [constr] masonry, o
mortared, headwall
— m — — f — **hormigón m** - [constr] concrete
headwall
— m — — f — — m **armado** - [constr] reinforced concrete headwall
— m — — f — **mampostería f** - [constr] masonry
headwall
— m — — f — — f **con roca f** - [constr]
(stone) masonry headwall
— m — — f — — f **reconstruida** - [constr] reconstructed stone headwall
— m — — f **de tablestacado m (de acero m)** -
[constr] (steel) sheet(ing) headwall
— m — — f **existente** - [constr] existing headwall
— m — — f **nuevo** - [constr] new headwall
— m — — f **opuesto** - [constr] opposite headwall
— m — — f **para alcantarilla f** - [constr] culvert headwall
— m — — f — **bóveda f** - [constr] arch headwall
— m — — f — **tubería f** - [constr] pipe headwall
— m — — f — — f **abovedada** - [constr] pipe-
-arch headwall
— m — — f **parcial** - [constr] partial headwall
— m — — f **proyectado** - [constr] designed
headwall
— m — — f **rústico** - [constr] rustic headwall
— m — — f **total** - [constr] full headwall
— m — **contención f** - [constr] bulkhead • [nav]
bulkhead
— m — **defensa f** - [constr] breast, o bulkhead,
wall • [hidr] sea wall
— m — **desviación f** - [constr] diversion, o di-
verting, wall
— m — **empotramiento m** - [constr] embedding
wall
— m — **estribo m** - [constr] bench wall
— m — — m **original** - [constr] original bench
wall
— m — **fachada f** - [constr] front wall
— m — **interceptación f** - [constr] cutoff wall
— m — **núcleo m** - [mec] core wall
— m — **protección f** - [constr] protection wall
— m — — f **contra metralla f** - [constr] splinter protection wall
— m — **refuerzo m** - [constr] reinforcing,cement
wall
— m — **retención f** - [constr] retaining, o
back, wall • [nav] bulkhead
— m — — f **curvado** - [constr] curved retaining
wall
— m — — f **de acero m** - [constr] steel retaining wall
— m — — f — — m **de tipo m con cajón(es) m** -
[constr] bin-type steel retaining wall
— m — — f — — m — — m **arcón(es) m** -
[constr] bin-type steel retaining wall
— m — — m — **encostillado m** - [constr] lagging retaining wall
— m — — f — — m **corrugado** - [constr] corrugated lagging retaining wall
— m — — f — — m **de acero m** - [constr]
corrugated steel lagging retaining wall
— m — — f — — m **de acero m** - [constr] steel
lagging retaining wall
— m — — f — **hormigón m** - [constr] concrete
retaining wall
— m — — f — — m **armado** - [constr] reinforced concrete retaining wall
— m — — f **de tipo m con, arcón(es) m, o cajo-
ne(s) m** - [constr] bin-type retaining wall
— m — — f — **por gravedad f** - [constr]
gravity type retaining wall
— m — — f **detrás de estribo m para puente m** -
[constr] back wall

muro m para retención f íntegramente de acero m
— [constr] all-steel Bin-Wall
— m —— f lateral - [constr] side retaining
wall
— m —— f tipo m cajón m - [constr] bin-type
retaining wall
— m —— f — m — de acero m - [constr]
steel Bin-Wall
— m —— f tipo m celular - [constr] bin-type
retaining wall
— m —— f vaciado - [constr] poured retaining
wall
— m — revestimiento m - [constr] breast wall
— m — riel(es) m - [grúas] leg wall
— m — m para descargadora f - [grúas] pier
leg wall
— m —— m para grúa f - [grúas] pier leg wall
— m —— m — puente m grúa - [grúas] pier leg
wall
— m — seguridad f - [constr] safety wall
— m — sostén(imiento) m - [constr] sustaining,
o retaining, o back, wall • (steel) bin wall •
revetement
— m —— m Bin-Wall (tipo m cajón m) -
[constr] Bin-Wall revetement
— m —— m de acero m - [constr] steel bin
wall revetement
— m —— m tipo m cajón - [constr] Bin-Wall
revetement
— m piñón - [constr] gable end
— m por gravedad f - [constr] gravity wall
— m posterior - [constr] back, o rear, wall
— m principal - [constr] main, o major, wall
— m protector - [constr] protection,tive wall
— m recargado - [constr] surcharged, o overloaded, wall
— m recio - [constr] tough, o strong, wall
— m soportado - [constr] supported wall
— m sostenido - [constr] supported wall
— m terminado - [constr] finished wall
— m terminal - [constr] end wall; endwall
— m — abocinado - [constr] flared endwall
— m — combado - [constr] warped endwall
— m — recto - [constr] straight endwall
— m tipo cajón m - [constr] Bin-Wall
— m —— m para retención f—[constr] Bin-Wall
retaining wall
— m —— m proyectado - [constr] designed
Bin-Wall
— m — jaula f - [constr] crib wall; cribbing
— m vertical - [constr] vertical wall
— m — curv(ad)o—[constr] vertical curved wall
— m — efectivo - [constr] effective vertical
wall
música f estridente - [mús] screeching music
— f tocada - [mús] played music
mutilación f de cadena f - [cadenas] chain mutilation
— f visible - visible mutilation
mutilado,da a - mutilated
— a visiblemente - mutilated visibly
mutilar v cadena f - [mec] to mutilate @ chain
— v visiblemente - to mutilate visibly
mutual f - [com] cooperative | a - . . .
— f de Fabricantes m - [seguros] Factory Mutual
mutualidad f de fabricantes m - [com] factory
mutual
mutuamente adv - . . .; to each other
muy adv - . . .; highly; well • too far • substantially • readily
— adv abrasivo,va - [mec] very abrasive
— adv accesible - very, o readily, accessible
— adv aceptable - very acceptable; very, o
pretty, good
— adv ácido - [quím] very acid(ic)
— adv adaptable - very, o highly, versatile
— adv afinado,da - [sold] washed up
— adv — en borde(s) m - [sold] washed up too
high
— adv agrietado,da - [mec] deeply checked
— adv alto - very, o extra, high
— adv ancho,cha - very, o extra, wide
— adv aparente - very, o readily, apparent

muy adv aplanado,da - [tub] severly elliptical
— adv atrás - well, o way, o very far, back
— adv bajo,ja - very, o extra, low
— m bituminoso,sa - very bituminous; rich in
bitumen
— adv blando,da - very, o extra, soft
— adv breve - very short • prohibitively short
— adv calificado,da - highly, o very, qualified
— adv capacitado - highly, qualified, o skilled
— adv coloidal - very, o highly, colloidal
— adv competidor,ra - very, o highly, competitive
— adv complacido,da - very well impressed
— adv comprometido,da - very, o heavily, involved, o committed
— adv común - very common
— adv confiable - highly reliable
— adv contaminado,da - very, o highly, contami-
nataed
— adv corriente - very, common, o current
— adv corrosivo,va - very, o highly, o badly,
corrosive
— adv corto,ta - very short • prohibitively
short
— adv desplazado,da - [mec] very, o widely,
offset
— adv deteriorado,da - very, o baldy, deteriorated
— adv duro,ra - very, o extremely, hard
— adv efectivo,va - very, o highly, effective
— adv endurecible - [metal-trat] high(ly)
hardenable,bility
— adv envuelto,ta - very, o heavily, involved
— adv especializado - very, o highly, especial-
ized
— adv estanco,ca - [tub] very, o extremely,
(water) tight
— adv experimentado,da - highly experienced •
highly accomplished
— adv flexible - very, o highly, flexible
— adv gastado,da - very, o badly, worn
— adv grande - very, o overly, large
— adv importante - very, o particularly, important • [ind] very large
— adv improbable - very, o highly, unlikely
— adv inclinado,da - steep (graded)
— adv — non armazón m rígido - [constr] rigid
frame high slope
— adv influyente - very influential • [fig]
leverage producing
— adv involucrado,da - substantially involved
— adv leve - very (s)light
— adv liviano,na - very, o unusually, light-
(weight)
— adv mejorado,da - very, o significantly, im-
provewd • [autom] highly revised
— adv merecido,da - very deserved; hard earned
— adv metalizado,da - [metal-prod] highly me-
talized
— adv móvil,vible - easily portable
— adv muelle - very, o deep(ly) cushioned
— adv peligroso,sa - very, o extremely, dangerous, o hazardous
— adv penetrante - very, o deeply, penetrating
— adv pequeño,ña - very small
— adv permeable - very, o highly, permeable
— adv portátil - very, o highly, portable
— adv prestigioso,sa - very prestigious
— adv quemado,da - very, o badly, burned
— adv raído,da - very, o badly, frayed
— adv ranurado,da - very, o severely, grooved
— adv rápido,da - very, o extra, fast
— adv resbaladizo,za - very, o extra, slippery
— adv resistente - very, o extra, strong
— adv sedimentado,da - very, o heavily, silted
— adv sensible - very, o extra, sensitive
— adv señor(es) mío(s)/nuestro(s) - dear sir(s)
— adv solicitado - [mec] highly stressed
— adv tóxico,ca - very, o highly, toxic
— adv transitado,da - [vial] very busy
— adv versátil - very, o highly, versatile
mV m - [electrón] véase milivoltio m

N

NAUCA f - [pol] véase Nomenclatura f Arancelaria Uniforme Centro Americana
NC a - [electrón] véase cerrado,da normalmente
NO a - [electrón] véase abierto,ta normalmente
NU f - [pol] véase Naciones f Unidas
nacer v - . . . • [astron] . . .; to come up
nacido,da a - born • [botán] sprouted • [astron] arisen; come up
naciente a - nascent • fledgling; developing • burgeoning | f - [hidr] headwater(s)
nacimiento m - . . . • [astron] . . .; coming up
— m de curva f - [arquit] beginning of @ curve
nación f - [pol] . . .; country
— f con déficit m de chatarra f - [metal-prod] scrap deficit, nation, o country
— f — exceso m de chatarra f - [metal-prod] scrap surplus, nation, o country
— f en vías f de desarrollo m - [econ] developing, nation, o country
— f iberoamericana - [pol] Latin American, nation, o country
— f latinoamericana - [geogr] Latin American, nation, o country
— f poco desarrollada - [econ] underdeveloped, nation, o country
nacional a - . . . • local • [transp] domestic • [pol] federal
nacionalidad f de firma f - [econ] firm's nationality
— f — — f consultora - consulting firm('s) nationality
nacionalismo m industirla—[econ] industrial nationalism
Naciones f Unidas - [pol] United Nations; U N
nada f - nil | [adv] . . .; none
— adv más - no(thing) more • as soon as • that's it
— adv menos - nothing, less, o short
— adv — que - nothing, less than, o short of
NAF f - [hidr] véase napa f acuífera
nafta f - [petról] . . .; véase también gasolina
— f desodorizada - [petról] deodorized naphtha
— f disolvente - [petról] solvent naphtha
naftalina f - [quím] . . .; naphthalene
— f cruda - [petról] crude naphthalene
napa f - [hidr] nappe; water sheet
— f acuífera - [hidr] water table
— f de agua n - [hidr] water nappe
— f — — f freática - [hidr] ground, o phreatic, water nappe
— f freática f - [hidr] (ground) water, o phreatic, table
— f — alta - [hidr] high water table
naranja f - [bot] . . . | a - véase anaranjado,da
nariz f de bóveda f - [arquit] roof knuckle
— f — husillo m - [tornos] spindle nose
narrado,da a - narrated
natura f - . . .; véase naturaleza f
natural n - [pol] . . . | a - . . . . undisturbed
naturaleza f - . . .; Mother Nature
— f alterada - altered, nature, o type
— f cambiada - changed nature
— f casual - casual nature
— f compacta - compact nature; compactness
— f de accidente m - [segurid] accident, type, o nature
— f — aplicación f - application nature
— f — — f de carga f - [mec] load application nature
— f — bien(es) m - good(s) nature
— f — carga f - [mec] load nature • [ind] charge nature
— f — cargo m - [admin] position nature
— f — cuenca f - [hidr] watershed nature
— f — defecto m - defect nature
— f — documento m - document nature

naturaleza f de falla f - vailure nature
— f — funcionamiento m inapropiado - malfunction nature
— f — invención f - invention nature
— f — invento m - invention nature
— f — investigación f - investigation nature
— f — — f de accidente(s) m - [segurid] accident(s) investigation nature
— f — lesión f - [med] injury nature
— f — pedido m - [com] order nature
— f — servicio(s) m - service(s) nature
— f — siniestro m - [segurid] damage nature
— f — terreno m - land, o terrain, nature
— f inalterada - unaltered, nature, o type
— f variada - varied nature
naturalmente adv - . . .; admittedly
nauseado,da a - nauseated; sick to @ stomach
naval a - [transp] . . .; ocean transportation
nave f - [constr] . . .; bay • [metal-prod] (blast furnace) tapping floor
— f adosada - [constr] lean-to bay
— f asociada - [náut] associated vessel
— f capitana - [milit] flagship
— f espacial - [astronáut] space, vessel, o ship, o vehicle • space shuttle
— f — reutilizable - [astronáut] space shuttle
— f exterior - [constr] outer,tside bay
— f industrial - [ind] industrial bay • high bay
— v para almacenamiento m - [ind] storage bay
— f — benzol m - [coque] brnzol building
— f — carga f - [metal-prod] charging, aisle, o bay
— f — clasificación f - [ind] (as)sorting bav
— f — colada f - [metal-prod] pouring, o casting, aisle, o bay • cast house; casting pit; teeming bay
— f — — f de lingote(s) m - [metal-prod] pig casting bay
— f — convertidor(es) m - [metal-prod] converter bay
— f — chatarra f - [metal-prod] scrap bay
— f — chimenea(s) f - [metal-prod] stack bay
— f —, deslingotado m, o desmoldeo m - [metal-prod] stripping, bay, o aisle
— f —, o — m, en horno m de fosa—[metal-prod] soaking pit stripping, aisle, o bay
— f escoria f - [metal-prod] slag bay
— f — fundición f - [metal-prod] cast house; melting bay
— f — fusión f - [metal-prod] melting bay
— f — horno(s) m - [metal-prod] charging floor; furnace(s) bay
— f — — m de fosa f - [metal-prod] soaking pit bay
— f — m Siemens-Martin - [metal-prod] open hearth, bay, o building
— f — laminación f - [metal-lam] rolling bay
— f — materia(s) f prima(s) - [ind] raw material(s) bay
— f — materiales m - [ind] materials bay
— m — mezcladora(s) f - [metal-prod] mixer bay
— f — moldeo m - [metal-prod] cast house
— f — palanquilla(s) f - [metal-prod] billet(s) bay
— f — producto(s) m - [ind] product(s) bay
— f — — m empaquetado(s) - [ind] packaged product(s) bay
— f — reparación(es) f - [ind] repair(s) bay
— f — seguridad f - [segurid] safety bay
— f — servicio(s) m - [ind] service(s) bay
— f — vaciado m - [metal-prod] pouring, aisle, o bay
— f Siemens-Martin - [metal-prod] open hearth bay
navecilla f - . . . • [combust] (small) boat
navegación f - [transp] . . .; navigating • shipping
— f aérea - [aeron] aerial navigating,tion
— f costera - [transp] coastal, navigation, o shipping
— f costanera - [transp] coastal shipping
— f cuidadosa - [transp] careful navigation

navegación f de cabotaje m - [nav] coastal shipping
— f — encontrado - [nav] two-way navigation
— f — recreo - [nav] (pleasure) boating
— f en dos direcciónes f - [nav] two-way navigation
— f para carrera f rally - [deport] rally navigation
— f por estima - [nav] dead reckoning
— f ultramarina - [nav] ocean navigation
navegado,da a - navigated
— a cuidadosamente - navigated carefully
navegar v cuidadosamente - to navigate carefully
— v por estima f - [nav] to dead reckon
naviero,ra a - [nav] . . .; ship building
neblina f de alborada f—[meteorol] pre-dawn fog
— f — partícula(s) f de tierra - [meteorol] particulate(s) fog
— f — pulverizador - spray mist
— f disipada - [meteorol] abated fog
— f, matinal, o matutina(l) f - [meteorl] (early) morning fog
— f polvorienta - [meteorol] dusty fog
nebulosidad f - . . .; fog(giness)
necesariamente adv - . . .; needfully; forcibly
necesario,ria a - . . .; needed; required
necesidad f,- . . .; requirement
— f actual - present, o today's, requirement, o need
— f — de agua m - present water requirement
— f — — m de pozo m - [hidr] present well water requirement
— f — — m de río m - [hidr] present river water requirement
— f agrícola - [agric] agriculture, need, o requirement
— f agrupada - grouped need
— f analizada - analyzed, need, o requirement
— f de agricultura f - [agric] agriculture,al, need, o requirement
— f — agua m - [hidr] water, requirement(s), o need
— f — — m de planta f - [hidr] plant water, need, o requirement
— f — — m de río m - [hidr] river water, need, o requirement
— f — aire m - air, need, o requirement • [metal-prod] blast requirement
— f — ajuste m - adjustment need
— f — aplicación f - application requirement
— f — — f final - final application requirement
— f — auxilio m - assistance need
— f — caja f - [fin] cash requirement
— f — capacitación f - [labor] training requirement
— f — cliente m - [com] customer('s) need
— f — — m satisfecho - satisfied customer's need
— f — conocer v - need to know
— f — crédito m - [fin] credit need
— f — determinación f - need determination
— f — determinar v - need to determine
— f — equipo m - [ind] equipment requirement • [labor] group's, requirement, o need
— f — estabilidad f - stability, need(s), o requirement(s)
— f — exportador(es) m - [com] exporter('s) need, o requirement
— f — fundente m - [sold] flux requirement(s)
— f — gravilla f - [constr] (pea) gravel requirement
— f — importación f - [com] import(ation) requierement
— f — importador(es) m - [com] importer('s), need, o requirement
— f — limpieza f - cleaning, o cleanness, need, o requirement
— f — mineral m - [miner] ore requirement(s)
— f — — m de hierro m - [miner] iron ore, requirement(s), o need(s)
— f — moldeador m - [plást] molder('s) need
— f — organización f - [admin] organiza-

necesidad f de personal m - [pers] personnel('s) requirement(s)
— f — población f - population('s) need(s)
— f — precalentamiento m - [sold] preheating, need, o requirement
— f — proyecto m - project('s), need(s), o requirement(s)
— f — reemplazo m - need of replacement
— f — reparación f - [ind] repair need
— f — — f de equipo m - [ind] equipment repair need
— f — — f — — m para amoladora f - [mec] grinding equipment repair need
— f — — — m — — f de rodillo(s) m - [metal-lam] roll grinding equipment repair need(s)
— f — — — m — granallado m - [mec] shot blasting equipment repair meed(s)
— f — — — m de rodillo(s) m - [metal-lam] roll shot blasting equipment repair need
— f — — — m — rectificación f - [mec] grinding equipment repair need
— f — — — m — f de rodillo(s) m - [metal-lam] roll grinding equipment repair need
— f — — — m — torneado m - [mec] turning equipment repair need
— f — — — m de rodillo(s) m - [metal-lam] roll turning equipment repair need
— f — saber v - need to know
— f — sínter m - [metal-prod] sinter requirement(s)
— f — subalterno m - [admin] subordinate's need(s)
— f — viento m - [metal-prod] blast requirement(s)
— f determinada - determined need
— f eléctrica - electrical, requirement, o need
— f eliminada - eliminated need
— f emocional - emotional, requirement, o need
— f esencial - essential need
— f — de población f - [econ] population('s) essential need(s)
— f especial - special need
— f especializada - specialized need
— f específica - specific, need, o requirement
— f — para capacitación f - [labor] specific training, requirement, o need
— f frecuente -frequent, requirement, o need
— f frustratoria - frustrating need
— f futura - future, need, o requirement
— f — de agua m - [hidr] future water requirement(s)
— f — — m de río n - [hidr] future river water requirement(s)
— f humana f - human need
— f importante - important, need, o want
— f imprevista - unforseen, need, o requirement
— f individual - individual need
— f industrial - [ind] industrial need
— — varia - [ind] various industrial needs
— f obviamente frustratoria - frustratingly obvious need
— f para capitalización f - [fin] capitalization need(s)
— f — financiamiento m - [fin] financing need
— f — preparación f de chatarra f - [metal-prod] scrap preparation requirement(s)
— f particular - particular, o specific, need, o requirement
— f personal - [ind] personal need
— f política - [pol] political need
— f posible - possible need • might need
— f primaria - primary need
— f probable - probable need
— f propia - own need
— f psicológica - psychological, need, o want
— — importante - important psychological, need, o want
— f real - real, o actual, need
— f singular - unique need

necesidad f suplida - supplied need
necesidades f varias - various, o sundry, needs
necesitado,da a - . . .; véase también necesario,ria
— a probablemente - probably needed
necesita v conocerse - must be known
necesitar v - . . .; to want • to require
— v ajuste m - to need @ adjustment
— v auxilio m - to need assistance
— v estabilidad f—to need @ stability
— v identificación f - to need identification
— v limpieza f - to, need, o require, cleaning
— v pieza f - [mec] to need @ part
— v probablemente - to probably need
— v reemplazo m - to need @ replacement
necesitarse v - to be necessary
negociable a - . . . • [com] marketable
— a contra factura(s) f - [com] negotiable a-
gainst invoice(s)
— a —— f mensual(es) - [com] negotiable a-
gainst monthly invoice(s)
negociación f - . . .; negotiating • [labor]
bargaining; dealing
— f con sindicato m - [labor] (labor) union
bargaining
— f de contrato m - [legal] contract negotiating,tion
— f — convenio - [labor] agreement, o contract, negotiating,tion
— f — oferta(s) f - [legal] offer(s), o propo-
sal(s), negotiating,tion
— f — propuesta(s) f - [legal] proposal(s), o
bid(s), negotiating,tion
— f — servicio(s) m - service(s) negotiation
— f —— m para consultoría* f - consulting
service(s) negotiating,tion
— f para convenio m - [labor] contract negotiating,tion
— f sindical-administrativa - [labor] (labor)
union-,management, o employer, bargaining
negociado m - [pol] . . .; office
— m de árbitro(s) m general(es) - [fisc] bureau
of general excise tax(es)
negociado,da a - negotiated; bargained
negociador m de gobierno m - [pol] government
negotiator
— m gubernamental - [pol] government negotiator
negociar v - . . .; to deal • [labor] to bargain
— v contrato m—[legal] to negotiate @ contract
— v directamente - to, negotiate, o deal, directly
— v letra f - [fin] to negotiate @ draft
— v mejor - [labor] to bargain better
negocio m - . . .; operation • affair
— m de arrendamiento(s) m - [com½] leasing business
— m — comprador m - [com] buyer('s) business
— m dirigido - [com] conducted business
— m independiente - independent business
— m privado - private business • [fin] individual, firm, o concern
negrita f - [imprent] véase negrilla f
negro m de carbón m - coal black
— m — carbono m - [quím] carbon black
— m —— seleccionado - chosen carbon black
— m — gas m - [petról] gas black
— m — humo m - [quím] lampblack; lamp, o carbon, black
— m —— m seleccionado - chosen, lamp, o carbon, black
— m grisáceo - gray-black
— m liso - [metal-lam] smooth black
negrura f - blackness
neoandínico,ca a - [geol] Neoandinic
neófito m - . . .; beginner • [deport] rookie
neopreno m - [quím] neoprene
— m hermético a aceite m - [electr-cond] oil
tight neoprene
— m para servicio m pesado - [electr-cond]
heavy duty neoprene
nervado,da* a - ribbed; véase también con nervadura f
nervadura f colada de biela f - [comb.int] con-

necting rod cast rib
— f con proyección f estándar - [autom-neumát]
standard rib design
— f de biela f - [comb.int] connecting rod rib
— f en plancha f metálica (para techo m) -
[constr] deck rib
— f estándar - [mec] standard rib
— f exterior - [mec] outisde, o exterior, rib
— f interior - [tub] internal, o inside, rib
— f moldeada f - [mec] cast rib
— f profunda - [mec] deep rib
nervio m - . . . [mec] . . .; stay • bead •
stiffener • [metal-lam] bulb
— m para refuerzo m - [ind] reinforcing, o supporting, o strengthening, rib, o web
neto,ta a - . . . • clean cut - [constr] lay
— a a . . . día(s) de fecha f de factura f -
[fin] . . . days net from invoice date
— a — . . . —— f — f pagadero en moneda de EE. UU. de N. A. - [fin] . . . days
net from invoice date payable in U.S.A. funds
— a contra efecto m por cobrar - [fin] net
against @ receivable (value)
— a de - net of
— a — amortización f - [fin] net of amortization
— a . . . días m - [fin] . . . days net
— a sin ganancia f - [contab] net of profit
— a —— f no realizada - [contab] net of, unrealized, o unearned, profit, o gain
neumático m - [autom-neumát] . . .; pneumatic
tire • skin; rubber; gumball
— m acepillado - [autom-neumát] shaved tire
— m actualizado - [autom-neumát] state-of-@-art
tire
— m adherente—[autom-neumát] sticking,cky tire
— m adheridor—[autom-neumát] sticking,cky tire
— m agresivo - [autom-neumát] aggressive tire
— m — para reemplazo m - [autom-neumát] aggressive replacement (tire)
— m alfa-numérico - [autom-neumát] alpha-numeric tire
— m ancho - [autom-neumát] wide tire
— m angosto - [autom-neumát] narrow tire
— m anterior - [autom-neumát] previous tire •
older tire
— m apareado - [autom-neumát] matched tire
— m aplicado - [autom-neumát] applied tire
— m apropiado - [autom-neumát] appropriate tire
— m arruinado - [autom-neumát] ruined tire
— m asombroso - [autom-neumát] wonderful tire
— m atendido - [autom-neumát] serviced tire
— m auxiliar - [autom-neumát] spare tire
— m balón m - [autom-neumát] balloon, o
do(ugh)nut, tire
— m barato - [autom-neumát] cheap, o inexpensive, tire
— m callejero m - [autom-neumát] street tire
— m cambiado - [autom-neumát] changed tire
— m cepillado - [autom-neumát] shaved tire
— m cinchado - [autom-neumát] belted tire
— m — a bies - [autom-neumát] belted bias tire
— m — con tela(s) f a bies - [autom-neumát]
belted bias tire
— m clasificado - [autom-neumát] classified, o
sorted, tire
— m colocado - [autom-neumát] installed tire
— m como equipo m original - [autom-neumát]
original equipment tire
— m compacto - [autom-neumát] compact tire
— m — con presión f alta - [autom-neumát] high
pressure compact tire
— m —— — f — para repuesto m - [autom-
-neumát] high pressure compact spare tire
— m — para repuesto m - [autom-neumát] compact
spare tire
— m comparable - [autom-neumát] comparable tire
— m competidor - [autom-neumát] competing,titor
tire
— m comprado - [autom-neumát] purchased, o
bought, tire
— m común - [autom-neumát] common tire • also
ran tire

neumático m con banda(s) f circunferencial(es) -
[autom-neumát] belted tire
— m — — f — ancha(s) - [autom-neumát] wide,
o full-width, belt(ed) tire
— f — — — f — de acero m - [autom-neumát]
wide, o full-width, steel belt(ed) tire
— m — — f — de acero m - [autom-neumát]
steel belted tire
— m — — f circunferencial múltiple ancha de
acero m - [autom-neumát] multiple full width
steel belt(ed) tire
— m — — f — múltiple de acero m - [autom-
-neumát] multiple steel belt(s) tire
— m — — f — para camión m liviano - [autom-
-neumát] light truck belted tire
— m — — f — plegada - [autom-neumát] folded
belt tire
— m — — f — de rayón m - [autom-neumát]
folded rayon belt tire
— m — cámara f - [autom-neumát] tube-type tire
— m — capacidad f para rendimiento m alto -
[autom-neumát] high performance capability
tire
— m — característica(s) f excepcional(es) para
conducción f - [autom-neumát] premium handling tire
— m — carcasa f con . . . tela(s) - [autom-
-neumát] . . .-ply carcass tire
— m — — . . . f de rayón m - [autom-
-neumát] . . .-ply rayon carcass tire
— m — — f de poliestero m - [autom-neumát]
polyester carcass tire
— m — cincha f circunferencial - [autom-
-neumát] belted tire
— m — — f — de acero m - [autom-neumát]
steel belted tire
— m — — f de acero m - [autom-neumát] steel
belted tire
— m con clasificación f nominal para velocidad
f . . . - [autom-neumát] . . .-speed rated
tire
— m — — f para velocidad f . . . - [autom-
-neumát] . . . .-speed rated tire
— m — — f . . . para velocidad f - [autom-
-neumát] . . . .-speed rated tire
— m — — — f alta—[autom-neumát] (high)
speed rated tire
— m — — f — — f elevada - [autom-neumát]
(high) speed rated tire
— m — — f H - [autom-neumát] H-rated tire
— m — costado m blanco - [autom-neumát] white
side wall tire
— m — — m negro - [autom-neumát] black side
wall tire
— m — cuerda(s) f a bies - [autom-neumát] bias
tire
— m — — f circunferencial(es)—[autom-neumát]
belted tire
— m — diámetro m grande - [autom-neumát] large
diameter tire
— m — — m reducido - [autom-neumát] small
diameter tire
— m — elemento(s) m profundo(s) - [autom-
-neumát] deep block tire
— m — malla f circunferencial - [autom-neumát]
belted tire
— m — marca f - [autom-neumát] brand tire
— n — medida(s) f distinta(s) - [autom-neumát]
different, tire size, o size tire
— m — — f mayor(es) - [autom-neumát] up-sized
tire
— m — — f menor(es) - [autom-neumát] down-
-sized tire
— m — — f P-métrica(s) - [autom-neumát]
P-metric size(d) tire
— m — nervadura(s) f - [autom-neumát] ribbed
tire
— m — — f profunda(s) - [autom-neumát] deep-
-rib(bed) tire
— m — pared(es) f lateral(es) alta(s)—[autom-
-neumát] high aspect ratio tire
— m — — f — baja(s) - [autom-neumát] low-
-aspect ratio tire

neumático m con pared f lateral blanca - [autom-
-neumát] white side wall tire
— m — — f — — resistente a intemperie f -
[autom-neumát] weather resistant white side
wall (tire)
— m — — f — con franja f blanca angosta -
[autom-neumát] narrow white side wall tire
— m — — f — negra - [autom-neumát] black
side wall tire
— m — — f — — resistente a intemperie f -
[autom-neumát] weather resistant black side
wall tire
— f — — f — resistente a intemperie f -
[autom-neumát] weather resistante side wall
(tire)
— m — presión f - [autom-neumát] pressurized
tire
— m — — f alta - [autom-neumát] high pressure tire
— m — — f baja - [autom-neumát] low pressure
tire
— m — proyección f con elemento(s) m profun-
do(s) - [autom-neumát] deep block design tire
— m — rendimiento m alto - [autom-neumát] high
performance tire
— m — — m, muy, o sumamente alto - [autom-
neumát] very, o ultra-, high performance tire
— m — — f cuesta f abajo fácil - [autom-
-neumát] easy rolling tire
— m — — f cuesta abajo más fácil - [autom-
-neumát] easiest rolling tire
— m — — f más fácil - [autom-neumát] easiest
rolling tire
— m — sensación f de tirantez f - [autom-
-neumát] taut feeling tire
— m — sobreprecio m - [autom-neumát] premium
priced tire
— m — talón m - [autom-neumát] beaded (edge)
tire
— m — tamaño m, diferente, o distinto -
[autom-neumát] different, tire size, o size
tire
— m — — excepcional - [autom-neumát] over-
size(d) tire
— m — tecnología f alta - [autom-neumát] high
technology tire
— m — tela f única - [autom-neumát] monoply
tire
— m — . . . tela(s) f - [autom-neumát] . . .-
ply tire
— m — — f a bies - [autom-neumát] bias(-ply)
tire
— m — — f — — para barro m - [autom-neumát]
bias ply mud tire
— m — — f — — camión(es) m - [autom-
neumát] bias ply truck tire
— m — — f — — — m liviano(s) m—[autom-
-neumát] bias (ply) light truck tire; light
truck bias (ply) tire
— m — — f circunferencial(es) - [autom-
-neumát] belted tire
— m — — f diagonal(es) - [autom-neumát] bias
(ply) tire
— m — — f — para barro m - [autom-neumát]
bias ply mud tire
— m — — f — camión(es) m—autom-neumát]
bias ply truck tire
— m — — f — — m liviano(s) - [autom-
neumát] bias ply light truck tire
— m — telas f múltiples - [autom-neumát] mul-
tiply* tire
— m — . . . tela(s) f - [autom-neumát] . . .-
-ply tire
— n confiable - [autom-neumát] dependable tire
— m — con calidad f alta - [autom-neumát] dependable high quality tire
— m convencional - [autom-neumát] conventional
tire
— m — para nieve f - [autom] conventional snow
tire
— m — — rendimiento m alto - [autom-neumát]
conventional high performance tire
— m correcto - [autom-neumát] right tire

neumático m̲ corrido - [autom-neumát] raced tire
— m̲ corriente - [autom-neumát] common, o̲ regular, o̲ conventional, tire
— m̲ — para nieve f̲ - [autom-neumát] conventional snow tire
— m̲ corto - [autom-neumát] short tire
— m̲ costoso - [autom-neumát] expensive tire
— m̲ dañado - [autom-neumát] damaged tire
— m̲ de calidad f̲ alta - [autom-neumát] high,
quality, o̲ performance, tire
— m̲ — caucho m̲ - [autom-neumát] rubber tire
— m̲ — competencia f̲ - [autom-neumát] compe-
tive,titor tire
— m̲ — compuesto m̲ - [autom-neumát] compound
tire
— m̲ — — m̲ suave - [autom-neumát] soft compound tire
— m̲ — planta f̲ piloto - [autom-neumát] pilot
plant tire
— m̲ — poliestero m̲ - [autom-neumát] polyester
tire
— m̲ — tipo m̲ para carretera f̲ - [autom-neumát]
highway type tire
— m̲ — — m̲ rendimiento m̲ alto - [autom-
-neumát] (high) performance type tire
— m̲ delantero - [autom-neumát] front tire
— m̲ derecho - [autom-neumát] straight tire •
right (hand) tire
— m̲ — para impulsión f̲ - [agric-equip] right
hand drive tire
— m̲ deseado - [autom-neumát] desired tire
— m̲ desgastado - [autom-neumát] worn tire
— m̲ desinflado - [autom-neumát] flat (tire)
— m̲ —, cambiado, o̲ permutado - [autom-neumát]
switched flat (tire) • changed flat (tire)
— m̲ deslizante - [autom-neumát] sliding tire
— m̲ despegado - [autom-neumát] unseated tire
— m̲ distintivo—[autom-neumát] distinctive tire
— m̲ Duplex - [autom-neumát] Duplex tire
— m̲ emparchado - [autom-neumát] patched tire
— m̲ en deslizamiento m̲ - [autom-neumát] sliding
tire
— m̲ en medida(s) f̲ para camión(es) m̲ - [autom-
neumát] truck size(d) tire
— m̲ — — f̲ — — m̲ liviano(s) m̲ - [autom-
neumát] light truck size(d) tire
— m̲ — — m̲ apropiado - [autom-neumát] appropriate (sized) tire
— m̲ — — m̲ nuevo - [autom-neumát] new,
size, o̲ size(d) tire
— m̲ ensayado - [autom-neumát] tested tire
— m̲ entizado - [autom-neumát] chalked tire
— m̲ equilibrado - [autom-neumát] balanced tire
— m̲ especial - [autom-neumát] special tire
— m̲ — para carrera(s) f̲ - [autom-neumát] special race,cing tire
— m̲ — — — f̲ Rally - [autom-neumát] special
rally (race,cing) tire
— m̲ estadounidense - [autom-neumát] American, o̲
United States, made tire
— m̲ estándar - [autom-neumát] standard tire
— m̲ estriado - [autom-neumát] rib(bed) tire
— m̲ europeo - [autom-neumát] European tire
— m̲ — con medida(s) f̲ métrica(s) - [autom-
-neumát] European metric size(d) tire
— m̲ excepcional - [autom-neumát] premium tire
— m̲ exclusivamente para carrera(s) f̲ - [autom-
-neumát] pure, o̲ real, racing (only) tire
— m̲ existente - [autom-neumát] existing tire
— m̲ experimental - [autom-neumát] experimental
tire
— m̲ exterior - [autom-neumát] outboard tire
— m̲ fabricado a̲ - [autom-neumát] manufactured
tire
— m̲ — de acuerdo con tecnología f̲ alta -
[autom-neumát] high technology tire
— m̲ — en Estados m̲ Unidos - [autom-neumát]
American, o̲ United States, -made tire
— m̲ favorito - [autom-neumát] favorite tire
— m̲ fiscalizado - [autom-neumát] checked, o̲r
tested, tire
— m̲ frontal - [autom-neumát] front tire
— m̲ gastado - [autom-neumát] worn tire

neumático(s) m̲ gemelo(s) - [autom-neumát] twin
tire(s)
— m̲ gobernado - [autom-neumát] steered tire
— m̲ grande - [autom-neumát] large tire
— m̲ — para camión m̲ - [autom-neumát] large
truck, o̲ truck large, tire
— m̲ — — — m̲ liviano - [autom-neumát] large
light truck, o̲ light truck large, tire
— m̲ — — camioneta f̲ - [autom-neumát] large
light truck, o̲ light truck large, tire
— m̲ guiador - [agric-equip] guide,ding tire
— m̲ para adentro - [autom-neumát] inboard tire
— m̲ hacia adentro - [autom-neumát] inboard tire
— m̲ hacia afuera - [autom-neumát] outboard tire
— m̲ hacia interior m̲ - [autom-neumát] inboard
tire
— m̲ inaceptable - [autom-neumát] unacceptable
tire
— m̲ incorrecto - [autom-neumát] wrong tire
— m̲ individual - [autom-neumát] individual tire
— m̲ industrial - [autom-neumát] industrial tire
— m̲ — con . . . tela(s) f̲ - [autom-neumát]
...-ply industrial tire
— m̲ — sin cámara f̲ - [autom-neumát] tubeless
industrial tire
— m̲ — — . . . — con . . . tela(s) f̲ -
[autom-nuemát] . . .-ply, tubeless industrial, o̲ industrial tubeless, tire
— m̲ inflable - [autom-neumát] inflatable tire
— m̲ inflado - [autom-neumát] inflated tire
— m̲ — en forma f̲ apropiada - [autom-neumát]
properly inflated tire
— m̲ — — — f̲ inapropiada - [autom-neumát]
improperly, o̲ inappropriately, inflated tire
— m̲ inspeccionado - [autom-neumát] inspected
tire
— m̲ instalado - [autom-neumát] installed, o̲
fitted, tire
— m̲ intercambiado - [autom-neumát] interchanged
tire
— m̲ interior - [autom-neumát] inboard tire
— m̲ izquierdo - [autom-neumát] left (hand) tire
— m̲ — para impulsión f̲ - [agric-equip] left
(hand) drive tire
— m̲ japonés - [autom-neumát] Japanese tire
— m̲ largo - [autom-neumát] long tire
— m̲ liso - [autom-neumát] slick, o̲ smooth, tire
— m̲ — para carrera(s) f̲ - [autom-neumát] slick
racing, o̲ racing slick, tire
— m̲ macizo m̲ - [autom-neumát] solid tire
— m̲ marcada - [autom-neumát] marked tire •
branded tire
— m̲ más largo - [autom-neumát] longer tire
— m̲ mejor - [autom-neumát] best tire • better
tire
— m̲ métrico - [autom-neumát] metric tire
— m̲ montado - [autom-neumát] mounted tire
— m̲ no convencional - [autom-neumát] unconventional tire
— m̲ — para rendimiento m̲ alto - [autom-
-neumát] unconventional high performance tire
— m̲ — director - [autom-neumát] non directional tire
— m̲ — radial - [autom-neumát] non-radial tire
— m̲ — — para automóvil(es) m̲ para pasajeros m̲
[autom-neumát] non-radial passenger tire
— m̲ — — — camión(es) m̲ - [autom-neumát] non-
-radial truck tire
— m̲ — — — — m̲ liviano(s) - [autom-neumát]
non-radial light truck tire
— m̲ normal para carretera f̲ - [autom-neumát]
(normal), street, o̲ highway, tire
— m̲ novedosísimo - [autom-neumát] hot tire
— m̲ nuevo - [autom-neumát] new tire
— m̲ — instalado - [autom-neumát] installed new
tire
— m̲ — para reemplazo - [autom-neumát] new replacement tire
— m̲ obtenible - [autom-neumát] available tire
— m̲ optativo - [autom-neumát] optional tire
— m̲ original - [autom-neumát] original tire
— m̲ P-métrico - [autom-neumát] P-metric tire
— m̲ — substitutivo - [autom-neumát] P-metric

substitute tire
neumático m pantanero - [autom-neumát] mud tire
— m — con tela(s) f a bies - [autom-neimát]
bias ply mud tire
— m —— f diagonal(es) f - [autom-neumát]
bias ply mud tire
— m para autobús m - [autom-neumát] bus tire
— m — automóvil m - [autom-neumát] automobile,
o car, tire
— m —— m deportivo - [autom-neumát] sports,
car tire, o rubber
— m —— para pasajero(s) m - [autom-neumát]
passenger (car, o automobile) tire
— m — auxilio m - [autom-neumát] spare tire
— m — barro m - [autom-neumát] mud tire
— m —— m y nieve f - [autom-neumát] mud and
snow tire
— m —— m y terreno(s) m abierto(s) - [autom-
-neumát] mud-terrain tire
— m — camión m - [autom-neumát] truck tire
— m —— m liviano - [autom-neumát] light
truck tire
— m —— m para servicio m pesado - [autom-
-neumát] heavy-duty truck tire
— m — camioneta f - [autom-neumát] light truck
tire
— m —— f para recreo m - [autom-neumát]
sport (light) truck tire
— m — carga f - [autom-neumát] load, o
freight, tire
— m —— f estándar - [autom-neumát] standard,
load, o freight, tire
— m —— f normal - [autom-neumát] standard
load tire
— m — carrera(s) f - [autom-neumát] race,cing,
o racy, tire, o rubber
— m —— f Rally - [autom-neumát] rally tire
— m — carretera f - [autom-neumát] street, o
highway, tire
— m — conducción f fuera de carretera f -
[autom-neumát] off-@-road tire
— m — eje m guía - [agric=equip] guide axle
tire
— m — equipo m para construcción f - [constr]
construction equipment tire
— m —— movimiento m de tierra f -
[autom-neumát] earth mover,ving (equipment)
tire
— m — fango m - [autom-neumát] mud tire
— m — flotabilidad f alta - [autom-neumát]
high flotation tire
— m —— f elevada - [autom-neumát] high flotation tire
— m — fuera de carretera f - [autom-neumát]
off-@-road tire
— m — impulsión f - [mec] drive,ving tire
— m — kilometraje m mayor - [autom-neumát] Ex-
tra-Miler tire; long mileage tire
— m — lado m derecho - [agric-equip] right
hand tire
— m —— m izquierdo m - [agric-equip] left
hand tire
— m ——m para impulsión f - [agric-equip]
left hand drive,ving tire
— m — mano f derecha - [autom-neumát] right
hand tire
— m —— f izquierda - [autom-neumát] left
hand tire
— m — motocicleta f - [autom-neumát] motorcycle
tire
— m — nieve f - [autom-neumát] snow tire
— m — pantano(s) m—[autom-neumát] mud-terrain
tire
— m — recambio m - [autom-neumát] replacement,
o spare, tire
— m —— m seleccionado - [autom-neumát] selected, o chosen, replacement tire
— m — reemplazar(se) v - [autom-neumát] tire
to be, replaced, o substituted
— m — reemplazo m - [autom-neumát] replacement
tire
— m —— m en medida(s) f más apropiada(s) -
[autom-neumát] closest replacement tire size

neumático m para reemplazo m para automóvil(es)
m para pasajero(s) m - [autom-neumát] substitute, o replacement, passenger (car) tire
— m —— m para camión(es) m - [autom-neumát]
truck replacement, o replacement truck, tire
— m —— m —— m liviano(s) - [autom-neumát]
replacement, o substitute, light truck tire
— m —— m —— camioneta(s) f - [autom-neumát]
substitute, o replacement, light truck tire
— m — rendimiento m alto - [autom-neumát]
(high) performance tire
— m —— m —, favorito, o preferido - [autom-
-neumát] favorite (high) performance tire
— m — reposición f - [autom-neumát] replacement tire
— m — repuesto m - [autom-neumát] spare (tire)
— m — rueda f - [autom-neumát] wheel tire
— m —— f de . . . pulgadas f—[autom-neumát]
. . .-inch (wheel) tire
— m — servicio m pesado m - [autom-neumát]
heavy duty tire
— m —— m para camión(es) m - [autom-
-neumát] heavy-duty truck tire
— m — substitución f - [autom-neumát] substitute tire
— m — terreno(s) m muy pantanoso(s) - [autom-
-neumát] high flotation tire
— m — toda(s) estación(es) f - [autom-neumát]
all-season tire
— m —— rueda(s) f - [autom-neumát] all-
-position tire
— m — todo terreno m - [autom-neumát] all-
-terrain tire
— m — tractor m - [agric-equip] tractor tire
— m — uso m fuera de carretera f - [autom-
-neumát] off @ highway tire
— m —— m sobre o fuera de carretera f -
[autom-neumát] on-off highway tire
— m — velocidad(es) f de hasta 90 millas (145
kms) por hora - [autom-neumát] S tire
— m —— f —— 115 millas (185 kms) por hora
- [autom-neumát] H tire
— m —— f —— 130 millas (210 kms) por hora
- [autom-neumát] V tire
— m patrón - [autom-neumát] control, o base-
line, tire
— m peor - [autom-neumát] worse tire • worst
tire
— m pequeño - [autom-neumát] small tire
— m permutado - [autom-neumát] switched, o
swapped, tire
— m pinchado - [autom-neumát] blown out tire •
punctured tire
— m posterior - [autom-neumát] rear tire
— m preferido - [autom-neumát] preferred, o favorite, tire
— m probado - [autom-neumát] tested tire
— m proyectado - [autom-neumát] designed tire
— m puesto a punto - [autom-neumát] tuned tire
— m que rueda libremente - [autom-neumát] free
rolling tire
— m quitado - [autom-neumát] removed tire
— m radial - [autom-neumát] radial (tire)
— m —— callejero - [autom-neumát] street radial
(tire)
— m —— para rendimiento m alto - [autom-
-neumát] (high) performance street radial
(tire)
— m —— con banda f circunferencial de acero m -
[autom-neumát] steel belted radial tire
— m —— f —— m para camión(es) m -
[autom-neumát] steel-belted truck radial tire
— m —— f —— m liviano(s) -
[autom-neumát] steel-belted light truck radial
(tire)
— m —— cincha f (circunferencial) de acero m
- [autom-neumát] steel belted radial (tire)
— m —— con perfil m bajo - [autom-neumát] low
profile radial (tire)
— m —— con tecnología f alta - [autom-neumát]
high technology radial (tire)
— m —— confiable - [autom-neumát] dependable
radial (tire)

neumático m radial confiable con calidad f alta
- [autom-neumát] dependable high quality radial (tire)
— m — **de producción** f - [autom-neumát] production radial (tire)
— m —— f **corriente** - [autom-neumát] regular-production radial (tire)
— m —— **serie** . . . - [autom-neumát] . . . -series, radial (tire)
— m — **fabricado** - [autom-neumát] manufactured, o constructed, radial (tire)
— m — **High-Tech** - [autom-neumát] High-Tech radial (tire)
— m — **iniqualable** - [autom-neumát] unique radial (tire)
— m — **métrico** - [autom-neumát] metric radial (tire)
— m — **nuevo** - [autom-neumát] new radial (tire)
— m — **para automóvil(es) m (para pasajeros m)** - [autom-neumát] passenger radial, o radial passenger, tire
— m —— **barro** m - [autom-neumát] radial mud tire
— m ——— m **y terreno(s) m abierto(s)** - [autom-neumát] radial Mud-Terrain tire
— m —— **camión(es)** m - [autom-neumát] radial truck tire
— m ——— m **liviano(s)** m - [autom-neumát] radial light truck, o light truck radial, tire
— m ——— m **para servicio m pesado** - [autom-neumát] heavy duty truck radial (tire)
— m —— **conducción** f **fuera de carretera** f - [autom-neumát] off-@-road radial (tire)
— m —— **cualquier terreno** m - [autom-neumát] all, o mud, terrain radial (tire)
— m —— **fuera de carretera** f - [autom-neumát] off-@-road radial (tire)
— m —— **kilometraje** m **mayor** - [autom-neumát] long-mileage radial (tire)
— m —— **lodo** m - [autom-neumát] radial mud tire
— m —— **rendimiento** m **alto** - [autom-neumát] (high) performance radial (tire)
— m ——— m **fuera de carretera** f - [autom-neumát] off-@-road (high) performance radial (tire)
— m ——— m **para camión(es)** m - [autom-neumát] high performance truck radial tire; truck (high) performance radial (tire)
— m ——— m ——— m **liviano(s)** m - [autom-neumát] light truck (high) performance radial (tire)
— m ——— m **muy elevado** - [autom-neumát] ultra-high-performance radial (tire)
— m ——— m **sumamente alto** - [autom-neumát] ultra-high-performance radial (tire)
— m —— **servicio** m **pesado** - [autom-neumát] heavy-duty radial (tire)
— m ——— **para camión(es)** - [autom-neumát] heavy-duty truck radial (tire)
— m —— **toda(s) estación(es)** f - [autom-neumát] all-season radial (tire)
— m —— **velocidad(es)** f **de hasta 130 millas** (209 kms) **por hora** f - [autom-neumát] VR tire
— m — **T/A High Tech** - [autom-neumát] T/A High Tech radial (tire)
— m **ranurado** - [autom-neumát] grooved tire
— m **recio** - [autom-neumát] strong, o tough, tire
— m **reemplazado** - [autom-neumát] replaced, o substituted, tire
— m **removido** - [autom-neumát] removed tire
— m **reparado** - [autom-neumát] repaired tire
— m **repuesto** - [autom-neumát] replaced tire
— m **resbalador** - [autom-neumát] slipping tire
— m **reventado** - [autom-neumát] blown out tire
— m **reversible** - [autom-neumát] reversible tire
— m **rodado** - [autom-neumát] rolled tire
— m **sacado** - [autom-neumát] removed, o pulled, tire
— m **satisfactorio** - [autom-neumát] satisfactory tire
— m **seleccionado** - [autom-neumát] selected tire

neumático m sin cámara f - [autom-neumát] tubeless tire
— m —— f **con . . . tela(s)** f - [autom-neumát] . . .-ply tubeless tire
— m —— f **para aire** m - [autom-neumát] tubeless tire
— m — **inflar** - [autom-neumát] uninflated tire
— m **substituido** - [autom-neumát] substituted tire
— m **substitutivo** - [autom-neumát] substitute tire
— m **trasero** m - [autom-neumát] rear tire
— m **triunfador** - [autom-neumát] winning tire
— m **triunfante** - [autom-neumát] winning tire
— m **vertical** - [autom-neumát] upright tire
— m **vendido** - [autom-neumát] sold tire
— m **viejo** - [autom-neumát] old tire
neumático,ca a - . . .; air (operated)
— a **únicamente** - [autom-neumát] straight air
neumáticos m - [autom-neumát] tire(s) equipment
— m **mezclados** - [autom-neumát] mixed tires
— m —, **inapropiadamente, o incorrectamente** - [autom-neumát] improperly mixed tire(s)
— m **gemelos** - [autom-neumát] twin tires
— m **posteriores dobles** - [autom-neumát] dual rear tires
neutral m - neutral | a - . . .
— m **hidrostático** - hydrostatic neutral
— m — **ajustado** - adjusted hydrostatic neutral
neutralización f **de, desecho(s)** m, o **residuo(s)** m - waste(s) neutralization
— f — **silicio** m - [metal-prod] silicon neutralization
neutralizado,da a - neutralized
neutralizador m - neutralizer
neutralizar v - . . . • to offset • to resist • [metal-prod] to, block, o kill
— v **desecho(s)** m - [nucl] to neutralize @ waste
— v **reacción** f - [metal-prod] to block
— v — f **de colada** f - [metal-prod] to block @ heat
— v **residuo(s)** - [nucl] to neutralize @ waste
neutro m - [electr-instal] neutral
— m **a tierra** f - [electr-instal] grounded neutral
— m **conectado con tierra** f - [electr-instal] grounded neutral
— m —— f **por medio de resistencia(s)** f - [electr-instal] resistance grounded neutral
— m **sólido** - [electr-instal] solid neutral
— m **sólido a tierra** f - [electr-instal] grounded solid neutral
neutro,tra a - . . . • [electr] no load
ni adv **aún** - not even
— adv — **necesario,ria** - not even necessary
— adv **cerca** - anywhere, o nowhere, near
— adv **por mucho** - not even close
nicho m - . . . • cell
niebla f **de ácido** - [quím] acid mist
— f **de alborada** f - [meteorol] pre-dawn mist
— f **disipada** - [meteorol] abated fog
nieve f **acumulada** - [meteorol] accumulated snow; snow drift
— f **artificial** - [deport] man-made, o artificial, snow
— f **compacta(da)** - compact(ed) snow
— f **derretida** - [meteorol] melted snow
— f **fresca** - [meteorol] fresh snow
— f **invernal** - [meteorol] winter('s) snow
— f, **quitada,** o **removida,** o **sacada** - removed snow
— f **semiderretida** - [meteorol] half melted, o melting, snow
ningún adv **metal** m **aportado** - [sold] no deposited metal
— adv **vacío** m - [fís] no vacuum
ninguna adv **de parte(s)** f - [legal] neither party
— adv **intensidad** f - no intensity
— adv **protección** f - no protection
— adv **velocidad** f - no speed • any speed
ninguno,na a - . . .; none; neither
— adv **de dos** - neither

ninguno(s),na(s) dos operan --no-two-operate
niña f - . . .; female child
niño m - . . . . • male child • [fam] kid
niño m talentoso - [educ] talented child
niños m - boys • children
niobio m - [metal] . . .; columbium
niple* m - [tub] nipple • véase entrerrosca f •
  [petról] véase rótula f para tubería f
— m terraja - [mec] die nipple
níquel m crómico - [metal] chrome-nickel; nickel
  and chrome,mium
— m en plancha(s) f - [metal-lam] sheet nickel
— m dúctil - [metal] ductile nickel
— m y cromo m - [metal] nickel and chrome,mium
nitidez f - [óptica] sharpness • definition; a-
  cuteness
— f adecuada - [óptica] adequate sharpness
— f de contorno m - contour, o outline, sharpness
— f — — m de imagen f - [óptica] image outline sharpness
— f de imagen f - [óptica] image sharpness
— f en película f - [fotogr] film definition
nitración f - [quím] nitration; nitriting
nitrado,da a - [quím] nitrated
nitrador,ra a - [quím] nitrating
nitrar v - [quím] to nitrate
nitrato m amónico - [quím] ammonium, o ammoniac,
  nitrate
— m aniónico - [quím] anionic nitrate
— m cálcico - [quím] calcium nitrate
— m de amonio - [quím] ammonium nitrate
— m calcio m - [quím] calcium nitrate
— m magnesio m - [quím] magnesium nitrate
— m mercurio m - [quím] mercury nitrate
— m monoetilo m - [quím] monoethyl nitrate
— m — m amina - [quím] amine monoethyl nitrate
— m plata f - [quím] silver nitrate
— m potasio m - [quím] potassium nitrate
— m sodio m - [quím] sodium nitrate
— m tolueno m - [quím] toluene nitrate
— m eliminado - [quím] eliminated nitrate •
  a - [quím] (con) denitrified
— m férrico - [quím] ferric nitrate
— m mercúrico - [quím] mercuric nitrate
nítrico,ca a fumante - [quím] fuming nitric
nitroalgodón* m - [quím] nitrocotton
nitrocelulosa f - [explos] nitrocellulose
nitrogelatina* - [quím] nitrogelatin
nitrógeno m amoniacal - [quím] ammonia nitrogen
— m atmosférico - [quím] atmospheric nitrogen
— m bombeado - pumped nitrogen
— m con agua m - water pumped nitrogen
— m residual - [quím] residual nitrogen
— m seco - [quím] dry nitrogen
nitronaftalina f - [quím] nitronaphthalene
nitruración f - [metal-trat] nitriding
nitrurado,da a - [quím] nitrided
nitrurar v - [quím] to nitride
nitruro m de aluminio m—[quím] aluminum nitride
— m boro m - [quím] boron nitride
— m — m cúbico m [quím] cubic boron nitride
nivel n - . . . . • grade; elevation • line; alti-
  tude; plateau • range • [econ] bracket •
  [constr] profile | a - (a) [constr] at grade
— m acostumbrado - customary level
— m acuífero - [hidr] water table
— m adecuado - adequate, o workable, level
— m para existencia(s) f - [ind] adequate, o
  workable, inventory level
— m administrativo - [admin] administration, o
  management, level
— m de empresa f - [admin] enterprise, o
  concern, management level
— m aguas f abajo - [hidr] downstream water
  level
— m f arriba - [hidr] upstream water level
— m ajustable - adjustable level
— m con potenciómetro m - [electrón] potentiometer adjustable level
— m alto - high level • high range | a - (de)
  elevated

nivel m alto de aceite m - high oil level
— m — — m hidráulico - hydraulic oil high
  level; high hydraulic oil level
— m — agua m - [hidr] high water (level)
— m — — m freática - [hidr] groundwater, o
  phreatic water, high level; high groundwater
  (level)
— m — carga f - [metal-prod] high charge
  level
— m — irradiación f - [nucl] high radiation
  level
— m — líquido m - [mec] high liquid level
— m — — m regulado - [mec] controlled high
  liquid level
— m — metalización f - [metal-trat] high,
  metallization level, o degree metallization
— m — producción f - [ind] high production,tivity level
— m — ruido(s) m - [acúst] high, noise, o
  sound, level
— m — sonido(s) m - [acúst] high, sound, o
  noise, level
— m — para líquido m - high liquid level
— m — regulado - [mec] controlled high level
— m apropiado - proper, o appropriate, level
— m — para abastecimiento m - proper, fill, o
  supply,lies, level
— m — lubricante m - [mec] proper, o appropriate, lubricant level
— m arancelario m - [fisc] tariff level
— m — alto - [fisc] high tariff level
— m — bajo - [fisc] low tariff level
— m — cero - [fisc] zero tariff level
— m bajo - low level • low range • [hidr] low
  flow level
— m — de aceite m - low oil, level, o supply
— m — — — m hidráulico - low hydraulic oil
  level
— m — — líquido m - low liquid level
— m — — — m regulado - [mec] controlled low
  liquid level
— m — irradiación f - [nucl] low radiation
  level
— m — metalización f - [metal-trat] low,
  metallization level, o degree metallization
— m — radiactividad f - [nucl] low radiation level
— m — — — f de desecho(s) m - [mucl] low
  (radiation) level waste
— m — — — f — residuo(s) m - [mic;] low radiation (level) waste
— m — ruido(s) m - [acúst] low noise level
— m — sonido(s) m - [acúst] low sound level
— m — regulado - [mec] controlled low level
— m básico para aislación f - [electr-instal]
  basic insulation level
— m — — f contra impulso(s) m - [electrón]
  basic impulse insulation level
— m cambiado - changed level
— m cambiario - [fin] exchange level
— m circular - [instrum] circular level
— m competitivo - competitive level
— m comprobado - checked, o proven, level •
  [constr] established grade
— m de planta f - [constr] established plant
  grade
— m compuesto - composite level
— m de voltaje(s) m - [electrón] composite
  voltage level
— m constante - constant, o steady, level
— m — para presión f - [mec] constant, o
  steady, pressure level
— m — — f para aceite m - [comb.int] constant, o steady, oil pressure level
— m — — f — m alcanzado - [comb.int]
  attained, o reached, steady oil pressure
  level
— m convenido - agreed (upon) level
— m cumplido - [constr] met, o attained, level,
  o grade
— m dado - given level
— m de Abney - [herram] Abney level
— m — aceite m - [mec] oil, level, o supply

nivel m de aceite m alto - high oil level
— m — — m **bajo** - low oil, level, o supply
— m — — m **comprobado** - [comb.int] proven, o checked, oil level
— m — — m **constatado** - [int.comb] gaged oil level
— m — — m **en bomba** f—[bombas] pump oil level
— m — — m — — f **con impulsión** f - [bombas] driven pump oil level
— m — — m — — f — — f **con cadena** f - [bombas] chain driven pump oil level
— m — — m — — f — — f **verificado** - [bombas] checked chain driven pump oil level
— m — — f **verificado** - [bombas] checked pump oil level
— m — — m **en caja** f • [comb.int] crank case oil level
— m — — m — — f **para cadena** f - [mec] chain case oil level
— m — — m — — f — — f **verificado** - [mec] checked chain case oil level
— m — — m — — f **para engranaje(s)** m - [mec] gear case oil level
— m — — n — — f **verificado** - [mec] checked case oil level
— m — — m **cambiador** m **para toma(s)** f - [electr-equip] tap changer oil level
— m — — m **cárter** m - [comb.int] crankcase oil level
— m — — m — — m, **comprobado,** o **verificado** - [comb.int] checked crankcase oil level
— m — — m **en conjunto** m - [mec] assembly oil level
— m — — m — — m **de árbol** m - [mec] shaft assembly oil level
— m — — m — — m — — m **para aguilón** m - [grúas] boom shaft assembly oil level
— m — — m — — m — — m **verificado** - [grúas] checked boom shaft assembly oil level
— m — — m — — m — — m **para cable** m a **gancho** m - [grúas| whip shaft assembly oil level
— m — — m — — m — — m **verificado** - [grúas] checked whip shaft assembly oil level
— m — — m — — m — — m a **motón** m - [grúas] main shaft assembly oil level
— m — — m — — m — — m — — m **verificado** - [grúas] checked main shaft assembly oil level
— m — — m — — m **verificado** - [mec] checked assembly oil level
— m — — m **en depósito** m - [mec] reservoir oil level
— m — — m — — **motor** m - [comb.int] engine oil level • [electr-mot] motor oil level
— m — — m — — m **verificado** - [comb.int] checked engine oil level • [electr-mot] checked motor oil level
— m — — m **hidráulico** - [mec] hydraulic oil level
— m — — m **medido** - [mec] measured, o gaged, oil level
— m — — m **para lubricador** m - [mec] lubricator oil level
— m — — m — — m **para línea** f - [grúas] line lubricator oil level
— m — — m — — m — — f **neumática** - [grúas] air line lubricator oil level
— m — — m — — m — — f — **verificado** - [grúas] checked air line lubricator oil level
— m — — m **verificado** - [comb.int] checked oil level
— m — **aceptación** f - acceptance level
— m — — f **de rodillo(s)** m - [metal-lam] roll acceptance level
— m — — f — — m **de acero** m - [metal-lam] steel roll(es) acceptance level
— m — — f — — m — — m **forjado** - [metal-lam] forged steel roll acceptance level
— m — — f — — m — — **fundido** - [metal-lam] cast steel roll acceptance level

nivel m de aceptación f **de rodillo(s)** m **para a- poyo** m - [metal-lam] back-up roll acceptance level
— m — f — — m — — m **de acero** m - [metal -lam] steel back-up roll acceptance level
— m — f — — m — — m — — m **fundido** - [metal-lam] cast steel back-up roll accep- tance level
— m — f — — m **para trabajo** m - [metal- -lam] work roll acceptance level
— m — f — — m — — m **de acero** m - [metal --lam] steel work roll acceptance level
— m — f — — m — — — — m **fundido** - [metal-lam] cast steel work roll acceptance level
— m — **acumulador** m - [electr-prod] battery level
— m — **administración** f - [admin] management level (line)
— f — — **en empresa** f - [Admin] concern, o enterprise, management level
— m — **agregado(s)** m - [ind] aggregate(s), o filler, level
— m — **agua** m - [hidr] water, level, o eleva- rion • water surface • streamline
— m — — m **aguas abajo** - [hidr| downstream water level
— m — — m — **arriba** - [hidr] upstream water level
— m — — m **alto** - [hidr] high water level
— m — — m **bajo** - [hidr] low water level
— m — — m **en acumulador** m - [electr-prod] battery water level
— m — — m — — m **verificado** - [electr-prod] checked battery water level
— m — — m **en caldera** f - [cald] boiler water level
— m — — m **fondo** m - bottom water level
— m — — m **de lavador** m - [metal-prod] scrubber water level
— m — — m — — m — — m **para venturi** m - [metal-pord] Venturi scrubber bottom water level
— m — — m — **lavador** m - [metal-prod] scrub- ber water level
— m — — m — **presa** f - [hidr| dam water level
— m — — m — **radiador** m - [comb.int] radiator water level
— m — — m — **Venturi** m - [metal-prod] Ventu- ri, water level, o flow water
— m — — m **freática** - [hidr] groundwater, lev- el, o elevation
— m — — m — **restaurado** - [hidr] restored groundwater level
— m — — m **para freno** m - [mec] brake water level
— m — — m **regulado** - [hidr] controlled water level
— m — — m **subterránea** - [hidr] ground, o un- derground, water level
— m — — m **verificado** - [hidr] checked water level
— m — **aislación** f - [electr-instal] insulation level
— m — — f **contra impulso(s)** m - [electrón] impulse insulation level
— m — **ajuste** m - adjustment level
— m — — m **preciso** - fine adjustment level
— m — **alumbrado** m - [electr-instal] lighting, o illuminating,tion, level
— m — **anteojo** m - [instrum] optical level • surveyor('s) level
— m — **aprovechamiento** m - utilization level
— m — **armónica** f - [electrón] harmonic(s) level
— m — **automatización** f - [ind] automation level
— m — **baño** m - [metal-prod] bath, level, o surface; liquid line
— m — **base** f - base level
— m — **burbuja** f - [herram] spirit, o air, level
— m — — f **de aire** m - [instrum] air level
— m — **calidad** f - quality level
— m — — f **aceptable** - [ind] acceptable quality

level
nivel m **de calidad** f **operativo** m - [ind] operational quality level
— m — **canaleta** f - [hidr] trough level
— f — **capacitación** f - [labor] qualification level
— m — — f **de personal** m - [labor] personnel qualification level
— m — **carbón** m - [metal-prod] coal line
— f — **carga** f - [ind] charge, o stockline, level
— m — — f **bajo** - [ind] low, charge level, o stockline
— m — — f **normal** - [ind] normal, charge level, o stockline
— m — — f **perdido** - [ind] lost, charge, o stockline, level
— m — **cesto** m **repartidor** - [col.cont] tundish level
— m — **colinesterase** m - [medic] cholinesterase level
— m — — m **en sangre** f - [medic] blood cholinesterase level
— m — **combustible** m - [comb.int] fuel level
— m — **comida** f - [avicult] feed level
— m — **compactación** f **de suelo** m - [constr] soil compacting,tion level
— m — **comparación** f - comparison level; datum
— m — — f **horizontal** - horizontal datum
— m — **concentración** f - concentration level
— m — — f **de ión(es)** m - [quím] ion(s) concentration level
— m — — f — m **de hidrógeno** - [quím] hydrogen ion(s) concentration level
— m — **corriente** f - [hidr] flow, o stream, level
— m — — f **alto** - [hidr] high, flow, o stream, level
— m — — f **bajo** - [hidr] low, flow, o stream. level
— m — — f **inferior** - [hidr] lower, flow, o stream, level
— m — — f **superior** - [hidr] higher, flow, o stream, level
— m — **corte** m - [constr-pil] cutoff (elevation, o level)
— m — **cortocircuito** m - [electr-oper] short circuit level
— m — **creciente** - [hidr] high water, o flood, stage, o mark
— m — **crisol** m - [metal-prod] hearth level
— m — **demanda** f - demand level
— m — — f **establecido** - established demand level
— m — **desarrollo** m - [econ] development level
— m — **desempeño** m - performance level
— m — **distribuidor(es)** m - [com] distributor level
— m — **eficiencia** f - [ind] efficiency level
— m — — f **internacional** - [econ] international efficiency level
— m — **electrolito** m - [electr-acumul] electrolyte level
— m — **energía** f - [electr-distrib] power level
— m — — f **para horno** m **(eléctrico)** - [metal-prod] arc, o electric, furnace power level
— m — — f **regulado** - [electr-distrib] controlled power level
— m — **enriquecimiento** m - enriching,chment level
— m — — m **con oxígeno** m - [metal-prod] oxygen enriching,chment level
— m — **envolvimiento** m - involvement level
— m — **escoria** f - [metal-prod] slag, level, o line
— m — **estudio** m - study level
— m — **existencia(s)** f - [ind] inventory level • [com] stocks, level, o quantity
— m — — f **adecuado** - [ind] adequate, o workable, inventory level
— m — — f **de repuesto(s)** m - [ind] adequate, o workable, parts inventory level
— m — — f **de repuesto(s)** m - [ind] parts inventory level

nivel m **de fluido** - fluid level
— m — — m **hidráulico** - [petról] hydraulic fluid level
— m — — m — **verificado** - [petról] checked hydrauylic fluid level
— m — — m **para accionamiento** m - [mec] drive fluid level
— m — — m — — m **para bomba** f - [bombas] pump drive fluid level
— m — — m — — m — — f **triple** - [bombas] triple pump drive fluid level
— m — — m — — m — — m — **verificado** = [petrólk] checked triple pump drive fluid level
— m — **fondo** m - [mec] bottom level
— m — — m **de lago** m - [hidr] lake bottom level
— m — — m — **tubería** f - [constr] pipe grade bottom • pipe bottom level
— m — **garganta** f - [metal-prod] throat level
— m — **humedad** f - moisture level
— m — — f **en cemento** m - [constr] cement moisture level
— m — **iluminación** f - [electr-instal] illuminating,tion level
— m — **impulso** m - [electrón] impulse level
— m — — m **básico** - [electrón] basic impulse level
— m — — m **de onda** f - [electrón] wave impulse level
— m — **incremento** m - increment level
— m — — m **de potencia** f - [electrón] power increment level
— m — **indiferencia** f - indifference level
— m — **ingreso(s)** m - [econ] income, level, o bracket
— m — — m **familiar(es)** - [econ] family income bracket
— m — **inspección** f - inspection level
— m — **interés** m - interest level
— m — **inyección** f - [combust] injection level
— m — — f **de fuel oil** m - [combust] fuel oil injection level
— m — **irradiación** f - [nucl] radiation level
— m — **laguna** f - [hidr] lake level
— m — **lecho** m - [hidr] (stream) bed level
— m — — m **para corriente** f - [hidr] streambed level
— m — **líquido** m - liquid level
— m — — m **refrigerante** - [comb.int] coolant level
— m — — m — **en motor** m - [comb.int] engine coolant level
— m — — m — — — m **verificado** - [comb.int] checked engine coolant level
— m — — m — **verificado** - [comb.int] checked coolant level
— m — — m **regulado** - [mec] controlled liquid level
— m — **lodo** m - [petról] mud level
— m — **lubricante** m - [lubric] lubricant level
— m — **llanta** f - [autom-neumát] rim level
— m — **metalización** f - [metal-prod] metallization, level, o degree
— m — **muelle** m - [constr] wharf level
— m — **norma** f **para aprobación** f - [ind] acceptance requirement level
— m — **ocupación** f - [labor] employment level
— m — **organización** f - [admin] organization level
— m — **orificio** m - hole level
— m — — m **para verificación** f - [mec] check-(ing) plug level
— m — **pared** f - [constr] wall level
— m — — f **para contención** f - [hidr] weir, o retaining wall, level
— m — **paridad** f - parity level
— m — — f **cambiaria** - [econ] exchange parity level
— m — **peligro** m - danger level
— m — **permeabilidad** f - [metal-prod] permeability level

nivel m de piquera f - [metal-prod] iron notch,
level, o elevation
— m — **piso** m - [constr] floor level
— m — — m **terminado** - [constr] finished
floor, level, o line
— m — **placa** f - [mec] plate level
— m — **planta** f - [ind] plant level • [constr]
plant grade
— m — **potencia** f - [electr-oper] power level
— m — **precio** m - price level
— m — **preparación** f - preparation level -
[labor] qualification level
— m — f **de personal** m - [labor] personnel
qualification level
— m — **presa** f - [hidr] dam level
— m — **presión** f - pressure level • [autom-
-neumát] inflation level
— m — f **de sonido** m - [acúst] sound pressure level; S P L
— f — f **inferior** - [autom-neumát], lower inflation level
— m — f **superior** - [autom-neumát] higher
inflation level
— m — **producción** f - [ind] production, o output, level
— m — f **alcanzado** - [ind] attained production level
— m — — f **de acero** m - [metal-prod] steel,
production, o output, level
— m — **profesionalismo** m - [admin] professionalism level
— m — **prosperidad** f - [econ] prosperity level
— m — **proyección** f - design level
— m — **punta** f - [constr-pil] tip elevation
— m — **radiactividad** f - [nucl] radioactivity,
o radiation, level
— m — **refrigerante** —[comb.int] coolant level
— m — — m **para motor** m - [comb.int] engine
coolant level
— m — — m — m **verificado** - [comb.int]
cnecked engine coolant level
— m — — m **verificado** - [comb.int] checked
coolant level
— m — **rendimiento** m - performance, o yield,
level
— f — **resistencia** f - [metal] strength level
— f — — f **específica** - [metal] specific
strength level
— m — **ruido** m - [acúst] noise, o sound, level
— m — — m **alto** - [acúst] high noise level
— m — — m — **en compresor** m - [ind] compressor high noise level
— m — — m — — m **para circulación** f -
[metal-prod] circulating compressor high
noise level
— m — — m — — m — — f **de gas** m -
[metal-prod] gas circulating compressor high
noise level
— m — — m — — m — — m **en tra-**
gante m - [metal-prod] throat gas circulating
compressor high noise level
— m — — m **bajo** - [acúst] low noise level
— m — — m — **en compresor** m - [ind] compressor low noise level
— m — — m — — m **para circulación** f -
[metal-prod] circulating compressor low noise
level
— m — — m — — m — — f **de gas** m -
[metal-prod] gas circulating compressor low
noise level
— m — — m — — m — — f — — m **en tra-**
gante m - [metal-prod] throat gas circulating
compressor low noise level
— m — — m **en compresor** m - [metal-prod] compressor noise level
— m — — m — — m — — f **para circulación** f
circulating compressor noise level
— m — — m — — m — — f **de gas** m - [metal-
-prod] gas circulating compressor noise level
— m — — m — — m — — f — — m **en tragante**
m - [metal-prod] throat gas circulating com-
poressor noise level
— m — — m **en salida** f - [comput] output noise
level
nivel m de ruta f - [metal-prod] runner elevation
— m **de salida** f - [electron] output level
— m — — f **ajustado** - [electrón] adjusted output level
— m — — f **dado**—[electrón] given output level
— m — — f **de tono** m - [electrón] tone output
level
— m — — f — — **audible** - [electrón] audio
tone output level
— m — — f **en decibelio(s)** m - [electrón] de-
cibel(s) output level
— m — — **indeseable** - [electrón] spurious output level
— m — **saturación** f - saturation level
— m — **separador** m - separator level
— m — — m **para venturi** - [metal-prod] venturi
separator level
— m — **sobrecarga** f - surcharge level
— m — **solicitación** f - [mec] stress level
— m — **sonido** m - [fís] noise, o sound, level
— m — **sueldo(s)** m - [labor] wage level
— m — — m **y salario(s)** m - [pers] payroll
level(s)
— m — **suelo** m - ground level • floor level
— m — — m **adyacente** - [constr] adjacent
ground level
— m — **suministro** m - [ind] supply level
— m — — m **de energía** f - [electr-distrib]
power supply level
— m — — m — — m — — **regulado** - [electr-distrib]
controlled power supply level
— m — **supervisión** f - [admin] supervision
level
— m — **supervisor** m - [admin] supervisor level
— m — — m **inmediato** - [labor] first-line supervisor level
— m — **temperatura** f - temperature level
— m — — f **elevado** - high temperature level
— m — — f **reducido** - low temperature level
— m — **tono** m - [electrón] tone level
— m — — m **audible** - [electron] audio tone
level
— m — — m **para salida** f - [electrón] tone
output, o output tone, level
— m — — m — — f **comprobado** - [electrón]
checked, tone output, o output tone, level
— m — — m — — f **verificado** - [electrón]
checked, tone output, o output tone, level
— m — **trabajo(s)** m - [ind] work(s) level
— m — — m **para investigación** f - [ind] research work level
— m — **tragante** m - [metal-prod] throat level
— m — **tubería** f—[tub] pipe level • pipe grade
— m — **umbral** m - [constr] sill level
— m — **venturi** m - [metal-prod] venturi level
— m — **vertedero** m - [hidr] spillway level
— m — — m **para emergencia** f - [hidr] emergency spillway level
— m — **vía** f - [constr] runway level • [ind]
track level
— m — — f **para grúa** f - [constr] runway level
— m — **viscosidad** f - [lubric] viscosity level
— m — **visibilidad** f - visibility level
— m **debido** - proper level
— m **demasiado alto** - too high level
— m — **bajo** - too low level
— m **deseado** - desired level
— m **desusado** - unusual level
— m **diferente** - different level
— m — **de sonido** m - [acúst] different sound
level
— m **directivo** - [admin] leadership level
— m **dirigente** - [admin] leadership level
— m **disponible** - available level
— m **distinto** - different level
— m — **de sonido** m - [electrón] different,
sound, o noise, level
— m **eficiente** - efficient level
— m — **para fabricación** f - [ind] efficient
manufacturing level
— m **elevado** - high level

nivel m elevado de líquido m - high liquid level
— m —— — m regulado - [mec] controlled high liquid level
— m —— — metalización f - [metal] high, metallization level, o degree metallization
— m —— — producción f - [ind] high productivity level
— m —— — productividad f - [ind] high productivity level
— m —— — radiación f - [nucl] high radiation level
— m ——, ruido(s) m, o sonido(s) m - [acúst] high, noise, o sound, level
— m — para líquido m - high liquid level
— m — regulado - [mec] controlled high level
— m en decibelio(s) m—[electrón] decibel level
— m — depósito m - tank level
— m —— — m para aceite m—[mec] oil tank level
— m —— — m —— — m hidráulico - [petról] hydraulic oil tank level
— m —— — m —— — m verificado - [mec] checked oil tank level
— m —— — m para alimentación f - [ind] feeding tank level
— m —— — — f de agua m - [cald] water feeding tank level
— m —— — m — combustible m - [comb.int] fuel tank level
— m —— — m — fluido m—[mec] fluid tank level
— m —— — m —— — hidráulico - [petról] hydraulic fluid tank level
— m —— — m —— — verificado - [petról] checked hydraulic fluid tank level
— m —— — m verificado - checked tank level
— m — embalse m - [hidr] reservoir level
— m — escalafón m - [admin] level
— m — organización f - [admin] organization level
— m — radiador m - [comb.int] radiator level
— m — salida f - [comput] output level
— m entre tono(s) m - [electrón] level between tonos m
— m específico - specific level
— m — de resistencia f - [metal] specific strength, o strength specific, level
— m establecido - etablished level
— m exacto - exact level
— m exigido - required level
— m exiguo - (too) low level
— m externo - external level
— m extremo - extreme level
— m final - final level • [constr] finished, o final, grade
— m freático - [hidr] phreatic, o groundwater, level, o table; water table; subsurface water level
— m — aislado - [hidr] perched water table
— m — alto - [hidr] high, water table, o subsurface water, level
— m — bajado - [hidr] lowered water table
— m — elevado - [hidr] raised water table • high water table
— m general - general level
— m habitual - customary, o habitual, level
— m hidráulico - [hidr] hydraulic level
— m — verificado - [hidr] checked hydraulic level
— m hidroestático - [hidr] ground water level
— m horizontal - horizontal level
— m — para comparación f - horizontal datum,ta
— m indicado - indicated level
— m inferior - lower, level, o range • bottom level • floor; bottom • underlevel
— m — de administración f - [admin] first line management (level)
— m —— — presión f - [autom-neumát] lower inflation level
— m —— — irradiación f - [nucl] lower radiation level
— m —— — radiación f - [nucl] lower radiation level
— m — que apropiado a - (a) below proper level
— m —— — mar a - (a) below sea level

nivel m inferior que rasante a - (a) below grade
— m —— — suelo a - (a) below grade
— m —— — declive m apropiado a - [constr] (a) below proper level grade
— m intermedio - intermediate level
— m internacional - international level
— m lógico - [electrón] logic(al) level
— m — cambiado - [electrón] changed logic(al) level
— m máximo - maximum level • top; ceiling
— m — de agua m - [hidr] maximum water level • high water level • surface level
— m —— — salida f - output maximum, o maximum output, level
— m —— — f indeseable - [electrón] spurious output maximum level
— m — para aislación f - [electr-instal] maximum insulation level
— m mayor - higher, o greater, level
— m mediano - intermediate level • average. o medium, level
— m medio - middle level • average level • mean level
— m — de bajamar - [hidr] mean low level
— m —— — mar - [hidr] mean sea level
— m —— — f alta - [hidr] mean high tide
— m —— — f baja - [hidr] mean low tide
— m menor - lower level
— m — de inspección f - [ind] lower inspection level
— m mínimo - minimum level • floor; bottom
— m — de salida f - output minimum, o minimum output, level
— m —— — f indeseable - [electrón] spurious output minimum level
— m — para agua m - [hidr] minimum water level
— m moderado - moderate(d) level
— m nacional - national level
— m natural - [constr] natural, level, o grade
— m — de suelo m - [hidr] natural ground level
— m normal - normal level
— m — de aceite m - [mec] normal oil level
— m —— — caudal m - [hidr] normal flow, level, o line
— m —— — colinesterase m - [medic] normal cholinesterase level
— m —— — en sangre f - [medic] normal blood cholinesterase level
— m —— — eficiencia f - [ind] normal efficiency level
— f operativo - [ind] operational level
— m original - original level
— m — de lecho m - [hidr] original bed level
— m —— — suelo m - [constr] original ground level
— m para abastecimiento m - [mec] fill level
— m — aislación f - [electr-instal] insulation level
— m para albañil m - [herram] mason's level
— m — asignación f - assignment,ning level
— m — cantero m - [herram] stone mason's level
— m — compactación f—[constr] compacting,tion level
— m —— — f de relleno m - [constr] backfill compacting,tion level
— m — circulación f - [constr-vial] traffic level
— m — ensayo(s) m - test(ing) level
— m — de impulso(s) m - [electrón] impulse test(ing) level
— m —— — f —— — m de onda f - [electrón] wave impulse test(ing) level
— m —— — m —— — m —— — f plena - [electrón] full wave impulse test(ing) level
— m — entrada f - [electrón] input level
— m —— — f de señal f - [electrón] signal input level
— m — fabricación f—[ind] manufacturing level
— m — lógica f - [electrón] logic(al) level
— m —— — f con frecuencia f - [electrón] frequency logic(al) level
— m —— — f —— — f alta - [electrón] high frequency logic(al) level

nivel m para lógica f con frecuencia f baja -
[electrón] low frequency logic level
— f — mano f - [herram] hand level
— n — operación f - operating level
— m — f de diafragma - [mec] diaphragm o-
perating,tion level
— m — prueba(s) f - test(ing) level
— m — radiación f - [nucl] radiation level
— m — rebalse m - [hidr] outflow level
— m — rechazo m - rejection level
— m — — m de rodillo(s) m - [metal-lam] roll
rejection level
— m — — m — — m de acero m - [metal-lam]
steel roll rejection level
— m — — m — — m — — n torjado - [metal-
-lam] forged steel roll rejection level
— m — — m — — m — — m fundido - [metal-
-lam] cast steel roll rejection level
— m — — m — — m para apoyo m - [metal-lam]
back-up roll rejection level
— m — — m — — m — — m de acero m—[metal-
-lam] steel back-up roll rejection level
— m — — m — — m — — m — — m forjado -
[metal-lam] forged steel back-up roll rejection level
— m — — m — — m — — m — — m fundido -
[metal-lam] cast steel back-up roll rejection
level
— m — — m — — m — — m trabajo m - [metal-lam]
work roll rejection level
— m — — m — — m — — m — — m de acero m—[metal-
-lam] steel work roll rejection level
— m — — m — — m — — m — — m forjado -
[metal-lam] forged steel work roll rejection
level
— m — — m — — m — — m — — m fundido -
[metal-lam] cast steel work roll rejection
level
— m — rigidez f - [mec] rigidity level
— m — — f dieléctrica - [fís] dielectric rigidity level
— m — — f — para devanado m - [electr-mot]
winding dielectric rigidity level
— m — — f — — — m para estator m—[electr-
-mot] stator winding dielectric rigidity
level
— m — — f — — — m — rotor m—[electr-mot]
rotor winding dielectric rigidity level
— m — salida f - [comput] output level
— m — — f de señal m - [comput] signal output
level
— m — seguridad f - [segur] safety level •
[ind] assurance level
— m — — f en calidad f - [ind] quality assurance level
— m — — f — confiabilidad f - [ind] reliability assurance level
— m — señal f - [comput] signal level
— m — supervisión f - [admin] supervision,vi-
sory level
— m — trabajo m - [ind] operating level
— m — — m seguro - [ind] safe operating level
— m — utilización f - utilization level
— m peligroso - [segur] danger(ous) level
— m preconvenido - agreed upon level
— m predeterminado - predetermined level
— m primero - first, level, o line
— m proyectado - designed level
— m razonable - reasonable level
— m reducido - low level • low profile •
[constr] low profile
— m — de (ir)radiación f - [nucl] low radiation level
— m — — metalización f - [metal] low, metalization level, o degree metallization
— m — — radiactividad f - [nucl] low radioactivity level
— m — — — f de, desecho(s) m, o residuo(s) m
- [nucl] low (radioactivity) level waste
— m — —, ruido(s) m, o sonido m - [acúst] low
noise level
— m regulado - controlled level
— m requerido - required level

nivel m salarial - [labor] salary, level, o zone
— m seguro - safe level
— m — para operación f - [ind] safe operating,tion level
— m — para trabajo m - [ind] safe work operating,tion level
— m sonoro m - [acust] sound, o noise, level
— m —, diferente, o distinto - [acúst] different, sound, o noise, level
— m superior - top, o upper, o higher, level, o
range • ceiling
— m — de carga f - [metal-prod] stock line
— m — — corriente f - [hidr] flow line
— m — — — f plano - [hidr] flat (grade) flow
line
— m — — irradiación f - [nucl] higher radiation level
— m — — material m (almacenado) - [ind] stock
line
— m — — presión f - [autom-neumát] high(er)
inflatioin level
— m — — radiación f - [nucl] higher radiation
level • top, o highest, radiation level
— m — — subsuelo m acuífero - [hidr] water
table
— m transmitido - [electrón] transmitted level
— m uniforme - uniform level
— m variado - varied level • different level
— m — de combustible m - [comb.int] varied
fuel level
— m verificado - checked level
— m — de aceite m - [mec] checked oil level
— m — — m hidráulico - [mec] checked hydraulic oil level
— m — — fluido m - checked fluid level
— m — — — m hidráulico - [petról] checked
hydraulic fluid level
— m — en depósito - [petról] checked tank
level
— m — — m para fluido m - checked fluid
tank level
— m — — m — — — m hidráulico - [petról]
checked hydraulic fluid tank level
niveles m y estructura(s) f de sueldo(s) m y
salario(s) m - [labor] payroll levels and
structure(s)
nivelación f - . . . . • leveling up • evening •
straightness • [constr] grading • cut and
fill balancing • [f.c.] surfacing
— f automática - automatic leveling
— f cuidadosa - [constr] careful, o close,
leveling
— f — árbol m - [mec] shaft leveling
— f — — m para transmisión f - [mec] line-
shaft leveling
— f — carril m - [mec] rail leveling
— f — metal m - [mec] metal leveling
— f — riel m - [mec] rail leveling
— f — sobrecarga f - surcharge leveling
— f — vía f - [f.c.] track, leveling, o grading
— f final - [constr] final grading
— f hidráulica - [constr] hydraulic leveling
— f longitudinal - lengthwise leveling
— f mejor - better leveling
— f transversal - crosswise leveling
nivelado* m - véase nivelación f
nivelado,da a - level(ed); even(ed); graded;
leveled up • on elevation • flush
— a cuidadosamente - [constr] leveled, carefully, o closely
nivelador m - . . . . • [topogr] levelman
niveladora f - [constr] grader; leveler
— f con cuchilla f - [constr] blade, o knife,
leveler
— f con rodillo(s) m - [metal-lam] roller leveler
— f — — m para apoyo m - [metal-lam] back(ed)
-up leveler
— f con rueda(s) f - [constr] wheel grader
— f — — f inclinada(s) - [constr] leaning
wheel grader
— f — — f neumática(s) f - [constr] rubber-

-tired grader
niveladora f **oscilante** - [constr] oscillating leveler
— f — **tensión** f **continua** - [metal-lam] continuous tension leveler
— f **elevadora** - [constr] elevating grader
— f **para empuje** m **angular** - [constr] bullgrader
— f — m **recto** - [constr] bulldozer
— f — **patrulla** f - [constr] patrol grader
— f — **placa(s)** f - [metal-lam] plate leveler
— f — **plancha(s)** f - [metal-lam] plate leveler
— f **para rodillo(s)** m - [metal-lam] roll(er) leveler
— f — **tijera** f - [metal-lam] shear leveler
— f **por estiramiento** m - [metal-lam] stretcher leveler
— f — **tensión** f—[metal-lam] stretcher leveler
— f **portátil** - portable leveler
nivelar v - . . .; to even; to smooth(en) • [hormig] to strike off • [constr] to grade
— v **árbol** m - [mec] to level @ shaft
— v — m **para transmisión** f - [mec] to level @, lineshaft, o transmission shaft
— v **carga** f - [transp] to level @ load • [metal-prod] to trim @ charge
— v **carril** m - [mec] to level @ rail
— v **cuidadosamente** - [constr] to level, closely, o carefully
— v **metal** m - [mec] to level @ metal
— v **riel** m - [mec] to level @ rail
— v **sobrecarga** f - to level @ surcharge
no <u>adv</u> . . . • non-
— <u>adv</u> **abrasivo,va** <u>a</u> - nonabrasive
— <u>adv</u> **abrir** v - to, not, o fail to, open
— <u>adv</u> **acelerar** v - [mec] to, fail to, o not, accelerate, o pick up speed
— <u>adv</u> **aceptación** f - non-acceptance • declination,ning
— <u>adv</u> **aceptado,da** <u>a</u> - not accepted; declined
— <u>adv</u> **aceptar** v - to, not accept, o decline
— <u>adv</u> **acumulable** - nonaccumulative
— <u>adv</u> **adjudicación** - not, o non, awarding
— <u>adv</u> — f **de oferta** f - offer non awarding
— <u>adv</u> — — **propuesta** f - proposal, o bid, non awarding
— <u>adv</u> **adjudicado,da** <u>a</u> - not awarded
— <u>adv</u> **adjudicar** v - to not award
— <u>adv</u> — v **oferta** f - to not award @, bid, o proposal
— <u>adv</u> — **propuesta** f - to not award @, proposal, o bid
— <u>adv</u> **administrativo,va** <u>a</u> - non administrative; nonmanagement
— <u>adv</u> **admitido,da** <u>a</u> - nonadmitted
— <u>adv</u> **admitir** v - [metal-prod] to not, admit, o tolerate; to reject
— <u>adv</u> — **soplado** m - [metal-prod] to not tolerate, o not take, @ blast; to reject @ blast
— <u>adv</u> — — m **horno** m - [metal-prod] to, not tolerate, o reject, @ blast
— <u>adv</u> **afectado,da** - unaffected • unassigned
— <u>adv</u> — **a explotación** f - non-operating
— <u>adv</u> **aglutinante** <u>a</u> - [miner] noncoking
— <u>adv</u> **aguantar** v - to not hold
— <u>adv</u> — **estufa** f - [metal-prod] to not hold @ heat (@, stove, o furnace)
— <u>adv</u> **ajustar** v - to not, adjust, o fit
— <u>adv</u> **alcanzar** v - to not, reach, o attain
— <u>adv</u> — v **temperatura** f - to not, reach @ temperature, o get hot enough
— <u>adv</u> — v — f **necesaria** - [ind] to not, get hot (enough)
— <u>adv</u> **alcohólico,ca** <u>a</u> - nonalcoholic
— <u>adv</u> **alterado,da** <u>a</u> - unaltered
— <u>adv</u> **andar** v **lejos** - to be close
— <u>adv</u> **apto,ta** - unfit
— <u>adv</u> — **para tratamiento** m - [metal non-treatable
— <u>adv</u> — — m **térmico** - [metal] unfit for heat treatment; not heat treatable
— <u>adv</u> **arrancado,da** <u>a</u> - [mec] not started
— <u>adv</u> **arrancar** v - to, not, o fail to, start
— <u>adv</u> **ascendente** - [válv] non-rising

no <u>adv</u> **asentar** v - to not fit
— <u>adv</u> — v **busa** f - [metal-prod] to not fit (@ blowpipe)
— <u>adv</u> **asignado,da** <u>a</u> - unassigned
— <u>adv</u> **autorizado,da** <u>a</u> - unauthorized
— <u>adv</u> **avanzar** v - to not advance • [mec] to not feed
— <u>adv</u> — v **alambre** m - [sold] to not feed @ wire
— <u>adv</u> **avenir(se)** v - to, decline, o not agree
— <u>adv</u> **bastar** - to not suffice • to not be sufficient
— <u>adv</u> **caber** v **duda** f - beyond @ doubt; certainly
— <u>adv</u> **calentar(se)** v - to not heat
— <u>adv</u> — v **suficientemente** - to not, heat sufficiently, o get hot enough
— <u>adv</u> **cambiar** v - to not change
— <u>adv</u> **centrado,da** - off-center; uncentered
— <u>adv</u> **coaxil** <u>a</u> - [electr-cond] non coaxial
— <u>adv</u> **cohesivo,va** <u>a</u> - non cohesive
— <u>adv</u> **comercial** <u>a</u> - noncommercial
— <u>adv</u> **comestible** <u>a</u> - nonedible
— <u>adv</u> **común** <u>a</u> - uncommon • noncurrent
— <u>adv</u> **comunista** <u>a</u> - noncommunist
— <u>adv</u> **concéntrico,ca** <u>a</u> - nonconcentric
— <u>adv</u> **condensante** <u>a</u> - noncondensing
— <u>adv</u> **conducir** v **corriente** f - [electr-oper] non current carrying
— <u>adv</u> **conductor** <u>a</u> - [electr-oper] nonconducting
— <u>adv</u> **conforme** <u>a</u> - unacceptable
— <u>adv</u> **conseguir** v - to not, obtain, o get
— <u>adv</u> — **eliminar fuga(s)** f - [metal-prod] inability to, seal, o close, o eliminate, @ leak(s)
— <u>adv</u> — **sacar** v **escoria** f - [metal-prod] inability to slag
— <u>adv</u> — v — f **por escorial** m—[metal-prod] inability to slag through @ slag notch
— <u>adv</u> — **sangrar** v - [metal-prod] inability to tap
— <u>adv</u> **consolidado,da** <u>a</u> - unconsolidated
— <u>adv</u> **contenido,da** <u>a</u> - [comb.int] unarrested
— <u>adv</u> **controlado,da** <u>a</u> - uncontrolled; noncontrolled
— <u>adv</u> **convencional** <u>a</u> - unconventional
— <u>adv</u> **coquizante** <u>a</u> - [combust] noncoking
— <u>adv</u> **corregido,da** <u>a</u> - uncorrected
— <u>adv</u> **corriente** <u>a</u> - noncurrent; nonstandard
— <u>adv</u> **corrosivo,va** <u>a</u> - noncorrosive • [petról] sweet
— <u>adv</u> **crítico,ca** <u>a</u> - noncritical
— <u>adv</u> **cumplir** v - to not comply; to fail to
— <u>adv</u> — v **con** - to fail to comply with
— <u>adv</u> **dañar** v - to not, damage, o harm, o hurt
— <u>adv</u> **dar** v - to not, give, o provide
— <u>adv</u> — v **más viento** m (**soplante** m) - [metal-prod] maximum blast
— <u>adv</u> **dar(se)** v **por vencido,da** - never give up; never say die
— <u>adv</u> — v **temperatura** f - [metal-prod] insufficient, temperature, o heat • temperature drop
— <u>adv</u> **de recomendar** v - not recommended
— <u>adv</u> **debe** . . . - do not . . .
— <u>adv</u> — . . .(se) - must not be . . .
— <u>adv</u> — **ensayar(se)** v - not to be tested • do not, attempt, o test, o try
— <u>adv</u> — **inhalar(se)** v - not to be inhaled; do not inhale
— <u>adv</u> — **instalar(se)** v - [ind] do not install
— <u>adv</u> — **intentar(se)** v - not to be attempted; do not attempt
— <u>adv</u> — **usar(se)** v - not to be used; do not use
— <u>adv</u> — **aplicar(se)** v - not to be applied; non pertinent • do not apply
— <u>adv</u> — **tragar(se)** v - not to be swallowed; do not swallow
— <u>adv</u> **desarmable** <u>a</u> - not disassembly,bling • no-take-apart
— <u>adv</u> **definitivo** <u>a</u> - non definitive • experimental

no adv denunciado,da - unreported; not reported
— desbaratable - [comput] undefeatable
— adv destructivo,va a - nondestructive
— adv detergente - [quim] nondetergent
— adv devengado,da a - [fin] unearned
— adv dibujar(se) v - [sold] to not give @ pattern
— adv director,ra a - [mec] non-directional
— adv disponer v - to not have available; to lack; to have no
— adv disponer(se) v - lacking
— adv disponible a - unavailable; non, o not, available • [contab] non-demand
— adv disputado,da a - undisputed; uncontested
— adv distribuido,da a - undistributed; undivided
— adv dividido,da a - undivided
— adv documentado,da - undocumented
— adv emitido,da a - unissued
— adv encender v - [combust] to not light
— adv — v horno m - [combust] to keep @ furnace unlit
— adv — v — m de fosa f - [metal-lam] to keep @ soaking pit unlit
— adv encogible a - non shrink(able)
— adv endurecible a - [metal] non-hardenable
— adv — a con tratamiento m térmico - [metal-trat] heat treatment non-hardenable
— adv envejecedor,ra - nonaging
— adv envejecimiento m - nonaging
— adv equidistante a - non equidistant; off-center
— adv — de extremo(s) m - off-center (lengthwise)
— adv escuadrado,da a - [mec] unsquared
— adv especificado,da a - unspecified
— adv — de otro modo m - not otherwise specified
— adv especificar(se) v de otro modo m - (de) unless otherwise specified
— adv especificar(se) v lo contrario m - (de) unless otherwise specified
— adv espectacular - nonspectacular
— adv estabilizado,da a - unstabilized; non stabilized
— adv estándar a - non-standard
— adv estar v en marcha f - not running
— adv — marchando - not running
— adv estropeador,ra a - non-marring
— adv exceder v - not to, o to not, exceed
— adv exclusivo,va a - nonexclusive
— adv exigible a - [contab] non-demand
— adv exigido,da a - not required
— adv — v engrase m - [mec] to not require @ grease,sing
— adv explosivo,va a - nonexplosive
— adv expuesto,ta a - not exposed; unexposed
— adv — a a fuego m - [cald] unfired
— adv factible a - unfeasible
— adv facturado,da a - [com] uninvoiced; not invoiced
— adv fermentado,da - [quím] unfermented; nonfermented
— adv ferroso,sa a - non-ferrous
— adv financiado,da a - unfinanced
— adv fiscalizado,da a - noncontrolled; uncontrolled
— adv fresable a - [metal] non-machinable
— adv funcionar v - to not, operate, o function
— adv — aún - [ind] not function still; still not function
— adv galvanizado,da a - ungalvanized
— adv garantizado,da a - [fin] unsecured • not, warranted, o guaranteed
— adv generar v - [electr-prod] to, not, o fail to, generate
— adv — corriente f - [electr-prod] to, not, o fail to, generate current
— adv giratorio,ria a - non-rotating
— adv haber v - to not have
— adv — f (en) existencia(s) f - [ind] to not have in stock
— adv hacer contacto - [electr-oper] to not,

close, o make contact
no adv hacer v falta - to not be necessary
— adv — v — mencionar v - needless to say
— adv — v funcionar v - to not operate
— adv — v — v grúa f - [grúas] to not operate @ crane
— adv — v nada malo - to do no wrong
— adv higroscópico,ca a - non-hygroscopic
— adv igualado,da a - nonequalled
— adv impedido,da a - unimpeded
— adv importante - unimportant; nonimportant
— adv imputable a - through no fault • [contab] nonchargeable
— adv imputado,da a - unassigned • [contab] uncharged
— adv incapacitante a - nondisabling
— adv incluido,da a - not included
— adv indicador,ra a - nonindicating
— adv indicar(se) v de otro modo - not indicated otherwise; except as noted
— adv inflamable - noninflammable
— adv informado,da - not informed • unreported
— adv inhalar v - to not inhale
— adv insertar v - to not insert
— adv — v nunca - to never insert
— adv integrado,da a - [fin] not integrated; unpaid (up)
— adv intencional a - unintentional
— adv interesar(se) v - to not care
— adv investigado,da a - uninvestigated
— adv labrable a - [metal] non-machinable
— adv lucrativo,va a - [fin] non-profit; unprofitable
— adv llegar v hasta - to not reach • to be clear of
— adv maduro,ra a - [fam] green • unripe
— adv magnético,ca a - nonmagnetic
— adv — a con cromo m níquel - [metal] nonmagnetic chromium-nickel
— adv maquinable* a - [metal] non-machinable
— adv marcador,ra a - nonmarking
— adv material a - nonmaterial; immaterial
— adv mayor a de - (of) not more than
— adv menor a de - (of) not less than
— adv metálico,ca a - nonmetallic
— adv miembro a - non-member
— adv modificar(se) a - without changing
— adv — variable(s) f restante(s) - (de) not, o without, changing @ other variable(s)
— adv muchos años - not many years
— adv necesariamente - not necessarily
— adv — labrable - [mec] not necessarily machinable
— adv — soldable - [sold] not necessarily weldable
— adv negociable a - [fin] nonnegotiable
— adv nivelado,da a - unlevel(ed)
— adv obstante - however • despite
— adv oficial - unofficial
— adv operar v - to not, run, o operate
— adv — v aguilón m - [grúas] to not boom
— adv — v excitador m - [electr-prod] to not operate @ exciter
— adv orientado,da a - nonoriented
— adv oxidante a - non-oxidizing
— adv pavimentado,da a - unpaved
— adv permanente a - nonpermanent
— adv permisible a - non permissible
— adv pesar v - underweight
— adv plano,na a - nonflat
— adv poder v - to be unable (to)
— adv — aventajar - to be no match for
— adv — colar v - [metal-prod] inability, o unable, to tap (@ blast furnace)
— adv — decir v - not able to say; can not be said; couldn't say
— adv — eliminar - inability to eliminate
— adv — — fuga(s) f - [ind] inability to, seal, o close, o eliminate, @ leak(s)
— adv — negar v - to be unable to deny; to be no getting around
— adv — parar - inability to stop
— adv — tapar v - inability to, stop, o plug

no adv poder tapar v escorial m - [metal-prod]
inability to plug @ slag notch
— adv —— v toberín m - [metal-prod] inability to plug a cooler
— adv —— v — m de escorial m - [metal-prod]
inability to plug @ slag notch cooler
— adv poder(se) aplicar v - to not be pertinent
— adv perforado,da a - [mec] unperforated
— adv preformado,da a - not preformed
— adv prestar(se) v - to not be suitable
— adv productivo,va a - nonproductive
— adv profesional a - nonprofessional; unpro-
fessional
— adv programado,da a - unscheduled
— adv próspero,ra a - non prosperous
— adv protegido,da a - unprotected • [electr-
-cond] unshielded; nonshielded
— adv provisto,ta a - nonprovided; unprovided
— adv — a separadamente - not provided separately • [ind] not serviced separately
— adv quemado,da a - unburned
— adv ranurado,da a - grooveless
— adv real - not, real, o true; untrue
— adv realizado,da a - unaccomplished; unrealized
— adv reconocer v - to not, acknowledge, o accept
— adv — cargo(s) m - [contab] to not accept @
charge(s)
— adv recubierto,ta a - uncovered; not covered
— adv redondo,da a - out-of-round
— adv regulable a - uncontrollable
— adv regulado,da a - non controlled; uncon-
trolable
— adv regular v - to not control
— adv — v venturi m - to not control @ venturi
— adv remitido,da a - unremitted; not sent
— adv remunerado,da a - non-salaried; unpaid
— adv renovable a - non renewable; expendable
— adv repetidor,ra a - non-repetitive; non repeating
— adv requerido,da a - not required; unrequired
— adv requerir v - to not require
— adv — v engrase m - [mec] to not require @
grease,sing
— adv — mantenimiento m - [mec] to not require
@ maintenance; maintenance free
— adv — fondo m - [mec] no bottom required
— adv residential a - nonresidential
— adv residente a - nonresident
— adv restringido,da - unrestricted
— adv retirar(se) v sonda f - [metal-prod] inability to remove @ stock rod
— adv reversible - non reversing; non reversible
— adv revisable - [legal] not subject to, escalation, o revision
— adv rígido,da a - non rigid
— adv rotante a - nonrotating
— adv rozar v - [mec] to clear
— adv salir v escoria f - [metal-prod] inability to slag
— adv — v — f por escorial m - [metal-prod]
inability to slag (through @ slag notch)
— adv — v — f por piquera f - [metal-prod]
inability to slag through @ iron notch
— adv saturado,da a - unsaturated
— adv se aprueba - not approved
— adv — . . . - do not . . .
— adv — exige - not required
— adv — ilustra - not illustrated
— adv — provee - not provided • not available
— adv — separadamente - [ind] not provided separately
— adv sensible a - non-sensitive
— adv — a detonador m - [explos] detonator
non-sensitive
— adv ser v así - (de) otherwise
— adv — de importancia f - to not be, important, o of importance
— adv — importante - to not be important
— adv — mayor - not major
— adv — necesario,ria - to not be necessary
— adv — otro,ra - to happen to be
— adv significativo,va - irrelevant

no adv solamente - not only • not alone
— adv soldable a - [sold] nonweldable
— adv sólido,da - not solid • unsound
— adv soplar v - [combust] to not blow
— adv — horno m - [metal-prod] to not blow (@
blast furnace)
— adv sumergido,da - non submerged; unsubmerged
— adv suplido,da a - not provided
— adv — separadamente - not provided separately
— adv sujeto,ta a multa f - not subject to fine
— adv surtido,da a - not provided
— adv — a separadamente - not provided separately
— adv tapar v - to not cover • to not plug
— adv — cañón m - [metal-prod] to not plug (@,
clay, o mud, gun)
— adv tener - to not have • to build without
— adv — aplicación f - to not apply
— adv — existencia(s) f - [com] to not have in
stock
— adv — fin m - to have no end; to be endless
— adv — nada malo - to do no wrong
— adv — regulación f - uncontrolled
— adv —— f de nivel - uncontrolled level
— adv —— f — — m de agua m - [hidr] uncontrolled water level
— adv —— f — — m en venturi m - [metal-
-porod] uncontrolled venturi water level
— adv —— f venturi m - [metal-prod] uncontrolled venturi
— adv — seguridad f - to be uncertain • to be
unsafe
— adv tocar - to not touch • [mec] to clear
— adv — v jamás - to never touch
— adv — v nunca - to never touch
— adv — tambor m - [mec] to clear @ drum • to
not play @ drum
— adv trabajable* - [metal] non-machinable
— adv trabajar v - to not, work, o operate
— adv — v excitador m - [electr-prod] non operating exciter
— adv tragar v - to not swallow
— adv transferible - nontransferable
— adv transferido,da - non transferred
— adv transponible a - nontransversable
— adv triturable a - uncrushable
— adv únicamente - not only • let alone
— adv uniforme m - nonuniform
— adv urbano,na a - nonurban
— adv utilizado,da a - unused
— adv variar - to not, vary, o change; invariable
— adv vencido,da - unconquered • [fin] unexpired
— adv vender(se) v - not, for sale, o available
— adv ventilado,da a - nonventilated
— adv zurdo,da - not left, o right, handed
nobiliario,ria a - . . . • equestrian
nocivo,va a - . . .; damaging
nocturno,na a - , , , • overnight
noche f helada - [meteorol] freezing, o frosty
(cold), night
— f íntegra - whole night | a - all night
— f polar - [geogr] polar night
nódulo m - . . . • [miner] lump • [metal-prod]
pellet; Renn-Krupp process nodule
— m metálico - [metal] metal, node, o lump
— m resistente - [suelos] resistant nodule
nogal m claro - [color] pecan; light walnut
— m obscuro - [color] dark walnut
nombrado,da a - named • appointed • commissioned
— a de nuevo - renamed • rebranded • relabeled
nombramiento m - . . . • commissioning
— m de garante m - guarantor naming
— m nuevo - renaming • rebranding • relabeling
nombrar v - . . . • to commission • to term
— v de nuevo - to reappoint • to rebrand • to
relabel
— v garante m - to name @ guarantor
nombre m - . . . • [fam] handle
— m apropiado - appropriate name
— m bien conocido - household name
— m original - original name

nombre m cambiado - changed name
— m comercial - [com] commercial, o business, o trade, name
— m — adquirido - [legal] acquired, commercial, o trade, name
— m — comprado - [legal] pur chased trade name
— m — empleado - [legal] trade name used
— m — gestionado - [legal] applied for trade name
— m — poseido - [legal] owned trade name
— m — registrado - [legal] registered trade name
— m — retenido - [legal] held trade name
— m — solicitado - [legal] applied for trade name
— m — usado - [legal] trade name used
— m — vendido - [legal] sold trade name
— m común - common name
— m de barco m - [nav] vessel, o steamer, name
— m — cliente m - customer('s) name
— m — comercio m - [legal] trade name
— m — — m registrado - [legal] registered trade name
— m — compañía f - [legal] company('s) name
— m — dueño m - owner('s) name
— m — electrodo m - [sold] electrode('s) name
— m — embarcación f - vessel, o ship('s), name
— m — empresa f - [legal] company, o corporate, name
— m — entidad f - concern, o company, name
— m — fabricante m - manufacturer('s) name
— m — firma f - firm('s) name
— m — — f consultora - consulting firm('s) name
— m — instalación f - [ind] installation name
— m — licitación f [com] tender('s) name
— m — modelo m - model name
— m — nave f - [transp] ship('s), o vessel('s) name
— m — obra f - [constr] job, o project, name
— m — pieza f - [mec] part('s) name
— m — pila f - . . .; given name
— m — propietario m - owner('s) name
— m — proponente m - bidder('s) name
— m — proveedor m - supplier('s) name
— m — punto m - [comput] site name
— m — — m distante—[comput] remote site name
— m — — m remoto - [comput] remote site name
— m — representante m - [legal] representative('s) name
— m — — m legal - [legal] legal representative('s) name
— m — repuesto m - [ind] (spare) part name
— m — sitio m - [comput] site name
— m — — m distante—[comput] remote site name
— m — — m remoto - [comput] remote site name
— m — sociedad f - [legal] company, o corporate, name
— m — socio m - [legal] partner('s) name
— m — vapor m - [transp] steamer, o vessel, o ship, name
— m familiar - family name • familiar name
— m muy familiar - household name
— m para alarma f - [comput] alarm name
— m — elemento m - element, o item, name
— m químico - [quím] chemical name
— m personal - personal name
— m segundo - middle, o second, name
— m societario - [legal] corporate, name, o title
nomenclador m - . . . .; véase nomenclatura f
nomenclatura f arancelaria - [fisc] tariff('s) nomenclature
— f — uniforme - [fisc] uniform tariff(s) nomenclature
— f — — Centro Americana - [fisc] Central American Uniform Tariff(s) Nomenclature
— f asignada - assigned nomenclature
— f de material(es) m - [ind] material(s) nomenclature
— f estándar - standard nomenclature
— f estandarizada - standardized nomenclature
— f normalizada - standardized nomenclature

nomenclatura f para planta f - [ind] plant('s) nomenclature
— f — — f para oxígeno m - [ind] oxygen plant nomenclature
— f para tamaño(s) m - size(s) nomenclature
— f — — m de neumático(s) m - [autom-neumát] tire size(s) nomenclature
— f según norma f - standard nomenclature
— f unificada - unified nomenclature
— f uniforme - uniform nomenclature
nómina f - . . .; listing
— f de conductor(es) m - [deport] driver('s) list
— f — directorio m - [legal] board of directors list
— f — ingeniero(s) m - engineer('s) list
— f — personal m - [labor] personnel list
— m — — m superior - [admin] management personnel list
— f — — m técnico - [labor] technical personnel list
— f — pieza(s) f - [ind] part(s), list(ing), o book
— f — sueldo(s) m - [labor] payroll
nominación f - nomination; naming; véase también nombramiento m
nominado,da a - nominated; named; véase también nombrado,da
nominal a - . . . • rated • [ind] trade
nominar v - . . .; véase también nombrar
nomo n - . . . . • véase también gnomo m
nomográfico,ca a - nomographic
nomograma* m - nomograph; véase también gráfico m
— m empleado - used nomogrpah
— m — regulación f - control nomograph
— m — — f para entrada f - [hidr] inlet control nomograph
— m — — f — salida f - [hidr] outlet control nomograph
— m usado - used nomograph
noray m - [náut] . . .; mooring post
norcentral a - northcentral
noreste m - véase nordeste m
noria f - [hidr] . . . . • [mec] crab
norma f - . . .; specification; requirement; criterion; method; regulation; policy; ruling • warrant; instruction • practice • rating • [herram] square rule
— f aceptada - accepted standard
— f — internacionalmente - internationally accepted standard
— f actual - present standard
— f — para seguridad f - [segurid] present safety standard
— f ad hoc - ad hoc standard
— f alta - high standard
— f aplicable - applicable, o related, standard
— f aplicada - applied standard
— f apropiada - appropriate, o adequate, standard
— f — para proyección f - adequate design(ing) standard
— f básica - basic standard
— f — para rosca(s) f - [mec] basic thread(s), o thread(s) basic, standard
— f — — — f para tubería f - [tub] pipe thread basic standard
— f — — f — — f de acero m - [tub] steel pipe thread basic standard
— f boliviana - Bolivian standard
— f británica - British standard
— f — para dureza f - [metal] British hardness standard; micum
— f — — Rosca(s) f Cónica(s) para Tubería(s) f - [tub] British Standard (for) Tapered Pipe Thread
— f contable - [contab] accounting, standard, o policy, o principle
— f — significativa - [contab] significant accounting policy
— f contra incendio(s) m - [segurid] fire protection standard
— f correspondiente - [ind] corresponding, o

respective, standard
norma f corriente - current standard
— f — para industria f - [ind] current industry standard
— f — — rueda(s) f - [mec-ruedas] current wheel(s) standard
— f de American Welding Society - [sold] American Welding Society, standard, o test requirement
— f dimensional - dimensional standard
— f — básica - basic dimensional standard
— f disponible - available standard
— f elemental - elementary, o basic, standard, o rule
— f — para circulación f - [vial] elementary traffic rule
— f — — tránsito m - [vial] elementary traffic rule
— f elevada - high standard
— f equivalente - equivalent standard
— f estadounidense - American, o United States, standard
— f — para inspección f de puente(s) m - [pol] National Bridge Inspection Standard
— f — — Seguridad f para vehículo(s) m Motorizado(s) - [autom] Federal Motor Vehicle Safety Standard
— f estatal - state standard
— f — corriente f - current state standard
— f estética - aesthetic standard
— f estricta - strict, o stringent, standard
— f estructural - [constr] structural, standard, o criterion,ria
— f europea - European standard
— f — para Cable(s) m - [electr-cond] European Cable Standard
— f — — ensayo(s) m - European test(ing) standard
— f — — prueba(s) f - European test(ing) standard
— f general - general, standard, o rule, o specification
— f implantada - implanted standard
— f indicada - indicated standard
— f inglesa - [tub] British standard
— f interna - internal standard
— f internacional - international standard
— f — equivalente - equivalent international standard
— f — para cable(s) m - [Cabl] international cable standard
— f legal - [legal] legal, standard, o requirement
— f máxima - maximum standard
— f métrica - [metric] metric standard
— f — o inglesa - [tub] metric or British standard
— f — — para rosca(s) f cónica(s) - [tub] metric or British standard for pipe thread(s)
— f — — — — — para tubería f—[tub] metric or British standard for taper pipe thread(s)
— f — para rosca f cónica para tubería f - [tub] metric standard for taper pipe thread
— f mínima - minimum standard
— f modificada - [admin] revised, o modified, standard, o policy
— f nacional - national standard
— f N E M A - [electr] National Electrical Manufacturer's Association, o N E M A, standard, o policy
— f operativa - operating, o working, standard, o policy
— f óptima - optimum standard
— f para aceptación f - acceptance standard
— f — actuación f - [admin] performance standard
— f — agua m - [hidr] water standard
— f — — m potable - [hidr] drinking water standard(s)
— f — almacenaje m—[com] warehousing practice
— f — alumbrado m - [electr-ilum] lighting standard

norma f para aplicación f - [applied, o applicable) standard
— f — aprobación f - acceptance, o approval, o qualification, standard, o requirement
— f — — para procedimiento m - [ind] procedure qualification standard
— f — — — m de soldadura f y soldadora(s) f para tubería(s) f - [sold] Standard for Qualification of Welding Procedures and Welders for Piping and Tubing
— f — — — m para soldadura f - [sold] welding procedures qualification standard
— f — — soldadora f - [sold] welder qualification standard
— f — aprovisionamiento m - [com] stocking, o supplying, practice
— f — auditoría f - [contab] auditing, standard, o practice
— f — — aceptada - [contab] accepted auditing, standard, o practice
— f — — con aceptación f general - [contab] generally accepted auditing, standard, o practice
— f — calidad f - quality standard
— f — — alta - high quality standard
— f — — f más alta - highest quality standard
— f — — f óptima - highest quality standard
— f — — f prescrito - prescribed quality standard
— f — capacidad f - capacity standard
— f — carga f - [constr] load standard
— f — colada f - [metal-prod] casting standard
— f — con máquina f para colar v - [metal-prod] pig casting machine casting standard; pigcaster casting standard
— f — con máquina f para moldeo m - [metal-prod] pig casting machine, o pigcaster, casting standard
— f — — f — — f — — v para arrabio m para afino m - [metal-prod] pig casting machine, o pigcaster, casting standard for refining pig iron
— f — color(es) m - color(s) standard(s)
— f — comparación f - comparison standard
— f — compra(s) f - [com] purchasing standard
— f — conducción f - [autom] driving standard
— f — — f de colada f - [metal-prod] tapping (handling) standard
— f — construcción f - [constr] construction standard
— f — — f vial - [vial] highway construction standard
— f — control m - control, standard, o practice, o procedure
— f — — m interno - [admin] internal control procedure
— f — corriente f alterna - [electr-prod] alternating current, standard, o rating
— f — — f continua - [elecrr-prod] direct current, standard, o rating
— f — cumplimiento m - [admin] compliance, o performance, standard
— f — — f establecida - curve set standard
— f — — f fijada - curve set standard
— f — deaprtamento m para vialidad f - [vial] highway department standard
— f — desempeño m - [admin] performance standard
— f — diseño m - design, standard, o criterion
— f — ecuanimidad f - fairness standard
— f — — f en empleo m - [pers] fair employment practice
— f — ejecución f - performance standard
— f — — f buena - good performance, o workmanship, standard
— f — embalaje m - [ind] pack(ag)ing standard
— f — empresa f - [admin] company policy,cies
— f — ensayo(s) m - [ind] test(ing) standard
— f — — m de comportamiento m - [ind] performance test(ing) standard
— f — — m — — m de caldera(s) f - [cald] boiler performance test(ing) standard
— f — fabricación f - [ind] manufacturing

standard
norma f para fabricación f de elemento m - [ind]
element(s) manufacturing standard(s)
— f — **grúa(s) f** - [grúas] crane(s) standard
— f — — f **según Instituto m Estadounidense
para Ingenieros m para Estructuras f** -
[constr] American Institute of Structural Engineers' Crane Standard
— f — **iluminación f** - [electr-instal] lighting
standard
— f — **industria f** - [ind] industry standard
— f — **instrumentación f** - [instrum] instrumentation standard
— f — **manejo n** - [ind] handling standard
— f — — m **de colada f** - [metal-prod] tapping
handling standard
— f — **operación f** - operating,tion standard
— f — — f **de máquina f** - [ind] machine operating standard
— f — **planta f** - plant standard
— f — **preparación f** - [ind] preparation standard
— f — — f **de junta(s) f** - [sold] joint preparation standard
— f — **procedimiento(s) m** - [ind] procedure(s)
standard
— f — — m **para soldadura f** - [sold] welding
procedure(s) standard
— f — **producción f** - production standard •
production policy
— f — **protección f** - [segurid] protection
standard
— f — — f **contra incendio(s) m** - [segurid]
fire protection standard
— f — **proyección f** - design, standard, o criterion,ria
— f — — f **estructural** - [constr] structural
design, standard, o criterion,ria
— f — **prueba(s) f** - [ind] test(ing) standard
— f — — f **para comportamiento m de caldera(s)
f** - [acold] boiler(s) performance test, standard, o criterion,ria
— f — **pureza f** - purity standard • [hidr]
quality standard
— f — — f **para agua m** - [hidr] water quelity
standard
— f — **recepción f** - [ind] acceptance standard
— f — **referencia f** - reference, o subject,
standard
— f — **rendimiento m** - yield, o capacity, standard • performance standard
— f — — m **para neumático m** - [autom-neumát]
tire performance standard
— f — **revisión f** - [contab] auditing standard
— f — **rosca(s) f** - [mec] thread standard
— f — — f **para tubería f** - [tib] pipe thread
standard
— f — — f — — f **de acero m** - [tub] steel
pipe thread standard
— f — **rueda(s) f** - [mec-ruedas] wheel standard
— f — **seguridad f** - [segurid] safety standard
— f — — f **implantada** - [segurid] implanted
safety standard
— f — — f **industrial** - [segurid] industrial
safety standard
— f — — f — **implantada** - [segurid] implanted industrial safety standard
— f — — f **vial** - [vial] highway safety standard
— f — **soldadora f** - [sold] welder standard
— f — **soldadura f** - [sold] welding standard
— f — — f **para tubería(s) f** - [tub] pipe
welding standard
— f — — f — — f **para conducción f** - [sold]
pipe line welding standard
— f — — f — — f — — f **e instalación(es),
anexa(s), o conexa(s)** - [sold] standard for
welding pipe lines and related facilities
— f — — f — — f — — f — — f **similares** -
[sold] standard for welding pipe lines and
related facility,ties
— f — **supercarretera(s) f** - [vial] interstate
(highway)   standard, o specification

norma f para terminación f - [mec] finish(ing)
standard
— f — **tolerancia f** - tolerance standard
— f — — f **para tubo(s) m** - [tub] pipe tolerance standard
— f — — f — — m **de acero m** - [tub] steel,
tube, o pipe, tolerance standard
— f — — f — — m — — m **soldado** - [tub]
welded steel, tube, o pipe, tolerance standard
— f — — f **según Sociedad f Estadounidense
para Ensayos y Materiales** - [metal-prod]
American Society for Testing and Materials
tolerance standard
— f — **tratamiento m** - treatment standard
— f — **tubería f** - [tub] pipe standard
— f — **tubo(s) m** - [tub] tube, o pipe, standard
— f — — m **de acero m** - [tub] steel tube standard
— f — **ventana(s) f con sección(es) f interme-
dias abisazgradas** - [constr] standard for intermediate casement section windows
— f **prescrita** - prescribed standard
— f **protectora** - [labor] protective standard
— f **publicada** - published standard
— f **racional** - rational, standard, o criterion
— f **recomendada** - recommended standard
— f — **por industria f** - [ind] industry recommended standard
— f **reconocida** - acknowledge, o accepted,
standard • recognized standard
— f — **internacionalmente** - internationally accepted standard
— f — **para ensayo(s) m** - [ind] recognized
testing standard
— f — — **prueba(s) f** - [ind] recognized testing standard
— f **reformada** - [admin] revised, o reformed,
standard, o policy
— f **representativa** - representative standard
— f — **para ejecución f buena** - representative
workmanship standard
— f **respectiva** - [ind] respective standard
— f **rígida** - rigid standard
— f — **para industria f** - [ind] industry rigid,
o rigid industry, standard
— f **según Asociación f Estadounidense de Fabricantes de Artículos m Eléctricos** - [electr]
National Electrical Manufacturers' Association, standard, o rating
— f — **Briggs** - [petról] Briggs' standard
— f — **Dirección f Estadounidense para Cons-
trucción(es) f Naval(es)** - [náut] American
Bureau of Shipping, standard, o rating
— f — — f — **Protección f de Ambiente m** -
[pol] Environmental Protection Administration
standard
— f — — f **para Vialidad f** - [pol] Bureau of
Public Roads standard
— f — **Instituto m Estadounidense para Hormigón
m** - [constr] American Concrete Institute
standard
— f — — m — — **Ingenieros m Eléctricos y E-
lectrónicos** - [electr] Institute of Electrical and Electronics Engineers' standard
— f — — m **Hidráulico** - [hidr] Hydraulic(s)
Institute standard
— f — **Junta f Estadounidense de Aseguradores m
contra Incendio(s) m** - [seguros] National
Board of Underwriters standard
— f — **Sociedad f Estadounidense para Ensayos m
y Material(es) m** - [ind] American Society for
testing   and Materials standard
— f — — f — — **Ingenieros m Mecánicos** -
American Society of Mechanical Engineers', o
A S M E, code, o standard
— f — — f — — **Soldadura f** - [sold] American
Welding Society, o A W S, test requirement
— f **significativa** - significant, standard, o
policy
— f **técnica** - technical standard
— f **variable** - variable standard

normal f - . . . . [metal-prod] dust catcher |
a - . . .; conventional • [labor] straight
time
— a a - normal to
— a — eje m - [mec] normal to @ centerline
— a — m de alma m - [metal-lam] normal to
@ web axis
— a — m longitudinal - normal to @ longitudinal axis
normalización f - normalization,zing; standardization,zing • conditioning
— f adicional - additional standardization
— f — en acería f - [metal-prod] additional
plant standardization
— f — misma acería f - [metal-prod] additional (in) plant standardization
— f — — planta f - [ind] additional (in)
plant standardization
— f antes de trefilería f - [trefil] normalization,zing before @ drawing
— f de acero n - [metal-prod] steel normalization,zing
— f — condición f - condition normalization,zing
— f — especificación f - specification normalization,zing
— f — — f técnica - technical specification
standardization,zing
— f — grúa(s) f - [grúas] crane standardization,zing
— f — método m - method standardization,zing
— f — — m para análisis m - [quím] analysis
method standardization,zing
— f — nomenclatura f - nomenclature standardization,zing
— f — perfil m - profile standardization,zing
• [metal-lam] shape standardization,zing
— f — — m estructural - [metal-lam] structural shape standardization,zing
— f — — m — de acero m - [metal-lam] steel
structural, o structural steel, shape standardization,zing
— f — placa(s) f - [metal-trat] plate normalization,zing
— f — plancha(s) f - [metal-trat] plate normalization,zing
— f — producto m - [ind] product normalization,zing, o standardization,zing
— f — respuesto(s) m - [ind] spare(s) (parts)
standardization,zing
— f — suministro(s) m - [ind] supply,plies
standardization,zing
— f — todas planta(s) f - [ind] interplant
standardization,zing
— f — trabajo(s) m - [ind] work standardization,zing
— f — — m, de repetición f, o repetidor -
[ind] repetitive work standardization,zing
— f interna - internal standardization,zing
— f siderúrgica - [metal-prod] steel, making, o
industry, standardization,zing
— f técnica - technical standardization,zing
— f — de perfil(es) m (estructural(es)) -
[metal-lam] structural shapes technical standardization,zing
normalizado* m - véase normalización f
normalizado,da a - normalized • standardized •
[metal-trat] normalized
— a antes de trefilería f - [trefil] normalized
before @ drawing
— a para resistencia f alta a tensión f - [metal-trat] normalized (for) high
strength
normalizador m - normalizer
normalizador,fa a - normalizer,zing
— m para fleje m—[metal-trat] strip normalizer
normalizar v acero m - [metal-trat] to normalize
@ steel
— v antes de trefilería f - [trefil] to normalize before @ drawing
— v condición f - to normalize @ condition
— v especificación f - to standardize @ specification

normalizar v especificación f técnica - [ind]
to standardize @ technical specification
— v producto m - [ind] to standardize @ product
— v trabajo m - to standardize @ work
normalmente adv - normally
— adv abierto,ta a - normally open(ed)
— adv cerrado,da a - normally closed
normativa* f - standard(s)
— f establecida - [ind] established standard(s)
normativo,va a - standard (setting)
noroccidental a - northwest(ern)
nororiental a - northeast(ern)
norpatagónico,ca a - [geogr] North Patagonian
norte m - . . . | a - (de) northern
— m a sur m - north to south; southbound
— m franco - due north
Norteamérica f - véase América f de Norte m
nota f - . . . . remark • [comunic] . . .; communication; message
— f a pie m - [public] footnote
— f aclaratoria - clarifying, o explanatory,
note, o letter
— f adicional - [comunic] additional, note, o
letter
— f correspondiente - corresponding note
— f cuidadosa - careful note
— f de alistamiento m - [transp] readiness
note,tice
— f — cargo m - [contab] debit, note, o memo-
(randum) ı
— f — crédito m - [contab] credit, note, o
memo(randum)
— f — débito - [contab] debit, note, o memo-
(randum)
— f — — mensual - [contab] monthly debit,
note, o memo(randum)
— f — empaque m - [transp] packing list
— f — entrega f - [transp] delivery slip
— f — gasto m - [contab] expense, note, o
memo(randum), o statement; bill of charge(s)
— f — recepción f - [com] acceptance, o warehouse, o material(s), receipt
— f — redacción f - [public] editor's note
— f — reunión f - [legal] minutes
— f — romaneo m - [transp] weight certificate
— f — venta(s) f - [com] sale(s) slip; invoice
• sale(s) receipt
— f especial - special note
— f explicativa - explanatory note
— f general - general, note, o remark
— f para estado m contable - [contab] accounting (statement) note
— f — remisión f - [transp] shipping slip •
delivery slip
— f — instalación f - installation note
— f primera - first note
— f sobre balance m (general) - [fin] (financial) statement note
— f — fabricación f - [ind] fabrication note
— f última - last note
notable a - . . .; outstanding
notación f - . . .; noting
— f de medida(s) f - [instrum] reading record
notado,da a - noted • detected
notar v - . . .; to detect • [instrum] to record
@ reading
— v lectura v - [instrum] to, note, o record, @
reading
— v medida f - [instrum] to, note, o record, @,
reading, o measurement
— v — f indicada - [instrum] to, note, o record, @ reading
— v — f que (se) indica - [instrum] to, note,
o record, @ reading
notario m - [legal] . . .; recording attorney
— m público - [legal] public, notary, o recording attorney; notary public
— m subscrito - [legal] undersigned,ning, notary (public), o recording attorney
noticia f buena - good news
— f mala - bad news
— f pronta - quick, o early, reply

noticioso m - [public] newsletter • [telecom]
  news cast; news bulletin
notificación f - . . .; notifying; report(ing);
  communication; notice; advice
— f a contratista(s) m - [constr] notice to @
  contractor(s)
— f anticipada - advance notice
— f de adjudicación f - award(ing) notification
— f — cambio m - change notice,tification
— f — — m de precio - [com] price change
  notice,tification
— f — iniciación f - [constr] start-up notice
— f — — f de trabajo(s) m - [constr] work
  start up notice • notice to proceed
— f — reglamento m - [pers] rule,ling, notification, o communication
— f — — m sobre seguridad f - [segurid] safety rule,ling, notification, o communication
— f escrita f - written notice
— f para cancelación f - [legal] termination
  notice
— f por escrito - written, notice, o advice
notificado,da a - . . .; advised • reported
— a por escrito - notified, o advised, in
  writing
notificar v - . . .; to advise • to report •
  [legal] to serve (@ notice)
— v adjudicación f - to notify @ award(ing)
— v anticipadamente - to notify beforehand
— v cambio m - no notify @ change
— v contracargo m - [contab] to notify @ counter charge
— v por escrito - to notify in writing
novato m - . . . • [petról] boll weevil •
  [fam] greenhorn
novato,ta a - . . .; new; recent
novedad f - . . .; news • (new) development
novedodísimo,ma a - very, new, o recent • hot
novedoso,sa a - . . .; innovative; recent • ima-
  ginative • hot
novel a - . . . • neophyte
novela f cinematográfica - [teatro] Hollywood
  script
novicio m - . . . • [deport] rookie
— m Pro Rally de año m - [deport] Pro Rally
  Rookie of the Year
noyero m - [mec] core maker
noyo m - [mec] core
— m macho - [metal-prod] core
nube f baja - [meteorol] low(-hanging) cloud
— f brumosa - [meteorol] misty cloud
— f de polvo m - dust cloud
— f — — m enceguecedora - blinding dust cloud
— f neblinosa - misty cloud
— f pequeña f - small cloud; cloudlet
— f — de polvo m - small dust cloud; dust puff
nublado,da a - . . .; clouded
nublamiento m - clouding
— m de responsabilidad f - [admin] responsibility, o accountability, clouding
nublar v - . . .; to cloud
— v responsabilidad f - [admin] to cloud @ responsibility
núcleo m - . . . • core; kern • [electr-mot]
  core • [petról] core section; drill core •
  core sample • [sold] core • [mec-ruedas] hub
  • [comb.int-radiador] core • [cabl] heart
— m ácido - [sold] acid core
— m (como) acabado de fundir - [metal-alambre]
  rough core
— m con alma m llena - [f.c.-ruedas] solid center
— m — — m ondulada - [f.c.]ruedas] dished
  center
— m — rayo(s) m - [f.c.-ruedas] spoked center
— m de acero m - [electr-cond] steel core
— m — — m galvanizado - [cabl] galvanized
  steel core
— m — aire m - [electr-mot] air core
— m — alambre m - [alambre] wire core
— m — — m de acero m - [sold] steel core wire
— m — — para electrodo m - [sold] electrode
  wire core

núcleo m de alambre m sólido - [sold] solid wire
  core
— m — — m tubular - [sold] tubular wire core
— m — cable m - [telecom-cond] cable, nucleus,
  o core • [cabl] rope, center, o core
— m — — m independiente - [cabl] independent
  rope core
— m — cáñamo m - [cabl] hemp core
— m — cordón m - [cabl] strand core
— m — cuerda f - [cabl] strand, o rope, core
— m — carbono m - [sold] carbon core
— m — electrodo m - [sold] electrode core
— m — electroimán m - (electro)magnet core
— m — fibra f - [cabl] fiber core
— m — hierro m - iron core
— m — instrucción f - [pers] training core
— m — programa m - program core
— m — — m para instrucción f - training program core
— m — — m — — f sobre seguridad f -
  [segurid] safety training program core
— m — — m — — f en trabajo m -
  [segurid] job safety training program core
— m — ramal m - [cabl] strand core
— m — rueda f - [f.c.-ruedas] wheel center
— m — tornillo m - screw nucleus • screw body
— m — torón* m - [cabl] strand core
— m doble - [petról] double core
— m ensanchado - expanded, nucleus, o core
— m ferruginoso - [quím] ferruginous nucleus
— m final - [alambre] finish core
— m formado - [telecom-cond] formed nucleus
— m fundente - [sold] flux-core; véase también
  alma m fundente
— m hincador - [constr-pil] drive,ving core
— m — Core-Drive - [constr-pil] Core-Drive
— m — con pared f gruesa - [constr-pil] heavy
  wall drive,ving core
— m independiente - [cabl] independent core
— m — para cable m de alambre m - [cabl] independent wire rope core
— m laminar - [electr-instal] lamination (assembly)
— m — fijo m - [electr-instal] stationary lamination (assembly)
— m — movible - [electr-instal] moving lamination (assembly)
— m — para estator m - [electr-mot] stator
  lamination (assembly)
— m limpiado - cleaned core
— m — con vapor m - [comb.int] steam cleaned
  core
— m limpio - clean core
— m lleno - [f.c.-ruedas] solid center
— m magnético - [electr-mot] magnetic frame
— m — laminado - [electr-mot] laminated frame
— m — — para generador m - [electr-prod] laminated generator, o generator laminated,
  frame
— m — — — motor m - [electr-mot] laminated
  motor frame
— m metálico - [cabl] metal(lic) core
— m — para electrodo m - [sold] electrode('s)
  metal core
— m movible - [electr-equip] movable core
— m móvil - [mec] plunger
— m para alimentación f - [sold] feed core
— m — — f de alambre m - [sold] wire feed
  core
— m — amplificación f - [sold] amplifier core
— m — — f magnética - [sold] magnetic amplifier core
— m — — f para alimentadora f (para alambre m) - [sold] wire feed(ing) magnetic amplifier core
— m — amplificador m - [sold] amplifier core
— m — — m magnético - [sold] magnetic amplifier core
— m — — m — para alimentadora f (para alambre m) - [sold] feed(ing) magnetic amplifier
  core
— m — bobina f - [electr-equip] coil core
— m — calentador m - heater core

núcleo m para condensador m - [electr-equip]
condenser core
— m — enfriador m - [comb.int] cooler core
— m — estator f - [electr-mot] stator, core, o
nucleus
— m — evaporador m - evaporator core
— m — inducido m - [electr-mot] armature core
— n — muro m central - [electr-transf] center
leg
— m — perforación f - [petról] drill(ing) core
— m — polo m - [electr-mot] pole, nucleus, o
core
— m — radiador m - [comb.int] radiator core
— m — — m exterior - [comb.int] external radiator core
— m — — m — comprobado - [comb.int] checked
external radiator core
— m — — m — limpiado - [comb.int] cleaned
external radiator core
— m — — m — limpio - [comb.int] clean external radiator core
— m — transformador m - [electr-prod] transformer coil
— m preliminar - [alambre] rough core
— m prolongado - [mec] dog point; extended core
— m — apuntado - [mec] pointed dog
point
— m — redondeado - [mec] rounded dog point
— m — y apuntado - - [mec] pointed dog point
— m — — redondeado - [mec] rounded dog point
— m regulado - [petról] controlled core
— m — uniformemente - [petról] uniformly controlled core
— m rotado - [mec] rotated core
— m salido - [cabl] popped core
— m saturado - [petról] bleeding core
— m — con petróleo m - [petról] bleeding core
— m sin rectificar - [alambre] rough core
— m sobresaliente m - [cabl] popped core
— m (tal) como acabado de fundir v - [alambre]
rough core; as welded core
— m testigo - [miner] boring core
— m tubular - [sold] tubular core
— m uniforme - uniform core
nudillo m con cicatriz,ces f - [anatom] scarred
(up) knuckle
nudo m - . . . • [mader] knot; knar • [mec]
knob; knurl • [constr] panel point
— m apretado - tight(ened) knot
— m atado - tied knot
— m con vuelta f redonda (con dos cotes m) -
[cabl] rolling hitch
— m corredizo - [cabl] slip(ping) knot
— m en forma de ocho m - [cabl] figure eight
knot
— m en viga f - [constr] panel point
— m — — f armada - [estruct] (truss) panel
point
— n — — f para cabriada f - [constr] truss
panel point
— m para escarpia f - [mec] pike knot
nudoso,sa a - . . .; knurled • [sold] knoby;
knotted,ty; ropy
nuestra adv referencia f - our reference
nuestro adv pedido - our order
— m plano m - our drawing
Nueva adv Inglaterra f - [geogr] New England
nuevamente adv - newly • (once) again; anew •
back
— adv a - back (in)to
— adv en estado m (completamente) fundido m -
(completely) remelted
Nuevas adv Lecciones para Soldadura f con Arco m
- [sold] New Lessons in Arc Welding
nuez f de nogal m - [botán] walnut
nulidad f - [legal] . . .; annuling; invalida-
tion; cancellation
nulo,la a - . . .; zero
numeración f consecutiva - consecutive, numera-
ation, o numbering
— f continua(da) - continuous numbering
— f correlativa - consecutive numbering
— f de agujero(s) m - [mec] hole(s) numbering

numeración f de partida f - item(s) number(ing)
— f en secuencia f - consecutive numbering
— f para inventario m - inventory numbering;
stock series
— f secuencial - consecutive numbering
numerado,da a - numbered
— a consecutivamente - numbered consecutively
— a correlativamente - numbered consecutively
— a en secuencia f - numbered consecutively
— a progresivamente - numbered, progressively,
o consecutively
numeral m - . . . • [legal] item; section
numerar v - . . .; to numerate
— v agujero - [mec] to number @ hole
— v consecutivamente - to number consecutively
— v correlativamente - to number consecutively
— v en forma f permanente - to number permanently
— v — secuencia f - to number consecutively
— v partida v - to number @ item
— v progresivamente - to number, progressively,
o consecutively
numéricamente adv - numerically
— adv correcto,ca - numerically correct
número m - . . .; digit • quantity • [public]
issue • [legal] item
— m cierto - certain number
— m clave - key number
— m codificado - code(d) number
— m consecutivo m- consecutive number
— m constante - constant number
— m correspondiente - corresponding number
— m — a oblicuidad f - [mec] skew number
— m dado - given, o certain, number
— m de accionista(s) m - [legal] number of,
shareholders, o stockholders
— f — aceptación f - acceptance number
— m — agujero m - hole number
— m — alambre m - [cabl] wire number
— m — albarán m - [transp] leading log number
— m — alimentadora f - [sold] feeder number
— m — — f para alambre m - [sold] wire feeder
number
— m — almacén m - [com] stock number
— m — aparato m - [ind] apparatus number
— m — arancel m - [fisc] tariff number
— m — bobina f - [metal-lam] coil number
— m — bulto m - [transp] pack(age) number
— m — caja f - case number - [metal-lam] stand
number
— m — catálogo m - catalog number
— m — cilindro m - [comb.int] cylinder number
— m — cilindros m - [comb.int] number of cylinders
— m — circuito m - [electr-instal] circuit
nuymber
— m — código - code number
— m — — m especificado—specified code number
— m — — m para cargo m - [admin] position
code number
— m — — m — máquina f - [ind] machine code
number
— m — — m para modelo m - model code number
— m — — m — — m de máquina f - [ind] machine model code number
— m — — m — m — soldadora f - [sold]
welder, o machine, model code number
— m — — m — serie f - serial,ies code number
— m — — m — soldadora f - [sold] welder, o
machine, code number
— m — — trabajo m - [ind] job code number
— m — — m entre . . . - code number between
. . .
— m — — m inferior a . . . - code number below; below code (number)
— m — — m sobre seguridad f - [segurid]
safety code number
— m — — m superior a . . . - code number a-
bove; above code number . . .
— m — colada f - [metal-prod] heat, o tapping,
o cast(ing), o pour(ing), number
— m — concordancia f - [sold] conformance number

número m de concordancia f según Sociedad f Estadounidense para Soldadura f - American Welding Society, o A W S, conformance number
— m — **conjunto** m - [mec] assembly number
— m —— m **envolvente** - [mec] involute assembly number
— m —— m **para producción** f - [ind] production assembly number
— m — **contrato** m - contract number
— m — **cordón(es)** m - [cabl] strand(s) number
— m — **cuenta** f - [contab] account number
— m — **desbaste** m - [metal-lam] bloom number
— m —— m **plano** - [metal-prod] flat bloom number
— m — **desemulsificación*** f - [petról] demulsification number
— m — **diagrama** m - diagram number
— m — **diente(s)** - [mec] tooth number • number of teeth
— m —— m **en corona** f **(dentada)** - [mec] ring gear teeth number
— m —— m **en piñón** m - [mec] pinion teeth number
— m — **eje** m - [mec] axle number
— m —— m **identificado** - [mec] identified axle number
— m — **elemento** m - [ind] element number
— m — **embolada(s)** f - [mec] stoke(s) number
— m —— f **para bomba** f - [bombas] pump stroke(s) number
— m — **envolvente** m - [mec] involute number
— m — **equipo** m - [ind] equipment number
— m — **especificación** f - specification number
— m —— f **completa** - [mec] complete specification number
— m —— f — **para eje** m - [mec] complete axle specification number
— f —— f **para eje** m - [mec] axle specification number
— m — **existencia** f - [com] stock number
— m — **expediente** m - [com] docket number
— m — **fabricación** f - [ind] manufacture,ring number
— m — **fabricante** m - [ind] manufacturer,ring number
— m — **fase(s)** f - [electr-prod] phase(s) number
— m — **ficha** f - [labor] check number
— m — **golpe(s)** - [mec] stroke(s), o blow(s), number
— m — **grupo** m - group number
— m —— m **de suelo** m - [suelos] soil group number
— m — **índice** m - index number
— m — **inscripción** f - inscription, o registration, number
— m — **instalación** f - [ind] installation number
— m — **inventario** m - [ind] inventory, o stock, number
— m —— m **para mordaza** f - [mec] clamp stock number
— m — **item** m - item number
— m — **juego** m - set number
— m —— m **apareado** - [mec] matched set number
— m —— m **de engranaje(s)** m - [mec] gear(s) set number
— m —— m —— m **apareado(s)** - [mec] matched gear set number
— m —— m —— m **hermanado(s)** — [mec] matched gear set number
— m —— m **hermanado** — [mec] matched set number
— m — **legajo** m - [com] docket number
— m — **licitación** f - [com] tender number
— m — **lingote** m - [metal-prod] ingot number
— m — **lote** m - lot, o batch, number
— m —— m **para producción** f - [ind] production assembly number
— m — **máquina** f - [ind] machine number
— m — **material** m - [ind] material(s) number
— m — **matrícula** f - license number
— m — **milla(s)** f - [transp] mileage
— m — **modelo** m - model number
— m —— m **de eje** m - [mec] axle model number
— m —— m **identificado** - identified model

número m de montaje m para producción f - [ind] production assembly number
— m — **obra** f — [constr] job, o project, number
— m — **orden** f - order number • chronological, o serial, number
— m —— f **para compra** f - [com] purchase order number
— m —— f — **fabricación** f - [ind] fabrication, o production, order number
— m —— f — **producción** f - [ind] production order number
— m — **paleta** f - [ventil] blade number
— m — **paquete** m - [ind] package number • [metal-lam] pack number
— m — **partida** f - lot, o batch, o item. number
— m —— f **para unidad** f - unit lot number
— m —— f —— f **motriz** - [agric-equip] power unit lot number
— m — **pasada** f - [sold] pass number
— m — **pasadas** f - [sold] number of passes
— m —— f **exigida(s)** f - [metal-lam] required number of passes
— m — **paso(s)** m - [mec] step(s) number
— m —— f **en trabajo** m - [ind] job step number
— m — **pedido** m - [com] order number
— m —— m **interno** - [ind] requisition number
— m —— m **para material(es)** m - [ind] material(s) requisition number
— m — **permiso** m - license number
— m — **pieza** f - [mec] part number
— m —— f **básica** - [ind] basic part number
— m —— f **en catálogo** m — catalog part number
— m —— f **para caja** f - [mec] housing part number
— m —— f — **carcasa** f - [mec] housing part number
— m —— f — **repuesto** m - [mec] spare part number
— m — **planchas** f - [mec] number of plates
— m —— f **en circunferencia** f - [tub] periphery number of plates
— m — **planchón** m - [metal-lam] slab number
— m — **plano** m - drawing number
— m — **proveedor** m - supplier('s) number
— m — **punto(s)** m - [mec] stitches number • [comput] site(s) number
— m — **rechazo** m - rejection number
— m — **regla** f - rule number
— m —— f **sobre seguridad** f - [segurid] safety rule number
— m — **referencia** f - reference number • stock number
— m — **repuesto** m - [mec] (spare) part number
— m — **rueda(s)** f - [mec] number of wheels
— m — **serie** f - serial number • code number
— m —— f **modificado** - [ind] changed serial number
— m —— f **para alimentadora** f — [sold] feeder serial number
— m —— f —— f **para alambre** m - [sold] wire feeder serial number
— m —— f — **equipo** m - [ind] unit serial number
— m —— m — **malacate** m - [petról] drawworks serial number
— m —— f — **máquina** f - [ind] machine serial number
— m —— f **para modelo** m - [ind] model serial number
— m —— f —— m **de máquina** f - [ind] machine model serial number
— m —— f — **motor** m - [electr-mot] motor serial number • [comb.int] engine serial number
— m —— f —— m — **soldadora** f - [sold] welder, o machine, model serial number
— m —— f — **unidad** f - unit serial number
— m —— f —— f **motriz** - [agric-equip] power unit serial number
— m — **sitio(s)** m - [comput] site(s) number
— m — **suelo** m - [suelos] soil number

número m de tamaño(s) m - size(s) number
— m — terminal m - terminal number
— m — tocho m - [metal-lam] bloom number
— m — — m plano - [metal-lam] flat bloom
number
— m — tubería(s) f - [tub] pipe(s) number
— m — tubo(s) m - [tub] pipe(s) number
— m — unidad f - unit number
— m — vaciado m - [metal-prod] pour number
— m — vehículo m - [electrón] vehicle number
— m — — m entrado - [comput] entered vehicle
number
— m — velocidad(es) f - [autom-mec] number pf
speed(s); gear(s) number
— m — f para avance m - [autom-mec] number
of forward speed(s)
— m — f — retroceso - [autom-mec] number
of reverse speed(s)
— m — viaje(s) m - trip(s) number
— m — visita(s) f - visit(ation)s number
— m deseado - desired number
— m obtenido - obtained desired number
— m determinado - determined number
— m en código m - code number
— m — m alfanumérico - alphanumerical code
number
— m — m para lote m - batch code number
— m — inventario m - stock number
— m — lista f - list number
— m — f de repuesto(s) m - part(s) list
number
— m — relieve - raised numeral
— m — tabla f - table number
— m — f de espesores f - thickness(es)
schedule number
— m — f — — según Instituto m Estadounidense para Normas f - [tub] American National Standards Institute, o A N S I, schedule
number
— m entero - whole number
— m entrado - [comput] entered number
— m entre - number between
— m formidable - formidable, o great, number
— m — de participantes m - [deport] formidable field (of entrants)
— m idéntico - identical number
— m identificado - identified number
— m igual - equal number
— m imaginado - imagined number
— m imaginario - imaginary number
— m impar - [matem] odd number
— m índice - index number
— m individual - individual number
— m — para paleta f - [ventil] blade individual number
— m inferior a - number below
— m limitado - limited number
— m — de medida(s) f - limited size(s) number
— m — — tamaño(s) m - limited size(s) number
— m mágico - magic number
— m marcado - marked number • [comput] entered
number
— m máximo - maximum number
— m — de perno(s) m - [mec] maximum bolt(s)
number
— m mínimo - minimum number
— m — de perno(s) m - [mec] minimum bolt(s)
number
— m mixto - [mat] mixed number
— m nutrido - [deport] large field
— m — de participante(s) n - [deport] large
field of entrants
— m obtenido - obtained number
— m par - [mat] even number
— m para especificación f - specification number
— m — — f para eje m - [mec] axle specification number
— m — identificación f - identification number • stock number
— m — — f para espesor m - thickness identification number • [mec] gage identification
number

número m para identificación f para espesor m
de penetrámetro m - [instrum] penetrameter
thickness identification number
— m — inventario m - stock number • stock
series
— m — lote m - [ind] batch, o assembly, o
lot, number
— m — — de producción f - [ind] production
assembly number
— m — — m — — f para portadiferencial m -
[autom-mec] differential carrier production
assembly number
— m — medida f - measurement, o size, number
— m — modelo m - [mec] model number
— m — — m para eje m - [mec] axle model number
— m — — m para mordaza f - [mec] clamp model
number
— m — mordaza f - [mec] clamp number
— m — — f para inventario m - [mec] clamp
stock number
— m — oblicuidad f - skew number
— m — pieza f - [mec] part number
— m — — f individual - [mec] individual part
number
— m — referencia f - reference, o key, number
— m — viscosidad f - [lubric] viscosity number
— m prerregulado - preset number
— m — para embolada(s) f - [bombas] stroke(s)
preset number
— m — — f para bomba f - [bombas] pump
stroke(s) preset number
— m primero - digit one; first number
— m progresivo - progressive, o consecutive,
number
— m promedio - average number
— m registrado - [comput] entered number
— m requerido - required number
— m segundo m - digit two; second number
— m superior a - number above
— m total - total number • entire amount
— m únicamente con . . . cifra(s) f - number
with only . . . digit(s); pure . . . digit
number
numerosos,sas adv - many
nunca adv - . . . . • exclusive
— adv creciente - never increasing
nutrido,da a - . . . . [fig] healthy

O

O A A f - véase Organización f para Alimento(s)
m y Agricultura f
o c f - véase orden f para compra f
O C D E f - véase Organización f para Coopera-
ción y Desarrollo m Económico
O D E C A f - véase Organización f de Estados m
Centro Americanos
O I E A f - véase Organización f, o Organismo
m para Energía f Atómica
O I T f - véase Organización f Internacional
para Trabajo m
O M M f - véase Organización f Meteorológica
Mundial; Organismo m Meteorológico Mundial
O N U f - [pol] véase Organización f de Naciones f Unidas
O N U D I a - véase Organización f de las Naciones Unidas para el Desarrollo Industrial
O S N f - véase Obras f Sanitarias de la Nación
o sea adv - [legal] to wit
O₂ - véase oxígeno m
objetivo m - . . . . • purpose; goal • target
— m actual - present, o current, objective
— m básico - basic, o prime, objective
— m conseguido - achieved objective
— m convenido - [admib] agreed (upon) objective

objetivo m̲ desarrollado - developed objective
— m̲ deseado - desired, o̲ sought for, objective
— m̲ — para empresa f̲ - [ind] desired, compa-ny, o̲ concern, objective
— m̲ establecido - established, o̲ set, objec-tive
— m̲ fijado - set, objective, o̲ goal
— m̲ fundamental - basic objective
— m̲ general - general objective
— m̲ logrado - attained, o̲ reached, o̲ achieved, o̲ accomplished, objective
— m̲ obtenido - achieved objective
— m̲ original - original objective
— m̲ para cargo m̲ - [admin] position objective
— m̲ para comportamiento m̲ - behavioral objec-tive
— m̲ — entidad f̲ - [admin] concern('s), o̲ or-ganization('s), objective
— m̲ — fideicomiso m̲ - [legal] trust objective
— m̲ — función f̲ - [admin] function objective
— m̲ — f̲ para control m̲ - [admin] control-ling function objective
— m̲ — f̲ para fiscalización f̲ - [admin] controlling function objective
— m̲ — f̲ — organización f̲ - [admin] orga-nizing,zation function objective
— m̲ — grupo m̲ - group('s) objective
— m̲ — instrucción f̲ - instruction objective
— m̲ — plazo m̲ largo - long, range, o̲ term, objective
— m̲ — rendimiento m̲ - yield, o̲ performance, objective
— m̲ — sección f̲ - [labor] unit('s) objective
— m̲ — subalterno m̲ - [admin] subordinate('s) objective
— m̲ — término(s) m̲ - term('s) objective
— m̲ — — m̲ para referencia f̲ - reference term(s) objective
— m̲ — trabajo m̲ - [labor] work objective
— m̲ — — m̲ administrativo - [admin] manage-ment work objective
— m̲ — — m̲ logrado - [admin] achieved management work objective
— m̲ — — m̲ para administración f̲ - [admin] management('s) work objective
— m̲ — venta(s) f̲ - [com] sales objective
— m̲ preconvenido - [admin] (pre)agreed (upon) objective
— m̲ preparado - [admin] prepared objective
— m̲ primario - prime,mary objective
— m̲ principal - main, objective, o̲ purpose
— m̲ secundario - secondary objective
— m̲ trascendente - [econ] transcendent objec-tive
objeto m̲ - . . . • item • scope
— m̲ agudo - sharp object
— m̲ cilíndrico - cylindrical object
— m̲ de arte m̲ - work of art • novelty
— m̲ — — m̲ colgado - [domést] hung, work of art, o̲ novelty
— m̲ — — m̲ suspendido - [domést] suspended, o̲ hung, work of art, o̲ novelty
— m̲ — compañía f̲ - company, purpose, o̲ object
— m̲ — contrato m̲ - [legal] contract object
— m̲ — cristal m̲ - [domést] crystal, o̲ glass, object
— m̲ — — m̲ para escritorio m̲ - [domést] of-fice glass
— m̲ — empresa f̲—company, o̲ corporate, object
— m̲ — invención f̲ - invention object
— m̲ — invento m̲ - invention object
— m̲ — licitación f̲—tender, object, o̲ purpose
— m̲ — presente m/f̲ - hereby covered
— m̲ — sociedad f̲ - corporate, object, o̲ pur-pose
— m̲ — viaje m̲ - trip object(ive)
— m̲ específico - specific object
— m̲ extraño - foreign object
— m̲ filoso - [segurid] sharp object
— m̲ llevado - carried object
— m̲ metálico - metal(lic) object
— m̲ — agudo - sharp metal(lic) object
— m̲ oscilante - [segurid] swinging object

objeto m̲ para reclamación f̲ - claim, object, o̲ purpose, o̲ reason
— m̲ — reclamo m̲ - claim, object, o̲ reason
— m̲ pesado - heavy object
— m̲ portado - carried object
— m̲ punzante - [segurid] piercing object
— m̲ que cae - [segurid] falling object
— m̲ suelto - loose object
— m̲ tirado - pulled object • thrown object
oblicuar v - . . .; to skew; to angle
oblicuidad f̲ - . . .; skew; bias; camber; bevel; slant
— f̲ — alcantarilla f̲ - [hidr] culvert skew
oblícuo,cua a̲ - . . .; beveled; skewed; askew; across
— a̲ escalonado,da - step beveled
— a̲ hacia derecha f̲ - skew right
— a̲ — izquierda f̲ - skew left
obligación f̲ - . . .; commitment; accountabi-lity • duty • forcing • responsibility • liability; (promisory) note • bond • deben-ture
— f̲ a su cargo m̲ - its responsibility
— f̲ activa - [fin] active, o̲ quick, liability
— f̲ adicional - additional obligation
— f̲ — de contratista m̲ - additional contrac-tor('s) obligation
— f̲ anterior - [legal] prior, o̲ previous, obli-gation
— f̲ asumida - assumed obligation
— f̲ bancaria - [fin] bank, obligation, o̲ note
— f̲ capitalizada - [contab] capitalized obliga-tion
— f̲ colocada - [fin] placed bond
— f̲ comercial - [com] commercial, o̲ business, obligation
— f̲ con plazo m̲ largo - [fin] long term note
— f̲ contingente - [contab] contingent liability
— f̲ contractual - [legal] contract(ual), obli-gation, o̲ liability
— f̲ — anterior - [legal] previous, o̲ prior, contract(ual) obligation
— f̲ — previa f̲ - [legal] prior contract obli-gation
— f̲ contraida - [com] contracted obligation
— f̲ cumplida - complied, o̲ met, obligation
— f̲ de afiliado m̲ - [fin] affiliate('s) note
— f̲ — contratista m̲—contractor('s) obligation
— f̲ — divisas f̲ - [fin] (foreign) exchange commitment
— f̲ — empresa f̲ - [fin] concern('s) note
— f̲ — — f̲ afiliada - [fin] affiliate('s), obligation, o̲ note
— f̲ — supervisor m̲ - supervisor('s) obligation
— f̲ exclusiva - [legal] exclusive, o̲ sole, obligation
— f̲ garantizada - [fin] guaranteed bond
— f̲ imperativa - imperative obligation
— f̲ para retroventa - [fin] recourse; repur-chase obligation
— f̲ por arriendo(s) m̲ - lease obligation
— f̲ — — m̲ capitalizado - [contab] capitalized lease obligation
— f̲ por cobrar v - [contab] note(s) receivable
— f̲ — — v en gestión f̲ - [contab] overdue note(s) receivable
— f̲ — pagar v - [contab] note payable
— f̲ previa - [legal] prior obligation
— f̲ pública - [fin] public debenture
— f̲ recibida - [fin] received note
— f̲ rentable de deuda f̲ pública - [fin] general obligation assessment bond
— f̲ suscri(p)ta - [fin] signed bond
— f̲ tomada - [fin] assumed bond
obligadamente - obligatorily; mandatorily
obligado,da a̲ - obliged; forced; bound; respon-siblke • [legal] liable
— a̲ a abandonar v - [deport] forced out
obligar f̲ - . . .; to force • to coerce
— v a abandonar v - [deport] to force out
obligarse v - to assume @ obligation; to bind oneself • to be(come) responsible
obligatoriamente a̲d̲v̲ - obligatorily; mandatorily

obligatoriedad f - obligation
obligatorio,ria a - . . . .; mandatory • binding
obra f - . . . . • [constr] job; project; job, o
project, site • installation, o construction, site; site; facility • scheme • struc-
ture • contract • operation | a - (en) field
— f audaz - challenging project
— f auxiliar - [constr] appurtenance; auxiliary project
— f civil - [constr] construction, o civil,
(engineering) work, o job; public construction • civil construction
— f — auxiliar - [constr] auxiliary civil,
work, o job, o project
— f — certificada - [constr] certified civil,
job, o project
— f cloacal - [constr] sewer project
— f colateral - [constr] collateral project
— f comenzada - [constr] begun, job, o project
— f comercial - commercial project
— f complementaria - [constr] related work
— f completa - [constr] complete, job, o project
— f completada - [constr] completed, o fin-
nished, job, o project
— f conexa - [constr] related, o connected, o
allied, project, o facility
— f contra inundación(es) f - [hidr] flood
(control) project
— f contratada - [constr] contracted, work, o
project
— f — en curso - [constr] contracted, work, o
contract, in progress
— f de categoría f - quality, work, o project
— f — — f primera - high(est) quality project (lavel)
— f de construcción f - [constr] construction,
project, o job • construction site
— f — — f relacionada - related construction
(work, o project)
— f — — f vial - [vial] highway construction, work, o project, o site
— f — horno m - [ind] furnace project
— f — — m alto - [metal-prod] blast furnace
project
— f — hormigón m - [constr] concrete, construction, o structure, o project
— f — ingeniería f - [constr] engineering,
project, o work
— f — ladrillo(s) m - [constr] brickwork
— f — referencia f - [public] reference
(work)
— f — tierra f - [constr] earthern structure
— f — vialidad f - [constr] highway (construction), project, o site
— f demorada - [constr] delayed project
— f deportiva - [constr] sports, project, o
structure
— f determinada - particular, job, o project
— f ejecutada - [constr] performed, project, o
job, o work • work performed
— f — bien - [ind] workmanlike, project, o
job
— f en cabecera f - [hidr] headworks
— f — construcción f - [constr] construction,
project, o job • construction in progress
— f — — f paralizada - [constr] paralized
construction, project, o job • paralized
construction in progress
— f — curso m - work in progress
— f — ejecución f - [constr] construction in,
progress, o process
— f — hierro m - [metal-fabr] iron work
— f — proceso m - [constr] work in process
— f — taller m - [mec] shop work
— f específica - specific, o particular, job
— f ferroviaria - [f.c.] railway structure
— f futura—[constr] future, project, o scheme
— f grande - [constr] large project
— f hidráulica - [hidr] hydraulic project
— f importante - [constr] important project
— f inconclusa f - [ind] incomplete project
— f iniciada - [constr] started (up) project

obra f metálica - [constr] metal, o steel, project, o (super)structure
— f minera-metalúrgica - [ind] mining-metallurgical project
— f nueva - [constr] new project
— f para ampliación f - expansion, o widening,
project
— f — — de calzada f - [constr] road
widening project
— f — — f — carretera f - [vial] road, o
highway, widening project
— f — — f vial - [vial] road widening project
— f — conservación f - [ind] maintenance work
• [constr] conservation, project, o work
— f — consolidación f - [constr] consolidation work
— f — — de suelo m - [constr] earth, o
soil, control project
— f — conservación f - conservation project
— f — — f de agua m - [hidr] water conservation project
— f — — f — suelo m - [suelos] soil conservation project
— f — construcción f - [constr] project, construction, o work • construction site
— f — contener v erosión f - [constr] erosion
control project
— f — defensa f - [hidr] protection project
— f — — f contra inundación(es) f - [hidr]
flood protection, project, o work
— f — — — f causada(s) f por hura-
cán(es) m - [hidr] hurricane-flood protection project
— m — dique m - [constr] dam project
— f — drenaje m - [hidr] drainage, project, o
work • drainage site
— f — — m para terraplén m - [constr] fill,
o embankment, drainage project
— f — ensanchamiento m de calzada f—[constr]
road widening project
— f — — m — carretera f - [constr] road, o
highway, widening project
— f — entretenimiento* m - [constr] maintenance work
— f — estabilización f - [constr] stabilization project
— f — — f de terraplén m - [constr] fill
stabilization project
— f — fomento m - [pol] improvement, o development, project, o program
— f — horadación f - [constr] tunnel(ing)
project
— f — infraestructura f - [constr] infra-
structure work
— f — instalación f para drenaje m - [constr]
drain(age) construction project
— f — — f — — dentro de terraplén m -
[constr] embankment drain(age) construction
project
— m — mantenimiento m - maintenance, project,
o work
— m — manutención f - maintenance work
— f — paso m a nivel - [constr] grade separation project
— f — prevención f - [constr] control structure
— f — prevenir inundación(es) f - [hidr]
flood (protection) project
— f — protección f contra inundación(es) f -
[hidr] flood protection work
— f — red f interestatal - [constr] Interstate System project
— f — regulación f - control project
— f — — f de contaminación f - [ambient]
pollution control project
— f — — f — — atmosférica - [coque] air
pollution control project
— f — — f contra contaminación f de aire m -
[ambient] air pollution control project
— f — — f — flujo m - [hidr] flow control
structure
— f — rehabilitación f—redevelopment project

obra f para rehabilitación f para, centro m urbano, o zona f céntrica - downtown redevelopment project
— f — reparación f - repair, project, o work
— f — separación f - separation project
— f — — f de nivel(es) m - [constr] grade
separation project
— f — toma f - [hidr] intake construction
— f portuaria - [constr] dock, o port, project
— f privada—[constr] private, project, o work
— f propuesta - proposed project • [constr]
construction project
— f pública - [constr] public, project, o work
— f — para condado m - [pol] county public,
project, o work
— f realizada - [constr] performed work
— f sanitaria - [sanit] sanitation (project)
— f simultánea - [constr] simultaneous, o collateral, project
— f vial - [constr] highway project
— f vial en construcción - [constr] highway
construction project
obrador m - . . . . • [ind] work, site, o yard •
[constr] construction, site, o field; shop
obrante a - . . .; appearing
obras f civiles - [constr] civil, projects, o
department • civil construction
— f Sanitarias de Nación f - [pol] (Argentina)
Waterworks Authority, o (National) Department of Sanitation
obrero m - [ind] . . .; workman; man
— m agrícola - [agric] farm, worker, o laborer
— m calificado—[labor] (qualified) journeyman
— m jornalizado - [labor] hourly, laborer, o
emloyee
— m municipal - [pol] municipal, o city, employee, o laborer
— m para conservación f—[ind] maintenance man
— m — cuadrilla f - crew member
— m — — f para construcción f - [constr]
construction crew, worker, o member
— m — entretenimiento m - [ind] maintenance,
worker, o man, o employee
— m — horno m - [metal-prod] furnace hand
— m — mantenimiento m - [ind] maintenance man
— m siderúrgico - [metal-prod] steel worker •
iron worker
— m trabajando - [ind] man at work
obscurecer v - . . .; to black(en)
obscurecido,da a - obscured; darkened; turned
dark; blackened
obscurecimiento m - . . .; obscuring; turning
dark(er)
obscuro,ra a - . . . . • black
obsequio m - . . . . • giveaway | a - [de] com-
plimentary
observación f - . . .; observing; monitoring;
watch(ing) • seeing; look(ing); looking at;
viewing • check(ing); inspecting,tion • detecting,tion • taking care of • comment
— f astronómica f - . . . .; stargazing
— f bajo luz f ultravioleta - viewing under
ultraviolet light
— f — microscopio m - observation under @ microscope
— f breve - brief viewing
— f celeste - [astron] stargazing
— f con cuidado - careful watching
— f constante - constant check(ing)
— f correspondiente - corresponding remark
— f cuidadosa - careful watching; close look
— f de astro(s) m - [astron] stargazing
— f — bomba f - [bombas] pump, viewing, o
watching, o monitoring
— f — calidad f - quality observation,ving
— f — carrera(s) f - [deport] race watching
— f — daño(s) - damage(s) observation,ving
— f — clavija f - [electrón] pin monitoring
— f — condición f - condition observing
— f — — f de camisa f - [mec] liner condition observing
— f — corrosión f - corrosion observation
— f — entrehierro m - [mec] gap observation

observación f de estrella(s) f - [astron]
stargazing
— f — manejo m - [ind] operation observation
— f — — m de bobinadora f - [metal-lam]
coiler operation observation
— f — — m de regulación f - [ind] automatic
control operation observation
— f — naturaleza f - [biol] natural observation, o study
— f — polaridad f - [electr-instal] polarity
observation,ving
— f — práctica f - [ind] practice observing •
practice following
— f — resultado(s) m - results observation
— f — — m de ensayo(s) m - [ind] test(s)
results observation
— f — — f — — m en laboratorio m - [ind]
laboratory test(s) results observation
— f — señal(es) f - signal(s) monitoring
— f — tablero m - [ind] panel observation
— f — — m para regulación f - [electr-oper]
control panel observation
— f — velocidad f - speed, watching, o monitoring
— f durante período m corto - short, time o
term, observation
— f — — m largo - long, time, o term, observation
— f en busca f de deformación f - watching for
@ deformation(s)
— f — obra f - field observation
— f escrita - written remark
— f especial - special, remark, o comment
— f fluoroscópica - [electrón] fluoroscopic
viewing
— f general - general remark
— f — sobre seguridad f - [segurid] general
safety, remark, o comment
— f metalográfica - [metal] metallographic,
observation, o check(ing)
— f oral - oral remark
— f perspicaz - perceptive, observation, o
remark
— f pertinente - pertinent, observation, o remark
— f planeada sobre seguridad f - [segurid]
planned safety, observation, o remark
— f posterior—later, o follow-up, observation
— f semanal - weekly observation
— f sísmica - [petról] seismic observation
— f — de pozo m - [petról] well shooting
— f sobre período m corto - short, time, o
term, observation
— f — — m largo m - long, time, o term. observation
— f — seguridad f - [segurid] safety, remark,
o comment
observado,da a - observed; seen; watched; monitored; looked at • inspected • detected •
taken care of • rejected
— a brevemente - looked at briefly
— a con cuidado - watched carefully
— a cuidadosamente - watched carefully
— a durante período m corto - short, time, o
term, observed
— a — — m largo - long, time, o term, observed
— a en busca de deformación f - watched for @
deformation
— a en obra - field observed
— a más cuidadosamente - watched, o observed,
more carefully, o (more) closely
— a sobre período m corto - short, time, o
term, observed, o watched, o monitored
— a — — m largo - long, time, o term, observed, o watched, o monitored
observador m - . . . .; onlooker • [deport] corner marshall
— m casual - casual, observer, o onlooker
— m deportivo - [autom] motor sports observer
— m metalúrgico - [metal-prod] metallurgical
observer
— m — para, control m, o fiscalización f, de

calidad f - [metal-lam] quality control metallurgical observer
observador m **metalúrgico para fiscalización** f **de
calidad** f - [metal-lam] quality control metallurgical observer
— m —— **verificación** f **de calidad** f - [metal-
-lam] quality control metallurgical observer
— m **para combustión** f - [combust] combustion
observer
observador,ra a - observing; observer
observar v - . . . .; to monitor; to look (at) •
to comply with; to take care of • to inspect
• to detect • to heed; to obey; to satisfy;
to follow • [com] to revise • to reject • to
appraise
— v **advertencia** f - [segurid] to, observe, o
take heed of, @ warning
— v **arco** m - [sold] to watch @ arc
— v — m **mismo** - to watch @ arc iteself
— v **astro(s)** m - [astron] to stargaze
— v **bajo luz** f **ultravioleta** - to view under @
ultraviolet light
— v **bomba** f - [bombas] to view @ pump
— v **brevemente**—to, observe, o look at, briefly
— v **calidad** f - to observe @ quality
— v **carrera** f - [deport] to watch @ race
— v **clavija** f - [electrón] to monitor @ pin
— v **con cuidado** - to watch carefully
— v — **vista** f **simple** - to detect visually
— v **condición** f - to, observe, o watch, @ condition
— v — f **de camisa** f - [mec] to observe @ liner
condition
— v — f — **émbolo** m - [mec] to observe @ pis-
ton('s) condition
— v **defecto** m - to observe @ defect; to detect
@ flaw
— v — **visible** - to observe @ visible defect
— v **corrosión** f - to observe @ corrosion
— v **cuidadosamente** - to, observe, o watch,
carefully
— v **daño** m - to observe @ damage
— v **diferencia** f - to observe @ difference • to
detect @ difference
— v — f **en espesor** m - to, observe, o detect,
@ difference in @ thickness
— v **durante período** m **corto** - to observe over @
short period
— v — m **largo** - to observe over @ long,
period, o time
— v **en busca** f **de deformación(es)** - to watch
for @ deformation(s)
— v — **forma** f **visual** - to detect visually
— v **en obra** f - to field observe
— v **entrehierro** m - [mec] to observe @ gap
— v **estrella(s)** f - [astron] to star gaze
— v **más cuidadosamente** - to observe more carefully; to look closer
— v **polaridad** f - [electr-instal] to observe @
polarity
— v **práctica** f - to observe @ practice
— v **precaución(es)** f - to observe @ precau-
tion(s)
— v **señal(es)** f - to, observe, o monitor, @
signal(s)
— v **sobre período** m **breve** - to observe for @
short time
— v **sobre período** m **corto** - to observe for @
short time
— v — m **largo** - to observe for @ long time
— v **velocidad** f - to, monitor, o check, @ speed
— v **tablero** - to, observe, o watch, @ panel
— v — m **para regulación** f - [electr-oper] to ,
observe, o watch, @ control panel
observatorio m - . . . . [topogr] overlook
obsesión f - haunting
obsesionado,da a - . . . .; haunted
obsesionar v - . . . .; to haunt
obsolescencia f - obsolescence
— f **de material(es)** f - [constr] material(s)
obsolescence
— f — **mercadería(s)** - goods obsolescence
obstaculizado,da a - obstructed

obstaculizar v - to be @ obstacle; to interfere
obstáculo m - . . . .; hurdle • break • [fig]
hitch
— m **afrontado** - confronted, o faced, obstacle
— m **confrontado** - confronted obstacle
— m **dañino** - damaging obstacle
— m **en vera** f **de camino** m - roadside obstacle
— m **encontrado** - encountered, o met, obstacle
— m **evitado** - avoided obstacle
— m **sobre superficie** f - surface obstacle
— m **subterráneo**—[constr] underground obstacle
obstétrica* f - **véase obstetricia** f
obstrucción f - . . . .; plug(ging)
— f **con aire** m - air obstruction
— f —— m **verificado** - [mec] checked air obstruction
— f — **vapor(es)** - [mec] vapor lock
— f **de conducto** m - conduit, o channel, obstruction, o clogging
— f —— m **para aire** m - air channel, obstruction, o clogging
— f **interna** - [tub] interior obstruction
— f **inusitada** - unusual obstruction
— f **para circulación** f - circulation obstruc=
tion
— f —— f **libre** - free circulation obstruction
— f —— f — **de aire** m - free air, flow, o
circulation, obstruction
— f — **drenaje** m - [hidr] drainage obstruction
— f **por aire** m - [neumát] air binding
— f **posible** - possible obstruction
— f **próxima** - neighboring obstruction
— f **superior** - overhead obstruction
— f **vecina** - neighboring obstruction
obstruido,da a - obstructed; clogged; plugged
— a **con salpicadura(s)** f - [sold] spatter
clogged
obstruir v - . . . .; to clog; to plug; to choke
(up)
— v **conducto** m - to, obstruct, o clog, @, conduit, o channel
— v — m **para aire** m - to clog @ air channel
obtención f - . . . .; obtaining; getting; acqui-
ring; attainment,ning; getting; reaching;
achieving,vment,shment •
capture,ring • granting • purchase,sing; re-
quisition(ing) • producing,ction; buildup •
[deport] collecting,tion; claiming; notching • snagging
— f **de acuerdo** m - agreement obtainment,ning
— f — **aprobación** f - approval, obtaining, o
getting
— f — **atención** f - attention getting
— f —— f **médica** - [medic] medical attention
getting
— f — **autorización**s f - authorization, getting, o obtaining
— f — **calidad** f - quality, getting, o obtaining
— f — **concesión** f - [com] concession, obtaining, o getting
— f — **derecho(s)** m - [legal] rights getting
— f — **literario(s)** - [legal] copyright,
obtaining, o getting
— f — **energía** f - [electr-distrib] power, o
energy, obtaining, o getting
— f —— f **de embalse** m - [hidr] energy obtaining from @ pond(ing)
— f — **estado** m - [comput] status attaining
— f — **hidrógeno** m - [quím] hydrogen production
— f — **patente** f - [legal] patenting
— f — **valor(es)** m - [labor] value(s) obtaining
— f **fácil** - easy, getting, o obtaining | adv -
(de) readily available
— f **por interpolación** f - obtaining by inter-
polation
— f — **memoria** f - [comput] calling from @
memory
— f — **nombre** m - name getting
— f —— m **comercial** - [legal] trade name ob-

tainment
obtención f de número m - number obtaining
— f —— m deseado—desired number obtaining
— f — objetivo m - [admin] objective attaining
— f — patente f - [legal] patent obtaining;
patenting
— f —— f de invención f - [legal] patent
obtaining
— f —— f para dispositivo m - device patenting
— f —— f para sistema m - [ind] system patenting
— f — permiso m - [fisc] permit getting
— f — presión f - pressure build(ing)up
— f —— f en aceite m - [comb.int] oil pressure buildup
— f — propiedad f - [legal] ownership obtaining
— f —— f intelectual - [legal] copyright
obtaining; copywriting
— f de rendimiento m - yield attaining
— f — respuesta f - answer getting
— f — resultado(s) m - result(s), obtaining,
o securing
— f — satisfacción f - satisfaction, obtaining, o deriving
— f — valor m - value obtaining
— f — victoria f - [deport] victory, obtaining, o claiming
— f — mercado m - market developing,pment
obtener v - . . .; to, attain, o achieve, o
secure, o reach, o accomplish • to build up
• to, acquire, o acquisition • to purchase •
to produce • to take • to succeed in •
[deport] to, collect, o capture, o claim, o
snag, o notch, o come away with
— v acuerdo m - tp obtain @ agreement
— v atención f - to get @ attention
— v — f médica - [medic] to get @ medidal attention
— v autorización f - to, get, o obtain, @ authorization
— v auxilio m - [segurid] to get @ help
— v campeonato m - [deport] to take @ championship
— f colocación f - [deport] to take @ place
— v — f primera - [deport] to take @ first
(place)
— v — f segunda - [deport] to take @ second
(place)
— v concesión f - [com] to obtain @ concession
— v cordón(es) m - [sold] to get @ bead
— v — m con tamaño m apropiado - [sold] to
obtain @ proper bead size
— v dato(s) m - to obtain @ datum,ta
— v de memoria f - [comput] to call from @
memory
— v derecho(s) m—[legal] to obtain @ right(s)
— v — m de patente f - [legal] to obtain @
patent right(s)
— v — m literario(s) - [legal] to obtain @
copyright
— v en mercado - to, obtain, o acquire, commercially
— v energía f - to obtain @ energy
— v — f de embalse m - [hidr] to obtain @
energy from @ pond(ing)
— v estado m - [comput] to, obtain, o get, @
status
— v éxito m - to succeed
— v experiencia f - to get, experience, o
practice
— v fijación f - [mec] to obtain @ fastening
— v — f de convertidor m - to obtain @ converter fastening
— v — f perfecta - [mec] to obtain @ perfect
fastening
— v — f — de convertidor m - [metal-prod] to
obtain @ perfect converter fastening
— v información f - to obtain @, information,
o datum,ta
— v marca f - [legal] to obtain @ (trade) mark

obtener v marca f de fábrica f - [legal] to obtain @ trade mark
— v — f industrial - [legal] to obtain @
trade name
— v — f registrada - [legal] to obtain @
trade mark
— v — m comercial - [legal] to obtain @
trade name
— v número m - to obtain @ number
— v — m deseado - to obtain @ desired number
— v objetivo - [admin] to achieve @ objective
— v patente f - [legal] to (obtain @) patent
— v — f de invención f - [legal] to obtain @
patent
— v patrón m - [mec] to obtain @ pattern
— v penetración f - to achieve @ penetration
— v — f buena - [sold] to get good penetration
— v — f completa - [sold] to get @ complete
penetration
— v — f total - [sold] to get @ total penetration
— v permiso m - [legal] to get @ permit
— v personería f - [pol] to incorporate
— v por interpolación f - to obtain by inter-
polation
— v práctica f - to get @ practice
— v presión f - to build up @ pressure
— v — f en aceite m - [comb.int] to build up
@ oil pressure
— v propiedad f intelectual - [legal] to obtain @ copyright
— v protección v - to obtain @ protection •
[sold] to rely on for @ shielding
— v — f de metal m fundido - [sold] to rely
on @ molten slag for shielding
— v rendimiento m - to obtain @ yield
— v respuesta f - to, obtain, o get, @ answer;
to be answered
— v resultado(s) m - to, obtain, o get, o accomplish, @ result(s)
— v — m deseado = [Admin] to accomplish @
desired result
— v triunfo m - [deport] to, obtain, o hammer
out, @ victory
— v valor m - to obtain @ value
— v ventaja f - [deport] to, obtain, o
squeeze out, @ lead
— v victoria f - [deport] to claim @ victory
obtenible a - . . . • available • on @ market
— a contra pedido m - available (up)on, request, o order
— f corrientemente - availably currently
— a de existencia(s) f available from stock(s)
— a en comercio m - [com] available commercially
— a — plaza f - commercial(y available)
— a fácilmente - obtainable, easily, o readily
— a normalmente - normally available
obtenido,da a - obtained; reached; gotten; accomplished; achieved; acquired • produced •
built up • purchased; requisitioned • captured • [deport] collected; claimed; snagged
• notched
— a como resultado de carrera f - [deport]
race-bred
— a de memoria f - [comput] called (up) from
memory
— a en mercado m - obtained commercially
— a por interpolación f - obtained by inter-
polation
obturación f - . . .; closure; clogging; plugging; blocking • [mec] capping off
— f con derivación f - [instrum] by-pass,
blocking, o blanking
— f con disco m de bronce m - [mec] bronze disk
plugging
— f — agujero m para colada f - [metal-prod]
taphole, plugging, o closing
— f de atomizador - [comb.int] atomizer plugging
— f — conducto m - conduit plugging
— f — entalladura f - [mec] notch clogging

obturación f de entalladura f en base f para a-
ceite m - [electr-mot] base oil notch clogging
— f — línea f para combustible m - [comb.int]
fuel line clogging
— f — ojo m para llave f - [sold] keyhole
clogging
— f — orificio m - orifice, o opening, o port,
clogging, o closing (off)
— f — — m para comprobación f - test opening
plugging
— f — — m — derivación f - [mec] by-pass
port closing (off)
— f — — — m para retorno m - [mec] return hole,
plugging, o clogging
— f — — — — m para aceite m - [mec] oil
return hole, plugging, o clogging
— f — pasaje m - passage clogging
— f — piquera f - [metal-prod] taphole closing
— f — retorno m para aceite m - [mec] oil return block(ing)
— f — soldadura f - [sold] weld seal(ing)
— f manual—[sold] manual, seal(ing), o closing
obturado,da a - closed; capped; plugged; clogged
• blocked • capped (off)
— a con salpicadura(s) f - [sold] spatter clogged
obturador m - . . . ; shutter; closure; closing
device • packer • plug; poppet • cutoff collar • [fotogr] shutter
— m anular - [petról] packer
— m con lodo m - [metal-prod] mud gun
— m con varilla f - [metal-prod] rod stopper
— m de bronce m - [válv] bronze plug
— m — piquera f - [metal-prod] notch stopper
— m — — f para escoria f - [metal-prod] slag
notch stopper
— m — tipo m renovable - [válv] remewable plug
type
— m flotante - [válv] floating poppet
— m — para válvula f - [válv] valve floating
poppet
— m para aceite m - [mec] oil seal
— m — aire m - [comb.int] choke
— m — anclaje m - [petról] anchor packer
— m — cabeza f - [petról] head packer
— m — — f para tubería f - [petról] (pipe)
head packer
— m — compresión f - [mec] pressure packer
— m — — f para conexión f - [mec] connection
compression packer
— m — — f para unión f - [mec] connection
compression packer
— m — conexión f - [mec] connection packer
— m — — f para salida f - [mec] outlet, o exhaust, connection packer
— m — empaque m - [petról] packer
— m — escorial m - [metal-prod] slag notch
stopper
— m — flujo m - [petról] flow packer
— m — orificio m - [metal-prod] hole stopper
— m — — m para escoria f - [metal-prod] slag
notch stopper
— m — retención f - [mec] retaining packer
— m — salida f - [mec] outlet, o exhaust,
packer
— m — surgencia f - [petról] surge, o flow,
packer
— n — unión f - [mec] connection packer
— m — válvula f - [válv] valve poppet
— m refrentable - [válv] regrindable plug
— m renovable - [válv] renewable plug
obturador,ra a - plugging
obturar v - . . . ; to, block, o clog, o choke
(up) • to cap (off)
— v abertura f - [mec] to, plug, o seal, @
opening
— f — en pasada f - [sold] to seal @ pass
opening
— v — — f para raíz f - [sold] to seal @
root pass opening
— v atomizador - [comb.int] to plug @ atomizer
— v con óxido m - [metal] to, plug, o rust, shut

obturar v entalladura f en base f - [electr-
-mot] to clog @ base notch
— v — f para aceite m en base f - [electr-mot]
to clog @ base oil notch
— v línea f - [comb.int] to clog @ line
— v — f para combustible m - [comb.int] to
clog @ fuel line
— v ojo m para llave f - [sold] to, close, o
plug, @ keyhole
— v orificio m - to, clog, o plug, o close, @,
orifice, o opening, o port
— v — m para comprobación f - to plug @ test
opening
— v — m — derivación f - [mec] to close off
@ by-pass port
— v — m para retorno m - [mec] to clog @ return hole
— v — m — — m para aceite m - [mec] to clog
@ oil return hole
— v — m para ventilación f - [comb.int] to
plug @ air vent hole
— v pasaje m - to clog @ passage
— v respiradero m - [comb.int] to plug @ air
vent hole
— v retorno m - [mec] to block @ return
— v — m para aceite m - [mec] to block @ oil
return
— v separación f - [sold] to fill @ gap
obviamente adv - obviously
— adv frustratorio,ria - frustratingly obvious
obviar v - . . . ; to avoid
obvio,via a - . . . ; apparent
ocasión f - . . . • incident; occurence; event •
time, instance
— f alguna - any (other) time
— f cualquiera - any (single) time
— f dada - given, occasion; chance
— f de incendio m - [segurid] chance of fire •
fire igniting,tion
— f descubierta - discovered, o found, occasion,
o event
— f determinada - specific, occasion, o event
— f inolvidable - unforgettable occasion •
[fig] bright spot
— f rara - rare occasion | adv - (en) very
rarely
— f única - single, occasion, o time
— f — cualquiera - any single time
ocasionado,da a - . . . . • caused; triggered •
incurred
ocasional a - . . . ; random • [labor] part time
ocasionalmente adv - . . . ; sometimes
ocasionar v - . . . ; to incur • to, produce, o
trigger; to be @ cause
— v incendio m - [segurid] to ignite @ fire
— v mella f - [mec] to, create, o cause, @ nick
— v movimiento m - to cause @, movement, o motion
— v — m ascendente - to cause @ upward, movement, o motion
— v — m — de arco m - [sold] to cause @ arc
upward, motion, o movement
— v — m de arco m - [sold] to cause @ arc,
movement, o motion
— v — m descendente - to cause @ downward,
movement, o motion
— v — m — de arco m - [sold] to cause @ arc
downward, movement, o motion
occidental a - . . . ; westerly
occipicio m - [anat] back of head
ocluido,da a - occluded
ocluidor,ra a - occluding
— a de burbuja(s) f - bubble occluding
ocluir v burbuja f - to occlude @ bubble
— v — f de aire m - [explos] to occlude @ air
bubble
— v gas m - to occlude @ gas
oclusión f - . . . ; occluding • stoppage • [sold]
occlusion
— f de burbuja(s) f - bubble(s) occlusion,uding
— f — — f de aire m - [explos] air bubble(s)
occlusion,uding
— f de gas m - gas occlusion,uding

oclusivo,va a de gas m — 1010 —

oclusivo,va a de gas m - gas occluding
octanaje m alto—[petról] high octane (rating)
— m bajo - [petról] low octane (rating)
octánico,ca* a - [petról] véase con octano(s) m
octavo m de galón m - [metric] pint
ocultar v - ...; to screen
— v cilindro m para pescante n - [grúas] to
  enclose @ jib cylinder
oculto,ta a - .... • unseen
ocupación f - .... engagement,ging; involvement,ving • occupation • spending •
  [deport] claiming
— f activa - active involvement,ving
— f básica - basic occupation
— f de tiempo m - time spending
ocupado,da a - occupied; busy; engaged; involved; employed • spent • [deport] claimed
— a activamente - involved actively
ocupante m - .... • tenant
— m — vehículo m - [transp] vehicle occupant
ocupar v - .... • to involve • to spend • to
  stay in • [deport] to claim
— v activamente - to involve actively
— v colocación f - [deport] to hold @ position
— v — f primera - [deport] to, take, o hold,
  @ first (position, o place)
— v — f segunda - [deport] to, take, o hold,
  @ second (position, o place)
— v habitación f - to stay, o live, in @ room
— v pieza f - to, stay, o live, in @ room
— v tiempo m - to, occupy, o spend, @ time
— v tres primeros puestos m - [deport] to be
  one-two-three
ocuparse v - to engage in; to be busy in
ocurrencia f de corrosión f - [quím] corrosion
  occurring,rence
— f de problema m - problem occurring,rence
— f — riesgo m comercial - [seguros] commercial risk occurrence,ring
— f — — m político - [seguros] political
  risk occurrence,ring
— f inconsciente - unwitting occurrence.ring
— f irregular - irregular ocurrence,ring |
  a - (de) occurring irregularly
— f real - actual occurrence,ring
— f repentina - sudden occurrence,ring
— f simultánea - simultanous occurrence,ring
ocurrido,da a - occurred; happened; took place
— a inconscientemente - occurred unwittingly
— a repentinamente - occurred suddenly
— a simultáneamente - occurred simultanously
ocurrir v - ...; to take place
— v corrosión f - [quím] to occur @ corrosion
— v inconscientemente - to occur unwittingly
— v repentinamente - to occur suddenly
— v simultáneamente - to occur simultanously
ocho adv cajas f - [metal-lam] eight stand(s)
— adv cilindros en V - [comb.int] V-eight
— adv en V - [int.comb] véase ocho cilindros m
  en V
odisea f en espacio m - space odyssey
odómetro m calibrado - [instrum] calibrated
  odometer
odontógrafo m - [instrum] odontograph
odorizado,da a - odorized
odorizar v - to odorize
oersted m - [magnet] oersted
oeste m—[geogr] ...; véase también occidente
  | a - véase occidental
— m a este m - [vial] east bound
— m Americano - [geogr] (American) West
— m franco - due west
ofensivo,va a - ...; abusive
ofensor m - offender
oferente m - [com] ...; bidder; proponent
oferta f - .... • [com] bid; proposal • quotation • supply; availability
— f aceptada - [com] accepted, offer, o bid, o
  proposal
— f ajustada - [com] adjusted, offer, o bid, o
  proposal
— f alternativa - [com] alternate,tive offer
— f base - [com] base offer

oferta f básica - basic, offer, o bid, o proposal
— f cerrada - [com] closed, offer, o bid
— f combinada - [com] combined proposal
— f comercial - [com] commercial, offer, o bid
— f comparada - compared, offer, o bid, o proposal
— f conveniente - [com] convenient, offer, o
  bid, o proposal
— f de condición f - condition offer(ing)
— f — equipo m - equipment offer(ing)
— f — precio m - price, offer(ing), o bid(ding)
— f — producto m - product offer • service offer
— f — proveedor m - [com] supplier('s), offer,
  o bid, o proposal
— f — servicio m - service offer(ing)
— f — ventaja f - advantage offer(ing)
— f descalificada - disqualified, offer, o bid,
  o proposal
— f económica - economic, o commercial, offer, o
  bid, o proposal • [fin] monetary offer
— f entregada - [com] delivered, o presented,
  offer, o bid, o proposal
— f especial - [com] special offer(ing) • package • bargain
— f evaluada - evaluated, offer, o bid, o proposal
— f final - [com] final, offer, o bid, o proposal
— f inicial - [com] initial, offer, o bid, o
  proposal
— f mejicana - [com] Mexican offer
— f monetaria - [fin] monetary offer
— f no adjudicada - [com] not awarded offer
— f original - original, offer, o bid, o proposal
— f para repuesto(s) m—[ind] spare(s) (parts),
  offer, o proposal
— f — sistema m - system offer(ing)
— f presentada - [com] presented, o submitted,
  offer, o proposal
— f primera - [com] first, offer, o bid. o proposal
— f rechazada - [com] rejected, offer, o bid, o
  proposal
— f retirada - [com] withdrawn, offer, o proposal
— f segunda - [com] second, offer, o bid, o proposal
— f siderúrgica - [metal] steel availability
— f técnica - technical, offer, o proposal
— f única - single, offer, o bid, o proposal
— f válida - valid, offer, o bid, o proposal
ofertado,da - offered; bid; proposed
ofertante* m - vease oferente m
ofertar v - [com] to, offer, o bid, o propose
— v equipo m - [ind] to, offer, o propose, @
  equipment
— v precio m - to, offer, o bid, o propose, @
  price
— v sistema m - to, offer, o propose, @ system
offset m - [imprent] offset
oficial m - .... • [labor] ...; craftsman;
  journeyman • [milit] ...; field grade offi-
  cer • [labor] assistant | a - .... • autho-
  rized • [pol] government(al)
— m certificado - [labor] certified craftsman
— m con certificado m de competencia f - [labor]
  certified craftsman
— f — campo m - [milit] field (grade) officer
— m de gobierno m - [pol] government official
— m — marina f - [milit] navy officer
— m — primera f - [ind] journeyman
— m — — ajustador - [tub] pipefitter journeyman
— m diestro - [labor] skilled craftsman
— m — con certificado m de competencia f -
  [labor] certified skilled craftsman
— m electricista - [electr-instal] journeyman, o
  certified, electrician • master electrician
— m mecánico - [mec] master mechanic
— f para combustión f - [combust] combustion,

officer, o supervisor
oficial m para costas,tes,tos - costs officer
— m — m para rodillo(s) - [metal-lam] rolls cost officer
— m para derrota f - [naút] navigator
— m — enlace m - liaison officer
— m — largada f - [deport] starter
— m — materia(s) f prima(s) - [ind] raw materials officer
— m — personal - [labor] personnel officer
— f — ruta f - [aeron] navigator
— m — seguridad f - [segurid] safety officer
— m — primero - first officer • [ind] journeyman; skilled workman • leader
— m soldador - [sold] welder (journeyman)
— m superior - [milit] general officer
oficialidad f alta - [milit] (top) brass
oficina f - . . . • [ind] section
— f central - [com] head office
— f Certificadora para Tiradas f - [public] Audit Bureau of Circulation(s)
— f cobradora f - [com] collecting,tion office
— f de correo(s) m - [comunic] post office
— f — Dirección f Impositiva - [pol] Internal Revenue office
— f en muelle m - [ind] dock office
— f — obra f - [ind] field office
— f — operación f - operating office
— f — planta f - plant office
— f Estadounidense para Investigación(es) f Camineras - [pol] United States Office of Road Inquiry
— f — para Normas f - [pol] United States Bureau of Standards
— f general - [com] general office
— f meteorológica - [meteorol] weather bureau
— f para cálculo m de costas,tes,tos - [ind] cost(s) computation office
— f — carrera(s) f - [deport] race, office, o headquarters
— f — combustión f - [combust] combustion office
— f — cómputo(s) m - [com] computation office
— f — contrato(s) m - [com] contract(s) office
— f — control m - control(ling) office
— f — — m de cambio(s) m - [fin] exchange control office
— f — — m — existencia(s) f - [ind] stock, section, o control office
— f — coordinación f - coordination office
— f — cronometraje m - time(keeping) office
— f — encaminamiento m - routing, office, o section
— f — enlace m - liaison office
— f — estudio(s) m - [admin] study,dies office
— f — impuesto m sobre, réditos, o renta f - [pol] income tax office
— f — información f - information office • hospitality, center, o desk
— f — órdenes f - [ind] order(s) section
— f — — f para producción f - [ind] production order(s) section
— f — personal - [admin] personnel (office)
— f — — m para reserva f - [ind] labor reserve office
— f — preparación f - preparation office
— f — — f para trabajo m - [labor] work preparation office
— f — — f y, fiscalización f, o verificación f - preparation and control office
— f — producción f - production office
— f — programación f - [ind] scheduling, office, o section, o division
— f — — f para producción f - [ind] production scheduling, office, o section
— f — registro m civil - [pol] vital statistics office
— f — regulación f - [ind] control(ling) office
— f — — f para combustión f - [combust] combustion (control,ling) office
— f — proyección f - design(ing) office
— f — — f carretera - [constr] highway design(ing) office

oficina f para proyección f carretera - [constr] highway design(ing) office
— f — — f ferroviaria - [constr] railway design(ing) office
— f — seguridad f - [segurid] safety office
— f — — f para planta f - [ind] plant safety office
— f — suministro(s) m - [ind] supply,lies office
— f — venta(s) f - [com] sale(s) office
— f — tipificación f - [pol] Bureau of Standards
— f — verificación f - [ind] checking, o control(ling) office
— f — viaje(s) m - [viajes] travel office
— f policial - [pol] police, o law enforcement, office, o agency
— f principal - [com] main office; headquarters
— f recaudadora f - [fisc] collection(s) office
— f regional - [com] regional office
— f — para recuperación f - [ind] reclamation, office, o district
— f sobre . . . rueda(s) f - . . .-wheel office
— f técnica - technical office
— f — en propiedad f industrial - [legal] trade mark, o patent, attorney('s) office
oficio m - . . . • [legal] legal notice; communiqué | a - [legal] (de) officially
ofrecedor m - [com] . . .; bidder
ofrecer v - . . . • to volunteer • to feature • to propose • to to come up with
— v aislamiento f térmico - to offer @ thermal insulation
— v cena f - [culin] to dine
— v — f opípara - [culin] to really dine
— v desgaste m - [mec] to offer wear
— v — m uniforme - to offer uniform wear
— v equipo m - [ind] to offer @ equipment
— v garantía f - to give @, guaranty, o warrant
— v precio m - to, offer, o bid, o propose, @ price
— v producto m - to offer @ product
— v — m nuevo - [com] to offer @ new product
— v protección f - to, offer, o provide, @ protection
— v — para regulador m - to offer @ control protection; to protect @ control
— v resistencia f - to offer, o provide, @ resistance
— v servicio m - to offer, o provide, @ service
— v sistema m - to offer @ system
— v ventaja(s) f - to offer @ advantage(s)
ofrecido,da a - offered; proposed • volunteered
ofrecimiento m - . . .; proposal,sing • volunteering
— m aislamiento m térmico - thermal insulation offer(ing)
— m — equipo m - [ind] equipment offer(ing)
— m — precio m - price, offering, o bidding
— m — producto m - product offer(ing)
— m — — m nuevo - new product offer(ing)
— m — servicio m - service offer(ing)
— m — ventaja(s) f - advantage(s) offering
ofuscado,da a - clouded; confused
ofuscamiento m - . . .; clouding; confusing
— m de responsabilidad f - responsibility, o accountability, clouding, o confusing
ofuscar v - . . .; to cloud
— v responsabilidad f - to, cloud, o confuse, @, responsability, o accountability
ohmímetro m conectado - [instrum] connected ohmmeter
— m desconectado - [electr-instal] disconnected ohmmeter
— m verificado - [instrum] checked ohmmeter
Ohmstone m - [electr-equip] Ohmstone
oído,da a - heard; listened (to)
Oilite m - Oilite
oir v sonido m - to hear @ sound
ojal m - . . . . • [mec] grommet; eyelet • [cabl] eye(let); loop
— m de cable m - [cabl] cable loop
— m — — m de alambre m - [cabl] wire rope eye
— m — caucho m - [mec] rubber grommet

ojal m de metal m - [cabl] (metal) grommet
— m —tornillo m - [mec] screw eye(let)
— m en tablilla f para circuito m - [electrón] circuit card eyelet
— m para cable m - [cabl] cable, o (wire) rope, grommet, o thimble
— m — elevación f - [mec] lifting eye
— m — — f para tapa f - [mec] cover lifting eye
ojalillo m - [mec] small eye
— m con diente(s) m - [mec] griplet
— m — — m para soldadura f - [electrón] griplet
— m — — m — — f posterior - [electrón] (rear) griplet
— m — — m para taladro m - [mec] drill grip-let
— m dentado - [mec] griplet
— m — para soldadura f - [electrón] griplet
— m — — f posterior - [electrón] (rear) griplet
ojeada f - . . .; superficiel look
— f ligera f - quick, o passing, glance
ojillo m - [mec] váse cáncamo m; perno m con ojo m
ojiva f - [arquit-arco] . . .; lanced arch
— f apuntada - [arquit-arco] pointed lancet
— f de lanceta f—[arquit-arco] pointed lancet
— f en punta f - [arquit-arco] pointed lancet
— f puntiaguda - [arquit-arco] pointed lancet
ojival a - . . .; lancet
— a apuntado,da - [arquit-arco] pointed lancet • ogee
— a en punta f - [archit-arco] pointed lancet
ojo m - . . . • [cabl] eye • [mec] eye(let) (hole) • [cadenas] eye; loop • [imprent] (type) face
— m abierto - open eye • opened eye
— m cerrado - [mec] closed eye
— m de buey - . . . • [náut] porthole
— m — impulsor m - [mec] driver eye
— m — pescado m - [sold] fish eye
— m — pez m - [sold] fish eye
— m desnudo - naked, o unaided, eye
— m eléctrico - [electrón] electric,ronic eye
— m elevador - [mec] lifting eye
— m en chaveta f - [mec] keyhole
— m en impulsor m - [mec] driver eye
— m giratorio - [mec] swivel eye
— m especial - [mec] special eye
— m fijo - [mec] fixed eye
— m lavado - [segurid] washed eye
— m nunca parpadeante - unblinking eye
— m para llave f - [mec] keyhole
— m — — f obturado - [sold] closed keyhole
— m parpadeante - blinking eye
— m protegido - [segurid] protected eye
— m roscado - [mec] threaded eye(let)
— m rotatorio - [mec] swivel eye
— m único - [cadenas] single, eye, o loop
ojuelo m - . . . • [mec] véase cáncamo m; perno m con ojo m
ola f de construcción(es) - [constr] construc-tion wave
— f — — f de horno(s) m alto(s) - [metal--prod] blast furnace(s) construction wave
— f — presión f - pressure wave
oleada f - . . .; burst • [fig] tempo
— f con presión f - pressure wave
— f — velocidad f - burst of speed
oleo m - [arte] oil painting; oil
oleoducto m troncal - [petról] trunk pipeline
oleoresina f - [metal-prod] runner clay
oligisto m - [miner] . . .; hematite • [metal--prod] iron glance
olor m desagradable - disagreeable, smell, o oder • noxious odor
— m pestilente - malodor; bad odor
olvidado,da a - . . . • forgotten; ignored
olvidar v - . . . • to ignore
olla f - [domést] . . . • [metal-prod] véase cuchara f
— f automática (crack pot)—[domést] crack pot

ollao m - [náut] . . .; grommet
ominoso,sa a - . . .; menacing
omisible* a - negligible; omittible
omisión f - . . .: omitting; deleting,tion • • dispensing with; neglect(ing) • passing failing,lure
— f anterior - prior omission
— f — de componente m - [electrón] component, omitting, o omission, o deleting,tion
— f — mano f - [pint] coat, omission, o omit-ting
— f — — f final - [pint] finish(ing), o fi-nal, coat, omission, o omitting
— f — — f — de pintura f - [pint] finish--(ing) paint coat, omission, o omitting
— f — interruptor m - [electr-instal] breaker omission
— f — perforación f - [mec] perforation, omis-sion, o omitting
— f — reóstato - [electr-instal] rheostat, omission, o omitting
— f en cumplimiento m - failure to, comply, o perform; noncompliance,plying
omitido,da a - omitted; deleted; neglected; dis-pensed with • passed
omitir v - . . .; to delete • to, neglect, o dispense with • to pass
— v componente m - [electrón] to, omit, o de-lete, @ component
— v etapa f - [deport] to omit @ lap
— v mano f - [pint] to omit @ coat
— v — f final - [pin] to omit @ finishing coat
— v — f — de pintura f - [pint] to omit @ finish(ing) paint coat
— v perforación(es) f - [mec] to omit @ perfo-ration(s)
— v reóstato - [electr-instal] to omit @ rheo-stat
— v velocidad f - [autom-mec] to skip @, gear, o shift
— v vuelta f - [deport] to skip @ lap
omnibus m de lujo - [autom] (de luxe) motorcoach
— m eléctrico - [transp] electric bus; trolley-bus
— m escolar - [transp] school bus
— m para transporte m de personal m - [transp] personnel (transportation) bus
once m - . . . • [culin] [Chi.] lunch; tea
onceavo,va a - eleventh
onda f - . . . • [electr] impulse • surge
— f casi sinusoidal - [electrón] near(ly) sine wave
— f de corriente f - [electr-oper] current wave
— f — — f armónica - [electr-oper] harmonic current wave
— f — — f — para alimentación f - [electr--oper] harmonic current feeding wave
— f — presión f - pressure wave
— f digital - [electrón] digital wave
— f hertziana - [electrón] Hertzian wave
— f llena f - [electrón] full wave
— f más corta f—[electrón] shorter wave
— f — — de rayo(s) m gamma f - [electrón] gamma ray(s) short(er,est) wave length
— f — larga f - [electrón] longer wave
— f — — f de rayo(s) gamma - [electrón] gamma ray(s) short(er,est) wave length
— f media - [electrón] medium, o middle, wave • véase media onda
— f plena - [electrón] full wave
— f portadora - [electrón] carrier wave
— f — mantenida - [electrón] maintained car-rier wave
— f presente - present wave
— f sinusoidal - [electrón] sine wave
— f — en salida f - [electrón] output sine, o sine output, wave
— f — presente - [electrón] present sine wave
— f vibratoria - [electrón] vibrating,tional wave
ondulación f - . . .; wave; waving; curl(ing) • corrugation • meandering • wave shape • [sold] ripple

ondulación f definida ~ [sold] distinct ripple
— f en fondo - [constr] invert corrugation
— f helicoidal - [hidr] helical corrugation;
Hel-Cor
— f igual - [sold] even, o smooth, ripple
— f leve - slight, ripple, o waviness
— f marcada - [sold] distinct, o deep, ripple
— f muy marcada - [sold] deep ripple(s)
— f pareja - [sold] even ripple
— f pronunciada - [metal-lam] pronounced waviness
ondulado,da a - . . .; wave shaped; véase también corrugado,da - [topogr] hilly; rolling
— a helicoidalmente - helically corrugated;
Hel-Cor
ondulante a - . . .; wavy • meandering
ondular v - . . . • to meander
ondulatorio,ria a - . . . • wavy,ving
oneroso,sa a - . . . . • consuming
— a en tiempo m - time consuming
opacidad f - . . .; opaqueness
opción f - . . .; choice; alternative; alter-
nate choice | adv - (a) at @ option
— f comprendida - understood option • included
option
— f de accionamiento m - [mec] power option
— f — m con motor m - [electr-mot] (electric) motor option • [comb.int] engine power
option
— f — m — m diesel - [mec] diesel (engine) power option
— m — m — m eléctrico - electric motor
(driving) option
— f — m eléctrico - [mec] electric power
option
— f — acoplador m - [comput] coupler option
— f — m para voz f - [comput] voice coupler option
— f — alarma f - [electrón] alarm option
— f — f con repetición f - [electrón] re-
peat(ing) alarm option
— f — f repetidora - [electrón] repeating
alarm option
— f — capacidad f para idioma m . . . -
(comput] . . . language capability option
— f — f — m español - [comput] Spanish
language capability option
— f — f — m inglés - [comput] English
language capability option
— f — circuito m - [electrón] circuit option
— f — m para amplificador m - [electrón]
amplifier circuit option
— f — m — m para medidor m—[electrón]
meter amplifier circuit option
— f de corriente f de entrada f - [comput] input current option
— f — decodificador m - [comput] decoder option
— f — m integral - [comput] integral decoder option
— f — dispositivo m para grúa f - crane, attachment, o device, option, o alternative
— f — duración sincronizada de tono m -
[comput] timed tone duration option
— f — embarque m - [transp] shipping option
— f — parcial(es) - [transp] partial ship-
ment(s) opotion
— f — extensión f hidráulica - [grúas] hydraulic extension option
— f — medida(s) - measurement(s), o size(s),
option
— f — f para reemplazo m - [autom-neumát]
replacement size(s) option
— f — motor(es) m - [comb.int] engine, option, o choice
— f — oscilador m - [electrón] oscillator option
— f — m externo - [electrón] external oscillator option
— f — prolongación f hidráulica - [grúas] hydraulic extension option
— f — prórroga - extension option
— f — f en período m para ejecución f

performance period extension option
opción f de retorno m audible - [electrón] audible feedback option
— f — sección f - section option
— f — f de decodificador m - [comput] decoder section option
— f — teclado m - [electrón] (key)board option
— f económica - economic, option, o choice
— f ejercida - excercised option
— f explorada - explored option
— f nueva - new, option, o choice
— f para compra f - [com] purchase,sing option
— f — f de edificio m - building purchase
option
— f — f — terreno m - land purchase option
— f — construcción f—[constr] building option
— f — dispositivo m - [mec] device, o attachment, option
— f — m para aguilón m - [grúas] boom attachment alternative
— f — m — m para grúa f - [grúas]
crane boom attachment, option, o alternative
— f — idioma n - [comput] language option
— f — operación f - operation option
— f — f con teclado m - [electrón] (key)-
board control option
— f — oruga f - [constr-equip] crawler, option, o alternative
— f — pescante m - [grúas] jib option
— f — prórroga f - [fin] extension option
— f — proyección f - design, option, o alternative
— f — f para reemplazo m - [constr] replacement design, option, o alternative
— f — m — m para puente m - [constr]
bridge replacement design, option, o alternative
— f — puente m - [constr] bridge alternative
— f — reducción f - reduction option
— f — f de corriente f - [sold] current
reduction option
— f — f — f de entrada f - [sold] input
current reduction option
— f — reemplazo m - replacement, option, o alternative
— f — m para puente m - [constr] bridge
replacement, option, o alternative
— f — regulación f - control option
— f — f con teclado n - [electrón] (key)-
board control option
— f — f de nivel m - [electrón] level control option
— f — f — m en tablero m - [electrón]
board level control option
— f para renovación f - [fin] renewal option
— f — sentido m - [electrón] sense option
— f — m en tablero m - [electrón] onboard
sense option
— f — trabajo m - job, o work, option
— f precisada - selected option
— f preferida - preferred, o number one, option
— f respectiva - respective option
— f retráctil - [grúas] telescoping option
— f señalada - outstanding option • indicated
option
— f sobre terreno m - land option
opcional* a - véase optativo,va
operabilidad f - operability
operación f - . . .; operating; running; functioning • transaction • project; job; process
• work; action • [electr-oper] switch(ing)
— f aceptable - acceptable, operation, o performance
— f adaptable - flexible operation
— f activada - [electrón] initiated operation
— f — por operador m - [electrón] dispatcher
initiated operation
— f actual - [ind] existing operation
— f amparada - covered operation
— f ampliada - [ind] expanded operation
— f apropiada - appropriate, o proper, operating,tion • proper running
— f — con amperaje(s) m bajo(s) - [sold] good

low current option
operación f apropiada de marcha f sin carga f -
[comb.int] proper idler operation
— f — — planta f - [ind] plant appropriate,
o appropriate plant, operation
— f — sin carga f - [comb.int] proper idler, o
idler proper, operation
— f atractiva - attractive operation
— f automática - automatic operation | a -
(con) automatically operated
— f baja condición(es) f anormal(es) - [ind]
opoerating,tion under abnormal condition(s)
— f — — f normal(es) - operating,tion under
normal condition(s)
— f bajo techo m - [ind] under roof, o indoor,
operation
— f básica - [ind] basic operation
— f bloqueada - blocked operation
— f breve - brief, o short, operation
— f cíclica - [ind] cycling
— f comercial - commercial operation • commercial start-up
— f como soldadora f - [sold] operation as @
welder
— f compleja - complex operation
— f completa - complete operating,tion
— f completada - completed operation
— f común - common operating,tion
— f con agua m - water, o hydraulic, operation
— f — — m con temperatura f (elevada) -
water (high) temperature operating,tion
— f — amperaje(s) m bajo(s) - [sold] low current operating,tion
— f — — m normal(es) - [sold] normal, amperage, o current, operating,tion
— f — balde(s) m—[ind] bucket operating,tion
— f — bomba f - [bombas] pump operating,tion
— f — — f manual - [mec] hand, o manual,
pump operating,tion
— f — cangilón(es) m - [ind] bucket operating,tion
— f — carga f - [ind] load operating,tion
— f — — f máxima - [mec] maximum load operating,tion
— f — — f mínima - [mec] minimum load operating,tion
— f — . . . cilindro(s) m - [comb.int] . . .
cylinder operation; running on . . . cyl-
inder(s)
— f — cilindro m neumático - [neumát] air
cylinder operating,tion | a - [neumát] (con)
air cylinder operated
— f — coque m - [coque] coke operating,tion
— f — corriente f - [sold] current operation
— f — — f alterna f - [sold] alternating
current operation
— f — — f — con . . . voltio(s) m - [sold]
. . . volt(s) alternating current operating,tion
— f — — f continua f - [sold] direct current operating,tion
— f — — f — con . . . voltio(s) m - [sold]
. . . volt(s) direct current, operating,tion
— f con cost m reducido - low cost operation
— f — crédito m - [fin] credit operation
— f — chatarra f—[metal-prod] scrap operation
— f — — f sola(mente) - [metal-prod] (all)
scrap (only) operation
— f — estado m sólido - [electrón] solid state
operating,tion
— f — factor m de utilización f adecuado (a
capacidad f) - [sold] operation at @ duty
cycle consistent with @ rating
— f — — m — — f normal - [sold] normal duty
cycle operation
— f — fuente f de energía f - [sold] power
source operating,tion
— f — — f — — f para corriente f alterna -
[sold] alternating current power source operating,tion
— f — — f — — f — — f continua - [sold]
direct current power source operating,tion

operación f con lanza f - [metal-prod] lance operating,tion
— f — — f para oxígeno - [metal-prod] oxygen
lance operating,tion
— f — marcha f lenta - [mec] (s)low speed operating,tion
— f — — f plena - [mec] full speed operating,tion • [comb.int] running at @ full
throttle
— f — oxígeno m - [metal-prod] oxygen operating,tion
— f — . . . período(s) m - [electr-oper] . . .
cycle operating,tion
— f — pie m - [mec] foot operating,tion
— f — m derecho - [mec] right foot operating,tion
— f — m izquierdo - [mec] left foot operating,tion
— f — presión f - pressure operating,tion
— f — — f alta - high pressure operating,tion
— f — — f baja - low pressure operating,tion
— f — — f reducida - reduced, o low, pressure
operating,tion • [metal-prod] fanning
— f — reculada - [mec] recoil operating,tion
— f — — f máxima - [mec] maximum, o high, recoil operating,tion
— f — — f mínima - [mec] minimum, o low, recoil operating,tion
— f — — f reducida - [mec] reduced, o low,
recoil operating,tion
— f — retorno - [comput] loopback operating,tion
— f — solenoide m - [electr-oper] solenoid operating,tion
— f — — m para corriente f - [electr-oper]
current solenoid operating,tion
— f — — m — f alterna - [electr-oper] alternating current solenoid operating,tion
— f — — m — — f continua - [electr-oper]
direct current solenoid operating,tion
— f — teclado m - [electrón] (key)board operating,tion
— f — temperatura f - [ind] temperature operating,tion
— f — — f elevada - [ind] high temperature
operating,tion
— f — — f — para agua m - [ind] high water
temperature operating,tion
— f — terminación f - [comput] terminating,tion operating,tion
— f — — f silenciosa - [comput] quiet terminating,tion operating,tion
— f — — f sin ruido(s) m - [comput] quiet, o
noiseless, terminating,tion operating,tion
— f — tiempo m caluroso m - hot weather operating,tion
— f — m frío m - cold weather operating,tion
— f — todos cilindros m - [comb.int] all cylinder operating,tion • running, with, o on,
all cylinders
— f — estado m sólido - [electrón] solid state
operating,tion
— f — transistor(es) m - [electrón] solid
state operating,tion
— f — tres pasos m - three-step operating,tion
— f — un turno m - [ind] one turn operating,tion
— f — vagón m convencional - [coque] multi-
-spot operating,tion
— f — — m detenido m - [coque] single spot
operating,tion
— f — — m en movimiento m - [coque] multispot
operating,tion
— f — — m single-spot - [coque] single-spot
operating,tion
— f — velocidad(es) f - [mec] . . .-
-speed operating,tion
— f — velocidad f alta - [mec] high, o fast, o
high range, speed operating,tion
— f — — f baja - [mec] (s)low, speed, o
range, operating,tion
— f — — f fija - [mec] fixed speed opera-

ting,tion
operación f con velocidad f lenta—[mec] (s)low
speed operating,tion
— f — — f media(na) - [mec] medium speed operating,tion
— f — — f variable - [mec] variable speed
operating,tion
— f — — m constante - [electr-oper] constant
speed operating,tion
— f — voltaje m constante - [electr-oper]
constant voltage operating,tion
— f — — m variable - [electr-oper] variable
voltage operating,tion
— f . . . voltio(s) m - [electr-oper] . . .
volt operating,tion
— f — — — en corriente f alterna -
[electr-oper] . . . volt(s) alternating current operating,tion
— f — . . . — — — f continua - [electr-
-oper] . . . volt(s) direct current operating,tion
— f confiable - reliable, o trustworthy, o dependable, operating,tion, o performance
— f — con costo m, bajo, o reducido - [ind]
reliable low cost operating,tion
— f — de motor m - [electr-mot] dependable
motor operating,tion • [comb.int] engine dependable operating,tion
— f — económica - [ind] dependable low cost
operating,tion
— f constante - constant, o continuous, operating,tion
— f continua - continuous operation • all-day
operating,tion
— f — de secador m - [ind] continuous drier
operating,tion
— f continuada - [ind] continuous operating,tion • continuity • sustained use
— f correcta - [ind] correct, o proper, operating,tion
— f corriente - [mec] current, o routine, operating,tion
— f — de motor m - [electr-mot] current, o
routine, motor operating,tion • [comb.int]
current, o routine, engine operating,tion
— f costa f adentro - [petról] on shore operating,tion
— f — f afuera - [petról] off-shore operating,tion
— f cotidiana - day-to-day operating,tion
— f crítica - critical operating,tion
— f — para fabricación f - [ind] critical
fabricating operating,tion
— f cronometrada - [electrón] timed operating,tion
— f cubierta - covered operation
— f cuidadosa - careful, o thoughtful, operating,tion
— f de acería f - [metal-prod] steel, mill, o
shop, operating,tion • steel mill practice
— f — acondicionador m para aire m—[ambient]
air conditioner operating,tion
— f — aguilón m - [grúas] boom, o crane, operating,tion; booming
— f — — m para grúa f - [grúas] crane boom
operating,tion
— f — alimentadora f - [avicult] feeder, running, o operating,tion
— f — — f para alambre m - [sold] wire feeder operating,tion
— f — alternador m - [electr-oper] alternator
operating,tion
— f — — m mantenida - [electr-oper] maintained alternator operating,tion
— f — Angledozer m - [constr] Angledozer operating,tion
— f — automóvil - [autom] automobile, o car,
operating,tion
— f — bocina f - [autom] horn operating,tion
— f — — f verificada - [segurid] checked
horn('s) operating,tion
— f — bomba f - [bombas] pump operating,tion
— f — — f manual - [bombas] hand, o manual,

pump operating,tion
operación f de bomba f para agua m - [bombas]
water pump operating,tion
— f — — f — — m para alimentación f -
[bombas] feed water pump operating,tion
— f — — f — combustible m - [comb.int] fuel
pump operating,tion
— f — — f — elevación f - [bombas] lift(ing)
pump operating,tion
— f — — f — — f de combustible m - [comb.-
int] fuel lift(ing) pump operating,tion
— f — — f — regulación f - [bombas] control
pump operation
— f — cabeza f para inyección f - [petról]
swivel operating,tion
— f — calentador m - heater operating,tion
— f — cambiador m - [mec] changer operating,tion
— f — — m para tope(s) m - [electr-oper]
tap changer operating,tion
— f — carburador m - [comb.int] carburetor operating,tion
— f — cilindro m - [mec] cylinder operating,tion
— f — — m con acción f doble - [mec] double
acting cylinder operating,tion
— f — circuito m - [comput] circuit operating,tion
— f — colocación f - placement operating.tion
— f — combustión f - [combust] combustion operating,tion
— f — compañía f - company operating,tion
— f — compresor m [mec] compressor operating,tion
— f — conducto m - [tub] conduit, o line, operating,tion
— f — — m para desecho(s) m - [nucl] waste,
conduit, o line, operating,tion
— f — — m — — m con radiactividad f—[nucl]
radioactive waste line operating,tion
— f — — m — — m — — f intermedia - [nucl]
intermediate level waste line operating,tion
— f — — m — — m — — f mediana - [nucl]
intermediate level waste line operating,tion
— f — conmutador m - [electr-oper] switch toggling
— f — convertidor m - [metal-prod] converter
operating,tion
— f — coquería f = [coque] coke plant operating,tion
— f — corona f - [mec] crown operating,tion
— f — — f perforadora - [petról] coring tool
operating,tion
— f — departamento m - department(al) operating,tion
— f — — m para rodillo(s) m - [metal-lam]
roll department operating,tion
— f — dispositivo m - [mec] device operating,tion
— f — — m para cambio(s) m - [autom-mec]
shift unit operating,tion
— f — elevador m—[grúas] hoist operating,tion
— f — empresa f - [admin] commpany, o concern,
o corporation, operating,tion
— f — equipo m—[ind] equipment operating,tion
— f — — m agrícola - [agric-equip] agricultural, o farm, equipment operating,tion
— f — espectrómetro m - [instrum] spectromoter
operating,tion
— f — estrangulador m - [mec] choke operating,tion • operated choke
— f — filtro m - [mec] filter operating,tion
— f — — m con tela(s) f - [coque] baghouse
operating,tion
— f — — m seco m - [coque] dry baghouse operating,tion
— f — — — con tela(s) f - [coque] dry baghouse (filter) operating,tion
— f — freno m - [mec] brake operating,tion
— f — función f - [comput] function operating,tion
— f — — f crítica - [comput] critical function operating,tion

operación f de grúa f - [grúas] crane, operating,tion, o function, o action
— f —— f sin carga f - [grúas] crane operation without @ load
— f — horno n - [ind] furnace operating,tion
— f —— m eléctrico - [metal-prod] electric
furnace operating,tion
— f — para recocido m - [metal-trat] annealing furnace operating,tion
— f — implemento m - [agric-equip] implement
operating,tion
— f — interruptor m - [electr-oper] breaker
operating,tion
— f —— m bloqueado - [electr-oper] blocked
breaker operating,tion
— f — laminador m - [metal-prod] mill operating,tion
— f — lavador m por vía f húmeda - [ind] wet
scrubber operating,tion
— f — línea f - [ind] line operating,tion
— f — manija f - [mec] handle operating,tion
— f —— f para bomba f - [mec] pump handle
operating,tion
— f — manipulador m - [electrón] keyer operating,tion
— f — m para tono(s) m - [electrón] tone
keyer operating,tion
— f — máquina f - [ind] machine, operating,tion, o running
— f — maquinaria f - [ind] machinery operating,tion
— f — marcha f lenta - [comb.int] idler operating,tion
— f — mecanismo m - [mec] mechanism operating,tion
— f — motor m - [electr-mot] motor operating,tion • [comb.int] engine operating,tion
— f —— m con marcha f lenta - [electr-mot]
motor, (s)low speed, operating,tion, o i-
dling • [comb.int] engine, (s)low speed o-
perating,tion, o idling
— f —— m —— f plena - [electr-mot] me-
tor full speed operating,tion • [comb.int]
engine full speed operating,tion
— f —— m con combustión f interna - [comb.-
int] internal combustion engine operating,tion
— f —— m hidráulico - [agric-equip] hydraulic motor, operting,tion, o running
— f — mototopadora - [constr-equip] bulldozer
operating,tion
— f — oficina f - office operating,tion
— f — palanca f - [mec] lever operating,tion
— f — planta f - [ind] plant operating,tion
— f —— f para ácido m - [ind] acid plant
operating,tion
— f —— f —— m sulfúrico - [coque] sulfuric acid plant operating,tion
— f —— f hidroeléctrica - [electr-prod]
hydroelectric plant operating,tion
— f — pozo m - [petról] well operating,tion
— f — prensa f - [mec] press operating,tion
— f — producto m - product operating,tion
— f — puesta f en marcha - [ind] start-up
operating,tion
— f — pulverización f - spraying operation
— f — quemador m - [combust] burner operating,tion
— f — red f - [ind] system operating,tion
— f —— f eléctrica - [electr-distrib] electric system operating,tion
— f —— f para aire m comprimido—[ind] compressed air system operating,tion
— f — regulador m - [mec] control operating,tion
— f —— m para marcha f - [ind] control
operating,tion
— f —— m —— f sin carga f - [comb.int]
idler operating,tion
— f — relé m - [electr-oper] relay, operating,tion • relay function(ing)
— f — rueda f - [mec] wheel operating,tion
— f — secador m [ind] drier operating,tion

operación f de sección f - [labor] unit, o section, operating,tion
— f —— f para laminación f - [metal-lam]
rolling section operating,tion
— f — sede f - [admin] home office operating,tion(s)
— f — seguro m - [mec] lock operating,tion
— f —— m contra rotación f - [grúas] rotation lock operating,tion
— f — selector m - [mec] selector, operating,tion, o switch(ing)
— f — m para estrangulador m - [mec] choke
selector, operating,tion, o switch(ing)
— f — sistema n - [ind] system operating,tion
• system driving
— f — sitio m - site operatng,tion
— f —— m para desecho(s) m - [ind] waste
site operating,tion
— f —— —— m radiactivo(s) - [nucl] radioactive waste site operating,tion
— f —— m para residuo(s) m - [ind] waste
site operating,tion
— f —— m —— m radiactivo(s) - [nucl] radioactive waste site operating,tion
— f — sociedad f - [admin] corporate operating,tion
— f — soldadora f - [sold] welder operating,tion
— f —— f con motor m con combustión f interna - [sold] engine welder operating,tion
— f — soldadura f - [sold] welding operating,tion
— f —— f en interior m - [sold] inside (diameter) weld(ing) operating,tion
— f —— f interior - [sold] inside (diameter)
weld(ing) operating,tion
— f — taller m - [ind] shop operating,tion
— f — tapa f - [mec] cover operating,tion
— f —— f para válvula f - [válv] valve cover
operating,tion
— f — tipo m - type operating,tion
— f —— m continuo - continuous (type) operating,tion
— f — topadora f - [constr-equip] bulldozer
operating,tion
— f — turboalimentador m - [comb.int] turbocharger operating,tion
— f — unidad f motriz - [electr-prod] power
unit operating,tion
— f — usina f - [electr-prod] power plant operating,tion
— f —— f industrial - [electr-prod] power
plant operating,tion (programming) • [comput]
power plant operating,tion programming
— f —— f — en planta f—[electr-prod] plant
industrial power plant operating,tion
— f —— f —— f siderúrgica - [electr-
-prod] steel plant industrial power plant operating,tion
— f — válvula f - [mec] valve switch(ing)
— f —— f para estrangulador m - [mec] choke
valve switching
— f — selectora - [mec] selector valve,
switch(ing), o operating,tion
— f — vehículo m - [transp] vehicle operating,tion
— f — ventilador m - [mec] fan operating,tion
— f —— m en frío - [mec] fan cold operating,tion
— f —— m para compresión f - [mec] pressurizer operating,tion
— f — voltímetro m - [electr-oper] voltmeter
operating,tion
— f — yacimiento m (petrolífero) - [petról]
(oil) field operating,tion
— f debida - [ind] proper, o correct, o smooth,
operating,tion
— f — de regulador m - [mec] control(ler), proper, o correct, operating,tion
— f — —— m para marcha f sin carga f—[comb.
-int] proper, o correct, o appropriate, idler
operating,tion
— f — verificada - checked proper operation

operación f defectuosa - defective, o faulty, operating,tion
— f desde, cabina f, o casilla f - [grúas] cab operating,tion
— f diaria - daily operating,tion • day-by-day operating,tion
— f difícil - hard, o difficult, operating,tion
— f distinta - distinct, o different, opera-ting,tion
— f diversa - different operation • pl multiple operations
— f durante invierno m - winter operating,tion
— f — verano m - summer operating,tion
— f económica - economical, o low cost, opera-ting,tion
— f — de grúa f - [grúas] low cost crane ope-rating,tion
— f efectiva - effective operating,tion
— f eficiente—[ind] efficient, operating,tion, o work(ing)
— f — de equipo m - [ind] efficient equipment operating,tion
— f — — grúa f - [grúas] efficient crane op-erating,tion
— f — desde punto m de vista f de costo m - [ind] cost efficient, operating,tion, o per-formance
— f ejecutada - performed operating,tion
— f eléctrica - [electr-oper] electric(al) op-erating,tion | a - (con) electrically opera-: ted
— f elemental - elementary, o basic, o simple, operating,tion
— f en campo m - [agric] field operating,tion
— f — carretera f - [vial] road operating,tion
— f — época(s) f calurosa(s) - hot, o warm, weather, operating,tion
— f — escala f comercial - [com] commercial scale operating,tion
— f — Estados m Unidos m - [com] United States operating,tion
— f — forma f continua(da) - continuous oper-ating,tion • day-in (and) day-out performance
— f — — f pareja - [mec] even, running, o op-erating,tion
— f — — f suave - smooth, operating,tion, o running
— f — frío m - [mec] cold operating,tion
— f — — de ventilador m - [mec] fan cold op-erating,tion
— f — línea f - [ind] line operating,tion
— f — país - in @ country operating,tion
— f — paralelo - [ind] parallel operating,tion
— f — seco - [mec] dry running
— f — sitio m - site operating,tion
— f — vacío - in @ vacuum operating,tion
— f errática - erratic operating,tion
— f específica - [ind] specific operating,tion
— f estable - stable operating,tion
— f estándar - [ind] standard operating,tion
— f excelente - excellent operating,tion
— f explicada - [ind] explained operating,tion
— f fabril - [ind] manufacturing operating,tion
— f — compleja - [ind] complex manufacturing operating,tion
— f fácil - easy, o simple, operating,tion • fast handling
— f flexible - flexible operating,tion
— f fría - [sold] cool operating,tion
— f funcionada - performed, function(ing), o operating,tion
— f fundamental - basic operating,tion
— f futura - future operating,tion
— f general - general operating,tion • overall operating,tion
— f — de departamento m - [ind] overall de-partment(al) operating,tion
— f — para conservación f - [mec] general maintenance operating,tion
— f — —, entretenimiento m, o conservación f - [ind] general maintenance operating,tion
— f — para mantenimiento m - [ind] general maintenance, operating,tion

operación f hermética - airtight operating,tion
— f hacia adelante - [mec] forward opera-ting,tion
— f — atrás - [mec] reverse operating,tion
— f hidráulica - [hidr] hydraulic opera-ting,tion
— f inapropiada - improper operating,tion
— f indebida - improper operating,tion
— f independiente - independent, opera-ting,tion, f function(ing)
— f industrial - industrial operating,tion
— f inferior - [mec] operating,tion from below | a - (con) bottom operated
— f iniciada f - initiated, o begun, o started, operating,tion
— f — por despachador m - [electrón] dispatch-er initiated operating,tion
— f inicial - [mec] break(ing)-in; start-up
— f ininterrumpida - [ind] continuous opera-ting,tion
— f intensa - [comb.int] hard running
— f interna - internal operating,tion
— f interrumpida - interrupted operating,tion
— f larga - long operating,tion
— f libre - free operating,tion
— f — de arena f - [hidr] sand-free opera-ting,tion
— f — — problema(s) m - trouble free, opera-ting,tion, o service
— f mantenida - [autom-neumát] maintained op-erating,tion • sustained use,sage
— f manual - manual, o hand, operating,tion | a - (con) hand, o manually, operated
— f — de aparejo m con cadena f - [mec] hand chain hoist operating,tion
— f — para elevación f con cadena f - [mec] hand chain hoist operating,tion
— f marina - [petról] offshore operating,tion
— f más difícil - harder operating,tion
— f — fácil - easier operating,tion • easiest operating,tion
— f — segura - [segurid] safer operating,tion • safest operating,tion
— f mayor para conservación f - [sold] major, o large, maintenance operation
— f — — entretenimiento m - [sold] major, o large, maintenance operation
— f — — mantenimiento m - [sold] major main-tenance operation
— f mecánica - mechanical operating,tion • [grúas] power function
— f — independiente - [mec] independent power function
— f — para limpieza f - [tub] mechanical cleaning operation
— f — — para tubería f - [tub] mechani-cal pipe cleaning operation
— f mecanizada - [grúas] power function
— f — independiente - [mec] independent power function
— f mejorada - improved operation
— f mezclada - mixed operating,tion
— f minera - [miner] mining operation
— f mixta - mixed operation
— f múltiple - multiple operation
— f neumática - [neumát] air operating,tion | a - (con) air operated
— f neutral - [electr-oper] neutral operation
— f — sin conexión f con tierra f - [electr--oper] ungrounded neutral operating,tion
— f nocturna - night(time) operation
— f normal - normal operating,tion
— f — de compañía f - [com] normal company operating,tion
— f — — empresa f - [com] normal, company, o corporation, o concern, operation
— f — — horno m - [combust] normal furnace operation
— f — — sociedad f - [com] normal corporation operating,tion
— f óptima - optimum, operating,tion, o per-formance
— f ordinaria - common operating,tion

operación f para acondicionamiento m - conditioning operation
— f — — m de cuchilla f - [mec] blade conditioning operation
— f — — — f para cizalla f—[metal-lam] shear blade conditioning operation
— f — — m de rodillo(s) m - [metal-lam] roll conditioning operation
— f — acoplamiento m—[mec] coupling operation
— f — alivio m - relief operation
— f — — m de tensión(es) f - [metal stress relief operation
— f — apilamiento m - [ind] piling operation
— f — armado m - assembling,bly operation
— f — arranque m - start-up operation
— f — aspiración f - [comb.int] intake, o suction, operation
— f — atención f - [ind] servicing (operation)
— f — — f de secador m - [ind] drier servicing (operation)
— f — bobinado m - [metal-lam] coiling operation
— f — caída f - [mec] fall(ing) operation
— f — — f libre - [grúas] free fall(ing) operation
— f — — f — de cable m - [grúas] line, o cable, free fall(ing) operation
— f — — f — — m para gancho m - [grúas] whip, line, o cable, free fall(ing) operation
— f — — f — — m — motón m - [grúas] main, line, o cable, free fall(ing) operation
— f — cálculo m - [mat] calculation, operation, o mechanics
— f — calentamiento m - heating operation
— f — canteado m—[metal-lam] edging operation
— f — carga f - [metal-prod] charging operation; filling • [transp] loading operation
— f — — f para carbón m - [miner] coal loading operation • [metal-prod] coal charging operation
— f — — f — — m en tren m (enterizo) - [miner] (unit) train coal loading operation
— f — — f en tren m - [f.c.] train loading operation
— f — colada f—[metal-prod] pouring operation
— f — — f continua f - [col.cont] continuous casting operation
— f — compactación f - [constr] compacting, o tamping, operation
— f — compresión f - [Comb.int] compression operation
— f — conformación f - [metal-fabr] forming operation
— f — — f en caliente - [metal-fabr] hot forming operation
— f — — f mediante rotación f - [metal-fabr] spinning operation
— f — conservación f - [ind] maintenance operation • service,cing operation
— f — — f de secador m - [ind] drier maintenance (operation)
— f — construcción f - [constr] construction operation
— f — coquización f - [coque] coking operation
— f — — f según norma f - [coque] standard coking operation
— f — corte m - [mec] cutting operation
— f — — m de borde(s) m - [metal-lam] edging operation
— f — decapado m - [metal-trat] pickling operation
— f — — m químico m - [metal-trat] chemical pickling operation
— f — departamento m para laminación f - [metal-lam] rolling department operation
— f — desarmado m - [mec] disassembly,bling operation
— f — desbaste m - [metal-lam] roughing operation
— f — descarga f—[transp] unloading operation
— f — desengrase m - degreasing operation
— f — desviación f - by-pass(ing) operation
— f — distribución f - [com] distribution op-

operación f para dragado m - [hidr] dredging operation
— f — elevación f - [grúas] hoisting operation
— f — eliminación f - elimination, o disposal, operation
— f — — f de, desecho(s) m, o residuo(s) m - waste(s) disposal operation
— f — embobinado* m - [metal-lam] coiling operation
— f — empalmado,me - [mec] splicing operation
— f — empernado m - [mec] bolting, operation, o procedure
— f — enfriamiento m - cooling operation
— f — ensayo m - trial, o test(ing), operation
— f — entretenimiento m - [ind] maintenance operation
— f — erección f - [constr] erecting,tion (operation)
— f — estirado,ramiento m - [trefil] wire-drawing operation
— f — excavación f - [constr] excavating (operation) • [miner] mining (operation)
— f — expandido*,nsión f - [metal-fabr] expanding,sion operation
— f — fabricación f - [ind] fabrication operation
— f — fundición f - [metal-prod] melt shop, o foundry, operation
— f — gunitado m - [metal-prod] guniting operation
— f — hincadura f - [constr-pil] driving operation
— f — — f de pilote(s) m - [constr-pil] pile driving operation
— f — horadación f - [constr] tunneling operation
— f — — f con inserción f con gato(s) m - [constr] jacked-in-place tunnel operation
— f — humectación f - [hormig] wetting operation
— f — inserción f (con gatos m) - [constr] jacking operation
— f — laminación f - [metal-lam] rolling operation
— f — — f en frío - [metal-lam] cold, rolling, o reduction, operation
— f — lavado m - washing operation
— f — limpieza f - cleaning operation
— f — — f para tubería f - [tub] pipe cleaning operation
— f — manipulación f - [electrón] keying operation
— f — mantenmiento m - maintenance operation
— f — perforación f - [petról] drilling operation
— f — picadura f - [sold] chipping operation
— f — — f para escoria f - [sold] slag chipping operation
— f — preparación f - preparation operation
— f — — f de chatarra f - [metal-prod] scrap preparation operation
— f — prerreducción f - [metal-prod] prereduction operation
— f — procesamiento m (de datos m) - [comput] (data) processing operation
— f — recalcadura f - [metal-fabr] forging operation
— f — recubrimiento m - coating operation
— f — — m exterior - external coating operation
— f — recuperación f - recovey,ring, o salvage,ging operation
— f — reducción f - [mec] reduction operation
— f — — f en caliente - [metal-lam] hot reduction operation
— f — — f — frío m - [metal-lam] cold reduction operation
— f — regulación f - [ind] control(ling) operation
— f — — f automática - [ind] automatic control(ling) operation
— f — relleno m - filling operation • [constr] backfilling operation

operación f para servicio m - service,cing operation
— f — terminación f—[ind] finishing operation
— f — torneado,dura - [metal-fabr] machining operation
— f — trefilería f - [alambre] wiredrawing operation
— f — vaciado m - [constr] pouring operation
— f — — m para lingote(s) m - [metal-prod] ingot pouring operation
— f pareja - [mec] smooth operation • even running
— f periódica - periodic(al) operation
— f permitida - permitted, o allowed, operation
— f piloto - [ind] pilot operation
— f plena - full operation
— f pobre - poor operation
— f polar - polar operation
— f primera - first operation
— f prolongada - extended operation
— f — de planta f - [ind] extended plant operation
— f — — — f industrial - [ind] industrial plant extended operation
— f química - [quím] chemical operation
— f — para limpieza f - [ind] chemical cleaning operation
— f — — — f de tubería f - [tub] chemical pipe cleaning operation
— f pulidora - [ind] polishing operation
— f rápida - fast, o quick, operation • fast handling
— f real - actual operation
— f — de rueda f - [mec-ruedas] actual wheel operation
— f regular - regular operation • smooth running
— f repetida—repeated, o repetitive, operation
— f — para vuelco m - [mec] repetitive dumping operation
— f restante - remaining, o other, operation
— f restringida - restricted operation—[metal-oper] fanning
— f ruda - rough, o rugged, operation
— f satisfactoria - satisfactory operation
— f según Asociación f para Industria f Electrónica - [electrón] Electronics Industry Association, o E I A, operation
— f — norma f - standard operating,tion
— f segura - safe, o reliable, operating,tion, o performance
— f — de grúa f - [grúas] crane safe, o safe crane, operation
— f explicada - [ind] explained safe operating,tion
— f semiintegrada - [ind] semi-integrated operating,tion
— f sencilla - simple operating,tion
— f separada - separate operating,tion
— f siguiente - next, o following, operation
— f silenciosa - quiet, o silent, operation
— f — confiable - quiet dependable operation
— f simultánea - simultaneous operation • multiple operation
— f — de soldadora f y producción f de, energía f, o fuerza f motriz - [sold] simultaneous welder and power operation
— f sin aire—[mec] operating without air • anaerobic operating,tion
— f — aumento m de temperatura f - [sold] cool operating,tion
— f — carga - no load, operating, o running
— f — conexión f con tierra f - [electr-oper] ungrounded operating,tion
— f — golpeteo(s) m - knockless, o knock-free, operation
— f — inconveniente(s) m - trouble-free operating,tion
— f — interrupción(es) f - interruption free operating,tion; operating,tion without @ interruption(s)
— f — preocupación(es) f—worry-free operation
— f — problema(s) m - trouble-, o problem-, free operating,tion

operación f sin problema(s) m para conservación f - maintenance-, o trouble-, free operation
— f sincronizada - [comput] synchronized,nous, o timed, operating,tion
— f sobre banda f - [culin] over @ band operating,tion
— f — término - [ind] term operating,tion
— f — — m corto—[ind] short term operation
— f — — m largo - [ind] long term operation
— f — vía f - track operating,tion
— f sobrecabeza f - [sold] overhead operating,tion
— f social,cietaria - [legal] company, o corporation, operating,tion
— f sólida - rugged operating,tion
— f sólo - operating,tion, alone, o only
— f suave - smooth, operating,tion, o running
— f sucesiva - sequential operation
— f superior - top, o overhead, operation
— f supervisada - supervised, o overseen, operation
— f suspendida - suspended operating,tion
— f técnica - [ind] technical operating,tion
— f — específica(da) - specific technical operating,tion
— f típica - typical operating,tion
— f — para inserción f (con gatos m) - typical jacking operation
— f total - total operating,tion
— f totalmente automática - fully automatic operating,tion
— f — con chatarra f - [metal-prod] all scrap operating,tion
— f ultramarina - overseas operating,tion
— f única - single operating • [mec] single stroke
— f uniforme—uniform, o smooth, operating,tion
— f — de horno m - [metal-prod] uniform furnace, o furnace uniform, operating,tion
— f unitaria - unit operating,tion
— f variable - variable operating,tion
— f veraniega - summer operating,tion
— f verificada - verified, o checked, o reviewed, operating,tion
— f vertical - [mec] vertical operating,tion
operacional a - operational; véase operativo,va
operado,da a - [mec] operated; run • [electr-oper] toggled; switched; moved
— a apropiadamente - operated, o run, properly, o appropriately
— a automáticamente - operated automatically
— a — con aire m - automatically air operated
— a bajo condición(es) f - operated under condition(s)
— a — — f anormal(es) - [ind] operated under abnormal condition(s)
— a — — f normal(es) - [ind] operated under normal condition(s)
— a con aire m—[ind] (compressed) air operated
— a — — m comprimido - [ind] (compressed) air operated
— a — amperaje m anormal - [sold] operated at @ abnormal, amperage, o current
— a — — m normal - [sold] operated at @ normal, amperage, o current
— a — bomba - [mec] pump operated
— a — — manual - [mec] hand pump operated
— a — cadena f - [mec] chain, operated, o driven
— a — carga f - operated with @ load
— a — — f máxima - [mec] maximum load operated
— a — — f mínima - [mec] minimum load operated
— a — . . . cilindro(s) m - [comb.int] run, on, o with, . . . cylinder(s)
— a — m neumático - [neumát] air, o pneumatic, cylinder operated
— a — corriente f alterna - [electr-oper] alternating current operated
— a — — f continua - [electr-oper] direct current opoerated
— a — engranaje(s) m - [mec] gear operated
— a — — m sin fin - [mec] worm gear operated

operado,da a con factor m de utilización f -
[sold] operated with @ duty cycle
— a —— m —— f anormal - [sold] operated
with @ abnormal duty cycle
— a —— m —— f normal - [sold] operated
with @ normal duty cycle
— a — marcha f lenta - operated at @ low speed
— a —— f plena - operated at @ full speed •
run at @ full throttle
— a — seguridad f - [ind] operated safely
— a con factor m de utilización f adecuado -
[sold] aperated at @ appropriate duty cycle
— a —— m —— f — a capacidad f - [sold]
operated at @ duty cycle consistent with @
rating
— a con llave f - [mec] wrench operated
— a — mano f - [mec] hand, operated, o controlled
— a — motor m - motor operated
— a — palanca f - [válv] lever operated
— a — pie m - [mec] foot, operated, o controlled
— a — m derecho - right foot operated
— a —— m izquierdo - left foot operated
— a — pulgar m - thumb operated
— a . . . cilindro(s) m - [comb.int] running
on . . . cylinder(s)
— a — tornillo m - [mec] screw operated
— a —— sin fin - [mec] worm gear operated
— a — velocidad f fija - [mec] fixed speed operated
— a —— f lenta - [mec] low speed operated
— a —— f plena - [mec] full speed operated
— a —— f variable - [mec] variable speed operated
— a continua(da)mente - operated continuously
— a debidamente - [mec] operated smoothly
— a económicamente—[mec] operated economically
— a eficientemente - operated efficiently
— a en base f a corriente f - [instrum] current operated
— f —— f — temperatura f - [instrum] temperature operated
— a en campo m - [agric] field operated
— a — forma f continua(da) - operated continuously
— a —— f pareja f - operated, o run, evenly
— a — forma f suave—operated, o run smoothly
— a — país m - operated in @ country
— a — seco - [mec] operated, o run, dry
— a — vacío - operated in @ vacuum
— a fácilmente - [mec] operated, o run, easily
— a inapropiadamente - operated improperly
— a independientemente - operated independently
— a intensamente - [comb.int] run hard
— a libremente - operated freely
— a manualmente - operated manually • hand held
— a neumáticamente - [neumát] air operated;
operated pneumatically
— a normalmente - operated normally
— a paralelamente - operated parallely
— a por, compañía f, o empresa f - [com] company operated
— a satisfactoriamente—operated satisfactorily
— a separadamente - operated separately
— a simultáneamente - operated, simultanously,
o at @ same time
— a sin aire m - [mec] operated without air
— a — carga f - operated, o run, empty, o idle
— a — interrupción(es) f - operated, without
interruption(s), o uninterruptedly
— a sobre banda f—[culin] operated over @ band
— a sólo - operated, only, o alone
operador m - operator; véase también conductor m
— m apilador - [ind] stacking operator
— m auxiliar - [ind] assistant operator • operator, assistant, o helper
— m para precintadora f - [ind] bander operator, assistant, o helper
— m —— quemador m - [combust] burner operator, assistant, o helper
— m —— zunchadora f - [ind] bander operator,
assistant, o helper

operador m ayudante para aplanadora f - [constr]
leveler operator, assistant, o helper
— m —— niveladora f - [constr] leveler operator, assistant, o helper
— m bien instruido - [labor] well-trained operator
— m competente - [labor] competent operator
— m con diafragma m - [válv] diaphragm operator
— m cortador - [ind] cutter,ting operator
— m costa f afuera - [petról] offshore operator
— m eficiente - efficient operator
— m en tierra f - [petról] land operator
— m individual - individual operator
— m laminador - [metal-lam] roller operator
— m nuevo - [ind] new operator
— m — usado - [ind] new operator used
— m — para molino m - [ind] new mill, o mill
new, operator
— m para aguilón m - [grúas] boom operator •
crane operator
— m — amasadora f - [metal-prod] pug mill operator
— m — apiladora f - [ind] piler, o stacker,
operator
— m — aplanadora f - [constr] leveler operator
— m — barra f (taponadora) - [metal-prod]
(stopper) rod operator; steel pourer
— m — báscula f - [ind] scale(s) operator;
weighmaster
— m —— f para, banda(s) f, o cinta(s) f, o
chapa(s) f, fina(s) - [metal-lam] thin strip
scales operator
— m —— f —, — f, o — f, o — f, gruesa(s) -
[metal-lam] thick, o heavy, strip scales operator
— m —— f para fleje m fino - [metal-lam]
thin strip scales operator
— m —— f —— m grueso - [metal-lam] thick
strip scales operator
— m —— f — línea f para corte m - [metal-
-lam] cutting line scales operator
— m —— f — tren m Steckel - [metal-lam]
Steckel mill scale(s) operator
— m —— f terminal - [ind] terminal scales
operator
— m — bobinadora f - [metal-lam] coiler operator
— m — bomba f para cascarilla f - [metal-lam]
scale pump operator
— m — brazo m - [grúas] boom operator
— m — caja f - [metal-lam] stand, operator, o
roller
— m — canteadora f - [metal-lam] edger operator
— f — carga f - [ind] load, o burden, operator
— m — cargadora f - [ind] loader operator
— m — cizalla f - [ind] shear operator
— m —— f para fleje(s) m - [metal-lam] band
shear operator
— m —— f en caliente - [metal-lam] hot shear
operator
— m —— f escuadradora - [mec] resquaring
shear operator
— m — cizalla f final - [metal-lam] final
shear operator
— m —— f para chatarra f - [metal-lam] scrap
shear operator
— m — clasificación f - [ind] grading operator
— m — combustión f - [combust] combustion operator
— m — compresora f - [ind] compressor operator
— m — convertidor(es) m - [metal-prod] converter operator
— m — cortadora f - [metal-lam] shear operator
— f —— f longitudinal - [metal-lam] slitter
(shear) operator
— m — corte m - [ind] cutting operator
— m — cuchara f - [metal-prod] ladle operator
— m — depósito m - [ind] tank operator
— m —— m para ácido(s) m - [metal-trat] acid
tank opoerator
— m — depurador m (para gases m) - [ind] gas
scrubber operator

operador m para descascarilladora f - [metal-
-lam] descaler operator
— m — desenrolladora f - [metal-trat] decoiler, o uncoiler, operator
— m — despuntadora f - [metal-lam] cropping
shear operator; end cut shearman
— m — empujadora f - [metal-lam] pusher operator
— f — — f y cargadora f - [metal-lam] pusher
and loader operator
— m — encabezadora f—[trefil] header operator
— m — enrolladora f - [metal-lam] coiler operator
— m — — f para chatarra f - [metal-lam] scrap
roller, o rolling, operator
— m — — f final - [metal-lam] final coiler
operator
— m — — f — para bobina(s) f - [metal-lam]
(coil) final coiler operator
— m — entrada f - [mec] intake operator
— m — equipo m - [ind] equipment operator
— m — — perforador - [petról] rig operator
— m — escarpadora f - [metal-lam] scarfer operator
— m — estación f - station operator
— m — — f de base - [electrón] base station
opoerator
— m — — f para fuel oil m - [ind] fuel oil
station operator
— m — estampadora f - [ind] stamper operator
— m — estufa f - [ind] stove operator
— m — filtro m - [ind] filter operator
— m — foso m - [ind] pit operator
— m — — m para aceite m - [ind] oil pit operator
— m — generador m - [electr-prod] generator
operator
— m — grúa f—[grúas] crane operator; craneman
— m — — f cargadora - [grúas] charging crane
operator • loading crane operator
— m — — f para almacén m - [ind] storage
crane operator
— m — — f — — m para sínter m - [metal-
-prod] sinter storage crane operator
— m — — f — carga f - [grúas] charging crane
operator • loading crane operator
— m — — f — colada f - [metal-prod] teeming
crane operator
— m — — f — mezcladora f - [grúas] mixer
crane operator
— m — — f — nave f—[ind] bay crane operator
— m — — f — — f para colada f—[metal-prod]
pouring bay crane operator
— m — — — f — — f — chatarra f—[metal-prod]
scrap bay crane operator
— m — — f — — f — servicio m - [ind] service bay crane operator
— m — — f — sínter m - [metal-prod] sinter
crane operator
— m — — f semipórtico - [grúas] semiportal
crane operator
— m — horno m - [combust] furnace, operator, o
tender
— m — — m alto - [metal-prod] blast furnace
operator
— m — — m basculante - [metal-prod] tilting
furnace operator
— m — — m eléctrico - [metal-prod] electric
furnace operator
— m — — m monopila - [metal-trat] single
stack furnace, operator, o tender
— m — — m — recocido m - [metal-trat] annealing furnace operator
— m — laminadora f - [metal-lam] mill operator
— m — — f para descascarillado m—[metal-lam]
descaling mill operator
— m — — f descascarilladora - [metal-lam] descaling mill operator
— m — — f terminadora - [metal-lam] speeder;
finishing mill operator
— m — lavador m para gas(es) m - [metal-prod]
gas scrubber operator
— m — luz,ces - [electr-oper] lights operator

operador m para manipulador m - manipulator operator
— m — máquina f - [mec] machine operator
— m — mesa f - [ind] table operator
— m — — f báscula - [ind] scale(s) table operator
— m — — f para inspección f - [ind] inspection table operator
— m — — f filtro - [ind] filter table operator
— m — mezcla - [ind] blending operator
— m — molino m - [miner] mill operator
— m — molino m nuevo - [ind] new mill operator
— m — motoestibadora f - [ind] fork lift truck
operator
— m — nave f para colada f - [metal-prod]
pouring, bay, o floor, operator
— m — niveladora f - [constr] leveler operator
— m — oxicorte m - [metal-lam] torch cutting
operator
— m — pértiga f - [grúas] boom operator
— m — pluma f - [grúas] boom operator
— m — precintadora f - [ind] bander operator
— m — prensa f - [ind] press operator
— m — proceso m - [ind] process operator
— m — pupitre m - [ind] pulpit operator
— m — quemador(es) m - [combust] burner(s) operator
— m — reclasificación f - [ind] reclassification,fying operator
— m — recogida f - gathering operator
— m — red f—[ind] network, o system, operator
— m — retardadora f - [f.c.] retarder operator
— m — salida f - [metal-lam] delivery operator
— m — sierra f - [mec] saw operator
— m — — f en caliente - [metal-lam] hot saw
operator
— m — — f — frío - [metal-lam] cold saw operator
— m — sistema m - [ind] system operator
— m — — m para descascarillado m—[metal-lam]
descaling system operator
— f — — m — — m hidráulico - [metal-lam]
hydraulic descaling system operator
— m — soldadora f - [sold] welder operator
— m — soplante m—[metal-prod] blower operator
— m — sótano m - [ind] pit operator
— m — — m para aceite m - [ind] oil pit operator
— m — tambor m - [mec] drum operator
— m — — m para peletización* f - [miner] pelletizing drum operator
— m — — m para mezcla f - [ind] mixing drum
operator
— m — — m mezclador - [ind] mixing drum operator
— m — teleproceso m - [ind] remote process operator
— m — terminal m - [ind] terminal operator
— m — — m para entrada f - [ind] intake terminal operator
— m — tijera f - [ind] shear operator
— m — — f para borde(s) m - [metal-lam] side
slitter, o edger, operator
— m — — f para fleje(s) m - [metal-lam] band
shear operator
— m — — f para forma(s) f - [metal-lam]
shape(s) shear operator
— m — — f despuntadora - [metal-lam] cropping
shear(s) operator
— m — torno m - [mec] lathe operator
— m — transferidor(ra) - [ind] transfer operator
— m — transportadora f - [ind] conveyor operator
— m — — f para cascarilla f - [metal-lam]
scale conveyor operator
— m — tren m - [metal-prod] mill operator •
[transp] train operator
— m — m para descascarillado m - [metal-
-lam] descaling mill operator
— n — trituración f - [miner] crusher operator
— m — vagoneta - [ind] (dump) car operator

operador m para vagoneta f para lingote(s) m

- 1022 -

operador m para vagoneta f para lingote(s) m -
[metal-prod] ingot car operator
— m — vástago(s) m - [metal-prod] stopper rod
operator
— m — vehículo m - [transp] vehicle operator
— m — volteadora f - [metal-lam] turner operator
— m — zunchadora f - [ind] bander operator
— m principal - main operator
— m reclasificador - [ind] reclassification, o
reclasifying, operator
— m sobre tierra f - [petról] land operator
— m soldador - [sold] welder,ding operator
— m sopletero m - [metal-lam] burner operator
— m — auxiliar - [metal-lam] assistant burner
operator
operante a - . . .; operative
operar v - . . . • to run; to perform; to function • to run • [electr-oper] to, switch, o
toggle, o (throw @) switch
— v acondicionador m (para aire m) - [ambient]
to operate @ (air) conditioner
— v aguilón m - [grúas] to (operate @) boom
— v — m para grúa f - [grúas] to operate @
crane boom
— v alimentadora f - [avicult] to, operate, o
run, @ feeder
— v alternador m - [electr-oper] to operate @
alternator
— v apropiadamente - to, operate, o run, properly, o appropriately
— v automáticamente
— v automóvil m - [autom] to, operate, o drive,
@, automobile, o car
— v bajo condición(es) f anormal(es) - [ind] to
operate under @ abnormal condition(s)
— v — f normal(es) - [ind] to operate under
@ normal condition(s)
— v bomba f - [bombas] to operate @ pump
— v — f para combustible m - [Comb.int] to operate @ fuel pump
— v — f — elevación f - [comb.int½ to operate
@ lift(ing) pump
— v — f — f para combustible m—[comb.int]
to operate @ fuel lift(ing) pump
— v calentador m - to operate @ heater
— v carburador m - [comb.int] to operate @ carburetor
— v circuito m - [comput] to operate @ circuit
— v como generador m para energía f - [sold] to
operate as @ power generator
— v — soldadora f - [sold] to operate as @
welder
— v con - [ind] to operate with • to handle
— v — aire m - [ind] to air operate
— v — — m comprimido - [ind] to compressed
air operate
— v — base f de pila(s) f - to operate with @
battery,ries | a - battery powered
— v — amperaje m anormal - [sold] to operate
with @ abnormal, amperage, o current
— v — — m normal - [sold] to operate with @
normal, amperage, o current
— f — bomba f manual - [mec] to hand pump operate
— v — capacidad f de hasta . . . - to operate
up to . . .
— v — carga f máxima - [mec] to operate with @
maximum load
— v — f mínima - [mec] to operate with @
minimum load
— v — . . . cilindro(s) m - [comb.int] to run,
on, o with, . . . cylinder(s)
— v — factor m de utilización f - [sold] to
operate at @ duty cycle
— v — — m — f adecuado a capacidad f -
[sold] to operate at @ duty cycle consistent
with @ rating
— v — — m — f anormal - [sold] to operate
with @ abnormal duty cycle
— v — — m — f normal - [sold] to operate
with @ normal duty cycle
— v — défecit m - [fin] to, be, o operate, in

@ red
operar v con marcha f lenta - to operate at @
low speed
— v — f plena - to, operate, o run, at @
full, speed, o throttle
— v — pie m - to foot operate
— v — pie m derecho - to right foot operate
— v — — m izquierdo - tp left foot operate
— v — resorte m - [mec] to, operate with @
spring, o be spring released
— v — seguridad f - [ind] to operate safely
— v — superávit m - [fin] to, operate, o be
in @ black, o with @ surplus
— v — todos cilindro(s) m - [comb.int] to run,
on, o with, all cylinder(s)
— f — velocidad f - to, operate, o run, at @
speed
— v — — f fija - to operate at @ fixed speed
— v — — f variable - to operate at @ variable
speed
— v conmutador m - [electr-oper] to, throw, o
toggle, @ switch
— v — m con palanca f - [electr-oper] to throw
@ toggle switch
— v — m para sentido m - [electr-oper] to
toggle @ sense switch
— v control m - [mec] to operate @ control
— v continuamente - to operate continuously
— v convertidor m - [metal-prod] to operate @
converter
— v debidamente - to operate, properly, o
smoothly
— v dentro de presupuesto m - [fin] to operate
within @ budget
— v económicamente - to operate economically
— v eficientemente - to operate efficiently
— v elevador m - [mec] to operate @ hoist
— v — m para aguilón m - [grúas] to operate @
boom hoist
— v en aceite m - [mec] to, operate, o run, in
@ oil
— v — campo m - [agric] to field operate
— v — forma automática - to operate automatically
— v — — f continua - to operate continuously
— v — — f instantánea - to operate instartly;
[autom] to shift like that
— v — — f pareja - to run, evenly, o smoothly
— v — — f suave - to run smoothly
— v — país m - to operate in @ country
— v — seco - [mec] to run dry
— v — vacío m - to operate in @ vacuum
— v equipo m - [ind] to, operate, o run, @
equipment
— v — m agrícola - [agric] to operate @, farm.
o agricultural, equipment
— v espectrómetro m - [instrum] to operate @
spectrometer
— v estrangulador m - [mec] to operate @ choke
— v fácilmente - to operate easily
— v función f - [comput] to operate @ function
— v — f crítica - [comput] to operate @ critical function
— v grúa f - [grúas] to operate @ crane
— v — f con carga f - [grúas] to operate @
crane with @ load
— v — f sin carga f - [grúas] to operate @
crane without @ load
— v horno m - [combust] to operate @ furnace
— v — m eléctrico - [metal-prod] to operate @
electric furnace
— v inapropiadamente - to operate improperly
— v independientemente - to operate independently
— v intensamente - [comb.int] to run hard
— v interruptor m - [electr-oper] to operate @
breaker • to throw @ switch
— v libremente - to operate freely
— v línea f - [ind] to operate @ line
— v llave f - [electr-oper] to throw @ switch
— v lleno,na - to run full
— v manija f - [mec] to operate @ handle
— v — f para bomba f - [bombas] to operate @

pump handle
operar v **manipulador** m - [electrón] to operate
@, keyer, o manipulator
— v — m **para tono(s)** m - [electrón] to operate
@ toner key
— v **máquina** f - [mec] to, run, o operate, @ machine
— v **manivela** f - to (operate @) crank
— v **manualmente** - to operate, manually, o by
hand
— f **mecanismo** m - [mec] to operate @ mechanism
— v **motor** m - [electr-mot] to, operate, o run,
@ motor; to start @ motor • [comb.int] to, o-
perate, o run, @ engine; to start @ engine
— v — m **con marcha** f **lenta** - [electr-mot] to
idle @ motor • [comb.int] to idle @ engine
— v — — f **plena** - [electr-mot] to, operate, o run, @ motor at full, speed, o throt-
tle• [comb.int] to, operate, o run, @ engine
at full, speed, o throttle
— v **hidráulico** - [agric-equip] to run @ hydraulic motor
— v **neumáticamente** - [neumát] to operate, pneumatically, o on air • to be air operated
— v **normalmente** - to operate normally
— v **oficina** f - to, operate, o run, @ office
— v **palanca** f - [mec] to operate @ lever •
[electr-oper] to throw @ switch
— v — f **para rotación** f - [grúas] to operate @
swing
— v **paralelamente** - to operate parallely
— v **prensa** f - [mec] to operate @ press
— v **producto** m - to operate @ product
— v **red** f - [ind] to operate @ system
— v — **para aire** m **comprimido** - [ind] to operate @ compressed air system
— v **regulador** m - to operate @ control
— v **rotación** f - [grúas] to operate @ swing
— v **satisfactoriamente** - to, operate, o perform, satisfactorily
— v **secador** m - [ind] to operate @ drier
— v — m **continua(da)mente** - [ind] to operate @
drier continuously
— v **seguro** m - [mec] to operate @, catch, o
lock
— v — m **contra rotación** f - [grúas] to operate
@ rotation, catch, o lock
— v **selector** m - [mec] to operate @ selector
— v — m **para estrangulador** m - [mec] to, operate, o switch, @ choke selector
— v **separadamente** - to operate separately
— v **simultáneamente** - [mec] to, operate, o
switch, simultaneously, o at @ same time
— v **sin aire** m - [mec] to operate without @ air
— v — **carga** f - to run empty
— v — **interrupción** f - [mec] to operate without @, breaker, o switch, o interruption
— v **sistema** m - [mec] to, operate, o drive, @
system
— v **sobre banda** f - [culin] to operate over @
band
— v — **rodaja** f - [mec] to operate over @, roller, o pulley
— v **soldadora** f - [sold] to operate @ welder
— v **sólo** - to operate, only, o alone
— v **tapa** f - [mec] to operate @ cover
— v — f **para válvula** f - [válv] to operate @
valve cover
— v **unidad** f - [ind] to operate @, unit, o machine
— v — f **motriz** - [mec] to operate @ power unit
— v **válvula** f - [mec] to, operate, o switch, @
valve
— v — f **para estrangulador** m - [mec] to, operate, o switch, @ choke valve
— v — f **selectora** - [mec] to operate @ selector valve
— v — f — **para estrangulador** m - [mec] to,
operate, o switch, @ choke selector valve
— v **vehículo** m - [autom] to operate @ vehicle
— v **ventilador** m - [mec] to operate @ fan
— v — m **para compresión** f - [mec] to operate @
pressurizer (fan)

operar v **voltímetro** m - [electr-oper] to operate
@ voltmeter
operario m - [ind] . . .; workman; man • personnel • [labor] labor
— m **asignado** - [labor] assigned workman
— m — **a equipo** m **(para perforación** f**)** -
[petról] (drilling) rig worker
— m — — **plataforma** f - [petról] (drilling)
rig worker
— m — — **pozo** m - [petról] well, o rig, (drilling) worker
— m **bien instruido** - [labor] well trained operator
— m **competente** - [labor] competent operator
— m **diestro** - [labor] skilled, operator, o help
— m **especializado** - [ind] skilled worker
— m **para conservación** f - [ind] maintenance man
— m **soldador** m - [sold] welder (craftsman)
— m **veterano** - [ind] veteran craftsman
— m — **experto** - [ind] veteran expert craftsman
operabilidad f - [ind] operability
operatividad* f - véase **operación** f
operativo,va a - . . .; operational
opinar v - . . .; to opine; to believe
opinión f - . . .; view • judgement • feeling
— f de **administración** f - [admin] management's
oipinion
— f — **dirección** f - [admin] management('s)
opinion
— f **expresada** - expressed opinion
— f **de sí mismo,ma** - self-opinion
— f **expuesta** - expounded, o expressed, opinion
— f **formada** - formed opinion
— f **general** - general opinion • overall rating
— f **personal** - personal, opinion, o feeling
— f **pertinente** - pertinent, o relevant, opinion
— f **popular** - popular opinion
— f **pública** - public opinion • consensus
oponer v - . . .; to counter
— v **resistencia** f—[mec] to oppose @ resistance
oponerse v - to oppose; to be opposed
— v **a resistencia** f - [mec] to oppose @ resistance
— v — — f **de resorte** m - [mec] to oppose @
spring('s) force
oportunamente adv - . . .; in due time; eventually • from time to time • previously
oportunidad f - . . .; chance • timing; timely-
ness
— f **comercial** - [com] commercial, o market, opportunity
— f **de accidente** m - [segurid] accident timing
— f — **inspección** f - [ind] inspection time
— f — **investigación** f - investigation timing
— f — — f **de accidente** m - [segurid] accident
investigation timing
— f **debida** - due, o proper, time
— f **para expansión** f - expansion opportunity
— f — — f **de exportación(es)** f - [com] export
expansion opportunity
— f **práctica** - practical opportunity
— f **primera** - first, o earliest (possible), opportunity, o chance
— f **singular** - unique opportunity
— f **única** - unique opportunity
oportuno,na a - . . .; well timed
oposición f - . . . • [mec] drag(ging)
— f **a resistencia** f - [mec] opposition to @,
force, o resistance
— f — — f **de resorte** m - [mec] spring force
opposition; opposition to @ spring force
opresión f - . . . • pushing; pressing
— f **de botón** m - [mec] button, pushing, o pressing
— f — — m **para encendido** m - [comb.int]
start(ing) button, pushing, o pressing
oprímase interj **para parar** v - push to stop
oprimir v **botón** m - [mec] to, press, o push, o
engage, @ button
— v — m **para arranque** m - [comb.int] to press
@ start(ing) button
— v — m — **encendido** m - comb.int] to, press,
o engage, @ starter button

oprimir v **gatillo** m - to, press, o hold, @ trigger
— v — **de pistola** f - [sold] to, press, o pull, @ gun trigger
— v **para, detener** v, o **parar** v = [mec] to, push, o press, to stop
optar v - . . .; to elect
optativamente adv - optionally
optativo,va a - . . .; optional • alternate
optimación f - optimization,zing; optimum operating,tion
— f **con computador** m - [comput] computer optimization,zing
— f — — m **de banda** f **para rodamiento** m - [autom-neumát] tread computer optimization
— f — — m — **configuración** f **de banda** f **para rodamiento** m - [autom-neumát] tread design computer optimization,zing
— f — — n — **elemento** m - [autom-neumát] element computer optimization,zing
— f **de ajuste** m - [mec] adjusting,tment optimization,zing
— f — — m **para máquina** f - [ind] (machine) tooling optimization,zing
— f — **duración** f **de ajuste** m - tooling life optimization,zing
— f — **largo(r)** m **de barra(s)** f - [cuerp-moled] bar length optimization,zing
— f — **razón** f **para engranaje** m - [autom-mec] gear ratio optimization,zing
— f — **temperatura** f - temperature optimization
— f — — f **de forja** f - [metal-prod] forge temperature optimization,zing
— f — **economía** f **de combustible** n - [combust] optimization,zing for fuel economy
optimado,da a - optimized
— a **con computador** n - [comput] computer optimized
— a **para economía** f **de combustible** - [combust] optimized for @ fuel economy
optimador,ra a - optimizer • optimizing
optimar v - to optimize
— v **bloque** n - [autom-beynát] to optimize @ element
— v — m **con computador** m - [autom-neumát] to computer optimize @ element
— v **combustión** f - [combust] to optimize @ combustion
optimar v **con computador** m - [comput] to computer optimize
— v — m **banda** f - [autom-neumát] to computer optimize @ tread
— v — n — f **para rodamiento** m - [autom-neumát] to computer optimize @ tread
— v — m **bloque** m - [autom-neumát] to computer optimize @ element
— v — m — m **para banda** f **(para rodamiento)** - [autom-neumát] to computer optimize @ tread element
— v — m **configuración** f - [autom-neumát] to computer optimize @ design
— v — m — f **de banda** f - [autom-neumát] to computer optimize @ tread design
— v — m — f — — f **para rodamiento** m - [autom-neumát] to computer optimize @ tread design
— v — m **elemento** m - [autom-neumát] to computer optimize @ element
— v — m — m **para banda** f **(para rodamiento** m) - [autom-neumát] to computer optimize @ tread element
— v **elemento** m - [autom-neumát] to optimize @ element
— v — m **con computador** m - [autom-neumát] to computer optimize @ element
— v **para economía** f - [ind] to optimize for economy
— v — — f **de combustible** m - [combust] to optimize for @ fuel economy
— v **razón** f - to optimize @ ratio
— v — f **para engranaje(s)** m - [autom-mec] to optimize @ gear ratio
optimista m - . . . . . a - . . . . • [fin] bull(ish)

óptimo,ma a - . . .; optimal,mum; best (choice); premium; maximum
opuesto,ta a - . . . . • alternative
— a **a resistencia** f - [mec] opposed to @ force
— a — — f **de resorte** m - [mec] opposed to @ spring force
— a **diametralmente** - diametrically opposite
oquedad f - . . .; void
oral a - . . .; verbal
oralmente adv - orally; by telling
orbe n - . . . . [astron] globe
orden m - . . . . • sequence • calling for • rank(ing); range • [comput] command; instruction
— m **ajustado** - adjusted sequence
— m — **para avance** m - [sold] adjusted travel sequence
— m **analizado** - analyzed order
— m **básico** - basic, order, o sequence
— m **científico** - scientific order
— f **colacionada**—(comput] reported back command
— f **complementaria** - [com] change order
— f **completa** - [com] complete order
— f **completada** - [com] completed order
— m **correcto** - correct, o proper, sequence
— m — **de cordón(es)** m - [sold] proper bead sequence • [cabl] proper strand order
— m **creciente** - increasing
— m **cronológico** - chronological order
— m **cultural** - cultural order
— f **cumplida** - [comput] performed command
— m **dado** - given, order, o sequence
— adv **de** - in the order of; approximately; approaching; roughly
— m — **actividad(es)** f - activity,ties sequence
— m — **adición(es)** f - [metal-prod] addition(s) sequence
— m — **avance** m - [sold] travel sequence
— f — **cambio** m - [com] change order
— f — — m **aprobada** - approved change order
— f — — **para equipo** m - [ind] equipment change order
— n — **carga** f - [ind] charge,ging, order, o sequence
— f — — f **para horno** m - [metal-prod] furnace charge,ging order
— f — — f — — m **de foso,sa** - [metal-prod] soaking pit charge,ging order
— f — — f **para lingote(s)** - [metal-lam] ingot charging order
— m — — f — m **en horno** m **de foso,sa** - [metal-lam] soaking pit ingot charging order
— m **ciclo** m - cycle sequence
— m — — m **de histéresis** f - [electr] hysteresis cycle sequence
— m **colocación** f - [sold] placement sequence
— n — — f **de cordón(es)** m - [sold] bead placement sequence • sample bead placement
— f **compra** f - [com] (purchase,sing) order
— f — — f **original** - [com] original purchase order • purchase order original
— f — — f **para bien(es)** m **durable(s)** - [econ] durable goods purchase order
— f **consideración** f - study, o discussion, order
— m — **cordón(es)** m - [sold] bead sequence • [cabl] strand(s), order, o sequence
— m — **descarga** f - [metal-lam] drawing order
— m — — f **para horno** m **de fosa** f - [metal-lam] soaking pit drawing order
— m — — f — **lingote(s)** m - [metal-lam] ingot drawing order
— f — **día** m - [legal] order of the day; agenda
— m — **embargo** m - [legal] attachment order
— m — **encendido** m - [comb.int] firing order
— m — **evento(s)** m - order, o sequence, of events • scenario
— m — — m **adicional** - additional event(s) sequence
— f — **exterior** m - [com] foreign order
— n — **fabricación** f - production, o manufacture,ring, order
— f — **instalación** f - installation sequence

orden m de llegada - arrival order • [deport]
finish(ing) order
— m — magnitud - size • magnitude order
— m — número(s) m - numerical sequence
— m — operación(es) - operations,ting, order, o
o sequence
— m — pago m - [admin] payment order
— m — m anticipado - [com] advance payment
order
— m — paridad - parity sequence
— m — paso(s) m - steps sequence • (operation)
sequence
— f — m eléctrico(s) m - [electr-oper] electrical operation(s) sequence
— m — m para operación f - operation(s) sequence
— m — m para trabajo m - job step sequence
— m — programa m - program sequence
— m — programa - program sequence
— f — recepción f - [com] materials receipt
— m — relés m - [electr-instal] relay sequence
— m — resistencia f decreciente a rotura f -
[mec] decreasing breaking strength sequence
— m — tabla f - table, order, o sequence
— m — taller m - [ind] shop order
— m — tiempo m time sequence
— m — título(s) m - [fotogr] title sequence
— m — trabajo m - [ind] work, o job, order
— m — m administrativo - [admin] management
work order
— m — m humano m - [admin] human work order
— m debido - proper order
— f educacional - [educ] educational order
— m en código m - [comput] code order
— m — m rechazado - [comput] rejected code
sequence
— m — que debe(n) considerar(se) - consideration, o attention, order
— m — tabla f - table, order, o sequence
— m estadístico - statistical sequence
— m — predeterminado - predetermined statistical sequence
— m financiero - [fin] financial order
— m general - [labor] general order
— m — sobre seguridad f - [segurid] general
safety order
— m inverso - reverse order | a - (en) in re-
verse
— f modificadora - [ind] change order
— m normal - normal, order, o sequence
— m numérico - numerical, order, o sequence
— m para adición f - [metal-prod] addition(s)
sequence
— m — f de ferroaleación(es) f - [metal-
-prod] ferroalloy addition(s) sequence
— f — compra f - [legal] purchase order
— f — f transferida - [legal] transferred
purchase order
— f — empernado m - [mec] bolting sequence
— m — material(es) m - [ind] material(s) order
— m — montaje m - [mec] assembling,bly, o e-
rection, o mounting, order
— m — muestreo m - [ind] sampling order
— f — multifrecuencia - [comput] multifrequency sequence
— m — f fon tomo m doble - [comput] dual
tone multifrequency sequence
— f — pago m - [fin] payment order
— f — m anticipado - [com] advance payment
order
— m — procesamiento m - [ind] processing order
— f — producción f - [ind] production order
— f — rearmado m - reassembly,ling order
— f — servicio m - [ind] service, o shop,
order
— f — taller m - [ind] shop order
— m — toma f de muestra(s) f - [ind] sampling
order
— f — trabajo m - [labor] work order
— f parcial - partial order
— m posible - possible, order, o sequence
— m predeterminado - predetermined sequence
— m preferido - preferred, order, o sequence

orden m probable - probable, order, o sequence
— m — de evento(s) m - probable event(s), order, o sequence
— f recuperada - [comput] reported back command
— f retornada - [comput] reported back command
— f sobre exterior - [fi] foreign order
— f sobre seguridad f - [segurid] safety order
— m sucesivo - sequence; sequential order
— m — para paso m - [mec] pitch sequencing
— m — — — m para banda f para rodamiento m -
[autom-neumat] tread pitch sequencing
— m tecnológico - technological order
— m válido - valid, order, o sequence
— m válido para dígito m - [comput] valid digit, o digit valid, sequence
ordenación f - . . .; arranging,gment • calling
for
— f, con cuidado, o cuidadosa - careful, i
thoughtful, ordering, o arranging,gment
— f de pieza(s) f - [ind] part(s) ordering
— f — — f para repuesto - [ind] (spare) parts
ordering
— f — respuesto(s) m - [ind] (spare) parts ordering
ordenada f - . . . • [topogr] offset
— f media - [geom] mid-ordinate
— f — T para curva T - [geom] curve mid-ordinate
ordenado,da a - . . .; orderly; (thoughtfully)
arranged; ranked; categorized • ordered; called for • in order • [comput] commanded; instructed
—,da a con cuidado - carefully, o thoughtfully,
arranged
— a eficientemente - [com] ordered efficiently
ordenadora f - [comput] computer
— f automática - [comput] automatic sequencer
— f — para secuencia f - [comput] automatic
sequential computer
— f de generación f cuarta - [comput] fourth
generation computer
— f — proceso(s) m - [comput] process(es)
computer
— f digital - [comput] digital computer
— f electrónica - [comput] (electronic) computer
— f — secuencia(s) f - [comput] sequential
computer
— f — verificar carga(s) f - [grúas] load
safety, o Lode-Safe-T, computer
— f pequeña - [comput] small computer
ordenamiento m - . . .; arranging,gment; sequence,cing • rank(ing); categorizing,zation
• [comput] commanding • [ind] housekeeping
— m bueno - [ind] good housekeeping
— m — de sitio m para trabajo m - [ind] (work-
place) good housekeeping
— m — paso(s) m - [ind] steps, ordering, o arranging
— m — — m para puesta f en marcha y detención
f - [electrón] stop and start sequence,cing
— m — sitio m para trabajo m - [ind] housekeeping
— m — tiempo m - time sequencing
— m — trabajo m - [admin] work, sequencing, o
arranging,gment
— m — trabajos m - [ind] jobs, arranging, o
sequencing
— m debido - [ind] (good) housekeeping • good,
arranging,gment, o sequencing
— m — de sitio m para trabajo m - [ind] (good)
housekeeping
— m eficiente - [com] efficient, ordering, o
arranging
— m — de repuesto(s) m - [ind] efficient parts
order(ing) • efficient part(s) arranging
— m para puesta f en marcha y detención f -
[electrón] start and stop sequencing
ordenante m - ordering party • purchaser
ordenanza f - [labor] orderly; janitor • [legal]
. . .; ruling; decree; edict
— f contra incendio(s) m - [pol] fire ordinance
— f — — de aplicación - [pol] applicable fire

ordinance
ordenanza f contra incendio m de aplicación f
  en localidad f - [segur] local fire ordinance
— f local - [legal] local, ordinance, o law
— f — de aplicación f - [legal] applicable
  local, ordinance, o law
— f municipal - [legal] city, o municipal, o
  local, ordinance, o law
— f Sobre Seguridad f y Salud Industrial -
  [legal] Occupational Safety and Health Act
ordenar v - . . . • to assort; to rank; to cate-
  gorize • to call for • [comput] to command;
  to instruct
— v, con cuidado m, o cuidadosamente - to arrange, carefully, o thoughtfully
— v eficientemente - to order efficiently
— v — respuesto(s) m - [ind] to order @ parts
  efficiently - to arrange part(s) efficiently
— f pieza f - [ind] to order @ part
— v — f para repuesto m - [ind] to order @
  part • to arrange @ part
— v repuesto(s) m - [ind] to order @ part(s)
ordene,no* m - ordering
ordeñadora f - [agric-equip] milking machine
ordinal a - . . . • [legal] section
ordinario,ria a - . . . • regular
ordoviciano m - [geol] Ordovician
ordoviciano,na a - [geol] Ordovician
oreado,da a - aired • [agric] tedded
oreadora f - [agric-equip] tedder
— f para, heno, o pasto m - [agric-equip] hay
  tedder
orear v - . . . • [agric] to ted
— v, heno m, o pasto m - [agric- to ted @ hay
oreja f - . . . • [mec] tab; lug; lobe • nose
— f con conformación f especial - [mec] specially designed, tab, o lug, o lobe
— f cuadrada - [mec] square, tab, o lug, o lobe
— f — de acero m - [mec] square steel, tab, o
  lug, o lobe
— f — acero m - [mec] steel, lug, o tab, o
  lobe
— f de placa f—[mec] plate, lug, o tab, o lobe
— f — — f para cierre m - [mec] lock(ing), o
  closing, plate, lug, o tab, o lobe
— f — — f sujetadora - [comb.int] lock(ing)
  plate, lug, o tab, o lobe
— f doblada - [mec] bent, lug, o tab, o lobe
— f en placa f—[mec] plate, lug, o tab, o lobe
— f — camino(s) m - [agric-equip] road lug
— f — palanca f - [mec] lever, ear, o tab
— f — — f para regulación f - control lever,
  ear, o tab
— f — f — — f hidrostática - hydrostatic
  control lever, ear, o tab
— f — petaca f - [metal-prod] cooling plate
  (lifting), ear, o tab, o lug
— f integral - [mec] integral, o built-in, lug
— f para anclaje m - [mec] anchor ear
— f — elevador m - [petról] lifter,ting, o
  elevator, bail, o lug
— f — empuje m - [mec] thrust tab
— f — leva f - [mec] cam lobe
— f — rueda f - [agric-equip] wheel lug
— f — — f para terreno(s) m arenoso(s) -
  [agric-equip] sand wheel lug
— f — suspensión f - [mec] hanging, o holding,
  ear, o lug
— f — tensor m—[mec] turnbuckle (lifting) lug
— f — terreno(s) m arenoso(s) - [agric-equip]
  sand wheel lug
— f plana f - [med] flat lug
orejeta f cuadrada - [mec] small square, lug, o
  tab
— f — de acero m - [mec] square steel, lug, o
  tab
— f de acero m - [mec] steel, lug, o tab
— f en ajustador m - [mec] adjuster lug
— f fiadora - [mec] retainer, lug, o tab
— f — para cojinete m - [med] bearing retainer
  tab
— f para ajuste m - [mec] adjusting tab
— f — cierre m - [mec] lock(ing) tab

orfebrería f - . . .; goldsmithing; silversmithing
organigrama m - [admin] organization(al), o personnel, chart
órganismo m - . . .; body • party • [pol] agency
  • deparment • véase también organización f
— m apropiado - [pol] proper, o appropriate, a-
  gency
— m competente - [pol] competent agency • respective agency
— m crediticio - [fin] credit organization
— m — internacional - [fin] international
  credit organization
— m directivo - [com] directive body • [pol]
  directive agency
— m extranjero - [pol] foreign agency
— m internacional - international, organization, o agency
— m — para Energía f Atómica - [pol] Atomic
  Energy International, o International Atomic
  Energy, Organization
— m — para Financiación f - [fin] International Financing Organization
— m — — Financiamiento m - [fin] International Financing Organization
— m — Normalización f - [pol] International
  Standards Organizaion; I S O
— m Meteorológico Mundial - [pol] World Meteorological Organization
— m — m para Salud f - [pol] World Health Organization
— m nacional—[pol] national, o federal, agency
— m — competente - [pol] competent national
  agency
— m — para Normalización f - [pol] National
  Standards,dization Organization
— m oficial - [pol] government agency; official
  organism
— m — Cooperación f Económica Europea - [pol]
  Organization for European Economic Cooperation
— m — normalización f - standardization organization
— m — registro m - [pol] recording, agency, o
  office
— m patrocinante - sponsoring body
— m público - public, organization, o agency
— m — descentralizado - [pol] decentralized
  public, agency, o organization
— m registrador - [pol] recording, agency, o
  office
— m respectivo - [pol] respective agency
— m técnico - technical, organization, o agency
organización f - . . .; organizing; orchestra-
  ting,tion • véase también distribución f -
  organismo m - [admin] staff • [legal] incorporation • format • gearing • engineering •
  [com] business
— f accidental - [legal] informal organization
— f administrativa - [admin] administrative, o
  management, organization
— f aerodinámica - [admin] streamlined organization
— f auxiliar - auxiliary organization
— f capacitada - [ind] skilled organization
— f capaz - capable organization
— f — para servicio m - capable service organization
— f central(izada) - central(ized) organization
— f — para venta(s) f - central(ized) sale(s)
  organization
— f confiable - dependable organization
— f Consultiva Marítima Intergubernamental -
  [pol] Intergovernmental Maritime Consultive
  Organization; I M C O
— f contable - [contab] accounting organization
— f de acería f - [metal-prod] steel, plant, o
  mill, organization
— f — administración f - [admin] management, o
  administrative, organization
— f — consumidor(es) m - [econ] consumer('s)
  organization
— f — distribuidores m - dealer's organization

organización f de distribuidor(es) m para motor(es) m - [electr-mot] motors distributors organization • [comb.int] engines distributors organization
— f — Estados m Centro Americanos - [pol] Organizatin of Central American States; O C A S
— f — Naciones f Unidas - [pol] United Nations Organization; U N O
— f — — f — para Actividades f Educativas, Científicas y Culturales - [pol] United Nations Educational, Scientific and Cultural Organization; U N E S C O
— f — planta f - [ind] plant organization
— f — programa n - program, organization, o planning
— f — — m para instrucción f - training program organization,zing
— f — — m — f sobre seguridad f - [segurid] safety training program organizing
— f — sección f - [ind] section organization,zing
— f — — operación f de horno(s) m - [metal-prod] furnace operating,tion section organization,zing
— f — servicio(s) m - [ind] service(s) organizing,zation
— f — — m administrativo(s) m - [admin] administration,tive service(s) organization
— f — — m auxiliar(es) - [ind] auxiliary service(s) organization,zing
— f — trabajo m - [admin] work organization,zing
— f — tren m - [f.c.] train make-up
— f demasiado extendida - [admin] over-expanded, o sprawled, organization
— f desparramada—[admin] sprawled organization
— f distribuidora - [com] distributing,tion organization
— f Educativa, Científica y Cultural de Naciones f Unidas - [pol] United Nations Educational, Scientific and Cultural Organization; U N E S C O
— f empresarial - [admin] corporate organization • internal organization
— f — interna - [admin] internal corporate organization
— f en torno m a personalidad(es) f - [admin] around @ personality,ties organization
— f escueta - [admin] streamlined organization
— f Financiera Internacional - [fin] International Financial Organization
— f industrial - [ind] industrial organization
— f — potencial - [ind] potential industrial organization
— f ineficaz - [admin] inefficient,fective organization
— f informal - informal organization,zing
— f Internacional para Aviación f Civil - [pol] International Civil Aviation Organization; I C A O
— f — — Energía f Atómica - [nucl] International Atomic Energy, Organization, o Agency
— f — — Normalización f - International Standards,dizing Organization
— f — — Normas f - International Standards Organization
— f — para Trabajo m - [pol] International Labor Organization; I L O
— f levantada - [admin] built, o raised, organization
— f local - local organization,zing
— f lógica - logical organization,zing
— f — de trabajo m - [admin] logical work organization,zing
— f meteorológica - [meteorol] meteorological organization
— f Mundial—[pol] World Meteorological Organization; W M O
— f moderna O [admin] modern organization,zing; streamlined organization,zing
— f mundial - world(wide) organization
— f — de distribuidores m - [com] worldwide dealer organization

organización f mundial para distribución f - worldwide distributor organization
— f — — Salud f - [pol] World Health Organization; W H O
— f — para servicio m - worldwide service organization
— f — — venta(s) f - [com] worldwide, dealer, o sales, organization
— f muy capacitada - [ind] highly skilled organization
— f no autorizada - unauthorize, o not authorized, organization
— f operativa - operating organization
— f Panamericana para Salud f - [pol] Panamerican Health Organization
— f para Alimentos m y Agricultura f - [pol] Food and Agriculture Organization; F A O
— f — carrera f - [deport] event format
— f — combustible(s) m - [ind] combustion, o fuels, setup
— f — — m en planta(s) f siderúrgica(s) - [metal-prod] steel plant combustion setup
— f — comercialización f - [com] marketing organization
— f — Cooperación f y Desarrollo m Económicos - pol] Organization for Economic Cooperation and Development
— f — distribución f - [com] distributing,tion organization; (distributing) channel
— f — energía f - [ind] power setup
— f — — f en planta(s) f siderúrgica(s) - [metal-prod] steel plant power setup
— f — ensayo(s) m - test(ing) engineering
— f — instrucción f - [ind] training organization,zing
— f — investigación(es) f - research organizing,zation
— f — logro m de objetivo(s) m - [admin] objective(s) achievement organization,zing
— f — mantenimiento m - [ind] maintenance organizing,zation
— f — — m asignado - [ind] assigned maintenance organizing,zation
— f — producción f - [ind] production setup
— f — productividad f - [ind] productivity organizing,zation
— f — — f de personal m - [ind] employee productivity organizing,zation
— f — salud f - health organizing,zation
— f — servicio m - service organizing,zation
— f — venta(s) f - [com] sales, o selling, organization • dealer('s) organizing,zation
— f por administración f - [admin] management organizing,zation
— f productiva—[admin] productive organization
— f profesional - professional organization
— f progresiva - progressive, o forward-looking, organization
— f responsable - responsible organization
— f — para servicio m - responsible, o responsive, service organization
— f responsive - responsive organization
— f sólida - [admin] solid, o sound, organization
— f técnica - technical organization
— f — Europea—European technical organization
— f — — para Neumáticos m y llantas f - [autom-neumát] European Tire and Rim Technical Organization
organizado,da a - organized • geared • orchestrated • engineered • [legal] incorporated
— a en torno a personalidad(es) f - [admin] organized around @ personality,ties
— a lógicamente - logically organized
organizador m - organizer • promoter
organizar v - . . . . • to plan • to gear • to orchestrate • to engineer • [legal] to incorporate
— v administración f - [admin] to organize @, administration, o management
— v consumidor(es) m - [econ] to organize @ consumer(s)
— v contabilidad f - [contab] to organize @ ac-

counting
organizar v̱ en torno a personalidad(es) f̱ - -
　[admin] to organize around @ personality,ties
— v̱ ensayo m̱ - to, organize, o̱ engineer, @ test
— v̱ lógicamente - to organize logically
— v̱ — trabajo m̱ - [admin] to organize @ work
　logically
— v̱ para lograr m̱ objetivo m̱ - [admin] to or-
　ganize to achieve (@ objective)
— v̱ productividad f̱ - [ind] to organize @ pro-
　ductivity
— v̱ trabajo m̱ - [admin] to organize @ work
organizativo,va* a̱ - organizational
órgano m̱ eléctrico - [mús] electric organ
orientación f̱ - . . . . • position • guidance;
　guiding
— f̱ correcta - accurate, orientation, o̱ guiding
— f̱ de conductor m̱ natural - [admin] natural
　leader('s) orientation
— f̱ — desviador m̱ - [petról] whipstock orien-
　tation
— f̱ — director m̱ natural - [admin] natural
　leader('s) orientation
— f̱ — grano m̱ - [metal- grain orientation
— f̱ — m̱ luego de laminación f̱ - [metal-lam]
　rolling oriented grain
— f̱ — guiasonda(s) m̱ - [petról] whipstock
　orientation
— f̱ — líder ṉ - [admin] leader('s) orientation
— f̱ — m̱ natural - [admin] natural leader's
　orientation
— f̱ — potencial m̱ - [con] potential orienta-
　tion
— f̱ egocéntrica—[admin] egocentric orientation
— f̱ — de líder m̱ - [admin] leader('s), egocen-
　tered, o̱ selfcentered, orientation
— f̱ hacia consumidor m̱ - [econ] consumer orien-
　tation
— f̱ — consumerismo* m̱ - [econ] consumerism
　orienting,tation
— f̱ — seguridad f̱ - [segurid] safety orienta-
　tion
— f̱ — — f̱ para, cargo m̱, o̱ puesto m̱ - [admin]
　position safety orienting,tation
— f̱ — servicio m̱ - [ind] service orientation
— f̱ — sistema m̱ - [ind] system orientation
— f̱ inicial - [segurid] initial orientation
— f̱ — hacia servicio m̱ - [segurid] initial
　safety orientation
— f̱ inteligente - intelligent orientation
— f̱ natural - natural orientation
— f̱ para cargo m̱ - [labor] position orientation
— f̱ preliminar - preliminary orientation
orientado,da a̱ - oriented
— a̱ hacia consumerismo* m̱ - [econ] consumerism
　oriented
— a̱ — consumidor(es) m̱ - [econ] consumer
　oriented
— a̱ — deporte(s) m̱ - [deport] sports oriented
— a̱ — — m̱ acuático(s) - [deport] water
　(sports) oriented
— a̱ — servicio m̱ - [ind] service oriented
— a̱ — sistema m̱ - [ind] system oriented
orientador,ra a̱ - orienting
— a̱ hacia consumerismo* m - [econ] consumerism
oriental m̱ - oriental • [Uru.] Uruguayan | a̱ -
　. . .; . . . • easterly
orientar v̱ - . . . • to guide • to swing
— v̱ hacia consumidor m̱ - [econ] to consumer
　orient(ate)
— v̱ — consumismo m̱ - [econ] to consumerism
　orient(ate)
— v̱ — servicio m̱ [ind] to service orient(ate)
— v̱ — sistema m̱ - [ind] to system orient(ate)
— v̱ potencial m̱ - [econ] to orient(ate) @ po-
　tential
orientativo,va a̱ - orien(ta)ting; guiding; sug-
　gestive
orificio m̱ - . . .; opening; aperture; eye; bore
　• [comb.int] well
— m̱ abocardado - [mec] counterbore; countersunk
　hole
— m̱ aforador - [mec] metering, orifice, o̱ hole

orificio m̱ agrandado - [mec] enlarged hole
— m̱ — para remache m̱ - [mec] enlarged rivet
　hole
— m̱ aliviador - [mec] weep hole
— m̱ — ovalado - oval weep hole
— m̱ atravesador - [mec] thruhole; through hole
— m̱ avellanado - [mec] countersunk hole
— m̱ blindado - [mec] grommet
— m̱ calibrado - [mec] metering orifice
— m̱ central - [mec] center hole
— m̱ — en múltiple m̱ - [comb.int] manifold cen-
　ter hole
— m̱ — — m̱ a bloque m̱ - [comb.int]
　manifold to block center hole
— m̱ en múliple m̱ - [comb.int] manifold hole
— m̱ — — m̱ a bloque ṉ - [comb.int] manifold to
　block hole
— m̱ defelctor - [mec] baffle(d) opening
— m̱ deformado - [mec] deformed, o̱ stretched,
　hole
— m̱ — para remache m̱ - [mec] stretched rivet
　hole
— m̱ delator - [petról] tell-tale hole
— m̱ — obstruido - [petról] clogged tell-tale
　hole
— m̱ en acelerador m̱ - [comb.int] throttle bore
— m̱ — base f̱ - [mec] base hole
— m̱ — basto - [mec] rough hole • rough core
— m̱ — bloque m̱ - [comb.int] block hole
— m̱ — — m̱ para bomba f̱ - [comb.int] block
　pump hole
— m̱ — — m̱ — — f̱ para combustible m̱ - [comb-
　-int] block fuel pump hole
— m̱ — caja f̱ - [mec] housing hole
— m̱ — — f̱ para montaje m̱ - [mec] housing
　mounting hole
— m̱ — — f̱ — motor m̱ - [electr-mot] motor
　housing hole • [comb.int] engine mounting
　hole
— m̱ — círculo m̱ - [mec] circumference hole
— m̱ — circunferencia f̱ - [mec] circumference
　hole
— m̱ — costado m̱ - [mec] side hole
— m̱ — — m̱ de cárter m̱ - [comb.int] crankcase
　side hole
— m̱ — cubierta f̱ - [mec] cover hole
— m̱ — — f̱ para caja f̱ - [autom-mec] housing
　cover hole
— m̱ — — f̱ — — f̱ para eje m̱ - [autom-mec]
　axle, o̱ shaft, housing cover (oil) filler
　hole
— m̱ — — f̱ — — f̱ — — m̱ para abastecimiento
　m̱ (de aceite m̱) - [autom-mec] axle housing
　cover (oil) filler hole
— m̱ — — f̱ — engranaje m̱ - [mec] gear cover
　hole
— m̱ — — f̱ posterior - [mec] rear cover hole
— m̱ — cubo m̱ - [mec] hub hole
— m̱ — — m̱ para carrete m̱ - [mec] reel hub
　hole
— m̱ — culata f̱ - [comb.int] head hole
— m̱ — — f̱ para cilindro m̱ - [comb.int] cylin-
　der head hole
— m̱ — — f̱ — — m̱ para manómetro m̱ (para tem-
　peratura f̱ - [comb.int] cylinder head tempe-
　rature gage hole
— m̱ — — f̱ — — m̱ — medidor m̱ para tempera-
　tura f̱ - [comb.int] cylinder head temperature
　gage hole
— m̱ — — f̱ — medidor m̱ para temperatura f̱ -
　[comb.int] head temperature gage hole
— m̱ — eje m̱ motor - [mec] output shaft cavity
— m̱ — émbolo m̱ - [mec] piston hole
— m̱ — — m̱ para pasador m̱ - [mec] piston pin
　hole
— m̱ — estrangulador m̱ - [mec] choke hole
— m̱ — extremo m̱ - [mec] end hole
— m̱ — — m̱ de múltiple - [comb.int] manifold
　end hole
— m̱ — — m̱ — — m̱ a bloque m̱ - [comb.int]
　manifold end to block hole
— m̱ — fuste m̱ - [mec] shank hole
— m̱ — — m̱ de piñón m̱—pinion shank hole

orificio m en grillete m - [mec] shackle hole
— m — m para remache m - rivet shackle hole
— m — motor m para aceite m - [comb.int] engine oil, hole, o filler
— m — ojal m - [electrón] eyelet hole
— m — m en tablilla f - [electrón] card
eyelet hole
— m — m — f para circuito m—[electrón]
circuit card eyelet hole
— m — pared f - [constr] wall opening
— m — placa f - [mec] plate orifice
— n — f para toma f de presión f - pressure
taking plate orifice
— m — plantilla f - [mec] templet hole
— m — quemador m - [combust] burner orifice
— m — respiradero m - [mec] air vent hole
— m — m para reductor m para velocidad f -
[mec] gear reducer air vent hole
— m — superficie f - surface opening
— m — tablero m - [electrón] board hole
— m — m para suministro m de energía f -
[electrón] power supply board hole
— m — m para circuito m - [electrón] circuit, card, o board, hole
— m — tambor m - [mec] drum hole
— m — m para balanceo m - [mec] rocking
drum hole
— m — tapa f - [mec] cover hole
— m — tubería f - [tub] pipe hole
— m — f para drenaje m - [tub] drain pipe
hole
— m — viga f - [mec] beam, o rail, hole
— m — f de base f - [mec] base rail hole
— m — viscosímetro m - [instrum] viscometer
orifice
— m escariado - [mec] reamed hole
— m.estirado - [mec] stretched hole
— m estrangulador - [petról] choke hole
— m — para surgencia f - [petról] flow bean
(choke) hole
— m faltante - [mec] missing hole
— m — para retorno m - [mec] missing return
hole
— m — — m para aceite m - [mec] missing
oil return hole
— m hacia abajo - [mec] downward(s) pointing,
port, o hole
— m — arriba - [mec] upward(s) pointing port
— m hacia costado m - [mec] sideways pointing,
port, o hole
— m horadado - [mec] drilled hole
— m — para chaveta f - [mec] cotter pin
drilled hole
— m inferior - [mec] lower, o bottom, hole, o
port
— m — en bloque m—[comb.int] block lower hole
— m — en palanca f - [bombas] lower lever, o
lever lower, hole
— m — — — f para bomba f - [bombas] pump
lever lower hole
— m — para espárrago m - [mec] stud lower hole
— m — — m roscado - [mec] stud lower hole
— m — — m — en tapa f - [comb.int] cover
stud lower hole
— m — — m — — f para engranaje m -
[comb.int] gear cover stud lower hole
— m — — — f — — m a bloque m -
[comb.int] gear cover to block stud lower
hole
— m limitador - [mec] limiting, orifice, o hole
— m limpiado - [mec] cleaned hole
— m limpio - [mec] clean hole
— m lubricado - [mec] lubricated hole
— m más alejado - [mec] farthest hole
— m — próximo - [mec] nearest, o closest, hole
— m obturado - [mec] plugged, o closed, hole, o
orifice, o opening · closed (off) port
— m — en quemador m - [combust] clogged burner
orifice
— m optativo para extracción f - [mec] optional
puller hole
— m para abastecimiento m - [mec] filler (hole,
o port)

orificio m para acceso m - [mec] access hole
— m — aceite m - [mec] oil (filler) hole
— m — m para retorno m - [mec] return oil
hole
— m — — m — m obturado - [mec] blocked
oil return hole
— m — — m en caja f - [mec] case oil hole
— m — — f para ventilador m -
[ventil] fan case oil hole
— m — — — m — cubo m - [mec] hub oil hole
— m — — m — m para palanca f - [mec]
lever hub oil hole
— m — — m — m — f para válvula f -
[mec] valve lever hub oil hole
— m — — m — — f — v neumática -
[mec] air valve lever hub oil hole
— m — — m — — f — f para aire -
[mec] air valve lever hub oil hole
— m — — m para motor m - [comb.int] engine
oil, hole, o filler
— m — — m — retorno m (para aceite m) -
[mec] oil return oil hole
— m — admisión f - inlet, port, o hole
— m — alivio m - [mec] relief, opening, o port
— m — aportación f - [mec] filler hole
— m — f para aceite m - [mec] oil fill(ing)
hole
— m — árbol m - [mec] shaft hole
— m — m para motor m - [electr-mot] motor
shaft hole
— m — bomba f - [bombas] pump hole
— m — f para combustible m - [comb.int]
fuel pump hole
— m — f — m en bloque m - [Comb.int]
block fuel pump hole
— m — bujía f - [comb.int] spark plug hole
— m — calentamiento m - heating opening
— m — cojinete m - [mec] bearing hole
— m — colada f - [metal-prod] taphole; (iron)
notch
— m — comprobación f - [mec] test(ing), o
check(ing), hole, o opening, o plug
— m — f obturado - [mec] plugged test(ing),
opening, o hole
— m — conmutador m - [electr-instal] switch, o
breaker, hole
— m — chaveta f - [mec] cotter pin hole
— m — derivación f - [mec] by-pass hole
— m — f abierto - [mec] open(ed) by-pass,
hole, o port
— m — f obturado - [mec] closed off by-pass
port
— m — descarga f - [mec] discharge, o exhaust,
hole, o port, o opening · [comb.int] (carburetor) bowl drain
— m — f en lado m para admisión f - [mec]
intake drip
— m — f — taza f - [comb.int] bowl drain
— m — f — — f para carburador m - [comb.-
int] carburetor bowl drain
— m — f para marcha f sin carga - [comb.-
int] idle discharge hole
— m — distribución f - distribution, hole, o
orifice
— m — drenaje m - [mec] drain(ing) hole
— m — drenar - [comb.int] drain(ing) hole
— m — v aceite m - [comb.int] oil drain-
(ing hole)
— m — v — m en motor m - [comb.int] engine oil drain(ing hole)
— m — eje m - [mec] axle, o shaft, hole
— m — enlechado m - [constr] grout(ing) hole
— m — entrada - [mec] inlet (port, o hole)
— m — escape m - [comb.int] exhaust, o outlet,
port, o vent; vent
— m — escoria f - [metal-prod] slag notch
— m — f con aire m comprimido m - [metal-
prod] air operated slag notch
— m — espárrago m (roscado) - [mec] stud hole
— m — — m (—) en tapa f - [comb.int] cover
stud hole
— m — — m (—) — — f para engranaje m -
[comb.int] gear cover stud hole

orificio m para espárrago m (roscado) en tapa f para engranaje m a bloque m - [comb.int] gear cover to block stud hole
— m — extracción f - [mec] puller hole
— m — fiador m - [mec] retainer opening
— m — flujo m - flow opening
— m — grasa f - [mec] grease hole
— m — guía f - [electr-instal] lead hole
— m — henchimiento m - [mec] filler (hole)
— m — inspección f - [constr] inspection, o lamp, hole, o port
— m — lanza f - [metal-prod] lance hole
— m — lubricación f - [mec] lubrication hole
— m — lubricante m - [mec] lubricant hole
— m — m en posición f horizontal - [mec] horizontal position lubricant hole
— m — llave f - [mec] wrench hole • [mec] keyhole
— m — mano f - [mec] hand hole
— m — f en cruceta f - [mec] crosshead hand hole
— m — manómetro m - [comb.int] (pressure) gage hole
— m — — m para temperatura f - [comb.int] temperature gage hole
— m — marcha f sin carga f - [comb.int] idle outlet
— m — medición f - [combust] measuring, o metering, hole, o orifice
— m — f de combustible m - [combust] fuel measurement,ring, hole, o orifice
— m — medidor m - [comb.int] gage hole
— m — m para temperatura f - [comb.int] temperature gage hole
— m — medir gasto m - [mec] metering orifice
— m — montaje m - [mec] mounting hole
— m — pasador m - [mec] pin hole
— m — m para émbolo m - [mec] piston pin hole
— m — perno m - [mec] bolt hole
— m — — m en círculo m - [mec] circumference bolt hole
— m — — m — circumferencia f - [mec] circumference bolt hole
— m — quemador m - [combust] burner, orifice, o hole
— m — rebalse m - [mec] overflow hole
— m — remache m - [mec] rivet hole
— m — respiradero m - [comb.int] (air) vent hole
— m — restricción f - restriction, o restraining, hole
— m — retén m—[mec] retainer, opening, o hole
— m — retorno m - [mec] return hole
— m — — m para aceite m - [mec] oil return hole
— m — — m — aceite m atascado - [mec] plugged, o clogged, oil return hole
— m — — m — m obturado - [mec] plugged, o clogged, oil return hole
— m — rociadura f - [mec] spray(ing) hole
— m — salida f - outlet, hole, o port
— m — — f para depósito m—[sold] tank outlet
— m — — d — m para fundente m - [sold] flux tank outlet
— m — — f — fundente m - [sold] flux outlet
— m — sangría f - [comb.int] bleed hole
— m — soldadura f - [sold] weld(ing) hole
— m — succión f - [bombas] suction port
— m — tapón m - [mec] plug, hole, o opening • [comb.int] filler plug, hole, o opening
— m — — m para aceite m - [comb.int] filler plug, hole, o opening
— m — — — nivel m - [comb.int] level plug, hole, o opening
— m — toma f para presión f - pressure taking, orifice, o hole
— m — tornillo m - [mec] screw hole
— m — — m fijador - [mec] setscrew hole
— m — — m para cubierta f - [mec] cover screw hole
— m — — f, fijador, o sujetador m - [mec] setscrew hole

orificio m para válvula f - [mec] valve hole
— m — varilla f - [mec] rod hole
— m — — f medidora - [mec] dipstick hole
— m — — f para medir v aceite m - [mec] dipstick hole
— m — ventilación f - [miner] vent(ing) shaft
— m — f obturado - [comb.int] plugged air, vent, o hole
— m — verificación f - [mec] check(ing), plug, o hole
— m perforado - [mec] drilled, o punched, hole
— m — con anticipación f - [mec] pre-punched hole
— m portante - [mec] bearing hole
— m — en cubierta f - [mec] cover bearing hole
— m — — f para engranaje(s) m - [mec] gear cover bearing hole
— m ranurado - [mec] slotted, hole, o opening
— m — en fiador m - [mec] retainer slotted opening
— m — — retén m - [mec] retainer slotted opening
— m — para soldadura f - [sold] slotted welding hole
— m reducido - [mec] undersized hole
— m roscado - [mec] tapped, o threaded, hole; tap
— m sucio - [mec] dirty, orifice, o hole
— m — en quemador m - [combust] dirty burner, orifice, o hole
— m superior - [mec] upper, o top, o higher, hole, o port
— m — para espárrago m (roscado) - [mec] stud upper hole
— m — — — m (—) en tapa f - [comb.int] cover stud upper hole
— m — — — m (—) — — f para engranaje(s) m - [comb.int] gear cover stud upper hole
— m — — — m (—) — — f — m a bloque m - [comb.int] gear cover to block stud upper hole
— m — en bloque m—[comb.int] block upper hole
— m — palanca f - [mec] lever upper hole
— m — — f para bomba f - [bombas] pump lever upper hole
— m variable - [mec] variable, orifice, o hole
— m ventilador - [mec] ventilating hole • [comb.int] air vent
— m — en depósito m - [comb.int] tank air vent
— m — — m para combustible m - [comb.int] fuel tank air vent(ing hole)
orificios m múltiples m - [mec] multiple, holes, o openings
origen m - . . . . • start(ing) • developing,pment • cause,sing • deriving; spawning
— m de caudal m - [hidr] flow source
— m — Estados m Unidos de América f de Norte - United States origin
— m — falla f - fault origin(ation)
— m — flujo m - [hidr] flow source
— m — fondo m - [fin] fund, origin, o source
— m — mercadería f - [com] merchandise, o goods, origin, o source
— m — mercancía f - [com] goods origin
— m — siniestro m - [seguros] loss origin
— m étnico - [demogr] ethnic, o national, origin
— m extranjero - foreign origin
— m mejicano - Mexican origin
— m mineral - [miner] mineral origin
— m nacional - [com] national, o local, origin
— m orgánico - [geol] organic origin
— m variable - variable, o varying, origin
origenación* f - origination • leading to • vease también origen m
— f de importe m - value origin(ation)
— f — residuo(s) m - waste(s) origin,nation
— f — valor m - value origin,nation
originado,da - originated • started; began • developed; spawned • caused; led to
original m - . . . . • [legal] reproducible; tracing • design | a - . . . ; unique
— m de documento m - [legal] document original
— m — — m para embarque m - [transp] shipping

document original
original m **de documento** m **para embarque** m **solicitado** - [transp] requested shipping document original
— m — **factura** f - [com] original invoice
— m — **póliza** f **de seguro** m - [seguros] original insurance policy
originar v - . . . • to, develop, o spawn, o derive, o cause • to lead to
— v **de** - to originate from
— v **desecho(s)** m - [nucl] to originate, o produce, @ waste(s)
— v **en** - to originate in
— v **falla** f - to originate @ fault
— v **importe** m - to originate @ value
— v **residuo(s)** m - [nucl] to originate, o produce, @ waste(s)
— v **valor** - to originate @ value
originario,ria a **de Estados** m **Unidos de América** f **de Norte** - United STates origin
orilla f - . . . ; rim
— f **de río** m - [hidr] river bank
orillado,da a - bordered; edged
orilladora f - [metal-lam] edger
orillar v - . . .; to edge
orín m - [metal] . . . .; véase también **oxido** m
oriundo,da a - . . . • hometown
ornamentación f **con ladrillo(s)** m - [constr] brick ornament(ation)
ornamental a - . . . • [imprent] display
orquesta f - [mús] . . .; band
orquestado,da a - [mús] orchestrated
orquestar v - [mús] to orchestrate
ortogeosinclinal m - [geol] orthogeosyncline • a - [geol] orthogeosyncline
ortogonal a - . . . • [sold] fillet
— a **en pasada** f **primera** - [sold] first pass fillet
oruga f - . . . • [constr-equip] . . .; crawler
— f **de pala** f **mecánica** - [constr-equip] shovel track
— f — **tipo** m **para tractor** m - [mec] tractor type crawler
— f **para tractor** m - [mec] tractor crawler
— f **tipo** m **tractor** m - [mec] tractor type crawler
osadía f - . . . • [fam] grit; gumption
osado,da a - . . . • challenging
oscilación f - . . .; oscillating; weave,ving; sway(ing); wobble,ling • vibrating,tion - [electr-oper] surge,ging; fluctuating,tion • [sold] side to side, o weaving, motion
— f **apropiada** - proper oscillating,tion
— f **de eje** m - [mec] axle oscillating,tion
— d — — m **hidráulico** - [grúas] hydraulic axle oscillating,tion
— f — **herramienta(s)** f - [petról] tool swing
— f — **tono** m—[electrón] tone oscillating,tion
— f **lateral** - [sold] side-to-side, o lateral, motion, o weave
oscilación f **posterior** - [grúas] tail swing
— f **en frecuencia** f - [electrón] fequency oscillation
oscilación(es) f **por minuto** m - oscillation(s) per minute
oscilado,da a - oscillated; swayed • fluctuated; wobbled
— a **apropiadamente** - oscillated properly
oscilador m - [electrón] . . . • [mec] rocker
— m **a cubo** m - [mec] rocker to @ rock
— m **de cristal** m—[electrón] crystal oscillator
— m **externo** m - [electrónl external oscillator
— m **para conmutador** m—[electrónl switch oscillator
— m — — m **para sentido** m - [electrón] sense switch oscillator
— m — — — m **en manipulador** m - [electrón] keyer sense swith oscillator
— m — **espacio(s)** m - [electrón] space oscillator
— m — **frcuencia** f **alta** - [electr-equip] high frequency oscillator
— m — **marca(s)** f - [electrón] mark oscillator

oscilador m **para marca(s)** f **y espacio(s)** m - [electrón] mark/space oscillator
— m — **salida** f - [electrón] output oscillator
— m — — f **seleccionado** - [electrón] selected output oscillator
— m **para tono** m - [electrón] tone oscillator
— m **principal** - [instrum] main, o master, oscillator
— m **seleccionado** - [electrón] selected oscillator
— m **tipo** m **disruptivo** - [comb.int] spark gap oscillator
oscilador,ra a - oscillating,tor
— m **para portaescobilla(s)** f - [electr-mot] brush oscillator
oscilante a - oscillating; side-to-side
oscilar v - . . .; to swing; to sway; to wobble; to fluctuate • to range
— v **apropiadamente** - to oscillate properly
— v **en ángulo** m **recto** - [sold] to oscillate at @ right angle(s)
— v — — m **con dirección** f **de avance** m - [sold] to oscillate at @ right angle(s) to @ direction of @ travel
— v **frecuencia** f - [electrón] to oscillate @ frequency
— v **tono** m - [electrón] to oscillate @ tone
osciloscópico,ca a - [electrón] oscilloscopic,pe
— m **vertical** - [instrum] vertical oscilloscopic,pe
oscilloscope m - [electrón] oscilloscope
— m **conectado** - [electrón] connected oscilloscope
— m **reeconectado** - [electrón] reconnected oscilloscope
— m **regulado** - [electron] set oscilloscope
oscuro,ra a - véase **obscuro,ra** a
ostentación f - . . .; show(ing); carrying
ostentado,da a - . . .; exhibited; shown • carried
ostentar v - . . .; to show; to present; to carry; to bear
otoñal a - . . .; fall; autumn
otorgado,da a - granted; assigned • extended; issued • [legal] executed
otorgamiento m - . . . • assigning • issuing; extending,sion • [legal] executing,tion
— m — **asesoría** f - advisorship, o consultantship, granting
— m — — f **técnica** - technical consultantship granting
— m — **beneficio(s)** m - benefit(s) granting
— m — — m **crediticio(s)** - [fin] credit, benefit(s), o term(s), granting, o extending
— m — **contrato** m - contract granting
— m — **crédito** m - [fin] credit granting
— m — **garantía** f - [fin] guaranty granting
— m — **poder** m - [legal] power (of attorney) granting
otorgante m - [legal] . . .; maker
otorgar v - . . .; to assign • to issue; to extend • [legal] to execute
— v **asesoría** f **técnica** - to grant @ technical, consultantship, o assistance,ntship
— v **beneficio(s)** m **crediticio(s)** - [fin] to, grant, o extend, @ credit, benefits, o terms
— v — n **de crédito** m - [fin] to, grant, o extend, credit, benefits, o terms
— v **contrato** m - to grant @ contract
— v **crédito** m - [fin] to grant @ credit
— v **garantía** f - [fin] to grant @ guaranty
otra adv **aplicación** f - (an)other application
— adv **calidad** - (an)other, quality, o grade
— adv **consideración** f - (an)other consideration
— adv **industria** f - (an)other industry
— adv **forma** - (an)other form • (de, o en) otherwise
— adv **función** f - (an)other function
— adv **laya** f - (an)other way
— adv **manera** - (an)other way • unless otherwise specified
— adv **mano** - (an)other hand
— adv **mina** f- [miner] (an)other mine

otra adv mina f propuesta - [miner] alternate
  mine site
— adv moneda f - [fin] (an)other currency
— adv partida f - [ind (an)other, batch, o run
— adv persona f - (an)other (person)
— adv pieza f - [mec] (an)other part
— adv pregunta f - (an)other, o further, question
— adv protección f - (an)other protection
— adv regulación f - [mec] (an)other setting
— adv venta f - [com] (an)other sale
— adv vez f - once, o over, again
— adv zona f - (an)other, zone, o area
otro m - (an)other • (an)other, o third, party
— adv activo m - [fin] (an)other asset
— adv código m - (an)other code
— adv color m - (an)other color
— adv comentario m - (an)other, remark, o com-
  mentary • further ado
— adv componente - (an)other component
— adv deudor m - [com] (an)other debtor
— adv dispositivo m - (an)other device
— adv equipo m - (an)other equipment
— adv estado m - [pol] (an)other state • (de) -
  out-of-state
— adv gasto m - (an)other expense
— adv extremo m - (an)other, end, o extreme
— adv — m de escala f - (an)other scale end
— adv fabricante m - [ind] (an)other manufacturer
— adv — para equipo m - [ind] (an)other equipment manufacturer
— adv ingreso m - [contab] (an)other income
— adv material n - (on)other, o differing, material
— adv medio m - (an)other mean(s)
— adv país m - [pol] (an)other country
— adv pasivo m - [fin] (an)other liability
— adv quebranto m - [fin] (an)other loss
— adv recubrimiento m - (an)other coat(ing)
— adv sitio m - (an)other, o alternate, site
— adv tipo - (an)other type
— adv — m de soldadura f - [sold] (an)other
  weld(ing) type
otro,tra a - (an)other; additional
— a que - other than
— a — estandar - other than standard
— a — según norma f - other than standard
ovado,da a - . . .; ovate
ovalamiento m - oval shaping
ovalidad f - ovalness • [metal-lam] out-of-round
  • [tub] hi-lo • [sold] hi-lo
— f de barra f - [metal-lam] bar ovalness
— f — diámetro m interior - [comb.int] bore
  out-of-round
ovalización f - ovalization
ovillo m - [textil] ball
— m de hilo m - [textil] twine ball
oxi-gas m - véase oxígeno m y gas m
oxiacetileno m - [quím] oxyacetylene
oxicorte m - [metal-fabr] torch cutting • [sold]
  oxyacetylene, o gas, o flame, cutting; lanc-
  ing
— m automático - [sold] automatic oxyacetylene
  cutting
oxidación f - . . .; oxidizing; rust; corrosion;
  rusting; corroding
— f anódica - [quím] anodic oxidation,dizing
— f atmosférica - [metal] atmosphere,ric oxidation
— f con gas n - [metal] gas oxidation,dizing
— f — — m de oxígeno m - [metal] oxygen gas
  oxidation,dizing
— f — — — m industrial - [metal] industrial oxygen gas oxidation,dizing
— f de eje m - [mec] axle oxidation,dizing
— f — líquido m cloacal - [sanit] sewage oxidation,dizing
— f — metal m - [metal] metal oxidation,dizing
— f — — m fundido - [metal-prod] molten metal
  oxidation,dizing
— f — sulfuro m - [quím] sulfide oxidation
— f — — m de hierro m - [quím] iron sulfide
  oxidation,dizing

oxidación f de superficie f - [metal-prod] surface oxidation,dizing
— f detenida - stopped oxidation,zing
— f evitada - [metal- prevented, o avoided,
  oxidation,dizing
— f futura - [metal] future, rust(ing), o oxidation,dizing
— f inhibida - [metal] inhibited, rust(ing), o
  oxidation,dizing
— f ligera - [metal] light, rust(ing), o oxidation,dizing
— f parcial - [metal] partial, rust(ing), o
  oxidation,dizing
— f — con gas m - [metal] gas partial oxidation,dizing
— f — — — m de oxígeno m - [metal] oxygen
  gas partial, rust(ing), o oxidation,dizing
— f — — — m — — m industrial - [metal] industrial oxygen gas partial, rust(ing), o
  oxidation,dizing
— f posterior - [quím] later, o future, oxidation,dizing
— f química - [quím] chemical oxidation,dizing
— f superficial - superficial oxidation,dizing
  • surface oxidation,dizing
— f total - [metal] total, rust(ing), o oxidation,dizing • rusting out
oxidado,da a - oxidized; corroded; rusty,ted
— a anódicamente - [metal] anodically oxidezed
— a completamente - completely, o thoroughly,
  oxidized, o corroded, o rusted
— a totalmente - [metal] totally, oxidized, o
  corroded, o rusted • rusted out
oxidante a en caliente - [sold] hot oxidizing
oxidar v - [metal] . . .; to corrode
— v anódicamente - [quím] to anodically oxidize
— v totalmente - [metal] to, oxidize, o rust. o
  corrode, totally • to rust out
óxido m - [metal] . . .; rust; corrosion
— m alto - [metal] high oxide
— m anódico - [metal] anodic oxide
— m bajo - [metal] low oxide
— m blanco - [quím] white, rust, o oxide
— m de aluminio m - [metal] aluminum oxide •
  alumina
— m — azufre m - [quím] sulfur oxide
— m — calcio m - [quím] calcium oxide • lime
— m — — m y cuarzo m - [quím] quartz/calcium,
  o calcium/quartz, oxide
— m — carbono m - [quím] carbon (mon)oxide
— m — cinc m - [metal] zinc oxide
— m — hierro m - [metal] iron oxide
— m — — m en escama(s) f - [metal-prod]
  (iron) scale
— m — — m — — f (en) seco - [metal-prod]
  dry (iron) scale
— m — — m en escoria f - [metal-prod] iron
  oxide in @ slag
— m — magnesio m - [quím] magnesium oxide •
  [miner] magnesite
— m — manganeso m - [quím] manganese oxide
— m — molíbdeno m - [miner] molybdenum oxide
— m — sodio m - [quím] sodium oxide
— m — titanio m - [metal] titanium oxide •
  [miner] titania
— m elevado - [metal] high oxide
— m excesivo - [metal] excess(ive), oxide, o
  rust
— m férrico - [miner] ferric oxide
— m — hidratado - [miner] hydrous ferric oxide
— m ferrohidroso - [miner] hydrous ferric oxide
  • limonite
— m ferroso - [miner] ferrous, o iron, oxide;
  hematite
— m —férrico - [miner] ferrous-ferric oxide;
  magnesite
— m globular - [metal] globular oxide
— m intermedio - [quím] intermediary oxide
— m laminado - [metal-lam] rolled-in scale
— m para planta f siderúrgica - [metal-prod]
  (steel) mill oxide
— m protector - [metal] protective oxide
— m puro m - [quím] pure oxide
— m — de hierro m - [quím] pure iron oxide

oxigenado,da a - [quím] oxygenated
oxígeno m atmosférico—[quím] atmospheric oxygen
— m básico - [quím] basic oxygen
— m biológico - [sanit] biological oxygen
— m — exigido - [sanit] biological oxygen de-
mand(ed)
— m bioquímico - [sanit] biochemical oxygen
— m — exigido - [sanit] biochemical oxygen de-
mand(ed)
— m disuelto - dissolved oxygen
— m — en electrolito m - [quím] electrolyte
disolved oxygen
— m en aire m - [combust] air oxygen
— m — m para tobera f - [metal-prod] tuyere
air oxygen
— m — baño m - [metal-prod] bath oxygen
— m — electrolito m—[quím] electrolyte oxygen
— m — hierro m - [metal-prod] iron oxygen
— m — m sólido - [metal] solid iron oxygen
— m gaseoso - [quím] gaseous oxygen; oxygen gas
— m — industrial - [metal-prod] industrial
oxygen gas
— m global - global oxygen
— m industrial - [ind] industrial oxygen
— m líquido - liquid oxygen
— m para acería f - [metal-prod] steel plant
oxygen
— m — f Linz-Donawitz - [metal-prod] basic
oxygen steel plant oxygen
— m — horno m - [metal-prod] furnace oxygen
— m — m eléctrico m - [metal-prod] electric
furnace oxygen
— m — lanza f - [metal-prod] lance oxygen
— m reducido - [metal-prod] reduced oxygen
— m y gas m - [combust] oxygen and gas
oxihidrogénico,ca a - véase oxhídrico,ca
oxoclorinación f - [quím] oxochlorinating,tion
oyente m - . . .; listener
ozoquerita f - [geol] ozokerite

P

P B m - véase punto Bedaux
P B I m - [econ] véase producto m bruto interno
P C - véase pérdida f por calcinación f
P E - véase peso m específico
p-H m - [hidr] p-H
P-métrico,ca a - [autom-neumát] P-metric
p p m - véase partes f por millón m
p p m m - véase partes f por millón m
pabellón f . . . • booth
— m en galería f - [com] mall booth
— m para feria f - fairgrounds pavilion
pábilo m para lámpara f - [combust] lampwick
paca f - [transp] . . .; bale
pactar v - . . .; to accord; to agree
Pacto m Andino m - [pol] Andean Pact
— m Subregional Andino - [pol] Subregional Andean Pact
padre m - . . . • parent
— m legítimo - legitimate parent
padrino m - . . . • sponsor
padrón m - . . .; list
— m de proveedor(es) m - supplier('s) list
pagadero,ra contra documento(s) m - [com] payable against @ document(s)
— a — m siguiente(s) - [com] payable a-
gainst @ following document(s)
— a — entrega f - [com] payable, against, o
(up)on, delivery
— a — f de documento(s) m - [com] payable
against delivery of document(s)
— a — f — m siguiente(s) - [com] payable against delivery of @ following documents
— a — evidencia(s) f de entrega f - [com] payable against proof of delivery

pagadero,ra a contra factura f - [com] payable
against @ invoice
— a — presentación f - [com] payable against @
presentation
— a — f de documento(s) m - [com] payable
against (presentation of) @ document(s)
— a — f — m para embarque m - [com]
payable against (@ presentation of) @ shipping document(s)
— a — f — evidencia f de entrega f—[com]
payable against @ presentation of proof of
delivery
— a — f — factura f—[com] payable against
(@ presentation of) @ invoice
— a — f primera - [com] payable against @
first presentation
— a en fondos m de . . . • [fin] payable in
. . . funds
— a por . . . a - [com] payable by . . .
— a — intermedio de . . . - [com] payable
through . . .
pagado,da a - paid (up)
— f anticipadamente - prepaid; paid in advance
— a contra - paid against
— f contra presentación f - [fin] paid ; upon,
o against, @ presentation
— a — f primera - [fin] paid, against, o
upon, first presentation
— a de más - overpaid
— a — menos - underpaid; paid partially
— a efectivamente - truly paid-up
— a directamente - paid directly
— a en efectivo - [fin] paid in cash
— a por adelantado - [fin] prepaid; paid in advance
— a realmente - truly paid(-up)
— a por transferencia f - [fin] paid by @
transfer
— a prontamente - [fin] paid promptly
pagar v - . . . • to pay off
— v anticipadamente - to, prepay, o pay in advance
— v cargo m - to pay @ charge
— v crédito m - [fin] to pay @ credit
— v de más - to overpay
— v — menos - to underpay
— v por etapa(s) f - to pay, by stages, o as
you go
— v derecho(s) m - [fin] to pay @ fee(s)
— v directamente - to pay directly
— v dividendo m - [fin] to pay @ dividend
— v honorario(s) m - [legal] to pay @ fee(s)
— v impuesto(s) - [fisc] to pay @ tax(es)
— v — m sobre, rédito(s), o renta f - [fisc]
to pay @ income tax
— v indemnización f - [seguros] to pay @ indemnity
— v interés v - [fin] to pay @ interest
— v pasivo m - [fin] to pay @ liability,ties
— v por transferencia f - [fin] to pay by @
transfer
— v prima f - [seguros] to pay @ premium
— v prontamente - [fin] to pay promptly
— v reclamo m - to pay @ claim
— v — prontamente - [seguros] to pay @ claim
promptly
pagaré m - [fin] (promissory) note
— m avalado - [fin] guaranteed promissory note
— m comercial - [fin] commercial (promissory)
note
— m — de primera (clase f) - [fin] prime promissory note
página f amarilla - [public] yellow page(s)
— f anterior - [public] previous page
— f contigua - [public] adjacent, o contiguous,
page
— f doble - [public] spread
— f — central - [public] center spread
— f indicada - [public] indicated page
— f interior - [public] inside page
— f opuesta - [public] opposite page
— f posterior - [public] back, o later, page
— m precedente - [public] preceeding page

página f que antecede - [public] preceding page
— f siguiente - [public] following page
— f verificada - [public] checked page
pago m - . . . ; pay • settlement
— m a cuenta f - [fin] payment on account
— m — realizar(se) - [fin] payment to be made
— m — vista f - [fin] (on) sight payment
— m adelantado - [fin] prepayment
— m adicional - [fin] additional payment
— m anticipado - [com] prepayment; advance(d) payment
— m — con pedido m - [com] advance payment with @ order
— m aplazado - [fin] delayed payment
— m asumido - [fin] assumed payment
— m bancario - [fin] bank payment
— m comprometido - [fin] committed, o pledged, payment
— m con orden f - [com] cash with @ order
— m contra entrega f - [com] cash on delivery • charge, o collect, on delivery
— m correspondiente - [fin] corresponding, o respective, payment
— m de almacenaje m—[transp] demurrage payment
— m — bodegaje m - [transp] demurrage payment
— m — cargo m - charge payment
— m — contrato m - contract payment
— m — crédito m - [fin] credit payment
— m — derecho(s) m - [fisc] fee(s) payment | a - fee basis
— m — interés(es) m - [fin] interest payment
— m — honorario(s) m - [legal] fee(s) payment
— m — impuesto(s) m - [fisc] tax payment
— m — — m sobre, réditos m, o renta f - [fisc] income tax payment
— m — indemnización f - [seguros] indemnity payment
— m — mismo(s) m - [fin] payment thereof
— m — pasivo(s) m - [fin] liability,ties payment
— m — pedido m - [com] order payment
— m — personal - [pers] personnel, payment, o compensation
— m — premio m - [seguros] premium payment
— m — prima f - [seguros] premium payment
— n — reclamación f - [seguros] claim payment
— m — reclamo m - [seguros] claim payment
— m — regalía f - [fin] royalty payment
— m demorado - [fin] delayed payment
— m diferido - [fin] deferred payment
— m directo - [fin] direct payment
— m efectivo - [fin] effective payment
— m efectuado - [fin] payment made
— m en compensación f - compensation payment • adjustment payment
— m en efectivo m - [fin] cash payment
— m — — contra orden f - cash with @ order
— m — exceso m - [fin] excess payment
— m — más - [fin] overpayment
— m — menos - [fin] underpayment
— m futuro - [fin] future payment
— m inicial - [fin] initial payment • down payment
— m máximo - [fin] maximum payment
— m — futuro - future maximum, o maximum future, payment
— m mínimo - [fin] minimum payment
— m — futuro - [fin] future minimum, o minimum future, payment
— m — para arriendo m - minimum, lease, o rent, payment
— m parcial - [fin] partial payment
— m — anticipado - [com] partial advance payment
— m por - [fin] payment by • payment for
— m — arriendo m - [fin] lease, o rent, payment
— m — intermedio de . . . - [fin] payment through . . .
— m — périda(s) f - [seguros] loss payment
— m — — f por cobrar v - [seguros] loss payment(s) receivable
— f — — f — — v por reaseguros m cedidos -

[seguros] loss payment(s) receivable from ceded reinsurance
pago m por pérdida(s) f por pagar v - [seguros] [seguros] loss payment(s) payable
— m — — f — por reaseguro(s) m cedidos - [seguros] loss payment(s) payable from ceded reinsurance
— m — — f — v — m cedido(s) - [segur] loss payment(s) payable from ceded insurance
— m — — f — reaseguro(s) m cedido(s) m - [seguros] loss payment(s) from ceded reinsurance
— m — — f — seguro(s) m cedido(s)—[seguros] loss payment(s) from ceded insurance
— m — trimestre m - [fin] quarterly payment • payment per quarter
— m por transferencia f - [fin] payment by @ transfer
— m previsto - [fin] foreseen payment
— m progresivo - [fin] progressive payment
— m pronto - [fin] ready, o prompt, payment
— m — de reclamo m - [seguros] claim, prompt, o ready, payment
— m realizado - [fin] payment, made, o making
— m según contrato m - [legal] contract payment • payment according to @ contract
— m suplementario - [fin] supplementary payment • bonus
— m trimestral - [fin] quarterly, payment, o instalment
— m último - [fin] last payment
— m — previsto - [fin] last foreseen payment
— m — recibido - [fin] last received payment
pago,ga a - véase pagado,da a
paila f - . . . ; [domést] pail • [ind] pan
— f de hojalata f - [comést] tin, pan, o pail
— f para miel f de caña f - [ind] sugar syrup kettle
— f redonda f - [mec] round pan
— f — de hojalata f - [mec] round tin pan
Paintgrip m - [metal-trat] Paintgrip
— m de calidad f comercial - [metal-trat] commercial quality Paintgrip
país m . . . | adv - (en) local(ly)
— m andino - [geogr] Andean country
— m aprobado - approved country
— m atractivo - [geogr] attractive country
— m atrasado - [econ] backward country
— m atrayente - [Geogr] attractive country
— m avanzado - [econ] advanced, o developed, country
— m bilingüe - [pol] bilingual country
— m caribe - [geogr] Caribbean country
— m científicamente poco desarrollado - [Econ] scientifically underdeveloped country
— m comunista - [pol] Communist(ic) country
— m con déficit m - [econ] deficit country
— m — — f de chatarra f - [metal-prod] scrap deficit, country, o nation
— m — desarrollo m incompleto - [econ] underdeveloped country
— m — — m industrial - [econ] industrially developed country
— m — — m relativo - [econ] developing country; véase también país m en desarrollo m
— m — — exceso m - [econ] surplus country
— m — — m de chatarra f - [metal-prod] scrap surplus country
— m de Caribe m - [geogr] Carribean country
— m — comprador m - [com] buyer('s), o purchaser('s), country
— f — Comunidad f (Británica) - [pol] (British) Commonwealth country
— m — destino m - [com] destination country
— m — exportación f - [com] exporting country
— m — origen m - [legal] country of origin
— m — — m Estados m Unidos de América f de Norte - (materials of), United States, o U S, origin
— m — — m para suministro , - [com] supply origin country
— m — — m de proveedor m - [econ] supplier('s) country of origin

país m de Pacto m Andino - [pol] Andean Pact
country
— m — procedencia f - [com] source country
— m — proviniencia f - [com] source country
— m — región f andina - [geogr] Andean region
country
— m desarrollado - [econ] developed country
— m — industrialmente - [econ] industrially
developed country
— f, diferente, o distinto - [pol] different
country
— m diverso - [pol] different country
— m en desarrollo m—[econ] developing country
— m — región f - [pol] region country
— m — vía(s) f de desarrollo m - [econ] developing country
— m exportador m - [econ] exporting country
— m iberoamericano - [geogr] Latinamerican,
country, o nation • Iberian-American nation
— m industrializado - [econ] industrialized
country
— m — altamente - [ind] highly industrialized
country
— m latinoamericano - [geogr] Latinamerican,
country, o nation
— m miembro - [pol] member country
— m — — Banco m Interamericano para Desarrollo - [pol] Interamerican Development
Bank member country
— m minero - [miner] mining country
— m multilingüe - [pol] multilingual country
— m no comunista - [pol] non Communist country
— m — miembro - [pol] non-member country
— m — — de Comunidad f Económica Europea -
[pol] non European Economic Community member
country
— m petrolero ⊦ [petról] oil country
— m pobre - [econ] poor country
— m — en recurso(s) m natural(es) - [econ]
natural resource(s) poor country
— m poco desarrollado - [econ] underdeveloped
country
— m proveedor - [econ] supplying country
— m — de carbón m - [com] coal supplying
country
— m regional - [geogr] regional country
— m respectivo - [pol] respective country
— m rico - [econ] rich country
— m — en mineral(es) m - [miner] mineral(s),
o ore, rich country
— m — — recurso(s) m natural(es) m - [econ]
natural resource(s) rich country
— m sede - [legal] country of incorporation
— m tercero - third country
— m trilingüe - trilingual country
— m vecino - [pol] neighbor(ing) country
paisaje m - . . . . • scene(ry)
— m rural - country landscape; countryside
— m — ondulado - [topogr] rolling, country
landscape, o countryside
— m subdesarrollado - [econ] underdeveloped
country
paja f - . . . . • [agric] mulch
pala f - [herram] . . .; spade; skid shovel;
scraper • [mec-ventil] blade
— f aporcadora - [agric] wing hiller; hiller
wing
— f carbonera f - [herram] (coal) scoop
— f cargadora f - [herram] loading shovel;
loader • power shovel
— f con cucharón m - [constr-equip] bucket
shovel
— f con punta f - [herram] spade
— f — soltador m - [herram] trip shovel
— f — — m con resorte m - [agric] spring
trip shovel
— f de vapor m - [constr] steam shovel
— f especial - [constr-equip] special shovel
— f hidráulica - [constr-equip] hydraulic
shovel
— f mecánica - [constr-equip] power, shovel, o
scraper; mechanical scraper
— f — con dos motores m - [constr-equip] twin
power scraper
pala f mecánica doble - [constr] twin scraper
— f minera f - [miner] mining shovel
— f neumática - [herram] air spade
— f niveladora f - [constr-equip] skimmer scoop
— f para arrastre m - [constr-equip] dragline,
o pull (shovel); scraper
— f — buey m - [constr-equip] scoop
— f — cereal(es) m - [herram] grain shovel
— f — grano(s) m - [herram] grain scoop
— f — irrigación f - [agric] irrigating shovel
— f — tiro m - [constr] draft, o pull, shovel
— f sin cucharón m - [constr-equip] bucketless
shovel
— f tapadora - [agric-equip] covering shovel;
shovel coverer
— f vertedora f - [agric-equip] turning shovel
palabra f clave - byword • key word
— f desusada - [filol] unusual word
— f para bienvenida f - welcome,ming word
— f — enlace m - [gram] function word
— f rara - [filol] unusual word
— f última - last word
palacio m municipal - [pol] town hall
palafrenero m - . . . . • [f.c.] brakeman
palanca f - [mec] . . .; handle; arm • link •
[autom] stick
— f accionadora - [mec] actuating lever
— f — para cambio(s) m - [mec] shift, actuating, o driving, lever
— f — — horquilla f - [mec] fork, actuating,
o driving, lever
— f — — — para cambio(s) m -
[autom-mec] shift fork, actuating, o driving,
lever
— f acodada - [mec] toggle (joint) • bearing
crank • swivel lever
— f — doble - [mec] double swivel lever
— f acodillada - [mec] toggle
— f activadora - [mec] activating, o driving,
lever
— f aflojada—[mec] released, o loosened, lever
— f ajustada - [mec] adjusted lever • set lever
— f angular - [mec] bearing crank • bell crank
— f articulada - [mec] hinged lever
— f — para acelerador m - [mec] hinged throttle lever
— f — — — m provista con resorte m - [mec]
spring loaded hinged throttle lever
— f auxiliar - [mec] auxiliary lever • relay
lever
— f — para regulación f - [mec] auxiliary control lever
— f avanzada - [mec] advanced, o forward,
moved, o pushed, lever
— f balancín - [mec] rocker lever
— f cambiada - [mec] shifted lever
— f colocada - [mec] positioned, o set, lever
— f de baquelita f - [mec] bakelite lever
— f dentada - [mec] geared, o rack, lever
— f — de baquelita f - [mec] bakelite geared
lever
— f derecha - [mec] right lever • right handle
— f — en consola f hacia derecha f - [mec]
right console outside lever
— f — — — f — izquierda f - [mec] left console inside lever
— f — — — f para mano f derecha - [grúas]
right console outside, lever, o handle
— f — — — f — derecha - [grúas] left
console inside, lever, o handle
— f desplazadora—[mec] shifter, lever, o shaft
— f empujada - [mec] pushed lever
— f en consola f - [mec] console, lever, o
handle
— f — — f derecha - [mec] right console, lever, o handle
— f — — f izquierda - [mec] left console, lever, o handle
— f engranada - [mec] engaged lever
— f equilibrada - [mec] balanced lever •
counterweight lever
— f exterior - [mec] outside, o external, lever

palanca f fijada - [mec] set lever • fastened
lever
— f fijadora - [mec] fastening lever
— f formadora - [mec] forming lever
— f girada - [mec] turned lever
— f golpeteada - [mec] struck lever
— f hacia derecha f - [mec] right, lever, o
handle
— f — izquierda f - [mec] left, lever, o
handle
— f hidráulica - [mec] hydraulic lever
— f — abierta - [mec] open(ed) hydraulic lever
— f — auxiliar - [mec] auxiliary hydraulic
lever
— f hidrostática - [mec] hydrostatic lever
— f — mantenida - [mec] held hydrostatic lever
— f inspeccionada - [mec] inspected lever
— f interior - [mec] inside, lever, o handle
— f izquierda - [mec] left, lever, o handle
— f — en consola f hacia derecha f - [mec]
right console inside, lever, o handle
— f — — — izquierda f - [grúas] left
console outside, lever, o handle
— f — — — f para mano f derecha - [grúas]
right console inside, lever, o handle
— f — — — f izquierda - [grúas] left
console outside, lever, o habdle
— f mantenida - [mec] held lever
— f manual - [mec] manual, o hand, lever
— f — para cebadura f - [Bombas] manual, o
hand, priming lever
— f — — consola f - [ind] console hand lever
— f — — f para perforador m - [petról]
driller('s) console hand lever
— f movida - [mec] moved lever
— f — hacia adelante - [mec] forward moved
lever
— f operada - [mec] operated lever
— f oscilante - [mec] rocker, o floating, lever, o arm
— f — para tubo(s) m - [herram] rocking pipe
lever
— f para accionamiento n - [mec] activating, o
driving, lever
— f — acelerador m - [comb.int] throttle lever
— f — — m fijada - [mec] set throttle lever
— f — — m floja - [comb.int] loose throttle
lever
— f — regulador m - [comb.int] throttle lever
— f — — m para admisión f - [comb.int] governor throttle lever
— f — ajustar v resorte m - [mec] spring winding lever
— f — ajuste m - [mec] setting, o winding,
lever
— f — altura f—[mec] height, lever, o control
— f — — f para cabezal m - [agric-equip]
header height lever
— f — — f espigadora f - [agric-equip]
header height lever
— f — apretar resorte m - [mec] spring winding
lever
— f — árbol m - [mec] shaft lever
— f — — m para desplazador m - [mec] shifter
shaft lever
— f — — m para freno m - [mec] brake shaft
lever
— f — arrollamiento m - [mec] winding lever
— f — — m para resorte m - [mec] spring winding lever
— f — avance m - [mec] forward, o feed, lever,
o rod; accelerator
— f — — m para carrillo m - [mec] carriage
feed lever
— f — — m para desplazador m - [mec] slide
feed lever
— f — — m — — m transversal - [mec] cross
slide feed lever
— f — — m y retroceso m - [mec] forward and
reverse lever
— f — banda f - [mec] band lever
— f — bomba f.- [bombas] pump lever
— f — — f para aspiración f - [comb.int]

lift(ing) pump lever
palanca f para bomba f para elevación f - [comb.
int] lift(ing) pump lever
— f — f— — f interna,rior - [bombas] internal lifting pump lever
— f — — — f para combustible m - [comb.
int] fuel lift(ing) pump lever
— f — botador m - [mec] kicker lever
— f — cable m - [grúas] cable handle
— f — — m para arrastre m - [grúas] pull
cable handle
— m — — m para gancho m - [grúas] whip handle
— f — — m — motón m - [grúas] main (cable)
handle
— f — — m — tracción f - [grúas] pull(ing)
cable handle
— f — cambio(s) m - [mec] shift(er), o change,
lever • [autom-mec] shift(ing) lever
— f — — m de marcha f - [autom] gear shift, o
change, lever; stick
— f — — m — f colocada - [mec] placed
(gear) shift lever
— f — — m — f engranada - [mec] engaged
gear shift lever
— f — — m — velocidad(es) f - [mec] gear, o
transmission, shift lever
— f — — m — f colocada f - [mec] placed
gear shift lever
— f — — m — — f movida - [autom-mec] moved,
gear shift lever, o stick
— f — — m para transmisión f - [mec] transmission shift handle
— f — carburador m - [comb.int] carburetor
link
— f — carrete m - [mec] spool lever
— f — — m para cuchareo m - [petról] sand
reel, handle, o lever
— f — — m trabada - [agric-equip] locked
spool lever
— f — carro,rillo m - [sold] carriage arm
— f — — m para avance m - [sold] travel carriage arm
— f — cebadura f - [bombas] priming lever
— f — — f manual - [bombas] hand, o manual,
priming lever
— f — — f para bomba f - [comb.int] pump
priming lever
— f — — f — — f para elevación f - [comb.-
int] lifting pump priming lever
— f — — f — — — f para combustible m
- [comb.int] fuel lifting pump priming lever
— f — cierre m - [mec] closing lever
— f — conexión f - [mec] link(ing) lever
— f — — f para regulador m - [mec] throttle,
o control, link lever
— f — — — m para marcha f lenta - [mec]
idler link lever
— f — — f — m — — f sin carga f - [mec]
idler link lever
— f — consola f hacia derecha f - [mec] right
console, lever, o handle
— f — — f hacia izquierda f - [mec] left console, lever, o handle
— f — contacto m - [mec] contact arm •
[electr-equip] trigger bar
— f — contramarcha f - [mec] reversing lever
— f — contrapeso m - [mec] counterweight lever
— f — control m - [mec] control lever
— f — charnela f - [mec] knuckle, arm, o bar
— f — — f para dirección f - [autom-mec]
steering knuckle arm
— f — chispa f - [comb.int] spark lever
— f — desacoplamiento m - [mec] release, o
throw-out, lever
— f — — m para embrague m - [mec] clutch,
release, o throw-out, lever
— f — desembragar v - [mec] release arm
— f — desembrague v - [mec] release arm
— f — — m para carro m - [sold] carriage release arm
— f — — m — — m para avance m - [sold]
travel carriage release arm
— f — desplazador - [mec] shifter lever

palanca f para dirección f - [autom-mec] steering lever
— f — eje m - [mec] axle, o shaft, lever
— m — m de botador m - [m3d] kicker shaft lever
— f — elevación f - [grúas] hoist(ing), o lift(ing) lever
— f — f manual - [mec] hand, o manual, lift(ing) lever
— f — f para aguilón m - [grúas] boom hoist, lever, o handle
— f — embrague m - [mec] clutch, lever, o finger
— f — m desengranado - [mec] disengaged clutch lever
— m — m para accionamiento m - [mec] main drive clutch
— f — m — m principal - [mec] main drive clutch lever
— f — m para implemento m - [agric-equip] implement clutch lever
— f — m — m desengranada - [agric--equip] disengaged implement clutch lever
— f — elevación f - [mec] lift)ing) lever
— f — estrangulación f - [comb.int-carburad] choke, lever, o rod
— f — estrangulador m - [comb.int-carburad] choke, lever, o rod
— f — expulsor m - [mec] kicker lever
— f — fiador m - [mec] tripper lever
— f — flotador m - [mec] float lever
— f — formación f - [mec] forming lever
— f — freno m - [mec] brake lever
— f — m ajustada - [mec] set brake lever
— f — m colocada - [mec] positioned brake lever
— f — m — en posición f - [mec] positioned brake lever
— f — m para árbol m - [mec] shaft brake lever
— f — m para carrete m - [mec] reel brake lever
— f — m — m para núcleo m - [petról] core reel brake lever
— f — m — malacate m - [petról] calf brake lever
— f — m — — m para cable m - [petról] calf wheel brake lever
— f — m — m — m para tubería f - [petról] calf wheel brake lever
— f — m — m — herramienta f—[petról] bull wheel brake lever
— f — m — — m para tubería f - [petról] calf wheel brake lever
— f — m para mano f - [mec] hand brake lever
— f — m — tambor m - [petról] wheel brake lever
— f — m — m para herramienta(s) f - [petról] bull wheel brake lever
— f — m para cable m - [grúas] cable brake lever
— f — m — — m a motón m - [grúas] main brake lever
— f — m — — m principal - [grúas] main brake lever
— f — principal - [mec] main brake lever
— f — m puesta a punto m - [mec] set brake lever
— f — m trabada - [mec] locked brake lever
— f — gancho m - [grúas] whip handle
— f — giro m—[grúas] swing, lever, o handle
— f — gobierno m - [mec] control, handle, o lever
— f — grillete m - [mec] clevis lever
— f — horquilla f - [mec] clevis lever
— f — f para cambio(s) m - [mec] shift fork lever
— f — inclinación f - [mec] tilt(ing) lever
— f — leva f - [mec] cam lever
— f — luz f (indicadora de virajes m) [autom] turn signal lever
— f — malacate m - [petról] reel, handle, o lever

palanca f para mando m - [mec] control, lever, o rod • activating lever
— f — m para regulador m - [comb.int] governor control lever
— f — manecilla f - [instrum] pointer activator
— f — mano f - [mec] hand lever
— f — f lubricada - [mec] lubricated hand lever
— f — f para consola f - [petról] console hand lever
— f — f — f para perforador m - [petról] driller('s) console hand lever
— f — f — f — m lubricada - [petról] lubricated driller('s) console hand lever
— f — marcha f atrás - [mec] reversing lever
— f — mariposa f - [comb.int-carburad] butterfly valve, lever, o rod
— f — motón m - [grúas] main handle
— f — operación f - [mec] operating rod
— f — pie m - [mec] treadle
— f — reactor m - [electr] reactor, handle, o lever
— f — regulación f - [mec] control, o governor, o throttle, control, o handle
— f — f colocada - [mec] positioned control lever
— f — f hidráulica - [mec] hydraulic control lever
— f — f hidrostática - [mec] hydrostatic control lever
— f — f — mantenida - [mec] held, o maintained, hydrostatic control lever
— f — f para velocidad f - [mec] hydrostatic speed control lever
— f — f para alimentación f - [mec] feed(ing) control lever
— f — f — altura f - [mec] height control lever
— f — f — f para espigadora f - [agric] header height control lever
— f — f — f — f cabezal m - [agric] header height control lever
— f — f para dirección f - [autom] steering control lever
— f — f para estrangulador m - [comb.int] choke, o throttle, control lever
— f — f — giro m - [grúas] swing control lever
— f — f — m para tornillo m sin fin - [agric] auger swing control lever
— f — f — rotación f - [grúas] swing control lever
— f — f — f para tornillo m sin fin - [agric] auger swing control lever
— f — f — tornillo m sin fin - [mec] auger control lever
— m — m — — para descarga f - [agric] unloading auger control lever
— f — f — trinquete m para traba f - pawl control lever
— f — f — — m para traba f - [grúas] lock pawl control (lever)
— f — f — — f desengranada - [grúas] disengaged lock pawl control lever
— f — f — válvula f - [válv] valve control lever
— f — f — f — f para acelerador m - [comb.-int] throttle control valve
— f — f — f — f mariposa - [válv] butterfly valve control, lever, o throttle
— f — f — f — para acelerador m - [comb.int] throttle (valve) control lever
— f — f — velocidad f - [mec] speed control lever
— f — f — — f para motor m - [comb.int] engine speed control lever
— f — f — — f variable - [mec] variable speed control lever
— f — regulador m - [mec] control, lever, o handle

palanca f para regulador m para alimentadora f -
[mec] feed(ing) control, lever, o handle
— f —— m — velocidad f - [comb.int] speed
throttle lever
— f —— m —— f alta - [comb.int] high
speed throttle lever
— f —— m —— f — sin carga f -
[comb.int] high idle speed throttle lever
— f —— m —— f baja - [comb.int] low speed
throttle lever
— f —— m —— f — sin carga f - [comb.int]
low idle speed throttle lever
— f —— m —— f variable - [mec] variable
speed control lever
— f —— m —— f para motor m - [mec]
motor variable speed control lever
— f —— f —— f —— m hidráulico -
- [mec] hydraulic motor variable speed control lever
— f —— m movida - [comb.int] moved throttle
lever
— f —— m regulado - [comb.int] positioned
throttle lever
— f — regular y profundidad f - depth lever
— f — resorte m - [mec] spring lever
— f —— m en espiral - [mec] coil, o winding,
spring lever
— f — retorno m - [mec] return lever
— f —— m de resorte m - [mec] spring return
lever
— f — retroceso m - [mec] reverse, o back-up,
lever
— f — rotación f - [grúas] swing, lever, o
handle
— f —— f para tornillo m sin fin - [mec]
auger swing lever
— f — soltador m - [mec] release lever
— f — soltar v - [mec] release lever
— f —— v sujetador m - [mec] grip release
lever
— f — suelta f - [mec] release lever
— f —— f de sujetador m - [mec] grip release
lever
— f — tambor m - [mec] reel, lever, o handle
— m —— m para cuchareo m - [petról] sand
reel, lever, o handle
— f — tirar v - [mec] pull, lever, o handle
— f — tornillo m - [mec] screw lever
— f —— m para marcha f lenta f - [mec] idle
speed screw lever
— f —— m —— f sin carga f - [mec] idle
speed screw lever
— f —— m sin fin - [mec] auger lever
— f —— m —— para descarga f - [agric-
-equip] unloading auger, o auger unloading,
lever
— f — torno m - [tornos] vise, lever, o
handle • [petról] cathead lever
— f —— m para banco m - [tornos] vise,
lever, o handle
— f — traba f - [mec] lock(ing) lever
— f — transferencia f - [mec] relay lever
— f — transmisión f - [mec] transmission,
lever, o handle • shaft handle
— f — trinquete m - [mec] pawl handle
— f —— m para traba f - [mec] lock pawl
handle
— f — tuerca f - [mec] nut lever
— f —— f hendida - [mec] split nut lever
— f —— f — para roscado m - [mec] thread
cutting split nut lever
— f —— f partida - [mec] split nut lever
— f —— f — para roscado m - [mec] thread
cutting split nut lever
— f — válvula f - [válv] valve lever
— f —— f mariposa - [válv] butterfly valve
lever
— f —— f neumática - [valv] air valve lever
— f —— f — para torno m - [petról] cathead
air valve lever
— f —— f para aire m - [neumát] air valve
lever
— f —— f —— m para torno m - [petról]
cathead air valve lever
palanca f para válvula f para aprovisionamiento
m - [ind] supply valve lever
— f —— f —— m de quemador m - [Combust]
burner supply valve lever
— f —— f para torno m - [petról] cathead
valve lever
— f — vástago m - [mec] rod, o shaft, lever
— f —— m para válvula f - [válv] valve shaft
lever
— f —— m —— f mariposa - [válv] butterfly
valve safety lever
— f —— m y grillete m - [válv] rod and clevis lever
— f —— m —— m para válvula t - [válv]
valve rod and clevis lever
— f —— m y horquilla f - [mec] rod and clevis lever
— f —— m —— f para válvula f - [mec]
valve rod and clevis lever
— f — velocidad f - [mec] speed lever
— f principal - [mec] main lever
— f proyectada - [mec] designed, lever, o handle
— f puesta - [mec] placed, o put, lever
— f — a punto m - [mec] set lever
— f — en marcha f atrás - [mec] lever put into
@ reverse
— f —— posición f para marcha f atrás -
[mec] lever put into @ reverse (position)
— f —— f para retroceso m - [mec] lever
put in(to) @ reverse
— f — retroceso m - [mec] lever put in(to)
@ reverse
— f regulada - [mec] positioned lever - controlled lever
— f reguladora - [mec] control(ling), o governor, lever, o handle
— f — para velocidad f - [mec] speed control
lever
— f sencilla - [mec] simple lever
— f soltada - [mec] released lever
— f soltadora - [mec] release,sing lever
— f — para sujetador m - [mec] grip release,sing lever
— f tercera - [mec] third lever
— f tirada - [mec] pulled lever
— f trabada - [mec] locked lever • stuck lever
— f trabadora - [mec] lock(ing) lever
palangre m - [pesca] . . .; setline
palanquilla f - [metal-prod] billet(s) • square
billet(s) • gad
— f austenítica—[metal-prod] austenitic billet
— f colada - [metal-prod] cast billet
— f común - [metal-prod] common billet
— f de acero m - [metal-prod] steel billet
— f —— m austenítico - [metal-prod] austenitic steel billet
— f —— m con carbono m - [metal-prod] carbon
steel billet
— f —— m —— m para alambrón m - [metal-
-lam] carbon steel wire rod billet
— f —— m no austenítico - [metal-prod] non-
-austenitic steel billet
— f fabricada por colada f continua - [metal-
-prod] continuous(ly) cast billet
— f fiscalizada - [metal-lam] controlled billet
— f no austenítica - [metal-prod] nonaustenitic
billet
— f para alambrón m - [metal-lam] wire rod billet
— f —— barra(s) f - [metal-prod] bar billet
— f —— para armadura f - [metal-lam] reinforcing bar billet
— f —— f —— m para hormigón m - [metal-
-lam] concrete reinforcing bar billet; rebar
billet
— f — perfil(es) m (de acero m) - [metal-lam]
profile, o shape, billet
— f —— m —— m con carbono m - [metal-lam]
carbon steel profile billet
— f plana f - [metal-prod] flat, o universal, billet
— f por colada f continua - [metal-col.cont]

continuous(ly) cast billet
palanquilla f primera - [metal-lam] first billet
— **f producida por tercero(s) m** - [metal-lam]
billet(s) produced by others/third party
— **f regulada** - [metal-lam] controlled billet
— **f Sivensa** - [metal-lam] Sivensa billet
— **f última** - [metal-lam] last billet
— **f universal** - [metal-lam] universal billet
palastro m - [metal-lam] . . .; slab; rolled,
steel, o iron, o sheet • plate iron
— **m de acero m** - [constr] steel sheet(ing)
— **m laminado** - [metal-lam] rolled, iron, o
steel
— **m — en caliente** - [metal-lam] hot rolled,
iron, o steel
— **m — — frío** - [metal-lam] cold rolled,
iron, o steel
paleado m - [metal-prod] working
paleado,da a - [metal-prod] worked
paleadora f - [constr-equip] power shovel
palear v - . . . • to shovel
paleta f - . . . [constr] (mason's) trowel;
(mortar) hoe • [metal-prod] skimmer; baffle
• [metal-prod-amasadora] (pug mill) blade •
[metal-prod-sinterizadora] sinter breaker
blade • [constr-equip] paddle • [mec-ventil]
blade • [mec] plow • flight • button •
[turb] vane
— **f adaptable** - [mec-ventil] adaptable blade;
Adaptair*
— **f Adaptair*** - [mec-ventil] Adaptair blade
— **f colada** - [mec] cast blade
— **f con inclinación f excesiva** - [mec-ventil]
overpitched blade
— **f — — f insuficiente** - [mec-ventil] under-
pitched blade
— **f de acero m** - [mec] steel blade • [herram]
steel trowel
— **f — — m austenítico** - [mec] austenitic
steel blade
— **f — — m con aleación f austenítica** - [mec]
austenitic alloy steel blade
— **f — — m inoxidable** - [mec] stainless steel
blade
— **f — — m — austenítico** - [mec] austenitic
stainless steel blade
— **f — — m martensítico** - [mec] martensitic stainless steel blade
— **f — — m martensítico** - [mec] martensitic
steel blade
— **f — — m para máquina(ria) f** - [herram]
machine steel trowel
— **f aleación f** - [herram] alloy blade
— **f — — f con níquel** - [mec] nickel base(d)
alloy blade
— **f — — f — titanio m** - [mec] titanium
base(d) alloy blade
— **f — — f ligera** - [mec] light alloy blade
— **f fabricación f especial** - [mec-ventil]
customized blade
— **f — — f propia** - [mec-fentil] customized,
o self manufactured, blade
— **f madera f** - [herram] wooden paddle
— **f desequilibrada** - [mec-ventil] unbalanced,
o out-of-balance, blade
— **f direccional** - [turb] directional vane
— **f — para flujo m** - [turb] flow directional
vane
— **f directriz** - [turb] guide,ding vane
— **f — completa** - [turb] complete guide,ding
vane
— **f — en posición f abierta** - [turb] open(ed)
position guide,ding vane
— **f — — — f cerrada** - [turb] closed position guide,ding vane
— **f en desequilibrio m** - [mec-ventil] unbalanced, o out-of-balance, blade
— **f — volante m** - [mec] flywheel wing
— **f equilibrada f** - [mec-ventil] balanced
blade
— **f — con pieza f metálica** - [mec-ventil]
balanced blade with @ iron

paleta f estándar - [mec-ventil] standard blade
• [turb] standard vane
— **f fija** - [mec-ventil] stationary blade •
[turb] stationary vane
— **f fijada** - [cabl] attached button
— **f floja** - [mec=ventil] loose blade
— **f forjada** - [mec-ventil] forged blade
— **f fuera de equilibrio m** - [mec-ventil]
out-of-balance blade
— **f giratoria** - [mec-ventil] rotating blade
— **f instalada** - [mec-ventil] installed blade
— **f — inapropiadamente** - [mec-ventil] improperly installed blade
— **f labrada** - [mec-ventil] machined blade
— **f mecánica f** - [herram] machine trowel
— **f — de acero m** - [hormig] machine steel
trowel
— **f mecanizada** - [mec-ventil] machined blade
— **f movible** - [mec] movable blade
— **f para agitador m** - [mec] agitator blade
— **f — albañil m** - [herram] mason('s) trowel
— **f — amasadora f** - [metal-prod] pug mill,
blade, o paddle
— **f — — f para colector m (para polvo m)** -
[metal-prod] dust catcher pug mill blade
— **f — — f — — m — primario** - [metal-
-prod] primary dust catcher pug mill blade
— **f — — f — — m — secundario** - [metal-
-prod] secondary dust catcher pug mill blade
— **f — — f — m primario** - [metal-prod]
primary dust catcher pug mill blade
— **f — — f — m secundario** - [metal-prod]
secondary dust catcher pug mill blade
— **f — — f — síntar m** - [metal-prod] sinter
pug mill blade
— **f — ceniza(s) f** - [combust] ash paddle
— **f — escoria f** - [metal-prod] slag skimmer
— **f — extracción f de ceniza(s) f** - [combust]
ash paddle
— **f — mano f** - [herram] hand trowel
— **f — — f de acero m** - [herram] steel hand, o
hand steel, trowel
— **f — purga f** - [metal-prod] skimmer (gate)
— **f — — f para sifón m** - [metal-prod] skimmer
gate; splasher
— **f — rompedora f** - [miner] breaker blade
— **f — — f para sínter m** - [metal-prod] sinter
breaker blade
— **f — roturadora f** - [metal-prod] breaker
blade
— **f — — f para sínter m** - [metal-prod] sinter
breaker blade
— **f — sifón m** - [metal-prod] splasher (blade)
— **f — sopladora f** - [mec] blower blade
— **f — turbina f** - [mec] turbine, vane, o blade
— **f — ventilador m** - [mec] ventil] fan blade
— **f — — m por aspiración f** - [mec] sucker fan
blade
— **f, quitada, o removida, o sacada** - [mec] removed blade
— **f separadora** - [metal-prod] splasher (blade)
— **f — para escoria f** - [metal-prod] slag, runner gate, o skimmer
— **f suelta** - [mec-ventil] loose blade •
[turb] loose vane
paleteado m - shoveling; spading
— **m manual** - [constr] hand spading
paliación f - . . .; reducing,ction
paliado,da a - palliated; reduced
paliar v - . . .; to reduce
paliativo m - . . . • recourse
palidecer v - . . . . • to dim
— **v luz f** - [ilumin] to dim @ light
palidecido,da a - paled • dimmed
palidecimiento m - paling • dimming
— **m de luz f** - [ilumin] light dimming
palier* n - [autom-mec] axle shaft
palillo m (envuelto) con algodón m - [medic]
Q-tip
palista* m - [ind] véase paleador m
palmitato* m - [quím] palmitate
— **m de aluminio m** - [quím] aluminum palmitate
palo m - . . . . • [constr] upright; support •

[náut] spar
palo m grueso - big, o thick, stick • [fam] top brass; big wig
— m para golf m - [deport] golf club
— m — moldear⁻ - [mec] dogtail
palote m - . . . • stroke; tally
palpado,da a - touched; felt • [penal] frisked
— a de armas f - [penal] frisked
palpadura f - . . . • [penal] frisking
— f de armas f - [penal] frisking
palpar v - . . . • [penal] to frisk
— v de armas f - [penal] to frisk
— v rejilla f - [mec] to feel @ screen
— v — f para colador m - [mec] to feel @ strainer screen
pallete f de madera f - [hidr] timber fender
— m — f sin tratar v - [hidr] untreated timber fender
panaché m de legumbre(s) f - [culin] vegetable plate
panal m para radiador m - [comb.int] radiator honeycomb
pandeado,da a - warped; buckled; bulged; bent
pandear v - to warp; to buckle; to bulge • to bend
pandearse v - to buckle
pandeo m - . . . ; (ring) buckling
— m anterior - [mec] previous buckling
— m crítico - critical buckling
— m de anillo m - [mec] ring buckling
— m — aro m - [mec] ring buckling
— m — columna f - [constr] column buckling
— m — flexión f - [mec] buckling
— m — pared f - [constr] wall buckling
— m — tipo m columnar - [constr] column type buckling
— m — punto m - [constr] local buckling
— m — — m evitado - [constr] prevented local buckling
— m — zona f - [constr] local buckling
— m — — f evitado - [constr] prevented, o avoided, area buckling
panel m - . . . • véase también tablero m
— m abisagrado - [mec] hinged panel
— m acústico - [constr] accoustical panel
— m aislador⁻[electr-instal] insulating panel
— m — de plomo m - [electr-instal] lead insulating panel
— m aluminizado - [constr] aluminized, panel,o panelling
— m — para pared f - [constr] aluminized wall panel(ling)
— m — — techo(s) m - [constr] aluminized roof panel(ling)
— m amovible - [mec] removable panel
— m anterior - [constr] front panel • former, o previous, panel
— m antisonoro - [constr] accoustical panel
— m con borne(s) m - [electr-instal] terminal strip • stud panel
— m — — m accesible - [sold] handy terminal strip
— m — — m con pieza f corrediza - [sold] fanning strip assembly
— m — — m para conductor(es) m - [electr-instal] lead(s) terminal strip
— m — — m — — m a alimentadora f - [sold] feeder lead(s) terminal strip
— m — — m — — m a alimentadora f para alambre m - [sold] wire feeder lead(s) terminal strip
— m — — m — — m para regulación f - [sold] control lead(s) terminal strip
— m — — m — — — f para alimentadora f - [sold] feeder control lead(s) terminal strip
— m — — m para conexión(es) f - [sold] connection terminal strip
— m — — m — — f para conductor m - [sold] lead(s) connection terminal strip
— m — — m — — f — — m para regulación f - [sold] control lead(s) connection terminal strip
— m — — m — — f — — m para ali-

mentadora f - [sold] feeder control lead(s) connection terminal strip
panel m con borne(s) m para conexión f para conductor m para regulación f para alimentadora f para alambre m - [sold] wire feeder control lead connection terminal strip
— m — — m — elemento m (portátil) - [sold] pod terminal strip
— m — — m para telemando n - [sold] (remote) control pod terminal strip
— m — — m — regulación f desde distancia f para voltaje m - [sold] remote voltage control pod terminal strip
— m — — m — — — f de voltaje m - [sold] voltage control pod terminal strip
— m — — — — m para telemando m—[sold] control pod terminal strip
— m — — m — equipo m motor/interruptor m automático - [sold] power pack/contactor kit terminal strip
— m — — m fuente f para energía f - [sold] power source terminal strip
— m — — m — generador m - [electr-prod] generator terminal strip
— m — — m selector m para voltaje m - [electr-equip] voltage selector, terminal strip, o stud panel
— m — — m numerado - [electr-instal] numbered terminal, block, o strip
— m — — m terminal(es) m - [sold] terminal strip
— m — — m para amplificador m - [electrón] amplifier terminal strip
— m — — m — — — m magnético - [electrón] magnetic amplifier terminal strip
— m — terminal(es) m - [sold] terminal strip
— m contra polvo m - [mec] dust (shield) panel
— m corriente - [constr] conventional panel
— m de acero m - [constr] conventional steel panel
— m — — — m común para pared f - [constr] plain steel conventional wall panel
— m — — — — m techo m - [constr] plain steel conventional roof panel
— m — — — m galvanizado - [constr] galvanized steel conventional panel
— m — — — m común - [constr] plain galvanized (steel) conventional panel
— m — — — — para pared f - [constr] plain galvanized conventional wall panel
— m — — — m — — techo m - [constr] plain galvanized conventional roof panel
— m corriente - [constr] conventional panel
— m — para pared(es) f - [constr] conventional wall panel
— m — techo(s) m - [constr] conventional roof panel
— m — — m de acero m - [constr] conventional steel roof panel
— m — — m — m galvanizado - [constr] plain galvanized steel conventional roof panel
— m — — m — — m — común - [constr] plain galvanized steel conventional roof panel
— m cortado - [mec] cut panel
— m — en obra f - [constr] field cut panel
— m — taller m - [constr] shop cut panel
— m cubrepuerta(s) f - [constr] door cover(ing) panel
— m de acero m - [metal-fabr] steel panel
— m — m de Armco - [metal-fabr] Armco, steel, o Steelox, panel
— f — m — — m para cartelera(s) f (para avisos) - [metal-fabr] Armco Steelox panel for poster panel(s)
— m — — m — — letreros m (para avisos) - [metal-fabr] Armco Steelox panel for poster panel(s)
— m — — m engatillado - [metal-fabr] interlockng steel panel
— m — — m — en (forma f de) U - [metal-fabr] channel (shaped) int erlocking steel panel
— m — — m galvanizado - [metal-fabr] galva-

nized steel panel
panel m de acero m galvanizado común - [metal-
-fabr] plain galvanized steel panel
— m — — m **engatillado m** - [metal-fabr]
interlocking galvanized steel panel
— m — — m **en (forma f de) U** - [metal-
-fabr] channel (shaped) interlocking galvanized steel panel
— m — — m **para pared(es) f** - [metal-fabr]
galvanized steel wall panel
— m — — m **techo m** - [constr] galvanized steel roof panel
— m — — m **para anuncio(s) m** - [metal-fabr]
Steelox poster panel
— m — — m **aviso(s) m** - [metal-fabr]
Steelox poster panel
— m — — m **cartelera(s) f** - [metal-fabr]
Steelox poster panel
— m — — m **para techo(s) m** - [constr] steel
roof panel
— m — — m **Steelox** - [metal-fabr] Steelox
panel
— m — — m — —, **aviso(s) m, o cartelera(s)**
f - [metal-fabr] Steelox poster panel
— **f** — **atrás** - back, o rear, panel
— m — **baquelita f** - [constr] bakelite panel
— m — **madera f** - [constr] wood panel
— m — **micarta f** - [electr-instal] micarta
panel
— m **de piedra f en seco*** - [constr] drywall
panel
— m — **plomo m** - [mec] lead panel
— m — — m **para aislación f** - [electr-instal]
lead insulating panel
— m — **yeso m** - [constr] gypsum panel
— m — — m **y madera f** - [constr] gypsum and
wood panel
— m **delantero** - [mec] front panel • [sold]
case wraparound
— m **deprimido** - [mec] recessed panel
— m **deslizable** - [constr] sliding panel
— m — **para ventilación f** - [constr] ventilation sliding, o sliding ventilation, panel
— m **dorsal** - [mec] back panel
— m — **deprimido** - [mec] recessed rear panel
— m — **hundido** - [mec] recessed rear panel
— m **empernado** - [mec] bolted panel
— m **embridado** - [mec] flanged panel
— m **en blanco** - blank panel
— m **(forma f de) U** - [constr] channel
(shaped) panel
— m **engatillado** - [mec] interlocking panel
— m — **en (forma f de) U** - [constr] channel
(shaped) interlocking panel
— m **estándar** - [constr] standard panel
— m — **para muro(s) m tipo cajón m** - [constr]
standard bin-wall panel
— m **frontal** - [mec] front panel
— m — **deprimido** - [mec] recessed front panel
— m — **hundido** - [mec] recessed front panel
— m — **para caja f** - [mec] case, o housing,
front panel
— m — **para manipulador m (para tonos m)** -
[electrón] (tone) keyer front panel
— m — — **máquina f** - [mec] machine front
(panel)
— m — **para parte f de centro m** - [mec] center
section front panel
— m — — **soldadora f** - [sold] welder front
(panel)
— m — **vertical** - [mec] vertical front panel
— m **horizontal** - [mec] horizontal panel
— m — **embridado** - [mec] horizontal flanged
panel
— m **hundido** - [mec] recessed panel
— m **inferior** - [mec] lower, o bottom, panel
— m — **deprimido** - [mec] recessed, lower, o
bottom, panel
— m — **hundido** - [mec] recessed, lower, o bottom, panel
— m — **para caja f** - [mec] case, o housing,
lower, o bottom, panel
— m — **para soldadora f** - [sold] welder, bot-

tom, panel
panel m insertado - inserted panel
— m **inserto** - inserted panel
— m **insonoro** - [constr] accoustical panel
— m **suspendido** - [constr] suspended accoustical panel
— m **izquierdo** - [mec] left (side) panel
— m **lateral** - [mec] side panel
— m — **deprimido** - [mec] recessed side panel
— m — **derecho m** - [mec] right side panel
— m — — **para caja f** - [mec] right side, case,
o housing, panel
— m — — **para soldadora f** - [sold] welder
right side panel
— m — **izquierdo** - [mec] left side panel
— **f** — — **para caja f** - [mec] left side, case,
o housing, panel
— m — — **para soldadora f** - [sold] welder left
side panel
— m — **para caja f** - [mec] case, o housing,
side panel
— m — — **cambiador m** - [mec] changer side panel
— m — — — m **para toma(s) f** - [electr-equip]
tap changer side panel
— m — — **motor** - [electr-mot] side motor, o
motor side, panel • [comb.int] side engine, o
engine side, panel
— m — **para soldadora f** - [sold] welder side
panel
— m **liso** - [constr] smooth panel • flush panel
— m **luminoso** - [electr-instal] luminous panel •
beacon
— m **para acceso m** - [mec] access panel
— m — — m **para caja f** - [mec] case, o housing, access panel
— m — — m **para respaldo m** - [mec] rear access
panel
— m — — m — — m **para caja f** - [mec] case,
o housing, rear access panel
— m — — m — — **soldadora f** - [sold]
welder, o case, rear access panel
— m — — m — **soldadora f** - [sold] case, o
housing, access panel
— m — — m **removido** - [mec] removed access
panel
— m — **aislación f** - [mec] insulating panel
— m — — f **para conductor(es) m** - [electr-
-instal] lead(s) insulting panel
— m — **alivio m** - [segurid] relief panel
— m — — m **contra explosión(es) f** - [segurid]
explosion relief panel
— m —, **anuncio(s) m, o aviso(s) m** - poster
panel
— m — **caja f** - [mec] case, o housing, panel
— m — **cambiador m** - [mec] changer panel
— m — — m **para toma(s) m** - [electr-equip]
tap changer panel
— m — **cartel(es) m** - poster panel
— m — **cierre m** - [sold] sealing panel
— m — **cobertura f** - cover(ing) panel
— m — **conductor(es) m** - [electr-instal]
lead(s) panel
— m — **cubierta f** - [mec] cover panel
— m — **edificio(s) m** - [constr] building panel
— m — **fondo m** - [mec] back, o rear, panel
— m — **frente m** - [mec] front panel
— m — **guardabarro(s) m** - [autom] fender panel
— m — — m **delantero** - [autom] front fender
panel
— m — — m **trasero** - [autom] rear fender panel
— m — **lectura f** - [instrum] reading panel
— m — **letrero(s) f** - [constr-vial] sign panel
— m — **manipulador m (para tonos m)** - [electrón]
(tone) keyer panel
— m — **montaje m** - [mec] mounting panel
— m — — m **para arrancador m** - [sold] starter
mounting panel
— m — — m **para caja f para telemando m** -
[sold] remote control box mounting panel
— m — — m **para pieza(s) f constitutiva(s)** -
[mec] component (parts) mounting panel
— m — **pared f** - [constr] wall panel • siding
— m — — f **deslizable** - [constr] sliding wall

panel

panel m para pared f deslizable para operación
f manual - [constr] manually operated sliding
wall panel
— m —— f — manualmente - [constr] manually
sliding wall panel
— m —— f — para ventilación f - [constr]
ventilation sliding wall panel
— m —— f para ventilación f - [constr] ventilation wall panel
— m — puerta(s) f - [constr] door panel
— m — f para automóvil m - [autom] automotive door panel
— m — regulación f - [instrum] console
— m —— f para dispositivo m - [sold] device
control panel
— m —— f —— m para fijación f - [sold]
fixture console
— m —— para instalación f soldadora - [sold]
welding fixture console
— m — respaldo m - [sold] welder back panel
— m — selección f - selector,ting panel
— m —— f para montaje m - [sold] voltage
selector panel
— m — techo m - [constr] roof panel
— m —— m de acero m - [constr] steel roof
panel
— m —— m —— m galvanizado - [constr] galvanized steel roof panel
— m —— m —— m galvanizado común—[constr]
plain galvanized steel roof panel
— m — terminal(es) m - [electr-instal] termi-
nal(s) strip
— m — ventilación f - [constr] ventilation
panel
— m posterior - [mec] back, o rear, panel
— m — interior - [mec] lower, rear, o back,
panel
— m — para caja f - [mec] rear, o back, case,
o housing, panel
— m —— manipulador m (para tonos m) -
[electrón] (tone) keyer rear panel
— m — soldadora f - [sold] welder, back, o
rear, panel
— m —, quitado, o removido, o sacado - [mec]
removed, back, o rear, panel
— m protector m - [mec] protective panel
— m — contra salpicadura(s) f - splash panel
— m, quitado, o removido, o sacado - [constr]
removed panel
— m Steelox m - [metal-fabr] Steelox panel
— m — de acero m - [metal-fabr] Steelox panel
— m —— m de Armco - [metal-fabr] Armco
Steelox panel
— m —— m —— para, cartelera(s) f, o
letrero(s) m, para, anuncio(s) m o aviso(s) m
- [metal-lam] Armco Steelox panel(s) for poster panel(s)
— m —— edificio(s) m - [constr] Steelox
building panel(s)
— m superior m - [mec] top, o upper, panel
— m — deprimido - [mec] recessed, top, o upper, panel
— m — hundido - [mec] recessed top panel
— m — para caja f - [mec] case, o housing,
top, o upper, panel
— m — soldadora f - [sold] welder, top, o
upper, panel
— m suspendido - [constr] suspended panel
— m trasero - rear, o back, panel
— m — para caja f - [sold] case, o housing,
back, o rear, panel
— m vertical - [mec] vertical panel
— m — embridado - [mec] vertical flanged panel
panelería f - [autom] panelling
— f interior - [autom] interior, o inside, panelling
— f —— para automóvil(es) m - [autom] automobile interior, o interior automobile, panelling (steel)
panfleto* m - [public] véase folleto m
panorma m - . . . . • setting • outlook
— m cambiante - changing panorama

panorama n escénico - scenic, panorama, o setting; panoramic scenery
— m estático - static panorama
— m exterior - exterior, o outdoor, panorama
pantalla f - . . . • [sold] shield; lead, block,
o shield • [fotogr] screen • [mec] baffle;
shield • [segurid] shield • [comput] screen;
display (unit); displaying • [sold] face-
shield
— f central - [mec] center,tral baffle
— f con tubo(s) m para rayo(s) m catódico(s) -
[comput] cathode ray tube, screen, o display;
C R T display
— f —— m —— m — optativa - [electrón]
optional cathode ray tube display
— f —— m —— m — para escritorio m -
[electrón] desktop cathode ray tube display
— f —— m —— m — representación f visual - [electrón] cathode ray tube, o C R T,
display
— f contra irradiación f - [electrón] radiation shield
— f de aluminio m - [telecom-cond] aluminum,
shield, o screen
— f — plomo m - lead shield
— f derecha f - [mec] right baffle
— f electrostática - [electrón] electrostatic
shield
— f enfriadora - [ind] cooling screen
— f — para gas m - [ind] gas cooling screen
— f facial - [sold] face shield; véase careta f
— f fluorescente - fluorescent screen
— f fluoroscópica - [electrón] fluoroscopic
screen
— f frontal - [mec] front baffle
— f grande - [electrón] large, o wide, screen
— f horizontal - [mec] horizontal, baffle, o
screen
— f húmeda - [mec] wet screen
— f — para lavado m - [ind] wet, wash(ing), o
scrubbing, screen
— f —— m de gas m - [metal-prod] wet gas,
o gas wet, scrubbing screen
— f inferior - [mec] lower, o bottom, screen, o
baffle
— f — posterior - [mec] bottom, o lower, rear
baffle
— f izquierda f - [mec] left, baffle, o screen
— f lateral - [mec] side baffle
— f — derecha f - [mec] right side baffle
— f — izquierda - [mec] left side baffle
— f limitada - [electrón] limited screen
— f no inflamable - [ind] nonflammable screen
— f optativa - [electrón] optional display
— f — con tubo m para rayo(s) m catódicos -
[electrón] optional cathode ray tube display
— f para alarma f - [electrón] alarm display
(unit, o screen)
— f — arco m - [sold] arc shield
— f — base f - [mec] base baffle
— f —— f central - [mec] center base baffle
— f —— f derecha - [mec] right base baffle
— f —— f izquierda - [mec] left base baffle
— f —— f posterior - [mec] rear base baffle
— f — bolsillo m - [sold] pocket shield
— f — condición f - [comput] status display
— f — difracción f - [instrum] diffraction
screen
— f — enfriamiento m - [mec] cooling screen
— f —— m para gas m - [metal-prod] gas cooling screen
— f — escritorio m - [comput] desktop display
— f —— m para condición f - [comput] desktop
status display
— f —— m — estado m - [comput] desktop status display
— f —— m — posición f - [comput] desktop
status display
— f —— m — representación f visual (de condición f) - [comput] desktop status display
— f —— m —— f — de estado m - [comput]
desktop status display
— f —— m —— f — de posición f - [comput]

desktop status display
pantalla f̲ para estado m̲ - [comput] status display (unit)
— f̲ — lámpara f̲ - [domést] lampshade
— f̲ — lavado m̲ - [ind] wash(ing), o̲ scrubbing, screen
— f̲ — — m̲ de gas m̲ - [metal-prod] gas scrubbing screen
— f̲ — mano(s) f̲ - [sold] handshield; hand guard
— f̲ — observación f̲ - [fotogr] monitoring, o̲ viewing, screen
— f̲ — posición f̲ - [comput] status display
— f̲ — proyección f̲ - [fotogr] viewing screen
— f̲ — refrigeración f̲ - [mec] cooling screen
— f̲ — representación f̲ visual - [comput] display; viewing screen
— f̲ — — f̲ — de condición f̲ - [comput] status display
— f̲ — — f̲ — — estado m̲ - [comput] status display
— f̲ — — f̲ — — posición f̲ - [comput] status display
— f̲ — — f̲ — para escritorio m̲ - [comput] desktop display
— f̲ — retención f̲ - [agric] check flap
— f̲ — sistema m̲ - [comput] system display
— f̲ — televisión f̲ - [electrón] television screen
— f̲ — televisor m̲ - [electrón] television screen
— f̲ — ventilador m̲ - [mec] fan screen
— f̲ posterior - [mec] rear baffle
— f̲ protectora - protective screen • [segurid] shield
— f̲ rectangular - [electrón] rectangular picture
— f̲ seca - [mec] dry screen
— f̲ superior - [mec] top, o̲ upper, baffle
— f̲ — frontal - [mec] top front baffle
— f̲ — posterior - [mec] top rear baffle
— f̲ vertical - [mec] vertical baffle
pantalón m̲ - . . . . - [metal-prod] (gas) uptake • (gas) down comer
— m̲ para horno - [metal-prod] furnace downcomer
— m̲ — tragante m̲ - [metal-prod] uptake lower end
pantalones m̲ - [vest] trousers; pants; jeans
— m̲ grueso(s) - [vest] heavy trousers
— m̲ para trabajo m̲ - [vest] work pants; overalls; jeans
— m̲ sin botamanga(s) f̲ - [vest] cuffless trousers
— m̲ — vuelta f̲ - [vest] cuffless trousers
pantanera f̲ - [autom-neumát] mud (and terrain) tire • véase también neumático m̲, pantanero, o̲ para barro m̲
— f̲ radial - [autom-neumát] radial mud tire
pantano m̲ - . . . .; swale
— m̲ salobre - [hidr] salt water, o̲ brackish, marsh
pantógrafo m̲ - . . . . • [metal-prod] (pantograph) trolley • bott mechanism
— m̲ para carro m̲ báscula - [metal-prod] scale car pantograph
— m̲ — corte m̲ - [tub] cutting pantograph
— m̲ — — de tubo(s) m̲ - [tub] tube cutting pantograph
paño m̲ - [textil] . . . • rag
— m̲ empapado - soaked rag
— m̲ con querosina - kerosene soaked rag
— m̲ húmedo - damp, cloth, o̲ rag
— m̲ limpiador - wiper; wiping, cloth, o̲ rag
— m̲ sin halacha(s) f̲ - lint, o̲ ravel, free rag
— m̲ suave - [textil] soft, cloth, o̲ rag
pañol m̲ para herramienta(s) f̲ - [ind] tool, shed, o̲ crib • [petról] doghouse
papel m̲ - . . . • role
— m̲ aislador - [electr-mat] insulating, o̲ duro, paper
— m̲ blanco - [papel] white paper • blank paper
— m̲ cambiado - changed role
— m̲ cambiante - changing role

papel (con) esmeril m̲ - [herram] emery paper
— m̲ (—) lija f̲ - [herram] sand paper
— m̲ (—) — fino m̲ - [herram] fine sand paper
— m̲ (—) — grueso - [herram] coarse sand paper
— m̲ (—) — mediano - [mec] medium sand paper
— m̲ — resina f̲ - [aislac] resin paper
— m̲ cuadriculado m̲ - [papel] quardilled paper • graph paper
— m̲ de conductor m̲ - [admin] leadership role
— m̲ — consumidor m̲ - [econ] consumer('s) role
— m̲ — estado m̲ - [pol] government('s) role
— m̲ — gobierno m̲ - [pol] government('s) role
— m̲ — hidrógeno m̲ - [quím] hydrogen('s) role
— m̲ — líder m̲ - [admin] leadership role
— m̲ — madera f̲ - [papel] Kraft paper
— m̲ — oficio - [papel] legal (size) paper
— m̲ — persona f̲ - [admin] person('s) role
— m̲ — tipo m̲ recubierto con aluminio m̲ - [comput] aluminum coated type paper
— m̲ desempeñado - role played
— m̲ duro - [papel] hard paper
— m̲ económico - [econ] economic role
— m̲ en blanco - [papel] blank paper
— m̲ — bobina(s) f̲ - [papel] web paper
— m̲ endurecido - [papel] hard(ened) paper
— m̲ engomado - [papel] adhesive paper
— m̲ — en un (sólo) lado m̲ - [papel] one side adhesive paper
— m̲ especial - [papel] special paper • [legal] documentary, paper, o̲ stationary
— m̲ exigido - [comput] required paper
— m̲ filtrador m̲ - [papel] filter(ing) paper
— m̲ impermeable - [papel] waterproof paper; (water) repellant paper
— m̲ — resistente - [papel] strong waterproof paper
— m̲ importante - important role
— m̲ impregnado - [electr-cond] impregnated paper
— m̲ — con aceite m̲ - [electr-cond] oil impregnated paper
— m̲ — — m̲ especial - [electr-cond] special oil impregnated paper
— m̲ — con viscosidad f̲ alta - [electr-cond] high viscosity impregnated paper
— m̲ — de tipo m̲ con viscosidad f̲ alta - [electr-cond] high viscosity type impregnated paper
— m̲ insubstituible - unsubstitutible role
— m̲ Kraft - [papel] Kraft paper
— m̲ — para bolsa(s) f̲ - [papel] Kraft bag paper
— m̲ — — f̲ de papel m̲ - [papel] Kraft bag paper
— m̲ metalizado - [electr-cond] metallized paper
— m̲ milimetrado - [papel] graph paper
— m̲ milimétrico - [papel] graph paper
— m̲ oro - [fin] paper gold
— m̲ para bolsa(s) f̲ - [papel] bag paper
— m̲ — embalaje m̲ - [papel] wrapping, o̲ Kraft, paper
— m̲ — envolver - [papel] wrapping paper
— m̲ — filtrar v̲ - [papel] filtering paper
— m̲ — filtro m̲ - [papel] filter paper
— m̲ — pago(s) m̲ - [fisc] revenue stamps
— m̲ picado - [papel] confetti
— m̲ protector - [papel] protective paper
— m̲ que no mancha - [papel] non-staining paper
— m̲ — repele agua m̲ - [transp] water repellent paper
— m̲ recubierto - [papel] coated paper
— m̲ — con aluminio m̲ - [comput] aluminum coated paper
— m̲ requerido - [comput] required paper
— m̲ resistente - [papel] tough, o̲ strong, paper
— m̲ — que no mancha - [constr] tough non--staining paper
— m̲ sellado - [legal] . . .; documentary, o̲ legal, paper, o̲ stationary
— m̲ — con valor m̲ mayor - [fisc] higher,est value documentary paper
— m̲ — — — m̲ menor - [fisc] lower,est value documentary paper

papel m sellado reintegrado - [fisc] reimbursed documentary paper • reimbursed documentary tax
— m — oficial - [legal] official documentary paper
— m sin carbonizar v - [papel] carbonless paper
— m tamaño m carta f - [papel] letter size(d) paper
— m — oficio m - [papel] legal size(d), o foolscap, paper
— m tornasol - [papel] litmus paper
papila f gustativa - [anat] taste bud
papilla f explosiva - [explos] explosive pap
páquer* m - véase empaquetador m
paquete m - .... • [metal-lam] pack
— m aéreo - [transp] air parcel
— m — asegurado - [transp] insured air parcel
— m almacenado - [transp] stored parcel
— m asegurado m - [transp] insured parcel
— m con cantidad f estándar - [ind] standard, package, o parcel, quantity
— m de barras f - [metal-lam] fagot
— m — chapas f - [metal-lam] sheet pack(age)
— m — chatarra f - [metal-prod] scrap bundle
— m — hojas f - [metal-lam] sheet pack(age)
— m — laminilla(s) f - [mec] shim pack
— m — — f nominal - [mec] nominal shim pack
— m — maquinaria f - [ind] machinery parcel
— m — — f definido - [ind] defined machine package
— m — perno(s) m - [mec] bolt(s) package
— m estándar - standard, package, o parcel
— m fácil para almacenar v - [ind] easy to store, o easily stored, package
— m individual - individual, parcel, o package
— m postal - [transp] parcel post
— m — aéreo m - [transp] air parcel post
— m — asegurado - [transp] insured air parcel post
— m — asegurado—[transp] insured parcel post
par m - .... • [constr] rafter; joist • [mec] torque
— m a tierra f - [telecom-cond] ground pair
— m cableado - [electr-cond] cabled pair
— m de armadura f - [constr] rafter
— m — conductores m - [electr-cond] conductors, o leads, pair
— m — contacto(s) m - [electr-equip] contact(s) pair
— m guante(s) m - [vest] glove(s) pair
— m — — m para trabajo m - [segurid] work glove(s) pair
— m — limatesa f - [constr] hip rafter; angle jack
— m — mordaza(s) f - [herram] jaw(s)
— m — pinza(s) f - [herram] (pair of) pliers
— m — — f grande - [herram] large pliers
— m — — f pequeñas - [herram] small pliers
— m — torsión f - [mec] torque (tightening)
— m — — f para tuerca(s) f - [mec] nut(s), torquing, o torque tightening
— m equilibrado - [mec] balanced pair
— m individual - [telecom-cond] individual pair
— m inferior - lower pair
— m — de tomacorriente(s) - [electr-instal] lower receptacle(s) pair
— m lubricado - [mec] lubricated pair
— m maximo - [mec] maximum torque
— m — para operación f - [mec] breakdown torque
— m mínimo - [mec] minimum torque
— m motor - [mec] torque; véase también momento torsional • motor torque
— m — alto - [mec] high torque
— m — bajo - [mec] low torque
— m — correcto m - [mec] correct torque
— m — desigual - [mec] unequal torque
— m — distribuido - [mec] distributed torque
— m — incorrecto - [mec] incorrect torque
— m — para bomba f - [bombas] pump (motor) torque
— m — para impulsión f - [mec] driving torque

par m motor para rotación f - [mec] rotation, o rolling, torque
— m — — f verificado - [mec] checked, rotation, o rolling, torque
— m — sujetador m - [mec] fastener torque
— m — tuerca f - [mec] nut torque
— m — recibido - [mec] accepted torque
— m — requerido - [mec] required torque
— m — simulado - [mec] simulated torque
— m — transmitido - [mec] transmitted torque
— m — verificado - [mec] checked torque
— m nominal - [mec] nominal, o rated, torque
— m — con carga f - [mec] nominal torque with @ load
— m para arranque m - [mec] starting torque
— m — elevación f - [mec] lifting, o elevating,tion torque
— m retorcido - twisted pair
— m superior - upper, o top, pair
— m — de tomacorrientes m - [electr-instal] upper receptacle(s) pair
— m torsional - [mec] torque
— m — de conjunto m - [mec] assembly torque
— m — — — m para girar - [mec] assembly rolling torque
— m trenzado - [telecom-cond] braided pair
para adv acceso m - [vial] on bound
— adv ajuste m - to fit
— adv — m con presión f - [mec] to press fit
— adv altura f variable - [agric-equip] (for) variable drop
— adv ambas direcciones f - [vial] two-way
— adv ambos sentidos m - [vial] two-way
— adv armar v - [mec] for assembly
— adv — f en obra f - [mec] for field assembly
— adv arranque m - [mec] (for) starting
— adv avión(es) m - [aeron] (for) aircraft
— adv cálculo m - (for) calculation • [instrum] slide
— adv carga f - (for), charge,ging, o load(ing)
— adv combustion f con - [combust] fired
— adv — — gas m - [combust] gas fired
— adv — — de coque m - [combust] coke (oven) gas fired
— adv con - to; towards
— adv conducción f - [tub] tubing; pipe line
— adv conexión f con - [electr-distrib] for connecting,tion with
— adv — f — línea f para abastecimiento m - [electr-instal] to @ suppply line
— adv — f — tierra f - [electr-instal] to @ ground
— adv conocimiento - for information
— adv — m y control m - for information and, control, o guidance
— adv constancia - [legal] in proof (whereof); in witness, thereof, o whereof; for @ record
— adv construcción f - [constr] for construction; constructional
— adv conveniencia f - for convenience (sake)
— adv — mayor - for (greater) convenience
— adv cuenta f de - [fin] for @ account of
— adv diario - [vest] casual • daytime
— adv dos direcciónes f - two-way
— adv — sentidos m - two-way
— adv durabilidad f - for durability
— adv efecto(s) m - [legal] for @ purpose
— adv ello - to this end
— adv encendido m - [sold] for start(ing)
— adv — m de frecuencia f alta (solamente) - [sold] for high frequency start(ing) only
— adv — m únicamente - [sold] for start(ing) only
— adv endurecimiento m - for hardening
— adv — de superficie f—for surface hardening
— adv — m superficial - [sold] hardsurfacing
— adv enfriamiento m - for @ cooling
— adv ensayo m - for testing; for practice
— adv entrada - incoming
— adv fin m - for, o to, @, purpose, o end • in regard to
— adv — m aduanero - [fisc] for custom(s) purpose(s)

- 1045 -

para adv **fin(es)** m **aduanero(s) únicamente** -
[fisc] for custom purpose(s) only
— adv — m **de referencia** f̱ - for reference
(purposes)
— adv **inserción** f - insertion type
— adv — f **forzada** - [mec] drive,ving type
— adv **introducción** f - insertion type
— adv **f forzada** - [mec] drive,ving type
— adv **largada** f̱ - for starting
— adv **operación** f **sin carga** f̱ - [comb.int] no
load (speed)
— adv **perforación** f̱ - [petról] for drilling •
drilling rig
— adv **propósito** m̱ - foe, o to, @, purpose, o
end
— adv **proyección** f̱ - for design
— adv **que** - in order for
— adv **recubrimiento** m̱ - for coating
— adv — **duro** - [metal- for hardsurfacing
— adv **referencia** f̱ - for reference
— adv **resistencia** - for strength
— adv **salida** f̱ - outgoing
— adv **siembra** f̱ - [agric-equip] lister
— adv — f̱ **variable** - [agric-equip] variable
drop
— adv **siempre** - forever
— adv **sorpresa** - to @ surprise
— adv — f̱ **grande** - to @ great surprise; much
to @ surprise
— adv **tarde** - [vest] (for) evening
— adv **todo efecto(s)** m̱ - for, o to, every, intent, o purpose • practically
— adv — **terreno** m̱ - all terrain
— adv **tonelaje** m̱ **reducido** - (for) small tonnage
— adv **trabajos** m̱ **(muy) pesados** - heavy duty
— adv **un hombre** m̱ - one-man
— adv — **(sólo) hombre** m̱ - one-man only
— adv **velocidad(es)** f̱ **de hasta 90 millas (145
kms) por hora** - [autom-neumát] S
— adv — — **115 millas (185 Kms) por hora** -
[autom-neumát] H
— adv — — **130 millas (210 kms) por hora** f̱
- [autom-neumát] V
— adv — f̱ **elevada** - [autom-neumát] speed
rated
— adv — **v** - [autom-neumát] V-speed rated
parabrisa(s) m̱ **agrietado** - [autom] cracked
windshield
— m **de cristal** m̱ **inastillable** - [autom] shatter proof, o safety, glass windshield
— m **en forma** f̱ **de V** - [autom] V-shaped windshield
— m — **V** - [autom] V-shaped windshield
— m **inastillable** - [autom] shatter proof, o
safety, windshield
— m **para recambio** m—[autom] replacement windshield
parada f̱ - . . . • [ind] shut down; banking;
stoppage; down time • [f.c.] stop • [vial]
rest station • [petról] stand • [deport]
service
— f̱ **comenzada** - [ind] begun, o started, shutdown
— f̱ **con pavimento** m̱ **mojado** - [autom] wet stopping
— f̱ — — m̱ **seco** - [autom] dry stopping
— f̱ **costosa** - [ind] costly, shut down, o down
time
— f̱ **de equipo** m̱ - [ind] equipment shutdown
— f̱ — **horno** m̱ - [ind] furnace shutdown
— f̱ — — m̱ **alto** - [metal-prod] blast furnace
shutdown
— f̱ — — m̱ **para reparación** f̱ - [ind] furnace
repair shutdown
— f̱ — **línea** f̱ - [ind] line shutdown
— f̱ — — **motor** m̱ - [electr-mot] motor, stopping,
o stalling • [comb.int] engine shut-off
— f̱ — **planta** f̱ - [ind] plant shut down
— f̱ — **sistema** m̱ **para incorporación** f̱ **en asfalto** m̱ - [nucl] asphalt incorporation system shutdown
— f̱ **definitiva** - [ind] final shutdown • [metal

-prod] dead banking
parada f̱ **en fosa** f̱ - [deport] pit stop
— f̱ — f̱ **programada** - [deport] scheduled pit
stop
— f̱ **importante** - [ind] long shutdown
— f̱ **imprevista** - unforseen, o unplanned, stop,
o downtime
— f̱ **imputable** - [labor] chargeable idle time
— f̱ **indefinida** - [ind] indefinite shutdown
— f̱ **inesperada** - unexpected stop(page)
— f̱ **innecesaria** - [ind] unnecessary, o needless, downtime, o shutdown
— f **instantánea** - [sold] clean stop
— f̱ **larga** - [ind] long, o lengthy shutdown
— f̱ **mecánica** f̱ - [mec] mechanical, stop(ping),
o shutdown
— f̱ **no imputable** - [labor] nonchargeable, idle
time, o hour(s)
— f **nocturna** - nightime stop • overnight rest
— f̱ **para emergencia** f̱ - [ind] emergency, stop-
-(ping), o shutdown, o shutoff
— f̱ — — f̱ **para motor** m̱ - [comb.int] emergency
engine shutoff • [electr-mot] emergency motor
shutoff
— f̱ — **limpieza** f̱ - [metal-prod] cleaning shutdown
— f̱ — **mantenimiento** m̱ - [ind] maintenance
shutdown
— f̱ — **reconstrucción** f̱ - [metal-prod] reline
shutdown
— f̱ — — f̱ **general** - [metal-prod] (total) reline,ning shutdown
— f̱ — **refrigerio** m̱ - [deport] meal, o lunch,
halt
— f̱ — — m̱ **vespertino** - [deport] evening meal
halt
— f̱ — **reparación(es)** f̱ - [ind] repair shutdown
— f̱ — — f̱ **para horno** m̱ - [ind] furnace repair
shutdown
— f̱ — — f̱ **limpieza** f̱ - [ind] repair and
cleaning shutdown
— f̱ — **servicio** m̱ - [ind] service, stop, o halt
• [deport] service break
— f̱ **por duración** f̱ **indefinida** - [ind] indefinite duration shutdown
— f̱ — **falta de arrabio** m̱ - [metal-prod] pig
iron shortage shutdown
— f̱ — **período** m̱ **indefinido** - [ind] indefinite
period shutdown
— f̱ — **tiempo** m̱ **indefinido** - [ind] indefinite
time shutdown
— f̱ **preparada** - [ind] prepared shutdown
— f̱ — **para horno** m̱ - [ind] prepared furnace
shutdown
— f̱ **prevista** - [ind] foreseen, o scheduled,
shutdown
— f̱ **programada** - [ind] scheduled shutdown •
scheduled, stop(page), o downtime
— f̱ — **en fosa** f̱ - [deport] scheduled pit stop
— f̱ **prolongada** - [ind] extended, o lengthy,
shutdown, o stoppage
— f̱ **semanal** - [ind] weekly shutdown
— f̱ — **para equipo** m̱ - [ind] weekly equipment
shutdown
— f̱ **temporaria** - [ind] temporary shutdown •
[metal-prod] hot banking
— f̱ **terminada** - [ind] finished shutdown
— f̱ **total** - total stop • [ind] total downtime
— f̱ **única** - single stop
paradas f̱ **frecuentes** - [ind] frequent shutdowns
— f̱ — **de horno** m̱ - [ind] frequent furnace
shutdowns
— f̱ **hasta fecha** f̱ - [ind] stoppages, o shutdowns, to date
— f̱ **para mes** m̱ - [ind] stoppages, o downtime,
this month
— f̱ — — m̱ **hasta fecha** f̱ - [ind] stoppages. o
downtime, this month to date
— f̱ **totales** - [ind] total, stoppages, o downtime
— f̱ — **para mes** m̱ - [ind] total, stoppages, o
downtime, this month
— f̱ — — — m̱ **hasta fecha** f̱ - [ind] total,

downtime, o stoppages, this month to date
paradero m - [vial] rest station
parado,da a - stopped; halted; stalled • inoperative; inactive • [ind] down • [electr--distrib] turned off
— a de punta f - standing up; upended
parador m - . . . • [vial] rest area • motel
parafina f - . . . • [petról] wax
parafinado,da a - [petról] waxed
parafinador m - [petról] waxer
parafinador,da a - [petról] waxer
parafínico,ca a - paraffinic
parafinoso,sa* a - paraffinic
parafraseado,da a - paraphrased
parafraseo* m - paraphrasing
paragolpe(s) m - [autom] . . .; bumper • [f.c.] bumper; buffer
— m de acero m - [autom] steel bumper
— m — madera f - [constr] wooden, o timber, bumper
— m delantero - [autom] front bomper
— m esquinero - [autom] corner, o side, bumper
— m — trasero - [autom] corner, o side, rear bumper
— m lateral m - [autom] side bumper
— m — trasero - [autom] side rear bumper; bumperette
— m móvil - [mec] disappearing, stop, o bumper
— m para capó m - [autom] hood bumper
— m — servicio m pesado - [f.c.] heavy duty, bumper, o coupling
— m — vagón(es) m - [f.c.] car, bumper, o coupling
— m por fricción f - [f.c.] friction (type), bumper, o coupling
— m — f para servicio m pesado - [f.c.] heavy duty friction (type) bumper
— m trasero - [autom] rear bumper
parágrafo m - . . . • [legal] section; clause
— m anterior - [legal] foregoing, paragraph, o clause, o section
paraje m - . . .; site • location; locale
— m completado - completed site
— m para vacación f - vacation spot
— m — — f en fin m de semana f - weekend vacation spot
paralela f - [mec] bar
paralelado,da a - paralleled
paralelamente adv - parallely; alongside
— adv a flujo m - [hidr] parallel(y) to @ flow
— adv — — m de material(es) m - parallel to @ material(s) flow
paralelamiento* m - véase puesta f en paralelo
paralelismo m - . . . • [electr-instal] parallelism; parallel circuit relation(ship)
— m de ala(s) f - [metal-lam] flange parallism
— m — perfil m - [metal-lam] shape parallelism
— m — — m soldado - [constr] welded shape parallelism
paralelo m - . . . | adv - [electr-instal] across @ line
paralelo,la a - . . .; allongside • [fin] unofficial • [abl] lang
— a sensiblemente - fairly parallel
paralización f - . . . • [ind] down time; closing, o shutting, down; down shutting
— f de compresor m - [mec] compressor shutting down
— f de grúa f - [grúas] crane shutting down
— f — horno m - [combust] furnace shut down
— f — obra f - [constr] project, o job, shutting, o closing, down
— f total - [ind] total, shutdown, o stoppage
paralizado,da a - paralyzed • [ind] shut down
paralizar v - . . . • [ind] to shut down • [metal-prod] to kill
— v compresor m - to shut down @ compressor
— v grúa f - [grúas] to shut down @ crane
— v malacate m - [petról] to shut down @ drawworks
paramento m - . . . • [arquit] . . .; wall face
— m exterior - [constr] (outer) apron
— m — de vertedero m - [constr] spillway apron

paramento m posterior - [arquit] back wall face
parámetro m - . . .; constant • range
— m anterior - previous, o old, parameter • [arquit] front wall face
— m básico - basic parameter
— m clave - key parameter
— m comprobado - proven, o checked, o verified, parameter
— m crítico - critical parameter
— m — para mecanización f - [mec] machining critical parameter
— m — — f por abrasión f - [mec] abrasion machining critical parameter
— m de corte m - [geol] shear parameter
— m — diseño m - design, parameter, o constant
— m — línea f - [ind] line parameter
— m — producción f - [ind] production parameter
— m específico - specific parameter
— m inherente - inherent parameter
— m nuevo - new parameter
— m para acero m - [metal-prod] steel parameter
— m — cálculo m - calculation parameter
— m — diseño m - design parameter
— m — equipo m - [ind] equipment parameter
— m — funcionamiento m - operating,tion parameter
— m — gobierno m - control(lability) parameter
— m — mecanización f - [mec] machining parameter
— m — — f por abrasión f - [mec] abrasion machining parameter
— m — operación f - operating,tion parameter
— m — proceso m para reducción f - [miner] reduction process parameter
— m — proyección f - design parameter
— m — reacción f - response parameter
— m — reducción f - [miner] reduction parameter
— m — suelo m - [suelos] soil parameter
— m que antecede - foregoing parameter
— m útil - useful parameter
— m verificado - verified, o checked, parameter
— m viejo - old parameter
parante* m - [mec] post; pillar • [constr] standard
— m para puerta f - [constr] door post; jamb
— m — — f para automóvil m - [autom] automotive door post; automotive jamb
parapeto m - . . . • [constr-puentes] railing
— m para puente m - [constr] bridge railing
parar v - . . . • [electr-oper] to turn off • [mec] to stand up • [ind] to, shut, o close, o go, o be, down • to take off
— v bomba f - [bombas] to stop @ pump
— v de punta f - to stand up; to upend
— v estufa f - [combust] to shut down @ stove
— v horno m - [constr] to shut down @ furnace
— v — m alto - [metal-prod] to shut down @ blast furnace
— v lentamente - [ind] to. stop, o shut down, slowly, o gradually
— v línea f - [ind] to stop @ line
— v — f para galvanización f - [metal-trat] to, stop, o shut down, @ galvanizing line
— v motor m - [electr-mot] to, stop, o turn off, @ motor • [comb.int] to, stop, o turn off, @ engine
— v soldadora f - to, stop, o turn off, @ welder
— v soldadura f - [sold] to stop @ weld(ing)
pararse v - to stop (itself); to stall
— v de cabeza - to, stand, o go, on @ head
— f — punta—to, stand, o go, on @, end, o nose
— v motor m - [electr-mot] to stall @ motor • [comb.int] to stall @ engine
pararrayos m - [electr-instal] . . .; rod; aerial terminal; lightning, arrester, o conductor, o protector
— m catódico - [electr] multigap arrester
— m dañado - [electr-instal] damaged (lightning), rod, o arrester

pararrayos n̲ con antena(s) f̲ - [electr-instal]
horn type lightning arrester
— m con cuerno(s) m̲ - [electr-instal] horn type
lightning arrester
— m — entrehierro m̲ múltiple - [electr-instal]
multigap (lightning) arrester
— m — retardo m̲ catódico - [elecrr-instal]
multiple lightning arrester
— m para transformador m̲ - [electr-instal]
transformer (lightning) arrester
— m — m dañado - [electr-instal] damaged
transformer (lightning) arrester
paraviento(s) m̲ - [autom] wind board
parcela f̲ de tierra f̲ - ground plot
parcial a̲ - . . . . random • interim • local •
[arquit] segmental
parcialmente adv̲ endurecido,da a̲ - [constr] partially hardened
— adv̲ reductor,ra - partially reducing
parchar v̲ - to patch; véase también emparchar v̲
parche m̲ - . . . . • patch; mend; repair
— m caliente - hot patch
— m de lardillo(s) m̲ - [metal-prod] brick(work)
patch
— m dorsal - back, o̲ rear, patch
— m en camino m̲ - [vial] road patch
— m en dorso m̲ - back, o̲ rear, patch
— m frío m̲ - cold patch
— m para neumático m̲ - [autom-neumát] tire
patch
— m permanente - permanent patch
— m temporario - temporary patch
pardo m̲ gris - gray-brown
!pare! - [interj] stop!
parecer m̲ - . . . . | v̲ - . . .; to resemble
— v̲ bien - to look good
— v̲ en principio m̲ - to appear at first
— v̲ fácil - to, look, o̲ seem, o̲ appear, easy
— m general - general opinion
— f̲ hecho,cha automáticamente - [sold] to, appear, o̲ look, o̲ seem, machine made
— m justificado,da - to, seem, o̲ appear, o̲
look, justified
— v̲ obvio,via - to, seem, o̲ appear, obvious
— v̲ significante - to appear significant
parecerse v̲ - to appear similar; to approximate
— v̲ poco - unlike
parecido m̲ - . . . .; resembling; appearing,rance
• approximating,tion
— a̲ en principio m̲ - appearance at first
parecido,da a̲ - . . . .; alike • appeared •
resembled; approximated
parecillo m̲ - [constr] counter rafter
pared f̲ a prueba de incendio(s) m̲ - [constr]
fire wall
pared f̲ acanalada - [constr] corrugated well
— f̲ agrietada - [constr] cracked wall
— f̲ alta - [constr] high wall
— f̲ anterior - [constr] front wall • previous
wall
— f̲ armada - [constr] assembled wall
— f̲ baja - [constr] low wall
— f̲ base - base wall
— f̲ básica - basic wall
— f̲ calentada - heated wall
— f̲ celular - [constr] cell(ular) wall
— f̲ con inclinación f̲ - [constr] slanted, o̲
leaning, wall
— f̲ — f de . . . - [constr] on . . . batter
wall
— f̲ conformada—[constr] formed, o̲ shaped, wall
— f̲ contra incendio(s) m̲ - [constr] fire wall
— f̲ correspondiente - corresponding wall
— f̲ corrugada - [constr] corrugated wall
— f̲ curva(da) - [constr] curved wall
— f̲ dañada - [constr] damaged wall
— f̲ de acero n̲ - steel wall
— f̲ — agitador m̲ - agitator wall
— f̲ — ala f̲ - [hidr] wing wall
— f̲ — alcantarilla f̲ - [constr] culvert wall
— f̲ — bloque(s) m̲ - [constr] block wall
— f̲ — bóveda f̲ - [constr] vault wall
— f̲ — f para transformador m̲ - [electr-

-instal] transformer vault wall
pared f̲ de cabecera f̲ - [constr] headwall
— f̲ — caja(s) f̲ - [constr] crib wall
— f̲ — — f de acero m̲ - [constr] steel crib
wall
— f̲ — cámara f̲ - [constr] chamber wall
— f̲ — — f subterránea - [electr-instal] manhole wall
— f̲ — camisa f̲ - [petról] liner wall
— f̲ — celda f̲ - cell well
— f̲ — cilindro m̲ - [comb.int] cylinder wall
— f̲ — compartimento m̲ - [constr] compartment,
wall, o̲ side
— f̲ — conducto m̲ - [tub] conduit wall
— f̲ — crisol m̲ - [metal-prod] hearth wall
— f̲ — cuba f̲ - [metal-prod] bosh wall
— f̲ — cubo m̲ - [mec=ruedas] hub wall
— f̲ — chaflán m̲ - [sold] bevel edge
— f̲ — depósito m̲ - tank wall
— f̲ — escorial m̲—[metal-prod] slag notch wall
— f̲ — estribo m̲ - [constr] abutment (side)wall
— f̲ — estructura f̲ - [constr] structure wall
— f̲ — — f para drenaje m̲ - [constr] drainage
structure wall
— f̲ — etalaje m̲ - [metal-prod] bosh wall
— f̲ — excavación f̲ - [constr] excavation, o̲
trench, (side) wall
— f̲ — fachada f̲ - [constr] front wall
— f̲ — fosa f̲ - [constr] pit wall
— f̲ — hormigón m̲ - [constr] concrete wall
— f̲ — — m conformado - [constr] formed, o̲
shaped, concrete wall
— f̲ — horno m̲ - [combust] furnace wall
— f̲ — — m reparada - [combust] repaired furnace wall
— f̲ — junta f̲ - [sold] joint wall
— f̲ — ladrillo(s) m̲ - [constr] brick wall
— f̲ — — m de arcilla f̲ - [refract] clay brick
wall
— f̲ — lingotera f̲ - [metal-prod] mold wall
— f̲ — madera f̲ - [constr] timber, o̲ wood, wall
— f̲ — — f sólida - [constr] solid timber wall
— f̲ — núcleo m̲ - [mec] core wall
— f̲ — pilote m̲ - [constr-pil] pile wall
— f̲ — — m exigida - [constr-pil] required
pile wall
— f̲ — plancha(s) f̲ - [constr] plate wall
— f̲ — — f para revestimiento m̲ - [constr]
liner plate wall
— f̲ — ranura f̲ - [sold] groove (side) wall
— f̲ — recipiente m̲ - [calderas] vessel wall
— f̲ — roca f̲ - [geol] rock wall
— f̲ — sala f̲ - [constr] (living) room wall
— f̲ — — f para regulación f̲ - [ind] control
room wall
— f̲ — — f — transformador m̲ - [electr-equip]
transformer room wall
— f̲ — tanque m̲ - tank wall
— f̲ — tipo m̲ de jaula f̲ - [constr] crib wall
— f̲ — — m entramado - [tub] truss shaped wall
— f̲ — transformador m̲ - [electr-equip] transformer wall
— f̲ — tubería f̲ - pipe, o̲ tube, wall • conduit
wall • [ambient] duct wall
— f̲ — f abovedada - [tub] pipe-arch wall
— f̲ — — f conformada - [tub] shaped pipe wall
— f̲ — — f de tipo m̲ entramado - [tub] truss
shaped pipe wall
— f̲ — — f entramada - [tub] truss pipe wall
— f̲ — — f helicoidal—[tub] helical pipe wall
— f̲ — — f ranurada - [tub] slotted pipe wall
— f̲ — — f remachada - [tub] riveted pipe wall
— f̲ — — f unión f̲ - [tub] coupling wall
— f̲ — — f sin roscar - [tub] unthreaded coupling wall
— f̲ — vagón m̲ - [f.c.] car wall
— f̲ deflectora - [hidr] diversion wall
— f̲ delgada - thin wall
— f̲ deslizable - [constr] sliding wall
— f̲ doble - double wall
— f̲ drenada - [constr] drained wall
— f̲ drenante - [constr] drain(ing), wall, o̲
curtain

pared f drenante de canto m rodado - [constr]
gravel drain, wall, o curtain
— f efectiva - [constr] effective wall
— f estable - [constr] stable wall
— f este - [constr] east(ern) wall
— f expuesta - [constr] exposed wall
— f exterior - [constr] exterior, o outer, o
outside, wall
— f — de tubería f - [tub] outer pipe wall
— f fea - [constr] ugly, o unsightly, wall
— f — agrietada - [constr] ugly, o unsightly,
cracked wall
— f fina - thin wall
— f frontal - [constr] front wall • [mec] front
panel
— f — inclinada - [constr] inclined, o slanting, front wall
— f gruesa - [constr] thick wall • heavy wall
— f — de tubería f - [tub] thick pipe wall
— f inatractiva - [constr] unsightly wall
— f — agrietada - [constr] unsightly cracked
wall
— f inclinada - [constr] inclined, o slanting,
o battered, wall
— f incombustible - [constr] fire wall
— f interceptora - [hidr] cutoff wall
— f interior - [constr] interior, o inner, o
inside, wall; inwall
— f — de tubería f - [tub] pipe, interior, o
inside, o inner, wall; interior pipewall
— f — inclinada - [constr] inclined, o sloping, interior, o inside wall; sloping inwall
— f — inferior - [constr] lower, interior, o
inside, wall; lower inwall
— f — media f - [constr] middle inwall
— f — superior - [constr] upper, interior, o
inside, wall; upper inwall
— f interna - [constr] inside wall; inwall
— f — inclinada - [constr] sloping inwall
— f intersectante - [constr] intersecting wall
— f lateral - [constr] side wall; sidewall;
siding • [mec] side panel
— f — alta - [constr] high side wall
— f — baja - [autom-neumát] low sidewall
— f — blanca - [autom-neumát] white sidewall
— f — — resistente a intemperie f - [autom-
-neumáta] weather resistant white sidewall
— f lateral con franja f blanca—[autom-neumát]
white sidewall
— f — — — f — angosta - [autom-neumát] narrow white sidewall
— f — con . . . tela(s) f - [autom-neumát]
. . .-ply sidewall
— f — contrapuesta - [autom-neumát] reverse
sidewall
— f — de ranura f - [sold] groove side wall
— f — de túnel - [constr] tunnel side wall
— f — de vagón m - [f.c.] car side wall
— f — estable - [constr] stable side wall
— f — flexible - [autom-neumát] flexible, o
flexed, sidewall
— f — flexionada - [autom-neumát] flexed sidewall
— f — inclinada - [constr] sloped side
wall
— f — inferior - [constr] lower side wall
— f — más baja - [autom-neumát] lower sidewall
— f — muy flexible - [autom-neumát] highly
flexible sidewall
— f — negra - [autom-neumát] black sidewall
— f — negra resistente a intemperie f—[autom-
-neumát] weather resistant black sidewall
— f — opuesta - opposing,site side wall
— f — rechoncha - [autom-neumát] squat(ty)
sidewall
— f — resistente a intemperie f - [autom-
neumát] weather resistant sidewall
— f luego de recalcada - [metal-fabr] swaged
wall
— f metálica - [constr] metal wall
— f norte - [constr] north(ern) wall
— oeste - west(ern) wall
— f ondulada - [constr] corrugated wall

pared f opuesta - opposing,site wall
— f para cámara f - [constr] chamber wall
— f — — f para, aspiración f, o succión f
[turb] suction chamber wall
— f — contención f - [constr] withholding wall
• [hidr] weir
— f — relleno m - [constr] curtain wall; spandrel
— f — seguridad f - [constr] safety wall •
[segurid] fire wall
— f — ventilación f—[constr] ventilation wall
— f perforada - [tub] perforated wall
— f porosa - [hidr] porous wall
— f posterior - [constr] back, o rear, wall •
[mec] rear panel
— f — inclinada - [constr] sloping back wall
— f principal - [constr] main, o major, wall
— f propuesta - [constr] proposed wall
— f recalcada - [metal-fabr] swaged wall
— f refractaria - [constr] fire wall
— f reparada - [constr] repaired wall
— f rocosa - [constr] rocky wall
— f rozada - [deport] tagged wall
— f sencilla - [tub] single wall
— f soldada por punto(s) m - [tub] spot welded
wall
— f sólida de madera f - soled timber wall
— f soportada - [constr] supported wall
— f sostenida - [constr] supported wall
— f sud - [constr] south(ern) wall
— f terminada - [constr] finished wall
— f terminal - [constr] end wall
— f — de vagón m - [f.c.] car end wall
— f trasera - [constr] back, o rear, wall
— f — inclinada - [constr] sloping, o slanting, backwall
— f variable - [constr] variable wall
— f vertical - [constr] vertical wall
— f — curva - [constr] vertical curved wall
— f — efectiva - [constr] effective vertical
wall
— f — exterior - [constr] exterior vertical
wall
pareja f - . . . . • [metal-lam] pack (before
rolling)
— f sobresaliente - [deport] outstanding team;
quite @ team
parejamente adv - evenly • smoothly
parejo,ja a - . . . . • level(ed) • flush • steady
• uniform • orderly • true
parentético,ca a - [gram] parenthetical
parhilera f - [constr] . . .; purlin
paridad f cambiaria—[econ] exchange parity
— f de dólar m - [fin] dollar parity
— f — moneda f - [fin] currency parity
— f mantenida - maintained parity
— f sostenida - sustained, o maintained, parity
paritario,ria a - [labor] labor-management
parlante m - [electrón] speaker | a - . . .
— m magnético - [electrón] magnetic speaker
parné m - . . . . • [metal] brass
paro m - [ind] . . .; shut-down; stoppage
— m clandestino - [labor] wildcat strike
— m de bomba f - [bombas] pump stopping
— m laboral - [labor] labor stoppage
— m no autorizado - [labor] wildcat strike •
industrial action
— m simultáneo - [labor] simultanoeus stoppage
parpadear v - [electr-ilum] to flash on and off
— v luz f - to flash @ light
parpadeo m - [electr-ilum] flashing on and off
— m — luz f - light flashing (on and off)
parque m - . . . . • [transp] pool; fleet • [ind]
(storage) yard • [deport] paddock
— m caminero - [vial] roadside park
— m con atracción(es) f - amusement park
— m de automotores m - [ind] automotive pool
— m — depósitos m - [petról] tank farm
— m — — m para clasificación f - [petról]
classification (tank) farm
— m — — m — f de desechos m - [mucl] was
waste classification tank farm
— m — — m — — f — — radiactivos m -

[nucl] radioactive waste classification tank
farm
parque m de depósitos m para clasificación f de
desecho(s) m radiactivo(s) líquido(s) m -
[nucl] liquid radioactive waste classification tank farm
— m —— m —— f de residuo(s) m - [nucl]
waste classification tank farm
— m —— f —— m radioactivo(s) m -
[nucl] radioactive waste classification tank
farm
— m —— m —— f —— m — líquido(s) m -
[nucl] liquid radioactive waste clarification
tank farm
— m —— m para recogida f - collection tank
farm
— m —— m —— f de desecho(s) m - [nucl]
waste collection tank farm
— m —— m —— f —— m radiactivo(s) -
[nucl] radioactive wate collection tank farm
— m —— m —— f —— m — líquido(s) -
[nucl] liquid radioactive waste collection
tank farm
— m —— m —— f — residuo(s) m - [nucl]
waste collection tank farm
— m —— m —— f —— m radioactivo(s) -
[nucl] radioactive waste collection tank farm
— m —— m —— f —— m — líquido(s) m -
[nucl] liquid radioactive waste collection
tank farm
— m — etanque(s) n - [petról] tank farm
— m — planta f - [ind] plant yard
— m estatal - state, park, o memorial
— m exterior - [ind] outside yard
— m — para almacenamiento m - [ind] outside
storage yard
— m —— lingote(s) m - [metal-prod] outisde
ingot (storage) yard
— m externo - [ind] outside yard
— m industrial - [ind] industrial, park, o yard
— m interior - [ind] inside yard
— m intermedio - [ind] intermediate yard
— m interno - [ind] inside yard
— m municipal - [pol] city park
— m nacional - [pol] national park
— m para almacenamiento m - [ind] storage, o
stock, yard; stockyard
— m —— m para coque m - [ind] coke storage
yard
— m —— m exterior - [ind] outside storage
yard
— m — almacenar v tocho(s) m - [metal-prod]
bloom storage yard
— m — aparcamiento m - [autom] parking lot
— m — apilamiento m - [ind] piling, yard, o
area
— m — camping(s) m - [deport] camping area
— m — carbón m - [miner] coal yard
— m — clasificación f - classification farm
— m — coque m - [metal-prod] coke yard
— m — chatarra f - [metal-prod] scrap yard
— m — depósito m - [ind] stockyard
— m — desbaste(s) m - [metal-lam] bloom yard
— m — descanso m - [vial] roadsie park
— m — diversión(es) m - amusement park; fair
— m — enfriamiento m - [ind] cooling yard
— m — equipo m - [ind] equipment yard
— m —— m móvil - [ind] mobile equipment pool
— f — escarpado m - [metal-lam] scarfing yard
— m — de desbaste(s) m - [metal-lam]
bloom scarfing yard
— m — escoria f - [metal-prod] slag yard
— m — estacionamiento m - [autom] parking lot
— m —— m drenado - [constr] drained parking
lot
— m —— m para hospital m - [constr] hospital
parking lot
— m —— m viejo - [constr] old parking lot
— m — expedición f - [ind] shipping yard
— m — lingote(s) m - [metal-prod] ingot yard
— m — lobo(s) m - [metal-prod] skull (cracker)
yard
— m — materia(s) f prima(s) - [ind] raw mate-

rial(s) yard; stockyard
parque m para material(es) m - [ind] material(s)
yard
— m —— m en planta f - [ind] plant materials
yard
— m — mineral(es) m - [metal-prod] ore yard
— m — molde(s) m - [metal-prod] mold storage
yard
— m — palanquilla f - [metal-lam] billet yard
— m — planchón(es) m - [metal-lam] slab yard
— m — preparación f para chatarra f - [metal-
-prod] scrap preparation yard
— m — recogida f - [ind] collection (tank)
farm
— m —— m recreo m - recreation, o amusement,
park
— m — reserva f - [metal-prod] reserve storage
yard
— m — soldadura f - [sold] welding yard
parquerización* f - [metal-prod] parkerizing
párrafo m anterior - [gram] preceding, o foregoing, paragraph
— m precedente - [Gram] preceding, o foregoing,
paragraph
— m que antecede - [gram] preceding paragraph;
paragraph above
— m precede - [gram] preceding paragraph;
paragraph above
— m — sigue m - [Gram] following paragraph;
paragraph below
parrilla f - [mec] . . .; grating; gridiron;
grill(s); véase también rejilla f • [miner]
grizzly
— f de cadena(s) f - [constr] chain grate
— f — hierro m - [mec] iron grating
— f eléctrica (pequeña) - [domést] hamburger
grill
— f para carbón m - [domést] coal grill
— f —— m de leña f - [domést] charcoal grill
— f —— m vegetal - [domést] charcoal grill
— f — decapado m - [metal-trat] pickling grate
— f — hogar m - [combust] boiler grate
— f — máquina f para sinterizar,zación f -
[metal-prod] sintering machine grate
— f pequeña - [domést] hamburger grill
— f sacudidora - [miner] shaking grate
parte m - . . .; section; element; component;
pieza • [ind] (daily) report • [legal] party
• [mec] véase pieza f | adv - (en) partially
— f abierta - [mec] open part
— f — de gancho m - [mec] hook open part
— f accesoria - appurtenance
— f activa - [mec] active, o moving, part
— f actora f - [legal] plaintiff; prosecution;
claimant
— f achaflanada - [sold] beveled, part, o area
— f — de superficie f - [sold] beveled surface
area
— f adyacente - adjacent part
— f — de plancha f - [sold] plate adjacent,
part, o area
— f afectada - affected part
— f ahusada - [mec] tapered part; taper
— f alguna - any part
— f alta f - upper, o high(er), part
— f — de escala f - range, top, o higher part
— f anterior - [mec] front (part) • previous
part
— f — de acoplado m - [mec] trailer front
(part)
— f — de base f - base front (part)
— f —— ranura f - [mec] groove front (part)
— f bajo tierra f - [constr] buried portion
— f — de cimiento m - [constr] foundation
buried, part, o portion, o section
— f —— f — poste m - [constr] post buried,
part, o portion, o section
— f básica - base,sic, part, o portion
— f buena - good part • major, part, o portion
— f buena de carrera f - [deport] much of @ way
— f — de corrida f - [deport] much of @ way
— f central - center,tral, part, o portion, o
section

parte f central de árbol m para leva(s) f -
[comb.int] camshaft center
— f —— Estados m Unidos (de América f del
Norte)—[geogr] Midwest | adv - [geogr] (de)
Midwesterner
— f —— neumático m - [autom-neumát] tire
center
— f céntrica - [pol] downtown
— f̄ — de ciudad f—[pol] city center; downtown
— f̄ —— Estados m Unidos - [geogr] Midwest
— f̄ cilíndrica f - [tub] barrel
— f̄ — de tubería f - [tub] pipe barrel
— f̄ clave - key part
— f̄ codemandada - [legal] codefendant
— f̄ comercial - [com] business, o commercial,
part, o portion; hub
— f compareciente - [legal] plaintiff
— f̄ componente - component part
— f con corriente f - [electr] live part
— f̄ cóncava - [sold] concave, part, o portion
— f̄ cónica - [mec] conical, part, o portion
— f̄ — de casquillo m - [mec] socket conical,
part, o portion
— f̄ considerable - considerable, o substantial,
o great(er), part, o portion
— f̄ constitutiva - component part
— f̄ — de equipo m para perforación f—[petról]
(drilling) rig component (part, o portion)
— f contratante - [legal] contracting party
— f̄ correspondiente - corresponding, part, o
portion
— f — de corona f - [constr] crown portion
— f̄ ——— f de bóveda f - [constr] arch
crown, part, o portion
— f crítica - critical, part, o portion
— f̄ — de terraplén m - [constr] embankment
critical portion
— f cualquiera - any part
— f̄ de accesorio m - fitting, o appurtenance,
part, o portion
— f̄ — acero m - [mec] steel, part, o component
— f̄ — alma m - [cabl] core, part, o member
— f̄ ——— m de cable m - [cabl] cable, o wire
rope, core, part, o member
— f ——— m de alambre m - [cabl] wire
rope core, part, o member
— f — aluminio m - [mec] aluminum, part, o
component
— f — atrás - back, part, o portion
— f̄ ——— de pista f - [deport] (race) track
back (part, o portion)
— f — caso m - case, part, o portion
— f̄ — caja f̄ - [mec] case, o housing, part, o
portion, o section
— f — centro m - center, tral, part, o portion
— f̄ ——— m de caja f - [mec] case, o housing,
center, part, o portion, o section
— f — cimentación f - [constr] foundation,
part, o portion, o section
— f — cimiento m - [constr] foundation portion
— f̄ — colada f - [metal-prod] cast(ing) report
— f̄ — conducto m - [hidr] waterway portion
— f̄ ——— m para agua m - [hidr] waterway portion
— f — cubierta f - [mec] cover part • [autom-
-neumát] tire part
— f — cuerpo m - [anatom] body, part, o area
— f̄ — defensa f - [mec] shroud(ing) part •
[legal] defendant part
— f — desagüe m - [hidr] drain(age) part
— f̄ ——— m inferior - [hidr] subdrain part
— f̄ — ejemplificación f - case, o example,
part, o portion
— f — equipo m - [ind] equipment part
— f̄ ——— m para torre f (para perforación f) -
[petról] derrick, o rig, component
— f̄ — espaciador m - [mec] spacer section
— f̄ — especificación f - specification, part,
o portion, o section
— f — espectro m - [fís] spectrum part
— f̄ — estructura f - [constr] structure, part,
o portion, o section
— f̄ ——— f que corresponde a corona f (de bó-

veda f) - [constr] structure arch crown
portion
parte f de fundamento m - [constr] foundation,
part, o portion
— f — garganta f - [sold] throat section
— f̄ — f total - [sold] full throat section
— f̄ — gasto(s) m - expense(s) share
— f̄ — hormigón m - [constr] concrete component
— f̄ — horno m - [combust] furnace part
— f̄ — línea f - line, part, o portion •
[electr-instal] line side
— m —— f de conductor(es) m eléctrico(s) -
[electr-distrib] power line side
— f —— f para energía f - [electr-distrib]
power line side
— f̄ — madera f - [constr] wood(en) part
— f̄ — neumático m - [autom-neumát] tire, part,
o component
— f̄ — obra f - [constr] job portion
— f̄ — país m - [pol] country, part, o section
— f̄ — pista f - [deport] course part
— f̄ — recepción f - [com] receipt, o acceptance, slip
— f — recorrido m - [deport] course, part, o
portion
— f — red f - [ind] system, part, o section
— f̄ —— f para aceite m - [mec] oil system,
part, o section
— f — rutina - routine part
— f̄ — separador m - [mec] spacer section
— f̄ — soldadura f̄ - [sold] weld portion
— f̄ — suministro m - supply part • [ind] sup-
ply(ing) report
— m —— m extranjero - foreign supply portion
— f̄ —— m nacional - domestic supply portion
— f̄ — terraplén m—[constr] embankment portion
• [f.c.] fill portion
— f — trabajo m [labor] work, o job, part •
work, o job, report
— f —— m completada - completed, work, o
job, o project, part
— f — tubería f - [tub] tube, o pipe, o line,
part, o portion
— f — urbanización f - [constr] subdivision
section
— f debilitada - weakened, portion, o part
— f̄ delantera - front, o forward, part, o end
— f̄ — a plomo - plumb front • flush front
— f̄ — de automóvil m - [autom] car front part
— f̄ —— bastidor m - [mec] frame front
— f̄ —— camión f - [autom] truck front
— f̄ —— carrocería f - [autom] front bodywork
— f̄ —— huella f - [autom-neumát] footprint
front
— f —— remolque m - [autom] trailer front
— f̄ delgada - thin part
— f̄ demandada - [legal] defendant
— f̄ demandante - [legal] claimant
— f̄ desgastada - worn, part, o portion
— f̄ — de eslabón m - [cadenas] link worn, portion, o part
— f deslizante - [mec] sliding part
— m diario - daily report; log
— f̄ dorsal - back, o rear, part
— f̄ económica - economic part
— f̄ eléctrica - electrical part • [sold] electrical end • véase también pieza f eléctrica
— f empotrada - [constr] embedded part
— f̄ en divisa(s) f - [fin] foreign currency,
part, o portion
— f — dólar(es) m - [fin] dollar, part, o portion
— f —— m canadiense(s) - [fin] Canadian
dollar(s), part, o portion
— f —— m estadounidense(s) - [fin] United
States, o American, dollar(s) part
— f — moneda f . . . - [fin] . . . currency
part
— f —— f extranjera - [fin] foregin currency
part
— f —— f nacional - [fin] local currency
part
— f — peso m - part of weight

parte f en peso(s) m - [fin] peso(s), part, o
portion • . . . . currency part
— f — — m argentino(s) - [fin] Argentina,ne
peso(s) part
— f — — m moneda f nacional - [fin] national
currency part • Argentine currency part
— f — sucre(s) m - [fin] sucre(s) part
— f ensanchada - expanded, part, o portion
— f específica - specific, part, o portion
— f — de obra f - [constr] project particular,
part, o portion • particular scheme
— f expuesta - exposed, part, o portion
— f exterior - outside, part, o portion; shell;
case
— f — de curva f - outside of @ curve
— f — — junta f - [sold] joint outside
— f — — plataforma f - platform outside
— f — — — f para descarga f - [ind] unloading platform outside
— f externa - external part; outside
— f extrema - end portion
— f — de estructura f - [constr] structure end
portion
— f fija - stationary part • [metal-prod]
port end
— f — de horno m basculante - [metal-prod]
tilting furnace port end
— f fina - thin, part, o portion, o section
— f final - final, o end, part • latter part
— f — de campaña f - [metal-prod] campaign
latter part
— f — — libro m - [public] book back (part)
— f — — manual m - [public] manual back
— f — — soldadura f - [sold] weld finish end
— f financiada - [fin] financed portion
— f financiera - [fin] financial, part, o section
— f frontal - front (part, o end)
— f — de horno m - [ind] furnace front (part)
— f grande - great portion • much
— f gruesa - thick part
— f hidráulica - [petról] fluid, o mud, end
— f — con presión f - [petról] pressurized
fluid end
— f — — — f alta - [petról] high pressure
fluid end
— f — — bomba f - [bombas] pump fluid end
— f — — — f para inyección f - [petról]
(slush) pump, mud, o fluid, end
— f identificada - identified, part, o portion
— f importada - [com] imported, part, o portion
— f importante - major, part, o portion, o item
— f indivisible - indivisible, part, o portion
— f inferior - lower part; underside • [constr-
-pil] bottom (part, o portion) • [metal-fabr]
invert
— f — de alcantarilla f - [constr] culvert invert
— f — — arado m - [agric-equip] plow bottom
— f — — banda f - [metal-lam] strip bottom
— f — — barra f - [metal-lam] bar bottom
— f — — bóveda f - [constr] arch leg
— f — — cabeza f - [mec] head bottom
— f — — — — f para inyección f - [petról]
swivel bottom
— f — — caja f - [mec] case, o housing, bottom, o lower part
— f — — carcasa f - [electr-mot] frame, bottom, o lower half
— f — — carrera f - [comb.int] stroke bottom
— f — — cartel m - chart bottom
— f — — cinta f - [metal-lam] strip bottom
— f — — chaflán m - [mec] chamfer bottom
— f — — chapa f - [metal-lam] strip bottom
— f — — chasis m - [mec] chassis bottom
— f — — escala f - range lower portion
— f — — — f para electrodo(s) m - [sold]
electrode(s) range lower portion
— f — — escalera,rilla f—[mec] ladder bottom
— f — — estructura f - [constr] structure
bottom
— f — — — f Super-Span - [constr] Super-Span
structure bottom

parte f inferior de falda f - [mec] skirt bottom
— f — — f de émbolo m - [comb.int] piston
skirt bottom
— f — — fleje m - [metal-lam] strip bottom
— f — — grúa f - [grúas] crane underneath
— f — — horno m - [combust] furnace bottom
— f — — — m para reducción f - [metal-prod]
reduction furnace bottom
— f — — junta f - [sold] joint bottom
— f — — larguero m - [constr] stringer bottom
— f — — losa f - [constr] slab bottom
— f — — — f para cubierta f - [constr-puent]
deck slab bottom
— f — — malacate m—[petról] drawworks bottom
— f — — neumático m - [autom-neumát] tire
bottom
— f — — núcleo m - [constr-pil] core bottom
— m — — — m hincador m - [constr-pil]
driving core bottom
— f — — orificio m - [mec] hole bottom
— f — — — m para abastecimiento m - [mec]
filler hole bottom
— f — — pared f - [constr] wall bottom
— f — — — f para celda f - cell wall bottom
— f — — — tabla f - chart bottom • [mec] board,
bottom, o underside
— f — — — f para túnel m - [constr] lower
tunnel, o tunnel lower, sidewall
— f — — — f lateral—[constr] lower sidewall
— f — — pendiente f - [topogr] slope bottom
— f — — pilote m - [constr=pil] pile bottom
— f — — rejilla f - [mec] screen bottom
— f — — soldadora f - [sold] welder bottom
— f — — soldadura f - [sold] weld bottom
— f — — talud m - [topogr] slope bottom
— f — — tapón m - [mec] plug bottom
— f — — — m de hormigón m - [constr-pil]
concrete plug bottom
— f — — tragante m - [metal-prod] lower shaft
— f — — tubería f - [tub] pipe, invert, o
bottom • [constr] conduit bottom • [ambient]
duct, bottom, o underside
— f — — — f abovedada - [metal-fabr] pipe
arch bottom
— f — — — f — corrugada - [metal-fabr] corrugated pipe arch bottom
— f — — — f — encajable - [tub] corrugated nestable pipe arch bottom
— f — — — f — galvanizada - [metal-fabr]
galvanized corrugated pipe arch bottom
— f — — — f — galvanizada - [metal-fabr]
galvanized pipe arch bottom
— f — — — f corrugada - [metal-fabr] corrugated pipe bottom
— f — — — f — galvanizada - [metal-fabr]
galvanized corrugated pipe bottom
— f — — — f encajable - [metal-fabr] nestable pipe bottom
— f — — — f — corrugada - [metal-fabr] corrugated nestable pipe bottom
— f — — — f — galvanizada - [metal-fabr]
galvanized corrugated nestable pipe bottom
— f — — — f galvanizada - [metal-fabr] galvanized pipe bottom
— f — — tubo m de pistola f - [sold] gun tube
bottom
— f — — viga f - [constr] beam, o stringer,
bottom
— f — — — f (en) U - [constr] channel bottom
— f — — — f (—) W - [constr] W-beam bottom
— f — derecha - lower right hand corner
— f — izquierda - lower left hand corner
— f — total - [constr] entire bottom
— f inicial - initial, o early, part
— f — de campaña f - [metal-prod] campaign
early part
— f inspeccionada - inspected part
— f integral - integral part
— f integrante - integrating, o integral, part
— f interesada - interested, o concerned, party
— f interior - inside • core • [autom-neumát]
understructure; undertread
— f — de banda f - [autom-neumát] (tread? un-

derstructure, o undertread
parte f interior de planchón m - [metal-lam] slab inside
— f —— **producto m** - [ind] product inside
— **f interna** - internal, part, o portion
— **f̄ — de soldadura f** - [sold] weld internal, part, o portion
— **f invitante** - [social] inviting party
— **f̄ más alta de escala f** - scale, o range, upper part
— f —— **f de electrodo m** - [sold] electrode range upper portion
— f — **amplia** - broadest part
— **f̄ — baja de escala f** - scale, o range, lower part
— f —— **f de electrodo m** - [sold] electrode range lower portion
— f **más intensa** - [mec] brunt
— **f̄ — saliente** - [mec] farthest extension
— **f̄ mayor** - great(er,est), part, o portion
— **f̄ mecánica** - [mec] mechanical part
— **f̄ media** - middle part; center
— **f̄ — a alta de escala f** - [sold] range middle to high portion
— f — **de cartel m** - chart middle (part)
— f̄ —— **concavidad f**—(con)cavity middle part
— f̄ —— **coraza f** - [metal-prod] shell center
— f̄ —— **escala f̄** - range middle (portion)
— f̄ —— **f para electrodo m** - [sold] electrode range middle (part)
— f̄ —— **tabla f** - chart middle (part)
— **f̄ menor** - small(er), part, o portion
— **f̄ mineral** - [miner] mineral, o ore, portion
— **f̄ — de mena f básica** - [miner] basic ore mineral portion
— f —— **básico** - [miner] basic ore mineral portion
— f **motriz** - [petról] (slush pump) power end
— **f̄ — de bomba f para inyección f** - [petról] slush pump power end
— f **nacional**—[ind] local(ly) manufactured part
— f **negativa** - negative part
— **f̄ no financiada** - [fin] unfinanced portion
— **f̄ occidental** - [geogr] western part
— **f̄ — de Estados m Unidos** - [geogr] Far West
— **f̄ óptica**—optical, part, o portion, o section
— **f̄ oriental** - [geogr] eastern part
— **f̄ — de Estados m Unidos** - [geogr] East
— **f̄ pagadera** - [fin] payable part
— **f̄ para repuesto m** - [mec] véase **pieza f para repuesto m**
— m — **trabajo m** - [labor] time, report, o card
— **f̄ plana** - flat (part)
— **f̄ — de arandela f** - [mec] washer flat part
— **f̄ —— cabeza f para perno n** - [mec] bolt head flat (part)
— f̄ —— **cubo m** - [mec] hub flat part
— **f̄ —— diámetro m** - [mec] diameter flat part
— **f̄ ——— m interior** - [mec] inside diameter flat part
— f ——— **m — de arandela f** - [mec] washer inside diameter flat part
— f —— **eje m** - [mec] axle, o shaft, flat
— **f̄ ——— m para bomba f** - [bombas] pump shaft flat
— f —— **émbolo m** - [mec] piston land
— **f̄ —— mazo m** - [mec] hub flat part
— **f̄ — para rodamiento m** - [mec-ruedas] float; clearance (para rieles)
— f **por millón m** - parts per million
— **f̄ por peso m** - part by weight
— **f̄ positiva** - positive part
— **f̄ posterior** - [mec] rear, o back, (part, o side, o end)
— f — **de árbol m** - [mec] shaft, back, o rear
— **f̄ ——— m para leva f** - [comb.int] camshaft, back, o rear
— f —— **automóvil m** - [autom] car('s) tail
— **f̄ —— base f** - base, back, o rear
— **f̄ —— camión m**—[autom] truck, back, o rear
— **f̄ —— compartimiento m** - [mec] compartment rear
— **f̄ ——— m para motor** - [electr-mot] motor

compartment, back, o rear • [comb.int] engine compartment, back, o rear
parte f posterior de pista f - [deport] track, o grid, back (part, o section)
— **f primera** - first, part, o portion, o section
— **f̄ principal** - principal, o main, o major, part, o section, o item
— **f̄ — de desagüe m** - [hidr] drain, main, o principal, part
— f ——— **m inferior** - [hidr] subdrain, main, o principal, part
— f —— **equipo m** - [ind] equipment main item
— **f̄ rebajada** - [mec] offset
— **f̄ removible** - [mec] removable part; knockout
— **f̄ ribera afuera** - [hidr] seaward end
— **f̄ segunda** - second, part, o portion
— **f̄ sellada f** - [mec] sealed, part, o portion
— **f̄ — de conducto m** - [electr-cond] conduit sealed, part, o portion
— f ——— **m para cable(s) m** - [electr-cond] cable conduit seal(ed part)
— **f sin chaflán m** - [sold] groove face
— **f̄ sobre tierra f** - [constr] above ground, part, o portion, o section
— f —— **f de cimiento m** - [constr] above ground foundation, part, o portion
— f —— **f — fundamento m** - [constr] above ground foundation, part, o portion
— f —— **poste m** - [constr] above ground post portion
— **f solapada** - [mec] lapped, part, o section
— **f̄ sujetadora** - [mec] gripping portion
— **f̄ — de bloque m** - [mec] block gripping portion
— **f superior** - upper, o top, part, o portion, o section, o piece │ adv - (por) overhead
— f — **de acumulador m** - [electr-acum] battery top
— f —— **armadura f** - [electr-mot] armature top
— **f̄ —— banco m** - [mec] (work) bench top
— **f̄ ——— m para trabajo m** - [mec] work bench top
— f —— **banda f** - [metal-lam] strip top
— **f̄ ——— f para rodamiento m** - [autom-neumát] tread surface • surface tread
— f —— **barra f** - [mec] bar top
— **f̄ —— boquilla f** - [mec] nozzle top
— **f̄ ——— f principal** - [comb.int] main nozzle top
— f —— **bóveda f** - [constr] arch top
— **f̄ —— cabeza f** - [mec] head top
— **f̄ ——— f para inyección f** - [petról] swivel top
— f —— **cabina f** - [mec] cab(in) top
— **f̄ —— carcasa f** - [electr-mot] frame, top, o upper, half
— **f̄ —— carrera f** - [comb.int] stroke top
— **f̄ —— cartel m** - chart top
— **f̄ —— cimiento m** - [constr] foundation top
— **f̄ —— cinta f** - ribbon top • [metal-lam] strip top
— f —— **cojinete m** - [cojin] bearing top
— **f̄ ——— m central** - [cojin] center bearing, top, o cap
— f ——— **m principal** - [mec] center main bearing, top o cap
— f ——— **m frontal** - [mec] front bearing, top, o cap
— f ——— **m — principal** - [mec] front main bearing, top, o cap
— f ——— **m principal** - main bearing top
— f ——— **m posterior** - [mec] rear bearing, top, o cap
— f ——— **m — principal** - [mec] rear main bearing, top, o cap
— f —— **chapa f̄** - [metal-lam] strip top
— **f̄ —— chasis m** - [mec] chassis top
— **f̄ —— depósito m** - [comb.int] tank top
— **f̄ —— émbolo m** - [mec] piston top
— **f̄ —— escala f̄** - range, top, o higher, portion, o part
— f —— **estructura f** - [constr] structure top
— f ——— **f Super-Span** - [constr] Super-Span

parte f superior de falda f para émbolo m - [comb.int] piston skirt top
— f — — — fleje m - [metal-lam] strip top
— f — — — fundamento m - [constr] fundation, top, o cap
— f — — — grúa f - [grúas] crane top
— f — — — horno m - [ind] furnace, upper part, o top
— f — — — m alto - [metal-prod] blast furnace, upper part, o top
— f — — — — m para reducción f - [metal-prod] reduction furnace, upper part, o top
— f — — indicador m—[instrum] indicator top (side)
— f — — — larguero m - [constr] stringer top
— f — — — losa f - [constr] slab top
— f — — — malacate m - [petról] drawworks top
— f — — — neumático m - [autom-neumát] tire top
— f — — — núcleo m - [mec] core, upper part, o top
— f — — — — m hincador - [constr-pil] driving core, upper part, o top
— f — — página f - [public] page top
— f — — — panel m - [mec] panel, upper part, o top
— f — — pared f - [constr] wall, upper part, o top
— f — — — — f de celda f - cell wall top
— f — — — pendiente f - [topogr] slope top
— f — — — pilote m - [constr-pil] pile top
— f — — — — m tubular - [constr-pil] pipe pile top
— f — — portadiferencial m - [autom-mec] differential carrier top
— f — — presa f - [hidr] dam top
— f — — ranura f - [mec] groove top
— f — — — f profunda—[sold] deep groove top
— f — — recubrimiento m de hormigón m - [constr] concrete envelope top
— f — — rejilla f - [mec] screen top
— f — — soldadora f - [sold] welder top
— f — — soldadura f - [sold] weld top
— f — — tabla f - chart top
— f — — talud m - [topogr] slope top
— f — — — f en ribera f - [hidr] upper bank slope
— f — — tapón m - [sold] plug top
— f — — — m de hormigón m - [constr-pil] concrete plug top
— f — — tobera f - [comb.int] nozzle top • [metal-prod] tuyere top
— f — — — f principal - [comb.int] main nozzle top
— f — — tragante m - [metal-prod] upper shaft
— f — — traviesa f - [f.c.] tie top
— f — — tubería f - [tub] pipe top • [constr] conduit top • [ambient] duct top
— f — — — f abovedada - [metal-fabr] pipe arch top
— f — — — — f — corrugada - [metal-fabr] corrugated pipe arach top
— f — — — — f — galvanizada - [metal-fabr] galvanized corrugated pipe arch top
— f — — — — f galvanizada - [metal-fabr] galvanized pipe arch top
— f — — — — f corrugada - [metal-fabr] corrugated pipe top
— f — — — — f galvanizada - [metal-fabr] galvanized corrugated pipe top
— f — — — — f encajable - [metal-fabr] nestable pipe top
— f — — — — f — corrugada - [metal-fabr] corrugated nestable pipe top
— f — — — — f — galvanizada - [metal-fabr] galvanized corrugated nestable pipe top
— f — — — — f galvanizada - [metal-fabr] galvanized pipe top
— f — — — — f para agua m - [petról] wash pipe top
— f — — — — f principal - [ambient] main (pipe) top
— f posterior - [mec] back, part, o side

parte f superior de viga f - [constr] beam. o stringer, top
— f — — — f (en) U - [constr] channel top
— f — — — f (—) W - [constr] W beam top
— f — derecha - upper right hand corner
— f — intercambiable - [tub] interchangeable top
— f — izquierda - upper left hand corner
— f — lisa f - [mec] even, o flush, top
— f — trasera - [mec] rear top deck
— f técnica - technical, part, o portion, o section
— f telescópica - [mec] telescopic,ping part
— f tierra f adentro - [hidr] land section
— f trasera - back, o rear, (part, o end)
— f — de automóvil - [autom-mec] car('s), tail, o rear (end)
— f — — bastidor m - [mec] frame rear (part)
— f — — huella f - [autom-neumát] footprint rear
— f — remolque m - [autom] trailer rear
— f traslapada - overlapped,ping part
— f tubular - [tub] tubular section; barrel
— f visible - visible part
— f — de electrodo m - [sold] visible electrode (part)
— f vulnerable - vulnerable part
partes f iguales - equal parts | adv - (en) half and half
partición f - . . . . • split(ting)
participación f - . . .; sharing; part taking • involvement,ving • role • pool • announcement • [fin] partnership • share • [legal] joint venture • [literat] contribution • [deport] campaign(ing) | a - (en) shared
— f accionaria - [fin] equity participation
— f acogida - welcomed participation
— f activa - active, participation, o role • [deport] active campaigning
— f amplia - ample, o broad, participation
— f aumentada - increased participation
— f de Banco m para Importación(es) f y Exportación(es) f - [fin] Eximbank participation
— f — capital m nacional - [econ] local, o domestic, capital participation
— f — empleado(s) m - [labor] employee involvement,ving
— f — Eximbank m - [fin] Eximbank participation
— f — grupo m - group participation • group activity
— f — industria f nacional - [ind] local industry paraticipation
— f — ingeniero(s) m - engineer('s) participation
— f — — m mecánico(s) - [mec] mechanical engineer('s) participation
— f — inversionista(s) m - [fin] investor('s) participation
— f — — m extranjero(s) m - [fin] foreign investor(s) participation
— f — — m nacional(es) - [fin] national, o domestic, investor(s) participation
— f — metalurgista(s) m - [metal] metallurgist('s) participation
— f — país m - country('s) participation
— f — público m - [legal] public participation
— f — personal m - [fin] employee(s) profit sharing
— f en actividad(es) f - activity,ties involvement,ving
— f — — f comunal(es) f - community involvement
— f — administración f - [admin] management, participation, o involvement, o sharing
— f — beneficio(s) m - profit share,ring
— f — capital m - [fin] capital, participation, ó share,ring
— f — carrera(s) f - [deport] racing
— f — — f para rendimiento m alto - [deport] performance, race, o event, involvement
— f — club(es) para servicio m - (service) club(s) involvement

participación f en exterior - [fin] foreign
joint venture
— f — ganancia(s) f - [fin] profit(s),
share,ring
— f — gestión f - [admin] management,
share,ring, o participation
— f — mercado m - [com] market, participation,
o share,ring, o participation
— f — patrimonio m - [fin] equity, participating,tion, o share,ring
— f — propiedad - [fin] ownership share,ring;
property ownership
— f — seguridad f - [segurid] safety, participating,tion, o share,ring
— f — utilidad(es) f - [contab] profit(s),
share,ring, o distribution
— f estatal - state, participation, o sharing
— f total - global, o total, share,ring
— f líquida - [econ] liquid, o net, share
— f — en efectivo m - [Econ] liquid, o net.
cash share
— f local - local, participation, o share,ring
— f máxima - maximum, participation, o share
— f — de industria f - [ind] maximum industry
participation
— f — — — f nacional m - [ind] maximum local
industry, participation, o share
— f mínima - minimum, participation, o share
— f — de industria f nacional - [ind] minimum
local industry, participation, o share
— f nacional - local, o national, o domestic,
participation, o share
— f neta - [econ] net, participation, o share
— f patrimonial - [fin] capital share,ring •
equity share,ring
— f pública - [legal] public participation
— f reducida - [fin] reduced, participation, o
share,ring
participado,da a - participated; shared; pooled
• involved • [deport] campaigned
— a activamente - participated actively •
[deport] campaigned actively
— a en administración f - [admin] shared in @
management
— a — gestión f - [admin] shared in @ management
— f — propiedad - shared in @ ownership
participante m . . . • [deport] entry; competi-
tor • pl field | a - . . . .; participant
— f — carrera f - [deport] race participant
— m — — f para rendimiento m alto - [deport]
performance event participant
— m — clínica f - clinic participant
— f — prueba f - [deport] trial, o event, participant
participantes m - [deport] field
participar v - . . .; to share; to take part •
to involve; to become involved • to pool •
. . .; to announce • [deport] to campaign
— v activamente - to participate actively •
[deport] to campaign actively
— v en gestión f - [admin] to, participate, o
to share, in @ management
— v — beneficio(s) - [fin] to share in @ prof-
it(s)
— v — carrera f - [deport] to, participate, o
enter, @ race, o in @ performance
— v — ganancia(s) f - to share in @ profit(s)
— v — gestión f - [admin] to share in @ management
— v — propiedad f - [econ] to share (in) @
ownership
— f — utilidad(es) f - [Econ] to share in @
profit(s)
participante m pleno - full, participant, o
partner
partícula f abrasiva - abrasive particle
— f arenosa - gritty particle
— f arrastrada - [Ambient] entrained particle
— f arrastrada de tamaño m mayor - [ambient]
large(r) entrained particle

partícula f arrastrada de tamaño m menor -
[ambient] small(er) entrained particle
— f — grande - [ambient] large entrained particle
— f — mayor m - [ambient] large(r) entrained
particle
— f — menor - [ambient] small(er) entrained
particle
— f — pequeña - [ambient] small entrained particle
— f caliente - hot particle
— f coloidal - [hidr] colloidal particle
— f con tamaño m mayor - [constr] oversized
particle
— f — — m menor - [constr] undersized particle
— f contenida - contained particle
— f cuarcítica - [hidr] quartzose particle
— f de aceite m - oil particle
— f — calidad f constante - [ind] consistent
quality particle
— f — carbono m - [comb.int] carbon particle
— f — composición f constante - [ind] consistent composition particle
— f disolución f - solution particle
— f escoria f - [sold] slag particle
— f líquido m - liquid particle
— f mineral - [miner] ore particle
— f — polvo m - powder particle • dust particle
— f — — m contenida - [sold] contained powder
particle
— f — rocío m - [ind] spray particle •
[meteorol] dew particle
— f — suelo m - [suelos] soil particle
— f en suspensión f - suspended particle
— f — — f en aire m - air suspended particle
— f — tamaño m mayor - oversize(d) particle
— f — — m menor - undersize(d) particle
— f extraña - foreign particle
— f filtrable - [ambient] filt(e)rable particle,culate
— f gaseosa - [quím] gaseous particle
— f — ionizada - [quím] ionized gaseous particle
— f gris - [miner] gray particle
— f — negruzca - [miner] black-gray particle
— f ionizada - [quím] ionized particle
— f lijada - [mec] sanded particle
— f líquida - liquid particle
— f liviana - light particle
— f magnética - magnetic particle • [sold] magnetic dirt
— f — fina - [electr] fine magnetic particle
— f — muy fina f - [electr] very fine, o divided, magnetic particle
— f menuda f - minute particle
— f metálica - [metal] metal(lic) particle
— f pesada - heavy particle
— f producida - [ind] produced particle
— f suspendida - suspended particle
— f — en aire m - air suspended particle
— f única - single particle
— f uniforme - uniform particle
— f volante - [segurid] flying particle
particulación* - particulating,tion*
particulado,da* a - particulated*
particulador,ra* a - particulating*
particular m - regard • item | a - . . . .; dis-
tinctive; specific • private; individual | _
v - to particulate*
particularmente adv - . . . . • uniquely
partículas f dispersas - distributed particles
— f en tamaños m diversos - [constr] several, o
varied, size(d) particles
partida f - . . . .; leaving; check-out • group-
ing; description • [transp] shipment; order;
lot • [com] item • [contab] entry • [ind]
batch; run • [legal] . . . . • item • [fin]
fund(s); amount; quantity • [grúas] lift
— f cargada f - [ind] charged batch • [transp]
loaded batch
— f efectiva - effective lot

partida f̲ de artículo m̲ - item batch
— f̲ — cable m̲ - [cabl] rope shipment
— m̲ — equipo m̲ - equipment batch
— f̲ — existencia(s) f̲ - [com] stock item
— f̲ — fundente m̲ - flux batch
— f̲ — metal m̲ - [metal- metal batch
— f̲ — — m̲ laminado - [metal-lam] metal stand
— f̲ — pella(s) f̲ - [miner] pellet(s) batch
— f̲ — — f̲ reducida(s) f̲ - [miner] reduced pellet(s) batch
— f̲ — sínter m̲ - [miner] sinter batch
— f̲ — trabajo m̲ - [ind] job run
— f̲ — unidad f̲ motriz - [mec] power unit lot
— f̲ desde país m̲ de origen - [legal] departure from @ country of origin
— f doble - [contab] double entry
— f en lote m̲ - item batch
— f grande - [ind] large batch
— f intermitente - [ind] intermittent run
— f mayor - [ind] large(r) batch
— f menor - [ind] small(er) batch
— f numerada - numbered item
— f número . . . - item number . . .
— f pequeña f̲ - [ind] small batch
— f prevista - foreseen item
— f separada - separate item
— f siguiente - following item
partido m̲ - . . . . • [deport] game • handicap • [pol] . . . • county
— m̲ final - [deport] final game • championship, o̲ all star, game
partido,da a̲ - departed; left • checked out • split; cloven • [mec] crushed • [agric] cracked
partidor m̲ - . . . . • [hidr] division, vider box
— m̲ principal - [hidr] main division, vider box
partir v̲ - . . . . • to, depart, o̲ leave • to check out • to crack; to crush • to split
parto m̲ [medic] . . . ; delivery
— m̲ normal - [medic] normal, delivery, o̲ childbirth
pasa adv̲ no pasa - go-no go
pasabajos m̲ - [electrón] low pass
pasabanda f̲ - [electrón] bandpass
— f̲ optativa - [electrón] optional bandpass
pasada f̲ - . . . . • [sold] pass • [constr-pil] pass • [metal-lam] pass • [hidr] drain(ing)
— f̲ adicional—[sold] additional, pass, o̲ layer
— f̲ ancha f̲ - [sold] wide pass
— f̲ — en ángulo m̲ (interior) - [sold] wide fillet pass
— f̲ — — interior m̲ de ángulo m̲ - [sold] wide fillet pass
— f̲ — ortogonal - [sold] wide fillet pass
— f̲ angosta - [sold] narrow pass
— f̲ aplicada - [sold] applied pass
— f̲ audaz f̲ - [deport] bold pass(ing)
— f̲ automática - [metal-lam] automatic pass
— f̲ con avance m̲ rápido—[sold] high speed pass
— f̲ con cuchara f̲ - [constr-hormig] troweling
— f̲ — electrodo m̲ - [sold] electrode pass(ing)
— f̲ — — m̲ con alma m̲ fundente - [sold] Innershield pass
— f̲ — — m̲ manual - [sold] stick pass
— f̲ — metal m̲ para relleno m̲ - [sold] filler metal, pass, o̲ layer
— f̲ — movimiento m̲ amplio (de tejido m̲) - [sold] wide weave, pass, o̲ weld
— f̲ — — m̲ de vaivén m̲ - [sold] weave pass
— f̲ — — m̲ de tejido m̲ - [sold] weave pass
— f̲ — — m̲ en zig-zag - [sold] weave pass
— f̲ — — m̲ zigzagueante* - [sold] weave pass
— f̲ — paleta f̲ - [constr] troweling
— f̲ — poco movimiento m̲ de tejido m̲ - [sold] small weave pass
— f̲ — tejido m̲ - [sold] weave pass
— m̲ — — m̲ angosto - [sold] small weave pass
— f̲ — — m̲ en caja f̲ - [sold] box weave pass
— f̲ — — m̲ triangular - [sold] triangular weave pass
— f̲ corriente - [sold] straight pass
— f̲ — para relleno m̲ - [sold] straight filler pass

pasada f̲ de banda f̲ - [metal-trat] strip passing
— f̲ — cinta f̲ - [metal-lam] strip pass(ing)
— f̲ — canto m̲ - [metal-lam] bullhead pass(ing)
— f̲ — corriente f̲ - [electr-oper] current passing
— f̲ — chapa f̲ - [metal-lam] strip pass(ing)
— f̲ — fleje m̲ - [metal-lam] strip pass(ing)
— f̲ — soldadura f̲ - [sold] weld pass
— f̲ . . . de soldadura f̲ circunferencial - [sold] . . . pass girth weld
— f̲ — soldadura f̲ con arco m̲ sumergido - [sold] submerged arc weld pass
— f̲, efectuada, o̲ ejecutada - [sold] applied, o̲ performed, pass
— f̲ — ángulo m̲ - [sold] fillet pass
— f̲ — — m̲ interior - [sold] fillet pass
— f̲ — — m̲ — vertical - [sold] vertical fillet pass
— f doble - [metal-lam] two-step rolling • [sold] double pass
— f — para obtener(se) espesor m̲ final - [metal-lam] two-step rolling to @ final gage
— f̲ — — m̲ interior - [sold] fillet pass
— f̲ — caliente - [sold] hot pass
— f̲ en interior m̲ de ángulo m̲ - [sold] fillet pass
— f̲ — — m̲ — — m̲ vertical - [sold] vertical fillet (fill) pass
— f̲ — laminador m̲ - [metal-lam] mill pass
— m̲ — — m̲ para planchón(es) m̲ - [metal-lam] slabbing mill pass
— f̲ — tejido m̲ triangular - [sold] triangular weave pass
— f̲ exigida - [metal-lam] required pass
— f̲ exterior - [sold] outside pass • [tub] outside diameter pass
— f final - [sold] finish, o̲ final, o̲ last, o̲ cover, pass
— f inicial - [sold] first, o̲ initial, pass
— f̲ Innershield - [sold] Innershield pass
— f̲ múltiple - [sold] multiple pass
— f̲ número . . . • [sold] pass number . . .
— f̲ para alisamiento m̲ - [sold] stripper, pass, o̲ bead
— f̲ — cierre m̲ - [sold] cover, o̲ last, pass
— f̲ — — m̲ en junta f̲ - [sold] joint cover pass
— f̲ — — m̲ — — f̲ de tubería f̲ - [sold] pipe joint cover pass
— f̲ — — m̲ tejida - [sold] weave cover pass
— f̲ — — m̲ — con ancho m̲ cabal - [sold] full width weave cover pass
— f̲ — cobertura f̲ - [sold] cover pass
— f̲ — endurecimiento m̲ - [metal-lam] hardening pass
— f̲ — — m̲ de superficie f̲ - [metal-lam] temper pass
— f̲ — — m̲ superficial - [metal-trat] temper pass
— f para raíz f̲ - [sold] root pass • stringer bead
— f̲ — — f̲ con electrodo m̲ Fleetweld - [sold] Fleetweld root pass
— f̲ — — f̲ depositada - [sold] deposited root pass
— f̲ — reducción f̲ reducida - [metal-lam] pinch pass
— f̲ — — f̲ para soldadura f̲ - [sold] weld root pass
— f̲ — — f̲ — — f̲ con pasada(s) f̲ múltiples - [sold] multiple pass weld root pass
— f̲ — — f̲ — — f̲ en ranura f̲ - [sold] groove weld root pass
— f̲ — recubrimiento m̲ - [sold] cap(ping) pass
— f̲ — relleno m̲ - [sold] fill(er) pass
— f̲ — — m̲ a tope - [sold] butt fill pass
— f̲ — — m̲ — con chaflan m̲ en V - [sold] V-butt fill pass
— f̲ — — m̲ — — m̲ — vertical - [sold] vertical V-butt fill pass
— f̲ — — m̲ — en V (vertical) - [sold] (vertical) V-butt fill pass
— f̲ — — m̲ — — m̲ (interior) vertical—[sold]

vertical V-butt fillet pass
pasada f para temple m—[metal-trat] temper pass
— f paralela - [metal-lam] parallel pass •
[sold] split weave (pattern)
— f — con movimiento m de tejido m - [sold]
weave weld
— f — — m reducido de tejido m - [sold]
split weave weld
— f penúltima f - [sold] second last pass
— f poco profunda - [mec] shallow pass
— f por - going, o passing, through
— f posterior - [sold] later pass
— f primera - [sold] first pass • sealing bead
— f — con cordón m recto - [sold] first pass
stringer bead
— f — con cuchara f - [constr] first troweling
— f — — paleta f - [constr] first troweling
— f — de soldadura f - [sold] first pass weld
— f — — f circunferencial - [sold] first
pass girth weld
— f — — f — exterior - [sold] first pass
outside diameter girth weld
— f — — f — interior - [sold] first pass
inside diameter girth weld
— f — — f en ángulo m (interior) - [sold]
fillet weld first pass
— f — — f en ranura f - [sold] joint weld
first pass
— f — — f — — f angosta f - [sold] narrow weld first pass
— f — — f — — f con chaflán m en V -
[sold] narrow V joint weld first pass
— f — — f vertical - [sold] first pass of
vertical weld; vertical weld first pass
— f — — f — ascendente - [sold] first
pass vertical-up weld(ing)
— f — — f — descendente - [sold] first
pass vertical-down weld(ing)
— f — únicamente - [sold] first pass only
— f — vertical ascendente - [sold] first vertical-up pass
— f — — descendente - [sold] first vertical-
-up pass
— f profunda - [mec] deep pass
— f restante - remaining pass
— f segunda - [sold] second pass; avoid hot
pass
— f — — soldadura f - [tub] second pass weld
— f — — — circunferencial - [tub] second
pass girth weld
— f — — f — exterior - [sold] second pass
outside diameter girth weld
— f — — f — interior - [sold] second pass
inside diameter girth weld
— f — en caja f - [sold] second pass box weave
— f siguiente - [metal-lam] next, o following,
pass
— f subsiguiente - [sold] succeding pass
— f sucesiva - [sold] successive pass
— f tejida - [sold] weave pass
— f — con ancho m cabal - [sold] full width
weave pass
— f triangular - [sold] trinagular pass
— f última - [sold] last, o finish, o cap, pass
— f única - [sold] single pass
— f — con avance m rápido - [sold] high speed
single pass
— f — pequeña - [sold] small single pass
— f veloz - [deport] scooting by
— f vertical - [sold] vertical pass
— f — ascendente - [sold] vertical-up pass
— f — descendente - [sold] vertical-down pass
— f — en ángulo m (interior) - [sold] vertical
fillet pass
— f — — interior m de ángulo m - [sold] vertical fillet pass
— f — para relleno m - [sold] vertical fill
pass
— f — — — m a tope - [sold] vertical butt
fill pass
— f — — — m — — con chaflán f en V—[sold]
vertical V-butt (fill) pass
— f — — — m en ángulo m (interior) - [sold]

vertical fillet fill pass
pasada f vertical para relleno en interior m de
ángulo m - [sold] vertical fillet fill pass
pasadera f - . . . . • [constr] catwalk; gangway
pasadizo m - passageway; gangway; catwalk
pasado,da a - past; beyond; former; last
— a mediodía m - past, noon, o midday, o meri-
dian
— a a comprador m - passed to @ purchaser
— a a cuarto m intermedio - [legal] recessed
— a a través de - passed, o shot, through
— a como exhalación f - [deport] blown, o shot,
o whizzed, by, o past
— a gradualmente - [mec] to inch through
— a — por hilera f - [trefil] inched through @
die
— a por - gone, o put, o led, o slipped,
through • [cabl] reeved
— a — alto - overlooked
— a — (inter)cambio m de ión(es) m - [nucl]
gone through @ ion exchange
— a — enfriador - passed through @ cooler
— a — interior m - routed, through, o inside
— a — rodillo(s) m alimentador(es) m - [mec]
passed through @ feeding, rolls, o mill
— a sobre - passed over
— a velozmente - [deport] scooted, o whizzed,
o raced, by
pasador m - [mec] . . .; lock; hook; dowel;
stud (bolt); drift, o split, pin; cotter;
spike; axis • forelock • [constr] plunger bar
— m accionador - [mec] driving ping
— m — para palanca f - [mec] lever driving pin
— m — — f trabadora - [mec] locking lever
driving pin
— m aceitado - [mec] oiled pin
— m acodado - [mec] toggle pin
— m aflojado - [mec] loosened, pin, o dowel
— m ahusado - [mec] spindle, o conical, o tapered, (drift) pin, o spike
— m — de acero m - [mec] spindle, o conical, o
tapered, steel, spike, o pin
— m — instalado - [mec] installed tapered
dowel
— m —, quitado, o removido, o sacado - [mec]
removed tapered dowel
— n central - [mec] kingbolt; kingpin; main
beam pin; center pin
— m — para viga f - [mec] main beam pin
— m con aleta(s) f - [mec] split, o cotter, pin
— m — arandela f - [mec] washered pin
— m — — f para empuje m - [mec] washered
thrust pin
— m — — f — — m de cigüeñal m a cárter m -
[mec] crankshaft to crankcase washered thrust
pin
— m con costo m bajo - [mec] low cost pin
— m — — m reducido - [mec] low cost pin
— m — — m — para, reemplazo m, o reposición
f - [mec] low cost replacement pin
— m — extremo m vivo - [mec] live end pin
— f — muesca f - [mec] hitch pin
— m — resorte m - [mec] spring pin
— m conectador - [mec] connecting,tor, o
linkage,king, pin
— m — colocado - [mec] inserted linkage pin
— m — para perno m - [mec] bolt link pin
— m — — m — m con ojo m - [mec] eye bolt
link pin
— m —, quitado, o removido, o quitado - [mec]
removed linkage pin
— m conformado - [mec] shaped, o formed, pin
— m cónico - [mec] dowel, o tapered, pin
— m — acero m - [mec] steel pin
— m — — m con aleación f - [mec] alloy (steel)
pin
— m de alambre m - [mec] wire pin • lockwire
— m — — m para fijación f - [mec] lockwire
— m — tipo m para rastreo(s) m - [mec] trawling type pin
— m — — m — seguridad f - [mec] safety style
pin
— m — — m redondo - [mec] round type pin

pasador m de tipo m roscado - [mec] screw type
pin; screw pin style
— m defectuoso - [mec] faulty, o defective, pin
— m desplazado - [mec] displaced, pin, o cotter
— m enclavado - [mec] interlocked pin
— m Esna - [mec] Esna pin
— m especial - [mec] special pin
— m espiga - [mec] dowel pin
— m — para jaula f - [mec] cage dowel pin
— f — — — f para cojinete m - [mec] bearing
cage dowel pin
— m estándar - [mec] standard pin
— m estriado - [mec] fluted, o grooved, pin
— m excéntrico - [mec] eccentric pin
— m expulsado - [mec] driven out pin
— m faltante - [mec] missing, pin, o cotter
— m — verificado - [mec] checked missing, pin,
o cotter
— m fijo - [mec] non-rotating pin
— m — para cojinete m - [mec] non-rotating
bearing pin
— m flojo - [mec] loose pin
— m girado - [mec] rotated pin
— m giratorio - [mec] rotating, o swivel, pin
— m — para horquilla f - [mec] fork swivel pin
— m — — m para cambio(s) m - [mec] shift-
-(ing) fork swivel pin
— m — — palanca f - [autom-mec] lever swivel
pin
— m — — — f accionadora - [autom-mec] activating lever swivel pin
— m — — perno m - [mec] bolt swivel pin
— m — — — m con ojo m - [mec] eye bolt
swivel pin
— m — que activa dispositivo m para cambio(s)
m - [autom-mec] shift unit activating swivel
pin
— m hembra - [mec] female pin
— m hidráulico - [mec] hydraulic pin
— m — para exclusión f - [mec] hydraulic lock
out pin
— m — — f de motor m - [agric-equip] hydraulic motor lock out pin
— m hincado - [mec] driven pin
— m horadado - [mec] drilled pin
— m hueco - [mec] hollow pin
— m ideal - [mec] ideal pin
— m impulsor - [mec] driving pin
— m — para eslabón m radial - [mec] radius
link drive,ving pin
— m — — palanca f—[mec] lever drive,ving pin
— m — — f trabadora - [mec] locking lever
drive,ving pin
— m inferior - [mec] lower, o bottom, pin
— m — para eslabón m - [mec] lower link pin
— m insertable* - [mec] insertable* pin
— m — con, golpes m, o presión f - [mec]
driving pin
— m insertado - [mec] inserted, o driven, pin
— m instalado - [mec] installed (roll) pin
— m intermedio - [mec] idler pin
— m lubricado - [mec] lubricated pin
— m macho - [mec] male pin
— m muescado - [mec] hitch pin
— m no rotante - [mec] non-rotating pin
— m — para cojinete m - [mec] non-rotating
bearing pin
— m para abrazadera f - clevis, o clamp, pin
— m — — f para camisa f - [mec] liner clamp
pin
— m — — f — vástago m - [mec] rod clamp pin
— m — afirmación f - [mec] steady pin
— m — — f para zapata f - [mec] shoe steady
pin
— m — — f — — f para freno m - [mec] brake
shoe steady pin
— m — aguilón m - [grúas] boom pin
— m — — m agrietado—[grúas] cracked boom pin
— m — — — m desgastado - [grúas] worn boom pin
— m — — — m desplazado - [grúas] dislocated
boom pin
— m — — m para servicio m pesado - [grúas]
heavy duty boom pin

pasador m para aguilón m torcido - [grúas] distorted boom pin
— m — — m verificado - [grúas] checked boom
pin
— m — alineación f - [mec] aligning,nment pin
— m — ancla f - [náut] anchor pin
— m — anclaje m - [mec] anchor(ing) pin
— m — — m para banda f - [mec] band anchor-
-(ing) pin
— m — — m — — f para freno m - [mec] brake
band anchor(ing) pin
— m — — — freno m - [mec] brake anchor-
-(ing) pin
— m — apoyo m - [mec] pad pin
— m — árbol m - [mec] shaft pin
— m — armella f - [mec] bail pin
— m — articulación f - [mec] linkage, o
jointer, pin
— m — banda f - [mec] band pin
— m — — f — freno m - [mec] brake band pin
— m — barra f - [mec] bar pin
— m — — f para tiro m - [mec] drawbar pin
— m — — f — tracción f - [mec] drawbar pin
— m — bisagra f - [mec] hinge, o knuckle, pin
— m — — f para flotador m - [comb.int] float
hinge pin
— m — — f — resguardo m - [mec] guard hinge
pin
— m — bloque m - [mec] block pin
— m — — m sujetador - [mec] grip block pin
— m — m botador m - kicker, o ejector, pin
— m — — m para estampa f - [mec] tool kicker
pin
— m — — m — troquel m - [mec] tool kicker
pin
— m — botalón m - [grúas] boom pin
— m — — m para cucharón m - [grúas] container
boom pin
— m — brazo m - [mec] arm pin • [grúas] boom
pin
— m — — m para devanado m protector m -
[electr-equip] release arm pin
— m — — — m — para tensión f nula -
[electr-equip] no voltage release arm pin
— m — — m — rueda f loca - [mec] idler arm
pin
— m — caballete m - [grúas] gantry pin
— m — — m verificado - [grúas] checked gantry
pin
— m — cabo m - [cabl] marline spike
— m — caja f - [mec] box pin
— m — — f sujetadora - [trevil] grip box pin
— m — cara f de engranaje m - [mec] gear face
pin
— m — carga f - [mec] load pin
— m — carraca f - [mec] ratchet pin
— m — cerrojo m - [mec] latch pin
— m — cierre m - [mec] latch(ing) pin
— m — cigüeñal m - [comb.int] crankshaft pin
— m — codo m - [mec] elbow, pin, o stud
— m — — m para salida f - [comb.int] outlet
elbow, pin, o stud
— m — — m — f de agua m - [comb.int] water
outlet elbow, pin, o stud
— m — — m — — f de agua m de cilindro(s) a
culata f - [comb.int] cylinder water outlet
elbow to head stud
— m — cojinete m - [mec] bearing pin
— m — conexión f - [mec] link, o connecting,
pin
— m — corte m - [mec] shear(ing) pin
— m — cruceta f - [mec] crosshead pin
— m — cubierta f - [mec] cover (dowel) pin
— m — — f para distribuidor m - [autom-mec]
divider cover dowel pin
— m — — f — — m para fuerza f - [autom-mec]
power divider cover dowel pin
— m — cucharón m - [constr-equip] ladle, o
bucket, pin
— m — — m para excavadora f - [constr-equip]
power shovel bucket pin
— m — — m — — f por arrastre m - [constr-
-equip] dragline (excavator) pin

range cam pin

pasador m para cuchilla f sujetadora - [mec]
  grip(per) blade pin
— m — charnela f - [mec] knuckle, o hinge, pin
— m — eje m - [mec] axle, o roll(ing) pin
— m — émbolo m - [mec] piston pin • [comb.int]
  wrist pin
— m — embrague m - [mec] clutch pin
— m — empuñadura f - [mec] grip handle, o hand
  grip, pin
— m — enclavamiento m - [mec] interlocking pin
— m — engranaje m - [mec] gear pin
— m — — m loco - [mec] idler (gear) pin
— m — — m para impulsión f - [mec] drive,ving
  gear pin
— m — — m — — f para bomba f - [bombas]
  pump drive,ving gear pin
— m — — m — — f — — f para aceite m
  - [comb.int] oil pump drive,ving gear pin
— m — — m — — f — — f — — m a árbol m -
  [comb.int] oil pump drive,ving gear to shaft
  pin
— m — eslabón m - [mec] link pin
— m — — m conectador - [mec] connecting link
  pin
— m — — m radial - [mec] radius link pin
— m — — m superior - [mec] upper link pin
— m — excavadora(s) f - [constr-equip] excavator pin
— m — — f por arrastre m - [constr-equip]
  dragline pin
— m — excéntrico m - [mec] eccentric pin
— m — exclusión f - [mec] lock-out pin
— m — — f de motor m - [mec] motor lock-out
  pin
— m — expulsor m - [mec] kicker, o excluder,
  pin
— m — extremo m - [mec] end pin
— m — — m fijo - [mec] dead end pin
— m — fijación f - [mec] fast, o attaching, o
  latch(ing), pin
— m — flotador m - [comb.int] float(er) pin
— m — freno m - [mec] brake pin
— m — — m compensador m - [mec] equalizer
  brake pin
— m — — m para árbol m—[mec] shaft brake pin
— m — fulcro m - [mec] fulcrum pin
— m — — m — brazo m - [mec] arm fulcrum pin
— m — — m — — m tirador - [mec] drawing arm
  fulcrum pin
— m — — m — palanca f - [mec] lever fulcrum
  pin
— m — — m — — f soltadora - [mec] release
  lever fulcrum pin
— m — — m — — f para sujetador m - [mec]
  grip release lever fulcrum pin
— m — gancho m - [mec] hook pin
— m — gorrón m - [mec] pivot pin
— m — gozne m - [mec] hinge pin
— m — — m para aguilón m - [grúas] boom hinge
  pin
— m — grillete m—[mec] clevis, o shackle, pin
— m — — m lubricado - [mec] lubricated, clevis, o shackle, pin
— m — — — m primero - [mec] first clevis pin
— m — — m, quitado, o removido, o sacado -
  [mec] removed clevis pin
— m — — m segundo - [mec] second clevis pin
— m — horquilla f - [mec] yoke, o clevis, pin
— m — — f lubricado - [mec] lubricated clevis
  pin
— m — — f primera - [mec] first clevis pin
— m — — f quitado - [mec] removed clevis pin
— m — — f removible - [mec] removable clevis
  pin
— m — — f, removido, o sacado - [mec] removed
  clevis pin
— m — — f segunda - [mec] second clevis pin
— m — impulsión f - [mec] drive,ving pin
— m — lengüeta f - [mec] tongue pin
— m — leva f - [mec] cam pin
— m — — f para velocidad(es) f alta(s)—[mec]
  high (speed) range cam pin
— m — — f — — baja(s) f - [mec] low (speed)

pasador m para mango m - [mec] handle pin
— m — — m para soltar - [mec] release handle
  pin
— m — — m — — v sujetador m - [mec] grip
  release handle pin
— m — manguito m - [mec] sleeve pin
— m — — m para regulador m - [comb.int] governor sleeve pin
— m — manivela f - [mec] crank, o wrist, pin
— m — ménsula f - [mec] bracket pin
— m — — f para posición f - [mec] placer
  bracket pin
— m — montaje m - [mec] mounting, pin, o stud
— m — montante m - [grúas] leg pin
— m — — m para caballete m - [grúas] gantry
  leg pin
— m — — — m verificado—[grúas] checked
  gantry leg pin
— m — — m — grúa f - [grúas] crane leg pin
— m — muñón m - [mec] trunnion pin
— m — operación f - [mec] operating,tion pin
— m — palanca f - [mec] lever, o handle, pin
— m — — f accionadora - [mec] activating lever pin
— f — — f activadora - [mec] activating lever
  pin
— m — — f para árbol m—[mec] shaft lever pin
— m — — f para freno m—[mec] brake lever pin
— f — — f — — m para árbol m - [mec] shaft
  brake lever pin
— m — — f para soltador m - [mec] release
  lever pin
— m — — f — suelta f - [mec] release lever
  pin
— m — — f para trinquete m - [mec] pawl, lever, o handle, pin
— m — — f — — m para traba f - [mec] lock
  pawl, lever, o handle, pin
— m — — f para válvula f - [mec] valve lever
  pin
— m — — f — — f neumática - [mec] air valve
  lever pin
— m — — f — — f por aire m - [mec] air
  valve lever pin
— m — — f soltadora - [mec] release lever pin
— m — perno m - [mec] bolt pin
— m — — m con ojo m - [mec] eye bolt pin
— m — pértiga f - [mec] boom pin
— m — pestillo m - [mec] latch pin
— m — — m para lengüeta f - [mec] tongue
  latch pin
— m — pierna f - [mec] leg, o shank, pin
— m — pieza f - [mec] part pin
— n — — f giratoria - [mec] pivot pin
— m — piñón m - [mec] pinion pin
— m — — m diferencial - [mec] differential,
  pinion, o gear, pin
— m — — m intermedio - [mec] idler pinion pin
— m — — m loco - [mec] idler pinion pin
— m — — m planetario - [mec] planetary pinion
  pin
— m — — m — loco - [mec] planetary idler
  pinion pin
— m — pistón m - [mec] piston, o wrist, pin
— m — pivote m - [mec] pivot pin
— m — placa f - [mec] plate pin
— m — — f para embrague m - [mec] clutch
  plate pin
— m — pluma* f - [grúas] boom pin
— m — polea f - [mec] pulley, o sheave, pin
— m — — f para cable m - [petról] (cat)line
  sheave pin
— m — — f — — m para cabrestante m -
  petról] catline, pulley, o sheave, pin
— m — — f — — m — torno m - [petról] catline, pulley, o sheave, pin
— m — — f — — m — maniobra(s) f - [petról]
  catline, pulley, o sheave, pin
— m — polín m - [mec] roller pin
— m — — m para leva f - [mec] cam roller pin
— m — — m levantador m - [mec] (hoisting) roller pin

pasador m para polín m levantador para leva f -
[mec] cam roller pin
— m — posición f - [mec] placer pin
— m — punto m para apoyo m - [mec] fulcrum pin
— m — resguardo m - [mec] guard pin
— m — resorte m - [mec] spring pin • [válv]
spring lock
— m — m para válvula f - [válv] valve
spring lock
— m — retención f - [constr] forelock • [mec]
retaining, o lock(ing), pin
— m — rodillo m - [mec] roll(er) pin
— m — m Esna - [mec] Esna roll(er) pin
— m — m para leva f - [mec] cam roller pin
— m — roldana f - [mec] sheave pin
— n — rueda f - [mec] wheel pin
— m — f dentada - [mec] sprocket pin
— m — f — loca - [mec] idler sprocket pin
— m — f loca - [mec] idler pin
— m — f — con rodillo(s) m - [mec] idler
roller pin
— m — seguridad f - [mec] safety pin • shear
pin • lock(ing) pin
— m — f de acero m - [mec] steel shear pin
— m — f para transformador m - [electr-
-transf] transformer locking pin
— m — f tronzable - [mec] shear pin
— m — f — de acero m - [mec] steel shear
pin
— m — f tronzado - [mec] sheared pin
— m — serie f . . . - [mec] . . . series pin
— m — servicio m - [mec] service pin
— m — pesado - [grúas] heavy duty pin
— m — m para aguilón m - [grúas] heavy
duty boom pin
— m — soltador m - [mec] release,sing pin
— m — m para bloque m - [mec] block release
pin
— m — m — m sujetador - [mec] grip
block release,sing pin
— m — m para caja f - [mec] box release pin
— m — m — f sujetadora - [trefil] grip
box release,sing pin
— m — soltar v sujetador m - [mec] grip release,sing pin
— m — suelta f - [mec] release,sing pin
— m — sujetador m - [mec] grip pin • cam pin
— m — m para camisa f - [mec] liner clamp
pin
— m — sujetar v horquilla f - [mec] clevis
holding pin
— m — tope m - [mec] bumper, o stop(ping), pin
— m — m para brazo m - [mec] arm stop pin
— m — m — m para palanca f,- [mec]
lever arm stop pin
— m — m — contacto m - [electr-equip] contact, bumper, o stop, pin
— m — m — m deformado - [mec] deformed
contact bumper pin
— m — m — m faltante - [mec] missing
contact bumper pin
— m — tornillo m - [mec] screw pin
— m — m hueco - [mec] hollow screw pin
— m — m para caja f - [mec] box hollow
screw pin
— m — m — f sujetadora - [trefil]
grip box screw pin
— m — traba f - [mec] lock(ing) pin
— m — f para ajustador m - [mec] adjuster
lock(ing) pin
— m — transformador m - [electr-transf] transformer pin
— m — trinquete m - [mec] pawl pin
— m — m para traba f - [mec] lock(ing) pawl
pin
— m — m — f engrasado - [mec] greased
lock(ing) pawl pin
— m — tuerca f - [mec] nut pin
— m — f ajustado - [mec] adjusted, o tightened, nut pin
— m — f para ajuste m - [mec] adjusting nut
pin
— m — f — m deprimido - [mec]

(de)pressed adjusting nut pin
pasador m para válvula - [mec] valve pin
— m — f neumática - [válv] air valve pin
— m — f para aire m - [válv] air valve pin
— m — varilla f - [mec] rod pin
— m — f para cambio m para marcha f - [mec]
reach rod pin
— m — viga f - [mec] beam pin
— m — vástago m - [mec] rod pin
— m — m para tiro,rar - [mec] pull rod pin
— m — velocidad f alta - [mec] high velocity
pin
— m — f baja - [mec] low velocity pin
— m — f reducida - [mec] low velocity pin
— m — vertedero m - [mec] chute, o spillway,
pin
— m — m para descarga f - [mec] discharge,
chute, o spillway, pin
— m — yugo m - [mec] yoke pin
— m — zapata f - [mec] shoe pin
— m — f para freno - [mec] brake shoe pin
— m pivote - [mec] pivot pin
— m primero - [mec] first pin
— m quitado - [mec] removed pin
— m ranurado - [mec] groove(d) pin
— m recio - [mec] tough, o rugged, pin
— m removible - [mec] removable pin
— m removido - [mec] removed pin
— m retenedor m - [mec] retaining pin
— m rompible m - [mec] shear pin
— m para seguridad f - [mec] shear pin
— m roscado - [mec] stud bolt; screw pin;
threaded pin
— m para cojinete m - [mec] rotating bearing
pin
— m rotatorio - [mec] rotating pin
— m para carga f - [mec] rotating load pin
— m sacado - [mec] removed pin
— m segundo - [mec] second pin
— m soltador m - [mec] release,sing pin
— m superior m - [mec] upper, o top, pin
— m para eslabón m - [mec] upper link pin
— m trabador - [mec] lock(ing) pin
— m tronzado - [mec] sheared, o shorn, pin
— m verificado - [mec] checked pin
— m vertical - [mec] vertical pin
pasador,ra a - passing • through
pasaje m - . . . ; pass; passageway; walkway •
[vial] throughway • [transp] . . . . véase
también billete m
— m aéreo m - [aeron] air transportation
— m bloqueado - blocked passage(way)
— m con columnas f - [arquit] arcade
— m — regulación f - [hidr] controlled pas-
sage(way)
— m — en entrada f - [hidr] inlet controlled passage(way)
— m correspondiente - [transp] respective fare
— m inferior - [constr] (lower level, o underground) passageway; underpass
— f — para bicicletas f - [constr] bicycle
underpass
— m — para peatón(es) m - [constr] pedestrian
underpass
— m libre - free passage
— m obturado - clogged passage
— m para aceite m - [mec] oil passage(way)
— m — aire m - air passage(way)
— m — m bloqueado - block air passage(way)
— m — distribución f - [hidr] header
— m — peatón(es) m - [constr] pedestrian pas-
sage(way)
— m restringido - restricted passage
pasajero m - [transp] . . . ; rider
pasamano(s) m - [constr] . . . ; railing; guardrail
— m con perfil m angular - [constr] angle hand
railing
pasamontaña m - [vest] ski mask
pasante a - . . . . . • through
pasante m en leye(s) m - [legal] lawyer's clerk
pasantía f - [labor] assistantship
pasaportar v - [pol] to issue @ passport

pasar v - . . . • to convert • to slip in; to thread • to hand • to happen • to spend • to elapse • to route • [legal] to route; to transfer • to proceed • [cabl] to reeve • [hidr] to drain • [deport] to, come, o go, o, get, by

pasar v a - to pass into • [hidr] to discharge into

— v a comprador m - to pass to @ purchaser

— v — conducir v - [autom] to get into @ driver's seat; to move into @ car

— v — cuarto m intermedio - [legal] to (go into) recess • to adjourn to @ call of @ chair

— v — través de - to, go, o shoot, through

— v agua m - [hidr] to cross @, water, o stream • [medic] to void

— v — n por alcantarilla f - [hidr] to pass @ water through @ culvert

— v alternativamente a y de - to flash in and out

— v — — — — realidad f - to flash into and out of reality

— v ante mi - [legal] to convey before me

— v alrededor - to pass around

— v cable m - [electr-instal] to, pass, o route, o thread, @ cable

— v como exhalación f - [deport] to blow, by, o past

— v con regulación f en salida - [hidr] to flow in @ outlet control

— v conductor m - [electr-instal] to run @ lead

— v — m por - [electr-instal] to run @ lead through

— v — m para regulación f - [electr-instal] to run @ control lead

— v — m — — f por - [electr-instal] to run @ control lead through

— v corriente f - [electr-oper] to pass @ current

— v — f por campo m - [electr-oper] to pass @ current through @ field

— v cuerda f - [cabl] to thread @ rope

— v de borde m de diente m - [mec] to run off (of) @ tooth

— v — caída f libre a elevación f mecánica - [grúas] to go from free fall to power (hoist)

— v — elevación f mecánica a caída f libre - [grúas] to go from power (hoist) to free fall

— v desapercibido - to go undetected

— v edad f - to outgrow

— v electrodo m - [sold] to feed @ electrode

— v — por prolongación f - [sold] to feed @ electrode through @ extension

— v — forma f audaz - [deport] to sweep by

— v gradualmente - [mec] to inch through

— v — por hilera f - [trefil] to inch through @ die

— v libremente - to pass freely

— v lista f - [legal] to call @ roll

— v manga,guera f - [mec] to thread @ hose

— v paño m - to wipe; to swab

— v — m por encima - to swab

— v por - to, put, o run, o go, o slip, o lead, through • [mec] to drive through • [cabl] to reeve

— v por alto - to skip; to overlook • to ignore

— v — bobina f - [electr-oper] to go through @ coil

— v caja f sin laminar v - [metal-lam] to dummy through

— v — cedazo m - to sieve

— v por encima - to pass over(head)

— v — de estructura f - [constr] to, pass over, o cross, @ structure

— v — enfriador m - to pass through @ cooler

— v — interior m - to, pass, o route, through

— v — rodillo(s) m alimentador(es) - [mec] to pass through @ feeder roll(s)

— v — sobrecarga f - [electr-oper] to, pass, o go, through @ overload

— v — soldadora f - [sold] to, run, o drive, through @ welder

— v propiedad f - [legal] to pass @, title, o ownership

pasar v rayos-X, por, o a través de - [electrón] to, pass, o shoot, @ X-rays through

— v sobre - to pass over

— v — roldana f - [cabl] to, pass, o go, o operate, over @, pulley, o sheave

— v soldando por (encima) de zona f para encendido m - [sold] to weld over @ striking area

— v tamiz m—to pass through @, mesh, o screen

— v tiempo m - to spend @ time

— v tiza f, por encima de, o sobre - to chalk (over)

— f unos a otros - to pass each other

— v — — — libremente - to pass each other freely

— v velozmente - [deport] to pass swiftly; to shoot by

— v vertiginosamente - to pass, swiftly, o speedily

— v — a y de - to flash in and out

— v — — — — realidad f - to flash in and out of reality

pasarela f - [constr] walkway; catwalk; footboard; gangway; crosswalk

— f adicional - [mec] additional walkway

— f armada - [mec] assembled walkway

— f con barand(ill)as f - [mec] catwalk with @, handrail(s), o railing(s)

— f enrejado m engatillado - [constr] Interlock Grating walkway

— f cubierta - [mec] covered walkway

— f elevada - [constr] skywalk; elevated walkway

— f en émbolo m - piston walk(way)

— f grúa f - [grúas] crane walkway

— f estándar - [grúas] standard walkway

— f exterior - outside walkway

— f integral - [petról] integral catwalk

— f interior - inside, walkway, o catwalk

— f metálica - metal walkway

— f montada - [mec] assembled walkway

— f normal - [grúas] standard, o normal, catwalk, o walkway

— f optativa - [constr] optional, crosswalk, o walkway, o catwalk

— f según norma f - [grúas] standard catwalk

— f y barandilla f - [mec] catwalk and railing

pasarlo v - to fare

pasatiempo m de contemplar sexo m bello - girl-watching

— m — otear sexo m bello - girl-watching

pasazanja(s) f - [agric-equip] ditch jumper

Pascua f de Resurrección f - [relig] Easter

pase m - [sold] véase pasada f

— m libre - free pass

pasear v - . . .; to walk; to stroll; to jaunt

paseo m - . . . • jaunt

— m con ambiente m regulado para peatón(es) m • [constr] all-weather pedestrian mall

— m cubierto m - [constr] mall

— m — para peatón(es) m - pedestrian mall

— m emocionante - [deport] thrilling ride

pasillo m - . . .; gangway; catwalk • hall • [ind] walkway

— m para cuba f - [metal-prod] stack walkway

— m — — f de horno m - [metal-prod] bosh walkway

— m — etalaje m - [metal-prod] bosh walkway

— m — — m de horno m - [metal-prod] bosh walkway

— m — horno m alto - [metal-prod] blast furnace walkway

— m superior m - [grúas] upper, catwalk, o platform

pasivación* f - passivation • [metal-prod| scaling • véase también inactivación f • [metal-prod] véase descascarillado m

— f de superficie f - [metal-trat] surface scaling

pasivado,da* a - passivated - [metal-prod] véase descascarillado,da a; scaled • vuelto,ta pasivo,va, o inactivo,va

pasivante* m - passivator*

pasivar* v - [metal-trat] to scale • véase tam-

pasivar v - to passivate • [metal-trat] to scale
• véase volver, o hacer, pasivo,va a, o inactivo,va a
— v superficie f - [metal-trat] to, passivate,
o scale, @ surface
pasivo m acumulado—[fin] accrued liability,ties
— m con plazo m corto - [contab] short term liability,ties
— m — — m largo - [contab] long term liability,ties
— m circulante m - [contab] current liability,ties
— m correspondiente a contrato m - contract
(related) liability,ties
— m corriente - [contab] current liability,ties
— m de empresa f - [fin] concern('s) liability,ties
— m — — f subsidiaria - [contab] subsidia-
ry(ries) (concern) liability,ties
— m devengado - [contab] accrued liability,ties
— m eventual - [contab] contingent liability,ties
— m exigible - [contab] demand liability,ties
— m neto - [contab] net liability,ties
— m — de empresa f - [contab] concern('s) net
liability,ties
— m — — — f subsidiaria - [contab] subsi-
diary('s) (concern's) net liability,ties
— m no exigible - [contab] non-demand liability,ties
— m pagado - [fin] paid (off) liability,ties
— m transitorio m - [contab] deferred liability,ties; deferred credits
— m vario - [fin] other liability,ties
pasmado,da a - stunned; astounded • [fam] open
mouthed
pasmo m - . . . .; stunning
paso m - . . . . • sequence • aperture • neck •
procedure • [mec] pitch; thread • [constr]
crossing • [geogr] pass • [electr- by-pass •
[hidr] flow • elapsing • [legal] transfer •
[vial] throughway • [deport] coming, o going, by
— m a comprador m - passing to @ purchaser
— m a desnivel - [f.c.] grade separation
— m a nivel m - [f.c.] grade crossing; road
crossing • [vial] level crossing
— m — — m carretero - [f.c.] roadway crossing
— m — — m ferroviario-carretero - [constr]
highway-railway grade crossing
— m a paso m - step by step
— m acelerado - rapid stride; fast stepping action
— m adecuado - [hidr] adequate flow
— m — de aceite m - adequate oil flow
— m alrededor - step, o passing, around
— m alternado - [mec] chordal pitch
— m apropiado - appropriate, o proper, step
— m ascendente - [sold] upward step; step-up
— m básico - basic step
— m — para trabajo m - basic job step
— m circular - [mec] arc pitch
— m clave - key step
— m — para creación f - key development step
— m — — desarrollo m - key development step
— m como exhalación f - blowing, by, o past
— m correcto - correct step
— m — para soldadura f - [sold] correct welding step
— m cortado - [transp] blocked way
— m dado - given step • taken step
— m de aceite m - [mec] oil, flow, o passage
— m — agravio m - [labor] grievance step
— m — agua m - [hidr] water pass(ing)
— m — — m por alcantarilla f - [hidr] water
flow through @ culvert
— m — banda f - [metal-lam] strip pass(ing) •
[electrón] bandpass
— m — — f optativo - [electrón] optional
bandpass
— m — cable m - [cabl] cable pitch; lay length
• [electr-instal] cable routing
— m — cadena f - [mec] chain pitch

paso m de cadena f para transmisión f - [mec]
(transmission) chain pitch
— m — calor m - [termol] heat, passage,sing, o
transfer
— m — cinta f - [metal-trat] strip pass(ing)
— m — corriente f - [electr-oper] current,
pass(ing), o flow
— m — — f por conducto m - [hidr] conduit
flow through
— m — corrugación(es) f - [metal-fabr] corrugation pitch
— m — — f en chapa(s) f para encofrado m -
[constr] form sheet corrugation pitch
— m — — f — — f para molde(s) m - [constr]
form sheet corrugation pitch
— m — cuerda f - [cabl] rope threading
— m — chapa f - [metal-trat] strip pass(ing)
— m — — m en engranaje m - [mec] (gear) pitch
— m — engranaje m - [mec] gear pitch, circle,
o line
— m — eslabón m - [mec] link pitch
— m — fleje m - [metal-trat] strip pass(ing)
— m — fluido m - [petról] fluid passage,sing
— m — lista f - [legal] to call @ roll
— m — montaña f - [geogr] mountain pass
— m — procedimiento m - procedure step
— m — proceso m - [ind] process(ing) step
— m — — m para fabricación f - manufacturing
(process) step
— m — — m — producción f - [ind] manufacturing (process) step
— m — programa - program step
— m — propiedad f—[legal] title, o ownership,
passing
— m — retroceso m - back step
— m — rosca f - [mec] (thread) pitch; furrow
— m — — f de tornillo m - [mec] screw thread
(pitch)
— m — — f incompleto - [mec] incomplete
thread
— m — — f por pulgada f - [metal-fabr]
thread(s) per inch
— m — soldadura f - [sold] weld(ing) sequence
— m — — f planeado - [sold] planned weld(ing)
sequence
— m — tambor m - [grúas] drum pitch
— m — tornillo m - [mec] screw pitch
— m — tortuga f - crawl(ing)
— m — trabajo m - work, o job, step. o rate
— m — trenzado m - [cabl] lay length
— m eléctrico - [electr-oper] electrical, operation, o step
— m — para operación f - [electr-oper] electrical operation(s) sequence
— m elevado - [vial] overpass
— m en fabricación f - [ind] manufacture,ring
step
— m — montaña f - [geogr] mountain pass
— m — proceso m - [ind] process step
— m — — m para fundición f de vidrio m -
[cerám] glass melting process step
— m — — m — producción f - [ind] production
process step
— m — producción f - [ind] production step
— m entre centros m - [mec] center-to-center
spacing; pitch
— m — — m en soldadura f intermitente -
[sold] intermittent weld(ing) center-to-center
spacing
— m específico - specific step
— m gigantesco - giant, step, o stride
— m gradual - [ind] inching
— m — por hilera f - [trefil] inching through @
die
— m hacia adelante - forward, step, o stride
— m — atrás - back(ward) step
— m inferior m - [constr] underpass; throughway;
(underground) subway
— m — amplio - [constr] roomy underpass
— m — — con . . . vías f - [constr] roomy
. . .-lane underpass
— m — carretero - [vial] highway underpass
— m — — gemelo - [vial] twin highway underpass

paso m inferior carretero gemelo hecho con plan-
cha(s) f estructural(es) - [vial] twin structural plate highway underpass
— m — construido - [constr] constructed, o
built, underpass
— m — de acero m - [constr] steel underpass
— m — — plancha(s) f estructural(es) - [vial]
structural plate underpass
— f — — f - de acero m - [constr] structural steel plate, o steel structural plate,
underpass
— m — — f para vehículo(s) m - [constr]
structural plate vehicular underpass
— f — — f para revestimiento m - [constr]
liner plate underpass
— m — — tubería f - [constr] pipe underpass
— m — — f de plancha(s) f múltiple(s) -
[constr] Multi-Plate pipe underpass
— m — — f — múltiple(s) Multi-Plate
- [constr] Multi-Plate pipe underpass
— m — en carretera f - [vial] highway underpass
— m — estabilizado - [constr] stabilized underpass
— m — estable - [constr] stable underpass
— m — ferroviario - [constr] railway underpass
— m — gemelo - [vial] twin underpass
— f — — de planchas f estructurales - [vial]
structural plate twin, o twin structural
plate, underpass
— m — gigantesco m - [vial] gigantic, o giant
size(d), underpass • Super-Span underpass
— m — — de acero m - [vial] giant size(d), o
Super-Span, steel underpass
— m — grande - [vial] large underpass
— m — instalado - [vial] installed underpass •
[G.B.] installed subway
— m — Multi-Plate - [constr] Multi-Plate underpass
— m — para (auto)camiones m - [vial truck underpass
— m — — automóviles m - [constr] automobile,
o car, underpass
— m — — carga(s) f - [constr] freight underpass
— m — — ganado m - [constr] cattle, o livestock, underpass
— m — — locomotora(s) f - [f.c.] locomotive,
o engine, underpass
— m — — peatón(es) m - [constr] pedestrian
underpass • [G.B.] (underground) pedestrian
subway
— m — — — m terminado - [vial] finished pedestrian, underpass, o (underground) subway
— m — — red f de servicio(s) m público(s) -
[constr] utility (lines) underpass
— m — — vehículo(s) m - [vial] vehicular underpass
— m — — . . . vía(s) f - [constr] . . .-lane
underpass
— m — pequeño - [vial] small underpass
— m — protegido - [constr] protected underpass
— m — Super-Span - [constr] Super-Span underpass
— m — terminado - [vial] finished, underpass,
o subway
— m — vehicular - [vial] vehicular underpass
— m — — de acero m - [vial] vehicular steel
underpass
— m — — gigantesco - [vial] giant size(d), o
Super-Span, vehicular underpass
— m — — — de acero m - [vial] giant size(d),
o Super-Span, vehicular steel underpass
— m — vial - [vial] highway underpass
— m internacional - [mec] international, o metric, thread
— m libre - [náut] free passage
— m mantenido - kept pace
— m métrico - [mec] international, o metric,
thread
— m montañoso - [geogr] mountain pass
— m para acción f - action step
— m — — f establecido - established action
step
— m — — f fijado - established action step

paso m para banda f para rodamiento m - [autom-
-neumát] tread pitch
— m — cadena f - [cadenas] chain pass
— m — conductor m - [electr-mat] lead grommet
— m — conversión f - change,ging procedure
— m — diseño m - design(ing) procedure
— m — embarcación(es) m - [nav] boat, o vessel, pass(ing)
— m — empaquetadura f - [mec] gasket, o packing, clearance
— m — fabricación f - [ind] manufacture,ring,
o fabricating,tion, step
— m — fundición f - [metal] melting step
— m — — f de vidrio m - [cerám] glass melting
step
— m — ganado m - [agric] cattle, o (live)-
stock pass
— m — — m menor - [f.c.] hogpass
— m — guranición f - [mec] gasket, o packing,
clearance
— m — inyección f - [int.comb] jet passage
— m — — f para acelerador m - [comb.int] accelerator jet passage
— m — operación f - operation, step, o sequence
— m — peatón(es) m - [vial] pedestrian pass-
-(ageway)
— m — preparación f - preparation step
— m — — f de herramental m - [herram] tooling, step, o procedure
— m — proceso m - [ind] process step
— m — producción f - [ind] production, o manufacture,ring, step
— m — — f de vidrio m - [cerám] glass, production, o manufacture,ring, step
— m — trabajo m - [ind] work, o job, step, o
procedure
— m por - pass(ing), o putting, o going, o
slipping, o leading, through • [cabl] reeving
— m — alto - overlooking
— m — encima - passing over; overhead passing
— m — enfriador m - [ind] pass(ing) through @
cooler
— m — interior - routing, through, o inside
— m — paso adv - step by step
— m — rodillo(s) m - [mec] pass(ing) through @
roll(s)
— m — — m alimentador(es) - [mec] pass(ing)
through @ feeding roll(s)
— m primero - first step
— m próximo - next step
— m ranurado - [mec] grooved pass • ground pass
— f rápido - quick pass(age,ing)
— m — de estado m líquido a sólido - [sold]
freezability
— m razonable - reasonable step
— m reducido - small step • [metal-prod] reduced, o small, section
— m repetido - repeated step
— m restante - remaining step
— m restringido - restricted passage
— m sencillo m - [mec] single thread
— m siguiente - following, o next, step
— m sobre - pass(ing) over
— m sobre arroyo m - [constr] creek crossing
— m superior - [vial] overpass • [mec] overcast
— m — antiguo - [vial] old overpass
— m — carretero - [constr] highway overpass
— m — para aire m - [mec] (air) overcast
— m — para carretera f - [vial] highway overpass
— m — viejo - [vial] old overpass
— m uniforme - [hidr] uniform flow
— m variable - [mec] variable pitch
— m veloz - rapid, o fast, pass(ing) • [deport]
shooting by
— m vehicular - [vial] vehicular crossing
— m — inferior - [vial] vehicular underpass
— m — superior - [vial] vehicular overpass
pasquín m - . . . . [public] tabloid
pasta f - . . . [culin] . . .; pasta • [papel]
pulp • [metal-prod] slurry; clay • dope
— f aguada - [papel] slurry; stock
— f aprobada - approved, paste, o compound

pasta f aprobada para junta(s) f—[tub] approved joint(ing) compound
— f celulósica* - [papel] cellulose paste
— f de asfalto m - [petról] asphalt paste
— f — carbonilla f - cinder paste
— f — — f y asfalto m - [constr] cinder-asphalt paste
— f — carbono m - [metal-prod] carbon paste; galipot
— f — — m con aceite m - [metal-prod] carbon and oil paste
— f — celulosa f - [papel] cellulose paste
— f — sulfato m - [metal-prod] sulfate slurry
— f dispersiva - [ind] dissipating paste
— f para cañón m - [metal-prod] (mud) gun clay
— f — correa(s) f - [mec] belt dressing
— f — junta(s) - [tub] joint(ing) compound
— f — relleno m - [ind] filling, o packing, paste
— f — — m para petaca(s) f - [metal-prod] cooling plate packing paste
— f — rosca(s) f - [mec] thread, filler, o paste
pasteca f - . . . . • [nav] (vertical) bull block
— f grande - [nav] bull block
— f vertical - [nav] vertical bull block
pasto m - [botán] grass
— m artificial - [constr] artificial grass
— m azulado - [botán] blue grass
— m crecido - [botán] grown grass
— m emparvado - [agric] stacked hay
— m oreado - [agric] tedded hay
— m sostenido - supported grass
pastón m - [hormig] batch
pastoril a - pastoral
pastoso,sa a - . . . . • grassy; grassed
pasturaje m - [agric] . . .; grazing land
pata f - . . . . [mec] jack • [sold] standing support
— f de acero m - [mec] steel, leg, o foot
— f — — m soldada - [mec] welded steel, leg, o foot
— f — — m con arco m - [mec] arc welded steel, leg, o foot
— f — araña f - [cojin] oil, o lubrication, groove; graphite groove; véase también raya f para engrase m
— f — cabra f - [constr-equip] sheepfoot (roller) • jenny; jimmy
— f — cordón m - véase cateto m de cordón m
— f — perro - . . . . • [cabl] dog leg
— f extendida - [mec] extended, leg, o foot; portruding, leg, o foot
— f para alimentadora f - [mec] feeder leg
— f — basculación f - [mec] tilting lug
— f — bastidor m - [mec] frame, foot, o leg
— f — — m para soldadora f - [electr-prod] (welder) generator frame, leg, o foot
— f — — m generadora f - [electr-prod] generator frame, foot, o leg
— f — — m — m para soldadora f - [sold] welder generator frame, foot, o leg
— f — compresor m - [mec] compressor foot
— f — depósito m - tank, o reservoir, leg
— m — — m para fundente m - [sold] flux tank, leg, o support
— f — desarraigadora f - [constr-equip] rooter standard
— f — elevación f - [mec] lifting lug
— f — escalera f - [constr] ladder, leg, o side piece
— f — generadora f - [electr-prod] generator, leg, o foot
— f — — f para soldadora f - [sold] welder generator, foot, o leg
— f — grúa f - [grúas] crane leg • gantry leg
— f — — f de pórtico m - [cranes] gantry leg
— f — mesa f - [domést] table leg
— f — — f para comedor m - [domést] dining (room) table leg
— f — silla f - [domést] chair leg
— f — soldadora f - [sold] welder, foot, o leg
— f — soporte m - [mec] support(ing) leg • drop-leg standing support

pata f para soporte m plegable - [mec] drop-leg standing support; standing support
— — — m plegadiza - [sold] standing support
— f, plegadiza, o plegable - [mec] drop-leg; standing support
— f soldada - [mec] welded, foot, o leg
— f — con arco m - [mec] arc welded foot
— f tipo gato - [mec] jack type leg
— f trasera f - [mec] rear, foot, o leg
patada f - . . . . • [electr-oper] shock
patear v - . . . . • [electr-oper] to shock
patentado,da a - [legal] patented • [metal-trat] recocido,da antes de trabajar v en frío
patentado* m - [metal-trat] heat treating,ment
patentar v conformación f - to patent @ pattern
— v dispositivo m - to patent @ device
— v sistema m - [ind] to patent @ system
patente f - . . . . • [autom] license plate | adv - [autom] (con) licensed
— f adquirida - [legal] acquired patent
— f afectada - affected patent
— f comprada - [legal] purchased patent
— f creada - [legal] developed patent
— f empleada - [legal] used patent; patent used
— f en trámite m - [legal] pending patent
— f gestionada - [legal] applied for patent
— f obtenida - [legal] obtained, o granted, patent
— f para conformación f - [ind] pattern patent
— f — invención f - [legal] (invention) patent
— f — — f adquirida - [legal] acquired patent
— f — — f comprada—[legal] purchased patent
— f — — f creada - [legal] developed patent
— f — — f empleada - [legal] used patent; patent used
— f — — f gestionada - [legal] applied for, o sought, patent
— f — — f obtenida - [legal] obtained, o granted, patent
— f — — f poseida - [legal] owned patent
— f — — f registrada - [legal] registered patent
— f — — f retenida - [legal] (with)held patent
— f — — f solicitada - [legal] applied for patent
— f — — f usada - [legal] used patent; patent used
— f — — f vendida - [legal] sold patent
— f — sistema m - [ind] system patent
— f poseida f - [legal] owned patent
— f registrada - [legal] registered patent
— f retenida - [legal] (with)held patent
— f solicitada - [legal] patent applied for
— f usada - [legal] used patent; patent used
— f vendida - [legal] sold patent; patent sold
— f violada - [legal] infringed patent
— f sobre - véase patente f para
patentización f - . . . . • [metal-trat] patenting • véase también recocido* m antes de trabajo m en frío
patín m - . . . . • [mec] skid
— m de madera f - [constr] wooden, slide, o skid
— m — perfil m - [metal-lam] shape base
— m — — m soldado - [metal-lam] welded shape base
— m — riel - [metal-lam] rail base
— m (— tipo m) para yacimiento m petrolífero - [petról] oil field type skid
— m — viga f - [metal-lam] beam base
— m delantero - [mec] front skid
— m para acondicionamiento m - [ind] conditioning skid
— m — árbol m - [mec] shaft skid
— m — bomba f - [bombas] pump skid
— m — carga f - [mec] load(ing) skid • [metal-prod] charging skid
— m — f para bobina(s) f - [metal-lam] coil loading skid
— m — — f para entrada f - [metal-lam] feeding end coil loading skid
— m — — f — salida f - [metal-lam]

delivery coil loading skid
patín m **para carga** f **para bobina(s)** f **para extremo** m **para entrada** f - [metal-lam] feeding end coil loading skid
— m —— f —— f —— m **para salida** f - [metal-lam] delivery coil loading skid
— m — **cojinete** n - [mec] bearing skid
— m —— m **para árbol** m - [mec] shaft bearing skid
— m —— m —— m **para transmisión** f - [mec] transmission shaft bearing skid
— m — **entrada** f **para bobina(s)** f - [metal-lam] coil entry skid
— m — **matriz** f - [mec] die shoe
— m —— m **inferior** - [mec] lower die shoe
— m —— f **maestra** - [mec] master die shoe
— m —— f — **inferior** - [mec] master lower die shoe
— m —— f — **superior** - [mec] master upper die shoe
— m —— f **superior** - [mec] upper die shoe
— m — **motor** m - [electr-mot] motor skid • [comb.int] engine skid
— m — **eléctrico** - [electr-mot] electric motor skid
— m — **palanquilla(s)** f - [metal-lam] billet(s) skid
— m — **prolongación** f - [mec] extension skid
— m —— f **para extremo** m - [petról] end extension skid
— m —— m —— m **hacia motor** m - [petról] power end extension skid
— n — **reparación(es)** f - [mec] repair(s) skid
— m — **tubo** m - [tub] pipe skid
— m — **tubería** f - [tub] piping skid
— m — **yacimiento** m **(petrolífero)** - [petról] (oil) field (type) skid
— m **posterior** - [mec] rear skid
— m **recto** - [metal-lam] straight, skid, o beam
— m **superior** - [mec] upper shoe
— m — **para matriz** f - [mec] upper die shoe
— m **trasero** - [mec] rear skid
patinado,da a - [mec] skidded
— a **circularmente** - [autom] spun
patinadura* f - véase **patinazo** m • slippage
patinaje m - slipping; skidding; skating; slip-(ping)
— m **circular** - circular, spinning, o skipping • [autom] (wheel) spinning
— m **de rueda** f - [autom] wheel spin(ning)
— m **circunferencial** - [autom] wheel spin(ning)
— m **de rueda** f - [autom] wheel spin(ning)
— m — **rueda** f - [mec] wheel, spinning, o skidding
— m **excesivo** - [mec] excessive, spinning, o slipping, o skidding
— m **sobre hielo** m - [deport] ice skating
patinar v - . . . .; to slip
— v **circularmente** - [autom] to wheel spin
— v **excesivamente** - [mec] to, slip, o skid, excessively
— m **incontrolable** - [vial] uncontrollable, skidding, o slipping
— v **incontrolablemente** - [autom] to, skid, o slip, uncontrollably
— v **circunferencialmente** - [autom] to wheelspin
patio m - . . . .; véase también **parque** m; playa f
— m **en subestación** f - [electr-distrib] substarion yard
— m **exterior** - [ind] outside yard
— m, **interior**, o **interno** - [ind] inside yard
— m **para enfriamiento** m - [ind] cooling yard
— m — **escurrimiento** m - [constr] drain(ing), yard, o board
— m — **expedición** f - [ind] shipping yard
— m — **maniobra(s)** f - drill(ing) yard
patrimonio m - . . . .; ownership • resources • value • [fin] (shareholders) equity • [deport] home • [contab] net worth
— m **accionario** - [fin] stock(s), share, o equity
— m **admitido** - [fin] admitted, o recognized, equity

patrimonio m **de, compañía** f, o **empresa** f - [fin] company('s) o corporation('s), net worth
— m — **sociedad** f - [fin] corporate net worth
— m **nacional** - [econ] national, resource(s), o net worth
— m **neto** m - [contab] net value; shareholders' equity
— m, **personal**, o **propio** - [fin] personal assets
— m **real** - [fin] true net worth
— m **social** - [fin] (shareholders') equity; net worth; company, o corporate, net worth
— m **admitido** - [fin] admitted, o acknowledged, (corporate) equity
— m — **de accionista(s)** m - [fin] stockholders, o shareholders, equity
— m — **declarado** - [fosc] declared, o reported, shareholders' equity
— m — **no admitido** - [fin] nonadmitted, o undeclared, equity
— m — **según declaración** f - [fisc] declared, o reported equity
— m —— f **sobre base** f **estatutaria** - [fisc] statutory basis reported shareholders' equity
— m — **total** - [fin] total (shareholders') equity
— m — **de accionistas** m - [fin] total shareholders' equity
— m **total** - [fin] total, equity, o net worth
patrocinado,da a - sponsered; backed; supported
— a **por fabricante** m - [ind] factory, o manufacturer, sponsored, o supported, o backed
patrocinador,da a - . . . .; patron
patrocinante a - sponsoring • patron
patrocinar v - . . . .; to back
patrocinio m - . . . .; sponsoring; backing
— m **para rendimiento** m - performance sponsoring,rship
patrola* f - [transp] véase **automóvil patrullero**
patrón m - . . . .; owner • [mec] standard; model • [sold] jig; template • [ind] pattern
— m **aceptable** - [mec] acceptable pattern
— m **agresivo** - aggressive pattern
— m — **para banda** f **(para rodamiento** m**)** - aggressive tread pattern
— m **ajustado** - adjusted, o tight, pattern
— m — **para contacto** m - [mec] adjusted contact pattern
— m —— m **para diente** m - [mec] adjusted tooth contact pattern
— m **Briggs** - [petról] Briggs standard
— m **comprobado** - checked, o proven, pattern
— m **concentrado** - [mec] concentrated pattern
— m **correcto** - [mec] correct pattern
— m **corregido** - [mec] corrected pattern
— m **cuadrado** - [mec] square pattern
— m **de tormenta(s)** f - [meteorol] storm(s) pattern
— m **desarrollado** - developed pattern
— m **evolutivo** - evolutionary pattern
— m **extendido** - extended, o broad, pattern
— m **incontrolable** - [mec] incorrect pattern
— m **obtenido** - [mec] obtained pattern
— m **para administración** f - [admin] management pattern
— m —— f **para trabajo** m - [admin] work management pattern
— m — **carga** f - [ind] charge,ging pattern • [transp] loading, order, o pattern
— m —— f **utilizado** - [ind] used, o followed, charging pattern • [transp] used, o followed, loading pattern
— m — **contacto** m - [mec] contact pattern
— m —— **ajustado** - [mec] adjusted, o tight, contact pattern
— m —— m **comprobado** - [mec] checked, o tested, contact pattern
— m —— m **cuadrado** - [mec] square contact pattern
— m —— m **desplazado** - [mec] displaced, o moved, contact pattern
— m —— m **movido** - [mec] moved contact pattern

patrón m para contacto m para corona f dentada
- [mec] ring gear contact pattern
— m —— m para diente m - [mec] tooth contact
pattern
— m —— m —— m ajustado - [mec] adjusted,
o tightened, tooth contact pattern
— m —— m —— m comprobado - [mec] checked
tooth contact pattern
— m —— m —— m verificado -[mec] checked
tooth contact pattern
— m —— m uniforme - [mec] uniform, o even,
contact pattern
— m —— m verificado - [mec] checked contact
pattern
— m — diente m - [mec] tooth pattern
— m — enfriamiento m - [ind] cooling pattern
— m — engranaje m - [mec] gear pattern
— m — estaquillado m—[mec] stake,king pattern
— m — llama f - [combust] flame pattern
— m — potabilidad f - [hidr] drinkability, o
drinking water, standard(s)
— m — precio(s) m - price(s) pattern
— m —— m para materia(s) f prima(s) - [ind]
raw material(s) price pattern
— m — regulador m - [ind] control pattern
— m —— m giratorio - [electr-oper] rototrol
pattern
— m —— m para antifluctuación* f -
[electr-oper] bias rototrol pattern
— m —— m —— m motor m - [electr-mot] motor
Rototrol pattern
— m —— m para generador m - [electr-prod]
generator Rototrol pattern
— m — Rototrol* m - [electr-oper] Rototrol
pattern
— m —— m para generador m - [electr-prod]
generator Rototrol pattern
— m satisfactorio - satisfactory pattern
— m uniforme - uniform, o even, pattern
— m variable - [mec] variable pattern
— m verificado - checked, o verified, pattern
patrona f - proprietress; (lady) owner
patrulla f - [milit] . . . . patrolling
— f caminera - [vial] road patrol
patrullado,da a - [segurid] patrolled
patrulladora f - [vial] patrol
— f para camino(s) m - [vial] road patrol
patrulla,lera f para camino(s) m - [vial] road
patrol
pausa f - . . . . • hesitation
— f marcada - distinct pause
pauta f - . . . . • course • policy • clue • hot
tip
— f cambiaria - [econ] exchange policy
— f operativa - operational pattern
— f para crecimiento m - growth pattern
pavimentación f - . . . . • surfacing • [vial]
road surface
— f apropiada - [constr] suitable paving
— f asfáltica - [constr] asphalt(ic), paving, o
pavement
— f — de fondo m - [constr] asphalt(ic) invert paving,vement
— f con hormigón m - [constr] concrete paving
— f de aeropuerto m - [aeron] airfield, paving,
o pavement
— f —— calzada f - [vial] road(way), paving, o
surfacing
— f — calle f - [constr] street paving
— f —— carretera f - [vial] highway paving
— f —— fondo m - [constr] invert paving,vement
— f —— m (colocada) en obra f - [constr]
field-applied invert paving,vement
— f — superficie f - [constr] surface paving
— f — talud m - slope paving
— f — tramo m - [vial] section, o stretch,
paving
— f evaluada - [constr] evaluated paving
— f exterior - outside, o outer, paving
— f — completa—[tub] complete exterior paving
— f interior - inside, o inner, paving
— f — completa - [tub] complete inside paving
— f —— de tubería f - [tub] complete pipe

interior paving
pavimentación f interior completa de tubería(s)
f con asfalto m - [tub] complete pipe interior paving,vement
pavimentado,da a - paved
— a con asfalto - [constr] asphalt paved;
blacktopped
— a totalmente - [constr] fully, o totally,
paved
pavimentador,ra a - [constr] paver,ving
pavimentadora f - [constr-equip] paver
pavimentar v superficie f - [constr] to pave @
surface
— v tramo m - [const] to pave @, section, o
stretch
pavimento m - [constr] . . .; road surface; pad
— m agrietado - [constr] cracked pavement
— m alisado - [constr] smoothed pavement
— m apropiado - [constr] suitable pavement
— m asfaltado m - [constr] asphalt pavement
— m — para pista f para despegue m - [constr]
asphalt runway pavement
— m asfáltico - [constr] asphalt(ic) pavement
— m — (em)parchado - [vial] patched asphalt
pavement
— m — en fondo m - [constr] asphalt invert
pavement
— m — mantenido - [constr] maintained asphalt
pavement
— m bituminoso - [constr] bituminous pavement
— m — pesado - [constr] heavy bituminous pavement
— m colocado - [constr] applied, o installed,
pavement
— m — en obra f - [constr] field-,applied, o
installed, pavement
— m completo m - [constr] full pavement | a -
[constr] (con) fully paved
— m con asfalto m - [constr] asphalt pavement
— m — espesor m de . . . metro(s) m - [constr]
. . . meter thick pavement
— m — ripio m - [constr] gravel pavement
— m construido - [constr] constructed, o built,
pavement
— m — apropiadamente - [constr] properly constructed pavement
— m costoso - [constr] costly, o expensive,
pavement
— m de asfalto - [constr] asphalt pavement
— m —— m colado - [constr] poured asphalt
pavement
— m —— m en lámina(s) f - [constr] sheet asphalt pavement
— m —— m vertido - [constr] poured asphalt
pavement
— m — bloque(s) m - [constr] block pavement
— m —— m de asfalto m - [constr] asphalt
block(s) pavement
— m — carpeta f asfáltica - [vial] sheet asphalt pavement
— m — fondo m - [tub] invert paving,vement
— m — hormigón m - [constr] concrete paving,vement
— m —— m armado - [vial] reinforced concrete paving,vement
— m —— m asfáltico - [constr] asphaltic concrete paving,vement
— m —— m bituminoso - [constr] bituminous
concrete paving,vement
— m —— m colocado - [constr] laid, o installed, concrete pavement
— m —— m en obra f - [constr] field-in-
stalled concrete pavement
— m —— m para pista f (para despegue m) -
[constr] concrete runway paving,vement
— m — lámina f - [constr] sheet paving,vement
— m —— f asfáltica - [constr] sheet asphalt
paving,vement
— m — macadán m - [constr] macadam pavement
— m — pista f - [constr] runway pavement,ving
— m — superficie f - [constr] surface paving
— m — talud m - [constr] slope paving,vement
— m — tierra f - [constr] earth paving,vement

pavimento m̲ de tierra f̲ asfáltica - [constr] asphaltic earth pavement
— m̲ — tipo m̲ bituminoso,- [constr] bituminous
type pavement
— m̲ — — m̲ rígido - [constr] rigid type pavement
— m̲ delgado - [constr] thin pavement
— m̲ deshecho - [constr] broken (up) pavement
— m̲ desigual - [constr] uneven, o̲ rough, pavement
— m̲ desnudo - [constr] bare pavement
— m̲ desparejo m̲ - [vial] uneven, o̲ rough, pavement
— m̲ duro - [vial] hard pavement
— m̲ destruido - [constr] destroyed pavement
— m̲ económico - [constr] economical, o̲ low
-cost, pavement
— m̲ emparejado - [constr] evened, o̲ smoothed,
pavement
— m̲ en declive - [constr] slope(d) pavement
— m̲ grueso - [constr] thick pavement
— m̲ liso - [constr] smooth pavement
— m̲ mojado - [vial] wet pavement
— m̲ parejo - [constr] even, o̲ smooth, pavement
— m̲ reemplazado - [constr] replaced pavement
— m̲ rígido - [constr] rigid pavement
— m̲ roto m̲ - [constr] broken pavement
— m̲ rupturado - [constr] ruptured pavement
— m̲ seco - [vial] dry pavement
— m̲ sobre talud m̲ - [constr] slope pavement
— m̲ soportado - [constr] supported pavement
— m̲ sostenido - [constr] supported pavement
pavor m̲ - . . .; awe
pavoroso,sa a - . . .; awesome
peaje m̲ - [vial] . . .; toll
peatón m̲ - . . .; pedestrian
— m̲ conducido - [vial] moved, person, o̲ people
pecar v - . . . • to be remiss
— v̲ de modesto - to understate
— v̲ — sencillo m̲ - over(ly) simplified
— v̲ — tonto - to be, foolish, o̲ crazy
peculiar a - . . .; odd; unusual
peculio m̲ - . . . • asset(s)
— m̲ propio - [fin] personal asset(s)
pechera f̲ - [vest] bib
pedacería f̲ - [metal-lam] scrap; véase también
chatarra f̲; recorte(s) m̲
— f̲ de acero m̲ - [metal-lam] steel scrap
— f̲ — — m̲ de laminadora f̲ (en caliente) -
[metal-lam] hot mill scrap
pedal m̲ - [mec] . . .; foot pedal
— m̲ acelerador m̲ - [comb.int] accelerator, o̲
gas, pedal
— m̲ ajustado - [mec] adjusted (foot) pedal
— m̲ alto - [mec] high pedal
— m̲ bajo - [mec] low pedal
— m̲ basculante - [mec] rocker,cking pedal
— m̲ deprimido - [mec] (de)pressed, o̲ pushed
(down), (foot) pedal
— m̲ inferior - [mec] low(er,est) pedal
— m̲ más alto - [mec] higher,est peda;
— m̲ — bajo - [mec] lower,est pedal
— m̲ operante - [mec] operating, o̲ working, pedal
— m̲ para acelerador m̲ - [comb.int] accelerator,
o̲ gas(oline) pedal
— m̲ — árbol m̲ - [mec] shaft (foot) pedal
— m̲ — — m̲ para desplazador m̲ - [mec] shifter
shaft (foot) pedal
— m̲ — desplazador m̲ - [mec] shifter (foot)
pedal
— m̲ — embrague m̲ - [mec] clutch (foot) pedal
— m̲ — — m̲ para emergencia f̲ - [mec] emergency
clutch (foot) pedal
— m̲ — — m̲ — — f̲ para mandíbula f̲ - [mec]
emergency jaw clutch pedal
— m̲ — — m̲ — — f̲ — — f̲ inferior - [mec]
emergency low jaw clutch pedal
— m̲ — — m̲ — — f̲ — — f̲ superior - [mec]
emergency high jaw clutch pedal
— f̲ — — f̲ — mandíbula f̲ - [mec] jaw clutch
pedal
— m̲ — — m̲ — — f̲ inferior - [mec] low jaw
clutch pedal

pedal m̲ para embrague m̲ para mandíbula f̲ superior - [mec] high jaw clutch pedal
— m̲ — freno m̲ - [mec] brake pedal
— m̲ — — m̲ deprimido - [mec] (de)pressed brake
(foot) pedal
— m̲ — inclinación f̲ - [mec] tilting pedal
— m̲ — — f̲ para volante m̲ - [autom] steering
wheel tilting level
— m̲ — — f̲ — — m̲ para dirección f̲ - [autom]
steering wheel tilting (position) pedal
— m̲ para pie m̲ - véase pedal m̲
— m̲ para regulación f̲ - [mec] control (foot),
pedal, o̲ throttle
— m̲ — — f̲ para motor m̲ - [comb.int] engine
control (foot) pedal
— m̲ regulado - [comb.int] controlled (foot)
pedal
— soltado - [mec] released (foot) pedal
— m̲ superior - [mec] high(er) (foot) pedal
— m̲ uniforme - [mec] uniform, o̲ even, pedal
pedazo m̲ de asfalto m̲ - [constr] asphalt, piece,
o̲ chunk
— m̲ — hierro m̲ - [sold] iron piece
pedernal m̲ - [geol] . . .; chert
pedestal m̲ - . . . • bracket • mounting; footing
• [metal-prod] (mud gun) swinging beam | a -
(de) floor type; vertical
— m̲ completo - [mec] complete pedestal
— m̲ — para maniobra(s) f̲ - [hidr] complete
operating pedestal
— m̲ de banco m̲ - [mec] bench pedestal
— m̲ fresado - [mec] machined pedestal
— m̲ para maniobra(s) f̲ - [hidr] operating pedestal floor stand
— m̲ — portadiferencial m̲ - [autom-mec] differential carrier pedestal
— m̲ — reparación f̲ - [mec] repair, pedestal, o̲
stand
— m̲ — soporte m̲ - [mec] support stand
— m̲ — tapa f̲ - [mec] cap, o̲ cover, pedestal
— f̲ — — f̲ para cojinete m̲ - [cojin] bearing
cap pedestal
— m̲ — trabajo m̲ - [mec] work stand
pedestrismo m̲ - [deport] . . .; walking; hiking;
track sports
pedido m̲ - . . . • request; requisition •
[legal] application for
— m̲ aceptado - accepted order
— m̲ adjudicado - [com] awarded order
— m̲ analizado - analyzed order
— m̲ cablegráfico - cable(d) order
— m̲ completado - [com] completed order
— m̲ completo - [com] complete order
— m̲ con pago m̲ anticipado - [com] advance payment order
— m̲ cursado - [com] issued, o̲ passed, order
— m̲ de cliente m - customer(‛s) order
— m̲ — emergencia f̲ - emergency order
— m̲ — exterior m̲ - foreign order
— m̲ — fabricación f̲ - [ind] production, o̲ fabrication, order
— m̲ efectuado - made order
— m̲ eficiente - [com] efficient ordering
— m̲ — por repuesto(s) m̲ - [ind] efficient
part(s) order(ing)
— m̲ emitido - issued order
— m̲ en trámite m - [legal] pending request
— m̲ especial - [com] special order
— m̲ eventual - eventual order
— m̲ facturado - [com] invoiced, o̲ billed, order
— m̲ final - [com] final order • [ind] production order
— m̲ finalizado - finalized order
— m̲ formulado - issued order
— m̲ interno - [com] requisition
— m̲ — de cliente m̲ - [com] customer('s) requisition
— m̲ nuevo - [com] new order
— m̲ original - [com] original order
— m̲ para compra f̲ - [com] purchase order
— m̲ — — f̲ de pieza(s) f̲ - [ind] parts purchase requisition
— m̲ — — f̲ — — f̲ para repuesto m̲ - [ind]

spare part(s) purchase requisition
pedido m **para compra** f **de repuesto(s)** m - [ind]
spare(s) (parts) purhase requisition
— m — **exportación** f - [com] export order
— m — **montaje** m - erection order
— m — **obra** f - [constr] work order
— f — — f **civil** - [constr] civil work order
— m — **trabajo** m - [ind] work, order, o requisition, o request
— m **parcial** - [com] partial order
— m **pendiente** - pending, order, o request
— m **por asesoramiento** m - assistance, o advice,
request
— m — **ayuda** f - assistance, o help, request
— m — **barra(s)** f - [metal-lam] bar(s) order
— m — — f **para puesta** f **en paralelo** - [ind]
paralleling assistance request
— m — **calidad** f - quality request(ing)
— m — **cambio** m - change request
— m — — m **para planta** f - [ind] plant, o
mill, change request
— m — **desistimiento** - desistance request •
waiver request
— m — **informe(s)** m - (information) inquiry
— m — — m **por cliente** m - costumer('s) information inquiry
— m — **material(es)** m - [ind] material(s) req-
uisition(ing) order
— m — **neumático(s)** m - [autom-neumát] tire(s)
order
— m — **modificación** - change request
— m — **pieza(s)** f - [ind] part(s) order(ing)
— m — — f **para repuesto** m - [ind] spare, o
replacement, part(s) order(ing)
— m — **precio(s)** - [com] price(s) request •
price tender
— m — **propuesta** f - [com] (proposal) inquiry
— m **por repuesto(s)** - [ind] part(s) order(ing)
• replacement(s) order(ing)
— m — **tolerancia** f - tolerance request
— m **principal** m - main order
— m **recibido** - received order
— m **seguido** - followed up order
— m **separado** - separate order
— m **siguiente** - next, o following, order
— m **sobre exterior** m - foreign order
— m **telefónico** - (tele)phone order
— m **telegráfico** - telegraph, o wire, order •
cable order
— m **total** - total, o entire, order
— m **viejo** - old order
pedido,da a - requested • requisitioned •
[legal] applied for
— a **eficientemente** - [com] ordered efficiently
— a **especialmente** - [com] ordered specially
— a **separadamente** - ordered separately
pedir v - . . . .; to order; to apply • [com] to
requisition
• [legal] to apply for
— v **calidad** f - to request @ quality
— v **cantidad** f - to order @ quantity
— v **eficientemente** - [com] to order efficiently
— v — **repuesto(s)** m - [ind] to order @ part(s)
efficiently
— v **en juicio** m - [legal] to claim (in @ court]
— v **especialmente** - [com] to order specially
— v **material(es)** m - [ind] to requisition (@
materials)
— v **pieza** f - [ind] to order @ part
— v — f **para repuesto** m - [ind] to order @,
spare, o replacement, part
— v **repuesto** m - to order @ (spare) part; to
order @ replacement
— v **separadamente** - to order separately
pedraplén* m - véase **enrocamiento** m
pedregoso,sa a - . . . .; gravelly
pedregullo m - [geol] gravel • [constr] aggregate; véase también **grava** f
pega f - . . . .; pasting • [f.c.] trouble
pegado* m - véase **pegadura** f; **pegamiento** m
pegado,da a - pasted • stuck; bound • hit]
[mec] frozen
— a **en cráter** m - [sold] stuck in @ crater •

pegado,da a **en frío** m - cold stuck
— a — **posición** f **abierta** - [electr-instal]
stuck (in @) open (position)
— a — — f **cerrada** - [electr-instal] stuck
(in @) closed (position)
— a — — f **conectada** - [electr-instal] stuck
(in @), closed, o connected, (position)
— a — — f **desconectada** - [electr-instal]
stuck (in @) open (position)
pegadura f - . . .; pasting; véase **pegamiento** m
— f **en cráter** m - [sold] sticking in @ puddle
pegajosamente adv - stickily • tackily
pegajosidad f - stickiness • tackiness
pegajoso,sa a - . . . . • tacky
pegamiento m - . . . . - binding • [sold] freezing
— m **de contacto** - [electr-oper] contact freezing
— m — — m **para interruptor** m - [electr-oper]
breaker, o contactor, freezing
— m — — m — — m **automático** - [electr-oper]
breaker, o contactor, contact freezing
— m — **de relé** m - [electr-instal] relay
contact, freezing, o sticking
— m — — m — — m **piloto** - [electr-oper]
pilot relay contact, freezing, o sticking
— m — **chapa** f - [metal-lam] strip sticking •
sheet sticking
— m — **electrodo** m - [sold] electrode sticking
— m — **interruptor** m - [electr-oper] breaker, -
contactor, freezing
— m — — m **automático** - [electr-oper] breaker,
o contactor, freezing
— m **en cráter** m - [sold] crater, o puddle,
sticking; sticking in @, crater, o puddle
— m — — m **evitado** - [sold] prevented, o a-
voided, crater sticking
— m — — m **prevenido** - [sold] prevented crater
sticking
— m — **frío** - [sold] cold sticking
— n **evitado** - prevented sticking
— m **prevenido** - prevented sticking
pegar v - . . . . • to paste • to bind
— v — **posición** f **abierta** - to stick (in @)
open (position)
— v — — f **cerrada** - to stick (in @) closed
(position)
— v — — f **conectada** - [electr-oper] to stick
(in @) closed (position)
— v — — f **desconectada** - [electr-oper] to
stick (in @) open (position)
pegarse v - to, stick, o bind • to freeze
— v a **camino** m - [autom-neumát] to, hug, o
stick to, @ road
— v **contacto** m - [electr-oper] to freeze @ contact
— v **electrodo** m - [sold] to stick @ electrode
— v **en cráter** m - [sold] to stick in @ puddle
— v — **frío** - [sold] to cold stick
— v **fundente** m - [sold] to stick @ flux
— v **interruptor** m - [electr-oper] to freeze @,
contactor, o breaker
— v — m **automático** - [electr-oper] to freeze @
contactor
pegmatítico,ca a - [geol] pegmatitic
pegote m - . . . . • [metal-prod] incrustation •
scaffolding
peine m - . . . . [mec] threading die; screw
tool
pelado,da a - . . . . • [constr] spalled (off) •
[geol] smooth
— a **hacia afuera** - [autom-neumát] peeled toward
@ outside
peladora f - [mec] peeler • [extrusión f de e-
lectrodos] stripper
— f **para electrodo(s)** m - [extrusión f de elec-
trodos] electrode stripper
— f — — m **defectuoso(s)** m - [extrusión f de
electrodos] defective electrode stripper
peladura f - . . . . • [constr] spalling (off)
— f **hacia afuera** - [autom-neumát] peeling toward @ outside
pelar v - . . . . • [sold] to peel off • [electr-
-instal] to peel; to remove • [constr] to

spall off
pelar v aislación f - [electr-instal] to remove
@ insulation
— v hacia afuera - [autom-neumát] to peel, toward @ outside, o outwards
peldaño m - [constr] . . .; rung; round; rundle;
crosspiece • tread
— alto - [constr] high step
— m bajo - [constr] low step
— m de acero m - [constr] steel, step, o rung
— m — m redondo - [constr] round steel rung
— m — supervisión f - supervision level
— m para acceso m - [constr] access step
— m — m para peatón(es) m - [constr] pedestrian access step
peleado,da a - fought
peletización f - [miner] pelletizing; véase también formación f de pella(s) f
peletizado,da a - pelletized
película f - [fotogr] . . .; movie
— f cinematográfica - [fotogr] movie picture
— f clásica - [fotogr] classic movie
— f continua - continuous film
— f de aceite m - oil film
— f — grasa f - [lubric] grease film
— f — humedad - [hidr] moisture film
— f — oxidación f - [metal-trat] oxidation
film
— f — f anódica - [metal-trat] anodic oxidation film
— f — óxido m - [metal] oxide film
— f — resina f - resin film
— f — f de epoxia f - epoxy resin film
— f delgada - thin, o light, film
— f — de aceite m - [mec] thin oil film
— f enfocada constantemente - [fotogr] zoom(ed)
film
— f examinada - examined film
— f gruesa - thick film
— f — de aceite m - [mec] thick oil film
— f impermeable - watertight film • vapor barrier
— f ligera - light film • sheen
— f metálica - [metal] metal(lic) film
— f — oxidante - [metal] oxidizing metallic, o
metallic oxidizing, film
— f para rayos-X - [electrón] X-ray film
— f preservadora - preserving, vative film
— f producida - [fotogr] produced film
— f radiográfica - [electrón] X-ray, o radiographic, film
— f revelada - [fotogr] developed film
— f — inmediatamente - [fotogr] quickly, o immediately, developed film
— f sensibilizada - sensitized film
— f sensible - sensitive film
— f sobre seguridad f - [segurid] safety film
— f tenue - tenuous, o (s)light, o thin, film
peligrado,da a - endangered • vulnerable
peligrar v - . . .; to endanger
peligro m - . . .; hazard • jeopardy
— m en vera f de camino m - [constr] roadside
hazard
— m atribuible a nieve f - [meteorol] snow hazard
— m aumentado - increased, o enhanced, danger
— m biológico - biological danger
— m común - common, danger, o hazard
— m — en zona f - [segurid] common area hazard
— m de agua m - water, danger, o hazard
— m — corto circuito m - [electr-instal] short
circuit danger
— m — choque(s) m - [electr-oper] shock danger
— m — daño m - damage danger
— m — deformación f - [mec] deflection distress
— m — derrubio m - [hidr] scour hazard
— m — m aumentado - [hidr] increased scour
hazard
— m — m reducido - [hidr] reduced scour
hazard
— m — incendio m - [segurid] fire hazard
— m — inundación f - [hidr] flood(ing) hazard
— m — recalentamiento m - [sold] overheating

danger
peligro m de resbalamiento m - [segurid] slipping hazard
— m — soltar v - [constr] loosening danger
— m — temporada f - [segurid] seasonal hazard
— m — traspie m - [segurid] tripping hazard
— m disminuido - [segurid] curtailed, o lessened, hazard, o danger
— m durante invierno m - [segurid] winter hazard
— m — verano m - [segurid] summer hazard
— m en planta f - [segurid] plant hazard
— m — f durante invierno m - [ind] winter
plant hazard
— m — f — verano m - [ind] summer plant
hazard
— m — suspensión f - [segurid] suspended hazard
— m — f en aire m - [segurid] airborne
(suspended) hazard
— m — zona f - [segurid] area hazard
— m erosivo - erosive hazard
— m evitado - [segurid] avoided, o prevented,
danger, o hazard
— m general - [segurid] general, hazard, o danger
— m — en zona f - [segurid] general area hazard
— m inusitado - [segurid] unusual hazard
— m minimizado - [segurid] minimized danger
— m mínimo - minimum, danger, o hazard
— m no transponible - [transp] nontransversable
hazard
— m para conducción f - [autom] driving hazard
— m — salud f - [segurid] health hazard
— m — seguridad f - [segurid] safety hazard
— m — tránsito m - [transp] traffic, danger, o
hazard
— m posible - [segurid] possible, o potential,
danger, o hazard
— m — de resbalamiento m - [segurid] potencial
slipping hazard
— m — — traspié(s) - [segurid] potential
tripping hazard
— m principal - [segurid] main, o major, hazard
— m — en zona f - [segurid] major area hazard
— m reducido - [segurid] reduced, o curtailed,
o diminshed, danger, o hazard
peligroso,sa a - . . .; hazardous; unsafe •
damaging
— a en invierno m - [segurid] hazardous in winter
— a — verano m - [segurid] hazardous in summer
— a para persona(s) f - [segurid] hazardous to
@ human(s)
pelo m de agua m - véase nivel m máximo de agua
pelota f - . . . • [miner] pellet
— f — barro m - [sanit] mud ball
— f — lodo m - [sanit] mud ball
— f para golf - [deport] golf ball
— f — tenis m - [deport] tennis ball
pelotilla f de escoria f fundida - [sold] molten
slag ball
pelotón m - [deport] . . .; pack; field
— m nutrido - [deport] strong field
peltre m - [metal] , , , • zinc
pelusa f - . . .; lint
pella f - . . . | a - [adv] - (en) pelletized
— f autofundente - [miner] self-fluxing pellet
— f cargada - [metal-prod] charged pellet
— f con mezcla f gaseosa - [metal-prod] gas
containing pellet
— f consolidada - [miner] burned pellet
— f de calidad f, diferente, o distinta -
[miner] varying quality pellet
— f — coque m - [coque] coke pellet
— f — horno m - [metal-prod] furnace pellet
— f — metalización f, alta, o elevada - [miner]
high(ly) metallized pellet
— f — mineral - [metal-prod] ore pellet
— f — m de hierro m - [metal-prod] iron ore
pellet
— f — óxido m - [miner] oxide pellet
— f — retorno m - [miner] return pellet
— f — m consolidada - [miner] burned return

pellet
pella f **metalizada** - [miner] metallized pellet
pella f **para horno** m **alto** - [metal-prod] blast
furnace pellet
— f **prerreducida** - [miner] prereduced pellet
— f — **con metalización** f, **alta**, o **elevada** -
[miner] high(ly) metallized prereduced pellet
— f **reducida** - [metal-prod] reduced pellet
— f **verde** - [miner] green pellet
— f **zarandeada** - [miner] screened pellet
pellas f **cargadas continuamente** - [metal-prod]
continuously charged pellets
pellet m - véase **pella** f
pena f **aceptada** - accepted penalty
— f **acumulada** - accumulated penalty
— f **adicional** - additional penalty
— f **admitida** - admitted, o accepted, penalty
— f **aplicable** - applicable penalty
— f **aplicada** - applied pendalty
— f **disciplinaria** - disciplinary penalty
— f **especificada** - specified penalty
— f **por demora** f - delay penalty
— f — — f **en entrega** f - delayed delivery
penalty
— f — **falta** f **en rendimiento** m - nonperform-
ance penalty
— f — **incumplimiento** m - noncompliance, o non-
performance, penalty
penalidad f **aceptada** - accepted penalty
— f **acumulada** - accumulated penalty
— f **adicional** - additional penalty
— f **admitida** - admitted penalty
— f **aplicable** - applicable penalty
— f **aplicada** - applied penalty
— f **esoecificada** - specified penalty
— f **plena** - full penalty
— f **por demora** f - delay penalty
— f — — f **en entrega** f - delivery delay, o
delayed delivery, penalty
— f **total** - total, o full, penalty
penalización f - penalizing,zation
— f **aduanera** - [fisc] customs penalization
pendiente f - . . .; descent; gradient; ascent;
sloping; batter; dip; scarp; inclination;
fall • [constr] gradation | adv - hanging;
dangling; descending • outstanding | adv -
(en) pending • [constr] sidehill
— f **aproximada** - [topogr] approximate grade
— f **ascendente** - [vial] up grade
— f **aumentada** - [constr] increased gradient
— f **con pasto** m - [constr] grassed (pasture)
backslope
— f **considerable** - steep grade,dient
— f **continua** - [topogr] continuous grade,dient
— f **cortada** - [constr] cut, slope, o face
— f **crítica** - [constr] critical, slope, o gradient
— f **de abismo** m - precipice slope
— f — **acequia** f - [constr] channel grade
— f — **acueducto** m - [constr] aqueduct gradient
— f — **alcantarilla** f - [hidr] culvert slope
— a — **aplicación** f - [fin] unallocated
— f — **cuenca** f - [hidr] (water)shed slope
— f — — f **hidrográfica** - [hidr] watershed
slope
— f — **entrega** - pending delivery
— f — **fondo** m **(de tubería** f) - [constr] invert
grade
— f — **liquidación** f - [com] pending, liquida-
tion, o settlement
— f — — f **final** - [com] pending final, liq-
uidation, o settlement
— f - . . . **por ciento** m—. . . per cent slope
— f — **talud** m - [constr] slope, face, o grade
— f — — m **cortado** - [constr] cut slope face
— f — **temperatura(s)** f - temperature(s)
gradient,dation
— f — **terreno** m - terrain slope
— f — **tubería** f - [constr] pipe slope
— f — **zanja** f - [hidr] ditch gradient
— f **descendente** - [constr] descending grade •
[vial] down grade
— f — **larga** - [vial] long descending grade

pendiente f **empinada** - [topogr] steep slope
— f **en ascenso** m - ascending, grade, o slope
— f — **aumento** - [topogr] increasing gradient
— f — **orilla** f - [hidr] side, o edge, slope, o
grade
— f — **zanja** f - [constr] ditch, slope, o grade
— f — — f **abierta** - [constr] open ditch,
slope, o grade
— f — **zigzag** - [constr] switchback (grade)
— f **estable** - stable, hill, o slope, o grade
— f **extendida** - [topogr] extended, o long,
slope, o grade
— f **fuerte** - [topogr] steep, slope, o grade
— f — **superada** - overcome steep grade
— f **hidráulica** - [topogr] hydraulic gradient •
[hidr] flow line
— f **ideal** - [constr] ideal, slope, o grade
— f — **para alcantarilla** f - [hidr] ideal culvert, slope, o grade
— f **igual** - equal, slope, o grade
— f **inclinada** - [topogr] steep, slope, o grade
• sloping incline
— f **inestable** - [topogr] unstable, hill, o
grade, o slope
— f **inusitada** - [vial] unusual, grade, o slope
— f **larga** - [topogr] long, o lengthy, slope, o
grade
— f **mantenida** - [constr] maintained grade
— f **más inclinada** - [topogr] steeper grade
— f — **plana** f - [topogr] flatter slope
— f **máxima** - [topogr] maximum, slope, o grade
— f **mayor** - [topogr] steeper. slope, o grade
— f **menor** - [topogr] flatter, slope, o grade
— f **mínima** - [topogr] minimum, slope, o grade,
o gradient
— f **muy inclinada** - [topogr] very steep, slope,
o grade, o approach
— f **no común** - [topogr] unusual grade
— f **para flotación** f - [hidr] flotation gradient
— f **plana** - [topogr] flat, slope, o grade
— f **pronunciada** - [topogr] pronounced, o steep.
slope, o grade
— f **reducida** - [constr] reduced grade
— f **relativamente inclinada** - [topogr] comparatively steep, slope, o grade
— f **satisfactoria** - [topogr] satisfactory gradient
— f — **establecida** - [constr] established satisfactory gradient
— f **segura** - [constr] safe, slope, o grade
— f **suave** - [topogr] slight, o flat, o easy,
slope, o grade, o gradient
— f **superada**—[topogr] overcome, slope, o grade
— f **única** - [constr] single slope
péndola f - . . . • [constr] strut • [puentes]
hanger • [mec] suspender
— f **extrema** - [puentes] end hanger
— f **intermedia** - [puentes] intermediate hanger
pendolón m - [constr] jamb; king post • [puentes]
hanger
pendular v - to swing | a - . . .
penetración f - . . . • piercing; lodging,gment;
entering; entrance • lodging,gment • [com]
keying in on • [hidr] seepage,ping • [sold]
penetration; fusion depth; fusing into •
[hidr] soaking into
— f **a través** - penetrating,tion through
— f **adecuada** - [sold] adequate penetration
— f **alta** - high penetration
— f **aumentada** - [sold] increased, o added, penetration
— f **baja** - low penetrating,tion
— f **buena** - [sold] good penetrating,tion
— f **completa** - [sold] complete, o full, penetration
— f **de aceite** m - oil penetration
— f — **área** m - area penetration
— f — **calor** m - heat penetration
— f — **congelación** f - [meteorol] frost penetration
— f — **formación** f **petrolífera** - [petról] oil
bearing formation, penetration, o drilling
— f — **humedad** f - moisture penetrating,tion

penetración f de material m - material penetra- -
ting,tion
— f —— m para relleno m - [hidr] backfill
(material) penetrating,tion
— f —— m en cloaca f - [sanit] backfill in-washing
— f — mercado m - [com] market penetrating,tion
— f — piedra f - stone penetrating,tion
— f — raíz f—[sold] root penetrating,tion
— f — roca f - tock, o stone, penetrating,tion
— f — sistema m - system, penetrating,tion, o
entering
— f — soldadura f - [sold] weld penetration
— f —— f más allá de ángulo de encuentro m
- [sold] beyond @ corner weld penetration
— f — techo m - [constr] roof penetrating,tion
— f — vehículo m—[transp] vehicle penetration
— f —— m sin gobierno m - [transp] errant
vehicle penetration
— f —— m —— m evitada - [transp] avoided,
o prevented, errant vehicle penetration
— f — zona f - area penetration
— f dentro de pieza(s) f soldada(s) - [sold]
penetration beyond @ corner
— f en acero m - [metal] steel penetrating,tion
— f — cubierta f - [mec] cover, entering, o
penetrating,tion
— f — horadación f - [constr] bore, entering,
o penetrating,tion
— f — hormigón m - [constr] concrete penetrating,tion
— f — pared f - [constr] wall penetrating,tion
— f —— f para conducto(s) m - [constr] conduit wall penetrating,tion
— f — piso m - [constr] floor penetrating,tion
— f —— m para conducto(s) m - [constr] floor
conduit penetration
— f — ranura f - [sold] groove penetration
— f —— f profunda - [sold] deep groove penetration
— f escasa - low penetrating,tion; lack of penetration • [sold] shallow penetration
— f estándar - standard penetrating,tion
— f excesiva—[sold] excessive penetrating,tion
— f exigida - [constr=pil] required bearing
— f fácil - easy penetrating,tion
— f hasta profundidad f considerable - [constr-
-pil] considerable depth penetrating,tion
— f incompleta f - incomplete penetrating,tion
— f —— de raíz f - [sold] incomplete root penetrating,tion
— f incrementada - [sold] added penetration
— f insuficiente - [sold] poor, o inadequate, o
insufficient, penetration
— f lograda - [sold] attained, o achieved, penetration
— f más allá de ángulo m de encuentro m—[sold]
beyond @ corner penetration
— f — difícil - more difficult penetration
— f — fácil - easier penetrating,tion
— f — profunda - [sold] deeper penetration
— f máxima - maximum, o highest, penetration
— f mayor - deeper, o added, penetration
— f media - average penetrating,tion
— f mediana - [sold] medium penetration
— f menor - shallower, o less, penetrating,tion
— f mínima - [sold] minimum, o minimized, penetration
— f no trabajada - [mec] unworked penetration
— f normal - normal penetration
— f obtenida - obtained penetration
— f para conducto m - [constr] conduit penetrating,tion
— f permitida - allowed penetration
— f plena - [sold] full penetration
— f pobre - [sold] poor penetration
— f poco profunda - [sold] shallow penetration
— f por medio de venta(s) f - [com] sale(s) penetrating,tion
— f — pared(es) f - [ambient] penetration
through @ wall(s)
— f por piso m - [constr] penetration through @

floor(s)
penetración f por techo m - [constr] penetration through @ roof
— f profunda - [sold] deep penetration
— f promedia - average penetration
— f — de congelación f - [meteorol] average
frost penetration
— f reducida - [sold] reduced, o (shal)low, penetration
— f reducida - [sold] (shal)low, o reduced, penetration
— f regulada - [sold] controlled penetration
— f resistida - resisted penetration
— f total - total, o full, o complete, penetration
— f a través de - penetrating,tion completely through
— f trabajada - worked penetration
— f y tracción f buenas - [autom-neumát] good
penetration and traction • get-in-there-and-
-dig-deep
penetrado,da a - penetrated • pierced • moved, o
gone, into • entered • lodged • [com] keyed
in on • [hidr] seeped, o soaked, into •
[sold] fused into
— a a través de - penetrated through
— a fácilmente - penetrated easily
— a normalmente - penetrated normally
— a por (ir)radiación f - [nucl] radiation penetrated
— a totalmente - penetrated, completely, o totally
— a — a través de - penetrated, totally, o
completely, through
penetrador m . . . • [mec] penetrator; piercer
penetrador,ra a - [mec] picercer; piercing
— a en mercado m - [com] market penetrating
penetrante m - penetrant | a - . . .; [metal]
penetrant,trating • piercing
— m fluorescente - fluorescent penetrant
— m líquido - [sold] liquid, o die, penetrant
— m teñidor - dye penetrant
penetrar v - . . . • to enter; to, flow, o go, o
move, o continue, into; to get through; to
dig in • to impress • to lodge • [com] to
key in • [hidr] to, seep, o soak, into •
[sold] to fuse into
— v a través de - to penetrate through
— v acero m - [metal] to penetrate @ steel
— v bien - to penetrate well
— v completamente - to penetrate completely
— v cubierta f - to, penetrate, o enter, @
cover
— v dentro de junta f - [sold] to flow into @
joint
— v directamente - [sold] to, penetrate, o
feed, directly
— v en alcantarilla f - [hidr] to flow into @
culvert
— v — costura f - [sold] to flow into @ seam
— v — horadación f - [constr] to enter @ bore
— v — suelo m - to, penetrate, o go into, @
ground
— v fácilmente - to, penetrate, o flow (into),
easily
— v — en junta f - [sold] to flow easily into
@ joint
— v formación f - to penetrate @ formation •
[petról] to drill in(to) (@) oil bearing
formation
— v — productiva - [petról] to drill in(to) @
oil bearing formation
— v fuga f - [mec] to penetrate @ leak
— v hormigón m - [constr] to penetrate @ concrete
— v junta f - [sold] to flow into @ joint
— v más allá de ángulo de encuentro m - [sold]
to penetrate beyond @ corner
— v mercado m - [com] to penetrate @ market
— v metal m - to penetrate @ metal
— v normalmente - to penetrate normally
— v profundamente - to, penetrate, o dig, deeply, o down

penetrar <u>v</u> **relleno** <u>m</u> - [constr] to move into @
backfill
— <u>v</u> **sistema** <u>m</u> - to, penetrate, <u>o</u> enter, @ system
— <u>v</u> **suelo** <u>m</u> - to, penetrate, <u>o</u> go into, @,
ground, <u>o</u> soil
— <u>v</u> **totalmente** <u>v</u> - to penetrate completely
— <u>v</u> — **a través** - to penetrate completely
through
penetrómetro <u>n</u> **típico** - [sold] typical penetro-
meter
península <u>f</u> **ibérica** - [geogr] Iberic, <u>o</u> Spanish,
peninsula
— <u>f</u> **inferior** - [geogr] lower peninsula
— <u>f</u> **superior** - [geogr] upper peninsula
paniplanicie* <u>f</u> - [geogr] peniplain*
pensado,da <u>a</u> - thought • expected; planned; contemplated; considered
pensamiento <u>m</u> - . . . . • planning; consideration
pensar <u>v</u> - . . . . • to, consider, <u>o</u> contemplate,
<u>o</u> have in mind; to plan; to expect
— <u>v</u> **detenidamente** - to think, through, <u>o</u> over
— <u>v</u> **ganar** - [deport] to plan on winning
— <u>v</u> **sobre** - to think, on, <u>o</u> about
pensión <u>f</u> - . . .; board
— <u>f</u> **completa** - [culin] full, <u>o</u> complete, board
pentalobulado,da <u>a</u> - [arq-arco] five foiled
pentóxido <u>m</u> - [quím] pentoxide
pentrita <u>f</u> - [explos] pentrite
penúltimo,ma <u>a</u> - . . .; second (from @) last
penumbra <u>f</u> - (semi-) dusk; . . .
peña <u>f</u> - [geol] . . .; cliff; tor
peñasco <u>m</u> - [geol] . . .; cliff; bluff; crag
— <u>m</u> **escarpado** - [geol] steep, bluff, <u>o</u> cliff
peón <u>m</u> - [labor] laborer; hand; man; labor; unskilled, workman, <u>o</u> labor
— <u>m</u> **para escritorio** <u>m</u> - office janitor
— <u>m</u> — **limpieza** <u>f</u> - [ind] janitor; clean-up man
— <u>m</u> — — <u>f</u> **para escritorio** <u>m</u> - office janitor
— <u>m</u> — — <u>f</u> **para oficina** <u>f</u> - office janitor
— <u>m</u> — **oficina** - [com] office janitor
— <u>m</u> — **vestuario** <u>m</u> - [ind] dressing room, at-
tendant, <u>o</u> janitor
peor <u>a</u> - . . . . • more
— <u>a</u> **de caso(s)** <u>m</u> - worst case
pepita <u>f</u> **técnica** - technical nugget • speaking
technically
pequeño,ña <u>a</u> - . . .; minute; exiguous • pinhead
— <u>a</u> **bien conformado,da** <u>a</u> - [sold] small well
shaped
— <u>a</u> **levemente cóncavo,va** - [sold] small (and)
slightly concave
— <u>a</u> — **convexo,xa** <u>a</u> - [sold] small slightly
convex
pera <u>f</u> - [botán] . . . . [mec] bulb • [medic]
syringe
— <u>f</u> **para espolvorear** <u>v</u> - [herram] spray(ing)
bulb
— <u>f</u> — **regulación** <u>f</u> - control bulb
— <u>f</u> — **romper** - [ind] drop ball
— <u>f</u> — — **chatarra** <u>f</u> - [metal-prod] scrap drop
ball
— <u>f</u> **rompedora** <u>f</u> - [constr] drop ball
peraltado,da <u>a</u> - [constr] raised; banked • elongated • [f.c.] superelevated • [geom] elongated vertically
peraltar <u>v</u> - [constr] to, raise, <u>o</u> elongate, <u>o</u>
camber vertically • [f.c.] to superelevate
peralte <u>m</u> - elongation • [constr] bank(ing) •
(girder) depth; rise • [tub] vertical elongation • [deport] bank(ing)
— <u>m</u> **alto** - [vial] high banking
— <u>m</u> **bajo** - [vial] low banking
— <u>m</u> **de perfil** <u>m</u> - [metal-lam] shape rise
— <u>m</u> — — <u>m</u> **soldado** - [metal-lam] welded shape
rise
— <u>m</u> **especificado** - [constr] specified, rise, <u>o</u>
vertical elongation
— <u>m</u> **mínimo** - [constr] minimum, depth, <u>o</u> vertical elongation
— <u>m</u> — **total** - [constr] minimum total, rise, <u>o</u>
depth
— <u>m</u> **nominal** - [metal-lam] nominal rise

percance <u>m</u> - . . . . • [segurid] accident • loss
percatado,da <u>a</u> - sensed; realized
percatamiento* <u>m</u> - sensing; realizing
percatar <u>v</u> - . . .; to sense; to realize
percatarse <u>v</u> - to, realize, <u>o</u> sense, <u>o</u> become,
aware
— <u>v</u> **de extensión** <u>f</u> **excesiva** - [admin] to sense
@ sprawl
— <u>v</u> — — <u>f</u> — **de organización** <u>f</u> - [admin] to
sense @ organization sprawl
— <u>v</u> — **desparramamiento** <u>m</u> - [admin] to sense @
sprawl
— <u>v</u> — — <u>m</u> **de organización** <u>f</u> - [admin] to
sense @ organizational sprawl
percepción <u>f</u> - . . .; perceiving • hearing; feel
• [fin] receipt; receiving
— <u>f</u> **de imagen** <u>f</u> - [óptica] image perception
— <u>f</u> — — <u>f</u> **fluoroscópica** - [electrón] fluoroscopic, image, <u>o</u> picture, perception, <u>o</u> perceiving
— <u>f</u> — **sonido** - [fís] sound, perceiving, <u>o</u>
hearing
— <u>f</u> — **valor(es)** <u>m</u> - value(s) perception
— <u>f</u> **fluoroscópica** - [electrón] fluoroscopic
perception,ceiving
perclorato <u>m</u> **amónico** - [quím] ammonic perchlorate
— <u>m</u> **de amonio** <u>m</u> - [quím] ammonia perchlorate
— <u>m</u> — **calcio** <u>m</u> - [quím] calcium perchlorate
— <u>m</u> — **magnesio** <u>m</u> - [quím] magnesium perchlorate
— <u>m</u> — **potasio** <u>m</u> - [quím] potassium perchlorate
— <u>m</u> — **sodio** <u>m</u> - [quím] sodium perchlorate
percibido,da <u>a</u> - perceived • heard • [fin] received
percibir <u>v</u> - . . .; to hear • to sense • to pick
up • [fin] to, receive, <u>o</u> collect; to draw
— <u>v</u> **imagen** <u>f</u> - [óptica] to perceive @ image
— <u>v</u> — <u>f</u> **fluoroscópica** - [electrón] to perceive @ fluoroscopic image
— <u>v</u> **señal** <u>f</u> - [electrón] to pick up @ signal
— <u>v</u> **sonido** <u>m</u> - to, perceive, <u>o</u> hear, @ sound
— <u>v</u> **valor(es)** <u>m</u> - to perceive @ value • [fin]
to collect @ value
percolante <u>a</u> - véase colante; rezumante
percolar <u>v</u> - véase colar <u>v</u>; rezumar <u>v</u>
percusión <u>f</u> - . . .; pounding
percusor <u>m</u> - . . . . • [petról] (fishing) jar •
véase también percutor <u>m</u>
— <u>m</u> **para equipo** <u>m</u> **para cable** <u>m</u> - [petról] cable
tool jar
— <u>m</u> — — <u>f</u> **para cable** <u>m</u> - [petról] cable tool
jar
percutor <u>m</u> - [mec] striker; hammer • [petról]
(fishing) jar
— <u>m</u> — **botador** <u>m</u> - [mec] kicker striker
— <u>m</u> — **eje** <u>m</u> - [mec] shaft striker
— <u>m</u> — — <u>m</u> **para botador** <u>m</u> - [mec] kicker shaft
striker
— <u>m</u> — — <u>m</u> **para expulsor** <u>m</u> - [mec] kicker
shaft striker
— <u>m</u> — **equipo** <u>m</u> **para cable** <u>m</u> - [petról] cable
tool jar
— <u>m</u> — **expulsor** <u>m</u> - [mec] kicker striker
— <u>m</u> — **herramienta** <u>f</u> **para cable** <u>m</u> - [petról]
cable tool jar
percha <u>f</u> - . . . . • [domést] (coat) rack • hanger
perchero <u>m</u> — [domést] clothes rack
perdedor,ra <u>a</u> **de tiempo** <u>m</u> - time losing
perder <u>v</u> - . . . . • to miss • to knock off •
[electr-instal] to knock out • [petról] to
by-pass
— <u>v</u> **a penas** - [deport] to miss narrowly
— <u>v</u> **agua** <u>m</u> - [hidr] to, lose, <u>o</u> leak, @ water
— <u>v</u> — <u>m</u> **para lubricación** <u>f</u> - to lose @ lubricating water
— <u>v</u> **aire** <u>m</u> - to, lose, <u>o</u> leak, @ air
— <u>v</u> **amortiguador** <u>m</u> - [autom-mec] to lose @
shock (absorber)
— <u>v</u> **área** <u>m</u> **en superficie** <u>f</u> - to lose @ surface
area
— <u>v</u> **carga** <u>f</u> - [ind] to lose @ charge • [transp]
to lose @ load

- 1072 -

perder v carrera f - [deport] to lose @ race
— v circuito m - [electrón] to lose @ circuit
— v colada f - [metal-prod] to loose @ cast •
@ wild cast
— v con frecuencia f - to lose frequently
— v contacto m - to lose, contact, o touch
— v en encuentro m - to lose in @ encounter
— v energía f - to lose @ energy
— v — f en tubería f - [hidr] to lose energy
in @ pipe
— v engranaje m - [mec] to lose @ gear
— v equilibrio m - to lose @ balance
— v equipo m - [seguros] to lose @ equipment
— v excitación f—[electr-oper] to lose @ excitation
— v faro m - [autom-electr] to lose @ headlight
• [nav] to lose @ lighthouse
— v fluido m - to lose @ fluid
— v frecuentemente - to lose frequently
— v freno m - [mec] to lose @ brake
— v fuerza v - [mec] to lose @ strength • to
lose @ power
— v gobierno m - to lose @ control • [autom-
-mec] to go out of control
— v junta f - [metal-prod] to leak (@ joint)
— v líquido m - to lose @, liquid, o fluid
— v — para perforación f - [petról] to by pass
@ drilling fluid
— v — m refrigerante - [comb.int] to lose @,
coolant, o refrigerant
— f máquina f - [seguros] to lose @ machine
— v maquinaria f—[seguros] to lose @ machinery
— v memoria f - to lose @ memory
— v por destilación f - [petról] to lose by
distillation
— v — evaporación f - [hidr] to lose by evaporation
— v por poco - to miss narrowly
— v potencia f - to lose @ power
— v presión f - to lose @ pressure
— f — de aceite m - [comb.int] to lose @ oil
pressure
— v — f hidrostática - [mec] to lose @ hydrostatic pressure
— v producción f - [ind] to lose @ production
— v refrigerante , m - [comb.int] to lose @
coolant
— v rendimiento m - to lose @ performance
— v resistencia f to lose @ strength
— v servodirección* f - [autom] to lose @ power
steering
— v terraplén m - [constr] to lose @ embankment
— v tiempo m - to lose @ time
— v tracción f - [autom] to, lose, o break, @
traction
— v transformador m - [electr-prod] to knock
out @ transformer
— v transmisión f - [autom] to lose @ transmission
— v velocidad f - [autom] to lose @ gear • to
lose @ speed
— v venta f - [com] to lose @ sale
— v ventilador m - [mec] to lose @ fan
— v voltaje m - [electr-oper] to lose @ voltage
— v — m de campo m - [electr-oper] to lose @
field voltage
— v volumen m - to lose @ volume
perderse v carga f - [metal-prod] to lose @,
charge, o stockline (level)
— v nivel m de carga f - [metal-prod] to lose @
stockline (level)
pérdida f - . . . . • leak • missing • droop •
knocking off • [hidr] leakage; seepage •
[petról] by-passing • [electr-oper] knocking
out
— f a tierra f - [electr-oper] ground, loss, o
fault • loss to @ ground
— f alta - high loss
— f antes de impuesto m sobre, réditos, o renta
f - [fin] loss before @ income tax(es)
— f aproximada - approximate loss
— f — de material m—approximate material loss
— f baja - low loss

pérdida f baja de calor m - [termic] low heat
loss
— f bruta - [fin] gross loss
— f catastrófica - [seguros] catastrophic loss
— f causada - caused loss
— f comercial - [comb] business, o commercial,
loss
— f computada - computed loss
— f — f de energía f - [hidr] computed energy
loss
— f considerable - considerable loss • considerable leakage
— f — de metal—[metal] significant metal loss
— f contabilizada - [contab] recognized loss
— f — para contrato m - [contab] recognized
contract loss
— f cubierta f - [seguros] covered loss
— f de aceite m - [mec] oil, loss, o escape
— f — en extremo m hacia motor m - [petról]
power end oil, loss, o escape
— f — m evitada - [mec] avoided, o prevented, oil, loss, o escape
— f — m prevenida - [mec] prevented oil,
loss, o escape
— f — m verificada - [mec] checked oil loss
— f — agua m - [hidr] water loss
— f — m para lubricación f - lubricating
water loss
— f — aire m - air, leak, o loss
— f — alineamiento m - loss of alignment
— f — aluminio m - [metal] aluminum loss
— f — amortiguador m - [autom-mec] (shock) absorber loss
— f — cabeza f - [hidr] head loss
— f — calor m - [termol] heat loss
— f — campo m - [electr-oper] field loss
— f — capital m - [fin] capital loss(es)
— f — — m no realizado - [fin] un paid-in
capital loss
— f — carbono m - [combust] carbon loss
— f — — m por azufre m en coque m - [combust]
carbon loss due to sulfur in @ coke
— f — — m escoria f en carbonato(s) m -
[combust] carbon loss due to slag in @ car-
bonate(s)
— f — — m — — f en ceniza f - [combust]
carbon loss due to slag in @ ash
— f — — m — humedad f en viento m—[combust]
carbon loss due to moisture in @ blast
— f — carga f - [metal-prod] charge, o burden,
loss • [hidr] head loss
— f — — f de energía f - [hidr] energy head
loss
— f — — f determinada - [hidr] determined
head loss
— f — — f por fricción f - [mec] friction
head loss
— f — carrera f - [deport] race, loss, o losing
— f — cascarilla f - [metal-lam] scale loss
— f — — f debida a calentamiento m de lingo-
te(s) en horno m de fosa - [metal-lam] scale
loss due to ingot(s) heating in @ soaking
pit
— f — cobre m - [metal-prod] copper loss
— f — consumo(s) m - [electr-oper] consumption
loss
— f — contacto m - contact, o touch, loss,sing
— f — chapa f magnética - [metal] magnetic
sheet loss
— f — depósito m - [ind] tank loss
— f — día(s) m - [segurid] days, o time, lost
— f — ductilidad f - [metal] ductility loss
— f — energía f - energy loss
— f — — f alta - high energy loss
— f — — baja - low energy loss
— f — — f computada - [hidr] computed energy
loss
— f — — f debido a fricción f - [hidr] energy
friction loss
— f — — f — — — f con tubería f - [hidr]
pipe friction energy loss
— f — — f en tubería f - [hidr] pipe energy
loss; loss of energy in @ pipe

pérdida f de energía f **en tubería debido a fricción** f̄ - [hidr] pipe friction energy loss
— f — — f **en unión(es)** f - [hidr] junction, o joint, energy loss
— f — — f **mayor** - high(er) energy loss
— f̄ — — f̄ **menor** - small(er) energy loss
— f̄ — — f̄ **por compresión** f **de pilote(s)** m - [constr-pil] energy loss through pile compression; pile compression energy loss
— f — — f **por compresión** f **elástica**—[constr-pil] energy loss through elastic compression
— f — — f — — **de pilote(s)** m—[constr-pil] energy loss through pile elastic compression
— f — — f — — f — — m **tubular(es)** - [constr-pil] energy loss through pipe pile elastic compression
— f — — f — — f — — **de tubería** f - [constr-pil] energy loss through pipe elastic compression
— f — **engranaje(s)** m - [mec] gear(s), loss, o losing
— f — **equilibrio** m - [segurid] loss of balance
— f̄ — **equipo** m - [seguros] equipment loss
— f̄ — **excitación** f - [electr-oper] excitation loss
— f — **faro(s)** m - [autom-electr] headlight(s) loss
— f — **fluido(s)** m - [hidr] fluid(s) loss
— f̄ — — m **para perforación** f - [petról] drilling fluid by-pass(ing)
— f — **fluido(s)** - (fluids) leakage
— f̄ — **freno(s)** m - [mec] brake(s) loss
— f̄ — **gas(es)** n - gas(es) loss
— f̄ — — m **en depósito** m - [ind] tank gas loss
— f̄ — — m — — m **para almacenamiento** m - [ind] stock tank, o reservoir, gas loss
— f — **humedad** f - moisture loss
— f̄ — **ingreso(s)** m - [com] revenue(s) loss(es)
— f̄ — **líquido(s)** m - [ind] liquid(s) loss
— f̄ — — m **refrigerante(s)** - [ind] coolant, o refrigerant, (liquids) loss
— f — — m — — **excesivo(s)** - [comb.int] excessive coolant loss
— f — **lodo** m - [petról] (drilling) mud loss
— f̄ — — m **para perforación** f - [petról] drilling mud, loss, o by-pass(ing)
— f — **máquina** f - [ind] machine loss
— f̄ — **maquinaria** f - [ind] machinery loss
— f̄ — **material(es)** m - [ind] material(s) loss • [metal-fabr] (materials) weight loss
— f — **memoria** f - memory loss; loss of memory
— f̄ — **metal** - [metal] metal loss
— f̄ — — m **considerable** - [metal] significant metal loss
— f — — m **de aleación** f - [sold] alloy (metal) loss
— f — **nivel** - [ind] level loss
— f̄ — — m **de carga** f - [ind] stockline level loss; loss of stockline level
— f — **núcleo** m - [magnet] core loss
— f̄ — **potencia** f - power loss
— f̄ — **precarga** f̄ - [mec] preload loss
— f̄ — — f **para cojinete** m - [mec] bearing preload loss
— f — **presión** f - pressure loss
— f̄ — — f **hidrostática** - [mec] hydrostatic pressure loss
— f — **producción** f - [ind] production loss • lost production
— f — **propiedad(es)** f - property,ties loss
— f̄ — — f **de asfalto** m - [constr] asphalt property,ties loss
— f — **radiador** - [comb.int] radiator leak
— f̄ — **rendimiento** m - yield loss
— f̄ — **resistencia** f - [mec] resistance, o strength, loss
— f — — f **a corrosión** f - [sold] corrosion resistance loss
— f — **servodirección** f - [autom] power steering loss
— f — **temperatura** f - temperature loss
— f̄ — **terraplén** m - [constr] embankment loss
— f̄ — **tiempo** m - time, loss, o losing

pérdida f **de título** m - [legal] certificate loss • [deport] title loss
— f — **tracción** f - [autom] traction loss
— f̄ — **transformador** m - [electr-oper] transformer, loss, o knocking out
— f — **transmisión** f - [autom-mec] transmission loss
— f — **venta(s)** f - [com] sale(s) loss
— f̄ — **velocidad** f - speed loss • [autom-mec] gear loss
— f — — f **deciba a patinaje** m - [autom] (speed loss due to) slippage
— f — — f — — **resbalamiento** m - [autom] (speed loss due to) slippage
— f — **ventilador** m - [mec] fan loss
— f̄ — **vida** f - life loss; loss of life
— f̄ **debida a fricción** f - friction loss
— f̄ — — f **con tubería** f - [hidr] pipe friction loss
— f — — f **con pared** f **de conducto** m - [tub] conduit wall friction loss
— f̄ — — f — — f **de tubería** f - [tub] conduit wall friction loss
— f — **interrupción** f - [com] interruption loss
— f — — f **comercial** - [com] business interruption loss
— f — — f **en negocio(s)** m - [com] business interruption loss
— f — **recalcadura*** f - [metal-fabr] loss by swaging
— f — **rotura(s)** f - [com] breakage
— f̄ **definitiva** - final, o permanent, loss
— f̄ **dieléctrica** - [metal] dielectric loss
— f̄ **durante transporte** m - [transp] loss during transportation
— f **eléctrica** f - [electr-oper] electrical loss
— f̄ **elevada** - high loss
— f̄ — **de calor** m - [termol] high heat loss
— f̄ — **en rendimiento** m - [ind] high yield loss
— f̄ **en altura** f - [hidr] head loss
— f̄ — — f **de fluido** m—[hidr] fluid head loss
— f̄ — **área** m - surface, o area, loss
— f̄ — — m **de superficie** f - surface area loss
— f̄ — **cambiador** m **para toma(s)** f - [electr-oper] tap changer loss
— f — **cambio** m - [fin] exchange loss
— f̄ — — m **de divisa(s)** f—[fin] exchange loss
— f̄ — — m — — **moneda(s)** f̄ **extranjera(s)** - [fin] foreign currency exchange loss
— f — **capacidad** f - capacity loss
— f̄ — — f **en entrada** f - [hidr] entrance (capacity) loss; inlet (capacity) loss
— f — — f — **salida** f - [hidr] outlet (capacity) loss
— f — — f **portadora** - [constr] bearing, capacity, o power, loss
— f — — f **resistente** - bearing power loss
— f̄ — **circuito** m - [electr] circuit leak • [electrón] circuit loss
— f — **compresión** f - [mec] compression loss
— f̄ — — f **para reducción** f - reduction compression loss
— f — — f — — f **de pella** f - [miner] pellet reduction compression loss
— f — **contrato** m - [legal] contract loss
— f̄ — — m **contabilizada** - [contab] recognized contract loss
— f — **conversión** f - [fin] translation loss
— f̄ — **entrada** f - [hidr] entrance, o inlet, loss • [turb] intake loss
— f — **fabricación** f—[contab] fabrication loss
— f̄ — **fuerza** f - [mec] power, o strength, loss
— f̄ — **línea** f - [electr-oper] line loss
— f̄ — **mordedura** f - [mec] bite, o pinch, loss
— f̄ — **muestra** f - sample loss
— f̄ — **núcleo** m - [magnet] core loss
— f̄ — **potencia** f - [mec] power loss
— f̄ — — f **de motor** m - [comb.int] engine power loss
— f — **presión** f - pressure, loss, o losing
— f̄ — — f **de aceite** m - [comb.int] oil pressure, loss, o lkoising
— f — **producción** f - [ind] production loss

pérdida f en producción f de mineral(es) m -
[miner] ore production loss
— f — — f — — m por volatilización f -
[miner] ore production loss by volatilization
— f — reducción f - reduction loss
— f — remache(s) m - [mec] loss at rivet(s)
— f — resistencia f - [mec] strength loss
— f — — f a compresión f - compression resistance loss
— f — — f de pella f - [miner] pellet resistance loss
— f — — f — — f a corrosión f - [miner]
pellet corrosion resistance loss
— f — salida f—[turb] outlet, o outflow, loss
— f — sección - section loss
— f — sello m - [mec] seal loss
— f — — m para árbol m—[mec] shaft seal loss
— f — transferencia f - [fin] translation loss
— f — transformador m - [electr-oper] transformer loss
— f — tránsito m - [transp] transit loss
— f — transporte m - [transp] transportation
loss; loss, in, o during, transportation
— f — tubería f - [hidr] pipe loss
— f — — f debido a fricción f - [hidr] pipe
friction loss
— f — vatios/kilogramo - [electr-oper] watt-
-/ kilogram loss
— f — velocidad f - [mec] speed loss •
[autom] gear, loss, o losing
— f — venta(s) f - [com] sale(s) loss
— f — voltaje m - [electr-oper] volt(age) loss
— f — — m en campo m—[electr-oper] field
voltage loss
— f — volumen m - volume loss
— f establecida - established loss
— f evitada - avoided, o prevented, loss, o
leak, o escape
— f excesiva - excessive loss
— f — de resistencia f - excessive resistance
loss
— f — — — f a compresión f - [mec] excessive
compression resistance loss
— f final - final loss
— f frecuente—frequent, loss, o losing
— f grande - great, o large, o high, loss
— f — de calor m - [termol] great, o high,
heat loss
— f hacia exterior m - [electr-oper] external,
leak(age), o loss
— f incurrida - incurred loss
— f magnética - [electr-oper] magnetic loss
— f máxima - maximum loss
— f — en núcleo m - [electr-magnet] maximum
core loss
— f mayor - major, o high(er), o heavy, o
great(er), loss
— f — de energía f - [hidr] major energy loss
— f — — metal m - [metal] major, o heavy, metal, loss
— f menor - minor, o lower, o light, loss
— f — de calor m - low(er) heat loss
— f — de energía f - [hidr] minor energy loss
— f — — metal m - [metal] minor, o light, metal loss
— f mínima - minimum loss
— f — de núcleo m - [electr-magnet] minimum
core loss
— f — por diafonía f - [telecom] minimum cross
talk loss
— f neta—[fin] net loss
— f — en cambio m - [fin] net exchange loss;
net loss on exchange,ging
— f — — — m de moneda(s) f extranjera(s) -
[fin] net foreign currency exchange loss
— f — no realizada - [fin] net unrealized loss
— f — por emisión f (de pólizas f) - [seguros]
net underwriting loss
— f — sobre seguro(s) m - [seguros] net underwriting loss(es)
— f — — — m emitido(s) - [seguros] net underwriting loss
— f — — venta f de valor(es) m - [fin] security,ties sale net loss
pérdida f no realizada - [contab] unrealized
loss
— f — — sobre cambio(s) m - [fin] unrealized
exchange loss
— f — — — — m extranjero(s) - [fin] unrealized foreign exchange loss
— f — — — capital m - [fin] unrealized capital loss
— f — — — divisa(s) f - [fin] unrealized
foreign exchange loss
— f normal - normal loss • [petról] normal
by-pass
— f — en lodo m para perforación f - [petról]
drilling mud normal by-pass
— f observada - observed loss
— f otra - other loss
— f parcial - partial loss
— f — de equipo m - [seguros] equipment partial, o partial equipment, loss
— f — — maquinaria f - [seguros] machinery
partial, o partial machinery, loss
— f pequeña f - small loss
— f — de calor m - [termol] small, o low, heat
loss
— f permanente - permanent loss
— f permitida - allowed loss • [ind] permissive waste
— f por alza f de temperatura f de horno m -
[metal] iron (heat) loss
— f — calcinación f - [refract] calcination
loss; loss by calcination
— f — colilla(s) f - [sold] stub loss
— f — conversión f - [fin] translation loss
— f — — f de divisa(s) f - [fin] currency, o
exchange, translation cost
— f — — f — — f en exterior - [fin] foreign
currency translation loss
— f — — — f — — f extranjera(s) - [fin] foreign currency translation loss
— f — corriente f de Foucault - [electr-oper]
Foucault current loss
— f — — — f parásita - [electr-oper] Foucault,
o parasite,tic, current loss
— f — chisporroteo m - [sold] spatter loss
— f — desgaste m - [mec] wear, o attrition,
loss
— f — destilación f - [petról] distillation
loss
— f — diafonía f - [telecom] cross talk loss
— f — disolución f - solution loss
— f — evaporación f - [hidr] evaporation loss
— f — fricción f - [mec] friction(al) loss
— f — — f con pared f - [tub] wall friction
loss
— f — — f en cojinete n - [cojin] bearing
friction loss
— f — fuego m - loss by fire • [metal-prod]
melting loss • descaling loss
— f — histérsis f - [electr] hysteretic,resis
loss - [metal] iron loss
— f — presión f - [mec] friction loss
— f — rotura f - [com] breakage
— f — procesamiento - [ind] process(ing), allowance, o loss
— f — recalcadura f - [metal-fabr] swaging
loss; loss by swaging
— f — salpicadura(s) f - [sold] spatter loss
— f — seguro(s) m - [seguros] underwriting
loss(es)
— f — transferencia f - transfer loss
— f — — f de divisa(s) f - [fin] currency,
transfer, o translation, loss
— f — volatilización f - [miner] volatilization loss; loss by volatilization
— f posible - possible loss
— f prevenida - prevented, loss, o escape
— f prevista - foreseen, o predicted, loss
— f — por fricción f - foreseen, o predicted,
friction loss
— f promedio - average loss
— f — de metal m - average metal loss
— f real - actual loss

pérdida f realizada - [contab] realized, o attained, loss
— f reclamada - [seguros] claimed loss
— f reconocida - [seguros] recognized loss
— f reducida - reduced, o low, loss • [segurid]
controlled loss
— f — de calor m - [termol] low heat loss
— f — en rendimiento m - [ind] low yield loss
— f restringida - restricted loss • restricted
leakage
— f severa - severe loss
— f sobre cambio m - [fin] exchange loss
— f — — m extranjero - [fin] foreign exchange
loss
— f — capital m - [fin] capital loss
— f — — m realizado - [fin] realized capital
loss
— f — divisa(s) f—[fin] foreign exchange loss
— f — moneda f extranjera - [fin] foreign exchange loss
— f sobre título(s) m—[fin] loss on securities
— f — venta(s) f - [com] sale(s) loss; loss on
@ sale(s)
— f — — f de valores m - [fin] securities
sale loss
— f — valor(es) m - [fin] loss on securities
— f total - total loss
— f — de freno(s) m - [autom] total brake loss
— f única - only, o single, loss
— f verificada - checked loss • [petról]
checked by-pass
perdidas f y ganancias f - [contab] véase ganancias f y pérdidas f
— f y gasto(s) m incurrido(s) - [seguros] incurred loss(es) and expense(s)
perdido,da a - . . . . • missed • knocked off •
[electr-oper] knocked out • [petról]
by-passed
— a con frecuencia f - lost frequently
— a frecuentemente - lost frequently
— a por amputación f - [medic] amputated; lost
by amputation
— a — destilación f - [petról] lost by distillation
— a — evaporación - evaporated; lost by evaporation
perdigón(es) m - . . . . • [metal-prod] pellet(s)
perezoso,sa a - . . . . • sold] sluggish
perfección f - . . . . • completion
— f de circuito m - [electr-instal] circuit
completion
perfeccionado,da a - perfected; improved; refined; advanced; developed
perfeccionador,ra a para producción f - [ind]
production perfecting
perfeccionamiento m - . . . .; improving,vement;
completing,tion; successful developing,pment
• [ind] final, development, o training
— m básico - basic, developing,pment, o improving,vement
— m con éxito m - successful developing,pment
— m de circuito m - [electr-instal] circuit
completion
— m — compuesto m - [quím] compound developing,pment
— m — disolución f - solution improvement
— m — — f para enfriamiento m - cooling solution improvement
— m — — f para refrigeración f - cooling solution improvement
— n — eficacia f - efficiency improvement
— m — — f de disolución f - solution efficiency improvement
— m — — f para cierre m - [mec] closure solution efficiency improvement
— m — fórmula f - formula developing,pment •
[quím] recipe developing,pment
— m — neumático(s) m - [autom-neumát] tire developing,pment
— m — procedimiento m - [ind] procedure, development, o improvement
— m — — m para ensayo(s) m - [ind] test procedure, development, o improvement

perfeccionamiento m de producción f - [ind]
production perfecting
— m — producto m - [ind] product developing,pment
— m — receta f - [quím] prescription, o formula, o recipe, developing,pment
— m — proyección f - advanced design design
improvement
— m exitoso - successful developing,pment
— m obtenido como resultado m de carrera(s) f
- [autom-neumát] race-bred development
— m — mediante fogueo m en competición(es) f -
[autom-neumát] race-bred development
— m patentado - patented improvement
— m producido como resultado m de carrera(s) f
- [autom-neumát] race-bred development   \ c
perfeccionar v - . . . .; to develop • to refine
— v compuesto m - [quím] to develop @ compound
— v fórmula v - [ind] to develop @ formula •
[quím] to develop @, prescription, o recipe
— v investigación f - [legal] to complete @,
investigation, o research
— v producción f—[ind] to perfect @ production
— v producto m - [ind] to, develop, o perfect,
@ product
— v receta f - [ind] to develop @ formula •
[quím] to develop @, recipe, o prescription
— v técnica f - [ind] to perfect @ technique
— v tecnología f—[ind] to develop @ technology
perfecto,ta a - . . . .; flawless
perfil m - . . . .; contour • section • shape •
skyline • [metal-lam] shape; profile; member
• (rail) section • [estruct] piece
— m alto - high profile
— m ancho - [metal-lam] wide shape
— m angular - [metal-lam] angle, shape, o profile; angle
— m — con nervio m - [metal-lam] bulb, o ship,
angle, o shape
— m — galvanizado - [metal-trat] galvanized,
angle, o shape
— m — laminado—[metal-lam] rolled angle
— m — de acero m - [metal-lam] rolled steel
angle
— m — — — m galvanizado - [metal-trat]
galvnized rolled steel angle
— m — para montaje m para cubierta f - [mec]
roof mounting angle
— m ángulo - [metal-lam] angle (shape, o profile)
— m — con alas f desiguales - [metal-lam] unequal flange angle beam
— m — — f iguales - [metal-lam] equal
flange angle (beam)
— m Armco de acero m - Armco steel shape
— m — — para calzada - [metal-fabr] Armco (steel floor) plank
— m — — m — f para puente m - [metal-
-fabr] Armco (steel) bridge plank
— m — para calzada f para puente m - [metal-
-fabr] Armco bridge plank
— m bajo - low profile
— m — y ancho - low side profile
— m cilíndrico - [geom] cylindrical, profile, o
shape
— m comercial - [metal-lam] commercial, o merchant, shape, o profile
— m con ala(s) f ancha(s) - [metal-lam] wide
flange(s), profile, o section
— m — — f desigual(es) - [metal-lam] unequal
flange(s), beam, o profile
— m — — f iguales - [metal-lam] equal flanges
beam
— m — — f inclinada(s) - [metal-lam] slanted
flange(s) beam
— m — — f — de acero m - [metal-lam] steel
slanted flange(s), beam, o shape
— m — — f — laminado en caliente - [metal-
-lam] hot rolled slanted flange beam
— m — — f — frío - [metal-lam] cold
rolled slanted flanges, beam, o shape
— m — patín m ancho - [metal-lam] wide base,
beam, o shape

perfil m con patín m ancho laminado en caliente
- [metal-lam] hot rolled wide base beam
— m — — m — — — **frío** - [metal-lam] cold rolled wide base beam
— m — — m — **(y) recto** - [metal-lam] wide (and) straight base, beam, o shape
— m — — m — **(—) laminado en caliente** - [metal-lam] hot rolled wide (and) straight base beam
— m — — m — **(—) — — frío** - [metal-lam] cold rolled wide (and) straight base beam
— m — — m **recto** - [metal-lam] straight base, beam, o shape
— m — — m — **laminado en caliente** - [metal-lam] hot rolled straight base beam
— m — — m — — — **frío** - [metal-lam] cold rolled straight base beam
— m **conformado en caliente** - [metal-lam] hot formed section
— m — — **frío** — [metal-lam] cold formed section
— m — **laminado** - [metal-lam] rolled formed section
— m **corriente** - [metal-lam] regular section
— m **cuadrado** - [metal-lam] square (section)
— m **con canto(s) m redondeados** - [metal-lam] round cornered square (section)
— m **de acero m** - [metal-lam] steel, profile, o shape, o member
— m — — m **con carbono m** - [metal-lam] carbon steel, profile, o shape
— m — — m — — m **laminado en caliente** - [metal-lam] hot rolled carbon steel shape
— m — — m — — — **frío** - [metal-lam] cold rolled carbon steel shape
— m — — m **en U** - [metal-lam] steel channel
— m — — — — — **con alas f desiguales** - [metal-lam] unbalanced steel channel
— m — — m — — — — f **iguales** - [metal-lam] balanced steel channel
— — — m **estructural** - [metal-lam] structural steel shape
— m — — m — **laminado** - [metal-lam] rolled structural steel shape
— m — — m — — **soldado** - [metal-lam] welded rolled structural steel shape
— m — — **soldado** - [metal-lam] welded structural steel shape
— m — — m **galvanizado** - [metal-trat] galvanized steel, shape, o member
— m — — m — **con laminación f profunda** - [metal-lam] deep(ly) formed galvanized steel, member, o shape
— n — — m **laminado** - [metal-lam] rolled steel, section, o shape
— m — — m — a **en caliente** - [metal-lam] hot rolled steel shape
— m — — m — — — **frío** - [metal-lam] hot rolled steel shape
— m — — m **soldado** - [estruct] welded steel shape
— m — **ángulo m** - [metal-lam] abgle, profile, o shape, o beam
— m — **asiento m** - [constr] bedding profile
— m — **banda f** — [metal-lam] strip cross section
— m — — f **para rodamiento m** - [autom-neumát] (tire) tread profile
— m — — — m **despezado** - [autom-neumát] tapered tread profile
— m — **base f** - [constr] base, o bedding, profile, o shape
— f — **calidad f alta** - [metal-lam] high quality, profile, o shape
— m — **canal** - [metal-lam] channel, beam, o profile, o shape
— m — **cauce m** - [hidr] bed profile
— m — — m **para curso m para agua m** - [hidr] streambed profile
— m — **columna f** - [metal-lam] column, o beam, profile, o shape
— m — **chapa f** - [metal-lam] sheet profile
— m — — f **en U** - [metal-lam] channel sheet, profile, o shape
— m — — f — — **con alas f desiguales** -

[metal-lam] unbalanced sheet channel
perfil m de diente m - [mec] tooth, profile, o outline
— m — — m **de engranaje m** - [mec] gear tooth, profile, o outline
— m — **engranaje m** - [mec] gear, profile, o outline
— m — **espesor m** - [mec] thickness profile
— m — — m **de banda f** - [metal-lam] strip thickness profile
— m — — m — — f **para rodamiento m** - [autom-meumát] tread depth profile
— m — **gola f** - [arquit] ogee
— m — **hierro m** - [metal] iron profile
— m — — m **en U** - [metal-lam] iron channel (profile)
— m — **laminación f** - [metal-lam] rolling shape
— m — — f **profunda** - [metal-lam] deep(ly) formed, profile, o member, o shape
— m — **lecho m** - [hidr] bed profile
— m — — m **de curso m de agua m** - [hidr] streambed profile
— m — **subyacente** - [constr] (underlying) bed(ding) profile
— m — **llanta f** - [f.c.-ruedas] rim contour
— m — **más de tres pulgadas f** - [metal-lam] structural size (shape, o profile)
— m — **menos de tres pulgadas f** - [metal-lam] bar size (profile, o shape)
— m — **neumático m** — [autom-neumát] tire profile
— m — **país m** - [econ] country('s) profile
— m — **paso m inferior** - [vial] underpass profile • [G.B.] subway profile
— m — **producción f normal** - [metal-lam] normal production shape
— m — **profundidad f** - [mec] depth profile
— m — — f **de banda f** - [metal-lam] strip depth profile
— m — — f — — f **para rodamiento m** - [autom-neumát] (tire) tread depth profile
— m — **río m** - [hidr] river profile
— m — **rodillo m** - [mec] roll, profile, o shape
— m — — m **para rotadora f** - [cuerp.moled] rotator roll profile
— m — **suelo m** - [suelos] soil profile
— m — **terreno m** - [geol] ground profile • [suelos] soil profile
— m — **tronco m de cono m** - [geom] truncated cone shape
— m — **viga f** - [metal-lam] beam, shape, o profile
— m **económico** - [econ] economic profile
— m — **de país m** - [econ] country('s) economic profile
— m **en barra f** - [metal-lam] bar shape
— m — **frío** - [metal-lam] cold (formed) section
— m **especial** - special profile • [metal-lam] special, section, o shape
— m — **de acero m inoxidable** - [metal-lam] stainless steel special shape
— m — **inoxidable** - [metal-prod] stainless special, o special stainless, shape
— m **esquinero** - [metal-lam] corner shape
— m — **para contención f** - [mec] corner hold down, rail, o shape
— m **estándar** - [metal-lam] standard, profile, o section
— m — **para paso m inferior** - [vial] standard underpass profile • [G.B.] standard subway profile
— m **estandardizado*** - [metal-lam] standardized profile
— m **estructural** - [metal-lam] structural, profile, o shape, o section, o member
— m — **con resistencia f alta (a tensión f)** - [metal-lam] high strength structural (shape)
— m — **de acero m** - [metal-lam] steel structural, profile, o shape
— m — — m **con carbono m** - [metal-lam] carbon steel structural shape
— m — — m — — m **laminado en caliente** - [metal-lam] hot rolled carbon steel structural shape

perfil m estructural estándar - [metal-lam]
standard structural, shape, o section
— m — hueco - [metal-lam] hollow structural,
shape, o section
— m — laminado - [metal-lam] rolled structural
shape
— m — en caliente - [metal-lam] hot rolled
structural shape
— m — — frío - [metal-lam] cold rolled
structural shape
— m — para barco(s) m - [metal-lam] ship
structural (shape)
— m — — embarcación(es) f - [metal-lam] ship
structural (shape)
— m — soldado - [metal-lam] welded structural
shape
— m formado - [metal-lam] formed shape
— m fuera de serie f—[metal-lam] out-of-series
shape
— m futuro - [econ] future profile
— m hidráulico - [hidr] hydraulic profile
— m — para caudal m - [hidr] flow hydraulic
profile
— m — — — m uniforme - [hidr] uniform flow
hydraulic profile
— m — — — m no uniforme - [hidr] non-uniform
flow hydraulic profile
— m homogéneo - [geol] homogenous profile
— m hueco - [metal-lam] hollow (section, o
shape)
— m I - [metal-lam] I, shape, o beam, o section
• I bar
— m — de acero m - [metal-lam] steel I, beam,
o shape
— m — — — m con carbono m - [metal-lam] carbon steel I shape
— m — — — m — — m laminado en caliente -
[metal-lam] hot rolled carbon steel I shape
— m — — — m — — m frío - [metal-lam]
cold rolled carbon steel I shape
— m — — — m laminado en caliente - [metal-
-lam] hot rolled steel I shape
— m — — — m — — en frío - [metal-lam] cold
rolled steel I shape
— m — laminado en caliente - [metal-lam] hot
rolled I beam
— m — — — frío - [metal-lam] cold rolled
I beam
— m industrial - [econ] industrial profile
— m L - [metal-lam] L, bar, o shape
— m laminado - [metal-lam] rolled, shape, o
profile, o (formed) section • stock
— m — con ala m ancha - [metal-lam] rolled
wide flange, section, o beam
— m — de acero m - [metal-lam] rolled steel
shape
— m — en caliente - [metal-lam] hot rolled,
shape, o profile
— m — — con patín m ancho - [metal-lam]
hot rolled wide, base, o flange, shape
— m — — — — m — y recto - [metal-lam]
hot rolled wide straight base shape
— m — — — — m recto - [metal-lam] hot
rolled straight base shape
— m — — frío - [metal-lam] cold, rolled, o
formed, section
— m — según norma f - [metal-lam] standard
rolled section
— m — — — f con ala m ancha - [metal-lam]
standard rolled wide flange section
— m ligero - [metal-lam] light profile
— m liviano - [metal-lam] light profile
— m longitudinal - longitudinal profile
— m — de terreno m - land longitudinal profile
— m mediano - [metal-lam] medium, section, o
profile
— m metálico - [metal-lam] metal(lic) shape
— m — para calzada(s) f - [constr] floor metal
plank
— m — — — f para puente m - [constr] bridge
metal plank
— m — — piso m para puente m - [constr]
bridge metal plank

perfil m no homogéneo - [geol] non-homogeneous
profile
— m normal - normal profile • [metal-lam] standard shape
— m — en I - [metal-lam] standard I shape
— m — — U - [metal-lam] standard U shape
— m normalizado - [metal-lam] standardized
shape
— m para automóviles m - [autom-neumát] automobile, o car, profile, o shape
— m — m en stock* - [autom-neumát] stock
profile
— m — — m — — en salón(es) m para venta(s)
m - [autom-neumát] showroom stock profile
— m — — m — — sin modificar—[autom-neumát]
showroom stock profile
— m — bóveda(s) f - [constr] arch channel
— m — calzada f - [constr] floor plank
— m — — f para puente(s) m - [constr] bridge
(floor) plank
— m — carrera(s) f - [autom-neumát] racing
profile
— m — — f entre pilón(es) m - [autom-neumát]
autocrossing profile
— m — contención f - [mec] restraining, o hold
down, shape, o rail
— m — marco m - [constr] frame bar
— m — piso(s) m - [constr] floor plank
— m — — m para puente(s) m - [constr] bridge
(*floor) plank
— m — rueda(s) f - [mec] wheel profile
— m — serie f - [autom-neumát] series profile
— m pesado - [metal-lam] heavy, shape, o beam
— m — y plano - [metal-lam] heavy and flat
shape
— m plano - [metal-lam] flat profile
— m — de rodillo m - [mec] roll flat profile
— m — — — m para botadora f - [cuerp.moled]
rotator roll flat profile
— m redondo - [metal-lam] round, o circular,
section, o shape
— m semiterminado - [metal-lam] semifinished,
shape, o form
— m soldado - [metal-lam] welded, shape, o profile
— m — columnar - [metal-lam] welded column(ar)
shape
— m — de acero m - [metal-lam] welded steel
shape
— m — — — m estructural - [metal-lam] welded
structural steel
— m — — — m — laminado - [metal-lam] welded
rolled structural steel
— m T - [metal-lam] tee section
— m terminado - [metal-lam] finished, shape, o
form
— m transversal - [metal-lam] cross section(al)
(profile)
— m tronco-cónico - [geom] truncated cone shape
— m U - [metal-lam] U shape; channel
— m — con alas f desiguales - [metal-lam]
unbalanced channel
— m — para apoyo m - [metal-lam] arch channel
— m — — bóveda(s) f - [constr] arch channel
— m Z - [metal-lam] Z (shape)
perfilado m - véase perfiladura f
perfilado,da a - shaped • streamlined • highlighted • shaped
perfilador m - [mec] shaper
— m para perfiles m comerciales - [metal-lam]
industrial shapes shaper
— m — secciónes f comerciales - [metal-lam]
commercial shapes shaper
perfiladura f - . . . . • shaping • streamlining •
highlighting • [petról] logging
— f eléctrica - [petról] electric logging
— f en entrada f - [hidr] entrance streamlining
— f — salida - [hidr] outlet streamlining
perfilar v - . . . . • to streamline • to highlight
— v entrada f - [hidr] to streamline @ entrance
— v salida f - [hidr] to streamline @ outlet
perfilómetro m - [instrum] perfilometer

perforación f - . . .; bore; slot; piercing •
boring through • [sold] burn(ing) through •
[petról] . . .; drill hole; borehole
— f **abocardada** - [mec] counterbore
— f **adecuada** - [constr] adequate bore,ring
— f **anticipada** - [mec] preboring; prepunching
— f — **de orificio** m - [mec] hole, preboring, o
prepunching
— f **bajo agua** m - [petról] underwater, o submarine, drilling
— f — **presión** f - [petról] pressure drilling
— f **completa** - complete, bore,ring, o drilling
— f **completada** - completed, boring, o drilling
— f **con cable** m - [petról] cable drilling
— f — m **con herramienta**(s) f - [petról]
cable tool drilling
— f — m **rígido** - [petról] pole dril(ling)
— f — **cilindro** m - [comb.int] cylinder boring
— f — **corona** f (sacatestigos)—[petról] coring
— f — **chorro** m - [mec] jet, boring, o drilling
— f — **diamante**(s) m - [miner] diamond drilling
— f **con diámetro** m **reducido** - [petról] slim, o
narrow, hole
— f — **liga** f **metálica** - [petról] metal bond
drilling
— f — **operación** f **única** - [mec] single stroke
punch(ing)
— f — **percusión** f - [petról] percussion drilling
— f — **poca profundidad** f - [petról] shallow
hole
— f — **precisión** f - [mec] precision, drilling,
o perforating,tíon
— f — **presión** f **baja** - [petról] low pressure
drilling
— f — **profundidad** f **reducida** - [petról] shallow hole
— f — **sacatestigo**(s) m—[petról] core drilling
— f — **sonda** f (**con diamantes** f) - [miner]
diamond drilling
— f — **varilla** f (**rígida**) - [petról] pole drilling
— f **costa afuera** - [petról] offshore drilling
— f **de acero** m - [sold] steel piercing
— f — **agujero** m - [mec] hole, boring, o drilling, o punching, o piercing
— f — m **con precisión** f - [mec] precision
hole boring
— f — m **para colada** f - [metal-prod] taphole drilling
— f — **cárter** m - [comb.int] oil pan puncture,ring
— f — **costura** f - [sold] seam burning through
— f — **chapa** f - [sold] sheet burn(ing) through
— f — — f **delgada** - [sold] thin plate burn-
-(ing) through
— f — — f **muy delgada** - [sold] very thin
sheet burn(ing) through
— f — **disco** m - [mec] disk drilling
— f — **estrato**(s) m - [petról] stratum,ta drilling
— f — **fondo** m - [constr] invert perforating
— f — **formación** f - [geol] formation drilling
— f — — f **petrolífera** - [petról] oil bearing
formation drilling
— f — **hilera** f - [mec] row perforating,tion
— f — — f **inferior** - [metal-fabr] bottom row
perforating,tion
— f — — f **superior** - [metal-fabr] upper, o
top, row perforating,tion
— f — **material** m - [mec] material perforating,tion
— f — **muestra** f - [mec] specimen, o sample,
drilling
— f — **orificio** m - [mec] hole, punching, o
piercing
— f — — m **para costura** f (**longitudinal**) -
[mec] longitudinal seam hole punching
— f — **plantilla** f - [mec] templet boring
— f — **pozo**(s) m - [petról] (well) drilling
— f — — m **con diámetro** m **reducido** - [petról]
slim, o narrow, hole drilling
— f **desde plataforma** f - platform drilling

perforación f **desde plataforma** f **apoyada sobre**
fondo m **de mar** m - [petról] drilling from @
bottom supported platform
—, **efectuada,** o **ejecutada** - performed, o made,
drilling
— f **en frío** - [mec] cold drilling
— f **estándar** - [mec] standard punching
— f — **con** . . ., **agujero**(s), o **orificio**(s) -
[mec] standard . . . hole punching
— f **hacia abajo** - [mec] down perforating,tion
— f **horizontal** - [mec] horizontal, drilling, o
punching
— f **mínima** - [mec] minimum perforating,tion •
[sold] minimum burn through
— f **omitida** - [mec] omitted perforating,tion
— f **para petróleo** m - [petról] oil drilling
— f — **carga**(s) f **explosiva**(s) - [explos]
blast(ing) hole
— f — **ensayo**(s) - [mec] test(ing) drilling •
[petról] wildcat, drill(ing), o well
— f — **voladura**(s) f - [explos] blast(ing) hole
— f **precisa** - [mec] precise perforating,tion
— f **primera** - first perforation
— f **profunda** f - [mec] deep, drilling, o perforating,tion • [petról] deep borehole
— f **rectificada** - [petról] reamed drill hole
— f **satisfactoria** - [constr] satisfactory, o a-
dequate, drilling, o boring, o perforating
— f **según norma** f - [metal-fabr] standard perforating,tion, o drilling, o boring
— f **submarina** - [petról] underwater, o submarine, drilling
— f **total** - [metal] rusting out
— f **vuelta hacia arriba** - [tub] turned up perforation
perforado,da a - perforated; drilled; punched;
slotted; pierced • bored through • [sold]
burned through
— a **bajo presión** f - [petról] pressure drilled
— a **completamente** - drilled completely
— a **con operación** f **única** - [mec] single stroke
punched
— a **totalmente** - [metal] rusted out
perforador m - . . . . • [petról] drilling rig
perforador,ra a - . . . .; perforator
perforadora f - [herram] drill press • [comput]
card punch(er)
— f **automática** - [metal-prod] automatic drill;
bott mechanism
— f **de balancín** f - [petról] spudder
— f **con cilindro**(s) m - [comb.int] cylinder
boring machine
— f **con chorro** m - [mec] jet drill
— f — **diamante**(s) m - [petról] diamond drill
— f — **proyectil** m - [petról] gun perforator
— f — **regulación** f **automática** - [petról] automatic drilling control unit
— f **de carretilla** f - [miner] wagon drill
— f — **pedestal** m - [mec] vertical drill
— f **costa** f **afuera** - [petról] offshore operator
— f **flotante** - [petról] floating, drill, o unit
— f **giratoria** - [petról] rotary drill
— f **horizontal** - [miner] horizontal drill;
drifter
— f **jackdrill*** - [miner] jackdrill
— f **manual** - [constr] jackhammer (drill)
— f **neumática**—[herram] pneumatic, o air, drill
— f — **manual** - [constr] jackhammer (drill)
— f **para agujero**(s) m - [mec] hole punch
— f — **exploración** f - [petról] core drill •
prospecting drill
— f — **poste**(s) m - [herram] post drill
— f — **pozo**(s) m - [constr] well driller
— f — — m **con diámetro** m **reducido** - [petról]
slim, o narrow, hole rig
— f — — m **para agua** m - [hidr] water well
drill
— f — **roca** f - [herram] rock drill
— f — **rueda**(s) f - [f.c.-ruedas] wheel boring
mill
— f — **tubería** f - [tub] piping drill
— f — — f **para entubación** f - [petról] casing
perforator

perforadora f por percusión f - [petról] percussion drill
— f portátil - [petról] portable drilling machine
— f radial - [herram] radial drill
— f rotativa - [petról] rotary (drilling) drill
— f — con punta f con diamante(s) m - [petról] diamond point rotary drill
— f vertical - [herram] vertical, o post, drill
perforar v - [mec] . . .; to pierce; to punch puncture; to bore through • to drift • [sold] . . .; to break, through, o out; to melt through; to burn through • [petról] to tap
— v acero m - [sold] to pierce @ steel
— v agujero m - [mec] to, drill, o bore, o make, o punch, @ hole
— v — m con precisión f - [mec] to precision bore @ hole
— v bajo presión f - [petról] to pressure drill
— v cárter - [comb.int] to puncture @ oil pan
— v codo m - [metal-prod] to pierce @ gooseneck
— v completamente - [mec] to drill completely
— v con operación f única - [mec] to single stroke punch
— v disco m - [mec] to drill @ disk
— v estrato m - [petról] to drill @ stratum,ta
— v fondo m - [metal-prod] to drill @ bottom
— v — m de horno m - [metal-prod] to drill @ furnace bottom
— v — m — m alto - [metal-prod] to drill @ blast furnace bottom
— v formación f - [petról] to drill (into) @ formation
— v — f petrolífera - [petról] to drill (into) @ oil bearing formation
— v — f productiva - [petról] to drill in(to) (@ oil bearing formation)
— v material - [mec] to perforate @ material
— v muestra f - [mec] to drill @, specimen, o sample
— v orificio m - [mec] to drill @ hole
— v — m con anticipación f - [mec] to pre-punch @ hole
— v plantilla f - [mec] to bore @ templet
— v pozo m - [constr] to drill @ well
— v totalmente - [mec] to drill completely • [metal] to rust out
— v busa f - [metal-prod] blowpipe breakthrough
perforarse v - to break through
— v busa f - [metal-prod] blowpipe breakthrough
— v cuchara f - [metal-prod] ladle breakout
— v piquera f - [metal-prod] iron notch breakout
performancia* f - véase rendimiento m; desempeño m; performance f
perfoverificación f - [petról] véase verificación f de perforación f
perfunctoriamente* adv - perfunctorily; carelessly; véase en forma f descuidada
perfunctorio,ria* a - perfunctory; careless
pericia f - . . .; expertise; craftsmanship
— f administrativa - [admin] management : expertise
— f de cliente m - [ind] customer's, skill, o expertise
— f para aplicación(es) f - [ind] application expertise
— f suplida - provided expertise
— f técnica - [ind] technical expertise
— f tecnológica - technological expertise
pericicloide* a - pericycloid*
periferia f - . . .; girth • [pol] edge; limit
— f de cauce m - [hidr] channel periphery
— f — engranaje m - [mec] gear, circle, o line
— f — m en base f - [mec] dedendum gear, circle, o line
— f — m — cabeza f - [mec] addendum gear, circle, o line
— f — m — pie m - [mec] dedendum gear, circle, o line
— f — petaca f - [metal-prod] cooler,ling plate edge
— f — polea f - [mec] sheave periphery
— f — tambor m - [mec] drum periphery

periferia f de tubería f - [tub] pipe, girth, o periphery
— f — — f de acero m corrugado - [tub] corrugated steel pipe periphery
— f equivalente - equal, o equivalent, periphery
— f exterior - outside, o exterior, periphery
— f igual - equal periphery
— f interior - inside, o interior, periphery
— f — total - [tub] complete, inside, o interior periphery
— f total - total, o complete, periphery
periféricamente adv - periferally
periférico,ca a - . . .; peripheral
periforme* a - pear shaped
perilla f . . . • button; knob; handle • [constr] doorknob
— f ajustada - [instrum] adjusted, o tightened, knob
— f con resorte m - [mec] loaded knob
— f — — m con tensión f - [mec] spring loaded knob
— f autoindicadora - [instrum] self-indicating dial
— f — para regulación f - [instrum] self-indicating control dial
— f girada - turned, o twisted, knob
— f para aire m fresco - fresh air knob
— f — ajuste m - [instrum] adjusting knob
— f — alineación f - [mec] alignment knob
— f — — f transversal - [sold] cross seam alignment, knob, o screw
— f — cambiador m - [mec] shifter knob
— f — arranque m - [comb.int] starter,ting, knob, o button
— f — cambio(s) m - [mec] shifter knob
— f — conmutador m - [electr-equip] switch, knob, o handle
— f — — f para cambio(s) m - [electr-equip] change switch knob
— f — — m — — m para polaridad f - [electr] polarity change switch knob
— f — — m para selección f - [electr-equip] selector switch, handle, o knob
— f — — m regulador m - [mec] control switch, knob, o handle
— f — — m selector - [electr-equip] selector switch, knob, o handle
— f — corredera f - [mec] slide, stud, o knob
— f — iniciación f - [comb.int] start(ing), button, o knob
— f — — f de precalentamiento m - [comb.int] preheater start(ing), button, o knob
— f — interruptor m - [electr-equip] breaker, o switch, handle
— f — palanca f - [mec] lever knob
— f — reactor m - [electr-equip] reactor, knob, o handle
— f — regulación f - [mec] control(ling) knob
— f — — f de aire m - air control knob
— f — — f para aire m fresco - [autom] fresh air control knob
— f — amperaje m - [sold] current, o amperage, control, knob, o handle
— f — reóstato m - [electr-equip] rheostat, knob, o handle
— f — selector m - [electr-equip] selector knob
— f — — m para amperaje(s) m - [electr-equip] ampere(s) selector knob
— f — sujeción f - [mec] holding, o locking, knob
— f — — f con tensión f - [mec] loaded holding knob
— f — — f — — f con resorte m - [mec] spring loaded holding knob
— f — tope m - [mec] stop, handle, o knob
— f — trabadura f - [mec] locking knob
— f — trinquete m - [mec] ratchet knob
perímetro m - . . .; girth
— m bañado - [hidr] wetted perimeter
— m de bola(s) f - [cuerp-moled] ball(s) perimeter
— m — círculo m - [geom] circle perimeter

perímetro m de círculo m medido - [geom] measured circle perimeter
— m — obra f - [constr] project limit(s)
— m — paquete m - [transp] package girth
— m — planta f - [ind] plant perimeter • plant
site
— m exterior - outer, o outside, perimeter
— m humedecido - wetted perimeter
— m interior - inner, o inside, perimeter
— m máximo - maximum, perimeter, o girth
— m — de paquete m - maximum package girth
— m medido - [geom] measured perimeter
— m — de círculo m - [geom] measured circle, o
circle measured, perimeter
— m mojado - wet(ted) perimeter
— m remojado - [hidr] wetted perimeter
— m total - total, o entire, perimeter
periodicidad f - . . . . • [electr-prod] véase
frecuencia f
periódico,ca a - . . .; regular
periodismo m - [public] . . .; media
periodista m - [public] . . .; reporter
— m independiente - [public] independent, o
free lance,cing, journalist, o reporter
— m profesional - [public] media professional
período m - . . . • stint • space • [pol] term •
[elecr-prod] cycle; hertz
— m acordado - agreed period
— m ajustable - adjustable period (of time)
— m ajustado - adjusted (time) period
— m anterior - prior, o former, period
— m — para recuperación f - former recovery
period
— m bajo bandera f amarilla - [deport] caution
period
— m certificado - certified period
— m completo - complete period
— m — para ejecución f - complete performance
period
— m con carga f - load(ed) period
— m — — f máxima - maximum load period
— m — — f mínima - minimum load period
— m — fósforo m alto - [metal-prod] high phosphorus period
— m considerable - lengthy, o substantial, period • time period
— m — de servicio m - extended service period
— m — — — m normal - [ind] extended normal
service (period)
— m corto - short (time) period
— m — de tiempo m - short time period
— m crítico - critical, period, o range
— m — año(s) m - year(s) period
— m de espera f - waiting period
— m exuberancia f - [econ] boom
— m — — f para bien(es) m raíz,ces - [econ]
real estate boom
— m — funcionamiento m - [ind] run
— m — garantía f - [com] guaranty, o warranty,
period
— f — — f eléctrica - [electr] electrical
guaranty period
— m — — f mecánica - [mec] mechanical guaranty period
— m — gracia f - [fin] grace period
— m — interrupción f - interruption period
— m — inundación f - [hidr] flood(ing) period
— m — operación f - operating,tion, o running,
period
— m — — f largo - [mec] long running period
— m — parada f - [ind] inactive, o shut down,
period
— m — permanencia f - stay (period)
— m — precipitación f - [meteorol] rainfall
perioood
— m — — f extraordinaria - [hidr] peak rainfall period
— m — punta f - [ind] peak period
— m — retención f - withholding, o detention,
period • [sanit] retention period
— m seca f - [meteorol] dry, period, o cycle
— m servicio m - service, period, o length
— m — suspensión f - suspension period

período m de tiempo m - time, period, o span
— m — — m ajustable - adjustable time period
— m — — m ajustado - adjusted time period
— m — — m regulado - controlled, o adjusted,
time period
— n — trabajo m - [ind] work period
— m — — m para grúa f - [grúas] crane work
period
— m — utilidad(es) f - profit period • performance (period)
— m — — f prolongado - prolonged, o extended,
performance (period)
— m diferente - different period
— m establecido - established period • time
limit
— m estipulado - stipulated, o set, period
— m experimental - experimental period
— m glacial - [geol] ice age
— m igual - equal, o same, period
— m indicado - indicated, o given, period, o
duration
— m inicial - initial period
— m largo - long (time) period
— m — de tiempo m - long time period
— m limitado - limited period
— m más largo - longer,gest period
— m — prolongado - longer,gest period
— m — admisión f - [comb.int] admission, o intake, stroke
— m para afino m - [metal-prod] refining period
— m — almacenaje,namiento - storage period
— m — amortización f - [fin] amortization period • [contab] depreciation period
— m — aprendizaje m - apprenticeship, o training, period
— m — aviso m - [legal] notice (period)
— m — calcinación f - [miner] calcining, o
burning, period
— m — capacitación f - training period
— m — cebadura f - [bombas] priming period
— m — — f inicial - [bombas] initial priming
period
— m — construcción f - [constr] construction
period
— m — contacto m - [mec] contact period
— m — — m para engranaje m - [mec] gear contact period
— m — coquización f - [coque] coking period
— m — corrección f - correcting,tion period
— m — — f de falla(s) f - [ind] debugging
period
— m — descanso m - [labor] rest(ing) period
— m — descuento(s) m - [com] discount period
— m — detención f - [ind] stop(ping) period •
[sanit] detention (period)
— m — ejecución f - performance period
— m — ensayo(s) m - test(ing) period
— m — entrega f - delivery period
— m — escurrimiento m - [hidr] runoff period
— m — establecimiento m - establishing period
— m — — m de llama f - [combust] flame establishing period
— m — fabricación f - [ind] manufacture, o
fabrication, period
— m — garantía f—guaranty, o warranty, period
— m — imposición f - [fisc] tax(ing) period
— m — marcha f - [ind] operation period
— m — — f sin carga f - [comb.int] idle
(operating) period
— m — — f — — f de motor m - [electr-oper]
motor idle period • [comb.int] engine idle
period
— m — parada f - [ind] shutdown period
— m — paralización f - [ind] shut down period
— m — — f para horno m - [ind] furnace shut
down period
— m — presentación f - offer presentation (period)
— m — promoción f - promotion period
— m — prueba(s) f - test(ing), o trial, period
— m — puesta f en marcha f - start-up period
— m — — f — — f de horno m - [ind] furnace
start-up period

período m **para purificación** f - purification, o
refining, period
— m — **recuperación** - recovery,ring period
— m — **reducción** f - [metal-prod] reduction, o
reducing, period
— m — — f **en estufa** f - [metal-prod] stove
reduction,cing period
— m — **reembolso** m - [fin] reimbursement, o
payback, period
— m — **refinación** f - [metal-prod] refining period
— m — **retorno** m - return period
— m — **secamiento** m - drying, period, o time
— m — **transporte** m - transportation period
— m — **validez** f - validity period
— m — — f **para propuesta** f—[com] proposal, o
offer, o bid, (guaranty) validity period
— m — **vigencia** f - validity period
— m **previsto** - foreseen period
— m **prolongado** - lengthy, o extended, period
— m — **de operación** f - [ind] lengthy, o extended, operating,tion period
— m — — f **para planta** f - [ind] lengthy, o
extended, plant operating,tion period
— m — — **seca** - [meteorol] extended dry, period, o cycle
— m — — **tiempo** m - extended time period
— m **que se establece de antemano** - adjustable
time • pre-established time period
— m **regulable** - adjustable (time) period
— m **regulado** - adjusted (time) period
— m **sucesivo** m - successive period
— m — **de un año** m - succesive one year period
⌐ adv - **(por)** from year to year
peritaje m - . . . . • arbitrating,tion
perito m - . . .; specialist; arbitrator • [ind]
supervisor
— m **calígrafo** - [legal] handwriting expert
— m **conjunto** - [legal] joint expert
— m **de, escritorio** m, o **oficina** f - office expert
— m — — f **técnica** - technical office expert
— m **en seguro(s)** m - [seguros] insurance expert
— m **financiero** - [fin] financial expert
— m **mercantil** - [contab] expert accountant •
master of business administration
— m — **conservación** f—[ind] maintenance expert
— m — **mantenimiento** m - [ind] maintenance expert
— m **técnico** - technical expert
perito,ta a - . . .; skilled
perjudicación f - [seguros] damaging; injuring
— f **de cobertura** f - [seguros] coverage prejudicing
perjudicado,da a - damaged • [deport] bit
perjudicar v - . . .; to harm; to have @ adverse
effect • to suffer • to prejudice • [deport]
to bite
— v **cobertura** f - [seguros] to prejudice @ coverage
— f **decisión** f - to prejudice @ decision
perjudicial a - . . .; detrimental
perjuicio m - . . .; harm; trouble • [seguros]
loss • [deport] biting
— m **grave** - grave, harm, o damage
perla f - . . . . • [quím] bead • [sold] bead
— f **aisladora** - [electr-instal] (insulating)
bead
perlita f - . . . . • [metal] perlite
— f **esferoidal** - [miner] granular perlite
— f **granular** - [miner] granular perlite
perlítico,ca a - [metal] perlitic
permanecer v - . . .; to abide • to stand
— v **abierto,ta** - to remain open
— v **apartado,da** - to stand clear
— v **cerrado,da** - to remain closed
— v **concentrado,ca** - [sold] to remain concentrated
— v **constante** - to remain constant
— v **dentro de** - to, remain, o be, (with)in
— v **en red** - [electr-instal] to remain in @
system
— v — **reposo** - to, hold, o stay • to set

permanecer v **en sistema** m—to remain in @ system
— v — **sitio** m - to remain in @ place
— v — **vigencia** - to remain in, force, o effect
— v — — f **plena** - to remain in @ full force
— v — **vigor** - to remain in force
— v — — m **pleno** - yo remain in @ full force
— v **estacionario** - to remain stationary; to
stand still
— v **fijo** - to remain stationary; to stay set
— v **fluido,da** - to remain fluid
— v **fresco,ca** - to remain fresh • to, keep, o
stay, cool
— v **firme** - to, remain, o stand, firm
— v **fuera** - to remain out(side) • [deport] to
sit out
— v — **de carrera** f - [deport] to sit out @
race
— v **inalterable** - to remain @ same
— v **intacto** - to remain, intact, o unbroken
— v **redondo,da** - to remain round
— v **relativamente fluido,da** - [sold] to remain
relatively fluid
— v **válido,da** - to remain valid
— v **visible** - to remain visible
permanecido,da a - remained
— a **abierto,ta** - remained open
— a **cerrado,da** - remained closed
— a **en sitio** m - remained in @ place
— a **intacto,da** - remained, intact, o unbroken
— a **válido,da** - remained valid
permanencia f - . . . . • durability • holding •
stint • remaining
— f **de acero** m - [metal] steel, permanence, o
holding
— f — **puente** m - [constr] bridge permanence
— f — **tubería** f - [tub] pipe durability
— f **en sitio** m - remaining in @ place
— f **excesiva** - [labor] overstay
— f **válido,da** - remaining valid
permanecido,da a - remained
permanecimiento* m - véase **permanencia** f
permanente a - . . .; continuous,nuing • year
round • [labor] full time • [mec] stay-, o
remain-, -in-place
— a **de reserva** - permanent reserve
— a **de respaldo** - permanent back-up
— a **parcial** - [seguros] partially permanent
— a **total** - [seguros] total permanent
permanentemente adv - . . .; indefinitely
— adv **a vista** f - permanently, in, o to, view
permeabilidad f **a inducción** f - [electr-oper]
induction permeability
— f **admisible** - [electr-oper] allowable permeability
— f **alta** - high permeability
— f **baja** - low permeability
— f **de acero** m - [metal] steel permeability
— f — — m **con silicio** m - [metal] silicon
steel permeability
— f — — m — — m **recocido** - [electr-
-oper] annealed silicon steel permeability
— f — **columna** f **(de carga** f**)** - [metal-prod]
burden permeability
— f — **lecho** m - [miner] bed permeability
— f — **material** - [metal] material permeability
— f — **muestra** - [electr] sample permeability
— f — **núcleo** m - [magnet] core permeability
— f — **suelo** m - [hidr] soil permeability
— f **elevada** - high permeability
— f **en corriente** f **alterna** - [electr] alternating current permeability
— f — — f **continua** - [electr] direct current
permeability
— f **extrema** - extreme permeability
— f **magnética** - [metal] magnetic permeability
— f — **alta**—[metal] high magnetic permeability
— f — **baja** - [metal] low magnetic permeability
— f **máxima** - maximum permeability
— f — **admisible** - [electr] maximum allowable
permeability
— f **mínima** - minimum permeability
— f — **admisible** - [electr] minimum allowable
permeability

permeabilidad f̲ **reducida** - low permeability
permeable a̲ - . . .; pervious; porous
permeámetro m̲ **absoluto** - [instrum] absolute permeameter
— m̲ **de tipo** m̲ **Metropolitan-Vickers** - [instrum] Metropolitan-Vickers type permeameter
— m̲ **estadounidense** - [instrum] American permeameter
— m̲ **Iliovici** - [instrum] Iliovici permeameter
— m̲ **Metropolitan-Vickers** - [instrum] Metropolitan-Vickers permeameter
— m̲ **N P L** - [instrum] N P L permeameter
— m̲ **Tinsley-Cambridge** - [instrum] Tinsley-Cambridge permeameter
— m̲ **tipo N P L** - [instrum] N P L type permeameter
— m̲ — **Tinsley-Cambridge** - [instrum] Tinsley-Cambridge type permeameter
— m̲ **Vickers** - [instrum] Vickers permeameter
permiano a̲ - [geol] Permian; véase también **pérmico,ca**
pérmico m̲ - [geol] Permian
permisible adv - . . .; allowable
permisión f̲ - . . .; permitting • accomodating; allowing; enabling; letting; making possible
— f̲ **de ajuste** m̲ - adjustment permitting
— f̲ — **aumento** m̲ - increase, permitting, o̲ allowing
— f̲ — **interrogación** f̲ - interrogating,tion, permitting, o̲ allowing
— f̲ — **operación** f̲ - operating,tion, permitting, o̲ allowing
— m̲ — **tiempo** m̲ - time allowance,wing
— f̲ — — m̲ **adecuado** - adequate time allowing
— f̲ — — m̲ **suficiente** - sufficient, o̲ adequate, time allowing
— f̲ — **transferencia** f̲ - [electr] transfer, allowing, o̲ permitting, o̲ permission
— f̲ — **visita** f̲ - visit, permitting, o̲ allowing
— f̲ **para desnegranaje** m̲ - [mec] disengagement, o̲ release, permitting, o̲ allowing
— f̲ — **engranaje** m̲ - [mec] engaging,gement, o̲ allowing
— f̲ — **funcionamiento** - [ind] operating,tion permitting, o̲ allowing
— f̲ — **operación** f̲ - [ind] operating,tion, permitting, o̲ allowing
— m̲ — **penetración** f̲ - penetration, allowing, o̲ permitting
permiso m̲ - . . .; permitting; allowing; letting • approval • véase también **permisión** f̲
— m̲ **cancelado** - [com] cancelled license
— m̲ **consular** - [pol] consular permit,mission
— m̲ **de cambio** m̲ - [pol] exchange permit
— m̲ — **control** m̲ **para cambio(s)** m̲ - [fin] exchange control permit
— m̲ — **oficina** f̲ **para cambio(s)** m̲ - [fin] exchange control permit
— m̲ **escrito** - written permit,mission
— m̲ **especial** - special permit,mission
— m̲ — **para importación** f̲ - [econ] special import(ation) permit,mission
— m̲ **exclusivo**—[legal] exclusive, permission, o̲ right(s)
— m̲ **general**—general, permit,mission, o̲ license
— m̲ **obtenido** - obtained, o̲ gotten, permit
— m̲ **para circulación** f̲ - circulation permission
— m̲ — — f̲ **libre** - free circulation permission
— m̲ — **conducción** f̲ - [autom] driver('s), o̲ operator('s), license
— m̲ — **conducir** v̲ **automóvil(es)** m̲ - [autom] (car) driver('s) license
— m̲ — **desplazamiento** m̲ - [tub] displacement permission
— m̲ — — m̲ **de tubería** f̲ - [tub] pipe displacement permission
— m̲ — **exportación** f̲ - [com] export(ation) permit
— m̲ — — f̲ **cancelado** - [com] cancelled export-(ation) license
— m̲ — **importación** f̲ - [com] import(ation), permit, o̲ license
— m̲ — — f̲ **anulado** - [com] annuled import-(ation) license

permiso m̲ **para importación** f̲ **cancelado** - [com] cancelled import(ation) license
— m̲ — **movimiento** m̲ - movement allowance,wing
— m̲ — **prospección** f̲ - [miner] prospecting permit,mission
— m̲ — **salida** f̲ - [legal] exit permit,mission
— m̲ — **trabajo** m̲ - [labor] work permit
— m̲ **precario** m̲ - tamporary permit • [autom] temporary license
— m̲ — **para conducción** f̲ - [autom] junior, o̲ temporary, operator('s) license
permitido,da a̲ - permitted; allowed; let • permissible; allowable • accomodated; enabled; made possible • provided
— a̲ **marchar** - allowed to, operate, u̲ run
permitir v̲ - . . .; to, let, o̲ make possible • to enable • to, provide, o̲ accomodate, o̲ afford
— v̲ **ajuste** m̲ - to, permit, o̲ allow, @ adjustment
— v̲ **aumento** m̲ - to allow @ increase
— v̲ **circulación** f̲ **libre** - to allow @ free circulation
— v̲ — f̲ — **de aire** m̲ - to allow @ free air circulation
— v̲ **continuación** f̲ to allow to continue
— v̲ **correr sin apresuramiento** m̲ - [deport] to loaf (home)
— v̲ **dar cumplimiento** m̲ - to enable to perform
— v̲ **desagüe** m̲ - [hidr] to, permit, o̲ allow, @ draining
— v̲ **desengranaje** m̲ - [mec] to allow @, disengagement, o̲ release
— v̲ **desplazamiento** m̲ - [mec] to allow @ @ displacement,cing
— m̲ — m̲ **de tubería** f̲ - [tub] to, permit, o̲ allow, @ pipe('s) displacement,cing
— v̲ **detección** f̲ - to, allow, o̲ permit, o̲ enable, @ detection
— v̲ **engranaje** m̲ - [mec] to permit @ engagement
— v̲ **entrega(s)** f̲ **parcial(es)** - [com] to, allow, o̲ permit, partial delivery,vies
— v̲ **escoger** v̲ - to, allow, o̲ permit, o̲ let, choose
— v̲ **funcionamiento** m̲ - [ind] to permit @ operation
— v̲ **funcionar** - to allow to, run, o̲ operate
— v̲ **holgura** f̲ - [mec] to allow @ clearance
— v̲ — f̲ **de aproximadamente** . . . - [mec] to, allow, o̲ permit, @ clerance of, about, o̲ approximately, . . .
— f̲ **interrogación** f̲ - to allow @ interrogation
— v̲ **lujo** m̲ - to afford (@ luxury)
— v̲ **marchar** - to allow to, run, o̲ operate
— v̲ **movimiento** m̲ - to allow @ movement
— v̲ — m̲ **de tubería** f̲ - [tub] to allow @ pipe's movement
— v̲ **operación** f̲ - to allow to, run, o̲ operate
— f̲ **operar** - to allow to, run, o̲ operate
— v̲ **penetración** f̲ - to allow @ penetration
— v̲ — f̲ **de grasa** f̲ - [mec] to allow @ grease penetration
— v̲ **que espartillo** m̲ **crezca debajo de rueda** f̲ - [deport] to let @ sagebrush to grow under @ wheel
— v̲ — **funcione** - to allow to, function, o̲ run, o̲ operate
— v̲ — **grasa** f̲ **penetre** - [mec] to allow @ grease to penetrate
— v̲ **reconocer** - to enable @ detection (of)
— v̲ **tiempo** m̲ - to allow @ time
— v̲ — m̲ **adecuado** - to allow @ adequate time
— v̲ — **suficiente** - to allow sufficient time
— v̲ **transferencia** - to, permit, o̲ allow, @ transfer
— v̲ **visita** f̲ - to, permit, o̲ allow, @ visit
permuta f̲ - . . .; exchanging; switch(ing); swap(ping); change-over • alternate
— f̲ **de automóvil** - [autom] car change,ging
— f̲ — **conductor(es)** m̲ - [autom] driver('s) (ex)change
— f̲ — **neumático(s)** m̲ - [autom-neumát] tire(s), switch(ing), o̲ swap(ping), o̲ exchange,ging
— f̲ — — m̲ **desinflado** - [autom-neumát] flat

tire switch(ing)
permuta f **de neumático(s)** m **para automóvil** m - [autom-neumát] car tire(s) switching
— f — **suspensión(es)** f - [autom-mec-] suspension(s) swap(ping)
permutación f - . . .; switch(ing); **véase también permuta** f
permutado,da a - interchanged; exchanged; swapped • changed over
permutar v - to switch; to swap; to change over
— v **automóvil** m - [autom] to (ex)change @ car
— v **conductor(es)** m - [autom] to change drivers
— v **neumático(s)** m - [autom-neumát] to, swap, o switch, o exchange, @ tire(s)
— v — m **desinflado(s)** - [autom-neumát] to switch @ flat tire(s)
— v **nombre(s)** - to change @ name(s); to rename
— v **suspensión(es)** f - [autom-mec] to swap @ suspension(s)
pernería f - [mec] bolts (and nuts); bolting • bolts, products, o assortment • fastener(s)
— f **floja** - [mec] loose bolts
— f **nueva** - [mec] new bolting
— f **suelta** - [mec] loose bolts,ting
pernil m - [zool] véase **anca** m
perno m - [mec] . . .; rivet • stud • dowel
— m **aceitado** - [mec] oiled, bolt, o rivet
— m **acodado** · [mec] toggle pin
— m **adyacente** - [mec] adjacent bolt
— m **aflojado** - [mec] loose(ned) bolt
— m **ahusado** - [mec] tapered pin
— m **ajustable** - [mec] adjustable, bolt, o pin
— m **ajustado** - [mec] tightened, o torqued, bolt, o pin
— m — **excesivamente** - [mec] overtightened bolt
— m **hasta momento** m **torsional de** . . . - [mec] bolt torqued to . . .
— m — **nuevamente** - [mec] retightened bolt
— m **alargado** - [mec] spreader bolt
— m **apareado** - [mec] matched bolt
— m — **con diámetro** m **de agujero** m - [mec] hole size matched bolt
— m **apretado** - [mec] tight(ened), o torqued, bolt
— m — **en menos** - [mec] undertightened bolt
— m — **excesivamente** - [mec] overtightened bolt
— m **arponado** - [med] rag-bolt
— m **atravesador** - [mec] through bolt
— m — **aflojado** - [mec] loose(ne)d through bolt
— m — **flojo** - [mec] loose through bolt
— m **cadmiado** - [mec] cadmium plated bolt
— m **capuchino** - [mec] cap bolt
— m **carbonatado** - [mec] carbonated bolt
— m — **con resistencia** f **alta** - [mec] high strength carbonated bolt
— m **central** - [mec] center, o central, o middle, bolt
— m **cincado** - [mec] zinc plated bolt
— m **colocado** - [mec] placed, o installed, bolt
— m — **en tresbolillo** - [mec] staggered bolt
— m — **nuevamente** - [mec] reinstalled bolt
— m **común** - [mec] machine bolt
— m — **con cabeza** f **cuadrada** - [mec] square (head) machine bolt
— m — **con gancho** m **y ojo** m - [mec] standard hook-and-eye bolt
— m — — m **y ojo** m **con tuerca** f - [mec] standard hook-and-eye bolt with @ nut
— m — **de acero** m - [mec] steel machine bolt
— m — — **con carbono** m - [mec] carbon steel machine bolt
— m — — — m — — m **con cabeza** f **cuadrada** - [mec] carbon steel square head machine bolt
— m — **para borne(s)** m - [mec] general stud bolt
— m **con aleación** f - [mec] alloy bolt
— m — — f **de cobre** m—[mec] copper alloy bolt
— m — — — f — m **para resistencia** f **alta** - [mec] high tensile copper alloy bolt
— m — **argolla** f - [mec] eye bolt; stirrup
— m — — f **con pasador** m - [mec] eyebolt and key
— m **con brida** f - [tub] flange bolt

perno m **con cabeza** f - [mec] cap bolt
— m — — f **avellanada** - [mec] countersunk head(ed) bolt
— m — — f **con talón** m - [mec] plow bolt
— f — — f **cónica** - [mec] conical head(ed) bolt
— m — — **embebida** - [mec] countersunk conical head(ed) bolt
— n — — f — **fresada** - [mec] countersunk (milled) conical head(ed) bolt
— m — — f **corriente** - [mec] standard head(ed) bolt
— m — — f **cuadrada** - [mec] square head(ed) bolt
— m — — f — **sin desbastar** v - [mec] rough square head(ed) bolt
— m — — f **chata** f - [mec] cheese headed bolt
— m — — f **de hongo y cuello** m **cuadrado** - [mec] carriage bolt
— m — — f **embebida** - [mec] countersunk bolt
— f — — — **con talón** m - [mec] countersunk plow bolt
— m — — f — **y cuello** m **cuadrado** - [mec] countersunk carriage bolt
— m — — f **embutida** - [mec] countersunk, o flush head(ed), bolt
— m — — f **esférica** - [mec] ball bolt
— m — — f **especial** - [mec] special head bolt
— m — — f — **para arado** - [mec] plow bolt
— f — — f **fresada** - [mec] countersunk bolt
— m — — f — **con talón** m - [mec] countersunk plow bolt
— m — — f — **y cuello** m **cuadrado** - [mec] countersunk carriage bolt
— m — — **hexagonal** - [mec] hex(agonal) headed bolt • hex(agonal) headed screw
— f — — f — **para placa** f **lubricadora** - [comb.int] oiler plate hexagonal head bolt
— m — — f — — f — **a bloque** m - [comb.-int] oiler plate to block hexagonal head bolt
— m — — f — — — f — **y (placa** f) **para empuje a bloque** m - [comb.int] oiler plate and thrust plate to block hexagonal head bolt
— m — — f — — — — f **para empuje** m **a bloque** m - [comb.int] thrust plate to block hexagonal head bolt
— m — — f — **tratado térmicamente** - [mec] heat treated hexagon(al) head bolt
— m — — f **movible** - [mec] movable head bolt
— m — — f **perdida** - [mec] countersunk bolt
— m — — f **plana** - [mec] cheese head(ed), o flat head, bolt
— m — — f **ranurada** - stove, o slotted head, bolt
— m — — f — **con chaveta** f - [mec] stove bolt with @ cotter pin
— m — — f **rasa** - [mec] flush head(ed) bolt
— m — — f **redonda** - [mec] round head(ed) bolt
— m — — f — **con talón** n - [mec] round head(ed) plow bolt
— m — — f **y cuello** m **cuadrado** - [mec] square neck carriage bolt; carriage bolt
— m — **cuello** m **cuadrado** - [mec] carriage, o square neck(ed), bolt
— m — — m **ovalado** - [mec] oval neck(ed) bolt
— m — **extremo** m **vivo** - [mec] live end pin
— m — **gancho** m - [mec] hook bolt
— m — — m **con tuerca** f - [mec] hook bolt and nut
— m — — m **galvanizado** - [mec] galvanized hook bolt
— m — — m **y ojo** m **con tuerca** f - [mec] hook and eye bolt and nut
— m — **mango** m **en T largo para elevar sujetador** m **para camisa** f - [herram] long tee handle liner clamp lifting bolt
— m — — m — **para elevar sujetador** m **para camisa** f - [herram] tee handle liner clamp lifting bolt
— m **con muesca** f - [mec] stove bolt
— m — — f **con chaveta** f - [mec] stove bolt with @ cotter pin

perno m con ojal m - [mec] eyebolt
— m — ojo m - [mec] eyebolt; eye bolt
— m — m ajustado - [mec] adjusted, o tightened, eyebolt
— m — m para ajuste m - [mec] adjusting eyebolt
— m — m — m de rueda f - [mec] wheel adjustment eyebolt
— m — m — m — f loca - [mec] idler adjustment eye bolt
— m para extremo m de banda f - [mec] band (dead) end eyebolt
— m — m — m — f ajustado - [mec] adjusted, o tightened, band (dead) end eyebolt
— m — m — m fijo de banda f - [mec] adjusted band dead end eyebolt
— m — n — m — f ajustado - [mec] adjusted band dead end eyebolt
— m — orejeta(s) f - [mec] step bolt
— m — resistencia f alta - [metal-fabr] high-strength bolt
— m — f alta para carga(s) f pesada(s) - [mec] bearing-type high-strength bolt
— m — rosca f - [mec] threaded bolt
— m — f fina - [mec] fine thread bolt
— m — f gruesa - [mec] coarse thread bolt
— m — f hacia derecha f - [mec] right hand thread(ed) bolt
— m — f — izquierda - [mec] left hand thread(ed) bolt
— m — sombrerete m - [mec] cap, bolt, o screw
— m — tapa f - [mec] cap bolt
— m — tornillo m - [mec] screw bolt
— m — tratamiento m térmico - [mec] heat treated bolt
— m — tuerca f - [mec] (nut) bolt
— m cónico - [mec] tapered pin • double pin
— m corriente - [mec] standard bolt • carriage bolt
— m corto - [mec] short bolt • short stud
— m chaveta - [mec] cotter pin • key bolt
— m de acero m - [mec] steel bolt
— m — m con carbono m - [mec] carbon steel bolt
— m — m — m con cabeza f cuadrada - [mec] carbon steel square head(ed) bolt
— m — n — resistencia f alta - [mec] high-strength bolt
— m — m con tratamiento m térmico - [mec] heat treated steel bolt
— m — cadmio m - [mec] cadmium bolt
— m — m plaqueado - [mec] plated cadmium, o cadmium plated, bolt
— m — m inoxidable - [mec] stainless steel bolt
— m — m para junta f estructural - [mec] structural joint steel bolt
— m — bronce m - [mec] bronze bolt
— m — latón m - [mec] brass bolt
— m — m con brida f - [mec] flanged brass bolt
— m — tipo m para carga(s) f pesada(s) - [mec] bearing-type bolt
— m disyuntor m - [electr-instal] breaker bolt
— m desajustado - [mec] loose(ned) bolt
— m elevador - [mec] lifting, o hoisting, bolt
— m — con mango m en T - [mec] tee handle lifting bolt
— m — m — para sujetador m - [mec] tee handle clamp lifting bolt
— m — m — m para camisa f - [mec] tee handle liner clamp lifting bolt
— m — para sujetador m - [mec] clamp lifting bolt
— m — m para camisa f - [mec] liner clamp lifting bolt
— m embridado - [mec] flanged bolt
— m — para tambor m - [mec] drum flanged bolt
— m — m para freno m - [mec] brake drum flange(d) bolt
— m emplomado - [mec] leaded(-in) bolt
— m en T - [mec] T bolt

perno m en U - [mec] U bolt
— m — — cadmiado - [mec] cadmium plated U bolt
— m — — galvanizado - [mec] galvanized cadmium plated U bolt
— m especial - [mec] special bolt
— m esquinero - [mec] corner bolt
— m estructural - [estruct] structural bolt
— m exterior - [mec] outside bolt
— m — para cojinete m - [cojin] outside bearing bolt
— m — m para rotación f - [grúas] outside swing bearing bolt
— m fijador - [mec] holding bolt • post bolt
— m flojo - [mec] loose bolt
— m — en apoyo m - [grúas] bearing loose bolt
— m — m para rotación f - [grúas] swing bearing loose bolt
— m — cojinete m - [mec] bearing loose bolt
— m — m para rotación f - [grúas] swing bearing loose bolt
— m — engranaje m - [mec] gear loose bolt
— m — m para reducción f - [mec] gear reducing loose bolt
— m — m — f para rotación f—[grúas[swing gear reducer loose bolt
— m — verificado - [mec] checked loose bolt
— m galvanizado - [mec] galvanized bolt
— m — para defensa(s) f (laterales) - [vial] guardrail galvanized, o galvanized guardrail, bolt
— m — poste(s) m - [mec] galvanized post, o post galvanized, bolt
— m — con resistencia f alta - [mec] galvanized high strength bolt
— m — por inmersión f - [mec] dip(ped) galvanized bolt
— m — — f en caliente - [mec] hot dip(ped) galvanized bolt
— m herrumbrado - [mec] rusted,ty bolt
— m —, quitado, o removido, o sacado - [mec] removed rusted,ty bolt
— m hexagonal - [mec] hex(agon(al) bolt
— m incluído - [mec] included bolt
— m insertado - [mec] inserted bolt
— m instalado - [mec] installed bolt
— m — en obra f - [constr] field (installed) bolt
— m — nuevamente - [mec] reinstalled bolt
— m interior - [mec] inside bolt
— m — para cojinete m - [mec] inside bearing, o bearing inside, bolt
— m — m para rotación f - [grúas] inside swing bearing, o bearing inside swing, bolt
— m invertido - inverted bolt
— m largo - [mec] long bolt • long stud
— m macho y hembra - [mec] hook and eye bolt
— m maestro - [mec] kingbolt • bolster bolt
— m mecánico - [mec] machine bolt
— m necesario - [mec] necessary, o required, bolt
— m negro - [mec] black bolt
— m — para tubería f - black pipe bolt
— m — — f de planchas f múltiples - [tub] black Multi-Plate, o Multi-Plate black, bolt
— m opuesto - [mec] opposite bolt
— m para acción f rápida - [mec] quick acting bolt
— m — acoplamiento m - [mec] coupling, bolt, o stud
— m — acumulador m - [comb.int] battery bolt
— n — adaptación f - [mec] fit(ting) bolt
— m — f para cuerpo m—[mec] body fit(ting) bolt
— m — t — m para acoplamiento m - [mec] coupling body fit(ting) bolt
— m — ajuste m - [mec] tightening, o adjusting, bolt • torque bolt
— m — m apretado - [mec] tightened hold-down bolt
— m — con cabeza f cuadrada - [mec] square heat set, bolt, o screw
— m — m — f hueca - [mec] socket set.

bolt, o screw
perno m **para ajuste** m **rápido** - quick acting, o speed, bolt
— m — **anclaje** m - [mec] anchor(ing), o hold--down, jack, o bolt • [constr] foundation bolt • structural bolt
— m — — m **colocado** - [mec] placed, o installed, anchor bolt
— m — — **con forma** f **de U** - [mec] U-shaped anchor bolt
— m — — m **de acero** m—[mec] steel anchor bolt
— m — — m — — m **inoxidable** - [mec] stainless steel anchor bolt
— m — — m **bronce** m - [mec] bronze anchor bolt
— m — — m **fijado** - [mec] fastened anchor bolt
— m — — m **galvanizado** - [mec] galvanized anchor bolt
— m — — m **instalado** - [mec] installed anchor bolt
— m — — m **modificado** - [constr] changed, o modified, anchor bolt
— m — — m **para apiladora** f - [mec] piler anchor bolt
— m — — **horno** m - [combust] furnace anchor bolt • [cement] kiln anchor bolt
— m — — m **para precalentador** m - [mec] preheater anchor bolt
— m — — m **recuperador** m - [ind] recuperator anchor bolt
— m — — m **para separador** m - [mec] spacer anchor bolt
— m — — m **removido**—[mec] removed anchor bolt
— m — **apoyo** m **para rotación** f - [grúas] swing bearing bolt
— m — **arado** m - [mec] plow bolt
— m — **asentamiento** m - [mec] fit(ting) bolt
— m — — m **para cuerpo** m—[mec] body fit(ting) bolt
— m — — m **para acoplamiento** m - [mec] coupling body fit(ting) bolt
— m — **base** f - [mec] base bolt
— m — **biela** f - [comb.int] (connecting) rod bolt
— m — **bomba** f - [bombas] pump bolt
— m — — f **hidráulica** - [bombas] hydraulic pump bolt
— m — **boquilla** f - [mec] nozzle bolt
— m — — f **para fiador** m - [mec] toggle nozzle bolt
— m — **brazo** m - [mec] arm bolt
— m — — m **para inclinación** f - [metal-prod] tilting arm bolt
— m — — m — — f **para cañón** m - [metal-prod] (mud, o clay) gun tilting arm bolt
— m — **brida** f - [mec] flange bolt
— m — — f **para tubería** f - [tub] tubing flange nut
— m — — f **hidráulica** - [tub] hydraulic tubing flange bolt
— m — **caballete** m - [grúas] gantry bolt
— m — — m **flojo** - [grúas] gantry loose bolt
— m — **camisa** f - [mec] lining, o sleeve. bolt
— m — **cangilón** m - [mec] bucket, o elevator, bolt
— m — **carcasa** f - [electr-mot] frame bolt
— m — **carrete** m - [mec] spool, o reel, bolt
— m — **carruaje** m - [mec] carriage bolt
— m — **cimentación** f - [constr] foundation bolt
— m — **cimiento** m - [constr] foundation bolt
— m — **coche** m - [mec] carriage bolt
— m — **cojinete** m - [cojin] bearing bolt
— m — — m **exterior** - [mec] outside bearing bolt
— m — — m — **para rotación** f - [grúas] outside swing bearing bolt
— m — — m **interior**—[mec] inside bearing bolt
— m — — m **para rotación** f - [grúas] swing bearing bolt
— m — **conectador** m **para línea** f **(para combustible** m**)** - [comb.int] (fuel) line connector bolt
— m — **conjunto** m - [mec] assembly bolt

perno m **para conjunto** m **para árbol** m - [mec] shaft assembly bolt
— m — **corona** f - [autom-mec] ring gear bolt
— m — — f **dentada** - [autom-mec] ring gear bolt
— m — **cubierta** f - [mec] cover bolt
— m — **cubo** m - [mec] hub bolt
— m — **cuchilla** f - [mec] blade bolt
— m — — f **sujetadora** - [mec] gripper blade, bolt, o rivet
— m — **chaveta** f - [mec] cotter, o safety, (key) bolt; key bolt
— m — **diferencial** m - [autom-mec] differential bolt
— n — m **entre eje(s)** m - [autom-mec] inter--axle differential bolt
— m — **dilatación** f - [mec] expansion bolt
— m — **defensa(s)** f **(laterales para caminos** m**)** - [vial] guardrail bolt
— m — **disyuntor** m—[electr-instal] breaker bolt
— m — **eclisa** f - [f.c.] rail, o track, bolt
— m — **eje** m - [mec] axle bolt • shaft bolt
— m — — m **anterior** - [autom-mec] forward, o front, axle bolt
— m — — m **posterior** - [autom-mec] rear axle bolt
— m — **elevación** f - [mec] lifting, o hoisting, bolt
— m — **émbolo** m - [mec] piston, bolt, o pin
— m — **embutir** v - [mec] countersunk (headed) bolt
— m — **empalme** m - [mec] splice, o junction, bolt
— m — **empuje** m - [mec] thrust bolt
— m — — m **para horquilla** f (toggle) clamp (thrust) bolt
— m — — m — — f **y boquilla** f - [mec] toggle clamp and nozzle (thrust) bolt
— m — **enganche** m - [mec] hook(ing) bolt
— m — **engranaje** m - [mec] gear bolt
— m — — m **para reducción** f - [mec] gear reducer,cing bolt
— m — — m — — f **para rotación** f - [grúas] swing gear reducer,cing bolt
— m — **erección** f - [estruct] erection bolt
— m — **eslabón** m - [mec] link, bolt, o rivet
— m — **expansión** f - [mec] expansion bolt
— m — **extremo** m - [mec] end, bolt, o pin
— m — — m **fijo** - [mec] dead end, bolt, o pin
— m — **fiador** m—[mec] toggle, o fastener, bolt
— m — **fijación** f - [mec] clamp(ing), o lock--(er), o locking, o place, o hold down, o post, bolt
— m — — f **a poste** m - [mec] post bolt
— m — — f **para biela** f - [comb.int] connecting rod place bolt
— m — **fundación** f - [ind] foundation bolt
— m — **horquilla** f - [mec] clevis bolt • [sold] toggle clamp bolt
— m — — f **para empuje** m - [mec] toggle clamp (thrust) bolt
— m — **fijación** f - [mec] place, o hold(ing), bolt
— m — — f **para montaje** m - [mec] mounting, place, o hold(ing), bolt
— m — — f — — m **para parte** f **superior de cojinete** n - [comb.int] bearing cap mounting, place, o hold(ing), bolt
— m — — f — — m — — f — — m **central** - [comb.int] center main bearing cap mounting place bolt
— m — — f — — m — — f — — m **principal** - [comb.int] center main bearing cap mounting place bolt
— m — — f — — m — — f — — m **posterior** - [comb.int] rear bearing cap mounting place bolt
— m — — f — — m — — f — — m **principal** - [comb.int] rear main bearing cap mounting place bolt
— m — — f — — m — — f — — m **principal** - [comb.int] main bearing cap mounting place bolt

perno m para fijación f para montaje m para parte superior de cojinete m principal frontal -
[comb.int] front main bearing cap mounting
place bolt
— m — freno m - [mec] brake bolt
— m — — m para llanta f—[mec] rim brake bolt
— m —, fundación f, o fundamento m - [constr]
foundation bolt
— m — gancho m - [mec] hook pin
— m — — m para retén m - [mec] retainer hook
bolt
— m — — m — — m enjugador - [mec] wiper retainer hook bolt
— m — grillete m - [mec] shackle bolt
— m — horquilla f - [mec] clevis bolt
— m — — f y boquilla f - [extrusión de elec-
trodos] toggle clamp and nozzle bolt
— m — inductor m - [electr-mot] stator bolt
— m — instalación f (en obra f) - [constr]
field (installed) bolt
— m — junta(s) f - [mec] splice, o joint, bolt
— m — — f estructural - [constr] structural
joint bolt
— m — leva f - [mec] cam bolt
— m — línea f - [mec] line bolt
— m — — f para combustible m - [comb.int]
fuel line bolt
— m — llanta f - [autom-mec] rim bolt
— m — máquina f - [mec] véase perno m común
— m — mazo m - [mec] hub bolt
— m — ménsula f - [mec] bracket bolt
— m — metal(es) f - [mec] machine bolt
— m — — m con cabeza f hexagonal - [mec]
hex(agon) head machine bolt
— m — — m — — — f redonda - round head machine bolt
— m — — m — — — f — sin ranurar - [mec] button head machine bolt
— m — — n sin tuerca f - [mec] cap screw
— m — montaje m - mounting, bolt, o stud •
assembly bolt
— m — — m de acero m - [mec] steel assembly
bolt
— m — — m — — m inoxidable - [mec] stainless steel assembly bolt
— m — m para parte f superior de cojinete m
- [comb.int] bearing cap mounting bolt
— m — — m — — — f — — m central - [mec]
center bearing cap mounting bolt
— m — — m — — — f — — m principal -
[comb.int] center main bearing cap mounting
bolt
— m — — m — — — f — — m frontal - [comb-
.int] front bearing cap mounting bolt
— n — — m — — — f — — m principal -
[comb.int] front main bearing cap mounting
bolt
— m — — m — — f — — m posterior -
[comb.int] rear bearing cap mounting bolt
— m — — m — — — f — — m principal -
[comb.int] rear main bearing cap mounting
bolt
— m — — m — — — f — — m principal -
[comb.int] main bearing cap mounting bolt
— m — — m — rejilla f - [mec] grid mounting
bolt
— m — — m — dispositivo m para cambio(s) -
[mec] shifting unit mounting, stud, o bolt
— m — motor m - [mec] motor bolt
— m — — m hidráulico - [mec] hydraulic motor
bolt
— m — parte f superior de cojinete m - [comb.-
int] bearing cap bolt
— m — — f — — — m central - [comb.int]
center,tral bearing cap bolt
— m — — f — — — m principal - [comb.-
int] center main bearing cap bolt
— m — — f — — — m frontal - [comb.int]
front bearing cap bolt
— m — — f — — m principal - [comb.-
int] front main bearing cap bolt
— m — — f — — — m posterior - [comb.int]
rear bearing cap bolt

perno m para parte f superior de cojinete m -
[comb.int] bearing cap bolt
— m — — f — — — m posterior - [comb.int]
rear bearing cap bolt
— m — — f — — — m — principal - [comb.-
int] rear main bearing cap bolt
— m — — f — — — m principal - [comb.int]
main bearing cap bolt
— m — pasador m - [mec] cotter key
— m — peldaño m - [mec] step bolt
— m — pivote m - [mec] kingpin
— m — plancha f - [mec] plate bolt
— m — — f estructural - [mec] structural
plate bolt
— m — portaescobilla(s) m - [mec] brushholder,
bolt, o stud
— m — poste m - [mec] post bolt
— m — precalentador m - [mec] preheater bolt
— m — ranura f - groove bolt • stove bolt
— m — — f con chaveta f - [mec] stove bolt
with @ cotter pin
— m — regulación f [mec] adjusting, o control-
-(ling) bolt
— m — — f con cabeza f - [mec] head(ed) control bolt
— m — — f — — f movible - [mec] movable
head control bolt
— m — rejilla f - [mec] grid (mounting) bolt
— m — retén m - [mec] retainer,ning bolt
— m — — m para carrete m - [mec] spool retainer,ning bolt
— m — retención f - [mec] holding, o lock-
-(ing), pin, o bolt
— m — roldana f - [mec] sheave, bolt, o pin
— m — rueda f - [mec] wheel bolt • lug bolt
— m — — f dentada - [mec] sprocket bolt
— m — seguridad f - [mec] safety bolt • locking pin; shake proof screw • linchpin
— m — sistema m - [mec] system bolt
— m — — m para inclinación f - [mec] tilting,
device, o system, bolt
— m — — m — — — f de cañón m - [metal-prod]
mud, o clay, gun tilting, device, o system,
bolt
— m — soporte m - [mec] support bolt
— m — — m para cojinete m - [mec] bearing
support bolt
— m — — m — — m para bancada f - [mec] main
bearing support bolt
— m — sujeción f - [mec] clamp(ing), o hold-
-down, o lock(ing), o keeper, bolt
— m — — f aflojado - [mec] loosened hold-down
bolt
— m — — f ajustado - [mec] adjusted, o
tightened, hold-down bolt
— m — — f para barra f - [mec] bar lock bolt
— m — — f — — f para tracción f - [mec]
drawbar lock bolt
— m — — f, quitado, o removido, o sacado -
[mec] removed hold-down bolt
— m — — f soltado - [mec] loosened hold-down
bolt
— m — sujetador m - [mec] holder bolt • cam
bolt
— m — — m para acumulador m - [comb.int] battery hold-down bolt
— m — — m — camisa f - [mec] liner clamp
bolt
— m — — m — — f colocado - [mec] installed
liner clamp(ing) bolt
— m — — m — — f instalado - [mec] installed
liner clamp(ing) bolt
— m — — m — — f, quitado, o removido, o
sacado - [mec] removed liner clamp(ing) bolt
— m — sujetar v acumulador m - [comb.int] battery holder,ding bolt
— m — tablero m - [mec] panel bolt
— m — — m para mando m - [sold] control panel
bolt
— m — tambor m - [mec] drum bolt
— m — — m para freno m - [mec] brake drum
bolt
— m — tapa f - [mec] cover bolt

perno m para traba f - [mec] lock(ing) bolt
— m — — f, quitado, o removido, o sacado -
[mec] removed lock(ing) bolt
— m — tracción f - [mec] traction bolt • head
screw • cap screw
— m — f con cabeza f hexagonal - [mec]
hex(agonal) head cap screw
— m — tubería f - [tub] pipe,ping bolt
— m — — f de plancha(s) f múltiple(s) - [tub]
Multi-Plate pipe bolt
— m — volante m - [mec] flywheel bolt •
[autom-mec] driving wheel bolt
— m pasante - [mec] through bolt
— m — aflojado - [mec] loose(ne)d through bolt
— m — flojo - [mec] loose through bolt
— m — para rotor m - [electr-mot] rotor
through bolt
— m — — m aflojado - [electr-mot]
loose,send motor through bolt
— m Penzote* - [mec] véase perno m pivote
— m portaescobilla(s) - [electr-mot] brushholder stud
— m prisionero m - [mec] stud (bolt) • steady
pin
— m que atraviesa v rotor m - [electr-mot]
rotor through bolt
— m — — v — aflojado - [electr-mot] loose-
-(ened) rotor through bolt
— m quitado - [mec] removed bolt
— m real - [mec] king bolt • [autom-mec] véase
pivote m para dirección f
— m recto - [mec] straight bolt
— m — para anclaje m - [mec] straight anchor
bolt
— m reforzado - [mec] reinforced, o heavy, bolt
— m remachado - [mec] riveted bolt; anchor bolt
— m removido - [mec] removed bolt
— m restante - [mec] remaining bolt
— m rompible - [mec] breakable, o shear, pin
— m — para seguridad f - [mec] shear pin
— m roscado - [mec] threaded bolt • dowel screw
— m sacado - [mec] removed bolt
— m semiterminado - [mec] semifinished bolt
— m sin ajustar - [mec] untightened bolt
— m — apretar - [mec] untightened bolt
— m sin cabeza f - [mec] headless bolt; stud
— m — desbastar - [mec] rough bolt
— m — pintar - [mec] unpainted bolt
— m — — para tablero - [mec] unpainted panel
bolt
— m — — — m para mando - [sold] unpainted
control panel bolt
— m — tuerca f - [mec] nutless bolt; cap screw
— m suelto - [mec] loose bolt
— m — para brida f - [mec] flange loose, o
loose flange, bolt
— m — — — f para tubería f - [tub] tubing,
flange loose, o loose flange, bolt
— m — — — f — f hidráulica - [tub] hydraulic tubing flange loose bolt
— m T - [mec] T bolt
— m tratado - [mec] treated bolt
— m — térmicamente - [mec] heat treated bolt
— m U - [mec] U bolt
— m U con ambos extremos roscados—[mec] U bolt
— m vertical - [mec] vertical bolt; kingpin
— m — para charnela f - [mec] knuckle (vertical) pin
— m — — — f para dirección f - [autom-mec]
kingpin; knuckle (vertical) pin
— m vuelto a ajustar v—[mec] retightened bolt
— m — — comprobar - [mec] (re)checked bolt
— m — — verificar - [mec] (re)checked bolt
— m y tuerca f - [mec] bolt and nut
— m — — f para poste(s) m - [mec] post bolt
and nut
pernos m por plancha f - [mec] bolts per plate
— m y tuercas f - [mec] bolts and nuts
pernoctación* f - overnight, rest, o stay
pernoctador* - overnighter*
pernoctar v - . . .; to overnight*
pero conj - . . .; however
peróxido m de hidrógeno m - [quím] (hydrogen)

peroxide; véase también agua m oxigenada
perpendicular v a - perpendicular to
— v — borde m - perpendicular to @ edge
— a — camino m—[vial] perpendicular to @ road
— a — eje m - [mec] perpendicular to @ axis
— a — m de tambor m - [mec] perpendicular
to @ drum axis
perpendicular(es) a entre sí - perpendicular to
each other
perpetuación f de error m - mistake perpetuation
perpetuado,da a - perpetuated
perpetuar v error m - to perpetuate @ mistake
perpiaño,ña - [arq-bóveda] superimposed
perro m caliente - [culin] hot dog
— m errante - [vial] errant, o stray, dog
— m vagabundo - [vial] errant, o stray, dog
persecución f - . . .; pursuing
— f activa - active, pursuing, o persecution •
[deport] active, following, o chasing
— f agresiva - [deport] aggressive, following,
o chasing
— f de cerca - [deport] close, o hot, pursuit
perseguido,da a - persecuted; pursued • [deport]
followed; chased
perseguido,da a - persecuted • pursued •
[deport] followed; chased
— a activamente - persecuted actively; [deport]
pursued, o followed, o chased, actively
— a agresivamente - persecuted actively;
[deport] pursued, o followed, o chased, ag-
gressively
— a de cerca - pursued, o followed, o chased,
closely, o hotly
perseguimiento m - . . . • [deport] pursuing •
[deport] following; chasing
perseguir v - . . . • [deport] to, pursue, o
follow, o chase
— v activamente - to persecute actively •
[deport] to, pursue, o follow, actively
— v agresivamente - to persecute agressively •
[deport] to, persue, o follow, o chase, ag-
gressively
— v de cerca - [deport] to pursue, o follow, o
chase, closely, o hotly; to be hot after
perseverado,da a - persevered
perseverancia f - perseverance,ring
persiana f - [mec] louver • [constr] jalousie;
shutter • [domést] screen
— f de tipo m estacionario - [carp] stationary
type louver
— f en V invertida - [Carp] inverted V louver
— f fija - [constr] fixed louver
— f graduable - [autom-radiador] shutter
— f — para radiador m - [int.comb] radiator
shutter
— f para radiador m - [autom] radiator shutter
— f — ventilador m - [electr-equip] fan louver
— f — — m para estufa f - [metal-prod] stove
fan louver
persistencia f - . . . • [fam] nagging
persistente a - . . .; persisting; continuing;
lingering • [fam] nagging • [deport] hard
charging
persistido,da a - persisted • [fam] nagged
persistir v - . . . • to hang on • to linger •
[fam] to nag
persona f - . . .; party; individual person •
human; one • [fam] guy; gal
— f activa - active person • [fam] live wire
— f afectada - affected, o involved, person
— f bien capacitada - [labor] well trained,
person, o party, o worker
— f buena - good person • pl good people
— f capacitada - [labor] qualified person • pl
qualified people • [artes] talented person •
pl talented people
— f con don m para música - [mús] musically,
talented, o gifted, person; natural musician
— f con llave(s) f - [admin] keyholder
— f — quien tratar v - [com] contact
— f dinámica - dynamic person
— f de empresa f - entrepreneur
— f — — capaz - [admin] capable entrepreneur

persona f empleada - employed, o used person
— f en cargo m directivo - [labor] leadership position person
— f — nivel m superior - [admin] top level person
— f enérgica - energetic person; live wire
— f especial - special, o unusual, person • pl special, o unusual, people
— f especializada - [ind] specialized, o trained, person
— f excepcional - exceptional, o unusual, person • pl exceptional, o unusual, people
— f física - [legal] physical person
— f idónea - [labor] qualified, person, o party
— f, influenciada*, o influida - influenced person
— f jurídica - [legal] juristic, o juridic(al) person • status of incorporation | adv - (con) incorporated
— f — de derecho m - [legal] rightful, o recognized, juristic person
— f — — — m privado - [legal] privately recognized juristic person
— f — extranjera - [legal] foreign juristic person
— f lesionada - [segurid] injured person
— f mayor - of age person • senior citizen
— f moral - moral person • [legal] juristic person
— f natural—[legal] natiral person; individual
— f — extranjera - [legal] foreign, natural person, o individual
— f no autorizada - unauthorized person
— f — zurda - right handed person
— f normal - normal person
— f — capacitada - [labor] normal qualified, o qualified normal, person
— f primera - [gram] first person
— f privada - [legal] private person
— f protegida - [segurid] protected person • pl protected people
— f pública - [legal] public person
— f que comete v infracción f - offending individual
— f — incurre en infracción f - offending individual
— f — emplea mano f derecha - [labor] right handed person
— f — — preferentemente mano f derecha - [labor] right handed person
— f segura de sí misma - self-reliant person
— f seria f - earnest person • dedicated person
— f sin autorización - unauthorized, person(s), o people
— f sin experiencia f - inexperienced, person, o people
— f talentosa - talented person
— f trabajadora - working, person, o people
— f usada - person used
— f zurda - left handed person
personas f - people
personaje m - . . . . • figure
— m científico - scientific figure
personal m - . . .; work force; manpower; employees
— m a capacitar v - [labor] trainee
— m — v para conservación f - [ind] maintenance trainee
— m — — para mantenimiento m eléctrico - [ind] electrical maintenance trainee
— m — — — — m mecánico - [ind] mechanical maintenance trainee
— m — — v para producción f - [ind] operating trainee
— n — instruir v - [labor] trainee
— m — jornal - [labor] hourly, o daily, worker • casual labor
— m — sueldo m - [labor] salaried, personnel, o employee(s)
— m — supervisar - [labor] personnel to be supervised • supervised staff
— m adicional - [labor] additional personnel
— m adiestrado—trained, o developed, personnel

personal m administrativo - [admin] administrative personnel; corporate staff; administration; management (personnel, o staff, o team)
— m — para escritorio m - [admin] administrative office personnel
— m — — oficina f - [admin] administrative office personnel
— m afectado - affected, o involved, personnel
— m agremiado - [labor] bargaining unit employee; labor union personnel
— m aplicado - [labor] dedicated personnel
— m apropiado - [ind] appropriate personnel • task force
— m asegurado - [seguros] insured personnel
— m asignado - [labor] assigned personnel
— m ausente - [labor] absent personnel ; absentee(s)
— m autorizado - [admin] authorized, personnel, o staff
— m auxiliar - [admin] auxiliary personnel; staff
— m buscado - [labor] sought for personnel • found, personnel, o people
— m calificado - [labor] qualified, o trained, personnel
— m — altamente - [labor] highly qualified personnel
— m — malamente - [labor] poorly, qualified, o trained, personnel
— m — para tratamiento m térmico - [metaltrat] qualified heat treating personnel
— m capacitado - qualified, o developed, o trained, personnel, o people, o staff
— m — altamente - [labor] highly, qualified, o trained, personnel
— m — en fábrica f - [ind] factory trained personnel
— m — malamente - [labor] poorly trained personnel
— m — para servicio m - [ind] trained service, o service trained, personnel, o staff
— m — técnicamente - [ind] technically qualified personnel
— m clave - [admin] key personnel
— m — para planta f - [ind] key plant, o plant key, personnel
— m competente - competent, personnel, o staff
— m con capacidad f técnica - [ind] technically qualified personnel
— m experiencia f—[labor] experienced staff
— m costoso - [labor] expensive, personnel, o manpower
— m de asegurado m - [seguros] insured('s) personnel
— m — civil - [pol] civilian personnel
— m — cliente m - [ind] customer('s) personnel
— m — condado m - [pol] county, personnel, o staff, o force
— m — empleado(s) m - [labor] non-production personnel
— m — empresa f - [labor] company, o enterprise, personnel
— f — entidad f - [com] company personnel
— m — fabricante m - [ind] manufacturer('s), personnel, o staff
— m — — de transformador(es) m - [electrequip] transformer manufacturer's personnel
— m — marina f - [milit] navy personnel
— m — técnico(s) m - [labor] technical personnel
— m destacado - [labor] assigned personnel
— m diestro - [labor] skilled, o qualified, personnel, o labor, o help
— m directivo - [labor] directing, o supervisory, personnel, o staff; management; supervision
— m diseñador - design(ing) staff
— m disponible - [labor] available personnel • labor pool
— m docente - [educ] teaching staff; faculty
— m empleado - employed personnel • people used
— m — innecesariamente - [labor] unnecessarily hired people

personal m encontrado - [labor] found personnel
— m escaso - [labor] scarce personnel
— m escogido - chosen, personnel, o people
— m especializado - [ind] specialized, o skilled, o trained, personnel, o labor, o men
— m — mayor - [ind] more skilled labor
— m — menor - [ind] less skilled labor
— m exigido - [labor] required personnel; force requirement
— m — normalmente - [labor] normally required personnel; standard force requirement
— m experimentado - [labor] experienced, personnel, o staff
— m experto - expert personnel
— m extranjero - foreign personnel
— m fuera de convenio m - [labor] exempt, personell, o employee(s)
— m geotécnico - [geol] geotechnical, staff, o personnel
— m hallado - [labor] found personnel
— m hotelero - hotel personnel
— m inexperto - [labor] unskilled labor
— m influenciado,uido - influenced personnel
— m jornalizado - salaried personnel • hourly worker • bargaining unit employee(s)
— m local - [pers] local personnel
— m mal calificado - [labor] poorly trained personnel
— m — capacitado - [labor] poorly trained personnel
— m médico - [medic] medical personnel
— m mensualizado - [labor] (salaried) employees
— m metalúrgico - [metal-prod] metallurgical personnel
— m minero - [miner] mining personnel
— m motivado - [labor] motivated personnel
— m muy calificado - [labor] highly, qualified, o trained, personnel
— m — capacitado - [labor] highly trained personnel
— m nacional - [labor] local, o national, personnel
— m no especializado - [labor] unskilled labor
— m normal - [labor] normal personnel
— m — para mantenimiento m - [labor] normal maintenance personnel
— m — — operación f - [labor] normal operating personnel
— m obrero - [labor] labor force • hourly personnel • [ind] production personnel
— m — productivo - [labor] hourly production personnel
— m ocasional - [labor] occasional, o casual, labor
— m operativo - [labor] operating, personnel, o staff
— m — para planta f - [ind] plant operating, personnel, o staff
— m para administración f - [labor] administration personnel; management; staff • [com] staff management
— m — almacén(es) m - [ind] storekeeping staff
— m — amoladura f - [mec] grinding personnel
— m — — de rodillo(s) m - [metal-lam] (roll) grinding personnel
— m — apoyo m - [labor] support personnel
— m — colada f - [metal-prod] (steel) pouring staff; steel pourer(s)
— m — compra(s) f - purchasing employee
— m — conservación f - [ind] maintenance, man, o person(nel)
— m — — f para división f de energía f [ind] Power Division maintenance personnel
— m — control m de calidad f - [labor] quality control personnel
— m — coquería f - [coque] coke plant personnel
— m — departamento m - [ind] department('s) personnel
— m — — m para laminación f - [metal-lam] rolling department personnel
— m — — m — procesamiento m - [ind] processing department personnel

personal m para departmento m para producción f - [ind] production department personnel
— m — — m — rodillo(s) m - [metal-lam] roll department personnel
— m — elaboración f - [ind] manufacturing, personnel, o staff
— m — entretenimiento m - [ind] maintenance personnel
— m — explotación f - [labor] operating personnel
— m — fiscalización f - [ind] control personnel
— m — — f para calidad f - [ind] quality control personnel
— m — granallado m - [metal-lam] shot blasting personnel
— m — — m para rodillo(s) m - [metal-lam] roll shot blasting personnel
— m — horno m - [ind] furnace personnel
— m — — m eléctrico - [metal-prod] electric furnace personnel
— m — inspección f - [ind] inspection personnel
— m — investigación(es) f - [ind] research, personnel, o staff • investigation personnel
— m — mando m - [labor] supervisory personnel
— m — mantenimiento m - [ind] maintenance personnel
— m — — m eléctrico para capacitar v - [ind] electrical maintenance trainee
— m — metalurgia f - [metal-prod] metallurgical personnel
— m — montaje m - [ind] mounting, o erecting, personnel
— m — oficina f - [labor] office personnel
— m — — f para viaje(s) m - [admin] travel (office) staff
— m — — f técnica - [ind] technical office, personnel, o staff
— m — operación(es) f - [ind] operation(s), personnel, o force, o staff
— m — — f de instalación(es) f - [ind] facility,ties operating personnel
— m — — f — planta f - [ind] plant operating personnel
— m — — f — propietario - [ind] owner('s) operating personnel
— m — planificación f - [admin] planning personnel
— m — planta f - [ind] plant, personnel, o staff
— m — — f para coquización f - [coque] coke plant personnel
— m — producción f - [ind] production personnel • operating, o manufacturing, personnel
— m — — f para división f para energía f - [ind] power division operating personnel
— m — rectificación f - [mec] grinding personnel
— m — — f de rodillo(s) m - [metal-lam] (roll) grinding personnel
— m para rendimiento m (alto) - [ind] (high) performance, personnel, o team
— m — reparación(es) f - [ind] repair(ing) personnel
— m — — f para horno m - [ind] furnace repair personnel
— m — reserva - [labor] labor reserve • stand-by personnel
— m — sección f - [labor] unit, o section, personnel
— m — seguridad f - [segurid] safety personnel
— m — servicio m - [ind] service, personnel, o staff, o manpower • serviceman,men
— m — — m para neumático(s) m - [autom--neumát] tire service personnel
— m — supervisión f - [admin] supervision,sory personnel
— f — — f para planta f - [ind] plant supervising personnel
— m — — f para producción f - [ind] production, o operating, supervisory personnel
— m — — f metalúrgica - [metal-prod] metal-

lurgical supervision personnel
personal m para taller - [ind] shop personnel
— m — **torneado** m - [metal-lam] turning personnel
— m — — m **de rodillo(s)** m - [metal-lam] roll
turning personnel
— m — **venta(s)** f - [com] sales, person(nel), o
people, o force, o staff
— m — — f **a detalle** - [com] retail (sales)
person(nel), o people
— m — — f **a por mayor** - [com] wholesale
(sales), person(nel), o people
— m — — f — **menor** - [com] retail (sales)
person(nel), o people
— m — — f **de fabricante** m - [ind] manufactu-
rer('s) sales, person(nel), o staff, o people
— m — — f — **productor** - [ind] producer('s),
o manufacturer('s), sales, personnel, o staff
— m — — f — **exportación** f - [com] export
sale(s), personnel, o staff, o people
— m — — f **para servicio** m **para neumático(s)** m
- [autom-neumát] tire center personnel
— m — **verificación** f - [ind] control personnel
— m — — f **de calidad** f - [ind] quality control personnel
— m **participante** - [labor] participating per-
son(nel)
— m **planificador** - [admin] planning personnel
— m **pleno** - [labor] full, personnel, o staff |
a - [labor] (con) fully staffed
— m **por hora** f - [labor] hourly worker
— m **presente** - [ind] personnel, present, o in
attendance
— m **productivo** - [labor] production,tive personnel
— m **propio** - [labor] company, o corporate, o
own, personnel
— m **próximo** - [labor] nearby personnel
— m **proyectista** - design(ing), personnel, o
staff
— m **que trabaja en forma** f **insegura** - [labor]
safety problem personnel
— m **según norma** f - [labor] standard staff
— m — — **autorizado** - [labor] standard authorized, personnel, o staff
— m **seleccionado** — [admin] selected, o screened,
personnel, o people, o staff
— m **semiespecializado** - [labor] semi-skilled
labor
— m **semiexperto** - [labor] semi-skilled labor
— m **societario** - [labor] corporate staff
— m **superior** - [admin] management (personnel) •
[labor] supervisory personnel
— m — **para administración** f - [admin] top, o
staff, management • higher supervision
— m **supervisor** - [com] line management; (line)
supervision • supervisory personnel
— m — **para operación(es)** f - [com] line operating management
— m — — **planta** f - [com] plant line supervision,sory personnel
— m — **superior** - [labor] higher supervision
— m **técnico** - [labor] technical, personnel, o
staff, o people, o team
— m **torneador** - [mec] lathe operator(s); turning personnel
— m **transportado** - [labor] transported personnel
— m **uniformado** - [pol-policía] sworn personnel
— m **usado** - [labor] personnel, o people, used
personalidad f - . . . . • mood • [legal] status
— f **distintiva** - distinct(ive) personality
— f **individual** - individual personality
personalmente adv - . . .; first hand
personas f - persons; people • [fam] folks
personería f - [legal] . . .; status
— f **societaria** - [legal] corporate status
personificación f **de cuenta** - [contab] account
personification
personificado,da a - personified
personificar v **cuenta** f - [contab] to personify
@ account
perspectiva f - . . . . • possibility | adv (en)

in prospect; prospective
perspectiva f **analizada** - analyzed perspective
— f **de ingeniería** f - engineering perspective
— f **diferente** - different perspective
— f **distinta** f - different perspective
— f **económica** - economic perspective
— f **futura** - future, perspective, o outlook
— f **minera** - [miner] mining perspective
— f **para fabricación** - [ind] manufacturing perspective; manufacuting possibility
— f — **minería** f - [miner] mining perspective
— f **técnica** - technical perspective
— f — **económica** - technical-economic perspective
perspicaz a - . . .; perceptive
persuadido,da a - persuaded; convinced
— a **a comprar** v - persuaded to buy
persuadir v **a cliente** - to persuade @ customer
persuasión f - persuasion; persuading
— f **a comprar** v - persuading,suasion to buy
— f **de cliente** n - customer persuading,suasion
— f **reforzada** - [labor] stepped-up persuasion
perteneciente a - . . . • pertinent • involved
pertenencia f **personal** - [legal] personal, be-
longing, o possession
pértiga f - [mec] . . .; boom
— f **para trole** m - [electr-equip] trolley pole
pertinaz a - . . .; dogged
pertinente a - . . .; related
— a **a contrato** m - contract pertinent
pertrechar v - . . . • [milit] to arm
pertrecho m - . . . • [milit] supplie(s); muni-
tion(s) • armature
— m **bélico** - [milit] ordnance
perturbación f - . . .; disturbing • [labor]
troublemaking; unsettling; upset(ting)
— f **activa** - [labor] active troublemaking
— f **de red** f **eléctrica** - [electr-oper] electrical system disturbance
— f — **circulación** f - [transp] traffic disturbing,bance
— f — **estructura** f - [constr] structure disturbing,bance
— f — — f **sobre superficie** f - [constr] surface structure disturbing,bance
— f — **superficie** f - surface disturbing,bance
— f — **tránsito** - [transp] traffic disturbing,bance
perturbado,da a - perturbed; (emotionally) upset; disturbed; unsettled
perturbador m - . . . • [labor] troublemaker
— m **activo** - [labor] active troublemaker
perturbador,ra a - . . .; unsettling; upsetting;
troublesome
perturbante a - perturbing; unsettling
perturbar v - . . .; to upset
— v **circulación** f - [transp] to disturb @
traffic
— v **estructura** f - [constr] to disturb @ structure
— v — f **sobre superficie** f - [constr] to disturb @ (on @) surface structure
— v **superficie** f - to distrub @ surface
— v **tránsito** m - [transp] to disturb @ traffic
peruviano,na* a - véase **peruano,na**
pesa f **para contrapeso** m - [mec]counterweight, o
counterbalance, weight
— f — **eje** m - [agric] axle weight
— f — — m **guía** - [agric] guide,ding axle
weight
— f — **eje** m **guía estándar** - [agric] standard
guide axle weight
— f — — m — **motriz** - [gric] power(ed) guide
axle weight
— f — **rueda** f - [mec] wheel weight
— f — — f **motriz** - [mec] drive wheel weight
— m — **unidad** f **motriz** - [agric] power unit
weight
— f **recomendada** - [mec] recommended weight
— v **volante** - [mec] flyweight
— f — **centrífuga** - [mec] centrifugal flyweight
pesada f **de bobina(s)** f - [metal-lam] coil
weighing

pesadez f . . . . • lingering
pesado,da a - . . . • weighed • heavyweight •
[lubric] thick
pesador m - . . . • weight recorder
— m para, cereal(es) m, o grano(s) - [agric]
grain weigher
pesaje m - weighing
pesar v m - . . . | v - . . . to measure by
weight
— v automóvil m - [autom] to weigh @ car
— v más - to weigh more
— v menos - to weigh less
— v plancha f - [mec] to weigh @ plate
— v — f individual - [mec] to weigh @ indivi-
cual plate
— v suelo m - [suelos] to weigh @ soil
— f — m húmedo - [suelos] to weigh @ a wet soil
— f — m seco - [suelos] to weigh @ dry soil
pesca f - . . . . • [deport] game fishing •
[petról] fishing
— f deportiva - [deport] game fish(ing)
— f por arrastre - [pesca] trawling
pescacable(s) m - [petról] rope spear
pescacasquillo(s) m - [petról] (combination)
socket
pescacuchara(s) m - [petról] boot, o latch,
jack, o socket
pescadespojo(s) m - [petról] mouse trap
pescado,da a - fished
— a por arrastre m - [pesca] trawled
pescador m - . . . • [petról] grab
— m con campana f - [petról] socket
— m — accionado con tijera(s) f - [petról]
jar socket
— m — cerrojo m - [petról] latch, o boot, jack
— m — cuello m - [petról] overshot
— m — enchufe m - [petról] overshot
— m — gancho m - [petról] grab
— m — garra f - [petról] latch jack
— m — mordaza f (recuperable) - [petról] releasing overshot
— m — — f excéntrico (recuperable) - [petról]
eccentric releasing overshot
— f — f para tubería f - [petról] pipe
(overshot) socket
— m — — f para entubación f—[petról]
casing (overshot) socket
— m — rosca f - [petról] threaded tap
— m — sopapa f - [petról] mouse trap
— m con trampa f - [petról] mouse trap
— m para tubería f - [petról] pipe catcher
— f — — f perdida - [petról] liner catcher
— f por fricción f - [petról] friction socket
— m — — f corrugado - [petról] corrugated
friction socket
pescaespigas m - [petról] pin socket
pescaherramientas f - [petról] tool catcher
— m abocinado - [petról] horn socket
pescante m - . . . . • [grúas] . . .; crane; (derrick) jib crane; jib; cathead; boom; mast arm
— m colgante - [grúas] underslung jib
— m — cable m contravienta(s) m - [gruas] jib
with @ guy line cable
— m — celosía f - [grúas] lattice jib
— m — pie m derecho m - [grúas] strutted jib
— m — — m — estándar - [grúas] standard
strutted jib
— m — plegamiento m lateral - [grúas] side
folding jib
— m — polea f - [grúas] sheave jib
— m — f única - [grúas] single sheave jib
— m — puntal m - [gruas] strutted jib
— m extraible* - [grúas] pull-out jib
— m — manualmente - [grúas] manual pull-out
jib
— m hidráulico - [grúas] hydraulic jib
— m manual - [grúas] manual jib
— m — único - [grúas] single manual jib
— m para torre f - [grúas] tower jib
— m plegable - [grúas] folding jib
— f — lateralmente - [grúas] side folding jib
— m recto - [grúas] straight jib
— m — transportable - [grúas] movable straight
jib

pescante m recto transportable en costado m (de
grúa) f - [grúas] side stowing straight jib
— m sobre carrillo m - [grúas] trolley boom
— m transportable - [grúas] stowing jib
— m — — costado m (de grúa f) - [grúas] side
stowing jib
— m único - [grúas] single jib
pescar v por arrastre m - [pesca] to trawl
pescasonda(s) m - [petról] socket
— f con enchufe m - [petról] slip socket
— m — media vuelta - [petról] half turn socket
— m hembra - [petról] slip socket
— f por fricción f - [petról] friction socket
— f — — corrugado - [petról] corrugated friction socket
pescatubos* m - [grúas] pipe grab(ber) •
[petról] collar socket
— m combinado - [petról] combination socket
— m universal - [petról] combination socket
pescaunión(es) m - [petról] collar socket
pesimista m - . . . | a - . . .; bear(ish)
pésimo,ma a - . . .; dismal
peso m adherente - adherance,ring weight
— m adicional - additional weight
— m aproximado - approximate, o rough, weight
— m — de estructura f - [constr] structure
approximate weight
— m — de plancha f - [mec] plate approximate,
o approximate plate, weight
— m — — f galvanizada - [mec] galvanized
plate approximate weight
— m — — — f individual - [mec| individual
galvanized plate approximate weight
— m — — f individual - [mec] individual
plate approximate weight
— m — en kilogramo(s) m - [mec] approximate,
kilogram weight, o weight in kilogram(s)
— m — — — m por metro m (linear) - approximate kilogram weight per (linear) meter
— argentino - [fin] Argentine peso
— m bruto - . . . . • [metal-trat] pot yield
— m — de eje m - [mec] axle gross weight
— m — nominal - [mec] nominal gross weight
— m — — de eje m - [autom] gross axle weight
rating
— m cabal - full, o heavy, weight
— m combinado - combined weight
— m computado - computed weight
— m — de carrillo m - [grúas] computed trolley
weight
— m — — puente m - [grúas] computed bridge
weight
— m con accesorio(s) m - weight with accessory,ries
— m empaque m doméstico - [transp] domestic
shipping weight
— m — — m marítimo - [transp] export shipping
weight
— considerable - considerable, o substantial,
weight
— m de aceite m - oil weight
— m — — m recomendado - [mec] recommended oil
weight
— m — acero m - steel weight
— m — — m en perfiles m - [metal-lam] steel
shapes weight
— m — agua m - water weight
— m — aire m - air weight
— m — aleación f - [metal] alloy weight
— m — aparejo m - [grúas] tackle weight
— m — alimentadora f - [mec] feeder weight
— m — — f para alambre m - [sold] wire feeder
weight
— m — aparejo m - [grúas] tackle weight
— m — — m para elevación f - [Grúas] hoisting
tackle weight
— m — automóvil m - [autom] car weight
— m — avión m - [aeron] airplane weight
— m — banda f - [mec] band weight
— m — — f para freno m - [mec] brake band
weight
— m — bobina f - [metal-lam] coil weight
— m — — f media - [metal-lam] average coil
weight

peso m de bola f - [cuerp-moled] ball weight
— m — — f individual - [cuerp.moled] individual ball weight
— m — — f mantenida - [cuerp-moled] maintained ball weight
— m — cable m - [cabl] cable weight
— m — capa f - [mec] layer weight
— m — — f de aleación f - [metal] alloy
layer weight
— m — carga f - [grúas] load weight
— m — — f para elevarse - [grúas] weight of
load to be lifted
— m — carrete m - [cabl] reel weight
— m — carrillo m - [grúas] trolley weight
— m — — m para montacarga(s) m - [metal-prod]
skip car weight
— m — cascarón m - [metal-prod] scab weight
— m — cilindro m - cylinder weight
— m — — m lleno - full cylinder weight
— m — — m vacío - empty cylinder weight
— m — cinc - [metal-trat] spelter weight
— m — — f para freno m - [mec] brake band
weight
— m — columna f - [constr] column weight
— m — — f para relleno m - [constr] fill column weight
— m — conjunto m - [mec] assembly weight
— m — — m de neumático m y rueda f - [autom]
tire-wheel assembly weight
— m — cubeta f - [metal-prod] (skip) car
weight
— m — — f para montacarga(s) m - [metal-prod]
skip car weight
— m — chapa f - [metal-lam] strip weight
— m — elemento m - element('s) weight
— m — — m de aleación f - [metal-prod] alloy
weight
— m — embarque m - [transp] shipping, o shipment, weight
— m — ensambladura f - [mec] assembly weight
— m — estructura f - [constr] structure weight
— m — fabricación f - [ind] fabrication weight
— m — fleje m - [metal-lam] strip weight
— m — fluido m - [mec] fluid weight
— m — — m en neumático m - [mec] tire fluid
weight
— m — grúa f - [grúas] crane weight
— m — — f con carga f - loaded crane weight
— m — — f sin carga f - [grúas] unloaded
crane weight
— m — lata f - tin weight • container weight
— m — lingote m - [metal-prod] ingot weight
— m — — m correspondiente - [metal-prod]
corresponding, o respective, ingot weight
— m — — m respectivo - [metal-prod] respective ingot weight
— m — lote m - [ind] lot weight
— m — material m - material weight • [constr]
backfill weight
— m — — m para relleno m compactado -
[constr] compacted back fill weight
— m — metal - [metal] metal weight
— m — — m para soldadura f - [sold] weld
metal weight
— m — mordaza f - [mec] clamp weight
— m — muestra f - sample weight
— m — neumático m - [autom-neumát] tire weight
— m — operador m - operator('s) weight
— m — — m individual - individual opera-
tor('s) weight
— m — peltre m - [metal] spelter weight
— m — perfil m - [metal-lam] shape weight
— m — — m de acero m - [metal-prod] steel
shape weight
— m — perno m - [mec] bolt weight
— m — pieza f - [mec] part weight
— m — — f más pesada - [mec] heavier,iest
part weight
— m — — f pesada - [mec] havy part weight
— m — placa f - [mec] plate weight
— m — — f para base f - [mec] base plate
weight
— m — plancha f - [metal-lam] plate weight

peso m de plancha f galvanizada - [metal-trat]
galvanized plate weight
— m — — f — individual - [mec] individual
galvanized plate weight
— m — plancha f - [metal-lam] plate weight
— m — — f estructural - [metal-lam] structural plate weight
— m — — f individual - [mec] individual plate
weight
— m — planchón m - [metal-lam] slab weight
— m — poste m - [electr-instal] pole weight
— m — protector m - protector weight
— m — — m para rodillo m - [mec] roller
protector weight
— m — puente m - [puentes] bridge weight
— m — — m para grúa f - [grúas] (crane)
bridge weight
— m — recubrimiento m - [metal-trat] coating
weight
— m — — m cinc m - zinc coating weight
— m — relleno m - [constr] fill weight
— m — revestimiento m - [mec] coating, o lining, weight
— m — — m de cinc m - zinc coating weight
— m — — m equivalente - [metal-trat] equivalent coating weight
— m — — m para cuchara f - [metal-prod] ladle
lining weight
— m — — m total - [metal-trat] total coating
weight
— m — riel(es) m - [f.c.] rail(s) weight
— m — rollo m - [metal-prod] roll, o coil,
weight
— m — — m de varilla f - [trefil] rod coil
weight
— m — sal(es) f fija(s) - [nucl] fixed tail-
ing(s) weight
— m — sección f - [mec] sction weight
— m — — f de plancha f - [mec] plate section
weight
— m — — f — — f estructural - [metal-lam]
structural plate section weight
— m — sistema m - system('s) weight
— m — — m hidráulico—hydraulic system weight
— m — soldadura f - [sold] weld weight
— m — suelo m - [suelos] soil weight
— m — — m en condición f natural - [constr]
undisturbed soil weight
— m — — m húmedo - [suelos] wet soil weight
— m — — m seco - [suelos] dry soil weight
— m — superstructura f - [constr] superstruc-
ture load
— m — tambor m - [mec] drum weight • [cabl]
reel weight
— m — tierra f—[constr] earth, o soil, weight
— m — — f excavada - [constr] excavated earth
weight
— m — tractor m - [mec] tractor('s) weight
— m — — m con accesorio(s) m - [mec] tractor
weight with accessories
— m — — m sin accesorio(s) m - [mec] trac-
tor('s) weight without accssories
— m — tubería f - [tub] pipe, ó tubing, weight
— m — — f corrugada - [tub] corrugated pipe
weight
— m — — f de acero m—[tub] steel pipe weight
— m — — — — m corrugado - [tub] corrugated steel pipe weight
— m — tubo m - [tub] tube('s) weight
— m — tuerca f - [mec] nut('s) weight
— m — vagón m - [f.c.] car weight
— m — vagoneta f - [metal-prod] (skip) car
weight
— m — — f para montacarga(s) m - [metal-prod]
skip car weight
— m decisivo - decisive weight
— m — vehículo m - [mec] vehicle('s) weight
— m dado - given weight
— m determinado - determined weight
— m — exactamente - exacly determined weight
— m en kilogramo(s) m - weight in kilogram(s)
— m — libra(s) - weight in pound(s)
— m — medida(s) f inglesa(s) - [metrіc] En-

glish, o̲ British, (system) weight
peso m̲ enorme—enormous weight • bruising weight
— m̲ equivalente - equivalent weight
— m̲ especificado - specified weight
— m̲ — para fluido m̲ - [mec] specified fluid weight
— m̲ — — — m̲ en neumático m̲ - [mec] tire specified fluid weight
— m̲ específico - specific, weight, o̲ gravity
— m̲ — alto - high specific weight
— m̲ — aparente - [constr] apparent specific weight
— m̲ — bajo - low specific weight
— m̲ — representativo - representative specific weight
— m̲ estimado - estimated weight
— m̲ — de petróleo m̲ - [petról] oil estimated, gravity, o̲ weight
— m̲ — — — m̲ crudo - [petról] crude oil estimated, gravity, o̲ weight
— m̲ — — — m̲ mejorado - [petról] improved oil estimated weight
— m̲ — de avión m̲ - [Aeron] estimated aircraft weight
— m̲ excesivo - excess(ive) weight
— m̲ ficticio - fictitious weight
— m̲ final - final weight
— m̲ individual - individual weight
— m̲ inicial - initial weight
— m̲ ligero - light weight
— m̲ — de pistola f̲ - [sold] gun light weight
— m̲ lineal - linear weight
— m̲ — efectivo - effective linear weight
— m̲ liviano - light weight
— m̲ — de pistola f̲ - [sold] gun light weight
— m̲ mantenido - maintained weight
— m̲ máximo - maximum weight
— m̲ — de bobina f̲ - [metal-lam] maximum coil weight
— m̲ mayor - greater, o̲ heavier, weight
— m̲ medio - average weight
— m̲ — de bobina f̲ - [metal-lam] average coil, o̲ coil average, weight
— m̲ menor - smaller, o̲ lighter, weight
— m̲ — de aguilón m̲ - [grúas] lower boom weight
— m̲ mexicano - [fin] Mexican peso
— m̲ minimizado - minimized weight
— m̲ mínimo - minimum weight
— m̲ — para operación f̲ - minimum operating weight
— m̲ molecular - [quím] molecular weight
— m̲ — alto - [quím] high molecular weight
— m̲ — bajo - [quím] low molecular weight
— m̲ muerto - dead, weight, o̲ load
— m̲ — de artefacto m̲ - [electr-instal] luminaire dead weight
— m̲ — de columna f̲ - [constr-vial] standard dead weight
— m̲ neto - net weight
— m̲ — aproximado - approximate net weight
— m̲ — total - total net weight
— m̲ nominal - nominal weight
— m̲ — bruto - gross net weight
— m̲ — — — para eje m̲ - [autom] gross axle weight rating; G A̲ W R
— m̲ — — — — m̲ delantero - [autom] front axle gross weight rating
— m̲ — — — — m̲ trasero - [autom] rear axle gross weight rating
— m̲ — de perfil m̲ - [metal-lam] shape nominal weight
— m̲ — — tubería f̲ - [tub] pipe, o̲ tube, nominal weight
— m̲ — para eje m̲ - [autom] axle weight rating
— m̲ — por metro m̲ - nominal weight per meter
— m̲ — — pie m̲ - nominal weight per foot
— m̲ normal - normal weight • standard weight
— m̲ para embarque m̲ - [transp] shipping weight
— m̲ — — m̲ para exportación f̲ - [transp] export shipping weight
— m̲ — — m̲ — exterior m̲ - [transp] export shipping weight
— m̲ — — m̲ — interior m̲ - [transp] domestic shipping weight

peso m̲ para fluido m̲ - [mec] fluid weight
peso m̲ — movimiento m̲ - moving weight
— m̲ — manipulación f̲ - handling weight
— m̲ — — f̲ de tubería f̲ - [tub] pipe handling weight
— m̲ — — f̲ — — f̲ corrugada - [tub] corrugated pipe handling weight
— m̲ — — f̲ — — f̲ de acero m̲ - [tub] steel pipe handling weight
— m̲ — — f̲ — — f̲ — — m̲ corrugado - [tub] corrugated steel pipe handling weight
— m̲ — manipuleo* m̲ - véase peso m̲ para manipulación f̲
— m̲ — operación f̲ - operating,tion weight
— m̲ — trabajo m̲ - work(ing) weight
— m̲ — transporte m̲ - [transp] transporation weight
— m̲ pesado - heavy weight | a̲ - heavyweight
— m̲ pluma a̲ - featherweight; lightweight
— m̲ por caballo m̲ de fuerza* - úsese peso m̲ por caballo m̲ de vapor m̲
— m̲ — — m̲ de vapor m̲ - weight per horsepower
— m̲ — metro m̲ - weight per meter
— m̲ — — m̲ de estructura f̲ - [constr] weight per meter of structure
— m̲ — — m̲ lineal - weight per linear meter
— m̲ — milla f̲ - weight per mile
— m̲ — par - weight per pair
— m̲ — pie - weight per foot
— m̲ — rollo m̲ - weight per, roll, o̲ coil • coil weight
— m̲ promedio - average weight
— m̲ — estimado - estimated average weight
— m̲ propio - own weight
— m̲ real - true weight
— m̲ recomendado - recommended weight
— m̲ reducido - reduced weight • light, o̲ low, weight • a̲ - lightweight
— m̲ — de pistola f̲ - [sold] gun light weight
— m̲ relativamente reducido - relatively light weight
— m̲ representativo - representative weight
— m̲ según análisis m̲ - analysis weight
— m̲ — micrografía f̲ - [metal] micrograph weight; weight according to @ micrograph
— m̲ — norma f̲ - [tub] standard weight
— m̲ sin abastecer v̲ - [int.comb] dry weight
— m̲ — accesorios m̲ - weight without accessories
— m̲ — acojinar v̲ - [autom] unsprung weight
— m̲ — carga f̲ - [grúas] unloaded weight; weight without @ load
— m̲, soportado, o̲ sostenido - [mec] supported weight
— m̲ tendiente m̲ - [autom] bias weight
— m̲ — hacia adelante - [autom] forward bias weight
— m̲ total - total weight
— m̲ — de revestimiento m̲ - [metal-trat] coating total weight
— m̲ — inicial - total initial weight
— m̲ — neto - total net weight
— m̲ — para transporte m̲ - [transp] total transportation weight
— m̲ transferido - transferred weight
— m̲ uniforme - uniform, o̲ even, weight
— m̲ unitario - unit weight; weight each
— m̲ — de suelo m̲ - [suelos] soil unit weight
— m̲ — — — m̲ húmedo - [suelos] wet soil unit weight
— m̲ — — — m̲ seco - [suelos] dry soil unit weight
— m̲ — — — m̲ sumergido - [suelos] submerged soil unit weight
— m̲ — — tierra f̲ - [constr] soil, o̲ dirt, unit weight
— m̲ — húmedo - [suelos] unit wet weight
— m̲ — seco - [suelos] unit dry weight
— m̲ — sumergido - [suelos] submerged unit weight
— m̲ variable - variable weight
— m̲ variado - varied weight
— m̲ vertical - vertical weight • [arq-bóveda] downward weight

peso m vivo ~ live weight
— m volumen - volume weight
pesos m y dimensiones f - weights and measure-
(ment)s
— m — medidas f - weights and measurements
pesquero m - . . . | a - fishing
pesquiza f - . . . . test • survey
pestaña f - [anat] . . . • [mec] . . .; rim •
rib; collar • offset • collar • bumper •
cleat • [autom] bead
— f cantonera - [mec] corner flange
— f de acero m - [mec] steel flange
— f — — m fundido - [mec] cast steel flange
— f — bronce m - [mec] bronze flange
— f — carburador m - [comb.int] carburetor
flange
— f — cárter m - [comb.int] crankcase flange
— f — cubo m - [mec] hub flange
— f — — m de engranaje m - [mec] gear hub,
— flange, o rib
— f — fregadero m - [sanit] sink flange
— f — freno m - [mec] brake, flange, o rim
— f — — m para carrete m - [mec] reel brake
rim
— f — — m — — m para núcleo m - [petról]
core reel brake rim
— f — hierro m - [mec] iron, flange, o rim
— f — — m fundido - [mec] cast iron flange
— f — llanta f para freno m - [mec] brake rim
flange
— f — maza f - [mec] hub flange
— f — neumático m - [autom-neumát] (tire) bead
— f — pileta f - [sanit] sink flange
— f — rueda f - [mec-ruedas] wheel flange
— f — — f para vía f - [autom] track wheel
flange
— f — silenciador m—[comb.int] muffler flange
— f — tambor m - [cabl] drum, o reel, chime
— f — — m para freno m - [petról] brake
(drum) rim
— f de taza f - [mec] cup flange • [cojin]
bearing flange
— f delgada - [mec] thin rim
— f derecha - [mec] right flange
— f — para montaje m - [comb.int] right mounting flange
— f — — — m en base f - [comb.int] base
right mounting flange
— f desgastada - [mec-ruedas] worn flange
— f doblada - [autom] bent, flange, o rim
— f doble - [mec-ruedas| double flange
— f en base f - [mec] base flange
— f — — f para montaje m - [mec] base mounting flange
— f — múltiple m - [int.comb] manifold flange
— f esquinera - [mec] corner flange
— f inferior - [mec] lower, o bottom, flange
— f — derecha f - [mec] lower, o bottom, right
flange
— f — izquierda f [mec] lower left flange
— f izquierda para montaje m - [comb.int] left
mounting flange
— f — — — m en base f - [comb.int] base left
mounting flange
— f — — — m — — f para cárter m - [comb.-
int] crankcase base left mounting flange
— f normal - [f.c.-ruedas] normal, o standard,
flange
— f para fricción f - friction flange
— f — neumático m - [autom] tire rim
— f — rueda f - [mec] wheel flange
— f — tapa f - [mec] cover flange
— f — — f delantera—[mec] front cover flange
— f — — f trasera - [mec] rear cover flange
— f rota - [mec] broken flange
— f superior - [mec] top, o upper, flange
— f única - [mec-ruedas] single flange
— f vertical - [mec] vertical flange
— f — sobre talón m - [autom-neumát] vertical
bead flange
pesticida m - [quím] pesticide
pestillo m - [mec] . . .; latch; catch; clamp;
dog • [constr] plunger

pestillo m en lengüeta f - [mec] tongue latch
— m — — f para grúa f - [grúas] crane tongue
latch
— m para asidero m - [mec] handle latch
— m — carrete m - [mec] reel latch
— m — — m para núcleo m - [petról] core reel
latch
— m — cerrojo m - [mec] latch, pin, o plunger
— m — fijación f - [mec] latch(ing) pin
— m — freno m - [mec] brake latch
— m — grúa f - [grúas] crane latch
— m — mesa f - [mec] table latch
— m — — f rotatoria - [petról] rotary latch
— m — palanca f - [mec] lever latch
— m — — f para freno m - [mec] brake lever
latch
— m — — f para mano f—[mec] hand lever latch
— m — soltar v - [mec] release,sing latch
— m — — v lengüeta f - [mec] tongue release
latch
— m — suelta f - [mec] release,sing latch
— m — sujeción f - holding, o release, latch
— m — — f de lengüeta f - [mec] tongue, holding, o release, latch
— m — torno m - [petról] cathead latch
— m por fricción f - [mec] friction latch
— m — para freno m - [mec] brake friction
latch
— m — — f — palanca f - [mec] lever friction
latch
— m — — f — — f para freno m - [mec] brake
lever friction latch
— m pasante m - [constr] sliding latch
— m tipo cerradura f - [mec] lock type latch
petaca f - . . . • [metal-prod] (water) (cooling) plate
— f aislada - [metal-prod] cut off, o isolated,
cooling plate
— f alimentada - [metal-prod] fed cooling plate
— f — independientemente - [metal-prod] independently fed cooling plate
— f condenada - [metal-prod] cut off, o
plugged, cooling plate
— f corta - [metal-prod] short cooling plate
— f de quemador m - [metal-prod] nose
— f para cañón m - [metal-prod] stack (cooling)
plate
— f — cuba f - [metal-prod] stack (cooling)
plate
— f — etalaje m - [metal-prod] bosh cooling
plate
— f — horno m - [metal-prod] furnace cooling
plate
— f — refrigeración f - [metal-prod] cooling
plate
— f — — f para zona f - [metal-prod] area, o
zone, cooling plate
— f — — f — — f de tobera(s) f - [metal-
-prod] tuyere (zone, o area) cooling plate
— f — tragante m - [metal-prod] furnace throat
cooling plate
— f fugada - [metal-prod] leaking cooling plate
— f primaria - [metal-prod] primary cooling
plate
— f quemada - [metal-prod] burned (out) cooling
plate
— f supletoria - [metal-prod] supplemental
cooling plate
— f tragada - [metal-prod] swallowed cooling
plate
petición f - . . . • [legal] applifcation for
— f de oferta f - [com] invitation to bid; bid,
o proposal, request
peticionado,da a - petitioned • applied for
peticionar v - to petition; to apply for
peto m - . . . . • [sold] protective, o safety,
apron
— m protector - [sold] protective, o safety,
apron
petróleo m asfáltico - [petról] asphalt(ic) base
petroleum
— m — parafínico - [petról] paraffin (and)
asphalt petroleum,

petróleo n azul - [petról] blue oil
— m barato - [petról] cheap, o inexpensive, oil
— m británico - [petról] British oil
— m bueno - [petról] good oil
— m compuesto m - [petról] compound(ed) oil
— m con base f de asfalto m - [petról] asphalt base(d) petroleum
— m — f mixta - [petról] paraffin (and) asphalt petroleum
— m — viscosidad f alta - [petról] high viscosity oil
— m — — f baja - [petról] low viscosity oil
— m crudo - [petról] crude oil • heavy oil
— m — dulce - [petról] sweet crude oil
— m — no corrosivo - [petról] non-corrosive, o sweet, crude oil
— m — parafina f - [petról] paraffin oil
— m — f y asfalto m - [petról] paraffin and asphalt, petroleum, o oil
— m destilado - [petról] distilled oil
— m estadounidense - [petról] American oil
— m graso - [petról] fatty oil
— m inglés - [petról] British oil
— m mezclado - [petról] mixed, o compounded, oil
— m motor - [petról] motor oil
— m para asentamiento m de polvo m - [petról] dust, laying, o abating, oil
— m — exportación f - [petról] export oil
— m parafinoso* estadounidense - [petról] American paraffin oil
— m producido - [petról] produced oil
— m — por pozo m - [petról] well produced oil
— m pirolizado - [petról] pyrolized oil
— m refinado - [petról] refined, oil, o petroleum
— m sintético - [petról] synthetic oil
— m soplado - [petról] blown oil
petrolero m - [naut] tanker
— m costa afuera - [petról] off-shore operator
petrolero,ra a - [petról] oil country • véase también petrolífero,ra a
petrolífero,ra a - [petról] petroliferous; oil (bearing); véase también petrolero,ra a
petrolina* f - [petról] petroline*
petrolización* f - [petról] petrolizing,zation
petrolizado,da a - [petról] petrolized
petrolizador,ra a - [petról] petrolizing
petrolizar v - [petról] to petrolize
pez f de carbón m - [miner] coal pitch
— f — hulla f - [miner] coal pitch
— f dura - [petról] hard pitch
— f negra - rosin
pez,ces m para pesca f deportiva - [deport] game fish
pezuña f - . . . .; (cloven) hoof
pH - pH; véase también concentración f (efectiva) de ion(es) m de hidrógeno m
— m de agua m - [hidr] véase concentración (efectiva) de ión(es) de hidrógeno en agua m
picacho m - [topogr] . . .; sharp peak
picado m - véase picadura f
picado,da a - . . . .; chopped
— a ligeramente - mild(ly), o slightly, chopped
— a con oxígeno m - [metal-prod] oxygen lanced
— a superficialmente - [sold] surface pock marked
picador m - . . . . • pickman • [constr] pneumatic drill operator
picadora f - [agric-equip] shredder
— f para ensilaje m - [agric-equip] silage cutter
— f — maíz m - [agric-equip] corn, o maize, shredder
— f — paja f - [agric-equip] straw shredder
picadura f - . . . .; chopping; shredding • [metal-prod] pit(ting) • [metal-lam] pick-up • [sold] pock (marking); pinhole,ling • clipping operation • [mec] chipping off
— f de escoria f - [sold] slag chip(ping)
— f — fibra f de vidrio - m - [plást] fiberglass chopping
— f — piedra f - [miner] stone hewing

picadura f de platino(s) m - [comb.int] point(s) pitting
— f — — m de magneto m - [comb.int] magneto point(s) pitting
— f — válvula f - [comb.int] valve pitting
— f en cojinete m - [mec] bearing pitting
— f — superficie f - [sold] surface pock marking
— f profunda - [metal] deep pitting
— f superficial - [sold] surface hole
picana f - . . .; prod
— f eléctrica - [penal] cattle, goad, o prod
picaporte m - [constr] . . .; latch
— m con resorte m - [constr] spring latch
picar v - . . . • [metal] to chip (away) • [constr] to roughen • [mec] to chip (off)
— v cojinete m - [cojin] to pit @ bearing
— v en demasía f - to chip (off), excessively, o too much
— v escoria f - [sold] to chip @ slag
— v fibra f de vidrio m - [plást] to chip @ fiber glass
— v piedra f - [miner] to hew @ stone
— v platino(s) m - [comb.int] to pit @ point(s)
— m — m de magneto m - [comb.int] to pit @ magneto point(s)
— v válvula f - [comb.int] to pit @ valve
pick-up m - [autom] véase camioneta f • [metal-lam] incorporación f; pick-up m
pico m - . . . . • nozzle; spout; tip • lip • [alambre] nib • [estruct] leader • [geogr] peak; alp; pike • [mec] (grease) fitting
— m como cabado de fundir - [alambre] rough cored, o as cast, nib
— m elevado - [ind] high peak
— m en demanda f - [electr-oper] demand peak
— m — — f elevado - [electr-oper] high demand peak; high peak demand
— m — — f reducido - [electr-oper] low demand peak; low peak demand
— m engrasado m - [mec] greased fitting
— m engrasador - [lubric] grease,sing fitting
— m extremo - end nozzle
— m guiador - [sold] guide,ding tip
— m obstruido - [sold] clogged tip
— m para boquilla f - [sold] nozzle tip
— m — — f para contacto m - [sold] contact nozzle tip
— m — calibrador m - [instrum] gage jaw
— m — conexión f con tubo m para bajada f - [constr] leader head
— m — — f de tubo(s) m para bajada f con desagüe(s) - [constr] stub-up
— m — cono m - [sold] cone tip
— m — — m para fundente m - [sold] flux cone tip
— m — contacto m - [sold] contact tip
— m — — m para boquilla f - [sold] nozzle contact tip
— m — — portador m de corriente f - [sold] current carrying contact tip
— m — corte m - [sold] cutting tip
— m — electrodo m - [sold] electrode tip
— m — — m que apunta hacia abajo - [sold] downwards pointing electrode tip
— m — engrase m - [lubric] (grease) fitting
— m — — Alemite - [mec] Alemite fitting
— m — — m — estándar - [mec] standard Alemite fitting
— m — — m — según norma f - [mec] standard Alemite fitting
— m — — m atendido - [mec] serviced fitting
— m — — m colocado - [mec] installed grease fitting
— m — — m con presión f - [mec] pressure fitting
— m — — m cubierto - [mec] covered grease fitting
— m — — — protector - [mec] guard covered grease fitting
— m — — m de acero m fundido - [mec] cast steel (grease) fitting
— m — — m de bronce m - [mec] bronze (grease)

fitting
— m **para engrase** m **de hierro** m - [mec] iron
(grease) fitting
— m — — — m **fundido** - [mec] cast iron
(grease) fitting
— m — — m **en extremo** m - [end] end grease
fitting
— m — — m — — m **de patín** m - [mec] skid end
(grease) fitting
— m — — m — — — m **para cojinete** m -
[mec] bearing skid end (grease) fitting
— m — — m — — — m — — m **para árbol**
m - [mec] shaft bearing skid end (grease)
fitting
— m — — m — — m — — m — — m **pa-**
ra transmisión f - [mec] transmission shaft
bearing skid end (grease) fitting
— m — — m **estándar** - [mec] standard fitting
— m — — m **expuesto** - [mec] exposed (grease)
fitting
— m — — m **instalado** - [mec] installed grease
fitting
— m — — m **lubricado** - [mec] lubricated grease
fitting
— m — — m **para accionamiento** m **de desplazador**
m - [mec] shifter linkage grease fitting
— m — — m — — m **para dispositivo** m **para**
cambio(s) m - [mec] shifter linkage (grease
fitting
— m — — m **araña** f **para tambor** m - [mec]
drum spider (grease) fitting
— m — — m — — f — — m **para embrague** m -
[mec] clutch drum spider (grease) fitting
— m — — m — — f — — m — — m **inferior** -
[mec] low(er) clutch drum spider (grease)
fitting
— m — — m — — f — — m — — m **superior** -
[mec] high, o upper, clutch drum spider
(grease) fitting
— m — — m **para arpón** m - [petról] spear
(grease) fitting
— m — — m — — m **para agua** m - [petról] water spear (grease) fitting
— m — — m **cabezal** m (**para torno** m) -
petról] catshaft (grease) fitting
— m — — m **caja** f - [mec] box (grease) fitting
— m — — m — — f **para cojinete** m - [cojin]
bearing box (grease) fitting
— m — — m **para carrete** m - [mec] reel
(grease) fitting
— m — — m — — m **para núcleo** m - [petról]
core reel (grease) fitting
— m — — m **cojinete** m - [cojin] bearing
(grease) fitting
— m — — m — — m **para cabezal** m (**para torno**
m) - [petról] catshaft bearing (grease) fitting
— m — — m — — m — — m — — m **para desco-**
nexión f - [petról] breakout catshaft bearing
(grease) fitting
— m — — m — — m — — m — — m **desenros-**
cadura f - [petról] breakout catshaft bearing
(grease) fitting
— m — — m **para conexión** f - [petról] connection (grease) fitting
— m — — m — — f **para agua** m - [petról] water connection (grease) fitting
— m — — m — — f — — m **para embrague** m -
[petról] clutch water connection (grease)
fitting
— m — — m — — f — — m — — m **inferior** -
[petról] low, o bottom, clutch water connection (grease) fitting
— m — — m — — f — — n **superior** -
[petról] high, o upper, clutch water connection (grease) fitting
— m — — m — — f — — m **y aire** m - [mec]
water and air connection (grease) fitting
— m — — m — — f — **aire** m - [mec] air connection (grease) fitting
— m — — m — — f — — m **para embrague** m -

clutch air connection (grease) fitting
pico m **para engrase** m **para conexión** f **para aire**
m **para embrague** m **inferior** - [petról] low, o
bottom, clutch air connection (grease) fitting
— m — — m — — f — — m — — m **superior** -
[petról] high, o upper, clutch air connection
(grease) fitting
— m — — m — — **contraárbol** m - [mec] jackshaft
(grease) fitting
— m — — m — — m — — **mesa** f (**rotatoria**) -
[petról] rotary (table) countershaft (grease)
fitting
— m — — m — — m **rotatorio** - [petról] rotary
countershaft (grease) fitting
— m — — m — — **contraeje** m - [mec] countershaft, o jackshaft, (grease) fitting
— m — — m — — m **inferior** - [mec] low, o
bottom, jackshaft (grease) fitting
— m — — m — — m **para mesa** f **rotatoria** -
[petról] rotary (table) countershaft (grease)
fitting
— m — — m — — m **rotatorio** - [mec] rotary
(table) countershaft (grease) fitting
— m — — m — — m **superior** - [mec] upper, o
top, o high, jackshaft (grease) fitting
— m — — m — — **cubierta** f - [mec] cover plate
(grease) fitting
— m — — m — — f **para caja** f - [mec] box
cover plate (grease) fitting
— m — — m — — f **para cojinete** m -
[mec] bearing box cover (grease) fitting
— m — — m **para desplazador** m - [mec] shifter
(grease) fitting
— m — — m — — m **para freno** m - [mec] brake
shifter (grease) fitting
— m — — m — — m — — m **auxiliar** - [mec]
auxiliary brake shifter (grease) fitting
— m — — m — **dispositivo** m **para cambio(s)** m -
[mec] shifter (grease) fitting
— m — — m — — m **para freno** m—[mec]
brake shifter (grease) fitting
— m — — m — — m — **cambio(s) para freno** m
auxiliar - [mec] auxiliary brake shifter
(grease) fitting
— m — — m — **embrague** m - [mec] clutch
(grease) fitting
— m — — m — — m **inferior** - [mec] low, o
bottom, clutch (grease) fitting
— m — — m — — m **superior** - [mec] high, o
top, o upper, clutch (grease) fitting
— m — — m — **eslabonamiento** m - [mec] linkage (grease) fitting
— m — — m — — m **para desplazador** m - [mec]
shifter linkage (grease) fitting
— m — — m — — m — **dispositivo** m **para cam-**
bio(s) m - [mec] shifter linkage (grease)
fitting
— m — — m — **extremo** m - [mec] end (grease)
fitting
— m — — m — — m **de contraárbol** m - [mec]
jackshaft end (grease) fitting
— m — — m — **contraeje** m - [mec]
jackshaft end (grease) fitting
— m — — m — **freno** m - [mec] brake (grease)
fitting
— m — — m — — m **para carrete** m - [petról]
reel brake (grease) fitting
— m — — m **para núcleo** m -
[petról] core reel brake (grease) fitting
— m — — m **para frente** m - [mec] front (side)
(grease) fitting
— m — — m — — m **para soporte** m - [mec]
support front (side) (grease) fitting
— m — — m — — m — — m **para desplazador** m
- [mec] shifter support front side (grease)
fitting
— m — — m — — m — — m — **dispositivo** m
para cambio(s) m - [mec] shifter support
front side (grease) fitting
— m — — m **para lanza** f - [petról] spear
(grease) fitting

pico m para engrase m para lanza f - [petról]
spear· (grease) fitting
— m — — m — — f para agua m - [petról] water spear (grease) fitting
— m — — m para malacate m - _petról] drawworks (grease) fitting
— m — — m — — m lubricado - [petról] lubricated drawworks (grease) fitting
— m — — m para palanca f - [mec] handle, o
lever, (grease) fitting
— m — — — f para trinquete m - [mec]
pawl, handle, o lever, (grease) fitting
— m — — m — — f — — m para traba f—[mec]
lock pawl, handle, o lever, (grease) fitting
— m — — m para patín m - [mec] skid (grease)
fitting
— m — — m — — m para árbol m - [mec] shaft
skid (grease) fitting
— m — — f para pestaña f - [mec] rim (grease)
fitting
— m — — m — — f para freno m - [mec] brake
rim (grease) fitting
— m — — m — — f — — m para carrete m -
[mec] reel brake rim (grease) fitting
— m — — m — — f — — m para núcleo
m - [petról] core reel brake rim (grease)
fitting
— m — — m — placa f - [mec] plate (grease)
fitting
— m — — m — — f para cubierta f - [mec]
cover plate (grease) fitting
— m — — m — — f — — f para caja f - [mec]
box cover plate (grease) fitting
— m — — m — — f — — f — — f para cojinete m - [mec] bearing box cover plate
(grease) fitting
— m — — m — reserva f - [mec] relief. Alemite, o (grease) fitting
— m — — m para respaldo m - [mec] back side
(grease) fitting
— m — — m — — m para soporte m - [mec] support back side (grease) fitting
— m — — m — — m — — m para desplazador m
- [mec] shifter support back side (grease)
fitting
— m — — m — — m — — m — dispositivo m
para cambio(s) m - [mec] shifter support back
side (grease) fitting
— m — — m — retén m—[mec] retainer (grease)
fitting
— m — — m — — m (para) enjugador m - [mec]
wiper retainer (grease(fitting
— m — — m — rodillo m - [mec] roll(er)
(grease) fitting
— m — — m — — m para cable m - [grúas] line
roller (grease) fitting
— m — — m — — m — — m de alambre m -
[grúas] wire line roller (grease) fitting
— m — — m — — m — — m para perforación f
- [petról] wire line roller (grease) fitting
— m — — m — soporte m - [mec] support
(grease) fitting
— m — — m — — m — desplazador m - [mec]
shifter support (grease) fitting
— m — — m — — m — — dispositivo m para cam-
bio(s) m - [mec] shifter support (grease)
fitting
— m — — m — torno m - [petról] cathead
(grease) fitting
— m — — m — — m para desconexión f -
[petról] breakout cathead (grease) fitting
— m — — m — — m — desenroscadura f -
[petról] breakout cathead (grease) fitting
— m — — m — — m para enroscadura f -
[petról] make-up cathead (grease) fitting
— m — — m — — m para reserva f - [petról]
reserve cathead (grease) fitting
— m — — m — trinquete m—[mec] pawl (grease)
fitting
— m — — m — — m para traba f - [mec] lock
pawl (grease) fitting
— m — — m — tubería f - [tub] pipe (grease)
fitting

pico m para engrase m para tubería f para agua
m - [petról] wash pipe (grease) fitting
— m — — m para viga f - [mec] beam (grease)
fitting
— f — — m — — f compensadora - [mec]
equalizer beam (grease) fitting
— m — — m — — f — para freno m - [mec]
brake equalizer beam (grease) fitting
— m — — m (puesto) a descubierto - [mec] exposed (grease) fitting
— m — — m, quitado, o removido - [mec] removed (grease) fitting
— m — — m roscado - [mec] screwed, o threaded, (grease) fitting
— m — — m de acero m fundido - [mec] cast
steel, screwed, o threaded, (grease) fitting
— m — — m sacado - [mec] removed (grease)
fitting
— m — — m según norma f standard, nib, o tip
— m — — m tipo m (con) botón m - [mec] button
head (type) (grease) fitting
— m — — m verificado - [mec] checked (grease)
fitting
— m — enlechado m - [constr] grouting, socket,
o plug, o hole
— m — — f guía f - [sold] guide,ding tip
— m — — f aislado - [sold] insulated guide
tip
— m — — f con largo(r) m intermedio - [sold]
medium length guide,ding tip
— m — — m corto - [sold] short guide,ding tip
— m — — m fino - [sold] small guide,ding tip
— m — — m largo - [sold] long guide,ding tip
— m — — m — para electrodo m electrizado -
[sold] long electrical stickout guide,ding
tip
— m — — m mediano - [sold] medium guide tip
— m — — m — para electrodo m electrizado -
[sold] electrical stickout medium guide tip
— m — — m para electrodo m electrizado -
[sold] electrical stickout guide tip
— m — guiadera f - véase pico m para guía
— m — introducción f de lechada f - [constr]
grout(ing), socket, o plug
— m — matriz f para trefilería f - [alambre]
wiredrawing die nib
— m — pistola f - [sold] gun tip
— m — — f obstruido - [sold] clogged gun tip
— m — — f — con salpicadura(s) f - [sold]
spatter clogged gun tip
— m — reserva f - [mec] relief fitting
— m — — f para engrase m - [mec] relief,
Alemite, o grease, fitting
— m — trefilería f - [alambre] wiredrawing nib
— m — — f según norma f - [alambre] standard
nib
— m — — f — — f sin rectificar v—[alambre]
rough standard (wiredrawing) nib
— m — — f — — f sin rectificar para diáme-
tro(s) m según norma f - [alambre] standard
nib rough cored to standard size
— m — — f sin montaje m - [alambre] nib without casing
— m — — f — rectificar - [alambre] rough
drawing nib
— m — tubo m - [tub] tube tip
— m — — m para fundente m - [sold] flux tube
tip
— m redondeado - [topogr] rounded peak
— m reducido - [ind] low peak
— m rociador - [hidr] spray(ing), tip, o nozzle
— m — con presión f - [hidr] pressure spray
tip
— m roscado - [hormig] threaded plug
— m — para enlechado m - [hormig] threaded
grouting plug
— m según norma f - [alambre] standard nib
— m — — f sin rectificar - [alambre] rough
standard nib
— m — montaje m - [alambre] nib without casing
— m — rectificar - [alambre] rough (cored) nib
— m mantenido - [ind] sustained peak
— m tal como acabado de fundir - [alambre]

rough cored nib
picofaradio m - [electrón] picofarad
picoso,sa a - . . . • [sold] pock marked
— a en superficie f—[sold] surface pock marked
pie n - . . . • [mec] leg • boot • stand • jack
• dedendum • [constr] footing; foothold
— m cuadrado - [metric] square foot
— m — de tabla f - [madera] board foot
— m cúbico - [metric] cubic foot
— m de aislador m - [electr-instal] insulator boot
— m — biela f - [mec] connecting rod, bottom, o small end
— m — bóveda f - [arquit] arch footing
— m — diente m - [mec] tooth, toe, o root
— m — falda f - [mec] skirt bottom
— m — f para émbolo m - [comb.int] piston skirt bottom
— m — hormigón m - [constr] concrete base
— m — — m colado - [constr] poured concrete base
— m — — m para, bordillo m, o cordón m, o encintado m - [constr] poured concrete curb base
— m — imprenta f - [imprent] dateline
— m — monte m - [topogr] piedmont
— m — muro m - [constr] (wall) toe
— m — obra f - work, site, o yard
— m — operador m - [ind] operator('s) foot
— m — pendiente f - [topogr] slope, o hill, toe, o bottom
— m — terraplén m - [constr] embankment toe
— m — talud m - [topogr] slope bottom • [constr] slope toe
— m — — de terraplén m - [constr] embankment slope toe
— m delantero - front foot
— m derecho m - [anat] right foot • [constr] stud; strut • standard • [herram-escalera] rail • [mec] post; mast
— m — de operador m - operator's right foot
— m — esquinero - [mec] corner (upright) post
— m — estándar - [mec] standard strut
— m — para escalera f - [herram] ladder rail
— m — grúa f - [grúas] crane post
— m — — f para torre f - [grúas] derrick crane post
— m — — f — f para perforación f - [petról] derrick crane post
— m — torre f - [petról] derrick leg
— m — — f para perforación f - [petról] derrick leg
— m — proyectado recientemente - [mec] recently, o newly, engineered mast
— m — según norma f - [mec] standard strut
— m igual - equal footing
— m inseguro - [segurid] dangerous footing
— m izquierdo - [anat] left foot
— m — de operador m - [ind] operator('s) left foot
—libra m - [metric] foot pound
— m libre - free foot
— m lubricado - lubricated foot
— m para aguilón m - [grúas] boom foot
— m — — m lubricado - [grúas] lubricated boom foot
— m — bordillo m - [constr] curb base
— m — — m de hormigón m - [constr] concrete curb base
— m — botalón m - [náut] boom foot
— m — encintado m - [constr] curb base
— m — — m de hormigón m - [constr] concrete curb base
— m — exhibición f - [com] stand
— m — — f de neumático(s) m - [autom-neumát] tire stand
— m — — f — — para camión(es) m - [autom--neumát] truck tire stand
— m — — f — m para vehículo(s) m para pasajero(s) m - [autom-neumát] passenger tire stand
— m — neumático(s) m - [autom-neumát] tire stand

pie m para neumático(s) m para camión(es) m - [autom-neumát] truck tire stand
— m recto - [constr] upright
— m seguro - safe footing
— m trasero - rear foot
piedra f - . . .; rock • [meteorol] hail
— f amoladera f - [mec] grindstone
— f arenisca - [geol] sandstone
— f — resistente a calor m - [geol] firestone
— f — — a fuego - [geol] firestone
— f asentadora - [herram] honing, o finishing, stone
— f — para escobilla(s) f - [electr-equip] brush stone
— f aserrada - [constr] sawn,wed stone
— f calcárea - [geol] limestone
— f caliza - [geol] limestone
— f — asfáltica - [geol] asphalt(ic) limestone
— f — calcinada - [miner] burned limestone
— f — consumida - [metal-prod] limestone consumed
— f — — en exceso m - [metal-prod] excess lime(stone) consumed
— f canchada* - [constr] crushed stone
— f colocada - [constr] laid, o placed, stone
— f — con mano f - [constr] hand, laid, o placed, stone
— f cortada - [constr] cut stone
— f cribada - [constr] screened, stone, o rock
— f de esmeril - [herram] emery wheel; grindstone
— f — voladura f - [miner] shot rock
— f en seco - [constr] drywall
— f enlechada - [constr] grouted stone
— f extraída - [miner] quarried, o extracted, stone
— f floja - loose stone
— f granítica - [geol] granite stone • [sanit] stone granite
— f labrada - [miner] hewn,wed stone
— f encachada - [constr] crushed stone
— f natural - [constr] natural stone
— f para afilar - [herram] grindstone; grinding stone; hone; honing stone
— f — — v sierra(s) f - [herram] swa grinding stone
— f — amolar v - [herram] grindstone • [mec] glazer
— f — asentar v - [herram] honing stone; hone
— f — — f escobilla(s) f - [electr-equip] brush stone
— f — colector(es) m - [electr-equip] commutator stone
— f — cortar,te - [herram] cutting stone
— f — escobilla(s) f - [electr-equip] commutator, o brush, stone
— f — — f con granulometría f de . . . • [electr-mot] . . . grit commutator stone
— f — muestra f - specimen, o sample, stone
— f — rectificar v - [electr-mot] commutator stone
— f — — v conmutadores m - [electr-mot] commutator stone
— f partida - [miner] crushed, stone, o rock
— f picada - [miner] hewn, o broken, o crushed, stone
— f pómez - [miner] pumice (stone, o material)
— f — pulverizada - [miner] powdered pumice stone
— f pulidora - [herram] polishing stone
— f — para escobilla(s) f - [electr-mot] commutator stone
— f rectificadora - [electr-mot] commutator stone
— f terminadora - [herram] finishing stone
— f triturada - [constr] crushed stone
— f — (bien) clasificada - [constr] (well) graded crushed stone
— f — con tamaño m uniforme - [constr] uniformly graded crushed stone
— f — cribada - [constr] screened crushed, stone, o rock
— f volcada - [constr] dumped, stone, o rock

piel f de lingote m - [metal-prod] ingot, o
chill, skin
— f delgada - thin skin
— f descubierta - [anatom] bare skin
— f desnuda - [anatom] bare skin
— f expuesta - [medic] exposed skin
— f gruesa - thick skin
— f lesionada - [medic] injured skin
— f más delgada - thinner skin
— f — gruesa - thicker skin
— f pálida - pale, skin, o complexion
— f protectora - [anatom] protective skin
— f protegida - [segurid] protected skin
pierna f cuadrada - [mec] square leg • [clavos]
square shank
— f redonda - [mec] round leg • [clavos] round
shank
pies m cúbicos por minuto m - cubic feet per
minute; C F M
— m — — segundo m - cubic feet per second;
C F S
— m — según norma f por minuto - standard cubic feet per minute; S C F M
— m — — — f — segundo m - standard cubic
feet per second; S C F S
— m — soldadura/hora - [sold] feet of weld
per hour
— m exactos - exact feet
— m por minuto m - feet per minute
— m — segundo m - feet per second
— m — tonelada f (de 2000 libras f) - feet
per ton
pieza f - . . . . • [mec] . . .; spare part • component • casting • fitting • unit • repair
part • [sold] work • [refract] shape •
[constr] . . .; chamber
— f accesoria - accessory part; appurtenance
— f — expuesta - exposed accessory (part)
— f acodada - [mec] dog leg
— f activada - [electr-oper] hot part
— f — eléctricamente - [electr-oper] electrically hot part
— f adyacente - [mec] adjacent, o adjoining,
part
— f agrietada - [mec] cracked, part, o member
— f aisladora - [mec] insulating, part, o pad
— f aislante - [electr-instal] insulating pad
— f alargada - [mec] elongated, o lengthened,
part
— f angular - [mec] angle • bracket
— f — de acero m - [mec] steel angle,gular
part
— f — galvanizada - [mec] galvanized angle
— f — — para aumentar rigidez f - [mec] galvanized stiffener,ning angle
— f — optativa - [mec] optional angle
— f — para montaje m - [mec] mounting bracket
— f — — refuerzo - [mec] angle reinforcement
— f — — sostén m - [mec] support angle
— f — pesada - [constr] heavy angle
— f — de acero m - [constr] heavy steel
angle
— f anticuada - [mec] obsolete part
— f apropiada - [mec] appropriate, o proper,
part
— f armada - [mec] assembled part • reinforced
part
— f auténtica - authentic, o genuine, part
— f — para reemplazo m - [mec] authentic, o
genuine, replacement part
— f autorizada - [ind] authorized part
— f — por fábrica - [ind] factory authorized
part
— f auxiliar - [mec] auxiliary part; accessory
— f averiada - [mec] damaged part
— f basta - [mec] blank
— f — para barreno m - [petról] drill blank
— f — herramienta f - [herram] tool blank
— f — — hilera f - [mec] die blank
— f — — matriz f - [mec] die blank
— f bruta - [mec] blank
— f — para herramienta f - [mec] tool blank
— f — — hilera f - [mec] die blank

pieza f bruta para matriz f - [mec] die blank
— f burda - [mec] rough part • blank
— f — para herramienta(s) f - [herram] tool
blank
— f — — hilera f - [mec] die blank
— f — — matriz f - [mec] die blank
— f buscada - [mec] needed, o sought, part
— f caliente - [mec] hot part
— f — para motor m - [electr-mot] hot motor
part • [comb.int] hot engine part
— f central - [mec] center,tral part
— f circular - [mec] round part
— f clave - [mec] key part
— f colada - [metal-prod] casting; cast part
— f — de hierro m - [metal-prod] iron casting
— f — — m gris - [metal-prod] gray iron
casting
— f — en coquilla f - [metal-prod] chill(ed)
iron casting
— f colocada - [mec] installed part
— f combada - [mec] warped, part, o piece
— f completa - [mec] complete part
— f completamente distinta - [mec] completely
different part
— f componente - [mec] component (part)
— f — de tipo m enchufable - [mec] plug-type
component
— f — — — m insertable - [mec] plug-type
component
— f — enchufable - [mec] plug-type component
— f — insertable - [mec] plug-in component
— f — interior - [mec] internal component
— f — original - [mec] original component
— f — — para eje m - [autom-mec] original
axle component
— f — para accionamiento m - [mec] drive,ving
component
— m — — cambio m entre eje(s) m - [autom-mec]
axle shifting component
— f — para eje m - [autom-mec] axle component
— f — — embrague m - [mec] clutch component
— f — — mecanismo m para cambio(s) entre ejes
m - [autom-mec] axle shift(ing) component
— f — — sistema m para cambio(s) - [autom-
-mec] shift system component
— f — suspensión f - [autom] suspension
component
— f — típica - [mec] typical component
— f compuesta f - [mec] built-up member
— f con calidad f - [ind] quality part
— f — carga f - [mec] loaded, part, o section
— f — corriente f - [electr-oper] (electrically) hot part
— f — dureza f (muy) elevada - [ind] (very)
high, o high-durometer, part
— f — inserción f metálica - [mec] metal insert component
— f — movimiento m - [mec] moving part
— f — — m alternativo - [mec] reciprocating
part
— f — — m crítica - [mec] critical moving
part
— f — — m inspeccionada - [mec] inspected
moving part
— f — — recíproco - [mec] reciprocating part
— f — pestaña f - [mec] flanged, part, o fitting
— f — — de hierro m (fundido) - [metal-prod]
(cast) iron flanged fitting
— f — prefijo m . . . - [mec] . . . prefix
part
— f — proyección f especial - [ind] specially
designed part
— f — refuerzo m - [mec] reinforced part
— f — — m metálico - [mec] metal reinforced,cement component
— f — tratamiento m - [mec] treated part
— f — — m térmico - [mec] heat treated part
— f cóncava - [mec] concave (part)
— f — bajada - [mec] lowered concave (part)
— f — elevada - [mec] raised concave (part)
— f conecta(da) - [mec] connected, o related,
part

pieza f conectadora - [mec] connecting part •
  connecting, strip, o channel • lug
— f — en U - [mec] connecting channel
— f — esquinera - [mec] corner connecting part
— f conexa - [mec] related part
— f — instalada - [mec] installed related part
— f — reemplazada—[mec] replaced related part
— f — verificada - [mec] checked related part
— f constitutiva - [mec] component (part)
— f — con transistor(es) m - [electrón] solid
  state (electronic) component
— f — débil - weak component (part)
— f — grande - [mec] large component
— f — incluida - [mec] enclosed, o included,
  component (part)
— f — interior - [mec] internal component
— f — más débil - weaker,kest component
— f — fuerte - stronger,gest component
— f — para malacate m - [petról] drawworks
  component
— f — soldadora f - [sold] welder component
— f corriente - [mec] current part • [cerám]
  standard shape; equivalent
— f corroída - [mec] corroded, part, o member
— f cortada - [mec] cut part • [metal-lam] cut
  length • [constr] sawed stone
— f crítica - [mec] critical part
— f curva(da) - [mec] curved part
— f — de madera f (para exterior de malacate
  m) - [petról] cant
— f dañada - [mec] damaged, part, o piece
— f de acero m - [mec] steel, part, o component, o member • [constr] steel
— f — — m alto en carbono m - [mec] high carbon steel part
— f — — m colado - [metal-prod] steel casting
— f — — m con calibre m delgado - [metal-lam]
  light gage steel, part, o member
— f — — m con carbono m - [mec] carbon steel
  part
— f — — m carbono m alto - [mec] high carbon steel part
— f — — m — — m mediano - [mec] medium carbon steel part
— f — — m manganeso m - [mec] manganese
  steel part
— f — — m — m (des)gastada - [sold] worn
  manganese steel part
— f — — m tenor m alto de carbono m -
  [sold] high carbon steel part
— f — m — m mediano de carbono m -
  [sold] medium carbon steel part
— f — — m gastada - [sold] worn steel part
— f — — m estructural - [metal-lam] structural member
— f — — m — con calibre m delgado - [metal-
  -lam] light gage steel structural member
— f — — m — conformada en frío m - [metal-
  lam] cold formed, steel structural, o structural steel, member
— f — — m fundida—[metal-prod] steel casting
— f — — m intemperizado - [constr] weathering
  steel, part, o member
— f — — m manganésico - [sold] manganese
  steel part
— f — — m moldeado - [metal-prod] molded
  steel part; steel casting
— f — m para conexión f - [tub] steel fitting
— f — m — extremo m - [constr] steel end
  part
— f — m soldada - [sold] welded steel part;
  steel weldment
— f — aleación f - [metal-prod] alloy casting
— f — — f porosa - [metal-prod] porous alloy
  casting
— f — — f refractaria - [mec] refractory alloy part
— f — aluminio m - [mec] aluminum, part, o
  component
— f — bronce m - [mec] bronze part
— f — caucho m - [mec] rubber part
— f — — m (muy) duro - [plást] very hard, o

high-durometer, rubber part
pieza f de caucho m para montaje m - [mec] rubber mounting
— f — — m — — m de chasis m - [mec] rubber
  chassis mounting
— f — cobre m - [mec] copper part
— f — — m soldada - [sold] welded copper part
— f — conjunto m - [mec] assembly part
— f — chapa f (metálica) - [mec] sheet metal
  part
— f — dos por cuatro (pulgadas f) - [maderas]
  two-by-four
— f — fundición f - [metal] iron casting
— f — — f gris - [metal] grey cast iron part
— f — hierro m - [mec] iron part • [sold]
  iron weld(ment)
— f — — m colado - [metal] cast iron part
— f — — m frío - [sold] cold iron part
— f — — m fundido - [metal-prod] iron casting
  • cast iron part • casting
— f — — m — caliente - [sold] hot cast iron
  part
— f — — m — delgada - [metal] thin iron
  casting
— f — — m — estanca a agua m - [sold] watertight casting
— f — — m — frío - [sold] cold cast iron
  part
— f — — m — gruesa - [metal] thick iron
  casting
— f — — m — protegida - [sold] protected
  (iron) casting
— f — — m U - [mec] U-iron
— f — hormigón m - [constr] concrete component
— f — inflexión f - [mec] offset
— f — junta f - [sold] joint member
— f — — f en lado m de flecha f—[sold] arrow
  side joint member
— f — latón m - [sold] brass part
— f — — m para conexión f - [sold] brass connection block
— f — — m — f de conductor m - [sold]
  cable brass connection block
— f — — m — f para conductor m a pistola
  f - [sold] gun cable brass connection block
— f — — m — — — — f a alimenta-
  dor m - [sold] feeder gun cable brass connection block
— f — — m — — m — — m pa-
  ra alambre m - [sold] wire feeder gun cable
  brass connection block
— f — linterna f (eléctrica) - [electr] flash-
  light part
— f — madera f - [constr] wood(en), component,
  o part
— f — — f de dos por cuatro (pulgadas f) -
  [mader] two-by-four
— f — material m plástico - [mec] plastic part
— f — mecanismo m - [mec] mechanism, o operating, part
— f — — m de máquina f para escritorio m -
  [com] business machine operating part
— f — metal m - [mec] metal part
— f — — m laminado - [mec] sheet metal part
— f — — m — suelto - [mec] loose sheet metal
  part
— f — motor m - [electr-mot] motor part •
  [comb.int] engine part
— f — pilote m - [constr-pil] pile, component,
  o part
— f — — m con núcleo m hincador - [constr-
  -pil] Core Drive pile component
— f — — m Core-Drive con núcleo m hincador -
  [constr-pil] Core Drive pile component
— f — polo m - [electr] pole piece
— f — producción f, limitada, o reducida -
  [ind] small production, item, o part
— f — puente m - [constr] bridge member
— f — refractario m - [metal-prod] refractory
  shape
— f — regulador m - [mec] control part
— f — — m para marcha f sin carga f - [mec]
  idler part

pieza f de relé m - [electr-equip] relay part
— f débil — [mec] weak part
— f decapada - [metal-trat] pickled part
— f defectuosa - [mec] defective, o faulty, part • [sold] defective work
— f — reemplazada - [mec] replaced defective part
— f — reparada - [mec] repaired defective part
— f deformada - [mec] deformed, part, o member
— f delgada - [mec] thin, o light gage, section, o part
— f descartable - [mec] expendable part
— f descartada - [ind] scrapped part
— f descri(p)ta - [ind] described part
— f desgastable - [mec] wear(ing) part
— f desgastada - [mec] worn part
— f — reemplazada - [mec] replaced worn part
— f desplazada - [mec] displaced, o dislocated. part
— f determinada - [ind] specific part
— f, diferente, o distinta—[mec] different part
— f divisoria - [mec] dividing part
— f — terminal - [mec] median terminal section
— f elaborada - fabricated part; fabrication
— f eléctrica - [ind] electrical part
— f electrizada - [electr-equip] (electrically) hot part
— f embridada - [mec] flanged, part, o fitting
— f empotrada - [constr] embedded part
— f en conjunto m - [mec] assembly breakdown
— f — forma f de T - [constr] tee
— f — — f — U - [mec] U-shaped member
— f — movimiento m - [mec] moving part
— f — S - [mec] offset
— f engrasada - [mec] greased part
— f entramada - [constr] truss member
— f entretejida - [constr] laced member
— f especial - [mec] special, part, o shape
— f específica - [mec] specific part
— f — para grúa f - [grúas] specific crane part
— f esquinera - [mec] corner, part, o member, o section
— f — para canaleta f—[mec] trough corner part
— f — — conexión f - [mec] corner connection, part, o piece
— f — vertical - [mec] vertical corner member
— f estampada - [metal-fabr] stamping
— f estándar - [mec] standard part
— f estañada - [metal-trat] tinned part
— f — soldada - [sold] soldered, o welded, tinned part
— f estructural - [metal-lam] structural (shape, o member)
— f — conformada - [metal-fabr] formed structural member
— f — — en frío [metal-fabr] cold-formed structural member
— f — de acero m - [constr] steel structural member
— f — — m con carbono m - [metal-lam] carbon steel structural shape
— f — — m inoxidable - [metal-lam] stainless (steel) structural shape
— f — hormigón m - [constr] concrete structural, member, o shape
— f — liviana - [constr] light structural shape
— f — conformada - light formed structural, member, o shape
— f — — en frío - [constr] light gage cold formed structural (member, o shape)
— f — menor - [constr] structural minor member
— f — normalizada - [metal-lam] normalized structural (member, o shape)
— f — con resistencia f alta (a tensión f) - [metal-lam] normalized high strength structural (member, o shape)
— f principal - [constr] main, o principal, structural (member, o shape)
— f exigida - [mec] required part
— f existente - [mec] existing part
— f fabricada - [metal-fabr] fabricated part; fabrication
— f fallada - [mec] failed part

pieza f faltante - [mec] missing part
— f fina - [sold] thin part
— f final - [mec] end fitting • [constr] end member
— f — de cobre m - [mec] copper end fitting
— f flotadora - [mec] floating part
— f — esférica - spherical floating part
— f flotante - [mec] floating part
— f forjada - [metal] forging; forged, piece, o part
— f — con estampa f - [metal-fabr] stamping
— f — martinete m - [mec] drop, forged part, o forging
— f — de acero m - [metal-prod] steel forging
— f — m de tamaño m grande - [metal-prod] large (size) steel forging
— f — grande - [metal-prod] large forging
— f — mayor - [metal-prod] larger forging
— f — menor - [metal-prod] smaller forging
— f — pequeña - [metal-prod] small forging
— f formada f - [metal-prod] formed part
— f — en prensa f - [metal-prod] pressed, in, o formed part
— f frontal - [constr] face,cing, member, o component
— f — para relleno m - [mec] front (filler) block
— f — — m para bloque m para motor m - [comb.int] front filler block to engine
— f — de acero m - [mec] steel face member
— f — — — m intemperizado - [constr] weathering steel face member
— f funcional - [mec] functional part
— f fundida - [metal-prod] (iron) casting
— f — con aleación f - [metal-prod] alloy casting
— f — de acero m - [metal-prod] steel casting
— f — — m con aleación f - [mec] alloy steel casting
— f — — m maleable - [mec] malleable steel casting
— f — hierro m - [metal-prod] iron casting
— f — — m gris - [metal-prod] grey iron casting
— f — metal - [metal-prod] (metal) casting
— f — — m maleable - [metal-prod] malleable (metal) casting
— f — delgada - [metal-prod] thin casting
— f — depósito m - in-stock casting
— f — en existencia f - in-stock forging
— f — grande - [metal-prod] large, forging, o casting
— f — gruesa - [metal-prod] thick casting
— f — hueca - [metal-prod] hollow casting
— f — maleable - [metal-prod] malleable casting
— f — más pequeña - [metal-prod] smaller casting
— f — mayor - [metal-prod] larger casting
— f — menor - [metal-prod] smaller casting
— f — para base f (con oruga f) - [constr] travel base casting
— f — — f para rotación f - [mec] rotating base casting
— f — — costado m - [metal-prod] side casting
— f — — estructura(s) f - [metal-prod] structural casting
— f — — portadiferencial - [autom-mec] (differential) carrier casting
— f — — puente m - [metal-prod] bridge casting
— f — — — m carretero - [metal-prod] highway bridge casting
— f — — quemador m - [combust] burner casting
— f — — revestimiento m - [mec] revetment casting
— f — — servicio m - [metal-prod] service casting
— f — — —m para presión f - [metal-prod] pressure service casting
— f — temperatura(s) f alta(s) - [metal-prod] high temperature cast
— f — — — f baja(s) - [metal-prod] low tem-

perature casting

pieza f fundida para uso m estructural - [metal-
-lam] casting for structural use
— f — turbina f (para vapor m) - [metal-
-prod] (steam) turbine casting
— f — pequeña - [metal-prod] small casting
— f — protegida - [sold] protected casting
— f — sin recocer v - [metal-prod] green, o
unannealed, casting
— f galvanizada - [mec] galvanized part
— f — soldada - [sold] welded, o soldered,
galvanized part
— f gastada - [mec] worn part
— f — con superficie f endurecida - [sold]
hardsurfaced worn part
— f — reconstruida - [sold] rebuilt worn part
— f — reemplazada - [mec] replaced worn part
— f — rellenada - [sold] rebuilt worn part
— f genuina - [mec] genuine, o original, part
— f — para reemplazo m - [mec] genuine, o original, replacement part
— f giratoria f - [mec] rotating part; pivot
(block)
— f — para asidero m - [sold] handle pivot
block
— f grande - [mec] large, part, o component •
[sold] heavy part
— f — fundida - [metal-prod] large casting
— f gruesa - [mec] heavy, o thick, part, o section • [sold] thick weldment
— f hallada - [mec] found part
— f hermanada - [mec] matched part
— f horizontal - [mec] horizontal, part, o member
— f identificada - [ind] identified, o disig-
nated, part
— f ilustrada - [mec] illustrated, o pictured,
part
— f importada - [ind] imported part
— f — para equipo m - [ind] imported equipment
part
— f inaceptable - [ind] unacceptable part
— f incluida - [mec] enclosed, o included, part
— f individual - [mec] individual part •
[hotel] single room
— f individualizada - [mec] designated part
— f inferior - [mec] bottom part • [tub] bottom
— f — de armadura f - [constr] chord
— f — — tubería f - [tub] pipe bottom (part)
— f — — abovedada - [tub] pipe arch bottom
(part)
— f — — f — corrugada - [tub] corrugated
pipe arch bottom
— f — — f — — encajable - [tub] corrugated nestable pipe arch bottom
— f — — f — encajable corrugada - [tub] corrugated nestable pipe arch bottom
— f — — — f — y galvanizada - [tub]
galvanized corrugated nestable pipe arch bottom
— f — — — f encajable - [tub] nestable pipe
bottom
— f — — — f — corrugada - [tub] corrugated
nestable pipe arch bottom
— f inspeccionada - [mec] inspected part
— f instalada - [mec] installed part
— f intercambiable - [mec] interchangeable part
— f — para matriz f - [mec] interchangeable
die, o die interchangeable, part
— f — — punzón f - [mec] interchangeable
punch, o punch interchangeable, part
— f interior - [mec] internal, part, o component; internal assembly; internal(s) •
[constr] inside room
— f — para caja f - [mec] case internal part
— f — — — f para engranaje(s) m - [mec] gear
case internal(s) (part)
— f — protegida—[mec] protected internal part
— f intermedia f - [mec] intermediate part
— f interna - [mec] internal part
— f — de motor m - [electr-mot] motor internal
part • [comb.int] internal engine part
— f — — rótula f de codo m - [metal-prod]
elbow hinge collar

pieza f interna para motor m - [electr] motor internal, o internal motor, part • [comb.int]
internal engine, o engine internal, part
— f invertida - [mec] inverted part
— f — para conexión f - [mec] inverted connector
— f laminada - [metal-lam] rolled section
— f lateral - [mec] side, part, o member
— f — frontal - [mec] front side member
— f — posterior - [mec] rear side member
— f — transversal - [mec] transverse side member
— f legítima - [mec] genuine part
— f — para reemplazo m - [mec] genuine replacement part
— f limpia(da) - [mec] clean(ed) part
— f longitudinal - [mec] longitudinal, part, o
member
— f — horizontal - [mec] horizontal longitudinal, part, o member
— f lubricada - [mec] lubricated, o oiled, part
— f mal cortada - [mec] improperly cut part
— f más débil - weakest, part, o component •
weaker, part, o component
— f — pesada - [mec] heavier, part, o component • heaviest, part, o component
— f — voluminosa - [mec] largest, part, o member
— f masiva - [mec] massive part
— f mecánica - [mec] mechanical part • operating part • mechanical(s)
— f — mayor m - [constr] heavy equipment
— f — para conversión f - [mec] conversion me-
chanical(s)
— f medida - [mec] measured, o sized, part, o
component
— f mejorada - [mec] improved, part, o component
— f menor - [estruct] minor member
— f — para esqueleto m estructural - [constr]
structural; frame,ming minor member
— f metálica - [mec] metal(lic), part, o piece
• [mec-ventil] metal; iron
— f — para paleta f - [ventil] blade iron
— f moldeada—[metal-prod] molded part; casting
— f montada - [mec] assembled part • mounted
part
— f movible - [mec] moving, o working, part
— f — para contacto m - moving contact
— f móvil - [mec] moving part
— f muy solicitada - [mec] highly stressed part
— f necesaria - [mec] necessary, o needed, o
required, part
— f no conductora f - [electr-equip] non-conducting part
— f normalizada - [mec] standardized, part, o
component • [metal-trat] normalized, part, o
component
— f nueva f - [mec] new part
— f — con superficie f endurecida - [sold]
hardsurfaced new part
— f — usada - [mec] used new part
— f número . . . - [mec] part number . . .
— f obsoleta - [mec] obsolete part
— f operativa - [mec] operating part
— f optativa - [mec] optional part
— f — para puerta f - [mec] optional door, o
door optional, part
— f ordenada - [ind] ordered part
— f original - [mec] original part
— f oxidada - [mec] osidized part
— f — anódicamente - [mec] anodically oxidized part
— f para accesorio m - [mec] fitting, member, o
part
— f — admisión f - [comb.int] intake part
— f — aislación f - [mec] insulating part
— f — ajuste m - [mec] fitting (part)
— f — — m aflojada - [mec] loose(ned) fitting
— f — — m, ajustado, o apretado - [mec]
thight(ened) fitting
— f — — m para bomba f—[bombas] pump fitting
— f — — m — línea f - line fitting
— f — — m — válvula f - [mec] valve fitting

pieza f para eje m - [mec] axle, part, o member
— f — — m con forma f de U - [mec] U-shaped axle, part, o member
— f — izaje* m - véase pieza f para elevación
— f — — m montaje m - [mec] mounting lock
— f — — m para sujetador m - [mec] clamp mounting block
— f — recambio m - [mec] replacement, o spare, part
— f — recambio m de estelita f - [mec] Stellite, replacement, o spare, part
— f — — m para soldadora f - [sold] welder replacement part
— f — relé m - [electr-equip] relay part • pl relay kit
— f — — m para campo m - [electr-equip] field relay part • pl field relay kit
— f — relleno m - [mec] filler block
— f — reposición f - [mec] replacement part
— f — repuesto m - [mec] spare, o replacement, part, o item; spare • véase también repuesto
— f — — m apropiado - [mec] appropriate spare, part, o component
— f — — m de estelita f - [mec] Stellite (spare, o replacement) part
— f — — m para motor m - [electr-mot] motor (spare) part • [comb.int] engine (spare) part
— f — — m normal - [ind] normal spare part
— f — — m para arrancador m - starter part
— f — — m — bomba f - [bombas] pump (spare) part
— f — — m — f para parte f hidráulica - [petról] fluid end pump part
— f — — m — grúa f - [grúas] crane spare (part)
— f — — m — f aérea - [grúas] overhead crane spare (part)
— f — — m — horno m - [ind] furnace spare (part)
— f — — m — m alto - [metal-prod] blast furnace spare (part)
— f — — m para rodillo m - [metal-lam] roll spare (part)
— f — — m para rotámetro* m - [mec[rotameter spare (part)
— f — — m — soldadora f - [sold] welder, replacement, o spare, part
— f — — m universal - [ind] universal spare part
— f — separador m - [mec] spacer,cing section
— f — — m para acoplamiento m - [mec] coupler spacer,cing section
— f — servicio m - [ind] service part • [constr] maid('s) room
— f — sistema m - [mec] system part
— m — — m de polea(s) f - [mec] pulley system part
— f — — m — f para carro m para montacarga(s) m para horno m alto - [metal-prod] blast furnace skip car pulley system part
— f — — m — f para montacarga(s) m - [metal-prod] skip pulley system part
— f — — m — f m para horno m - [metal-prod] furnace skip pulley system part
— f — — m — m — m para horno m alto - [metal-prod] blast furnace skip pulley system part
— f — — m — f — vagoneta f—[metal-prod] car pulley system part
— f — — m — f — f para montacarga(s) m - [metal-prod] skip car pulley system part
— f — — m — f — f — m para horno m - [metal-prod] furnace skip car pulley system part
— f — — m — f — f — m — m alto - [metal-prod] blast furnace skip car pulley system part
— f — soporte m - [tub] supporting, o hanger, part
— f — — m de acero m - [constr] steel plate cradle • steel supporting part
— f — — m de plancha(s) f de acero m - [constr] steel plate (supporting) cradle

pieza f para sostén m - [estruct] supporting member
— f — — m entramada - [constr] supporting truss member
— f para substitución f - [estruct] substitution (member)
— f — suspensión f - [metal-prod| holding part
— f — — f para codo m (portavientos) - [metal-prod] gooseneck holding part
— f — tapa f - [mec] cover part
— f — — f optativa—[mec] optional cover part
— f — — f para carrete m - [mec] reel optional cover part
— f — — f — m para alambre m - [sold] optional wire reel cover part
— f — — f para carrete m - [mec] reel cover part
— f — — f — m para alambre m - [sold] wire reel cover part
— f — tolva f - [mec] hopper part
— f — trabajar(se) v - [mec] work
— f — trabajo m - [mec] working part
— f — — m protegida - [mec] protected working part
— f — transportador m - [mec] carrier part
— f — — m diferencial - [mec] differential carrier part
— f — turista(s) - [domést] tourist('s) room
— f — vaivén m - [mec] reciprocating part
— f — válvula f - [válv] valve part
— f — — f armada - [válv] assembled valve part
— f — — f colocada - [válv] installed valve part
— f — — f instalada - [válv] installed valve part
— f — — f montada - [válv] assembled valve part
— f — ventilador m - [mec] fan part
— f — — m para techo m - [mec-ventil] ceiling fan part
— f — yugo m - [mec] yoke part • yoke block
— f pedida - [ind] ordered part
— f pequeña - [mec] small part
— f — de hierro m fundido - [metal-prod] small casting
— f — fundida - [metal-prod] small casting
— f pesada - [mec] heavy part
— f plana - [mec] flat (part)
— f — de acero m - [metal-lam] steel flat
— f polar - [electr-equip] polar piece • pole (piece)
— f — para campo m - [electr-equip] field pole piece
— f — — — m para generador m - [electr-prod] generator field polar,le piece
— f — — excitador m - [electr-mot] exciter polar,le piece
— f — — generador m - [electr-prod] generator polar,le piece
— f — — polo m auxiliar - [electr-equip] interpole pole piece
— f — laminada - [electr-prod] pole lamination
— f — — principal - [electr-prod] main pole lamination
— f — principal - [electr] main pole piece
— f portadora de corriente f - [electr-equip] current carrying part
— f postal - [com] mail out; piece of mail
— f posterior - [mec] rear, block, o part
— f — para relleno m - [mec] rear (filler) block
— f precisa - [mec] precise, o accurate, part
— f prefabricada - [mec] prefabricated part | a - (de) preengineered
— f primaria - [constr] primary member
— f principal - [mec] principal, o main, part • [constr] major member
— f — de esqueleto m (estructural) - [constr] structural frame,ming major member
— f protegida - [mec] protected part
— f prototipo - [mec] prototype part
— f provista - [ind] provided part

pieza f proyectada - [ind] designed part
— f — específicamente - [ind] specially, o specifically, designed part
— f pulida - [mec] polished part
— f — electrolíticamente - [mec] electrolytically polished part
— f que conduce corriente f - [electr-instal] current carrying part
— f — conecta f - [mec] connecting,tion part
— f — — con pistola f - [sold] gun connection block
— f — no conduce corriente f - [electr-equip] non current carrying part
— f — proteje - [mec] protecting,tion part
— f — — caja f para refrigeración f - [metal-prod] cooling plate protecting,tion part
— f — — — f — — f de petaca f - [metal-prod] iron notch cooling plate protection part
— f quitada - [mec] removed part
— f rearmada - [mec] reassembled part
— f recalcadora f - [mec] swager
— f — alternante fijada a plato m rotante - [herram] reciprocating rotary swager
— f reconstruible - [mec] rebuildable part
— f recubierta - [mec] coated part
— f recuperada - [ind] recovered part
— f rechazada - [ind] reject(ed part)
— f redonda - [mec] round part
— f Reed - [mec] Reed part
— f reemplazada - [mec] replaced part
— f refractaria - [refract] refractory shape
— f reinstalada - [ind] reinstalled part
— f relacionada - [mec] related part
— f rellenable - [sold] rebuildable, o refillable, part
— f rellenadora - [mec] filler block
— f removida - [mec] removed part
— f reparada - [mec] repaired part
— f repetida - [ind] duplicate part
— f reprocesada - [sold] rework(ed part)
— f repuesta - [mec] replaced part
— f requerida - [ind] required part
— f resistente - [ind] resisting,tant part
— f — a calor m - [ind] heat resistant part
— f restante - [mec] remaining part • other part
— f retráctil - [mec] telescoping part
— f — a prueba f de falla(s) f - [mec] fool proof telescoping part
— f rígida - [sold] rigid part
— f rotante - [mec] rotating part
— f rotativa - [mec] rotating part • [electr-mot] véase inducido m
— f sacada - [mec] removed part
— f separada - [mec] separate(d) part
— f serializada* - [ind] serially numbered part
— f sin armar - unassembled, o knocked-down, section, o part
— f — corriente f - [electr-oper] cold part
— f — formar - [mec] blank
— f — terminar - unfinished part; blank
— f soldada - [sold] weldment; work piece • welded part
— f — de acero m - [sold] steel weldment
— f — — m inoxidable - [sold] stainless steel weldment
— f — en taller m - [mec] shop welded part
— f — grande - [sold] large, o heavy, weldment
— f — liviana - [sold] light weldment
— f — pequeña - [sold] small weldment
— f solicitada - [mec] stressed part
— f — altamente - [mec] highly stressed part
— f subcontratada - [ind] subcontracted part
— f suelta - [mec] loose part
— f superior - [mec] cap • [constr] top (piece, o part)
— f — de tubería f - [tub] pipe top
— f — — — f abovedada - [tub] pipe arch top
— f — — — f — corrugada encajable - [tub] corrugated nestable pipe arch top
— f — — — f — encajable - [tub] nestable pipe arch top

pieza f superior de tubería f abovedada encajable corrugada (y) galvanizada - [tub] galvanized corrugated nestable pipe arch top
— f — — — f encajable - [tub] nuestable pipe top
— f terminal - [metal-fabr] end, o terminal, section • [tub] end fitting
— f — mal colocada - [cabl] poorly, o improperly, attached end fitting
— f — metálica - [metal-fabr] (metallic) end section
— f — para defensa f lateral - [vial] Flex-Beam terminal
— f termoformada* - [plást] thermoformed part
— f típica - [mec] typical part
— f torcida - [mec] twisted, o distorted, member, o part • crooked part
— f totalmente de hierro m - [metal] entire iron part
— f — — — m colado - [metal] entire cast iron part
— f — — — m fundido - [metal] entire cast iron part
— f transversal - [constr] transverse, o cross, member, o piece
— f — especial - [mec] special transverse member
— f — inferior - [mec] bottom transverse member
— f — — especial - [mec] special bottom transverse member
— f — premontada - [constr] subassembled, o premounted, transverse member
— f — superior - [mec] top transverse member
— f — — especial - [mec] special top transverse member
— f traslapada - [mec] overlapping, part, o piece
— f tratada - [metal-trat] treated part
— f — térmicamente - [metal-trat] heat treated part
— f trazada - [mec] traced part
— f única - single, piece, o part
— f usada - [mec] used, part, o piece
— f — nuevamente - [mec] reused part
— f vertical - [mec] vertical member
— f — acanalada - [mec] U shaped, o channel, vertical member
— f — en forma f de U - [mec] U-shaped vertical member
— f vinculada - [mec] related, o connected, part
— f vital - [mec] vital part
— f vuelta a armar - [mec] reassembled part
— f viva - [mec] moving part
piezas f apareadas - [mec] matched parts
— f para - [ind] kit
— f por ficha f - pieces per card
— f varias - miscellaneous, o sundry, parts
piezoeléctrico,ca a - piezoelectric
pifia f - . . .; goof
pifiar v - . . .; to goof
pila-estribo f - [constr] aburment pier
— f para almacenamiento m - [ind] storage pile; stockpile
— f — — m de agregado(s) m - [ind] aggregate storage pile
— f — — m — carbón m - [miner] coal stockpile; coal storage pile
— f — — m — — m lavado - [miner] washed coal, stockpile, o storage pile
— f — — m para material(es) m - [ind] material(s) storage pile
pigmento m de titanio m y calcio m - [pint] titanium-calcium pigment
— m metálico - [constr] metallic pigment
— m molido - [pint] ground pigment
— m — para pintura(s) f - [pint] ground paint pigment
— m para pintura(s) f - [pint] paint pigment
— m pulverizado - [pint] pulverized pigment
— m — para pintura(s) f - [pint] pulverized paint pigment

pila f - . . . . • [ind] (stock)pile • [constr]
pier; bent • [electr] battery; cell • [miner]
bed • stack
— f ácida - [electr] acid, battery, o cell
— f — con plomo m - [electr] lead-acid, battery, o cell
— agotada - [coque] finished pile • [electr]
discharged battery
— f de . . . .-high (stack)
— f — agregado(s) m - [ind] aggregate(s) pile
— f — caballete m - [constr] bent
— f — carbón m - [min] coal, pile, o stack
— f — fundente m - [sold] flux pile
— f — gruesos m - [metal-prod] lump(s) pile
— f — hormigón m - [constr] concrete pier
— f — material(es) m - [ind] stockpile
— f — neumáticos m - [autom-neumát] tire stack
— f — tres - three-high stack
— f descargada - [electr- discharged, o shot,
battery
— f en declive - [ind] sloped stack
— f — empalme m - [vial] interchange pier
— f para mezcla(dura) f y homogeneización f -
[miner] blending pile
— f — paso m inferior—[constr] underpass pier
— f — pilote m - [constr] (pile) bent
— f — puente m - [constr] bridge, bent, o pier
— f — recocido m - [metal-trat] annealing box
— f — sostén m - [constr] support(ing) pier
— f — — m de hormigón m - [constr] concrete
support(ing) pier
— f — radio f - [telecom] radio battery
— f rectificadora - [electr] rectifier stack
— f — en corto circuito m - [electr-oper]
shorted rectifier stack
— f seca - [electe] dry (cell) battery
— f termoeléctrica - [instrum] thermocouple
— f tipo - [miner] typical stack
— f vertical - [ind] vertical stack
pilar m - . . .; standard • [constr] pier; bent;
buttress; buckstay
— m armado - [constr] reinforced pier
— m de hormigón m - [constr] concrete pier
— m — — m armado - [constr] reinforced (concrete) pier
— m en celosía f - [constr] lattice column
— m — río m - [constr] river pier
— m metálico - [constr] (metallic) bent
— m — para puente m - [constr] bridge bent
— m para barrera f - [constr] barrier pillar
— m — puente m - [constr] bridge pier
— m — soporte m - [constr] support(ing) bent
— m — torre f - [constr] tower, pier, o leg
— m — — f para perforación f - [petról] derrick leg
— m reforzado - [constr] reinforced pier
— m sobre tierra f - [constr] land (based) pier
pilastra f - [constr] . . .; pylon
— f central - [constr] center,tral, o middle,
pylon
— f — con crucero m (en T) - [constr] T-head
center,tral pylon
— f — — dos cruceros m en T - [constr] double
T-head center,tral pylon
— f — para soporte m - [constr] center,tral, o
middle, (support) pylon
— m con crucero m (en T)—[constr] T-head pylon
— f para soporte m—[constr] support(ing) pylon
pileta f blindada - [hidr] armor-plated basin
— f para, sedimentación f, o tranquilización
f - [hidr] armor-plate, settling, o stilling,
basin
— f con arenilla f - [sanit] grit tank
— f — calefacción f - [deport] heated pool
— f — — f para natación f - [deport] heated
swimming pool
— f interior f—[deport] indoor (swimming) pool
— f — con calefacción f - [deport] indoor
heated (swimming) pool
— f — — f para natación f—[deport] heated
indoor swimming pool
— f — para natación f - [deport] indoor swimming pool

pileta f interior para natación f con calefacción f - [deport] heated indoor swimming pool
— f lavatorio - [sanit] wash, fountain, o stand
— f para aireación f - [sanit] aeration tank
— f — cloro m - [sanit] chlorine tank
— f — decantación f - [sanit] settling tank
— f — — f final - [sanit] final settling tank
— f — — f primaria - [sanit] primary settling
tank
— f — desborde m - [sanit] overflow basin
— f — enjuague m - [sanit] rinse,sing tank
— f — escoria f - [metal-prod] slag pit
— f — — f para horno m alto - [metal-prod]
blast furnace slag pit
— f — lavado,dura - wash(ing), o rinse,sing,
tank
— f — — en caliente m - hot rinse tank
— f — natación f - [deport] swimming pool
— f — — f con calefacción f - [deport] heated
swimming pool
— f — piso m - [sanit] floor drain; sump
— f — rebalse m - [hidr] overflow basin
— f — — m para usina f - [electr-prod] power
plant overflow basin
— f — sedimentación f - [sanit] sedimentation,
o settling, tank
— f — tranquilización f - [hidr] stilling, o
settling, basin
— f — vuelco m - [ind] dump(ing) basin
pilón m - . . . . • [constr-equip] hammer; striking, part, o head • [vial] pylon
— m para tránsito m - [vial] traffic, cone, o
pyloin
piloncillo m - véase pilón m
pilotado,da a - véase piloteado,da a
pilotaje m - . . . . • [náut] piloting
pilote m - . . .; piling • caisson
— m abandonado - [constr-pil] abandoned pile
— m ahusado - [constr-pil] tapered pile,ling
— m aplastado - [constr-pil] collapsed pile
— m Armco - [constr-pil] Armco pile,ling
— m — de tubería f - [constr-pil] Armco pipe
pile,ling
— m colado - [constr-pil] cast pile
— m — en obra f - [constr-pil] cast-in-place
pile,ling
— m — — f para sostener cimiento m -
[constr-pil] foundation supporting cast-in-
-place pile,ling
— m — — f — fundamento m - [constr-
-pil] foundation supporting cast-in-place
pile,ling
— m compuesto - [constr-pil] composite pile
— m — para cimiento(s) m - [constr-pil] composite (foundation) pile,ling
— m con bulbo m - [constr-pil] bulb (type) pile
— m — cabeza f - [constr]pil] head pile
— m — — f con rebajo m - [constr-pil] offset
head pile
— m — — f — retallo m - [constr-pil] offset
head pile
— f — capacidad f alta - [constr-pil] high capacity pile
— m — — f mayor - [constr-pil] higher capacity pile
— m — — f menor - [constr-pil] lower capacity
pile
— m — — f reducida - [constr-pil] low capacity pile
— m — diámetro m cabal - [constr-pil] full
size pile
— m — — m deseado - [constr-pil] desired diameter pile
— m — — m grande - [constr-pil] large diameter pile
— m — — m mayor - [constr-pil] larger diameter pile
— m — — m menor - [constr-pil] smaller diameter pile
— m — — m uniforme - [constr-pil] uniform
diameter pile
— m — extremo m abierto - [constr-pil] open
end pile

pilote m con núcleo m hincador - [constr-pil]
core drive pile
— m — pared f delgada - [constr-pil] thin wall
pile
— m — pedestal m - [constr-pil] bulb (type)
pile
— m — tapa f - [constr-pil] capped pile
— m cónico - [constr-pil] tapered pile
— m cónicoescalonado - [constr-pil] step-ta-
pered pile
— m continuo - [constr-pil] continuous pile
— m Core-Drive - [constr-pil] core-Drive pile
— m — — con núcleo m hincador - [constr-pil]
Core-Drive pile
— m — — para hincadura f con núcleo m -
[constr-pil] Core-Drive pile
— m corrugado - [constr-pil] corrugated pile
— m — horizontalmente - [constr-pil] horizontally corrugated pile
— m dañado - [constr-pil] damaged pile
— m de acero m - [constr-pil] steel pile
— m — con pared f gruesa - [constr-pil]
heavy wall steel pile
— m — — m corrugado - [constr-pil] corrugated
steel pile
— m — — m estructural - [constr-pil] structural steel pile
— m — — m hincado - [constr-pil] driven steel
pile
— m — — m recubierto con hormigón m con barreno m vertical - [constr-pil] soldier pile
— m — — m rellenado - [constr-pil] filled
steel pile
— m — — m — con hormigón m - [constr-pil]
concrete filled steel pile
— m — Armco para hincadura f con núcleo m -
[constr-pil] Armco Core-Drive pile
— m — camisa f perdida - [constr-pil] pile
shell
— m — — f de acero m - [constr-pil] steel
pile shell
— m — — f de acero m corrugado - [constr-
-pil] corrugated steel pile shell
— m — — f — Armco - [constr-pil] Armco
pile shell
— f — f — — soldado helicoidalmente -
[constr-pil] Armco Hel-Cor pile shell
— m — — f — hincado - [constr-pil] driven
pile shell
— f — — f — — con mandril m - [constr-pil]
mandrel driven pile shell
— f — — f — soldado helicoidalmente -
[constr-pil] Hel-Cor pile shell
— m — columna - [constr-pil] end bearing pile
— m — hormigón m - [constr-pil] concrete pile
— m — colado en obra - [constr-pil] cast-
-in-place (concrete) pile
— m — — m moldeado en sitio m - [constr-pil]
cast-in-place concrete pile
— m — — m vaciado en obra f - [constr-pil]
cast-in-place (concrete) pile
— f — madera f - [constr-pil] lumber, o wood,
o timber, pile
— m — — f sin tratar - [constr-pil] untreated
timber pile
— m — tubería f - [constr-pol] pipe piling
— m — — f ahusada - [constr-pil] tapered pipe
piling
— m — — f de acero m - [constr-pil] steel
pipe pile,ling
— m — — f de Armco m - [metal-fabr] Armco
pipe pile,ling
— m — tubo m - [constr-pil] pipe pile,ling
— m delgado - [constr-pil] thin pile,ling
— m en H - [constr-pil] H pile,ling
— m estriado - [constr-pil] fluted pile,ling
— m — verticalmente - [constr] vertically
fluted pile,ling
— m existente - [constr-pil] existing pile,ling
— m falso - [constr-pil] follower
— m grande - [constr-pil] large pile
— m hincado - [constr-pil] driven pile; drive
shell

pilote m hincado con mandril m - [constr-pil]
mandrel driven pile
— m — en forma f corriente - [constr-pil] conventionally driven pile
— m — hasta capacidad f portante exigida -
[constr-pil] pile driven to @ required bearing
— m — — penetración f exigida - [constr-pil]
pile driven to @ required penetratión
— m — mediante chorro m de agua m - [constr-
-pil] jet set pile,ling
— m hueco - [constr-pil] hollow, o empty,
shell, o pile
— m inclinado - [constr-pil] batter pile
— m mayor - [constr-pil] larger pile
— m individual - [constr-pil] individual pile
— m metálico - [constr-pil] metal(lic) pile
— m — delgado - [constr-pil] thin metal(lic)
pile
— m menor - [constr-pil] smaller pile
— m moldeado (en lugar m) - [constr-pil] cast-
-in-place pile
— m para apoyo m - [constr] bearing pile
— m — cabecera f (para puente m) - [constr-
-pil] abutment pile
— m — carga f - [constr-pil] bearing pile
— m — cimentación f - [constr-pil] foundation
pile,ling
— m — — f para pila f - [constr-pil] pier
foundation pile
— m — cimiento(s) m - [constr] foundation
pile,ling
— m — ensayo(s) m - [constr-pil] test(ing)
pile,ling
— m — fricción f - [constr-pil] friction pile
— m — fundación f - [constr-pil] foundation
pile,ling
— m — máquina f - [ind] machine pile,ling
— m — — f lingoteadora - [metal-prod] pig,
casting machine, o caster, pile,ling
— m — pila f - [constr-pil] pier pile
— m — recalzo m - [constr-pil] underpinning
pile,ling
— m — sostén m - [constr-pil] supporting pile
— m — sostener cimiento m - [constr-pil] foundation supporting pile
— m — — v fundamento m - [constr-pil] foundation supporting pile
— m pequeño - [constr-pil] small pile
— m prefabricado - [constr-pil] precast pile
— m — de hormigón m - [constr-pil] precast
concrete pile
— m próximo - [constr-pil] near(-by), o adjacent, pile
— m rechazado - [constr-pil] rejected pile
— m redondo - [constr] round pile
— m reforzado - [constr-pil] reinforced pile
— m rellenado - [constr-pil] filled pile
— m — con hormigón m - [constr-pil] concrete
filled pile
— m revestido - [constr-pil] encased pile
— m — con tubería f - [constr-pil] pipe encased pile
— m — — f de acero m - [constr-pil] steel
pipe encased pile
— m soldado - [constr-pil] welded pile
— m sustentador - [constr-pil] bearing pile
— m tubular - [constr-pil] (tubular) (pipe)
pile; pipe, piling, o caisson
— m — ahusado - [constr-pil] tapered pipe pile
— m — de acero m - [constr-pil] steel pipe,
pile, o caisson
— m — — Armco - [constr-pil] Armco (pipe)
caisson
— m — colado - [constr-pil] cast pipe pile
— m — — en obra f - [constr-pil] cast-in-
-place pipe pile
— m — — — sitio m - [constr-pil] cast-in-
-place pipe pile
— m — — — — m para sostener cimiento m -
[constr-pil] foundation supporting cast-in-
-place pipe pile
— f — — — m con pared f gruesa - [constr-

-pil] heavy weall   steel pipe pile
pilote m **tubular de acero** m **rellenado con hormi-gón** m̄ - [constr-pil] concrete filled steel pipe pile
— m — **de Armco** - [constr-pil] Armco pipe pile,ling
— m — **grande** - [constr-pil] large pipe pile
— m̄ — **hincado** - [constr] driven pipe pile
— m̄ — **en forma** f **corriente** - [constr-pil] conventionally driven pipe pile
— m — **hasta capacidad** f **portante exigida** - [constr-pil] pipe pile driven to @ required bearing
— m — — **penetración** f **exigida** - [constr--pil] pipe pile driven to @ required penetration
— m — **mayor** - [constr-pil] larger pipe pile
— m̄ — **para sostén** m - [constr-pil] supporting pipe pile
— m — **pequeño** - [constr-pil] small pipe pile
— m — **rellenado**—[constr-pil] filled pipe pile
— m — **con hormigón** m - [constr-pil] concrete filled pipe pile
— m — **soldado** - [constr-pil] welded pipe pile
— m̄ — **de acero** m - [metal-fabr] welded steel pipe, pile, o caisson
— m — **uniforme**—[constr-pil] uniform pipe pile
— m̄ **uniforme** - [constr-pil] uniform pile
— m̄ **vaciado** - [constr-pil] cast pile
— m̄ — **en obra** f - [constr-pil] cast-in-place pile
— m — — f **completado** - [constr-pil] completed cast-in-place pile
— m̄ **vertical** - [constr-pil] soldier pile
— m̄ — f **en fila** f - [constr-pil] aligned soldier pile
piloteado m - piloting
— m **de conexión** f - [mec] connection piloting
piloteado,da a - piloted
piloto m - . . . . • [cabl] core • [combust] pilot (flame) (unit)
— m **aprendiz** - [aeron] apprentice, o student, pilot
— m̄ **encendido** - [combust] lit pilot • pilot on
— m̄ **para acelerador** m - [comb.int] throttle pilot
— m — **cable** - [electr-instal] cable pilot
— m̄ — **caja** f - [mec] case pilot
— m̄ — m **para sostén** m - [mec] support case pilot
— m — — f — — m **para asiento** m - [mec] seat support case pilot
— m — **camisa** f - [mec] liner pilot
— m̄ — **línea** f **aérea** - [aeron] airline pilot
— m̄ — **sangrador** m - [metal-prod] bleeder, igniter, o lighting device
— m̄ — **soldadora** f - [sold] welder pilot
— m̄ — **válvula** f - [tub] valve pilot
— m̄ — — f **reguladora** - [tub] control valve pilot
pincel m - . . . . • brush
— m **con cerdas** f - [herram] bristle brush
— m̄ — — f **medianas** - [herram] medium bristles brush
pincelada f - . . . . • brushing
pincelado,da a - brushed
pincelar v - . . . ; to brush
pinchado,da a - [autom-neumát] punctured
pinchadura f - . . . . • [autom-neumát] puncture; blowout; puncturing
— f **de cámara** f - [autom-neumát] tire, o tube, puncture,ring
— f — **neumático** m - [autom-neumát] tire, puncture,ring, o blowout
pinchar v - . . . . • [metal-prod] to drill • [autom-neumát] to puncture
— v **cámara** f—[autom-neumát] to puncture @ tube
— v **cuchara** f - [metal-prod] to (oxygen) lance @ ladle
— v **escorial** m - [metal-prod] to drill @ slag notch
— v **horno** m - [metal-prod] to tap @ furnace; to drill @ taphole

pinchar v **neumático** m - [autom-neumát] to puncture @ tire
— v **piquera** f - [metal-prod] to drill @ taphole
pino m **Brasil** - [mader] Brazilian pine
— m **con calidad** f **para construcción** f - [mader] construction grade fir
— m **de sur** m - [mader] southern, pine, o fir
— m̄ **Douglass** - [mader] Douglass, pine, o fir
— m̄ — **con calidad** f **para construcción** f̄ - [maderas] construction grade Douglass fir
— m̄ — **para construcción** f - [maderas] construction grade Douglass fir
— m **para construcción** f - [maderas] construction grade fir
— m **spruce** - [mader] spruce (pine)
pintado* m - véase **pintura** f
pintado,da a - painted
— a **con brea** f - [constr] pitch painted
— a — **cal** f̄ - [constr] whitewashed; calcimined
— a — **compuesto** m **asfáltico** - [constr] asphalt compound painted
— a — **soplete** m - [pint] spray painted
— a **en planta** f - plant painted
— a — **obra** f - [pint] (job) site painted
— a **finalmente** - [pint] finish painted
pintar v - . . . . • to coat
— v **bien** - to paint well • to look, encouraging, o rosy
— v **color de rosa** f - to be rosy
— v **con cal** - [pint] to, whitewash, o calcimine
— v **con soplete** m - [pint] to spray paint
— v — **yeso** m - [constr] to cacimine • to gypsum paint
— v **corona** f **dentada** - [mec] to paint @ ring gear
— v **diente** m - [mec] to paint @ tooth
— v — m **de corona** f **dentada** - [mec] to paint @ ring gear tooth
— v — m — **engranaje** m - [mec] to paint @ gear tooth
— v **en planta** f - [ind] to paint at @ plant
— v **equipo** m - [ind] to paint @ equipment
— v **finalmente** - [pint] to finish paint
— v **frente** m - [pint] to paint @ face
— v — m **de indicador** m - [vial] to paint @ sign face
— v — m — **letrero** m - [vial] to paint @ sign face
— v **material** m - [ind] to paint @ material
— v **superficie** f - to paint @ surface
pintoresco,ca a - . . . . • colorful • scenic
pintura f **acrílica** - [pint] acrylic paint
— f **anaforésica** - [metal-trat] anaphoretic paint(ing)
— f **anticorrosiva** - [pint] corrosion, o rust, resisting,tant, paint(ing); anticorrosive paint(ing)
— f **antioxidante** - [pint] rust resisting paint
— f **aplicada** - [pint] applied paint
— f **aprobada** - [pint] approved paint(ing)
— f **con aceite** m - [pint] oil paint(ing)
— f **bituminosa** - [pint] bituminous paint(ing)
— f **con base** f **alquitranada** - [pint] tar base(d) bituminous paint(ing)
— f — — — f **asfáltica** - [pint] asphalt(ic) base(d) bituminuous paint(ing)
— f **blanca** - [pint] white paint(ing)
— f — **de aluminio** m - [pint] white aluminum paint(ing)
— f **cataforésica** - [metal-trat] cataphoretic paint(ing)
— f **con base** f **alquitranada** - [pint] tar base(d) paint(ing)
— f — — f **asfáltica** - [pint] asphalt(ic) base(d) paint(ing)
— f — **cal** f - [constr] whitewash(ing); calcimine,ing
— f — **soplete** m - [pint] paint spraying • spray paint(ing)
— f **corriente** - [pint] standard paint(ing)
— f **de aluminio** m - [pint] aluminum paint
— f — — m **blanco** - [pint] white aluminum paint(ing)

pintura f de base f - [metal-prod] stool coating
— f — **corona f dentada** - [mec] ring gear paint(ing)
— f — **diente m** - [mec] tooth painting
— f — — m **de corona f dentada** - [mec] ring gear tooth painting
— f — — m — **engranaje m** - [mec] gear tooth painting
— f — **epoxia f** - [pint] epoxy paint
— f — **equipo m** - [ind] equipment painting
— f — **esmalte m** - [pint] enamel paint(ing)
— f — **fondo m** - [pint] prime (painting)
— f — **frente m** - [vial] sign face paint(ing)
— f — — m **de letrero m** - [vial] sign face paint(ing)
— f — **lingotera(s) f** - [metal-prod] mold, painting, o coating
— f — **material m** - [ind] material painting
— f — **placa(s) f** - [metal-prod] stool painting
— f — **superficie f** - surface painting
— f — **tipo m epoxídico** - [pint] epoxy type paint(ing)
— f — — m **lavable**—[pint] washable type paint
— f **en planta f** - [ind] plant painting
— f — **relieve** - [pint] relief paint(ing)
— f — — **de superficie(s) f plana(s)** - [pint] texture paint(ing)
— f **especial** - [pint] special paint(ing)
— f **espesa** - [pint] thick paint
— f **fácilmente desprendible** - [pint] easily removable paint
— f **final** - [pint] final, o finish, paint(ing)
— f **fraguada** - [pint] cured paint
— f **fresca** - [pint] fresh paint
— f **interior** - [pint] interior paint(ing)
— f **lavable** - [pint] washable paint
— f **líquida** - [pint] liquid, o can, paint
— f **marina** - [pint] deck paint
— f **metálica** - [pint] metal(lic) paint
— f **para acabado** - [pint] finishing, paint, o coat
— f — **cubierta(s) f** - [nav] deck paint • [autom-neumát] tire paint
— f — **diente(s) m** - tooth paint
— f — **estructura(s) f** - [pint] structure,ral paint
— f — — f **de acero m** - [pint] steel structure,ral paint
— f — **exterior(es) m** - [pint] outside paint
— f — **imprimación f** - [pint] priming paint; prime(r)
— f — — f **aplicada en fábrica f** - [metal-fabr] mill primer
— f — — f **roja** - [pint] red primer
— f — — f — **aplicada en fábrica f** - [metal-fabr] red mill primer
— f — **preparación f** - [pint] undercoat(ing)
— f — **recubrimiento m** - [pint] sealer
— f — **tubería(s) f** - [tub] pipe, paint, o coating
— f **pulverizada** - [pint] spray(ed) paint
— f **reflectora** - [pint] reflecting, o (light) reflective, paint
— f **según norma f** - [pint] standard paint(ing)
— f **vinílica** - [pint] vinyl paint
pinza(s) f - [mec] . . .; pinch; clip; tong(s) • [herram] pliers; forceps; nippers
— f **con lagarto m** - [herram] alligator grab
— f — **mandíbula f** - [herram] alligator, jaw, o grab
— f **con punta(s) f chata(s)** - [herram] duck bill pliers
— f — — f **larga(s)** - [herram] long nose pliers
— f **giratoria** - [metal-prod] rotating tongs
— f **grande(s)** - [herram] large pliers
— f **motorizada** - [metal-lam] mechanical tongs
— f **para forja f** - [herram] forge tongs
— f **sin giro m** - [metal-prod] nonturning tongs
— f **trabadora** - [herram] locking pliers
— f **volteadora(s)** - [mec] turning tongs
pinzas f - [mec] tongs
piñata f - pinata • jackpot

piñón m - [mec] . . . • cog wheel; gear
— m **ajustado** - [mec] tight(ened), o adjusted, pinion
— m **biselado** - [mec] bevel(ed) pinion
— m **cambiado** - [mec] changed pinion
— m **con engranaje m (recto)** - [mec] spur gear
— m **cónico** - [mec] bevel(ed), gear, o pinion
— f **diferencial** - [mec] pinion gear • differential pinion
— m **hacia derecha f** - [mec] right hand pinion
— m — **izquierda f** - [mec] left hand pinion
— m **identificado** - [mec] identified pinion
— m **impulsor** - [mec] drive,ving pinion
— m — **armado** - [mec] assembled drive,ving pinion
— f **desarmado** - [autom-mec] disassembled drive,ving pinion
— m — **desplazado** - [mec] displaced, o moved, drive,ving pinion
— f — **movido** - [mec] moved drive,ving pinion
— m — **para rodillo m** - [mec] roll drive,ving pinion
— m — — — m **para impulsión f** - [mec] feed-(ing) roll drive,ving pinion
— m **instalado** - [mec] installed pinion
— m **interior** - [mec] internal pinion
— m **intermediario** - [mec] idler pinion
— m **intermedio** - [mec] idler pinion
— m **lateral** - [mec] lateral, o side, pinion
— m — **instalado** - [mec] installed, lateral, o side, pinion
— m — **removido** - [mec] removed, lateral, o side, pinion
— m **loco** - [mec] idler pinion • sliding pinion
— m **movido** - [mec] moved pinion
— m **para arrancador m** - [comb.int] starter pinion
— m — **ataque m** - [mec] attack pinion • rack pinion
— m — — m **para cremallera f** - [metal-prod] rack pinion
— m — **cigüeñal m** - [comb.int] crankshaft pinion
— m — **cuadrante m** - [instrum] quadrant pinion
— m — — m **para reactor m** - [instrum] reactor quadrant pinion
— m — **eje m** - [autom-mec] axle pinion
— m — — m, **anterior, o frontal** - [autom-mec] forward, o front, axle pinion
— m — — m **posterior** - [autom-mec] rear axle pinion
— m — **empuje m** - [mec] crowding, o thrust, pinion
— m — **engranaje m** - [mec] gear pinion
— m — **émbolo m** - [mec] piston pinion
— m — — m **interior** - [mec] internal gear pinion
— m — **impulsión f** - [mec] drive,ving pin(ion)
— m — — f **armado** - [mec] assembled drive,ving pin(ion)
— m — — f **desarmado** - [autom-mec] desassembled, o unassembled, drive,ving pinion
— m — — f **para alimentación f** - [mec] feed drive,ving pinion
— m — **mando m** - [mec] drive,ving pinion
— m — **oscilación f** - [grúas] swing(ing) pinion
— f — **rotación f** - [mec] rotating pinion • [grúas] swing(ing) pinion
— m — — f **lubricado** - [grúas] lubricated swing(ing) pinion
— m — **selector m** - [mec] selector pinion
— m — — m **para amperaje(s) m** - [electr-equip] ampere(s) selector pinion
— m **planetario** - [mec] planetary pinion
— m — **loco** - [mec] planetary idler, o idler planetary, pinion
— m **principal** - [mec] main pinion
— m — **para engranaje m** - [mec] main gear pinion
— m — — — m **interior** - [mec] internal gear main pinion
— m **recto** - [mec] spur pinion
— m **removido** - [mec] removed pinion
— m **rotado** - [mec] rotated pinion

piñón m rotante - [mec] rotating pinion
— m sesgado - [mec] bevel(ed) pinion
— m — para impulsión f - bevel(ed) drive,-
ving pinion
— m — — f de rodillo m - [mec] roll bevel
drive,ving pinion
— m — — — f — — m para impulsión f - [mec]
feed(ing) roll bevel drive,ving pinion
— m y cremallera f - [mec] rack and pinion
— m — cruceta(s) f - [bombas] pinion and
crosshead(s)
piquera f - [metal-prod] taphole; (iron) notch;
tapping spout • véase también orificio m para
colada • spout; pouring lip • steel notch
— f corta - [metal-prod] short taphole
— f de horno m - [metal-prod] furnace, iron
notch, o taphole
— f larga - [metal-prod] long taphole • iron
too far in @ taphole
— f mal tapada - [metal-prod] poorly plugged,
taphole, o iron notch
— f para arrabio m - [metal-prod] iron notch
— f — desescoriado m - [metal-prod] slag notch
— f — escoria f - [metal-prod] slag notch
— f — por aire m comprimido—[metal-prod]
air operated slag notch
— f — hierro m - [metal-prod] iron notch
piqueta f - . . . • [sold] chipping hammer
piramidación* f - pyramidation,dizing
— m de diamante m - diamond pyramid
— f truncada - truncated pyramid
piramidizado,da a - pyramidized
piramidizar v - to pyramid
pirateado,da a - . . . • [transp] hijacked •
[aeron] skyjacked
piratear v - . . . . • [transp] to hijack •
[aeron] to skyjack
piratería f - . . . . • [transp] hijacking •
[aeron] skyjacking
— f aérea - [aeron] skyjacking
pirita f - [metal] . . .; iron disulfide
— f calcinada - [miner] calcined, o roasted,
pyrite
— f de hierro m - [miner] iron pyrite
piritoso,sa a - [geol] pyritic
piroclástica f - [geol] pyroclastic
piroclástico,ca a - [geol] pyroclastic
— a volcánico,ca - [Geol] volcanic pyroclastic
piroconsolidación f - [miner] burning (period)
pirolización f - [quím] pyrolization,zing
pirolizado,da* a - [quím] pyrolized
pirolizar* v - [quím] to pyrolize
pirómetro m de aire m - [instrum] air pyrometer
— m para bóveda f - [metal-prod] roof pyrometer
— m radiomático - [instrum] radiomatic pyrometer
— m registrador - [instrum] recording pyrometer
pirotecnia f - [explos] . . .; fireworks
pisado,da a - stepped on; treaded; trod
pisando adv talón(es) - hot on @ heel(s)
piscifactoría f - [ictiol] (fish) hatchery
piscina f - . . . . • [metal-prod] pool; tank
— f con calefacción f - [deport] heated swimming pool
— f interior - [deport] indoor swimming pool
— f — con calefacción f - [deport] heated indoor, o indoor heated, swimming pool
— f para natación f - [deport] swimming pool
piso m - . . . . • [autom] deck • [metal-prod]
course • [autom] floorboard; footboard
— m alfombrado - [constr] carpeted floor
— m alto - [constr] upper, floor, o stor(e)y
— m asegurado - [constr] secured floor
—cielo raso - [constr] floor-ceiling
— m conformado - [constr] formed floor(ing)
— m — para puente m - [constr] formed bridge
floor(ing)
— m de acero m - [constr] steel floor(ing)
— m — — m para puente m - [constr] bridge
steel, o steel bridge, flooring
— m — adoquines m - [constr] block floor(ing)
— m — — m de madera f - [constr] wood block
floor(ing)

piso m de aminato m - [constr] asbestos floor-
-(ing)
— m — — m vinílico - [constr] vinyl asbestos
floor(ing)
— m — baldosa(s) f - [constr] tile floor(ing)
— m — — f de amianto - [constr] asbestos tile
floor(ing)
— m — — f — — m vinílico - [constr] vinyl
asbestos tile floor(ing)
— m — — f elástica(s) - [constr] resilient
tile floor(ing)
— m — bloques m - [constr] block floor(ing)
— m — — m de madera f - [constr] wood(en)
block(s) floor(ing)
— m — — — f de contrahilo m - [constr]
end grain wood(en) block(s) floor(ing)
— m — — — m — — f para uso m interior -
[constr] interior use wood block floor(ing)
— m — cámara f subterránea - [electr-instal]
manhole floor(ing)
— m — emparrillado m - [constr] grid flooring
— m — — m de acero m - [constr] steel grid
floor(ing)
— m — hormigón m - [constr] concrete flooring
— m — para puente(s) m - [constr] concrete bridge deck(ing)
— m — — m prevaciado - [constr] precast concrete deck(ing)
— m — ladrillo(s) m - [constr] brick flooring
— m — — m refractario(s) - [constr] fire-
brick floor(ing)
— m — madera f - [constr] wood (block) floor-
-(ing)
— m — — f de contrahilo m - [constr] end
grain wood floor(ing)
— m — — f desgastado - [constr] worn wood
floor(ing)
— m — — f dura - [constr] hardwood timber,
floor(ing), o deck(ing)
— m — — f gastado - [constr] worn wood
floor(ing)
— m — — f para uso m interior - [constr]
interior use wood floor(ing)
— m — metal - [constr] metal floor(ing)
— m — muelle m - [constr] deck, floor(ing), o
surface
— m — plancha(s) f - [constr] plate(s) floor-
-(ing)
— m — — f estriada(s) f - [constr] checkered
plate floor(ing)
— m — rejilla f - [constr] (steel) grid
floor(ing)
— m — — f de acero m - [constr] steel grid
floor(ing)
— m — tabla(s) f - [constr] board, o plank,
floor(ing)
— m — tablón(es) m - [constr] plank floor(ing)
— m — — m de madera f - [constr] wood planks
floor(ing)
— m — — m — — f desgastado - [constr] worn
wood plank(s) floor(ing)
— m — valle - [topogr] bottom land
— m — velódromo m - [deport] cycling track
— m desgastado - [constr] worn floor(ing)
— m detrás de puerta f - [constr] floor(ing),
inside, o behind, @ door
— m existente - [constr] existing floor(ing)
— m fijado - [constr] secured floor(ing)
— m gastado - [constr] worn floor(ing)
— m horizontal - [constr] horizontal floor(ing)
• [autom] floor board
— m inferior - [constr] lower floor
— m instalado - [constr] installed floor(ing) •
[puentes] installed deck(ing)
— m metálico - [constr] metal(lic) floor(ing)
— m oblicuo - [autom] toe board
— m para cabina f - [autom] cab floor
— m — camioneta f (rural) - [autom] station
wagon, floor, o deck(ing)
— m — carga f - [ind] charging, floor, o aisle
— m — colada f - [metal-prod] casting floor
— m — distribuidor m - [metal-prod] larry (car)
floor

piso m para enganche m - [petról] lazy board
— m — fosa f - [metal-lam] pit bottom
— m — f plano - [metal-lam] flat pit bottom
— m — f razonablemente plano - [metal-lam] reasonably flat pit bottom
— m — horno m - [combust] furnace floor
— m — m de fosa f - [metal-lam] soaking pit bottom
— m — m — f plano - [metal-lam] flat soaking pit bottom
— m — m — f razonablemente plano - [metal-lam] rasonably flat soaking pit bottom
— m — plano - [metal-lam] flat furnace bottom
— m — m razonablemente plano - [metal-lam] reasonably flat furnace bottom
— m — lavador m - [ind] washer, o scrubber, floor
— m — m para gas m - [metal-prod] gas, scrubber, o washer, floor
— m — moldeo m - [metal-prod] casting floor
— m — operación(es) f - [ind] operating, deck, o floor, o platform
— m — f para bomba f - [bombas] pump operating, floor, o platform
— m — plataforma f - [constr] platform floor • [ind] bench floor
— m — puente m - [constr] bridge, floor(ing), o deck
— m — m asegurado - [constr] secure bridge, floor(ing), o deck
— m — m de acero m - [constr] steel bridge floor(ing) • bridge steel floor(ing)
— m — m fijado - [constr] scured bridge floor(ing)
— m — torre f — f para perforación f - [petról] (drilling) derrick floor(ing)
— m — trabajo m - [ind] work, o operating, floor
— m — uso m interior - [constr] interior use floor(ing)
— m — vagón m - [f.c.] car floor(ing)
— m — m distribuidor - [metal-prod] larry (car) floor(ing)
— m — m para carga f - [f.c.] freight car floor(ing)
— m — vía f - [f.c.] road, o track, bed
— m plano - [constr] flat floor
— m — para horno m - [ind] flat, furnace, o pit, floor, o bottom
— m — m de fosa f - [metal-lam] soaking pit flat, o flat soaking pit, bottom
— quitado - [constr] removed floor
— m removible - [constr] removable floor
— m, removido, o sacado - [constr] removed floor
— m sólido - [constr] solid floor • [puentes] solid deck
— m superior - [constr] upper, floor, o story
— m terminado - [constr] finished floor
— m típico - [constr] typical floor(ing)
— m — para puente m - [constr] typical bridge floor(ing)
— m vaciado - [constr] poured, o cast, floor
— m vinílico - [constr] vinyl floor(ing)
pisón m - [herram] . . .; tamper; tamping pad • [mec] ram
— m corriente - [herram] conventional, o sidewalk, tamper
— m neumático m - [constr-equip] pneumatic tamper
— m — copa f - [extrusión de electrodos] cup tamper
pisonador m - [mec] holder
— m para chapa(s) f - [sold] sheet holder
pista f - . . . • [miner] road • [cojin] race; cup • [aeron] tarmac • [deport] (running) track; course; rink • slope
— f alpina - [deport] Alpine, slope, o slide
— f asentada - [cojin] seated, race, o cup
— f áspera - [deport] rough course • [vial] rugged course

pista f atlética - [deport] (race) track
— f barrosa - [deport] muddy, o soggy, course
— f caminera - [deport] road course
— f cerrada - [deport] closed course
— f con cambio m doble de vía f - [autom] double lane change course
— f — figura f de lazo m - [deport] loop type, course, o track
— f — . . . milla(s) f - [deport] . . .-mile track
— f obstáculo(s) - [deport] steeplechase, o obstacle, track • [autom] pylon, o cone, course
— f — valla(s) f - [deport] obstacle course
— f conservada - [miner] maintained road
— f corta - [deport] short, track, o course
— f de aeropuerto m - [aeron] airport runway
— f — asfalto m - [aeron] asphalt runway
— f — m para despegue m - [aeron] (takeoff) asphalt, o asphalt (takeoff), runway
— f — hormigón m (para despegue m) - [aeron] concrete (takeoff) runway
— f — una milla f - [deport] (one) mile track
— f diferencial - [cojin] differential cup
— f difícil - [deport] difficult, o hard, course
— f — para carrera(s) f - [deport] difficult race course
— f en mina f - [miner] mine road
— f entre montaña(s) f - [deport] mountain course
— f exigente - [deport] demanding course
— f exterior - [deport] outside course • [cojin] outer race
— f — en cono m - [cojin] cone outer race
— f — — m para cojinete m - [cojin] bearing cone outer race
— f — golpe(te)ada - [cojin] tapped outer race
— f — para cojinete m - [cojin] bearing outer, cup, o race
— f — — rodamiento m - [cojin] outer ring
— f fuera de carretera f - [deport] off-road course
— f golpe(te)ada - [cojin] tapped race
— f instalada - [cojin] installed cup
— f — correctamente - [cojin] correctly installed cup
— f interior - [deport] inside course • [cojin] inner race
— f — en cono m • [cojin] cone inner race
— m — — o para cojinete m - [cojin] bearing cone inner race
— f — golpe(te)ada - [cojin] tapped inner race
— f — para cojinete m - [mec] inner bearing, o bearing inner, cup
— f — para rodamiento m - [cojin] inner ring
— f intermedia f - [deport-ski] intermediate slope
— f larga - [deport] long, track, o course
— f lenta - [deport] slow course
— f limpia f - [miner] clean road • [deport] clean track
— f limpiada - [miner] cleaned road • [deport] cleaned track
— f mayor - [deport-ski] advanced slope
— f mojada - [vial] wet, o sloppy, track, o course
— f para aparcamiento m (para avións m) - [aeron] apron
— f — aterrizar,zaje - [aeron] landing strip; runway
— f — — m con instrumento(s) m - [aeron] instrument (landing), runway, o strip
— f — — m existente - [aeron] existing (landing) runway
— f — — m para avión(es) m comercial(es) - [aeron] commercial use runway
— f — — m pavimentada - [aeron] paved, o surfaced, runway
— f — — m — para uso m comercial - [aeron] commercial use, paved, o surfaced, runway
— f — bola(s) f - [cojin] ball race
— f — — f para cojinete(s) m - [cojin] bear-

ing ball race

pista f para carrera(s) f - [deport] racetrack; race course

— f —— f para velocidad f alta - [deport] speedway; high speed race course

— f —— f —— f — en carretera(s) f - [autom] high spoeed road racing course

— f —— f pedestre(s)—[deport] running track

— f — carreteo m - [aeron] taxiway

— f — cojinete m - [cojin] bearing, race, o cup

— f —— m apretada dentro de portadiferencial m - [autom-mec] bearing cup pressed in(to) @ differential carrier

— f —— m asentado - [mec] seated bearing cup

— f —— m diferencial - [cojin] differential bearing cup

— f —— m exterior - [cojin] outer bearing cup

— f —— m golpe(te)ada - [cojin] tapped bearing cup

— f —— m instalada - [mec] installed bearing cup

— f —— m interior - [cojin] inner bearing cup

— f —— m para diferencial - [mec] differential bearing cup

— f —— m — engranaje m - [mec] gear bearing cup

— f —— m —— m lateral - [autom-mec] side gear bearing cup

— f —— m —— m para árbol m - [autom--mec] shaft side gear bearing cup

— f —— m —— m —— m para salida f (de fuerza f) - [autom-mec] output shaft side gear bearing cup

— f —— m — piñón m - [mec] pinion bearing cup

— f — cono m - [cojin] bearing cup ring

— f —— m apretada - [mec] pressed bearing cup

— f — dato(s) m - [comput] data track

— f — despegue m o aterrizaje m - [aeron] runway; take-off or landing strip

— f — diferencial m - [mec] differential cup

— f — enfriadora f - [mec] cooler track

— f — engranaje m - [cojin] gear cup

— f —— m lateral - [cutom-mec] side gear cup

— f —— m — para salida f (de fuerza f) - [autom-mec] output side gear cup

— f — ensayo(s) m - testing, o proving, grounds, o track

— f — entrenamiento m—[deport] training track

— f —— m con obstáculo(s) m - [deport] steeplechase training track

— f — instrucción f - [deport] teaching slope

— f — maniobra(s) f - [aeron] taxiway

— m — patinaje m - [deport] skating rink - [autom] skidpad

— f — principiante(s) m - [deport] beginner's, o teaching, slope, o track

— f — prueba(s) f - [autom] proving grounds

— f — quemador m - [combust] burner trail

— f —— m principal - [combust] main burner trail

— f — recorrido m - [deport] road course

— f — rodadura f - [mec] roller,ling track

— f —— f para enfriadora f - [mec] cooler rolling track

— f —— f —— f rotativa - [mec] rotary cooler roller track

— f —— f —— f — para sínter m - [metal--prod] sinter rotary cooler roller track

— f — rodamiento m - [cojin] ring

— f — slalom m - [deport] slalom course

— f — viento(s) m atravesado(s) - [aeron] cross-wind runway

— f paralela - [aeron] parallel runway

— f — para aterrizaje m - parallel landing, strip, o runway

— f —— m con instrumento(s) m - [aeron] parallel instrument, strip, o runway

— f pavimentada - [aeron] paved, o hardsur-

faced, runway

pista f peraltada - [constr] banked track

— f piloto - pilot trail

— f rápida - [deport] fast track

— f ruda - [deport] tough course

— f sembrada con despojo(s) m - [deport] littered course

— f torcida f - [deport] crooked track • [cojin] cocked cup

pistola f - . . . . • [pint] sprayer; spray gun • [sold] (welding) gun; torch

—- f adaptable - [sold] flexible gun

— f alimentadora - [sold] feeder,ding gun

— f — para alambre m - [sold] wire feeder,ding gun

— f con peso m reducido - [sold] light weight gun

— f — prolongación f - [sold] stickout gun

— f —— f para electrodo m sobresaliente - [sold] extended stickout gun

— f — soplete m - [mec] spray gun • [sold] torch gun

— f — suplemento m - [sold] stickout gun

— f —— m para prolongación f para electrodo m sobresaliente - [sold] extended stickout gun

— f eléctrica - [sold] electric gun

— f en operación f - [sold] operating, o in action, gun

— f engrasadora - [lubric] grease,sing gun

— f equipada para avance m mecanizado - [sold] travel unit equipped gun

— f Innershield - [sold] Innershield gun

— f — liviana - [sold] light Innershield gun

— f —— (y) felxible - [sold] light (and) flexible Innershield gun

— f — maniobrable - [sold] flexible Innershield gun

— f — para alambre m con alma m fundente - [sold] Innershield gun

— f Lincoln - [sold] Lincoln (Squirt)gun

— f — electrodo m de alambre - [sold] Lincoln Squirtgun

— f liviana - [milit] light gun • [sold] light gun

— f — y maniobrable - [sold] light (and) flexible gun

— f manual - [sold] hand (held) gun

— f — para alimentación f automática de electrodo m - [sold] hand-held Squirt gun

— f — Squirt - [sold] hand-held Squirt gun

— f obstruida - [sold] clogged gun

— f — con escoria f - [sold] spatter, o slag, clogged gun

— f para alambre - [sold] wire gun

— f ···—— m antiguo - [sold] old wire gun

— f —— m con alma m fundente - [sold] Innershield gun

— f —— m —— m — liviana - [sold] light Innershield gun

— f —— m —— m —— maniobrable - [sold] light (and) flexible Innershield gun

— f — alimentación f automática - [sold] semiautomatic gun; Squirtgun

— f —— f de, alambre m, o electrodo m - [sold] Squirtgun; semiautomatic gun; Squirt gun

— f — arco m sumergido - [sold] submerged arc gun

— f — electrodo m de alambre m - [sold] Squirtgun

— f —— m —— m con alma m fundente - [sold] Innershield Squirtgun

— f —— m —— m — para arco m sumergido - [sold] submerged arc Squirtgun

— f —— m —— m — soldadura f por arco m sumergido - [Sold] submerged arc Squirtgun

— f — engrase m - [mec] grease gun

— f — enjuague m (con chorro m de agua m) - [hidr] flushing gun

— f — de radiador m con chorro m de agua m - [autom] (radiator) flushing gun

— f — grasa f - [lubric] grease,sing gun

pistola f para metalización f - [metal-fabr]
metallizing gun
— f —— f con gas m de acetileno m - [metal-
-fabr] acetylene gas metallizing gun
— f — perforación f - [petról] gun perforator
— f — pintar v - [pint] painting, o paint
spraying, gun
— f — polvo(s) m - powder gun
— f — pulverización f - [pint] pulverizing, o
spray(ing), gun
— f — soldadura f - [sold] welding gun
— f —— f con arco m sumergido - [sold] submerged arc welding gun
— f — trabajo(s) m - [sold] work(ing) gun
— f —— m liviano - [sold] light-duty gun
— f —— m mediano(s) m - [sold] medium-duty
gun
— f —— m pesado(s) m - [sold] heavy-duty gun
— f — voltaje m alto - [sold] high voltage gun
— f —— m bajo - [sold] low voltage gun
— f recomendada - [sold] recommended gun
— f rociadora - [pint] spray(ing) gun
— f sin prolongación f - [sold] no stickout gun
— f —— f para electrodo m sobresaliente -
[sold] gun without @ extended stickout
— f — suplemento m - [sold] no stickout gun
— f —— m para prolongación f para electrodo
m sobresaliente - [sold] gun without extended
stickout
— f singular - [sold] unique gun
— f — con alimentación f automática - [sold]
semiautomatic weldinggun
— f soplete - [sold] torch gun
— Squirt - [sold] Squirt (welding) gun
— f — para alimentación f de, alambre, o electrodo - [sold] Squirt welding gun
— f Squirtgun - [sold] (innershield) Squirtgun
— f — de Lincoln - [sold] Lincoln Squirtgun
— f — Innershield - [sold] Innershield Squirtgun
— f — para arco m sumergido - [sold] submerged
arc Squirtgun
— f —— soldadura f con arco m sumergido -
[sold] submerged arc Squirtgun
— f taponada - [sold] clogged gun
— f — con escoria f - [sold] spatter clogged
gun
pistolete m - . . . . • [dib] French curve
pistón m - . . . . • prefiérase émbolo m
pistonear v - [petról] to swab
pistoneo* m - [mec] swabbing
pito m de alarma - [segurid] alarm whistle
pitón m - [mec] . . . . • screw eye
pivote m - [mec] . . .; pivot • spindle; gudgeon
• axis; knuckle; pintle
— m central - [mec] center pivot; véase también
pasador m central
— m con charnela f - [autom] knuckle pivot
— m —— f para dirección f - [autom-mec]
steering knuckle pivot
— m guía - [mec] véase pivote m
— m inferior - [mec] lower pivot
— m — para dirección f - [autom] lower steering pivot
— m para dirección f - [autom-mec] kingpin;
kingbolt; steering pivot
— m — mecanismo m para dirección f - [autom-
-mec] steering pivot
— m — rueda f - [autom] wheel spindle
— m — sujetador m - [mec] fastener pivot
— m sujetador - [mec] fastener,ning pivot
— m superior - [autom-mec] upper pivot
— m — para dirección f - [autom-mec] upper
steering pivot
pizarra f arcillosa - [geol] argillaceous slate
— f extraída - [miner] quarried slate
— f micácea - [geol] micaceous slate
pizarroso,sa a - [geol] shal(e)y
placa f - [metal-prod] (ingot) mold stool -
[metal-lam] sheet; plate; faceplate • [metal-
-prod] cooling plate • [electrón] shield •
• [deport] plaque • [mec] platen • [pol] com-
memorative, plaque, o brass

placa f aforadora f - [comb.int] throttle plate
— f aisladora - [metal-prod] insulating,tion
plate
— f — para carburador m - [comb.int] carburetor insulating plate
— aislante - [metal-prod] side board; insulating plate
— f — exotérmica - [metal-prod] exothermic
(insulating) side board
— f angular - [mec] angle plate
— f — para refuerzo m - [mec] gusset plate
— f apropiada - [mec] proper, o appropriate,
plate
— f — para filtro m - [segurid] proper safety
plate
— f —— tapar careta f - [segurid] proper, o
appropriate, cover(ing) (shield) cover plate
— f calibrada - [mec] orifice plate
— f — para medición f - [combust] measurement
orifice plate
— f —— f para combustible m - [combust]
fuel measurement orifice plate
— f central - [mec] center,tral plate
— f — para embrague m - [mec] clutch center
plate
— f —— m inferior - [mec] low clutch center plate
— f —— m superior - [mec] high clutch
center plate
— f cobertora* - [mec] cover(ing) plate
— f compactadora - [mec] compacting, plate, o
beam
— f compensadora - [mec] compensating plate;
wearing plate
— f complementaria - [com] complementary plate
— f completa - [mec] complete plate
— f con aleta(s) f - [mec] ear, o lug, plate
— f — borne(s) m - [electr-mot] terminal
plate
— f — característica(s) f - characteristic(s)
plate; nameplate
— f — calidad f estructural - [metal-lam]
structural quality plate
— f — dato(s) m - information plate
— f — dureza f brinell elevada - [mec] high
Brinell plate
— f — información f - information plate
— f —— f en vehículo m -
[autom] vehicle information, plate, o placard
— f —— f sobre neumático m - [autom-neumát]
tire information placard
— f — instrucción(es) f - instruction plate
— f —— f para cubierta f - [mec] cover instruction plate
— f —— f — operación f - [mec] operation
instruction plate
— f —— f —— f de tapa f - [válv] cover
operation instruction plate
— f —— f —— f —— f para válvula f -
[válv] valve cover operation instruction
plate
— f —— f — tapa f - [mec] cover instruction
plate
— f —— f —— f para válvula f - [válv]
valve cover instruction plate
— f — ojo m - [mec] eye plate
— f —— m con resorte m - [mec] spring eye
plate
— f — orificio m - [mec] orifice plate
— f —— m para medición f - [combust] measurement orifice plate
— f —— m —— f para combustible m -
[combust] fuel measurement orifice plate
— f —— m —— toma f de presión f - [mec]
pressure taking orifice plate
— f —— m variable - [mec] variable orifice
plate
— f — resorte m - [mec] spring plate
— f corrediza - [mec] sliding plate
— f corta - [mec] short plate
— f criba - [mec] screen plate
— f cuadriculada - [mec] checker(ed) plate
— f de acero m - [metal-prod] steel plate

placa f de acero m con manganeso m - [metal-lam] manganese steel plate
— f —— m —— m para mandíbula f - [mec] manganese steel jaw plate
— f —— m para mandíbula f - [mec] steel jaw plate
— f —— m para refuerzo m - [mec] steel reinforcing plate
— f —— m —— m exterior - [mec] external steel plate reinforcement,cing
— f — cinc m - [metal-lam] zinc plate
— f — cobre m - [metal-lam] copper plate
— f — hierro m - [metal-lam] iron plate
— f —— m magnético - [metal-lam] magnetic iron plate
— f deflectora - [mec] baffle plate
— f — contra remolino(s) m [hidr] (anti-vortex) baffle, plate, o board
— f delantera - [mec] front (end) plate
— f — para bloque m - [mec] block front (end) plate
— f —— m de cilindro(s) m - [comb.int] cylinder block front end plate
— f descargadora f - [mec] stripper plate
— f descartada - [metal-prod] rejected stool
— f desconectadora - [petról] breakout plate
— f deslizante - [mec] sliding plate
— f desmontable - [mec] removable plate
— f — para tapa f - removable cover plate
— f —— f inferior - [mec] bottom cover removable plate
— f —— f superior - [mec] top cover removable plate
— f desplazadora - [mec] shifter plate
— f despojadora - [agric-equip] stripper plate
— f desviadora - [constr] deflection vane • véase también placa f deflectora
— f difusora - [mec] diffuser (plate)
— f en extremo m - [mec] end plate
— f —— m para accionamiento m - [mec] end drive,ing plate
— f en vehículo m - [autom] vehicle, placard, o plate
— f —— m con información f - [autom] vehicle information, placard, o plate
— f —— m —— f sobre neumático(s) m - [autom] vehicle tire information placard
— f enfriada - [metal-lam] cooled plate
— f espaciadora f - [mec] spacer plate
— f estriada - [metal-lam] rippled plate
— f exotérmica - [metal-prod] exothermic, side board, o plate
— f fija - [mec] fixed, o stationary, plate
— f final - [mec] final plate • end plate
— f frontal - [torno] front plate
— f fusible - [mec] fuse plate
— f — débil - [mec] weak fuse plate
— f giratoria f - [mec] turntable; swivel
— f — principal - [grúas] main swivel
— f horizontal - [suelos] horizontal slab
— f — para base f - [constr] horizontal base plate
— f horquilla - [mec] yoke, o clevis, plate
— f identificadora - identifying, o number, plate; nameplate
— f identificadora para motor m - [electr-mot] motor nameplate • [comb.int] engine nameplate
— f — para voltaje(s) m - [electr-equip] voltage, number, o name, plate
— f inclinada - [mec] slanting, o baffle, plate
— f — para base f - [constr-vial] inclined, o slanting, base plate
— f indicadora - nameplate
— f — Luz-No - [electr-instal] On-Off face plate
— f — para capacidad f nominal - [mec] load rating plate
— f —— conmutador m - [sold] switch nameplate
— f —— fuente f para energía f - [sold] voltage power source nameplate
— f —— f —— m en voltaje m constante - [sold] constant voltage power source nameplate

placa f indicadora f para fuente f para energía en voltaje m variable - [sold] variable voltage power source nameplate
— f —— regulación f - control plate
— f —— f para velocidad(es) f - [mec] speed(s) control plate
— f — para reóstato m - [electr-equip] rheostat nameplate
— f —— voltaje m - [electr-instal] voltage number plate
— f —— m para entrada f - [electr-instal] input voltage number plate
— f —— m — salida f - [sold] output voltage number plate
— f —— m constante - [electr-instal] constant voltage nameplate
— f —— m variable - [electr-instal] variable voltage namelate
— f —— base f para enchufe m - [electr-instal] plug, o receptacle, nameplate
— f —— fusible(s) m - [electr-instal] fuse nameplate
— f individual - [metal-lam] individual plate
— f inferior - [mec] lower plate
— f — para soporte m - [mec] lower, o bottom, support plate
— f —— sujeción f - [mec] lower face plate
— f informativa - information, plate, o placard
— f — sobre automóvil m - [autom] automobile information, plate, o placard
— f — neumático m - [autom-neumát] tire information, plate, o placard
— f inmovilizada - [mec] fixed, o unmovable, plate
— f laminada - [metal-lam] rolled plate
— f larga - [metal-lam] long plate
— f lateral - [mec] side plate
— f — para embrague m - [mec] clutch side plate
— f limitadora - [mec] limit(ing) plate • top plate
— f limpia - [mec] clean plate
— f limpiada - [metal-lam] cleaned plate • [metal-lam] cleaned (ingot) stool
— f lubricadora - [comb.int] oiler plate
— f — a bloque m - [comb.int] oiler plate to @ block
— f — para engranaje m - [mec] gear oiler plate
— f —— m para sincronización f - [comb.-int] timing gear oiler plate
— f — y (placa f) para empuje m a bloque m - [comb.int] oiler (plate) and thrust plate to block
— f matriz - [mec] stamp(ing) plate
— f nervada* - [metal-lam] ribbed plate
— f orificio - véase placa f con orificio m
— f para accionamiento m - [mec] drive,ving plate
— f —— m en extremo m - [mec] end drive,ving plate
— f — acelerador m - [comb.int] throttle plate
— f — acumulador m - [electr-acum] battery plate
— f — adaptación f - [mec] adapter plate
— f —— f para motor m - [int.comb] engine adapter plate • [electr-mot] motor adapter plate
— f — aislación f - [electr-instal] insulating plate
— f — ajuste m - [sold] adapter plate
— f — anclaje m - [constr] anchor(ing) plate
— f — anillo m - [mec] ring, platen, o plate
— f —— m para compresión f - [constr] compression ring plate(n)
— f — aplicación f para presión f - [constr] plat(t)en
— f — apoyo m - [constr] bearing plate • [mec] bed plate • rocket plate • [f.c.] level plate
— f — árbol m - [comb.int] shaft plate
— f —— m para leva(s) f - [comb.int] camshaft plate
— f — arrancar buje(s) m - bushing puller

plate
placa f para arrastre m - [tornos] dog driver
plate
— f — asiento m - [constr] bearing, o base,
plate • [mec] bedplate; soleplate • [f.c.]
tie plate
— f — — m modificada - [mec] changed, o modified, seat plate
— f — — m para busa f - [metal-prod] blow
pipe seat
— f — m — — f (para) codo m portavientos
- [metal-prod] blowpipe and gooseneck adaptor
(to make ENSIDESA blast furnace elements interchangeable)
— f — m — codo m - [metal-prod] gooseneck seat plate
— f — m — columna f - [constr] column
bearing plate
— f — banda f articulada - [ind] apron plate
— f — base f - [metal-prod] stool • [mec] bed
plate; ground plate • [constr] base plate
— f — f horizontal - [constr] horizontal
base plate
— f — f inclinada - [constr] inclined, o
slanted, base plate
— f — f para columna f - [constr] column
base plate
— f — f para soporte m - [mec] support base
plate
— f — f para sostén m - [mec] support base
plate
— f — f laminada - [metal-lam] rolled base
plate
— f — f para lingotera(s) f - [metal-prod]
stool; ingot, mold, o stool
— f — f — molde(s) m - [metal-prod] stool
— f — f proyectada - [constr] designed
base plate
— f — biela f - [mec] crankshaft plate
— f — f con sujeción f con perno(s) m -
[comb.int] rod bolt lock plate
— f — f para sujeción f de perno(s) m -
[combs.int] rod bolt lock plate
— f — caja f - [mec] box, o housing, plate
— f — f para cojinete m - [mec] bearing box
plate
— f — f — desplazador m - [mec] shifter
box plate
— f — f transportadora - [mec] apron plate
— f — cambiador m - [mec] shifter plate
— f — cambio(s) m - [mec] shifter plate
— f — cierre m - [mec] seal, o lock, plate •
end dam
— f — m para aceite m—[mec] oil seal plate
— f — cigüeñal m - [comb.int] crankshaft plate
— f — cimentación f - [constr] ground plate
— f — cimiento m - [mec] ground plate
— f — cobertura f - [mec] cover plate
— f — conexión(es) f - [sold] lead block •
[electr-instal] connection plate
— f — f con tierra f - [electr-instal]
ground(ing) plate
— f — f de cobre m - [electr-instal] copper
connection plate
— f — f triangular - [electr-instal] triangle, o triangular, connection plate
— f — f — de cobre m - [electr-instal]
triangular copper connection plate
— f — conmutador m - [electr-equip] switch
plate
— f — m para polaridad f - [sold] polarity
switch plate
— f — m para régimen m para carga f -
[electr-instal] charging rate switch plate
— f — crisol m - [metal-prod] hearth, stave, o
plate
— f — cuadrante m - [instrum] dial plate
— f — m para amperaje m - [instrum] amperage, o current, dial plate
— f — cubierta f - [mec] cover plate; crown
sheet
— f — f para caja f - [mec] box, o cabinet,
cover plate

placa f para cubierta f para caja f auxiliar
para regulación f - [mec] auxiliary cabinet
cover plate
— f — — f — caja f para cojinete m - [mec]
bearing box cover plate
— f — — f para regulación f - [mec]
(control) cabinet cover plate
— f — cubo m - [mec-ruedas] hub plate
— f — — para rueda f - wheel hub plate
— f — — para extremidad f - [mec] end hub
plate
— f — cubrir v - [mec] cover(ing) plate
— f — v cursor m - [mec] slide cover plate
— f — cuña f - [mec] wedge plate
— f — desconexión f - [petról] breakout plate
— f — f para barreno,na - [petról] bit
breakout plate
— f — f — con cola f de pescado m -
[petról] fish tail bit breakout plate
— f — f para trépano m - [petról] bit
breakout plate
— f — desenroscadura f - [petról] breakout
plate
— f — f de barreno,na - [petról] bit
breakout plate
— f — f — trépano m - [petról] bit breakout plate
— f — desenroscar v trépano(s) m - [petról]
bit breaker
— f — desgaste m - [mec] wear plate
— f — — m para cojinete m - [mec] bearing
wear plate
— f — deslizamiento m - [mec] slide,ding plate
— f — — m para rodillo m - [metal-lam] roll
sliding plate
— f — desobstrucción f - [mec] clean-out plate
— f — desplazador m - [mec] shifter plate
— f — desviación f - [mec] baffle plate
— f — diafragma m - [mec] diaphragm plate
— f — embrague m - [mec] clutch plate
— f — — m inferior - [mec] low(er) clutch
plate
— f — — m para automóvil m - [autom-mec]
automotive clutch plate
— f — — m — desenroscadura f - [petról]
breakout clutch plate
— f — — m — enroscadura f - [petról] make-up
clutch plate
— f — — m velocidad f alta - [autom-mec]
high speed clutch plate
— f — — m — f baja - [autom-mec] low
speed clutch plate
— f — m superior - [mec] high, o top, o
upper, clutch plate
— m — — m rápido - [autom-mec] quick clutch
plate
— f — — m ultrarrápido - [autom-mec] (very)
high speed clutch plate
— f — empalme m - [mec] gusset plate
— f — empuje m - [mec] thrust, o push(er),
plate
— f — — m para árbol m - [mec] shaft thrust
plate
— f — — m — — m para leva(s) f - [comb.int]
camshaft thrust plate
— f — — m — — m — — f a bloque m - [int.-
-comb] camshaft thrust plate to block
— f — — m para bloque m - [int.comb] thrust
plate to block
— f — — m para cigüeñal m - [comb.int] crankshaft thrust plate
— f — — m — eje m - [mec] shaft thrust plate
— f — — m — — m para leva(s) f - [mec] camshaft thrust plate
— f — enfriamiento m - [ind] cooling plate
— f — — m para cuba f - [metal-prod] stack
cooling plate
— f — — m — etalaje m - [metal-prod] bosh
cooling plate
— f — — m — solera f - [metal-prod] hearth
cooling plate
— f — enroscadura f - [petról] make-up plate
— f — escobilla f - [electr-mot] brush plate

placa f para esfera f - [instrum] dial plate
— f — — f para amperaje m - [instrum] amperage dial plate
— f — — f — voltaje m - [instrum] voltage
dial plate
— f — estrangulador m - [comb.int] choke plate
— f — etalaje m - [metal-prod] bosh plate
— f — extremo m - [mec] end, plate, o cover
— f — — m para árbol m - [mec] shaft end
plate
— f — — — m para tambor m - [mec] drum
shaft end plate
— f — — m para contraárbol m - [mec] jackshaft end plate
— f — — m — contraeje m - [mec] jackshaft
end plate
— f — — m — impulsión f - [mec] end drive
plate
— f — filtrar v - [mec] filter(ing) plate
— f — fondo m - [mec] bottom, plate, o sheet •
[metal-prod] stool
— f — — m para lingotera(s) f - [metal-prod]
(ingot) stool
— f — — m — molde(s) m - [metal-prod] ingot
mold stool
— f — forro m - [mec] filler bar
— f — fricción f - [mec] friction plate
— f — — f (des)gastada - [mec] worn friction
plate
— f — — f para embrague m - [mec] clutch
friction plate
— f — — — m para desenroscadura f -
[petról] breakout clutch friction plate
— f — — — m para enroscadura f -
[petról] make-up clutch friction plate
— f — — f para torno m - [petról] cathead
friction plate
— f — gato(s) m - [mec] jack(ing) plate
— f — identificación f - identification, o
number, plate; nameplate
— f — — f con diagrama m—[ind] diagram(atic)
nameplate
— f — — f diagramático - [ind] diagram(atic)
nameplate
— f — — f mostrada - shown identification
plate
— f — — f para conmutador m - [electr-instal]
switch nameplate
— f — — — — m para polaridad f - [electr-
-instal] polarity switch nameplate
— f — — f — — m — — f para medidor m -
[electr-instal] meter polarity switch nameplate
— f — — f — corriente f alterna - [sold]
alternating current nameplate
— f — — — — f continua - [sold] direct
current nameplate
— f — — f — embrague m - [mec] clutch nameplate
— f — — f — — m con rueda f libre - [mec]
overrunning clutch nameplate
— f — — f — fusible m - [electr-instal] fuse
nameplate
— f — — f — interruptor m - [electr-instal]
switch nameplate
— f — — — m para línea f - [electr-
-instal] line switch nameplate
— f — — f — malacate m - [petról] drawworks,
nameplate, o identification plate
— f — — f — máquina f - [mec] machine nameplate
— f — — f — puerta f - [mec] door, number
plate, o nameplate
— f — — f — — f para tablero m - [sold]
panel door, number plate, o nameplate
— f — — — — — m para selección f
para voltaje m - [sold] - [sold] voltage selector panel door, number plate, o nameplate
— f — — f — soldadora f - [sold] welder
nameplate
— f — — f — tablero m para selección f -
[electr-instal] selector panel nameplate
— f — — f — torno m - [petról] cathead name-

plate
placa f para identificación f para transformador
m - [electr-instal] transformer nameplate
— f — impulsión f - [mec] drive,ving plate
— f — — f para extremo m - [mec] end drive,-
ving plate
— f — inspección f - [mec] inspection plate
— f — interruptor m - [electr-instal] switch,
o breaker, plate
— f — — m para magneto m - [comb.int] magneto
switch plate
— f — jaula f - [mec] cage plate
— f — — f para cojinete m - [cojin] bearing
cage plate
— f — — f — — m para piñón m - [cojin] pinion bearing cage plate
— f — limpieza f - [mec] clean-out plate
— f — lingotera f - [metal-prod] ingot mold
stool
— f — — f descartada - [metal-prod] rejected
(ingot) mold stool
— f — — f limpia - [metal-prod] clean (ingot)
mold sstool
— f — — f limpiada - [metal-prod] cleaned
(ingot) mold stool
— f — — f recubierta - [metal-prod] coated
(ingot) mold stool
— f — lubricante(s) m - [mec] lubricant plate
— f — llave f - [electr-instal] switch plate
— f — — f para encendido m - [mec] ignition
switch plate
— f — madrastra f - [metal-prod] mantle plate
— f — mandíbula f - [mec] jaw plate
— f — marca f - [ind] (mark) nameplate
— f — montaje m - [mec] mounting plate
— f — — m para arrancador m - [sold] starter
mounting plate
— f — — m — caja f - [mec] box mounting
plate
— f — — m — — f para engranaje(s) m -
[sold] gears box mounting plate
— f — — m para enchufe m - [electr-instal]
receptacle mounting plate
— f — — m — — m para conversión f—[electr]
conversion receptacle mounting plate
— f — — m — interruptor m - [electr-instal]
switch, o breaker, mounting plate
— f — motor m - [electr-mot] motor plate •
[comb.int] engine plate
— f — — m con borne(s) m - [electr-mot]
motor terminal(s) plate
— f — nivel m para aceite m - [mec] oil level
plate
— f — nivelación f - [mec] leveling plate
— f — número(s) m - number(s) plate
— f — — m para puerta f - [constr] door
number(s) plate
— f — — m — — f para tablero m - [ind]
panel door number(s) plate
— f — — m — — f — — m para selección f -
[sold] selector panel door number plate
— f — — m — — f — — m — — f de voltajes
m - [sold] voltage selector panel door number
plate
— f — obturador m - [mec] packer plate
— f — — m para retén m - [mec] retainer,ning
packer plate
— f — perno m - [mec] bolt plate
— f — piquera f - [metal-prod] (iron) notch
cooling plate, o cooler
— f — portaescobilla(s) m - [electr-mot]
brushholder plate
— f — presión f - [mec] pressure plate
— f — — f para embrague m - [mec] clutch
pressure plate
— f — protección f - [mec] protection plate
— f — — f para nivel m - [ind] level protection plate
— m — — f — de carga f - [metal-prod]
stock lining protection plate; stockline ring
— f — punta f - [mec] point, o end, plate
— f — reconocimiento m - (recognition) plaque
— f — — m por operación f sin accidentes m -

placa f para recubrimiento m

     [ind] safety plaque
placa f para recubrimiento m - [mec] cover(ing)
plate
— f — — m para tapa f - [ind] lid covering
plate
— f — — m — — f inferior - [mec] lower, o
bottom, lid covering plate
— f — — m — — f superior - [mec] top, o upper, lid covering plate
— f — referencia f - [ind[reference plate
— f — refrigeración f - [ind] cooling, o
chill(ing), plate
— f — — f para etalaje m - [metal-prod] bosh
cooling plate
— f — — f — piquera f - [metal-prod] (iron)
notch cooling plate
— f — — f para zona f - [ind] zone, o area,
cooling plate
— f — refuerzo m - [constr] stiffening plate;
stiffener • [mec] gusset plate
— f — regulación f - [ind] control nameplate •
[comb.int] throttle plate
— f — regulador m - [comb.int] governor, o
throttle, plate
— f— rejilla f - [metal-prod] checker plate
— f — respaldo m - [mec] back-up, o back(ing),
plate
— f — — m para disco m - [instrum] disk backing plate
— f — — m — embrague m - [mec] clutch back
plate
— f — — m — posicionador m - [mec] placer, o
positioner, back(-up), o backing, plate
— f — — m para regulador m - [instrum] dial
plate
— f — respiración f - [válv] breather plate
— f — retén m - [mec] retainer plate
— f — — m para cojinete m - [cojin] bearing
retainer plate
— f — — m — m para bancada f - [mec] main
bearing retainer plate
— f — retención f - [mec] retaining plate
— f — — t corta - [mec] short retaining plate
— f — — f larga - [mec] long retaining plate
— f — revestimiento m - [mec] lining plate
— f — — m de tapa f - [mec] lid, o cover,
lining plate
— f — — m — — f inferior - [mec] lower, o
bottom, lid, o cover, lining plate
— f — — m — — f superior - [mec] top, o upper. lid, o cover, lining plate
— f — roce m - [mec] wear(ing) plate
— f — salpicadura(s) f - [metal-prod] splash(er)
plate
— f — salpique m - [metal-prod] splash(er)
plate
— f — sello m - [mec] seal(ing) plate
— f — — m para aceite m - [mec] oil seal(ing)
plate
— f — señal(es) f - signal(ling) plate •
[f.c.] target
— f — soporte m - [mec] support(ing) plate •
[sold] back-up strip • [metal-trat] thrust
plate • [constr] bearing plate
— f — — m para enclavamiento m - [electr-mot]
interlock(ing) support plate
— f — sostén m - [mec] support(ing) plate
— f — — m para contacto m - [comb.int] contact support(ing) plate
— f — sujeción f - [mec-tornos] faceplate •
[constr] anchor plate
— f — — f con perno(s) m - [mec] bolt locking
plate
— f — sujetar perno(s) m - [mec] bolt locking
plate
— f — tapa f - [mec] crown, plate, o sheet
— f — tapón m - [mec] packer plate
— f — — m para retención f - [mec] retaining
packer plate
— f — tierra f - [electr-instal] ground(ing),
rod, o plate
— f — tiro m - [mec] pull(ing) plate
— f — tope - [mec] bumper, o bomp(ing) plate

placa f para tobera f - [metal-prod] tuyere
plate
— f — tope m [mec] thrust plate
— f — — m para eje m - [mec] shaft thrust
plate
— f — — m — — m para leva(s) f - [mec] camshaft thrust plate
— f — — m para cursor m - [mec] slide bump-
ing plate
— f — — m — — m para hilera f - [mec] die
slide bump(ing) plate
— f — tracción f - [mec] pull(ing) plate
— f — tránsito m sobre pavimento m - [constr]
street plate
— f — transportador m (con banda t articulada)
- [ind] apron plate
— f — — m con cajas f - [ind] apron plate
— f — — m — mandril m - [ind] apron plate
— f — tubo(s) m - [tub] pipe, o flue, plate
— f — — m para humo(s) m - [combust] flue
plate
— f — — m — caldera(s) f - [cald] flue plate
— f — válvula f - [valv] cover plate
— f — varilla f medidora - [mec] dipstick
plate
— f — — f para nivel m para aceite m -
[mec] oil level dipstick plate
— f — velocidad f alta - [mec] high speed
plate
— f — — f para embrague m - [mec] high
speed clutch plate
— f — ventilación f - [mec] ventilation plate
• baffle plate
— f — zona f - [mec] zone, o area, plate
— f portadora - [constr] bearing, plate, o pad
— f — de neopreno m - [constr] neoprene bearing pad
— f portante - [constr] bearing plate
— f — de acero m - [constr] steel bearing
plate
— f preformada - [metal-fabr] preshaped plate
— f — de acero m - [metal-fabr] preshaped
steel plate
— f protectora - [mec] protection,tive plate •
face platae
— f — para cubierta f - [mec] cover(ing) plate
— f — — — f para horno m - [metal-lam] furnace covering plate
— f — — — f — — m de fosa f - [metal-lam]
soaking pit covering plate
— f quitada - [mec] removed plate
— f ranurada - [mec] grooved, o slotted, plate
— f recubierta - [metal-prod] covered (ingot)
stool
— f removible - [mec] removable plate
— f removida - [mec] removed plate
— f respiradora - [válv] breather plate
— f retenedora - [mec] retainer,ning plate
— f — para rodillo m - [mec] roller retainer
plate
— f — — m enderezador - [mec] straightening roller retainer plate
— f rotulada - (inscribed) nameplate
—rótulo - nameplate; véase también placa f para identificación f
— sacada - [mec] removed plate
— f semillera - [agric-equip] seed plate
— f separada - [mec] separated plate
— f sujetadora - [comb.int] holding, o lock,
plate
— f superior - [mec] upper plate • [cald] crown
sheet
— f — para hogar m - [cald] hearth crown sheet
— f — — — m para caldera f - [cald] (bolier)
hearth crown sheet
— f — — soporte m - [mec] upper support plate
— f — — sujeción f - [mec] upper face plate
— f terminal - [mec] end plate
— f — para conjunto m - [mec] assembly end
plate
— f — — — m de tambor m - [petról] drum assembly end plate
— f — — — m — — m para sacasonda(s) m -

[petról] coring reel drum end plate
placa f terminal y cubo m - [mec] end plate and hub
— f **tope** - [mec] stop plate
— f **trinagular** - [mec] triangular plate
— f **trituradora** - [mec] crushing, o grinding, plate
— f **tubular** - crown sheet
— f **única** - [mec] single plate
— f **variable** - [mec] variable plate
— f **y retén m** - [mec] plate and retainer
placé* adv - [deport] in @ money
placentero,ra a - . . .; enjoyable
placer m - . . .; enjoyment
plafón m - [arq] . . .; soffit
plafond m - véase techo m
plagado,da a con duende(s) m - gremlin-plagued
plagiopórfido m - [geol] plagio porphyry
plan m - . . . • program • project • scope
— m **aceptado** - accepted plan
— m **amplio** - broad, o comprehensive, plan, o project
— m — **para regulación f** - comprehensive control, plan, o project
— m — — — f **de inundación(es) f** - [hidr] broad, o comprehensive, flood control, plan, o project
— m **con aporte(s) m de empleado(s) m** - [labor] contributory plan
— m **con muestra f única** - single sample plan
— m **conocido** - known plan
— m **convenido** - agreed (upon) plan
— m **de escalafón m** - [labor] seniority promotion plan
— m **económico** - economic(al) plan
— m **enmendado** - amended, o revised, plan
— m — **para jubilación f** - [labor] revised, retirement, o pensions, plan
— m **extenso** - comprehensive, plan, o project
— m — **para regulación f** - [hidr] comprehensive control, plan, o project
— m — — — f **de inundación(es) f** - [hidr] comprehensive flood control, project, o plan
— m **funcional** - [ind] functional, plan, o pattern
— m — **para alimentación f** - [ind] funciontal feed(ing), plan, o pattern
— m — — — f **para horno m** - [ind] functional furnace feed(ing), plan, o pattern
— m **general** - general, o overall, o master, plan
— m — **para urbanización f** - [pol] master (development) plan
— m **impositivo** - [fisc[tax(ing) plan
— m **inmediato** - immediate plan
— m **maestro** - master, o key, plan
— m **Nacional para Desarrollo m** - [pol] (Argentina's) National Development Plan
— m — — — — m **Urbano m** - [pol] National Urban Development Plan
— m **original** - original plan
— m **para acción f** - action plan; policy
— m — — f **aceptado** - accepted action plan
— m — — f **comprendido** - understood action plan
— m — — **creado** - created, o established, o developed, action plan, o policy
— m — — f **entendido** - understood action plan
— m **acopio m** - gathering plan
— m — — m **de material(es) m** - [ind] materials gathering plan
— m — **alimentación f** - [ind] feeding pattern
— m — — f **para horno m** - [ind] furnace feeding, plan, o pattern
— m — **ampliación f** - expansion project
— m — **calidad f** - [ind] quality plan
— m — **capacitación f** - [labor] training, o qualifying, plan, o program
— m — **carga f** - [transp] loading schedule
— m — **comercialización f** - [com] marketing plan(ning)
— m — **construcción f** - [constr] construction plan

plan m para construcción f de central f nuclear (para energía f) - [nucl] nuclear power plant construction plan
— m — — f — **planta f nuclear (para energía f)** - [nucl] nuclear power plant construction plan
— m — **contingencia(s) f** - contingency plan
— m — **contrato m** - contract, plan, o scheme
— m — — m **sobre mejora(s) f** - [constr] improvement contract, plan, o scheme
— m — **control m** - [ind] control plan
— m — — m **para calidad f** - [ind] quality control plan
— m — **crédito m** - [fin] credit plan
— m — **cuidado m** - care plan
— m — — m **perpetuo** - perpetual care plan
— m — **entrega(s) f** - delivery,ries schedule
— m — **entrenamiento m** - [pers] training plan
— m — **estudio(s) m** - [educ] curriculum
— m — — m **para seminario m** - [labor] seminar, curriculum, o agenda
— m — **expansión f** - expansion plan
— m — **fabricación f** - [ind] manufacturing plan • manufacturing schedule
— m — **instalación f** - [ind] installation plan
— m — **inversión f** - [fin] investment plan
— m — **inyección f de agua m** - [petról] water-flooding plan
— m — **jubilación(es) f** - [labor] pension plan
— m — — f **con aporte(s) m de empleado(s) m** - [labor] contributory pension plan
— m — — f **sin aporte(s) m de empleado(s) m** - [labor] non-contributory pension plan
— m — **mercadeo m** - [com] marketing plan(ning)
— f — **montaje m** - mounting plate
— f — — m **en obra f** - [ind] field erection plan
— m — **muestra(s) f** - [com] sampling plan
— m — — f **doble(s)** - [com] double sample(s) plan
— m — — f **múltiple(s)** - [com] multiple samples plan
— m — **muestreo m** - [ind] sampling plan
— m — **pago m** - payment plan
— m — **promoción f** - [labor] promotion plan
— m — — f **por antiguedad f** - [labor] seniority promotion plan
— m — **protección f** - [ind] protection,tive, plan, o scheme
— m — — f **para red f** - [ind] system('s) protection,tive, plan o scheme
— m — **regulación f** - control, plan, o scheme, o project
— m — — f **de contaminación f** - [ambient] pollution control project|
— m — — — f **atmosférica** - [ambient] air pollution control project
— m — — f — — f **de aire m** - [ambient] air pollution control project
— m — **renovación f** - [pol] renewal project
— m — — f **urbana** - [pol] urban renewal project
— m — **seguimiento m** - follow-up plan
— m — **seguridad f** - [segurid] safety plan
— m — — f **radiológica** - [nucl] radiological safety plan
— m — **trabajo m** - work plan
— m — **verificación f** - control plan
— f — — f **no destructiva** - nondestructive control plan
— m — **viaje m** - travel plan
— m **perfecto** - perfect, plan, o scheme
— m — — m **de emisión(es) f** - [cerám] perfect emission(s) control scheme
— m — **para emisión(es) m** - [combust] perfect emission(s) scheme
— m — — **regulación f de emisión(es) f** - [cerám] perfect emission(s) control scheme
— m **perpetuo** - perpetual plan
— m — **para cuidado m** - perpetual care plan
— m **preconvenido** - (previously) agreed (on) plan
— m **preliminar** - preliminary plan

plan m principal - main, o key, o principal, plan
— m sin aporte(s) m de empleado(s) m - [labor] non-contributory plan
— m tributario - [fisc] tax(ing) plan
plana f - . . . . [herram] board • [imprent] spread
plancha f - . . . . • plate, o panel, section • platen • [tub] section • [sold] material • [domést] iron
— f aceitosa - [mec] oily plate
— f achaflanada - [metal-fabr] beveled plate
— f adyacente - [mec] adjacent plate
— f agrietada - [metal-lam] cracked plate • [constr] cracked plank
— f aisladora - [electr-mat] insulating plate
— f angular f - [metal-lam] angle plate
— f — a tope - [sold] square edge butt
— f apilada - [metal-lam] stacked plate
— f armada - [constr] assembled plate
— f Armco para revestimiento m - [metal-fabr] Arcmo liner plate
— f — — — m de túnel(es) m - [metal-fabr] Armco tunnel liner plate
— f Asbestos-Bonded - [metal-trat] Asbestos--Bonded, plate, o strip
— m — — de Armco - [metal-trat] Armco Asbestos-Bonded, plate, o strip
— f atiesadora f - [metal-fabr] battening member
— f azulada - [metal-prod] blued plate
— f calentada - [metal-trat] heated plate
— f — con soplete m—[sold] torch heated plate
— f central - [tub] heated panel
— f colocada - [mec] installed, o fitted, plate • [sold] positioned plate • [mec] placed plate
— f completa - [constr] full, o complete, plate
— f con aleación f - [metal-prod] alloy plate
— f — — f alta—[metal-prod] high alloy plate
— f — — f baja - [metal-prod] low alloy plate
— f — — f laminada en caliente - [metal-prod] hot rolled alloy steel plate
— f — — f — — frío - [metal-prod] cold rolled alloy steel plate
— f — borde(s) m recortado(s) - [metal-lam] sheared edge plate
— f — calibre m delgado - [metal-prod] light gage plate
— f — — m grueso - [metal-prod] heavy gage plate
— f — — m liviano - [metal-prod] light gage plate
— f — — m pesado - [metal-prod] heavy gage plate
— f — calidad f comercial - [metal-lam] commercial quality plate
— f — — f estructural - [metal-lam] structural quality plate
— f con carbono m - [metal-prod] carbon plate
— f — — m alto - [metal-prod] high carbon plate
— f — — m bajo—[metal-prod] low carbon plate
— f — — laminada en caliente - [metal-prod] hot rolled carbon plate
— f — — m — frío - [metal-prod] cold rolled carbon plate
— f — —manganeso m - [metal-lam] carbon-manganese plate
— f — — — m para recipientes m para presión f - [metal-lam] carbon-manganese pressure vessel plate
— f — dureza f Brinell alto - [metal-prod] high Brinell plate
— f — — f — bajo - [metal-prod] low Brinell plate
— f — — f — elevado - [metal-prod] high Brinnel plate
— f — enrejado m engatillado (interlock grating) - [metal-fabr] Interlock Grating plank
— f — espesor m diferente - [sold] different, plate, o material, thickness

plancha f con espesor m distinto.- [sold] different, plate, o material, thickness
— f — — m diverso - [sold] different, plate, o material, thickness
— f — — m igual - [metal-lam] same thickness plate
— f — — m ilimitado - [metal-lam] unlimited thickness plate
— f — forma f irregular - [metal-lam] sketch plate
— f — junta f traslapada - [constr] lap(ped) joint liner plate
— f — liga f de asbesto m - [metal-trat] Asbestos-Bonded, plate, o sheet, u strip
— f — — f — m de Armco - [metal-trat] Armco Asbestos-Bonded strip
— f — mismo calibre m - [metal-lam] same thickness plate
— f — — espesor m - [metal-lam] same thickness plate
— f — pestaña(s) f para revestimiento m - [constr] flanged liner plate
— f — f — — m para túnel(es) m - [constr] flanged tunnel liner plate
— f — propiedad(es) f para transición f - [metal-lam] transition property plate
— f — — f — — f mejorada(s) - [metal-lam] improved transition property plate
— f — radio m corto - [constr] short radius plate
— f — — m largo - [constr] long radius plate; large radius plate
— f — ranura f profunda a tope - [sold] deep groove butt
— f — tenor m alto de azufre m - [metal-lam] high sulfur plate
— f — vapor m - [domést] steam iron
— f conformada - [metal-lam] shaped, o formed, plate
— f — en U - [metal-lam] U, formed, o shaped, plate
— f cónica - [metal-fabr] conic(al) plate
— f — para techo m - [constr] conic(al) roof plate
— f contigua - adjacent plate
— f corrediza - [mec] sliding plate
— f corroída - [metal-lam] corroded plate
— f corrugada - [metal-fabr] corrugated, plate, o sheet
— f — curva(da) - [tub] curved corrugated plate
— f — de acero m - [constr] corrugated steel, section, o plate
— f — — — m remachada - [constr] riveted corrugated steel, section, o plate
— f — liviana - [metal-lam] light(weight) corrugated sheet
— f — Multi-Plate - [metal-fabr] Multi-Plate corrugated, o corrugated Multi-Plate, sheet
— f — pesada - [metal-fabr] heavy corrugated, o corrugated heavy, plate, o sheet
— f — para revestimiento m - [constr] corrugated liner plate
— f corta - [metal-fabr] short, plate, o panel
— f cortada - [metal-lam] sheared plate
— f — en obra f - [constr] field, sheared, o cut, plate
— f costilla f - [constr] poling plate
— f curva(da) - [tub] curved, section, o plate
— f chapada - [metal-fabr] clad plate
— f chapeada - [metal-fabr] clad plate
— f — acero m - [metal-lam] steel, plate, o sheet • plate steel
— f — — m con aleación f - [metal-prod] alloy steel plate
— f — — m — — f alta - [sold] high alloy steel plate
— f — — m — — f baja - [sold] low alloy steel plate
— f — — — — f laminada en caliente - [metal-lam] hot rolled alloy steel plate
— f — — m — f — — frío - [metal-lam]

cold rolled alloy steel plate
— f de acero m con borde m, laminado, o de laminación f - [metal-lam] mill edge steel
plate
— f —— m — calidad f comercial - [metal-
-lam] commercial quality steel plate
— f —— m —— f estructural - [metal-lam]
structural quality steel plate
— f —— m — canto m de laminación f -
[metal-lam] mill edge steel plate
— f —— m colocada - [mec] placed steel plate
— f —— m con carbono m - [metal-lam] carbon
(steel) plate
— f —— m —— m alto - [sold] high carbon
steel plate
— f —— m —— m bajo - [sold] low carbon
steel plate
— f —— m —— m con calidad f comercial -
[metal-lam] commercial quality carbon steel
plate
— f —— m —— f estructural - [metal-lam]
structural quality carbon steel plate
— f —— m —— laminada en caliente -
[metal-lam] hot rolled carbon plate
— f —— m —— frío - [metal-lam]
cold rolled carbon plate
— f —— m —— para anillo m para base f -
[petról] carbon steel base ring plate
— f —— m para fondo m - [petról]
carbon steel bottom plate
— f —— m —— m para estanque m -
[petról] tank bottom carbon steel plate
— f —— m —— m para manto m - [petról]
carbon steel mantle plate
— f —— m —— m para estanque m -
[petról] tank mantle carbon steel plate
— f —— m —— m techo m - [petról] carbon steel roof plate
— f —— m —— m para estanque m -
[petról] tank roof carbon steel plate
— f —— m — carbono-silicio m - [metal-lam]
carbon-silicon steel plate
— f —— m — cromo-molíbdeno m - [metal-lam]
chromium-molybdenum steel plate
— f —— m — junta(s) f traslapada(s) -
lap(ped) joint steel plate
— f —— m —— f para revestimiento m -
[constr] lap(ped) joint steel liner plate
— f —— m — molibdeno m - [metal-prod] molybdenum steel plate
— f —— m con pestaña(s) - [metal-lam]
flanged steel plate
— f —— m —— f para revestimiento m -
[constr] flanged steel liner plate
— f —— m —— f —— m para túnel(es) m -
[constr] flanged steel tunnel liner plate
— f —— m conformada - [metal-lam] formed, o
shaped steel plate
— f —— m — para piso m - [constr] formed
steel flooring
— f —— m —— m para puente(s) m -
[constr] formed steel bridge flooring
— f —— m corrugada - [metal-fabr] corrugated
steel, plate, o section
— f —— m — y curva(da) - [metal-lam] curved
(and) corrugated steel sheet
— f —— m de Armco - [metal-lam] Armco steel
plate
— f —— m —— m con borde(s) m de laminación f - [metal-lam] Armco mill edge steel
— f —— m —— con canto(s) m de laminación
f - [metal-lam] Armco mill edge steel plate
— f —— m decapada - [metal-trat] pickled
steel plate
— f —— m dulce—[metal-lam] mild steel plate
— f —— m especial - [metal-lam] special(ty)
steel plate
— f —— m galvanizada - [metal-trat] galvanized steel plate
— f —— m — corrugada - [tub] corrugated
galvanized steel plate
— f —— m inoxidable - [metal-prod] stainless
steel, plate, o sheet

plancha f de acero m inoxidable con carbono m
muy bajo - [sold] extra-low carbon stainless
(steel) plate
— f —— m — decapada - [metal-trat] pickled
stainless steel plate
— f —— m — laminada en caliente - [metal-
-lam] hot rolled stainless steel plate
— f —— m —— decapada - [metal-trat]
pickled hot rolled stainless steel plate
— f —— m —— recocida - [metal-trat]
annealed hot rolled stainless steel plate
— f —— m — recocida - [metal-trat] annealed
stainless steel plate
— f —— m laminada - [metal-lam] rolled steel
plate
— f —— m con calidad f estructural - [metal-
-lam] structural quality rolled steel plate
— f —— m — en caliente - [metal-lam] hot
rolled steel, sheet, o plate
— f —— m —— frío - [metal-lam] cold
rolled steel, sheet, o plate
— f —— m —— plano m - [metal-lam] flat
rolled steel plate
— f —— m —— con calidad f comercial -
[metal-lam] commercial quality flat rolled
steel plate
— f —— m —— —— f estructural -
[metal-lam] structural quality flat rolled
steel plate
— f —— m normalizado - [metal-prod] normalized steel plate
— f —— m para piso m - steel flooring
— f —— m para fondo m - [petról] bottom
steel plate
— f —— m —— m para estanque m - [petról]
tank bottom steel plate
— f —— m —— m puente(s) m - [constr]
steel bridge plank
— f —— m para avance m - [constr] steel poling place
— f —— m — piso m para puente(s) m -
[constr] steel bridge, floor(ing), o plank
— f —— m para revestimiento m - [constr]
steel liner plate
— f —— m —— m instalada - [constr] installed steel liner plate
— f —— m — techo m - [constr] steel roof
deck(ing)
— f —— m —— m para estanque m - [petról]
tank roof steel plate
— f —— m perforada - [metal-lam] perforated
steel plate
— f —— m plaqueado - [metal-trat] clad steel
plate
— f —— m prensada - [metal-lam] pressed
steel plate
— f —— m — para revestimiento m - [metal-
lam] pressed steel liner plate
— f —— m soldada - [tub] welded, o fabricated, steel plate
— f —— m templado - [metal=trat] tempered
steel plate
— f — aluminio m - [metal-lam] aluminum plate
— f — caucho m - [mec] sheet rubber
— f — cobre m - [metal-lam] copper plate
— f — cromo-molíbdeno m - [metal-lam] chromium-molybdenum plate
— f —— m para caldera(s) f - [metal-lam]
chromium-molybdenum boiler plate
— f — fundición f - [metal-prod] sole plate
— f — hierro m - [metal-lam] iron plate
— f —— m fundido - [metal-lam] cast iron
plate
— f —— m — preparada - [sold] prepared
cast iron plate
— f —— m magnético - [metal-prod] magnetic
iron plate
— f —— m para esmaltar v - [metal-lam] enameling iron, sheet, o plate
— m —— v con calidad f para embutición f - [metal-lam] drawing quality enameling iron sheet
— f —— m —— v —— f para estampar v -

[metal-lam] drawing quality enameling iron
sheet
plancha f de hierro m puro - [metal-lam] ingot
iron, sheet, o plate
— f — m **galvanizado** - [metal-trat] galvanized ingot iron, sheet, o plate
— f — **laminador m** - [metal-lam] mill plate
— f — — m **universal** - [metal-lam] universal
mill plate
— f — **madera f** - [maderas] wood(en) sheet
— f — — f **laminada** - [maderas] plywood sheet
— f — — f **terciada** - [mader] plywood sheet
— f — **metal m** - [metal-lam] metal plate
— f — **níquel m** - [metal-lam] nickel, sheet, o
plate
— f — **sección f** - [hidr] sectional plate
— f — **tosca f** - [suelos] hardpan slab
— f **delgada** - [metal-lam] thin sheet • thin
metal
— f — **de acero m** - [metal-lam] light steel
plate
— f — — — m **con carbono m** - [metal-lam]
carbon steel light plate
— f — — — m — — m **con calidad f comercial**
- [metal-lam] commercial quality carbon steel
light plate
— f — — — m — **calidad f comercial** - [metal-
-lam] commercial quality light steel plate
— f — **soldada** - [sold] welded thin plate
— f **deslizante** - [mec] sliding plate
— f **dura** - [metal-lam] hard plate
— f **eléctrica** - [domest] electric iron
— f **elevada** - [grúas] lifted, o raised, o
hoisted, plate, o sheet
— f **emplomada** - [metal-prod] leaded, o terne,
plate, o sheet
— f — **para recipiente(s) m para presión f** -
[metal-lam] pressure vessel leaded plate
— f **empotrada** - [constr] embedded plate
— f — **colocada** - [constr] placed embedded plate
— f — **en bobina(s) f** - [metal-lam] coiled, sheet,
o plate
— f — **rollo(s) m** - [metal-lam] coiled, sheet,
o plate
— f **en sección(es) f** - [metal-lam] sectional,
sheet, o plate
— f **enrejada** - [metal-lam] grating, sheet, o
plank
— f — **engatillada** - [metal-fabr] interlocking
grating, sheet, o plank
— f **escuadrada** - [metal-lam] squared plate
— f **especificada** - [metal-lam] specified plate
— f **esquinera** - [constr] corner plate
— f **estampada** - [metal-lam] stamped, plate, o
blank, o sheet
— f **estándar** - [metal-lam] standard plate
— f **estriada** - [metal-lam] riffled, o rippled,
plate • [constr] checkered plate
— f — **para piso m** - [metal-lam] checkered
floor plate
— f **estructural** - [metal-lam] structural plate
• [constr-puentes] structural deck(ing)
— f — **armada f** - [constr] assembled structural
plate
— f — — **en obra f** - [constr] field assembled
structural plate
— f — **conformada** - [metal-lam] formed, o
shaped (steel) structural plate
— f — **corrugada** - [metal-fabr] corrugated
structural plate
— f — — **curva(da)** - [constr] curved corrugated structural plate
— f — **curva(da)** - [metal-fabr] curved structural plate
— f — **de acero m** - [metal-lam] steel structural plate
— f — — — m **conformada** - [metal-fabr]
formed, o shaped, steel structural plate
— f — **para bóveda f** - [tub] arch structural
plate
— f — — **piso(s) n** - [constr] structural plate flooring
— f — — — m **para puente(s) m** - [constr]

structural plate bridge flooring
plancha f estructural para puente(s) m - [metal-
-fabr] structural plate bridge plank(ing)
— f — **para tubería f** - [tub] pipe, o skelp,
structural plate
— f — — — f **abovedada** - [tub] pipe arch
structural plate
— f — **perforada** - [metal-fabr] perforated
structural plate
— f — **sin curvar** - [metal-lam] uncurved structural plate
— f **fija(da)** - [mec] fixed, o attached, plate
— f **fina** - [metal-lam] light gage plate
— f **final** - [mec] end plate
— f **frontal** - [mec] front, plate, o panel, o
face
— f **fusible** - [metal-lam] fuse plate
— f — **débil** - [metal-prod] weak fuse plate
— f **galvanizada** - [metal-trat] galvanized plate
— f — **con magnesio m** - [metal-trat] magnesium,
zinc coated plate, o Zinc-Grip
— f — **para guardia f** - [metal-trat] galvanized
toe plate
— f **gruesa** - [metal-lam] thick, o heavy (gage),
plate • heavy metal • [sold] thick section
— f — **con calidad f comercial** - [metal-lam]
commercial quality, heavy, o thick, plate
— f — **de acero m** - [metal-lam] heavy steel
plate
— f — — — m **con calidad f comercial**—[metal-
-lam] commercial quality heavy steel plate
— f — — — m — **carbono m** - [metal-lam] commercial quality heavy carbon plate
— f — — — m — — m **con calidad f comercial**
- [metal-lam] commercial quality carbon steel
heavy plate
— f — **soldada** - [sold] welded heavy plate
— f **horizontal** - [sold] horizontal plate •
[suelos] horizontal slab
— f — **de tosca f** - [suelos] horizontal hardpan slab
— f — **para base f** - [constr] horizontal base
plate
— f **húmeda** - [sold] wet plate
— f **inclinada** - [sold] inclined, o slanted,
plate
— f — **para base f** - [vial] inclined base plate
— f **individual** - [mec] individual plate
— f — **pesada** - [mec] heavy individual plate •
weighed individual plate
— f **inferior** - [sold] bottom plate • invert
plate
— f — **para bóveda f** - [tub] bottom arch plate
— f **inmovilizada** - [sold] fixed plate
— f — **rígidamente** - [sold] rigidly fixed plate
— f **inoxidable** - [metal-lam] stainless plate
— f — **con carbono m** - [metal-lam] stainless
carbon (steel) plate
— f — — — m **muy alto** - [sold] extra high
carbon stainless (steel) plate
— f — — — m — **bajo** - [sold] extra low carbon stainless (steel) plate
— f — **laminada en caliente** - [metal-lam] hot
rolled stainless (steel) plate
— f — — — — **decapada** - [metal-trat] pickled
hot rolled stainless (steel) plate
— f — — — **recocida** - [metal-trat] annealed hot rolled stainless (steel) plate
— f — **recocida** - [metal-trat] annealed stainless (steel) plate
— f **interior** - [mec] inside plate
— f — **para alma m** - [metal-lam] inside web
plate
— f **Interlocking Grating** - [metal-fabr] Interlocking Grating plate
— f — **enrejada engatillada** - [metal-fabr]
Interlocking Grating, plate, o plank
— f **laminada** - [metal-lam] rolled plate
— f — **en caliente** - [metal-lam] hot rolled
plate
— f — — — **decapada** - [metal-trat] pickled
hot rolled plate
— f — — — **recocida** - [metal-trat] annealed

hot rolled plate

plancha f laminada en caliente sin aceitar - [metal-lam] hot rolled dry sheet
— f — **en plano** - [metal-lam] flat rolled plate
— f **larga** - [metal-lam] long plate • [mec] long panel
— f **lateral** - [metal-fabr] side plate • [tub] side panel
— f — **con contacto m** - [hidr] side-rubbing plate
— f — **para base f** - [mec] base side plate
— f — — f **rotante** - [mec] rotating base side plate
— f **limpia** - [constr] clean plate
— f **limpiada** - [constr] cleaned plate
— f **lisa** - [metal-lam] smooth, plate, o strip
— f **liviana** - [metal-lam] light plate
— f — **con calidad f comercial** - [metal-lam] commercial quality light plate
— f — **de acero m** - [metal-lam] light steel plate
— f — — m **con calidad f comercial** - [metal-lam] commercial quality light steel plate
— f — — m — **carbono m con calidad f comercial** - [metal-lam] commercial quality carbon steel light plate
— f — — — m **prensado** - [constr] lightweight pressed steel plate
— f **manipulada** - [mec] handled plate
— f **más gruesa** - [metal-lam] thicker, o heavier, plate
— f — **pesada** - [metal-lam] heavier plate
— f **mediana** - [metal-lam] medium plate
— f **metálica** - [metal-lam] metal plate
— f — **para piso m** - [metal-fabr] floor plate
— f — — m **para puente(s) m** - [metal-fabr] bridge (floor), plate, o plank
— f **modificada** - [mec] modified, o changed, plate
— f **Multi-Plate** - [tub] Multi-Plate plate
— f — **de acero m corrugado** - [tub] Multi-Plate corrugated steel plate
— f **múltiple** - [tub] multiple, o sectional, plate
— f **muy delgada** - [metal-lam] very thin plate
— f **necesaria** - [constr] necessary plate
— f **negra** - [metal-lam] black, plate, o strip
— f — **lisa** - [metal-lam] smooth black, strip, o plate
— f — **para revestimiento m** - [metal-lam] black liner plate
— f — — m **para túnel(es) m** - [metal-lam] black (tunnel) liner plate
— f **nervada** - [metal-lam] ribbed plate
— f **normalizada** - [metal-trat] normalized plate
— f **optativa** - [mec] optional plate
— f **oxidada** - [metal-lam] rusty plate
— f — **calentada** - [sold] heated rusty plate
— f — — **con soplete m** - [sold] torch heated rusty plate
— f — **limpia** - [metal-lam] clean rusty plate
— f — **limpiada** - [metal-lam] cleaned rusty plate
— f — **sin limpiar** - [metal-lam] uncleaned rusty plate
— f **para aislación f** - [constr] insulation board
— f — **ala m** - [metal-lam] flange plate
— f — **alma m** - [metal-lam] web plate
— f — **anillo(s) m** - [mec] ring plate
— f — — m **para base f** - [metal-fabr] base ring plate
— f — **asiento m** - [mec] seat, o base, plate
— f — — m **para compresor** - [mec] compressor seat plate
— f — — m — m **modificada** - [mec] changed compressor seat plate
— f — — m — **ventilador m** - [mec] fan seat plate
— f — — m — m **modificada** - [mec] changed fan seat plate
— f — **avance m** - [constr] poling plate
— f — **base f** - [mec] base plate; bedplate;

sub-plate • [constr] toe plate

plancha f para base f horizontal - [constr] horizontal base plate
— f — — f **inclinada** - [vial] slanted, o inclined, base plate
— f — **blindaje** - [metal-prod] armor plate • [constr] poling plate
— f — **bóveda f** - [tub] arch, plate, o sheet
— f — — f **de plancha(s) f estructural(es)** - [constr] structural plate arch plate
— f — **brida(s) f** - [metal-lam] flange plate
— f — **buque(s) m** - [metal-lam] ship plate
— f — **caldera(s) f** - [metal-lam] boiler plate
— f — — f **resistente a fatiga f** - [metal-lam] fatigue resisting boiler plate
— f — **cilindro(s) m** - [mec] cylinder plate
— f — **cubierta f** - [mec] cover(ing) plate • [nav] deck plate
— f — — f **modificada** - [mec] modified, o changed, cover(ing) plate • [nav] modified, o changed, deck plate
— f — **cumbre(ra) f** - [metal-fabr] roof plate
— f — **desgaste m** - [mec] wear plate
— f — — m **de acero m** - [mec] steel wear plate
— f — — m **de acero m inoxidable** - [mec] stainless steel wear plate
— f — — m **para carro m** - [metal-prod] car wear plate
— f — — m — — m **para montacarga(s) m** - [metal-prod] skip car wear plate
— f — — m — — m **para horno m alto** - [metal-prod] blast furnace skip car wear plate
— f — — m — **cilindro m** - [petról] cylinder wear plate
— f — — m — **circuito m** - [mec] circuit wear plate
— f — — m — **cubeta f** - [ind] hopper wear plate
— f — — m — **mesa f** - [mec] table wear plate
— f — — m — — f **alimentadora** - [ind] feeder table wear plate
— f — — m — — f **para planta f para sínter m** - [metal-prod] sinter(ing) plant feeder table wear plate
— f — — m **para tolva f** - [mec] hopper wear plate
— f — — m — — f **para horno m** - [metal-prod] furnace hopper wear plate
— f — — m — — f **tragante m** - [metal-prod] top hopper wear plate
— f — — m — — m **de horno m** - [metal-prod] furnace top hopper wear plate
— f — — m — — f — — m — m **alto** - [metal-prod] blast furnace top hopper wear plate
— f — — m — **tragante m para horno m** - [metal-prod] furnace top wear plate
— f — — m — **para horno m alto** - [metal-prod] blast furnace top wear plate
— f — — m — **vagoneta f** - [ind] car wear plate
— f — — m — — f **para montacarga(s) m** - [metal-prod] skip car wear plate
— f — — m — — m — — m **para horno m alto** - [metal-prod] blast furnace skip car wear plate
— f — **deslizamiento m** - [mec] sliding plate
— f — **edificio(s) m** - [constr] building, plate, o panel
— f — **ensayo(s) m** - [sold] test(ing) plate
— f — **escurrimiento m** - [constr] (roof) flashing
— f — — m **para techo(s) m** - [constr] roof flashing
— f — **estampado m** - [metal-fabr] drawing sheet
— f — — m **profundo** - [metal-prod] deep drawing sheet
— f — **fondo m** - [mec] bottom, sheet, o plate • [tub] bottom, o invert, plate
— f — — m **de acero m** - [petról] bottom steel, o steel bottom, plate
— f — — m — — m **con carbono m** - [petról]

carbon steel bottom plate
plancha f para fondo m para estanque m—[petról]
tank bottom plate
— f — **guardia** f - [mec] toe plate
— f —— f **optativa** - [mec] optional toe plate
— f — **hogar(es)** m - [ind] firebox plate
— f — **frotamiento** m - [mec] wear plate
— f — **indicador(es)** m - [vial] sign blank
— f — **instalación** f - [mec] installation plate
— f — **junta(s)** f - [mec] splice, o joint,
plate
— f — **lastre** m - [nav] ballast plate
— f — **letrero(s)** m - [vial] sign, plate, o
blank
— f — **manto** m - [petrol] mantle plate
— f —— m **para estanque** m - [petról] tank
mantle plate
— f — **nivel** m **para carga** f - [metal-prod]
stockline plate
— f — **perfil** m - [metal-lam] shape plate
— f —— m **soldado** - [metal-lam] welded shape
plate
— f — **perforar(se)** - [sold] plate to be
pierced
— f — **pico** m **para engrase** m - [mec] (grease)
fitting plate
— f — **piso** m - [constr] floor plate •
[puentes] flooring
— f —— m **asegurado** - [constr] secured
flooring
— f —— m **fijada** - [puentes] secured, o attached, flooring
— **f** —— m **para puente** m - [constr-puentes]
plate gage, o bridge, flooring; bridge floor-
-(ing) plank
— f ——— m **de acero** m - [constr-puent]
steel bridge, o bridge steel, plank
— f —— m —— m **fijada** - [constr] secured
bridge flooring
— f —— m **quitada** - [constr] removed floor
plate
— f —— m **removible** - [constr] removable
floor plate
— f —— m **removida** - [constr] removed floor
plate
— f —— m **sacada** - [constr] removed floor
plate
— f — **puente** m - [constr] bridge plank
— f — **recipiente(s)** m **para presión** f - [metal-
lam] pressure vessel plate
— f —— f —— f **para temperatura(s)** f **al-
ta(s)** - [metal-lam] high temperature pressure
vessel plate
— f —— m —— f —— f **baja(s)** - [metal-
-lam] low temperature pressure vessel plate
— f — **refuerzo** m - reinforcement,cing, o backing, o back-up • [constr-cabeza f de viga f]
cover plate
— f —— m **de acero** m - [metal-lam] steel reinforcing plate
— f —— m —— m **con carbono** m - [metal-lam]
carbon steel reinforcing plate
— f —— m **traslapada** - [constr] lapped cover
plate; cover plate lap
— f — **respaldo** m - [mec] backing, o back-up,
plate; backup
— f —— m **de acero** m - [sold] steel back-up
(plate)
— f — **revestimiento** m - liner plate
— f —— m **de acero** m - [metal-lam] steel
liner plate
— f —— m —— m **acanalado** - [metal-lam]
corrugated steel liner plate
— f —— m —— m **corrugado** - [metal-lam]
corrugated steel liner plate
— f —— m —— m **encarrujado** - [metal-lam]
corrugated steel liner plate
— f —— m —— m **ondulado** - [metal-lam] corrugated steel liner plate
— f —— m — **Armco** - [metal-fabr] Armco liner
plate
— f —— m —— m **con borde(s)** m **pestañado(s)**
y extremo(s) m **traslapado(s)** m - [metal-fabr]

offset type Armco liner plate
**plancha f para revestimiento m de acero m acana-
lado** - [metal-fabr] corrugated steel liner
plate
— f —— m **de acero** m **corrugado** - [metal-fabr]
corrugated steel liner plate
— f —— m —— m **encarrujado** - [metal-fabr]
corrugated steel liner plate
— f —— m —— m **ondulado** - [metal-fabr]
corrugated steel liner plate
— f —— m **instalada** - [constr] installed
liner plate
— f —— m **Liner Plate** - [metal-fabr] Armco
liner plate
— f —— m **para túnel(es)** m - [constr] (tunnel) liner plate
— f — **roce** m - [mec] wear plate
— f — **ropa** f - [domest] (flat)iron
— f — **separación** f - véase **salpicadero** m
— f — **solera** f - [tub] invert (plate)
— f — **tapa** f - [cald] coverplate
— f —— f **para válvula** f - [petról] valve
cover plate
— f — **techo** m - [metal-lam] roof, sheet, o
deck, o plate
— f —— m **de acero** m **con carbono** m - [petról]
carbon steel roof plate
— f —— m **para estanque** m - [petról] tank
roof plate
— f — **temperatura(s)** f **alta(s)** - [metal-prod]
high temperature plate
— f —— f **baja(s)** - [metal-prod] low temperature plate
— f — **tolva(s)** f - [metal-prod] hopper strip
— f — **tubería(s)** f - [metal-lam] pipe plate;
skelp
— f —— f **abovedada(s)** f - [metal-lam] pipe-
-arch plate
— f —— f — **de plancha(s)** f **estructural(es)**
- [metal-lam] structural plate pipe-arch
plate
— f —— f **circunferencial** - [metal-lam] circular pipe plate
— f —— f — **Multi-Plate** - [tub] Multi-Plate
circular pipe plate
— f —— f **de plancha(s)** f **estructural(es)** -
[metal-lam] structural plate pipe plate
— f —— f **elíptica** - [metal-lam] elliptical
pipe plate
— f — **tubo(s)** m - [metal-lam] skelp
— f **perforada** - [metal-lam] perforated plate
— f **pesada** - [metal-lam] heavy, o thick, plate
• weighed plate
— f — **con calidad** f - [metal-lam] quality
heavy (steel) plate
— f — **corrugada** - [metal-lam] corrugated
(steel) plate
— f —— **curva(da)** - [metal-fabr] curved corrugated steel plate
— f ——— **de acero** m - [metal-fabr] curved
corrugated heavy steel plate
— f —— **de acero** m - [metal-fabr] heavy corrugated steel plate
— f — **para fondo** m - [tub] heavy invert plate
— f **plana** - [metal-lam] flat plate
— f — **con cruceta** f **(fijada)** - [constr] flat
plate with (attached) rock cross
— f —— **acero** - [metal-lam] flat steel
plate
— f **plaqué** - [metal-trat] clad plate
— f **plaqueada** - [metal-trat] clad plate
— f **posterior** - [constr] back, o rear, plate, o
panel
— f **preformada** - [metal-fabr] preformed, o preshaped, plate
— f — **de acero** m - [metal-fabr] preshaped, o
preformed, steel plate
— f **prensada** - [metal-fabr] pressed plate
— f — **liviana** - [metal-lam] light-weight
pressed (steel) plate
— f **preparada** - [metal-fabr] prepared plate
— f **protectora** - protective plate
— f **quitada** - [mec] removed plate

plancha f ranurada - [metal-fabr] grooved, o slotted, plate
— f recubierta - [metal-trat] coated plate
— f reflectante - [metal-trat] reflecting, o reflectorized, plate, o sheeting
— f removible - [mec] removable plate
— f removida - [mec] removed plate
— f resistente - [metal-prod] strong plate
— f — a fatiga f - [metal-prod] fatigue resisting plate
— f — — — f para caldera(s) f - [metal-prod] fatigue resisting boiler plate
— f restante - [mec] remaining plate
— f rota - [metal-lam] broken plate • [constr] broken plank
— f sacada - [mec] removed plate
— f seca - [constr] dry plate
— f según norma f - [metal-lam] standard plate
— f sin aceitar - [mec] dry, plate, o sheet
— f — — laminada en caliente - [metal-lam] dry hot, o hot dry, rolled, plate, o sheet
— f curvar - [metal-fabr] uncurved plate
— f — limpiar - [metal-trat] uncleaned plate
— f — recubrir - [metal-trat] uncoated plate
— f sobrecabeza - [sold] overhead plate
— f soldada - [sold] welded plate
— f — a tope - [sold] butt welded plate
— f sucia - [sold] dirty plate
— f suelta - [mec] loose plate
— f superior - [mec] top, o upper, plate
— f templada - [metal-trat] tempered plate
— f — de acero m - [metal-trat] tempered steel plate
— f terne - [metal-treat] terne sheet; véase también plancha f emplomada
— f transversal - [constr] transverse, plate, o section
— f variable - [mec] variable plate
— f vertical - [mec] vertical plate
— f Zinc-Grip - [metal-trat] Zinc-Grip (plate, o sheet)
— f — con magnesio m - [metal-trat] magnesium Zinc-Grip (sheet, o plate)
planchada f - [ind] platform • [náut] gangplank
— f guía - [metal-lam] ramp-guide
— f — para entrada f - [metal-lam] feed(ing) end ramp-guide
— f — — salida f - [metal-lam] delivery end, (end), ramp, o platform
— f — — f para laminador m - [metal-lam] mill delivery ramp
— f para soporte m - [metal-lam] supporting, platform, o ramp
planchadora f - [metal-lam] leveller
— f en caliente - [metal-lam] hot leveller
. . . planchas f contiguas a - [constr] . . . planks wide
— f en todos espesores m - [sold] all plate thicknesses
— f por anillo(s) m - [tub] plates per ring
— f sin limitación f de espesor(es) m - [metal-lam] unlimited thickness plate(s)
plancheado* m - [metal-trat] plating
— m con cobre m - [metal-trat] copper plating
— m — cromo m - [metal-trat] chrome plating
— m electrolítico - [metal-trat] electrolytic plating
plancheado,da a - [metal-trat] plated
— a con plomo m - [metal] lead plated
planchón m - [metal-lam] . . .l slab; flat billet
— m caliente m - [metal-lam] hot slab
— m calmado - [metal-lam] killed slab
— m con aluminio - [metal-lam] aluminum killed slab
— m colado - [col.cont] cast slab
— m con cola f de pescado m - [metal-lam] fish tail(ed) slab
— m de acero m - [metal-prod] steel slab
— m — — m efervescente - [metal-prod] rimming steel slab
— m escarpado - [metal-lam] scarfed slab
— m por colada f continua - [col.cont] continuous cast slab

planchón m procesado - [col.cont] processed slab
— m — por colada f continua - [col.cont] continuous cast processed slab
— m sin escarpar - [metal-lam] unscarfed slab
planchuela f - [metal-lam] flat, (fishplate, o billet, o bar);fishplate; flat; fish joint; flat bar; thick heavy strip; strip; steel flat; heavy, o thick, strip • [herram] trowel
— f de acero m - [metal-lam] steel flat, o flat steel, bar; bar
— f de acero m dulce - [metal-lam] mild steel flat (bar)
— f — hierro m - [metal-lam] flat iron
— f laminada - [metal-lam] rolled flat (bar)
— f — en caliente - [metal-lam] hot rolled flat (bar)
— f para ajuste m - [mec] adjusting strap
— f — m para generador m - [electr-prod] generator adjusting strap
— f — condensador m - [electr-equip] condenser strap
— f — contacto m - [electr-instal] contact strap
— f — — m con tierra f - [electr-instal] ground(ing) strap
— f — derrame m - [sold] run-off, o run-out, tab
— f — — m de acero m - [sold] steel run-off tab
— f — descarte m - [mec] run-out tab
— f — guía f - [sold] clamp-on strip
— f — respaldo m - [sold] backing, o back-up, strip
— f — montaje m - [mec] mounting, strap, o strip
— f — — m para condensador m - [electr-equip] condenser (mounting), strap, o strip
— f — puesta f a tierra f - [electr-instal] ground(ing) strip
— f — sujeción f - [mec] clamp(ing), o adjusting, strap
— f — sujetar acumulador m - [mec] battery holding, flat, o strap
— f — traba(dura) f - [hormig] kick strip
— f — — f vertical - [constr] vertical kick strip
— f supletoria - [mec] run-out, o additional, tab, o strap
— f vertical - [constr] vertical strip
planeado,da a - planned; designed
— a en vacío m - vacuum planned
planeador m - . . . . - planner
planeadora f - [metal-lam] leveller
planeamiento* m - planning
— m (por) anticipado - preplanning
— m de actividad(es) m - activity,ties planning
— m — incremento(s) m - [mec] increment(s) planning
— m — instrucción f - [labor] instruction, o training, planning
— m — investigación f - [ind] research planning
— m — programa - program planning
— m — — m para instrucción f - training program planning
— m — — — f sobre seguridad f - [segurid] safety training program planning
— m — — m sobre seguridad f - [segurid] safety program planning
— m — trabajo(s) m - [ind] work(s) planning
— m — — m para investigación(es) f - [ind] research (work) planning
— m en vacío m - vacuum planning
— m estratégico - [admin] strategic planning
— m general - general planning
— m para beneficiación f - [miner] beneficiation planning
— m pobre - poor planning
— m urbano - [pol] urban planning
planear v con anticipación f - to preplan
— v en vacío m - to vacuum plan
— v ensayo m - to plan @ test(ing)
— v incremento m - [ind] to plan @ increment

planear v instalación f - to plan @ installation
planeidad* f - levelness
planes m - planning
planicidad* f - véase llanura f
planificación f - planning; projecting
— f adelantada - forward planning; preplanning
— f con miras f de utilidad(es) f [admin] profit oriented planning
— f de actividad(es) f - activity,ties planning
— f —— f para departamento m - [ind] departmental activity,ties planning
— f —— f general(es) - [ind] overall, o general, activity,ties planning
— f — administración f - [admin] management
planning
— f — ensayo(s) m - test(ing) planning
— f — estimación(es) f - [admin] budgeting
planning
— f económica - economic planning
— f en vacío m - vacuum planning
— f estratégica - [admin] strategic planning
— f para conservación f - [ind] maintenance
planning
— f — expansión f - [econ] expansion planning
— f — instalación f - [ind] installation planning
— f — investigación f - [ind] research planning • control work planning
— f — mantenimiento m - [ind] maintenance
planning
— f —— m para equipo m - [ind] equipment
maintenance planning
— f —— m —— m para producción f - [ind]
production, o operating, equipment maintenance planning
— f —— m —— m para trabajo m - [ind]
work, o operating, equipment maintenance
planning
— f — materia(s) f prima(s) - [ind] raw materials planning
— f — plazo m corto - short range planning
— f —— largo - long range planning
— f — producción f - [ind] production planning
— f — producto m - [ind] product, planning, o
designing
— f — reparación(es) f - [ind] repair(s) planning
— f —— f para equipo m - [ind] equipment
repair planning
— f —— f —— m para producción f - [ind]
production, o operating, equipment repair
planning
— f —— f —— m para trabajo m - [ind]
operating equipment repair planning
— f — trabajo m - [admin] work planning
— f —— m para evaluación f - [ind] evaluation work planning
— f —— m — investigación(es) f [ind] research, o investigation, (work) planning
— f — transmisión f - [electr-distrib] transmission planning
— f por administración f - [admin] management
planning
— f y control m - [admin] planning and control
— f — fiscalización f - [admin] planning and
control(ling)
planificado,da a - planned; projected
planificador m - planner
planificar v - to, plan, o project
— v expansión f - [econ] to plan @ expansion
planilla f - . . . . • [com] form; sheet; record;
report sheet • [com] . . .; tabulation; list;
schedule; exhibit; table • [dib] chart •
[deport] score sheet
— f comparativa - comparative, chart, o table
— f con dato(s) m - data, o information, sheet
— f —— m sobre rueda(s) f - [mec-ruedas)
wheel data sheet
— f —— m ténico(s) m - [ind] technical, o
engineering, data sheet
— f — especificación(es) f - specification(s)
sheet
— f — información f - information, o data,
sheet, o page
planilla f con información f sobre rueda(s) f -
[mec-ruedas] wheel data sheet
— g —— f —— f para riel(es) m - [mec=
-ruedas] track wheel data sheet
— f — informe m - report, form, o sheet
— f — carga(s) - [metal-prod] burden sheet •
[transp] load(ing sheet
— f individual m - individual, sheet, o record
— f — para rodillo m - [metal-lam] individual
roll record
— f — resumen m - summary statement
— f —— m de oferta(s) f - [com] proposal(s)
summary statement
— f — tiempo(s) m para descarga f - [transp]
lay day statement
— f en forma f de gráfico m - plotted record
— f llenada - filled form
— f para análisis - analysis sheet
— f — cable(s) m - [electr-cond] cable
schedule
— f — cálculo(s) m - [matem] computation sheet
— f — cambio(s) m - exchange form
— f —— m para carga(s) f - [metal-prod] burden change(s) sheet
— f — circulación f - [ind] circulation, o
flow, sheet, o chart
— f — cobertura f - [com] coverage form •
[fin] routing sheet
— f — colada(s) - [metal-prod] batch report
sheet; heat log
— f — cómputo(s) m - calculation (sheet) •
[deport] score sheet
— f — conducto(s) m - [electr-oper] conduit
schedule
— f — conversaciones f sobre seguridad f -
[segurid] safety contact record
— f — cotización(es) f - quotation(s), o pro-
posal(s), sheet, o form
— f — cronometraje m - [labor] time study
sheet
— f — dato(s) m - data, o information, sheet
— f — descarga f - [transp] lay day statement
— f — escrutinio m - tally, o analysis, sheet
— f — flujo m - [ind] flow chart
— f — funcionamiento m - [ind] flow, chart, o
sheet
— f — gasto(s) m - [com] expense(s) schedule
— f — fosa f - [combust] soaking pit sheet
— f — información f - data sheet
— f —— m para inspección f - [ind] inspection report form
— f — instrucción(es) - instruction sheet
— f — oferta(s) f - proposal, o quotation,
sheet, o form
— f — permuta(s) f - [com] exchange(s) form
— f — personal m - [labor] organization chart
— f — reajuste m de precio(s) m - price escalation form
— f — recuento m - tally sheet
— f — registro m - [instrum] recording chart
— f —— m para inspección f - [ind] inspection checklist record
— f — regulación f - control(ling) sheet
— f — ruta f - [transp] log • [deport] scoring
sheet
— f — tiempo(s) m - [labor] time sheet •
[deport] score sheet
— f — venta(s) f - [com] sales form
— f —— f con cambio(s) - [com] exchange
sales form
— f —— f — permuta(s) f - [com] exchange
sales form
— f —— f para cambio(s) m - [fin] exchange
sales form
— f —— f para divisa(s) - [fin] exchange
sales form
— f — verificación f - tally(ing) sheet
planimetría f - . . . .; survey(ing)
planitud f - flatness; levelness • [constr] flat
grade • [topogr] véase llanura f
— f — cordón m dentro de ranura f - [sold]
flat inside bead

planitud f de plancha f - [metal-lam] plate,
  levelness, o flatness
— f extrema - [constr] extreme flatness;
  tremely flat grade
— f lograda - attained, o accomplished, flat-
  tness
plano m - plane • [dib] drawing; design; blueprint; layout; projection; print; plan •
  [metal-lam] flat; plate • [topogr] level
— m acotado - [dib] dimensioned drawing
— m adicional - [dib] additional drawing
— m adjunto - [dib] attached, o enclosed, drawing
— m analógico - analogical drawing
— m ancho m - [metal-lam] wide flat
— m angosto - [metal-lam] narrow flat
— m aprobado - [dib] approved, o certified,
  drawing
— m básico - [dib] basic drawing
— m certificado - [dib] certified drawing
— m comentado - commented (on) drawing
— m completo - [dib] complete drawing
— m con dimensión(es) f - [dib] dimension(ed),
  drawing, o print
— m confeccionado - [dib] prepared drawing
— m constructivo - [mec] construction, drawing,
  o plan
— m de ala m - [metal-lam] flange plane
— m — alma m - [metal-lam] web plane
— m — anillo m - [mec] ring drawing
— m — cara f - face, drawing, o plane
— m — — f externa - [f.c.-ruedas] front face
  plane
— m — — interna - [f.c.-ruedas] back face
  plane
— m — cliente m - customer('s), o your, drawing
— m — concepto(s) m - concept(s) chart
— m — — m para ingeniería f - engineering
  concept(s) chart
— m — conjunto m - [dib] general (assembly),
  drawing, o plan
— m — contorno(s) m - [dib] outline drawing
— m — detalle m - [dib] detailed drawing
— f — dimensión(es) f - [dib] dimension(s)
  drawing
— m — diseño m - [dob] design drawing
— m — — — m aprobado - [dib] certified design
  drawing
— m — disposición f - [dib] layout, o arrangement, drawing; plan view
— m — distribución f - [dib] arrangement, o
  layout, o outline, drawing • plan drawing •
  ground plan
— m — fabricante m - [dib] manufacturer('s)
  drawing
— m — fachada - [arq] (building) front drawing
— m — — m de edificio m - [arq] (building)
  front drawing
— m — falla f - [geol] fault plane
— m — patinadura f - [f.c.-ruedas] slide flat
— m — piso m - véase dibujo m, o plano m, para
  distribución f
— m — rodadura f - [f.c.-ruedas] tread
— m — — f cilíndrico - [f.c.-ruedas] cylind-
  drical tread
— m — — f endurecido - [ruedas] toughened
  tread
— m — — f normal - [f.c.-ruedas] normal tread
— m — — f para rueda f - [f.c.-ruedas] wheel
  tread
— m — ruptura f - [constr] rupture plane
— m — — f para material m para relleno m -
  [constr] fill mterial rupture plane
— m — situación f - location plan
— m — taller m - [dib] shop drawing
— m — ubicación f - [dib] location, o ground,
  plan; setting plan
— detallado m - [dib] detailed drawing
— m — según norma f - [dib] standard detailed
  drawing
— m diferente - [geom] different plane
— m digital - [dib] digital drawing

plano m distinto - [Geom] different plane
— m doble - [dib] double plane
— m eléctrico - [dib] electrical drawing
— m elemental - [dib] elementary drawing
— m en detalle m - detailed drawing
— m escuadrado - [metal-lam] (re)squared flat
— m final - [dib] final drawing
— m frontal - [constr] front view
— m general - [dib] general, drawing, o layout,
  o assembly drawing • plot plan
— m — para distribución f - [dib] general arrangement drawing
— m — — montaje - [dib] general assembly
  drawing
— m hasta ejecución f - [ind] from blueprint to
  hardware
— m horizontal - horizontal plane
— m inclinado - inclined plane; ramp; incline
— m incorrecto - [dib] incorrect drawing
— m industrial - [dib] industrial drawing
— m internacional - international level
— m lógico - [dib] logical drawing
— m —digital - [dib] logical-digital drawing
— m mecánico - [dib] mechanical drawing
— m normalizado - [dib] standardized, o normalized, drawing
— m para árbol m - [mec] shaft drawing
— m — armado m - [mec] assembly,bling drawing
— m — — m de plancha(s) f - [constr] plate
  assembly drawing
— m — colocación f - [mec] installation, o
  location, drawing
— m — comparación f - [ing] datum,ta
— m — construcción f - [constr] construction
  plan
— m — — f de planta f - [ind] plant construction plan
— m — — f — — f nuclear (para energía f) -
  [nucl] nuclear (power) plant construction
  plan
— m — eje m - [mec] axle drawing
— m — equipo m - [dib] equipment drawing
— m — exfoliación f - [metal] grain boundary
— m — — f de grano(s) m - [metal] grain
  boundary (plane)
— m — fabricación f - [ind] fabrication, o
  manufacturing, drawing
— m — fundamento m - [constr] foundation,
  drawing, o plan, o print
— m — instalación f - [ind] installation, o
  erection, drawing
— m — — f para estructura f - [constr] structure erection drawing
— m — perfil m - [dib] profile, drawing, o map
— m — planta f - [dib] plant, o layout, drawing
— m — — f para edificio m - [arq] building
  layout drawing
— m — — f y corte m - [dib] plan and section
  drawing
— m — programación f - [comput] programming
  board
— m — quemador m - [combust] burner drawing
— m — recuperador m - [metal-lam] recuperator
  drawing
— m — referencia f - reference drawing • datum
— m — regulación f - control plan
— m — — f para desecho(s) m (radiactivos) -
  [nucl] (radioactive) waste control plan
— m parcial - [geogr] partial map
— m preliminar - [dib] preliminary drawing;
  draft
— m primero - foreground
— m — a derecha f - right foreground
— m — a izquierda f - left foreground
— m principal - [dib] key plan
— m tal como construido,da - as built drawing
  m último - background
— m — en centro m - center background
— m único - single plane
— m vertical - vertical plane
plano,na a - . . . .; level
— a a tope m - [sold] flat butt

plano,na a̱ a tope con chaflán m̱ - [sold] flat
   grooved butt
—,— a con paso(s) múltiple(s) m̱ - [sold] multiple pass downhand
—,— a̱ con solapo m - [sold] flat lap
—,— a̱ —— m̱ con pasada f̱ única - [sold] single pass flat lap
—,— a̱ en ángulo m̱ - [sold] flat fillet
—,— a̱ —— m̱ interior - [sold] flat fillet
—,— a̱ —— m̱ — con pasada f̱ única - [sold]
   single pass flat fillet
—,— a̱ — interior m̱ de ángulo m̱ - flat fillet:
—,— a̱ — ranura f̱ profunda - [sold] flat deep
   groove
—,— a̱ ortogonal - [sold] flat fillet
—,— a̱ redondeado,da - flat round(ed)
planos m̱ - [metal-lam] flat products; véase
   también producto(s) m plano(s)
planta f̱ - . . . • [ind] plant; mill • [constr]
   story; level • [ind] factory; works • layout;
   plant view; floor plan - [fotogr] top view •
   [dib] (ground) plan • [ind] facility | [adv]
   - (en) in plant; mill
— f aceptada - [ind] accepted plant
— f̱ — finalmente - [ind] finally accepted
   plant
— f — provisoriamente - [ind] plant accepted
   provisionally
— f alta - [constr] upper, floor, o̱ stor(e)y
— f̱ ambulante - [ind] ambulatory, o̱ traveling,
   plant
— f anual - [botán] annual
— f̱ apropiada - [ind] appropriate plant
— f̱ automatizada - [ind] automated plant
— f̱ — altamente - [ind] highly automated plant
— f̱ — para acero m̱ - [metal-prod] automated
   steel plant
— f baja - [constr] ground, o̱ first, o̱ main,
   floor, o̱ level, o̱ story
— f Bessemer - [metal-prod] Bessemer plant
— f̱ — duplex - [metal-prod] Bessemer du-
   plex(ing) plant
— f — para duplex - [metal-prod] duplex(ing)
   Bessemer plant
— f caliente - [ind] hot plant
— f̱ cloacal - [sanit] sewage,wer plant
— f colgada - [domest] hung plant
— f̱ comercial - [ind] commercial plant
— f — para asfalto - [constr] commercial asphalt plant
— f —— m caliente - [constr] commercial
   hot (asphalt) plant
— f con atmósfera f̱ neutra - neutral atmosphere
   plant
— f — horno m̱ de fosa - [metal-prod] soaking
   pit, plant, o̱ shop
— f — horno(s) m̱ - [ind] furnace(s) plant
— f̱ —— m̱ para recocido m̱ - [metal-prod] annealing furnace plant
— f —— m̱ eléctrico(s) m̱ - [metal-prod] electric furnace(s), plant, o̱ shop
— f —— m̱ Siemens (Martín) - [metal-prod]
   open hearth, plant, o̱ shop
— f construida - [ind] built plant
— f̱ contaminada - contaminated plant
— f̱ cribadora - [constr] screening plant
— f̱ de alambique(s) m̱ - [petról] still(s) plant
— f̱ —— m̱ tubular(es) - [petról] pipe still
   (distillation) unit
— f — cliente m̱ - [ind] customer('s) plant;
   your plant
— f — comprador m̱ - [ind] buyer(s), o̱ custo-
   mer(s), plant
— f — convertidor(es) m̱ - [metal-prod] converter shop
— f —— m con oxígeno m̱ - [metal-prod] oxygen
   converter shop
— f — horno(s) m̱ - [ind] furnace(s) plant
— f̱ —— m̱ para coque m̱ - [coque] coke oven(s)
   plant
— f — origen - [ind] manufacturing plant
— f̱ — soplantes m̱ - [metal-prod] blower plant
— f̱ — turbosoplantes m̱ - [metal-prod] turbo-

   -blower plant
planta f̱ de vendedor m̱ - [ind] seller('s), o̱
   vendor('s), shop, o̱ plant, o̱ factory
— f — zonificación f̱ - [ind] zoning layout
— f̱ depuradora - purification,fying, o̱ treatment, plant
— f — para agua m̱ - [sanit] water treatment
   plant
— f —— líquido(s) m̱ cloacal(es) - [sanit]
   sewage treatment plant
— f desintegradora - [petról] cracking plant
— f̱ determinada - [ind] specified plant
— f̱ dosificadora - [ind] batching plant
— f̱ — para hormigón m̱ - [constr] concrete
   batching plant
— f — flotante - [hormig] floating batching
   plant
— f duplex - [metal-prod] duplex(ing) plant
— f̱ elaboradora - [ind] manufacturing, o̱ fabricating, o̱ producing,ction, plant
— f eléctrica - [electr-prod] electric (light,
   o̱ power) plant
— f̱ emisora - [ind] issuing plant
— f̱ en aire libre - [ind] open air plant
— f̱ — base f̱ a chatarra f̱ - [metal-prod] scrap
   based plant
— f en caliente - [ind] hot plant
— f̱ — cantera f̱ - [miner] quarry plant
— f̱ — frío - [ind] cold plant
— f̱ — ultramar - [ind] overseas plant
— f̱ —— para fabricación f̱ - [ind] overseas
   manufacturing plant
— f ———— f̱ de alcantarilla(s) f̱ -
   [metal-fabr] overseas culvert plant
— f — vacío m̱ - [ind] vacuum plant
— f̱ especificada - [ind] specified plant
— f̱ externa - external plant
— f̱ fabril - [ind] manufacturing plant
— f̱ — para artículo(s) m̱ liviano(s) -
   [ind] light manufacturing plant
— m ——— m mediano(s) - [ind] medium manufacturing plant
— f ——— m pesado(s) - [ind] heavy manufacturing plant
— f farmacéutica - [quím] pharmaceutical plant
— f̱ fría f̱ - [ind] cold plant
— f̱ generadora - [electr-prod] (electric) power, plant, o̱ station; light plant
— f general - [int] general plant • [dib]
   general drawing
— f — de ubicación f̱ - [ind] general layout •
   [dib] general location drawing
— f — de zonificación f̱ - [ind] general zoning
   layout
— f grande - [ind] large, o̱ full sized, plant,
   o̱ mill; maximill
— f̱ hidroeléctrica - [electr-prod] hydraulic
   (power) plant
— f hidrosulfurizadora - [quím] hydrosulfuriz-
   ing plant
— f independiente - [ind] independent plant •
   package(d) plant
— f — para alumbrado m̱ - [electr-prod] independent light(ing) plant
— f —— depuración f̱ - [ind] treatment plant
— f ———— f̱ para líquido(s) m̱ cloacal(es) -
   [sanit] independent, o̱ package, sewage treatment plant
— f —— iluminación f̱ - [electr-prod] independent light(ing) plant
— f individual - [ind] individual plant
— f̱ — pequeña - small, individual, o̱ package,
   plant
— f —— para depuración f̱ de líquido(s) m̱
   cloacal(es) - [sanit] small package sewage
   treatment plant
— f industrial - [ind] industrial, o̱ manufacturing, plant, o̱ mill • plant property
— f integrada - [ind] integrated, plant, o̱ mill
   • integrated works
— f — grande - [ind] large integrated plant
— f̱ — pequeña - [ind] small integrated plant
— f̱ — totalmente - [ind] totally integrated

plant

planta f integral - [ind] integrated, plant, o
works • package plant
— f interna - [ind] internal plant
— f invernal - [bot] winter plant
— f L D - [metal-prod] véase planta f Linz
Donawitz
— f laminadora f - [metal-lam] (rolling) mill
— f — para acero m - [metal-prod] steel (rolling) mill
— f lavadora f - [ind] washing plant; washer
— f Linz Donawitz - [metal-prod] basic oxygen,
(steel), plant, o shop
— f — — con inyección f inferior - [metal-
-prod] quiet basic oxygen plant
— f — — — f por fondo m - [metal-prod]
quiet basic oxygen plant
— f local - [ind] local plant
— f lubricadora - [ind] lubricating plant
— f manufacturera - [ind] manufacturing plant
— f mediana - [ind] midi mill
— f metalista* —[metal-fabr] metalworking plant
— f metalúrgica - [metal-prod] metallurgical
plant
— f mezcladora - [ind] mixer,xing plant; mixer
building
— f moderna - [ind] modern, plant, o mill
— f — integrada - [ind] integrated modern,
plant, o mill
— f — — totalmente - [ind] totally, o completely, integrated modern, plant, o mill
— f montada - [ind] erected plant
— f motriz - [electr-prod] power plant
— f no integrada - [metal-prod] nonintegrated,
plant, o mill
— f normal - [ind] normal plant
— f nuclear - [nucl] nuclear plant
— f — para energía f - [electr-prod] nuclear
power plant
— f — prototipo m - [nucl] prototype nuclear
power plant
— f nueva - [ind] new, plant, o mill
— f papelera - [papel] paper, mill, o plant
— f para abono(s) m - [quím] fertilizer plant
— f — absorción f - absorbing,btion plant
— f — — f para amoníaco - [subprod] ammonia
absorbing plant
— f — acabado m - [ind] fnishing plant
— f — — m de plancha(s) f - [metal-lam] plate
finishing, plant, o shop
— f — acero m - [metal-prod] steel, plant, o
mill
— f — — m automatizada - [metal-prod] automated steel, plant, o mill
— f — — m — altamente - [metal-prod] highly
automated steel, plant, o mill
— f — — m con colada f continua - [metal-
-prod] continuous casting steel plant
— f — — m especial - [metal-prod] special
steel(s) plant
— f — — m no integrada - [metal-prod] nonintegrated steel, plant, o mill
— f — ácido(s) m - [quím] acid plant
— f — — m nítrico - [quím] nitric acid plant
— f — — m sulfhídrico - [coque] hydrosulfuric
acid plant
— f — — m sulfúrico - [quím] sulfuric acid
plant
— f — — m — en planta f para coque m -
[quím] coke plant sulfuric acid plant
— f — aglomeración f - [metal-prod] sintering
plant
— f — aglutinación f - [metal-prod] sintering
plant
— f — agregado(s) m - [constr] aggregate(s), o
gravel, plant
— f — agua m - [hidr] water plant
— f — aire m (comprimido) - [ind] (compressed)
air plant
— f — alimentación f - [ind] feed(ing) plant
— f — almacenamiento m - [ind] storage,ring
plant • [petról] tank farm
— f — alquilación f - [petról] alkylation,

plant, o unit

planta f para alquilación f con ácido m sulfrico
- [petról] sulfuric acid alkylation plant
— f — alumbrado m - [electr-prod] light(ing)
plant
— f — aluminio m - [metal- aluminum plant
— f — ánodo(s) m - anode plant
— f — arrabio m - [metal-prod] pig iron plant
— f — aserrado f - [ind] sawing plant
— f — — m de piedra(s) f - [miner] stone
sawing plant
— f — asfalto m - [constr] asphalt plant
— f — automóvil(es) m - [autom] automobile
plant
— f — beneficiación f - [miner] benefitting, o
beneficiation, plant
— f — — f de mineral m—[miner] ore beneficiation plant
— f — beneficio m - [miner] beneficiation
plant
— f — benzol(io) m - [subprod] benzol plant
— f — bombeo m - [bombas] pumping plant
— f — calcinación f - [miner] calcination,ning
plant
— f — camión(es) m - [autom] truck plant
— f — cicloexano - [subprod] cyclohexene plant
— f — clarificación f - [hidr] clarification,
o clarifying, plant
— f — clasificación f - classification plant •
[miner] grading, o sizing, plant
— f — — f para mineral(es) m - [miner] ore,
classification, o classifying, o grading,
plant
— f — — f — — m de hierro m - [miner] iron
ore, classification,fying, o grading, plant
— f — — f y sinterización f - [metal-prod]
classification and sintering plant
— f — cloruro m de polivinilo m - [quím] polyvinyl chloride plant
— f — colada f continua - [col.cont] continuous casting, plant, o shop
— f — — f con . . . línea(s) f - [col-
-cont] . . . strand continuous casting plant
— f — — f continua para palanquilla f - [col-
-cont] continuous billet, casting plant, o
caster
— f — — f para palanquilla f - [col.cont]
billet, casting, plant, o caster
— f — concentración f - [miner] concentration
plant
— f — congelación f de carne(s) f - [ind]
packing, house, o plant
— f — coque m - [coque] coke plant
— f — coquización f - [coque] coke,king plant
— f — corte m - [ind] shearing, o sawing,
plant
— f — — m para piedra(s) f - [miner] stone
sawing plant
— f — caucho m - [ind] rubber plant
— f — craqueo m - [petról] cracking plant
— f — cribado m - [ind] screening, plant, o
station
— f — — m para coque m - [coke] coke screening plant
— f — — m — — m metalúrgico - [coque] metallurgical coke screening, plant, o station
— f — — m para horno m alto - [coque]
(blast) furnace coke screening (plant)
— f — depuración f - [sanit] purification, o
treatment, plant
— f — — f para agua m - [hidr] water, purification, o treatment, plant
— f — — f — — m corriente - [hidr] (drinking) water treatment plant
— f — — f — — m servida - [constr] waste
water, o (sanitary) sewage, treatment plant
— f — — f — — m servida en base f a lodo m
activado - [sanit] activated sludge sewage
treatment plant
— f — — m — gas(es) m - [ind] gas, cleaning,
o scrubbing, plant
— f — — f — líquido(s) m cloacal(es) -
[sanit] sewage treatment plant

planta f para depuración f para líquido(s) n
cloacal(es) con base f de lodo(s) m activados
- [sanit] activated sludge sewage treatment
plant
— f —— f —— m — de tipo m con laguna f -
[sanit] lagoon type sewage treatment plant
— f —— f para residuo(s) m - [sanit] waste
treatment plant
— f —— f primaria - [sanit] primary treatment plant
— f —— f secundaria - [sanit] secondary
treatment plant
— f — desbenzolación f - [coque] debenzolizing
plant
— f — desgasificación f - [metal-prod] degassing plant
— f —— f para acero m - [metal-prod] steel
degassing plant
— f —— f —— m por vacío m - [metal-prod]
vacuum steel degassing plant
— f — desintegración f - [petról] cracking
plant
— f — destilación f - [ind] distillation plant
— f —— f con alambique(s) m tubular(es) -
[petról] pipe still distillation plant
— m —— m en vacío m - [petról] vacuum distillation plant
— f —— f para alquitrán m - [petról] tar
distillation,ling plant
— f — desulfuración f de gas(es) m en horno m
para coque m - [coque] coke oven gas desulfurization plant
— f — dolomí(t)a f - [miner] dolomite plant
— f — dosificación f - [constr] batching plant
• [quím] formulating,tion plant
— f —— f para hormigón m - [constr] concrete
batching plant
— f — elaboración f - [ind] fabricating plant
— f —— f para alcantarilla(s) f - [metal-
-culvert fabricating plant
— f —— f para metal(es) m - [metal-fabr]
metalworking plant
— f — eliminación f de líquido(s) m cloa-
cal(es) - [sanit] sewage disposal plant
— f — encendido m con gas m - [combust] gas
fueled furnace
— f — energía f - [electr-prod] power, plant,
o station, o unit
— f —— f eléctrica - [electr-prod] electric,
power, o light, plant
— f —— f nuclear—[nucl] nuclear power plant
— f —— f —— f — primitiva - [nucl] early nuclear
power plant
— f —— f — prototipo - [nucl] prototype nuclear power plant
— f —— f —— f —— primitiva - [nucl] early prototype nuclear power plant
— f —— f primitiva - [electr-prod] early, o
primitive, power plant
— f —— f prototipo primitiva - [electr-prod]
early prototype power plant
— f —— f para universidad f - [electr-prod]
university power station
— f —— f temprana - [electr-prod] early power
plant
— f —— f y soldadora f combinada - [sold]
combined power plant and welder
— f —— f y soldadora f portátil combinada -
[sold] combined power plant and portable
welder
— f para enriquecimiento m - [combust] enriching plant
— f — escoria f - [metal-prod] slag plant
— f — estañadura f - [metal-trat] tinning
plant
— f para evacuación f - [sanit] disposal plant
— f —— f de líquido(s) m cloacal(es) -
[sanit] sewage disposal plant
— f — extrusión f - [metal-fabr] extrusion,
plant, o mill
— f — fabricación f—[ind] manufacturing plant
— f —— f en escala f (grande) - [ind] (large

scale) production plant
planta f para fabricación f en escala f industrial - [sold] production plant
— f —— serie - [sold] production plant
— f —— f para alcantarilla(s) f - [metal-
-fabr] culvert plant
— f —— f — camión(es) f - [autom] truck
(manufacturing) plant
— f — fertilizante m - [ind] fertilizer plant
— f — filtración f - [hidr] filtering,tration
plant
— f — finos m - [miner] fines plant
— f — formulación f - [quím] formulating,tion
plant
— f — fraccionamiento m - [ind] fractioning
plant
— f —— m para aire m - [ind] air fractiona-
ting,tion plant
— f — fundición f - [metal-prod] foundry;
melt(ing) shop
— f —— f con horno(s) m eléctrico(s) m -
[metal-prod] electric furnace melt(ing) shop
— f — galvanización f - [metal-trat] galvanizing plant
— f — gas m - [ind] gas plant
— f —— m para horno m - [metal-prod] furnace
gas plant
— f —— m —— m purgada - [metal-prod]
blown down, o bled, furnace gas plant
— f —— m purgada - [metal-prod] blown down,
o bled, gas plant
— f —— m reformado - [metal-prod] reformed
gas plant
— f — iluminación f - [electr-prod] light(ing)
plant
— f — industria f - [ind] industry,trial plant
— f —— f química - [ind] chemical industry
plant
— f — laminación f - [metal-lam] rolling,
mill, o plant
— f —— f para fleje m - [metal-lam] strip
mill, plant, o mill
— f —— f —— en caliente - [metal-lam] hot
strip (rolling), plant, o mill
— f —— f —— m — frío - [metal-lam] cold
strip (rolling), plant, o mill
— f —— f — tubo(s) m - [tub] pipe mill
— f — lavado m - [ind] washing plant; washery*
— f —— m para mineral(es) m - [miner] ore
washing plant
— f — limpieza f - [ind] cleaning plant
— f —— f para gas m - [combust] gas cleaning
plant
— f — metalistería f - [metal-fabr] metal
working plant
— f — mezcla(dura) f - [quím] formulating,tion
plant • [constr] aggregate(s) plant
— f — molienda f - [miner] grinding, o crushing, plant
— f — munición(es) f - [milit] munition(s)
plant
— f — neumático(s) m - [autom-neumát] tire
plant
— f — operación(es) f - [ind] operating,tion
plant
— f — óxido(s) m - [metal-prod] oxide plant
— f —— m para planta(s) f siderúrgica(s) -
[metal-prod] (steel) mill oxide plant
— f — oxígeno m - [metal-prod] oxygen plant
— f —— m líquido - [ind] liquid oxygen plant
— f —— m básico - [metal-prod] basic oxygen
plant
— f —— m — con inyección f inferior -
[metal-prod] quiet basic oxygen plant
— f —— m —— f por fondo m - [metal-
-prod] quiet basic oxygen plant
— f — peletización f - [miner] pelletizing
plant
— f — perfil(es) m - [metal-lam] shapes plant
— f — pertrecho(s) m - [milit] munition(s)
plant
— f — pieza(s) f de hormigón m prefabricado -

[constr] (prefabricated beams) casting plant

planta f **para piso(s)** m **para puente(s)** m - [constr] bridge floor(ing) plant

— f — **plancha(s)** f **cizallada(s)** - [metal-fabr] sheared plate, plant, o mill

— f — **preparación** f - [ind] preparation, yard, o plant

— f — — f **para carbón** m - [miner] coal preparation, yard, o plant

— f — — f **escoria** f - [metal-prod] slag preparation, plant, o yard

— f — — f **mineral(es)** m - [metal-prod] ore preparation, plant, o yard

— f — **prerreducción** f - [metal-prod] prereduction, plant, o yard

— m — **procesamiento** m - [ind] processing, yard, o plant

— f — — m **para caucho** m - [ind] rubber plant

— f — **producción** f - [ind] production, o manufacturing, plant, o facility,ties • [sold] production plant

— f — — f **para ácido** m **sulfúrico** - [quím] sulfuric acid production plant

— f — — f **arrabio** m - [metal-prod] (pig) iron production plant

— f — — f **oxígeno** m - [ind] oxygen generating plant

— f — — f **tubería** f - [tub] pipe fabricating plant

— f — — — f **de acero** m - [tub] steel pipe fabricating plant

— f — **reducida** - [ind] small output plant

— f — **producto(s)** m **liviano(s)** - [petról] light, o bright, stock plant

— f — — m **para drenaje** m - [metal-fabr] drainage plant

— f — **purificación** f - [ind] purification, o treatment, plant

— f — — f **para agua** m - [hidr] water treatment plant

— f — **recuperación** f - [ind] recovery,ring plant • [extrusión f de electrodos] recuperating mill

— f — — f **para ácido** m - [ind] acids, recovery, o recuperating,tion, plant

— f — — f — m **residual** - [ind] residual acid recovery plant

— f — — f — **subproducto(s)** m - [subprod] by-products (recovery) plant

— f — — f **vapor** m - [petról] vapor recovery plant

— f — **redestilación** f - [petról] distillate rerun(ning) unit

— f — — f **de destilado*** m **bruto de craqueo** m - [petról] pressure distillate rerun(ning) unit

— f — **reducción** f - [metal-prod] reducing, o reduction, plant

— f — — f **con gas** m - [metal-prod] gaseous reduction plant

— f — — f — m **reformado** - [metal-prod] reformed gas reduction plant

— f — — f **directa** - [metal-prod] direct reduction plant

— f — — f — **para óxido** m - [metal-prod] oxide direct reduction plant

— f — — f — **en planta(s)** f **siderúrgica(s)** - [metal-prod] mill oxide direct reduction plant

— f — — f **en caliente** - [metal-prod] hot reduction mill

— f — — f **en frío** - [metal-prod] cold reduction mill

— f — — f **para óxido(s)** m [metal-prod] oxide reduction plant

— f — — f — — m **en planta(s)** f **siderúrgica(s)** - mill oxide reduction plant

— f — **refinación** f - [ind] refining plant

— f — **refrigeración** f - [ind] cooling, o refrigerating,tion plant

— f — — f **con agua** m - [ind] water cooling plant

— f — — f **para agua** m - [ind] water cooling plant

planta f **para regulación** f **primaria** - [petról] primary, control, o regulating, plant

— f — **reprocesamiento** m - [ind] reprocessing plant

— f — — m **para combustible** m - [nucl] fuel reprocessing plant

— f — — m — m **agotado** - [nucl] spent fuel reprocessing plant

— f — **sinter(ización)** - [metal-prod] sinter--(ing) plant

— f — **síntesis** m - [quím] synthesis plant

— f — — m **para amoníaco** m - [quím] ammonia synthesis plant

— f — **soldadura** f - [sold] welding plant

— f — — f **para tubo(s)** m - [tub] pipe welding mill

— f — — — m **por fusión** f - [tub] fusion weld(ed) pipe mill

— f — **subproducto(s)** m - [metal-prod] by-products plant

— f — — m **para coquería** f - [metal-prod] coke by-products plant

— f — **sulfato** m - [quím] sulfate, plant, o building

— f — — m **de amonio** m - [coque] ammonium sulfate, plant, o building

— f — **sulfuro** m - [quím] sulfide plant

— f — — m **de hidrógeno** m - [quím] hydrogen sulfide plant

— f — **terminación** f - [ind] finishing, plant, o shop

— f — — f **para plancha(s)** f - [metal-trat] plate finishing shop

— f — **tonelaje** m **nuevo** - [ind] new tonnage, plant, o mill

— f — **trabajo** m **en escala** f **grande** - [sold] large scale, o production, plant

— f — — m — f **industrial** - [sold] production plant

— f — **trabajo(s)** m **en serie** - [sold] production plant

— f — **tratamiento** m - [ind] treating,tment plant

— f — — m **continuo** - [ind] continuous treating plant

— f — — m — **para gasolina** f - [petról] continuous gasoline treating plant

— f — — m **para ácido** m **sulfúrico** - [quím] sulfuric acid contact plant

— f — — m — **combustible** m - [combust] fuel treatment plant

— f — — — m **irradiado** - [nucl] irradiated fuel treatment plant

— f — — m — **desecho(s)** m - [nucl] waste treatment plant

— f — — — m **radiactivo(s)** - [nucl] radioactive waste treatment plant

— f — — — m — **líquido(s)** - [nucl] liquid radioactive waste treatment plant

— f — — m — **escoria** f - [metal-prod] slag treatment plant

— f — — m — **residuo(s)** m - [nucl] waste treatment plant

— f — — m — m **radiactivo(s)** - [nucl] radioactive waste treatment plant

— f — — m — m **líquido(s)** - [nucl] liquid radioactive waste treatment plant

— f — **trituración** f - [miner] crushing plant

— f — — f **para carbón** m - [miner] coal crushing plant

— f — — f — **mineral** m - [miner] ore crushing plant

— f — — f — — m **de hierro** m - [miner] iron ore crushing plant

— f — — f **primaria** - [miner] primary crushing plant

— f — — f **secundaria** - [miner] secondary crushing plant

— f — — f **y clasificación** f - [miner] crushing and, classification, o grading, plant

— f — — f — **cribado** m - [miner] crushing and screening plant

planta f para trituración f y cribado m para mi-
neral(es) m - [miner] ore crushing and
screening plant
— f — **tubería** f̱ - [tub] pipe mill
— f̱ — — f **con costura** f̱ - [tub] seam(ed) pipe
plant
— f — — f — — f **recta** - [tub] straight seam
pipe plant
— f — — f **soldada** - [tub] welded pipe mill
— f̱ — **tubo(s)** m̱ - [tub] pipe, o̱ tube, plant
— f̱ — **zarandeo** m̱ - [miner] screening plant
— f̱ **pequeña** - [ind] small, plant, o̱ operation •
mini mill
— f **perenne** - [botán] perennial
— f̱ **piloto** - [ind] pilot plant
— f̱ — **grande** - [ind] large pilot plant
— f̱ — **para ensayo(s)** m̱ - [ind] pilot test(ing)
plant
— f — **pequeña** - [ind] small pilot plant
— f̱ — **neumático(s)** m̱ - [autom-neumát] tire
pilot, o̱ pilot tire, plant
— f **portátil** - [ind] portable plant
— f̱ — **para agregado(s)** m̱ - [constr] portable
aggregate plant
— f — — **bombeo** m̱ - [petról] portable pumping
plant
— f — — **fuerza** f̱ **motriz** - [electr-prod] portable power plant
— f **previsible** - foreseeable plant
— f̱ **principal** - [ind] main plant • [constr]
main, o̱ ground, floor
— f **productora** - [ind] producing plant
— f̱ — **para abono(s)** m̱ - [ind] fertilizer plant
— f̱ — — **fertilizante(s)** m̱ - [ind] fertilizer
plant
— f **prototipo** - [ind] prototype plant
— f̱ — **para energía** f̱ - [electr-prod] prototype
power plant
— f **proyectada** - [ind] projected mill • designed mill
— f **química** - [quím] chemical plant
— f̱ **radiactiva** - [nucl] radioactive, o̱ hot,
plant
— f **recuperadora** - [ind] recovery plant
— f̱ — **para ácido** m̱ - [quím] acid, recovery, o̱
restoring, plant
— f **reductora** - [ind] reducing,ction plant
— f̱ **reguladora** - regulating, o̱ controlling,
plant
— f **semi-integrada**—[ind] semi-integrated plant
— f̱ **sencilla** - simple plant
— f̱ **siderúrgica** - [metal-prod] steel, plant, o̱
mill • iron works
— f — **integrada** - [metal-prod] integrated
steel, plant, o̱ mill
— f **Siemens Martín** - [metal-prod] open hearth,
plant, o̱ mill, o̱ shop
— f **superior** - superior plant • [constr] upper,
floor, o̱ story
— f **termoeléctrica** - [electr-prod] thermoelectric, o̱ thermal, o̱ steam, power, plant, o̱
house, o̱ station
— f — **convencional** - [electr-prod] conventional thermal,moelectrical plant
— f **textil** - [textil] textile plant
— f̱ **típica** - [ind] typical, plant, o̱ mill
— f̱ **tipo** m̱ - [ind] typical, o̱ standard, plant
— f̱ — **sencilla** - [ind] simple standard plant
— f̱ **totalmente moderna** - [ind] completely, o̱
totally, o̱ fully, modern, plant, o̱ mill
— f **ubicada** - [ind] located plant
— f **urbana** - [pol] townsite
— f̱ **vieja** - [ind] old plant
— f̱ **vivaz** - [botán] perennial
planteado,da a̱ - proposed; set forth; stated •
plotted • approached
planteamiento m̱ - . . . . • approach; proposal;
suggestion; design; planning • statement;
stating • outline • scenario • position
— m **actual** - present, o̱ current, outline,ning
— m̱ **con** . . . **paso(s)** m̱ - . . . step approach
plantear v̱ - . . . • to, outline, o̱ expound •
to, frame, o̱ pose, o̱ present, o̱ propose • to

set, forth, o̱ up; to establish; to plot • to
found
plantel m̱ - . . . • [labor] force; staff; organization • operating, o̱ work, force • team
— m **obrero** - [labor] operating, o̱ labor, force
— m̱ **para rendimiento** m̱ **(alto)** - [labor] performance team
planteo m̱ - . . .; stating; approach; plot(ting)
— m **actual** - present outline,ning
— m **indirecto** - indirect approach
plantilla f̱ - . . . • [mec] hard die; mold; jig
• [dib] curvature guide • [labor] personnel;
roster; crew; staff • [sold] gage • subject
— f **adecuada** - [dib] suitable, o̱ adequate, template
— f **armada** - [mec] assembled template,plet
— f̱ **autorizada** - [admin] authorized staff
— f̱ **cambiable** - [mec] changeable (hard) die
— f̱ **colocada en sitio** m̱ - [mec] positioned
template
— f **con** . . . **agujero(s)** m̱ - [mec] . . .-hole
templet,late
— f **con escala** f̱ - [mec] templet with @ gage
— f̱ **con matríz,ces** f̱ - [mec] die, section, o̱
assembly
— f — **punzón(es)** m̱ - [mec] punch, assembly, o̱
section
— f **de acero** m̱ - [mec] steel templet,late
— f̱ — **personal** - [labor] staff; personnel •
[admin] organization chart
— f̱ — **radios** m̱ - [sold] fillet, guide, o̱ gage;
gage; radius gage
— f **estándar** - [instrum] standard gage
— f̱ **idéntica** - [mec] identical templet,late
— f̱ — **de acero** m̱ - [mec] identical steel templet,late
— f **modular** - [mec] modular, o̱ hard, die
— f — **cambiable** - [mec] changeable, modular, o̱
hard, die
— f **normal** - [labor] normal, o̱ standard, staff,
o̱ personnel • [instrum] standard gage
— f̱ **nueva** - [mec] new templet,late
— f̱ **para cordón(es)** m̱ **cóncavo(s)** - [sold] concave (fillet) gage
— f — — m — **en ángulo** m̱ **interior** - [sold]
concave fillet gage
— f — — m **convexo(s)** - [sold] convex (fillet)
gage
— f — — m — **en ángulo** m̱ **interior** - [sold]
convex fillet gage
— f — **medir** - measuring, o̱ size, gage
— f̱ — — v̱ **cordón(es)** m̱ - [sold] weld size
gage
— f — — v — m **cóncavo(s)** m̱ - [sold] concave
(fillet) gage
— f — — v — m — **en ángulo** m̱ **interior** -
[sold] concave fillet gage
— f — — v — m **convexo(s)** - [sold] convex
(fillet) gage
— f — — v — m — **en ángulo** m̱ **interior** -
[sold] convex fillet gage
— f — — v — m **en ángulo** m̱ **interior** - [sold]
fillet gage
— f — **producción** f̱ - [admin] operating, o̱ production, personnel
— f — **soldador** m - [sold] welding jig
— f̱ — **soldadura(s)** f̱ - [sold] weld(ing) gage
— f̱ — **supervisión** f—[admin] supervisory staff
— f̱ — — f **para planta** f̱ - [ind] plant supervisory staff
— f — **taller** m̱ - [ind] shop crew
— f̱ **perforada** - [mec] punched templet,late
— f̱ **precisa** - [mec] accurate templet,late
— f̱ **preparada** - [mec] prepared, o̱ assembled, o̱
set up, templet,late
— f **según norma** f̱ - [mec] standard gage •
[labor] standard staff
— f — — f **autorizada** - [labor] standard authorized staff
— f **taladrada** - [mec] punched, o̱ bored, templet
— f̱ **vieja** - [mec] old templet
planura* f̱ - flatness
plaqué m̱ . . . | a̱ - [metal-trat] clad

plaqueado* m̲ - [metal-trat] cladding
plaqueado,da a̲ - [metal-trat] clad
— a̲ con cadmio m̲ - [metal-trat] cadmium plated
— a̲ con cinc m̲ - [metal-trat] zinc plated
— a̲ — cromo m̲ - [metal-trat] chrome plated
plaquear* v̲ - [metal-trat] to plate; véase también planchear* v̲
plaqueta f̲ - [autom] placard
— f̄ para vehículo m̲ - [autom] vehicle placard
plaquita f̲ compacta - [quím] compact product
plasmadar* v̲ - véase plasmar
plasmado,da a̲ - molded; shaped • established
plasmar v̲ - . . . . • to establish
plasticidad f̲ alta - [suelos] high plasticity
— f̲ baja - [suelos] low plasticity
— f̄ elevada - [suelos] high plasticity
— f̄ media(na) - [suelos] medium plasticity
— f̄ reducida - [suelos] low plasticity
plástico m̲ flexible - [plást] flexible plastic
— m̲ liso - [plást] smooth plastic
— m̲ moldeado - [plást] molded plastic
— f̄ para uso(s) m̲ múltiple(s) - [plást] multi-use, o̲ multiple use, plastic
— m̲ procesado - [plást] processed plastic
— f̄ reforzado - [plást] reinforced plastic
— m̲ — con fibra(s) f̲ de vidrio m̲ - [plást] fiber glass reinforced plastic
— m̲ tenaz - [plást] tough plastic
— m̲ vinílico - [plást] vinyl plastic
plastificado,da a̲ - plasticized
platabanda f̲ - [metal-lam] flat sheet • [constr] cover plate
plataforma f̲ - . . .; bench • deck; porch; floor • trolley • [transp] pallet • flat bed • [ind] dock • [geogr] shelf • [petról] drilling rig • [f.c.] . . .; platform • [autom] flat bed truck | a̲ - [transp] flat bed
— f̲ adicional - [mec] additional platform
— f̲ apoyada - supported platform
— f̄ — sobre fondo m̲ (de mar m̲) - [petról] bottom supported platform
— f̲ armada - [mec] assembled platform
— f̄ arrastrada - [f.c.] drawn platform
— f̄ — por locomotora f̲ - [f.c.] locomotive drawn platform
— f̲ astillero - [petról] fourble board
— f̄ auxiliar - [mec] auxiliary platform
— f̄ con rodillo(s) m̲ - [transp] dolly
— f̄ — — f̲ para giro m̲ - [metal-prod] bull wheel platform
— f̲ continental - [geol] continental shelf
— f̄ costa afuera - [petról] offshore, platform, o̲ rig
— f̄ — — para perforación f̲ - [petról] off-shore (drilling) rig
— f̲ — — — — f̲ para gas m̲ - [petról] offshore gas (drilling) rig
— f̲ — — — — f̲ — petróleo n̲ - [petról] offshore oil (drilling) rig
— f̲ de acero m̲ - [mec] steel platform
— f̄ — camión f̲ - [autom] truck bed
— f̄ — madera f̄ - [transp] wooden pallet
— f̄ — patín(es) f̲ - [autom] skid
— f̄ — — n̲ para camión f̲ - [transp] truck skid
— f̄ — plancha(s) f̲ - [constr] plate platform
— f̄ — — f̲ estriada(s) - [constr] checkered plate platform
— f̲ — tipo m̲ giratorio - [mec] turntable type platform
— f̲ elástica - [mec] elastic platform
— f̄ — integral - [mec] integral elastic platform
— f̲ — solidaria - [mec] integral elastic platform
— f̲ en construcción f̲ para perforación f̲ - [petról] under construction oil rig
— f̲ epicontinental - [geol] continental shelf
— f̄ fija - [mec] stationary, o̲ fixed, platform
— f̄ — para enfriamiento m̲ - stationary, o̲ fixed, cooling platform
— f̲ final - [mec] final platform
— f̄ — para enfriamiento m̲ - final cooling platform

plataforma f̲ frontal - front, platform, o̲ porch
— f̲ giratoria - [mec] revolving, o̲ rotating, platform, o̲ table; turntable (type) platform; swivel platform
— f̲ — superior - [mec] upper turntable
— f̄ — para plancha(s) f̲ - [mec] plate turntable
— f̲ inferior - bottom, o̲ lower, platform
— f̄ integral - [mec] integral platform
— f̄ lateral - [mec] side, o̲ lateral, platform; sidestand
— f̲ — en costado m̲ derecho - [mec] right hand, platform, o̲ sidestand
— f̲ — — — m̲ izquierdo - [mec] left hand, platform, o̲ sidestand
— f̲ metálica - [ind] metal platform
— f̄ montada - [mec] mounted, o̲ erected, platform
— f̲ movible - [mec] movable platform • trolley
— f̄ — para enfriamiento m̲ - [mec] movable cooling platform
— f̲ para acceso m̲ - [ind] access platform
— f̄ — — m̲ para mantenimiento m̲ - [ind] maintenance access platform
— f̲ — alubias f̲ - [agric-equip] bean platform
— f̄ — amplificador m̲ - [mec] amplifier platform
— f̄ — — m̲ magnético - [sold] magnetic amplifier, o̲ mag amp, platform
— f̲ — aparcamiento m̲ - parking platform
— f̄ — — m̲ para avión(es) m̲ - [aeron] air, field, o̲ airplane, (parking) apron
— f̲ — apromtamiento m̲ - [constr] ready apron
— f̄ — apronte m̲ - warm up apron
— f̄ — atadora f̄ - [agric-equip] binder deck
— f̄ — balancín(es) m̲ - [metal-prod] bell beam platform
— f̲ — — m̲ para campana f̲ - [metal-prod] bell beam platform
— f̲ — para carga f̲ - [ind] loading rack • charging floor
— f̲ — cereal(es) m̲ - [agric] grain platform
— f̄ — colada - [metal-prod] pouring, o̲ teeming, platform
— f̲ — comando m̲ - [ind] pulpit
— f̄ — cuba f̲ - [metal-prod] mantle platform
— f̄ — cuchara(s) f̲ - [metal-prod] ladle stand
— f̄ — chapín(es) m̲ - [metal-prod] bleeder platform
— f̲ — descarga f̲ - [Transp] unloading platform
— f̄ — embolso m̲ - [agric] bagging platform
— f̄ — enfriamiento m̲ - [ind] cooling platform
— f̄ — — m̲ final - [ind] final cooling platform
— f̄ — enganchador m̲ - [petról] finger board
— f̄ — enganche m̲ - [petról] finger, o̲ lazy, board
— f̲ — — m̲ — tiro(s) cuádruple(s) - [petról] fourble board
— f̲ — — m̲ — tramo(s) m̲ cuádruple(s) - [petról] fourble board
— f̲ — ensacado* m̲ - [agric] bagging platform
— f̄ — ensayo(s) m̲ - [ind] test(ing) program
— f̄ — espigadora f̲ - [agric-equip] header platform
— f̲ — ferrocarril m̲ - [f.c.] (railway) flat (bed) car
— f̲ — fréjoles m̲ - [agric] bean platform
— f̄ — grano(s) m̲ - [agric] grain platform
— f̄ — grúa f̲ - [grúas] crane platform
— f̄ — — f̲ para servicio m̲ - [grúas] service crane platform
— f̲ — horno m̲ - [ind] furnace base
— f̄ — — m̲ para recocido m̲ - [metal-trat] annealing furnace base
— f̲ — limpieza f̲ - [ind] cleaning pulpit
— f̄ — malacate m̲ - [petról] drawworks porch
— f̄ — mando m̲ - [ind] pulpit; control platform
— f̄ — mantenimiento m̲ - [ind] maintenance platform
— f̲ — montaje m̲ - [mec] mounting, o̲ erecting, platform
— f̲ — motor - [electr-mot] motor platform • [comb.int] engine platform

plataforma f para natación f - [deport] swimming platform
— f — operador m - [ind] operator('s) platform
— f — perforación f - [petról] (drilling) platform; oil rig
— f — — costa afuera - [petról] offshore (drilling), rig, o platform
— f — — f marina - [petról] offshore (drilling) rig
— f — — f para gas m - [petról] gas (drilling) rig
— f — — f — petróleo m - [petról] oil drilling, platform, o rig
— f — — pozo(s) m para petróleo n - [petról] oil (drilling) rig
— f — — f submarina - [petról] offshore drilling platform
— f — personal m - [grúas] personnel platform
— f — preparación f de cabeza f caliente - [metal-prod] véase f plataforma f para preparación f de mazarota(s) f
— f — — f de mazarota(s) f - [metal-prod] hot top(ping) (preparation) platform
— f — recocido m - [metal-trat] annealing (furnace) base
— f — remolque m - [ind] undercarriage
— f — reparación(es) f - [ind] repair platform
— f — — f para grúa f - [grúas] crane repair platform
— f — — f — — f puente - [grúas] (bridge) crane repair platform
— f — rodillo(s) m - [metal-lam] roll(er) stand
— f — — m para perfil(es) m (estructurales) - [metal-lam] structural (mill) roll stand
— f — salida f - [metal-lam] delivery ramp
— f — sangría f—[metal-prod] tapping platform
— f — seguridad f - [segurid] escape platform
— f — servicio m - [ind] service platform
— f — — m para grúa f - [grúas] crane service platform
— f — taladrado m - [petról] drilling rig
— f — termopar(es) m - [metal-prod] thermocouple(s) platform
— f — tolva f para carga f - [metal-prod] receiving hopper platform
— f — trabajo m - [ind] working platform; deck
— f — tragante m - [metal-prod] main, o top, o throat, platform
— f — transporte m - [transportation platform
— f — — m para carga f - [mec] load transporting platform
— f — vaciado m—[metal-prod] pouring platform
— f perforadora - [petról] drilling rig
— f — costa afuera - [petról] off-shore drilling rig
— f portalingoteras - [metal-prod] véase vagón m portalingoteras
— f posterior - [mec] back, o rear, platform, o porch
— f — para malacate m - [petról] drawworks back, platform, o porch
— f rodante - [ind] gantry
— f solidaria - [ind] integral platform
— f superior - [ind] upper, o top, platform
— f — para termopar(es) m - [metal-prod] upper thermcouple(s) platform
— f — — torre f - [petról] crow's nest
platasoldadura f - [sold] silver (alloy) brazing
platea f - . . . • [constr] mat; apron
— f de hormigón m - [constr] concrete apron
— f — tierra f - [constr] dirt, o mud, mat
— f para cimentación f—[constr] foundation mat
— f — — f para equipo m - [constr] equipment foundation mat
— f — — f — — m mayor - [constr] major equipment foundation mat
platear v - . . . • véase también planchear* v
platería f - [metal] . . .; silversmithing
plática f - . . . • discussion
— f planeada - [labor] planned discussion
— f sobre seguridad f - [segurid] safety talk
— f técnica - technical discussion

platicado,da a - discussed
platicar v - . . .; to discuss
platillo m - . . . . • [válv] saucer
— m desviador - véase placa f desviadora
— m para acoplamiento m - [metal-prod| coupling flange
platina f - . . . . • [metal-lam] flat
— f reforzada - [mec] ribbed platen
platino m - [metal] . . . . • [comb.int] (breaker) point
— m abierto - [comb.int] open(ed) point
— m ajustado - [comb.int] adjusted point
— m cerrado - [comb.int] closed point
— m con metal m adherido - [comb.int] fused point
— m con picadura f - [comb.int] pitted point
— m fundido - [comb.int] fused point
— m para interruptor m - [comb.int] breaker point
— m — — m para encendido m - [comb.int] ignation breaker point
— m — magneto - [comb.int] magneto point
— m — — m ajustado - [comb.int] adjusted magneto point
— m — — m con metal m adherido - [comb.int] fused magneto point
— m — — m con picadura(s) f - [comb.int] pitted magneto point
— m — fundido - [comb.int] fused magneto point
— m — — m picado - [comb.int] pitted magneto point
— m — — m rectificado - [comb.int] dressed magneto point
— m — rotor m - [comb.int] rotor point
— m — — m ajustado - adjusted breaker point
— m — — m reemplazado - [comb.int] replaced breaker point
— m — ruptor m - [comb.int] breaker point
— m picado - [comb.int] pitted point
— m rectificado - [comb.int] dressed point
— m reemplazado - [comb.int] replaced point
plato m - . . . . • [mec] chuck; spindle • . . .; base
— m adaptador - [tornos] spindle nose
— m barbotador - [metal-prod] splasher; baffle
— m frontal - [comb.int] front (end) plate
— m inferior - [mec] bottom, o lower, plate
— m lateral - [mec] side plate
— m para embrague m - [mec] clutch side plate
— m magnético - [grúas] magnetic plate • [tornos] magnetic chuck
— m — para manipuleo m de tocho(s) m - [metal-prod] bloom handling magnetic plate
— m para embrague m - [mec] clutch plate
— m — — m gastado - [mec] worn clutch plate
— m — — m inferior - [mec] low(er) clutch plate
— m — — m reemplazado - [mec] replaced clutch plate
— m — — m revestido - [mec] (re)lined clutch plate
— m — — m superior - [mec] high, o upper, clutch plate
— m — extracción f - [mec] extraction baffle
— m — fijación f - [mec] fastening plate • angle iron
— m — retén m - [mec] retainer plate
— m — sistema m planetario - [mec] planetary system plate
— m — torno m - [herram] lathe chuck
— m revestido - [mec] (re)lined plate
— m rotante - [herram] revolving head; spinning spindle
— m único - [mec] single plate
playa f - [hidr] . . . • [ind] yard
— f aérea - [vial] parking deck
— f arenosa - [hidr] sandy beach
— f con lomo m de asno m - [f.c.] hump yard
— f — regulación f electrónica - [f.c.] electronically controlled railroad yard
— f de estanque(s) m - [petról] tank farm

playa <u>f</u> de hormigón <u>m</u> - [constr] concrete apron
— <u>f</u> electrónica - [<u>f</u>.c.] electronic yard
— <u>f</u> — automática - [f.c.] automatic electronic yard
— <u>f</u> ferroviaria - [f.c.] railway, <u>o</u> railroad, yard
— <u>f</u> — para clasificación <u>f</u> - [f.c.] railroad (classification) yard
— <u>f</u> — — — <u>f</u> con regulación <u>f</u> electrónica - [f.c.] electronically controlled railroad classification yard
— <u>f</u> interior - [ind] inside yard
— <u>f</u> intermedia - [ind] intermediate yard
— <u>f</u> interna - [ind] inside yard
— <u>f</u> para almacenaje <u>m</u> - [ind] storage yard; stockpile
— <u>f</u> — — <u>m</u> para contratista <u>m</u> - [constr] contractor's, storage yard, <u>o</u> stockpile
— <u>f</u> — — <u>m</u> en planta <u>f</u> - [ind] plant storage yard
— <u>m</u> — almacenamiento <u>m</u> - [ind] storage yard
— <u>m</u> — — <u>m</u> para carbón <u>m</u> - [ind] coal storage yard
— <u>f</u> — — <u>m</u> — coque <u>m</u>—[ind] coke storage yard
— <u>f</u> — — — palanquilla <u>f</u> - [metal-lam] billet(s) storage yard
— <u>f</u> — — <u>m</u> para planta <u>f</u> - [ind] plant storage yard
— <u>f</u> — — almacenar - [ind] storage yard
— <u>f</u> — — <u>v</u> producto(s) <u>m</u> - [ind] product(s) storage yard
— <u>f</u> — — tocho(s) <u>m</u> - [metal-lam] bloom storage yard
— <u>f</u> — aparcamiento <u>m</u> - [autom] parking, lot, <u>o</u> pad, <u>o</u> yard
— <u>f</u> — — <u>m</u> de equipo(s) <u>m</u> - [vial] equipment (storage) yard
— <u>f</u> — — <u>m</u> — — <u>m</u> mecanizado(s) - [vial] mechanical equipment (storage) yard
— <u>f</u> — apronte <u>m</u> - warm up yard
— <u>f</u> — avión(es) <u>m</u> - [aeron] apron
— <u>f</u> — carbón <u>m</u> - [ind] coal yard
— <u>f</u> — clasificación <u>f</u> - [f.c.] (railroad) classification, yard, <u>o</u> facility
— <u>f</u> — — <u>f</u> con lomo <u>m</u> de asno - [f.c.] hump yard
— <u>f</u> — — <u>f</u> con regulación <u>f</u> electrónica - [f.c.] electronically controlled railroad yard
— <u>f</u> — — <u>f</u> electrónica - [f.c.] electronic (classification) yard
— <u>f</u> — — <u>f</u> automátizada - [f.c.] automated electronic (classification) yard
— <u>f</u> — — <u>f</u> por gravedad <u>f</u> - [f.c.] humping (classification) yard
— <u>f</u> — coque <u>m</u> - [coque] coke yard
— <u>f</u> — chatarra <u>f</u> - [metal-prod] scrap, <u>o</u> junk, yard
— <u>f</u> — depósito <u>m</u> - storage yard; stockyard
— <u>f</u> — — <u>m</u> para palanquilla <u>f</u> - [metal-lam] billet storage yard
— <u>f</u> — — <u>m</u> — tocho(s) <u>m</u> - [metal-lam] bloom storage yard
— <u>f</u> — desbaste(s) <u>m</u> - [metal-lam] bloom yard
— <u>f</u> — deslingoteado <u>m</u> - [metal-lam] stripping yarad
— <u>f</u> — desmoldeo <u>m</u> - [metal-lam] stripping yard
— <u>f</u> — escarpado <u>m</u> - [metal-lam] scarfing yard
— <u>f</u> — — <u>m</u> para desbaste(s) <u>m</u> - [metal-lam] bloom scarfing yard
— <u>f</u> — escoria <u>f</u> - [metal-prod] slag yard
— <u>f</u> — estacionamiento <u>m</u> - [autom] parking lot
— <u>f</u> — — <u>m</u> drenado—[hidr] drained parking lot
— <u>f</u> — — <u>m</u> para hospital <u>m</u> - [constr] hospital parking lot
— <u>f</u> — hierro <u>m</u> viejo - [metal-prod] junk yard
— <u>f</u> — homogeneización <u>f</u>—[miner] blending yard
— <u>f</u> — lanzamiento <u>m</u>—[astronáut] launching pad
— <u>f</u> — lingoteras <u>f</u> - [metal-prod] mold (storage) yard
— <u>f</u> — lingote(s) <u>m</u> - [metal-prod] ingot yard
— <u>f</u> — — <u>m</u> frío(s) - [metal-prod] cold ingot yard

playa <u>f</u> para maniobra(s) - drill(ing) yard
— <u>f</u> — materia(s) <u>f</u> prima(s) - [ind] raw materials, <u>o</u> stock, yard
— <u>f</u> — mezcla(dura) <u>f</u> - [ind] mixing floor
— <u>f</u> — — <u>f</u> y homogeneización <u>f</u> - [miner] blending yard
— <u>f</u> — mezclar <u>v</u> - [ind] mixing yard
— <u>f</u> — mineral(es) <u>m</u> - [metal-prod] ore (stock) yard
— <u>f</u> — molde(s) <u>m</u> - [metal-prod] mold yard
— <u>f</u> — moldeo <u>m</u> - [metal-prod] molding, <u>o</u> casting, yard
— <u>f</u> — moldura <u>f</u> - [metal-prod] molding yard
— <u>f</u> — palanquilla <u>f</u> - [metal-prod] billet yard
— <u>f</u> — planchón(es) <u>m</u> - [metal-lam] slab yard
— <u>f</u> — prefabricación <u>f</u> - [constr] (pre)casting yard
— <u>f</u> — preparación <u>f</u> - [ind] preparation yard
— <u>f</u> — — <u>f</u> de chatarra <u>f</u> - [metal-prod] scrap preparation yard
— <u>f</u> — salamandra <u>f</u> - [metal-prod] polo field
— <u>f</u> — remolque(s) <u>m</u> (carretero(s) - [transp] piggyback yard
— <u>f</u> — — semirremolque(s) <u>m</u> (carreteros) para cargar sobre vagon(es) <u>m</u> plataforma—[transp] piggyback yard
— <u>f</u> — soldadura <u>f</u> - [sold] welding yard
— <u>f</u> — — <u>f</u> circunferencial - [sold] girth welding yard
— <u>f</u> — — <u>f</u> perimétrica - [sold] girth welding yard
— <u>f</u> — trabajo <u>m</u> - [ind] work(ing) yard
— <u>f</u> por gravedad <u>f</u> - [f.c.] gravity (classification) yard
playo,ya <u>a</u> - shallow
plaza <u>f</u> - . . . • seat • [com] area • unit • [vial] rotary intersection; [G.B.] roundabout; circus | <u>adv</u> - (en) commercially
— <u>f</u> central - [pol] village green
— <u>f</u> con jardín(es) <u>m</u> - landscaped square
— <u>f</u> en, población <u>f</u>, <u>o</u> pueblo <u>m</u> - [pol] village green
— <u>f</u> para acompañante <u>m</u> - [autom] navigator('s) seat
— <u>f</u> — almacenamiento <u>m</u> - [ind] stockyard
— <u>f</u> — conductor <u>m</u> - [autom] driver('s) seat
— <u>f</u> principal - [com] main market area
— <u>f</u> única - [autom] single seat | <u>adv</u> - [autom] (para) single seater
— <u>f</u> y jardín(es) - landscaped area
— <u>f</u> — — <u>m</u> elevado(s) - [constr] elevated, <u>o</u> raised, landscaped area
plazas <u>f</u> principales - <u>adv</u> - (en) commercially available
plazo <u>m</u> - . . . .; period; deadline; closing date; termination; performance period; time, period, <u>o</u> length • space
— <u>m</u> acordado - agreed, period, <u>o</u> term
— <u>m</u> adicional - additional (time) period
— <u>m</u> ampliado - [com] extended time (period)
— <u>m</u> anual - [lega;] yearly term
— <u>f</u> contractual - [legal] contract period
— <u>m</u> contratado - [legal] contracted period
— <u>m</u> corto - short, term, <u>o</u> period | <u>adv</u> - (a) short range
— <u>m</u> crediticio - [fin] credit, term, <u>o</u> period
— <u>f</u> determinado - established period
— <u>m</u> especificado - specified time period
— <u>m</u> establecido - established time (period); alloted time; time limit
— <u>m</u> estipulado - stipulated, <u>o</u> set, period
— <u>m</u> exigido - required (time) period
— <u>m</u> extendido - [legal] extended, time, <u>o</u> term
— <u>m</u> — para pago <u>m</u> - [fin] extended payment, time, <u>o</u> term
— <u>m</u> faltante - remaining, time, <u>o</u> period
— <u>m</u> fijado - established, <u>o</u> set, (time) period
— <u>m</u> fijo - [fin] fixed, term, <u>o</u> time
— <u>m</u> intermedio - intermediate term
— <u>m</u> largo - long, time, <u>o</u> term, <u>o</u> period • long range
— <u>m</u> mayor - longer (time) period
— <u>m</u> — para entrega <u>f</u> - longer delivery period

plazo m mediano - [com] medium term
— m medio - [com] medium term
— m menor - shorter time period
— m — para entrega f - [com] shorter delivery period
— m no mayor que - [legal] period not in excess of
— m original - [legal] original, term, o time period
— m para amortización f - [fin] amortization period
— m — — f de crédito m - [fin] credit amortization period
— m — armado m - [mec] assembly,bling time
— m — — m de plancha(s) f - [mec] plate assembly time
— f para armadura f - [mec] assembly time
— m — contrato m - [legal] contract, time, o period
— m — crédito m - [fin] credit term
— m — descuento m - [fin] discount period
— m — — m sobre factura f - [com] invoice discount period
— m — ejecución f - [com] performance time
— m — entrega f - [com] delivery, period, o term, o date; deadline
— m — — f contractual - [com] contract(ual) delivery, period, o date
— m — — f estipulado - [com] stipulated delivery, date, o period
— m — erección f - [mec] assembly time
— m — — f de plancha(s) f - [mec] plate(s) assembly, time, o period
— m — fabricación f - [ind] manufacture,ring period
— m garantía f - guaranty, o warranty, period
— m — mantenimiento m - validity period • maintenance period
— m — montaje m - [mec] assembly time
— m — pago m - [fin] payment, date, o term, o period
— m —, reembolso m, o reintegro m - [fin] repayment, time, o term
— m — presentación f - presentation, period, o date, o term, o time
— m — — f de oferta f - [com] offer presentation, period, o term, o date, o time
— m — respuesta f - answer(ing) period
— m — revisión f - check(ing) (time) period
— m — validez f - [legal] validity, time, o term, o period
— m — verificación f - check(ing) (time) period
— m — vigencia f - validity period
— m previsto - foreseen period | adv - (en) on schedule
— m regular - regular period | adv - (en) periodic
plazoleta f con jardín(es) m - [pol] landscaped plaza
plegadizo,za - . . . .; collapsible
plegado n - . . . . • fold; bend
— m alternante - [mec] alternating bend(ing)
— m excesivo - [mec] excessive bending
— m con cara f de raíz en parte f exterior - [mec] root bending | adv - [mec] root bent
— m — — f — — f hacia afuera - [mec] root bending | adv - root bent
— m — — f — soldadura f en parte exterior - [mec] face bending | adv - face bent
— n — extremo(s) m apoyado(s) - [sold] free bent | adv - [mec] free bent
— m — mandril m - [mec] guided bending | adv - [mec] guided bending
— m — raíz f sometida a tracción f - [sold] root bending | adv - root bent
— m — soldadura f sometida a tracción f - [mec] face bent | adv - [mec] face bent
— m guiado - [mec] guided bend(ing)
— m lateral - [mec] side bending | adv - [mec] side bent
— m libre - [mec] free bending | adv - [mec] free bending

plegado m sin mandril m - [mec] free bending | adv - free bent
— m sobre conformador m - [mec] guided bending | adv - mec] guided bent
— m — mandril m - [mec] guided bending | adv - (con- guided bent
plegado,da a - [mec] folded; bent • telescoped
plegadura f - [mec] crimp(ing)
plegamiento m - [geol] fold(ing)
— m intensivo - [geol] intensive folding
plegar v - . . .; to double; to kink; to crimp
plegarse v - to fold • to collapse
— v en forma aguda - [mec] to bend sharply
pletina f - [metal-lam] flat; small plate; shim; filler; strap; stock; plate • sheet billet
— f de acero m - [metal-lam] steel shim
— f — — m galvanizado - [metal-trat] galvanized steel, strap, o shim
— f para contención f - [metal-lam] retaining plate
— f — estirado m - [alambre] stretcher bar
pliego m - . . . . • [constr] specification(s)
— m adyacente - [mec] adjacent sheet
— m — de malla f de alambre m - [constr] adjacent wire mesh sheet
— m de características f - specification(s) sheet
— m — — f — f técnica(s) - technical specification(s) sheet
— m — condición(es) f - specification(s); specification, o term, sheet
— m — — f técnica(s) - technical specification(s) sheet
— m — especificación(es) f - specification(s) sheet
— m — — f técnica(s) f - technical specification(s) sheet
— m — malla f - [metal-fabr] mesh sheet
— m — — f de alambre m - [constr] wire mesh sheet
— m — ruta - release sheet • route,ting sheet
— m — — f para orden f - [ind] order, release, o route,ting, sheet
pliegue m - . . .; bend; kink; crimp • [metal-lam] bent edge
— m agudo - [mec] sharp bend
— m . . . grado(s) m - [mec] . . . degree bend
— m . . . — sin falla(r) - [sold] . . . degree bend without failure
— m en V - [mec] (V) crimp
— m sin falla f - [sold] bend without failure
plomo m - . . . | adv - (a) plumb
— m en chapa(s) f - [metal-lam] milled lead
— m endurecido - [metal] hardened lead
— m reforzado - [electr-cond] compounded lead
— m vertido - [metal-prod] poured, o cast, lead
pluma f - . . . • [grúas] boom; jib
— f básica - [grúas] basic boom
— f con . . . sección(es) f - [grúas] . . . section boom
— f en voladizo - [grúas] projecting boom
— f extendida - [grúas] extended boom
— f — totalmente - [grúas] fully, o totally, extended boom
— f hidráulica - [grúas] hydraulic boom
— f — telescópica - [grúas] hydraulic telescopic boom
— f montada - [grúas] mounted boom
— f — sobre cojinete(s) m - [grúas] bearing(s) mounted boom
— f — — m con rodillo(s) m - [grúas] roller bearing mounted boom
— f para carga f - [grúas] load, o Chicago, boom
— f — grúa f - [grúas] crane boom
— f — — f con mástil m arriostrada - [constr] stiff leg crane boom
— f retraída - [grúas] retracted boom
— f — totalmente - [grúas] totally, o fully, retracted boom
— f telescópica - [grúas] telescopic boom

pluma f **telescópica hidráulica** - [grúas] teles-
cópic hydraulic boom
plusvalía f - [fisc] unearned, increment, o increase, o value • [contab] added value; good
will
— f **revaluada** - [fisc] revalued unearned, increase, o increment
poa f - . . . • [botán] (Kentucky) bluegrass
— f **kentuckiensis** - [botán] Kentucky bluegrass
población f - [pol] . . .; community; settlement
— f **agrícola** - [pol] farm(ing) town
— f **amistosa** - [pol] friendly town
— f **apreciable** - [pol] substantial population
— f **considerable** - [pol] substantial population
— f **creciente** - [pol] growing, o expanding,
population, o town
— f **con casas** f **quinta** - [pol] second home (recreational) community
— f — **residencias** f - [pol] home(s), o residential, community, o town
— f — **villa(s)** f - [pol] second home (recreational) community
— f **de zona** f **para venta(s)** f - [com] trading
area population
— f **en expansión** f - [pol] growing, o expanding, town, o population
— f **existente** - [pol] existing community
— f **fantasma** - [pol] ghost town
— f **fronteriza** - [pol] pioneer settlement
— f **grande** - [pol] large town
— f **industrial** - [ind] industrial, town, o city
— f **íntegra** - [pol] entire community
— f **mayor** - [pol] larger, o bigger, town
— f **menor** - [pol] smaller town
— f **minera** - [miner] mining town
— f **mundial** - [pol] world population
— f **para recreo** m - [pol] recreational, town, o
community
— f **pequeña** - [pol] small town
— f **recreativa** - [pol] recreational community
— f **residencial** - [pol] residential community
— f — **sobre lago** m - [pol] lakeside residential community
— f **ribereña** f - [pol] waterfront community
— f **rural** - [pol] rural community • rural population
— f **sobre lago** m - [pol] lakeside community
— f **urbana** - [pol] urban population
poblador m - [pol] resident; settler
— m **primero** - [pol] pioneer; founder
pobre a - . . .; lean; meager; weak
— a **en recurso(s)** m **natural(es)** - [econ] poor
in natural resource(s)
poca adv **absorción** f - [quím] poor absorption
— adv **agua** m - [hidr] meager water (supply)
— adv **conservación** f - [ind] little, o poor,
maintenance
— adv **intensidad** f - little intensity
— adv **protección** f - little protection
— adv **significación** f - little significance
pocillo m - . . . • [domést] cup
poco,ca adv - . . .; some; somewhat • slightly
— adv **a poco** - gradually
— adv **apto,ta** - unfit
— adv **común** - unusual; uncommon
— adv **desarrollado,da** - underdeveloped
— adv **espesor de ladrillo** m - [constr] thin
(brick) wall
— adv **frecuente** - infrequent • occasional
— adv **inclinado,da** - uninclined
— adv **mantenimiento** m—[ind] little maintenance
— adv **parecido,da** - un(a)like
— adv **metal** m **aportado** - [sold] little deposited weld(ing metal)
— adv **movimiento** m **(en tejido** m**)** - [sold]
small weave,ving
— adv **paso** - [mec] reduced section
— adv **peso** - light weight
— adv **impráctico,ca** - unpractical
— adv **profundo,da** - shallow
— adv **tiempo** m **después** - soon after
— adv **vacío** - [neumát] little vacuum
— adv **viscoso,sa** - low viscous

pocero m - [hidr] well, digger, o driller •
[ind] pitman
poceta f - [mec] well
— f **en guardabarro(s)** n - [autom] fenderwell
— f — , **delantero,** o **frontal** - [autom] front
fenderwell
— f — — m, **posterior,** o **trasero** - [autom]
rear fenderwell
— f **frontal** - [autom] front fenderwell
— f **trasera** - [autom] rear fenderwell
pocos,cas adv - few
poder m - power • [legal] power (of attorney) •
[electr-oper] véase **energía** • [mec] leverage
— m **adquisitivo** - purchasing power
— m — **de dinero** m - [fin] currency('s) purchasing power
— m **amplio** - [legal] broad, o ample, o full,
power (of attorney)
— m **calorífico** - [combust] caloric, o heating,
power, o value, o rating, o content
— m — **alto** - [termol] high heat(ing) value
— m — **bajo** - [combust] low heat(ing) value
— m — **de carbón** m - [combust] coal heat(ing)
value
— m — — m **vegetal** - [combust] charcoal
heat(ing) value
— m — — **combustible** m - [combust] fuel
heat(ing) value
— m — — **fuel oil** m - [combust] fuel oil
heat(ing) value
— m — — **gas** m - [combust] gas heat(ing) value
— m — — — **de petróleo** m - [combust] petroleum gas heat(ing) value
— m — — — m **de horno** m - [combust] furnace
gas heat(ing) value
— m — — — — m **alto** - [combust] blast
furnace gas heat(ing) value
— m — — — m **licuado** - [combust] liquefied
gas heat(ing) value
— m — — — m — **de petróleo** m - [combust]
liquefied petroleum gas heat(ing) value
— m — **disponible** - [combust] available heat-
(ing), value, o content
— m — **inferior** - [combust] lower,west heat-
(ing) value
— m — — **de fuel oil** m - [combust] fuel oil
lower,west heat(ing) value
— m — — **gas** m - [combust] gas lower,west
heat(ing) value
— m — — — m **de horno** m - [combust] furnace gas lower,west heat(ing) value
— m — — — m — m **alto** - [combust]
blast furnace gas low(er,est) heat(ing) value
— m — **medio,diano de gas** m - [combust] gas average heat(ing) value
— m — — — m **en horno** m - [combust] furnace gas average heat(ing) value
— m — — — m — m **alto**—[combust] blast
furnace gas average heat(ing) value
— m — **superior** - [combust] higher,hest hat-
-(ing) value
— m — — **de fuel oil** m - [combust] fuel oil
higher,hest heat(ing) value
— m — — **gas** m - [combust] gas higher,hest
heat(ing) value
— m — — — m **en horno** m - [combust]
furnace gas higher,hest heat(ing) value
— m — — — m — m **alto** - [metal-prod]
blast furnace higher,hest heat(ing) value
— v **colar** v - [metal-prod] to be ready for tapping
— v — **rápidamente** - [metal-prod] to be able to
tap quickly
— v — **horno** m - [metal-prod] (furnace) ready
for tapping
— v — — m **alto** - [metal-prod] blast furnace
ready for tapping
— m **de elemento** m **para recuperación** f - [quím]
element, recouping, o recovery, power
— m **ejecutivo** - [pol] executive power; adminis-
tration | adv - (de) presidential
— m — **nacional** - [pol] national administration
— m **especial**—[legal] special power of attorney

— v **exigir** - to have @ right to require
— m **general** - [legal] general, o ample, power of attorney
— m **judicial** - [pol] judiciary; judicial power
— m **legal** - [legal] legal power of attorney
— m **legislativo** - [pol] legislature; legislative power
— m **otorgado** - [legal] granted power of attorney)
— m **para administración** f - [legal] management, authority, o power (of attorney)
— m — **representación** f - [legal] power of representation
— m — **resolución** f - [optic] resolving, o resolution, power
— m — **sustentación** f - [constr] support(ing)
— v **rescindir** v - [legal] to have @ power to rescind
— v **sangrar** v - [metal-prod] to be, able, o ready, for tapping
— m **suficiente** - [legal] sufficient, o broad, o ample, power (of attorney)
— v **tapar** v - [metal-prod] to be able to plug
poderse v **aplicar** - to be, pertinent, o applicable
— v **leer** - to be, legible, o readable
— v **decir** v - (that) can be said; sayable
— v — v **que** - literally
— v **emplear** v - (that) can be, used, o employed
⌐ adv - (de) where applicable
— v — v **con herramienta** f - [mec] to fit, o that can be used with, @ tool
— v **obtener** v - to be available
— v — v **con cargo** m **adicional** - available, o obtainable, at @ extra cost
— v — **con recargo** m **en precio** m - available at @ extra cost
— v — **sin recargo** m **en precio** m - available at no extra charge
— v — **para voltaje(s)** m **distinto(s)** - other voltage(s) available
— v — — **con recargo** m **en precio** m - other voltages available at extra cost
— v — — — m **distinto(s) sin recargo** m **en precio** m - other voltages available at no extra cost
— v **ofrecer** v - (to be able) to offer
— v **regular** v - to be able to adjust
— v — **separadamente** - to be able to adjust separately; separately adjustable
— v **tolerar** v - to be, tolerable, o permissible
podría v **estar** - (that) might be
— v **necesitar(se)** - might need
— v **ser** - might be
poise* m - [petról] poise
polaina f **de cuero** m - [segurid] leather legging
— f — — m **con cromo** m - [segurid] chrome leather legging
— f — — m — — m **sin resorte** m - [segurid] springless chrome leather legging
— m — — m **sin resorte** m - [segurid] springless leather legging
— f **sin resorte** m—[segurid] springless legging
polar a - . . .; pole
— m **de campo** m - [electr-instal] field pole
polaridad f **apropiada** - [electrón] proper, o appropriate, polarity
— f **cambiada** - [sold] changed polarity
— f **contraria** - [electr] opposite polarity
— f **con corriente** f **continua** - [sold] direct current, o D C, polarity
— f — — f — **negativa** - [sold] negative, current, o D C negative, polarity
— f — — f — **positiva** - [sold] direct current, o D C, positive polarity
— f **de electrodo** m - [sold] electrode polarity
— f — — m **deseada** - [sold] desired electrode polarity
— f — — m **negativa** - [sold] negative electrode polarity
— f — — m **positiva** - [sold] positive electrode polarity
— f — **energía** f - [electr-prod] power polarity

polaridad f **de energía** f **de salida** f **en generador** m - [electr-prod] generator output polarity
— f — f — — — m **de soldadora** f - [sold] welder generator output polarity
— f — **fuente** f **de energía** f - source polarity
— f — — f **para energía** f - [sold] power source polarity
— f — **salida** f - [sold] output polarity
— f — **señal** f **en entrada** f - [electrón] input signal polarity
— f — **voltímetro** m - [electr] voltmeter polarity
— f **deseada** - [sold] desired polarity
— f **directa** - [sold] direct polarity • straight polarity
— f **en entrada** - [sold] input polarity
— f **equivocada** - [electr] wrong polarity
— f **inversa** - [sold] reverse polarity
— f **invertida** - [sold] reverse(d) polarity • changed polarity
— f — **energía** f - [sold] reversed output polarity
— f — — — f **en salida** f - [sold] reversed output polarity
— f — — — f — — f **de generador** m - [sold] reversed generator output polarity
— f — — — f — — f — — m **en soldadora** f - [sold] reversed welder generator output polarity
— f **negativa** - [sold] negative, o minus, polarity
— f — **en corriente** f **continua** - [electr-oper] direct current negative polarity
— f **normal** - [sold] normal, o straight, polarity
— f **observada** - [electr-instal] observed polarity
— f **opuesta** - [electr-oner] opposite polarity
— f **para electrodo** m - [sold] electrode polarity
— f — — m **negativo** - [sold] electrode negative polarity
— f — — m **positivo** - [sold] electrode positive polarity
— f — **marca/espacio** - [electrón] mark/space polarity
— f — **señal** f **para entrada** f - [electrón] input signal polarity
— f — — f — — f **para marca/espacio** - [electrón] mark/space input signal polarity
— f **positiva** - [electr-prod] positive, o plus, polarity
— f — **en corriente** f **continua** - [electr-prod] direct current positive polarity
— f **regulada** - [sold] controlled polarity
— f **seleccionada** - [electrón] selected, o chosen, polarity
— f **verificada** - [electr-oper] checked polarity
polarización f - . . .; polarizing • [electrón] bias(ing)
— f — **voltaje** m - [electrón] voltage biasing
— f — — m **para bloqueo** m - [electrón] blocking voltage biasing
— f **directa** - [electrón] forward bias(ing)
— f **inversa** - [electrón] reverse bias(ing)
— f **magnética** - [electr] magnetic polarization
polarizado,da a - polarized • [electrón] biased
polarizar v - . . . • [electrón] to bias
— v **voltaje** m - [electr-oper] to polarize @ voltage • [electrón] to bias @ voltage
— v — m **para bloqueo** m - [electr-oper] to polarize @ blocking voltage • [electrón] to bias @ blocking voltage
polea f - [mec] . . .; belt pulley; sheave; tackle
— f **acanalada** - [mec] grooved, o corrugated, sheave • sheave (wheel)
— f — **templada** - [mec-ruedas] hardened sheave wheel
— f **agrietada** - [cabl] cracked sheave
— f **alineada** - [mec] aligned pulley
— f **ancha**—[mec] wide (faced), sheave, o pulley

polea f̄ angosta - [mec] narrow (faced), sheave,
o pulley
— f̄ auxiliar - [grúas] auxiliary sheave
— f̄ con acanaladura ancha - [mec] wide throat,
sheave, o pulley
— f — — f̄ para cable m para arrastre m -
[constr] wide throat dragline sheave
— f — — f única - [mec] single groove sheave
— f̄ — — acanaladura(s) f - [mec] . . .
groove sheave
— f — garganta f - [mec] (grooved) sheave
— f̄ — brida(s) f̄ - [mec] flange(d) sheave
— f̄ — cuatro tiros m - [cabl] four-part pulley
— f̄ — diámetro m reducido - [cabl] small diameter sheave
— f — gancho m - [petról] hook block
— f̄ — ojo m fijo - [mec] fixed eye, sheave, o
pulley
— f — — m giratorio - [mec] swivel eye, pulley, o sheave
— f — paso m variable - [mec] variable pitch
pulley
— f — . . . ranura(s) f - [cabl]
. . . groove pulley
— f — . . . escalón(es) m - [cabl] . . .-step
pulley
— f cónica - [cabl] step pulley
— f̄ — con . . . escalón(es) m - [cabl] . . .-
step pulley
— f corriente - [grúas] standard sheave
— f̄ corrugada - [cabl] corrugated sheave
— f̄ desgastada - [cabl] worn, sheave, o drum
— f̄ — excesivamente - [mec] excessively worn
sheave
— f doble - [mec] double, block, o pulley
— m — con ojo m fijo - [mec] fixed eye double
pulley
— f en extremo m de aguilón m - [grúas] boom
point sheave
— f — — — m para cable m a motón m -
boom point main (load) line sheave
— f — m — — m — — m para elevación f -
[grúas] boom point load line sheave
— f en medio m - [mec] center sheave
— f̄ — m para extremo m para cucharón m -
[grúas] container tip center sheave
— f escalonada - [cabl] stepped pulley • three-
-step pulley
— f estándar - [grúas] standard sheave
— f̄ fija - [grúas] fixed, o fast, sheave, o
pulley
— f — primera - [mec] first fixed pulley
— f̄ flotante - [grúas] floating sheave
— f̄ — para aguilón m - [grúas] boom floating
sheave
— f — — elevación f - [grúas] hoist(ing)
floating sheavew
— f — — — f para aguilón m - [grúas] boom
hoist floating sheave
— f giratoria - [cabl] turn(ing), o swivel,
sheave
— f — para cable m - [grúas] line turn(ing)
sheave
— f — — m para elevación f - [grúas] hoist
line turn(ing) sheave
— f guía - [mec] mule pulley • [cabl] idler
— f̄ — giratoria - [grúas] swivel(ing), o revolving, sheave, o fairleader
— f — oscilante - [grúas] swing fairlead(er)
— f̄ — para cable m para montacarga(s) m -
[metal-prod] skip cable guide pulley
— f inspeccionada - [cabl] inspected sheave
— f̄ loca - [mec] idler pulley
— f̄ — ajustada - [mec] adjusted idler
— f — alineada - [mec] aligned idler
— f̄ — completa - [mec] complete idler
— f̄ — — para motor m - [comb.int] complete
engine idler • [electr-mot] complete motor
idler
— f — con acanaladura f - [cabl] troughing
idler
— f — para accionamiento m - [mec] drive,ving
idler pulley

polea f loca para correa f - [mec] belt idler
— f — — mando m - [mec] drive,ving idler
pulley
— f — — para motor m - [electr-mot] motor idler
• [comb.int] engine idler
— f — — retorno m - [mec] return idler pulley
— f̄ — — retroceso m - [mec] return idler
— f̄ — — tensión f - [mec] tension idler
— f̄ — — transporte m - [mec] carrying idler
— f lubricada - [grúas] lubricated, sheave, o
pulley
— f magnética - magnetic pulley
— f̄ mecanizada - [mec] machined pulley
— f̄ — reforzada - [mec] reinforced, o heavy
duty, machined pulley
— f montada - [mec] mounted sheave
— f̄ — sobre cojinete(s) m - [grúas] bearing
mounted sheave
— f — — m con rodillo(s) m - [cojin] roller bearing mounted sheave
— f motriz - [mec] tug wheel • [sold] power
take-off
— f — para malacate m - [petról] (pulley)
tug wheel
— f — — m para herramienta(s) f - [petról]
bull wheel tug wheel
— f — — f — — m para cable m para tubería f - [petról] calf wheel tug wheel
— f — — tambor m para herramienta(s) f -
[petról] bull wheel tug wheel
— f movible - [grúas] traveling sheave
— f̄ — lubricada - [grúas] lubricated traveling
sheave
— f — para caballete m - [grúas] gantry traveling sheave
— f — — — m lubricada - [grúas] lubricated
gantry, o gantry lubricated, traveling sheave
— f móvil - [cabl] movable, o traveling, block
— f̄ múltiple - [cabl] three-step pulley
— f̄ para aguilón m - [grúas] boom sheave
— f̄ — — m flotante - [grúas] floating boom
sheave
— f — — m lubricada - [grúas] lubricated
floating boom sheave
— f — — m para elevación f - [grúas] boom
hoisting sheave
— f para arrancador m (con cuerda f) - [comb.-
-int] (rope) starter,ting pulley
— f — arranque m - [comb.int] starting pulley
— f — arrastre m - [petról] tug(ging) pulley
— f̄ — bomba f - [bombas] pump pulley
— f̄ — — f para cuchara f - [petról] sand
pump pulley
— f — — — cuchareo m - [petról] sand pump
pulley
— f — — f rociadora - [petról] spray(ing)
pump sheave
— f — — f — para camisa f - [petról] liner
spray(ing) pump sheave
— f — — f — para camisa f - [petról] liner
spray(ing) pump sheave
— f — — f — — — — f y émbolo m - [petról]
piston-liner spray(ing) pump sheave
— f — caballete m - [grúas] gantry sheave
— f̄ — — m lubricada - [grúas] lubricated
gantry sheave
— f — cable m - [cabl] (wire) rope, o line,
sheave, o pulley
— f — — m de alambre m - [cabl] wire rope
pulley
— f — — m para cabrestante - [petról] catline
sheave — f — — f — llave f para enroscadura f - [petról] tong(s) line pulley
— f — — m para maniobra(s) f - [petról] catline sheave
— f — — m — motón m - [grúas] main (load)
line sheave
— f — — m — elevación f - [grúas] hoist, o
load, line sheave
— f — — m — tenaza(s) f - [petról] tong line
pulley
— m — — — f̄ — enroscadura f - [petról]
tong line pulley

polea f **para cable** m **para torno** m - [petról]
catline sheave
— f —— m **principal**—[grúas] main line sheave
— f —— m — **para carga** f - [grúas] main load
sheave
— f —— m —— **elevación** f - [grúas] main
load line sheave
— f —— f **en extremo** m **de aguilón**
m - [grúas] boom point main load line sheave
— f —— m —— f —— m —, **pértiga** f,
o **pluma** f - [grúas] boom point main load line
sheave
— f — **cadena** f - [mec] sprocket
— f — **campana** f - [metal-prod] bell pulley
— f — **carrete** m - [mec] reel pulley
— f — **carrillo** m - [grúas] trolley sheave
— f — **carro** m - [metal-prod] car pulley
— f — m **para montacarga(s)** f - [metal-prod]
skip car pulley
— f —— m — f **para horno** m **alto** - [metal-
-prod] blast furnace skip car pulley
— f — **chapín** m - [metal-prod] bleeder pulley
— f — **cola** f - [mec] tail, o rear, pulley
— f — **corona** f **(para torre** f) - [petról] (rig)
crown pulley
— f — **correa** f - [mec] belt pulley
— f — **cuerda** f - [mec] cord pulley
— f —— f **redonda** - [mec] round cord pulley
— f — **desvío** m - [cabl] guide,ding pulley
— f — **elevación** f - [grúas] hoist sheave
— f —— f **para aguilón** m - [grúas] boom hoist
sheave
— f —— f —— m **flotante** - [grúas] floating
boom hoist(ing) sheave
— f —— f —— m — **lubricada** - [grúas] lubricated floating boom hoist(ing) sheave
— f —— f —— m — **caballete** n - [grúas]
gantry boom hoist(ing) sheave
— f —— f —— m **lubricada** - [grúas]
lubricated gantry boom hoist(ing) sheave
— f —— f — **vagoneta** f - [metal-prod] skip
sheave
— f — **extremo** m - [grúas] point, o end, sheave
— f —— m **para aguilón** m - [grúas] boom, end,
o head, o point, sheave
— f —— m — **botalón** m - [grúas] boom, head,
o end, o point, sheave
— f —— m — **cucharón** m - [grúas] container
tip sheave
— f — m **superior** m - [petról] crown sheave
— f — **gancho** m - [grúas] hook sheave
— f —— m **auxiliar** - [grúas] auxiliary hook
sheave
— f —— m **principal**—[grúas] main hook sheave
— f — **guía** f - [mec] guide, o jockey, pulley,
o sheave • idler
— f —— f **para cable** m - cable guide sheave
— f —— f —— m **para cabrestante** m - [mec]
catline guide sheave
— f —— f —— m **maniobra(s)** f - [petról]
catline guide,ding sheave
— m —— f —— m **para montacarga(s)** m -
[metal-prod] skip cable guide,ding pulley
— f —— f —— m — **torno** m - [petról] catline guide,ding sheave
— f — **herramienta(s)** f - [petról] crown pulley
— f — **impulsión** f - [mec] drive,ving sheave
— f —— f **para ventilador** m - [mec] fan
drive,ving pulley
— f — **malacate** m - [grúas] winch sheave
— f —— m **auxiliar** - [grúas] auxiliary winch
sheave
— f —— m **para herramienta(s)** f - [petról]
bull wheel tug (side)
— f — **montacarga(s)** m - [metal-prod] skip,
pulley, o sheave
— f —— m **para horno** m - [metal-prod] furnace
skip pulley
— f —— m —— m **alto** - [metal-prod] blast
furnace skip pulley
— f — **motón** m—[grúas] block, sheave, o pulley
— f —— m **para extremo** m - [grúas] point, o
tip, main (load) line sheave

polea f **para motón** m **para extremo** m **de aguilón** m
[grúas] boom point main (load) line sheave
— f —— m **para gancho** m - [grúas] hook block
sheave
— f —— m —— m **para grúa** f - [grúas] crane
hook block sheave
— f — **motor** m - [electr-mot] motor pulley •
[comb.int] engine pulley
— f —— m **lubricada** - [comb.int] lubricated
engine pulley • [electr-mot] lubricated motor
pulley
— f —— m **principal** - [electr-mot] main motor
pulley • [comb.int] main engine pulley
— f — **pescante** m - [grúas] jib sheave
— f — **prolongación** f—[grúas] estension sheave
— f —— f **para aguilón** m - [grúas] boom extension sheave
— f —— f — **cable** m - [grúas] line extension sheave
— f —— f —— m **para gancho** m - [grúas]
whip line extension sheave
— f —— f — **trozo** m **final** - [grúas] end extension sheave
— f —— f —— m — **de cable** m - [grúas]
whip line end extension sheave
— f — **puesta** f **en marcha** - [comb.int] starter,
o starting, pulley
— f — **punta** f - [grúas] point, o tip, sheave
— f —— f **de aguilón** m - [grúas] boom, point,
o tip, sheave
— f —— f —— m **auxiliar** - [grúas] auxiliary boom, point, o tip, sheave
— f —— f **prolongada** - [grúas] extended,
point, o tip, sheave
— f —— f — **de aguilón** m - [grúas] extended
boom, point, o tip, sheave
— f —— f —— m **auxiliar** - [grúas] extended auxiliary boom, point, o tip, sheave
— f — **reducción** f **de velocidad** f - [mec] gear
reducer pulley
— f — **reductor** m—[mec] reducer sheave
— f — **reenvío** m - [mec] countershaft pulley
• guide pulley
— f — **remolque** m - [petról] tug(ging) pulley
— f — **salida** f - [mec] output sheave
— f — **skip** m - [metal-prod] skip, sheave, o
pulley
— f — **sonda** f - [metal-prod] stock rod pulley
— f — **soporte** m - [mec] support sheave
— f —— m **para cable** - [grúas] cable, o support, sheave
— f —— m —— m **para elevación** f - [grúas]
hoist line support sheave
— f — **tensión** f - [mec] tension, o jockey,
pulley
— f — **tornillo** m - [mec] screw pulley
— f —— m **alimentador** - [mec] feeder screw
pulley • [petról] temper screw pulley
— f —— m **(pequeño)** - [tornos] mandrel
— f — **tragante** m - [metal-prod] throat pulley
— f —— m **para cargador** m - [metal-prod] skip
hoist throat pulley
— f —— m — **elevador** m - [metal-prod] hoist
throat pulley
— f —— m —— m **para montacarga(s)** m -
[metal-prod] skip hoist throat pulley
— f —— m — **grúa** f **(skip)** - [metal-prod]
skip hoist throat pulley
— f — **trozo** m **final de cable** m - [cabl] whip
line sheave
— f — **ventilador** n - [mec] fan pulley
— f —— m **para motor** m - [electr-mot] motor
fan pulley • [comb.int] engine fan pulley
— f —— m **lubricada** - [mec] lubricated fan
pulley
— f — **vagoneta** f - [ind] car pulley
— f —— f **para carga** f - [metal-prod] skip
pulley
— f —— f **para montacarga(s)** m - [metal-prod]
skip car pulley
— f —— f —— m **para horno** m - [metal-prod]
furnace skip car pulley
— f —— f —— m —— m **alto** - [metal-prod]

blast furnace skip car pulley
polea f para válvula f - [valv] valve pulley
— f — — f **para descarga** f - [metal-prod] discharge valve pulley
— f — — f **igualadora** - [metal-prod] equalizer
valve pulley
— f — **volante** m - [mec] flywheel pulley
— f **pequeña** - [mec] small, pulley, o sheave
— f **primera** f - [mec] first pulley • [grúas]
first sheave
— f **principal** - [mec] head pulley • [petról]
crown pulley
— f **ranurada** - [mec] grooved sheave
— f **reforzada** - [mec] heavy duty pulley
— f **rota** - [mec] broken, pulley, o sheave
— f **rotatoria** - [cabl] turn(ing) sheave
— f — **para cable** m - [grúas] line turn(ing)
sheave
— f — — — m **para elevación** f - [grúas] hoist
line turn(ing) sheave
— f **sencilla** - [mec] single, pulley, o sheave,
o block
— f — **con ojo** m - [mec] eye single pulley
— f — — — m **fijo** - [mec] fixed eye single
pulley
— f — — — m **giratorio** - [mec] single eye
swivel(ling) pulley
— f **suelta** - [mec] loose pulley
— f **superior** - [mec] top, o upper, pulley
— f **tensora** - [mec] idler (pulley)
— f — **ajustada** - [mec] adjusted idler (pulley)
— f — **colocada** - [mec] positioned idler
— f — **para accionamiento** m - [mec] drive(r)
idler (pulley)
— f — — **correa** f - [mec] belt idler
— f — — — f **para impulsión** f - [mec]
drive,ving belt idler
— f — — — m **para implemento** m—[agric-equip]
implement drive,ving idler (pulley)
— f — — **correa** f - [mec] belt idler
— f — — — f **para implemento** m—[agric-equip]
implement belt idler
— f — — — f — — f **para impulsión** f -
[agric-equip] implement drive,ving belt idler
— f — — **implemento** m - [agric-equip] implement idler (pulley)
— f — — — m **ajustada** - [agric-equip] adjust-
ted implement idler (pulley)
— f — — — f **de implemento** m - [agric-equip]
implement drive,ving idler
— f — — **mando** m - [mec] drive,ving idler pulley
— f **triple** - [cabl] triple block
— f **única** - [grúas] single, sheave, o block
— f — **para cable** m - [grúas] single sheave
— f — — m **en** . . . **sección(es)** f - [grúas]
single . . . part line sheave
— f — — **motón** m - [grúas] block single sheave
— f — — — m **para gancho** m - [grúas] hook
block single sheave
— f — — — m — — m **para cable** m - [grúas]
cable hook block single sheave
— f — — — — m **para cable** m **en** . . .
sección(es) f - [grúas] . . . part line block
single sheave
— f **verificada** - [mec] checked sheave
— f —, **con vista** f, o **visualmente** - [mec] visually checked sheave
— f **viajera** - [grúas] traveling sheave
— f — **lubricada** - [grúas] lubricated traveling
sheave
— f — **para caballete** m - [grúas] gantry traveling sheave
— f — **para caballete** m - [grúas] gantry traveling sheave
— f — — — m **lubricada** - [grúas] lubricated
gantry traveling sheave
policía f - . . . . • [ind] security personnel
— f **caminera** - [pol] highway patrol(man)
— f — **Estatal** - [pol] State Highway Patrol
— f — - [pol] Federal Bureau of Investigation:
F B I
— f **interna** - [ind] police department; plant

protection
policía f para tránsito m - [pol] traffic police
— f **privada** - [ind] (private) security personnel
policíclico,da a - polycyclic
policloruro m - [quím] polychloride
— m de **vinilo** m - [quím] polyvynil chloride
poliducto* m - [petról] polyduct*
polielectrolito m - [sanit] polyelectrolyte
poliéster* m - [plást] **véase poliestero** m
— m **armado** - [plást] reinforced polyester
— m — **con fibra(s)** f **de vidrio** m - [plást] (fiberglass) reinforced polyester
polietileno m **sólido** - [quím] solid polyethylene
polífono,na a - polyphonic
polígono m - . . . • [milit] . . .; proving
ground(s); rifle range
— m **de fuerza(s)** f - force(s) polygon
— m **para tiro** m - [deport] rifle range
polímero m - [quím] polymer
— m **líquido** m - [quím] liquid polymer
— m — **de polisulfuro** m - [quím] polysulfide
liquid polymer
— m **natural** - [quím] natural polymer
— m **neutro** - [quím] neutral polymer
polín m - [mec] . . .; placer; roller
— m **levantador** - [mec] lifting cam
— m — **para leva** f - [mec] cam roller bracket
— m **para leva** f - [mec] cam roller
— m — **dedo** m - [mec] cam roller
— m — — m **levantador** - [mec] placer cam roller
polio m - [medic] **véase poliomielitis** f
poliomielitis m - [medic] . . .; infantile paral-
ysis
polipasto m - [cabl] tackle; hoist; block; rigging
polipropilene m - [quím] polypropylene
polirrevestido,da a - polycoated
polirrevestir v - to polycoat
polirrevestimiento m - polycoating
polispasto m - [cabl] **véase polipasto** m
política f - . . . • [pol] politics • approach
— f **ágil** m - agile policy
— f **anterior** - former, o previous, policy
— f **antidumping** - [econ] antidumping policy
— f **buena** - good policy
— f **cambiaria** - [fin] exchange policy
— f — **estable** - [fin] stable exchange policy
— f **con efecto** m **neutro** - [econ] neutral effect
policy
— f **contable** - [contab] accounting policy
— f **contra dumping** m - [econ] antidumping policy
— f **crediticia** - [fin] credit policy
— f **económica** - [econ] economic policy
— f — **minera** - [miner] economic mining policy
— f **empresarial** - corporate policy
— f **estable** - stable policy
— f **fiscal** - [econ] fiscal policy
— f **flexible** - flexible policy
— f **fronteriza** - [econ] frontier policy
— f **general** - general policy
— f **implantada** - implemented policy
— f **implementada** - implemented policy
— f **industrial** - [ind] industrial policy
— f — **general**—[ind] general industrial policy
— f **metalúrgica** - [metal] metallurgical policy
— f **minera** - [miner] mining policy
— f **modificada** - [admin] revised, o changed,
policy
— f **monetaria** - [fin] monetary policy
— f **nacional** - national policy
— f **para apertura** f - [econ] opening (up) policy
— f **defensa** f - defense policy
— f — **estímulo(s)** m - [labor] incentive(s) policy
— f — **exportación(es)** f - [econ] export(s)
policy
— f — **frontera** f - [econ] frontier policy
— f — **importación(es)** f - [econ] import(s) policy
— f — **incentivo(s)** m - [labor] incentive(s)
policy
— f **preventiva** - preventive policy
— f **reformada** - [admin] revised policy

política f siderúrgica - [metal] steel(making) policy
— f — nacional - [metal] national steel(making) policy
— f social - social policy
polivalencia f - [quím] . . .; multivalence
polivinilo m - [quím] polyvinyl
— m rígido - [plást] rigid polyvinyl
poliuretano m - [quím] polyurethane
póliza f administrada - [seguros] administered policy
— f contra accidente(s) m - [legal] accident (insurance) policy
— f de asegurado m - [seguros] insured('s) policy
— f de Corporación f Estadounidense para Seguro m sobre Crédito(s) m - [seguros] Federal Credit Insurance Corporation policy
— f flotante - [seguros] floating policy
— f — para transporte m - [seguros] floating transportation policy
— f interpretada - [seguros] interpreted policy
— f mantenida - [seguros] maintained policy
— f para importación f - [com] importation certificate; certificate of importation
— f — garantía f - [instrum] guaranty policy
— f — incendio m - [seguros] fire policy
— f — seguro(s) m - [seguros] insurance policy
pólizas f emitidas - [seguros] issued policies • (premiums) written
polo m a caja f - [electr-prod] pole to frame
— m adicional - [electr-instal] additional, o extra, pole
— m auxiliar - [electr-instal] auxiliary pole • commutating pole • interpole
— m — en generador m - [electr-prod] generator interpole
— m cilíndrico - [electr-instal] cylindrical pole
— m doble - [electr-instal] double pole
— m eléctrico - [electr-instal] electric(al) pole
— m en campo m - [electr-instal] field pole
— m — — m para excitador m - [electr-instal] exciter field pole
— m laminado - [electr-instal] laminated pole
— m laminar - [electr-equip] laminated pole (piece)
— m magnético - [electr-equip] magnetic pole
— m negativo - [electr-instal] negative pole
— m para alternador m - [electr-equip] alternator pole
— m — campo m - [electr-equip] field pole
— m — — m para excitador m - [electr-equip] exciter field pole
— m — — m — generador m - [electr-equip] generator field pole
— m — conmutación f - [electr-equip] commutating pole • interpole
— m — estator m - [electr-mot] stator pole
— m — excitador m—[electr-equip] exciter pole
— m — generador m - [electr-prod] generator pole
— m positivo - [electr-equip] positive pole
— m principal - [electr-equip] main pole
— m rectangular - [electr-equip] rectangular pole
— m vivo - [sold] hot lead
poltrona f - [domést] easy chair
polución f térmica - thermal pollution
polvareda f enceguecedora—blinding dust (cloud)
— f espesa - thick, o heavy, dust
— f persistente - lingering dust
— f rojiza - [vial] red dust
polvillo m en colector m - [metal-prod] dust catcher dust
— m — horno m - [metal-prod] furnace dust
— m — — m alto - [metal-prod] blast furnace dust
polvo m - . . . . • [miner] fines • [constr] dry mix • [miner] breeze | a - (en) powdered
— m abrasivo - abrasive dust
— m acumulado - accumulated, dust, o soot, o dirt • dust slug

polvo m adherido - [electr-oper] clinging, dust, o powder
— m asentado - settled dust
— m asfáltico - asphalt(ic) dust, o powder
— m conductor - conductive,ting dust
— m dañino - [segurid] harmful dust
— m de antracita f - [miner] anthracite, dust, o culm
— m — asfalto m - asphalt, powder, o dust
— m — azufre m - [quím] sulfur, powder, o dust
— m — carbón m - [miner] coal, fines, o dust
— m — carbono m - carbon, dust, o powder
— m — caucho m - [plást] rubber, dust, o powder
— m — corcho m - [mader] cork, dust, o powder
— m — hierro m - [metal] iron, powder, o dust
— m — — m adherido - [metal] clinging iron, dust, o powder
— m — — m que (se) adhiere - [metal] clinging iron, dust, o powder
— m — limadura(s) f - [mec] file, powder, o dust
— m — magnesio m - [miner] magnesium powder
— m — óxido m - [miner] oxide, dust, o powder
— m — — m de hierro m - [metal] crocus
— m — proceso m - [ind] process, dust, o powder
— m — — m L-D - [metal-prod] basic oxygen process, dust, o powder
— m — roca f - [constr] rock flour
— m depurador - clean(s)ing powder
— m desértico - [ambient] desert dust
— m en atmósfera f - atmosphere,ric, o airborne, dust
— m — colector m (para polvo m) - [metal-prod] dust catcher dust
— m — condición f húmeda - wet form, dust, o powder
— m — — f seca - dry form, dust, o powder
— m — chimenea f - [combust] flue dust
— m — horno m - [ind] furnace dust • flue dust
— m — — m alto - [metal-prod] blast furnace dust • flue dust
— m — — m L. D. - [metal-prod] basic oxygen furnace dust
— m —, red f, o sistema m - [ind] system gas
— m —, — f, o — m, para purificación f (de gas m) - [metal-prod] (gas) scrubbing system gas
— m — tragante m - [metal-prod] throat, o flue, dust, o fines
— m — — m en horno m - [metal-prod] furnace flue dust
— m — — m — — m alto - [metal-prod] blast furnace flue dust
— m — trampa f - [metal-prod] trap, o dust catcher, dust
— m enceguecedor - blinding dust
— m excesivo - excessive dust
— m — acumulado - [ind] accumulated excessive dust
— m exotérmico - [metal-prod] exothermic, dust, o powder
— m extrudido - [mec] extruded powder
— m fino - fine, dust, o powder
— m fluorescente - fluorescent powder
— m fundente - [sold] fluxing, o welding, powder; flux
— m húmedo - wet, dust, o powder
— m ingerido - ingested, dust, o powder
— m levantado - billowed dust
— m magnético - magnetic, dust, o powder
— m — en condición f húmeda - wet form magnetic, dust, o powder
— m — — — f seca f - dry form magnetic powder
— m — fluorescente - fluorescente magnetic magnetic powder
— m — húmedo - wet magnetic powder
— m — seco - dry magnetic powder
— m micrométrico - micron,crometric powder
— m — sin clasificar - ungraded miacron,crometric powder
— m moderado - moderate, dust, o powder

polvo m negro - black, dust, o powder
— m nocivo - [segurid] noxious, o poisonous, o
harmful, dust, o powder
— m ordinario - ordinary, dust, o powder
— m — removido - removed ordinary, dust, o
powder
— m para recubrimiento m - [sold] coating dry
mix
— m que (se) adhiere - clinging, dust, o powder
— m químico - [quím] chemical powder
— m — mojado - [quím] wet chemical powder
— m — seco - [quím] dry chemical powder
— m quitado - removed dust • wiped dust
— m recolectado - gathered, o collected, dust
— m rojizo - [vial] red(dish) dust
— m sacado - removed dust • wiped dust
— m seco - dry, dust, o powder
— m sin clasificar - ungraded, dust, o powder
pólvora f negra - [explos] black (gun)powder
— f sin humo m - [explos] smokeless (gun)powder
polvoriento,ta a - . . .; dirty
polla f - . . . . • [deport] sweepstakes; pool
pollo m y costilla(s) f - [culin] chicken and
rib(s)
ponchada f - . . . . • [fam] [Arg.] bundle
ponderación f - . . .; weighing
— f de índice(s) f - index weigh(t)ing
ponderado,da a - pendered • considered •
[contab] weigh(t)ed
ponderar v - . . . . to give weight
poner v - to, put, o place • to, set, o lay,
down • to provide; establish • to apply
— v a consideración f - to present for study
— v — descubierto - to expose
— v — en abertura f - [mec] to expose in @
opening
— v — pico m para engrase m - [mec] to expose @ (grease) fitting
— v — ranura f - [mec] to expose @ spline
— v — día - to update
— v — disposición f - to make available
— v — plomo - to plumb
— v — punto m - to, set, o adjust - [mec] to
jog • [ind] to, time, o tune (up), o set @
control
— v — con precisión f - [ind] to fine tune
— v — m encendido m - [comb.int] to time @
ignition
— v — m freno m - [mec] to set @ brake
— v — matriz f - [mec] to set @ die
— v — m neumático m - [autom-neumát] to
tune @ tire
— v — m nuevamente - [instrum] to reset
— v — m — arrancador m - [electr] to reset
@ starter
— v — m — m para motor m - [electr-mot]
to reset @ motor starter
— v — m palanca f - [mec] to set @ lever
— v — m — f para freno m - [mec] to set @
brake lever
— v — resguardo m - [ind] to store
— v — m unidad f - [ind] to store @ unit
— v — m — f motriz - [agric] to store @
power unit
— v — rojo m - to turn red hot • [metal-prod]
to develop @ hot spot
— v — tierra f - [electr-oper] to ground
— v — f banco m - [sold] to ground @ bench
— v — f blindaje m - [electr-cond] to
ground @ shield
— v — f caja f - [mec] to ground @ body
— f — f chasis m - [electrón] to ground @,
chassis, o frame
— v — f mesa f - [sold] to ground @ table
— v — f soldadora f - [sold] to ground @
welder
— v — f trabajo m - [sold] to ground @,
weld, o job, o work
— v aparte - to set, apart, o aside
— v atención f - [ind] to pay attention; to exercise, care, o attention • to perform maintenance
— v — f normal - [ind] to pay normal attention
• to perform normal maintenance

poner v confianza f (en) - to, rely on, o trust
in
— v conmutador m - [electr-oper] to, place, o
set, @ switch
— v contrasoplado m - [metal-prod] to apply @
counterblast
— v cuidado m - to exercise care
— v cuidado m - to exercise care
— v — normal - to exercise normal care
— v de manifiesto m - to, show, o bring out;
to evidence
— v empaquetadura f - véase empaquetar v
— v en automático - to put on automatic
— v — boca f - to, put, o place, in(to) @
mouth
— v — clave f - to code
— v — código m - to code
— v — contacto m (con) - to contact
— v — cortocircuito m - [electr-oper] to
short(circuit)
— v — m consigo mismo - [electr-oper] to
short on itself
— v — m entre sí - [electr-oper] to short
together
— v — estado m bueno - [ind] to overhaul
— v — función f - [admin] to install • [mec]
to start (up)
— v — funcionamiento m - [mec] to make operative • [comb.int] to start (up) • to come on
— v — gas f (estufa f) - [combust] véase poner gas a estufa f
— v — juego m - to bring into play
— v — marcha f - to, start, o turn, o switch,
on • [metal-prod] to start up; to restart;
to blow in
— v — f acondicionador m - [ambient] to
start (up) @ conditioner
— v — f bomba f - [bombas] to start (up) @
pump
— v — f compresor m - [ind] to start (up) @
compressor
— v — f en emergencia f - to emergency
start (up)
— v — f equipo m - [ind] to start (up) @,
project, o equipment
— v — f instalación f - [ind] to start (up)
@, project, o installation
— v — m motor m - [electr-mot] to start
(up) @ motor • [comb.int] to start (up) @ engine • [electr-mot] to turn @ motor • [comb.-
int] to turn @ engine
— v — f normalmente - to start normally
— v — f nuevamente - [mec] to restart
— v — f separadamente - to turn on separately
— v — f soldadora f - [sold] to start @
welder
— v — f soplador m—[mec] to start @ blower
— v — f — m para combustión f - [combust]
to start @ combustion blower
— v — operación f - [mec] to put into operation; to turn on - [comb.int] to crank
— v — f motor m - [electr-mot] to start @
motor - [comb.int] to crank @ engine
— v — paralelo m - [electr-instal] to parallel
— v — m soldadora f - [sold] to parallel @
welder
— v — m — f con motor m con combustión f
interna - [sold] to parallel @ engine driven
welder
— v — peligro m - to endanger • to jeopardize
— v — m calzada f - [constr] to jeoparize @
roadway
— v — m cimiento(s) m - [constr] to endanger @ foundation(s)
— v — m estructura f - [constr] to endanger
@ structure
— v — posición f - [mec] to, set, o put in, o
place in, @ position - [comb.int] to operate
(@ switch)
— v — f de conectado - [electr-oper] to
turn on
— v — f — desconectado - [electr-oper] to
turn off

poner v̱ en posición f̱ electrodo m̱ - [sold] to position @ electrode
— v̱ ——— f̱ **neutral** - [mec] to, put, o̱ place, ı̱n @ neutral position
— v̱ ——— f̱ **para aparcamiento** m̱ - [autom] to, put, o̱ place, in @ parking position
— v̱ ——— f̱ — **estacionamiento** m̱ - [autom] to, put, o̱ place, in @ parking position
— v̱ ——— f̱ — **vertical** - to, upright, o̱ put, o̱ place, in @ vertical position
— v̱ — **práctica** f̱ - to put in(to) practice; to implement; to make (it) mature
— v̱ — **relación** f̱ - to contact
— v̱ — **relieve** m̱ - to emphasize
— v̱ — **servicio** m̱ - to, put, o̱ place, in(to) service • [ind] to start up
— v̱ — **sitio** m̱ - to put in(to) place
— v̱ — **vigor** m̱ - to put in(to) force
— v̱ — **voladizo** - [mec] to cantilever
— v̱ ——— m̱ **indicador** m̱ - [constr] to cantilever @ sign
— v̱ **entre   paréntesis** m̱ - [gram] to enclose, in, o̱ between, parentheses
— v̱ **fin** m̱ - to, put, o̱ make, @ end • [fam] to snap
— v̱ **fuera de carrera** f̱ - [deport] to put out of @ race
— v̱ **fuera de circuito** m̱ - [electr-oper] to short
— v̱ **gas** m̱ - [combust] to (connect @) gas
— v̱ — m̱ **a estufa** f̱ - [ind] to, start, o̱ connect, @ gas to @ stove
— v̱ **guillotina** f̱ - [mec] to place @ slide plate
— v̱ — f̱ **en salida** f̱ - [metal-prod] to place @ slide plate in @, outlet, o̱ discharge
— v̱ ——— f̱ **para polvo** m̱ **(en botellón** m̱**)** - [metal-prod] to place @ slide plate in @ dust catcher discharge valve
— v̱ **junto(s) en cortocircuito** m̱ - [electr-oper] to short together
— v̱ **límite(s)** m̱ - to, set, o̱ place, @ limit(s)
— v̱ **marca** f̱ - to check
— v̱ **mira** f̱ - to set @ sight
— v̱ **nuevamente** - to reset
— v̱ — **en servicio** m̱ - to restore @ service
— v̱ **oblicuo,cua** - véase **oblicuar***
— v̱ **palanca** f̱ - [mec] to, put, o̱ place, o̱ set, @ lever
— v̱ — f̱ **en marcha** f̱ **atrás** - [mec] to, put, o̱ place, @ lever in reverse
— v̱ — f̱ — **posición** f̱ - [mec] to position @ lever
— v̱ — f̱ ——— f̱ **para marcha** f̱ **atrás** - [mec] to put @ lever in @ reverse (position)
— v̱ ——— f̱ — **retroceso** m̱ - [mec] to put @ lever in(to) @ reverse (position)
— v̱ — f̱ — **retroceso** m̱ - [mec] to put @ lever in(to) @ reverse (position)
— v̱ **presión** f̱ **alta** - [ind] to apply @ high pressure
— v̱ **puente** m̱ - [electr-instal] to, jump, o̱ put @ jumper
— v̱ **punta arriba** - to upend
— v̱ **refractario(s)** m̱ - [metal-prod] to insert @ refractory,ries
— v̱ **regulador** m̱ - [instrum] to set (@, control, o̱ throttle)
— v̱ — m̱ **para velocidad** f̱ - [mec] to, set, o̱ place, o̱ put, @ speed control lever
— v̱ **sitio** m̱ - [milit] to, besiege, o̱ lay siege
— v̱ **sobre rueda(s)** f̱ - [mec] to put on wheel(s)
— v̱ **transmisión** f̱ - [mec] to, put, o̱ place, @ transmission
— v̱ — f̱ **en posición** f̱ **neutral** - to, put, o̱ place, @ transmission in @ neutral (position)
— v̱ **tubería** f̱ - [tub] to connect @ pipe
— v̱ — f̱ **a petaca** f̱ - [metal-prod] to connect @ pipe to @ cooling plate
— v̱ **vapor** m̱ - [vapor] to apply steam
— v̱ **viento** m̱ - [metal-prod] to blow in
ponerse v̱ - [astron] to, set, o̱ go down
— v̱ **a par** - [deport] to catch up
— v̱ — **trabajo** m̱ - to, go, o̱ get, to work

ponerse v̱ **bota(s)** f̱ **alta(s)** - to put on @ hip-wader(s)
— v̱ **casco** m̱ - [segurid] to wear @ helmet
— v̱ **cómodo** - to relax
— v̱ **de acuerdo** - to, agree, o̱ get together
— v̱ — **pie** - to stand (up)
— v̱ **en carrera** f̱ - [deport] to get on @ way
— v̱ — **comunicación** f̱ - to contact
— v̱ — **marcha** f̱ - to come on • [deport] to get under @ way
— v̱ **nervioso,sa** - to get, nervous, o̱ edgy, o̱ wary
ponteado m̱ - [metal-prod] bridging • scaffolding • [constr] bridge job
ponteado,da a̱ - [constr] bridged; gapped
pontear v̱ - . . .; to bridge; to span
— v̱ **corriente** f̱ - [constr] to span @ stream
— v̱ **separación** f̱ - to, bridge, o̱ span, @ gap
— v̱ — f̱ **grande** - [constr] to bridge @ wide, gap, o̱ opening
ponteo m̱ - bridging; spanning
— m̱ **de corriente** f̱ - [constr] stream bridging
ponzoñoso,sa a̱ - . . .; toxic
pool* m̱ - [transp] véase **parque** m̱
popular a̱ - . . .; crowd pleasing
popularísimo,ma a̱ - (very) crowd pleasing
por prep - . . .; per • due to | [adv] over • hovering (around)
— adv **acta** f̱ - [legal] at @ . . . meeting
— adv **administración** f̱ - self, o̱ company, performed • [constr] by, o̱ under, plant management
— adv **delegada** - [com] cost plus
— adv **añadidura** — in addition • [fam] for spice
— adv **ante mi** - [legal] before me; in my presence
— adv **apariencia** f̱ - for @ look(s)
— adv **aplicar** - [fin] unallocated
— adv **avión** m̱ - [transp] (by) air mail
— adv **bajo cuerda** f̱ - underhandedly • through @ grapevine
— adv **bobina** f̱ - by @, reel, o̱ coil
— adv **cabeza** f̱ - per capita
— adv **ciento** - [fin] per cent
— adv — m̱ **de compactación** f̱ - [constr] compaction percentage; percent compaction
— adv — m̱ **de flecha** f̱ - rise per cent
— adv — m̱ ——— f̱ **total** - total rise per cent
— adv — m̱ — **henchimiento** m̱ - per cent fill
— adv —— m̱ **de conductor** m̱ - [electr-instal] per cent conductor fill
— adv — m̱ **más fino** - [constr] per cent finer
— adv — m̱ —— **por peso** m̱ - [constr] per cent finer by weight
— adv — m̱ **máximo de carbono** m̱ - [quím] per cent maximum carbon
— adv — m̱ **mínimo** - minimum per cent
— adv — m̱ **para impacto** m̱ - per cent for impact
— adv — m̱ **pavimentado** - per cent paved
— adv **cobrar** v̱ - [constab] receivable
— adv — v̱ **de empresa(s)** f̱ **afiliada(s)** - [fin] receivable(s) from affiliate(s)
— adv **comprador** m̱ **(individualmente)** - per buyer
— adv **compuerta** f̱ - [hidr] per gate
— adv **común** m̱ - ordinarily; usually
— adv **consiguiente** - therefore; thereafter; consequently
— adv **contraste** m̱ - by contrast • unlike
— adv **correo** m̱ - by mail • mail order
— adv **cuanto** - whereas; inasmuch • because
— adv **cuenta** f̱ **(de)** - for @ account (of)
— adv **debajo** - [mec] undershot
— adv — **de mitad** f̱ **de velocidad** f̱ **(de régimen** m̱**)** - below half speed
— adv —— **nominal** - [mec] below rating
— adv —— **plataforma** f̱ **(de camión** f̱**)** - [autom] below @ (truck) frame
— adv — **velocidad** f̱ - below @ speed
— adv **delante** - ahead
— adv **demás** ṉ - furthermore • besides
— adv **desgaste** m̱ - [mec] by wear
— adv — m̱ **sin lubricar** - [mec] unlubricated frictional

por adv despachar(se) - [transp] to be shipped
— adv diferencia(l) m - [transp] differential
— adv — m de flete(s) m - [transp] plus freight differential
— adv doquiera - everywhere • across @ board
— adv ejemplo m - for, example, o instance • that is • i. e.
— adv ello - therefore; so
— adv embarcar(se) v - [transp] to be shipped
— adv encima - overshot
— adv — de - over; above; over and above
— adv — — calzada f—[constr] above @ roadway
— adv — — nominal n - above @ rating
— adv — — plataforma f (de camión m) - [autom] above @ (truck) frame
— adv — — rectificador m - over @ rectifier
— adv — — suelo m - above @ ground
— adv ende - therefore; consequently
— adv entre - through
— adv — rodillo(s) m - [metal-lam] through @ roll(s)
— adv — — m enderezador(es) (para varilla f) - [trefil] through @ rod straightening rolls
— adv escrito - in writing; written
— adv esta vez f - for, once, o this time
— adv estructura f - per structure
— adv excepción f - by exception; exceptionally
— adv — f solamente - by exception (only)
— adv experiencia f - by experience
— adv — propia - by self experience; first hand
— adv general n - generally; typically
— adv hora f - per, o by @, hour
— adv inhabitante m - per, inhabitant, o capita
— adv individuo m - per. individual, o capita
— adv instalar(se) v - to be installed
— adv instinto m - by instinct; instinctive(ly)
— adv intermedio - through
— adv — de otra(s) persona(s) f - through, others, o other people
— adv — — tercero(s) m - through other, person(s), o people
— adv libra f - per pound
— adv mayor - [com] wholesale
— adv medio - every other
— adv — de - by, means, o way, of; through
— adv menor - [com] retail
— adv menos n - at least
— adv metro m - per meter
— adv — de longitud f - per, meter of length, o linear meter
— adv — lineal - per linear meter
— adv mucho - by far • far and away
— adv no pesar - [transp] not weighing
— adv orden de - by order of
— adv otra parte f - on @ other hand; otherwise • converesely • furthermore • however • also • contrary(ly) • on @ plus side • [legal] as @ other party
— adv otro motivo m - unrelated
— adv pagar - [com] payable; owed
— adv par - per pair
— adv pared f - through @ wall
— adv — f lateral - through @ sidewall
— adv parte de - by
— adv partida(s) f - by batch(es)
— adv pedido m de - by @ request, o at @ order of
— adv partida f doble - [contab] double entry
— adv persona - per, o capita
— adv peso m - by, o according to, weight
— adv pie - per, o by, @ foot
— adv presente m/f - hereby; by these presents
— adv qué - why • therefore
— adv razón f de apariencia f - for, look(s), o appearance(s) sake
— adv de conveniencia f - for convenience
— adv regla f general - in general; by large
— adv retraso(s) m - due to @ delay(s)
— adv rollo m - per, o by, roll
— adv rueda f - by, o per, wheel
— adv separado - separately
— adv sí - per se • (by) itself

por adv si acaso m - if by chance; just in case
— adv — mismo,ma - (by) itself
— adv — sóla - by herself
— adv — sólo - by himself • by itself
— adv sobre - over; across • [electrón] úsese entre borne(s) m
— adv — . . . fase(s) f - [electr-instal] across . . . phase(s)
— adv — ojal m - [electrón] cross @, grommet, o eyelet
— adv — ranura f - [mec] across @ spline
— adv — f en árbol m - [mec] across @ shaft spline
— adv — f — — m exterior - [mec] over @ outer shaft spline
— adv — fase f - [electr-instal] across @ phase
— adv su cuenta f - for, his, o her, o its, account • at, his, o her, o its, expense
— adv — intermedio m - thereby
— adv suerte - luckily • hopefully
— adv supuesto - admittedly; of course; surely
— adv tanda(s) f - by @ batch(es)
— adv tanteo(s) m - by trial and error
— adv tanto - [legal] therefore
— adv todo,da - throughout
— adv — país m - [geogr] throughout @ country
— adv tonelada f - per (metric) ton
— adv — f de dos mil libras f - per ton
— adv — métrica - per metric ton
— adv neta - per net ton
— adv trozo m - per length
— adv tubo m - through @ pipe
— adv una parte f - [legal] as one party
— adv — vez - for once
— adv unidad f - (per) unit
— adv vez f - at @ time
— adv — f primera - for @ first time • first of @ kind
— adv vía f de - via; by way of
— adv — f húmeda - [ind] wet process(ing)
— adv — f seca - [ind] dry process(ing)
— adv yunta f - per pair
porcelana f - [domést] fine china
— f vítrea - [ceram] vitreous china
porcelanizado,da a - [ceram] percelanized
porcentaje m - . . . .; per cent • content
— m actual - present, o current, percentage
— m alto - high, percentage, o content
— m apreciable - substantial percentage
— m aproximado - approximate percentage
— m aumentado - increased percentage
— m bajo - low, percentage, o content
— m con bate* m - [deport] batting average
— m considerable - substantial, o large, o considerable, percentage
— m correspondiente - corresponding percentage
— m de abandono(s) m - [deport] attrition rate
— m — ajuste(s) m - adjustment(s) rate
— m — alargamiento m - per cent of elongation
— m — aleación f - [metal] alloy percentage
— m — asociación f accidental - [legal] joint venture percentage
— m — aumento m - percentage increase
— m — — en superficie f - surface increase percentage
— m — carbón m - [miner] coal content
— m — carbono m - [metal] carbon content
— m — carga f - load percentage
— m — para transformador m - [electr-prod] transformer load percentage
— m — compactación f - [constr] compaction percent(age)
— m — — f de relleno m - [constr] backfill compaction percent(age)
— m — — f — — m para tubería f - [constr] pipe backfill compaction percent(age)
— m — cromo m - [metal-prod] chromium, percentage, o content
— m — densidad f - density percent(age)
— m — — f estándar - [constr] standard density percent(age)
— m — — f — según A A S H T O - [constr]

percent(age) standard A A S H T O density
porcentaje m de desgaste m - wear percent(age)
— m — eliminación f - elimination percent(age)
• [deport] attrition rate
— m — elongación f - [metal-fabr] elongation
percentage
— m —— f de metal m - [metal-fabr] metal
elongation percentage
— m — estímulo m—[labor] incentive percentage
— m — expansión f - [metal] expansion percentage
— m — gas m - [ind] gas percentage
— m —— m para reducción f - [metal-prod]
reducing,ction gas percentage
— m —— m reductor - [metal-prod] reducing,ction gas percentage
— m — humedad f - moisture percentage
— m — huella f - [autom-neumát] footprint percentage
— m — impuesto m - [fisc] tax, percentage, o
rate
— m — incentivo m - [labor] incentive percentage
— m — incremento m - increase percentage
— m —— m en metal m crudo - [ind] raw material increased percentage
— m —— m —— m máximo - [ind] maximum
raw material increased percentage
— m — instalación(es) f - installation(s) percentage
— m — níquel m - [metal] nickel percentage
— m — premio(s) m - [labor] premium(s) percentage
— m —— m obtenido(s) - [labor] premium(s)
percentage obtained
— m — prima f - [labor] premium percentage
— m —— f obtenida - [labor] premium(s) percentage obtained; obtained premium percentage
— m — rapidez f - speed percentage
— m —— f para avance m - [sold] procedure
speed percentage
— m —— f —— m correcta - [sold] correct
procedure speed percentage
— m —— f —— m para procedimiento m -
[sold] correct procedure speed percent(age)
— m — reducción f - reduction percentage; per
cent reduction
— m —— f de área m - [metal-fabr] per cent
reduction in @ area • [alambre] cross area
reduction percentage
— m — saturación f - saturation percentage
— m — terminación f - [legal] completion percentage
— m —— f en base f a costo m - [legal] cost
completion percentage
— m — pérdida f - loss percentage
— m — peso m - weight percentage • percentage
of @ weight
— m — reducción f - reduction percentage
— m — trabajo m - [labor] work percentage
— m —— m completado - [admin] labor, o work,
completion percentage
— m —— m — en oficina f (matriz) - [admin]
(home) office , labor, o work, completion
percentage
— m elevado - high percentage • high content
— m — de carbono m - [metal-prod] high carbon
content
— m grande - large, o high, percentage
— m máximo - maximum percentage
— m — actual - maximum current percentage
— m —— de impuesto m - [fisc] maximum current tax rate
— m — de carbono m - [metal] maximum carbon,
percentage, o content
— m medio - average percentage
— m menor - smaller percentage
— m — en aumento m en superficie f - smaller, o
smallest, surface increase percentage
— m mínimo - minimum percentage
— m — de alargamiento m - [metal] minimum
elongation percentage • [metal] minimum percent ultimate elongation

porcentaje m mínimo de alargamiento m hasta ruptura f - [metal] minimum percent ultimate
elongation
— m obtenido - obtained percentage
— m para facturación f - invoicing percentage
— m pequeño - low percentage • few percent
— m por producto m - [com] product mix
— m promedio - average percentage
— m — de hierro m - [miner] average iron percentage
— m —— m total - [miner] iron average percentage
— m —— de hierro m - [miner] total iron
average percentage
— m reducidísimo - minute percentage
— m reducido - reduced, o low, percentage, o
content
— m — de carbono m - [metal-prod] low carbon
content
— m regulado - [quím] controlled, percentage,
o analysis
— m relativo - relative percentage
porcentual* a - véase de porcentaje m
porcino m - [ganad] . . .; pork
— m alimentado con grano(s) m - [ganad] grain
fed pork
porción f - . . . • section • batch • [deport]
leg
— f bifurcada - [electr-distrib] tapped off
portion
— f — de energía f - [electr-distrib] power
tapped off portion
— f —— — f en línea f - [electr-distrib]
line power tapped off portion
— f cerámica - [cerám] ceramic portion
— f cónica - [mec] conical, portion, o section
— f de cerámica f - [cerám] ceramic portion
— f — energía f - [electr-distrib] power portion
— f —— f en línea f - [electr-distrib] line
power portion
— f — mercado m - [com] market portion
— f — peso m - weight portion
— f —— m de vagón m - [f.c.] car weight portion
— f — recorrido m - [deport] course leg
— f derivada - [electr-distrib] tapped off
portion
— f — de energía f - [electr-distrib] power
tapped off portion
— f —— — f en línea f - [electr-distrib]
line power tapped off portion
— f desgastada - [mec] worn (off) portion
— f — de eslabón m - [cadenas] link worn portion
— f en dólares m - [fin] dollar portion
— f —— m canadienses - [fin] Canadian dol-
lars portion
— f — sucres m - [fin] sucre(s) portion
— f pequeña - small portion
— f primera - first portion
— f reducida - small portion
— f — de mercado m - [com] small market portion
— f según especificación f - specification portion
— f segunda - second portion
porfirita m - [geol] porphyrite
— f hornabléndica - [geol] hornblend(ic) porphyrite
porfirítico,ca a - [geol] porphyritic
pórlan(d) m - véase cemento m Portland
pormenor m - . . . • information; datum,ta •
intricacy
pormenores m - ins and outs
poro m de gas m - [metal] pinhole
— m en superficie f - [sold] (surface) pinhole
— m grande - [sold] large, pore, o hole
— m pequeño - [metal-prod] pinhole; small hole
— m superficial - [sold] surface pore
pororó m - [culin] . . .; popcorn
porosidad f - [sold] . . .; pinhole porosity
— f causada por soplo m de arco m - [sold]

arc blow porosity
porosidad f **causada por soplo** m̲ **magnético de arco** m̲ - [sold] arc blow porosity
— f **de lecho** m̲ - [miner] bed voidage
— f̲ — **metal** m̲ - [metal] metal porosity
— f̲ — — m̲ **para soldadura** f̲ - [sold] weld metal porosity
— f — **raíz** f̲ - [sold] root porosity
— f̲ — **soldadura** f̲ - [sold] weld porosity
— f̲ **debajo de superficie** f̲ - [sold] subsurface
porosity
— f̲ **debida a contaminación** f̲ - [metal] contamination porosity; porosity due to a contamination
- f — — — f **orgánica** - [metal] organic contamination porosity; porosity due to @ organic contamination
— f̲ - — **oxidación** f̲ - [metal] rust porosity
— f̲ — — **óxido** m̲ - [metal] rust porosity; porosity due to @ rust
— f̲ — — **soplo** m̲ (**magnético**)-·[sold] (magnetic) blow porosity; porosity due to @ (magnetic) blow
— f̲ — — — m̲ (—) **de arco** m̲ - [sold] (magnetic) arc porosity; porosity due to @ (magnetic) arc blow
— f̲ **dispersa** - [sold] scattered, o̲ disperse,
porosity
— f̲ — **limitada** - [sold] limited, scattered, o̲
disperse, porosity
— f̲ **en acero** m̲ - [metal] steel porosity
— f̲ — — m̲ **sulfuroso** - [metal] sulfur bearing
steel porosity
— f̲ — **superficie** f̲ - [sold] surface (pinhole)
porosity
— f̲ **intermitente** - [sold] scattered, o̲ intermittent, porosity
— f̲ — **en superficie** f̲ - [sold] scattered, o̲
intermittent, surface porosity
— f̲ **interna** - [sold] internal porosity
— f̲ **limitada** - [sold] limited porosity
— f̲ **reducida** - [sold] reduced porosity
— f̲ **severa** - [sold] severe, porosity, o̲ (pin)-
holes
poroto m̲ - [botán] bean; **véase fréjol** m̲
— m̲ **frutilla** f̲ - [botán] cowpea
— m̲ **soja,ya** - [botán] soy bean
porra f̲ - [herram] . . .; about sledge
porta. . . a - [mec] . . . holder; . . . carrier
portaacumulador m̲ - [electr] battery carrier
portaavión(es) m̲ - [milit] . . .; carrier
portabagazo(s) m̲ - [agric] bagasse carrier
portabarras m̲ - [electr-instal] bus bar bracing
portabarrenos,nas m̲ - [petról] drill, chuck, o̲
bit, holder
portabilidad* t̲ - **véase calidad** f̲ **de portátil**
portabolsa(s) m̲ - [agric=equip] bag holder
portaboquilla(s) m̲ - [mec] nozzle, holder, o̲
support • [comb.int] jet holder
— m̲ **para pulverizador** m̲ - [comb.int] nozzle
support
— m̲ **principal** - [comb.int] main jet holder
portabroca(s) m̲ - [herram] (Jacob's) chuck; pad
— m̲ **electromagnético** - [mec] electromagnetic
chuck
— m̲ — **según norma** f̲ - [herram] standard electromagnetic chuck
— m̲ **según norma** f̲ - [herram] standard chuck
portaburil(es) m̲ - [tornos] (checkering) tool
rest
portacabeza(s) m̲ - [mec] head carrier
portacable(s) m̲ - [electr-instal] cable support •
[petról] wire, line, o̲ rope, socket
— m̲ **con mandril** m̲ - [petról] Prosser swivel
socket
— m̲ **fijo** - [cabl] stiff neck(ed) socket
— m̲ **giratorio** - [cabl] swivel rope socket
— m̲ **liviano** - [sold] light(weight) mounting
— m̲ — **para rollo** m̲ (**de alambre** m̲) - [sold]
light(weight) wire reel mounting
— m̲ **para alambre** m̲ - [sold] wire reel mounting
portacadena(s) m̲ - [mec] chain, container, o̲
carrier

portacarbón(es) * m̲ - [electr-mot] **véase porta-
escobilla(s)** m̲
portacarrete(s) m̲ - [mec] reel, rack, o̲ support,
o̲ mounting • [sold] (wire) reel mounting
— m̲ **abierto** - [sold] open frame (wire) reel
mounting
— m̲ — **para rollo** m̲ (**de alambre** m̲) - [sold]
open frame wire reel mounting
— m̲ **liviano** - [sold] light(weight) mounting
— m̲ — **abierto** - [sold] light(weight) open
frame mounting
— m̲ — — **para rollo** m̲ (**de alambre** m̲) - [sold]
light(weight) open frame wire reel mounting
— m̲ **tipo jaula** - [sold] cage type (wire) reel, ·
support, o̲ mounting
portacojinete(s) m̲ - [cojin] bearing, carrier, o̲
support, o̲ block
— m̲ **de caucho** m̲ - [cojin] bearing rubber support
portacortadora m̲ - [mec] cutter holder
portacristal(es) m̲ - [mec] glass holder •
[óptica] lens holder
— m̲ **abisagrado** - [segurid] hinged lens holder
— m̲ — **levadizo** - [segurid] flip-front lens
holder
— m̲ **de fibra** f̲ **de vidrio** m̲ - [óptica] fiberglass lens holder
— m̲ — — f̲ — m̲ **fundido** - [óptica] molded
fiberglass lens holder
— m̲ — **metal** m̲ - [óptica] metal lens holder
— m̲ — — m̲ **fundido** - [óptica] cast metal lens
holder
— m̲ — — m̲ — **en matriz** f̲ - [óptica] die cast
metal lens holder
— m̲ **integrado** - [óptica] integral lens holder
— m̲ **liviano** - [óptica] light(weight) lens
holder
— m̲ **recio** - [óptica] rugged lens holder
portacuchilla(s) m̲ - [torno] cutter head
— m̲ **anterior** - [tornos] front cutter head
— m̲ **posterior** - [tornos] back, o̲ rear, cutter
head
portachumacera(s) m̲ - [cojin] bearing support
— m̲ **de caucho** m̲ - [cojin] rubber bearing support
portada f̲ - . . . . [constr] doorway • [public]
cover painting
portadado(s) m̲ - [mec] die holder
portadiferencial m̲ - [autom-mec] differential
carrier
— m̲ **armado** - [autom-mec] assembled differential
carrier
— m̲ **para eje** m̲ - [autom-mec] axle differential
carrier
— m̲ — — m̲, **anterior,** o̲ **delantero** - [autom-
-mec] front, o̲ forward, axle differential
carrier
— m̲ **removido** - [autom-mec] removed differential
carrier
— m̲ — **reparado** - [autom-mec] overhauled, o̲ repaired, differential carrier
— m̲ **sujetado** - [autom-mec] fastened differential carrier
portado,da a̲ - . . . . • carried; conveyed
— a̲ **en parte** - carried in part
— a̲ **normalmente** - carried normally
portador m̲ **para cabeza** f̲ - [sold] (weld)ing
head carrier
— m̲ — **rueda** f̲ - [mec] wheel, o̲ roll, carrier,
o̲ holder
— m̲ — — f̲ **para guía** f̲ - [mec] guide roll(er)
holder
portaeje(s) m̲ - [mec] axle carrier • spindle
carrier
portaelectrodo(s) m̲ - [sold] electrode, holder,
o̲ clamp
— m̲ **aislado** - [sold] insulated (electrode)
holder
— m̲ — **cabalmente** - [sold] well, o̲ fully, insulated (electrode) holder
— m̲ — **totalmente** - [sold] fully, o̲ totally,
insulated (electrode) holder
— m̲ **autoenfriante** - [sold] self-cooling (elec-

trode) holder
portaelectrodo(s) m colgado - [sold] hung (electrode) holder
— m **colgante**—[sold] hanging (electrode) holder
— m **con enfriamiento** m - [sold] cooled (electrode) holder
— m —— m **para electrodo(s)** m **con carbono** m -
[sold] cooled carbon electrode holder
— m **Cooltong** - [sold] Cooltong electrode holder
— m — **con enfriamiento** m - [sold] Cooltong
[electrode holder]
— m **corriente**—[sold] standard electrode holder
— m **de metal** m - [sold] metal electrode holder
— m **defectuoso** - [sold] defective electrode
holder
— m **enfriado** - [sold] cooled electrode holder
— m — **con agua** m - [sold] water cooled electrode holder
— m **estándar** - [sold] standard electrode holder
— m **metálico** - [sold] metal(lic) electrode
holder
— m **normal** - [sold] notmal, o standard, (electrode) holder
— m **para carbón(es)** m - [sold] carbon electrode
holder
— m —— m **enfriado (con agua** m) - [sold]
(water) cooled carbon electrode holder
— m — **electrodo(s)** m - [sold] electrode holder
— m —— m **con carbono** m - [sold] carbon elec-
trode(s) holder
— m **según norma** f - [sold] standard (electrode)
holder
portaenchufe(s) m - [electr-equip] plug holder
portaequipage(s) m - [autom] (baggage, o lug-
gage,) carrier
portaescariadora f - [mec] end mill holder
portaescobilla(s) m - [electr-mot] brushholder
— m **a caja** f - [electr-prod] brushholder to @
frame
— m **completo**—[electr-mot] complete brushholder
— m **con anillo** m **para ajuste** m - [electr-mot]
slip ring brushholder
— m **dextrorso** - [electr-mot] clockwise brushholder
— m **fuera de ajuste** m - [electr-mot] out-of-ad-
justment (brush)holder
— m **hacia derecha** f - [electr-mot] right hand
brushholder
— m — **izquierda** f - [electr-mot] left hand
brushholder
— m **negativo** - [electr-mot] negative brushholder
— m **para alternador** m—[electr-prod] alternator
brushholder
— m — **bobina** f - [electr-equip] coil brushholder
— m —— f **para reacción** f - [electr-equip]
reactor brushholder
— m **excitador** m - [electr-equip] exciter
brushholder
— m — **generador** m - [electr-prod] generator
brushholder
— m —— m **para soldadora** f - [sold] welder
generator brushholder
— m — **reactor** m - [electr-equip] reactor
brushholder
— m — **reóstato** m - [electr-equip] rheostat
brushholder
— m — **trabajo(s)** m **liviano(s)** - [electr-mot]
light duty brushholder
— m **positivo** - [electr-equip] positive brushholder
— m **principal** - [electr-mot] main brushholder
— m **roto** - [electr-mot] broken (brush)holder
— m **según norma** f - [electr-mot] standard
brushholder
— m **siniestrórsum**—[electr-mot] counterlockwise
brushholder
portaestampa(s) m - [mec] die, o tool, o insert,
holder • die block • die insert
portaestribo(s) m - [autom] step hanger
portafiltro(s) m - [mec] filter, holder, o can-
nister
portafresa(s) m - [tornos] (milling) tool holder

portafusible(s) m - [electr-instal] fuse, holder, o block
— m **para dos fusibles** m - [electr-equip] two-,
o double, fuse block
portagatillo(s) m - [sold] trigger pad
portagavilla(s) m - [agric-equip] bundle, o
sheave, carrier
portaguía(s) m - [mec] guide holder • [sold]
faceplate • (nozzle extension) insert
— m **para prolongación** f - [mec] extension insert
— m —— f **para boquilla** f - [sold] nozzle extension insert
portahacina(s) m - [agric-equip] sheave, o bundle, carrier
portaherramienta(s) m - [herram] tool, carrier,
o holder • chuck • arbor
— m **flotante** - [tornos] floating tool holder
— m **para barreno,na** - [herram] boring chuck
— m — **máquina** f **para barrenar** v - [herram]
boring chuck
— m — **taladro** m - [herram] boring chuck
portahilera(s) m - [cabl] block
portahusillo(s) m - [tornos] spindle head
portal m - [constr] . . .; portal; gateway
portalámpara(s) m - [electr-instal] socket • receptacle • lamp holder
portamandril(es) m - [herram] chuck
— m **para conformación** f - [herram] forming mandrel holder
portamatriz,ces m - [mec] die holder • die set
portamecha(s) m - [petról] drill collar
— m **con unión(es)** f **integral(es) (cambiables)** -
[petról] drill collar
portamuestra(s) m - [mec] (sample) carrier
portaneumático(s) m - [autom] tire, holder, o
carrier
— m **auxiliar** - [autom] spare tire carrier
— m **trasero** - [autom] rear tire carrier
— m **lateral** - [autom] side tire carrier
portante a - conveying; carrying • [geol] bearing
portapatente(s) m - [autom] license, bracket, o
plate frame
portapeine(s) m - [mec] (threading) die holder
portapico(s) m - [alambre] casing
portapieza(s) m - [torno] work holder
portapistola(s) m - [sold] gun, holder, o carriage • travel carriage
— m **abisagrado** - [sold] hinged gun holder
— m **automotriz** - [sold] self-propelled carriage
— m — **sin carril(es)** m - [sold] self-propelled trackless carriage
— m — **corriente** - [sold] standard gun holder
— m — **para avance** m **mecanizado** - [sold] standard Squirtmobile gun holder
— m **mecanizado** - [sold] mechanical(ly carried)
gun
— m **para avance** m **automotriz** - [sold] self-propelled travel carriage
— m —— m **mecanizado** - [sold] Squirtmobile
gun holder
— m **sin carril(es)** m - [sold] trackless (gun)
carriage
— m **Squirtmobile** - [sold] Squirtmobile gun
holder
— m — **para avance** m **mecanizado** - [sold]
Squirtmobile gun holder
portapunzón(es) m - [mec] punch holder
portar v - . . .; to convey • [constr] to bear •
[constr] to, carry, o support
— v **acarreo(s)** m - [hidr] to carry @ silt
— v **carga** f - to, carry, o bear, @ load | adv -
(para) for bearing purpose(s)
— v — f **directa** - [constr] to carry @ direct
load
— v — f — **pesada** - [constr] to, bear, o carry, @ direct heavy load
— v — f **viva** - [constr] to carry @ live load
— v **con seguridad** f - [mec] to carry safely
— v **directamente** - to carry directly
— v — **carga** f - to carry directly @ load
— v —— f **pesada** - [constr] to carry directly
@ heavy load

portar v económicamente - [mec] to carry economically
— v en parte - to carry in part
— v normalmente - to carry normally
— v objeto m - to carry @ object
portarodillo(s) m - [mec] roll, carrier, o stand
  • [metal-lam] roll stand
portarresistor m - [electr-instal] resistor
holder
portarrodamiento(s) m - [cojin] cup follower
portarrueda(s) m - [mec] wheel holder • [autom]
wheel carrier
portasaco(s) m - [agric-equip] bag holder
portasurtidor(es) m - [comb.int] jet holder
— m principal - [mec] main jet holder
portataladro(s) m - [petról] bit holder
portataza(s) m - [mec] cup carrier
portaterraja(s) m - [mec] die head
portatrépano(s) m - [petról] bit holder
portatroquel(es) m - [mec] die, o insert, head,
o holder
portavástago(s) m - [mec] rod holder
— m para llama(s) f—[combust] flame rod holder
portavidrio(s) m - glass holder
portaviento(s) m - [metal-prod] blow pipe; tuyere stock • gooseneck
— m perforado - [metal-prod] burned (through),
blow pipe, o gooseneck
porte m - . . . • span • [constr] importance •
[comunic] postage
— m amplio - wide span
— m de acarreo(s) m - [hidr] silt carrying
— m — carga f - [transp] load carrying
— m — — f viva - [constr] live load carrying
— m — estructura f - [constr] structure carrying
— m — objeto m - object carrying
— m en parte f - carrying in part
— m grande - [constr] large size; Super-Span •
king size
— m máximo - maximum, size, o load • [grúas]
working load limit
— m normal - normal carrying
— m para producción f - [ind] production size
— m suficiente - sufficient size
— m — para producción f - (sufficient) production size
portería f - . . . . • [ind] gate house
portero m - . . .; doorman
portezuela f - [mec] grid door
— f lateral - [mec] side door
— f — para descarga f - [mec] side discharge
door
— f para descarga f de tambor(es) m - [mec]
drum discharge door
— f — seguridad f - [ind] explosion, o safety,
door
— f — — f contra explosión(es) f - [segurid]
explosion (safety) door
portezuelo m - [geogr] mountain pass • neck
pórtico m - portal • [ind] portal crane; gantry;
ore bridge • reserve coke stock • [constr]
portal; bent • [electr-instal] tower
— m de reserva(s) f - [metal-prod] reserve coke
stock • reserve bridge • reserve store yard
ore bridge
— m — — f para mineral(es) m - [metal-prod]
reserve ore bridge
— m en parque m - [metal-prod] yard portal
(crane); ore bridge
— m para indicador(es) m - [vial] sign bridge
— m — — m con voladizo m - [vial] cantilever
sign bridge
— m — — m elevado(s) - [vial] overhead sign
bridge
— m — recogida f - [metal-prod] ore bridge
portillo m - . . . . • port
— m auxiliar - [electrón] auxiliary port
Portlan(d) m - [constr] véase cemento m Portland
portón m - [constr] gate
— m con una hoja f - [constr] single gate
— m — — sóla hoja f - [constr] single gate
— m corredizo - [constr] sliding gate

portón m para cerca f - [constr] fence gate
— m para planta f - [ind] plant gate
— m pendular - [constr] swing(ing) gate
posaneumático(s) m - [autom] fender well
posar v - . . . • [grúas] to, lower, o let down
posarse v - [mec] to ride
— v sobre rodillo m - [mec] to ride (on) @ roller
— v — — m loco - [mec] to ride (on) @ idler
— v — rueda f - [mec] to ride (on) @ wheel
— v — — f loca - [mec] to ride (on) @ idler
poscalentamiento* m - [sold] postheating
poscalentar* v - [sold] to postheat
poseedor m de marca f - [deport] record holder
— m de record* m - [deport] record holder
poseer v - . . . .; to have; to hold
— v derecho m - [legal] to, have, o own, @
right
— v — m literario(s) - [legal] to own @ copyright
— v — m para patente f (de invención f) -
[legal] to own @ patent (rights)
— v marca f de fábrica f - [legal] to own @
trade mark
— v — f industrial - [legal] to own @ trade
mark
— v — f registrada - [legal] to own @ (registered) trade mark
— v nombre m comercial - [legal] to own @ trade
name
— v patente f (para invención f) - [legal] to
own @ patent
poseído,da a - possessed • owned
posesión f - owning; ownership
— f de cargo m - incumbency | adv - [con] in-
cumbent
— f — Estados m Unidos - [pol] United States
possession
— f de nombre m comercial - [legal] trade name,
owning, o ownership
posguerra f - [hist] post war
posibilidad f - . . .; capability • chance;
prospect • scope; potential
— f aminorada - reduced, o lessened, possibility
— f aumentada - increased possibility
— f — de separación f - increased separation
possibility
— f — aceptación f - acceptance, chance, o
poissibility
— f de agrietamiento m - [metal] cracking,
chance, o possibility
— f — m de reja f - [agric-equip] share
cracking, chance, o possibility
— f — — m — — f para arado m - [agric-
-equipo] plow share cracking, chance, o possibility
— f — apilamiento m - [ind] piling possibility
— f — — m estratificado - [ind] bedding possibility
— f — armado m - assembly,bling possibility
— f — — m previo - [constr] pre-assembly possibility
— f — contacto(s) m - [segurid] contact potential
— f — — m personal(es) - personal contact(s),
possibility, o potential
— f — corregir(se) adv - (con) repairable
— f — corrosión f - corrosion potential
— f — deslizamiento m - [autom-neumát] hydroplaning possibility
— f — — m aumentada - [autom-neumát] increased hydroplaning possibility
— f — — m sobre agua m - [autom-neumát] hydroplaning possibility
— f — — m — — m aumentada - [autom-neumát]
increased hydroplaning possibility
— f — — m — — m reducida - [autom-neumát]
decreased, o reduced, hydroplaning possibility
— f — exportación f - [com] export(ing), potential, o possibility
— f — interconexión f - interconnection possi-

bility
posibilidad f **de interrupción** f - interruption,
possibility, o capacity
— f — **lesión(es)** f - [segurid] injury possibility
— f — — f **personal(es)** - [segurid] personal
injury,ries possibility
— f — **máquina** f - [ind] machine('s) capability
— f — **muerte** f - [segurid] death possibility
— f — **penetración** f - penetration possibility
— f — **prearmado** m - [ind] pre-assembly, possibility, o feature
— f — **premontaje** m - [constr] pre-assembly,
possibility, o feature
— f — **representación** f - [com] dealer(ship),
opportunity, o possibility
— f — **separación** f - separation possibility
— f — — f **aumentada** - increased separation
possibility
— f — — f **reducida** - reduced separation possibility
— — **urbanización** f - [constr] development possibility
— f —, **uso** m, o **utilización** f—use possibility
— f **disminuida** - decreased possibility
— f — **de separación** f - decreased separation
possibility
— f **económica** - [econ] economic possibility
— f **excepcional** - exceptional possibility
— f **matemática** - mathmatical, possibility, o
chance
— f **merecedora** - deserving possibility
— f **nacional** - [econ] national possibility
— f **para almacenamiento** m - [ind] storage, o
storing, possibility
— f — **especialización** f - specialization,zing
possibility
— f — **rendimiento** m - performance potential •
yield possibility
— f — **soldadura** f - [sold] weldability
— f — — f **con electrodo(s)** m **manual(es)** -
[sold] stick electrode welding capability
— f **recreativa** - [deport] recreational, possibility, o potential
— f **reducida** - reduced, o lessened, possibility, o chance, o potential
— f **remota** - remote possibility
— f **técnica** - technical possibility
posibilidades f **de acería** f - [metal-prod] steel
plant, possibilities, o potential
— f **para electrodo** m - [sold] electrode potential
— f — **empleo** m - operability
— f — **mercado** - [com] market possibilities;
marketability
— f — **planta** f **para acero** m - [metal-prod]
steel plant, possibilities, o potential
— f — **promoción** f - [labor] promotional, possibilities, o potential
— f — **soldadura** f - [sold] weldability; welding capability
— f — — f **Linc-Fill** - [sold] Linc-Fill, welding capability, o weldability
— f — — f **con electrodo** m **muy sobresaliente** - [sold] Linc-Fill long stickout welding capability
— f — **tracción** f - [autom] traction potential
— f — **viraje(s)** m—[autom] cornering potential
posibilitación f - making possible; enabling
posibilitado,da a - made possible; enabled
posibilitar v - to make possible; to enable
posible a - . . . • probable • available • potential • practiable; feasible • prospective
— a **beneficiario** m - possible beneficiary -
a - eligible
— a **empleo** m **de tierra(s)** f - probable land use
— a **tránsito** m - [vial] possible traffic
— a — m a **preverse** - [vial] possible traffic
expectation
posiblemente adv - . . . ; probably • maybe
posición f - . . . • ranking • item • positioning • [instrum] setting • [fin] holding(s) -
[deport] slot; spot | adv - [en] in position

posición f **abierta** - open position
— f — **para estrangulador** m - [mec] choke
open(ed) position; open choke position
— f — **totalmente** - [mec] full(y) open(ed) position
— f **acostada** - lying position
— f **actual** - present position
— f **adelantada** - forward position
— f **ajustada** - adjusted position
— f **alta** - high, o up, position • [ind] high
setting
— f **apropiada** - proper, o appropriate, position
| adv - (en) properly set
— f — **para operación** f - proper operating position
— f — — **rodillo** m - [metal-lam] proper roll
position
— f — — **soldar** v - [sold] proper welding position
— f **aproximada** - approximate position
— f **arancelaria** - [fisc] tariff, position, o
classification, o number
— f **ascendente** - [sold] uphill position | adv -
[sold] (en) uphill positioned
— f **avanzada** - advanced, o forward, position
— f **baja** - low position • down position • [mec]
low • [electr-oper] low, position, o setting
— f **bajada** - lowered position; lowered location
— f **buena** - good position • [deport] good
showing
— f **cambiada** - changed position
— f **central** - [instrum] range center
— f — **en escala** f **de amperaje(s)** m - [sold]
current range center
— f **cero** - [instrum] zero position
— f — **para traslación** f - [mec] zero translation position
— f — — f **longitudinal** - [mec] zero
lengthwise translation position
— f — — f **transversal** - [mec] zero, crosswise, o transverse, translation position
— f **cerrada** - [mec] closed position
— f — **para estrangulador** - [mec] choke closed,
o closed choke, position
— f — **totalmente** - [mec] totally, o fully,
closed position
— f **cómoda** - comfortable position • convenient
position
— f — **para trabajo** m - convenient work(ing)
position
— f **comparativa** - comparative position; ranking
— f — **para contaminación** f **de aire** m - [ecol]
air pollution ranking
— f **con traba** f **abierta** - [mec] lock open position
— f — — f **cerrada** - [mec] locked closed position
— f **conectada** - [electr-oper] closed position
— f **contra palo(s)** m - [deport] pole position
— f **conveniente** - convenient position
— f — **para trabajo** m - convenient work(ing)
position
— f **correcta** - correct position | adv - (en)
properly set
— f — **para soldadura** f - [sold] correct welding position
— f **correspondiente** - corresponding, o respective position
— f — a . . . f **en reloj** m - . . . . o'clock position
— f — **para máquina** f - [mec] machine position
— f **cualquiera** - any position
— f **de apertura** f **total**—[mec] full(y) open(ed)
position
— f — **árbol** m - [mec] shaft position
— f — — m **cambiada** - [comb.int] changed shaft
position
— f — — m **para toma** f **para fuerza** f - [mec]
take-off shaft position
— f — **arco** m—[sold] arc, location, o position
— f — **avance** m - [mec] forward. o advance, position
— f — **borne** m - [sold] stud, position, o set-

ting
posición f de cabeza f - [mec] head·position •
[sold] (welding) head position
— f — cable m - [cabl] cable, o rope, position
— f — cabo n = [cabl] rope position
— f — canaleta f - [sold] trough position
— f — carga f - [transp] load(ing) position
— f — cerrado,da a - [mec] closed position
— f — cierre m - [mec] close(d) position
— f — — m total - [mec] full(y) close(d) position
— f — cilindro m - [mec] cylinder position
— f — — m para elevación f - [agric-equip]
lift(ing) cylinder, position, o location
— f — conectado,da a - [electr-oper] on (position)
— f — conmutador m - [electr-oper] switch position
— f — — m para selección f - [electr-oper]
control switch position
— f — — m para escala f de voltaje(s) m -
[sold] voltage range switch position
— f — — m — límite(s) m para voltaje m -
[sold] voltage range switch position
— f — — m para selección f - [sold] selector
(switch) position
— f — — m — — f para amperaje(s) - [sold]
current selector switch position
— f — — m — sentido m - [sold] sense switch
position
— f — — m — — m para manipulador m -
[electrón] keyer sense switch position
— f — corona f dentada - [mec] ring gear position
— f — desconectado,da - [electr-oper] off (position)
— f — detenido.da a - [mec] stopped position
— f — eje m - axis position • axle position
— f — — m para toma f para fuerza f - [mec]
take-off shaft position
— f — electrodo m - [sold] electrode position
— f — estrangulador m - [mec] choke position
— f — flotador m - float position
— f — gancho m - [mec] hook position
— f — horquilla f - [mec] yoke, o clevis, position, o setting
— f — interruptor m - [electr-oper] switch position
— f — junta f - [sold] joint position
— f — leva f - [mec] cam position
— f — metal m en lingote m - [metal-lam] ingot
traceability; traceability to @ ingot
— f — mordaza f - [mec] clamp('s) position
— f — ojal m - [mec] eyelet position • grommet
position
— f — parado,da a - [mec] stopped position
— f — piñón m - [mec] pinion position
— f — — m ajustada - [mec] adjusted pinion
position
— f — — m impulsor - [mec] drive,ving pinion
position
— f — pistola f - [sold] gun position
— f — puerta f - door position
— f — puntero m - [deport] lead
— f — — m asegurada—[deport] solidified lead
— f — regulador m - [mec] control, position, o
setting • [comb.int] throttle position •
[electr-oper] switch position
— f — — m para operación f sin carga f -
[comb.int] no load speed throttle position
— f — reóstato m - [electr-equip] rheostat position, o setting
— f — — m para voltaje m - [electr-oper]
voltage rheostat, position, o setting
— f — — m — — m en circuito m abierto -
[electr-oper] open circuit voltage rheostat,
position , o setting
— f — rodillo m - [metal-lam] roll position
— f — rueda f - [mec] wheel position
— f — — f, anterior, o delantera - [autom]
front wheel position
— f — — f, posterior, o trasera - [autom]
rear wheel position

posición f de sistema m - [comput] system status
— f — — m informada - [comput] reported system status
— f — soldadura - [sold] weld(ing) position
— f — sujetador m - [mec] cam, o holder, position
— f — tablilla f - [mec] board position
— f — — f con circuito m (estampado) -
[electrón] (printed) circuit board position
— f — — f — — m impreso - [electrón] printed circuit board position
— f — tornillo m - [mec] screw position •
auger position
— f — — m sin fin m - [mec] auger position
— f — — m — — m en giro m - [mec] auger
swing position
— f — — m — — m — rotación f - [mec] auger
swing position
— f — — m — — m para descarga f - [agric-
-equip] unloading auger position
— f — — m — — m — — f en giro m - [mec]
unloading auger swing position
— f — — m — — m — — f en rotación f -
[mec] unloading auger swing position
— f — tubería f - [tub] pipe position
— f — — f para rotación f - [petról] drill-
(-ing) pipe position
— f — tubo m - [tub] pipe position
— f — — m para fundente m - [sold] flux tube
position
— f — — m para salida f para fundente m -
[sold] flux tube outlet position
— f — válvula f - [válv] valve position
— f — — f verificada - [válv] checked valve
position
— f — vanguardia f - leadership position
— f — — f mundial - world(wide) leadership
position
— f — vástago m - [mec] rod position
— f — — m para llama f - [combust] flame rod
position
— f — — m — — f para quemador m - [combust]
burner flame rod position
— f — volante m - [mec] flywheel position •
[autom-mec] steering wheel position
— f desconectada - [electr-oper] open position
— f deseada - desired position
— f descendente - [sold] downhill position |
adv - (en) downhill positioned
— f desengranada - [mec] disengaged position
— f destacada - outstanding position
— f detenida - stopped, o static, position
— f determinada - [mec] determined, o established, position
— f para piñón m - [mec] pinion('s) determined, o determined pinion('s), position
— f disparada - [electr-oper] tripped position
— f elevada - [mec] raised position
— f embutida - [mec] (plugged) in position
— f en calificación f - [deport] qualifying position
— f campeonato m - [deport] championship,
position, o standing
— f categoría f - category, position, o
standing
— f — f (para) producción f - [deport]
production standing
— f — f general - [deport] overall standing
— f esfera f - [instrum] dial setting
— f giro m - [mec] swing position
— f — m para tornillo m (sin fin m) - [mec]
auger swing position
— f — m — — m sin fin m para descarga f -
[mec] unloading auger swing position
— f que se embarca - [transp] shipping position
— f — — expide - [transp] shipping position
— f rotación f - [mec] swing position
— f — f para tornillo m (sin fin m) - [mec]
auger swing position
— f — f — m — — m para descarga f -

[mec] unloading auger swing position
posición f encontrada - conflicting position
— f **engranada** - [mec] engaged position
— f̄ **equivocada** - wrong position
— f̄ **erguida** - upright, o standing, position
— f̄ **establecida** - established position | adv -
[con] positioned
— f̄ — **automáticamente** - automatically estab-
lished position | adv - (con) positioned au-
tomatically
— f **estática** - static position
— f̄ **estratigráfica** - [geol] stratigraphic position
— f **extendida** - extended, o out, position
— f̄ **final** - final position • erect(ed) position
— f̄ — . . . - [deport] . . . place finish
— f̄ — **primera** - [deport] first place finish
— f̄ **financiera** - [fin] financial position
— f̄ **floja** - [mec] released position
— f̄ **flotadora** - [agric-equip] floating position
— f̄ — **para cabezal m** - [agric-equip] header
float(ing) position
— f — **espigadora f** - [agric-equip] header
float(ing) position
— f **geográfica** - [geogr] geographic position
— f̄ **hacia abajo** - down(wards) position
— f̄ — **arriba** - up(wards) position
— f̄ — **derecha f** - right hand position
— f̄ — **izquierda f** - [mec] left hand position
— f̄ **horizontal** - horizontal position • [sold]
flat, o level, position | adv - [sold] (en)
flat positioned
— f **inclinada** - [sold] inclined, o slanting,
position | adv - [sold] (en) inclined, o
slanting, position(ed)
— f — **ascendente** - [sold] uphill position |
adv - (en) uphill positioned
— f̄ — **descendente** - [sold] downhill position |
adv - (en) downhill positioned
— f̄ **incorrecta** - incorrect, o wrong, position
— f̄ **indebida** - improper, o wrong, position
— f̄ **indicada** - indicated position
— f̄ — **para abrazadera f** - [mec] clamp location
— f̄ **inferior** - low(er), o down, position
— f̄ **intermedia** - intermediate position
— f̄ **invertida** - reverse(d), o inverted, o
wrong, position • adv - upside down
— f **levantada** - raised position
— f̄ **levemente descendente** - [sold] slightly
downhill position
— f **longitudinal** - [mec] longitudinal, o
lengthwise, position
— f — **cero** - [mec] zero lengthwise position
— f̄ **mala** - [deport] bad showing
— f̄ **más adelantada** - [mec] full forward position
— f̄ — — **posible** - [sold] extreme forward position
— f **más alta** - highest, position, o setting
— f̄ — **baja** - lowest, position, o setting
— f̄ — **atrasada** - [mec] full rear position
— f̄ — **avanzada** - [mec] extreme, o most, for-
ward(most), position
— f — **baja** - lowest position
— f̄ — **retrasada** - [mec] extreme rear(most) position
— f **máxima** - [instrum] high setting | adv -
(en) on maximum
— f — **para retroceso m** - [mec] farthest back
position
— f **medio abierta** - [mec] half open position
— f̄ — **cerrada** - [mec] half closed position
— f̄ **mejor** - best position
— f̄ **mejorada** - improved position
— f̄ **menos favorable** - less favorable position |
adv - [deport] (en) down (in @ standing)
— f̄ **mínima** - [mec] low setting | adv - (en) on
minimum
— f **misma** - same position
— f̄ **natural** - natural position
— f̄ **negativa** - [instrum] (electrode) negative
(setting)
— f **neta** - [fin] net holding
— f̄ **neutra(l)** - neutral position

posición f neutral para palanca f - [mec] lever
neutral position
— f — — — **f para cambio(s) m para marcha f** -
[mec] gear shift lever neutral position
— f **nivelada** - [mec] level(ed) position •
[sold] | adv - [sold] level positioned
— f — **final** - [ind] final level(ed) position
— f̄ **normal** - normal position
— f̄ — **para operación f** - [mec] normal running
position
— f **original** - original position
— f̄ **otra que plana f** - [sold] out of position
— f̄ **para abrazadera f** - [mec] clamp, position,
o location
— f̄ — **aceleración f** - [mec- grúas] acceleration setting
— f̄ — **aparcamiento m** - [autom] park(ing) position
— f̄ — **arranque m** - [comb.int] start(ing), o
start up, position
— f̄ — — **m únicamente** - [sold] start(ing) only
position
— f̄ — **ascenso m o descenso m de pasajero(s) m**
[aeron] gate position
— f̄ — **avance m** - [mec] forward position
— f̄ — **bloqueo m** - [mec] blocking position
— f̄ — — **m completo** - [mec] complete locking,
o full locked, position
— f̄ — **caída f** - [mec] fall(ing) position
— f̄ — — **f libre** - [grúas] free fall(ing) position
— f̄ — — **f para avión(es) m** - [aeron] aircraft loading position
— f̄ — — **f y descarga f para avión(es) m** -
[aeron] aircraft loading (and unloading) position
— f̄ — — **f** — — **f de, mercadería(s), o mer-
cancía(s) f** - [aeron] cargo gate
— f̄ — **cilindro m** - [mec] cylinder position
— f̄ — — **m para elevación f** - [mec] lift(ing)
cylinder position
— f̄ — **combustible m** - [comb.int] fuel(ing) position
— f̄ — — **m pleno** - [comb.int] full fuel position
— f̄ — **conmutador m** [electr-instal] switch position
— f̄ — **corona f** - [mec] ring gear position
— f̄ — — **f dentada** - [mec] ring gear position
— f̄ — **detención f** - [mec] stop(ping) position
— f̄ — **electrodo m** - [sold] electrode position
— f̄ — — **m negativo** - [sold] electrode negative, o negative electrode, position
— f̄ — — **m positivo** - [sold] electrode positive, o positive electrode, position
— f̄ — **elevación f** - [mec] lifting, o hoisting,
position
— f̄ — **embarque m** - [transp] shipping position
— f̄ — **estacionamiento m** - [autom] park(ing)
position
— f̄ — **estrangulación f** - [comb.int] choke,king
position
— f̄ — **estrangulador m abierto** - [mec] choke
open(ed) position
— f̄ — — **m contenido** - [mec] choke holding
position
— f̄ — **flotación f** - float(ing) position
— f̄ — **frío m** - cold position
— f̄ — — **m mayor** - [mec] coldest position
— f̄ — **funcionamiento m** - [mec] run(ning), o
operating, position
— f̄ — **gas m** - [sold] gas position
— f̄ — — **m inerte** - [sold] inert gas position
— f̄ — . . . **hora(s) f** - . . . o'clock position
— f̄ — **impulsión f** - [mec] drive,ving position
— f̄ — **inserción f** - [tub-instal] stabbing position
— f̄ — **largada f** - [deport] starting position •
grid, position, o spot
— f̄ — — **f sacada** - [deport] drawn starting
spot
— f̄ — **marcha f** - [mec] run(ning) position
— f̄ — — **f atrás** - [autom] reverse, o back-up,

position
posición f para marcha f lenta - [mec] idle, ⌣
slow, speed position
— f — — f **sin carga f** - [mec] (s)low idle
position
— f — — f **plena** - full, speed, o load, position
— f — — f **sin carga f** - no, charge, o
load, position
— f — — f **rápida** - [mec] fast idle position
— f — — f **sin carga f** - idle speed position
— f — **montaje m** - [mec] mounting position •
mounted attitude
— f — **mordedura f** - [mec] gripping position
— f — **operación f** - [mec] operating, o running, position • setting
— f — — f **sin carga f** - [comb.int] no load
speed position
— f — **operar v** - [ind] operating position
— f — — v **pistola** - [sold] gun operating position
— f — **palanca f** - [mec] lever position
— f — — f **para cambio(s) m en marcha f** -
[mec] gear shift lever position
— f — **parada f** - [mec] stop(ping) position
— f — **prueba(s)** - test(ing) position
— f — **regulación f** - control(ling) position
— f — — f **para máquina f** - [sold] machine position
— f — **pera f** - bulb position
— f — **plancha f** - [mec] plate position
— f — **polaridad f** - polarity position
— f — — f **deseada** - [electrón] desired polarity position
— f — **puesta f en marcha f** - [comb.int] start-
(ing), position, o office
— f — **regulación f** - [electr-instal] control-
-(ling) position
— f — — f **alta** - [comb.int] high (control)
position
— f — — f — **sin carga f** - [comb.int] high
idle position
— f — — f **automática** - [comb.int] automatic
position
— f — — f — **sin carga f** - [comb.int] automatic idle position
— f — — f **baja** - [comb.int] low idle position
— f — — f — **sin carga f** - [comb.int] low
idle (control) position
— f — **trabajo m** - [sold] work(ing), o operating, position
— f — — m **para pistola f** - [sold] gun operating position
— f — **traslación f** — [mec] translation position
— f — — f **cero** - [mec] zero translation position
— f — — f **longitudinal** - [mec] lengthwide, o
longitudinal, translation position
— f — — f **transversal** - [mec] crosswise, o
transverse, translation position
— f — **velocidad f** - speed position
— f — — f **alta** - full, o high, speed position
— f — — f **baja** - [mec] low, speed, o range,
position
— f — — f **para máquina f** - machine
speed position
— f — — f — **motor m** - [ind] engine
speed position • [electr-mot] motor speed position
— f — — f **lenta f** - [mec] slow speed position
— f — — f — **de motor m** - [omb.int] slow
engine speed position • [electr-mot] motor
engine speed position
— f — — f — **sin carga f** - [comb.int] low
idle position
— f — — f **reducida** - [comb.int] slow engine
speed position
— f — **peor** - worse position • worst position
— f — **plana** - [sold] downhand position • level, o
flat, position
— f — **planeada** - planned position
— f — **positiva** - [instrum] positive position •
[sold] electrode positive (setting)

posición f posterior - [mec] rear position
— f **precisa** - precise, o particular, setting, o
position
— f **predeterminada** - predetermined position
— f **preeminente** - preeminent, o leadership, position
— f **preferida** - preferred position
— f **prefijada** - preset, o preestablished, position
— f **prevista** - [constr-pil] design position
— f **primera** - first position • [deport] top
slot; fastest position; lead
— f — **retenida** - [deport] retained lead (position)
— f **que corresponde a desde cinco f hasta siete
f en reloj m** - [instrum] five to seven
o'clock position
— f **reciente** - [comput] recent status
— f **reclinada,nante** - reclining, o lying, position
— f **regulada** - controlled position | a - (con)
positioned
— f **remota** - remote, position, o location
— f **rígida** - rigid position
— f **segura** - safe, position, o location
— f **semiabierta** - half open position
— f **sentada** - seated,ting position
— f **sobrecabeza** - [sold] overhead position
— f **sólida** - solid position • strong position
— f **suelta** - loose position | adv - [en] swing-
ing freely
— f **superior** - [mec] up(per) position
— f **típica** - typical position
— f — **para sostén m** - [constr] typical sup-
port(ing) position
— f **totalmente abierta** - [mec] full(y) open(ed)
position
— f — **aplicada** - [mec] full(y) on position
— f — **bajada** - [mec] full(y), o completely,
lowered position
— f — **cerrada** - full(y) closed position
— f — **desaplicada** - [mec] full(y) off position
— f — **(a)floja(da)** - [mec] full(y) released
position
— f **transversal** - [mec] crosswise, o transverse, position
— f — **cero** - [mec] zero crosswise position
— f **verificada** - checked position
— f **vertical** - vertical, o up(right), position
| adv - (en) upright; vertical
— f — **ascendente** - [sold] vertical up position
— f — **de cabeza f** - [sold] vertical head position
— f — **descendente** - [sold] vertical down position
— f **yaciente** - lying position
— f **zaguera** - [deport] back, pack, o position
posicionador m - [mec] positioner; placer •
[sold] jig • [metal-lam] (coil) positioner
— m **electroneumático** - electropneumatic positioner
— m **para bobina(s) f** - [metal-lam] (coil) positioner
— m — **chapa(s) f** - [sold] sheet positioner
— m — **soldadura f** - [sold] welding jig
posicionamiento* m - [mec] positioning
— m **de rodillo m** - [metal-lam] roll positioning
posscionar* v - [mec] to position • [electr-
-opler] to operate (@ switch) • **[sold] to set
up** • véase **poner v en posición f**
positivo,va a - . . . . • concrete • innovative
posponer v - . . . . • to delay
pospuesto,ta a - postponed; delayed
post venta a - after @ sale
posta f - [transp] stop
postcalentamiento m - véase **poscalentamiento m**
postcomprimido,da a - [constr] poststressed
poste m - [constr] pillar; support; post; pole •
shank; pillar; support • [constr] upright;
vertical member • standard
— m **abatible** - [constr] breakaway post
— m **circular** - [constr] circular, pole, o post
— m **colocado** - [constr] placed post

poste m creosotado - [constr] creosoted, o creosote treated, post
— m cuadrado - [constr] square, post, o pole
— m de acero m - [constr] steel, post, o pole
— m — m para alumbrado m - [constr] steel light(ing) pole
— m — cepa f - [constr] stub post
— m — hormigón m - [constr] concrete post • [telecom-instal] concrete pole
— m — — m armado - [constr] reinforced concrete post
— m — madera f - [constr] wood(en), post, o pole
— m   tipo m abatible - [constr] breakaway (design) post
— m — — m no abatible - [constr] nonbreakaway design post
— m — — m — — para alumbrado m - [constr] nonbreakaway design light pole
— m — tubería f - [metal-fabr] pipe post
— m — — f con corrugación f helicoidal - [constr] Hel-Cor pipe post
— m — — f Hel-Cor—[constr] Hel-Cor pipe post
— m — — f para indicador(es m - [constr] (standard) pipe post
— m — — f para indicador(es) m - [constr] standard pipe post
— m débil - [constr] weak post
— m doble - [constr] double post
— m esquinero - [constr] corner, post, o pole
— m esquinero de hormigón m armado - [constr] reinforced concrete corner post
— m estándar - [constr] standard post
— m — de tubería f - [constr] standard pipe post
— m extremo - [constr] tail, o end, post
— m final - [constr] end, o terminal, o tail, post
— m frontal - [constr] front post
— m — para rueda f motriz - [mec] drive,ving wheel front post • [petról] front jack post
— m fuerte - [constr] strong post
— m — con voladizo m - [constr] blocked-out, o cantilevered, strong post
— m — de acero m - [constr] strong steel post
— m gemelo - [constr] double, o twin, post
— m — de tubería f - [constr] double pipe post
— m — — — f para indicador(es) m - [constr] double pipe sign post
— m — — de tubería f - [constr] double standard pipe post
— m grúa - [grúas] pedestal crane
— m — hidráulico - [grúas] hydraulic pedestal crane
— m imprimado - [constr] primed, post, o pole
— m — con pintura f - paint-primed pole
— m independiente - [constr] independent pole
— m maestro - [mec] master, o Samson, post
— m magnético - [constr] magnetic, post, o pole
— m metálico - [constr] metal, post, o pole
— m — para alumbrado m - [constr] metal light pole
— m hexagonal - [constr] hexagon(al) pole
— m octogonal - [constr] octagonal pole
— m para alumbrado m - [constr] light, pole, o post, o standard
— m para amarre m - [constr] mooring post • buckstay - [petról] anchor
— m — anclaje m - [constr] anchor post
— m — apoyo m - [constr] support (pole) • [petról] headache post
— m — baranda f - [constr] rail post
— m — — f para puente m - [constr] bridge rail post
— m — barrera f - [constr] barrier post
— m — cabeza f - [constr] head, o terminal, o dead end, post, o pole
— m — cerca f - [constr] fence post
— m — cimiento(s) m - [constr] foundation post
— m — — m para torre f - [petról] derrick foundation post
— m — — m — — f para perforación f - [petról] (drilling) derrick foundation post

poste m para defensa f - [vial] guardrail post
— m — desconexión f - [petról] breakout post
— m — desenroscadura f - [petról] breakout post
— m — desenroscamiento m - [petról] breakout post
— m — desvío m - [constr] corner post
— m — detener balancín m - [petról] headache post
— m — enroscadura,camiento - [petról] make-up post
— m — fijación f - [constr] fastening pole
— m — guía f - [constr] guide, o locator, pole, o post
— m — indicador(es) m - [constr] sign, o standard, post
— m — letrero(s) m - [constr] sign post
— m — línea f - [constr] line post
— m — — f de hormigón f - [constr] [constr| reinforced concrete line post
— m — malacate m - [petról] wheel post
— m — m
— m — — m para herramienta(s) f - [petról] bull wheel post
— m — — m — tubería f - [petról] calf wheel post
— m — puente m - [constr] bridge post
— m — retención f - [mec] backup post
— m — rueda f - [mec] wheel, post, o jack
— m — — f fija - [mec] fixed wheel post
— m — — f motriz - [mec] drive,ving, jack, o post
— m — sujeción f - fastening, pole, o post
— m — tope m - [petról] headache post
— m — torno m - [tornos] heel post • [petról] wheel post
— m — viga f - [vial] rail post
— m — — f para defensa f - [vial] guardrail post
— m — pluma - [grúas] jib pole
— m portaherramientas - [torno] tool post
— m posterior - [mec] rear post
— m — para rueda f - [mec] wheel rear post
— m — — — f motriz - [mec] drive wheel rear post • [petról] rear jack post
— m primero - [mec] first post
— m — en línea f - [mec] first line post
— m rígido - [constr] rigid post
— m situador - [mec] locator,ting post
— m terminal - [constr] (dead) end, o terminal, post, o pole
— m tronchable - [constr] brekaway post
— m único - [constr] single post
— m — de tubería f - [constr] single pipe post
— m — — — f para indicadore(s) m - [constr] single standard sign post
— m — estándar - [constr] single standard post
— m — — de tubería f - [constr] single standard pipe post
— m vertical - [constr] vertical post
— m — de acero m - [constr] vertical steel post
postear* v = [contab] véase asentar v; entrar v
postemulsificador* m - post emulsifier
postensado,da a - [constr] poststressed
postensar* v - [constr] to post stress
postergación f - . . .; deferring; postponement
postergado,da a - deferred
postergar v - . . .; to defer
posterior a - . . .; further; later
— a a embutición f - [metal-fabr] after drawing
— adv a estampación,ado - [metal-fabr] after drawing
posteriormente adv - . . .; later (on) thereafter; afterwards
postigo m para ventanilla f - [constr] window shutter
postizo m - . . . - [mec] insert
postizo,za a - . . .; slip-on
postor m - [com] . . .; proponent
postrefrigerador m - [ind] aftercooler
postrimería(s) f - . . . | adv - (en) late in
postulante m - [labor] candidate
— m favorecido - [com] prime candidate

postura f aceptada - [com] successful bid
— f favorecida - [com] succesful, o favored, bid
— f final - final position
— f seria - hard-line stance
potabilización f - [hidr] . . .; purification
— f de agua m - [hidr] water purification
potable a - [hidr] . . .; drinking
potasa f cáustica - [quím] caustic potash
pote m - . . . • [metal-prod] thimble - [metal-prod] pipe box
— m ahusado - [válv] tapered pot
— m alimentador - [ind] feeding pot
— m — con centrífuga(dora) f - [ind] cyclone feed(ing) pot
— m — — f con velocidad f variable - [ind] variable speed cyclone feed(ing) pot
— m armado - [válv] assembled pot
— m limpio - [válv] clean pot
— m montado - [válv] assembled pot
— m para aleación f - [sold] solder pot
— m — alimentación f - [ind] feed(ing) pot
— m para cinc m - [metal-trat] (zinc) pot
— m — escoria f - [metal-prod] slag, thimble, o pot
— m — estañadura f - [metal-trat] tinning pot
— m — galvanización f - [metal-trat] galvanizing, pot, o kettle
— m — nivel m para regulación f - [metal-prod] bottom water level regulator
— m — — m y regulación f para venturi m - [metal-prod] Venturi scrubber bottom water level regulator
— m — sellar,llado - seal(ing) pot
— m — soldadura f - [sold] solder(ing) pot
— m — válvula f - [válv] valve pot
— m — — f ahusado - [válv] tapered valve pot
— m — — f armado - [válv] assembled valve pot
— m — — f colocado - [válv] installed valve pot
— m — — f desarmado - [válv] disassembled, o unassembled, valve pot
— m — — f para aspiración f - [válv] suction valve pot
— m — — f instalado - [válv] installed valve pot
— m — — f limpiado - [válv] cleaned valve pot
— m — — f limpio - [válv] clean valve pot
— m — — f montado - [válv] assembled valve pot
— m — — f purgado - [válv] drained valve pot
— m — venturi m - [metal-prod] Venturi pot
— m — volteo m - [ind] tilting pot
— m — — m frontal - [ind] front tilting pot
— m purgado - [válv] drained pot
potencia f - . . .; strength; force; energy; power • degree • [electr] size; rating • [ind] horsepower; output • [electr] rating
— f absorbida - [electr-distrib] absorbed power • [ind] power required,ement(s) • [sold] input current
— f — para, estirado m, o trefilería f - [alambre] wiredrawing power requirement
— f adicional - additional, o extra, power
— f alta - [ind] high, power, o rating
— f amplia - [ind] ample power
— f aportada - [ind] (power) input
— f apropiada - [electr-mot] correct, o appropriate, o sufficient, power, o size
— f asignada - [electr-oper] rated, capacity, o current, o energy, o power
— f asignada para entrada f - [sold] rated input current
— f auxiliar - [electr-oper] auxiliary power
— f — disponible - [electr-oper] available auxiliary power; auxiliary power available
— f — en corriente f alterna - [electr-oper] alternating current auxiliary power
— f — — — f continua - [electr-oper] direct current auxiliary power
— f baja - [electr-oper] low, power, o rating
— f bruta—[electr-oper] gross, rating, o power
— f calorífica - [combust] heat(ing) value

potencia f centralizada - [ind] centralized (horse)power
— f comparable - [electr-mot] comparable size
— f comprobada - [electr-oper] checked power
— f considerable - [mec] significant power
— f constante - [electr-prod] constant potential
— f contra freno m - [comb.int] brake,king horsepower
— f correspondida - [mec] matched horsepower
— f de compresor m - [mec] compressor rating
— f — automóvil m - [autom] car('s) power
— f — generador m - [electr-prod] generator's power
— f — máquina f - [mec] machine, size, o power
— f — motor m - [electr-mot] motor, rating, o size, o (horse)power • [comb.int] engine, rating, o (horse)power
— f — — m contra freno m - [comb.int] engine braking power • [electr-mot] motor braking power
— f — — m para elevación f - [grúas] hoisting motor rating
— f — — m en H P - [electr-mot] motor horse power rating • [comb.int] engine horsepower rating
— f — — m — — — para funcionamiento m intermitente - [electr-mot] motor horsepower intermittent rating
— f — — m — — — según N E M A - [electr-mot] N E M A, o National Electrical Manufacturer's Association, motor horsepower rating
— f — — m según N E M A - [electr-mot] N E M A, o National Electrical Manufacturer's Asociation, motor rating
— f — régimen - [mec] horsepower rating
— f — — m estipulada - [mec] given rating
— f — reserva - [mec] reserve horsepower
— f — resistencia f - [constr] bearing power
— f — soldadora f - [sold] welder, o machine, sized, o rating
— f — — f en amperio(s) m - [sold] machine, o welder, size, o rating, in amperes
— f — turbina f - [turb] turbine rating
— f desarrollada - [mec] developed horsepower
— f disponible - [ind] available, capacity, o rating, o (horse)power
— f efectiva - [mec] brake,king horsepower • [electr-mot] rated output
— f elevada - high rating
— f emésima - [mat] mth power
— f en amperio(s) m - [electr-mot] current rating
— f — barra f para tiro m - [constr] draw bar horsepower
— f — — f tracción f - [constr] draw bar horsepower
— f — H P para funcionamiento m intermitente - [electr-oper] horsepower intermittent rating
— f enésima - [mat] nth power
— f establecida - [electr-mot] rated current
— f especificada - [electr-oper] specified power
— f excéntrica - [mec] eccentric power
— f exigida - [mec] required horsepower • [electr-prod] required power system
— f — desarrollada - [mec] developed required horsepower
— f fijada - [electr-prod] rated current
— f garantizada - [electr-oper] rated output
— f grande - high power
— f hidrostática - [mec] hydrostatic power
— f indicada - [electr-oper] current rating
— f — para trabajo m continuo - [sold] continuous rating
— f máxima - [mec] maximum, energy, o power • [electr-oper] maximum rating
— f — contra freno m - [mec] maximum braking horsepower
— f — en borne m - [electr-instal] maximum terminal(s) rating
— f — instantánea - [electr-prod] maximum surge power
— f — para entrada f - [electr-distrib] maxi-

potencia f máxima para salida f   - 1154 -

mum, input, o incoming, power
potencia f máxima para salida f - [electr-
distrib] maximum output horsepower
— f **mayor** - [mec] greater, o higher, o increased, rating, o (horse)power
— f — **de motor m** - [comb.int] higher engine
horsepower
— f **menor** - [mec] lower, o decreased, rating, o
horsepower
— f — **de motor m** - [comb.int] lower engine
horsepower
— f **mermada** - [mec] decreased, o lost, power
— f **mínima** - [electr-mot] minimum, rating, o
energy, o (horse)power
— f — **contra freno m** - minimum brake,king
horsepower
— f — **en borne(s) m** - [electr-oper] minimum
terminal(s) rating
— f **neta** - [mec] (net) output • net rating
— f — **en volante m** - [mec] net flywheel rating
— f **nominal** - [mec] nominal energy • rated energy • nominal rating • stated horsepower •
[sold] rated output
— f — **contra freno m** - [mec] rated brake,king
horsepower
— f — **de motor m** - [electr-mot] stated motor
horsepower
— f — — **soldadora f** - [sold] welder rating
— f — **en borne(s) m** - [electr-oper] terminals
nominal rating
— f — — **exceso m** de . . . - [mec] rating over
. . . | adv - (con) rated at over . . .
— f — **para entrada f** - [sold] rated input current
— f — — — f **para perforación f** - [petról]
rated drilling input horsepower
— f **para acoplamiento m hidráulico contra freno**
m - [mec] hydraulic coupling braking power
— f — **funcionamiento m continuo** - [electr]
continuous rating
— f — — m **intermitente** - [electr-oper] intermittent rating
— f — — — m **para operación f** - [petrol]
rated drilling horsepower
— f — — **salida f** - [sold] rated output
— f — — — f **para fuerza f motriz** - [sold]
rated power output
— f — — — — f — **soldadura f** - [sold] rated
welder output
— f — **alimentación f** - [electr-oper] input
current
— f — **entrada f** - [electr-oper] input (horse)-
power • input (current)
— f — — f **para perforación f** - [petról] drilling input horsepower
— f — **estirado m** - [trefil] wiredrawing power
— f — **excavación f** - [constr] digging power
— f — **impulsión f** - [mec] drive,ving power
— f — — f **requerida** - [sold] required
drive,ving power
— f — **perforación f** - [petról] drilling horsepower
— f — **salida f** - [mec] output (horse)power
— f — **sustentación f** - [constr] support(ing)
power
— f **perdida** - lost power
— f **requerida** - required power
— f — **para estiramiento m de alambre n** -
[alambre] wiredrawing power requirement
— f **resistente** - [constr] bearing power
— f **según N E M A** - [electr-oper] N E M A, o
National Electrical Manufacturer's Association, rating
— f **suficiente** - [electr-oper] sufficient,
power, o size
— f **superior** - extra power
— f — **a** . . . - power above . . .
— f **ultraelevada** - [electr-oper] ultrahigh
power
potencial m - . . . . • rationale • [hidr] force
— m **a tierra f** - [electr-instal] ground potential
— m **anódico** - [tub] anode potential

potencial m catódico - [tub] cathode potential
— m **cinético** - [fís] kinetic potential
— m **constante** - [electr-oper] constant potential
— m — **regulado** - [sold] controlled constant
potential
— m — — **electrónicamente** - [sold] electronically controlled constant potential
— m **efectivo** - [electr-oper] effective, o actual, potential:
— m **eléctrico** - [electr-oper] electric potential
— m **en corriente f alterna** - [electr-oper] alternating current potential
— m — — f **continua** - [electr-oper] direct
current potential
— m — **más** - [electrón] plus potential
— m — **superficie f** - [electrón] surface potential
— m **exportador** - [com] exporting potential
— m **humano** - [econ] human potential
— m **neutral** - [electr-instal] neutral potential
— m **orientado** - [econ] oriented potential
— m **real** - [electr-oper] actual potential
— m **térmico** - heat, o thermal, potential
— m **variable** - [electrón] variable potential
— m **variante** - [electrón] varying potential
— m — **en superficie f** - [electrón] varying
surface potential
potenciómetro m ajustado - [electrón] adjusted
potentiometer
— m **conectado** - [electr-oper] closed potentiometer
— m **desconectado** - [electr-oper] open potentiometer
— m **electrónico** - [instrum] electronic potentiometer
— m — **registrador** - [instrum] recording electronic potentiometer
— m — — **para velocidad f alta** - [instrum]
high speed recording electronic potentiometer
— m **magnético** - [instrum] magnetic potentiometer
— m **para amplitud f** - [electrón] amplitude potentiometer
— m — — f **ajustado** - [electrón] adjusted amplitude potentiometer
— m — **nivel m de tono(s) m** - [electrón] tone
level potentiometer
— m — **regulación f** - [instrum] control potentiometer
— m — — f **para voltaje m** - [electrón] voltage
control potentiometer
— m — — — m **conectado** - [electr-oper]
closed voltage control potentiometer
— m — — f — — m **desconectado** - [electrón]
open voltage control potentiometer
— m — **voltaje m** - [instrum] voltage potentiometer
— m **registrador** - [instrum] recording potentiometer
— m — **con pluma f** - [instrum] pen recording
potentiometer
pozo m - . . . . • perforation - [metal-lam] low
spot • pit | adv - [de] well (produced)
— m **abajo** - [petrol] downhole
— m **abandonado** - [petról] junked hole
— m **activo** - [petról] operating well
— m **artesiano** - [hidr] artesian well
— m **brotante** - [hidr] flowing well
— m **colector** - [tub] catch basin
— m **(con) compensador** - [comb.int] accelerating
well
— m — **diámetro m reducido** - [petról] slim hole
— m — **profundidad f grande** - [constr] deep
well
— m — — f **reducida** - [constr] shallow well
— m **contenido** - [petról] capped well
— m **corriente** - [petról] conventional well
— m **de alquitrán m** - [petról] tar well
— m **costa afuera** - [petról] offshore well
— m **descontrolado** - [petról] wild well
— m **descubridor** - [petról] pilot well

pozo m en erupción f - [petról] wild well
— m — escoria f - [sold] slag hole
— m — producción f - flowing well
— m — tierra f - [petról] land well • conventional well
— m fuera de gobierno m - [petról] wild well
— m gasífero - [petról] gas well
— m hincado - [constr] driven well
— m húmedo - [constr] wet pit
— m improductivo - [petról] dry well
— m iniciado - started well
— m marino - [petról] offshore well
— m múltiple - [petról] multiple well(s)
— m negro - [sanit] cesspool; cesspit
— m para acceso m - [constr] access, o entry,
well, o shaft
— m — m para cloaca f - [constr] sewer entry shaft
— m — m — mina f - [constr] mine entry
shaft
— m — m — túnel m - [constr] tunnel access shaft
— m — agua m - [hidr] water well
— m — alivio m - [hidr] relief well
— m — m de tubería f perforada - [hidr]
perforated pipe relief well
— m — ascensor m - [constr] elevator shaft
— m — aspiración f - [metal-prod] sump •
[hidr] wet well
— m — f para bomba f - [metal-prod] pump
sump
— m — f — f para depuración f (de gases
m) - [metal-prod] gas washer pump sump
— m — barrenado m - [constr] shaft
— m — m circular - [constr] circular shaft
— m — bombeo m - [bombas] pump(er) (well)
— m — carga f - [transp] loading pit
— m — cargar montacarga(s) m - [metal-prod]
skip pit
— m — carro m - [metal-prod] car pit
— m — m para montacarga(s) m - [metal-prod]
skip car pit
— m — ceniza(s) f - [metal-prod] cinder pit
— m — colada f - [metal-prod] casting, o pouring, pit
— m — combustión f - [metal-prod] (stove) combustion, chamber, o pit
— m — f para estufa f - [metal-prod]
(stove) combustion chamber
— m — condensado(s) m - [ind] condensate sump
— m — confluencia f — [hidr] junction structure
— m — conexión f con tierra f - [electr-inst]
ground(ing) well
— m — cubeta f - [metal-prod] skip car pit
— m — drenaje m - [hidr] drain(age), pit, o
pocket
— m — escoria f - [metal-prod] slag, o cinder,
pit
— m — exploración f - [petról] wildcat well •
— m — explosión f - [explos] blast(ing) hole
— m — gas m - [petról] gas well
— m — inspección f - floor pit
— m — petróleo m - [petról] oil well
— m — purga f - [metal-prod] draining pit •
slag pit • blow off well
— m — recogida f - [hidr] sump
— m — separación f - drop(ping) pit
— m — skip m - [metal-prod] skip pit
— m — sondaje,deo - [mec] trial bore pit
— m — temperatura f - temperature pit
— m — f para cojinete m - [mec] bearing
temperature pit
— m — termómetro m - [hormig] thermometer well
— m — tubería f - [tub] pipe,ping well
— m — f perforada - [hidr] perforated pipe
well
— m — f — para alivio m - [hidr] perforated pipe relief well
— m — m para expulsión f - [cald] blowoff
pit
— m — vagoneta f - [ind] car pit
— m — f para montacarga(s) m - [metal-prod]
skip car pit

pozo m para ventilación f - [miner] air shaft •
[constr] shaft
— m — f ascendente - [miner] upcast
— m perforado - [constr] drilled well
— m — para alivio m - [hidr] perforated relief well
— m petrolífero - [petról] oil well
— m profundo - [constr] deep well
— m seco - [hidr] dry well • [constr] dry pit
— m submarino - [petról] subsea well
— m terrestre - [petról] land well
— m torpedeado - [petról] shot well
— m único - single well
pozos m varios - miscellaneous, o multiple, well
práctica f - . . . • way; manner • experience •
performance • training • [com] policy
— f a seguir(se) v - [ind] practice to be followed
— f acostumbrada - customary practice
— f actual - current, o present, practice
— f — en taller m - [ind] modern shop practice
— f — más correcta - best current practice
— f abnormal - abnormal practice
— f anterior - earlier, o prior, o previous,
practice, o experience
— f aplicable - [ind] applicable practice
— f aprobada - approved practice
— f apropiada - [ind] appropriate, o good,
practice
— f — para instalación f - proper installation
practice
— f — laminación f - [metal-lam] proper
rolling practice
— f avanzada - [ind] advanced practice
— f buena - good practice
— f buena para horno m - [ind] good furnace
practice
— f — — instalación f - [ind] good installation practice
— f — — f de conductor(es) m - [electr-
-instal] good wiring practice
— f — — f eléctrica - [electr-instal] good
electrical (installation) practice
— f — para línea f para conducción f [tub]
good pipeline practice
— f carente de seguridad f - [segurid] unsafe
practice
— f competente - good practice
— f común - common practice
— f con horno m eléctrico - [metal-prod] electric furnace practice
— f conocida - known practice
— f conservadora - conservative practice
— f contable - [contab] accounting, practice, o
policy
— f — conformada - [contab] approved, o conformed, accounting practice
— f — permitida - [contab] allowed accounting
practice
— f continuada - continued, o repeated, practice, o experience
— f convencional - conventional practice
— f — para proyección f - conventional de-
sign(ing) practice
— f conveniente - convenient, o good, practice
— f correcta - correct practice
— f — para ingeniería f - sound engineering
practice
— f correspondiente - corresponding, o respective, practice
— f corriente - [com] standard, practice, o
policy • [ind] common, o normal, practice
— f — actual - current common practice
— f — para taller m - [ind] common shop practice
— f crediticia - [fin] credit practice
— f de dumping m - [pol] dumping practice
— f — empresa f - [com] company, practice, o
policy,cies
— f deficiente - [ind] defective practice
— f desconocida - [ind] unknown practice
— f en nave f para colada f - [metal-prod] cast
house practice

práctica f especial - [ind] special practice
— f **establecida** f - [ind] established practice
— f̄ — **de empresa** f - [ind] established company practice
— f **estadounidense** f - [ind] American practice
— f̄ — **actual** - [ind] current American practice
— f̄ **estándar** - [ind] standard practice
— f̄ — **para fabricación** f - [ind] standard manufacturing practice
— f — — — f **de arrabio** m - [metal-prod] standard pig iron production practice
— f — **para producción** f - [ind] standard production practice
— f — — — f **de arrabio** m - [metal-prod] standard pig iron production practice
— f **exigida** - [labor] required practice
— f̄ **falto de seguridad** f - [segurid] unsafe practice
— f — — — f **probable** - [segurid] likely unsafe practice
— f **fiscalizada** - [ind] controlled practice
— f̄ — **con precisión** f - [ind] closely controlled practice
— f — **para producción** f **de acero** m - [metal-prod] controlled steel (production) practice
— f **general** - [ind] general practice
— f̄ **inapropiada para instalación** f - improper installation practice
— f **industrial** - [ind] industrial, o shop, practice
— f **limitada** - limited, o restricted, practice
— f̄ **local** - local practice
— f̄ — **satisfactoria** - satisfactory local practice
— f **mala** - [ind] bad practice
— f̄ **metalúrgica** —[metal] metallurgical practice
— f̄ — **establecida** - [metal] established metallurgical practice
— f **moderna** - [ind] modern practice
— f̄ **normal** - [ind] normal practice
— f̄ — **para fabricación** f - [ind] normal, o standard, manufacturing practice
— f — **para producción** f - [ind] normal production practice
— f **observada** - observed practice
— f̄ **operativa** - [ind] operating practice
— f̄ — **para cambio** m **de rodillo(s)** m - [metal-lam] roll change operating practice
— f — — **laminador** m - [metal-lam] mill operating practice
— f — — — m **para temple** m - [metal-lam] temper mill operating practice
— f **para aceración** f - [metal-prod] steel making practice
— f — — f **fiscalizada** - [metal-prod] controlled steel making practice
— f — **acería** f - [metal-prod] steel plant practice • open hearth practice • melting practice
— f — **acero** m **caliente** - [metal-lam] hot steel practice
— f — **adición(es)** f - [metal-prod] addition(s) practice
— f — — f **de aluminio** m - [metal-prod] aluminum addition(s) practice
— f — — f **exotérmica** - [metal-prod] exothermic addition(s) practice
— f — — f **estimulador** m **para efervescencia** f - [metal-prod] rimming stimulator additions practice
— f — **ajuste(s)** m - [mec] fitting practice • [seguros] adjustmkent making
— f — — m **de cono(s)** m - [cojin] cone(s) fitting practice
— f — — m **para taza** f - [cojin] cup fitting practice
— f̄ — **alcantarilla(s)** f - [constr] culvert practice
— f — **apareamiento** m - [mec] fitting practice
— f̄ — **auditoría** f - [contab] auditing practice
— f̄ — **bobinado** m—[metal-lam] coiling practice
— f̄ — **calentamiento** m - heating practice
— f̄ — — m **aprobada** - [metal-trat] approved heating practice

práctica f̄ **para calentamiento** m **de lingote(s)** m - [metal-lam] ingot heating practice
— f — **cambio** m **de rodillo(s)** m - [metal-lam] roll change,ging practice
— f̄ — **carga** f - loading practice
— f̄ — — f **para torpedo(s)** m - [metal-prod] hot metal car loading practice
— f — **colada** f - [metal-prod] pouring practice
— f̄ — — f **anormal** - [metal-prod] abnormal casting practice
— f — — f — f **continua** - [col.cont] continuous abnormal casting practice
— f — **colocación** f **de mazarota(s)** f - [metal-prod] hot top placing practice
— f — **combustión** f - [combust] combustion practice
— f — **configuración** f - layout practice
— f̄ — — f **para instalación** f - installation layout practice
— f — **compra** - buying, o purchasing, practice
— f̄ — **construcción** f - [constr] construction practice
— f — **descascarillado** m - [metal-lam] descaling practice
— f — **eliminación** f - elimination, o removal, practice
— f̄ — **escarpado** m - [metal-trat] scarfing practice
— f — **estimulador** m **para efervescencia** f - [metal-prod] rimming stimulator practice
— f — **fabricación** f - [ind] fabrication, o manufacturing, practice, o procedure
— f — — f **de arrabio** m - [metal-prod] pig iron production practice
— f — **fosa(s)** f **(para calentamiento** m**)** - [metal-prod] (soaking) pit practice
— f — — f **cerrada** - [metal-lam] closed pit practice
— f — **gunitado** m - [metal-prod] guniting practice
— f — **horno(s)** m - [combust] furnace practice
— f̄ — — m **alto** - [metal-prod] blast furnace practice
— f — — m **buena** - [combust] good furnace practice
— f — — m **mala** - [combust] bad furnace practice
— f — **ingeniería** f - engineering practice
— f̄ — **instalación** f - installation practice
— f̄ — — f **para conductor(es)** m - [electr-instal] wiring practice
— f — **laminación** f - [metal-lam] (rolling) mill practice
— f — **línea** f **para conducción** f - [tub] pipeline practice
— f — **lingote(s)** m - [metal-lam] ingot practice
— f — — m **caliente(s)** - [metal-lam] hot ingot practice
— f — — m **frío(s)** m - [metal-lam] cold ingot practice
— f — — m **semifrío(s)** - [metal-lam] semi-cold ingot practice
— f — **manejo** m - [ind] operation practice
— f̄ — — m **para bobinadora** f - [metal-lam] coiler operation practice
— f — — m — **regulación** f **automática** - [ind] automatic control operation practice
— f — **mantenimiento** m - [ind] maintenance practice
— f — **mazarota(s)** f - [metal-prod] hot top practice
— f — **muestreo** m - [ind] sampling practice
— f̄ — **operación** f - [ind] operating practice
— f̄ — — f **establecida** - [ind] established operating practice
— f — — f **para laminación** f - [metal-lam] rolling operation practice
— f — **preparación** f - preparation practice
— f̄ — — f **para muestra(s)** f - [ind] sample preparation practice
— f — **prerreducción** f - [miner] prereduction

practice
práctica f **para proceso** m - [ind] process practice
— f — **producción** f - [ind] production practice
• production, o operating, policy
— f — — f **de acero** m - [metal-prod] steel-(making) practice
— f — — f **de arrabio** m - [metal-prod] pig iron production practice
— f — **proyección** f - design practice
— f — — f **de alcantarilla(s)** f - [constr] culvert design practice
— f — **reducción** f - [metal-lam] reduction practice
— f — — f **directa** - [metal-prod] direct reduction practice
— f — **relleno** m - [constr] backfilling practice
— f — **remoción** f - [ind] removal practice
— f — — f **de cascarón** m - [metal-lam] scab removal practice
— f — **remover cascarón** m - [metal-lam] scab removal practice
— f — **rociadura** f - [ind] spray practice
— f — **sacar cascarón** m - [metal-lam] scab removal practice
— f — **seguridad** f - [segurid] safety practice
— f — — f **exigida** - [segurid] required safety practice
— f — — f **perturbadora** - [segurid] troublesome safety practice
— f — — f **problemática** - [segurid] troublesome safety practice
— f — **soldadura** f - [sold] welding practice
— f — — f **buena para línea** f **para conducción** f - [sold] good pipeline practice
— f — — f **con arco** m - [sold] arc welding pracice
— f — — f **para línea(s)** f **para conducción** f - [sold] pipeline practice
— f — **taller** m - [ind] shop practice
— f — **toma** f **de muestra(s)** f - [ind] sampling practice
— f — **trabajo** m - [ind] work, o operating,tion practice
— f — — m **establecida** - [inf] established, work, o operating,tion, practice
— f — **tratamiento** m **térmico** - [metal-trat] heat treatment practice
— f — **trazado** m - layout practice
— f — — m **para instalación** f - installation leyout practice
— f — **zunchado** m—[metal-lam] banding practice
— f **peligrosa** - [segurid] dangerous practice
— f **permitida** - allowed practice
— f **perturbadora** - troublesome practice
— f **pobre** - [ind] poor practice
— f — **para relleno** m - [constr] poor backfilling practice
— f **previa** - prior, practice, o experience
— f **probable** - likely practice
— f **profesional** - professional practice
— f **real** - actual practice
— f **recomendada** - recommended practice
— f — **para ensayo(s)** m - recommended testing practice
— f — — m **para aislación** f - [electr-inst] Recommended Practice for Testing Insulation
— f — — m **para resistencia** f **de aislación** f **para máquinas** f **con movimiento** m **rotativo** - [electr-instal] recommended practice for testing insulation resistance of rotating machinery
— f — **para instalación** f - recommended installation practice
— f — — **relleno** m - [constr] recommended backfilling practice
— f **respectiva** - [ind] respective practice
— f **restringida** - restricted practice
— f **sana** - [ind] sound practice
— f — **para combustión** f - [combust] sound combustion practice
— f **según norma** f - [ind] standard practice

práctica f **según norma** f **para fabricación** f - [ind] standard production practice
— f — — f **para arrabio** m - [metal-prod] standard pig iron production practice
— f — — f **para operación** f - [ind] standard operating practice
— f — — f **recomendada** - [ind] recommended standard practice
— f — — f **para diseño** m **de mezcla(s) de hormigón** m - [constr] Standard Recommendation Practice for the Design of Concrete Mixes
— f **segura para operación** f - [segurid] safe operating practice
— f **uniforme** - [com] uniform practice
— f **usual** - [ind] (standard) practice
— f — **para taller** m - [ind] (modern) shop practice
practicado,da a - practiced • made
prácticamente adv - . . . .; practically
— adv **nulo,la** - practically none; nil
practicar v **agujero** m - [mec] to, make, o drill, @ hole
— v **ajuste** m - to make @ adjustment
práctico,ca a - . . . .; handy; convenient; realistic; practicable • utilitarian
— a **desde punto** m **de vista económico** - economically practical
— a **económicamente** - economically practical
pradera f - . . . • turf
preagujereado,da a - [mec] prebored; predrilled
preagujerear v - [mec] to prebore; to predrill
preaireación f - [sanit] preaeration
prearmado m - [mec] preassembly,bling
— a **antes de instalación** f - [constr] preassembly,bling before @ installation
— m **de estructura** f - [constr] structure preassembly,bling
— m — **extremo** m - [mec] end preassembling,bly
— m — — m **achaflanado** - [constr] bevel(led) end preassembling,bly
— m **en fábrica** - [ind] factory preassembling
prearmado,da a - preassembled • preerected
— a **antes de instalación** f - [constr] preassembled before @ installation
— a **en fábrica** - [ind] factory preassembled
prearmar v - to preassemble
— v **antes de instalación** f - [constr] to preassemble before @ installation
— v **en fábrica** - [ind] to factory preassemble
— v **extremo** m - [mec] to preassemble @ end
— v — m **achaflanado** - [constr] to preassemble @ bevel(led) end
— v — **planta** f - to shop assemble
preavisado,da a - preadvised; prenotified
preavisar v - to notify (in advance)
preaviso m - (advance) notice
— m **breve** - [labor] short notice
— m **escrito** - [legal] written notice
precalentado,da a - preheated
— a **antes de conformación** f - [sold] preheated before forming
— a — **plegar** v - [sold] preheated before bending
— a — **soldar(se)** - [sold] preheated before @ welding
precalentador m - [ind] preheater
— m **gas-eléctrico** - [metal-trat] gas-electric prehater
— m **no oxidante** - [metal-trat] non-oxidizing preheater
— m **oxidante** - [metal-trat] oxidizing preheater
— m **para arranque** m - [comb.int] starter preheater
— m — **artesa** f - [col.cont] tundish preheater
— m — **cuchara** f - [metal-prod] ladle preheater
— m — **gas** m - [combust] gas preheater
— m — — m **para artesa** f - [col.cont] tundish gas preheater
— m — — m — **cuchara** f - [metal-prod] ladle gas preheater
— m **totalmente eléctrico** - [metal-trat] all, o fully, electric preheater
precalentamiento m - [ind] preheat(ing)

precalentamiento m afectado - affected preheating
— m antes de conformación f - [sold] preheating
before @ forming
— m —— plegadura f - [sold] preheating before bending
— m —— soldadura f - [sold] preheating before welding
— m con gas m - [combust] gas preheating
— m controlado - [sold] controlled preheat(ing)
— m correcto - [sold] correct preheating
— m para soldadura f - [sold] welding,dment
correct preheating
— m controlado - [sold] controlled preheating
— m —— cuidadosamente - [sold] carefully controlled preheating
— m de aire m - aire preheating
— m —— alambre m - [sold] wire preheat(ing)
— m —— artesa f - [col.cont] tundish preheating
— m —— gas m - [combust] gas preheating
— m —— m natural - [combust] natural gas
preheating
— m —— pieza f fundida - [sold] casting preheating
— m —— rodillo(s) m - [metal-lam] roll(s) preheating
— m evitado - [ind] avoided preheating
— m excelente - excellent preheating
— m insuficiente - insufficient preheating
— m máximo - [sold] maximum preheat(ing)
— m mayor - [sold] greater, o higher, preheating
— m menor - [sold] less preheat(ing)
— m mínimo - [sold] minimum preheat(ing)
— m —— recomendado - [sold] minimum recommended
preheat(ing)
— m necesario - [sold] needed preheating
— m para emergencia f - [ind] emergency preheating
— m para soldadura f - [sold] weld(ment) pre-
heat(ing)
— m regulado - [sold] controlled preheat(ing)
— m suficiente - [sold] sufficient preheat(ing)
— m uniforme - [sold] uniform preheating
precalentar v - to preheat
— v alambre m - [sold] to preheat @ wire
— v antes de conformación f - [sold] to preheat
before @ forming
— v —— plegadura f - [sold] tp preheat before @ bending
— v —— soldar,dadura f - [sold] to preheat
before @ weld(ing)
— v electrodo m - [sold] to preheat @ wire
— v gas m - [combust] to preheat @ gas
— v —— m natural - [combust] to preheat @ natural gas
— v tubería f - [tub] to preheat @ pipe
precalificación f - prequalifying,fication
precalificado,da a - prequalified
precalificar v - to prequalify
precarga f - preload(ing)
— f ajustada - [mec] adjusted preload(ing)
— f apreciable - appreciable preload(ing)
— f apropiada - proper preload(ing)
— f aumentada - increased preload(ing)
— f correcta - correct preload(ing)
— f —— para cojinete m - [mec] correct bearing
preload(ing)
— f —— —— m para diferencial m - [autom-mec]
correct differential bearing preload(ing)
— f corregida - correct preload(ing)
— f —— para cojinete m - corrected bearing pre-
load(ing)
— f disminuida - [mec] decreased, o reduced,
preload(ing)
— f final - final preload(ing)
— f —— para cojinete m - [mec] final bearing
preload(ing)
— f incorrecta - [mec] incorrect preload(ing)
— f —— para cojinete m - [mec] incorrect bearing preload(ing)
— f inicial - [mec] initial preload(ing)
— f medida - [mec] measured preload(ing)

precarga f para armado m - [mec] build-up pre-
load(ing)
— f —— m para ensayo(s) m - [mec] trial
build-up preload(ing)
— f —— cojinete m - [mec] bearing preload(ing)
— f —— m ajustada - [mec] adjusted bearing
preload(ing)
— f —— m para diferencial m - [autom-mec]
differential bearing preload(ing)
— f —— m —— m ajustada - [autom-mec] adjusted differential bearing preload(ing)
— f —— m —— piñón m - [mec] pinion bearing
preload(ing)
— f —— m —— m ajustada - [mec] adjusted
pinion bearing preload(ing)
— f —— m verificada - [mec] checked bearing
preeload(ing)
precargado,da a - preloaded
— a apropiadamente - [mec] preloaded properly
precargar v - [mec] to preload
— v apropiadamente - [mec] to preload properly
— v cojinete m - [mec] to preload @ bearing
precaución f - . . . • caution(ing) • effort
— f acostumbrada - [segurid] customary, o common, precaution, o safety
— f —— para seguridad f - [segurid] common
safety precaution
— f adicional - additional, o extra, precaution
— f contra lesión(es) f - [segurid] precaution,
o guarding, against @ injury,ries
— f cuidadosa - [segurid] extreme (pre)caution
— f especial - special precaution
— f excesiva - excessive precaution
— f necesaria - necessary precaution
— f observada - [segurid] observed precaution
— f para armado m - [mec] assembling precaution
— f —— manejo m - [mec] handling, o operating,
precaution
— f —— m de máquina f - [segurid] machine
operating,tion precaution
— f —— m —— soldadora f - [sold] welder operating,tion precaution
— f —— operación f - [mec] handling, o operating, precaution
— f —— f máquina f - [segurid] machine
operating,tion precaution
— f —— f —— soldadora f - [sold] welder operating,tion precaution
— f —— relleno m - [constr] backfilling precaution
— f —— preparación f - preparation precaution
— f —— f de base f - [constr] base preparation precaution
— f —— seguridad f - [segurid] safety precaution
— f —— f para soldadura f - [sold] welding
safety precaution
— f razonable - reasonable precaution
— f recomendada - recommended precaution
— f siguiente - [segurid] following precaution
— f técnica - technical precaution
— f —— especial - special technical precaution
— f tomada - taken precaution
— f usual - usual precaution
— f vinculada - related precaution
precaucional* a - véase precautorio,ria; preventivo,va
precaver v - . . . • to caution
— v contra lesión(es) f - [segurid] to guard
against @ injury,ries
— v —— porosidad f - [sold] to, resist, o
guard against, porosity
— v —— f en soldadura f - [sold] to, resist,
o guard against, @ weld porosity
precaverse v - to beware; to guard against
precavido,da - . . . .; cautioned
— a contra lesión(es) f - [segurid] beware of,
o guarded against, @ injury,ries
precedente m - . . . | a - . . .; former; above
preceder v - to precede; to be ahead
pecedentemente adv - previously • earlier; above
precedido,da a - preceded

preceptivamente - . . . . • mandatorily
preceptivo,va a - . . . . • mandatory
preciarse v - to, (take) pride, o treasure
precintador m - [ind] bander
precintadora f - [ind] bander
precinto m - [mec] . . .; strap
precio m - . . . . • price tag • outlay
— m a por menor - [com] retail price
— m actual - [com] actual, o real, price
— m — de costo m - [com] actual, o current,
cost price
— m — de chatarra f - [metal-prod] scrap, actual, o current, price
— m ajustado - adjusted price • escalated price
— m alto - high price
— m — para petróleo m - [petról] high oil, o
oil high, price
— m anterior - former, o previous, price
— m aproximado - approximate price
— m — en mercado m - [com] approximate market,
price, o cost
— m astronómico - [com] astronimic(al), o sky
high, price
— m — para chatarra f - [metal-prod] astrono-
mic(al), o sky high, scrap price
— m aumentado - increased, o escalated, price
— m bajo - low, price, o cost
— m — de petróleo m - [petról] low oil price
— m base - base price
— m básico - basic price
— m — para máquina - basic machine price
— m — — propuesta f - [com] basic, proposal,
estimated, price
— m bueno - [com] good price
— m — para chatarra f - [metal-prod] good
scrap price
— m combinado - combined, o composite, price
— m — para chatarra f - [metal-prod] composite
scrap, o scrap composite, price
— m — — — f comprada - [metal-prod] composite purchased scrap price
— m competidor - [com] competitive price
— m competitivo - competitive price
— m — en mercado m internacional - [econ] international level competitive price
— m compuesto - composite price
— m — para chatarra f - [metal-prod] composite
scrap, o scrap composite, price
— m — — — f comprada - [metal-prod] composite purchased scrap price
— m — promedio - composite average price
— m — — anual - composite average annual
price
— m — — — para chatarra f - [metal-prod]
composite average annual scrap price
— m — — para chatarra f - [metal-prod] composite average scrap price
— m consignado - given, o quoted, price
— m contractual - contract price
— m corriente - current price • market cost
— m — aproximado - [com] approximate current
cost
— m — — mercado m - [com] market current cost
— m — — — m exterior - [com] foreign market
current price
— m — — — m extranjero - [com] foreign market current price
— m — — — m interior - [econ] domestic market current price
— m — para chatarra f - [metal-prod] scrap
current price
— m — — — f comprada - [metal-prod] purchased scrap current price
— m costo m, seguro m y flete m - [com] cost,
insurance and freight price
— m — y flete m - [com] cost and freight price
— m cotizado - quoted price
— m definido - defined,nite price
— m definitivo - definitive price
— m desglosado - broken down price
— m detallado - detailed price
— m determinado - determined price
— m diario - daily price • daily rate

precio m difícil para igualar - hard to beat, o
rock bottom low, price
— m ecuánime - fair price
— m en curso m - current price
— m — en mercado m interior - [econ] de-
mestic market current price
— m — — m — — m mundial - [com] world market current price
— m — fábrica f - [ind] factory price
— m — fecha f para embarque - [com] shipping
date price
— m — mercado m exterior - [com] foreign market price
— m — — m extranjero - [com] foreign market
price
— m — — m interior - [econ] domestic market
price
— m — — m mundial - [com] world market price
— m — nivel m competitivo - competitive price
level
— m — — internacional - [econ] international level price
— m — obra f - price at @ (project) site
— m — planta f - [ind] plant, o mill, price
— m — — f principal - [ind] main, plant, o
mill, price
— m — plaza f - market, price, o value
— m especial - special price
— m extraordinariamente alto - exceptionally, o
surprisingly, high, price, o cost
— m — bajo - exceptionally, o surprisingly,
low, price, o cost
— F O B - véase precio m libre a bordo m
— m facturado - [com] invoiced price
— m favorable - favorable price • value pricing
— m fijo - fixed, o firm, price
— m — diario - fixed daily, price, o rate
— m — e inamovible - [com] fixed and firm
price
— m firme - fixed, o firm, price
— m global - global, o overall, price
— m implícito - [econ] implicit price
— m inferior - lower price • lowest price
— m — a todos menos uno - second lowest price
— m interno - internal price
— m — único - single internal price
— m libre a bordo m - free on board price
— m — — costado m - free along side, o F A S,
price
— m más alto - higher price • highest price
— m — bajo - lower price • lowest price
— m — reducido - lower price • lowest price
— m máximo - maximum price
— m mayorista - [com] wholesale price
— m mínimo - minimum price
— m minorista - [com] retail price
— m módico - modest price
— m no válido para venta f por separado - price
not for separate sale
— m nuevo - new price
— m ofertado - offered, o proposed, o bid,
price
— m oficial - official price
— m — en fábrica f - official factory price;
factory list price
— m ofrecido - offered, o bid, price
— m optativo - optional price
— m para caliza f - [miner] limestone price
— m — — f por tonelada - [miner] limestone
price per ton
— m — compra f - purchase price
— m — — f considerablemente inferior para
equipo m - [ind] substantially lower equipment, price, o cost
— m — — f inferior - lower (purchase) cost
— m — — f — para equipo m - [ind] lower
equipment cost
— m — — f invertido - [fin] invested purchase
price
— m — — menor - lower (purchase) cost
— m — consumidor m - consumer, o retail,
cost, o price
— m — coque m - [coque] coke price

precio m para coque m por tonelada f - [coque]
coke price per ton
— m — coste m y costas f - [constr] cost plus
price
— m — costo m - [com] cost price
— m — m actual - [com] current cost price
— m — chatarra f - [metal-prod] scrap price
— m — f comprada - [metal-prod] purchased
scrap price
— m — f marginal - [metal-prod] marginal
scrap price
— m — dólar m - [fin] dollar price
— m — economía f abierta - [econ] open economy
price
— m — f cerrada - [econ] closed economy
price
— m — equipo m - [ind] equipment price • machine price
— m — m nuevo - [ind] new equipment price
— m — exportación f - [com] export price
— m — factura f - [com] invoice price
— m — facturación f - [com] invoice,cing price
— m — hierro m - [metal] iron price
— m — m esponja - [metal] sponge iron price
— m — máquina f - [mec] machine price
— m — f básica - [mec] basic machine price
— m — f nueva - [mec] new machine price
— m — materia(s) f prima(s) - [ind] raw mate-
rial(s) price
— m — material(es) m - [ind] material(s) price
— m — mercadería f - [com] merchandise, o
goods, price
— m — mercado m - [com] market price
— m — m consumidor - [econ] consumer market
price
— m — mercancía f - [com] goods price
— m — mineral m - [miner] ore price
— m — m de hierro m—[miner] iron ore price
— m — m prerreducido - [metal-prod] prereduced ore price
— m — m reducido - [metal-prod] reduced ore
price
— m — neumático(s) m - [autom-neumát] tire(s)
price
— m — oferta f - offer, o bid(ding) price
— m — petróleo m - [petról] oil price
— m — piedra f - [constr] stone price
— m — f caliza - [miner] limestone price
— m — f — por tonelada - [miner] limestone
price per ton
— m — propuesta f - proposal, o bidding, price
— m — reposición f - replacement price
— m — reventa f - resale price
— m — seguro f - [seguros] insurance price
— m — soldadora f - [sold] welder price
— m — suministro(s) m - supply,lies price
— m — transporte m - [transp] transportation
price
— m — vendedor m - [com] vendor's, o seller's,
price
— m — venta f - [com] sales, o selling, price
— m — f de equipo m - [ind] equipment sale
price
— m parcial - partial price
— m por separado - separate price; priced sepa-
- rately
— m — tonelada f - price per ton
— m — trabajo m - job, o work, price
— m — unidad f - unit price
— m previo - previous, o prior, price
— m promedio - average price
— m propuesto - proposed, o quoted, price
— m razonavle - reasonable price | adv - (con)
reasonably priced
— m reajustable - [com] escalatable price
— m reajustado - escalated price
— m real - actual price
— m reducido - reduced price • low price
— m — para chatarra f - [metal-prod] low scrap
price • scrap low price
— m — mineral m - [miner] reduced ore, o
ore reduced, price
— m representativo - representative price

precio m revisable* - escalatable price
— m según lista f - [com] list price
— m sobre vagón m [ferroviario) - [transp]
price on (board) @ (railway) car
— m — m (—) en fábrica f - [transp] price
on board @ (railway) car at @ plant
— m sombra - shadow price
— m suelto—[com] single price • price separately
— m sugerido - suggested price
— m — para reventa f - suggested resale price
— m sujeto a reajuste m - [com] escalatable
price; price subject to escalation
— m superior - higher price • highest price
— m — a todos menos uno - second highest price
— m total - total price
— m — convenido - [com] total agreed price
— m — para equipo m - [ind] equipment total, o
total equipment, price
— m — oferta f - total offer price
— m — propuesta f - total, proposal, o bid,
price
— m — según lista f - total list price
— m único - single price
— m unitario - unit price
— m — optativo - optional unit price
— m variable - variable price
— m variado - varied price
— m vendedor - seller's price
— m viejo - old price
— m vigente - [com] current, o in vigor, price
— m — en mercado m - current market price
— m — m mundial - [com] world market
current,o current world market, price
precipcio m - [topogr] . . .; canyon • ragged
edge
— m peligroso - dangerous, o perilous, precipice, o drop off
precipitación f - . . . • cascading; falling •
[sanit] settling • [meteorol] precipitation;
rainfall; rain (water) • [nucl] fallout
— f abundante - [meteorol] abundant, o heavy,
rainfall
— f acumulada - accumulated precipitation
— f acuosa - [meteorol] liquid sunshine
— f anual - [meteorol] yearly rainfall
— f — media - [meteorol] average annual rainfall
— f calculada - [meteorol] estimated, o cálcu-
lated, rainfall
— f cresta - [meteorol] peak rainfall
— f de carburo m - [sold] carbide precipitation
— f — m de cromo m - [metal] chromium carbide precipitation
— f — m intercristalino - [metal] inter-
crystaline carbide precipitation
— f — cobre m - [miner] copper precipitation
— f — lluvia f - showring; rainfall
— f — metal - [metal-prod] metal precipitarion
— f en . . . minuto(s) m - [meteorl] . . . minute rainfall
— f escasa - [meteorol] scant precipitation
— f excesiva - [hidr] excess precipitation •
excess runoff
— f extraordinaria - [hidr] extraordinary
rainfall • peak rainfall
— f fuerte - [meteorol] heavy rainfall
— f horaria - [hidr] hourly rainfall; rainfall
per hour
— f — máxima - [meteorol] maximum rainfall per
hour
— f intensa - [meteorol] intense precipitation;
heavy rainfall
— f máxima - [meteorol] maximum rainfall
— f media - [meteorol] average rainfall • mean
precipitation
— f — para estación f - [meteorol] mean seasonal precipitation
— f para estación f - [meteorol] seasonal precipitation
— f perjudicial - damaging precipitation
— f — de carburo(s) m - [metal-prod] damaging
carbide precipitation

precipitación f pluvial - [meteorol] rainfall;
rain (water); storm water
— f — anual - [meteorol] annual rainfall
— f — — media - [meteorol] average annual
rainfall
— f — en localidad f - [meteorol] local rainfall
— f — escasa - [meteorol] light rainfall
— f — estimada—[meteorol] calculated rainfall
• design rainfall intensity
— f — media - [meteorol] average rainfall
— f prevista - [meteorol] expected rainfall
— f promedia - [meteorol] average rainfall
— f total - [meteorol] total rainfall
precipitado m - [quím] precipitate
precipitado,da a - . . .; precipitated • cas-
caded; fallen
precipitador m - [metal-prod] precipitator
— m (de) Cotrell - [electr-equip] Cottrell, o
electrostatic, precipitator
— m eléctrico - [electr-equip] electrical precipitator
— m electrostático - [electr-equip] electrostatic, o Cottrell, precipitator
— m por vía f húmeda - [ind] wet precipitator
— m — — f seca - [ind] dry precipitator
precipitante a - . . .; cascading
precipitar v - . . . • to cascade
— v lluvia f - [metal-prod] to shower
— v metal m - [metal-prod] to precipitate @
metal
precipitarse v - to, fall, o tumble
precipitoso,sa a - . . . • [topogr] steep
precisado,da a - specified; determined; indicated • pointed out; instructed • required
precisamente adv - . . . • right
precisar v - . . . • to, point out, o indicate •
to require
precisamente adv - . . .; accurately
precisar v - . . .; to, specify, o establish, o
point out • to instruct
— v material m - [ind] to require @ material
— v opción f - to select @ option
precisión f - . . . • specification,fying; determination,ning • specific information • re-
quirement | adv - (con) precisely; accura-
tely; fine
— f asegurada - [ind] assured accuracy
— f contractual - [legal] contract precision
— f — para báscula f - [ind] scales contract
precision
— f dimensional - dimensional precision
— f — alta - [metal-lam] high dimensional precision
— f inherente - [ind] built in accuracy
— f máxima - [mec] maximum precision • maximum
consistency
— f para balanza f - [ind] scales precision
— f — báscula f - [ind] scales precision
— f — frecuencia f - [electrón] frequency, accuracy, o precision
— f — método m - method precision
— f — opción f - option selection
— f — permeámetro m - [instrum] permeameter
accuracy
— f — — Iliovici - [instrum] Iliovici permeameter accuracy
— f — plantilla f - [mec] templet accuracy
— f — reproducibilidad f - [electrón] reproducibility precision
— f positiva - positive accuracy
— f razonable - reasonable precision
— f relativa - relative preciseness; closeness
| adv - (con) precisely; closely
— f total - complete, o total, accuracy
preciso,sa a - . . .; certain • particular •
actual • refined • fixed • required • sophisticated
precolado,da a - [metal-prod] precast
precolar v - [metal-prod] to precast
precoladura f - [metal-prod] precasting
precomprimido,da a - precompressed • prestressed
precomprimir v - to precompress • to prestress

preconizado,da a - favored; commended
preconizar v - . . .; to favor; to commend
preconvenido,da a - (previously) agreed upon
preconvenir v - (previously) agree on
precursor m - . . .; pioneer • prerunner
precurvado m - precurving
precurvado,da a - precurved
precurvar v - to precurve
predecible a - véase previsible
predecir v acertadamente - to predict accurately
— v aumento m de población f - [pol] to predict
@ population increase
— v carga f externa - [constr] to predict @ external load
— v con precisión f relativa - to predict
closely
— v deformación f - to predict @, deformation,
o deflection
— v efecto m - to predict @ effect
— v precisamente - to predict, precisely, o
accurately
— v ritmo m - to predict @ rate
— v — m de desgaste m - [mec] to predict @
wear rate
— v tasa f de desgaste m - [mec] to predict @
wear rate
predefinido,da - predefined; predetermined
predefinir v - . . .; to predefine
predepuración f - [mec] precleaning • [sanit]
pretreatment
predepurado,da a - [mec] precleaned
predepurador m - [mec] precleaner
predepurador,ra a - [mec| precleaner,ning
predepurar v - [mec] to preclean
predeterminación f - . . .; preselecting,tion
— f de curso m para acción f - action course
predeterminating,tion
— f — nivel m - level predetermination,ting
— f — precio m - price predetermining,tion
— f — velocidad f - [mec] speed predetermination,ting
predeterminado,da a - predetermined; predefined;
preselected • specified
predeterminar v - . . .; to preselect
— v curso m para acción f - to predetermine @
action course
— v nivel - to predetermine @ level
— v precio m - to predetermine @ price
— v velocidad f - [mec] to predetermine @ speed
predicado,da a - preached; expounded
predicador m - . . .; expounder
predicar v - . . .; to expound
predicción f - . . .; forecast(ing) • anticipation
— f acertada - accurate prediction
— f confirmada - confirmed prediction
— f de aumento n - increase prediction
— f — — en población f - [pol] population increase prediction
— f — carga f externa - [constr] external load
predicting,tion
— f — deformación f - [constr] deformation, o
deflection, predicting,tion
— f — incremento m - increment forecasting
— m — — m en potencia f - [electr-ioer] power
increment forecasting
— f — nivel m - [electr] power forecasting
— f — — m — — m en potencia f - [electr-
-oper] power increment level forecast(ing)
— f — m de potencia f - [electr] power
level forecast(ing)
— f — potencia f - [electr-oper] power fore-
cast(ing)
— f — ritmo m - rate prediction
— f — — m para desgaste m - [mec] wear rate
predicting,tion
— f — tasa f - rate predicting,tion
— f — — f para desgaste m - [mec] wear rate
predicting,tion
— f precisa - precise, o accurate, prediction
predictible* a - véase previsible a
predictivo,va* a - predictive
predicho,cha a - predicted; forecast

predicho,cha a acertadamente - predicted accurately
— a con precisión f relativa—predicted closely
— a precisamente - predicted, accurately, o precisely
predio m - . . .; land; real estate; premise(s) • property
— m colindante - adjoining property
— m industrial - [ind] industrial property
— m rural - rural property
prediseñado,da a - preengineered
predistribuidor m - [turb] predistributor
predominante a - . . .; dominant • pervasive
predominantemente adv - predominantly
— adv arcilloso,sa a - predominantly clayey
preeminencia f - . . .; leadership • status
preeminente a - . . . • leadership
preencendido m - [comb.int] preignition
preensamblado,da a - [mec] preassembled
preensambladura f - [mec] preassembly
preensamblar v - to preassemble
preensayar v - to pretest
preensayo m - pretest(ing)
— m de soldadura f - [sold] weld pretest(ing)
— m — — — f en serie - [sold] production weld pretest(ing)
— m significativo - meaningful pretest(ing)
— m vigoroso - vigorous pretest(ing)
preestablecer v tiempo m - to, preestablish, o set, o adjust, @ time,ming
preestablecido,da a - preestablished; predefined
preestudiado,da* a - véase estudiado,da de antemano
preestudio m - véase estudio m previo
prefabricado,da a - prefabricated • precast
— a en taller m - [ind] shop prefabricated
prefabricante m - [hormig] precaster
prefabricar v - to prefabricate • [hormig] to precast
prefactibilidad* f - prefeasibility*
prefacio m - [imprent] . . .; foreword
preferencia f - . . .; choice
— f de material m - material preferring,rence
— f — operador m - [ind] operator appeal
— f — soldador m - [sold] welder, o operator, preference
— f estudiantil - [educ] student preference
— f personal - personal, o own, preference
preferencialmente adv - preferentially
preferido,da a - preferred; chosen • favored
— a más a menudo - most often preferred
— a con más frecuencia - most often preferred
preferir v - . . .; to choose
— v material m - to prefer @ material
prefijación f - presetting; preestablishing
prefijado,da a - preset; preestablished
prefijar v - to preset; to preestablish
prefiltro m - [mec] prefilter
— m instalado - installed prefilter
preflexión f - [mec] prebending
— f de plancha f - [sold] plate prebending
prefloculación f - [hidr] preflocculation
preformación f - . . .; preforming
— f de cordón m - [cabl] strand preforming
preformado,da a - [mec] preformed; preshaped • [cabl] union formed
— a en obra - [hormig] site preformed
preformar v - to perform; to preshape
— v cordón m - [cabl] to preform @ strand
pregonero m - . . . .; announcer
pregunta f adicional - additional, o further, question
— f de cliente m - client('s), o customer('s), question
— f específica - specific question
— f importante - important question
— f provocativa - dynamic, o provocative, question
— f recibida - received question
— f repetida - repeated, o repetetive, question
— f repetidora - repetitive question
— f sobre aplicación(es) f - [ind] application question

pregunta f sobre dinámica f - [autom] dynamics question
— f — — f para vehículo m - [autom] vehicle dynamics question
— f — neumático(s) m - [autom-neumát] tire question
— f — — m para rendimiento m alto - [autom--neumát] high performance tire question
— f — pedido(s) m - [com] order(ing) question
— f — rueda(s) f - [autom] wheel question
— f — técnica f - [ind] technique question
— f — — f para aplicación f - [ind] application technique question
prehomogeneización f - prehomogenizing,zation
prehomogeneizado,da a - prehomogenized
prehomogeneizar v - to prehomogenize
prejuicio m - prejudging,gment; advance judging
— m de decisión f - decision prejudicing
— m sin fundamento m - unfounded prejudice
— m personal - personal prejudice
prejuzgado,da a - prejudged; judged in advance
prejuzgar v - . . .; to judge in advance
preliminar,rio,ria a - . . .; primary • prestart(ing)
— a a uso - prestart(ing); prior to @ use
prelimpiado,da a - [mec] precleaned
prelimpiador m - [mec] precleaner
— m para motor m - [comb.int] engine precleaner
prelimpiador,ra a - [mec] precleaner,ning
prelimpiar v - [mec] to preclean
prelimpieza f - precleaning
prelimpio,pia a - [mec] preclean
prelubricado,da a - prelubricated
prelubricar v - to prelubricate
preludiado,da a - preluded
preludiar v - to prelude
prematuro,ra a - . . .; early
premezcla f - [constr] premix(ing)
— f de arcilla(s) f - [constr] clays premixing
premzclado,da a - premixed; ready-mixed
premezclador m - premixer
premzclar v - to premix
— v arcilla(s) f - [cement] to premix @ clay(s)
premio m - . . . • award • [seguros] premium
— m a iniciativa f - suggestion award
— m — inventiva f - suggestion award
— m — personal m horario - [labor] hourly personnel incentive(s)
— m — — m obrero - [labor] hourly personnel incentive(s)
— m — todo personal m - [labor] general bonus
— m colectivo - [labor] colective, bonus, o premium
— m devengado - [seguros] earned premium
— m especial - special, award, o prize
— m general - [deport] overall honor(s)
— m gordo - grand prize
— m grande - grand prize; grand prix
— m — para resistencia f - [deport] grand endurance prize; endurance grand prix
— m habitual - [labor] habitual incentive
— m indirecto - [labor] indirect, premium o bonus
— m individual - [labor] individual, premium, o bonus
— m mayor - [deport] véase premio m grande
— m mensual - [labor] monthly bonus - [seguros] monthly premium
— m negativo - [labor] negative premium
— m neto - [seguros] net premium
— m no devengado - [seguros] unearned premium
— m obtenido - obtained premium
— m pagado - [seguros] paid premium
— m por cobrar - [seguros] premium receivable
— m — pagar - [seguros] premium payable
— m sobre póliza(s) f emitida(s) - [seguros] premium(s) written
premisa f básica - basic premise
— f — para diseño m - basic design premise
— f — — proyección f - basic design premise
— f original - original premise
— f para diseño m - design premise
— f seguida - followed premise

premoldeado,da a - premolded; preformed; precast
premoldear v - to, premold, o preform, o precast
premontado,da a - preassembled
— a antes de instalación f - [constr] preassembled before @ installation
premontaje m - [mec] premounting • preassembling,bly • sumbassembling,bly
— m antes de instalación f - [constr] preassembling,bly before @ installation
— m de pieza(s) f - [constr] part(s) preassembling,bly
— m —— f transversal(es) - [constr] transverse member(s) subassembling,bly
premontar v - to preassemble
— v antes de instalación f - [constr] to preassemble before @ installation
— v pieza(s) f - to preassemble @ part(s)
— v — transversal(es) - [constr] to premount @ transverse member(s)
premura f - . . . ; hurry(ing)
prenda f - . . . • [legal] chattel mortgage; security interest • [vest] garment
— f agraria - [legal] chattel mortgage
— f contaminada - [vest] contaminated garment
— f gruesa - [vest] thick, o heavy, garment
— f industrial - [fin] industrial chattel mortgage
— f libre de aceite m - [segurid] oil free garment
— f protectora - [segurid] protective garment
— f — libre de aceite m - [segurid] oil free protective garment
— f quitada - [vest] removed garment
— f suelta - [vest] loose, garment, o clothing
prender v - . . . ; to grip • to button • [metal-prod] to blow in • [combust] to, light, o ignite • [electr-oper] to turn on
— v firmemente - to hold, securely, o tightly
— v fuego m - [combust] to light @ fire
— v luz f - [electr-oper] to light @ lamp; to turn on @ light
— v — f indicadora - to light @ signal lamp
— v — para llamada f - [telecom] to light @ call light
prenderse v - [electr-oper] to come on • [mec] to snap on • [comput] to light up
— v gas m - [combut] to light @ gas
— v — m en tragante m - [ind] to ignite @ furnace top gas
prendido,da a - gripped; seized • [electr-oper] turned on • [comput] lit up
— m firmemente - held, o seized, securely
prendimiento m - . . . ; gripping • [comput] lighting up
— m de luz f para llamada f - [telecom] call light lighting (up)
— m seguro - secure holding
prensa f - . . . . • [mec] drop press • jack; gag • vise • [metal-fabr] brake • [metal-lam] baler
— f con tornillo m - [herram] vise; C clamp
— f — . . . unidad(es) f - [mec] . . .-unit press
— f conformadora - [ind] forming press
— f — en O - [tub] O forming press
— f — U - [mec] U forming oress
— f cortadora - [mec] cutting press
— f — muescadora - [mec] notch cutting press
— f dobladora - [mec] brake press
— f especializada - [imprent] trade press
— f estampadora - [ind] stamping, o forming, press
— f forjadora - [metal-fabr] forging press
— f hidráulica - [metal-fabr] hydraulic press
— f — para chatarra f - [metal-prod] hydraulic scrap press
— f hidrostática - hydrostatic press
— f inactiva - [ind] down, o inactive, press
— f manual - hand, o manual, press
— f mecánica - [mec] (mechanical) press
— f muescadora - [mec] notch(ing) press
— f operada - [mec] operated press

prensa f para alinear - [mec] aligning, o gag, press
— f para banco m - [mec] machine,nist's vise
— f — copero m - [metal-prod] dishing press
— f — corte m - [tub] cutoff press
— f — costado m - [mec; side press • check screw
— f — chatarra f - [metal-lam] scrap, press, o baler
— f — dos copas f - [extrusion] two-cup press
— f — embutir - [f.c.-ruedas] dishing, o conning, press
— f — extrusión f - [metal-fabr] extruding, o extrusion, press
— f —— f de electrodo(s) m - [extrusion] electrode(s) extruding press
— f —— en planta f para electrodo(s) m - [sold] electrode(s) plant extruding press
— f — filtrar - filer(ing) press
— f — forja f - [metal-fabr] hammer press
— f — forjadura f - [mec] forging press
— f — forjar v - [metal-fabr] forging press
— f —— v en caliente - [metal-fabr] drop forge,(ging press)
— f —— v rueda(s) f - [metal-fabr] wheel (forging) press
— f — guarnición(es) f - [mec] trim press
— f — heno m - [agric-equip] hay, press, o baler
— f —— m con impulsión f mecánica - [agric-equip] power hay, baler, o press
— f — industria f - [ind] industrial press
— f —— f para neumático(s) m - [autom-neum] tire trade press
— f — papel - [papel] paper press
— f —— m en bobina(s) f - [imprent] web press
— f — redondear v - [mec] rounding,dness press
— f —— v extremo(s) m - [tub] end roundness press
— f —— v — m de tubo(s) m - [tub] pipe end roundness press
— f — redondo(s) m - [mec] roundness press
— f — tubería f - [tub] pipe,ping press
— f — tubo(s) m - [mec] pipe press
— f preparada - [mec] set up press • [ind] tooled press
— f primitiva - [mec] early press
— f punzonadora - [ind] punch(ing) press • drop press
— f taladradora - [mec] drill(ing) press
— f técnica - technical press
— f troqueladora - [metal-fabr] punch(ing) press
— f U - [mec] U-press
— f usada - [mec] used press
— f vertical - [mec] vertical press
prensado* - m véase prensadura f
prensado,da a - [mec] pressed
— a bien - hard packed
— a cabalmente - hard packed
— a en frío - [ind] cold pressed
prensadura f cabal - hard packing
— f en frío - [mec] cold pressing
— f rápida - [mec] rapid pressing
prensaestopa(s) m - [mec] . . . ; (packing) gland
— m colocado - [mec] installed packing box
— m completo - [mec] complete packing box
— m con aceite m - [lubric] oil gland
— m deslizado - [mec] slipped packing box
— m destornillado - [mec] unscrewed packing box
— m empujado - [mec] pushed packing box
— m engrasado - [mec] greased packing box
— m instalado - [mec] installed packing box
— m integral - [mec] built-in, o complete, packing box
— m invertido - [mec] inverted packing box
— m para inundación f con aceite m - [petról] oil-flood stuffing box
— m — tubería f - [tub] pipe packing box
— m —— f para agua m - [petról] wash pipe packing box
— f — válvula f - [válv] valve packing box
— m —— f para regulación f - [metal-prod]

control valve packing box
prensaestopa(s) m̲ **para válvula** f̲ **para regulación**
f̲ **de descarga** f̲ - [metal-prod] discharge control valve packing box
— m̲ — **vástago** m̲ - [mec] rod, gland, o̲ packing
box
— m — — m **para campana** f̲ - [metal-prod] bell,
rod gland, o̲ packing box
— m̲, **quitado,** o̲ **removido,** o̲ **sacado** - [mec] removed packing box
prensar v̲ **bien** - [mec] to hard pack
— v̲ **cabalmente** - [mec] to hard pack
— v̲ **en frío** - to cold press
prensista m̲ - [ind] . . .; press operator
preocupación f̲ - . . .; concern
— f̲ **apropiada** - appropriate concern
— f̲ **causada** - caused concern
preocupado,da a̲ - . . . . • involved
preocupar v̲ - . . .; to concern • to involve
preoperativo,va* a̲ - **véase anticipado,da** a̲
preparación f̲ - . . . . • readiness,dying • developing,pment • training • [sold] preparation;
set-up; conditioning • [mec] warm-up •
dressing; tooling; set(ting) up; makeup;
making up • [miner] dressing • [ind] changeover • design(ing); engineering • [quím]
formulating,tion • [legal] execution •
[domést] packing • [comput] writing
— f̲ **aceptable** - [sold] acceptable preparation
— f̲ — **para junta** f̲ **a tope** - [sold] acceptable
butt joint preparation
— f̲ **apresurada** - hurried preparation
— f̲ **atrasada** - delayed preparation
— f̲ **avezada** - advanced preparation
— f̲ **básica** - basic preparation
— f̲ **bélica** - [milit] war preparation
— f̲ **beneficiosa** - beneficial preparation
— f̲ **buena** - good preparation
— f̲ — **para automóvil** - [autom] good, automobile, o̲ car, preparation
— f̲ **correcta** - correct preparation
— f̲ — **para alcantarilla** f̲ - [tub] correct culvert preparation
— f̲ — — f̲ **corrugada** - [tub] corrugated
culvert correct preparation
— f̲ — — f̲ **de acero** m̲ - [tub] steel culvert
correct preparation
— f̲ — — f̲ — — m **corrugado** - [tub] corrugated steel culvert correct preparation
— f̲ **cuidadosa** - careful preparation
— f̲ **de accesorio** m̲ - [tub] fitting fabrication
— f̲ — — m **en taller** m̲ - [mec] shop fitting
fabrication
— f̲ — **acta** m̲ - report, o̲ minutes, preparation
— f̲ — **alcantarilla** f̲ - [constr] culvert preparation
— f̲ — — f̲ **corrugada** - [constr] corrugated
culvert, preparation, o̲ fabrication
— f̲ — — f̲ **de acero** m̲ - [constr] steel culvert
fabrication
— f̲ — — f̲ — — m **corrugado** - [constr] corrugated steel culvert preparation
— f̲ — **arista(s)** f̲ - [sold] edge preparation
— f̲ — **automóvil** m̲ - [autom] car preparation
— f̲ — **base** f̲ - base preparation • [constr]
bedding
— f̲ — **boquilla** f̲ - [mec] nozzle preparation
— f̲ — **borde(s)** m̲ - [sold] edge preparation
— f̲ — — m **en V** - [sold] V-edge preparation
— f̲ — **cabeza** f̲ - [mec] head preparation
— f̲ — — f̲ **caliente** - [metal-prod] hot topping
— f̲ — **canal** m̲ - [metal-prod] spout preparation
— f̲ — — m **para sangría** f̲ - [metal-prod] tapping spout preparation
— f̲ — **canto** m̲ - [sold] edge preparation
— f̲ — — m **vivo** - [sold] feather edge preparation
— f̲ — **carbón** m̲ - [miner] coal preparation
— f̲ — **carga** f̲ - [Transp] load preparation •
[metal-prod] burden, o̲ charge, preparation
— f̲ — **cimentación** f̲ - [constr] foundation preparation

preparación f̲ **de cimiento(s)** m̲ - [constr] foundation preparation
— f̲ — **colada** [metal-prod] casting preparation
— f̲ — **componente** m̲ - component preparation
— f̲ — **compuesto** m̲ - [quím] compound preparation
— f̲ — — m **explosivo** - [explos] explosive
compound preparation
— f̲ — **cuchara(s)** f̲ - [metal-prod] ladle preparation
— f̲ — **cucharón(es)** - [metal-prod] ladle(s)
preparation
— f̲ — **chatarra** f̲ - [metal-prod] scrap preparation
— f̲ — **desecho(s)** m̲ - [nucl] waste preparation
— f̲ — — m **radiactivo(s)** m̲ - [nucl] radioactive waste(s) preparation
— f̲ — — m — **sólido(s)** - [nucl] solid radioactive waste preparation
— f̲ — **diseño(s)** m - drawing(s) preparation
— f̲ — — m **detallado(s)** - [dib] detailing
— f̲ — **disolución** f̲ - [quím] solution preparation
— f̲ — — f̲ **oxidante** - [quím] oxidizing solution preparation
— f̲ — **ensayo** m̲ - test, preparation, o̲ engineering
— f̲ — **equipo** m̲ — [ind] equipment preparation
— f̲ — **escenario** - [teatr] stage setting
— f̲ — **escoria** f̲ - [metal-prod] slag preparation
— f̲ — — f̲ **para horno** m̲ - [metal-prod] furnace
slag preparation
— f̲ — — f̲ — — m **Siemens-Martin** - [metal-
prod] open hearth slag preparation
— f̲ — **especificación** f̲ - specification preparing,ration
— f̲ — **estudio** m̲ - study preparation
— f̲ — — m **semejante** - similar study preparation
— f̲ — — m **similar** - similar study preparation
— f̲ — **extremo** m̲ - [mec] end preparation
— f̲ — — m **de tubería** f̲ - [tub] pipe end preparation
— f̲ — **floculante** m̲ - [hidr] flocculant preparation
— f̲ — **fluido** m̲ - fluid preparation
— f̲ — — m **regenerador** - regenerating fluid
preparation
— f̲ — **fundamento** m̲ - [constr] foundation preparation
— f̲ — **gráfico(s)** m̲ - graph(s) preparation
— f̲ — **horno** m̲ - [ind] furnace preparation
— f̲ — — m **alto** - [metal-prod] blast furnace
preparation
— f̲ — — m — **para detención** f̲ - [metal-prod]
blast furnace shutdown preparation
— f̲ — — m — — **prolongada** - [metal-prod]
blast furnace preparation for @ lengthy shutdown
— f̲ — — m — **parada** - [ind] furnace shutdown
preparation
— f̲ — — m — **puesta** f̲ **en marcha** - [ind] furnace start-up preparation
— f̲ — **hueco** m̲ - [mec] hole preparation
— f̲ — **informe** m̲ - [ind] report preparation
— f̲ — **junta** f̲ - [sold] joint preparation
— f̲ — — f̲ **a tope** - [sold] butt joint preparation
— f̲ — — f̲ **en canal** m̲ - [metal-prod] runner, o̲
spout, joint preparation
— f̲ — **lecho** m̲ - [constr] bed preparation; bedding
— f̲ — — m **para estructura** f̲ - [constr] structure bedding
— f̲ — **lingotera(s)** f̲ - [metal-prod] (ingot)
mold preparation
— f̲ — **máquina** f̲ - [ind] (machine) retooling •
[deport] race(r) (car) preparation
— f̲ — **material(es)** n̲ - material(s) preparation
— f̲ — **mazarota(s)** f̲ - [metal-prod] hot topping
— f̲ — **mezcla** f̲ - blend grading

preparación f de mina f - [miner] mine preparation
— f — mineral - [miner] ore, preparation, o dressing
— f interna - [mec] internal preparation
— f más fácil - easier preparation
— f — — de borde m - [sold] easier edge preparation
— f para colada f - [metal-prod] casting preparation
— f — embarque m— - [Transp] shipment preparation
— f — entrevista f - [labor] interview, o contact, preparation
— f — operación f - [ind] operation preparation
— f — parada f - [ind] shutdown preparation
— f — — f para horno m - [ind] furnace shutdown preparation
— f — pieza f - [ind] part(s) preparation
— f — — f nueva - [ind] new part preparation
— f — — f producción f - [ind] changeover
— f — — f de pieza(s) f nueva(s) - [ind] new part(s) changeover
— f — puesta f en marcha f - [ind] start-up, o starting, preparation
— f — — t — — f de horno m - [ind] furnace start-up preparation
— f — remesa,misión f - [transp] shipment preparation
— f — soldadura f - [sold] weld(ing) preparation
— f por separado - separate preparation
— f previa - previous, o prior, preparation • preplanning
— f prescri(p)ta—[sold] prescribed preparation
— f primorosa - elegant, o careful, preparation
— f rápida - [mec] quick, o fast, preparation, o set(ting) up
— f recomendada - recommended preparation
— f — para borde m - recommended edge preparation
— f — para canto m - [sold] recommended edge preparation
— f según norma f - standard(ized) preparation
— f separada - separate preparation
— f típica - [sold] typical preparation
— f y presentación f - [sold] preparation and fit-up
preparado,da a - prepared; readied; conditioned; ready • [legal] drafted • [quím] formulated • [mec] machined • set up; made up; (re)tooled • [ind] engineered • [domést] packed • [comput] written
— a bien - well prepared
— a cuidadosamente - prepared carefully
— a en laboratorio m - laboratory prepared
— a — taller m - [mec] shop fabricated
— a especialmente - prepared especially
— a fácilmente - [mec] set up easily
— a para carrera f - [deport] race prepared
— a — exportación f - [com] export prepared
— a — operación f - [ind] operation prepared
— a para puesta f en marcha f - [ind] start-up prepared
— a por separado - prepared separately
— a rápidamente - [mec] set up quickly
— a separadamente - prepared separately
preparador m - [ind] preparer • fixer
— m para barra(s) f taponadora(s) f - [metal-prod] stopper rods, preparer, o fixer
— m — bobina(s) f - [metal-lam] coil preparer
— m — cuchara(s) f - [metal-prod] ladle preparer
— m — — f de primera - [metal-prod] ladle preparation first helper
— f — — f segunda - [metal-prod] ladle preparation second helper
— m — equipo(s) m - [ind] equipment preparer
— m — lingotera(s) f - [metal-prod] ingot mold preparer
— m — palanquilla(s) f - [metal-prod] billet preparer

preparador m para parte m - report preparer
— m — placa(s) f - [metal-prod] stool fixer
— m — solera(s) f - [metal-lam] (soaking pit) bottom(s) fixer
— m — f para horno(s) m de fosa f - [metal-lam] soaking pit bottom(s) fixer
— m — tren(es) m - [f.c.] train preparer • [metal-lam] mill preparer
— m — vástago(s) m - [metal-prod] (stopper) rod(s) fixer
preparador,ra a - preparer; preparing
preparar v - . . .; to ready • to, finish, o terminate • to install • [legal] to draft • [mec] to, make, o set, up • [ind] to design; to engineer • [quím] to formulate • [domést] to pack • [comput] to write
— v alimentadora f - [mec] to prepare @ feeder
— v — f para alambre m - [sold] to prepare @ wire feeder
— v — f — — m con diámetro m distinto - [sold] to change @ wire size
— v automóvil m - [autom] to prepare @, automobile, o car
— v bártulo(s) m - [fam] to pack @ bag(s)
— v base f - [constr] to prepare @ base; to bed
— v — f para estructura f - [constr] to bed @ structure
— v bien - to prepare well
— v — automóvil m - [autom] to prepare well @, automobile, o car
— v borrador m - to (prepare @) draft
— v busa f - [metal-prod] to prepare @ blowpipe
— v carrera f - [deport] to prepare @ race
— v cimentación f - [constr] to prepare @ foundation
— v cimiento(s) m - [constr] to prepare @ foundation(s)
— v colada f - [metal-prod] to prepare @ cast(ing)
— v componente m - to prepare @ component
— v compuesto m - [quím] to prepare @ compound
— v — m explosivo - [explos] to prepare @ explosive compound
— v con escoria f - [metal-prod] to prepare with slag
— v cuidadosamente - to prepare carefully
— v cucharón m - [metal-prod] to prepare @ ladle
— v cuidadosamente - to prepare carefully
— v disolución f - [quím] to prepare @ solution
— v — f oxidante - [quím] to prepare @ oxidizing solution
— v embarque m - [transp] to prepare @ shipment
— v en forma f especificada - to prepare as specified
— v — taller m - [ind] to shop fabricate
— v — — m accesorio m - [mec] to shop fabricate @ fitting
— v equipo m - [ind] to prepare @ equipment
— v escenario m - [teatr] to set @ stage
— v especialmente - to prepare especially
— v específicamente - to, prepare, o write, @ specification, o specifically
— v estudio m - to prepare @ study
— v — m semejante - to prepare @ similar study
— v — m similar 0 to prepare @ similar study
— v extremo m - [mec] to prepare @ end
— v — m de tubo m - [tub] to prepare @ pipe end
— v fácilmente - to, prepare, o set up, easily
— m fundamento m - [constr] to prepare @ foundation
— v horno m - [combust] to prepare @ furnace
— v hueco m - [mec] to prepare @ hole
— v informe m - to prepare @ report
— v lecho m - [constr] to prepare @ bed
— v — m para estructura f - [constr] to bed @ structure
— v lingotera f - [metal-prod] to prepare @ ingot mold
— v máquina f - [ind] to (re)tool @ machine • to set up @ tooling • [deport] to prepare @ racer
— v modelo m - to prepare @ model

preparar v modelo m matemático - [matem] to prepare @ mathematical model
— v muela v - [medic] to prepare @ (molar) tooth • [med] to dress @ grinding wheel
— v nuevamente - to prepare again • [legal] to redraft
— v objetivo m - [admin] to prepare @ objective
— v para alambre m con diámetro m distinto - [sold] to change @ wire size
— v — carrera f - [deport] to prepare for @ race
— v — carretera f - [autom] to get ready to roll
— v — operación f - [ind] to prepare for @ operation
— v — puesta f en marcha - [ind] to prepare for @, start-up, o starting
— v — parada f - [ind] to prepare for @ shutdown
— v — f para horno m - [ind] to prepare @ furnace shutdown
— v parte f - to prepare @, part, o portion
— v — f de suelo m - [constr] to prepare @ soil (structure) portion
— v plancha f - [sold] to prepare @ plate • to use @ plate preparation
— v — f de hierro m fundido - [sold] to prepare @ cast iron plate
— v — f en forma f especificada - [sold] to prepare @ plate as specified • to use @ plate preparation as specified
— v plantilla f - [mec] to, set up, o assemble, @ templet
— v pliego m - [legal] to prepare @ sheet
— v — m con condición(es) f - [legal] to write @ specification(s)
— v por separado - to prepare separately
— v prensa f - [ind] to, prepare, o tool, @ press • to set up @ press
— v primorosamente - to prepare carefully
— v propuesta f - to prepare @ proposal
— v puesta f en marcha de horno m - [ind] to prepare @ furnace start-up
— v programa m - [comput] to write @ program
— v rápidamente - [mec] to, prepare, o set up, quickly
— v reclamo m - to prepare @ claim
— f remesa f - [fin] to prepare @ remittance • [transp] to prepare @ shipment
— v remisión f - [transp] to prepare @ shipping
— v secador m - [ind] to prepare @ drier
— separadamente - to prepare separately
— v tapa f - to, prepare, o make up, @ cover
— v tobera f - [metal-prod] to prepare @ tuyere
preparativo m para cimentación f - [constr] foundation preparation
preplancheado,da a - [metal-trat] preplated
preplanchear v - [metal-trat] to preplate
preplaneado,da a - preplanned
preplaneamiento m - proplanning
preplanear v to preplan
prepolímero m - [quím] prepolymer
— m formador de celdilla(s) f - [quím] cell forming prepolymer
— m — — — f cerrada(s) - [quím] closed cells forming prepolymer
preponderante a - . . . . • outstanding
preposicionamiento* m - prepositioning
— m de guía(s) f lateral(es) - [metal-lam] side guide(s) prepositioning
— m futuro - [ind] future prepositioning
prepotente a - . . . . • arrogant
preprocesado,da a - preprocessed
preprocesamiento m - preprocessing
— m de risiduo(s) m - [nucl] waste(s) preprocessing
— m — — m sólido(s) m - [nucl] solid waste(s) preprocessing
preprocesar v - to preprocess
— v residuo(s) m - [nucl] to preprocess @ waste
— v — m sólido(s) - [nucl] to preprocess @ solid waste(s)
preproceso m - preprocess(ing)

prerreducción f - [miner] prereduction,cing
— f de mineral m - [miner] ore prereduction,cing
— f — — m de hierro m - [miner] iron ore prereduction,cing
prerreducido m - [miner] prereduced iron; véase también mineral m prerreducido
— m con proporción f alta de hierro m - [metal-prod] high content prereduced ore
— m — ley f alta de hierro m - [metal-prod] high conent prereduced ore
prerreducido,da a - [metal-prod] prereduced
prerregulación f - [mec] presetting
— f de embolada(s) f - [mec] stroke presetting
— f — número m - number presetting
prerregulado,da a - preset
prerregular v embolada(s) - to preset @ stroke(s)
— v número m - to preset @ number
presa f - . . . . • [hidr] . . .; reservoir; embankment • farm pond • weir section
— f con altura f escasa - [hidr] low head dam
— f construida - [constr] built, o constructed, dam
— f de roca f - [hidr] rock (fill) dam
— f — terraplén m - [hidr] earth fill dam
— f — — m de tierra f - [hidr] eath fill dam
— f — tierra f - earth-fill, o earthen, dam • sump(le) hole
— f — — f encespedada - [hidr] sodded earth dam
— f drenada - [hidr] drained dam
— f encespedada - [hidr] sodded dam
— f grande - [hidr] large dam
— f móvil - [hidr] movable dam
— f para conservación f - conservation dam
— f — — f de suelo(s) m - [hidr] soil saving dam
— f — derivación f - [hidr] diversion dam
— f — desviación f - [hidr] diversion, dam, o reservoir
— f — lodo(s) m - [petról] sump hole
— f — regulación f - [hidr] check(ing) dam
— m — — f para erosión f - [hidr] erosion check(ing) dam
— f — retención f - [hidr] nonoverflow dam • nonoverflow section
— f pequeña f - [hidr] small dam
— f principal - [hidr] main dam
— f sumergible - [hidr] weir
— f permanente - [hidr] permanent dam
— vertedero f - [hidr] drop structure
prescindencia f - abstention; abstaining • waiving
prescindido,da - abstained • waived
prescindir v - . . .; to prescind • to, abstain, o waive • to cancel
prescribir v procedimiento m - to prescribe @ procedure
— v — m para muestreo m - [ind] to prescribe @ sampling procedure
prescripción f - . . . . • [legal] (statute of limitations)
prescri(p)to,ta a - prescribed
presecado m - predrying
presecado,da a - predried
presecarse v - to predry
preselección f - preselecting,tion
preseleccionado,da a - preselected
preseleccionar v - to preselect
presencia f - . . . . • attendance • [legal] witnessing
— f de aceite m - oil presence
— f — actividad f - activity presence • [nucl] radioactivity presence
— f — agua m - [hidr] water presence
— f — — freática - [hidr] groundwater presence
— f — ánimo m - . . . • alertness • confidence
— f — — m demandada - tested confidence
— f — — m exigida - tested confidence
— f — condición(es) f - condition(s) presence
— f — — anormal(es) - abnormal condition(s) presence
— f — defecto m - defect, presence, o existence

presencia f de estado - government's presence
— f — gobierno m - government's presence
— f — humedad f - moisture presence
— f — nitrato m - [quím] nitrate's presence
— f — polvo m - dust presence
— f — — m conductor m - conductive dust presence
— f — punto m - point witnessing
— f — radiactividad f - [nucl] radioactivity presence
— f — ruido(s) m - [comput] noise presence
— f en inspección f - [ind] inspection witnessing
— f industrial - [ind] industrial presence
— f orientadora - orienting presence
presenciación f - witnessing
— f de ensayo(s) m - test(s) witnessing
— f — prueba(s) f - test(s), o trial(s), witnessing
presenciado,da a - witnessed; watched; observed
presenciar v - to witness; to watch; to observe
— v carrera f - [deport] to watch @ race
v ensayo m - to witness @ test
— v inspección f - to witness @ inspection
— v prueba f - to witness @, test, o trial
presentación f - . . .; presenting; introducing; introduction; discussing; discussion - producing; production • submitting; submittal • filing - feature,ring • display,ing; format • producing,ction • submitting; submittal • appearance • report; paper • popping up • [sold] fit-up; position(ing) • [com] launching • [imprent] layout
— f a cliente(s) m - [com] customer introduction
— f actualizada - [ind] updated presentation
— f ajustada - [sold] tight fit-up
— f — necesaria - [sold] required tight fit-up
— f apropiada - [sold] proper fit-up
— f audiovisual - audiovisual presentation
— f comercial - commercial introduction
— f de característica f - chareceritic, o feature, introduction
— f — certificado n - [legal] certificate presenting,tation
— f — componente m - component presentation
— f — documentación f documentation presentation
— f — documento m - document, o statement, presentation, o rendering
— f — — m para embarque m - [transp] shipping document presentation
— f — factura f - [com] invoice presentation
— f — informe m - report, presentation, o filing, o submitting, o rendering
— f — junta f - [sold] joint fit-up
— f — oferta f - [legal] proposal, o bid, presentation, o submitting; bidding
— f — pieza(s) f - [sold] part(s) fit-up
— f — producto m - [com] product, presentation, o introduction
— f — — m nuevo - [com] new product, presentation, o introduction
— f — propuesta(s) f - [com] bidding; proposal(s), o bid(s), presentation, o submittal
— f — reclamación f - claim filing
— f — — f escrita - written claim filing
— f — reclamo m - claim filing
— f — — m escrito - written claim filing
— f — saludo(s) m - greeting(s) presentation
— f debida - [sold] proper fit-up
— f detallada - detailed, presentation, o discussion
— f feliz - successful presentation
— f final - [deport] final appearance
— f ideal - [sold] ideal fit-up
— f impropia - [sold] improper fit-up
— f inapropiada - [sold] improper fit-up
— f mayor - [com] large(est) launch(ing)
— f mejor - better presentation • [sold] better fit-up
— f normal - [sold] normal fit-up
— f para aprobación f - approval presentation

presentación f para prueba(s) f - test(ing) presentation
— f — soldadura f - [sold] weld(ing) position
— f — — f plana - [sold] downhand weld(ing) position
— f pobre - [sold] poor fit-up
— f previa - prior, o previous, presentation
— f primera - first presentation
— f técnica - technical presentation
— f variable - [sold] varying fit-up
— f visual - visual presentation
— f y preparación f - [sold] fit-up and presentation
presentado,da - presented; introduced • produced • appeared • featured; displayed; discussed; submitted • filed; reported • afforded • popped up • [com] launched ´sold] positioned
— a completamente - presented fully
— a con poca frecuencia f - rarely attained
— a para aprobación f - presented for approval
— a — prueba(s) f - presented for tests,ting
— f — soldadura f - [sold] positioned for @ weld(ing)
— a — — f plana - [sold] positioned for @ downhand weld(ing)
— a por escrito m - rendered, o submitted, in writing
— a previamente - presented priorly
— a visualmente - presented visually
presentador m de componente m - component presenter
presentante m - presenter; bearer | a - . . . . • submitting
presentar v - . . .; to submit; to inroduce; to produce • to occur • to take @ form; to appear • to set • to, file, o report • to discuss • to display • [com] to, feature, launch, o introduce • to announce @ release • [deport] to field • [sold] to position
— v bien ajustado,da - [sold½ to butt tight (together)
— v camino m - to, present, o provide, @ path
— v característica(s) - to, present, o introduce, @ characteristic(s)
— v certificado m - [legal] to present @ certificate
— v completamente - to present, fully, o completely
— v documento m - to, render, o present, o submit, @, document, o statement
— v ejemplo m - to, set, o present, @ example
— v factura f - to present @ invoice
— v informe - to, submit, o file, @ report
— v oferta f - [legal] to, submit, o present, o offer, @ proposal
— v para aprobación f - to, present, o submit, for approval
— v — prueba(s) - to present for @ test(ing)
— v — soldadura f - [sold] to present for @ welding
— v — — f plana f - [sold] to position for @ downhand weld(ing)
— v problema m - to present @ problem
— v producto m - [com] to, present, o introduce, @ product
— v — m nuevo - [com] to, present, o introduce, @ new product
— v propuesta f - [legal] to, present, o submit, @ proposal, o bid; to bid
— v reclamación f - file @ claim
— v — f escrita - to file @ wrriten claim
— v — reclamo m - to file @ claim
— v — — m escrito - to file @ written claim
— v saludo m - to present @ greeting
— v visualmente - to present visually
presentarse v - to appear; to, pop up, o arise
— v ante - [legal] to, appear, o come, before
— v bien - to appear (all) rosy
— v con poca frecuencia f - to appear (only) infrequently
— v mal - to appear poor(ly)
— v — trabajo m - [sold] work poor preparation

presente   - 1168 -

presentes adv - this; these • [legal] present(s)
preserva v madera f - [maderas] (que) wood preserving
preservación f de ambiente m - [ambient] environmental protection
— f — borde m - [sold] edge preservation
— f — competividad* f - [econ] competitiveness, preseving,vation
— f — madera f - [mader] wood, preservation, o
preserving
— f — medio ambiente m - [ambient] environmental protection
preservador m de ambiente m - [ambient] environ-
mentalist
— m — naturaleza f - [ambient] environmental-
ist
preservar v competitividad* v - [econ] to preserve @ competitiveness
preservativo m en forma f de película f - film
preservative
— m — película f - film preservative
presidente m - . . . . • [legal] (board) chairman
— m de compañía f - [legal] company president
— m — concejo m - [pol] commission, president,
o chairman
— m — m para condado m - [pol] county commission, chairman, o president
— m — consejo m - [legal] commission, president, o chairman; chief executive
officer
— m — directorio m - [legal] (board) chairman
— m — empresa f - [legal] company president
— m — sociedad f - [legal] corporation, o company, president
— m para división f—[legal] division president
presidir v reunión f - to, preside at, o conduct, @ meeting
presilla f - [mec] . . .; clip
presión f - . . . .; pressing • [mec] torque;
véase también empuje m • [hidr] head | [adv]
(con) pressurized
— f activa - [suelos] active pressure
— f — de suelo m - [suelos] active soil pressure
— f — — tierra f - [suelos] earth active, o
active earth, weight, o pressure
— f — determinada - [suelos] determined active
pressure
— f — lateral - [suelos] active lateral pressure
— f — — de suelo m - [suelos] active lateral
soil pressure
— f adecuada - adequate, o proper, pressure
— f — de aceite m - adquate, o proper, oil
pressure
— f — mantenida - [mec] maintained adequate
pressure
— f adicional - additional pressure
— f — para inflación f - [autom-neumát] additional inflation pressure
— f admisible - [suelos] allowable pressure
— f — aproximada - [suelos] approximate allowable pressure
— f — para suelo m - [beol] allowable soil
pressure
— f ajustada - adjusted pressure
— f alcanzada - reached pressure
— f aliviada - relieved pressure
— f alta - high pressure
— f — de aceite m - [mec] high oil pressure
— f — — agua m - [hidr] high water pressure
— f — — aire m - [neumát] high air pressure
— f — en fondo m - [ind] high bottom pressure
— f — — horno m - [ind] furnace high pressure
— f — — — m de cuba f - [metal-prod] shaft
furnace high pressure
— f — — tope m - [ind] top high, o high top,
pressure
— f — — tragante m - [metal-prod] high throat
pressure
— f — para descarga f - [ind] high, exhaust, o
discharge, pressure
— f — para gato(s) m - [constr] high jacking

pressure
presión f alta para operación f - [petról] high
operating pressure
— f — peligrosa . f - dangerous, o unsafe, high
pressure
— f — — de gas m - [combust] unsafe high
gas pressure
— f aplicada - applied, o exerted, pressure
— f — uniformemente - [mec] uniformly, applied, o distributed, pressure
— f apropiada - appropriate, o proper, o correct, pressure
— f — para aceite m - proper oil pressure
— f — — inflación f - [autom-neumát] proper
inflation pressure
— f — producida - [mec] proper exerted pressure
— f apropiada - appropriate pressure
— f aproximada - approximate pressure
— f — para aplastamiento m - [constr] approximate collapse,sing pressure
— f ascendente m - upward, o rising, pressure
— f atmosférica - [meteorol] atmospheric pressure
— f — en poro(s) m en relleno m - [constr]
pore(s) (atmospheric) pressure
— f computada - computed, o calculated, pressure
— f con ariete m hidráulico - [constr] hydraulic ram pressure
— f — — m — aumentada - [constr] increased
hyraulic ram pressure
— f — — m — disminuida - [constr] decreased
hydraulic ram pressure
— f — dedo m - finger pressure
— f — límite m para fluencia f - [tub] yield
strength pressure
— f concentrada - concentrated pressure
— f constante - constant, o steady, pressure •
standing pressure
— f — — para aceite m - [comb.int] steady oil
pressure
— f — — — m alcanzada - [comb.int] attained
steady oil pressure
— f correcta - correct pressure • [autom-
-neumát] correct (inflation) (pressure)
— f — neumático m - [autom-neumát] correct
tire inflation
— f correspondiente - corresponding pressure
— f crítica - critical pressure
— f de aceite m - [mec] oil pressure
— f — — m alta - [comb.int] high oil pressure
— f — — m baja - [comb.int] low oil pressure
— f — — m comprobada - [comb.int] checked oil
pressure
— f — — m en caja f para cambio(s) m - [mec]
gear box oil pressure
— f — — m — — f — velocidad(es) f -
[mec] gear box oil pressure
— f — — m — embrague m - [mec] clutch oil
pressure
— f — — m — depósito m comprobada - [comb.-
-int] checked reservoir oil pressure
— f — — m — m verificada - [comb.int]
checked reservoir oil pressure
— f — — m en motor m - [comb.int] engine oil
pressure • [electr-mot] motor oil pressure
— f — — m indicada - [mec] shown oil pressure
— f — — m lograda - [mec] attained, o built
up, oil pressure
— f — — m mostrada - [mec] shown oil pressure
— f — — m obtenida - [comb.int] built up oil
pressure
— f — — m para trabajo m - working oil pressure
— f — — m perdida - [comb.int] lost oil pressure
— f — — m señalada - [mec] shown oil pressure
— f — — m verificada - [comb.int] checked oil
pressure
— f — agua m - [hidr] water pressure
— f — aire m - [neumát] air pressure
— f — — m para entrada - inlet air pressure

presión f de aire m para salida f - oulet air
pressure
— f — — m soplado - [ind] air blast pressure
— f̄ — — m̄ verificada - checked air pressure
— f — arena f - [constr] sand pressure
— f̄ — — f acuífera - [hidr] water bearing
sand pressure
— f — cable m - [cabl] cable, o wire rope,
pressure
— f — — m de alambre m - [cabl] wire rope
pressure
— f — carga f - load pressure
— f̄ — — f muerta f - [constr] dead load pressure
— f — — f viva - [constr] live load pressure
— f̄ — circuito m - circuit pressure
— f̄ — — m para regulación f - control circuit
pressure
— m — — f para hoja f - [constr] blade control circuit pressure
— f — — m — f — — f (para topadora f) -
[constr] (dozer) blade control circuit pressure
— f — columna f de agua m - [hidr] water
column pressure
— f — — — m de . . .—[hidr] . . . head
— f̄ — combustible m - [combust] fuel pressure
— f̄ — — m reducida - reduced, o low, fuel
pressure
— f — confinamiento m - [geol] confining pressure • [suelos] confinement pressure
— f — contacto m - [mec] contact pressure •
engagement (pressure)
— f — falla f - fault pressure
— f̄ — fluencia f - yield pressure
— f̄ — fluido(s) m - [petról] fluid(s) pressure
— f̄ — — m para perforación f - [petról] drilling fluid pressure
— f — forja f - [metal-fabr] forging pressure
— f̄ — freno m - [mec] brake pressure
— m — — en línea f a motón m - [grúas] main
line brake pressure
— f — — — — f principal - [grúas] main
line brake pressure
— f — — m para pie m - [mec] foot brake pressure
— f — — m — — en línea f principal - [mec]
main line foot brake pressure
— m — — — m — — f — verificada - [mec]
cheked main line foot brake pressure
— f — — m para pie m sobre cable m a motón m
• [grúas] main line foot brake pressure
— f — — — — m verificada - [mec] checked
foot brake pressure
— f — — m — tambor m - [mec] drum brake
pressure
— f — — m — — m para cable m - [grúas] drum
brake pressure
— f — — m — — m a gancho m - [grúas] whip
drum brake pressure
— f — — m — — m — — m verificada - [grúas]
checked whip drum brake pressure
— m — — m — — m — — n a motón m - [grúas]
hoist, o main, drum brake pressure
— m — — m — — m — — m — — m verificado
- [grúas] checked main drum brake pressure
— f — — m verificada - [mec] checked brake
pressure
— f — gas m - gas pressure
— f̄ — — m combustible - [combust] fuel gas
pressure
— f — — m en horno m alto - [metal-prod]
blast furnace gas pressure
— f — — — m — — en línea f - [metal-
-prod] blast furnace gas line pressure
— f — — m en línea f - [metal-prod] gas line
pressure
— f — — m en tragante m - [metal-prod] throat
gas pressure
— f — — m inerte - inert gas pressure
— f̄ — gato(s) m - [const] jacking pressure
— f̄ — golpe m - [mec] hammer pressure
— f̄ — — m de ariete m - [hidr] (water) hammer
pressure

presión f de humo m - [combust] fume(s), o
smoke, pressure
— f — líquido m - liquid pressure
— f̄ — mechero m - [combust] burner pressure
— f̄ — mordaza f̄ - [mec] clamp(ing) pressure
— f̄ — mordedura f - [papel] nip, o pinching,
pressure
— f — muelle m - [mec] spring pressure
— m — — m para rodillo m - [mec] roll spring
pressure
— f — — m — — m loco - [mec] idle roll
spring pressure
— f — neumático m - [autom-neumát] tire, pressure, o inflation • inflation pressure
— f — — m delantero - [autom-neumát] front
tire pressure
— f — — m posterior - [autom-neumát] rear
tire pressure
— f — oxígeno m - [ind] oxygen pressure
— f̄ — prisma m - [constr] prism pressure
— f̄ — — m de suelo m - [constr] soil prism
pressure
— f — pulgar m - thumb pressure
— f̄ — punta f̄ - [petról] pressure peak
— f̄ — — f para bombeo m - [petról] peak pump
pressure
— f — reacción f - reaction pressure
— f̄ — — f de suelo m - [suelos] soil reaction
pressure • soil support pressure
— f — refrigerante m - refrigerant, o coolant,
pressure
— f — relleno m - [suelos] fill pressure
— f̄ — resistencia f - [ambient] resistance
pressure
— f — resorte m - [mec] spring pressure
— f̄ — — m para rodillo m - [mec] roll spring
pressure
— m — — m — — m impulsor - [sold] drive,ving
roll spring pressure
— m — riego m - [hidr] circulating pressure
— f̄ — — m de agua m - [metal-prod] circulating
(water) pressure
— f — rodillo(s) m - [mec] roll(s) pressure
— f̄ — — m impulsor - [sold] drive,ving roll
pressure
— f — — m — regulable - [mec] adjustable
drive,ving roll pressure
— f — — m — con . . . posición(es) f - [mec]
. . . position adjustable drive,ving roll
pressure
— f — — m — superior - [mec] upper drive,ving
roll pressure
— f — — m para impulsión f - [mec] drive,ving
roll pressure
— f — — m — — f regulable - [mec] adjustable
drive,ving roll pressure
— f — — m — — f con . . . posición(es) -
[mec] . . . position drive,ving roll pressure
— f — — m — — f regulable - [mec] adjustable
drive,ving roll pressure
— f — — m superior - [mec] upper roll pressure
— f̄ — sobrecarga f - [constr-pil] overload
pressure
— f — suelo m - [constr] soil, o earth, pressure
— f — — m en sitio m - [suelos] soil at @
site pressure
— f — sobre esquina(s) f - [constr] corner
soil pressure
— f — — m sobre tubería f - [constr] soil
pressure on @ pipe
— f — suministro m para aire m - [mec] air supply pressure
— f — tafilete m - [vest] around @ head, o
headband, pressure
— f — tubería f - [tub] pipe pressure
— f̄ — vacío m - vacuum pressure; pressure vacuum
— f — vapor m - [cald] steam, o vapor, pressure
— f̄ — — m en depósito m - tank vapor pressure
— f̄ — — m saturado - [cald] saturated steam
pressure

presión f de vapor m supercalentado ·   - 1170 -

presión f de vapor m supercalentado - [cald]
superheated steam pressure
— f —— m vivo - [cald] live steam pressure
— f̄ — ventilador m - [mec] fan pressure
— f̄ —— m para aire m - [combust] air fan
pressure
— f —— m — gas m - [combust] gas fan pressure
— f — viento m - wind pressure • [metal-prod]
(air) blast pressure
— f —— m soplado - [metal-prod] blast pressure
— f descendente - descending, o downward, pressure
— f deseada - desired pressure
— f̄ — en tubería f - [petról] casing pressure
— f̄ ——— f para entubación f - [petról] desired casing presssure
— f — f —— f lograda nuevamente -
[petról] regained desired casing pressure
— f — marcada - [mec] dialed desired pressure
— f̄ determinada - determined pressure
— f̄ diferencial - differential pressure
— f̄ — transmitida - [instrum] transmitted differential pressure
— f diferente - different pressure
— f̄ digital - finger pressure
— f̄ — leve - slight finger pressure
— f̄ disminuida - decreased, o lower(ed), pressure
— f distinta - different pressure
— f̄ distribuida - distributed pressure
— f̄ — transferida - transferred distributed
pressure
— f ejercida - exerted, o developed, pressure
— f̄ — por bomba f - [bombas] pump pressure
— f̄ elevada - high pressure • raised presssure •
[comb.int] boost(ed) pressure
— f — de aceite m - [mec] high oil pressure
— f̄ —— agua m - [hidr] high water pressure
— f̄ —— rodillo m - [mec] high roll pressure
— f̄ ——— m impulsor - [mec] high drive,ving
roll pressure
— f ——— m para impulsión f - [mec] high
drive,ving roll pressure
— f — ejercida - great exerted pressure
— f̄ — máxima - [comb.int] maximum boost(ing)
pressure
— f — para descarga f - [mec] high discharge
pressure
— f —— forja f - [mec] high forging pressure
— f̄ en acumulador m - accumulator pressure
— f̄ —— m, indicada, o mostrada - [mec] shown
accumulator pressure
— f — boca f para aspiración f - [mec] air intake, o intake opening, pressure
— f — bomba f - [bombas] pump pressure
— f̄ —— f para inyección f - [petról] mud
pump pressure
— f —— f — lodo m - [petról] mud pump pressure
— f̄ —— f verificada - [bombas] checked pump
pressure
— f — broca f - [petról] drill pressure
— f̄ — cimentación f - [constr] foundation
pressure
— f —— f concentrada - [constr] concentrated foundation pressure
— f — cimiento m—[constr½ foundation pressure
— f̄ — depósito m - tank pressure
— f̄ —— m inferior - lower tank pressure
— f̄ —— m separador - separator tank pressure
— f̄ — esquina f - corner pressure; pressure at
@ corner
— f — etapa f - stage pressure
— f̄ —— f final - final stage pressure
— f̄ — evaporador m - [ind] evaporator pressure
— f̄ —— m para refrigerante m - [ind] refrigerant, o coolant, evaporator pressure
— f —— m verificada - [ind] checked evaporator pressure
— f — filtro m - filter pressure
— f̄ —— m inferior - bottom filter pressure

presión f en filtro m superior - upper, o top,
filter pressure
— f — hogar m - [combust] hearth pressure
— f̄ — horno m - [ind] furnace pressure
— f̄ —— m con cuba f - [metal-prod] shaft
furnace pressure
— f —— m — fosa f - [metal-lam] soaking pit
pressure
— f — línea f - line pressure
— f̄ —— f para aceite m - [ind] oil line
pressure
— f —— f — combustible m - [ind] fuel line
pressure
— f —— f — fuel oil m - [comb.int] fuel
oil line pressure
— f —— f principal - [tub] main line pressure
— f — neumático m - [autom-neumát] tire pressure
— f — planta f - [ind] plant pressure
— f̄ — recipiente m - [cald] vessel pressure •
[ind] container pressure
— f — red f - system pressure
— f̄ —— f general - [ind] plant system pressure
— f —— f mantenida - [mec] maintained system
pressure
— f —— f para agua m - [hidr] water system
pressure
— f — regulador m - regulator pressure
— f̄ — salida f - output pressure
— f̄ — sistema m - [mec] system pressure
— f̄ —— m mantenida - [mec] maintained system
pressure
— f — tope m - [metal-prod] top pressure
— f̄ — tragante m - [metal-prod] throat, o furnace top, pressure
— f —— m de horno m - [metal-prod] furnace
top pressure
— f — tubería f - [tub] line pressure
— f̄ —— f abovedada - [constr] pipe arch
pressure
— f —— f para aceite m - [ind] oil line
pressure
— f —— f — combustible m - [comb.int] fuel
line pressure
— f —— f — entubación f - [petról] drill
casing, o casing (pipe), pressure
— f —— f —— f fiscalizada - [petról] monitored (drill) casing pressure
— f —— f —— f lograda nuevamente -
[petról] regained casing pressure
— f —— f — fuel oil m - [comb.int] fuel oil
line pressure
— f —— f — perforación f - [petról] drill-
-(ing) pipe pressure
— f —— f —— f fiscalizada - [petról] monitored drill(ing) pipe pressure
— f —— f en salida f—(pipe) outlet pressure
— f̄ —— f en salida f para agua m - [hidr]
water outlet pressure
— f —— f principal - [tub] main pressure
— f̄ —— f — para gas m - [combust] gas main
pressure
— f — unidad f - unit pressure
— f̄ entre bomba f y válvula f - pressure between @ pump and @ valve
— f equivalente - equivalent pressure
— f̄ — de líquido m - equivalent liquid, o liquid equivalent pressure
— f especial - special pressure
— f̄ — para inflación f - [autom-neumát] special inflation pressure
— f específica - specific pressure
— f̄ — de engranaje m - [mec] gear specific
pressure
— f — para inflación f - specific inflation
pressure
— f — por unidad f - [mec] unit specific
pressure
— f —— f de superficie f - [mec] pitch
line specific pressure
— f ——— f —— f sobre diente m - [mec]

gear pitch line specific pressure
presión f esquinera - corner pressure
— f — **calculada** - [constr] calculated, o computed, corner pressure
— f — **computada** - [constr] computed, o calculated, corner pressure
— f — **de suelo m** - [constr] corner soil, o soil corner, pressure
— f **estabilizada** - stabilized pressure
— f **estándar** - [ind] standard pressure
— f — **para ensayo(s) m** - [ind] standard testing) pressure
— f **estática** - static pressure
— f — **comprobada** - [electr-mot] checked static pressure
— f **estimada** - estimated pressure
— f — **para aplastamiento m** - estimated collapse,sing pressure
— f **excedida** - exceeded, o surpassed, pressure
— f **excesiva** - excessive, o high wide, pressure
— f — **en esquina f** - [constr] excessive corner pressure
— f — — **horno m** - [metal-prod] furnace excessive, o excessive corner, pressure
— f — — **horno m con cuba f** - [metal-prod] shaft furnace excessive pressure
— f — — **tragante m** - [metal-prod] excessive furnace pressure
— f **excesivamente alta** - excessively high pressure
— f — — **en horno m con cuba f** - [metal-prod] shaft furnace excessively high pressure
— f — **elevada** - excessively high pressure
— f **exterior** - external, o outside, pressure
— f **externa** - external pressure
— f — **uniforme** - uniform external pressure
— f **extrema** - extreme pressure
— f **final** - final pressure
— f **fiscalizada**—monitored, o checked, pressure
— f **fluida** - [petról] fluid pressure
— f **forjadora** - [metal-fabr] forging pressure
— f — **elevada** - [mec] high forging pressure
— f **freática** - [hidr] ground water pressure
— f — **aliviada** - [hidr] relieved ground water pressure
— f **generada** - generated, o built up, pressure
— f **grande** - great pressure
— f — **ejercida** - exerted great, o great exerted, pressure
— f **hacia abajo** - [mec] downward(s) pressure
— f — **arriba** - upward(s) pressure
— f **hidráulica** - [hidr] hydraulic pressure
— f — **apropiada** - proper, o appropriate, hydraulic pressure
— f — — **comprobada** - [mec] proven, o checked, proper, o appropriate, hydraulic pressure
— f — — f **verificada** - [mec] checked, proper, o appropriate, hydraulic pressure
— f — **comprobada** - checked hydraulic pressure
— f — **correcta** - [hidr] correct hydraulic pressure
— f — — **comprobada** - [hidr] checked correct hydraulic pressure
— f — — **verificada** - [mec] checked correct hydraulic pressure
— f — **para trabajo m** - [hidr] hydraulic working, o working hydraulic, pressure
— t — **verificada** - checked hydraulic pressure
— f **hidrostática** - [hidr] hydrostatic pressure
— f — **de arena f** - [hidr] sand hydrostatic pressure
— f — — — f **acuífera** - [hidr] water bearing sand hydrostatic pressure
— f — **externa** - [hidr] external hydrostatic pressure
— f — — f **alta** - [hidr] high external hydrostatic pressure
— f — — **baja** - [hidr] low external hydrostatic pressure
— f — **interna** - [hidr] internal hydrostatic pressure
— f — — **alta** - [hidr] high internal hydrostatic pressure

presión f hidrostática interna baja - [hidr] low internal hyudrostatic pressure
— f — **perdida** - [hidr] lost hydrostatic pressure
— f **horizontal** - horizontal pressure
— f **igualada** - equalized pressure
— f **indicada** - shown pressure
— f **inferior** - lower pressure
— f — **a nominal f** - pressure lower than rated
— f **inicial** - initial pressure
— f **insuficiente** - insufficient pressure | adv - [autom-neumát] (con) underinflated
— f **interior** - internal pressure
— f — **proyectada** - internal design pressure
— f **intermedia** - [mec] medium pressure
— f — **con límite m mínimo de fluencia f** - [tub] minimum yield strength internal pressure
— f — — **m de fluencia f** - [tub] yield strength internal pressure
— f — **de fluencia f** - [tub] internal yield pressure
— f **ininterrumpida** - uninterrupted pressure
— f **interrumpida** - interrupted pressure
— f **interior** - inside, o inner, pressure
— f **intermedia** - [cald] intermediate pressure
— f **interna** - internal pressure
— f — **de agua m** - [hidr] internal water, o water internal, pressure
— f **lateral** - lateral, o side, pressure
— f — **de confinamiento m** - [geol] lateral confinement,ning pressure
— f — **generada** - generated lateral pressure
— f — **liviana** - light side pressure
— f — **pesada** - heavy side pressure
— f — **total** - [constr] total lateral pressure
— f — **unitaria** - [constr] unit lateral pressure
— f **leída** - [instrum] read pressure
— f **leve** - slight pressure
— f — **con dedo m** - slight finger pressure
— f — **hacia abajo** - [mec] slight downward(s) pressure
— f — — **y (hacia) adelante** - [mec] slight down(wards) and forward(s) pressure
— f — — **adelante** - [mec] slight foward pressure
— f **limitada** - limited pressure
— f **liviana** - light pressure
— f **lograda** - attained, o built up, pressure
— f — **nuevamente** - [mec] regained pressure
— f **manométrica** - [instrum] gage pressure
— f **mantenida** - [autom-neumát] held, o maintained, pressure, o air
— f **máxima** - maximum pressure • [petról] pressure peak
— f — **de agua m** - [hidr] water maximum, o maximum water, pressure
— f — — **m industrial** - [hidr] industrial water maximum pressure
— f — — **m potable** - [hidr] drinking water maximum pressure
— f — **aire m** - maximum air pressure
— f — — **m comprimido** - [ind] maximum compressed air pressure
— f — **en circuito m** - circuit maximum pressure
— f — — **m para regulación f** - control circuit maximum pressure
— f — — **m — f de hoja f** - [constr] blade control circuit maximum pressure
— f — — **m — — f para topadora f** - [constr] dozer blade control circuit maximum pressure
— f — **red f** - maximum system pressure
— f — **tubería f** - [petról] maximum pipe pressure
— f — — **f para entubación f** - [petról] maximum casing pressure
— f — **exigida** - [tub] maximum required pressure
— f — **garantizada** - maximum guaranteed pressure
— f — **normal** - maximum normal pressure

presión f máxima para aceite m — 1172

presión f máxima para aceite m - maximum oil pressure
— f — — agua n - maximum water pressure
— f — — m para enfriamiento m - cooling water maximum pressure
— f — — bombeo m - [petról] maximum, o peak, pump pressure
— f — — ensayo(s) m - maximum test(ing) pressure
— f — — — m exigida - maximum required test(ing) pressure
— f — condensación f - [ind] refrigerant, o coolant, condensation,sing pressure
— f — — gas m - [combust] maximum gas pressure
— f — — inflación f - [autom-neumát] maximum inflation pressure
— f — — inyección f - [constr] maximum injecting pressure
— f — — — f de lechada f - [constr] maximum grouting pressure
— f — — operación f - [ind] maximum operating pressure
— f — — servomotor m - [ind] servomotor maximum pressure
— f — — trabajo m - [ind] maximum operating pressure
— f — — — m permitida - [cald] maximum allowable working pressure
— f — permitida - maximum allowable pressure
— f — — para trabajo m - [ind] maximum allowable, working, o operating, pressure
— f — segura - [ind] maximum safe pressure
— f — — para operación f - [ind] maximum safe operating pressure
— f mayor - higher, o greater, pressure
— f — para agua m - greater water pressure
— f media - average pressure
— f — — m industrial - [hidr] industrial water average pressure
— f — — m potable - [hidr] drinking water average pressure
— f — — aire m - average air pressure
— f — — m comprimido - [ind] average compressed air pressure
— f — — trabajo m - [cald] average operating pressure
— f mediana - medium pressure
— f menor - lower, o less, pressure
— f — para aire m - [autom-neumát] lower air pressure
— f mínima - minimum pressure
— f — para agua m - [hidr] minimum water pressure
— f — — — m depurada - [sanit] treated water minimum pressure
— f — — — m industrial - [hidr] industrial water minimum pressure
— f — — — m potable - [hidr] drinking water minimum pressure
— f — — aire m - minimum air pressure
— f — — — m comprimido - [ind] compressed air minimum pressure
— f — — circuito m - circuit minimum pressure
— f — — ensayo(s) m - [tub] minimum test(ing) pressure
— f — — — exigida - [ind] minimum required test(ing) pressure
— f — — inyección f [constr] minimum injecting pressure
— f — — — f de lechada f - [constr] minimum grouting pressure
— f — — red f - minimum system pressure
— f — — trabajo m - minimum working pressure
— f — — tubería f - [tub] minimum pipe pressure
— f — — — f para entubación f - [petról] minimum casing pressure
— f — segura - [ind] minimum safe pressure
— f — — para operación f - [ind] minimum safe operating pressure
— f mostrada - [instrum] shown pressure
— f negativa - negative pressure

presión f negativa mantenida - maintained negative pressure
— f neumática - [neum] air pressure
— f nominal - nominal, o rated, pressure
— f — en salida f - rated output pressure
— f — mínima - mínimum nominal pressure
— f — — para aceite m - minimum, nominal, o rated, oil pressure
— f — para aceite m - nominal oil pressure
— f normal - normal pressure
— f — para trabajo m - normal, work(ing), o operating, o service, pressure
— f — — — m para manómetro m - [instrum] gage normal work(ing) pressure
— f — — viento m - normal wind pressure • [metal-prod] normal blast pressure
— f — máxima - normal maximum pressure
— f — — para condensación f - [ind] normal maximum condensation,sing pressure
— f — — — f para refrigerante - normal maximum, refrigerant, o coolant, condensing pressure
— f — — — para refrigerante m - [ind] normal, refrigerant, o coolant, condensing pressure
— f nueva - new pressure
— f obtenida - obtained, o built up, pressure
— operativa - operating pressure
— f óptima - imptimum, o ideal, pressure
— f osmótica - [fís] osmotic pressure
— f para agua m - water pressure
— f — — m para enfriamiento m - [ind] cooling water pressure
— f — — m — freno m - [mec] brake water pressure
— f — — m — llanta f - [mec] rim water pressure
— f — — m — — f para freno m - [mec] brake rim water pressure
— f — — aire m - [neumát] air pressure
— f — — m recomendada - [neumát] recommended air pressure
— f — — m — por fabricante m - [neumát] manufacturer('s) recommended air pressure
— f — aplastamiento m - collapse,sing pressure
— f — — m para suelo m - [suelos] soil collapse,sing pressure
— f — apoyo m - [constr] bearing, o support-(ing) pressure
— f — — m esquinera - [constr] corner bearing pressure
— f — — m — sobre suelo m - [constr] soil corner bearing pressure
— f — — m sobre esquina f - [constr] corner bearing pressure
— f — — m plancha f esquinera - [constr] corner plate bearing pressure
— f — — m — suelo m - [constr] soil bearing pressure
— f — aspiración f - [mec] suction pressure • intake pressure
— f — bombeo m - [mec] pump(ing) pressure
— f — condensación f - condensation,sing pressure
— f — — f para refrigerante m - [ind] refrigerant, o coolant, condensation,sing pressure
— f — conformación f - [metal-fabr] forming pressure
— f — contacto m - [mec] contact pressure
— f — — m para leva f - [mec] cam engagement (pressure)
— f — — m para sujetador m - [mec] cam (contact) engagement
— f — derrumbamiento m - [suelos] collapse,-sing pressure
— f — — m para suelo m - [suelos] soil collapse,sing pressure
— f — desasentamiento m - [hidr] unseating pressure
— f — descarga f - discharge,ging, o exhaust-(ing) pressure
— f — — f alta - high discharge,ging pressure
— f — — f baja - low discharge,ging pressure

presión f para descarga f elevada - high discharge,ging pressure
— f — — f reducida - low discharge,ging pressure
— f — desmoronamiento m - [constr] collapsing
pressure
— f — — m de suelo m - [suelos] soil collapsing pressure
— f — — m previsto - [suelos] anticipated col
lapse pressure
— f — — m — para suelo m - [suelos] anticipated soil collapse,sing pressure
— f — diseño m - [cald] design pressure
— f — — m para recipiente m - [cald] vessel
design pressure
— f — empuje m - [constr] jacking pressure
— f — ensayo(s) m - test(ing) pressure
— f — — m para recipiente(s) m - [cald] vessel test(ing) pressure
— f — — m hidrostático(s) - [cald] hydrostati
test(ing) pressure
— f — entrada f - input, o incoming, o intake,
pressure
— f — — f para agua m - water intake pressure
— f — formación f - [metal-fabr] forming pressure
— f — impulsión f - [constr] jacking pressure
— f — inflación f - [autom-neumát] inflation
pressure
— f — — f adicional - [autom-neumát] addition
al inflation pressure
— f — — f específica - specific inflation
pressure
— f — — f mantenida - maintained inflation
pressure
— f — inyección f - [bombas] charge,ging pressure
— f — — f para lechada f - [constr] grouting
pressure
— f — — f verificada - [bombas] checked, o
verified, charge,ging pressure
— f — operación f - operating pressure
— f — prueba(s) f - test(ing) pressure
— f — — f de ensayo m para recipiente(s) m ·
[cald] vessel proof testing pressure
— f — — f hidrostática(s) - [tub] hydrostatic
test(ing) pressure
— f — — f — con agua m - [tub] water hydrostatic test(ing) pressure
— f — quemador m - [combust] burner pressure
— f — retroceso m - back-off pressure
— f — servicio m - [ind] service pressure
— f — trabajo m - [ind] work(ing), o operating
o service, pressure
— f — — m para agua m - [hidr] water working
pressure
— f — — m para manómetro m - [instrum] gage
work(ing) pressure
— f — — m exigida - design working pressure
— f — — m permitida - [cald] allowable working pressure
— f — — m prevista - design working pressure
— f — verificar - pressure to be checked
— f parcial - partial pressure
— f pasiva - passive pressure
— f — de suelo m - [suelos] passive soil pressure • variable soil pressure
— f — — tierra f - [suelos] earth passive
pressure • earth passive weight
— f — determinada - [suelos] determined passive pressure
— f — variable - variable passive pressure
— f — — de suelo m - [constr] variable passive soil pressure
— f peligrosa - dangerous, o unsafe, pressure
— f — para gas m - [combust] unsafe gas, o gas
unsafe, pressure
— f perdida - lost pressure
— f permitida - allowed,wable pressure
— f pesada - heavy, o weighty, pressure
— f planeada - planned pressure
— f — para inflación f - [autom-neumát]
planned inflation pressure
presión f plena - full pressure
— f política - [pol] political pressure
— f positiva - positive pressure
— f — manetnida - maintained positive pressure
— f prevista - [constr] design pressure
— f — para viento m - [constr] design wind
pressure
— f producida - produced, o exerted, pressure
— f que actúa sobre tubería f - [constr] pressure exerted on @ pipe
— f radial - [constr] radial pressure
— f — exterior - [constr] external radial
pressure
— f — igualada - equalized radial pressure
— f — para conformación f - [metal-fabr] radial, forming, o shaping, pressure
— f recomendada - recommended pressure
— f — para inflación f - [autom-neumát] recommended inflation pressure
— f — oxígeno m - [ind] recommended oxygen
pressure
— f reducida - reduced, o decreased, o low,
pressure ∩ back-off pressure
— f — de aceite m - [mec] reduced, o low, oil
pressure
— f — — agua m - [mec] low water pressure
— f — para descarga f - low discharge pressure
— f registrada - [instrum] recorded pressure
— f regulable - [mec] adjustable pressure
— f — de rodillo m - [sold] adjustable roll
pressure
— f — — m impulsor - [sold] adjustable
drive,ving roll pressure
— f regulada - regulated, o controlled, pressure
— f — con tapa f - [comb.int] cap regulated
pressure
— f — — — f para radiador m - [comb.int] radiator cap regulated pressure
— f — de gas m - controlled gas pressure
— f según norma f - [ind] standard pressure
— f — — f para ensayo(s) m - [ind] standard
test(ing) pressure
— f — plaqueta f - [autom-neumát] placcard
pressure
— f segura - [ind] safe pressure
— f — para operación f - [ind] safe operating
pressure
— f sentida - sensed, o felt, pressure
— f señalada - [mec] signalled, o shown, o
sensed, pressure
— f severa - severe, o extreme, pressure
— f sobre cable m - [grúas] main line pressure
— f — diente m - [mec] gear pitch line pressure
— f — esquina,nera f - [constr] corner pressure
— f — fondo m - [constr] bottom pressure
— f — pedal m - [mec] pedal pressure
— f — — m para freno m - [mec] brake pedal
pressure
— f — plancha f - [mec] plate pressure
— f — — f esquinera - [constr] corner plate
pressure
— f — revestimiento m - [constr] pressure on
@ lining
— f — suelo m - [constr] pressure on @ soil
— f — tambor m - [mec] drum pressure
— f — — m para freno m - [mec] brake drum
pressure
— f — tapa f - cover, o top, pressure
— f — tubería f - [constr] pressure on @ pipe
— f — — f abovedada - [constr] pressure on @
pipe-arch
— f sobrepasada - surpassed, o overriden,
pressure
— f soportada - supported, o held, pressure
— f sostenida - [mec] held pressure
— f súbita - sudden pressure
— f substitutiva - alternate pressure
— f suficiente - sufficient, o enough, pressure

presión f típica - typical pressure
— f — para neumático m - [autom-neumát] typical tire pressure
— f transferida - transferred pressure
— f transmitida - transmitted pressure
— f transversal - transverse pressure
— f uniforme - uniform pressure
— f — aplicada - applied uniform pressure
— f unitaria - unit pressure
— f — de prisma m - [constr] prism unit pressure
— f — — — m de suelo m - [constr] soil prism unit pressure
— f variable - variable pressure
— f variable de agua m - [hidr] variable water pressure
— f — — — m en planta f - [hidr] variable plant water pressure
— f — — suelo m - variable soil pressure
— f — pasiva - variable passive pressure
— f — — de suelo m - [constr] variable passive soil pressure
— f variada - varied, o changed, pressure
— f verificada - checked pressure • [autom-neumát] checked inflation
— f — de freno m - [mec] chedked brake pressure
— f — — — m para pie m - [mec] checked foot brake pressure
— f — — — m — — m para línea f principal - [mec] checked main line foot brake pressure
— f vertical - vertical pressure
— f — de suelo m - [suelos] vertical soil pressure
— f — unitaria - [constr] unit vertical, o veretical unit, pressure
— f Zerk - Zerk pressure
presionado,da a - pressured • pressed
prestación f - . . . . • provision,viding • service; provided service • furnishing; granting
— f con compensación f - providing,vision with compensation
— f de asesoramiento m - advice, providing,vision, o supply(ing), o furnishing
— f — — m técnico - [ind] technical aid, providing,vision, o furnishing, o supply(ing)
— f — asesoría f - consultation providing,vision
— f — atención f - attention paying; servicing
— f — — f para bomba f - [bombas] pump servicing
— f — — f — filtro m - [mec] filter servicing
— f — fianza f - [vin] bond grant(ing)
— f — garantía f - guaranty, o bond, posting
— f — personal m - personnel service(s)
— f — servicio(s) m - service(s), providing, o offering, o rendering, o supplying; servicing • performance,ming • personnel services • field service(s) • [legal] service(s) term • [labor] service(s), rendering, o rendered
— f — — m de consulta,toría - [ind] consultation furnishing
— f — — para filtro m - [med] filter servicing
— f — — m — motor m - [electr-mot] motor servicing • [comb.int] engine servicing
— f — — m — unidad f motriz - [agric-equip] power unit service,cing
— f — — m para bomba(s) f - [bombas] pump servicing
— f sin compensación f - providing,vision without compensation
prestaciónes f sociales - [labor] social charges • fringe benefits
prestado,da a - loaned • provided; furnished; granted • borrowed
— a con compensación f - loaned, o provided, with compensation
— a sin compensación f - loaned, o provided, without compensation
préstamo m - . . . .; loaning • borrowing
— m con aval m - [fin] guaranteed loan

préstamo m con fianza f - [fin] guaranteed loan
— m de casa f matriz - [fin] home office loan
— m directo - [fin] direct loan
— m — exterior m - [fin] direct foreign loan
— m exterior - [fin] foreign loan
— m — directo - [fin] direct foreign loan
— m garantizado m - [fin] guaranteed, o secured, loan
— m pendiente - [fin] outstanding loan
— m prendario - [fin] secured loan
— m privado - [fin] private loan
— m sin aval m - [fin] unsecured loan
— f — fianza f - [fin] unguaranteed loan
— m — garantía f - [fin] unguaranteed, o unsecured, loan
prestar v - . . . . • to borrow • to provide; to grant
— v asesoramiento m - to provide @ advice
— v — m técnico - [ind] to, provide, o furnish, @ technical aid
— v asesoría m - to provide @ consultation
— v asistencia f - to, aid, o assist
— v atención f - to pay @ attention • to give consideration • to service
— v — para filtro m - to service @ filter
— v ayuda - to, provide help, o assist
— v con compensación f - to provide with compensation
— v conformidad f - to, approve, o give approval
— v fianza f - [fin] to grant @ bond
— v garantía f - to post @ bond
— v servicio m - to, perform, o give, o supply, o offer, o provide, @ service • to perform • [mec] to service • [legal] to work
— m — de consulta,toría f - [ind] to furnish @ consultation
— m — para filtro m - to service @ filter
— m — — motor m - [electr-mot] to service @ motor • [comb.int] to service @ engine
— m — — neumático m - [autom-neumát] to service @ tire
— m — — unidad f motriz - [agric-equip] to service @ power unit
— v — — zapata f - [mec] to service @ shoe
— v — total - to provide total service
— v — m — para todo mundo m - [com] to be all things to all @ people
— v validez - to lend @ validity
prestarse v - to lend, one, o it, self
— v bien - to lend itself well
— v especialmente - to be particularly appropriate
— v para - to be suitable for
prestó v juramento y firmó v ante mí - [legal] subscribed and sworn to before me
presumido,da a - . . . . • presumed • assumed • design • hypothesized
presumir v - . . . . • to assune • to hypothesize • to assume
presunción f - . . . . • assuming,mption • hypothesizing
— f de diseño m - [constr] design assumption
— f para proyección f - design assumption
presupuestado,da a - [fin] budgeted • projected
presupuestar v - . . . .; to project
presupuestario,ria a - budget(ary)
presupuesto m - . . . .; budget • project
— m de inversión f - [fin] investment budget
— m excedido - overrun budget
— m normal - normal budget
— m — para operación(es) f - normal operating budget
— m para costo(s) m para estimación f - engineering estimate
— m reajustado - escalated budget
— m reducido - [fin] limited budget
presupuestos y normas - [fin] budgets and standards
presurestato* m - véase manostato m - [metal-prod] pressurestat
pretaladrado,da a - [mec] predrilled
pretaladrar v - [mec] to predrill

pretender v - . . . . • to contend • to claim • to
attempt • to, represent, o make @ represemta-
tion
— v contratar - to pretend to contract
pretendido,da a - pretended; claimed • contended
• represented; attempted
pretensado m - [metal] prestressing
pretensado,da a - [metal] prestressed
— a circunferencialmente - prestressed circumferentially; circumferentially prestressed
pretensión f - . . .; pretending; contending;
contention • representing,tation • attempt •
[labor] application
— f demostrada - pretended chance
— f substancial - substantial claim
pretil m - [constr] . . .; low wall • guardrail
— m de hormigón m - [constr] (low) concrete
wall
preunido,da a - preassembled; preunited
preunir v - to preassemble; to preunite
prevaciado,da a - [metal-prod] precast
prevalecer v - . . . . • to take precedence
prevalecido,da a - prevailed
prevalecimiento m - prevailing
prevalencia* f - véase prevalecimiento m
prevalente* a - prevalent; véase prevaleciente @
prevención f - . . .; preventing; avoiding,dance
• control • caution; warning; alarm
— f contra corrosión f - corrosion, prevention,
o avoidance
— f — choque(s) - [electr-oper] shock prevention
— f — daño(s) m - damage prevention
— v — — m a ojo(s) m - [segurid] eye injury
prevention
— v — — m — vista f - [segurid] eye(sight),
o vision, injury prevention
— f — entrada f - entering,trance prevention
— f — f de aire m - air entering,trance
preventing,tion
— f — erosión f - [constr] erosion prevention
— f — evaporación f - evaporation prevention
— f — — f de fluido(s) m - [petról] fluid(s)
evaporation prevention
— f — — f — — m volátil(es) - [petról] vol-
átile fluid(s) evaporation prevention
— f — ludimiento m - [mec] galling prevention
— f — oxidación f - oxidation preventing,tion
— f — presión f alta - [mec] high pressure
warning • high pressure preventing,tion
— f — — t de aceite m - [mec] high oil
pressure warning
— f — — f — agua m - [mec] high water
pressure warning
— f — — f baja - [mec] low pressure warning
— f — — f de aceite m - [mec] low oil
pressure warning
— f — — — de agua m - [mec] low water
pressure warning
— f — — f elevada - [mec] high pressure warning
— f — — f — de aceite m - [mec] high oil
pressure warning
— f — — f — agua m - [mec] high water
pressure warning
— f — — f reducida - [mec] low pressure warning
— f — — f — de aceite m - [mec] low oil
pressure warning
— f — — f — agua m - [mec] low water
pressure warning
— f — temperatura f alta - [mec] high temperature warning
— f — — f — de aceite m - [mec] high oil
temperature warning
— f — — f — — — m hidráulico - [mec] high
hydraulic oil temperature warning
— f — — f — — m para convertidor m -
[mec] converter oil high temperature warning
— f — — f — — — m — m para torsión f -
[mec] torque converter oil high temperature
warning
— f — trabamiento m - [mec] seizure prevention

prevención f de accidente(s) f - [segurid] accident, avoidance,ding, o prevention • safety
— f — erosión f - erosion, preventing,tion, o
control
— f — agrietamiento m - cracking preventing
— f — contracción f - contracting,tion, preventing,tion, o avoidance,ding
— f — corriente f - [electr-oper] current a-
voidance,ding, o preventing,tion
— f — — f en árbol - [electr-oper] current a-
voidance,ding, in @ shaft
— f — daño m - damage, preventing,tion, o a-
voidance,ding
— f — erosión f - [hidr] erosion, control, o
preventing,tion
— f — grieta(s) f - [metal] crack(s) preventing,tion
— f — inundación(es) f - [hidr] flood control
— f — lesión(es) f - [segurid] injury preventing,tion
— f — — a piel f - [segurid] skin injury preventing,tion
— f — perforación(es) f - [sold] burn through
preventing,tion
— f — problema(s) m - problem(s), o trouble,
avoidance,ding, o prevention
— f — rotación f - [mec] rotation prevention
— f — sedimentación f - [constr] sedimentation,
preventing,tion, o anticipating
— f — socavación f - undermining, preventing,tion, o avoidance,ding
— f — vibración f - vibration, avoidance,ding,
o preventing,tion
prevenido,da a - . . . . • prevented; avoided •
warned
prevenir v - . . . . • to prevent; to avoid • to
resist
— v accidente(s) m - [segurid] to avoid @ acci-
dent(s)
— v adhesión(es) v - to, avoid, o prevent, @,
bonding, o adhesion(s)
— v aflojamiento m - [mec] to prevent @ loosening
— v — de conexión f - [electr-instal] to prevent @ connection loosening
— v — — f de conductor m - [electr-instal]
to prevent @ cable connection loosening
— v agrietamiento m - [sold] to prevent @
cracking
— v cambio m - to prevent @ change
— v — m en procedimiento m - to prevent @ procedure change
— v — m inesperado - to prevent @ unexpected
change
— v — m — en procedimiento m - to prevent @
unexpected procedure change
— v contra corrosión f - to prevent @ corrosion
— v — daño(s) m - to prevent @ damage(s)
— v — detención f - [mec] to prevent stopping
— v — — f de compresor m - [mec] to prevent @
compessor stopping
— v — entrada f - to prevent, entrance, o entering
— v — evaporación f - to prevent @ evaporation
— v — ludimiento m - [mec] to prevent @ galling
— v socavación f - [constr] to prevent @ undermining
— v — trabamiento m - [mec] to prevent @ seizure m
— v corriente f - to avoid @ current
— v — f en árbol m - [electr-oper] to, prevent,
o avoid, @ current in @ shaft
— v corrosión f - to prevent @ corrosion
— v cumplimiento m - to prevent from performing
— v daño m - to prevent @ damage
— v — m a borne m - [sold] to prevent @ stud
damage
— v — m — rectificador m - [electr-oper] to
prevent @ rectifier damage
— v — — m con silicio m - [electr-oper]
to prevent @ silicon rectifier damage
— v deformación(es) f - to prevent @, deformation, o deflection

prevenir v desenrollamiento m - [mec] to prevent
@, uncoiling, o unwinding
— v — m excesivo - [mec] to prevent @ overrun
— v — m — de carrete m - [mec] to prevent @
reel overrun
— v desviación f - to prevent @ deflection
— v encorvadura f - [mec] to prevent @ bending
— v entrada f - to prevent @ entering,trance
— v — f de aire m - [neumát] to prevent @ air
entering,trance
— v erosión f - [hidr] to prevent @ erosion
— v ludimiento f - [mec] to prevent @ galling
— v movimiento m - to prevent @ movement,ving
— v pegamiento m - to, prevent, o avoid, @
sticking
— v — m en cráter m - [sold] to prevent @ crater sticking
— v pérdida f - to prevent @, loss, o escape
— v — f de aceite m - [mec] to prevent @ oil,
loss, o escape
— v problema m - to prevent @, problem, o trouble
— v recalentamiento m - to avoid @ overheating
— v — m de rectificador m - [electr-oper] to
prevent @ rectifier overheating
— v — m — transformador m - [electr-oper] to
prevent @ transformer overheating
— v sobrecalentamiento m - [electr-oper] to
avoid @ overheating
— v — m de rectificador n - [electr-oper] to
prevent @ rectifier overheating
— v — m — transformador m - [electr-oper] to
prevent @ transformer overheating
— v socavación f - [constr] to, prevent, o
avoid, @ undermining
— v trabamiento m - [mec] to prevent @ seizure
— v vibración f - to prevent @ vibration
preventivo m - . . . .; deterrent; véase también
medida f preventiva
— m contra óxido - [metal] rust preventive
— m mejor - best preventive • better preventive
preventivo,va a - . . . . • guardian
preventorio m - . . . . • [penal] jail
— m infantil - [penal] juvenile jail
prever v - . . . . • to predict • to project •
to plan • to forecast • to expect
• to schedule • to contemplate
— v carga f - to predict @ load
— v — f externa - [constr] to predict @ external load
— v circulación f - to predict, @ circulation,
o flow
— v — f de aire m - to, predict, o design, @
air flow
— v exportación f - [com] to foresee @ exports
— v fecha f - to foresee @ date
— v pago m - [fin] to foresee @ payment
— f pérdida f - to predict @ loss
— v — f debida a fricción f - to predict @
friction loss
— v posibilidad f - to, foresee, o anticipate,
@ possibility
— v proveedor m - to foresee @ supplier
previo,via adv autorización f - when authorized
— adv precalentamiento m - [sold] after preheating
previo adv desengrasamiento,grase - after degreasing
— adv engrasamiento,grase m - after greasing
— adv pago m - after payment • fee basis
— adv — m de derecho(s) m - on @ fee basis
previsibilidad* f - predictability
previsible a - foreseeable; conceivable; predic-
table • expected
previsión f - . . . . • anticipation • [accntg] reserve • [fin] budget • [com] projecting,tion;
scheduling; foreseeing • possibility; contem-
plation; planning
— f de carga f - load predicting,tion
— f — exportación f - [com] export(s) foreseeing
— f — pago m - [fin] payment foresseeing
— f — pérdida f - loss predicting,tion

previsión f de proceso m - [ind] process fore-
cast(ing)
— f — producción f - [ind] production fore-
cast(ing)
— f — proveedor m - [com] supplier foreseeing
— f — proyecto m - project specification
— f deducible - [seguros] deductible provision
— f para aislación f - [ambient] insulation
provision
— f — ajuste(s) m - [contab] adjustment reserve
— f — conservación f - prospective maintenance
— f — cuenta(s) f de cobro m dudoso - [com]
doubtful accounts reserve
— f — f dudosa(s) - [com] doubtful accounts
reserve
— f — desvalorización f - [fin] devaluation
reserve
— f — f de inversión(es) f - [fin] invest-
ment(s), devaluation, o impairment, reserve
— f — deudor(es) m moroso(s) - [contab] bad
debt(s) reserve
— f — imprevisto(s) m - [fin] contingency reserve
— f — m diverso(s) - [fin] miscellaneous
contingency,cies reserve
— f — m vario(s) - [fin] miscellaneous
contingency,cies reserve
— f — impuesto(s) m - [fin] tax, reserve, o
provision
— f — m demorado(s) - [fin] deferred tax,
provision, o reserve
— f — m diferido(s) - [fisc] deferred tax,
provision, o reserve
— f — m federal(es) m - [fisc] federal tax,
provision, o reserve
— f — m federal sobre, renta f, o réditos
m - [fisc] federal income tax, reserve, o
provision
— f — m nacional(es) - [fisc] federal tax,
reserve. o provision
— f — m — corriente - [fisc] current federal tax, reserve, o provision
— f — m — sobre, rédito(s) m, o renta f
- [fisc] current federal income tax, reserve, o provision
— f — m nacional - [fisc] federal income
tax, reserve, o provision
— f — m — diferido - [fisc] deferred federal tax, reserve, o provision
— f — m — sobre, réditos, o renta f -
[fisc] deferred federal income tax reserve
— f — m sobre, rédito(s) m, o renta f -
income tax, reserve, o provision
— f — m —, renta f, o réditos m, demorado,
o diferido - [fin] deferred income tax, reserve, o provision
— f — indemnización f - indemnity reserve
— f — f para personal - [labor] personnel
indemnity reserve
— f — inversión(es) f - [fin] investment(s),
reserve, o provision
— f — menoscabo m - [fin] impairment, reserve,
o provision
— f — m de inversión(es) f - [fin] invest-
ment(s) impairment, provision, o reserve
— f — obra f en construcción f - [contab] construction in progress, reserve, o provision
— f — f — f paralizada - [contab] paralyzed construction in progress reserve
— f — obsolescencia f - [contab] obsolescence
reserve
— f — f de material(es) f - [contab] materials obsolescence reserve; obsolete materials reserve
— f — f — mercadería(s) f - obselete
goods reserve
— f — participación f en utilidad(es) f -
[contab] profit, distribution, o sharing,
reserve
— f — pérdida(s) f - [fin] reserve for losses
— f — — debida(s) a fricción f - friction
loss(es), reserve, o provision

previsión f para preparación f - [miner] preparation reserve
— f — quebranto(s) m - [fin] reserve for @
loss(es)
— v — reconocimiento m - [miner] exploration
reserve
— f — reserva(s) f - [legal] reserve provision
— f — — f legal - [legal] legal reserve provision
— f social - [labor] social security
previsor,ra a - . . .; foreseeing • smart
previsoramente adv - foreseeingly • smartly
previsto,ta a - foreseen; expected • anticipated
• prepared; provided; forecast(ed); scheduled
• projected; planned; predicted • design •
contemplated • forthcoming • limiting • intended
— a para carga f - [electr-oper] load rated
— a — — f plena - [electr-oper] full load
rated
— a — — f — continua - [electr-oper] full
load continuous rated
prima f - . . . . • . . .; bonus • amenity | a -
[ind] raw
— f de seguro(s) m - [seguros] insurance premium
— f — — m marítimo - [seguros] maritime insurance premium
— f debida - [seguros] owed premium; premium
due
— f devengada - [seguros] earned premium
— f diaria - [seguros] daily premium • [labor]
daily bonus
— f directa f - direct, premium, o bonus
— f indirecta - indirect, premium, o bonus
— f legal - [contab] legal reserve
— f neta - [seguros] net primium
— f — sobre póliza(s) f emitida(s) - [seguros]
net premium(s) written
— f — — seguro(s) m emitidos - [seguros] net
premiums written
— f no devengada - [seguros] unearned premium
— f normal - [seguros] normal premium
— f obtenida - [labor] obtained, premium, o
bonus
— f pagada - [seguros] paid premium
— f — descarga f rápida f (de buque m) •
[transp] dispatch money
— f por antigüedad f - [labor] seniority premium
— f — cobrar v - [seguros] premium receivable
— f sobre póliza(s) f emitida(s) - [seguros]
premium(s) written
— m — producción f - [ind] production bonus
— f — tonelaje m - [miner] tonnage bonus
— f tal - [seguros] such premium
primado,da a - . . . • prevailed
primar v - to prevail • to govern; to rule
primariamente adv - . . .; basically
primario m - [metal-prod] primary dust catcher •
primary cleaner
— m purgado - [metal-prod] flushed primary dust
catcher
primario,ria a - . . .; prime • basic • coarse •
[mec] input
primas f sobre seguro(s), interés(es), y varios
m - [contab] insurance, premiums, interest
and miscellaneous
— f — —, y varios, diferido(s) - [contab]
deferred insurance, premiums, interest and
miscellaneous
primavera f - [meteorol] . . .; springtime
— f húmeda - [meteorol] wet spring(time)
— f seca - [meteorol] dry spring(time)
primer adv - véase primero,ra • (después de la
labra f modificada)
— arco m - [sold] véase arco m primero
— lugar - [deport] first, o (place)
— nivel m - first level • [admin] first line
— m — de administración f - [admin] first line
management
— relleno m - [constr] first fill
— tercio m - first third

primer usuario m - first user
primera adv - (de) first, o prime. quality, o
class • senior
— instancia f - [legal] first appeal(s)
— intención f - véase Intención f primera
— mitad f - first half
— pasada f - [sold] véase pasada a primera
primeras tres colocaciones f - [deport] first, o
top, three (positions)
primerísimo,ma adv - premium; very first
primero m - [ind] first helper
— m absoluto - [deport] first over all finish
— m de dos - first (of two); former
— m para barra f taponadora - [metal-prod] stopper rod first helper
— m — convertidor(es) m - [metal-prod] converter first helper
— m — cuchara f - [metal-prod] ladleman first
helper
— m — horno(s) m - [metal-prod] (furnace) first
helper; mixer
— m — horno m basculante - [metal-prod] tilting furnace first helper
— m — — m con fosa f - [metal-lam] soaking
pit, first helper, o operator
— m — — m fijo - [combust] stationary furnace
first helper
— m — — m para mezcla f - [metal-prod] mixer
furnace first helper
— m — — nave f para colada f - [metal-prod]
casting, o pouring, bay first helper
— m — máquina f - [ind] machine first helper
— m — — f para escoria f - [metal-prod] slag
bay first helper
— m — vástago m - [metal-prod] stopper rod
first helper
— a y segundo a - [Deport] one-two; first and
second
primero,ra a - . . . • [ind] raw • early • prime
primero,ra en (su) clase f - first of @ kind
— a en clasificación f general - [deport] first
overall
primitivo,va a - . . . • early
primo m hermano - [social] first cousin
primor m - . . . • care
primordial a - . . . • primal; prime; primary •
basic • critical • utmost
primorosamente adv - . . .; carefully
primoroso,sa a - . . .; careful
principal m - manager| a - . . .; chief; master •
primary; prime; premier; major; main; leading
• main; outstanding
— m para horno m - [metal-prod] melter
principalmente adv - . . .; basically
principiante m no profesional - nonprofessional
beginner
— m profesional - [labor] professional beginner
principio m - . . . . • principle; criterion,ria •
fundamental
— m aceptado - accepted principle
— m administrativo - [admin] management principle
— m apareado - matched principle
— m aplicado - applied principle
— m básico - basic (principle); fundamental
— m — para prevención f contra accidente(s) m -
[segurid] accident prevention, principle, o
fundamental
— m comercial - [com] business principle
— m conceptual - conceptual principle
— m contable - [contab] accounting, principle, o
policy
— m — aceptado - [contab] accepted accounting
principle
— m — con aceptación f general - [contab]
generally accepted accounting principle
— m — importante - [contab] important, o significant, accounting principle
— m — prescrip(t)o - [contab] prescribed accounting principle
— m de abatibilidad f - breakaway principle
— m — año - year, start, o beginning
— m — arco m - [arquit] arch principle

principio m de bóveda f - [constr] arch principle
— m —— f distribuidora - [constr] spreader arch principle
— m —— f — para muro m para retención f - [constr] retaining wall spreader arch principle
— m — capítulo m - chapter beginning
— m — causa(s) f múltiple(s) - multiple causes principle
— m para combustión f - [combust] combustion principle
— m — consolidación f—consolidation principle
— m — contabilidad f - [contab] accounting principle
— m — corrosión f - [quím] corrosion principle
— m — desmoldeo m - [metal-lam] stripping beginning
— m — desviación f - [vial] detour beginning
— m — dispositivo m - [mec] device's principle
— m — engranaje(s) m interno(s) - [mec] internal gear(s) principle
— m — estrategia f - strategy principle
— m — fuerza(s) f viva(s) - [mec] vis viva principle
— m — fulcro m - [mec] fulcrum principle
— m — funcionamiento m - operating principle
— m —— m alternativo - [mec] alternating operation principle
— m —— m continuo - [ind] continuous operation principle
— m — ingeniería f - engineering principle
— m —— m establecido - established engineering principle
— m — medición f - [instrum] measuring principle
— m —— f con integrador m electrónico - [instrum] electronic integrator (assisted) measuring principle
— m — nivel(es) m de organización f - [admin] organization level(s) principle
— m — operación f continua - [ind] continuous operation principle
— m — orden m - order, o sequence, principle
— m —— m normal - normal sequence principle
— m — organización f - [admin] organization principle
— m — permeámetro m - [instrum] permeameter principle
— m — prioridad(es) f - [admin] priority,ties principle
— m —— f técnica - [admin] technical priority principle
— m — relación(es) humana(s) - [labor] human relation(s) principle
— m — resultado(s) m para administración f - [admin] management results principle
— m — secuencia f - sequence principle
— m —— f normal - normal sequence principle
— m — seguridad f - [segurid] safety, principle(s), o policy
— m — rejilla f de Davy - [miner] Davy screen principle
— —— m tratamiento m - [mader] treatment, principle, o fundamental
— m —— m para madera(s) f - [mader] wood, o lumber, treatment, principle, o fundamental
— m —— m y preservación f - [mader] treatment and preservation, principle, o fundamental
— m —— m —— f de madera f - [mader] wood treatment and preservation fundamental
— m — ubicación f - [constr] location principle
— m —— f para alcantarilla f - [constr] culvert location principle
— m prescri(p)to - prescribed principle
— m sano - sound principle
— m — para combustión f - [combust] sound combustion principle
— m técnico - technical principle
prioridad f - . . . . • [vial] right of way
— f alta - high priority

prioridad f asignada - assigned, o given, priority
— f baja - low priority
— f dada - given priority
— f para carga f - [ind] charge,ging priority
— f — dirección f - [mec] steering priority
— f — investigación f - [ind] research priority
— f — paso m - [vial] right of way
— f — rechazo m primero - [legal] right of first refusal
— f — servicio m - service priority
— f técnica - [admin] technical priority
prioritario,ria* a - prioritary*
priorización* f - prioritization,zing
prioritizado,da* a - prioritized
prioritizar* v - to prioritize
prisa f - . . .; hurry | adv - (de) in @ hurry
prisionero m - . . . . • [mec] stud, o screw, bolt; pin; headless (set) screw • cap screw
— m ahusado - [mec] taper(ed), stud, o set screw
— m — inferior - [mec] bottom, o lower, stud. o set screw
— m — superior - [mec] top, o upper, stud, o set screw
— m con cabeza f hueca - [mec] socket, stud, o set screw
— m —— f — ranurada - [mec] fluted socket, stud, o head screw
— m — concavidad f hexagonal - [mec] Allen screw
— m corto - [mec] short, stud, o set screw
— m fijador - [mec] set screw
— m hueco - [mec] hollow, stud, o set screw
— m inferior - [mec] lower, o bottom, stud, o set screw
— m largo - [mec] long, stud, o set screw
— m para acoplamiento m - [mec] coupling stud
— m — afirmación f - [mec] steady pin
— m —— f para zapata f - [mec] shoe steady, stud, o set screw, o pin
— m —— f —— f para freno m - [mec] brake shoe steady pin
— m — ajuste m - [mec] (headless) stud, o set screw, o socket (screw)
— m —— m con filo m anular - [mec] headless, stud, o set screw
— m —— m corto - [mec] short (headless) set screw
— m —— hueco m - [mec] hollow set screw
— m —— m con macho m corto - [mec] hollow half dog point set screw
— m —— m ranurado - [mec] slotted headless, stud, o set screw
— m — caja f deslizadora - [mec] sliding box, stud, o set screw
— m — casquete m - [mec] cap, stud, o set screw
— m —— m para eje m - [mec] shaft cap, stud, o set screw
— m —— m — eje m - [mec] shaft cap, stud, o set screw
— m —— m —— m principal - [mec] main shaft cap, stud, o set screw
— m — cierre m - [mec] lock(ing), stud, o set screw
— m —— m ahusado - [mec] taper(ed) lock, stud, o set screw
— m —— m — inferior - [mec] lower, o bottom, taper(ewd) lock, stud, o set screw
— m —— m — superior - [mec] top, o upper, taper(ed) lock, stud, o set screw
— m —— m inferior - [mec] bottom, o lower, lock, stud, o set screw
— m —— m superior - [mec] top, o upper, lock, stud, o set screw
— m — culata f - [comb.int] head, stud, o set screw
— m —— f a bloque m - [comb.int] head to block, stud, o set screw
— m —— f para cilindro(s) m - [comb.int] cylinder head, stud, o set screw
— m —— f —— m a bloque m - [comb.int] cylinder head to block, stud, o set screw
— m — cuña f - [mec] wedge, stud, o set screw

prisionero m para cursor m - [mec] slide, stud,
u set screw
— m — — m enderezador - [mec] straightening
slide, stud, o set screw
— m — engranaje m - [mec] gear, stud, o set
screw
— m — — m para marcha f sin carga f—[bombas]
idler gear, stud, o set screw
— m — m — — f — — f de bomba f -
[bombas] pump idler gear, stud, o set screw
— m — — m — — f — — f — — f para aceite
m - [comb.int] oil pump idler gear, stud, o
set screw
— m — fijación f - [mec] headless set screw
— m — — f ranurado - [mec] slotted headless,
stud, o set screw
— m — montaje m - [mec] mounting, stud, o set
screw
— m — — m para soporte m - [mec] bracket
mounting, stud, o set screw
— m — — — m — — m para alternador m - [comb.
-int] alternator bracket mounting, stud, o
set screw
— m ranurado - [mec] slotted, stud, o set screw
— m — para ajuste m - [mec] slotted headless
set screw
— m — roscado m - [mec] slotted headless set
screw
— m — — para ajuste m - [mec] slotted headless set screw
— m — recto - [mec] straight stud
— m — — enteramente - [mec] straight all-
-thread(ed) stud
— m — — íntegramente - [mec] straight all-
-thread(ed) stud
— m roscado - [mec] stud • pin drive
— m — enteramente - [mec] all-thread(ed) stud
— m — íntegramente - [mec] all-thread(ed) stud
— m sujetador - [mec] set screw
— m superior - top, o upper, stud, o set screw
prisma m de material (para relleno m) - [constr]
fill prism
— m — relleno m - [constr] fill prism
— m — suelo m - [constr] soil prism
— m exterior - exterior, o outside, prism
— m — de relleno m - [constr] exterior, o outside, fill prism
prístino,na a - . . . .; undisturbed
privación f - depriving,vation
— f de corriente f - [electr-oper] de-energizing
privado,da a - . . . . • particular
privar v de corriente f - [electr-oper] to de-
-energize
privativo,va a - . . . . • private preserve
privilegio m - , , , , • [legal] franchise
— m de salario m - [labor] wage privilege
pro. . . - [legal] assistant; under
—forma a - pro-forma
— rally a - [deport] pro rally
— — internacional - [deport] international pro
rally
— — nacional - [deport] national pro rally
— rata a - véase prorrata
probabilidad f pluvial - [meteorol] rainfall expectancy
— f pluviométrica - [meteorol] rainfall expectancy
probado,da a - . . . .; proven; tested; probed
— a durante año(s) a - time, proved, o tested
— a en servicio m - service, proved, o tested
— a hidrostáticamente - tested hydrostatically
— a ya - already, o previously, tested
probador m - tester
— m de tipo m sonoro - sound type tester
— m — — m — para regulación f para encendido
m - [comb.int] sound type timing tester
— m para absorción f - [petról] absorption tester
— m — aislación f - [instrum] insulation tester
— m — aislamiento m - [instrum] insulation
tester

probador m para amperio(s) - [instrum] amprobe*;
ampere tester
— m — inducido(s) m - [instrum] growler
— m — regulación f para encendido m - [comb.-
-int] timing tester
— m — tubería(s) f - [tub] pipe tester
— m — — f para entubación f - [petról] casing tester
— m sonoro - [instrum] sound tester
probar v - . . . . • to assay • to probe
— v electrodo m - [instrum] to test @ electrode
— v en fábrica f - [ind] to factory test
— v — planta f - to plant test
— v equipo m - [ind] to test @ equipment
— v hidrostáticamente - to test hydrostatically
— v material m - [ind] to test @ material
— v muestra f - to test @ sample
— v resistencia f - to test @ resistance
— v — f a humedad f - to test @ moisture resistance
— v rueda f - [mec-ruedas] to test @ wheel(s)
— v — f para riel(es) m - [mec-ruedas] to
test @ track wheel(s)
— v transformador m - [electr-oper] to test @
transformer
— v tubería f - [tub] to test @, pipe, o main
— v — f para agua m - [tub] to test @ water,
pipe, o main
— v — f — — m para venturi m - [metal-prod]
to test @ Venturi water main
— v ventilador m - to, test, o try, @ fan
probeta f - [quím] . . .; flask • [sold] specimen • [metal-prod] test(ing), piece, o specimen; sample
— f aluminizada - [metal-trat] aluminized specimen
— f con sección f circular - [mec] circular, o
round, section specimen
— f — borde m no protegido - [metal] unprotected edge specimen
— f — — m protegido - [metal] protected edge
specimen
— v con entalle m - [mec] notch(ed) test piece
— f — — m en V—[mec] V-notch(ed) test piece
— f cuadrada - [mec] square, test piece, o
specimen
— f de banda f - [metal-lam] strip specimen
— f — — f aluminizada - [metal-trat] aluminized strip specimen
— f — cinta f - [metal-lam] strip specimen
— f — — f aluminizada - [metal-trat] aluminized strip specimen
— f — chapa f - [metal-lam] strip specimen
— f — — f aluminizada - [metal-trat] aluminized strip specimen
— f — fleje m - [metal-lam] strip specimen
— f — — m aluminizado - [metal-trat] aluminized strip specimen
— f — imán m - [electr] magnet specimen
— f — metal m de aportación f - [sold] all
weld metal specimen
— f — — m soldador - [sold] all-weld-metal
test specimen
— f de tubería f - [tub] pipe, test piece, o
specimen
— f graduada - [labp graduate (specimen)
— f Iliovici - [instrum] Iliovici specimen
— f longitudinal - [metal-lam] longitudinal, o
lengthwise, test piece, o sample
— f no protegida - unprotected specimen
— f para ensayo(s) m - [metal] (test) specimen
— f — — m de entalle m - [metal] nick break
specimen
— f — — m de muesca(s) f - [metal] nick break
specimen
— f — — m para tracción f - [metal-prod] tensile, o traction, test piece, o specimen
— f — — m graduada - [lab] graduate
— f — — m de corte m (con cizalla f) - [mec]
shearing, test piece, o specimen
— f — tracción f - [metal-lam] véase probeta f
para ensayo(s) m para tracción f
— f protegida - protected specimen

probeta f̲ recocida - [metal-trat] annealed, test
piece, o̲ specimen
— f̲ **representativa** - [ind] representative, test
p̲iece, o̲ specimen, o̲ sample
— f̲ **significativa** - [ind] significant specimen
— f̲ **transversal** - [metal-lam] transverse, test
piece, o̲ specimen, o̲ sample
problema m̲ - . . .; trouble • worry; question;
woe • factor • bottleneck
— m̲ **actual** - present, o̲ today's, problem
— m̲ **adicional** - additional problem
— m̲ **aislado** - isolated problem
— m̲ **altamente técnico**—highly technical problem
— m̲ **anterior** - above, o̲ previous, problem
— m̲ **aparente** - apparent, problem, o̲ trouble
— m̲ **aparte** - separate problem
— m̲ **causado por agua m̲ freática** - [hidr]
groundwater problem
— m̲ — **por cerca f̲** - [comput] fence problem
— m̲ — — f̲ **de estaca(s) f̲** - [comput] picket
fence problem
— m̲ — — **suelo m̲** - [suelos] soil problem
— m̲ **comenzado** - started problem
— m̲ **comprendido** - [admin] understood problem
— m̲ **común** - common problem
— m̲ **con abastecimiento** - supply(ing) problem
— m̲ — — m̲ **de agua m̲** - [hidr] water supply
problem
— m̲ — **acero m̲** - [metal-prod] steel problem
— m̲ — **adherencia f̲** - adherence problem
— m̲ — **agotamiento** - [constr] drainage problem
• depletion problem
— m̲ — **agrietamiento** m̲—[sold] cracking problem
— m̲ — **agua m̲** - [hidr] water problem
— m̲ — **alabeo** m - warpage problem
— m̲ — — m̲ **debido a calor m̲** - heat warpage
problem
— m̲ — **alimentación f̲** - feed(ing) problem
— m̲ — — f̲ **de alambre m̲** - [sold] wire feeding
problem
— m̲ — — f̲ — **fundente m̲** - [sold] flux feeding
problem
— m̲ — — f̲ **deficiente de combustible** - [comb.-
-int] fuel starvation problem
— m̲ — — f̲ **insuficiente de combustible m̲** -
[comb.int] fuel starvation problem
— m̲ — **almacenamiento** m̲ - storage problem
— m̲ — **altura f̲ limitada** - [constr] limited
headroom problem
— m̲ — **amoladura f̲** - [mec] grinding problem
— m̲ — **anegación f̲** - [hidr] flooding problem
— m̲ — **aparcamiento** m̲ - [vial] parking problem
— m̲ — **arrollamiento** m̲ - roll(ing)-up problem
— m̲ — **aserrado** m̲ - [mec] sawing problem
— m̲ — **bobina(s) f̲ caliente(s)** - [metal-lam]
hot coil problem
— m̲ — **borde(s) m̲** - [metal-lam] edge(s) problem
— m̲ — **cámara f̲** - [autom-neumát] tube problem
— m̲ — **cambio(s) m̲** - [mec] shift(ing) problem •
[fin] exchange problem
— m̲ — — m̲ **para marcha f̲** - [mec] shift(ing)
problem
— m̲ — **carburación f̲** - [comb.int] carburation
problem
— m̲ — **carburador m** - [comb.int] carburetor,
problem, o̲ trouble
— m̲ — **carga f̲** - load(ing) problem
— m̲ — — f̲ **excesiva** - overload, o̲ excess load,
problem
— m̲ — **cascarilla f̲** - [metal-lam] scale problem
— m̲ — — f̲ **en banda f̲** - [metal-lam] strip
scale problem
— m̲ — — f̲ — — f̲ **gruesa** - [metal-lam] thick
strip scale problem
— f̲ — — f̲ — **bobina(s) f̲** - [metal-lam] coil
scale problem
— m̲ — — f̲ — — f̲ **caliente(s)** - [metal-lam]
hot coil scale problem
— m̲ — — f̲ — **cinta f̲** - [metal-lam] thick
strip scale problem
— m̲ — — f̲ — **cinta f̲ gruesa** - [metal-lam]
thick strip scale problem
— m̲ — — f̲ — **chapa f̲** - [metal-lam] strip

scale problem
peoblema m̲ **con cascarilla f̲ en chapa f̲ gruesa** -
[metal-lam] thick strip scale problem
— f̲ — — f̲ **en fleje m̲** - [metal-lam] strip
scale problem
— m̲ — — — m̲ **grueso** - [metal-lam] thick
strip scale problem
— m̲ — **cimentación f̲** - [constr] foundation
problem
— m̲ — **circuito m̲** - circuit problem
— m̲ — — m̲ **para carga f̲** - [electr-instal]
charging circuit problem
— m̲ — **combustión f̲** - [combust] combustion
problem
— m̲ — — f̲ **en horno m̲** - [combust] furnace combustion problem
— m̲ — — f̲ — — m̲ **para recalentamiento m̲** -
[metal-lam] reheating furnace combustion
problem
— m̲ — **conducción f̲** - [autom] handling, o̲ driving, problem
— m̲ — **conservación f̲** - [ind] maintenance
problem | a - [ind] (con) maintenance prone
— m̲ — **consolidación f̲** - [suelos] consolidation problem
— m̲ — — f̲ **de suelo(s) m̲** - [constr] soil, o̲
earth, consolidation, o̲ retention, problem
— m̲ — **construcción f̲** - [constr] construction,
o̲ building, problem
— m̲ — **contaminación f̲** - pollution problem
— m̲ — — f̲ **de agua m̲** - [hidr] water pollution
problem
— m̲ — — f̲ — **aire m̲** - [ambient] air pollution problem
— m̲ — **control m̲** - [ind] control problem
— m̲ — — f̲ — **calidad f̲** - [ind] quality control problem
— m̲ — **corrosión f̲** - [metal] corrosion problem
— m̲ — **crecimiento m̲** - growing, problem, o̲ pain
— m̲ — **dentición f̲** - [medic] teething problem
— m̲ — **desagüe(s) m̲** - [sanit] sewer problem
— m̲ — — m̲ **exiguo(s)** - [sanit] inadequate
sewer (system) problem
— m̲ — — m̲ **inadecuado(s)** - [sanit] inadequate
sewer (system) problem
— m̲ — — m̲ **pluvial(es)** - [hidr] storm, sewer,
o̲ drainage, problem
— m̲ — **desgaste m̲** - [mec] wear problem
— m̲ — — m̲ **de borde(s) m̲** - [autom-neumát]
edge wear problem
— m̲ — **deslizamiento(s) m̲** - [constr] landslide
problem
— m̲ — **deterioro,ración** - deterioration problem
— m̲∴ — **diferencial m̲** - [autom-mec] differential
problem
— m̲ — **diseño m̲** - design problem
— m̲ — **distorsión f̲** - [mec] distortion problem
— m̲ — **distribución** f- distribution problem
— m̲ — **distribuidor** - [comb.int] distributor
problem
— m̲ — **drenaje m̲** - [hidr] drainage problem
— m̲ — **elaboración f̲** - fabrication problem
— m̲ — **eliminación f̲** - [sanit] disposal problem
— m̲ — **embalse m̲** - [hidr] flooding problem
— m̲ — **empaquetadura f̲** - [mec] packing problem
— m̲ — **encendido m̲** - [comb.int] ignition, problem, o̲ trouble
— m̲ — **encharcamiento m̲** - [hidr] ponding problem
— m̲ — **equipo m̲** - [ind] equipment problem
— m̲ — **erosión f̲** - erosion, problem, o̲ trouble
— m̲ — — f̲ **resuelto** - (re)solved erosion
problem
— m̲ — — f̲ **solucionado** - solved erosion problem
— m̲ — **escamas f̲** - [metal-lam] scale problem
— m̲ — **estanqueidad f̲** - sealing problem
— m̲ — — f̲ **en punto m̲ de extracción f̲** - [min]
extraction point sealing problem
— m̲ — **etalaje m̲** - [metal-prod] bosh problem
— m̲ — **evacuación f̲** - [sanit] disposal problem
— m̲ — — f̲ **de líquido(s) m̲** - [sanit] liquid(s)
disposal problem

problema m con evacuación f de líquido(s) cloa-
cal(es) - [sanit] sewage disposal problem
— m — explotación f minera - [miner] mine exploitation problem
— m — extracción f minera - [miner] mining extraction problem
— m — fabricación f - fabrication, o manufacturing, problem
— m — extremo m - end problem
— m — falseo m - falsing problem
— m — m causado por cerca f de estaca(s) f
- [comput] picket fence falsing problem
— m — fijación f - holding, o restraint,ning,
problem
— m — filtración f - [constr] filtering, o
leakage, problem
— m — f en obra f - [constr]½ field leakage
problem
— m — fuente - source problem
— m — f para energía f - [electr-distrib]
power source problem
— m — fuga(s) f - leakage problem
— m — f de aire m - air leakage problem
— m — henchimiento m por succión f - [petról]
suction filling problem
— m — hierro m metalizado - [metal-hierro]
metallized iron problem
— m — holgura f - [mec] clearance problem
— m — horno m - [combust] furnace problem
— m — m alto - [metal-prod] blast furnace
problem
— m — m para recalentamiento m - [metal-
-lam] reheating furnace problem
— m — inundación f - [hidr] flooding problem
— m —, manejo m, o maniobra(s) f, o manipuleo
m - handling problem
— m — mantenimiento m - [ind] maintenance
problem
— m — m de rodillo(s) m - [metal-lam] roll
maintenance problem
— m — m — m especial - [metal-lam]
special roll maintenance problem
— m — material(es) m - [ind] material(s) problem
— m — motivación f - [labor] motivation problem
— m — motor m - [electr-mot] motor problem •
[comb.int] engine problem
— m — m para ventilador m - [mec] fan motor
problem
— f — navegación f - [nav] navigation(al)
problem
— m — neumático(s) m - [autom-neumát] tire
problem
— m — óxido m - [metal] oxide problem
— m — pandeo m - [constr] buckling problem
— m — pegadura(s) f - sticking problem
— m — pegamiento m - sticking problem
— m — m en cráter m - [sold] crater sticking problem
— m — perforación f - [mec] drilling problem •
[sold] burn through problem
— m — pistón m - [petról] piston problem
— m — puente m - [constr] bridge,ging problem
— m — ranuración f - [mec] slotting, o grooving, problem
— m — recalentamiento m - overheating problem
— m — m de motor m - [electr-mot] motor
overheating problem • [comb.int] engine overheating problem
— m — rectificador m - [electr-equip] rectifier, problem, o trouble
— m — m con silicio m - [electr-equip] silicon rectifier, problem, o trouble
— m — reemplazo m - [mec] change,ging problem
— m — m de rodillo(s) m - [metal-lam] roll
change,ging problem
— m — refractario(s) m - [cerám] refractory,ries problem
— m — refrigeración f - cooling problem
— m — regulación f [ind] control(ling) problem
— m — f de (medio) ambiente - [ambient] environmental control problem

problema m con renovación f- renewal problem
— m — f urbana = [constr] urban renewal
problem
— m — reparación f - [ind] repair problem
— m — f de rodillo m - [metal-lam] roll
repair problem
— m — f de rodillo m especial - [metal-lam]
special roll repair problem
— m — rigidez f - [mec] restraint, o rigidity,
problem
— m — seguridad f - [segurid] safety problem
— m — sobrecalentamiento m - overheating
problem
— m — m de motor m - [electr-mot] motor
overheating problem • [comb.int] engine
overheating problem
— m — sobrecarga f - overload(ing) problem
— m — sobrevoltaje m - [electr-oper] overvoltage, o excess voltage, problem
— f — socavación f - undercutting problem
— m — f interna - [sold] internal under-
cut(ting) problem
— m — soldadura f - [sold] welding problem
— m — subdrenaje m - [hidr] subdrainage problem
— m — suelo m - [suelos] soil problem
— m — m fluido - [suelos] fluid soil problem
— m — supervisión f - [admin] supervision
problem
— m — texturación* f - [constr] texturing
problem
— m — tiro,raje m - [combust] draft problem
— m — tracción f - [autom-neumát] traction
problem
— m — transmisión f - [mec] transmission,
problem, o trouble
— m — vaciado m - pouring problem
— m — válvula f - [válvl] valve problem
— m — velocidad f - [mec] speed problem
— m — ventilador m - [mec] fan problem
— m — voltaje m - [electr-oper] voltage problem
— m — m deficiente - [electr-oper] under-
voltage problem
— m — m excesivo - [electr-oper] overvoltage problem
— m conocido - known problem
— m continuado - continuing, o ongoing, problem
— m con transmisión f - [mec] ongoing transmission, problem, o trouble
— m corregido - corrected, o solved, problem
— m cotidiano - daily, o every-day, problem
— m para elevación f - [grúas] daily, o
every-day, lifting, o hoisting, problem
— m de importancia f - major problem
— m ingeniería f - [constr] engineering
problem
— m — f difícil - [constr] tough, o hard,
o difficult, engineering problem
— m rutina - routine problem
— m tránsito - [transp] traffic problem
— m deparado - caused, o problem, o trouble
— m diario - daily, o every-day, problem
— m diferente - different problem
— m difícil - difficult, o tough, problem
— m de ingeniería f - [constr] tough engineering problem
— m disuasivo - deterrent,ring problem
— m eléctrico - [electr-oper] electric(al)
problem
— m con motor m - [electr-mot] motor electrical problem
— m — m para ventilador m - [mec] fan
motor electric(al) problem
— m — ventilador m - [mec] fan electric(al)
problem
— m eliminado - eliminated problem
— m elusivo - elusive, problem, o matter
— m empezado - started problem
— m en alimentación f de alambre m - [sold]
wire feeding problem
— m — banda f gruesa - thick strip problem

problema m̲ en caja f̲ para cambios m̲ - [autom] gearbox, problem, o̲ trouble, o̲ woe
— m̲ — cambio m̲ de rodillo(s) m̲ - [metal-lam] roll, changing, o̲ replacement, problem
— m̲ — carrera f̲ - [deport] race problem
— m̲ — cinta f̲ - [metal-lam] strip problem
— m̲ — — f̲ gruesa - [metal-lam] thick strip problem
— m̲ — circuito m̲ - [electrón] circuit problem
— m̲ — circulación f̲ - [vial] traffic problem
— n̲ — comunicación(es) f̲ - [labor] communica-tion(s) problem
— m̲ — día m̲ de carrera f̲ - [deport] race day problem
— f̲ — chapa f̲ - [metal-lam] strip problem
— m̲ — — f̲ gruesa - [metal-lam] thick strip problem
— m̲ — diseño m̲ - [ind] design problem
— m̲ — — m̲ de material(es) m̲ - [ind] materials design problem
— m̲ — fleje m̲ - [metal-lam] strip problem
— m̲ — — m̲ grueso - [metal-lam] thick strip problem
— m̲ — industria f̲ - [ind] industry problem
— m̲ — — f̲ para construcción f̲ - [constr] con-struction industry problem
— m̲ — ingeniería f̲ - [ind] engineering problem
— m̲ — instalación f̲ - [ind] installation, pro-blem, o̲ difficulty
— m̲ — motor m̲ - [electr-mot] motor problem • [comb.int] engine problem
— m̲ — — m̲ para ventilador m̲ - [mec] fan motor problem
— m̲ — obra f̲ - [constr] field problem
— n̲ — operación f̲ - [ind] operation(al), pro-blem, o̲ trouble
— m̲ — planta f̲ - [ind] plant problem
— m̲ — — f̲ para ácido m̲ - [quím] acid plant problem
— m̲ — f̲ — — m̲ sulfúrico - [quím] sulfuric acid plant problem
— m̲ — producción f̲ - [ind] production problem
— m̲ — — f̲ de acero m̲ - [metal-prod] steel production, problem, o̲ bottleneck
— m̲ — red f̲ - [ind] network, o̲ system, problem
— m̲ — — f̲ para telecomunicación(es) f̲ - [telecom] telecommunications network problem
— m̲ — reemplazo m̲ - [ind] replacement problem
— m̲ — — m̲ de rodillo(s) m̲ - [metal-lam] roll replacement problem
— m̲ — sitio m̲ para aparcamiento m̲ - [ind] parking (area) problem
— m̲ — soldadura f̲ - [sold] welding problem
— m̲ — superficie f̲ - [metal-prod] surface pro-blem
— m̲ — telecomunicación(es) f̲ - [telecom] tele-communication(s) problem
— m̲ — trabajo m̲ - [labor] work, o̲ job, problem
— m̲ — transmisión f̲ - [autom-mec] transmission problem
— m̲ — transporte m̲ - [transp] transportation, o̲ shipping, problem
— m̲ encontrado - encountered problem
— m̲ energético - [energ] energy problem
— m̲ escolar - [educ] schools, o̲ educational, problem
— m̲ — en centro(s) m̲ urbano(s) - [educ] inner--city education problem
— m̲ especial - special problem
— m̲ — para drenaje m̲ - [hidr] special drainage problem
— m̲ — — proyección f̲ - special design problem
— m̲ específico - specific problem
— m̲ evaluado - evaluated problem
— m̲ evitado - avoided, o̲ prevented, problem
— m̲ existente - existing problem
— m̲ experimentado - experienced problem
— m̲ experimental - experimental problem
— m̲ fácil - easy, o̲ simple, problem
— m̲ físico - [medic] physical problem
— m̲ frecuente - frequent problem
— m̲ futuro - future problem
— m̲ general - general problem

problema m gremial - [labor] labor problem
— m̲ hidráulico - [hidr] hydraulic problem
— m̲ hipotético - hypothetical problem
— m̲ idéntico - identical problem
— m̲ iniciado - started, o̲ initiated, problem
— m̲ intermitente - intermittent problem
— m̲ laboral - [labor] labor problem
— m̲ legal - [legal] legal problem
— m̲ mayor - major problem • main, o̲ principal, problem • bigger, o̲ greater, problem • big-gest, o̲ greatest, problem
— m̲ mecánico - [mec] mechanical problem
— m̲ — con motor m̲ - [electr-mot] motor mechan-ical problem • [comb.int] engine mechanical problem
— m̲ — — m̲ para ventilador m̲ - [mec] fan motor mechanical problem
— m̲ — — ventilador m̲ - [mec] fan mechanical problem
— m̲ — persistente - [mec] nagging mechanical problem
— m̲ menor - minor problem • smaller problem • smallest problem
— m̲ minero - [miner] mining problem
— m̲ nuevo - new problem
— m̲ ocurrido - ocurred problem
— m̲ operativo - [ind] operating problem
— m̲ para alineación f̲ - [mec] alignment,gning problem
— m̲ — balanceo m̲ - [autom-neumát] (tire) balance,cing problem
— m̲ — conservación f̲ - [ind] maintenance pro-blem • service problem
— m̲ — drenaje m̲ - [hidr] drainage problem
— m̲ — — m̲ agravado - [hidr] aggravated, o̲ worsened, drainage problem • complicated drainage problem
— m̲ — — m̲ complicado - [hidr] complicated drainage problem
— m̲ — equilibrio m̲ - balance,cing problem
— m̲ — explotación f̲ - [miner] exploitation problem
— f̲ — fiscalización f̲ - [ind] control problem
— m̲ — — f̲ de calidad f̲ - [ind] quality con-trol problem
— m̲ — funcionamiento m̲ - [ind] operational problem
— m̲ — operación f̲ - [ind] operating problem
— m̲ — pontear,teo - [constr] bridging problem
— m̲ — procesamiento,so m̲ - [ind] processing problem
— m̲ — proyección f̲ - design(ing) problem
— m̲ — reemplazo m̲ - replacement problem
— m̲ — — m̲ prematuro - [ind] early replacement problem
— m̲ — soldadura f̲ - [sold] welding problem
— m̲ — subdrenaje m̲ ferroviario - [f.c.] rail-way subdrainage problem
— m̲ — — m̲ para carretera f̲ - [constr] highway subdrainage problem
— m̲ — verificación f̲ - [ind] control problem
— m̲ — — f̲ para calidad f̲ - [ind] quality control problem
— m̲ persistente - nagging, o̲ continuing, pro-blem
— m̲ poco común - uncommon, o̲ unusual, problem
— m̲ posible - possible, o̲ potential, problem
— m̲ prematuro - premature, o̲ early, problem
— m̲ — para reemplazo m̲ - premature, o̲ early, replacement problem
— m̲ previsto - foreseen problem • prevented problem
— m̲ principal - principal, o̲ main, problem
— m̲ progresivo - progressive, problem, o̲ trou-ble
— m̲ — con erosión f̲ - progressive erosion, problem, o̲ trouble
— m̲ propio - own problem
— m̲ repetido - repeated,petitive problem
— m̲ repetidor - repetitive problem
— m̲ repetitivo* - repetitive problem
— m̲ resuelto - (re)solved problem
— m̲ sanitario - [medic] health problem

problema m serio - serious problem
— m — con material(es) m - [ind] serious mate-
rial(s) problem
— m — para conservación f - [ind] serious
maintenance problem
— m — para mantenimiento m - [ind] serious
maintenance problem
— m — diseño m - [ind] serious design problem
— m — — m de material(es) m - [ind] serious material(s) design problem
— m simple - simple problem
— m sindical-administrativa - [labor] (labor)
union-, management, o employer, problem
— m singular - unique problem
— m solucionado - solved problem
— m técnico - technical problem
— m temporario - temporary problem
— m — de seguridad - [segurid] temporary safety problem
— m tenaz - tough problem
— m urbano - [pol] urban problem
— m variado - varied, o different, problem
— m viejo - old problem
problemas m múltiples para conservación f - mul-
tiple maintenance problems | adv - (con)
maintenance prone
— m y soluciones f - [fam] tricks and treats
problemático,ca a - . . . ; troublesome
procedencia f - . . . • propriety | adv - (de)
source • [nav] home port
— m de fondo(s) m - [fin] fund(s) source
— f mexicana - Mexican origin
— f nacional - [com] local origin
procedente de adv - proceeding, o coming, from
proceder v - . . . . • to be in order • to play @
thing
— v con cautela f - to, exercise, o proceed
with, caution
— v — cuidado - to, take care, o use caution
— v de - to, come, o originate, from
— v — acuerdo m - to act accordingly
— v en forma f contraria - to do @ opposite
— v — — f correcta - to act correctly
— v — — f inversa - to do @ opposite
— v — otra forma - to do otherwise; to fail to
do
— v para arrancar - [comb.int] to proceed to
start
— v — puesta f en marcha f - [comb.int] to
proceed to start
— v sin premura v - to take @ time
procedido,da a - proceeded; gone ahead; acted
— a de acuerdo - acted accordingly
— a para arrancar - [comb.int] to proceed to
start
— a — puesta f en marcha - [comb.int] proceeded to start
procedimiento m - . . . . • method • approach |
adv - (de) procedural
— m actual - present procedure
— m — para proyección f - present. o actual,
design procedure
— m adecuado - adequate, o proper, procedure
— m administrativo - [admin] administrative, o
management, procedure
— m — para procedimiento m para solidificación
f - [nucl] solidification processing management procedure
— m ajustado - adjusted procedure
— m anual - yearly procedure
— m — para conservación f - [ind] yearly maintenance procedure
— m aplicable - applicable procedure
— m aprobado - approved procedure
— m apropiado - appropriate, o proper, procedure
— m — para puesta f a tierra f - [electr-inst]
appropriate, o proper, grounding procedure
— m — soldadura f - [sold] appropriate, o
proper, welding procedure
— m — — f para línea(s) f para conducción
- [sold] appropriate, o proper, pipeline
(welding) procedure

procedimiento m apropiado para soldadura f de
tubería(s) f para conducción f - [sold] proper pipeline (welding) procedure
— m — seguido - followed proper procedure
— m básico - basic procedure
— m bueno - [sold] good, o sound, procedure
— m — para mantenimiento m - [ind] good maintenance procedure
— m — — soldadura f - [sold] good welding
procedure
— m cambiado - changed procedure
— m comentado - discussed procedure
— m completo - complete procedure
— m — para sangría f - [comb.int] complete
bleeding procedure
— m complicado - involved procedure
— m comprobado - proven procedure
— m con alineación f - [ind] alignment,gning
procedue
— m con arco m abierto - [sold] open arc procedure
— m — — m — con alimentación f (Squirt) automática de electrodo m - [sold] open arc
semi-automatic Squirt process; semiautomatic
Squirt process
— m — — m — con voltaje m bajo - [sold] low
voltage open arc procedure
— m — — autoprotegido - [sold] self-shielded arc, procedure, o process
— m — — m en tándem - [sold] tandem arc procedure
— m — — m sumergido - [sold] submerged arc,
procedure, o process
— m — baño m de escoria f - [metal-prod] slag
bottom process
— m — convertidor m - [metal-prod] coverter,
procedure, o process
— m electrodo m con alma m fundente - [sold]
controlled Innershield, procedure, o process
— m — — m — — m — — autoprotegido -
[sold] Innershield self-shielded process
— m — — muy sobresaliente - [sold] long
stickout, o Linc-Fill, procedure
— n — — m poco sobresaliente - [sold] short
stickout procedure
— m — — m muy sobresaliente - [sold] long
stickout procedure
— m — — m sobresaliente - [sold] Linc-Fill
procedure
— m — hidrógeno m bajo - [sold] low hydrogen,
process, o procedure
— m — manipulador m - [electrón] keyer processing
— m — — m para tono(s) m - [electrón] tone
keyer processing
— m — pasada(s) f múltiple(s) - [sold] multiple pass(es) procedure(s)
— m — — f única(s) - [sold] single pass(es)
procedure
— m — . . . paso(s) m - . . . .-step procedure
— m — punta f coladora - [hidr] well point
procedure
— m — retroceso m - [sold] back-step procedure
— m — — m salteado - [sold] skip back-step
procedure
— m — tren m mandril - [tub] plug mill process
— m — vacío m - [sold] vacuum process
— m — voltaje m constante - [sold] constant
voltage, process, o procedure
— m conciso - concise procedure
— m confiable - reliable, o trustworthy, procedure
— m contable - [contab] accounting procedure
— m continuo - continuous procedure
— m contradictorio - contradictory, o inconsistent, procedure
— m — para ensayo(s) m - [ind] constadictory,
o inconsistent, test(ing) procedure
— m correctivo - [labor] corrective procedure
— m — normal - [labor] normal corrective procedure
— m correcto - correct procedure
— m — seguido - followed correct procedure
— m corriente - common, o standard, procedure

procedimiento m̲ corriente para empaque m̲ -
[ind] current, o̲ standard, packing procedure
— m̲ —— **proyección** f̲ - [ind] present design
procedure
— m̲ **creado** - created, o̲ developed, procedure
— m̲ **dado** - given procedure
— m̲ — **para laminación** f̲ - [metal-lam] given
rolling procedure
— m **de acuerdo con** - acting accordingly
— m̲ — **Burton** - [petról] Burton, process, o̲
procedure
— m̲ — **paso(s)** m̲ **para agravio(s)** m̲ - [labor]
grievance step(s) procedure
— m̲ — **práctica** f̲ - practice procedure
— m̲ —— f̲ **estándar** - standard practice procedure
— m̲ — **rutina** - routine procedure
— m̲ — **para, conservación** f̲, o̲ **mantenimiento**
m̲ - [ind] routine maintenance procedure
— m̲ — **tipo** m̲ **continuo** - [ind] continuous type
procedure
— m̲ **debajo de grúa** f̲ - [grúas] underneath @
crane procedure
— m̲ **destructivo** - destructive procedure
— m̲ **desulfurante** - [metal-prod] desulfurizing
procedure
— m̲ **determinado** - determined, o̲ given, procedure
— m̲ — **para laminación** f̲ - [metal-lam] given
rolling procedure
— m̲ **diario** - [ind] daily, o̲ day-by-day, procedure
— m̲ — **para, conservación** f̲, o̲ **mantenimiento** m̲
- [ind] daily, o̲ day-by-day, maintenance procedure
— m̲ **distinto** - different, o̲ other, procedure
— m̲ **divulgado** - [ind] published, o̲ divulged,
procedure
— m̲ **efectivo** - effective procedure
— m̲ — **para, conservación** f̲, o̲ **mantenimiento** n̲
- [ind] effective maintenance procedure
— m̲ **efectuado** - performed procedure
— m̲ **ejecutado** - performed procedure
— m̲ **escogido** - [ind] chosen procedure
— m̲ **especial** - special procedure
— m̲ — **para laminación** f̲ - [metal-lam] special
rolling procedure
— m̲ —— **soldadura** f̲ - [sold] special welding
procedure
— m̲ **específico** - specific procedure
— m̲ — **para ensayo(s)** m̲ - specific test(ing)
procedure
— m̲ **establecido** - established procedure • outlined procedure
— m̲ **estándar** - standard procedure
— m̲ — **de práctica** - standard practice procedure
— m̲ —— **empaque** m̲ - standard packing procedure
— m̲ **exacto** - exact, o̲ accurate, procedure
— m̲ **exigido** - required procedure
— m̲ — **para trabajo** m̲ - required job procedure
— m̲ **expuesto** - presented, o̲ above, procedure
— m̲ **general** - general procedure
— m̲ **idéntico** - identical procedure
— m̲ **importante** - important procedure
— m̲ — **para, conservación** f̲, o̲ **mantenimiento** n̲
- [ind] important maintenance procedure
— m̲ **impráctico** - impractical procedure
— m̲ **inconsecuente** - inconsistent procedure
— m̲ — **para ensayo(s)** m̲ - [ind] inconsistent
test(ing) procedure
— m̲ **incorrecto** - [ind] incorrect procedure
— m̲ **inesperado** - unexpected procedure
— m̲ **Innershield (autoprotegido)** - [sold| Inner-
shield self-shielded process
— n̲ — **con electrodo** m **con alma** m **fundente au-**
toprotegido - [sold] Innershield self-shielded process
— m̲ — **fijado** - [sold] set Innershield process
— m̲ — **regulado** - [sold] controlled Innershield
procedure
— m̲ **íntegro** - entire procedure

procedimiento m̲ **intermedio** - [ind] intermediate,
process(ing), o̲ procedure
— m̲ **judicial** - [legal] judicial proceeding •
court action
— m̲ **largo** - long, o̲ lengthy, process, o̲ procedure
— m̲ **Linc-Fill** - [sold] Linc-Fill procedure
— m̲ —— **con electrodo** m̲ **muy sobresaliente** -
[sold] Linc-Fill long stickout procedure
— m̲ **Kaldo** - [metal-prod] Kaldo process
— m̲ **mecánico** - [ind] mechanical process
— m̲ **mejor** - [sold] best procedure • better procedure
— m̲ **mejorado** - [ind] improved procedure
— m̲ **mensual** - monthly procedure
— m̲ — **para, conservación** f̲, o̲ **mantenimiento** -
[ind] mopnthly maintenance procedure
— m̲ —— **inspección** f̲ - [ind] monthly inspection procedure
— m̲ **múltiple** - [ind] multiprocess
— m̲ **necesario** - necessary procedure
— m̲ **normal** - normal, o̲ standard, procedure
— m̲ — **para arranque** m̲ - [comb.int] normal
starting procedure
— m̲ —— **industria** f̲ - [ind] normal industry
practice
— m̲ —— **puesta** f̲ **en marcha** f̲ - [comb.int]
normal starting procedure
— m̲ —— **relleno** m̲ - [constr] normal (back)-
-filling procedure
— m̲ **nuevo** - new, procedure, o̲ process
— m̲ **operativo** - operating,tional procedure
— m̲ — **para procesamiento** m̲ - [nucl] processing operating procedure
— m̲ —— m̲ **para solidificación** f̲ - [nucl]
solidification processing operating procedure
— m̲ **para ablandamiento** m̲ -softening process
— m̲ —— m̲ **para cubo** m̲ - [f.c.-ruedas] hub
softening process
— m̲ — **accionamiento** m̲ - [mec] drive,ving procedure
— m̲ —— m̲ **con cadena** f̲ - [mec] chain drive
procedure
— m̲ — **agravio** m̲ - [labor] grievance procedure
— m̲ — **aguilón** m̲ - [grúas] boom procedure
— m̲ — **ajuste** m̲ - [mec] adjustment, procedure,
o̲ process
— m̲ —— m̲ **para cojinete** m̲ - [mec] bearing adjustment procedure
— m̲ — **almacenamiento** m̲ - [ind] storage,ring
procedure
— m̲ — **aplicación** f̲ - applicable procedure
— m̲ —— f̲ **amplia** - widely applicable procedure
— f̲ —— f̲ **para trabajo** m̲ - [labor] applicable
job procedure
— m̲ — **armado** m̲ - [mec] assembly,bling procedure
— m̲ —— m̲ **para ensayo(s)** m̲ - [mec] trial,
assembly,bling, o̲ build-up, procedure
— m̲ —— m̲ — **dispositivo** m̲ - [mec] device
assembly,bling procedure
— m̲ —— m̲ **para cambio(s)** m̲ - [autom-
-mec] shift unit assembly,bling procedure
— m̲ — **arranque** m̲ - [comb.int] starting, o̲
start-up, procedure
— m̲ — **auditoría** f̲ - [contab] auditing procedure
— m̲ — **beneficio,ciación** f̲ - [miner] beneficiation procedure
— m̲ — **calificación** f̲ - qualifying procedure
— m̲ — **combinación** f̲ - [quím] compounding procedure
— m̲ — **composición** f̲ - [quím] compounding procedure
— m̲ — **comprobación** f̲ - test(ing) procedure
— m̲ —— f̲ **para fuga(s)** f̲ - leak test(ing)
procedure
— m̲ —— f̲ **previa** - pretesting procedure
— m̲ —— f̲ **en caliente** - [metal-fabr] hot
forming procedure
— m̲ —— f̲   **frío** - [metal-fabr] cold forming
procedure

procedimiento m para conservación f - [ind] maintenance procedure
— m — — f comentado - [ind] discussed maintenance procedure
— m — — f de rutina - [ind] routine maintenance procedure
— m — — f — — f para bomba f - [bombas] pump routine maintenance procedure
— f — — f — — f — — f para inyección f - [petról] mud pump routine maintenance procedure
— m — — f — — f — — f — lodo m - [petról] mud pump routine maintenance procedure
— — — f — — f para malacate m - [petról] drawworks routine maintenance procedure
— m — — en parte f superior de grúa f - [grúas] crane top maintenance procedure
— m — — f para aguilón m - [grúas] boom maintenance procedure
— m — — f para bomba f - [bombas] pump maintenance procedure
— m — — f — — f para inyección f - [petról] mud pump maintenance procedure
— m — — f — — lodo(s) m - [petról] mud pump maintenance procedure
— m — — f — malacate m - [petról] drawworks maintenance procedure
— m — — f recomendado - [mec] recommended maintenance procedure
— m — control* m - véase procedimiento m para fiscalización f
— m — — m de calidad f - [ind] véase procedimiento m para fiscalización f de calidad f
— m — — m proceso m - véase procedimiento para fiscalización f de proceso m
— m — corte m - [sold] cutting procedure
— m — desarmado m - diassembling,bly procedure
— m — desarme m - disassembling,bly procedure
— m — descongelación f - [sold] thawing procedure
— m — f de tubería(s) f - [sold] pipe thawing procedure
— m — detección f de falla(s) f - troubleshooting procedure
— m — detención f - stopping procedure
— m — determinación f de falla(s) f - troubleshooting procedure
— m — embalaje m - packaging procedure
— m — emergencia(s) f - emergency,cies procedure
— m — empaquetado m - packing procedure
— m — empernado m - [mec] bolting procedure
— m — ensayo(s) m - test(ing) procedure
— m — según Sociedad f Estadounidense para Soldadura f - [sold] American Welding Society, o A W S, testing procedure
— m — — m — — f - para Ensayo m de Material(es) m - [sold] American Society for Testing Materials, o A S T M, testing procedures
— m — excoriación f - skinning procedure
— m — explotación f - [miner] extraction procedure
— m para extremo(s) m - end(s) treatment
— m — m para entrada f - [constr] inlet end(s) treatment
— m — — m — salida f - [constr] outlet end(s) treatment
— m — fabricación f - [ind] manufacturing procedure
— m — fiscalización f - control procedure
— m — — f para instrucción f - [labor] training control procedure
— m — flotación f - flotation procedure
— m — fusión f - [metal-prod] melting down process
— m — — única - [metal-prod] single melting (down) process
— m — industria f - industry practice
— m — — f para rendimiento , - [ind] performance guaranty procedure
— m — industrialización f - [ind] industriali-

zation process
procedimiento m para ingeniería f - engineering procedure
— m — inserción f [mec] inserting procedure
— m — — f con gato(s) m - [constr] jacking procedure
— m — inspección f - inspection procedure
— m — — f para calidad f - [ind] quality control procedure
— m — m para motón m - [grúas] block inspection procedure
— m — — m — — m para gancho m—[grúas] hook block inspection procedure
— m — — f destructiva - [ind] destructive inspection procedure
— m — — f mensual - [mec] monthly inspection procedure
— m — — f no destructiva - nondestructive inspection procedure
— m — — f por característica(s) f - [ind] characteristic(s) inspection procedure
— m — — f posterior - [ind] subsequent inspection procedure
— m — — f radiográfica - [ind] radiographic inspection procedure
— m — — f ultrasónica - [electrón] ultrasonic inspection procedure
— m — instalación f installation procedure
— m — instrucción f - instruction, o training, procedure
— m — laminación f - [metal-lam] rolling procedure
— m — limpieza f - cleaning procedure
— m — lubricación f - [mec] lubrication procedure
— m — — f con aceite m - [mec] oil lubrication procedure
— m — — f con grasa f - [mec] grease lubrication procedure
— m — — f recomendado - [mec] recommended lubrication procedure
— m — manejo m - handling procedure
— m — manipulación,leo - handling procedure
— m — mantenimiento m - [ind] maintenance procedure
— m — — m preventivo - [ind] preventive maintenance procedure
— m — marcación f - marking procedure
— m — — f entre operación(es) f - [ind] between operations marking procedure
— m — mecanización f - [mec] machining procedure
— m — — f por abrasión f - [mec] abrasion machining procedure
— m — muestreo m - [ind] sampling procedure
— m — — m prescri(p)to - [ind] prescribed sampling procedure
— m — neumático(s) m - [autom-neumát] tire procedure
— m — nivelación f - [ind] leveling procedure
— m — operación f - [ind] operating,tor procedure
— m — ordenar v - [com] ordering procedure
— m — pedir,dido - ordering procedure
— m — penetración f - [mec] penetrating,tion, process, o procedure
— m — — f profunda - [sold] deep penetrating,tion process
— m — perforación f - [mec] perforating,tion, ptovrdd, o procedure
— m — — f automática - [mec] plug mill process • [tub] véase procedimiento m con tren m mandril
— m —pintura f - [pint] painting procedure • paint(ing) system
— m — — f con . . . capa(s) f - [pint] . . . coat paint(ing) system
— m — producción f - [ind] production, o manufacturing, o operating, procedure, o process
— m — posición f - position procedure
— m — — f horizontal - [sold] horizontal position procedure
— m — — f plana - [sold] flat position proce-

procedimiento m para posición f sobrecabeza -
[sold] overhead position procedure
— m — — f vertical - [sold] vertical position
procedure
— m — producción f - production procedure
— m — — f de acero m - [metal-prod] steel,
making, o producing, process, o procedure
— m — protección f - protection procedure
— m — proyección f - design(ing) procedure
— m — — f aplicado - [constr] applied de-
sign(ing) procedure
— m — — f contra abrasión f - [constr] abrasion design procedure
— m — f corrosión f - [constr] corrosion
design procedure
— m — prueba(s) f - test(ing), o trial, procedure
— m — — f de garantía f - [ind] guaranty
test(ing) procedure
— m — — f — — f para rendimiento m - [ind]
performance guaranty test(ing) procedure
— m — puesta f a tierra f - [electr-instal]
grounding procedure
— m — — f en marcha f - [ind] starting, o
start-up, procedure
— m — — f — — f de motor m - [electr-mot]
motor, starting, o start-up, procedure •
[comb.int] engine, starting, o start-up, procedure
— m — — f — operación f - [ind] turn-on procedure
— m — purga f - bleeding procedure
— m — rearmado m - reassembling,bly procedure
— m — recepción f - [ind] receiving, o acceptance, procedure
— m — recocido m - [metal-trat] annealing procedure
— m — reconstrucción f - [mec] reconstruction,
o rebuilding, procedure, o process
— m — reducción f - [metal-prod] reduction,
procedure, o process
— m — — f directa - [metal-prod] direct reduction, procedure, o process
— m — — f — de mineral m - [metal-prod] ore
direct reduction, procedure, o process
— m — — f — — — m de hierro m - [metal-
-prod] iron ore direct reduction, procedure,
o process
— m — registro m - registration procedure
— m — — m en hotel m - [transp] hotel registration procedure
— m — regresión f - regression process
— m — — f múltiple - multiple regression process
— m — regulación f - control procedure
— m — — f para proceso m - process control
procedure
— m — reιleno m - [constr] backfilling procedure
— m — remoción f - [mec] removal procedure
— m — — f para horquilla f - [mec] yoke removal poroced ure
— m — revelación f - [fotogr] developing,pment
procedure
— m — — f para película f - [fotogr] film
develping,pment procedure
— m — revisión f - revision procedure •
[contab] auditing procedure
— m — — f analítica - [contab] analytical
auditing procedure
— m — sangría f - bleeding procedure
— m — secado,camiento m - drying procedure
— m — seguridad f - [segurid] safety procedure
— m — — f según norma f - [segurid] standard
safety procedure
— m — — f recomendado - [segurid] recommended
safety procedure
— m — soldadura - [sold] welding, procedure, o
process, o practice
— m — — f adecuado - [sold] adequate, o proper, welding procedure
— m — — f apropiado - [sold] appropriate, o
proper, welding procedure

procedimiento m para soldadura f comprobado -
[sold] proven, o checked, welding procedure
— m — — f confiable - [sold] reliable welding
procedure
— m — — f de chapa(s) f (metálicas) - [dolf]
sheet metal (welding) procedure
— m — — f divulgado - [sold] known, o published, welding procedure
— m — — f en ángulo m (interior) - [sold]
fillet weld(ing) procedure
— m — — f — posición f horizontal - [sold]
horizontal (welding) position procedure
— m — — f — — f plana - [sold] flat position
(welding) procedure
— m — — f — — f sobrecabeza - [sold] overhead position (welding) procedure
— m — — f — — f vertical - [sold] vertical
position (welding) procedure
— m — — f mecanizada - [sold] mechanized
welding, process, o procedure
— m — — f para línea(s) f para conduciión f -
[sold] pipeline (welding) procedure
— m — — f — tubería(s) f para conducción f -
[sold] pipeline (welding) procedure
— m — — f con arco m - [sold] arc welding
procedure
— m — — f — — m abierto - [sold] open arc
welding, process, o procedure
— m — — f preplaneado* - [sold] pre-planned
welding procedure
— m — — f probado - [sold] proven, o tested,
welding procedure
— m — — f recomendado - [sold] recommended
welding procedure
— m — — f regulado - [sold] controlled welding
procedure
— m — — f según norma f - [sold] standard
welding procedure
— m — — f — — f divulgado - [sold] published standard welding procedure
— m — — f semiautomática - [sold] semi-automatic welding, process, o procedure
— m — — f (manual) con arco m sumergido -
[sold] squirt (welding) process
— m — — f sobrecabeza - [sold] overhead welding, procedure, o process
— m — substitución f - alternate procedure
— m — toma f de muestra(s) f - [ind] sampling
procedure
— m — — f — temperatura f - [instrum] temperature taking procedure
— m — trabajo m - work(ing), o job, procedure
— m — — m completo - entire job procedure
— m — — m para aplicación f amplia - [segur]
widely applicable job procedure
— m — — m para turno m - [ind] turn work(ing)
procedure
— m — transmisión f - transmittal procedure
— m — — f de dibujo(s) m - drawing(s) transmittal procedure
— m — — f — plano(s) m - drawing(s) transmittal procedure
— m — tubería(s) f con pared(es) f delgada(s) -
[sold] thin wall pipe procedure
— m — turno m - [labor] turn procedure
— m — unión f - [sold] joint procedure
— m — — f doble - [tub] double joint welding
procedure
— m — verificación f - checking, o verification, procedure
— m pobre - [ind] poor, process, o procedure
— f — para, conservación f, o mantenimiento m -
[ind] poor maintenance procedure
— m por vía f química - [quím] chemical, process, o procedure
— m posterior - later, o subsequent, procedure
— m preciso - precise, o specific, procedure
— m preplaneado - pre-planned procedure
— m preventivo - preventive procedure
— m — recomendado - recommended preventive
procedure
— m probado - proven, o tested, procedure
— m que produce cortocircuito(s) m - [sold]

short circuiting procedure
**procedimiento m que produce corte(s) m en cir-
cuito m** - [sold] short circuiting procedure
— **m que sigue** - following procedure
— **m químico** - [quím] chemical process,cedure
— **m radiográfico** - [electrón] radiographic procedure
— **m — correcto** - [electrón] correct radiographic procedure
— **m recomendado** - recommended procedure
— **m — para, conservación f, o mantenimiento m**
- recommended maintenance procedure
— **m — f preventiva f** - [ind] preventive
maintenance procedure
— **m — mantenimiento m preventivo** - [ind]
recommended preventive maintenance procedure
— **m — para trabajo m** - [ind] recommended, job,
o work, procedure
— **m seguro** - [segurid] safe job procedure
— **m — para trabajo m** - [segurid] recommended safe job procedure
— **m regulado** - controlled procedure
— **m regular** - regular procedure
— **m repetido** - repeated procedure
— **m rutinario** - routine procedure
— **m satisfactorio** - satisfactory, procedure, o
treatment
— **m — extremo m** - [constr] satisfactory end
treatment
— **m secador** - drying procedure
— **m seguido m** - taken, o followed, procedure
— **m según Instituto m Estadounidense para Petróleo m** - [petról] Anerican Petroleum Institute procedure
— **m — norma f**—standard, procedure, o practice
— **m — f divulgado** - published standard procedure
— **m — f para soldadura f en serie f** -
[sold] published standard production welding
procedure
— **m — f para soldadura f** - [sold] standard
welding procedure
— **m — f — f en serie f** - [sold] standard production welding procedure
— **m — f seguro** - [segurid] standard safe, o
safe standard, procedure
— **m — f recomendado** - [segurid] recommended safe job procedure
— **m — f — para trabajo m** - [segurid]
standard recommended safe job procedure
— **m seguro** - [segurid] safe procedure
— **m — aplicable** - [segurid] applicable safe, o
safe applicable, procedure
— **m — para trabajo m** - [segurid] applicable
safe, o safe applicable, job procedure
— **m — para manejo m** - [segurid] safe handling
procedure
— **f — manipulación,leo** - [segirod] safe
handling procedure
— **m — operación f** - [segurid] safe operating procedure
— **m — exigido** - required safe procedure
— **m — para trabajo m** - [segurid] required
safe job procedure
— **m — para aplicación f** - [segurid] applicable
safe procedure
— **m — f para trabajo m** - [segurid] applicable safe job procedure
— **m — f amplia** - [segurid] safe job procedure
— **m — f — para trabajo m** - [segurid]
widely applicable safe job procedure
— **m — trabajo m** - [segurid] safe job procedure
— **f — m con aplicación f amplia** -]
[segurid] widely applicable safe job procedure
— **m — recomendado** - [segurid] recommended
safe procedure
— **m semanal** - weekly procedure
— **m — para, conservación f, o mantenimiento m**
- weekly maintenance procedure
— **m sencillo** - simple procedure

procedimiento m sencillo para soldadura f -
[sold] simple welding procedure
— **m siderúrgico** - [metal-prod] steel (making),
procedure, o process
— **m siguiente** - next, o following, procedure
— **m similar** - [sold] similar procedure
— **m simplificado** - simplified procedure
— **m sistemático** - systematic procedure
— **m típico** - typical procedure
— **m — para extremo(s) m** - [constr] typical
end, procedure, o treatment
— **m — soldadura f** - [sold] typical welding
procedure
— **m último** - last, o latter, procedure
— **m ultrasónico** - ultrasonic procedure
— **m usual** - usual, o standard, procedure
— **m vertical**--[sold] vertical procedure
— **m — ascendente** - [sold] vertical-up procedure
— **m — descendente** - [sold] vertical-down procedure
procesado m - processing; véase procesamiento m
procesado,da a - processed • [comput] handled
— **a por colada f continua** - [col.cont] continuous cast processed
procesador(a) m/f - [ind] processor; processing
unit
— **m/f para alimento(s) m** - [domést] food processor
procesamiento m - . . . . • processing • [comput]
véase también **proceso m**
— **m con calor m** - [ind] heat processing
— **m — corte m igual** - [metal-lam] equal section processing
— **m confiable** - [ind] reliable processing
— **m de alimento(s) m** - [culin] food processing
— **m — compuesto m** - [quím] compound processing
— **m — concentrado(s) m** - [ind] concentrate(s)
processing
— **m — chapa(s) f** - [metal-lam] strip processing
— **m — — f en caliente** - [metal-lam] hot
strip processing
— **m — — f — frío** - [metal-lam] cold strip
processing
— **m — dato(s) m** - [comput] data processing
— **m — material(es) m** - [ind] material(s) processing
— **m — — m degradado(s) m** - [ind] downgraded
material(s) processing
— **m — — m electrohipobárico(s)** - [electrón]
electrohypobaric materials processing
— **m — — m nuevo(s)** - [ind] new material(s)
processing
— **m — — m sin analizar** - [ind] unanalyzed
material(s) processing
— **m — metal(es) m** - [metal] metal(s) processing
— **m — — m fundido(s)** - [metal-prod] molten
metal(s) processing
— **m — mineral(es) m** - [miner] ore, o miner-
al(s), processing
— **m — órden(es) f** - [ind] order(s) processing
— **m — papel m** - paper processing
— **m — planchón(es) m** - [metal-lam] slab(s)
processing
— **m — plástico(s) m** - [plást] plastic(s) processing
— **m — producción f** - [ind] production processing
— **m — producto(s) m** - [ind] product(s) processing
— **m — rodillo(s) m** - [mec] roll(s) processing
— **m — — m de acero m** - [metal-lam] steel
roll(s) processing
— **m — — m para laminación f** - [metal-lam]
rolling roll(s) processing
— **m difícil** - hard, o difficult, processing
— **m eficiente** - [ind] efficient processing
— **m para concentración f** - [miner] concentration processing
— **m — solidificación f** - [nucl] solidifica-

procesamiento m para sustancia(s) f química(s) -   - 1188 -

tion, o solidifying, process(ing)
procesamiento m de sustancia(s) f química(s) -
[quím] chemical(s) processing
— m electrolítico m - electrolytic processing
— m electrónico - [comput] electronic processing
— m — de dato(s) m - [comput] electronic data
processing
— m en horno m - [combust] furnace processing
— m — — m Siemens-Martin - [metal-prod] open
hearth furnace processing
— m futuro - [ind] future processing
— m por laminación f en frío - [metal-lam] cold
rolled processing
— m Siemens-Martin - [metal-prod] open hearth
processing
— m térmico - [ind] heat, o thermal, processing
procesar v - . . . . • to, process, o handle
— v alimento(s) m - [alim] to process @ food(s)
— v compuesto m - [quím] to process @ compound
— m concentrado m - [ind] to process @ concentrate
— v íntegramente - [ind] to process totally
— v material - [ind] to process @ material
— v — m degradado - [ind] to process @ downgraded material
— v metal m - [metal] to process @ metal
— v — m fundido - [metal-prod] to process @
molten metal
— v material m plástico - [plást] to process @
plastic(s)
— v plástico(s) m - [plást] to process @ plas-
tic(s)
proceso m - . . . . • process; procedure; processing; handling • treatment
— m ácido - [metal-prod] acid process(ing)
— m actual - [ind] current, o existing, process
— m — de planta f - [quím] existing plant
processing
— m — — — f para ácido(s) m - [quím] existing acid plant proces(sing)
— m — para pintura f - [ind] current painting
process
— m — — — f de lingotera(s) f - [metal-prod]
current (ingot) mold painting process
— m — — — f — placa(s) f - [metal-prod]
current (ingot) stool painting process
— m actualizado - [ind] updated process
— m aplicado - applied process)ing)
— m apropiado - appropriate, o suitable, pro-
cess(ing)
— m Asbestos-Bonded - [metal-trat] Asbestos-
-Bonded process(ing)
— m avanzado - [ind] advanced process(ing)
— m — para fabricación f - [ind] advanced, o
sophisticated, manufacturing process
— m bacteriano - [miner] bacterial process
— m básico - [metal-prod] basic process
— m — con oxígeno m - [metal-prod] basic oxygen process
— m Bessemer - [metal-prod] Bessemer, o acid,
process
— m — básico - [metal-prod] basic Bessemer
process
— m — — calmado - [metal-prod] killed basic
Bessemer process
— m — desoxidado - [metal-prod] deoxidized
basic Bessemer process
— m — — — y calmado - [metal-prod] killed
deoxidized basic Bessemer process
— m biológico - [biol] biological process
— m cal-soda - [sanit] cal-soda process
— m completo - complete process
— m complicado - complicated process
— m comprobado - [ind] proven process
— m con alimentación f automática (de electrodo m) - [sold] semiautomatic process
— m con arco m - [sold] arc process
— m — — m abierto - [sold] open arc process
— m — — m — con alimentación f automática de
electrodo m - [sold] open arc semiautomatic
process
— m — — m protegido - [sold] shielded arc

process
proceso m con arco m protegido entre metal(es) m
- [sold] shielded metal arc process
— m — arco m sumergido - [sold] submerged arc
process
— m — cámara f fluidizada - [metal-prod]
fluidized chamber process
— m — — f — con reducción f gaseosa—[metal-
-prod] fluidized chamber gas reduction process
— m con capacidad f grande - [ind] large capacity process:
— m — — — pequeñs -ssmall capacity processs
— m — — — f reducida - [ind] small capacity
process
— m — circuito m- [mec] circuit process
— m — — m para regulación f - [mec] control
circuit process
— m — convertidor m - [metal-prod] converter
process
— m — — m con oxígeno m - [metal-prod] oxygen
converter process
— m — costo m alto - [ind] high cost process
— m — — m bajo - low cost process
— m — — m elevado - high cost process
— m — — m más alto - higher cost process •
highest cost process
— m — — m — elevado - [ind] higher cost process • highest cost process
— m — — m — reducido - lower cost process •
lowest cost process
— m — — m mayor - higher cost process •
highest cost process
— m — — m menor - lower cost process • lowest
cost process
— m — — m reducido - [ind] low cost process
— m — electrodo m con alma m fundente - [sold]
flux cored (electrode) process
— m — — — m fundente con alimentación f
automática (de electrodo m) - [sold] flux
cored semiautomatic process
— m — electrodo m manual - [sold] manual, o
stick, electrode process
— m — — m con alma m fundente - [sold] Innershield (electrode) process
— m — horno m alto - [metal-prod] blast furnace process
— m — — m alto y (con) convertidor m para o-
xígeno m - [metal-prod] blast furnace-oxygen
converter process
— m — — m bajo con reducción f gaseosa -
[metal-prod] low furnace gas reduction process
— m — — m Bessemer - [metal-prod] Bessemer
furnace process
— m — — m eléctrico - [metal-prod] electric
furnace process
— m — — m rotatorio - [metal-prod] rotary
furnace process
— m — — — m con combustible m sólido -
[metal-prod] solid fuel rotary furnace process
— m — — m Siemens-Martin - [metal-prod] open
hearth furnace process
— m — — m sobre eficiencia f - [comput] efficiency report process(ing)
— m — intercambio m de ión(es) m - [nucl]
ion(s) exchange process
— m — lecho m fuidificado - [metal-prod]
fluidized bed process
— m — liga f de asbesto m - [metal-trat] Asbestos-Bonded process
— m — oxígeno m - [metal-prod] oxygen process
— m continuado - continuing,nuous process
— m continuo - [ind] continuous process
— m de absorción f - absorption process
— m — catálisis f - catalysis process
— m — — f con gas m - gas catalyis process
— m — dato(s) m - [electrón] data processing
— m — — m sobre consumo m de oxígeno m -
[combust] oxygen consumption data processing
— m — — — — m de combustible m - [comput]
fuel consumption data processing

proceso m de dato(s) m sobre producción f -
[comput] production data processing
— m — informe(s) m—[comput] report processing
— m — — m combustión f - [combust] combustion report processing
— m — — m eficiencia f - [combput] efficiency report processing
— m — flotación f - flotation processing
— m — lingote(s) a banda(s) f en caliente -
[metal-lam] hot strip rolling ingot process
— m — material m - [ind] material processing
— m — partícula(s) f - particle(s) processing
— m — — f magnética(s) - [ind] magnetic par-
ticle(s) processing
— m — planchón(es) m - [metal-lam] slab processing
— m — — f con arco m - [sold] arc welding
process(ing)
— m diferente - different process(ing)
— m distinto - different process
— m doble - [ind] duplex process
— m eléctrico - [ind] electric process(ing)
— n electroquímico - electrochemical process
— m elegido - chosen process
— m — para tratamiento m - chosen treatment
process
— m escogido - selected, o chosen, process
— m especial - [ind] special process
— m específico - specific, o particular, process
— m — para montaje m - [mec] particular assembling,bly process
— m Esso-Fior - [metal-prod] Esso-Fior process
— m estructural - [constr] structural process
— m exclusivo - exclusive process
— m fabril - [ind] manufacturing process
— m — progresivo - [ind] progressive, o aggressive, manufacturing process
— m Frasch - [petról] Frasch process
— m Hojalata f y Lámina f - [metal-trat] Hojalata y Lámina process
— m Innershield - [sold] Innershield, process,
o system
— m ininterrumpido - [ind] uninterrupted, o
continuous, process
— m integracionista - integrationist process
— m íntegro - entire, o integral, process
— m Krupp-Renn - [metal-prod] Krupp-Renn process
— m L D - [metal-prod] basic oxygen process
— m ligante - bonding process
— m Linz-Donawitz - [metal-prod] basic oxygen
process
— m magnético - magnetic process
— m mecánico - [ind] mechanical process
— m — para soldadura f - [sold] mechanized
welding process
— m — práctico - [sold] practical mechanized
process
— m — — para soldadura f - [sold] practical
mechanized welding process
— m mejor - better process • best process
— m — para soldadura f - [ind] better mechanical process • best mechanical process
— m — — — f — f con arco m - [sold] best
arc welding process
— m metalúrgico - [metal-prod] metallurgical
process
— m — para recuperación f - [miner] metallurgical recovery process
— m Midrex - [metal-prod] Midrex process
— m mejor - better process • best process
— m moderno - modern process
— m — para proyección f—modern design process
— m múltiple - [ind] multiple process
— m muy complicado - very complicated process
— m operativo - [ind] operating process
— m para acepilladura - shaving process
— m — aceración f - [metal-prod] steel making
process
— m — — m para herramienta(s) f - [metal-
-fabr] tool steel process
— m — acopio m - [ind] gathering process

proceso m para afinar v - [metal-prod] refining
process
— n — aglomeración f - [miner] agglomeration
process
— m — aumento m de capital m - [fin] capital
increase process
— m — austemplado* m - [metal-trat] austem-
pering* process
— m — avance - advance,cing process
— m — — m gradual - [mec] inching process
— m — cálculo m - calculation, o computation,
process
— m — catálisis f - catalysis process
— m — — f con gas m húmedo - wet gas catalysis process
— m — cementación f - [metal-trat] cementation
process
— m — cepilladura f - [mec] shaving process
— m — circuito m - circuit process
— f — colada f - [metal-prod] casting process
— m — — f continua - [metal-prod] continuous
casting, o concasting, process
— m — combustión f - [combust] combustion
process
— m — — f en caldera f - [cald] boiler combustion process
— m — — f en horno(s) m - [combust] furnace
combustion process
— m — — f — — m con fosa f - [metal-lam]
soaking pit combustion process
— m — — f — — m para coque m - [coque] coke
oven combustion process
— m — — f — — m igualación f de tempe-
ratura(s) f - [combust] soaking pit combustion process
— m — — f — — m recalentamiento m -
[combust] reheating furnace combustion process
— m — — f — — m — — m de tocho(s) m -
[combust] bloom reheating furnace combustion
process
— m — — f — — m recocido m - [metal-
-trat] annealing furnace combustion process
— m — — f — planta(s) f para calcinación f -
[metal-prod] calcination plant combustion
process
— m — cómputo(s) m - computation process
— m — comunicación(es) f - [labor] communica-
tion(s) process
— m — concentración f - [miner] concentration
process
— m — — f magnética - [miner] magnetic concentration process
— m — conformación f - [mec] shaping, o forming, process
— m — — f de bola(s) f - [cuerp.moled] ball,
shaping, o formation,ming process
— m — — f en caliente - [metal-fabr] hot
forming process
— m — — f mediante rotación f - [metal-fabr]
spinning process
— m — coquización f - [coque] coking process
— m — conversión f - conversion process
— m — — f para metal m blanco - [metal-prod]
Babbitt metal conversion process
— m — corrugación f - [metal-fabr] corrugating,tion process
— m — corte m - [mec] cutting process
— m — craqueo m - [petról] cracking process
— m — decapado m - [metal-trat] pickling process
— m — deformación f - [metal-fabr] deformation
process
— m — descomposición f - decomposing process
— m — — f de hidrocarburo(s) m - [petról]
cracking process
— m — desgasificación f - [metal-prod] degassing process
— m — descontaminación f - decontamination, o
decontaminating, process
— m — diseño m de estructura f - [constr]
structure design process
— m — distribución f - distribution process

proceso m para ejecución f - implementation process
— m — elaboración f - [ind] fabricating,tion, manufacturing, process
— m — endurecimiento m - hardening process
— m — entrega f - [ind] delivery process
— m — fabricación f - [ind] manufacturing, o fabricating,tion, o production, process
— m — — f para pella(s) f - [coque] pellet(s) production process
— m — — f — — f de coque m - [coque] coke pellet(s) production process
— m — flotación f - [miner] flotation process
— m — forja(dura) f - [metal-prod] forging process
— m — formación f - [metal-lam] forming, o rolling, process
— m — — f para pella(s) f - [miner] pellet(s) forming, o pelletizing, process
— m — — f — — f en caliente - [miner] hot pelletizing process
— m — — f — — f frío - [miner] cold pelletizing process
— m — fundición f - [metal-prod] (s)melting process
— m — — f con oxígeno m - [metal-prod] oxygen smelting process
— m — — f de vidrio m - [cerám] glass melting process
— m — fusión f - [metal-prod] (s)melting process
— m — — f con oxígeno m - [metal-prod] oxygen (s)melting process
— m — galvanización f - [metal-trat] galvanizing process
— m — industrialización f - [ind] industrialization,zing process
— m — ingeniería f - engineering process
— m — integración f - integrating,tion process
— m — laminación f - [metal-lam] rolling process
— m — — f en caliente - [metal-lam] hot rolling process
— m — — f en frío - [metal-lam] cold rolling process
— m — limpieza f - cleaning process
— m — manufactura f - manufacturing process
— m — mecanización f - [mec] machining process
— m — — f abrasiva f - [mec] abrasive machining process
— m — — f por abrasión f - [mec] abrasion machining process
— m — montaje m - [ind] assembling,bly process
— m — metalización f - [metal] metallizing process
— m — operación f - [ind] operating process
— m — peletización f - [miner] pelletizing process
— m — pintura f - painting process
— m — — f de lingotera(s) f - [metal-prod] (ingot) mold painting process
— m — — f de placa(s) f - [metal-prod] (ingot) stool painting process
— m — planta f - [ind] plant process
— m — — f para ácido(s) m - [ind] acid plant process
— m — producción f - [ind] production, o fabrication, process
— m — — f para acero m - [metal-prod] steel making process
— m — — f — — m Siemens-Martin - [metal-prod] open hearth steel production process
— m — — f — pella(s) f - [coque] pellet(s) production process
— m — — f — — f de coque m - [coque] coke pellet(s) production process
— m — — f — vidrio m - [cerám] glass production process
— m — proyección f - [ind] design process
— m — — f de alcantarilla(s) f - [constr] culvert design(ing) process
— m — — f escultural - [constr] structural design(ing) process

proceso m para reconstrucción f - [sold] rebuilding process
— m — recuperación f - [miner] recovering,ry process
— m — reducción f - [ind] reducing,ction process
— m — — f directa f - [metal-prod] direct reducing,ction process
— m — — f — de mineral m - [metal-prod] ore direct reducing,ction process
— m — — f gaseosa - [metal-prod] gas reducing,duction process
— m — — f en caliente - [metal-lam] hot reducing,ction process
— m — — f en frío - [metal-lam] cold reducing,ction process
— m — — f gaseosa - [metal-prod] gas reducing,ction process
— m — — f — con lecho m estático - [metal-prod] static bed gas reducing,ction process
— m — refinación f - [metal-prod] refining process
— m — secado,camiento m - drying process
— m — separación f - separation process
— m — — f magnética - [miner] magnetic separation process
— m — selección f - selecting,tion process
— m — sinterización f - [metal-prod] sintering process
— m — soldadura f - [sold] welding, process, o procedure, o system
— m — — f con arco m - [sold] arc welding process
— m — — f — — m sumergido - [sold] submerged arc welding process
— m — — f — pasada f única - [sold] single pass welding process
— f — — f — — f — con costo m reducido - [sold] low cost single pass welding process
— m — — f con transferencia f de metal - [sold] metal transfer welding process
— m — — f — — f — — mediante corte(s) m en circuito m - [sold] short circuiting metal transfer welding process
— m — — f en frío - [sold] cold welding process
— m — solidificación f - [sold] freezing action
— m — — f de soldadura f - [sold] weld(ing) freezing action
— m — — f — — f a tope - [sold] butt weld freezing action
— m — — f — — f en ángulo m (interior) - [sold] fillet weld freezing action
— m — — f — — f en ranura f - [sold] groove weld freezing action
— m — temple,plado m - [metal-trat] tempering process
— m — terminación f - [ind] finishing process
— m — torcedura f - [mec] twisting process
— m — — f de par(es) m - [telecom-cond] pair(s) twisting process
— m — tratamiento m - treating,tment process
— m — — m en vacío m - [metal-prod] vacuum treatment process
— m — trefilación,lería en frío - [alambre] cold drawing process
— m práctico - practical process
— m — para soldadura f - [sold] practical welding process
— m principal - [ind] main process
— m — para producción f - [ind] main production process
— m productivo - [ind] production,tive process
— m progresivo - [ind] progressive process
— m real - actual process
— m — para selección f - actual selection process
— m recomendado - recommended process
— m — para fabricación f - [ind] recommended manufacturing process
— m repetido - repeated, o repetitive, process
— m S L R N - S L R N process
— m secuencial - [comput] sequential process

proceso m **semiautomático** - [sold] semiautomatic process
— m **sencillo** - simple process
— m **siderúrgico** - [metal-prod] steelmaking process
— m **Siemens-Martin** - [metal-prod] open hearth process
— m **simple** - simple process
— m — **para soldadura** f - simple welding process
— m **simulado** - [ind] simulated process
— m **sin limitación** f - unlimited process
— m **tecnológico** - technological process(ing)
— m — **para peletización** f - [miner] pelletizing technological process(ing)
— m **Teniente** - [metal-cobre] Teniente process
— m **tradicional** - traditional, o conventional, process
proclamación f - . . .; proclaiming
proclamado,da a - proclaimed
procura f - search; quest; attempt(ing)
procurado,da a - tried; attempted; searched • endeavored
procurador m - [legal] solicitor; attorney (at law) • [pol] . . .; attorney (at law)
— m **general** - [pol] attorney general
— m — **para estado** m - [pol] (state's) attorney general
procurar v - . . .; to try; to procure; to (make @) attempt; to search; to endeavor
— v **complacer** v - to try to please • to curry @ favor
producción f - . . .; developing,pment; putting, o turning, out • manufacture,ring; fabricating,tion • making • yield • exerting,tion • output (rating • productivity • incurring; taking place; giving [petról] . . .; lift(ing) • [ind] output • operation; tonnage • throughput • [sold] mileage • [agric] growing • [ind] production department; operation(s)
— f **actual** - [ind] current production
— f **acumulada** - accumulated, o cumulative, production, o output
— f **adicional** - additional production
— f **agrícola** - [agric] agricultural production
— f **alcanzada** - attained production
— f **alentada** - encouraged production
— f **alta** - high production • [electr-prod] high output • [sold] high, otuput, o mileage
— f **ampliada** - [ind] expanded output
— f — **en horno** m - [metal-prod] expanded furnace output
— f — — m **eléctrico** - [metal-prod] expanded electric furnace output
— f **anual** - annual, o yearly, production, o output
— f — **de cobre** m - [metal-prod] annual copper, ouput, o production
— f — **estimada** - estimated yearly production
— f **artesana** - [ind] handicraft production
— f **artificial** - [ind] artificial production • [petról] artificial lift
— f **económica** - [petról] economical, o low cost, artificial lift(ing)
— f **aumentada** - [ind] increased, production, o output
— f **aumentada** - increased, production, o output
— f — **de operario** m - [ind] increased operator('s), production, o output
— f — — **soldador** m - [sold] increased, operator's, o weldor's, production, o output
— f **automatizada** - [ind] automated, production, o output
— f — **compleja** - [ind] complex automated, producion, o output
— f **baja** - [labor] low, production, o output
— f **carbonera** - [miner] coal, production, o output
— f — **nueva** - [miner] mew coal production
— f **casera** - home(made) production
— f **competitiva** - [econ] competetive production
— f **compleja** - [ind] complex production

producción f **Comunal** - [econ] Community, production, o output
— f **con granulometría** f **constante** - [miner] constant size, production, o throughput
— f — — f **irregular** - [miner] irregular size production
— f — **temperatura** f **alta** - [ind] high temperature production
— f — — f **elevada** - [ind] high temperature production
— f **con velocidad** f **de avance** m **grande** - [sold] high speed production
— f **confiable** - [ind] dependable production
— f **constante** - [electr-prod] continuous, o steady, output
— f — **de energía** f - [electr-prod] continuous power, production, o output, o operation
— f — — **soldadura** f - [sold] constant welding output
— f **continua** - [ind] continuous production
— f **corriente** - [ind] current, o common, production, o output
— f **costa afuera** - [petról] off-shore production
— f **chilena** - [ind] Chile(an), production, o output
— f — **de acero** m - [metal-prod] Chile(an) steel production
— f **de acería** f - [metal-prod] steel plant production
— f — **acero** m - [metal-prod] steel, production, o making, o manufacture, o output
— f — — m **bruto** - [metal-prod] raw steel production
— f — — m **con boro** m - [metal-prod] boron steel production
— f — — m **con carbono** m - [metal-prod] carbon steel production
— f — — m **crudo** - [metal-prod] raw steel, production, o output
— f — — m **efervescente** - [metal-prod] rimming steel proucion
— f — — m **en molde(s)** m **tipo** m **botella** - [metal-prod] bottle top mold rimming steel production
— f — — m **estabilizado** - [metal-prod] stabilized steel production
— f — — m **para hojalata** f - [metal-prod] tinning steel production
— f — — m **por sistema** m L D - [metal-prod] basic oxygen steel, production, o making
— f — — m **según especificación(es)** f **de cliente** m - [metal-prod] [ind] customizing; customized production; production customizing
— f — — m **Siemens-Martin** - [metal-prod| open hearth steel production
— f — **ácido** m - [quím] acid production
— f — — m **sulfúrico** - [quím] sulfuric acid production
— f — **alimento(s)** m - [econ] food(s) production
— f — **América** f **Latina** a- [econ] Latin American, production, o output
— f — **arco** m - [sold] véase formación f de arco
— f - [electr-oper] arc developing,pment
— f — — m **con magnitud** f **grande** - [electr-oper] high magnitude arc(ing)
— f — — m **dentro de aceite** m - [electr-oper] arcing, in, o under, @ oil
— f — — m **entre contacto(s)** m - [electr-oper] contact arcing
— f — **arrabio** m - [metal-prod] pig iron, production, o making, o output
— f — — m **básico** - [metal-prod] basic pig iron production
— f — — m **de tipo** m **para fundición** f - [metal-prod] foundry type pig iron production
— f — — m **especial** - [metal-prod] special (quality) pig iron production
— f — — m **para fundición** f - [metal-prod] foundry pig iron production
— f — **avión(es)** m - [aeron] aircraft manufacture
— f — **banda** f - [metal-lam] strip production
— f — — f **gruesa** - [metal-lam] thick strip

production
producción f **de barra(s)** f **laminadas en caliente**
m - [metal-lam] hot rolled bar production
— f — — f — **en caliente** - [metal-lam] hot
rolled bar production
— f — — f — — **para armadura(s)** f -
[metal-lam] hot rolled reinforcing bar production
— f — — f — — — — — f **para hormigón** m
armado - [metal-lam] hot rolled concrete reinforcing bar production
— f — . . . **barril(es)** m - [petról] . . . bar-
rel(s) output
— f — — m **por día** f - [petról] . . . barrels
per day output
— f — **barro(s)** m - [metal-trat] sludge formation
— f — **bien(es)** m - [econ] good(s) production
— f — — m **de capital** - [econ] capital good(s)
production
— f — **bobina(s)** m - [metal-lam] coil(s) production
— f — — f **laminada(s)** - [metal-lam] rolled
coil(s) production
— f — — f — **en caliente** - [metal-lam] hot
rolled coil(s) production
— f — — f — — **frío** - [metal-lam] cold
rolled coil(s) production
— f — **bola(s)** f - [cuerp.moled] ball(s) production
— f — — f **de acero** m - [cuerp-moled] steel
ball(s), production, o manufacture
— f — — f — — m **para molienda** f - [cuerp.-
-moled] steel grinding ball(s), production, o
manufacture
— f — — f **para molienda** f - [cuerp.moled]
grinding ball(s), production, o manufacture
— f — **calidad** f - [ind] quality production
— f — — f **en soldadura** f - [sold] weld quality production
— f — **calor** m - heat, production, o building
up
— f — **carbón** m - [miner] coal production
— f — **cañería(s)** f - [tub] pipe making
— f — **cemento** m - [constr] cement production
— f — **cereal(es)** m - [agric] grain(s) production,cing
— f — **cinta** f - [metal-lam] strip production
— f — — f **gruesa** - [metal-lam] thick strip
production
— f — **cobre** m - [metal-prod] copper production
— f — — m **aumentada** - [metal-prod] increased
copper production
— f — — m **incrementada** - [metal-prod] increased copper production
— f — **combustible(s)** m - [combust] fuel(s)
production
— f — — m **fósil(es)** - [petról] fossil fuel(s)
production
— f — **compañía** f - company('s) production
— f — **comunidad** f - [econ] Community('s),
production, o output
— f — **conjunto(s)** m **soldado(s)** - [sold] weld-
ment(s), production, o manufacture
— f — **convertidor** m - [metal-prod] converter
output
— f — — m **amplificada** - [metal-prod] increased converter output
— f — **coque** m - [coque] coke, production, o
manufacturing; cokemaking
— f — **corriente** f - [electr-prod] current,
production, o output
— f — — f **alterna** - [electr-prod] alternating
current, production, o output
— f — — f **continua** - [electr-prod] direct
current, production, o output
— f — **combustible(s)** m - [combust] fuel production
— f — **cuerpo(s)** m **moledor(es)** - [cuerp-moled]
grinding media production
— f — **cortocircuito(s)** m - [electr-oper] short
circuiting; shorting
— f — **chapa** f - [metal-lam] strip production •
sheet(s) production

producción f **de chapa** f **gruesa** f - [metal-lam]
thick strip production
— f — — f **laminada** - [metal-lam] rolled
sheet production
— f — — f — **en caliente** - [metal-lam] hot
rolled sheet production
— f — — f — **frío** - [metal-lam] cold rolled sheet
production
— f — **desecho(s)** m - waste(s) production
— f — **desgaste** m - [mec] wear production
— f — **diamante(s)** f - diamond production
— f — **disco(s)** m - [metal-fabr] circling
— f — **emanación(es)** f - fume(s) production
— f — — f **tóxica(s)** f - [ind] toxic fume(s),
production, o poroducing
— f — **empresa** f - corporate, o company, production
— f — **energía** f - [electr-prod] power •
[sold] power generator output
— f — — f **eléctrica** - [electr-prod] electric, energy, o power, production
— f — — f **en corriente** f **alterna** - [electr-
-prod] alternating current power output
— f — — f — — f **continua** - [electr-prod]
direct current power output
— f — **escarpado** m - [metal-trat] scarfing,
production, o output
— f — **estañadura** f - [metal-trat] tinning production
— f — **excitación** f - [electr-mot] exciting
production
— f — **excitador** m - [electr-mot] exciter,
production, o output
— f — — f **alta** - [electr-mot] high exciter
output
— f — — m **baja** - [electr-mot] low exciter
output
— f — **planta** f - [ind] plant, o factory, output, o production
— f — **fleje** m - [metal-lam] strip production
— f — — m **grueso** - [metal-lam] thick strip
production
— f — **fuente** f - source, production, o output
— f — — f **de energía** f - [electr-prod] power
source, output, o production
— f — — — f **motogeneradora** - [sold]
motor-generator power source, production, o
output
— f — **fuerza** f **motriz** - [electr-prod] electric, energy, o power, production
— f — **fundición** f - [metal-prod] melt shop
production
— f — **gas** m - [metal-prod] gas, production, o
output
— f — **gel** m - [quím] gel production,cing
— f — **generador** m - [electr] generator output
— f — — m **para energía** f - [electr-prod]
power generator output
— f — **grano(s)** m - [agric] grain production
— f — **grieta(s)** - crack(s), production, o
development
— f — **hierro** m - [miner] iron (ore) production
• [metal-prod] iron production
— f — — m **esponja** - [metal-prod] sponge iron
production
— f — — m **y coque** m - [metal-prod] iron and
coke poroduction
— f — **hollín** m - [combust] soot production
— f — **horno** m - [metal-prod] furnace production
— f — — m **alto** - [metal-prod] blast furnace,
production, o output
— f — — m **eléctrico** - [metal-prod] electric
furnace, production, o output
— f — **hulla** f - [miner] coal production
— f — **imagen** f - [electrón] image production
— f — **incendio** m - [segurid] fire ignition
— f — **laminador** m - [metal-lam] mill, production, o throughput
— f — — m **en unidad** f **de tiempo** m - [metal-
-lam] mill throughput in @ time unit
— f — **línea** f - [ind] line, production, o
output
— f — — f **para galvanización** f - [metal-

galvanizing line production
producción f de lodo m - [ind] sludge output
— f — mineral(es) m - [miner] ore, production, o output
— f — — m de hierro m - [miner] iron ore output
— f — miniplanta f - [metal-prod] miniplant, production, o output
— f — motogenerador m - [sold] motor-generator output
— f — nave(s) f - [nav] shipbuilding
— f — neumático(s) m - [autom-neumát] tire production
— f — nieve f - [meteorol] snowmaking
— f — operario m - [ind] operator('s) output
— f — palanquilla f - [metal-prod] billet(s) production
— f — — f de acero m - [metal-lam] steel billet production - [metal-lam] steel billet(s) production
— f — — f — — m con carbono m - [metal-lam] carbon steel billet(s) production
— f — — f — — m — — m para alambrón m - [metal-lam] carbon steel wire rod billet(s) poroduction
— f — — f para alambrón m - [metal-lam] wire rod billet production
— f — — f — — barra(s) f - [metal-lam] bar(s) billet production
— f — — f — — f para armadura f para hormigón m - [metal-lam] concrete reinforcing bar, o rebar, billet production
— f — — f — perfil(es) m de acero m - [lam] (steel), profile, o shape, billet production
— f — — f — — m — — m con carbono m - [metal-lam] carbon steel, profile, o shape, billet production
— f — partícula(s) f - [ind] particle(s) production
— f — película(s) f - [fotogr] film production
— f — pella(s) f - [metal-prod] pellet(s) production
— f — picadura(s) f - [metal] pinholing
— f — planta f - [ind] plant production
— f — portadiferencial m - [autom-mec] differential carrier production
— f — prensa f - [mec] press production
— f — presión f - pressure exerting,tion
— f — — f apropiada - [mec] proper pressure exerting
— f — — f en radiador m - [comb.int] radiator pressurizing,ization
— f — producto(s) m - [ind] product(s), production, o manufacture,ring
— f — — m de acero m - [metal-fabr] steel product(s) manufacture,ring
— f — — m siderúrgico(s) - [metal-fabr] steel products manufacture,ring
— f — punzonado m - [mec] punching production
— f — — n de agujero(s) m - [mec] hole punching production
— m — recalentamiento m - overheating causing
— f — redondo(s) m - [metal-lam] round(s) production
— f — — m de acero m - [metal-lam] steel round(s) production
— f — — m — — m con aleación f baja - [metal-lam] low alloy steel rounds production
— f — reformación f - [metal-prod] reforming production
— f — — f con temperatura(s) f alta(s) - [metal-prod] high temperatura reforming production
— f — régimen m - [sold] rated output
— f — residuo(s) m - [ind] waste(s) production
— f — sínter m — [metal-prod] sinter production
— f — sociedad f - [ind] corporate production
— f — soldador m - [sold] welder('s), o operator('s), production, o output
— f de soldadora f - [sold] welding machine output
— f — — f con calidad f alta - [sold] high

quality weld(s) production
producción f de tecnología f - [ind] technology production
— f — trabajo m - work production
— f — transformador m - [electr-oper] transformer output
— f — tren m - [metal-lam] mill production
— f — tubería f - [tub] piping production
— f — tubo(s) m - [tub] tube, o pipe, production
— f — — m sin costura f - [tub] seamless, tube(s), o pipe(s), production
— f — vidrio m - [cerám] glass production
— f — . . . vatio(s) m - [electr-prod] . . . watt(s) production
— f — voltaje m - [electr-prod] voltage production
— f — — m requerido - [electr-prod] required voltage production
— f desalentada - [econ] discouraged production
— f detenida - [ind] stopped production
— f diaria - daily, production, o output
— f — de arrabio m - [metal-prod] daily pig iron, production, o output
— f diaria inicial - [ind] initial daily production
— f disminuida - [ind] decreased output
— f — de operario m - [ind] operator('s) decreased output
— f — — soldador m - [sold] weldor('s), o operator('s) output
— f disponible - [ind] available output
— f doméstica - [ind] domestic, production, o output
— f echada a perder - [ind] spoilage
— f eficiente - [ind] efficient production
— f — desde punto m de vista de costo(s) m - [ind] cost efficient performance
— f elevada - [ind] high, production, o output • raised, o increased, production
— f en barril(es) m por día m - [petról] barrel(s) per day output; B P D O
— f — cantidad f - [ind] mass production
— f — escala f grande - [ind] large scale, o high volume, production
— f — — f industrial - [ind] industrial, o full scale, production
— f — general - [ind] general production
— f — hercio(s) m - [electr-prod] hertz output
— f — masa f - [ind] mass production
— f — serie - [ind] mass, o industrial scale, production, o fabrication
— f — tiempo m - [ind] general production
— f — unidad f de tiempo m - [ind] time unit throughput
— f — . . . voltio(s) m - [electr-prod] . . . volt(s) output
— f especial - [ind] special production
— f especializada - [ind] specialized, production, o output
— f establecida - [ind] settled production
— f estándar - [ind] standard production
— f estimada - [ind] estimated, production, o output
— f — acumulada - [ind] estimated accumulated, production, o output
— f exigida - [ind] required, production, o output
— f experta - expert, o experienced, production, o manufacturing
— f extranjera - foreign production
— f ferrífera - [miner] iron ore production
— f forzada - [ind] forced production
— f hidráulica - [hidr] hydraulic, production, o output • [petról] artificial lift
— f horaria - [ind] hourly, production, o output
— f hullera - [miner] coal production
— f — nueva - [miner] new coal production
— f incrementada - [ind] increased production
— f industrial - [ind] industrial production
— f — manufacturera - [ind] industrial manu-

facturing production
producción f iniciada - [ind] started, o begun, **production**
— f **inicial** - [ind] initial production
— f **ininterrumpida** - uninterrupted production
— f **insuficiente**—[ind] insufficient production
— f **interrumpida** - [ind] interrupted production
— f **laminación f y tratamiento m de acero m** - [metal-prod] making, shaping and treating of steel
— f **laminada** - [metal-lam] rolled production
— f **— en caliente** - [metal-lam] hot rolled production
— f **— — frío** - [metal-lam] cold rolled production
— f **local** - [ind] local production
— f **manufacturera** - [ind] manufacturing production
— f **máxima** - [ind] maximum, production,tivity, o capacity, o output
— f **— asegurada** - [ind] maximum assured production • maximized throughput
— f **mensual** - [ind] monthly, production, o output
— f **— de acero m** - [metal-prod] monthly steel, production, o output
— f **— — — m bruto** - [metal-prod] monthly raw steel, production, o output
— f **— — arrabio m** - [metal-prod] monthly pig iron production
— f **— — hierro m** - [metal-prod] monthly iron production
— f **— — — m primario** - [metal-prod] monthly primary iron production
— f **metálica** - [metal-prod] metallic, production, o output
— f **minera** - [miner] mining production
— f **mínima** - [ind] minimum, production,tivity, o output
— f **mundial** - [ind] world(wide), production, o output
— f **— de acero m** - [metal-prod] world(wide) steel, production, o output
— f **nacional** - [ind] local, o domestic, manufacture, o production, o output
— f **no metálica** - [miner] nonmetallic, production, o output
— f **nominal** - [ind] nominal, o rated, production, o output
— f **— disponible** - [metal-prod] available rated, production, o output
— f **— total** - [ind] total available rated, production, o output
— f **— total** - [ind] total rated output
— f **normal** - [ind] normal, production, o output
— f **nueva** - [ind] new production
— f **— de carbón m**—[miner] new coal production
— f **— — hulla f** - [miner] new coal production
— f **obtenida** - [ind] attained production
— f **para fin(es) m general(es)** - [ind] general, o all purpose, production, o fabrication
— f **— uso(s) m general(es)** - [ind] all purpose production
— f **parcial** - [ind] partial production
— f **perdida** - [ind] lost production
— f **— recuperada** - [ind] recovered lost production
— f **perfeccionada** - [ind] perfected production
— f **por centro m para producción f** - [ind] production center, production, o output
— f **— operación f** - [ind] operation output
— f **— producto m** - [ind] product output
— f **primaria** - [ind] primary production
— f **principal** - [ind] main production
— f **programada** - [ind] scheduled, production, o output
— f **propia** - [ind] own, o generated, production
— f **racional** - [ind] rational, production, o output
— f **real** - [ind] actual production
— f **recuperada** - [ind] recovered production
— f **reducida** - [ind] reduced, o decreased, o fallen, o low, production, o output

producción f regulada - [ind] controlled production
— f **— cuidadosamente** - carefully controlled production
— f **— de tubería(s) f** - [tub] controlled pipe production
— f **regular** - [ind] regular, o settled, production
— f **requerida** - [ind] required, production, o output
— f **satisfactoria** - [ind] satisfactory, o healthy, production, o output
— f **secundaria** - [ind] secondary production
— f **según norma** - [ind] standard production
— f **siderúrgica** - [metal-prod] steel production
— f **— mundial** - [metal-prod] world(wide) steel, production, o output
— f **superior** - [ind] superior production
— f **típica** - [ind] typical, production, o output
— f **total** - [ind] total, production, o output • [electr-prod] full output
— f **— de acero m** - [metal-prod] total steel, production, o output
— f **— — energía f** - [electr-prod] total, o full, power, production, o output
— f **uniforme** - [ind] uniform, o steady, production, o output
— f **— de corriente f** - [electr-prod] steady, o uniform, power, production, o output
— f **— f alterna** - [electr-prod] uniform, o steady, alternating current, production, o output
— f **— — f continua** - [electr-prod] uniform, o steady, direct current, production, o output
— f **— de soldadura f** - [sold] uniform, o steady, weld(ing) output
— f **útil** - useful, o effective, production
— f **y venta f de material(es) m especial(es)** - [com] special(ty) material(s), production and sales, o enterprise
— f **verificada** - [ind] checked output
— f **— con estadística(s) f** - [ind] statistically controlled production
— f **— permanentemente** - [ind] permanently controlled production
— f **— — con estadística(s) f** - [ind] permanently statistically controlled production
producido m - product • proceeds
producido,da a - produced; turned, o put, out • evolved; developed • incurred • exerted; given; yielded • taken place
— a **adicionalmente** - additionally produced
— a **como resultado m de carrera f** - [deport] race-bred
— a **en acería f** - [metal-prod] steel plant produced
— a **— — f eléctrica** - [metal-prod] electric steel plant produced
— a **— base f a** - [ind] produced from
— a **— — f petróleo m** - [ind] produced from oil
— a **— — f — — m crudo** - [ind] produced from crude oil
— a **— horno m** - [ind] furnace produced
— a **— — m eléctrico** - [metal-prod] electric furnace produced
— a **— masa f** - [ind] mass produced
— a **— planta f** - [ind] shop fabricated
— a **— serie** - [ind] series, o mass, produced
— a **ininterrumpidamente** - produced uninterruptedly
— a **localmente** - [ind] produced locally
— a **por otro(s)** - [ind] produced by (an)others
— a **— tercero(s)** - [ind] produced by, @ third party,ties, o other(s)
producir v - . . . . to, fabricate, o manufacture, o make, o develop • to, evolve, o generate • to, perform, o put out • to give • to turn out • to take place • to exert • to, incur, o take place • to arise; to account

for
producir v̲ **acero** m̲ - [metal-prod] to make @ steel
— v̲ — m̲ **con carbono** m̲ - [metal-prod] to produce @ carbon steel
— v̲ **adicionalmente** - to produce additionally
— v̲ **alimento(s)** m̲ - [econ] to produce @ food(s)
— v̲ **arco** m̲ - [electr-oper] to develop @ arc • [sold] v̲éase también **formar** v̲ **arco** m̲
— v̲ — m̲ **con magnitud** f̲ **grande** - [electr-oper] to high magnitude arc
— v̲ — m̲ **dentro de aceite** m̲ - [electr-oper] to arc under @ oil
— v̲ **barra(s)** f̲ - [metal-lam] to produce @ bar(s)
— v̲ — f̲ **laminada(s)** - [metal-lam] to produce @ rolled bar(s)
— v̲ — f̲ — **en caliente** - [metal-lam] to produce @ hot rolled bar(s)
— v̲ — f̲ — — — **para armadura(s)** f̲ - [metal-lam] to produce hot rolled reinforcing bar(s)
— v̲ — f̲ — — — f̲ **para hormigón** m̲ **armado** - [metal-lam] to produce @ hot rolled concrete reinforcing bar(s)
— v̲ **bien(es)** m̲ - [ind] to produce @ good(s)
— v̲ — m̲ **de capital** - [econ] to produce @ capital good(s)
— v̲ **bola(s)** f̲ - [cuerp-moled] to produce @ ball(s)
— v̲ **calidad** f̲ - to produce @ quality
— v̲ **de soldadura** f̲ - [sold] to produce @ weld quality
— v̲ **calor** m̲ - [sold] to, produce, o̲ build up, @ heat
— v̲ **carbonización** f̲ - to produce @ carbonization
— v̲ — f̲ **parcial** - [sold] to produce @ partial carbonization
— v̲ **cobre** m̲ - [metal-prod] to produce @ copper
— v̲ **combustible(s)** m̲ - [combust] to produce @ fuel(s)
— v̲ — m̲ **fósil(es)** - [combust] to produce @ fossil fuel(s)
— v̲ **conjunto** m̲ - [ind] to produce @ assembly
— v̲ — m̲ **soldado** - [sold] to, produce, o̲ manufacture, @ weldment
— v̲ **cordón(es)** m̲ - [sold] to make @ bead(s)
— v̲ — m̲ **pequeño(s)** - [sold] to make @ small bead(s)
— v̲ **cortocircuito** m̲ - [electr-oper] to short-(circuit)
— v̲ — m̲ **en interruptor** m̲ - [electr-oper] to short(circuit) @ switch
— v̲ — m̲ — — m̲ **en encendido** m̲ - [comb.int] to short(circuit) @ ignition switch
— v̲ — — **rectificador** m̲ - [electr-oper] to short(circuit) @ rectifier
— v̲ **chispa(s)** f̲ to, spark, o̲ cause @ sparking
— v̲ **de acuerdo con especificación(es)** f̲ - [ind] to customize
— v̲ — — — — f̲ **de cliente** m̲ - [ind] to customize (to @ customer's specifications)
— v̲ **desecho(s)** m̲ - [nucl] to produce @ waste(s)
— v̲ **desgaste** m̲ - [mec] to produce @ wear
— v̲ — m̲ **en motor** m̲ - [electr-mot] to produce @ motor wear • [comb.int] to produce @ engine wear
— v̲ — m̲ **mínimo** - [electr-mot] to produce @ minimum motor wear • [comb.int] to produce @ minimum engine wear
— v̲ **descontinuidad** f̲ - [sold] to skip
— v̲ **económicamente** - to produce economically
— v̲ **emanación(es)** f̲ - to produce @ fume(s)
— v̲ — f̲ **tóxica(s)** - [ind] to produce @ toxic fume(s)
— v̲ **en base a** - [ind] to produce from
— v̲ — — — f̲ **a petróleo** m̲ - [ind] to produce from @ oil
— v̲ — — f̲ — — m̲ **crudo** - [ind] to produce from @ crude oil
— v̲ **en masa** f̲ - [ind] to mass produce
— v̲ — **serie** - [ind] to mass produce
— v̲ **energía** f̲ - [electr-prod] to produce @ power
— v̲ **eléctrica** - [electr-prod] to produce @ electric, energy, o̲ power
— v̲ **fuerza** f̲ **motriz** - [electr-prod] to, produce, o̲ generate, @ electric, energy, o̲ power
producir v̲ **gel** m̲ - [quím] to produce @ gel
— v̲ **gas** - to, produce, o̲ generate, @ gas
— v̲ **imagen** f̲ - [electrón] to produce @ image
— v̲ **incendio** m̲ - [segurid] to ignite @ fire
— v̲ **ininterrumpidamente** - to produce uninterruptedly
— v̲ **localmente** - to produce locally|
— v̲ **magnetización** f̲ - [electr] to produce magnetization
— v̲ **martensita** f̲ - [metal] to develop @ martensite
— v̲ **palanquilla** f̲ - [metal-prod] to produce @ billet(s)
— v̲ — f̲ **de acero** m̲ - [metal-lam] to produce @ steel billet(s)
— v̲ — f̲ — — m̲ **con carbono** m̲ - [metal-lam] to produce @ carbon steel billet(s)
— v̲ — f̲ — — m̲ — — m̲ **para alambrón** m̲ - [metal-lam] to produce @ carbon steel wire rod billet(s)
— v̲ — f̲ — **alambrón** m̲ - [metal-lam] to produce @ wire rod billet(s)
— v̲ — f̲ — **barra(s)** f̲ - [metal-lam] to produce @ bar billet(s)
— v̲ — f̲ — — f̲ **para armadura** f̲ - [metal-lam] to produce @ reinforcing bar(s) billet(s)
— v̲ — f̲ — — f̲ — — f̲ **para hormigón** m̲ - [metal-lam] to produce @, concrete reinforcing bar, o̲ rebar, billet(s)
— v̲ — f̲ — **perfil(es)** m̲ **de acero** m̲ - [metal-lam] to produce @ profile billet(s)
— v̲ — f̲ — — m̲ — — m̲ **con carbono** m̲ - [metal-lam] to produce @ carbon steel profile billet(s)
— v̲ **partícula(s)** v̲ - [ind] to produce @ particle(s)
— v̲ — f̲ **con calidad** f̲ **constante** - [ind] to produce @ consistent quality particle(s)
— v̲ — f̲ **con composición** f̲ **constante** - [ind] to produce @ consistent composition particle
— v̲ **película** f̲ - [fotogr] to produce @ film
— v̲ **potencia** f̲ - [electr-prod] to, produce, o̲ generate, @ output
— v̲ — f̲ **nominal (indicada)** - [electr-prod] to, produce, o̲ generate, @ (rated) output
— v̲ **presión** f̲ - [mec] to, exert, o̲ put out, @ pressure
— v̲ — f̲ **apropiada** - [mec] to put out @ proper pressure
— v̲ — f̲ — **en radiador** m̲ - [comb.int] to pressurize @ radiator
— v̲ **residuo(s)** m̲ - [nucl] to produce @ waste(s)
— v̲ **resultado(s)** m̲ - to produce @ result(s)
— v̲ — m̲ **bueno(s)** - to produce @ good result(s)
— v̲ — m̲ **malo(s)** - to produce @ poor result(s)
— v̲ — m̲ **mejor(es)** - to produce @ better result(s) • to produce @ best result(s)
— v̲ — m̲ **peor(es)** - to produce @ worse results • to produce @ worst results
— v̲ — m̲ **pobre(s)** - to produce @ poor results
— v̲ **salpicadura(s)** f̲ - [sold] to produce @ spatter
— v̲ — f̲ **excesiva(s)** - [sold] to produce @ excessive spatter
— v̲ **soldadura** f̲ - [sold] to produce @ weld
— v̲ — **con calidad** f̲ **alta** - [sold] to produce @ high quality weld
— v̲ **tecnología** f̲ — [ind] to produce @ technology
— v̲ **trabajo** n̲ - to produce @ work
producirse v̲ - to occur
— v̲ **cuelgue** m̲ - [metal-prod] to hang
— v̲ **fuga** f̲ - to leak
— v̲ **grieta** v̲ - to (develop @) crack
productividad f̲ - . . . . • [ind] output (rate)
— f̲ **alta** - [ind] high productivity
— f̲ — **de horno** m̲ **alto** - [metal-prod] blast furnace high, o̲ high blast furnace, productivity
— f̲ **aumentada** - [ind] increased, o̲ improved, o̲ upped, productivity
— f̲ **de horno** m̲ - [ind] furnace productivity

productividad f̲ de horno m̲ alto - [metal-prod]
blast furnace productivity
— f̲ — minería f̲ - [miner] mining productivity
— f̲ — — f̲ de hierro m̲ - [miner] iron mining
productivity
— f̲ — personal m̲ - [ind] personnel, o̲ employee, productivity
— f̲ efectiva - [labor] effective productivity
— f̲ en tonelada(s)/hora(s) - [labor] ton/hour
productivity
— f̲ — —/— efectiva(s) f̲ - [labor] effective
ton/hour productivity
— f̲ máxima - maximum productivity
— f̲ material - material productivity
— f̲ mayor - [ind] greater, o̲ more, productivity
• greatest, o̲ most, productivity
— f̲ mejorada - [ind] improved productivity
— f̲ menor - [ind] lesser producvivity • least
productivity
— f̲ mínima - [ind] minimum productivity
— f̲ organizada - [ind] organized productivity
producto m̲ - . . . . • output • supply,lies •
[ind] material(s) • [com] proceeds • commodity • véase también producción f̲
— m̲ abrasivo - [mec] abrasive product
— m̲ — con liga f̲ - [ind] bonded abrasive product
— m̲ acabado - finished product
— m̲ — en caliente - [metal-lam] hot finished
product
— m̲ — — frío - [metal-lam] cold finished
product
— m̲ accesorio - [ind] by-product
— m̲ adaptado - [admin] adapted product
— m̲ adicional - additional product
— m̲ adquirido - acquired, o̲ purchased, o̲
bought, product
— m̲ — para distribución f̲ - product, bought, o̲
purchased, for distribution
— m̲ — venta f̲ - product, bought, o̲ pur-
cchased, for sale
— m̲ aglomerado - [miner] agglomerated product
— m̲ ajustado - [ind] adjusted product
— m̲ algo similar - [ind] somewhat similar product
— m̲ analizado - [ind] analyzed product
— m̲ aplicado - applied product
— m̲ asfáltico - [nucl] asphalted,tic product
— m̲ bruto - [Econ] gross product
— m̲ — interno - [com] gross national product;
G N P
— m̲ cementado - [nucl] cemented product
— m̲ centroamericano - [geogr] Central American
product
— m̲ comercial - commercial product
— m̲ compacto - compact product
— m̲ competidor - [com] competing,titive product
— m̲ comprado - purchased, o̲ bought, product
— m̲ — para distribución f̲ - product, bought, o̲
purchased, for distribution
— m̲ — para venta f̲ - product, bought, o̲ purchased, for @ sale
— m̲ con costo m̲ alto - high cost product
— m̲ — — m̲ bajo - low cost product
— m̲ — — m̲ elevado - high cost product
— m̲ — — m̲ reducido - low cost product
— m̲ — fibra(s) f̲ corta(s) - [textil] short fi-
ber(ed) product
— m̲ — temperatura f̲ alta - [ind] high temperature product
— m̲ — — f̲ baja—[ind] low temperature product
— m̲ — — f̲ elevada - [ind] high temperature
product
— m̲ — — f̲ reducida - [ind] low temperature
product
— m̲ corrosivo - corrosive product
— m̲ corrugado - [constr] corrugated product
— m̲ — de acero m̲ - [constr] corrugated steel
product
— m̲ — — — m̲ para drenaje m̲ - [constr] corrugated steel drainage product
— m̲ — helicoidalmente - [constr] helically
corrugated product

producto m̲ costoso - expensive product
— m̲ cúbico - [ind] cubical product
— m̲ de acero m̲ - [metal] steel product
— m̲ — — m̲ con carbono m̲ - [metal] carbon
(steel) product
— m̲ — — m̲ corrugado - [metal-fabr] corrugated
steel product
— m̲ — — m̲ dulce—[metal-lam] mild steel plate
stock
— m̲ — — m̲ elaborado - [metal-fabr] wrought, o̲
fabricated, steel product
— m̲ — — m̲ — para drenaje m̲ - [metal-fabr]
wrought, o̲ fabricated, steel drainage product
— m̲ — — m̲ laminado - [metal-lam] rolled steel
product
— m̲ — — m̲ — en caliente - [metal-lam] hot
rolled steel product
— m̲ — — m̲ — en frío - [metal-lam] cold
rolled steel product
— m̲ — — m̲ para carretera(s) f̲ - [metal-fabr]
highway steel product
— m̲ — — m̲ — construcción f̲ - [constr] construction steel, o̲ steel construction, product
— m̲ — — m̲ — — f̲ vial - [vial] highway construction steel product
— m̲ — — m̲ para drenaje m̲ - [metal-fabr] steel
drainage, o̲ drainage steel, product
— m̲ — — m̲ — — m̲ instalado - [constr] installed steel drainage product
— m̲ — — m̲ — — — permanentemente - [constr]
permanently installed steel drainage product
— m̲ — — m̲ para herramienta(s) f̲ - [herram]
tool steel product
— m̲ — alambre m̲ - [alambre] wire product
— m̲ — baldosería f̲ - [cerám] tile product
— m̲ — banda f̲ - [metal-lam] strip product
— m̲ — — f̲ de acero m̲ - [metal-lam] steel
strip product
— m̲ — — f̲ — — m̲ especial - [metal-lam]
specialty steel strip product
— m̲ — barra(s) f̲ - [metal-lam] bar(s) product
— m̲ — — f̲ acabaa(s) en caliente - [metal-lam]
hot finished bar product
— m̲ — — — — frío - [metal-lam] cold finished bar product
— m̲ — — f̲ terminada(s) en caliente - [metal-
-lam] hot finished bar product
— m̲ — — f̲ — en frío - [metal-lam] cold finished bar product
— m̲ — broca(s) f̲ - [herram] bit product
— m̲ — cadena(s) f̲ - [cadenas] chain product(s)
— m̲ — calidad f̲ - [ind] quality product
— m̲ — — f̲ primera - [ind] first quality product; first(s)
— f̲ — carbono m̲ - [quím] carbon product
— m̲ — cemento m̲ - [constr] cement product
— m̲ — cinta f̲ - [metal-lam] strip product
— m̲ — — f̲ de acero m̲ - [metal-lam] steel strip
product
— m̲ — — — — m̲ especial - [metal-lam] specialty steel strip product
— m̲ — cliente m̲ - customer('s) product
— m̲ — combustión f̲ - [combust] combustion pro-
duct(s) • flue gas(es)
— m̲ — compañía f̲ - company product
— m̲ — competencia f̲ - [com] competitive product
— m̲ — corrosión f̲ - [quím] corrosion product
— m̲ — chapa f̲ - [metal-lam] strip product
— m̲ — — f̲ de acero m̲ - [metal-lam] steel strip
product
— m̲ — — f̲ — — m̲ especial - [metal-lam] specialty steel strip product
— m̲ — destilación f̲ - [quím] (di)still(ation)
product
— m̲ — diamante m̲ - diamond product
— m̲ — empresa f̲ - company, o̲ corporate, product
— m̲ — ferretería f̲ - hardware product
— m̲ — fisión f̲ - [nucl] fision product
— m̲ — fleje m̲ - [metal-lam] strip product
— m̲ — — m̲ de acero m̲ - [metal-lam] steel strip
product
— m̲ — — m̲ — — m̲ especial - [metal-lam] specialty strip steel product

producto m de hierro m - [miner] iron product
— m — hormigón m - [constr] concrete product
— m —— m pretensado - [constr] prestressed concrete product
— m — jabón m—[quím] soap product
— m ——— m de hidroestearato m de litio m - [lubric] lithium hydroxy stearate soap product
— m — plaquita f - [quím] compact product
— m — proceso(s) m siderúrgico(s) - [metal-prod] steel process(es) product
— m — segunda (calidad) - [ind] second quality good; seconds
— m — sociedad f - [legal] corporate product
— m — tubería f - [tub] tubing product
— m ——— f de acero m - [tub] steel tubing product
— m ——— f soldada - [tub] welded tubing product
— m ——— f — de acero m - [tub] welded steel tubing product(s)
— m — venta(s) f - [contab] sales product
— m — derivado(s) m - derived product
— m — de (piedra) caliza f - [miner] limestone product(s)
— m desactualizado - [ind] outdated product
— m descartable - disposable product
— m desconcentrado - [ind] deconcentrated product
— m disponible - available product
— m económico - inexpensive product
— m efectivo - effective product
— m eficaz - effective, o efficient, product
— m eficiente - effective product
— m elaborado - [ind] fabricated, o manufactured, o produced, product • finished goods
— m — en consignación f - [contab] on consignment finished product
— m — por cliente m - customer's product
— m eléctrico - electrical product
— m en barra(s) f - [metal-lam] bar product
— m ——— f terminado - [metal-lam] finished bar porbduct
— m — bruto - [ind] crude, o unfinished, product
— m — curso m de elaboración f - [ind] good(s) in process (of manufacture, ring)
— m — neumático(s) m - [autom-neumát] tire(s) product(s)
— m esférico - spherical product
— m especial - [ind] special product
— m — de banda f - [metal-lam] specialty strip product
— m —— cinta f - [metal-lam] specialty strip product
— m —— chapa f - [metal-lam] specialty strip product
— m —— fleje m - [metal-lam] specialty strip product
— m ——— m de acero m - [metal-lam] specilaty steel strip product
— m específico - specific product
— m estándar - standard, product, o material
— m estructural - [constr] structural product
— m evolucionado - [ind] developed product
— m exhibido - [com] displayed product
— m existente - existing product
— m explosivo - [explos] explosive product
— m extranjero - foreign product
— m fabricado - [ind] manufactured, o fabricated, product
— m — en cantidad(es) f grande(s) - [ind] high volume product
— m farmacéutico - [medic] pharmaceutical (product)
— m final - [ind] end product
— m — terminado - [ind] finished end product
— m fisionable - [nucl] fissionable product
— m — no recobrable - [nucl] nonretrievable, o nonrecoverable, fissionable product
— m ——— recuperado - [nucl] unretrieved, o unrecovered, fissionable product
— m — recuperable - [nucl] retrievable, o recoverable, fisionable product
producto m fisionable recuperado - [nucl] retrieved, o recovered, fisionable product
— m forjado - [metal-trat] forged product
— m — con precisión f - [metal-fabr] precision forged product
— m garantizado - guaranteed, o warranted, product
— m identificado - identified product
— m industrial - [ind] industrial product
— m inicial - initial product
— m insecticida - [quím] insecticie (product)
— m — formulado - [quím] formulated insecticide (product)
— m instalado - [constr] installed product
— m — permanentemente - [constr] permanently installed product
— m intermedio - [ind] intermediate product
— m — de destilación f - [ind] intermediate (di)still(ation) product
— m interno - [econ] internal, o national, product
— m — bruto - [econ] gross, internal, o national, product
— m irritante - irritating product
— m lácteo - [agric] dairy, o milk, product
— m laminado - [metal-prod] rolled,ling product • stock
— m — de acero m - [metal-lam] rolled steel product
— m — en caliente - [metal-lam] hot rolled product
— m —— frío - [metal-lam] cold rolled product
— m —— no plano - [metal-lam] non-flat rolled product
— m —— pesado - [metal-lam] heavy rolled product
— m —— plano - [metal-lam] flat rolled product
— m ——— liviano - [metal-lam] light flat rolled product
— m ——— pesado - [metal-lam] heavy flat rolled product
— m — sencillo - [metal-lam] simple, rolled, o merchant, product
— m — terminado - [metal-lam] finished rolled product
— m lapidario - [miner] lapidary product
— m largo - [metal-lam] long product
— m listado - listed product
— m manufacturado - [metal-fabr] manufactured product
— m — de acero m - [metal-fabr] manufactured steel product
— m —— hierro m - [metal-fabr] manufactured iron product
— m ——— m y acero m - [metal-fabr] manufactured iron and steel product
— m más costoso - costlier, o more expensive, product
— m masivo - [ind] massive product
— m mecánico - [mec] mechanical product
— m mejorado - [ind] improved product • [ganad] improved breed
— m menos costoso - less, costly, o expensive, product
— m metálico - metal(lic) product
— m metalizado - [metal] metallized product
— m modificado - [ind] modified, o changed, product
— m monolítico - [constr] monolithic product
— m — pretensado - [constr] prestressed monolithic product
— m movido - moved product
— m no plano - [metal-lam] non-flat product
— m — recubierto - [metal-trat] uncoated product
— m recuperable - unretrievable, o nonrecoverable, product
— m recuperado - unretrieved, o unrecovered, product
— m normal - normal product
— m normalizado - [ind] normalized product

producto m nuevo - [ind] new product • entry
— m — **en neumático(s)** m - [autom-neumát] new tire product
— m — **renglón m de neumático(s)** m - [autom--neumát] new tire product
— m — **ofrecido** - [com] offered new product
— m — **presentado** - [com] presented, o introduced, new product
— m **ofrecido** - [com] offered product
— m **Oilmaster** - [petról] Oilmaster product
— m — **para industria f petrolera** - [petról] Oilmaster (oil industry) product
— m **operado** - operated product
— m — **activación** f - [nucl] activation product
— m **para cimentación** f - [constr] foundation poroduct
— m — **construcción** f - [constr] construction product
— m — — **f vial** - [vial] highway construction product
— m — **consumidor m** - [ind] consumer product
— m — **consumo m** - consumable,mption product
— m — — **m primario** - primary consumable product
— m — **drenaje m** - [hidr] drainage product
— m — — **m instalado** - [constr] installed drainage product
— m — **fabricar(se)** - [ind] product to be manufactured
— m — **fundación(es)** f - [constr] foundation product
— m — **ingeniería** f - [constr] engineering product
— m — — **f y construcción** f - [constr] engineering and construction product
— m — **mantenimiento m** - maintenance product
— m — — **m industrial** - [ind] industrial maintenance product
— m — **molienda** f - [metal-prod] grinding media
— m — **motor m** - [electr-mot] motor product • [comb.int] engine product
— m — **procesamiento m** - processing product
— m — **soldadura** f - [sold] welding product
— m — — **f con arco m** - [sold] arc welding, product, o supply,lies
— m — **transmisión** f - [austom-mec] transmission product
— m — **yacimiento(s) m** - [petrol] field product
— m — — **para gas m** - [petról] gas field product
— m — — m — **petróleo m** - [petról] oil field product(s)
— n — — m — — **m y gas m** - [petról] oil and gas field product(s)
— m **patentado** - [ind] patented product
— m **perfeccionado** - [ind] developed product • perfected product
— m **permanente** - permanent product
— m **pesado** - heavy product
— m **plano** - [metal-lam] flat product
— m — **laminado** - [metal-lam] flat (rolled), o rolled flat, product
— m — — **en caliente** - [metal-lam] hot rolled flat product
— m — — — **frío** - [metal-lam] cold rolled flat product
— m **plaqueado** - [metal-trat] clad product
— m **prerreducido** - [metal-prod] prereduced product
— m **presentado** [com] presented, o introduced, product
— m **procesado** - [ind] processed product
— m **protegido** - protected product
— m **provisto** - [ind] provided, o supplied, product
— m **proyectado** - [ind] designed product • upcoming, o future, product
— m **puro** - [ind] pure product
— m **químico** - [quím] chemical (product)
— m — **cáustico** - [segurid] caustic chemical product
— m — **contrarrestante** - [quím] counteracting chemical (product)

producto m químico corrosivo - [quím] corrosive chemical (product)
— m — **derivado de carbón m** - [coque] coal chemical (product)
— m — **tóxico** - [segurid] toxic chemical (product)
— m — **valioso** - [quím] valuable chemical (product)
— m **reconocido** - [ind] acknowledged product
— m **recubierto** - coated product
— m **recuperable** - retrievable, o recoverable, product
— m **recuperado** - retrieved, o recovered, product
— m **refinado** - refined product
— m **relaminado** - [metal-lam] rerolled product
— m **relativamente nuevo** - [ind] relatively new product
— m **sano** - [ind] sound product
— m **secundario** - [ind] secondary product • by--product
— m **seleccionado** - selected product
— m **semiacabado** - [ind] semifinished product
— m **semielaborado** - [ind] semifinished product
— m — **de hierro m** - [metal-prod] semifinished iron product
— m — **para relaminación** f - [metal-lam] rerolling semifinished, o semifinished rerolling, product
— m **semiterminado** - [ind] semifinished product
— m — **para relaminación** f - [metal-lam] rerolling semifinished product
— m **sencillo** - [ind] simple product
— m **siderúrgico** - [metal-prod] steel (mill) product
— m **similar** - similar product
— m **soldado** - [sold] wlded,ding product
— m **superabrasivo** - [quím] superabrasive product
— m **suplido** - supplied, o provided, o furnished, product
— m **terminado** - [ind] finished product
— m — **de acero m** - [metal-lam] finished steel product
— m — — **m laminado** - [metal-lam] finished rolled steel product
— m — **en caliente** - [metal-lam] hot finished product
— m — — **frío** - [metal-lam] cold finished poroduct
— m — — **tránsito** - [com] finished in transit product
— m — **total** - [ind] total(ly) finished product
— m **total** - [ind] total product
— m **tóxico** - toxic product
— m **tubular** - [tub] tubular, o pipe, product
— m — **de acero m** - [tub] steel pipe product
— m — — **Armco** - [tub] Armco tubular product
— m **usado** - used product
— m **vencido** - outdated product
— m **vendido** - [com] sold product
— m **versátil** - versatile product
productor m - . . .; manufacturer • [ind] operating,tor personnel
— m **asociado** - [ind] associate producer
— m **de acero** - [metal-prod] steel, producer, o maker, o manufacturer
— m — **alambre m** - [alambre] wire producer
— m — — **m para resorte(s) m** - [alambre] spring wire producer
— m — **fleje m** - [metal-lam] strip producer
— m — — **m de acero m** - [metal-lam] steel strip producer
— m — — m — — **m especial** - [metal-lam] specialty steel strip producer
— m — **mineral** - [miner] ore producer
— m — — **m prerreducido** - [metal-prod] prereduced ore producer
— m — **resorte(s) m** - [metal-fabr] spring producer
— m — **cereal(es) m** - [agric| grain producer
— m — **grano(s) m** - [agric] grain producer
— f — **lámina f** - [metal-lam] strip producer

productor m de lámina(s) m y fleje m - [metal-lam] sheet and strip producer
— m —— f —— laminado(s) m - [metal-lam] rolled sheet and strip producer
— m —— f —— en caliente - [metal--lam] hot rolled sheet and strip producer
— m —— f —— m — en frío - [metal-lam] cold rolled sheet and strip producer
— m — plancha(s) f - [metal-lam] sheet producer
— m —— f galvanizada(s) - [metal-lam] galvanized sheet producer
— m — trigo m - [agric] wheat producer
— m ejecutivo - [teatro] executive producer
— m en comunidad f - [poll] community producer
— m estadounidense - United States producer
— m local - [econ] local producer
— m más adelantado - [ind] foremost producer
— m mayor - [ind] largest, o greatest, producer
  • larger, o greater, producer
— m minero - [miner] mining producer
— m — grande - [miner] large mining producer
— m — pequeño - [miner] small mining producer
— m regional - [ind] regional producer
— m siderúrgico - [metal-prod] steel producer
productor,ra a de alimento(s) m - [econ] food producing
— a — calor m - heat producing
— a — desgaste - [mec] wear producing
— a — dólar(es) m - [econ] dollar, producing, o generating
— a — gas m - gas producing
— a — gel m - [quím] gel producing
— a — trigo m - [agric] wheat producing
productos m y servicios m en curso - [contab] products, work and services in process
proemio m - [public] . . .; prologue
profesión f - . . .; career
— f de dirigente—[admin] management profession
— f — ingeniería f - engineering profesion
— f —— f minera - [miner] mining engineering profession
— f minera - [miner] mining profession
— f para caracterización f - identification profession
profesional m para difusión f electrónica - [telecom] electronic media professional
— m — televisión f - [telecom] electronic media, o television, professional
— m principiante - [labor] professional beginner; beginning professional
profesionalsmo m - professionalism
profesor m de ciencias f - [educ] science, professor, o teacher
— m destacado - [educ] outstanding, professor, o teacher
profundidad f - . . . . • elevation • [nav] draft
— f aceptable - acceptable depth
— f — de agua m - acceptable water depth
— f —— m en entrada f - [hidr] acceptable headwater depth
— f adicional - [constr] additional depth
— f alcanzada - reached, o attained, depth
— f apropiada - suitable, o appropriate, depth
— f bajo tierra f - [constr] cover (height)
— f considerable - considerable depth
— f controlada - controlled depth
— f — de imperfección f - flaw, o imperfection, controlled depth
— f correcta - correct depth
— f corriente - ordinary depth
— f — para zanja f - [constr] ordinary trench depth
— f crítica - critical depth
— f — para caudal - [hidr] critical flow depth
— f de chaflán m - [sold] chamfer depth
— f — agua m - [hidr] water depth
— f —— m en entrada f - [hidr] headwater depth
— f —— f en diámetro(s) m - [hidr] headwater depth in diameter(s)
— f —— m —— f para regulación f - [hidr] control headwater (depth)

profunidad f de agua m en entrada f para regulación f en entrada f - [hidr] inlet control headwater depth
— f —— m —— f para regulación f en entrada f - [hidr] inlet control headwater depth
— f —— m en entrada f - [hidr] headwater, height, o depth
— f —— m —— salida f - [hidr] tailwater, height, o depth
— f — arcón m - [constr] caisson depth
— f —— m para excavación f - [constr] (excavation) caisson depth
— f — banda f - [autom-neumát] skid depth
— f —— f para rodamiento m - [autom-neumát] tire, o skid, depth
— f — baño m - [metal-prod] bath depth
— f — canal - canal, o channel, depth
— f — corte m - cut depth
— f — cauce m - [hidr] channel, o canal, depth
— f — caudal - [hidr] flow depth
— f — cobertura f - [constr] cover depth
— f —— f para pista f - [constr] runway cover depth
— f —— f —— f para aeropuerto m - [constr] airport runway cover depth
— f — congelación f - [meteorol] frost depth
— f — contacto m - contact depth
— f —— m de diente - tooth depth contact
— f — corriente f—[hidr] stream, o flow, depth
— f —— f normal - [hidr] normal, stream, o flow, depth
— f — corrugación(es) - [mec] corrugation depth
— f — corte m - cut depth - [metal-lam] scarfing depth
— f — costura f - [sold] seam depth
— f — curso m de agua m - [hidr] flow depth
— f — chavetero m - [mec] keyway depth
— f — defecto m - depth, o flaw, depth
— f — diente m - [mec] tooth depth
— f — descontinuidad f - [metal-lam] discontinuity depth
— f — distribución f - spread depth
— f — embutido m - [metal-fabr] draw(ing) depth
— f — empotramiento m - [constr-pil] embedding depth
— f —— m para tablestacado m - [constr-pil] sheetpiling embedding depth
— f — enfriamiento m - [metal-prod] chill, o cooling, depth
— f — escotadura f - [mec] notch depth
— f —— f en azud m - [hidr] weir notch depth
— f — estrato m - [geol] stratum,ta depth
— f —— m impermeable - [geol] impervious stratum,ta depth
— f —— m permeable - [geol] pervious stratum,ta depth
— f — excavación f - [constr] excavation depth; trench depth
— f — flujo m - [jodr] flow depth
— f — fundente m - [sold] flux depth
— f — fusión f - [sold] fusion depth
— f —— f de soldadura f - [sold] weld fusion depth
— f —— f razonable - [sold] reasonable fusion depth
— f — grieta f - crack depth
— f — helada f - [meteorol] frost depth
— f — imperfección f - imperfection, o flaw, depth
— f — línea f - line depth
— f —— f para dragado m - [constr] dredging line depth
— f — muestra f - [suelos] sample depth
— f — pasada f - [sold] pass depth
— f — piñón m - [mec] pinion depth
— f —— m ajustada - [mec] adjusted pinion depth
— f — pista f - [aeron] runway depth
— f — penetración f - penetration depth
— f —— f de congelación f - [meteorol] frost penetration depth
— f — pista f - [aeron] runway depth
— f —— f para aeropuerto - [aeron] airport

profundidad f de relleno m - [sold] filling
- depth
— f —— m —— en ranura f - [sold] slot
weld filling depth
— f —— —— f para tapón m - [sold] plug
weld(ing) filling depth
— f — remanso m - [hidr] backwater depth
— f — resalto m - [mec] thread depth
— f —— m para tornillo m - [mec] screw
thread depth
— f — salamandra f - [metal-prod] salamander
depth
— f — soldadura f - [sold] weld depth
— f — tragante m - [metal-prod] throat depth
— f — tres decímetros m - foot depth
— f — zanja f - [constr] trench depth
— f debajo de superficie f - [constr] bearing
depth
— f —— —— f natural - [constr-pil] depth
below @ natural grade
— f diversa - different depth
— f en centro m - center depth
— f —— m de capa f para patinaje m——[autom-
-neumát] center skid depth
— f escasa - [hidr] small depth
— f — de napa f - [hidr] shallow water table
— f —— — f freática - [hidr] high water
table
— f específica - specific depth
— f especificada - specified depth
— f estándar - standard depth; véase profundidad f según norma f
— f estipulada - stipulated, o design, depth
— f excesiva - excessive depth
— f — de fundente m - [sold] excessive flux
depth
— f exigida - required depth
— f final - final, o finished, depth
— f — para balasto m - [f.c.] finished ballast depth
— f grande - great depth
— f — de banda f para deslizamiento m -
[autom-neumát] deep skid depth
— f —— —— f — resbalamiento m - [autom-
-neumá] deep skid depth
— f igual - equal depth
— f incorrecta f - incorrect depth
— f indicada - indicated, o design, depth
— f máxima - maximum depth
— f máxima para grieta f - maximum crack depth
— f mayor - greater, o additional, depth •
greatest depth
— f — de - depth greater than
— f media - average, o mean, depth
— f mínima - minimum depth
— f — de cobertura f - [constr] cover minimum
depth
— f —— grieta f - minimum crack depth
— f nominal - nominal depth
— f — de perforación f - [petról] nominal
depth rating
— f notable - [petról] great depth
— f para descenso m de nivel m - [hidr] drawdown depth
— f pareja - equal, o even, depth
— f — de banda f (para rodamiento m)——[autom-
-neumát] equal tread depth
— f prevista - design depth
— f razonable - reasonable depth
— f reducida - reduced, o shallow, depth
— f según norma f - standard depth
— f —— — f en centro m (de banda f - [autom-
-neumát] center standard depth
— f —— — f para excavación f - [constr]
trench standard depth
— f siguiente - following depth
— f suficiente - sufficient depth
— f total - total depth
— f uniforme - uniform depth
— f varia - various depths
— f variable - variable depth
— f variada - varied depth
profundización f deepening

profundización f de canal m - [hidr] channel
deepening
programa m - . . . . • schedule • calendar • a-
genda • policy • format • [comput] software
— m abandonado - abandoned program
— m académico - academic program
— m actual - [ind] current program
— m para ampliación f - [ind] current expansion program
— m adecuado - adequate, schedule, o program
— m administrativo - management program
— m ajustado - tight schedule
— m — para entrega(s) f - tight delivery
schedule
— m ambicioso - ambitious program
— m aplicado - applied program
— m aprobado - approved, program, o scheulde
— m apropiado - proper, o appropriate, program
— m arancelario - [fisc] tariff schedule
— m automovilístico - [telecom] motorsport(s)
program
— m completo - complete, schedule, o program
— m — para ampliación f - [ind] complete expansion program
— m —— — f de planta f - [ind] complete
plant expansion program
— m computerizado - [comput] computerized program
— m para destilación f - [petról] distillation computer program
— m —— —— f con evaporación f instantánea -
[petról] flash distillation computer program
— m con coparticipación f - [com] co-op program
— m — . . . etapa(s) f - . . .-step program
— m — plazo m largo - [fin] long term program
— m conectado - [comput] program in
— m constatado - checked program
— m contractual - [legal] contractual program
— m cooperativo - cooperative program
— m — estadounidense - [constr] national co-
operataive program
— m — para investigación(es) f sobre ca-
rretera(s) f - [constr] National Cooperative
Highway Research Program
— m crediticio - [fin] credit program
— m especial - [fin] special credit program
— m crítico - critical, program, o schedule
— m cultural - cultural program
— m de Allen m - [admin] Allen program
— m — para acción f administrativa -
[admin] Allen management action program
— m —— —— f para administradores m -
[admin] Allen management action program
— m — fecha(s) f - date(s) schedule
— m —— — f para envío(s) m - shipping date(s)
schedule
— m — operabilidad f - [ind] operability program
— m — plazo(s) m - date(s) schedule
— m —— m para envío m - shipping date(s)
schedule
— m — punto(s) m - point(s) program
— m —— — n para inspección f - inspection
point(s) program
— m — urgencia f - crash program
— m departamental - [ind] departmental schedule • [pol] county-(wide) program
— m — para inspección f - [pol] county-wide
inspection program
— m —— —— f de puente(s) m - [pol] county-
-wide bridge inspection program
— m — para instrucción f - [ind] departmental
training program
— m —— —— f sobre seguridad f - [segurid]
departmental safety training program
— m deportivo - [deport] sports program
— m — automovilístico - [deport] motorsports
program
— m desconectado - [comput] program out
— m educacional - [educ] educational program
— m ejecutivo - executive program
— m — para control m - [comput] executive
control program

programa m en principio m - tentative schedule
— m espacial - [astronáut] space program
— m — tripulado - [astronáut] manned space program
— m especial - special program
— m establecido - established program
— m — para mantenimiento m - [ind] established maintenance program
— m experimental - experimental program
— m extenso - extensive, o massive, program
— m — nuevo - new extensive program
— m — para reemplazo m de puentes - [constr] massive bridge replacement program
— m flexible - [comput] flexible program
— m gastronómico - gastronomic schedule
— m general - general, program, o schedule
— m hidráulico - [hidr] hydraulic, o water. program
— m impedido - impeded, o hindered, program
— m integrado - integrated program
— m integral - integral program
— m masivo - massive program
— m — para reemplazo m de puentes m - [constr] massive bridge replacement program
— m mixto - mixed schedule
m modificado - [ind] modified, o revised, o changed, schedule
— m muy ajustado - critical, o very tight, schedule
— m nacional - [pol] national program
— m naval - [nav] naval, o shipbuilding, program
— m nuevo - new, program, o schedule
— m — para inversión f - [fin] new investment program
— m — — f para producción f carbonera - [miner] new coal production investment program
— m — — f — — f hullera - [miner] new coal production investment program
— m — producción f carbonera - [miner] new coal production program
— m — — — f hullera - [miner] new coal production program
— m operacional - [comput] operating program
— m operativo - [ind] operating schedule
— m orientado hacia servicio m - [ind] service oriented program
— m para acción f - [admin] action program
— m — — f administrativa - [admin] management action program
— m — — f para administrador(es) m - [admin] management action program
— m — actividad(es) f - activity,ties schedule
— m — — f — amoladura f - [metal-lam] grinding activity schedule
— m — — f — — f para rodillo(s) m - [metal-lam] roll grinding acivity schedule
— m — — f — granallado m - [metal-lam] shot blasting activity schedule
— m — — f — — m de rodillo(s) m - [metal-lam] roll shot blasting activity schedule
— m — — f para rectificación f - [metal-lam] grinding activity schedule
— m — — f — — f de rodillo(s) m - [metal-lam] roll grinding activity schedule
— m — — f para torneado m - [metal-lam] turning activity schedule
— m — — f — — m de rodillo(s) m - [metal-lam] roll turning activity schedule
— m — administración f - [admin] management program
— m — adquisición(es) f - acquisition program
— m para alerta(r) - [comput] alert(ing) program
— m — — m audible - [comput] audible alert program
— m — alimentación f - [mec] feeding schedule
— m — ampliación f - expansion program
— m — aprendizaje m - apprenticeship, o training, program
— m — aprobación f - approval, o certification, program

programa m para aprovechamiento m - improvement program - utilization program
— m — — m de agua m - [hidr] water utilization program
— m — — m de recurso(s) m - resource(s) utilization program
— m — ayuda f - aid, o support, program
— m — f para comercialización f - [com] marketing support program
— m — f — venta(s) f - [com] marketing support program
— m — f — desarrollo m - [pol] development aid program
— m — f — — m de aeropuerto(s) m - [pol] Airport Development Aid Program
— m — barco m - [nav] ship('s) schedule
— m — calidad f - quality program
— m — calificación f - certification program
— m — cambio m - [com] changeover program • [fin] exchange program
— m — — m de rodillo(s) m - [metal-lam] roll change, program, o plan
— m — campana f - [metal-prod] bell program
— m — — f grande m - [metal-prod] large bell program
— m — — f pequeña - [metal-prod] small bell program
— m — capacitación f - [labor] training program
— m — — f para personal m - [ind] personnel, o employee, training program
— m — carga f - [transp] loading schedule • [ind] charging schedule
— m — — f para barco m - [nav] ship('s) loading schedule
— m — cargo(s) m - [contab] charge(s) schedule
— m — carrera(s) f - [deport] races,cing program
— m — certificación f - certification program
— m — coladas f - [metal-prod] tapping schdule
— m — compañías f varias - multicompany program
— m — comprobación f [ind] control program
— m — — f para calidad f - [ind] quality control program
— m — computador m - [comput] computer program
— m — conductor(es) m - [deport] driver schedule
— m — conservación f - [ind] maintenance, program, o schedule
— m — — f para cuchara(s) f - [metal-prod] ladle, maintenance, o inspection, schedule
— m — construcción f - [constr] construction, program, o schedule
— m — contrato m - contract schedule
— m — control m - [comput] control program
— m — — m ejecutivo - [comput] executive control program
— m — — m para calidad f - [ind] quality control program
— m para corrida(s) f para ensayo m - [ind] trial run schedule
— m — costo(s) m - [contab] charges schedule
— m — cualificación* f - [labor] qualification program
— m — — f para personal m - [labor] personnel qualification program
— m — departamento m - [ind] department(al) schedule
— m — desarrollo m - development program
— m — — m nacional - [econ] national development program
— m — despacho(s) m - [ind] shipping schedule
— m — distribución f - distribution, schedule, o program
— m — — f para combustible(s) m - [ind] fuels distribution schedule
— m — — f para fluido(s) m - [ind] fluid(s) distribution schedule
— m — — f para servicio(s) m - [ind] service(s) distribution schedule
— m — documentación f - documentation schedule

programa m para documentación f para ingeniería
  f - engineering documentation schedule
— m — embarque(s) m - [transp] shipping,pments schedule
— m — — m parcial(es) - [transp] partial shipping,pments schedule
— m — ensayo(s) m - [ind] test(ing), o trials, program, o schedule
— m — — m para neumático(s) m - [autom-neum] tire trial(s), program, o schedule
— m — entrega(s) f - [trans] delivery schedule
— m — — f de dibujo(s) m - [dib] drawing(s) delivery schedule
— m — — f — plano(s) m - [dib] drawing(s) delivery schedule
— m — entrenamiento m - [ind] training program
— m — esímulo m - [econ] incentive(s) program
— m — estudio(s) m - [educ] curriculum
— m — — m aprobado - [educ] approved curriculum
— m — — m para seminario m - [labor] seminar curriculum
— m — evento(s) m - event(s), calendar, o format, o program
— m — expansión f - expansion program
— m — — f completo - [ind] complete expansion program
— m — — f para empresa f - [ind] corporation('s) expansion program
— m — — f — planta f - [ind] plant expansions program
— m — exploración(es) f - exploration program
— m — — f de subsuelo m - [suelos] subsurface exploration program
— f — fabricación f - [ind] fabrication,ting, o manufacturing, program, o schedule
— m — financiación f - [fin] financing program
— m — fomento m - development, o improvement, program
— m — — m Vecinal - [pol] Neighborhood Development Program
— m — formación f - [labor] training program
— m — garantía f - [ind] guaranty program
— n — — f de calidad f - [ind] quality guaranty program
— m — incentivo(s) m - [labor] incentive(s) program
— m — ingeniería f - [ind] engineering program
— m — iniciación f - starting, program, o schedule
— m — inspección f - [ind] inspection, program, o schedule
— m — — f de cucharón(es) m - [metal-prod] ladle(s) inspection, program, o schedule
— m — — f — puente(s) m - [constr] bridge(s) inspection program
— m — instrucción f - instruction, o training, program
— m — f para personal m - [labor] employee safety training program
— m — — f — — f em trabajo m - [segurid] job safety training program
— m — intercambio m - changeover program
— m — inversión(es) f - [fin] investment(s), program, o schedule
— m — — f con plazo m largo - [fin] long term investment program
— m — — f para producción f - [ind] production investment program
— f — — f — — f, carbonera, o hullera - [miner] coal production investment program
— m — — f — — f de, carbón m, o hulla f - [miner] coal production investment program
— m — investigación(es) f - research program
— m — — f aplicada - applied research program
— m — laminación f - [metal-lam] rolling schedule
— m — lubricación f - [mec] lubrication, program, o schedule
— m — mantenimiento m - [ind] maintenance, program, o schedule,ling
— m — — m preventivo - [ind] preventive maintenance program

programa m para mejora(s) f - improvement(s) program
— m — mejor aprovechamiento m - [hidr] imporovement program
— m — — m de agua(s) m natural(es) - [hidr] water resource(s) improvement program
— m — mejoramiento mpara activo m fijo - [ind] capital improvement program
— m — montaje m - [ind] erection schedule
— m — — m en obra - [ind] field erection schedule
— m — motivación f—motivational progrem
— m — normalización f - [ind] standardization program
— m — — f para grúa(s) f - [grúas] crane(s) standardization program
— m — — f para (todas) planta(s) f - [ind] interplant standization program
— m — obra(s) - [ind] project, o work, o job, schedule • improvement program
— m — — f bajo empréstito m - [constr] bond imrovement program
— m — operación f - [comput] operating program
— m — — f para distribución f - [ind] distribution operating schedule
— m — — — f de servicio(s) m - [ind] service(s) distribution operating schedule
— m — parada(s) f - [ind] shutdown(s), program, o schedule
— m — — f semanal(es) - [ind] weekly shutdown, program, o schedule
— m — — f semanal(es) para equipo m - [ind] weekly equipment shutdown, program, o schedule
— m — permuta(s) f - [com] changeover program
— m — personal m - [ind] manpower, o personnel, schedule
— m — — m asignado - [labor] assigned personnel schedule
— m — — m para departamento m - [ind] departmental, manpower, o personnel, schedule
— m — — m destacado - [labpr] assigned, manpower, o personnel, schedule
— m — planificación f - planning program
— m — plano(s) m - [dib] drawing(s) schedule
— m — plazo m corto - short term schedule
— m — — m largo - long term schedule
— m — prevención f - [ind] prevention program
— m — — f de accidente(s) m - [segurid] accident prevention program
— m — producción f - [ind] production, program, o schedule
— m — — f carbonera - [miner] coal production program
— m — — f para carbón m - [miner] coal production prgram
— m — — f — hulla f - [miner] coal production program
— m — — f hullera - [miner] coal production program
— m — — f por departamento m - [ind] departmental production schedule
— m — propaganda - [com] advertising program
— m — — f con coparticipación f - [com] co-op advertising program
— m — prueba f - trial program
— m — publicidad f - [com] advertising program
— m — puesta f en marcha - [ind] start-up schedule
— m — reconstrucción f - [ind] reconstruction program • [metal-prod] relining program
— m — — f para horno m - [metal-prod] furnace relining, program, o schedule
— m — reemplazo m - replacement program
— m — — m para puente(s) m - [constr] bridge replacement program
— m — — preventivo - [ind] preventive replacement program
— m — regulación f - [comput] control program
— m — rehabilitación f - rehabilitation program
— m — rendimiento m alto - [autom-neumát] performance program

programa m para reparación(es) f - [ind] repair schedule
— m — seguridad f - [segurid] safety program
— m — f para planta f - [segurid] plant safety program
— m — f — trabajo m - [segurid] work safety schedule
— m — f — — m en obra f - [segurid] project site work safety schedule
— m — seminario m - seminar program
— m — servicio(s) m - [ind] service(s), program, o schedule
— m — — m exigido(s) - [ind] required service(s), program, o schedule
— m — subsidio(s) m - [pol] subsidy,dies, o aid, program
— m — — m para desarrollo m - [pol] Development Aid Program
— m — — m — m de Aeropuerto(s) m - [pol] Airport Development Aid Program
— m — sugerencia(s) f - [labor] suggestion(s) program
— m — f de personal m - [labor] employee suggestion(s) program
— m — suministro(s) m - [ind] supply,lies, schedule, o program
— m — supercarretera(s) f [vial] superhighway program
— m — f intersestatales f - [vial] interstate highway program
— m — tarea f - [labor] job schedule
— n — televisión f - [electrón] television, program, o entertainment
— m — terminación f - completion, program, o schedule
— m — todas planta(s) f - [ind] interplant program
— m — torneado m - [mec] turning schedule
— m — trabajo m - [ind] work, o operating, schedule • [constr] project schedule
— m — transmisión f - [electr-distrib] transmission, schedule, o program
— m — transporte m - [transp] transportation schedule
— m — trueque m - [com] changeover program • barter(ing) program
— m — unidad f para producción f - [ind] operating unit schedule
— m — — f funcional - [ind] operating, o functional, unit schedule
— m — — f operativa - [ind] operating unit schedule
— m — urbanización f - [pol] development program
— m — venta f de acción(es) f - [legal] shares sale program
— m — verificación f - [ind] control program
— m — — f de calidad f - [ind] quality control porogram
— m — visita(s) f - visitor('s) program; open house
— m patentado - [electrón] proprietary software
— m permanente - permanent, o continuing, program
— m — para instrucción f - continuing training program
— m — — f sobre seguridad f - [segurid] continuing safety training program
— m planeado - planned program
— m posible - possible, program, o schedule
— m — para entrega(s) f - possible, o probable, delivery schedule
— m predeterminado - [ind] predetermined, program, o schedule
—· m preestablecido - preestablished, o planned, program, o schedule
— m preliminar - [ind] preliminary, program, o schedule
— m — para fabricación f - [ind] preliminary manufacturing schedule
— m preparado - prepared program • [comput] written program
— m propuesto - [ind] proposed schedule

programa m propuesto para fabricación f - [inc] proposed fabrication schedule
— m publicitario - [com] advertising program
— m recomendado - recommended, program, o schedule
— m — para mantenimiento m - [ind] recommended maintenance schedule
— m registrado - registered program • [comput] proprietary software
— m revisado - revised program
— m seminario m - seminar program
— m sencillo - simple program
— m — para sitio m - [comput] simple field program
— m similar--similar, program, o schedule
— m — para fabricación f - [ind] similar fabrication schedule
— m tentativo - tentative, program, o schedule
— m trimestral - quarterly, program, o schedule
— m urgente - urgent, o crash, program
— m valorado - valu(at)ed, program, o schedule
— m vivo - [teatr] live, show, o program
— m — para televisión f - [teatro] live television, show, o program
programable a - programmable
— m en sitio m - [comput] field programmable
— a individualmente - [comput] individually programmable, o programmable individually
programación f - programming; scheduling • engineering
— f adecuada - adequate scheduling
— f anticipada - [electrón] in-programming
— m — para de-énfasis* - [electrón] de-emphasis in-programming
— f bead-chain - [electrón] bead change programming
— f de actividad(es) f - activity,ties scheduling
— f — — f para amoladura f - [metal-lam] grinding activity,ties scheduling
— f — — f — — f de rodillo(s) m - [metal-lam] roll grinding activity,ties scheduling
— f — — f — granallado m - [metal-fabr] shot blasting activity,ties scheduling
— f — — f — m de rodillo(s) m - [metal-lam] roll shot blasting activity,ties scheduling
— f — — f — rectificación f - [metal-lam] roll grinding activity,ties scheduling
— f — — f — torneado m - [metal-lam] turning activity,ties scheduling
— f — — f — m de rodillo(s) m - [metal-lam] roll turning activity,ties scheduling
— f — característica(s) f - [electrón] characteristic(s) programming
— f — — f seleccionable(s) - [electrón] selectable characteristic(s) programming
— m — código m - [comput] code programming
— f — — m específico - [comput] specific code programming
— f — corrida f - [ind] run scheduling
— f — — f para ensayo(s) m - [ind] trial run, programming, o scheduling
— f — exigencia(s) f - [ind] requirement(s), scheduling, o programming
— f — — f para producto(s) m - [metal-fabr] product(s) requirement(s) scheduling
— f — — f — — m laminado(s) - [metal-lam] rolled products requirement(s) scheduling
— f — fecha(s) f - date(s) scheduling
— f — lingote(s) m a planchón(es) m y palanquilla(s) f - [metal-lam] ingot(s) to slab(s) and billet(s) schduling
— f — microplaqueta(s) f - [comput] chip microprogramming
— f dinámica - dynamic, scheduling, o programming
— f en fábrica f - [comput] factory programming
— f — — f para microplaqueta(s) f - [comput] chip factory programming
— f — entrada f - [comput] input programming
— f — salida - [comput] output programming
— f — sistema m - [comput] programming in(to)

@ system
programación f en sitio m - field programming
— f **individual**—[comput] individual programming
— f **industrial** - [ind] industrial scheduling
— f **lineal** - linear scheduling
— f **para conductores** m - [deport] driver scheduling
— f — **chapa(s)** f - [metal-lam] strip scheduling
— f — — f **en caliente** - [metal-lam] hot strip scheduling
— f — — f **en frío** - [metal-lam] cold strip scheduling
— f — **ensayo(s)** m - test(s), scheduling, o engineering, o programming
— f — **estimación(es)** f - [admin] budgeting programming
— f — **mano** f **de obra** - [ind] manpower scheduling
— f — **mantenimiento** m - [ind] maintenance scheduling
— f — — m **para usina** f - [electr-prod] power plant maintenance scheduling
— f — — m **eléctrico** - [electr-prod] electical maintenance scheduling
— f — — m **para usina** f - [electr-prod] power plant electrical maintenance scheduling
— f — — — — f **en planta** f siderúrgicas - [electr-prod] steel power plant electrical maintenance scheduling
— f — — m — — — f **industrial en planta(s)** f **siderúrgica(s)** - [electr-prod] steel plant industrial power plant electrical maintenance scheduling
— f — — m **mecánico** - [ind] mechanical maintenance scheduling
— f — — m — **para usina** f - [electr-prod] power plant mechanical maintenance scheduling
— f — — — — f **industrial** - [electr-prod] industrial power plant mechanical maintenance, programming, o scheduling
— f — — — — f **en planta(s)** f **siderúrgica(s)** - [electr-prod] steel plant industrial power plant mechanical maintenance, programming, o scheduling
— f — m — — f — f **siderúrgica(s)** - [electr-prod] steel plant power plant mechanical maintenance scheduling
— m — **operación** f - [comput] operation programming
— f — — f **para usina** f - [comput] power plant operation programming
— f — f — — f **industrial** - [comput] industrial power plant operation programming
— f — — f — — f — **en planta(s)** f **siderúrgica(s)** - [comput] steel plant industrial power plant operation programming
— f — **palanquilla(s)** f - [metal-lam] billet(s) scheduling
— f — **personal** m - [labor] personnel, o manpower, o force, scheduling
— f — **planchón(es)** m - [metal-lam] slab(s) scheduling
— f — **planificación** f - [admin] planning programming
— f — — f **para estimación(es)** f - [admin] budgeting planning programming
— f — **producción** f - [ind] production, programming, o scheduling
— f — **producto(s)** m - [ind] product scheduling
— f — — m **laminado(s)** - [metal-lam] rolled product(s) scheduling
— f — **reparación** f - [ind] repair scheduling
— f — — f **para convertidor(es)** m - [metal-prod] converter repair scheduling
— f — **riel(es)** m **y perfil(es)** m - [metal-lam] rail(s) and structural(s) scheduling
— f — **tarea(s)** f - [labor] job(s) scheduling
— f — **torneado** m - [mec] turning scheduling
— f — **trabajo** m - [labor] work scheduling
— f — **transpondedor** m - [telecom] transponder, programming, o scheduling
— f **sencilla** - [comput] simple programming

programación f **sencilla en sitio** m - [comput] simple field programming
— f **variable** - [electrón] variable programming
— f — **con tope(s)** m - [electrón] bead chain programming
programado,da a - programmed; scheduled • planned • engineered
— a **adecuadamente** - adequately scheduled
— a **anticipadamente** - [electrón] in-programmed
— a **en fábrica** f - [comput] factory programmed
— a — **planta** f - [comput] plant programmed
— a — **sistema** m - programmed into @ system
— a — **sitio** m - [comput] field programmed
— a **sencillamente** - programmed simply
— a — **en sitio** m - [comput] simple,ply field programmed
programador m - [electrón] programmer, scheduler • schedule man
— m **para distribuidor** m - [metal-prod] distributor programmer
— m — **laminador** m—[metal-lam] mill programmer
— m — m **para desbaste(s)** m **(planos)** - [metal-lam] slabbing mill programmer
— m — m — **tocho(s)** m **y planchón(es)** m - [metal-lam] blooming and slabbing mill programmer
— m — — m — — m **y desbaste(s)** m **(planos)** - [metal-lam] blooming and slabbing mill programmer
— m — **parada(s)** f - [comput] shutdown programmer
— m — **pasada(s)** f - [metal-lam] passes programmer • preset screwdown control
— m — — f **para laminador** m - [metal-lam] mill pass(es) programmer
— n — — f — m **para desbaste(s)** m **(planos)** - [metal-lam] slabbing mill pass(es) programmer
— m — — f — — m — **planchónes** m - [metal-lam] slabbing mill pass(es) programmer
— m — — f — — m — **tocho(s)** m - [metal-lam] blooming mill pass(es) programmer
— m — — f — — — m **y desbaste(s)** m **(planos)** - [metal-lam] blooming and slabbing mill pass(es) programmer
— m — — f — — — m **y planchón(es)** m - [metal-lam] blooming (and slabbing) mill, passes programmer, o preset screwdown control
— m — **tornillo** m **para ajuste(s)** m - [metal-lam] screwdown programmer
programar v - . . .; to schedule • to engineer
— v **adecuadamente** - to schedule adequately
— v **anticipadamente** - [electrón] to in-program
— v — **de-énfasis** m - [electrón] to in-program @ de-emphasis
— f **asignación(es)** f - [ind] to schedule @ assignment(s)
— v — f **para personal** - [ind] to schedule @ personnel assignment(s)
— v — f **para trabajo** m - [ind] to schedule @ work assignment(s)
— v **característica(s)** - [electrón] to program @ characteristic(s)
— v — f **seleccionable(s)** - [electrón] to program @ selectable characteristic(s)
— v **código** m - [comput] to program @ code
— v — m **específico** - [comput] to program @ specific code
— v **conductor(es)** m - [deport] to schedule @ driver(s)
— v **destino** m - [ind] to schedule @ assignment
— v — m **para personal** m - [ind] to schedule @ personnel assignment(s)
— v — m — **trabajo** m - [ind] to schedule @ work assignment(s)
— **en fábrica** f - [comput] to factory program
— v — **planta** f - [comput- to plant program
— v — **sistema** m - [comput] to program into @ system
— v — **sitio** m - [comput] to field program
— v **ensayo(s)** m - [ind] to, program, o engineer, @ test(s)
— v **exigencia(s)** f - to schedule @ requirements

programar v exigencia(s) f para producción f -
[ind] to schedule @ production program(s)
— v fecha(s) f - to schedule @ date(s)
— v individualmente - [comput] to program individually
— v interrogación f - [comput] to program @ interrogate,tion
— v — f en sitio m - [comput] to field program
@ interrogate,tion
— v mantenimiento m - [ind] to schedule @ maintenance
— v material(es) m - [ind] to schedule @ supply,plies, o material(s)
— v microplaqueta(s) f - [comput] to program @
chip(s)
— v — f en fábrica f - [comput] to factory
program @ chip(s)
— v personal m—[labor] to schedule @ personnel
— v reparación(es) f - [ind] to schedule @ re-
pair(s)
— v — f para convertidor m - [metal-prod] to
schedule @ converter repair(s)
— v salida f - [comput] to program @ output
— v sencillamente - to program simply
— v — en sitio m - [comput] to simple field
program
— v supervisión f - [admin] to schedule @ supervision
— v — f para personal m - [labor] to schedule
@ personnel supervision
— v tarea(s) f - [labor] to schedule @ job(s)
— v trabajo m - [labor] to schedule @ work
— v transpondedor m - [telecom] to program @
transponder
programista* m - programmer; scheduler; planner
progresado,da a - progressed; advanced; proceded
— a espectacularmente - [deport] rocketed (forward)
progresar v - . . .; to proceed • to grow
— v en dirección f ascendente - [sold] to progress upward(s)
— v — m — por junta f - [sold] to progress up @ joint
— v hacia arriba - [sold] to progress upward(s)
progresión f - . . .; advancement • proceeding
— f actualizada - updated progression
— f analítica - analytic progression
— f aritmética - [mat] arithmetic(al) progression • arithmetic(al) ratio
— f directa - [sold] straight progression;
stringer bead
— f aritmética - multiplier effect • [mat] geo-
metrical ratio • arithmetic progression
progresista f - . . .; forward-looking
progresiva f - [topogr] station • [vial] station
progresivamente adv - . . . • increasingly •
consecutively • back-to-back
progresivo,va a - . . . . • sequential • growing •
aggressive • advanced; forward looking • consecutive
— a moderadamente - moderately progressive
progreso m - . . . . • growth • proceeding
— m acelerado - sped up progress
— m científico - scientific progress
— m con dibujo(s) m - drawing(s) progress(ing)
— m de avance m - [ind] completion progress
— m — m de obra f - [constr] work completion, progress, o percentage
— m — trabajo m acelerado - [ind] sped up work
progress
— m económico - [econ] economic progress
— m en fabricación f - [ind] manufacture,ring
progress
— f — obra f - [constr] work, o job, progress
— m — tecnología f - technology advancement
— m — — f para proyección f - design technology, progress, o advancement
— m — trabajo m - [ind] work progress
— m espectacular - [deport] rocketing forward
— m hacia cuota f - progress toward @ quota
— m mayor - [deport] major gain
— m mediante investigación f - progress through
research

progreso m menor - [deport] minor gain
— satisfactorio - satisfactory progress
— m significativo - significant progress
— m técnico - technical progress
— m tecnológico - technological, progress, o
advancement
— m tremendo - tremendous progress
— m verificado - checked progress
prohibición f de renuncia f - [legal] unre-
nounceability
— f — traspaso m - nontransferability
prohibido,da a - forbidden; prohibited
— a aparcar - [vial] no parking
— a detenerse v - [vial] no stopping
— a estacionar(se) - [vial] no parking
prohibitivamente adv - prohibitively
prohibitorio,ria a - prohibitory; forbidding
proliferado,da a - proliferated
proliferar v - to proliferate
prólogo m - [publ] . . .; foreword
prolongación f - . . .; extending,nsion • projecting,tion; elongating,tion • [electr-cond]
extension (cord) • [sold] stickout • [constr]
splice • [herram] cheater
— f abovedada - [constr] arched extension
— f adicional - [sold] added stickout
— f ahusada - [mec] tapered extension
— f —para árbol m - [mec] shaft tapered extension
— f — — m para generador m - [comb.int]
generator shaft tapered extension
— f aislada - [sold] insulated extension
— f — Linc-Fill - [sold] Linc-Fill insulated
extension
— f — para electrodo m - [sold] insulated extension
— f — — m sobresaliente - [sold] Linc-Fill
insulated extension
— f ajustable - [mec] adjustable extension
— f apropiada - [sold] proper, o appropriate,
extension, o stickout
— f — para electrodo m - [sold] proper, o appropriate, (electrode) stickout
— f auxiliar - [grúas] auxiliary extension
— f con polea f - [grúas] sheave extension
— f — — f única - [grúas] single sheave extension
— f — — f — para punta f - [grúas] single
sheave point extension
— f — — f — — — f para aguilón m - [grúas]
single sheave boom point extension
— f — resalto m - [mec] offset extension
— f — — m para polea f - [grúas] offset
sheave extension
— f considerable - material lengthening
— f — de vida f (útil) - [mec] service life
material lengthening
— f de alambre m - [sold] wire stickout
— f — árbol m - [mec] shaft, extending,sion,
o stickout
— f — — alcantarilla f - [constr] culvert extension,nding
— f — — — abovedada - [constr] arched culvert
extension
— f — — m para bomba f - [comb.int] pump
shaft extension
— f — — m — generador m - [comb.int] generator shaft extension
— f — arco m (en cráter m) - [sold] crater
control
— f — cordón m - [electr-cond] cord lengthening
— f — cubo m - [mec-ruedas] hub extension
— f — desagüe m - [hidr] sewer extension
— f — — m pluvial - [hidr] storm sewer extension
— f — electrodo m - [sold] (electrode) stickout
— f — — m de carbono m - [sold] carbon stickout
— f — — m — — m ajustada - [sold] adjusted
carbon stickout
— f — ferrocarril m - [f.c.] railway extension

prolongación f̲ de ferrocarril m̲ urbano - [f.c.]
rapid transit extension
— f̲ —— m̲ a aeropuerto m̲ - [f.c.] airport
rapid transit extension
— f̲ — hora(s) f̲ programada(s) - [ind] scheduled time extension
— f̲ —— f̲ — para mantenimiento m̲ - [ind]
excess scheduled maintenance downtime
— f̲ — llanta f̲ - [autom-neumát] rim extension
— f̲ — servicio(s) m̲ técnico(s) - [ind] technical service(s) extension
— f̲ — utilidad f̲ - usefulness, prolonging, o̲
extension
— f̲ — vida f̲ - life, prolongation, o̲ stretching, o̲ extending
— f̲ —— f̲ útil - useful life, prolongation, o̲
extending
— f̲ —— f̲ — de motor m̲ - [comb.int] engine
life, stretching, o̲ extending
— f̲ desplazable - [grúas] swing away extension
— f̲ — para aguilón m̲ - [grúas] swing away boom
extension
— f̲ desplazada - [mec] offset extension
— f̲ electrizada - [sold] electrical stickout
— f̲ — de electrodo m̲ - [sold] electrode (electrical) stickout
— f̲ en voladizo - [mec] cantilever extension
— f̲ fácil - [constr] easy extension
— f̲ — de alcantarilla f̲ - [constr] culvert
extension ease • easy culvert extension
— f̲ hasta aeropuerto m̲ - [transp] airport extension
— f̲ hidráulica - [grúas] hydraulic extension
— f̲ insertable - [mec] insert(able) extension
— f̲ — para aguilón m̲ - [grúas] boom insertable
extension
— f̲ Linc-Fill - [sold] Linc-Fill extension
— f̲ —— para electrodo(s) m̲ muy sobresalien-
tes - Linc-Fill long stickout extension
— f̲ normal - normal extension • [sold] standard, o̲ normal, stickout
— f̲ — de electrodo m̲ - [sold] normal electrode
stickout; electrode normal stickout
— f̲ — aguilón m̲ - [grúas] boom extension
— f̲ para alcantarilla f̲ - [constr] culvert ex-
ension
— f̲ — alimentadora f̲ - [sold] feed(er) extension
— f̲ —— f̲ para alambre m̲ - [sold] wire feed
extension
— f̲ para apoyo m̲ - [mec] support, o̲ rest, extension
— f̲ —— m̲ para aguilón m̲ - [grúas] boom rest
extension
— f̲ — árbol m̲ - [mec] shaft extension
— f̲ — barra f̲ - [mec] bar extension
— f̲ —— f̲ para derivación f̲ - [electr-instal]
shunt bar extension
— f̲ — boquilla f̲ - [sold] nozzle extension
— f̲ —— f̲ para electrodo m̲ (sobresaliente) -
[sold] stickout nozzle extension
— f̲ —— f̲ —— m̲ muy sobresaliente - [sold]
long stickout nozzle extension
— f̲ — cable m̲ - [electr-cond] cable extension
— f̲ —— m̲ a gancho m̲ - [grúas] whip line
extension
— f̲ —— m̲ con camisa f̲ de alambre - [sold]
extension wire sheath cable
— f̲ —— m̲ —— f̲ —— m̲ en espiral - [sold]
extension wire sheath cable
— f̲ —— m̲ para pistola f̲ - [sold] gun cable
extension
— f̲ —— m̲ — regulación f̲ - [e;ectr-instal]
control cable extension
— f̲ — carrete m̲ - [mec] reel extension
— f̲ —— m̲ para alambre m̲ - [sold] wire reel
extension
— f̲ — cloaca f̲ - [snit] sewer extension
— f̲ —— f̲ sanitaria - [sanit] sanitary sewer
extension
— f̲ — cubo m̲ - [mec-ruedas] hub extension
— f̲ — descanso m̲ - [mec] rest extension
— f̲ —— m̲ para aguilón m̲ - [grúas] boom rest
extension

prolongación f̲ para eje m̲ - [mec] shaft extension
— f̲ —— m̲ motor - [mec] drive,ving axle extension
— f̲ —— m̲ para impulsión f̲ - [mec] drive,ving
axle extension
— f̲ — electrodo(s) m̲ - [sold] electrode extension
— f̲ —— m̲ muy sobresaliente - [sold] long
stickout extension
— f̲ —— m̲ sobresaliente - [sold] stickout
extension • Linc-Fill extension
— f̲ — escalerilla f̲ - [agric-equip] ladder extension
— f̲ — extremo m̲ - [mec] end extension
— f̲ —— m̲ hacia motor m̲ - [petról] power
end extension
— f̲ — garrucha f̲ - [grúas] sheave extension
— f̲ — herramienta f̲ - [mec] tool extension
— f̲ —— f̲ para corte m̲ - [mec] (cutting)
tool extension
— f̲ — llave f̲ - [herram] (wrench) cheater
— f̲ — motón m̲ - [grúas] sheave extension
— f̲ —— m̲ — punta f̲ de aguilón m̲ - [grúas]
boom point sheave extension
— f̲ — paragolpe(s) m̲ - [autom] búmper extension
— f̲ —— m̲ delantero - [autom] front bumper
extension
— f̲ —— m̲ frontal - [autom] front bumper extension
— f̲ —— m̲ posterior - [autom] rear bumper extension
— f̲ —— m̲ trasero - [autom] rear bumper extension
— f̲ — pata f̲ - [mec] leg extension
— f̲ — pico m̲ para engrase m̲ - [mec] Alemite, o̲
grease fitting, extension
— f̲ — plancha f̲ - [mec] plate extension
— f̲ —— f̲ para guardia f̲ - [mec] toe plate
extension
— f̲ — pluma f̲ - [grúas] boom extension
— f̲ — polea f̲ - [grúas] sheave extension
— f̲ —— f̲ desplazada - [grúas] offset sheave
extension
— f̲ — punta f̲ - [grúas] point extension
— f̲ —— f̲ de aguilón m̲ - [grúas] boom point
extension
— f̲ —— f̲ —— m̲ con polea f̲ - [grúas]
sheave boom point extension
— f̲ —— f̲ —— m̲ —— f̲ única - [grúas]
single sheave boom point extension
— f̲ — roldana f̲ - [Grúas] sheave extension
— f̲ — suplemento m̲ - [grúas] insert extension
— f̲ —— m̲ para aguilón m̲ - [grúas] boom insert extension
— f̲ — tambor m̲ ⊷ [grúas] barrel extension
— f̲ — transmisión f̲ - [mec] transmission extension
— f̲ — válvula f̲ - [mec] valve extension
— f̲ —— f̲ ajustable - [mec] adjustable valve
extension
— f̲ —— f̲ regulable - [mec] adjustable valve
extension
— f̲ — vástago m̲ - [mec] stem, o̲ rod, extension
— f̲ —— m̲ ajustable - [mec] adjustable, stem,
o̲ rod, extension
— f̲ —— m̲ para cilindro m̲ - [mec] cylinder
rod extension
— f̲ —— m̲ — válvula f̲ - [mec] valve stem
extension
— f̲ —— m̲ para válvula f̲ ajustable - [mec]
adjustable valve, stem, o̲ rod, extension
— f̲ —— m̲ —— f̲ regulable - [mec] adjustable valve, stem, o̲ rod, extension
— f̲ —— m̲ regulable - [mec] adjustable, stem,
o̲ rod, extension
— f̲ reajustada - [mec] readjusted stickout
— f̲ — de electrodo m̲ de carbono m̲ - [sold]
readjusted carbon stickout
— f̲ recta - [mec] straight, extension, o̲ stickout
— f̲ — roscada - [mec] straight threaded, extension, o̲ stickout

prolongación f̱ regulable - [mec] adjustable extension
— f̱ roscada - [mec] threaded extension
— f̱ — para árbol m̱ - [mec] threaded shaft extension
— f̱ — — — m̱ para bomba f̱ - [bombas] pump threaded shaft extension
— f̱ totalmente hidráulica - [grúas] all hydraulic extension
— f̱ usada - [mec] used extension
— f̱ visible - [sold] visible, extension, o̱ stickout
— f̱ — de electrodo m̱ - [sold] visible electrode, extension, o̱ stickout
prolongado,da a̱ - . . ..; prolonged • lengthy; extended; lengthened • lasting
— a̱ anteriormente - previously extended
— a̱ considerablemente - lengthened materially
prolongador,ra a̱ para utilidad f̱ - usefulness prolonging
— m̱ — arco m̱ (en cráter m̱) - [sold] crater control
prolongar v̱ - . . .; to, lengthen, o̱ extend • [constr-pil] to splice • [fam] to breathe @ longevity
— v̱ alcantarilla f̱ - [constr] to, extend, o̱ suplement, @ culvert
— v̱ árbol m̱ - [mec] to extend @ shaft
— v̱ considerablemente - to lengthen materially
— v̱ — vida f̱ - to materially lengthen @ life
— v̱ — — f̱ útil - to materially lengthen @ service life
— v̱ parada f̱ - [ind] to extend @ shut down
— v̱ utilidad f̱ - to prolong @ usefulness
— v̱ vida f̱ - to, extend, o̱ increase, o̱ stretch, @ life
— v̱ — f̱ de herramienta(s) f̱ - [herram] to, extend, o̱ prolong, @ tool('s) life
— v̱ — f̱ — motor m̱ - [electr-mot] to, extend, o̱ increase, o̱ stretch, @ motor('s) life • [comb.int] to, extend, o̱ increase, o̱ stretch, @ engine('s) life
— v̱ — f̱ útil - to extend @ useful life
— v̱ — f̱ — de herramienta(s) f̱ - [herram] to, extend, o̱ prolong, @ tool('s) useful life
— v̱ — f̱ — — motor m̱ - [comb.int] to, extend, o̱ stretch, @ engine('s) useful life • [electmot] to, extend, o̱ stretch, @ motor('s) useful life
promedio m̱ a bate m̱ - [deport] batting average
— m̱ anual - annual average
— m̱ aproximado - approximate average
— m̱ de acierto(s) m̱ - [deport] batting average
— m̱ — atenuación f̱ - [telecom-cond] average attenuation
— m̱ — colada(s) f̱ - [metal-prod] average tappings
— m̱ — — f̱ — por campaña f̱ - [metal-prod] campaign average tappings
— m̱ — consumo m̱ - [ind] average requirement
— m̱ — — m̱ de coque m̱ - [metal-prod] average coke requirement
— m̱ — — m̱ específico - [ind] average specific requirement
— m̱ — — m̱ — de combustible m̱ - [ind] average specific fuel requirement
— m̱ — contenido m̱ - [ind] content average
— m̱ — — m̱ de hierro m̱ - [miner] average iron content
— m̱ — determinación(es) f̱ - determination(s) average
— m̱ — distribución f̱ - [vial] distribution average
— m̱ — — f̱ unitaria—unit distribution average
— m̱ — estímulo(s) m̱ - [labor] incentive(s) average
— m̱ — repuesto(s) m̱ - [ind] average spare parts
— m̱ — tiempo m̱ - average time
— m̱ — — m̱ en mano f̱ de obra - [labor] average labor time
— m̱ diario - daily average; average per day
— m̱ estimado - estimated average

promedio m̱ estimado - estimated average
— m̱ global - overall average
— m̱ industrial - [ind] industrial average
— m̱ máximo - maximum average
— m̱ — para atenuación f̱ - [telecom-cond] maximum average attentuation
— m̱ mínimo - minimum average
— m̱ para industria f̱ - [ind] industry average
— m̱ — — f̱ estadounidense - [ind] American, o̱ United States, industry average
— m̱ ponderado - weighted, mean, o̱ average
promesa f̱ buena - [legal] good promise
— f̱ valiosa - [legal] valuable promise
prometedor,ra a̱ - . . . . • trick(y)
prominencia f̱ - . . .; high spot • [mec] boss
— f̱ elevada - [topogr] high knoll
prominente a̱ - . . . . • bold
promoción f̱ - . . .; promoting; sponsoring,rship • development • advertising • [labor] advancement; furthering • [legal] initiation,ting; encouragement,ging
— f̱ conjunta - [com] (joint) (changeover) program
— f̱ considerable - sizable promotion
— f̱ de comprensión f̱ - comprehension, o̱ understanding, creating,tion
— f̱ — entendimiento m̱ - understanding creating,tion
— f̱ — gobierno m̱ - [pol] government promotion
— f̱ — industria f̱ - [ind] industry promotion
— f̱ — — f̱ nacional - [ind] national, o̱ local, industry promotion
— f̱ — inversión(es) f̱ - [fin] investment(s) promotion
— f̱ — minería f̱ - [miner] mining promotion
— f̱ — — f̱ en escala f̱, pequeña, o̱ reducida - [miner] small scale mining promotion
— f̱ — — f̱ grande - [miner] large scale mining promotion
— f̱ — — f̱ — bajo contrato m̱ - [miner] large scale contract mining promotion
— f̱ — — f̱ pequeña - [miner] small scale mining promotion
— f̱ — — f̱ por contrato m̱ - [miner] contract mining promotion
— m̱ — reducción f̱ - [metal-prod] production average
— f̱ — rendimiento m̱ - [ind] yield, o̱ performance, promotion, o̱ development
— f̱ — — m̱ de neumático(s) m̱ - [autom-neumát] performance tire development
— f̱ — seguridad f̱ - [segurid] safety, promotion, o̱ campaign
— f̱ — — f̱ para personal m̱ - [segurid] personnel, o̱ employee, safety promotion
— f̱ — utilización f̱ - use promotion
— f̱ — — f̱ máxima - maximum use promotion
— f̱ — venta(s) f̱ - [com] sale(s) promotion
— f̱ — — f̱ de neumático(s) - [autom-neumát] tire sales promotion
— f̱ — — f̱ — — m̱ para rendimiento m̱ alto - [autom-neumát] (high) performance tire sales promotion
— f̱ emprendedora - [com] aggressive promotion
— f̱ en general - [econ] general promotion
— f̱ estatal - [pol] government(al) promotion
— f̱ exitosa - successful promotion
— f̱ general - [econ] general promotion
— f̱ indirecta - [com] indirect promotion • soft selling
— f̱ industrial - [econ] industrial promotion
— f̱ social - social promotion • [pol] social uplift(ing)
— f̱ para inversión f̱ - [econ] investment promotion
— f̱ — reinversión f̱ - [Econ] reinvestment promotion
— f̱ por antigüedad f̱ - [labor] seniority promotion
— f̱ por distribuidor(es) m̱ - [com] dealer promotion
— f̱ regional - regional promotion
— f̱ tecnológica - [ind] technological promotion

promocional* a - véase promotor,ra; promotivo,va
promotivo,va - promotional
promotor m - . . . . • pioneer; developer; initia-
tor • proponent; sponsor • incorporator
— m de empresa f - project sponsor
— m — obra f - project sponsor
— m — proceso m - [ind] process sponsor
— m — proyecto m - project sponsor
promotor,ra a - promotional
— a de comprensión f - understanding, promoting, o creating
— a — entendimiento m - understanding, promoting, o creating
— a — retorno m - return promoting
promover v - . . .; to encourage1 to sponsor •
to advertise • [legal] to promote; to ini-
tiate
— f calidad f - to, promote, o advertise, @
quality
— v distorsión f - [sold] to promote @ distortion
— v eficiencia f - to promote @ efficiency
— v empleo m - to promote @ use
— v — máximo - to promote @ maximum use
— v generación f - [ind] to, promote, o develop, @ generating,tion
— v precio m - to, promote, o advertise, @
price
promovido,da a - promoted; sponsored; encouraged
• advertised • [legal] initiated
peomulgación f - [legal] . . .; enacting,tion;
passage,sing
promulgado,da a - promuulgated • [legal] enacted
promulgar v - . . . • [legal] to enact
pronosticabilidad f - predictability
pronosticación f - . . . . • forecasting
pronosticado,da a - prognosticated; forecast(ed)
• predicted
pronosticar v - . . .' to predict; to forecast
— v aumento - to, predict, o forecast, @ increase
— v efecto m - to predict @ effect
pronóstico m - . . .; prognostication; forecast;
prediction; projection • prophecy
— m para aumento m de población f - [pol] population increase, prediction, o forecast
— m — producción f - [ind] production forecast
pronto,ta a - . . . . • willing
prontuario m - . . . . • dossier; docket • individual record; personnel file • [labor] record
pronunciado,da a - . . . . • bold • declared
pronunciamiento m - . . . . • declaration
pronunciar v - . . . • to declare
propagación f de incendio m - [segurid] fire,
propagation, o spreading
propagado,da a - propagated • spread
propaganda f - . . . . • [com] advertising
— f con coparticipación f - [com] co-op advertising
— f con precio(s) m - [com] price advertising
— f impresa - [com] print(ed) advertising
— f indirecta - [com] indirect, o low key, advertising
— f por televisión f - [telecom] television advertising
propano m licuado - [combust] liquefied propane
— m líquido m - [combust] liquid propane
— m sin usar - [combust] unused propane
propenso,sa a adv - subject to; prone to
— a a deslizar(se) - [topogr] prone, o subject,
to (land)slide
— a filtración(es) f - [hidr] prone, o subject,
to @ seepage
propiamente dicho,cha - properly said • proper •
itself
propiedad f - . . . . • propriety; appropriateness
• feature • [legal] title; ownership • [com]
real estate
— f absoluta - absolute ownership | adv - (en)
wholly owned
— f adhesiva - adhesive property
— f adyacente - adjacent property
— f anestésica - [medic] anesthetic property

propiedad f apropiada - approved, adequacy, o
propriety • satisfactory property
— f austenítica - [metal-prod] austenitic property
— f buena - good property
— f — contra impacto(s) m - [sold] good impact
property
— f calculada - calculated property
— f — con computador m - [comput] computer
calculated property
— f coligante - [miner] binding property
— f colindante - abutting property
— f como acabado,da de soldar - [sold] as-weld-
ed property
— f compresiva - compressive property
— f computada - [comput] computed property
— f — de corte m transversal - computed
(cross)sectional property
— f con alivio m de tensión(es) f - stress relieved property
— f conglomerante - [miner] binding property
— f contra impacto(s) m - [metal] impact property
— f — — m con entalladura f - notch impact
property
— f — — m — entalladura f Charpy - [metal]
Charpy notch impact property
— f — — f — — f — — en V - [metal] Charpy
V-notch impact property
— f — pandeo m - [metal] non-sagging feature
— f dañada - damaged property
— f dañosa - damaging property
— f de acero m - [metal-prod] steel property
— f — — m austenítico - [metal-prod] austenitic steel property
— f — aportación f - [sold] deposit property
— f — asfalto m - [petról] asphalt property
— f — cable m - [telecom-cond] cable property
— f — — m telefónico - [telecom-cond] telephone cable property
— f — cauce - [hidr] channel property
— f — condado m - county property | a - (de)
county owned
— f — corrugación f - [mec] corrugation property
— f — — f de alcantarilla f - [metal-fabr]
culvert corrugation property
— f — — f — — f de acero m - [metal-fabr]
steel culvert corrugation property
— f — — f en arco m - [mec] arch corrugation
property
— f — — f tangencial - [mec] tangent(ial)
corrugation property
— f — corte m transversal - sectional property
— f — — m — de chapa f corrugada - [metal-
-fabr] corrugated sheet sectional property
— f — — m — — f de acero m corrugada -
[metal-fabr] corrugated steel sheet(s) sectional property,ties
— f — chapa f corrugada - [metal-fabr] corrugated sheet property
— f — — f de acero m corrugada - [metal-fabr]
corrugated steel sheet property,ties
— f — derecho(s) m literario(s) - [legal]
copyright ownership
— f — descascarillado m - [metal] scaling
property
— f — deshidratación f - [sanit] dewatering
characteristic
— f — electrodo m - [sold] electrode property
— f — endurecimiento m - [metal-fabr] hardening property
— f — escoria f - [metal-prod] slag property
— f — expansión f - [fís] expansion property
— f — — f térmica - [metal] thermal, o heat,
expansion property
— f — familia f - family property; family's
— f — gas m - [fís] gas property
— f — hidrógeno m - [quím] hydrogen property
— f — — m bsjo - low hydrogen property
— f — imán m - magnet('s) property,ties
— f — m permanente - permanent magnet property,ties

propiedad f de ingrediente(s) m

propiedad f de ingrediente(s) m - ingredient('s)
property,ties
— f — limpieza f - [sold] cleaning property
— f — marca f de fábrica f - [legal] trade
mark ownership
— f — — f industrial - [legal] trade mark
ownership
— f — — f registrada - [legal] trade mark
ownership
— f — material m - material('s) property,ties
— f — — m duro - hard material property,ties
— f — — m magnéticamente duro - [electr]
magnetically hard material property,ties
— f — m para relleno m - [constr] backfill
property,ties
— f — metal m - [metal] metal('s) property
— f — m aportado - [sold] (weld) deposit
property
— f — m de aportación f - [sold] weld metal
property
— f — m — — f sin diluir - [sold] undiluted weld(ing) metal
— f — m — soldadura f aportado - [sold]
weld metal deposit property
— f — m sin diluir - [sold] undiluted metal
property
— f — nombre m comercial - [legal] trade name
ownership
— f — patente f - [legal] patent ownership
— f — — f de invención f - [legal] patent
ownership
— f — pilote m - [constr-pil] pile property
— f — — m tubular - [constr-pil] pipe pile
property
— f — plancha f - [metal-lam] plate property
— f — — f para revestimiento m - [const]
liner plate property
— f — planta f - [ind] plant property • plant
ownership
— f — sección f - [constr] section property
— f — — f transversal - (cross)sectional
property,ties
— f — — f — calculada(s) - calculated
(cross)sectional property,ties
— f — — — f con computador m -computer
calculated (cross)sectional property,ties
— f — — f — de corrugación f - [mec] corrugation (cross)sectional property
— f — — — f de alcantarilla f -
[metal-fabr] culvert corrugation (cross)sec-
tional property
— f — — f — — f — — f de acero m -
[metal-fabr] steel culvert corrugation
(cross)sectional property,ties
— f — — — f de arco m - [mec] arc(h)
corrugation (cross)sectional property
— f — — f — de corrugación f tangencial -
[mec] tengent(ial) corrugation (cross)sectio-
nal property,ties
— f — — f — chapa(s) f de acero m -
[metal-fabr] steel sheet (cross)sectional
property,ties
— f — — f — — — f — — corrugado m -
[metal-fabr] corrugated steel sheet(s)
(cross)sectional property,ties
— f — sistema m magnético - [electr- magnetic
system property,ties
— f — soldadura f - [sold] weld property,ties
— f — suelo m - [suelos] soil property,ties
— f — tenor m bajo de hidrógeno m - [quím] low
hydrogen property,ties
— f — transición f - [metal] transition property,ties
— f — — f mejorada(s) - [metal] improved
transition property,ties
— f — tubería f - [tub] pipe property,ties
— f deseada - desired property
— f dimensional - [mec] dimension(al) property
— f diversa - different, o diverse, properties
— f en estado m soldado - [sold] as welded property

propiedad f esfuerzo-tensión f - [suelos] ·
stress-strain property,ties
— f — de suelo m - [suelos] soil stress-
-strain property,ties
— f excelente - excellent property,ties
— f exclusiva - [legal] exclusive, property, o
ownership | adv - (de) wholly owned
— f exigida - required property,ties
— f familiar - family property
— f física - [fís] physical property,ties
— f — de soldadura f - [sold] weld physical
property,ties
— f — de suelo m - [suelos] soil physical property,ties
— f — importante - special physical property
— f — necesaria - necessary, o needed, physical property,ties
— f — típica - typical physical property
— f geométrica - [geom] geometric property
— f homogénea - homogenous property
— f horizontal - horizontal property • [constr]
cooperative apartment • condominium
— f inigualada - unequalled property
— f inmueble - inmovable property • [legal]
real estate
— f intelectual - [legal] copyright
— f — adquirida - [legal] acquired, o purchased, o acquired, copyright
— f — comprada - [legal] purchased copyright
— f — creada - [legal] developed copyright
— f — ejercida - [legal] used copyright
— f — gestionada - [legal] applied for copyright
— f — obtenida - [legal] obtained, o acquired,
copyright
— f — poseída - [legal] owned copyright
— f — registrada - [legal] registered copyright
— f — retenida - [legal] held, o kept, copyright
— f — solicitada - [legal] applied for copyright
— f — vendida - [legal] sold copyright
— f ligante - [miner] binding property
— f literaria - [legal] copyright
— f — registrada - [legal] copyright(ed)
— f magnética - [metal] magnetic property
— f — de acero - [metal-prod] steel magnetic
property,ties
— f — imán m - [electr] magnet('s) magnetic property,ties
— f — — m permanente - [electr] permanent
magnet('s) magnetic property,ties
— f — exigida - [electr] required magnetic
property,ties
— f — requerida - [electr] required magnetic
property,ties
— f mecánica - mechanical property,ties
— f — alta - [metal] high mechanical property
— f — buena - good mechanical property,ties
— f — como acabado,da de soldar v - [sold] as
welded mechanical property,ties
— f — de acero m - [metal-prod] steel mechanical property,ties
— f — — aportación f - [sold] deposit mechanical property,ties
— f — — aportación f sin diluir - [sold] all-
-weld metal (deposit) mechanical property;
undiluted weld metal deposit mechanical property,ties
— f — de metal m - [metal] metal mechanical
property,ties
— f — — m aportado - [sold] metal deposit
mechanical property,ties
— f — — m — sin diluir - [sold] undiluted
weld metal mechanical property,ties
— f — — m para soldadura f aportado -
[sold] weld metal deposit mechanical property
— f — — m — — f — sin diluir - [sold]
undiluted weld metal deposit mechanical property,ties
— f — — m sin diluir - [sold] undiluted

metal mechanical property,ties
propiedad f mecánica de soldadura f - [sold]
weld mechanical property,ties
— f — — **suelo** m -]suelos] soil(s) mechanical
property,ties
— f — **destacada** - outstanding mechanical property,ties
— f — **específica** - [sold] specific mechanical
property,ties
— f — **sobresaliente** - [sold] outstanding mechanical property,ties
— f — **como acabado,da de soldar** - [sold] as
welded mechanical property,ties
— f — **típica** - [sold] typical mechanical property,ties
— f — — **como acabado,da de soldar** - [sold]
typical as welded mechanical property,ties
— f — — — **de metal** m **aportado sin diluir** -
[sold] typical all-weld metal mechanical property,ties
— f — — — — **de aportación** f - [sold] typical weld metal mechanical property,ties
— f — — — m — — f **sin diluir** - [sold] typical undiluted, o all, weld metal mechanical
property,ties
— f — — **tal como acabado,da de soldar**—[sold]
typical as welded mechanical property,ties
— f **metálica** - [metal] metal(lic) property
— f **metalúrgica** - [metal] metallurgical property,ties
— f **movible** - [fin] movable asset
— f **necesaria** - necessary property
— f **para conformación** f - [metal-trat] necessary, forming, o shaping, property
— f **para encendido** m - [sold] striking property
— f — — m **de arco** m - [sold] (arc) striking
property,ties
— f — **formación** f **de escoria** f - [sold] slag
forming property,ties
— f — **fraguado** m - [constr] setting property
— f — — m **de hormigón** m - [constr] concrete
setting property
— f — **manipulación** f - [constr] working property
— f — — f **de hormigón** m - [constr] concrete
working property
— f — **manipuleo** m - [constr] working, o handling, property
— f — — m **de hormigón** m - [constr] concrete,
working, o handling, property
— f — **reencendido** m - [sold] restriking property
— f — — m **de arco** m - [sold] (arc) restriking
property
— f — **relleno** m—[constr] backfilling property
— f — **resistencia** f **a impacto(s)** m - [metal]
impact (resistance) property
— f — — f **alta a impacto(s)** m - [sold] high
impact (resistance) property
— f — **temperatura(s)** f, **alta(s)**, o **elevada(s)** -
[metal] high temperature property
— f **pasada** - [legal] passed title
— f **permanente** - permanent property
— f **personal** - [legal] personal, property, o
possession(s)
— f **privada** - private property
— f **próxima** - near, o adjacent, property
— f **química** - [quím] chemical property
— f — **específica** - [sold] specific chemical
property
— f **real** - actual property
— f **requerida** - required property
— f **satisfactoria** - satisfactory property
— f — **para manipulación,leo** - [constr] satisfactory working property
— f **tal como soldado,da** - [sold] as welded property
— f **tecnológica** - technological property
— f — **de acero** m - [metal-prod] steel technological property
— f — — m **austenítico** - [metal-prod] austenitic steel technological property
— f **típica** - typical property

propiedad f típica como acabado,da de soldar -
[sold] typical as welded property
— f — **con alivio** m **de tensión(es)** f - [metal]
typical stress relieved property,ties
— f — **de metal** m, **aportado**, o **de aportación** f -
[sold] typical weld metal property
— f — **tal como soldado,da** - [sold] typical as
welded property,ties
— f **transferida** - [legal] transferred property
propietario,ria de acción(es) f - [legal] share
owner
— m **de automóvil** - [autom] automobile, o car,
owner
— m — — m **de lujo** - [autom] luxury car owner
— m — **camión** m - [autom] truck owner
— m — — m **liviano** - [autom] light truck
owner
— m — **derecho(s)** m - [legal] right(s) owner
— m — — m **literario(s)** - [legal] copyright
owner
— m — **obra** f - project owner
— m — **vehículo** m - [autom] vehicle owner
— m — — m **deportivo** - [autom] sport(s) vehicle owner
— m — — m **para carga(s)** f **liviana(s)** -
[autom] utility vehicle owner
— m — — m — **deporte(s)** m **y carga(s)** f **li-
viana(s)** - [autom] sport(s)-utility vehicle
owner
— m — — m — — m **servicio** m **general** -
[autom] sport(s)-utility vehicle owner
— m — — m **para servicio** m **general** - [autom]
utility vehicle owner
— m — **vivienda** f - home owner
propietario,ria a - proprietary
propileno m - [quím] propylene
propina f amplia - [viajes] ample gratuity,ties
— f **liberal** - [viajes] liberal, o ample, gratuity,ties
— f **suficiente** - [viajes] ample gratuity,ties
propio suministrador m - supplier himself
propio,pia a - . . . • self-; individual • apropriate • peculiar • it,him,her self
proponente m - . . . . • bidder • proposing, o
possible, bidder
— m **adjudicado** - awarded bidder
— m **extranjero** - foreign bidder
— m **favorecido** - awardee; favored bidder
— m **nacional** - domestic bidder
— m **para fabricación** f **de grúa(s)** f - [grúas]
crane bidder
proponer v -- . . . • to intend • to bid
— v **afinamiento** m - to propose @ refinement
— v **condición** f - to propose @ condition
— v **equipo** m - to, propose, o offer, @ equipment
— v **programa** m - [ind] to propose @ schedule
— v — m **para fabricación** f - [ind] to propose
@ fabrication schedule
— v **red** f - to propose @ system
— v **servicio(s)** m - to propose @ service(s)
— v — m **de personal** m - [labor] to propose @
personnel service(s)
— v **sistema** m - to, propose, o offer, @ system
— v **solución** f - to, propose, o offer, @ solution
— v **valor** m - to propose @ value
proporción f - . . . . • rate; ratio; percentage
• extent; quantity • [ind] giving; servicing
• delivery • providing
— f **ácido-agua** - [quím] acid-water, proportion,
o ratio
— f — — **incorrecta** - [quím] incorrect acid-
-water ratio
— f **aire-gas** - [combust] air-gas blend
— f **algo menor** - somewhat less
— f — — **de molibdeno** m - [metal] somewhat
less, o lower, in molybdenum
— f **alta** - high proportion
— f — **de aleación** f - [metal] high alloy (content)
— f — — **azufre** m - [metal] high sulfur con-
tent | adv - [metal] (con) high sulfur con-

tent
proporción f **alta de carbono** m - [quím] high
carbon content
— f —— **ferrita** f - [quím] high ferrite | a -
[metal] (con) high ferrite
— f —— —— f **libre**—[metal] high free ferrite
— f —— **manganeso** m - [metal] high manganese
content | adv - [metal] (con) high manganese
— f —— **polvo** m **de hierro** m - [sold] high
iron powder
— f —— **sílice** m—[metal] high silica content
— f —— **silicio** m - [metal] high silicon con-
tent | adv - [metal] (con) high silicon
— f **apreciable** - appreciable, o substantial,
proportion, o percentage
— f **apropiada en mezcla** f - [comb.int] proper
mix(ing) ratio
— f **baja** - low proportion
— f —— **de carbono** m - [metal] low carbon con-
tent | adv - [metal] low carbon
— f —— **hidrógeno** m - [sold] low hydrogen
conent | adv - [metal] (con) low hydrogen
— f **considerable** - substantial, o sizeable,
percentage, o proportion; considerable, o
large, amount, o share • substantial margin
— f **de manganeso** n - [metal] (fairly) high
manganese content | adv - [sold] (con) (fair-
ly) high manganese content
— f **de abandono(s)** m - [deport] attrition rate
— f —— **aceite(s)** m - [quím] oil(s) content
— f —— m **graso(s)** - [petról] fatty oil(s)
content
— f —— **agua** m - water content
— f —— **aleación** f - [metal] alloy content
— f —— —— f **en plancha** f - [metal-prod] plate
alloy content
— f —— —— f **requerida** - [metal] required alloy
content
— f —— **aluminio** m - [metal] aluminum content
— f —— —— m **en acero** m - [metal-prod] aluminum
content in @ steel
— f —— **anhidrido** m - [quím] hydrogen content
— f —— —— m **sulfuroso** - [quím] hydrogen sulfide
content
— f —— **azufre** - [quím] sulfur content
— f —— **banda** f - [metal-lam] strip proportion
— f —— —— f **para rodamiento** m - [autom-neumát]
tread (pro)portion
— f —— **carbón** m - [miner] coal content
— f —— **carbonato(s)** m - [miner] carbonate(s)
content
— f —— m **de calcio** m - [miner] calcium carbonate content
— f —— **carbono** m - [metal] carbon content
— f —— —— m **en acero** m - [metal-prod] steel
carbon content
— f —— —— m —— m **para tubería** f - [tub] pipe
(steel) carbon content
— f —— —— m **en plancha** f - [metal-prod] plate
carbon content
— f —— —— m —— **tubería** f - [tub] pipe carbon
content
— f —— **carga** f - [metal-prod] charge proportion
— f —— —— f **para horno** m - [metal-prod] furnace
charge proportion
— f —— —— f —— m **alto** - [metal-prod] blast
furnace charage proportion
— f —— **cemento** m - [petról] cement content
— f —— —— m **asfáltico** - [petról] asphalt cement
content
— f —— **cobre** m - [metal-prod] copper content
— f —— —— m **en escoria** f - [metal-prod] slag
copper content
— f —— **cromo** m - [metal-prod] chromium content
— f —— **chatarra** f - [metal-prod] scrap percentage
— f —— **diámetro** m - diameter proportion
— f —— **elemento(s)** m **de aleación** f - [metal]
alloy(ing elements) content
— f —— —— m —— f **en metal** m, **aportado, o de
aportación** f - [sold] deposit alloy content
— f —— **ferrita** f - [metal- ferrite content

proporción f **de ferrita** f **delta** - [metal] delta
ferrite content
— f —— **fósforo** m - [metal] phosphorus content
— f —— m **en arrabio** m - [metal-prod] pig
iron phosphorus content
— f —— m —— **metal** m **caliente** - [metal-prod]
hot metal phosphorus content
— f —— **gas** m - [combust] gas proportion
— f —— **hierro** m - [miner] iron, proportion, o
content
— f —— m **en agua** m - [hidr] water iron content
— f —— **huella** f - [autom-neumát] footprint percentage
— f —— **humedad** f - moisture content
— f —— **manganeso** m - [metal] manganese content
— f —— m **en acero** m - [metal-prod] steel
manganese content
— f —— m —— m **en tubería** f - [tub] pipe
steel manganese content
— f —— m **en tubería** f - [tub] pipe manganese
content
— f —— **material(es)** m - material(s) proportion
— f —— m **sin quemar** - [combust] unburned ma-
terial(s) proportion
— f —— **mezcla** f - mix(ture) proportion
— f —— —— f **por peso** m - mix(ture) proportion
by weight
— f —— —— f **por volumen** m - mix(ture) proportion by volume
— f —— **molibdeno** m - [metal] molybdenum content
— f —— **niobio (columbio)** m - [metal] columbium
content
— f —— **oxígeno** m - [quím] oxygen content
— f —— m **en acero** m - [metal-prod] steel
oxygen content
— f —— **polvo** m - dust content • powder content
— f —— m **de hierro** m - [sold] iron powder
content
— f —— **producto(s)** m - [com] commodity mix
— f —— **producción** f **siderúrgica** - [metal-prod]
steelmaking share
— f —— **recomendación** f - recommendation giving
— f —— **sílice** m - [miner] silica content
— f —— **sínter** - [metal-prod] sinter, content, o
percentage
— f —— **solución*** f - solution yield(ing)
— f —— **tiempo** m - [admin] time proportion
— f **definitiva** - definite proportion
— f **directa** - direct proportion
— f **elevada** - high proportion
— f —— **de ganga** f - [miner] high gangue (proportion)
— f **en mezcla** f - [comb.int] mix(ing) ratio
— f —— **porcentaje(s)** m - [quím] percentage(s)
proportion
— f **equilibrada** - balanced proportion
— f **establecida** - established, o stated, proportion
— f **final** - final proportion • final analysis
— f —— **de fósforo** m - [quím] final phosphorus
analysis
— f **inapropiada** - [comb.int] improper mix(ing
ratio)
— f **incorrecta** - [quím] incorrect ratio
— f **indemnizable** - [seguros] indemnifiable proportion
— f **indemnizada** - [seguros] indemnified proportion
— f **inversa** - inverse, o reverse, proportion
— f **lineal** - linear proportion(ing)
— f **máxima** - maximum, proportion, o content
— f —— **de aleación** f - [metal-prod] maximum alloy content
— f —— —— **carbono** m - [metal] maximum carbon
content
— f —— —— **elemento** m **de aleación** f - [metal-
-prod] maximum alloy(ing element) content
— f —— —— **hierro** m - [metal] maximum iron content
— f —— —— m **en agua** m - [hidr] maximum water
iron content

proporción f máxima permitida - [metal-prod]
maximum allowed content
— f — — de aleación f - [metal-prod] maximum
allowed alloy content
— f — — elemento m de aleación f - [metal-
-prod] maximum allowed alloy(ing element)
content
— f mayor - greater, o higher, o larger, proportion, o share • [quím] higher content
— f — de aleación f - [metal] higher alloy
content
— f — — carbono m - [metal-prod] higher carbon content
— f — — manganeso m - [metal-prod] higher
manganese content
— f — — polvo m de hierro m - [Sold½ higher
iron powder content
— f — — silicio m - [metal-prod] higher sili-
con content | adv - (con) higher silicon
— f — — tiempo m - [admin] higher time proportion
— f media - average, proportion, o content
— f — de hierro m—average iron content
— f — — — m en agua m - [hidr] average water
iron content
— f menor - lower, o smaller, proportion, o
share
— f — de carbono m - [metal-prod] lower carbon
content
— f — — molibdeno m—lower molybdenum content
— f — de tiempo n - [admin] lower time proportion
— f mínima - minimum, o smallest, content
— f — de aleación f - [metal-prod] minimum alloy content
— f — — carbono m - [metal] minimum, o lowest, carbon content
— f — de elemento m de aleación f - [metal-
-prod] minimum alloy content
— f — — hierro m - [metal] minimum iron content
— f — — — m en agua n - [hidr] water minimum
iron content
— f — permitida - [metal-prod] minimum allowed
content
— f — — de aleación f - [metal-prod] minimum
allowed alloy content
— f — — — elemento m de aleación f - [metal-
-prod] minimum allowed alloy content
— f muy alta - [quím] very high content
— f — — f de carbono m - [metal-prod] very
high carbon content
— f — reducida de carbono m - [metal-prod]
very low carbon content
— f óptima - optimum, proportion, o content
— f — de humedad f - optimum moisture content
— f original - original, proportion, o content
— f para mezcla f - [constr] mix proportion(s)
— f preferida - [quím] preferred analysis
— f reducida - reduced, o small, proportion
— f — de carbono m - [metal-prod] low carbon
content
— f variable - variable,rying, proportion, o a-
mount
— f variante - varying,riable amount
proporcionado,da a - . . . .; proportional; com-
mensurate • given; furnished; provided; afforded; delivered • permitted • [ind] serviced
— a separadamente - [com] provided, o serviced,
separately
— a simultáneamente - provided simultaneously
— a sin cargo - supplied without @ charge
— a tardíamente - provided late(ly)
proporcionador m - proportioner; supplier
— m para ácido(s) m - [quím] acids proportioner
proporcional a - . . . .; ratable
proporcional a directamente - directly proportional
— a inversamente - inversely proportional
proporcionamiento* giving; affording
— m de asesoramiento m - advice, providing, o
supplying

proporcionamiento m de imagen f - [electrón] i-
mage, o picture, providing
proporcionar v - . . . .; to afford; to make a-
vailable • to give, to permit • to deliver •
[ind] to service
— v ajuste m - to provide @ adjustment
— v — m aproximado - [sold] to provide @
broad adjustment
— v — m amplio - [Sold] to provide @ broad adjustment
— v — m preciso - [sold] to provide @ precise
adjustment
— v — m — para voltaje m - [sold] to provide
@ precise voltage adjustment
— v asesoramiento m - to, provide, o supply, @
advice
— v — m técnico - [ind] to, provide, o furnish, o supply, technical, aid, o advice
— v calidad f - to, provide, o supply, quality
— v característica(s) f - to provide @ charac-
teristic(s)
— v — f excelente(s) - to provide excellent
characteristic(s)
— f — f para soldadura f - [sold] to provide
@ welding characteristic(s)
— v — f — — excelente(s) - [sold] to provide excellent welding characteristics
— v contragolpe m - [mec] to provide @ backlash
— v cordón m parejo - [sold] to give @ smooth
bead
— v desempeño m destacado - to give @ outstanding performance
— v — m — costa afuera - [petról] to, give, o
provide, @ outstanding offshore performance
— v dureza f - to provide @ hardness
— v — f nominal - to provide @ nominal hardness
— v exceso m de combustible m - [comb.int] to
overfuel
— v frecuencia f alta - [sold] to provide @
high frequency
— v imagen f - [electrón] to provide @, image,
o picture
— v — en televisión f - [electrón] to provide
@ television, image, o picture
— v incentivo m - to provide @ incentive
— v información f - to, provide, o supply, o
feed, @ information
— v protección f - to, provide, o give, @ protection
— v — f adicional - to, give, o provide, @
added protection
— v — f — contra atmósfera f - [ambient] to
give @ added protection against @ atmosphere
— v — f — — — f corrosiva - to give @ added
protection against @ corrosive atmosphere
— v — f — — humedad f - to give @ added protection against @ moisture
— v — f — contra atmósfera f - to give protection
against @ atmosphere
— v — — — f corrosiva - to give protection
against @ corrosive atmosphere
— v — f — contra humedad f - to give protection
against @ moisture
— v recomendación f - to give @ recommendation
— v regulación f - to provide @ adjustment
— v — f aproximada - [sold] to provide @ broad
adjustment
— v — f precisa - [sold] to provide @ precise
adjustment
— v — f — para voltaje m - [sold] to provide
@ precise voltage adjustment
— v rendimiento m - to give @ performance
— v resistencia f - to provide @ resistance
— v separadamente - to provide separately
— v servicio m - to supply @ service
— v — m de consulta f - [ind] to furnich @
consultation
— v simultáneamente - to provide simultaneously
— v sin cargo - to supply without @ charge
— v sitio m - to provide room
— v soldadura f - [sold] to provide @ welding
— v solución f - to provide @ solution

proporcionar v sostén m - to provide @ support
— v tardíamente - to provide late(ly)
— v traslapo m - [mec] to provide @ (over)lap
— v ventaja(s) f - to provide @ advantage(s)
— v vida f (útil) larga - to, provide, o give, @ long (useful) life
— v — f —— a banda f para rodamiento m - [autom-neumát] to, provide, o give, @ long tread life
— v voltaje m - [sold] to provide @ voltage⸱
— v — m preciso - [sold] to provide @ precise voltage
proposición f para afinamiento m - refinement proposal,sing
propósito m - . . . .; goal • objective • intent • scope • use
— m administrativo - [admin] management purpose
— m apropiado - appropriate, o good, purpose
— m combinado - combined purpose
— m de administración f - [admin] management's purpose
— m — advertencia f - warning purpose
— m — baranda(s) f para puente(s) m - [constr] bridge rail purpose(s)
— m — compañía f - company's purpose
— m — defraudar v - [legal] intent to defraud
— m — empresa f - company, o corporation, o concern, purpose, o end
— m — inspección f - inspection purpose
— m — instalación f - installation purpose
— m — instrucción(es) f - instruction(s) purpose(s)
— m — invención f - invention purpose
— m — investigación f - investigation purpose
— m — f de accidente m - [segurid] accident investigation purpose
— m — sociedad f - [legal] corporate,tion purpose(s)
— m — soldadura f - [sold] weld purpose
— m — trabajo m - work purpose
— m — m administrativo - [admin] management work purpose
— m doble - double, o dual, purpose • combined purpose
— m ejemplificativo - illustrative purpose
— m especializado - specialized purpose
— m específico - specific purpose
— m general - general purpose
— m guiador - guiding purpose
— m ilustrativo - illustrative purpose
— m para regulación f - control purpose(s)
— m parecido - similar purpose(s)
— m principal - principal, o main, purpose
— m regulatorio - regulatory purpose
— m similar - similar purpose
— m único - single purpose
propuesta f - [com] . . .; bid • project • draft • proposition • suggestion • intent(ion) • [legal] motion
— f abierta - [com] open(ed), proposal, o bid
— f aceptada - [legal] accepted, proposal, o bid • successful, o winning, o low, bid, o proposal, o offer
— f alternativa - alternate, proposal, o bid
— f baja - [com] low bid
— f base,sica - base,sic proposal
— f calificada - qualified proposal
— f comparada - compared, proposal, o offer, o bid
— f conveniente - convenient, proposal, o bid
— f de condición f - condition proposal,sing
— f — contratista m - [com] contractor's bid
— f — equipo m - [ind] equipment proposal
— f — m eléctrico - [ind] electrical equipment proposal
— f — m mecánico - [ind] mechanical equipment proposal
— f — programa - [ind] schedule proposal
— f — proveedor m - [com] supplier's, proposal, o bid
— f — seguro m - [seguros] insurance proposal
— f — solución f - solution proposal,sing
— f — valor m - value proposal

propuesta f descalificada - disqualified proposal
— f económica - economic(al), o commercial, proposal • valued proposal
— f — final - final economic(al) proposal
— f elaborada - prepared proposal
— f en consideración f - [com] subject proposal
— f entregada - delivered proposal
— f evaluada - evaluated, proposal, o offer
— f favorecida - [legal] favored, o successful, o winning, bid
— f habilitada - accepted, o approved, proposal
— f inicial - initial, proposal, o bid
— f jerarquizada - graded proposal
— f más baja - [com] lowest bid
— f mejor - [legal] best, o successful, bid
— f mexicana - Mexican, proposal, o bid
— f monetaria - [fin] monetary proposal
— f no adjudicada - [com] not awarded, proposal, o bid
— f optativa - [com] optional, proposal, o bid
— f original - [com] original proposal
— f para contrato - [legal] contract, proposal, o draft
— f — costo(s) m - cost(s) proposal
— f — m para servicio(s) m - [legal] service cost(s) proposal
— f — programa m - [ind] schedule proposal
— f — m para fabricación f - [ind] fabrication schedule proposal
— f — realización f - performance proposal
— f — f de estudio m - study performance proposal
— f — regulación f - [ind] control proposal
— f — repuesto(s) m - [ind] spare(s) (parts) proposal
— f — sistema m - system. proposal, o offer
— f — solución f - solution proposal
— f por equipo m - [ind] equipment bid
— f servicio(s) m - service(s) proposal
— f — m de personal m - [labor] personnel service(s) proposal
— f preparada - prepared proposal
— f presentada - [legal] presented, o submitted, proposal, bid
— f primera - [com] first, proposal, o bid
— f rápida - quick proposal • [legal] rapid action
— f recibida - received, proposal, o bid
— f rechazada - [com] rejected, proposal, o bid
— f retirada - [com] withdrawn, proposal, o bid
— f segunda - [com] second, proposal, o bid
— f substitutiva - [com] alternate, proposal, o bid
— f — optativa - [com] optional alternate bid
— f técnica - [ind] technical proposal
— f — devuelta - returned technical proposal
— f — final - final technical, proposal, o bid
— f tramitada - handled, proposal, o bid
propuesto,ta a - . . . .; propounded; suggested • intended
propugnado,da a - advocated; propounded
propugnar v - . . . .; to propound
propulsado,da - powered • pioneered
— a con motor m rotativo - [comb.int] rotary engine powered
propulsar v - . . . . • to impel • to power • to pioneer
propulsión f - . . . .; driving; propelling; véase también impulsión f
— f automática - [mec] self propelling | a - [sold] (con) self propelled
— f autónoma - self propelling | a - self-propelling
— f con aire m - [neumát] air drive,ving
— f con cable(s) m - [mec] cable, o rope, drive
— f — m Texrope - [mec] Texrope drive
— v correa f en V - [mec] V-belt drive
— f — f trapezoidal - [mec] V-belt drive
— f — engranaje(s) m - [mec] gear drive,ving
— f — gas m - [mec] gas drive,ving
— f de gancho m giratorio - [grúas] swivel drive,ving

propulsión f delantera - [autom- front-wheel
  drive,ving
— f directa - [mec] direct drive
— f eléctrica - [ind] electrical drive
— f frontal - [autom] front wheel drive
— f independiente - independent, drive,ving, o
  propelling, o propulsion
— f para carretera f - [autom] highway propulsion
— f — vía f - [autom] roadway propulsion
— f propia - [mec] self propelling | a - [mec]
  - (con) self-propelled
— f uniforme - [mec] uniform drive,ving
— f — directa biselada   [mec] bevel(ed) uniform drive,ving
— f variable - [mec] variable drive,ving
propulsor m - impeller • [pol] activist • [com]
  promoter; pioneer • [mec] driver • [legal] organizer; developer
propulsor,ra a - propelling; driver,ving • [pol]
  activist
prorrata f - . . . | adv - (a) prorated; propor-
  tionately; pro rata
— f mensual - monthly pro rata
— f por fracción f de día - prorrated for @
  fraction of @ day
— a igualmente - equally prorated
prorratear v - . . .; to pro rata
— v valor m - to pro rata @ value
prorrateo m - . . .; prorating • assesment
— m de valor m - value prorating
— m mensual - monthly pro rata
prórroga f - . . . • amplification
— f de carta f de crédito - [fin] credit letter
  extension
— f de período m para ejecución f - performance
  period extension
— f — plazo m - time extension
prorrogable adv - extensible • renewable
prorrogar v carta f de crédito - [fin] to extend
  @ credit letter
proscribir v - . . . • to rule out
proscripción f - . . .; proscribing • ruling out
proscri(p)to,ta a - . . . • ruled out
prosecretario m - assistant, o under-,secretary
prosecución f - . . .; continuation,nuing; going
  on
— f afanosa - [deport] pressing on relentlessly
— f implacable - [deport] pressing on relentlessly
proseguido,da - continued; gone on
— a afanosamente - [deport] pressed on relentlessly
— a implacablemente - [deport] pressed on relentlessly
proseguir v afanosamente - [deport] to press on
  relentlessly
— v así - to continue thus
— v en forma determinada - to press on dtermi-
  nedly
— v hasta ganar - to go on to win
— v — triunfar - to go on to win
— v implacablemente - [deport] to press (on) relentlessly
prospección f geológica - [geol] geological
  prospecting,tion
— f y exploración f - prospecting and exploring
  • [petról] shot survey(ing)
prospecto m - [public] . . .; brochure
prosperado,da a - prospered; flourished
prosperar v - . . .; to flourish
prosperidad f - . . . • [econ] boom
próspero,ra a - . . .; successful
protección f - . . . • armor; defense; safety •
  encasement • [mec] guard; protective device •
  [sold] shield(ing) • [constr] envelope •
  [electr] relaying
— f adecuada - [constr] adequate, o suitable,
  protection
— f a prueba de agua - waterproof protection
— f adicional - additional, o added, o extra,
  protection
— f — contra corrosión f - extra corrosion
  protection

protección f aduanera - [fisc] protective tariff
— f ambiental - environmental protection
— f amplia - ample protection
— f — para rectificador m - [electr-equip] ample rectifier protection
— f — — m con silicio m - [electr-equip]
  ample silicon rectifier protection
— f anticorrosiva - [metal] anticorrosive pro-
  textion
— f aprobada - approved protection
— f apropiada - appropriate, o proper, protecting,tion
— f Arcair - [sold] Arcair protection
— f automática - automatic protection
— f — contra sobrecarga(s) f - [electr-equip]
  automatic overload protection
— f catódica - [electr] cathode protection
— f completa - complete protection
— f con aislación f - insulation protection
— f — — f doble - double insulation protection
— f — arpillera f - [transp] burlap protection; burlapping
— f — asbesto m - asbestos protection
— f — empedrado - [constr] riprap protection
— f — (materail) plástico m - plastic, protection, o shielding
— f — revestimiento m con roca(s) f - [constr]
  riprap protection
— f contra - protection against
— f — acción f de remolino(s) m [hidr] protection against @ eddy action
— f — atmósfera f - atmosphere protection
— f — — f corrosiva - corrosive atmosphere
  protection
— f — caída(s) f - protection from @ fall(s)
— f — calor m - protection against @ heat
— f — — m excesivo - protection against @ excessive heat
— f — carga(s) f - protection against @ load(s)
— f — — f directa(s) - protection against @
  direct load(ing)
— f — conexión(es) f con tierra f - [electr-
  instal] protection against @ ground(ing)
— f contra posible(s) conexión(es) f con tierra
  f - [electr-instal] protection against @ possible ground(ing)
— f — contaminación f - [segurid] contamination
  protection
— f — corrosión f - corrosion protection; protection, from, o against, corrosion
— f — cortocircuito(s) m - [electr-instal] pro-
  textion against @ short circuit(s); short circuit protection
— f — chispa(s) f - [segurid] spark(s) protection; protection, from, o against, @ spark(s)
— f — daño(s) m - damage protection
— f — derrubio(s) m - [hidr] scouring protection
— f — destello(s) m - [sold] flash protection
— f — enfriamiento m - [sold] protection a-
  gainst cooling
— f — — m rápido - [sold] protection against @
  fast cooling
— f — — m súbito - protection against @ sudden
  cooling
— f — erosión f - erosion protection; protection, against, o from, erosion
— f — error(es) m - protection against, mis-
  take(s), o error(s)
— f — escopleadura(s) f - [sold] gouging protection
— f — — f con arco - [sold] arc gouging protection
— f — estallido(s) m - [combust] blowout protection
— f — — m en horno m - [combust] furnace, o
  oven, blowout protection
— f — — m equilibrada - [combust] balanced
  blowout protection
— f — extremo(s) m - protection against @ ex-
  treme(s)
— f — — m en temperatura(s) f - protection a-
  gainst @ temperature extreme(s)

protección f contra frente(s) f de onda(s) -
[electr-oper] protection against @ wave(s)
front; wave front protection
— f — humedad f - moisture protection; protection, from, o against, @ moisture
— f — incendio(s) m—[segurid] fire protection
— f — m con espuma f - [segurid] foam, o
Foamite, fire protection
— f — m — fomita - [segurid] Foamite fire
proection
— f — intemperie f - [ambient] weather protection; weatherproofing • [constr] weather-
tightness
— f — impacto(s) m - protection against @ im-
pact(s)
— f — m directo(s) - [nec] protection a-
gainst @ direct impact(s)
— f — inundación(es) f - [hidr] flood protection
— f — pérdida(s) f - loss protection
— f — f en obra - [electr-equip] field loss
protection
— f — perturbación(es) f - [electr-instal]
perturbation protection
— f — radiación(es) f - [electrón] radiation,
protection, o shield
— f — rayo(s) - [segurid] ray(s) protection
— f — m de arco m - [segurid] arc ray protection; protection, from, o against, @ arc
ray(s)
— f — sabotaje m - [segurid] protección f contra sabotaje m
— f — sequía(s) f—drought protection
— f — sobrecarga(s) f - [electr-instal] overload, protection, o heater link
— f — sobrecorriente(s) f - [electr-instal]
overcurrent protection
— f — f con curva f de tiempo m inversa -
[electr-instal] inverse time overload protection
— f — f para fase f a fase f - [electr-
-instal] phase-to-phase overcurrent protection
— f — f — f a tierra f - [electr-
-instal] phase to ground overcurrent protection
— f — f en elemento m térmico - [electr-
-instal] thermal element overcurrent protection
— f — f — falla f a tierra f - [electr-
-instal] ground fault overcurrent protection
— f — socavación f - [constr] protection a-
gainst @ undermining
— f — suelo(s) m corrosivo(s) - [suelos] protection against @ corrosive soil(s)
— f — temperatura(s) f - protection against
temperature(s)
— f — f extrema(s) - protection against @
extreme temperature(s)
— f — tierra f - [mec] dirt, o earth, guard
— f — vandalismo m - protection against @ vandalism
— f — voltaje(s) m - [electr-instal] voltage
protection
— f — m alto(s) - [electr-instal] high
voltage protection
— f — m bajo(s) - [electr-instal] low
voltage protection
— f cuidadosa - [segurid] careful protection
— f debida - due, o proper, protecting,tion
— f densa - [sold] dense shield
— f — para arco m - [sold] dense arc shielding
— f diferencial - [electr-instal] differential
protection
— f — para circuito(s) m - [electr-instal]
circuit differential, o differential circuit,
protection
— f doble - double protection | adv - (con)
double shielded
— f económica - economical, o low cost, protection
— f eficaz - efficient protection
— f — contra humedad f - efficient protection

against @ moisture
protección f eléctrica - [electr-instal] elec-
tric(al) protection • [ind] electrical safety device
— f industrial - [ind] industrial protection
— f — f para piel f - [segurid] occupational skin protection
— f — f —, ojo(s) m, o vista f—segurid]
occupational, eye, o sight, protection
— f especial - [segurid] special protection •
[mec] special shielding
— f — escopleadura f - [sold] special
gouging protection
— f — — f con arco m - [sold] special
arc gouging protection
— f eventual - eventual protection
— f exterior - exterior, o outside, protection • [mec] external shielding
— f externa - [mec] external shielding
— f facial - [segurid] face shield(ing)
— f fácil - [segurid] easy shielding
— f galvánica - [electr-oper] galvanic protection
— f gaseosa - [sold] gaseous shield(ing)
— f — desapareciente - [sold] disappearing
gaseous shield(ing)
— f inferior - lower, o bottom, shield(ing)
— f inherente - inherent protection
— f — contra cortocircuito(s) m - [electr-
-instal] inherent short circuit protection
— f — para circuito m - [electr-instal] inherent circuit protection
— f — — m para energía f - [electr-
-instal] inherent power circuit protection
— f — — m — f auxiliar - [electr-
-instal] inherent auxiliary power circuit
protection
— f — — m — f — contra cortocircui-
to(s) m - [electr-instal] inherent auxiliary
power circuit short circuit protection
— f — — energía f - [electr-instal] inherent power protection
— f — — f auxiliar - [electr-instal] inherent auxiliary power protection
— f — — m — f contra cortocircuitos m
- [electr-instal] inherent power circuit
short circuit protection
— f instantánea - instant(aneous) protection
— f — contra cortocircuito(s) m - [electr-
-instal] instantaneous short circuit protection
— f — sobrecarga(s) f - [electr-instal]
instantaneous overload protection
— f interior - interior, o inside, protection
— f lateral - (anteojos[side shield
— f — de malla f de (material) plástico m -
[segurid] plastic wire screen sideshield
— f — de (material) plástico m - [segurid]
plastic sideshield
— f — para, ojo(s) m, o vista f - [segurid]
sideshield eye protection
— f — perforada - [segurid] perforated sideshield
— f — — de (material) plástico m - [segurid]
perforated plastic sideshield
— f máxima - maximum protection
— f mecánica - mechanical protection • [ind]
safety, guard, o shield
— f medianera - [vial] median protection
— f mediante subdrenaje m - [constr] sub-
drain(age) protection
— f metálica - [electr-cond] metal(lic)
shield(ing)
— f — continua - [electr-cond] continuous
metallic shield(ing)
— f — ininterrumpida - [electr-cond] continuous metallic guard
— f momentánea - transient protection
— f óptima - optimum protection
— f para acero m - steel protection
— f para acoplamiento m - [mec] coupling guard
— f — aislación f - insulation protection
— f — alcantarilla f - [constr] culvert pro-

tection • embankment protection
protección f **para alumbrado** m - [electr-instal]
lighting protection
— f — **ambiente** m - [ambient] environmental
protection
— f — **arco** m - [sold] arc, protection, o
shielding
— f — — m **entre metal(es)** m - [sold] metal
arc, protection, o shielding
— f — — **para volver(lo) estable** - [sold]
stable arc shielding
— f — **asbesto** m - asbestos protection
— f — **barranca(s)** f - [hidr] bank protection
— f — **borde** m - [mec] edge protection
— f — — m **de cauce** m - [hidr] channel slope
protection
— f — — m **para cierre** m - [mec] seal lip protection
— f — — **cable(s)** f - [electr-instal] cable protection
— f — **cara** f - [segurid] face protection
— f — **cauce** m - [hidr] channel protection
— f — — m **contra erosión** f - [hidr] channel
erosion protection
— f — **cepa** f - [mec] stub protection
— f — **cierre** m - [mec] seal protection
— f — **circuito** m - [electr] circuit protection
— f — **cobertura** f - [mec] cover guard
— f — — f **total** - [mec] full cover guard
— f — — f — **para motor** m - [mec] motor
full cover guard
— f — **cojinete** m - [mec] bearing protection
— m — **colador** m - [mec] strainer protection
— m — **componente** - [mec] component protection
— f — — m **eléctrico** - [electr-instal] electrical component protection
— f — **conductor(es)** m - [electr-instal] cable,
o lead, protection
— f — **conexión** f - [electr-instal] connection
protection
— f — **cristal** - [óptica] lens protection
— f — **chapa** f - [metal-lam] strip protection
— f — **chorro** m - stream protection
— f — **elemento** m **térmico** - [sold] thermal, o
heat, element protection
— m — **enganche** m - [mec] coupling, protection,
o guard
— f — **enlechado** m - [hormig] grout protection
— f — **entrada** - [hidr] inlet protection
— f — **equipo** m - [ind] equipment protection
— f — **estructura** f - [constr] structure protection
— f — — f **para drenaje** m - [constr] drainage
structure protection
— f — **excavación** f - [constr] excavation, o
ditch, protection
— f — **exterior** m - [mec] exterior protection
— f — — m **de tubería** f - [tub] pipe exterior
protection
— f — **extremo** m - [mec] end protection
— f — **fabricación** f **nacional** - [econ] local,
production, o manufacture, protection
— f — **fabricante** m - [ind] manufacturer('s)
protection
— f — — f **nacional** - econ] local manufactu-
rer('s) protection
— f — **fleje** m - [metal-lam] strip protection
— f — **fondo** m - [constr] invert protection
— f — **garantía** f - [ind] warranty protection
— f — **generador** m - [electr-prod] generator
protection
— f — **hormigón** m - [hormig] concrete protection • concrete envelope
— f — — m **armadura** f - [hormig] reinforcement concrete protection
— f — **horno** m - [ind] furnace, o oven, o kiln,
protection
— f — — m **rotatorio** - [cerám] kiln protection
— f — **indicador** m - [mec] meter guard
— m — **interior** m - interior protection
— f — — m **de tubería** f - [tub] pipe interior
protection

protección f **para inversión** f - investment protecting,tion
— f — **lecho** m - [hidr] bed, o channel, protection
— m — **lente** m - [óptica] lens protection
— f — **losa** f - [constr] slab protection
— f — — f **de hormigón** m - [electr-instal]
concrete slab protection
— f — **llanta** f - [mec] rim protection •
[autom-neumát] tire protection
— f — — f **para horno** m **rotatorio** - [cerám]
kiln tire protection
— f — **manómetro** m - [instrum] gage protection
— f — **máquina** f - [ind] machine protection
— f — **maquinaria** f - [ind] machinery protection
— f — **margen** m - [hidr] shore, o bank, protection
— f — — m **de cauce** m - [hidr] channel slope
protection
— f — **material(es)** m - [ind] material(s) protection
— f — **medidor(es)** m - [sold] meter(s), protection, o guard
— f — **medio ambiente** m - [ambient] environmental protection
— f — **metal** m - [sold] metal protection
— f — **motor** m - [electr-mot] motor protection
• [comb.int] engine protection
— f — — m **principal** - [electr-mot] main motor
protection
— f — **nivel** m **de carga** f - [metal-prod] stock
lining protection
— f — **ojos(s)** m - [segurid] eye protection
— f — **paleta** f - [turb] vane protection
— f — — f **directriz** - [turb] guiding vane
protection
— f — **parque** m - [ind] yard protection
— f — **paso** m **inferior** - [constr] underpass
protection
— f — **piel** f - [segurid] skin protection
— f — **pieza** f - [mec] part protection
— f — — f **de hierro** m **fundido** - [sold] casting protection
— f — — f **en movimiento** m - [mec] working
part protection
— f — — f **fundida** - [sold] casting protection
— f — — f **interior** - [mec] internal part
protection
— f — — f **para trabajo** m - [mec] working
part protection
— f — **planta** f - [ind] plant protection
— f — **playa** f - [hidr] shore protection
— f — **polea** f - [grúas] sheave protection
— f — **puente** m - [constr] bridge protection
— f — **persona(s)** f - [segurid] (persons) protection
— f — **producto** m - [ind] product protection
— f — **público** m - [segurid] public protection
— f — **puente** m - [constr] bridge protection
— f — **punta** f - [mec] point protection
— f — — f **de aguilón** m - [grúas] point sheave
protection
— f — **rectificador** m - [electr-equip] rectifier protection
— f — — m **con silicio** m - [sold] silicon rectifier protection
— f — **regulador** - [mec] control protection
— f — **ribera** f - [hidr] shore protection
— f — **rosca** f - [mec] thread protection
— f — **salida** f - [hidr] outlet protection
— f — **sello** m - [mec] seal protection
— f — **separación** f **de nivel(es)** m - [constr]
grade elevation protection
— f — **soldador** m - [segurid] weldor protection
— f — **soldadora** f - [sold] welder protection •
welder protecting fuse
— f — **soldadura** f - [sold] weld protection
— f — — f **contra corrosión** f - [sold] weld
corrosion protection
— f — **sobrecarga(s)** f - [electr-oper] overload

protection

protección f **para subdrenaje** - [hidr] subdrainage protection
— f — **superficie** f - surface protection
— f — — f **decapada** - [metal-trat] pickled surface protection
— f — **talud** m - [topogr] slope protection
— f — — m **para cauce** m - [hidr] channel slope protection
— f — **tensión** f **nula** - [electr] non-voltage protection
— f — **termopar** m - [combust] thermocouple protection
— f — **tobera** f - [ind] tuyere protection
— f — **tubería** f - [tub] pipe,ping protection
— f — — f **contra corrosión** f - [tub] pipe corrosion protection
— f — — f **para conducción** f - [tub] pipeline protection
— f — **vista** f - [segurid] eye, o sight, protection
— f — **voltaje** m - [electr-oper] voltage protection
— f — — m **bajo** - low voltage protection
— f — — m **momentáneo** - [electr-oper] transient voltage protection
— f **plena** - full, o ample, protection | adv - (con) fully protected
— f **para rectificador** m - [electr-equip] ample rectifier protection
— f — — — m **con silicio** m - [electr-equip] silicon rectifier ample protection
— f — **detención** f - [mec] shutdown protection
— f **por detención** f **automática** - [mec] automatic shutdown protection
— f — **medio de relés** m - [electr-oper] protective relaying
— f **propia** - [segurid] own protection
— f **proporcionada** - provided protection
— f **provista** - provided protection
— f — **para arco** m - [sold] arc shielding
— f **química** - [quím] chemical protection
— f **regional** - [pol] regional protection
— f **respiratoria** - [segurid] respiratory protection
— f **superior** - top shield(ing)
— f **temporaria** - temporary protection
— f **térmica** - thermal protection
— f — **contra sobrecarga(s)** f - [electr] thermostatic overload protection
— f — — **sobrecorriente(s)** f - [sold] thermal overload protection
— f **para generador** m - [electr-prod] generator thermal protection
— f — — **motor** m - [electr-mot] motor thermal protection • [comb.int] engine thermal protection
— f — — — m **para alimentación** f - [sold] feed(ing) motor thermal protection
— f — — **soldadura** f - [sold] weld(ing) thermal protection
— f **termostática** - thermostatic protection
— f — **contra sobrecarga(s)** f - [electr-equip] thermostatic overload protection
— f **total** - [mec] full cover guard • [sold] total shield(ing)
— f — **para arco** m - [sold] total arc protection
— f — — **motor** m - [mec] full (cover) motor guard
protectivo,va* a - véase **protector,ra** a
protector m - . . . . • shield; (safety) guard; saver • [electr-oper] release • [mec] guard • [autom-neumát] flap
— m **a prueba de polvo** m - [electr-mot] dust tight case
— m **con resorte** m - [mec] steel spring guard
— m — — m **de acero** m - [mec] steel spring guard
— m **contra corriente(s)** f **de aire** - [sold] wind, o draft, shield(ing)
— m — **falta** f **de tensión** f - [electr-equip] circuit breaker

protección f **contra falta** f **de tensión** f - [electr-equip] circuit breaker
— f — **retorno** m **de llama(s)** f - [combust] backfire preventor
protector m **contra sobrecarga(s)** f - [electr-instal] overload protector; circuit breaker
— m — — f **defectivo** - [electr-equip] defective overload protector
— m — **tierra** f - dirt, o dust, shield
— m **de latón** m - [herram] brass protector
— m — — m **para mandíbula(s)** f - [herram] brass jaw protector
— m — — m — f **para tornillo** m **(para banco** m - [herram] brass vise jaw protector
— m — **tensión** f **nula** - [sold] no voltage release
— m — — m **nulo** - [sold] no voltage release
— m **defectivo** - defective protector
— m **lateral** - [mec] side protector • [anteojos] side shield
— m **momentáneo** - [electr] transient protector
— m **para aguilón** m - [grúas] boss saver
— m — **alambre** m - [mec] wire shield
— m **para cable** m - [petról] line saver
— m — **correa** f - [mec] belt guard
— m — **chapa** f - [metal-lam] strip protector
— m — **indicador** m - [mec] meter guard
— m — **lente** f - [óptica] lens protector
— m — **mandíbula** f - [mec] jaw protector
— m — — f **para tornillo** m **(para banco** m) - [herram] vise jaw protector
— m — **manómetro** m - [instrum] gage protector
— m — **mazo** m **(de conductores** m) - [electr-instal] harness protector • loom
— m — **pestaña** f - [mec] flange protector
— m — — f **inferior** - [mec] bottom flange protector
— m — — f **superior** - [mec] top, o upper, flange protector
— m — **radiador** m - [comb.int] radiator guard
— m — **rodillo** m - [mec] roll(er) protector
— m — **rosca** f - [mec] thread protector
— m — **seguridad** f - [mec] safety, protector, o cover
— m — **terminal** m - [mec] terminal protector
— m — — m **para conectador** m - [mec] connector terminal cover
— m — **tobera** f - [mec] tuyere protector
— m — **tubería** f - [tub] pipe,ping protector
— m — — f **para entubación** f - [petról] casing protector
— m **rotativo** - [segurid] rotary, protector, o shield
— m **térmico** - [electr-instal] thermal protector
— m — **contra sobrecarga(s)** f - [electr-instal] thermal overload, protector, o link
— m — **para motor** m - [electr-mot] motor thermal protector,tion
— m — — m **para alimentación** f - [sold] feed(ing) motor thermal protection
protector,ra a **para inversión** f - [fin] investment protecting
— a — **tensión** f **nula** - [electr-instal] no voltage release
— m — **voltaje** m **nulo** - [electr-instal] no voltage release
— a — **mano(s)** f - [segurid] handshield
proteger v - . . . .; to defend; to guard • to save
— v **a, otros, o otras personas** f - [segurid] to protect, others, o other person(s)
— v — **sí mismo** - to protect, one-, o him-, o her-self
— v — **tercero(s)** m - [segurid] to protect, others, o @ third party,ties
— v **acero** m - [mec] to protect @ steel
— v **alcantarilla** f - [constr] to protect @, culvert, o embankment
— v **apropiadamente** - to protect properly
— v **arco** m - [sold] to shield @ arc
— v — m **entre metales** m - [sold] to shield @ metal arc
— v **borde** m - to protect @, edge, o lip

proteger v borde m de cauce m—[hidr] to protect @ channel slope
— v — m — **cierre** m - [mec] to protect @ seal lip
— v — m — **sello** m - [mec] to protect @ seal lip
— v **borne** m - [electr-instal] to protect @ stud
— v **cable** m - [electr-cond] to protect @ cable
— v **cauce** m - [hidr] to protect @ channel
— v — m **contra** erosión f - [hidr] to erosion protect @ channel
— v **cigarrillo(s)** m - [segurid] to protect @ cigarette(s)
— v **colador** m - [mec] to pprotect @ strainer
— v **componente** m - [mec] to protect @ component
— v — m **eléctrico** - [electr-instal] to protect @ electrical component
— v **con aislación** v - [mec] to insulation protect
— v — — f **doble** - to double insulation protect
— v — **asbesto** m - to asbestos protect
— v — **capa** f **de material** m **orgánico** - [agric] to mulch
— v — **empedrado** m—[constr] to riprap protect
— v — **pavimento** m - [constr] to pavement protect
— v — — m **sobre talud** m - [constr] to slope pavement protect
— v — **revestimiento** m **de roca(s)** f - [constr] to riprap protect
— v **conexión** f - [electr-instal] to protect @ connection
— v **contra** - to protect against
— v — **acción** f - to protect against @ action
— v — — f **de remolino** m - [hidr] to protect against @ eddy action
— f — **caída(s)** f—to protect against @ fall(s)
— v — **calor** m - to protect against @ heat
— v — — m **excesivo** - to protect against @ excessive heat
— v — **carga(s)** f to protect against @ load(s)
— v — — f **directa(s)** - to protect against @ direct load(s)
— v — **contaminación** f - [segurid] to protect against @ contamination
— v — **corrosión** f - to protect against @ corrosion
— v — **chispa(s)** f - [segurid] to protect against @ spark(s)
— v — **daño(s)** m - to protect against @ damage
— v — **derrubio(s)** m - [hidr] to protect against @ scour(ing)
— v — **enfriamiento** m - [sold] to protect against @ cooling
— v — — m **rápido** - [sold] to protect against @, fast, o rapid, cooling
— v — **erosión** f - to protect against @ erosion
— v — **extremo(s)** m - to protect against extremes
— v — — m **en temperatura(s)** f to protect against @ temperature(s) extreme(s)
— v — **humedad** f—to protect against @ moisture
— v — **impacto(s)** m - to protect against @ impact(s)
— v — — m **directo(s)** - to protect against @ direct impact(s)
— v — **intemperie** f - to protect against @ weather; to weatherproof
— v — **rayo(s)** m - [segurid] to protect against @ ray(s)
— v — — m **de arco** m - [segurid] to protect against @ arc ray(s)
— v — **sabotaje** m - to protect against sabotage
— v — **sobrecarga(s)** f - [electr-instal] to protect against @ overload(ing)
— v — **socavación** f - [constr] to protect against @ undermining
— v — **temperatura(s)** f - to protect against @ temperature(s)
v — **temperatura(s)** f **extrema(s)** - to protect against temperature extreme(s)
— f — **vandalismo** n - to protect against @ vandalism

proteger v cristal m - [optica] to protect @ lens
— v **cuidadosamente** - to protect carefully
— v **de elemento(s)** m - to protect, from, o against,@ element(s)
— v **debidamente** - to protect properly
— v **económicamente** - to protect economically
— v **eficazmente** - to protect efficiently
— f — **contra humedad** f - to protect efficiently against @ moisture
— f **en ruta** f - to protect (while) en route
— v **entrada** f - [hidr] to protect @ inlet
— v **equipo** m - [ind] to protect @ equipment
— v **estructura** f - [constr] to protect @ structure
— v — f **para drenaje** m - [constr] to protect @ drainage structure
— f **horno** m - [ind] to protect @, furnace, o oven, o kiln
— v — m **rotativo** - [cerám] to protect @ kiln
— v **inversión** f - to protect @ investment
— v **lecho** m - [hidr] to protect @ bed
— v **lente** m - óptica] to protect @ lens
— v **llanta** f - [mec] to protect @ tire
— v — f **para horno** m **rotativo** - [cerám] to protect @ kiln tire
— v **manómetro** m - [instrum] to protect @ gage
— v **máquina** f - [mec] to protect @ machine
— v **maquinaria** f - [mec] to protect @ machinery
— v **márgen** m - [hidr] to protect @ bank
— v — m **de cauce** m - [jodr] to protect @ channel bank
— v **material** - [ind] to protect @ material
— v **motor** m- [electr-mot] to protect @ motor • [comb.int] to protect @ engine
— v — m **contra sobrecarga(s)** f - [electr-mot] to protect @ motor against @ overload(s)
— v **ojo(s)** m - [segurid] to protect @ eye(s)
— v **paso** m - to protect @ pass
— v — m **inferior** - [constr] to protect @ underpass
— v — n **superior** - [constr] to protect @ overpass
— v **persona(s)** f - [segurid] to protect @, person(s), o human (being), o people
— v **piel** f - [segurid] to protect @ skin
— v **pieza** f - [mec] to protect @ part
— v — f **de hierro** m **fundido** - [sold] to protect @ (iron) casting
— f — f **fundida** - [sold] to protect @ casting
— v — f **interior** - [mec] to protect @ internal part(s)
— v — f **para trabajo** m - [mec] to protect @ working part
— v **plenamente** - to protect fully
— v **producto** m - to protect @ product
— v **público** m - [segurid] to protect @ public
— v **puente** m - [constr] to protect @ bridge
— v **regulador** m - to protect @ control
— v **salida** f - [hidr] to protect @ outlet
— v **sello** m - [mec] to protect @ seal
— v **separación** f **de nivel** - [constr] to protect @ grade elevation
— v **soldadura** f - [sold] to protect @ weld
— v — f **contra corrosión** f - [sold] to corrosion protect @ weld
— v **subdrenaje** m - [constr] to protect @ subdrainage
— v **talud** m - [hidr] to protect @ slope
— v — m **de cauce** m - [hidr] to protect @ channel slope
— v **tobera** f - [constr] to protect @ tuyere
— f **transferencia** v - to protect @ transfer
— v — **de metal** m - [sold] to protect @ metal transfer
— v — f — — m **a través de arco** m - [sold] to protect @ metal transfer through @ arc
— v **tubería** f - [tub] to protect @ pipe,ping
— v — **contra corrosión** f - [tub] to protect @ pipe against @ corrosion
— v — f **para conducción** f - [tub] to protect @ pipeline
— v **vista** f - [segurid] to protect @, eye(s), o

eyesight
protegido,da a - protected • saved • enclosed; guarded; sheltered; shielded
— a **apropiadamente** - properly protected
— a **con asbesto** m - asbestos protected
— a — **aislación** f - [ind] insulation protected
— a — — f **doble** - double insulation protected
— a — **alambre** m - [mec] wire protected
— m **empedrado** m - [constr] riprap protected
— a — **losa** f **de hormigón** m - [constr] concrete slab protected
— a — **pavimento** m **sobre talud** m - [constr] slope pavement protected
— a — — m **de roca(s)** f - [constr] riprap protected
— a **contra** - protected against
— a — **acción** f - protected against @ action
— a — — f **de remolino(s)** m - [hidr] protected against @ eddy,dies action
— f — **caída** f - protected against @ fall
— a — **calor** m - protected against @ heat
— a — — m **excesivo** - protected against @ excessive heat
— a — **carga** f - protected against @ load(ing)
— a — — f **directa** - protected against @ direct load(ing)
— a — **contaminación** f - [segurid] protected against @ contamination
— a — **corrosión** f - protected, against, o from, @ corrosion
— a — **chispa(s)** f - [segurid] protected, from, o against, @ spark(s)
— a — **daño(s)** m - protected against @ damage
— a — **derrubio(s)** m - [hidr] protected against @ scour(ing)
— a — **enfriamiento** m - protected against @ cooling
— a — — m **rápido** - [sold] protected against @ fast cooling
— a — — m **súbito** - [sold] protected against @ sudden cooling
— a — **erosión** f - protected against @ erosion
— a — **extremo(s)** a - protected against @ extremes
— a — — m **en temperatura** f - protected against @ temperature extreme(s)
— a — **humedad** f - protected against @ moisture
— a — **impacto(s)** m - protected against @ impact(s)
— a — — m **directo(s)** - protected against @ direct impact(s)
— a — **intemperie** f - protected against @ weather; weather protected • weather proofed
— a — **rayo(s)** m **de arco** m - [segurid] protected, from, o against, @ arc ray(s)
— a — **sobrecarga(s)** f - [electr-instal] overload protected; protected from @ overload(s)
— a — **socavación(es)** f - [constr] protected against @ undermining
— a — **temperatura** f - temperature protected
— a — — f **extrema(s)** - protected against extreme temperature(s)
— a — **sabotaje** m—protected against @ sabotage
— a — **vandalismo** m - protected against @ vandalism
— a **cuidadosamente** - protected carefully
— m — **contra daño(s)** m - protected carefully against @ damage(s)
— a **debidamente**—properly, o appropriately, protected
— a **económicamente** - protected economically
— a **eficazmente** - efficiently protected
— a **plenamente** - fully protected
protestar v - . . . ; to object
— v **bandera** f - [pol] to swear @ allegiance
protocolización f - [legal] recording; registering • protocolizing
protocolizado,da a -[leg] (officially), recorded, o registered; protocolized
prototipo m - . . . ; standard
protuberancia f - . . . ; protrusion; bulge; node • [hormig] arise; high spot • mec; boss; nipple; knurl • [vial] bump; hump

protuberancia f **encontrada** - [vial] met, o encountered, bump, o hump
— f **en carretera** f - [vial] highway bump
— f — **cojinete** m - [cojin] bearing protrusion
— f — — m **en terraja** f - [mec] die nipple
— f — **perno** m - [mec] bolt, boss, o protrusion
— f — **aportación** f **de aceite** m - [mec] (oil) filler boss
protuberante a **largo,ga** - [mec] long protruding
proveedor m - . . . ; supplier • vendor
— m **capaz** - capable supplier
— m **cercano** - nearby supplier
— m **confiable** - [com] dependable supplier
— m — **para equipo** m - [ind] dependable equipment supplier
— m — — m **para soldadura** f - [sold] dependable welding equipment supplier
— m **de acero** m - [metal] steel supplier
— m — **artículo** m **automovilístico(s)** - [autom] automotive (service) supplier
— m — **bobinadora** f - [metal-lam] coiler supplier
— m — **combustible** m - fuel supplier; fueler
— m — **equipo** m - [ind] equipment supplier
— m — — m **confiable** - [ind] dependable equipment supplier
— m — — m **para soldadura** f - [sold] welding equipment supplier
— m — **grúas** f - [grúas] crane(s) supplier
— m — **máquina(s)** f - [ind] machine(s) supplier
— m — **maquinaria(s)** f - [ind] machinery, o equipment, supplier
— m — **neumático(s)** m - [autom-neumát] tire supplier
— m — **pieza(s)** f - [ind] part(s) supplier
— m — — f **para automóvil(es)** m - [autom] automotive, supplier, o supply source
— m — **preferencia** f - preferred supplier; source of (your) choice
— m — **rodillo(s)** m - [metal-lam] roll supplier
— m — **superabrasivo(s)** m - [ind] superabrasives supplier
— m **dedicado** - dedicated, o committed, supplier
— m **doméstico** - [com] domestic, o national, supplier
— m **en exterior** - [com] foreign supplier
— m **experto** - [com] expert supplier; supply,ies expert
— m **extranjero** - foreign supplier
— m **habitual** - [com] habitual supplier
— m **local** - [com] local supplier
— m **nacional** - [com] national, o local, supplier
— m **preferido** - [com] preferred supplier; choice source
— m **previsto** - foreseen, o contemplated, supplier
— m **principal** 0 [com] main, o leading, supplier
— m **próximo** - nearby supplier • next supplier
— m **seguro** - [com] safe, o solid, supplier
— m — **de equipo(s)** m - [com] solid equipment supplier
— m — — m **para soldadura** f - [sold] solid welding equipment supplier
— m **sugerido** - suggested, supplier, o source
— m **tradicional** - [com] traditional supplier
proveer v - . . . ; to furnish; to purvey • to render • to insure • to service • to, fit, o equip, o outfit • to give; to include • to stock
— v **acero** m - [metal-prod] to supply @ steel
— v **adecuadamente** - to supply adequately
— v — **aire** m - to supply @ air
— v — m **fresco** - to, provide, o supply, fresh air
— v — m **limpio** - [ind] to, provide, o supply, @ clean air
— v **ajuste** m - to, provide, o supply, @ adjustment
— v — m **aproximado** - [ind] to, provide, o supply, @, broad, o approximate, adjustment
— v — m **preciso** - [ind] to, provide, o supply,

@ precise adjustment

proveer v con arco m - [sold] to arc(h)

— v **asesoramiento** m - to, supply, o provide, @ advice

— v **asiento** m - to provide @ seat • [constr] to provide @ bed(ding)

— v — m **apropiado** - [ind] to provide @ proper bed(ding)

— v — m — **para tubería** f - [tub] to provide @ proper pipe bed(ding)

— v — m **para tubería** f - [tub] to provide @ pipe bed(ding)

— v **atención** f - [ind] to service

— v — **para filtro** m—[mec] to service @ filter

— v **base** f - to provide @ base,sis

— v **borne**(s) m - [sold] to provide @ stud(s)

— v — m **para electrodo**(s) m - [sold] to provide @ electrode stud(s)

— v — **separado**(s) - [sold] to provide @ separate stud(s)

— v — m — **para electrodo**(s) m - [sold] to provide @ separate electrode stud(s)

— v **cilindro** - [mec] to furnish @ cylinder

— v **combustible** m - to supply @ fuel

— v **compuesto** m - to provide @ compound

— v **con** - to, supply, o furnish, o fit (up) with

— v — **aleta**(s) - [mec] to (provide with @) fin

— v — **arrancador**(es) m - [comb.int] to, equip, o provide, with @ starter(s)

— v — — m **eléctrico**(s) - [comb.int] to, equip, o provide, with @ electric starter(s)

— v — **bóveda** f - [constr] to arch

— v — **cadena**(s) f - [mec] to fit with @ chain

— v — **cilindro** m - [mec] to furnish with @ cylinder

— v — — m **para éter** m - [comb.int] to furnish with @ ether cylinder(s)

— v — **compensación** f - to provide with @ compensation

— v — **resorte**(s) m - [mec] to spring load

— v — **tratamiento** m **con gas**(es) m - to gas treatment provide

— v — **ventilación** f - to provide with @ ventilation

— v **contra pedido** m - to furnish on request

— v **contragolpe** m - [mec] to provide @ backlash

— v **corriente** f - [electr-oper] to provide with electricity

— v — — f **de salida** f - [electr-oper] to provide @ output

— v — — f — — f **uniforme** - [electr-oper] to provide (with) @ steady output

— v **dureza** f - to provide @ hardness

— v — f **nominal**—to provide @ nominal hardness

— v **e instalar** v - to furnish and install

— v **electrodo** m - [sold] to, provide, o supply, @ electrode

— v **energía** f—[electr-oper] to provide @ power

— v — f **para regulación** f - [sold] to provide @ control power

— v **equipo** m - [ind] to, provide, o supply, (with) equipment

— v **espacio** m t- provide @ space • to provide @ clearance

— v **estabilidad** f - to provide @ stability

— v — f **motriz** - [electr-oper] to provide @ power

— v **fusión** f—[metal-prod] to provide @ melting

— v — f **inmediata** - [metal-prod] to provide @ inmediate melting

— v **herramienta**(s) f - to, furnish, o provide, (with) tool(s); to tool

— v — f **para producción** f - [ind] to provide @ production tool(s)

— v **información** f - to provide @ information

— v **instalación**(es) f - to provide @ facilities

— v **instantáneamente** - to provide instantly

— v **instrucción** f - to provide @ instruction

— v **lecho** m - [constr] to provide @ bed(ding)

— v — m **apropiado** - [ind] to provide @ proper bedding

— v — m — **para tubería** f - [tub] to provide @

proveer v lecho m para tubería f - [tub] to provide @ pipe bed(ding)

— f **leva** f - [mec] to provide @ cam

— v — f **con resorte** m - [mec] to spring load @ cam

— v **lubricación** f - [lubric] to provide @ lubrication

— v — f **adicional** - [lubric] to provide @ additional lubrication

— v **magnetización** f - [electr] to provide @ magnetizatioṇ,ziṇg

— v **material**(es) m - [ind] to provide @, material(s), o supplies

— v — m **para producción** f - [ind] to provide @ production supplies

— v **modelo** m - to provide @, model, o pattern

— v **normalmente** - to, provide, o supply, normally

— v **originalmente** - to, provide, o furnish, originally

— v **para costo**(s) m - to provide @ cost(s)

— m — — m **para jubilación**(es) f - [labor] to fund @ pension(s) cost(s)

— v **penetración** f - [mec] to provide @ penetration

— v — f **más profunda** - [sold] to provide @ deeper penetration

— v **pieza** f - [ind] to provide @ part

— v **potencia** f **auxiliar** - [electr-prod] to provide @ auxiliary power

— v **producto** m - [ind] to, provide, o supply, @ product

— v **protección** f - to provide @ protection

— v — f **contra cortocircuito**(s) m - [electr-instal] to provide @ short circuit protection

— v — f **inherente** - to provide @ inherent protection

— v — f — **contra cortocircuito**(s) m—[electr] to provide @ inherent short circuit protection

— v **regulación** f - to provide @ adjustment

— f — f **amplia** - [sold] to provide @ broad adjustment

— v — f **aproximada** - [sold] to provide @ broad adjustment

— v **resistencia** f - to provide @ resistance • to provide @ strength

— v **seguridad** f - [segurid] to provide @ assurance

— v — f **máxima** - [segurid] to provide a maximum, assurance, o safety

— v — f — **en calidad** f - to provide @ maximum quality assurance

— v **selección** f - to provide @ selection

— v **separación** f - [mec] to provide @ clearance

— v **separadamente** - to provide separately

— v **servicio** m - [ind] to, (provide, o supply @) service

— v — m **para bomba** f - [bombas] to service @ pump

— v — — **filtro** m - [mec] to service @ filter

— v — m **para freno**(s) m - [mec] to service @ brake(s)

— v — m **para motor** m - [comb.int] to service @ engine • [electr-mot] to service @ motor

— v — m **para unidad** f **motriz** - [agric-equip] to service @ power unit

— v **servicio**(s) m **público**(s) - to provide @ utility,ties

— v **sin compensación** f - to provide without @ compensation

— v **solidez** f - to provide @ strength

— v **sostén** m - to provide @ support

— v **sujetador** m - [mec] to provide @ fastener

— v — m **con resorte** m - [mec] to spring load @ cam

— v **supervisión** f - [ind] to provide @ supervision

— v **tope** m - [mec] to provide @ stop

— v **totalmente** - to provide, totally, o fully, o completely

proveer v transferencia f - [electrón] to provide (for) @ transfer
— v ubicación f - to provide @ location
— v — f substitutiva - to provide @ alternate
location
— v umbral m - [electrón] to provide @ threshold
— v velocidad f - [mec] to provide @ speed
— v — f correcta - [mec] to provide @ correct
speed
— v voltaje m - [electr-oper] to provide @
voltage
— v — m en salida f - [electr-distrib] to provide @ output voltage
proveniencia f - source; origin
proveniente a - proceeding; coming
proveniente a de - coming, o arising, o proced-
ing, from
— a de exterior m - foreign origin
provenir v - . . . . to, proceed, o come
— v de - to, come, o proceed, o originate, from
• to accrue
providencia f - . . .; measure • facility
— f para servicio m - service facility,ties
provincia f - [pol] . . . . county
— f de registro m - [legal] home province
provisión f - . . .; providing; equipping; rendering; (out)fitting • [fin] reserve • [com]
stocking • pl - supply(ing); supplies; furnishing; giving; purveying,yance • allowance
• including,usion • service,cing
— f abundante - abundant, o adequate, supply
— f adecuada - adequate supply(ing)
— f — de energía f - [electr-distrib] adequate
power supply
— f — de líquido m refrigerante m - [ind] adequate coolant supply
— f — — — m — para camisa f - [mec] adequate liner coolant supply
— f apropiada - proper supply
— f — de aire m - proper, o appropriate, air
supply
— f — — líquido - adequate liquid supply
— f — — — m refrigerante - [mec] adequate
coolant supply
— f — — — para émbolo m - [mec] adequate piston coolant supply
— f — de energía f - [electr-distrib] adequate
power supply
— f asegurada - insured supply
— f con - provision, o providing, o furnishing,
o fitting, o supplying, with
— f — aleta(s) f - [mec] finning
— f — arrancador m - starter equipping
— f — — m eléctrico - electric starter equipping
— f — cadena f - [mec] fitting with @ chain
— f — cilindro m - [mec] furnishing, o fitting, with @ cylinder
— f — — m para éter m - [comb.int] furnishing
with @ ether cylinder
— f con compensación f - provision, o providing, with @ compensation
— f — instrumento(s) m - instrumenting,tation
— f — motor m - [comb.int] engine equipping •
[electr-mot] motor equipping
— f — — m diesel - [comb.int] diesel engine
equipping
— f — resorte m - [mec] spring loading
— f — — m integral - [mec] spring loading
— f — ventilación f - [mec] ventilation providing,vision
— f contra pedido m - furnishing on request
— f de acero m - [metal-prod] steel supplying
— f — agua m - [hidr] water supplying
— f — — m auxiliar - [mec] auxiliary water
supply(ing)
— f — — m para freno m - [mec] brake water
supply(ing)
— f — — m — auxiliar - [mec] auxiliary
brake water supply(ing)
— f — — m para alimentación f - [cald] feed-
-(ing) water supply(ing)

provisión f de agua m para llanta f - [mec] rim
water supply(ing)
— f — — m — — f para freno m - [mec] brake
rim water supply(ing)
— f — aire m - [neumát] air supply(ing)
— f — — m fresco - fresh air, provision, o
supply(ing)
— f — — m para equipo m - [ind] equipment air
supply(ing)
— f — — m — — m perforador - [petról] rig
air supply(ing)
— f — — m — quemador m - [combust] burner air
supply(ing)
— f — — m para trefilador m - [trefil] wire
drawer air supply(ing)
— f — asesoramiento m - advice supply(ing)
— f — atención f - [ind] servicing
— f — — f para bomba f - [bombas] pump servicing
— f — — f — filtro m - filter service,cing
— f — base f - base,sis supply(ing)
— f — calefacción f - [ambient] heat supplying
— f — calor m - heat providing,vision
— f — cilindro m - [mec] cylinder furnishing
— f — combustible m - fuel supply(ing)
— f — — m cortada - [combust] cut off fuel
supply(ing)
— f — — m para red f - [combust] system fuel
supply(ing)
— f — — m regulada - [comb.int] controlled
fuel supply(ing)
— f — compuesto m - compound providing,vision
— f — contragolpe m - [mec] backlash provision
— f — corriente f - [electr-distrib] electricity, providing,vision, o supply(ing)
— f — — f alterna - [electr-distric] alternating current supply(ing)
— f — — f continua - [electr-distrib] direct
current supply(ing)
— f — chatarra f - [metal-prod] scrap sup-
ply(ing)
— f — — f disponible - [metal-prod] available
scrap supply(ing)
— f — — f vieja - [metal-prod] old scrap sup-
ply(ing)
— f — dureza f - hardness providing,vision
— f — — f nominal - nominal hardness providing,sion
— f — electrodo(s) m - [sold] electrode(s)
supply(ing)
— f — energía f - [electr-distrib] power, sup-
ply(ing), o providing,vision; electric service
— f — — f con transistor(es) m - [sold] solid
state power supply(ing)
— f — — f cortada - [electr-distrib] turned on
power supply
— f — — f correcta - [electr-distrib] correct
power supply(ing) • correct power requirement
— f — — f desconectada - [electr-distrib]
turned off power supply
— f — — f para impresora f - [electrón] printer power supply(ing) • printer power requirement
— f — — f transistorizado,da - [sold] solid
state power supply(ing)
— f — equipo(s) m - equipment supply(ing)
— f — espacio m - [ind] space providing,vision
• [mec] cleranace providing,vision
— f — estabilidad f - stability providing,vi-
sion
— f — existencia(s) f - [ind] inventory stocking
— f — fluido m refrigerante - [mec] coolant
(fluid) supply,ing
— f — fuerza f motriz - [electr-distrib] power
providing,vision
— f — fundente m - [sold] flux supply(ing)
— f — gas m - [combust] gas supply(ing)
— f — — m para quemador m - [combust] burner
gas supply(ing)
— f — — m — horno m - [combut] furnace gas
supply(ing)

provisión f de gas m para horno m para combustion m - [combust] combustion furnace gas supply(ing)
— f — herramental m - [herram] tooling
— f — herramienta(s) f - [ind] tool(s) furnishing
— f — holgura f - [mec] clearance providing
— f — información f - information providing
— f — instalación(es) f [ind] facility,ties providing,vision
— f — instrucción f - instruction providing
— f — instrumento(s) m - [instrum] instrumentation
— f — — m para detección f - [instrum] detection instrumentation
— f — líquido m - liquid providing,vision
— f — — m refrigerante - [mec] coolant providing,sion
— f — — m — para camisa f - [mec] liner coolant supply(ing)
— f — — m — émbolo m - [mec] piston coolant supply(ing)
— f — lubricación f - [lubric] lubrication providing,vision
— f — material(es) m - material(s) supply(ing)
— f — modelo m - pattern providing,vision
— f — motor m - [comb.int] engine, supply(ing) o providing,vision
— f — pieza(s) f - [ind] part(s) providing,vision
— f — — f para repuesto m - [ind] spare parts supply(ing)
— f — procedimiento(s) m - procedure(s), supply(ing), o providing,vision
— f — — m para determinación f de falla(s) f - troubleshooting procedure providing,vision
— f — producto(s) m - [ind] product(s), supply(ing), o providing,vision
— f — protección f - [ind] protection providing,vision
— f — refrigerante m - [mec] coolant supply(ing)
— f — — m para camisa f - [mec] liner coolant supply(ing)
— f — — m — émbolo m - [mec] piston coolant supply(ing)
— f — repuesto(s) m - [ind] spare(s) (parts), providing,vision, o supply(ing)
— f — — m crítico(s) m - [ind] critical spare(s) (parts) providing,vision
— f — — m para motor m - [comb.int] engine spare(s) (parts), providing,vision, o supply(ing) • [electr-mot] motor spare(s) (parts), providing,vision, o supply(ing)
— f — resistencia(s) f resistance(s) providing,vision • [mec] strength providing,vision
— f — rodillo(s) m - [metal-lam] roll(s), providing,vision, o supply(ing)
— f — seguridad f - [segurid] safety, o assurance, providing,vision, o supply(ing)
— f — — f máxima - [segurid] maximum, safety, o assurance, providing,vision, o supply(ing)
— f — — f — en calidad f - maximum quality assurance, providing,vision, o supply(ing)
— f — selección f - selection providing,vision
— f — separación f - [mec] clearance providing
— f — servicio m - [ind] servicing; service, providing,vision, o supply(ing)
— f — — m completo - [ind] complete, o comprehensive, service providing,vision
— f — — m — para motor m - [electr-mot] complete, o comprehensive, motor service providing,vision • [comb.int] engine, complete, o comprehensive, motor service, supply(ing), o providing,vision
— f — — m — bomba f—[bombas] pump servicing
— f — — m — filtro m—[mec] filter servicing
— f — — m — motor m - [comb.int] engine servicing • [electr-mot] motor servicing
— f — m público(s) - utility,ties providing
— f — sostén m - [mec] support providing
— m — tope m - [mec] stop, providing,vision, o supply(ing)

provisión f de transferencia f - [electrón] transfer providing,vision
— f — tubería f - [tub] pipe supply(ing)
— f — ubicación f - location providing,vision
— f — — f substitutiva - alternate location providing,vision
— f — umbral m - [electrón] threshold providing,vision
— f — velocidad f - [mec] speed providing
— f — — f correcta f - [mec] correct speed providing,vision
— f debida f - proper providing,vision
— f — de aire m - proper air supply(ing)
— f deficiente - short supply(ing)
— f disponible - available supply(ing)
— f e instalación f - furnishing and installation,ling
— f económica - economic(al), o low cost, providing,vision, o supply(ing)
— f en mercado m interior - domestic (market) supply(ing)
— f — plaza f - [com] market supply(ing)
— f excesiva - oversupply
— f igual - equal supply(ing)
— f inadecuada - inadequate, o short, supply(ing)
— f mercadeada - [com] marketed supply,plies
— f normal - [electr-distrib] normal, supply, o service
— f — de energía f - [electr-distrib] normal electric supply
— f original - original, supply(ing), o furnishing
— f para amplificador m - [electrón] amplifier supply(ing)
— f — — m magnético - [electrón] magnetic amplifier supply(ing)
— f — costo(s) m - cost(s) funding
— f — — m para jubilación(es) f - [fin] pension(s) cost(s) funding
— f — equipo m - [mec] equipment supply(ing)
— f — — m perforador m - [petról] rig supply(ing)
— f — fusión f - [metal-prod] melting providing,sion
— f — — f inmediata - [metal-prod] inmediate melting providing,vision
— f — gasto(s) m - [fin] expense(s), reserve, o providing,vision
— f — tratamiento m - treatment provision
— f — — m con gas m - [metal-trat] gas treatment providing,vision
— f pequeña - small, o limited, supply(ing)
— f regulada - controlled supply(ing)
— f separada - separate, supply(ing), o service,cing
— f — de agua m - [hidr] separate water supply
— f sin compensación f - compensationless, o unreimbursed, providing,vision
— f suficiente - sufficient supply(ing)
— f tardía - late, supply(ing), o providing
— f total - total, o full, providing,vision, o supply(ing)
provisoriamente adv - provisionally; temporarily
provisto,ta a - . . . . • given; purveyed • (out)fitted; equipped; furnished; with • included • [com] stocked • [ind] serviced • rendered
— a adecuadamente - adequately supplied
— a con aleta(s) f - [mec] finned
— a con arrancador m - [mec] starter equipped
— a — — m eléctrico - [comb.int] electric starter equipped
— a con cadena f - [,ec] fitted, o equipped, with @ chain; chain equipped
— a — cilindro m - [mec] furnished with @ cylinder; cylinder equipped
— a — — m para éter m - [comb.int] furnished with @ ether cylinder; ether cylinder equipped
— a — compensación f - compensation provided; provided with @ compensation
— a — empaquetadura f - [tub] gasketed
— a — fusible(s) m [electr-instal] fused

provisto,ta con imán m—[electr] magnet equipped
— a — lengüeta f - [electr-instal] lugged
— a — neumático(s) m - [autom] tired; shod
— a — perno(s) m - [mec] bolt equipped
— a — — m ajustable(s) - [mec] adjustable bolt equipped
— a — resorte(s) m - [mec] spring, equipped, o loaded
— a — rodillo(s) m - [mec] roller equipped
— a — tratamiento m con gases m - gas treatment equipped
— a — varilla(s) f - [mec] rodded
— a — — f redonda(s) - [tub] rodded
— a — — f — para ajuste m - [tub] rodded
— a — ventilación f - ventilation equipped
— a contra pedido - furnished on request
— a e instalado,da - furnished and installed
— a en construcción f - built in
— a instantáneamente - provided instantly
— a normalmente - supplied normally
— a originalmente - furnished originally
— a por tercero(s) m - furnished by others
— a sin compensación f - furnished without compensation
— a totalmente - provided, o furnished, totally, o fully, o completely
provocación f de incendio m - [segurid] fire, setting, o starting
provocador m - provoker; provocateur
provocante a - . . .; thought provoking
provocar v - . . .; to cause • to start
— v incendio m - [segurid] to start @ fire
próximo,ma a - . . .; adjoining; near(by) • upcoming
— a a - near, o close to
— a a conmutador m - [electr-instal] close to, o near, o next to, @ switch
— a a máximo,ma - near @ maximum
— a — — admitido,da - near @ allowable maximum
— a — núcleo m laminar - near @ lamination
— a — superficie f - near, o close to, @ surface
— a — estructura - near, o close to, @ structure
proximidad f - . . .; nearness; closeness
proyección f - . . . • developing; engineering; tailoring • conditioning • rating • [com] forecast • [constr] . . . • rendering • [mec] tang • protrusion
— f de pared f de tubería f - [tub] pipe wall design(ing)
— f a escala f - projecting to @ scale
— f abarcada - covered design(ing)
— f abocada - approached design
— f actual - [mec] current design
— f actualizada - updated design
— f adecuada - adequate, o good, design(ing)
— f — de tubería f - [constr] barrel adequate design(ing)
— f — — f para estructura f - [constr] structure barrel adequate design(ing)
— f ajustada - [constr] close design(ing)
— f — a medida(s) f - [constr] measurement(s) adjusted design(ing)
— f — a medida(s) f de alcantarilla f - [constr] culvert measurements adjusted design
— f alternativa - alternative design
— f anterior - [mec] previous design • discontinued design
— f anticipada - pre-engineering; predesigning
— f antesísmica - antiseismic design
— f anual - annual, projection, o forecast
— f aprobada - approved, o sound, design
— f apropiada - proper, o good, o suitable, o adequate, design(ing)
— f artística - [constr] artist('s) rendering
— f avanzada - advanced, o aggressive. design
— f — para banda f para rodamiento m - [autom-neumát] advanced tread design
— f — extremo m hidráulico m - [petról] fluid end advanced design
— f basada en resistencia f estructural - structural strength based design

proyección f buena - good, design(ing), o engineering
— f carretera - [constr] highway design(ing)
— f civil - [constr] civil design(ing)
— f comercial - [com] commercial projection
— f compacta - compact design(ing)
— f comprobada - proven, o sound, design(ing)
— f competidora - competing design
— f compuesta - composite design(ing)
— f — para estructura f - [constr] structure composite design(ing)
— f — — f subterránea - [constr] underground structure composite design(ing)
— f — — — f de acero m - [constr] steel structure composite design(ing)
— f común - common design(ing)
— f con larguero(s) m - [constr] stringer design(ing)
— f con porte m grande - [constr] Super-Span design(ing)
— f contra - design(ing) against
— f — abrasión f - [hidr] design against @ abrasion; abrasion design
— f — corrosión f - [hidr] design against @ corrosion; (anti)corrosion design(ing)
— v convencional - conventional design(ing)
— f correctiva - corrective, o remedial, design(ing)
— f cubierta f - [constr] covered design(ing)
— f cuidadosa - careful, design(ing), o developing,pment
— f de acequia f - [constr] ditch design(ing)
— f — acuerdo con especificación(es) f - designing, o engineering, to @ specifications
— f — característica(s) f - characteristic(s), projecting,tion, o design(ing)
— f — carretera f - [constr] highway, o roadway, design(ing)
— f — — f rural(es) - [constr] rural, o country, highway design(ing)
— f — cauce(s) m - channel(s) design(ing)
— f — — m abierto(s) - [hidr] open channel designing
— f — — m estable(s) - [hidr] stable channel design(ing)
— f — — m para drenaje m - [constr] drainage channel design(ing)
— f — — m — — m junto a camino m - [constr] roadside drainage channel m - roadside drainage channel design(ing)
— f — cloaca(s) f - [sanit] (sanitary) sewer design(ing)
— f — — f pluvial - [hidr] storm sewer design
— f — — f — sanitaria - [sanit] sanitary storm sewer design(ing)
— f — — f sanitaria - [sanit] sanitary sewer design(ing)
— f — conducto m - [constr] conduit designing
— f — — m subterráneo m - [constr] underground conduit design(ing)
— f — conductor(es) m - [electr-cond] conductor(s) design(ing)
— f — conformación f - shape design(ing)
— f — consumo(s) m - [com] requirements forecast(ing)
— f — corriente f - [hidr] current, o flow, design(ing)
— f — — f bajo presión f - [hidr] pressure flow design(ing)
— f — demanda f - demand forecast(ing)
— f — diámetro m - diameter, o size, design
— f — — m de tubería f - [tub] pipe,ping size design(ing)
— f — ensayo m - test engineering
— f — fabricante m - [ind] manufacturer('s) design
— f — — m de equipo m - [ind] equipment manufacturer('s) design(ing)
— f — ferrocarril m - [constr] railway designing
— f — franja f divisoria - [constr] median design(ing)
— f — — f — hundida - [constr] median swale design(ing)
— f — imagen f - image projecting,tion

proyección f̲ de instalación(es) f̲ - [ind] system, o facility,ties design(ing)
— f̲ de junta f̲ - [mec] joint design(ing)
— f̲ — — f̲ empernada - [mec] bolted joint design(ing)
— f̲ — larguero m̲ - [constr] stringer design
— f̲ — letra(s)· f̲ - letter(ing) design(ing)
— f̲ — línea f̲ - [ind] line design(ing)
— f̲ — manual - [public] manual design(ing)
— f̲ — material(es) m̲ - material(s) design(ing)
— f̲ — — m̲ para relleno m̲ - [constr] backfill design(ing)
— f̲ — — m̲ para servicio m̲ - service material design(ing)
— f̲ — metal m̲ corrugado - [metal-fabr] corrugated metal design
— f̲ — mordaza f̲ - [mec] clamp design(ing)
— f̲ — motor m̲ - [electr-mot] motor design(ing) • [comb.int] engine design(ing)
— f̲ — muro m̲ - [constr] wall design(ing)
— f̲ — — m̲ para cabecera f̲ - [constr] headwall design(ing)
— f̲ — — m̲ tipo cajón - [constr] bin-wall design(ing)
— f̲ — neumático m̲ - [autom-neumát] tire design(ing)
— f̲ — nivel - level design(ing)
— f̲ — obra f̲ - [constr] project design(ing)
— f̲ — pared f̲ - [constr] wall design(ing)
— f̲ — — f̲ para conducto m̲ - [tub] conduit wall design(ing)
— f̲ — placa f̲ - [mec] plate design(ing)
— f̲ — — f̲ para base f̲ - [constr] base plate design(ing)
— f̲ — plancha f̲ - plate design(ing)
— f̲ — — f̲ para revestimiento m̲ - [constr] liner plate design(ing)
— f̲ — política f̲ - [admin] policy creation
— f̲ — proceso m̲ - process, o system, designing
— f̲ — producto m̲ - product, design(ing), o plan(ning)
— f̲ — — m̲ bruto interno - [econ] gross national product forecast
— f̲ — proyecto m̲ - project design(ing)
— f̲ — ranura(s) f̲ - [mec] spline design(ing)
— f̲ — — f̲ en árbol m̲ - [mec] shaft spline(s) design(ing)
— f̲ — — f̲ — eje m̲ - [mec] shaft spline(s) design(ing)
— f̲ — red f̲ - system design(ing)
— f̲ — — f̲ cloacal - [sanit] sewer system design(ing)
— f̲ — red f̲ de tubería(s) f̲ - [tub] piping system design(ing)
— f̲ — — f̲ — — f̲ con presión f̲ - [tub] pressure piping system design(ing)
— f̲ — — f̲ eléctrica - [electr-instal] electrical system design(ing)
— f̲ — relleno m̲ - [constr] backfill(ing), design(ing), o engineering
— f̲ — — m̲ de tierra f̲ - [constr] earth backfill(ing) design(ing)
— f̲ — renglón m̲ - [ind] line design(ing)
— f̲ — resultado(s) m̲ - result(s) forecast(ing) • [fin] profit and loss forecast(ing)
— f̲ — revestimiento m̲ - [constr] liner design(ing)
— f̲ — rodillo m̲ - [mec] roll(er) design(ing)
— f̲ — — m̲ para alimentación f̲ - [mec] feed-(ing) roll design(ing)
— f̲ — — f̲ entrada - [constr] inlet (cross section) design(ing)
— f̲ — — f̲ — salida f̲ - [constr] output (cross section) design(ing)
— f̲ — sección f̲ terminal - [tub] end section design(ing)
— f̲ — servicio m̲ para material(es) m̲ - [ind] material(s) service design(ing)
— f̲ — sistema m̲ - system design(ing)
— f̲ — — m̲ hidráulico - [hidr] hydraulic system design(ing)
— f̲ — — m̲ movible - [electrón] mobile system design(ing)

proyección f̲ de sistema m̲ móvil - [electrón] mobile system design(ing)
— f̲ — — m̲ para fundición f̲ - [metal-prod] melting system design(ing)
— f̲ — — m̲ — — f̲ eléctrica - [metal-prod] electric melting system design(ing)
— f̲ — — m̲ para manipulación f̲ - [ind] handling system design(ing)
— f̲ — — m̲ — — f̲ de material m̲ - [ind] material(s) handling system design(ing)
— f̲ — — m̲ transpondedor - [electrón] transponder system design(ing)
— f̲ — — m̲ — movible - [electrón] mobile transponder system design(ing)
— f̲ — — m̲ — móvil - [electrón] mobile transponder system design(ing)
— f̲ — soporte m̲ - [mec] support design(ing)
— f̲ — — m̲ estructural - [vial] structural support design(ing)
— f̲ — subestación f̲ - [electr-prod] substation design(ing)
— f̲ — sumidero m̲ - [hidr] catchbasin design
— m̲ — — m̲ típico - [hidr] typical catchbasin design(ing)
— f̲ — suspensión f̲ - suspension design(ing)
— f̲ — tablestacado m̲ - [constr-pil] sheetpiling design(ing)
— f̲ — tecnología f̲ - [ind] technology design
— f̲ — tubería f̲ - [tub] pipe design(ing) • [constr] barrel design(ing)
— f̲ — — f̲ abovedada - [tub] pipe-arch design
— f̲ — — f̲ — corrugada - [tub] corrugated pipe-arch design(ing)
— f̲ — — f̲ — — de acero m̲ - [tub] corrugated steel pipe-arch design(ing)
— f̲ — — f̲ corrugada - [tub] corrugated pipe design(ing)
— f̲ — — f̲ — — de acero m̲ - [tub] corrugated steel pipe design(ing)
— f̲ — — f̲ de acero m̲ - [tub] steel pipe design(ing)
— f̲ — — f̲ de plancha(s) f̲ - [tub] plate pipe design(ing)
— f̲ — — f̲ — — f̲ estructurales - [metal-fabr] structural plate pipe design(ing)
— f̲ — — f̲ para estructura f̲ - [constr] structure barrel design(ing)
— f̲ — válvula f̲ - [válv] valve design(ing)
— f̲ — venta(s) f̲ - [com] sale(s) forecast(ing)
— f̲ — vertedero m̲ - [constr] chute, o spillway, design(ing)
— f̲ deficiente - deficient design(ing); underdesign(ing)
— f̲ definitiva - final design(ing)
— f̲ — para tubería f̲ - [constr] final pipe design(ing)
— f̲ detallada - detailed design(ing)
— f̲ discontinuada - [mec] discontinued design
— f̲ distinta - alternate design(ing)
— f̲ — para alcantarilla f̲ - [constr] alternate culvert design(ing)
— f̲ económica - economical, o low-cost, design
— f̲ eficiente - efficient design(ing)
— f̲ eléctrica - [electr] electrical design(ing)
— f̲ — para turbogenerador m̲ - [electr-prod] turbogenerator electrical design(ing)
— f̲ emprendida - undertaken design(ing)
— f̲ en defecto m̲ - underdesign(ing)
— f̲ — exceso m̲ - overdesign(ing)
— f̲ general - general design(ing)
— f̲ enmendada - [ind] amended design(ing)
— f̲ equilibrada - balanced design(ing)
— f̲ — para alcantarilla f̲ - [constr] balanced culvert design(ing)
— f̲ escogida - chosen design
— f̲ especial - special, o specific, o custom, design(ing), o tailoring
— f̲ establecida - established design(ing)
— f̲ estándar - standard design(ing)
— f̲ — para nervadura f̲ - [autom-neumát] standard rib design(ing)
— f̲ estrella triángulo - [electr-equip] star-delta design(ing)

proyección f estructural - [constr] structural design(ing)
— f — clave - [constr] key structural design
— f — para alcantarilla f - [constr] culvert structural design(ing)
— f — — — f de tubería f - [constr] pipe culvert strutural design(ing)
— f — — — f abovedada - [constr] pipe-arch culvert structural design(ing)
— f — mínima - [constr] minimum structural design(ing)
— f — para bóveda f - [constr] arch structural design(ing)
— f — — estructura f bajo tierra f - [constr] underground, o buried, structure structural design(ing)
— f — — — f subterránea - [constr] buried, o underground, structure structural design(ing)
— f excelente - excelent, design(ing), o engineering
— f excesiva - overdesign(ing)
— f exigida - [ind] required design(ing)
— f fácil - easy design(ing)
— f familiar - familiar design(ing)
— f ferroviaria - [f.c.] railway design(ing)
— f final - final design(ing)
— f — para red f - system final design(ing)
— f — — sistema m - system final design(ing)
— f general - general design(ing)
— f geométrica - geometric design(ing)
— f — para carretera f - [constr] highway geometric design(ing)
— f — — — f rural - [constr] rural highway geometric design(ing)
— f global - [com] overall forecast
— f hidráulica - hydraulic design(ing)
— f — para cloaca(s) f - [constr] sewer hydraulic design(ing)
— f horizontal - horizontal projection
— f ilustrada - illustrated design(ing)
— f inaceptable - unacceptable design(ing)
— f individual - individual, design(ing), o engineering
— f — de estructura f - [constr] structure individual design(ing)
— f — justificada - [constr] justified individual design(ing)
— f interior - interior, o internal, design
— f justificada - justified design(ing)
— f macroeconómica - [com] macroeconomic forecast(ing)
— f más aproximada - [constr] closest, o nearest, design(ing)
— f mecánica - mechanical design(ing)
— f — para turbogenerador m - [electr-prod] turbogenerator mechanical design(ing)
— f moderna - modern design(ing)
— f — para aeropuerto m - [aeron] modern airport design(ing)
— f modificada - modified design(ing)
— f modular - modular design(ing)
— f — para sistema m hidráulico - hydraulic system modular design(ing)
— f necesaria - [ind] necessary, o required, design(ing)
— f negativa - negative projection
— f nueva - new design(ing)
— f original - original design(ing)
— f para acero m - [metal] steel design(ing)
— f — — m corrugado - [metal-fabr] corrugated steel design(ing)
— f — aeropuerto m - [aeron] airport design
— f — alcantarilla f - [constr] culvert design(ing)
— f — — — f basada en resistencia f estructural - [metal-fabr] structural strength based culvert design(ing)
— f — — — f de tubería f - [constr] pipe culvert design(ing)
— f — — — f metálica - [constr] metal culvert design(ing)
— f — — — f corrugada - [constr] corrugated metal culvert design(ing)

proyección f para alcantarilla f para dique m - [constr] levee culvert design(ing)
— f — arcón m - [constr] caisson design(ing)
— f — — m para excavación f - [constr] caisson design(ing)
— f — asiento m - [mec] seat design(ing)
— f — — m para válvula f - [válv] valve seat design(ing)
— f — automóvil m - [autom] automobile design(ing)
— f — avión m - [aeron] aircraft design(ing)
— f — banda f para rodamiento m - [autom-neumát] tread design(ing)
— f — baranda f - [constr] rail design(ing)
— f — — f para puente m - [constr] bridge rail design(ing)
— f — bloque(s) m - [mec] block design(ing)
— f — — m para tracción f - [autom-neumát] traction block design(ing)
— f — — m — — f para banda f para rodamiento m - [autom-neumát] traction block tread design(ing)
— f — boca f para registro m - [constr] manhole design(ing)
— f — — f — — m de acero m - [constr] steel manhole design(ing)
— f — — f — — m — — m corrugado - [tub] corrugated steel manhole design(ing)
— f — bomba f - [bombas] pump design(ing)
— f — bóveda f - [constr] arch design(ing)
— f — — f con porte m grande - [constr] Super-Span (arch) design(ing)
— f — cable - [cabl] cable, o rope, design
— f — — m de alambre m - [cabl] wire rope, design(ing), o construction
— m — cabo m - [cabl] rope design(ing)
— f — calandria f - [papel] calender design(ing)
— f — calor m - [sold] heat driving
— f — calzada - [constr] roadway design(ing)
— f — cámara f - [autom-neumát] tube design
— f — camisa f - [mec] sleeve, o liner, design(ing)
— f — canal m - [hidr] channel design(ing) • [constr] gutter design(ing)
— f — — m abierto - [hidr] open channel design(ing)
— f — capacidad f - capacity design(ing)
— f — carga f - [ind] load design(ing)
— f — — f viva - [constr] live load design
— f — cloaca f - [sanit] sewer design(ing)
— f — — f pluvial - [hidr] storm sewer design(ing)
— f — — f sanitaria - [sanit] sanitary sewer design(ing)
— f — compacidad f - compactness design(ing)
— m — conjunto m - [mec] assembly design(ing)
— f — — m para descargador m - [mec] dump(ing) assembly design(ing)
— f — — m — — m para depurador m - [ind] scrubber dump(ing) assembly design(ing)
— f — — m — — m automático - [ind] automatic scrubber dump(ing) assembly design(ing)
— f — corte m transversal - [mec] cross-section design(ing)
— f — costura f - [mec] seam design(ing)
— f — — f empernada - [mec] bolted seam design(ing)
— f — cuerpo m - [mec] body design • barrel design(ing)
— f — — m de estructura f - [constr] barrel design(ing)
— f — demanda f - demand forecast(ing)
— f — desagüe m - [hidr] drainage design(ing)
— f — — m pluvial - [hidr] storm sewer design(ing)
— f — — m — sanitario - [sanit] sanitary storm sewer design(ing)
— f — descarga f - [hidr] discharge design
— f — descargador m - [mec] dump(er) design
— f — dispositivo m para cambio(s) m — [autom-

-mec] shift(ing) unit design(ing)
proyección f para durabilidad f - durability design(ing)
— f — **electrodo** m - [sold] electrode design
— f — **entrada** f - [hidr] inlet design(ing)
— f — **equipo** m - [ind] equipment design(ing)
— f — — m **básico** - [ind] basic equipment design(ing)
— f — **estructura** f - [constr] structure design(ing)
— f — — f **bajo tierra** f - [constr] buried, o underground, structure design
— f — — f **compuesta** - [constr] structure composite design(ing)
— f — — f — **de acero** m - [constr] steel structure composite design(ing)
— f — — f **de acero** m - [constr] steel structure design(ing)
— f — — f **individual** - [constr] individual structure design(ing)
— f — — f — **justificada** - [constr] justified individual structure design(ing)
— f — — f **subterránea** - [constr] underground structure design(ing)
— f — — f — **de acero** m - [constr] underground steel structure design(ing)
— f — **extremo** m - [constr] end design(ing)
— f — — m **de alcantarilla** f - [constr] culvert end design(ing)
— f — — m — — f **flexible** - [constr] flexible culvert end design(ing)
— f — — m **hidráulico** m - [petról] fluid end design(ing)
— f — **fabricante** m - [ind] manufacturer('s) design(ing)
— f — — m **de equipo** m - [ind] equipment manufacturer('s) design(ing)
— f — **ferrocarril** m - [f.c.] railway design
— f — **franja** f - [mec] strip design(ing)
— f — — f **divisoria** - [constr] median strip design(ing)
— f — — f — **hundida** - [constr] median swale design(ing)
— f — **grúa(s)** - [grúas] crane, design(ing), o engineering
— f — — f **sobre pedestal** m - [grúas] pedestal crane design(ing)
— f — **hilera** f - [trefil] die design(ing)
— f — **horno** m - [combust] furnace design(ing)
— f — — m **eléctrico** - [combust] electric furnace design(ing)
— f — **inundación** f mayor (previsible) en . . . año(s) m - [hidr] design for @ . . . year storm
— m — **inyector** m - [mec] injector design
— f — — m **para ácido** m - [petról] blow case design
— f — **manguito** m - [mec] sleeve design(ing)
— f — — m **para barra** f **taponadora** - [metal-prod] stopper rod sleeve design(ing)
— m — **manipulador** m **para tono(s)** m—[electrón] tone keyer design(ing)
— f — **maquinaria** - [ind] machinery design(ing)
— f — — f **moderna** - [ind] modern machinery design(ing)
— f — **matriz** f - [mec] die design(ing)
— f — **mercado** m - [econ] market forecast(ing)
— f — **molino** m - [ind] mill design(ing)
— f — **mordaza** f - [mec] clamp design(ing)
— f — **nervadura** f - [autom-neumát] rib design
— f — **neumático** m - [autom-neumát] tire design
— f — **oferta** f - supply forecast(ing) • proposal design(ing)
— f — **operar** v - design(ing) to operate
— f — **palanca** f - [mec] handle, o lever, design(ing)
— f — **pandeo** m - [constr] buckling design(ing)
— f — **pared** f - [constr] wall design(ing)
— f — **pavimento** m - [constr] pavement design
— f — **pieza** f - [mec] part design(ing)
— f — **pista** f - [aeron] pavement, o runway, design(ing)
— f — **planta** f - [ind] plant design(ing)

proyección f para portar - design(ing) to carry
— f — **producto** m - [ind] product design(ing)
— f — **provisión** f - design(ing) to provide
— f — **reducción** f - reducing,ction design(ing)
— f — — f **directa** - [metal-prod] direct reducing,ction design(ing)
— f — **rendimiento** m - [ind] performance design(ing)
— f — **resistencia** f - [constr] resistance, o strength, design(ing)
— f — **resistir** v **aplastamiento** m - [constr] buckling resistance design(ing)
— f — **servicio** m - [ind] service design(ing)
— f — **sitio** m - [constr] site, o location, design(ing)
— f — — m **para obra** f - [constr] site design
— f — **tecnología** f - [ind] technology design(ing)
— f — **terminación** f - [mec] end(ing) design
— f — — f **de extremo** m - end treatment design(ing)
— f — — f — — m **de alcantarilla** f - [constr] culvert end treatment design(ing)
— f — **transmisión** f - [electr-distrib] transmission design(ing)
— f — **troquel** m - [mec] die design(ing)
— f — **turbogenerador** m - [electr-prod] turbogenerator design(ing)
— f **preferida** - preferred, o chosen, design
— f **pronta** - ready design(ing)
— f **recomendada** - recommended design(ing)
— f **requerida** - required design(ing)
— f **revelada** - revealed design(ing)
— f **satisfactoria** - satisfactory design(ing)
— f **según pedido** m - custom design(ing)
— f **segura** - safe, o confident, design(ing)
— f **sencilla** - [ind] simple design(ing)
— f **similar** - similar design(ing)
— f **sísmica** - [geol] seismic design(ing)
— f **sobre pantalla** f - [fotogr] screen projecting,tion
— f — **telón** m - [fotogr] screen projecting
— f **sólida** - solid design(ing)
— f **substitutiva** - alternate design(ing)
— f **sugerida** - suggested design(ing)
— f **Super-Span** - [constr] Super-Span design(ing)
— f **total** - overall design(ing)
— f **verificada** - verified design(ing)
— f **vertical** - vertical, projecting,tion, o design(ing)
proyecciones f varias - various designs
proyectación* f - véase **proyección f**
proyectado,da - projected • planned; developed; designed; engineered; tailored • conditioned • expected • scaled • rated
— a **anticipadamente** - pre-engineered; predesigned
— a **apropiadamente** - properly, o suitably, designed
— a **bien** - well, designed, o engineered
— a **compactamente** - compactly designed
— a **convencionalmente** - designed conventionally
— a **cuidadosamente** - carefully, designed, o developed
— a **de acuerdo con especificación(es)** f - engineered to @ specification(s)
— a **debidamente** - properly designed
— a **deficientemente** - designed deficiently; underdesigned
— a **económicamente** - designed economically
— a **en defecto** - underdesigned; underdeveloped
— a — **escala** f - projected to @ scale
— a — **exceso** - overdesigned; overdeveloped
— a — **forma compacta** - compactly designed
— a — **plantilla** f - [mec] designed into @ template
— a **especialmente** - specially, o custom, designed
— a **excesivamente** - overdesigned
— a **fácilmente** - easily designed
— a **específicamente** - specifically, o specially, designed; specifically tailored

proyectado,da a horizontalmente - projected horizontally
— a individualmente - individually, designed, o engineered
— a originalmente - designed originally
— a para - designed, to, o for
— a para compacidad f - designed, compactly, o for compactness
— a — durabilidad f—designed for @ durability
— a — operar v - designed to operate
— a para inundación f mayor (previsible) en . . . año(s) m - [hidr] designed for @ . . . year storm
— a — portar v - designed to carry
— a — proveer - designed to provide
— a — resistencia f - designed for @ strength
— a — resistir v aplastamiento m - [constr] designed for @ buckling
— a prontamente - designed readily
— a recientemente - recently, o newly, designed, o engineered
— a según especificación(es) g de Asociación f Estadounidense de Fabricantes de Productos m Eléctricos - [electr-equip] designed to @, National Electrical Manufacturers Asocation, o N E M A, specification(s)
— a — norma(s) f de Asociación f Estadounidense de Fabricantes m de Artículos m Eléctricos - [electr-equip] designed to, National Electrical Manufacturers Association, o N E M A, Standards
— a — pedido m - custom designed
— a sencillamente - [ind] designed simply
— a sobre pantalla f—[fotogr] screen projected
— a — telón m - [fotogr] screen projected
— a teniéndo(se) en cuenta seguridad f - [segurid] designed with safety in mind
— a verticalmente - projected vertically
proyectar v - . . . . ; to, engineer, o tailor; to develop • to plan • to expect • to condition • to rate
— v acequia f - [constr] to design @ ditch
— v alcantarilla f - [constr] to design @ culvert
— v — f metálica - [constr] to design @ metal culvert
— v — f — corrugada - [constr] to design @ corrugated metal culvert
— v — f para dique m - [constr] to design @ levee culvert
— v anticipadamente - to, predesign, o pre-engineer
— f apropiadamente - to design, properly, o suitably
— v asiento m - [mec] to design @ seat
— v — m para válvula f - [válv] to design @ valve seat
— v automóvil m—[autom] to design @ automobile
— v avión m - [aeron] to design @ aircraft
— v baranda f - [constr] to design @ rail(ing)
— v — f para puente m - [constr] to design @ bridge rail(ing)
— v bien - to, design, o engineer, well
— v calandria f - [papel] to design @ calender
— v calor m - [sold] to drive @ heat
— v calzada f - [constr] to design @ roadway
— v calle f - [vial] to design @ street
— v cámara f - [autom-neumát] to design @ tube
— v canal m - [hidr] to design @ channel • [constr] to design @ gutter
— v — m abierto - [hidr] to design @ open channel
— v característica f - to project @ characteristic
— v carretera f - [constr] to design @, highway, o roadway
— v — rural - [constr] to design @ rural highway
— v cauce m - [hidr] to design @ channel
— v — m abierto - [hidr] to design @ open channel
— v — m estable - [hidr] to design @ stable channel

proyectar v cauce m para drenaje m junto a camino m - [constr] to design @ roadsie drainage channel
— v circulación f - to design @ circulation
— v — para aire m - to design @ air flow
— v cloaca f - [sanit] to design @ sewer
— v — f sanitaria - [sanit] to design @ sanitary sewer
— v compactamente - to design compactly
— v conducto m - [constr] to design @ conduit
— f conformación f - to design @ shape
— v convencionalmente - to design conventionally
— v corriente f bajo presión f - [hidr] to design @ pressure flow
— v corte m transversal - to design @ cross-section
— v cuerpo m - to design @ body
— v — v de estructura f - [constr] to design @ structure body • to design @ barrel
— v cuidadosamente - to, design, o develop, carefully
— v — neumático m - [autom-neumát] to carefully develop @ tire
— v de acuerdo con especificación(es) f - to engineer to @ specification(s)
— f deficientemente - to design dificiently; to underdesign
— v descarga f - [hidr] to design @ discharge
— v diámetro m - to design @, diameter, o size
— v — m para tubería f - [tub] to design @ pipe, diameter, o size
— v directamente hacia adelante - to stick straight out
— v — afuera - to stick straight out
— v económicamente - to design economically
— v electrodo m - [sold] to design @ electrode
— v en defecto - to underdesign
— v — escala f - to project to @ scale
— v — plantilla f - [mec] to design into @ template
— v — exceso - to overdesign
— v — forma f compacta - to design compactly
— v ensayo m - to engineer @ test
— v especialmente - to design specially, to custom design
— v específicamente - to, design, o tailor, especially, o specifically
— v estructura f - [constr] to design @ structure
— v excesivamente - to overdesign
— v extremo m - [constr] to design @ end
— v — m de alcantarilla f - [constr] to design @ culvert end
— v — m — f flexible - [constr] to design @ flexible culvert end
— v fácilmente - to design easily
— v ferrocarril m - [f.c.] to design @ railway
— v franja f - [constr] to design @ strip
— v — f divisoria - [Constr] to design @ median (strip)
— v — f — hundida - [constr] to design @ median swale
— v grúa f - [grúas] to, design, o engineer, @ crane
— v hacia adelante - to, stick out, o protrude
— v — afuera - to, stick out, o protrude
— v herramienta f - [herram] to design @ tool
— v horizontalmente - to project horizontally
— v horno m - [combust] to design @ furnace
— v — m eléctrico - [combust] to design @ elecric furnace
— v imagen f - [fotogr] to project @ image
— f individualmente - to, project, o engineer, individually
— v instalación f - [ind] to design @ system • to design @ facility
— v larguero m - [constr] to design @ stringer
— v línea f - [ind] to design @ line
— v manipulador m para tono(s) m - [electrón] to design @ tone keyer
— v manual - to design @ manual
— v máquina f - [ind] to design @ machine

proyectar v material m - to design @ material
— v — m para relleno m - [constr] to design @ backfill
— v molino m - [ind] to design @ mill
— v mordaza f - [mec] to design @ clamp
— v motor m - [comb.int] to design @ engine • [electr-mot] to design @ motor
— v muro m - [constr] to design @ wall
— v — m para cabecera f - [constr] to design @ headwall
— v — m tipo cajón m - [constr] to design @ bin wall
— v neumático m - [autom-neumát] to design @ tire
— v nivel m - to design @ level
— v obra f - to design @ project
— v originalmente - to design originally
— v palanca f - [mec] to design @, handle, o lever
— v para compacidad f - to design compactly
— v para inundación f mayor previsible en . . . año(s) m - [hidr] to design for @ . . . year storm
— v — operar v - to design to operate
— v — portar v - to design to carry
— v — proveer v - to design to provide
— v — resistencia f - to design for @ strength
— v — resistir aplastamiento m - [constr] to design for @ buckling
— v placa f - [mec] to design @ plate
— v — f para base f - [constr] to design @ base plate
— v planta f - [ind] to design @, mill, o plant
— v proceso m - to design @, system, o process
— v producto m - to design @ product
— v prontamente - to design readily
— v red f - to design @ system
— v — f cloacal - [sanit] to design @ sewer system
— v relleno m - [constr] to, design, o engineer, @ backfill
— v — m de tierra f - [constr] to design @ earth backfill
— v renglón m - [ind] to design @ line
— v corte m (transversal) para entrada f - [constr] to design @ inlet (size)
— v — m (—) para salida f - [constr] to design @ outlet (size)
— v según pedido m - to custom design
— v sencillamente - [ind] to design simply
— v sistema m - to design @ system
— v — m móvil - [electrón] to design @ mobile system
— v — m para fundición f - [metla-prod] to design @ melting system
— v — m — — f eléctrica - [metal-prod] to design @ electric melting system
— v — m transpondedor* m - [electrón] to design @ transponder system
— v — m — móvil - [electrón] to design @ mobile transponder system
— v sobre pantalla f - [fotogr] to screen project
— v sobre telón m - [fotogr] to screen project
— v soldadora f - [sold] to design @ welder
— v soporte m estructural - [constr] to design @ structural support
— v subestación f - [electr-distrib] to design @ substation
— v Super-Span m - [constr] to design @ Super-Span
— v teniendo en cuenta seguridad f - to design with @ safety in mind
— v transición f - to design @ transition
— v troquel m - [mec] to design @ die
— v tubería f - [constr] to design @ barrel
— v — f de acero m - [tub] to design @ steel pipe
— v válvula f - [válv] to design @ valve
— v vertedero m - [constr] to design @ chute
— v verticalmente - to project vertically
proyectil m con reacción f - [milit] missile
proyectista m - . . . .; draftsman; planner; layout man

proyectista m jefe - [constr] chief draftsman
— m novel - [constr] neophyte designer
— m para puente(s) m—[constr] bridge designer
— m — tubería f - [constr] pipe designer
proyecto m - . . . .; proposal; planning • engineering • expectation • need • development • rendering • layout
— m ambicioso - ambituous project
— m cambiado - changed design
— m caminero - [vial] highway project
— m carretero - [vial] highway project
— m cívico - [constr] civil project
— m comercial - [constr] commercial, project, o development
— m completo - complete, o total, project
— m de carácter m nacional - national scope project
— m definitivo - final project
— m demorado - delayed project
— m efectuado - implemented project
— m especial - [ind] special project
— m final - final project • final design
— m hidroeléctrico - [constr] hydroelectric project
— m implementado - implemented project
— m importante - important project
— m industrial - [constr] industrial project
— m integral - integral, o complete, project
— m minero - [miner] mining project
— m —-metalúrgico - mining-metallurgical project
— m modificado - modified, o changed, project
— m para amplicación f - expansion project
— m — calle f - [vial] street planning
— m — estatuto(s) m - [legal] by-laws, o regulation(s) (code), draft
— m — explotación f - [miner] development project
— m — canalización f—[hidr] waterway project
— m — caracterización f - [com] identificacion project
— m — construcción f - [constr] building project • building plans
— m — contener erosión f - [constr] erosion control project
— m — contrato m - [legal] contract draft; proposed contract
— m — decreto m - [pol] decree, o executive order, draft
— m — entubamiento m—[hidr] enclosure project
— m — — m para curso m de agua m - [hidr] stream enclosure project
— m — identificación f—identification project
— m — inversión f - [fin] investment project
— m — modificación f - [legal] amendment project
— m — — f para carta f de crédito - [fin] credit letter, amendment, o change, proposal
— m — obra(s) m - project plan(ning)
— m — reemplazo m para puente m - [constr] bridge replacement project
— m — resultado(s) m - [fin] profit and loss forecast
— m — sitio m - site, project, o plan
— m — — m para recreo m - resort project
— m — — m — — m, de montaña f, o serrano - [constr] mountain resort project
— m — venta(s) f - [com] sales forecast
— m residencial - [constr] residential development
— m según norma - [legal] standard draft
— m siderúrgico - [metal-prod] steel plant, o steelmaking, project
— m total - total project
— m vial - [vial] highway project
— m — grande - [vial] major, highway project, o road scheme
proyector m cinematográfico - [fotngr] motion picture, o movie, projector
prudencia f - . . . • judgement; sense
prudencial a - . . .; reasonable
prudente a - . . .; smart • [pol] conservative
prudentemente adv - . . .; smartly; prudently; sagely

prueba f - . . . . • attempt; experiment; assay •
probe,bing • assurance • [deport] véase tam-
bien ensayo m - [deport] session | [adv]
(para) test(ing)
— f a distancia f - [comput] remote test(ing)
— f — efectuar(se) v - test to be conducted
— f — realizar(se) v - test to be conducted
— f — rendir(se) v - testing procedure
— f — tensión f - [metal- tensile test
— f ácida - [quím] acid test
— f — en caliente - [quím] acid heat test
— f agotadora - grueling, test, o contest
— f atestiguada - witnessed, o certified, test
— f — en fábrica f - [ind] certified plant
test
— f bajo presión f - pressurization test
— f — techo m - indoor test(ing)
— f balística - [metal] ballistic test(ing)
— f certificada - certified test(ing)
— f con ácido m - [quím] acid test(ing)
— d — ataque m con ácido m - etch(ing) test
— f — carbonómetro m - [instrum] carbometer
test(ing)
— f — extensómetro m - [instrum] extensometer
test(ing)
— f — impacto(s) m - impact(s) test(ing)
— f — — m térmico(s) - thermal impact(s) test
— f — medidor m para carbono m - [instrum]
carbometer test(ing)
— f — sometimiento m a presión f - pressurization test(ing)
— f — voltaje m alto - [electr-oper] high
voltage test(ing)
— f corta - short test(ing)
— f de absorción f - absorption test(ing)
— f — aceite m - oil test(ing) | adv - (a) oil
tight
— f — aceptación f - acceptance test(ing)
— f — ácido m - acid test(ing) | adv - (a)
acid proof
— f — adherencia f - [metal-trat] adherence
test(ing)
— f — agua m - water test(ing) | adv - (a)
waterproof; hermetic
— f — aire m - air test(ing) | adv - (a) air
proof
— f — asentamiento m—[constr] slump(ing) test
— f — burla(s) - scoffing test | adv - (a)
scoff-proof
— f — calidad f - quality test(ing)
— f — campo m - [electr-oper] field probe,bing
— f — capacidad f - capacity test(ing[|
adv - (a) capacity proof
— f — carga f - [ind] load test(ing)
— f — — f muerta - dead load test(ing)
— f — — f viva - live load test(ing)
— f — cohesión f - [coque] shatter test(ing)
— f — combustión f - [combust] fire,ring test
— f — comportamiento m - performance test(ing)
— f — — m para caldera(s) f - [cald] boiler
performance test(ing)
— f — conducción f - [autom] drive,ving test-
-(ing); test drive,ving
— f — conformidad f - [legal] agreement proof
— f — congelación f - cold test; freezing test
— f — corrosión f - corrosion test(ing)
— f — deterioro,ración - deterioration test |
adv - (a) non-deteriorating
— f — doblado m - véase prueba f de plegado m
— f — doctor m - [petról] doctor('s) test
— f — durabilidad f - durability test(ing)
— f — — f estructural - structural durability
test(ing)
— f — electrodo m - [sold] electrode test(ing)
— f — eliminación f - elimination test(ing) •
[autom] qualifying test(ing)
— f — emulsificación* f - [petról] emulsifica-
tion test(ing)
— f — emulsionamiento* m - [petról] emulsifi-
cation test(ing)
— f — ensayo n - proof test(ing)
— f — — m para recipiente(s) m - [cald] vessel proof test(ing)

prueba f de equipo m - [ind] equipment test(ing)
— f — — m eléctrico - [ind] electric(al)
equuipment test(ing)
— f — estanqueidad f - tightness test(ing)
— f — evaporación f - evaporation test(ing)
— f — explosión(es) - [explos] explosion
test(ing) | adv - (a)explosion proof
— f — extracción f con sifón m - siphon ex-
traction proof | adv - (a) siphon-proof
— f — fatiga - [metal] fatigue test(ing)
— f — filtración f - filtration test(ing) |
adv - (a) watertight; airtight
— f — — f — aire - air filtration test |
adv - (a) airtight
— f — — f — gas m - gas filtration proof |
adv - (a) gas tight
— f — flexión f - [mec] bending test | adv -
véase también prueba f de plegado m
— f — flotación - floating, o flotation, test
— f — fluidez f - fluidity test(ing)
— f — — f de compuesto m - compound fluidity test(ing)
— f — forja f - [metal] forge,ging test
— f — forjabilidad f - [metal] foregeability,
o forging, test
— f — fractura f - break(ing) test
— f — — f por plegado m - [sold] bending
breaking test
— f — — f — — m con entalla f - [sold]
nick break(ing) test
— f — — f — — — f de cordón m (de soldadura f) - [sold] nick breaking test
— f — fuego m - [combust] fire,ring test •
[fam] acid test(ing)
— f — fuerza(s) f - stress(es) test
— f — — f para chaqueta f - telecom-cond]
jacket stress test
— f — fuga(s) f - leak(age) test(ing) | adv -
(a) leakproof
— f — funcionamiento m - [ind] operating, o
running, test
— f — garantía f - guaranty test(ing)
— f — — en rendimiento m - [ind] performance
guaranty test(ing)
— f — goteo m - drip test | adv - (a) drip
proof; non leaking • [petról] leak proof
— f — hermeticidad f - airtightness test(ing)
— f — holgura f - [mec] play test
— f — humedad f - humidity test(ing) | adv -
(a) humidity, o moisture, o damp, proof •
dampproof(ing[waterproof(ing)
— f — — f y agua m - moisture and water test
| adv - (a) dampproof and waterproof
— f — inclemencia(s) f de mal tiempo - weather
proof
— f — infiltración(es) - infiltration test |
adv - (a) infiltration, free, o proof
— f — intemperie f - weather proof | adv -
(a) weatherproof
— f — laboratorio m - laboratory test
— f — — m de aseguradores m contra incendios
m - [segurid] Underwriter Laboratory('s) Test
— f — — m en escala f piloto - pilot scale
laboratory test
— f — luz f - light test | adv - light proof
— f — mal tiempo m - bad weather test | adv -
(s) weatherproof
— f — materia(s) f prima(s) - [ind] raw ma-
terial(s) test(ing)
— f — material(es) m - material(s) test(ing)
— f — mineral(es) m - [miner] ore test(ing)
— f — montaje m - erection test(ing)
— f — muestra(s) f - sample test(ing)
— f — niebla f - cloud, o fog, test(ing)
— f — opacidad f - cloud test
— f — ovalización f - ovalization test(ing)
— f — penetración f - penetration test(ing)
— f — — f de humedad f - moisture penetration
test(ing)
— f — pérdida(s) f - leak test(ing) | adv - (a)
leakproof
— f — plegado m - [sold] bend)ing) test
— f — — m con cara f de raíz f hacia afuera -
[sold] root bend test

[sold] root bend test

prueba f de plegado m con cara f de soldadura f hacia afuera - [sold] face bend test
— f —— m con extremo(s) m apoyado(s)—[sold] free bend test
— f —— m — mandril m - [sold] guided bend test
— f —— m con plantilla f - [metal-fabr] (guided) root bend test(ing)
— f —— m — raíz f sometida a tracción f - [sold] root bend test
— f —— m — soldadura f sometida a tracción f - [sold] face bend test
— f —— — m de costado m - [sold] side bend test
— f —— m guiado - [sold] guided bend test
— f —— m lateral - [sold] side bend test
— f —— m libre - [sold] free bend test
— f —— m sobre conformador m - [sold] guided bend test
— f —— m — mandril m - [sold] guided bend test
— f — pliegue m - véase prueba f de plegado m
— f — polvo - dust test | adv - (a) dust proof
— f — precarga f - [mec] preload(ing) test
— f —— f para armado m - [mec] build-up preload(ing) test
— f —— — m para ensayo m - [mec] trial build-up preload(ing) test
— f — prehomogeneización f - prehomogenizing test
— f — presurización* f - véase prueba f con sometimiento m a presión f
— f — putrefacción f - rotting test | adv - (a) rot(ting) proof
— f — (ir)radiación f [electrón] radiation test | adv - [electrón] (a) radiation shielded
— f — recepción f - acceptance test
— f — recipiente m - [cald] vessel test(ing)
— f — reducibilidad f - [miner] reducibility test
— f —, remesa f, o remisión f - [fin] remittance proof
— f —— f original - [fin] original remittance proof
— f — rendimiento m - performance, o yield, test, o proof
— f —— m para apilador m - [ind] piler performance test
— f —— m —— m para arcilla f - [cement] clay piler performance test
— f —— m — rascador m - [cement] scraper performance test
— f —— — m para arcilla f - [cement] clay scraper performance test
— f — resistencia f - [deport] endurance test
— f —— f a choque(s) m - [mec] drop test
— f —— f — humedad f - moisture resistance test(ing)
— f — roedor(es) adv - (a) rodent proof; vermin resistant
— f — rueda f - [mec] wheel test(ing)
— f —— f para riel(es) m - [mec-ruedas] track wheel test(ing)
— f — sacudida(s) f - [mec] shaking test | adv - (a) shakeproof
— f — soldadura f - [sold] weld, testing, o trying
— f — tensión f - [metal] tensile test
— f — tracción f - [mec] traction, o pulling, test
— f — transformador m - [electr-equip] transformer test(ing)
— f — turbiedad f - clouding test
— f — turbieza f - clouding test
— f dieléctrica - dielectric test
— f dura - hard test • [fam] acid test(ing)
— f efectuada - performed test
— f en caliente - [quím] hot, o heat, test(ing)
— f — carretera f - [transp] road test(ing)
— f — escala f - scale test(ing)
— f —— f de laboratorio m - laboratory scale

test(ing)

prueba f en escala f de planta f piloto - [ind] pilot plant scale test(ing)
— f —— f piloto m - pilot scale test(ing)
— f —— fábrica f - [ind] factory, o manufacturing plant, test
— f —— f de choque m termico - [ind] manufacturing plant thermal shock test
— f — medio ambiente m - environment(al) test
— f — obra f - field test(ing)
— f — planta f - [ind] plant test(ing)
— f — taller m - [ind] shop test(ing)
— f — vacío m - vacuum test(ing) • [ind] cold run
— f estándar - standard test(ing)
— f final - final test(ing)
— f fiscalizada - controlled test(ing)
— f hidráulica - [hidr] hydraulic, o water, test(ing)
— f hidrostática - [hidr] hydrostatic test(ing)
— f — con agua m - [tub] water hydrostatic test(ing)
— f internacional - international test(ing)
— f — en Baja California f - [deport] Baja International (test, o race)
— f larga - long test(ing)
— f mayor - [deport] major test • great(er)test
— f no destructiva - nondestructive test(ing)
— f para aceptación f [ind] acceptance test(ing
— f —— f en obra - field acceptance test
— f —— aplastamiento m - [mec] collapse,sing test
— f — bola(s) f - [cuerp.moled] ball(s) test(ing)
— f — cable m - [cabl] cable test(ing)
— f —— m telefónico - [telecom-cond] telephone cable test(ing)
— f — calificación f - [deport] qualifying test(ing)
— f — comportamiento m - performance test(ing)
— f —— m de soplador m - [combust] blower performance test(ing)
— f —— m —— m para hollín m - [combust] soot blower performance test(ing)
— f —— m para ventilador m - [mec] fan performance test(ing)
— f —— m —— m centrífugo - [mec] centrifugal fan performance test(ing)
— f — comprobación f - checking, o verification, trial, o test(ing)
— f — conducción f - [autom] driving test
— f — conductor - [autom] driver('s) test
— f — control m - control(ling) test(ing)
— f —— m de calidad f - [ind] quality control test(ing)
— f — inspección f - inspection test(ing)
— f — propiedad(es) f - property,ties test
— f —— f de material(es) m - material(s) poroperty,ties test(ing)
— f positiva - positive proof
— f posterior - later test • later proof • [deport] later session
— f preceptiva - mandatory test
— f — para control m de calidad f - [ind] mandatory quality control test(ing)
— f Preece - [cabl] Preece test
— f presenciada - witnessed, test, o trial
— f realizada - performed test(ing)
— f reconocida - recognized test(ing)
— f remota - [comput] remote test
— f respectiva - respective test(ing)
— f rigurosa - rigorous, o tough, test(ing)
— f — para conducción f - [autom] tough driving test
— f —— conductor(es) m - [autom] tough driving test
— f simulada - simulated test(ing)
— f satisfactoria - satisfactory test(ing)
— f según norma f - standard test(ing)
— f severa - severe test(ing) • [fam] acid test
— f solicitada - requested test(ing)
— f subjetiva - subjective test(ing)
— f supletoria - substitutive, test, o proof

prueba f suprema

prueba f suprema - supreme test(ing)
pruebas f considerables - considerable trials
pruebatubos* m - [petról] casing tester • pipe
tester
psicólogo m industrial - [medic] industrial psy-
chologist
psicópata m - [medic] . . .; psychopath
púa f - . . . . • [alambre[barb
— f chata - [alambre] flat barb
— f de cuchilla f - [mec] knife guard
— f redonda - [alambre] round barb
— f reemplazable - [mec] replaceable, barb, o
point
— f semiredonda - [alambre] half round barb
— f entralazada - [alambre] interlocked barb
publicación f - . . .; publishing • release •
advertisement,sing • [public] literature •
[legal] publicity - pl media
— f abarcante - [public] comprehensive publication
— f automovilista - [public] automotive, o motoring, publication
— f de fabricante m - [ind] manufacturer('s),
publication(s), o literature
— f — — m de camión(es) m - [autom] truck
manufacturer('s) literature
— f — información f - information, publication,lishing
— f — valor m - value publication,lishing
— f (más) destacada - [public] leading publication
— f pertinente - [public] pertinent literature
— f prestigiosa - [public] prestigious publication
— f primera - [public] first publication
— f según norma f - [public] standard publication,lishing
— f — pieza(s) f - [mec] part(s) publication
— f técnica - [ind] technical publication
— f última - last, o most recent, publication
publicaciones f varias - [public] miscellaneous
literature
publicar v - . . . . • to issue • to expose
— v información f - to publish @ information
— v norma f - to publish @ standard
— v valor m - to publish @ value
publicidad f - . . . . • advertising • exposure
publicitario,ria a - advertising
público m automovilista - [transp] motoring public
— m comprador - [com] buying, o consuming, public; consumers
— m (en) general - general public
— m motorista - [transp] motoring public
— m protegido - [segurid] protected public
— m usuario - [com] consuming public; consumers
• [transp] traveling public
público,ca a - . . . . • [legal] publicly, held, o
owned
pudelado m - [metal-prod] véase pudelación f
pudelado,da a - puddled
pudiente a - . . .; wealthy; upper class
pudinga f - [miner] conglomerate
puede omitir(se) v - can, o may be, omitted
— ser v - can, o may, be
puente m - . . . .; bridging structure • [constr]
(roof) tie • [electr-instal] jumper (connec-
tion); jumper cable | adv - bridging
— m actual - [constr] existing bridge
— m aéreo • [transp] . . . . • [constr] aerial
bridge
— m — para acceso m - [constr] aerial access
bridge
— m alto - [constr] high bridge
— m angosto - [constr] narrow bridge
— m antiguo - [constr] old, o ancient, bridge
— m Armstrong - [vial] Armstrong bridge
— m atractivo - [constr] attractive bridge
— m bajo - [constr] low bridge
— m basculante m - [constr] drawbridge; bascule
bridge

puente m bóveda - [constr] arch bridge
— f — con plancha(s) f múltiple(s) - [constr]
Multi-Plate arch bridge
— m cansado - [constr] tired bridge
— m carretero m - [constr] highway bridge
— m — atractivo - [Constr] attractive highway
bridge
— m — económico - [constr] economical, o low
cost, highway bridge
— m — reemplazado - [constr] replaced highway
bridge
— m colgado - [constr] véase puente m colgante
— m comprobado - [constr] tested bridge •
[electr.instal] checked, jumper, o bridge
— m común - [constr] common, o conventional,
bridge
— m — de hormigón m - [constr] common, o conventional, concrete bridge
— m con alma m llena — [constr] solid web bridge
— m — arco m - [constr] arch bridge
— m — armadura f - [constr] truss, o open web,
bridge • truss construction
— m — — f con tablero m inferior - [constr]
through truss bridge
— m — — f de acero m - [constr] steel truss
bridge
— m — — f — — m con tablero m inferior -
[constr] steel through truss bridge
— m — — con cuerda f convexa - [constr]
curved chord truss bridge
— m — — f inferior convexa - [constr]
curved chord truss bridge
— m — — f sin arriostramiento m superior -
[constr] pony truss bridge
— m — bóveda f - [constr] arch(ed) bridge
— m — — f metálica — [constr] Armstrong bridge
— m — . . . bóveda(s) f - [constr] . . .-arch
bridge
— m — caballete(s) m - [constr] trestle (type
bridge)
— m con celosía f - [constr] truss, o open web,
bridge
— m — disyuntor m - [electr-instal] contactor
jumper; breaker
— f — interruptor m automático - [electr-
instal] contact, jumper, o breaker
— m — nivel m elevado - [constr] high level
bridge
— m — onda(s) f - [constr] wave bridge
— m — — f plena - [electrón] full wave
bridge
— m — peaje m - [vial] toll bridge
— m — porte m grande - [constr] large span, o
Super-Span, bridge
— m — tablero m - [constr] half through bridge
— m — — m inferior - [constr] floor, o
through, bridge
— m — — m superior - [constr] deck bridge
— m — tramo(s) m - [constr] span bridge
— m — — m con viga f maestra - [constr] girder span bridge
— m — — — f — de hormigón m - [constr]
concrete girder span bridge
— m — — m corto(s) - [constr] short span
bridge
— m — — m único - [constr] single span bridge
— m — un sólo tramo m - [constr] single span
bridge
— m — — tramo m - [constr] single span bridge
— m — una sóla vía f - [constr] one, lane, o
way, bridge
— m — vía f - [constr] one lane bridge; one
way bridge
— m — . . . vía(s) f - [constr] . . .-lane
bridge
— m — vía f superior - [constr] deck bridge
— m — f — de hormigón m - [constr] concrete deck bridge
— m — viga(s) f compuesta(s) - [constr] girder bridge
— m — — f de acero m - [constr] steel, girder, o beam, bridge
— m — — f con alma m llena - [constr] solid

web bridge
puente m **con viga(s)** f **de hormigón** m **pretensado**
  - [constr] prestressed concrete girder bridge
— m —— f **en caja** f **de hormigón** m - [constr]
concrete box girder bridge
— m —— f —— f —— m **pretensado** -
[constr] prestressed concrete box girder
bridge
— n — **viga** f **maestra** - [constr] girder bridge
— m —— f — **de hormigón** m - [constr] concrete girder bridge
— m **conservado** - [constr] maintained bridge
— m **construido** - [constr] constructed, o built,
bridge
— m **convencional** - [constr] conventional bridge
— m — **de hormigón** m - [constr] conventional
concrete bridge
— m **corriente** - [constr] conventional bridge
— m — **de hormigón** m - [constr] conventional
concrete bridge
— m **cortado** - [electrón] cut jumper
— m **cruzado** - [constr] crossed bridge
— m **cubierto** - [constr] covered bridge
— m **de acero** m - [constr] steel bridge
— m —— m **con una (sóla) vía** f - [constr] one
lane (only) steel bridge
— m —— m **estructural** - [constr] structural
steel bridge
— m —— f **galvanizado** - [constr] galvanized
steel bridge
— m — **alcantarilla** f - [constr] culvert bridge
— m — **cobre** m - [electr-instal] copper jumper
— m — **condado** m - [constr] county bridge
— m — **fratás** m - [herram] float bridge
— m — **Graetz** - [sold] Graetz bridge
— m — **hormigón** m - [constr] concrete bridge
— m —— m **armado** - [constr] reinforced concrete bridge
— m —— m — **con tramo** m **único** - [constr]
single span reinforced concrete bridge
— m —— m —— **tramos** m **múltiples** - [constr]
multi-span reinforced concrete bridge
— m —— m —— **un (sólo) tramo** m - [constr]
single span reinforced concrete bridge
— m —— m **deteriorado** - [constr] deteriorated
concrete bridge
— m —— m **pretensado** - [constr] prestressed
concrete bridge
— f — **madera** f - [constr] wood(en), o timber,
bridge
— m — **ojal** - [mec] eyelet jumper
— m — **plana** f - [herram] float bridge
— m — **plancha(s)** f - [constr] plate bridge
— m —— f **estructural(es)** - [constr] structural plate(s) bridge
— m — **suspensión** f - [constr] suspension bridge
— m —— m **con caballete(s)** m - [constr]
trestle-type bridge
    — m —— m **convencional** - [constr] conven-
tional type bridge
— m — **tubería** f - [constr] culvert bridge
— m —— f **gemela** - [constr] twin culvert
bridge
— m **débil** - [constr] weak bridge
— m **decrépito** - [constr] decrepit bridge
— m **defectuoso** - [constr] defective bridge •
substandard bridge
— m **departamental** m - [constr] county bridge
— m **desmontable** - [constr] dismountable bridge;
Bailey bridge
— m — **tipo** m **Bailey** - [constr] dismountable,
o Bailey, (type) bridge
— m **deteriorado** - [constr] deteriorated bridge
— m — **de hormigón** m - [constr] deteriorated
concrete bridge
— m **doble** - [constr] double bridge
— m **ducha** - [metal-prod] spray(er) bank
— m **económico** - [constr] economical, o low
cost, bridge
— m — **para reemplazo** m - [constr] economical,
o low cost, bridge replacement
— m **elevado** - [constr] high (level), o aerial,
bridge

puente m **elevado de acero** m **corrugado** - [constr]
corrugated steel aerial bridge
— m **elevado de plancha(s)** f - [constr] plates
aerial bridge
— m —— f **estructural(es)** - [constr]
structural plate(s) aerial bridge
— m — **para acceso** m - [constr] aerial access
bridge
— m **en voladizo** - [constr] cantilever bridge
— m **estabilizado** - [constr] stabilized bridge
— m **estable** - [constr] stable bridge
— m **estándar** - [constr] standard, o conventional, bridge
— m — **de hormigón** m - [constr] standard, o
conventional, concrete bridge
— m **estrecho** - [constr] narrow bridge
— m **estructural** - [constr] structural bridge
— m **existente** - [constr] existing bridge
— m **ferroviario** - [f.c.] railaroad, o railway,
bridge
— m — **con cubierta** f - [f.c.] deck, railway, o
railroad, bridge
— m —— f **sólida** - [f.c.] closed deck,
railroad, o railway, bridge
— m —— **piso** m - [f.c] deck railway bridge
— m —— m **sólido** - [f.c.] closed deck,
railroad, o railway, bridge
— m **soldado** - [sold] welded railway bridge
— m **fuerte** - [constr] strong bridge
— m **gemelo** - [constr] twin bridge
— m **grande** - [constr] large bridge
— m **grúa** - [constr] bridge crane • overhead
crane; véase también **grúa** f **puente**
— m — **con mando** m, **a distancia, o remoto** -
[grúas] remote(ly) controlled bridge crane
— m —— **telemando** m - [grúas] remote(ly)
controlled bridge crane
— m — **para horno** m - [ind] furnace bridge
crane
— m —— m **con fosa** f - [metal-prod] soaking pit (cover) crane
— m — **móvil** - [grúas] overhead crane
— m —— **cucharón** f **(para colada** f**)** - [metal-
-prod] (teeming, o pouring), ladle crane
— m —— **mineral(es)** m - [metal-prod] ore
bridge (crane)
— m — **portacaldero(s)** m - [metal-prod] ladle
crane
— m — **portacuchara(s)** - [metal-prod] ladle
crane
— m **inseguro** - [constr] unsafe bridge
— m **instalado** - [electrón] installed bridge
— m **íntegro** - [constr] entire, o whole, bridge
— m **levadizo** - [constr] drawbridge; bascule
bridge
— m **más seguro** - [constr] safer bridge
— m **Multi-Plate de planchas** f **múltiples** -
[constr] Multi-Plate bridge
— f **normalizado** - [grúas] standardized bridge
— m **nuevo** - [constr] new bridge
— m **para acceso** m - [constr] access bridge
— m **para camino** m - [constr] road bridge
— m —— m **para acarreo** m - [constr] haul(ing)
road bridge
— m — **chispa** f - [comb.int] véase **entrehierro** m
— m — **indicador(es)** m - [vial] sign bridge
— m — **montacarga(s)** m - [ind] skip bridge
— m — **ojal** m - [electrón] eyelet jumper
— m — **paso** m - [constr] footbridge
— m — **petaca** f - [metal-prod] cooling plate
jumper
— m —— f **sin refrigeración** f - [metal-prod]
cooling plate jumper without @ refrigeration
— m — **rectificación** f - [electrón] rectifier,fying bridge
— f —— f **con diodo** m **(con silicio** m**)** -
[electrón] (silicon) diode rectifier bridge
— m —— f —— m —— m **semiconductor** -
[electrón] semiconductor silicon dioe rectifier bridge
— m — f —— m —— m **semiconductor con
onda** f **plena** - [electrón] full wave semiconductor silicon diode rectifier bridge

puente m para rectificación f de silicio m -
[electrón] rectifier bridge
— m —— f semiconductor - [electrón] semiconductor rectifier bridge
— m — rectificador m - [electrón] rectifier
bridge
— m —— m comprobado - [electrón] checked
rectifier bride
— m — reemplazo m - [const] replacement bridge
— m —— m económico - [constr] economical, o
low cost, replacement bridge
— m — servicio m - [constr] temporary, o service, bridge
— m — skip m - [metal-prod] skip bridge
— m — tensor m - tension member bracket
— m — transportador m - [ind] conveyor bridge
— m — viaducto m - [constr] viaduct bridge
— m peligroso - [constr] dangerous, o unsafe,
bridge
— m pequeño - [constr] small bridge
— m permanente - [constr] permanent bridge
— m poco resistente - [constr] weak bridge
— m práctico - [constr] practical bridge
— m protegido - [constr] protected bridge
— m — con onda f plena - [electrón] full wave,
bridge rectifier, o rectifier bridge
— m reemplazado - [constr] replaced bridge •
bridge replacement
— m repavimentado - [constr] repaved, o resurfaced, o refloored, bridge
— m resistente - [constr] strong bridge
— m retráctil - [grúas] telescoping bridge
— m según norma f - [constr] standard, o conventional, bridge
— m —— f de hormigón m - [constr] standard,
o conventional, concrete bridge
— m seguro - [constr] safe bridge
— m soldado - [sold] welded bridge
— m Super-Span - [constr] Super-Span bridge
— m —— con porte m grande - [constr] Super-
-Span bridge
— m —— m — económico - - [constr] economical, o low cost, Super-Span bridge
— m —— económico - [constr] economical, o
low cost, Super-Span bridge
— m —— — para reemplazo m - [constr] economical, o low cost, Super-Span bridge replacement
— m sepultado - [constr] buried bridge
— m suspendido—[constr] suspended,nsion bridge
— m techado - [constr] covered bridge
— m transportador - [mec] conveyor bridge
— m vetusto - (very) old, o obsolete, bridge
— m vial - [constr] highway bridge
— m — soldado - [sold] welded highway bridge
— m viejo - [constr] old bridge
pueraria f thunbergiana - [botán] kudau
puerta f abierta - open door
— f abisagrada - [constr] hinged door
— f accionada - operated door
— f — con aire m - air operated door
— f arrollable - [constr] roll-up door
— f arrollada - [constr] rolled up door
— f automática - [constr] automatic door
— f — contra incendio(s) m - [constr] automatic fire door
— f bajada - [mec] lowered door
— f cerrada - closed door
— f — con . . . candado(s) m - [securid] 'door
closed with . . . padlock(s)
— f con panel(es) m liso(s) - [constr] flush
panelled door
— f — persiana(s) f - [constr] louvered door
— f — rejilla(s) f - [constr] grid door
— f — resorte(s) m - [constr] spring loaded
door
— f — tablero(s) m liso(s) - [constr] flush
panelled door
— f — vidrio(s) m - [constr] glazed door
— f — explosión(es) f [constr] explosion door;
blast gate
— f —— f accionada con resorte(s) m - spring
loaded explosion door

puerta f contra incendio(s) m - [constr] fire
(resisting) door
— f corrediza - [constr] slide,ding door
— f — con dos hojas f - [constr| double leaf
sliding door
— f ———— f operadas con motor m -
[constr] double leaf motor operated sliding
door
— f — doble - [constr] double sliding door
— f — operada con motor m - [constr] motor
operated sliding door
— f — vertical - [Constr] overhead (sliding)
door
— f de acero m - [constr] steel door
— f —— m contra incendio m - [constr] steel
fire door
— f — aluminio m - [constr] aluminum door
— f — metal m - [constr] metal door
— f —— m hueca - [constr] hollow metal door
— f — acceso m - [constr] access door
— f —— m a rodillo(s) m impulsor(es) - [ind]
drive roll(s) door
— f — batería f para coque m - [metal-coque]
coke battery door
— f para caja f - [mec] case, o housing, o enclosure, door
— f — carrete m - [mec] reel door
— f —— f para carrete m - [mec] reel, o
housing, o enclosure, door
— f —— f para carrete m para alambre m -
[sold] wire reel housing door
— f — cierre m - [mec] seal(ing) door
— f —— m para gas m - [metal-prod] gas seal
door
— f — circuito m - [electrón] circuit gate
— f — descarga - discharge,ging door
— f —— f accionada con aire m - air operated
discharge,ging door
— f — entrada f - [constr] entrance door
— f — escotilla f - [nab] hatch door
— f — extremo m - end door
— f — hogar m - [combust] hearth, o fire, door
— f — horno m - [combust] furnace door • oven
door
— f —— m Siemens-Martin - [metal-prod] open
hearth furnace door
— f — inspección f - [ind] inspection door
— f — limpieza f - cleaning door • [petról]
cleanout door
— f —— f tipo m liso - [petról] flush type
cleanout (door)
— f —— f derecha f - right hand (swing door
— f —— f izquierda - [mec] left hand (swing)
door
— f — panel m - [constr] panel door
— f —— m para selección f de voltaje m -
[sold] voltage selector panel door
— f — paso m para personal m - [ind] pedestrian door
— f — quemador m - [combust] burner door
— f —— m para estufa f - [metal-prod] stove
burner door
— f — recambio m - [coque] spare door
— f — respuesto m - [coque] spare door
— f — rodillo(s) m - [mec] roll(s) door
— f —— m impulsor(es) - [sold] drive roll(s)
door
— f — sección f de precipitador m - [metal-
-prod] precipitator(s) section door
— f — tablero m - [ind] panel door
— f —— m para selección f - [sold] selector
panel door
— f —— m — f de voltaje m - [sold]
voltage selector panel door
— f plegable - [constr] folding door
— f plegadiza - [constr] folding door
— f posterior - [mec] rear door
— f para caja f - [mec] case back door
— f refractaria—[constr] fire resisting door
— f superior - [mec] top door • [coque] chuck
door
— f trasera - [mec] back door
— f — para caja f - [mec] case back door
— f única - [Constr] single door

puertas f abiertas adv - (de) open house
— f adentro - indoors; inside
— f afuera - outdoors • outside
— f — en zona f polvorienta - outside in @ dusty area
puerto m artificial - [nav] artificial harbor
— m de destino m - [transp] port of destination
— m — escala f - [transp] port of call
— M — origen m - [transp] port of origin
— n estadounidense - [transp] U.S.A., o American, port
— m interior - [transp] inland port
— m mejicano - [transp] Mexican port
— m natural - [transp] natural harbor
— m para descarga - [transp] port of discharge
— m — embarque m - [transp] port of, embarkation, o loading, o shipping, o shipment
— m — exportación f - [transp] port of exportation • port of embarkation
— m — importation - [nav] port of importation
— m — materia(s) f prima(s) - [ind] raw material(s) dock
— m sobre Atlántico m - [transp] Atlantic port
— m — — m Norte - North Atlantic port
— m — — m Sur - [transp] South Atlantic port
— m — costa f de Atlántico - [transp] Atlantic seaboard port
— m — f de Golfo m (de México) - [transp] Gulf (of Mexico(seaboard port
— m — f de Pacífico - [transp] Pacific seaboard port
— m — Golfo m (de México) [transp] Gulf (of Mexico) port
— m — Pacífico m - [transp] Pacific port
pues . . .; well
puesta f - . . .; placing • [astron] . . .; setting (down) • providing • bid • [vest] donning
— f a descubierto - uncovering; exposing
— f — — m de pico m para engrase m - [mec] (grease) fitting exposing,sure
— f — — m — ranura f - [mec] spline exposing
— f — — m en abertura f - [mec] exposing,sure in @ opening
— f — día - updating
— f — disposición f - making available
— f — fiscalización f - [labor] control start-up
— f — punto - tuning (up); adjusting • tune-up • setting; adjustment; control setting • [comb.int] timing
— f — — m con precisión f - fine tuning
— a — — m para encendido m - [comb.int] ignition timing
— f — — m — forjadora f - [metal-fabr] forging machine timing
— f — — m — freno m - [mec] brake setting
— f — — m — línea f de cizalla(s) f - [metal-lam] shear line tune,ning up
— f — — m de máquina f forjadora - [mec] forging machine timing
— f — — m — matriz f - [mec] die setting
— f — — m — motor m - [electr-mot] motor tune-up • [comb.int] engine tune-up
— f — — m — neumático m - [autom-neumát] tire tuning
— f — — m — palanca f - [mec] lever setting
— f — — m — — f para freno m - [mec] brake lever setting
— F — — m de proceso m - [ind] process, adjusting, o tuning
— f — — m definitiva - [mec] final adjustment
— f — — m final - final tune up
— f — — m incorrecta - [comb.int] incorrect timing
— f — — m nueva - [instrum] retiming
— f — — m para encendido m - [comb.int] ignition retiming
— f — — m nuevamente - [comb.int] retiming
— f — — m — para arrancador m - [comb.int] starter, retiming, o resetting
— f — — m — — m para motor m - [electr-mot] motor starter, resetting, o retiming •

[comb.int] engine starter, resetting, o retiming
puesta f a régimen m - adjusting • setting
— f — resguardo m - storage,ring
— f — m de unidad f motriz - [agric-equip] power unit storage,ring
— f — tierra f - [electr-instal] ground(ing)
— f — — f accidental - [sold] accidental grounding
— f — — f apropiada - [electr-instal] proper, o suitable, o appropriate, ground(ing)
— f — — f dentro de soldadora f [sold] suitable ground(ing) inside @ welder
— f — — f — para soldadora f - [sold] suitable welder ground(ing)
— f — — f de armadura f—[mec] body grounding
— f — — f — banco m - [sold] bench grounding
— f — — f — bastidor m para soldadora f - [sold] welder frame grounding
— f — — f — blindaje m - [sold] shield grounding
— f — — f — caja f - [mec] body grounding
— f — — f — chasis m - [electrón] chassis grounding
— f — — f — lampar(ill)a f - [electr-instal] lamp grounding
— f — — — f para ensayo(s) m - [electr-instal] test lamp grounding
— f — — f — máquina f - [mec] machine frame grounding
— f — — f — mesa f - [sold] table grounding
— f — — — f rotativa - [petról] rotary table grounding
— f — — f — soldadora f - [sold] welder (frame) grounding
— f — — f — trabajo m—[sold] work grounding
— f — — f dentro de soldadora f - [sold] grounding inside @ welder
— f — — f para chasis - [Electrón] chassis ground(ing)
— f — — f pobre - [electr-instal] poor, grounding, o ground connection
— f — — trabajo m - [labor] going to work
— f — una fase f - [electr-instal] single phasing
— f accidental a tierra f - [sold] accidental grounding
— f aparte - setting apart
— f apropiada a tierra f - [sold] proper grounding (method)
— f de atención f - attention paying • [ind] maintenance providing
— f — — f normal - [ind] normal maintenance providing
— f — cuidado - care exercise,sing
— f — — m normal - normal caretaking
— f — manifiesto - show(ing); evidencing; bringing out
— f — marca f - marking • checking
— f — mira(s) f - sight(s) setting
— f — palanca f - [mec] lever setting
— f — — f en marcha f atrás - [mec] setting @ lever in @ reverse
— f — — f — posición f para marcha f atrás - [mec] setting @ lever in @ reverse (position)
— f — — f — — f — retroceso m - [mec] setting @ lever in @ reverse (position)
— f — — f en retroceso - [mec] setting @ lever in @ reverse
— f — regulador m - [mec] control setting
— f — — m para velocidad f - [mec] speed control lever setting
— f — sol m - sunset; sundown
— f debida - proper setting
— f — a tierra f - [electr-instal] proper grounding
— f en automático - put(ting) into automatic
— f — boca f - putting in(to) @ mouth
— f — comunicación f - contacting; putting into contact
— f — cortocircuito - [electr-oper] shorting; short circuiting
— f — — m consigo mismo,ma - [shorting on,

itself, o oneself
puesta f en cortocircuito m entre sí - [electr-
-oper] shorting together
— f — **ejecución** f - implementation
— f — f **de convenio m** - [labor] contract
implementation
— f — **existencia(s)** - [com] stocking
— f — **funcionamiento m** - [ind] start(ing)-up;
making operative
— f — **lista** f - [com] listing
— f — f **de producto m** - [com] product listing
— f — **marcha** f - start(ing) up • turning on
— f — f **accidental** - [mec] accidental
start(ing) up
— f — f **alentadora** - [ind] encouraging
start-up
— f — f **automática** - automatic start(-up)
— f — f **de carro m** - [sold] automatic
travel start(-up)
— f — f **con corriente f para soldadura f** -
[sold] hot start (ing)
— f — f **con telemando m** - [comb.int] remote
start(ing)
— f — f **con temperatura(s) f bajs(s)** -
[comb.int] cold weather start(ing)
— f — f **reducida(s)** - [comb.int]
cold weather start(ing)
— f — f **tiempo m frío** - [comb.int] cold
start(ing) up
— f — f — m **normal** - [comb.int] normal
weather start(ing) up
— f — f **de acondicionador m (para aire m)** -
[ambient] (air) conditioner, start(ing), o
start-up
— f — f **bomba** f - [bombas] pump, start-
-up, o start(ing)
— f — f **caja** f **de generador m** - [electr-
-prod] generator building start-up
— f — f **compresor m** - [ind] compressor
start(ing)-up
— f — f **equipo m** - [ind] equipment.
start(ing) up • project start(ing) up
— f — f — m **eléctrico** - [electr-oper]
electrical equipment start(ing)-up
— f — f — m **para laminador m para
palanquilla f** - [metal-lam] billet mill electrical equipment start(ing)-up
— f — f — m — m **tren m para palanquilla f** - [metal-lam] billet mill electrical
equipment start(ing)-up
— f — f **galvanización f** - [metal-trat]
galvanization,zing start(ing)-up
— f — f **grúa f** - [grúas] crane start-up
— f — f **horno m** - [metal-prod] furnace,
start(ing) up, o blowing in
— f — f **instalación f** - [ind] project, o
installation, o facility, start(ing)-up
— f — f **en época(s) f fría(s)** - [comb.int]
cold weather start(ing) up
— f — f **fácil** - [comb.int] easy start(ing)
up
— f — f **inicial** - [ind] initial start(ing)
up
— f — f **normal** - [comb.int] normal starting
— f — f **nueva** - [comb.int] restart(ing)
— f — f **preparada** - [ind] prepared start-up
— f — f **para horno m** - [ind] prepared
furnace start-up
— f — f **separada** - separate, start=up, o
turning on
— f — f **separadamente** - separate, turning
on, o start(ing)-up
— f — f **sin problema(s)** - [ind] smooth
start(ing) up
— f — f **supervisada** - supervised startup
— f — f **de bomba f** - [bombas] pump start-
-up
— f — f — f **supervisada** - [bombas] supervised pump start-up
— f — f **en emergencia f** - emergency start-
(ing) up
— f — f **para ensayo m** - [mec] running in,

o test(ing), start-up
puesta f en marcha f para ensayo m de equipo m -
[ind] equipment, running, o testing, start-up
— f — f **y detención f** = [ind] start(ing)
and stopping
— f — **operacion f** - [mec] putting into operation; start(ing) up; turn(ing) on
— f — **paralelo** - [ind] paralleling
— f — m **de conmutador(es) m**—[sold] starter
paralleling
— f — m — **soldadora f** - [sold] welder
paralleling
— f — m — f **motogeneradora** - [sold] motor-
-generator welder paralleling
— f — m — f **con motor m con combustión
f interna** - [sold] engine driven welder paralleling
— f — m —**soldadora/rectificadora f** -
[sold] welder rectifier paralleling
— f — m —/— **transformadora** - [sold]
transformer/rectifier welder paralleling
— f — m — **soldadora/transformadora f** -
[sold] transformer-welder paralleling
— f — m **defectuosa** - [sold] improper paralleling
— f — m **inapropiada** - [sold] improper paralleling
— f — m **incorrecta** - [sold] improper paralleling
— f — m **indebida** - [sold] improper paralleling
— f — **peligro m** - endangering; jeopardizing
— f — m **de calzada f** - [constr] roadway,
endangering, o jeopardizing
— f — m — **cimiento(s) m** - [constr] foundation endangering
— f — m — **estructura f** - [constr] structure endangering
— f — **posición f** - positioning; putting in(to)
position
— f — f **de conectado,a** - [electr-oper]
turning on
— f — f **desconectado,da** - [electr-oper]
turning off
— f — f **electrodo m** - [sold] electrode
positioning
— f — **posición f neutral** - [mec] putting, o
placing, in (a) neutral (position)
— f — f **para aparcamiento m** - [autom] putting in @ parking position
— f — f **para estacionamiento m** - [autom]
putting in @ park(ing) position
— f — **relación f** - placing in relation; contacting
— f — **servicio m** - [ind] putting, o placing,
in service • starting; start-up
— f — m **nuevamente** - [ind] restoration to
@ service
— f — **sitio m** - putting in @ place
— f — **vigor m** - [legal] enforcement,cing; putting in force
— f — **voladizo m** - cantilevering
— f — m **de indicador m** - [constr] sign
cantilevering
— f **fuera de carrera f** - [deport] putting out
of @ race
— f **junto,ta en corto circuito m**—[electr-oper]
shorting together
— f **tensa** - [mec] tensioning
— f **de cadena f** - [mec] chain tensioning
puesto n - . . . . • [ind] station • [labor] . . .
• job; post; office • [deport] slot
— m **aéreo m** - [ind] overhead position
— m — **para transformación f** - [electr-instal]
platform, o aerial, transformer bank
— m **contra palo(s) m** - [deport] pole position
— m **en mercado m** - [com] market, booth, o
stand | adv - [com] on @ market
— m **final** - final position • [sold] final station
— m **para abastecimiento m** - supply station
— m — **auxilio(s) m** - [segurid] aid station
— m — —(s) m **primero(s)** - first aid station

puesto m para bebida(s) f gaseosa(s) - [culin]
soft drink stand
— m — cambio(s) m - [f.c.] switch(ing) stand;
(reversing) switchstand • exchange teller
— m — — m para ariete m - [mec] ram changing
station
— m — — m — émbolo(s) m [mec] ram changing
station
— m — — m de tipo m bajo - [f.c.] low type
switch(ing) stand
— m — catadura f - [culin] tasting booth
— m — catar v - [culin] tasting booth
— m — comando m - [ind] pulpit
— m — conducción f - [autom] steering position
— m — control m - [deport] checkpoint; control
— m — — m de carrera f - [deport] race headquarters
— m — — m — — f rally - [deport] rally
(race) headquarters
— m — desenrollamiento m - [mec] unwinding
stand
— m — emergencia f - [ind] emergency station
— m — encalado m - [metal-prod] limewashing
station
— m — — m de cono(s) m -[metal-prod] thimble
lime washing station
— m — enclavamiento m - block(ing) system
station
— m — enfriamiento m - cooling station
— m — — m para cono(s) m - [metal-prod]
thimble cooling station
— m — escucha(r) - listening post
— m — espectador(es) m - [deport] spectator(s)
stage
— m — expendio m - [com] selling, booth, o
stand
— m — — m de bebida(s) f - drink(s) stand
— m — — m — — f gaseosa(s) - [culin] soft
drink(s) stand
— m — gaseosa(s) - [culin] soft drink(s) stand
— m — instalación f - [ind] set-up stand
— m — — f de rodillo(s) m - [metal-lam] roll
see-up, stand, o station
— m — mando m - [mec] control(s) • command
post • [sold] control station • [ind] oykout;
control balcony; govern house; bench; cab'
operator('s) cab(inet)
— m — — m con botonera f - [ind] push button,
cabinet, o station
— m — — m para laminador m - [metal-lam] mill
opoerator('s) pulpit
— m — manejo m - goern(ing) house
— m — marcación f - [ind] marking station
— m — — f para lata(s) f - [ind] can marking
station
— m — observación f - [ind] observation, bay,
o post
— m — operación f - [ind] operating,tion, station, o post
— m — operador m - [ind] operator('s), station, o pulpit, o cab(inet)
— m — peaje m - [vial] toll, booth, o plaza
— m para refección f - [culin] feed, o snack,
booth, o stand
— m — regulación f - [ind] control balcony •
bench • [sold] control station
— m — — f para motor m - [electr-mot] motor
control, stand, o center
— m — reparación f - repair stand
— m — — f para cubierta f - [ind] cover repair stand
— m — — f — — f para horno m - [metal-lam]
furnace cover repair stand
— m — — f — — f — — m de fosa f - [metal-
-lam] soaking pit cover repair stand
— m — — f para tapa f - [ind] cover repair
stand
— m — — f — — f para horno m - [metal-lam]
furnace cover repair stand
— m — — f — — f — — m de fosa f - [metal-
-lam] soaking pit cover repair stand
— m — separador m—[ind] separator('s) station
— m — servicio m - [com] service bay

punesto m para socorro m - [ind] emergency station
— m — soldador m [sold] weldor('s), o opera-
tor('s), station
— m — soldadura f - [sold] welding station
— m — soldadura f - [sold] welding station
— m — — f de oreja(s) f - [sold] tab welding
station
— m — trabajo m - [ind] work site • [labor]
job
— m — transformación f -[electr] transformer
bank
— m — — f en aire m libre - [electr-instal]
open air, o ground level, transformer bank
— m — transformador(es) m - [electr-instal]
(transformer) bank
— m — — m aéreo - [electr-instal] platform
transformer bank
— m — venta(s) f - sales, o selling, station
— m — — f de neumático(s) m - [autom-neumát]
tire selling, station, o stand
— m — vigilancia f - [ind] observation post •
pulpit
— m primero - [ind] first station • [deport]
first, o top, place
— m principal - [ind] main location
— m — para control m - [deport] headquarters
— m — — mando m - [electr-distrib] main controller
— que - [conj] inasmuch; because
— m segundo - [ind] second station • [deport]
second place
— m subterráneo m - (underground) chamber
— m — para transformación f - [electr-distr]
underground transformer station
— m — para transformador(es) m - [electr-
-distrib] (underground) transformer chamber
— m tercero - [ind] third station • [deport]
third place
puesto,ta a - put; placed; set (down) • provided • [astron] set; gone down
— a a descubierto - exposed • uncovered
— a — — m en abertura f - exposed in @ opening
— a — día a - updated
— a — disposición f - made available
— a — punto m [comb.int] timed; tuned
— a — — m con precisión f - [ind] fine tuned
— a — — nuevamente - [ind] reset; retimed
— a — resguardo - stored (away
— a — tierra f - [electr-instal] grounded
— a — — f eléctricamente - [sold] electrically grounded
— a aparte - set apart
— a de manifiesto - evidenced; brought out;
showed; shown
— a en automático - [ind] put on automatic
— a — boca f - put into @ mouth
— a — comunicación f - put in contact; contacted
— a — corto circuito m - shortcircuited;
shorted
— a — — — m consigo mismo,ma - [electr-oper]
shorted on itself
— a — — a — entre sí - [electr-oper] shorted
together
— a — existencia f - [com] stocked
— a — funcionamiento m - made operative
— a — marcha f - started; turned, o switched,
on
— a — — f en emergencia f - emergency started
— a — — f nuevamente - [comb.int] restarted
— a — — f separadamente - restarted, o turned
on, separately
— a — operación f—[ind] turned on; put into
operation
— a — paralelo - paralleled
— a — peligro m - endangered; jeopardized
— a — posición f - positioned, put, o placed,
in @ position
— a — — f de conectado,da - [electr-oper]
turned on
— a — — f — desconectado,da - [electr-oper]

turned off
puesto,ta en posición f neutral - [ind] put, o
  placed, in @ neutral (position)
— f —— f para aparcamiento m—[autom] put
  in @ park(ing) (position)
— a — f — estacionamiento - [autom] put in
  @ park(ing) position
— a — f para soldadura f plana - [sold] in
  flat weld(ing) position
— a — práctica - put in(to) practice
— a — relación f - contacted; put in contact
— a — servicio m - placed, o put, in service
— a — sitio m - put in place
— a — vigor m - enforced; put into force
— a — voladizo - cantilevered
— a fuera de carrera f - put out of @ race
— a junto en corto circuito m - [electr-oper]
  shorted together
— a nuevamente en servicio m - restored to service
— a punta arriba - upended
puja f - . . .; bid •[deport] game • bash
— f ardua - [deport] rough rally
pujante a - . . . • thriving; booming; growing •
  pulsating
pujanza f - . . . • boom
pulcritud f - . . . • spit and polish
pulcro,ra a—. . .; tidy • spit and polish
pulg/min t—- véase pulgada(s) f por minuto
pulgada f - [metric] inch
— f cúbica - [metric] cubic inch
pulgadas f a . . . potencia f por pulgada f -
  inches to @ . . . power per inch
— f por minuto m - inch(es) per minute
pulido,da a - . . .; burnished • smooth; bright
  • ground • planished
— a duro,ra - hard polished
— a electrolíticamente - [metal-trat] polished
  electrolytically
— a en ambos lados m - polished on both sides
— a — un (sólo) lado - polished on (only) one
  side
— a ligeramente - [mec] polished lightly
— a mecánicamente - [mec] polished mechanically
pulidor m - . . . • [metal-trat] brannerman
— m con afrecho m - [metal-trat] brannerman
— m — salvado m - [metal-trat] brannerman
pulidora f - polisher; buffer • branner • hone
— f con afrecho m - [metal-trat] branner
— f — salvado m - [metal-trat] branner
— f para banda f - [metal-trat] band polisher
— f — cinta(s) f - [metal-trat] sheet polisher
pulimentado,da a - polished; burnished • bright
pulimentador,ra - véase pulidor,ra
pulimento m - [mec] polish(ing); buffing; planishing • lapping • grinding
— m con banda f - [mec] band polishing
— m con anillo(s) m - [mec] ring polishing
— m — aro(s) m - [mec] ring polishing
— m de contorno m - [mec] contour polishing
— m — ranura f - [mec] groove polishing
— m — superficie f - surface polishing
— m — vástago m - [mec] rod polishing
— m electrolítico - [metal-trat] electrolytic
  polish(ing)
— m ligero - [mec] light polishing
— m mecánico - [mec] mechanical polishing
pulir v - . . . • to face • [metal-prod] to,
  mill, o grind, o planish • to lap
— v ligeramente - [mec] to polish lightly
— v anillo m - [mec] to polish @ ring
— v aro m - [mec] to polish @ ring
— v electrolíticamente - [metal-trat] to polish
  electrolytically
— v mecánicamente - [mec] to polish mechanically
— v ranura f - [mec] to polish @ groove
— v superficie f - to polish @ surface
— v vástago m - [mec] to polish @ rod
pulpa f - . . . • [miner] slurry
— f alcalina - [miner] alkaline pulp
— f de mineral - [miner] ore slurry
— f — petróleo m - [petról] oil pulp
— f lavada - [miner] washed pulp

pulpadora f - [miner] pulper
pulpería f - [com] . . . ; post exchange
púlpito m - . . . . • [ind] véase pupitre m;
  puesto m para mando m; console
— m existente - [ind] existing console
pulsación-eco f - [electrón] pulse-echo
— f en línea f - [electrón] line pulsation
— f —— f para aspiración f - [mec] suction
  line pulsation
— f — tubería f - [tub] line pulsation
pulsador m para avance m gradual - [sold]
  inch(ing) pushbutton
— m — corte m - [sold] stop, o cutoff, pushbutton
— f — espacio m - [electrón] space pulse
— f — marca f - [electrón] mark pulse
pulsador m - . . . .; pulsator • plunger • jog •
  [sanit-enfriador para agua] bubbler
— m con pedal m - [sanit-enfriador para agua]
  foot operated bubbler
— m — rototrol m - [electr-prod] rototrol
  jog
— m —— m para generador m - [electr-prod]
  generator rototrol jog
— m manual - [mec] jog • [sanit-enfriador para
  agua] hand (operated) bubbler
— m para arranque m - [sold] start(ing) pushbutton
— m —— m y corte - [sold] start and stop
  pushbutton
— m — arranque m - [electr-oper] start(ing)
  pushbutton
— m —— m y parada f - [electr-oper] start
  and stop pushbutton
— m — detención f - [electr-oper] stop button
— m — parada f - [electr-oper] stop (push)-
  -button
— m — puesta f en marcha f y detención ᶠ -
  [electr-oper] start and stop pushbutton(s)
pulsadora f para carbón m - [metal-prod] coal
  shaker
pulsar v - . . . • to beat • to press • to jog
pulsera f - . . . | a - (de) wrist
pulsómetro* m - [instrum] véase pulsímetro m
pulverizable a - . . . . • grindable
pulverización f - . . .; pulverizing; mulling •
  atomization,zing; powdering • mist • spray-
  -(ing) mist
— f con presión f alta - high pressure spraying
— f —— baja - low pressure spray(ing)
— f de agua m - water, spray, o mist
— f — azufre m - sulfur powdering
— f — emulsión f - [pint] emulsion spray(ing)
— f —— f para pintura f - [pint] paint emulsion spray(ing)
— f — hojalata f - [metal-prod] tin pest
— f — mineral(es) m - [miner] ore pulverizing
— f — pigmento(s) m - [pint] pigment(s) pulverizing,zation
— f —— m para pintura(s) f - [pint] paint
  pigment pulverizing,zation
— f — pintura f - [pint] paint, pulverizing, o
  spray(ing)
— f fina - fine mist
— f sin aire m - [pint] airless, pulverizing, o
  spraying
— f tenue - fine mist
— f — de agua m - fine water mist
pulverizado,da a - pulverized • powdered •
  sprayed
pulverizador m - [mec] . . .; jet; sprayer; nozzle; spray gun (head, o injector) • [comb.-
  -int] (carburetor), jet, o nozzle • muller
— m con aire m comprimido - [pint] air brush
— m economizador - [mec] economizer jet
— m para carbón m - [ind] coal pulverizer
— m —— enfriador m para rodillo(s) m - [metal-
  .-lam] roll cooling spray
— m — suelo(s) m - [agric] soil pulverizer
pulverizador,ra a - pulverizing
pulverizar v - . . . .; to mull
— v agua m - to pulverize @ water
— v azufre m - to powder @ sulfur

pulverizar v emulsión f - [pint] to spray @
emulsion
— v — f para pintura f - [pint] to spray @
paint emulsion
— v magnesio m - [quím] to powder @ magnesium
— v pintura f - [pint] to spray @ paint
pulverulento,ta a - pulverulent; véase también
polvoriento,ta
punta f - . . . • toe • prong • comb* • (clavos)
brad; nail • [ind] peak • [tornos] center;
chuck point; stock • [deport] lead • [f.c.]
point • [metal-lam] crop(ping) • [electrón-
-clavija] (jack) tip
— f achaflanada - [constr-pil] beveled tip
— f — para hincadura f - [constr-pil] beveled
driving tip
— f aguda - [mec] sharp point
— f ahusada - [mec] cup point • tapered tip
— f — con peso m reducido - [grúas] light
weight tapered tip
— f balística - [mec] ballistic point
— f — con precisión f - [mec] precision ballistic point
— f — precisa—[mec] precision ballistic point
— f coladora f - [dren] well point | a [constr]
(de) wellpointing
— f con cabeza f cónica - [clavos] brad; casing
nail
— f — — f perdida - [clavos] finishing nail
— f — — f redonda - [clavos] oval head nail
— f — . . . polea(s) f - [grúas] . . . sheave
point
— f — precisión f - [mec] precision point
— f cónica - [mec] cone point
— f cortada - [mec] clipped end
— f chata - [mec] dull (bill) point
— f de acero m - [mec] steel point
— f — — m prensado - [constr] pressed steel
pile point
— f — aguilón m - [grúas] boom, point, o top;
véase también extremo m de aguilón m
— f — — m auxiliar - [grúas] auxiliary boom
point
— f — — m con altura f reducida - [grúas]
reduced height boom point
— f — — m con . . . polea(s) f - [grúas] . .
. . . sheave boom point
— f — aguja f - [clavos] needle point
— f — alambre m con núcleo m - [sold] cored
wire tip
— f — barra f - [mec] bar, tip, o end
— f — bobina f - [metal-lam] coil end
— f — botalón m - [náut] véase extremo m de
botalón m
— f — brazo m - [anat] arm, tip, o end •
[grúas] boom point
— f — cable m - [cabl] (wire) rope, o cable,
end
— f — — m de alambre m - [cabl] wire rope end
— f — cabo m - [cabl] rope end
— f — cincel m - [clavos] chisel point
— f — clavija f - [electrón] jack tip
— f — clavo m - [clavos] nail point
— f — cuchilla f - [mec] blade tip
— f — cuerda f - [cabl] rope end
— f — dedo m - [anat] finger tip
— f — diente m - tooth, tip, o point
— f — eje m - [mec] axle tip • trunnion; journal • spindle • [autom-mec] wheel spindle
— f — electrodo m - [sold] electrode end •
electrode butt
— f — — m con carbono m - [sold] carbon tip
— f — — m para aportación f - [sold] filler
rod, end, o tip
— f — — m que apunta hacia arriba - [sold]
upwards pointing electrode tip
— f — lingote m - [metal-lam] ingot, end, o
tip • crop(ping)
— f — llave f - [mec] key, end, o tip •
[herram] wrench, end, o tip
— f — parís - [clavos] wire nail
— f — pértiga f - [mec] boom point
— f — pilote m - [constr-pil] pile, tip, o
end, o bottom

punta f de pilote m tubular - [constr-pil] pipe
pile, tip, o end, o bottom
— f — pluma f - [grúas] boom point
— f — riel m - [f.c.] rail end
— f — riel(es) m - [f.c.] railhead
— f — tornillo m - [mec] screw, point, o end
— f — torno m - [tornos] chuck, point, o center
— f — tubería f - [tub] pipe,ping end
— f — — f para agua m - [petról] wash pipe
end
— f — varilla f (para aportación f) - [sold]
(welding) rod, end, o tip
— f en producción f - [ind] production peak
— f mocha f - [mec] dull, o blunt, point
— f negativa - [electrón] negative peak
— f para ajuste m - [mec] (pin) tight side
— f — grúa f - [grúas] crane tip
— f — para servicio m pesado - [grúas]
heavy duty crane tip
— f — receptor m - [electrón-clavija] receiver
(jack) ring
— f — — m para estación f—[electrón-clavija]
field receive (jack) tip
— f — transmisión f - [electrón-clavija]
transmit (jack) tip
— f — transmisor m para estación f—[electrón-
-clavija] field transmitter (jack) tip
— f — trazar v - [herram] graver
— f positiva - [electrón] positive peak
— f precisa - [mec] precision point
— f reducida - reduced tip
— f — de aguilón m - [grúas] reduced boom tip
— f reversible - [mec] reversible point
— f saliente - [mec] spit; protruding point
— f suplementaria - [grúas] tip section
— f — tipo m con cabeza f de martillo m -
[herram] hammerhead tip
— f diagonal - [clavos] side point
— f diamante - [clavos] diamond point
— f esférica - [mec] oval point
— f excéntrica - [clavos] side point
— f fija - [tornos] dead center
— f — en torno m - [tornos] lathe dead center
— f hacia bobinadora f - [metal-lam] coiler end
— f — enrolladora f - [metal-lam] coiler end
— f hembra - [mec] female end
— f inferior - [mec] lower end
— f — de lanza f - [metal-prod-LD] lance lower
end
— f machimbre - [clavos] flooring brad
— f mala - bad end
— f máxima - [ind] maximum peak
— f movible - [tornos] live center
— f — de torno m - [tornos] lathe live center
— f para aguilón m - [grúas] single sheave
boom point
— f — alimentación f - [ind] feeding end
— f — apoyo m - [mec] trunnion; journal • fulcrum
— f — desabobinamiento m - [metal-lam] uncoiling end
— f — desenrollado m - [metal-lam] decoiling
end
— f — desenrollamiento m - [metal-lam] uncoiling end
— f — encendido m - [comb.int] firing tip
— f — entrada f - [ind] entry end
— f — hincadura f - [constr-pil] driving tip
— f — mandril m - [tornos] live center
— f — — m para torno m - [tornos] lathe live
center
— f — martillo m neumático - [constr] jackbit
— f — pilote m - [constr-pil] pile tip; ferrule
• véase también azuche m
— f — rebobinado m - [mec] recoiling end
— f — reenrollado m - [mec] recoiling end
— f — salida f - exit end
— f parís - [clavos] wire nail
— f postiza - [constr-pil] slip-on point
— f — de acero m - [constr-pil] slip-on steel
point
— f — — — m para pilote(s) m - [constr-pil]
slip-on steel pile point

punta f postiza para pilote(s) m - [constr-pil]
  slip-on pile point
— f roma - [clavos] blunt point
puntada f - [mec] stitch • [medic] (sharp) pain
puntaje* m - [deport] score; véase también
  tanteo m
puntal m - [constr] strut; brace; bracing; shore
  • [mec]; post; spit • truss
— m central - [mec] center post
— m de madera f - [constr] timber strut
— m delantero - [mec] front strut
— m frontal - [mec] front strut
— m para soporte m - [hidr] supporting strut
— m roto - [mec] broken strut
— m según norma f - [mec] standard strut
— m sólido m - [constr] solid strut
— m temporario - [constr] temporary. brace, o
  strut
— m trasero - [mec] rear strut
puntas f y cruces f - [f.c.] points and cross-
  ings
punteado m - . . . . • dotting
punteado,da a - dotted
punteador,ra a - dotting
puntear v - [contab] to check • [deport] to lead
— v trabajo m - [sold] to tack @ weld
punteo m - [contab] check(ing); dotting
puntera f - [vest] (toe) cap • [constr] blade
  tip
— f de cuchilla f - [herram] blade tip
— f — embudo m - véase pico m de cono m
puntero m - [instrum] hand; indicator • [deport]
  front runner; leader; first place (car);
  front running; pace setter
— m — reloj m - [instrum] clock hand
— m en industria f - [ind] industry leader
— m — número m de punto(s) m - [deport] points
  leader
— m — plazo(s) m - compliance leader
— m — producción f - [ind] production leader
— m — — f de neumático(s) m - [autom-neumát]
  tire production leader
— m — — m para rendimiento m (alto) -
  [autom-neumát] performance tire leader
— m — punto(s) m en categoría f - [deport]
  class point leader
— m — tecnología f—[ind] technological leader
— m — amperaje m para salida f - [sold] output
  (current) pointer
— m — categoría f - [deport] class leader
— m — clasificación f - [deport] class leader
puntero,ra a - top • lead
puntilla f - [clavos] tack
punto m - . . . .; spot; location; area • dègree •
  item • [sold] tack; spot • [instrum] setting
  • [legal] matter • [comput] site
— m a favor - point in favor; favorable point
— m a rojo m - [ind] hot spot
— m aceitado - [mec] oiled, point, o spot
— m adjudicado - awarded point
— m aislado - [electr-cond] insulated spot
— m ajustado - [mec] adjusted spot • tight spot
— m alejado—distant, o outlying, point, o spot
— m alguno - any, point, o location
— m alto - high, point, o spot • [ind] high
  setting
— m — brillante - [mec] bright, o polished,
  high spot
— m — de corrugación f - [metal-lam] corrugation, high point, o ridge
— m — para fusión f - [metal-prod] high melting point
— m — pulido - [mec] polished high spot
— m aparte - [gram] period (and) paragraph
— m áspero m - rough spot
— m atendido - [ind] attended, o serviced, spot
— m — regularmente - [ind] regularly serviced,
  site, o spot
— m atribuido - [labor] attributed, o credited.
  point
— m B - véase punto Bedaux
— m bajo - [mec] low, spot, o point
— m — de corrugación f - [metal-lam] corruga-

tion, low point, o valley
punto m bajo en línea f - line low point
— m — tubería f - [tub] line low point
— m básico - basic point
— m Bedaux - [labor] Bedaux point
— m blando m - soft spot
— m — eliminado—[constr] eliminated soft spot
— m brillante - bright, o shiny, spot
— m en árbol m - [electr-mot] shaft polished
  spot
— m cálido - hot spot
— m caliente - hot, point, o spot
— m — creado - created hot spot
— m — en devanado m - [electr-equip] winding
  hot spot
— m cedente - [mec] yielding spot—[metal-prod]
  véase límite m de fluencia f
— m central - central point
— m — de matriz f - [mec] die center (spot)
— m — — f para prensa f en O - [mec]
  O press die center (spot)
— m — para descarga f - central dumping station
— m ciego - [sold] blind spot
— m común - common point
— m — para reflexión f - common reflection
  point
— m con aislación f - [electr-cond] insulated
  section
— m — — f de conductor m - [electr-cond]
  lead insulated section
— n — atención f - [ind] serviced site •
  [electrón] attended site
— m — — frecuente - [electrón] frequently attended site
— m — — f infrecuente - [electrón½ infrequently attended site
— m — — f regular - [electrón] regularly attended site
— m — desgaste m - [mec] wear(ing) point
— m — — m aceitado - [mec] oiled wear(ing)
  point
— m — — m limpiado - [mec] cleaned wearing
  point
— m — hielo m - [vial] icy spot
— m — óxido m - rust(y) spot
— m — tramo m de tubería f - [ambient] point
  within @ duct run
— m concedido - [labor] allowed point
— m considerado - point under considerations •
  point under design
— m convenido - agreed point • agreed location
— m crítico - critical point - [ind] halt •
  [metal-prod] critical temperature
— m cualquiera - any point
— m culminante - high point
— m dado - given point
— m de atracción f - [electr-oper] attraction
  point
— m — ciclo m - [electr-oper] cycle point
— m — — m para histéresis f - [electr] hysteresis cycle point
— m — consumo m - consumption location
— m — contrapunta f - [herram] tailstock center
— m — deformación f - [metal-prod] yield point
— m — demora f - [ind] holding point
— m — derrumbe m - failing point
— m — destino m - [transp] destination point
— m — dureza f - hardness point
— m — — f alto - high hardness point
— m — — f bajo - low hardness point
— m — escape - breakthrough (point)
— m — falla f - failing point
— m — fluidez f - pour(ing) point
— m — funcionamiento m - operation point
— m — garantía f - guaranty point
— m — importancia f - important point
— m — interés m primario - primary interest
  area
— m — origen m - point, o place, of origin
— m — punta f - [tornos] headstock center
— m — rocío m- dew point

punto m de ruptor m - [comb.int] breaker point
— m —— **soldadura f** - [sold] welding point • tack (weld); spot (weld); weld spot
— m —— f **aislado** - [sold] isolated weld spot • pl scattered weld(s)
— m —— **soldadura f inicial** - [sold] initial tack weld
— m —— f **por proyección f** - [sold] projection weld
— m —— **sutura f** - [medic] suture stitch
— m —— **tiempo m** - time, spot, o point
— n —— m **dado** - [cronol] given time point
— m —— m **determinado** - [cronol] given time point
— m —— **tierra f** - [electr-instal] grounding point
— m —— **trocha f** - [f.c.-ruedas] gage,ging point
— m —— **tubería f** - [constr] pipe point
— m —— f **abovedada** [constr] pipe-arch point
— m —— **vacilación f** - [sold] hesitation point
— m —— **valor m** - [labor] value point
— m —— **vista f** - viewpoint; standpoint; view; side
— m —— f **de administración f** - [admin] management('s) viewpoint
— m —— f — **administrador m** - [admin] management('s) viewpoint
— m —— f **básico** - basic viewpoint
— m —— f **de circulación f** - [transp] traffic standpoint
— m —— f — **conservación f** - [ind] maintenance standpoint
— m —— f — **contaminación f** - [ambient] pollution, viewpoint, o aspect
— m —— f — **enseñanza f** - training, viewpoint, o aspect
— m —— f **hidráulica f** - [hidr] hydraulics standpoint
— m —— f **ingeniería f** - engineering viewpoint
— m —— f — **mantenimiento m** - [ind] maintenance standpoint
— m —— f — **operador m** - [ind] operator('s) standpoint
— m —— f — **seguridad f** - [segurid] safety standpoint
— m —— f — **soldador m** - [sold] weldor('s) standpoint
— m —— f — **soldadura f** - [sold] weld(ing) standpoint
— m —— f — **tránsito m** - [transp] traffic standpoint
— m —— f **educativo** - educational standpoint
— m —— f **financiero** - [fin] financial viewpoint
— m —— f **inseguro** - dangerous footing aspect
— m —— f **metalúrgico** - [metal-prod] metalurgical viewpoint
— m —— f **subjetivo** - subjective viewpoint
— m —— f **técnico** - technical viewpoint • engineering viewpoint
— m —— f **tecnológico** - technological viewpoint
— m **debido a porosidad f** - porosity spot
— m **débil** - weak, point, o spot
— m **deficitario** - [labor] deficit point
— m **demasiado ajustado** - [mec] (too) tight spot
— m **desfavorable** - unfavorable point
— m **destacado** - outstanding point • highlight
— m **determinado** - given, site, o location • specific point
— m **disperso(s)** - pl - scattered location(s)
— m **distante** - distant, o remote, o outlying, spot, o point, o location
— m **duro** - hard spot
— m **elevado** - high, o elevated, spot, o point
— m **emisor** - [electrón] beacon location
— m **en canal m** - [hidr] channel point
— m —— m **para curso m de agua m** - [hidr] waterway channel point
— m — **centro** - central point
— m — **cielo m raso** - [constr] ceiling spot
— m — **concurso m** - prize point

punto m en contra - point against • unfavorable point
— m — **periferia** - periphery,ral point
— m — **que** - point at which; at which point
— m — — **hace contacto m** - operating point • [electr-oper] point at which contact is made
— m **ensayado** - tested, point, o location
— m **equidistante** - equidistant point • compromise
— m **específico** - specific point
— m — **remoto** - specific remote location
— m **exigible** - [labor] requirable point
— m **extremo** - extreme, place, o location
— m — **de carrera f** - [comb.int] stroke extreme point
— m **favorable** - favorable point
— m **fijo** - fixed point • anchor position
— m **final** - [electr-instal] (final) terminating point
— m **financiero** - [fin] financial point
— m **focal** - focal point
— m — **para corrosión f** - [sold] corrosion focal point
— m — — **fatiga f** - fatigue focal point
— m — **posible** - possible, o potential, focal point
— m — — **de corrosión f** - [sold] possible, o potential, corrosion focal point
— m — — **fatiga f** - [sold] possible, o potential, fatigue focal point
— m **frío m** - cold, spot, o point
— m **fuerte** - strong point • [sold] heavy tack
— m — **de soldadura f** - [sold] heavy tack weld
— m **fundamental** - fundamental (point)
— m **generador** - generating point
— m **grueso** - [sold] heavy tack
— m **histórico** - historial site
— m **hora** - [labor] point hour; hour point
— m — **atribuido** - [labpr] attributed point hour
— m — **pagado** - [labor] paid point hour
— m — **real** - [labor] actual, o true, hour point, o point hour, o value
— m **importante** - important point
— m **inicial** - initial, o starting, point, o location • [sold] initial tack
— m — **de ebullición f** - initial, o starting, boiling point
— m **internacional** - [transp] international, point, o location
— m — **para entrada f** [transp] international entry point
— m — — **salida f** - [transp] international, embarkation, o shipping, point
— m **interrogado** - [comput] interrogate site
— m **más bajo de fondo m** - [constr] invert
— m — **remoto** - remotest, point, o place
— m **máximo** - maximum (point); peak • [sold] highest range
— m — **de rotación f a derecha f** - [mec] fully clockwise
— m — — f **a izquierda f** - [mec] fully counter clockwise
— m — **para operación f** - maximum operating, weight, o point
— m **medio** - medium, o mid(dle), point; half way; halfway point; midpoint
— m — **en carrera f** - [deport] race midpoint • [mec] stroke midpoint
— m — — **corredera f para grúa** - [grúas] crane runway midpoint
— m — — **escala f** - [sold] (current) range, center, o midpoint
— m — — f **para amperaje m** - [sold] current range midpoint
— m — — — f —— m **para electrodo m** [sold] elecrode range midpoint
— m — **en mirilla f** - [instrum] glass midpoint
— m — — **vano m** - [mec] span midway
— m **mínimo** - [sold] lowest, point, o range
— m — **para fluencia f** - [sold] minimum yield point
— m — **especificado** - specified minimum point

punto m **mínimo especificado para fluencia** f - minimum specified yield point
— m —————— f **para acero** m - [metal] steel minimum specified yield point
— m — **para fluencia** f **para acero** m [metal] steel minimum yield point
— m **muerto** - dead, o blind, point; neutral, position, o point • [mec] dead center
— m — **inferior** - [mec] low(er), o bottom dead center
— f — **superior** - [mec] top, o upper, dead center
— m **muy apartado** - distant location | adv —(en) in @ middle of nowhere
— m **negativo** - negative point
— m — **para transición** f - [electrón] negative transition point
— m **neutral,tro** - [mec] neutral (point)
— m **óptimo** - [instrum] optimum setting
— m **oxidado** - rust(y), o rusted, spot
— m — **limpiado** - [mec] cleaned rusty spot
— m — **limpio** - [mec] clean rusty spot
— m **para ablandamiento** - softening point
— m — **activación** f - [mec] activating point
— m — **ajuste** m - adjustment point
— m ——— m **para instrumento(s)** m - [instrum] instrument(s) adjustment point
— m — **almacenamiento** m - storage,ring point
— m ——— m **permanente** - permanent storage,ring place
— m — **amarre** m—[transp] mooring point
— m — **anclaje** m - [constr] anchor(ing) (point)
— m — **aplicación** f - [mec] application point
— m — **apoyo** m - [fís] fulcrum • [mec] bearing; support(ing) point
— m ——— m **para desplazador** m - [mec] shifter fulcrum
— m ——— m — **palanca** f - [mec] lever fulcrum
— m ——— m — **pasador** m - [mec] pin fulcrum
— m — **arranque** m - starting point
— m — **captación** f - pick-up point
— m — **carga** f - load(ing) point
— m ——— f **para cinta** f **transbordadora** f - [mec] conveyor load(ing) point
— m ——— f **para correa** f—[mec] belt load(ing) point
— m — **combustión** f—[combust] combustion point
— m — **congelación** f - freezing (point)
— m — **consideración,rar** - item for study
— m — **contacto** m **para ruptor** m - [comb.int] breaker point
— m — **cristalización** f - [quím] crystalization point
— m — **descanso** m - [vial] rest(ing) area
— m — **descarga** f - unloading, o dumping, o discharage, o off-loading, point
— m ——— f **para cinta** f **transportadora** - [mec] conveyor (belt) discharge,ging point
— m — **desgaste** m - [mec] wear(ing) point
— m ——— m **para empaquetadura** f - [mec] packing wear(ing) area
— m — **diseño** m - design point
— m ——— f **para bomba** f - [bombas] pump design point
— m — **ebullición** f - boiling point
— m — **encendido** m - firing, o ignition, point
— m — **enganche** m - [mec] hook-up location
— m — **engrase** m - [mec] greasing point
— m ——— m **para cojinete** m - [mec] bearing greasing point
— m — **enlace** m - [constr] tie(-in) point
— m ——— m **con roca** f - [constr] rock tie-in point
— m — **ensayo(s)** m - test(ing) point
— m — **entrada** f - intake, o inlet, point
— m ——— f **para aire** m - air inlet point
— m — **entrega** f - [transp] delivery, point, o location
— m ——— f **para tubería** f - [tub] pipe delivery, point, o station
— m — **espera** f - waiting, o holding, point

punto m **para estrobo,bado** m - [transp] grommet location
— m — **exportación** f - [transp] exit point
— m — **extracción** f - extraction point
— m — **fluencia** f - [metal] yield(ing) point
— m — **fusión** f - [metal-prod] fusion, o melting, point
— m ——— f **alto** - high fusion point
— m ——— f **bajo** - low fusion point
— m ——— f **elevado** - high melting point
— m ——— f **reducido** - low melting point
— m — **generación** f - generating,tion point
— m — **giro** m—[mec] pivoting, o turning, point
— m — **ignición** f - ignition, o firing, o flashing, point
— m — **inflamación** f - [combust] flash point
— m ——— f **en vasija** f **abierta** - open vessel flash point
— m — **inflexión** f - [mec] inflection point; break(ing) point
— m — **inspección** f - inspection point
— m ——— f **típica** - typical inspection point
— m — **intercambio** m - exchange, o transfer, point
— m — **interés** m - interest area
— m — **interrogación** f - [comput] interrogate site
— m — **inversión** f - [mec] inversion point • [fin] investment point
— m — **invitación** f - invitation, point, o location
— m — **izamiento** m - [transp] hoisting point
— m — **largada** f - [deport] starting point
— m — **llegada** f - arrival point
— m — **medición** f - measuring, point, o position
— m ——— f **para horno** m - [combust] furnace measuring, point, o position
— m — **operación** f - operating,tion point
— m — **partida** f - [deport] starting point • departure
— m — **premio** m - [labor] premium point
— m ——— m **negativo** - [labor] negative premium point
— m ——— m **positivo** - [labor] positive premium point
— m — **presenciar** v - [ind] witness(ing) point • point to be witnessed
— m — **presión** f - pressure point
— m ——— f **en tubería** f - [tub] pipe pressure point
— m ——— f ——— f **para entubación** f [petról] casing pressure point
— m ——— f ——— f ——— f **regulado** - [mec] set casing pressure point
— m ——— f **regulado** - [mec] set pressure point
— m — **proyección** f - design point
— m ——— f **para bomba** f - [bombas] pump design point
— m — **rechupe** m - [metal-lam] sagging point
— m — **referencia** f - reference point • monument • [mec] bench mark
— m ——— f **náutico** - [náut] nautical reference point
— m — **reflexión** f - [óptica] reflection point
— m — **regulación** f - [mec] set(ting) point
— m ——— f **para admisión** m - [comb.int] intake set(ting) point
— m ——— f — **caudal** m - [hidr] flow setting point
— m ——— f **para presión** f - [tub] pressure set(ting) point
— m ——— f ——— f **en tubería** f - [tun] pipe pressure set(ting) point
— m ——— f ——— f ——— f **para entubación** f - [petról] casing pressure set(ting) point
— m — **rendimiento** m **disminuido** - diminished, o diminishing, return point
— m — **repetición** f - [comput] repeat(ing) site
— m — **resbalamiento** m - slippage point
— m — **restricción** f - restriction point
— m — **rotación** f - [mec] rotation point
— m — **separación** f separation point • [electr-

-instal] fanout point
punto m para soldar(se) v - [sold] welding point
— m — solidificación f - solidification, o
setting, point
— m — suspensión f - [mec] suspension point •
support
— m — tracción f - [mec] traction, o pulling,
point
— m — transbordo m - [transp] transfer, o
transshipment, point
— m — m para cinta f transportadora - [ind]
conveyor belt transfer point
— m — transferencia f - transfer point
— m — transición f - transition point
— m — vaporización f - vaporization,zing point
— m — venta f - [com] sales, point, o tool
— m — venteo m - venting point
— m — vigilancia - observation point
— m — vuelco m - [ind] dumping, point, o location
— m particular - specific, point, o location
— m — remoto - specific remote (place)
— m permanente - permanent, point, o location
— m positivo - positive point
— m — para transición f - [electrón] positive
transition point
— m preciso - precise, o specific, location
— m predeterminado - predetermined, o preselected, point, o location
— m preseleccionado - preselected point
— m pulido - polished, point, o spot
— m — en árbol m - [electr-mot] shaft polished spot
— m real - [labor] actual, o true, point
— m regulado - [mec] set point
— m remoto - remote, point, o place, o site;
remote
— m repetidor - [comput] repeater site
— m rojo - red spot
— m seguido—[gram] period and no new paragraph
— m sin recubrir v - uncoated spot
— m soldado - [sold] weld(ed) spot
— m terminal - terminal (point); terminus •
extremity
— m — eléctrico - [electr=instal] electrical,
terminal (point), o terminus
— m — mecánico - [mec] mechanical, terminal
(point), o terminus
— m único - single point
puntos m de destino m y origen m - [transp] destination and origin places
— m — m y origen m iniciales - [transp]
initial destination and origin places
— m diacríticos - [gram] diacritical marks •
umlaut(s)
— m múltiples - multiple points | adv - multi-
-point
puntual a - . . .; short, term, o duration
puntualización f - punctualization; pinpointing
puntualizar v - to pinpoint • to emphasize
puntualizado,da a - pinpointed • emphasized
puntualmente adv - . . .; on schedule
punzado m - [mec] punching; véase punzadura f
punzado,da a - [mec] punched
punzadora f - [metal-fabr] punch press
— f de extremo(s) m - [mec] end punch
punzadura f - [mec] punching
— f de caja f—[transp] container puncture,ring
— f para contrapeso(s) m - [mec] counterweight
punching
— f — m — . . . kilogramos - [mec] . . .
pounds counterweight punching (use pounds)
— f — marcación f - [mec] mark punching
punzante a - . . .; puncturing
punzar v abertura f - [mec] to punch @ opening
— v cajón m - [transp] to puncture @ container
— v extremo m - [mec] to end punch
— v hueco m - [f.c.-ruedas] to punch @ bore
punzatubos m - [mec] pipe punch(er) • [petról]
casing perforator
punzón m - [herram] . . .; point; drift; spike;
gab; graver; trebler • center punch • die •
piercing ram

punzón m armado - [mec] assembled punch
— m extractor - véase ariete m (extractor)
— m individual - [mec] individual punch
— m intercambiable - [mec] interchangeable
punch
— m macho - [mec] male punch
— m maestro m - [mec] master punch
— m — con adaptador m - [mec] male punch with
@ adapter
— m — macho - [mec] master male punch
— m Magna Die - [mec] Magna Die punch
— m — — de Whistler - [mec] Whistler Magna
Die (punch)
— m modular - [mec] modular punch
— m para clavo(s) m - [herram] nail punch
— m — conjunto m maestro - [mec] master set
punch
— m — centrar f - [herram] center(ing) punch
— m — empalmar v cable(s) m - [cabl] marline
spike
— m — marcar,cación f - [mec] marking die
— f — muesca f - [mec] notch punch
— m según medida f - [mec] custom punch
— m troquelador - [mec] die punch
— m — central - [mec] center,tral die punch
— m y matriz,ces f - [mec] punch and die(s)
punzonado m - [mec] punch(ing); véase también
punzadura f
punzonado,da a - [mec] punched
puño m - . . . • [mec] handle bar
— m abocinado - [seguird] gauntlet
— m — contra calor m - [seguird] heat breaker
gauntlet
— m abocinado para guante m - [seguird] glove
gauntlet
— m — — — f de lana f - [seguird] wool
glove gauntlet
— m — — — m — — f contra calor m - [segu-
rid] wool heat breaker glove gauntlet
— m — — — m industrial - [seguird] industrial glove gauntlet
— m — — — m — contra calor m - [seguird]
industrial glove heat breaker gauntlet
— m — — m — de lana f - [seguird] industrial wool glove gauntlet
— m — — m — — f contra calor m -
[seguird] industrial wool heat breaker glove
gauntlet
— m — industrial - [seguird] industrial gauntlet
— m — — contra calor m - [seguird] industrial
heat breaker gauntlet
— m — — de lana f - [seguird] industrial wool
gauntlet
— m — — f de lana f contra calor m—[seguird]
industrial wool heat breaker gauntlet
pupinización f - pupinization
pupinizado,da a - pupinized
pupinizar v - to pupinize
pupitre m - . . .; desk; pulpit • cabinet
— m eléctrico - [electr-instal] electrical console
— m — línea f para galvanización f - [metal-
-trat] galvanizing line pulpit
— m para mando m - [ind] control pulpit
— m — maniobra(s) f - [ind] bench board
— m — operador m - [ind] operator('s) desk
— m — regulación f - [ind] control desk
pureza f - . . . • [fig] white(ness)
— f alta - high purity
— f de agua m - [hidr] water, purity, o quality
— f — hierro m - [metal-prod] iron purity
— f — — m reducida - [metal-prod] reduced
iron purity
— f metálica - [metal-prod] metal(lic) purity
purga f - . . .; purge • discharge,ging; drain-
-(ing); flush(ing); bleed(ing); exhaust(ing) •
[válv] blowdown
— f anual - [comb.int] annual draining
— f completa - complete draining
— f con contracorriente f - back flush(ing)
— f — — de aire m - air back flush(ing)

purga f con nitrógeno m - nitrogen purge,ging
— f — vapor m - [cald] steam purge,ging
— f de aceite m - [comb.int] oil draining
— f — — m en colador m - [mec] sump oil drain(ing)
— f — acumulación f - accumulation, drainage, o draining
— f — — f de humedad f - moisture accumulation drainage,ning
— f — — f de lodo m - [petról] sludge accumulation drainage,ning
— f — aire m - air, purge,ging, o bleeding
— f — — m en red f - system air bleeding
— f — — m en sistema m - system air bleeding
— f — bajada f - [hidr] drop (leg) draining
— f — bomba f - [bombas] pump release,sing
— f — — f para inyección f - [petról] mud pump release,sing
— f — — f para lodo(s) m - [petról] mud pump release,sing
— f — cámara f - [mec] chamber drain(ing)
— f — — f para filtro m - [mec] filter chamber drain(ing)
— f — — f — — m para aire m - [mec] air filter chamber drain(ing)
— f — colector m - [mec] sump drain(ing)
— f — — m para aspiración f - [mec] suction manifold drain(ing)
— f — — m primario m (para polvo m) - [metal-prod] primary (dust) catcher flushing
— f — condensación f - condensation discharge,ging
— f — condensador m - condenser, exhaust, o blowdown
— f — cono m - cone discharge,ging
— f — — m de Venturi m - [metal-prod] Venturi cone (emergency) discharge,ging
— f — depósito m - [mec] tank drainage,ning
— f — — m para abastecimiento m - [comb.int] supply(ing) tank drainage,ning
— f — — m — reserva f - [mec] reservoir drainage,ning
— f — — m — — f para aceite m - [mec] oil reservoir drainage,ning
— f — eje m - [mec] axle draining
— f — escoria f - [metal-prod] slag flush-off
— f — filtro m - [mec] filter, draining, o flushing
— f — — m para aceite m - [comb.int] oil filter draining
— f — — m — agua m - [comb.int] water filter, draining, o flushing
— f — — m — aire m - [mec] air filter drain-(ing)
— f — — m — combustible m - [comb.int] fuel filter, drainage,ning, o bleeding
— f — — m por contraflujo m - [comb.int] (fuel) filter back flushing
— f — freno m - [mec] brake, drainage,ning, o discharge
— f — humedad f - [mec] moisture, bleeding, o drainage,ning
— f — lavador - [metal-prod] (gas) scrubber, o washer, flushing
— f — — m para gas m - [metal-prod] gas, scrubber, o washer, flushing
— f — línea f - [mec] line discharge,ging
— f — lodo m - [petról] sludge drainage,ning
— f — lubricante m - [mec] lubricant, drainage, o draining
— f — manguera f - hose purge,ging
— f — planta f para gas m - [metal-prod] gas plant, blowdown, o bleed(ing)
— f — — f — m para horno m - [metal-prod] furnace gas plant, blowdown, o bleeding
— f — pote m - [válv] pot drainage,ning
— f — — m para válvula f - [válv] valve pot drainage,ning
— f — primario m - [metal-prod] primary dust catcher flushing
— f — red f - [ind] system, purge,ging, o bleeding
— f — — f — combustible m - [comb.int] fuel system bleeding
purga f de red f para enfriamiento m - [comb.-int] cooling system draining
— f — separador m - [ind] separator bleeding
— f — — m para agua m - [comb.int] water separator, bleeding, o drainage,ning
— f — sincronizador m - timer purge,ging
— f — sistema m - system, purge,ging, o bleeding
— f — suciedad f - dirt, cleaning, o draining
— f — — f acumulada - accumulated dirt, cleaning, o draining
— f — tubería f - [mec] line discharge,ging
— f — tubo m - [tub] pipe, draining, o cleaning
— f — — m para refrigeración f - cooling pipe drain(ing)
— f — — m — — f exterior - [ind] external cooling pipe drain(ing)
— f — — m — — m de coraza f - [metal-prod] shell external cooling plate draining
— f — válvula f - [válv] valve draining
— f — vapor m - vapor purge,ging • steam bleed(ing)
— f — — m de propano m - [combust] propane vapor purge,ging
— f — zona f - [ind] area drainage,ning
— f — — f para asentamiento m - [comb.int] settling area drainage,ning
— f — — f — decantación f - [comb.int] settling area drainage,ning
— f inferior m - [mec] bottom, o low(er), outlet, o drain
— f para aire m - air vent
— f — combustible m - [comb.int] fuel drain
— f — suciedad f dirt drain
— f — tambor m - [mec] drum, drain, o outlet
— f — tubería tubería f - [mec] line drain
— f — — f para colector m - [mec] sump line drain
— f periódica - periodic(al) flush(ing)
— f superior - top, o upper, outlet, o drain
— f — para tambor m - [mec] drum, top, o upper, outlet
— f total - full, o total, draining
purgado,da a - purged; flushed; drained; bled; discharged
— a anualmente - [comb.int] drained annually
— a completamente - drained completely
— a con contracorriente f - back flushed
— a — — f de aire m - backflushed with air
— a periódicamente - drained, o flushed, periodically|
— a totalmente - fully, o totally, drained
purgador m - . . . . • bleeder • drain cock; drip pot • flushing pump • [metal-prod] exhaust
— m magnético - magnetic drain
— m — gas m - [ind] gas bleeder
— m — horno m - [metal-prod] furnace bleeder
— m — — m alto - [metal-prod] blast furnace bleeder
— m — presión f alta - high pressure flushing, pump, o bleeder
— m — quemador m - [ind] burner exhaust
— f — — m para estufa f - [metal-prod] stove burner exhaust
purgar v - . . . . to drain; to bleed; to flush • to exhaust; to blow off; to discharge • [metal-prod-salamandra] to pour out • [combust] to bleed
— v aceite m - [mec] to drain @ oil
— v — m en colador m - [mec] to drain @ sump oil
— v aceite m - [mec] to drain @ oil
— v — m en colador f - [mec] to drain @ sump oil
— v acumulación f - to drain @ accumulation
— v — f de humedad f - to drain @ moisture accumulation
— v — f de lodo m - [petról] to drain @ sludge accumulation
— v aire m - to purge, o drain, @ air
— v — m en red f - to bleed @ air from @ system bleeding

tem
purgar v **aire** m **en sistema** m - to bleed @ air from @ system
— v **anualmente** - [comb.int] to drain annually
— v **bajada** f - [hidr] to drain @ drop leg
— v **cámara** f - [mec] to drain @ chamber
— v — f **para filtro** m - [mec] to drain @ filter chamber
— v — f —— m **para aire** m - [mec] to drain @ air filter chamber
— v **colector** m - [mec] to drain @ sump
— v — m **primario** - [metal-prod] to flush @ dust catcher
— v — m — **para polvo** m - [metal-prod] to flush @ dust catcher
— v **completamente** - to drain completely
— v **con contracorriente** f - to back flush
— v —— f **de aire** m - to back flush with air
— v **condensación** f - to, drain, o discharge, @ condensation
— v **depósito** m - to drain @ reservoir
— v — **para abastecimiento** m - [comb.int] to drain @ supply tank
— v — m — **reserva** [mec] to drain @ reservoir
— v — m — f **para aceite** m - [mec] to drain @ oil reservoir
— v **eje** m - [mec] to drain @ axle
— v **filtro** m - to, drain, o flush, @ filter
— v — m **para aire** m - [mec] to drain @ air filter
— v — m **para combustible** m - [comb.int] to drain @ fuel filter
— v — m **con contraflujo** m - [comb.int] to back flush @ filter
— v **gas** m - [metal-prod] to purge @ gas
— v **horno** m - [combust] to bleed @ furnace
— v **humedad** f - [ind] to, drain, o bleed, @ moisture
— v **lavador** m - [metal-prod] to purge @ gas, washer, o scrubber
— v — m **para gas** m - [metal-prod] to flush @ gas, washer, o scrubber
— v **línea** f - [mec] to, flush, o discharge, @ line
— v **lodo** m - [petról] to drain @ sludge
— v **lubricante** m - [mec] to drain @ lubricant
— v **manguera** f - to purge @ hose
— v **periódicamente** - to, drain, o flush, periodically
— v **planta** f - [ind] to, bleed, o blow down, @ plant
— v — f **para gas** m - [metal-prod] to, bleed, o blow down, @ gas plant
— v — f —— m **para horno** m - [metal-prod] to, bleed, o blow down, @ furnace gas plant
— v **por contraflujo** m - to back flush
— v —— m **filtro** m - [mec] to back flush @ filter
— v —— m — m **para combustible** m [comb.int] to back flush @ fuel filter
— v **pote** m - [válv] to drain @ pot
— v — m **para válvula** f - [válv] to drain @ valve pot
— v **primario** m (**para polvo** m) - [metal-prod] to purge @ primary dust catcher
— v **red** f - [ind] to purge @ system
— v — f **para enfriamiento** m - [comb.int] to drain @ cooling system
— v **separador** m - to drain @ separator
— v — m **para agua** m - [comb.int] to, drain, o bleed, @ water separator
— v **sincronizador** m - to purge @ timer
— v **sistema** m - to purge @ system
— v — m **para combustible** m - [comb.int] to bleed @ fuel system
— v — m — **enfriamiento** m - [comb.int] to, flush, o drain, @ cooling system
— v — m — **refrigeración** f - [comb.int] to flush @ cooling system
— v **suciedad** f - to drain @ dirt
— v — f **acumulada** - to drain @ accumulated dirt
— v **totalmente** - to drain, fully, o completely

purgar v **tubería** f - [mec] to discharge @ line
— v — f **para agua** m - [ind] to purge @ water pipe
— v — f —— m **para Venturi** m - [metal-prod] to purge @ Venturi water pipe
— v **válvula** f - [válv] to drain @ valve
— v **vapor** m - to purge @, vapor, o steam
— v — m **de propano** m - [combust] to purge @ propane vapor
— v **zona** f - to drain @, zone, o area
— v — f **para asentamiento** m - [comb.int] to drain @ settling area
— v — f — **decantación** f - [comb.int] to drain @ settling area
purgas f - drips
purificación f - . . .; purifying • [metal-prod] scrubbing
— f — **gas** m - [metal-prod] gas, cleaning, o scrubbing
— f —— m **en horno** m **alto** - [metal-prod] blast furnace gas scrubbing
— f —— m **en coquería** f - [coque] coke gas scrubbing
— f — **líquido(s)** m - [sanit] liquid(s) purification
— f —— m **cloacal(es)** - [sanit] sewage purification,fying
purificador m **para aire** m - [comb.int] air, cleaner, o filter • [metal-prod] air scrubber • véase también **depurador** m **para aire** m
— v **líquido** m - [ind] to purify @ liquid
— v — m **cloacal** - [sanit] to purify @ sewage
puro,ra a - . . . . • [fig] white
— a **comercial(mente)** - commercially pure
puzolana f - [miner] puzzuolana

Q

que adv - . . . . • than
— **abarca** f **empresa** f **íntegra** adv - organization wide
— — f **organización** f **íntegra** - organization wide
— **ahorra** v **tiempo** m - time saving
— **antecede** v - foregoing; above
— **aparece a pie** m below
— **arrastra** v **aire** m -[hormig] air entraining
— **asigna** v **tarea(s)** f - job assigning
— **avanza** f **por sí mismo** - self propelled
— **causa inquietud** f - concern causing
— **preocupación** f worry, o concern, causing
— **circunscribe** - circumscribing
— **compite** - competing; competitive
— **concuerda** - agreeing; concordant
— **contiene** - containing
— **corre** - current • this
— **en oleada(s)** f - running in wave(s)
— **corresponde** - involved
— **cruza por encima** - crossing above
— **deja huella(s)** f - marring; leaving tracks
— — **marca(s)** f - marring; leaving tracks
— **demanda presencia** f **de ánimo** - confidence testing
— **depende** adv - depending
— **figura(n)** adv - appearing; listed
— **fluye(n)** adv - flowing
— — adv **lleno,na** - full flowing
— **ha demostrado bondad(es) en uso** m - field tested
— **hace(n) contacto** - contacting • [electr-oper] closing
— **(se) hincha con congelación** f - [constr] frost heaving
— — — **helada(s)** f - frost heaving
— — **ilustra(n)** - illustrated
— — **indica(n)** - indicated; listed

que (se) **intenta** ~ intended
— — **levanta con congelación** f̲ - [constr] frost heaving
— — — — **helada(s)** f̲ - [constr] frost heaving
— — **mueve por sí mismo,ma** - self propelled
— **no deja huella(s)** f̲ - non-marring
— — — **marca(s)** f̲ - non-marring
— — **devenga interés** - [fin] non interest bearing
— — **requiere conservación** f̲ - maintenance free
— — **se produce ya** - no longer in production
— — — **surte(n) separadamente** - not serviced separately
— — **tiene dedo(s)** m̲ - fingerless
— — **venga a(l) caso** m̲ - irrelevant
— **opera eficientemente** - operating efficiently
— **pasa por zaranda** f̲ - passing @ sieve
— — — — f̲ **estándar** - passing @ standard sieve
— **pone pelo(s)** m̲ **de punta** - hair raising
— **(se) presentará en breve** adv - soon-to-arrive
— **prueba confianza** adv - confidence proving
— **puede girar** v̲ - [mec] free to pivot
— **puede interpretar(se)** - interpretable
— — **obtener(se)** - available; obtainable • on @ market
— — **presentar(se)** - available
— **reacciona apropiadamente a maniobra(s) de volante** m̲ - [autom-neumát] responsive
— **requiere (mucho) tiempo** adv - time consuming
— **reside** adv - residing
— **resulta de ello** adv - resulting therefrom
— **rueda** v̲ **libremente** - [autom-neumát] free rolling
— **se requiere** - required
— **sea** adv - that is
— **sigue** adv - below; following
— **soporte** v̲ **tensión(es)** f̲ tension carrying
— **suele(n) llamar(se) (acostumbradamente)** • (usually, o commonly) called
— **suscribe** adv - [legal] undersigned; who signs below
— **trabaja adecuadamente** - hard working
— — **libremente** - free working
— **vale** adv - that count(s)
— **(se) vende en juego(s) únicamente** - sold in sets only
quebrada f̲ - . . . . • [topogr] . . .; draw
quebradizo,za **en caliente** - [metal-prod] hot short
quebrado m̲ - [matem] fraction
quebrado,da a̲ - broken • [topogr] hilly; rough; rugged(ness)
quebrador,ra a̲ - . . . . • [miner] véase **triturador,ra**
quebradora f̲ - [miner] crusher
— f̲ **primaria** - [coque] primary crusher
— f̲ **secundaria** - [coque] secondary crusher
quebradura f̲ - . . . • [metal-lam] break
— f̲ **de bobina** f̲ - [metal-lam] coil break
quebrantador m̲ - [metal-lam] scale breaker
quebrantadora f̲ **neumática** - [miner] pneumatic breaker
quebranto m̲ - . . . . • [fin] bankruptcy
— m̲ **antes de impuesto(s)** m̲ - [fisc] loss before tax(es)
— m̲ — — m̲ **sobre, rédito(s)** m̲, o **renta** f̲ - [fisc] loss before @ income tax(es)
— m̲ **comercial** - [fin] (commercial) bankruptcy
— m̲ **contabilizado** - [contab] accounting, o recognized, loss
— m̲ — **para contrato** m̲ - [contab] accounted, o recognized, contract loss
— m̲ **incurrido** - [fin] incurred loss
— m̲ **neto** - [fin] net loss
— m̲ — **no realizado** - [contab] net unrealized loss
— m̲ — — — **sobre valor(es)** m̲ **especulativo(s)** - [fin] net unrealized loss on @ marketable (equity) security,ties
— — — m̲ — — — **negociable(s)** - [fin] net unrealized loss on marketable (equity) security,ties

quebranto m̲ **neto por emisión** f̲ - [seguros] net underwriting loss
— m̲ — — — f̲ **de póliza(s)** f̲ - [seguros] net underwriting loss
— m̲ — **sobre seguro(s)** m̲ **emitido(s)** - [seguros] net underwriting loss
— m̲ **no realizado** - [seguros] unrealized loss
— m̲ — **sobre capital** m̲ - [fin] unrealized capital loss
— n̲ **sobre contrato** m̲ - [legal] contract loss
— m̲ — — m̲ **contabilizado** - [contab] recognized contract loss
queda hecho depósito m̲ que marca f̲ ley - [legal] copyright(ed)
quedado,da a̲ - remained
quedar v̲ - . . . . • to be • to continue
— **aislado,da** - [ind] to, become, o remain, isolated, o cut off
— v̲ — **petaca** f̲ - [metal-prod] to, remain, o become, isolated (a cooling plate)
— v̲ **apartado,da** - to, become, o stand, separated, o clear
— v̲ **atrapado,da** - to become trapped
— v̲ **atrás** - to stay behind • to retrench
— v̲ **boquiabierto,ta** - to become openmouthed
— v̲ **con** - to be up to
— v̲ — **poca agua** m̲ - [metal-prod] to receive insufficient water
— v̲ — **vapor** m̲ - to remain with steam
— v̲ — — m̲ **precipitador** m̲ - [metal-prod] to leave @ precipitator with steam
— m̲ **convenido,da** - to remain agreed
— v̲ **en** - to remain in; to stand at
— v̲ — **observación** f̲ - to, remain, o hold, under observation
— v̲ — **red** f̲ - [comb.int] to remain in @ system
— v̲ — **sitio** m̲ - to remain in @ place
— v̲ **engranado,da** - [mec] to remain engaged
— v̲ **entendido** - to be understood
— v̲ **escoria en ruta** f̲ - [metal-prod] slag remaining in @ runner
— v̲ **flojo,ja** - to, be, o remain, loose, o slack
— v̲ **fuera de combate** - [deport] to quit
— v̲ — **servicio** m̲ - to be (put) out of service; to fail
— v̲ **inutilizado,da** - to be left useless
— v̲ **lleno,na** - to, remain, o become, full
— v̲ — **codo** m̲ - [metal-prod] to fill @ gooseneck
— v̲ **pegado,da** - to remain stuck
— v̲ **sin** - to run out; to deplete
— v̲ — **agua** m̲ - to be left without @ water
— v̲ — **petaca** f̲ - [metal-prod] cooling plate without water
— v̲ — **serie** f̲ - [metal-prod] circuit without water
— v̲ — **venturi** m̲ - [metal-prod] Venturi left without water
— v̲ **sin energía** f̲ - [electr-distrib] to be left without power
— v̲ — **actuación** f̲ - to remain motionless; to not operate
— v̲ — **descubrir** - to, go, o remain, undetected
— v̲ — **neumático(s)** m̲ - [autom-neumát] to run out of tire(s)
— v̲ — **tensión** f̲ [electr-oper] power failure; to remain without power
— v̲ **sumergido,da** - to, be, o remain, submerged
— v̲ **tabicado,da** - [metal-prod] to be bricked up (a slag notch)
— v̲ **terminado,da** - to have been completed
queja f̲ - . . . .; claim
— f̲ **de cliente** m̲ - customer('s) complaint
— f̲ — **empleado** m̲ - [labor] employee complaint
— f̲ — **personal** m̲ - [labor] personnel complaint
quejado,da a̲ - complained
quema f̲ - burn(ing) (out, o up)
— f̲ **completa** - (complete) burning up
— f̲ **de campo** m̲ - [electr-equipo] field burning out
— f̲ — — m̲ **para embrague** m̲ - [mec] clutch field burn(ing) out

quema f de conmutador m - [electr-oper] switch,
o breaker, burning out
— f — contacto m - [electr-oper] contact burning out
— f — devanado m - [electr-equip] coil burning
out
— f — embrague m - [mec] clutch burning out
— f — etapa f - stage burning up
— f — excitador m - [electr-instal] exciter
burning out
— f — fusible - [electr-oper] fuse, burning
out, o blowing
— f — — m para circuito m - [electr-oper]
circuit fuse, burning out, o blowing
— f — — — m para regulación f -
[electr-oper] control circuit fuse, burning out, o blowing
— f — — m para energía f - [electr-oper]
power fuse, burning out, o blowing
— f — petaca f - [metal-prod] cooling plate,
burning up, o burnout
— f — válvula f - [comb.int] valve burn(ing)
quemado m - burning
quemado,da a - burned (up, o out) • [electr]
blown • [comb.int] blown • [color] brown
— a muerte - véase calcinado,da totalmente
— a completamente - burned up; consumed
— a lentamente - [combust] burned slowly
— a más lentamente - burned slower
— a — rápidamente - burned faster
— a rápidamente - [combust] burned fast(ly)
quemador m - [metal-prod] burner; G.B. • port
— m ajustado - [combust] adjusted burner
— m con aspiración f natural - [combust] natural draft, o naturally aspirated, burner
— m convencional—[combust] conventional burner
— m encendido - [combust] lit, o ignited, burner • burner on
— m — rápidamente - [combust] quickly ignited
burner
— m fabricado - [combust] manufactured burner
— m final - [combust] end, o final, burner
— m inferior - [combust] bottom, o lower, burn-
er | adv - [combust] (con) bottom fired
— m lateral - [combust] side burner
— m para aire m y petróleo m - [combust] air-
-oil burner
— m lateral - [combust] lateral, o side. burner
— m — para fuel oil m - [combust] side fuel
oil burner
— m — — gas m - [combust] lateral, o side,
gas burner
— m — azufre m - sulfur burner
— m — bóveda f - [ind] roof burner
— m — caldera f - [combust] boiler burner
— m para encendido m—[combust] ignition burner
— m — estufa f - [combust] stove burner
— m — — flujo m máximo m [metal-prod] maximum
flow burner
— m — — m mínimo - [metal-prod] minimum flow
burner
— m — fuel oil m - [combust] fuel oil burner
— m — gas m - [combust] gas burner
— m — — m en bóveda f - [combust] roof gas
burner
— m — — m residual - [combust] waste gas burner
— m — llama   f larga - [combust] long flame
burner
— m — — f piloto - [combust] pilot (flame)
burner
— m — máxima f - [combust] maximum flow burner
— m — mínima f - [combust] minimum flow burner
— m — oxígeno m y gas m - [combust] oxygen
(and) gas burner
— m — — m y petróleo m - [combust] oxygen-oil
burner
— m — petróleo m - [combust] oil burner
— m — salida f - [combust] screening burner
— m — techo m - véase quemador m para bóveda f
— m individual - [combust] individual burner
— m principal - [combust] main burner
— m reencendido - [combust] relit burner

quemador m regulado - [combust] regulated, o
controlled, burner
— m y cabeza f - [combust] burner and head
quemadura f - . . .; burning • [deport] tan
— f de arco m - [sold] arc burn; arcburn
— f — contacto m - [electr-oper] contact
burn(ing)
— f — fusible m - [electr-oper] fuse blowing
— f — — m en línea f - [electr-oper] line
fuse blowing
— f — — m — — f para entrada f - [electr-
-oper] incoming, o supply, line fuse blowing
— f — — — — f — suministro m - [electr-
-instal] supply line fuse blowing
— f — . . . grado m—[medic] . . . degree burn
— f — rayo(s) m - [medic] ray(s) burn
— f — sol m - [medic] sunburn
— f — válvula f - [comb.int] valve burn(ing)
— f leve - [medic] slight burn
— f — con arco m - [sold] slight arc burn
— f química - [medic] chemical burn
— f severa - [medic] severe burn
— f — con arco m - [sold] severe arc burn
— f térmica - [segurid] heat burn(ing)
quemar v - . . . . to burn, in, o out, o up •
to frit • to tan • to blaze
— v busa f - [metal-prod] to burn @ blowpipe
— v campo m—[electr-equip] to burn out @ field
— v — m para embrague m - [mec] to burn (up)
@ clutch field
— v carbono m - [combust] to burn @ carbon
— v completamente - to burn (up) completely
— v con explosión f - [metal-prod] to burn out
with @ explsion; to blow out; to explode
— v — — f toberín m - [metal-prod] to burn
out with @ explosion @ slag notch nozzle
— v — — f toberón m - [metal-prod] to burn
(out) with @ explosion @ tuyere cooler
— v con oxígeno m - [metal-fabr] to burn (out)
with oxygen
— v — — m burro m (de hierro m) - [metal-
-prod] to burn out @ frozen pig with oxygen
— v — — m tobera f - [metal-prod] to burn
(out) @ tuyere with oxygen
— v conmutador m - [electr-oper] to burn out @
switch
— v contacto m—[electr-oper] to burn @ contact
— v devanado m - [electr-oper] to burn (out) @
coil
— v electrodo m - [sold] to burn @ electrode
— v embrague m - [mec] to burn (out) @ clutch
— v etapa f - to burn up @ stage
— v excitador m - [electr-equip] to burn (out)
@ exciter
— v fusible m - [electr-oper] to, burn (out), o
blow, @ fuse
— v — m para circuito m - [electr-oper] to
blow @ circuit fuse
— v — m — — m para regulación f - [electr-
-oper] to blow @ control circuit fuse
— v — para energía f - [Electr-oper] to blow @
power fuse
— v — m — línea f - [electr-oper] to blow @
line fuse
— v — m — — f para entrada f - [electr-oper]
to blow @ supply line fuse
— v — m — — — f — suministro m - [electr-
-instal] to blow @ supply line fuse
— v hasta consumir - [sold] to burn, up o away
— v hierro m - [metal-prod] to burn @ iron
— v lentamente - [combust] to burn slowly
— v más lentamente - to burn more slowly
— v — rápidamente - to burn more rapidly
quemar v motor m - [electr-mot] to burn out @
motor - [comb.int] to burn out @ engine
— v petaca f - [metal-prod] to burn (out) @
cooling plate
— rápidamente - to burn, rapidly, o quickly
— v soldadora f - [sold] to burn (up) @ welder
— v tobera f - [metal-prod] to burn out @ tuyere
— v — por entrarle hierro m - [metal-prod] to
burn (out) @ tuyere with @ hot metal

quemar $\underline{v}$ toberín $\underline{m}$ - [metal-prod] to turn (out)
@ slag notch
— $\underline{v}$ — $\underline{m}$ para escorial $\underline{m}$ - [metal-prod] to burn
out @ slag notch nozzle
— $\underline{v}$ toberón $\underline{m}$ - [metal-prod] to burn out @
tuyere cooler
— $\underline{v}$ trozo $\underline{m}$ de tobera $\underline{f}$ para colocar otro -
[metal-prod] to burn (out) @ part of @ tuyere
to make room for another
— $\underline{v}$ válvula $\underline{f}$—[comb.int] to burn (out) @ valve
quemarse $\underline{v}$ - véase quemar $\underline{v}$
querellar $\underline{v}$ - . . . • [legal] to sue
querer $\underline{v}$ - . . . • to hope (for)
— $\underline{v}$ decir - to intend to say
querógeno $\underline{m}$ - [petról] kerogen
querosén(o) $\underline{m}$ - [petról] kerosene; coal oil
— $\underline{m}$ frío - [petról] cold kerosene
querosina $\underline{f}$ - [petrol] kerosene
quid $\underline{m}$ - . . .; substance
quiebra $\underline{f}$ inminente - [legal] inminent, bankruptcy, $\underline{o}$ receivership
quien pron corresponda - whom it may concern
— pron diseña soldadura $\underline{f}$ - [sold] weld designer
— pron ejecuta soldadura $\underline{f}$ - [sold] weld fabri-
cator
quienquiera pron - . . .; anyone
quijada $\underline{f}$ abierta - open(ed) jaw
— $\underline{f}$ de tornillo $\underline{m}$ - [herram] chap
— $\overline{f}$ — — $\underline{m}$ para banco $\underline{m}$ - [herram] (vise) jaw
— $\overline{f}$ separada - [mec] separate, $\underline{o}$ open, jaw
química $\underline{f}$ de escoria $\underline{f}$ - [metal-prod] slag chemistry
— $\underline{f}$ liviana - [quím] light chemistry
— $\overline{f}$ — ensayo(s) $\underline{m}$ - [quím] test(s) chemistry
— $\overline{f}$ pesada - [quím] heavy chemistry
químico $\underline{m}$ capacitado - [quím] qualified, $\underline{o}$ knowledgeable, chemist
— $\underline{m}$ conocedor - [quím] knowledgeable chemist
— $\underline{m}$ entendido - [quím] knowledgeable chemist
— $\underline{m}$ industrial - [quím] industrial chemist
— $\underline{m}$ — ensayo(s) $\underline{m}$ - [quím] test(ing) chemist
— $\underline{m}$ — perfeccionamiento $\underline{m}$ de neumático(s) $\underline{m}$ -
[autom- neumát] (advanced) tire development
chemist
quinceavo,va $\underline{a}$ - fifteenth
quincuagésimo,ma - fiftieth
quincena $\underline{f}$ - [cron] . . .; half month
quincenalmente adv - . . .; on @ fortnightly
basis; (once) each fortnight
quinelas $\underline{f}$ - [juego] numbers (racket)
quinta $\underline{f}$ - . . . • [constr] cottage
quintal $\underline{m}$ - [metric] . . .; hundredweight
quintuplicado $\underline{m}$ - quintuplicate
quintuple $\underline{a}$ - . . .; five pronged
quita $\underline{f}$ - [com] rebate; discount
quitado,da $\underline{a}$ - removed; taken away; lifted •
[mec] chipped (off)
— $\underline{a}$ completamente - removed completely
— $\overline{a}$ con cuidado - removed carefully
— $\overline{a}$ — esmeriladora $\underline{f}$ - [mec] grinder removed
— $\overline{a}$ — — $\underline{f}$ mecánica - [mec] electric grinder
removed
— $\underline{a}$ con extractor $\underline{m}$ - extractor removed • [mec]
puller removed
— $\underline{a}$ — lima $\underline{f}$ - [mec] file removed
— $\overline{a}$ — $\underline{f}$ manual - [mec] hand file removed
— $\overline{a}$ — lubricante - lubricant removed
— $\overline{a}$ — martillo $\underline{m}$ - hammer removed
— $\overline{a}$ — pincel $\underline{m}$ - brushed off
— $\overline{a}$ de cajón $\underline{m}$ - [transp] removed from @, case,
$\underline{o}$ container
— $\underline{f}$ — cruceta $\underline{f}$ - [mec] removed from @ crosshead
— $\underline{a}$ — eje $\underline{m}$ - [mec] removed from @ axle
— $\overline{a}$ — encima - lifted off of
— $\overline{a}$ — motor $\underline{m}$ - [comb.int] removed from @ engine • [electr-mot] removed from @ motor
— $\overline{a}$ — prensaestopa(s) $\underline{m}$ - [mec] removed from @
packing box
— $\underline{a}$ — separador $\underline{m}$—[mec] removed from @ spacer
— $\overline{a}$ fácilmente - easily, $\underline{o}$ readily, removed
— $\overline{a}$ lentamente - removed slowly

quitado,da $\underline{a}$ para inspección $\underline{f}$ - removed for @
inspection
— $\underline{a}$ — limpieza $\underline{f}$ - [mec] removed for cleaning
— $\overline{a}$ periódicamente - removed periodically
— $\overline{a}$ rápidamente - removed quickly
quitaherrumbre $\underline{m}$ - [metal] rust remover
quitanieves $\underline{n}$ - [vial] snowblower; snowplow
— $\underline{m}$ rotativo - [vial] (rotary) snowblower
quitapuerta(s) $\underline{m}$ - véase sacapuerta(s) $\underline{m}$
quitar $\underline{v}$ - . . .; to lift off; to take away •
to seal @ leak • [mec] to chip off
— $\underline{v}$ abrazadera $\underline{f}$ - [mec] to remove @ clamp
— $\overline{v}$ aceite $\underline{m}$ - to remove @ oil
— $\overline{v}$ acoplamiento $\underline{m}$ - [mec] to remove @, coupling, $\underline{o}$ linkage
— $\underline{v}$ acumulador $\underline{v}$ - [autom-electr] to remove @
battery
— $\underline{v}$ agua $\underline{m}$ - to remove @ water
— $\overline{v}$ aire $\underline{m}$ - to remove @ air
— $\overline{v}$ anillo $\underline{m}$ - [mec] to remove @ ring
— $\overline{v}$ — $\underline{m}$ para compresión $\underline{f}$ - [mec] to remove @
compression ring
— $\underline{v}$ — $\underline{m}$ — empaquetadura $\underline{f}$ - [mec] to remove @
packing ring
— $\underline{v}$ arandela $\underline{f}$ - [mec] to remove @ washer
— $\overline{v}$ — $\underline{f}$ de bronce $\underline{m}$ - [mec] to remove @ bronze
washer
— $\underline{v}$ — $\underline{f}$ para seguridad $\underline{f}$ - [mec] to remove @
lockwasher
— $\underline{v}$ árbol $\underline{m}$ - [botán] to remove @ shaft • [bot]
to remove @ tree
— $\underline{v}$ armadura $\underline{f}$ - [mec] to remove @ armature
— $\overline{v}$ aro $\underline{m}$ - [mec] to remove @ ring
— $\overline{v}$ — $\underline{m}$ para compresión $\underline{f}$ - [mec] to remove @
compression ring
— $\underline{v}$ — $\underline{m}$ — empaquetadura $\underline{f}$ - [mec] to remove @
packing ring
— $\underline{v}$ articulación $\underline{f}$ - [mec] to remove @ linkage
— $\overline{v}$ artículo $\underline{m}$ - [gram] to remove @ article
— $\overline{v}$ asiento $\underline{m}$ - [mec] to remove @ seat
— $\overline{v}$ — $\underline{m}$ para válvula $\underline{f}$ - [válv] to remove @
valve seat
— $\underline{v}$ astilla $\underline{f}$ - [mec] to remove @, splinter, $\underline{o}$
chip
— $\underline{v}$ atomizador $\underline{m}$ - [comb.int] to remove @ atomizer
— $\underline{v}$ banda $\underline{f}$ - [mec] to remove @ band
— $\overline{v}$ — $\underline{f}$ para freno $\underline{m}$ - [mec] to remove @ brake
band
— $\underline{v}$ barra $\underline{f}$ - [mec] to remove @ bar
— $\overline{v}$ barra $\underline{f}$ conformada (para perforación $\underline{f}$) -
[petról] to remove @ Kelly
— $\underline{m}$ barreno $\underline{m}$ - [petról] to remove @ drill bit
— $\overline{v}$ — $\underline{m}$ para perforación $\underline{f}$ - [petról] to re-
moe @ drill bit
— $\underline{v}$ bloque $\underline{m}$ - [mec] to remove @ block
— $\overline{v}$ bomba $\underline{f}$ - [bombas] to remove @ pump
— $\overline{v}$ boquilla $\underline{f}$ - [mec] to remove @ nozzle
— $\overline{v}$ — $\underline{f}$ para rociadura $\underline{f}$ - [mec] to remove @
spray nozzle
— $\underline{v}$ broca $\underline{f}$ - [petról] to remove @ bit
— $\overline{v}$ — $\underline{f}$ para perforación $\underline{f}$ - [petról] to remove @ drill bit
— $\underline{v}$ buje $\underline{m}$ - [mec] to remove @ bushing
— $\overline{v}$ — $\underline{m}$ para camisa $\underline{f}$ - [mec] to remove @
liner bushing
— $\underline{v}$ — $\underline{m}$ piloto - [mec] to remove @ pilot bushing
— $\underline{v}$ — $\underline{m}$ — para camisa $\underline{f}$ - [mec] to remove @
liner pilot bushing
— $\underline{v}$ bujía $\underline{f}$ - [comb.int] to remove @ spark plug
— $\overline{v}$ burro $\underline{m}$ - [metal-prod] to remove @ runner
scrap
— $\underline{v}$ cabeza $\underline{f}$ para inyección $\underline{f}$ - [petról] to remove @ swivel
— $\underline{v}$ cadena $\underline{f}$ - [mec] to remove @ chain
— $\overline{v}$ caja $\underline{f}$ - [mec] to remove @, box, $\underline{o}$ housing
— $\overline{v}$ — $\underline{f}$ para conmutador $\underline{m}$ - [mec-ventil] to
remove @ switch housing
— $\underline{v}$ calce $\underline{m}$ - [mec] to remove @ shim
— $\overline{v}$ calor $\underline{m}$ - to, remove, $\underline{o}$ take away, @ heat
— $\overline{v}$ calza $\overline{f}$ - [mec] to remove @ shim
— $\overline{v}$ calza $\overline{f}$ para piñón $\underline{m}$ - [mec] to remove @

pinion shim
quitar v cámara f - [mec] to remove @ chamber •
[autom-neumát] to remove @ tube
— **f — f para filtro m** - [mec] to remove @
filter chamber
— **v — f — — m para aire m** - [mec] to remove
@ air filter chamber
— **v camisa f** - to remove @ shirt • [mec] to remove @ liner
— **f campana f** - [mec] to remove @ bell • to remove @ canopy
— **v carga f** - [transp] to remove @ load
— **v cartucho m** - [mec] to remove @ cartridge
— **v casquete m** - [comb.int] to remove @ cap
— **v cierre m** - [mec] to remove @ seal
— **v cilindro m**—[comb.int] to remove @ cylinder
— **v cinta f** - [mec-frenos] to remove @ lining
— **v — f para freno m** - [mec] to remove @ brake
lining
— **v cojinete m** - [mec] to remove @ bearing
— **v — m con grasa f** - [mec] to remove @ grease
bearing
— **v colador m** - [mec] to remove @ strainer
— **m colector m** - [mec] to remove @ sump
— **v — para purga f** - [mec] to remove @ drain
sump
— **v combustible m** - to remove @ fuel
— **v completamente** - to remove completely
— **v con cuidado m** - to remove carefully
— **v — esmeriladora f** - [mec] to grinder remove
— **v — — f mecánica** - [mec] to remove with @
electrical grinder
— **v — extractor m** - [mec] to remove with @ extractor • to remove with @ puller
— **v — lima f** - [mec] to remove with @ file
— **v — — f manual** - [mec] to hand file remove
— **v — paño m** - to wipe (off)
— **v — pincel m** - to brush (off)
— **v conductor n** - [electr-instal] to remove @,
cable, o lead
— **v conjunto m** - [mec] to remove @ assembly
— **v — m de cabeza f** - [mec] to remove @ head
assembly
— **v — — — f para inyección f** - [petról] to
remove @ swivel assembly
— **v — m de distribuidor m para fuerza f** -
[autom-mec] to remove @ power divider assembly
— **v — m — llanta f** - [autom] to remove @
rim assembly
— **v — m — neumático n** - [autom-neumát] to remove @ tire assembly
— **v — m — portadiferencial** - [autom-mec] to
remove a differential carrier asembly
— **v contrapeso m** - [grúas] to remove @ counterweight
— **v cortocircuito m** - [electr-oper] to eliminate @ short circuit
— **v cruceta f** - [mec] to remove @ crosshead
— **v cubierta f** - [mec] to remove @ cover
— **v — f para mesa f** - [mec] to remove @ table
cover
— **v — f para transmisión f** - [mec] to remove @
transmission cover
— **v — principal** - [mec] to remove @ main cover
— **v cuerpo m** - [mec] to remove @ body
— **v — m extraño** - [medic] to remove @ foreign
object
— **v cuidadosamente** - to remove carefully
— **v chapa f** - [mec] to remove @ sheet
— **v — f para cubierta f** - [mec] to remove @
cover sheet
— **v de cajón m** - [transp] to remove from @ container • [mec] to remove from @, box, o case
— **v — eje m** - [mec] to remove from @ axle
— **v — encima** - to lift off of
— **v — motor m** - [comb.int] to remove from @
engine • [electr-mot] to remove from @ motor
— **v — orificio m** - [mec] to remove from @ hole
— **v — prensaestopa(s) m** - [mec] to remove from
@ packing box
— **v — separador m** - [mec] to remove from @
spacer
— **v defensa f** - [mec] to remove @ guard

quitar v defensa f para seguridad f - [segurid]
to remove @ safety guard
— **v demasiado poco** - to remove too little
— **v derivación f** - [electr-oper] to remove @,
bias, o shunt
— **v distribuidor m** - [mec] to remove @ manifold
— **v — m rociador** - [mec] to remove @ spray
manifold
— **v eje m** - [mec] to remove @, axle, o shaft
— **v elemento m** - [quím] to remove @ element •
[mec] to remove @ canister
— **v embalaje m** - [transp] to remove @ packing
— **v émbolo m** - [comb.int] to remove @ piston
— **v embrague m** - [mec] to remove @ clutch
— **v — m para elevación f** - [grúas] to remove
@ hoist clutch
— **v — m — mecanismo m para elevación f** -
[grúas] to remove @ hoist mechanism clutch
— **v — m — m — rotación f** - [grúas] to remove @ swing clutch
— **v empaquetadura f** - [mec] to remove @ packing
• [mec] to remove @ gasket
— **v en defecto m** - to remove too little
— **v — menos** - to remove too little
— **v engranaje m** - [mec] to remove @ gear
— **v escoria f** - [sold] to, remove, o clean, @
slag
— **v eslabón m** - [mec] to remove @ link
— **v — m de cadena f** - [mec] to remove @ chain
link
— **v esqueleto m** - [transp] to remove @ crate
— **v esquina f** - [mec] to remove @ corner
— **v fácilmente** - to remove readily
— **v fiador m** - [mec] to remove @ retainer
— **v — m para trinquete m** - [mec] to remove @
ratchet retainer
— **v filtro m** - [mec] to remove @ filter
— **v — m para aire m** - [mec] to remove @ air
filter
— **v freno m** - [mec] to remove @ brake
— **v fuga f** - [mec] to remove @ leak
— **v — f de agua m** - [mec] to stop @ water leak
— **v — f — gas m** - [metal-prod] to seal @ gas
leak
— **v — f — viento m** - [metal-prod] to, seal, o
stop, @ blast leak
— **v grasa f** - [mec] to remove @ grease; to degrease
— **v grillete m** - [mec] to remove @ clevis
— **v guarnición f** - [mec] to remove @ gasket
— **v — f circular** - [mec] to remove @ O ring
— **v guía(dera)f** - to remove @ guide
— **v — f para vástago m** - [válv] to remove @
stem guide
— **v — f — — m para válvula f** - [válv] to remove @ valve stem guide
— **v — f — — m — — f inferior** - [válv] to
remove @ lower valve stem guide
— **v — f — — m — — f superior** - [válv] to
remove @ upper valve stem guide
— **v horquilla f** - [mec] to remove @ clevis
— **v inhibición f** - to uninhibit
— **v injerto m** - [mec] to remove @ insert
— **v — m para válvula f** - [válv] to remove @
valve insert
— **v laminilla f** - [mec] to remove @ shim
— **v — f para piñón m** - [mec] to remove @ pinion shim
— **v lodo m** - to remove @ mud • [petról] to remove @ sludge
— **v material m** - to remove @ material
— **v — m extraño** - to remove @ foreign material
— **f mezcla f** - to remove @ mixture • [metal-
prod] to kish
— **v montaje m** - [mec] to remove @ mounting
— **v mordaza f** - [mec] to remove @, muzzle, o
clamp
— **v neumatico m** - [autom-neumát] to remove @
tire
— **v nieve f** - to remove @ snow
— **v paleta f** - [mec] to remove @ blade
— **v panel m** - [mec] to remove @ panel
— **v — m posterior** - [mec] to remove @ rear
panel

- 1249 -

quitar v para inspección f - to remove for inspection
— v — limpieza f—[mec] to remove for cleaning
— v pasador m - [mec] to remove @ pin
— v — m conectador - [mec] to remove @, connecting, o linkage, pin
— v — m para grillete m - [mec] to remove @ clevis pin
— v — m — horquilla f - [mec] to remove @ clevis pin
— v periódicamente - to remove periodically
— v perno m - [mec] to remove @ bolt
— v — m herrumbrado - [mec] to remove @ rusty bolt
— v — m para sujeción f - [mec] to remove @ hold-down bolt
— v — m — m para camisa f - [mec] to remove @ clamp(ing) bolt
— v — m para traba f - [mec] to remove @ lock bolt
— v pico m - [mec] to remove @ tip
— v — m para engrase m - [mec] to remove @ grease fitting
— v pieza f - [mec] to remove @ part
— v — f para bomba f - [bombas] to remove @ pump part
— v piso m - [constr] to remove @ floor
— v placa v - [mec] to remove @ plate
— v plancha f - [mec] to remove @ plate
— v — f para piso m - [mec] to remove @ floor plate
— v polvo m - to (remove @) dust
— v prenda f - [vest] to remove @ garment
— v prensaestopa(s) m - [mec] to remove @ packing box
— v presión f - [mec] to remove @ pressure
— v — f alta - [mec] to remove @ high pressure • [metal-prod] to cut off @ high pressure
— v — f — en tragante m - [metal-prod] to cut off @ furnace top high pressure
— v puente m - [constr] to remove @ bridge • [electr-instal] to remove @ jumper
— v rápidamente - to remove quickly
— v rebaba(s) f - [mec] to remove @ burr(s)
— v recalcadura f - [mec] to remove @ upsetting
— v — f metálica - [mec] to remove @ metal upsetting
— v rejilla f - [mec] to remove @, grill, o screen
— v resguardo m - [mec] to remove @ guard
— v resistencia f - [electrón] to remove @ resistance
— v resorte m - [mec] to remove @ spring
— v restricción f - to remove @ restriction
— v retén m - [mec] to remove @ retainer
— v — m para cierre m - [mec] to remove @ seal retainer
— v — m — sello,lladura - [mec] to remove @ seal retainer
— v retenedor m - [mec] to remove @ retainer
— v rotor m - [electr-mot] to remove @ rotor
— v rueda f - [mec] to remove @ wheel
— v — f motriz - [mec] to remove @ drive wheel
— v sección f - [mec] to remove @ section
— v — f de plancha f - [mec] to remove @ plate section
— v — f — f para piso m - [mec] to remove @ floor plate section
— v — para piso m - [constr] to remove @ floor section
— v segmento m - [mec] to remove @ segment
— v — m para vástago m - [mec] to remove @ subrod
— v sello m - [mec] to remove @ seal
— v semieje m - [autom-mec] to remove @ axle shaft
— v separador m - [mec] to remove @ separator • to remove @ spacer
— v — m para armado m - [mec] to remove @ assembly,bling spacer
— v solenoide m - [electr-instal] to remove @ solenoid
— v soplado m - [metal-prod] to cut off @ blast

quitar v soporte m - [mec] to remove @ holder
— v suciedad f - to remove @, dirt, o filth
— v suelo m - to remove @ soil
— v sujetador m - [mec] to remove @, fastener, o clamp
— v — m para camisa f - [mec] to remove @ liner clamp
— v — m — conductor m - [electr-instal] to remove @ cable clamp
— v — m — conexión f - [electr-instal] to remove @ connecting clamp
— v — m — f para conductor m - [sold] to remove @ cable connecting clamp
— v suplemento m - [mec] to remove @ insert • to remove @ supplement
— v — m para valvula f - [válv] to remove @ valve insert
— v tapa f - [mec] to remove @, cover, o cap
— v — f para colector m - [mec] to remove @ sump cover
— v — f — mesa f - [mec] to remove @ table, cover, o top
— v — f — transmisión f - [mec] to remove @ transmission cover
— v — f — válvula f - [válv] to remove @ valve cover
— v — f principal—[mec] to remove @ main cover
— v tapón m - [mec] to remove @ plug
— v — m para drenaje m - [mec] to remove @ drain(age) cover
— v — m — filtro m - [mec] to remove @ filter plug
— v — m para freno m - [mec] to remove @ brake plug
— v — m — nivel m - [mec] to remove @ brake plug
— v — m — m de aceite m - [mec] to remove @ oil level plug
— v — m — — m en caja f - [mec] to remove @ case oil level plug
— v — m — m — m — f para cadena f.- [mec] to remove @ chain case oil level plug
— v — m para purga f - [mec] to remove @ drain plug
— v — m — f para colector m - [mec] to remove @ manifold drain plug
— v — m — f — m para aspiración f - [mec] to remove @ suction manifold drain plug
— v — m para tubo(s) m - [mec] to remove @ pipe plug
— v tierra f - to remove @ dirt
— v toberón m - [metal-prod] to remove @ cooler
— v — m para escorial m - [metal-prod] to remove @ slag notch cooler
— v tornillo m - [mec] to remove @ screw
— v — m con casquete m - [mec] to remove @ cap screw
— v trinquete m - [mec] to remove @ ratchet
— v tubería f - [tub] to remove @ tube,bing
— v tubo m - [tub] to remove @, tube, o pipe
— v tuerca f - [mec] to remove @ nut
— v — f herrumbrada - [mec] to remove @ rusty nut
— v unidad f - to remove @ unit
— v válvula f - [válv] to remove @ valve
— v — f inferior - [válv] to remove @ lower valve
— v — f superior - [válv] to remove @ upper valve
— v vástago m - [mec] to remove @ rod • [válv] to remove @ stem
— v — m para válvula f - [válv] to remove @ valve stem
— v — m — f inferior - [válv] to remove @ lower valve stem
— v — m — f superior - [válv] to remove @ upper valve stem
— v — m intermedio - [mec] to remove @ intermediate rod
quorum m reglamentario m - [legal] required quorum

R

rabo m - . . . . • [mec] tang
racionalización f - rationalization,zing
— f de marca(s) f - [ind] make(s) rationalization,zing
— f — modelo(s) m - [ind] model(s) rationalization,zing
— f — producción f - [ind] production rationalization,zing
— f económica - economic rationalization,zing
racionalizado,da a - rationalized
racionalizar v - to rationalize
racha v - . . . . • rash
radiación f - . . . . • irradiation
— f alta - [nucl] high (level) (ir)radiation
— f baja - [nucl] low (level) (ir)radiation
— f de desecho(s) m - [nucl] waste(s) (ir)ra-
diating,tion
— f — intensidad f alta - [nucl] high intensity (ir)radiation
— f — — f baja - [nucl] low intensity (ir)ra-
diation
— f — — f mayor - [nucl] higher intensity
(ir)radiation
— f — — f menor - [nucl] lower intensity
(ir)radiation
— f — residuo(s) m - [nucl] waste(s) (ir)ra-
diating,tion
— f — — m sólido(s) m - [nucl] wold waste(s).
(ir)radiating,tion
— f — máquina f para rayos-X - [electrón] X-ray
machine radiation
— f elevada - [nucl] high (level) radiation
— f lumínica f - [segurid] light radiation
— f mayor - greater, o higher, radiation
— f menor - lower, o lesser, radiation
— f penetrante - [electrón] penetrating radiation
— f — gamma - [electrón] penetrating gamma
radiation
— f — X - [electrón] penetrating X radiation
— f reducida - [nucl] low level radiation
— f resistida - [nucl] resisted radiation
— f superior - [nucl] high(er) radiation
radiactividad f - . . . .; fallout • radiation
level
— f acumulada—[nucl] accumulated radioactivity
— f alfa - [nucl] alpha radioactivity
— f alta - [nucl] high radioactivity
— f baja - [nucl] low radioactivity
— f beta - [nucl] beta radioactivity
— f elevada - [nucl] high radioactivity
— f empleada - [nulc] used radioactivity
— f ensayada - [nucl] tested radioactivity
— f gamma - [nucl] gamma radioactivity
— f intermedia - [nucl] intermediate radioactivity
— f mediana - [nucl] intermediate radioactivity
— f reducida - [nucl] low radioactivity
— f total - [nucl] total radioactivity
— f — almacenada - [nucl] total stored radioactivity
radioactivo,va a - [nucl] . . .; hot
radiador m - . . . .; irradiator • [mec] gill
— m atascado - [comb.int] clogged radiator
— m comprobado - [comb.int] checked radiator
— m con frente f en V - [comb.int] V front radiator
— m con presión f - [comb.int] pressurized radiator
— m — tubo(s) m - [comb.int] tube radiator
— m — — m achatado(s) n - [comb.int] flat
tube radiator
— m corriente - [int.comb] standard radiator
— m con aceite m - [mec] standard oil radiator
— m de aceite m - [comb.int] oil radiator
— m — — m de red f hidráulica - [mec] hydrau-

lic system oil radiator
radiador m de calor m - heat radiator
— m — panal m - [comb.int] honeycomb radiator
— m — presión f - [comb.int] pressure,rized
radiator
— m — — f positiva - [comb.int] positive
pressure type radiator
— m — tipo m con presión f - [comb.int] pressure type radiator
— m — — m — — f positiva - [comb.int] positive pressure type radiator
— m enjuagado - [comb.int] flushed radiator
— m estándar - [mec] standard radiator
— m para aceite m - [mec] standard oil radiator
— m exterior - [comb.int] external radiator
— m — limpiado - [comb.int] cleaned external
radiator
— m — limpio - [comb.int] clean external radiator
— m — comprobado - [comb.int] checked external
radiator
— m — verificado - [comb.int] checked external
radiator
— m grande - [comb.int] large, o big, radiator
— m — de tipo m con presión f - [comb.int]
large, pressurized, o pressure type, radiator
— m limpiado - [comb.int] cleaned radiator
— m — con vapor m - [comb.int] steam cleaned
radiator
— m limpio - [comb.int] clean radiator
— m llenado - [comb.int] filled radiator
— m para motor m - [comb.int] engine radiator
— m regulado - [comb.int] controlled radiator
— m — con termostato m - [comb.int] thermostat
controlled radiator
— m — termostáticamente - [comb.int] thermostatically controled radiator
— m sin agua m - [comb.int] dry radiator
— m tabicado - [comb.int] partitioned radiator
— m tipo vertical - [comb.int] vertical (type)
radiator
— m — m — con tubo(s) m - [comb.int] vertical
type tube radiator
— m — m — — m achatado(s) - [comb.int]
vertical type flat tube radiator
— m tubular - [comb.int] tube,bular radiator
— m verificado - [comb.int] checked radiator
— m vertical - [comb.int] vertical radiator
— m — achatado - [comb.int] vertical flat radiator
— m — con tubo(s) m - [comb.int] vertical tube
radiator
— m — — — m achatado(s) m - [comb.int] vertical flat tube radiator
radial a - radial • radius
— a callejero,ra a - [autom-neumát] street radial
— a con rendimiento m sumamente alto - [autom-
-neumát] ultra-high performance radial
radio m de acción f corto - [aeron] short range
— m — f largo - [aeron] long range
— m — f mediano - [aeron] medium range
— m — acordamiento m - transition radius
— m — alcantarilla f - [tub] culvert, radius,
o one-half diameter
— m — borde m para ataque m - [mec] attack
edge radius
— m — cabeza f - [mec] head radius
— m — f de riel m - [metal-lam] rail head
radius
— m — cara f - face radius
— m — f frontal—[cojin] front face radius
— m — — f de cono m - [cojin] cone front
face radius
— m — — f de taza f - [cojin] cup front
face radius   ius
— m — — f posterior - [cojin] back, o rear,
face radius
— m — — — de cono m - [cojin| cone, back,
o rear, face radius
— m — — f — de taza f - [cojin] cup, back, o
rear, face radius

radio m de curvatura f - curvature, o bending, radius
— m — diámetro m - [tub] diameter radius
— m — dobladura f - bend(ing) radius
— m — eje m - [mec] axle radius
— m — — m longitudinal - center line radius
— m — engranaje m - [mec] gear radius
— m — esquina f - [constr] corner radius
— m — esquinero m - [constr] corner radius
— m — fondo m - [constr] bottom radius
— m — formación f - [geom] forming radius
— m — — f interior - [geom] inside corner forming radius
— m — giro m - [mec] turning, o swing, radius; gyration radius
— m — — m pequeño—[grúas] small swing radius
— m — línea f central - centerline radius
— m — — f para curvatura f - curvature centerline radius
— m — mandril m - [cojin] plug radius
— m — nudillo(s) m - [cald] knuckle radius
— m — pared f - [tub] wall radius
— m — — f de conducto m - [tub] conduit wall radius
— m — plancha f - [metal-lam] plate radius
— m — — f esquinera - [constr] corner plate radius
— m — raíz f - [sold| root radius
— m — rueda f - [mec-ruedas] wheel radius
— m — superficie f - [geom] surface radius
— m — — f para rodadura f - [ruedas] wheel radius
— m — tapa f - [constr] cover radius
— m — taza f - [cojin] cup radius
— m — tiro m - [milit] fire,ring radius
— m — tope m - [constr] top radius
— m — transición f - [cald] knuckle radius
— m — tubería f - [tub] pipe radius
— m — vértice f de cabeza f de riel m - [metal-lam] rail head corner radius
— m — viraje m - [mec] turning radius
— m esquinero - [mec] corner radius
— f estereofónica - [telecom] stereo(phonic) radio
— m exterior - external, o outer, o outside, radius
— m — de curva f - curve outside radius
— m — — tubería f - [tub] pipe outside radius
— m externo - external radius
— m exterior para acordamiento m - transition, outer, o external, radius
— m grande - large radius
— m hidráulico - [hidr] hydraulic radius
— m interior - internal, o inside, o inner, radius
— m — de curva f - curve inside radius
— m — — tubería f - [tub] pipe inside radius
— m interno - internal radius
— m — de acordamiento m - transition, inner, o internal, radius
— m largo - long radius
— m lateral - side, o lateral, radius
— m liso - smooth radius
— m máximo - maximum radius
— m — de curvatura f—maximum bend(ing) radius
— m — — transición f - maximum knuckle radius
— m — m para doblamiento m - [tub] maximum bend(ing) radius
— m mayor - greater, o longer, radius
— m mínimo - [mec] minimum radius
— m — curvatura f - [tub| minimum bending radius
— m — — transición f - minimum knuckle radius
— m — para doblamiento m - minimum bend(ing) radius
— m para descarga f - [constr] dumping radius
— m — evacuación f - [constr] dumping radius
— m — operación f - operating radius
— m para sección f trasera - [mec] rear end radius
— m — viraje(s) m - [autom] turning radius
— m pequeño - small radius
— m reducido - small radius
— m superior - top, o upper, radius

radio m superior - top, o upper, radius
— m urbano m - [pol] city limit(s)
radioactividad* f - véase radiactividad f
radioactivo,va a - véase radiactivo,va a
radioenlace m - [telecom] radio linkage
radioelemento m - [nucl] radioactive element
— m almacenado - [nucl] stored radioactive element
radiofrecuencia f - [electrón] radio frequency
— f con calidad f para voz f - [electrón] voice grade radio frequency
radiografía f - [electrón] radiograph(y); X-ray picture
— f de defecto m - [electrón] defect radiograph
— f — m en soldadura f - [sold] weld defect radiograph
— f — soldadura f - [sold] weld, X-ray, o radiograph(y)
radiomático m - [instrum] radiomatic pyrometer
radiooyente m - [electrón] radio listener
radiotelefonear v - [telecom] to radiotelephone
radiotelefonía f - [telecom] . . .; radiotele-phone; radio; wireless
— f bidireccional - [telecom] two-way radio
radiotelefónico,ca a - [telecom] . . .; wireless
radioteléfono m - [telecom] . . .; wireless
radiotelegrafiar v - [electrón] to radio(tele-graph)
raedera f - . . . • (agric-equip) jointer
ráfaga f - . . .; burst
— f automática - [electrón] automatic burst
— f de impulso(s) m - [electrón] impulse(s) burst
— f — — m transmitida - [electrón] transmit-ted impulse(s) burst
— f — — m — automáticamente - [electrón] automatically transmitted impulse(s) burst
— f transmitida - [electrón] transmitted burst
— f — automáticamente - [electrón] automat-ically transmitted burst
raído,da a - . . .; abraded
raíz f - [botán] . . . • [sold] root; stringer bead
— f de árbol m - [botán] tree root
— f — cordón m - [sold] bead root
— f — diente m - tooth root
— f — — m de engranaje m - [mec] gear tooth root
— f — junta f - [sold] root joint
— f — soldadura f - [sold] weld root
raja f - . . . • [mec] véase grieta f
rajatubo(s) m - [petról] casing ripper
rajado,da a - - - . . .; ripped • cloven
rajadura f - [mec] . . .; cut; split; rend; flaw; chap
— f en cigüeñal m - [comb.int] crankshaft crack
rajar v cigüeñal v - [comb.int] to crack @ crankshaft
ralentí* - [mec] véase marcha f sin carga f
ralentir* v - véase marchar v sin carga f
rallado,da a - [culin] grated
Rally m - [deport] Rally
— m en Reino m Unido - [deport] (United King-dom) Royal Automobile Club Rally
— m entre pilón(es) m - [deport] rally cross
rama f - . . .; limb
— f de eslinga f - [cabl] sling, branch, o leg
— f ejecutiva - [pol] executive branch
— f judicial - [pol] judicial branch
— f legislativa - [pol] legislative branch
— f para investigación(es) f - investigating, o research, branch, o arm
— f — — f y perfeccionamiento m - [ind] re-search and development. branch, o arm
— f — perfeccionamiento m - development, arm, o branch
— f sumergida - submerged branch
ramal m - . . . • [vial] extension; spur route • [cabl] (each) pulley (in @ block) • [electr-instal] branch circuit; offset; Y • [f.c.] branch line • [tub] lateral • [cadenas] leg
— m aferente - [hidr] inlet
— m central - [cabl] central strand
— m cloacal - [sanit] branch sewer

ramal m con . . . hilo(s) m - [cabl] . . .-wire
  strand
— m con porte m grande—[tub] king size lateral
— m — sección f triangular - [cabl] triangular
  section strand
— m conectador - [ambient] connecting branch
— m continuo - [cabl] continuous strand
— m de alambre m - [cabl] wire strand
— m — cable m - [cabl] (wire) rope strand
— m — de desagüe m - [sanit] sewer branch
— m — — m pluvial - [constr] storm sewer
  feeder
— m — fibra f - [cabl] fiber strand
— m — — f vegetal - [cabl] vegetable fiber
  strand
— m domiciliario - [tub] house lateral • véase
  también acometida f
— m envuelto m - [cabl] wrapped strand
— m — separadamente—[cabl] separately wrapped
  strand
— m ferroviario m—[f.c.] branch rail(way) line
— m galvanizado - [cabl] galvanized strand
— m — simple - [cabl] single galvanized strand
— m — único m—[cabl] single galvanized strand
— m liso - [cabl] smooth strand
— m para acortamiento m - [f.c.] cutoff
— m — alimentación f - [f.c.] feeder line
— m — barbetar v - [cabl] seizing strand
— m — cablecarril m - [cabl] track strand
— m — cortar - [cabl] cutting strand
— m — — f piedra f - [cabl] stone sawing
  strand
— m — corte m - [cabl] cutting, o sawing
  strand
— m — drenaje m - [tub] flushing line
— m para enrollamiento m - [cabl] coil(ing)
  strand
— m — — m liso - [cabl] smooth coil(ing)
  strand
— m — eslinga f - [cadenas] sling leg
— m — inyección f de agua m - [tub] water (injecting) line
— m — — f — — m para lavado m - [tub] washing water line
— m — ligar v - [cabl] seizing strand
— m — purga f - [tub] flushing, o purging,
  line; drip leg
— m — sujeción f - [cabl] seizing strand
— m — tranvía m - [cabl] track strand
— m — vapor m - [tub] steam line
— m — — m para purga f - [tub] steam purging,
  o purging steam, line
— m revestido - [cabl] clad strand
— m — separadamente - [cabl] separately clad
  strand
— m — — con merlín m - [cabl] separately marline clad strand
— m trenzado - [cabl] braided strand
— m — en hélice m - [cabl] helically braided
  strand
— m triangular - [cabl] triangular strand
— m único - [cabl] single strand
ramificación f - remification; branch
ramo m - . . . • [com] . . .; line • area •
  business
— m de seguros m - [seguros] insurance line
rampa f - [mec] . . .; inclined plane; chute •
  [vial] grade; incline; ascent • [ind] wharf •
  bridge
— f con declive - [constr] sloping ramp
— f de tipo m para interconexión f de superca-
  rretera(s) f - [vial] interchange type ramp
— f doble - [constr] double ramp
— m — para, acceso m, o entrada f - [constr]
  double entry ramp
— f — — entrada f para semirremolques carreteros para transporte m sobre vagón m plataforma - [f.c.] double entry piggyback ramp
— f húmeda - [ind] wet, o slippery, ramp, o
  dump
— f inclinada - [constr] sloping ramp
— f para acceso m - [vial] access, o on-bound,
  o approach, o on, ramp; ramp

rampa f para acceso m de tipo m para intercone-
  xion f de supercarretera(s) f - [vial] interchange-type on ramp
— f — — m para peatón(es) m - [constr] pedestrian approach ramp
— f — aceleración f—[vial] acceleration ramp
— m — almacenamiento m de bobina(s) f—[metal-
  -lam] coil storage ramp
— f — — — — f para entrada f - [metal-
  -lam] (mill) feed end coil storage ramp
— f — botadura f - [naút] slipway
— f — — f para barco(s) m - [naut] boat
  launching ramp; slipway
— f   caída f - [ind] discharge chute
— m — — f para máquina f para colar v -
  [metal-prod] pig casting machine discharge
  chute
— f — carga f - [ind] loading ramp
— f — cinta f transportadora - [mec] conveyor
  ramp
— f — — f — para soldadora f - [mec] conveyor to @ welder ramp
— f — coque m - [coque] coke ramp • coke wharf
— f — egreso m - [vial] off ramp
— f — — m de tipo m para interconexión f de
  supercarretera(s) f - [vial] interchange-type
  off ramp
— f — embarcación(es) f - [nav] boat ramp •
  slipway
— f — encaladora f - [metal-prod] lime, machine, o mixing vat, chute
— f para engrasar v - [autom-mec] grease,sing
  rack, o ramp
— f para horno m - [metal-prod] furnace ramp
— f — lanzamiento m - [naút] slipway; boat
  ramp • [f.c.] humping ramp
— f — leva f - [mec] cam ramp
— f — lubricar v - [autom-mec] grease,sing
  rack
— f — maíz m -   agric-equip] corn, o maize,
  chute
— f — mineral m - [metal-prod] ore chute
— f — retardación f - [vial] deceleration ramp
— f — salida f de horno m - [metal-lam] furnace exit ramp
— f — semirremolgue(s) m (carreteros) - [vial]
  piggyback ramp
— f — — m para transporte m sobre vagón(es)
  m plataforma - [f.c.] piggyback ramp
— f para velocidad f - [vial] retarder
— f salmonera - [piscicultura] fish ladder
— f seca - dry, ramp, o dump
— f suave - [constr] long approach
rampollo m - [botán] . . .; sprig
rancho m - . . . • [constr] shack • [RP1] hut
rango m - . . . • rating • véase también límite
  m; escala f; gama* f
rangua f - . . . • [metal] brass
ranura f - [mec] . . .; channel • gain • notch;
  keyhole notch • rabbet • pad way • furrow;
  fold; trough • open, groove, o slot • spline
  • [cojin] race
— f abierta - [mec] open groove
— f — en centro m - [autom-neumát] open center
  groove
— f achaflanada - [sold] bevel(ed) groove
— f alineada - [mec] aligned slot
— f angosta - [mec] narrow groove
— f — y profunda - [sold] deep narrow groove
— f atascada - [mec] clogged groove
— f biselada - bevel(led) groove
— f calentada - [sold] heated, crack, o groove
— f cerrada - [mec] closed, groove, o slot
— f central - [mec] central groove
— f — abierta - [autom-neumát] open central
  groove
— f colocada en tresbolillo m - [mec] staggered
  groove
— f comparada - compared groove
— f con ángulo(s) m recto(s) - [mec] square
  groove
— f — chaflán m - [sold] V-groove; V-butt
— f — — m doble - [sold] double bevel groove

ranura f con chaflán m único - [sold] single
  bevel groove
— f — rejilla f - [mec] slotted grate opening
— f — solapa f - [mec] lap(ped) groove
— f — — — con diámetro m de alambre m - [mec]
  wire diameter lap(ped) groove
— f continua - [mec] continuous, groove, o slot
— f contrastada - [mec] gaged groove
— f cortada f - [mec] cut groove
— f correcta - [mec] correct groove
— f correspondiente - [mec] matching, groove, o
  slot
— f de adaptador m - [mec] adapter slot
— f — fundición f - [mec] cast groove
— f desgastada - [mec] worn groove • worn
  spline
— f — excesivamente - [mec] excessively worn
  spline
— f diagonal - [mec] diagonal groove
— f en árbol m - [mec] shaft, slot, o spline
— f — — m engranada - [auto-mec] engaged
  shaft spline
— f — — m — para, entrada f, o aportación f,
  de fuerza f engranada - [autom-mec] engaged
  input shaft spline
— f en aro m - [cojin] ring (ball) race
— f — — m exterior m - [cojin] outer ring
  ball race
— f — m interior - [cojin] inner ring ball
  race
— f — bloque m - [mec] block groove
— f — borde m - [mec] edge groove
— f — — m de banda f - [autom-neumát] shoulder groove
— f — — m — — f para rodamiento m - [autom-
  -neumát] (tread) shoulder groove
— f — m de mesa f - [mec] table rim slot
— f — carrete m - [mec] reel groove
— f — clavija f - [mec] key, seat, o slot
— f — cojinete m - [cojin] bearing race
— f — — m para espiga f - [mec] shank bearing
  race
— f — chapa f - [mec] sheet slot
— f — — f para cubierta f - [mec] cover(ing)
  sheet slot
— f — eje m - [mec] axle, o shaft, spline
— f — — m para entrada f (de fuerza f)—[mec]
  input shaft spline
— f — — salida f (de fuerza f) - [mec]
  output shaft spline
— f — émbolo m - [comb.int] piston groove
— f — embrague m - [mec] clutch spline
— f — — m deslizante - [mec] sliding clutch
  spline
— f — engranaje m - [mec] gear spline
— f — — m lateral - [mec] side gear spline
— f — espiga f - [mec] shank race • [tub]
  spigot groove
— f — espiral - [mec] spiral(led), o helical,
  groove
— f — fiador m - [mec] retainer slot
— f — — m para trinquete m - [mec] ratchet
  retainer slot
— f — fondo m - [mec] bottom slot
— f — — de gamella f - [miner] riffle
— f — garrucha f - [mec] sheave groove
— f — — f verificada - [mec] checked sheave
  groove
— f — — f —, de vista, o visualmente - [mec]
  visually checked sheave groove
— f — hélice - [mec] helical, o spiral(led),
  groove
— f — J - [sold] J groove
— f — neumático m - [autom-neumát] tire groove
— f — pestaña f - [mec] flange groove
— f — piñón m - [mec] pinion, groove, o spline
— f — — m impulsor m - [mec] drive,ving pinion spline
— f — — m para impulsión f - [mec] drive,ving
  pinion spline
— f — plato m - [mec] plate slot • spindle
  slot
— f — polea f - [mec] sheave groove
— f — — f verificada - [mec] checked sheave

groove
ranura f en polea f verificada, de vista, o visualmente - [mec] visually checked sheave
  groove
— f — retén m - [mec] retainer slot
— f — roldana f - [mec] sheave groove
— f — — f verificada - [mec] checked sheave
  groove
— f — — f —, de vista, o visualmente - [mec]
  visually checked sheave groove
— f — semieje m—[autom-mec] axle shaft spline
— f — superficie f - [mec] surface slot
— f — tambor m - [cabl] drum groove
— f — tresbolillo - [mec] staggered groove
— f — tuerca f - [mec] nut slot
— f — — f alineada - [mec] aligned nut slot
— f — U - [mec] U groove
— f — — doble - [sold] double U groove
— f — — sencilla - [sold] single U groove
— f — V - [sold] V groove
— f — — doble - [sold] double V groove
— f — — sencilla - [sold] single V groove
— f — válvula f - [válv] valve groove
— f endurecida - [cojin] hardened race
— f engranada - [mec] engaged spline
— f escopleada - [mec] slotted groove • cut
  keyway
— f — con arco m - [sold] arc gouged groove
— f — para chaveta f - [mec] slotted key
  groove
— f esmerilada - [mec] ground groove
— f especial - [mec] special, slot, o groove
— f espiralada - [mec] involute spline
— f — en engranaje m - [mec] involute gear
  spline
— f — — m lateral - [mec] involute side
  gear spline
— f expuesta f - [mec] exposed spline
— f exterior - [mec] outside groove • [cojin]
  outer race
— f — en cojinete m - [cojin] outer bearing,
  o bearing outer, race
— f — para bola(s) f - [cojin] outer ball race
— f faltante - [mec] missing groove
— f — para aceite m - [mec] missing oil groove
— f fresada - [mec] cut groove
— f hacia borde m (exterior) - [autom-neumát]
  outside shoulder groove
— f — m interior - [autom-neumát] inside
  shoulder groove
— f inferior - [cojin] lower race
— f — en cojinete m - [cojin] bearing lower
  race
— f interior - [mec] inside, o inner, groove •
  [cojin] inner race
— f — en cojinete m - [cojin] inner bearing, o
  bearing inner, race
— f — para bola(s) f - [cojin] inner ball race
— f intermedia f - [cojin] middle race
— f — en cojinete m - [cojin] bearing middle
  race
— f labrada - [mec] wrought, o opened, groove
— f lateral - [mec] lateral groove
— f — en borde m (de banda f) - [autom-neumát]
  lateral shoulder groove
— f — — m — — f para rodamiento m -
  [autom-neumát] shoulder lateral groove
— f — profunda - [autom-neumát] deep lateral
  groove
— f — — en borde m de banda f (para rodamiento) - [autom-neumát] deep lateral soulder, o
  shoulder deep lateral, groove
— f mecanizada - [mec] machined groove
— f para aceite m - [mec] oil(ing) groove
— f monobiselada - [sold] single bevel groove
— f original - [sold] original groove
— f para aceite m faltante - [mec] missing oil
  groove
— f — — m atascada - [mec] clogged oil groove
— f — — m lubricante - [mec] (lubricating)
  oil groove
— f — — m para lubricación f - [mec] lubricating oil, groove, o trough
— f — apoyo m - [mec] pad way

ranura f para aro m - [comb.int] ring groove
— f —— m para émbolo m - [mec] piston ring
groove
— f — bola(s) f - [cojin] (ball, o bearing,)
race
— f —— f endurecida - [cojin] hardened ball
race
— f —— f para aro m - [cojin] ring ball race
— f —— f —— m exterior - [cojin] outer
ring ball race
— f —— f —— m interior - [cojin] inner
ring ball rac
— f —— f para cojinete m - [cojin] bearing
ball race
— f —— f refrentada - [cojin] ground ball
race
— f — bolilla(s) f - [cojin] ball race
— f — circulación f de aceite m lubricante -
[mec] (lubricating) oil groove
— f — cuña f - [mec] key seat; keyway
— f — chaveta f - [mec] key groove; keyway
— f —— f escopleada - [mec] cut keyway
— f —— f falsa—[mec] dummy, o false, keyway
— f — eliminación f - [mec] gutter groove
— f — enganche m - [mec] lock(ing) slot
— f —— m en borde m de mesa f - [mec] table
rim lock(ing) slot
— f — impulsión f - [mec] drive,ving slot
— f — lubricación f - [mec] oil(ing) groove
— f — perno m - [mec] bolt slot
— m —— m para sujeción f - [mec] bolt holding slot
— m —— m —— f a poste m - [mec] post bolt
holding slot
— f — planta(s) f - [agric] planter,ting slot
— f — solapa f - [mec] lap groove
— f —— f con diámetro m de alambre m - [mec]
wire diameter lap groove
— f — soldadura f - [sold] welding slot
— f — sujetar v - [mec] gripping, o holding,
groove, o slot
— f —— v alambre m - [mec] wire, grip(ping),
o holding, groove, o slot
— f profunda - [sold] deep, groove, o slot
— f — con chaflán m - [sold] bevel(led) deep
groove
— f —— — m pequeño - [sold] small
bevel(led) deep groove
— f — en borde m de banda f (para rodamiento
m) - [autom-neumát] shoulder deep groove
— f — plana f - [sold] flat deep groove
— f puesta a descubierto - [mec] exposed spline
— f pulida - [mec] polished groove
— f raspada - [mec] scraped groove • lapped
groove
— f reconstruida f—[mec] rebuilt, o redressed,
groove
— f recta - [mec] straight, spline, o groove
— f — en engranaje m - [autom-mec] straight
gear spline
— f —— — m lateral - [autom-mec] straight
side gear spline
— f refrentada - [cojin] ground race
— f superior - [cojin] upper race
— f — en cojinete m - [cojin] bearing upper
race
— f transversal - [mec] transverse, o cross,
groove
— f — esmerilada—[mec] ground cross groove
— f verificada - [mec] checked groove
— f —, de vista, o visualmente - [mec] visually checked groove
ranuración f - [mec] grooving; slotting
— f de extremo m - [mec] end, grooving, o slotting • end slot
— f — neumático m - [autom-neumát] tire
grooving
— f — tambor m - [cabl] drum grooving
ranurado m - grooving
ranurado,da a - grooved; slotted; fluted; castellated; notched
ranurador m - [herram] grooving tool; groover
— m mecánico - [herram] mechanical groover;

jumper
ranuradora f - [mec] slotter; key seater •
[herram] groover
— f para asfalto m - [herram] asphalt groover
— f — hormigón m - [herram] concrete groover
ranuramiento* m - [mec] véase ranuración f;
splining
ranurar v - to groove; to notch; to slot; to
castellate
— v extremo m - [mec] to slot @ end
— v neumático m - [autom-neumát'] to groove @
tire
— v tubería f - [tub] to groove @ pipe
— v — f para agua m - [tub] to groove @ water
pipe • [petról] to groove @ wash pipe
rapel* m - véase descuento m
rápidamente adv - . . .; quickly
rapidez f - . . .; speed • rate
— f aceptable - acceptable speed
— f alta - high rate
— f — para aportación f - [sold] high deposition rate
— f buena - good speed
— f crítica - [sold] critical, speed, o rate
— f — para emfriamiento m - [sold] critical
cooling rate
— f de aceleración f - [mec] acceleration rate
— f —— f de, carro m, o carrillo m - [grúas[
trolley acceleration (rate)
— f —— f de puente m - [grúas] bridge acceleration (rate)
— f —— f —— m (de) grúa f - [grúas]
bridge acceleration (rate)
— f — agotamiento m - depletion, speed, o
rate • [hidr] draining, speed, o rate
— f — alimentación f - [mec] feed(ing), speed,
o rate
— f —— f de alambre m - [sold] wire feed-
-(ing), speed, o rate
— f —— f — electrodo m - [sold] electrode,
o wire, feed(ing), speed, o rate
— f — aportación f - [sold] deposit(ion) rate
— f —— f considerable - [sold] fast depo-
sit(ion) rate
— f —— f de metal m para soldadura f -
[sold] weld metal deposit(ion) rate
— f — que puede emplear(se) - [sold] usable
deposit(ion) rate
— f —— f utilizable - [sold] usable depo-
sit(ion) rate
— f — aporte m - [sold] deposit(ion) rate
— f — arco m - [sold] arc speed
— f — alta - [sold] high arc speed
— f — avance m - [sold] (travel) speed
— f —— m con separación f de . . . - [sold]
speed with @ . . . gap
— f —— m prefijada - [sold] preset travel
speed
— f —— m recomendado - [sold] recommended
travel speed
— f —— m — con separación f de . . . -
[sold] recommended travel speed with @ . . .
gap
— f —— m sin separación f - [sold] speed
without @ gap
— f — bajada f - [mec] drop(ping) speed
— f — caída f - fall(ing) speed
— f — corrosión f - [metal] corrosion speed
— f — cumplimiento m - compliance, o accomplishment, speed, o rate
— f — deposición f - [sold] deposit(ion) rate
— f —— f uniforme - [sold] uniform depo-
sit(ion) rate
— f de desbaste,tado m - [metal-lam] slabbing
speed
— f — desplazamiento m - [mec] travel, o displacement, o line, speed
— f — difusión f - diffusion rate
— f — elevación f - [grúas] hoisting speed
— f — empuje m - [mec] pushing, o crowding,
speed
— f — enfriamiento m - cooling rate
— f —— m rápido - rapid, o fast, cooling

rapidez f de enfriamiento m de aportación f para
endurecimiento m de superficie f - [sold]
hardsurfacing deposit cooling rate
— f — m reducida - [sold] slow cooling rate
— f — m relativamente reducida - [sold]
relatively slow cooling rate
— f — ensayo m - test(ing) speed
— f — equipo m para recuperación f - [ind] recovery unit rate
— f — m — f de vapor(es) m - [petról]
vapor recovery unit rate
— f — erección f - erection speed
— f — escurrimiento m - [hidr] runoff rate
— f — evaporación f - evaporation, speed, o
rate
— f — excavación f - [constr] excavation rate
— f — extensión f - [mec] extension speed
— f — función f - function speed
— f — funcionamiento m - [mec] operation speed
— f — fusión f - [metal] fusion speed • [sold]
melt-off, o melting, rate
— f — f de electrodo m - [sold] electrode,
melt-off, o melting, rate
— f — giro m - [grúas] swing(ing) speed
— f — lingoteado m—[metal-lam] slabbing speed
— f — marcha f - [ind] operating,tion speed
— f — movimiento m - movement speed
— f — operación f - operation, o function,
speed
— f — oscilación f - [grúas] swing(ing) speed
— f — prensadura f - [mec] pressing speed
— f — proceso m - [ind] process speed
— f — m para formación f - [metal-lam]
rolling process speed
— f — reacción f - reaction rate
— f — f para combinación f neumático-vehículo m—[autom] tire-vehicle system quickness
— f — rotación f - [grúas] swing speed
— f — f de grúa f—[grúas] crane swing(ing)
speed
— f — soldadura f - [sold] welding speed
— f — f (blanda) - [sold] soldering speed
— f — tendido m - [tub] laying speed
— f — m de tubería f - [tub] pipe laying
speed
— f elevada de aportación f - [sold] high deposition rate
— f grande de aportación f - [sold] high, o
excellent, deposit(ion) rate
— f máxima de aportación f - [sold] maximum deposition rate
— f mayor de aportación f - [sold] higher deposition rate
— F — — f que puede emplearse - [sold]
highest usable deposit(ion) rate
— f mediana f - medium rate
— f — de aportación f - [sold] medium depo-
sit(ion) rate
— f para derretimiento m - melting speed
— f — fresado,adura f - [mec] machining speed
— f — hincadura f - [constr-pil] driving speed
— f — instalación f - [ind] installation, o
erection, speed
— f labrado m - [mec] machining speed
— f — mecanización f - [mec] machining speed
— f — recuperación f recovery,ring rate
— f — f de vapor m - [petról] vapor recovery rate
— f relativamente reducida - relatively slow
rate
— f suficiente - sufficient speed
— f típica - typical speed
— f uniforme - uniform, speed, o rate
— f utilizable - usable, speed, o rate
rápido,da a - . . .; quick; fast; speedy
raqueta f para tenis m - [deport] tennis racket
rara vez adv - rarely; seldom(ly)
rarificación* f - véase enrarecimiento m
raro,ra a - . . .; unusual; uncommon; freak(ish)
ras adv - . . .; flat
— adv con extremo n - flush with @ end
rasante f - [constr] . . .; ground, line o level
— f adversa - [constr] adverse gradient
— f establecida - [constr] established gradient

rasante f excavada - [constr] excavated grade
line
— f favorable - [constr] favorable gradient
— f final - [constr] final grade,dient
— f — luego de asentamiento m - [constr] final,
grade after settlement, o settlement final
grade
— f hidráulica - [hidr] hydraulic grade,dient
— f — adversa f - [hidr] adverse hydraulic
gradient
— f hidráulica favorable - [hidr] favorable
hydraulic gradient
— f ideal - [constr] ideal, gradient, o grade
line
— f — para alcantarilla f - [constr] ideal
culvert, gradient, o grade line
— f inferior - [constr] lower grade (line)
— f — de camino m - [constr] lower road, o
road lower, grade (line)
— f menos inclinada - [hidr] flatter grade line
— f original - [constr] original, ground level,
o grade line
— f — para alcantarilla f - [constr] culvert
grade line
— f — para camino m - [constr] road grade (line)
— f superior - [constr] higher grade (line)
— f — de camino m - [constr] higher road grade
rascado,da a - [mec] scraped; scratched
rascador m - [herram] scratcher
— m para aceite m - [mec] oil scraper
— m — arcilla f - [herram] clay scraper
rasgado,da a - ripped; torn
rasgo m - . . .; outline • feature • trait
— m característico - personality trait
— m común - common characteristic
— m de personalidad f - personality trait
— m grande - great stroke • broad stroke
— m individual - personality trait
— m mecánico - [mec] mechanical feature
— m particular - particular, o special, trait
— m personal - personal(ity) trait
raso m - . . . | adv - out-of-doors
raspa f - . . . • [herram] véase escofina f
raspa f lateral - [herram] side rasp
raspado m - scrape,ping; scratch(ing) • lapping
— m a través de fundente m - [sold] scratching
through @ flux
— m de pared(es) f - [constr] wall scraping
— m manual - manual, o hand, scratch(ing)
raspado,da a - scratched; scraped • lapped •
[legal] erased
— a excesivamente - [mec] lapped excessively
raspador n - [mec] . . .; scratcher • [petról]
stripper • [metal-lam] friction piece
— m ajustado - [mec] adjusted scraper
— m con . . . aleta(s) f - [herram] . . .-wing
rasp
— m con criba f - [herram] scraping grizzle
— m de media caña - [herram] half-round file •
fluted scraper
— m fijo - [mec] stationary scraper
— n oscilante - [herram] rocking scraper
— m para aceite m - [comb.int] oil scraper
— m — colector m para ceniza(s) f - [mec] ash
pan rim scraper
— m — cuchilla f - [mec] blade scraper
— m — fondo m - [petról] bottom scraper
— m — m para perforación f - [petról] hole
bottom, o bottom hole, scraper
— m — — m — pozo m - [petról] hole bottom, o
bottom hole, scraper
— m — pared(es) f - [herram] wall scraper
— m — perforación f - [petról] hole scraper
— m — fondo m - [petról] hole scraper
— m — rueda f - [autom] wheel scraper
— m — tubería f - [tub] pipe scraper • [petról]
stripper
— m — f para entubación f - [petról] casing
stripper
raspador,ra a - scraping • scalping
raspadura f - . . .; scrape; scratch(ing) • lapping • scalp(ing)
— f de arena f - [mec] sand scratch(ing)
— f — metal m - [mec] metal scratch(ing)

raspadura f de pared f - [constr] wall scraping ﾐ
— f — ranura f - [mec] groove scaping • groove lapping
— f excesiva - [mec] excessive scraping • excessive lapping
raspaparedes m - [herram] wall scraper
raspar v - . . .; to scratch • to lap • [legal] to erase
— v a través de fundente m - [sold] to scratch through @ flux
— v electrodo m - [sold] to scratch @ electrode
— v — m lentamente - [sold] to scratch @ electrode slowly
— v — m sobre plancha f - [sold] to scratch @ electrode over @ plate
— v excesivamente—[mec] to, scratch, o scrape. excessively • to lap excessively
— v lentamente - [mec] to, scratch, o scrape, slowly
— v — electrodo m - [sold] to scratch @ electrode slowly
— v metal - to, scratch, o scrape, @ metal
— v pared f - to, scratch, o scrape, @ wall
— v ranura f - [mec] to scratch, o scrape, @ groove • to lap @ groove
— v sobre - to, scratch, o rub, on
raspatubos m - [tub] pipe, scratcher, o scraper • [petról] (casing) scraper; go devil
rasqueta f para colector m de ceniza(s) f - ash pan rim scraper
rasqueteado,da a - scraped - [ganad] curried
rasquetear v—[mec] to scrape • [ganad] to curry
rastra f - [transp] . . .; drag • [agric-equip] harrow • [mec] cradle
— f con cepillo m—[vial] brush, drag, o harrow
— f — m liviana - [vial] light brush drag
— f — diente(s) m - [agric-equip] tooth harrow
— f — m con resorte(s) m - [agric-equip] spring tooth harrow
— f — m — tiro m excéntrico—[agric-equip] offset disk harrow
— f — m excéntrico(s) - [agric-equip] offset disk harrow
— f — m rígido(s) - [agric-equip] peg tooth harrow
— f — disco(s) m - [agric-equip] disk harrow
— f — m a tope - [agric-equip] bumper disk harrow
— f — m de tipo m ancho - [agric-equip] wide (type) disk harrow
— f — m para, maleza(s) f, o monte m - [agric-equip] bush disk harrow
— f — m — monte(s) m y pantano(s) n - [agric-equip] bush and bog disk harrow
— f — m para pantano(s) m - [agric-equip] bog disk harrow
— f — m para huerto(s) m - [agric-equip] orchard harrow|
— f — m reversible(s) m - [agric-equip] reversible disk harrow
— f — m sin palanca f - [agric-equip] leverless disk harrow
— f — m — f tractor m - [agric-equip] leverless tractor disk harrow
— f — tandem - [agric-equip] tandem disk harrow
— f — extremo(s) m abierto(s) - [agric-equip] open end harrow
— f — m cerrado(s) - [agric-equip] closed end harrow
— f de tipo m ancho - [agric-equip] wide type harrow
— f liviana - [constr] light drag
— f — huerto(s) m - [agric-equip] orchard harrow
— f reversible—[agric-equip] reversible harrow
— f sin palanca f - [agric-equip] leverless harrow
rastreado,da a - [agric-equip] harrowed • [hidr] trawled • dredged
rastreador m - [nav] trawler • mine sweeper
rastrear v - . . . • [nav] to trawl
rastrel m - [constr] ground

rastreo m - . . . • [hidr] trawling
rastrillado m - [agric] raking • [metal-prod] rabbling
rastrillado,da a - [agric] raked • [metal-prod] rabbled
rastrillar v - [agric] to rake • [metal-prod] to rabble
rastrillo m - [agric] . . . • [metal-prod] rabble
— m amolador - [agric-equip] buck rake
— m autodescargador - [agric-equip] self dump(ing) rake
— m con descarga f automática - [agric-equip] self dump(ing) rake
— m — f lateral - [agric-equip] side rake
— f — f y oreadora- [agric-equip] side rake and tedder
— m — entrega f lateral - [agric-equip] side rake
— m emparvador - [agric-equip] rake stacker; stacking rake
— m hacinador - [agric-equip] rake stacker; stacking rake
— m para empuje m - [agric-equip] sweep rake
— m — heno m - [agric-equip] hay rake
— m — m con descarga f lateral - [agric-equip] side(-delivery) hay rake
— m — m — f y oreador m (combinado) - [agric-equip] (combined) side-delivery rake and tedder
— m — pasto m - [agric-equip] hay rake
rata f - . . . • rate*; véase razón f; gasto m; índice m - rapidez f • canon m
ratificación f - . . .; ratifying; confirming • approval,ving
ratificado,da a - ratified; confirmed; approved
ratificador,ra a - ratifying; ratifier
ratificar v - . . .; to approve
ratonera f - [zool] rat hole • [petról] rat hole
raudal m - . . . • jet
raudo,da a - . . .; fast
raya f - . . .; scratch • [metal-trat] polishing line • [fam] line on @ ground
Raya f Azul - Blue Stripe
— f causada por rodillo m para salida f - [metal-trat] exit roll scratch
— f de engrase m - [mec] lubrication groove
— f — laminación f - [metal-lam] rolling streak
— f — lubricación f - [mec] lubrication groove
— f en sentido m de laminación f - [metal-lam] lengthwise streak
— f fuerte - heavy scratch
— f — causada por rodillo m para salida f - [metal-trat] heavy exit roll scratch
— f para personal m - [labor] personnel dues
rayado m - [mec] hatching; scoring • [imprent] ruling
rayado,da a - [mec] scores • [imprent] ruled
rayadura f - [mec] score,ring • . . .
— f ligera - [mec] (light) scratch
rayar v - [mec] . . .; to score • [imprent] to rule
rayante* - véase abrasivo,va a
rayo m - [mec] . . . . • [óptica] . . .; beam
— m catódico - [electrón] cathode ray
— m — para televisión f - [electrón] video, o television, cathode ray
— m de antorcha f de arco m eléctrico - [sold] arc torch ray
— m — f entre carbón(es) m - [sold] carbon arc torch ray
— m — f con dos electrodos m (de carbón m) - [sold] arc torch ray
— m — arco m - [sold] arc ray
— m — electrodo m - [electrón] electrode beam
— m electrónico - [electrón] electron(ic) beam
— m gamma - [electrón] gamma ray
— m — con onda f corta - short wave gamma ray
— m — — f más corta - [electrón] shorter wave gamma ray
— m — — más larga - [electrón] longer wave gamma ray
— m — producido por radioisotopo(s) m—[nucl]

radioisotope (produced) gamma ray
— m incidente - [nucl] incident ray
— m infrarrojo m - [electrón] infrared ray
— m Láser - [Electrón] Laser ray
— m luminoso - [fís] light ray
— m — amplificado - [electrón] amplified light ray
— m — electrónicamente - [electrón] electronically amplified light ray
— m — convertido - [electrón] converted light ray
— m para bicicleta f - [mec] bicycle spoke
— m — rueda f - [mec] wheel spoke
  • wheel arm
— f — f para malacate m - [petról] bull wheel arm
— m — f — — m para cable m para herramienta f - [petról] bull wheel arm
— m — f — — m — — m — tubería f - [petról] calf wheel arm
— m — f — — m — herramienta(s) f - [petról] bull wheel arm
— m — f — — m — tubería f - [petról] calf wheel arm
— m solar - [astron] sun ray
— m ultravioleta - [electrón] ultraviolet ray
— X m - [electrón] X-ray
— — m comercial - [electrón] commercial X-ray
— — m convertido - [electrón] converted X-ray
— — m — en rayo m luminoso - [electrón] X-ray converted (in)to @ light ray
— — m fluoroscópico m - [electrón] fluoroscopic X-ray
rayón m - streak • [textil) . . .
— m resistente - [textil] tough rayon
razón f - . . . .; rationale • cause • consideration • [com] concern
— f alta - high ratio
— f — entre caucho m y oquedad(es) f - [autom-neumát] high rubber-to-void ratio
— f — — m y vacío m - [autom-neumát] high rubber-to-void ratio
— f asignada - assigned rate,tio
— f atendible - sound reason
— f aumentada - greater ratio
— f baja - [mat] low ratio
— f — entre caucho m y oquedad(es) f - [autom-neumát] low rubber-to-void ratio
— f — — m y vacío m - [autom-neumát] low rubber-to-void ratio
— f básica - basic reason - basic rate,tio
— f buena - good ratio
— f — entre espesor m útil de cordón m y espesor m de plancha f - [sold] good weld throat to plate thickness ratio
— f comercial - commercial reason • commercial concern
— f costo-beneficio - [com] cost-benefit ratio
— f de acero m - [metal] steel ratio
— f — — m y hormigón m - [constr] steel to concrete ratio
— f — altura f - [constr] depth ratio - headroom reason(s)
— f — apariencia f - appearance, sake, o reason
— f — área f - [constr] area ratio
— f — f de acero m - [constr] steel area ratio
— f — — f — — m a hormigón m - [Constr] steel to concrete area ratio
— f — bonificación f - [labor] bonus rate
— f — calor m - [termol] heat ratio
— f — cambio m - [fin] exchange rate,tio
— f — carga f - [constr] load ratio
— f — f nueva - [curp.moled] recharge,ging ratio
— f — cobertura f - [fin] coverage ratio
— f — — f con interés(es) m - [fin] interest coverage ratio
— f — compresión f - [mec] compression ratio
— f — delgadez f - [constr] thinness, o slenderness, ratio
— f — desgaste m - [mec] wear ratio
— f — deuda f - [fin] debt ratio

razón f de engranaje(s) m - [mec] gear ratio
— f — — m alta - [mec] high gear ratio
— f — — m apropiada - [mec] appropriate, o right, gear ratio
— f — — m baja - [mec] low gear ratio
— f — — m inapropiada - [mec] inappropriate, o wrong, gear ratio
— f — escurrimiento m - [hidr] runoff rate
— f — espacio m - space, limitation, o reason
— f — expansión f - [mec] expansion ratio
— f — inducción f - [electr-oper] induction ratio
— f — masa f - [mec] mass ratio
— f — multiplicación f - [mec] multiplication, gear, o ratio
— f — — f alta - [mec] high gear ratio
— f — — f baja - [mec] low gear ratio
— f — — f (de) par - [mec] par multiplication ratio
— f — — f de torsión f - [mec] torque multiplication ratio
— f — otra laya - other, o different, reason
— f — pago m - [labor] pay rate
— f — — m para dividendo(s) m - [fin] dividend(s) pay(out) ratio
— f — pendiente f - [topogr] slope ratio
— f — pérdida f - loss rate,tio
— f — — f de metal m - metal loss rate
— f — Poisson - Poisson('s) ratio
— f — producción f - [ind] production rate
— f — profundidad f - depth ratio
— f — — f crítica - [hidr] critical depth ratio
— f — realización f - [labor] time rate
— f — — f normal m - [labor] standard time rate
— f — regulación f - control reason
— f — f de descarga f - [hidr] discharge control, reason, o rate
— f — resistencia f - [constr] strength, reason, o rate
— f — — f a rotura f - [mec] breaking strength ratio
— f — — f de tubería f - [tub] pipe strength ratio
— f — sincronización f - timing consideration
— f — superficie f de acero m - [constr] steel area ratio
— f — — m a la de hormigón m - [constr] steel to concrete area ratio
— f — f — — — — — m en corona f - [constr] crown steel to concrete area ratio
— f — torsión f - [mec] torque ratio
— f — velocidad f - [mec] speed ratio
— f — — f para accionamiento m - [mec] drive speed ratio
— f — f — — m para cabezal m para torno m - catshaft drive,ving speed ratio
— f — f — — m — mesa f rotative - [petról] rotary drive,ving speed ratio
— f — f — — m para rotación f - [petról] rotary drive,ving speed ratio
— f — f — cabezal m para torno m - [petról] catshaft speed ratio
— f diámetro-espesor - [tub] diameter-thickness ratio
— f disminuida - smaller, o reduced, ratio
— f económica - economic reason • economics
— f efectiva - [mec] effective ratio
— f — de engranaje(s) m - [mec] effective gear(s) ratio
— f — para impulsión f - [autom-mec] effective drive,ving ratio
— f elevada - [matem] high ratio
— f en baúd(es) m - [electrón] baud rate
— f — caja f con engranaje(s) m - [mec] gear box ratio
— f — eje trasero - [autom-mec] rear axle ratio
— f — porcentaje m - percentage ratio
— f — tanto por ciento m - percentage ratio
— f entre aire m y combustible m - [combust] fuel-air ratio
— f — altura f de pared f (lateral) y ancho m

de banda f̲ para rodamiento m̲ ~ [autom-neumát]
aspect ratio
razón f̲ entre altura f̲ y ancho(r) m̲ - [autom-
neumát] aspect ratio
— f̲ — f̲ — m̲ de neumático m̲ - [autom-
neumát] tire aspect ratio
— f̲ — ancho(r) m̲ y altura f̲ - width-height
ratio
— f̲ — beneficio m̲ y costo m̲ - profit/cost
ratio
— f̲ — carga f̲ sólida y líquida - [metal-prod]
solid/liquid charge ratio
— f̲ — carga f̲ e inflación f̲ - [autom-neumát]
load/inflation ratio
— f̲ — caucho f̲ y oquedad(es) f̲—[autom-neumát]
rubber-to-void ratio
— f̲ — diámetro m̲ interior y (diámetro m̲) exterior · [tub] inner (diameter)/
outer diameter ratio
— f̲ — espesor m̲ de cordón m̲ y (espesor m̲) de
plancha f̲ - [sold] weld bead to plate thickness ratio
— f̲ — m̲ útil de cordón m̲ y espesor m̲ de
plancha f̲ - [sold] weld throat to plate
thickness ratio
— f̲ — intensidad f̲ de escurrimiento m̲ e intensidad f̲ de precipitación f̲ - rate of runoff
to rate of rainfall ratio
— f̲ — relleno m̲ y cuerda f̲ - [constr] fill to
span ratio
— f̲ — resistencia f̲ y peso m̲ - [mec] strength
to weight ratio
— f̲ — señal f̲ y ruido m̲ - [comput] signal to @
noise ratio; sinad
— f̲ especial - special reason • [mec] special
ratio
— f̲ específica - specific reason • [mec] specific ratio
— f̲ — de calor m̲—[termol] specific heat ratio
— f̲ — carga f̲ - [hidr] specific head ratio
— f̲ estándar - standard ratio
— f̲ estimada - estimated rate
— f̲ — para emanación(es) f̲ - [petról] estimated vapor(ization) rate
— m̲ — — f̲ en depósito m̲ - [petról] estimated tank vapor(ization) rate
— f̲ — — m̲ — m̲ para almacenamiento m̲ -
[petról] estimated stock tank vapor(ization)
rate
— f̲ — vapor(es) m̲ en depósito m̲ para almacenamiento m̲ - [petról] estimated stock tank
vapor(ization) rate
— f̲ estratégica - strategic reason
— f̲ expresa - express reason
— f̲ expresada - expressed reason
— f̲ final - final reason • final ratio
— f̲ — para impulsión f̲ - [autom-mec] final
drive,ving ratio
— f̲ general - general reason • general, o overall, ratio
— f̲ — de engranaje(s) m̲ - [mec] overall gear,
o gear overall, ratio
— f̲ geopolítica - [pol] geopolitical reason
— f̲ hidráulica - [hidr] hydraulic reason
— f̲ horaria - [labor] hourly rate
— f̲ idiomática - [filol] idiomatic reason
— f̲ importante - important, o big, reason
— f̲ individual - [ind] individual rate • [mec]
individual ratio
— f̲ — para engranaje(s) m̲ - [mec] individual
gear ratio
— f̲ inicial - initial reason
— f̲ legal - legal reason
— f̲ límite - limiting ratio
— f̲ — de carga f̲ - [constr] maximum load ratio
— f̲ — entre altura f̲ y largo(r) m̲ - [constr]
depth to length limiting ratio
— f̲ normal - standard rate • standard ratio
— f̲ para accionamiento m̲ con cadena(s) f̲ -
[mec] chain drive,ving consideration(s)
— f̲ — complementariedad* f̲ - complimentary
status reason
— f̲ — construcción f̲ - [constr] construction

reason(s)
razón f̲ para emanación(es) f̲ - [petról] vaporization rate
— f̲ — — f̲ en depósito m̲ - [petról] tank va-
por(ization) rate
— f̲ — — f̲ — m̲ para almacenamiento m̲ -
[petról] stock tank vapor(ization) rate
— f̲ — engranaje(s) m̲ - [mec] gear ratio
— f̲ — m̲ optimizada* - [autom-mec] optimized gear ratio
— f̲ — falla f̲ - failure, reason, o cause
— f̲ — f̲ estructural - [constr] structural
failure, reason, o cause
— f̲ — garantía f̲ - guaranty reason
— f̲ — impulsión f̲ - [autom-mec] drive,ving
ratio
— f̲ — excepción f̲ - exception reason
— f̲ — quemadura f̲ - burn(ing) reason
— f̲ — — f̲ de fusible - [electr-oper] fuse,
burning, o blowing, reason
— f̲ — reclamación f̲ - claim reason
— f̲ — reclamo m̲ - claim reason
— f̲ — resistencia f̲ - strength ratio
— f̲ — seguridad f̲ - [segurid] safety, consideration, o reason
— f̲ — transferencia f̲ - transfer reason
— f̲ — — f̲ tecnológica - technological transfer reason
— f̲ — vapor(es) f̲ - [petról] vapor(ization)
rate
— f̲ — — f̲ en depósito m̲ - [petról] tank vaporization rate
— f̲ — — f̲ — m̲ para almacenamiento m̲ -
stock tank vapor(ization) rate
— f̲ pobre - poor rate • poor ratio
— m̲ — entre espesor m̲ útil de cordón y (el) de
plancha f̲ - [sold] poor weld throat to plate
thickness ratio
— por (la) cual adv - so it is (that)
— f̲ positiva - positive ratio • soundness
— f̲ primaria - primary reason
— f̲ principal - principal, o main, o primary, o
greatest, reason
— f̲ reducida - reduced ratio
— f̲ según norma f̲ - standard rate,tio
— f̲ siguiente - following reason
— f̲ sobrada - (very) good reason
— f̲ social - [legal] corporation; concern
— f̲ sólida - sound reason
— f̲ total - [mec] total, o overall, ratio
— f̲ — de engranaje(s) m̲ - [mec] overall gear
ratio
— f̲ verificada - checked reason • checked ratio
— f̲ vinculada f̲ - related reason
— f̲ — con regulación f̲ - control (related)
reason
— f̲ vigente - current rate • current ratio
— f̲ vinculada f̲ - related reason
— f̲ — con regulación f̲ - [hidr] control (related) reason
— f̲ — — f̲ de descarga f̲ - [hidr] discharge
control (related) reason
razonable a - reasonable • fair
razonablemente adv - reasonably • fairly
— adv bueno,na - reasonably good
— adv firme - reasonably, firm, o steady
— adv igual - reasonably equal
— adv plano,na - reasonably flat
— adv uniforme - reasonably uniform
razonamiento m̲ - . . . .; rationale
reabastecedor m̲ - resupplier
reabastecedor,ra adv - resupplying
reabastecer v - to, resupply, o refill, o re-
plenish • to take on • [comb.int] to refuel
— v aceite m̲ - [lubric] to resupply @ oil • to
refill with oil
— v — m̲ en extremo m̲ hacia motor m̲ - [petról]
to refill @ power end oil
— f̲ armadura f̲ - [mec] to refill @ body
— v bomba f̲ - [bombas] to refill @ pump
— v — f̲ con aceite m̲ - [bombas] to refill @
pump with oil
— f̲ caja f̲ - [mec] to refill @ case

reabastecer v caja f para cadena f - [mec] to refill @ chain case
— v combustible m - [comb.int] to refuel
— v con aceite m - [lubric] to refill with oil
— v — combustible m - [comb.int] to refuel
— v depósito m - [comb.int] to refill @ tank
— v — m para combustible m - [comb.int] to refill @ fuel tank
— v — m para gasolina f - [Comb.int] to refill @ gasoline tank
— v extremo m hacia motor m - [petró] to refill @ power end
reabastecido,da a - refilled; resupplied; replenished • taken on • [comb.int] refueled
— a con aceite m - [ind] oil refilled
— a — combustible m - [comb.int] refueled
reabastecimiento m - resupply(ing) • taking on • refilling • refueling • replenishing
— n con aceite m - [mec] oil refilling
— m — — m de caja f para cadena f - [mec] chain case oil refill(ing)
— m — — m extremo m hacia motor m - [petról] power end oil refill(ing)
— m — combustible m - [comb.int] refueling
— m de armadura f - [mec] body refilling
— m — bomba f - [bombas] pump refilling
— m — caja f - [mec] case refill(ing)
— m — — f para cadena f - [mec] chain case refill(ing)
— m — colector m - [mec] sump refill(ing)
— m — — m para aceite m - [mec] oil sump refill(ing)
— m — depósito m - [ind] tank refill(ing)
— m — — m para combustible m—[comb.int] fuel tank refill(ing)
— m — — m — gasolina f - [comb.int] gasoline tank refill(ing)
— m — extremo m hacia motor m - [petról] power end refill(ing)
reabierto,ta a - reopened
reacción f - . . .; reacting; response • feedback • responsiveness • boost • backlash | a - [aeron] (de) jet
— f a acelerador m - [autom-mot] response to @ throttle
— f — cambio(s) m - [mec] reaction to changes
— f — — m en dirección f - [autom] steering response
— f — caracterización f - [com] identification response
— f — deformación f - deflection response
— f — dirección f - [autom-mec] steering response
— f — maniobra(s) f - [autom-mec] steering response
— f — — f con volante m - [autom-mec] steering response • handling performance
— f — — f mejorada - [autom-mec] improved steering response
— f — movimiento(s) m de dirección f - [autom-mec] steering response
— f — — m — volante m - [autom] steering response
— f a viraje(s) m - [autom-neumát] cornering response
— f — volante m - [autom] steering response; response,siveness to @ steering wheel
— f ácida - [quím] acid reaction
— f afectada - affected reaction
— f ajustada - [mec] tight(ened) response
— f aluminotérmica - [sold] thermit reaction
— f apropiada - appropriate response; responsiveness
— f — a maniobra(s) f (con volante m) - [autom-neumát] responsiveness
— f básica - [quím] basic reaction
— f buena - good response • good reaction
— f — a dirección f - [autom] good steering response
— f — — movimiento(s) m de volante m - [autom-mec] good steering response
— f causada - caused reaction
— f con aire m - reaction with @ air
— f con vaho(s) m - [quím] reaction to @ vapor

reacción f con vaho(s) m disolvente(s) - [quím] reaction with @ solvent vapor(s).
— f de apatía f - [labor] apathy,thetic reaction
— f — asiento m - [constr] bedding reaction
— f — automóvil m - [autom-mec] car('s), o automobile('s), reaction, o attitude
— f — Boudouard—[quím] Boudouard('s) reaction
— f — cascarilla f - [metal-lam] scale reaction
— f — conductor m - [autom] driver reaction
— f — conjunto m neumático-vehículo m - [autom-mec] tire-vehicle system response,siveness
— f — desoxidación f - [metal-prod] deoxidation reaction
— f — generador m - [electr-prod] generator response
— f — indiferencia f - [labor] indifference, reaction, o attitude
— f — lecho m - [constr] bed(ding) reaction
— f — mercado m - [com] market(place) response
— f — motorista m - [autom] motorist('s) response
— f — neumático m - [autom-neumát] tire('s) response
— f — — m y vehículo m - [autom-neumát] tire-vehicle response
— f — — m — a maniobra(s) f (con volante m) - [autom-mec] tire-vehicle handling
— f — refinación f - [metal-prod] refining reaction
— f — suelo m - [suelos] soil reaction
— f — usuario m - user reaction
— f — voltaje m - [electr-oper] voltage reaction
— f directa - direct reaction • [mec] direct response
— f distribuida - distributed reaction
— f electroquímica - [quím] electrochemical reaction
— f en ánodo m - [quím] anode reaction
— f — cadena f - chain reaction
— f — cátodo n - [quím] cathode reaction
— f — curva f - [autom-neumát] cornering reaction
— f — fondo m - [constr] bottom reaction
— f entre sí - reaction with each other; mutual reaction
— f favorable - favorable reaction
— f fundamental - [quím] fundamental, o basic, reaction
— f individual - individual reaction
— f inmediata - immediate, o instantaneous, reaction, o response
— f lenta - slow, reaction, o response
— f — a maniobra(s) f - [autom] slow steering response
— f — — movimiento m de dirección f - [autom] slow steering response
— f más directa - [mec] more direct response
— f más lenta f - slower response
— f — — a maniobra(s) f - [autom] slower steering response
— f — — — movimiento m de dirección f - [autom] slower steering response
— f rápida - quicker response
— f — — maniobra(s) f - [autom] quicker steering response
— f — — movimiento m de dirección f - [autom] quicker steering response
— f suave - [mec] smoother response
— f máxima f - maximum response
— f — a maniobra(s) f—[autom] maximum steering response • handling capability
— f — — f con volante m - [autom] maximum handling capability
— f — de neumático m a maniobra(s) f (con volante m) - [autom] tire handling capability
— f para reducción f - [metal-prod] reducing, o reduction, reaction
— f positiva - positive, o sharp, reaction
— f — a movimiento(s) m de dirección f—[autom] positive steering response
— f — de conductor m - [autom] sharp driver reaction
— f precisa - [mec] precise, o tight, response

reacción f química - [quím] chemical reaction
— f rápida - quick, reaction, o response; responsiveness
— f — a maniobra(s) f - [autom] quick steering response
— f — — movimiento m de dirección f - [autom] quick steering response
— f reductora - [metal-prod] reducing,ction reaction
— f refleja f - reflex (reaction)
— f — rápida - quick reflex (reaction)
— f secundaria - [quím] secondary reaction
— f suave - [mec] smooth, reaction, o response
⌐ adv - [mec] (con) smooth responding
— f (de) termita f - [sold] thermit reaction
— f típica - typical, reaction, o response
— f transitoria - transient, reaction, o response
— f ultrarrápida - ultraquick response; snap action
— f — a dirección f - [autom] ultraquick steering response
reaccionado,da a - reacted • responded
— a con aire m - reacted with @ air
— a — vaho(s) m - [quím] reacted with @ vapor
— a — m disolvente(s) m - [quím] reacted with @ solvent vapor(s)
— a directamente - reacted, o responded, directly
— a entre sí - reacted, with each other, o mutually
— a más suavemente - responded more smoothly
— a suavemente - reacted, o responded, smoothly
reaccionante a - reacting
reaccionar v - . . .; to respond
— v a volante m - [autom] to respond to @ steering wheel
— v con aire m - to react with @ air
— v — vaho(s) m - [quím] to react with @ vapor
— v — — disolvente(s) m - [quím] to react with @ solvent vapor
— v directamente - [mec] to, react, o respond, directly
— v entre sí - to react, with each other, o mutually
— v instantáneamente - to respond instantly
— v más directamente - to, react, o respond, more directly
— v — suavemente - to, react, o respond, more smoothly
— v suavemente - to, react, o respond, smoothly
reacio,cia a - . . . . • balky
reacondicionado,da - [mec] reconditioned; overhauled
reacondicionamiento m - . . . .; overhauling; reconditioning
reactancia f - [electr-equip] . . .; reactor
— f básica - [electr-oper] basic, o fundamental, reactance
— f con secuencia f cero - [electr-oper] zero sequence reactance
— f — — f nula - [electr-oper] zero sequence reactance
— f — — f negativa - [electr-oper] negative sequence reactance
— f — — f positiva - [electr-equip] positive sequence reactance
— f inductiva - [electr-oper] inductive reactance
— f para cortocircuito m - [electr-equip] short circuit reactance
— f saturada - [electr-equip] saturated reactance
— f sincrónica - [electr-equip] synchronous, o synchronic, reactance
— f — no saturada - [electr-equip] unsaturated synchronous,nic reactance
— f — saturada - [electr-equip] saturated synchronous,nic reactance
— f subtransitoria - [electr-equip] subtransitory reactance
— f — no saturada - [electr-equip] unsaturated subtransitory reactance

reactancia f subtransitoria saturada - [electr-equip] saturated subtransitory reactance
— f transitoria - [electr-equip] transitory reactance
— f — no saturada - [electr-equip] unsaturated transitory reactance
— f — saturada - [electr-equip] saturated transitory reactance
reactivación f - [ind] (re)cycling • [electr-oper] re-energizing
— f de transformador m - [electr-oper] transformer reenergizing,zation
— f rápida - [mec] fast (re)cycling
reactivado,da a - reactivated • [electr-oper] re-energized
reactivar v - to reactivate • [electr-oper] to re-energize
— v transformador m - [electr-oper] to re-energize @ transformer
reactivo,va a - reactive
— a químicamente - [quím] reactive chemically
reactor m - . . .; mixer • [electr-equip] choke • [aeron] jet • [nucl] evaporator
— m actualizado - [nucl] advanced reactor
— m — con enfriamiento m - [nucl] advanced cooled reactor
— m — — — con gas m - [nucl] advanced gas cooled reactor
— m con agua m - [nucl] water reactor
— m — — m bajo presión f - [nucl] pressurized water reactor
— m concentrador m - [nucl] reactor, o mixer, concentrator
— m de tipo m con rotor m - [electr-equip] rotor type reactor
— m destapado - [nucl] uncovered reactor
— m evaporador - [nucl] reactor-evaporator; mixer-evaporator
— m — por motor m - [electr-equip] motor driven reactor
— m limpio - [nucl] clean reactor
— m mecánico - [electr-equip] mechanical reactor
— m — de tipo m con rotor m - [electr-equip] rotor type mechanical reactor
— m — — — m — — m impulsado por motor m - [electr-equip] motor driven rotor-type mechanical reactor
— m — impulsado por motor - [electr-equip] motor driven mechanical reactor
— m nuclear - [nucl] nuclear reactor
— m — para energía f - [nucl] nuclear power reactor
— m para arranque m - [electr-equip] starting reactor
— m — gancho m - [grúas] hook reactor
— m — lecho m - [metal-prod] bed reactor
— m — — m fluido - [metal-prod] fluid bed reactor
— m — puente m - [grúas] bridge reactor
— m — regulación f - control(ling) reactor • [electr-equipo] ballast
— m — radiofrecuencia f - [electr-equip] radio frequency choke
— m — reducción f - [metal-prod] reducing,ction reactor
— m — — f directa - [metal-prod] direct reduction reactor
— m — regulación f - [electr-equip] control reactor
— m — — f para amperaje m - [electr-equip] ampere(s) control reactor
— m — — f estabilizadora - [electr-equip] stabilizing ballast
— m — sobrelevación f (de corriente f) - [electr-equip] surge reactor
— m regenerador - [nucl] breeder reactor
— m saturable - [electr-equip] saturable reactor
— m tapado - [nucl] covered reactor
— m tubular - tubular reactor
reactualizado,da - véase actualizado,da a
reactualizar v - véase actualizar v
reafilación f - [mec] resharpening
reafilado,da a - resharpened

reafilar <u>v</u> - [mec] to resharpen
reafirmación <u>f</u> - . . .; reaffirming; repeating; repetition • reiterating,tion
reafirmado,da - reaffirmed; repeated;reiterated
reafirmar <u>v</u> - . . .; to reiterate; to repeat
reagrupación <u>f</u> - regrouping • [milit] rallying
reagrupado,da <u>a</u> - regrouped • [milit] rallied
reagrupar <u>v</u> - to regroup • [milit] to rally
reaguzado,da <u>a</u> - [mec] resharpened
reaguzamiento <u>m</u> - [mec] resharpening
reaguzar <u>v</u> - [mec] to resharpen
reajustabilidad <u>f</u> - readjustability • [com] escalatability
reajustable <u>a</u> - readjustable • [com] escalatable • subject to escalation
reajustado,da <u>a</u> - readjusted • [com] escalated • [comput] reset
— <u>a</u> independientemente - [comput] reset independently
reajustar <u>v</u> - to readjust • to re-tighten • to reset • [com] to escalate • [comput] to reset
— <u>v</u> accesorio <u>m</u> - [mec] to retighten @ fitting
— <u>v</u> amperaje <u>m</u> - [sold] to readjust @, output, <u>o</u> amperage, <u>o</u> amperes
— <u>v</u> botón <u>m</u> - [electrón] to reset @ button
— <u>v</u> conexión <u>f</u> - [comb.int] to readjust @ linkage
— <u>v</u> — <u>f</u> para acelerador <u>m</u> - [comb.int] to readjust @ throttle linkage
— <u>v</u> — <u>f</u> — regulador <u>m</u> - [comb.int] to readjust @ throttle linkage
— <u>v</u> entrehierro <u>m</u> - [mec] to readjust @ gap
— <u>v</u> independientemente - [comput] to reset independently
— <u>v</u> modalidad <u>f</u> - [electrón] to reset @ mode
— <u>v</u> presupuesto <u>m</u> [fin] to escalate @ budget
— <u>v</u> prolongación <u>f</u> - [sold] to readjust @ stickout
— <u>v</u> — <u>f</u> de electrodo <u>m</u> - [sold] to readjust @ (electrode) stickout
— <u>v</u> — <u>f</u> — <u>m</u> de carbono <u>m</u> - [sold] to readjust <u>a</u> carbon (electrode) stickout
— <u>v</u> resorte <u>m</u> - [mec] to readjust @ spring
— <u>v</u> — <u>m</u> para retorno <u>m</u> - [mec] to readjust @ return spring
reajuste <u>m</u> - readjustment • [com] escalation • [mec] overhaul(ing); resetting
— <u>m</u> alzado - [com] escalation
— <u>m</u> común - [comput] common reset(ting)
— <u>m</u> corriente - [comput] common reset(ting)
— <u>m</u> de accesorio <u>m</u> - [mec] fitting retightening
— <u>m</u> — amperaje <u>m</u> - [sold] output readjustment
— <u>m</u> — botón <u>m</u> - [electrón] button reset(ting)
— <u>m</u> — conexión <u>f</u> - [mec] linkage readjusting
— <u>m</u> — — <u>f</u> para acelerador <u>m</u> - [comb.int] throttle linkage readjusting,tment
— <u>m</u> — — <u>f</u> — regulador <u>m</u> - [comb.int] throttle linkage readjusting,tment
— <u>m</u> — entrehierro <u>m</u> - [mec] gap readjusting,tment
— <u>m</u> — modalidad <u>f</u> [electrón] mode reset(ting)
— <u>m</u> — precio(s) <u>m</u> - [com] (price) escalation
— <u>m</u> presupuesto <u>m</u> - budget escalation
— <u>m</u> — prolongación <u>f</u> - [electr] stickout readjustment
— <u>m</u> — — <u>f</u> de electrodo <u>m</u> - [sold] electrode stickout escalation
— <u>m</u> — — <u>f</u> — — <u>m</u> de carbono <u>m</u> - [sold] carbon (electrode) stickout readjustment
— <u>m</u> — resorte <u>m</u> - spring readjusting,tment
— <u>m</u> — — <u>m</u> para retorno <u>m</u> - [mec] return spring readjusting,tment
— <u>m</u> — tren <u>m</u> - [metal-lam] mill resetting
— <u>m</u> general - [ind] general, <u>o</u> major, overhaul
— <u>m</u> independiente - [comput] independent reset(ting)
— <u>m</u> manual - [ind] manual reset(ting)
— <u>m</u> para carácter <u>m</u> - [comput] character reset
— <u>m</u> — — <u>m</u> equivocado - [comput] wrong character reset(ting)
— <u>m</u> por costo(s) <u>m</u> mayor(es) - [com] escalation
—.<u>m</u> provisorio - [com] temporary, adjustment, <u>o</u> escalation

real <u>a</u> - . . .; true; virtual
— <u>a</u> en obra <u>f</u> - [constr] actual field
realce <u>m</u> - . . . • enhancing • compounding • [mec] boss
— <u>m</u> de rendimiento <u>m</u> - performance enhancing
— <u>m</u> — — <u>m</u> administrativo - [admin] management work emphasis
realidad <u>f</u> - . . .; reality • fact-• position
— <u>f</u> financiera - [fin] financial, reality, <u>o</u> position
realimentación <u>f</u> - . . . • [electrón] refeeding
— <u>f</u> por derivada - [electrón] derived refeeding
— <u>f</u> — integral - [electrón] integral refeeding
realineación <u>f</u> - . . .; realigning
realineado,da <u>a</u> - realigned
realinear <u>v</u> - to realign
realizable <u>a</u> - [fin] realizable; collectable; convertible • attainable
realización <u>f</u> - . . .; realizing; performance; performing; accomplishing,shment • meeting • attainment • design • preparation • implementation; handling; conducting • making • holding
— <u>f</u> adecuada - adequate, attaining, <u>o</u> meeting
— <u>f</u> conveniente - convenient, attainment, <u>o</u> achieving,vement
— <u>f</u> de cambio <u>m</u> - change(over) achievement
— <u>f</u> — ensayo(s) <u>m</u> - test(s), performance, <u>o</u> performing, <u>o</u> conducting, <u>o</u> achieving,vement
— <u>f</u> — entrega <u>f</u> - delivery, making, <u>o</u> performance,ming
— <u>f</u> — inspección <u>f</u> - inspection, performance, <u>o</u> performing
— <u>f</u> — inversión <u>f</u> - [fin] investment making
— <u>f</u> — pago <u>m</u> - [fin] payment making
— <u>f</u> económica - economical, attaining, <u>o</u> accomplishing,shment
— <u>f</u> — prueba(s) <u>f</u> - test(s) performing,mance
— <u>f</u> — trabajo <u>m</u> - work performance,ming • getting @ job done
— <u>f</u> hidráulica - [hidr] hydraulic performing
— <u>f</u> nacional - national accomplishment
— <u>f</u> normal - normal accomplishment
— <u>f</u> rápida - rapid achievement,ving
— <u>f</u> simultánea - simultaneous, achievement,ving • simulteneous performance,ming
realizado,da <u>a</u> - accomplished; performed; conducted; attained; realized • made; handled; carried out • held; met • [fin] paid up
— <u>a</u> adecuadamente - performed, <u>o</u> met, <u>o</u> accomplished, adequately
— <u>a</u> convenientemente - achieved conveniently
— <u>a</u> económicamente - accomplished economically
— <u>a</u> hidráulicamente - handled hydraulically
— <u>a</u> rápidamente—achieved, rapidly, <u>o</u> quickly
— <u>a</u> realmente - truly performed
— <u>a</u> simultáneamente - performed simultaneously
realización <u>f</u> - accomplishing,shment; carrying out; conducting; making; performing,mance
— <u>f</u> — estudio <u>m</u> - study performance
— <u>f</u> — investigación <u>f</u> - investigation conducting
realizar <u>v</u> - to realize • to accomplish; to perform; to realize; to carry out; to conduct • to fulfill; to attain; to obtain; to make • to handle; to hold; to meet • [com] to sell out
— <u>v</u> adecuadamente - to meet, <u>o</u> accomplish, adequately, <u>o</u> properly
— <u>v</u> cambio <u>m</u> - to, achieve, <u>o</u> accomplish, @ change(over)
— <u>v</u> por muestreo <u>m</u> - to, accomplish, <u>o</u> perform, by sampling
— <u>v</u> convenientemente - to achieve conveniently
— <u>v</u> económicamente - to accomplish economically
— <u>v</u> ensayo <u>m</u> - to, perform, <u>o</u> conduct, @ test
— <u>v</u> — <u>m</u> en laboratorio <u>m</u> - [ind] to conduct @ laboratory test
— <u>v</u> entrega <u>f</u> - to perform, <u>o</u> make, @ delivery
— <u>v</u> estudio <u>m</u> - to, make, <u>o</u> perform, <u>o</u> carry out, @ study
— <u>v</u> hidráulicamente - to perform hydraulically
— <u>v</u> inspección <u>f</u> - to perform @ inspection

realizar v inversión f - [fin] to make @ investment
— v investigación f—to conduct @ investigacion
— v mucho trabajo m - to perform much work • to work productively
— v pago m - [fin] to make @ payment
— v prueba f - to perform @ test
— v rápidamente - to perform quickly
— v simultáneamente - to perform simultaneously
— v trabajo m - to preform @, work, o task; to get @, job, o work, done
realmente adv - really; actually; truly; virtually • nothing short of
— adv automático,ca - truly automatic
realzado,da a - emphasized; stressed - [arquit] stilted
realzar v - to enhance; to emphasize • to compound
— v rendimiento m - to enhance @ performance
— v trabajo m administrativo - [admin] to emphasize @ management work
reallanamiento m - [metal-lam] rerolling
— m de soldadura f - [metal-lam] weld rerolling
reallanar v - [metal-lam] to reroll
— v soldadura f - [sold] to reroll @ weld
reanimación f - reviving • restarting
reanimado,da a - revived; restarted
reanimación f - . . .; resurrecting,tion
reanudado,da a - . . . • [fig] resurrected
reanudar v - . . . • [fig] to resurrect
— v cordón m - [sold] to restart @ bead
— v soplado m - [metal-prod] to restart, @ (air) blast, o @ wind
reaparecido,da a - reappeared
reaparición f - reappearance,ring
reaplicación f - reapplication,lying; véase también aplicación f nueva
— f — órden f—[ind] order reapplication,lying
— f de tablero m para distribución f - [electr-instal] distribution panel reapplication
reaplicado,da a - reapplied
reaplicar v - to reapply
reapretado,da a - . . .; retightened
— a ligeramente - [mec] retightened lightly
reapretamiento m - [mec] retightening
— m de tuerca f - [mec] nut retightening
— m ligero - [mec] slight retightening
reapretar v - . . . • [mec] to retighten
— v accesorio m - [mec] to retighten @ fitting
— v ligeramente - [mec] to retighten slightly
— v tuerca f - [mec] to retighten @ nut
reaprovisionado,da a - resupplied; taken on
reaprovisionamiento m - resupplying; taking on • [comb.int] refueling
reaprovisionar v - to resupply; to take on • [comb.int] to refuel
— v con combustible m - [comb.int] to refuel
rearmado m - [mec] reassembly,ling
— m de bomba f - [bombas] pump reassembly,ling
— m — carburador m - [comb.int] carburetor reassembly,ling
— m — motor n - [comb.int] engine reassembly,ling • [electr-mot] motor reassembly,ling
rearmado,da a - [mec] reassembled
rearmar v - . . . • [mec] to reassemble
— v árbol m - [mec] to reassemble @ shaft
— v — para entrada f (de fuerza f) - [mec] to reassemble @ input shaft
— v bomba f - [bombas] to reassemble @ pump
— v carburador m - [comb.int] to reassemble @ carburetor
— v distribuidor m - [mec] to reassemble @, divider, o distributor
— v — m para fuerza f - [autom-mec] to reassemble @ power, divider, o distributor
— v motor m - [electr-mot] to reassemble @ motor • [comb.int] to reassemble @ engine
— v pieza f - [mec] to reassemble @ part
— v — f para ajuste m - [mec] to reassemble @, adjustment part, o fitting
— v — f — m para válvula f - [válv] to reassemble @ valve fitting
rearme m - . . . • [mec] reassembly,ling

rearme m de árbol m—[mec] shaft reassembly,ling
— m — — m para entrada de fuerza f - [mec] input shaft reassembly,ling
— m — — f — ajuste m - [válv] valve fitting assembly,ling
— m — — f — — m para válvula f - [válv] valve fitting reassembly,ling
reasegurado,da a - [seguros] reinsured
reasegurdador m - [seguros] reinsurer
— m autorizado - [seguros] authorized reinsurer
— m cedente - [seguros] ceding reinsurer
— m no autorizado - [seguros] unauthorized reinsurer
reasegurador,ra a - [seguros] reinsuring
reaseguro(s) a prorrata - [seguros] pro rata reinsurance
— m asumido - [insur] assumed reinsurance
— m autorizado - [seguros] authorized reinsurance
— m cedido - [seguros] ceded reinsurance
— m compartido - [seguros] shared reinsurance
— m con otra(s) compañía(s) de seguro(s) m - [seguros] reinsurance with (an)other insurance company,nies
— m contratado(s) - [seguros] reinsurance written
— m en participación f - [seguros] shared reinsurance
— m — — f dentro de cuota f - [seguros] quota share reinsurance
— m — — en exceso m de cuota f - [seguros] excess quota share reinsurance
— m excedente - [seguros] excess reinsurance
— m no autorizado - [seguros] unauthorized reinsurance
— m por pagar - [seguros] reinsurance payable
— m recuperable - [seguros] recoverable reinsurance
reavalúo m - véase revaluación f
reavivamiento m - [relig] revival • reviving
rebaba f - [metal] . . .; chip; beard; chipping • [metal-lam] trim stock
— f controlada - [sold] controlled flash
— f creada - [mec] created burr
— f de fabricación f - [metal-fabr] fabrication burr
— f — soldadura f - [sold] weld(ing) flash
— f en barra f - [mec] bar burr
— f — buza f - [metal-prod] nozzle burr
— f exterior - [sold] outside flash
— f interior - [tub] inside flash
— f quitada - [mec] removed burr(s)
— f removida - [mec] removed, burr(s), o flash
— f sacada - [mec] removed burr(s)
rebabado m - [metal-fabr] burring • deseaming • [metal-prod] fettling; gas cutting; chipping; scarfing
— m con llama(s) f - [metal-prod] (flame), burring, o scarfing
— m — soplete m - [metal-lam] (torch) scarfing • [metal-fabr] deseaming
rebabado,da a - [metal-fabr] (de)burred
— a con llama(s) f - [metal-prod] flame scarfed
rebabadora f - [metal-lam] scarfer; flash trimmer • [mec] trimmer • [herram] burr
— f de carburo m - [herram] carbide burr
— f para palanquilla f - [metal-lam] billet scarfer
rebabadura f - [metal] fettling
rebabar v - [metal] to, chip, o scarf. o burr. o fettle
— v con llama(s) f - [metal-prod] to flame scarf
rebaja f eventual - [com] eventual rebate
rebajado,da a - lowered • offset • [arquit-arco] depressed
rebajador m - [cald] desuperheater
rebajadora f - [constr] earth mover; scraper
rebajar v - . . . • [com] to rebate • [mec] to, offset, o thin • [carp] to rabbet
— v prensaestopa(s) m - [metal-prod] to machine @ gland
rebaje m - [metal-lam] cutdown
rebajo m - offset • . . .; recess • lap • relief

• bearding • counterbore ◄ [carp] rabbet
rebajo m de tipo m naval - [constr] ship lap
— n en cordón m - [mec] bead recess
— m — reborde m - [mec] bead recess
— m — rueda f - [autom] wheel offset
— m — trinquete m - [mec] ratchet recess
— m integral - [carp] integral rabbet
rebalsado,da a.- overflowed; run over
rebalsar v - [hidr] . . .; to overflow; to run
over
rebalse m - [hidr] running over; overtopping;
overflowing • outflow; pool
— f de aceite m - [mec] oil overrun
— m — agua m - [hidr] water, outflow, o overflow
— f — m regulado - [hidr] controlled water,
outflow, o overflow
— f — condensación f - [cald] condensate overflow
— m — desaereador m - [cald] deaereator overflow
— m — emergencia f - [hidr] emergency overflow
— m empedrado - [hidr] riprapped pool
— m regulado - [hidr] controlled overflow
rebanadora f - [metal-lam] véase cortadora f
longitudinal
rebarba f - véase rebaba f
rebasadero m - . . . • [sanit-lavabo] overflow
— m integral - [sanit-lavabo] integral overflow
rebasado,da a - overflowed
rebase m - overflow
rebatido,da a - rejected • questioned • refuted
rebatimiento m - . . . .; rejection • questioning
rebatir v - . . .; to reject • to question
rebelde a - . . . . • balky
rebobinado m - re-coiling
— m — bobina f - [lam] coil re-coiling
— m — — f laminada - [metal-lam] rolled coil,
re-coiling, o re-rolling
— m — bobina(s) f laminada(s) en caliente -
[metal-lam] hot roll coil re-coiling
— m — f — en frío - [metal-lam] cold roll
coil re-coiling
rebobinado,da a - [metal-lam] re-coiled •
[electr-equipo] rewound
rebobinadora f - [metal-lam] recoiler •[electr-
-equipo] rewinder
rebobinar v - [metal-lam] to re-coil • [electr-
equip] to rewind
— v inducido m - [electr-mot] to rewind @ armature
reborde m - . . . . • [mec] raised face • [autom-
-neumát] (tire), head, o shoulder • [metal-
-lam] side trimming • [mec] shoulder; lip,
boss; bead; border; rib; ledge
— m ancho m - [mec] wide flange
— m angosto m - [mec] narrow flange
— m corriente - [metal-fabr] standard flange
— m dañado - [mec] damaged, bead, o flange
— m — banda f para rodamiento m - [autom-neum]
shoulder
— m — f — — m escuadrado - [autom-neumát]
square shoulder
— m de llanta f - [mec] dish
— m — sello m - [mec] seal lip
— m — — m para aceite m - [mec] oil seal lip
— m — tambor m - [mec] drum flange
— m — — m resellable - [mec] resealable drum
bead
— m — tornillo m - [tornos] screw collar
— m — — m de tornillo m para banco m -
[herram] vise screw collar
— m en cara f frontal de cono m - [cojin] cone
front face rib
— m — — f posterior - [cojin] back face rib
— m — — f — de cono m - [cojin] cone back
face rib
— m — cono m - [cojin] cone rib
— m — cordón m - [mec] bead recess
— m — extremo m - [mec] end flange
— m — — m de estructura f - [constr] structure end flange
— m exterior - [autom-neumát] outer shoulder

reborde m interior - [autom-neumát] inner shoulder
— m para conexión f - [mec] connection shoulder
— m — guía f - [cojin] shoulder
— m recubierto - [mec] coated shoulder
— m según norma f - [metal-fabr] standard flange
— m superior - [constr] wale
rebordeado m - [mec] shouldering; flanging
— m de acero m - [metal-fabr] steel flanging
— m — extremo m - [tub] end flanging
— m — — m de tubo m - [tib] tube end flanging
— m — tubería f - [metal-fabr] tube flanging
— m — — f de acero m - [tub] steel tube flanging
— m — — f soldada - [tub] welded tube flanging
— m — — f — de acero m - [metal-fabr] welded
steel tube,bing flanging
— m — tubo m - [metal-fabr] tube flanging
— m — — soldado - [tub] welded tube flanging
— m en caliente - [metal-fabr] hot flanging
— m — dos etapas f - [metal-fabr] double lap
flanging
— m — frío - [metal-fabr] cold flanging
— m — una (sola) etapa f - [tub] single lap
flanging
— m — — — f de extremo m de tubo m - [tub]
tube end single lap flanging
— m para conexión f - [mec] connection shouldering
— m por método m Van Stone - [metal-fabr] Van
Stone flanging
— m — — — en frío - [metal-fabr] cold
Van Stone, o Van Stone cold, flanging
— m Van Stone en frío - [metal-fabr] cold Van
Stone, o Van Stone cold, flanging
rebordeado,da a - [mec] flanged; shouldered
— a en ambos extremos m - [tub] flanged at both
ends
— a — un extremo m - [tub] flanged at one end
— a — lado m - [tub] flanged at one side
— a — — sólo extremo m - [tub] flanged at one
end only
rebordeador m - [mec] beader • side cutter
— m para tubo m - [metal-fabr] tube beader
— m — — m para caldera f - [cald] flue beader
— m — — humo(s) m - [cald] flue beader
rebordeadora f - [metal-lam] edger
— v vertical - [mec] vertical edger
rebordear v - to edge • to shoulder
— v conexión f - [mec] to shoulder @ connection
rebordeo m - véase bordeado m
rebosamiento m - [hidr] overflow(ing) • spill
spillway
rebotado,da a - bounced; rebounded
rebotadura f - . . .; bouncing
— f de suspensión f—[autom] suspension bouncing
— f — — f de vehículo m - [autom] vehicle('s)
suspension bouncing
rebotar v - . . .; to bounce
rebote m - . . .; bounce,cing • resilience,cy;
elasticity; back-spring(ing)
— m — — f de vehículo m - [autom] vehicle('s)
suspension bouncing
recabado,da a - requested; entreated
recabar v - . . .; to, elicit, o request • to
gather
— f información f - to request @ information; to
gather @ data
recadista m - messenger; errand boy
recalcado m - véase recalcadura f
recalcado,da a - [mec] undercut • calked • upset;
swaged • [náut] listed • packed • stressed;
(re)emphasized; pointed out
— a en caliente - [metal-trat] hot headed
— a — frío - [metal-trat] cold headed
recalcador,ra - [mec] upsetter,ting
recalcadora f - [mec] upsetter; upsetting, o
swaging, machine; heading tool
— f alternante - [herram] reciprocating swager
— f con cabezal m fijo - [herram] stationary
spindle swager
— f con movimiento m alternante - [herram] re-
cioprocating swager

recalcadora f corriente—[mec] standard upsetter
— f horizontal - [mec] horizontal upsetter
— f — corriente - [mec] standard horizontal upsetter
— f para barreno,na m/f - [petról] bit ram
— f — trépano(s) m - [petról] bit ram
— f — troquel(es) m - [metal-fabr] die swager
— f — — m para cierre m - [metal-fabr] die closing swager
— f rotativa - [metal-fabr] rotary swager
— f vertical - [mec] vertical upsetter
recalcadura f - [mec] . . .; packing; calking; upsetting; swaging • [sold] setting • [náut] list(ing) - pointing out
— f de cabeza f - [metal-fabr] heading
— f — — f en caliente - [metal-fabr] hot heading
— f — — f — frío - [metal-fabr] cold heading
— f — junta f - [náut] joint packing
— f — remache m - [mec] rivet heading
— f en caliente - [metal-fabr] hot, heading, o swaging
— f — frío - [metal-fabr] cold, heading, o swaging
— f — — de clavija f - [metal-trat] pin cold heading
— f horizontal - [mec] horizontal upsetting
— f metálica - [mec] metal upsetting
— f — quitada - [mec] removed metal upsetting
— f — removida - [mec] removed metal upsetting
— f — sacada - [mec] removed metal upsetting
— f quitada - [mec] removed upsetting
— f removida - [mec] removed upsetting
— f sacada - [mec] removed upsetting
— f sobre mandril m - [metal-fabr] swaging over @ mandrel; mandrel swaging
recalcante a - véase recalcador,ra a
recalcar v - . . .; to calk; to swage; to upset; véase también calafatear - to (re)emphasize; to point out
— v en caliente - [metal-trat] to hot head
— v — frío - [metal-trat] to cold head
— v — — clavija f - [mec] to cold head @ pin
recalculado,da a - recalculated
recálculo m - recalculating,tion
recalcular v - to recalculate
recalentado,da a - reheated • overheated • [metal-lam] soaked
— a localmente - reheated locally • overheated locally
recalentador f - reheater • superheater
recalentamiento m - reheating ʼoverheating • [sold] high heat(ing) • [metal-prod] soaking • [ind] hot spot
— m causado - caused reheating • caused overheating
— m de bisagra f - [mec] hinge overheating
— m — cojinete m - [mec] bearing overheating
— m — condensador m - [electr-equip] condenser overheating
— m — lingote m - [metal-lam] ingot reheating
— m — motor m - [electr-mot] motor overheating • [comb.int] engine overheating
— m — palanquilla(s) f - [metal-lam] billet reheating
— m — planchón(es) m - [metal-lam] slab reheating
— m — portaelectrodo(s) m - [sold] (electrode) holder overheating
— m — rectificador m - [electr-equip] rectifier overheating
— m — reóstato - [electr-equip] rheostat overheating
— m — — m para campo m - [electr-instal] field rheostat overheating
— f — soldadora f - [sold] welder overheating
— m — tomacorriente m - [electr-instal] receptacle overheating
— m — transformador m - [electr-equip] transformer overheating
— m — vapor m - [cald] steam reheating
— m interior - [electr-oper] internal overheating

recalentamiento m intermedio - [metal-prod] intermediate reheating
— m local - local reheating • local overheating
recalentar v - . . . . • . . . • [mec] to run hot • [metal-prod] to soak
— v bisagra f - to overheat @ hinge
— v cojinete m - [cojin] to overheat @ bearing
— v condensador m - [electr-equip] to overheat @ condenser
— v localmente - to reheat locally • to overheat locally
— v motor m - [electr-mot] to overheat @ motor • [comb.int] to overheat @ engine
— v reóstato m - [electr-oper] to overheat @ rheostat
— v — m para campo m - [electr-oper] to overheat @ field rheostat
— v soldadora f - [sold] to overheat @ welder
— v tomacorriente m - [electr-oper] to overheat @ receptacle
— v vapor m - [cald] to reheat @ steam
recalescente a - [metal] recalescent
recalque m - véase recalcadura f
recalzado,da a - [constr] underpinned
recalzar v - [constr] to underpin
recalzo m - [constr] underpinning
recambiado,da - rechanged; refilled; replaced
— a frecuentemente - replaced frequently
recambiar v - . . .; to refill
— v frecuentemente - to replace frequently
recambio m - . . . . • replacement,cing • [mec] refill(ing) | a - (para) replacement; spare; swing
— m frecuente - frequent replacement
— m de eje m - [autom-mec] axle replacement
recapacitando adv - coming down to
recapacitar v - to think, over, o through; to mull
recarbonatación f - recarbonation
recarbonatar v - to recarbonate
recarburación f - [metal-prod] recarburizing; recarbureting; recarburation,rization
— f en cuchara f - [metal-prod] ladle recarburizing
recarburante m - [metal-prod] recarburizer
recarburar v - [metal-prod] to recarburize,rate
recarburización f - véase recarburación f
recargable a - rechargeable
recargado,da a - recharged • overloaded • weighted • [sold] built up • [com] marked up • [mec] heavied up
recargar v - to overload; to overburden • to surcharge • [sold] to build up; to crown (a fillet) • [mec] to heavy up • [com] to mark up
— v muro m - [constr] to surcharge @ wall
recargo m - [mec] build up; heavying up • [sold] (weld) build up • [fisc] surcharge; extra • [com] marking up
— m cambiario - [fin] exchange surcharge
— m en precio m - [com] extra charge
— m para importación f - [com] import surcharge
— m por calidad f - [ind] quality surcharge
— m — dimensión(es) f - [transp] dimension(s) surcharge,ging
— m — enchapado m - [metal-trat] plating surcharge
— m — espesor m - [metal-lam] thickness surcharge
— m — galvanización f - [metal-trat] galvanizing (sur)charge
— m — — f en caliente - [metal-trat] hot galvanizing (sur)charge
— sobre muro m - [constr] wall surcharge,ging
recargue m - [sold] build-up; reinforcement
— m máximo - [sold] maximum build-up
— m mayor - [sold] greater build-up
recata f - . . . . • [constr] recess(ing)
recatado,da a - [constr] recessed
recatar v - . . . . • [constr] to recess
recauchutado m - [autom-neumát] recap(ping)
recauchutaje m - [autom-neumát] . . .; retreading
recauchutar v - [autom-neumát] to, recap, o retread

recaudación f - . . . . • [fisc] intake
— f bruta f - [fisc] gross intake
— f fiscal - [fisc] fiscal collection
— f por exportación(es) f - [fisc] export(s),
intake, o collection
— f — importación(es) f - [fisc] import(s),
intake, o collection
recebo m - . . . . • [geol] hoggin
receloso,sa a - . . .; hesitant; apprehensive •
touchy
recepción f - receipt; receiving • [com] acceptance,ting
— f con coctel(es) m - [social] cocktail reception
— m de bidón(es) f - [ind] drum(s) receiving
— f — bien(es) m - good(s) acceptance
— f — bobina(s) f - [metal-prod] coil(s) receiving
— f — — f caliente(s) - [metal-lam] hot coils
receiving
— m — código m - code receiving,ceipt
— f — condición f - [comput] status receipt
— f — conformidad f - conformity, o acceptability, receiving,ceipt
— f — coque m - [metal-prod] coke receiving
— f — embarque m - [transp] shipment receipt
— f — empuje m - thrust receiving
— f — equipo m - [ind] equipment, receipt, o
receiving, o reception, o acceptance
— f — estado m - [comput] status receiving
— f — fuerza f - [mec] force receiving
— f — impulso m - boost receiving • [electrón]
impulse receiving
— f — información f - information, receiving,
o receipt
— f — — f sobre, condición f, o estado m -
[comput] status information receiving,ceipt
— f — — f — posición f - [comput] status, o
position, information receiving,ceipt
— f — lingote(s) m - [metal-prod] ingot receiving,ceipt
— f — manual - [ind] manual receiving,ceipt
— f — materia(s) f prima(s) - [ind] raw materials, receiving,ceipt, o acceptance,ting
— f — material(es) m - material(s), receipt, o
receiving, o reception
— f — obligación f—[fin] note receiving,ceipt
— f — par m motor - [mec] torque, acceptance,
o accepting
— f — posición f - [comput] status receiving
— f — propuesta(s) f - [com] proposal(s), o
bid(s), receiving,ceipt; receipt of bid(s)
— f — recibo m - receipt receiving,ceipt
— f — solicitud - application reception
— f — señal f - [electrón] signal receiving
— f — tambor(es) m - [ind] drum receiving
— f — torsión f - [mec] torque input
— f — vuelta f - getting, o accepting, back
— f definitiva - final acceptance
— f física - physical acceptance
— f parcial - [com] partial acceptance
— f provisional - provisional, o temporary, acceptance
— f provisoria - [com] temporary acceptance
— f — parcial - [com] temporary partial acceptance
— f rápida - rapid, o quick, receiving,ceipt, o
acceptance
— f real - [transp] true acceptance
recepcionado,da - received • accepted
recepcionar v - to receive • to accept
receptáculo m - . . . . • [electr-equip] socket •
[mec] holder; bin
— m — metal m - [mec] metal(lic) container
— m — — m laminar - [mec] laminar, o laminated, metal(lic) receptacle
— m instalado m - [electr-instal] installed
socket
— m — en fábrica f - [electr-instal] factory
insalled socket
— m para neumático n (en guardabarros m) -
véase posaneumático m
— m — percusor m - [petról] jar socket

receptáculo m para relé m - [electr-instal]
relay, socket, o receptacle
receptor m - . . . . • [legal] licensee
— m con retorno m - [comput] loopback receiver
— m de aire m - air receiver
— m encaminable - [comput] addressable receiver
— m — con retorno m - [comput] addressable
loopback receiver
— m heterodino - [electrón] heterodyne receiver
— m indicador de posición f - [instrum] Selsyn
receiver
— m instalado - [electrón] installed receiver
— m para estación f - [comput] field receiver
— m — remache(s) m - [constr] rivet catcher
— m radiotelefónico—[electrón] radio (receiver}
— m remoto - [electrón] remote receiver
— m Selsyn - [electrón] Selsyn receiver
— m sincrónico - [e;ectrón] Selsyn, o synchronous, receiver
— m y transmisor m remoto(s) - [electrón] remote receiver and transmitter
receptor,ra a - . . . . • [metal-lam] catcher
receso m - . . . . • intermission
receta f afinada - fine tuned recipe
— f complicada - complicated, o compound, recipe
— f compuesta - compound recipe
— f formulada - [quím] developed, o formulated,
recipe
— f modificada - modified, recipe, o prescription
— f perfeccionada - [quím] developed, o perfected, recipe
— f prometedora - [quím] promising recipe
recetado,da a - [medic] prescribed
recibido,da a - received • accepted • incoming •
[educ] graduate(d)
— a a cuenta f - [fin] received on account;
payment received
— a de vuelta - received back
— a rápidamente - received, quickly, o rapidly
recibidor-secador m - [agric-equip] receiver-
-drier
recibir v - . . . . • to accept
— v adelanto m - [fin] to receive @ advance
— v anticipo m - [fin] to receive @ advance
— v bidón m - [ind] to receive @ drum
— v bien(es) m - to receive @ good(s) • to accept @ good(s)
— v código m - to receive @ code
— v conformidad f - to receive @ conformity • to
receive @ receipt
— v choque m - [electr-oper] to get @ shock
— v de vuelta f - to, receive, o get, back
— v embarque m - [transp] to receive @ shipment
— v empuje m - [mec] to receive @ thrust
— v equipo m - [ind] to receive @ equipment
— v fuerza f - [mec] to receive @ force • to
receive @ strength
— v impacto m - [mec] to receive @ impact
— v — m en servicio m - in service pounding
— v impulso m - to receive @, impulse, o boost
— v manual m - [ind] to receive @ manual
— v materia(s) f prima(s) - [ind] to, receive, o
accept, @ raw material(s)
— v material(es) f - [ind] to receive @ mate-
rial(s)
— v obligación f - [fin] to receive @ note
— v par m motor - [mec] to accept @ torque
— v pedido m - to receive @ order
— v propuesta f - to receive @, proposal, o bid
— v provisoriamente - [com] to accept temporarily
— v rápidamente - to receive, rapidly, o quickly
— v recibo m - [fin] to receive @ receipt
— f respuesta f - to, receive, o get, @ answer
— v señal f - [electrón] to receive @ signal
— v solicitud f - to receive @ application
— v tambor m - [ind] to receive @ drum
recibir(se) v - [educ] to graduate
recibo m acusado - (receipt) acknowledged
— m de adelanto m - advance receiving,ceipt
— m — almacén m - [ind] warehouse receipt
— m — anticipo m - advance receiving,ceipt

recibo m de embarque m - [transp] shipment receipt
— m de factura f - [com] invoice receipt
— m —manual m - [ind] manual receiving,ceipt
— m —materia(s) f prima(s) — [ind] raw material(s) receiving,ceipt, o reception
— m —paquete m postal - [transp] parcel post receiving,ceipt
— m — vuelta - getting, o receiving, back
— m original - [com] original receipt
— m por encomienda f por ferrocarril - [transp] railway express parcel receipt
— m — impuesto(s) m - [fisc] tax receipt
— m recibido - [com] received receipt
reciclación* f - véase recirculación f
reciclado,da* a - véase recirculado,da; reprocesado,da
reciclador* m - [ind] véase recirculador m; reprocesador m
reciclar* v - véase recircular v; reprocesar v
recién adv casado,da - newlywed
— adv cribado,da - [ind] freshly screened
— adv entonces - only then
— adv instalado,da - newly installed
— adv llegado - new comer; new arrival
— adv vaciado - [constr] newly poured
reciente a - recent; new • late
recientemente adv - . . . . • not long ago
recilindrado m - [metal-lam] rerolling; véase relaminación f
— m de soldadura f - [sold] weld rerolling
cilindrar v - [sold] to reroll @ weld
— v soldadura f - [sold] to reroll @ weld
recinto m - . . .; enclosure; room; chamber
— alambrado - [constr] wire enclosure
— m con alambre m tejido - [constr] wire mesh enclosure
— m con tejido m (de alambre m) - [constr] wire mesh enclosure
— cerrado m - [constr] (en)closed space
— m con corriente f de aire m - [constr] drafty location
— m de acero m - [ind] steel enclosure
— m — alambre m tejido - [constr] (wire) mesh enclosure
— m estrecho - close quarter(s)
— m metálico - [constr] metal enclosure
— m para motor(es) m - [electr-mot] motor room • [comb.int] engine, enclosure, o room
— m — práctica f de tiro m - [deport-golv] driving range
— m para seguridad f - [ind] safety enclosure
— m reducido - [ind] small, o confined, space, o place, o area, o enclosure; close quarters
— m para transformador(es) - [electr-distrib] transformer(s) enclosure
— m ventilado - ventilated, space, o enclosure
— m adecuadamente - [ind] adequately ventilated space
— m — inadecuadamente - [ind] inadequately ventilated space
recio,cia a - . . .; rugged; hefty
recipiente m - . . .; basin; reservoir; vessel; pan; can; chamber; tank; sump; bowl; holder • receiver • [autom] cell
— m abastecedor - supply(ing), container, o can
— m abierto - open(ed), container, o vessel
— n apropiado - appropriate, o suitable, container
— m autosoportante - [cald] self-supporting, container, o vessel
— m — instalado - [cald] installed self-supporting vessel
— — — verticalmente - [cald] vertically installed self-supporting vessel
— m blindado - shielded, o armored, container
— m calentado - [ind] heated container
— m cerrado - closed, container, o vessel
— m cilíndrico - [cald] cylindrical, container, o vessel
— m — autosoportante - [cald] self-supporting cylindrical, container, o vessel
— m — — instalado - [cald] installed self-supporting cylindrical vessel

recipiente m cilíndrico instalado - [cald] installed cylindrical vessel
— m con arena f - [ind] sand (storage) drum
— m — — f seca - [ind] dry sand (storage) drum
— m con boca f ancha - wide-mouth container
— m — fuga(s) f - leaking container
— m — oxígeno m - [metal-acero] oxygen, vessel, o container
— f — pared(es) f delgada(s) - [cald] thin walled vessel
— m — presión f - [ind] pressure vessel
— m — — f autosoportante - [cald] self-supporting pressure vessel
— m — — f — instalado - [cald] installed self-supporting pressure vessel
— m — — f — — verticalmente - [cald] vertically installed self-supporting pressure vessel
— m — — f cilíndrico - [cald] cylindrical pressure vessel
— m — — f — autosoportante - [cald] self-supporting cylindrical pressure vessel
— m — — f — — instalado - [cald] installed self-supporting cylindrical pressure vessel
— m — — f — instalado - [cald] installed cylindrical pressure vessel
— m — — f — — verticalmente - [cald] vertically installed cylindrical pressure vessel
— m — — f — autosoportante m - [cald] self-supporting cylindrical pressure vessel
— f — — f — — instalado verticalmente - [cald] vertically installed self-supporting pressure vessel
— m — — f instalado - [cald] installed pressure vessel
— m — — f — — verticalmente - [cald] vertically installed pressure vessel
— m — — f expuesto a fuego m - [cald] fired pressure vessel
— m — — f no expuesto a fuego m - [cald] unfired pressure vessel
— m — — f para temperatura(s) f alta(s) f - [cald] high temperature pressure vessel
— m — — f — — f baja(s) - [cald] low temperature pressure vessel
— m — — f según código(s) f de Sociedad f Estadounidense de Ingenieros m Mecánicos - [cald] American Society of Mechanical Engineers' Pressure Vessels
— m — — f sin fuego m - [cald] unfired pressure vessel
— m contaminado - contaminated container
— m cortado - [mec] cut container
— m de acero m - [ind] steel container
— m — — m dulce - [ind] mild steel container
— m — — m para almacenamiento m - [ind] mild steel storage container
— m — — m — elaboración f -[ind] mild steel processing container
— m — cinc m - [mec] zinc, container, o can
— m doble - [petról] dual, vessel, o container
— m hueco - [mec] hollow container
— m instalado - [cald] installed vessel • [ind] installed container
— m — verticalmente - [cald] vertically installed vessel
— m libre de polvo m - dust free container
— m limpio m - clean, container, o can
— m metálico - [ind] metal, container, o can
— m no expuesto a fuego m - [cald] unfired vessel
— m para aceite m - [comb.int] (oil) sump
— m — — m usado - [ind] used oil container
— m — almacenamiento m - [ind] storage, container, o equipment
— m — calefacción f - [cald] heating vessel
— m — — f con circulación f de vapor m - [cald] steam circulation heating vessel
— m para calentamiento m - heating vessel

recipiente m para cereales m - [agric-equip] grain, box, o tank
— m —— m para caja f para carro m - [agric--equip] wagon box grain tank
— m — combustible m - fuel container * [autom] fuel cell
— m —— m apoyado - [comb.int] grounded, o supported, fuel container
— m —— m conectado con tierra f - [comb.int] grounded, fuel container
— m para elaboración f - [ind] processing, container, o equipment
— m — entrada f - entry vessel
— m — fluido(s) m - fluid(s) container
— m — insecticida m - [quím] insecticide container
— m — líquido(s) m - [ind] liquid(s) container
— m — mezcla f - [ind] mixing chamber
— m — purga f - [autom-mec] drain(ing) pan
— m — retorno m - [mec] return pan
— m — transvase m - [mec] filler can
— m que haya contenido combustible m - container that has held @, combustible, o fuel
— m sin fuego m - [cald] fireless, o unfired, vessel
— m sin óxido m - rust free, container, o can
— m soldado - [sold] welded container • [cald] welded vessel
recirocante* a - véase recíproco,ca; alternante
recíproco,ca a - . . . . • reciprocating
recirculación f - recirculation • recycling • [hidr] reflux
— f con bomba f para alimentación f - [bombas] feed pump recirculation
— f —— f para alimentación f de caldera f - [cold] boiler feed pump recirculation
— f continua - continual,nuous recirculation
— f de agua m - water recirculation
— f — aire m - air recirculating,tion
— f —— m interior - inside air recirculation
— f —— m para enfriamiento m - cooling air recirculating,tion
— f asfalto m - [constr] asphalt recycling
— f — fundente m - [sold] flux, recyling, o recirculating,tion
— f — gas m - [combust] gas recirculating,tion
— f — material m - [ind] material recycling
— f — pella(s) f - [miner] pellet(s) recycling
recirculado,da a - recirculated • recycled
— a continuamente - recirculated continuously
recirculador m - recirculator • recycler
recirculadora f para asfalto m—[constr] asphalt recycler
recirculante a - recirculating
recircular v - to recirculate • to recycle
— v aire m - to recirculate @ air
— v — m exterior - to recirculate @ outside air
— v — m interior - to recirculate @ inside air
— v — m para enfriamiento m - to recirculate @ cooling air
— v asfalto m - [constr] to recycle @ asphalt
— v continuamente - to recirculate continuosly • to recycle continuously
— v fundente m - [sold] to recirculate @ flux
— v — m para arco m sumergido - [sold] to recirculate @ submerged arc flux
— v material m - [ind] to recycle @ material
recirura f - . . . .; ruggedness; strength
— f excepcional - exceptional strength
reclamación f de cliente m - [com] customer('s) claim
— f de zona f explotada - [miner] mined area reclamation
— f —— f minada - [miner] mined area reclamation
— f efectuada - filed claim
— f emergente - arising claim
— f escrita - written claim
— f — presentada - filed written claim
— f especial - special claim
— f neta - net claim
— f pendiente - pending claim
— f por daño(s) m - damage(s) claim

reclamación f por defecto(s) m - defect(s) claim(ing)
— f preparada - prepared claim
— f presentada - filed claim
reclamado,da a (re)claimed • objected
reclamante m - [legal] claimant
reclamar v - . . . . • to object • to claim • to reclaim
— v cantidad f - to claim @ amount
— v infracción f - to claim @ infringement
— v pérdida f - [seguros] to claim @ loss
reclamo m - . . . . • objecting,tion
— m bajo garantía f - [com] warranty claim
— m compensable - [seguros] compensable claim
— m compensado - [seguros] compensated claim
— m efectuado - claim, filed, o made
— m emergente - arising claim
— m — presentado - filed arising claim
— m purga f - [labor] grievance
— m pagado - paid claim
— m pagado - [seguros] paid claim
— m — prontamente - [seguros] promptly paid claim
— m personal - personal claim
— m por compensación f - [labor] compensation claim
— m — daño(s) m - [transp] damage claim
— m — equipage m - [transp] baggage claim(ing)
— m — infracción f - infringement claim
— m — merma(s) f - [seguros] shortage(s) claim
— m — mora f en pago m - [fin] default of payment claim
— m — pérdida f - [seguros] loss claim
— m presentado - [seguros] filed claim
— m preparado - prepared claim
reclasificación f - reclassification,fying
reclasificado,da a - reclassified
reclasificar v - to reclassify
reclutamiento m de personal m - [labor] personnel recruitment
recobrado,da a - recovered • recuperated • recouped; regained; made up
recobrar v - . . .; to recoup; to make up
recobro m - recovery; recouping; making up
— v gobierno m - [autom] to regain @ control • [pol] to recover @ power
— m de gobierno m - [autom] control regaining • [pol] power regaining
recocción* f - véase recocido m
recocer v - to recook • to reheat • to overboil • [metal-trat] to anneal • to draw
— v en caliente - [metal-trat] to heat anneal
— f en forma f incompleta - [metal-trat] to underanneal
recocido m - . . . . • (over)firing
— m abierto - [metal-trat] open annealing
— m anterior - [metal-trat] preannealing
— m — a laminación f en frío - [metal-trat] patenting
— m antes de trabajo m en frío - [metal-trat] patenting
— m azul - [metal-trat] blue annealing
— m blanco - [metal-trat] bright annealing
— m brillante - [metal-trat] bright annealing
— m completo m - [metal-trat] full annealing
— m con gas m - [metal-trat] gas annealing
— m con llama f - [metal-trat] flame annealing
— m —— f abierta - [metal-trat] open flame annealing
— m — tubo m radiante - [metal-trat] radiant tube annealing
— m continuo - [metal-trat] continuous annealing
— m de banda f - [metal-trat] strip annealing
— m —— f comercial - [metal-trat] commercial strip annealing
— m — bobina(s) f - [metal-trat] coil annealing
— m —— f abierta(s) f - [metal-trat] open coil annealing
— m —— f cerrada(s) - [metal-trat] closed coil annealing
— m —— f con llama f abierta - [metal-trat]

open flame coil annealing
recocido m de cinta f - [metal-trat] strip annealing
— **m —— f comercial** - [metal-trat] commercial strip annealing
— **m — chapa f** - [metal-trat] strip, o sheet, annealing
— **m —— f comercial f** - [metal-trat] commercial, strip, o sheet, annealing
— **m —— f en continuo** - [metal-trat] continuous strip annealing
— **m — fleje m** - [metal-trat] strip annealing
— **m —— m comercial** - [metal-trat] commercial strip annealing
— **m — hojalata f** - [metal-trat] tin (strip) annealing
— **n — soldadura f**—[sold] weld(ment) annealing
— **m — tipo m con campana f** - [metal-trat] bell type annealing
— **m —— m —— f convectora** - [metal-trat] convector bell type annealing
— **m —— m —— f — radiante** - [metal-trat] radiant convector type bell annealing
— **m —— m con convector m** - [metal-trat] convector type annealing
— **m —— m —— m radiante** - [metal-trat] radiant type convector annealing
— **m desoxidante** - [metal-trat] deoxidizing annealing
— **m después de corte m** - [metal-trat] annealing after shearing
— **m en caja f** - [metal-trat] box, o pot, o closed, annealing
— **m — campana f** - [metal-trat] bell annealing
— **m — cofre m** - véase **recocido m en caja f**
— **m — continuo** - [metal-trat] continuous annealing
— **m —— de chapa f** - [metal-trat] continuous strip annealing
— **m — descubierto m** - [metal-trat] open annealing
— **m — forma f incompleta** - [metal-treat] underannealing
— **m — horno m** - [metal-trat] furnace annealing
— **m — laboratorio m** - [metal] laboratory annealing
— **m — tanda(s) f**—[metal-trat] batch annealing
— **m estabilizador** - [metal-trat] stabilizing annealing
— **m intermedio** - [metal-trat] intermediate annealing • batch annealing
— **m monopila** - [metal-trat] single stack annealing
— **m — en caja f** - [metal-trat] single stack (box) anannealing
— **m negro** - [metal-trat] black annealing
— **m para eliminar esfuerzo(s) m** - [metal-trat] stress free annealing
— **m —— v tensión(es) f** - [metal-trat] stress free annealing
— **m — estabilización f** - [metal-trat] stabilizing annealing
— **m — suprimir v esfuerzo(s) m** - [metal-trat] stress free annealing
— **m —— v tensión(es) f** - [metal-trat] stress free annealing
— **m — tratamiento m** - [metal-trat] process, o treating,tment, annealing
— **m posterior** - [metal-trat] postannealing • galvannealing
— **m — a galvanización f** - [metal-trat] galvannealing
— **m subcrítico** - [metal-trat] subcritical annealing
— **m total** - [metal-trat] full annealing
recocido,da - [metal-trat] annealed
— **a en caliente** - [metal-trat] heat annealed
— **a — forma f incompleta** - [metal-treat] underannealed
— **a posteriormente** - [metal-trat] postannealed • galvananealed
recodo m - . . . . • [vial] corner • [topogr] oxbow

recogedora f - [metal-prod] - (ore yard) reclaimer
— **f — playa f para mineral m** - [metal-prod] ore yard reclaimer
recoger v - . . .; to collect; to actach; to include • to, draw, o take, o pick, o scoop, up • [metal-fabr] to gather •[hidr] to intercept
— **v agua m** - [hidr] to, collect, o intercept, o gather (un), @ water
— **v — en zona f, divisoria, o medianera** - [hidr] to intercept @ median water
— **v — sobre superficie f** - [hidr] to intercept @ surface water
— **v cable m** - [grúas] to pick up @ cable
— **v emanación(es) f** - to collect @ emission(s)
— **v información f** - to gather @ information
— **v líquido m** - to catch @ liquid
recogida f - gathering; collecting,tion; scooping, o drawing, o picking, o taking, up • [ind] removing; handling • [hidr] catching; catchment; capture; holding
— **f de agua m** - water collecting
— **f — cable m** - [grúas] cable pick(ing)-up
— **f — chatarra f** - [metal-prod] scrap collecting
— **f — desecho(s) m** - [nucl] waste collection
— **f —— m— m radiactivo(s) m** - [nucl] radioactive waste(s) collecting,tion
— **f —— m — líquido(s)** - [nucl] liquid radioactive waste(s) collecting,tion
— **f — emanación(es) f** - [ind] emission(s) collecting,tion
— **f — líquido(s) m** - liquid(s) catching,chment
— **f —— m separada** - [nucl] separated liquid catching
— **f — residuo(s) m** - [nucl] waste(s) collecting,tion
— **f —— m radiactivo(s)** - [nucl] nuclear waste(s) collecting,tion
— **f —— m — líquido(s)** - [nucl] liquid radioactive waste collecting,tion
recogido,da a - . . . . • gathered; collected • included • picked, o taken, o scooped, o drawn, up - [hidr] intercepted; caught
recogimiento m - . . . . • collecting,tion; picking up
— **m de agua n** - [hidr] water collecting,tion
recolección f - collecting,tion; gathering • [transp] pick(ing) up • [hidr] intercepting,tion
— **f con (auto)camión m** - [transp] truck pick(ing)-up
— **f de agua m** - [hidr] water, collecting,tion, o gathering, o interception,ting
— **f —— m en franja f, divisoria, o medianera** - [hidr] median water intercepting,tion
— **f —— m sobre superficie f** - [hidr] surface water intercepting,tion
— **f — basura(s) f** - [sanit] trash, o garbage, o refuse, pick(ing)-up, o collecting,tion
— **f — dato(s) m** - [comput] data gathering
— **f —— m sobre consumo m de oxígeno m** - [combust] oxygen consumption data gathering
— **f —— m — producción f** - [ind] production data gathering
— **f — desperdicio(s) m** - [sanit] trash, o garbage, o refuse, collecting,tion
— **f — información f** - [ind] data gathering
— **f —— f sobre producción f** - [ind] production data gathering
— **f — gas m** - gas collecting,tion
— **f — polvo m** - dust, collecting,tion, o gathering
— **f en lado m de afuera** - collecting on @ outside
— **f — obra f** - field collecting,tion
— **f —— f de información f** - field data collecting,tion
recolectado,da a - collected; gathered; hoarded
— **a de lado m de afuera** - collected on @ outside
recolectar v agua m - to collect @ water
— **v de lado de afuera** - to collect on @ outside

recolectar v polvo m - to, gather, o collect, @ dust
recolector,ra a - gathering
recomendable a - . . .; recommendable; advisable
recomendación f - . . .; recommending; suggesting,tion; advocating • guideline - pl • policy
— f adicional - additional recommendation
— f buena - good, recommendation, o guideline
— f completa - complete recommendation
— f completa para lubricante n - [lubric] complete lubricant recommendation
— f corregida - corrected, o revised, recommendation
— f correspondiente - corresponding, note, o recommen dation
— f dada - given recommendation
— f de código m - code recommendation
— f escala f - range recommending,mendation
— f - fabricante m - manufacturer('s) recommendation
— f — m de electrodo(s) m - [sold] electrode(s) manufacturer's recommendation
— f ingeniero(s) m - engineer('s) recommendation(s)
— f precaución f - precaution recommendation
— f enfática - strong recommendation
— f específica - specific recommendation
— f sobre electrodo m - [sold] specific elecrode recommendation
— f general - general recommendation
— f máxima - maximum recommendation
— f para precalentamiento m - [sold] maximum preheat(ing) recommendation
— f mínima - minimum recommendation
— f para recalentamiento m - [sold] minimum preheat(ing) recommendation
— v oral - oral, o verbal, recommendation
— f aceite m - [mec] oil recommendation
— f para alimentadora f - [ind] feeder recommendation
— f — f para alambre m - [sold] wire feeder recommendation
— f amperaje m - [sold] amperage, o current, recommendation
— f calidad f - quality, o grade, recommendation
— f cantidad f - quantity recommendation
— f densidad f - density recommendation
— f — f para aceite m - [comb.int] oil, density, o weight, recommendation
— f para diámetro m - diameter, o size, recommendation
— f — m de cable m - [electr-cond] cable size recommendation
— f ejecución f de medición(es) - [electr-instal] recommended guide for making measurements
— f — f — f dieléctrica(s) - [electr-instal] recommended guide for making dielectric measurement(s)
— f — f — f en obra f - [electr-instal] recommended guide for making dielectric measurements in the field
— f filtro m - [mec] filter recommendation
— f invierno m - winter recommendation
— f lubricante m—[lubric] lubricant recommendation
— f llanta f - [autom] rim recommendation
— f mantenimiento m - [ind] maintenance recommendation
— f mejora(s) f - improvement recommendation
— f en capacitación f - [labor] training improvement recommendation
— f — f en costo(s) m - [ind] cost(s) improvement(s) recommendation
— f momento m torsional - [mec] torque value recommendation
— f par m motor - [mec] torque value recommendation
— f perfil(es) m estructural(es) m - [metal-lam] structural(s) (shapes) recommendation
— f — m — de acero m - [metal-lam] steel structural shape recommendation
recomendación(es) f para perfil(es) estructurales de acero m con carbono m - [metal-lam] carbon steel structural shape(s) recommendation
— f — m — m de acero m laminado(s) - [metal-lam] carbon steel structural shape recommendation(s)
— f — m — m en caliente - [metal-lam] hot rolled carbon steel structural shape recommendation(s)
— f — peso m - weight recommendation
— f — m para aceite m - [comb.int] oil weight recommendation
— f — pieza f - [ind] part recommendation
— f — f para repuesto m - [ind] spare part recommendation
— f — pistola f - [sold] gun recommendation
— f — recalentamiento m - [sold] reheat(ing) recommendation
— f — proyección f - design recommendation
— f — repuesto(s) m - [ind] spare part(s) recommendation
— f — rueda(s) f y neumático(s) m - [autom-neumát] tire/wheel recommendation
— f — seguridad f - [segurid] safety recommendation
— f — sistema m - system recommendation
— f — tamaño m - size recommendation
— f — tipo m - type recommendation
— f — valor m - value recommendation
— f — verano m - summer recommendation
— f particular - specific recommendation
— f proporcionada - given recommendation
— f revisada - revised recommendation
— f sobre procedimiento(s) m - procedure(s) recommendation
— f — m específico(s) - [ind] specific procedure(s) recommendation
— f típica - typical recommendation
— f verbal - verbal, o oral, recommendation
— f y advertencia f - do's and dont's
recomendado,da a - recommended; advocated
— a enfáticamente - strongly recommended
— a para acero(s) m A I S I (número m) . . . [sold] recommended for American Iron and Steel Institute type . . . (steel)
— a por industria f - [ind] industry recommended
recomendar v - . . .; to suggest; to advocate
| a - (de) recommended
— v alimentadora f - [mec] to recommend @ feeder
— v — f para alambre m - [sold] to recommend @ wire feeder
— v amperaje m - [sold] to recommend @, amperage, o current
— v calidad f - to recommend @, quality, o grade
— v cantidad f - to recommend @ quantity
— v — f de existencia(s) f - [ind] to recommend @ stocking quantity
— v diámetro m - to recommend @ diameter
— v — m para cable m - [sold] to recommend @ cable size
— v enfáticamente - to recommend, emphatically, o strongly
— v escala f - to recommend @, scale, o range
— v existencia(s) - [ind] to recommend @ stocking quantity
— v filtro m - [mec] to recommend @ filter
— v llanta f - [autom] to recommend @ rim
— v mantenimiento m - [ind] to recommend @ maintenance
— v medida(s) - to recommend @ measurement(s)
• to recommend @ action
— v — f correctiva(s) - to recommend @, corrective, o remedial, action
— v peso m - to recommend @ weight
— v — m para aceite m - [mec] to recommend @ oil weight
— v pieza f - [ind] to recommend @ part
— v — f para repuesto m - [ind] to recommend

recomendar v pistola f — 1270 —

@ spare part
recomendar v pistola f - to recommend @ gun
— v precaución f - to recommend @ precaution
— f proyección f to recommend @ design(ing)
— v reconocimiento m - to recommend @ recognition
— v — m apropiado - to recommend @ appropriate recognition
— m régimen m para motor m - [comb.int] to recommend @ engine speed • [electr-mot] to recommend @ motor speed
— v — m sin carga f - [mec] to recommend @ idle speed
— v — m —— f para motor m - [electr motɔ̌ to recommend @ motor idle speed • [int.comb] to recommend @ engine idle speed
— v repuesto(s) m - [ind] to recommend @ spare part(s)
— v sistema m - to recommend @ system
— v tamaño m - to recommend @ size
— v tipo m - to recommend @ type
— v valor m - to recommend @ value
— v — m para momento m torsional - [mec] to recommend @ torque value
— v — m — por m motor - [mec] to recommend @ torque value
— v velocidad f - [mec] to recommend @ speed
— v — f para motor m - [electr-mot] to recommend @ motor speed • [comb.int] to recommend @ engine speed
— v — f sin carga f - [comb.int] to recommend @ idle speed
— v — f —— f para motor m - [comb.int] to recommend @ engine idle speed
recomenzado,da a - started over again
recomenzar v - to start over again
recomienzo m - starting over again
recompactación f - [constr] backpacking
— f demorada - [constr] delayed backpacking
— f inadecuada—[constr] inadequate backpacking
recompactado,da a - [constr] backpacked
— a inadecuadamente - inadequately backpacked
recompactar v - [constr] to backpack
— v inadecuadamente - [constr] to backpack inadequately
recompensa f - . . .; reward(ing) • pay off
— f material - material reward
recompensado,da a - rewarded • paid off
recompensar v - . . . • to pay off
reconciliado,da a - reconciled
reconciliar v - . . .; to compromise
reconducción f - . . . • rerun(ning)
reconducido,da a - rerun
reconducir v - . . .; to rerun
reconectado,da - reconnected • [instrum] reset
— a automáticamente - [electr-oper] reset automatically • reconnected automatically
— a de un voltaje m para otro - [electr-oper] reconnected from one voltage to another
— a incorrectamente - [electr-instal] reconnected incorrectly
reconectador,ra a - reconnecting
reconectar v - to reconnect; to reset • to reassemble
— v acoplamiento m - [mec] to recoonect @ linkage
— v acumulador m - [comb.int] to reconnect @ battery
— v automáticamente - [electr-oper] to reset automatically
— v borne m - [electr-instal] to reconnect @ terminal
— v conductor m - [electr-instal] to reconnect @ lead
— v conexión f - [electr-instal] to rewire @, connection, o wiring
— v de un voltaje m para otro - [electr-instal] to reconnect from one voltage to another
— v disyuntor m - [electr-opr] to, reconnect, o reset, @, breaker, o switch
— v incorrectamente - [electr-instal] to reconnect incorrectly
— m interruptor m - [electr-oper] to reset @

breaker
reconectar v osciloscopio m - [electrón] to reconnect @ oscilloscope
— v paso(s) m - [electrón] to reconnect @ sequence
— v — m para puesta f en marcha y detención f - [electrón] to reconnect @ stop and start sequence
reconexión f - reconnecting,tion • [instrum] reset(ting)
— f automática - [electr-oper] automatic resetting
— f de acoplamiento m - [mec] linkage, reconnecting,tion, o linkage,ing
— f — acumulador m - [comb.int] battery reconnecting,tion
— f — conductor m - [electr-instal] lead reconnecting,tion
— f — conexión(es) f - [electr-instal] wiring rewiring • lead(s) reconnecting,tion
— f — disyuntor m - [electr-oper] switch, reconnecting,tion, o resetting
— f — entrada f - [electr-instal] input reconnecting,tion
— f — interruptor m - [electr-oper] breaker, reconnecting,tion, o resetting
— f — osciloscopio m - [electrón] oscilloscope, reconnecting,tion, o resetting
— f — un voltaje m para otro - [electr-oper] reconnecting,tion from one voltage to another
— f incorrecta - [electr-instal] incorrect reconnecting,tion
reconfiguración f - reconfiguring,ration
reconfigurado,da a - reconfigured
reconfigurar v - to reconfigure
reconfortante a - recomforting • refreshing
reconocer v - . . . • to accept; to acknowledge to repute • to explore • to identify • to detect • to note • to distinguish • [legal] to grant • [com] to accept @ charge(s) • [miner] to prospect
— v beneficio(s) v - [contab] to recognize @ profit(s)
— v característica f - to recognize @ characteristic
— v comercio m - to identify @ business
— v de vista - to inspect visually
— v edificio m - to identify @ building
— v establecimiento m - to identify @, business, o establishment
— v formalmente - to recognize formally
— v información f - [comput] to recognize @ data
— v insolvencia f - to recognize, o admit, @ insolvency
— v instantáneamente - [comput] to recognize instantaneously
— v — información f - [comput] to instantaneously recognize @ data
— v internacionalmente - to, recognize, o accept, internationally
— v norma f - to accept @ standard
— v pérdida f - [insur] to recognize @ loss
— v terreno m - [deport] to read @ dirt
— v universalmente - to recognize universally
— v utilidad f - [contab] to recognize @ profit
reconocible adv - recognizable; identifiable • legible
reconocido,da a - . . .; established • reputed; reputable; famous; outstanding; authoritative • established; big name ••reconnoitered • inspected • noted • indebted • [miner] prospected
— a como válido,da - recognized (as) valid
— a desde hace mucho (tiempo m) - long recognized • time-tested
— a de vista - inspected visually
— a formalmente - recognized formally
— a internacionalmente - accepted internationally
— a universalmente - recognized universally
— a válido,da - recognized (as) valid
reconocimiento m - . . . • reconnaisance • re-

puting • approval • confession • survey •
award • noting • awareness; realization •
commendation • excellence award • indebted-
ness • identification • exposure • [miner]
development; prospecting
reconocimiento m **apropiado** - appropriate, o proper, recognition
— m **como estado** m - [pol] statehood
— m **de beneficio(s)** m - [contab] profit recognition
— m — **edificio** m - building identification
— m — **información** f—[comput] data recognition
— m — **insolvencia** f - insolvency recognition
— m — **pérdida** f - [seguros] loss recognition
— m — **público** m - [admin] public's recognition
— m — **subsuelo** m - [miner] underground survey
— m — **utilidad** f - [contab] profit recognition
— m — **vista** - visual inspection
— m **eficiente** - [com] effective identification
— m **especial** - special, o particular, recognition, o appreciation
— m **formal** - formal recognition
— m **geológico** - [geol] geological, reconnaisance, o survey
— m **instantáneo** - [comput] instant recognition
— m — **de información** f - [comput] instant data
recognition
— m **internacional** - international recognition •
international acceptance
— m **mineralógico** - [miner] mineralogical study
— m **mundial** - worldwide recognition
— m **para selección** f **de sitio** m - location survey
— m **por asistencia** f - [deport] assist
— m — **escrito** m - [legal] written acknowledge-
ment
— m **preliminar** - [geol] preliminary survey(ing)
— m **público** - [admin] public recognition
— m **según norma** f - standard acceptance
— m **topográfico** - (topographical) survey(ing)
— m **transpondido** - [electrón] transponded ac-
knowledgement
— m **universal** - univerdal recognition
reconstitución f—. . . • [legal] reorganization
— f **de sociedad** f - [legal] corporate reorganization
reconstituido,da a - [legal] reorganized
reconstituir v - . . . • [legal] to reorganize
reconstrucción f - [constr] . . .; rebuilding •
build-up • [metal-prod] relining • [legal]
reenactment • [mec] (re)dressing
— a **a condición** f **nueva** - [mec] rebuilding to @
new condition
— f — **nuevo** - rebuilding to new
— f **con pieza(s)** f **nueva(s)** - [mec] rebuilding
with @ new part(s)
— f — **trazado** m **nuevo** - [vial] relocation
— f **de accidente** m - [segurid] accident reen-
acting,tment
— n — **acero** m - [sold] steel rebuilding
— f — — m **con carbono** m - [sold] carbon steel
rebuilding
— f — — m — **manganeso** m - [sold] manganese
steel rebuilding
— f — — m **manganésico** - [sold] manganese
steel rebuilding
— f — **batería** f **para coque** m - [coque] coke
battery rebuilding
— m — **bloque** m - [mec] block, rebuilding, o
redressing
— f — — m **para sujeción** f - [mec] grip(ping)
block, rebuilding, o redressing
— f — — m **sujetador** m - [mec] grip(ping)
block, rebuilding, o redressing
— f — **cambiador** m **para toma(s)** m - [electr-
-equip] tap changer rebuilding
— f — **cauce** m - [hidr] channel rebuilding
— f — — m **de hormigón** m - [hidr] concrete
channel rebuilding
— f — — m — **para arroyo** m - [hidr]
creek('s) concrete channel rebuilding
— f — **eje** m - [sold] axle, o shaft, rebuilding, o build-up

reconstrucción f **de embrague** m - [mec] clutch
rebuilding
— f — — m **mecánico** - [mec] mechanical clutch
rebuilding
— f — **horno** m - [metal-prod] furnace, rebuilding, o relining
— f — m **alto** - [metal-prod] blast furnace,
rebuilding, o relining
— f — — m **para acería** f - [metal-prod] steel
plant furnace, rebuilding, o relining
— f — **motor** m - [electr-mot] motor rebuilding
• [comb.int] engine rebuilding
— f — **pieza(s)** f - [ind] part(s) rebuilding
— f — — f **de acero** m - [sold] steel part rebuilding
— f — — f — — m **con carbono** m - [sold]
carbon steel part(s) rebuilding
— f — — f — — m — **manganeso** m - [sold]
manganese steel part(s) rebuilding
— f — — f — — m — — m **desgastada(s)** -
[sold] worn manganese steel parts rebuilding
— f — — f — — m **desgastada(s)** - [sold]
worn steel part(s) rebuilding
— f — — f **(des)gastada(s)** - [sold] worn
part(s) rebuilding
— f — — **piquera** f - [metal-prod] tap hole, rebuilding, o relining
— f — **ranura** f - [mec] groove redressing
— f — **refractario(s)** m - [metal-prod] refractory,ries rebuilding
— f — — m **para horno** m - [metal-prod] furnace refractory,ries rebuilding
— f — **revestimiento** m - [metal-prod] lining
rebuilding; relining
— f — **rodillo(s)** m - [sold] roll(s) rebuilding
— f — **rueda** f - [mec] dressed wheel
— f — **zona** f - [ind] area rebuilding
— f — f **desgastada** - [sold] worn area rebuilding
— f **estándar** - [constr] standard rebuilding •
[metal-prod] standard relining
— f **general** - [metal-prod] total reline,ning
— f **hasta tamaño** m **original** - [sold] rebuilding to @ original size
— f **próxima** - [metal-prod] next, o coming,
reline,ning
— f **según norma** f - [constr] standard rebuilding • [metal-prod] standard reline,ning
— f **última** - [metal-prod] last reline,ning
— f **única** f - [ind] single, o only, relining
reconstruible adv - rebuildable
reconstruido,da a - reconstructed; rebuilt;
remade; reworked • [hormig] built up • [mec]
(re)dressed
— a **condición** f **nueva** - [sold] rebuilt to @
new condition
— a **a nuevo,va** - rebuilt to new
— a **con pieza(s)** f **nueva(s)** - [ind] rebuilt
with @ new part(s)
— a — — f **usada(s)** - [ind] rebuilt with @
used part(s)
— a **en tamaño** m **insuficiente** - [petról] undersize, rebuilt, o reworked
— a **hasta medida(s)** f **deseada(s)** - [sold]
(re)built up to @, dimension(s), o size
— a — **tamaño** m **deseado** - [sold] (re)built
(up) to @, (desired), dimension, o size
reconstruir v - [mec] . . . • to reenact •
[sold] to build up • [metal-prod] to reline
• [mec] to (re)dress
— v **a condición** f **nueva** - [mec] to rebuild to
@ new condition
— v — **nuevo** - to rebuild to new
— v **accidente** m - [segurid] to reenact @ ac-
cident|
— v **bloque** m -[mec] to, rebuild, o redress, @
block
— v — m **para sujeción** f - [mec] to, rebuild,
o redress, @ grip block
— v — m **sujetador** m - [mec] to, rebuild, o
redress, @ grip block
— v **cambiador** m - [mec] to rebuild @ changer

reconstruir v cambiador m para toma(s) m -
[electr-equip] to rebuild @ tap changer
— v con pasta f - to rebuild with paste
— v — — f de carbono m y ladrillo m - [metal-
-prod] to rebuild with carbon paste and brick
— v con pieza(s) f nueva(s) - [mec] to rebuild
with @ new part(s)
— v — — f usada(s) - [mec] to rebuild with @
used part(s)
— v embrague m - [mec] to rebuild @ clutch
— v — m mecánico - [mec] to rebuild @ mechanical clutch
— v hasta medida f deseada - [sold] to rebuild
to @ (desired), dimension, o size
— v — — f estipulada - [sold] to build up to
@ stipulated, dimension, o size
— v — tamaño m deseado - [sold] to build up to
@ desired, dimension, o size
— v — — m estipulado - [sold] to build up to
@ stipulated, dimension, o size
— v horno m - [ind] to, rebuild, o reline, @
furnace
— v motor m - [electr-mot] to rebuild @ motor •
[comb.int] to rebuild @ engine
— v pieza f - [sold] to rebuild @ part
— v — f gastada - [sold] to rebuild @ worn
part
— v polea f - [sold] to build up @ pulley
— v — f loca - [sold] to build up @ idler
— v ranura f - [mec] to redress @ groove
— v revestimiento m - [combust] to, rebuild @
lining, o reline
— v rodillo m - [sold] to build up @ roller
— v rueda f - [sold] to build up @ wheel •
[mec] to dress @ wheel
— v zona f - [sold] to rebuild @ area
— v — f desgastada - [sold] to rebuild @ worn
area
recontado,da a - recounted; related; tallied
recontar v - . . .; to tally
recopilación f - . . . • gathering • guide
— f de anal(es) - [ind] history compilation
— f — — m de horno m - [ind] furnace history
compilation
— f — dato(s) m - data, o fact(s), compilation
— f — — m sobre consumo m de combustible(s) m
- [combust] fuel consumption data gathering
— f — historia f - history compilation
recopilado,da a - compiled; gathered
recopilador m - . . .; collector
recopilar v - . . .; to gather
— v dato(s) m - to compile @, data, o facts
— v historia f - to compile @ history
— v — f de producto m - [ind] to compile @
product('s) history
— v información f - to, compile, o gather, data
— v — f existente - to compile @ existing data
record* m - [deport] . . .; véase también marca
— m de durabilidad f - durability record
— m — producción f [ind] production record
— m mensual - monthly record
— m — de producción f - [ind] monthly production record
— m para categoría f - [deport] class, o category, record
— m para etapa f - [deport] lap record
recordación f - . . .; remembering; reminding;
recalling
recordado,da a - remembered; reminded; recalled
recordar v - . . .; to recall • to bear in mind
recorrer v - . . .; to travel; to tour •
[public] to page (through)
— v distancia f - [deport] to go @ distance
— v — f total - to go @ (whole) distance
— v recorrido m - [deport] to navigate @ course
recorrida* f - véase recorrido m
recorrido m - travel; route,ting; trip; throw •
run; stroke; length • [constr] traverse •
[electr-oper] path • [hidr] flow • [deport]
course • [vial] stretch; mileage driven •
path • [mec] track
— m áspero - [deport] rough, o rugged, course
— m aumentado - [mec] increased travel

recorrido m barroso - [deport] soggy course
— m brutal - [deport] very difficult course
— m cargado - [transp] when loaded (course)
— m completado - [deport] completed course
— m con conformación f de óvalo m alargado -
[deport] dog-bone-shape(d) course
— m con figura f de lazo m - [deport] loop
type, course, o race
— m de cable m - [electr-instal] cable routing
— m — — largo - [electr-instal] long cable
run
— m — — m — en exceso m de . . . . • [electr-
-instal] long cable run over . . .
— m — camino m - [vial] road course
— m — caudal m - [hidr] flow, o course, path
— m — circuito m - [electr-instal] circuit
path
— m — — m a tierra f - [sold] ground circuit
path
— m — — m de retorno m a tierra f - [electr-
-instal] ground current return path
— m — conducto m - [electr-instal] conduit
route,ting
— m — corriente f - current flow • [hidr] flow
path
— m — cubeta f - [metal-prod] skip travel
— m — diente m - [mec] tooth travel
— m — — m de engranaje m - [mec] gear tooth
travel
— f — — m — — m durante contacto m - [mec]
gear tooth travel during contact
— m — — m durante contacto m - [mec] tooth
travel during contact
— m — disco m - disk, o plate, travel
— m — — m aumentado - [mec] increased, disk,
o plate, travel
— m — elemento m de banda f para rodamiento m
- [autom-neumát] tread block path
— m — engranaje m - [mec] gear travel
— m — gancho m - [grúas] hook travel
— m — montacarga(s) m - [ind] skip travel
— m — placa f - [mec] plate travel
— m — placa f aumentado - [mec] increased
plate travel
— m — — f incrementado - [mec] increased
plate travel
— m — plato m - [mec] plate travel
— m — resorte m - [mec] spring travel
— m — — m verificado - [mec] checked spring
travel
— m — skip m - [metal-prod] skip travel
— m — tubería f - [tub] pipe course
— m — válvula f - [válv] valve travel
— m durante contacto m - [mec] travel during
@ contact
— m eléctrico - [electr-oper] electrical path
— m en helicóptero m - [transp] helicopter tour
— m entre cono(s) m - [deport] pilon gauntlet
— m — montaña(s) f - [deport] mountain course
— m — pilón(es) m - [deport] pilon gauntlet
— m excéntrico - [mec] eccentric throw
— m — máximo - [mec] maximum eccentric throw
— m — mínimo - [mrv] minimum eccentric throw
— m excesivo - [mec] overtravel
— m excesivo de contacto(s) m - [electr-oper]
contact overtravel
— m fuera de carrera f - [deport] transit
— m fuera de carretera f - [deport] off-road
course
— m hecho - [deport] navigated course
— m incrementado - [mec] increased travel
— m máximo - [mec] maximum travel
— m — de gancho m - [grúas] maximum hook, o
hook maximum, travel
— m mínimo - [mec] minimum travel
— m — de gancho m - [grúas] hook minimum, o
minimum hook, travel
— m para conducción f (eléctrica) - [sold]
electrically conductive path
— m — ómnibus f - [transp] bus route
— m — — m escolar - [transp] school bus route
— m raudo - [Deport] fast course
— m recto - [autom] straight, path, o course

recorrido m sinuoso - [deport-esquí] slalom
course
— m traicionero - treacherous course
— m verificado - [mec] checked travel
— m vertical - vertical, course, o run
recorrido,da a - traveled • toured
recortado m - véase recortadura f
recortado,da a - resheared; trimmed; cropped •
[dib] cutaway • [agric] mowed; clipped
recortadora f - [agric-equip] trimmer; clipper
— f para ligustro m - [herram] hedge trimmer
recortadura f - . . .; trimming; cropping; cutting; shearing • [sold] cutout • [dib] cutaway • [agric] mowing
— f — alambre m - [mec] wire clipping
— f — borde(s) m - [mec] edge, o side, trimming, o shearing
— f — césped m - [agric] grass mowing
— f — extremo(s) m - [mec] end clipping
— f — — m de alambre m - [mec] wire end(s)
clipping
— f — sobrante(s) m - trimming
— f y corte m - [metal-lam] trimming and shearing
recortar v - [mec] . . . • to blank • [metal-
-lam] to crop; to reshear • to chip (out) •
[agric] to mow
— v alambre m - [mec] to clip @ wire(s)
— v césped m - to mow @, grass, o lawn
— v extremo m - [mec] to clip @ end
— v — m de alambre m—[mec] to clip @ wire end
— v rebaba(s) f - [metal-fabr] to clip (@,
burrs, o flash)
— v sobrante(s) m - [mec] to trim
recorte m - [mec] scrap; cutting; crop(ping);
(re)shearing; trim(ming) • cut out • offset •
[agric] mowing • clipping
— m de alambre m - [mec] wire, clipping, o cutting
— m — barra f - [metal-lam] bar crop(ping)
— m — — f en laminador m de barra(s) f -
[metal-lam] bar mill (bar) crop(ping)
— m — borde(s) m -[mec] side, shearing, o
trimming
— m — catálogo m—catalog, clipping, o cutting
— m — césped m - [agric] grass mowing
— m — extremo(s) m - [mec] end clipping
— m — — m de alambre m - [mec] wire end clipping
recortes m de laminación f en caliente - [metal-
-lam] hot metal scrap
— m — — f — frío - [metal-lam] cold metal
scrap
— m — laminador m - [metal-lam] mill croppings
— m — — m para barra(s) f - [metal-lam] bar
mill (bar) crop(pings)
recostar(se) sobre rueda f - [autom] to lean on
@ wheel
recreación f activa - [deport] active recreation
— f — en aire m libre - [deport] active outdoor recreation
— f en aire m libre - [deport] outdoor recreation
— f familiar - [deport] family recreation
recreamiento* m - véase recreación f
recrear v - [deport] to recreate
recreativo,va a - recreative; recreational
— a familiar.- [deport] family recreation(al)
recrecer v - . . .; to build up • to regrow
recrecimiento m - . . . .; building up
recreo m - [deport] pleasure; play • [educ] recess • [labor] coffee break
recristalización f - recrystalization
— f parcial - [quím] partial recristalization
recristalizado,da a - recrystalized
— a parcialmente - [quím] partially recrystal-
lized
recristalizador,ra a parcial - [quím] partial
recrystalizing
recristalizar v - [quím] to recrystalize
— v parcialmente - [quím] to partially recrystalize
recta f - [dib] straight line • [deport]

straight (away
recta f con fosa(s) f - [deport] pit straight
— f crítica - [deport] critical straight
— f principal - [deport] main straightway
— f para correlación f - correlation line
rectangular a - . . . . • oblong
— f con borde m recto - straight edge rectangular
— a — — m redondeado - rounded edge rectangular
rectángulo m con borde(s) m recto(s) - straight
edge(s) rectangle
— m — — m redondeado(s) - rounded edge(s)
rectangle
rectificación f - . . .; rectifying • correction • honing; truing (up) • [comb.int] re-
boring • grinding • [mec] dressing • [electr-
-mot] (motor brush(es) stoning
— f cilíndrica - [mec] cylindrical rectification,fying
— f con muela f - [herram] stoning; dressing
— f — — f abrasiva - [herram] abrasive grinding wheel, stoning, o dressing
— f — . . . pasada(s) - [mec] . . . pass(es)
rectifying
— f — — f poco profunda(s) - [mec] shallow
pass(es) rectifying
— f — — f profunda(s) - [mec] deep pass(es)
rectifying
— f de aceite m - [petról] oil(s) rectifying
— f — — m ligero - [petról] light oil rectifying
— f — — m liviano m—[petról] light oil rectifying
— f — benzol(es) m - [coque] benzol rectifying
— f — cilindro m - [comb.int] cylinder, re-
boring, o honing
— f — colector m - [electr-mot] commutator,
honing, o truing
— f — conmutador m - [electr-mot] commutator
stoning
— f — curso m - [hidr] channel cut(ting)-off
— f — chapa(s) f - [metal-lam] hot strip
grinding
— f — deficiencia(s) f - deficiency,cies rectifying,fication
— f — motor m - [electr-mot] motor rectifying,fication • [comb.int] engine re-boring
— f — muela(s) f - [mec] grinding wheel dressing
— f — platino(s) m - [comb.int] point(s)
dressing
— f — — m para magneto(s) m - [comb.int] magneto point dressing
— f — perforación f - [petrol] drill hole
reaming
— f — rodillo(s) m - [metal-lam] roll grinding
— f — — m para trabajo m - [metal-lam] work
roll(s) grinding
— f — trabajo m - [labor] work rectifying,fi-
cation
— f — — m en ejecución f - [labor] work in
process rectification,fying
— f en seco - [mec] dry grinding
— f final - [mec] final, o finish(ing), machining, o grinding, o dressing
— f plana f - [mec] flat rectifying
rectificado,da a - rectified • [mec] rebored;
honed; trued (up) • [electr-mot] stoned
— a hasta, dimensión(es) f, o medida(s) f, e-
xacta(s) - [f.c.-ruedas] finished to exact
dimension(s)
rectificador m - . . . . • [herram] reamer; hone
— m al, bies m, o sesgo m - [electr] bias rectifier
rectificador m - [electr-equip] rectifier
— m abierto - [electr-equip] open rectifier
— m bifásico - [electr-equip] two-phase rectifier
— m — para corriente f alterna - [electr-equip]
alternating current two-phase rectifier
— m — — — f continua - [electr-equip] direct
current two-phase rectifier

rectificador m cerrado - [electr-equip] closed rectifier
— m comprobado - [electr-equip] checked rectifier
— m con arco m - [electr-equip] arc rectifier
— m — m con mercurio m - [electr-equip] mercury arc rectifier
— m — cortocircuito m - [electr-oper] short circuited, o shorted, rectifier
— m — diodo m - [electr-equip] diode rectifier
— m — puente m - [electr-equip] bridge rectifier
— m — — m para onda f plena - [electrón] full wave bridge rectifier
— m — selenio m - [electr-equip] selenium rectifier
— m — silicio m - [electr-equip] silicon rectifier
— m — — m dañado - [electr-equip] damaged silicon rectifier
— m — tres fases f - [sold] three-phase rectifier
— f — — — f manual - [sold] three-phase manual rectifier
— m — vapor m de mercurio m - [electr-equip] mercury, vapor, o arc, rectifier
— m conectado - [electr-instal] connected rectifier
— m desconectado - [electr-instal] disconnected rectifier
— m dañado - [electr-equip] damaged rectifier
— m en corto circuito - [electr-instal] short circuited, o shorted, rectifier
— m fallado - [electr-oper] failed rectifier
— n ...fásico - [electr-instal] . . .-phase rectifier
— m monofásico - [electr-equip] single-phase rectifier
— m — para corriente f alterna—[electr-equip] alternating current single phase rectifier
— m — — — f continua - [electr-equip] direct current single-phase rectifier
— m multifásico - [electr-equip] multi-phase rectifier
— m — para corriente f alterna - [electr-equip] alternating current multi-phase rectifier
— m — — — f continua - [electr-equip] direct current multi-phase rectifier
— m para aceite(s) m - [petról] oil rectifier
— m — — m liviano(s) - [petról] light oil(s) rectifier
— m — — m ligero(s) - [petról] light oil(s) rectifier
— m — — m pesado(s) - [petról] heavy oil(s) rectifier
— m — . . .amperio(s) m - [electr-equip] . . . ampere(s) rectifier
— m — amplificación f - [electr-equip] amplifier rectifier
— m — — f magnética - [electr-equip] magnetic amplifier rectifier
— m — avance m - [sold] travel rectifier
— m — carga f para acumulador m - [electr-equip] battery charge rectifier
— m — cargadora f - [electr-equip] charger rectifier
— m — — f para acumulador(es) m - [electr-equip] battery charger rectifier
— m —:corriente f (para entrada f) - [electr-instal] power rectifier
— m — — f alterna - [electr-equip] alternating current rectifier
— m — — f continua - [electr-equip] direct current rectifier
— m — — f — para soldadura f - [sold] direct current welding rectifier
— m — desviación f - [electr-equip] bias rectifier
— m — diodo m - [electr-equip] diode rectifier
— m — dispositivo m para avance m - [sold] travel rectifier
— m — (electro)imán m para grúa f - [grúas]

[grúas[crane (electro)magnet rectifier
rectificador m para motor m - [electr-mot] motor rectifier
— m — onda f plena - [electrón] full wave rectifier
— m — polaridad f - [electr-equip] polarity rectifier
— m — potencia f - [electr-equip] power rectifier
— m — regulación f - [electr-equip] control rectifier
— m — f con silicio m - [electrón] silicon control rectifier
— m — soldadura f - [sold] welding rectifier
— m — — f con corriente f alterna - [sold] alternating current welding rectifier
— m — — f — — f continua - [sold] direct current welding rectifier
— m — — f con arco m - [sold] arc welding rectifier
— m — válvula(s) f - [petról] valve grinder
— m polarizado - [electr-equip] bias rectifier
— m principal - [electr-equip] main rectifier
— m — para energía f - [electr-equip] main power rectifier
— m trifásico - [electr-equip] three-phase rectifier
— m — manual - [sold] three phase manual rectifier
— m — para corriente f alterna - [electr-equip] alternating current three-phase rectifier
— m — — — f continua - [electr-equip] direct current three-phase rectifier
rectificadora f - [mec] grinder; grinding machine; dresser; honing machine • [metal-lam] (roll) grinder
— f con pedestal - [herram] upright grinder
— f — plato m magnético - [grúas] magnetic plate rectifier
— f en seco - [herram] dry grinder
— f para benzol(es) n - [subprod] benzol rectifier
— f — cilindro(s) m - [mec] cylinder grinder
— f — dado(s) m - [extrusión f de electrodos] die rectifier
— f — superficie(s) f - [mec] grinder wheel
— f recta - [petról] straight reamer
— f sesgada - [herram] bias rectifier
— f soldadora - [herram] welding rectifier
— f — para . . . amperios m - [sold] . . . amperes welding rectifier
— f — . . . m en corriente f alterna - [electr-equip] . . . amperes alternating current welding rectifier
— f — . . .— m — f continua f - [electr-equip] . . . amperes direct current welding rectifier
rectificar v - . . .; to true; to do up • [mec] to grind; to mill • to re-bore; to hone • [metal-fabr] to ream • [electr-mot] to stone
— v cilindro m - [mec] to rebore @ cylinder
— v colector m - [electr-mot] to true @ commutator
— v conmutador m - [electr-mot] to stone @ commutator
— v deficiencia f - to rectify @ deficiency
— v motor m - [comb.int] to re-bore @ engine
— v muela f - [herram] to dress @ grinding wheel
— v perforación f - [petról] to ream (a drill hole)
— v platino m - [comb.int] to dress @ point
— v — m para magneto m - [comb.int] to dress @ magneto point
— v trabajo m—[mec] to rectify @, work, o job
rectitud f de borde m - [metal-lam] edge staightness
— f — canto(s) m - [metal-lam] edge straightness
— m — perfil m - profile straightness • [metal-lam] shape straightness
— f — — m soldado - [metal-lam] welded shape

straightness

recto m - . . . . • [mec] spur

rector m - [educ] . . .; headmaster

recubierto,ta a - covered • coated; plated • enveloped • surface • [grúas] lagged • [culin] topped; iced

— a **adicionalmente** - additionally coated

— a **completamente** - completely covered • completely coated

— a **con aluminio** m - alumin(i)um coated

— a — **alquitrán** f (**de hulla** f) - (metal-trat) (coal) tar coated

— a — **arena** f - sand covered

— a — **asfalto** m - asphalt covered

— a — **cadmio** m - [metal-trat] cadmium plated

— a — **cal** f - lime, covered, o coated

— a — **calzada** f - [constr] pavement covered

— a — **caucho** m - rubber covered

— v — **cinc** - [metal-trat] zinc coated

— a — **cobre** m - [metal-trat] copper coated

— a — **esmalte** m - [metal-trat] enamel coated

— a — — m **de alquitrán** m (**de hulla** f) - [tub] (coal) tar enamel coated

— a — **estaño** m - [metal-trat] tin, covered, o coated; tinned

— a — **fosfato(s)** m - [metal-trat] phosphate(s) coated

— a — **fundente** m - [sold] flux covered

— a — **imprimación** f—[metal-trat] prime coated

— a — — f **por inmersión** f - dip-prime coated

— a — **ladrillo(s)** m - [constr] brick covered

— a — — m **refractario(s)** - [refract] refractory brick(s) covered

— a — **lubricante** m - [mec] lubricant, covered, o coated

— a — **material** m - material covered

— a — — m **duro** - hard material covered

— a — **óxido** m **de titanio** m - [metal-trat] titania, o titanium oxide, coated, o covered

— a — **pavimento** m - [vial] pavement covered

— a — **plástico** m - plastic coated

— a — **plomo** m - lead, coated, o plated

— a — **pieza(s)** f **angular(es)** - [mec] angle(s) covered

— a — — f — **de acero** m - [mec] steel angles covered

— a — **plomo** m - [metal-trat] lead, covered, o coated

— a — — m **y estaño** m - [metal-trat] terne (coated)

— a — **porcelana** f - porcelain coated

— a **en obra** f - field coated

— a — — f **con asfalto** m—field asphalt coated

— a — — f **exteriormente** - field outer coated

— a — — f **exterior(mente) e interior(mente)** - field (outer and inner) coated

— a — — f — — **con asfalto** m - field asphalt coated

— a — **planta** f - [metal-trat] shop coated

— a — **enteramente** - [metal-trat] coated entirely

— a **exteriormente** - [tub] (outside) coated

— a — **con asfalto** m - [tub] (outside) asphalt coated

— a **interiormente** - véase **revestido,da interiormente**

— a **liberalmente** - liberally coated

— a — **con grasa** f - [mec] liberally grease coated

— a **parcialmente** - partially coated

— a — **por inmersión** f - partially dip coated

— a — — f **en asfalto** m - [tub] partially asphalt dip coated

— m — — — f — — m **en caliente** - [tub] partially heat dip(ped) asphalt coated

— a — — — f **en caliente** - [metal-trat] partially heat dip coated

— a **por inmersión** f - dip coated

— a — — f **en caliente** - hot dip coated

— a **totalmente** - fully, o entirely, o completely, coated, o covered

recubrimiento m - covering; coating; overlay • wrapping; enveloping,pment; cover(ing); facing • [metal-trat] coat(ing) • [sold] surfacing; coat(ing) • [pint] finishing coat • [electr-cond] sheath(ing) • [mec] lagging; reeving • [culin] topping; icing • véase también **revestimiento** m

recubrimiento m **ácido** - acid coating • véase también **revestimiento** m **ácido**

— m **adecuado** - adequate cover(ing) • [tub] adequate coating

— m — **especial** - [tub] adequate special coating

— m **adicional** - [sold] additional coating

— m **aislante** - insulating covering • lagging

— m **ajustado** - [tub] tight coating

— m **anaranjado** - orange coating

— m **anticorrosivo** - [metal-trat] anticorrosive, o corrosion resistant, coating

— m — **con liga** f **térmica** - [metal-trat] thermally bonded corrosion resistant coating

— m — **ligado térmicamente** - [metal-trat] thermically bonded corrosion resistant coating

— m — **tenaz** - [metal-trat] tough corrosion resistant coating

— m **antifricción** f - [mec] Babbitt metal coating

— m — **para cojinete** m - [turb] bearing Babbitt metal coating

— m — — m **para empuje** m - [turb] thrust bearing Babbitt metal coating

— m — — **zapata** f - [turb] shoe Babbitt metal coating

— m **aplicado** - applied coating

— m **asfáltico** - [tub] asphalt(ic), coating, o covering

— m — **doble** - [tub] double asphalt coating

— m — **en obra** f - field asphalt(ic) coating

— m — — **taller** m - shop asphalt(ic) coating

— m — **exterior** - [constr] asphalt(ic) exterior coating

— m **atractivo** - attractive coating

— m **azul** - blue coating

— m **bajo en hidrógeno** m - [sold] low hydrogen coating

— m **bitumástico** - bitumastic coating

— m **bituminoso** - bituminous, o bitumen, o asphalt(ic), coat(ing), o cover(ing)

— m — **dañado** - [constr] damaged bituminous coating

— m — **doble** - double bituminous, o bituminous double, coat(ing)

— m — — **total** - bituminous double full coat

— m — **en obra** f - [tub] bituminous, o bitumen, field coating

— m — **total** - bituminous full, o full bituminous, coat(ing)

— m **blanco** - white coat(ing)

— m **bueno** - good, coat(ing), o cover(ing) • [sold] good coverage • [tub] good, o right. coat(ing)

— m — **con escoria** f - [sold] good slag coverage

— m **calcáreo** m - lime coating

— m **calizo** - lime coating

— m **canela** - tan coating

— m — **claro** - light tan coating

— m — **obscuro** - dark tan coating

— m **castaño** - brown coating

— m — **claro** - light brown, o tan, coating

— m — **gris(áceo)** - gray(ish) brown(ish) coating

— m — **rojizo** - red(ish) brown(ish) coating

— m **color durazno** - peach (color) coating

— m **completo** - full coverage - [sold] complete (slag) coverage

— m — **con capa** f **bitumástica** - [metal-fabr] full bituminous coating

— m — **de escoria** f - [sold] complete slag coverage

— m **con acero** m - [metal] steel coating

— m — — m **inoxidable** - [sold] stainless steel coating

— m — **aluminio** m - [metal] aluminum coating

— m — **soga** f - [grúas] rope lagging

— m — **calzada** f - [vial] pavement covering

— m — **cinc** m **sobre cobre** m - [metal-trat] zinc over copper coating

recubrimiento m con costo m bajo - low cost, o inexpensive, coating
— m — pavimento m - [vial] pavement covering
— m — pistola f - [pint] spray coating
— m — soga f - [grúas] rope lagging
— m — caucho m - rubber covering
— m — espesor, estándar, o según norma f - standard thickness coating
— m — fosfato(s) m - [alambre] phosphate coat(ing)
— m — hidrógeno m bajo - [sold] low hydrogen covering
— m — imprimación f - prime coating
— m — — f por inmersión f - dip-prime coating
— m — látex m - latex coating
— m — liga f térmica - [metal-trat] thermally bonded coating
— m — lubricante - [mec] lubricant coating
— m — materia(s) f orgánica(s) - [sold] organic (matter) covering
— m — metal m - metal, coating, o covering
— m — — m antifricción f - [mec] Babbitt metal coating
— m — plástico m - plastic coating
— m — plomo m - lead, coating, o plating
— m — pulverización f - [metal-trat] jet coating
— m — stellite m - stellite coating
— m — tierra f - [constr] earth cover(ing)
— m consumido - [sold] burned away coat(ing)
— m continuo - [pint] continuous coating
— m corriente - [grúas] standard lagging
— m dañado - damaged coat(ing)
— m de acero m - [metal] steel covering
— m — — m líquido - [metal-trat] molten steel cover(ing)
— m — alquitrán m - [tub] tar coating
— m — — m de hulla f - [tub] coal tar (enamel) coat(ing)
— m — asfalto m - [tub] asphalt coat(ing)
— m — banda f - [metal-trat] strip coat(ing)
— m — cal f - lime coat(ing)
— m — caucho m - rubber coat(ing)
— m — cemento m - [constr] cement coat(ing)
— m — — m de látex - [constr] latex cement coating
— m — cinc m - [metal-trat] zinc coat(ing)
— m — — m fundido - [metal-trat] molten zinc coating
— m — — m puro - [metal-trat] pure zinc coat(ing)
— m — cloruro m—chloride, coating, o covering
— m — — m de polivinilo m - [electr-cond] polivynil chloride, covering, o coating
— m — cobre m - [sold] copper coating
— m — — m consumido - [sold] burned (away) coppoer coating
— m — coco m (rallado) - [culin] (grated) coconut topping
— m — compuerta f - [hidr] gate coat(ing)
— m — — f rodante - [hidr] rolling gate coat(ing)
— m — chapa f - [metal-trat] strip coat(ing)
— m — electrodo m - [sold] electrode coat(ing)
— m — — m para soldadura f - [sold] welding electrode coating
— m — escoria f - [sold] slag, coating, o cover(ing)
— m — esmalte m - [tub] enamel coat(ing)
— m — — m de alquitrán m - [tub] tar enamel coating
— m — — m — — m de hulla f - [tub] coal tar enamel coating
— m — — m — — m — — f protegido - [tub] shielded coal tar enamel coating
— m — estaño m - [metal-trat] tin coating
— m — fleje m - [metal-trat] strip coating
— m — fundente m - [sold] flux coat(ing)
— m — hidrógeno m - hydrogen coat(ing)
— m — — m bajo - [sold] low hydrogen coating
— m — — m — con polvo m de hierro m - [sold] iron powder low hydrogen coating
— m — hormigón m - [constr] concrete, encase-

ment, o envelope
recubrimiento m de irregularidad(es) f en corrugación(es) f - [tub] corrugation covering
— m — lingotera f - [metal-prod] ingot mold coating
— m — material m plástico - [electr-cond] plastic covering
— m — — m de cloruro m de polivinilo m - [electr-cond] polyvinyl chloride plastic covering
— m — — m — — — m polivinílico - [electr-cond] polyvinyl chloride plastic covering
— m — mineral m - mineral covering
— m — níquel m - [metal-trat] nickel coating
— m — óxido m - [metal-trat] oxide skin
— m — — m de titanio m - [sold] titania coating
— m — papel - paper coating
— m — — m con aluminio m - paper aluminum coating
— m — placa f - [metal-prod] (ingot) stool coating
— m — plástico m - plastic coating
— m — plomo m - [metal-trat] lead coating
— m — — m y estaño m - [metal-trat] terne coat(ing)
— m — porcelana f - porcelain coating
— m — f dañado - damaged porcelain coating
— m — resina f - resin coating
— m — — f con base f de epoxia f - epoxy resin coating
— m — tipo m básico - basic (type) coating
— m — — m con óxido m de titanio m - titanium oxide coating
— m — — m con hidrógeno m bajo - [sold] low hydrogen type coating
— m — — — m bajo con polvo m de hierro m - [sold] iron powder low hydrogen type coating
— m — yute m - [electr-cond] jute covering
— m — — m impregnado - [electr-cond] impregnated jute covering
— m delgado - thin coating
— m — de asfalto m - [constr] thin asphalt coating
— m — escoria f - [sold] thin slag coating
— m distinto - different, o other, coating
— m doble - double coating
— m — con asfalto m - [tub] double asphalt coating
— m — — cinc m - [metal-trat] double zinc, o zinc double, coating
— m — — — m sobre cobre m - [metal-trat] zinc over copper double coating
— m — total - [metal-fabr] double full coating
— m durable - durable coating
— m duro - [sold] hard surface | a - [sold] hardsurfacing
— m — m de horno m para calcinación f - [sold] kiln hardsurfacing
— m — resistente a abrasión f - [sold] hardsurfacing to resist @ abrasion • abrasion resisting hardsurfacing
— m económico - low cost, o inexpensive, coating
— m en obra f - [tub] field coating
— m — — f con cemento m - [constr] cement field coating
— m — — f — — m de látex m - [constr] latex cement field coating
— m — — f — látex m - [constr] latex field coating
— m — — f de compuerta f - [hidr] gate field coating
— m — — f — — f rodante - [hidr] roller gate field coating
— m — planta f - [ind] shop coat(ing)
— m . . . sección(es) f - [grúas] . . .-part reeving
— m . . . sección(es) f para aguilón m - [grúas] . . . part hoist reeving
— m . . . — f para elevador m - [grúas] . . . — hoist reeving
— m . . . — f — — m para aguilón m -

[grúas] . . .-part boom hoist reeving
recubrimiento m **en taller** m **con cemento** m **de**
látex - [constr] latex cement shop coating
— m — — m **con látex** m - latex shop coating
— m **entero** m - whole coating
— m **especial** m - [tub] special coating
— m — **adecuado**—[tub] adequate special coating
— m **estándar** - standard coating • [grúas]
standard lagging
— m **expuesto** - exposed coat(ing)
— m **exterior** - outer coating
— m — **aplicado en obra** f - [tub] field applied
external coating
— m — **asfáltico** - asphalt exterior coating
— m — **bituminoso** - [metal-fabr] bitumen (ex÷
terior) coating
— m — **de asfalto** m - [constr] asphalt exterior
coating
— m — — **hormigón** m - [hormig] concrete encasement
— m — — **neopreno** m - [electr-cond] outer
neoprene, coating, o sheeting
— m **fuerte** - [metal-trat] heavy coating
— m **fuertemente serpentado** - [metal-trat] heavy
snaky coating
— m **galvanizado** - [metal-trat] galvanized coating
— m — **normal** - [metal-trat] normal galvanized
coating
— m — **para alambre** m - [alambre] wire galvanized coating
— m — **uniforme** - [metal-fabr] uniform galvanized coating
— m **gris** - gray coating
— m **obscuro** - dark gray coating
— m **grueso** - [sold] thick, o heavy, coating
— m — **de asfalto** m - [constr] thick asphalt
coating
— m — — **cinc** m - [metal-trat] thick, o heavy,
zinc coating
— m **intemperizado** - weathered coating
— m **liberal** - liberal coating
— m — **con grasa** f - [mec] liberal grease coating
— m **ligado térmicamente** - [metal-trat] thermally bonded coating
— m **liso** - [pint] smooth coating
— m **liviano** - light coat(ing)
— m — **con cal** f - [sold] light lime coating
— m **marcadamente serpenteado** - [metal-trat]
heavy snaky coating
— m **más grueso** - heavier coating
— m — **delgado** - lighter coating
— m **mediante extrusión** f - [sold] extruded,
coating, o covering
— m **mediante inmersión** f - [sold] dipped, coating, o covering
— m **metálico** - [metal-trat] metal(lic) coating
— m **mineral** - [miner] mineral covering
— m **mínimo** - [constr] minimum cover(ing)
— m **múltiple** - multiple coating | a - (con)
polycoated
— m **muy uniforme** - very, o highly, uniform
coating
— m **negro** - black coat(ing)
— m **normal** - [metal-fabr] normal coating
— m **obtenido mediante extrusión** f - [sold] extruded, cover(ing), o coat(ing)
— m — — **inmersión** f - [sold] dipped, coating,
o covering
— m **para abrazadera** f - [mec] clamp coat(ing)
— m — — f **para camisa** f - [mec] liner clamp
coating
— m — **apoyo** m - [mec] pad cover(ing)
— m — **base** f - [metal-prod] stool coating
— m — **buje** m - [mec] bushing coating
— m — — m **piloto** m - [mec] pilot bushing
coating
— m — **camisa** f - [mec] lining coating
— m — **can** m - [mec] shoulder coating
— m — — m **para vástago** m - [mec] rod shoulder
coating
— m — **conducto** m - [electr-instal] conduit envelope,pment

recubrimiento m **para corrugación(es)** f - [tub]
corrugation(s) covering
— m — **edificio** m - [constr] building, covering,
o facing
— m — **electrodo** m - [sold] electrode covering
— m — **elevador** m - [grúas] hoist reeving
— m — — m **para aguilón** m - [grúas] boom hoist
reeving
— m — **endurecimiento** m - [sold] hardening deposit
— m — — m **de superficie** f - [sold] hardsurfacing deposit
— m — **lingotera** f - [metal-prod] mold coating
— m — **pieza** f - [mec] part coating
— m — **placa** f - [metal-prod] stool, lining, o
coating
— m — — f **para asiento** m - [metal-prod] stool
coating
— m — **quemador** m - [metal-prod] burner cover-
(ing)
— m — — m **para encendido** m - [sold] ignition
burner cover(ing)
— m — **reborde** m - [mec] shoulder coating
— m — **sujetador** m - [mec] clamp, o holder,
coat(ing)
— m — **superficie** f - [mec] surface coat(ing)
— m — — f **expuesta** - [mec] exposed surface
coat(ing)
— m — **tambor** m - [cabl] drum lagging
— m — **tapa** f - [mec] cover coat(ing)
— m — — f **para válvula** f - [válv] valve cover
coat(ing)
— m — **válvula** f - [válv] valve coat(ing)
— m — **viga** f - [constr] rail coating
— m **parcial** - [mec] partial coat(ing)
— m **por inmersión** f - partial dip coat(ing)
— m — — f **en asfalto** m - [constr] partial
dip(ping) asphalt coat(ing)
— m — — f **en caliente** - [constr]
partial heat dip(ping) asphalt coat(ing)
— m — — f **en caliente** - partial heat dip
coat(ing)
— m **pintado** - [pint] painted coating
— m **plástico** - plastic coat(ing)
— m **plateado** - [metal] silver coat(ing)
— m **por inmersión** f - dip(ping), o immersion,
coating
— m — — f **en caliente** - [metal-trat] hot
dip(ping) coat(ing)
— m **posterior** - [metal-trat] later coat(ing) •
[grúas] rear lagging
— m **protector** - protective, coat(ing), o cover,
o skin
— m — **con calidad** f **alta** - [mec] high quality
protective coating
— m — **de cloruro** m - [metal-trat] chloride protective, coating, o cover
— m — — m **de polivinilo** m - [electr-cond]
polivinyl (chloride) protective cover
— m — **de compuesto** m **para rosca(s)** f - [mec]
threading compound protective coating
— m — — m — — f **de calidad** f **alta** - [mec]
high quality threading compound protective
coating
— m — — **hormigón** m - [hormig] concrete protective coating
— m — **para hormigón** m - [constr] concrete protective, coating, o covering
— m — **de óxido** m - [metal] protective oxide,
coat(ing), o skin
— m **refractario** - [refract] refractory, coating,
o covering
— m **rojo ladrillo** - brick red coat(ing)
— m — **obscuro** m - [pint] dark red coating
— m **rosa(do)** - pink coating
— m **saturante** - saturating coat(ing)
— m — **bituminoso** - bituminous saturating coat
— m **según norma** f - standard coat(ing)
— m **separado** - separate coating • separated
coat(ing) • adv - (con) stripped
— m **serpenteado** - [metal-trat] snaky, o wavy,
coat(ing)
— m **suave** - smooth, o soft, coating
— m **tal** - such (a) coating

recubrimiento m̲ tenaz - tough coating
— m̲ tenue - (s̄)light, coating, o̲ covering
— m̲ — con escoria f̲ - [sold] (s̄)light slag,
coating, o̲ covering
— m̲ terne - [metal-trat] terne coating
— m̲ tipo - coat(ing), class, o̲ type
— m̲ total - total, o̲ complete, o̲ entire, o̲
full, coat(ing), o̲ cover(ing)
— m̲ — con escoria f̄ - [sold] complete slag
coverage,ring
— m̲ uniforme - uniform coat(ing)
— m̲ — ajustado - [tub] uniform tight coat(ing)
— m̲ — con cinc m̲ - [metal-trat] uniform zinc
coat(ing)
— m̲ verde - green coat(ing)
recubrir v̲ - . . . • to coat; to surface; to
overlay; to overlap; to surface • [constr] to
backfill; to do over • [sold] to surface; to
deposit (over) • [mec] to face • [grúas] to
lag • [culin] to top; to ice; to white
— v̲ abrazadera f̲ - [mec] to coat @ clamp
— f̄ — f̲ para camisa f̲ - [mec] to coat @ liner
clamp
— v̲ apoyo m̲ - [mec] to cover @ pad
— v̲ buje m̲ - [mec] to coat @ bushing
— v̲ — m̲ piloto - [mec] to coat @ pilot bushing
— v̲ camisa f̲ - [mec] to coat @ liner,ning
— v̲ can m̲ - [mec] to coat @ shoulder
— v̲ — m̲ para vástago m̲ - [mec] to coat @ rod
shoulder
— v̲ con aluminio m̲ - [metal-trat] to aluminum
coat
— v̲ — asfalto m̲ - [constr] to asphalt coat
— v̲ — calzada f̄ - [vial] to pavement cover
— v̲ — caucho m̲ - to rubber cover
— v̲ — cinc m̲ - [metal-trat] to zinc coat
— v̲ — fosfato(s) m̲ - [alambre] to phosphate
coat
— v̲ — fundente m̲ - [sold] to flux coat
— v̲ — imprimación f̲ - [pint] to prime(r) coat
— v̲ — — f̲ por inmersión f̲ - [metal-trat] to
dip prime(r) coat
— v̲ — lubricante m̲ - [mec] to lubricant coat
— v̲ — (otro) material m̲ - to, do over, o̲ re-do
— v̲ — pavimento m̲ - [vial] to cover with @
pavemento
— v̲ — plástico m̲ - to plastic coat
— v̲ — plomo m̲ - [metal-trat] to lead plate
— v̲ — porcelana f̲ - to percelain coat
— v̲ en planta f̲ - [ind] to shop coat
— v̲ enteramente - to coat entirely
— v̲ liberalmente - to coat liberally
— v̲ — con grasa f̲ - to liberally grease coat
— v̲ lingotera f̲ - [metal-prod] to coat @ ingot
mold
— v̲ papel m̲ - [papel] to coat @ paper
— v̲ — m̲ con aluminio m̲ - [comput] to aluminum
coat @ paper
— v̲ pieza v̲ - [mec] to coat @ part
— v̲ placa f̄ - [metal-trat] to coat @ (ingot)
stool
— v̲ — f̲ para lingotera f̲ - [metal-prod] to
coat @ ingot stool
— v̲ por inmersión v̲ - [metal-trat] to dip coat
— v̲ — — f̲ en caliente - [metal-trat] to hot
dip coat
— v̲ reborde m̲ - [mec] to coat @ shoulder
— v̲ sujetador m̲ - [mec] to coat @ clamp
— v̲ superficie f̲ - [metal-trat] to coat @ surface • [sold] to hardsurface
— v̲ — f̲ expuesta - [mec] to coat @ exposed
surface
— v̲ — f̲ para rodadura f̲ - [vial] to resurface
(@ tread)
— v̲ tapa f̲ - [mec] to coat @ cover
— v̲ — f̲ para válvula f̲ - [válv] to coat @
valve cover
— v̲ totalmente - to, coat, o̲ cover, completely
— v̲ válvula f̲ - [válv] to coat @ valve
recubrir(se) v̲ - to be coated; to coat itself
recuento m - . . . ; to count • to tally
— m̲ de embarcador m̲ - [transp] shipper's count
recuperabilidad f̲ - recuperability; salvability

recuperabilidad f̲ total - total, o̲ complete,
recoverability, o̲ recuperability, o̲ salva=
bility
recuperable adv - . . .; recuperable; salvable;
salvageable • reusable
recuperación f̲ - . . .; recouping; making up;
retrieval,ving; salvage,ging • comeback •
[hidr] reclamation • [cald] recuperation •
[comput] recal(ling); reporting back • [fin]
offsetting • [ind] recycling • [medic] con-
valescing,scence
— f̲ anterior - former recovery
— f̄ cíclica - [econ] cyclic recovery
— f̲ de ácido m̲ [quím] acid, recovery, o̲ restoration,ring
— f̲ — — m̲ residual - [ind] residual acid,
recovery, o̲ restoration,ring
— f̲ — agregado m̲ - [constr] aggregate, re-
claim(ing), o̲ recovery
— f̲ — argón m̲ - [quím] argon recovery
— f̄ — baritina f̲ - [miner] barityne recovery
— f̄ — — f̲ por flotación f̲ - [miner] flotation barytine recovery
— f̲ — — f̲ — — f̲ selectiva - [miner] selective flotation barytine recovery
— f̲ — bobina(s) f̲ - [metal-lam] coil recovery
— f̄ — boro m̲ - [metal-prod] boron recovery
— f̄ — calor m̲ - heat, recovery, o̲ recuperation • heat, use, o̲ utilization
— f̲ — m̲ con eficiencia f̲ alta - [cerám]
high efficiency heat recovery
— f̲ — — m̲ de gas(es) m̲ - [combust] (gas)
heat, recovery, o̲ utilization
— f̲ — — m̲ — — m̲ residual(es) - [combust]
waste heat, recovery, o̲ utilization
— f̲ — — m̲ residual - [combust] waste heat,
recovery, o̲ use, o̲ utilization
— f̲ — capital m̲ - [fin] capital recovery
— f̄ — característica(s) f̲ - characteristic(s)
recovery
— f̲ — — f̲ magnética(s) - magnetic characte-
ristic(s) recovery
— f̲ — carbón m̲ - [ind] coal recovery,ring
— f̄ — cobre m̲ - [miner] copper recovery
— f̄ — concentrado m̲ - [miner] concentrate
recovery
— f̲ — chatarra f̲ - [metal-prod] scrap recovery
— f̄ — demora f̲ - [ind] delay make,king up
— f̄ — derrame m̲ - [ind] spillage recovery
— f̄ — desperdicio(s) m̲ - waste(s) recovery
— f̄ — destilado(s) m̲ - [quím] distillate(s)
recovery
— f̲ — economía f̲ - [econ] economic recovery
— f̄ — fundente m̲ - [sold] flux, recovery, o̲
pick-up
— f̲ — gas(es) m̲ - [ind] gas(es) recovery
— f̄ — — m̲ en tragante m̲ - [metal-prod]
throat gas, recovery, o̲ recuperation
— f̲ — información f̲ - [comput] information,
retrieval, o̲ recovery
— f̲ — lodo(s) m̲ - [ind] sludge recovery
— f̄ — mandato(s) m̲ - [comput] command back
reporting
— f̲ — material(es) m̲ - [ind] material(s), recovery,ring, o̲ reclamation,claiming - mate-
rial(s) recycling
— f̲ — — m̲ reducido(s) m̲ - [metal-prod] reduced material(s) recovering,ry
— f̲ — — m̲ — parcialmente - [metal-prod]
paratially reduced material(s) recovery
— m̲ — nitrógeno m̲ - [quím] nitrogen recovery
— f̄ — orden m̲ - order recovery,ring
— f̄ — — f̲ - [comput] command back reporting
— f̄ — — f̲ pieza(s) f̲ - [mec] part(s) recovery,ring
— f̄ — — f̲ para recambio m̲ - [mec] spare
part(s), recovery, o̲ recuperation
— f̲ — — f̲ usada(s) - [ind] used part(s), recovery, o̲ recuperation; cannibalizing
— f̲ — posición(es) f̲ - position(s) recovery •
[deport] come-from-behind performance
— f̲ — producción f̲ - [ind] production recovery
— f̄ — — f̲ perdida - [ind] lost production recovery

recuperación f de reserva(s) f - [miner] reserve
  recovering,ery
— f — residuo(s) m - [ind] waste(s) recovery •
  waste(s) utilization
— f — rodillo(s) m - [metal-lam] roll(s), recovery, o salvaging
— f — m usado(s) - [metal-lam] used roll(s)
  recovery,ring
— f — señal(es) f—[electrón] signal recalling
— f — — f anterior(es) - [electrón] previous
  signal(s) recall(ing)
— f — — f previa(s) - [electrón] previous
  signal(s) recall(ing)
— f — subproducto(s) m - [ind] by-product(s),
  recovery, o recuperation
— f — suelo(s) m - [suelos] soil(s) reclamation, o recovery
— f — terreno(s) m - [topogr] land reclamation
— f — — m para minería f - [miner] mining
  land reclamation
— f — — — f bajo cielo m abierto -
  [miner] open pit mining land reclamation
— f — tubo(s) m - [cald] tube(s) salvage,ging
— f — — m para caldera(s) f - [cald] flue(s)
  salvage,ging
— f — vapor(es) m vapor(s) recovery • [cald]
  steam recovery
— f — zona f - [miner] area, recovery, o reclamation
— f — — f explotada - [miner] mined area re-
  lamation
— f — — f minada - [miner] mined area reclamation
— f económica - [econ] economic recovery
— f fácil - easy, o ready, salvage,ging
— f industrial - [ind] industrial recovery
— f — de baritina f - [miner] barytine industrial recovery,ring
— f metalúrgica—[miner] metallurgical recovery
— f óptima - [quím] optimum recovery
— f parcial - partial recovery • [fin] partial
  offset(ting)
— f — de material(es) m - [ind] partial mate-
  rial(s), o material(s) partial, recovery
— f por flotación f—[miner] flotation recovery
— f — — f selectiva - [miner] selective flotation recovery
— f — vacío m - [mec] vacuum, recovery, o
  pick-up
— f rápida - [comput] fast, o quick, retrieval
— f reducida de material(es) m - [ind] reduced
  material(s) recovery
— f secundaria - [petról] secondary recovery
— f simple - [metal-prod] simple recovery
— f — de gas m - [ind] gas simple, o simple
  gas, recovery
— f — — m en tragante m - [metal-prod]
  throat gas simple recovery
— f total - [ind] complete, salvage,ging, o recovery,ring
recuperado,da a - recovered; recuperated; retrieved; reclaimed; recouped • recycled •
  made up [ind] reused; salvaged • [fim] offset • [comput] recalled; reported back
— a con facilidad - easily, o readily, recalled, o salvaged
— a facilmente - easily, o readily, salvaged
— a parcialmente - partially recovered • [fin]
  partially offset
— a rápidamente - quickly, o rapidly, retrieved
— a totalmente - salvaged completely
recuperador m - . . . . • [metal-lam] reheating
  furnace • checker
— m automático - [mec] automatic recuperator
— m — para cuerda f para arrancador m - [comb.
  -int] recoil starter
— m para aire m caliente - [metal-prod] hot
  blast stove
— m — arcilla f - [cerám] clay recuperator
— m — calor m - [ind] heat exchanger
— m — cuerda f para arrancador m - [comb.int]
  recoil starter
— m — fundente m - [sold] flux recovery unit

recuperador m para inyección f - [petról] mud,
  saver, o wiper
— m — lodo m - [petról] mud, saver, o wiper
recuperar v - . . .; to retrieve, to reclaim; to
  recoup • to make up • to salvage • [ind] to
  salvage • [ind] to recycle • [comput] to recall, to report back • [fin] to offset
— v capital m - [fin] to recover @ capital
— v completamente - to, recover, o salvage,
  completely
— v con facilidad f - to salvage readily
— v chatarra f - [metal-prod] to recover @
  scrap
— v fácilmente - to, recover, o salvage, easily, o readily
— v información f - [comput] to retrieve @, information, o data
— v mandato m - [comput] to report back @ command
— v material m - [ind] to, reclaim, o recover,
  @ material • to recycle @ material
— v — m reducido - [metal-prod] to recover @
  reduced material
— v — m — parcialmente - [metal-prod] to recover @ partially reduced material
— n nivel m de carga f - [metal-prod] to recover @ stockline level
— v orden m - to recover @ order
— v — f - [comput] to report back @ command
— v parcialmente - to recover partially •
  [fin] to partially offset
— v posición f - to recover @ position •
  [deport] to play catch-up
— v producción f—[ind] to recover @ production
— v — f perdida - [ind] to recover @ lost
  production
— v rápidamente - to recover quickly •
  [comput] to retrieve quickly
— v reserva(s) f - [miner] to recover @ reserve
— v señal f - [electrón] to recall @ signal
— v — f anterior - [electrón] to recall @ previous signal
— v — f previa - [electrón] to recall @ previous signal
— v soplado m - [metal-prod] to, recover, o
  build, o normalize, @ blast
— v terreno m - [topogr] to reclaim @ land
— v totalmente - to, recover, o salvage, completely
— v zona f - [topogr] to reclaim @ area
— v — f explotada - [miner] to reclaim @ mined
  area
— v — f minada - [miner] to reclaim @ mined
  area
— v ventaja f - [deport] to make up (@ gap)
recurrencia f - [véase] repetición f
recurrente a - . . .; recurring • periodic
recurrido,da a - resorted;ting
recurrir v - . . .; to recur
recurso m - resorting • [legal] . . .; remedy
— m administrativo - [admin] administrative recourse
— m agotado - exhausted resource
— m aplicado - [admin] applied recourse • [fin]
  applied resource
— m asignado - [econ] assigned, o allocated,
  resource
— m autogenerado - [econ] self-generating resource
— m cercano - nearby resource
— m con derecho m de reexportación f - [fin]
  resource with re-exportation right(s)
— m — ahorro(s) m - [fin] saving(s) resources
— m — — m de país m - [econ] country('s)
  savings resources
— m de estado m - [econ] state's resource
— m — laboratorio m - [ind] laboratory, resource, o facility
— m económico - [econ] economic resource
— m — exterior - [fin] outside economic resource
— m ejercido - [legal] taken resource
— m estatal - [econ] state('s) resource

recurso m exterior - [fin] foreign resource
— f financiero - [econ] financial resource
— m físico - physical resource
— m hidráulico - [hidr] hydarulic, o water, resource
— m hídrico - [hidr] water resource
— m humano - [Econ] human resource
— m — empleado - [admin] used human resource
— m — no numeroso - [econ] non-plentiful human resouce
— m — valioso—[econ] valuable human resource
— m — empleado - [admin] used human resource
— m — usado - [admin] used human resource
— m — utilizado - [admin] used human resource
— m intersectorial - [econ] inter-sector resource
— m limitado - [fin] limited resource | adv - (con) budget conscious
— m material - [ind] material resource
— m mecánico - [ind] mechanical resource
— m mejorado - [econ] improved resource
— m mineral - [miner] mineral resource
— m minero - [miner] mining resource
— m natural - [econ] natural resource
— m — no renovable - [écon] expendable, o non-renewable, natural resource
— m necesario - [econ] needed resource
— m no numeroso—[econ] non-plentiful resource
— m — renovable - [econ] non-renewable, o expendable, resource
— m para coarrera(s) f - [deport] racing resource
— m propio - [fin] company's, o own, resource
— m real - [econ] true resource
— m reinvertido - [econ] reinvested resource
— m técnico - technical resource
— m transferido - [econ] transferred resource
— m último - last, recourse, o resort
— m usado - [econ] used resource
— m útil - [econ] useful resource
recursos m materiales—[ind] material resources
— m de mineral m de hierro m - [miner] iron ore resources
— m — país m - [econ] country's, o national, resources
— m — préstamo(s) m - [fin] loan resources
— m descartables - [econ] expendable resources
— m disponibles - [econ] available resources
rechazado,da a - rejected; discarded; declined; turned down
rechazado,da a definitivamente - rejected definitely
— a parcialmente - partially rejected
— a totalmente - rejected totally
rechazamiento m - . . .; discarding
rechazar v - . . .; to discard; to turn down; to decline
— v artículo m - to reject @ article
— v carga f - [transp] to reject @ load
— v definitivamente - to reject definitely
— v equipo m - [ind] to reject @ equipment
— v material(es) m - to reject @ material(s)
— v oferta f - [com] to reject @, offer, o proposal
— v orden m en código m - [comput] to reject @ code sequence
— v parcialmente - to reject partially
— v pedido m - to reject @ order
— v producto m - [ind] to reject @ product
— v propuesta f - [com] to reject @, proposal, o bid
— v repuesto(s) m - [ind] to reject @ spare(s) (parts)
— v secuencia f en código m - [comput] to reject @ code sequence
— v señal f - [comput] to reject @ signal
— v señalización f - [comput] to reject @ signalling
— v soldadura f - to reject @ weld
— v totalmente - to reject totally
rechazo m - . . .; refusal; turning down; non-acceptance; declining • discarding • [ind] salvage

rechazo m de artículo m - article, rejecting,tion, o refusal
— m — carga f - [transp load rejecting,tion
— m — equipo m - [ind] equipment rejecting,tion
— m — material(es) m - material(s) rejecting,tion
— m — oferta f - [com] offer, o proposal, rejecting,tion
— m — orden m en código m - [comput] code sequence rejection
— m — propuesta f - [com] proposal, o bid, rejecting,tion, o refusal
— m — respuesto(s) m - [ind] spare(s), o spare parts, rejecting,tion, o refusal
— m — rodillo(s) m - [metal-lam] roll(s) rejecting,tion
— m — — m de acero m - [metal-lam] steel roll rejecting,tion
— m — — — m forjado - [metal-lam] forged steel roll rejecting,tion
— m — — — m fundido - [metal-lam] cast steel roll rejecting,tion
— m — — m para apoyo m - [metal-lam] back-up roll(s) rejecting,tion
— m — — m — — m de acero m - [metal-lam] steel back-up roll rejecting,tion
— m — — m — — m fundido - [metal-lam] cast steel back-up roll rejecting,tion
— m — — m — trabajo m - [metal-lam] work roll rejecting,tion
— m — — m — — m de acero m - [metal-lam] steel work roll rejecting,tion
— m — — m — — m fundido - [metal-lam] cast steel work roll rejecting,tion
— m — secuencia f en código m - [comput] code sequence rejecting,tion
— m — señal f - [comput] signal rejecting,tion
— m — señalización f - [comput] signalling rejecting,tion
— m — sistema m - [electrón] system rejecting,tion
— m — — m silenciador m - [electrón] squelch system rejecting,tion
— m — — m — regulado con tono m continuo - [electrón] continuous tone controlled squelch system rejecting,tion
— m — soldadura f - [sold] weld rejecting,tion
— m definitivo - definite rejecting,tion
— m especificado - [ind] specified discard
— m excepcional - [metal-lam] special discard
— m final - ultimate rejecting,tion
— m parcial m - partial rejecting,tion
— m primero - first, refusal, o rejecting,tion
— m total - total rejecting,tion
rechinamiento m - . . .; creaking
rechoncho,cha a - . . .; squatty
rechupe m - [metal-prod] pipe • [metal-lam] shrinkage (cavity)• (coil) telescoping
— m de lingote m - [metal-prod] ingot pipe
— m a tierra f - [electr-instal] ground(ing) system
— m excesivo - [metal-prod] excessive pipe
red f - . . .; gauze • [tub] utility • [constr] matting • [serv.publ] utility; service; line
— f acondicionadora - [ambien] conditioning system
— f — para aire m - [ambient] air conditioning system
— f adecuada - suitable system)
— f apropiada - [constr] proper, o appropriate, system
— f — para combustible m - [comb.int] good, o appropriate, fuel system
— f — — drenaje m - [constr] proper, o appropriate, drainage system
— f — — subdrenaje m - [sanit] proper, o appropriate, subdrainage system
— f automática - [mrv] automatic system
— m — para lubricación f - [mec] automatic lubricating,tion system

red f **auxiliar** = auxiliary, system, o network
— f **bifásica** - [electr-instal] two-phase system
— f **caminera**—[vial] roadway, network, o system
— f — **departamental** - [vial] county road system
— f **carretera** - [vial] highway, o road, system
— f **cebada** f - [bombas] primed system
— f — **automáticamente** - [bombas] automatically primed system
— f — **manualmente** - [bombas] manually primed system
— f **central** - central, system, o network
— f — **para lubricación** f - [mec] central lubricating system
— f **centralizada** - central(ized) system
— f — **automática** - [mec] automatic centralized system
— f — — **para lubricación** f [mec] automatic central(ized) lubrication system
— f **cloacal** - [sanit] sewage,wer, system, o line
— f — **colectora** - [sanit] interceptor sewer system
— f — — **con infiltración** f **reducida** - [sanit] low infiltration sewer system
— f **cloacal para urbanización** f - [constr] development, o subdivision, sewer line
— f — **proyectada** - [sanit] designed sewer system
— f **colectora** f - [sanit] collection system
— f **con presión** f - [mec] pressure,rized system
— f — — f **alta** - high pressure system
— m — **tres conductores** m - [electr-instal] three-wire system
— f **costosa** f - [electr-instal] costly system
— f — **para servicio** m **público** - [constr] expensive utility
— f **de alambre** m - [alambre] netting; wire net; fence,cing
— f — — m **galvanizado** - [alambre] galvanized fence,cing
— f — — m **hexagonal** - [alambre] hexagon, netting, o fencing
— f — **alcantarillado** m - [sanit] sewer system
— f — **alcantarilla(s)** f—[sanit] sewer system
— f — **altavoces** m - [electrón] public address system
— f — **altoparlante(s)** m - [electrón] public address system
— f — **bomba(s)** f - [bombas] pump(ing) system
— f — — f **para lubricación** f - [lubric] pumps lubricating system
— f — **calzada(s)** - [vial] road(s), o sidewalks, network
— f — — f **para interconección** f - [vial] interchange complex
— f — **canal(es)** f - [tub] raceway system
— f — — m **conectada a tierra** f - [constr] grounded raceway system
— f — — m **de acero** m - [tub] steel raceway system
— f — — m — — **totalmente puesto(s) a tierra** f - [electr-instal] completely grounded steel raceway system
— f — — m — — m **conectada a tierra** f—[tub] grounded steel raceway system
— f — — m **totalmente conectada a tierra** f - [electr-instal] completely grounded raceway system
— f — **carretera(s)** f - [vial] highway system
— f — — f **interestatal(es)** - [vial] interstate highway(s) system
— f **de cloaca(s)** f - [sanit] sewer(age) system, o line
— f — — f **domiciliaria(s)** f - [sanit] sewage collection system
— f — **conducto(s)** n - [electr-instal] conduit system • [tub] raceway system
— f — — m **conectada con tierra** f - [constr] grounded raceawy system
— f — — m **de acero** m - [tub] steel raceway system
— f — — m — — **conectada con tierra** f—[tub] grounded steel raceway system

red f **de conducto(s)** m **separado(s)** m - [electr-instal] separate conduit systems
— f — — **totalmente puestos a tierra** f - [electr-instal] completely grounded raceway system
— f — **conductor(es)** f - [electr-instal] cable, o lead, system
— f **de Davy** - [miner] Davy screen
— f — **desagüe(s)** m - [sanit] drain(age), o sewer, system
— f — — m **cloacal(es)** - [sanit] sanitary sewer(s) (system)
— f — — m **pluvial(es)** - [constr] storm drainage system; storm sewer (system)
— f — **distribuidor(es)** m - dealer network
— f — — m **asociado(s)** - [com] associate dealer(s) network
— f — **malla** f - mesh net
— f — **petaca(s)** f - [metal-prod] cooling plate circuit
— f — **precisión** f **para división** f **de voltaje** m - [electrón] precision voltage divider network
— f — **presión** f - pressure system
— f — **rociadores** m - [segurid] sprinkler system
— f — — m **instalada** - [segurid] installed sprinkler system
— f — **subestación(es)** - [electr-distrib] súbstation(s) system
— f — **supercarretera(s)** f - [vial] Interstate network; [G.B.] motorway network
— f — **tomacorriente(s)** f - [electr-instal] outlet(s) system
— f — — **para uso** m **general** - [electr-instal] convenience outlet(s) system
— f — **tubería(s)** f - [tub] tube,bing, o pipe, system • [petról] piping system • [ambient] duct(work) system
— f — — f **aérea** - [tub] aerial, o overhead, tube,bing, o pipe,ping, system
— f — — f **con presión** f - [tub] pressure piping system
— f — — f **hermética(s)** - [ambient] airtight duct system
— f — — f **para refinería** f - [petról] refinery piping system
— f — — f — — f **para petróleo** m - [petról] petroleum refinery piping system
— f — — f — **transporte** m **para petróleo** m - [petról] petroleum transportation piping system
— f — **tubo(s)** m - [tub] tube, o pipe, system
— f — **ventilador(es)** m - [ind] fan system
— f — — m **para recogida** f - [ind] collection fan system
— f — — m — — f **de polvo** m - [ind] dust collection fan(s) system
— f **desenfatizadora** f - [comput] de-emphasis, o de-emphasizing, network
— f **debida** - good, o proper, system
— f — **para combustible(s)** m - [comb.int] good fuel system
— f **departamental** - [vial] county system
— f — **de puente(s)** - [vial] county bridge system
— f **distribuidora** - distributing,tion system
— f **económica** - economical system
— f — **para servicio** m **público** - [constr] economical utility
— f **eléctrica** - [electr-distrib] electric, o power, o wiring, o utility, system, o network, o line
— f — **comprobada** - [electr-oper] checked electrical system
— f — **domiciliaria** - [electr-instal] house, o premises, wiring system
— f — **para cambio(s)** m - [autom-electr] electric shift(ing) system
— f — **para casa** f - [electr-instal] house wiring system
— f — **para . . . voltio(s)** m - [electr-instal] . . . volt electric(al) system
— f — **simétrica** - [electr-instal] symmetric

electric system
red f eléctrica verificada - [electr-oper]
checked electrical system
— f **elevada** - [electr-instal] overhead utility
— f **enfriada** - cooled (off) system
— f **estatal** - [vial] state system
— f **extintora** - [segurid] extinguishing system
— f **existente** - existing system
— f — **para aire m comprimido** - [ind] existing
compressed air system
— f . . .**-fásica** - [electr-distrib] . . .-phase
system
— f **federal** - [vial] federal system
— f — **de Carreteras** f - [vial] Federal Highway system
— f **general** - [electr-distrib] general, system,
o network - [ind] plant system
— f **hidráulica** - [hidr] hydraulic system
— f — **completa** - [hidr] complete hydraulic
system
— f **hidrostática** f - [hidr] hydrostatic system
— f **inadecuada** - inadequate system
— f — **para drenaje m** - [hidr] inadequate
drainage system
— f — **para subdrenaje m** - [hidr] inadequate
subdrainage system
— f **inoperante** - inoperative system
— f **íntegra** - entire system
— f **integrada** - integrated system
— f — **para lubricación** f - integrated lubricating,tion system
— f **interestatal** - [vial] Insterstate System
— f **limpia** - clean system
— f **limpiada** - cleaned system
— f **monofásica** - [electr-distrib] single phase
system
— f — **con tres conductores m** - [electr-instal]
three-wire single-phase system
— f **nacional de carretera(s)** f - [vial] National, o Federal, Highway System
— f **neumática** - [neumát] pneumatic system •
[ind] (air) pumping system
— f — **para cambios m** - [autom-mec] air shift-
(ing) system
— f — **para distribución** f - [ind] pneumatic
distribution system
— f **operante** - operative system
— f **para abastecimiento** m - supply(ing) system
— f — — **m de agua m** [hidr] water supply(ing)
system
— f — **aceite m** - [lubric] oil system
— f — — **m comprobada** - [mec] checked, o prov-
en, oil system
— f — — **m con presión** f - [comb.int] pressure
oil system
— f — — **limpia** - [mec] clean oil system
— f — — **m limpiada** → [mec] cleaned oil system
— f — — **m lubricante** - [comb.int] lubricating oil system
— f — **m para regulación** f - regulating, o
control, oil system
— f — **sucia** f - [mec] dirty oil system
— f — — **m verificada** - [mec] checked oil system
— f — **agua m** - [hidr] water system
— f — — **m corriente** - [hidr] (drinking) water, line, o system - public water system
— f — — **m — fría** f - [hidr] cold water,
line, o system
— f — — **f — y cloaca(s)** f - [hidr] water and
sewer system
— f — — **m de pozo m** - [hidr] well water system
— f — — **m de río m** - [hidr] river water system
— f — — **m industrial** - [hidr] industrial water system
— f — — **m — de río m** - [ind] industrial
river water system
— f — — **m para circulación** f - [hidr] circulating water system
— f — — **m — — f para reemplazo m** - [hidr]
replacement circulating water system

red f para agua m para enfriamiento m - [hidr]
cooling water system
— f — — **m — — m para rodillo(s) m** - [metal-
-lam] roll cooling water system
— f — — **m — laminadora(s)** f - [metal-lam]
mill water system
— f — — **m para refrigeración** f - [metal-lam]
cooling water system
— f — — **m — — f para rodillo(s) m** - [metal-
-lam] roll cooling water system
— f — — **m para sellar v** - [hidr] sealing water system
— f — — **m — tobera(s)** f - [metal-prod] tuyere water system
— f — — **m potable** - [jodr] (drinking) water,
facility, o system
— f — — **m sucia** - [ind] dirty water system
— f — **aire m** - [neumát] air system
— f — **m acondicionado** - [ambient] air conditioning system
— f — — **m comprimido** - [neumát] compressed
air system
— f — — **m — con presión** f - neumát] pressure,rized air system
— f — — **m — — f baja** - [neumát] low
pressure compressed air system
— f — — **m — existente** - [neumát] existing
compressed air system
— f — — **m — operada** - [neumát] operated compressed air system
— f — — **m — para planta** f - [ind] plant compressed air system
— f — — **m — depósito m** - [sold] tank air
system
— f — — **m — — m para fundente m** - [sold]
(flux) tank air system
— f — — **m para enfriamiento m** - [neumát]
cooling air system
— f — **alarma** f - [segurid] alarm, o warning,
system
— f — — **f contra carga(s)** f **excesiva(s)** -
[grúas] overload warning system
— f — **alimentación** f - [electr-prod] power
source
— f — **aprovisionamiento m** - supply system
— f — — **m para agua m** - [hidr] water sup-
ply(ing) system
— f — **aprovechamiento m** - [ind] utilization
system
— f — — **m para gas m** - [combust] gas utilization system
— f — **asfalto m** - [nucl] asphalt system
— f — — **m bombeo m** - [petról] pumping system
— f — — **m subterráneo** - [petról] subsurface,
o underground, pumping system
— f — **circulación** f - circulating,tion system
— f — — **f para agua m** - [hidr] water circulating system
— f — — **f — accionamiento m hidráulico** -
[ind] hydraulic circulating system
— f — — **f hidráulica** - [hidr] hydraulic circulating system
— f — **combustible m** - [comb.int] fuel system
— f — — **m mantenida** - [comb.int] maintained
fuel system
— f — — **m sangrada** - [comb.int] bled fuel
system
— f — — **m taponada** - [comb.int] clogged fuel
system
— f — **compresión** f - compressing system
— f — — **f de aire m** - [neumát] air compressing system
— f — **comunicación(es)** f - [electrón] commu-
nication(s) system
— f — — **f adecuada** - [comunic] suitable com-
munication(s) system
— f — — **f para planta** f - [ind] plant commu-
nication(s) system
— f — **condensado m** - [ind] condensate system
— f — **conexión** f - connecting,tion system
— f — — **f con tierra** f - [electr-instal]
grounding, o ground wire, system
— f — — **m de combustión** f - combustion con-

trol system
red f̲ para dato(s) m̲ - [comput] data network
— f̲ — desenfatización* f̲ - [comput] de-emphasis, o̲ de-emphasizing, network
— f̲ — distribución f̲ - [ind] distributing,
system, o̲ network, o̲ channel
— f̲ — f̲ dentro de planta f̲ - [electr-
distrib] plant distribution system
— f̲ — eléctrica - [electr-distrib] elec-
tric(al) distribution, network, o̲ system
— f̲ — f̲ neumática - [ind] pneumatic distribution system
— f̲ — f̲ para agua m̲ - [hidr] water distribution system
— f̲ — f̲ — m̲ de río m̲ - [hidr] river
water distribution system
— f̲ — f̲ — para agua m̲ potable - [hidr]
drinking water distributing,tion system
— f̲ — f̲ combustible - [comb.int] fuel
distribution system
— f̲ — f̲ para energía f̲ - [electr-distrib]
power distribution, network, o̲ system
— f̲ — f̲ — f̲ eléctrica en abanico -
[electr-distrib] radial (distribution) system
— f̲ — f̲ para combustible m̲ - [comb.int]
fuel distribution system
— f̲ — división f̲ - [electrón] divider system
— f̲ — f̲ para voltaje m̲ - [electrón] voltage
divider network
— f̲ — drenaje m̲ - [hidr] drainage system
— f̲ — encendido m̲ - [electr-instal] ignition
system
— f̲ — energía f̲ - [electr-distrib] power system
— f̲ — enfriamiento m̲ - [comb.int] cooling system
— f̲ — m̲ atascada - [comb.int] clogged cooling system
— f̲ — m̲ enjuagada - flushed cooling system
— f̲ — m̲ purgada - [comb.int] drained cooling system
— f̲ — m̲ verificada - [Comb.int] checked
cooling system
— f̲ engrase m̲ - [mec] lubricating,tion system
— f̲ — m̲ para distribuidor m̲ - [mec] distributor, greasing, o̲ lubrication, system
— f̲ — excitación f̲ - [electr-mot] exciting
system
— f̲ — freno m̲ - [mec] brake, line, o̲ system
— f̲ — m̲ para camión m̲ - [mec] truck('s)
brake,king, line, o̲ system
— f̲ — frío m̲ - [ind] cold, o̲ cooling, system
— f̲ — gas(es) m̲ - [ind] gas(es) system
— f̲ — fuel m̲ oil - [combust] fuel oil system
— f̲ — información f̲ - [comput] data network
— f̲ — irrigación f̲ - [hidr] irrigation system
— f̲ — lavado m̲ - washing system
— f̲ — lubricación f̲ - [mec] lubricating,tion
system
— f̲ — f̲ cebada - [ind] primed lubricating
system
— f̲ — f̲ con aceite m̲ - [ind] oil lubricating,tion system
— f̲ — f̲ con presión f̲ - [mec] pressure lubricating,tion system
— f̲ — f̲ forzada - [mec] forced lubricating,tion system
— f̲ — f̲ para bomba f̲ - [lubric] pump lubricating,tion system
— f̲ — oxígeno m̲ - [ind] oxygen (distribution,
system, o̲ service
— f̲ — pesca f̲ - [pesca] fishing net
— f̲ — protección f̲ - [segurid] protection system
— f̲ — f̲ para alumbrado m̲ - [electr-instal]
electric, o̲ lighting, protection system
— f̲ — purga f̲ - [ind] bleeding system
— f̲ — recolección f̲ - gathering, o̲ collecting,
system
— f̲ — f̲ para agua m̲ - [ind] water, gathering, o̲ collecting, system
— f̲ — recirculación f̲ - [ind] recirculating,tion system

red f̲ para recirculación f̲ para flujo m̲ - [ind]
flow recirculation system
— f̲ — f̲ — m̲ mínimo - [ind] minimum flow
recirculating,tion system
— f̲ — f̲ para gas m̲ - [combust] gas recirculating,tion system
— f̲ — refrigeración f̲ - [ind] cooling, o̲ refrigerating,tion, system
— f̲ — f̲ inoperante - [ind] inoperative refrigerating,tion system
— f̲ — f̲ operante - [ind] operating refrigerating,tion system
— f̲ — f̲ verificada - [comb.int] checked
cooling system
— f̲ — sangría f̲ - [ind] bleeding system
— f̲ — servicio m̲ - service system
— f̲ — servicio m̲ pesado - [ind] heavy duty
system
— f̲ — m̲ público - [serv.publ] (public), u-
tility, o̲ system, o̲ line; service
— f̲ — m̲ — destruida—[serv.publ] destroyed
utility
— f̲ — m̲ — existente - [serv.públ] existing utility,ties
— f̲ — m̲ — reemplazada - [serv.publ] replaced utility,ties
— f̲ — subdrenaje m̲ - [hidr] subdrainage system
— f̲ — suministro(s) - supply(ing) system
— f̲ — m̲ para agua m̲ - [hidr] water sup-
ply(ing) system
— f̲ — telecomunicación(es) f̲ - [telecom] te-
lecommunication(s) network
— f̲ — transmisión f̲ - [electr-distrib] transmission system
— f̲ — transporte(s) m̲ - [transp] transportation system
— m̲ — m̲ acelerado(s) m̲ - [trans] rapid
transit system
— m̲ — m̲ metropolitano(s) - [transp] (metropolitan) transit system
— f̲ — m̲ — acelerado(s) m̲ - [transp] (metropolitan) rapid transit system
— f̲ — vapor m̲ - [capor] steam, system, o̲ line
— f̲ — m̲ para secador m̲ - [papel] drier
steam, o̲ steam drier, system
— f̲ — . . . voltio(s) m̲ - [electr-distrib]
. . . volt, system, o̲ network, o̲ line
— f̲ pluvial - [hidr] storm system
— f̲ primaria - [electr-distrib] primary, system. o̲ network
— f̲ para distribución f̲ - primary distributing,tion system
— f̲ principal - [ind] main, system, o̲ network
— f̲ — para distribución f̲ - [electr-distrib]
main distributing,tion, system, o̲ network
— f̲ propuesta - proposed, system, o̲ network
— f̲ proyectada - projected, o̲ designed, system, o̲ network
— f̲ purgada - [ind] purged system
— f̲ sangrada - [ind] bled system
— f̲ secundaria - [ind] scondary, network, o̲
system
— f̲ para Ayuda f̲ Federal - [vial] Federal
Aid Secondary System
— f̲ — para distribución f̲ - [electr-distrib]
secondary distribution system
— f̲ separada - separate, system, o̲ network
— f̲ — de conducto(s) m̲ - [electr-instal] separate conduit, system, o̲ network
— f̲ séptica f̲ - [sanit] septic, system, o̲ network
— f̲ simétrica - symmetric system
— f̲ sonora - [electrón] sound system
— f̲ — electrónica - [electrón] electronic
sound system
— f̲ subterránea - [ind] underground, system, o̲
network, o̲ utility • [petról] subsurface
system
— f̲ — para distribución f̲ - [electr-instal]
underground distribution, system, o̲ network
— f̲ — para servicio m̲ público - [constr] underground utility
— f̲ taponada - clogged, system, o̲ network

red f telefónica - [telecom] (tele)phone system, o network
— f total - total, network, o system
— totalmente puesta a tierra f—[electr-instal]
completely grounded, network, o system
— f trifásica - [electr-instal] three-phase
system
— f trifilar - [electr-instal] three-wire, system, o network
— f — monofásica - [electr-instal] three-wire
single phase, system, o network
— f troncal - trunk, network, o system
— f vial - [vial] road, o highway, network
— f — departamental - [vial] county highway
system
— f — estatal - [vial] state highway system
— f — nacional - [vial] national, o federal,
highway system
— f — provincial - [vial] provincial, o state,
highway system
redacción f - . . . • writing
— f de informe(s) m - report writing
— f siguiente - [gram] following wording
redactado,da a - edited; worded • drafted
redactar v - to edit; to write • to word •
[legal] to draft
redactor m - . . . . • [legal] drafter
— m asociado - [public] associate editor
— m automovilista—[public] auto(mobile) editor
— m en jefe - [public] editor-in-chief
— m técnico - [public] technical editor •
engineering editor
redestilación f - redistilling • [petrol] rerun
— f de destilado m - [petról] distillate rerun
— f — — m bruto (de craqueo m) - [petról]
pressure distillate rerun
redestilado,da a - [petról] rerun
redestilar v - [petról] to rerun
redimensión* f - redimensioning
redimensionado,da a - redimensioned
redimensionamiento* m - redimensioning
redimensionar v - to redimension
redirección f - redirecting • redirection
— f de vehículo m - [transp] vehicle redirecting
— f — — m sin gobierno m - [transp] errant
vehicle redirecting,tion
— f severa - [transp] severe redirecting,tion
redirigido,da a - redirected
redirigir v - to redirect
— v vehículo m - [transp] to redirect @ vehicle
rediseñado,da a - redesigned
rediseñar v - to redesign
rediseño m - redesign(ing)
— m de sistema m - system redesign(ing)
rédito m - . . . . • return
— m disminuido - [fin] diminished,ining return
redituable a - . . .; income producing
redituado,da a - [fin] returned
redituar v - . . . . • to return
redondeado,da a - rounded (off • to @ nearest
whole number
redondear v - . . . . • [vial] to crown
— v arista f - to round @ (sharp) corner
— v banda f para rodamiento m - [autom-neumát]
to round off @ tread
— v borde m - [autom-neumát] to round (off)
— v — m de banda f para rodamiento m - [autom-
-neumát] to round off @ (tire) tread
— v canto m - [mec] to cant
— v esquina f - [mec] to round @ corner
redondeo m - rounding (off)
— m de arista f - [mec] corner rounding (off)
— m — banda f (para rodamiento m) - [autom-
-neumát] tread rounding (off)
— m — esquina f - [mec] corner rounding (off)
redondez f - . . . . • round shape • [f.c.-ruedas]
rotundity
— f de extremo m - end roundness
— f — — m de tubería f - [tub] pipe end
roundness
— f — tubería f - [tub] pipe roundness
— f original - original, roundness, o round

shape
redondez f uniforme - uniform roundness
redondo m - . . . . [metal-lam] round; rod
— m de acero m - [metal-lam] steel round
— m — — con aleación f baja - [metal-lam]
low alloy steel round
— m — — m — calidad f alta - [metal-lam]
high quality steel round
— m — — m — — baja - [metal-lam] low
quality steel round
— m desbastado - [metal-lam] roughed round
— m para armadura f—[constr] reinforcing round
— m — hormigón m - [constr] reinforcing round
— m — — m armado - [constr] reinforced concrete round
— m para trefilación f - [metal-lam] wiredrawing round
redondo,da á uniformemente - uniformly round
— a estriado,da - [clavos] round serrated
reducción f - . . .; reducing; lessening; constraint; abatement; curtailment • palliation
• [tub-fitting] reduction • [metal-prod] cut
back • [metal-lam] draft • [mec] lopping off
• [ind] bleeding off • [deport] slowing down
— f a absurdo m - reduction ad absurdum
— f — hierro m esponja - [metal-prod] reduction to sponge iron
— f — mínimo m - minimizing,zation
— f — — m de agrietamiento m - cracking minimizing,zation
— f — — m — — m de calzada f - [constr]
pavement crack(ing) minimizing,zation
— f — — m — — pavimento m - [constr]
pavement crack(ing) minimizing,zation
— f — — m — ciclo(s) m - [mec] cycl(s) minimizing,zation
— f — — m — ciclo m para marcha/detención -
[mec] on/off cycle minimization
— f — — m — — —/— f de compresor m -
[mec] compressor on/off cycling minimization
— f — — m — corrosión f - [mec] corrosion
minimizing,zation
— f — — m — costo(s) m—[ind] cost(s) minimizing,zation
— f — — m — desgaste m - [mec] wear minimizing,zation
— f — — m — distorsión f - [mec] distortion
minimizing,zation
— f — — m — efecto(s) m - [mec] effect(s)
minimizing,zation
— f — — m — — m de corrosión f - [mec] corrosion effect(s) minimizing,zation
— f — — m — existencia(s) f - [ind] inventory minimizing,zation
— f — — m — fuga f - leak(ing,age) minimizing,zation
— f — — m — grieta f - crack minimizing,zation
— f — — m — — f en calzada f - [constr]
pavement crack minimizing,zation
— f — — m — — f — pavimento m - [constr]
pavement crack minimizing,zation
— f — — m — inversión f - [ind] investment
minimizing,zation
— f — — m — — f en material(es) m - [ind]
inventory investment minimizing,zation
— f — — m — material(es) m - [ind] inventory minimizing,zation
— f — — m — problema m - [mec] problem minimizing,zation
— f — — m — — m con distorsión f - [mec]
distortion problem minimizing,zation
— f — partícula(s) f - particulating,tion
— f adecuada - [mec] adequate reduction
— f adicional - additional, o further, reducing,duction
— f alta - [mec] high reducing,ction
— f apropiada - [mec] appropriate, o adequate,
reducing,duction
— f arancelaria - [fisc] tariff reduction
— f automática - automatic reducing,duction
— f — de velocidad f - automatic speed, reducing,dcution, o decrease,sing

reducción f B S R - [metal-prod] Boehler strand reducing,duction
— f baja - [metal-prod] low reduction
— f̄ Boehler - [metal-prod] Boehler strand reducing,ction
— f calculada - [metal-lam] calculated draft
— f̄ con contracorriente f - [metal-prod] countercurrent reducing,ction
— f — engranaje(s) m - [mec] gear reducing,ction
— f — — m doble(s) - [mec] double gear reducing,ction
— f — — m helicoidal(es) - [mec] worm gear reducing,ction
— f — — m sin fin - [mec] worm gear reducing,duction
— f — — m — — en hélice f - [mec] helical gear reducing,ction
— f — — — — — — f de correa f articulada - [mec] link(ed) belt helical worm gear reducing,duction
— f — gas m - [metal-prod] gas(eous) reduction
— f̄ — hidrógeno m - [metal-prod] hydrogen reducing,duction
— f — óxido m - [metal-prod] oxide reduction
— f̄ — m de carbono m - [metal-prod] carbon (mon)oxide reduction
— f — — m — — m e hidrógeno m—[metal-prod] carbon (mon)oxide plus hydrogen reduction
— f — oxígeno m—[metal-prod] oxygen reduction
— f̄ concéntrica - [tub] concentric reducer,cing
— f — embridada - [tub] concentric flanged reducer,cing
— f considerable - [mec] significant reduction
— f̄ — de carga f - significant load reduction
— f̄ continua - [metal-lam] continuous reduction
— f̄ de agua m - [hidr] water reduction
— f̄ — — m en entrada f - [hidr] headwater reduction
— f — alambre m - [trefil] wire, draft, o drawing, o reducing,duction
— f — amplitud - [electrón] amplitude decrease,sing
— f — área m - [alambre] cross area reduction
— f̄ — — m mínima - minimum area reduction
— f̄ — — m transversal - crosswise area reduction
— f — — m — de alcantarilla f - [hidr] culvert (crosswise) area reduction
— f — beneficio m - [contab] earning(s), o profit(s), decrease
— f — — m retenido - [contab] retained earning(s) decrease
— f — capacidad f - capacity reduction
— f̄ — capital m - [fin] capital, reduction, o decrease
— f — — m de compañía f - [fin] company capital, reduction, o decrease
— f — — m — empresa f - [fin] company, o corporate, capital decrease
— f — — m — sociedad f - [fin] corporate capital, decrease, o reduction
— f — capitalización f - [fin] capitalization, decrease, o reduction
— f — carga f - load, reduction, o decrease, o lessening
— f — sobre conexión f - [electr-equip] connecting,tion load reducing,ction
— f — coercividad f - coercivity reduction
— f̄ — contaminación f - [ambient] pollution abatement
— f — — f de agua m - [hidr] water pollution abatement
— f — contragolpe m - [mec] backlash reducing,ction, o removal,ving
— f — corriente f - [electr-oper] current reducing,ction
— f — — f en entrada f - [electr-oper] input current reducing,ction
— f — costo(s) m - cost reducing,ction
— f — — m de soldadura f - [sold] welding cost, reducing,ction, o decrease
— f — cuadrilla f - [labor] crew reduction

reducción f de cuadrilla f para reemplazo m - [ind] replacement crew reduction
— f — choque m - transp] collision, o impact, reduction
— f — daño m - damage, reduction, o minimizing,zation
— f — deformación f - deformation, o deflection, reducing,ction
— f — diámetro m - diameter reduction
— f̄ — — m de cable m - [cabl] cable diameter reducing,ction
— f — — m durante recalcadura f - [metal-fabr] diameter reduction during swaging
— f — — m exterior - [metal-fabr] outside diameter reducing,ction
— f — — m — durante recalcadura f - [metal-fabr] outside diameter reducing,ction during swaging
— f — — m interior - [metal-fabr] inside diameter reducing,ction
— f — — m — durante recalcadura f - [metal-fabr] inside diameter reducing,ction during swaging
— f — dispersión f - dispersion reducing,ction
— f̄ — — f de aluminio m - [metal-prod] aluminum dispersion reducing,ction
— f — dureza f - hardness reducing,duction
— f̄ — energía f - [electr-distrib] energy reducing,ction
— f — erosión f - [hidr] erosion, reduction, o control
— f — facilidad f para traslado m - [mec] portability, reduction, o decrease
— f — fatiga f - [segurid] fatigue reduction
— f̄ — filtración f - [hidr] filtration, o seepage, reducing,ction
— f — flojedad f - slack reducing,ction
— f — flujo m - flow, decrease, o reduction
— f̄ — — f habilidad f - impairment
— f̄ — — f física - [labor] physical impairment
— f — hierro m - [metal-prod] iron reduction
— f̄ — — m con contracorriente f - [metal-prod] countercurrent iron reduction
— f — infiltración f - [sanit] infiltration reduction
— f — intensidad f - [electr-oper] intensity reduction
— f — — f de corriente f - [electr-oper] current (intensity) reducing,ction
— f — — f — — f en entrada f - [electr-oper] input current intensity
— f — juego m longitudinal - [mec] end play, reducing,ction, o decrease,sing
— f — marcha f - [metal-prod] blast reducing,ction
— f — mineral m - [metal-prod] ore reduction
— f̄ — — m de hierro m - [metal-prod] iron ore reducing,ction
— f — participación f - [fin] participation, o share, reducing,ction
— f — partida f - [ind] batch reducing,ction
— f̄ — — f de pella(s) f - [miner] pellet(s) batch reducing,ction
— f — — f de terrón(es) m - [metal-prod] lump batch reducing,ction
— f — peligro m - [segurid] danger, o hazard, reducing,ction
— f — — m de derrubio(s) m - [hidr] scour hazard reducing,ction
— f — pella(s) f - [metal-prod] pellet(s) reducing,ction
— f — pendiente f - [constr] gradient, o grade, reduction • slope reduction
— f̄ — pérdida f - loss reducing,ction
— f — personal m - [labor] crew, reduction, o cut(ting) back
— f — — m en cuadrilla f - [labor] crew cut(ting) back
— f — plazo m - time, o lapse, reduction
— f̄ — porosidad f - porosity reducing,ction
— f̄ — precio m - price reducing,ction
— f̄ — presión f - pressure reducing,ction •

  - [ind] pressure bleeding

reducción f de producto m - [ind] product reducing,duction

— f — **producto m plano m** - [metal-lam] flat product reducing,ction

— f —— m — **laminado** - [metal-lam] flat rolled product reducing,ction

— f — **rapidez f** - speed, o rate, decrease, o reducing,ction

— f —— f **de difusión f** - diffusion rate decrease,sing

— f — **remanso m** - [hidr] backwater, o eddy, decrease,sing

— f — **recurso(s)** - resource(s) reducing,ction

— f —— m **mineral(es)** - [miner] mineral resource(s) reducing,ction

— f — **residuo(s) m** - waste reducing,ction

— f — **resistencia f** - resistance, reduction, o decrease

— f —— f **a daño(s) m** - damage(s) resistance, reducing,ction, o decrease

— f — **ruido m** - noise, reduction, o abatement

— f —— m **de motor m** - [comb.int] engine noise reducion

— f — **salpicadura(s) f** - [sold] spatter, reducing,ction, o minimizing

— f —— f **fina(s)** - [sold] fine spatter minimizing

— f — **separación f** - (gap) decrease,sing

— f — **sólido(s) m** - solid(s) reducing,ction

— f — **soplado m** - [metal-prod] blast reduction

— f — **tamaño m** - size, reducing,ction, o decrease,sing

— f —— m **de mineral** - [miner] ore size reducing,ction

— f — **temperatura f** - temperature, decrease, o reducing,ction, o lowering

— f — **tendencia f** - tendency reducing,ction

— f — **tensión f** - tension reducing,ction • [electr-distrib] voltage reducing,ction • under-voltage; voltage drop

— f — **terrón(es) f** - [metal-prod] lump(s) reducing,ction

— f — **velocidad f** - speed, reducing,ction, o decrease

— f —— f **de banda f** - [metal-lam] strip speed reducing,ction

— f —— f — **chapa f** - [metal-lam] strip speed reducing,ction

— f —— f — **fleje m** - [metal-lam] strip speed reducing,ction

— f — **viscosidad f** - viscosity lowering

— f — **voltaje m** - [electr-distrib] voltage reducing,ction • under-voltage

— f **directa f** - [metal-prod] direct reduction

— f — **de hierro m** - [metal-prod] direct iron, o iron direct, reducing,ction

— f —— **mineral m** - [metal-prod] ore direct, o direct ore, reducing,ction

— f ——— m **de hierro m** - [metal-prod] iron ore direct, o direct iron ore, reducing,ction

— f —— **partida f** - [miner] batch direct reducing,ction

— f ——— f **de pella(s) f** - [miner] pellet batch direct reducing,ction

— f —— f **en terrón(es) m** - [metal-prod] lump batch direct reducing,ction

— f — **según Armco** - [metal-prod] Armco direct reducing,duction

— f **doble** - [mec] double reducing,ction

— f — **planetaria f** - [mec] double planetary, o planetary double, reducing,ction

— f **efectiva** - effective reducing,ction • effective control

— f **elevada** - [mec] high reducing,ction

— f **embridada** - [tub] flanged reducer

— f **en activo m** - [contab] asset(s) decrease

— f —— m **fijo** - [contab] fixed asset(s) decrease

— f — **alabeo m** - [mec] warpage reducing,ction

— f **en amperaje m** - [electr-oper] amperage reducing,duction, o lowering

— f ——— m **para entrada f** - [electr-mot] input current reducing,duction

reducción f **en anchura f** - width reducing,ction

— f — **ángulo m** - angle, reducing,ction, o decrease,sing

— f —— m **de corrimiento m** - [autom-neumát] slip angle decrease,sing

— f — **caliente** - [metal-lam] hot reduction

— f — **capacidad f** - capacity reducing,ction; impairing,rment

— f —— f **física** - [labor] physical impairment

— f —— f **mental** - [labor] mental impairment

— f —— f **portadora** - [constr] bearing power loss

— f —— f **resistente** - [constr] bearing power loss

— f — **capital m** - [fin] capital, loss, o decrease

— f —— m **operativo** - [fin] working capital loss

— f **en carga f** - load reducing,ction

— f — **circulación f** - flow decrease

— f — **consumo m** - consumption reducing,ction

— f —— m **de combustible m** - fuel conservation

— f —— m **de energía f** - [electr-oper] energy (consumption) reduction

— f — **contaminación f** - [ambient] pollution, reduction, o abatement

— f —— f **de agua m** - [hidr] water pollution, reduction, o abatement

— f — **continuo** - [metal-lam] continuous reducing,ction

— f — **corriente f** - [electr-oper] current reducing,ction

— f —— f **de entrada f** - [electr-oper] input current reducing,ction

— f **en costo m** - cost, reducing,ction, o decrease,sing, o lowering, o shaving

— f —— m **a máximo m** - [ind] cost shaving

— f —— m **de comisión(es) f** - [fin] commission(s) cost(s), reducing,ction, o decrease

— f —— m —— f **no amortizada(s)** - [fin] unamortized commission(s) cost(s) decrease

— f —— m **para operación f** - operating costs reduction

— f —— m —— f **para soldadora f** - [sold] welder operating cost(s) reducing,ction

— f —— m **para reemplazo m** - replacement cost(s) reducing,ction

— f — **deformación f** - [mec] deflection, constraint,ning

— f — **desecho(s) m** - [ind] waste reduction

— f — **desgaste m** - [ind] wear reducing,ction

— f —— m **para banda f (para rodamiento m)** - [autom-neumát] tread wear reducing,ction

— f —— m **para motor m** - [comb.int] engine wear reducing,ction

— f — **deslizamiento m** - [autom-neumát] hydroplaning reducing,ction

— f —— **sobre agua m** - [autom-neumát] hydroplaning reducing,ction

— f — **habilidad f física** - [labor] physical impairment

— f —— f **mental** - [labor] mental impairment

— f — **deslumbramiento m** - [vial] glare reducing,ction

— f — **desperdicio(s) f** - [ind] waste(s) reducing,ction

— f — **diámetro m** - diameter reducing,ction • [tub] size reducing,ction

— f —— m **de cable m** - [cabl] cable, o rope, diameter reducing,ction

— f —— m **de extremo m** - [cabl] end diameter reducing,ction

— f —— — m **de cable m (de alambre m)** - [cabl] wire rope end diameter reducing,ction

— f — **eje m trasero** - [autom] rear axle reducing,ction

— f **en emisión(es) f** - [petról] emission(s) reducing,ction

— f —— f **de hidrocarburo(s) m** [petról] hydrocarbon emission reducing,ction

reducción f en emisión(es) f de hidrocarburo(s)
m a atmósfera f - [petról] hydrocarbon(s)
atmosphere emission(s) reducing,ction
— f — endurecimiento m - hardening decrease
— f — erogación(es) f—[fin] spending cut-back
— f — escoria f - [metal-prod] slag reduction
— f — fatiga f - fatigue reducing,ction
— f — f de soldador m - [sold] welder's, o
operator's, fatigue reducing,ction
— f — formación f de arco m - [sold] arcing
minimizing,zation
— f — frío m - [metal-lam] cold reducing,ction
— f — habilidad f - [labor] impairing,rment
— f — hélice f - [mec] helical reducing,ction
— f — doble - [mec] double helical, o helical double, reducing,ction
— f — f triple - [mec] helical triple, o
triple helical, reducing,ction
— f — horno m con cuba f - [metal-prod] shaft
furnace reducing,ction
— f — ingreso(s) m - [fin] income reduction
— f — m neto(s) f—[fin] net income reduction
— f — impacto(s) m - [mec] impact(s), reducing, o reduction
— f — impuesto(s) m - [fisc] tax, reducing, o
reduction
— f — f federal(es) - [fisc] federal
tax(es), decrease, o reducing,ction
— f — m sobre, rédito(s), o renta f -
[fisc] federal income tax, decrease, o reducing,ction
— f — m —, — m o renta f, consolidado
m - [fisc] consolidated federal income tax
reducing,duction
— f — m nacional - [fisc] federal, o national, tax, decrease, o reducing,ction
— f — intensidad f - intensity reducing,ction
— f — f de amperaje m - [electr-oper] current, o amperage, intensity reducing,ction
— f — f — m de entrada f - [electr-mot]
input current reducing,ction
— f — corriente f - [electr-mot] current (intensity) reducing,ction
— f — f — f de entrada f - [electr-mot]
input current (intensity) reducing,ction
— f — juego m - slack taking up
— f — laminación f - [metal-lam] mill draft
— f — laminador m - [metal-lam] mill draft
— f — marcha f - [ind] slowing (down)
— f — óxido m - [metal] oxide reducing,ction
— f — m de hierro m - [metal-prod] iron oxide reducing,ction
— f — m — m con óxido de carbono m e
hidrógeno - [metal-prod] carbon (mon)oxide
plus hydrogen iron oxide reducing,ction
— f — pasada f - [metal-lam] pass draft
— f — f por laminador m - [metal-lam] mill
pass draft
— m — patrimonio m (social) - [fin] shareholder's equity reducing,ction
— f — pérdida(s) f - [segurid] loss control
— f — porcentaje m de reducción f - reduction
percentage, reduction, o drop, o decrease
— f — porosidad f - porosity reducing,ction
— f — posibilidad f - possibility, o chance,
reducing,ction, o lessening
— f — f de deslizamiento m (sobre agua n) -
[autom-neumát] hydroplaning possibility reducing,ction
— f — f de separación f - separation possibility reducing,ction
— f — precio(s) f - [com] price cutting
— f — presión f - pressure, reducing,ction, o
decrease,sing
— f — f de combustible m - [combust] fuel
pressure reducing,ction
— f — f en red f - [ind] system pressure,
lowering, o decrease,sing
— f — producción f - [ind] production, drop, o
fall, o reducing,ction
— f — provisión f - supply(ing) decrease
— f — regulación f - [ind] setting decrease
— f — f para caudal m - flow setting, lowering, o decrease

reducción f en resistencia f - [mec] resistance
reducing,ction
— f — f eléctrica f - [electr-oper] electric resistance, o resistivity, decrease
— f — f efectiva - [electr-oper] resistivity decrease,sing
— f — riesgo m—[segurid] risk reducing,ction
— f — m de incendio(s) m - [segurid] fire,
risk, o hazard, reducing,ction
— f — m — m inundación f - [hidr] flood(ing)
risk, o hazard, reducing,ction
— f — ritmo m - [mec] speed reducing,ction
— f — m — movimiento m - [mec] movement
speed, reducing,ction, o decrease,sing
— f — m — m de estrangulador m - [mec]
choke movement speed, reducing,ction, o decrease,sing
— f — saturación f - saturation reducing,ction
— f — f de relleno m - [constr] (back)fill
saturation reducing,ction
— f — sección f - [constr] section reduction
— f — f hidráulica - [hidr] waterway reducing,ction
— f — f transversal (de conducto m)—[hidr]
waterway, area, o crossection, reducing,ction
— f — seguridad f - [segurid] safety, reducing,ction, o decrease,sing
— f — f en general - [segurid] overall
safety, reducing,tion, o decrease,sing
— f — separación f - separation reducing,ction
— f — serpenteo m - [autom-neumát] squirm(ing)
reducing,ction
— f — severidad f - [segurid] severity reducing,ction
— f — f de accidente(s) m - [segurid] accident severity reducing,ction
— f — tamaño m - size reducing,ction
— f — temperatura f - temperature, lowering, o
reducing,ction, o decrease,sing • cooling
down
— f — f de horno m - [ind] furnace temperature, reducing,ction, o lowering
— f — f — motor m - [comb.int] engine temperature, reducing,ction, o decrease,sing •
[comb.int] motor temperature, reducing,ction,
o decrease,sing
— f — tiempo m - time reducing,ction
— f — m de detención f - [ind] downtime
reducing,duction
— f — m — inactividad f - [ind] downtime
reducing,ction
— f — m inactivo - [ind] downtime reducing,ction
— f — trabajo m - work reducing,ction
— f — tren m laminador—[metal-lam] mill draft
— f — utilidad f - [contab] earning(s), reducing,ction, o decrease,sing
— f — f retenida - [contab] retained earn-
ing(s) decrease
— f — velocidad f - velocity, o speed, reducing,ction, o decrease,sing • speed dropping • [autom] down(shifting) • [deport]
slowing (down)
— f — f de agua m - [hidr] water velocity
reducing,ction, o slowing (down)
— f — f — carga f - [fís] load(ing), velocity, o speed, decrease,sing
— f — f — motor m - [electr-mot] motor
speed, reducing,ction, o decrease,sing •
[comb.int] engine speed, reducing,ction, o
decrease,sing
— f — f — movimiento m - [mec] movement
speed, reducing,ction, o decrease,sing
— f — f — m de estrangulador m - [mec]
choke movement speed decrease,sing
— f — f efectuada - [autom] made down
shift(ing)
— f — f para entrada f - [hidr] inlet, velocity, o speed. reducing,ction, o decrease
— f — f — salida f - [hidr] outlet, speed,
o velocity, reducing,ction, o decrease,sing
— f — versatilidad f - versatility decrease
— f — voltaje m - [electr-oper] voltage, reducing,ction, o cut(ting)

reducción f̱ en voltaje m̱ en circuito m̱ abierto -
[electr-oper] open circuit voltage cut(ting)
— f̱ — **volumen** m̱ - volume reducing,ction
— f̱ — — m̱ **de escoria** f̱ - [metal-prod] slag
volume reducing,ction
— f̱ — — m̱ **medio**—average volume reduction
— f̱ **endotérmica** - [metal] endothermic reduction
— f̱ **excéntrica** - [tub] eccentric reducing
— f̱ — **embridada** - [tub] eccentric flanged reduction
— f̱ **gaseosa** - [metal-prod] gas(eous) reduction
— f̱ **con cama f̱ estática** - [metal-prod] static bed gas reduction
— f̱ **hasta tamaño** m **apropiado** - reducing,ction
to @ proper size
— f̱ **helicoidal** - [mec] helical reducing,ction
— f̱ — **doble** - [mec] helical double, o̱ double
helical, reducing,ction
— f̱ — **triple** - [mec] helical triple, o̱ triple
helical, reducing,ction
— f̱ **inicial** - [metal-prod] initial reduction
— f̱ **leve** - slight reducing,ction
— f̱ **macho y hembra** - [mec] box and pin substitute
— f̱ **marcada** - marked, o̱ dramatic, reduction
— f̱ — **de diámetro** m—marked diameter reduction
— f̱ **media** - average reducing,ction
— f̱ — **de volumen** m̱ - volume average, o̱ average
volume, reducing,ction
— f̱ **normal** - [metal-lam] normal reducing,ction
— f̱ **notoria** - noticeable, reduction, o̱ decrease
— f̱ **para ajuste** m̱ - [mec] bushing
— f̱ — **rotación** f̱ - [grúas] swing reduction
— f̱ — **sedimentación** f̱ - [hidr] silting reduction
— f̱ — **tubería** f̱ - tubing substitute
— f̱ — — f̱ **para bombeo** m̱ - [petról] tubing
substitute
— f̱ — — f̱ — **entubación** f̱ - [petról] casing
substitute
— f̱ **parcial** - partial reducing,ction
— f̱ — **de material** m̱ - partial material reducing,ction
— f̱ **pequeña** - small, reducing,ction, o̱ decrease
— f̱ — **en provisión** f̱ - small supply decrease
— f̱ **planetaria** - [mec] planetary reduction
— f̱ — **doble** - [mec] planetary double, o̱ double
planetary, reducing,ction
— f̱ — — . . . **etapa(s)** f̱ - [mec] . . .-stage
planetary reduction
— f̱ **por estiramiento** m̱ - [tub] stretch(ing)
reducing,ction
— f̱ — — m̱ **en caliente** - [metal-trat] hot
stretch(ing) reducing,ction
— f̱ — m̱ — **soldada** - [tub] welded hot
stretch(ing) reducing,ction
— f̱ — m̱ — f̱ — — **por resistencia** f̱
eléctrica - [tub] electric resistance welded
hot-stretch reducing,ction
— f̱ — m̱ — **continuo** - [metal-lam] continuous stretch(ing) reducing,ction
— f̱ — **laminación** f̱ - [metal-lam] mill reducing,ction
— f̱ **previsible** - [metal-lam] foreseeable reducing,ction, o̱ decrease
— f̱ **reducida** - [metal-lam] low, o̱ limited, o̱
small, reducing,ction
— f̱ **seria** - serious, o̱ large, reducing,ction
— f̱ **simple** - [mec] single reduction
— f̱ **temporaria** - temporary, reduction, o̱ decrease
— f̱ — f̱ **en capacidad** f̱ - [labor] temporary impairment
— f̱ — — **habilidad** f̱ - [labor] temporary impairment
— f̱ — — — f̱ **física** - [labor] temporary physical impairment
— f̱ — — — f̱ **mental** - [labor] temporary mental impairment
— f̱ **total en cada caja f̱ de tren** m—[metal-lam]
drafting practice
— f̱ **triple** - [metal-lam] triple reduction
— f̱ **y regulación** f̱—[mec] reduction and control

reducción f̱ y regulación f̱ de presión f̱ - pressure reduction and control
reducidísimo,ma a̱ - very reduced; minute
reducido,da a̱ - . . .; decreased; lessened; low;
slowed (down); confined • [ind] bled off •
[metal-prod] plugged (tuyere) • palliated;
constrained • impaired • fallen • small •
[mec] lopped off
— a̱ **a partículas** f̱ - particulated
— a̱ **adicionalmente** - reduced, further, o̱ additionally
— a̱ **directamente** - reduced directly
— a̱ **efectivamente** - controlled effectively
— a̱ **en continuo** - [metal-lam] reduced continuously
— a̱ **en frío** - [metal-fabr] cold reduced
— a̱ **hasta tamaño** m̱ **apropiado** - reduced to @
proper size
— a̱ **levemente** - reduced slightly
— a̱ **marcadamente** - reduced dramatically
— a̱ **mediante esmerilado** m̱ - [mec] reduced by
grinding
— a̱ — **recalcadura** f̱ - [metal-fabr] reduced by swaging
— a̱ **parcialmente** - reduced partially
— a̱ **por estiramiento** m̱ - [tub] stretch reduced
— a̱ — — m̱ **en caliente** - [metal-trat] hot
stretch(ed) reduced
— a̱ — — m̱ — **continuo** - [tub] continuously
stretch reduced
— a̱ — — m̱ — — **para industria** f **petrolera** -
[tub] continuous(ly) stretch(ed) reduced oil
country
— a̱ **seriamente** - seriously reduced
reducir v̱ - . . .; to curb; to cut, down, o̱ back;
to curtail; to minimize • to alleviate • to
damp(en) • to impair • to constrain; to slow
(down) • to discourage • to palliate • to fall
• [ind] to bleed off • [mec] to lop off
— v̱ **a mínimo** m̱ - to minimize; to, hold, o̱ reduce, to @ minimum
— v̱ — — m̱ **conservación** f̱ - [ind] to minimize
@ maintenance
— v̱ — — m̱ **costo** m̱ - to minimize @ cost(s)
— v̱ — — m̱ — m̱ **para conservación** f̱ - [ind] to
minimize @ maintenance cost(s)
— v̱ — — m̱ — m̱ **mantenimiento** m̱ - [ind] to
minimize @ maintenance cost(s)
— v̱ — — m̱ — m̱ **de tiempo** m̱ **inactivo** - [ind] to
minimize @ downtime cost(s)
— v̱ — — m̱ **derrame** m̱ - [sold] to minimize @
spilling
— v̱ — — m̱ **desgaste** m—[mec] to minimize @ wear
— v̱ — — m̱ **destellado** m̱ - [sold] to minimize @
flash through
— v̱ — — m̱ **inversión** f̱ **en material(es)** m̱ -
[ind] to minimize @ inventory (costs)
— v̱ — — m̱ **mantenimiento** m̱ - [ind] to minimize
@ maintenance
— v̱ — — m̱ **parada(s)** f̱ - [ind] to minimize @
downtime
— v̱ — — m̱ **penetración** f̱ - [sold] to minimize
@ penetration
— v̱ — — m̱ **socavación** f̱ - [sold] to minimize @
undercut(ting)
— v̱ — — m̱ **tiempo** m̱ **de detención(es)** f̱ - [mec]
to minimize @ downtime
— v̱ — **partícula(s)** f̱ - to particulate
— v̱ **adaptabilidad** f̱ - to decrease @, adaptabi-
lity, o̱ versatility
— v̱ **adicionalmente** - to reduce, additionally, o̱
further
— v̱ **agrietamiento** m—[sold] to reduce @ cracking
— v̱ **agua** m̱ - [hidr] to reduce @ water
— v̱ — m̱ **en entrada** f̱ - [hidr] to reduce @ headwater
— v̱ **alabeo** m̱ - [mec] to reduce @ warpage,ping
— v̱ **amperaje** m̱ - [electr-oper] to reduce @ amperage • [sold] to, reduce, o̱ lower, @ current
— v̱ **ángulo** m̱ - to, reduce, o̱ decrease, @ angle
— v̱ — m̱ **para corrimiento** m̱ - [autom-neumát]
to, reduce, o̱ decrease, @ slip angle
— v̱ **aportación** f̱ - [ind] to reduce @ input

reducir v @ mínimo m - to minimize
— v — m aportación v - to minimize @ input
— v — de calor m - [sold] to minimize @ heat input
— v área m - to reduce @ area
— v — m transversal - to reduce @ cross area
— v — m — de alcantarilla f - [hidr] to reduce @ culvert area
— v calentamiento m—[sold] to reduce @ heating
— v cantidad f—to reduce @, quantity, o amount
— v — f de arco m - [sold] to reduce @ arc, quantity, o amount
— v capacidad f - to reduce @ capacity; to impair
— v capital m - [fin] to, reduce, o decrease, @ capital
— v — m operativo - [fin] to, reduce, o decrease, @ working capital
— v capitalización f - [fin] to, reduce, o decrease, @ capitalization
— v carga f - to, reduce, o decrease, o lessen, @ load
— v — f sobre conexión f - [electr-equip] to reduce @ connection load
— v caudal m - to, reduce, o decrease, @ flow
— v ciclo m - to reduce @ cycling
— v circulación f - to decrease @ flow
— v — (líquido) refrigerante - [comb.int] to decrease @ coolant (liquid) flow
— v coercividad f - to reduce @ coercivity
— v consumption m—to, reduce, o cut @ consumption
— v — m de combustible m - to conserve @ fuel
— v — m — fundente m - [sold] to reduce @ flux consumption
— v — m — a mínimo m - [sold] to minimize @ flux consumption
— v contragolpe m - [mec] to reduce @ backlash
— v costo m - to, reduce, o cut, o lower, o shave, @ cost(s)
— v — m para operación f - to, reduce, o lower, @ operating cost(s)
— v — m — f de soldadora f - [sold] to, reduce, o lower, @ welder operating cost(s)
— v choque(s) m - [electr-oper] to reduce @ shock • [transp] to reduce @ collision(s)
— v deformación f - [mec] to, reduce, o constrain, @, deflection, o deformation
— v desgaste m - [mec] to, reduce, o decrease, @ wear
— v — m de banda f - [mec] to reduce @ band wear
— v — m — — f para rodamiento m - [aitp,-neumát] to reduce @ tread wear
— v — m de motor m - [electr-mot] to reduce @ motor wear • [comb.int] to reduce @ engine wear
— v deslizamiento m - [autom-neumát] to reduce @ hydroplaning
— v — m sobre agua m - [autom-neumátl to, reduce, o decrease, @ hydroplaning
— v desperdicio(s) m - to reduce @ waste(s)
— v diámetro m - to, reduce, o decrease, @, diameter, o size
— v — m de cable m - [cabl] to reduce @ rope diameter
— v directamente - to reduce directly
— v — mineral m - [miner] to directly reduce @ ore
— v — — m de hierro m - [metal-prod] to directly reduce @ iron ore
— v dispersión f - to reduce @ dispersion
— v — f de aluminio m - [metal-prod] to reduce @ aluminum dispersion
— v distorsión f - [sold] to, reduce, o minimize, @ distortion
— v dureza f - to reduce @ hardness
— v efectivamente - to, reduce, o control, effectively
— v emisión f - to reduce @ emission(s)
— v — — hidrocarburo(s) m - [petról] to reduce @ hydrocarbon(s) emission(s)

reducir v emisión f de hidrocarburo(s) m a atmósfera f - [petról] to reduce @ hydrocarbon(s) atmosphere emission(s)
— v en continuo - [metal-lam] to reduce continuously
— v — frío - [metal-lam] to cold reduce
— v erosión f - [hidr] to, reduce, o control, @ erosion
— v escoria f - [metal-prod] to reduce @ slag
— v facilidad f para traslado m - [mec] to, reduce, o decrease, @ portability
— v fatiga f - [segurid] to reduce @ fatigue
— v — f de soldador m - [sold] to reduce @, weldor, o operator, fatigue
— v filtración f - [hidr] to, reduce, o discourage, @, filtration, o seepage
— v flojedad f - to take up @ slack
— v flujo m - to decrease @ flow
— v gasto(s) m - to reduce @ expense(s)
— v — m para reemplazo m - to reduce @ replacement cost
— v habilidad f - to impair
— v impacto(s) m - [mec] to reduce @ impact(s)
— v incorporación f - [sold] to reduce @ admixture
— v juego m - to take up @ slack
— v — m longitudinal - [mec] to decrease @ end play
— v largo(r) m - to reduce @ length
— v — m de arco m - [sold] to reduce @ arc length
— v levemente - to reduce slightly
— v marcadamente - to reduce significantly
— v marcha v - to slow down
— v mediante esmerilado m - [mec] to reduce by grinding
— v — recalcadura f - [metal-fabr] to reduce by swaging
— v mezcla f de metal(es) m - [sold] to reduce @ admixture
— v mineral m de hierro m - [metal-prod] to reduce @ iron ore
— v mordedura f - [mec] to lose @ pinch
— v notablemente - to slash
— v oxígeno m - [metal-prod] to reduce @ oxygen
— v parcialmente - to reduce partially
— v — material m - to partially reduce @ material
— v participación f - [fin] to reduce @, participation, o share
— v partida f - [miner] to reduce @ batch
— v — f de pella(s) f - [miner] to reduce @ pellet(s) batch
— v peligro m - [segurid] to, reduce, o curtail, @, danger, o hazard
— v — m de derrubio m - [hidr] to reduce @ scour(ing) hazard
— v pella(s) f - [metal-prod] to reduce @ pellet(s)
— v pendiente f - [constr] to reduce @ gradient
— v penetración f - [sold] to, reduce, o minimize, @ penetration
— v — f dentro de metal m de base f - [sold] to minimize @ penetration into @ base metal
— v pérdida(s) f—[segurid] to, reduce, o control, @ loss(es)
— v penetración f - [sold] to reduce @, penetration, o burn-through
— v peso m - to reduce @ weight; to lighten
— v por estiramiento m - [tub] to reduce by stretching
— v — — m en caliente - [metal-trat] to hot stretch reduce
— v — — m — continuo - [metal-lam] to stretch reduce continuously
— v porosidad f - [metal-prod] to, reduce, o decrease, @ porosity
— v posibilidad f - to, reduce, o lessen, @, possibility, o chance
— v — f de deslizamiento m (sobre agua m) - [autom-neumát] to reduce @ hydroplaning possibility
— f — f de separación f - to reduce @ separa-

tion possibility
— v **precipitación** f - [metal-prod] to reduce @ precipitation
— v — f **de carburo(s)** m - [metal-prod] to reduce @ carbide(s) precipitation
— v **presión** f - to, reduce, o decrease, @ pressure • [ind] to bleed off @ pressure
— v — f **de combustible** m - [combust] to reduce @ fuel pressure
— v **remanso** m - [hidr] to decrease @ backwater
— v **rendimiento** m - [ind] to reduce @ yield • to lose @ performance
— v **resistencia** f - to decrease @ strength
— v — f **a rotura** f - [mec] to decrease @ breaking strength
— v **riesgo** m - to reduce @, risk, o hazard
— v **riesgo** m **de incendio** m - [segurid] to reduce @ fire hazard
— v — m **de inundación** f - [hidr] to, reduce, o alleviate, @ flood(ing) hazard
— v **ritmo** m - [mec] to reduce @ speed
— v — m — **movimiento** m - [mec] to decrease @ movement speed
— v — m — — m **de estrangulador** m - [mec] to decrease @ choke movemen speed
— v **ruido** m - [comb.int] to reduce @ noise
— v — m — **motor** m - [comb.int] to reduce @ engine noise
— v **salpicadura(s)** f - [sold] to reduce @ spatter
— v **saturación** f - to reduce @ saturation
— v — f **de relleno** m - [constr] to reduce @ backfill saturation
— v **sección** f **hidráulica** - [hidr] to reduce @ waterway
— v **sedimentación** f—[hidr] to reduce @ silting
— v **seguridad** f—[segurid] to decrease @ safety
— v — f **en general** - [segurid] to, reduce, o decrease, @ overall safety
— v **separación** f - to reduce @ separation • [mec] to decrease @ gap
— v **seriamente** - to reduce considerably
— v **serpenteo** m - [autom-neumát] to reduce @ squirm(ing)
— v **severidad** f—[segurid] to reduce @ severity
— v — f **de accidente** m - [segurid] to reduce @ accident severity
— v **soplado** m - [metal-prod] to, reduce, o cut back, @, blast, o wind
— m **soplo** m - [sold] to reduce @ blow
— v — m **magnético** - [sold] to reduce @ magnetic blow
— v — m — **de arco** m - [sold] to reduce @ (magnetic) arc blow
— v **tamaño** m - to, reduce, o decrease, @ size
— v **temperatura** f - to, reduce, o decrease, o lower, @ temperature • to cool down
— v — f **en horno** m - [metal-prod] to lower @ furnace temperature
— v — f **de motor** m - [electr-mot] to, reduce, o decrease, @ motor temperature • [comb.int] to, reduce, o decrease, @ engine temperature
— v — f **de soplado** m - [metal-prod] to lower @ blast temperature
— v **tendencia** f - to reduce @ tendency
— v — f **a agrietamiento** m - [sold] to reduce @ cracking tendency
— d — f **a distorsión** f - [sold] to reduce @ distortion tendency
— v **terrón** m - [metal-prod] to reduce @ lump
— v **tiempo** m - to, reduce, o shorten, @ time
— v — m **de inactividad** f - [ind] to, reduce, o shorten, o minimize, @ downtime
— v — **parada** f - [ind] to, reduce, o shorten, o minimize, @ downtime
— v — m — **soldadura** f - [sold] to, reduce, o shorten, o minimize, @, arc, o welding, time
— v — m **inactivo** - [ind] to reduce @ downtime
— v **trabajo** m - to reduce @ work
— v **velocidad** f - to, reduce, o decrease, o drop, @, speed, o velocity • to slow down • [autom-mec] to downshift
— f — f **a marcha** f **lenta** - [autom-mec] to re-

duce to @ idle speed
reducir v **velocidad** f **de agua** m - [hidr] to reduce @ water velocity
— v — f — **entrada** f - [hidr] to reduce @ inlet velocity
— v — f — **motor** m - [electr-mot] to reduce @ motor speed • [comb.int] to reduce @ engine speed
— v — f — **movimiento** m - [mec] to reduce @ movement speed
— v — f — — m **de estrangulador** m - [mec] to decrease @ choke movement speed
— v — — **salida** f - [hidr] to reduce @ outlet velocity
— v **viento** m - [metal-prod] to, reduce. o cut back, @, blast, o wind
— v **viscosidad** f - to, reduce, o lower, o decrease, @ viscosity
— v **voltaje** m - [electr-oper] to, reduce, o decrease, o lower, @ voltage
— v — m **en circuito** m **abierto** - [electr-oper] to, reduce, o cut, @ open circuit voltage
— v **volumen** m - to reduce @ volume
— v — **de escoria** f - [metal-prod] to reduce @ slag volume
reductibilidad* f - reduc(t)ibility* m
— f **alta** - [metal-prod] high reduc(t)ibility
— f — **de mineral** m - [metal-prod] high ore reduc(t)ibility
— f **buena** - [metal-prod] good reduc(t)ibility
— f **de mineral** m - [metal-prod] ore reduc(t)ibility
— f — — m **de hierro** m - [metal-prod] iron ore reduc(t)ibility
— f **de sínter** m - [metal-prod] sinter reduc(t)ibility
— f **pobre** - [metal-prod] poor reduc(t)ibility
reductor m - reducer
— m **compatible** - [quím] compatible reducer
— m — **preferentemente soluble** - [quím] preferentially, o preferently, compatible reducer
— m **común** - common, o regular, o standard, reducer
— m **con engranaje(s)** m - [mec] gear reducer
— m — — m **helicoidal** - [mec] worm gear reducer
— m — — m **sin fin** - [mec] worm gear reducer
— m **con . . . velocidad(es)** f - [mec] . . . speed reducer
— m **concéntrico** - [tub] concentric reducer
— m **cónico** - [mec] cone shaped, o tapered, reducer
— m — **excéntrico** - [petról] eccentric tapered reducer
— m **conocido** - [quím] known reducer
— m **estándar** - [tub] standard reducer
— m **excéntrico** - [mec] eccentric reducer
— m — **embridado** - [tub] excentric flanged reducer
— m **general** - [mec] general, o over-all, reducer, o gear box
— m **para longotera,teadora** f - [metal-prod] pig casting machine general gear box
— m — — **máquina** f **para colar** v - [metal-prod] pig casting machine general gear box
— m **para amperaje** m - [electr-mot] amperage reducer • [sold] current reducer
— m — — m **para entrada** f - [electr-mot] input current, reducer, o reducing starter; G A C starter (box)
— m — **deformación** f - deformation,ming reducer
— m — **distribuidor** m - [metal-prod] distributor reducer
— m — **émbolo** m - [mec] piston reducer
— m — — m **para cañón** m - [metal-prod] (mud) gun piston reducer
— m — **engranaje(s)** m - [mec] gear reducer
— m — — m **para rotación** f - [grúas] swing gear reducer
— m — **marcha(s)** f - [autom-mec] gear reducer
— m — **mecanismo** - [mec] mechanism reducer
— m — — m **para giro** m - [grúas] rotation, o revolving, mechanism reducer

reductor m para mecanismo m para rotación f - [grúas] rotation mechanism reducer
— m para motor m - [electr-mot] motor reducer • [comb.int] engine reducer
— m — velocidad(es) f - [mec] speed(s) reducer • [autom-mec] gear reducer
— m — — f alineado - [mec] aligned speed reducer
— m — — f alta - [mec] high speed reducer
— m — — f con engranaje m sin fin - [mec] worm gear speed reducer
— m — f — m — — planetario - [mec] planetary gear speed reducer
— m preferentemente soluble - [quím] preferable, o preferentially, solluble reducer
— m reforzado - [mec] heavy duty reducer
— m — para velocidad f alta - [mec] heavy duty high speed reducer
— m regular - [mec] regular reducer
— m según norma f - [tub] standard reducer
— m sólido - [quím] solid reducer
— m soluble - [quím] solluble reducer
reductor,ra a - reducer; reducing
— a para corriente f - current reducing
— a — deformación f - deformation, o deflection, reducing
— a — desecho(s) m - waste reducing
— a — desperdicio(s) m - waste reducing
— a — intensidad f de corriente f - [sold] current, o amperage, (intensity) reducing
— a — presión f - [tub] pressure reducing
— a — residuo(s) m - waste reducing
reductora f - [extrusión de electrodos] reducer
redundado,da a - redounded; added up
redundar v - to redound; to add up
reedificado,da a - rebuilt
reelaboración f - reworking; rehandling; reprocessing; repreparation
— f comercial - [mec] commercial, reworking, o regeneration
— f de combustible m - [nucl] fuel, rehandling, o reworking
— f — material(es) m - [ind] material(s), reworking, o rehandling, o reprocessing, o recycling
— f en escala f comercial - [mec] commercial scale, reworking, o rehandling, o regenerating,tion
reelaborado,da a - [ind] reworked; rehandled; reprocessed; reprepared
reelaborar v - [ind] to rework; to rehandle; to reprocess; to reprepare
— v combustible m - [nucl] to, rework, o rehandle, @ fuel
Reelite* m - [gruas] Reelite
reembalado,da para almacenamiento m—repackaged for storage,ring
— a para expedición f - [ind] repackaged for shipment,pping
— a — reexpedición f - repackaged for reshipment,pping
reembalaje m para almacenamiento m - repackaging for storage
— m — expedición f - repackaging for shipment
— a — reexpedición f - repackaging for reshipment
reembalar v para almacenamiento m - repackaging for storage,ring
— v — expedición f - to repackage for shipment
— v — reexpedición v - to repackage for reshipment,pping
reembolsable a - . . . .; reimbursable
reembolsado,da a - reimbursed • paid
reembolsar v costo m - to reimburse @ cost
reembolso m - . . . .; reimbursing • payback; paying back; refund(ing)
— n de costo m - cost reimbursement
— m máximo - maximum. reimbursement, o refund
— m mínimo - minimum, reimbursement, o refund
reemisión f - [fin] reissue,uing
— f de acción(es) f - [fin] share(s), o stocks, reissue,uing
reemitir v - [fin] to reissue

reemitir v acción(es) f - [fin] to reissue @, share(s), o stock
reempaquetado* m - repacking
— m — cojinete m - [mec] bearing repacking
reempaquetado,da a - repacked
— a anualmente - [mec] repacked annually
reempaquetador m - repacker
reempaquetador,ra a - repacking
reempaquetadura f - repacking
reempaquetamiento* m - repacking
— m anual - annual repacking
reempaquetar v - to repack
— v anualmente - to repack annually
— v cojinete m - [mec] to repack @ bearing
reemplazable adv - . . . .; changeable; removable
reemplazado,da a - replaced; substituted; superseeded; swapped • change, out, o over; removed in favor of ; taken over
— a completamente - replaced completely
— a con relleno m - [mec] replaced with @ fill
— a — — m apropiado - [constr] replaced with @ suitable fill
— a convenientemente - replaced conveniently
— a económicamente - replaced economically
— a en fábrica - factory replaced
— a en juego(s) - [mec] replaced in @ set(s)
— a expresamente - replaced expressly
— a frecuentemente - replaced frequently
— a inmediatamente - replaced immediately
— a rápidamente - replaced quickly
— a sin cargo m - reeplaced, free(ly), o at no cost
— a totalmente - replaced totally
reemplazante m - replacement; alternate
reemplazar v - to, replace, o substitute • to superseed • to, change, o take, over; to change out • to swap; to remove in favor of
— v aceite m - [mec] to replace @ oil
— v — m en cárter m - [comb.int] to replace @ crankcase oil
— v — m — — m de motor m - [comb.int] to, replace, o change, @ engine crankcase oil
— v — m hidráulico - [comb.int] to, change, o replace, @ hydraulic oil
— v agua m - [hidr] to replace @ water
— v apoyo m - [mec] to replace @ pad
— v árbol m - [agric] to replace @ tree - [mec] to replace @ shaft
— v armadura f - [mec] to replace @ armature
— v aro m - [mec] to replace @ ring • [comb.int] to re-ring
— v — m para émbolo m - [comb.int] to re-ring
— v — m con resorte m - [mec] to replace @ snap ring
— v — m para cambiador m - [mec] to replace @ shifter ring
— v — m desplazador m - [mec] to replace @ shifter ring
— v — m — mecanismo m para cambop(s) m - [mec] to replace @ shifter ring
— v atomizador m - [comb.int] to replace @ atomizer
— f barra f - [mec] to replace @ bar
— v — f para perforación f (conformada) - [petról] to replace @ kelly
— v bobina f - [electr-instal] to replace @ coil
— v bomba f - [bombas] to replace @ pump
— v — f para inyección f - [comb.int] to replace @ injection pump
— v bombilla f - [electr-ilum] to replace @ bulb
— v — f para luz f - [electr-ilum] to replace @, light, o lamp, bulb
— v bujía f - [comb.int] to replace @ spark plug
— v cabeza f - [mec] to replace @ head
— v — f para inyección f - [petról] to replace @ swivel
— v cadena f - [mec] to replace @ chain
— v caja f - [mec] to replace @ housing
— v cambiador m - [mec] to replace @ shifter
— v camisa f - [mec] to, replace, o change(out), @, liner,ning
— f carcasa f - [mec] to replace @ housing
— v cartucho m - [mec] to replace @ cartridge

reemplazar v cartucho m para filtro m - [mec] to replace @ filter cartridge
— v — m — — m hidráulico - [mec] to replace @ hydraulic filter cartridge
— v cilindro m - [mec] to replace @ cylinder
— v — m para éter m - [comb.int] to replace @ ether cylinder
— v cojinete m - [mec] to replace @ bearing
— v completamente - to replace completely
— v componente m - tp replace @ component
— v compresor m - [ind] to replace @ compressor
— v con relleno m - to replace with @ fill
— v — — m apropiado - [constr] to replace with @ suitable fill
— v conductor m - [electr-instal] to replace @ lead
— v conexión f - [electr-instal] to replace @, connection, o wiring
— v conmutador m - [electr-instal] to replace @ switch
— v — m defectuoso - [electr-pper] to replace @ faulty switch
— v contacto m - [electr-instal] to replace @ contact
— v — m para relé m - [electr-instal] to replace @ relay contact
— v — m — — m piloto - [electr-instal] to replace @ pilot relay contact
— m contador v—[electrón] to replace @ counter
— v — m para frecuencia(s) f - [electrón] to replace @ frequency counter
— v convenientemente - to replace conveniently
— v — empaquetadura f - to conveniently replace @ packing
— v convertidor m - [metal-prod] to replace @ converter
— v correa f - [mec] to replace @ belt
— v — f para impulsión f - [mec] to replace @ drive,ving belt
— v chatarra f—[metal-prod] to replace @ scrap
— v cristal m - to replace @ crystal
— v cubierta f - [constr] to replace @ decking
— v — f de hormigón m - [constr] to replace @ concrete deck(ing)
— v cuerda f - [mec] to replace @ rope
— v chaveta f - [mec] to replace @ key
— v defecto m - [mec] to replace @ defect
— v desconexión f - [electr-instal] to replace @, disconnection, o open
— v desplazador m - [mec] to replace @ shifter
— v detrás de volante m - [autom] to take over (@ wheel)
— v devanado m - [electr-equip] to replace @ coil
— v diagrama m - to, replace, o supersede, @ diagram
— v dibujo m - to replace @ drawing
— v — m de pieza f - to replace @ part drawing
— v diodo m - [electrón] to replace @ diode
— v — m emisor (de luz f) - [electrón] to replace @ (light) emitting diode
— v disco m - [mec] to replace @, disk, o plate
— v distribuidor m - to replace @ distributor • to replace @ divider
— v — n para fuerza f - [autom-mec] to replace @ power divider
— v económicamente - to replace economically
— v electrodo m - [sold] to, replace, o change, @, electrode, o rod, o stick
— v — manual - [sold] to, replace, o change, @, stick, o manual electrode
— v elemento m - to replace @ element • [mec] to replace @, cartridge, o canister
— v — m para filtro m - [mec] to replace @ filter, element, o cartridge
— v émbolo m - [mec] to, replace, o change out, @ piston
— v embrague m - [mec] to replace @ clutch
— v — m (para mecanismo m) para elevación f - [grúas] to replace @ hoist clutch
— v — — (— — m) — rotación f - [grúas] to replace @ swing clutch
— v enjugador m - [mec] to replace @ wiper

reemplazar v empaquetadura f - [mec] to, replace, o change, @ packing
— v en fábrica f - tp replace at @ factory
— v — juego(s) m - to replace in sets
— v engranaje m - [mec] to replace @ gear
— v equipo m - [ind] to replace @ equipment
— v — m para rueda f - [autom-mec] to replace @ wheel('s) equipment
— v escobilla f - [electr-mot] to replace @ brush
— v estrangulador m - [mec] to, replace, o change, @ choke
— v estructura f - [constr] to replace @ structure
— v — f existente - [constr] to replace @ existing structure
— v expresamente - to expressly replace
— v fiador m - [mec] to replace @ toggle
— v filtro m - [mec] to replace @ filter
— v — m doble - [comb.int] to replace @ dual filter
— v — m para aceite m - [mec] to replace @ oil filter
— v — m — — m hidráulico - [petról] to, replace, o change, @ hydraulic oil filter
— v — m — combustible m - [comb.int] to, replace, o change, @ fuel filter
— v forro m - [mec] to replace @ liner,ning
— v frecuencímetro m - [electrón] to replace @ frquency counter
— v frecuentemente - to replace frequently
— v fusible m - [electr-equip] to, replace, o change, @ fuse
— v — m en circuito m - [electr-oper] to, replace, o change, @ circuit fuse
— v — m — — m para regulación f - [electr-oper] to replace @ control circuit fuse
— v guarnición f - [autom] to replace @ trim
— v — f anular - [mec] to replace @ O-ring
— v — f circular - [mec] to replace @ O-ring
— v guía(dera) f - [mec] to replace @ guide
— f — f para válvula f - [comb.int] to reploace @ valve guide
— v inmediatamente - to replace immediately
— v instrumento m - [instrum] to replace @ instrument
— v — interruptor m - [electr-oper] to replace @ breaker
— v — m automático - [electr-instal] to replace @ contactor
— v leva f - [mec] to replace @ cam
— v línea f - to replace @ line
— v — f para filtro m - [comb.int] to replace @ filter line
— v material(es) m - to replace @ material(s)
— v mecanismo m - [mec] to replace @ mechanism
— v — m para cambio(s) m - [mec] to replace @ shifter
— v montaje m - [mec] to replace @ mounting
— v motor m - [electr-mot] to replace @ motor • [comb.int] to replace @ engine
— v — m para ventilador m - [ventil] to replace @ fan motor
— v neumático m - [autom-neumát] to, replace, o change, o substitute, @ tire
— v pavimento m - [constr] to replace @ pavement
— v pieza f - [mec] to replace @ part
— v — f conexa - [mec] to, replace, o change, @, related, o connected, part
— v — f defectuosa - [mec] to replace @ defective part
— v — f (des)gastada - [mec] to replace @ worn part
— v — f para cojinete m - [mec] to replace @ bearing part
— v platino m - [comb.int] to replace @ point
— v — m para rotor m - [comb.int] to replace @ breaker point
— v plato m - [mec] to replace @, disk, o plate
— v — m para embrague m - [mec] to replace @ clutch plate
— v puente m - [constr] to replace @ bridge

reemplazar v puente m carretero - [vial] to replace @ highway bridge
— v — m departamental - [constr] to replace @ county bridge
— v rápidamente - to replace quickly
— v red f - to replace @ system
— v — f para servicio m público - [constr] to replace @ utility
— v relé m - [electr-equip] to replace @ relay
— v — m piloto - [electr-instal] to, replace, o change, @ pilot relay
— v remache m - [mec] to replace @ rivet
— v resorte m - [mec] to, replace, o change, @ spring
— v — m para escobilla f - [electr-mot] to, replace, o change, @ brush spring
— v revestimiento m - [mec] to replace @ lining
— v rueda f - [mec] to replace @ wheel
— v — f — dentada - [mec] to, replace, o change, @ sprocket
— v — f — loca - [mec] to replace @ idler sprocket
— v — f loca - [mec] to replace @ idler
— v sello m - to replace @ seal
— v — m contra polvo m - [mec] to replace @ dust seal
— v — m para enjugador m - [mec] to replace @ wiper seal
— v semieje m - [autom-mec] to replace @ axle shaft
— v símbolo m - to replace @ symbol
— v sin cargo m - to replace, free(ly), o at no cost
— v sincronizador m - [instrum] to replace @ timer
— v — m para purga f - [Combust] to replace @ purge timer
— v solenoide m - [electrón] to replace @ solenoid
— v — m para detención f - [comb.int] to replace @ shut-down solenoid
— v — — f de combustible m - [comb.int] to replace @ fuel shut-down solenoid
— v suelo m - [suelos] to replace @ soil
— v — m para cimentación f - [constr] to replace @ foundation soil
— v sujetador m - [mec] to replace @ cam
— v tablero m - to replace @ board
— v — con circuito m - [electrón] to replace @ circuit board
— v — m — m, estampado, o impreso - to replace @ printed circuit board
— v tamaño m—to, replace, o substitute, @ size
— v — m de neumático m - [autom-neumát] to, replace, o substitute, @ tire size
— v tapa f - [mec] to replace @ cover
— v tapón m - [mec] to replace @ plug
— v termóstato m - [instrum] to replace @ thermostat
— v tomacorriente m - [electr-instal] to replace @ receptacle
— v totalmente - to replace totally
— v trampa f - [mec] to replace @ trap
— v transformador m - [electr-equip] to replace @ transformer
— v transmisión f - [mec] to replace @ transmission
— f tubería f - [tub] to replace @, tube,bing, o pipe,ping, o line
— v — para agua m - [hidr] to replace @ water pipe • [petrol] to replace @ wash pipe
— v tubo m - [tub] to replace @, tube, o pipe
— v — a atomizador m - [comb.int] to replace @, tube, o pipe, to @ atomizer
— v — m — válvula f - [tub] to replace @ tube to @ valve
— v varilla f - [mec] to, replace, o change, @ rod
— v ventilador m - [ventil] to replace @ fan
— v viaducto m - [constr] to replace @ trestle
— v — m ferroviario - [constr-f.c.] to replace @ railway trestle
reemplazo m - . . . .; replacing; substituting •

removal,ving • remedy; change,ging out • swapping • superseeding
reemplazo m apropiado - [mec] proper replacement
— m — para neumático m - [autom-neumát] proper tire replacement
— m completo - complete, replacement, o change
— m — de juego m - [mec] set complete, replacement, o change
— m — — — m de malla(s) f - [mec] mat set complete, replacement, o change
— m — — — — f para criba f - [mec] screen mat set complete, replacement, o change
— m con relleno m - [constr] fill replacement
— m — — m apropiado - [constr] suitable fill replacement
— m conveniente - convenient, replacement, o change
— m — para empaquetadura f - [mec] convenient packing replacement,cing
— m de accesorio m - [mec] accessory replacement
— m — — m para criba f - [metal-prod] screen accessory replacement
— m — — — m — — f para sínter m - [metal-prod] sinter screen accessory change,ging
— m — — — — f — sinterización f - [miner] sinter(ing) screen accessory change
— m — — — — f — — f en caliente—[miner] hot sinter(ing) screen accessory change,ging
— m — — — m para vagoneta f - [ind] car accessory, change, o replacement
— m — — — — f para montacarga(s) f - [metal-prod] skip car accessory change
— m — — — m para zaranda f - [metal-prod] screen accessory change
— m — — — — f para sínter m - [metal-prod] sinter screen accessory change
— m — aceite m - [mec] oil replacement,cing
— m — en cárter m - [comb.int] crankcase oil, replacement, o change
— m — — m — — m de motor m - [comb.int] engine crankcase oil, replacing, o changing
— m — — m hidráulico - [petról] hydraulic oil, replacement, o change,ging
— m — agua m water, replacement,cing, o change
— m — anillo m - [mec] ring, replacement, o change,ging
— m — — m para desgaste m - [mec] wear ring, replacement, o change,ging
— m — — m — — m para vástago m - [metal-prod] rod wear ring, replacement, o change
— m — — m — — — m para campana f - [metal-prod] bell rod wear ring change,ging
— m — — m — — — — f grande - [metal-prod] large bell rod wear ring change
— m — — m — — — — f pequeña - [metal-prod] small bell rod wear ring change
— m — — m — válvula f - [metal-prod] valve ring, replacement, o change,ging
— m — — m — — para viento m (caliente) - [metal-prod] (hot) blast valve ring, replacement, o change,ging
— m — — m — vástago m - [metal-prod] rod ring change,ging
— m — — m — — m para campana f - [metal-prod] bell rod ring change,ging
— m — — m — — — f grande - [metal-prod] large bell rod ring change,ging
— m — — m — — — — f pequeña - [metal-prod] small bell rod ring change(ging)
— m — apoyo m - [mec] pad replacing, o change
— m — árbol m - [botan] tree replacing • [mec] shaft, replacing, o change
— m — armadura f - [mec] armature replacing
— m — aro m - [mec] ring replacement
— m — — m desplazador - [mec] shifter ring replacement
— m — — m para émbolo m - [comb.int] re-ringing
— m — — m — mecanismo m para cambio(s) m - [mec] shifter ring replacement
— m — aspirador m - [metal-prod] (extractor) fan change,ging
— m — — m para sínter m - [metal-prod] sinter

sinter fan, change,ging, o replacement,cing
reemplazo m **de aspirador** m **principal** - [mec]
main (extractor) fan change,ging
— m — **atomizador** m - [comb.int] atomizer replacing,cement
— m — **barra** f - [mec] bar replacement,cing
— m — — f **para perforación** f - [petról] drilling bar relacement,cing
— m — — f — — f **conformada** - [petról]
kelly replacement
— m — — f **para zapata** f **(de oruga** f**)** -
[constr] grouser bar replacement,cing
— m — — f — — f — — f **para tractor** m -
[constr] tractor grouser bar replacement
— m — **bobina** f - [electr-equip] coil replacement,cing
— m — — f **para reactor** m - [electr-equip] reactor coil replacement,cing
— m — **bomba** f - [bombas] pump replacement,cing
— m — — f **para inyección** f - [comb.int] injection pump replacement,cing
— m — **bombilla** f - [electr-ilum] bulb replacement,cing
— m — — f **para luz** f - [electr-ilum] light, o
lamp, bulb replacement,cing
— m — **cabeza** f - [mec] head replacement,cing
— m — — f **para inyección** f - [petról] swivel
replacement,cing
— n — **cable** m - [cabl] cable, replacement, o
change
— m — — m **para campana** f - [metal-prod] bell
cable change,ging
— m — — m — — f **para horno** m—[ind] furnace
bell cable, change,ging, o replacement,cing
— m — — m — — f — — m **alto** - [metal-prod]
blast furnace bell cable replacement,cing
— m — — m — — f **grande** - [metal-prod] large
bell cable, change,ging, o replacement,cing
— m — — m — — f **pequeña** - [metal-prod] small
bell cable, change,ging, o replacement,cing
— m — — m — **grúa** f - [cranes] crane cable,
change,ging, o replacement,cing
— m — — m — **montacarga(s)** m - [metal-prod]
skip cable, change,ging, o replacement,cing
— m — — m — — m **para horno** m - [ind] furnace skip cable, change,ging, o replacement
— m — — m — — m — — m **alto** - [metal-prod]
blast furnace skip cable, change,ging, o replacement,cing
— m — — m — **sonda** f - [metal-prod] stock
rod cable, change,ging, o replacement,cing
— m — — m — **vagoneta** f - [metal-prod]
(skip) car cable, change,ging, o replacement
— m — — m — — f **para montacarga(s)** m -
[metal-prod] skip car cable, change,ging, o
replacement,cing
— m — — m — — f — — m **para horno** m -
[metal-prod] furnace skip car cable, change,
o replacement,cing
— m — — m — — f — — m — — m **alto** m -
[metal-prod] blast furnace skip car cable,
change,ging, o replacement,cing
— m — **cadena** f - [mec] chain replacement,cing
— m — **caja** f - [mec] housing replacement,cing
— m — **camisa** f - [mec] liner, change,ging, o
replacement,cing
— m — **campana** f - [ind] bell, change,ging, o
replacement,cing
— m — — f **para horno** m - [metal-prod] furnace
bell, change,ging, o replacement,cing
— m — — f — — m **alto** - [metal-prod] blast
furnace bell, change,ging, o replacement,cing
— m — **candado** m - [segurid] padlock,
change,ging, o replacement,cing
— m — **canón** m - [metal-prod] mud gun,
change,ging, o replacement,cing
— m — **carrete** m - [mec] reel, change,ging, o
replacement,cing
— m — — m **para alambre** m - [sold] wire reel,
change,ging, o replacement,cing
— m — **carro** m - [ind] car replacement,cing
— m — — m **para montacarga(s)** m - [metal-prod]
skip car, change,ging, o replacement,cing

reemplazo m **de cierre** m - seal replacement,cing
— m — — m **hermético** - [combust] (airtight)
seal replacement,cing
— m — — m — **para gas** m - [combust] gas seal,
change,ging, o replacement,cing
— m — — m — — m **en distribuidor** m - [metal-
-prod] distributor gas seal, change,ging, o
replacement,cing
— m — — m — — m — — m **para horno** m -
[metal-prod] furnace distributor gas seal,
change,ging, o replacement,cing
— m — — m — — m — — m — — m **alto** -
[metal-prod] blast furnace distributor gas
seal, change,ging, o replacement,cing
— m — — m **para enjugador** m - [mec] wiper seal
replacement
— m — **cilindro** m - [mec] cylinder
replacement,cing
— m — — m **para éter** - [comb.int] ether cylinder, change,ging, o replacement,cing
— m — **cinta** f - [mec] belt change,ging
— m — — f **para transportadora** f - [mec] conveyor belt, change,ging, o replacement,cing
— m — **cojinete** m - [mec] bearing, change,ging,
o replacement,cing
— m — **componente** m - component, change,ging, o
replacement,cing
— m — **compresor** m—compressor replacement,cing
— m — **conductor** m - [electr-instal] lead replacement,cing
— m — **conexión(es)** m - connection(s) replacement,cing • [electr-instal] wiring rewiring
— m — **conmutador** m - [electr-instal] switch,
replacement,cing, ó change,ging
— m — — m **defectuoso** - [electr-instal] defective, o faulty, switch, change,ging, o replacement,cing
— m — **contacto** m - [electr-equip] contact replacement,cing
— m — — m **para relé** m - [electr-instal] relay
contact, change,ging, o replacement,cing
— m — **contador** m - [electrón] counter, change,
o replacement,cing
— m — — m **para frecuencia(s)** f - [electrón]
frequency counter replacement,cing
— m — **convertidor** m [metal-prod] converter replacement,cing
— m — **correa** f - [mec] belt, change,ging, o
replacement,cing
— m — — f **para impulsión** f - [mec] drive,ving
belt, change,ging, o replacement,cing
— m — **criba** f - [metal-prod] screen, replacement,cing, o change,ging
— m — — f **para sínter** - [metal-prod] sinter
screen, replacement,cing, o change,ging
— m — — f **para sinterización** f - [miner]
sinter(ing) screen, change,ging, o replacing
— m — **crisol** m - [metal-prod] hearth replacement,cing, o change,ging
— m — — m **de horno** m - [metal-prod] furnace
hearth, replacement,cing, o change,ging
— m — — m — — m **alto** - [metal-prod] blast
furnace hearth, change,ging, o replacement
— m — **cristal** m - crystal replacement,cing
— m — **cubierta** f - [constr-puentes] deck(ing)
replacement,cing
— m — — f **de hormigón** m - [constr-puentes]
concrete deck(ing) replacement,cing
— m — — f **obsoleta** - [constr-puentes] obsolete deck(ing) replacement,cing
— m — **cuerda** f - [comb.int] rope replacement,cing
— m — **cuerpo** m - body, replacement, o change
— m — — m **de válvula** f - [valv] valve body,
change,ging, o replacement,cing
— m — **chapín** m - [metal-prod] bleeder, replacement,cing, o change,ging
— m — — m **para horno** m - [metal-prod] furnace bleeder, change,ging, o replacement,cing
— m — — m — — m **alto** - [metal-prod] blast
furnace bleeder, replacement,cing, o
change,ging

reemplazo m de chatarra f - scrap replacing
— m — defecto m - defect replacement,cing
— m — desconexión f—[electr-instal] open(ing) replacement,cing
— m — desplazador m—[mec] shifter replacement
— m — devanado m - [electr-install] coil re- placement,cing
— m — diagrama m - diagram, replacement, o superseding
— m — dibujo m drawing replacement,cing
— m — diodo m - [electrón] diode replacement
— m — m emisor m - [electrón] emitting diode replacement,cing
— m — — m - de luz f - [electrón] light emitting diode replacement,cing
— m — disco m - [petról] disk replacement
— m — — m para embrague m - [mec] clutch plate replacement,cing
— m — distribuidor m - distributor replacement • [autom-mec] divider replacement,cing
— m — — m para fuerza f - [autom-mec] power divider replacement,cing
— m — — m potencia f - [autom-mec] power divider replacement,cing
— m — eje m - [mec] axle replacement,cing
— m — electrodo m - [sold] electrode, repla- cement,cing, o change,ging
— m — elemento m - element replacement,cing • [segurid] canister replacement,cing
— m — — m defectuoso - [sold] defective elec- trode replacement,cing
— m — — m para filtro m - [mec] filter ele- ment replacement,cing
— m — — m refrigerado m - [combust] cooled part, change,ging, o replacement,cing
— m — émbolo m - [mec] piston, change,ging, o replacement,cing
— m — embrague m - [mec] clutch, change,ging, o replacement,cing
— m — — m para elevación f - [grúas] hoisting clutch replacement
— m — — m rotación f - [grúas] swing cluch replacement,cing
— m — empaquetadura f - [mec] packing, re- placement,cing, o change,ging
— m — — f para cierre m - [mec] seal packing, replacement,cing, o change,ging
— m — — m — — m para distribuidor m - [metal-prod] distribution seal packing, re- placement,cing, o change,ging
— m — — f — — m para horno m - [metal-prod] furnace distributor seal pack- ing, replacement,cing, o change,ging
— m — — — — — m — — m alto - [metal-prod] blast furnace distributor seal packing, replacement,cing, o change,ging
— m — — f — m hermético m - [combust] seal packing, replacement,cing, o change,ging
— m — — f — m para gas m - [combust] gas seal packing, replacement,cing, o change
— m — — f — m — en distribuidor m para horno m alto - [metal-prod] blast fur- nace distributor gas seal packing change,ging
— m — enjugador m - [mec] wiper replacement
— m — equipo m - [ind] equipment, change,ging, o replacement,cing
— m — — m para armadura f - [metal-lam] grinding equipment replacement,cing
— m — — m — amoladura f - [metal-lam] grinding equipment replacement,cing
— m — — m — f de rodillo(s) m - [meta- -lam] roll grinding equipment replacement
— m — — m — granallado m - [metal-lam] shot blasting equipment replacement,cing
— m — — m — — m para rodillo(s) m - [metal- -lam] roll shot blasting equipment re- plaement,cing
— m — — m para lingoteadora f - [metal-prod] pig casting machine equipment replacement
— m — — m — máquina f para colar v - [metal- -prod] pig casting machine equipment, re- placement,cing, o change,ging

reemplazo m de equipo m para rectificación f de rodillo(s) m - [metal-lam] roll grinding equipment replacement,cing
— m — — m para rueda f - [autom-mec] wheel equipment replacement,cing
— m — — m para torneado m - [metal-lam] turning equipment replacement,cing
— m — — m — de rodillo(s) m - [metal- -lam] roll turning equipment replacement
— m de escobilla(s) f - [electr-mot] brush(es) replacement,cing
— m — estrangulador m - [mec] choke, re- placement,cing, o change,ging
— m — estructura f - [constr] structure re- placement,cing, o change,ging
— m — — f existente - [constr] existing structure replacement,cing
— m — filtro m—[mec] filter replacement,cing
— m — — m doble - [comb.int] double filter replacement,cing
— m — — m para aceite m - [lubric] oil fil- ter, replacement,cing, o change,ging
— m — — m — — m hidráulico - [petról] hy- draulic oil filter, replacement,cing, o change,ging
— m — — m — combustible m - [comb.int] fuel filter, replacement,cing, o change,ging
— m — fondo m - [mec] bottom, change,ging, o replacement,cing
— m — — m de crisol m - [metal-prod] hearth bottom, replacement,cing, o change,ging
— m — — m — de horno m - [metal-prod] furnace hearth bottom replacement,cing
— m — — m — — m — — m alto—[metal-prod] blast furnace hearth bottom replacement,cing
— m — forro m - [mec] lining replacement,cing
— m — fusible m - [electr-oper] fuse, re- placement,cing, o change,ging
— m — — m para circuito m - [electr-oper] circuit fuse, replacement, o change,ging
— m — guarnición f - [mec] trim replacement
— m — — f anular - [mec] O-ring replacement
— m — — f circular - [mec] O-ring replacement
— m — guía(dera) f - [mec] guide replacement
— m — — f para válvula f - [comb.int] valve guide replacement,cing
— m — herramienta f - [mec] tool replacement • tool substitute
— m — — f para perforación f - [petról] drilling tool, replacement, o substitute
— m — impulsor m - [mec] impeller replacement
— m — — m para aspirador m - [metal-prod] exhaust(ing) fan impeller, replacement,cing, o change,ging
— m — — m — — m para sínter m—[metal-prod] sinter fan impeller, replacement, o change
— m — — m — — m principal - [mec] main fan impeller, replacement,cing, o change,ging
— m — — m para sínter m - [miner] main sinter fan impeller change,ging
— m — instrumento m - [instrum] instrument replacement,cing
— m — interruptor m - [electr-instal] switch replacement,cing
— n — — m automático - [electr-instal] con- tactor, replacement,cing, o change,ging
— m — juego m - set replacement,cing
— m — — n de malla(s) f - [mec] mat set, re- placement,cing, o change,ging
— m — lenteja f - [válv] body change,ging
— m — — f de válvula f - [válv] valve body, replacement,cing, o change,ging
— m — — f — f para viento m - [metal- -prod] blast valve body, replacement,cing, o change,ging
— m — — f — m caliente - [metal-prod] hot blast valve body, replacement,cing, o change,ging
— m — leva f - [mec] cam replacement,cing
— m — línea f - [tub] line replacement,cing
— m — — f para filtro m - [comb.int] filter line, replacement,cing, o change,ging

reemplazo m de malla f - [mec] mat change,ging
— m — — f para criba f - [mec] screen mat, replacement,cing, o change,ging
— m — material(es) m - [ind] material(s), replacement,cing, o change,ging
— m — mecanismo m - [mec] mechanism replacement
— m — — m para cambio(s) - [mec] shifter replacement,cing
— m — pieza f - [ind] part replacement,cing
— m — molde m - [ind] mold, replacement,cing, o change,ging
— m — — m para lingoteadora f - [metal-prod] pig casting mechine mold, replacement,cing, o change,ging
— m — n para máquina f para colar v—[metal] prod] pig casting machine, change,ging, o replacement,cing
— m — montaje m - [mec] mounting, change,ging, o replacement,cing
— m — motor m—[electr-mot] motor, change,ging, o replacement,cing • [comb.int] engine, change,ging, o replacement,cing, o transplanting
— m — — m para ventilador m - [ventil] fan replacement,cing
— m — neumático m - [autom-neumát] tire replcement,cing
— m — paleta f - [mec] blade, replacement,cing, o change,ging
— m — — f para rompedora f - [mec] breaker blade, replacement,cing, o change,ging
— m — — f — — f para sínter m - [metal-prod] sinter breaker blade, replacement,cing, o change,ging
— m — — f para roturadora f para sínter m - [metal-prod] sinter breaker blade, replacement,cing, o change,ging
— m — pavimento m - [constr] pavement, replacement,cing, o change,ging
— m — pieza f - [mec] part, replacement,cing, o change,ging
— m — — f conexa - [mec] related part replacement,cing, o change,ging
— m — — f defectuosa - [ind] defective part, replacement,cing, o change,ging
— m — — f (des)gastada - [mec] worn part, replacement,cing, o change,ging
— m — — f para cojinete m - [cojin] bearing part, relacement,cing, o change,ging
— m — placa f - [mec] plate, replacement,cing, o change,ging
— m — plancha f—[mec] plate, replacement,cing, o change,ging
— m — — f para desgaste m - [mec] wear plate, replacement,cing, o change,ging
— m — platino m—[comb.int] point, change,ging, o replacement,cing
— m — — m para rotor m - [comb.int] breaker point, replacement,cing, o change,ging
— m — plato m - [mec] plate, replacement,cing, o change,ging
— m — polea f - [mec] pulley, replacement,cing, o change,ging
— m — — f para montacarga(s) m - [metal-prod] skip pulley, replacement,cing, o change,ging
— m — — f — — m para horno m - [metal-prod] furnace skip pulley, replacement,cing, o chane,ging
— m — — f — — m — — m alto - [metal-prod] blast furnace skip pulley, replacement,cing, o change,ging
— m — — f para vagoneta f - [metal-prod] car pulley, replacement,cing, o change,ging
— m — — f — — f para montacarga(s) m—[metal-prod] skip car pulley, replacement,cing, o change,ging
— m — — f — — m para horno m [metal-prod] furnace skip car pulley, replacement,-cing, o change,ging
— m — — f — — m — — m alto - [metal-prod] blast furnace skip car pulley, replacement,cing, o change,ging
— m — portadiferencial m - [autom-mec] differ-

ential carrier, replacement,cing, o change
reemplazo m de portadiferencial para eje m - [autom-mec] axle differential carrier, replacement,cing, o change,ging
— m — — m, anterior, o delantero - [autom-mec] forward axle differential carrier, replacement,cing, o change,ging
— m — puente m - [constr] bridge replacing
— m — — m carretero - [const] highway bridge replacing
— m — — m de hormigón m - [constr] concrete bridge replacing
— m — — — m deteriorado - [constr] deteriorated, concrete bridge replacing
— m — — m deteriorado - [constr] deteriorated bridge replacing
— m — — m Super-Span - [Constr] Super-Span bridge replacing
— m — purgador m - [mec] bleeder replacing
— m — — m para horno m - [metal-prod] furnace bleeder, replacement,cing, o change,ging
— m — — — m alto - [metal-prod] blast furnace bleeder, replacement,cing, o change
— m — red f - system replacement,cing
— m — — f para servicio m público - [constr] utility replacement,cing
— m — relé m - [electr-equip] relay replacing
— m — resorte m - [mec] spring replacing
— m — revestimiento m - [mec] lining replacing
— m — rodete m - [ventil] impeller, change, o replacement,cing
— m — — m para ventilador m - [mec] fan impeller, replacement,cing, o change,ging
— m — rodillo m - [mec] roll, change,ging, o replacement,cing
— m — — m para cinta f - [mec] belt roller, replacement,cing, o change,ging
— m — — m — — f transportadora f - [mec] conveyor belt roller, replacement,cing, o change,ging
— m — rollo m - [mec] roll change,ging • [sold] coil change,ging
— m — — m de alambre m - [sold] wire reel change,ging
— m — rueda f - [mec] wheel, change,ging, o replacement,cing
— m — — f dentada - [mec] sprocket, replacement,cing, o change,ging
— m — — f para grúa f - [Grúas] crane wheel, replacement,cing, o change,ging
— m — — f — lingoteadora f - [metal-prod] pig casting machine wheel, change,ging, o replacement,cing
— m — — f — máquina f para colar - [metal-prod] pig casting machine wheel change,ging
— m — sello m - [mec] seal replacement,cing
— m — — m contra polvo m - [mec] dust seal, replacement,cing, o change,ging
— m — semieje m - [autom-mec] axle shaft, replacement,cing, o change,ging
— m — servicio(s) m técnico(s) - [ind] technical service(s), replacement,cing, o substitution, o change,ging
— m — símbolo m - symbol replacement,cing
— m — sincronizador m - [instrum] timer replacement,cing
— m — — m para purga f - [combust] purge,ging timer, replacement,cing, o change,ging
— m — soldadora f - [sold] welder replacement
— m — solenoide m - [electrón] solenoid replacement,cing
— m — sonda f - [metal-prod] stock rod change
— m — — f para horno m - [metal-prod] furnace stock rod, replacement,cing, o change,ging
— m — — f — — m alto - [metal-prod] blast furnace stock rod, replacement,cing, o change
— m — suelo m - soil replacement,cing
— m — — m para cimentación f - [constr] foundation soil replacement,cing
— m — sujetador m - [mec] cam replacement,cing
— m — tapón m - [mec] plug replacement,cing
— m — termóstato m - [instrum] thermostat, replacement,cing, o change,ging

reemplazo m de tolva f - [mec] hopper, replacement,cing, o change,ging
— m — f para horno m - [metal-prod] furnace
hopper, replacement,cing, o change,ging
— m — f — — m alto - [metal-prod] blast
furnace hopper, replacement,cing, o change
— m — tomacorriente m - [electr-instal] receptacle, replacement,cing, o change,ging
— m — trampa f - [mec] trap replacement,cing
— m — transformador m - [electr-equip] transformer, replacement,cing, o change,ging
— m — transmisión f - [mec] transmission, replacement,cing, o change,ging
— m — tubería f - [tub] tube,bing, o pipe
line, replacement,cing, o change,ging
— m — tubo m - [tub] tube, o pipe, replacing
— m — — m a atomizador m - [comb.int] tube
to @ atomizer, replacement,cing, o change
— m — — válvula f - [tub] tube to @
valve, replacement,cing, o change,ging
— m — vagoneta f - [metal-prod] car, replacement,cing, o change,ging
— m — — f para montacarga(s) m - [metal-prod]
skip car, replacement,cing, o change,ging
— m — — f — — m para horno m (alto) -
[metal-prod] (blast) furnace skip car replacement,cing, o change,ging
— m — varilla f - [sold] rod change,ging
— m — ventilador m - [ventil] fan replacement
— m — viaducto m - [constr] trestle, replacement, o changing
— m — — m ferroviario - [constr-f.c.] railway
trestle, replacement,cing, o change,ging
— m — zaranda f - [miner] screen replacement
— m — f para sínter(ización) - [miner] sin-
ter(ing) screen, replacement,cing, o change
— m económico - inexpensive, o low cost, replacement, o substitute
— m en fábrica f - replacement at @ factory
— m — juego m—[emc] replacement,cing in @ set
— m — obra f - replacement,cing on @ job
— m — sitio m - at @ site replacement,cing
— m exigido - required replacement,cing
— m expreso - express replacement,cing
— m frecuente - frequent replacement
— m inmediato - immediate replacement
— m más sencillo - simpler replacement • sim-
plest replacement
— m necesario—necessary, o needed, replacement
— m para alcantarilla f - [constr] culvert replacement
— m para árbol m - [botán] tree replacement •
[mec] shaft replacement
— m — armadura f—[mec] armature replacement
— m — aspirador m - [mec] (extractor) fan replacement
— m — atomizador m - [comb.int] atomizer replacement
— m — bobina f - [electr-equip] coil replacement
— m — bomba f - [bombas] pump replacement
— m — — f para inyección f - [comb.int] injection pump replacement
— m — bombilla f - [electr-ilum] bulb replacement
— m — — f para luz f - [electr-ilum] light, o
lamp, bulb replacement
— m — bujía f - [comb.int] spark plug replacement
— m — caja f - [mec] house replacement; replacement housing
— m — componente m - component replacement
— m — cartucho m - [mec] cartridge replacement
— m — — f para filtro m - [mec] filter cartridge replacement
— m — — m — — m hidráulico - [mec] hydraulic filter cartridge replacement
— m — cilindro m - [mec] cylinder replacement
— m — — m para éter m - [comb.int] ether cylinder replacement
— m — cojinete m - [mec] bearing replacement
— m — compresor m - [ind] compressor replacement

reemplazo m para conmutador m - [electr-instal]
switch replacement
— m — conservación f - [constr] maintenance
replacement
— m — contacto m - [electr-instal] contact
replacement
— m — — m para relé m - [electr-instal] relay
contact replacement
— m — — m — — m piloto - [electr-instal]
pilot relay contact replacement
— m — contador m - [electrón] counter replacement
— m — — m para frecuencia f - [electrón] frequency counter replacement
— m — conmutador m - [electr-instal] switch
replacement
— m — — m defectuoso - [electr-instal] defective, o faulty, switch replacement
— m para construcción f - [constr] construction, replacement, o substitute
— m — f corriente - [constr] standard
construction, replacement, o substitute
— m — — f estándar - [constr] standard construction, substitute, o replacement
— m — convertidor m - [metal-prod] converter
replacement
— m — cristal m - crystal replacement
— m — chatarra f - [metal-prod] scrap replacement
— m — dibujo m - [dib] replacement drawing
— m — diodo m - [electrón] diode replacement
— m — — m emisor (de luz f) - [electrón]
(light) emitting diode replacement
— m — dispositivo m - attachment, o device,
replacement, o substitute, o alternate
— m — — m para cambio(s) m - [autom-mec]
shift(ing) unit replacement
— m — — m — — grúa f - [grúas] crane. attachment, o device, replacement, o alternate
— m — eje m - [mec] axle replacement
— m — electrodo m - [sold] electrode replacement
— m — elemento m - [mec] element replacement •
cartridge replacement
— m — — m para filtro m - [mec] filter cartridge replacement
— m — embrague m - [mec] clutch replacement
— m — engranaje m - [mec] gear replacement
— m — equipo m - [ind] equipment replacement
— m — — m para rueda f - [autom-mec] wheel
equipment replacement
— m — escobilla f - [electr-mot] brush replacement
— m — estructura f - [constr] structure replacement
— m — — f existente - [constr] existing
structure replacement
— m — fiador m - [mec] toggle, o holder, replacement
— m — filtro m - [mec] filter replacement
— m — — m para combustible m - [comb.int]
fuel filter replacement
— m — frecuencímetro m - [electrón] frequency
counter replacement
— m — guarnición f - [mec] gasket replacement
— m — — f anular - [mec] O-ring replacement
— m — — f circular - [mec] O-ring replacement
— m — instrumento m - [instrum] instrument replacement
— m — interruptor m - /[electr-instal] switch,
o breaker, replacement
— m — leva f - [mec] cam replacement
— m — montaje m - [mec] mounting replacement
— m — motor m - [electr-mot] motor replacement • [comb.int] engine replacement
— m — — m para ventilador m - [ventil] fan
motor replacement
— m — — m inferior - [grúas] lower, motor, o
engine, replacement, o alternate
— m — — m superior - [grúas] upper, motor, o
engine, replacement, o alternate
— m — neumático m - [autom-neumát] tire replacement

reemplazo m para pavimento m - [constr] pavement replacement
— m — **pieza f** - [mec] part replacement
— m — — f **conexa** - [mec] related part replacement
— m — — f **defectuosa** - [mec] defective part replacement
— f — — f **(des)gastada** - [mec] worn part replacement
— m — **puente m** - [constr] bridge replacement
— m — — m **carretero m** - [vial] highway bridge replacement
— m — — m de **condado m** - [vial] county bridge replacement
— m — — m **departamental**—[vial] county bridge replacement
— m — **red f** - system replacement
— m — — f **para servicio m público** - [constr] utility replacement
— m — **relé m**—[electr-equip] relay replacement
— m — — m **piloto**- [electr-instal] pilot relay replacement
— m — **remache m** - [mec] rivet replacement
— m — **resorte m** - [mec] spring replacement
— m — — m **para escobilla f** - [electr-mot] brush spring replacement
— m — **rodillo m** - [metal-lam] roll replacement
— m — **rueda f** - [mec] wheel replacement
— m — — f **dentada**—[mec] sprocket replacement
— m — — f — **loca** - [mec] idler sprocket replacement
— m **para rueda f** - [mec] wheel replacement
— m — — f **loca** - [mec] idler replacement
— m — **sello m** - [mec] seal replacement
— m — — m **contra polvo m** - [mec] dust seal replacement
— m — — m **enjugador m** - [mec] wiper replacement
— m — **sincronizador m** - [instrum] timer replacement
— m — — m **para purga f** - [combust] purge timer replacement
— m — **solenoide m** - [electrón] solenoid replacement
— m — **sujetador m** - [mec] cam replacement
— m — **tablilla f**—[electrón] board replacement
— m — — f **con circuito m** - [electrón] circuit board replacement
— m — — f — — m, **estampado**, o **impreso** - [electrón] printed circuit board replacement
— m — **tapón m** - [mec] plug replacement
— m — **transformador m** - [electr-equip] transformer replacement
— m — **transmisión f** - [mec] transmission replacement
— m — **tubería f** - [tub] tubing replacement
— n — **tubo m** - [tub] tube, o pipe, replacement
— m — — m a **atomizador m** - [comb.int] tube to @ atomizer replacement
— m — — m a **válvula f** - [tub] tube to @ valve replacement
— m — **vagoneta f** - [ind] car replacement
— m — **vehículo m transportador** - [grúas] carrier, replacement, o alternate
— m — **ventilador m** - [ventil] fan replacement
— m **posible** - possible replacement
— m **prematuro** - early replacement
— m **rápido** - [mec] quick, o fast, replacement
— m **sencillo** - simple replacement
— m **sin cargo m** - free replacement,cing
— m **substitutivo** - alternate replacement
— m **total** - total replacement
reencaminado,da a - rerouted • switched back
— a a **estrangulador m** - [mec] switched to @ choke
— a — — m **nuevo** - [mec] switched to a new choke
reencaminamiento m - rereouting • switching
— m a **estrangulador m** O [mec] rerouting, o switching back, to @ choke
— m — — m **nuevo** - [mec] rerouting, o switching (back) to @ new choke
— m de **caudal** - [hidr] flow switching

reencaminamiento m de caudal m @ estrangulador m - [mec] flow switch(ing) to @ choke
— m — — m — m **nuevo** - [mec] flow switching to @ new choke
— m — **circulación f** - flow switching
— m — — f a **estrangulador m** - [mec] circulation switching to @ flow
— m — — m — m **nuevo** = [mec] circulation switch(ing) to @ new flow
— m — — f de **lodo m** --[petról] mud, circularion, o flow, switch(ing)
— m — — f — — m a **estrangulador m** - [petról] mud circulation switch(ing) to @ choke
— m — — f — — m — — m **nuevo** - [petról] mud circulation switch(ing) to @ new choke
reencaminar v - to reroute • to switch (back)
— v a **estrangulador m** - [mec] to switch to @, flow, o choke
— v — — m **nuevo** - [mec] to switch to @ new, flow, o choke
— v **caudal m** - to switch @ flow
— v — m a **estrangulador m** - [mec] to switch @ flow to @ choke
— v — m — — m **nuevo** - [mec] to switch @ flow to @ new choke
— v **circulación f** - to switch @ flow
— v — f a **estrangulador m** - [petról] to switch @ circulation to @, flow, o choke
— v — — m **nuevo** - [petról] to switch @ circulation to @ new, flow, o choke
— v — f de **lodo m** - [petról] to switch @ mud, circulation, o flow
— v — f — — m a **estrangulador m** - [petról] to switch @ mud circulation to a, flow, o choke
— v — f — — m — — m **nuevo** - [petról] to switch @ mud circulation to a new choke
reencarrilar v - to get back on @ track
reencauzado,da a - [transp] rerouted
reencauzamiento m - [transp] rerouting
— m de **circulación f**—[transp] traffic rerouting
— m — **tránsito m** - [transp traffic rerouting
— m — **vehículo m** - [transp] vehicle redirecting
— m **severo m** - [transp] severe rerouting
reencauzar v - [transp] to reroute
— v **circulación f**—[transp] to reroute @ traffic
— v **tránsito m** - [treansp] to reroute @ traffic
— v **vehículo m** - [transp] to reroute @ vehicle
reencender v - to relight • [sold] to restart
— v **arco m** - [sold] to restart @ arc
— v **quemador m** - [combust] to relight @ burner
reencendido,da a - [combut] relit • [sold] restarted
reengranado,da a - [mec] re-engaged
reengranaje m - [mec] re-engaging
— m de **arrancador m** - [comb.int] starter re-engagement,gaging
— m — **motor m** - [comb.int] starter motor re-engagement,ging
— m — — m **para arrancador m** - [comb.int] starter motor re-engagement,ging
reengranar v - [mec] to re-engage
— v **arrancador m** - [comb.int] to re-engage @ starter
— v **motor m** - [electr-mot] to reengage @ motor • [comb.int] to re-engage @ engine
— v — m **para arrancador m** - [comb.int] to re-engage @ starter motor
reengrase m - [mec] regreasing
— m **fácil** - easy regreasing
reenhebrado m - [mec] rethreading
reenhebrado,da a - [mec] rethreaded
reenhebrar v - [mec] to rethread
reenrollado m - [mec] re-coiling
reenrollado,da a - [mec] - rerolled; recoiled; rewound
reenrollar v - [mec] to rewind; to re-coil
reenrolladora f - [metal-lam] rewinder; recoiler
reescuadrado,da a - resquared
reescuadrar v - [mec] to resquare
reesmerilado m - [mec] regrinding
reescuadrado,da a - [mec] reground
reesmerilar v - [mec] to regrind
reestructuración f - restructuring
reestructurado,da a - restructured

restructurar v - to restructure
reexpandido,da a - re-expanded
reexpandir v - to re-expand
reexpandir v - to re-ecpand
reexpansión f - re-expansion; re-expanding
reexpedición f - [transp] reshipping • forwarding
— f por carga f—[transp] forwarding by freight
reexpedido,da a - reshipped - forwarded
reexpedir v - [transp] to reship; to forward
reexportable adv - [fin] reexportable
— adv en moneda f convertible (libremente) - [fin] reexportable in freely convertible currency
reexportación f - reexporting,tation
— f de capital m - [fin] capital reexporting,tation
— f — mercadería f - [fisc] good, o merchandise, reexporting,tation
— f — valor(es) m - [fin] value(s) reexporting,tation
reexportado,da - [fisc] reexported
reexportar v - [com] to reexport
— v capital m - [fin] to reexport @ capital
refacción f . . . . • repair • [Mex.] véase pieza
f para repuesto m
— f especial - [mec] special repair(ing)
— f identificada - [ind] identified repair
refaccionario,ria a - repairing
refección f - [culin] snack
— f de medianoche f - [culin] midnight snack
referencia f - . . . . • narrating • relation •
aplication • [ind] datum,ta | adv - [con]
with reference, o referring, to; referenced;
in regard to • subject
— f a tierra f - [electrón] reference,cing to @
ground
— f bancaria - [fin] bank reference
— f comercial - [fin] commercial reference
— f — cliente m - [com] customer('s), o your,
reference
— f — soldadura f - [sold] welding reference
— f detallada - [com] detail(ed) reference
— f directa - direct reference
— f fácil - easy reference
— f final - final reference
— f futura - future reference
— f hecha - made reference
— f indirecta - indirect reference
— f para lote m - [ind] batch reference
— f — partida f - [ind] batch reference
— f — pieza f - [ind] part('s) reference
— f personal - [fin] personal reference
— f recíproca - [dib] reciprocal reference
— f vertical - vertical reference
referenciación f - referencing
— f de documentación f - documentation referec-
ing
— f — documento m - document referencing
referenciado,da a - referenced
— a a cristal m - [electrón] crystal referenced
referenciar v - to reference
— v documento m - to reference @ document
referente a - . . .; referring; relating
— a a - regarding; concerning
— a a construcción f - constructional
— a — textura - textural; texture relate
referido,da a - . . . .; referred • narrated • applied
— a a tierra f - [electrón] referenced to @
ground
— a directamente - referenced,rred directly
referir v - . . .; to apply
referirse a - to refer to
— v a guía f - [public] to refer to @ directory
refinación f - . . . . • [metal-prod] . . .; puddling; véase también afino m
— f de acero m - [metal-prod] steel refining
— f — m especial - [metal-prod] specialty
steel refining
— f — cobre m - [metal-prod] copper refining
— f — estructura f—[metal] structure refining
— f — f granular - [metal] grain structure

refinación f
refinación f de grano(s)—[metal] grain refining
— f — metal(es) m - [metal] metal(s) refining
— f — petróleo m - [petról] petroleum, o oil,
refining
— f — sulfuro m - [quím] sulfide refining
— f — — m de cobre m - [quím] copper sulfide
refining
— f primaria - primary refining • [petról]
topping; skimming; stripping
refinado,da a altamente - highly refined
— a en aire m - air, refined, o blown
— a sin concentrar - [nucl] unconcentrated refined
refinador m para grano m - grain refiner
refinamiento m en cálculo m - [mat] computing
refinement
— m — estructura f - [metal] structure refinement
— m — — f granular - [metal] grain, o granular, structure refinement
refinanciado,da a - [fin] refinanced
refinanciamiento m - [fin] refinancing
refinanciar v - [fin] to refinance
refinar v cobre m - [metal-prod] to refine @
copper
— v chatarra f - [metal-prod] to refine @ scrap
— v metal m - [metal-prod] to refine @ metal
— v petróleo m - [petról] to refine @, oil, o
petroleum
reflector m - . . .; floodlight • [autom] spot-
light
— m ambar - [autom-electr] amber reflector
— m de aluminio m - [electr-ilum] aluminum reflector
— m — material m plástico - [electr-ilum]
plastic reflector
— m — vibración(es) m - vibration reflector
— m — — f con frecuencia f alta - [electrón]
high frequency vibration reflector
— m delantero - [autom] front floodlight
— m esférico - spheric reflector
— m frontal - [autom] front floodlight
— m — para cabina f - [autom] cab front floodlight
— m para aguilón m - [grúas] boom floodlight
— m — cabina f - [autom] cab floodlight
— m — frente m de cabina f - [autom] cab front
floodlight
— m plano - flat, o plane, reflector
— m posterior - [autom] rear floodlight
— m — para cabina f - [autom] cab rear floodlight
— m rojo - [autom-electr- red reflector
— m trasero - [autom] rear floodlight
reflector,ra a - reflector; reflecting
reflejante a - véase refelctor,ra; reflectante
reflejar v - . . .; to mirror
— v en utilidad(es) f - [fin] to reflect in,
earnings, o profits
— v hacia atrás - to reflect back(wards)
— v variable - to reflect @ variable
— v — f desconocida - to reflect @ unknown
variable
reflejo m - . . .; reflecting
reflexión f - . . .; reflecting
— f de variable f - variable reflecting,tion
— f — — f desconocida - unknown variable reflecting,tion
reflexionado,da a - reflected
reflujo m - . . . . • flowing back • overflow
reforma f - . . . .; correction; véase también reformación f - [ind] rework(ing) • [legal]
revising,sion; amending,dment; change,ging
— f de contrato m - [legal] contract change
— f — m social - [legal] social contract
change
— f — escritura f social - [legal] social contract change
— f de política f - [admin] policy revision
— f, educacional, o educativa - [educ] educational reform

reforma f de toma f - [hidr] intake change
— f —— f para agua m - water intake change
— f —— f —— m de río m - [hidr] river
water intake change
— f posterior - [legal] later change
— f — de contrato m - [legal] contract later
change; later contract change
reformación f- . . . .; reforming - [legal] revising; revision
— f con base f de vapor m - [metal-prod] steam
base reforming
— f — catalizador m - [metal-prod] catalyzer
reforming
— f — gas m para oxidación f - [metal-prod]
oxidation gas reforming
— f —— m — f con oxígeno m—[metal-prod]
oxygen oxidation gas reforming
— f — temperatura(s) f alta(s) - [metal-prod]
high temperature reforming
— f —— f elevada(s) - [metal-prod] high temperature reforming
— f — vapor m - [metal-prod] steam reforming
— f —— m e hidrocarburo(s) m - [metal-prod]
steam-hydrocarbon(s) reforming
— f —— m por oxidación f - [metal-prod] oxidation gas reforming
— f —— m — f parcial - [metal-prod] oxygen oxidation gas partial reforming
— f —— m —— f con oxígeno m - [metal-
-prod] partial oxygen oxidation gas reforming
— f —— m —— f parcial - [metal-prod] partial oxidation gas reforming
— f en base f de oxidación f - [metal-prod]
base reforming
— f —— f —— f con gas m - [metal-prod]
gas oxidation base reforming
— f —— f —— f —— m de oxígeno m -
[metal-prod] oxygen gas oxidation base reforming
— f —— f —— f —— m —— m industrial
- [metal-prod] industrial oxygen gas oxidation base reforming
— f —— f —— f parcial - [metal-prod]
partial oxidation base reforming
— f —— f —— f — con gas m - [metal-prod]
gas partial oxidation base reforming
— f —— f —— f —— m de oxígeno m -
[metal-prod] oxygen gas partial oxidation
base reforming
— f —— f —— f —— m —— m industrial - [metal-prod] industrial oxygen gas
partial oxidation base reforming
— f en estufa f - [metal-prod] stove reforming
— f —— f cargada - [metal-prod] charged
stove reforming
— f —— f — con bola(s) f - [metal-prod]
ball(s) charged stove reforming
— f —— f —— f cerámica)s f - [metal-
-prod] ceramic balls charged stove reforming
— f —— f —— f con contenido m alto en
alúmina f - [metal-prod] high alumina ball
charged stove reforming
— f singular - [metal-prod] unique reforming
— f —— con vapor m - [metal-prod] unique steam
reforming
— f —— m e hidrocarburo(s) m - [metal-
-prod] unique steam-hydrocarbon(s) reforming
— f térmica - [metal-prod] thermal reforming
— f — en estufa f - [metal-prod] stove thermal
reforming
— f —— f cargada - [metal-prod] charged
stove thermal reforming
— f —— f — con bola(s) f - [metal-prod]
ball(s) charged stove thermal reforming
— f —— f —— f cerámica(s) - [metal-
-prod] ceramic ball(s) charged stove thermal
reforming
— f —— f —— f — con alúmina f -
[metal-prod] alumina ceramic balls charged
stove thermal reforming
— f —— f —— f —— contenido m
alto en alúmina f - [metal-prod] high alumina
ceramic ball(s) charged stove thermal refor-
mins
reformación f térmica en estufa f cargada con
bola(s) f con contenido m alto en alúmina f -
[metal-prod] high alumina ball charged stove
thermal refining
reformado,da a - reformed · [mec] reworked ·
[legal] revised; amended
reformador m catalítico - [petról] catalytic
reformer
— m — para gasolina f - [petról] gasoline catalytic reformer
— m catalizador - [petról] catalytic reformer
— m con vapor m - [metal-prod] steam reformer
— m —— m e hidrocarburos m - [metal-prod]
steam, hydrocarbons, o methane, reformer
— m —— m y metano m - [metal-prod] steam-
-methane reformer
reformar v - . . . . • [mec] to rework · [legal]
. . .; to revise
— v norma f - [admin] to, reform, o revise, @,
standard, o policy
— v política f - [admin] to, reform, o revise,
@ policy
reformulación f - reformulation
reformulado,da a - reformulated
reformular v - to reformulate
reforzado,da a - . . . . • stepped up • rugged •
[constr] buttressed; braced; trussed; built
up • haunched • [mec] banded • heavy duty
— a apropiadamente - appropriately, o properly,
reinforced
— a con fibra(s) f de vidrio m - [plást] fiber
glass reinforced
— a con hormigón m - [constr] concrete reinforced
— a — malla f - [constr] mesh reinforced
— a —— f de alambre m - [constr] wire mesh reinforced
— a — mampostería f - [constr] masonry, o
brickwork, reinforced
— a — varilla(s) f - [tub] rod reinforced
— a debidamente - properly reinforced
— a transversalmente - [mec] cross braced
reforzador m - reinforcer; booster; strengthener
— m para presión f - véase compresor m
reforzamiento - reinforcement,cing; strengthening; véase también refuerzo m • [mader]
blocking
— m para muro(s) m - [constr] wall reinforcing
reforzar v - . . .; to fortify • to back • to
boost • [mec] to, brace, o truss, o butress
— v apropiadamente - to reinforce properly
— v bastidor m - [mec] to truss @ frame
— v borde m - [mec] to reinforce @ edge
— v bóveda f - [constr] to reinforce @ (top)
arch·
— v — f superior - [constr] to reinforce @ top
arch
— v cabrestante m - [mec] to reinforce @ winch
— v celda f - to, butress, o rinforce, @ cell
— v con fibra de vidrio m - [plást] to, fiber
glass reinforce, o reinforce with fiber glass
— v — mampostería f [constr] to reinforce
with, masonry, o brickwork
— v — varilla(s) f - [tub] to rod reinforce
— v debidamente - to reinforce properly
— v eje m - [mec] to reinforce @ axle
— v extremo m - to reinforce @ end
— v — m cortado - to reinforce @ cut end
— v suspensión f - [autom] to stiffen @ suspension
— v tramo m - [mec] to strengthen @ span
refractario m - [refract] refractory
— m ácido - [refract] acid refractory • siliceous refractory
— m aluminoso - [refract] alumina, o aluminous,
refractory
— m básico - [refract] basic refractory
— m con alúmina f - [refract] alumina refractory
— m — contenido m alto en alúmina f - [refract]
high alumina refractory
— m enfriado - [combust] cooled refractory
— m de alúmina f - [refract] alumina refractory
— m de sílice m - [refract] silica refractory

refractario m **deteriorado** - [metal-prod] deteriorated refractory (brick)
— m **en, muro** m, o **pared** f - [constr] wall refractory (brick)
— m — **pila** f - [metal-prod] checker work refractory [brick]
— m **enfriado** - [combust] cooled refractory
— m — **con agua** m - [combust] water cooled refractory (brick)
— m **para acería** f - [metal-prod] steel plant refractory (brick)
— m — — f **eléctrica** - [metal-prod] electric (steel) plant refractory (brick)
— m — — f **Linz-Donawitz** - [metal-prod] basic oxygen steel plant refractory (brick)
— m — **barra** f **taponadora** - [metal-prod] stopper rod refractory (brick)
— m — **bóveda** f - [combust] refractory arch
— m — **colada** f - [metal-prod] casting refractory (brick)
— m — **etalaje** m - [metal-prod] bosh refractory
— m — **fraguado** m - [metal-prod] setting refractory • forging refractory (brick)
— m — **horno** m - [metal-prod] furnace refractory
— m — — m **alto** m - [metal-prod] blast furnace refractory (brick)
— m — **horno** m—[metal-prod] furnace refractory
— m — — m **para encendido** m - [metal-prod] ignition, o lighting, furnace refractory brick
— m — — m — m **de sínter** m - [metal-prod] sinter, ignition, o lighting, furnace refractory (brick)
— m — — m **eléctrico** - [metal-prod] electric furnace refractory (brick, o shape)
— m — **industria** f - [ind] industry refractory
— m — — f **siderúrgica** - [metal-prod] steel industry refractory (brick, o shape)
— m — **muro** m - [constr] wall refractory
— m — — m **anterior** - [combust] front wall refractory, brick, o shape
— m — — f **frontal** - [combust] front wall refractory, brick, o shape
— m — **nivel** m **superior de carga** f - [metal-prod] stock line level refractory (brick)
— m — **pared** f - [constr] wall refractory
— n — — f **anterior** - [combust] front wall refractory, brick, o shape
— m — — f **frontal** - [combust] front wall refractory, brick, o shape
— m — — **posterior** - [combust] rear wall refractory, brick, o shape
— m — **piquera** f - [metal-prod] (iron) notch refractory, brick, o shape
— m — **quemador** m - [metal-prod] burner refractory, brick, o shape
— m — — m **para encendido** m - [metal-prod] ignition, o lighting, burner refractory
— m — — m **para estufa** f - [metal-prod] stove burner refractory, brick, o shape
— m — **recubrimiento** m - [refract] covering refractory, brick, o shape
— m — — m **para quemador** m - [metal-prod] burner covering refractory, brick, o shape
— m — — m — m **para encendido** m—[combust] ignition burner covering refractory
— m — **ruta** f **para arrabio** m - [metal-prod] pig iron runner refractory, brick, o shape
— m — — f **para escoria** f - [metal-prod] slag runner refractory, brick, o shape
— m **perfilado** - [cerám] shaped refractory
— m **quemado** - [metal-prod] burned refractory
— m **refrigerado** - [combust] cooled refractory
— m — **con agua** m - [combust½ water cooled refractory, brick, o shape
— m **silícico** - [cerám] siliceous refractory
— m **silicoso** - [refract] acid refractory
refractario,ria a - refractory • ceramic • fire (resisting)
refractarista m - [refract] refractory, bricklayer, o mason, o repairman
— m **ayudante** m - [ind] assistant bricklayer

refractarista m **ayudante para cuchara(s)** f - assistant ladleman
— m **para cuchara(s)** f - [refract] ladleman; ladle.liner, o bricklayer
refrán m **ingenioso** - wise, o smart, saying
refregadora f - [miner] scrubber
refrenar v . . .; to refrain • to coerce
refrendación f - . . .; attestation
refrendado,da a - [legal] attested; countersigned; authenticated • endorsed • visaed
refrendar v - [legal] . . .; to attest
refrentable a - [mec] refaceable; regrindable
— a **de tipo** m **obturador** - [válv] regrindable plug type
refrentación f - [mec] (spot) facing
— f **de ranura** f - [cojin] race grinding
— f — — f **para bola(s)** f - [cojin] ball race grinding
refrentado* m - véase **refrentación** f] facing; dressing
— m **de rueda** f - [mec] wheel, facing, o dressing
— m **automático** - [mec] automatic facing
refrentado,da a - [mec] faced; milled; ground; correct(ed); dressed; spot faced
— a **con bronce** m - [metal-prod] bronze faced
— a — **estelita** f - [válv] stellite face(d)
refrentador,ra a - [mec] facer,cing
refrentadora f - [mec] facing, mill, o machine
— f **para extremo(s)** m - [mec] end facer
— f — — m **único** - [mec] single end facing machine
refrentar v - [mec] . . .; to dress; to reface; to spot face • to correct
— v **ranura** f - [cojin] to grind @ race
— f — **para bola(s)** f - [cojin] to grind @ ball race
— v **rueda** f - [mec] to, grind, o dress, @ wheel
refrescante a - . . . . • [meteorol] crisp
refrigeración f - . . .; cooling
— f **adecuada** - [sold] adequate cooling
— f **con agua** m - water cooling
— f — **aire** m - [ambient] air cooling
— f **efectiva** - [ind] effective cooling
— f **eficiente** - [comb.int] efficient cooling
— f **exterior** - [ind] outside, o exterior, cooling
— f — **de coraza** f - [metal-prod] shaft outisde cooling
— f **forzada** - [ind] forced cooling
— f **hidráulica** f - hydraulic, o water, cooling
— f **insuficiente** - [ind] insufficient cooling
— f **intermedia** - [ind] intermediate cooling
— f **mecánica** - [ind] mechanical cooling • mechanical refrigeration
— f **para coraza** f - [metal-prod] shaft water cooling
— f — **fondo** m - [ind] bottom cooling
— f — — m **de crisol** m - [metal-prod] hearth bottom cooling
— f — — m — **horno** m - [metal-prod] furnace bottom cooling
— f — — m — — m **alto** - [metal-prod] blast furnace bottom cooling
— f — — m **de solera** f **de horno** m **alto** - [metal-prod] blast furnace hearth bottom cooling
— f — **horno** m - [metal-prod] furnace cooling
— f — **planta** f - [ind] plant cooling
— f — — f **generadora** - [electr-prod] power plant cooling
— f — **refractario(s)** m - [combust] refractory cooling
— f **regulada** - [ind] controlled, cooling, o refrigeration
— f — **rodillo** m - [metal-lam] roll cooling
— f — **tapa** f - [ind] cover, o cap, cooling
— f — — f **para lingotera** f - [metal-prod] mold cap cooling
— f — — f — **molde** m - [metal-prod] mold cap cooling
— f — **tope** m - [metal-prod] top cooling
— f **primaria** - primary cooling

refrigeración f rápida - quick cooling
— f secundaria f - [ind] secondary cooling
refrigerado,da a - [ind] refrigerated; cooled
— a con agua m - [ind] water cooled
— a — líquido m - [ind] liquid cooled
— a hidráulicamente - [ind] hydraulically, o water, cooled
refrigerador m con agua m - [ind] water cooler • heat exchanger
— m intermedio m - [ind] intercooler
— m para puerta(s) - [metal-prod] door, o port, chiller, o cooler
— m rotativo - [ind] rotary cooler
— m secundario - [ind] secondary cooler • aftercooler
refrigerante m agregado—[ind] added refrigerant
— m de tipo m permanente - [comb.int] permanent type coolant
— m ensuciado - dirtied coolant
— m limpiado - cleaned coolant
— m limpio - [ind] clean coolant
— m para camisa f - [mec] liner coolant
— m — f verificado - [ind] checked liner coolant
— m — convertidor m - [metal-prod] converter coolant
— m — émbolo m - [mec] piston coolant
— m — m para camisa f - [petról] liner piston coolant
— m — m — f verificado - [petról] checked liner piston coolant
— m — m verificado - [petról] checked piston coolant
— m — motor m - [electr-mot] motor coolant • [comb.int] engine coolant
— m perdido - [comb.int] lost coolant
— m permanente - [comb.int] permanent coolant
— m rociado - [mec] sprayed coolant; coolant spray
— m sucio - dirty coolant
— m verificado - [ind] checked, coolant, o refrigerant
refrigerar v - . . .; to chill
— v exteriormente - [metal-prod] to cool externally
— f refractario(s) m - [combust] to cool @ refractory,ries
refrigerio m vespertino m - [culin] evening meal
refuerzo m - . . .; reinforcing; strengthening • [carp] haunch • [sold] head excess thickness • [constr] bracing; brace; butress(ing); abutment; backing; truss(ing)
— m angular - [constr] angle brace
— m apropiado - appropriate, o proper, reinforcement,cing
— m central - [mec] center,tral brace,cing
— m circunferencial - [tub] circumferential reinforcement,cing
— m — de acero m - [tub] circumferential steel reinforcement,cing
— m con alma m - [metal-lam] web stiffener
— m — cordón m - [sold] bead reinforcement
— m — fibra f de vidrio m - [plást] fiber glass reinforcement
— m — hormigón m - [constr] concrete reincement,cing
— m — mampostería f - [constr] masonry reinforcement,cing
— m — tela f - fabric, o cloth, reinforcement
— m — varilla(s) f - [tub] rod reinforcement
— m corrido - [constr] running reinforcement
— m cuadrado - [mec] square reinforcement • [carp] haunch
— m de acero m - [mec] steel reinforcement
— m — m circunferencial - [tub] circumferential steel reinforcement
— m — bóveda f - [constr] (top) arch reinforcement,cing
— m — f superior - [constr] top arch reinforcement,cing
— m — tipo entramado - [constr] truss type reinforcement,cing
— m debido - proper reinforcement,cing

refuerzo m dorsal - [mec] back brace,cing
— m esquinero - [constr] corner reinforcement,cing
— m exterior - [mec] external, o outside, reinforcement,cing
— m — con placa(s) f de acero m - [mec] external, o outside, steel plate reinforcement,cing
— m horizontal - [constr] horizontal reinforcement,cing
— m longitudinal - longitudinal reinforcement
— m — inferior - [constr-cercas] bottom rail
— m — superior - [constr-cercas] top rail
— m mediante pieza f angular - [mec] angle reinforcement
— m metálico - [mec] metal(lic) reinforcement
— m para ala m - [constr] flange, reinforcement, o stiffener
— m — apertura f - [cald] opening reinforcement
— m — bastidor m - [autom] frame, reinforcement, o reinforcing, o trussing
— m — — m para automóvil m - [autom] automobile, o car, frame reinforcement,cing
— m — — camión m - [autom] truck frame reinforcement,cing
— m — borde m - [mec] edge reinforcement,cing
— m — brida f - flange, reinforcement,cing, o stiffener
— m — cabrestante m - [mec] winch reinforcement
— m — celda f - cell, reinforcement,cing, o butressing
— m — coraza f - [ind] shell stiffener
— m — cordón m - [sold] bead reinforcement
— m — eje m - [mec] axle trussing
— m — entramado m - truss(ing) reinforcement
— m — eslabón m - [cadenas] (link) stud
— m — esquina f - [constr] corner reinforcement
— m — extremo m - [mec] end reinforcement,cing
— m — — m cortado - cut end reinforcement,cing
— m — hormigón m - [constr] concrete reinforcement,cing
— m — mampostería f - [constr] masonry reinforcement,cing
— m — muro m - [constr] wall reinforcement,cing
— m — — m de tipo m entramado - [constr] truss type reinforcement,cing
— m — — m entramado - [constr] truss wall reinforcement,cing
— m — — m de tipo m entramado - [constr] truss type wall reinforcement,cing
— m — — — m prefabricado - [constr] prefabricated truss type wall reinforcing
— m — soldadura f - [sold] weld reinforcement,cing
— m — suspensión f - [autom-mec] suspension, reinforcement,cing, o stiffening
— m — tramo m - [mec] span strengthening
— m prefabricado - [constr] prefabricated reinforcement,cing
— m — para esquina f - [constr] prefabricated corner reinforcement,cing
— m soldado - [constr] welded, attachment, o reinforcement,cing
— m sólido - [mec] solid reinforcement,cing
— m total - total reinforcement,cing
— m transversal - [mec] transverse, o cross, reinforcement,cing, o brace,cing
— m usado - [mec] used, reinforcement, o brace
— m vertical - [constr] vertical reinforcement
refugio m - . . . . • safety • [vial] isle,land
— m colector - [hidr] drainage island
— m contra radiactividad f - [constr] fallout shelter
refulado m - [constr] hydraulic fill
refulgencia f - . . . .; glitter(ing)
refulgente a - glittering
refulgido,da a - shined; glittered
refulgir v - . . .; to glitter
refundición f - [metal-prod] . . .; remelting
— f de arco m - [sold] arc remelting
— f — soldadura ı - [sold] weld remelting
— f — f por punto(s) m - [sold] tack weld remelting
— f — zona f - [sold] area, o zone, remelting

refundición <u>f</u> **de zona** <u>f</u> **endurecida** - [sold] hardened zone remelting
— <u>f</u> —— <u>f</u> — **de plancha** <u>f</u> - [sold] plate hardened zone remelting
— <u>f</u> **en vacío** <u>m</u> - [metal-prod] vacuum remelting
— <u>f</u> **por arco** <u>m</u> - [metal-prod] arc remelting
— <u>f</u> —— <u>m</u> **en vacío** <u>m</u> - [metal-prod] vacuum arc remelting; V A R
refundido,da <u>a</u> - remelted
refundir <u>v</u> **soldadura** <u>f</u> **por puntos** <u>m</u> - [sold] to remelt @ tack weld
— <u>v</u> **zona** <u>f</u> **endurecida** - [sold] to remelt @ hardened zone
— <u>v</u> — <u>f</u> — **de plancha** <u>f</u> - [sold] to remelt @ plate hardened zone
regado,da <u>a</u> - [hidr] watered; sprinkled; irrigated
— <u>a</u> **con aspersión** <u>f</u> - [hidr] spray sprinkled
regalado,da <u>a</u> - given
regalar <u>v</u> - . . .; to give
regalo <u>n</u> - . . . | <u>adv</u> - (de) free; at no charge
regalía <u>f</u> - [com] bonus • [legal] royalty
regañadiente(s) <u>adv</u> - (a) . . .; begrudgingly
regar <u>v</u> **con aspersión** <u>f</u> - [hidr] to spray (sprinkle)
regencia <u>f</u> - . . .; governing; governance
— <u>f</u> **de adquisición** <u>f</u> - procurement governing
regeneración <u>f</u> - . . .; regenerating • revivification,fying
— <u>f</u> **catiónica** - [nucl] cationic regeneration
— <u>f</u> **de arcilla** <u>f</u> - [petról] clay rivivification,fying (system)
— <u>f</u> — **combustible** - [nucl] fuel regeneration
— <u>f</u> —— **resina** <u>f</u> - [nucl] resin, <u>o</u> rosin, regenerating,tion
— <u>f</u> —— <u>f</u> **aniónica** <u>f</u> - [nucl] anionic, resin, <u>o</u> rosin, regenerating,tion
— <u>f</u> —— <u>f</u> **catiónica** - [nucl] cationic, resin, <u>o</u> rosin, regenerating,tion
regenerado,da <u>a</u> - regenerated • revivified
regenerador <u>m</u> - . . . • [metal-prod] checker, work, <u>o</u> chamber
— <u>m</u> **en canal** <u>m</u> - [metal-prod] duct regenerator
— <u>m</u> —— <u>m</u> **de solera** <u>f</u> - [metal-prod] hearth duct regenerator
— <u>m</u> **Siemens-Martin** - [metal-prod] open hearth checker
regenerar <u>v</u> - . . .; to refivify
— <u>v</u> **arcilla** <u>f</u> - [petról] to revivify @ clay
— <u>v</u> **combustible** <u>m</u> - [nucl] to regenerate @ fuel
regido,da <u>a</u> - governed; ruled
régimen <u>m</u> - . . .; rate; load; rating; condition • plan • range • speed • [hidr] . . .; flow • [medic] . . .; diet • [com] . . .; conduct; policy; system
— <u>m</u> **bajo** - (s)low speed
— <u>m</u> — **de velocidad** <u>f</u> - slow speed rate
— <u>m</u> **calorífico** - [ind] heat(ing) schedule
— <u>m</u> **común** - [pol] common rule,ling
— <u>m</u> — **para capital** - [pol] capital common rule,ling
— <u>m</u> — **para capital(es)** <u>m</u> **extranjero(s)** - [pol] foreign capital(s) common rule,ling
— <u>m</u> —— —— <u>m</u> **según Acuerdo** <u>m</u> **de Cartagena** [pol] Cartagena Agreement Foreign Capital Common Rules,lings system
— <u>m</u> **de aire** <u>m</u> - [metal-prod] air rate
— <u>m</u> — **concesión(es)** <u>m</u> - [miner] concession(s) system
— <u>m</u> —— <u>f</u> **para exploración** <u>f</u> - [miner] exploration concession(s) system
— <u>m</u> — **desplazamiento** <u>m</u> - displacement rate
— <u>m</u> — **franquicia(s)** <u>f</u> - [fisc] exemption(s) schedule
— <u>m</u> — **impulso(s)** <u>m</u> - [electrón] pulse rate
— <u>m</u> —— <u>m</u> **para dato(s)** <u>m</u> - [electrón] data pulse rate
— <u>m</u> —— <u>m</u> — **entrada** <u>f</u> - [electrón] input pulse rate
— <u>m</u> —— **espacio(s)** <u>m</u> - [electrón] space pulse rate
— <u>m</u> —— <u>m</u> — **marca(s)** <u>f</u> - [electrón] mark(s) pulse rate

régimen <u>m</u> **económico** - [econ] economic plan • [fin] payment plan
— <u>m</u> **en vacío** <u>m</u> - [mrv] idle speed
— <u>m</u> **para motor** <u>m</u> - [electr-mot] motor idle speed • [comb.int] engine idle speed
— <u>m</u> **establecido** - established policy
— <u>m</u> **impositivo** - [fisc] tax plan
— <u>m</u> **individual** - individual rate
— <u>m</u> **lento** - slow rate • slow speed
— <u>m</u> — **para motor** <u>m</u> - [electr-mot] motor slow, <u>o</u> slow motor, speed • [comb.int] engine slow, <u>o</u> slow engine, speed
— <u>m</u> **necesario** - [mec] required, speed, <u>o</u> rate, <u>o</u> value
— <u>m</u> **nominal** - nominal, <u>o</u> rated, speed • [sold] welder rating
— <u>m</u> — **en H P** - [electr-mot] horse power rating
— <u>m</u> —— — **para funcionamiento** <u>m</u> **continuo** - [electr-mot] horse power continuous rating
— <u>m</u> —— **para regulación** <u>f</u> - [electr-mot] controller rating
— <u>m</u> —— <u>f</u> **en H P** - [electr-mot] controller horsepower rating
— <u>m</u> —— <u>f</u> —— — **para funcionamiento** <u>m</u> **continuo** - [electr-mot] controller horsepower continuous rating
— <u>m</u> —— <u>f</u> —— — **motor** <u>m</u> - controller horsepower continuous rating
— <u>m</u> —— <u>f</u> —— <u>m</u> — **para funcionamiento** <u>m</u> **continuo** - [electr-mot] controller horsepower continuous rating
— <u>m</u> —— <u>f</u> —— —— <u>m</u> **continuo según N E M A** - [electr-mot] National Electrical Manufacturer's Association, <u>o</u> N E M A, controller horsepower continuous rating
— <u>m</u> —— <u>f</u> **según N E M A** - [electr-mot] National Electrical Manufacturer's Association controller rating
— <u>m</u> **según N E M A** - [electr-mot] National Electrical Manufacturer's Association ruling
— <u>m</u> **normal** - [sold] welder rating
— <u>m</u> — **para realización** <u>f</u> - normal accomplishment rate
— <u>m</u> — **soplado** <u>m</u> - [metal-prod] normal blast rate
— <u>m</u> **operativo** - [ind] operating, schedule, <u>o</u> rate
— <u>m</u> **para alimentación** <u>f</u> - [ind] feed(ing) rate
— <u>m</u> — <u>f</u> **regulado** - [ind] controlled feeding rate— <u>m</u> — <u>f</u> **regulado** <u>m</u> - [ind] controlled
— <u>m</u> —— <u>f</u> — **automáticamente** - [ind] automatically regulated feed(ing) rate
— <u>m</u> — **bomba** <u>f</u> - [bombas] pump rate
— <u>m</u> — **entrada** <u>f</u> - [ind] input rate
— <u>m</u> — **filtración** <u>f</u> - [hidr] filtering, rate, <u>o</u> regimen
— <u>m</u> — **marcha** <u>f</u> - [ind] operating rate
— <u>m</u> —— <u>f</u> **variable** - variable operating rate
— <u>m</u> — **operación** <u>f</u> - [ind] operating rate
— <u>m</u> — **producción** <u>f</u> - [ind] production rate
— <u>m</u> — **realización** <u>f</u> - accomplishment rate
— <u>m</u> —— <u>f</u> **normal** - normal accomplishment rate • [labor] standard time rate
— <u>m</u> — **salida** <u>f</u> - [electrón] output rate
— <u>m</u> — **soplado** <u>m</u> - [metal-prod] blast rate
— <u>m</u> — **temperatura** <u>f</u> - temperature(s) range
— <u>m</u> — **trabajo** <u>m</u> - operating rate
— <u>m</u> — **velocidad** <u>f</u> - speed rate; rate of speed
— <u>m</u> — **carga** <u>f</u> - [ind] charge,ging, <u>o</u> feed(ing) rate
— <u>m</u> —— <u>f</u> **regulado** - controlled feeding, <u>o</u> loading, rate
— <u>m</u> —— <u>f</u> — **automáticamente** - [ind] automatically controlled, loading, o feeding, rate
— <u>m</u> — **grúa** <u>f</u> - [grúas] crane rating
— <u>m</u> — **marcha** <u>f</u> - [ind] operating speed
— <u>m</u> —— <u>f</u> **aminorado** - [ind] reduced operating speed
— <u>m</u> — **motor** <u>m</u> - [electr-mot] motor speed • [comb.int] engine speed
— <u>m</u> — **movimiento** <u>m</u> - [mec] movement rate
— <u>m</u> —— <u>m</u> **para tapón** <u>m</u> - [mec] plug movement rate

régimen m para movimiento m de tapón m - ⌈mec⌉
plug movement rate
— m —— m ——— m **para estrangulador** m -
⌈mec⌉ choke plug movement rate
— m **operación** f **de motor** m - [electr-mot]
motor operating speed • ⌈comb.int⌉ engine
operating speed
— m **pleno** m - full rate (speed) • [sold] full
rated output
— m **recomendado** - recommended speed
— m — **para motor** m - [comb.int] recommended
engine speed • ⌈electr-mot⌉ recommended motor
speed
— m ——— m **sin carga** f - [electr-mot] recommended motor idle speed • [comb.int] recommended engine idle speed
— m **regulado** - controlled rate
— m **sin carga** f - idle speed
— m **tributario** - [fisc] tax plan
— m **variable** - variable rate
región f - . . . .; area
— f **andina** - [eogr] Andean, o Andes, region
— f **árida** - [geogr] arid region
— f **central** - [geogr] central, region, o area
• [EE.UU] Midwest
— f **de Grandes Lagos** m - ⌈geogr⌉ Great Lakes,
area, o region
— f **desarrollada** - developed, region, o area
— f **desolada** - ⌈geogr⌉ desolate, area, o region
— f **en conjunto** m - ⌈geogr⌉ region as ? whole
— f — **desarrollo** m - [econ] developing, area,
o region
— f **literal** - [geogr] coast(al), area, o region
— f **metropolitana** - ⌈pol⌉ metropolitan, area, o
region • urban complex
— f **nororiental** - Northeast; north eastern region; [EE.UU.] New England
— f **pantanosa** - [geogr] swampy, region, o area
— f **para proyecto** n - project, region, o area
— f — **recreo** m - ⌈deport⌉ playground (region)
— f **petrolera** - [petról] oil country
— f **poco desarrollada** - [econ] underdeveloped,
area, o region
— f **por, avenar, o desaguar, o drenar** v -
⌈hidr⌉ drainage,nable area
— f **remota** - [geogr] remote, region, o area
— f **rural** - [pol] rural area
— f **sur central** - southcentral, area, o region
— f — **de Estados Unidos** - [pol] Southwest
regir v **adquisición** f - to govern @ procurement
— v **materia** f - to govern @ subject
registración f - registration; registering; recording • record keeping • processing • **véase**
también **registro** m - [instrum] reading •
⌈cronol⌉ clocking
registrado,da a - . . . .; recorded • shown
checked in • ⌊instrum⌋ read • [cronol]
clocked • [legal] proprietary
— a **con impresor** m - [electrón] printer recorded
— a —— m **para escritorio** m - [electrón]
desktop printer recorder
— a **en ensayo** m - recorded in @ test
— a — **esfera** f—[instrum] registered on @ dial
— a — **hotel** m—⌈transp⌉ registered, o checked
in, in @ hotel
— a **entre** - registered between
registrador m **automático** - [instrum] automatic,
o strip-chart, recorder
— m **con banda** f - [instrum] strip-chart recorder
— m — **cinta** f - [instrum] strip-chart recorder
— m — f **con información** f **(aportada)** -
⌈electrón⌉ fed strip-chart recorder
— m — **gráfico** m - [instrum] wound recorder
— m —— m **con cuerda** f - [instrum] wound
chart recorder
— m — **información** f **(aportada)** - [electrón]
fed recorder
— m — . . . **punto(s)** m - [instrum] . . .-
point recorder
— m **integrador** - [instrum] integrating recorder
— m — **para caudal(es)** m - [instrum] integrat-

ing flow recorder
registrador m **para documento(s)** m **público(s)** -
[legal] public documents recorder
— m — **laminación** f - [metal-lam] rolling
recorder
— m — **línea** f **para carga** f - [metal-prod]
stock line recorder
— m — **pared** f - [electrón] wall recorder
— m — **pesadas** f - [instrum] weight recorder
— m — **pesos** m - [ind] weight recorder
— m — **potencial** m - [ind] potential recorder
— m — **presión** f - [instrum] pressure recorder
— m — **producción** f - [ind] production recorder
• productimeter
— m — **(de) Selsyn** - [instrum] Selsyn productimeter
— m — **temperatura(s)** f - [instrum] temperature
recorder
— m **potenciómetro electrónico** - [instrum] **véase**
potenciómetro m **electrónico registrador**
—/**regulador** m - ⌈instrum⌉ recording controller
—/— m **para temperatura(s)** f - [instrum] temperature recorder/controller
— m **sobre banda** f - [instrum] strip-chart recorder
— m — **cinta** f - [instrum] strip-chart recorder
registrador,ra a - recording
registrar v - . . . • to list • to show • to
check in • [instrum] to read; to clock
— v **actividad** f - to record @ activity
— v **actuación** f - to record @ performance
— v **aumento** m - to show @ increase
— v **carga** f - to record @ load
— v **con impresor** m - [electrón] to printer record
— v —— m **para escritorio** m - [electrón] to
desktop printer record
— v **contrato** m - [legal] to, register, o record, @ contract
— v **cumplimiento** m - to record @ performance
— v **convenio** m—⌈legal⌉ to register @ agreement
— v **daño** m - to record @ damage
— v **derecho** m - [legal] to register @ right
— v — m **de patente** f **(de invención** f**)** -
⌈legal⌉ to register @ patent right(s)
— v **derecho(s)** m **literario(s)** - [legal] to
register @ copyright
— v **desempeño** m - to record @ performance
— v **en dirección** f **ascendente** - [sold] to read
up scale
— v —— f **descendente** - [sold] to read down
scale
— v — **ensayo** m - to record in @ test
— v — **esfera** f - [instrum] to register on @
dial
— v **ensanchamiento** m - [instrum] to record @
expansion
— v **ensayo** m - to record @ test
— v **entre** - to register between
— v **expansión** f - [instrum] to record @ expansion
— v **factura** f - [com] to, register, o record, @
invoice
— v **falla** f - to record @ failure
— v **horas** f **de operación** f **de motor** m - [instr]
to register, o record, @, motor, o engine, o-
peration hour(s)
— v **incremento** m - to, register, o show, @ increase
— v **información** f - to record @ datum,ta
— v — f **sobre ensayo** m - to record @ test
datum,ta
— v **invención** f - [legal] to register @ invention
— v **marca** f **de fabricación** f - [legal] to register @ trade mark
— v — **industrial** - [legal] to register @ trade
mark
— v — f **registrada** - [legal] to register @
trade mark
— v **nombre** m **comercial** - [legal] to register @
trade name
— v **nombre** m **de comercio** m - [legal] to regis-

ter @ trade name
registrar v̲ nombre de comercio m̲ - [legal] to
register @ trade name
— v̲ número m̲ - [comput] to enter @ number
— v̲ patente f̲ - [legal] to register @ patent
— v̲ — f̲ de invención f̲ - [legal] to register @
patent
— v̲ peso - [ind] to record @ weight
— v̲ presión f̲ - [instrum] to record @ pressure
— v̲ — f̲ para ensanchamiento m̲ - [instrum] to
record @ expansion pressure
— v̲ — f̲ para expansión f̲ - [instrum] to record
@ expansion pressure
— v̲ propiedad f̲ intelectual - [legal] to register @ copyright
— v̲ — f̲ literaria f̲ - [legal] to (register @)
copyright
— v̲ tiempo m̲ - [cronol] to, record, o̲ clock, @
time
— v̲ velocidad f̲ - to, record, o̲ clock, @ speed
registrarse v̲ (en hotel m̲) - to, register, o̲
check, in(to) @ hotel
regístrese v̲ y comuníquese v̲ - [legal] let it be
recorded and communicated
registro m̲ - . . . .; register; log; record(ing);
booking • checking in • [metal-prod] manhole;
[combust] damper • [ind] inspection hole;
cleanout • [pol] registry • [instrum] reading
• clocking • [legal] patent, o̲ trade mark,
office • [hidr] gate • [ind] processing
— m̲ centralizado - [ind] central control
— m̲ civil - [pol] vital statistics office
— m̲ comercial - [com] commercial reqistration
— m̲ con impresor m̲ - [electrón] printer recording
— m̲ — m̲ para escritorio m̲ - [electrón]
desktop printer recording
— m̲ — información f̲ - [ind] data control
— m̲ — f̲ sobre rueda(s) f̲ - [mec]ruedas]
wheel data control
— m̲ constante - [instrum] constant, o̲ continuous, record(ing)
— m̲ contable - [contab] accounting record •
book(s) of record
— m̲ corredizo - [combust] slide,ding damper
— m̲ de accidente(s) m̲ - [sequrid] accident(s)
record(s)
— m̲ — acción(es) f̲ - [legal] share, registry,
o̲ record
— m̲ — actividad(es) f̲ - activity,ties, re-
cord(s), o̲ recording
— m̲ — — f̲ industrial(es) - [pol] industrial
activity,ties, record, o̲ registry
— m̲ — actuación f̲ - [admin] performance record
— m̲ — aumento(s) m̲ - increase(s), record, o̲
showing
— m̲ — capital(es) m̲ - [fin] capital, recording, o̲ registration
— m̲ — — m̲ contable - [fin] book capital,
record(s), o̲ registration
— m̲ — carga f̲ - [ind] load record(ing)
— m̲ — Comercio m̲ - [legal] Commercial Registry
— m̲ — Compromiso(s) m̲ para Inversión(es) f̲
Extranjera(s) f̲ - [pol] Foreign Investment(s)
Commitment Registry,ration
— m̲ — constitución f̲ - [legal] incorporation
registration
— m̲ — contrato(s) m̲ - [pol] contract(s) registry • [legal] contract(s), recording, o̲ registering, o̲ registration
— m̲ — convenio(s) m̲ - [legal] agreement(s) registration
— m̲ — corrimiento m̲ - [electrón] shift(ing),
recording, o̲ registration
— m̲ — cumplimiento m̲ - [admin] compliance, o̲
performance, record(ing)
— m̲ — departamento m̲ - [ind] department record
— m̲ — derecho(s) m̲ - [legal] right(s), registration, o̲ registering
— m̲ — — m̲ sobre patente f̲ (de invención f̲) -
[legal] patent right(s) registering,tration
— m̲ — — m̲ literario(s) - [legal] copyright,
registry,tering

registo m̲ de desempeño m̲ - [admin] performance
record(ing)
— m̲ — desplazamiento(s) m̲ - [electrón] shift
register(ing)
— m̲ — desempeño m̲ - [admin] performance re-
cord(ing)
— m̲ — documento(s) m̲ público(s) - [legal]
public document(s) registry,ration
— m̲ — enmienda(s) f̲ - [legal] amendment(s) re-
cord(ing)
— m̲ — ensanchamiento m̲ - [instrum] expansion
record(ing)
— m̲ — ensanche m̲ - [instrum] expansion, re-
cord(ing), o̲ registration
— m̲ — ensayo(s) m̲ - test(s) record(ing)
— m̲ — entrevista(s) f̲ - [labor] contact(s)
record(ing)
— m̲ — — f̲ sobre seguridad f̲ - [sequrid] safety contact(s) record(ing)
— m̲ — escrituras f̲ (públicas) - [legal] public
instrument(s), registry, o̲ registration, o̲
recording
— m̲ — estatuto(s) m̲ - [legal] incorporation(s)
registry,ration
— m̲ — expansión f̲ - [instrum] expansion record
— m̲ — factura(s) f̲ - [com] invoice(s), recording, o̲ registering,tration
— m̲ — falla(s) f̲ - [ind] fault(s), o̲ failures,
record(ing), o̲ registering,tration
— m̲ — firma(s) f̲ - [legal] signature(s), registry, o̲ record(ing)
— m̲ — firma(s) f̲ interesada(s) - [legal] registry, o̲ record, of interested parties
— m̲ — garantía(s) f̲ - warranty,ties registration, o̲ register
— m̲ — f̲ de concesionario - [com] dealer('s)
warranty,ties registration, o̲ register,try
m̲ — importación(es) f̲ - [com] import(s)
(permits) record
— m̲ — incremento(s) m̲ - increase(s) record(s)
— m̲ — información f̲ - information, o̲ datum,ta,
record(ing), o̲ registration
— m̲ — — f̲ sobre ensayo(s) m̲ - test(ing), information, o̲ data,tum, recording
— m̲ — inspección(es) f̲ - [ind] inspection(s)
record • [constr] lamp hole
— m̲ — instrumento(s) m̲ - [legal] document(s)
registry,tration
— m̲ — — m̲ público(s) - [legal] public in-
strument(s) registry,ration
— m̲ — inversión(es) m̲ - [fin] investment(s),
registration, o̲ record
— m̲ — — f̲ extranjera(s) - [fin] foreign in-
vestments registry,ration
— m̲ — — f̲ nacional(es) f̲ - [fin] local in-
vestment(s) registry,tration
— m̲ — laminación f—[metal-lam] rolling record
— m̲ — mantenimiento m̲ - [ind] maintenance re-
cord(s)
— m̲ — marca(s) f̲ - [legal] (trade)mark(s),
registry, o̲ record
— m̲ — — f̲ de fabricación f̲ - [legal] trade
mark(s), registry,tration, o̲ registering
— m̲ — — f̲ industrial(es) - [legal] industrial, o̲ trade, mark(s), registry, o̲ registering, o̲ registration
— m̲ — — f̲ registrada(s) - [legal] trade
mark(s), registry,tering,tration
— m̲ — Minería f̲ - [miner] Mining Registry
— m̲ — modelo(s) m̲ industrial(es) - [legal]
industrial model(s) registry,tration; trademark
— m̲ — modificación(es) f̲ - [legal] change(s),
o̲ amendment(s), registry,tration
— m̲ — nombre(s) m̲ comercial(es) - [legal]
trade name registry.tration
— m̲ — — para comercio m̲ - [legal] business, o̲
trade, name registry,tration
— m̲ — número m̲ - [comput] number entering
— m̲ — operación(es) f̲ - [ind] operating,tion
record(ing)
— m̲ — patente(s) f̲ - [legal] patent(s) registry, o̲ registry,tration

registro m de patente f de invención f - [legal] patent registry,trating,tration
— m — **patrimonio m** - [legal] net worth registration
— m — **perforación(es) m** - [suelos] drilling record
— m — — f **analizado(s)** - [geol] analyzed perforation record(s)
— m — **pesada(s) f**—[ind] weight(s) record(ing)
— m — **planta f** - [ind] plant record(s)
— m — **presión(es) f** - [instrum] pressure(s) records,ding
— m — — m **para ensanchamiento m** - [instrum] expansion pressure record(ing)
— m — — f — **ensanche m** - [instrum] expansion pressure(s) record(ing)
— m — — f — **ensayo m** - **test(ing) pressure** record(ing)
— m — — f — **expansion f** - [instrum] expansion pressure(s) record(ing)
— m — **producción f** - [ind] production, o operating, record(ing)
— m — **producto(s) m** - [ind] product(s) record
— m — — m **elaborado(s)** - [ind] manufactured product(s) record(s)
— m — **propiedad f** - property record(s)
— m — — f **intelectual** - [legal] copyright(s), records, o registry,tration, o registering
— m — — f **literaria** - [legal] copyrighting
— m — **proveedor(es) m** - [com] supplier('s), registry, o list
— m — — m **de estado m** - [pol] government('s) supplier(s) registry
— m — **reinversión(es) f** - [fin] reinvestments registry,tration
— m — **reparación(es) f** - [ind] repair(s) record(s)
— m — **salida(s) f** - [metal-prod] manhole
— m — — f **para polvo m** - [metal-prod] dust catcher manhole
— m — — f — — m **cerrado** - closed manhole
— m — — f — — m **en colector m primario** - [metal-prod] dust catcher (outlet) manhole
— m — **servicio(s) m** - [ind] service(s) record
— m — — m **documentado(s)** - [labor] documented service(s) record
— m — **serie m** - [ind] series registry
— m — — f **a paralelo m** - [electrón] serial to parallel register
— m — — f — **en . . . etapa(s) f** - [electrón] . . .-stage serial to parallel register
— m — **tiempo(s) m**—[ind] time record(s)
— m — **transacción f** - [legal] transaction registration
— m — — f **de tecnología f** - [legal] technology transfer, record, o registration
— m — **valor m neto** - [fin] net value registration
— f — **velocidad f** - speed record(ing)
— m — **venta(s) f** - [com] sales, record, o register
— m **departamental** - [ind] department record(s)
— m **doble para limpieza f** - [válv] double cleanout
— m **documentado** - documented record
— m — **de servicio m** - documented service record
— m **en chimenea f** - [combust] stack damper • chimney valve
— m — **colector m para polvo m** - [metal-prod] dust catcher manhole
— m — **estado m** - test record(ing)
— m — **esfera f** - [instrum] registering on @ dial
— m — **hotel m** - [trans] hotel registering,tration, o checking in
— m — **trámite m** - [legal] requested registration
— m **entre** - registering between
— m **escrito** - written record
— m **general** - general registry
— m **gráfico** - [instrum] recording chart

registro m igual - [instrum| same reading
— m **impreso** - printed record • [electrón] hard copy record
— n **individual** - individual record
— m — **para rodillo m** - [metal-lam] individual roll record
— m **ininterrumpido** - [instrum] uninterrupted, o continuous, record
— m **litológico** - [geol] lithological record
— m **manual** - manual record(ing) • [mec] hand hole
— m — **en crucela f**—[mec] crosshead hand hole
— m **mayor** - [instrum] higher,est reading
— m **menor** - [instrum] lower,west reading
— m **Mercantil** - [pol] Commercial, Registry, o Record
— m **Nacional de Firma(s) f Consultora(s)**— [pol] National Consulting Firm(s) Registry
— m **para contrato(s) m** - [pol] contracts registry
— m — — m **para transferencia f de tecnología f** - [legal] technology transfer contracts registry
— m — **control m de calidad f** - [ind] quality control(s) record
— m **para corte m** - [válv] cut-off, register, o opening
— m — **entrada f** - [ind] inlet damper
— m — **inspección f** - [mec] inspection, hole, o port
— m — **limpieza f** - [mec] clean out cover • [válv] cleanout
— m — **padrón m de proveedores m** - supplier(s) (list) registry
— m — **tiraje m** - [combust] draft (regulating) damper
— m — **tiro m** - [combust] damper
— m — **ventilador m** - [mec] fan damper
— m **patronal** - employer(s) registry
— m **permanente** - permanente record • [electrón] hard copy record
— m **público** - [legal] public, registration, o recording • public registry
— m — **para comercio m** - [com] (public) commercial, register, o registry
— m — — **minería f** - [miner] public mining, o mining public, registry
— m — — **propiedad f** - [pol] public property, o property public, registry
— m **regulado** - [ambient] controlled damper
— m — **automáticamente** - [ambient] automatically, o instrument, controlled damper
— m **serie-paralelo** - [electrón] serial to parallel register
— m **sobre servicio m** - service record
— m **tributario m** - [fisc] tax(ation) registry
— m — **nacional** - [fisc] national taxation registry
— m **solicitado** - [legal] requested, o pending, registration, o recording
— m **vigente** - [legal] existing (valid) registration

regla f - . . . • standard; policy • card • [herram] straight edge
— f **aplicable** - applicable rule
— f **básica** - basic rule
— f **crítica** - critical rule
— f — **para seguridad f** - [segurid] critical safety rule
— f **de aplicación f** - applicable rule
— f — — f **amplia** - widely applicable rule
— f — **arte m** - standard procedure
— f — **bolsillo m** - [herram] pocket rule(r)
— f — **tres** - [matem] (simple) proportion
— f — — **sencilla** - [matem] simple proportion
— f **departamental** - departmental rule
— f — **para seguridad f** - [segurid] departmental safety rule
— f **elemental** - elementary rule
— f **empírica** - empirical rule • rule of thumb
— f **esencial** - essential rule
— f **específica** - specific rule
— f **establecida** - established rule

regla f exacta - exact rule
— f general - general rule • rule of thumb |
adv - (por) usually; as @ general rule
— f — sobre seguridad f - [segurid] general
safety rule
— f graduada - [instrum] scale(d ruler)
— f inamovible - immovable, o hard and fast,
rule
— f obligatoria - mandatory rule
— f — sobre seguridad f - [segurid] mandatory
safety rule
— f para apareamiento m - [autom-neumát] tire
fitment rule
— f — — m de neumático m - [autom-neumát]
tire firment rule
— f — bolsillo m - pocket rule(r)
— f — cálculo m - slide rule(r); calculator
— f — m para drenaje m - [hidr] drainage
calculator
— f — m — — m con tubo(s) m abovedado(s)
- [hidr] pipe-arch drainage calculator
— f — m — — m — — m circular(es) -
[hidr] round pipe drainage calculator
— f — calibre(s) m - gage
— f — circulación f - [vial] traffic rule
— f — conciliación f - [labor] conciliation
rule
— f — — f y arbitraje m - [labor] conciliation and arbitration rule
— f — enrasar v - [hormig] straightedge
— f — instalación f - [mec] installation,
rule, o practice
— f — — f apropiada - [ind] good installation
practice
— f — — f eléctrica - [electr-instal] electric, installation rule, o practice
— f — nivelación f - [constr] grade bar
— f — reemplazo m - replacement rule
— f — m de neumático(s) m - [autom-neumát]
tire(s) replacement rule
— f — seguridad f - [segurid] safety rule
— f — — f aplicable - [segurid] applicable
safety rule
— f — — f para aplicación f - [segurid] applicable safety rule
— f — — f — planta f - [segurid] plant
safety rule
— f — tránsito m - [transp] traffic rule
— f para trazar (líneas) - [dib] straightedge
— f — código m - [legal] code, ruling, o regulation
— f sencilla - simple rule
— f simple - simple rule
— f sobre seguridad f - [segurid] safety rule
— f — — f de aplicación f amplia -
[segurid] broad, o widely, applicable safety
rule
reglado,da* a - controlled
reglaje m - adjusting; setting • control(ling)
reglamentación f - regulating,tion; ruling(s);
requirement(s) • coding • [legal; law • implementation
— f aplicable - applicable, ruling(s), o regu-
lation(s)
— f de actuación f - [admin] performance regulating,tion
— f aplicación f - applicable ruling(s)
— f — — f amplia - widely applicable rulings
— f — código m - [legal] code, ruling(s), o
regulation(s)
— f — cumplimiento m - [admin] compliance, o
performance, regulating,tion
— f — desempeño m - [admin] performance regulating,tion
— f — Dirección f para Protección f de Ambiente m - [ambient] Environmental Protection
Administration, regulation, o ruling
— f — gobierno m - government, regulating,tion
— f — ley f - [legal] coding of @ law
— f — Pacto m Andino - [pol] Andean Pact ruling
— f establecida - established ruling
f estricta f - strict, o stringent, regula-

ting,tion, o requirement
reglamentación f federal - [legal] federal regulating,tion
— f interna - internal ruling(s)
— f local - [legal] local, regulation, o law
— f municipal - [pol] city, code, o ruling(s)
— f para adquisición f - [com] procurement regulating
— f para salud f - [medic] health regulations
— f — seguridad f - [segurid] safety regula-
tion(s)
— f — trabajo m en curso - [ind] work in progress regulation(s)
— f — — m — ejecución f - [labor] work in
progress regulation(s)
— f radiológica - [nucl] radiological regula-
tion(s)
— f — aplicable - [segurid] applicable radiological regulation(s)
— f — para salud f - [nucl] radiological
health regulation(s)
— f radiológica para salud f - [segurid] applicable radiological health regulation(s)
— f vigente - [pol] existing regulation(s)
reglamentar v adquisición f - to regulate @
procurement
reglamentario,ria a - . . .; statutory
reglamento m - . . .; rule; ruling
— m de departamento m de transporte(s) m -
[pol] Department of Transportation Regula-
tion(s)
— m de empresa f - [com] company rule
— m establecido - established rule,ling
— m — de empresa f - established company rule
— m gubernamental - [pol] government, rule, o
regulation, o ruling
— m local - [pol] local requirement; city ordinance
— m oficial - official ruling
— m para salud f - [segurid] health regulation
— m — seguridad f - [segurid] safety, rule, o
regulation
— m — — f estadounidense para transporte(s) m
automotor(es) - [transp] Federal Motor Car-
rier(s) Safety Regulation(s)
— m — tránsito m - [transp] traffic rule(s)
reglar v - . . .; to control
reglas f - rulings; good practice
— f para instalación(es) f eléctrica(s) -
[electr-instal] good electrical practice
— f y Reglamento(s) de Comisión f Reguladora Nuclear (Estadounidense) - [nucl] Nuclear Regulatory Commission Rules and Regulations
regleta f - [imprent] . . .; separator
reglilla f - [regla de cálculo] slide
regodear v - . . .; to luxuriate
regresar v - . . .; to come back
reguera f - . . . • [metal-prod] trough; runner
— f para horno m alto - [metal-prod] blast furnace runner
regulable a - controllable; governable • adjustable
— a con mano f - hand controllable
regulación f - . . .; control; controlling; adjusting,tment; control; setting; timing; regulating; set(ting) up • sanctioning • [electr-
-oper] switching; toggling
— f desde distancia f - [electr-oper] remote
control(ling)
— f — — f de amperaje m - [electr-oper] remote current control
— f — — f — interruptor m - [electrón] remote breaker control
— f — — f — polaridad f - [sold] remote
polarity control
— f abierta f - [electr-oper] open control
— f adicional - additional, o extra, control
— f ajustada - adjusted control • tight control
— f alta - [ind] high setting
— f — de amperaje m - [sold] high current
setting
— f — de voltaje m - [sold] high voltage setting

regulación f alta de voltaje m en circuito a-
bierto - [sold] high open circuit voltage
setting
— f — en vacío m - [sold] high idle
— f — sin carga f - [sold] high idle
— f amplia - [ind] broad adjustment
— f — de voltaje m - [sold] broad voltage adjustment
— f analógica - [comput] analogical control
— f — electrónica - [electrón] electronic analogical control
— f — — f para caldera f - [cald] boiler
electronic analogical control
— f — — para sistema m - [ind] system('s)
electronic analogical control
— f — — — m auxiliar - [ind] auxiliary
system('s) electronic analogical control
— f — para caldera f - [cald] boiler analogical control
— f — — sistema m - [ind] system('s) analogical control
— f — — — m auxiliar - [ind] auxiliary sys-
tem('s) analogical control
— f anticipada - [ind] (pre)setting
— f apropiada - appropriate, o proper, setting, o control(ling)
— f aproximada - approximate, o coarse, o
rough, o broad, setting, o adjusting,tment
— f — de característica(s) f - [sold] broad
characteristic(s), setting, o adjusting,tment
— f — — — f para arco m - [sold] broad arc
characteristic(s), setting, o adjusting,tment
— f automática - automatic, adjustment, o control; fully automatic, control, o regulation
— f — de altura f - [mec] automatic height
control
— f — — — f para cabezal m - [agric-equip]
automatic header height control
— f — para agua m - [ind] water automatic, o
automatic water, control
— f — — m en molde m - [metal-prod] mold
water automatic control
— f — — carga f - [ind] automatic charge,ging
control
— f — — — f de coque m - [metal-prod] automatic coke charge,ging control
— f — combustión f - [combust] automatic
combustion control
— f — espesor m - [metal-lam] automatic,
gage, o thickness, control
— f — — posición f - automatic position control
— f — — temperatura f - [ind] automatic temperature control
— f — — — f para viento m - [metal-prod]
blast temperature automatic, o automatic
blast temperature, control
— f — — tiro,raje - [combust] automatic
draft, o draft automatic, control
— f — viento m caliente - [metal-prod] hot
blast automatic, o automatic hot blast, control
— f — sin carga f - [ind] automatic idle
— f automatizada - [ind] automated control
— f — para espesor m - [metal-lam] automated
gage control
— f auxiliar - [ind] auxiliary control
— f baja - [ind] low control | adv - (de) low
idle
— f — para amperaje m - [sold] low current
setting
— f — — voltaje m—[sold] low voltage setting
— f — — m en circuito m abierto - [sold]
low open circuit voltage setting
— f — sin carga f - [mec] low idle speed
— f bipolar - [electr-oper] two-pole control
— f cambiada - [mec] changed setting
— f compleja - complex control
— f completa - complete control
— f — para amperaje m - [sold] complete current control
— f — — encendido m - [comb.int] complete
(spark) timing

regulación f completa para gas m - [combust]
complete gas control
— f comprensiva - [ind] comprehensive control
— f — para inundación(es) f - [hidr] comprehensive flood control
— f comprobada - checked, setting, o control
— f común - regular, o common, control, o setting
— f con botonera f - [ind] push button control
— f — cable m - [mec] cable control
— f — . . . circuito(s) m - [ind] . . .-loop
control(ling)
— f — computador m - [comput] computer con-
trol(ling)
— f — cuadrante m—[ind] dial control(ling)
— f — dos circuito(s) - [ind] two loop con-
trol(ling)
— f — esipigón(es) m - [hidr] jetty con-
trol(ling)
— f — muro(s) m - [hidr] wall control(ling)
— f — — m para, contención f, o retención f -
[hidr] retaining wall control(ling)
— f — pedal m - [mec] (foot) pedal con-
trol(ling)
— f — pulsador(es) m - [ind] push button
control(ling)
— f con regulador m - [mec] governor con-
trol(ling)
— f — — m automático - [mec] (automatic)
governor control(ling)
— f — tablestaca(s) f - [hidr] sheet(ing)
control
— f — — f de acero m - [hidr] steel sheeting
control(ling)
— f — tapa f - [mec] cap control(ling)
— f — — f para regulador m - [int.comb] radiator cap, control(ling), o regulating
— f — teclado m - [electrón] (key)board con-
trol(ling)
— f constante - constant, o continuous, control
— f continua - continuous control(ling)
— f — de amperaje m - [electr-oper] continuous, amperage, o current, control(ling)
— f — — — m mediante manivela f - [sold]
crank-type continuous current control(ling)
— f — — — m — manubrio m - [sold] crank
type continuous current control(ling)
— f — — — m y voltaje m - [sold] dual continuous control(ling)
— f — — potencia f - [electr] power continuous control(ling)
— f — — tipo m con cuadrante m - [ind] continuous dial-type control(ling)
— f — — voltaje m - [electr-oper] continuous
volt(age) control(ling)
— f — doble - [sold] dual continuous control
— f — mediante dos reguladores m - [sold] dual
continuous control
— f — para voltaje m - [sold] continuous voltage, control, o adjustment
— f — — — m en circuito m abierto - [sold]
continuous open circuit control
— f continuada - [ind] continued control
— f contra (re)torcedura(s) f - [mec] twist
control(ling)
— f coordenada - coordinated control
— f conveniente - convenient control
— f correcta - [comb.int] correct, o accurate,
timing
— f — para encendido m - [comb.int] accurate,
o correct, timing
— f cuidadosa - carful control(ling)
— f de abastecimiento m - supply(ing) control
— f — — m de combustible m - [comb.int] fuel
supply control
— f — aberutra f - [mec] opening control(ling)
— f — — f en estrangulador m - [comb.int]
choke opening control(ling)
— f — accionamiento m - [mec] drive control
— f — — m para mesa f rotativa - [petról]
rotary drive control
— f — acelerador m - [comb.int] throttle con-
trol(ling)

regulación f de adherencia f

regulación f de adherencia f - sticking control
— admisión f - [comb.int| inlet control
— f — agua m - [hidr] water, regulating,tion, o control
— f — — m freática - [constr] groundwater control
— f — — m para enfriamiento m - [ind] cooling water control
— f — — m — freno m - [mec] brake cooling water control
— f — — m — — m — — m auxiliar - [mec] auxiliary brake cooling water control
— f — — m — — — llanta f - [mec] rim cooling water control
— f — — m — — — m — — f para freno m—[mec] brake rim cooling water control
— f — — m pluvial - [hidr| rain, o storm, o drainage, (water) control
— f — — m sobre superficie f - [hidr] surface water control
— f — aguilón m - [grúas] boom control
— f — aire m - [neumát] air control
— f — — m fresco - [ambient] fresh air control(ling)
— f — ajuste m - [mec] setting control(ling)
— f — — m para amperaje m - [sold] current setting, control(ling), o change
— f — alimentación f - [ind] feed(ing) control
— f — — f para mesa f rotativa - [petról] rotary (table) feed control(ling)
— f — alimentador m para alambre m - [sold] wire feeder,ding control(ling)
— f — altura f - [mec] height, control(ling), o setting
— f — — f para cabezal m - [agric] header height control(ling)
— f — espigadora f - [agric-equip] header height control(ling)
— f — ambiente m - [ambient] environmental, o climate, management, o control, o engineering, control
— f — amperaje m - [electr-oper] amperage,res control • [sold] current, o heat, adjustment, o control, o setting
— f — — m ajustada - [sold] adjusted current control
— f — — m aproximado - [sold] coarse current adjustment
— f — — m cambiada - [sold] changed current, setting, o control
— f — — (mediante regulador m) de tipo m con cuadrante - [electr-oper] dial-type current control
— f — — m elegido - [sold] chosen current setting
— f — — m escogido - [sold] chosen current setting
— f — — m para encendido m - [sold] starting current adjustment
— f — — m — soldadura f - [sold] welding, current control, o heat adjustment
— f — — m seleccionado - [sold] chosen current setting
— f — — m y de voltaje m - [sold] current and voltage control
— f — — m por escalón(es) m - [sold] step type current control
— f — acho(r) m - [mec] width setting
— f — — m para laminación f - [metal-lam] mill width setting
— f — antepozo m - [petról] cellar control
— f — aportación f - [ind] input control(ling)
— f — — f de calor m - [sold] heat input control(ling)
— f de arco m - [sold] arc control(ling)
— f — — m excelente - [sold] excellent arc control(ling)
— f — — m optativa - [sold] optional arc control • arc control option
— f — — m variable - [sold] variable arc control(ling)

regulación f de artefacto m - [electr-oper] appliance, o convenience, control(ling)
— f — Askania m - [metal-prod] Askania control(ling)
— f — atmósfera f - [ambient] atmosphere, o climate, control
— f — — f con argón m - [sold] argon atmosphere control
— f — — f con gas m de argón m - [sold] argon gas atmosphere control
— f — — f de argón m gaseoso - [sold] argon gas atmosphere control
— f — — f — gas m de argón m - [sold] argon gas atmosphere control
— f — — f protectora f - [metal-trat] protective atmosphere control(ling)
— f — avance m de barreno,na - [petról] rotary feed control
— f — — m gradual - [sold] inching control
— f — baño m - [metal-prod] bath control • [sold] puddle control
— f — — m con gráfico m para carbono-temperatura - [metal-prod] carbon-temperature, graph, o chart, bath control
— f — — m para soldadura f - [sold] weld(ing) puddle control
— f — basicidad f de escoria f - [metal-prod] slag basicity control
— f — batería f de bomba(s) f - [bombas] pump(s) battery control
— f — cabeza f - head control
— f — cabezal m - [sold] head control • [agric-equip] header control
— f — cable m - [mec] cable control
— f — — m a gancho m - [grúas] whip (hoist) control(ling)
— f — — m — — m verificada—[grúas] checked whip (hoist) control
— f — — m a motón m - [grúas] main (hoist) control(ling)
— f — — m — — m verificada - [grúas] checked main hoist control
— f — caída f - [mec] fall, o drop, control
— f — — f de cabezal m - [agric-equip] header drop control
— f — — f de espigadora f - [agric-equip] header drop control
— f — — f de regulador m giratorio - [electr] Rototrol drop control
— f — — f — — m giratorio para motor m - [electr-mot] motor Rototrol drop control
— f — calidad f - [ind] quality control
— f — — f de coque m - [coque] coke quality control
— f — calorímetro m - [metal-prod] calorimeter control
— f — — f — superficie f final - [metal-lam] finish(ed) surface quality control
— f — calor m - heat control(ling)
— f — cambiador m - [ind] changer setting
— f — — m para toma(s) m - [electr-oper] tap changer setting
— f — campo m - [electr-oper] field control
— f — (cantidad f de) escoria f en cucharón m - [metal-prod] ladle slag quality control
— f — capacidad f - capacity control(ling)
— f — característica(s) f - characteristic(s), control, o adjustment
— f — — f de arco m - [sold] arc characteristic(s), control, o adjustment, o tuning
— f — — f — — m para (cada) aplicación f - [sold] arc characteristic(s) tuning for each application
— f — — f física(s) f - physical characteristic(s) control(ling)
— f — — f para encendido m - [sold] striking characteristic(s) control
— f — — f química(s) - [quím] chemical characteristic(s) control
— f — carburador m - [comb.int] carburetor, setting, o adjusting
— f — — m para mezcla f (muy) pobre - [comb.-int] carburetor (very) lean setting

regulación f de carburador m para mezcla f (muy) rica - [comb.int] carburetor (very) rich set-setting
— f — carga f - [constr] load control • [metal-prod] charge,ging control • skip house
— f — f automática - [metal-prod] automatic charge,ging control
— f — f — de coque m - [metal-prod] automatic coke charging control
— f — f de coque m - [metal-prod] coke charge,ging control
— f — f de horno m - [metal-prod] furnace charge,ging control
— f — f para alternador m - [electr-prod] alternator load control(ling)
— f — carrera f - [mec] stroke control(ling)
— f — carrete m - [mec] reel, o spool, control
— f — m para núcleo(s) m - [petról] core reel control(ling)
— f — caudal m - flow control(ling)
— f — m de agua m - [hidr] water flow control(ling
— f — m — retorno m - [combust] return flow control(ling)
— f — m variable - variable flow control
— f — ciclo m - cycle control(ling)
— f — m para avance m - [sold] travel cycle control(ling)
— f — m para desviación f - by-pass cycle control(ling)
— f — m — soldadura f - [sold] weld(ing) cycle control(ling)
— f — f y avance m - [sold] welding and travel cycle control(ling)
— f — cierre m - [mec] closing control • closure control(ling)
— f — m para estrangulador m - [mec] choke closing,sure control
— f — circuito m - [electrón] circuit control(ling)
— f — cizalla f - [mec] shear control(ling)
— f — f para planchón(es) m - [metal-lam] slab shear control(ling)
— f — combustible m - [combust] fuel, control, o regulating
— f — combustión f - [ind] combustion control
— f — compuerta f - [hidr] gate control(ling)
— f — f exclusiva OR - [electrón] exclusive OR gate control(ling)
— f — f OR - [electrón] OR gate control
— f — f para sótano m - [petról] cellar gate control(ling)
— f — conformación f - shape control(ling)
— f — f de cordón m - [sold] bead shape control(ling)
— f — conmutador m - [electr-oper] switch, setting, o positioning
— f — m para selección f - [electr-oper] selector switch setting
— f — m — — f de amperaje m - [sold] current selector switch, setting, o position
— f — m para sentido(s) m - [electrón] sense switch, setting, o toggling
— f — contaminación f - pollution control
— f — f ambiental - [ambient] environment, o air, pollution control
— f — f atmosférica - [ambient] air pollution control
— f — f de agua m - [hidr] water pollution control
— f — f — aire m - [ambient] air pollution control
— f — f — ambiente m - [ambient] environment pollution control
— f — f — medio m ambiente - [ambient] environmental pollution control
— f — contenido m - [ind] content control
— f — m de oxígeno m—oxygen content control
— f — m — m en acero m - [metal-prod] steel oxygen content control
— f — cordón m - [sold] bead control
— f — corriente f - current control

regulación f de corriente f alterna - [electr-distrib] alternating current control
— f — f — con . . . voltio(s) m - [electr-distrib] . . . volt alternating current control
— f — f continua - [electr-distrib] direct current control
— f — f — en . . . voltio(s) m - [electr] . . .-volt direct current control
— f — f imantadora - [electr] magnetizing current control
— f — f para magnetización f - [electr] magnetizing current control
— f — f — salida f - [sold] output (current), adjustment, o control
— f — f — f constante - [sold] continuous output (current) control
— f — f para soldadura f - [sold] welding current, setting, o control, o adjustment
— f — corte m - [sold] stop(ping) control
— f — cráter m - [sold] crater, o molten puddle, control
— f — m para soldadura f - [sold] weld(ing) puddle control
— f — creciente(s) f - [hidr] flood control
— f — cuadrante m - [instrum] dial control
— f — chatarra f - [metal-prod] scrap control
— f — chispa f - [comb.int] spark control
— f — f para encendido m - [comb.int] ignition, spark control, o timing
— f — declive m - [topogr] slope control(ling)
— f — deformación f - [ind] deformation control
— f — demanda f - demand control
— f — f de usina f - [Electr-prod] power plant demand control
— f — densidad f - density control
— f — descarga f - [hidr] discharge control
— f — en entrada f - [constr] inlet discharge control(ling)
— f — f — separador m - [metal-prod] separator discharge control
— f — f — m de Venturi m - [metal-prod] scrubber separator discharge control
— f — f para drenaje m - [hidr] drainage discharge control
— f — descongelación f - [criog] defrosting control
— f — descongelador m - [refrig] defroster control
— f — desempañador m - [autom] defroster control; windshield wiper control
— f — dirección f - [autom] steering, o directional, control
— f — f mecanizada - [autom] power steering control
— f — disminución f - [electr] taper(ing) control
— f — f de amperaje m - [sold] current taper(ing) control
— f — f — f — m para relleno m de cráter m - [sold] crater filling current taper control
— f — dispositivo m - [ind] device control • [electr-oper] appliance, o convenience, control
— f — m para avance m - [sold] travel unit, adjusting,tment, o control
— f — distancia f - distance control
— f — f entre cordón(es) m - [sold] step-over control
— f — distorsión f - [mec] distortion control
— f — f de chapa f (de acero m) - [metal-fabr] (steel) sheet metal distortion control
— f — distribución f - [ind] distribution control
— f — disyuntor m - [electr-oper] circuit breaker control
— f — dosificación f - [quím] dosis control • [combust] proportional control
— f — efervescencia f - effervescence control • [metal-prod] rimming (action) control
— f — elevación f - [grúas] hoist(ing) control
— f — f de aguilón m - [grúas] boom hoist-

-(ing) control
regulación f de elevación f de aguilón m verificada—[grúas] checked boom hoist(ing) control
— f —— f verificada - [grúas] checked hoist-
-(ing) control;
— f — elevador m - [grúas] hoist control
— f —— m principal - [grúas] main hoist control
— f —— m — verificada - [grúas] checked
main hoist control
— f — emanación(es) f - [ind] emission(s) control
— f —— f en deshornamiento m - [coque]
pushing emission(s) control
— f — enbrague m - [mec] clutch control
— f —— m para mesa f (rotativa) - [petról]
rotary (table) clutch control
— f — emisión(es) f - [ind] emission(s), o
discharge, control
— f — encendido m - [comb.int] ignition, o
spark, control; timing • [sold] strike,king
control
— f —— m completada - [comb.int] completed
timing
— f —— m comprobada - [comb.int] checked
timing
— f —— m — nuevamente - [comb.int] rechecked timing
— f —— m en motor m - [comb.int] engine timing
— f —— m marcada - [comb.int] marked timing
— f —— m y corte m - [sold] start(ing) and
stop(ping) control
— f — enfriamiento m - [ind] cooling control
— f —— m primario - [ind] primary cooling
control(ling)
— f —— m secundario - [ind] secondary cooling control(ling)
— f — engargantado - [mec] meshing control
— f —— m de engranaje m - [mec] gear meshing
control
— f — engranaje m - [mec] gage control • engagement control
— f — entrada f - [hidr] inlet control
— f —— f de aire m - air intake control
— f —— f —— m tipo m celosía f - [comb.-
-int] louver type air intake control
— f — equilibrio m de temperatura f - [ind]
balanced temperature control
— f — equipo m - [ind] equipment control •
equipment setting
— f —— m magnético - [electr] magnetic
equipment control
— f — erosión f - [hidr] erosion control
— f — escombro(s) m - [hidr] debris control
— f — escoria f - [sold] slag control
— f —— f por ruido m de lanza f - [metal-
-prod] lance noise slag control
— f — espesor m - [metal-lam] thickness, o
gage, control, o setting
— f —— m con rayos-X - [metal-lam] X-ray
gage control
— f —— m de laminación f - [metal-lam] mill
gage setting
— f — espigadora f - [agric-equip] header
control
— f — estabilidad f - stability control
— f —— f de arco m - [sold] arc stability
control
— f — estrangulador m - [comb.int] throttle, o
choke, control, o setting
— f —— m por operador m - [mec] operator
choke control(ling)
— f — exportación(es) f - [com] export(s) con-
trol(ling)
— f —— f de chatarra f - [metal-prod] scrap
export(s) control
— f — factor m - factor control(ling)
— f — filete m - [sold] flash control
— f —— m interior - [sold] flash control
— f — filtración f - [hidr] filtration, o
seepage, control
— f — flotador m - [comb.int] float setting

regulación f de fluido(s) m - fluid(s) control
— f — flujo m - flow control
— f —— m de agua m - [hidr] water flow control
— f —— m — combustible m - [combust.int]
fuel flow control
— f — freno m - [mec] brake, control(ling), o
adjusting,tment
— f —— m contra rotación f - [grúas] swing
brake control(ling)
— f —— m — accionamiento m - [mec] drive
brake control(ling)
— f —— m — m para mesa f rotativa -
[petról] rotary drive brake control(ling)
— f —— m por inercia f - [mec] inertia brake
control(ling)
— f — fuente f para energía f - [electr-oper]
power source, control, o setting
— f — gas m - [combust] gas control
— f —— m de argón m - argon gas control
— f —— m inerte - inert gas control(ling)
— f — gasto m - expense control • [hidr] flow
control
— f —— m de agua m - [hidr] water flow control
— f — generador m - [electr-prod] generator
control
— f — giro m - [grúas] swing control
— f —— m de tornillo m sin fin - [mec] auger
swing control
— f — gobierno m - [pol] government control
— f — grano m - [metal] grain control
— f — granulometría f - [metal-prod] grain
(size) control
— f — grieta(s) f - [sold] crack(ing) control
— f —— f interior(es) - [sold] internal
crack(s) control
— f — grúa f - [grúas] crane control
— f — hoja f - [mec] blade control
— f —— f para topadora f - [mec] (bull)dozer
blade control
— f — horno m - [combust] furnace control •
oven control
— f —— m eléctrico - [metal-prod] electric,
o arc, furnace control
— f — humedad f - humidity, o moisture, control
— f —— f de viento m - [metal-prod] (air)
blast moisture control
— f — importación f - import(s) control(ling)
— f —— f de chatarra f - [metal-prod] scrap
(iron) import(ation) control
— f — indicador m - signal(ling) control
— f —— m para viraje(s) m - [autom] turn(s)
signal control
— f — inductancia f - [electr-oper] inductance
control
— f —— f variable - [electr-oper] variable
inductance control
— f — inmisión f - [ind] intake control
— f — intensidad f - intensity control
— f —— f de calor m - [sold] heat intensity
control
— f — interruptor m - [electr-oper] breaker, o
switch, setting, o control
— f —— m en vacío m - [electr-oper] vacuum
breaker, o switch. control
— f —— m para circuito m - [electr-oper]
circuit breaker control
— f — inundación(es) f - [hidr] flood control
— f — inyección f - [metal-prod] injection
control
— f —— f de vapor m - [metal-prod] steam injection control
— f —— f —— m en viento m - [metal-prod]
blast steam injection control
— f — laminador m - [metal-lam] (rolling)
mill control(ling)
— f — largo(r) m - [mec] length, setting, o
control
— f —— m para laminación f - [metal-lam]
(rolling) mill length setting
— f — leva f - [mec] cam, setting, o control

regulación f de línea f - [tub] line control
— f — — f para aceite m - [ind] oil line con-trol
— f — — f — agua m - [combust] water line control
— f — — f — fuel oil m - [comb.int] fuel oil line control
— f — líquido m - liquid control
— f — magnetización f - [electr-oper] mag-ntization,zing control
— f — máquina f - [mec] machine control
— f — marcha f - [mec] operation control • speed control
— f — — f sin carga f - [comb.int] idler,ling control
— f — — f — — f lenta f - [comb.int] slow idling, setting, o control(ling)
— f — — f — — f rápida - [comb.int] fast idling, setting, o control
— f — medio ambiente m - [ambient] environ-mental control
— f — metal m - [metal] metal control
— f — — m fundido - [metal-prod] molten metal control • [sold] molten puddle control
— f — método m - [ind] method control
— f — mezcla f - mixture, o blend, control
— f — — f aire m y gas m - [metal-prod] air--gas blend control
— f — motor m - [electr-mot] motor control--(ling) • [comb.int] engine, setting, o con-trol(ling)
— f — — m con amplificación f magnética - [electr-oper] magnetic amplifier motor con-trol(ling)
— f — — m hidráulico - [hidr] hydraulic motor control(ling)
— f — — m monofásico - [electr-oper] single phase motor control(ling)
— f — — m para fuerza f motriz - [electr-mot] power motor control • [sold] engine, o motor, speed set for power
— f — — m para soldadura f - [sold] motor, o engine, speed set for welding
— f — movimiento m - [mec] movement control
— f — nivel m - [electrón] level control(ling)
— f — — m de agua m - [hidr] water level con-trol(ling)
— f — — m — — m para freno m - [mec] brake motor water level control(ling)
— f — — m — energía f - [electr-distrib] power level control(ling)
— f — — m — líquido - liquid level control
— f — — m en separador m - [metal-prod] sepa-rator level control(ling)
— f — — — — m para Venturi m - [metal--prod] Venturi separator level control(ling)
— f — — m para suministro m de energía f - [electr-distrib] power level supply con-trol(ling)
— f — operación f - [ind] operating,tion con-trol(ling)
— f — — f de equipo m - [ind] equipment ope-ration(al) control
— f — orificio m - [mec] opening control(ling)
— f — — m en estrangulador m - [mec] choke opening control(ling)
— f — oxidación f - [metal] oxidation,dizing control(ling)
— f — palanca f - [mec] lever, control(ling), o positioning, o setting
— f — — f para regulador m—[comb.int] throt-tle lever, setting, o positioning
— f — palanquilla f - [metal-lam] billet con-trol(ling)
— f — pasada(s) f - [metal-lam] pass(es) set-ting
— f — pedal m - [mec] (foot) pedal controlling
— f — penetración f - [sold] penetration con-trol(ling)
— f — perforación f - [petról] drilling con-trol(ling)
— f — perilla f—[mec] bulb, o handle, setting
— f — — f para control m para amperaje m -
[sold] current control, knob, o handle, set
regulación f de perilla f - knob, o handle, setting
— f — — f para regulación f - [sold] control, knob, o handle, setting
— f — — f — — f para amperaje m - [sold] current control, knob, o handle, setting
— f — período m (de tiempo m) - (time) period adjusting,tment
— f — pistola f - [sold] gun operating control
— f — polaridad f - [sold] polarity control
— f — posición f - position control(ling) • positioning
— f — — f de cabeza f - [sold] head position, adjusting,tment, o control
— f — potencia f - power control
— f — potencial - [electr] potential control
— f — — m constante - [electr] constant po-tential control
— f — pozo m - [petról] well control
— f — — m corriente - [petról] conventional well control
— f — — m en tierra f - [petról] conven-tional, o on land, well control
— f — — m submarino - [petról] subsea well control
— f — precipitación f - [metal-prod] precipi-tation control
— f — — f de carburo(s) m - [metal-prod] carbide(s) precipitation control
— f — presión f - pressure, control,ling, o regulating,tion
— f — — f con tapa f para radiador m - [comb.int] pressure regulating with @ radia-tor cap
— f — — f de combustible m - [combust] fuel pressure control
— f — — f en hogar m - [combust] hearth pres-sure control
— f — — f en línea f - line pressure control
— f — — f — — f para aceite m - [ind] oil line pressure control
— f — — f — — f para combustible m - [combust] fuel line pressure control
— f — — f — — f — — fuel oil m - [combust] fuel oil line pressure control
— f — — f en tubería f - [tub] line pressure control
— f — — f — — f para aceite m - [ind] oil line pressure control
— f — — f — — f — combustible m - [combust] fuel line pressure control
— f — — f — — f — — fuel oil m - [combust] fuel oil line pressure control
— f — problema(s) m - problem(s) control
— f — — m de contaminación f (de ambiente m) - [ambient] (environment) pollution problems control
— f — procedimiento m—procedure(s) control
— f — — m con electrodo(s) m - [sold] elec-trode(s) procedure control(ling)
— f — — m — — m con alma m fundente—[sold] Innershield procedure control(ling)
— f — — m Innershield - [sold] Innershield procedure control(ling)
— f — — m para soldadura f - [sold] welding procedure control(ling)
— f — proceso(s) m - [ind] process(es) con-trol(ling)
— f — — m en circuito m - [cuerp.moled] cir-cuit process(es) control
— f — — m — — m para regulación f - [cuerp--moled] grinding circuit process control
— f — — m para minería f - [miner] mining process(es) control
— f — producción f - production control
— f — producto(s) m - [ind] product(s) control
— f — — m de acero m - [metal-prod] steel product(s) control(ling)
— f — — m siderúrgico(s) - [metal-prod] steel product(s) control(ling)
— f — proporción f - [quím] content control
— f — — f aire-gas - air-gas blend control

regulación f de proporción f de oxígeno m -
[metal-prod] oxygen content control
— f —— f —— m en acero m - [metal-prod]
steel oxygen content control
— f — provisión f - [ind] supply(ing) control
— f —— f de combustible m - [comb.int] fuel
supply(ing) control
— f — puesta f en marcha f - [ind] start(ing)
control
— f —— f —— f y detención f - [ind]
start(ing) and stop(ping) control
— f — punto(s) m - [mec] point(s) controlling
— f —— m de presión f - [petról] pressure
point setting
— f —— m —— f en tubería f - [petról]
pipe pressure point setting
— f —— m —— f —— f para entubación f -
[petról] casing pressure point setting
— f —— m para regulación f - [mec] set(ting)
point control(ling)
— f —— m —— f para presión f - [mec]
pressure set(ting) point control(ling)
— f —— m —— f —— f en tubería f -
[petról] pipe pressure point set(ting) con-
trol(ling)
— f —— m —— f —— f —— f para entubación f - [petról] casing pressure point
set(ting) control(ling)
— f — rayos-X - [electrón] X-ray control(ling)
— f — rapidez f - speed, o rate, control(ling)
— f —— f para enfriamiento m - [sold] cooling rate control(ling)
— f — razón f - [ind] ratio control(ling)
— f —— f entre aire m y combustible m -
[combust] fuel-air ratio control(ling)
— f — rebalse m - [hidr] outflow, o overflow,
control
— f —— m de agua m - [hidr] water, outflow,
o overflow, control
— f —— recorrido m - [ind] travel control(ling)
— f —— m para cubeta f - [metal-prod] skip
(car) travel control(ling)
— f —— m para montacarga(s) m - [metal-prod]
skip travel control(ling)
— f —— m de skip m - [metal-prod] skip travel control(ling)
— f — red f - [ind] system, o network, control
— f —— f auxiliar - [ind] auxiliary, system,
o network, control(ling)
— f —— régimen m - [mec] rate control
— f —— m para carga f - [electr-oper] charging rate control(ling)
— f —— m — movimiento m - [mec] movement
rate control(ling)
— f —— m —— m de tapón n - [mec] plug
movement rate control(ling)
— f —— m —— m —— m hacia adentro -
[mec] inwar plug movement rate control(ling)
— f —— m —— m —— m —— en asiento m
- [mec] plug movement into @ seat rate con-
trol(ling)
— f —— m —— m —— m para estrangulador
m - [mec] choke plug movement rate control
— f —— m —— m —— m —— m hacia adentro en asiento m - [mec] choke plug movement
into @ seat rate control(ling)
— f —— m —— m hacia adentro en asiento m
- [mec] movement into @ seat rate control
— f — regulador m - [mec] governor, control, o
regulating,tion • [combust] throttle control
— f —— m para aire m - [neumát] air regulator
setting
— f — relé m para sobrecarga f - [electr-oper]
overload relay control(ling)
— f —— m interruptor m - [electr-oper] trip-
-(ping) relay control
— f —— m por falla(s) f - [electr-oper]
fault trip(ping) relay control
— f — relé m maestro - [electr-equip] master
relay control(ling)
— f —— m para fase(s) f - [electr-equip]
phase relay setting
— f —— m principal - [electr-oper] master
relay control(ling)

regulación f de resorte m = [mec] spring adjusting,tment
— f — río(s) m - [hidr] river(s) control
— f — rotación f - [grúas] swing, o rotation,
control
— f —— f de tornillo m sin fin - [mec] auger
swing control
— f — rueda f - [mec] wheel control(ling)
— f —— f directriz - [agric-equip] guide
wheel control(ling)
— f — salida f - [hidr] outlet control
— f — sensor m - [instrum] sensor control
— f —— m para calor m - [instrum] heat sensor control
— f —— m — rotadora/alimentadora f -
[cuerp-moled] rotator/feeder heat sensor
control
— f de séptum m - [metal-prod] septum valve
control
— f — servodirección f - [autom] power
steering control
— f — sistema f - system, control(ling), o
setting up
— f —— m auxiliar - [ind] auxiliary system
control(ling)
— f —— m para encendido m - [combust] igni-
rion system control(ling)
— f —— m — enfriamiento m - [metal-lam]
cooling system control(ling)
— f —— m —— m de rodillo m - [metal-lam]
roll(er) cooling system control(ling)
— f — soldadora f - [sold] welder, setting, o
controlling
— f — soldadura f - [sold] weld(ing) control
— f —— f soplador m - [mec] blower, setting, o
controlling
— f — sótano m - [petról] cellar control(Ling)
— f — suministro m - [ind] supply(ing) con-
trol(ling)
— f —— m de combustible m - [ind] fuel sup-
ply(ing) control(ling)
— f —— m — energía f - [electr-distrib]
power supply control(ling)
— f —— m para gas m - [combust] gas supply
control(ling)
— f — tablero m - [electr-oper] panel control
— f — tamaño m - size control(ling)
— f —— m de cordón m - [sold] bead size con-
trol(ling)
— f —— m — gota(s) f - [sold] drop size
control(ling)
— f —— m — gotita(s) f - [sold] droplet
size control(ling)
— f — temperatura f - [ind] temperature, con-
trol(ling), o setting - [sold] heat control
— f —— f de banda f - [metal-lam] strip temperature control(ling)
— f —— f — calentador m - [ambient] heater
temperature control(ling)
— f —— f — horno m - [metal-prod] furnace
temperature control(ling)
— f —— f — fin m de soplado m - [metal-
-prod] blow end temperature control(ling)
— f —— f — viento m - [metal-prod] blast
temperature control(ling)
— f —— f entre pasada(s) f - [sold] interpass temperature control(ling)
— f —— f equilibrada - [ind] balanced temperature control(ling)
— f —— f para acondicionador m para aire m -
[ambient] air conditioner temperature control
— f —— f — calefacción f - heating temperature control(ling)
— f —— f — soplado m - [metal-prod] blow,
o blast, temperature control(ling)
— f —— f — recalentamiento m - [metal-trat]
reheat(ing) temperature control(ling)
— f — tensión f - [mec] tension control(ling)
• [electr-prod] véase regulación f de voltaje
— f — tiempo m - time, adjustment, o control
— f — tipo m con botonera f - [electr-instal]
push button type control
— f —— m con cuadrante m - [sold] dial type
control(ling)

regulación f de tipo m con disco m - [sold] dial
  type control(ling)
— f — — m — esfera f - [sold] dial type con-
  trol(ling)
— m — — m con pulsador(es) m - [electr-oper]
  push button type control
— f — tiraje,ro - [combust] draft, regulating,
  o control
— f — toma m - [electr-oper] tap setting
— f — tornillo m sin fin - [mec] auger control
— f — — m para descarga f - [agric-
  -equip] unloading auger control
— f — transformador m - [electr-equip] transformer control
— f — transmisión f - [mec] transmission control
— f — — f para malacate m - [petról] drawworks transmission control
— f — tratamiento m - treatment control
— m — — m térmico - [metal-trat] heat, o
  thermal, treatment control(ling)
— f — — m por regulador(es) m - [metal-lam]
  regulator(s) mill control(ling)
— f — — m — m para velocidad f - [metal-
  -lam] speed regulator mill control
— f — tubería f—[tub] tube,bing, o pipe,ping,
  o line, control
— f — — f para aceite m - [ind] oil, tube, o
  pipe, o line, control
— f — — f — combustible - [conbust] fuel
  line control
— f — — f — fuel oil m - [comb.int] fuel oil
  line control
— f — unidad f - unit, control, o adjusting
— f — válvula f - [válv] valve control •
  valve timing
— f — — f para agua m - [mec] water valve
  control(ling)
— f — — f — bloque m - [electrón] block
  valve control
— f — — f — solenoide m - [electr-equip]
  solenoid valve control
— f — — f septum - [metal-prod] septum valve
  control
— f — variable f - variable control(ling)
— f — — f identificable - identifiable variable control(ling)
— f — velocidad f - [ind] speed, control(ling)
  o governing
— f — — f para alimentación f - [ind] feed-
  -(ing) speed control
— f — — f — f para alambre m - [Sold]
  wire feed(ing) speed control(ling)
— f — — f para estrangulador m - [mec] choke
  speed, control(ling), o setting
— f — — f deseada - [mec] desired speed, con-
  trol(ling), o setting
— f — — f — para estrangulador m - [mec] desired choke speed, control(ling), o setting
— f — — f — elevación f - [mec] lifting, o
  hoisting, speed control(ling)
— f — — f — elevador m - [mec] elevator
  speed control(ling)
— f — — f — — m para vagón(es) m - [agric-
  -equip] wagon elevator speed control(ling)
— f — — f para estrangulador m - [mec] choke
  speed control(ling)
— f — — f para motor m - [electr-mot] motor
  speed control(ling) • [comb.int] engine speed
  control(ling)
— f — — f — — m hidráulico - [mec] hydraulic motor speed control(ling)
— f — — f para traslación f - [mec] ground
  speed control(ling)
— f — — f variable - [mec] variable speed
  control(ling)
— f — — f — para motor m - [electr-mot] motor variable speed control(ling) • [comb.int]
  engine variable speed control(ling)
— f — — f — — m hidráulico - [mec] hydraulic motor variable speed control(ling)
— f — ventilador m - [mec] fan, o blower, con-
  trol(ling), o setting

regulación f de vertedero m - [agric-equip]
  spout control(ling)
— f — — m para tajadora f - [àgric-equip]
  chopper spout control(ling)
— f — voltaje m - [electr-prod] voltage, regulating,tion, o contol(ling), o setting, o adjusting,tment
— f — — m bajo - [electr-oper] low voltage
  control(ling)
— f — — m en fuente f para energía f - [sold]
  power source voltage control(ling)
— f — — m de tipo m con cuadrante m—[electr-
  -prod] dial-type voltage control(ling)
— f — — m en arco m - [sold] arc voltage
— f — — m en circuito m - [sold] circuit
  voltage, control(ling), o setting
— f — — m en circuito m abierto - [sold] open
  circuit voltage, control(ling), o setting
— f — — m para alimentador m - [ind] feeder
  voltage control(ling)
— f — — m — — m para alambre m - [sold]
  wire feeder voltage control(ling)
— f — — m para corriente f alterna - [electr-
  -prod] alternating current voltage, regulating,tion, o control(ling)
— f — — m — — f continua - [electr-prod]
  direct current voltage, regulating,tion, o
  control(ling)
— f — — m para encendido m - [sold] starting
  voltage, control(ling), o adjusting,tment
— f — — m preciso - [sold] fine voltage, con-
  trol(ling), o setting
— f — — m variable - [electr-prod] variable
  voltage, control(ling), o setting
— f deficiente - poor control(ling)
— f — de velocidad f—poor speed control(ling)
— f — — — f para alimentación f - [sold]
  poor feed(ing) speed control(ling)
— f demasiado alta - [electr-prod] too high
  setting
— f — — para amperaje m - [sold] too high
  current setting
— f — baja - [electr-prod] too low setting
— f — — para amperaje m - [sold] too low current setting
— f deseada - [ind] desired setting
— f digital - [comput] digital control
— f dimensional - [metal-lam] dimensional, control, o setting
— f — en caliente - [metal-lam] hot dimensional, control(ling), o setting
— f dinámica - [comput] dynamic control • on
  line control
— f doble - double, o dual, control • two loop
  control
— f — continua - [electr-oper] dual continuous, o continuous dual, control
— f — — para amperaje m - [sold] dual continuous current control
— f — — — voltaje m - [sold] dual continuous
  voltage control
— f — de amperaje m - [sold] dual current control
— f — — voltaje m - [sold] dual voltage control
— f — permanente - [sold] permanent dual control
— m — — de amperaje m - [sold] dual continuous current control(ling)
— f — — — m para soldadura f - [sold]
  dual continuous welding output control
— f — — — corriente f producida - [electr-
  -prod] dual continuous output control
— f eléctrica - [electr-oper] electric(al) control
— f — automática - [electr-oper] automatic
  electric control
— f electrónica - [electrón] electronic control
— f — analógica - [electrón] analogical electronic control
— f — de espesor m - [metal-lam] electronic,
  thickness, o gage, control
— f — — mezcla f - [comb.int] electronic,

blend, o mixture, control
regulación f electrónica para caldera f - [cald]
boiler electronic control
— f — **para red f** - [ind] system electronic
control
— f — — **sistema m** - [ind] system electronic
control
— f — — — **m auxiliar** - [ind] auxiliary system electronic control
— f **en admisión f** - inlet control
— f — **entrada f** - [hidr] inlet control
— f — **esfera f** - [instrum] dial control
— f — **fábrica f** - [mec] factory (re)setting
— f — **frío** - [mec] cold setting
— f — **salida f** - [hidr] outlet control(ling)
— f — **más** - [ind] turning up
— f — **menos** - [ind] turning, down, o back
— f — **panel m** - panel control(ling)
— f — — **m delantero** - [ind] front panel con-
trol(ling)
— f — — **m frontal** - front panel control(ling)
— f **energética** - [metal-prod] power control
— f **errática** - [ind] erratic control
— f **errática de velocidad f** - erratic speed
control
— f **especial** - [ind] special control(ling)
— f — **para deformación f** - [mec] special deformation control
— f **específica** - [ind] specific setting
— f — **para toma f** - [electr-oper] specific tap
setting
— f **estadística** - statistical control
— f **estándar** - standard setting
— f — **para flujo m** - standard flow setting
— f **estática** - [comput] static control • off
line control
— f **estricta** - strict control
— f **exacta** - exact, o accurate, o precise, setting, o control • fine tuning
— f — **de amperaje m** - [sold] precise current
setting
— f — — **arco m** - [sold] precise arc control
— f — **característica(s) f de arco m**—[sold]
arc characteristic(s) fine tuning
— f — — — **f** — — **m para cada aplicación f** -
[sold] arc characteristic(s) fine tuning for
each application
— f **excelente** - excellent control
— f — **de arco m** - [sold] excellent arc control
— f **excesiva** - too high setting
— f — **de amperaje m** - [sold] too high current
setting
— f **exigida** - required control(ling)
— f **exigua** - too, o very, low setting
— f — **de amperaje m** - [sold] too low current
setting
— f **explicada** - [ind] explained adjustment
— f **fácil** - easy, control, o adjustment
— f — **para arco m** - [sold] easy arc control
— f — — **corriente f** - [electr-oper] easy current control
— f — — — **f de salida f** - [sold] easy output
control
— f **final** - [electr-oper] final, control, o adjustment
— f — **de velocidad f** - final speed control
— f **graduable** - [mec] step control
— f — **de tipo m con cuadrante m** - [sold-oper]
dial type step control
— f — — **m con esfera f** - [sold-oper] dial
type step control
— f **graduada** - [mec] step control(ling)
— f— **conveniente** - [sold] convenient step control
— f — — **con cuadrante m** - [sold] convenient
dial step control
— f — — **m de tipo m con cuadrante m** - [sold]
convenient dial type step control
— f **hidráulica** - [hidr] hydraulic, o water,
control
— f — **automática** - automatic hydraulic control
— f **hidrostática** - [mec] hydrostatic control
— f **ideal** - [sold] ideal control

regulación f ideal para encendido m - [sold]
ideal starting control
— f — — — **m para arco m** - [sold] ideal arc
starting control
— f **importante** - important control
— f **imprevisible** - unforeseeable control • erratic control
— f **incierta f** - uncertain, o erratic, control
— f **inconstante** - erratic control
— f — **de velocidad f** - erratic speed control
— f **independiente** - independent control(ling)
— f **indicadora** - indicating control
— f **inferior** - [ind] lower, o lowest, setting
— f **inmediata** - immediate, control, o setting •
next setting
— f **inmediatamente inferior** - [mec] next lower
setting
— f — **superior** - [mec] next higher setting
— f **limitada** - [ind] limited control
— f — **para velocidad f** - [sold] speed limited,
o limited speed, control
— f **local** - local control
— f **maestra** - master control
— f **magnética** - [electr-oper] magnetic control
— f **manual** - [ind] manual, o hand, operation, o
control | adv - (con) manually controlled
— f **manual de temperatura f** - [ind] manual temperature control
— f — — **f de viento m** - [metal-prod] manual blast temperature control
— f — — **transmisión f para malacate m** - [ind]
drawworks transmission manual control
— f **más alta** - [ind] higher setting • highest
setting
— f — **baja** - lower setting • lowest setting
— f **máxima** - [ind] maximum, control, o setting
— f — **para perilla f** - [mec] maximum knob
setting
— f — — — **f para (regulación f de) amperaje
m** - [sold] current control, handle, o knob,
maximum setting
— f — **para conmutador m** - [sold] switch maximum, o maximum switch, setting
— f **mecánica** - [mec] mechanical control | adv -
(con) mechanically controlled
— f **mediante inspección f** - [admin] inspection
control(ling)
— f **mínima** - minimum, control, o setting
— f — **para perilla f** - minimum knob setting
— f — — — **f para amperaje m** - [sold] current
control, knob, o handle, minimum setting
— f **modificada** - modified, o changed, setting
— f **motorizada** - [sold] motorized control
— f — **para voltaje m** - [sold] motorized voltage control
— f **motriz** - [mec] power control
— f **muy precisa** - [mec] outstanding, o very
precise, control
— f **necesaria** - necessary control
— f **neumática** - pneumatic control
— f **normal** - normal control
— f **nueva** - new ruling • new setting; resetting
— f — **común** - [comput] common reset(ting)
— f — **corriente** - [comput] common reset(ting)
— f — **para encendido m** - [comb.int] retiming
— f — — — **m de motor** - [comb.int] engine retiming
— f — **repetida** - [electr-oper] repeat(ed) resetting
— f — — **arrancador m para motor m** - [electr-
-mot] motor starter reset(ting)
— f **opcional*** - véase **regulación f optativa**
— f **optativa** - optional control
— f — **para arco m** - [sold] optional arc control
— f — — **arranque m** - optional starter control
— f — — **voltaje m** - [sold] optional voltage
control
— f **para** - setting for
— f — **activación f** - [ind] activating setting
— f — **agua m** - [hidr] water control
— f — **ajuste m** - adjusting,tment setting
— f — **amperaje m**—amperage, o current, control

regulación f para amperaje m de tipo m con esfera f - [electr] dial-type current control
— f — . . . amperio(s) m - [electr-instal] . . . amperes setting
— f — avance m - [mec] forward setting
— f — — m gradual - [mec] jog(ging) control • inch(ing) setting
— f — cable m - [cabl] cable control
— f — caudal m - [hidr] flow setting
— f — combustión f alta - [combust] high fire setting
— f — — f baja - [combust] low fire setting
— f — conectado,da - [electr-oper] switching on; on setting
— f — conexión f - [electr-oper] switching on; on setting
— f — descongelación f—[sold] thawing setting
— f — desecho(s) m radiactivo(s) - [nucl] radioactive waste, control, o setting
— f — embrague m - [mec] clutch control(ling)
— f — — m para tambor m - [mec] drum clutch control
— f — escala f - [ind] range setting
— f — — f alta—[electr-oper] high (range) setting
— f — especialidad(es) f - [ind] special-ty,ties, setting, o control
— f — espita f - [mec] spout, control, o setting
— m — excéntrico m - [mec] eccentric control
— f — — m ajustable - [mec] adjustable eccen-tric control
— f — flujo m—[hidr] flow, setting, o control
— f — — m de oxígeno m - [combust] oxygen flow, control, o setting
— f — fuerza f motriz - [electr-oper] power, operation, o setting
— f — función(es) f - [comput] function(s) control
— f — — f múltiple(s) m - [comput] multi--function control
— f — mezcla f - [comb.in] mixture, o blend, control, o setting
— f — — f (muy) pobre - [comb.int] (very) lean setting
— f — — f (—) rica f - [comb.int] (very) rich setting
— f — minería f - [miner] mining control
— f — osciloscopio m - [electrón] oscillo-scope setting
— f — presión f - [mec] pressure set(ting)
— f — — f en tubería f - [petról] pipe,ping pressure set(ting)
— f — — f — — f para entubación f—[petról] casing pressure set(ting)
— f — puesta f en marcha - [ind] starting con-trol
— f — punto(s) m - [mec] point(s) setting
— f — — m para presión f - [mec] pressure point(s) setting
— f — refrigeración f - [ind] refrigeration, control, o setting
— f — — f ajustada - [ind] adjusted refrig-eration, control, o setting
— f — residuo(s) m - [nucl] waste(s) control
— f — — m radiactivo(s) - [nucl] radioactive waste(s) control
— f — resorte m - [mec] spring setting
— f — rueda f - [mec] wheel control
— f — — f directriz - [agric-mec] guide wheel control(ling)
— f — — f — mecanizada - [agric-equip] power guide wheel control
— f — soldadora f - [sold] welder setting
— f — soldadura f - weld(ing) setting
— f — suministro m - setting to supply
— f — suplir v - setting to supply
— f — tanda f - [ind] set(ting) up for @, run, o batch
— f — velocidad f - [comb.int] speed control
— f — — f alta - [mec] high (speed) setting • [autom-mec] high range (position)
— f — — f — o baja - [comb.int] high-low

(speed), control, o setting
regulación f para vertedero m - [mec] spout con-trol
— f perfecta - perfect control
— f — para emisión(es) f - [combust] perfect emission(s) control
— f permanente - permanent, o continuous, con-trol
— f — para amperaje m - [sold] permanent, o continuous, amperage, o current, control
— f — — m y voltaje m - [sold] dual con-tinuous control
— f — — corriente f - [electr-distrib] con-tinuous current control • [sold] continuous, current, o amperage, control
— f — — — f en salida f - [electr-prod] con-tinuous output control
— f — — — para soldadura f - [sold] con-tinuous (welding) output control
— f — — — f producida - [electr-prod] con-tinuous output control
— f — doble - [sold] continuous dual control
— f — mediante dos reguladores m - [sold] con-tinuous dual control
— f permisiva - permissive control
— f plena - [mec] full speed
— f — sin carga f - [mec] full idle speed
— f pobre - poor control
— f — de velocidad f - poor speed control
— f — — f para alimentación f - [sold] poor feed speed control
— f por computador m - [comput] computer con-trol(ling)
— m — — m con ciclo(s) m cerrado(s) - [metal--pro] closed loop computer control
— f — operador m—[mec] operator control(ling)
— f portátil m - [electr] portable control
— f — de voltaje m - [electr] portable volt-age control
— f posible - possible control
— f práctica - practical, o convenient, control
— f — para motor m - [comb.int] practical, o convenient, engine control • [electr-mot] practical, o convenient, motor control
— f — para operario m - [ind] practical, o conveniente, operator control
— f precisa f - precise, o accurate, control; fine tuning • particular setting • [electr-oper] fine current control
— f — para amperaje m - [sold] precise, o fine, current, o amperage, setting, o ad-justment, o control
— f — — arco m - [sold] arc precise, set-ting, o control
— f — — característica(s) f - characteris-tic(s) precise control
— f — — — f para arco m - [sold] precise, o fine, arc characteristic(s), setting, o con-trol, o adjustment, o (fine) tuning
— f — — — — m para cada aplicación f - [sold] arc characteristic(s) fine tuning for each application
— f — — — f para encendido m - [sold] strik-ing characteristic(s) precise control
— f — para conformación f - shape precise con-trol(ling)
— f — — — f de cordón m - [sold] bead shape precise control(ling)
— f — — encendido m - [sold] striking precise control(ling)
— f — — — m y corte m - [sold] start(ing) and stop(ping) precise control
— f — estabilidad f - stability precise con-trol(ling)
— f — — — f para arco m - [sold] arc stabi-lity precise control(ling)
— f — — procedimiento m - procedure precise control(ling)
— f — — — m para soldadura f - [sold] weld-ing procedure precise control(ling)
— f — temperatura f - [sold] accurate heat control(ling)
— f — — puesta f en marcha - [mec] start(ing)

and stop(ping) precise control
regulación f precisa para tamaño m - size precise control
— f — f — — m de cordón m - [sold] bead size
precise control
— f — — voltaje m - [electr-oper] precise, o
fine, voltage, control, o adjustment
— f previa f - [ind] setting • prior control
— f primaria - primary, control, o regulating
— f principal - [mec] main control(ling)
— f racional - rational control(ling)
— f rápida—quick, control(ling), o adjustment
— f — para amperaje m - [sold] quick heat adjustment; quick current adjustment
— f — — — m para soldadura f - [sold] quick
(welding) heat, setting, o adjustment
— f — — soldadura f - [sold] quick welding,
control(ling), o adjustment
— f recomendada - recommended, setting, o adjusting,tment
— f — para amperaje m - [sold] recommended,
current, o amperage, setting, o adjustment
— f remota - [electrón] remote control
— f — para interruptor m - [electrón] remote
breaker control
— f requerida - required, control, o setting
— f restante—[mec] remaining, o other, control, o setting
— f simple - single, o simple, control
— f sin funcionar v—[sold] nonoperating control
— f — — v para arco m - [sold] nonoperating
arc control
— f sumamente ligera—[mec] finger tip control
— f superior - [mec] higher setting
— f térmica - [metal-prod] thermal control -
[metal-prod] heat, o temperature, control, o
adjusting,tment • [sold] heat setting
— f — aproximada - [sold] approximate, o
coarse, heat setting
— f — de soldadura f - [sold] welding heat,
control, o adjusting,tment
— f — exacta - [sold] exact, o accurate, heat,
control, o adjusting,tment
— f — precisa - [sold] accurate, o precise.
heat control
— f típica - typical, control, o setting
— f — para osciloscopio m - [electrón] typical
oscilloscope setting
— f tipo manubrio m - [mec] crank type control
— f total - total, o complete, control
— f — de cráter m - [sold] complete, o total,
puddle control
— f totalmente automática - [electr-oper] fully, o totally, automatic control
— f utilizable - usable, o useful, control
— f variable - [instrum] variable control
— f — en circuito m - [electrón] circuit variable control
— f variada - varied, o changed, setting
— f verificada - changed setting
regulado,da a - . . . • regulated; controlled •
adjusted; set (up) • [electr-oper] toggled
— a a distancia f - remotely controlled
— a desde distancia f - remotely controlled
— a a entrada f - inlet controlled
— a — salida f - [hidr] outlet controlled
— a a nivel m inferior - set, o adjusted, to @
lower level
— a — — m — que para procedimiento(s) m para
soldadura f - [sold] set, o adjusted, lower
than for @ welding procedure(s)
— a — — m superior - set, o adjusted, to @
higher level
— a — — — que para procedimiento(s) m para
soldadura f - [sold] set, o adjusted, higher
than for @ welding procedure(s)
— a apropiadamente - set, appropriately, o properly
— a automáticamente - controlled, o regulated,
automatically; instrument controlled
— a completamente - controlled completely
— a con computador m - [comput] computer controlled
— a — conmutador m - [electr-instal] switch

controlled
regulado,da con conmutador m para selección f -
[sold] selector switch controlled
— a — — m — — f con . . . posición(es) f -
[sold] . . .-position selector switch controlled
— a — — m selector - [electr-instal] selector
switch controlled
— a — espigón m - [hidr] jetty controlled
— a — flotador m—[comb.int] float controlled
— a — muro m - [constr] wall controlled
— a — — para contención f - [hidr] retaining wall controlled
— a — precisión f - precision, o accurately, o
closely, controlled
— a — regulador m - [mec] governor controlled
— a — relé m - [electr-oper] relay controlled
— a — — m para sobrecarga f - [electr-oper]
overload relay controlled
— a — resorte m - [mec] spring controlled
— a — tablestacado m - [hidr] sheet piling
controlled
— a — tablestacas f de acero m - [hidr] steel
sheeting controlled
— a — tapa f - cover, o cap, controlled
— a — — f para radiador m - [comb.int] radiator cap controlled
— a cuidadosamente - controlled carefully
— a electrónicamente - [electrón] electronically controlled
— a en fábrica f - factory (pre)set
— a — más - turned up
— a — menos - turned down
— a estrictamente - closely, o strictly, controlled
— a exactamente - exactly, o closely, controlled
— a fácilmente - easily controlled
— a hidráulicamente - [hidr] hydraulically controlled
— a impropiamente - set, o controlled, improperly
— a independientemente - controlled independently
— a mecánicamente - [mec] mechanically, o machine, controlled
— a mediante energía f - [electr-oper] power
controlled
— a mediante inspección f - [admin] inspection
controlled
— a — termóstato - [instrum] thermo-
stat(ically) controlled
— a nuevamente - [ind] reset
— a para conectado,da - [electr-oper] switched
on
— a — desconectado,da - [electr-oper] switched
off
— a — fuerza f motriz - [sold] power setting
— a — mezcla f (muy) pobre - [comb.int] set
(very) lean
— a — — f (—) rica - [comb.int] set for
(very) rich
— a — producción f de energía f - [comb.int]
power setting
— a — sin carga f - [sold] on no load
— a — soldadura f - [sold] weld(ing) setting
— a — solenoide m - [electr-oper] solenoid,
controlled, o set
— a — suministro,trar - [electr- set to supply
— a — suplir - [electr] set to supply
— a — tanda f - [ind] set (up) for @ run
— a — telemando - remotely controlled
— a — válvula f - valve controlled
— a — velocidad f - [mec] speed set
— a — — f alta - [mec] set for @ high speed
— a — — f baja - [mec] set for @ low speed
— a perfectamente - set, o controlled, perfectly
— a por fase(s) f - phase controlled
— a — medio de relé(s) m - [electr-oper] relay
controlled
— a — operador m - [mec] operator controlled
— a — relé(s) m - relay controlled
— a — telemando m - remotely controlled

regulado,da a por válvula f - valve controlled
— a rápidamente - adjusted quickly
— a termostáticamente - thermostatically controlled
— a uniformemente - controlled uniformly
regulador m - . . . . • [comb.int] throttle •
[electr-equip] slope selector • [sanit] buffer • [mec] adjuster • equalizer
— m a distancia f - [electr] remote control
— m —— f con pedal m - [sold] foot operated
remote control
— m —— f —— m para amperaje m - [sold]
foot operated remote current control
— m —— f para amperaje m - [electr-equip]
remote amperage control • [sold] remote, current, o output, control
— m —— f —— m impulsado por motor m eléctrico - [sold] motor driven remote, current,
o output, control
— m —— f — polaridad f - [sold] remote polarity control
— m abierto - open(ed) control(ler); expanded
regulator
— m accionado por pulgar - [ind] thumb control
— m adaptable - versatile control(ler)
— m adicional - additional control(ler)
— m ajustado - adjusted, o set, control(ler), o
governor, o throttle
— m automático - (fully) automatic control(ler)
— m — para altura f para cabezal m - [agric-
-equip] automatic header height control
— m —— f de espigadora f - [agric-equip]
automatic header height control
— m —— combustión f - [combust] automatic
combustion control(ler)
— m —— energía f - [electr-equip] automatic
power control
— m —— f para velocidad f - [sold] automatic speed power control
— m —— marcha f, en vacío m, o sin carga f -
[comb.int] (automatic) engine, idler, o idling device
— m —— f lenta (en vacío m, o sin carga))
- [sold] automatic engine idler
— m —— f sin carga f - [comb.int] (automatic) engine, idler, o idling device
— m — para soldadora f - [sold] automatic
welder control(ler)
— m —— soldadura f - [sold] automatic
welder,ding, control(ler)
— m —— tiro,raje m - [combust] automatic
draft contol(ler)
— m auxiliar - [mec] auxiliary control(ler)
— m avanzado - [comb.int] advanced throttle
— m calibrado - calibrated, control(ler), o
throttle, o governor
— m centrífugo - [mec] centrifugal, o ball
type, control(ler), o governor
— m — con bola(s) - [vapor] centrifugal
(fly)ball (type) governor
— m —— contrapesos m - [mec] centrifugal
type governor
— m — de tipo m con bola(s) f (volantes) -
[comb.int] centrifugal flyball type governor
— m —— m — pesa(s) f volante(s) - [mec]
centrifugal flyweight type governor
— m cero - [ind] zero regulator
— m climatérico - climate control(ler)
— m — individual - [constr] individual climate control(ler)
— m colocado - [comb.int] positioned throttle
— m completo - [ind] complete control(ler)
— m — para agua m - [hidr] complete water con-
trol(ler)
— m —— gas m - [combust] complete gas con-
trol(ler)
— m — para marcha f lenta - [comb.int] complete idler
— m —— f — para motor m - [comb.int]
complete engine idler • [electr-mot] complete
motor idler
— m con amplidino m - [electr-equip] amplidyne,
regulator, o contol(ler)

regulador m con amplidino m para horno m -
[electr-equip] amplidyne furnace regulator
— m — bola(s) m - [mec] flyball governor
— m —— f centrífugo - [mec] centrifugal flyball control
— m —— f volantes - [comb.int] fly-weight
governor
— m — calibrador m - [mec] regulator with @
gage
— m —— m para presión f - [mec] pressure
regulator (equipped) with @ gage
— m — cuadrante m - [instrum] dial control
— m — esfera f - [instrum] dial control
— m — frecuencia f alta - [electr-oper] high
frequency control
— n —— f alta para intensidad f - [electr-
-oper] high frequency intensity control
— m — fusible(s) m - [electr-equip] fused con-
trol(ler)
— m — lazo m - [electrón] loop control
— m —— m abierto - [electrón] open loop control
— f —— m cerrado - [electrón] closed loop
control
— m — manómetro m - [instrum] gage equipped,
control, o regulator
— m —— m con presión - f -[instrum] gage
equipped pressure regulator
— m — palanca f - [mec] lever control
— m — pedal - [mec] (foot) pedal, o throttle,
control
— m —— m para amperaje m - [electr-oper]
(foot), pedal, o throttle, control • [sold]
foot operated current control
— m — pesa(s) f volante(s) - [mec] fly weight
governor
— f — . . . posición(es) f - [electr-instal]
. . .-position switch
— m — pulsador(es) m - [electr-oper] push button control
— m — reóstato m - [electr-oper] rheostat
control
— m —— m para campo m - [sold] field rheostat control
— m — telemando m - [ind] remote control
— m —— m optativo - [sold] optional remote
control
— m —— m — para amperaje m - [sold] optional remote current control
— m — transistor(es) m - [electrón] solid
state, o transistorized, control
— m —— m para amperaje m - [electr-oper]
transistorized, amperage, o current, control
— m —— m — voltaje m - [electr-oper] transistorized voltage, control, o regulator
— m conectado - connected, o linked, control, o
throttle
— m — con bomba f inyectora - [comb.int] con-
trol(ler), o throttle, linked to @ injection
pump
— f confiable - [instrum] dependable control
— m — con transistor(es) m - [instrum] dependable solid state control
— m constante - [mec] constant, control, o governor
— m — para velocidad f - [mec] constant speed,
control, o governor
— m continuo - [ind] continuous control
— m — de tipo m con cuadrante m - [ind] continuous dial-type control
— m — para amperaje m - [electr-prod] continuous amperage control • [sold] continuous
current control
— m —— voltaje m - [electr-prod] continuous
voltaje, control, o adjustor
— m —— m en circuito m abierto - [sold]
continuous open circuit voltage, control, o
adjustor
— m — tipo m manubrio m - [electr-equip] crank
type control
— m —— m — para corriente f - [electr-
-equip] continuous crank-type current control
— m —— m —— voltaje m - [electr-equip]

continuous voltage control
regulador m **contra torcedura(s)** f - [mec] twist control(ling)
— m **conveniente** - convenient, o handy, control
— m **dañado** - damaged control
— f **de baja** f - [mec] idler
— m — **Reeves** - [electr-equip] Reeves control
— m — — **para corriente** f **alterna** - [electr-instal] Reeves alternating current control
— n — **precisión** f - [electr-oper] precision control(ling)
— m — **reserva** f - [ind] spare, control(ler), o regulator
— m — **silicio** m - [electrón] silicon control
— m — **tipo** m **centrífugo** - [mec] centrifugal, o flyball, (type), governor, o regulator
— m — — m **con bola(s)** (**volantes**) - [mec] flyweight type governor
— f — — m — **botonera** f - [electr-instal] push button type control(ler)
— m — — m — **cuadrante** m - [ind] dial type control(ler)
— m — — m — — m **calibrado** - [instrum] calibrated dial type control(ler)
— n — — m — **esfera** f - [instrum] dial type control(ler)
— m — — m — **manubrio** m - [electr--equip] crank type control
— m — — m — — m **para corriente** f - [electr--equip] crank type current control(ler)
— m — — m — **palanca** f - [mec] lever type control(ler)
— m — — m — **pesa(s)** f (**volantes**)—[comb.int] flyweight type governor
— m — — m **con reóstato** m - [electr-equip] rheostat type control(ler
— m — — m — **pulsador(es)** m - [electr-equip] push button type control(ler)
— m — — m **rotativo** - [electr-equip] rotary type control(ler)
— m — — m — **para bobina** f - [electr-equip] rotary type coil control(ler)
— m — — — — f **para reactancia** f - [electr-equip] rotary type reactor control
— m — — m — **devanado** m - [electr-equip] rotary type coil control(ler)
— m — — m — **devanado** m - [electr-equip] rotary type coil control(ler)
— m — — m — **para reactancia** f - [electr-equip] rotary type reactor control
— m **dentro de cabina** f - [agric-equip] in-cab control
— m **doble** - [electr-equip] dual, o double, control(ler)
— m — **continuo** - [electr-equip] dual continuous control(ler)
— m — **permanente** - [electr-equip] dual continuous control(ler)
— m — — m **para amperaje** m - [sold] dual continuous output (welding) control(ler)
— m — — — — m **para soldadura** f - [sold] dual continuous (welding) output control(ler)
— m — — — — **corriente de salida** f - [electr--prod] dual continuous output control(ler)
— m — — — f **producida** f - [electr-prod] dual continuous (current) output control(ler)
— m **eléctrico** - [electr-equip] electric(al) control(ler)
— m — **automático** - [electr-equip] automatic electric(al) control(ler)
— m — — — **marcha** f **sin carga** f - [comb.int] electric automatic idler
— m — — — f — — f **para motor** m - [comb--int] electric automatic engine idler
— m — — — f **sin carga** f - [Comb.int] electric engine idler
— m — **manual** - [instrum] manual electric control(ler)
— m **electrónico** - [electrón] electronic control
— m — **con transistor(es)** m - [electrón] transistorized, o solid state, electronic control
— m — **panel** m **deprimido** - recessed control
— m **escalonado** - [sold-equip] step control(ler)

regulador m **escalonado de tipo** m **con cuadrante** m - [sold] dial type step control(ler)
— m — — — m — **esfera** f - [sold] dial type step control
— m — — **para amperaje(s)** m - [sold] step type current control
— m **especial** - [ind] special control(ler)
— m **estrangulado** - [mec] choked control(ler)
— m **exacto** - [ind] exact, o precise, control
— m **expandido** - [mec] expanded, regulator, o control(ler)
— m ...**fásico** - [electr-equip] ...-phase, control(ler). o regulator
— m ...— **para inducción** f - [electr-equip] ...-phase induction, control(ler), o regulator
— m **fijado** - [mec] locked lever • [comb.int] set throttle
— m **final** - [mec] final control(ler) • [electr--oper] final adjustment switch
— m **fuera de uso** - [mec] inoperative control
— m **giratorio** - [electr-equip] rototrol; rotary control(ler)
— m — **a bies** - [electr-equip] bias control
— m — **antifluctuante** - [electr-equip] antihunt rototrol
— m — — **para equilibrio** m **para carga** f - [metal-prod] load balance antihunt rototrol
— m — — — **motor** m - [electr-equip] motor antihunt rototrol
— m — **para equilibrio** m [metal-prod] balance rototrol
— m — — — m **de carga** f - [metal-prod] load balance rototrol
— m — **para laminador** m - [metal-lam] (rolling) mill rototrol
— m — — **motor** m - [electr-equip] motor rototrol
— m **graduable** - [sold] step, o adjustable, control(ler)
— m — **de tipo** m **con cuadrante** - [sold] dial type step control
— m — — — m — **esfera** f - [sold] dial type step control
— m **graduado** - [sold] step control
— m — **conveniente** - [sold] convenient step control
— m — — **con cuadrante** m - [sold] convenient dial step control
— m — — **de tipo** m **con cuadrante** m - [sold] convenient dial type step control
— m **hidráulico** - [mec] hydraulic control(ler), o governor
— m — **automático** - automatic hydraulic control(ler)
— m — **con válvula** f - [válv] valve hydraulic control
— m — — — f **con . . . vía(s)** f - [válv] ...-way valve hydraulic control
— m — **manual** - [ind] manual hydraulic control
— m **horizontal** - [mec] horizontal, regulator, o adjuster
— m — **para cabeza** f - [sold] horizontal head adjuster
— m **hundido** - [ind] recessed control
— m **inactivado** - [mec] inactivated, o inoperative, control
— m **independiente** - [ind] independent control
— f — **para freno** m - [mec] independent brake control
— m — — — m **para accionamiento** m - [mec] independent drive,ving brake control
— m — — — m **para mesa** f (**rotativa**) - [petról] independent rotary drive brake control
— m **indicador** - [mec] indicating control
— m — **calibrado** - indicating calibrated control(ler)
— m — **de tipo** m **con cuadrante** m - [mec] indicating calibrated dial type control(ler)
— m — **de tipo** m **con cuadrante** m - [instrum] indicating dial type control(ler)
— m — — — m — — m **calibrado** - [instrum]

calibrated dial type indicating controller
regulador m **indicador para temperatura** f -
[instrum] indicating temperature control
— m **isocrónico** - [instrum] isochronous governor
— m **isócrono** - [instrum] isochronous governor
— m **limitador** - [ind] limiting control(ler)
— m — **para temperatura** f - [combust] temperature limit control
— m — — — f **alta** - [Combust] temperature
high limit control(ler)
— m **limpiado** - [comb.int] cleaned, governor, o
control(ler)
— m **limpio** m - [comb.int] clean, governor, o
control(ler)
— m **local** - [electr-equip] local control(ler)
— m **lubricado** - [comb.int] lubricated governor
— m **mal ajustado** - [mec] poorly adjusted, o
misadjusted, control(ler), o governor
— m **manual** - [sold] manual control(ler)
— m — **para marcha** f **lenta** - [comb.int] manual
engine idler
— m — — — f — **sin carga** f - [sold] manual
idler
— m — — — f — — — f **para motor** m - [comb-
-int] manual engine idler
— m — — — — **válvula** f—[mec] valve manual control
— m — — — — f **para fundente** m - [sold] flux
valve manual control
— m **mecánico** - [mec] mechanical control(ler)
— m **movido** - [ind] moved, throttle, o control
— m **neumático** - [neumát] pneumatic, o air, operated, regulator, o control(ler)
— m — **y mecánico** - [ind] pneumatic, o air, and
mechanical control
— m **no indicador** - [ind] nonindicating controller
— m — — — m **calibrado** -[ind] un-, o non-
calibrated control(ler)
— m — — **de tipo** m **con cuadrante** m - [instrum]
nonindicating calibrated dial type controller
— m — — — — m — — — **calibrado**—[instrum]
calibrated dial type non-indicating control-
-(ler)
— m — **registrador** m - [instrum] non-recording
control(ler)
— m — — — **de tipo** m **con cuadrante** m - [instr]
dial type non-recording control(ler)
— m — — — — m — **esfera** f - [instrum] dial
type non-recording control(ler)
— m **normal** - [instrum] normal control(ler)
— m — **para amperaje** m - [electr-equip] normal
amperage control(ler) • [sold] normal current
control(ler)
— m **Oilmaster** - [petról] Oilmaster wellhead
— m **operado** - [mec] operated control
— m — **con mano** f - hand operated control
— m — — **pie** m - foot operated control
— m — **manualmente** - [mec] manually, o hand,
operated control
— m — — m **para transmisión** f—[petról] transmission manually operated control
— m — — — — f **para malacate** m - [petról]
drawworks transmission manually operated control
— m **optativo** - [ind] optional control
— m — **para arco** m - [sold] optional arc control
— m — — **arranque** m - [sold] optional starter
control
— m — **para marcha** f **sin carga** f - [comb.int]
optional engine idler
— m — **para voltaje** m - [sold] optional voltage
control
— m **para apertura** f - [mec] opening control
— m — — f **en estrangulador** m - [mec] choke
opening control
— m **acción** f - [mec] action control
— m — — f **inversa** - [mec] reverse action,
control(ler), o regulator
— m **accionazmiento** n - [mec] drive,ving con-
trol(ling)
— m — — m **para mesa** f **rotativa** - [petról[rotary (table) drive control

regulador m **para aceleración** f - [comb.int]
throttling, governor, o control(ler)
— m — — m **con pedal** m - [mec] foot pedal
throttle control(ler)
— m — **adaptabilidad**—[mec] versatility control
— m — — f **inigualada** - unequalled, o unparalleled, versatility control
— m — **agua** m - water, regulator, o control
— m — — m **para alimentación** f - feed water
controller
— m — — — m **para enfriamiento** m - cooling water
control(ling)
— m — — — — — m **para freno** m - [mec] brake
cooling water control(ling)
— m — — — — — — m **auxiliar** - [mec]
auxiliary brake cooling water control
— m — — f — — — m — **llanta** f - [mec] rim
cooling water control
— m — — — — — — — f **para freno** m—[mec]
brake rim cooling wter control
— m — **aguilón** m - [grúas] boom control
— m — **aire** m - [comb.int] choke; air regulator
— m — — **fresco** - [ambient] fresh air control
— m — — m **regulado** - [neumát] set, o controlled, air regulator
— m — **ajuste** m **preciso** - [sold] fine adjustment control
— m — **alarma** f - [segurid] alarm control
— m — **alimentación** f - [ind] feed)ing control
— m — **alimentadora** f - [ind] feeder control
— m — — f **para mesa** f (**rotativa**) - [petról]
rotary (table) feed(er) control
— m — **altura** f - [mec] height control(ler)
— m — — f **para cabezal** m - [agric-equip]
header height control
— m — — f — **espigadora** f - [agric-equip]
header height control
— m **para amperaje** m - [electr-oper] amperage,
o amperes, control • [sold] cuurent, control,
o regulator, o adjuster; Currentrol
— m — — **ajustado** - [electr-oper] adjusted, o
set, amperes, o current, control
— m — — m **aproximado** - [sold] coarse current,
control, o adjustment
— m — — — m **con regulador** m **de tipo** m **con, cua-**
drante m, o **esfera** f - [electr-oper] dial
type amperage control
— m — **amperaje** m — — m **con cuadrante** m -
[electr-oper] dial-type current control
— m — — — — m **con reactor** m - reactor type
current control
— m — — — m — — — m **con núcleo variable**
- [electro-equip] saturable (type) reactor,
design, o type, current control
— m — — — m — — — m **reactor** m **saturable** -
[electr-equip] saturable reactor, design, o
type, current control
— m — — — m **fijo** - [sold] fixed current control
— m — — — m **movido** - [sold] adjusted current
control
— m — — — m **normal** - [electr-equip] normal amperage control • [sold] normal current control
— m — — — m **para ajuste** m **preciso** - [sold] fine
-tuning current control
— m — — — m **para carga** f - [electr-equip]
charging rate switch
— m — — — m **para fuente** f **para energía** f -
- [electr-equip] power source current control
— m — — — m **y voltaje** m - [sold] current and
voltage control
— m — **amplidino** n - [electr-equip] amplidyne
reulator
— m — **arco** m - [sold] arc control (switch)
— m — — m **con voltaje** m **fijo** - [sold] fixed
voltage arc control
— m — — — m — — — m **variable** - [sold] variable
voltage arc control
— m — — — **variable** - [sold] variable arc control
— m —**argón** m - [sold] argon regulator

regulador m para arrancador m - [electr-equip]
starter control
— m — — m con pulsador m—[electr-equip] push
button starter control
— m — Askania m - [metal-prod] Askania control
— m — avance m - [sold] travel control
— m — — m para barreno m - [petról] rotary
feed control
— m — — m gradual - [sold] inch(ing) speed
control
— m — bobina f - [electr-equip] coil control
— f — — f para reactancia f - [electr-equip]
reactor (coil) control
— m — cabeza f - [sold] head ajuster
— m — — f para pozo m - [petról] wellhead
control
— m — cable m - [grúas] cable control
— m — — m a gancho m - [grúas] whip (hoist)
control
— m — — m — — m verificado - [grúas]
checked whip hoist control
— m — — m — motón m - [grúas] main hoist
control
— m — — m — — m verificado - [grúas]
checked main hoist control
— m — caída f - [electr- droop control
— m — — f para regulador m giratorio -
[electr-equip] Rototrol droop control
— m — — — m — para motor m - [electr-
-equip] motor Rototrol droop control
— m — campo m - [electr-equip] field, control,
o regulador
— m — carburante m - [comb.int] fuel throttle
— m — carga f para horno m - [metal-prod]
furnace charging control
— m — carrete m - [mec] reel control
— m — — m — núcleo m - [petról] core reel
control(ler)
— m — caudal m - [hidr] flow control
— m — — m de agua m [hidr] water flow
control
— m — ciclo m - cycle control(ler)
— m — — m abierto - [electrón] open loop
control
— m — — m cerrado - [electrón] closed loop
control(ler)
— m — — m para avance m - [sold] travel cycle
control(ler)
— m — — — soldadura f - [sold] welding
cycle control(ler)
— m — — m — — f y avance m - [sold] welding
and travel cycle control(ler)
— m — cierre m - [mec] closing,sure control
— m — — m de estrangulador m - [mec] choke
closing control(ler)
— m — cizalla f - [metal-lam] shear control
— m — — f para planchón(es) m - [metal-lam]
slab shear control
— m — combustible m - [combust] fuel control
— m — — m para caldera f - [cald] boiler
fuel, governor, o control(ler)
— m — combustión f - [combust] combustion con-
trol(ler)
— m — compuerta f - [hidr] gate control(ler) •
sluice gate
— m — contrapeso m - [cald] loaded governor
— m — control m - [ind] control regulator
— m — corriente f - [electr-equip] current
control • reactor
— m — — f para carga f - [electr] charging
rate switch
— m — — f — entrada f - [electr] input (power) control
— m — — f — salida f - [electr] output
(power) control
— m — — f — soldadura f - [sold] (welding)
output control
— m — curva f - [sold] curve control
— m — — f para amperios m - [sold] ampere(s)
curve control(ler)
— m — — f — voltamperios m - [sold] volt-em-
- pere(s) curve control • slope selector switch
— m — chispa f - [comb] spark control(ler)

regulador m para desempañador m - [autom] defroster control(ler)
— m — desplazamiento m - [mec] displacement,
control(ler, o regulator
— m — dispositivo m - device control(ler) •
[sold] attachment (speed) control
— m — devanado m - [electr-mot] winding control
— m — — m para reactancia f - [electr-equip]
reactor control
— m — dosificación f - [combust] proportional
control
— m — electrodo m - [electr-instal] electrode,
regulator, o control(ler)
— m — — m amplidino m - [electr-instal] amplidyne electrode, regulator, o control(ler)
— m — — m para amplidino m - [electr-instal]
amplidyne electrode regulator
— m — elevación f - [mec] lift(ing), o hoist-
-(ing), control - [grúas] boom hoist control
— m — — f de aguilón m - [grúas] boom hoist
control
— m — — f — — m verificado - [grúas]
checked (boom) hoist control(ler)
— m — — f vertical - [mec| vertical lift(ing),
adjuster, o control(ler)
— m — — m para cable m - [grúas] cable hoist
control
— m — — — — m a gancho m - [grúas] whip
hoist control
— m — elevador m principal - [grúas] main hoist
control
— m — — m — verificado - [grúas] checked main
hoist control
— m — embalse m - [hidr] reservoir equalizer
— m — embrague m - [mec] clutch control(ler)
— m — — m para carrete m - [petról] reel
clutch control(ler)
— m — — m para núcleo m - [petról] core reel
clutch control(ler)
— m — — m — mesa f rotatoria - [petról]
rotary (table) clutch control(ler)
— m — encendido m [comb.int] ignition, o
spark, control • [sold] start, o striking,
control
— m — — m y corte m - [sold] start(ing) and
stop(ping) control
— m — energía f - [electr-mot] power control
— m — — f para velocidad f - [sold] speed
power control
— m — enfriamiento m - [ind] cooling control
— m — — m primario - [metal-col.cont] primary
cooling control(ler)
— m — — m secundario - [metal-col.cont] secondary cooling control(ler)
— m — entrada f - [comput] input controller
— m — — f de aire m - air intake control •
[comb.int] choke
— m — equipo m - [ind] equipment control(ler)
— m — — m para soldadura f - [sold] welding
equipment control(ler)
— m — — m — — f automática - [sold] automatic welding equipment control(ler)
— m — espita f - [mec] spout control
— m — estrangulador m - [comb.int] choke control
— m — excéntrico m - [mec] eccentric control
— m — — m ajustable - [mec] adjustable eccentric control
— m — flujo m - [hidr] flow control
— m — — m de combustible - [combust] fuel flow
control
— m — — m — oxígeno m - [combust] oxygen
flow control
— m — — m mínimo - [instrum] minimum flow regulator
— m — frecuencia - [electr-oper] frequency
control
— m — — f alta - [electr-oper] high frequency
control
— m — — f alta para intensidad f - [electr-
-oper] high frequency instensity control
— m — freno m - [mec] brake control(ler)

regulador m para freno m independiente m - [mec]
independente brake control
— m —— m — para accionamiento m - [petról]
independent drive,ving brake control
— f —————— m para mesa f rotatoria -
[petról] independent rotary drive brake con-
trol(ler)
— m —— m para accionamiento m - [mec] drive
brake control
— m —— m ——— m para mesa f rotatoria -
[petról] rotary drive brake control
— m —— m por inercia f - [mec] inertia brake
control
— m — fuente m - [mec] source control
— m —— f para energía f - [electr-equip]
power source control
— m —— f contraelectromotriz - [electrón]
counter electromotive power control
— m —— f electromotriz - [electrón] electromotive power control
— m — función(es) f - [comput] function(s)
controller
— m —— f múltiple(s) - [comput] multi-function controller
— m —— f única - [comput] single function
controller
— m — gas m - [combust] gas, control(ler), o
regulator
— m —— m inerte - [ind] inert gas control
— m —— m — de tungsteno m - [sold] tungsten
inert gas control
— m — generador m - [electr-prod] generator
control
— m —— m para imán m - [grúas] magnet generator control(ler)
— m — giro m - [grúas] swing, o rotation, con-
trol(ler)
— m — gobierno m - [comb.int] governor,ning
control(ler)
— m — grúa f - [grúas] crane control(ler)
— m — horno m - [ind] furnace control(ler)
— m — indicador m - [autom] signal control
— m —— para viraje(s) m - [autom] turn
signal control(ler)
— m — índice m - index, o rate. control(ler)
— m —— m para carga f - [electr-oper]
charge,ging rate switch
— m — inducción f - [electr-oper] induction
regulator
— m — inercia f - inertia, governor, o control
— m — intensidad f - intensity control(ler)
— m —— f para luz f - [electr-instal] dimming switch
— m — intervalo m—[instrum] (circuit) timer
— m — laguna f - [hidr] lagoon equalizer
— m — laminador m—[metal-lam] mill controller
— m — líquido m - liquid control(ler)
— m — lodo m - [petról] véase acondicionador m
para lodo m
— m — malacate m - [petról] drawworks throttle
— m —— m auxiliar - [grúas] auxiliary winch
control(ler)
— m —— m principal - [grúas] auxiliary winch
control(ler)
— m — mando m - control(ling) regulator; operator control
— m — mano f - [mec] manual control(ler)
— m — máquina f - [mec] machine control(ler)
— m — marcha f - [comb.int] speed control(ler)
— m —— f con régimen m bajo - [comb.int]
idler; idling device
— m —— f en vacío m - [comb.int] idler; idling device
— m —— f lenta - [comb.int] idler
— m —— f — automática - [comb.int] automatic idler, control, o lever
— m —— f — de motor m - [comb.int] automatic engine idler, control, o lever
— m —— f sin carga f - [comb] idler; idling,
device, o control, o lever; engine idler
— m —— f automática - [comb.int]
automatic idler, control(ler), o lever
— m —— f — f de motor m - [comb.int]
automatic engine idler, control(ler, o lever

regulador m para marcha f sin carga f verificado - [mec] checked idler
— m para mezcla f - [ind] blend control
— m —— f aire-gas - [ind] air-gas blend con-
trol(ler)
— m — mordaza f - [mec] grip control
— m —— f contra (re)torcedura(s) f - [mec]
twist control
— m —— f — f en palanca f para rotación
f - [grúas] swing lever twist grip control
— m — motor m - [electr-mot] motor control •
[comb.int] engine, governor, o control, o
regulator
— m —— m conectado - [electr-instal] wired,
o connected, motor control
— m —— m hidráulico - [mec] hydraulic motor
control
— m —— m para corriente f alterna - [electr-
-mot] alternating current motor control
— m —— m —— continua - [electr-mot] direct current motor control
— m — nivel m alto - [mec] high level control
— m —— m para líquido m - [mec] high
liquid level control
— m —— m bajo - [mec] low level control
— m —— m —— líquido m - [mec] low liquid
level control
— m —— m para agua m - [hidr] water level,
control, o regulator
— m —— m —— m en fondo m - bottom water
level, control, o regulator
— m —— m —— m — de lavador m -
[ind] scrubber bottom water level regulator
— m —— m —— m —— m — Venturi m -
[metal-prod] (Venturi) (scrubber) bottom
water level regulator
— m —— m —— m en freno m - [mec] brake
water level control
— m —— m —— lavador m - [ind] scrubber water level regulator
— m —— m elevado - [hidr] high level control
— m —— m de líquido m - liquid level con-
trol(ler)
— m — operación f - [sold] operator control
— m —— f manual - [mec] manually operated
control
— m — orificio m - [mec] opening control(ler)
— m —— m en estrangulador m - [mec] choke
opening control(ler)
— m — palanca f - [mec] lever control(ler)
— m — perforación f - [petról] drilling con-
trol(ler)
— m — pie m - [mec] foot control(ler)
— m — posición f - positioner
— m — potencial m - potential control(ler)
— m —— f constante - [electr-oper] constant
potential control(ler)
— m —— m para laminación f—[electr-oper]
mill constant potential control(ler
— m — presión f - pressure, control(ler), o
regulator
— m — presión f - [neumat] air pressure regulator
— m —— f — gas m - [combust] gas pressure
regulator
— m —— f diferencial - differential pressure
regulator; Askania
— m — proceso m - [ind] process control(ler)
— m — propano m - [combust] propane regulator
— m — proporción f - blend control(ler)
— m —— f aire-gas m - [ind] air-gas blend
control
— m — puesta f en marcha f - [sold] start(ing)
control
— m —— f —— f y detención f - [ind]
start(ing) and stop(ping) control
— m — pulgar m - [mec] thumb control(ler)
— m — quemador m - [combust] burner regulator
— m — refrigeración f - [ind] refrigeration, o
cooling, control(ler)
— m — régimen m - rate,tio control(er)
— m —— m para carga f - [electr-acumul]
charging rate switch
— m — regulación f - control regulator

regulador m — regulación f de velocidad(es) f alta y baja para motor m - [comb.int] high-low engine speed control lever
— m — — f — — f de motor m - [electr-mot] mot or speeds control lever • [comb.int] engine speed control lever
— m — relación f de gas m natural y aire m - [combust] natural gas/air ratio regulator
— f — relé m - [electr-oper] relay control
— m — — m para sobrecarga f - [electr-oper] overload relay control
— m — — m interruptor - [electr-equip] trip relay control
— m — — m interruptor por falla(s) f - [electr-oper] fault trip(ping) relay control
— m — relé m maestro - [electr-oper] master relay control
— m — — m principal - [electr-oper] master relay control(ler)
— m — reloj m - [electr-instal] clock control
— m — — m conectado - [electr-instal] wired clock control(ler)
— m — reóstato m - [electr] rheostat control
— m — — m para voltaje m - [sold] job selector control(ler)
— m — resorte m - [mec] spring adjuster
— m — ritmo m - [mec] rate control(ling)
— m — — para alimentación f - [ind] feed-(ing) rate control(er)
— m — rodillo m - [metal-lam] rolling speed
— m — rotación f - [grúas] swing, o rotation, control
— f — rueda f - [mec] wheel control
— m — — f directriz - [autom-mec] guide,ding wheel control
— m — — f motriz f - [autom] driv,ving wheel control
— m — seguridad f - [segurid] safety control
— m — — f para temperatura f - [segurid] safety temperature control(ler)
— m — — f — — f alta - [combust] safety high temperature controller
— m — — f — — f límite - [combust] safety high limit temperature control(ler)
— m — — f — — f baja - [combust] safety low temperature control(ler)
— m — — f — — f límite - [combust] safety low limit temperature controller
— m — — f — — límite inferior - [combust] safety low limit temperature controller
— m — — f — — f superior - [combust] safety high limit temperature controller
— m — sensor m - [instrum] sensor control
— m — — m para calor m - [instrum] heat sensor control
— m — — m — en rotadora f - [cuerp-moled] rotator heater sensor control
— m — — m — /alimentadora f - [cuerp-moled] rotator/feeder heater sensor control
— m para Septum m - [metal-prod] septum valve control
— m para servicio m - [ind] service regulator
— m — — m fallado - [ind] faulty service regulator
— m — soldadora f - [sold] welder,ding control
— m — — f automática - [sold] (fully) automatic welder control
— m — — f semiautomática - [sold] semiautomatic welder control
— m — — f totalmente automática - [sold] fully automatic welder control
— m — soldadura f - [sold] weld(ing) control
— n — f manual - [sold] control for manual welding; manual welding control
— m — suministro m - [ind] supply,ing control
— m — — m de gas m - [combust] gas supply,ing control
— m — tambor m - [mec] drum control(ler)
— m — telegobierno m - [ind] remote control
— f — m operado con mano f - [ind] remote control operated
— m — — — pie m - [ind] remote foot operated control

regulador m para temperatura f - temperature control(ler) • [sold] heat control(ler)
— m — — f de viento m - [metal-prod] blast temperature control
— m — — f elevada - [ind] high temperature control(ler)
— m — — f límite - [combust] high limit temperature control(ler)
— m — — f — alta - [combust] high limit temperature control(ler)
— m — — f — baja - [combust] low limit temperature control(ler)
— m — — f — inferior - [combust] low limit temperature control(ler)
— m — — f — superior - [combust] high limit temperature control(ler)
— m — — f para acondicionador m (para aire) m - [ambient] (air) conditioner temperature control(ler)
— m — — f — calentador m - [ambient] heater temperature control(ler)
— m — tensión f - [electr-distrib] véase regulador para voltaje m
— m para terraja f - [mec] stock gage
— m — tiro,raje m - [ambient] draft regulating damper; draft control; damper
— m — tiro m contra llama(s) f - [ambient] fire damper
— m — transformador m - [electr-prod] transformer control(ler)
— m — — m para bobinadora f - [metal-lam] coil(er) transformer control(ler)
— m — transmisión f - [mec] transmission control(ler)
— m — — f para malacate m - [petról] drawworks transmission control(ler)
— m — trinquete m - [mec] pawl control(er)
— m — — m para traba f - [grúas] lock pawl control (handle)
— m — — — f desengranado - [grúas] disengaged lock pawl control (handle)
— m — unidad f para alimentación f - [sold] feeding equipment control(ler)
— m — — f — para alambre m - [sold] wire feeding equipment control(ler)
— m — válvula f - [válv] valve control(ler)
— m — — f para agua m - [mec] water valve control(ler)
— m — — f — fundente m - [sold] flux valve control(ler)
— m — — f Septum - [metal-prod] septum valve control(ler)
— m — velocidad f - [ind] speed, control, o governor, (lever); speed, operator, o regulator; throttle control
— f — — f alta - [comb.int] high speed throttle
— m — — f alta en vacío - [comb.int] high idle speed throttle
— f — — f alta o baja - [comb.int] high-low control (lever)
— m — — — — de motor m - [comb.int] high-low engine speed control lever
— m — — f — en vacío - [comb.int] low idle speed throttle
— m — — f constante - [comb.int] constant speed governor
— f — — f para avance m - [sold] (travel) speed control
— m — — f — — m para tractor m - [sold] tractor travel speed control
— m — — f — — m gradual - [sold] inch(ing) speed control
— m — — f para carga f - [electr-acumul] charging rate switch
— m — — f — dispositivo m - [sold] attachment speed control
— m — — f para elevador m - [mec] elevator speed control
— m — — f — — m para vagón(es) m - [agric-equip] wagon elevator speed control
— m — — f para motor m - [comb.int] engine (speed), governor, o control

regulador m para velocidad f para motor m hidráulico—[mec] hydraulic motor speed control
— m —— f — tornillo m para ajuste m -
[metal-lam] screw-down speed control
— m —— f, ajustado, o puesto - [mec] set
speed control lever
— m —— variable - [mec] variable speed con-
trol(ler)
— m —— f variable para motor m hidráulico -
[mec] hydraulic motor variable speed control
— m —— vertedero m - [mec] spout control
— m —— m para tajadora f - [agric] chopper
spout control(ler)
— m —— voltaje m - [electr-oper] volts,tage,
control(ler), o regulator • [sold] job selector
— m —— m alto - high voltage control(ler)
— m —— m, bajo, o reducido - [electr-oper]
low voltage control(ler)
— m —— m en circuito m abierto - [sold] open
circuit voltage, control(er), o adjuster
— m —— de tipo m con cuadrante m—[electr]
dial-type voltage control(ler)
— m —— m fuera de uso m - [electr-equip]
inoperative voltage control(ler)
— m —— m inactivado - [electr-equip] inoperative, o deactivated, voltage control(ler)
— m —— para alimentador m para alambre
m - [sold] inoperative, o deactivated, wire
feeder voltage control(ler)
— m —— m optativo - [sold] optional voltage
control(ler)
— m —— m para alimentador m - [sold] wire
feeder voltage control(ler)
— m — m —— m para alambre m - [sold]
wire feeder voltage control(ler)
— m — m — m —— m fuera de uso m -
[sold] inoperative, o deactivated, wire feeder voltage control(ler)
— m — m — m — m inactivado - [sold]
inoperative wire feeder voltage control(ler)
— m —— — arco m - [sold] arc voltage con-
trol(ler)
— m —— m para corriente f - [electr-equip]
current voltage control(ler)
— m —— fuente f para energía f - [sold]
power source voltage control(ler)
— m —— m — inducción f - [electr-equip] induction voltage, control(ler), o regulator
— m —— — manipulador m - [electrón] keyer
voltage, control(ler), o regulator
— m — m — m para tono(s) m - [electrón]
tone keyer voltage, control(ler), o regulator
— m —— m variable - [electr] variable voltage, control(ler), o regulator
— m —— m verificado - [electrón] checked
voltage, control(ler), o regulator
— m paralelo m - [comput] parallel, o compan-
ion, control(ler), o regulator
— m Partlon m - [ind] Partlon controller
— m — para temperatura f - [ind] Partlon temperature control(ler)
— m permanente - [electr-equip] permanent, o
continuous, control(ler)
— m —— corriente f - [electr-equip] continuous current control(ler)
— m —— — m tipo manubrio m - [electr-equip]
continuous crank-type current control(ler)
— m — de tipo m con cuadrante m - [electr-
-equip] continuous dial-type control(ler)
— m por inducción f - [electr-equip] induction,
regulator, o control(ler)
— m —— f para, tensión f, o voltaje m -
[electr-equip] induction voltage regulator
— m portátil - [electr-equip] portable control
— m —— optativo - [sold] optional portable con-
trol(ler)
— m —— m para campo m - [sold] optional portable field control(ler)
— m — para campo m - [metal-sold] portable
field control(ler)
— m —— voltaje m - [electr-equip] portable
voltage control(ler)

regulador m práctico - [inc] convenient, o
handy, o practical, control(ler)
— m — para operación f - [ind] convenient, o
handy, o practial, operating,tion control
— m preciso - [electr-oper] fine adjustment,
switch, o control(ler)
— m — para amperaje m - [sold] fine current,
adjuster, o control(ler)
— m —— puesta f en marcha f - [ind] start-
-(ing) precise control(ler)
— m —— f — f y detención f - [mec]
start(ing) and stop(ping) precise control
— m —— encendido m y corte m - [sold]
start(ing) and stop(ping) precise control
— m — para voltaje m - [sold] fine voltage
control(ler)
— m principal - [electr-oper] main, o important, (operator) control(ler)
— m — para amperaje m - [sold] main current,
adjuster, o control(ler)
— m —— mando m - [sold] main, o important,
operator control(ler)
— m protegido - [ind] protected control(ler)
— m que se describe v a continuación f - [ind]
following control(er)
— m reforzado - [electr-equip] heavy duty con-
trol(ler)
— m — con potencial m constante - [electr]
heavy duty constant potential control(er)
— m —— — m para laminación f - [electr]
heavy duty mill constant potential control
— m —— de tipo m para laminación f
[electr] heavy duty mill type constant potential control
— m registrador m - [instrum] recording, con-
trol(ler), o regulator
— m — neumático - [instrum] air operated recording, control(ler), o regulator
— m regulado - [comb.int] positioned throttle
— m remoto - [ind] remote control(ler)
— m rotativo - [instrum] rotary control(ler)
— m — para bobina f - [instrum] rotary coil
control
— m —— — f para reactancia f - [electr]
rotary reactor control(ler)
— m —— devanado m - [electr] rotary coil
control
— m —— — m para reactancia f - [electr]
rotary reactor control(ler)
— m rotatorio - [electr] Rototrol
— m selectivo para amperaje(s) m - [electr]
selective current control(ler)
— m sin funcionar v - [sold] non functioning
control(ler)
— m —— v para arco m - [sold] nonfunctioning
arc control(ler)
— m soplado - [electr-oper] blown out control
— m tipo manubrio m - [mec] crank-type control
— m tirado hacia afuera - pulled out throttle
— m totalmente automático - [electr-equip]
fully automatic control(ler)
— m variable - [electr-instal] selector switch
— m — para amperaje m - [sold] current selector switch
— m — para avance m - [sold] variable speed
control(ler)
— m — para motor m hidráulico - [mec] hydraulic motor variable control(ler)
— m ——, tensión f, o voltaje m - [electrón]
variable voltage, control(ler), o regulator;
Variac
— m verificado - [mec] checked control(ler)
— m vertical - [mec] vertical, adjuster, o
control(ler), o regulator
regulador,ra a - regulating; control(ling) •
[deport] sanctioning
regular a - . . . . - fair • ordinary; consistent;
. standard • faithful • ordinary | v - . . .;
to (set @) control • to tune (up) • to check
• [electr-oper] to toggle
— v a conectado,da - [electr-oper] to switch on
— v — nivel m inferior - to adjust, down, o
lower

regular <u>v</u> a nivel <u>m</u> inferior a(1) para procedimientos para soldadura <u>f</u> - [sold] to adjust lower than for @ welding procedure(s)
— <u>v</u> — nivel <u>m</u> superior - to adjust, up, <u>o</u> higher
— <u>v</u> — — <u>m</u> — a(1) para procedimiento(s) <u>m</u> para soldadura <u>f</u> - [sold] to adjust higher than for @ welding procedure(s)
— <u>v</u> abastecimiento <u>m</u>—[ind] to control @ supply
— <u>v</u> — <u>m</u> de combustible <u>m</u> - [comb.int] to control @ fuel supply
— <u>v</u> actuación <u>f</u> - to regulate @ performance
— <u>v</u> admisión <u>f</u> - to control @ intake
— <u>v</u> agua <u>m</u> freática - [hidr] to control @ ground water
— <u>v</u> — <u>m</u> para enfriamiento <u>m</u> - [ind] to control @ cooling water
— <u>v</u> ajustadamente - to control tightly • [sold] to fine tune
— <u>v</u> ajuste <u>m</u> - to change @ setting; to set @ adjustment
— <u>v</u> — para amperaje <u>m</u> - [sold] to change @ current setting
— <u>v</u> ajuste <u>m</u> preciso - [sold] to set @ fine adjustment; to fine tune
— <u>v</u> amperaje <u>m</u> - [sold] to adjust @, current, <u>o</u> amperage
— <u>v</u> — <u>m</u> para encendido <u>m</u> - [sold] to adjust @ starting, current, <u>o</u> amperage
— <u>v</u> — <u>m</u> — soldadura <u>f</u> - [sold] to set @ welding, current, <u>o</u> amperage
— <u>v</u> aportación <u>f</u> - to control @ input
— <u>v</u> — de calor <u>m</u> - [sold] to control @ heat input
— <u>v</u> apropiadamente - to, set, <u>o</u> control, appropriately
— <u>v</u> — encendido <u>m</u> - [comb.int] to time properly; to set @ spark appropriately
— <u>v</u> — <u>m</u> y voltaje <u>m</u> - [sold] to, set, <u>o</u> adjust, @, current, <u>o</u> amperage, and @ voltage
— <u>v</u> — <u>m</u> — — <u>m</u> para encendido <u>m</u> - [sold] to, adjust, <u>o</u> set, @ starting, current, <u>o</u> amperage, and voltage
— <u>v</u> anticipadamente - [mec] to set (beforehand)
— <u>v</u> apropiadamente - to control, properly, <u>o</u> appropriately
— <u>v</u> atmósfera <u>f</u> - [ambient] to control @ atmosphere
— <u>v</u> — <u>f</u> con argón <u>m</u> - [sold] to control @ argon atmosphere
— <u>v</u> — <u>f</u> — — <u>m</u> gaseoso - [sold] to control @ argon gas atmosphere
— <u>v</u> — <u>f</u> — gas <u>m</u> de argón <u>m</u> - [sold] to control @ argon gas atmosphere
— <u>v</u> — <u>f</u> protectora - [metal-trat] to control @ protective atmosphere
— <u>v</u> — — gas <u>m</u> de argón <u>m</u> - [sold] to control @ argon gas atmosphere
— <u>v</u> automáticamente - [mec] to, set, <u>o</u> control, automatically
— <u>v</u> avance <u>m</u> - [sold] to control @ travel
— <u>v</u> baño <u>m</u> - [sold] to control @ puddle
— <u>v</u> cable <u>m</u> - [cabl] to control @ cable
— <u>v</u> — a gancho <u>m</u> - [grúas] to control @ whip
— <u>v</u> — — motón <u>m</u> - [grúas] to control @ main
— <u>v</u> calidad <u>f</u> - [ind] to control @ quality
— <u>v</u> calor <u>m</u> - to control @ heat
— <u>v</u> cambiador <u>m</u>—[electr-oper] to set @ changer
— <u>v</u> — <u>m</u> para toma <u>f</u> - [electr-oper] to set @ tap changer
— <u>v</u> capacidad <u>f</u> - to control @ capacity
— <u>v</u> característica <u>f</u> - to control @ characteristic
— <u>v</u> — <u>f</u> de arco <u>m</u> - [sold] to control @ arc(s) characteristic(s)
— <u>v</u> carburador <u>m</u> - [comb.int] to set @ carburetor
— <u>v</u> — <u>m</u> para mezcla <u>f</u> muy pobre - [comb.int] to set @ carburetor (very) lean
— <u>v</u> — <u>m</u> — — <u>f</u> — rica - [comb.int] to set @ carburetor (very) rich
— <u>v</u> carga <u>f</u> - to control @, load, <u>o</u> charge
— <u>v</u> — <u>f</u> para alternador <u>m</u> - [instrum] to control @ alternator charge

regular <u>v</u> carrera <u>f</u> - [mec] to control @ stroke
— <u>v</u> ciclo <u>m</u> - [electr] to control @ cycle
— <u>v</u> — <u>m</u> para avance <u>m</u> - [sold] to control @ travel cycle
— <u>v</u> — <u>m</u> — derivación <u>f</u> - [electr-oper] to control @ bypass cycle
— <u>v</u> — <u>m</u> — desviación <u>f</u> - [electr-oper] to control @ shunt cycle
— <u>v</u> — <u>m</u> — soldadura <u>f</u> - [sold] to control @ weld(ing) cycle
— <u>v</u> circuito <u>m</u> - [comput] to control @ circuit
— <u>v</u> completamente - to control completely
— <u>v</u> componente(s) <u>m</u> - to control @ component(s)
— <u>v</u> compresor <u>m</u> - [mec] to control @ compressor
— <u>v</u> compuerta <u>f</u> - [hidr] to control @ gate
— <u>v</u> — <u>f</u> O - [electrón] to control @ OR gate
— <u>v</u> — <u>f</u> — exclusiva - [electrón] to control @ exclusive OR gate
— <u>v</u> — <u>f</u> para antepozo <u>m</u> - [petról] to control @ cellar gate
— <u>v</u> — <u>f</u> — sótano <u>m</u> - [petról] to control @ cellar gate
— <u>v</u> con espigón <u>m</u> - [hidr] to jetty control
— <u>v</u> — muro <u>m</u> - [hidr] to control with @ wall
— <u>v</u> — — <u>m</u> para retención <u>f</u> - [hidr] to control with @ retaining wall
— <u>v</u> precisión <u>f</u> - to control, precisely, <u>o</u> accurately
— <u>v</u> — arco <u>m</u> - [sold] to fine tune @ arc
— <u>v</u> — — <u>f</u> característica(s) <u>f</u> - [sold] to fine tune @ characteristic(s)
— <u>v</u> — — <u>f</u> — <u>f</u> de arco <u>m</u> - [sold] to fine tune @ arc characteristic(s)
— <u>v</u> — regulador <u>m</u> - [mec] to governor control
— <u>v</u> — — automático - [mec] to (automatic) governor control
— <u>v</u> — tablestacas,cado - [hidr] to sheeting control
— <u>v</u> — — <u>f</u> de acero <u>m</u> - [hidr] to steel sheeting control
— <u>v</u> — tapa <u>f</u> - to, cap, <u>o</u> cover, control
— <u>v</u> — — <u>f</u> para radiador <u>m</u> - to, control, <u>o</u> regulate, with @ radiator cap
— <u>v</u> conmutador <u>m</u> - [electr-oper] to, position, <u>o</u> set, @ switch
— <u>v</u> — <u>m</u> para sentido <u>m</u> - [electrón] to control with @ sesnse switch
— <u>v</u> — <u>m</u> para velocidad <u>f</u> alta - [electr-oper] to set @ switch on high
— <u>v</u> — — — <u>f</u> baja - [electr-oper] to set @ switch on low
— <u>v</u> contaminación <u>f</u> - to control @ pollution
— <u>v</u> cordón <u>m</u> - [sold] to control @ bead
— <u>v</u> corriente <u>f</u> - [electr-oper] to control @ current
— <u>v</u> — <u>f</u> para soldadura <u>f</u> - [sold] to control @ weld(ing) current
— <u>v</u> costo(s) <u>m</u> - [ind] to control @ cost(s)
— <u>v</u> cráter <u>m</u> - [Sold] to control @ puddle
— <u>v</u> — <u>m</u> de soldadura <u>f</u> - [sold] to control @ weld(ing) puddle
— <u>v</u> cuidadosamente - to control carefully
— <u>v</u> cumplimiento <u>m</u> - to, control, <u>o</u> regulate, @ performance
— <u>v</u> declive <u>m</u> - [topogr] to control @ slope
— <u>v</u> descarga <u>f</u> - [hidr] to control @ discharge
— <u>v</u> desempeño <u>m</u> - to, control, <u>o</u> regulate, @ performance
— <u>v</u> dispositivo <u>m</u> - [ind] to control @ device
— <u>v</u> — <u>m</u> para avance <u>m</u> - [sold] to, control, <u>o</u> adjust, @ travel unit
— <u>v</u> emanación(es) <u>f</u> - [ind] to control @ emission(s)
— <u>v</u> embrague <u>m</u> - [mec] to control @ clutch
— <u>v</u> en fábrica <u>f</u> - [ind] to factory (pre)set
— <u>v</u> — más - to turn up • to over control
— <u>v</u> — menos - to turn down • to under control
— <u>v</u> encendido <u>m</u> - [comb.int] to time @, spark, <u>o</u> engine
— <u>v</u> engargantado <u>m</u> - [mec] to control @ meshing
— <u>v</u> — <u>m</u> de engranaje <u>m</u> - [mec] to control @

gear mesh(ing)
— m engranaje m - [mec] to control @ engagement
  • to control @ gear
— v erosión f - [hidr] to control @ erosion
— v escombro(s) m - [hidr] to control @ debris
— v estrangulador m - [mec] to, control, o set,
  @ choke, o throttle
— v estratificación f - [indo control @ strati-
  fication
— v exportación(es) f - [fisc] to control @ ex-
  port(s)
— v fácilmente - to, control, o adjust, easily
— v factor m - to control @ factor
— v filtración f - [hidr] to control @ seepage
— v flujo m - to, control, o regulate, @ flow
— v — m de agua m - [hidr] to control @ water
  flow
— v freno m - [mec] to control @ brake
— v — m contra rotación f - [grúas] to control
  @ swing brake
— v funcionamiento m - [mec] to control @ operation
— m — m de compresor m - [mec] to control @
  compressor operation
— v gas m de argón m - [sold] to control @ argon gas
— v gasto m de agua m - [hidr] to control @
  water flow
— v grieta(s) v - [sold] to control @ crack(s)
— v — f interior(es) - [sold] to control @ internal crack(s)
— v hidráulicamente - to control hydraulically
— v horno m - [combust] to control @ furnace •
  [domest] to control @ oven
— v — m eléctrico - [metal-prod] to control @
  arc furnace
— v humedad f - to control @ moisture
— v independientemente - to control independently
— v intensidad f - to control @ intensity
— v — f de calor m - [sold] to control @ heat
  intensity
— v interruptor m - [electr-oper] to position @
  switch
— v — m a posición f de conectado - [electr-
  -oper] to switch on
— v — m por vacío m - [electr-oper] to control
  @ vacuum breaker
— v inundación(es) f - [hidr] to control @
  flood(s)
— v leva f - [mec] to set @ cam
— v máquina f - [ind] to, control, o set, @
  machine
— v mecanismo m - [mec] to control @ mechanism
— v — m para avance m - [sold] to control @
  travel mechanism
— v mediante inspección f - [admin] to inspection control
— v — mezcla (muy) rica - [comb.int] to set
  (very) rich
— v motor m - [electr-mot] to control @ motor •
  [int.comb] to control @ engine
— v — m hidráulico - [mec] to control @ hydraulic motor
— v nivel m - to control @ level
— v — m alto - [mec] to control @ high level
— v — — m de líquido m - [mec] to control @
  high liquid level
— v — bajo - [mec] to control @ low level
— v — m — de líquido m - [mec] to control @
  low liquid level
— v — m elevado—[mec] to control @ high level
— v — m — de líquido m - [mec] to control @
  high liquid level
— v nivel m elevado - [mec] to control @ high
  level
— v — n — de líquido m - [mec] to control @
  high liquid level
— v — m para agua m - [hidr] to control @
  water level
— v — m — energía f - [electr-distrib] to
  control @ power level
— v — m — suministro m - [electr-distrib] to

control @ supply level
regular v nivel m para suministro m de energía f
  [electr-distrib] to contol @ power supply
  level
— v nuevamente - [instrum] to reset
— v — encendido m - [comb.int] to retime
— v — — m en motor v - [comb.int] to retime @
  engine
— v operación f - [ind] to control @ operation
— v — f de compresor m - [ind] to control @
  compressor operation
— v osciloscopio m - [electrón] to control @
  oscilloscope
— v palanquilla f - [metal-lam] to control @
  billet(s)
— v para mezcla f (muy) pobre - [int.comb] to
  set (very) lean
— v — — f — rica - [comb.int] to set (very)
  rich
— v — suministrar v - [ind] to set to supply
— v — suplir v - [ind] to set to supply
— f — tanda f - [ind] to set up for @ run
— v pedal m- [mec] to control @ (foot) pedal
— v perfectamente - to control perfectly
— v período m - to, control, o adjust, @ time
— v — m de tiempo m - to, control, o adjust,
  @ time period
— v polaridad f - [electr] to control @ polarity
— v precisamente v - to control precisely •
  [sold] to fine tune
— v — arco m - [sold] to fine tune @ arc
— v presión f - to, control, o regulate, @
  pressure
— v — con tapa f - [comb.int] to, control, o
  regulate, @ pressure with @ cap
— v — — f — f para radiador m - [comb.-
  -int] to, control, o regulate, @ pressure
  with @ radiator cap
— v procedimiento m - to control @ procedure
— v — m — electrodo m - [sold] to control @
  procedure with @ electrode
— v — — m con alma m fundente - [sold] to
  control @ Innershield procedure
— v — m Innershield - [sold] to control @ Innershield procedure
— v provisión f - to control @ supply
— v — f de combustible m - [comb.int] to control @ fuel supply
— v punto m - [mec] to control @ point
— v — m para presión f - [mec] to control @
  pressure point
— v — m — — f en tubería f - [tub] to, control, o set, @ pressure point
— v — m — f — f para entubación f -
  [tub] to, control, o set @ casing pressure
  point
— v quemador m - [combust] to, control, o regulate, @ burner
— v rápidamente - to, control, o adjust, rapidly, o quickly
— v rapidez f - [mec] to, control, o regulate,
  o adjust, @ speed
— v — f para bajada f - [mec] to control @
  drop(ping) speed
— v rebalse m - [hidr] to control @, outflow,
  o overflow, o spillage
— v — m de agua m - [hidr] to control @ water
  outflow
— v refrigeración f - [ind] to control @, refrigeration, o cooling
— v régimen m - to control @, rate, o speed
— v — m para carga f - [electr-acumul] to
  control @ charge,ging rate
— v regulador m - [comb.int] to, control, o
  position, @ throttle, o control
— v — m para aire m - [neumát] to set @ air
  regulator
— v resorte m - [mec] to, control, o set, o
  adjust, @ spring
— v rotación f - [grúas] to control @ swing
— v salida f - [hidr] to control @, outlet, o
  outflow

regular v **salpicadura** f - [sold] to control @ spatter
— f — f **excesiva** - [sold] to control @ excessive spatter
— v **sistema** m - [ind] to set (up) @ system
— v **soldadora** f - [sold] to set @ welder
— v **soplador** m - [ind] to set @ blower
— v **suministro** m - to control @ supply

— v — m de **combustible** m - [comb.int] to control @ fuel supply
— v — m de **energía** f - [electr-distrib] to control @ power supply
— m **tamaño** m - to control @ size
— v — m de **cordón** m - [sold] to control @ bead size
— v — — **gota** f—[hidr] to control @ drop size
— v — m — **gotita** f - [sold] to control @ droplet size
— v **temperatura** f - to, control, o set, @ temperature
— v **tiempo** m - to, control, o adjust, @ time
— v **toma** f - [electr-oper] to set @ tap
— v **tornillo** m - [mec] to adjust @ screw
— v — m para **ajuste** m - [mec] to set @ adjusting screw
— v — m — — m para **velocidad** f - [mec] to adjust @ speed adjusting screw
— v — m — **resorte** m - [mec] to adjust @ spring screw
— v **trabajo** m - [ind] to, control, o regulate, @, work, o job
— v — m en **ejecución** f - [ind] to, control, o regulate, @ work in progress
— v **tratamiento** m - to control @ treatment
— v — m **térmico** - [metal-trat] to control @ heat treatment
— v **válvula** f - [ind] to operate @ valve
— v — f para **estrangulación** f - [mec] to control @ choke valve
— v **variable** f - to control @ variable
— v — f **identificable** - to control @ identifiable variable
— v **velocidad** f - to, control, o govern, o adjust, o set, @ speed
— v — f **alta** - [ind] to set @ high speed
— v — f **baja** - [ind] to set @ low speed
— v — f **deseada** - [ind] to set (for) @ desired speed
— v — f — para **estrangulador** m - [ind] to set (for) @ desired choke speed
— v — f para **bajada** f - [ind] to control @ drop(ping) speed
— v — f — **elevación** f - [grúas] to control @ hoist-ing speed • [agric-equip] to control @ lifting speed
— v — f — **elevador** m - [mec] to control @ elevator speed
— v — f — m para **vagón(es)** m - [agric] to control @ wagon elevator speed
— v — f — **estrangulador** m - [ind] to, control, o set, @ choke speed
— v **ventilador** m - [mec] to set @, fan, o blower
— v **voltaje** m - [electr-oper] to, control, o set, o regulate, @ voltage
— v — m en **corriente** f **alterna** - [electr-oper] to control @ alternating current voltage
— v — m — — f **continua** - [electr-oper] to control @ direct current voltage
— v — m para **arco** m - [sold] to control @ arc voltage
— v — m — **encendido** m - [sold] to control @ starting voltage
— v — m — **soldadura** f - [sold] to, control, o set, @ weld(ing) voltage
regularidad f = ... • consistency • reliability • [deport] rally
— f — **análisis** - [quím] analysis regularity
— f — — m **químico** - [quím] chemical analysis regularity
— f en **espesor** m - [mec] thickness, o gage, regularity

regularidad f en **funcionamiento** m - reliability
regularización f de **emanación(es)** f - emissions control • [sanit] odor control
regulatorio,ria - regulatory
regularizar v - . . . ; to normalize
regularmente adv - . . .; consistently; fairly
regurgitar v - . . . • [fig] to cough up
rehabilitación f - . . . . • redeovelopment • reopening • comeback
— f de **camino** m - [vial] road reopening
— f — **centro** m **urbano** - [pol] downtown redevelopment
— f — **terreno(s)** m - [constr] land reclamation
— f — — m para **minería** f - [miner] mining land reclamation
— m — — m — — f **bajo cielo** m **abierto** - [miner] open pit (mining) land reclamation
— f — **zona** f - [pol] area redevelopment
— f — — f **céntrica** - [pol] downtown redevelopment
— f **mediante revestimiento** m **nuevo** - [constr] rehabilitation through relining
rehabilitado,da a - rehabilitated • reopened
— a **mediante revestimiento** m **nuevo** - [constr] rehabilitated through (@) relining
rehabilitar v - . . . ; to reestablish; to resore to service • to reopen
— v **camino** m - [vial] to reopen @ road
— v **mediante revestimiento** m **nuevo** - [constr] to rehabilitate through @ relining
rehacer v **boca** f - [metal-prod] to, repair, o rebuild, @ iron notch
— v — f **con ladrillo(s)** m - [metal-prod] to rebuild @ (iron) notch (with bricks)
— v **con ladrillo(s)** v - [constr] to rebuild with brick(s)
— v **piquera** f - [metal-prod] to rebuild @, taphole, o iron notch
— v **refractario(s)** m - [metal-prod] to rebuild @ refractories
rehusado,da a - refused • withheld • declined
rehusamiento* m - refusal • declination,ning • withholding
rehusado,da a - refused • declined • withheld
— a **irrazonablemente** - unreasonably withheld
rehusar f - . . . . to withhold
— v **irrazonablemente** - to unreasonably, withhold, o reject
reimpresión f **directa** - [imprent] direct reprint(ing)
reimpreso,sa a - [imprent] reprinted
— a **directamente** - [imprent] reprinted directly
reimprimir v **directamente** - [imprent] to reprint directly
reindustrialización f - [econ] reindustrialization,lizing
reindustrializado,da a - [econ] reindustrialized
reindustrializar v - [econ] to reindustrialize
reingresado,da a - reentered
— v en **carrera** f - [deport] to reenter @ race
reingreso n - reentry;tering
— m en **carrera** f - [deport] race reentry,tering
reiniciación f - reinitiating,tion; restarting; resuming,mption
— f de **consumo** m de **energía** f - [electr-oper] power load restarting
— t — **soldadura** f - [sold] weld(ing) restarting
reiniciado,da a - reinitiated; restarted; resumed
reiniciar v - to, reinitiate, o restart, o resume
— v **consumo** m de **energía** f - [electr-oper] to restart @ power load
— v **soldadura** f - [sold] to restart @ weld(ing)
reino m - [pol] . . . ; realm
reinserción f - [mec] reinserting,tion; véase también **vuelta** f **a insertar** v
reinsertado,da* a - [mec] reinserted; véase también **vuelto,ta a insertar**
reinsertar* v - [mec] to reinsert; véase también **volver** v **a insertar** v
reinstalación f - . . . . • replacement,cing

reinstalación f̲ apropiada ~ [mec] proper reinstalling,lation
— **f̲ de interruptor m̲** - [electr-instal] breaker, o̲ switch, reinstalling,lation
— **f̲ — pieza f̲** - [ind] part reinstalling,lation
— **f̲ inapropiada** - [ind] improper reinstalling,lation
reinstalado,da a̲ - [mec] reinstalled
— **a̲ apropiadamente** - [mec] reinstalled properly
— **a̲ inapropiadamente** - [mec] improperly reinstalled
reinstalar v̲ - [mec] to reinstall
— **v̲ apropiadamente**—[mec] to reinstall properly
— **v̲ arandela f̲** - [mec] to reinstall @ washer
— **v̲ conmutador m̲** - [electr-instal] to reinstall @ switch
— **v̲ inapropiadamente** - [mec] to reinstall improperly
— **v̲ interruptor n̲** - [electr-instal] to reinstall @, breaker, o̲ switch
— **v̲ pieza f̲** - [ind] to reinstall @ part
reinstrucción* f̲ - véase **instrucción f̲ adicional**
reintegración f̲ - . . .; reintegrating,tion; return(ing) • [transp] reshipment
— **f̲ de sistema m̲** - system reintegrating,tion
reintegrado,da a̲ - reintegrated
reintegrar v̲ - to reintegrate • [fin] to refund
— **v̲ papel m̲ sellado m̲** - [fisc] to reimburse @ documentary tax
reintegro n̲ - [fin] reimbursement; restitution • return(ing) • [fin] refund(ing) • [transp] reshipment,pping
— **m̲ creciente** - [fisc] increasing reimbursement
— **m̲ de impuesto(s)** - [fisc| tax, refund(ing), o̲ reimbursement,sing
— **m̲ papel m̲ sellado** - [fisc] documentary tax reimbursement,sing
— **m̲ máximo** - [fin] maximum, refund, o̲ reimbursement
— **m̲ mínimo** - [fin] minimum, refund, o̲ reimbursement
— **m̲ por exportación(es) f̲** - [fisc] export(s) reimbursement
— **m̲ según ley f̲** - [fisc] legal reimbursement
— **m̲ único** - [fisc] single reimbursement
reintroducción f̲ - reintroducing,ction
— **f̲ de gas m̲** - [combust] gas, reintroduction, o̲ reinjection
reintroducido,da a̲ - reintroduced; reinjected
reintroducir v̲ - [combust] to, reintroduce, o̲ reinject, @ gas
reinversión f̲ consecutiva - [fin] consecutive reinvesting,tment
— **f̲ de utilidad(s) f̲** - [econ] profit(s) reinvesting,tment
reinvertido,da a̲ - [fin] reinvested
reinvertir v̲ - [fin] to reinvest
— **v̲ utilidad(es) f̲** - [fin] to reinvest @ profit
reiteración f̲ - . . .; reiterating; repeating; repetition; reaffirmation,ming • assuring
reiterado,da a̲ - reiterated; restated • assured
reiterar v̲ - . . .; to, repeat, o̲ reaffirm, o̲ restate • to assure
reivindicación f̲ - [legal] . . .; replevying • claim
— **f̲ precedente** - [legal] foregoing, replevin, o̲ claim
— **f̲ siguiente** - [legal] following, replevin, o̲ claim
reivindicado,da a̲ - [legal] replevied
reja f̲ - [mec] . . .; grating • rack • [agric-equip] (plow)share; bottom
— **f̲ angosta (para escardillar v̲)**—[agric-equip] deer tongue
— **f̲ con pasador m̲ rompible** - [agric-equip] pin break shovel
— **f̲ — punta f̲ movible** - [agric-equip] bar point share
— **f̲ esmerilada** - [agric-equip] ground share
— **f̲ para arado m̲** - [agric-equip] plowshare
— **f̲ — — m̲ esmerilada** - [agric-equip] ground plowshare
— **f̲ — desmontaje m̲** - [agric-equip] detachable share

reja f̲ para desmontaje m̲ rápido - [agric-equip] quick(ly) detachable share
— **f̲ — escardadora f̲** - [agric-equip] cultivator share
— **f̲ — escarificadora f̲** - [agric] cultivator share
— **f̲ — resistencia f̲** - resistance grid
— **f̲ — subsuelo m̲** - [agric-equip] subsoiler
— **f̲ — — m̲ para implemento m̲ Lister** - [agric-equip] Lister subsoiler
— **f̲ triangular (para escardillar v̲)** - [agric-equip] duck feet
rejilla f̲ - [mec] . . .; screen(ing); grid; grate • [autom-radiador] grill
— **f̲ abierta** - [mec] open slot
— **f̲ apropiada** - [constr] suitable screen
— **f̲ cerrada** - [mec] closed slot • [electr-instal] closed grid
— **f̲ — para conexión f̲ a tierra f̲** - [electr-instal] closed gounding, grid, o̲ rod
— **f̲ coladora** - [mec] strainer screen
— **f̲ — magnética** - [mec] magnetic strainer screen
— **f̲ de acero m̲** - [mec] steel, grid, o̲ grate
— **f̲ — alambre m̲** - [mec] wire screen(ing)
— **f̲ — — m̲ de acero m̲** - [constr] steel wire screen(ing)
— **f̲ — — m̲ grueso** - [constr] coarse wire screen(ing)
— **f̲ — — m̲ — de acero m̲** - [constr] coarse steel wire screen(ing)
— **f̲ de barra(s) f̲** - [sanit] bar screen; trash rack
— **f̲ — bronce m̲** - [sanit-mingitorio) bronze strainer
— **f̲ — — m̲ tipo panal** - [sanit-mingitorio] beehive bronze strainer
— **f̲ — fieltro m̲** - [mec] felt screen
— **f̲ — — m̲ para colador m̲** - [mec] felt strainer screen
— **f̲ — hierro m̲** - [mec] iron grating
— **f̲ — latón m̲** - [mec] brass screen • [sanit-mingitorio] brass strainer
— **f̲ — metal m̲** - [mec] metal grating
— **f̲ — — m̲ desplegado** - [mec] expanded metal grid
— **f̲ — pletina(s) f̲** - [mec] bar grating
— **f̲ filtro m̲** - [mec] filter screen
— **f̲ — magnética** - [mec] magnetic filter screen
— **f̲ incombustible** - [segurid] nonflammable screen(ing)
— **f̲ — apropiada f̲** - [segurid] suitable nonflammable screen(ing)
— **f̲ inspeccionada f̲** - [mec] inspected screen
— **f̲ limpia f̲** - [mec] clean screen
— **f̲ limpiada f̲** - [mec] cleaned screen
— **f̲ magnética** - [mec] magnetic screen
— **f̲ — f̲ armada** - [mec] assembled magnetic screen
— **f̲ — instalada** - [mec] installed magnetic screen
— **f̲ metálica** - [mec] metal(lic), grate,ting, o̲ screen(ing)
— **f̲ palpada f̲** - [mec] felt, o̲ touched, screen
— **f̲ para aceite m̲** - [comb.int] oil screen
— **f̲ — admisión f̲** - [comb.int] intake screen
— **f̲ — — f̲ para aire m̲** - [comb.int] air intake screen
— **f̲ — aire m̲** - [comb.int] air screen
— **f̲ — aspiración f̲** - [mec] intake, o̲ suction, screen
— **f̲ — colador m̲** - [comb.int] strainer screen
— **f̲ — — m̲ palpada** - [mec] felt, o̲ touched, strainer screen
— **f̲ — — m̲ para aceite m̲** - [comb.int] oil strainer screen
— **f̲ — conducto m̲** - [mec] conduit screen
— **f̲ — — m̲ vertical** - [combust] stack screen
— **f̲ — — m̲ — para aire m̲** - [combust] air stack screen
— **f̲ — conexión f̲ con tierra f̲**—[electr-instal] ground(ing) grid
— **f̲ — desagüe m̲** - [hidr] dewatering screen

rejilla f para devanado m - [electr] coil grid
— f — — m para campo m - [electr-instal]
field coil grid
— t — embudo m - [mec] funnel screen
— f — enfriamiento m - [mec] cooling, grid, o
screen
— f — fondo m - [mec] bottom screen
— f — lavado m - [mec] washing, o scrubbing,
screen
— f — lavador m - [metal-prod] washer, screen,
o grid
— f — máquina f - [mec] machine, grid, o grate
— f — — f para sínter(ización) - [metal-prod]
sinter(ing) machine grate
— f — prelimpieza f - [comb.int] pre-cleaner
screen
— f — — f para motor m - [comb.int] engine
pre-cleaner screen
— f — protección f - [mec] protector screen
— f — — f para indicador m—[mec] meter guard
— f — radiador m - [comb.int] radiator, grill,
o screen • radiator shield
— f — refrigeración f - [mec] cooling screen
— f — regulación f—[mec] control(ling) screen
— f — — f para aire m - [ambient] air control
louver
— f — resistencia f - [electrón] resistance
grid
— f — toma f - [mec] intake screen
— f — — f para aire m - [comb.int] air intake
screen
— f — — f — — m para volante m - [comb.int]
flywheel air intake screen
— f — tubería f—[ind] pipe, grating, o screen
— f — — f para descarga f - [ind] discharge
pipe grating
— f — — f — — f para presión f alta -
[metal-prod] high pressure discharge pipe
grating
— f — volante m - [comb.int] flywheel screen
— f prelimpiadora f - [comb.int] pre-cleaner
screen
— f protectora f - [mec] protector screen
— f — para radiador m - [comb.int] radiator
screen
— f quitada - [mec] removed screen
— f removida - [mec] removed screen
— f rotativa - [avicult] rotary, o roto, screen
— f sacada - [mec] removed screen
— f tipo panal - [sanit-mingitorios] beehive
strainer
— f soldada - [sold] welded, grate, o screen
— f — de acero m - [sold] welded steel grate
rejuntado m - [constr] pointing
rejuntante a - [constr] pointing
relación f - . . .; relationship; liaison; association • rate; ratio; relative rate -
list(ing) • narrating,tion; communicating •
recording • area | adv - (con) with regard to
— f alta - high ratio
— f — entre caucho m y oquedad(es) f - [autom-
-neumát] high rubber-to-void ratio
— f — — m y vacío m - [autom-neumát] high
rubber to void ratio
— f amigable - [fam] rapport
— f amistosa - rapport; friendly relation(ship)
— f ancho/espesor - width/thickness ratio
— f apropiada - proper, o correct, relationship
— f aproximativa - approximative ratio
— f baja - low ratio
— f — entre altura f y anchura f - [cutom-
-neumát] low aspect (ratio)
— f — — caucho m y vacío m - [autom-neumát]
low rubber to void ratio
— f bases/ácidos - [metal-prod] base(s)/acid(s)
ratio
— f buena f - [labor] good relation(ship)
— f clara - [admin] good relationship
— f — con otros niveles m de administración f
- [admin] clear line relationship(s)
— f — — personal m subalterno - [admin] clear
staff relationship(s)
— f con otros niveles m de administración f -
[admin] line relationship(s)

relación r con personal m subalterno - [admin]
staff relationship(s)
— f confusa - [admin] confused relationship •
noncordial relationiship
— f — con personal m subalterno - [admin]
confused staff relationship(s)
— f — — otro(s) nivel(es) m de administración f—[admin] confused line relationship(s)
— f cordial - cordial relation(ship); rapport
— f correcta - correct relation(ship)
— f costo-beneficio - [com] cost-benefit ratio
— f de cambio m - [fin] exchange rate,tio
— f — compresión f - [mec] compression ratio
— f — cortocircuito m - [electr-oper] short
circuit reactance
— f — dependencia f -[labor] employee status •
employee relationship
— f — empleo m - [labor] employee, relation, o
status
— f — engranaje m - [mec] gear ratio
— f — expansión f - [mec] expansion ratio
— f — fabricante(s) m - manufacturers list
— f — fusión f - [sold] fusion (relative) rate
— f — intercambio m - interchange relationship
— f — — m entre capa(s) f - [geol] layer
interchange relation(ship)
— f — multiplicación f - [mec] multiplication
ratio
— f — — f de torsión f - [mec] torque multiplication ratio
— f — precio(s) m - [com] price(s), relation-
-(ship), o ratio • price(s) listing
— f — resistencia f a rotura f - [mec] breaking strength ratio
— f — tarea(s) f - [admin] task(s), o work.
relation(ship)
— f — — f entre sí - [admin] task(s), o work,
(inter)relationship
— f — torsión f - [mec] torque ratio
— f — trabajo m - [admin] work relation(ship)
— f — — m entre sí - [admin] work interrela-
tion(ship)
— f — velocidad(es) f - [mec] speed(s) ratio
— f diámetro-espesor m - [tub] diameter-thickness ratio
— f directa - direct relation(ship)
— f elevada - [matem] high ratio
— f en eje m delantero - [mec] front axle ratio
— f — — m trasero - [mec] rear axle ratio
— f — porcentaje m - percentage ratio
— f — tanto por ciento m - percentage ratio
— f — altura f de pared(es) f lateral(es) y
ancho(r) m de banda f para rodamiento m -
[autom-neumát] aspect ratio
— f — — f y ancho(r) m - [autom-neumát] aspect ratio
— f — — f y anchura f - height-to-width ratio
• [autom-neumát] aspect ratio
— f — carga f sólida y líquida - [metal-prod]
solid/liquid charge ratio
— f — — f e inflación f - [autom-neumát]
load/inflation, ratio, o relation(ship)
— f — caucho m y oquedad(es) f—[autom-neumát]
rubber-to-void ratio
— f — — m — vacío m - [autom-neumát] rubber-
-to-void ratio
— f — corrosión f y resistencia f eléctrica -
[quím] corrosion-resistivity, relation, o
ratio
— f — diámetro m interior y (el) exterior m -
[tub] inner (diameter) outer diameter ratio
— f — engranaje(s) m - [mec] gear ratio
— f — esfuerzo m y tensión f - [suelos] stress
-strain, relationship, o ratio
— f — — m — de suelo m - [suelos] soil
stress-strain relationship
— f — gas m y aire m - [combust] gas/air ratio
— f — — m natural y aire m - [combust] natural gas/air ratio
— f — mineral m y coque m - [metal-prod] ore/
coke ratio
— f — resistencia f y peso m - [mec] strength-
-to-weight ratio
— f — sí - interrelation(ship)

relación f entre resistencia f y peso m
[metal] strength/weight ratio
— f —— f —— m muy alta - [metal] (very)
high strength/weight ratio
— f —— m —— m baja - [metal] (very)
low strength/weight ratio
— f — tiempo m en fosa f y en vía f - [metal-
[prod] pit time/track time ratio
— f especial - special ratio
— f estándar - standard ratio
— f estrecha - close relation(ship) • close
liaison
— f evidente - evident relation(ship)
— f favorable - favorable relation(ship)
— f futura f - future relation(ship)
— f general - general relation(ship)
— f interna - internal relation; interrelation
— f reducida - [matem] low ratio
— f según norma f - standard ratio
— f sólida - solid, o sound, relation(ship)
— v tiempo m en vía/tiempo m en fosa - [metal-
-prod] track time/pit time ratio
relaciones f exteriores - [pol] foreign affairs
— f humanas f - [labor] human relations
— f industriales - [ind] industrial relations
— f indepartamentales f - [admin] interdepart-
ment(al) relations
— f laborales - [labor] labor relations
— f —— de empresa f - [labor] corporate labor
relations
— f —— de sociedad f - [labor] corporate labor
relations
— f —— malas - [labor] poor labor relations
— f —— pobres - [labor] poor labor relations
— f públicas - [labor] public relations
— f —— empresa f - [labor] corporate public
relations
— f — favorables - favorable public relations
— f malas - [labor] poor labor relations
— f satisfactorias - satisfactory relations
— f sindicales - [labor] labor, o union, rela-
tionship(s)
— f — administrativas - [labor] union-manage-
ment relations
relacionado,da a - . . . .; associated • [com] associated • affiliated • listed
— a con - related, o associated, with
— a — trabajo m - job-, o work-, related
— a directamente - related directly
— a entre sí - interrelated
relacionar v - . . . .; to associate • to list
— v directamente - to relate directly
— v entre sí - to (inter)relate
— v trabajo m - [admin] to relate @ work
— v trabajos m entre sí - [admin] to interre-
late a, work, o jobs
relay* m - [electr-equip] véase relé m
relaminación f - [metal-lam] rerolling
— f de cordón m - [metal-lam] seam rerolling
— f —— costura f - [metal-lam] seam rerolling
— f de soldadura f - [sold] weld rerolling
— f en caliente - [metal-lam] hot rerolling
— f —— frío m - [metal-lam] cold rerolling
— f exterior - [metal-lam] outside rerolling
— f interior - [metal-lam] inside rerolling
relaminado,da a - [metal-lam] rerolled
— a en caliente - [metal-lam] hot rerolled
— a —— frío - [metal-lam] cold rerolled
— a exteriormente—[metal-lam] outside rerolled
— a interiormente - [metal-lam] inside rerolled
relaminar v - [metal-lam] to reroll
— v cordón m - [metal-lam] to reroll @ seam
— v costura f - [metal-lam] to reroll @ seam
— v en caliente - [metal-lam] to hot reroll
— v — frío - [metal-lam] to hot reroll
— v exteriormente - [metal-lam] to outside reroll; to reroll outside
— v interiormente - [metal-lam] to inside reroll; to reroll inside
— v soldadura f - [wold] to reroll @ weld
relatado,da a - related • narrated; told
relatar v - to relate; to narrate; to tell
relativamente adv - . . . .; comparatively •
rat her; fairly

relativamente adv alto,ta - relatively high •
upper stage
— adv bajo,ja - relatively low • lower stage
— adv breve - relatively, brief, o short
— adv brillante - relatively, bright, o shining
— adv cedente - relatively yielding
— adv constante - relatively, o fairly, constant
— adv corto,ta - relatively, short, o brief
— adv costoso,sa - relatively, costly, o expensive
— adv débil - relatively weak
— adv delgado - relatively thin
— adv fluido,da - [sold] relatively fluid
— adv grande - relatively, o fairly, large
— adv húmedo,da - relatively, o fairly, moist
— adv inclinado,da - relatively slanted; comparatively steep
— adv largo,ga - relatively, short, o brief
— adv lento,ta - relatively, o rather, slow
— adv liviano,na - relatively, o comparatively,
light(weight)
— adv liso,sa - relatively, smooth, o even
— adv nuevo,va - relatively new
— adv pequeño,ña - relatively, o comparatively,
o fairly, small
— adv pesado,da - relatively heavy
— adv rápido,da - relatively, o rather, fast
— adv reducido,da - relatively small • comparatively shallow
— adv regular - fairly regular
— adv suelto,ta - relatively loose
— adv uniforme - relatively, o fairly, uniform,
o regular
relativo,va adv - . . . . • related; pertinent •
proportional • fair(ly)
— adv a - related, o corresponding, to
— a a productividad f - [ind] productivity related
relave m - [miner] . . .; tailing(s)
— m con contenido m alto de sal(es) f - [miner]
high content tailing
— m —— m bajo de sal(es) f - [miner] low
content tailing
— m —— porcentaje m alto de sal(es) f - [miner]
high (salts) content tailing
— m —— m bajo de sal(es) f - [miner] low
(salts) content tailing
— f de flotación f - [miner] flotation tailing
relé m - [electr-equip] . . .; actuator
— m ajustable - [electr-equip] adjustable relay
— m apropiado - [electr-instal] proper relay
— m auxiliar - [electr-equip] auxiliary relay
— m — con reposición f eléctrica - [electr-
-equip] electric reset auxiliary relay
— m C - [electrón] C relay
— m con pestilo m - [comput] latch(ing) relay
— m —— m magnético - [comput] magnetic
latch(ing) relay
— m — velocidad f alta - [electr] high speed
relay
— m —— f alta para protección f diferencial
de corriente f - [electr] high speed differential current relay
— m contrarrestador - [electr-equip] overriding
relay
— m de tiempo m - [electr-equip] time relay
— m — tipo magnético - [electr-equip] magnetic
type relay
— m —— m — para sobrecarga f - [electr-
-equip] magnetic type overload relay
— m —— m reversible - [electr-equip] reversible,sing type relay
— m —— m térmico - [electr-equip] thermal
type relay
— m desconectado - [electr-oper] tripped relay
— m detector - [electr-equip] detecting,tor
relay
— m — para temperatura(s) f - [instrum] temperature (detecting,tor) relay
— m —— — f para cojinete(s) m - [electr-
-equip] bearing temperature (detecting,tor)
relay
— m diferencial - [electr] differential relay

relé eléctrico -- [electr-equip] electrical relay
— m, equivocado, o errado - [electr-equip] wrong relay
— m extrarrápido - [electr-equip] very quick, o instantaneous, relay
— m inapropiado - [electr-instal] wrong relay
— m instantáneo m - [electr-equip] instantaneous relay
— m — para falta f a tierra f - [electr-equip] instananeous ground fault relay
— m — — f para sobrecorriente(s) f - [electr-equiip] instantaneous overcurrent ground fault relay
— m — — pérdida f a tierra f.- [electr-equip] instantaneous ground fault relay - [electr-equip] instantaneous ground fault relay
— m — — — f — — f para sobrecorriente(s) f - [electr-equip] instantaneous overcurrent ground fault relay
— m — — sobrecarga f - [electr-equip] instantaneous (overcharge) relay
— m intermedio - [electr-equip] intermediate, o transfer, relay
— m interruptor - [electr-equip] trip(ping) relay
— m — para falla(s) f - [electr-equip] fault trip(ping) relay
— m inversor - [electr-equip] reversing relay
— m — para campo m - [electr-equip] field reversing relay
— m limitador - [electr-equip] limit(ing) relay
— m — con transistor(es) m - [electr-equip] transistorized limit(ing) relay
— m — — — m para corriente f—[electr-equip] transistorized current limit(ing) relay
— m — para corriente f - [electr-equip] current limit(ing) relay
— m maestro - [electr-equip] master relay
— m magnético - [electr-equip] magnetic relay
— m — para retraso m - [electr-equip] time delay magnetic relay
— m — — — m con cubeta f para aceite m - [electr-equip] oil cup time delay magnetic relay
— m — — sobrecarga f - [electr-equip] magnetic overload, o overload magnetic, relay
— m — — — f con recargo m - [electr-equip] time delay magnetic overload relay
— m — — — f para retardo m con cubeta f para aceite m - [electr-equip] oil cup time delay magnetic overload relay
— m motriz - [electr-equip] power relay
— m operado con onda f portadora - [electrón] carrier (wave) operated relay
— m optativo - [electr-equip] optional relay
— m — Linc-Fill—[sold] Linc-Fill relay option
— m — —·— para encendido m [con electrodo m muy soresaliente} - [sold] Linc-Fill starting relay option
— m para activación f - [comput] keying relay
— m — amperaje m - [electr-equip] amp(eres) relay
— m — arranque m - [comb.int] start(ing) relay
— m — bloqueo m - [electr-equip] (b)locking relay
— m — cambiador m - [electr-equipo (tap) changer,ging relay
— m — cambio m de fase f - [electr-equip] phase sequence relay
— m — campo m - [sold] field relay
— m — — m de tipo m reversible - [electr-equip] reversing type field relay
— m — — m reversible - [electr-equip] reversing field relay
— m — clave f - [electrón] keying relay
— m — cojinete m - [electr-equip] bearing relay • [instrum] bearing thermocouple relay
— m — contacto m—[electr-equip] contact relay
— m — control m - [comput] control relay
— m — — m para función f—[comput] function control relay
— f — — m — — f retenida magnéticamente - [comput] (magnetically) latched function con-

trol relay
relé m para control m para función f sujetado magnéticamente - [comput] latched function control relay
— m — corriente f - [electr-equip] current relay
— m — — f con transistores m - [electrón] transistorized current relay
— m — detención f - [electr-equip] stop(ping) relay
— m — — f automática - [sold] automatic stop(ping) relay
— m — encendido m - [sold] starting relay
— m — — con electrodo m muy sobresaliente - [sold] Linc-Fill starting relay
— m — enclavamiento m - [electr-equip] interlocking, o lock-out, relay
— m — excitación f total - [electr-equip] full field relay
— m — falla f a tierra f - [electr-equip] ground fault relay
— m — — f de sobrecorriente f a tierra f - [electr-equip] overcurrent ground fault relay
— m — falta f de voltaje m - [electr-equip] no voltage relay
— m — fase f - [electr-equip] phase relay
— f — — f para sobrecorriente f - [electr-equipo] phase overcurrent relay
— m — — f — — f para inducción f - [electr-equip] overcurrent relay induction
— f — — — f — — f monofásica - [electr-equyip] single phase overcurrent relay induction
— m — frequencia f - [electr-equip] frequency relay
— m — frenado m - [electr-equip] brake relay
— m — gatillo m - [electr-equip] trip relay
— m — línea f - [electr-instal] line relay
— m — llama f - [combust] flame relay
— m — pérdida - [electr-equip] fault relay
— m — — f a tierra f - [electr-equip] ground fault relay
— m — — f para campo m - [electr-equip] field loss relay
— m — — f sobrecorriente f - [electr-equip] overcurrent fault relay
— m — — f — — f a tierra f - [electr-equip] overcurrent ground fault relay
— m — presión f - [instrum] pressure relay
— f — — f para falla(s) f - [electr-equip] fault pressure relay
— m — — f gas m - [electr-equip] gas pressure relay
— m — — f súbita - [electr-equip] sudden pressure relay
— m — protección f - [electr-equip] protective relay
— m — — f para campo m - [electr-equip] field protective relay
— m — — f diferencial - [electr-equip] differential (protective) relay
— m — — f para corriente f - [electr-equip] differential current relay
— f — regulación f - [electr-equip] control-(ling) relay
— m — retardación,do - [sold] time delay relay
— m — — m para regulación f de tiempo m - [electr-equip] control(ling) delay relay
— m — — — m — — f de tiempo m - [electr-equip] time control delay relay
— m — — — m — — f — — m para sobreelevación f de corriente f - [electr-oper] surge time control delay relay
— m — — — m con cubeta f de aceite m - [electr-equip] oil cup time delay relay
— m — retención f - [comput] latching relay
— m — . . . segundo(s) m - [comput] . . . second(s) time relay
— m — . . . — m para entrada f - [comput] . . . second(s) input time relay
— m — sobrecarga f - [electr-equip] overload relay
— m — — f de retardo m - [electr-equip] time delay overload relay

delay overload relay
relé m **para sobrecarga** f **de retardo** m **con cubeta**
  f **de aceite** m - [electr-equip] oil cup time
  delay overload relay
— m —— f — **tiempo** m - [electr] inverse time
  overload relay
— m —— f— **tipo** m **térmico** - [electr-equip]
  thermal type overload relay
— m —— f **instantáneo** m - [electr] instantaneous overload relay
— m —— f **magnética** - [electr] magnetic overload relay
— m —— f **de tipo** m **magnético**—[electr-equip]
  magnetic type overload relay
— m —— **sobrecorriente** f - [electr-equip] overcurrent relay
— m —— f **a tierra** f - [electr-equip] ground
  overcurrent relay
— m —— f —— f **monofásica** - [electr-equip]
  single phase ground overcurrent relay
— m —— m —— f — **conectado directamente** -
  [electr-equip] single phase ground overcurrent residually coonected relay
— m —— f **monofásica** - [electr-equip] single
  phase overcurrent relay
— m —— **soldadora** f - [sold] welder relay
— m —— **termocupla** f - [electr-equip] thermocouple relay
— f —— f **para cojinete** m - [instrum] bearing
  (thermocouple) relay
— m —— **tiempo** m - [electr-equip] time relay
— m —— m **de** . . . **segundo(s)** m - [electrón]
  . . . second(s) time relay
— m —— m — . . . m **para entrada** f -
  [electrón] . . . second(s) input time relay
— m —— **velocidad(es)** f **alta(s)** - [electr-equip]
  high speed(s) relay
— m —— f **alta(s) y baja(s)** - [electr-equip]
  high-low speed(s) relay
— m —— f **baja(s)** - [electr-equip] low speeds
  relay
— m —— **voltaje** m - [electr-equip] voltage relay
— m —— m **alto(s)** - [electr-equip] overvolt-
  age(s) relay
— m —— m **bajo(s)** - [electr-equip] undervolt-
  age(s) relay
— m —— **voltaje** m **mínimo** - [electr-equip] under-
  voltage relay
— m —— f — **con** . . . **fase(s)** f - [electr-
  -equip] . . . phase(s) undervoltage relay
— m **piloto** m—[electr-equip] (wire) pilot relay
— m — **para conductor** m - [electr-equip] wire,
  o lead, pilot relay
— m —— — m **nuevo** m - new cable voltage relay
— m —— — m **para protección** f - [electr-
  -equip] protection wire pilot relay
— m —— m —— f **para cable** m - [electr-
  -equip] cable protection wire pilot relay
— m — — **soldadora** f - [sold] welder pilot
  relay
— m **principal** - [electr-equip] main, o master,
  relay • supervisory relay
— m **protector** - [electr-equip] protector, o
  Protecto, relay
— m **quemado** - [electr-equip] burned (out) relay
— m **reemplazado** - [electr-equip] replaced relay
— m **regulador** - [electr-equip] control relay
— m — **principal** - [electr-equip] main control
  relay
— m **retardador**—[electr-equip] delay(ing) relay
— m **retenido** - [electrón] latched relay
— m **reversible** - [electr-equip] reversible,sing
  relay
— m **sujetado** - [electrón] latched relay
— m — **magnéticamente** - [electrón] magnetically
  latched relay
— m **supervisor** - [electr-equip] supervisory
  relay
— m **térmico** - [electr-equip] thermal relay
— m — **ajustable** - [electr-equip] adjustable
  thermal relay|
— m — **para sobrecarga** f - [electr-equip] thermal overload relay

relé m **termosensible** - [electr-equip] temperature responsive relay
— m **vinculado** - [electr-equip] related, o associated, relay
relegación f - . . .; relegating • holding back
relegado,da a - relegated; held back • banished
relegar v - . . .; to banish • to hold back
relevación f - . . . • release,sing
— f **de esfuerzo(s)** m - [mec] véase **alivio** ₥ de
  tensión(es) f
— f **total** - [legal] full release,sing
relevado,da a - released • relieved
— a **totalmente** - [legal] fully released
relevamiento m - release,sing • survey,ing
— m **aéreo** - [topogr] aer(ial) survey
— m **completo** m - complete survey(ing)
— m — **de suelo** m - [suelos] complete soil sur-
  vey(ing)
— m — **subsuelo** m - [topogr] subsoil, o underground, survey(ing)
— m — **suelo** m - [suelos] soil survey(ing)
— m — **zona** f - area survey(ing)
— m **fotogramétrico** - [topogr] aerial survey
— m **geológico** - [geol] geological survay
— m — **Estadounidense** - [geol] United States
  Geological Survey
— m **topográfico** - [topogr] (topographic) sur-
  vey(ing)
relevancia* f - véase **importancia** f
relevante a - . . .; outstanding; significant;
  noticeable; important • pertinent
relevar v - . . . • to not, injure, o hurt •
  [electr-oper] to relay
— v **totalmente** - [legal] to release fully
— v **zona** f - [topogr] to survey @ area
relevo m - . . . • [labor] new, o change of,
  shift • [electr-equip] véase **relé** m • [legal]
  véase también **dar realce** m ⎸ a - [labor]
  spell; relief; shift • step-up
— m **para tensión** f - tension release
relieve m - . . . ⎸ a - (en) relief; embossed
— m **bajo** - bas-relief; véase **bajo relieve** m
relimpiador m - recleaner
reliquia f - . . .; memento
reloj m **calibrador** - [instrum] gage,ging clock
— m **con cuerda** f - [mec] spring clock
— m **conectado** - [electr-instal] wired clock
— m **de segundo(s)** m **muerto(s)** - [instrum]
  stop watch
— m **despertador** m - [instrum] alarm clock
— m **eléctrico** - [instrum] electric clock
— m **indicador** m - [instrum] gage,ging clock
— m **integral** - [instrum] integral clock
— m **inteligente** - [instrum] smart clock
— m — **integral** - [instrum] integral smart
  clock
— m **maestro** m - [instrum] master clock
— m **para bolsillo** m - [instrum] (pocket) watch
— m — . . . **día(s)** m - [instrum] . . . day
  clock
— m — **división** f **por cincuenta** - [electrón]
  divide-by-fifty clock
— m — **registro** m - [electrón] register, o recording, clock
— m **principal** - [instrum] master, o main, clock
— m **pulsera** - [instrum] (wrist) watch
— m **registrador** - [instrum] recorder clock
— m — **con cuerda** f - [mec] wound, o spring,
  recorder clock
— m — — **gráfico** m - [instrum] chart recorder
  clock
— m — — m **con cuerda** f - [instrum] wound,
  o spring, chart recorder clock
— m **sincronizador** - [instrum] time,ming clock
— m **suizo** - [instrum] Swiss, watch, o clock
reluctancia f - [electr] reluctance
rellenable a - refillable
rellenado m - [domést] filling; stuffing;
  véase **relleno** m
rellenado,da a - refilled; filled in • [domést]
  overstuffed
— a **alrededor de tubería** f - [constr] backfilled around @ pipe

rellenado,da a **con aluminio** m - [autom-carroc]
aluminum filled
— a **con diorita** f - [miner] diorite, stuffed, o
filled
— a —— **espuma** f - foam filled
— a —— f **de poliestireno** m - [quím] poly-
styrene foam filled
— a — **fibra** f **de vidrio** m - [cerám] fiberglass
filled
— a — **gelatina** f - gelatin filled
— a — **grasa** f - [lubric] grease, filled, o
packed
— a — **hormigón** m - [constr] concrete filled
— a —— m **con peso** m **reducido** - [constr]
lightweight concrete filled
— a — **material** m **granular** - [constr] granular
material filled
— a — **suelo** m - [constr] soil (back)filled
— a **cuidadosamente** - [constr] backfilled carefully
— a **prontamente** - [constr] backfilled promptly
— a **rápidamente** - [constr] backfilled quickly
— a **temporariamente** - packed temporarily
rellenador* m - [autom-carroc] véase **relleno** m
rellenamiento* n - véase **relleno** m
rellenar v - . . .; to backfill; to refill; to
fill in • [metal-prod] to pack • [constr] to
grout • [sold] to flush up • [constr] to
slush; to chink
— v **alrededor de** [constr] to (back)fill around
— v — **de costado(s)** m - [constr] to (back)fill
around @, side(s), o edge(s)
— v ——— f **de estructura** f - [constr] to
(back)fill around @ structure side(s)
— v —— **tubería** f - [constr] to (back)fill
around @ pipe
— v **bóveda** f - [constr] to (back)fill @ arch
— v **caja** f **para refrigeración** f - [metal-prod]
to grout @ cooling plate
— v **cajón** m - [constr] to backfill @, box, o
bin
— v **colector** m - [mec] to refill @ sump
— v **con aluminio** m - [autom-carroc] to fill
with aluminum
— v — **fibra** f **de vidrio** m - [cerám] to fill
with fiberglass
— v — **gelatina** f - to fill with gelatin
— v — **grasa** f - [lubric] to, fill, o pack,
with grease
— v **con hierro** m - [metal-prod] to, fill, o
plug, with iron
— v — **material** m **granular** - [constr] to
(re)fill with granular material
— v — **mortero** m **blando** - [constr] to slush
with mortar
— v — **pasta** f - [metal-prod] to pack with @
paste • to grout (@ cooling plate)
— v — **suelo** m - [constr] to backfill with soil
— v **cordón** m - [sold] to flush up @ weld
— v **cráter** m - [sold] to fill @ crater
— v **depósito** m - [comb.int] to refill @ tank
— v — m **para combustible** m - [comb.int] to refill @ fuel tank
— v **depresión** f - [sold] to fill @ depression
— v — f **en superficie** f - [sold] to fill @
surface depression
— v **estructura** f - [constr] to backfill @
structure
— v **grieta** f - [sold] to fill @ crack
— v **hueco(s)** m - [sold] to bridge @ gap
— v **parte** f **cóncava** - [sold] to flush @ concave
weld portion
— v — f — **de soldadura** f - [sold] to flush up
@ weld concave portion
— f **petaca** f - [metal-prod] to, pack, o plug, @
cooling plate
— f **pieza** f - [sold] to rebuild @ part
— v — f **gastada** [sold] to rebuild @ worn part
— v **separación** f - [sold] to, bridge, o fill, @
gap
— v **prontamente** - [constr] to (back)fill
promptly
— v **rápidamente** [constr] to (back)fill quickly

rellenar v **soldadura** f - [sold] to flush up @
weld
— v **temporariamente** - to, fill, o pack, temporarily
— f **tubería** f - [constr] to backfill @ pipe
— v — f **abovedada** - [constr] to backfill @
pipe-arch
relleno m - fill(ing); padding; filler; cover •
[sold] build-up; rebuilding • [constr] filler
• backfill(ing) fill work; backfiller; cover
• filling (in) • [mec] pad; véase también
empaquetadura f
— m a **costado** m - [constr] side fill(ing)
— m **adecuado** - [constr] adequate, o proper, o
good, backfill
— m **aislante** - [constr] insulating (back)fill
— m — **para cable(s)** m - [electr-cond] insulating cable filling
— m **alrededor de conducto** m - [constr] backfill around @ conduit
— m —— **tubería** f - [constr] backfill around
@, conduit, o pipe
— m **amarillo** - [autom-carroc] yellow filler
— m — **para carrocería** f **(para automóvil)** m -
[autom-carroc] yellow autobody filler
— m —— **chapistería** f - [autom-carroc] yellow
autobody filler
— m **apisonado** - [constr] tamped (back)fill
— m — **cabalmente** - [constr] thoroughly tamped
backfill
— m — **cuidadosamente** - [constr] carefully
tamped backfill
— m — **levemente** - [constr] lightly tamped
(back)fill
— m **apropiado** - [constr] appropriate, o proper, o suitable, (back)fill(ing)
— m **asentado** - [constr] settled (back)fill
— m **atascado** - [hidr] clogged backfill
— m **bien compactado** - [constr] well compacted
backfill
— m **drenado** - [constr] well drained backfill
— m **blanco** - [autom-carroc] white filler
— m — **para carrocería(s)** f **(para automóvil** m**)**
[autom-carroc] white autobody filler
— m —— **chapistería** f - [autom-carroc] white
autobody filler
— m **blando** - [constr] soft (back)fill
— m **circundante** - [constr] surrounding (back)-
fill
— m — **de tierra** f - [constr] surrounding,
soil, o earth, (back)fill
— m **clasificado** - [constr] graded backfill
— m — **cuidadosamente** - [constr] finely, o
carefully, graded backfill
— m **colocado** - [constr] placed backfill
— m **compactado** - [constr] compacted (back)fill
— m — **acabadamente** - [constr] thoroughly compacted (back)fill
— m — **cabalmente** - [constr½] thoroughly tamped
backfill
— m — **cuidadosamente** - [constr] carefully compacted backfill
— m — **levemente** - [constr] lightly compacted
(back)fill
— m — **moderadamente** - [constr] moderately compacted backfill
— m **completado** - [constr] completed (back)fill
— m **completo** - [constr] entire fill | adv -
(de) [sold] full size
— m — **de cráter** m - [sold] crater complete
filling
— m **comprimido** - [constr] compressed backfill
— m **con altura** f **escasa** - [constr] shallow fill
— m —— f **reducida** - [constr] shallow fill
— m — **aluminio** m - [autom-carroc] aluminum
filler,ling
— m — **calidad** f - [autom-carroc] premium filler,ling
— m ——— **alta** - [constr] high grade backfill
— m ——— f **buena** - [constr] high grade back-
fill(ing)
— m ——— f **máxima** - [autom-carroc] premium
quality autobody filler

relleno m̲ con calidad f̲ máxima para chapistería
(para automóvil(es) m̲ - [autom-carroc] premium quality autobody filler
— m̲ — f̲ mejor - [autom-carroc] premium filler
— m̲ — f̲ pobre - [constr] low grade backfill
— m̲ — f̲ primer(ísim)a - [constr] premium
(quality) filler
— m̲ — f̲ — para chapistería f̲ para automó-
vil(es) m̲ - [autom-carroc] premium quality
autobody filler
— m̲ — — f̲ superior - [constr] superior quality backfill
— m̲ — fibra f̲ de vidrio m̲ - [cerám] fiberglass
filling
— m̲ — gelatina f̲ - gelatin filling
— m̲ — grasa f̲ - [lubric] grease pack(ing)
— m̲ — nitrógeno m̲ - [electr] nitrogen filling
— m̲ — pasta f̲ - [metal-prod] (paste) grouting
— m̲ — peso m̲ reducido - light-weight filler
— m̲ — — para carrocería f̲ - [autom-
-carroc] lightweight autobody filler
— m̲ — — m̲ — — f̲ para automóvil m̲ -
[autom-carroc] lightweight automobile filler
— m̲ — — chapistería f̲ - [autom-
-carroc] lightweight autobody filler
— m̲ — poca altura f̲ -[constr] shallow fill |
adv̲ - [con] shallow buried
— m̲ — soldadura f̲ vertical en ángulo m̲ interior - [sold] vertical fillet fill
— m̲ — suelo m̲ - [constr] soil backfill(ing)
— m̲ — superficie f̲ horizontal • [hidr] horizontal surface fill
— m̲ conformado - [constr] shaped fill(ing)
— m̲ consolidado - [constr] consolidated fill
— m̲ contenido - [constr] contained, o̲ held
back, fill
— m̲ contra cabezal m̲ - [constr] spandrel fill
— m̲ convencional - [autom-carroc] conventional
fill(er)
— m̲ — con peso m̲ reducido - [autom-carroc]
conventional lightweight filler
— m̲ corriente m̲ - [autom-carroc] conventional,
o̲ ordinary, fill(er)
— m̲ dañado - [constr] damaged fill
— m̲ de aluminio m̲ - [autom-carroc] aluminum
filler
— m̲ — balasto m̲ - [f.c.] ballast fill
— m̲ — cajón m̲ - [constr] bin (back)filling
— m̲ — canto(s) m̲ rodado(s) - [constr] cobbles
backfill(ing)
— m̲ — colector m̲ - [hidr] sump refilling
— m̲ — — m̲ para aceite m̲ - [mec] oil sump refilling
— m̲ — cráter - [sold] crater filling
— m̲ — depósito m̲ - [comb.int] tank refilling
— m̲ — — m̲ para combustible m̲ - [comb.int]
fuel tank (re)filling
— m̲ — estructura f̲ - [constr] structure back-
fill(ing)
— m̲ — estufa f̲ - [combust] stove filling
— m̲ — fibra f̲ de vidrio m̲ - [cerám] fiberglass
filler
— m̲ — — f̲ de vidrio m̲ con hebra(s) f̲ larga(s)
- [cerám] long strand fiberglass filler
— m̲ — fragmento(s) m̲ de roca f̲ - [constr] rock
fragment(s) fill(er)
— m̲ — hormigón m̲ - [hormig] concrete filling
— m̲ — — con peso m̲ reducido - [constr]
lightweight concrete fill(ing)
— m̲ — hueco(s) m̲ - void(s) filling
— m̲ — piedra f̲ - [constr] stone fill(ing)
— m̲ — material m̲ granular - [constr] granular
material backfill(ing)
— m̲ — roca f̲ - [constr] rock, o̲ stone, fill
— m̲ — separación f̲ - [sold] gap, bridge,ging,
o̲ fill(ing)
— m̲ — suelo m̲ - [constr] soil backfill(ing0
— m̲ — — m̲ inestable - [constr] unstable soil
(back)fill(ing)
— m̲ — tierra f̲ - [constr] earth, o̲ soil,
back(fill)
— m̲ — — f̲ con calidad f̲ pobre - [constr] low

grade soil backfill(ing)
relleno m de tierra f proyectado - [constr] designed earth backfill
— m̲ — yute m̲ - [electr-cond] jute filler
— m̲ — zanja f̲ - [constr] trench backfill(ing)
— m̲ desaguado - [constr] drained fill • [hidr]
dewatered fill
— m detrás de cabezal m̲ - [constr] spandrel
fill
— m̲ distribuido - [constr] distributed, o̲
spread, fill
— m̲ drenado - [constr] drained (back)fill
— m̲ eficiente - [constr] efficient backfill
— m̲ elástico - [constr] elastic filler
— m̲ especial - [constr] special backfill •
[autom-carroc] specialty filler
— m̲ especificado - [constr] specified backfill
— m̲ estándar - [constr] standard backfill(ing)
— m̲ evaluado - [constr] evaluated backfill
— m̲ excesivo - [constr] excess(ive) fill(er)
— m̲ exterior - [constr] exterior fill
— m̲ fino - [constr] fine fill(er)
— m̲ granular - [constr] granular (back)fill
— m̲ — bien compactado - [constr] well tamped
granular, (back)fill, o̲ material
— m̲ — fino - [constr] fine granular fill
— m̲ — seleccionado - [constr] selected, o̲
chosen, granular backfill
— m̲ heterogéneo - [geol] heterogenous, o̲ uneven, fill
— m̲ hidráulico - [constr] hydraulic fill
— m̲ impermeable - [autom-carroc] waterproof
filler • [constr] impervious backfill
— m̲ — para carrocería f̲ - [autom-carroc] waterproof body filler
— m̲ inapropiado - [constr] inapropiate, o̲ improper, backfill
— m̲ inicial - [suelos] initial fill
— m̲ inoxidable - [autom-carroc] rustproof
filler
— m̲ labrable - [autom-carroc] workable filler
— m̲ lateral - [constr] side fill(ing)
— m̲ liso - [autom-carroc] smooth filler
— m̲ lubricado - [constr] lubricated (back)fill
— m̲ — por lluvia f̲ - [constr] rain-lubricated
(back)fill
— m̲ mantenido - [constr] maintained, o̲ kept,
fill
— m̲ más alto - [constr] higher fill
— m̲ mecanizable - [autom-carroc] workable filler
— m mejor para carrocería(s) f̲ - [autom-
-carroc] premium body filler
— m̲ — para cráter m̲ - [sold] best crater filling
— m̲ metálico - [autom-carroc] metal(lic) filler
— m̲ no metálico - [autom-carroc] nonmetal(lic)
filler
— m̲ normal - [constr] normal (back)fill(ing)
— m̲ óptimo - [autom-carroc] premium filler
— m̲ — para carrocería(s) f̲ - [autom-carroc]
premium body filler
— m para aislación f̲ - [constr] isolation, o̲
insulation, filler
— m̲ — bóveda f̲ - [constr] arch backfill(ing)
— m̲ — cable(s) m̲ - [electr-cond] cable, filler, o̲ compound
— m̲ — carrocería f̲—[autom-carroc] body filler
— m̲ — — f̲ para automóvil(es) m̲ - [autom-
-carroc] autobody filler
— m̲ — costado(s) m̲ - [constr] side backfill
— m̲ — chapistería f̲ - [autom-carroc] (auto)-
-body filler
— m̲ — — f̲ para automóvil(es) m̲ - [autom-
-carroc] autobody filler
— m̲ — esquina(s) f̲ - [constr] corner backfill
— m̲ — estructura f̲ - [constr] structural backfill • structure backfill
— m̲ — junta f̲ - [constr] joint filler
— m̲ — — f̲ para aislación f̲ - [constr] insulating,tion, o̲ isolating,tion, joint filler
— m̲ — — f̲ para dilatación f̲ - [constr] expansion joint filler

relleno m para junta f para regulación f - [mec]
    control joint filler
— m — muro m - [constr] wall backfill
— m — m para retención f - [constr] retain-
    ing wall backfill
— m — pestaña f - [autom-neumát] bead filler
— m — plancha f - [constr] plate backfill
— m — f esquinera - [constr] corner plate
    backfill
— n — retención f—[constr] retaining backfill
— m — rosca(s) f - [mec] thread filler
— m — sandwich(es) m - [culin] sandwich, fill-
    ing, o cream
— m — subdrenaje m—[constr] subdrain backfill
— m — terraplén m - [constr] embankment back-
    fill
— m — tubería f - [constr] pipe backfill(ing)
— m — — f abovedada - [constr] pipe arch
    backfill(ing)
— m — zanja f - [constr] trench backfill(ing)
— m permeable - [constr] pervious, o permeable,
    backfill
— m plástico - [autom-carroc] plastic filler
— m poroso - [constr] porous backfill
— m preformado - [constr] preformed filler
— m — para junta f para dilatación f—[constr]
    preformed expansion joint filler
— m — — f para construcción f es-
    tructural - [constr] preformed expansion
    joint filler for structural construction
— m — — f — pavimentación f con
    hormigón m - [constr] preformed expansion
    joint filler for concrete paving,vement
— m protector - [constr] protective fill
— m próximo - [constr] adjacent fill
— m — a estructura f - [constr] structure ad-
    jacent fill
— m proyectado - [constr] designed, o engi-
    neered, backfill
— m — debidamente - [constr] well engineered
    backfill
— m rápido - [constr] fast, o prompt, backfil-
    ling • [sold] fast fill(ing) | adv - [sold]
    fast fill
— m reforzado - [cerám] reinforced filler
— m — de fibra f de vidrio m - [cerám] fiber-
    glass reinforced, o reinforced fiberglass,
    filler
— m regulado m - [sold] controlled fill(ing)
— m — de cráter - [sold] controlled crater
    fill(ing)
— m restante - [constr] remaining backfill
— m rodeante - [constr] surrounding (back)fill
— m satisfactorio - [constr] satisfactory back-
    fill(ing)
— m saturado - [constr] saturated backfill
— m seguro - [constr] safe backfill(ing)
— m seleccionado - [constr] selected backfill
— m sin desaguar v - [constr] undrained fill
— m — drenar v - [constr] undrained fill
— m singular - [autom-carroc] unique filler
— m — para chapistería f - [autom-carroc]
    unique autobody filler
— m sobre estructura f - [constr] over struc-
    ture fill
— m — soporte m - [constr] backfill above @
    cradle
— m — tubería f - [constr] pipe, backfill, o
    cover
— m suelto - [constr] loose (back)fill
— m temporario - [constr] temporary packing
— m total—[constr] total, o entire, (back)fill
— m uniforme - [autom-carroc] smooth filler
— m vertical - [sold] vertical fill
— m y avance m - [sold] fill-follow
— m — — m rápido(s) m - [sold] (fast) fill-
    -follow(ing)
— m y solidificación f - [sold] fill-freeze
relleno,na a - véase rellenado,da
remachado m - [mec] riveting • gripping
— m bueno - [mec] good, o close, riveting
— m con punto(s) m - [mec] stitching • spot
    welding

remachado m de costura f - [mec] seam riveting
— m — — f longitudinal - [mec] longitudinal
    seam riveting
— m — solapo m - [mec] lap riveting
— m — traslapo m - [mec] lap riveting
— m — unión(es) f - [constr] joint riveting
— m en caliente - [mec] hot riveting
— m — fábrica f - [mec] shop riveting
— m — frío m - [mec] cold riveting
— m — obra f - [mec] field riveting
— m — planta f - [mec] shop riveting
— m — taller m - [mec] shop riveting
— m exigido - [mec] required riveting
— m por punto(s) m - [mec] spot riveting
remachado,da a - riveted • clinched
— a bien - [mec] close, o well, riveted
— a en caliente - [mec] hot riveted
— a — fábrica f - [mec] shop riveted
— a — frío - [mec] cold riveted
— a — obra f - [mec] field riveted
— a — planta f - [mec] shop riveted
— a — taller m - [mec] shop riveted
remachador m - [mec] . . . . • dolly
— m por punto(s) m - [mec] stitcher
remachar v bien - [mec] to close rivet • to
    rivet well
— v con punto(s) m - [mec] to (rivet) stitch
— v costura f - [mec] to rivet @ seam
— v — f longitudinal - [mec] to rivet @ longi-
    tudinal seam
— v en caliente - [mec] to hot rivet
— v — fábrica f - [mec] to shop rivet
— v — frío - [mec] to cold rivet
— v — obra f - [mec] to field rivet
— v — planta f - [mec] to shop rivet
— v — taller m - [mec] to shop rivet
— v junta f - [constr] to rivet @ joint
— v por punto(s) m - [sold] to stitch
— v unión f - [constr] to rivet @ joint
remache m a colocarse v en obra f - [estruct]
    field rivet
— m aceitado - [mec] oiled rivet
— m acoplador - [mec] linkage,king rivet
— m apropiado - [mec] proper, o appropriate,
    rivet
— m caliente - [mec] hot rivet
— m ciego - [mec] blind rivet
— m circunferencial - [mec] circumferential
    rivet
— m — apropiado - [mec] proper circumferential
    rivet
— m — en tamaño m apropiado - [mec] proper
    size(d) circumferential rivet
— m colocado—[mec] placed, o installed, rivet
— m — en caliente m - [mec] hot driven rivet
— m — — frío - [mec] cold driven rivet
— m con cabeza f embutida - [mec] countersunk,
    o flush head(ed), rivet
— m — — f plana - [mec] flathead, o flat
    headed, rivet
— m — — f rasa - [mec] flush head(ed) rivet
— m — — f redonda - [mec] round headed rivet
— m — tamaño m apropiado - [mec] proper size
    rivet
— m conectador - [mec] connecting, o linkage,
    rivet
— m de acero m - [mec] steel rivet
— m — — m con cabeza f redonda - [mec] round
    head(ed) steel rivet
— m — latón m - [mec] brass rivet
— m — — m con cabeza f plana - [mec] brass
    flat headed rivet
— m doblado - [mec] bent rivet
— m dúctil - [mec] ductile rivet
— m en caliente - [mec] hot rivet(ing)
— m — frío - [mec] cold rivet(ing)
— m expulsado - [mec] driven out rivet
— m frío - [mec] cold rivet
— m galvanizado - [mec] galvanized rivet
— m hendido - [mec] split rivet
— m hueco - [mec] hollow rivet
— m inspeccionado - [mec] inspected rivet
— m instalado en obra f - [mec] field rivet

remache m instalado mecánicamente - [constr] power driven rivet
— m **para acoplamiento** m - [mec] linking rivet
— m — **apoyo** m - [mec] pad rivet
— m — **cinta** f - [mec] lining rivet
— m — — f **para freno** m - [mec] brake lining rivet
— m — **conexión** f - [mec] linkage,king rivet
— m — **conjunto** m - [mec] assembly rivet
— m — — m **de contacto** m - [electr-instal] contact assembly rivet
— m — **correa** f - [mec] belt rivet
— m — **costura** f - [mec] seam rivet
— m — **cuchilla** f - [mec] blade rivet
— m — — f **sujetadora** - [mec] gripper blade rivet
— m — **eslabón** m - [mec] link rivet
— m — — m **conectador** - [mec] connecting link rivet
— m — **forro** m - [mec] lining rivet
— m — — m **para freno** m - [mec] brake lining rivet
— m — **grillete** m - [mec] shackle rivet
— m — **junta** f - [mec] joint rivet
— m — **leva** f - [mec] cam rivet
— m — **manivela** f - [mec] crank rivet
— m — **pasador** m - [mec] pin rivet
— m — — m **para manivela** f - [mec] crank pin rivet
— m — **plancha** f - [mec] plate rivet
— m — — f **lateral** - [mec] side plate rivet
— m — **retén** m - stop, o retainer, rivet
— m — **rodillo** m - [mec] roller rivet
— m — **sujetador** m - [mec] holder, o cam, rivet
— m — **tope** m - [mec] stop rivet
— m **partido** m - [mec] split rivet
— m **por punto(s)** m - [mec] stitch rivet(ing)
— m **reemplazado** - [mec] replaced rivet
— m **semitubular** - [mec] semi-tubular rivet
— m — **con cabeza** f **plana** - [mec] semi-tubular flat head(ed) rivet
— m — **de latón** m - [mec] semitubular bronze, o bronze semitubular, rivet
— m — — — m **con cabeza** f **plana** - [mec] semi--tubular bronze flat head(ed) rivet
— m **tubular** - [mec] tubular rivet
— m — **de latón** m - [mec] bronze tubular rivet
remanencia* f - véase **remanente** m
remanente m - remanent • remanence • remainder • [contab] balance | a - remaining; residual
— m **de aportación(es)** f - [sold] deposit remain(s)
— m — — m **anterior(es)** - [sold] old deposit remains
— m — — f — **con aleación** f - [sold] old alloy deposit remains
— m — — f — **con aleación** f **alta** - [sold] old high alloy deposit remains
— m — — m — — f **para endurecimiento** m **de superficie** f - [sold] old high alloy hardsurfacing deposit remains
— m — — m — **para endurecimiento** m **de superficie(s)** f - [sold] old hardsurfacing deposit remains
— m — — m **con aleación** f **alta** - [sold] hardsurfacing high alloy deposit remains
— m — — — — f **alta para endurecimiento** m **de superficie(s)** - [sold] high alloy hardsurfacing deposit remains
— m **determinado** - determined, o specified, remnant, o remainder
remansarse v - [hidr] . . .; to back up
remanso n **en entrada** f **a alcantarilla** f - [hidr] above @ culvert backwater
— m **reducido** - [hidr] diminished, o decreased, backwater
rematado,da—. . . . • [constr] capped; topped off
rematar v - . . .; to terminate • [constr] to abut • to top off
— v **refractario(s)** m - [constr] to finish off @ refractory,ries
— v — m **para cúpula** f - [metal-prod] to finish off @ refractory bricks in @ stove dome

remate m - . . .; top; termination; finale • [constr] . . .; crown; top line; cap • topping off
— m **atractivo** - [tub] attractive (end) finish
— m **cónico** - [mec] cone,nical point
— m **chato** - [mec] flat, o blunt, point
— m **de tornillo** m - [mec] screw end
— m **eficiente** - [tub] efficient end finish(ing)
— m **en púas** f - [alambre] barbed finish
— m **esférico** - [mec] oval point
— m **superior** - upper crest
— m **terminal** - [metal-fabr] end finish
remediar v - . . .; to cure
remedio m - . . .; cure • resource • fix • [medic] . . .; drug; medication
— m **económico** - economical remedy
— m **efectivo** - effective remedy
— m **mejor** - better remedy • best remedy
— m **milagroso** - [medic] miracle drug
— m **para erosión** f - erosion remedy
— m — **plazo** m **corto** - short term remedy
— m **rápido** - quick remedy • [fam] quick fix
— m **somero** • [fam] quick fix • [medic] quick remedy
remendar v **metal** m - [med] to mend @ metal
remesa f **aérea** - [transp] air shipment
— f **comprobatoria** - [transp] trial shipment
— f **de dividendo(s)** m - [fin] dividend remittance
— f — **divisa(s)** f - [fin] foreign currence remittance
— f — — f **a exterior** m - [fin] foreign currency remittance out of @ country
— f **en tránsito** m **y con corresponsales** m - [contab] remittance(s) in transit and with correspondents
— f **original** - [fin] original remittance
— f **por vía** f **aérea** - [transp] air shipment
— f **preparada** - [transp] prepared shipment
remezcla f - remix(ing)
remezclar v - [constr] to remix
remiendo m - . . .; mend
— m **de metal** - [mec] metal mend(ing)
remisión f - . . . • [fin] remittance; transfer • [comunic] mailing • [com] delivery slip
— f — **dividendo** m - [fin] dividend remittance
— f **original** - [fin] original remittance
— f **por avión** m - [transp] sending by air
— f **preparada** - [transp] prepared shipment
— f **total** - [fin] total, transfer, o remittance
remitente m - . . .; shipper
remitido,da a - [fin] remitted; transferred • [comunic] mailed
— a **por avión** m - [transp] shipped, o sent, by air
— a — **vía** f **aérea** - [Transp] shipped by air
remitir v - . . .; to send; to transmit; to convey; to forward • [fin] to transfer • [pol] to refer • to abide • to depend • [comunic] to mail
— v **a exterior** m - [fin] to send out (of @ country)
— v **por avión** m - to send by air
— v — **vía** f **aérea** - [transp] to ship by air; to air freight
remitirse v - to refer to
remito m - [com] delivery slip; packing list
remoción f - . . . • disposal • cleanout • slipping off • [metal-prod] rabbling
— f **algo mejor** - [ind] somewhat better removal • slightly better removal
— f — — **de escoria** f - [sold] slightly, o somewhat, better slag removal
— f **automática** - [ind] automatic removal • self removal
— f **de contrapeso** m - [grúas] counterweight, self, o automatic, removal
— f **completa** - complete removal
— f **con deslizamiento** m - slipping off
— f — **esmeriladora** f - [mec] grinder removal
— f — — f **mecánica** - [mec] electric grinder removal

remoción f con lima f - [mec] file removal
— f — —f manual - [mec] hand file removal
— f cuidadosa - [mec] careful removal
— f — aceitador m - [mec] oiler removal
— f — — m con fieltro m - [mec] felt oiler
removal,ving
— f — aceite m - [mec] oil removal,ving
— f — acoplamiento m - [mec] coupling, o linkage, removal,ving
— f — acumulador n - [electr-acumul] battery
removal,ving
— f — agua m - [mec] water removal,ving •
[miner] dewatering
— f — — m sobre superficie f - [hidr] surface
water removal,ving
— f — aire m - [neumát] air removal ving
— f — anhidrido m - [quím] anhydride removal
— f — — m sulfuroso - [quím] sulfur dioxide
removal,ing
— f — anillo m - [mec] spring removal,ving
— f — — m circular—[mec] O-ring removal,ving
— f — — m — para árbol m - [autom-mec] shaft
O-ring removal,ving
— f — — m — — — m para salida f (de fuerza f) - [autom-mec] output shaft O-ring removal,ving
— f — — m para compresión f - [mec] compression ring removal,ving
— f — — m — empaquetadura f - [mec| packing
ring removal,ving
— f — — m con resorte m - [mec] snap ring removal,ving
— f — arandela f - [mec] washer removal,ving
— f — — f con orificio m en forma de D—[mec]
D-washer removal,ving
— f — — f de bronce m - [mec] bronze washer
removal,ving
— f — — f para árbol m - [autom-mec] shaft
washer removal,ving
— f — — f — — m para entrada f (de fuerza
f) - [autom-mec] input shaft washer removal,ving
— f — — f de bronce m - [mec] bronze washer
removal,ving
— f — — f — — m para empuje m—[mec] bronze
thrust washer removal,ving
— f — — f para empuje m - [mec[thrust washer
removal,ving
— f — — f para seguridad f - [mec] lockwasher
removal,ving
— f — — f — — f para cubierta f - [autom-
-mec] cover lockwasher removal,ving
— f — — f — — — f para cojinete m -
[autom-mec] bearing cover lockwasher removal
— f — — f — — f — — f — — m para entrada (de fuerza f) - [autom-mec] input bearing
cover lockwasher removal,ving
— f — — f plana - [mec] flat washer removal,ving
— f — — f — para árbol m - [autom-mec] shaft
flat washer removal,ving
— f — — f — — — m para entrada f (de
fuerza f)—[auto-mec] input shaft flat washer
removal,ving
— f — araña f - [mec] spider removal,ving
— f — árbol m - [botán] tree removal,ving •
[mec] shaft removal,ving
— f — — m para salida f (de fuerza f) - [mec]
output shaft removal,ving
— f — armadura f - [mec] armature removal,ving
— f — aro m - [mec] ring removal,ving
— m — — m para compresión f - [mec] compression ring removal,ving
— f — — empaquetadura f - [mec] packing ring removal,ving
— f — articulación f - [mec] link(age) removal,ving
— f — atomizador m - [comb.int] atomizer removal,ving
— f — asiento m - [mec] seat removal,ving
— f — — m para válvula f - [válv] valve seat
removal,ving
— f — astilla(s) f - [mec] chip removal,ving

remoción f de banda f - [mec] band removal,ving
— f — — f para freno m - [mec] brake band
removal,ving
— f — bloque m - [mec] block removal,ving
— f — — m deslizable - [autom-mec] sliding
block removal,ving
— f — bomba f - [bombas] pump removal,ving
— f — boquilla f - [mec] nozzle removal,ving
— f — — f para rociadura f - [mec] spray
nozzle removal,ving
— f — — f para perforación f - [petról]
drill(ing) bit removal,ving
— f — buje m - [mec] bushing removal,ving
— f — — m para camisa f - [mec] liner bushing removal,ving
— f — — m piloto - [mec] pilot bushing removal,ving
— f — — m — para camisa f - [mec] liner
pilot bushing removal,ving
— f — bujía f - [comb.int] spark plug removal,ving
— f — cabeza f - [mec] head removal,ving
— f — — f para inyección f - [petról] swivel
removal,ving
— f — cadena f - [mec] chain removal,ving
— f — caja f - [mec] housing, o body. removal,ving
— f — — f para conmutador m - [mec-ventil]
switch housing removal,ving
— f — — f para eje m - [autom-mec] axle
housing removal,ving
— f — cajón m - [transp] container removal,ving • removal from @ container
— f — calce m - [mec] shim removal,ving
— f — calor m - heat removal,ving
— f — calza f - [mec] shim removal,ving
— f — — f para piñón m - [mec] pinion shim
removal,ving
— f — cámara f - [mec] chamber removal,ving
— f — — f para filtro m - [mec] filter chamber removal,ving
— f — — f — — m para aire m - [mec] air
filter chamber removal,ving
— f — camisa f - [mec] liner removal
— f — campana f - [mec] bell, o canopy, removal,ving
— f — carga f - [mec] load removal,ving
— f — carrillo m - [mec] trolley removal,ving
— f — cartucho m - [mec] cartridge removal,ving
— f — cascarilla f - [metal-lam] scale, flushing, o removal,ving
— f — cascarón m - [metal-prod] scab removal,ving
— f — casquete m - [comb.int] cap removal,ving
— f — causa f - cause removal,ving
— f — — f de error m - error cause, removal,ving
— f — ceniza f - [combust] ash removal,ving
— m — cierre m - [mec] closure, o seal, removal,ving
— f — cilindro m - [comb.int] cylinder removal,ving
— f — cinta f - [mec] lining removal,ving
— f — — f para freno m - [mec] brake lining
removal,ving
— f — cojinete m - [mec] bearing removal,ving
— f — — m con grasa f - [mec] grease bearing
removal,ving
— f — colador m - [mec] strainer removal,ving
— f — colector m - [mec] sump removal
— f — — m para purga f - [mec] drain(age)
sump removal,ving
— f — combustible m - fuel removal,ving
— f — conexión f - [mec] linkage removal,ving
• [electr-instal] connection removal,ving
— f — conjunto m - [mec] assembly removal,ving
— f — — m de árbol m - [mec] shaft assembly
removal,ving
— f — — m — — m para entrada f (de fuerza f)
- [autom-mec] input shaft assembly removal,ving
— f — — m — cabeza f - [mec] head assembly

removal,ving • [petról] swivel assembly removal,ving
remoción f **de conjunto** m **de cabeza** f **para inyección** f—[petról] swivel assembly removal,ving
— f — — m — **calce** m - [mec] shim pack removal,ving
— m — — m — **distribuidor** m - [autom-mec] divider assembly removal,ving
— f — — m — — m **para fuerza** f - [autom-mec]
power divider assembly removal,ving
— f — — m — **llanta** f - [autom] rim assembly
removal,ving
— f — — m — **neumático** m - [autom-neumát]
tire assembly removal,ving
— f — — m — **portadiferencial** m - [autom-mec]
differential carrier assembly removal,ving
— f — **cono** m - [mec] cone removal,ving
— f — — m **para cojinete** m - [mec] bearing
cone removal,ving
— m — **contragolpe** m - [mec] backlash, removal,ving, o reducing,ction
— m — **contrapeso** m - [grúas] counterweight removal,ving
— f — **contratuerca** f - locknut removal,ving
— f — **cruceta** f - [mec] crosshead removal,ving
• removal,ving from @ crosshead
— f — **cubierta** f - [mec] cover removal,ving •
[autom-neum•at] tire removal,ving
— f — — f **grande** - [mec] large cover removal,ving
— f — — f **para bomba** f - [combas] pump cover
removal,ving
— f — — f — **caja** f - [mec] housing cover
removal,ving
— f — — f — — f **para eje** m - [autom-mec]
axle housing cover removal,ving
— f — — f — **cojinete** m - [autom-mec] bearing
cover removal,ving
— f — — — m **para entrada** f **(de fuerza**
f**)** - [autom-mec] input bearing cover removal,ving
— f — — f — **mesa** f - [mec] table top removal,ving
— f — — f — **transmisión** f - [mec] transmission cover removal,ving
— f — — f **principal** - [mec] main cover removal,ving
— f — **cuerpo** m - [mec] body removal,ving
— f — — m **extraño** - [medic] foreign, body, o
object, removal,ving
— f — **chapa** f - [mec] sheet removal,ving
— f — — f **para cubierta** f - [mec] cover(ing)
sheet removal,ving
— f — **defecto(s)** m - [metal-trat] defect(s)
removal,ving
— f — — m **mediante escarpado** m - [metal-trat]
scarfing defect(s) removal,ving
— f — **defensa** f - [segurid] protection removal
— f — — f **para seguridad** f - [segurid] safety
guard removal,ving
— f — **derivación** f - [electr-instal] bias removal,ving
— f — **despunte(s)** m - [metal-lam] cropping(s)
removal,ving
— f — — m **de laminación** f - [metal-lam] mill,
o plant, croppings, o scrap, removal,ving
— n — **dispositivo** m **para cambio(s)** m - [autom-
-mec] shift(ing) unit removal,ving
— f — **diferencial** m **entre eje(s)** m - [autom-
-mec] interaxle differential removal,ving
— f — **distribuidor** m - [mec] divider removal,ving
— f — — m **para fuerza** f - [mec] power divider
removal,ving
— f — — m **rociador** - [mec] spray manifold removal,ving
— f — **eje** m—[mec] axle, o shaft, removal,ving
• removal,ving from @, axle, o shaft
— f — **elemento** m - [mec] element removal,ving
• [segurid] canister removal,ving
— f — **embalaje** m - packing removal,ving
— f — **émbolo** m - [mec] piston removal,ving
— f — **embrague** m - [mec] clutch removal,ving

remoción f **de embrague** m **para (mecanismo** m **para)
elevación** f - [grúas] hoist(ing) clutch re-
moval,ving
— f — — m **(— m — m (— m —) rotación** f -
[grúas] swing clutch removal,ving
— f — **empaquetadura** f - [mec] packing removal
— f — **encofrado** m - [constr] form, break(ing)-
-up, o removal,ving
— f — **engranaje** m - [mec] gear removalv,ving
— f — **escamilla** f - [sold] scale removal,ving
— v — — f **de laminación** f - [sold] mill
scale removal,ving
— f — **escoria** f - [sold] slag removal.ving •
[metal-prod] slag removal,ving
— f — — f **en ranura(s)** f **(profunda(s)**—[sold]
deep groove slag removal,ving
— f — **eslabón** m - [mec] link removal,ving
— f — — m **de cadena** f - [mec] chain link removal,ving
— v — **esmeriladora** f - [mec] grinder removal
— f — — f **mecánica** - [mec] electric grinder
removal,ving
— f — **esqueleto** m - [transp] crate,ting removal,ving
— f — **esquina** f - [mec] corner removal,ving
— f — **exceso** m - [sold] excess removal,ving
— f — — m **de fundente** - [sold] excess flux
removal,ving
— f — **fiador** m - [mec] retainer removal,ving
— f — — m **para cierre** m - [mec] seal retainer
removal,ving
— f — — m — **trinquete** m - [mec] ratchet
retainer removal,ving
— f — **fijación** f—[mec] fastening removal,ving
— f — **filtro** m - [mec] filter removalv,ving
— f — — m **para aire** m - [ind] air filter removal,ving
— f — **freno** m - [mec] brake removal,ving
— f — **fundente** m - [sold] flux cleanout,ning
— f — **grillete** m - [mec] clevis removal,ving
— f — **guia(dera)** f - [mec] guide removal,ving
— f — — f **para vástago** m - [válv] stem guide
removal,ving
— f — — f — — m **para válvula** f - [válv]
valve stem guide removal,ving
— f — — f — — m — f **inferior** - [válv]
lower valve stem guide removal,ving
— f — — f — — m — f **superior** - [válv]
upper valve stem guide removal,ving
— f — **horquilla** f - [mec] clevis, o yoke, o
fork, removal,ving
— f — — f **para árbol** m - [autom-mec] input
shaft yoke removal,ving
— f — — f — — m **salida** f **(de fuerza** f**)** -
[autom-mec] output shaft yoke removal,ving
— f — — f — **cambio(s)** m - [autom-mec] shift
fork removal,ving
— f — **injerto** m - [mec] insert removal,ving
— f — — m **para válvula** f - [válv] valve insert removal,ving
— f — **inyección** f - [petról] mud removal.ving
— f — **jaula** f - [transp] crate,ting removal
— f — **laminilla(s)** f - [mec] shim removal,ving
— f — **lodo** m - [petról] mud removal
— f — **material(es)** m - [ind] material(s) removal,ving
— f — — m **excavado(s)** m - [constr] excavated
material(s) removal,ving
— f — — m **extraño(s)** - [ind] foreign mate-
rial(s) removal,ving
— f — **molde** m - [constr½ form removal,ving
— f — **montaje** m - [ind] mounting removal,ving
— f — **mordaza** f - [mec] clamp removal,ving
— f — **motor** m - [comb.int] engine removal,ving
— f — **neumático** m - [autom-neumát] tire removal,ving
— f — **nieve** f - [vial] snow removal,ving
— f — **orificio** m - [mec] removing from @ hole
— f — **oxígeno** m - [metal-prod] oxygen removal,ving
— f — — m **en hierro** m **sólido** - [metal-prod]
solid iron oxygen removal,ving
— f — **paleta** f - [ventil] blade removal,ving

remoción f de panel m - [mec] panel removal,ving
— f — — m para acceso m - [mec] access panel removal,ving
— f — — m posterior - [mec] back panel removal,ving
— f de pasador m - [mec] pin removal,ving
— f — — m ahusado - [mec] tapered, pin, o dowel, removal,ving
— f — — m conectador - [mec] connecting, o linkage,king, pin removal,ving
— f — — m para grillete m - [mec] clevis pin removal,ving
— f — — m — horquilla f - [mec] clevis pin removal,ving
— f — perno m - [mec] bolt removal,ving
— f — — m para anclaje m - [mec] anchor bolt removal,ving
— f — — m para sujeción f - [mec] fastening, o hold-down, bolt removal,ving
— f — — m — sujetador m - [mec] clamp(ing) pin removal,ving
— f — — — m para camisa f - [mec] liner clamp(ing) bolt removal,ving
— f — — m para traba(r) - [mec] lock(ing) bolt removal,ving
— f — — m sujetador - [mec] clamp(ing) bolt removal,ving
— f — — m — para camisa f - [mec] liner clamp(ing) pin removal,ving
— f — pico m - [mec] (grease) fitting removal
— f — — m para engrase m - [mec] grease fitting removal,ving
— f — pieza f - [mec] part removal,ving
— f — — f para bomba f - [bombas] pump part removal,ving
— f — piñón m - [mec] pinion removal,ving
— f — — m lateral - [mec] side pinion removal,ving
— f — piso m - [constr] floor removal,ving
— f — placa f - [mec] plate removal,ving
— f — plancha f - [mec] plate removal,ving
— f — — f para piso m - [mec] floor plate removal,ving
— f — polvo m - dust, removal,ving, o wiping
— f — — m ordinario - [mec] ordinary dust removal,ving
— f — portadiferencial m - [autom-mec] differential carrier removal,ving
— f — prenda f - [vest] article removal,ving
— f — prensaestopa(s) m - [mec] packing box removal,ving
— f — rebaba(s) f - [sold] flash, o burr, removal,ving
— f — recalcadura f - [mec] upsetting removal,ving
— f — — f metálica - [mec] metal upsetting removal,ving
— f — rejilla f - [mec] screen removal,ving
— f — resguardo m - [mec] guard removal,ving
— f — resistencia f - [electrón] resistor removal,ving
— f — resorte m - [mec] spring removal,ving
— f — restricción f - restriction removal,ving
— f — retén m - [mec] retainer removal,ving
— f — — m para cierre m - [mec] seal(ing) retainer removal,ving
— f — retenedor m—[mec] retainer removal,ving
— f — rotor m - [mec] rotor removal,ving
— f — rueda f - [mec] wheel removal,ving
— f — — f motriz - [mec] drive wheel removal,ving
— f — sección f - [mec] section removal,ving
— f — — f de piso m - [constr] floor section removal,ving
— f — — f — plancha f - [mec] plate section removal,ving
— f — — f — — f para piso m - [mec] floor plate section removal,ving
— f — segmento m - [mec] segment removal,ving
— f — — m de vástago m - [mec] subrod removal,ving
— f — sello m - [mec] seal removal,ving

remoción f de sello m para aceite m - [mec] oil seal removal,ving
— f — semieje m - [autom-mec] axle shaft removal,ving
— f — separador m - [mec] spacer removal,ving
— f — — m para armado m - [mec] assembling spacer removal,ving
— f — — m cojinete m - [mec] bearing spacer removal,ving
— f — sobre - [mec] raising off of
— f — solenoide m - [electr-instal] solenoid removal,ving
— f — sonda f - probe removal,ving
— f — — f para empleo m en obra f - field probe removal,ving
— f — soporte m - [mec] holder removal,ving
— f — suciedad f - dirt removal,ving
— f — suelo m - [suelos] soil removal,ving
— f — — m superficial - [constr] stripping
— f — sujetador m—[mec] fastener removal,ving
— f — — m para camisa f - [mec] liner clamp removal,ving
— f — suplemento m - [mec] supplement, o insert, removal,ving
— f — — m para válvula f - [válv] valve insert removal,ving
— f — tapa f - [mec] cover, o cap, o top, removal,ving
— f — — f para colector m - [mec] sump cover removal,ving
— f — — f — mesa f - [mec] table cover removal,ving
— f — — f — transmisión f - [mec] transmission cover removal,ving
— f — — f — válvula - [válv] valve cover removal,ving
— f — — f principal - [mec] main cover removal,ving
— f — tapón m - [mec] plug removal,ving
— f — — m mostrada - [mec] shown plug removal,ving
— f — — m para drenaje m - [mec] drain plug removal,ving
— f — — m para filtro m - [mec] filter plug removal,ving
— f — — m — — m mostrada - [petról] shown filter plug removal,ving
— f — — m — freno m - [mec] brake plug removal,ving
— f — tapón m para nivel m de aceite m - [mec] oil level plug removal,ving
— f — — m — m — m en caja f - [mec] case oil level plug removal,ving
— f — — m — m — m — f para cadena - [mec] chain case oil level plug removal,ving
— f — — m para purga f - [mec] drain plug removal,ving
— f — — — f de colector m para aspiración f - [mec] suction manifold drain plug removal,ving
— f — — m — tubo(s) m - [mec] pipe plug removal,ving
— f — tierra f - dirt, o ground, removal,ving
— f — — f ordinaria - [constr] ordinary, dirt, o ground, removal,ving
— f — tornillo m - [mec] screw removal,ving
— m — — m con casquete m - [mec] cap screw removal,ving
— f — — m — — m para cubierta f - [mec] cover cap screw removal,ving
— f — — m — — m — — f para cojinete m - [mec] bearing cover cap screw removal,ving
— f — — m — — m — — f — m para aportación f (de fuerza f) - [autom-mec] input bearing cover cap screw removal,ving
— f — trampa f - [mec] trap removal,ving
— f — tubería f - [tub] pipe,ping removal,ving
— f — tubo m - [tub] tube, o pipe, removal,ving
— f — tuerca f - [mec] nut removal,ving
— f — — f herrumbrada - [mec] rusty nut re-

moval,ving
remoción f de tuerca f para árbol m—[autom-mec]
shaft nut removal,ving
— f — f — m — **aportación f (para fuerza f)** - [autom-mec] (power) input shaft nut
removal,ving
— f — f — **engranaje m** - [mec] gear nut removal,ving
— f — f — m **para impulsión f** - [mec]
drive,ving gear nut removal,ving
— f — f — **espárrago m** - [mec] stud nut removal,ving
— f — f — **horquilla f** - [autom-mec] yoke
nut removal,ving
— f — f — f **para aportación f (de fuerza f)** - [autom-mec] (power) input yoke nut
removal,ving
— f — **unidad f** - [mec] unit removal,ving
— f — **válvula f** - [válv] valve removal,ving
— f — f **inferior** - [válv] lower, o bottom,
valve removal,ving
— f — f **superior** - [válv] upper, o top,
valve removal,ving
— f — **vástago m** - [mec] rod removal • [válv]
stem removal
— f — m **intermedio** - [mec] intermediate
rod removal,ving
— f — m **para válvula f** - [válv] valve stem
removal,ving
— f — m — f **inferior** - [válv] lower, o
bottom, valve stem removal,ving
— f — m — f **superior** - [válv] upper, o
top, valve stem removal,ving
— f **desde árbol m** - [mec] removal,ving from
(off) @ shaft
— f — **eje n** - [mec] removal,ving from @, axle,
o shaft
— f — m **para aportación f (de fuerza f)** -
[auom-mec] removal,ving from @ input shaft
— f — **base f** - [mec] removal,ving from @ base
— f — **caja f** - [mec] removal,ving from @,
case, o housing
— f — f **para eje m** - [autom-mec] removal,ving from @ axle, case, o housing
— f — **émbolo m** - [mec] removal,ving from @
piston
— f — **varilla f**—[mec] removal,ving from @ rod
— f — f **para empuje m** - [mec] removal,ving
grom @ push(ing) rod
— f **deslizando** - slipping off
— f **difícil** - [sold] hard, o poor. removal,ving
— f — **de escoria f** - [sold] poor slag removal
— f **fácil** - easy, o ready, removal,ving
— f — **de escoria f** - [sold] easy, o good, slag
removal,ving
— f **lenta** - slow removal,ving
— f **mejor** - better removal,ving; best removal,ving
— f — **de escoria f** - [sold] better slag removal,ving • best slag removal,ving
— f — f **en ranura(s) f (profunda(s)** -
[sold] best deep groove slag removal,ving
— f **mejorada** - improved removal,ving
— f — **de escoria f** - [sold] improved slag removal,ving
— f **mostrada** - shown removal,ving
— f **muy fácil** - [sold] very easy removal,ving
— f — **de escoria f** - [sold] very, o extremely,
easy slag removal,ving
— f **para inspección f** - removal,ving for @ inspection
— f — **limpieza f** - [mec] removal,ving for
cleaning
— f **periódica** - periodic removal,ving
— f **rápida** - rapid, o quick, removal,ving
remojado,da a - soaked • wetted; drenched
— a **en agua m** - water soaked
remojar v aceitador m - to soak @ oiler
— v — m **para fieltro m** - to soak @ felt oiler
— v — m — m **para émbolo m** - [mec] to soak
@ piston felt oiler
— v — **agua m** - to soak in water; to water soak
remojo m de aceitador m - [mec] oiler soaking

remojo m de aceitador m para fieltro m - [mec]
felt oiler soaking
— m **en agua m** - water soaking
remolcado,da a - towed
remolcador m para empuje m - [nav] pusher tug
remolcar v embarcación f - [náut] to tow @ vessel
— v **equipo m** - to tow @, equipment, o machine
— v **implemento m** - [agric] to tow @ implement
— v **máquina f** - [ind] to tow @ machine
— v **unidad f motriz** - [Agric] to tow @ power
unit
remoldeado,da a - remolded; reshaped
remolque m - . . . . • [transp] trailer; véase
también acoplado m
— m **carretero** - [transp] road towing
— m **con descarga f por arriba** - [transp] top
dump trailer
— m — f **por fondo m** - [transp] bottom dump
trailer
— m — f **trasera** - [transp] rear (end) dump
trailer
— m **con plataforma f**—[transp] platform trailer
— m — f **baja** - [transp] low-boy (trailer)
— f **de embarcación(es) f** - [náut] vessel towing
— m — **equipo m** - [ind] equipment, o machinery,
towing
— m — **implemento(s) m** - [agric-equip] implement towing
— m — **máquina f** - [ind] machine towing
— m — **unidad f motriz** - [agric-equip] power
unit towing
— m **en planta f** - [ind] in-plant towing
— m — **parque m** - [ind] yard towing
— m — **playa f** - [ind] yard towing
— m — **taller m** - [ind] shop, o plant, towing
— m **inclinado** - [transp] tilted, o slanting,
trailer
— m **manual** - [ind] hand, towing, o pulling
— m **para transporte m** - [transp] transporting
trailer
— m **por carretera f** - [transp] highway towing
— n **rápido** - [transp] fast, o high speed, towing
— m — **por carretera f** - [transp] high speed
road towing
remontado,da a - remounted • [metal-lam] riding
remontarse v - to go back
remonte m - [metal-lam] riding
removedor m - [ind] remover
— m **mecanizado** - [grúas] power remover
— m — **para contrapeso m** - [grúas] power(ed)
counterweight remover
— m — **puerta(s) f** - [coque] door, remover, o
exractoe, o machine
— m — f **para horno(s) m (para coque m)** -
[coque] coke oven door, remover, o extractor
— m — **residuo(s)** - trash remover • scavenger
— m **para tierra f** - [constr] earth, o dirt, remover •
remover v - . . . .; to shovel out • to slip off •
[sold] to chip out • [metal-prod] to rabble
— v **abrazadera f** - [mec] to remove @ clamp
— v **aceitador m** - to remove @ oiler
— v — m **con fieltro m** - [mec] to remove @
felt oiler
— v **aceite m** - to remove @ oil
— v **acoplamiento m** - [mec] to remove @, linkage, o coupling
— v **acumulador m** - [electr-acumul] to remove @
battery
— v **agua m** - to remove @ water; to dewater
— v — m **sobre superficie f** - [hidr] to remove
@ surface water
— v **aire m** - [neumát] to remove @ air
— v **anillo m** - [mec] to remove @ ring
— v — m **circular** - [mec] to remove @ O ring
— v — m **para árbol m para salida f (para
fuerza f)** - [autom-mec] to remove @ output
shaft O ring
— v — m **para compresión f** - [mec] to remove @
compression ring
— v — m — **empaquetadura f** - [mec] to remove @
packing ring

remover v arandela f - [mec] to remove @ washer
— v — f con orificio m en forma de D - [mec] to remove @ D washer
— v — f de bronce m - [mec] to remove @ bronze washer
— v — f —— m para empuje m - [mec] to remove @ bronze thrust washer
— v — f para árbol m - [mec] to remove @ shaft washer
— v — f —— m para aporte m (de fuerza f) - [autom-mec] to remove @ input shaft washer
— v arandela f para empuje m - [mec] to remove @ thrust washer
— v — f para seguridad f - [mec] to remove @ lockwasher
— v — f —— f para cubierta f - [autom-mec] to remove @ cover lockwasher
— v — f —— f para cojinete m - [auto-mec] to remove @ bearing cover lockwasher
— v — f —— f —— m para aportación f (de fuerza f) - [autom-mec] to remove @ input bearing cover lockwasher
— v arandela f plana - [mec] to remove a flat washer
— v — f — para árbol m - [autom-mec] to remove @ shaft flat washer
— v — f —— m para aportación f (de fuerza f) - [autom-mec] to remove @ input shaft flat washer
— v — t para seguridad f - [mec] to remove @ lockwasher
— v araña f - to remove @ spider
— v árbol m - [botán] to remove @ tree - [mec] to remove @ shaft
— v — m para aportación f (de fuerza f) - [autom-mec] to remove @ input shaft
— v — m — salida f (de fuerza f) - [autom-mec] to remove @ output shaft
— v armadura f - [mec] to remove @ armature
— v aro m - [mec] to remove @ ring
— v — m para compresión f - [mec] to remove @ compression ring
— v — m — empaquetadura f - [mec] to remove @ packing ring
— v articulación f - [mec] to remove @, joint, o linkage
— v asiento m - [mec] to remove @ seat
— v — m para válvula f - [válv] to remove @ valve seat
— v astilla f - [mec] to remove @, splinter, o chip
— v atomizador m - [comb.int] to remove @ atomizer
— v banda f - [mec] to remove @ band
— v — f para freno m - [mec] to remove @ brake band
— v barra f - to remove @ bar
— f — conformada - [mec] to remove @ shaped bar • [petról] to remove @ Kelly
— v — f para perforación f - [petról] to remove @ Kelly
— v barreno,na - [petról] to remove @ bit
— v — para perforación f - [petról] to remove @ drill(ing) bit
— v bloque m - [mec] to remove @ block
— v bomba f - [bombas] to remove @ pump
— v boquilla f - [mec] to remove @ nozzle
— v — f para rociadura f - [mec] to remove @ spray nozzle
— v broca f - [mec] to remove @ bit
— v — f para perforación f - [petról] to remove @ drill(ing) bit
— v buje m - [mec] to remove @ bushing
— v — m para camisa f - [mec] to remove @ liner bushing
— v — m piloto - [mec] to remove @ pilot bushing
— v — m — para camisa f - [mec] to remove @ liner pilot bushing
— v bujía f - [comb.int] to remove @ spark plug
— v cabeza f - [mec] to remove @ head
— v — f para inyección f - [petról] to remove

@ swivel
remover v cadena f - [mec] to remove @ chain
— v caja f - [mec] to remove @ box • to remove @, housing, o body
— v — f para conmutador m - [electr-instal] to remove @ switch housing
— v — f para eje m - [autom-mec] to remove @ axle housing
— v cajón m - [transp] to remove @ container
— v, calce m, o calza f - [mec] to remove @ shim
— v — para piñón m - [mec] to remove @ pinion shim
— v calor m - to, remove, o take away, @ heat
— v cámara f - [mec] to remove @ chamber
— v — f para filtro m - [mec] to remove @ filter chamber
— v — m para aire m - [mec] to remove @ air filter chamber
— v camisa f - [vest] to remove @ shirt • [mec] to remove @ liner
— v campana f - [mec] to remove @, bell, o canopy
— v carga f - to remove @ load
— v carrillo m - [mec] to remove @ trolley
— v cartucho m - [mec] to remove @ cartridge
— v casquete m - [comb.int] to remove @ cap
— v causa f - to remove @ cause
— v — f de error m - to remove @ error cause
— v conexión f - [electr-instal] to remove @ connection
— v cierre m - [mec] to remove @ seal
— v cilindro m - [comb.int] to remove @ cylinder
— v cinta f - [mec=frenos] to remove @ lining
— v — f para freno m - [mec] to remove @ brake lining
— v cojinete m - [mec] to remove @ bearing
— v — m con grasa f - [mec] to remove @ grease bearing
— v colador m - [mec] to remove @ strainer
— v colector m - [mec] to remove @ sump
— v — m para purga f - [mec] to remove @ drain sump
— v completamente - to remove completely
— v combustible m - to remove @ fuel
— v con cuidado - to remove carefully
— v — esmeriladora f - [mec] to remove with @ grinder
— v — — f mecánica - [mec] to remove with @ electric grinder
— v — extractor m - [mec] to remove with @, extractor, o puller
— v — lima f - [mec] to file (away)
— v — — f manual - [mec] to remove with @ hand file
— v paño m - to wipe
— v conductor m - [electr-instal] to remove @ lead
— v — de(sde) borne m - [electr-instal] to remove @ lead from @ terminal
— v conexión f - [mec] to remove @ linkage
— v conjunto m - [mec] to remove @ assembly
— v — m de árbol m - [mec] to remove @ shaft assembly
— v — m —— m para aportación f (de fuerza f) - [autom-mec] to remove @ input shaft assembly
— v — m ——, calce, o calza - [mec] to remove @ shim pack
— v — m — distribuidor - [autom-mec] to remove @ divider assembly
— v — m —— para fuerza f - [autom-mec] to remove @ power divider assembly
— v — m — cabeza f - [mec] to remove @ head assembly
— v — m —— f para inyección f - [petról] to remove @ swivel assembly
— v — m — llanta f - [autom-ruedas] to remove @ rim assembly
— v — m — neumático m - [autom-neumát] to remove @ tire assembly
— v — m — portadiferencial m - [autom-mec] to remove @ differential carrier assembly

remover v cono m - [mec] to remove @ cone
— v — de cojinete m - [mec] to remove @ bearing cone
— v continuamente - to remove continuously
— v contragolpe m - [mec] to remove @ backlash
— v contrapeso m - [grúas] to remove @ counterweight
— v contratuerca f - [mec] to remove @ locknut
— v convenientemente - to remove conveniently
— v cruceta f - [mec] to remove @ crosshead
— v cubierta f - [mec] to remove @ cover • [autom-neumát] to remove @ (outer) tire
— v — f grande - [mec] to remove @ large cover
— v — f para bomba f - [bombas] to remove @ pump cover
— v — f — caja f - [mec] to remove @ housing cover
— v — f — — f para eje m - [aitom-mec] to remove @, axle, o shaft, housing cover
— v — f para cojinete m - [autom-mec] to remove @ bearing cover
— v — f — — m para aportación f (de fuerza f) [autom-mec] to remove @ input bearing cover
— v — f — mesa f - [mec] to remove @ table cover
— v — f — transmisión f - [mec] to remove @ transmission cover
— v — f principal - [mec] to remove @ main cover
— v cuerpo m - [mec] to remove @ body
— v — m extraño - (medic) to remove @ foreign, body, o object
— v cuidadosamente - to remove carefully
— v chapa f - [mec] to remove @ sheet
— v — f para cubierta f - [mec] to remove @ cover sheet
— v de cajón m - [ind] to remove from @ box • [transp] to remove from @ container
— v — eje m - [mec] to remove from @ axle
— v — motor m - [electr-mot] to remove from @ motor • [comb.int] to remove from @ engine
— f — orificio m - [mec] to remove from @ hole
— v — prensaestopa(s) m - [mec] to remove from @ packing box
— v — separador m - [mec] to remove from @, spacer, o separator
— v — sobre m - to raise off of • to remove from @ envelope
— f — válvula f - [mec] to remove from @ valve • [comb.int] valve removal
— v demasiado - to remove too much
— v — poco - to remove too little
— v desde - to remove from
— v — árbol m - [botán] to remove from @ tree • [mec] to remove from @ shaft
— v — — — m para aportación f de fuerza f - [auto-mec] to remove from @ input shaft
— v — base f - [mec] to remove from @ base
— v — caja f - [mec] to remove from @, case, o housing
— v — — f para eje m - [autom-mec] to remove rom @ axle housing
— v — émbolo m - [mec] to remove from @ piston
— v — varilla f - [mec] to remove from @ rod
— v — — f para empuje m - [mec] to remove rom @ push rod
— v deslizando - to, slide, o slip, off
— v diferencial m entre eje(s) m - [auto-mec] to remove @ inter-axle differential
— v distribuidor m - [mec] to remove @ manifold
— v — m rociador - [mec] to remove @ spray manifold
— v eje m - [mec] to remove @, axle, o shaft
— v elemento m - [mec] to remove @ element • [segurid] to remove @ canister
— v embalaje m - to remove @ packing
— v émbolo m - [comb.int] to remove @ piston
— v embrague m - [mec] to remove @ clutch
— v — m para (mecanismo m para) elevación f - [grúas] to remove @ hoist(ing) clutch
— v — m para rotación v - [grúas] to remove @ swing clutch

remover v enpaquetadura f - [mec] to remove @ packing
— v en defecto - to remove too little
— v — menos - to remove too little
— v encofrado m - [constr] to remove @ form
— v engranaje m - [mec] to remove @ gear
— v escoria f - [metal-prod] to flush (@ slag)
— v eslabón m - [mec] to remove @ link
— v — para cadena f - [cadenas] to remove @ chain link
— v esqueleto m - [transp] to remove @ crate
— v esquina f - [mec] to remove @ corner
— v excedente m - to remove @ excess
— v — m de fundente m - [sold] to remove @ excess flux
— v fácilmente - to remove easily
— v fiador m - [mec] to remove @ retainer
— v — m para trinquete m - [mec] to remove @ ratchet retainer
— v fijación f - [mec] to remove @ fastening
— v fijador - [mec] to remove @ fastener
— v filtro m - [mec to remove @ filter
— v — m para aire m - [mec] to remove @ air filter
— v freno m - [mec] to remove @ brake
— v grasa f - [mec] to remove @ grease
— v grillete m - [mec] to remove @ clevis
— v guarnición f - [mec] to remove @ trim(ming)
— v — f circular - [mec] to remove @ O ring
— v guía(dera) f - [mec] to remove @ guide
— v — f para vástago m - [válv] to remove @ stem guide
— v — f — — m para válvula f - [válv] to remove @ valve stem guide
— v — f — — m — — f inferior - [válv] to remove @ lower valve stem guide
— v — f — — m — — f inferior - [válv] to remove @ upper valve stem guide
— v horquilla f - [mec] to remove @, clevis, o yoke
— v — f para árbol m - [mec] to remove @ shaft clevis, o yoke
— v — — m para aportación f (de fuerza f) - [autom-mec] to remove @ input shaft yoke
— v — f — — m — salida f (de fuerza f) - [autom-mec] to remove @ output shaft yoke
— v — f para cambio(s) m - [autom-mec] to remove a shift(ing) fork
— v material m - to remove @ material
— v — m extraño - [medic] to remove @ foreign, material, o substance
— v mejor - to remove better • to remove best
— v molde m - [constr] to remove @ form
— v montaje m - [mec] to remove @ mounting
— v mordaza f - [mec] to remove @ clamp
— v neumático m - [autom-neumát] to remove @ tire
— v nieve f - [vial] to remove @ snow
— v paleta f - [mec] to remove @ blade
— v panel m - [mec] to remove @ panel
— v — m, anterior, o delantero n - [mec] to remove @ front panel
— v — m posterior - [mec] to remove @ rear panel
— v para inspección f - [mec] to remove for inspection
— v limpieza f - [mec] to remove for cleaning
— v pasador m - [mec] to remove @ pin
— v — m ahusado - [mec] to remove @ tapered, pin, o dowel
— v — — conectador - [mec] to remove @ linking pin
— v — m para grillete m - [mec] to remove @ clevis pin
— v — m — horquilla f - [mec] to remove @ clevis pin
— v periódicamente - to remove periodically
— v perno m - [mec] to remove @ bolt
— v — m herrumbrado - [mec] to remove @ rusty bolt
— v — m para anclaje m - [mec] to remove @ anchor bolt

remover v **perno** m **para sujeción** f - [mec] to remove a hold-down pin
— v — m — — f **para camisa** f - [mec] to remove @ liner clamnp(ing) bolt
— v — m **para traba** f - [mec] to remove @ lock(ing) bolt
— v **pico** m - [mec] to remove @ nozzle
— v — m **para engrase** m - [mec] to remove @ grease fitting
— v **pieza** f - [mec] to remove @ part
— v — f **para bomba** f - [bombas] to remove @ pump part
— v **piñón** m - [mec] to remove @ pinion
— v — m **lateral** — [mec] to remove @ side pinion
— v **piso** m - [constr] to remove @ floor
— v **placa** f - [mec] to remove @ plate
— v **plancha** f - [mec] to remove @ plate
— v — f **para piso** m - [mec] to remove @ floor plate
— v **polvo** m - to (remove @) dust; to brush
— v — m **ordinario** - to remove @ ordinary dust
— v **portadiferencial** m - [autom-mec] to remove @ differential carrier
— v **prensaestopa(s)** m - [mec] to remove @ packing box
— v **rápidamente** - to remove, quickly, o rapidly, o speedily
— v **rebaba(s)** f - [metal-prod] to remove @ burr • [sold] to remove @ flash
— v **recalcadura** f - [mec] to remove @ upsetting
— v — f **metálica** - [mec] to remove @ metal upsetting
— v **rejilla** f - [mec] to remove @ screen
— v **resguardo** m - [mec] to remove @ guard
— v **resistencia** f - [mec] to remove @ resistance • [electrón] to remove @ resistor
— v **resorte** m - [mec] to remove @ spring
— v **restricción** f - to remove @ restriction
— v **retén** m - [mec] to remove @ retainer
— v — m **para cierre** m - [mec] to remove @ seal retainer
— v — m — **sello** m - [mec] to remove @ seal retainer
— v **retenedor** m - [mec] to remove @ retainer
— v **rotor** m - [electr-mot] to remove @ rotor
— v **rueda** f - [mec] to remove @ wheel
— v — f **motriz** - [mec] to remove @ drive,ving wheel
— v **sección** f - [mec] to remove @ section
— v — f **de pieza** f - [mec] to remove @ part section
— v — f — **piso** m - [constr] to remove @ floor section
— v — f — **plancha** f - [mec] to remove @ plate section
— v — f — — f **para piso** n - [mec] to remove @ floor plate section
— v **segmento** m - to remove @ segment • [mec] to remove @ subrod
— v **sello** m - [mec] to remove @ seal
— v — m **para aceite** m - [mec] to remove @ oil, seal, o ring
— v **semieje** m - [autom-mec] to remove @ axle shaft
— v **separador** m - [mec] to remove @, spacer, o separator
— v — m **para armado,dura** - [mec] to remove @ assembling spacer
— v — m **para cojinete** m - [mec] to remove @ bearing spacer
— v **solenoide** m - [electr-instal] to remove @ solenoid
— v **sonda** f - [mec] to remove @ probe
— v — f **para uso** m **en obra** f - to remove @ field probe
— v **soporte** m - [mec] to remove @, support, o holder
— v **suciedad** f - to remove @ dirt
— v **suelo** m - [suelos] to remove @ soil
— v **sujetador** m - [mec] to remove @, fastener, o clamp
— v — m **para camisa** f - [mec] to remove @ liner clamp

remover v **suplemento** m - [mec] to remove @, insert, o supplement
— v — m **para válvula** f - [válv] to remove @ valve insert
— v **tapa** f - to remove @, cover, o top
— v — f **para colector** m - to remove @ sump cover
— v — f — **mesa** f - [mec] to remove @ table cover
— v — f — **transmisión** f - [mec] to remove @ transmission cover
— v — f — **válvula** f - [válv] to remove @ valve cover
— v — f **principal** - [mec] to remove @ main cover
— v **tapón** m - [mec] to remove @ plug
— v — m **para purga** f - [mec] to remove @ drain plug
— v — m — — f **para colector** m - [mec] to remove @ manifold drain plug
— v — m — f — — m **para aspiración** f - [mec] to remove @ suction manifold drain plug
— v — m — **drenaje** m - [mec] to remove @ drain plug
— v — m — **filtro** m - [mec] to remove @ filter plug
— v — m — **freno** m - [mec] to remove @ brake plug
— v — m — **nivel** m **de aceite** m - [mec] to remove @ oil level plug
— v — m — — m — — **en caja** f - [mec] to remove @ case oil level plug
— v — m — — m — — f — — f **para cadena** f - [mec] to remove @ chain case oil level plug
— v — m — **tubo(s)** m - [tub] to remove @ pipe plug
— v **tierra** f - [constr] to (re)move @, earth, o ground, o dirt
— v — f **ordinaria** - to remove @ ordinary dirt
— v **tornillo** m - [mec] to remove @ screw
— v — m **con casquete** m - [mec] to remove @ cap screw
— v — m — — m **para cubierta** f - [autom-mec] to remove @ cover cap screw
— v — n — — m — — f **para cojinete** m - [autom-mec] to remove @ bearing cover cap screw
— v — m — — m — — f — — m **para entrada** f **(de fuerza** f) - [autom-mec] to remove @ input bearing cover cap screw
— v **trampa** f - [mec] to remove @ trap
— v **trinquete** m - [mec] to remove @ ratchet
— v **tubería** f - [tub] to remove @ tubing
— v **tubo** m - [tub] to remove @ tube
— v **tuerca** f - [mec] to remove @ nut
— v — f **herrumbrada** - [mec] to remove @ rusty nut
— v — f **para árbol** m - [autom-mec] to remove @ shaft nut
— v — f — — m **para aportación** f **(de fuerza** f) - [autom-mec] to remove @ input shaft nut
— v — f — **engranaje** m - [mec] to remove @ gear nut
— v — f — — m **para impulsión** f - [mec] to remove @ drive,ving gear nut
— v — f — **espárrago** m - [mec] to remove @ stud nut
— v — f — **horquilla** f - [mec] to remove @ yoke nut
— v — f — — f **para aportación** f **(de fuerza** f) - [autom-mec] to remove @ input yoke nut
— v **unidad** f - [mec] to remove @ unit
— v **válvula** f - [válv] to remove @ valve
— v — f **inferior** - [válv] to remove @ lower valve
— v — f **superior** - [válv] to remove @ upper valve
— v **vástago** m - [mec] to remove @ rod • [válv] to remove @ stem
— v — m **intermedio** m - [mec] to remove @ intermediate rod
— v — m **para válvula** f - [válv] to remove @ valve stem

remover v vástago m para válvula f inferior -
[válv] to remove @ lower valve stem
— v — m — — f superior - [válv] to remove @
upper valve stem
removible a fácilmente - easily removable
— a mecánicamente - power removable
— a para limpieza f - [mec] removable for
cleaning
removido,da a - removed; slipped off; taken
away • [metal-prod] rabbled
— a completamente - removed completely
— a con cuidado - removed carefully
— a con esmeriladora f - [mec] grinder removed
— a — — f mecánica - [mec] electric, o power,
grinder removed
— a — extractor m - [mec] extractor, o puller,
removed
— f — lima f - [mec] file removed
— a — — f manual - [mec] hand file removed
— a — paño m - wiped (off)
— a continuamente - removed continually
— a convenientemente - removed conveniently
— a cuidadosamente - removed carefully
— a de cruceta f - [mec] removed from @ crosshead
— a — cajón m - [transp] removed from @, box,
o case, o container
— a — eje m - [mec] removed from @ axle
— a — motor m - [electr-mot] removed from @
motor • [comb.int] removed from @ engine
— a — orificio m - [mec] removed from @ hole
— a — prensaestopa(s) m - [mec] removed from @
packing box
— a — separador m—[mec] removed from @ spacer
— a — sobre - [mec] raised off of - removed
from @ envelope
— a de(sde) árbol m - [botán] removed from @
tree • [mec] removed from @ shaft
— a — m — aportación f (de fuerza f) -
removed from @ input shaft
— a — base f - [mec] removed from @ base
— a — caja f - [mec] removed from @, box, o
housing
— a — — f para eje m - [autom-mec] removed
from @ axle housing
— a — émbolo m - [mec] removed from @ piston
— a — varilla f - [mec] removed from @ rod
— a — — f para empuje m - [mec] removed from
@ push(ing) rod
— a deslizando - slipped, o slid, off
— a fácilmente - easily, o readily. removed
— a lentamente - removed slowly
— a mejor - removed, better, o best
— a para inspección f - removed for inspection
— a — limpieza f - [mec] removed for cleaning
— a periódicamente - removed periodically
— a rápidamente - removed, quickly, o speedily
remozado,da a - rejuvenated; freshened
remozamiento m - . . .; freshening
remozar v - . . .; to freshen
remuneración f - . . . . • [admin] employment
cost(s) • [labor] compensation
— f comparativa - [labor] comparative remunera-
tion
renaciente a - . . . . • [fig] burgeoning
rendición f - . . . . • [filol] translation
rendido,da a - . . . . • (sur)rendered; yielded •
[filol] translated • [ffs] returned; given
rendidor,ra a - [ind] hard working; yielding
rendija f - . . .; opening
— f grande - large crack
— f pequeña - small crack
rendimiento m - . . . . • rendering; giving •
[ind] output (rate); production; rendering •
giving • recovery • capacity • [fin] return •
[combusr] fuel rate
— m afectado - affected, yield, o performance
— m adecuado - adequate, yield, o performance
— m alto - high, yield, o performance • high
efficiency • [sold] high, efficiency rate, o
operating factor
— m — logrado - attained high yield
— m — — en procesamiento m - [ind] high yield

attained in @ processing
rendimiento m alto logrado en procesamiento m
de metal m - [metal-prod] high yield attained
in @ metal processing
— m — — m — — m fundido - [metalprod]
high yield attained in @ molten metal processing
— m apropiado - [ind] proper, o appropriate,
performance
— m aproximado - approximate yield
— m articulado - articulated yield
— m asegurado - assured, yield, o performance
— m atribuido - [labor] attributed yield
— m aumentado - increased, yield, o performance
— m bajo - low, yield, o efficiency
— m bruto - gross yield
— m bueno - good performance
— m con enfriamiento m natural - [electr] self-
-cooled rating
— m — tracción f delantera - [autom] front
drive performance
— m — velocidad f grande - high speed performance
— m confiable - reliable, o dependable, performance
— m continuo - [electr] continuous rating
— m contratado - contracted performance
— m de acero m fundido - [metal-prod] cast
steel yield
— n — artículo m - product performance
— m — bomba f - [bombas] pump, performance, o
yield, o duty
— m — carga f - [ind] burden yield
— m — cobre m - [miner] copper yield
— m — combustible - [combust] fuel yield
— n — combustión f - [combust] combustion
yield
— m — coque m - [coque] coke yield
— m — desbaste m - [metal-prod] ingot yield
— m — — m a lingote m - [metal-prod] bloom/
/ingot yield
— m — — m plano a lingote m - [metal-prod]
slab to ingot yield
— m — diamante m - diamond performance
— m — excitador,tatriz - [electr] exciter output
— m — fábrica f - [ind] plant production
— m — ferroaleación f - [metal-prod] ferralloy
yield
— m — gas m - [combust] gas yield
— m — — m de coque m - [metal-prod] coke gas
yield
— m — — m de coque m por tonelada f de carbón m - [coque] coke gas yield per ton of
coal
— m — generador m - [electr-prod] generator
yield
— m — grúa f - [grúas] crane performance
— m — — f sobre pedestal m - [grúas] pedestal crane, performance, o yield
— m — horno m - [ind] furnace, yield, o performance
— m — — m basculante - [metal-prod] tilting
furnace, yield, o performance
— m — — m eléctrico - [metal-prod] electric
furnace, yield, o performance
— m — — m fijo - [metal-prod] stationary furnace yield
— m — intercambiador m (para calor m) - heat
exchanger performance
— m — intercambio m sólido-gaseoso - [metal-
-prod] solid-gaseous interchange yield
— m — lingote(s) m - [metal-prod] ingot yield
— m — — m a desgaste m - [metal-lam] ingot-
-bloom yield
— m — máquina f - [ind] machine performance
— m — molino m - [miner] mill, o grinder, output, o production
— m — motón m - [grúas] block performance
— m — — para gancho m - [grúas] hook-block
performance
— m — motor m - [electr-mot] motor output •
[comb.int] engine output

rendimiento m de motor m diesel - [comb.int]
diesel engine performance
— m — — m unidad f motriz - [agric-equip]
power unit engine, yield, o performance
— m — neumático m - [autom-neumát] tire performance
— m — — m en carrera f - [autom-neumát]
tire('s) racing performance
— m — plancha f a lingotte m - [metal-prod]
slab to ingot yield
— m — planta f - [ind] plant, production, o
yield, o output
— m — pote m - [metalprod] pot yield
— m — — en caja f básica - [metal-prod] base
box pot yield
— m — producto m - [ind] product performance
— m — quemador m - [combust] burner output
— m — reducción f - [metal-prod] reduction
yield
— m — rodillo m - [metal-lam] roll life
— m — soldadora f - [sold] welder output
— m — soldadura f - [sold] welding output
— m — transformador m - [electr-prod] transformer rating
— m — tubería f - [tub] pipe,ping yield
— m — turboalimentador m - [comb.int] turbocharger performance
— m — vehículo m - [autom] vehicle performance
— m — ventilador m - [ind] fan, performance, o
yield
— m desbaste m a lingote m - [metal-lam] slab
to ingot yield
— m destacado - [ind] outstanding performance
— m — costa afuera - [petról] outstanding offshore poerformance
— m disminuido - [ind] diminishing,shed return
— m disponible - available output
— m — total - [ind] total available output
— m doble - [ind] double yield
— m eficiente - [ind] efficient performance
— m — desde punto m de vista f de costo(s) m -
[ind] cost efficient performance
— m elevado,dísimo - [ind] (very, o ultra),
high, yield, o performance, o efficiency
— m en carrera f - [autom-neumat] race,cing
performance
— m — coque m - [coque] coke yield
— m — escarpado m - [metal-lam] scarfing yield
— m — — m automático - [metal-lam] automatic
scarfing yield
— m — flujo m - [ind] flow yield
— m — general - overall, yield, o performance
— m — tijera f de desbastadora f - [metal-lam]
blooming shear yield
— m — trabajo m - [labor] job performance
— m — uso m - on-@-job performance
— m equitativo - [com] equitable return
— m especificado m - specified yield
— m específico - [ind] specified yield ·
[combut] specific fuel rate
— m establecido - established, yield, o performance, o rating
— m estimado - estimated, o assumed, yield
— m excelente - excellent, yield, o performance
— m — de motor m - [electr-mot] excellent motor performance · [comb.int] excellent engine
performance
— m — de motor m diesel - [comb.int] excellent
diesel engine performance
— m exceptional - exceptional, o outstanding, o
all-out, performance
— m exigible - requirable yield
— m fuera de planta f - [ind] outdoor, o out-
-of-plant, performance
— m fundición f a palanquilla f - [metal-prod]
melt to billet yield
— m importante - [ind] important, o optimum,
performance
— m incomparable - incomparable performance
— m intrínseco - intrinsic yield
— m ininterrumpido - [ind] uninterrupted, o
continuous, performance, o rating, o yield
— m interrumpido - interrupted, o discontinuous, yield, o production

rendimiento m intrínseco - [ind] intrinsic yield
— m — en reducción f - [metal-prod] reduction
intrinsic yield
— m máximo - maximum, yield, o performance ·
peak performance · maximum, duty, o rating ·
maximum output
— m — de bujía f (para encendido m) - [comb.-
-int] spark plug peak performance
— m — — f para chispa f - [comb.int] spark
plug performance
— m — — f — motor m - [comb.int] maximum
performance · [electr-mot] maximum motor performance
— m — — — m diesel - [comb.int] diesel engine maximum performance
— m — de quemador m - [combust] maximum burner
output; burner maximum output
— m — de ventilador m - [combust] maximum fan
yield
— m — garantizado - [sold] full rated output
— m — nominal - [sold] full rated output
— m mayor - greater, o higher, o better, output, o performance
— m medio - average efficiency · all day efficiency
— m medido - measured, yield, o performance
— m mejor - better performance · best performance
— m mejorado - improved performance
— m — de horno m - [ind] improved furnace performance
— m — — m eléctrico - [metal-prod] improved electric furmace performance
— m menor - less performance
— m metálico - [metal-prod] metallic yield
— m mínimo - minimum, yield, o output, o performance, o rating
— m — de motor m - [comb.int] minimum engine
performance · [electr-mot] minimum motor performance
— m — — — m diesel - [comb.int] minimum
diesel engine performance
— m — — quemador m - [combust] minimum burner
output
— m — — ventilador m - [combust] minimum fan
yield
— m muy elevado - very, o ultra-, high performance
— m neto - [ind] net, output, o yield
— m nominal - [electr-oper] rated, output, o
performance, o rating
— m — de soldadora f - [sold] welder rating
— m — disponible - available rated output
— m — — total - total avilable rated output
— m — establecido - established performance
rating
— m — total - total rated output
— m normal - normal yield
— m obtenido - obtained yield
— m óptimo - optimum, yield, o performance
— m — de intercambiador m de calor m -
heat exchanger optimum performance
— m — para neumático m - [autom-neumát] optimum tire performance
— m para acoplamiento m - [mec] coupling yield
— m — — m hidráulico - [mec] hydraulic coupling yield
— m — bomba f - [bombas] pump yield
— m — motor m - [electr-mot] motor yield ·
[comb.int] engine yield
— m — oferta f - [com] offer yield
— m — turbina f - [turb] turbine yield
— m — turbogenerador m - [electr-prod] turbogenerator yield
— m perdido - [ind] lost performance
— m pleno - full performance
— m ponderado - weighted yield
— m — para oferta f - offer weighted yield
— m — — turbina f - [turb] turbine weighted
yield
— m — para prueba f - test weighted yield
— m presumido - presumed, o assumed, yield
— m previsto - foreseen, o predicted, yield, o
performance

rendimiento m productivo - productive output
— m razonable - reasonable, o equitable, return
— m real - actual, o true, yield, o performance
— m real de tubería f - [tub] pipe actual yield
— — — — f para ensayo m - [tub] test pipe
actual yield
— m realzado - enhanced performance
— m reducido - reduced, o low, yield, o output,
o efficiency
— m relativo - relative, yield, o performance
— m sacrificado - [ind] sacrificed performance
— m satisfactorio - satisfactory performance
— m — para neumático m - [autom-neumát] satisfactory tire performance
— m sólido - solid, o rugged, performance
— m sumamente alto - ultra-high performance
— m superior - superior performance
— m volumétrico - volumetric, performance, o
efficiency
— m térmico - [combust] thermal, efficiency, o
yield, o performance
— n — alto - [termol] high thermal, efficiency
o yield
— m — bajo - [termol] low thermal, efficiency,
o yield
— m — bueno - [termol] good thermal, efficiency, o yield
— m — malo - [termol] bad, o poor, thermal
yield
— m — — para proceso m - process, bad,
o poor, thermal yield
— m — pobre - [termol] poor thermal yield
— m — — para proceso m - [ind] process poor
thermal, efficiency, o yield
— n total - total, yield, o output
rendir v - . . . . [filol] to translate • [fin]
to return • [educ] to take (@ exam)
— v beneficio m - [to render @ benefit • [fin]
to pay @ dividend
— v examen - [educ] to take @ test
— v servicio m - [mec] to (render @) service
renglón m - . . . .; row; range • area • [com]
commodity
— m amplio - [comb] broad line
— m completo - [com] full, o complete, o broad,
line, o assortment
— m — de accesorio(s) m - [ind] complete accessory,ries line
— m — para compañía f - [com] company complete
line
— m — — empresa f - [com] corporate, o company, complete line
— m — — sociedad f - corporate complete line
— m — clasificación H - [autom-neumát]
H-rated line
— m de accesorio(s) m - accesory,ries line
— m — camión(es) m - [autom] truck line
— f — — m liviano(s) m - [autom] light truck
line
— m — especialización f - [ind] specialization
line
— m — grúas f - [grúas] crane line
— m — máquinas f - [mec] machine(ry) line
— m — matriz,ces - [mec] die(s) line
— m — neumático(s) m - [autom-neumát] tire(s)
line
— m — pago m - [contab] pay item
— m — producción f - [ind] product(ion), line,
o item
— f — — — f de fabricante m - [ind] manifactu-
rer('s) product(ion) line
— m — producto(s) m - [ind] product(s) line •
product(s) scope
— m distintivo - distinctive line
— m especializado - [ind] specialized line
— m inferior - [imprent] bottom line
— m lucrativo - [com] profitable line
— m proyectado - [ind] designed line
— m superior - upper, o superior, line •
[imprent] top line
rengo,ga a - . . . .; lame
renguear v - to limp; to hobble
renguera f - limp(ing)

renivelación f - [constr] releveling
— f de, carril, o riel m - [f.c.] rail releveling
renivelado,da - [constr] releveled
renivelar v - [constr] to relevel
— v carril, o riel m - to relevel @ rail
renovación f - . . . .; renewal
— f de aceite m - [mec] oil, renewal, o change
— f — alcantarilla f - [constr] culvert, renewal,wing
— f — — f abovedada - [constr] arch culvert
renewal,wing
— f — asiento m - [mec] seat renewal,wing
— f — buje m - [mec] bushing renewal,wing
— f — — m para válvula f - [válv] valve bushing renewal,wing
— f — — m para vástago m - [petról] stem
bushing renewal,wing
— f — — m — m para válvula f - [petról]
valve stem bushing renewal,wing
— f — cierre m - [mec] lock renewal,wing
— f — — m para válvula f - [petról] valve
lock renewal,wing
— f — contrato m - [legal] contract renewal,wing
— f — disco m - [mec] disl renewal,wing
— f — personal m - [pers] (personnel) turnover
— f — resorte m - [mec] spring renewal,wing
— f — — m para válvula f - [mec] valve spring
renewal,wing
— f — válvula f - [válv] valve renewal,wing
— f — vástago m - [mec] stem, o rod, renewal
— f — — m para válvula f - [petról] valve
stem renewal,wing
— f — vía(s) f - [f.c.] track(s) renewal,wing
— f periódica - [periodic(al), renewal,wing, o
change,ging
— f prematura - premature, o early, renewal,wing
— f urbana - urban, renewal, o (red)development
renovado,da a - renewed; changed
— a periódicamente - changed periodically
renovar v - . . . .; to renew
— v aceite m - [mec] to renew @ oil
— v alcantarilla f - [constr] to renew, o replace, @ culvert
— v — f abovedada - [constr] to renew, o replace, @ arch culvert
— v contrato m - [legal] tp renew @ contract
— v periódicamente - to, renew, o change, periodically, o every so often
renta f bruta - [econ] gross income
— f de empresa f - [fin] corporate,tion income
— f declarada - [fisc] declared income
— f diversa - [contab] sundry income
— f interna - [pol] internal revenue
— f neta f - [econ] net income • net profit
— f sobre activo m neto - [fin] return on net
asset(s)
rentabilidad f - [econ] profitability
— f alentada - [econ] encouraged profitability
— f alta - [econ] high profitability
— f aumentada - increased profitability
— f baja - [econ] low profitability
— f de, compañía f, o empresa f, o sociedad f -
[fin] corporate, o company, profitability
— f posible - possible profitability • payback
potential
— f restituida - restored profitability
rentable a - [econ] revenue producing
rentas f generales - [fin] general income •
[fisc] operating budget
— f diversas - [contab] sundry income
renuevo m - [botán] . . .; sprig
renuncia f - . . . .; resigning; renunciation; re-
nouncement • relinquishing,shment • waiving;
waivering
— f de derecho(s) m - [legal] right(s) waiver
— f — fuero m - [judicial] jurisdiction relinquishing,shment
— f explícita - [legal] explicit waiver(ing)
— f gratuita - [legal] gratuitious waver(ing)
— f implícita - [legal] implicit waiver(ing)

renunciación f - resignation; resigning; relinquishing,shment; resignation
renunciado,da a - resigned; renounced; relin-
quished • waived; waivered
— a gratuitamente - gratuitiously waivered
renunciamiento m - renouncing,cement • relinquishing,shment
renunciar v - . . . . • to waive(r) • to part
— v derecho m - [legal] to waive @ right
— v fuero m - [judicial] to relinquish @ jurisdiction
— v gratuitamente - [legal] to gratuitiously
waiver
reñidamente adv - . . . . • [deport] hotly
reñido,da a - . . . . • [deport] heated
reolavador m - rheolaveur
reología f - rheology
reológico,ca a - rheological
reordenado,da - rearranged • realigned
reordenamiento m - rearranging • realigning •
realignment
— m de término(s) m - term(s) rearranging,gment
reordenar v . . .; to realign; to transpose
— v programa n - to rearrange @ schedule
— v término(s) m - to rearrange @ term(s)
reorganizado,da - reorganized • rearranged
reostato m distante - [electr-equip] remote
rheostat
— m eliminado - [electr-instal] eliminated, o
omitted, rheostat
— m inferior - [electr-equip] lower rheostat •
inferior rheostat
— m inspeccionado - [electr-instal] inspected
rheostat
— m limpiado - [electr-instal] cleaned rheostat
— m omitido - [electr-instal] omitted rheostat
— m para ajuste m - [electr-equip] adjustment
rheostat
— m — m de voltaje m - [electr-equip] voltage adjustment rheostat
— m — m — m en circuito m abierto -
[electr-equip] open voltage adjustment rheostat
— m — m, exacto, o preciso - [sold] fine
tuning rheostat; fine adjusting rheostat
— m — m — de amperaje m - [electr-equip]
current fine, o fine current, adjustment
rheostat
— m — m, — de voltaje m - [electr-equip]
voltage fine adjustment rheostat
— m — m — — m en circuito m abierto .
[electr-equip] open circuit voltage fine adjustment rheostat
— m para amperaje m - [electr-equip] amperage,
o current, rheostat
— m — avance m - [sold] travel rheostat
— m — gradual - [sold] inch rheostat
— m — campo m - [electr-equip] field rheostat
— m — — m para soldadora f - [sold] welder
field rheostat
— m — — m magnético - [electr-magnet] (magnetic field rheostat
— m — — m recalentado - [electr-oper] overheated field rheostat
— m — compensador m - [electr-equip] compensator rheostat
— m — — m para voltaje m - [electr-equip]
voltage compensator rheostat
— m — — m — — m en línea f - [electr-equip]
line voltage compensator rheostat
— m — criba f - [metal-prod] screen rheostat
— m — — f para coque m - [metal-prod] coke
screen rheostat
— m — fuente f (para energía f) - [electr-
-equip] power source rheostat
— m — regulación f - [electr-equip] control-
-(ling) rheostat
— m — — f continua (de corriente f) - [elect-
-equip] continuous, amperage, o current, current control rheostat
— m — — f de voltaje m - [electr-equip]
continuous voltage control rheostat
— m — — — f de amperaje m - [electr-equip] cur-

rent, control, o adjustment, rheostat
reostato m para regulación f de arco m - [sold]
arc control rheostat
— m — — f — velocidad f - [electr-equup]
speed (control) rheostat
— m — f — — f para avance m - [sold]
speed control rheostat
— m — selección f - [electr-equip] selector
rheostat
— m — soldadora f - [sold] welder rheostat
— m — soldadura f - [sold] welding rheostat
— m — soldar v - [sold] weld(ing) rheostat
— m — velocidad f - [electr-equip] speed
rheostat
— m — — f para alimentación f - [sold]
feed(ing) speed rheostat
— m — — f para alimentación f [de alambre m)
[sold[(wire) feed(ing) speed rheostat
— m — — — m gradual - [sold] inch(ing)
speed rheostat
— m — — f para dispositivo m para avance m -
[sold] travel speed rheostat
— m — voltaje m - [electr-equip] voltage rheostat • [sold] output control; job selector
— m — — m en circuito m abierto - [electr-
-equip] open circuit voltage rheostat
— m — — m — m en fuente f para energía f
[electr-equip] power source open circuit
voltage rheostat
— m — — m preciso - [sold] fine voltage rheostat
— m por agua m - [instrum] water rheostat
— m recalentado - [electr-oper] overheated
rheostat
— m regulador - [electr-oper] control(ling)
rheostat
— m — para voltaje m - [sold] job selector
rheostat
— m selector - [electr-equip] selector rheostat
— m — para voltaje(s) m - [sold] job selector
rheostat
— m superior - [electr-equip] upper rheostat •
superior rheostat
— m suprimido - [electr-instal] eliminated
rheostat
reoxidación f - reoxidation,dizing
reoxidado,da a - reoxidized
reoxidar v - to reoxidize
reparable a - . . .; repairable
reparación f - . . . • mend(ing) • repairing •
fixing • overhauling • remedy • piecing together • [metal-prod] relining
— f a nuevo - [ind] overhauling
— f — — de eje m - [mec] axle, o shaft, over-
haul(ing)
— f acelerada - [ind] speedy,ded repair
— f anormal - [ind] abnormal repair
— f con acero m - [metal-prod] steel abnormal, o abnormal steel, repair
— f básica - [mec] basic repair(ing) • [ind]
basic overhaul(ing)
— f de transportador m - [transp] carrier
basic overhaul(ing)
— f completa - [mec] complete repair(ing) •
[ind] complete overhaul(ing)
— f — de empaquetadura(s) f - [mec] gasket(s)
complete overhaul(ing)
— f — — — f para motor m - [comb.int] engine
gasket('s) complete overhaul(ing)
— f — — juego de empaquetaduras f - [ind]
gasket set complete overhaul(ing)
— f — — — m — — f para motor m - [comb.-
-int] engine gasket set complete overhaul(ing)
— f — — artesa f - [col.cont] tundish complete
repair,ing
— f — — transportador m - [mec] carrier complete overhaul(ing)
— f completada - [ind] completed, o final, re-
pair(ing)
— f con parch(es) m - patching; patch repairing
— f considerable - [ind] major, o extensive,
repair(ing)
— f de agujero(s) - [mec] hole(s) repair(ing)

reparación f de arrancador m - [comb.int] starter repair(ing)
— f — — m con arrollamiento m automático - [comb.int] rewind(ing) starter repair(ing)
— f — artesa f - [col.cont] tundish repair
— f — — f para colada f - [col.cont] tundish repair
— f — atracadero m - [naút] wharf, o mooring site, repair
— f — balanza f - [ind] scales repair
— f — bomba f - [bombas] pump repair
— f — — f para agua m - [bombas] water pump repair
— f — — — combustible m - [comb.int] fuel pump repair
— f — cable - [cabl] cable repair(ing)
— f — — m subterráneo - [electr-instal] underground cable repair(ing)
— f — canal m - [hidr] channel repair(ing) • [metal-prod] runner repair(ing)
— f — — m para colada f - [metal-prod] (tapping) runner repair(ing)
— f — carburador m - [comb.int] carburetor repair(ing)
— f — carretera f - [vial] road repair(ing)
— f — carrocería f - [autom] body repair(ing)
— f — cinta f - [metal-lam] strip repair
— f — — f transportadora - [ind] conveyor repair(ing)
— f — — f — con bandeja(s) f - [ind] skid conveyor repair(ing)
— f — — f — — patín(es) m - [ind] skid conveyor repair(ing)
— f — cojinete(s) m - [mec] bearing repair
— f — — m para motor m - [mec] motor bearing repair • [comb.int] engine bearing repair
— f — conductor m - [electr-instal] lead repair(ing)
— f — — m desconectado - [electr-instal] disconnected, o open(ed), lead repair(ing)
— f — conexión f - [electr-instal] connection repair(ing) • wiring repair(ing)
— f — contacto m - [electr-instal] contact repair(ing)
— f — — m para relé m - [electr-instal] relay contact repair(ing)
— f — convertidor m - [metal-prod] converter repair(ing)
— f — — m efectuado - [ind] performed converter repair(ing)
— f — — m programada m - [metal-prod] scheduled converter repair(ing)
— f — coraza f—[metal-prod] shell repair(ing)
— f — — f para crisol m - [metal-prod] hearth shell repair(ing)
— f — — f — cuba f - [metal-prod] hearth shell repair(ing)
— f — cubierta f - [mec] cover repair(ing)
— f → cuchara(s) f - [metal-prod] ladle repair
— f — chapa f - [mec] plate repair(ing)
— f — — f para crisol m - [metal-prod] hearth plate repair(ing)
— f — — f para crisól m - [metal-prod] hearth plate repair(ing)
— f — chapistería f - [autom-carroc] body repair(ing)
— f — — f para carrocería f - [autom-carroc] body repair(ing)
— f — defecto(s) m - defect(s) repair(ing)
— f — desacoplador m - [autom-mec] lockout repair(ing)
— f — desconexión f—[electr-instal] open(ing) repair(ing)
— f — dispositivo m - [mec] device, o unit, repair(ing), o overhaul(ing)
— f — — m eléctrico - [autom-mec] electric, device, o unit, repair(ing), o overhaul(ing)
— f — — m — para cambio(s) m - [autom-mec] electric shift(ing) unit, repair(ing), o overhaul(ing)
— f — — m neumático - [autom-mec] air, o pneumatic, device repair(ing)
— f — — m — para cambio(s) m - [autom-mec] air, o pneumatic, shift(ing) unit, repair-

-(ing) unit, repair(ing), o overhaul(ing)
reparación f de dispositivo m para cambio(s) m - [autom-mec] shift(ing) unit, repair(ing), o overhaul(ing)
— f — distribuidor m - [autom-mec] divider, repair(ing), o overhaul(ing)
— f — distribuidor m - [autom-mec] divider, repair(ing), o overhaul(ing)
— f — — m para potencia f - [autom-mec] power divider, repair(ing), o overhaul(ing)
— f — edificio(s) m - [constr] building repair(ing)
— f — eje m - [mec] axle, repair(ing), o overhaul(ing)
— f — — m motor - [autom-mec] drive,ving axle, o axle drive,ving, overhaul(ing)
— f — — m para impulsión f - [autom-mec] drive,ving axle overhaul(ing)
— f — emergencia f - [ind] emergency repair
— f — empaquetadura(s) f - [mec] gasket(s) overhaul(ing)
— f — — f para motor m - [electr-mot] motor gasket overhaul(ing) • [comb.int] engine gasket overhaul(ing)
— f — — f para prensa f para vástago m para campana f - [metal-prod] bell rod packing box repair
— f — engranaje(s) m - [mec] gear(s) repair
— f — equipo m—[ind] equipment, repair(ing), o overhaul(ing), o repair(ing)
— f — — m para amoladura f - [mec] grinding equipment repair(ing)
— f — — m — — f para rodillo(s) n—[metal-lam] roll grinding equipment repair(ing)
— f — — m — combustión f - [combust] combustion equipment repair(ing)
— f — — m — granallado m - [mec] shot blasting equipment repair(ing)
— f — — m — — m de rodillo(s) m - [metal-lam] roll shot blasting equipment repair
— f — — m para procesamiento m - [ind] processing equipment repair(ing)
— f — — m — — m térmico - [ind] heat processing equipment repair(ing)
— f — — m — producción f - [ind] production, o operating, equipment repair(ing)
— f — — m — rectificación f - [metal-lam] grinding equipment repair(ing)
— f — — m — — f de rodillo(s) m - [metal-lam] roll grinding equipment repair(ing)
— f — — m — torneado m - [mec] turning equipment repair(ing)
— f — — m — — m de rodillo(s) m - [metal-lam] roll turning equipment repair(ing)
— f — — m — trabajo m - [ind] work, o operating, equipment repair(ing)
— f — — m pesado - [ind] heavy equipment repair(ing)
— f — espectrómetro m - [instrum] spectrometer repair(ing)
— f — extremo(s) m - [mec] end(s) repair(ing)
— f — fuente f para abastecimiento m - [electr-distrib] power wupply repair(ing)
— f — — f — suministro m - [ind] power supply repair(ing)
— f — fuga f - leak repair(ing)
— f — grúa(s) f - [grúas] grane(s) repair
— f — herramienta(s) f - [ind] tool repair
— f — hierro m fundido - [sold] cast iron repair(ing)
— f — hormigón m - [constr] concrete repair
— f — horno m - [combust] furnace repair(ing)
— f — — m para acero m - [metal-prod] steel plant furnace repair(ing)
— f — impulsión f para eje m - [autom-mec] axle drive, repair, o overhaul(ing)
— f — instrumento m - [ind] instrument repair
— f — — m para combustión f - [combust] combustion instrument repair(ing)
— f — interruptor m - [electr-oper] breaker repair(ing)
— f — — m automático - [electr-instal] contactor repair(ing)
— f — — m — para relé m - [electr-equip]

relay contactor repair(ing)
reparación f **de juego** m - [mec] set repair(Ing)
— f —— m **de empaquetadura(s)** f - [mec] gasket set, repair(ing), o overhaul(ing)
— f —— m **para motor** m - [comb.int] engine gasket set, repair(ing), o over(hauling) • [electr-mot] motor gasket set repair(ing), o overhaul(ing)
— f — **línea** f - [electr-instal] line repair
— f —— f **conectada** - [electr-instal] closed line repair(ing)
— f —— f **desconectada** - [electr-instal] open line repair(ing)
— f — **lingotera(s)** f - [metal-prod] ingot mold repair(ing)
— f — **locomotora(s)** f - [f.c.] locomotive, repair(ing), o overhaul(ing)
— f — **mecanismo** m - [mec] mechanism, o unit, repair(ing)
— f —— m **eléctrico** - [mec] electric, mechanism, o unit, repair(ing)
— f —— m — **para cambio(s)** m - [mec] electric shift(ing) unit repair(ing)
— f — **motor** m - [electr-mot] motor repair(ing) • [comb.int] engine repair(ing)
— f — **neumático** m - [autom-neumát] tire repair(ing)
— f — **pared** f - [constr] wall repair(ing)
— f —— f **para horno** m - [combust] furnace wall repair(ing)
— f — **parte** f **oxidada** - [autom-carroc] rust repair(ing)
— f — **pieza** f - [mec] part repair(ing)
— f —— f **de aluminio** m - [mec] aluminum part repair(ing)
— f —— f **defectuosa** - [mec] defective part repair(ing)
— f —— f **para equipo** m - [ind] equipment part repair(ing)
— f — **piquera** t - [metal-prod] tap hole repair(ing) • iron notch repair(ing)
— f — **plancha** f - [mec] plate repair(ing)
— f — **portadiferencial** - [autom-mec] differential carrier, repair(ing), o overhaul(ing)
— f — **pozo** m - [hidr] well, repair(ing), o servicing
— f — **refractario(s)** m - [metal-prod] refractory,ries repair(ing)
— f — **rotura(s)** f - break(s) repair(ing)
— f — **solera** f - [metal-prod] floor repair
— f —— f **y cordón(es)** m - [metal-prod] fettling
— f — **soplante** m - [metal-prod] blower, repair(ing), o overhaul(ing)
— f — **superficie** f - surface repair(ing)
— f — **tajo** m - [miner] cut repair(ing)
— f — **tapa** f - [mec] cover repair(ing)
— f — **transportador** m - [transp] carrier, repair(ing), o overhaul(ing)
— f —— m **diferencial** - [mec] differential carrier, repair(ing), o overhaul(ing)
— f — m **con bandeja(s)** f - [ind] skid conveyor repair(ing)
— f —— m **con patín(es)** m - [ind] skid conveyor repair(ing)
— f — **vagón(es)** m **(ferroviario(s)** - [f.c.] (railway) car repair(ing)
— f — **vagoneta(s)** f - [ind] buggy repair(ing)
— f — **vagón(es)** m **ferroviario(s)** - [f.c.] (railway) car repair(ing)
— f — **válvula(s)** f - [válv] valve repair(ing)
— f —— f **neumática** - [válv] air valve repair(ing)
— f —— f **para aire** m - [valv] air valve repair(ing)
— f **después de colada** f - [metal-prod] after tapping repair(ing)
— f **efectiva** - effective repair(ing)
— f **efectuada** - [ind] made repair
— f **ejecutada** - [ind] performed repair
— f **electromecánica** - [ind] electro-mechanical
— f **en obra** f - [ind] on-the-job repair(ing)
— f — **caliente** - [ind] hot repair • repair without cooling @ furnace

reparación f **especial** - special repair(ing)
— f **eventual** - [mec] eventual repair(ing)
— f **extensiva** - [mec] extensive repair(ing)   •
— f **fácil** - easy repair(ing)
— f **ferroviaria** - [f.c.] railway repair(ing)
— f **fuera de garantía** f - out-of-guaranty, o out-of-warranty, repair(ing)
— f **general** - [ind] general repair(ing); major, o overhaul, repair(ing) • [metal-prod] general reline,ning
— f — **mayor** - [ind] major (general), repair, o overhaul(ing)
— f —— **para motor** m - [electr-mot] major motor overhaul(ing) • [comb.int] major engine overhaul(ing)
— f **grande** - [ind] large, o major, repair(ing), o overhaul(ing)
— f **hecha** - [ind] performed repair
— f **identificada** - identified repair
— f **libre de picadura(s)** f - [metal] pinhole-free repair(ing)
— f **mayor** - [ind] major, repair(ing), o overhaul(ing)
— f **mecánica** - [ind] mechanical repair(ing)
— f **mejor** - [ind] better, o improved, repair
— f **naval** - [nav] ship, o vessel, repair
— f **normal** - [ind] normal repair(ing)
— f — **después de colada** f - [metal-prod] normal after tapping repair
— f **óptima** - optimum, o premium, repair
— f **para eje** m **en tándem** - [autom-mec] tandem drive axle overhaul(ing)
— f — **mantenimiento** m - [ind] maintenance repair(ing)
— f — **servicio** m - [mec] routine service,cing
— f **parcial** - [ind] partial repair(ing)
— f **pequeña** - small, o minor, repair(ing)
— f **permanente** - permanent repair(ing)
— f — **para cadena** f - [cadenas] permanent chain repair(ing)
— f **programada** - [ind] scheduled repair • scheduled shutdown
— f **provisoria** - temporary repair)ing)
— f **rápida** - fast, o wuick, o speedy, repair
— f **sencilla** - [ind] simple repair(ing)
— f **temporaria** - temporary repair(ing)
— f — **para cadena** f - [cadenas] temporary chain repoair(ing)
— f **y mantenimiento** m - [ind] repair and maintenance
— f —— m **para edificio(s)** m - [ind] building(s) repair and maintenance
reparado,da a - repaired • restored • [metal-prod] relined • fixed • overhauled • pieced together
— a **a nuevo** - [mec] overhauled; repaired like new
— a **con parche(s)** m - repaired by patching; patch repair(ing)
— a **eventualmente** - repaired eventually
— a **fácilmente** - repaired easily
— a **fuera de garantía** f - repaired out of, warranty, o guaranty
— a **permanentemente** - repaired permanently
— a **provisoriamente** - repaired temporarily
— a **rápidamente** - repaired quickly
reparador m - . . . . • [ind] repairman; menderman
— m **para barrenos,nas** - [petról] bit dresser
— m — **cuchara(s)** f - [metal] ladle repairman
— m — **equipo(s)** m - [ind] equipment repairman
— m — **herramienta(s)** f - [mec] tool dresser
— f — **pieza(s)** f **para equipo** m - [ind] equipment part repairman
— m — **trépano(s)** m - [petról] bit dresser
reparar v - . . . ; to, mend, o fix, o overhaul • to patch; to piece together • to undergo @ repair • to service • [metal-prod] to reline
— v **a nuevo** - [ind] to overhaul
— v **agujero** m - to, repaint, o cover, @ hole
— v **arrancador** m - [mec] to repair @ starter
— v — m **con arrollamiento** m **automático** - [comb.int] to repair @ rewind starter
— v **atracadero** m - [náut] to repair @ wharf
— v **cable** m - [cabl] to repair @ cable

reparar v **cable** m **subterráneo** - [electr-instal]
to repair @ underground cable
— v **cadena** f - [cadenas] to repair @ chain
— v **carburador** m - [comb.int] to repair @ carburetor
— v **con parche(s)** m - to repair by patching
— v **conductor** m - [electr-instal] to repair @ lead
— v — m **conectado** - [electr-instal] to repair @, connected, o closed, lead
— v — m **desconectado** - [electr-instal] to repair a, disconnected, o open, lead
— v **conexión(es)** f - [electr-instal] to repair @, connection, o wiring
— v **contacto** m - [electr-instal] to repair @ contact
— v — m **para relé** m - [electr-instal] to repair @ relay switch
— v **convertidor** m - [metal-prod] to repair @ converter
— v **cucharón** m - [metal-prod] to repair @ ladle
— v **daño** m - to repair @ damage
— v **desconexión** f - [electr-instal] to repair @ open (connection)
— v **dispositivo** m - [mec] to repair @, device, o unit
— v — m **eléctrico** - [electr-instal] to repair @ electric, device, o unit
— v — m — **para cambio(s)** m - [mec] to repair @ electric shift(ing), device, o unit
— v **distribuidor** m - [autom-mec] to, repair, o overhaul, @ divider
— v — m **para fuerza** f - [autom-mec] to, repair, o overhaul, @ power divider
— v **eje** m—[mec] to, repair, o overhaul, @ axle
— v — m **motor** - [autom-mec] to overhaul @ drive,ving axle
— v — m — **en tándem** - [autom-mec] to, repair, o overhaul, @ tandem drive,ving axle
— v — m **para impulsión** f - [autom-mec] to, repair, o overhaul, @ drive,ving axle
— v **equipo** m - [ind] to repair @ equipment
— v — m **para procesamiento** m - [ind] to repair @ processing equipment
— v — m — m **térmico** - [ind] to repair @ heat processing equipment
— v **espectrómetro** m - [instrum] to repair @ spectrometer
— v **eventualmente** - [ind] tp repair eventually
— v **fácilmente** - [mec] to repair easily
— v **fisura** f - to repair @, fissure, o crack
— v **fuente** f **para suministro** m - to repair @ power, source, o supply
— v **fuera de garantía** f - to repair out of @, guaranty, o warranty
— v **fuga** f - to repair @ leak
— v **grieta** f - [mec] ro repair @ crack
— v **hierro** m - [sold] to repair @ iron
— v — m **fundido** - [sold] to repair @ cast iron
— v **horno** m - [combust] to repair @ furnace
— v **instrumento** m—[ind] to repair @ instrument
— v **interruptor** v - [electr-instal] to repair @ breaker
— v — m **automático** - [electr-instal] to repair @ contactor
— v — m — **para relé** m - [electr-instal] to repair @ relay contactor
— v **línea** f - [electr-instal] to repair @ line
— v — f **conectada** - [electr-instal] to repair @ closed line
— v — f **desconectada** - [electr-instal] to repair @ open line
— v **mecánicamente** - [ind] to make @ mechanical repair
— v **mecanismo** m - [electr-oper] to repair @ mechanism
— v — m **eléctrico** - [electr-instal] to repair @ electric mechanism
— v — m — **para cambio(s)** m - [autom-mec] to repair @ electric shift, unit, o mechanism
— v **motor** m - [electr-mot] to repair @ motor • [comb.int] to repair @ engine
— v **neumático** m - [autom-neumát] to repair @ tire

reparar v **pared** f - [constr] to repair @ wall
— f — f **para horno** m - [combust] to repair @ furnace wall
— v **parte** f - to repair @, part, o section
— v — f **oxidada** - [autom-carrocl] to repair @ rusty, part, o section, o area
— v **permanentemente** - to repair permanently
— v **pieza** f [mec] to repair @ part
— v — f **defectuosa** - [mec] to repair @ defective part
— v — f **para equipo** m - [ind] to repair @ equipment part
— v **portadiferencial** m - [autom-mec] to repair @ differential carrier
— v **rápidamente** - [ind] to repair quickly
— v **rotura** f - to repair @ break
— v **ruta** f - [metal-prod] to repair @ runner • [vial] to repair @ route
— v **sobre marcha** f - [ind] to repair (while) in operation
— v **soldadora** f - [sold] to repair @ welder
— v **soldadura** f - [sold] to repair @ weld
— v — f **defectuosa** - [sold] to repair @, defective, o damaged, weld
— v **solera** f - [metal-prod] to repair @ sill
— v — f **y cordón(es)** m—[metal-prod] to fettle
— v **soplante** f - [combut] to repair @ blower
— v **superficie** f - to repair @ surface
— v **tajo** m - [miner] to repair @ cut
— v **válvula** f - [válv] to repair @ valve
— v — f **neumática** - [válv] to repair @ air valve
— v — f **para aire** m - [válv] to repair @ air valve
reparo m - . . . . • [constr] protection; defense • comfort
— n de **cobertizo** m - [autom] garage('s) comfort
repartición f - . . . . • spreading • splitting • giving out • [pol] department; division; agency, branch • authority • véase también **organismo** m
— f de **carga** f - load distribution
— f **estatal** - [pol] state department
— f **federal** - [pol] federal department
— f **gubernamental** - [pol] government department • government authority
— f **municipal** - [pol] city department
— f **nacional** - [pol] federal, o national, department
— f **provincial** - [pol] provincial, o state, department
repartido,da a - distributed • spread • given out
repartir v - . . .; to, give, o spread, (out); to go around • to split
— v **carga** f - to distribute @ load
reparto m - . . . . • distributing,tion; giving out • [com] delivery
repasado,da a - gone over (again)
repasar v - . . .; to review; to go over (again)
— v a **fondo** m - to review in depth
repaso m - . . .; going over—[ind] overhaul(ing)
— m a **fondo** - in depth review
repatriación f de **capital** m - [fin] capital repatriation
repavimentación f - [constr] repaving; reflooring; resurfacing
— f de **fondo** m - [tub] invert repaving
— f — **puente** m - [constr] bridge, repaving, o reflooring, o resurfacing
repavimentado,da - a - [constr] repaved; resurfaced; reflooored
repavimentar v - [constr] to, repave, o refloor, o resurface
— v **fondo** m - [tub] to repave @ invert
— v **puente** m - [constr] to, repave. o refloor, o resurface, @ bridge (floor)
repecho m - . . .; incline; slope; ascending, o up grade
repelar v - . . .; to remove; to take out
— v **agua** m - to repel @ water
repelente a - . . .; repelling

repentino,na - . . .; abrupt
repercucisón f económica - [econ] economic, re-
percussion, o effect
repercutir v - . . .; to react; to reflect; to
re-echo
repertorio m - . . . • catalog; listing; inventory
repetibilidad f - repetitivity
repetición f - . . .; repeat(ing); doing again;
reiterating,tion; reaffirmation,ming; recurrence,cy; recurring
— f → alarma f - [segurid] alarm, repeat(ing),
o repetition
— f de condición f - condition, repetition, o
repeating
— f de operación f - [ind] operation, repeating, o repetition
— f de paso m - [ind] step, repeating, o repetition
— f — procedimiento m - [ind] procedure, repeating, o repetition
— f — proceso m - [ind] process, repeating, o
repetition
— f exacta - [ind] exact, repeating, o repetition, o duplicate,tion
— f promedia - average, repeating, o repetition
• mean recurrence,ring
repetidamente adv - . . .; continuously
repetido,da a - repeated • repetitive; reiterated; reaffirmed; done again • recurred,ring;
recurrent
— a frecuentemente - recurred frequently
repetidor m - [mec] repeater
— m para laminador m—[metal-lam] mill repeater
— m — m para barra(s) f - [metal-lam] rod,
o bar, mill repeater
repetidor,ra a - repeater,ting; repetitive
repetir v - . . .; to reiterate; to reaffirm •
to do again • to recur
— v condición f - to repeat @ condition
— v conservación f - to repeat @ maintenance
— v frecuentemente - to repeat frequently
— v operación f - to repeat @ operation • to
start over again
— v — f íntegra - to repeat a total operation;
to do it all over again
— v paso m - to repeat @ step
— v procedimiento m - to repeat @ procedure
— v proceso m - to repeat @ process
repetirse v frecuentemente - to recur frequently
repetitivo,va* a - véase repetido(s),da(s)
repillado* n - [constr] plaster
repisa f para sombrero(s) m - [domést] hat shelf
replanchar* v - [metal-lam] to level (again)
replanteado,da a - restated • reestablished
replanteador m - [constr] layout man
replantear v - . . . • to plan • to reestablish
replanteo m - . . . • reestablishing,shment •
[constr] layout; plotting
repleto,ta a . . . . • chock(ed) • full house
— a de holgorio m - fun filled
repliegue m de estrato m - [geol] anticline
reponer v - . . .; to, add, o refill • to put
back • to reinstate • to reposition • to, re-
plenish, o refill
— v aceite m - [mec] to replace @ oil; to refill with oil
— m cañón m - [armas] to replace @ barrel
— v cable m - [electr-instal] to replace @,
cable, o conductor, o lead
— v combustible m - [ind] to refuel
— v eje m - [mec] to replace @ axle • to reposition @ axle
— v electrodo m - [sold] to, replace, o change,
@, electrode, o rod
— v émbolo m - [mec] to replace @ piston
— v en servicio m - [ind] to put back into service
— v filtro m - [mec] to replace @ filter
— v guía(dera) f - [mec] to replace @ guide
— f — f para válvula f - [válv] to replace @
valve guide
— v mercadería f - [com] to replace @ merchandise

reponer v neumático - [autom-neumát] to replace @ tire
— v rápidamente - to replace quickly
— v tapón m - [mec] to replace @ plug
— v — m para drenaje m - [mec] to replace @
drain(age) plug
— v varilla f - [mec] to replace @ rod
report* m - . . .; véase informe m
reportero m - [public] journalist; reporter
— m escéptico - [public] skeptical reporter
reposadero m - . . . • [hidr] véase cámara f
desarenadora
reposado,da a - . . . • [metal-prod] killed
reposar v - . . . • [metal-prod] to kill
reposición f - . . .; replacing • replenishing
• reinstatement • resetting • repositioning
• adding • putting back • [electr-oper] re-
set(ting)
— f con costo m reducido - low cost replacement,
— f de aceite m - [mec] oil replacement,cing
— f — aro m - [mec] ring replacement,cing
— f — m con resorte m - [mec] snap ring
replacement,cing
— f — cañón m - [armas] barrel replacement •
barrel replacing
— f — eje m - [mec] axle replacement,cing •
axle repositioning
— f — electrodo m - [sold] electrode, o
stick, change,ging, o replacement,cing
— f — m manual - [sold] stick replacement,cing
— f — émbolo m - [mec] piston, replacement, o
replacing
— f — filtro m - [mec] filter replacing
— f — gas m - [combust] gas restoring,ration
— f — guía(dera) f - [válv] guide replacement,cing
— f — f para válvula f - [válv] valve
guide replacement,cing
— f — mercadería f - [com] merchandise replacement,cing
— f — pieza f - [mec] part replacement,cing
— f — tapón m - [mec] plug replacement,cing
— f — m para drenaje m - [mec] drain(age)
plug replacement,cing
— f — m — purga f - [mec] drain plug replacement,cing
— f económica - inexpensive replacement,cing
— f eléctrica - [electr-oper] electric re-
set(ting)
— f en servicio m - [ind] replacing, o putting
back, into service
— f fácil - [mec] easy replacement,cing
— f para varilla f - [sold] rod replacement
— f — tapón m - [mec] plug replacement
reposicionamiento m - [electr-oper] reset(ting)
reposo m - . . .; resting; leisure • [ind]
holding
— m de acero m - [metal-prod] (steel) holding
— m — suelo m - [suelos] soil repose,sing
— m — tren m - [metal-prod] track time
— m natural - natural, repose, o rest(ing)
— m — de suelo m - [suelos] natural soil, o
soil natural, repose
repostado m - [comb.int] refueling
repostado,da a - [comb.int] refueled
reprender v - . . .; to rebuke
represión f - . . .; rebuke
represa f - [hidr] . . .; reservoir • damming
— f de tierra f - [hidr] earth(en) dam •
[petról] earth(en) sump
— f — f para petróleo m - [petról] earthen
sump
represado,da a - [hidr] dammed
representación f - . . .; representing • [com]
dealership; agency • [pol] serving
— f adecuada - adequate representation
— f con éxito m - [com] profitable, o successful, dealership
— f esquemática - schematic representation
— f gráfica - graphic representation; photograph • [mat] plot(ting) • [dib] curve

representación f ininterrumpida - uninterrupted, o continuous, picture
— f — de zona f - area continuous picture
— f — — f para soldadura f - [sold] welding area continuous picture
— f laboral - [labor] labor representation
— f legal - [legal] legal representation; counsel
— f obrera - [labor] labor representation
— f política - [pol] political representation (allowance)
— f profesional - [legal] professional representation
— f sobre tablado m - [teatro] stage show
— f visual - visual representation • [comput] display
— f — para sistema n - [comput] system display
representado m - [legal] principal(s) - [com] supplier
representado,da a - represented • [pol] served
— a adecuadamente - adequately represented
— a gráficamente - [mat] plotted
— a legalmente - [legal] legally represented
— a predominantemente - predominantly represented
— a visualmente - represented visually; pictured - [comput] displayed
representante m - . . . . • [com] agent; dealer; (local) representative
— n autorizado—[com] authorized representative
— m calificado - [com] qualified, representative, o dealer
— m capacitado - [com] qualified, representative, o dealer
— m competente - [com] competent, o qualified, representative, o dealer
— m común - common representative • joint representative
— f de comprador m - purchaser('s) representative
— m — contratista m - contractor('s) representative
— m — — m en obra f - [constr] job superintendent
— m — departamento m - department representative
— m — — m de refractario(s) m - [ind] refractory,ries department representative
— m — distribuidor m - distributor('s) representative
— m — — m para servicio m - distributor('s) service representative
— m — empleado(s) m - [labor] employee(s) representativew
— m — empleador(es) m - [labor] employer(s) representative
— m — empresa f - [legal] corporation('s) representative
— f — — f siderúrgica - [metal-prod] steel company representative
— m — medio(s) m informativo(s) - [public] media, representative, o member
— m — personal m - [labor] employee(s), o personnel, representative
— m — planta f - [metal-prod] mill representative • [ind] plant representative
— m — prensa f - [public] press (representative)
— f — propietario m - proprietor('s) representative
— m — — m a cargo (directamente) de obra f - [ind] resident engineer
— m exclusivo - exclusive representative
— m — para exportación f - [com] exclusive export representative
— m — — importación f - [com] exclusive import representative
— m legal - [legal] legal representative
— m — para proponente m - [legal] bidder('s) legal representative
— m local - [com] local, representative, o dealer
— m — para servicio m - [ind] local service

representative
representante m para inspección f - inspection representative
— m — servicio m - [ind] service, representative, o personnel • [com] service dealer
— m — — m para inspección f - [ind] inspection service representative
— m — — m — — f para comprador m - [com] purchaser('s) inspection service representative
— m — venta(s) f - [com] sale(s), representative, o person(nel)
— m — zona f - [com] area, o local, representative
— m patronal m - [labor] employer('s) representative
— m posible - [com] prospective, o possible, dealer, o representative
— m regional - [com] regional representative
— m titular - [com] titular representative
representar v - . . . • [pol] to serve
— v adecuadamente - to represent adequately
— v graficamente - to represent graphically • to photograph • [mat] to plot
— v legalmente - to represent legally
— v visualmente - to represent visually • [comput] to display
represión f - . . . • containing,nment
— f de contaminación f - [ambient] contamination containment
— f — — f de ambiente m - [ambient] environmental contamination containment
— f — — f — medio m ambiente m - [ambient] environmental contamination containment
reprimido,da a - repressed; contained
reprimir v - . . .; to contain
— v contaminación f - to, repress, o contain, @ contamination
— v — f de ambiente m - [ambient] to, repress, o contain, @ environmental contamination
— v — f de medio m ambiente m - [ambient] to, repress, o contain, @ environmental contamination
reprobar v - . . . . [educ] to, fail, o flunk
— v examen m - [educ] to, fail, o flunk, @ test
reprocesado,da a - [ind] reprocessed; reworked recycled
reprocesadora f - [ind] reprocessor • recycler
— f para asfalto m - [constr] asphalt recycler
reprocesamiento m - . . . . • rework(ing); recycling
— m de acero m - [metal-prod] steel recycling
— m asfalto m - [constr] asphalt recycling
— m — combustible m - [nucl] fuel recycling
— m — — m agotado - [nucl] spent fuel recycling
— m — material(es) m - [ind] material(s), reprocessing, o recirculating,tion, o recycling
reprocesar v - [ind] to reprocess • to rework • to recycle
— v acero m - [metal-prod] to recycle @ steel
— v asfalto m - [constr] to recycle @ asphalt
— v material(es) m - [ind] to, reprocess, o recirculate, o recycle, @ material(s)
reproducción f - . . .; reproducing; duplicating; duplication • replica • [imprent] reprint--(ing) • [electrón] displaying
— f anterior - previous reproducing,ction • [electrón] previous display(ing)
— f con precisión f - [electrón] precision reproducing,ction
— f continua - [electrón] continuous, reproducing,ction, o display(ing)
— f de condición f - condition, reproducing, o reproduction, o duplicating, o duplication
— f — — f de carga f - [constr] load condition reproducing,ction
— f — — f — — f en obra f - [constr] field load condition reproducing,ction
— f — — f real - actual, o true, condition reproducing,ction
— f — — f — de carga f - [constr] actual load condition reproducing,ction

reproducción f **de condición** f **real de carga** f - [constr] actual load condition reproduction

reproducción f **de condición** f **real de carga** f **en obra** f - [constr] actual field load condition reproducing,ction

— f —— f — **en obra** f - [constr] actual field condition reproducing,ction

— f — **estado** m - [electrón] status display

— f —— m **de alarma** f - [electrón] alarm status display(ing)

— f — **señal** f - [electrón] signal display(ing)

— f —— f **actual** - [electrón] current signal display(ing)

— f —— f **anterior** - [electrón] previous signal display(ing)

— f —— f **corriente** - [electrón] current signal display(ing)

— f —— f **previa** - [électrón] previous signal display(ing)

— f —— f **reciente** - [Electrón] recent signal display(ing)

— f **en miniatura** - (miniature) replica

— f — **tamaño** m **menor** - (small scale) replica

— f — **pantalla** f—[electrón] tube display(ing)

— f —— **con tubo** m **para rayo(s)** m **catódico(s)** m - [electrón] cathode ray(s) tube display(ing)

— f —— m **para rayo(s)** m **catódico(s)** - [electrón] cathode ray tube display(ing)

— f **exacta** - exact reproduction • duplication

— f **previa** - [electrón] previous display(ing)

— f **radiográfica** - [electrón] radiogrpahic, reproduction, o picture

reproducibilidad f - reproducibility

reproducible a - [dib] reproducible

reproducido,da a - reproduced • duplicated • [electrón] displayed

— a **anteriormente** - [electrón] displayed previously

— a **con precisión** f - [electrón] reproduced precisely

— a **continuamente** - reproduced continuously • [electrón] displayed continuously

— a **en pantalla** f - [electrón] displayed

— a —— f **con tubo** m **para rayo(s)** m **catódicos** - [electrón] displayed on @ cathode ray tube

— a — **tubo** m **para rayo(s)** m **catódico(s)** - [electrón] displayed on @ cathode ray tube

— a **previamente** - previously reproduced • [electrón] displayed previously

reproducir v - . . . • to duplicate • [imprent] to reprint • [electrón] to display

— v **con precisión** f - to reproduce precisely • [electrón] to display precisely

— v **condición** f - to, reproduce, o duplicate, @ condition

— f — f **de carga** f - [constr] to reproduce @ load condition

— v — f —— f **en obra** f - [constr] to reproduce @ field load condition

— f — **real** - to reproduce @ actual condition

— v — f — **de carga** f - [constr] to reproduce @ actual load condition

— v — f —— f **en obra** f - [constr] to reproduce @ actual field load condition

— f — f — **en obra** f - [constr] to reproduce @ actual field condition

— v **continuamente** - [electrón] to reproduce continuously

— v **en pantalla** f - [electrón] to display on @ screen

— v —— f **con tubo** m **para rayo(s)** m **catódicos** - [electrón] to display on @ cathode ray tube

— v — **tubo** m - [electrón] to display on @ tube

— v —— m **para rayo(s)** m **catódico(s)** - [electrón] to display on @ cathode ray tube

— v **señal** f - [electrón] to display @ signal

— v — f **actual** - [electrón] to display @, present, o current, signal

— v — f **anterior** - [électrón] to display @ previous signal

— f — f **corriente** - [electrón] to display @ current signal

reproducir v **señal** f **previa** - [electrón] to, reproduce, o display, @ previous signal

— v — f **reciente** - [electrón] to, reproduce, o display, @ recent signal

reproductor m **de plano(s)** m - [dib] blueprint reproducer

reprogramación f - [comput] reprogramming

— f **de computador** m - [comput] computer reprogramming

reprogramar v - [comput] to reprogram

— v **computador** m - [comput] to re-program @ computer

reproyección * f - redesigning

reproyectado,da a - redesigned

reproyectar v - to redesign

República f **de Africa** m **de Sur** - [pol] Republic of South Africa

repudiado,da a - repudiated

repudiar v **contrato** m - [legal] to repudiate @ contract

repudio m **de contrato** m - [legal] contract repudiation

repuesto m - [mec] (spare, o repair,) part; replacement • véase también **reemplazo** m | a - (de) spare; reserve; replacement; stand-by

— m **adquirido** - [ind] acquired, o purchased, spare (part)

— m **aprobado** - [mec] approved spare (part)

— m **apropiado** - appropriate, o suitable, spare (part), o component

— m **autorizado** - [ind] authorized (spare) (part)

— m — **por fábrica(nte)** - [ind] factory authorized (spare) part

— m **cargado** - [transp] loaded spare (part)

— m **codificado** - [ind] coded spare (part)

— m **crítico** - [ind] critial spare (part)

— m **para grúa** f - [grúas] crane critical spare (part)

— m —— **laminador** m - [metal-lam] mill critical spare (part)

— m —— — m **para chapa(s)** f - [metal-lam] strip mill critical spare (part)

— m —— — m —— f **en caliente** - [metal-lam] hot strip mill critical spare (part)

— m —— — m —— f **frío** - [metal-lam] cold strip mill critical spare (part)

— m —— **puesta** f **en marcha** f - [ind] critical start-up spare (part)

— m —— f —— f **de grúa(s)** f - [grúas] crane(s) start-up critical spare (part)

— m —— —— **laminador** m - [metal-prod] mill start-up critical spare (part)

— m —— f —— f—— m **para chapas** f - [metal-lam] strip mill start-up critical spare (part)

— m —— f —— f —— m —— f **en caliente** - [metal-lam] hot strip mill start-up critical spare (part)

— m —— f —— f —— f — **frío** - [metal-lam] cold strip mill start-up critical spare (part)

— m **de acero** m - [ind] steel (spare) (part)

— m —— m **con manganeso** m - [mec] manganese steel spare (part)

— m —— m **manganésico** - [mec] manganese steel spare (part)

— m — **estelita** f - [ind] Stellite spare (part)

— m —— f **reconstruido** - [ind] Stellite rebuilt spare (part)

— m — **importación** f - [ind] imported spare (part)

— m —— f **adquirido** - [ind] acquired, o purchased, imported spare (part)

— m **descargado** - [transp] unloaded spare (part)

— m **eléctrico** - [electr-equip] electrical spare (part)

— m **específico** - [ind] specific spare (part)

— m **forjado** - [ind] forged (spare) part

— m **hallado** - [ind] found spare (part)

— m **ilustrado** - [ind] illustrated spare (part)

— m **importado** - [ind] imported spare (part

— m — **adquirido** m - [ind] purchased imported

spare (part)
repuesto m intercambiable - [mec] interchangeable
spare (part)
— m mecánico - [ind] mechanical spare (part)
— m — específico - specific mechanical spare
(part)
— m — universal - [ind] universal mechanical
spare (part)
— m nacional - [ind] local(ly manufactured)
spare (part)
— m normal - [ind] normal spare (part)
— m ordenado - [ind] ordered spare (part)
— m — eficientemente - [ind] efficiently ordered spare (part)
— m para acoplado m - [autom-mec] trailer spare
(part)
— m — automotor m - [autom] automotive spare
(part)
— m — automóvil m - [autom] automobile spare
(part)
— m — bomba f - [bombas] pump spare (part)
— m — — f con engranaje(s) m - [bombas] gear
pump spare (part)
— m — — f — — m para aceite m - [bombas]
oil gear pump spare (parat)
— m — — f para extremo m hidráulico—[bombas]
fluid end pump spare (part)
— m — — f para inyección f - [petról] slush
pump (spare) part
— m — — m — lodo m - [petról] slush pump
(spare) part
— m — boquilla f - [sold] nozzle (spare) part
— m — borne m - [electr-instal] stud (spare)
part
— m — — m para salida f - [electr-instal]out-
put stud (spare) part
— m — centro m para regulación f - [electr-
instal] control center spare (part)
— m — — — — f para motor(es) m - [electr-
-instal] motor control center spare (part)
— m — conmutador m - [electr-instal] switch
(spare) part
— m — — m para polaridad f (de arco m) -
[electr-instal] arc polarity switch (spare)
part
— m — computador m - [comput] computer spare
(part)
— m — conservación f - [ind] maintenance spare
(part)
— m — — f de equipo m - [ind] equipment maintenance spare (part)
— m — distribuidor m - [metal-prod] distributor spare (part)
— m — — m para horno m alto - [metal-prod]
blast furnace distributor spare (part)
— m — — rotativo - [metal-prod] rotary distributor spare part
— m — — m rotativo para horno m alto - [metal
-prod] blast furnace rotary distributor spare
(part)
— m — equipo m - [ind] equipment spare (part)
— m — extremo m hidráulico - [petról] fluid
end (spare) part
— m — grúa f - [grúas] crane spare (part)
— m — — f aérea - [grúas] overhead crane
spare (part)
— m — horno m - [combust] furnace spare (part)
— m — — m alto - [metal-prod] blast furnace
spare (part)
— m — instrumental - [instrum] instrument(s)
spare(s)
— m — interruptor m - [electr-instal] switch
(spare) part
— m — — m automático - [electr-instal] contactor (spare) part
— m — laminador m - [metal-lam] mill spare
(part)
— m — — m para chapa(s) f - [metal-lam] strip
mill spare (part)
— m — — f — — f en caliente - [metal-lam]
hot strip mill spare (part)
— m — — m — — f — frío - [metal-lam] cold
strip mill spare (part)

repuesto m para laminador m en frío - [metal-
-lam] cold mill spare (part)
— m — línea f - [ind] line spare (part
— m — — f para estañadura f - [metal-trat]
tinning line spare (part)
— m — — f — — f electrolítica - [metal-
-trat] electrolytic tinning line spare (part)
— m — mantenimiento m - [ind] maintenance
spare (part)
— m — — m para equipo m - [ind] equipment
maintenance spare (part)
— m — máquina f - [ind] machine spare (part)
— m — — f calibradora - [ind] sizing machine
spare (part)
— m — — f forjadora f - [mec] forging machine spare (part)
— m — medidor m para rayos-X - [electrón]
X-ray meter spare (part)
— m — motor m - [electr-mot] motor spare
(part) • [comb.int] engine spare (part)
— m — operación f - [ind] operating,tion
spare (part)
— m — — f para equipo m - [ind] equipment
operating,tion spare (part)
— m — parte f hidráulica - [petról] fluir end
spare (part)
— m — portaescobilla(s) m - [electr-mot]
brushholder spare (part)
— m — — m con anillo m para ajuste m -
[electr-mot] slip ring brushholder (spare)
part
— m — prensa f - [mec] press spare (part)
— m — — f para extrusión f - [mec] extruding
press spare (part)
— m — — — — m para electrodo(s) m -
[sold] electrode extruding press spare (part)
— m — — f — planta f para electrodo(s) m -
[sold] electrode plant extruding press spare
(part)
— m — programador m - [metal-lam] programmer
spare (part)
— m — — m para pasada(s) f - [metal-lam]
pass(es) programmer spare (part)
— m — puerta f - [mec] door spare (part)
— m — puesta f en marcha f - [ind] start-up
sapre (part)
— m — — f — — f para laminador m - [metal-
-lam] mill start-up spare (part)
— m — — f — — f — — m para chapa(s) f -
[metal-lam] strip mill start-up spare (part)
— m — — f — — f — — m — — f en caliente
- [metal-lam] hot strip mill start=up spare
(part)
— m — — f — — f — — m — — f en frío -
[metal-lam] cold strip mill start-up spare
(part)
— m — regulador m - [mec] controller spare
(part)
— m — — m para marcha f sin carga f - [mec]
idler (spare) part
— m — remolque m - [autom-mec] trailer (spare)
part
— m — rodillo(s) m - [metal-lam] roll(s)
spare (part)
— m — — m deflector(es) - [metal-lam] deflector roll(s) spare (part)
— m — — m — en línea f para estañadura f -
[metal-trata] tinning line defelctor roll
spare (part)
— m — — — — — f — — f electrolítica -
[metal-trat] electrolytic tinning line deflector roll spare (part)
— m — rotametro m - [instrum] rotameter spare
— m — sistema m - [ind| sustem spare (part)
— m — — m de polea(s) f - [mec] pulley system
spare (part)
— m — — m — — f para carro m - [metal-prod]
car puelly system spare (part)
— m — — m — — f — — m para montacarga(s)
m - [metal-prod] skip car pulley system spare
(part)
— m — — m — — f — — m — — m para horno
m - [metal-prod] skip car pulley system spare

(part)

repuesto m **para sistema** m **de poleas** f **para carro**
m **para montacarga(s)** m **para horno** m **alto** -
[metal-prod] blast furnace skip car pulley
system spare (part)
— m —— m —— f **para montacarga(s)** m -
[metal-prod] skip pulley system spare (part)
— m —— m —— f **para horno** m -
[metal-prod] furnace skip pulley system spare
(part)
— m —— m —— f —— m —— m **alto** -
[metal-prod] blast furnace skip pulley system
spare (part)
— m —— m —— f **para vagoneta** f - [metal-
-prod] car pulley system spare (part)
— m —— m —— f —— f **para montacarga(s)**
m - [metal-prod] skip car pulley system spare
(part)
— m —— m —— f —— f —— m **para horno**
- [metal-prod] furnace skip car pulley system
spare (part)
— m —— m —— f —— f —— m —— m **al-
to** - [metal-prod] blast furnace skip car pulley system spare (part)
— m —— m **para regulación** f - [ind] control
system spare (part)
— m —— m —— f **de demanda** f—[electr-prod]
demand control system spare (part)
— m —— m —— f —— f **de usina** f— -
[electr-prod] power plant demand control system spare (part)
— m — **soldadora** f - [sold] welder (spare) part
— m — **soldadura** f—[sold] welding (spare) part
— m —— f **por arco** m - [sold] arc welding
spare (part)
— m — **tablero** m - [electr-instal] panel spare
(part)
— m —— m — **distribución** f - [electr-instal]
distribution panel spare (part)
— m —— m —— f **de tensión** f **alta**—[electr-
-prod] high voltage distribution panel spare
(part)
— m —— m —— f —— f **baja** - [electr-
-prod] low voltage distribution panel spare
(part)
— m —— m —— f — **tensión** f **mediana** -
[electr-prod] medium voltage distribution
panel spare (part)
— m —— m —— f **de voltaje** m—[electr-prod]
high voltage distribution panel spare (part)
— m —— m —— f —— m **bajo** -
[electr-prod] low voltage distribution panel
— m —— m —— d —— m **medio,diano** -
[electr-prod] medium voltage distribution panel
— m — **tren** m **para laminación** f - [metal-lam]
rolling mill spare (part)
— m —— m —— f **de chapa(s)** - [metal-lam]
stip mill spare (part)
— m —— m —— f —— f **en caliente** -
[metal-lam] hot strip mill spare (part)
— m —— m —— f **en frío** - [metal-prod] cold
strip mill spare (part)
— m **pedido** m-[ind] ordered, spare (part), o replacement (part)
— m — **eficientemente** - [ind] efficiently ordered spare (part)
— m **recomendado** - [ind] recommended spare part
— m **rechazado** - [ind] rejected spare (part)
— m **sangrado** - [imprent] indented, o bled,
spare (part)
— m **solicitado** - [mec] requested spare (part)
— m **suministrado** - [mec] provided, o supplied,
spare (part)
— m **universal** - [mec] universal spare (part)
— m **usado** - [ind] used spare (part)
— m **vario** m - [ind] miscellaneous, o sundry,
spare (part)
repuesto,ta a - replaced • replenished • reposi-
tioned • put back • added • refilled
— a **en servicio** m - [ind] replaced, o put back,
into service
— a **rápidamente** - replaced quickly

repujado m - [mec] embossing • hammering •
drawing
— m **de indicador** m - [constr-vial] sign embossing
— m — **letrero** m - [constr-vial] sign embossing
— m **profundo** - [metal-fabr] deep drawing
repujado,da a - [mec] embossed • hammered
repujar v - [mec] to emboss • to hammer
— v **indicador** m - [constr-vial] to emboss @
sign
— v **letrero** - [constr-vial] to emboss @ sign
repulpadora f - [miner] repulper
repuntado,da a - . . . . • spurted
repuntar v - . . . . • to spurt
repunte m - . . . . • spurt(ing)
reputación f **considerable** - fair reputation
— f **granjeada** - earned, o won, reputation
— f **inigualada** - unequalled, o unmatched, reputation
— f **prestigiosa** - prestigious reputation
— f **prohibitoria** - forbidding, o prohibiting,
reputation
requebrado,da a - complimented
requerido,da a - required • needed; called for •
taken
— a **acostumbradamente** - customarily, o usually,
required
— a **para todo equipo** m - [mec] required for
all, equipment, o machine(s)
— a **por ley** f - required by law
— a **previamente** - previously required
requerimiento m - . . . .; requirement • need •
request • calling for • taking • véase también **requisito** m • **exigencia** f
— m **calórico** - [termol] caloric requirement
— m **cumplido** - met requirement
— m **de agua** m - [hidr] water requirement
— m —— m **de río** m - [hidr] river water requirement
— m —— m **industrial** - [hidr] industrial
water requirement
— m **asesoramiento** m - advice requirement
— m — **atención** f - [ind] servicing requirement
— m — **demanda** f - [com] demand requirement
— m —— f **interna** - [com] internal demand
requirement
— m —— f **nacional** - [com] national, o internal, o local, demand requirement
— m — **energía** f - [electr-prod] energy, o
power, demand requirement
— m — **exportador** m - exporter('s), need, o requirement
— m — **equipo** m - [ind] equipment requirement
— m — **fuerza** f - [ind] energy, o power, re=
quirement
— m — **grasa** f—[mec] grease requirement
— m — **inspección** f inspection requirement
— m —— f **periódica** - periodic inspection requirement
— m — **limpieza** f - cleaning requirement
— m — **lubricación** f - [ind] lubrication requirement
— m —— f **periódica** - [ind] periodic lubrication requirement
— m — **papel** m - [ind] paper requirement
— m — **par motor** m - [mec] torque requirement
— m — **servicio** m - [ind] service, requirement,
o need
— m — **tiempo** m - time requirement
— m **específico** - specific requirement
— m — **cumplido** - met specific requirement
— m — **de servicio** m - specific service, requirement, o need
— m **exigente** - demanding requirement
— m **general** - general requirement
— m **magnético** - magnetic requirement
— m **máximo** - maximum requirement
— m **mínimo** - minimum requirement
— m **municipal** - [pol] municipal, o city, requirement
— m **para conmutador** m - [electr-oper] switching
requirement

requerimiento m para derivación f - [electr-
-oper] shunting requirement
— **m** — **dimensión(es) f** - size requirement
— **m** — **entrada f** - input requirement
— **m** — **interruptor m** - [electr-oper] switching
requirement
— **m** — **salida f** - output requirement
— **m previo** - previous, o prior requirement
— **m químico** - [quím] chemical requirement
requerimientos m desglosados - broken down requirements
requerir v - . . . . • to, claim, o call for • to
take
— **v acostumbradamente** - to usually require
— **v ajuste m** - to require @ adjustment
— **v asesoramiento m** - to require @ advice
— **v atención f** - to require @ attention • [mec]
to require @, maintenance, o servicing
— **v cinta f** - to require @, ribbon, o strip
— **v corrección f** - [sold] to require @, repair,
o correction
— **v energía f** - to require @, energy, o power
— **v engrase m** - [mec] to require grease,sing
— **v equipo m** - to require @ equipment
— **v fuerza f**—[electr-oper] to require @ energy
— **v generalmente** - to usually require
— **v información f** - to require @, information,
o data
— **v** — **f para proyección f** - [ind] to require @
design data
— **v inspección f** - to require @ inspection
— **v** — **f periódica** - to require periodic(al)
inspecion
— **v limpieza f** - to require @ cleaning
— **v lubricación f** - [mec] to require @, oiling,
o lubrication
— **v** — **f periódica** - [lubric] to require @ periodic lubrication
— **v mantenimiento m** - to require @ maintenance
— **v más tiempo m** - to require more time; to be
more time consuming
— **v menos tiempo m** - to require less time; to
be less time consuming
— **v mucho tiempo n** - to require much time; to
be (very) time consuming
— **v papel m** - [comput] to require @ paper
— **v par m motor** - [mec] to require @ torque
— **v previamente** - to require previously
— **v proyección f** - to require @ design(ing)
— **v solución f** - to require @ solution
— **v tiempo m** - to require @ time
requiebro m - . . . . • [fam] pass
requisado,da a - requisitioned
requisar v - . . . .; to acquire
requisición f - . . . . • request; véase también
pedido m; orden f
— **f de equipo m** - equipment requisitioning
— **f** — — **m nuevo** - [ind] new equipment requisitioning
— **f** — **material(es) m** - material(s) requisitioning
— **f** — **materia(s) f prima(s)** - [ind] raw mate-
rial(s) requisitioning
requisito m - . . . .; qualification - need(ing)
— **m absoluto** - mandatory requirement
— **m de durabilidad f** - durability requirement
— **m** — **experiencia f** - experience requirement
— **n** — **mantenimiento m**—maintenance requirement
— **m** — **proyecto m** - project requirement
— **m** — **resistencia f** - [sold] strength requirement
— **m** — — **f a tensión f** - [sold] tensile
strength requirement
— **m detallado** - detailed requirement
— **m especial** - special requirement
— **m especificado** - specified requirement
— **m establecido** - established requirement
— **m exigido** - required requirement
— **m fiscal** - [fisc] fiscal requirement
— **m fundamental** - fundamental, o basic, requirement
— **m general** - general requirement
— **m indispensable** - indispensable, o mandatory,

requirement
requisito m legal - [legal] legal requirement
— **m máximo** - maximum requirement
— **m mínimo** - minimum requirement
— **m necesario** - necessary requirement
— **m para acondicionamiento m** - conditioning
requirement
— **m** — **calidad f** - quality requirement
— **m** — **capacitación f** - [ind] qualification
requirement
— **m** — — **f técnica** - [sold] technical qualification requirement
— **m** — — **f** — **y práctica f de operario m** -
[sold] procedure and qualification requirement
— **m** — **control m de calidad f** - quality control
requirement
— **m** — **ejecución f** - performance requirement
— **m** — **importación f** - [com] importation, re-
quirement, o formality|
— **m** — **estanqueidad f** - watertightness requirement
— **m** — **procedimiento m** - [ind] procedure, requirement, o qualification
— **m** — — **m para soldadura f** - [sold] welding
procedure, requirement, o qualification
— **m** — **prueba f** - [ind] testing requirement
— **m** — **resistencia f** - strength requirement
— **m** — — **f a tensión f** - [metal] tensile
(strength) requirement
— **m** — **seguimiento m** - follow-up requirement
— **m** — **soldador(es) m** - [sold] weldor, o operator, requirement, o qualification
— **m** — **soldadura f** - [sold] welding, requirement, o qualification
— **m previo** - previous, o prior, requirement
— **m químico** - [quím] chemical requirement
— **m** — **detallado** - [quím] detailed chemical
requirement
— **m técnico** - technical requirement
**Requisitos m para Procedimientos m para Solda-
dura f y Soldadores m** - [sold] Qualification
of Welding Procedures and Operators
rerregulación* f - [mec] resetting
rerregulado,da* a - reset
rerregular* v - [mec] to reset
— **v botón m** - [mec] to reset @ button
resaca f - [hidr] . . .; countercurrent
resaltado,da a - stressed
resaltar v - . . .; to stress
resalte* m - véase **resalto m**
resalto m - [mec] . . .; shoulder; offset; notch
• lug • thread • [constr] land; ledge •
[autom-neumát] cleat • rib • [sold] projection (in spark welding)
— **m cuadrado** - [mec] square, rib, o thread
— **m en borde m** - [autom-neumát] shoulder rib
— **m** — **neumático m** - [autom-neumát] shoulder rib
— **m** — **soldadura** - [sold] weld rib
— **m** — **tornillo m** - [mec] screw thread
— **m exterior** - [autom-neumát] shoulder rib •
outer rib
— **m** — **en banda f para rodamiento m** - [autom-
-neumát] tread shoulder rib
— **m** — — **neumático m** - [autom-neumát] tire
shoulder rib
— **m fino** - [mec] fine thread
— **m triangular** - [mec] (tri)angular thread
— **m macho** - [mec] male thread
— **m matriz** - [mec] female thread
— **m ovalado** - [mec] oval shoulder
— **m redondo,deado** - [mec] rounded thread
— **m trapezoidal** - [mec] buttress thread
resarcido,da - compensated; indemnified
resarcimiento m - indemnity; compensation
resbalada f - slip(ping)
resbaladera f - [instrum] véase **corredera f;
cursor m**
— **m de arcilla f** - clay slip
— **m seco** - [deport] dry skidpad
resbaladizo,za a - . . . .; slick
resbalado,da a - slipped; slid
resbaladura f - slippage

resbalamiento m̲ - . . .; slippage; sliding
— m̲ de alambre m̲ - [mec] wire slipping,page
— m̲ de eje m̲ [mec] axle, o̲ shaft, slipping,page
— m̲ —— m̲ **para impulsión** f̲ - [electr-mot]
    drive,ving shaft slippage
— m̲ — . . . m̲ **para motor** m̲ - [electr-mot]
    motor drive,ving, axle, o̲ shaft, slipping
— m̲ — **rueda** f̲ - [mec] wheel slipping,page
— m̲ **evitado** - [constr] prevented slipping
— n̲ **excesiivo** - [mec] excess(ive) slippage,ping
resbalar v̲ **alambre** m̲ - [mec] to slip @ wire
— v̲ **excesivamente** - [mec] to, slip, o̲ slide,
    excessively, o̲ too much
— v̲ **sobre garrucha** f̲ - [mec] to slide, o̲ slip,
    on @ sheave
rescatado,da a̲ - rescued
rescatador n̲ - . . .; rescuer
rescatador,ra - . . .; rescuer • rescuing
rescate m̲ - . . .; rescue,cuing
rescindido,da a̲ - rescinded • [legal] terminated
— a̲ **parcialmente** - [legal] partially rescinded
— a̲ **totalmente** - [legal] totally rescinded
— a̲ **unilateralmente** - [legal] rescinded unila-
    terally
rescindir v̲ - . . . • to terminate
— v̲ **adjudicación** f̲ - [legal] to rescind @ award
— v̲ **contrato** m̲ - [legal] to rescind @ contract
— v̲ **parcialmente** - [legal] to rescind partially
— v̲ **totalmente** - [legal] to rescind totally
— v̲ **unilateralmente** - [legal] to rescind uni-
    laterally
rescisión f̲ - . . .; rescinding,dment • [legal]
    termination
— f̲ **de adjudicación** f̲ - [legal] award(ing)
    rescinding,cission
— f̲ — **contrato** m̲ - [legal] contract res-
    cinding,cission
— f̲ —— m̲ **presente** - [legal] rescinding,cis-
    sion hereof
— f̲ **parcial**—[legal] partial rescinding,cission
— f̲ **total** - [legal] total rescinding,cission
— f̲ **unilateral** - [legal] unilateral rescinding
    • unilateral termination
rescuadrar* v̲ - véase **escuadrar**
reseco,ca a̲ - . . . • [fam] bone dry
resellable* a̲ - resealable
resentido,da a̲ - . . .; resented • annoyed
resentimiento m̲ - . . . • annoyance
— m̲ **de personal** m̲ - [labor] personnel, o̲ em-
    ployee, resentment
resentir v̲ - . . . • to annoy
reseña f̲ - . . .; report; outline,ning • high-
    lighting
reseñado,da a̲ - outlined; highlighted
reseñar v̲ - . . .; to highlight
reserva f̲ - . . . • reservoir • [fin] reserve •
    holding; conserving; saving • [contab] re-
    serve, véase también **previsión** f̲ - [mec] back-
    -up • [legal] confidential status | a̲ - (de)
    stand-by, spare; back-up
— f̲ **adjudicada** - [miner] awarded, o̲ assigned,
    reserve
— f̲ **beneficiada** - [miner] benefitted reserve
— f̲ **calculada** - [miner] calculated, o̲ esti-
    mated, reserve
— f̲ **comprobada** - [miner] proven reserve
— f̲ **conocida** f̲ - known reserve
— f̲ **constituída** - [fin] established reserve
— f̲ **cubicada** - [miner] measured reserve
— f̲ **de acero** m̲ - [metal-prod] steel reserve
— f̲ — m̲ **recuperable** - [metal-prod] recover-
    able steel reserve
— f̲ — **agua** m̲ - [hidr] water, storage, o̲ supply
— f̲ — **carbón** n̲ - [miner] coal reserve(s)
— f̲ — **derecho(s)** m̲ - [legal] right(s), reser-
    ving, o̲ reservation
— f̲ —— m̲ **exclusivo(s)** - [legal] exclusive
    right(s) reserving,vation
— f̲ — **esfuerzo** m̲ - effort reservation
— f̲ —— m̲ **personal** - (personal) effort re-
    serving,vation
— f̲ — **hermatita** f̲ - [miner] hematite reserves
— f̲ —— f̲ **pulverulenta** - [miner] pulverulent

hematite reserve(s)
reserva f̲ **de hierro** m̲ - [miner] iron reserve(s)
— f̲ **de hulla** f̲ - [miner] coal reserve(s)
— f̲ — **mineral** - [miner] ore reserve(s)
— f̲ —— m̲ **adjudicada** - [miner] awarded, o̲ as-
    signed, ore reserve(s)
— f̲ —— m̲ **de hierro** m̲ - [miner] iron ore re-
    serve(s)
— f̲ —— m̲ —— **adjudicada** - [miner] awarded,
    o̲ assigned, iron ore reserve(s)
— f̲ — **piedra** f̲ **caliza** - [miner] limestone re-
    serve(s)
— f̲ — **yacimiento** m̲ - [miner] deposit reserves
— f̲ **en dólar(es)** m̲ - [fin] dollar reserve(s)
— f̲ — **oro** m̲ - [fin] gold reserve(s)
— f̲ **especial** - [contab] special reserve(s)
— f̲ **estatutaria** - [legal] statutory reserve(s)
— f̲ **explorada** - [miner] explored reserve(s)
— f̲ **explotada** - [miner] worked reserve(s)
— f̲ **extranjera** - [econ] foreign reserve(s)
— f̲ **facultativa** - [rin] optional reserve
— f̲ **ferrífera** - [miner] iron bearing reserve(s)
— f̲ **legal** - [contab] legal reserve(s)
— f̲ **local** - [miner] local reserve(s)
— f̲ **mental** - mental reserve • [fam] grain of
    salt
— f̲ **minera** - [miner] mining reserve(s)
— f̲ **mineral** - [miner] mineral reserve(s)
— f̲ **mundial** - world(wide) reserve(s)
— f̲ **neta** f̲ - [fin] net, reserve, o̲ holding
— f̲ **oficial** - official reserve(s)
— f̲ — **en dólares** m̲ - [fin] official dollar
    reserve(s)
— f̲ — **acumulada** - [econ] accumulated official
    reserve(s); official reserve(s) accumulation
— f̲ **para amortización** f̲ - [contab] depreciation
    reserve
— f̲ — **cuenta(s)** f̲ **de cobro** m̲ **dudoso** - [contab]
    doubtful account(s) reserve(s)
— f̲ —— f̲ **mala(s)** - [contab] bad account(s)
    reserve
— f̲ — **depreciación** f̲ - [contab] depreciation
    reserve
— f̲ — **imprevisto(s)** m̲ - [contab] contingency
    reserve(s)
— f̲ —— m̲ **diverso(s)** - [contab] sundry, o̲
    miscellaneous, contingency reserve(s)
— f̲ —— m̲ **vario(s)** - [contab] sundry, o̲ mis-
    cellaneous, reserve(s)
— f̲ — **inversión(es)** m̲ - [contab] investment(s)
    reserve
— f̲ **permanente** - permanent, reserve, o̲ back-up
— f̲ **posible** - [miner] possible, o̲ potential,
    reserve(s)
— f̲ **positiva(s)** f̲ - [miner] positive reserve(s)
— f̲ — **de hierro** m̲ - [miner] positive iron re-
    serve(s)
— f̲ —— **mineral** m̲ - [miner] positive ore re-
    serve(s)
— f̲ —— m̲ **de hierro** m̲ - [miner] positive
    iron ore reserve(s)
— f̲ **privada(s)** - [econ] private reserve(s)
— f̲ — **en dólar(es)** m̲ - [fin] private dollar
    reserve(s)
— f̲ — **extranjera(s)** - [econ] private foreign
    reserve(s)
— f̲ — **en dólares** m̲ - [fin] private foreign
    dollar reserve(s)
— f̲ —— **en oro** m̲ - [fin] private foreign gold
    reserve(s)
— f̲ **probable(s)** - [miner] probable reserve(s)
— f̲ **recuperable(s)** - recoverable reserve(s)
— f̲ — **de acero** m̲ - [metal-prod] steel recove-
    rable, o̲ recoverable steel, reserve(s)
— f̲ — **chatarra** f̲ - [metal-prod] recoverable
    scrap reserve(s)
— v̲ **recuperada** - [miner] recovered reserve(s)
— f̲ **resultante** - [fin] resulting reserve(s)
— f̲ **social(es)** - [contab] corporate reserve(s)
— f̲ **total(es)** f̲ - total reserve(s)
— f̲ **trabajada** - [miner] worked reserve(s)
reservación f̲ - . . . • conserving,vation; sav-
    ing • accomodation • holding

reservación f de automóvil m - [autom] car reserving
— f — autoridad f - [admin] authority reservation
— f — derecho(s) m - [legal] right(s) reservation,ving
— f — — m exclusivo(s) m - [legal] exclusive right(s) reservation,ving
reservado,da a - . . . . saved; conserved • held
— a para sí mismo,ma - reserved, to, o for, itself,ves, o oneself,ves, o himself,ves, o herself,ves
reservar v - . . . . • to, keep, o save, o conserve, o hold
— v automóvil m - [autom] to reserve @ car
— v autoridad f—[admin] to reserve @ authority
— v — f para sí mismo,ma - [admin] to reserve @ authority for, itself,ves, o oneself, o himself, o herself, o themselves
— v derecho(s) m—[legal] to reserve @ right(s)
— v — m exclusivo(s) - [legal] to reserve @ exclusive right(s)
— v esfuerzo m - to reserve @ effort
— v — m personal - to reserve @ personal effort
— v para sí mismo,ma - to reserve for, itself, o oneself, o himself, o herself
reservarse f autoridad - to reserve @ authority for, itself, o oneself, o himself, o herself
reservorio* m - véase depósito m; reserva f
resguardado,da a - guarded; sheltered
resguardo m - . . .; shield • [mec] kicker
— m de alambre m - [mec] wire shield
— m delantero - front guard
— m inferior - [mec] low(er), o bottom, guard
— m para accionamiento m - [mec] drive guard
— m — — m inferior - [mec] low(er), o bottom, drive guard
— m — — m para mesa f rotatoria - [petról] rotary drive guard
— m — — m superior - [mec] top, o upper, drive guard
— m — acoplamiento m - [mec] coupling guard
— m — — m para freno m - [mec] brake coupling guard
— m — cable m - [electr-instal] cable guard • [mec] [cabl] cable guard
— m — — m para llave f para enroscadura f - [petról] tong line guard
— m — — m tenaza f para enroscadura f - [petról] tong line guard
— m — carrete m - [mec] reel guard
— m — — m para cucharreo m - [petról] sand reel guard
— m — conexión f - [mec] connection guard
— m — correa f - [mec] belt guard
— m — — f en V - [mwx] V belt guard
— m — embrague m - [mec] clutch guard
— m — — m para mesa f rotativa - [petról] rotary clutch guard
— m — — m inferior - [mec] low(er), o bottom, clutch guard
— m — — m superior - [mec] high, o top, o upper, clutch guard
— m — freno m - [mec] brake guard
— m — — m por inercia f - [mec] inertia brake guard
— m — tubo m - [mec] tube guard
— m — válvula f - [válv] valve guard
— m — válvula f neumática - [mec] air valve guard
— m — — f para aire m - [mec] air valve guard
— m posterior - [mec] rear, o back, guard
— m quitado - [mec] removed, o withdrawn, guard
— m removido—[mec] removed, o withdrawn, guard
— m sacado - [mec] removed, o withdrawn, guard
— m superior—[mec] top, o high, o upper, guard
— m tipo m paraguas - [mec] umbrella (type) (guard)
— m tubular - [mec] tube (shaped) guard
residencia f - . . .; housing; cottage • [pol] impeachment
— f cercana - [constr] nearby, residence, o

home, o house
residencia f individual - [constr] individual, o single-family, home
— f invernal - winter, residence, o home
— f para solaz m para fin(es) m de semana f - [constr] weekend retreat
— f — veraneo m - [constr] summer(ing) home
— f permanente - [constr] permanent, residence, o home; year-round dwelling
— f privada - [constr] private, residence, o home
— f ribereña f - [constr] waterfront porperty
— f sobre curso m de agua m - [constr] waterfront property
— f suburbana - [constr] suburban residence • townhouse— [constr] city dwelling
— f urbana -[constr] city dwelling • townhouse
— f veraniega - [constr] summer, residence, o cottage
residencial a - . . .; residential
residente m - . . . . • [constr] project supervisor
— m de condado m - [pol] county resident
— m — fin m de semana f - [pol] weekender
— m — Pensilvania descendente de (inmigrantes) alemanes - [pol] Pennsylvania Dutch
— m — zona f - resident; local
— m en ejido m - [pol] township resident
— m — perspectiva - [pol] prospective resident
— m permanente - permanent resident
— m posible - [pol] prospective resident
residual a - . . . . - [petról] bottom settling
residualmente adv - residually
residuo m - residue; debris • [constr] muck • [miner] tailing(s)• [petról] residue bottom • foot(s) • [agric] tailing(s) • [nucl] waste (materials)
— m aceitoso de exudación f - [petról] foots
— m almacenado - [constr] stored waste
— m con actividad f baja - [nucl] low activity waste
— m — — f intermedia - [nucl] intermediate activity waste
— m — — f mediana - [nucl] intermediate activity waste
— m — nivel m de (ir)radiación f alto - [nucl] high level radiation waste
— m — — m — — f bajo - [nucl] low level radiation waste
— m — — m — — f reducido - [nucl] low level radiation waste
— m — (ir)radiación f alta - [nucl] high level radiation waste
— m — — f elevada - [nucl] high level radiation waste
— m — — f baja - [nucl] low level radiation waste
— m — — f reducida - [nucl] low level radiation waste
— m — radiactividad [nucl] radioactive waste
— m — — f alta - [nucl] high (level) radioactive,vity waste
— m — — f — almacenado - [nucl] stored high level waste
— m — — f baja - [nucl] low (level) radioactivity waste
— f — — f baja almacenado - [nucl] stored low (level) radioactivity waste
— m — — f elevada - [nucl] high level radioactivity waste
— m — — f intermedia - [nucl] intermediate level waste
— m — — f mediana _ [nucl] intermediate level waste
— m — — reducida - [nucl] low radioactivity (level) waste
— m concentrado - [nucl] concentrated waste
— m crudo - [petról] crude residue
— m de afino m de cobre m - [metal] copper refining residue
— m — — m de sulfuro m de cobre m - [miner] copper sulfide refining residue
— m — categoría f . . . . - . . . category, re-

sidue, o waste
residuo m de ceniza(s) f—[combust] residual ash
— m — concentrado* m—[nucl] concentrate waste
— m — cribadura f - [miner] tailing(s) •
screening waste
— m — destilación f - distillation residue
— m — edificio(s) m público(s) - [sanit] in-
stitutional waste
— m — escoria f - [sold] remaining slag
— m — hierro m - iron residue
— m — lavado m - wash(ing) waste
— m — perforación f - [petról] drill cuttings
— m — f de pozo(s) m - [petról] well drilling cutting(s)
— m — petróleo m - [petról] bottom settling(s)
— m — m en fondo m - [petról] bottom set-
tling(s)
— m — m — m de depósito m - [petról]
(tank) bottom settling(s)
— m — refinación f - [metal-prod] refining
residue
— m — f de cobre m - [metal] copper refining residue
— m — f — sulfuro m de cobre m - [quím]
copper sulfide refining residue
— m elevado - [ind] high residue
— en fondo m de depósito m - [petról] bottom
settling(s)
— m existente - existing waste
— m generado - [nucl] generated, o produced,
waste
— m libre - [sanit] free residue
— m líquido - [nucl] liquid waste
— m — con radiactividad - [nucl] radioactive
liquid waste
— f — — f alta - [nucl] high level radioactive liquid waste
— m — — f baja - [nucl] low level liquid
radioactive waste
— m — — f elevada - [nucl] high level radioactive liquid waste
— m — — f intermedia - [nucl] intermediate
level (radioactive) liquid waste
— m — — f mediana - [nucl] intermediate
level radioactive liquid waste
— m negro - [sold] black residue
— m neutralizado - [nucl] neutralized waste
— m no radiactivo—[nucl] non radioactive waste
hazardous waste
— m — — peligroso - [nucl] nonradioacative
hazardous waste
— m originado - [nucl] originated, o produced,
waste
— m peligroso m - [segurid] hazardous waste
— m pesado - heavy residue • [petról] crude
residue
— m producido - [nucl] produced, o generated,
residue
— m radiactivo - [nucl] radioactive waste
— m — de categoría f . . . - [nucl] . . . category radioactive waste
— m — existente - [nucl] existing radioactive
waste
— m — líquido—[nucl] liquid radioactive waste
— m — sólido - [nucl] solid radioactive waste
— m reducido - [ind] low residue,dual
— m sedimentado - [hidr] sedimented residue
— m sólido - sold waste; waste solid(s)
— m — con radiactividad f alta - [nucl] high
level solid radioactive waste
— m — — f baja - [nucl] low level solid
radioactive waste
— m — — f elevada - [nucl] high level solid radioactive waste; solid high level waste
— m — — f intermedia - [nucl] intermediate
level (radioactivity) solid waste
— m — — f mediana - [nucl] intermediate
level (radioactivity) solid waste
— m — — f reducida - [nucl] low level solid radioactive waste; solid low level waste
— m — pre-procesado - [nucl] pre-processed
solid waste
— m tolerado - [ind] tolerated residue,dual

residuo m transferido - [nucl] transferred waste
— m tratado - [nucl] treated waste
residuos m y desechos m—[ind] scrap and rejects
resignación f - . . . . • settling
— f a hecho m - living with, o accepting, @
fact
resignarse v - to resign to • to settle
— v a hecho m - to live with @ fact
resiliencia* f - véase elasticidad f; rebote m
resiliente* a - véase elástico,ca
resina f - . . . . • pitch
— f catiónica - [nucl] cationic, resin, o rosin
— f con base f de epoxia f - epoxy, resin, o
rosin
— f acronitrilo-butadieno-estirénica - [quím]
acronitrile-butadiene-styrene resin
— f con calidad f alta - high-grade resin
— f de acrilonitrilo-butadineo-estireno m -
[plást] acrilonitrile-butadiene-styrene resin
— f — epoxia f - [quím] epoxy resin
— f poliestérica - [quím] polyester resin
— f sintética - [quím] synthetic resin
resinoide* m - resinoid
resinoso,sa a - . . . .; resinoid
resistencia f . . . . - stamina; endurance] withstanding • dragging • working strength •
[electr-equip] resistor • grid • reactance
— f a abrasión f - [mec] abrasion, o wear, resistance
— f — ácido(s) m - acid resistance
— f — adherencia f—[sold] sticking resistance
— f — f de salpicadura(s) f - [sold] spatter sticking resistance
— f — agrietamiento m - [sold] crack(ing) resistance
— f — m debida a inmovilización f (total) -
[sold] high restraint cracking resistance;
resistance to high restraint cracking
— f — m — proporción f, alta, o elevada, de azufre m - [sold] resistance to cracking due to high sulfur
— f — m — f, _, o —, de carbono m
- [sold] resistance to cracking due to high
carbon
— f — m longitudinal - [sold] longitudinal
cracking resistance
— f — agua m - water resistance
— f — aislamiento m - resistance to @ insulation
— f — aplastamiento m - [mec] crushing, resistance, o strength • collapse resistance •
[cabl] resistance to crushing
— f — m por impacto(s) m - [cabl] resistance to crushing
— f — asentamiento m - resistance to settling
— f — caldeo m - heating resistance
— f — m de motor m - [electr-mot] motor
heating resistance • [comb.int] engine heating resistance
— f — calor m - heat resistance; resistance to
@ heat
— f — carga(s) f - [mec] load(ing) resistance
— f — f con impacto(s) m - [mec] shock
loading resistance
— f — f por aplastamiento m - [cabl] resistance to @ crushing load
— f — f — m contra tambor m - [cabl]
resistance to crushing drum load
— f — cizalla(do) - [mec] véase resistencia f
a corte m
— f — cizalleo m - véase resistencia f a corte
— f — compresión f - compression resistance;
resistance to @ compression • [constr] compressive strength
— f — f admisible - [constr] allowable compressive strength
— f — f anular - [constr] ring compression.
resistance, o resisting
— f — concusión* f - [explos] concussion resistance
— f — condición f de suelo m - [suelos] resistance to @ soil condition
— f — f química - [quím] resistance to @

@ soil('s) chemical condition
resistencia f a condición f química de suelo m -
[suelos] resistance to @ soil('s) chemical
condition
— f — **corrosión f** - corrosion, o rust, resistance,ting
— f — — f **aceptable** - [metal] acceptable corrosion resistance
— f — — f — **generalmente** - [metal] generally
acceptable corrosion resistance
— f — — f **aceptada** - [metal] accepted corrosion resistance
— f — — f — **generalmente** - [metal] generally
accepted corrosion resistance
— f — — f **algo reducida** - [metal] lower corrosion resistance
— f — — f **alta de ácido**—high acid corrosion
resistance
— f — — f **atmosférica** - [metal] resistance to
@ atmospheric corrosion; atmospheric corrosion resistance
— f — — f **de ácido m** - acid corrosion resistance
— f — — f **necesaria** - [sold] needed corrosion
resistance
— f — — f **reducida** - [metal] low(er) corros-
sion resistance
— f — **corrosividad f alta de ácido m** - high
acid corrosion resistance
— f — — f **de ácido m** - acid corrosion resistance
— f — **corte m** - [suelos] shear(ing), strength,
o resistance
— f — **crecimiento m** - growth resistance
— f — — m **de corte(s) m** - [autom-neumát] cut
growth resistance
— f — **choque(s) m** - [mec] shock resistance
— f — — m **térmico(s)** - thermal shock resistance
— f — — m — **severo(s) m** - severe thermal
shock resistance
— f — **daño(s) m** - damage(s) resisting,tance
— f — — m **en borde(s) m** - [constr] edge damage resistance
— f — — m — — m **para, hincadura f, o penetración f** - [constr] leading edge damage resistance
— f — **derrumbamiento m** - [hidr] washout resistance,ting
— f — **desastre(s) m** - disaster resistance
— f — **descascarillado m** - [metal] scaling resistance
— f — **desgaste m** - [metal] wear resistance;
wearability • wear(ing)
— f — — m **aumentada** - [autom-beumát] added
wearability
— f — — m **de superficie f** - [mec] surface
wear resistance
— f — — m **por abrasión f** - [sold] abrasion
wear resistance
— f — — m — **impacto(s) m** - [sold] impact
wear resistance
— f — **desmenuzamiento m** - [sold-electrodos]
breakdown resistance
— f — **desplazamiento m** - [autom-neumát] push-
-off resistance
— f — — m **de talón m** - [autom-neumát] bead
push-off resistance
— f — **destello(s) m** - [sold] flash through resistance
— f — — m **a través de fundente m** - [sold]
flash through resistance
— f — **destrucción f** - destruction resistance
— f — **deterioro,ración** - deterioration resistance
— f — **deslocación f** - disjointing resistance
— f — **elemento(s) m** - element(s) resistance
— f — — m **corrosivo(s)** - [metal] corrosive
element(s) resistance
— f — **empuje m** - [mec] thrust withstanding
— f — — m **oblicuo** - [archit-arco] thrust resistance
— f — **entalle(s) m** - [metal] notch toughness

resistencia f a erosión f - erosion resistance
— f — — f **severa** - severe erosion resistance
— f — **escurrimiento m** - [metal] creep, resistance, o strength
— f — **esfuerzo m** - stress resistance,ting •
stress taking
— f — — m **cortante m** - [mec] shear strength;
véase también **resistencia f a corte m**
— f — **estallido(s) m** - [metal] bursting
strength
— f — **exfiltración f** - [hidr] exfiltration
resistance
— f — **extrusión f** - [mec] extrusion resistance,ting
— f — **fallo m** - failure resistance,ting
— f — **fallo m** - véase **resistencia f a rotura f**
— f — **fatiga f** - [metal] fatigue, resistance,
o strength, o limit • fatigue (useful) life
— f — **filtración f** - [hidr] filtration resistance
— f — **flambeo m** - [mec] column strength
— f — **flexión f** - [constr] bending, resistance, o strength; moment strength; flexing
withstanding
— f — — f **de acero m** - [metal] steel bending
strength
— f — — f — — m **corrugado** - [metal] corrugated steel bending strength
— f — — f **de pared f** - [constr] wall bending
strength
— f — — f — — f **de tubería f** - [constr]
pipe wall bending strength
— f — — f — — f — — f **abovedada** -
[constr] pipe arch wall bending strength
— f — — f — **tubería f** - [constr] pipe bending strength
— f — — f — — f **abovedada** - [constr] pipe-
-arch bending strength
— f — **fluencia f** - [metal] creep strength •
yield strength
— f — **flujo m** - flow resistance • [petról]
flow resistance
— f — **fractura f** - fracture resistance
— f — **fricción f** - [mec] friction resistance
— f — **fuego m** - fire, resistance, o endurance
— f — **fuerza f** - [tub] force resistance
— f — — f **de aplastamiento m** - [tub] collapsing force resistance
— f — **golpe(s) m** - [mec] shock resistance
— f — **hincadura f** - [constr-pil] driving resistance
— f — **humedad f** - moisture resistance
— f — — f **probada**—tested moisture resistance
— f — **impacto(s) m** - impact, resistance, o
strength, o value • notch toughness
— f — — m **con temperatura(s) f baja(s)** -
[metal] low temperature notch toughness
— f — — m **reducido(s)** - [mec] low impact resistance
— f — — m **repetido(s)** - [metal] repeated im-
pact(s) resistance
— f — **infiltración f** - [hidr] infiltration resistance
— f — **intemperie f** - weather resistance
— f — **interrupción f** - interruption resistance
— f — — f **de arco m** - [sold] pop-out resistance
— f — **inserción f con gato(s) m** - [constr]
jacking resistance
— f — **irradiación f** - [electrón] radiation
resistance
— f — **magulladura(s) f** - bruise resistance
— f — **pandeo m** - buckling, strength, o stress
— f — **paso de corriente f eléctrica** - [metal]
electrical resistance
— f — **penetración f** - [mec] penetration resistance
— f — — f **de piedra(s) f** - [autom-neumát]
stone penetration resistance
— f — — f — **roca(s) f** - [autom-neumát]
rock penetration resistance
— f — **perforación f** - [suelos] resistance to
drilling

resistencia f a picadura(s) f - [sold] pock marking resistance
— f — porosidad f - [sold] porosity resistance
• resistance to porosity
— f — — f debida a contaminación f - [metal]
contamination porosity resistance
— f — — f — — — f orgánica - [metal] organic contamination porosity resistance; resistance to @ organic contamination porosity
— f — — f — — óxido m - [metal] resistance
to @ rust porosity; rust porosity resistance
— f — — f — — soplo m (magnético) de arco m
- arc blow porosity resistance; resistance to
@ arc blow porosity
— f — precipitación f - [metal- precipitation
resistance
— f — — f de carburo(s) m - [metal] carbide
precipitation resistance
— f — radiación f - [nucl] radiation
resistance
— f — retirada - withdrawing resistance
— f — rodaje m - [autom-neumát] rolling resistance
— f — — m de neumático(s) m - [autom-neumát]
tire rolling resistance
— f — rotura f - [metal] breaking, o rupture,
strength, o resistance
— f — — f de alambre m - [trefil] wire breaking strength
— f — — f — cable m - [cabl] wire rope
breaking strength
— f — — — m de alambre m - [cabl] wire
rope breaking strength
— f — — f aumentada - [mec] increased breaking strength
— f — — f disminuida - [mec] decreased breaking strength
— f — — f por impacto(s) m - [mec] impact(s)
breakage,king resistance
— f — — f — tracción f - [cabl] tension, o
pulling, breaking strength
— f — — f — — f de cable m - [cabl] (wire)
rope tension breaking strength
— f — — f — — — f de cabo m - [cabl] rope
tension breaking strength
— f — rozamiento m - [mec] friction resistance
— f — ruptura f - [metal] bursting strength
— f — socavación f - [hidr] undermining resistance
— f — soldadura f - [sold] welding resistance
— f — — f con pasadas f múltiples - [sold]
multiple pass welding resistance
— f — soplo m - [sold] (arc) blow resistance
— f — — m magnético (de arco m) - [sold] arc
blow resistance;resistance to @ arc blow
— f — temperatura f - temperature resistance
— f — — f alta - high temperature resistance
— f — — f baja - low temperature resistance
— f — — f mayor - greater, o higher, temperature resistance
— f — — f menor - low(er) temperature resistance
— f — tensión f - [metal] (tensile) strength
— f — — f aliviada - [sold] stress relieved
tensile strength
— f — — f con alivio m de tensión f - [sold]
stress relieved tensile strength
— f — — f de aportación f - [sold] tensile
strength deposit
— f — — f — metal m de base - [metal] base
metal (tensile) strength
— f — — f — — m de aportación f - [sold]
weld metal tensile strength
— f — — f observada - [sold] observed (tensile) strength
— f — termofluencia f - [electr-cond] creep
resistance
— f — torsión f - [metal] torsion(al), resistance, o strength • [mec] torque
— f — — f a(l), girar, o rotar - [mec] rolling torque
— f — tracción f - [metal] tensile strength •
[mec] tractive resistance

resistencia f a tracción f con alivio m de ten-
sión(es) f - [metal] stress relieved tensile
strength
— f a vibración f - [mec] vibration, resistance, o stress • vibratory stress
— f — voltaje(s) m - [sold] voltage resistance
— f ← — m alto(s) - [sold] high voltage resistance; resistance to @ high voltage(s)
— f — — m bajo(s) - [sold] low voltage(s) resistance; resistance to low voltage(s)
— f abierta - open resistance • [electr-instal]
open(ed) resistor
— f aceptable - acceptable resistance
— f — generalmente - generally acceptable resistance
— f aceptada - accepted resistance
— f — generalmente - generally accepted resistance
— f adecuada - adequate, resistance, o strength
— f adicional - additional, o extra, resistance, o strength
— f admisible - [mec] allowable strength
— f — contra compresión f - [constr] allowable
compressive strenth
— f — para costura f - [mec] allowable seam
strength
— f — — — f empernada - [mec] allowable
bolted seam strength
— f — — junta f - [mec] allowable joint
strength
— f aerodinámica - aerodynamic, resistance, o
drag
— f agregada - added, resistance, o strength
— f algo mayor - [sold] somewhat, o slightly,
greater strength
— f — a tensión f - [sold] somewhat, o
slightly, greater (tensile) strength
— f — menor - somewhat, o slightly, less, o
lower, strength
— f — — a tensión f - [sold] somewhat, o
slightly, lesser, o lower, strength
— f alta - high, resistance, o strength •
[metal] high toughness
— f — a abrasión f - [autom-neumát] high abrasion resistance
— f — — ácido(s) m - [quim] high acid, resistance, o corrosiveness
— f — — aplastamiento m - [mec] high collapse
resistance
— f — — calor m - high heat resistance
— f — — corrosión f - [metal] high corrosion
resistance
— f — — — f por ácido(s) m - high acid corrosion resistance
— f — — corrosividad f - [metal] high corrosion resistance
— f — — — f por ácido(s) m - [metal] high
acid corrosiveness resistance
— f — — fatiga f - high fatigue resistance •
high yield stress point
— f — — impacto(s) m - high impact resistance
• [metal] high notch toughness
— f — — paso m de corriente f - [electr] high
electrical resistance
— f — — temperatura f - high temperature resistance
— f — — tensión f - [metal] high tensile
strength
— f — — torsión f - [mec] high torsion resis-
ance
— f — de soldadura f - [sold] high weld, o
weld high, strength
— f amplia - ample, strength, o resistance
— f apropiada - proper, o appropriate, strength
— f — asegurada - assured proper strength
— f aproximada - [metal] approximate strength
— f — a tensión f - [metal] approximate tensile strength
— f asociada - [electrón] associated resistor
— f aumentada - increased strength • improved
strength
— f baja a abrasión f - low abrasion resistance
— f — — aplastamiento m - [mec] low collapse

resistance

resistencia f baja a calor m · low heat resistance
— f — — corrosión f—low corrosion resistance
— f — — — f de ácido(s) m - [metal] low acid corrosion resistance
— f — — impacto(s) m - low impact resistance
— f — — paso m de corriente f - [electr] low electrical resistance
— f — — rodadura f - [autom-neumát] low resistance rolling
— f — — temperatura f - low temperature resistance
— f — — tensión f - [metal] low tensile strength
— f brindada - provided strength
— f buena - good resistance - [metal] good strength
— f — a corrosión f - [metal] good corrosion resistance
— f — a descascarillado m - [metal] good scaling resistance
— f — — impacto(s) m - [sold] good impact resistance
— f calculada - [electr-instal] calculated resistance
— f — total - [electr] total calculated resistance
— f cerrada f - [electr-oper] closed resistor
— f columnar - [mec] column strength
— f como columna f - [mec] column strength
— f comparable - [sold] comparable, o equal, strength
— f completa f - [electr-equip] complete resistor
— f — para desexcitación f - [electr-equip] complete de-energizing resistor
— f compresiva - compressive strength
— f — admisible - [constr] allowable compressive strength
— f — de pared f - [constr] wall, compression, o compressive, strength
— f — final - [metal] ultimate compressive strength
— f — máxima - ultimate compressive strength
— f — unitaria - unit compressive strength
— f — — máxima - [constr] ultimate unit compressive strength
— f comprobada - proved reistance • [electrón] checked resistor
— f con alivio m para tensión f -   [metal] stress relieved strength
— f — entalladura f en V - [metal] V-notch strength
— f — rejilla f - [electrón] grid resistance
— f — seguridad f - [segurid] safe resisting
— f concéntrica - concentric strength
— f conectada - [electr-oper] closed resistor • [electrón] connected resistor
— f contra corrosión f - corrosion fighting
— f — corte m - [mec] shear(ing), resistance, o value
— f — choque(s) m - [mec] shock resistance
— f — deslizamiento m - sliding resistance
— f — deformación f - deflection resistance
— f — descamación f - [metal] spalling resistance
— f — descascarillado m - [metal] shelling resistance
— f — desgaste m - [mec] wear resistance
— f — desviación f - deflection resistance
— f — erosión f - erosion resistance
— f — fatiga f - [metal] fatigue resistance
— f — — de superficie f para rodadura f - [mec-ruedas] (rolling) contact fatigue (life)
— f — — f — — f — — f de rueda f - [mec-ruedas] wheel rolling contact fatigue life
— f — — f — — f — — f — — f para rieles m - track wheel rolling contact fatigue life
— f — golpe(s) m - [mec] shock resistance
— f — incendio(s) m - fire resistance
— f — pandeo m - [mec] beam strength
— f — picadura(s) f - [metal] pitting resis-

tance

resistencia f contra porosidad f - [metal] porosity resistance
— f — — f en soldadura f - [sold] weld porosity resistance
— f — rodadura f - [autom-neumát] rolling resistance
— f — — f cuesta f abajo - [autom-neumát] (downhill) rolling resistance
— f corriente - standard resistance
— f crítica - [constr] critical resistance
— f — a pandeo m - [constr] critical buckling, stress, o resistance
— f — — tensión f - [metal] critical high (tension) strength
— f cúbica - [constr-hormig] cubic strength
— f Charpy - [metal] Charpy strength
— f — contra impacto(s) m - [metal] Charpy impact strength
— f — — — m con entalladura f - [metal] Charpy notch impact strength
— f — — — v en V - [metal] Charpy V-notch impact strength
— f de acero m - [metal] steel strength
— f — — m a corrosión f - [metal] steel corrosion resistance
— f — — m a flexión f - [metal] steel bending strength
— f — — m corrugado - [metal] corrugated steel strength
— f — — m — a flexión f - [metal] corrugated steel bending strength
— f — aislación f - [electr-instal] insulation resistance; insulation level (strength)
— f — — f en devanado m - [electr-instal] winding insulation resistance
— f — aislamiento m - [electr-cond] insulation resistance
— f — alambre m - [trefil] wire strength
— f — — m a deformación f - [trefil] wire, stability, o deformation resistance
— f — — m a rotura f - [alambre] wire breaking strength
— f — — m — — f por tracción f - [trefil] wire traction breaking strength
— f — — m interior - [cabl] inner, o inside, wire strength
— f — — m — en cable m - [cabl] rope, inner, o inside, wire strength
— f — alma m - [cabl] core strength
— f — cable m - [electr-cond] cable drag • [cabl] rope strength
— f — — m a rotura f - [cabl] (wire) rope breaking strength
— f — — m de alambre m - [cabl] wire rope strength
— f — — m — — m a rotura f - [cabl] wire rope breaking strength
— f — — m a rotura f - [cabl] (wire) rope breaking strength
— f — — m — — f por tracción f - [cabl] wire rope traction breaking strength
— f — campo m - [electr-instal] field resistance
— f — ceniza f - [combust] ash resistance
— f — — f a compresión f - [combust] ash compression resistance
— f — circuito m - [electr-oper] circuit resistance
— f — — m para campo m - [electr-instal] field circuit resistance
— f — — f de pilote m - [constr-pil] pile shaft strength
— f — — f — — m tubular - [constr-pil] pipe pile shaft strength
— f — conducto m - [constr] conduit strength
— f — — m rígido - [constr] rigid conduit strength
— f — conexión f - [mec] connection strength • connection force
— f — — f con tierra f - [electr-equip] ground(ing) resistance
— f — conjunto m - [mec] assembly strength

resistencia f de conjunto m a rotura f - [metal]
  aggregate, o assembly, breaking strength
— f — cordón m - [cabl] strand strength
— f — — con punto(s) - spot weld(ed) strength
— f — — m doble - [tub] double spot weld,
  resistance, o strength
— f — costura f - [mec] seam strength
— f — — f con una (sóla) hildera de remaches
  m - [mec] single rivet(s) seam strength
— f — — f doble - double seam strength
— f — — f empernada - [mec] bolted seam
  strength
— f — — m — en plancha(s) f estructural(es)
  - [mec] bolted structural plate seam strength
— f — — f en tubería f - [metal-fabr] pipe
  seam strength
— f — — f — — f Multi-Plate - [metal-fabr]
  Multi-Plate (pipe) seam strength
— f — — f longitudinal - [mec] longitudinal
  seam strength
— f — — f — empernada - [mec] bolted longitudinal seam strength
— f — — f — — en planchas f estructurales -
  [mec] bolted structural plate longitudinal
  seam strength
— f — entalle,lladura - [metal] notch strength
— f — estructura f - [constr] structure, resistance, o strength
— f — con carga(s) f vertical(es) -
  [constr] vertical load structure resistance
— f — hierro m fundido - [electr-equip] cast
  iron resistor
— f — hormigón m - [constr] concrete strength
— f — — a compresión f - [constr] concrete
  compressive strength
— f — junta f - [mec] joint strength • [metal-
  -fabr] seam strength
— f — — f traslapada - [mec] lap(ped) joint
  strength
— f — lecho m - [constr] bedding strength
— f — material(es) m - [ind] material(s)
  strength
— f — metal m - [metal] metal strength
— f — — a tensión f - [metal] metal tensile
  strength
— f — momento m - moment(um) strength
— f — muestra f - sample strength
— f — — f ensayada - tested sample strength
— f — — m para aportación f - [sold] weld
  metal strength
— f — — m — base f - [metal] base metal
  strength
— f — neumático m - [autom-neumát] tire, resistance, o strength
— f — — m a rodadura f - [autom-neumát] tire
  rolling, resistance, o strength
— f — pared f - wall strength
— f — — f de tubería f - [tub] pipe wall
  strength
— f — — f — — f abovedada - [tub] pipe-arch
  wall strength
— f — — f — — f — a flexión f - [constr]
  pipe-arch wall bending strength
— f — — f — — f — circular - [constr]
  circular pipe-arch wall bending strength
— f — — f — — f helicoidal - [tub] helical
  pipe wall strength
— f — — f — — f remachada - [tub] riveted
  pipe wall strength
— f — — f — — f soldada - [tub] welded pipe
  wall strength
— f — — f — — f — por punto(s) m - [tub]
  spot welded pipe wall strength
— f — pared f soldada - [sold] welded wall
  strength
— f — — f soldada con punto(s) m - [tub] spot
  welded wall strength
— f — pella f - [miner] pellet resistance
— f — — f a compresión f - [miner] pellet
  compression resistance
— f — perno m - [mec] bolt, strength, o force
— f — piloto m - [ind] pilot strength
— f — — m para cable m - [electr-cond] cable

pilot, resistance, o strength
resistencia f de plancha(s) f - [electr-equip]
  plate(s) resistor
— f — — f estampada(s) - [electr-equip]
  stamped plate(s) resistor
— f — reja(s) f - grid resistance
— f — resorte m - [mec] spring force
— f — sínter m - [metal-prod] sinter resistance
— f — sistema m - [electr] system resistance
— f — — m para conexión f con tierra f -
  [electr-instal] grounding system resistance
— f — — m — puesta f a tierra f - [electr-
  -instal] grounding system resistance
— f — soldadura f - [sold] weld, resistance,
  o strength; joint strength resistance
— f — — f a agrietamiento m - [sold] weld
  crack(ing) resistance
— f — — f a tensión f - [sold] weld tensile
  strength
— f — suelo m - [suelos] soil resistance •
  [electr-oper] earth resistance
— f — — m a corte m - [suelos] soil shearing
  strength
— f — terraplén m - [constr] embankment, o
  fill, strength
— f — — m para sostén m - [constr] embankment
  supporting strength
— f — terreno m - [suelos] soil bearing power
— f — tierra f - [electr-oper] ground(ing)
  resistance
— f — tipo m para plancha(s) f estampada(s) -
  [electr-equip] stamped plate type resistor
— f — toma f a tierra f - [electr-instal]
  ground resistance
— f — tubería f - [tub] pipe strength
— f — — f de acero m - [tub] steel pipe
  strength
— f — — f para sustentación f - [tub] pipe
  supporting strength
— f — — f para terraplén m - [constr] embankment conduit supporting strength
— f — tubería f helicoidal - [tub] helical
  pipe strength
— f — — f Multi-Plate - [tub] Multi-Plate
  (pipe) strength
— f — — f remachada - [tub] riveted pipe
  strength
— f — — f rígida - [tub] rigid pipe strength
— f — — f — para sustentación f - [constr]
  rigid pipe supporting strength
— f — — f — — f en obra f - [constr]
  rigid pipe field supporting strength
— f — — soldada - [tub] welded pipe strength
— f — — f — por punto(s) m - [tub] spot
  welded pipe strength
— f — unión f - [mec] joint strength
— f — viento m - [constr] wind resistance
— f — viga f - [constr] beam strength
— f — — f a tensión f - [constr] rail, o
  beam, tensile strength
— f — — f continua - [metal-fabr] continuous
  beam strength
— f debida a fricción f - frictional resistance
— f decreciente - decreasing strength
— f — a rotura f - [mec] decreasing breaking
  strength
— f definitiva - ultimate, o definite, strength
— f destacada - outstanding, strength, o resistance
— f — a agrietamiento m - [sold] outstanding
  crack resistance
— f — a impacto(s) m - [sold] excellent impact
  resistance
— f determinada - specified,fic strength
— f — a compresión f - [constr] specified
  compressive strength
— f dieléctrica - [electr] dielectric, resistance, o strength, o rigidity • electric, resistance, o strength • insulation resistance
— f — f, alta, o elevada - [electr] high dielectric strength
— f — baja - [electr] low dielectric strength

resistencia f dieléctrica de líquido m aislador
- [electr] insulating liquid dielectric
strength
— f —— m — para transformador m -
[electr] transformer insulating liquid dielectric strength
— f — entre conductores m - [telecom-cond]
dielectric resistance between conductor(s)
— f disminuida - decreased resistance
— f — a rotura f - [mec] decreased breaking
strength
— f —— crecimiento m - [autom-neumát] cut
growth resistance
— f disminuida @ rotura f - [mec] decreasing
breaking strength
— f divisora - [electrón] divider,ding resistor
— f — con precisión f - [electrón] precision
divider,ding resistor
— f —— voltaje m preciso - [electrón] precise voltage divider,ding resistor
— f — para voltaje m - [electrón] voltage divider,ding resistor
— f efectiva - effective resistance • [electr]
resistivity
— f eléctrica - [electr] electric(al) resistance; electric resistor • resistivity
— f — alta - [quím] high resistivity
— f — baja - [quím] low resistivity
— f — de agua m - [quím] water (electrical)
resistivity
— f —— suelo m - [suelos] soil, electrical
resistance, o resistivity
— f — efectiva - [quím] effective (electric)
resistivity
— f —— alta - [quím] high (effective electric) resistivity
— f —— baja - [quím] low (effective electric) resistivity
— f —— de agua m - [quím] water (electric)
resistivity
— f eléctrica (efectiva) de agua m
[quím] determined water resistivity
— f — (—) —— m medida - [quím] measured
water resistivity
— f — (—) — muestra - [quím] sample resistivity
— f — (—) — f de agua m - [quím] water sample resistivity
— f — (—) — f —— m determinada - [quím]
determined water sample resistivity
— f — (—) — muestra f de agua m medida -
[quím] measured water sample resistivity
— f — (—) — suelo m - [suelos] soil
sample resistivity
— f — (—) — f —— m determinada -
[suelos] determined soil sample resistivity
— f ——— suelo m—[suelos] soil resistivity
— f — (—) determinada - [quím] determined resistivity
— f — (—) en obra f - [quím] field resistivity
— f — (—) infinita - [quím] infinite resistivity
— f — (—) — medida - [quím] measured infinite resistivity
— f — (—) máxima de suelo m - [suelos] maximum soil resistivity
— f — máxima - [quím] maximum restivity
— f —— de suelo m - [suelos] maximum soil
resistivity
— f — mayor - [quím] higher resistivity •
[electr] higher, o greater, electric resistance
— f — menor - [quím] lower resistivity •
[sold] lower electrical resistance
— f — mínima - [quím] minimum resistivity
— f —— de suelo m - [suelos] soil minimum
resistivity
— f elevada—[mec] high, resistance, o strength
— f — a abrasión f - high abrasion resistance
— f —— corrosión f - [metal] high corrosion
resistance
— f ——— f de ácido(s) m - [metal] high
acid corrosion resistance

resistencia f elevada a corrosividad f de ácido
m - [quím] high acid corrosiveness resistance
— f —— temperatura f - high temperature resistance
— f — de soldadura f - [sold] high weld, resistance, o strength
— f en canto m - [mec] edge strength
— f — corriente f—[electr] current resistance
— f —— f alterna - [electr-oper] alternating
current resistance
— f —— f continua - [electr-oper] direct
current resistance
— f —— f — en arollamiento - [electr-equip]
winding direct current resistance
— f —— f —— m de bobina f - [electr-
-equip] coil winding direct current resistance
— f —— f —— m —— f patrón - [electr
-equip] base coil winding direct current
resistance
— f en ensayo m - test strength, o resistance
— f —— m con tres cantos m - [constr] three
edge test strength
— f —— m —— m para sustentación f -
[constr] three edge bearing test strength
— f — flujo m - [petról] flow resistance
— f — junta f - [mec] joint strength
— f — laboratorio m - laboratory strength
— f — obra f - [constr] field resistance
— f — serie f - [electr] series resistor
— f — viga f - [constr] beam strength
— f — voladizo m - [constr] cantilever resistance
— f equivalente - equivalent strength
— f específica - [electr] resistivity
— f —a compresión f - [constr] specified compressive strength
— f especificada - specified strenqth
— f — máxima—[mec] maximum specified strength
— f —— a compresión f - [mec] maximum specified compressive strength
— f — mínima—[mec] minimum specified strenqth
— f —— a compresión f - [mec] minimum specified compressive strength
— f — para fluencia f - [metal] specified
yield strength
— f estructural - [constr] structural strength
— f — adecuada - [constr] adequate structural
strength
— f — destacada - outstanding structural
strength
— f — real - [constr] actual structural
strength
— f excelente - excellent, resistance, o
strength
— f — a agrietamiento m - [sold] excellent
crack(ing) strength
— f —— corrosión f - excellent corrosion resistance
— f —— soplo m magnético (de arco m)—[sold]
excellent resistance to @ arc blow
— f excepcional - [mec] exceptional, o outstanding, o super, o extra high. resistance,
o strength
— f — a agrietamiento m - [sold] exceptional,
o outstanding, crack(ing) resistance
— f —— corrosión f - [metal-prod] exceptional corrosion resistance
— f —— desgaste m - [metal] outstanding wear
resistance
— f ——— m debido a impacto(s) m - [metal]
outstanding impact wear resistance
— f — a impacto(s) m - [metal] outstanding impact resistance
— f — contra deformación f - exceptional deflection resistance
— f —— desviación(es) f - [metal] exceptional deflection resistance
— f — de soldadura f contra agrietamiento m -
[sold] exceptional weld cracking resistance
— f externa - external resistance
— f extraordinaria - extraordinary resistance;
super strength
— f final - ultimate strength

resistencia f final de acero m - [metal] ultimate steel, strength, o stress
— f — de costura f - [mec] seam ultimate
strength
— f — — — f longitudinal - [mec] longitudinal seam ultimate, strength, o stress
— f — — — f — empernada - [mec] bolted longitudinal seam ultimate strength
— f — — — f — en plancha(s) f estructu-
ral(es) - [mec] bolted structural plate longitudinal seam ultimate strength
— f — — — tubería f - [tub] ultimate pipe, o
pipe ultimate, strength
— f graduable - [electr] adjustable resistance
— f grande—high, o great, strength • véase
también resistencia f alta
— f — a flambeo m—[mec] great column strength
— f — como columna f - [mec] great column
strength
— f hidráulica - [hidr] hydraulic, o water, resistance
— f in-situ - in-sito, o on site, resistance
— f — — de suelo m - suelos] in-situ soil resistance
— f inductiva - [electr] inductive resistance
— f inherente—inherent, resistance, o strength
— f — de conducto m - [tub] conduit inherent
stength
— f — — tubería f - [tub] pipe inherent, o
inherent pipe, strength
— f inicial - initital, resistance, o force •
early strength
— f — alta - [constr] high early strength
— f — — con fraguado m rápido - [constr] high
early stength
— f — de perno m - [mec] initial bolt force
— f insertada - [electrón] inserted resistor
— f interna - internal resistance
— f lateral - lateral, resistance, o strength
— f — de suelo m - [suelos] lateral soil, o
soil lateral, resistance
— f — — viga f - [constr] beam lateral
strength
— f — — — f maestra - [constr] girder lateral strength
— f — — — f para grúa f - [grúas] crane
girder lateral strength
— f longitudinal - [mec] longitudinal strength
— f — máxima - [mec] ultimate longitudinal
strength
— f — — — f de costura f - [mec] ultimate, o
maximum, longitudinal seam strength
— f magnética - [electr] magnetic resistance
— f mala - bad, o poor, resistance
— f — a corrosión f - [metal] bad, o poor,
corrosion resistance
— f máxima - maximum, o best, o ultimate. o
upper, resistance, o strength
— f a abrasión f - [sold] maximum abrasion resistance
— f — — — agrietamiento m - [sold] maximum, crack(ing) resistance
— f — — — m debido a inmovilización f total
- [sold] best resistance to high restraint
cracking; high restraint cracking best resistance
— f — — — m — — proporción f alta de azufre m - [sold] best resistance to cracking
due to high sulfur (content)
— f — — — m — — — f — carbono m -
[sold] best resistance to cracking due to
high carbon (content)
— f — — — m longitudinal - [sold] longitudinal cracking best resistance
— f — — — m transversal - [sold] cross
cracking best resistance
— f — — compresión f - [metal] ultimate compressive strength
— f — corrosión f - [metal] maximum, o
highest, corrosion resistance
— f — — destello(s) m a través de fundente m
- [sold] best, flash through resistance, o
resistance to flash through

resistencia f máxima a extrusión f - [mec] maximum extrusion resistance
— f — — fluencia f - [metal] maximum yield,
strength, o resistence
— f — — impacto(s) m - [metal] highest impact
value • [sold] maximum impact
— f — — picadura(s) f - [sold] best, pock
marking resistance, o. resistance to pock
marking
— f — a porosidad f - [metal] best, porosity
resistance, o resistance to porosity
— f — — — f debida a contaminación f orgánica - [metal] best organic contamination porosity resistance; best resistance to organic
contamination porosity
— f — — — f — óxido m - [metal] best
rust porosity resistance
— f — — — f — soplo m (magnético) de arco m - [sold] best resistance to arc (back)
blow porosity
— f — — presión f - maximum resistance to,
pressure, o compression
— f — — soldadura f con pasadas f múltiples -
[sold] best resistance to multiple pass weld-
(ing); best multiple pass welding resistance
— f — — tensión f - [sold] ultimate, upper, o
maximum tensile strength
— f — — — f en límite m de fluencia f -
[metal] maximum yield(ing) strength
— f — — tracción f - [mec] maximum tensile
strength
— f — — voltaje(s) m - [sold] best, o maximum, voltage resistance
— f — — — m alto(s) - [sold] best. high
volate resistance, o resistance to high volt-
age(es)
— f — admisible - [mec] ultimate, o maximum,
allowable strength
— f — contra desgaste m - [mec] maximum wear
resistance
— f — de acero m - [metal] ultimate, o maximum, steel, stress, o resistance
— f — — conexión f - [mec] maximum connection
force
— f — — costura f - [mec] ultimate, o maximum, seam strength
— f — — — f longitudinal - [mec] ultimate, o
maximum, longitudinal seam strength
— f — — — f transversal - [mec] ultimate, o
maximum, crosswise seam strength
— f — — junta f - [mec] ultimate joint
strength
— f — — — f traslapada - [constr] ultimate
lap joint strength
— f — — tubería f - [tub] ultimate pipe, o
pipe ultimate, strength
— f — exigida - maximum required strength
— f mayor - higher, o improved, resistance, o
strength • greatest, o highest, resistance
— f — a abrasión f - [sold] improved, o higher, abrasion resistance
— f — — calor m - higher heat resistance
— f — — corrosión f - better, o higher, corrosion resistance
— f — — corrosividad f - higher corrosiveness
resistance
— f — — — f de ácido - higher acid corrosion
resistance
— f — — estallido(s) m - [mec] greater
burst(ing), resistance, o strength
— f — — impacto(s) m - [metal] higher impact,
resistance, o property • highest impact property
— f — — paso m de corriente f - [electr-equip]
higher electrical resistance
— f — — tensión f - [metal] greater, o higher, tensile strength
— f — de viga f—[constr] greater beam strength
— f mecánica - mechanical, resistance, o
strength
— f — alta - [mec] high mechanical strength
— f — buena - [metal] high mechanical resistance

resistencia f mecánica de sínter m—[metal-prod]
  sinter mechanical resistance
— f — débil - [mec] low mechanical resistance
— f — equivalente - equivalent mechanical
  strength
— f — mayor - higher mechanical resistance •
  highest mechanical resistance
— f media - average resistance | a - [metal]
  (con) medium tensile
— f — a tensión f - [metal] medium tensile
  strength | a - [metal] (con) medium tensile
— f mejor - best resistance • better resistance
— f — a abrasión f - [metal] better abrasion
  resistance • best abrasion resistance
— f — — agrietamiento m - [sold] best cracking resistance
— f — — — m longitudinal - [sold] best long-
  gitudinal crack(ing) resistance
— f — — impacto(s) m - [metal] better impact
  resistance • better notch toughness • best
  impact resistance
— f — — porosidad f - [sold] better porosity
  resistance • best porosity resistance
— f mejorada - improved, o better, resistance
  • [metal] improved toughness
— f — a impacto(s) m - [metal] improved notch
  toughness
— f — — — m con temperatura(s) f baja(s) -
  [metal] improved, o better, low temperature
  notch toughness
— f menor - lower resistance
— f — a calor m - lower heat resistance
— f — — corrosión f - lower corrosion resistance
— f — — corrosividad f - lower corrosiveness
  resistance
— f — — estallido(s) m - [metal] lower, o
  smaller, burst(ing) strength
— f — paso m de corriente f (eléctrica) -
  [electr-oper] lower electrical resistance
— f — — temperatura f - lower temperature resistance
— f — — tensión f - [metal] lower, o smaller,
  (tensile) strength
— f mediocre - mediocre resistance
— f — — corrosión f - [metal] mediocre corrosion resistance
— f mínima - minimum, resistance, o strength
— f — a compresión f - [constr-píl] minimum,
  compressive strength, o resistance to compression
— f — — fluencia f - [metal] minimum (creep)
  yield, limit, o strength
— f — — tensión f - [metal] minimum yield
  strength • [sold] minimal, o minimum, tensile
  strength
— f — — — f determinada - [sold] specified
  minimum (tensile) strength
— f — — — f en límite m para fluencia f -
  [metal] minimum yield strength
— f — — — f especificada - [sold] specified
  minimum (tensile) strength
— f — — tracción f - [metal] minimum tensile
  strength
— f — contra desgaste m - [mec] minimum wear
  resistance
— f — de aislamiento m - [electr-cond] minimum
  insulation, o insulation minimum, resistance
— f — especificada - [mec] specified minimum,
  o minimum specified, strength
— f — — contra tensión f - [mec] minimum specified tensile strength
— f — — — fluencia f - [metal] specified
  minimum yield strength
— f — — de material m - [sold] minimum material strength
— f — exigida - minimum required strength
— f — requerida - minimum required strength
— f minimizada - minimized resistance
— f Mullen - [papel] Mullen (resistance)
— f muy alta - [mec] very, o extra high, resistance, o strength
— f necesaria - necessary, resistance, o

strength
resistencia f necesaria como columna f -
  [constr] necessary column strength
— f negativa - negative resistance
— f nominal - nominal resistance • resistance
  rating
— f — a abrasión f - [metal] abrasion resistance rating
— f — — impacto(s) m - [metal] impact resistance rating
— f normal - normal resistance • normal
  strength
— f observada - [sold] observed strength
— f óhmica - [electr-oper] ohmic resistance
— f óptima - optimum, resistance, o strength
— f — a presión f - optimum resistance to a
  compression
— f original - original resistance • original
  strength
— f para aceleración f - [electr] accelerating
  resistor
— f — arranque m - [mec] starting resistor
— f — compensación f - [electr-equip] ballast
  resistor
— f — — f estabilizadora - [electr-instal]
  stabilizing ballast resistor
— f — fundición f - [electr] cast iron resistor
— f — descarga f - [electr] discharge resistor
— f — — f para campo m - [electr] field discharge resistor
— f — — — m para derivación f -
  [electr] shunt field discharge resistor
— f — desexcitación f - [electr-equip] disen-
  ergizing resistor
— f — devanado m - [electr-equip] winding resistance
— f — . . . día(s) f - [constr-hormig] . . .
  day(s) strength
— f — ensayo(s) m - test(ing) strength
— f — — m en laboratorio m - laboratory test-
  (ing) strength
— f — frenado m - [mec] braking strength
— f — — m de contramarcha f - [electr-oper]
  plugging resistor
— f — freno m - [electr-equip] brake resistor
— f — — m dinámico - [electr-equip] dynamic
  brake resistor
— f — instalación f - [electr-instal] installation resistance • installation strength
— f — — f para conexión f con tierra f -
  [electr-instal] grounding installation resistance
— f — — f — puesta f a tierra f - [electr-
  instal] grounding installation resistance
— f — limitación f - [electr-equip] limiting,
  o limitation, resistor
— f — — f de corriente f - [electr-equip]
  current limiting,tation resistor
— f — — — f para carga f - [electr-
  equip] charging current limiting,tation resistor
— f — manipuleo m - [constr] handling strength
— f — protector m - protector resistance
— f — — m para voltaje m - [electr-equip]
  voltage release resistor
— f — — m — m nulo - [electr-equip] no
  voltage release resistor
— f — recubrimiento m - [sold] coating resistance
— m — — m para desmenuzamiento m - [sold]
  coating breakdown resistance
— f — soporte m - [constr] support(ing)
  strength
— f — — m exigida - [constr] required sup-
  port(ing) strength
— f — — m requerida - [constr] required sup-
  port(ing) strength
— f — sostén m - [constr] supporting strength
— f — sustentación f - supporting strength
— f — — f determinada - [constr] determined
  supporting strength
— f — — f en obra f - [constr] field sup-

port(ing) strength
resistencia f **para trabajo** m - [constr] working
strength
— f **parcial** - partial strength
— f **perdida** - lost, strength, o resistance
— f **permisible** - [electrón] allowable resistance
— f — **en circuito** m - [electrón] circuit allowable resistance
— f — — — m **de conmutador** m - [electrón]
switch circuit allowable resistance
— f **pertinaz** - [fam] dogged resistance
— f **plena** - [constr] full strength
— f — **a aplastamiento** m - [constr] full crushing strength
— f — — **pandeo** m - full buckling strength
— f **pobre** - poor, o bad, o weak, resistance
— f — **conversión** f - poor conversion resistance
— f — — **corrosión** f - [metal] poor, o bad,
corrosion resistance
— f **portante** - [constr] supporting, o bearing,
resistance, o strength
— f — **de conducto** m - [constr] conduit supporting strength
— f — — — m **rígido** - [constr] rigid conduit
supporting strength
— f — **exigida** - [constr] required supporting
strength
— f — — **segura** - [constr] required safe supporting strength
— f — **requerida** - [constr] required supporting
strength
— f — — **segura** - [constr] required safe supporting strength
— f **positiva** - positive resistance
— f **prevista** - [mec] foreseen, o anticipated,
force, o resistance
— f **probada** - proven, o tested, resistance
— f **proporcionada** - provided resistance
— f **provista** - [mec] provided, resistante, o
strength
— f **quitada** - [electrón] removed resistor
— f **real** - actual, resistance, o strength
— f **reducida** - reduced, o decreased, resistance
— f — **a abrasión** f - [mec] low abrasion resistance
— f — — **corrosión** f - [metal] low corrosion
resistance
— f — — — f **de ácido(s)** m - [quím] low acid
corrosion resistance
— f — — **corrosividad** f **de ácido(s)** m - [quím]
low acid corrosiveness resistance
— f — — **rodadura** f - [autom-neumát] low rolling resistance
— f **torsional** - [mec] rolling torque
— f — **de conjunto** m (**a girar**) - [mec] assembly
rolling torque
— f **total** - total resistance • aggregate
strength
— f — **a rotura** f - [metal] aggregated breaking
strength
— f — **calculada** - calculated, o estimated,
total resistance
— f — **de alambre** m - [alambre] aggregate, o
total, wire strength
— f — — — m **a rotura** f - [alambre] total, o
aggregate, wire breaking strength
— f — — — **cable** m - [cabl] aggregate, o total,
rope strength
— f — — — — m **a rotura** f - [cabl] aggregate, o total, wire rope breaking strength
— f **tubular** - tubular resistance
— f **unitaria** - unit strength • unit stress
— f **variable** - variable resistance • [electr-
-equip] variable resistor • rheostat
— f — **abierta** - [electr-instal] open(ed) variable resistor
— f — **cerrada** - [electr-instal] closed variable resistor
— f — **conectada** - [electr-oper] closed vari-

able resistor
resistencia f **variable de campo** m - [electr-
-instal] field variable resistance
— f — — **circuito** m - [electr-instal] circuit
variable resistance
— f — — — m **para campo** m - [electr-instal]
field circuit variable resistance
— f — **conectada** - [electr-instal] connected
variable resistance
— f **verificada** - [electr-oper] checked resistance
resistente a - . . . .; resistant; strong; sturdy;
husky; rugged; heavy; substantial • hard
— a — **abrasión** f - abrasion, o wear, resistant
— a — **aceite** m - oil resisting,ta
— a — **ácido** - acid resisting,tant
— a — **agrietamiento** m - crack resisting,tant
— a — **agua** m - water resisting,tant
— a — **aplastamiento** m - [cabl] crush resistant
— a — **asentamiento** m - [suelos] settlement resisting,tant
— a — **ataque(s)** m **químico(s)** - [quím] chemical
attack resisting,tant
— a — **calor** m - heat resisting,tant
— a — — m, **humedad** f, **y aceite** m - heat,
moisture, and oil resisting,tant
— a **carga(s)** f - load(ing) resisting,tant
— a — — f **con impacto** m - shock, o impact,
load(ing) resisting,tant
— a — **condición(es)** f **atmosférica(s)** f -
[electr] atmospheric condition(s) resistant
— a — **corrosión** f - corrosion, o rust, resisting,tant
— a — **choque(s)** m - [mec] shock resisting,tant
— a — **daño(s)** m - damage resisting,tant
— m — **desastres(s)** - disaster resisting,tant
— a — **desgaste** m - [mec] wear resisting,tant
— a — **deslizamiento(s)** m - [metal] creep resisting,tant
— a — **dislocación(es)** f - disjointing resisting,tant
— a — **elemento(s)** m - [meteorol] weather resisting,tant
— a — — m **corrosivo(s)** - corrosive element(s)
resisting,tant
— a — **envejecimiento** m - age resisting,tant
— a — **esfuerzo(s)** m - stress resisting,tant
— a — **fatiga** f - fatigue resisting,tant
— a — **fuego** m - [seguríd] fire resisting,tant
— a — **golpe(s)** m - [mec] shock(s), o blows(s),
resisting,tant
— a — **herrumbre** m - [metal] rust, o corrosion,
resisting,tant
— a — **humedad** f - moisture resisting,tant
— a — **impacto(s)** m - impact resisting,tant
— a — **inclemencia(s)** f **de tiempo** m [**malo**] -
weatherproof
— a — **infiltración** f - infiltration resisting,tant
— a — **intemperie** f - weatherproof; outside
weather resisting,tant
— a — **magulladura(s)** f - bruise resisting,tant
— a — **mal tiempo** m - weatherproof
— a — **oxidación** f - rust resisting,tant
— a — **ozono** m - ozone resisting,tant
— a — **rotura(s)** f - break(age) resisting,tant
— f — — f **por impacto(s)** m - impact breakage
resisting,tant
— a — **sacudida(s)** f - [mec] shakeproof
— a — **tensión** f - [mec] stress resisting,tant
— a — — f **interna** - [mec] stress resisting,tant
— a — **tiempo** m - weather resisting,tant • age
resisting,tant
— a — **vibración(es)** f - [mec] vibration(s) resisting,tant
— a **contra fuego** m - fire resisting,tant
resistido,da a - resisted • withstood • [mec]
dragged
— a **satisfactoriamente**—resisted satisfactorily
resistimiento* m - resisting • withstanding
resistir v - . . . .; to withstand; to stand up;
to combat; to last • [mec] to drag

resistir v abrasión f - [metal] to resist @ a-
brasion
— v — f moderada f - [mec] to resist @ moderate abrasion
— v acción f - to resist @ action
— v — f de viento m - to resist @ wind action •
to not blow away
— v acumulación f - [sold] to resist @ build-up
— v — f de (elementos m de) aleación f -
[sold] to resist @ alloy build-up
— f — f — silicio m - [sold] to resist @ silicon build-up
— f adherencia(s) f - [sold] to resist @ sticking
— v — f de salpicadura(s) f - [sold] to resist
@ spatter sticking
— v agrietamiento m - [sold] to resist @ cracking
— v — m debido a deformación f - [metal] to
resist @ cracking from deformation
— v — — impacto(s) m - [metal] to resist
@ cracking from impact(s)
— v — m longitudinal - [sold] to resist @ longitudinal cracking
— v — m transversal - [sold] to resist @
crosswise cracking
— v aplastamiento m - [mec] to resist @, crushing, o collapsing
— v — m de suelo m - [suelos] to resist @ soil
collapsing
— v asentamiento m - [suelos] to resist @ settling,tlement
— v calor m - to resist @ heat
— v compresión f - to resist @ compression
— v — f elevada—to resist @ heavy compression
— v bajo condición(es) f de economía f - [mec]
to, resist, o carry, economically
— v — f — seguridad f - [mec] to carry
safely
— v bien v - to, resist, o wear, well
— v — a desgaste m - [mec] to wear well
— v carga(s) f - [mec] to resist @ load(s)
— v — f con impacto(s) m - to resist @, impact, o shock, load(ing)
— v — f desequilibrada - to resist @ unbalanced load(ing)
— v — f estática - [constr] to resist @ static
load(ing)
— v compresión f - to resist @ compression
— v — f anular - [constr] to resist @ ring
compression
— v con seguridad f—[segurid] to resist safely
— v concusión f - [explos] to resist @ concussion
— v corriente f - [electr] to withstand @ current
— v corrosión f - to, resist, o fight, @, corrosion, o rust
— v — f leve - [metal] to resist @ mild corrosion
— v choque(s) v - [mec] to resist @ shock(s)
— v — m térmico - to, resist, o withstand, @
thermal shock
— v — m — severo - [mec] to, resist, o withstand, @ severe thermal shock
— v daño m - to resist @ damage
— v derrumbamiento m - [hidr] to resist @ washout • [constr] to resist @ collapse,sing
— v desastre m - to resist a disaster
— v desgaste m - to resist @ wear • to wear
— v — m como si fuera(n) de acero m - [mec] to
wear like, steel, o iron
— v —m de metal m contra metal m - [mec] to
resist @ metal-to-metal wear
— v desmoronamiento m - [constr] to resist @
collapse,sing
— v destrucción f - to resist @ destruction
— v dislocación f - [mec] to resist @ disjointing
— v doblamiento m - [mec] to resist @ bending
— v elemento(s) m - to resist @ element(s)
— v — m corrosivo(s) - to resist @ corrosive
element(s)

resistir v empuje m - [mec] to withstand @
thrust
— v esfuerzo m - to resist @ stress
— v extrusión f - [mec] to resist @ extrusion
— v falla f - to resist @ failure
— v flexión f - to withstand @ bending
— v fuego m - [segurid] to resist @ fire
— v golpe(s) m - [mec] to resist @, shock(s), o
blow(s)
— v impacto(s) m - [mec] to resist @ impact(s)
• to support @ impact
— v — m de tránsito m - [vial] to, support, o
withstand, @ traffic impact(s)
— v — m moderado(s) - [sold] to resist @ moderate impact(s)
— v — m severo(s) - [sold] to resist @ severe impact(s)
— v infiltración f - to resist @ infiltration
— v intemperie f - [meterol] to resist @ weather
— v irradiación f—[nucl] to resist @ radiation
— v más que - to, outlast, o out tough
— v momento m - to resist @ moment(um)
— v — m de flexión f - [mec] to resist @
bending moment(um)
— v oxidación f - [metal] to resist @, oxidation, o corrosion, o rust
— v pegamiento m - [mec] to resist @ sticking
— v penetración f - to resist @ penetration
— v presión f - [constr] to, resist, o withstand, @ pressure
— v — f de aplastamiento m - [constr] to resist, o withstand, @ collapse,sing pressure
— v — f — — m de suelo m - [suelos] to resist @ soil collapse,sing pressure
— v — f — derrumbamiento m - [constr] to resist @ collapse,sing pressure
— v — f — desmoronamiento m - [constr] to resist @ collapse,sing pressure
— v — f — suelo m - [suelos] to resist @ soil
pressure
— f radiación f - [nucl] to resist @ radiation
— v rotura f - [mec] to resist @ breakage,king
— v — f por impacto(s) m - [mec] to resist @
impact(s) breaking,kage
— v satisfactoriamente - to resist satisfactorily
— v sin deformar(se) v - to, resist, o withstand, without deformation,ming
— v socavación f - [hidr] to resist @, undercutting, o undermining
— v suelo m - [suelos] to resist @ soil
— v torcimiento m - [mec] to resist @, bending,
o twisting
— v tratamiento m - to, resist, o withstand, @
treatment
— v — m rudo - to withstand @ rough treatment
— v vibración(es) f - [mec] to resist @ vibra-
tion(s)
— v volteo m - [mec] to resist @ overturning
resistividad f - [electr] resistivity • véase
también resistencia f, efectiva, o eléctrica
— f máxima f - [electr] maximum resistivity
— f — de cobre m - [metal] copper maximum, o
maximum copper, resistivity
resistivo,va* a - [electr] resistive
resistor* m - [electr-equip] véase resistencia f
resma f de . . . hoja(s) f - [papel] . . . sheet
ream
— f . . . pliego(s) m - [papel] . . . sheet
ream
resoldado,da a - [sold] rewelded; resoldered •
reworked
resoldadura f - [sold] reweld(ing); resol-
der(ing) • rework(ing)
— f costosa - [sold] expensive, reweld(ing), o
resolder(ing), o rework(ing)
— f de conductor m - [electr-instal] wire, o
conductor, o lead, rewelding, o resoldering
— f — — m para puente m - [electrón] jumper
wire resoldering
resoldar v - [sold] to, reweld, o resolder • to
rework

resoldar v conductor m para puente m - [electr]
   to resolder @ jumper wire
resolución f - . . .; solving; working out •
   (ac)commodating,tion • [pol] order; decision
   • [legal] action; vote
— f de ecuación f - [matem] equation, solution,
   o solving
— v — empate m - tie breaking
— f — problema m—[matem] problem (re)solving
— f fácil - [matem] easy solution • easy ac-
   comadating,tion
— f — — m de erosión f - erosion problem
   resolution,lving
— f — adjudicación f - award(ing) decision
— f particular - specific decision
— f suprema f - [pol] executive decision
resolvedor m - solver
— m de problema(s) - problem solver
resolver v - . . .; to solve • to decide • to
   handle • to work out; to accomodate • to
   overcome • to analyze • to dissolve • [legal]
   to take action
— v adjusicación f - to decide @ award(ing)
— v atascamiento m - [ind] to dislodge @ jam
— v — m de chatarra f - [metal-prod] to dislodge @, cobble, o scrap jam
— v equation - [matem] to solve @ equation
— v empate m - to break @ tie
— v fácilmente - to (re)solve easily • to accomodate easily
— v parcialmente - to solve partially
— v problema m - to (re)solve @ problem • to
   overcome • [int] to trouble shoot
— v — m de erosión f - to (re)solve @ erosion
   problem
resoplido m - . . . . • blowing
— m de aire m - [ind] air puf(fing)
resorte m ajustado - [mec] tight(ened), o set, o
   adjusted, spring
— m alzador - [mec] lift(ing) spring
— m amortiguador m - [mec] snubber spring
— m anular - [mec] coil spring
— m apretado - [mec] tight(ened) spring
— m arrollado - [mec] wound spring
— m capilar - [mec] hair spring
— m — roto - [mec] broken hair spring
— m cilíndrico - [mec] cylinder,drical spring •
   coil spring
— m compensador - [mec] balance,cing spring
— m compresible—[mec] compressible,sion spring
— m comprimible - [mec] compression spring
— m comprimido - [mec] compressed spring
— m con hoja(s) f - [mec] leaf spring
— m — f roto - [mec] broken leaf spring
— m — lámina(s) f - [mec] leaf spring
— m — ojo(s) m - [mec] spring with @ eye(s)
— m cónico - [mec] conical spring
— m — mueblero - conical upholstering spring
— m — para tapicería f - conical upholstering
   spring
— m corto - [mec] short spring
— m — para tapicero,ría - [mec] short upholstering spring
— m cuadrado - [mec] square spring •
— m — para tapicero,ría - square upholstering
   spring
— m cuartoelíptico - [mec] quarter elliptical
   spring
— m de acero m - [mec] steel spring
— m — — m apretado - [mec] tight steel spring
— m — — m arrollado—[mec] wound steel spring
— m — — m muy apretado - [mec] tightly
   wound steel spring
— m — — m inoxidable - [mec] stainless
   steel spring
— m defectuoso - [mec] defective, o faulty,
   spring
— m delantero - [autom] front spring
— m doble - [mec] dual spring
— m en espiral m - [mec] coil, o helical,
   spring • wound spring • spring winding
— m — voladizo—[mec] cantilever (type) spring
— m — Z - [mec] Z spring
— m enganchado - [mec] hooked spring

resorte m especial - [mec] special spring
— m — para válvula f - [válv] valve special,
   o special valve, spring
— m exterior - [mec] outer spring
— m extrarresistente - [mec] heavy duty spring
— m — especial - [mec] special heavy duty
   spring
— m — para válvula(s) f - [válv] special
   heavy duty valve spring
— f — para válvula(s) f - [válv] heavy duty
   valve spring
— m fiador - [mec] detent spring
— m — con tornillo m - [mec] screw detent
   spring
— m — — m con bola f - [mec] ball screw
   detent spring
— m — para eje m—[mec] shaft retainer spring
— m — — para estrangulador m - [comb.-
   -int] choke shaft retainer spring
— m fijado - [mec] set spring
— m fuera de ajuste m - [mec] out of adjustment, o unadjusted, spring
— m fuerte m - [mec] strong, o heavy, spring
— m helicoidal - [mec] helical spring
— m igualador - [mec] equalizing spring
— m impulsor - [mec] drive,ving spring
— m inoxidable - [mec] stainless spring
— m inspeccionado - [mec] inspected spring
— m interior - [mec] inner, o internal, o inside, spring
— m — para espiga f—[mec] shank inner spring
— m laminado - [mec] laminated spring
— m liviano - [mec] light (series) spring
— m — para locomotora f - [mec] locomotive
   light (series) spring
— m mal ajustado - [mec] misadjusted, o poorly
   adjusted, spring; loose spring
— m mecánico - [mec] mechanical spring
— m mueblero - upholstering spring
— m — cónico - conical upholstering spring
— m — corto - [mec] short upholstering spring
— m — cuadrado - square upholstering spring
— m para alivio m - [mec] relief spring
— m — — m para presión f - [mec] pressure
   relief spring
— m — — — f de aceite m - [mec] oil
   pressure relief spring
— m — anclaje m - [mec] anchor spring
— m — asidero m - [mec] handle spring
— m — — m para tope m - [mec] stop handle
   spring
— m — bloque m - [mec] block spring
— m — — m para sujetador m - [trefil] grip-
   (ping) block spring
— m — sujetador m - [mec] grip(ping) block
   spring
— m — — m para avance m gradual—[trefil]
   inching grip block spring
— m — bolilla f - [mec] ball spring
— m — — f para retención f - [mec] poppet
   (ball) spring
— m — borne m - [mec] stud spring
— m — — m para portaescobilla(s) f—[electr-
   -mot] brushholder stud spring
— m — brazo m - [mec] arm spring
— m — — m tensor - [mec] tension arm spring
— m — can m - [mec] dog spring
— m — cierre m - [mec] lock(ing) spring
— m — compresión f - [mec] compression spring
   • compression ring
— m — comprimir v - [mec] compression spring
— m — conmutador m - [electr-equip] switch
   spring
— m — contacto m - [electr-equip] contact(or)
   spring
— m — — m principal - [electr-equip] main
   contactor spring
— m — cuña f - [mec] wedge spring
— m — — f para cierre m - [mec] locking
   wedge spring
— m — dedo m - [electr-equip] finger spring
— m — — m para contactador m—[electr-equip]
   contact(or) finger spring
— m — dispositivo m - [mec] device spring

resorte m̲ para dispositivo m̲ para cambio(s) m̲ -
[autom-mec] shift(ing) unit spring
— m̲ — eje m̲ - [mec] shaft, o̲ axle, spring
— m̲ — — m̲ para estrangulador m̲ - [comb.int]
choke shaft spring
— m̲ — émbolo m̲ - [mec] piston spring
— m̲ — embrague m̲ - [mec] clutch spring
— m̲ — empuñadura f̲ - [mec] handle spring
— m̲ — escobilla f̲ - [electr-mot] brush spring
— m̲ — — f̲ reemplazado - [electr-nit] replaced
brush spring
— m̲ — — f̲ roto - [electr-mot] broken brush
spring
— m̲ — eslabón m̲ - [mec] link spring
— m̲ — — m̲ para regulador m̲ - [comb.int] governor link spring
— m̲ — espiga f̲ - [mec] shank spring
— m̲ — fiador m̲ - [mec] retainer spring
— m̲ — — m̲ para estrangulador m̲ - [comb.int]
choke retainer spring
— m̲ — freno m̲ - [mec] brake spring
— m̲ — gatillo m̲ - [mec] trigger, o̲ dog, spring
— m̲ — gobierno m̲ - [mec] control spring
— m̲ — leva f̲ - [mec] cam spring
— m̲ — limitador m̲ - [mec] stop, o̲ limit, spring
— m̲ — para acelerador m̲ - [comb.int]
throttle stop spring
— m̲ — mando m̲ - [mec] control spring
— m̲ — mango m̲ - [herram] handle spring
— m̲ — manija f̲ - [mec] handle spring
— m̲ — montaje m̲ - [mec] mounting spring
— m̲ — palanca f̲ - [mec] lever spring
— m̲ — — m̲ para válvula f̲ - [mec] valve lever
spring
— m̲ — — f̲ — — f̲ neumática - [mec] air valve
lever spring
— m̲ — — f̲ — — f̲ — para torno m̲ - [petról]
cathead air valve lever spring
— m̲ — — f̲ — — f̲ para aire m̲ - [mec] air
valve lever spring
— m̲ — — f̲ — — f̲ para torno m̲ - [petról]
cathead air valve lever spring
— m̲ — perilla f̲ - [mec] knob spring
— m̲ — — f̲ — tope m̲ - [mec] stop handle spring
— m̲ — pestillo m̲ - [mec] latch spring
— m̲ — — m̲ para lengüeta f̲ - [mec] tongue
latch spring
— m̲ — portaescobilla(s) m̲ - [electr-mot]
brushholder spring
— m̲ — posicionador* - [mec] positioner spring
— m̲ — presión f̲ - [mec] pressure spring
— m̲ — — f̲ para rodillo m̲ - [mec] roll pressure spring
— m̲ — — f̲ — — m̲ impulsor - [sold] driving
roll pressure spring
— m̲ — — f̲ — — m̲ — exterior - [sold] outside drive,ving roll pressure spring
— f̲ — — f̲ — — m̲ — superior - [sold] upper
drive,ving roll pressure spring
— m̲ — regulador m̲ - [mec] governor, o̲ control,
spring
— m̲ — — m̲ ajustado - [mec] adjusted tovernor
spring
— m̲ — — m̲ mal ajustado - [mec] misadjusted
governor spring
— m̲ — retorno m̲ - [mec] return(ing) spring
— m̲ — — m̲ para impulsor m̲ - [electr-mot]
drive,ving return spring
— m̲ — — m̲ — — m̲ para piñón m̲ - [electr-mot]
pinion drive,ving return spring
— m̲ — — m̲ reajustado - [mec] readjusted return spring
— m̲ — rodillo m̲ - [mec] roll spring
— m̲ — — m̲ impulsor - [mec] drive,ving roll
spring
— m̲ — — m̲ — inferior - [mec] lower
drive,ving roll spring
— m̲ — — m̲ — superior - [mec] upper
drive,ving roll spring
— m̲ — sello m̲ - [mec] seal(ing) spring
— m̲ — servicio m̲ - [mec] service spring
— m̲ — — m̲ pesado - [mec] heavy duty spring
— m̲ — sujetador m̲ - [mec] cam, o̲ holder, spring

resorte m̲ para tapicería f̲ - upholstering spring
— m̲ — tensión f̲ - [mec] tension spring
— m̲ — — f̲ roto - [mec] broken tension spring
— m̲ — tope m̲ - [mec] stop spring
— m̲ — tornillo m̲ - [mec] screw spring
— n̲ — — m̲ hueco - [mec] hollow screw spring
— m̲ — — m̲ — para caja f̲ - [mec] box hollow
screw spring
— m̲ — — m̲ — — f̲ sujetadora - [mec] grip
box hollow screw spring
— m̲ — — m̲ limitador - [mec] stop screw spring
— m̲ — — m̲ — para acelerador m̲ - [comb.int]
throttle stop screw spring
— m̲ — — m̲ — caja f̲ - [mec] box screw spring
— m̲ — — m̲ — — f̲ sujetadora - [mec] grip box
screw spring
— m̲ — — m̲ piloto - [mec] pilot screw spring
— m̲ — — m̲ — para acelerador m̲ - [comb.int]
throttle pilot screw spring
— m̲ — — m̲ limitador - [mec] stop screw spring
— m̲ — torsión r̲ - [mec] torsion spring
— m̲ — trinquete m̲ - [mec] pawl spring
— m̲ — válvula f̲ - [mec] valve spring
— m̲ — — f̲ neumática - [mec] air valve spring
— m̲ — — f̲ — para torno m̲ - [petról] cathead
air valve spring|
— m̲ — válvula f̲ - [válv] valve spring
— m̲ — — f̲ para aire m̲ - [válv] air valve
spring
— m̲ — — f̲ — — m̲ para torno m̲ - [petról]
cathead air valve spring
— m̲ — — f̲ — alivio m̲ - [mec] relief valve
spring
— m̲ — — f̲ — — m̲ para presión f̲ - [mec]
pressure relief valve spring
— m̲ — — f̲ para servicio m̲ - [válv] service
valve spring
— m̲ — — f̲ — — m̲ pesado - [válv] heavy duty
valve spring
— m̲ — — f̲ — — m̲ para presión f̲ de aceite m̲
- [mec] oil pressure relief valve spring
— m̲ — varilla f̲ - [mec] rod spring
— m̲ — — f̲ para empuje m̲ - [mec] thrust, o̲
push, rod spring
— m̲ — vástago m̲ - [mec] rod spring
— m̲ — — m̲ para desplazador m̲ - [mec] shifter
rod spring
— m̲ — — m̲ — — m̲ para embrague m̲ - [mec]
clutch shifter rod spring
— m̲ — — m̲ — — m̲ — — m̲ inferior - [mec]
lower clutch shifter rod spring
— m̲ — — m̲ — — m̲ — — m̲ superior - [mec]
upper clutch shifter rod spring
— m̲ — — m̲ empuje m̲ - [mec] push rod spring
— m̲ — vida f̲ útil (prolongada) - [mec] long
life spring
— m̲ patentado - [mec] patented spring
— m̲ principal - [mec] main spring
— m̲ protector - [mec] spring guard; guard(ing)
spring
— m̲ quitado - [mec] removed spring
— m̲ reajustado - [mec] readjusted spring
— m̲ recuperador - [mec] recovering, o̲ return,
spring
— m̲ reemplazado - [mec] replaced spring
— m̲ reforzado - [mec] heavy duty spring • reinforced spring
— m̲ regulado - [mec] adjusted, o̲ set, spring
— m̲ removido - [mec] removed spring
— m̲ retornado - [mec] returned spring
— m̲ roto - [mec] broken spring
— m̲ sacado - [mec] removed spring
— m̲ semielíptico - [mec] semielliptical spring
— m̲ transversal - [mec] cross(wise) spring
— m̲ trasero - [mec] rear spring
— m̲ tres cuartos elíptico - [mec] three quar-
ter(s) elliptic(al) spring
— m̲ y sello m̲ - [mec] spring and seal
respaldado,da a - backed up • supported
resaldar v̲ - . . .; to back up; to stand behind
• to support
— v̲ seguro m̲ - [seguros] to back up @ insurance
— v̲ soldadura f̲ - [sold] to back up @ weld

respaldar v venta f - [com] to, back up, o stand behind, @ sale
respaldo m - . . . .; back, side, o face; rear (face) • back(ing) up; standing behind • support(ing) • backing up • [domést] back rest • [imprent] back cover • [mec] shoulder • véase también apoyo m
— m ajustado - [autom] adjusted back rest
— m con brida(s) f - [hidr] flanged back
— m corrugado - [mec] corrugated back
— m de acero m - [sold] steel, back-up, o back--(ing)
— m — acoplamiento m - [mec] coupling back (side)
— m — — m para freno m - [mec] brake coupling back side
— m — — m — — m auxiliar - [mec] auxiliary brake coupling back side
— m — bastidor m - [mec] frame back
— m — bomba f - [bombas] pump rear
— m — caja f - [mec] case back
— m — — f para regulación f - [sold] control box, back, o rear
— m — cono m - [mec] cone back
— m — — m exterior - [mec] outer cone back face
— m — — m — de cojinete m - [mec] outer bearing cone back face
— m — corona f dentada - [mec] ring gear back
— m — consola f - [mec] console, back, o rear
— m — — f para perforador m - [petról] driller's console back
— m — cordón m - [sold] bead back(side)
— m — émbolo m - [mec] piston back (side)
— m — hierro m - [sold] iron backing
— m — hormigón m - [constr] concrete backing
— m — horquilla f - [mec] yoke back(-face)
— m — — f para aporte m [de fuerza f) - [autom-mec] input yoke back-face
— m — junta f - [sold] joint back side
— m — — f con chaflán m doble - [sold] double V joint back side
— m — letrero m - [constr] sign back
— m — — m de acero m - [constr] steel sign back
— m — — — m para carretera f - [constr] steel highway sign back (side)
— m — — m para carretera f - [constr-vial] highway sign back
— m — malacate m - [petról] drawworks back
— m — máquina f - [ind] machine, back, o rear
— m — sección f - section, back, o rear
— m — — f para regulación f - [sold] control section, back, o rear
— m — soldadora f - [sold] welder, o case. back. o rear
— m — soporte m - [mec] support back(side)
— m — — m para desplazador m - [mec] shifter support back(side)
— m — — — m para dispositivo m (para cambios m) - [mec] shifter support back(side)
— m — t(e) - [mec] tee back side
— m efectivo - effective, o concrete, backing
— m — contra reacción f - effective, o concrete, reaction backing
— m golpeable - [mec] rear striking face
— m golpeteado - [mec] tapped back-face
— m integral - [mec] integral back-up
— m — de acero m—[mec] integral steel back-up
— m para caja f - [mec] box, o case, back
— m — horquilla f - [mec] yoke back-face
— m — — f para aportación f (de fuerza f) golpeado - [autom-mec] tapped input yoke back-face
— m — producto(s) m - [ind] product back-up
— m — seguro m - [seguros] insurance backup
— m — venta(s) f - [com] sale(s) support
— m permanente - permanent back-up
— m técnico - [ind] technical, support, o back--up
— m — fuerte - [ind] strong technical, support, o back-up
— m — sólido - [ind] strong technical, support

port, o back-up
respectivo,va a . . . . • applicable • concerned • thereof
— a tierra f - [electrón] referenced to @ ground
respecto adv de sí mismo,ma - self respect
respeto m máximo - maximum, o utmost, respect, o attention
— m mutuo - mutual respect
— m propio - self respect
respiración f - . . . .; breath
respiradero m - breather; outlet; vent (hole) • respirator • [constr] atmospheric vent; riser • [combust] flue • [miner] air shaft; chimney • [ind] breather; vent • [comb.int] breather • [metal-prod] air port • [fam] loophole
— m comprobado - [ind] checked breather
— m contra explosión(es) f - [ambient] blast gate
— m en tapa f - [comb.int] cap vent hole
— m — — f para depósito m - [comb.int] tank cap (air) vent hole
— m — — f — m para combustible m - [comb.-int] fuel tank cap (air) vent hole
— m estándar - [tub] standard riser
— m limpiado - cleaned breather
— m limpio - clean breather
— m obturado - [comb.int] plugged (air) vent hole
— m para aire m - [constr] air vent • [miner] air shaft
— m — — m para depósito m - [comb.int] tank air vent
— m — — m — — m para combustible m - [comb.-int] fuel tank air vent
— m — caja f - [mec] housing breather
— m — calefacción f - [ambient] heating, vent, o outlet
— m — cárter m - [comb.int] crankcase, vent, o breather
— m — depósito m - [comb.int] tank (air) vent
— m — — m para combustible m - [comb.int] fuel tank (air) vent (hole)
— m — reductor m - [mec] reducer air vent hole
— m — — m para engranaje(s) m - [mec] gear reducer air vent (hole)
— m — — m para velocidad f - [mec] gear reducer (air) vent (hole)
— m — ventilación f -ventilating,tion outlet
— m verificado - checked breather
respirador m - [segurid] respirator • [mec] véase también respiradero m
— m aprobado - [segurid] approved respirator
— m apropiado - [segurid] appropriate, o suitable, respirator
— m con obturador m - [válv] poppet breather
— m — — m flotante - [válv] floating poppet breather
— m empleado - [segurid] used respirator
— m guardado - [segurid] stored respirator
— m limpiado - [segurid] cleaned respirator
— m limpio - [segurid] clean respirator
— m llevado - [segurid] worn respirator
— m usado - [segurid] used respirator
resplendente a - resplendent; brilliant
respondedor m - . . . • [electrón] responder
— m en base f - [electrón] base responder
— m para base f - [electrón] base responder
respondedor,ra a - responsive • responding
responder v - . . . . • to be, answerable, o obliged,gated, o bound • to, meet, o comply • [fin] to pay
— v a circuito m - [electron] to respond to @ circuit
— v — especificación f - to meet, o correspond to, @ specification
— v — exigencia f - to meet @ requirement
— v — frecuencia f - [electrón] to (cor)respond to @ frequency
— v — teléfono m - [telecom] to answer @ (telé)phone
— v — tono m - [electrón] to respond to @ tone
— v — variación(es) f - to respond to @ varia-

tion(s)
responder v **instantáneamente** - to respond instantly,taneously
— v **prontamente** - to respond promptly
— v **rápidamente** - to, respond, o answer, quickly, o promptly
respondido,da a - responded; answered
— a **prontamente** - responded, o answered, quickly, o promptly
responsabilidad f - . . . . • task; function; duty • accountability (objective) • role •
[legal] liability
— f **absoluta** - absolute responsibility
— g **aceptada** - accepted responsibility • accepted liability
— f **actual** - current responsibility
— f **administrativa** - [admin] management responsibility
— f — **crítica** - [admin] critical management
responsibility
— f **asignada** - assigned responsibility:
— f — **en materia** f **de seguridad** f - [segurid]
assigned safety responsibility
— f **asumida** - assumed responsibility
— f — **para cumplimiento** m **de contrato** m -
[legal] contractually, assumed, o stated,
performance liability
— f **completa** - complete responsibility
— f **confiada** - entrusted responsibility
— f **crítica** - critical responsibility
— f — **accionista** m - [legal] stockholder's, o
shareholder's, liability
— f — **administración** f - [com] management('s)
responsibility
— f — **analista** m - [ind] analyst('s) responsibility
— f — **cargo** m - [ind] accountability objective
• position('s) responsibility
— f — **componedor** m - [quím] acompounder('s).
job, o responsibility|
— f — **comprador** m - buyer('s) responsibility
— f — **concesionario** - [com] dealer('s), o con-
cessionaire('s), responsibility
— f — **contratista** m - contractor('s) responsibility
— f — **limpieza** f - [ind] clean-up responsibility
— f — **distribuidor** m - [com] distributor('s),
o dealer('s), responsibility
— f — **ordenamiento** m - [ind] housekeeping responsibility
— f — — m **en sitio** m **de trabajo** m - [ind]
housekeeping responsibility
— f — **personal** m - [com] staff responsibility
— f — — m **de administración** f - [com] staff
management('s) responsibility
— f — — m **superior (de administración** f) -
[com] staff management responsibility
— f — — m **supervisor** - [com] line management
responsibility
— f — **proponente** m - bidder('s) responsibility
— f — **representante** m - [com] dealer('s) o
representative('s), responsibility
— f — **supervisor** m - supervisor('s) responsibility
— v — **vendedor** m - [com] vendor('s), o sel-
ler('s), responsibility, o liability
— f **delegada** - delegated responsibility
— f **denegada** - denied, responsibility, o liability
— f **desconocida** - unknown responsibility • denied responsibility
— f **diluida** - diluted, o clouded, responsibility
— f **económica** - economic responsibility
— f **en materia** f **de seguridad** f - [segurid]
safety responsibility
— f **establecida** - established responsibility
— f **eventual** .- [fin] eventual, o contingent,
liability
— f **exclusiva** [legal] exclusive, o sole, responsibility, o obligation
— f **final** - [fin] final, o ultimate, responsi-

bility
responsabilidad f **financiera** - [fin] financial,
responsibility, o liability
— f **forzosa** - mandatory, o necessary, responsi-
bilty, o function
— f **fundamental** - fundamental responsibility
— f **ilimitada**—[legal] unlimited responsibility
— f **impositiva** - [fisc] tax liability
— f **indivisa,sible** - [legal] indivisible responsibility
— f **legal** - [legal] legal responsibility
— f **limitada** - limited responsibility • [legal]
limited liability
— f **llevada** - [admin] carried, o borne, responsibility
— f **nublada** - clouded responsibility
— f **ofuscada** - clouded responsibility
— f **para con comprador** m - responsibility to-
-(wards) @ purchaser
— f — **proyección** f - design responsibility
— f **plena** - full responsibility
— f **por cumplimiento** m - performance liability
— f — **planura,nitud*** f - [metal-lam] flatness
responsibility
— f — **resultado(s)** m - [admin] result(s) accountability
— f — **superficie** f - [metal-lam] surface responsibility
— f **posterior** - later responsibility
— f **presente** - present, o current, responsibility
— f **sobre reaseugor(s)** m - [seguros] reinsurance liability
— f — — m **autorizado(s)** - [seguros] authorized reinsurance liability
— f — — m **no autorizado(s)** - [seguros] unauthorized reinsurance liability
— f — — m **autorizado(s)** m - [seguros] authorized insurance liability
— f — — m **no autorizado(s)** - [seguros] unauthorized insurance liability
— f **soldaria** - [legal] solidary responsibility
— f **técnica** - [ind] technical responsibility
— f — **crítica** - [admin] critical technical
responsibility
— f **total** - total, o full, responsibility
— f **tributaria** - [fisc] tax liability
responsabilizar v - to make responsible
responsabilizarse v - to be responsible
responsable m - responsible party • incumbent •
[ind] supervisor |
a - . . .; accountable • responsive
— a **contingentemente** - [legal] contingently
liable
— m **determinado** - given responsible party
— a **personalmente** - personally, responsible, o
accountable
— m — **compromiso(s)** m - [fin] commitment(s)
liable
responsablemente adv - responsibly
responsivo,va a - responsive
respuesta f - . . . . • feedback • solution •
[electrón] response
— f **a circuito** m - [comput] response to @ circuit
— f **adaptable** - versatile, response, o answer
— f **afirmativa** - affirmative, answer, o reply
— f **arribada** - arrived at answer
— f **automática** - automatic answer • [comput]
auto-answer
— f **de contratista** m - contractor('s), answer,
o reply
— f — **voltaje** m - [electr-oper] voltage, reaction, o response
— f **de frecuencia** f - [comput] frequency response
— f — **realimentación** f - [electrón] refeeding
response
— f **desfavorable** - unfavorable answer
— f **diferente** - different answer • [comput]
different response
— f **distinta** - different answer • [comput]
different response

respuesta f favorable - favorable, answer, o reply, o response
— f inmediata - immediate, o instantaneous, reply, o answer, o response
— f instantánea - instantaneous response
— f más frecuente - more frequent, o predominant, answer • most frequent answer
— f más rápida - [electrón] quicker response
— f negativa - negative, answer, o reply
— f obtenida - received, o gotten answer
— f óptima - [electrón] optimum response
— f posible - possible answer
— f positiva - positive, answer, o reply
— f probada - proven, o tested, answer
— f — en producción f - [ind] production tested, answer, o reply • tested response
— f pronta - prompt, o quick, o early, reply
— f rápida - quick, answer, o reply • [electr] quick, o prompt, response
— f sencilla - simple answer
— f sensata - sensible answer
— f sugerida - suggested, answer, o response
— f típica - typical response
— f transitoria—[electrón] transitory response
— f — de realimentación f - [electrón] transitory refeeding response
— f única - single, answer, o reply • [electr] single response
resquebrado,da a - cracked
resquebradura f - crack(ing)
— f por fatiga - [metal] fatigue crack(ing)
restablecer v - . . .; to reset
— v automáticamente - to reset automatically
— v contacto m - [electr-oper] to open @ short; to reestablish @ contact
— v — m a través de reóstato m para avance m gradual - [sold] to open @ short across @ inch(ing) rheostat
— v soplado m - [metal-prod] to restart @, (air) blast, o wind; to recover @ blast; to begin @ blast recovery
— m automático - automatic reset(ting)
restablecido,da a - restablished; restarted; reset
— a automáticamente - [ind] restarted, o reset, automatically
— m automático - automatic reset(ting)
restado a - [matem] subtracted
restante a - . . . . • other
restar v válido - to remain valid
restauración f de compresión f - compression restoration,ring
— f de rentabilidad f - profitability restoration,ring
— f — utilidad f - usefulness restoration,ring
— f — zona f explotada - [miner] mined area restoration,ring
— f — — f minada - [miner] mined area restoration,ring
— f para poder volver a emplear - [sold] restoring for service
— f — rentabilidad f - [econ] profitability restoration,ting
restaurante m - [culin] . . .; dining, o eating, facility, o establishment
— f carretero - [transp] roadside eating establishment
— m flotante - [culin] floating restaurant
— m gastronómico - [culin] gourmet, dining, o eating, facility, o establishment
restaurar v compresión f - [comb.int] to restore @ compression
— v nivel m - [hidr] to restore @ level
— v — m de agua - [hidr] to restore @ water level
— v — m — — m freática - [hidr] to restore @ ground water level
— v para poder volver a emplear v - [sold] to restore to @ service
— v resistencia f - to restore @ resistance
— v — f alta - to restore @ high resistance
— v — f a abrasión f - [sold] to restore @ abrasion resistance
— f — f a corrosión f - [sold] to restore @ corrosion resistance
restaurar v utilidad f - to restore @ usefulness
— v zona f explotada - [miner] to restore @ mined area
— v — f minada—[miner] to restore @ mined area
restitución f - . . . •[fin] reimbursement; repaying,yment
restituido,da a - restored • [fin] repaid
restituir v - . . . • [fin] to, reimburse, o repay
— v rentabilidad f - to restore @ profitability
resto m - . . . • [mec] cut-off • [miner] tailing
— m de camino m - [vial] road remainder
— m — colada f - [metal-prod] casting residue
— m — instalación f - [tub] line remainder
— m — planta f - [ind] plant remainder
— m — red f - [tub] line, o system, remainder
— m — temporada f - [deport] season remainder
— m — tubería f - [constr] pipe,ping remainder
restregado,da a - scoured
restregadura f - . . .; scouring • scraping
restregamiento m - . . .; scouring • scraping
restregar v - . . .; to scour; to brush
restricción f - . . .; restraint; curtailing • confinement; confining • [petról] capping • [comb.int] choking
— f aumentada - increased restraint
— f cualitativa - qualitative restriction
— f cuantitativa - [com] quantitative restriction
— f de aire m - air restricting,tion
— f — ancho(r) m - width restricting,tion
— f — — m de calle f - [vial] street width restricting,tion
— f — circulación f - circulation restricting,tion
— f — componente(s) m - component(s) restricting,tion
— f — compresión f - compression confining
— f — depurador m (para aire m) - [comb.int] air cleaner restricting,tion
— f — desplazamiento - displacement,cing restriction • [hidr] flow restriction
— f — — m de aceite m - oil flow restriction
— f — — m — petróleo m - [petról] oil flow restriction
— f — escape m - [comb.int] exhaust restriction
— f — filtración f - filtration, o leakage, restriction
— f — filtro m - filter restriction
— f — — m para aire m - [comb.int] air filter restricting,tion
— f — — — m para motor m - [comb.int] motor air filter restricting,tion
— f — — m doble - [comb.int] double air filter restricting,tion
— f — — — m para motor m - [comb.int] motor air double filter restricting,tion
— f — junta f - [sold] joint restraint
— f — pérdida f - loss. o leakage, restricting,tion
— f — pasaje m - passage restricting,tion
— f — paso m passage restricting,tion
— f dentro de núcleo m - restricting,tion (with)in @, core, o nucleus
— f elevada - [metal] high, restraint, o restricting,tion
— f en circulación f - circulation restricting,tion
— f — — de aire m - air ciculating,tion restricting,tion
— f — disponibilidad f - availably, o supply, restricting,tion, o tightening
— f — — f de electrodo(s) m - [sold] electrode supply tightening
— f — práctica f - practice,cal restriction
— f — suministro m - supply tightening
— f — — m de electrodo(s) m - [sold] electrode supply tightening
— f — uso m - use restricting,tion
— f excesiva - [sold] excessive, restraint, o restricting,tion
— f impuesta por embalaje m - packaging restricting,tion

restricción f lateral - [mec| side restraint;
  lateral restricting,tion
— f mayor - increased, restricting,tion. o restraint
— f monetaria—[fin] monetary restricting,tion
— f para altura f - height restricting,tion
— f — embarque m - [transp] shipping restriction • shipping clearance
— f — profundidad f - [constr] depth, limit, o
  restriction
— f — — f de zanja f - [constr] trench depth,
  limit, o restriction
— f práctica - practical restriction
— f quitada - removed restriction
— f removida - removed restriction
— f sacada - removed restriction
— f severa - severe, o stiff, restriction
restringido,da a - restricted; restrained; confined; constrained; curtailed • tight(ened) •
  [petról] capped • [comb.int] choked
— a dentro de núcleo m - restricted within @,
  core, o nucleus
restringimiento* m - véase restricción f
restringir v - . . .; to refrain; to constrict;
  to curtail; to hold to @ minimum; to confine;
  to constrain • [petról] to cap • [comb.int]
  to choke
restringirse v - [constr] to underdesign
— v ancho(r) m - to restrict @ width
— v — m de calle f - [vial] to restrict @
  street width
— v circulación v - to restrict @ circulation
— v — f de aire m - to restrict @ air circulation
— v componente m - to restrict @ component
— v compresión f - to, restrict, o confine, @
  compression
— v dentro de núcleo m - to restrict within @,
  nucleus, o core
— v depurador m - [comb.int] to restrict @
  cleaner
— v — m para aire m - [comb.int] to restrict @
  air cleaner
— v desplazamiento m - to restrict @ displacement • [hidr] to restrict @ flow
— v — m de aceite m - to restrict @ oil flow
— v — m petróleo m - [petról] to restrict @
  oil flow
— v disponibilidad f - to, restrict, o tighten,
  @, availability, o supply
— v escape m - [comb.int] to restrict @ exhaust
— v filtración f - to restrict @, filtration, o
  leakage
— v pasaje m - to restrict @ passage
— v paso m - to restrict @ pass(age)
— v pérdida f - to restrict @, loss, o leakage
— v suministro m - to, restrict, o tighten, @
  supply
resuelto,ta a - . . . . • resolved • decided; settled; assured; solved; worked out • accomodated
— a fácilmente—easily, resolved, o accomodated
resultado m - . . . . • consequence; effect •
  performance • accomplishment • product; outcome • turning out • finding • outgrowth •
  ending up • [contab] profit and loss
— m acumulado - accumulated result(s)
— m ajustado - adjusted result
— m aplicado - applied result
— m asegurado - assured result
— m a lograr(se) v - result to be, reached, o
  established, o accomplished
— m bueno - good result • good advantage
— m aceptable - acceptable result
— m compatible - compatible result
— m con voltaje m constante - [sold] constant
  voltage performance
— m — — m variable - [sold] variable voltage
  performance
— m confiable - reliable result
— m constante - cons(is)tant result
— m correcto - correct result
— m correlacionado - correlated result

resultado m correspondiente - corresponding, o
  related, result
— m dañino - harmful, o damaging, result
— m de accidente m - [segurid] accident result
— m — administración f - [admin] management
  result
— m — análisis m - analysis result • survey
  outcome
— m — conjunto m - [labor] assembly result
— m — — m de taller m - [labor] combined
  shop result(s)
— m — desempeño - performance result(s)
— m — — m real - actual performance result(s)
— m — ensayo m - test, result, o outcome
— m — — m de banda f - [metal-prod] strip
  test(s) result(s)
— m — — m — — f magnética - [metal-prod]
  magnetic strip test result(s)
— n — — — carga f - load(ing) test(s)
  result(s)
— m — — m — cinta f - [metal-lam] strip test
  result(s)
— m — — m — — f magnética - [metal-lam]
  magnetic strip test result(s)
— m — — m — chapa f - [metal-prod] strip
  test result(s)
— m — — — — f magnética - [metal-lam]
  magnetic strip test result(s)
— m — — m — empalme,madura - [electr-instal]
  splice,cing test result(s)
— m — — — fleje m - [metal-lam] strip test
  result(s)
— m — — m — m magnético - [metal-lam] mag-
  magnetic strip stest result(s)
— m — — m — propiedad(es) f - property,ties
  test(ing) result(s)
— m — — m — — f mecánica(s) - [metal] mechanical property,ties test(ing) result(s)
— m — — m — suelo m - [suelos] soil test-
  -(ing) result(s)
— m — escala f industrial - [ind] industrial
  scale result(s)
— m — explotación f - [miner] development, o
  operating,tion, result(s)
— m — exposición f - exposing,sition result(s)
— m — investigación f - investigation, o survey, result
— f — operación f - operating,tion result(s)
— m — principio m - [admin] principle result
— m — producción f - [ind] production result
— m — prueba f - test result(s)
— m — — f en laboratorio m - laboratory test
  result(s)
— m — soldadura f - [sold] weld(ing), performance, o result(s)
— m — taller m - [ind] shop result(s)
— m — transporte m - [transp] shipping re-
  sult(s)
— m — uso m - use result(s)
— m desastroso - disastrous result
— m deseado - desired result
— m — determinado - determined desired result
— m — logrado m - [admin] accomplished desired
  result
— m determinado - determined result
— m diferente - different result
— m directo - direct result
— m — de accidente m - [segurid] direct accident, o accident direct, result
— m distinto - different result
— m duradero - (long) lasting result
— m — asegurado - assured (long) lasting result
— m efectivo - effective result
— m equilibrado - balanced result
— m establecido - established result
— m estético - aesthetic result
— m exact - exact, o accurate, result
— m experimental - experimental result
— m favorable - favorable result
— m final - final, o end, result
— m gratificador,cante - gratifying result
— m impresionante - impressive result

resultado m indirecto - indirect result
— m — m de accidente m - [segurid] accident indirect, o indirect accident, result
— m individual - individual result
— m — de taller m - individual shop result
— m inmediato - immediate result
— m inteligible - clear cut result
— m logrado - [admin] obtained, o achieved, o accomplished, result
— m más efectivo - [admin] more effective result • most effective result
— m mejor - best result • better result
— m — asegurado - assured best result • assured better result
— m numéricamente correcto - numerically correct result
— m numérico - numerical result
— m obtenido - [admin] accomplished, o secured, o attained, o obtained, result
— m operativo - [contab] operating result
— m optimista - optimistic result
— m óptimo - optimum result
— m para curva f = curve result
— m — ejercicio m - [contab] fiscal year('s), result, o profit or loss
— m — gobierno m - handling result
— m peor - worse result • worst result
— m pobre - poor result
— m prometedor - promising result
— m real - actual, o true, result, o outcome
— m respectivo - respective result • related result
— m satisfactorio - satisfactory result • [educ] satisfactory test
— m — de prueba f - satisfactory test result
— m significativo - meaningful result
— m sin afectación f - [contab] unassigned profit and loss
— m sorprendente - surprising result
— m técnico - [ind] technical result
— m — buscado - sought technical result
— m útil - useful result • usable result
— m verificado - checked, result, o finding
— m vinculado a operación(es) f - [admin] operation(s) related result
— m valorado - [admin] assessed result
— m valuado—[admin] assessed, o valued, result
resultado,da a - resulted • turned out
resultados m - [contab] profit and loss
resultando adv - resulting
— adv en - resulting in • operating through
resultante a - . . . .; arising • spinoff
resultar v - . . . . • to, add, o end, up; to work • to arise • to prove • [deport] to come away with
— v agradable - to be agreeable • to be fun
— v atrayente - to be attractive; to have (operator) appeal
— v — a operador m - to have operator appeal
— v cierto,ta - to, be, o hold, true
— v claro - clearly; to be clear
— v como mandado hacer - [fam] to fit; to not hurt
— v contraproducente - to defeat @ purpose
— v conveniente - to be convenient; to pay off
— v corroído,da - to become corroded
— v dañado,da - to become damaged
— v de - to, result, o stem, from
— v — interés m - interestingly
— v divertido,da - to be fun
— v en - to result in • to cause
— v evitable - to become avoidable
— v factible - to become feasible
— v ganador - [deport] to, become, o come away, @ winner
— v interesante—to be, interesting, o exciting
— v inevitable - to become unavoidable
— v mejor - to become best • to work best
— v necesario - to be(come), necessary, o required | adv - (de) if necessary
— v posible - to be(come) possible
— a segundo - [deport] to take @ second
— v triunfante - [deport] to come away @ winner

resultar valioso,sa - to pay off
resumen m actualizado - [contab] updated, summary, o statement, o report
— m audiovisual - [electrón] audiovisual review
— m cualitativo - qualitative summary
— m cuantitativo - quantitative summary
— m de cuenta f - [com] statement (of account)
— m — f actualizado - [contab] updated, report, o statement (of account)
— m — f para propaganda - [com] advertising statement
— f — f con coparticipación f—[com] co-op advertising statement
— m — ensayo(s) m - [ind] test(s) summary
— f — m en laboratorio - [ind] laboratory tests summary
— m — expresión f - [gram] expression abbreviation
— m — investigación f - investigation summary
— m — norma(s) f - policy,cies summary
— m — — f contable(s) - [contab] accounting policy,cies summary
— m — — f — significativa(s) - [contab] significant accounting policy,cies summary
— m — observación(es) f - observations, o remarks, summary
— m de oferta(s) f - [com] proposal(s), o bid(s), summary
— m — reclamación(es) f - [com] claims summary
— m — — f de cliente(s) m - [com] customer claims summary
— m — valor(es) m - [fin] security,ties summary
— m — venta(s) f - [com] sales summary
— m escrito - written summary
— m general - general summary
— m histórico - [hist] historic(al) summary
— m semanal - weekly summary
— m — de observación(es) f - weekly, remarks, o observations, summary
resumidero m - [sanit] sump; véase también rezumadero m
resumido,da a summarized; abbreviated
resumir v - to summarize; to sum up
— v expresión f - to summarize @ expression
— v investigación f - to summarize @ investigation
resurgido,da a - resurged
resurgimiento m - . . . .; resurgence
resurgir v - . . . .; to resurge
retacado* m - [náut] caulking; packing • véase también recalcado m; recalcadura f
retacado m - [náut] caulking
retacado,da a - caulked
retacador m - caulker
retacamiento* m - [náut] packing
— m de junta f - [mec] joint packing
retacar v - [mec] to, plug, o pack, o calk, o tamp • to clinch (a rivet) • véase también recalcar; martillar
— v con amianto m - [ind] to pack with asbestos
— v — — contorno m de placa f - [metal-prod] to pack around @ cooling plate with asbestos
— v — asbesto m - [metal-prod] to pack with asbestos
— v — mortero m - [metal-prod] to pack with mortar
— v — — m refractario - [metal-prod] to pack with refractory mortar
— v junta f - [metal-prod] to pack @ joint
— v — f de petaca f - [metal-prod] to pack @ cooling plate joint
— v periféricamente - to pack around @ edge
— v — petaca f - [metal-prod] to pack around @ cooling plate
— v petaca f - [metal-prod] to pack @ cooling plate
retahila f - . . . .; system
retal m - . . . . • [metal] scrap
retallar v - . . . . • [mec] to offset
retallo m - . . . . • [arquit] offset; step
— m para derrame m - [arquit] water table
retardación f - . . . .; retarding; slowing; slow down • lag • [electr-oper] time delay

retardación f de árbol m - [mec] shaft slow-down
— f — arco m - [sold] arc slow-down
— f — — m en cráter m - [sold] crater (arc), control, o slow-down
— f — eje m - [mec] axle, o shaft, slow-down
— f — encendido m - [comb.int] ignition retardation,ding
— f — flujo m - flow retardation,ding
— f — fuga f - leak(age) retardation,ding
— f — unidad f motriz - [agric] power unit slowing (down)
retardado,da a - slowed down; delayed
retardador m - retarder • [sold] timer • [f.c.] retarder • [instrum] time delay
— n automático - automatic retarder • [sold] automatic time delay
— m integral - [electr] built-in time delay
— m para alimentación f - [sold] feed(ing) retarder
— m — — f gradual - [sold] inch(ing) time delay
— m — arco m - [sold] arc, retarder, o control
— m — — m en cráter - [sold] crater control
— m — frecuencia f alta - [sold] pre-flow timer
— m — interrupción f - [sold] after-flow timer
— m — — f de gas m - [sold] after-flow timer
— m — suministro m posterior - [sold] after--flow timer
— m — — m — de gas m - [sold] gas after-flow timer; after-flow gas timer
— m — — m — — m inerte - [sold] inert gas after-flow timer
— m — vagón(es) m - [f.c.] car retarder
retardante a - retarder,ding
retardar v - . . .; to slow down
— v agrietamiento m - [metal] to retard @ crack(ing)
— v chispa f - [comb.int] to, retard, o delay, @, spark, o ignition
— v flujo m - to retard @ flow
— v fuga f - to retard @, flow, o leak(age)
— v hincadura f - [constr] to slow down @ driving
— v oxidación f - [metal] to retard @ oxidation
— v rapidez f - to slow down
— v — de soldadura f - [sold] to slow (down) @ welding speed
— v unidad f motriz - [agric] to slow (down) @ power unit
retardo m - . . .; retarding • lag • slow(ing) (down) • time delay • drag
— m de aceite m - oil lag
— m de chispa f - [comb.int] spark, delay(ing), o retardation,ding
— m — encendido m - [comb.int] ignition, o spark, retardation,ding
— m — flujo m - flow retardation,ding
— m — fuga f - leak(age) retardation,ding
— m — mesa f - [mec] table retardation,ding
— m — — f para enfriamiento m - [metal-lam] cooling table retardation,ding
— m — unidad f motriz - [agric] power unit, slowing, o retardation,ding
— m en desconexión f eléctrica - [sold] burn--back
— m — flujo m - flow lag
— m — interrupción f automática de corriente f a electrodo m - [sold] burn-back
— m preestablecido - [sold] preset time delay
— m prerregulado - [sold] preset time delay
retazo m - [metal-lam] scrap; cut off
— m de acero m - [metal-lam] steel scrap
— m — electrodo m - [sold] electrode butt
— m — hierro m - [metal-prod] iron scrap; scrap iron
— m — material m - scrap material
retemplado,da a - [metal-prod] retempered
retemplar v - [metal-prod] to retemper
retemple m - [metal-prod] retempering
retén m - . . . . • [mec] pawl; retainer; clip; hook; lock; backstope; detent; keeper; bridle; snap lock; stop; tumbler; check; clamp;

dog; holdback • pallet • véase también retención f
retén m colocado - [mec] installed retainer
— m con aro m - [mec] ring retainer
— m — — m para cojinete m - [mec] bearing ring retainer
— m de fibra f - [mec] fiber retainer
— m — — f para tornillo m - [mec] fiber screw retainer
— m — fieltro m - [mec] felt retainer
— m — enderezador - [mec] straightening retainer
— m exterior - [mec] outer,tside retainer
— m flotante - [mec] floating, retainer, o bridle
— m frontal - [mec] front, retainer, o guard
— m — para aceite - [mec] front oil guard
— m instalado - [mec] installed retainer
— m intercambiable - [mec] interchangeable retainer • modular retainer
— m interior - [mec] inner, o inside, retainer
— m invertido - [mec] inverted, o reversed, retainer
— m modular - [mec] modular retainer
— m montado - [mec] mounted retainer
— m para aceite m - [mec] oil, retainer, o guard
— m — — m para cojinete m - [mec] bearing oil guard
— m — — m — — m frontal - [mec] front bearing oil guard
— m — — m — — m posterior - [mec] rear bearing oil guard
— m — — m para jaula f - [mec] cage oil retainer
— m — — m — jaula f para cojinete m - [mec] bearing cage oil retainer
— m — — m — — f — m para bancada f - [mec] main bearing cage oil retainer
— m — agua m - [constr] waterstop
— m — amortiguador m - [mec] dampener retainer
— m — — m para línea f - [mec] line dampener retainer
— m — buje m - [mec] (slide) bushing, retainer, o holder
— m — — m para cursor m - [mec] slide bushing holder
— m — — m — — m para hilera f - [trefil] die slide bushing holder
— m — canaleta f - [mec] trough stop
— m — carrete m - [mec] spool, o reel, retainer
— m — cemento m - [constr] cement retainer
— m — cierre m - [mec] seal, retainer, o holder
— m — — m cargado - [válv] loaded seal retainer
— m — — n instalado m - [mec] installed seal retainer
— m — — m manipulado - [mec] handled seal retainer
— m — — m para aceite m - [mec] oil seal retainer
— m — — m para tapa f - [mec] cover seal retainer
— m — — m — — f para válvula f - [mec] valve cover seal retainer
— m — — f — válvula f - [mec] valve seal retainer
— m — — m, quitado, o removido, o sacado - [mec] removed seal retainer
— m — — m tirado - [mec] pulled seal retainer
— m — cojinete m - [mec] bearing retainer
— m — — m con rodillo(s) m - [mec] roller bearing retainer
— m — — m para bancada f - [mec] main bearing retainer
— m — — m — biela f - [mec] connecting rod bearing retainer
— m — — m — — piñón m - [mec] pinion bearing retainer
— m — — m — vástago m - [mec] rod bearing retainaer
— m — cruceta f - [mec] crosshead retainer
— m — cursor m - [mec] slide holder

retén m para cursor m enderezador - [mec]
  straightening slide retainer
— m para diafragma m - [mec] diaphragm retainer
— m — — m para amortiguador m - [mec] dampener diaphragm retainer
— m — — m — — m para línea f - [mec] line
  dampener diaphragm retainer
— m — — m — — f para aspiración f -
  [mec] suction line dampener diaphragm
  retainer
— m — embrague m - [mec] clutch retainer
— m — engranaje m - [mec] gear retainer
— m — — m planetario - [mec] sun gear retainer
— m — — m solar - [mec] sun gear retainer
— m — enjugador m - [mec] wiper retainer
— m — — m para aceite m - [mec] oil wiper retainer
— m — — m — — m para vástago m - [mec] rod
  oil wiper retainer
— m — — m — — m — — m intermedio - [mec]
  intermediate rod oil wiper retainer
— m — fijación f - [mec] fastening retainer
— m — — f para cojinete m - [mec] bearing
  fastening, retainer, o shim
— m — grasa f - [mec] grease, retainer, o seal
— m — — f en cojinete m—[mec] bearing grease
  seal
— m — guarnición f - [mec] gasket retainer
— m — junta f - [mec] joint retainer
— m — — f para aceite m - [mec] oil joint retainer
— m para llanta f - [mec] rim retainer
— m — — f para rueda f - [mec] wheel rim retainer
— m — — f — — f dentada - [mec] sprocket
  rim retainer
— m para mandil m - [constr] apron step
— m — — m frontal - [constr] (front) apron
  step
— m — matriz f - [mec] die retainer
— m — — f modular - [mec] modular die retainer
— m — muelle m - [mec] spring retainer
— m — — m para válvula f - [comb.int] valve
  spring retainer
— m — pasador m - [mec] pin retainer
— m — — m para cruceta f - [mec] crosshead
  pin retainer
— m — perno m - [mec] locking clip
— m — — m para canjilón n - [constr] stud
  clamp
— m — — m — — m para draga f - [constr]
  dredge stud clamp
— m — petaca f - [metal-prod] cooling plate
  fastener
— f — placa f - [mec] plate retainer
— m — — f para embrague m - [mec] clutch
  plate retainer
— m — — f para fricción f - [mec] friction
  plate retainer
— m — — f — — f para embrague m - [mec]
  clutch friction plate retainer
— m — puerta f - [constr] door, stop, o holdback, o hook
— m — punzón m - [mec] punch retainer
— m — — m modular - [mec] modular punch retainer
— m — resorte m - [mec] spring, clip, o retainer
— m — rodillo m - [mec] roller retainer
— m — — m enderezador - [mec] straightening
  roller retainer
— m — rueda f - [mec] wheel retainer
— m — — f dentada - [mec] sprocket retainer
— m — — f — loca - [mec] idler sprocket retainer
— m — — f loca - [mec] idler retainer
— m — — f — con rodillo(s) m - [mec] idler
  roller retainer
— m — sello m - [mec] seal retainer
— m — — m para aceite m - [mec] oil seal retainer

retén m para sello m para tapa f - [mec] cover
  seal retainer
— m — — m — — f para válvula f - [mec]
  vale cover seal retainer
— m — — m para válvula f - [mec] valve seal
  retainer
— m — — m, quitado, o removido, o sacado -
  [med] removed seal retainer
— m — — m tirado - [mec] pulled seal retainer
— m — soporte m - [mec] holder retainer
— m — tapa f - [mec] cover retainer
— m — tornillo m - [mec] screw retainer
— m — válvula f - [válv] valve retainer
— m posterior - [mec] backstop; back stop •
  rear, guard, o retainer
— † — para aceite m - [mec] rear oil guard
— m, quitado, o removido, o sacado - [mec] removed retainer
retención f - . . .; retaining; holding • entrainment,ning • [fisc] withholding • [hidr]
  storage; collecting,tion • containing,tention
  • [comput] latching • [fotogr] freezing
— f con relé m = [electrón] relay latching
— f — aire m - air entrainment,ning
— f — calor m - [sold] heat retention
— f de camisa f - [petról] liner retention
— f — — f metal contra metal - [petról]
  metal-to-metal retention
— f — carga f - [grúas] load holding
— f — cierre m - seal retaining
— f — conformación f - shape, retaining, o
  holding
— f — cuadro m - [fotogr] frame holding
— f — charnela f - [válv] swing check(ing)
— f — derecho(s) n - [legal] right(s) holding
— f — — n literario(s) m - [legal] copyright holding
— f — empaquetadura f—[mec] packing retaining
— f — — f para camisa f- [petról] liner packing retaining
— f — escurrimiento m - [hidr] flow detaining
— f — — m sobre superficie f - [hidr] overland flow detaining
— f — ingreso(s) m - [fin] earning(s), holding, o retaining
— f — humedad f - [hidr] moisture, holding, o
  retaining
— f — — f en suelo m - [hidr] soil moisture,
  holding, o retaining
— f — impuesto(s) m - [fisc] tax withholding
— f — — m sobre, rédito(s) m, o renta f -
  [fisc] income tax withholding
— f — lama(s) f - [miner] slime, holding, o retaining
— f — marca f - [legal] mark holding
— f — — f de fábrica(ción) f - [legal] trade
  mark holding
— f — † industrial - [legal] trade mark
  holding
— f — — f registrada - [legal] trade mark
  holding
— f — nombre m comercial - [legal] trade name
  holding
— f — pago(s) m - [fin] payment(s) withholding
— f — patente f - [legal] patent holding
— f — — f de invención f - [legal] patent
  holding
— f — posición f - [deport] place holding
— f — — f primera - [deport] first place,
  holding, o retaining
— f — propiedad f intelectual - [legal] copyright hold(ing)
— f → sello m - [mec] seal withholding
— f — tierra f - [constr] earth retaining
— f — tubería f - [tub] piping holding
— f — utilidad(es) - [fin] profit(s), o earn-
  ing(s), withholding, o retaining
— f en cloaca f - [sanit] sewer, holding, o
  retaining
— f — — f lateral - [sanit] lateral sewer,
  holding, o retaining

retención f en cuneta f - [hidr] gutter holding
— f — poder m - retaining in @ power
— f — soporte m - [mec] retaining in @ holder
— f — sumidero m - [hidr] catch basin, detention, o holding
— f fuerte m - tight holding
— f lateral - lateral, o side, holding
— f neta - [maderas] net, retention, o holding
— f nominal - nominal, retention, o holding
— f momentánea - [comput] momentary latch(ing)
— f parcial - partial withholding
— f positiva - [mec] positive, retaining, o
holding
— f substitutiva—[fin] substituted withholding
— f total - total withholding
retenedor m - retainer • holder • [comput] latch
— m común - [comput] common latch
— m — para reajuste m - [comput] common reset
latch
retenedor m - retainer
— m de fieltro m - [mec] felt retainer
— m independiente - [comput] independent retainer
— m — para conjunto m - [comput] independent
group latch
— f — — m para ajuste m - [comput] independent reset group latch
— m — — reajuste m - [comput] independent reset latch
— m — — m para reajuste m - [comput] reset
group latch
— m momentáneo - [comput] momentary latch
— m para cojinete m - [mec] bearing retainer
— m — conjunto m - [comput] group latch
— m — — m para reajuste m - [comput] reset
group latch
— m — — m — — m independiente - [comput]
independent reset group latch
— m para derrumbe(s) m - [petról] cave in
catcher
— m — empaquetadura f - [petról] packing retainer
— m — engranaje m - [mec] gear retainer
— m — — m planetario - [mec] sun, o planetary, gear retainer
— m — — m solar - [mec] sun gear retainer
— m — grava f - [constr] gravel stop(per)
— m para grupo m - [comput] group latch
— m — — m para reajuste m - [comput] reset
group latch
— m — — m — — m independiente - [comput]
independent reset group latch
— m — — reajuste m - [comput] reset latch
— m — — m independiente - [comput] independent reset latch
— m — resorte m - [mec] spring holder
— m — suplemento m - [sold] insert retainer
— m — — m para boquilla f - [mec] nozzle insert retainer
retenedor m, quitado, o removido, o sacado -
[mec] removed retainer
retener v - . . .; to hold (back) • [hidr] to,
collect, o remove • [fisc] to withhold •
[mec] to entrain • [comput] to latch •
[fotogr] to freeze
— v acción(es) f - [fin] to (with)hold @,
share(s), o stock
— v aire m - to entrain @ air
— v calor m - to retain @ heat
— v carga f - [grúas] to hold @ load • [transp]
to hold back @ load
— v cierre n - [mec] to retain @ seal
— v cliente m - [com] to keep @ customer • to
keep @ business
— v conformación f - to retain @ shape
— v — f circular - to, remain round, o to hold
@ circular shape
— v cuadro m - [fotogr] to preeze @ frame
— v derecho(s) v - to hold @ right
— v — m literario(s) - [legal] to hold @ copyright
— v en mente f - to hold in mind; to reflect
— v — poder m - to retain in @ power

retener v ingreso(s) v - to. retain, o hold
back, @, income, o earning(s)
— v magnéticamente - [comput] to latch magnetically
— v marca f - [legal] to, hold, o keep, @ name
— v — f de fabricación f - [legal] to hold @
trade, mark, o name
— v — f industrial - [legal] to hold @ trade,
mark, o name
— v — f registrada - [legal] to hold @ trade,
mark, o name
— v nombre m comercial - [legal] to hold @
trade name
— v parcialmente - to withhold partially
— v patente f - [legal] to hold @ patent
— v — f de invención f - [legal] to hold @
patent
— v polvo m de hierro m - [electr] to hold @
iron powder
— v posición f primera - [deport] to retain @,
first position, o lead
— v propiedad(es) f - to retain @ property,ties
— v — f intelectual - [legal] to hold @ copyright
— v relé m - [electrón] to latch @ relay
— v sello m - [mec] to retain @ seal
— v totalmente - to withhold totally
— v tuerca f - [mec] to retain @ nut
— v utilidad(es) f - [fin] to retain @, pro-
fit(s), o earning(s)
retenida f - . . . • [petról] tie down
— f para poste m de hormigón m - [constr] concrete pole guy
— f — — m — — m con . . . lado(s) m -
[electr-instal] . . .-side concrete pole guy
retenido,da a - retained • (with)held • entrained • [hidr] collected • [fotogr] frozen
— a en poder m - retained, o held, in @ power
— a — soporte m - [mec] retained in @ holder
— m fuertemente - held tightly
— a parcialmente - withheld partially
— a totalmente - withheld totally
retentiva f - retentiveness
retentivamente adv - retentively
retentividad f - . . .; retentiveness; véase
también retentiva f
— v magnética - [electr] magentic retentivity
retícula f - . . .; reticula
reticulado m - [imprent] screen
retiración f - [imprent] back side; inside
— f de contratapa f - [imprent] back cover inside; inside of @ back cover
— f — tapa f - [imprent] (front) cover inside
retirado,da a - . . . • removed • withdrawn • departed • [fisc] withheld • [ind] dropped •
[labor] retired • véase también removido,da
— a de servicio m - withdrawn from @ service
— a fácilmente - removed easily
— a lentamente - drawn away slowly
retirar v - . . .; to remove; to withdraw • to
depart • [ind] to drop • [milit] to draw back
— v colada f - [metal-prod] to interrupt @,
pour, o casting
— f de circulación f - to, remove, o withdraw,
from circulation; to retire
— v — servicio m - to withdraw from @ service
— v — uso - to retire
— v electrodo m - [sold] to, withdraw, o take
out, @ electrode
— v fácilmente - to, remove, o withdraw, easily
— v lentamente - to, remove, o withdraw, o draw
away, slowly
— v oferta f - to withdraw @ offer
— v propuesta f - [com] to withdraw @ proposal
— v sonda f - [ind] to withdraw @ probe
— v sonda f para uso en obra f - [ind] to withdraw @ field probe
retiro m - . . . • withdrawal • removal • departure • [rel] retreat • [ind] cleanout •
véase también remoción f
— m de cuenta corriente - [fin] checking account, withdrawal, o closure
— m — — f de ahorro(s) m - [fin] savings ac-

count, withdrawal, o closure
retiro m de despunte(s) m de laminación f -
mill croppings, o plant scrap, removal
— m de escoria f - [metal-prod] slag removal
— m — oferta(s) f - [com] proposal(s), o of-
fer(s), withdrawal
— m — propuesta(s) f - [com] proposal(s), o
bid(s), withdrawal
— v — sonda f - [ind] probe withdrawal,wing
— m — f para uso m en obra f - [ind] field
probe withdrawal,wing
— m fácil - easy, removal, o withdrawal,wing
retocado n - [ind] retouching; finishing; véase
también **retoque** m
— m manual - [metal-fabr] hand finishing
— m — después de embutición f - [metal-fabr]
after-drawing hand finishing
— m — posterior a estampado m - [metal-fabr]
after-drawing hand finishing
— m posterior a embutición f - [metal-fabr] after-drawing finishing
— m — estampado m - [metal-fabr] after-
-drawing finishing
retocado,da a - touched up; finished
— a con abrasión f - [mec] abrasion retouched
— a — pintura f - [pint] paint retouched
— a — f pulverizada - [pint] spray paint
retouched
— a — f sopleteada - [pint] spray paint
retouched
retocar v anillo m para ajuste m - [electr-mot]
to touch up @ slip ring
— v con abrasión f - to abrasion touch up
retoma f - retaking • regaining • resumption;
resuming
— f de rumbo m - direction resumption,ming
retomado,da a - retaken • regained • resumed
retomador,ra a - resuming,mer
retomar v - to retake • to resume • to regain
— v delantera f - [deport] to, regain, o take
again, @ lead
— v rumbo m - to resume @ direction
retoño m - [botán] . . .; burgeon
retoque m - . . .; touching up • [pint] finishing
— m con abrasión f - [mec] abrasion retouching
— m — anillo m - [mec] ring touch(ing) up
— m — m para ajuste m - [electr-mot] slip
ring touching up
retorcedura f - . . .; twitch(ing) • [cabl]
kink(ing) • [segurid] sprain(ing)
retorcer v - . . . • to squirm • to twitch • to
twist together
— v eslabón m - [cadenas] to twist @ link
retorcido m - . . . • [alambre] twisted finish •
[cabl] lay
— m Lang - [cabl] Lang lay
retorcido,da a - twisted (together) • squirmed;
twitched
• [cabl] kinked
retorcijo m - twitch(ing)
retorcijón m - twitch(ing)
— m en vientre m - [medic] bowel twitch(ing)
retorcimiento m - . . . • twitch(ing); squirming
• twisting together
— m — banda f para rodamiento m - [autom-neum]
tread squirming
— m — eslabón m - [cadenas] link twisting
retornado,da a - returned; brought, o switched,
o come, back
— a elevación f mecánica - [grúas] returned
to power hoist(ing)
— a — nivel m normal - returned to @ normal
level
— a — posición f - returned to @ position
— a — f neutral - [mec] returned to @ neutral (position)
— a — f original - returned to @ original
position
— a — servicio m - returned to @ service
— a automáticamente - [mec] returned automati-
calle; sprung back
— a — a posición f neutral - [mec] sprung back

to @ neutral position
retornado,da de por sí - returned by itself •
[mec] sprung back
retornar v - . . .; to, bring, o switch, back •
[deport] to climb back • véase **volver** v
— v a elevación f mecánica - [grúas] to return
to @ power hoist(ing)
— v — nivel m - to return to @ level
— v — m normal - to return to @ normal
(level)
— v a posición f - to return to @ position
— v — f neutral - [mec] to, return, o
spring back, to @ neutral (position)
— v — operación f - to return to @ operation
— f — servicio m - to return (in)to @ service
— v automáticamente - to return automatically;
to spring back
— v — estrangulador m - [mec] to return, automatically @ choke, o @ choke automatically
— v — — m a posición f - [mec] to automatically return @ choke to @ position
— v — — m — f contenida - [mec] to automatically return @ choke to @ hold position
— v de por sí - [mec] to, return by itself, o
spring back
— v escoria f - [metal-prod] to return @ slag
— v informe m - [comput] to report back
— v mandato m - [comput] to report back @ command
— v orden f - [comput] to report back @ command
— v palanca f - [mec] to return @ lever
— v — f a posición f - [comb.int] to return @
lever to @ position
— v — — f — f lenta sin carga f -
[comb.int] to return @ lever to @ low idle
(position)
— v regulador m - [mec] to return @ control
— m — m para estrangulador m - [mec] to return
@ choke control
— v — m — m a operador m - [mec] to return
@ choke control to @ operator
— v resorte m - [mec] to return @ spring
— v voltaje m - [electr-oper] to return @ voltage
— v — de energía v - [electr-oper] to return @
power voltage
— v — — f a nivel m normal - [electr-oper-
to return @ power voltage to normal (level)
retorno m - . . .; switching back • [comput]
feedback • loopback - [deport] climb(ing)
back
— m a elevación f - [grúas] to return to hoist
— m — f mecánica - [grúas] return to @ power hoist
— v — nivel m - to return to @ level
— m — m normal—to return to @ normal level
— m — operación f - return(ing) to operation
— m — posición f - return)ing) to @ position
— n — — f neutral - [mec] return to @ neutral
(position)
— m — — f original - return(ing) to @ original position
— m — servicio m - return(ing), o restoring,
to service
— m ajustado - adjusted return
— m audible - [electr) audible feedback
— m automático - automatic return - swing(ing)
back
— m — — a neutral - to, return automatically,
o swing back, to @ neutral (position)
— m — — posición f neutral - [mec] automatic
return, o spring(ing) back, to @ neutral position
— m de aceite m - [mec] oil return
— m — agua m - [hidr] water return
— m — aire m - [combust] air return
— m — comida f - food return • [avicult] feed,
return, o intake
— m — condensador m - [ind] condenser return
— m — m enfriado (con agua m) - [ind]
(water) cooled condenser return
— m — corriente f - [electr] current return
— m — drenaje m - [cald] drip return

retorno m de drenaje m de calentador n - [cald]
heater drip return
— m — — m — — m para presión f (inter)me-
dia - [cald] intermediate pressure heater
drip return
— m — émbolo m - [mec] piston return
— m — — m accionado con resorte m - [mec]
piston spring return
— m — escoria f - [metal-prod] slag return
— m — — f de convertidor m - [metal-prod]
converter slag return
— m — freno m - [mec] brake return
— m — gas m - [combust] gas return - gas
back-up
— m — leva f - [mec] cam return
— m — — f cortadora - [mec] cutter cam return
— m — — f para cortadora f - [mec] cutter
cam return
— m — llama f - [combust] return flame -
flareback; backfire,ring
— m — mandato m - [comput] command back reporting
— m — marcha f sin carga f - [mec] idler
spring
— m — orden f - [comput] command back reporting
— m — por sí - [mec] springing back
— m — regulador m - [mec] control return
— m — — m para marcha f sin carga f - [mec]
idler return
— m — resorte m - [mec] spring return
— m — tambor m - [mec] drum return
— m — tobera f - [metal-prod] tuyere return
— m — tope m - [mec] top return
— m — trampa f - [mec] trap return
— m — — f para condensación f - [cald] condensate trap return
— m — voltaje m - [electr] voltage return
— m — — m de energía f - [electr] power
voltage return
— m — — m — — f a nivel m normal -
[electr-oper] power voltage return to normal
— m externo - [comput] external loopback
— m movido por resorte m - [mec] spring return
— m para aceite m - [mec] oil return
— m — — m obturado—[mec] blocked oil return
— m — impulsor m - [mec] drive return
— m — — m para piñon m - [electr-mot] pinion
drive return
— m positivamente sensible - [comput] positive
feedback
— m — — a tacto m - [comput] positive tactile feedback
— m positivo - [comput] positive feedback
— m promovido - promoted return
— m rápido - [mec] quick, o fast, return
— m sano y salvo - safe return
— m sensible - [comput] sensitive feedback
— m sensible a tacto m - [comput] tactile
feedback
— m sobre activo m - [fin] return on @ assets
— m — — m neto m - [fin] return on @ net
assets
— m — capital m - [fin] return on @ capital;
capital return(s)
retorsión f - . . .; twitch(ing)
retortijado,da a - twitched
retortijar v - . . .; to twitch
retortijón m - . . .; twitching
retozado,da a - romped
retozar v - . . . - [labor] to horseplay
retozo m - . . .; romping
retracción f automática - [mec] automatoc retraction
— f — de electrodo m - [sold] automatic electrode retraction
— f — cilindro m - cylinder retracting,tion
— f — — m para dirección f - [autom] steering cylinder retracting,tion
— f — — m — — f mecanizada - [autom] power
steering cylinder retracting,tion
— f — — m para elevación f - [mec] lift(ing)
cylinder retracting,tion

retraccion f de economía f - [econ] economic
retrenchment
— f de electrodo m - [sold] electrode retracting,tion
— f — rodillo m - [mec] roller retracting,tion
— f económica - [econ] economic retrenchment
retractación f - . . .; retracting,tion
retractado,da a - retracted
retráctil a - . . .; retractable [mec] disappearing; telescopic,ping
— a prueba de falla(s) f - fool proof telescoping
— m simple - [mec] single telescopic
retraer v - to, retract, o retreat, o retrench
— v apropiadamente - to retract properly
— v cilindro m - [mec] to retract @ cylinder
— v — — m para dirección f mecanizada -
. [autom] to retract @ power steering cylinder
— v — m — elevación f - [mec] to retract @,
lift(ing), o hoist(ing), cylinder
— v electrodo m - [sold] to retract @ electrode
— v — m de cráter m - [sold] ro retract @
electrode from @ crater
— v rodillo m - [mec] to retract @ roller
retraerse v - to, withdraw, o back away; to
move in
retraído,da a - . . . - [mec] retracted; re-
trenched; withdrawn; retreated · moved in
shy · collapsed
— a totalmente - [mec] totally retracted
retraimiento m - . . .; retraction; moving in ·
[econ] retrenchment; withdrawal; curtailment
— m significativo - [econ] significant, re-
trenching,chment, o curtailing,lment
retransmisión f - [electr] retransmitting; relaying
retrasado,da a - delayed; deferred; lagged;
slowed down; set back; fallen behind ﹥ late;
back
— a considerablemente - considerably, o materially, delayed, o slowed down
retrasar v - to, delay, o slow down, o lag, o
defer · to retard; to set back; to fall behind
— v conjunto m - to delay @ assembly
— v considerablemente - to, delay, o slow down,
considerably, o materially
— v chispa f - [comb.int] to, delay, o retard,
@ spark
— v desconexión f - [sold] to delay @ dropout
— v — f de interruptor m - [sold] to delay @
breaker dropout
— v — f — — m automático - [sold] to delay
@ contactor dropout
— v embarque m - [transp] to delay @
shipment,pping
— v entrega f - [transp] to delay @ delivery
— v trabajo m - to delay @, work, o job, o task
retrasarse v - to fall behind
— v arranque m - [ind] to delay @ restarting
— v colada f - [metal-prod] to delay @ tapping
— v parada f - [ind] to delay @ shutdown
retraso m - . . .; falling behind; deferring;
setback; lagging · arrears ، [deport] deficit
— m costoso - costly delay
— m de aceite m - [comb.int] oil lag
— m — conjunto m - assembly delay(ing)
— m — chispa f - [comb.int] spark, delay(ing),
o retardation,ding, o delay(ing)
— m — embarque m - [transp] shipment,pping
delay(ing); delayed shipment,pping
— m en entrega f - [transp] delivery delay(ing)
— m — trabajo m - work delay(ing)
— m normal - normal delay(ing)
— m preestablecido - preset time delay
— m prerregulado - preset time delay
retribución f - . . .; compensation; fee(s) ·
consideration
— f básica - [labor] basic, o base, rate
— f para director(es) m - [legal; director('s)
fee(s)

retribución f para estímulo m - [pers] incentive
pay
— m — asesoramiento m - [ind] consultant('s)
fee(s)
— f por conocimiento(s) m - [ind] compensation
for @ knowledge
retribuido,da - [labor] compensated
retribuir v - . . .; to compensate
retroactivación f - back dating
— f de cobertura f - [seguros] coverage back-
dating
retroactivo,va a - . . . .; back dated
retroajustado,da* a - retrofitted
retroajustar* v - to retrofit
retroajuste* m - retrofit(ting)
— m de dispositivo m - [mec] device, o unit,
retrofitting
— m — — m para cambio(s) m - [autom-mec]
shift, unit, o device, retrofit(ting)
retroceder v - . . .; to move back; to back,
away, o off - [sold] to back step - [milit]
to fall back
— v a cráter m - [sold] to, back up, o move
back, (in)to @ crater
— v gradualmente - [sold] to inch back(wards)
— v lentamente - [mec] to ease backwards slowly
— v rápidamente - to move back quickly
— m — a cráter m - [old] to move back quickly
into @ crater
— v sobre cordón m (terminado) - [sold] to move
back over @ (finished) weld
retrocedido,da a - backed, off, o up, o away;
moved back
— a lentamente - [mec] eased backwards slowly
retrocesión f - . . . .; backing away
retroceso m - back(ing) up; pull back; back(ing)
off; retrogression • [sold] back step(ping)
[metal-prod] return • [autom] reverse; driving backwards • [mec] back, travel, o pull;
reversing; backing away; reverse gear
— m de electrodo m - [sold] electrode back-up
— m — explosión f - [comb.int] backfire,ring
— m — llama f - [combust] flame back-up
— m — unidad f - [autom] unit reversing
— m gradual - backward(s) inching; inching,
away, o back(wards)
— m — de electrodo m - [sold] inching away
— m — — — m desde trabajo m - [sold] inching
away from @ work
— m máximo - [mec] maximum back-up; extreme
back travel
— m mínimo - [mec] minimum back-up
— m rápido - fast, o quick, return, o back-up
retrodescarga f - [electr] back discharge,ging
— f a través de generador m - [electr] back
discharge,ging through @ generator
retrodescargado,da a - [electr] back discharged
— a a través de generador m - [electr] discharged (back) through @ generator
retrodescargar v - [electr] to discharge back
— v a través de generador m - [electr] to discharge back through @ generator
retroexcavadora f - [constr-equip] (back)hoe;
back digger; trench hoe
— f hidráulica - [constr-equip] hydraulic back-
hoe
— f con cucharón m - [constr-equip] hoe with @
bucket
— f para . . . yarda(s) f cúbica(s) - [constr-
-equip] . . . cubic yard(s) (back)hoe
— f sin cucharón m - [constr-equip] (back)hoe
without @ bucket
retrogradación f-- . . .   [mec] setback
retroimpulso,sión - véase tambien reacción f
retrointegración f - retrointegrating,tion
retrointegrado,da a - retrointegrated
retrointegrar v - to retrointegrate
— v sistema n - to retrointegrate @ system
retroquemado,da a - [sold] burnbacked*
retroquemadura f - [sold] back burning; burning
back; burnback
retrosoplo* m - [sold] backward blow
— m de arco m - [sold] backward arc, o arc
backward, blow

retrosoplo m magnético - [sold] (magnetic)
backward blow
— m — de arco m - [sold] backward arc, o arc
backward, blow
retrospección f - . . .; soul search(ing)
retrospectivamente adv - retrospectively; in retrospect
retrotapón m - [mec] backplug
retrotaponado,da a - [petrol] retroplugged
retrotaponamiento m - [petrol] retroplugging
retrotaponar v - to backplug; to plug back
retrotope m - [mec] backstop
— m con resorte m - [mec] spring loaded backstop
— m — para aguilón m - [grúas] spring loaded boom backstop
— m — amortiguador m - [mec] buffer backstop
— m — — m con resorte m - [mec] spring
(loaded) buffer backstop
— m para aguilón m - [gruas] boom backstop
— m — — m con amortiguador m - [grúas] buffer
boom backstop
— m — — m — — m con resorte m - [gruas]
spring (loaded) buffer boom backstop
— m — torre f - [mec] tower backstop
— m retráctil - [gruas] telescopic backstop
— m — para aguilón m - [gruas] telescopic
boom, o boom telescopic, backstop
— m — simple - [gruas] single telescopic backstop
— m — — para aguilón m - [gruas] single telescopic boom backstop
retrover* v - to flash back
retrovisión f - flashback • [autom] rear view
retrovisor m - [autom] rear view mirror
— m derecho m - [autom] right, hand, o side,
rear view mirror
— m izquierdo - [autom] left, hand, o side,
rear view mirror
retrovisto,ta a - flashed back
retumbar v - . . .; to rumble
reubicación f - relocating,tion
— f de descarga f - discharge relocating,tion
— f — estación f station relocating,tion
— f — subestación f - [electr-distrib] substation relocating,tion
— f — tubería f - [constr] pipe, o tube, relocating,tion
— f — — f de cemento m - [constr] cement,
tube, o pipe, relocating,tion
reubicado,da a - relocated
reubicar v - to relocate
— v descarga f - to relocate @ discharge
— v estación f - to relocate @ station
— v subestación f - [electr-distrib] to relocate @ substation
reunido,da a - gathered; collected; called (together); met; brought together • provided;
supplied
— a de lado m de adentro—collected on @ inside
— a — — m — afuera - collected on @ outside
— a simultáneamente - gathered simultanously
reunión f - . . .; session • calling, o bringing, together; collecting,tion • providing;
supplying
— f administrativa - [legal] management meeting
— f — sobre seguridad f - [ind] management
safety meeting
— f anual - annual, o yearly, meeting
— f — de comisión f - [admin] yearly committee
meeting
— f comercial - business, o commercial, meeting
— f — importante - [com] important business
meeting
— f de accionista(s) m - [legal] stockholders,
o shareholders, meeting
— f — club automovilístico - [autom] car club
meeting
— f — comisión f - [admin] (sub)committee,
meeting, o session
— f — — f paritaria - [labor] bargaining
(session)
— f — directores m - [legal] board of directors meeting

reunión f de directorio m - [legal] board of directors meeting
— f — fin m de semana f - weekend meeting
— f — prensa f - [public] press conference
— f — trabajo m - work(ing) meeting
— f — vendedor(es) m - [com] salesmen meeting
— f diaria - daily meeting
— f — de comision f - [admin] daily committee
meeting
— f efectuada - meeting held; conducted meeting
— f en lado m de afuera - meeting, o collecting, on @ outside
— f — obra f - field collecting
— f —— f de informacion f - field data collecting
— f extraordinaria - [legal] special meeting
— f general de accionistas m - [legal] stockholders' meeting
— f — sobre seguridad f - [segur] general
safety meeting
— f importante - important meeting
— f mensual - monthly meeting
— f — de comision f - [admin] monthly committee meeting
— f — regular - regular monthly meeting
— f —— sobre seguridad f - [segurid] regular
monthly safety meeting
— f — sobre seguridad f - [segurid] monthly
safety meeting
— f —— f para personal m - [segurid]
monthly emplyee(s) safety meeting
— f nacional - national meet(ing)
— f ordinaria f - [legal] regular meeting
— f para negocio(s) m - [com] business meeting
— f — promoción f - promotion(al) meeting
— f —— f de venta(s) f - [com] sales, promotional, o incentive(s), meeting
— f plenaria - [legal] plenary, o full, meeting
— f previa - previous, o prior, meeting
— f próxima - coming, o future, meeting
— f pública - public, meeting, o gathering
— f regular - [legal] regular meeting
— f sectorial - [legal] sectional meeting
— f semanal - weekly, meeting, o gathering
— f — de comisión f - [admin] weekly committee
meeting
— f simultánea - simultaneous, gathering, o
meeting
— f sobre seguridad f - [ind] safety meeting
— f —— f para personal m - [segurid] employees safety meeting
— f — venta(s) f - [com] sales meeting
— f técnica - [ind] technical meeting
reunir v - . . .; to, assemble, o bring, o call,
together • to, provide, o supply
— v en lado m de afuera - to collect on @ outside
— v información f to, gather, o collect, @ information, o datum,ta
— v — f obtenida - to collect, obtained,
datum,ta, o information
— v — f — en obra f - to, collect, o gather,
@ field, information, o dateum,ta
— v — f en obra f - tom gather, o collect, @
field, information, o datum,ta
— v simultáneamente v - to, gather, o collect,
simultaneously
reurbanización f - (urban) redevelopment
reubanizado,da a - [constr] redeveloped
reurbanizar v - [constr] to redevelop
revalidado,da a - ratified • confirmed
revaloración f - véase revalorización f
revalorado,da a - véase revalorizado,da a
revalorar v - véase revalorizar v
revalorización f - revaluation
revalorizado,da a - revalued
revalorizar v - [fin] to revalue
revaluación f contable - [contab] accounting revaluation
— f de activo m - [contab] asset(s) revaluation
— f —— m fijo - [contab] fixed asset(s) revaluation
— f — bien(es) m - [contab] good(s) revaluation,ling

revaluación f de bien(es) m raíz,ces - [contab]
real estate revaluation
— f — equipo m - [fin] equipment revaluation
— f — inmueble(s) m - real estate revaluation
— f — máquina(s) f - [ind] machine(s),ery revaluation
— f — maquinaria(s) f - [ind] machinery revaluation
— f — plusvalía f - [fisc] unearned increment
revaluation
— f — terreno(s) m - [fin] land revaluation
— f hecha debidamente - properly made revaluation
— f — indebidamente - improperly made revaluation
— f posible - possible revaluation
revaluado,da a - revalued
revaluar v - [fin] to revalue
— v activo m - [fin] to revalue @ asset(s)
— v — fijo - [fin] to revalue @ fixed asset(s)
— v terreno(s) m - [fin] to revalue @, land, o
property
— v bien(es) m - [contab] to revalue @ good(s)
revegetación f - revegetation
revegetado,da - revegetated
revegetar v - to revegetate
revelación f - . . .: disclosing,sure • [quím]
developing,pment
— f — placa f - [fotogr] plate develping,pment
— f — película f - [fotogr] film developing,pment
— f —— f radiográfica - [fotogr] X-ray film
developing,pment
— f — proyección f - design, revealing, o revelation
— f — técnica f - technique, revealing, o revelation
— f —— f para material m - material technique, revealing, o revelation
— f —— f proyección f - design technique,
revealing, o revelation
— f —— f tratamiento m - [metal-trat]
treating,tment technique revealing,velation
— f —— f — m térmico - [metal-trat]
heat treating,tment technique revealing,ve-
lation
— f rápida - [fotogr] fast, o quick, develping,pment
— f — de película f - [fotorg] film, quick, o
fast, developing,pment
revelado,da a - revealed; disclosed
— a de inmediato - [fotogr] developed immediately
— a por ensayo(s) m destructivo(s) - destruc-
tively revealed
— a —— m no destructivo(s) - nondestructi-
vely revealed
— a rápidamente - [fotogr] developed quickly
revelador m - . . . . • indicator • [quím] developer
— m con anilina f - [quím] dye developer
— m de falla(s) f - fault finder
— v — película(s) f - [fotogr] film developer
— m fluorescente - [quím] fluorescent developer
— m líquido - wet, o liquid, developer
— m seco -dry developer
revelar v - . . .; to disclose • to detect; to
expose; to tell • [fotogr] to develop
— v de inmediato - to reveal immediately •
[fotogr] to develop quickly
— v defecto(s) m - to reveal @ defect(s)
— v película f - [fotogr] to develop @ film
— v proyección f - to reveal @ design
— v rápidamente - [fotogr] to develop quickly
— v — película f - [fotorg] to develop @ film
quickly
— v técnica f - to reveal @ technique
— v — f para material m - to reveal @ material
technique
— v — f — proyección f - to reveal @ design
technique
— v — f para tratamiento m - [metal-trat] to
reveal @ treating,tment technique

revelar v técnica f para tratamiento m térmico - [metal-trat] to reveal @ heat treating technique
revendedor m - [com] . . .; dealer
— m por correo m - [com] mail, retailer, o order dealer
revender v acción(es) f - [fin] to resell @, stock, o share(s)
— v para exportación f - [com] to resell for export(ation)
revendido,da a - resold
— a para exportación f - [com] resold for export(ation)(
revenido m - [metal-trat] annealing; véase también revenimiento m
— m negro - [metal-trat] black annealing
— m intermedio - [metal-trat] intermediate annealing
— m posterior - [metal-trat] subsequent, o later, annealing • subsequent, o later, drawing
— m subcrítico - [metal-trat] subcritical annealing
— m subsiguiente - [metal-trat] subsequent, o later, annealing • subsequent, o later, drawing
revenido,da a - [metal-trat] annealed
— a en aceite m - [metal-trat] oil tempered
revenir v - [metal-trat] to anneal
— v en forma f incompleta - [metal-trat] to underanneal
reventa f - . . .; reselling
— f de acción(es) f - [fin] stock, o share(s), resale, o reselling
— f para exportación f - [com] resale for export(ation)
reventado,da a - blown, out, o open • broken • [fam] busted
reventamiento m - breaking • [fam] busting
reventar v - . . .; to blow, out, o open • [fam] to bust
— v busa f - [metal-prod] to, blow, o break, out (suddenly) @ blowpipe
— v guarnición f - [mec] to blow @ gasket
— v — f para culata f - [comb.int] to blow @ head gasket
— v refractario(s) m - [metal-prod] refractory,ries breakout
reventarse v busa f - [metal-prod] to, blow, o break, out (suddenly) @ blowpipe
— v junta f - [ind] to blow out @ joint
— v tobera f - [metal-prod] to, blow, o break, out (suddenly) @ tuyere
— v — f y (su) toberón m - [metal-prod] to, blow, o break, out (suddenly) @ tuyere and (its) cooler
— v toberón m - [metal-prod] to, blow, o break, out (suddenly) @ cooler
reventazón m - véase reventón m
reventón m - breakout; breakthrough; blow out; burst(ing); blowing
— f de busa f - [metal-prod] blowpipe blowout
— m — guarnición f - [mec] gasket blowout
— m — — f para culata f - [comb.int] head gasket blow(ing) out
— m — neumático m - [autom-neumát] tire blow-out
— m — refractario(s) m - [metal-prod] refractories breakout
rever v continuamente - to review continually
— v desempeño m - [labor] to review @ performance
— v programa m - to review @, program, o schedule
— v texto m - [Gram] to review @ wording
— v — m propuesto - [gram] to review @ proposed wording
reverberante a - reverberatory
reverberatorio m - [metal-prod] reverberatory
reverificación f - double check(ing)
reverificado,da a - double checked
reverificar v - to double check
reversible a - . . .; reversing

reversible a por aire m - air reversing
reversión f de ciclo m - [electr] cycle reversal
— f — — m de corriente f alterna - [electr] alternating current cycle reversal
— f — material m - [electr] material recoil
— f — — m duro - [electr-oper] hard material recoil
— f — — m magnéticamente duro - [electr] magnetically hard material recoil
— f — — m — en circuito m magnético - [electr] magnetic circuit magnetically hard material recoil
— f — presión t - pressure reversal
reverso m - back, o rear, face, o side • [imprent] reverse side
— m de árbol m - [mec] shaft rear
— m — — m para aportación f (de fuerza f) - [autom-med] input shaft rear
— m — cono m - [cojin] cone back face
— m — cordón m - [sold] head back
— m — pestaña f - [cojin] flange back face
revés m - back (side) • [mec] set back
— m — junta f - joint back side • [sold] back-weld
— m — relé m - [electr] relay back side
revestido,da a - [mec] coated; lined; plated • encased • coated internally • [tub] lined; inside coated
— a con acero m - [metal-trat] steel, clad, o encased, o lined
— a — asfalto m - aspjalt lined
— a — cascote(s) m - [constr] rubble lined
— a — caucho m - rubber lined
— a — cobre m - [metal] copper lined • copper coated
— a — empedrado m - [constr] riprap lined
— a — filástica f - [cabl] marline clad
— a — fundente m - [sold] flux lined • flux coated
— a — hormigón m - [constr] concrete lined
— a — ladrillo(s) m - [constr] brick lined
— a — material m plástico - [plast] plastic lined • plastic coated
— a — merlín m - [cabl] marline clad
— a — metal m - metal lined • metal coated
— a — — m blanco - [metal] Babbit lined • Babbit coated
— a con pasto m - [hidr] grass lined
— a — plástico m - [plást] plastic lined • plastic coated
— a — plomo m - lead, coated, o plated • lead lined
— a — poliestero m - polyester coated • polyester lined
— a — roca(s) f - [constr] riprapped; rip rap lined
— a — tubería f - [constr-pil] pipe encased
— a — — f de acero m - [constr-pil] steel pipe encased
— a — vegetación f - [hidr] vegetation lined
— a electrolíticamente - [metal] electrolytically coated
— a — con cobre m - [metal] electrolytically copper coated
— a en hélice - spirally, coated, o wrapped
— a — — con filástica f - [cabl] marlin(e) spirally wrapped
— a interiormente - [tub] internally, o inside, lined, o coated
— a — con asfalto m - asphalt (inside) lined
— a — — plástico m - [tub] internally plastic lined
— a separadamente - [cabl] separately clad
— a — con merlín m - [cabl] separately marlin(e) clad
— a totalmente - totally coated • [tub] fully, lined, o paved
— a y recubierto,ta a - inside and outside coated • coated and lined
revestidor m - [petról] liner
— m para fondo m - [petról] bottom liner
revestidor,ra a - véase para revestir,timiento m
revestimiento m - lining; relining • coating;

revetment • casing • relining • [electr-cond]
sheathing • [válv] trim • [hormig] veneer •
[constr-túnel] lagging • [vial] surfacing •
[constr] facing • [cald] lagging • [tub] inside coating; paving; pavement • [mec] dressing • véase también **recubrimiento** m
revestimiento m **ácido** -[metal-prod] acid lining
— m **adecuado** - [tub] adequate lining
— m — **especial** - [tub] adequate special lining
— m **adicional** - additional coat(ing) • additional lining
— m **aleado** - [metal-trat] alloyed coat(ing)
— m **aluminoso** - [metal-prod] aluminous lining
— m **aplicado** - applied lining • applied coating
— m — **en obra** f—[constr] field applied lining
— m **apropiado** - [constr] appropriate, o proper,
lining o revetment
— m **asfáltico** - asphalt lining
— m **básico** - [metal-prod] basic lining
— m **bituminoso** - [tub] bituminous lining •
[constr] bituminous pavement,ving
— m — **centrifugado** - [tub] bituminous spun
lining
— m **blanco** - white lining
— m **centrifugado** - [tub] spun lining
— m — **totalmente** - [tub] fully spun lining
— m **circular** - [constr] circular liner,ning
— m — **para túnel** m - [constr] circular tunnel
liner,ning
— m **colocado** - placed liner,ning
— m **completo** - complete, coating, o lining
— m **con asfalto** m - [constr] asphalt lining
— m — **bóveda** f - [constr] arch lining
— m — f **de planchas** f **múltiples** - [constr]
Multi-Plate arch lining
— m — f **Multi-Plate** - [constr] Multi-Plate
(arch) lining
— m — **cascote(s)** r - [constr] rubble lining
— m — **cobre** m - [metal] copper coating • copper lining
— m **costo** m, **bajo**, o **reducido** - inexpensive
lining
— m **empedrado** m - [constr] riprap lining
— m — — m **de cauce** m - [hidr] channel riprap
lining
— m — — m — — m **abierto** - [hidr] open channel riprap lining
— m — **hormigón** m - [constr] concrete lining
— m — — m **de cauce** m - [hidr] channel concrete lining
— m — — m — m **abierto** m - [hidr] open
channel concrete lining
— m — **inyector(es)** m - [metal-lam] jet coating
— m — **ladrillo(s)** m - [constr] brick lining •
brick facing
— m — **metal** m - metal surfacing
— m — — m **duro** - [sold] hard surfacing
— m — **pasto** m - [hidr] grass lining
— m — **poliéstero** m - polyester coating; polycoating
— m — **roca(s)** f - [constr] riprap(ping)
— m — **tira(s)** f - [cald] strip lining
— m — **vegetación** f - [hidr] vegetation lining
— m — — f **de cauce** m - [hidr] channel vegetation lining
— m — — f — — m **abierto** - [hidr] open chen-
nel vegetation lining
— m **continuo** m—[metal-trat] continuous coating
— m **de acero** - steel, coating, o encasement
— m — — m **inoxidable** - (stainless) claddings•
[válv] stainless steel trim
— m — **alcantarilla** f - [constr] culvert lining
— m — — f **a punto** m **de fallar** v - [constr]
failing culvert lining
— m — **alquitrán** m - [constr] tar lining
— m — — m **de hulla** f - [tub] coal tar enamel
lining
— m — **asfalto** m - asphalt coating • asphalt
lining
— m — — m **de cauce** m - [hidr] channel asphalt
lining
— m — — m — — m **abierto** - [hidr] open channel asphalt lining

revestimiento m **de bronce** m - bronze lining •
[válv] bronze trim(ming
— m — **carbono** m - [metal-prod] carbon lining
— m — m **de crisol** m - [metal-prod] hearth
carbon lining
— m — m — m **de horno** m **alto** - [metal-
-prod] blast furnace hearth carbon lining
— m — m — **fondo** m **de crisol** m - [metal-
-prod] hearth bottom carbon lining
— m — m — m — — m **de horno** m -
[metal-prod] furnace hearth bottom carbon
lining
— m — m — m — — m — — m **alto** -
[metal-prod] blast furnace hearth bottom
carbon lining
— m — m — m — m — **horno** m -
[metal-prod] furnace hearth carbon lining
— m — **cascote(s)** m - [constr] rubble lining
— m — **cauce** m - [hidr] channel lining
— m — m **abierto** - [hidr] open channel lining
— m — m **abierto con cascote(s)** m - [hidr]
open channel rubble lining
— m — m **con cascote(s)** m -[hidr] channel
rubble lining
— m — m — **empedrado** m - [hidr] channel
riprap lining
— m — m — **hormigón** m - [hidr] channel
concrete lining
— m — m — **vegetación** f - [hidr] channel
vegetation lining
— m — **caucho** m - rubber lining
— m — **cinc** m = zinc, coating, o lining
— m — **cloaca** f - [sanit] sewer coating
— m — f **a punto** m **de fallar** v - [constr]
failing sewer lining
— m — **cobaltocromo** - cobalt-chrome facing
— m — **cobre** m - [metal] copper coating
— m — **conducto** m - [tub] conduit lining
— m — m **a punto** m **de fallar** v—[tub] failing conduit lining
— m — **convertidor** m - [metal-prod] converter
lining
— m — m **Linz-Donawitz** - [metal-prod] basic
oxygen converter (re)lining
— m — **corrugación** f - corrugation lining
— m — f **interior** - [metal-fabr] smooth-Flo
lining
— m — **crisol** m - [metal-prod] hearth lining
— m — m **de horno** m - [metal-prod] furnace
hearth lining
— m — m — m **alto** - [metal-prod] blast
furnace hearth lining
— m — **cuba** f - [metal-prod] bosh lining
— m — **cubilote** m - [refract] ganister lining
— m — **cuchara** f - [metal-prod] ladle relining
— m — f **para acero** m - [metal-prod] steel
ladle lining
— m — **cuero** m - leather, lining, o facing
— m — **chapa** f - [metal-trat] strip coating
— m — **desagüe** m - [hidr] drain lining
— m — m **a punto** m **de fallar** - [hidr] failing drain lining
— m — **dique** m - [hidr] dike shell
— m — **disco** m - [mec] disk facing
— m — **embrague** m - [mec] clutch facing
— m — m **con disco(s)** m - [mec] disk clutch
facing
— m — **esmalte** m - [metal-trat] enamel lining
— m — m **de alquitrán** m - [metal-trat] tar
enamel lining
— m — m — m **de hulla** f - [metal-trat]
coal tar enamel lining
— m — **estelita** f - [válv] stellite trim
— m — **estructura** f - [constr] structure lining
— m — **etalaje** m - [metal-prod] bosh lining
— m — **fondo** m - [metal-prod] bottom lining
— m — m **de crisol** m - [metal-prod] hearth
bottom lining
— m — m — — m **de horno** m - [metal-prod]
furnace hearth bottom lining
— m — m — m — — m **alto** - [metal-prod]
blast furnace hearth bottom lining

revestimiento m de fondo m de horno - m [metal-
-prod] furnace hearth lining
— m — freno m - [mec] brake, lining, o facing
— m — grasa f - grease coating
— m — hierro m - [metal-prod] iron liner,ning
— m — — m fundido - [mec] cast iron lining
— m — hogar m - [combust] hearth lining
— m — hormigón m - [constr] concrete lining
— m — — m armado m - [hidr] reinforced concrete lining
— m — — para cauce m - [hidr] channel concrete lining
— m — — m — — m abierto - [hidr] open channel concrete lining
— m — horno m - [metal-prod] furnace lining
— m — ladrillo(s) m - [constr] brick lining
— m — — m antiácido(s) - [tub] acid brick
lining
— m — — m para cauce m - [hidr] channel brick
lining
— m — — m — — m abierto - [hidr] open
channel brick lining
— m — laja(s) f - [constr] (slab) riprap
— m — morcilla f - [metal-prod] bustle pipe
lining
— m — mortero m - [constr] mortar lining
— m — — m de cemento m - [metal-fabr] cement
mortar lining
— m — muro m - [constr] wall facing
— m — — m para cabecera f - [constr] headwall
facing
— m — neopreno m - [electr-instal] neoprene
sheeting
— m — óxido m - [metal] oxide coating
— m — piedra f - [constr] stone, lining, o
facing; riprap
— m — — f colocada a mano f - [hidr] hand
placed stone lining
— m — — f enlechada - [hidr] grouted stone
lining
— m — — f volcada—[hidr] dumped stone lining
— m — plancha(s) f - [cald] plate lining
— m — — f plaqueada(s) - [cald] clad plate
lining
— m — plomo m - lead coating
— m — plástico m - [electr-cond] plastic
sheathing
— m — portaviento(s) m - [metal-prod] bustle
pipe lining
— m — refuerzo m - [mec] reinforcing liner
— m — — m de acero m - [mec] reinforcing
steel liner
— m — — — plancha(s) f de acero m - [mec]
reinforcing steel liner
— m — roca f - [constr] rock face
— m — ruta f - [metal-prod] runner lining
— m — tabla(s) f - board lining • [cabl]
(cable reel) lagging
— m — — f para carrete(s) m - [cabl] reel
lagging
— m — — f — — m para cable m - [cabl] cable
reel lagging
— m — tapón m - [metal-prod| stopper head
— m — tubería f - [tub] pipe lining
— m — — f de acero m - [constr-pil] steel
pipe encasement
— m — — f fallada - [constr] failed pipe lining
— m — tubo m - [tub] pipe lining • pipe casing
— m — vegetación f - vegetation lining
— m — zanja f - [constr] trench lining
— m — — f con plancha(s) f - [constr] trench
sheeting
— m — zapata f - [mec] shoe lining
— m — — f para freno m - [mec] brake shoe
lining
— m desgarrado - [mec] torn lining
— m diferente m - different lining
— m distinto - different, o other, lining
— m diverso - different lining
—- m duro m - [sold] hard facing; hardsurfacing;
véase también recubrimiento m duro
— m económico—inexpensive, o low cost, lining

revestimiento m en mal estado m - [metal-prod]
poor condition lining
— m equivalente - equivalent, coating, o lining
— m especial - [tub] special lining
— m — adecuado - adequate special lining
— m exterior - outside coating; covering •
wrapping • [electr-cond] outer sheathing
— m — completo - complete outer coating
— m extraído - [mec] pulled, lining, o coating
— m flojo - loose lining
— m grueso - heavy, o thick, lining • [sold]
heavy coating
— m — de cinc m - [metal-trat] heavy zinc
lining • heavy zinc coating
— m instalado - [constr] installed liner,ning
— m interior - inside, coating, o lining • internal coating • [tub] internal, o inside,
coating, o paving, o liner
— m — centrifugado - [tub] spun interior lining
— m — con mortero m - [tub] inner mortar lining
— m — — — m con cemento m - [tub] inner cement mortar, o cement mortar inner, lining
— m — con plástico m - internal plastic coating
— m — de alquitrán m - [tub] internal tar
coating
— m — — m de hulla f - [tub] internal
coal tar coating
— m — — asfalto m - [tub] asphalt lining
— m — — esmalte m - [tub] enamel lining
— m — — m de alquitrán m - [tub] internal
enmamel lining
— m — — m — — m de hulla f - [tub] internal coal tar enamel lining
— m — morcilla f - [metal-prod] bustle pipe
(inner) lining
— m — portaviento(s) m - [metal-prod] bustle pipe (inner) lining
— m — refuerzo m - [metal-fabr] reinforcing
steel liner,ning
— m — — m de plancha(s) f - [metal-fabr]
plate reinforcing (inner) lining
— m — — — — f de acero m - [metal-
-fabr] liner plate reinforcing steel liner
— m — inspeccionado - [tub] inspected internal
coating
— m — para cloaca f - [sanit] sewer interior
coating
— m — — tubería f - [tub] internal pipe liner
— m — total - [tub] full inside, liner,ning,
o paving,vement
— m — totalmente centrifugado - [tub] spun
full interior, lining, o paving,vement
— m — — de asfalto m centrifugado - [tub]
spun full interior asphalt lining
— m liso - [tub] smooth lining
— m moldeado - [mec] molded facing
— m no rígido - [constr] non-rigid liner,ning
— m — para túnel(es) m - [constr] non rigid
tunnel, liner,ning
— m nuevo - new liner,ning; relining
— m para alcantarilla f - [constr] culvert
liner,ning
— m — — f a punto m de fallar v - [constr]
failing culvert lining
— m — canal m - [hidr] channel lining
— m — — m para colada f - [metal-prod] runner
lining
— m — cauce m - [constr] channel surfacing
— m — cilindro m - [bombas] véase camisa f
véase también camisa f para cilindro m
— m — cloaca f - [sanit] sewer liner
— m — — f a punto m de fallar v - [constr]
failing sewer lining
— m — — f instalada - [constr] installed
sewer liner
— m — conducto m - [tub] conduit, o duct, lining
— m — — m a punto m de fallar v - [tub]
failing conduit lining
— m — desagüe m - [hidr] drain lining

revestimiento m para desagüe m a punto m de fallar v - [hidr] failing drain lining
— m — disco m - [mec] disk plate relining
— m — — m para embrague m - [mec] clutch
plate relining
— m — edificio(s) m - [metal-fabr] building
facing
— m — embrague m - [mec] clutch lining
— m — — m neumático - [mec] air clutch lining
— m — — m para aire m - [mec] air clutch lining
— m — estructura f - [constr] structure lining
— m — — f vieja - [tub] old structure lining
— m — freno m - [mec] brake lining
— m — fricción f - [mec] friction facing
— m — lingotera f - [metal-prod] (ingot) mold
lining
— m — molino m - [mec] mill lining
— m — — m triturador - [mec] crushing, o
grinding, mill lining
— m — placa f - [metal-trat] plate lining
— m — — f para fricción f - [mec] friction
plate lining
— m — — f — — f para embrague m - [mec]
clutch friction plate lining
— m — — f — — f — — m para desenroscadura
f - [petról] breakout clutch friction plate
lining
— m — — f — — f — — m — enroscadura f -
[petról] make-up clutch friction plate lining
— m — — f — — m para desenroscadura f -
[petról] breakout clutch plate lining
— m — — f — — m — enroscadura f - [petról]
make-up clutch friction plate lining
— m — — f — lingotera f - [metal-prod]
(ingot) mold stool lining
— m — plato m - [mec-frenos] plate (re)lining
— m — — m embrague m - [mec] clutch
plate (re)lining
— m — tubería f - [tub] pipe liner
— m — — f fallada - [constr] failed pipe lining
— m — túnel(es) m - tunnel liner,ning
— m — — m cargado - [constr] loaded tunnel
liner
— m — zanja f - [constr] (trench) sheeting
— m permanente - [constr] permanent lining
— m protector - protective lining
— m pesado - [metal-trat] heavy coating
— m por inmersión f - immersion coating
— m protector - protective, lining, o coating
— m reconstruido - rebuilt lining
— m reemplazado - [mec] replaced lining • replaced coating
— m reforzado - [mec] heavy duty, o rinforced,
liner
— m — de hierro m fundido - [comb.int] heavy
duty, o reinforced, cast iron liner
— m refractario—[metal-prod] refractory lining
— m — básico - [metal-prod] basic refractory
lining
— m — para horno m - [metal-prod] furnace refractory, o refractory furnace, lining
— m reversible - reversible lining
— m — de caucho m - reversible rubber lining
— m rígido - [hidr] rigid lining
— m roto - torn, o broken, lining
— m sin estrellas f - [metal-trat] unspangled
coating
— m Smooth-Flo - Smooth-Flo lining
— m sobre fondo m - [tub] invert coating
— m soldado - [tub] welded casing
— m — para pozo m - [tub] welded well casing
— m suelto - loose lining
— m tejido - [vest] woven facing
— m total - [metal-trat] total coating
— m totalmente centrifugado - [tub] spun full
lining;full spun coating
— m uniforme - uniform lining
— m vario - different lining
— m y recubrimiento m de liga f de asbesto m -
[metal-trat] asbestos bonded Smooth-Flo
lining

revestir v - . . .; to coat internally; to face
— v adicionalmente - to coat additionally
— v alcantarilla f - [tub] to line @ culvert
— v — f a punto m de fallar v - [constr] to
line @ failing culvert
— v cauce m - [hidr] to coat @ channel
— v — m abierto - [constr] to line @ open
channel
— v — m — con cascote(s) m - [hidr] to rubble
line @ open channel
— v — m — — empedrado m - [hidr] to riprap @
open channel
— v — m — — hormigón m - [hidr] to concrete
line @ open channel
— v — m — — ladrillo(s) m - [constr] to
brick line @ open channel
— v — m asfalto m - [hidr] to asphalt line
@ channel
— v — m con cascote(s) m - [hidr] to rubble
line @ channel
— v — m — empedrado m - [hidr] to riprap line
@ channel
— v — m — hormigón m - [hidr] to concrete
line @ channel
— v — m — ladrillo(s) m - [hidr] to brick
line @ channel
— v — m — vegetación f - [hidr] to vegetation
line @ channel
— v con - to line with • to encase
— v — asfalto m - [constr] to asphalt line
— v — — m cauce m - [constr] to asphalt line
@ channel
— v — — m cauce m abierto - [hidr] to asphalt
line @ open channel
— v — — m — m abierto - [hidr] to asphalt
line @ open channel
— v — cascote(s) m - [constr] to rabble line
— v — — m cauce m - [hidr] to rabble line @
channel
— v — — m — m abierto - [hidr] to rabble
line @ open channel
— v — empedrado m - [constr] to riprap line
— v — — m cauce m - [hidr] to riprap @ channel
— v — — m — m abierto - [hidr] to riprap @
open channel
— v — hormigón m - [constr] to concrete line
— v — — m cauce m - [hidr] to concrete line
@ channel
— v — — m — m abierto - [hidr] to concrete
line @ open channel
— v — ladrillo(s) m - [constr] to brick line
— v — — m cauce m - [constr] to brick line @
channel
— v — — m — m abierto - [constr] to brick
line @ open channel
— v — pasto m - [hidr] to grass line
— v — — m cauce m - [hidr] to grass line @
channel
— v — — m — m abierto - [hidr] to grass line
@ open channel
— v — poliestero m - to polyester coat; to
polycoat
— v — roca(s) f - [constr] to riprap; to line
with rock(s)
— v — vegetación f - [hidr] to line with vegetation
— v — — f cauce m - [hidr] to line @ channel
with vegetation
— v — — f — m abierto - [hidr] to line @
open channel with vegetation
— v conducto m - [tub] to line @ conduit
— v — m a punto m de fallar v - [tub] to line
@ failing conduit
— v desagüe m - [hidr] to line @ drain
— v — m a punto m de fallar v - [hidr] to line
a failing drain
— v disco m - [mec-frenos] to (re)line @ plate
— v — m — embrague m - [mec] to (re)line @
clutch plate
— v estructura f - [constr] to line @ structure
— v forma f - to take @, shape, o form

revestir v importancia f - to be important
— v interiormente—to, coat, o line, internally
— v — con plástico n - to internally plastic coat
— v nuevamente - to reline; to coat anew
— v plato m - [mec-frenos] to (re)line @ plate
— v — m para embrague m - [mec] to reline @ clutch plate
— v ruta f - [metal-prod] to (re)line @ runner
— v tubería f - [tub] to (re)line @ pipe
— v — fallada - [constr] to (re)line @ failed pipe
— v zanja f - [constr] to, line, o sheet, @ trench
— v — f con plancha(s) f - [constr] to sheet (line) @ trench
revezar v - to reverse • to alternate
revirar v - [mec] to lap • [náut] . . .
revisable a - subject to @ revision • [legal] subject to @ escalation
revisación f - revision; inspection; review(ing) • going over • overhauling • [med] checkup • véase también revisión f
— f médica - [med] medical checkup
revisado,da a - . . .; inspected; checked; reviewed • gone over • updated • reviewed • [fin] escalated • [contab] audited
— a analíticamente - [contab] audited analytically
— a periódicamente - revised, o reviewed, periodically
revisador m - checker • reviewer • auditor
— m de cuenta(s) f - [contab] auditor
revisador,ra a - checker • reviewer • auditor
revisar v - . . .; to, check, o inspect, o go over • to update • [contab] to audit • [fin] to escalate
— v analíticamente - [contab] to audit analytically
— v estado m - to check @ condition
— v excitatriz f - [electr-] to check @ exciter
— v expediente m - to check @ dossier • to revise @ dossier
— v legajo m - to revise @ dossier • to check @ dossier
— v periódicamente - to, revise, o check, periodically
— v precio m - to revise @ price • to check @ price
— v previamente - to preview
— v programa m - to revise @ program • to revise @ schedule
— v recomendación f - to, review, o revise, @ recommendation
— v texto m - [gram] to review @ wording
— v — m propuesto - [gram] to review @ proposed wording
revisión f - . . .; inspecting,tion; check(ing); review • [com] change order • updating • [med] checkup • [fin] escalation • [contab] audit(ing)
— f analítica - analytic(al) revision • [contab] analytic(al) audit(ing)
— f correspondiente - corresponding revision
— f de cuenta(s) f - [contab] audit(ing)
— f — equipo m - [ind] equipment, revision, o check(ing)
— f — — m nuevo - [ind] new equipment, revision, o check(ing)
— f — — m viejo - [ind] old equipment, revision, o check(ing)
— f — legajo m - dossier, revision, o checking
— f — expediente m - dossier, revision, o check(ing)
— f — precio m - price revision • price escalation
— f — — m detallada - [com] detailed price escalation
— f — programa m - program revision • program reviewing
— f — recomendación f - recommendation, reviewing, o revising,sion·
— f estándar - standard, review, o escalation

revisión f general - [ind] major overhaul(ing)
— f médica - [med] medical checkup
— f medicopsicológica - [mec] medical (and) psychological checkup
— f para fin(es) m fiscal(es) - [fisc] tax audit
— f — — m impositivo(s) - [fisc] tax auditing
— f periódica - periodical revision
— f programada - [ind] programed, o scheduled, revision, o checkup • scheduled repair
— f psicológica - [med] psychological checkup
— f última - latest revision
revisor m - reviewer
— m de cuentas f - [contab] auditor
revista f - . . . • profile • [public] review; magazine; journal
— f a fondo - in depth review
— f automovilística - [public] car magazine
— f de informe m - report review
— f — ingeniería f - engineering, review, o magazine, o journal
— f — programa m - program review(ing)
— f independiente - independent review
— f para aficionados m - [public] amateurs magazine
— f — — m a automóvil(es) m - [autom] car buff magazine
— f — caballero(s) m - [public] men's magazine
— f — consumidor(es) m - [public] consumer('s) magazine
— f — dama(s) f - [public] women('s) magazine
— f para evaluación f - [admin] evaluation review
— f — — f de programa m - [admin] program evaluation review
— f técnica - [public] technical magazine
revistado,da a - reviewed
revisto,ta a - reviewed
revivido,da a - relived • revived
revivificación f - . . .; revivifying
revivificado,da a - revivified
revivificar f - . . .; to revive
revivir v - . . . • to relive
revocación f - . . . • [legal] to repeal
revocado,da a - revoked • repealed
revocadura f - . . . • [constr] whitewash(ing)
revocar v - . . .; to revoke • to repeal
revolución f - . . . • [mec] revolving • turn
— f completa - [mec] complete, o full, revolution, o turn
— f de cigüeñal m - [mec] crankshaft revolution
— f — eje m - [mec] shaft revolution • spindle revolution
— f — motor m - [electr-mot] motor revolution • [comb.int] engine revolution
— f — tambor m - [mec] drum, turn, o revolution
— f — — m para malacate m - [grúas] winch drum, turn, o revolution
— f — — m — — m auxiliar - [grúas] auxiliary winch drum, turn, o revolution
— f — — m — — m principal - [grúas] main winch drum, turn, o revolution
— f industrial - [ind] industrial revolution
— f social - social revolution
— f única - [mec] somgle revolution
revoluciones f por kilómetro m - [mec] revolution(s) per kilometer (use miles)
— f — minuto m - [mec] revolutions per minute
— f — — m ajustadas - [mec] adjusted revolutions per minute
— f — — m comprobadas - [mec] proven, o checked, revolutions per minute
— f — — m de cilindro m - [mec] cylinder('s) revolutions per minute
— f — — m — — m de segadora f trilladora - [agric] combine('s) cylinder revolutions per minute
— f — — m — motor m - [electr-mot] motor('s) revolutions per minute • [comb.int] emgine('s) revolutions per minute
— f — — m — — m fiscalizadas - [electr-mot] monitored motor('s) revolutions per minute • [comb.int] monitored engine('s) revolutions per minute

revoluciones f por minuto m fiscalizadas - [mec] monitored revolutions per minute
— f —— m verificadas - [mec] checked revolutions per minute
revolvedora f - [ind] mixer
revolvente* a - véase rotativo,va; circulante
revoque m - [constr] . . .; plaster; stucco
— m acústico - [constr] accoustic plaster
— m antisonoro - [constr] accoustic plaster
— m de pared f - [constr] wall plaster(ing)
— m —— f exterior - [constr] outer surface plaster(ing)
— m exterior - [constr] outside (surface) plaster(ing)
— m grueso - [constr] rough plaster(ing)
— m preliminar - [constr] rough plaster(ing)
revuelto,ta a - stirred; agitated
rezagarse v - [deport] to, lag, o fall behind
rezagado,da a - . . . ; held back; fallen behind
rezagar v - . . . • to hold back
rezago m - . . . • [deport] falling behind
— m neumático - [autom-neumát] pneumatic trail
rezumadero m - . . .; pool • seepage,ping
— m de petróleo m - [petról] oil seepage
rezumado m - oozing • seepage
rezumado,da a - seeped
rezumante a - seeping • trickling
rezumar v - to, trickle, o seep
ría f - [hidr] tidal river; inlet; waterway • [nav] roads; véase también rada f; ensenada f
riachuelo m - [hidr] . . .; small river
riacho m - [hidr] . . .; small stream
riada f - [hidr] flash flood; flood flow
ribera f - [hidr] . . . • shore line; waterfront
— f de arroyo m - [hidr] stream, o creek, bank
— f — canal m - [hidr] channel, bank, o edge
— f — lago m - [hidr] lakeshore; lake bank
— f desbordada - [hidr] overflowed bank
— f erosionada - [hidr] eroded, bank, o shore
— f irregular - [topogr] irregular shoreline
— f meridional - [hidr] south, side, o shore
— f — de lago m - [hidr] lake south, side, o shore
— f inclinada - [hidr] steep bank
— f norte - [hidr] north, side, o shore
— f — de lago m - [hidr] lake north side
— f occidental - [hidr] west side
— f — de lago m - [hidr] lake west side
— f oeste - [hidr] west, side, o shore
— f oriental - [hidr] east, side, o shore
— m — de lago m - [hidr] lake east side
— f septentrional - [hidr] north, side, o shore
— f — de lago m - [hidr] lake north, side, o shore
— f sud - [hidr] south, side, o shore
— f — de lago m - [hidr] lake south side
ribereño,ña a - . . . ; water, o lake, front
ribete m - [vest] . . .; border • [calzado] welt
ribeteadora f—[sold] véase soldadora f por puntos
ricino m - [botán] . . .; castor
rico,ca a en cobre m - [miner] copper rich
— a — mineral(es) m - mineral, o ore, rich
— a — recurso(s) m natural(es) - [econ] natural resource(s) rich
riego m - [hidr] . . .; sprinkle,ling
— m con aspersión f - [hidr] spray sprinkling
— m tercero - [constr] third spray
— m — ligante - [vial] third binding spray
riel m - . . . • [mec] runner
— m americano - [metal-lam] T-rail
— m con cremallera f - [mec] rack; cog rail
— m — resistencia, grande, o mayor - [f.c.] high strength rail
— m —— superficie f endurecida - [sold] hard-surfaced rail
— m — tratamiento m térmico - [metal-trat] heat treated rail
— m de acero m - [metal-lam] steel rail
— m estándar - [f.c.] standard rail
— m — con tratamiento m térmico - [metal-lam] standard heat treated; o heat treated standard, rail

riel m liviano - [metal-lam] light rail
— m más resistente - [f.c.] higher strength rail
— m muy resistente - [f.c.] high strength rail
— m para depósito m - [mec] tank rail
— m —— m para gasolina f - [comb.int] gasoline tank rail
— m — grúa f - [grúas] crane runway
— m para montaje m - [mec] mounting rail
— m —— m para depósito m (para gasolina f) - [comb.int] (gasoline) tank mounting rail
— m — mandil m - [mec] gate rail
— m —— m trasero - [mec] tailgate rail
— m patín m - [mec] T-rail
— m renivelado - [f.c.] releveled rail
— m según norma f - [metal-lam] standard rail
— m —— f con tratamiento m térmico - [metal-lam] standard heat treated rail
— m tratado térmicamente - [metal-trat] heat treated rail
— m vignole - [metal-lam] T-rail
rienda f - . . .; guy; bridle • [cabl] bridle cable
— f de cable m metálico - [constr] cable, guy, o tie back
riesgo m - . . . . • jeopardy
— m analizado - [fin] analyzed risk
— m asumido - assumed risk; taken chance
— m cambiario - [fin] exchange risk
— m comercial - [seguros] commercial risk
— m — cubierto m - [seguros] covered commercial risk
— m — de Corporación f Estadounidense para Seguro m sobre Crédito(s) m - [seguros] Federal Credit Insurance Corporation commercial risk
— m corrido - risk run • [seguridad] chance taken
— m cualquiera - any risk
— m cubierto - [seguros] covered risk
— m de accidente(s) m - [seguridad] accident risk
— m — Corporación f Estadounidense para Seguros m sobre Crédito(s) m - [seguros] Federal Credit Insurance Corporation risk
— m — choque(s) m - [electr-oper] shock danger
— m — daño(s) - [seguridad] damage risk
— m — desmoronamiento(s) m - [constr] (trench) collapse hazard
— m —— m de excavación f - [constr] trench collapse hazard
— m — explosión f - [seguridad] explosion risk
— m — exportador m - [seguros] exporter('s) risk
— m — guerra f - [seguros] war risk
— m — incendio m - [seguridad] fire hazard
— m —— m aumentado - [seguridad] increased fire hazard
— m —— m eliminado - [seguridad] eliminated, o removed, fire hazard
— m —— m implicado - [seguridad] involved fire hazard
— m — inundación f - [hidr] flood hazard
— m —— f reducido - [hidr] alleviated flood hazard
— m — pérdida(s) f - loss(es) risk
— m — rotura f - breakage risk
— m — tiempo m malo - [constr] bad weather hazard
— m — torpedeamiento m - [seguros] torpedo risk
— m — torpedo(s) m - [seguros] torpedo risk
— m definido - [seguros] defined risk
— m extremadamente limitado - [seguros] extremely limited risk
— m implicado - [seguridad] implied, o involved, hazard
— m imprevisto - unforeseen, o unexpected, hazard, o risk
— m — para tránsito m - unforseen, o unexpected, traffic, risk, o hazard
— m incurrido - incurred risk
— m inesperado - unexpected, risk, o hazard
— m — para tránsito m - unexpected traffic hazard
— m limitado - limited risk
— m marítimo - [seguros] maritime risk

riesgo m̲ **minimizado** - [segurid] minimized, risk,
o̲ hazard
— m̲ **mínimo** - minimum, risk, o̲ hazard
— m̲ — **para tránsito** m̲ - [vial] minimum traffic
hazard
— m̲ — **construcción** f̲ ≑ [constr] construction
hazard
— n̲ — **seguridad** f̲ - [segurid] safety hazard
— m̲ — **tránsito** m̲ - [transp] traffic hazard
— m̲ **político** - [seguros] political risk
— m̲ — **cubierto** - [seguros] covered political
risk
— m̲ — **definido** - [seguros] defined political
risk
— m̲ **reducido** m̲ - [segurid] reduced, hazard, o̲
risk, o̲ danger
— m̲ **tomado** - [segurid] taken, risk, o̲ chance
rigidez f̲ - . . .; stiffness • restraint •
strength • [tub] beam strength
— f̲ **a rotación** f̲ - [mec] restraint stiffness
— f̲ **aumentada**—increased, rigidity, o̲ stiffness
— f̲ **contra bamboleo** m̲ - [mec] roll stiffness
— f̲ — **rodadura** f̲ - [mec] roll stiffness
— f̲ — **anillo** m̲ - [mec] ring stiffness
— f̲ — **aro** m̲ - [mec] ring stiffness
— m̲ — **costado** m̲ - [autom-neumát] sidewall
stiffness
— f̲ — **de neumático** m̲ - [autom-neumát]
tire sidewall stiffness
— f̲ — — m̲ — — m̲ **aumentada** - [autom-neumát]
stiffened tire sidewall
— f̲ — **junta** f̲ - [sold] joint, restraint, o̲
stiffness
— f̲ — **pared** f̲ - [constr] wall stiffness
— f̲ — — f̲ **de alcantarilla** f̲ - [constr] culvert wall stiffness
— f̲ — **suelo** m̲ - [suelos] soil stiffness
— f̲ — **tubería** f̲ - [tub] pipe, stiffness, o̲
strength • [tub] beam strength
— f̲ — — f̲ **para manejo** m̲ - [tub] pipe handling
stiffness
— f̲ **dieléctrica** - [electr] dielectric rigidity
— f̲ **graduada** - [mec] graduated stiffness
— f̲ **infinita** - infinite, rigidity, o̲ stiffness
— f̲ **lateral** - [constr] lateral, rigidity, o̲
stiffness
— f̲ **máxima** - [mec] maximum, rigidity, o̲ stiffness
— f̲ — **de tubería** f̲ - [tub] pipe maximum, rigidity, o̲ stiffness
— f̲ **medida** - measured, rigidity, o̲ stiffness
— f̲ **mayor** - increased, rigidity, o̲ stiffness
— f̲ **mecánica** - mechanical rigidity
— f̲ **mínima** - minimum, rigidity, o̲ stiffness
— f̲ — **de tubería** f̲ - [tub] minimum pipe, rigidity, o̲ stiffness
— f̲ **para manejo** m̲ - handling stiffness
— f̲ — **manipulación** f̲—[mec] handling stiffness
— f̲ — — f̲ **verificada** - [mec] checked handling
stiffness
rígido,da a̲ - . . .; unyielding • restrained •
[fam] stiff necked
rigor m̲ - . . . • harshness
— m̲ **de invierno** m̲ - [meteorol] winter rigor •
harsh winter weather
— m̲ — **tiempo** m̲ - [meteorol] weather (rigor)
riguroso,sa a̲ - rigorous • stringent • tough;
harsh
rijan adv - (que) applying
rimero m̲ - . . . • [public] signature
rincón m̲ **calentado** - [sold] heated corner
— m̲ **de piso** m̲ - [constr] floor corner
— m̲ **opuesto** - opposite corner
— m̲ **perpendicular** - perpendicular corner
ringlera f̲ - tier
riñón m̲ - . . . • [arquit] haunch
río m̲ **abajo** adv - downstream; down river
— m̲ **adentro** adv - offshore
— m̲ **arriba** adv - upstream; up river
— n̲ **impetuoso** - [hidr] raging river
— m̲ **navegable** - [nav] navigable river
— m̲ **por medio** - across @, river, o̲ stream

río m̲ **rugiente** - [hidr] raging river
— m̲ **selvático** - [hidr] jungle river
— m̲ **subterráneo** - [hidr] underground river
riolita f̲ - [miner] rhyolite
rioplatense a̲ - (from @) River Plate
riostra f̲ - [constr] . . .; bracing member; angle brace; strut; tie back • tie wire
— f̲ **ajustable** - [mec] adjustable brace
— f̲ **cruzada** - [constr] cross bracing; waler
— f̲ **estándar** - [mec] standard brace
— f̲ **para caldera(s)** f̲ - [cald] belly brace
— f̲ — **escaler(ill)a** f̲ - [mec] ladder brace
— f̲ — **prolongación** f̲ - [mec] extension brace
— f̲ — **torre** f̲ - [petról] derrick brace
— f̲ — — f̲ **para perforación** f̲ - [petról]
derrick brace
— f̲ **temporaria** - [constr] temporary tie wire
ripar v̲ - [mec] to, lathe, o̲ chip; véase también
laminar v̲
ripia f̲ - [constr] . . .; batten; lath
ripiar v̲ - [constr] . . .; to shingle
ripio m̲ - . . .; gravel • cuttings • aggregate
• [geol] shingle(s)
— m̲ **de pedernal** m̲ - [constr] flint gravel
riqueza f̲ - . . . • [quím] purity
— f̲ **garantizada** - [quím] guaranteed purity
risco m̲ - [topogr] . . .; peak
— m̲ **escarpado** - [topogr] craggy peak
ristra f̲ **de** . . . - row of . . . in @ row
ritmo m̲ - . . . • rate; pace; tempo • speed
— m̲ **acelerado** - fast, o̲ high, rate
— m̲ — **de inflación** f̲ - high infiltration rate
— m̲ **acompasado** - steady pace
— m̲ **actual** - current, o̲ present, rate
— m̲ — **de producción** f̲ - [ind] current production rate
— m̲ **aumentado** - increased, rate, o̲ speed
— m̲ — **en movimiento** m̲ - [mec] increased movement speed
— m̲ — — m̲ **de estrangulador** m̲ - [mec] increased choke movement speed
— m̲ **cómodo** - [deport] comfortable pace
— m̲ **crítico** - [ind] critical rate
— m̲ — **para enfriamiento** m̲ - critical cooling,
rate, o̲ speed
— m̲ **de crecimiento** m̲ - [ind] growth rate
— m̲ — — m̲ **en producción** f̲ - [ind] production
growth rate
— m̲ — **desgaste** m̲ - [mec] wear(ing) rate
— m̲ — — **de bola** f̲ - [cuerp.moled] ball
wear(ing) rate
— m̲ — — m̲ — — f̲, **moledora**, o̲ **para molienda**
f̲ - [cuerp.moled] grinding ball wear(ing)
rate
— m̲ — — m̲ **predicho** - [cuerp.moled] predicted
wear rate
— m̲ — **economía** f̲ - [econ] economy rhythm
— m̲ — **enfriamiento** m̲ - cooling rate • [sold]
quenching rate
— m̲ — **movimiento** m̲ - [mec] movement speed
— m̲ — — m̲ **de estrangulador** m̲ - [mec] choke
movement speed
— m̲ — **pella(s)** f̲ - [miner] pellet(s) rate
— m̲ — **producción** f̲ - [ind] production rate
— m̲ — — f̲ **actual** - [ind] current production
rate
— m̲ — — f̲ **lento** - [ind] slow production rate
— m̲ — — f̲ **rápido** - [ind] high production rate
— m̲ — **salida** f̲ - [mec] discharge rate
— m̲ — — f̲ **de pella(s)** f̲ - [miner] pellet(s)
discharge rate
— m̲ **esclofriante** - dizzying rate
— m̲ **desusado** - startling rate
— m̲ **discreto** - [deport] discreet, o̲ comfortable, rate, o̲ pace
— m̲ **disminuido** - decreased, rate, o̲ speed
— m̲ — **de movimiento** m̲ - [mec] diminished, o̲
decreased, movement speed
— m̲ — — — m̲ **de estrangulador** m̲ - [mec] decreased, o̲ diminished, choke movement speed
— m̲ **económico** - [econ] economic rhythm
— m̲ **exigido** - required, speed, o̲ rate
— m̲ **menor que máximo** - [ind] slower than maxi-

mum speed

ritmo m **para avance** m - [sold] travel speed
— m —— m **parejo** - [sold] steady travel speed
— m — **carga** f - [electr-oper] charging rate
— m — **colocación** f - [constr] laydown rate
— m — **crecimiento** m - [econ] growth rate
— m — **cumplimiento** m - accomplishment rate
— m — **descarga** f - [mec] discharge rate
— m —— f **de pella(s)** f - [miner] pellet(s) discharge rate
— m — **inversión** f - [fin] investment rate
— m — **marcha** f - [ind] operating rate
— m — **operación** f - [ind] operating rate
— m — **producción** f - [ind] production rate
— m — **soldadura** f - [sold] welding rate
— m —— **mayor que máximo** - [sold] faster than maximum welding speed
— m —— f **menor que máximo** - [sold] slower than maximum welding speed
— m — **trabajo** m - [ind] operating, rate, o speed
— m **impreso** - [deport] set, pace, o speed
— m **incrementado** - [mec] increased speed
— m — **para movimiento** m - [mec] increased movement speed
— m —— m **de estrangulador** m - [mec] increased choke movement speed
— m **industrial** - [ind] industrial, speed, o tempo
— m **inigualable** - [deport] unequalled pace • blistering pace
— m **irregular** - irregular, speed, o rate
— m — **para descarga** f - [ind] irregular discharge, speed, o rate
— m —— f **de pella(s)** f - [miner] irregular pellet discharge,ging, speed, o rate
— m — **para pella(s)** f - [miner] irregular pellet rate
— m — **salida** f - irregular discharge,ging rate
— m —— f **de pella(s)** f - [miner] irregular pellet(s) discharge,ging rate
— m **lento** - (s)low rate
— m — **para enfriamiento** m - [sold] slow quenching rate
— m —— **infiltración** f - [hidr] low infiltration rate
— m —— **producción** f - [ind] (s)low production rate
— m **mantenido** - [deport] maintained, o kept up, pace
— m **mayor** - faster, pace, o rate
— m — **para enfriamiento** m - [sold] faster quenching rate
— m **menor** - slower rate
— m — **para enfriamiento** m [sold] slower quenching rate
— m **mundial** - world(wide) rate
— m **para descarga** f - discharge,ging rate
— m —— f **de pella(s)** f - [miner] pellet(s) discharge,ging rate
— m — **enfriamiento** m - cooling rate
— m —— m **de pella(s)** f - [miner] pellet(s) cooling rate
— m **infiltración** f—[hidr] infiltration rate
— m — **marcha** f - [ind] operating speed
— m — **operación** f - [mec] operating speed
— m — **salida** f - [mec] discharge,ging rate
— m —— f **de pella(s)** f - [miner] pellet(s) discharge,ging rate
— m **rápido** - fast, o high, rate, o pace
— m **parejo** - even, o steady, rate, o pace
— m — **para enfriamiento** m - [sold] fast quenching , rate, o speed
— m —— **producción** f - [ind] high, o fast, production, rate, o speed, o pace
— m **reducido** - slow, rate, o pace, o speed
— m —— m **para movimiento** m - [mec] decreased movement, rate, o speed
— m ———— m **para estrangulador** m - [ind] decreased choke movement speed
— m **uniforme** - uniform, speed, o pace
— m — **para avance** m - [sold] uniform travel

soeed

ritmo m **vertiginoso** - [deport] torrid, o neck-breaking, speed
rival m - . . .; competitor | a - rival; competive
— m **más próximo** - [deport] closest, o nearest, rival, o competitor
— m **principal** - [deport] main, o principal, rival, o competitor
— m **próximo** - [deport] close, rival, o competitor
robado,da a - robbed; stolen • pilfered
robar v - to rob; to steal • to pilfer
robinete m - [tub] cock • [válv] véase **grifo** m
roble m **blanco** - [mader] white oak
— m **rojo** - [mader] red oak
roblón m **con cabeza** f **plana** - [mec] flathead rivet
— m **dúctil** - [mec] ductile rivet
— m **hendido** - [mec] split rivet
— m **para junta** f - [mec] joint rivet
— m **partido** - [mec] split rivet
roblonado,da* a - [mec] véase **remachado,da** a
— a **bien** - [mec] close riveted
roblonar v **bien** - [mec] to close rivet
robo m - . . . . - pilerage,ring
robot m **parlante** - [electrón] talking robot
robustez f - . . .; ruggedness
robusto,ta a - . . .; rugged • [fam] beefy
roca f **aguda** - [geol] sharp rock
— f —— **como navaja** f - [suelos] razor sharp rock
— f **almacenadora** - [petról] reservoir rock
— f **alterada** - [geol] altered rock
— f —— **por carbonatación** f - [Geol] carbonation altered rock
— f **andesítica** - [geol] Andesitic rock
— f **arenisca** - [geol] sandstone rock
— f **artificial** - [constr] artificial, rock, o stone
— f **asfáltica** - [petról] asphalt rock
— f **blanca** - [geol] white rock
— f **blanda** - [geol] soft rock
— f —— **(color) verdoso,sa** - [geol] greenish (colored) soft rock
— f **calcárea** - [miner] calcareous rock
— f **calibrada** - [constr] sized, o gaged, rock
— f **caliza** - [geol] limestone rock
— f **color** m **verdoso** - [geol] greenigh colored rock
— f **compuesta** - [geol] compound rock
— f **contaminada** - [constr] contaminated rock
— f **cortada** - [constr] cut. rock, o stone
— f **cristalina** - [geol] crystallyne rock
— f **de cubierta** f - [geol] cap. o cover, rock
— f —— **cuerda** f - [geol] ledge rock
— f **desintegrada** - [geol] decomposed rock
— f **desmenuzada** - [constr] crushed rock
— f **desproporcionada** - oversized rock
— f **determinante** - [petról] key rock
— f **disgregada** - [geol] weathered rock
— f **dura** - [geol] hard rock
— f **en base** f **de estribo** m - [constr] abutment rock
— f —— **cimiento** m - [constr] foundation rock
— f —— **fundación** f - [constr] foundation rock
— f —— **polvo** m - [constr] powdered rock; rock flour
— f **encajonante superior** - [geol] cap rock
— f **errante** - [geol] errant rock
— f **esquisto** - [geol] schist rock
— f **estratificada** - [geol] stratified rock
— f **floja** - [constr] loose, rock, o stone
— f **fracturada** - [geol] fractured rock
— f **grande** - [constr] large rock; boulder
— f **granítica** - [geol] granite,tic rock
— f **ígnea** - [geol] igneous rock • eruptive rock
— f **intrusiva** - [geol] intrusive rock
— f **madre** - [geol] country rock
— f **menuda** - [constr] small rock
— f —— **fracturada** - [constr] small fractured rock
— f **micácea** - [miner] micaceous rock

roca f petrolífera - [petról] oil bearing rock
— f — porosa—[petról] porous oil bearing rock
  • sand
— f piroclástica - [geol] pyroclastic rock
— f reconstruída - [constr] reconstructed,
  rock, o stone
— f rosada - [geol] pink rock
— f sedimentaria - [geol] sedimentary rock
— f sepultada - [geol] buried, o subsurface,
  rock
— f subyacente - [geol] bedrock; underground
  boulder
— f suelta - [geol] loose rock
— f trapeana - [miner] trap rock
— f triturada - [constr] crushed rock • fractured rock • broken stone
— f viva - [geol] ledge rock • solid rock
— f volada - [miner] blasted, o shot, rock
— f voladiza - [geol] overhanging rock
— f volcánica - [geol] volcanic rock
rocallera f - [miner] (stone) quarry • [constr]
  bulldozer type wheel
rocalloso,sa a - . . . ; rock filled
roce m - . . . • scratching
— m de ladera f - [autom] bank tagging
rociado m - . . . ; spray(ing) • véase también
  rociadura f; rociamiento m
— m bajo presión f - pressure spraying
— m — f alta - high pressure spraying
rociado,da a - . . . ; sprinkled; sprayed
— a bajo presión f - pressure sprayed
— a — f alta - high pressure sprayed
— a con aceite m - oil, sprinkled, o sprayed
— a — alquitrán m - tar, sprinkled, o sprayed
— a — asfalto m - asphalt sprayed
rociador m - [hidr] . . . ; spray head
— m automático - [hidr] automatic sprinkler
— m con campana f - [ind] bell spray(er)
— m — presión f - [vial] pressure sprayer
— m — soplete m - spray gun
— m exterior - [ind] outside spray(er)
— m — para blindaje m - [metal-prod] shell
  outside spray
— f — — m para crisol m - [metal-prod]
  hearth shell outside spray(er)
— m — para coraza f - [metal-prod] external
  shell spray(er)
— m hidráulico - [hidr] hydraulic spray(er)
— m inferior - [hidr] lower spray(er)
— m limpiador - [metal-lam] cleaning spray(er)
— m para agua m - [hidr] water sprayer
— m — campana f - [metal-prod] bell spray(er)
— m — — f grande m - [metal-prod] large bell
  spray(er)
— m — — f pequeña - [metal-prod] small bell
  spray(er)
— m — camisa f - [mec] liner spray(er)
— m — — f y émbolo m - [mec] piston-liner
  spray(er)
— m — — — f accionado con motor m elec-
  trico - [ind] electric motor driven piston-
  -liner spray(er)
— m — — f — — m — montado en parte f
  posterior - [ind] rear mounted electric motor
  driven piston liner spray(er)
— m — enfriamiento m - [metal-lam] cooling
  spray(er)
— m — — m de rodillo(s) m - [metal-lam] roll
  cooling spray(er)
— m — horno m - [ind] oven spray(er)
— m — — m para coque m - [coque] coke oven
  spray(er)
— m — rodillo m - [metal-lam] roll spray(er)
— m — solución f - [ind] solution spray(er)
— m — — f refrigerante - [metal-lam] cooling
  solution spray(er)
— m — — f — para rodillo(s) m - [metal-lam]
  roll cooling solution spray(er)
— m purgador - [tub] flush spray(er)
— m superior - top, o upper, spray(er)
rociadora f - [hidr] spray(er)
rociadura f - spray(ing)
— f de aceite m - [mec] oil spray(ing)

rociadura f de agua m - water spray(ing)
— f — — m para enfriamiento m - [ind] cooling
  water spray(ing)
— f — lavadora f = [ind] washer spray(er)
— f — líquido m - [hidr] liquid spray(ing)
— f — — m refrigerante - [mec] coolant liquid
  spray(ing)
— f — refrigerante m—[mec] coolant spray(ing)
— f — vapor m - vapor spray(ing)
— f singular - unique spray(ing)
rociamiento m - . . . • sprinkling; spraying
— m con aceite m - oil, sprinkling, o spraying
— m — tubo m - [tub] pipe spraying
— m de tubo m - [tub] pipe spraying
rociar v bajo presión f - to pressure spray
— v — — f alta - to high pressure spray
— m con aceite m - to oil, sprinkle, o spray
— v tubo m - [tub] to spray @ pipe
rocío m - . . . • [segurid] spray mist
— m de aceite m - oil spray
— m inhalado - [segurid] inhaled spray
— m para enfriamiento m - cooling spray
rodado m - [mec] wheel • [geol] boulder
— m grande - [geol] large boulder
— m para cojinete m - [mec] véase guiadera f
  para cojinete n
— m pequeño m - [geol] small boulder
— m silícico - [geol] solicic boulder
rodado,da a - rolled; rotated • [miner] scattered • [geol] rounded • [mec] run in
— a libremente - rolled freely
rodadura f - [autom] roll over • véase también
  rodaje m
— f con carril(es) m - véase oruga f
— f — engranaje m - [mec] gear rolling
— f — tambor(es) m - [mec] drum rolling
— f — — m de acero m - [mec] steel drum
  rolling
— f rápida - [mec] fast rolling • [autom]
  fast rollover
— f violenta - [autom] hard rollover
rodaja f - [mec] . . . ; pulley; (block) sheave;
  caster
— f acanalada - [mec] split roll
— f con bola(s) - [mec] ball caster
— f — — f accionada a con cilindro m neumático - [mec] air cylinder operated ball
  caster
— f impulsora f - [mec] drive,ving roll
— f — acanalada - [mec] split drive,ving roll
— f — con canto m acanalado - [mec] split
  drive,ving roll
— f para mesa f - [mec] table caster
— f — — f para cepilladora f - [herram] plane
  table caster
rodaje m - . . . • [autom] run(ning) in• [mec]
  undercarriage • [sold] (under)carriage • [mec]
  runway • rolling • run in • [autom] break-in
— m con cuatro ruedas f - [sold] four-wheel(ed)
  undercarriage
— f — — — f para taller m - [sold] four
  wheel(ed) shop running gear
— f — dos ruedas f - [sold] two-wheel(ed), undercarriage, o running gear
— n — — f para taller m - [sold] two-
  -wheeled shop, undercarriage, o running gear
— m — tres ruedas f - [sold] three-wheel(ed)
  undercarriage
— m — — — f para taller m - [sold] three-
  -wheel(ed) shop, undercarriage, o running
  gear
— m — neumático(s) m - [autom-neumát] tire
  rolling
— m de tipo m para taller m - [sold] shop type
  undercarriage
— m optativo - [sold] optional undercarriage
— m para armario m - [ind] cabinet truck
— m — remolque m - [sold] towing under carriage
— m — — m carretero - [sold] road towing undercarriage
— m — taller m - [sold] shop, undercarriage, o
  running gear

rodaje m práctico - [sold] practical, o convenient, undercarriage
— m sólido - [sold] sturdy undercarriage
rodamiento m - rolling motion • [coque] tumbler
test • [coque] bearing; véase también cojinete m
— m ajustado - [cojin] set bearing
— m — lateralmente - [cojin] set bearing
— m antefricción f - [cojin] anti-friction
bearing
— m armado - [cojin] assembled bearing
— m central - [cojin] center,tral bearing
— m — para caja f - [cojin] center housing
bearing
— m clásico - [cojin] standard bearing
— m — con hilera f única (de rodillos m) -
[cojin] standard single row bearing
— m — una sóla hilera f de rodillos m -
[cojin] standard single row bearing
— m con contacto m angular - [cojin] véase cojinete con rodillos m cónicos
— m — cuatro hileras f (de rodillos) - [cojin]
four row bearing
— m — dos hileras f (de rodillos m) - [cojin]
two, o double, row bearing
— m — hilera f cuádruple (de rodillos m) -
[cojin] four row bearing
— m — — f doble (de rodillos f) - [cojin]
double row bearing
— m — — f sencilla (de rodillos m) •
[cojin] single row bearing
— m — — f única (de rodillos m) - [cojin]
single row bearing
— m — rodillos m ahusados m - [cojin] tapered
rolls bearing
— m — — m cónicos - [cojin] tapered roller
bearing
— m — una (sola) hilera f (de rodillos m) -
one, o single, row bearing
— m de acero m - [cojin] steel bearing
— m desgastado - [cojin] worn bearing
— m estándar - [cojin] standard bearing
— m flotante - [mec] floating bearing
— m para caja f - [mec] housing bearing
— m para diferencial m - [mec] differential
bearing
— m — embrague m - [mec] clutch bearing
— m — rodillo m - [mec] roll bearing
— m — — m laminador m - [metal-lam] (rolling)
roll bearing; mill bearing
— m — rueda f - [mec] wheel bearing
— m — ventilador m - [mec] fan bearing
— m precargado - [cojin] preloaded bearing
— m según sistema m métrico - [cojin] metric
(system) bearing
— m — turboalimentador(es) m - [comb.int]
turbocharger bearing
— m unitario - [cojin] unit bearing
rodante a - . . .; traveling
rodapié m - [constr] . . .; baseboard
rodar v - . . .; to rotate; to roll • [mec] to
run in • [autom] to roll over
— v adelante - to roll ahead (of)
— v delante - to roll ahead
— v desde acantilado m - to roll, over, o off,
@ cliff
— v engranaje m—[mec] to, roll, o turn, @ gear
— v libremente - to, roll, o turn, freely
— v neumático m - [autom-neumát] to roll @ tire
— v sin lubricación f - [mec] to, roll, o turn,
without lubricating,tion
— v tambor m - [mec] to, roll, o turn, @ drum
— v uno contra otro - [mec] to roll against
each other
— v — — — sin lubricación f - [mec] to roll
against each other without lubricating,tion
rodeaban adv - (que) environmental
rodeado,da a - surrounded
— a de - sorrounded by • encased
— a completamente - completely, surrounded, o
encased
— a — con hormigón m - [constr] completely
concrete encased

rodear v estructura f - [constr] to surround @
structure
— f tubería f - [constr] to surround @ pipe
rodeo m - . . . . • surrounding • [vial] detour
— m de estructura f - [constr] structure surrounding
rodete m - . . . . • [electr-mot] rotor • [ventil]
impeller • [mec] shoulder • véase también
aspirador m
— m para aspirador m - [ventil] fan impeller
— m — ventilador m - [ventil] fan impeller
rodillo m - . . . .; [Esp.] coating roll • [mec]
wheel • [textil] cot
— m acanalado - [mec] split roll
— m activo - [mec] active, o live, roller
— m ahusado - [mec] tapered, o conical, roller
— m albardillado m - [mec] hourglass roller
— m ajustable - [mec] adjustable roll(er)
— m ajustado - [mec] adjusted roll
— m ajustador - [mec] adjusting roll(er)
— m alimentador m - [mec] feed(ing) roll(er)
— m — abierto - [mec] open feed(ing) roll(er)
— m — mecanizado - [ind] power feed(ing) roll
— m — para encabezadora f - [trefil] header
feed(ing) roll(er)
— m — — máquina f - [mec] machine feeding roll
— m altamente pulido - [mec] high(ly) polished
roll(er)
— m anterior - [mec] front roll(er) • previous
roll(er)
— m apisonador - [constr] tamping roll(er)
— m — con neumático(s) m de caucho m - [constr]
rubber-tired tamping roller
— m — — pata(s) f de cabra f - [constr]
sheepsfoot tamping roll(er)
— m aplanador - [metal-lam] leveling roll(er)
— m aspirador - [papel] suction roll(er)
— m aspirante - [papel] suction roll(er)
— m automotor - [metal-lam] self-driven roll(er)
— m auxiliar - [mec] auxiliary roll(er)
— m — mecanizado - [mec] auxiliary power roll
— m — — para iniciar marcha f -[mec] auxiliary power starting roll
— m — — — v operación f - [mec] auxiliary
power starting roll
— m — — v marcha f - [mec] auxiliary power
staring roll(er)
— m — — — v operación f - [mec] auxiliary
power starting roll(er)
— m "Bridle" - [metal-lam] bridle roll
— m calibrador - [mec] sizing roll(er)
— m canteador - [metal-lam] edge,ging roll(er) •
[f.c.-ruedas] edging roll
— m captador - [mec] pickup roll
— m centrifugador - [mec] spinner,ning roll
— m compresor - [metal-lam] press roll
— m con diámetro m grande - [mec] large diameter roll
— m — — m pequeño - [mec] small diameter roll
— m — — m reducido - [mec] small diameter roll
— m — diseño m singular - [mec] unique design
roll
— m — imán - [sold] magnet(ic) roll
— m — — m permanente - [sold] permanent magnet roll
— m — — m — grande - [sold] large permanent
magnet roll
— m — largo m de . . . pulgada(s) f - [papel]
. . .-inch face roll
— m — neumático(s) m de caucho m - [constr]
rubber-tired roller
— m — pata(s) f de cabra f - [constr] sheepsfoot roller
— m con presión f - [mec] pressure roll
— m — reborde m - [mec] hook roller
— m — superficie f endurecida - [sold] hard-
surfaced roller
— m conformado - [mec] shaped roll(er)
— m conformador - [mec] shaper,ping, o forming,
roll
— m — para borde(s) m - [metal-lam] edge, forming, o shaping, roll
— m cónico - [mec] tapered roll(er)

rodillo m contorneado—[mec] contour shaped roll
— m — de acero m - [mec] contoured steel roll
— m — — m endurecido - [mec] contour shaped hardened steel roll
— m — — m templado - [mec] contour shaped hardened steel roll
— m controlador - [mec] controlling roll
— m — para conformación f - [mec] shape controlling roll
— m — — forma f - [mec] shape controlling roll
— m corrugador - [mec] corrugating roll
— m corto ~ [mec] short roller
— m — tipo m carrete—[mec] short spool roller
— m cromado - [metal-trat] chrome,mium plated roll
— m de acero m - [mec] steel roll(er)
— m — — m contorneado - [mec] contoured, o contour shaped steel roll(er)
— m — — m — endurecido - [mec] contoured, o contour shaped, hardened steel roll
— m — — m — templado - [mec] contour shaped, o contoured, hardened steel roll(er)
— m — — m endurecido - [mec] hardened steel roll(er)
— m — — m forjado - [metal-lam] forged steel roll(er)
— m — — m fundido m - [metal-prod] cast steel roll(er)
— m — — m para apoyo m - [metal-lam] cast steel back-up roll
— m — — m — trabajo m - [metal-lam] cast steel work roll
— m — — m para apoyo m - [metal-lam] steel back-up roll
— m — — m — trabajo m - [metal-lam] steel work roll
— m — — m recubierto - [mec] covered steel roll
— m — — m — con caucho m - [metal-trat] rubber covered steel roll
— m — — m templado - [mec] tempered steel roll
— m — aleación f - [metal-lam] alloy roll
— m — caucho m - [mec] rubber roll(er)
— m — — m para remover agua m - [mec] squeegee roll
— m — coquilla f - [metal-lam] chilled iron roll
— m — — fundida - [metal-lam] cast chilled iron roll(er)
— m — — fundición f - [lam] cast (iron) roll(er)
— m — — f modular - [metal-lam] modular cast roll
— m deficiente - [metal-lam] substandard roll
— m defectuoso - [metal-lam] defective roll(er)
— m deflector - [metal-lam] deflecting,tor roll • [mec] baffle roll
— m — en entrada f - [metal-lam] feed end deflection roll
— m — para línea f - [metal-trat] line deflector roll
— m — — — f estañadora - [metal-trat] tinning line deflector roll
— m — — — f — electrolítica - [metal-trat] elecrolytic tinning line deflector roll
— m — — salida f - [metal-trat] delivery end deflector roll
— m — — — — f para laminador m - [metal-lam] mill delivery deflecting roll
— m delantero - [mec] frontal roll(er)
— m derecho m - [metal-lam] right roll(er)
— m — para laminación f de alma m - [f.c.-ruedas] right web roll
— m desarmado - [mec] disassembled roll
— m desbastador - [metal-lam] puddler('s) roll
— m deschalador - [agric] husking roll
— m despinochador - [agric] husking roll
— m despojador m - [agric-equip] snapping roll
— m detenido - [mec] stopped, o dead, roll(er)
— m embocador - [agric-equip] paddle roll
— m en cuna f - [metal-lam] roll in @ cradle
— m en tránsito m - [metal-lam] roll in transit

rodillo m enderezador - [mec] straightening roll - [metal-lam] unfolding roll
— m — ajustado - [mec] adjusted straightening roll
— m — para varilla(s) f - [trefil] rod straightening roll
— m endurecido - [mec] hardened roll(er)
— m — contorneado - [mec] countour shaped hardened roll
— m — — de acero m - [mec] contour shaped hardened steel roll
— m — de acero m - [mec] hardened steel roll
— m — — m contorneado - [mec] contour shaped hardened steel roll
— m — para muñón m - [mec] hardened trunnion roll(er)
— m enorme - [mec] huge roller
— m entorneador - [mec] turn roll
— m — para borde(s) m - [mec] edge turn roll
— m escurridor m - [mec] squeegee, o wringer, roll
— m Esna - [mec] Esna roll
— m especial - [metal-lam] special roll
— m — para línea f - [metal-trat] special line, o line special, roll
— m — — f para estañadura f - [metal-trat] tinning line special roll
— m — — — f — galvanización f - [metal-trat] galvanizing line special roll
— m estabilizador—[metal-lam] stabilizing roll
— m estructural - [metal-lam] véase rodillo m para perfil(es) m estructural(es)
— m exterior - [metal-lam] outside roll
— m extractor - [col.cont] withdrawal, o pinch, o supporting, roll
— m forjador - [metal-fabr] forging roll
— m formador - [metal-lam] forming, o shaping, roll
— m frontal - [mec] front roll(er)
— m galvanizador—[metal-trat] galvanizing roll
— m grande - [mec] large roll(er)
— m — ajustable - [mec] large adjustable, o adjustable large, roll
— m — regulable - [mec] large adjustable, o adjustable large, roll
— m guía - [mec] guide,ding roll • [Esp.] pinch roll
— m — inferior - [metal-lam] bottom pinch roll
— m — superior - [metal-lam] upper pinch roll
— m guiacable(s) - [cabl] cable guide roll(er)
— m guiador - [mec] guide roller
— m — con tensión f - [mec] (spring) loaded guide roll
— m — — — f con resorte m - [mec] spring loaded guide roll
— m — lateral - [mec] side guide roll
— m — — con tensión f (con resorte m) - [mec] (spring) loaded side guide roll
— m igualador - equalizing roller
— m imantado - [sold] magnet roll
— m impresor - [imprent] printing roll
— m — para filigrana f - [papel] watermark printing roll
— m — marca f de agua m - [papel] watermark printing roll
— m impulsado - [mec] driven roll
— m impulsor - [mec] drive,ving roll
— m — para alambre m - [sold] wire drive,ving roll
— m — acanalado - [mec] split drive,ving roll
— m — con canto m acanalado - [mec] split drive,ving roll
— m — exterior—[mec] outside drive,ving roll
— m — inferior - [sold] lower, o bottom, drive,ving roll
— m — interior - [mec] inner, o inside, drive,ving roll
— m — para marcha f sin carga f - [sold] idle drive,ving roll
— m — superior - [sold] top, o upper, drive,ving roll
— m individual - [mec] individual roll
— m — para apoyo m - [metal-lam] individual

back-up roll
rodillo m inerte - [ind] dead roll(er)
— **m inferior** - [mec] lower, o lowest, o bottom, roll(er), o mill
— **m — de carril m** - [constr-equip] belt bottom roller
— **m — para rotadora f** - [cuerp.moled] lower rotator roll
— **m — — cable m** - [grúas] lower cable roller
— **m iniciador** - [mec] starting roll(er)
— **m inmenso** - [mec] huge roll(er)
— **m intermediario** - [mec] idler roller
— **m intermedio** - [mec] middle roll(er)
— **m — para cable m** - [grúas] middle cable roller
— **m izquierdo m** - [mec] left roller
— **m — para laminación f** - [metal-lam] left rolling roll(er)
— **m — — — f de alma m** - [f.c.-ruedas] left web roll(er)
— **m laminador** - [metal-lam] rolling, o press, o laminating, roll(er)
— **m largo** - [mec] long roll(er)
— **m — de tipo m de carrete m** - [mec] long spool roll(er)
— **m lateral** - [mec] side roll(er)
— **m — para guía f** - [mec] side guide,ding roll
— **m — — f con tensión f (con resorte m)** - [mec] (spring) loaded side guide,ding roll
— **m liso m** - [mec] smooth roll(er)
— **m loco** - [mec] idler, o dancer, roll(er); (carrying) idler
— **m — endurecido** - [sold] hardened, o hard-surfaced, roll(er)
— **m — para alimentación f** - [metal-lam] feed-(ing) idling roll(er)
— **m — — — f para bobinadora f** - [metal-lam] decoiler feed(ing) roll(er)
— **m — — rotación f** - [mec] idler rotation roll(er)
— **m — — transportadora f** - [mec] conveyor (carrying) idler
— **m — rotante** - [mec] idle(r) rotation roll
— **m — tipo carrete m** - [mec] spool idle roll
— **m ludido** - [mec] galled roll(er)
— **m mecanizado** - [mec] power(ed) roll(er)
— **m — iniciador** - [mec] starting power roll
— **m — para iniciar marcha f** - [mec] power(ed) starting roll
— **m — — — f operación f** - [mec] power(ed) starting roll
— **m modular** - [mec] modular roll(er)
— **m moleteado** - [mec] knurled roll(er)
— **m Morgoil** - [metal-lam] Morgoil roll(er)
— **m motor** - [mec] drive,ving, o pinch, roll(er) • live roll(er)
— **m — para rotación f** - [mec] drive(n) rotation roll(er)
— **m — superior** - [mec] upper drive,ving roll
— **m motorizado** - [metal-lam] drive,ving roll
— **m neumático** - [constr] pneumatic, o rubber tire, roller
— **m nivelador** - [metal-lam] leveling roll(er)
— **m nuevo** - [metal-lam] new roll(er)
— **m — para calandria f** - [papel] new calender roll(er)
— **m — accionamiento m** - drive,ving roll(er)
— **m — — m para alimentador m** - [mec] drive(r) feed(ing) roll(er)
— **m — agarre m** - [metal-lam] pinch roll
— **m — — m para entrada f** - [metal-lam] feed-(ing) and pinch(ing) roll
— **m — ajuste m** - [mec] adjusting roll
— **m — alimentación f** - [mec] feed(ing) roller • feed (ing) throat
— **m — — f automática** - [mec] automatic feed-(ing) roll
— **m — — f para encabezadora f** - [trefil] header feed(ing) roll
— **m — — f mecanizada** - [mec] power(ed) feed-(ing) roll
— **m — — f para alambre m** - [sold] (wire), feeder, o feed(ing), roll

rodillo m para alimentación f para máquina f. para bola(s) f - [cuerp.moled] ball machine feed(ing) roll
— **m — — f delantera** - [mec] front feed roller
— **m — — f trasera** - [mec] back feed roller
— **m — apiladora f** - [mec] piler roller
— **m — aplanamiento m** - [metal-lam] leveling roll(er)
— **m — apoje* m** - [metal-lam] véase rodillo m para apoyo m
— **m — apoyo m** - [metal-lam] back-up, o support, roll(er) • idler roll
— **m — — m defectuoso** - [metal-lam] defective back-up roll
— **m — — m para laminador m para temple m** - [metal-lam] temper mill back-up roll
— **m — — m — tren m para temple m** - [metal-lam] temper mill back-up roll
— **m — — m — terminador** - [metal-lam] finishing mill back-up roll
— **m — — m — m desbastador** - [metal-lam] roughing mill back-up roll
— **m — — m — m continuo** - [metal-lam] continuous mill back-up roll
— **m — — m — m — para laminación f** - [metal-lam] - continuous rolling mill back-up roll(er)
— **m — — m — m — — f en caliente** - [metal-lam] continuous hot strip mill back-up roll(er)
— **m — — m Morgoil** - [metal-lam] Morgoil back-up roll(er)
— **m — — m roto** - [metal-lam] brocken back-up roll(er)
— **m — arrastre m** - [mec] pinch roll
— **m — — m inferior** - [mec] lower pinch roll
— **m — — m para entrada f** - [metal-lam] feed-(ing) end pinch roll
— **m — — — f para foso m para bucle(s) m** - [metal-lam] looping pit feeding end pinch roll
— **m — — m para foso m para bucle(s) m** - [metal-lam] looping pit pinch roll
— **m — — m — transferencia f** - [metal-lam] transfer pinch roll
— **m — — m superior** - [mec] top pinch roll
— **m — avance m** - [mec] (forward) advance roll
— **m — banda f para oruga f** - [constr-equip] crawler track roller
— **m — basculación f** - [metal-lam] tilting roll
— **m — biela f** - [mec] connecting rod roller
— **m — cable m** - [mec] cable, o rope, o line, roller
— **m — — m de alambre m** - [grúas] wire, cable, o line, roller
— **m — — m inferior** - [mec] lower cable roller
— **m — — m para perforación f** - [petról] (drilling) wire line roller
— **m — — m superior** - [mec] upper cable roller
— **m — calandria f** - [papel] calender roll • stack(ing) roll
— **m — — f nuevo** - [papel] new calender roll
— **m — — f usado** - [papel] used calender roll
— **m — canto(s) m** - [metal-lam] edger,ging roll
— **m — carril m** - [constr-equip] belt roller
— **m — casilla f** - [constr-equip] cab, o house, roller
— **m — cinta f** - [mec] belt roller
— **m — — f transportadora** - [mec] conveyor belt roll(er)
— **m — cojinete m** - [cojin] bearing roller
— **m — colocación f en posición f** - [mec] positioning roll(er)
— **m — — — f de bobina f** - [metal-lam] coil positioning roll(er)
— **m — conformación f** - [metal-fabr] forming roll
— **m — conformar v** - [metal-fabr] forming roll
— **m — — v borde(s) m** - [metal-lam] edge forming roll
— **m — — f progresiva** - [mec] progressive forming roll
— **m — — f — de borde(s) m** - [mec] progres-

sive edge forming roll
rodillo m̲ **para curvar** v̲ - [mec] curving roll(er)
• bending roll(er)
— m̲ — **desbastar,te** - [metal-lam] blooming, o̲ roughing, roll(er)
— m̲ — **embrague** m̲ - [mec] clutch roll(er)
— m̲ — — m̲ **con** **rueda** f̲ **libre** - [mec] overrunning clutch roller
— m̲ — **empuje** m̲ - [mec] thrust roller
— m̲ — **encorvar** v̲ - [metal-lam] bending roll
— m̲ — **enderezamiento** m̲ - [mec] straightener roller
— m̲ — — m̲ **de varilla(s)** f̲ - [trefil] rod straightening roll(er)
— m̲ — **enderezar** v̲ - [trefil] straightening roll(er)
— m̲ — — v̲ **varilla(s)** f̲ - [trefil] rod straightening roll(er)
— m̲ — **engrasamiento,se** - [mec] greasing, o̲ oiling, roll
— m̲ — **entrada** f̲ - [mec] entrance roll(er)
— m̲ — **para bobinadora** f̲ - [metal-prod] coiler turndown roll(er)
— m̲ — **escuadrar** v̲ - [metal-lam] edging roll • squaring roll(er)
— m̲ — **estañadura** f̲ - [metal-trat] tinning roll
— m̲ — **expulsor** m̲ - [metal-lam] ejecting roll
— m̲ — — m̲ **para plancha(s)** f̲ - [metal-lam] plate ejecting roll(er)
— m̲ — **forjar** m̲ - [metal-trat] forging roll(er)
— m̲ — **formación** f̲ - [metal-lam] forming roll
— m̲ — **galletita(s)** f̲ - [culin] cookie roll(er)
— m̲ — **giro** m̲ - [mec] turning roll
— m̲ — **guía** f̲ - [mec] guide pulley; jockey roller • [metal-lam] guide roll(er) • [sold] lead roll(er)
— m̲ — — f̲ **con tensión** f̲ **(con resorte** m̲**)** - [mec] (spring) loaded guide roll
— m̲ — **guiar cable** m̲ - [cabl] cable guide roll
— m̲ — **impulsión** f̲ - [mec] drive,ving roll
— m̲ — — f̲ **con diseño** m̲ **singular** - [sold] unique design drive,ving roll
— m̲ — — **con tensión** f̲ **(con resorte** m̲**)** - [sold] (spring) loaded drive,ving roll
— m̲ — — f̲ **singular** - [sold] unique drive,ving roll
— m̲ — **iniciar marcha** f̲ - [mec] starting roll
— m̲ — — v̲ **operación** f̲ - [mec] starting roll
— m̲ — **inmersión** f̲ - [mec] immersing roll
— m̲ — **laminación** f̲ - [metal-lam] work(ing), o̲ rolling, roll, o̲ mill
— m̲ — — f̲ **de acero** m̲ - [metal-lam] steel, rolling, o̲ mill, roll
— m̲ — — f̲ — **alma** m̲ - [f.c.-ruedas] web roll
— m̲ — — f̲ **de plano** m̲ **para rodadura** f̲ - [f.c.-ruedas] tread roll(er)
— m̲ — — f̲ **final** - [metal-lam] finishing roll
— f̲ — — f̲ **nuevo** - [metal-lam] new mill roll
— m̲ — — f̲ **usado** - [metal-lam] used mill roll
— m̲ — — f̲ **viejo** - [metal-lam] old mill roll
— m̲ — **laminador** m̲ - [metal-lam] mill roll
— m̲ — — m̲ **para temple** m̲ - [metal-lam] temper mill roll
— m̲ — — m̲ **viejo** - [metal-lam] old mill roll
— m̲ — **leva** f̲ - [mec] cam roller
— m̲ — **línea** f̲ - [metal-lam] line roll
— m̲ — **para estañadura** f̲ - [metal-trat] tinning line roll
— m̲ — — f̲ — — f̲ **electrolítica** - [metal-lam] electrolytic tinning line roll
— m̲ — — f̲ **para galvanización** f̲ - [metal-trat] galvanizing line roll
— m̲ — **mecanismo** m̲ - [mec] mechanism roll
— m̲ — — m̲ **alimentador** - [sold] feed(ing) mechanism roll
— m̲ — — m̲ — **para alambre** m̲ - [sold] wire feed(ing) mechanism roll
— m̲ — **mesa** f̲ - [mec] table roll(er)
— m̲ — — f̲ **auxiliar** - [mec] trailer table roll
— m̲ — — f̲ —, **corrediza,** o̲ **deslizante,** o̲ **desplazable** - [ind] traveling trailer table roller
— m̲ — — f̲ **basculante** - [mec] tilting table roller

rodillo m̲ **para mesa** f̲ **basculante, corrediza,** o̲ **delizante** - [mec] traveling tilting table roller
— m̲ — — f̲ **para retroceso** m̲ - [mec] pull back table roll
— m̲ — — f̲ **suplementaria** - [mec] trailer, o̲ tail, table roller
— m̲ — — f̲ —, **corrediza,** o̲ **delizante** - [mec] traveling trailer table roller
— m̲ — **mezcladora** f̲ - [ind] mixer roll
— m̲ — **muñón** m̲ - [mec] trunnion roller
— m̲ — **nivelación** f̲ - [metal-lam] leveling roll
— m̲ — **oruga** f̲ - [constr-equip] crawler roller • [mec] tractor roller
— m̲ — **palastro** m̲ - [metal-lam] plate roll
— m̲ — **perfil(es)** m̲ **estructural(es)** - [metal-lam] structural roll
— m̲ — **plancha** f̲ - [metal-lam] plate roll(er)
— m̲ — — f̲ **vertical** - [mec] vertical plate roll(er)
— m̲ — **polín** m̲ - [mec] (placer) cam roller
— m̲ — — m̲ **elevador** m̲ - [mec] placer cam roll
— m̲ — **posicionamiento*** m̲ **para bobina** f̲ - [metal-lam] coil positioning roll
— m̲ — **pote** m̲ - [metal-trat] pot roll(er)
— m̲ — **presión** f̲ - [metal-lam] pressure roll • [mec] pinch roll • [f.c.-ruedas] back pressure roll
— m̲ — **quebrantadora** f̲ - [metal-lam] (scale) breaker roll
— m̲ — **repuesto** m̲ - [metal-lam] spare roll
— m̲ — **respaldo** m̲ — **véase rodillo** m̲ **para apoyo**
— m̲ — **revestimiento** m̲ - [mec] coating roll
— m̲ — **rizado** m̲ - [metal-lam] curling roll
— m̲ — **rotación** f̲ - rotation roll • [metal-lam] turn roll
— m̲ — **rotadora** f̲ - [cuerp.moled] rotator roll
— m̲ — **rueda** f̲ - [mec] wheel roller
— m̲ — — f̲ **loca** - [mec] idler roller
— m̲ — **salida** f̲ - [metal-trat] exit roll
— m̲ — **soldadura** f̲ - [sold] weld(ing) roll
— m̲ — **soporte** m̲ - [mec] support(ing) roll
— m̲ — — m̲ **para tubo(s)** m̲ - [tub] pipe support(ing) roll
— m̲ — **sostén** m̲ - [mec] support(ing) roller
— m̲ — — m̲ **detrás de banda** f̲ - [petról-equip] band (back) roller
— m̲ — **succión** f̲ - [papel] suction roll
— m̲ — **tensión** f̲ - [mec] tension, o̲ jockey, roll • [metal-lam] looper
— m̲ — — f̲ **para accionamiento** m̲ - [metal-lam] drive,ving looper
— m̲ — — f̲ — — m̲ **eléctrico** - [metal-lam] electric drive,ving looper
— m̲ — — f̲ — — m̲ **hidráulico** - [metal-lam] hydraulic drive,ving looper
— m̲ — — f̲ **para tren** m̲ - [metal-lam] strip mill looper
— m̲ — — f̲ — — m̲ **semicontinuo** - [metal-lam] semicontinuous strip mill looper
— m̲ — — f̲ — — m̲ — **para laminación** f̲ **en caliente** - [metal-lam] semicontinuous hot strip mill looper
— m̲ — — f̲ **loco** - [metal-lam] tension idling roll
— m̲ — **terminación** f̲ - [metal-lam] finishing roll
— m̲ — **tiro** m̲ - [mec] pinch, o̲ pulling, roll
— m̲ — **tocho(s)** m̲ - [metal-lam] blooming roll
— m̲ — **trabajo** m̲ - [metal-lam] work(ing) roll
— m̲ — — m̲ **para tren** m̲ - [metal-lam] mill work(ing) roll
— m̲ — — m̲ — — m̲ **terminador** - [metal-lam] finishing mill work(ing) roll
— m̲ — — m̲ — — m̲ **desbastador** - [metal-lam] roughing mill work(ing) roll
— m̲ — — m̲ — **roto** - [metal-lam] broken work(ing) roll
— m̲ — **tracción** f̲ - [mec] pulling roll • [metal-lam] tension, o̲ pinch, roll
— m̲ — — f̲ **defectuoso** - [metal-lam] defective pinch roll
— m̲ — — f̲ **roto** - [metal-lam] broken pinch roll

rodillo m para tracción f para expulsor m -
[metal-lam] (r)ejector pinch roll
— m —— f —— m para plancha(s) f - [metal-
-lam] plate (r)ejector pinch roll
— m — transferencia f - [metal-lam] transfer
roll
— m — transportador m - [ind] conveyor roll
— m — tren m - [metal-lam] mill roll
— m —— m para temple m - [metal-lam] temper
mill roll
— m —— m continuo - [metal-lam] continuous
mill roll
— m —— m — para laminación f - [metal-lam]
continuous strip mill roll
— m —— m —— f en caliente - [metal-
-lam] continuous hot strip mill roll
— m —— m —— f — frío - [metal-lam]
continuous cold strip mill roll
— m —— m — f de acero m - [metal-lam]
steel (rolling) mill roll
— m —— m desbastador - [metal-lam] roughing
mill roll
— m —— m terminador - [metal-lam] finishing
mill roll
— m — trituradora f - [constr] crusher roll
— m — unión(es) f - [metal-lam] véase aplanadora f para costuras f
— m — válvula f - [mec] valve roller
— m —— f neumática - [mec] air valve roller
— m —— f para aire m—[mec] air valve roller
— m — vástago m - [mec] rod roller
— m pequeño m - [mec] small roll
— m — ajustable - [mec] small adjustable, o
adjustable small, roll
— m posterior - [mec] back, o rear, roller
— m principal - [metal-lam] main roll
— m progresivo - [mec] progressive roll
— m quebrantador - [miner] breaker roll
— m ranurado - [metal-lam] groove(d) roll(er)
— m recubierto - [mec] covered roll
— m — con fibra f - [mec] fiber covered roll
— m —— (material) plástico - [ind] plastic
covered roll
— m regulable - [mec] adjustable roll
— m regulador - [mec] controlling roll
— m — para conformación f - [mec] shape controlling roll
— m —— forma f—[mec] shape controlling roll
— m retraído - [mec] retracted roll(er)
— m rompedor m - [miner] breaker,king roll
— m rotante - [mec] rotating roll(er)
— m roto - [metal-lam] broken roll(er)
— m rugoso - [metal-lam] rough roll • Pangborn
roll
— m secador - [mec] drier, o wringer, roll
— m sin convexidad f - [mec] crownless roll
— m — costura f - [mec] seamless roll(er)
— m singular - [mec] unique roller
— m sumergido - [metal-trat] submerged roller •
sink roll
— m superior - [mec] upper roll(er); top roller
— m — aspirante - [papel] upper suction roll
— m — para cable m - [cabl] upper cable roller
— m —— carril m - [constr-equip] belt, top,
o upper, roller
— m — para guía f - [mec] top pinch roll
— m —— rotadora f - [cuerp.moled] upper rotator roll
— m tableado - [metal-lam] shattered roll
— m templado - [mec] tempered roll(er) • hardened roll(er)
— m — contorneado - [mec] countour shaped, o
contoured, hardened roll
— m — de acero m - [mec] contour shaped, o
contoured, hardened steel roll
— m — de acero m contorneado - [mec] contour
shaped, o contoured, hardened steel roll
— m — para muñón m - [mec] hardened trunnion
roller
— m tensor - [mec] tension roll(er)
— m — loco - [mec] tension idling roll(er)
— m Timken - [mec] Timken roller
— m tipo carrete n - [mec] spool roller

rodillo m tipo reloj m de arena - [mec] hourglass roll
— m tornado - [mec] turned roll(er)
— m torneado - [mec] ground roll
— m tractor - [mec] traction,tor roller
— m transportador - [mec] carrying roller •
[metal-prod] conveyor roll
— m trasero - [mec] back, o rear, roller
— m usado - [mec] used roll
— m vertical - [metal-lam] vertical roll
— m — tipo m de guía f vertical - [metal-lam]
side guide type vertical roll
— m vibratorio - [constr] vibrating roll(er)
— m — de acero m - [constr] vibrating steel
roller
— m —— m arrastrado - [constr] towed vibrating steel roller
— m —— m — por tractor m - [constr]
tractor-towed vibrating steel roller
— m viejo - [metal-lam] old roll
roedor m excavador - [zool] burrowing rodent
rojo m - . . . • [metal] red hot
— m ladrillo - [color] brick red
— m mate - [colores] matte, o dull, red
— m para pulir v - [mec] crocus
rol m - véase papel m
rolado,da* a - [metal-lam] véase laminado,da
rolar* v - [metal-lam] véase laminar v
roldana f - [mec] . . . • washer
— f agrietada - [mec] cracked roller
— f — para cable m - [cabl] cracked wire rope
roller
— f con espiral - [cabl] spiral sheave
— f de acero m - [mec] steel pulley
— f —— m galvanizado - [mec] galvanized
(steel) pulley
— f desgastada - [mec] worn sheave
— f para cable m - [cabl] wire rope, pulley, o
sheave, o roll(er)
— f —— m desgastada - [Cabl] worn wire rope,
pulley, o sheave, o roller
— f desplazada - [cabl] dislocated sheave
— f giratoria - [mec] swivel sheave
— f inferior - [mec] bottom, o lower, shave, o
pulley • [petról] heel sheave
— f libre - [cabl] free, o dangling, sheave
— f loca f - [constr] padlock sheave
— f lubricada - [Grúas] lubricated sheave
— f — cable m - [cabl] wire rope, roller, o
sheave
— f — cangilón m - [constr] padlock
— f — corona f - [petról] crown sheave
— f — cucharón m - [constr] padlock
— f — cuerda f - [mec] rope, roller, o sheave
— f — elevación f - [cabl] hoist(ing) sheave
— f —— m para vagoneta f - [metal-prod] skip
sheave
— f — guía f - [cabl] guide sheave
— f — horquilla f - [cabl] yoke sheave
— f — soga f - [mec] rope, sheave, o roller
— f posterior m - [mec] rear, o back. sheave •
[petról] heel sheave
— f ranurada - [mec] sheave
— f roscada - [mec] threaded washer
— f torcida - [cabl] distorted, sheave, o roller
— f — para cable m - [cabl] distorted wire
rope roller
— f verificada - [mec] checked sheave
rollo m - . . . . • [transp] circular bundle
— m concéntrico - [cabl] concentric roll
— m de cadena f - [mec] chain, roll, o bundle
— m . . . kilogramo(s) m - [mec] . . . kilogram, coil, o roll
— m — varilla f - [trefil] rod coil
— m estándar - [mec] standard, roll, o coil
— m iniciado - [trefil] started coil
— m nuevo - [trefil] new coil
— m según norma f - standard, coil, o roll
— m soldado - [trefil] welded coil
rollos m por bandeja f - coils per pallet
— m unidos mediante soldadura f - [trefil]

'(butt) welded coil(s)
romaneado,da a - weighed
romanear v - to weigh
romaneo m - weighing
rombal a · . . . .; diamond shape(d)
rómbico,ca a - diamond (shaped)
rombo m **invertido** - reverse diamond
rompeespatas m - [agric] cob breaker
rompefarfollas m - [agric] cob breaker
rompemarlos m - [agric] cob breaker
rompetusas m - [agric] cob breaker
rompecascarilla(s) m - [metal-lam] scale breaker
• véase también **rompedora** f **para cascarilla**
rompecollares m - [mec] collar buster
rompecoplas m - [mec] **véase rompecollares; rom-**
peuniones m
rompedora f - [mec] crusher; breaker
— f **con mandíbula(s)** f - [mec] jaw breaker
— f **para cascarilla** f - [metal-lam] scale
breaker
— f — **escama** f - [metal-lam] (roughing) scale
breaker
— f — **escoria** f - [metal-prod] slag breaker;
skull breaker
— f — **espuma** f - [hidr] scum breaker
— f — **farfolla(s)** f - [agric] cob breaker
— f — **espato(s)** m - [agric] cob breaker
— f — **lobo(s)** m - [metal-prod] skull, cracker,
o breaker
— f — **loma(s)** f - [agric] ridge buster
— f — **marlo(s)** m - [agric] cob breaker
— f — **sínter** m - [metal-prod] sinter breaker
— f — **tusa(s)** f - [agric] cob breaker
rompeespaldas a - backbreaking
rompehuesos m - bone crusher | a - bonecrushing
rompeolas m - [constr] jetty • [hidr] breakwall
romper v - . . . .; to snap • to crack through •
to break away
— v **alambre** m - [mec] to break @ wire
— v **amortiguador** m - [autom] to break @ shock
(absorber)
— v **apoyadero** m - [mec] to break @ strut
— v **apoyo** m - [mec] to break @ strut
— v **barra** f - [mec] to break @, bar, o rod
— v — f **para acoplamiento** m - [autom-mec] to
break @ tie rod
— v **cable** m - [cabl] to break @, rope, o cable
— v — m **con tensión** f - [cabl] to tension
brake @, rope, o cable
— v — m **de alambre** m - [cabl] to break @,
cable, o wire rope
— v — m — — m **con tensión** f - [cabl] to tension break @, wire rope, o cable
— v **cabo** m - [cabl] to break @ rope
— v — m **con tensión** f - [cabl] to tension
break @ rope
— v **cadena** f - [cadenas] to break @ chain
— v **caja** f - [mec] to break @ case
— v — f **para transferencia** f - [autom-mec] to
break @ transfer case
— v **cigüeñal** m - [comb.int] to break @ crankshaft
— v **con tracción** f - to tension braek
— v **conductor** m - [electr-instal] to break @,
wire, o lead
— v — m **a regulador** m **para voltaje** m—[electr-
-instal] to break @ voltage control lead
— v **columna** f - [constr] to break @ column •
[electr-instal] to break @ pole
— v **conductor** m - [electr-instal] to break @
lead
— v — m **para energía** f - [electr-instal] to
break @ power lead
— v — m — **entrada** f - [electr-instal] to
break @ power lead
— v **conexión** f - to break @ connection
— v **contacto** m - [electr-equip] to break @ contact
— v **costra** f - to, break, o crack through, @
crust
— v **depósito** m - [mec] to rupture @ tank
— m **disco** m - [mec] to break @ disk • [música]
to break @ record

romper v **eje** m - [mec] to break @ axle
— v — m **de papel** m - [mec] to break @ paper
shaft
— v — m — — m **endurecido** - [mec] to break @
hard(ened) paper shaft
— v **escoria** f - [metal-prod] to (break @) slag
— v **escorial** - [metal-prod] to break through
(@ slag notch)
— v **eslabón** m - [cadenas] to break @ link
— v **espalda(s)** f - to break @ back
— v **esquina** f - [mec] to break @ corner
— v **guarnición** f - [mec] to break @ gasket
— v — f **para culata** f - [comb.int½] to blow @
head gasket
— v **hierro** m - [metal] to break @ iron
— v — m **fundido** - [metal] to break @ cast
iron
— v **hoja** f - [mec] to break @ leaf
— v — f **de resorte** m - [mec] to break @ spring
leaf
— v **huso,sillo** m - [mec] to break @ spindle
— v **letrero** m - [com] to break @ sign
— v **pestaña** f - [mec] to break @ flange
— v **portaescobilla(s)** m - [electr-mot] tp break
@ (brush)holder
— v **puntal** m - [mec] to break @ strut
— v **resorte** m - [mec] tp break @ spring
— v — m **con hoja(s)** f - [mec] to break @ leaf
spring
— v — m **para portaescobilla(s)** m - [electr-
-mot] to break @ brush spring
— v **sangría** f - [metal-prod] to tap @ furnace
— v **sello** m - [mec] to break @ seal
— v — m **para vacío** m - [neumát] tp break @
vacuum seal
— v **semieje** m - [autom-mec] to break @ axle
shaft
— v **soldadura** f - [sold] to break @ weld
— v — f **en hierro** m - [sold] to break @ iron
weld
— v — f **en hierro** m **fundido** - [sold] to break
@ cast iron weld
— v **superficie** f - [mec] to break @ (sur)face
— v — f **de letrero** m - [com] to break @ sign
face
— v **tapón** m - [mec] to break @ plug
— v **tubo** m - [tub] to break @, tube, o pipe
— v **ventilador** m - [ventil] to break @ fan
romperse v - to break • to rip • to disrupt
— v **sangría** f - [metal-prod] to burst @ iron
notch
— v **súbitamente** - to, snap, o break, suddenly
rompeunión(es) m - [mec] collar buster
rompeviento(s) m - [vest] windbreaker; jacket
roncador,ra a - snoring • [autom] thundering
ronda f - . . . . • stalking; prowl(ing)
— f **en torno** m - stalking (around)
rondado,da a - stalked; prowled
— a **en torno** a - stalked (around)
rondar v - . . . . • to circle • to, stalk, o
prowl
— v **en torno** - to stalk (around)
rondear* v - véase **rondar**
rondpoint m - [vial] circle; rotary intersec-
tion • [G.B.] roundabout; circus
ropa f - [vest] clothing; clothes • attire
— f **apropiada** - [vest] suitable clothing
— f **blanca** - [domést] . . . ; bedding
— f **cambiada** - [vest] changed clothing,thes
— f **cómoda** - [vest] comfortable clothes,thing
— f — **para viajar** v - [Vest] comfortable traveling, clothes,thing, o attire
— f **contaminada** - [segurid] contaminated
clothing
— f **corriente** - [vest] every day, o casual, attire, o clothes, o dress
— f — **para diario** - [vest] every day casual
attire
— f — — **tarde(cita)** f - [vest] casual evening
attire
— f — — **velada** f - [vest] casual evening, attire, o dress
— f **de civil** - [vest] civilian, dress, o attire

ropa f empapada - [Segurid] soaked, o wet(ted). clothing
— f floja - [vest] loose clothes,thing
— f gruesa - [vest] thick, o heavy, clothes
— f húmeda - [vest] damp, o humid, o wet
— clothing
— f interior - [vest] underwear; underclothing
— f — limpia - [vest] clean, underwear, o underclothing
— f — sucia - [vest] dirty, o soiled, underwear, o underclothing
— f lavada - [domést] washed, o laundered, clothes,thing
— f libre de aceite m - [segurid] oil free, garment(s), o clothes,thing
— f limpia - [vest] clean clothes,thing
— f limpiada - [vest] cleaned clothes,thing
— f mojada - [vest] wet clothes,thing
— f para caballero m - [vest] men's clothing
— f — calle f - [vest] street clothes,thing
— f — cama f - [domést] bedding; bed clothing
— f — dama f - [vest] woman's clothing
— f — deporte(s) m - [vest] sports clothing
— f — — m invernal(es) - [vest] winter sports clothes,thing • snowmobile, suit, o clothing
— f — diario - [vest] daytime, clothing, o attire
— f — golf - [vest] golf clothes
— f — hombre m - [vest] men's clothing
— f — mujer(es) f - women's, o ladies', clothing
— f — niña f - [vest] girl's clothing
— f — niño m - [vest] boy's clothing
— f — niños m - [vest] children's clothing
— f — tarde(cita) f - [vest] evening attire
— f — trabajo m - [vest] work(ing) clothes
— f — velada f - [vest] evening attire
— f personal - [vest] personal clothing
— f — nueva - [vest] new personal clothing
— f — usada - [vest] used personal clothing
— f protectora - [segurid] protective. clothes, o clothing, o garment(s)
— f — libre de aceite m - [segurid] oil free protective, clothes,thing, o garment(s)
— f resistente a humedad f - [segurid] moisture resisting clothing
— f — a líquido(s) m - [segurid] liquid(s) resisting clothing
— f suelta - [vest] loose clothing
— f terminada - [vest] finished clothing
— f usada - [vest] used clothes,thing
— f — para caballero m - [vest] men's used clothing
— f — — cama f - [domést] used, bedding, o bed clothes,thing
— f — — dama f - [vest] women's used clothing
— f — — hombre m - [vest] men's used clothing
— f — — mujer f - [vest] women's used clothing
— m con tapa f inclinada - [domést] sloping lid locker
rosa f - . . . . • [colores] pink; rose
rosado,da a - [color] pink • rosy
rosario m de soldadura f - [sold] (weld) beads
rosca f - [mec] . . . . • [culin] . . .; doughnut; donut
— f ahusada - [mec] taper tapping
— f alineada - [mec] aligned thread
— f áspera - [mec] coarse thread
— f — brillante - [mec] bright coarse thread
— f brillante - [mec] bright thread
— f de tipo m estadounidense - [tub] national pipe thread
— f ahusada - [mec] tapered thread
— f alterada - [mec] upset thread
— f aplicada con rodillo(s) m - [metal-fabr] rolled thread
— f cilíndrica - [mec] running thread
— f con extremo m no desgastado - [mec] straight thread
— f — largo m especial - [mec] special thread length
— f — paso m derecho - [mec] right hand thread

rosca f con paso m izquierdo - [mec] left hand thread
— f cónica - [mec] tapered thread
— f — en tubo m - [tub] pipe taper thread
— f cortada - [mec] cut thread
— f dañada - [mec] damaged, o crossed, thread
— f de Arquímedes - [hidr] Archimedes screw; screw conveyor
— f de tornillo m - [mec] screw thread
— f derecha f - [mec] right hand thread
— f diestrogira - [mec] right hand thread
— f en ajustador m - [mec] adjuster thread
— f — — m para cojinete m - [cojin] bearing adjuster thread
— f — árbol m - [mec] shaft, o axle, thread
— f — eje m - [mec] axle, o shaft, thread
— f — pistola f - [sold] gun thread
— f — portadiferencial m - [autom-mec] (differential) carrier thread
— f estadounidense - [tub] national pipe thread
— f — común - [tub] National Pipe Straight Thread
— f — para tubería f - [tub] National Pipe Thread; American National Taper Pipe Thread
— f estropeada - [mec] stripped thread
— f estándar - [mec] standard thread
— f exterior - [mec] external, o outside, o male, thread; outside screw; threading
— f — aplicada con rodillo(s) m - [metal-fabr] external rolled thread
— f — laminada - [metal-fabr] external rolled thread
— f — y caballete m - [válv] outside screw and yoke; O S & Y
— f fina - [mec] fine thread
— f gruesa - [mec] coarse thread
— f — hacia derecha - [mec] right hand thread
— f — izquierda f - [mec] left hand thread
— f hembra - [mec] tapping; inside, o female, thread • box
— f — con ojal m - [mec] box with @ eye
— f incompleta - [mec] incomplete thread
— f interior - [mec] tapping; inside, o female, thread
— f — descubierta - [tub] running thread
— f internacional - [mec] international, o metric, thread
— f izquierda - [mec] left hand thread
— f limpia - [mec] clean thread
— f lubricada - [mec] lubricated thread
— f macho - [mec] male, o outside, thread
— f métrica - [mec] metric, o international, thread
— f múltiple - [mec] multiple thread
— f para, anillo m, o aro m - [mec] ring thread
— f — tapa f - [mec] cover thread
— f — — f para válvula f - [válv] valve cover thread
— f — tubería f - [tub] pipe,ping thread
— f — — f de acero m - [tub] steel pipe,ping thread
— f — tubo(s) m - [tub] pipe thread
— f — válvula f - [válv] valve thread
— f primera - [mec] first thread
— f protegida - [mec] protected thread
— f redonda,deada - [mec] round(ed) thread
— f sencilla - [mec] simple thread • single thread
— f siniestrogira - [mec] left hand thread
— f sobresaliente - [mec] protruding thread
— f y unión f - [tub] thread and coupling
— f — — f estadounidense(s) - [tub] American thread and coupling
roscado m - [mec] threading; thread cutting • [carp] tapping
— m de agujero m - [mec] hole tapping
— m — rosca f interna - [mec] véase roscado m interior
— m — válvula f - [mec] valve thread(ing)
— m — — f hacia adentro - [mec] valve in threading
— m — — f — afuera - [mec] valve out threading

roscado m̱ exterior - [mec] threading
— m̱ hembra - [mec] tapping
— m̱ interior - [mec] tapping
— m̱ — de agujero m̱ - [mec] hole tapping
— m̱ mediante laminación f̱—[mec] roll threading
roscado,da a̱ - [mec] . . .; tapped; with thread
— a̱ con cara f̱ plana - [mec] threaded flat
faced
— a̱ enteramente - [mec] all-, o̱ fully, threaded
— a̱ exteriormente - [mec] externally threaded
— a̱ íntegramente - [mec] all-threaded
— a̱ interiormente - [mec] tapped; internally
threaded
— a̱ . . . paso(s) m̱ - [mec] . . . treads
tapped
— a̱ mediante laminación f̱ - [mec] roll threaded
roscadora f̱ - [mec] bolt cutter
— f̱ para tubo(s) m̱ - [tub] pipe threader
roscar v̱ - [mec] to thread • to tap
— v̱ agujero m̱ - [mec] to tap @ hole
— v̱, anillo m̱, o̱ aro m̱ - [mec] to thread @ ring
— v̱ borne m̱ - [mec] to thread @ stud
— v̱ interiormente - [mec] to tap
— v̱ — agujero m̱ - [mec] to tap @ hole
— v̱ manguito m̱ - [mec] to tap @ coupling
— v̱ mediante laminación f̱ - [mec] to roll @
thread
— v̱ orificio m̱ - [mec] to tap @ hole
— v̱ tubo m̱—[tub] to thread @ pipe
— v̱ — m̱ cónicamente—[tub] to taper tap @ pipe
— v̱ válvula f̱ - [mec] to thread @ valve
— v̱ — f̱ hacia adentro - [mec] to thread @ pipe
in
— v̱ — f̱ afuera—[mec] to thread @ valve out
roseta f̱ - [constr] estutcheon • [metal-trat]
spangle
roseta f̱ de maíz m̱ - [culin] pop corn
rostro m̱ - . . .; face
— m̱ lavado - washed face
rotación f̱ - . . . • turn(ing); spin(ning) •
tolling (motion) • turning, around, o̱ over •
[grúas] swing • [comb.int] turning over
— f̱ apropiada - proper, rotation, o̱ turning
— f̱ bajo carga f̱ - [mec] rotation, o̱ turning,
under @ load
— f̱ centrífuga - [mec] centrifugal, rotation, o̱
spinning
— f̱ completa - [mec] complete, o̱ full, rotation
• [grúas] complete swing(ing)
— f̱ con cremallera f̱ y piñón m̱ - [mec] rack and
pinion rotation
— f̱ con velocidad f̱ alta - [mec] high speed rotation
— f̱ — f̱ baja - [mec] low speed rotation
— f̱ — f̱ plena - [mec] full speed rotation
— f̱ — f̱ reducida—[mec] slow speed rotation
— f̱ continua - [mec] continuous rotation
— f̱ correcta - [mec] correct, o̱ right, rotation
— f̱ — verificada - [mec] checked correct rotation
— f̱ de aguilón m̱ - [grúas] booming; boom rotation
— f̱ — araña f̱ - [mec] spider rotation
— f̱ — árbol m̱ - [mec] shaft rotating,tion
— f̱ — — m̱ para aportación f̱ (de fuerza f̱) -
[autom-mec] input shaft rotating,tion
— f̱ — m̱ para motor m̱ - [electr-mot] motor
shaft rotation • [comb.int] engine shaft rotation
— f̱ — — m̱ — piñón m̱ - [mec] pinion shaft rotation
— f̱ — — m̱ — salida f̱ (de fuerza f̱) - [autom-
-mec] output shaft, rotation, o̱ turning
— f̱ — armadura f̱ - [mec] body rotation
— f̱ — bomba f̱ - [bombas] pump rotation
— f̱ — brazo m̱ - [mec] arm rotation
— f̱ — — m̱ para soporte m̱ - [mec] overarm rotation
— f̱ — carga f̱ - [mec] load rotation
— f̱ — carrete m̱ - [mec] reel, rotation, o̱
turning
— f̱ — — m̱ para núcleo m̱ - [petról] core reel

rotation
rotación f̱ de cigüeñal m̱ - [comb.int] crankshaft
rotating,tion
— f̱ — compresor m̱ - [ind] compressor rotation
— f̱ — conjunto m̱ - [mec] assembly, o̱ unit,
rotating,tion
— f̱ — corona f̱ - [mec] ring rotating,tion
— f̱ — — f̱ dentada - [mec] ring gear rotation
— f̱ — cultivo(s) m̱ - [agric] crop rotation
— f̱ — depósito m̱ - [mec] tank, turning, o̱
rolling
— f̱ — — m̱ de acero m̱ - [mec] steel tank, rotating,tion, o̱ rolling, o̱ turning
— f̱ — devanado m̱ - [electr-mot] armature rotating,tion
— f̱ — diferencial m̱ - [mec] differential rotating,tion
— f̱ — electrodo m̱ - [sold] electrode rotating,tion
— f̱ — — m̱ de carbono m̱ - [sold] carbon (electrode) rotating,tion
— f̱ — elemento m̱ - [mec] element, o̱ unit, rotating,tion
— f̱ — embrague m̱ - [mec] clutch rotating,tion
— f̱ — empleado(s) m̱ - [labor] employee rotating,tion
— f̱ — engranaje m̱ - [mec] gear rotating,tion
— f̱ — gancho m̱ - [mec] hook rotating,tion
— f̱ — generador m̱ - [electr-prod] generator,
rotating,tion, o̱ turning
— f̱ — — m̱ para soldadura f̱ - [sold] welding
generator, rotating,tion, o̱ turning
— f̱ — grúa f̱ - [grúas] crane rotating,tion
— f̱ — inducido m̱ - [electr-mot] armature rotating,tion
— f̱ — jaula f̱ - [mec] cage rotating,tion
— f̱ — — f̱ para cojinete m̱ - [cojin] bearing
cage rotating,tion
— f̱ — — m̱ para piñón m̱ - [mec] pinion
bearing cage rotating,tion
— f̱ — máquina f̱ - [mec] machine rotating,tion
— f̱ — motor m̱ - [electr-mot] motor rotating,tion • [comb.int] engine, rotating,tion,
o̱ turning, o̱ cranking
— f̱ — núcleo m̱ - [mec] core rotating,tion
— f̱ — pasador m̱ - [mec] pin rotating,tion
— f̱ — piñón m̱ - [mec] pinion rotating,tion
— f̱ — rodaja f̱ - [mec] caster rotating,tion
— f̱ — — f̱ con bola(s) f̱ - [mec] ball caster
rotating,tion
— f̱ — tambor m̱ - [mec] drum rotating,tion
— f̱ — — m̱ de acero m̱ - [mec] steel drum rolling
— f̱ — tornillo m̱ - [mec] screw turning
— f̱ — — m̱ sin fin - [mec] auger, swing, o̱
turn(ing)
— f̱ — — m̱ — — m̱ para descarga f̱ - [mec] unloading auger, swing, o̱ turn(ing)
— f̱ — torrecilla f̱ - [tornos] turret rotation
— f̱ — tubería f̱ - [tub] pipe rotating,tion
— f̱ — tubo m̱ - [tub] tube, o̱ pipe, rotating,tion
— f̱ — — con velocidad f̱ elevada - [metal-
fabr] high speed tube rotating,tion
— f̱ — ventilador m̱ - [mec] fan rotating,tion
— f̱ — vertedero m̱ - [mec] spout swing(ing)
— f̱ — volante m̱ - [comb.int] flywheel, rotating,tion
— f̱ dextrogira - [mec] clockwise rotating,tion
— f̱ dextrorsa - [mec] clockwise rotating,tion •
clockwise motion
— f̱ direccional - [mec] directional, rotating,tion, o̱ turning
— f̱ en sentido m̱ contrario a de agujas f̱ de reloj m̱ - [mec] counterclockwise rotating,tion
— f̱ — — m̱ apropiado - [electr-mot] proper rotating,tion
— f̱ — — m̱ de aguja(s) f̱ de reloj m̱ - [mec]
clockwise rotating,tion
— f̱ — — m̱ en que gira motor m̱ - [comb.int]
enginewise rotating,tion
— f̱ fuera de posición f̱ - [mec] out-of-position

rotating,tion
rotación f hacia derecha f - [grúas] right,
swing(ing), o turning
— f — **izquierda** f - [Grúas] left swing(ing)
— f **horizontal** - horizontal rotating,tion
— f **incorrecta** - incorrect, o wrong, rotation
— f **independiente** - [mec] independent rotating,tion
— f **libre** - [mec] free rotating,tion
— f — **bajo carga** f - [mec] free rotation under
@ load
— f — **de diferencial** m - [mec] free differential rotating,tion
— f **Magnetorque** - [grúas] Mangetorque swing
— f **manual**—[mec] manual, o hand, rotating,tion
— f — **de bomba** f - [bombas] manual pump rotating,tion
— f **normal** - [mec] normal rotating,tion
— f **para arranque** m - [comb.int] start(ing)
cranking
— f **para tornar accesible** - [mec] rotating,tion
to make accessible
— f **periódica** - [mec] periodic rotating,tion
— f **precisa** - [mec] precise rotating,tion
— f — **con cremallera** f **y piñón** m - [grúas]
precise rack and pinion rotating,tion
— f **pura** - [mec] true rolling motion
— f **regulada** - [grúas] controlled swing(ing)
— f **siniestrogira** - [grúas] counterclockwise,
rotating,tion, o swing
— f **siniestorsa** - [mec] counterclockwise rotating,tion
— f **sobre rodaja(s)** f - [mec] caster rotation
— f — — f **con bola(s)** f - [mec] ball caster
rotation
— f **suave** - [Grúas] smooth rotation
— f — **con cremallera** f **y piñón** m - [grúas]
smooth rack and pinion rotation
— f **verificada** - [mec] checked rotation •
[grúas] checked swing
rotado,da a - rotated • rolled; spun • turned
around • [comb.int] turned over
— a **centrífugamente** - [mec] rotated, o spun,
centrifugally
— a **continuamente** - [mec] rotated continuously
— a **fuera de posición** f - [mec] rotated out of
position
— a **hacia derecha** f - [grúas] swung right
— a — **izquierda** f - [grúas] swung left
— a **independientemente** - [mec] rotated independently • [grúas] swung independently
— a **libremente** - [mec] rotated freely
— a **manualmente** - [mec] rotated manually
— a **para arranque** m - [comb.int] cranked for @
start(ing)
— a — **tornar** v **accesible** - [mec] rotated to
make accessible
— a **precisamente** - rotated precisely
rotador m - [mec] rotator
— m **de vástago** m **para perforadora** f - [petról]
Kelly spinner
— m — **espita** f - [mec] spout rotator
— m — — f **para segadora** f - [agric-equip]
harvester spout rotator
— m — — f — f **para forraje** m - [agric-
-equip] forage harvester spout rotator
— f — **válvula** f - [válv] valve rotator
— m — f **para descarga** f - [válv] exhaust
valve rotator
— m — — — f **refrentado con estelita** f -
[comb.int] stellite faced exhaust valve roto-
tor
— m — **vertedero** m - [mec] spout rotator
— m — — m **para segadora** f - [agric-equip]
harvester spout rotator
— m — — m — — f **para forraje** m - [agric-
-equip] forage harvester spout rotator
rotadora f - [mec] rotator
— f/**alimentadora** - [cuerp.moled] rotator/feeder
rotámetro m - [instrum] rotameter • véase también **medidor** m **para caudal(es)** m; **fluidímetro**
rotar v - . . .; to rotate; to spin; to turn
around • [comb.int] to turn over

rotar v **aprendiz,ces** m - [labor] to rotate @,
apprentice(s), o trainee(s)
— v **araña** f - [mec] to rotate @ spider
— v **árbol** m - [mec] to rotate @ shaft
— v — m **para aportación** f **(de fuerza** f**)** -
[autom-mec] to rotate a input shaft
— v — m **para piñón** m - [mec] to rotate @ pinion shaft
— v **armadura** f - [mec] to rotate @ body
— v **bomba** f - [bombas] to rotate @ pump
— v **carrete** m - [mec] to rotate @ reel
— v — m **para núcleo** m - [mec] to rotate @ core
reel
— v **centrífugamente** - [mec] to, rotate, o spin,
centrifugally
— v **cigüeñal** m - [comb.int] to rotate @ crankshaft
— v **conjunto** m - [mec] to rotate @ assembly
— v **continuamente** - [mec] to rotate continuously
— v **corona** f - [mec] to rotate @ ring
— v — **dentada** f - [mec] to rotate @ ring gear
— v **cultivo(s)** m - [agric] to rotate @ crop(s)
— v **depósito** m - [mec] to roll @ tank
— v — m **de acero** m - [mec] to roll @ steel
tank
— v **electrodo** m - [sold] to rotate @ electrode
— v — m **de carbono** m - [sold] to rotate @
carbon (electrode)
— m **elemento** m - [mec] to rotate @, element, o
unit
— v **embrague** m - [mec] to rotate @ clutch
— v **empleado(s)** m - [labor] to rotate @ em-
ployee(s)
— v **engranaje(s)** m - [mec] to rotate @ gear
— v **fuera de posición** f - [mec] to rotate out
of position
— v **gancho** m - [mec] to rotate @ hook
— v **hacia derecha** f - [grúas] to swing right
— v — **izquierda** f - [grúas] to swing left
— v **independientemente** - [mec] to rotate independently
— v **jaula** f - [mec] to rotate @ cage
— v — f **para cojinete** m - [mec] to rotate @
bearing cage
— v — f — — m **para piñón** m - [mec] to rotate
@ pinion bearing cage
— v **libremente** - [mec] to rotate freely
— v **manualmente** - [mec] to rotate manually
— v — **bomba** f - [bombas] to rotate @ pump manually
— v **motor** m - [comb.int] to, rotate, o turn, o
crank, @ engine • [electr-mot] to, rotate, o
turn, @ motor
— v **núcleo** m - [mec] to rotate @ core
— v **nuevamente** - [mec] to rotate again
— v **para arranque** m - [comb.int] to crank to
start
— v **piñón** m - to rotate @ pinion
— v **precisamente** - [mec] to rotate precisely
— v **tambor** m - [mec] to rotate @ drum
— v — m **de acero** m - [mec] to roll @ steel
drum
— v **volante** m - [comb.int] to rotate @ flywheel
rotatoria f - [metal-prod] distributor •
[petról] véase **mesa** f **rotativa**
rotatorio,ria a - rotating; revolving • véase
también rotativo,va
roto,ta a - . . . . • torn • snapped
— a **súbitamente** - broken, o snapped, suddenly
rotor m - [electr-mot] rotor - [comb.int] rotor,
• [comb.int½] rotor; breaker
— m **ajustado** - [comb.int] adjusted breaker
— m **bloqueado** - [electr-mot] blocked rotor
— m **bobinado** - [electr] wound rotor
— m **colado** - [electr-mot] cast, o poured, rotor
— m **completo** - [electr-prod] complete rotor
— m — **para generador** m - [electr-prod] complete generator rotor
— m **con** . . . **lóbulo(s)** m - [comb.int] . . .-
-lobe rotor
— m — **tres lóbulos** m - [comb.int] three-lobe
rotor

rotor m de aluminio m - [electr-mot] aluminum rotor
— m — — m fundido - [electr-mot] cast aluminum rotor
— m devanado m - [electr-mot] wound rotor
— m equilibrado m - [electr-mot] balanced rotor
— m fundido - [electr-mot] cast rotor
— m magnético - [electr-mot] magnetic rotor
— m para generador m - [electr-prod] generator rotor
— m — — m con imán(es) m permanente(s) - [electr-prod] permanent magnet generator rotor
— m — motor m - [electr-mot] motor rotor
— m — skip m - [metal-prod] skip rotor
— m quitado - [electr-mot] removed rotor
— m removido - [Electr-mot] removed rotor
— m sacado - [electr-mot] removed rotor
— m trabado - [electr-mot] locked rotor
rototrol - [electr-equip] rototrol • véase también regulador m giratorio
— m antifluctuante - [metal-prod] antihunt rototrol
— m — para equilibrio m - [metal-prod] balance antihunt rototrol
— m — — m para carga f - [metal-prod] load balance antihunt rototrol
— m para equilibrio m - [metal-prod] balance rototrol
— m — — m para carga f - [metal-prod] load balance rototrol
— m — generador m - [electr-prod] generator rototrol
— m — laminador m - [metal-lam] mill rototrol
rótula f - . . . • [mec] swivel • ball • [metal-trat] thrust ball
— f para codo m - [metal-prod] ball pin
— f — prolongación f - [metal-lam] extension segment
— f — — f para tren m para chapa(s) f - [metal-lam] sheet mill extension segment
— f — tubería f - [petról] casing swivel
— f — — f para entubación f - [petról] casing swivel
rotulación f - tagging; labeling; marking
— f de categoría f - [ind] quality labeling
— f — — f de calidad f - [ind] quality grade labeling
— f — — f — f de neumático m - [autom-neumát] tire quality grade labeling
— f — sobre m - envelope, addressing, o labeling, o marking
— f según Asociación f Estadounidense de Fabricantes de Artículos m Eléctricos - [electr-equip] National Electrical Manufacturer's Association, o N E M A, tagging
— f uniforme - [ind] uniform labeling
rotulado,da a - labeled; tagged; marked
rotular v - [ind] . . .; to tag; to mark
— v sobre m - to, label, o address, o mark, @ envelope
rótulo m - . . .; mark(r); nameplate; (price) tag; sticker mark(ing)
— m con información f - information placard
— m — — f sobre neumático(s) m - [autom-neumát] tire information placard
— m con precio m - [com] price tag
— m informativo - (information) placard
— m — en vehículo m - [autom] vehicle placard
— m — sobre neumático m - [autom-neumát] tire information placard
— m — — — para vehículo m - [autom-neumát] vehicle tire (information) placard
— m — conmutador m - [electr-instal] switch nameplate
— m — equipo m automático - [ind] automatic equipment marker
— m — fuente f para energía f - [sold] power source market
— m — identificación f - identification label; identity tag
— m — — f para carcasa f - [ind] housing identification tag

rótulo m para identificación f para carcasa f para eje m - [ind] axle housing identification tag
— f — — f — conductor(es) m - [electr-cond] cable tag
— m — interruptor m - [electr-equip] switch, identification tag, o nameplate
— m para interruptor m para línea f - [electr-instal] line switch nameplate
— m — motor n - [electr-mot] motor, tag, o nameplate - [comb.int] engine, tag, o nameplate
— m — nivel m - [ind] level, label, o sticker
— m — m de aceite m - [comb.int] oil level label
— m — orificio m - [mec] hole label
— m — — m para aceite m - [comb.int] oil fill label
— m — prevención f - caution, tag, o label
rotura f - . . .; break; fracture • break up • failure • tear • snap(ping)
— f de alambre m - [mec] wire break(ing)
— f — — m exterior - [cabl] outside wire break(ing)
— f — alma m - [cabl] core collapse,sing
— f — amortiguador m - [autom-mec] shock (absorber) break(ing)
— f — apoyadero m - [mec] strut break(ing)
— f — apoyo m - [mec] strut break(ing)
— f — banda f - [metal-trat] strip break(ing)
— f — — f para acoplamiento m - [autom-mec] tie rod break(ing)
— f — cable m - [cabl] cable, o (wire) rope, break(ing)
— f — — m de alambre m - [cabl] wire rope break(ing)
— f — cadena f - [cadenas] chain break(ing)
— f — caja f - [autom] case break(ing)
— f — — f para transferencia f - [autom] transfer case break(ing)
— f — cañería f - [tub] pipe break(age)
— f — — f de arcilla f - [tub] clay pipe break(age)
— f — cigüeñal f - [comb.int] crankshaft break(ing)
— f — conductor m - [electr-instal] wire, o lead, break(ing)
— f — — m a electrodo m - [sold] electrode lead break(ing)
— f — — m para entrada f - [electr-instal] power lead break(ing)
— f — — m para regulador m - [electr-instal] control lead break(ing)
— f — — m — m para voltaje m - [sold] voltage control lead break(ing)
— f — — m para tierra f - [sold] ground lead break(ing)
— f — conexión f - [electr-instal] connection break(ing)
— f — contacto m - [electr-oper] contact break(ing)
— f — chapa f - [metal-trat] strip break(age)
— f — depósito m - tank, rupture,ring, o break(ing)
— f — disco m - [mec] disk break(ing) • [mús] record break(ing)
— f — eje m - [mec] axle break(ing)
— f — — m de papel m - [mec] paper shaft break(ing)
— f — — m — — m endurecido - [mec] hard paper shaft break(ing)
— f — equipo m - [ind] equipment, break(ing), o breakdown
— f — — m para amoladura f - [metal-lam] grinding equipment brakdown
— f — — m — — f de rodillo(s) m - [metal-lam] roll grinding equipment breakdown
— f — — m para granallado m - [metal-lam] shot blasting equipment breakdown
— f — — m — — f de rodillo(s) m - [metal-lam] roll shot blasting equipment breakdown
— f — — m para rectificación f - [metal-lam] grinding equipment breakdown

rotura f de equipo m para rectificación f de ro-
   dillo(s) m - [metal-lam] roll grinding equipment breakdown
— f —— m — torneado m - [metal-lam] turning
   equipment breakdown
— f —— m —— m de rodillo(s) m - [metal-
   -lam] roll turning equipment breakdown
— f — eslabón m - [cadenas] link break(ing)
— f — esquina f - [mec] corner breaking
— f — fleje m - [metal-lam] strip break(age)
— f — guarnición f - [comb.int] gasket blowing
— f —— f para culata f - [comb.int] head
   gasket blowing
— f — hoja f - [mec] leaf breaking
— f —— f para resorte m - [mec] spring leaf
   break(ing)
— f — husillo m - [mec] spindle breaking
— f — letrero m - [com] sign breaking
— f — liga f - [hormig] bond break(ing)
— f — montacarga(s) m - [metal-prod] skip
   break(down)
— f — pestaña f - [mec-ruedas] flange breaking
— f — portaescobilla(s) m - [electr-mot]
   (brush)holder break(ing)
— f — puntal m - [mec] strut breaking
— f — resorte m - [mec] spring break(ing)
— f —— m con hoja(s) f - [mec] leaf spring
   break(ing)
— f —— m para escobilla f - [electr-mot]
   brush spring break(ing)
— f — rodillo(s) m - [metal-lam] roll breakage
— f — sello m - [mec] seal break(ing)
— f —— m para vacío m - [neumát] vacuum seal
   break(ing)
— f — semieje m - [autom-mec] halfshaft break-
   -(ing)
— f — soldadura f - [sold] weld break(ing)
— f —— f en hierro m - [sold] iron weld
   break(ing)
— f —— f —— m fundido - [sold] cast iron
   weld break(ing)
— f — superficie f - (sur)face break(ing)
— f —— f de letrero m - [com] sign face
   breaking
— f — tapón m - [mec] plug break(ing)
— f — tubería f - [tub] pipe break(age)
— f —— f de arcilla f - [tub] clay pipe
   breakage
— f — tubo m - [tub] tube, o pipe, break(ing)
— f —— m de arcilla f - [tub] clay pipe
   break(age)
— f — unidad f - unit break(ing)
— f — ventilador m - [mec] fan break(ing)
— f en cigüeñal m - [comb.int] cranksahft break
— f — conductor m - [electr-instal] lead, o
   wire, break
— f — esquina f - [mec] corner break
— f — pestaña f - [mec] flange break
— f — soldadura f - [sold] weld break
— f evitada - avoided, o prevented, break(age)
— f frágil - [metal] fragile, o brittle, break,
   o fracture
— f — con entalla f - [metal] fragile notch
   break(ing)
— f junto a casquillo m - [Cabl] failure at @
   socket
— f máxima - maximum break(age)
— f mayor - [sold] major break
— f mínima - minimum, o small, break(age)
— f por deformación f - deflection failure,ling
— f — desviación f - [mec] deflection failure
— f — flexión f - [mec] bending failure
— f — impacto(s) m - impact breakage
— f — tracción f - traction break(ing); tensile fracture
— f reducida - small, o little, break(age)
— f reparada - repaired break
— f súbita - sudden break(ing)
— f transversal - cross, o transverse, break
— f unitaria - unit break(ing)
— f violenta - [mec] blowout
— f — por presión f - [mec] blowout
roturador m - turner • breaker

roturador m para sínter - [metal-prod] sinter
   breaker
round m - [deport-box] round; véase vuelta f
roya f - [agric] . . .; blight
— f urbana - [pol] urban blight
rozadera* f - [mec] friction plate
rozado,da a - . . .; rubbed • [transp] tagged
rozadura f - . . .; rubbing; scraping; chaffing
   • [transp] tagging
rozamiento m - . . .; rub(bing) • galling; a-
   brasion • [transp] tagging
— m con pared f - [deport] wall tagging
— m de cable m - [cabl] cable, o (wire) rope,
   rubbing, o friction
— m —— m de alambre m - [cabl] wire rope,
   friction, o rubbing
rozar v - . . . • [transp] to tag
— v ladera f - [autom] to tag @ bank
— v ligeramente - to brush (against)
— v pared f - [deport] to tag @ wall
ruberoide* m - [constr] Rubberoid
rubio claro a - (light) blond
— rojizo,za a - sandy
rúbrica f - [legal] . . .; initial(s) • [public]
   heading
rubricación f - [legal] signing • initialling •
   [contab] validating,tion; registration
rubricado,da a - [legal] signed; initialled •
   [contab] (government), approved, o validated;
   registered
rubricar v - [legal] to sign (and seal) •
   [contab] to, validate, o register
rubro m - . . . • item; heading • label •
   [public] subject
— m aceptable - acceptable, o suitable, label
— m apropiado - suitable, heading, o label
— m semántico - semantic label
— m — apropiado - appropriate, o suitable,
   semantic label
rudo,da a - . . .; rugged; tough; husky; harsh •
   demanding
rueda f abierta - [mec] open wheel
— f accionada - [mec] driven whee;
— f — mecánicamente - [mec] motor driven wheel
— f acoplada - [mec] coupled wheel
— f aflojada - [mec] loosened wheel
— f agrícola - [agric] agricultural wheel
— f ahusada - [mec] tapered wheel
— f alineada - [autom-mec] aligned wheel
— f Allen m - [admin] Allen wheel
— f — para administración f - [admin] Allen
   management wheel
— f amoladora - [mec] grinding wheel
— f — motorizada - [mec] (electric) power
   driven grinding wheel
— f ancha - [mec] wide wheel
— f angosta - [mec] narrow wheel
— f anterior - [mec-ruedas] front wheel • previous wheel
— f apareada - [mec] matched, o mated, wheel
— f atascada - [mec] blocked wheel
— f cabilla - [mec] véase rueda f dentada
— f caliente - [mec] hot wheel
— f cambiada - [mec] changed wheel
— f cargada - [mec] loaded wheel
— f colada - [mec] cast wheel
— f — de aluminio m - [mec] cast aluminum wheel
— f colectora f - [electr-oper] collecting, o
   trolley, wheel
— f — para trole m - [electr-transp] trolley
   wheel
— f compacta - [mec] compact wheel
— f — con presión f - [mec] compact pressure
   wheel
— f —— f alta - [autom-neumát] high pressure compact wheel
— f —— f — para repuesto - [autom-neumát]
   high pressure compact spare wheel
— f — para repuesto m - [autom-neumát] compact
   spare wheel
— f competidora - [mec-ruedas] competetive wheel
— f común - [mec] common, o ordinary, wheel
— f con alma m embutida - [f.c.ruedas] dished

wheel • dowel wheel
rueda f **con cabilla(s)** f - [mec] _véase_ **rueda** f **dentada**
— f — f **empernada(s)** - [mec] bolted dowel wheel
— f — **calidad** f **de acuerdo con pedido** m **de cliente** m - [mec-ruedas] custom quality wheel
— f — **cangilon(es)** m - [mec] ferris wheel • [constr] wheel cutter
— f — **cojinete** m - [mec] bearing wheel
— f — m **con bola(s)** f - [mec] ball bearing wheel
— f — m — **rodillo(s)** m - [mec] roller bearing wheel
— f — **diámetro** m **de . . .** - [mec] . . . diameter wheel
— f — **disco** m - [mec] disk wheel
— f — **dos vidas** f - [f.c.-ruedas] two wear wheel
— f — **engranaje** m **interior** - [mec] internal gear wheel
— f — m **recto** - [mec] spur gear wheel
— f — **espiral** - [mec] spiral wheel
— f — **freno** m - [f.c.-ruedas] braking wheel
— f — **garganta** f **radial** - [grúas] radial tread wheel
— f — **llanta** f **abierta** - [mec] open tire wheel
— f — f **con tratamiento** m **térmico** - [f.c.-ruedas] rim treated wheel
— f — f **de acero** m - [mec] steel (tired, o rimmed] wheel
— f — f — **caucho** m - [mec] rubber tire(d) wheel
— f — f — m **macizo** - [mec] solid rubber tire wheel
— f — f — 1½ **pulgadas** f (o menos) - [f.c.-ruedas] one-wear wheel
— f — f — f (**o menos) para vagón(es)** m - [f.c.-ruedas] one wear car wheel
— m — f — f (— —) — — m **para carga** f - [f.c.-ruedas] one wear freight car wheel
— f — f — **dos pulgadas** f - [f.c.-ruedas] two-wear wheel
— f — f — 2½ **pulgadas** f (o más) • [f.c.-ruedas] multiple wear wheel
— f — f **delgada** - [f.c.-ruedas] one-wear wheel
— f — f **para vagón(es)** m - [f.c.-ruedas] one wear car wheel
— f — f — — m **para carga** f - [f.c.-ruedas] one wear freight car wheel
— f — f **gruesa** - [f.c.-ruedas] multiple wear wheel
— f — f **mediana** - [f.c.-ruedas] two-wear wheel
— f — f **templada** - [f.c.-ruedas] rim quenched, o quenched rim, wheel
— f — **pestaña** f - [mec] flanged wheel
— f — f **de acero** m - [mec] steel flanged, o flanged steel, wheel
— f — f **doble** - [mec] double flanged wheel
— f — f **única** - [mec] single flanged wheel
— f — **presión** f - [autom-neumát] pressure(d) wheel
— f — f **alta** - [autom-neumát] high pressure wheel
— f — **sin fin** - [mec] worm wheel
— f — **superficie** f **para rodadura** f **radial** + [mec] radial tread wheel
— f — **tratamiento** m **térmico** - [f.c.-ruedas] heat-treated wheel
— f — m **integral** - [f.c.-ruedas] entirely treated wheel
— f **con trinquete** m - [mec] ratchet wheel
— f — m **con accionamiento** m **doble** - [mec] double drive ratchet wheel
— f — **una (sóla) vida** f - [f.c.-ruedas] one--wear wheel
— f — (—) — f **para vagón(es)** m - [f,c,-ruedas] one wear car wheel
— f — (—) — f — m **para carga** f - [f.c.-ruedas- one wear freight car wheel

rueda f **con vida(s)** f **múltiple(s)**—[f.c.-ruedas] multiple wear wheel
— f **confiable** - [mec] dependable wheel
— f **cónica en escudo** m - [tornos] plate bevel wheel
— f **cóncava** - [f.c.-ruedas] dished wheel
— f **conductora** - [f.c.] leading wheel
— f **contra pedido** m - [mec] custom wheel
— f **convergente,gida** - [autom] toed-in wheel
— f **correcta** - [mec] correct, o right, wheel
— f **corriente** - [mec] ordinary wheel
— f **de acero** m - [mec] (bare) steel wheel
— f — m **con dos vidas** f - [f.c.-ruedad] two-wear steel wheel
— f — m — **llanta de** 1½ **pulgadas** (o menos) - [f.c.-ruedas] one-wear steel wheel
— f — m — f **de dos pulgadas** - [f.c.--ruedas] two-wear steel wheel
— f — m — f **de** 2½ **pulgadas** (o más) - [f.c.-ruedas] multiple-wear steel wheel
— f — m — **llanta** f **delgada** - [f.c.-ruedas] one-wear steel wheel
— f — m — f **gruesa** - [f.c.-ruedas] multiple wear steel wheel
— f — m — f **mediana** - [f.c.-ruedas] two-wear steel wheel
— f — m — **una (sóla) vida** f - [f.c.--ruedas] one-wear steel wheel
— f — m — **vida** f **múltiple** - [f.c.-ruedas] multiple wear steel wheel
— f — m **forjado** - [metal-fabr] forged steel wheel • [f.c.-ruedas] wrought steel wheel
— f — m **con tratamiento** m **térmico** - [f.c.-ruedas] heat treated wrought steel wheel
— f — m — **dos vidas** f - [f.c.-ruedas] two-wear wrought steel wheel
— f — m — **llanta de** 1½ **pulgadas** (o menos) - [f.c.-ruedas] one wear wrought steel wheel
— f — m — f **de dos pulgadas** f - [f.c.-ruedas] two-wear wrought steel wheel
— f — m — f — 2½ **pulgadas** (o más) - [f.c.-ruedas] multiple wear wrought steel wheel
— f — m — f **delgada** - [f.c.-ruedas] one-wear wrought steel wheel
— f — m — f **gruesa** - [f.c.-ruedas] multiple wear wrought steel wheel
— f — m — f **mediana** - [f.c.-ruedas] two-wear wrought steel wheel
— f — m — **una (sóla) vida** f - [f.c.--ruedas] one wear wrought steel wheel
— f — m — **vida** f **múltiple** - [f.c.--ruedas] multiple wear wrought steel wheel
— f — m **no rectificable** - [f.c.-ruedas] one-wear wrought steel wheel
— f — m **para grúa** f - [grúas] forged steel crane wheel
— f — m — **varias rectificaciones** f - [f.c.-ruedas] multiple wear wrought steel wheel
— f — m — **rectificable una sóla vez** - [f.c.-ruedas] two-wear wrought steel wheel
— f — m **no rectificable** - [f.c.-ruedas] one-wear steel wheel
— f — m **para ferrocarril** m - [f.c.-ruedas] steel railway wheel
— f — m — **varias rectificaciones** f - [f.c.-ruedas] multiple wear steel wheel
— f — f **rectificable una (sóla) vez** - [f.c.-ruedas] two-wear steel wheel
— f — m **sin llanta** f - [mec] bare, o rimless, steel wheel
— f — m **soldada** - [mec] welded steel wheel
— f — m — **por arco** m - [mec] arc welded steel wheel
— f **acuerdo con pedido** m **de cliente** m - [mec] custom wheel
— f **alambre** m - [mec] wire wheel
— f **aluminio** m - [mec] aluminum wheel
— f — m **colado** - [mec] cast aluminum wheel
— f — m **fundido** - [mec] cast aluminum wheel

rueda f de aluminio m fundido de equipo m original - [autom-mec] original equipment cast aluminum wheel
— f — carburo m - [mec] carbide wheel
— f — m con silicio m - [herram] silicon carbide wheel
— f — m — m para amoladura f - [herram] silicon carbide grinding wheel
— f — equipo m - [mec] equipment wheel
— f — — m original - [autom] original equipment wheel
— f — m — para automóvil m - [autom-mec] original equipment car wheel
     f   m   camión m [autom mec] original equipment truck wheel
— f — m — — m liviano - [autom-mec] original equipment light truck wheel
— f — esmeril - [herram] emery wheel
— f — fundición f - [autom-mec] cast (metal) wheel
— f — madera f - [mec] wood(en) wheel
— f — metal - [mec] metal wheel
— f — m fundido - [autom-mtec] cast metal wheel
— f — óxido m de aluminio m - [herram] aluminum oxide wheel
— f — m — m para amoladura f - [herram] aluminum oxide grinding wheel
— f — presos - [legal] (prisoner) lineup
— f — tipo m con borne m único - [autom] uni-, o single, lug wheel
— f — m — perno m único - [autom] uni-, o single, lug wheel
— f delantera - [mec] front end • [autom-mec] front wheel
— f — doble - [autom] dual front wheel
— f — normal - standard front wheel
— f — para repuesto m - [autom] spare front wheel
— f dentada - [mec] . . .; sprocket; gear, o cog, o mortised, wheel (gear); toothed wheel; spur gear • ratchet
— f — accionada - [mec] driven sprocket
— f — accionadora - [mec] driver,ving sprocket
— f — ajustable - [mec] adjustable sprocket
— f — alineada - [mec] aligned sprocket
— f — angosta - [mec] narrow sprocket
— f — con impulsión f - [mec] drive,ving sprocket
— f — — f con cadena f - [mec] chain drive sprocket
— f — — rodillo m - [mec] roller sprocket
— f — cónica - [mec] miter wheel
— f — dañada - [mec] damaged sprocket
— f — delgada - [mec] narrow sprocket
— f — doble - [mec] double sprocket
— f — helicoidal - [mec] helical, sprocket, o gear wheel
— f — impulsada - [mec] driven sprocket
— f — — inferior - [petról] low, o bottom, driven sprocket
— f — — superior - [petról] top, o high, o upper, driven sprocket
— f — impulsora - [mec] driver,ving sprocket
— f — inferior - [mec] low(er) sprocket
— f — — para accionamiento m - [mec] low(er) drive,ving sprocket
— f — — — impulsión f - [mec] lower drive,-ving sprocket
— f — inspeccionada - [mec] inspected sprocket
— f — intermedia - [mec] intermediate sprocket
— f — intermediaria f - [mec] idler sprocket
— f — loca - [mec] idler sprocket
— f — — reemplazada - [mec] replaced idler sprocket
— f — motriz f - [mec] drive,ving, o power, sprocket
— f — — angosta - [mec] narrow, drive,ving, o power, sprocket
— f — — delgada - [mec] narrow, drive,ving, o power, sprocket
— f — — para oruga f - [constr-equip] tractor sprocket; drive,ving sprocket

rueda f dentada motriz para oruga f para pala f mecánica f - [constr] (power shovel) tumbler
— f — para accionamiento m - [mec] drive,ving sprocket
— f — — — m optativo m - [mec] optional drive,ving sprocket
— f — — — m para cabezal m - [petról] catshaft drive,ving sprocket
— f — — m — — m para torno m - [petról] catshaft drive,ving sprocket
— f — — — m emergencia f - [mec] emergency drive,ving sprocket
— f — — — m — f optativa - [constr] optional emergency drive,ving sprocket
— f — — — m para mesa f - [mec] (rotary) table drive,ving sprocket
— f — — — m — f rotativa - [mec] rotary (table) drive,ving sprocket
— f — — árbol m - [mec] shaft sprocket
— f — — — m con leva(s) f - [mec] camshaft sprocket
— f — — cadena f - [mec] chain sprocket
— f — — — f verificada - [mec] checked chain sprocket
— f — — cigüeñal m - [comb.int] crankshaft sprocket
— f — — encabezadora - [Trefil] header sprocket
— f — — impulsión f - [mec drive,ving sprocket
— f — — — m para cabezal m (para torno m) - [petról] catshaft drive sprocket
— f — — — f — oruga f - [constr-equip] crawler drive,ving sprocket
— f — — mando m - [mec] drive,ving sprocket
— f — — oruga f - [mec] tractor sprocket
— f — — transmisión f - [mec] transmission sprocket
— f — — trefilador(a) - [trefil] wire drawing sprocket
— f — — válvula f - [mec] valve sprocket
— f — — velocidad f alta - [mec] high speed sprocket
— f — — — f baja - [mec] low speed sprocket
— f — propulsora - [mec] drive,ving sprocket
— f — recta - [mec] spur, sprocket, o gear
— f — reemplazada - [mec] replaced sprocket
— f — reforzada - [mec] heavy duty, o reinforced, sprocket
— f — reversible - [mec] reversible sprocket
— f — superior - [mec] top, o upper, o high(er), sprocket
— f — — para accionamiento m - [mec] top, o upper, o high(er), drive,ving sprocket
— f — — — impulsión f - [mec] top, o upper, o high(er), drive,ving sprocket
— f derecha f - [mec] right wheel
— f descargada - [mec] unloaded wheel
— f desgastada - [mec] worn wheel
— f desplazada - [autom-mec] offset wheel
— f despojadora - [agric-equip] picker wheel
— f desviada - [autom-mec] offset wheel
— f directriz - [mec] guide,ding wheel • [f c -ruedas] leading wheel
— f — mecanizada - [agric-equip] power(ed) guide,ding wheel
— f — motorizada - [agric-equip] powered guide,ding wheel
— f — motriz - [agric-equip] power(ed) guide,-ding wheel
— f divergente - [autom] toed out wheel
— f doble - [mec] double, o dual, wheel
— f doble doble - [mec] dual-dual, o double-double, wheel
— f embridada - [mec] flanged wheel
— f — para cable m - [mec] flanged cable wheel • [petról] flanged calf wheel
— f — — tubería f - [petról] flanged calf wheel
— f empernada - [mec] bolted wheel
— f en bruto m - [mec-ruedas] (wheel) blank
— f — — de acero m - [mec-ruedas] forged steel blank

rueda f en bruto de acero m forjado - [mec-
-ruedas] forged steel (wheel) blank
— f — — estándar - [mec-ruedas] standard
(wheel) blank
— f — — forjada - [f.c.-ruedas] forged
(wheel) blank
— f — descubierto - [autom] open wheel
— f — existencia - [ind] stock wheel
— f — una (sóla) pieza f - [f.c.-ruedas] one
piece wheel
— f enganchada - [constr] snagged wheel
— f ensayada - [mec-ruedas] tested wheel
— f esmeril(adora) - [mec] grinding wheel
— f — conformada - [mec] shaped grinding wheel
— f especial - [mec] special wheel • custom
wheel
— f esquinera - [mec] corner wheel
— f estándar - [mec] standard wheel
— f — de acero m - [mec] standard steel wheel
— f — en equipo m - [autom] standard equipment
wheel
— f — — — m original - [autom] standard original equipment wheel
— f excéntrica - [mec] eccentric wheel
— f extraída - [autom] pulled wheel
— f floja - [mec] loose wheel
— f forjada - [mec] forged wheel; forging
— f — en bruto - [mec-ruedas] forged blank •
forging
— f — — estándar - [mec-ruedas] standard
forged blank
— f — — — para riel(es) m - [mec-ruedas]
(forged) track wheel blank
— f — en una (sóla) pieza f - [f.c.-ruedas]
one piece forged wheel
— f — estándar - [f.c.-wheels] standard,
forged, o wrought, wheel
— f — normal - [f.c.-ruedas] standard, forged,
o wrought, wheel
— f — para grúa f - [grúas] forged crane wheel
— f — — mecanización f - [mec-ruedas] available forging
— f — — — f inmediata - [mec-ruedas] readily
available forging
— f fría - cold wheel • [turb] cold wheel
— f frontal - [autom] front wheel
— f giratoria - [mec] caster (wheel) • pinwheel
— f — frontal - [mec] front caster
— f guía - [mec] idler • guide wheel
— f — mecanizada - [agric-equip] power guide
wheel
— f — motriz - [agric] powered guide wheel
— f — — optativa - [agric-equip] optional
guide wheel
— f — según especificación(es) f - [autom]
custom (made) wheel
— f hidráulica - [hidr] hydraulic wheel
— f — Pelton - [hidr] Pelton hydraulic wheel
— f hiperbólica - [mec] hyperbolic(al) wheel •
skew(ed), o bevel(ed), wheel
— f impulsora - [mec] drive,ving wheel
— f — derecha - [autom] right drive wheel
— f — izquierda - [mec] left drive wheel
— f inclinada—[mec] slanting, o leaning, wheel
— f incorrecta - [autom-neumát] wrong wheel
— f individual - [mec-ruedas] individual wheel
— f inferior - [mec] bottom, o lower, wheel
— f — para skip m - [metal-prod] lower skip, o
skip lower, wheel
— f inspeccionada—[mec-ruedas] inspected wheel
— f interior - [mec] inner, o inside, o center,
wheel
— f intermedia - [mec] middle wheel
— f intermediaria - [mec] idler
— f invertida - [autom] reversed wheel
— f izquierda - [mec] left wheel
— f laminada - [f.c.-ruedas] rolled wheel
— f — de una (sóla) pieza f - [f.c.-ruedas]
one piece rolled wheel
— f libre - [mec] idler
— f limitadora - [mec] limiting wheel
— f — para profundidad f - [agric-equip] gage,
o depth, wheel

rueda f loca - [mec] idler
— f — ajustada - [mec] adjusted idler
— f — dentada - [mec] idler sprocket
— f — — con rodillo(s) m - [mec] idler roller
sprocket
— f — endurecida - [sold] hardsurfaced idler
— f — para eje m central - [mec] center line
sprocket
— f — — — m — excéntrica = [mec] eccentric
center line sprocket
— f — para oruga f - [constr-equip] caterpillar idler
— f — reemplazada - [mec] replaced idler
— f lubricada - [mec] lubricated wheel
— f lubricadora - [mec] lubricating wheel
— f, llanta f y eje m - [mec] wheel, rim and
axle
— f maciza - [mec] solid wheel
— f — para presión f - [agric] solid pressing
wheel
— f manual - [mec] handwheel
— f — para avance m - [herram] carriage handwheel
— f — de contrapunta f - [tornos] tailstock
handwheel
— f — — soporte m - [tornos] rest handwheel
— f — — — m combinado - [tornos] compound,
o combined, rest handwheel
— f — — — m compuesto - [mec] compound rest
handwheel
— f — para desplazamiento m - [mec] slide
handwheel
— f — — — m transversal - [mec] cross slide
handwheel
— f — — deslizamiento m - [mec] slide handwheel
— f — — — m transversal - [mec] cross slide
handwheel
— f mezcladora f - [miner] mulling, o mixing,
wheel
— f moledora - [miner] mulling, o grinding,
wheel
— f motriz - [mec] drive,ving, o power(ed),
wheel • band wheel • [autom] traction wheel •
[grúas] driver
— f — ahusada - [mec] tapered drive,ving wheel
— f — con garganta f ahusada - [grúas] tapered
tread driver,ving (wheel)
— f — — superficie f para rodadura f ahusada
- [grúas] tapered tread driver,ving (wheel)
— f — delantera - [autom] front traction wheel
— f — derecha - [autom] right traction wheel
— f — instalada - [autom] installed traction
wheel
— f — izquierda - [autom] left traction wheel
— f — optativa - [mec] optional, traction, o
powered, wheel
— f — para avance m - [sold] travel drive,ving
wheel
— f — — dispositivo m para avance - [sold]
travel unit drive,ving wheel
— f — — guía f - [autom] powered guide,ding
wheel
— f — — — f optativa - [autom] optional
powered guide,ding wheel
— f — — oruga f - [constr=equip] (caterpillar (tractor) sprocket; drive,ving sprocket
— f — — — f para pala f mecánica - [constr-
-equip] (power shovel) sprocket, o tumbler
— f — — tracción f - [autom-mec] traction
drive,ving wheel
— f — — — f delantera - [autom-mec] front
traction drive,ving wheel
— f — — — f trasera - [autom] rear traction
drive,ving wheel
— f — — unidad f - unit drive,ving wheel
— f —, quitada, o removida, o sacada - [autom]
removed drive,ving wheel
— f — trasera - [autom] rear traction wheel
— f muy desplazada - [autom] deeply, o widely,
offset, wheel
— f — — hacia afuera - [autom] widely offset
wheel

rueda f neumática - [autom-neum•at] pneumatic
wheel; rubber tired wheel
— f — doble - [autom-mec] dual, o double,
pneumatic, o rubber tired, wheel
— f — gemela - [autom] dual, o twin, pneumatic
wheel
— f no rectificable - [f.c.-ruedas] one-wear
wheel
— f — para vagón(es) m - [f.c.-ruedas] one
wear car wheel
— f — — m para carga f - [f.c.-ruedas]
one wear freight car wheel
— f nueva - [mec] new wheel
— f optativa - [mec] optional wheel
— f original - [mec] original wheel
— f para accionamiento m - [mec] driving wheel
— f — — m para traslación f - [mec] travel
drive,ving wheel
— f — ajuste m - [mec] adjustment wheel
— f — alimentación f - [mec] feeding wheel
— f — aporte m de comida f - [avicult] feed
intake wheel
— f — automotor(es) m - [autom] automotive
wheel
— f — autocamión m - [autom] truck wheel
— f — avance m - [sold] travel (drive), wheel,
o roll
— f — bogie m - [f.c.-ruedas] bogie, o truck,
wheel
— f — — m distinto - [f.c.-ruedas] different,
truck, o bogie, wheel
— f — — m para arrastre m - [f.c.-ruedas]
trailing, bogie, o truck, wheel
— f — — m para locomotora f - [f.c.-ruedas]
locomotive, o engine, truck wheel
— f — — m piloto - [f.c.-ruedas] pilot, bogie, o truck. wheel
— m — — m posterior - [f.c.-ruedas] trailer,
bogie, o truck, wheel
— f — — m para locomotora f - [f.c.-
-ruedas] locomotive, o engine, trailer,
truck, o bogie, whee;
— f — — m trasero - [f.c.-ruedas] trailer,
bogie, o truck, wheel
— f — camión m - [autom] truck wheel
— f — — m liviano - [autom] light truck wheel
— f — cangilón(es) m - [constr] shovel wheel
— f — carretera f - [autom] highway wheel
— f — carretón m - [f.c.-ruedas] bogie, o
truck, wheel
— f — — m para locomotora f - [f.c.-ruedas]
locomotive, o engine, bogie, o truck, wheel
— f — — m delantero - [f.c.-ruedas] (locomotive) front, bogie, o truck, wheel
— f — — m posterior - [f.c.-ruedas] (locomotive) trailer, bogie, o truck, wheel
— m — — m para locomotora f - [f.c.-
-ruedas] locomotive trailer, bogie, o truck,
wheel
— f — carrillo m - [ind] trolley wheel
— f — carro m - [mec] wagon, o cart, wheel
— f — — m para montacargas f - [metal-prod]
skip car wheel
— f — — m — — m para horno m (alto) -
[metal-prod] (blast) furnace skip car wheel
— f — centro m abierto - [mec] open center
wheel
— f — coche m - [f.c.-ruedas] (passenger) car
wheel
— f — — m para pasajero(s) m - [f.c.-ruedas]
passenger car wheel
— f — competencia f - [mec-ruedas] competetive
wheel
— f — contacto - [grúas] trolley wheel
— f — cubeta f - [metal-prod] (skip) car wheel
— f — — f para skip m - [metal-prod] skip car
wheel
— f — dispositivo - [mec] device wheel
— f — — m para avance m - [sold] travel drive
wheel
— f — embrague m - [mec] clutch wheel
— f — ferrocarril m - [f.c.-ruedas] railway
wheel

rueda f para ferrocarril m eléctrico - [f.c.-
-ruedas] electric railway wheel—
— f — — m elevado - [f.c.] elevated railway
wheel
— f — — m subterráneo - [f.c.] subway wheel
— f — frenado m - [petról] braking wheel
— f — — m para malacate m - [petról] calf
braking wheel
— f — — m — — m para tubería f - [petról]
calf wheel brake side
— f — freno m - [mec] brake king wheel
— f — garrucha f - [mec] sheave wheel
— f — giro m - [metal-prod] bull wheel
— f — grano m - [agric] grain wheel
— f — grúa f - [grúas] crane wheel
— f — guía f - [mec] guide, wheel, o lead;
idler • véase también rueda f guía • [sold]
- guide roll
— f — — f ahusada - [mec] tapered idler
— f — — f anterior - [sold] front guide wheel
— f — — f con garganta f ahusada - [grúas]
tapered tread idler
— f — — f — superficie f para rodadura f
ahusada - [grúas] tapered tread idler
— f — — f para rodadura f ahusada -
[grúas] tapered tread idler
— f — guía f motriz - [autom] powered guide
wheel
— f — — f optativa - [autom] optional powered
guide wheel
— f — — posterior - [sold] rear guide wheel
— f — lingoteadora f - [metal-prod] pig casting machine wheel
— f — locomotora f - [f.c.-ruedas] locomotive,
o engine, wheel
— f — — f ferroviaria - [f.c.-ruedas] railway
locomotive wheel
— f para mando m - [mec] control wheel •
drive,ving wheel
— f — manejo m - [válv] operating wheel
— f — máquina f para colar - [metal-prod] pig
casting machine wheel
— f — — f para sinterización f - [metal-prod]
sintering machine wheel
— f — mezclado m - [mec] mulling wheel
— f — mezcladora f - [miner] muller wheel
— f — mismo bogie m - [f.c.-ruedas] same bogie
wheel
— f — moledora f - [miner] mullor wheel
— f — moledura f - [mec] mulling wheel
— f — motón m - [mec] sheave wheel
— f — pala f - [constr] shovel wheel
— f — polea f - [mec] sheave
— f — presión f - [mec] press(ure) wheel
— f — — f con centro m abierto - [agric]
open center press wheel
— f — — f maciza - [mec] solid press wheel
— f — puente m - [grúas] bridge wheel
— f — — m para grúa f - [grúas] (crane)
bridge wheel
— f — pulverización f - [mec] mulling wheel
— f — recambio m - [autom] spare wheel
— f — remolque m - [mec] tug, o towing, wheel
— f — repuesto m - [autom] spare wheel
— f — retorno m - [mec] return wheel
— f — — m de comida f - [avicult] feed, return, o intake, wheel
— f — riel(es) m - [mec-ruedas] track wheel
— f — — m forjada en bruto - [mec-ruedas]
forged track wheel blank
— f — — m probada - [mec] rails tested track
wheel
— f — — m con pestaña f - [mec-ruedas] flange
track wheel
— f — — m — — f doble - [mec-ruedas] double
flange track wheel
— f — — m templada - [mec-ruedas] hardened
track wheel
— f — — m — con pestaña f doble - [mec-
-ruedas] hardened (double) flange track wheel
— f — roldana f - [mec] sheave wheel
— f — sin fin m - [mec] worm wheel
— f — tambor m - [mec] drum, wheel, o reel

rueda f para ténder m - [f.c.-ruedas] tender wheel
— f — — m ferroviario - [f.c.-ruedas] railroad tender wheel
— f — .terreno m arenoso - [agric-equip] sand wheel
— f — tierra f - [mec] land wheel
— f — tracción f - [autom] traction,tive wheel
— f — — f delantera - [autom] front traction wheel
— f — — f trasera - [autom] rear traction wheel
— f — tractor m - [autom] tractor wheel
— f — transporte m - [transp] transport(ation) wheel
— f — trinquete m - [mec] ratchet (wheel)
— f — turbina f - [turb] turbine wheel
— f — unidad f - [mec] unit wheel
— f — vagón m - [f.c.] railway car whee;
— f — — m para carga f - [f.c.-ruedas] freight car wheel
— f — — m — pasajeros m - [f.c.-ruedas] passenger car wheel
— f — — m ferroviario - [f.c.-ruedas] railway car wheel
— f — — m — para pasajeros m - [f.c.-ruedas] railway passenger car wheel
— f — vagoneta f - [metal-prod] car wheel
— f — — f minera - [miner] mine car wheel
— f — — f para montacarga(s) m - [metal-prod] skip car wheel
— f — — — — m para horno m (alto) - [metal-prod] (blast) furnace skip car wheel
— f — válvula f - [válv] valve wheel
— f — varias rectificaciones f - [f.c.-ruedas] multiple-wear wheel
— f — vehículo m - [transp] vehicle wheel
— f — vía f - [mec] track wheel
— f — volcador m - [mec] dumpster wheel • [ind] tilter gear wheel
— f para zurco m - [agric-equip] furrow wheel
— f Pelton - [hidr] Pelton wheel
— f planetaria - [mec] planetary wheel • idler
— f posterior - [mec] rear, o back, wheel
— f — para recambio m - [autom] spare rear wheel
— f — — repuesto m - [autom] spare rear wheel
— f primera - [mec] first wheel
— f principal - [mec] principal, o main, wheel
— f probada - [mec] tested wheel
— f propulsora - [mec] drive,ving wheel
— f prototipo - [mec] prototype wheel
— f — especial - [mec] special prototype wheel
— f reconstruida - [mec] reconstructed, o rebuilt, wheel • dressed wheel
— f rectificable una sóla vez - [f.c.-ruedas] two-wear wheel
— f reemplazada - [mec] replaced wheel
— f refrentada - [mec] faced, o dressed, wheel
— f reguladora f - [mec] control(ling) wheel
— f — para profundidad - [agric-equip] gage, o depth, wheel
— f resistente - [mec] rugged wheel
— f reversible - [autom] reversing,sible wheel
— f sacada - [mec] removed wheel
— f secundaria - [mec] secondary wheel • idler; follower
— f según concepto m de más tres pulgadas f - [autom] plus three concept wheel
— f sin fin - [mec] worm wheel
— f soldada - [mec] welded wheel
— f — por arco m - [mec] arc welded wheel
— f sólida - [mec] solid wheel
— f — de acero m - [mec] solid steel wheel
— f T S P - [mec] T S P wheel
— f — — para riel(es) m - [mec-ruedas] T S P track wheel
— f tapadora - [agric-equip] covering wheel

rueda f templada - [mec-ruedas] tempered, o hardened, o quenched, wheel
— f — íntegramente - [f.c.-ruedas] entirely, tempered, o quenched, wheel
— f — para riel(es) m - [mec-ruedas] hardened track wheel
— f — — vía f - [mec-ruedas] hardened track wheel
— f tensora - [mec] tension wheel; idler
— f terminada - [mec-ruedas] finished wheel
— f típica - [mec-ruedas] typical wheel
— f trabada - [mec] locked, o stuck, wheel
— f trasera - [mec] rear wheel
— f — doble - [mec] rear dual, o dual rear, wheel
— f — para repuesto m - [autom] spare rear wheel
— f vieja - [mec] old wheel
— f y eje m - [mec-ruedas] wheel and axle
ruedas f apareadas perfectamente - [grúas] perfectly matched wheels
— f armónicas - [mec] interchangeable wheels • change gears
— f con distancia(s) f igual(es) entre sí - [mec] evenly spaced wheels
— f gemelas - [autom] dual, o twin, wheels
ruedecita f - [mec] small wheel • caster; roller
— f con bolas f - [mec] ball caster
ruedo m - . . . • [mec] rolling
— m libre - [mec] free rolling
rugido m - . . .; roaring; rumble; rumbling
rugido,da a - roared • rumbled
rugidor,ra a - roaring • rumbling
rugiente a - . . .; rumbling
rugir v - . . . • to rumble
rugosidad f - . . . • roughness
— f absoluta - absolute roughness
— f de acequia f - [constr] channel roughness
— f — alcantarilla f—[hidr] culvert roughness
— f — cauce m - [hidr] channel roughness
— f — conducto m - conduit roughness • [constr] barrel roughness
— f — — m tubular - [constr] culvert barrel roughness
— f — — m — para alcantarilla f - [constr] culvert barrel roughness
— f — pared f - [constr] wall roughness
— f — rodillo m - [mec] roll(er) roughness
— f efectiva - [mec] effective roughness
— f — absoluta - [hidr] effective absolute roughness
— f interior - interior, o inside, roughness
— f — de alcantarilla f - [hidr] culvert, interior, o inside, roughness
— f superficial - [mec] surface roughness
— f — máxima - maximum surface roughness
— f — mínima - minimum surface roughness
— f — en álabe m - [turb] vane surface roughness
— f — en anillo m laberinto - [turb] laberynth ring surface roughness
— f — en paleta f - [turb] vane surface roughness
— f — — vástago m - [turb] rod surface roughness
— f — — cara f - [turb] face surface roughness
— f — — — f de paleta f - [turb] vane face surface roughness
rugoso,sa a - . . .; rough; knurled,ly • ragged
ruido m ambiental - environment(al), o ambient, noise
— m audible - audible noise
— m buscado - sought, o checked for, noise
— m comprobado - proven, o checked, noise
— m crujiente - grinding noise
— m de compresor m - compressor noise
— m — — m por circulación f de gas m - [ind] gas circulation compressor noise
— m — — m — — f — — m en tragante m - [metal-prod] throat gas circulation compressor noise

ruido m de golpes,peteo m - tapping, o knocking, o striking, noise
— m — **lanza** f - [metal-prod] lance noise
— m — **motor** m - [electr-mot] motor noise • [comb.int] engine noise
— f — **percusión** f - percussion, o striking, noise
— n — **sacudida(s)** f - [mec] rattling noise
— m **en salida** f - [comput] output noise
— m **excesivo** - excessive noise
— m **fuerte** - loud noise
— m **reducido** - low noise
— m — **de motor** m - [electr-mot] motor low noise • [comb.int] engine low noise
— m **sibilante** - hissing noise
— m **verificado** - checked noise
ruidoso,sa a - noisy
ruina f - . . .; ruining
rulemán* m - [mec] véase **cojinete** m (con, bolas f, o rodillos)
ruleta f - [juego] . . .; wheel • gambling
— f **girante**—[juego] turning, o whirling, wheel
rumbo m - . . . • objective; trend
— m **aclarado** - clarified direction
rumbo m **de falla** f - [geol] fault strike
— m **errático** - erratic course
— m **estimado** - [nav] dead reckoning
— m **retomado** - resumed, direction, o course
rumorar v - to rumor
rumoreado,da a - rumored
rumorear v - to rumor
ruptor m - [electr-instal] breaker; véase también **interruptor** m; **disyuntor** m
— m **para circuito** m [electr-instal] circuit breaker; véase también **disyuntor**; **interruptor** m **para circuito** m
— m — **contacto** m - [comb.int] contact breaker
ruptura f - . . . • [mec] parting
— f **de contrato** m—[legal] contract termination
— f — — m **laboral** - [labor] labor contract termination
— f — — m **para trabajo** m - [labor] labor contract termination
— f — **material** m - [constr] material rupture
— f — — m **para relleno** m - [constr] fill material rupture
— f — **pavimento** m - [constr] pavement, breaking, o rupture,ring
rupturar v - to rupture
— v **pavimento** m—[constr] to rupture @ pavement
rural f - [autom] station wagon | a - . . .; non urban
ruta f - . . .; routing • [electr-instal] path • [metal-prod] runner
— f **bloqueada** - [metal-prod] plugged, o clogged, runner
— f **brutal** - [deport] brutal, route, o course
— f **con etapa(s)** f - [deport] stage route
— f **dorsal** - [vial] ridge route
— f **exterior** - [vial] outer, o outside, route, o loop
— f — **para circunvalación** f - [vial] outer, loop, o belt(way)
— f **general** - [metal-prod] main runner
— f **interestatal** - [vial] Interstate (route)
— f **más corta** - [vial] shortest route
— f **para arrabio** m - [metal-prod] iron runner
— f — **carrera** f - [deport] race route
— f — — f **Rally** - [deport] Rally (race) route
— f — **circunvalación** f - [vial] loop; beltway
— f — **escoria** f - [metal-prod] slag runner
— f — **fundición** f - [metal-prod] iron runner
— f — **nave** f **para colada** f - [metal-prod] casting bay runner
— f — **navegación** f - [nav] navigation route
— f — — f **ultramarina** - [nav] overseas, o ocean, navigation route
— f — **ómnibus** - [transp] (omni)bus route
— f — — m **escolar** - [transp] school bus (omni)bus route
— f — **piquera** f - [metal-prod] iron notch runner; iron, runner, o trough
— f **para purga** f - [metal-prod] purge runner

ruta f **para trasponer** v **sierra** f - [vial] trans-mountain road
— f — **tubería** f - [ambient] duct path
— f **principal** - [vial] main route • [metal-prod] main runner
— f **serrana** f - [vial] mountain route • ridge route
— f **substitutiva** - [vial] alternate route
— f **subterránea** - [vial] underground, o subway, route
rutear* v - véase **encaminar**
rutero m - [metal-prod] runner repairman
rutina f - . . . | a - (de) rutinary
— f **de inspección** f - [ind] inspection routine
— f **mecánica** - [mec] mechanical routine
— f **para lubricación** f - [mec] lubrication routine
— f — f **de malacate** m - [petról] routine drawworks routine
rutinario,ria a - routine; routinary*

S

S A - [legal] véase **sociedad** f **anónima**
S A de C V - [legal] véase **sociedad** f **anónima con capital variable**
S H2 - [quím] véase **anhidrido** m **sulfuroso**
S E I C I f - [pol] véase **Secretaría** f **de Estado para Industria** f **y Comercio** m **Interior**
s n m adv - véase **sobre nivel** m **de mar**
S U S - [lubric] véase **Saybolt Universal en Segundo(s)**
sábado m - [chronol] Saturday • Sabbath
sable m - [milit] . . . • [metal-lam] saber • (strip) shape
— m **máximo** - [metal-lam] maximum saber
— m — **admisible** - [metal-lam] maximum allowable saber
— m **mínimo** - [metal-lam] minimum saber
sabor m **objetable** - [culin] objectionable, savo(u)r, o taste
saboreado,da a - savo(u)red
— a **anticipadamente** - savo(u)red beforehand; anticipated
— a **desde ya** - anticipated already
saborear v - to, savo(u)r, o taste • to get @ flavor • to anticipate
— v **anticipadamente** - to taste beforehand; to anticipate
— v **desde ya** - taste, o anticipated, already
saboreo m - getting @ taste
— m **anticipado** - anticipation
— m **desde ya** - anticipation
saca f - . . .; pulling, off, o out of; taking, o lifting, out; removal; moving, out, o from; extricating,tion
— f **a licitación** f - [com] tender issue,suing
— f **completa** - complete removal
— f **con deslizamiento** - slipping off; sliding out
— f **con esfuerzo** m - eking out
— f — **esmeriladora** - [mec] grinder removal
— f — — f **mecánica** - [mec] electric grinder removal
— f — **lima** f - [mec] filing off
— f — — f **manual** - [mec] hand file removal
— f — **mucho esfuerzo** m - eking out
— f — **paño** m - wiping (off)
— f — **puntapiés** m - kicking out
— f **cuidadosa** - careful removal,ving
— f **de aceite** m - oil removal,ving
— f — **acoplamiento** m - [mec] coupling, o linkage, removal,ving
— f — **acumulador** m - [autom-electr] battery removal,ving
— f — **agua** m - [hidr] water removal,ving

saca f de anillo m - [mec] ring removal,ving
— f — — m con resorte m - [mec] snap ring removal,ving
— f — — m para compresión f - [mec] compression ring removal,ving
— f — — m para empaquetadura f - [mec] packing ring removal,ving
— f — — m para retención f - [mec] holding ring removal,ving
— f — arandela f - [mec] washer removal,ving
— f — árbol m - [botán] tree removal,ving • [mec] shaft removal,ving
— f — aro m - [mec] ring removal,ving
— f — — m para seguridad f - [mec] lockwasher removal,ving
— f — armadura f - [mec] armature removal,ving
— f — aro m - [mec] ring removal,ving
— f — — m con resorte m - [mec] snap ring removal,ving
— f — — m para compresión f - [mec] compression ring removal,ving
— f — — m para empaquetadura f - [mec] packing ring removal,ving
— f — — m — retención f - [mec] holding ring removal,ving
— f — articulación f - [mec] linkage removal,ving
— f — asiento m - [mec] seat, removal,ving, o pulling
— f — — m para válvula f - [válv] valve seat, removal,ving, o pulling
— f — — m — — f hidráulica - [válv] hydraulic valve seat, removal,ving, o pulling
— f — atomizador m - [comb.int] atomizer removal,ving
— f — banda f - [mec] band removal,ving
— f — — f para freno m - [mec] brake band removal,ving
— f — — f — perforación f - [petról] Kelly removal,ving
— f — bloqueo m - [mec] block removal,ving
— f — bomba f - [bombas] pump removal,ving
— f — boquilla f - [mec] nozzle removal,ving
— f — — f — rociadura f - [mec] spray nozzle removal,ving
— f — broca f - [herram] bit removal,ving
— f — — f para perforación f - [petról] drill bit removal,ving
— f — buje m - [mec] bushing removal,ving
— f — — m para camisa f - [mec] liner bushing removal,ving
— f — — m piloto - [mec] pilot bushing removal,ving
— f — — m — para camisa f - [mec] liner pilot bushing removal,ving
— f — bujía f - [comb.int] spark plug removal,ving
— f — cabeza f - [mec] head removal,ving
— f — — f para inyección f - [petról] swivel removal,ving
— f — cadena f - [mec] chain removal,ving
— f — cadenero m - chainman removal,ving
— f — caja f - [mec] box, o case, o housing, removal,ving
— f — cajón m - [transp] removal,ving from @, container, o box
— f — calce m - [mec] shim removal,ving
— f — calor m - heat, removal,ving, o extracting,tion
— f — calza f - [mec] shim removal,ving
— f — — f para piñón m - [mec] pinion shim removal,ving
— f — cámara f - [mec] chamber removal,ving
— f — — f para filtro m - [mec] filter chamber removal,ving
— f — — f — — m para aire m - [mec] air filter chamber removal,ving
— f — cambiador m - [mec] changer removal,ving
— f — — m para toma f - [electr-equip] tap changer removal,ving
— f — camisa f - [mec] liner removal,ving
— f — campana f - [mec] bell removal,ving • canopy removal,ving

saca f de carga f - [transp] load removal,ving
— f — cartucho m - cartridge removal,ving
— f — cascarón m - [metal-prod] scab removal
— f — casquete m - [comb.int] cap removal,ving
— f — cierre m - [mec] closure, o seal, removal,ving
— f — cilindro m - [comb.int] cylinder removal,ving
— f — cinta f - [mec] (brake) lining removal
— f — — f para freno m - [mec] brake lining removal,ving
— f — cojinete m - [mec] bearing removal,ving
— f — — m con grasa f - [mec] grease bearing removal,ving
— f — colador m - [mec] strainer removal,ving
— f — colector n - [sanit] sump removal,ving
— f — — m para purga f - [mec] drain sump removal,ving
— f — combustible m - [combust] fuel removal,ving
— f — conclusión f - conclusion drawing
— f — conductor m - [electr-instal] lead removal,ving
— f — — m de(sde) borne m - [electr-instal] lead removal,ving from @ terminal
— f — conexión f - [mec] connection, o linkage, removal,ving
— f — conjunto m - [mec] assembly removal,ving
— f — — m de cabeza f - [mec] head assembly removal,ving
— f — — m — f para inyección f - [petról] swivel assembly removal,ving
— f — — m de distribuidor m - [mec] divider, o distributor, assembly removal,ving
— f — — m para fuerza f - [autom-mec] power divider assembly removal,ving
— m — — m — llanta f - [autom] rim assembly removal,ving
— f — — m — neumático m - [autom-neumát] tire assembly removal,ving
— f — — m portadiferencial m - [autom-mec] differential carrier assembly removal,ving
— f — contrapeso m - [grúas] counterweight removal,ving
— f — cruceta f - [mec] crosshead removal,ving
— f — cubierta f - [mec] cover removal,ving
— f — — f grande - [mec] large cover removal,ving
— f — — f para mesa f - [mec] table cover(ing) removal,ving
— f — — f para transmisión f - [mec] transmission cover(ing) removal,ving
— f — — f principal - [mec] main cover(ing) removal,ving
— f — cuerpo m - dodging
— f — — m extraño - [med] foreign object removal,ving
— f — chapa f - [mec] sheet removal,ving
— f — — f para cubierta f - [mec] cover(ing) sheet removal,ving
— f — defensa f - [segurid] guard removal
— f — — f para seguridad f - [segurid] safety guard removal,ving
— f — distribuidor m - [mec] distributor removal,ving
— f — — m rociador m - [mec] spray(ing) manifold removal,ving
— f — eje m - [mec] axle removal,ving • removal,ving from @ axle
— f — eje m - [mec] axle, o shaft, removal,ving
— f — elemento m - [mec] element removal,ving • canister removal,ving
— f — embalaje m - packing removal,ving • removal,ving from @ packing
— f — émbolo m - [comb.int] piston removal,ving
— f — embrague m - [mec] clutch removal,ving
— f — — m para mecanismo m - [mec] mechanism clutch removal,ving
— f — — — — m para elevación f - [grúas] hoist(ing) clutch (mechanism) removal,ving
— f — — m — m — rotación f - [grúas] swing cluth (mechanism) removal,ving
— f — empaquetadura f - packing removal,ving

saca f de enchufe m - [electr-instal] plug removal,ving • [electr-oper] plug pulling
— f — engranaje m - [mec] gear removal,ving
— f — eslabón m - [mec] link removal,ving
— f — — m de cadena f - [mec] chain link removal,ving
— f — esmeriladora f - [mec] grinder removal,ving
— f — — f mecánica - [mec] electric, o power, grinder, removal,ving
— f — esquina f - [mec] corner removal,ving • removal,ving from @ corner
— f — fiador m - [mec] retainer removal,ving
— f — — m para cierre m - [mec] seal retainer removal,ving
— f — — m — trinquete m - [mec] ratchet retainer removal,ving
— f — filtro m - [mec] filter removal,ving
— f — freno m - [mec] brake removal,ving • removal,ving from @ brake
— f — grillete m - [mec] clevis removal,ving
— f — guarnición f - [mec] gasket removal,ving
— f — — f circular - [mec] O-ring removal
— f — guía(dera) f - [mec] removal,ving
— f — — f para vástago m - [válv] stem guide removal,ving
— f — — f — — m para válvula f - [válv] valve stem guide removal,ving
— f — — f — — m — f inferior - [válv] lower valve stem guide removal,ving
— f — — f — — m — f superior - [válv] upper valve stem guide removal
— f — horquilla f - [mec] clevis removal,ving
— f — injerto m - [mec] insert removal,ving
— f — — m para válvula f - [válv] valve insert removal,ving
— f — inyección f - [petról] mud removal,ving
— f — laminilla f - [mec] shim removal,ving
— f — — f para piñón m - [mec] pinion shim removal,ving
— f — lodo m - [petról] mud removal,ving
— f — material m - material removal,ving
— f — — m extraño - foreign materail removal
— f — montaje m - [mec] mounting removal,ving
— f — mordaza f - [mec] clamp removal,ving
— f — motor m - [electr-mot] removal,ving from @ motor • [comb.int] removal from @ engine • [electr-mot] motor removal,ving • [comb.int] engine removal,ving
— f — neumático m - [autom-neumát] tire, removal,ving, o peeling
— f — nieve f - [vial] snow removal,ving
— f — paleta f - [mec] blade removal,ving
— f — panel m - [mec] panel removal,ving
— f — — m inferior - [mec] lower panel removal,ving
— f — — posterior - [mec] rear panel removal,ving
— f — — m superior - [mec] upper panel removal,ving
— f — pasador m - [mec] pin removal,ving
— f — — n conectador - [mec] connecting, o linkage, kingpin , removal,ving
— f — — m para grillete m - [mec] clevis pin removal,ving
— f — — m — horquilla f - [mec] clevis pin removal,ving
— f — perno m - [mec] bolt removal,ving
— f — — m para sujeción f - [mec] hold-down, o fastening, bolt removal,ving
— f — — m — sujetador m para camisa f - [mec] liner clamp bolt removal,ving
— f — — m para traba(r) - [mec] lock bolt removal,ving
— f — pico m para engrase m - [mec] grease fitting removal,ving
— f — pieza f - [mec] part removal,ving
— f — — f para bomba f - [bombas] pump part removal,ving
— f — placa f - [mec] plate removal,ving
— f — plancha f - [mec] plate removal,ving
— f — polvo m - dust, removal,ving, o wiping; dusting

saca f de prensaestopa(s) - [mec] removal,ving from @ packing box • packing box removal,ving
— f — recalcadura f - [mec] upsetting removal
— f — — f metálica - [mec] metal upsetting removal,ving
— f — rejilla f - [mec] screen removal,ving
— f — resguardo m - [mec] guard removal,ving
— f — resistencia f - [electrón] resistor removal,ving
— f — resorte m - [mec] spring removal,ving
— f — restricción f - restriction removal,ving
— f — retén m - [mec] retainer removal,ving
— f — retenedor m - [mec] retainer removal,ving
— f — — m para cierre m - [mec] seal retainer removal,ving
— f — rueda f - [mec] wheel, removal,ving, o pulling
— f — — f motriz - [autom] drive wheel removal,ving
— f — sección f - [mec] section removal,ving
— f — — f de piso m - [constr] floor section removal,ving
— f — — f — plancha f - [mec] plate section removal,ving
— f — segmento m - [mec] subrod, o segment, removal,ving
— f — sello m - [mec] seal removal,ving
— f — — m para aceite m - [mec] oil seal removal,ving
— f — semieje m - [auto-mec] axle shaft removal,ving
— f — separador m - [mec] spacer removal,ving • removal,ving from @ spacer
— f — — m para armado m - [mec] assembly,bling spacer removal,ving
— f — solenoide m - [electr-instal] solenoid removal,ving
— f — soporte m - [mec] support removal,ving
— f — suciedad f - dirt removal,ving
— f — suelo m - soil removal,ving
— f — sujetador m - [mec] fastener removal,ving
— f — — m para camisa f - [mec] liner clamp removal,ving
— f — suplemento m - [mec] supplement, o insert, removal,ving
— f — — m para válvula f - [válv] valve insert removal,ving
— f — tapa f - cover, o cap, o top, removal
— f — — f para colector m - [mec] sump cover removal,ving
— f — — f — mesa f - [mec] table cover removal,ving
— f — — f para transmisión f - [mec] transmission cover removal,ving
— f — — f para válvula f - [válv] valve cover removal,ving
— f — — f principal - [mec] main cover removal,ving
— f — tapón m - [mec] plug removal,ving
— f — — m mostrada - [mec] shown plug removal
— f — — m para drenaje m - [mec] drain plug removal,ving
— f — — m — filtro m - [mec] filter plug removal,ving
— f — — m — — m mostrada - [petról] shown filter plug removal,ving
— f — — m — nivel m - [mec] level plug removal,ving
— f — — m — — m de aceite m - [mec] oil level plug removal,ving
— f — — m — — m — — m en caja f - [mec] case oil level plug removal,ving
— f — — m — — m — — m — — f para cadena f - [mec] chain case oil level plug removal
— f — — m — purga f - [mec] drain plug removal,ving
— f — — m — — f de colector m - [mec] drain plug removal,ving
— f — — m — — f — — m para aspiración f - [mec] suction manifold drain plug removal,ving
— f — — m — tubo(s) m - [mec] pipe plug removal,ving
— f — testigo m - [petról] sample taking

saca f de testigo m cilíndrico - [petról] core sample taking
— f — tierra f - earth, o dirt, removal,ving
— f — tomacorriente m - [electr-instal] receptacle, removal,ving, o pulling
— f — tornillo m - [mec] screw removal,ving
— f — — m con casquete m - [mec] capscrew removal,ving
— f — traba f - [mec] catch removal,ving • lock release,sing
— f — — f para freno m - [mec] brake lock release,sing
— f — trinquete m - [mec] ratchet removal,ving
— f — tubería f - [tub] tubing removal,ving
— f — tubo m—[tub] tube, o pipe, removal,ving
— f — tuerca f - [mec] nut removal,ving
— f — — f herrumbrada - [mec] rusty nut removal,ving
— f — unidad f - [mec] unit removal,ving
— f — válvula f - [válv] valve removal,ving
— f — — f inferior - [válv] lower valve removal,ving
— f — — f superior - [válv] upper valve removal,ving
— f — vástago m - [mec] rod removal,ving • [válv] stem removal,ving
— f — — m intermedio - [mec] intermediate rod removal,ving
— f — — m para válvula f - [válv] valve stem removal,ving
— f — — m — — f inferior - [válv] lower, o bottom, valve stem removal,ving
— f — — m — — f superior - [válv] upper, o top, valve stem removal,ving
— f — ventaja f - advantage gaining
— f — — f mayor - getting @ most
— f deslizando - slipping off
— f — con facilidad f - easy slipping off
— f lenta - slow removal,ving
— f mediante soplado,dura - [mec] blowing out
— f mostrada - shown removal,ving
— f — inspección f - removal for inspection
— f — limpieza f - removal,ving for cleaning
— f periódica - periodic removal,ving
— f rápida - quick removal,ving
sacaasientos m - [herram] seat puller
— m hidráulico - [herram] hydraulic seat puller
— m — válvula f - [herram] valve seat puller
— m — — f hidráulica - [herram] hydraulic valve seat puller
sacabarba(s) m - [agric-equip] (barley) awner (attachment)
— f para cebada f - [agric-equip] barley awner (attachment)
sacabuje(s) m - [herram] bushing extractor
sacaclavo(s) m - [herram] nail puller
sacado,da a - removed; withdrwn; drawn; lifted, o taken, out; pulled, from, o out, of; extricated • moved out • [metal] slipped (off)
— a completamente - removed completely
— a con cuidado - removed carefully
— a con esfuerzo - eked out
— a — esmeriladora f - [mec] grinder removed
— a — — f mecánica - [mec] electric grinder removed
— a — extractor m - [mec] puller removed
— a — lima f - [mec] file removed
— a — — f manual - [mec] hand file removed
— a — mucho esfuerzo - eked out
— a — paño m - wiped (off, o away)
— a — puntapiés m - kicked out
— a cuidadosamente - removed carefully
— a de cajón m - removed from @ box • [transp] removed from @ container
— a — carrera f - [deport] retired from @ race
— a — cruceta f - [mec] removed from @ crosshead
— a — eje m - [mec] removed from @, axle, o shaft
— a — motor m - [electr-mot] removed from @ motor • [comb.int] removed from @ engine
— a — orificio m - [mec] removed from @ hole
— a — prensaestopa(s) f - [mec] removed from @ packing box
sacado,da de separador m - [mec] removed from @ spacer
— a deslizando - slipped off
— a — con facilidad f - slipped off easily
— a en sorteo m - [juego] drawn
— f fácilmente - removed, readily, o easily
— a lentamente - removed slowly
— a mediante soplo m - blown out
— a — inspección f - removed for inspection
— a para limpieza f - removed for cleaning
— a periódicamente - removed periodically
— a rápidamente - removed quickly
sacador m para asiento(s) m - [mec] seat puller
— m — — m para válvula f - [herram] valve seat puller
— m — núcleo(s) m - [petról] core extractor
— m — testigo(s) m - [petról] core extractor
sacaestopa(s) m - [mec] packworm
sacamancha(s) - [domést] spot remover; benzene; benzine
sacamuestra(s) m - sampler • [petról] core barrel • thief
sacanúcleo(s) m - [petról] core, drill, o barrel
— m para cable m para perforación(es) f - [petról] wire line core barrel
sacapajas f - [agric-equip] straw rack
sacapuerta(s) m - [coque] door, remover, o extractor; door machine
— m convencional - [coque] conventional, o multi-spot, door machine
— m single-spot - [coque] single-spot door machine
sacar v - . . .; to take • to remove • to withdraw • to, lift, o take, o move, o bring, out; to extricate • [electr-oper] to draw • [metal-prod] to chip off • [deport] to bring in
— v a licitación f - [com] to issue @ tender
— v — superficie f - [sold] to float (to @ surface)
— v — — f mediante flotación f - [sold] to float to @ surface
— v abrazadera f - [mec] to remove @ clamp
— v — f para brida f - [mec] to remove @ flange clamp
— v — f — pestaña f - [mec] to remove @ flange clamp
— v aceite m - [comb.int] to, remove, o drain, @ oil
— v acoplamiento m - [mec] to remove @, coupling, o linkage
— v acumulador m - [autom-electr] to remove @ battery
— v agua m - [hidr] to remove @ water
— v aire m - to remove @ air
— v amperaje m - [electr-oper] to draw current
— v . . . amperio(s) m - [electr-oper] to draw . . . amperes
— v anillo m - [mec] to remove @ ring
— v — m con resorte m - [mec] to remove @ snap ring
— v — m para aceite m - [mec] to remove @ oil, seal, o ring
— v — m para empaquetadura f - [mec] to remove @ packing ring
— v — m — retención f - [mec] to remove @ holding ring
— v arandela f - [mec] to remove @ washer
— v — f de bronce m - [mec] to remove @ bronze washer
— v — f para seguridad f - [mec] to remove @ lockwasher
— v árbol m - [botán] to remove @ tree • [mec] to remove @ shaft
— v armadura f - [mec] to remove @ armature
— m aro m - [mec] to remove @ ring
— v — m con resorte m - [mec] to remove @ snap ring
— v aro m para empaquetadura f - [mec] to remove @ packing ring
— v — m — retención f - [mec] to remove @ holding ring
— v articulación f - [mec] to remove @ linkage

sacar v asiento m - to, remove, o pull, @ seat
— v — m para válvula f - [válv] to remove, o
to pull, @ valve seat
— v — — f hidráulica - [válv] to pull @
hydraulic valve seat
— v astilla f - [mec] to remove @ chip
— v atomizador m - [comb.int] to remove @ atomizer
— v automóvil m - [autom] to roll out @ car
— v — m de cobertizo m - [autom] to roll @ car
out of @ garage
— v banda f - [mec] to remove, f lift, @ band
— v — f para freno m - [mec] to remove @ brake
band
— v barra f - [mec] to remove @ bar
— v — f conformada (para perforación f) -
[petról] to remove @ Kelly
— v barreno m - [petról] to remove @ bit
— v — m para perforación f - [petról] to remove @ drill bit
— v bloque m - [mec] to remove @ block
— v bomba f - [bombas] to remove @ pump
— v boquilla f - [mec] to remove @ nozzle
— v — f para rociadura f - [mec] to remove @
spray(ing) nozzle
— v broca f - [mec] to remove @ bit
— v — f para perforación f - [petról] to remove @ (drill) bit
— v buje m - [mec] to remove @ bushing
— v — m para camisa f - [mec] to remove @
liner bushing
— v — m piloto - [mec] to remove @ pilot bushing
— v bujía f - [comb.int] to remove @ spark plug
— v burro m - [metal-prod] to remove @ (frozen)
pig iron
— v — m de ruta f - [metal-prod] to remove @
(frozen) pig iron from @ runner
— v cabeza f - to remove @ head
— v — f para inyección f - [petról] to remove
@ swivel
— v cadena f - [mec] to remove @ chain
— v caja f - to remove @ box • [ind] to remove
@ housing
— v — f para conmutador m - [ventil] to remove
@ switch housing
— v calce m - [mec] to remove @ shim
— v calor m - to remove @ heat
— v calza f - [mec] to remove @ shim
— v — f para piñón m - [mec] to remove @ piston shim
— v cámara f - [mec] to remove @ chamber
— v — f para filtro m - [mec] to remove @ filter chamber
— v — f — — m para aire m - [mec] to remove
@ air filter chamber
— v cambiador m - [mec] to remove @ changer
— v — m para toma(s) f - [electr-equip] to remove @ tap changer
— v camisa f - [vest] to remove @ shirt • [mec]
to remove @ liner
— v campana f - [mec] to remove @ bell • to remove @ canopy
— v cañón m - [metal-prod] to remove @ mud gun
— v — m en piquera f - [metal-prod] to remove
@, clay, o mud, gun from @ iron notch
— v carga f - [transp] to remove @ load
— v cartucho m - to remove @ cartridge
— v casquete m - [comb.int] to remove @ cap
— v cierre m - [mec] to remove @ seal
— v cilindro m - [ind] to remove @ cylinder
— v cinta f - [mec-frenos] to remove @ liner
— v — f para freno m - [mec] to remove @ brake
lining
— v cojinete m - [mec] to remove @ bearing
— v — m con grasa f - [mec] to remove @ grease
bearing
— v colador m - [mec] to remove @ strainer
— v colector m - [sanit] to remove @ sump
— v — m para purga f - [sanit] to remove @
drain sump
— v combustible m - [combust] to remove @ fuel

sacar v completamente - to remove completely
— v con esfuerzo m - to eke out
— v — esmeriladora f - [mec] to grinder remove
— v — — f mecánica - [mec] to remove with @
electric grinder
— v — extractor m - [mec] to puller remove
— v — lima f - [mec] to file away
— v — — f manual - [mec] to hind file (away)
— v — mucho esfuerzo m - to eke out
— v — paño m - to wipe (with @ cloth)
— v — puntapies m - to kick out
— v conclusión f - to draw @ conclusion
— v conductor m - [electr-instal] to remove @
lead
— v — m de borne m - [electr-instal] to remove
@ lead from @ terminal
— v conexión f - [mec] to remove @, connection,
o linkage
— v conjunto m - [mec] to remove @ assembly
— v — m de cabeza f - [mec] to remove @ head
assembly
— v — — — f para inyección f - [petról] to
remove @ swivel assem
— v — m — distribuidor m - [mec] to remove @,
divider, o distributor, assembly
— v — — — para fuerza f - [autom-mec] to
remove @ power divider assembly
— v — m de llanta f - [autom] to remove @ rim
assembly
— v — m — neumático m - [autom] to remove @
tire assembly
— v — m — portadiferencial m - [autom-mec] to
remove @ differential carrier
— v conmoción f - to remove @ commotion
— v contrapeso m - [grúas] to remove @ counterweight
— v corriente f - [electr-oper] to draw @, current, o power
— v cruceta f - [mec] to remove @ crosshead
— v cubierta f - [mec] to remove @ cover
— v — f para mesa f - [mec] to remove @ table
cover
— v — f — transmisión f - [mec] to remove @
transmission cover
— v — principal - [mec] to remove @ main cover
— v chapa f - [mec] to remove @ sheet
— v — f para cubierta f - [mec] to remove @
cover sheet
— v cubierta f - to remove @ cover
— v cuerpo m - [fam] to get out of @ way
— v extraño - [med] to remove @ foreign object
— v de cajón m - [ind] to remove from @ box •
[ind] to remove from @ container
— v — eje m - [mec] to remove from @ axle
— v — motor n - [electr-mot] to remove from @
motor • [comb.int] to remove from @ engine
— v — orificio m - [mec] to remove from @ hole
— v — pista f - [deport] to, remove, o tow off
of, @ course
— v — prensaestopas m - [mec] to remove from @
packing box
— v — separador m - [mec] to remove from @
spacer
— v — servicio m - to take out of service
— v defensa f - [mec] to remove @ guard
— v — f para seguridad f - [segurid] to remove
@ safety guard
— v demasiado poco - to remove too little
— v desde atrás - [mec] to pull back out
— v — extremo m - to pull back from @ end
— v deslizando - to, slip out, o slide off
— v — con facilidad f - to slip off easily
— v distribuidor m - [mec] to remove @ manifold
— v — m rociador - [mec] to remove @ spray manifold
— v eje m - [mec] to remove @ axle
— v elemento m - [mec] to remove @ element •
[segurid] to remove @ canister
— v embalaje m - to remove @ packing
— v émbolo m - [mec] to remove @ piston
— v embrague m - [mec] to remove @ clutch

sacar v embrague m para (mecanismo m para) elevación f - [grúas] to remove @ hoist clutch
— v —— —(— m —) rotación f - [grúas] to remove @ swing clutch
— v empaquetadura f - [mec] to remove @ packing
— v en defecto - to remove too little
— v — menos - to remove too little
— v — sorteo m - (lotería) to draw
— v enchufe m - [electr-oper] to pull @ plug
— v energía f - [electr-oper] to draw @ power
— v engranaje m - [mec] to remove @ gear
— v escoria f - [metal-prod] to remove @ slag
— v — f mediante flotación - [sold] to float
(away) @ slag
— v eslabón m - [mec] to remove @ link
— v — m de cadena f - [mec] to remove @ chain
link
— v esqueleto m - [transp] to remove @ crate
— v esquina f - [mec] to remove @ corner
— v fácilmente - to remove easily
— v fiador m - [mec] to remove @ retainer
— v — m para trinquete m - [mec] to remove @
ratchet retainer
— v filtro m - [mec] to remove @ filter
— v — m para aceite m - [mec] to remove @ oil
filter
— m — m — agua m - [mec] to remove @ water
filter
— v — m — aire m - [mec] to remove @ air
filter
— v freno m - [mec] to remove @ brake
— v grasa f - [mec] to remove @ grease
— v grillete m - [mec] to remove @ clevis
— v guarnición f - [mec] to remove @ gasket
— v — f circular - [mec] to remove @ O ring
— v guía(dera) f - [mec] to remove @ guide
— v — f para vástago m - [válv] to remove @
stem guide
— v — f — — m para válvula f - [válv] to remove @ valve stem guide
— v — f — — m para válvula f inferior - [válv] to
remove @, lower, o bottom, valve stem guide
— v — f — — m — f superior - [válv] to
remove @, top, o upper, valve stem guide
— v horquilla f - [mec] to remove @ clevis
— v injerto m - [mec] to remove @ insert
— v — m para válvula f - [válv] to remove @
valve insert
— v laminilla f - [mec] to remove @ shim
— v — f para émbolo m - [mec] to remove @ piston shim
— v lentamente - to remove slowly
— v lodo m - [petról] to remove @ mud
— v material m - to remove @ material
— v — m extraño - to remove @ foreign material
— v mediante flotación f - [sold] to float
(away)
— v — soplado m - to blow out
— v montaje n - [mec] to remove @ mounting
— v mordaza f - [mec] to remove @ clamp
— v muestra f - [int] to, take, o obtain, @
sample(s)
— v — f de arrabio m - [metal-prod] to obtain
@ pig iron sample(s)
— v — limpieza f—[mec] to remove for cleaning
— v neumático m - [autom-neumát] to, remove, o
peel, @ tire
— v nieve f - [meteorol] to remove @ snow
— v paleta f - [mec] to remove @ blade
— v panel m - [mec] to remove @ panel
— v — delantero—[mec] to remove @ front panel
— v — m posterior - [mec] to remove @ rear
panel
— v para inspección f - [ind] to remove for @
inspection
— v pasador m - [mec] to remove @ pin
— v — m conectador - [mec] to remove @ linkage
pin
— v — m para grillete m - [mec] to remove @
clevis pin
— v — m — horquilla f - [mec] to remove @
clevis pin
— v periódicamente - to remove periodically

sacar v perno m - [mec] to remove @ bolt
— v — m herrumbrado - [mec] to remove @ rusty
bolt
— v — m para sujeción f - [mec] to remove @,
holding, o hold-down, bolt
— v — m — sujetador m para camisa f - to remove @ liner clamp
— v — m para traba(r) - [mec] to remove @
lock(ing) bolt
— v petaca f - [metal-prod] to remove @ cooling
plate
— v — con extractor m - [metal-prod] to remove
@ cooling plate with @ extractor
— v pico m para engrase m - [mec] to remove @
grease fitting
— v pieza f - [mec] to remove @ part
— v — f para bomba f - [bombas] to remove #
pump part
— v piso m - [constr] to remove @ floor
— v placa f - [mec] to remove @ plate
— v plancha f - [mec] to remove @ plate
— v — f para piso m - [mec] to remove @ floor
plate
— v polvo m - to (remove @) dust
— v posición f - [deport] to draw @ spot
— v — f para largada f - [deport] to draw @
starting spot
— v prensaestopa(s) m - [mec] to remove @
packing box
— v punta f - [mec] to sharpen
— v — a lápiz - to sharpen @ pencil
— v rápidamente - to remove quickly
— v rebaba(s) f - [mec] to remove @ burr(s)
— v recalcadura f - [mec] to remove @ upsetting
— v — f metálica - [mec] to remove @ metal upsetting
— v rejilla f - [mec] to remove @ screen
— v resguardo m - [mec] to remove @ guard
— v resistencia f - [electrón] to remove @ resistor
— v resorte m - [mec] to remove @ spring
— v restricción f - to remove @ restriction
— v retén m - to remove @ retainer
— v — m para cierre m - [mec] to, remove, o
pull, @ seal retainer
— v — m — sello m - [mec] to, remove, o pull,
@ seal retainer
— v retenedor m - [mec] to remove @ retainer
— v rotor m - [electr-mot] to remove @ rotor
— v rueda f - [autom] to, remove @ wheel
— v — f motriz - [autom] to remove @ drive
wheel
— v sección f - [mec] to remove @ section
— v — f de piso m - [constr] to remove @ floor
section
— v — f — placa f - [mec] to remove @ plate
section
— v — f — plancha f - [mec] to remove @ floor
plate section
— v segmento m - [mec] to remove @, segment, o
subrod
— v sello m - [mec] to remove @ seal
— v semieje m - [autom-mec] to remove @ axle
shaft
— v separador m - [mec] to remove @ spacer
— v — armado m - [mec] to remove @ assembling spacer
— m solenoide m - [electr-instal] to remove @
solenoid
— v soporte m - [mec] to remove @, support, o
holder
— v suciedad f - to remove @ dirt(iness)
— v suelo m - [suelos] to remove @ soil
— v sujetador m - [mec] to remove @, clasp, o
fastener
— v — m para camisa f - [mec] to remove @
liner clamp
— v suplemento m - [mec] to remove @, supplement, o insert
— v — m para válvula f - [válv] to remove @
valve insert
— v tapa f - to, remove, o take off, @, cover,
o top, o cap

sacar v̱ t̲apa f̱ para colector m̱ - [mec] to remove @ sump cover
— v̱ — f̱ — **mesa f̱** - [mec] to remove @ table cover
— v̱ — f̱ — **transmisión f̱** - [mec] to remove @ transmission cover
— v̱ — f̱ — **válvula f̱** - [válv] to remove @ valve cover
— v̱ — f̱ **principal** - [mec] to remove @ main cover
— v̱ **tapón m̱** - [mec] to remove @ plug
— v̱ — m̱ — **drenaje m̱** - [mec] to remove @ drain plug
— v̱ — m̱ — **filtro m̱** - [mec] to remove @ filter plug
— v̱ — m̱ — **freno m̱** - [mec] to remove @ brake plug
— v̱ — m̱ — **nivel m̱** - [mec] to remove @ level plug
— v̱ — m̱ — m̱ **de aceite m̱** - [mec] to remove @ oil level plug
— v̱ — m̱ — — — m̱ **en caja f̱** - [mec] to remove @ case oil level plug
— v̱ — **m̱** — m̱ — m̱ — — f̱ **para cadena f̱** - to remove @ chain case oil level plut
— v̱ — m̱ — **purga** f̱ - [mec] to remove @ drain plug
— v̱ — m̱ — — f̱ **para colector m̱** - [mec] to remove @ manifold drain plug
— m̱ — m̱ — — f̱ — — m̱ **para aspiración f̱** - [mec] to remove @ suction manifold drain plug
— v̱ — m̱ **para tubo(s) m̱** - [mec] to remove @ pipe plug
— v̱ **testigo m̱** - [petról] to take @ sample
— v̱ — m̱ **cilíndrico** - [petról] to take @ core sample
— v̱ **tierra f̱** - to remove @, dirt, o earth
— f̱ **tobera f̱** - [metal-prod] to remove @ tuyere
— v̱ **tomacorriente m̱** - [electr-oper] to pull @ socket
— v̱ **tornillo m̱** - [mec] to remove @ screw
— v̱ — m̱ **con casquete m̱** - [mec] to remove @ cap screw
— v̱ **traba f̱** - [mec] to release @ lock
— v̱ — f̱ **para freno m̱** - [mec] to release @ brake lock
— v̱ **trinquete m̱** - [mec] to release @, ratchet, o pawl
— v̱ **tubería f̱** - [tub] to remove @ tubing
— v̱ **tubo m̱** - [tub] to remove @, tube, o pipe
— v̱ **tuerca f̱** - [mec] to remove @ nut
— v̱ — f̱ **herrumbrada** - [mec] to remove @ rusty nut
— v̱ **unidad f̱** - [mec] to remove @ unit
— v̱ **válvula f̱** - [mec] to remove @ valve
— v̱ — f̱ **inferior** - [válv] to remove @ lower valve
— v̱ — f̱ **superior** - [válv] to remove @ upper valve
— v̱ **vástago m̱** - [mec] to remove @ rod • [válv] to remove @ stem
— v̱ — m̱ **intermedio** - [mec] to remove @ intermediate rod
— v̱ — m̱ **para válvula f̱** - [válv] to remove @ valve stem
— v̱ — m̱ — — f̱ **inferior** - [válv] to remove @ lower valve stem
— v̱ — m̱ — — f̱ **superior** - [válv] to remove @ upper valve stem
— v̱ **ventaja f̱** - to advantage
— v̱ — f̱ **mayor** - to get @ most
sacarruedas m̱ - [autom] wheel puller
sacatestigo(s) m̱ - [petról] core, barrel, o extractor; sampler
— m̱ — **cable m̱ para perforación f̱** - [petról] wire line core barrel
saco m̱ de arena f̱ - [constr] sand bag
— m̱ — **arpillera f̱** - [textil] jute bag
— m̱ — **cemento m̱** - [constr] cement bag
— m̱ — **fundente m̱** - [sold] flux bag
— m̱ — **con (elementos de) aleación f̱** - [sold] alloy flux bag
— ṉ — — m̱ **para endurecimiento m̱ de superficie** - [sold] hardsurfacing flux bag

saco m̱ de papel - [papel] (paper), bag, o sack
— m̱ — **yute** - [textil] jute bag
— m̱ **impermeable** - water, o moisture, proof bag
sacrificación* f̱ - véase **sacrificio m̱**
sacrificadamente adv - sacrificially
sacrificado,da a - sacrificed
sacrificar v̱ rendimiento m̱ - to sacrifice @, yield, o performance
sacrificio m̱ - . . . . • sacrificing
— m̱ **de rendimiento m̱** - yield, o performance. sacrificing
— m̱ **excesivo** - excessive sacrifice
sacude huesos a - bone jarring
sacudida f̱ - . . .; jerk(ing); jar(ring) • rattle,tling
sacudido,da a - . . .; rattled
sacudidor m̱ para vagón(es) m̱ - [metal-prod] car, shaker, o vibrator
sacudimiento m̱ - . . .; jar(ring)
sacudir v̱ - . . .; to jar; to rock; to rattle
saetilla f̱ de reloj m̱ - [instrum| watch, o clock, hand
saetín m̱ de tipo m̱ Lennon - [hidr] Lennon type flume
sagacidad f̱ - . . . . • wisdom
sagita f̱ - . . . . [arquit] rise
sal f̱ - [quím] . . .; sodium chloride • [nucl] tailing(s)
— f̱ **ácida** - [quím] acid salt
— f̱ **común** - [culin] common, o table salt; sodium chloride
— f̱ **corriente** - [quím] common salt
— f̱ **de ácido m̱** - [quím] acid salt
— f̱ — — m̱ **inorgánico** - [quím] inorganic acid salt
— f̱ — — m̱ — **oxidante** - [quím] oxidizing inorganic acid salt; inorganic oxidizing acid salt
— f̱ — **catión m̱** - [quím] cation salt
— f̱ — — m̱ **metálico** - [quím] metallic cation salt
— f̱ — — m̱ **polivalente** - [quím] polyvalent cation salt
— f̱ — — m̱ — **metálico** - [quím] metallic polyvalent cation salt; polyvalent metallic cation salt
— f̱ — **nitrato m̱** - [quím] nitrate salt
— f̱ — — m̱ **de amonio m̱** - [quím] ammonium nitrate salt
— f̱ — — m̱ — **sodio m̱** - [quím] sodium nitrate salt
— f̱ **disuelta** - [quím] dissolved salt
— f̱ — **en electrolito m̱** - [quím] electrolyte dissolved salt
— f̱ **en agua m̱** - [hidr] water salt
— f̱ — **electrolito m̱** - [quím] electrolyte salt
— f̱ — **suelo m̱** - [suelos] soil salt
— f̱ **fijada** - fixed tailing(s)
— f̱ **inorgánica** - [quím] inorganic salt
— f̱ — **oxidante** - [quím] oxidizing inorganic, o inorganic oxidixing, salt
— f̱ **orgánica** - [quím] organic salt
— f̱ — **oxidante** - [quím] oxidizing organic salt
— f̱ **oxidante** - [quím] oxidizing salt
— f̱ **polivalente** - [quím] polyvalent salt
— f̱ **resultante** - [quím] resulting salt
— f̱ **soluble** - [hidr] soluble salt
— f̱ — **en agua m̱** - [hidr] water soluble salt
— f̱ — — **suelo m̱** - [suelos] soil soluble salt
— f̱ **yodada** - [quím] iodized salt
sala f̱ - [constr] . . .; room; [ind] bay; aisle
— f̱ **para computador(es) m̱** - [comput] computer room
— f̱ — **disyuntor m̱** - [electr-instal] breaker, o interruptor, room
— f̱ — — m̱ **en vacío m̱** - [electr-instal] vacuum, breaker, o interruptor, room
— f̱ — **interruptor m̱** - [electr-instal] breaker, o interruptor, room
— f̱ — — m̱ **en vacío m̱** - [electr-instal] vacuum, breaker, o interruptor, room
— f̱ — **mando m̱** - [ind] control, room, o house

sala f comedor m - [constr] (formal) dining room
— f con tablero(s) m (para regulación f eléc-trica) - [electr-instal] electrical switch-board room
— f para acumulador(es) m - [ind] battery room
— f — auxilio m - [mec] aid station
— f — ₥ primero - [mec] first aid station; infirmary; dispensary; hospital
— f — bomba(s) - [ind] pump, room, o house
— f — caldera(s) f - [ind] boiler, room, o house
— f — cañón m - [metal-prod] clay, o mud, gun control room
— f — carga(r) - [metal-prod] charging aisle
— f — clasificación f - [ind] (as)sorting room
— f — colada f - [metal-prod] pouring aisle; pit side
— f — computación f - [comput] computer room
— f — computador(es) m - [comput] computer room
— f — descanso m - [ind] lounge
— f — enfermero(s) m - [mec] nurse(s) station
— f — estar - [constr] family, o living, room
— f — extractor(es) m - [ind] extractor room
— f — gobierno m - [ind] control, house, o room
— f — granallado m - [metal-lam] pangborn shop
— f — instrumento(s) m - [ind] instrument(s) room
— f — mando m - [ind] control room
— f — — cerrada - [ind] (en)closed control room
— f — ₥ encerrada - [ind] enclosed control room
— f — máquina(s) - [ind] machine room • [metal-prod] hoist, o skip, house
— f — f para montacargas m - [metal-prod] skip house
— f — f herramientas f - [ind] machine tool(s) shop
— f — motor(es) m - [ind] motor room
— f — ₥ diesel - [ind] diesel engine room
— f — — ₥ modificada - [ind] changed die-sel engine room
— f — ₥ para laminación f en caliente - [metal-lam] hot strip motor room
— f — — f — frío - [metal-lam] cold stip motor room
— f — mufa(s) f - [telecom-instal] muff room
— f — presión f - [ind] pressure room
— f — f alta f - [metal-prod] (furnace throat) high pressure control room
— f — regulación f - [ind] control room
— f — regulador m - [electr] regulator room
— f — — ₥ para electrodo m - [metal-prod] electrode regulator room
— f — reunión(es) f meeting room
— f — soplador(es) m - [metal-prod] blast room
— f — soplante(s) m - [metal-prod] blast room • booster building
— f — transformador m - [electr-distrib] transformer room
— f principal m - [constr] main room
— f — para regulación f (eléctrica) - [electr-instal] main electrical switchboard room
— f recreativa - [constr] recreation room
salamandra f - [metal-prod] salamander • [biol] salamander
salario m anual—[labor] yearly salary • payroll
— f devengado - [labor] accrued wage
— ₥ industrial - [labor] industrial wage
— ₥ pagado - [labor] paid wage(s)
— ₥ para calificación f - [labor] rated wage
— ₥ personal m - [labor] personnel salaries
— ₥ total - [labor] total wage(s) • payroll
saldo m - . . . ; remainder; rest • [contab] due from • due to
— acumulado - accumulated balance
— ₥ con - [contab] due, to, o from
— ₥ de balance m comercial - [econ] trade balance
— ₥ — costado m - [constr] side remainder

saldo m de patrimonio m social (de accionistas) - [fin] shareholders' equity balance
— ₥ — planta f - [ind] remainder of @ plant
— ₥ — prima(s) f - [seguros] premium(s) balance
— ₥ — — f estimada(s) incobrable(s) - [seguros] premiums balance(s) deemed uncollectible
— f — — f incobrable(s) - [seguros] uncollectible premium(s) balance
— ₥ — reaseguro(s) m - [seguros] reinsurance balance
— ₥ — — m cedido(s) - [seguros] ceded reinsurance balance
— ₥ — — m — por cobrar - [seguros] ceded reinsurance balance(s) receivable
— ₥ — — m — — pagar - [seguros] ceded reinsurance balance(s) payable
— ₥ — relleno m - [constr] (bacl)fill, balance, o remainder
— ₥ — — m lateral - [constr] side fill, balance, o remainder
— ₥ — reserva(s) f - reserve(s) balance
— ₥ — resultado(s) m - [fin] result(s) balance
— ₥ — — m acumulado(s) m [fin] accumulated result(s) balance
— ₥ — temporada f - sason('s), balance, o remainder
— ₥ — tubería f - [constr] pipe,ping balance
— ₥ disponible m - available balance
— ₥ en caja f - [fin] cash, balance, o in hand
— ₥ fin de año m - [contab] year-end balance
— ₥ — fondo m - [fin] fund('s) balance
— ₥ entre empresa(s) f - [fin] intercompany balance
— ₥ — — f afiliada(s) - [fin] (affiliated) intercompany balance
— ₥ estimado - [fin] deemed balance
— ₥ — incobrable - [fin] balance deemed uncollectible
— ₥ impagado - [com] unpaid balance
— ₥ incobrable - [fin] uncollectible balance
— ₥ — de prima f - [seguros] uncollectible premium(s) balance
— ₥ insoluto - [fin] unpaid balance
— ₥ no utilizado - [fin] unused balance
— ₥ pagado - [com] paid balance
— ₥ por cobrar - [contab] balance receivable
— ₥ — — v de agente m - [seguros] agent('s) balance receivable
— ₥ — — v por reaseguro(s) m cedido(s) - [seguros] ceded reinsurance balance receivable
— ₥ — — v por seguro(s) m cedido(s) - [seguros] ceded insurance balance receivable
— ₥ — giro m a vista f - [com] balance (in) sight draft
— ₥ — — m — — f con documento(s) m adjunto(s) - [com] balance (in) sight draft documents attached
— ₥ — pagar - [contab] balance payable
— ₥ — — por reaseguro(s) m cedido(s) m - [seguros] ceded reinsurance balance payable
— ₥ — — — seguro(s) m - [seguros] ceded insurance balance payable
— ₥ — — — — m cedido(s) - [seguros] ceded insurance balance payable
— ₥ — utilizar - [fin] unused balance
— ₥ significativo - [com] significant balance
— ₥ — entre empresa(s) f - [com] significant intercompany balance
— ₥ sin utilizar v - [fin] unused balance
— ₥ sobre prima(s) f - [seguros] premium(s) balance
saledizo,za - [arquit] . . .; offset
salida f - . . .; outfall; ourflow payout • output • departure • going, o getting, out • [fam] loophole • [mec] discharge (end, o side); delivery end • comput] output • [hidr] flow • [tub] take off • check(ing) out • emergence • extricating,tion • [mec] run-out
— f acampanada f - [mec] flared outlet
— f achaflanada - [constr] tapered outlet

salida f achaflanada para alcantarilla f - [tub]
culvert tapered outlet
— f **aislada** - [comput] insulated output
— f — **ópticamente** - [comput] optically insulated output
— f **analógica** - [comput] analogue,gic output
— f **audible** - [comput] audio,dible output
— f — **transmitida** - [comput] transmitted audio
output
— f **aumentada** - boosted output
— f **auxiliar** - auxiliary outlet
— f **básica** - [comput] basic output
— f **brida(da)** - [mec] flange(d) outlet
— f **cambiada** - [electrón] changed output
— f **común** - [electrón] common output
— f **con control** m—[comput] control(led) output
— f — — m **bi-estable** - [comput] bi-stable
control(led) output
— f — **manguera** f - [mec] hose outlet
— f **con tono** m **manipulado** - [electrón] tone
keyed output
— f **continua** - [comput] continuous output
— f — — **de tono** m - [comput] continuous tone, o
tone continuous, output
— f — **fijable** - [comput] strappable continuous
output
— f — — **de tono** m - [comput] continuous
strappable tone output
— f — **mantenible** - [comput] strappable continuous, o continuous strappable, output
— f — — **de tono** m - [electrón] continuous
strappable tone output
— f **cronometrada** - [comput] timed output
— f **de acequia** f - [hidr] channel outlet
— f — **acería** f - [metal-prod] steel plant outlet
— t — **acumulador** m - [electr] battery output
— f — **agua** m - [tub] water, exfiltration, o
leak(ing)
— f — — m **de cilindro(s)** m - [comb.int] cylinder water outlet
— f — **alcantarilla** f - [constr] culvert, outlet, o exit
— f — **boquilla** f - [mec] nozzle outlet
— f — **cable** m - [grúas] cable pay-out
— f — **caja** f - [metal-lam] stand, exit, o delivery
— f — — f **reductora** - [mec] reduction stand
output • output shaft
— f — **camino** - [vial] road exit • [autom]
going off @ road
— f — **canal** m - [hidr] channel exit
— f — **carrera** f - [deport] parking
— f — **cauce** m - [hidr] channel exit
— f — **ciclón** m - [coque] cyclone (separator)
exit (end)
— f — **cilindro** m - [comb.int] cylinder outlet
— f — **circuito** m - [electrón] circuit output
— f — **colector** m - [metal-prod] dust catcher,
discharge, o exit
— f — — m **para polvillo** m - [metal-prod] dust
catcher, chute, o discharge
— f — **convertidor** m - [electrón] converter
output
— f — — **amplificada**—amplified converter output
— f — **cuba** f - [metal-trat] vat exit, o outlet
— f — **depósito** m - [ind] (receiver) tank outlet
— f — **desagüe** m - [hidr] drain outlet
— f — — m **para planta** f **para oxígeno** m -
[ind] oxygen plant drain outlet
— f — — m **para refrigeración** f - [ind] cooling drain outlet
— f — — m — — f **para horno** m - [ind] furnace cooling drain outlet
— f — **detector** m **para umbral** m - [electrón]
threshold detector output
— f — **enderezadora** f - [mec] straightener
delivery
— f — **energía** f - [sold] power output
— f — — f **de fuente** f - [sold] power source
output
— f — **estructura** f - [hidr] structure outlet

salida f **de filtro** m - [mec] filter output
— f — — m **para paso** m **de banda** f - [electrón]
bandpass filter output
— F — — m **pasabanda** - [electrón] bandpass
filter output
— f — **flujo** m - [papel] flow exit
— f — **fotodisyuntor** - [electrón] photo-insu-
lator output
— f — **fuente** f - [electr] source output
— f — — f **para energía** f - [electr] power
source output
— f — **fuerza** f - [mec] (power) output
— f — **gas** m - [coque] gas outlet
— f — **granulometría** f - [comput] particle
size output
— f — **horno** m - furnace outlet
— f — — m **en acería** f - [metal-prod] steel
plant furnace outlet
— f — **laminación** f **en caliente** - [metal-prod]
hot strip mill outlet
— f — **laminador** m - [metal-lam] mill delivery
(end)
— f — **lavadora** f - [ind] washer outlet
— f — **material(es)** m - [contab] outgoing ma-
terial(s)
— f — **oscilador** m - [electrón] oscillator
output
— f — — m **para espacio(s)** m - [electrón]
space oscillator output
— f — — m — **marca(s)** f - [electrón] mark
oscillator output
— f — — m **seleccionada** - [electrón] selector
oscillator output
— f — **palanquilla(s)** f - [metal-prod] billet
delivery
— f — **pella(s)** f - [miner] pellet(s) discharge
— f — **petaca** f - [metal-prod] cooling plate
outflow
— f — **pista** f - [deport] spin(ning) out
— f — **planchón** f - [metal-lam] slab exit
— f — **planta** f - [ind] plant outlet
— f — — f **para oxígeno** m - oxygen plant outlet
— f — **refrigeración** f - [ind] cooling drain
outlet
— f — — f **para horno** m - [ind] furnace cooling drain outlet
— f — **secador** m - [ind] drier outlet
— f — — m **con aire** m - [mec] air drier outlet
— f — **señal** f - [comput] signal output
— f — — f **a red** f **telefónica** - [comput] signal output to @ telephone network
— f — **separador** m **ciclónico** - [coque] cyclone
separator exit
— f — **tablero** m - [electrón] board output
— f — — m **para manipulador** m - [electrón]
keyer board output
— m — — m — — m **para tono(s)** m - [electrón]
tone keyer board output
— f — **teclado** m - [electrón] keyer board output
— f — — m **para manipulador** n - [electrón]
keyer board output
— f — — m — — m **para tono(s)** m - [electrón]
tone keyer board output
— f — **tono** m - [electrón] tone output
— f — — m **audible** - [electrón] audio,dible
tone output
— f — — m — **equilibrada** - [electrón] balanced audio tone output
— f — — m **continua** - [comput] continuous
tone output
— f — — m **cronometrado** - [comput] timed tone
output
— f — — m **de codificador** m - [comput] encoder
tone output
— f — — m **continua** - [comput] continuous
tone output
— f — — m **equilibrada** - [electrón] balanced
tone output
— f — — m **paralela** - [comput] sidetone output
— f — — m **fijable** - [comput] strappable tone
output

salida f de tono m lateral - [comput] side tone output
— f —— m mantenible - [comput] strappable tone output
— f —— m sincronizada - [comput] timed tone output
— f —— m fijada - [comput] strappable timed tone output
— f —— m mantenido - [comput] strappable timed tone output
— f — tubería f - [tub] pipe,ping outlet
— f —— f para gas m - [combust] gas pipe outlet
— f —— f —— m limpio m - [metal-prod] clean gas pipe outlet
— f —— f vertical - [tub] riser outlet
— f —— m para fundente m - [sold] flux tube outlet
— f — tuerca f - [mec] bolt removal,ving
— f — válvula f - [comb.int] valve exhaust
— f —— f piloto - [mec] pilot valve outlet
— v —— solenoide - [mec] solenoid pilot valve outlet
— f —— f solenoide - [válv] solenoid valve outlet
— f — ventilador m - fan, exhaust, o outlet
— f digital - [comput] digital output
— f directa - [hidr] direct, exit, o outlet
— f en decibelios m - [electrón] decibel output
— f — manipulador m - [electrón] keyer output
— f — nivel m lógico - [electrón] logic level output
— f — relé m - [comput] relay output
— f —— m auxiliar - [comput] auxiliary relay output
— f — voladizo m - [constr] cantilever(ed) outlet
— f equilibrada - [electrón] balanced output
— f — con tono(s) m - [electrón] balanced tone output
— f —— m audible(s) - [electrón] balanced audio tone output
— f espúrea - [electrón] spurious output
— f fijable - [comput] strappable output
— f impresa - printed output • [comput] print-out
— f — adaptada - [comput] tailored printout
— f — adecuada - [comput] tailored printout
— f — en español m - [comput] Spanish printout
— f —— inglés m - [comput] English printout
— f — simple - [comput] plain printout
— f — en idioma m . . . - [comput] . . . language printout
— f —— m español - [comput] Spanish (language) printout
— f —— m inglés - [comput] English (language) printout
— f — para alarma f - [comput] alarm printout
— f — para condición f - [comput] status printout
— f —— estado m - [comput] status printout
— f indeseable - [electrón] spurious, o undesirable, output
— f inferior - [mec] bottom outlet
— f ingleteada - [constr] mitered outlet
— f — no sumergida - [constr] non-submerged mitered outlet
— f — sumergida - [constr] submerged mitered outlet
— f libre - [hidr] free outlet
— f — no sumergida - [hidr] non-submerged free outlet
— f manipulada - [electrón] keyed output
— f mantenible - [comput] strappable output
— f momentánea f - [comput] momentary output
— f —— de conmutador m - [comput] momentary switch, o switch momentary, output
— f no sumergida - [hidr] non-submerged outlet
— f —— para alcantarilla f - [constr] non--submerged culvert outlet
— f nominal - nominal, o rated, output
— f para agua m - water out(let)
— f — aire m - [mec] air outlet
— f — cámara f espiral - [turb] scroll case outlet

salida f para emergencia f - [segurid] escape output
— f — espacio(s) m - [electrón] space output
— f — humedad f - [ind] moisture discharge
— f —— f condensada - [ind] condensed moisture (discharge) outlet (connection)
— f — luz f - [electrón] light output
— f —— f exterior - [electrón] external light output
— f — manguera f - [mec] hose outlet
— f —— f para depósito m - [mec] tank hose outlet
— f — marca(s) f - [electrón] mark(s) output
— f — nivel m - [electrón] level output
— f —— m lógico - [electrón] logic level output
— f —— m — cambiado - [electrón| changed logic level output
— f — rociador m - [mec] sprinkler outlet
— f — salida f - [electrón] output connector
— f —— f para conectador m - [electrón] signal output connector
— f —— f toma f - [electrón] signal output connector
— f — soldadora f - [sold] welder output
— f — tono m - [comput] tone output
— f — turbina f - [turb] turbine outlet
— f programable - [comput] programmable output
— f — en sitio m - [comput] field programmable output
— f programada - [comput] programmed output
— f protegida - [hidr] protected outlet
— f regulada - [electr-prod] regulated, o controlled, output
— f requerida - [electrón] required output
— f seleccionada - [electrón] selected output
— f — para luz f - [electrón] light selected output
— f —— f externa - [electrón] external light selected output
— f —— sonar bocina f - [electrón] horn blow(ing) selected output
— f similar - [electrón] similar output
— f sincronizada - [comput] timed output
— f — fijable - [comput] strappable timed output
— f — mantenible - [comput] strappable timed output
— f sobresaliente - [constr] projecting outlet
— f — no sumergida - [constr] unsubmerged projecting outlet
— f — sumergida - [constr] submerged projecting outlet
— f sumergida - [constr] submerged outlet
— f — para alcantarilla f - [constr] submerged culvert outlet
— f superior - top outlet
— f tope - top outlet
— f transformada - [electrón] transformed output
salidero× m - véase fuga f; pérdida f; escape m
salidizo m - [arquit] offset
salido,da a - emerged • extricated • gotten, o gone, o checked, out
— a de camino m - [autom] gone off @ road
— a — carrera f - [ceport] parked; eliminated
— a — pista f - [deport] spun out
saliente m - . . . .; salient • extension; shoulder; portrusion; offset • boss; jut • véase también horquilla f - [constr] land | @ - protruding • outbound • [arquit] offset
— f —— entrada f - [mec] véase horquilla f de entrada
— f — ladrillo(s) m - [constr] protruding brick(s)
— f —— f de codo m portavientos - [metal-prod] protruding gooseneck bricks
— f — soldadura f - [sold] weld protrusion
— f en carrete n - [mec] reel tang
— f — interior m - [sold] interior protrusion
— f — leva f - [mec] cam lobe
— f interior - [mec] inner portrusion
— f — de soldadura f - inner weld protrusion
salimiento m - extricating,tion

salir v - . . .; to exit; to come out; to emerge
— v brincando - to jump out
— v a fosa f - [deport] to retire to @ pit
— v coque m por piquera f - [metal-prod] coke flow(ing) from @ iron notch
— v de - to go out • to check out
— v — baño m - [sold] to move out of @ puddle
— v — cabina f - [autom] to get out of @ cab
— v — encaje m - [metal-prod] to not fit in @ place
— v — entre junta(s) f - [constr] to come out of @ joint
— f — fábrica f - [ind] to leave @, factory, o plant
— v — fosa f - [deport] to leave @ pit
salirse v - to come, off, o out; to leak; to, get, o go, o slip, out • to extricate oneslef
— v de camino m - [vial] to, run, o go, off @, course, o road
— v — carrera f - [deport] to park
— v — cráter m - [sold] to leave @ crater
— v — pista f - [deport] to, run, o go, off @ course; to spin out • to retire to @ pit
— v — vía f - [f.c.] to leave @ track
— v directamente - [sold] to come straight out
— v en forma f de burbuja(s) f - to bubble out
— v hacia afuera - to pull out
— v hierro m por escorial - [metal-prod] to flow iron through @ slag notch
— v líquido m - [metal-prod] to drain out
— v nuevamente - to get back out
— v por tangente f - to digress
— v tobera f de (su) encaje m - [mec] to come out of its hole
salón f de automóvil m - [autom] car show
— m para belleza f - beauty, parlor, o shop
— m — conferencia(s) f - lecture hall
— m — entrevista(s) f - conference room
— m — estar v - [constr] lounge • living room
— m — exhibición f - [com] showroom
— m — junta(s) f - conference, o meeting, room
— m — muestra(s) f - [com] showroom
— m — promoción f de salud f - [medic] health club
— f — reunión(es) f - [constr] meeting, o conference, room
— m — venta(s) f - [com] sales room; showroom
— m — — f abarrotado - [com] cluttered sales-room
— m — — f bien iluminado - [com] well-lit, o well lighted, showroom
— m — — f ordenado - [com] orderly showroom
— m social - lounge
salpicadero m - [autom-mec] splashboard; splash-guard; dashboard; splasher
salpicado,da a - splattered; splashed
— a excesivamente - spattered excessively
salpicadura f - . . .; splashing
— f adherida - [sold] stuck spatter
— f caliente - [sold] hot spatter(ing)
— f de aceite m - [mec] oil, splash(ing), o spatter(ing)
— f — camino m - road splash(ing)
— f — líquido m - liquid(s) splashing
— f — metal m - [sold] metal spatter(ing)
— f — — m fundido - [sold] weld spatter(ing)
— f — rebote m - back-spring(ing)
— f — — m atrapada - [mec] trapped back--spring(ing)
— f — — m desde émbolo m -[mec] piston back-spring(ing
— f evitada - avoided, spatter, o splash(ing)
— f excesiva - [sold] excess(ive) spatter
— f fina - [sold] fine spatter(ing)
— f — minimizada - [sold] minimized fine spatter
— f insignificanate - [sold] negligible spatter
— f mayor - [sold] more, o greater, spattering
— f reducida - [sold] small, o reduced, spatter
salpicar v aceite m - [lubric] to spatter @ oil
— v de rebote m - to back-spring
— v en demasía f—[sold] to spatter excessively
— v excesivamente - [sold] to spatter excessi-vely

salpicar v líquido m - to splash @ liquid
salpicón m - . . .; splash(ing)
salpique n - spatter(ing)
— m y goteo m - [mec] spatter and drip
salsa f blanca f - [culin[(white) gravy
saltado,da a - jumped; vaulted; gone over • [mec] popped off • [electr-oper] blown; trip-ped; popped out • [comb.int] blown
— a hacia atrás - [mec] snapped back
saltadura f - [metal] . . .; spalling
saltar v - . . .; to vault; to go over • to bounce • to skip; to by-pass • [sold] to pop out • [mec] to, trip, o pop off • [electr-oper] to pop out
— v afuera - [mec] to pop out
— v arco m - [sold] to strike @ arc
— v circuito m - [electr-oper] to pop out @ circuit
— v dentro de - to jump into
— v diente m - [mec] to, jump, o skip, @ tooth
— v disyuntor m - [electr-oper] to trip @ (circuit) braker
— v — m para circuito m - [electr-oper] to trip @ ciurcuit breaker
— v en escama(s) f - [mec] to flake
— v fusible m - [electr-oper] to blow @ fuse
— v hacia afuera - [mec] to pop out
— v — atrás v - [mec] to snap back
— v muro m - to, jump, o go over, @ wall
— v velocidad f - [autom-mec] to skip @ shift
salteado,da a - every other
salto m - . . .; vault(ing); going over • [sold] skip(ping) • [electr-oper] surg(ing); popping out • [mec] popping off • [hidr] fall(s); cataract • drop • head
— m de agua m - [hidr] (water) falls
— m — arco m - [sold] véase encendido m de arco m
— m — circuito m - [electr-oper] circuit popping out
— m — cubierta f - [mec] cover blow(ing) off
— m — diente m - [mec] tooth. jumping, o skipping
— m estático - [hidr] static head
— m — máximo - [hidr] maximum static head
— m — mínimo - [hidr] minimum static head
— m hacia atrás - [mec] snapping back
— m hidráulico - [hidr] hydraulic fall(s)
— m máximo - maximum. jump, o leap • [hidr] maximum head
— m mínimo - [hidr] minimum, jump, o leap • minimum head
— m neto - [turb] net head
— m nominal - [hidr] nominal head
salubridad f - . . . • sanitation | a - health; medical
salud f en hogar m - [medic] home health
— f mundial - [medic] world health
saludo m presentado - presented greeting
salvado,da a - . . . . • [public] clarified; cor-rected
salvaguardado,da a - safeguarded • [legal] held, o saved, harmless
salvaguardar v - . . .; to, hold, o save, harm-less
salvaguardia f - . . . . • [legal] to, hold, o save, harmless
salvamento m - . . . . • [seguros] salvageability • salvage,ging
salvar v - . . . . • to, bridge, o span • [legal] to, amend, o correct, o clarify
— v cambio m - [mec] to safe @ shift
— v — m sincrónico - [mec] to save @, shift, o synchromesh
— v de cualquier pérdida - [legal] to save harmless
— v inconveniente - to solve @ problem
— v separación f - [constr] to bridge @, gap, o opening
— v — f grande - [constr] to bridge @ wide, gap, o opening
— v error m - [legal] to correct @ error

salvedad f - . . .; proviso; clarifying,fication
— f hecha adv - excepting
salvo adv - . . .; unless; except
— adv especificación f contraria - unless
  otherwise speciwied
— adv indicación f contraria - unless otherwise, specified, o indicated
— adv venta f previa - subject to prior sale
salvo,va a - . . .; exclusive
sanción f - . . .; penalty
— f adicional - additional, sanction, o penalty
— f pecuniaria - [legal] monetary penalty
— f por conducción f - [deport] driving, o nav-
  igational, penalty
sancionado,da a - sanctioned; penalized
sancionador,ra a - sanctioning; penalizing
sancionamiento m - sanctioning; penalizing
sancionar v - . . .; to penalize • [legal] to,
  approve, o enact
sandwich n - [culin] sandwich
saneado,da a - . . . • [ind] finished
saneamiento m - . . . • [hidr] reclamation •
  [legal] adjustment; reparation(s) • [constr]
  . . .; draining • [medic] sanitizing
— m ambiental - [segurid] environmental sanitation,tizing
— m de coraza f - [metal-prod] shell repair(ing)
— m — raíz f - [sold] root cleaning
sanear v - . . . • [ind] to repair • [legal] to
  adjust • [metal-prod] to chip
— v boquete m - [ind] to repair @ hole
— v raíz f - [sold] to clean @ root
— v coraza f - [metal-prod] to cut out @ bad
  part(s) of @ shell
saneo* m - véase saneamiento m
sangrador m - [ind] bleeder • [metal-prod] teemer
— m para horno m alto - [metal-prod] blast furnace bleeder
sangrar v - . . . • [metal-prod] to, tap, o
  pour, o cast, @ furnace; to teem • [medic]
  to bleed • [imprent] to bleed
— v aire m - [ind] to bleed @ air
— v bien - [metal-prod] to bleed well
— v bien horno m - [metal-prod] to cast well @
  furnace
— v con oxígeno m - [metal-prod] to tap with
  oxytgen
— v filtro m - [comb.int] to bleed @ filter
— v horno m - [metal-prod] to tap @ furnace
— v humedad f - [mec] to bleed @ moisture
— v red f - [ind] to bleed @ system
— v — f para combustible m - [comb.int] to
  bleed @ fuel system
sangre f - . . . | a - [transp] horse drawn
— f normal - [medic] normal blood
sangría f - . . . • [metal-prod] casting; tapping; teeming • [imprent] indentation • [med]
  bleeding
— f de aire m - [ind] air bleeding
— f — energía f - energy drain(ing)
— f — escoria f - [metal-prod] slagging; slag
  tapping • slag, notch, o hole
— f — filtro m - [comb.int] filter bleeding
— f — — para combustible m - [comb.int]
  fuel filter bleeding
— f — humedad f - [mec] moisture bleeding
— f — presión f - [ind] pressure bleeding off
— f — red f - [ind] system bleeding
— f — — f para combustibleem - [ind] fuel system bleeding
— f desde exterior m - foreign drain
— f en lingote(s) m - [metal-prod] ingot bleeding
— f mala - [metal-prod] poor, bleeding, o casting
— f rota - [metal-prod] iron thorough iron notch
  flow(ing) without tapping
sanitarista m - [medic] sanitarian
sanseacabó adv - that's, it, o all; it's that
  simple
sapo m - [zool] . . . • [f.c.] frog
saponificado,da a - saponified
sarta f - [mec] string • [petról] string; stem

sarta f para perforación f - [petról] drill
  string
— f — — f rotativa - [petról] rotary drilling line
— f perforadora - [petról] drill(ing) system
— f rotativa - [petról] rotary line
— f vástago - [petról] drill(ing) stem
sartén f - [domést] . . .; skillet; frypan
— f eléctrica - [domést] electric, fry(ing),
  pan, o skillet
satén m - [textil] . . .; satin
satinado,da a - [textil] sateening • [metal-
  -lam] temper rolling | a - glossy; satin
satinador m - [papel] calender
satinador,ra a - [papel] sateening • calendering
satinar v - [textil] to sateen • [metal-lam] to
  temper roll
satisfacción f - . . .; satisfying; gratifying;
  gratification • meeting; providing • accomodating,tion • apology • success; performance
— f ajena a trabajo m - [labor] off-@-job satisfaction
— f cabal - complete satisfaction • clear-cut
  performance
— f — anhelo(s) m - want(s) satisfaction
— f — — m de cliente m - customer('s) want(s)
  satisfaction
— f — cliente m - customer('s) satisfaction
— f — especificación f - specification, meeting, o satisfaction
— f — — f radiográfica - [sold] X-ray, o radiographic, specification, satisfaction, o
  meeting
— f — exigencia f - requirement, o demand,
  satisfaction, o meeting; need supplying
— f — — f de carga f - [mec] load requirement
  satisfaction
— f — — f — exportador m - exporter('s)
  need(s), satisfying, o supplying
— f — necesidad(es) f - need(s) supplying
— f — — — cliente m - customer('s) need,
  satisfying, o meeting
— f — — f — exportador m - [com] exporters
  need(s), satisfying, o supplying
— f — requerimiento(s) m - requirement(s),
  meeting. o satisfying, o supplying
— f — — m — exportador m - [com] exporters
  need(s), satisfying, o supplying
— f — soldador m - [sold] welder, o operator,
  appeal
— f derivada - derived satisfaction
— f dudosa - doubtul, o questionable, satisfaction
— f obtenida - derived, o obtained, satisfaction
— f plena - full satisfaction
— f de soldador(es) m - [sold] full, o excellent, operator, o welder, appeal
satisfacer v - . . . • to, meet, o fulfill • to,
  provide, o supply • to, accomodate, o suit, o
  conform • to perform
— v anhelo(s) m - to meet @, want(s), o need(s)
— v — m de cliente m - to satisfy @ custom-
  er(s), want(s), o need(s)
— v cliente m - to satisfy @ customer
— v condición f - to meet @ requirement
— v demanda f - to meet @ demand
— v especificación f - to meet @ specification
— v — f radiográfica - [electrón] to meet @,
  radiographic, o X-ray. specification
— v establecido m - to conform
— v exigencia f - to, meet, o supply, o satisfy, @ demand(s), o requirement(s), o need(s)
— f — f de carga f - [mec] to, meet, o satisfy, @ load('s) requirement(s)
— v — f de propiedad(es) f - to meet @ property,ties requirement(s)
— v — f mínima - to meet @ minimum requirement
— v — — de propiedad(es) f - to meet @ minimum property,ties requirement
— f necesidad f - to supply @ need
— v — f de cliente m - to satisfy @ customer's
  need(s)

satisfacer v norma f - to meet @ standard
— v — f para aceptabilidad f - [ind] to meet @
acceptability standard
— v — f rígida - to meet @ rigid standard
— v — f — para aceptabilidad f - [ind] to
meet @ rigid acceptability standard
satisfaciente a - . . .; satisfaction providing
satisfactorio,ria a - . . .; satisfying; agreeable; suitable; adequate; enjoyable; healthy;
practical; succesful; O.K.
satisfecho,cha • accomodated • provided; met;
supplied
— a generalmente - generally, o usually, satisfied
saturación f - . . .; saturating • soaking
— f de cimiento(s) m - [constr] foundation saturating,tion
— f — elemento m - element saturating,tion
— f — fundación f - [constr] foundation saturating,tion
— f — material - material saturating,tion
— f — m para relleno m - [constr] fill (material) saturating,tion
— f — relleno m - [constr] (back)fill saturating,tion
— f — m reducido - [constr] reduced, o low,
backfill saturating,tion
— f — subbase* f - [hidr] subbase saturation
— f — subrasante* f - [hidr] subgrade saturating,tion
— f — suelo m - [suelos] soil, o ground, saturating,tion
— f en caliente - hot saturating,tion
— f — frío - cold saturating,tion
— f posterior - subsquent, o later, saturating,tion
— f reducida - reduced, o low, saturating,tion
saturado,da a en caliente - hot saturated
— a — frío - cold saturated
saturador m con inyección -- [metal-prod] injection saturator
— m — — f de agua m - [metal-prod] water injection saturator
— m — — f — — m salada - [metal-prod] salt
water injection saturator
— m para planta f - [ind] plant saturator
— m — — f para sulfato m (de amonio f) -
[coque] ammonium sulfate plant saturator
saturador,ra a - . . .; saturating
saturante a - saturating; saturator
saturar v - . . . • to soak
— v cimiento(s) m - [constr] to saturate @
foundation
— v completamente - to saturate, completely, o
thoroughly
— v elemento - to saturate @ element
— v en caliente - to hot saturate
— v — frío - to cold saturate
— v fundación f - [constr] to saturate @ foundation
— v relleno m - [constr] to saturate @ backfill
— v subbase* - [hidr] to saturate @ subbase
— v subrasante f—[hidr] to saturate @ subgrade
— v suelo m - [suelos] to saturate @, soil, o
ground
sauna f - [medic] sauna
sauzal m denso - [botán] dense willow grove
Saybolt m universal en segundo(s) m - [lubric]
Saybolt universal second(s) • S U S
se v acuerda - [legal] it is resolved
— v provee - it is available
— v recomienda - it is recommended
— v — para - recommended for
— v requiere - (it is) required
— v — para todo(s) equipo(s) m - (it is) required for all, machine(s), o equipment
— v resolvió - [legal] it was resolved
sea v como fuere - at any rate; whatever
secado m - drying; véase también secamiento m
— m acabado - thorough drying
— m de armadura f - [electr-mot] armature, baking, o drying
— m — artesa f - [col.cont] tundish drying

secado m de artesa f para colada - [col.cont]
tundish drying
— m — banda f - [metal-lam] strip drying
— m — baritina f - [miner] barytine drying
— m — barra f - [metal-lam] rod drying
— m — — f de acero m - steel rod drying
— m — — f para taponamiento m - [metal-prod]
stopper rod drying
— m — canal(es) m - [metal-prod] runner(s)
drying
— m — cilindro(s) m - [comb.int] cylinder(s),
drying, o wiping
— m — cuchara f - [metal-prod] ladle drying
— m — electrodo m - [sold] electrode drying
— m — horno m - [metal-prod] furnace drying
— m — inducido m - [electr-mot] armature drying
— m — muestra(s) f - sample(s) drying
— m — plancha(s) f - [constr] plate(s) drying
— m — tubería f - [tub] pipe,ping drying
— m — tubo m - [tub] tube, o pipe, drying
— m en aire m - [ind] air drying
— m — galpón m - [agric] barn drying
— m — granja f - [agric] farm, o barn, drying
— m — vacío m - [ind] vacuum drying
secado,da a - dried (off)
— a acabadamente - dried thoroughly
— a con aire m - air, dry, o dried
— a en aire m - air, dry, o dried
— a — vacío - [ind] vacuum dried
secador m - drier
— m centrífugo - [ind] centrifugal drier
— m desconectado - [ind] disconnected, o
turned off, drier
— m enjuagado - [ind] flushed, o rinsed, drier
— m — en sentido m inverso - [ind] flushed re-
versely drier
— m giratorio - [ind] rotary drier
— m operado - [ind] operated drier
— m — continuamente - [ind] continuosly operated drier
— m para cabello m - [domést] hair (blower)
drier
— m — sal f - [ind] salt drier
— m — sulfato m de amonio m - [coque] ammonium
sulfate drier
— m por aire m - [mec] air drier
— m — — comprimido - [mec] compressed air
drier
— m — — m — enfriado por agua m - [ind]
water cooled compressed air drier
— m — — m — refrigerado - [mec] refrigerated
compressed air drier
— m — — refrigerado - [ind] refrigerated air
drier
— m preparado - [ind] prepared drier
— m rotativo - [ind] rotary drier
— m sobrecargado - [int] overloaded drier
— m vibratorio - [ind] vibrating,tory drier
— m — para sulfato m de amonio - [coque] vibrating,tory ammonia sulfate drier
secamiento m - . . .; drying off; véase también
secado m • drying time
— m con aire - [ind] air drying
— m de alambre m - wire drying
— m — barra f - [metal-lam] rod drying
— m — electrodo m - [sold] electrode drying
— m — tubería f - [tub] pipe,ping drying
— m — tubo m - [tub] tube, o pipe, drying
— m — zanja f - [constr] trench drying
— m en aire - [ind] air drying
— m — galpón m - [agric] barn drying
— m — granja f - farm, o barn, drying
— m — vacío m - [ind] vacuum drying
secar v acabadamente - to dry thoroughly
— v alambre m - to dry @ wire
— v cilindro m - [comb.int] to, dry, o wipe, @
cylinder
— v con aire m - to air dry
— v electrodo m - [Sold] to dry @ electrode
— v en vacío m - to vacuum dry
— v muestra f - [ind] to dry @ sample
— v nuevamente - to, redry, o dry again

secar v plancha f - [constr] to dry @ plate
— v tubería f - [tub] to dry @ pipe,ping
— v tubo m - [tub] to dry @, tube, o pipe
— v zanja f - [constr] to dry (up) @, trench, o
ditch
secarse v - to dry oneself • to dry off
sección f -. . . • [mec] length; shape • sector
• stage - area - [dib] (cross) section •
[hidr] waterway area • [petról-destil] leg •
[constr] stretch; span; (clear) opening; end
area • project • [electr-cond] size • line •
[constr] plank • [admin] component • [labor]
unit • [instrum] range • [legal] section •
[com] office • [ind] line • vease también
corte m
sección f abovedada - [tub] arch section
— f acampanada - [constr] flared section
— f acodada - [tub] elbow section
— f adecuada - adequate stction • [hidr] adequate waterway area • [electr-cond] proper
size
— f adelgazada - [trefil] (pre)pointed section
— f adyacente - adjacent section
— f agrietada - cracked section
— f ahusada - tapered section
— f — con peso m reducido - [grúas] light
weight tapered section
— f — — m con . . . polea(s) f—[grúas]
. . . sheave light weight tapered section
— f — — m . . . — f para punta f -
[grúas] . . . sheave light weight tapered tip
section
— f — para punta f - [grúas] tapered tip section
— f aislada - isolated section • [electr-cond]
insulated section
— f — de conductor m - [electr-cond] lead insulated section
— f aluvional - [hidr] alluvium area
— f anclada - [constr] anchored section
— f anterior - front section
— f — de conmutador m - [electr-instal] switch
front section; front switch section
— f arbolada f - [topogr] wooded section
— f armada - [constr] built-up, o assembled, o
erected, section
— f básica - basic section
— f blanda - soft, section, o spot
— f boscosa - [topogr] forestry section
— f carcomida - [metal] rusted out section
— f central f termoeléctrica - [electr-prod]
thermoelectric power plant section
— f circular - circular, o round, section
— f combustión f - [combust] combustion section
— f completa - complete, section, o portion
— f completada - completed, section, o portion
— f compuesta—compound(ed), o built-up section
— f — normal - [metal-lam] standard compound
section
— f con aislación f - [electr-cond] insulated
section
— f con ancho,chura de . . . - [mec] . . . wide
section
— f con cal f - [destil] lime leg (distillation
column)
— f — corte(s) m y terraplén(es) f - [vial]
cut-and -fill section
— f con fotografías f - [public] photographic
section
— f — guía f - [public] quide section
— f — — f para substitución f - substitution
guide,ding section
— f — — f — f de neumático(s) m - [autom-
-neumát] tire substitution guide section
— f — largo m doble - double length section
— f — — variable - [tub] random length(s)
— v nivel alto - [vial] high level span
— f — peso m reducido - light weight section
— f — — n — con . . . polea(s) f - [grúas]
. . . sheave light weight section
— f — — m — — — f para punta f -
[grúas] . . . sheave light weight tip section
— f — — m — para punta f - [grúas] light
weight tip section

sección f con . . . polea(s) f - [grúas] . . .-
sheave section
— f — . . . — f para extremo m - [grúas]
. . . sheave tip section
— f — . . . — f — — m para cucharón m -
[grúas] . . . sheave container tip section
— f — . . . — f — punta f - [grúas] . . .
sheave tip section
— f — . . . — f — servicio m pesado - •
[grúas] . . . sheave heavy duty crane tip
section
— f con riel m - [grúas] rail(ed) section
— f — — m colector - [grúas] collector rail
section
— f con riel m colector principal - [grúas]
main collector rail section
— f — m — — aislado - [grúas] insulated
main collector rail section • isolated main
collector rail section
— f cónica - [mec] conical section
— f — para transición f - [cald] conical
transition section
— f cóncava f - [mec] concave section
— f — dentada - [mec] toothed concave section
— f conformada - [mec] formed, o shaped, section
— f — en caliente - [constr] hot shaped section
— f — — frío - [constr] cold formed section
— f cónica - [mec] conical, o tapered, section
— f constante - [mec] constant section
— f contigua - contiguous, o adjacent, o adjoining, o abutting, sectión
— f — de tubería f - [tub] adjacent, o adjoining, o contiguous, o abutting, section
— f — — — f con peso m mayor - [ambient]
heavier,viest adjacent pipe section
— f continua - continuous, section, o length
— f — de cordón m - [cabl] continuous strand
section
— f — — ramal m - [cabl] continuous strand
section
— f corriente - [vial] standard length section
— f corta f - short, section, o length
— f cortada - cut section
— f — para camino(s) m - [constr] road cut
section
— f — — canal m - [constr] channel cut section
— f — — carretera f - [constr] highway cut
section
— f crítica - critical, section, o portion
— f — de terraplén m - [constr] embankment
critical section
— f cuadrada - square section • [metal-lam]
square
— f cualquiera - any, o random, section
— f curva(da) - curve(d) section
— f — en fábrica f - [metal-fabr] precurved
section
— f de abajo - [mec] bottom section
— f — abertura f - [constr] cross sectional
area
— f — acero m - [mec] steel section
— f — — fabricada en taller m - [tub] shop
fabricated steel section
— f — aguilón m - [grúas] boom section
— f — — m inferior - [grúas] lower boom section
— f — — m superior - [grúas] upper boom section
— f — ala f - [metal-lam] flange section
— f — — f ancha - [metal-lam] wide flange
section
— f — alambre m - [alambre] wire section
— f — — alcantarilla f - [constr] culvert length
— f — — f con rueda(s) f - [constr] wheeled
culvert length
— f — aliviadero m - [hidr] spillway, o overflow, section
— f — alma m - [cabbl] core (cross) section
— f — cartelera f - [metal-fabr] poster panel,
section, o unit
— f — anillo m - [mec] ring sector,tion

sección f de anillo m para muñón m - [mec] trunnion ring sector
— f — base f - [mec] base section
— f — — f para aguilón m - [mec- boom base
section
— f — bóveda f - [tub] arch section
— f — cable m - [grúas] line, section, o part
— f — — m contraviento(s) m - [grúas] guy
cable section
— f — — m de alambre m - [cabl] wire rope,
part, o section
— f — — m para elevación f - [grúas] hoisting
line, part, o section
— f — cadena f - [cadenas] chain section
— f — — f con rodillo(s) - [cadenas] roller
chain section
— f — camino m - [constr] road section
— f — canal m - [hidr] channel section •
[metal-prod] runner (cross) section
— f — — m para colada f - [metal-prod] runner
(cross) section
— f — canalón m - [constr] flume section
— f — codificador m - [comput] encoder section
— f — conducto m - [tub] conduit section •
[hidr] waterway, area, o section
— f — conducto m - [tub] conduit section •
[ambient] ductwork section
— f — — m vertical - [ambient] vertical ductwork section
— f — conjunto m - [mec] assembly section
— f — — m soldado - [sold] weldment section
— f — contrato(s) m - [com] contracts section
— f — convertidores m - [metal-prod] conver-
ter(s) section
— f — corte m - [constr] cut section
— f — decodificación,cadores - [comput] decoder,ding section
— f — — m optativo - [comput] optional decoder,ding section
— f — defensa f (lateral) - [vial] guardrail,
section, o element
— f — electrodo m - [sold] electrode section
— f — — m único - [sold] single electrode
section
— f — extremo n - [grúas] tip section
— f — — m de cucharón m - [grúas] container,
o bucket, tip section
— f — fondo m - [tub] bottom section
— f — hormigón m - [constr] concrete section
— f — — m armado - [constr] reinforced concrete, section, o member
— f — — m prefabricado - [constr] precast
(reinforced (concrete) structural section
— f — — m en forma f de mediacaña f -
[constr] concrete shell
— f — — m para techo m - [constr] concrete
roof plank
— f — horno m - [metal-prod] furnace (cross)
section
— f — malla f - [hormig] mesh section
— f — manual m - [public] manual section
— f — — m sobre armado m - [ind] manual assembly section
— f — nave f - [ind] bay section
— f — — f para colada f - [metal-prod] casting, o pouring, bay section
— f — — f — convertidor(es) m - [metal-prod]
converter bay section
— f — — f — deslingot(e)ado m - [metal-prod]
stripping bay section
— f — — f — desmoldeo m - [metal-prod]
stripping bay section
— f — — m horno(s) m - [metal-prod] fur-
nace(s) bay section
— f — neumático m - [autom-neumát] tire component
— f — núcleo m - [petról] core section
— f — — m regulado - [petról] controlled core
section
— f — — m — uniformemente - [petról] uni-
formely controlled core section
— f — órdenes f para producción f - [ind] production order section

sección f de panel m - [mec] panel, section, o
unit
— f — — m para anuncio(s) m - poster panel,
section, o unit
— f — — m — aviso(s) m - [com] poster
panel, section, o unit
— f — — m — cartel(es) m - [com] poster
panel, section, o unit
— f — pared f - [constr] wall section
— f — — f para contención f - [hidr] weir
— f — película f - [fotogr] film section
— f — — f fotográfica - [fotogr] photographic film section
— f — — f radiográfica - [electrón] X-ray
film section
— f — perfil m - [metal-lam] shape section
— f — — m de acero m - [metal-lam] steel
shape (cross) section
— f — pescante m - [grúas] jib, section, o
part, o piece
— f — pestaña f - [mec-ruedas] rim section
— f — pilote m - [constr-pil] pile section
— f — — m tubular - [constr-pil] pipe pile
section
— f — piso m - [constr] floor section
— f — — m quitada - [constr] removed floor
section
— f — — m removible - [constr] removable
floor section
— f — — m, removida, o sacada - [constr] removed floor section
— f — plancha f - [mec] plate section
— f — — f de acero m - [constr] steel plate
section
— f — — f — m corrugada - [constr] corrugated steel plate section
— f — — estructural - [mec] structural plate
section
— f — planchas f quitada - [mec] removed plate
section
— f — — f para piso m - [mec] floor plate(s)
section
— f — — f — quitada - [mec] removed
floor plate(s) section
— f — — f — removible - [mec] removable
floor plate(s) section
— f — — f — m, removida(s), o sacada(s) -
[mec] removed floor plate(s) section
— f — — f removible(s) - [mec] removable
plate(s) section
— f — — f, removida(s), o sacada(s) - [mec]
removed floor plate(s) section
— f — pluma f - [grúas] boom section
— f — precipitador m - [metal-prod] precipitator section
— f — presa f - [hidr] weir section
— f — probeta f - test piece (cross) section
— f — punta f - [grúas] point, o tip, section
• [herram] tip section
— f — punzón m - [mec] punch section
— f — riel m - [f.c.] rail, section, o element
— f — — m estándar - [f.c.] standard rail
section
— f — — m según norma f - [f.c.] standard
rail section
— f — rodillo(s) m - [mec] roll section
— f — — m impulsores - [sold] drive,ving
roll section
— f — — m para impulsión f - [mec] drive,ving
roll section
— f — rueda f - [mec-ruedas] wheel section
— f — ruedas f - [mec-ruedas] wheels section
— f — — f común - [mec-ruedas] ordinary
wheel(s) section
— f — soldadura f - [sold] weld section
— f — sufridera(s) f - [mec] die section
— f — tapa f - [tub] top section
— f — terminal - [transp] terminal, section, o
unit
— f — tolva f - [hidr] bin, o trough, section
— f — tren(es) m desbastador(es) - [metal-lam]
blooming mill(s) section
— f — tubería f - [tub] pipe,ping, section, o

o length • tubing, o conduit, section •
[ambient] duct, section, o length
sección f de tubería f abovedada - [tub] pipe-
-arch section
— f — f **circular** - [tub] round pipe section
— f̄ — f̄ **con ángulo m sesgado** - [constr]
skewed angle pipe length
— f — f **largo(r) m doble** - [tub] double
length pipe section
— f̄ — f̄ **contigua** - [tub] contiguous, o adjacent, o abutting, pipe section
— f — f̄ **de acero m** - [tub] steel pipe section
— f — — m **de plancha(s) f** - [tub]
plate steel pipe section
— f — — m — f **estructurales** -
[tub] structural plate steel pipe section
— f — — m — **para ensayo(s) m**
- [tub] structural plate steel pipe test(ing)
section
— f — f — **hormigón m** - [tub] concrete pipe
section
— f — f **especial** - [tub] special steel
pipe
— f — f **para alcantarilla f** - [tub] culvert
pipe section
— f — f — f **de acero m** - [tub] steel
culvert pipe section
— f — f — f — m **armada en fábrica f**
- [metal-fabr] shop-fabricated steel culvert
pipe section
— f — f **para subdrenaje m** - [tub] subdrainage pipe length
— f — f **pavimentada** - [constr] paved pipe
section
— f — f **perforada** - [tub] perforated pipe
section
— f — f — **de acero m** - [tub] perforated
steel pipe section
— f — f **rellenada** - [constr-pil] filled
pipe section
— f — — **con hormigón m** - [constr-pil]
concrete filled pipe section
— f — f **vertical** - [tub] riser section •
[ambient] vertical ductwork section
— f — **tubo m** - [tub] pipe, section, o length;
length of, pipe, o conduit
— f — m **abovedado** - [metal-fabr] pipe-arch
section
— f — m — **encajable** - [metal-fabr] nestable pipe-arch section
— f — **túnel m** —[constr] tunnel (cross) section
— f̄ — — m **aplastada** - [constr] flattened tunnel section
— f — — m **de acero m** - [metal-fabr] steel
tunnel (cross) section
— f — — m — m **corrugado** - [metal-fabr]
corrugated steel tunnel (cross) section
— f — — m — m — **para servicio m** —[metal-
-fabr] corrugated steel service tunnel
(cross) section
— f — — m — m — — m **interior** -
[metal-fabr] corrugated steel inside service
tunnel (cross) section
— f — — m **deformada** - [constr] deformed, o
deflected, tunnel (cross) section
— f — m **para servicio m** - [ind] service
tunnel (cross) section
— f — m — m **interior** - [ind] (inside)
service tunnel (cross) section
— f — **urbanización f** - [constr] subdivision
section
— f — **varilla f** - [trefil] rod, section, o
length
— f — **vía f** - [f.c.] track, section, o stretch
— f̄ — f̄ **recta** - [f.c.] straight (track)
section
— f — **viga f** - [constr] beam, o girder, section • rail, element, o span
— f — **con ala(s) f ancha(s)** • [constr]
wide flange beam section
— f — f **con conformación f en W** - [constr]
W-beam rail, section, o span

sección f de zubia f - [hidr] flume section
— f **débil** - weak section
— f̄ — **de aguilón m** - [grúas] weak boom, o boom
weak, section
— f **deformada** - [constr] deformed, o deflected,
section
— f **dentada** - [mec] toothed section
— f̄ **derecha** - straight section • right (side)
section
— f — **de varilla f** - [metal-lam] straight rod,
o rod straight, section
— f̄ **despacho(s) m** - [com] mailing department
— f̄ **divisoria** - [vial] median section
— f̄ — **terminal** - median terminal, o terminal
median, section
— f **doble** - [tub] double, section, o length •
[sold] double ending
— f — **con largo(r) m variable** - [tub] double
random (length)
— f — **de tubería f** - [tub] double pipe length
— v̄ — **variable** - [tub] double random (length)
— f̄ **dura** - hard, section, o spot
— f̄ **empernada** - [tub] bolted, section, o length
• bolted-together section
— f **empotrada** - [mec] recessed section
— f **en caliente m** - [metal-lam] hot (formed)
section
— f — **frío** - [metal-lam] cold (formed) section
— f̄ — **T** - [metal-lam] T, section, o shape
— f̄ — **X** - [metal-lam] X, section, o shape
— f **enana** - dwarf(ed) section
— f̄ **encajable** - [metal-lam] nestable section
— f̄ — **semicircular** - [tub] nestable half circle
section
— f **enderezadora** - [trefil] straightening section
— f **especial** - [metal-lam] special shape
— f̄ **especificada** - specified section
— f̄ — **de tubería f** - [tub] specified pipe
section
— f **estándar** - [mec] standard section • standard shape
— f — **conformada en W** - [metal-lam] standard
W-section
— f — **para extremo(s)** - [tub] standard end
finish
— f **este** - east section
— f̄ **estructural** - [constr] structural shape
— f̄ — **de hormigón m** - [constr] concrete
structural member
— f — — — m **prefabricada** - [constr] precast
concrete structural member
— f **exigüa** - small (cross) section
— f̄ **expandida** - [metal-fabr] expanded section
— f̄ — **ahusada** - [metal-fabr] tapered expanded
section
— f **extrema** - [grúas] tip, o end, section
— f̄ **fabricada** - [estruct] built-up, o fabricated, section
— f **final** - end section • [grúas] tip section •
[mec] final section; cap section • [electr-
-mot] end bracket
— f — **de cucharón m** - [grúas] container, o
bucket, tip section
— f **final para torre f** - [grúas] tower cap section
— f **fotográfica** - [public] photographic section
— f̄ **frontal** - front section
— f̄ — **de conmutador m** - [electr-instal]
switch(board) front section
— f **fuerte para aguilón m** - [grúas] strong boom
section
— f **grande** - large section
— f̄ — **para aguilón m** - [grúas] large boom
section; boom large section
— f **hidráulica** - [hidr] hydraulic section •
waterway, area, o opening
— f — **reducida** - [hidr] reduced waterway (area)
— I - [metal-lam] I section
— f **igual** - equal (size) (cross) section
— f̄ **inadecuada** - inadequate, o small, cross section
— f **individual** - individual section

sección f individual de tubería f - [constr] individual pipe section
— f — semicircular - [metal-fabr] individual helf-circle section
— f —— encajable - [metal-fabr] individual half-circle nestable section
— f inferior - lower, o bottom, section
— f — de aguilón m - [grúas] lower boom, o boom lower, section
— f inicial - [tub] initial section
— f insertable - [grúas] insert(able) section
— f intercambiable - [metal-fabr] interchangeable section
— f intermedia - intermediate section • [cald] transition section
— f inundada - [petról] flooded section
— f invariable - invariable, o constant, section
— f irregular - irregular section
— f laminación - [metal-lam] rolling section
— f laminada - [metal-lam] rolled section
— f — con ala f ancha - [metal-lam] rolled wide flange(d) section
— f — en caliente - [metal-lam] hot rolled section
— f —— frío - [metal-lam] cold rolled section
— f — según norma - [metal-lam] standard rolled section
— f ——— f con ala m ancha - [metal-lam] standard rolled wide flange(d) section
— f larga - [mec] long section
— f limitada - limited, o finite, section
— f liviana - [mec] light(weight) section
— f — de tubería f - [tub] light(weight) pipe section
— f más débil - weaker,kest section
— f ——— de aguilón m - [grúas] boom, weaker, o weakest, section
— f — fuerte - stronger,gest section
— f ——— de aguilón m—[grúas] boom, stronger, o strongest, section
— f máxima f - [cabl] maximum diameter
— f — de cable m - [cabl] cable maximum diameter • maximum line part
— f mayor - [mec] larger,gest section • [tub] longer,gest length
— f — para aguilón m - [grúas] larger,gest boom section
— f media - middle section • average section
— f menor - [mec] smaller,lest section • [tub] shorter length
— f — para aguilón m - [grúas] smaller,lest boom section
— y movimiento m - [pol] traffic department
— f Multi-Plate - [metal-fabr] Multi-Plate section
— f no rebosable - [hidr] no overflow section
— f normal - [metal-lam] normal, o standard. section
— f oblicua - [mec] oblique section
— f — escalonada - step-bevel(ed) section
— f ——— f estándar - [tub] standard step-bevel(ed) section
— f ——— para extremo m - [tub] standard step-bevel(ed) end finish
— f —— para extremo m - [tub] step-bevel(ed) end finish
— f oeste - west section
— f ojival - Gothic section
— f óptica - [instrum] optical section
— f oval(ada) - oval(led) section
— f oxidada - [metal] rusted area
— f — totalmente - [metal] rusted out area
— f para acabado m - [metal-lam] finishing section
— f ——— m de, banda f, o cinta f, o chapa f, o fleje m - [metal-trat] strip finishing section
— f — acondicionador m (para aire m) - [ind] air conditioner section
— f — acondicionamiento m para aire m - [ind] air conditioning sector

sección f para aguilón m - [grúas] boom section
— f — alcantarilla f - [constr] culvert section
— f — base f - [constr] base section
— f — calcinación f - [miner] calcination section
— f — caldera(s) f - [cald] boiler section
— f — clasificación f - [ind] classification, o (as)sorting, house, o area
— f — colada(s) f - [metal-prod] casting, o tapping, section, o area
— f — conmutador(es) m - [electr-instal] switch(ing) section
— f — conservación f - [ind] maintenance section
— f —— f eléctrica - [ind] electric maintenance section
— f —— f mecánica - [ind] mechanical maintenance section
— f — construcción(es) f - [ind] construction section
— f ——— f civil(es) f - [constr] civil construction(s) section
— f — contrato(s) m - [com] contracts section
— f — corte m - [ind] cutting section
— f ——— m de borde(s) m - [metal-lam] side trimming, o edging, section, o line
— m —— m longitudinal - slitting (section, o department)
— f — diseño(s) m - [ind] design(ing) section
— f — empalme(s) m - [f.c.] connecting section
— f — enfriamiento m - [ind] cooling section
— f — ensayo(s) m - [ind] test(ing) section
— f ——— m no destructivo(s) - [ind] nondestructive test(s), section, o laboratory
— f — entrada f - [ind] inlet, o input, o feeding, section, o end
— f —— f proyectada - [constr] designed inlet section, o area
— m — extremidad f - [grúas] tip section
— f ——— f para cucharón m - [grúas] container tip section
— f — extremo m - [ind] end finish(ing) section
— f ——— m para tubería f - [tub] pipe end(s) finishing section
— f ——— m —— f abovedada - [tub] pipe arch end finishing section
— f ——— m ——— f— de plancha(s) f estructural(es) - [tub] structural plate pipe-arch end finish(ing) section
— f ——— m —— f de plancha(s) f estructural(es) - [tub] structural plate pipe end(s) finish(ing section)
— f — fiscalización f - [ind] control section
— f —— f de calidad f - [ind] quality control section
— f — fosa(s) f - [ind] pit(s) section
— f —— f para colada f - [metal-prod] casting pit(s) section
— f — grúa(s) f - [grúas] crane(s), section, o area
— f ——— f para servicio m - [grúas] service crane section
— f ——— f —— m pesado - [grúas] heavy duty crane(s) section
— f — horno(s) m - [ind] furnace(s) section
— f —— m de fosa f - [metal-prod] soaking pit(s) section
— f — laminador m - [metal-lam] mill section
— f — mantenimiento m - [ind] maintenance section
— f ——— m asignado - [ind] assigned maintenance section
— f — materia(s) f prima(s) - [ind] raw materials section
— f — matriz,ces - [mec] die(s) section
— f — movimiento m - [transp] traffic department
— f — operación(es) f - [ind] operation(s) section
— f —— f de horno(s) m - [ind] furnace(s) operating section

sección f̱ para oxicorte m̱ - [metal-lam] torch
cutting section
— f̱ — plancheado m̱ - [metal-trat] plating
(department)
— f̱ — producción f̱ - [ind] production section
— m̱ — producto(s) m̱ - products section
— f̱ — m̱ con diamantes m̱ - [ind] diamond
product(s) section
— f̱ — programación f—[ind] scheduling section
— f̱ — f̱ para producción f̱ - [ind] (production) scheduling section
— f̱ — prolongación f̱ - [grúas] extending section
— f̱ — f̱ insertable - [grúas] insert extending,sión section
— f̱ — para aguilón m̱ - [grúas] boom
insert extension section
— f̱ — f̱ suplementaria - [grúas] insert extension section
— f̱ — f̱ para suplemento m̱ para aguilón m̱ -
[grúas] boom insert extension section
— f̱ — prueba f̱ para resistencia f̱ - [sold]
resistance section test(ing)
— f̱ — f̱ — f̱ a tensión f̱ - [sold] tensile (testing) section
— f̱ — punta f̱ - [mec] point, o̱ tip, section
— f̱ — f̱ (de) tipo m̱ con cabeza f̱ de martillo m̱ - [grúas] hammerhead tip section
— f̱ — f̱ para grúa f̱ - [grúas] crane tip
section
— f̱ — f̱ — f̱ para servicio m̱ - [grúas]
service crane tip section
— f̱ — f̱ — f̱ — m̱ pesado - [grúas]
heavy duty crane tip section
— f̱ — punzón(es) m̱ - [mec] punch(es) section
— f̱ — rebobinado m—[metal-lam] recoiling line
— f̱ — recocido m̱ - [metal-trat] annealing (department)
— f̱ — recorte,tado m̱ - [mec] trimming line
— f̱ — m̱ — m̱ de borde(s) m̱ - [metal-lam]
side trimming line
— f̱ — m̱ — m̱ y división f̱ en banda(s) f̱ -
[metal-lam] side trimming and slitting line
— f̱ — m̱ lateral - [metal-lam] side trimming
line
— f̱ — recubrimiento m̱ - [metal-trat] coating
(section, o̱ department)
— f̱ — reducción f̱ en caliente - [metal-lam]
hot reduction, mill, o̱ department
— f̱ — f̱ — frío - [metal-lam] cold reduction, mill, o̱ department
— f̱ — reenrollado m̱ - [metal-lam] recoiling,
line, o̱ section
— f̱ — regulación f̱ - [electr-instal] con-
trol(ling) section
— f̱ — reinspección f̱ - [metal-trat] (re)ins-
pection, section, o̱ line, o̱ stand
— f̱ — repuesto(s) m̱ - [ind] parts department
— f̱ — salida f̱ - [ind] outlet; exit - pay-off
section; delivery end
— f̱ — f̱ proyectada - [constr] designed outlet
— f̱ — secado,camiento m̱ - [ind] drying section
— f̱ — ser calentada - section to be heated
— f̱ — soldadura f̱ - [sold] weld(ing) section
— f̱ — sufridera(s) f̱ - [mec] die section
— f̱ — temple,plado m̱ - [metal-trat] tempering.
section, o̱ department
— f̱ — terminación f̱ - [metal-lam] finishing
section
— f̱ — f̱ de, banda f̱, - cinta f̱, o̱ chapa f̱,
o̱ fleje m̱ - [metal-trat] strip finishing section
— f̱ — f̱ en caliente m̱ - [metal-lam] hot
finishing section
— f̱ — f̱ — f̱ frío - [metal-lam] cold finishing section
— f̱ — transición f̱ - [cald] transition section
— f̱ — venta(s) f̱ - [com] sales, office, o̱ section
— f̱ — verificación f̱ - [ind] control section
— f̱ — f̱ para calidad f̱ - [ind] quality control section
— f̱ pavimentada - [constr] paved section

sección f̱ pequeña - small section
— f̱ — para aguilón f̱ - [grúas] small boom, o̱
boom small, section
— f̱ peraltada - [vial] superelevated section
— f̱ perforada - [mec] perforated section
— f̱ — de tubería f̱ - [tub] tube perforated
section
— f̱ — — f̱ vertical - [tub] vertical pipe
perforated section • [hidr] riser perforated
section
— f̱ perpendicular - [tub] perpendicular, o̱ tee,
section|
— f̱ — soldada - [tub] welded tee section
— f̱ poligonal - poligonal (cross) section
— f̱ por sección f̱ - section by section
— f̱ posterior - rear section
— f̱ — de conmutador m̱ - [electr-equip] switch
rear, o̱ rear switch, section
— f̱ predeterminada - [mec] predetermined, section, o̱ length • [trefil] predetermined
length feed
— t — de varilla f̱ - [trefil] rod predetermined, o̱ predetermined rod, length
— f̱ preparada - [ind] prepared section
— f̱ — en taller m̱ - [tub] shop prepared section
— f̱ primera - first section
— f̱ próxima - near, o̱ close, section
— f̱ quitada - [mec] removed section
— f̱ recta - straight section
— f̱ — de conducto m̱ - [ambient] conduit
straight section
— f̱ — varilla f̱ - rod straight section
— f̱ rectangular - rectangular section
— f̱ redonda f̱ - [metal-lam] round (section)
— f̱ reforzada - [mec] reinforced section
— f̱ regular - regular section
— f̱ removible - [mec] removable section
— f̱ — de plancha f̱ - [mec] removable plate
section
— f̱ removida - [mec] removed section
— f̱ respectiva - respective section • [admin]
respective unit
— f̱ rígida - [mec] rigid, o̱ unyielding, section
— f̱ rota - broken section • [constr] broken
plank
— f̱ rural - [geogr] country, o̱ rural, section
— f̱ sacada - [mec] removed section
— f̱ segunda - second section
— f̱ semicilíndrica - [mec] half round section
— f̱ — de troquel m̱ - [mec] half die section
— f̱ semicircular - [tub] half-circle section
— f̱ — de acero m̱ - [tub] half-circle steel
section
— f̱ — — bóveda f̱ - [tub] half-circle arch
section
— f̱ — — f̱ de acero m̱ - [tub] half-circle
steel arch section
— f̱ — — plancha(s) f̱ para revestimiento m̱ -
[constr] liner plate half circle
— f̱ — encajable - [metal-fabr] half circle
nestable section
— f̱ semirrígida - [metal-fabr] semirigid section
— f̱ sin curvar - [metal-fabr] uncurved section
— f̱ — — de plancha f̱ - [metal-fabr] uncurved
plate section
— f̱ — — — f̱ estructural - [metal-fabr]
uncurved structural plate section
— f̱ sin perforar - [mec] unperforated section
— f̱ sobre ajuste(s) m̱ - [public] adjustment(s)
section
— f̱ armado m̱ - [public] assembly section
— f̱ soldada - [sold] welded section
— f̱ — con refuerzo m̱ - [sold] reinforced
welded section
— f̱ — — — m̱ doble - [Sold] double reinforced weld(ed) section
— f̱ — reforzada - [sold] reinforced weld(ed)
section
— f̱ sujetadora - [mec] gripping, section, o̱
portion
— f̱ superior - top, o̱ upper, section
— f̱ — de aguilón ṉ - [grúas] boom upper sec-

tion
sección f suplementaria - [grúas] insert section
— f T - [metal-lam] T section
— f̄ tan corta como posible - short as possible length
— f telescópica - [mec] telescopic,ping section
— f̄ terminal - [metal-fabr] terminal, o end, section • [tub] last section
— f — abocinada - [tub] flared end section
— f̄ — — preparada en planta f - [tub] flared prefabricated end section
— f — corriente - [tub] standard terminal section
— f — de acero m - [tub] steel end section
— f̄ — — m galvanizado - [tub] galvanized steel end section
— f — para tubería f - [tub] pipe end section
— f̄ — — f abovedada - [tub] steel pipe-arch end section
— f — — m para tubería f - [constr] pipe steel end section
— f — — m — f abovedada - [constr] pipe-arch steel end section
— f — — m — f circular - [tub] round pipe steel end section
— f — — m galvanizado - [tub] galvanized steel end section
— f — — m para tubería f - [tub] pipe galvanized steel end section
— f — — m — f abovedada - [tub] pipe-arch galvanized steel end section
— f — — m — f circular - [tub] round pipe galvanized steel end section
— f — — m — f abovedada - [tub] pipe-arch steel end section
— f — — m preparada en taller m - [tub] shop fabricataed steel end section
— f — — defensa f lateral - [constr-vial] [constr] guardrail terminal section
— f — metálica - [metal-fabr] (metallic) end section
— f — optativa - alternate terminal secion
— f̄ — para alcantarilla f - [constr] culvert end section
— f — — tubería f - [constr] pipe end section
— f̄ — — f abovedada - [constr] pipe-arch end section
— f — — — f de acero m - [tub] pipe-arch steel end section
— f — — — f — — m galvanizado - [tub] pipe-arch galvanized steel end section
— ⌐ — — f circular - [tub] round pipe end section
— f — — — f de acero m - [tub] steel round pipe end section
— f — — — m galvanizado - [tub] round pipoe galvanized steel end section
— f — — — f de acero m - [tub] steel pipe end section
— f — — — f — — m galvanizado - [tub] pipe galvanized steel end section
— f — prefabricada - [metal-fabr] prefabricated end section
— f — preparada en planta f - [tub] prefabricated end secion
— f — — — taller m - [tub] shop fabricated end section
— f — substitutiva - [mec] alternate terminal section
— f tierra f adentro - [hidr] land section
— f̄ típica f̄ - typical section • typical cross section
— f̄ — de película f - [fotogr] film typical section
— f — — — f radiográfica - [electrón] X-ray film typical section
— f tipo de cabeza f de martillo m - hammerhead section
— f total - total section • [tub] total end area
— f transversal - cross section • [constr] transverse section • [tub] opening size • véase también corte m transversal

sección f̄ transversal de acero m - [metal] steel cross section
— f — de bóveda f - [tub] arch cross section
— f̄ — — canal m - [metal-prod] runner cross section
— f — — — m para colada f - [metal-prod] runner cross section
— f — — conducto m - [hidr] waterway area
— f̄ — — perfil m - [metal-lam] shape cross section
— f — — m de acero m - [metal-lam] steel shape cross section
— f — — m soldado - [metal-lam] welded shape cross section
— f — — pieza f - [mec] member cross section
— f̄ — — polo m - [electr] pole cross section
— f̄ — en exceso m de . . . - ⍟ area greater than . . .
— v — típica - [dib] typical cross section
— f̄ — — de bóveda f - [tub] arch typical cross section
— f — — tubería f - [tub] pipe typical cross section
— f — — — — f abovedada - [tub] pipe-arch typical cross section
— f trapecial - [hidr] trapeze section
— f trasera - rear, section, o end
— f̄ trasera de tubo m - [autom] véase tubo m para escape m
— f trinagular - traingular (cross) section
— f̄ tubular - [tub] tubular section
— f̄ última - last section
— f̄ única - single section
— f̄ — con largo(r) m variable - [tub] single random length
— f única variable—[tub] single random length
— f̄ uniforme - uniform section
— f̄ unitaria - unit section
— f̄ urbana - [pol] urban, section, o area
— f̄ útil - [tub] pay length
— f̄ — de tubería f - [tub] pipe pay length
— f̄ variable - variable section
— f̄ vertical - vertical section
— f̄ X - [metal-lam] X, section, o shape
seccionador m - [electr-equip] section, o isolating, switch
seccional* a - sectional • véase también en secciones
seccionamiento m - sectioning
seccionar v - . . . . • [alambre] to cut through
secciones a - (en) sectional
— f unida(s) - [mec] joined sections
seco,ca a - . . . . • sharp
— a en aire m - air dry
secreción f de adrenalina - [biol] adrenaline flow
secretaría f - . . . . • office • [pol] secretariat • department
— f de Agricultura f - [pol] Agriculture Department
— m — — f — Recursos m Naturales - [pol] Department of Agriculture and Natural Resources
— f — Comercio m - [pol] Department of Commerce
— f — — m e Industria f - [pol] Department of Industry and Commerce
— f̄ — Estado m - [pol] Department of State • government, department, o secretariat
— f — — m para Industria f y Comercio m - [pol] Department of Industry and Commerce
— f — — m — — f — — m Interior - [pol] Department of Industry and Interior Commerce
— f — Hacienda f - [pol] Treasury Department
— m — — f y Crédito m Público - [pol] [Department of ⍟ Treasury and Public Credit
— m — Industria f y Comercio m - [pol] Department of Industry and Commerce
— f — Patrimonio m Nacional - [pol] National Resources Department
— f — Recursos m Nacionales - [pol] Natural Resources Department
— f — Salud f Pública f - [pol] Public Health

Department
secretaría f **estatal** - [pol] state department
— f **general** - [pol] general secretariat
— <u>f</u> **técnica** - [pol] technical secretariat
secretario m - . . . • [com] clerk • [pol] secretary; minister
— m **de administración** f - [adm] executive secretary
— m — **gerencia** f - [adm] executive secretary
— m — **condado** m - [pol] county clerk
— m — **estado** m - [pol] secretary of @ state •
secretary; minister
— m — **juzgado** m - [legal] court clerk
— m — — m **para condado** - [pol] county court
clerk
— m — **Obras** f **Públicas y Transportes** - [pol]
Public Works and Transportation scretary
— m — **Transportes** m - [pol] Transportation Secretary
— m — **tribunal** - [pol] court clerk
— m — — m **para condado** m - [pol] county court
clerk
— m **municipal** - [pol] city, o town, clerk, o
registrar • county, clerk, o registrar
secreto m - . . . . • [fam] hot tip • [legal] confidentiality
sector m - . . . . • sector; area; zone; quadrant;
segment; department; division; field; activity; banch • [sold] step • [econ] enterprise •
[ind] department • [econ] activity
— m **agrícola** - [econ] agricultural, activity, o
sector
— m — **construcción** f - [constr] construction
sector
— m — **estudio** - study sector
— m **dentado** - [mec] toothed sector
— m — **para freno** m - [mec] brake quadrant
— m **industrial** ,- [ind] industrial, sector, o
field, o activity,ties
— m **metalmecánico** - [ind] metallurgical and
mechanical, sector, o activity; metal-mechanical, sector, o field, o activity
— m **minero** - [miner] mining activity,ties
— m **mayorista** - [com] wholesale business
— m **minorista** - [com] retail business
— m **para fabricación** f - [ind] manufacturing
sector
— f — **producción** f - [econ] production sector
— m **parcialmente superpuesto** - [sold] overlapping step
— m — — **de selector** m **para gamas** f - [sold]
range selector overlapping step
— m — — — m **para amperaje** m -
[sold] current range selector overlapping
step(s)
— m **privado** - [econ] private, sector, o acti-
vity.ties, o enterprise
— m **productivo** - [econ] production,tive sector
— m **público** - [econ] public, sector, o activity,ties
— m — **nacional** - [econ] national public sector
— m **superpuesto** - [sold] overlapping step
— m — **parcialmente** - [sold] partially overlapping step
— m — — **en selector** m **de límite(s)** m **para am-**
peraje m - [sold] current range selector
overlapping step
— m **usuario** - using sector; user
secuencia f - . . . . • <u>véase también</u> **orden** m
— f **básica** - basic sequence
— <u>f</u> — **para operación** f - [ind] operation basic
sequence
— f **cero** - [electr-oper] zero sequence
— <u>f</u> **de actividad(es)** f activity,ties sequence
— <u>f</u> — **adición(es)** <u>f</u> - [metal-prod] addition(s)
sequence
— f — **carácter(es)** m - [electrón] character(s)
sequence
— f — **operación(es)** f - [ind] operation(s) sequence
— f — **señal(es)** f - [electrón] signal(s) sequence
— <u>f</u> — **tiempo(s)** m - time(s) sequence

secuencia f **de tiempo(s)** m **establecida** - established time sequence
— f — — m **fijada** - fixed, o established, time
sequence
— f — **título(s)** m - [comput] title(s) sequence
— <u>f</u> **en código** m - [comput] code sequence
— <u>f</u> — — m **rechazada** - [comput] rejected code
sequence
— f — **conmutador** m - [electr-oper] switch sequence
— f — **orden** m - order sequence
— <u>f</u> — — m **de tiempo(s)** m - time sequence
— <u>f</u> **en tabla** f - table sequence
— <u>f</u> **negativa** - negative sequence
— f **normal** - normal sequence
— <u>f</u> **nula** - [electr-oper] zero sequence
— f **para carga** f - [ind] charge,ging sequence
• [transp] load(ing) sequence
— f — **muestreo** m - [ind] sampling sequence
— <u>f</u> — **multifrecuencia(s)** f - [comput] multifrequency sequence
— f — — f **con tono** m - [comput] tone multifrequency sequence
— f — — f — — m **doble** - [comput] dual tone
multifrequency sequence
— f — **operación** f - [ind] operating,tion sequence
— f— **señalización** f - [electrón] signalling
sequence
— f — **toma** f **de muestra(s)** f - [ind½ sampling
sequence
— f **posible** - possible sequence
— <u>f</u> **positiva** - positive sequence
— <u>f</u> **válida** - valid sequence
— <u>f</u> — **para dígito(s)** m - [comput] valid digit,
o digit valid, sequence
secundaria f - secondary
— f **provista con fusible(s)** m - [electr-equip]
fused secondary
secundario m - secondary • [metal-prod] secondary, cleaner, o scrubber, o washer • [mec]
idler
— m **para transformador** m - [electr-equip]
transformer secondary
secundario,ria a - secondary • accessory
sed f **excesiva** - excessive thirst
— f **mostrada** - shown, o developed, thirst
— <u>f</u> **transformada** - transformed thirst
seda f **artificial** - [textil] artificial, o synthetic, silk
-**sedán** m **de lujo** - [autom] luxury sedan; limou-
sine; limo
— m — — **importado** - [autom] imported luxury,
sedan, o limo(usine)
— m **deportivo** - [autom] sport sedan
— m **para carrera(s)** f - [autom] racing sedan
sede f - . . . .; seat • [legal] head, o home,
office
— f **oficial** - [legal] official, o registered,
headquarters, o (home, o head) office
— f **para carrera** f - [deport] race headquarters
— <u>f</u> — — f **Rally** - [deport] Rally headquarters
— <u>f</u> — **personal** m **para rendimiento** m **alto** -
[ind] performance team workhouse
— <u>f</u> — **Rally** † - [deport] Rally headquarters
— <u>f</u> — **sociedad** f - [legal] corporate headquarters
— f **social** - [legal] head, o corporate, office,
o headquarters
sedimentación f - . . . .; settling; silting;
siltation • puddling
— f **aguas** f **abajo** - [hidr] downstream silting
— <u>f</u> — **arriba** - [hidr] upstream silting
— <u>f</u> **causada** - [hidr] caused sedimentation
— <u>f</u> **de aceite** m - oil, sedimenting,tation, o
settling, o silting,tation
— <u>f</u> **en alcantarilla** f - [hidr] culvert silting
— <u>f</u> **evitada** - [hidr] avoided sedimentation
— <u>f</u> **profunda** - [geol] deep silt(ing)
— <u>f</u> **rápida** - [hidr] rapid, sedimentation, o
silting, o settling
— f **reducida** - [hidr] reduced silt(ing)
sedimentado,da <u>a</u> - sedimented; silted; settled

sedimentador m - [sanit½ settler; settling,
 tank, o basin
— m en entrada f - [hidr] intake settling basin
sedimentar v - to, sediment, o settle, o silt •
 [constr] to puddle
sedimento m - . . .; silt; lees • [metal-prod]
 scruff; scale • [petról] slush
— m acarreado - [jodr] carried silt
— m arcilloso - [geol] clay(ey), sediment, o
 silt
— m asentado - [hidr] deposited, o settled,
 sediment
— m atrapado - [hidr] trapped silt
— m de agua m - [hidr] water sediment
— m — arena f arcillosa - [geol] clayey sand
 sediment
— m — petróleo m - [petról] bottom settling(s)
— m — — m en fondo m - [petról] bottom set-
 tling(s)
— m — m — — m de depósito m - [petról]
 (container) bottom settling(s)
— m — m de depósito m - [petról] bottom
 settling(s)
— m en suspensión f - [hidr] suspended sediment
— m polvoriento - dusty silt(ing)
— m profundo - [geol] deep silt
— m saturado - [constr] saturated silt
segadora f - [agric-equip] reaper
— f gavilladora f - [agric-equip] (binder)
 reaper
— f — forraje(s) m - [agric-equip] forage
 reaper
— f para heno m - [agric-equip] hay mower
— f trilladora - [agric-equip] combine
— f — de modelo m reciente - [agric-equip]
 late model combine
— f — montada - [agric-equip] mounted combine
— f — reciente - [agric-equip] late (model)
 combine
segatrilladora* f - [agric-equip] combine
segmentado,da a - segmented
segmental a - [arquit] segmental
segmento m - . . . • [alambre] picket • [comb.-
 int] piston ring
— m ahusado - [mec] tapered segment
— m — para cuña f - [mec] tapered wedge segment
— m ampliado - broadened segment
— m amplio—broad, o extensive, o wide, segment
— m comercial - [com] commercial segment
— m — amplio - [com] broad, o extensive, commercial segment
— m de acero m - [mec] steel segment
— m — m galvanizado - [mec] galvanized
 steel segment
— m — alambre m - [alambre] wire picket
— m — alambrera f - [constr] fence picket
— m — bobina f - [metal-lam] coil strip; tail
 end • head end
— m — bobinadora f—[metal-lam] coiler segment
— m — cadena f - [cadenas] chain, segment, o
 section
— m — — f con rodillo m - [cadenas] roller
 chain, segment, o section
— m — círculo m—[mec] circle, part, o segment
— m — cordón m - [sold] bead segment
— m — — m de soldadura f - [sold] weld bead
 segment
— m — cuña f - [mec] wedge segment
— m — espectro m - [fís] spectrum segment
— m — instrucción f - [educ] instructional
 segment
— m — mandril m - [mec] mandrel segment
— m — — m para bobinadora f - [metal-lam]
 coiler mandrel segment
— m — mercado m - [com] market segment
— m — soldadura f - [sold] weld segment
— m — f intermitente - [sold] intermittent
 weld segment
— m — soplador m - [mec] blower segment
— m — tubería f - [constr] pipe,ping segment
— m — f para conducción f - [hidr] transmission line segment

segmento m de tubería f para conducción f de
 agua m - [hidr] water transmission line segment
— m — — f — — f — — m sin tratar - [hidr]
 raw water transmission line segment
— m — ventilador m - [mec] fan segment
— m — — m para generador m - [electr-prod]
 generator fan segment
— m — — m para soldadura f - [sold] welder
 generator fan segment
— m — — m para soldadora f - [sold] welder
 fan segment
— m específico - [mec] specific segment
— m — de mercado m - [com] specific market
 segment
— m galvanizado - [mec] galvanized segment
— m individual - [mec] individual segment
— m — de alambre m - [alambre] individual wire
 picket
— m intermitente - intermittent segment
— m para agua m - [hidr] water segment
— m — — m sin tratar v - [hidr] raw water
 segment
— m — conducción f - transmission segment
— m — — f de agua m - [hidr] water transmission segment
— m — — f — — m sin tratar v - [hidr] raw
 water transmission segment
— m — molienda f - [metal-prod] grinding slug
— m — vástago m - [petról] subrod
— m — — m colocado - [mec] installed subrod
— m — — m conectado - [mec] connected subrod
— m — — m para émbolo m - [petról] piston
 subrod
— m — — m, quitado, o removido, o sacado -
 [med] removed subrod
— m rentable - [com] revenue, o profit, producing segment
— m semitubular - [tub] half-round segment
— m vigoroso - strong segment
segregación f de azufre m - [miner] sulfur segregation
— f — — m en acero m - [metal- steel sulfur
 segregation
— f — centro m - [metal-prod] center segregation
— f — material(es) m - [constr] material(s)
 segregation
— f — peatón(es) m - [vial] pedestrian, separation, o segregation
— f — sulfuro(s) m - [metal] sulfide(s), segregation, o separation
— f severa - [metal- severe segregation
segregado,da - . . .; segregated
segueta f - [herram] . . .; keyhole, o saber,
 saw; hacksaw (blade)
— f eléctrica - [herram] saber saw
seguido,da a - . . .; followed (up) • in @ row
— a cuidadosamente - followed carefully
— a de cerca - followed closely
— a estrictamente - followed strictly
— a explícitamente - followed explicitly
seguidor m para caja f - [mec] cup follower
— m — empaquetadura f - [mec] packing follower
seguimiento m - . . .; follow-up • tracking
— m cuidadoso - careful following
— m de artículo m - item follow-up
— m — cerca - close following
— m — contrato m - contract follow-up
— m — costo(s) m - cost(s) follow-up
— m — diagrama m - diagram following
— m — socavación f - [constr] excavation following
— m — fabricación f - [ind] manufacture,ring
 follow-up
— m — instrucción(es) f - instruction(s) following
— m — medida(s) f - measure(s) following
— m — — f precaucionaria(s) - precautionary
 measure(s) following
— m — metal m de fusión f - [sold] follow(ing)
— m — — m — — f detrás de electrodo m -
 [sold] follow

seguimiento m de metal m en fusión f detrás de electrodo m mientras se suelda - [sold] fol-low(ing)
— m — órden(es) - [ind] order(s) following
— m — pedido m - order follow-up
— m — pieza f - [ind] part follow-up
— m — — f forjada - [ind] forged part follow--up
— m — premisa f - premise following
— m — procedimiento m - procedure following
— m — — m apropiado - proper procedure fol-lowing
— m — — m correcto - correct procedure fol-lowing
— m — servicio(s) m - service(s) follow-up
— m — — m contratado(s) m - contracted ser-vice(s) follow-up
— m — tiempo(s) m - time follow-up
— m estricto - strict, following, o follow-up
— m explícito - explicit following
— m rápido - [sold] fast follow(ing)
seguir v - . . . • to trail; to track (along) • to follow up
— v adelante - to follow on; to keep on going
— v avanzando - to keep (ón) going
— v carrera f - [deport] to, cover, o follow, @ race • [educ] to follow @ carreer
— v con poca agua m - [ind] continue(d) poor water supply
— v consejo m - to follow @ advice
— v contrato m - to follow up @ contract
— v costura f - [sold] to follow @ seam
— v cuidadosamente - to follow carefully
— v diagrama m - to follow @ diagram
— v estrictamente - to follow strictly
— v excavación f - [constr] to follow @ excava-tion
— v explícitamente - to follow explicitly
— v extendiendo - to extend further
— v fabricación f - [ind] to follow (up) @ man-ufacture
— v fielmente - to follow faithfully
— v instrucción(es) f - to follow @ instruc-tion(s)
— v indicación f - to follow @, indication, o advice
— v instrucción(es) f - to follow @ instruc-tion(s)
— v junta f - [sold] to, follow, o track along, @ joint
— v — irregular - [sold] to, follow, o track along, @ irregular joint
— v medida(s) f - to follow @ measure(s)
— v — f de precaución f - follow @ precaution-ary measure(s)
— v — f precautoria(s) - to follow @ precau-tionary measure(s)
— v pedido m - to follow (up) @ order
— v plan m - to follow @ plan
— v premisa f - to follow @ premise
— v procedimiento m - to follow @ procedure
— v — m apropiado - to, follow, o take, @ proper procedure
— m — m correcto - to follow @ correct proce-dure
— v relación f - [fotogr] to follow @ narration
— v soldando - to continue welding
— v trabajando - to continue working
seguirse v - (de) when following
según adv - . . .; as per; according to; in ac-cordance (with)
— adv colilla(s) f que se deje(n) - [sold] stub loss practice
— adv concepción f - according to @ design
— adv corresponda - as, appropriate, o may cor-respond
— adv cual sea - depending on
— adv — posición f—depending on @ position
— adv — — punto m de vista - depending on @ viewpoint
— adv diseño m - according to @ design
— adv elección f - according to @ chice
— adv estimación f - as projected

según adv exigencia(s) f - as required
— adv haga falta - ase, needed, o required • depending on
— adv instrucción(es) f - as per, o according to, @ instruction(s)
— adv — f recibida(s) - as per, o according to, @ instruction(s) received
— adv lista f de empaque(tamiento) m - [com] as per, o according to, @ packing list
— adv — f — m adjunta - [com] as per, o according to, @ attached packing list
— adv Manning - [ind] (as per) Manning('s)
— adv norma f - (according to @) standard
— adv proceda - as may be in order
— adv proyección f - (according to @) design
— adv se ilustra - as shown
— adv — especifica - as specified
— adv — ilustra - as shown
— adv — indica - as, indicated, o directed
— adv — refiere - as narrated
— adv — requiera - as (may be) required
— adv — ve - as, seen, o shown
— adv sea caso m - as may be @ case
— adv — necesario - as (may be) required
segunda adv - . . . • junior
— f colocación f - [deport] second place
— f — meritoria - [deport] fine second
— mitad f - second half
— pasada f - [sold] véase pasada f segunda
— potencia f - [mat] squared
— f soldadura f - [sold] second weld
— f — circunferencial - [sold] second girth weld
segundo m - . . . • [ind] assistant; second helper • [instrum] tick
— m absoluto - [deport] second over all
— m en clasificación f general - [deport] sec-ond overall
— lugar n - [deport] second (place)
— — m en clasificación f final - [deport] sec-ond place finish(ing)
— m para convertidor(es) m - [metal-prod] con-verter(s) second helper
— m — cuchara(s) f - [metal-prod] ladleman second helper
— m — horno m - [metal-prod] furnace second helper
— m — — m basculante - [metal-prod] tilting furnace second helper
— m — — m de fosa f - [metal-lam] soaking pit second helper; assistant soaking pit operator
— m — — m fijo - [metal-prod] stationary fur-nace second helper
— m — — m mezclador - [metal-prod] mixing furnace second helper
— m — — m para mezcla f - [metal-prod| mixing furnace second helper
— m — máquina f - [ind] machine second heller
— m — nave f - [ind] bay second helper
— m — — f para colada f - [metal-prod] cast-ing bay second helper
— m — — f — escoria(s) f - [metal-prod] slag bay second helper
— plano m - background
— — m a derecha f - right background
— — m — izquierda f - left background
— tercio - second third
segundo,da en categoría f - [deport] second in @ class
seguridad f - . . . • reliability • assurance; insurance
— f adecuada - [segurid| adequate safety
— f adicional - [segurid] additional, o added, o extra, security, o safety
— f apropiada - appropriate, o suitable, safety
— f — de armadura f—[mec] proper assembly as-surance
— f aumentada - increased, o improved, safety
— f comprobada - [segurid] checked, o proven, safety
— f de adherencia f - [pint] adherence assur-ance
— f — — f buena f - [pint] good adherence as-

surance

seguridad f de adherencia f buena de pintura f - [pint] good paint adherence assurance
— f — cañón m - [tub] barrel safety
— f — carretera f - [segurid] highway safety
— f — contacto - contact assurance
— f — — m eléctrico - [electr-instal] electrical contact assurance
— f — continuidad f - continuity assurance
— f — comportamiento m - performance assurance
— f — desempeño m - performance assurance
— f — cuadrilla f - [segurid] crew safety
— f — departamento m - [segurid] departmental safety
— f — equipo m - equipment safety
— f — firme m - [constr] roadbed safety
— f — funcionamiento m - dependability; reliability; performance assurance
— f — — m confiable - dependable performance assurance
— f — gancho m - [grúas] hook safety
— f — junta f - [mec] joint assuring
— f — — f ajustada - [mec] tight joint assuring
— f — motorista m - [transp] motorist safety
— f — operación f - [segurid] operating,tion safety
— f — operario m - [segurid] operator safety
— f — personal - [segurid] personnel, o employee, o worker, safety
— f — planta f - [ind] plant safety
— f — precisión f - [ind] accuracy assurance
— f — puente m - [constr] bridge safety
— f — quemador m - [combust] burner safety
— f — resistencia f - [mec] strength assurance
— f — — f apropiada - appropriate, o proper, strength assurance
— f — resultado(s) m - result(s) assurance
— f — — m duradero(s) - (long) lasting result(s) assurance
— f — servicio m - uninterrupted service
— f — soldadora f - [sold] welder safety
— f — soldadura f - [sold] welding safety
— f — — f por arco m - [sold] arc welding safety
— f — tránsito m - [vial] traffic safety
— f en carretera f - [segurid] highway safety
— f — circulación f - [vial] safe travel(ing)
— f — corte m - [segurid] cutting safety
— f — general - [segurid] overall safety
— f — — aumentada - [segurid] increased overall safety
— f — reducida - [segurid] decreased, o diminshed, o reduced, overall safety
— f — industria f - [segurid] industry,trial safety
— f — — f de acero m - [metal-prod] steel industry safety
— f — mantenimiento m - [segurid] maintenance safety
— f — — m de soldadora f - [sold] welder maintenance safety
— f — — m — — f con motor m con combustión f interna - [sold] engine welder maintenance safety
— f — marcha f - (operating) dependability
— f — mina f - [min] mine safety
— f — operación f - operating,tion safety
— f — — f de soldadora f - [sold] welder operating,tion safety
— f — — f — — f con motor m con combustión f interna - [sold] engine welder operation safety
— f — planta f - [segurid] plant safety
— f — producción f - [ind] production, o operating,tion, safety
— f — sí mismo,ma - [labor] self reliance
— f — soldadura f - [sold] welding safety
— f — — f y corte m - [sold] welding and cutting safety
— f — trabajo m - [segurid] work safety
— f — — m en obra f - [segurid] project site work safety

seguridad f en viaje(s) m - [transp] travel(ing) safety
— f exigida - required assurance
— f final - final assurance
— f — de calidad f - [ind] final quality assurance
— f — — — f de tubería f - [tub] final pipe quality, o pipe quality final, assurance
— f fiscalizada - [segurid] monitored safety
— f general - [segurid] general safety
— f industrial - [ind] industrial safety
— f integral - [segurid] built-in safety
— f máxima - [segurid] maximum, safety, o security, o assurance
— f — de calidad f - [segurid] maximum quality provided
— f — en operación f - [segurid] maximum operating safety
— f — provista - [segurid] maximum provided, security, o assurance
— f — — en calidad f - [ind] maximum provided quality assurance
— f mejorada - [segurid] improved safety
— f nacional - [pol] national, security, o safety
— f óptima - [segurid] optimum safety
— f — para neumático(s) m - [autom-neumát] optimum tire safety
— f para cargo m - [labor] position safety
— f — código m con . . . dígito(s) m - [comput] . . . digit code security
— f — conservación f - [ind] maintenance safety
— f — física f - [segurid] physics safety
— f — neumático m - [autom-neumát] tire safety
— f — puesto m - [segurid] position safety
— f — personal m - [segurid] personnel safety • protection
— f — soldadura f - [sold] welding safety
— f personal - [segurid] personal safety
— f radiológica - [nucl] radiological safety
— f reducida - [segurid] decreased safety
— f según norma f - [segurid] standard safety
— f sobre calidad f - [ind] quality assurance
— f social - [labor] social security
— f verificada - [segurid] checked safety
seguro m - . . . . • insuring • underwriting • [electr-equip] breaker block • [mec] lock; lock; click • safety device • tumbler
— m a prorrata - [seguros] pro rata insurance
— m — puerto . . . - [seguros] insurance to . . . port
— m asumido - [seguros] assumed insurance
— m autorizado - [seguros] authorized insurance
— m barato - [seguros] cheap insurance
— m cedido - [seguros] ceded insurance
— m con tornillo m - [mec] screw latch
— m contra accidente(s) m - [seguros] accident, o casualty, insurance
— m — catástrofe(s) f - [seguros] catastrophe (re)insurance
— m — riesgo m cualquiera - [seguros] any risk insurance • all risk insurance
— m — responsabilidad f civil - [seguros] liability insurance
— m — riesgo(s) m de guerra f - [seguros] war risk insurance
— m — rotación f - [mec] rotation lock
— m — — f de grúa f - [grúas] crane rotation lock
— m — todo riesgo m - [seguros] all risk insurance; insurance against all risks
— m contratado - [seguros] written insurance
— m de responsabilidad f - [seguros] liability insurance
— v — validez - [seguros] validity insurance
— m económico - [seguros] economical, o low cost, o cheap, insurance
— m emitido - [seguros] insurance written
— m en participación f - [seguros] shared insurance
— m — — f dentro de cuota f - [seguros]

quota share insurance
seguro m **en participación** f **en exceso** m **de cuota**
f - [seguros] excess quota share insurance
— m — **vigencia** f - [seguros] in force, o existing, insurance
— m — **vigor** - [seguros] in force insurance
— m **excedente** - [seguros] excess insurance
— m **exigido** - [seguros] required insurance
— m **económico** - [seguros] economical, o cheap,
insurance3
— m **hasta obra** f - [seguros] insurance to @,
job, o project, site
— m — **sitio** m **de obra** f - [seguros] insurance
to @, job, o project, site
— m **hospitalario** m - [seguros] hospital(ization)
insurance
— m **marítimo** - [seguros] marine, o maritime, insurance
— m **médico** - [seguros] medical, o health, insurance • Blue Shield (insurance)
— m **no autorizado** - [seguros] unauthorized insurance
— m **operado** - [mec] operated lock
— m — **para evitar rotación** f - [grúas] operated
rotation lock
— m — **caja** f **para transmisión** f - [mec] transmission case insurance
— m — **evitar rotación** f - [grúas] rotation,
lock, o latch
— m — — — f **de grúa** f - [grúas] crane rotation lock
— m **para fidelidad** f - [seguros] fidelity, o security, insurance
— m — **fijación** f - [mec] lock(ing) latch
— m — **permitir** v **rotación** f - [grúas] (rotation unlock(ing) latch
— m — **provisión** f - supply insurance
— m — **transmisión** f - [mec] transmission lock
— m — **varilla** f - [mec] rod latch
— m **respaldado** - [seguros] backed up insurance
— **servidor** a - [com] yours truly; sincerely
yours
— m **sobre bien(es)** m **raíz,ces** - [seguros] real
estate, o property, insurance
— m — **crédito(s)** m - [seguros] credit insurance
— m — — m **para exportación** f - [seguros] export credit insurance
— m — **equipo** m - [seguros] equipment insurance
— m — **exportación(es)** f - [seguros] export insurance
— m — **inmueble(s)** m - [seguros] real estate, o
property, insurance
— m — **material(es)** m - [seguros] material(s)
insurance
— m — **personal** - [seguros] personnel insurance
— m — **propiedad(es)** f - [seguros] property insurance
— m — **transporte(s)** m - [seguros] transportation insurance
— m — — m **terrestre(s)** - [transp] inland
transportation insurance
— m — **vehículo(s)** m - [seguros] vehicle(s) insurance
— m **social** - [seguros] social security
— m **total** - [seguros] total insurance
— m **vigente** - in force, o unexpired, o valid,
insurance
seguro,ra a - . . . . • accurate; insured • reliable; foolproof • confident
— a **a tierra** f - [electr-instal] tight ground
— a **de sí mismo,ma** - self-reliant
selección f - . . . .; selecting; choosing; picking
• determination • sorting (out) • range •
[admin] screening
— f **afectada** - affected, choice, o selection
— f **alternativa** - alternate, selection, o choice
— f **amplia** - ample, o broad, o wide, selection,
o choice, o range
— f — **de producto(s)** m—broad product selection
— f **amplísima** - huge, o very broad, selection
— f **apropiada** - appropriate, o proper, o right,
o good, selection, o choice
— f — **de electrodo** m - [sold] appropriate, o

proper, electrode selection
selección f **buena** - good, selection, o choice
— f **con confianza** f - confident, selection, o
choice
— f — **conmutador** m - [somput] switch, selecting,tion, o choice
— f **correcta** - correct, o right, selection, o
choice
— f **corriente** - current, o common, choice
— f **cuidadosa** - careful, selection, o choice
— f — **de materia(s)** f **prima(s)** - [ind] careful raw material(s), selection, o choice
— f **de alambre** m - [sold] wire selection
— f — **alineación** f - [mec] alignment, selection, o choice
— f — — f **apropiada** - [constr] appropriate,
o proper, alignment, selection, o choice
— f — **amperaje** m - [sold] current selection
— f — — m **para soldadura** f - [sold] welding
current selection
— f — **ancho(s)** m - width(s) range
— f — **antemano** - preselecting,tion
— f — **borne(s)** m - [electr-equip] stud selection
— f — — m **para salida** f - [electr-instal]
output stud selection
— f — **cable(s)** m - [cabl] cable, selection, o
choice • [electr-cond] cable, selection, o
choice
— f — **coeficiente** m - coefficient choice
— f — **colador** m - [mec] strainer, selection,
o choice
— f — **color(es)** m - color(s), selection, o
choice
— f — **condición** f - [electrón] condition, selection, o choice
— f — — f **de estado** m - [electrón] status
condition, selection, o choice
— f — **conformación** f - [mec] shape, selection, o choice
— f — **cono** m - [sold] cone selection
— f — **contratista** m - [constr] contractor,
selection, o choice
— f — **diámetro** m - diameter, selection, o
choice
— f — — m **de electrodo** m - [sold] electrode
diameter, selection, o choice
— f — — m — — m **de carbono** m - [sold] carbon (electrode) size, selection, o choice
— f — — m — **tubería** f - [tub] pipe, size, o
diameter, selection, o choice
— f — **diseño** m - design, selection, o choice,
o determination
— f — **electrodo** m - [sold] electrode, selection, o choice
— f — — m **apropiado** - [sold] appropriate, o
proper, electrode, selection, o choice
— f — **estado** m - [electrón] status, selecting,tion, o choice
— f — **estrangulador** m - [mec] choke, selection, o choice
— f — **estructura** f - [constr] structure, selection, o choice
— f — **filtro** m - [mec] filter, selection, o
choice
— f — **firma** f - firm, selection, o choice
— f — — f **consultora** - consulting firm, selection, o choice
— f — **frecuencia** f - [electrón] frequence,
selection, o choice
— f — — f **para tono** m - [electrón] tone frequency, selection, o choice
— f — — f — — m **marca-espacio** - [electrón]
mark-space tone frequency, selection, o
choice
— f — **fundente** m - [sold] flux, selection, o
choice
— f — **grúa** f - [grúas] crane, selection, o
choice
— f — **materia(s)** f **prima(s)** - [ind] raw mate-
rial(s), selection, o choice
— f — **material** m - [ind] material, selection,
o choice

selección f de material m específico - [ind]
  specific material, selection, o choice
— f —— m para relleno m - [constr] backfill
  material, selection, o choice
— f — modalidad f - [comput] node selection
— f —— f para ensayo(s) m - [comput] test
  mode(s), selection, o choice
— f —— f — prueba(s) f - [comput] test(ing)
  mode, selection, o choice
— f — motor(es) m - [electr-mot] motor, selection, o choice • [comb.int] engine, selection, o choice
— f — muestra(s) f - [ind] sample(s), selection, o choice
— f —— f de suelo m - [suelos] soil sample,
  selection, o choice
— f — negro m de carbono m - [autom-neumát]
  carbon black, selection, o choice
— f — m — humo m - [autom-neumát] carbon,
  o lamp, black, selection, o choice
— f — neumático m - [autom-neumát] tire, selection, o choice
— f —— m para recambio m - [autom-neumát]
  replacement tire selection
— f — ofertante m - bidder, selection, o
  choice
— f — oscilador m - [electrón] oscillator,
  selection, o choice
— f —— m para salida f - [electrón] output
  oscillator, selection, o choice
— f — personal - [admin] personnel, o people,
  selection, o choice, o screening
— f — polaridad f - [electrón] polarity, selection, o choice
— f — poste m - [constr] post, selection, o
  choice
— f — procedimiento(s) m - [ind] procedure(s),
  selection, o choice
— f — proceso m - [ind] process, selection, o
  choice
— f — producto m—product, selection, o choice
— f — proponente m - bidder, selection, o
  choice
— f — proveedor m - supplier, selection, o
  choice
— f — proyección f - projection, o design, selection, o choice
— f —— f final - final, design, o projection, selection, o choice
— f — regulación f - [ind] control(ling), o
  setting, selection, o choice
— f —— f para amperaje m - [sold] amperage,
  o current, setting, selection, o choice
— f —— salida f - [electrón] otuput, selection,
  o choice
— f —— f para oscilador m - [electrón] oscillator output, selection, o choice
— f — sitio m - site, o location, selection, o
  choice
— f — temperatura f - temperature. selection,
  o choice, o pick(ing)
— f —— f para precalentamiento m - [sold]
  preheat(ing) temperature, selection, o
  choice, o pick(ing)
— f — torre f - tower, selection, o choice
— f — velocidad f - [mec] speed, selection, o
  choice
— f —— f apropiada - [mec] appropriate, o
  proper, speed, selection, o choice
— f —— f — para tambor m - [mec] appropriate, o proper, drum speed, selection, o
  choice
— f —— f para tambor m - [mec] drum speed,
  selection, o choice
— f — viscosidad f - [lubric] viscosity, selection, o choice
— f —— f específica - [lubric] specific, o
  particular, viscosity, selection, o choice
— f —— f — para aceite m - [lubric] specific, o particular, oil viscosity, selection,
  o choice
— f — voltaje(s) m - [sold] voltage(s), selection, o choice

selección f e incorporación f - [pers] selection and induction
— f económica - economic(al) selection
— f final - final, selection, o choice
— f física - physical, selection, o choice
— f definitiva - final, selection, o choice
— f ilimitada - unlimited, selection, o choice
— f — de procedimiento(s) m - [ind] unlimited
  procedure(s), selection, o choice
— f inapropiada - inappropriate, o improper, o
  poor, selection
— f — de procedimiento(s) m - inappropriate,
  o improper, o poor, procedure(s) selection
— f manual - manual, o hand, selection, o
  choice, o pick(ing
— f más corriente - most common, selection, o
  choice
— f posible - possible, selection, o choice
— f práctica - practical, selection, o choice
— f preliminar - preliminary selection
— f previa - previous selection
— f primera - first, selection, o choice
— f provista - provided, selection, o choice
— f real - actual, selection, o choice
— f segunda - second choice
— f substitutiva - alternate, selection, o
  choice
— f — apropiada - appropriate alternate, selection, o choice
— f verificada - checked, selection, o choice
seleccionable a - selectable
— a con conmutador m - [comput] switch selectable
seleccionado,da a - selected; chosen • sorted
  out • [admin] screened
— a con confianza f - selected confidently
— a — conmutador m - [comput] switch selected
— a de antemano - preselected
— a por operador m - [electrón] operator selected
— a previamente - selected previously
seleccionador m - selector
— m de límite(s) para velocidad f - [instrum]
  speed limit selector
seleccionar v - . . . .; to select • to sort out •
  [admin] to screen
— v amperaje m - [sold] to select @, amperage,
  o current
— v — m para soldadura f - [sold] to select @
  welding, amperage, o current
— v coeficiente m - to, select, o choose, @
  coefficient
— v colador m - [mec] to select @ strainer
— v condición f - to, select, o choose, @ condition
— v — f de estado m - [electrón] to select @
  status condition
— v con confianza f - to select, confidently, o
  with confidence
— v — conmutador m - [comput] to switch select
— v de antemano - to preselect
— v diámetro m de electrodo m - to, select, o
  choose, @ electrode diameter
— v — m — m de carbono m - [sold] to, select, o choose, @ carbon (diameter) size
— v — m — tubería f - [tub] to, select, o
  choose, @ pipe, diameter, o size
— v estado m - [electrón] to, select, o choose,
  @ status
— v estrangulador m - [mec] tp, select, o
  choose, @ choke
— v — m apropiado - [mec] to, select, o
  choose, @ appropriate choke
— v filtro m - [mec] to, select, o choose, @
  filter
— v frecuencia f - [electrón] to, select, o
  choose, @ frequency
— v material m - [ind] to, select, o choose, @
  material
— v — m específico - [ind] to, select, o
  choose, @ specific material
— v — m para relleno m - [constr] to, select,
  o choose, @ backfill material

seleccionar v modalidad f - [comput] to, select, o choose, @ mode
— v — f para ensayo m - [comput] to, select, o choose, @ test mode
— v — f prueba f - [comput] to, select, o choose, @ test mode
— v muestra f - [ind] to, select, o choose, @ sample
— v — f de suelo(s) m - [suelos] to, select, o choose, @ soil sample
— v negro m - to, select, o choose, @ black
— v — m de carbono m - to, select, o choose, @ carbon black
— v m — humo m - to, select, o choose, @ lamp black
— v neumático - [autom-neumát] to, select, o choose, @ tire
— v — m para recambio m - [autom-neumát] to, select, o choose, @ replacement tire
— v oscilador m - [electrón] to, select, o choose, @ oscillator
— v — m para salida f - [electrón] to, select, o choose, @ output oscillator
— v personal m - [admin] to, select, o choose, o screen, @ personnel • to select @ organization
— v plantel m - [admin] to, select, o choose, @ organization
— v producto m—to, select, o choose, @ product
— v polaridad f - [electrón] to, select, o choose, @ polarity
— v — f apropiada - [electrón] to, select, o choose, @, appropriate, o proper, polarity
— v regulación f - to, select, o choose, @, control, o setting
— v — f para amperaje m - [sold] to, select, o choose, @ current setting
— v previamente - to, select, o choose, previously, o in advance
— v salida f - [electrón] to, select, o choose, @ output
— v — f para oscilador m - [electrón] to, select, o choose, @ oscillator output
— v torre f - to, select, o choose, @ tower
— v velocidad f - [mec] to, select, o choose, @ speed • [autom-mec] to, select, o choose, @ gear
— v — f apropiada - [mec] to, select, o choose, @ proper speed
— v — f para tambor m - [mec] to, select, o choose, @, appropriate, o proper, drum speed
selectivamente adv - selectively
selectividad f - selectivity
selector m con cuadrante m - [electr] dial selector
— m — esfera f - [electr] dial selector
— m girado - [electr-oper] turned switch
— m para amperaje(s) m - [electr-oper] amperes selector
— m — — m de tipo m con, cuadrante, o esfera • [sold] dial-type, amperes, o current, selector
— m — m — m — esfera f - [sold] dial--type, amperes, o current, selector
— m circuito(s) m - [electrón] circuit selector
— m — m para nivel m alto - [electrón] high level circuit selector
— m — — m — — m — para teleimpresor(es) - [electrón] high level teleprinter circuit selector
— m — escala(s) f - scale, o range, selector
— m — f para amperaje(s) m - [electr] amperage, o current, range selector
— m — — f — voltaje m - [sold] voltage range selector
— m — estrangulador m - [mec] choke selector
— m — — m operado - [mec] switched choke selector
— m — operación f - [mec] operation selector • function switch
— m — — f para laminador m - [metal-lam] mill operation selector

selector m para operación(es) f girado - [mec] turned function switch
— m — — f girado a "contención" - [mec] function switch turned to "Hold"
— m — — f "Prerregulación f hacia abajo" - [mec] function switch turned to "Preset down"
— m — — f — "— f — arriba" - [mec] function switch turned to "Preset up"
— m — trabajo m - work, o job, selector
— m — velocidad(es) m - [mec] speed selector
— m — — f para medidor m - [instrum] meter speed selector
— m — — f — m registrador - [instrum] recording meter speed selector
— m — voltaje m - [electr-oper] voltage selector • [sold] job selector
selectora f - véase selector m
Self* m - véase transformador m para filtración
selva f - [topogr] jungle
— f tropical - [geogr] tropical jungle
sellado m - [fisc] documentary, tax, o stamp • documentary, o legal, paper • [ind] capping • véase también selladura f
— m final - [ind] final capping
— m hermético - tight sealing
— m oficial - [legal] official documentary paper
— m — seguridad f - [legal] safety documentary paper
— m para contrato m - [com] contract documentary fee
sellado,da a - sealed • [legal] set seal
— a apropiadamente - sealed properly
— a con agua m - water sealed • [ind] water cap(ping)
— a — nitrógeno m - nitrogen sealed
— a contra - sealed against
— a — filtración f - [hidr] sealed against @ seepage
— a — humedad f - moisture sealed
— a en cemento m - cement sealed
— a — cilindro f - cylinder sealed
— a herméticamente - sealed, hermetically, o airtight
— a mecánicamente - [mec] machine, o mechanically, sealed
— a para toda vida f (útil) - sealed for life
— a permanentemente - sealed permanently
sellador m - . . . . - sealant
— m aprobado - [constr] approved sealer,lant
— m de caucho m - rubber sealant
— m — m silicónico - silicone rubber sealant
— v — plomo m - [constr] lead sealant
— m — m vertido - [constr] poured lead sealant
— m líquido - [constr] liquid sealant
— m — que no mancha - [constr] non staining liquid sealant
— m mineral - [constr] mineral seal(ant)
— m para empaquetadura f - [mec] gasket sealant
— m — f conformado en sitio m - [mec] formed-in-place gasket sealer,lant
— m silicónico - [quím] silicone sealant
sellador,ra a - . . . .; sealing; sealant
selladora f con calor m - heat sealer
— f con polietileno m - [extrusión f de electrodos] polyethylene sealer • heat sealer
selladura f apropiada - proper sealing
— f contra filtración f - [hidr] seal(ing against @ seepage
— f de acumulador m - [electr-acumul] battery sealing
— f — batería f - battery sealing
— f — cilindro m - [mec] cylinder sealing
— f — conjunto m - [mec] assembly seal(ing)
— f — — m de calce(s) m - [mec] shim pack seal(ing)
— f — extremo m - [mec] end seal(ing)
— f — — m de tubería f - [tub] pipe end sealing
— f — — m — — f para entubación f - [tub]

casement end sealing
selladura f **de grieta(s)** f - crack sealing
— f — **sobre** m - envelope sealing
— f **efectiva** - effective seal(ing)
— f — **zanja** f - [constr] trench sealing
— f **en cemento** m - (in) cement sealing
— f **hermética** - hermetical, o air tight, sealing
— f **para tapón** m - [valv] plug sealing
— f — — m **para tapa** f - [válv] valve cover plug sealing
— f — m — f **para válvula** f - [válv] valve cover plug sealing
— f — m — f — f **para cilindro** m - [válv] cylinder valve cover plug sealing
— v **para (toda) vida** f **(útil)** - [mec] sealing for @ (whole) life
sellamiento* m - véase **selladura** f
sellante a - sealer; véase también **sellador,ra**
sellar v - . . . . • [legal] to, affix, o set, @ seal
— v **acumulador** m - [autom-electr] to seal @ battery
— v **apropiadamente** - to seal properly
— v **batería** f - [electr] to seal @ battery
— v **con sello** m **oficial** - [legal] to affix @ official seal
— v **conjunto** m - [mec] to seal @ assembly
— v — m **de calce** m - [mec] to seal @ shim, assembly, o pack
— v **contra** - [mec] to seal against
— v — **filtración** f - [hidr] to seal against @ seepage
— v — **humedad** f - to seal against @, moisture, o humidity
— v **empaquetadura** f - [mec] to seal @ packing
— v — f **para árbol** m - [mec] to seal @ shaft packing
— v — f — m **para turbina** f - [turb] to seal @ turbine shaft packing
— v **en cemento** m - [Constr] to seal in cement
— v — **cilindro** m - [mec] to seal in @ cylinder
— v **enjugador** m - [mec] to seal @ wiper
— v **extremo** m - to seal @ end
— v **grieta** f - to seal @ crack
— v **herméticamente** - to seal, hermetically, o airtight
— v **para (toda) vida** f **(útil)** - [mec] to seal for @ (whole) life(time)
— v **permanentemente** - to seal permanently
— v **sobre** m - to seal @ envelope
— v **tapón** m - [mec] to seal @ plug
— v — m **para tapa** f - [válv] to seal @ cover plug
— v — m — f **para válvula** f - [válv] to seal @ valve cover plug
— v — m — f — f **en cilindro** m - [válv] to seal @ valve cover plug in @ cylinder
— v **zanja** f - [constr] to seal @ trench
sello m **accionado por presión** f - [mec] pressure driven seal
— m **anular** - [mec] ("O") ring seal
— m **apropiado** - appropriate, o proper, seal
— m **atrapado** - [mec] trapped seal
— m **buscado** - sought, o looked for, seal
— m **con anillo** m - [tub] ring seal
— m **con fuga** f - [mec] leaking seal
— m **cónico** m - [mec] conical, o tapered, seal
— m — **metal** m **contra metal** m - [petról] conical, o tapered, metal to metal seal
— m **contra** - [mec] seal against
— m — **polvo** m - [mec] dust seal
— m — — m **reemplazado** - [mec] replaced dust seal
— m **corredizo** - [mec] sliding, o moving, seal
— m **Chicago** - [mec] Chicago seal
— m — **de cuero** m **crudo** - [mec] Chicago rawhide seal
— m — — — m **sin curtir** v - [mec] Chicago rawhide seal
— m **de caucho** m - [tub] rubber seal
— m — **cuero** m **crudo** - [mec] rawhide seal
— m — — m **sin curtir** - [mec] rawhide seal

sello m **de oferente** m - bidder('s) seal
— m — **proponente** m - bidder('s) seal
— m **desgastado** - [mec] worn (out) seal
— m **gastado** - [mec] worn seal
— m — **árbol** m - [mec] shaft seal
— m — — m **para aportación** f **de fuerza** f - [autom-mec] output shaft seal
— m **encerrado** - [mec] trapped, o enclosed, seal
— m **estanco** m - watertight seal
— m **Fawick** m - [mec] Fawick seal
— m — **para rotor** m - [mec] Fawick rotor seal
— m **guardapolvo(s)** m - [mec] dust seal
— m **hermético** - hermetic, o airtight, seal(ing)
— m **inferior** - [mec] bottom, o lower, seal
— m — **para aceite** m - [mec] lower, o bottom, oil seal
— m **instalado** - [mec] installed seal
— m — **correctamente** - [mec] correctly installed seal
— m **lubricado** - [mec] lubricated seal
— m **mecánico** - [mec] mechanical seal
— m **mejorado** - [mec] improved seal
— m — **para aceite** m - [mec] improved oil seal
— m **mineral** - mineral seal
— m **nuevo** - [mec] new seal
— m — **para aceite** m - [mec] new oil seal
— m **oficial** - [legal] official seal
— m **para aceite** m - [mec] oil seal
— m — — m **aflojado** - [mec] loosened oil seal
— m — — m **apretado** - [mec] tightened oil seal
— m — — m **dañado** - [mec] damaged oil seal
— m — — m **en extremo** m - [mec] end oil seal
— m — — m — m **hacia compresor** m - [comb.-int] compressor end oil seal
— m — — m **instalado** - [mec] installed oil seal
— m — — m **mejorado** - [mec] improved oil seal
— m — — m **nuevo** - [mec] mew improved oil seal
— m — — m **para adaptador** m - [mec] adapter oil seal
— n — — m **biela** f - [mec] pitman, o connecting rod, oil seal
— m — — m **caja** f **para volante** m - [comb.-int] flywheel, o bell, housing oil seal
— m — — m **para cárter** m - [comb.int] crankcase oil seal
— m — — m **cigüeñal** m - [comb.int] crankcase oil seal
— m — — m **para cojinete** m - [mec] bearing oil seal
— m — — m — m **principal** - [mec] main bearing oil seal
— m — — m **cubierta** f - [mec] cover oil seal
— m — — m — f **para cojinete** m - [autom-mec] bearing cover oil seal
— m — — m — f — m **para aportación** f **de fuerza** f - [autom-mec] input bearing cover oil seal
— m — — m — f — m — f — f **instalado** - [autom-mec] installed input bearing cover oil seal
— m — — m — f **engranaje** m **para motor** m - [comb.int] engine gear cover oil seal
— m — — m — **chaveta** f - [mec] [mec] key oil seal
— m — — m — f **para polea** f - [autom-mec] pulley key oil seal
— m — — m — f **para impulsión** f - drive,ving pulley key oil seal
— m — — m — f — f — f **de ventilador** m - [autom-mec] fan drive,ving pulley key oil seal
— m — — m — **extremo** m - [comb.int] end oil seal
— m — — m — m **para aportación** f **de fuerza** f - [comb.int] take-off end oil seal
— m — — m — **grúa** f - [grúas] crane oil seal
— m — — m — f **aérea** - [grúas] overhead crane oil seal
— m — — m **para piñón** m - [mec] pinion oil seal

sello m para aceite m, removido, o sacado -
[mec] removed oil seal; removed oil ring
— m — agua m - [mec] water (shedder) seal
— m — árbol m - [mec] shaft seal
— m — m para bomba f - [bombas] pump shaft
seal
— m — — m para motor m - [electr-mot] motor
shaft seal • [comb.int] engine shaft seal
— m — armella f - [mec] bale seal
— m — f para elevación f - [mec] lift(ing)
bale seal
— m — barra f - [mec] bar, o rod, seal
— m — f para conexión f - [mec] connecting
rod, o Pitman, arm seal
— m — base f - [mec] base, o bottom, seal
— m — f para cabeza f - [petról] swivel
bottom seal
— m — — f para inyección f - [petról]
swivel bottom seal
— m — biela f - [mec] connecting rod, o Pitman, seal
— m — brazo m - [mec] arm seal
— m — — m para biela f - [mec] connecting
rod, o Pitman, arm seal
— m — caja f - [mec] housing seal
— m — f para volante m - [comb.int] bell
housing seal
— m — camisa f - [mec] liner seal
— m — carcamo m - sump seal
— m — cilindro m - [mec] cylinder seal
— m — — m neumático - [neumát] air cylinder
seal
— m — cobertura f - [mec] cover(ing) seal
— m — columna f - [mec] column seal
— m — — f para dirección f - [autom-mec]
steering column seal
— m — compresor m - [mec] compressor seal
— m — conducto m—[electr-instal] conduit seal
— m — — m para cable(s) m - [electr-instal]
cable(s) conduit seal
— n — conductor m - [electr-instal] lead seal
— m — conjunto m - [mec] assembly, o pack,
seal(ing)
— m — — m para calce m - [mec] shim pack seal
— m — costura f - [mec] seam seal(ant)
— m — cubierta f - [mec] cover seal
— f — f para tapa f para caja f - [mec]
case top cover(ing) seal
— m — chaveta f - [mec] key seal
— m — f para polea f—[mec] pulley key seal
— m — — f — — f para impulsión f de ventilador m—[mec] fan drive,ving pulley key seal
— m — dispositivo m - [autom-mec] unit seal
— m — — m para cambio(s) m - [autom-mec]
shift(ing) unit seal
— m — émbolo m - [mec] piston seal
— m — empaquetadura f - [mec] packing seal
— m — f formado en sitio m - [mec] formed-
-in-place gasket seal(er)
— m — f para árbol m - [mec] shaft packing
seal
— m — — f — — m para turbina f - [turb]
turbine shaft packing seal(ing)
— m — — f — camisa f - [mec] liner packing
seal
— m — f — eje m - [mec] axle, o shaft,
packing seal
— m — enjugador m - [mec] wiper seal
— m — — m desgastado - [mec] worn (out) wiper
seal
— m — — m gastado - [mec] worn wiper seal
— m — — m reemplazado - [mec] replaced wiper
seal
— m — envase m - [transp] container seal
— m — flecha* f - véase sello m para árbol m
— m — fondo m - [mec] bottom seal
— m — — m para cabeza f para inyección f -
[petról] swivel bottom seal
— m — horquilla f - [mec] fork seal
— m — — f para cambio(s) m - [autom-mec]
shift fork sealk
— m — jubilación f - [fisc] retirement (fund)
stamp

sello m para jubilación f notarial - [fisc]
notary,ries' retirement fund stamp
— m — junta f - [carp] joint seal
— m — — f ensamblada - [carp] mortised joint
seal
— m para motor m - [electr-mot] motor seal •
[comb.int] engine seal
— m — obturador m [mec] plug seal
— m — pieza f - [mec] part seal
— m — f frontal - [mec] front block seal
— m — f — para relleno m - [mec] front
filler block seal
— m — f para relleno m - [mec] filler block
seal
— m — f posterior para relleno m - [mec]
rear filler block seal
— m — piñón m - [mec] pinion seal
— m — presión f - [mec] pressure seal
— m — puerta f - [mec] door seal
— m — regulador m - [mec] regulator seal
— m — m para presión f - [mec] pressure
regulator seal
— m — rotor m - [mec] rotor seal
— m — sumidero m - [sanit] sump seal
— m — tapa f - [mec] cover seal
— m — f para válvula f - [mec] valve cover
seal
— m — tapón m - [mec] plug seal
— m — — m para caja f - [bombas] housing
plug seal
— m — transmisión f - [mec] transmission seal
— m — f con fuga f - [mec] leaky,king
transmission seal
— m — traslapo m - [maec] lap(ping) seal
— m — unidad f - [mec] unit seal
— m — f para cambio(s) m - [autom-mec]
shift(ing) unit seal
— m — vacío m - [neumát] vacuum seal
— m — — m roto - [neumát] broken vacuum seal
— m — vástago m - [válv] stem seal
— m — zanja f - [constr] trench seal
— m permanente - permanent seal(ing)
— m positivo - [mec] positive seal
— m protegido - [mec] protected seal
— m quitado - [mec] removed seal
— m reemplazable - [mec] replaceable seal(ing)
— m reemplazado - [mec] replaced seal
— m removido - [mec] removed seal
— m retenido - [mec] retained seal
— m roto - [mec] broken seal
— m sacado - [mec] removed seal
— m superior - [mec] top, o upper, seal
— m — para aceite m - [mec] top, o upper, oil
seal
— m tipo m campana - [mec] bell type seal
— m — — para aceite m - [mec] bell, type,
o housing, oil seal
— m verificado - [mec] checked, o tested, seal
— m y cambio m - [mec] seal and shift
— m — — m para horquilla f - [mec] fork seal
and shift
— m y resorte m - [mec] seal and spring
— m — — m para horquilla f - [mec] fork seal
and spring
— m — — m — — f para cambio(s) m - [mec]
shift fork seal and spring
semáforo m - . . . . • [vial] traffic, o signal,
light; signal post • [f.c.] signal
— m — tránsito m - [vial] traffic, signal, o
light
semana f de retraso - week('s) delay
— f — trabajo m - [labor] work week
— f laborable - [labor] work week
— f natural - [cronol] calendar week
semanalmente adv - . . . .; once each week; on @
weekly basis
semántica f - . . . . • terminology
— f adecuada - adequate semantics
— f apropiada - appropriate semantics
— f inadecuada - inadequate semantics
— f inapropiada - inapropiate semantics
semántico,ca a - . . . .; semantical
sembrado m - . . . .; field

sembrado,da a - [agric] sown; seeded • [ictiol] stocked
— a **con despojo(s)** m - littered; strewn
— a — **roca(s)** f - rock, littered, o strewn
sembrador m - [agric] . . . • drill
sembradora f **a voleo** m—[agric] broadcast seeder
— f —— m **con azadón(es)** m - [agric] hoe broadcast seeder
— f — **vuelo** m - [agric] broadcast seeder
— f —— m **con azadón(er)** m - [agric] hoe broadcast seeder
— f **con azadón(es)** m - [agric] hoe drill
— f —— m **para, cereal(es), o grano(s)** m - [agric] hoe drill
— f —— **caída** f - [agric] drop planter
— f —— f **de canto** m - [agric] edge drop planter
— f — **corredera** f - [agric] runner planter
— f — **disco(s)** m - [agric] disk planter
— f —— m **para maíz** m - [agric] disk, corn, o maize, planter
— f — **elevación** f **mecánica** - [agric] power lift, planter, o drill
— f —— f **para tractor** m - [agric] power lift tractor, planter, o drill
— f — **esteva(s)** f - [agric] walking planter
— f — **mancera(s)** f - [agric] walking planter
— f — **ruedas** f - [agric] wheel(ed) drill
— f —— f **para presión** f—[agric] press drill
— f — **trocha** f **ancha** - [agric] wide track, o tread, seeder
— f —— f **angosta**—[agric] narrow gage seeder
— f — **vía** f **ancha** - [agric] wide track seeder
— f —— f **angosta**—[agric] narrow gage seeder
— f **con zapata(s)** f **(para granos** m**)** - [agric] shoe drill
— f **entre zurco(s)** m - [agric] drill
— f **lister** - [agric] lister planter
— f **para alfalfa** f - [agric] alfalfa drill
— f —— f **y, hierba** f, o **pasto** m - [agric] alfalfa and grass drill
— f — **algodón** m - [agric] cotton, planter, o drill
— f —— m **con asiento** m - [agric] riding cotton planter
— f — m **y maíz** n - [agric] cotton and, corn, o maize, planter, o drill
— f — **betarraga** f - [agric] beet, drill, o seeder
— f —— f **y judía(s)** f - [agric] beet and bean drill
— f — **cereal(es)** f - [agric] grain drill
— f — **culata** f **de carro,ruaje** m - [agric] end gate seeder
— f — **fertilizante** m—[agric] fertilizer sower
— f — **grano(s)** m - [agric] grain drill
— f —— m **con elevación** f **mecánica** - [agric] power lift drill
— f —— m —— **para tractor** m - [agric] power lift tractor drill
— f —— m **con fertilizante** m - [agric] fertilizer grain drill
— f —— m **con rueda** f **para presión** f - [agric] press drill
— f — **hierba** f - [agric] grass drill
— f — **judías** f - [agric] bean planter
— f — **maíz** m - [agric] corn, o maize, planter, o drill
— f —— m **con caída** f **de canto** m - [agric] edge drop, corn, o maize, planter
— f —— m —— m **en montecillo(s)** m—[agric] hill drop, corn, o maize, planter
— f —— m —— f **plana** - [agric] flat drop, corn, o maize, planter
— f —— m **con corredera** f - [agric] runner (corn, o maize) planter
— f —— m **para altura** f **variable** - [agric] variable drop, corn, o maize, planter
— f —— m — **tierra** f **suelta** - [agric] loose ground (lister) planter
— f — **tipo** m **lister con cultivador** m - [agric] combined lister
— f — **pasto** m - [agric] grass drill

sembradora f **para porotos** m - [agric] bean planter
— f — **remolacha** f - [agric] beet, drill, o seeder
— f — f **y poroto(s)** m - [agric] beet and bean planter
— f — **semilla(s)** f - [agric] drill; seed, o lister, planter
— f — f **lister** - [agric] lister (seed) planter
— f — f — **para tractor** m - [agric] tractor lister
— f — f **para tractor** m - [agric] tractor lister
— f — **siembra** f **en cuadro** m - [agric] check-row planter
— f — f **variable** - [agric] variable drop planter
— f — **tierra** f **suelta** - [agric] loose grond planter
— f — **trigo** m - [agric] wheat drill
— f **sencilla** f - [agric] plain drill
— f — **para grano(s)** m - [agric] plain grain drill
sembrado,da a - sown • [fig] dotted
— a **con roca(s)** f - rock strewn
sembrar v - . . . • [ictiol] to stock
— v **con despojo(s)** m - to litter; to strew
— v — **roca(s)** f - to litter with rocks
— v **despojo(s)** m - to litter
— v **pista** f **(con despojos** m**)** - [deport] to litter @ course
semejante a - . . .; fellow, man, o being | a - . . .; alike
semestre m **primero** - first semester
— m **segundo** - second semester
— m **último** - last semester
semi m - [transp] véase **semiacoplado** m; **semi-rremolque** m
semiabierto,ta a - half open(ed)
semiacabado,da a - véase **semiterminado,da**
semiacero m - [metal-prod] semisteel
semiacoplado m - [transp] semitrailer; véase también **semirremolque** m
— m **para transporte** m **sobre vagón** m **(ferro-viario)** - [transp] piggyback trailer
semiacoplamiento m - [mec] half coupling
— m **con rosca** f **(de tipo** m**) estadounidense** - [tub] National Pipe Thread half coupling
— m — f **estadounidense** - [tub] National Pipe Thread half coupling
semianillo m - [mec] half ring; ring half
semianual a - semi-annual; half yearly
semiárido,da a - semi-arid
semiaro m - [mec] half arch • half ring
— m **para repuesto** m - [mec] backup half ring
semiarrastrado,da a - [mec] semi-trailing
semiatascado,da a - half, o partially, clogged
semibrillante a - [alambre] semiglossy
semicalmado,da a - [metal-prod] semikilled
semicerrado,da a - partially closed
semicircular a - [geom] half circle
semicírculo m - [geom] semicircle; half circle
semicircunferencial a - semicircular
semicojinete m - [cojin] half bearing
— m **para biela** f - [comb.in] connecting rod half bearing
semiconductor n - [electrón] semiconductor
semiconductor,ra a - [electr-cond] semiconducting,tor
semicontinuo m - [metal-lam] véase **tren**, o la-minador, **semicontinuo**
semicontinuo,nua a - semi-continuous
— m **cristalino,na** a - hemicrystaline; semi-crystaline
semidiagrama m - half diaphragm
semidiámetro m - half diameter; semidiameter
semiduro,ra a - half hard
semieje m - [autom] half (axle) shaft
— m **derecho** - [autom-mec] right half shaft
— m **estriado** - [autom] axle half shaft
— m **izquierdo** - [autom-mec] left half shaft

semieje m reemplazado - [autom] replaced axle shaft
— m roto - [autom] broken axle shaft
— m̄ quitado - [autom-mec] removed axle shaft
— m̄ ranurado - [autom] splined axle shaft
— m̄ removido - [autom-mec] removed axle shaft
— m̄ sacado - [autom-mec] removed axle shaft
semielaborado m - [metal-prod] semi-finished product
semielaborado,da a - semifinished; semiprocessed
semielaborar v - [ind] to semifinish
semielíptico,ca a - semielliptical
semiembutido,da a - semiflush
semiemptorado,da - [mec] semiflush
semienfriado,da a - semi-cooled
semihemisférico,ca a - semihemispherical
semiflecha f - [alambre] small rod • [autom-mec] [Méx.] véase semieje m
semiflexible a - semiflexible; semi-spring quality
semiflotación f - semifloating,tation
semiflotante a - semifloating
semifrío,ría a - semicold
semiintegrado,da a - semiintegrated
semillero m - [agric] . . .; seed hopper
semimate a - semimatte
semimuro m - [constr] half wall
— m para cabecera f - [constr] half headwall
seminario m - . . . • seminar
— m sobre juego m - gambling, o gaming, seminar
— m̄ — venta(s) f̄ - [com] sale(s) seminar
— m̄ técnico - [ind] technical seminar
semiordenada f - [matem] semi-ordinate
— f determinada—[matem] determined semi-ordinate
— f para curva f - [matem] curve semi-ordinate
semiportátil a - semiportable
semipórtico m - [grúas] semigantry • véase también grúa f semipórtico
semipórtico,ca a - [grúas] semiportal
— f — para chatarra f - [metal-prod] scrap semiportal (crane)
semiprocesado,da a - semiprocessed
— a laminado,da en frío - [metal-lam] cold rolled semiprocessed
semiproducto m - [ind] semiproduct; semifinished product; véase también producto m semielaborado, o semiterminado
— m para relaminación f - [metal-lam] rerolling semifinished product
semirremolcado,da a - [mec] semitrailer,ling
— m carretero - [transp] truck trailer
— m̄ — para transporte m sobre vagón m plataforma - [transp] piggyback trailer
— m con plataforma f baja - [transp] low-boy (trailer)
— m para transporte m sobre vagón m ferroviario - [transp] piggyback trailer
semirrígido,da a - semirigid
semisección f - [mec] half section
semisólido,da a - semisolid
semisuperimpuesto,ta a - semisuperimposed; half lap(ped)
semitécnico,ca a - semitechnical
semiterminado,da a - semifinished; semiprocessed
— a con sección f pequeña - [metal-lam] small section semifinished
— a laminado,da en frío - [metal-lam] cold rolled semifinished
semiterminar v - [ind] to semifinish
semitragado,da a - half swallowed
semitramo m - [mec] half section
— m con conformación f elíptica - elliptically shaped half section
semitroquel m - [mec] half die
semiunión f - [mec] half coupling
— f de latón m - [mec] brass half coupling
— f̄ imperial - [mec] imperial half coupling
— f̄ para tubería f - [mec] tube, o pipe, half coupling
— f̄ — f de latón m - [mec] brass tube half coupling
semivivo,va a - [metal-prod] semikilled

semoviente a - [mec] self propelled
sencillo,la a - . . . • single • uncomplicated; unsophisticated • straightforward • véase también único,ca a
senda f - . . .; trail
— f ecuestre - [deport] riding trail
— f̄ para acceso m - [constr] access trail
— f̄ equitación f - [deport] riding trail
sendero m - . . .; trail; track; cowpath
— m para acceso m - [constr] access trail
— m̄ — caminata(s) f̄ - [deport] hiking, path, o trail
— m̄ — pedestrismo - [deport] hiking path
sendos,das a - . . . • several
seno m - . . . • hollow; cavity • [cabl] slack
sensación f - . . . • feeling • experience • aura
— f buena - good feeling
— f̄ con velocidad f̄ grande - [autom] high speed feel(ing)
— f de aire m - air feel(ing)
— f̄ — aprendizaje m - learning experience
— f̄ — arco m - [sold] arc feel(ing)
— f̄ — confianza f - sense of confidence
— f̄ — control m - [admin] control feel(ing)
— f̄ — equilibrio m - balance, sense,sation, o feel(ing) • on-center feel(ing)
— f̄ — flojedad f - loose feel(ing)
— f̄ — grasitud f̄ - greasy feel(ing)
— f̄ — neutral - neutral feel(ing)
— f̄ — placer - pleasure feel(ing); enjoyment, sensation, o feel(ing)
— f̄ — poder m - [admin] power feel(ing)
— f̄ — reacción f - reaction feel(ing)
— f̄ — — f rápida f - quick response feeling
— f̄ — para seguridad f - safe(ty) feeling
— f̄ — tirantez f - taut feeling
— f̄ especial - special feel(ing)
— f̄ experimentada - felt, o experienced, sensation
— f satisfaciente - satisfying, sensation, o feeling
— f satisfactoria - satisfactory,fying feeling
— f̄ sentida - felt sensation
— f̄ subjetiva - subjective, sensation, o feel(ing)
sensacional adv - . . .; impressive • major
sensato,ta a - . . .; sane
sensibilidad f - . . .; responsiveness
— f analizada - analyzed sensitivity
— f̄ básica - basic sensitivity
— f de balanza f - [instrum] scales sensitivity • balance sensitivity
— f — — f para torsión f - [instrum] torsion balance sensitivity
— f̄ — empresa f - [com] business, o concern, responsiveness
— f — — f pequeña - [com] small concern responsiveness
— f̄ — material(es) m - [ind] material(s), responsiveness, o sensitivity
— f̄ — método - method sensitivity
— f̄ de papilla f - [explos] pap sensitivity
— f̄ deseada - desired sensitvity
— f̄ exigida - required sensitivity
— f̄ final - final sensitivity
— f̄ — de papilla f - [explos] pap final sensibility
— f̄ — deseada - desired final sensitivity
— f̄ fundamental - basic sensitivity
— f̄ máxima - maximum sensitivity
— f̄ mayor - greater sensitivity
— f̄ mínima - minimum sensitivity
— f̄ de entalladura f - [metal-mec] motch sensitivity
— f̄ para, solidificación f, o congelación f - [sold] freezability
— f̄ regulable - [instrum] controllable sensitivity
— f̄ controlada - [instrum] controlled sensitivity
sensibilización f de película f - [fotogr] film sensitization

sensible adv   - 1438 -

sensible adv - responsive • appreciable; notable
  • considerable
— adv a - sensitive to • operated by
— adv a absorción f - [sold] absorption, o
  pick-up, sensitive
— adv a — f de elemento(s) m - [quím] absorption, o pick-up, sensitive
— adv a — f — f de aleación f - [sold] alloy pick-up sensitive
— adv a acelerador m - [autom-mec] throttle
  sensitive
— adv a corriente f - [electr-oper| current
  sensitive
— adv a detonador m - [explos] detonator sensitive
— adv a falla(s) - [ind] failure sensitive
— adv a halógeno(s) m - [quím] halogen(s) sensitive
— adv a mecanización f - [mec] machining sensitive
— adv a — f por abrasión f - [mec] abrasion
  machining sensitive
— adv a precio m - price sensitive
— adv a tacto m touch sensitive • [comput]
  tactile
— adv a temperatura f - temperature, sensitive,
  o operated
— adv especialmente - especially sensitive
sensiblemente adv - . . .; notably; considerably
  • fairly
sensitivo,va a - . . . . • sensitized
sensor m abierto—[electr-equip] open(ed) sensor
— m externo - [electr-equip] external sensor
— m — abierto - [electrón] open(ed) external
  sensor
— m — cerrado - [electrón] closed external
  sensor
— m para calor m - [instrum] heat sensor
— m — m para rotadora f - [cuerp-moled]
  rotator heat sensor
— m — m — f alimentadora—[cuerp-moled]
  rotator/feeder heat sensor
— m — falla f - [instrum] failure sensor
— m — f a tierra f - [electr-equip] ground
  (failure) sensor
— m — nivel m - [mec] level sensor
— m — rotadora f—[cuerp-moled] rotator sensor
— f — — f alimentadora - [cuerp-moled] rota-
  tor/feeder sensor
— m — vataje n - [electr-instrum] wattage sensor
— m — m alto - [electr-instrum] high wattage sensor
— m — m bajo - [electr-instrum] low wattage
  sensor
— m — m elevado - [electr-instrum] high
  wattage sensor
— m — m reducido - [electr-instrum] low
  wattage sensor
— m térmico - [instrum] thermal, o heat, sensor
sentar v real(es) m - to make @ claim
sentencia f - . . . . • [comput] statement
— f diferente - [comput] different statement
— f — para alarma f - [comput] different alarm
  statement
— f — — condición f - [comput] different status statement
— f — — estado m - [comput] different status
  statement
— f judicial - [legal] judicial sentence
— f para alarma f - [comput] alarm statement
— f — condición f - [comput] status statement
— f — estado m - [comput] status statement
— f suprema - supreme sentence
sentido m - . . . . • area • direction • [electr-
  -oper] flow • [mec] pitch
— m amplio - broad sense
— m anticronométrico - véase sentido m contrario a de aguja(s) f de reloj
— m cambiado - changed direction
— m — para flujo m - changed flow direction
— m contrario - opposite direction
— m contrario a de aguja(s) f de reloj • [mec]

counterclockwise direction
sentido m contrario a de manecilla(s) f de reloj m - [mec] counterclockwise direction
— m correcto - [mec] correct, o proper, direction
— m — para alimentación f - [mec] right feed-
  -(ing) direction
— m — — f de alambre m - [sold] right
  wire feed(ing) direction
— m — — rotación f - [mec] right (rotation)
  direction
— m cronométrico - véase sentido m de aguja(s)
  f de reloj m
— m cualquiera - either, o any, direction
— m para ahusamiento m - [mec-ruedas] taper
  direction
— m — alambre m - [mec] wire direction
— m — alimentación f - [mec] feed(ing) direction
— m — — f para alambre m - [mec] wire feed-
  -(ing) direction
— m — apertura f - [mec] opening direction
— m — avance m - [mec] advance direction
— m — — m gradual - [sold] inching direction
— m — — m guiado - [sold] led travel direction
— m — — m para tubo m - [mec] pipe travel
  (direction)
— m — carrera f - [mec] stroke direction
— m — — f para cilindro m - [mec] cylinder
  stroke direction
— m — — f invertido - [mec] reversed stroke
  direction
— m — caudal - [hidr] flow direction
— m — corriente f - [hidr] flow direction •
  [electr-oper] current flow
— m — chavetero - [mec] keyway direction
— m — flecha f - arrow direction
— m — flujo m - [hidr] flow (direction)
— m — — indicado - designated flow (direction)
— m — giro m - [mec] rotation, o turn(ing),
  direction
— m — — m para motor m - [electr-mot] motor
  rotation direction • [comb.int] engine rotation direction
— m — laminación f - [metal-lam] rolling direction • lengthwise (direction)
— m — manipulador m - [electrón] keyer sense
— m — marcha f - [vial] travel direction
— m — mecanización f - [mec] machining direction
— m — procesamiento m - [ind] processing direction
— m — rendimiento m - performance, o yield,
  direction
— m — rotación f - [mec] rotation (direction)
— m — — f cambiado - [mec] changed rotation
  direction
— m — — f comprobado - [mec] proven, o
  checked, rotation (direction)
— m — — f correcto - [mec] correct, o right,
  rotation (direction)
— m — — f verificado - [mec] checked correct rotation (direction)
— m — — f para bomba f - [bombas] pump rotation direction
— m — — f — — f invertido - [bombas] reversed pump rotation (direction)
— m — — f para compresor m - [ind] compressor
  rotation (direction)
— m — — f — — m comprobado - [mec] checked
  compressor rotation (direction)
— m — — f — verificador m - [mec] fan rotation direction
— m — — f incorrecto - wrong rotation (direction)
— m — — f invertido - [mec] reversed, o
  changed, rotation (direction)
— m — — f verificado - [mec] checked rotation (direction)
— m — tránsito m - [transp] traffic direction
— m — trenzado - [cabl] lay direction

sentido m para veta f - [electr-magnet] seam direction
— m práctico - practical sense
— m transversal - [mec] crosswise direction
— m — a flujo m—[mec] across @ flow direction
— m — — — m de material(es) m - [mec] across
@ material(s) flow (direction
— m — unión f - [sold] across @ joint (direction)
— m único - [mec] single direction • [vial] one
way
sentido,da a - sensed; felt • heard
sentir v - . . . . • to sense • to get @ feeling •
to hear
— v aire m - to feel @ air
— v malestar m - to feel, ill, o sick
— v — m cualquiera - [medic] to be, sick, o
ill, in any way
— v (para) adentro - to get @ feeling
— v presión f - to feel @ pressure
— v sensación f - to feel @ sensation; to get
@ feeling
— v simpatía f - to sympathize
sentirse v enfermo - [medic] to feel sick
— v muerto - to feel dead
— v obligado - to feel, obliged, o you must
— v seguro - to feel safe
— v vivo,va - to feel alive
seña f - . . . . • trait
— f particular - special trait
— f vital - [biol] vital sign
— f vital para empresa f - [com] business' vital sign
señal f - . . . . • [mec] furrow • [deport] cue •
[comput] beep • [f.c.] signal
— f actual - [electrón] current signal
— f — reproducida - [electrón] displayed current signal
— f acústica - [electrón] acoustic signal
— f anterior - [electrón] previous signal
— v — recuperada - [electrón] recalled previous signal • recovered previous signal
— f audible - audible signal; audio; beep
— f — alta - [electrón] audio hi
— f — — de salida f - [electron] audio out hi
— f — baja - [electrón] audible lo
— f — — de salida f - [electrón] audio lo
— f — de entrada f - [electrón] audio in
— f — — salida f - [electrón] audible signal,
o audio, out
— f — normal - [electrón] normal, audible signal, o audio
— f — para entrada f - [electrón] input audio
— f — — salida f - [electrón] output audio
— f — recibida - [electrón] received audio
signal
— f auditiva - [comput] audible signal; audio
— f — alta - [electrón] hi, audio signal, o
audio
— f — — de salida f - [electrón] audible signal, o audio, out hi
— f — baja - [electrón] lo, audio signal, o
audio
— f — — de salida f - [electrón] audible signal, o audio, out lo
— f auditiva de entrada f - [electrón] audio in
(signal)
— f — — salida f — [electrón] audio out signal
— f cambiada - [electrón] changed signal
— f cambiante - [electrón] changing signal
— f carretera - [vial] road(side), o highway,
sign
— f con destello(s) m - [ind] flashing signal
— f conectada - [electrón] connected signal
— f corriente - [electrón] current signal
— f — reproducida f - [electrón] reproduced, o
displayed, current signal
— f de acuse m (de) recibo - [electrón] ac-
knowledge(ment) signal; handshake
— f — ajuste m - [mec] matchmark
— f — alarma f - [electrón] alarm signal
— f - alto - [vial] stop signal

señal f de amplitud f - [electrón] amplitude
signal
— f — — f insuficiente - [electrón] insufficient amplitude signal
— f — — f suficiente - [electrón] sufficient
amplitude signal
— f — debilitamiento m - sign of weakening
— f — desgaste m - [mec] wear sign
— f — deterioración f - deterioration sign
— f — deterioro m - véase señal f de deterioración f
— f — entrada f - [electróm] input, o incoming, signal
— f — — f audible - [electrón] input audio
— f — — f conectada - [electrón] connected
input signal
— f — — f válida - [electrón] incoming valid,
o valid incoming, signal
— f — frecuencia f - [electrón] frequency
signal
— f — — f apropiada - [electrón] apropriate,
o proper, frequency signal
— f funcionamiento - functioning signal
— f — — m apropiado - [electrón] appropriate
functioning signal
— f — — m inapropiado - malfunction(ing)
sign(al)
— f — oscilador m - [electrón] oscillator
signal
— f — peligro m - danger sign(al)
— f — presión f - pressure signal
— f — — f de gas m - [combust] gas pressure
signal
— f — — f — — m en horno m - [ind] furnace
gas pressure signal
— f — — f — — m — — m alto - [metal-
-prod] blast furnace gas pressure signal
— f eléctrica - [electr] electric signal
— f — para regulador m - [mec] throttle, o
control(ler), electrical signal
— f — — m para malacate m - [petról] drawworks, throttle, o control(ler), electrical
signal
— f en carretera f - [vial] highway sign
— f enana - dwarf signal
— f errónea - wrong signal; miscue
— f fiscalizada - monitored signal
— f externa - [electrón] external signal
— f generada - [electrón] generated signal
— f hidráulica - hydraulic signal
— f indicadora - indicating signal • [f.c.]
target
— f — de posición f de aguja(s) f - [f.c.]
(signal) target • points position sign
— f insuficiente - [electrón] insufficient
signal
— f interpretada - interpreted signal
— f luminosa - [vial] luminous signal
— f manipulada - [electrón] keyed signal
— f — para desplazamiento m - [electrón] shift
keyed signal
— f — — m de audiofrecuencia f—[electrón]
audio frequencey shift keyed, o A F S K,
signal
— f múltiple - [electrón] multiple signal
— f necesaria - necessary, o required. signal
— f — conectada - [electrón] connected necessary signal
— f neumática - [instrum] pneumatic signal
— f normal - normal signal
— f observada - monitored signal
— f para advertencia f - warning sign
— f — apareamiento m - [mec] matchmark
— f — desconexión f - [electrón] disconnecting, o strip(ping), signal
— f — espacio m - [electrón] space signal
— f — marca f - [electrón] mark signal
— f — parada f - [vial] stop signal
— f — polaridad f - [electrón] polarity signal
— f — — f apropiada - [electrón] appropriate,
o proper, polarity signal
— f — prevención f - warning, o alarm, o caution, signal

señal f para regulación f - [instrum] control
  signal
— f —— f para sensor m - [instrum] sensor
  control signal
— f —— f —— m para calor m - [instrum]
  heat sensor control signal
— f —— f —— m —— m para rotadora f a-
  limentadora - [cuerp.moled] rotator/feeder
  heat sensor control signal
— f — regulador m - [mec] throttle signal
— m —— m para malacate m - [petról] drawworks throttle signal
— f — relé m - [electrón] relay signal
— f — para sobrecorriente f - [electr-
  -oper] overcurrent relay signal
— f — salida f - [electrón] output signal •
  outgoing signal
— f —— f a red f telefónica - [comput] signal output to @ telephone network
— f —— f audible - [electrón] output audio
— f —— f conectada - [electrón] connected
  output signal
— f —— f válida - [electrón] outgoing valid,
  o valid outgoing, signal
— f — tránsito m - [vial] traffic, o highway,
  signal
— f —— m para carretera f - [vial] highway
  sign
— f presente - present signal
— f previa - [electrón] previous signal
— f — recuperada - [electrón] recalled previous signal
— f — reproducida - [electrón] displayed previous signal
— f recibida - [electrón] received signal
— f reciente - [electrón] recent signal
— f reclamada - [electrón] recalled signal
— f rechazada - [electrón] rejected signal
— f reflectora - reflective signal
— f reflejada - reflected signal
— f reproducida - [electrón] displayed signal
— f similar - similar signal
— f sobre velocidad f - [vial] speed signal
— f sonora - audible signal
— f válida - [electrón] valid signal
— f vigilada - monitored signal
— f visible - visible sign(al)
señalado,da a - signalled; indicated; pointed
  (out); designated; identified • marked; shown
  • displayed • remarkable; wonderful
— a claramente - signalled clearly
— a específicamente - pointed out specifically
— a expresamente - pointed out, expressly, o
  specifically
— a hacia - pointed, towards, o to
señalador m - signaler,ling • signal
— m para archivo m - [com] file signal(er)
— m — carretilla f - [ind] truck signaler
— m — ficha f - [com] file signal(er)
señalamiento m - signalizing,lization; signaling,
  showing; indicating,tion; pointing (out); de-
  signating,tion; identififcation,fying
— m a - pointing to
— m de presión f - [mec] pressure showing
— m —— f de aceite m - [mec] oil pressure
  showing
— m —— f — aire m - [mec] air pressure
  showing
— m en entrada f - [electrón] input signalling
— m — salida f - [electrón] output signalling
— m específico - specific, signaling, o pointing out
— m expreso - express, o specific, signaling, o
  pointing out
— m hacia - pointing to(wards)
— m polar - [electrón] polar signalling
— m — para entrada f - [electrón] polar input,
  o input polar, signalling
— m —— salida f - [electrón] polar output
  signalling
señalar v - . . . .; to, signal, o specify, o
  point out, o pinpoint, o identify, o desig-
  nate, o show, o indicate, o direct • to emphasize • to display • to dial

señalar v a rojo m - [instrum] to, be, o point
  to, @ red
— v amperaje m - [sold] to dial @, amperage, o
  current
— v claramente - to, signal, o mark, clearly
— v defecto m - to show @ defect
— v entrada f - [electrón] to signal @ input
— v — f polar - [electrón] to signal @ polar
  input
— v específicamente - to point out specifically
— v expresamente - to point out, expressly, o
  specifically
— v hacia - to point to(wards)
— v largada f - [deport] to flag off
— v límite m - to mark @ limit
— v presión f - [mec] to show @ pressure
— v — f de aceite m - [mec] to show @ oil
  pressure
— v — f — aire m - [mec] to show @ air pressure
— v rojo m - [instrum] to show @ red
— v rumbo(s) m - [fig] to set @ pace
señalero m - [f.c.] flagman
señalización f - signaling,lization; indicating
  • [vial] signing; signage
— f automática - [electrón] automatic signaling
— f completa - [vial] complete signaling
— f, equivocada, o errónea - miscue(ing)
— f incorrecta - [electrón] incorrect signaling
— f manual - [electrón] manual signaling
— f para agua m - [hidr] water signaling
— f —— m en Venturi m - [metal-prod] Venturi
  (flow) water, signaling, o indicating
— f — multifrecuencia f - [electrón] multifrequency signaling
— f —— f con tono m doble - [electrón] dual
  tone multifrequency signaling
— f — nivel m - level, indicating, o signaling
— f —— m para agua m - [metal-prod] water
  level, signaling, o indicating
— f —— m —— m para Venturi m - [metal-
  -prod] Venturi (flow) water, signaling, o indicating
— f — posición f - [transp] position marking
— f — prueba(s) f - [electrón] test signaling
— f — seguridad f - [segurid] safety, signaling, o marking
— f rechazada - [electrón] rejected signaling
señalizado,da a - posted
señalizar v - to post
señorita f - . . . • [petról] [Ven.] gin pole
separabilidad f - separability; severability
separable adv - . . . . • disconnectable
separación f - . . . - dividing; division; pulling apart • gapping; opening; spacing
  space • barrier • sorting; screening out •
  [mec] gap; space,cing; screening; clearance •
  loosening • [sold] gap; spacing; screen;
  clearance •[tub] pitch • [electr-instal] fanout • quartering, o splitting, out • crack;
  parting
— f accidental - accidental pulling apart
— f adecuada - suitable spacing
— f ajustada - [mec] adjusted gap
— f angosta - [sold] narrow gap
— f apropiada - [mec] suitable spacing
— f aumentada - increased, separation, o gap
— f con ángulo m inapropiado - improper angle
  separation
— f —— m de paleta(s) f - [ventil] improper blade angle separation
— f — bloque(s) m - [mec] blocking
— f — calce(s) m - [mec] shimming; blocking
— f — forma f de media luna - [mec] crescent
  shaped partition
— f conservada - [mec] maintained gap
— f correcta f - [mec] correct spacing
— f — para árbol m - [mec] correct shaft
  separation
— f costosa - [miner] high cost separation
— f de aceite m - oil separation
— f —— m para lavado m - wash oil separation

separación f de agua m - water separation
— f — — m para lavado m - wash water separation
— f — m — — m y aceite m - wash water and oil separation
— f — — m y aceite m - water and oil separation
— f — alambre m - [cabl] wire spacing
— f — conductor(es) m - [electr-instal] lead, o conductor, separating,tion, o spreading
— f — espira(s) f - [mec-resortes] coil separation
— f — gota(s) f - drop(s) separation
— f — grasa f - [sanit] degreasing
— f — labio(s) m de boquilla f - [mec] nozzle tip opening
— f — material m - material screening out
— f — mitad(es) f - [mec] half,ves separation
— f — — f de caja f - [mec] case half,ves separating,tion
— f — nivel(es) m - [constr] grade(s), separationg, o elevation
— f — — m estable - [constr] stable grade elevation
— f — — m protegida - [constr] protected grade elevation
— f — niebla f - [quím] mist separation
— f — partícula(s) f - particle(s) separation
— f — — f de líquido m - liquid particle(s) separation
— f — pavimento m - [autom-neumát] breaking loose
— f — pieza(s) f - [mec] part(s) separation
— f — placa(s) f - [mec] plate(s) separation
— f — plancha(s) f - [mec] plate(s) separation
— f — quijada f - [mec] jaw separation
— f — recubrimiento m - [metal-trat] stripping
— f — renta(s) f - [fin] income(s) separation
— f — rueda(s) f - [mec] wheel(s) separation
— f — — f guía - [agric-equip] guide wheel(s) spacing
— f — — f directriz,ces - [agric-equip] guide wheel(s) spacing
— f — tránsito m - [transp] traffic separation
— f — soldadura(s) f - [sold] weld separation; increment(s) pitch
— f — tensión(es) f - [mec] tension(s), o stress(es), separation
— f — voltaje(s) m - [electr-instal] voltages separation
— f difícil - difficult, o hard, separation
— f disminuida - [mec] decreased gap
— f económica - [miner] low cost separation
— f eliminada - eliminatws, o closed, gap
— f en aro m - [comb.int] ring gap
— f — — m para émbolo m - [comb.int] piston ring gap
— f — bujía f - [comb.int] spark plug gap
— f — — f ajustada - [comb.int] adjusted spark plug gap
— f — medio m líquido - [miner] liquid medium separation
— f — raíz f - [sold] root, opening, o spacing
— f — válvula f - [comb.int] valve clearance
— f entre barra(s) f - [mec] bar(s) spacing
— f — — f para armadura f - [constr] reinforcing bar(s) spacing
— f — — — f para losa f - [constr] slab reinforcing bar(s) spacing
— f — bloque(s) m - [mec] gap between blocks
— f — — m separador(es) - [mec] gap between @ grip block(s)
— f — centro(s) m - [mec] center-to-center spacing
— f — cimiento(s) m - [constr] fundation(s) spacing
— f entre corrugación(es) f - [mec] corrugation(s) pitch
— f — chapa(s) f - [sold] plate(s) separation
— f — espira(s) f - [mec-resortes] separation between coils
— f — larguero(s) m - [constr] stringer(s), separation, o spacing

separación f entre mitad(es) - half,lves separation
— f — — m de acoplamiento m - [mec] coupling half,lves separation
— f — pieza(s) f - [sold] gap (between parts)
— f — placa(s) f - [mec] plate separation
— f — plancha(s) - [mec] plate separation • [sold] gap between @ plate(s); gap
— f — poste(s) m - [constr] post spacing
— f — púas f - [alambre] barb(s) spacing
— f — ranura(s) f - [mec] spacing between slots
— f — recubrimiento m de electrodo m y plancha f - gap between @ electrode coating and @ plate
— f — remache(s) m - [mec] rivet(s) spacing
— f — — m circunferencial(es) - [mec] circumferential rivet(s) spacing
— f — rueda(s) f - [mec] wheel(s) spacing
— f excesiva - [sold] excessive gap
— f fácil - easy separating,tion
— f final - [mec] final, separation, o gap
— f física - [mec] physical separation
— f graduada - [constr] graduated spacing
— f — entre poste(s) m - [constr] graduated post(s) spacing
— f grande - [constr] wide, o large, opening
— f — en raíz f - [sold] large root opening
— f gravimétrica - pravity,vimetric separation
— f — en medio m líquido - [miner] liquid medium gravity,vimetric separation
— f igual - equal, separation, o space,cing
— f inapropiada - [mec] improper clearance
— f — de paleta(s) f - [ventil] blade improper, o improper blade, separation
— f — en válvula f - [comb.int] improper valve clearance
— f inicial - [mec] initial, separation, o gap
— f insuficiente - insufficient separation • [sold] insufficient gap
— f intermedia - intermediate, separation, o gap
— f — de partícula(s) f - particle(s) intermediate separation
— f — — f de líquido m - [ind] liquid particle intermediate separation
— f irregular - [sold] irregular gap
— f lógica - logical separation
— f magnética - magnetic separation
— f mantenida - [mec] maintained gap
— f máxima - [sold] maximum gap
— f mayor—[sold] greater, separation, o spacing
— f — en raíz f - [sold] larger root space,cing
— f menor - [sold] smaller, o narrower, gap
— f — en raíz f - [sold] smaller root, gap, o space,cing
— f mínima - [mec] minimum, space,cing, o gap
— f — entre brazo(s) m - [mec] minimum space between @ fork(s)
— f para árbol(es) m - [mec] shaft spacing
— f — peatón(es) m - [vial] pedestrian separation
— f pequeña - [sold] small, o narrow, gap
— f — en raíz f - [sold] small root spacing
— f plena - [mec] full separation
— f por etapa(s) f - [petról] stage, o phase, separation
— f por fracción f gravimétrica - gravimetric fraction separating,tion
— f — gravedad f - gravity separating,tion
— f — hundimiento m - [miner] sink(ing) separation
— f — — m y flotación f - [miner] sink(ing) and float(ing) separating,tion
— f preliminar - preliminary separation; [miner] scalping
— f provista - [mec] provided clearance
— f recomendada - [sold] recommended gap
— f reducida - [mec] reduced, separation, o gap
— f rellenada - [sold] bridged, o filled, gap
— f suficiente - sufficient, separation, o gap, o clearance
— f uniforme - uniform, separation, o clearance
— f variable - [constr] variable,rying spacing
— f — entre larguero(s) m - [constr] varying

stringer spacing
separación f visual,- visual separation
separadamente - separately • individually
separado,da a -separated; pulled apart; divided;
moved (farther) apart; split; quartered out;
spaced clear • loose; free standing • sorted
• screened out • gapped
— a accidentalmente - accidentally pulled apart
— a a penas - just off
— a con bloque(s) m - [mec] blocked (away)
— a — calce(s) m - [mec] shimmed; blocked away
— a de pavimento m—[autom-neumát] broken loose
— a fácilmente - separated easily
— a gravimétricamente - separated gravimetri-
cally
— a logicamente - separated logically
— a magnéticamente - separated magnetically
— a por . . . minuto(s) - . . . minute(s) apart
— a suficientemente - separated sufficiently
separador m - . . . . • spacer •ind] separation
tank • [petról] separator; scrubber • [mec]
brace • stripper • set • [cojin] cage; separator • [metal-prod] véase separador m para
agua m para Venturi • [electrón] standoff |
a - [sanit] screening
— m a múltiple m - [mec] spacer to manifold
— m ciclónico - [coque] cyclone separator
— m con espesor m apropiado - [mec] correctly
sized spacer
— m — — m original - [mec] original size(d)
spacer
— m — — f nominal - [mec] nominal spacer
— m de acero m - steel, separator, o set
— m — — m dulce - [mec] soft steel set
— m — caucho m - [mec] rubber spacer
— m — cobre m - [sold] copper, spacer, o block
— m — fibra f - [mec] fiber spacer
— m — — f cerámica - [mec] ceramic fiber
spacer
— f — latón m - [mec] brass, set, o spacer
— m — madera f - [mec] wood(en) spacer
— m — metal - [mec] metal spacer
— m delgado - [mec] thin spacer
— f dispuesto - [mec] arranged spacer
— m electrostático - [constr] electrostatic
separator
— m exterior - [mec] outer spacer
— m — para cojinete(s) m - [cojin] outer bearing spacer
— m grande - [mec] large spacer
— m — para cojinete m - [mec] large bearing, o
bearing large, spacer
— m grueso - [mec] thick spacer
— m inferior - [mec] lower, o bottom, spacer
— m insertado - mec] inserted, set, o spacer
— m interior - [mec] inner, o inside, spacer
— m — para cojinete m - [cojin] inner bearing,
o bearing inner, spacer
— m intermedio - [mec] middle, o intermediate,
spacer
— m magnético - magnetic separator
— m martillado - [mec] driven, set, o spacer
— m más delgado - [mec] thinner spacer
— m — grueso - [mec] thicker spacer
— m mayor - [mec] larger spacer
— m menor - [mec] smaller spacer
— m metálico - [mec] metal(lic) spacer
— m normal - [mec] normal, spacer, o separator
— m — en Venturi - [metal-prod] (Venturi)
scrubber normal (water) separator
— m original - [mec] original spacer
— m para abrasivo(s) m - [mec] abrasive(s) separator
— m — aceite m - [ind] oil, separating,tion, o
separator, tank
— m — — m ligero - [coque] light oil(s) separator
— m — — m liviano m - [petról] light oil separator
— m — — m para lavado m - wash(ing) oil, separator, o separating,tion, tank
— m — ácido m - [coque] acid separator
— m — acoplamiento m - [mec] coupler spacer

separador m para agua m - [mec] water separator
(tank)
— m — — m para lavado m y aceite m - [ind]
wash(ing) water and oil separation tank
— m — — Venturi m - [metal-prod] (scrubber, o Venturi) water separator
— m — — m purgado - [comb.int] drained, o
bled, water separator
— m — — m verificado - [mec] checked water
separator
— m — aire m - air separator
— m — amolar,ladura f - [mec] grind(ing) spacer
— m — amortiguador m - [mec] damp(en)er spacer
— m — — m para línea f - [mec] line damp(en)er
spacer
— m — — m — — f para aspiración f - [mec]
suction line damp(en)er spacer
— m — arandela f - [mec] washer spacer; spacer
washer
— m — árbol m - [mec] shaft spacer
— m — — m para palanca f - [mec] lever shaft
spacer
— m — armado m - [mec] assembling spacer
— m — — m colocado - [mec] installed assembling spacer
— m — — m instalado - [mec] installed assembling spacer
— m — — m, quitado, o removido, o quitado -
[mec] removed assembling spacer
— m — arranque m - [comb.int] starting spacer
— m — — m con manija f - [comb.int] crank
starting jaw spacer
— m — bomba f - [comb.int] pump spacer
— m — — f para aceite m - [comb.int] oil pump
spacer
— m — cierre m - [mec] seal, spacer, o retainer
— m — para aceite m - [mec] oil seal
spacer
— m — cojinete m - [mec] bearing spacer
— m — — m cambiado - [mec] changed bearing
spacer
— m — — m grande - [mec] large bearing spacer
— m — — m instalado - [mec] installed bearing
spacer
— m — — m para árbol m - [mec] shaft bearing
spacer
— m — — m — — m para tambor m - [mec] drum
shaft bearing spacer
— m — — m — biela f - [mec] connecting rod, o
pitman, spacer
— m — — m para piñón m - [mec] pinion bearing
spacer
— m — — m — — m cambiado - [mec] changed
pinion bearing spacer
— m — — m — — m con espesor m nominal—[mec]
nominal pinion bearing spacer
— m — — m pequeño - [mec] small bearing
spacer
— m — — m removido - [mev] removed bearing
spacer
— m — colador m - [comb.int] strainer spacer
— m — cono m - [cojin] cone spacer
— m — condensación f - [petról] drip
— m — descarga f - [mec] discharge spacer
— m — — f con perno(s) m prisionero(s) - [mec]
studded discharge spacer
— m — — f — prisioneros m - [mec] studded
discharge spacer
— m — disco m - [mec] disk spacer
— m — — m ranurado - [mec] slotted disk spacer
— m — empuje m - [mec] thrust spacer
— m — emergencia f - [mec] emergency spacer
— m — — f para Venturi m - [metal-prod]
scrubber emergency (water) separator
— m — encabezadora f—[mec] heading tool spacer
— m — enjugador m - [mec] wiper spacer
— m — — m para aceite m - [mec] oil wiper
spacer
— m — — m — — m para vástago m - [mec] rod
oil wiper spacer
— m — — m — — m — — m intermedio - [mec]
intermediate rod oil wiper spacer
— m — estampa f - [mec] insert spacer

ser v̲ dueño m̲ de patente f̲ (de invención f̲)

separador m̲ para estampa f̲ para troquel m̲—[mec]
die insert spacer
— m̲ — gas m̲ - [petról] gas separator
— m̲ — — m̲ y petróleo m̲ - [petról] gas and oil
separator
— m̲ — gota(s) f̲ - [mec] drop separator
— m̲ — hormigón m̲ - [hormig] concrete spacer
— m̲ — junta f̲ - [mec] joint separator • seal
spacer
— m̲ — leva f̲ - [mec] cam spacer
— m̲ — montaje m̲ - [mec] mounting spacer
— m̲ — — m̲ para separador m̲ - [mec] stabilizer
mounting spacer
— m̲ — palanca f̲ - [mec] lever spacer
— m̲ — perno m̲ - [mec] bolt spacer
— m̲ — — m̲ con ojo m̲ - [mec] eye bolt spacer
— m̲ — petróleo m̲ - [mec] oil separator
— m̲ — placa f̲ - [electr-instal] plate, separator, o̲ spacer
— m̲ — — f̲ para acumulador m̲ - [electr-instal]
battery plate separator
— m̲ — — f̲ retenedora - [mec] retainer plate
spacer
— m̲ — polvo m̲ - [sold] dirt separator •
[metal-prod] dust catcher
— m̲ — prensaestopa(s) f̲ - [mec] packing box
spacer
— m̲ — sello m̲ - [mec] seal spacer
— m̲ — — m̲ para aceite m̲ - [mec] oil seal
spacer
— m̲ — soporte m̲ - [mec] support spacer
— m̲ — — m̲ para cojinete m̲ - [mec] bearing
support spacer
— m̲ — — m̲ — m̲ para bancada f̲ - [mec] main
bearing support spacer
— m̲ — transmisión f̲ - [mec] transmission, separator, o̲ spacer
— m̲ — troquel m̲ - [mec] die spacer
— m̲ — — m̲ para recalcadora f̲ - [mec] heading
tool spacer
— m̲ — válvula f̲ - [mec] valve spacer
— m̲ — — f̲ neumática - [mec] air valve spacer
— m̲ — — f̲ para aire m̲—[mec] air valve spacer
— m̲ — ventilador m̲ - [ventil] fan spacer
— m̲ — Venturi m̲ - [metal-prod] Venturi, o̲
scrubber, separator; véase separador m̲ para
agua m̲ para Venturi
— m̲ pequeño m̲ - [mec] small spacer
— m̲ — para cojinete m̲ - [mec] small bearing, o̲
bearing small, spacer
— m̲, quitado, o̲, removido, o̲ sacado - [mec]
removed spacer
— m̲ superior - [mec] upper, o̲ top, spacer
— m̲ transversal - [constr] transverse spacer
— m̲ variable - [mec] variable spacer
— m̲ — para cojinete m̲ - [mec] variable bearing
spacer
separar v̲ - . . . ; to space; to move (farther),
away, o̲ apart; to pull away; to quarter; to
split out; to pull apart • to cut off • to
loosen • to screen out
— v̲ accidentalmente - to, separate, o̲ pull
apart, acidentally
— v̲ apropiadamente - to space suitably
— v̲ con bloque(s) v̲ - [mec] to block (apart)
— v̲ — calce(s) m̲ - to shim • to block
— v̲ — facilidad f̲ - to separate easily
— v̲ conductor(es) m̲ - [electr-instal] to, separate, o̲ spread, @, lead(s), o̲ conductor(s)
— v̲ electrodo m̲ - [sold] to space @ electrode
— v̲ espira(s) f̲ - [mec-resortes] to separate @
coil(s)
— v̲ fácilmente - to separate easily
— v̲ fracción f̲ - to separate @ fraction
— v̲ — f̲ gravimétrica - to separate @ gravimetric fraction
— v̲ grasa f̲ - [sanit] to degrease • [mec] to
remove @ grease
— v̲ gravimétricamente - to separate gravime-
trically
— v̲ horno m̲ - [metal-prod] to, separate, o̲ cut
off, o̲ isolate, @ furnace

separar v̲ impureza(s) f̲ - [metal-prod] to separate @ impurity,ties
— v̲ lógicamente - to separate logically
— v̲ magnéticamente - to separate magnetically
— v̲ material m̲ - [ind] to, separate, o̲ screen
out, @ material
— v̲ mitad(es) f̲ - [mec] to separate @
half,lves
— v̲ — f̲ de caja f̲ - [mec] to separate @ case
half,lves
— v̲ pieza f̲ - [mec] to separate @ part
— v̲ plancha f̲ - [mec] to. separate, o̲ move
apart, @ plate
— v̲ por etapa(s) f̲ - [petról] to,
stage-, o̲ phase-, separate
— v̲ quijada(s) f̲ - [mec] to, separate, o̲ open,
@ jaw(s)
— v̲ recubrimiento m̲ - [metal-trat] to strip
— v̲ suficientemente - to separate sufficiently
— v̲ torón(es) m̲ - [cable] to, separate, o̲ push
apart, @ strand(s)
— v̲ tránsito m̲ - [transp] to separate @ traffic
— v̲ verticalmente - to space vertically
separarse v̲ - to, pull, o̲ break, apart, o̲ loose
— v̲ de pavimento m̲ - [autom-neumát] to break
loose (from @ pavement)
separata f̲ - [public] reprint
séptum m̲ - [válv] septum
sepultado,da a̲ - buried
sepultamiento* m̲ - véase sepultura f̲
sepultar v̲ puente m̲ - [constr] to bury @ bridge
sepultura f̲ - . . . . burying
— f̲ de puente m̲ - [constr] bridge burying
sequía f̲ severa - [meteorol] severe drought
ser v̲ agradable - to be agreeable • to be fun
— v̲ aparente - to be apparent
— v̲ así que - so
— v̲ asunto m̲ importante - to be @ big deal
— v̲ capaz - to be capable
— v̲ — de - to be, capable, o̲ able, to, o̲ of
— v̲ causa f̲ - to be @ cause
— v̲ centro m̲ - to be @ center; to centralize
— v̲ — m̲ de autoridad f̲ - [admin] to be @ center of @ authority; to centralize
— v̲ conveniente - to be convenient
— v̲ cuidadoso,sa - [segurid] to be careful
— v̲ de aplicación f̲ - to apply
— v̲ — desear - to be desirable
— v̲ — despreciar v̲ - [fam] to be nothing to
sneeze at
— v̲ — esperar v̲ - to be expected
— v̲ — extrañar v̲ - to be no wonder
— v̲ — importancia - to be important
— v̲ — — f̲ primordial - to be, essential, o̲ of
primary importance; to be @ must
— v̲ — interés - interestingly
— v̲ — — notar - interestingly
— f̲ — notar - interestingly; notably
— v̲ — rigor - to be required
— v̲ deseable - to be desirable
— v̲ diferente - to be different; to differ
— v̲ divertido - to be fun
— v̲ dueño m̲ - to, own, o̲ be @ owner
— v̲ — m̲ de derecho(s) m̲ - [legal] to own @
right(s)
— v̲ — m̲ — — de patente f̲ (de invención f̲) -
to, own, o̲ be @ owner, of @ patent right(s)
— v̲ — m̲ — — m̲ literario(s) - [legal]
to, own, o̲ be @ owner, of @ copyright
— v̲ — m̲ — marca f̲ - [legal] to, own, o̲ be @
owner, of @ mark
— v̲ — m̲ — — f̲ de fabricación f̲ - [legal] to,
own, o̲ be @ owner, of @ trade mark
— v̲ — m̲ — — f̲ industrial - [legal] to, own,
o̲ be @ owner, of @ trade mark
— v̲ — m̲ — — f̲ registrada - [legal] to,
own, o̲ be @ owner, of @ trade mark
— v̲ — m̲ de movimiento m̲ de vehículo m̲ - to
control @ vehicle
— v̲ — m̲ — nombre m̲ comercial - [legal] to,
own, o̲ be @ owner, of @ trade name
— v̲ — m̲ — patente f̲ (de invención f̲) -[legal]
to, own, o̲ be @ owner, of @ patent

ser v dueño m de propiedad f intelectual -
[legal] to, own, o be @ owner, of @ copyright
— v **esencial** - to be essential
— v **evidente** - to be evident • to show
— v **exponente** m - to be @ exponent • to personify
— v **factible** - to be feasible
— v **halagüeño** - to be (all) rosy
— v **hipócrita** m - to be @ hypocrite,tic(al)
— v **humano** - human being • [admin] human
— v **importante** - to be important
— v **elegible** - to be eligible
— v **legible** - to be, legible, o readable
— v **más (de)moroso,sa** - to be (more) time consuming
— v **mayor** - to, exceed, o be, greater, o more
— v **menor** - to be, smaller, o less
— v **menos** - to be less
— v **— moroso,sa** - to be less time consuming
— v **motivo m para reclamación f** - to be claim-
able • (@ claim) to be in order
— v **necesario,ria** - to be, necessary, o essential ⌈ adv - (de) if necessary
— v **posible** - to be possible
— v **propietario m** - [legal] to be @ owner
— v **— m de derecho(s) m literario(s)** - [legal]
to, own, o be @ owner, of @ copyright
— v **— m — marca f de fabrica(ción) f**—[legal]
to, own, o be @ owner, of @ trade mark
— v **— — — f industrial** - [legal] to, own, o
be @ owner of, @ trade mark
— v **— — — f registrada** - [legal] to, own, o
be @ owner, of @ (registered) trade mark
— v **— nombre m comercial** - [legal] to, own, o
be @ owner, of @ trade, mark, o name
— v **— m de patente f** - [legal] to own @ patent
— v **primordial** - to be @ must
— v **puntero m** - [deport] to, be first, o lead @
way
— v **puntual** - to be punctual
— v **responsable** - [admin] to be, responsible, o
accountable
— v **— directamente** - to be directly responsible • to report directly
— v **— por** - to, account, o be responsible, for
— v **ruidoso,sa** - to be, o sound, noisy
— v **según norma f** - to be standard
— v **susceptible**—to be, susceptible, o eligible
— v **víctima** - [deport] to, be, o fall, @ victim
seriado,da* a - serial; produced in @ series;
véase también **producido,da en serie f**
— **a paralelo m** - [electrón] serial-to-parallel
serialización f - serialization,zing; serial
numbering • mass production
serializado,da* a - serialized; numbered serially
serializador,ra a - serializing
serializar f - to serialize; to number serially
sericite m - [miner] sericite
serie f - . . .; set • range • system - [electr]
serie(s); in @ line - [electrón] burst •
[fin] family • [deport] streak; sweep
— f **anterior** - former, o old series
— f **comanche** - [petról] Comanche series
— f **completa** - complete series • [ind] whole
family • [com] full line
— f **común** - common series
— f **con borde(s) no protegido(s)** - unprotected
edge series
— — f — — m **protegido(s)** - protected edge(s)
series
— f **— circuito m** - [electr-instal] circuited
series
— f **de . . . .** in @ row
— f **— borne(s) m** - [electr-instal] stud series
— f **— carga(s) f explosiva(s)** - [explos] powder load(s) series
— f **— carrera(s) f** - [deport] racing series
— f **— — f fuera de carretera f** - [deport]
off-road (race) series
— f **— — f Pro Rally** - [deport] Pro Rally
(races) series
— f **— eje(s) m** - [mec] axle(s) series

serie f de ensayo(s) m - test(s) series
— f **— equipo(s) m elevador(es)** - [grúas] lifting system
— f **— . . . evento(s) m** - [deport] . . .
event(s) series
— f **— fórmula(s) f** - [deport] formulas series
— f **— impulso(s) m** - [electrón] burst
— f **— máquina(s) f** - [mec] machine(s), series,
o family
— f **— modelo(s) m** - model(s) series
— f **— — m identificado** - identified model
series
— f **— neumático(s) m** [autom-neumát] tire(s)
series
— f **— m radial(es)** - [autom-neumát] radial
(tires) series
— f **— película(s) f** - [fotogr] film(s) series
— f **— perfil(es) m** - [metal-lam] shape(s)
series
— f **— petaca(s) f** - [metal-prod] cooling plate
circuit • plate(s) circuit
— f **— — f en cuba f** - [metal-prod] bosh (cooling) plate(s) circuit
— f **— — f etalaje m** - [metal-prod] bosh
(cooling) plate(s) circuit
— f **— pieza(s) f** - [mec] part(s) serie(s)
— f **— propiedad(es) f** - property,ties set
— f **— prueba(s) f** - test(s) series
— f **— radial(es) m** - [autom-neumát] radial
series
— f **— rodillo(s) m** - [metal-lam] roll(s)
series
— f **— — m conformador(es)** - [metal-lam]
shaping roll(s) series
— f **— tiempo(s) m** - [ind] time(s) series
— f **— tres posición(es) f mejor(es)** - [deport]
one-two-three sweep
— f **— triunfo(s) m** - [deport] winning streak
— f **— tubería(s) f** - [constr] pipe(s) series •
multiple pipe(s)
— f **— vista(s) f** - [fotogr] shot(s) series
— f **electromotriz** - electromotive series
— f **en etalaje m** - [metal-prod] bosh circuit
— f **tándem** - tandem series
— f **fuera de carretera f** - [deport] off-road
series
— f **homóloga** - homologous series
— f **identificada** - identified series
— f **ininterrumpida** - uninterrupted series •
[deport] streak
— f **interrumpida** - interrupted series
— f **larga** - [deport] long series
— f **— de triunfo(s) m** - [deport] long winning
streak
— f **liviana** - light series
— f **media(na)** - medium series
— f **nueva** - new series
— f **numérica** - numerical series
— f **óptima** - optimum, series, o set
— f **para inventario m** - [contab] stock series
— f **pesada** - [metal-lam] heavy series
— f **Pro Rally** - [autom-neumát] Pro Rally series
— f **tronchada** - broken, series, o streak
— f **vieja** - old series
seriedad f de propuesta f - [legal] proposal
seriousness
serpenteado,da a - [metal-lam] snake,ky •
[autom-neumát] squirmed
serpenteante a - snaking • squirming
serpentear v - . . .; to curve
serpenteo m - [autom-neumát] squirm(ing)
— m **aumentado** - [autom-neumát] increased squirm
— m **controlado** - [autom-neumát] controlled
squirm(ing)
serpenteo m - snaking • [neumát] squirming
— m **de elemento m** - [autom-neumát] block
squirm(ing)
— m **— — m de banda f para rodamiento m** -
[autom-neumát] tread block squirm(ing)
— m **gobernado** - [autom-neumát] controlled
squirm(ing)
— m **reducido** - [autom-neumát] reduced squirming
serpentina f - . . . . • [tub] . . .; heating tube

serpentina f de tubo m de cobre m - [tub] coiled copper tube
— f para aceite m - [ind] oil coil
— f — agitación f - [tub] agitating coil
— f — calefacción f - [cald] heating coil
— f — calentamiento m - [tub] heating coil
— f — — m de tubería f - heating pipe coil
— m — — m y agitación f - heating and agitating coil
— f para tubería f - [tub] pipe,ping coil
serrano,na a - . . .; mountain • transmountain
serrucho m con punta f - [herram] keyhole saw
— m para mano f - [herram] hand saw
— m — trozar v - [herram] (hand) crosscut saw
servicio m - . . . • performance • utility • [ind] sanitary building (and locker rooms) • utility • [ind] department
— m activo - [pers] active service
— n adicional - additional service
— m adjudicado - [ind] awarded service
— m administrativo - [admin] management service
— m — m para fiscalización f - [admin] management control(ling) service
— m — — organización f - [admin] management organizing service
— m — — planificación f - [admin] management planning service
— m — para procesamiento m de dato(s) m - [admin] management data processing service
— m adquirido - acquired, o procured, service
— m afable m - friendly service
— m alimentario - [culin] food service
— m alimenticio - [culin] food service
— m analítico - [ind] analytical service
— m análogo - analogous, o similar, service
— m antes de compra f - [ind] prepurchase service
— m apropiado - appropriate, o proper, service
— m atendido - supplied, o provided, service
— m autorizado - authorized service
— m — en fábrica f - [ind] factory authorized service
— m auxiliar - [admin] auxiliary, o staff, service
— m — para administración f - [admin] auxiliary administrative service
— m bajo tierra f—[constr] underground service
— m brindado - provided, o supplied, service
— m capaz - capable service
— m civil - [pol] civil service
— m — de marina f - [milit] navy civil service
— m comercial - commercial service
— m completado - completed service
— m completo - complete, o comprehensive, service
— m — brindado - provided, complete, o comprehensive, service
— m — para cliente m - total customer service
— m — para motor m - [electr-mot] complete, o comprehensive, motor service,cing • [comb.-int] complete, o comprehensive, engine service,cing
— m — — — m brindado - [electr-mot] complete, o comprehensive, motor service, provided, o supplied • [comb.int] complete, o comprehensive, engine service, provided, o supplied
— m con duración f larga - [ind] long service
— m — velocidad f alta - [f.c.] high-speed service
— m confiable - dependable service
— m consultivo - [ind] consultation service
— m — para determinación f de falla(s) f - [ind] trouble-shooting consultation service
— m continuado - continuous service
— m continuo - continuous, service, o duty
— m contratado - [ind] contracted service
— m — empleado - [ind] used contracted service
— m — usado - [ind] used contracted service
— m correspondiente - [legal] corresponding, o respective, service
— m corriente - [ind] routine service
— m — para conservación f - [ind] routine maintenance service,cing
servicio m de aceite m - [ind] oil(ing) service
— m — — m motriz - [petról] power oil service
— m — balsa(s) f - [transp] ferry service
— f — calidad f - quality service
— f — — f más alta - top quality service
— m — — f óptima - top quality service
— m — Camino(s) m - [pol] Roads Department
— m — consulta f - [ind] consultation service
— m — — f prestado - [ind] consultation, furnished, o provided
— m — — f proporcionado - [ind] furnished consultation
— m — consultoría f - [ind] consulting service
— m — — f adjudicado - [ind] awarded consulting service
— m — contacto,tador - [sold] contact life
— m — correo(s) m - [pol] postal service
— m — departamento m - [ind] department service
— m — — m para combustión f - [combust] combustion department service
— m — hombre(s) hora - [labor] man/hour service
— m — importación f - importation service • imported, services, o supplies
— m — ingeniería f - [ind] engineering service
— m — — f profesional - [ind] professional engineering service
— m — — m supervisora - [ind] supervisory engineering service
— m — lanzadera f - [transp] shuttle service
— m — mensajeros m y correspondencia f - [com] (messenger and) mail service
— m — Muestreos de Suelo m - [pol] Soil Sampling Service
— m — oxígeno m - [ind] oxygen service
— m — personal m - [labor] personnel service
— m — — m propuesto - [labor] proposed personnel service
— m — petróleo m motriz - [petról] power oil service
— m — proveedor m - [ind] supplier('s) service
— m — — m con calidad f - [ind] supplier('s) quality service
— m — rutina f - [ind] routine service,vicing
— m — — f sencillo - [mec] routine simple, o simple routine, service
— m — — f simple - [mec] simple routine service
— m — semirremolques m - [transp] trailer(s) service
— m — — m carreteros - [transp] trailer service
— m — — m — sobre vagón(es) m plataforma - [transp] piggyback (service)
— m — subcontratista f - [ind] subcontractor(s) service
— m — supervisor m - [ind] supervisor('s) service
— m — — m para montaje m - erection supervisor(s) service
— m — técnico m - [ind] technicians service
— m — — m supervisor(es) - [ind] supervising technician(s) services
— m — Transportes m Metropolitanos - [transp] Rapid Transit, System, o District
— m — — m — de Zona f de (la Ciudad f de) San Francisco - [transp] Bay Area Rapid Transit District
— m en aeropuerto m - [transp] airport service
— m — carretera f - [transp] road(side) service
— m descri(p)to - described service
— m desinteresado - missionary work
— m después de compra f - [com] postpurchase, o after purchase, service
— m diagnóstico - [ind] diagnostic service
— m — en laboratorio m - [ind] laboratory diagnostic service
— m directo - direct service
— m — de personal m - [labor] direct personnel service
— m disponible - available service
— m doméstico - [labor] domestic service
— m durable,radero • durable service

servicio m económico - economical service
— f **efectivo** - effective service,cing • field service
— m **eficiente** - [ind] efficient service
— m **eléctrico** - [electr-oper] electric(al) service
— m — **para horno** m - [ind] furnace electrical service
— m — — — m **eléctrico** - [metal-prod] arc furnace electric service
— m **electrónico** - [electrón] electronic service
— m **en fábrica** f - [ind] factory service
— m — **hotel** m - [viajes] hotel service
— m — **laboratorio** m - [ind] laboratory service
— m — **obra** f - [ind] field service
— m — **tierra** f - [transp] ground, service, o arrangement
— m **entre conductores** m - [electr-instal] service between conductor(s)
— m **especializado** - [ind] specialized service
— m **específico** - specific service
— m — **propuesto** - specific proposed service
— m **Estadounidense de Guardacostas** m - [milit] United States Coast Guard
— m — **para Conservación** f **de Suelo(s)** m - [pol] Soil Conservation Service
— m **eventual** - eventual, o standby, service
— m **exigido** - required service
— m **experto** - [ind] expert service,cing
— m **externo** - external service
— m — **de mensajeros** m **y correspondencia** f - [com] external mail service
— m **extranjero** - foreign service
— m **extremo** - véase **servicio** m **pesado**
— m **fácil** - [ind] easy service
— m **forestal** - [pol] forest(ry) service
— m — **(Estadounidense)** - [pol] (United States) Forest Service
— m **fuera de carretera** f - [autom] off-road service
— m — — **taller** m - [ind] field service
— m **garantizado** - guaranteed service
— m **general** - general service • utility,ties (and services)
— m **geológico** m - [geol] geological service
— m — **estadounidense** - [pol] United States Geological Service
— m **gubernamental** - [pol] government service
— m **hidráulico** - [ind] hydraulic service(s) (department)
— m **hotelero** - [viajes] hotel service(s)
— m **igual** - equal service
— m **importado** - [ind] imported service
— m **industrial** - [ind] industrial service
— m — **de Dow** m - [ind] Dow Industrial Service
— m — — — m **para Sistemas** m **para Lubricación** f **e Hidráulico** - [ind] Dow Industrial Service Specifications for Lube,brication and Hydraulic Systems
— m **ininterrumpido** - uninterrupted service
— m — **de grúa(s)** f - [grúas] uninterrupted crane(s) service
— m **intentado** - attempted service
— m **intermitente**—intermittent, service, o duty
— m **interno** - [ind] internal service • yard service
— m — **de mensajeros y correspondencia** f—[com] internal mail (and messenger) service
— m — **en planta** f - [ind] plant, internal, o yard, service
— m **libre** - free service • free performance
— m — **de problemas** m - trouble-free service
— m **liviano** - light, service, o duty
— m — **corriente** - conventional light duty
— m — **intermitente**—light intermittent service
— m **local** - [com] local service
— m **más rápido** - faster service
— m **máximo** - maximum service
— m **mecánico** - [mec] mechanical service • routine servicing
— m **mediano** - [ind] medium duty
— m **médico** - [ind] medical service(s)
— m **medido** - [telecom] measured service

servicio m mundial - [ind] world(wide) service
— m **municipal** - [fin] municipal service • municipal utility
— m **nacional** - national, o domestic, o local, service
— m — **de Caminos** m - [vial] National, o Federal, Roads Department
— m **necesario** - necessary service
— m **normal** - normal service
— m **ofrecido** - offered service • tendered service
— m **óptimo** - optimum, o top quality, service
— m **para administración** f - administration service • [admin] management service(s)
— m **para administración** f **de crédito** -[fin] credit administration service
— m — **aplicación(es)** f **para cliente(s)** - customer application service
— m — **árbol** m **impulsor** - [autom-mec] drive shaft service,cing
— m — — m **para impulsión** f - [autom-mec] drive shaft service,cing
— m — **asesoramiento** m - consulting,tant service
— m **para avalúo(s)** m - appraisal service
— m — **bomba** f - [bombas] pump service
— m — — f **para inyección** f - [comb.int] injection pump service
— m — **capital** m - [fin] capital service,cing
— m — **caracterización** f - [com] identification service
— m — **carcasa** f - [mec] housing service,cing
— m — **cliente(s)** m - [com] customer service • [ind] field service,cing
— m — **cojinete** m - [cojin] bearing service
— m — — m **planetario** - [mec] sun gear service
— m — **compra(s)** f - purchasing service(s)
— m — **conservación** f - [ind] maintenance service,cing • routine service,cing
— m — **Conservación** f **de Suelo(s)** m - [pol] Soil Conservation Service
— m — — f **preventiva** - [ind] preventive maintenance service
— m — **construcción** f - [constr] construction service
— m — **control** m - [admin] control service
— m — **deuda** f - [fin] debt service,cing
— m — **dirección** f - supervision service
— m — — f **para construcción** f - [constr] construction supervision service
— m **para eje(s)** m - [autom-mec] axle service
— m — — m **impulsor** - [autom-mec] drive,ving axle service,cing
— m — — m **para impulsión** f - [autom-mec] drive,ving axle service
— m — **embrague** m - [mec] clutch service
— m — — m **corredizo** - [mec] sliding clutch service
— m — — m **deslizante** - [mec] sliding clutch service
— m — **emergencia(s)** f - [mec] emergency service,cing
— m — **entrega** f - delivery service
— m — — f **de paquete** m - [transp] United Parcel service
— m — **fabricación** f - manufacturing service • fabrication service
— m — **filtro** m - [mec] filter service
— m — **fiscalización** f - [ad., control(ling) service
— m — **gobierno** m - [pol] government service
— m — **grúa(s)** f - [grúas] crane service
— m — **horno(s)** m - [ind] furnace(s) service
— m — — **eléctrico** - arc furnace service
— m — **información** f - information service
— m — — f **técnica** - [ind] technical information service
— m — **inspección** f - [ind] inspection service
— m — — f **de comprador** m - purchaser('s) inspection service
— m — — f **por representante** m **de comprador** m - véase **representante** m **de servicio** m **para inspección** f **por comprador** m

servicio m para instalación f - [ind] installation, o facility,ties, service
— m — investigación f - research service
— m — — f y desarrollo m - [ind] research and
development service
— m — lubricación f - [lubric] lubrication
service
— m — magneto m - [comb.int] magneto service
— m — mantenimiento m - [ind] maintenance service
— m — — m de pozo(s) m - [petról] well service,cing operation
— m — motor m - [electr-mot] motor service •
[comb.int] engine service,cing
— m — neumático m - [autom=neumát] tire service,cing
— m — organización f - [admin] organization, o
organizing, service
— n — planeamiento m - planning service
— m — — m general - general planning service
— m — planificación f - planning service
— m — — m y fiscalización f - [admin] planning and control service
— m — planta f - [ind] plant, o mill, service
— m — — f caliente - [ind] hot plant service
— m — — f para acero m - [metal-prod] steel,
plant, o mill, service,cing
— m — presión f - [ind] pressure service,cing
— m — procesamiento m - [electrón] processing
service
— m — — m para dato(s) m - [electrón] data
processing service
— m — proyección f - [ind] design(ing) service
— m — proyecto m - project service,cing
— m — puesta f en marcha f - [ind] start-up
service,cing
— m — rendimiento m - [ind] performance service,cing
— m — — m alto - [ind] (high) performance
service,cing
— m — reparación f - [ind] repair service,cing
— m — — f para pozo(s) m - [petról] well repair servicing
— m — respaldo m - support service,cing
— m — — m técnico - technical support service
— m — rodillo(s) m - [metal-lam] roll(s) service,cing
— m — rueda f - [mec] wheel(s) service,cing
— m — secador m - [ind] drier service,cing
— m — supervisión f - supervision,sory service
— f — — f para montaje m - [ind] erection
supervision,sory service(s)
— m — — — f y puesta f en marcha f -
[ind] erection and start-up supervision service
— m — — f para puesta f en marcha f - [ind]
start-up supervisory service,cing
— m — — f técnica - [ind] technical supervisory,sing,sion service,cing
— m — transmisión f - [ind] transmission service,cing
— m — vigilancia f - [pol] security service
— m personal - personal service
— m pesado - heavy, service, o duty
— m corriente - conventional heavy duty
— m por distribuidor m - distributor('s) service
— m — personal m auxiliar - [admin] staff('s)
service
— m — supervisor m - supervisor('s) service(s)
— m — teléfono m - [ind] (tele)phone service
— m portuario - [náut] port service(s)
— m positivo - positive, service, o performance
— m postal.- [comunic] postal service
— m posventa - [com] after-@-sale service
— m prescri(p)to - prescribed service
— m prestado - service, rendered, o provided, o
supplied, o performed, o offered
— m preventivo - preventive service
— m — de inspección f - [ind] guardian inspection service
— m privado - private service
— m profesional - professional service

servicio m profesional especializado - specialized professional service
— m prolongado - extended, o lasting, service
— m proporcionado - provided, o supplied, service
— m propuesto - proposed, o intended, service
— m provisto - service, provided, o supplied
— m público - public service • pl - utilities
— m — apropiado - [serv.públ] proper, o appropriate, public service, o utility,ties
— m — existente - [serv.públ] existing utility,ties
— m — provisto - [serv.públ] provided utility,ties
— m puntual - punctual, o certain, service
— m — adicional - additional certain service
— m rápido - rapid, o quick, o prompt, o fast,
service
— m — autorizado - authorized quick service
— m — — por fabricante m - quick, factory, o
manufacturer, authorized service
— m real - true, o actual, service
— m regular - [ind] regular, o routine, service
— m sencillo - [mec] simple routine service
— m — simple - [mec] simple routine service
— m relacionado - related service
— m responsable - responsible service • [ind]
responsible department
— m rudo - rugged service
— m sanitario - [sanit] sanitation • [med]
dispensary; hospital; medical service
— m satisfactorio - satisfactory, service, o
performance
— m — de neumático m - [autom-neumát] satisfactory tire performance
— m según Instituto m Estadounidense para Petróleo m - [petról] American Petroleum Institute, o A P I, service
— m según norma f - [ind] standard service •
standard duty
— m sencillo - [ind] simple service
— m — de rutina - [ind] simple routine service
— m severo - severe service • rugged service
— m similar - similar service
— m sin inconveniente(s) - trouble-free, performance, o service
— m — problema(s) m para conservación f -
[ind] trouble-free (maintenance) service
— m — — m — entretenimiento m - [ind] trouble-free maintenance service
— m — — m — mantenimiento m - trouble-free
(maintenance) service
— m social - [labor] social service • general
welfare
— m subcontratado - [labor] subcontracted service
— m suministrado—provided, o supplied, service
— m superior - [ind] total service
— m — para cliente m - [ind] total customer
service
— m suplido - supplied, o provided, service
— m vinculado - related service
— m técnico - technical, service, o assistance
— m — para puesta f en marcha f - [ind] technical start-up service
— m — — supervisión f - supervision, o supervisory, technical service
— m — — f para montaje m - [ind] erection
supervisory technical service
— m — — f — puesta f en marcha f - [ind]
start-up supervisory technical service
— m — extranjero - [ind] foreign technical
service
— m — nacional - [ind] local, o national,
technical service
— m — a cliente m - [com] complete customer
service
servido,da a - served
— a como viga f - [mec] served as a beam
— a rápidamente - served, rapidly, o quickly
servidumbre f - . . . . • [legal] easement; right
of way
— f de paso m - [legal] right of way

servilleta f de papel m - [domést] paper napkin
servir v - . . . • to service • to, fit, o work
— v como viga f - [mec] to serve as @ beam
— v deuda f - [fin] to service @ debt
— v igualmente - to serve,vice, equally, o also
— v para - to serve,vice; to handle • to, fit, or be good for • to rate • to house | adv - suitable
— v — aplicación f—to serve for @ application
— v — — f con amperaje m, alto, o elevado - [sold] to handle @ high current application
— f — — m, elevado, o reducido - [sold] to handle high and low current appli- cation(s)
— v para carcasa f—[mec] to serve as @ housing
— v — fin m - to serve @, purpose, o end
— v por sí sólo,la - to serve alone
— v primariamente - to, serve, o intend, prima- rily
— v rápidamente - to serve, fast, o quickly
— v también - to serv(ic)e also
servodirección f - [autom-mec] power steering; servodrive
— f hidráulica - [mec] servo fluid steering
— f perdida - [autom-mec] lost power steering
servomotor m completo - complete servomotor
— m neumático - pneumatic servomotor
— m para impulsión f - drive,ving servomotor
— m — mecanismo m para regulación f control mechanism servomotor
— m piloto - pilot servomotor
servotransmisión f - [mec] servotransmission
sesenta grados m debajo de corona f - ten and two o'clock position(s)
— m — — — f de estructura f - [constr] structure ten and two o'clock position(s)
sesgamiento m - . . .; angling
sesgado,da a - . . .; (a)skew; biased • across
— a horizontalmente [constr] beveled horizon- tally
— a verticalmente - beveled vertically; skewed
— a y achaflanado,da - skewed and beveled
sesgadura f - . . .; bias(ing); angling; de- flecting • skewing
— f de extremo n - [mec] end beveling
— f extrema - extreme skew(ing)
— f y achaflanadura a—skew(ing) and bevel(ing)
— f — — f combinada(s) - [constr] combined skew(ing) and bevel(ing)
sesgar v - . . .; to bezel; to angle
— v extremo m - [mec] to skew @ end
— v voltaje m - [electrón] to bias @ voltage
— v — m para bloqueo m - [electrón] to bias @ blocking voltage
— v y achaflanar v - to skew and bevel
sesgo m - . . .; slant | adv - (a) bias; slanted
— m de alcantarilla f - [hidr] culvert skew
sesión f - . . . • stage • [legal] meeting
— f con cronistas m y fotógrafos m - [public] press and photo stage
— f — fotógrafos m - [fotogr] photo stage
— f de . . . día(s) m - . . . day session
— f extraordinaria - [legal] special meeting
— f ordinaria - [legal] regular, session, o meeting
— f para calificación f - [deport] qualifying session
— f — clasificación f - [deport] qualifying sessio
— f — determinar posición(es) f - [deport] qualifying session
— f — ensayo(s) m - test(ing) session
— f — práctica - [deport] practice session
— f — — f para corrida(s) f entre pilón(es) m - [deport] autocross practice session
— f — prensa f - [public] press stage
— f práctica - [labor] workshop session
— f técnica - technical session
sesor m - véase asesor m
sesquióxido m férrico - [miner] ferric sesqui- oxide
severamente adv - . . .; badly
— adv corrosivo,va - badly corrosive

severidad f de accidente m - [segurid] accident severity
— f — — m reducida - [Segurid] reduced acci- dent severity
— f evaluada - [segurid] evaluated severity
severo,ra a - . . .; serious; stringent; stiff; tough
shieldarc m - [sold] Shieldarc
shunt n - [electr-instal] véase derivación f
si algo va a fallar, fallará - [fig] Murphy's Law
si bogas porque bogas, si no bogas porque no bo- gas - damned if you do (and) damned if you don't
— adv coloidal - if, o when, colloidal
— adv los,as hay - if any
— adv se mira(n) desde arriba - looking down (from above)
— pron mismo,ma -shimself; herself; itself; yourself • (a) - themself,ves • (de) - self
— adv no coloidal - if, o when, noncolloidal
— adv tan sólo - if only
— adv vamos a caso m - for that matter
siderita f - [miner] . . .; siderite
siderometalmecánico,ca a - [metal-prod] steel, metallurgical, and mechanical
siderometalúrgico,da a - [metal-prod] steel and metallurgical
Siderperú m - [metal-prod] véase Empresa f Side- rúrgica de Perú m
siderurgia f - [metal-prod] steel industry • steel, production, o making
— f en escala f grande - [metal-prod] large scale steel production
— f integrada - [metal-prod] integrated steel (making) industry
— f por reducción f directa - [metal-prod] di- rect reduction steel making (industry)
— f tradicional - [metal-prod] traditional steel making (industry)
siderúrgico m - [metal-prod] steel maker • véase también técnico m en siderurgia f
siderúrgico,ca a - [metal-prod] steel; steel industry, making, o production
siembra f - [agric] . . . • dotting • [ictiol] stocking
— f con despojo(s) m - littering
— f — roca(s) f - rock, scattering, o litter- ing
— f de pista f con despojo(s) m - [deport] course littering
— f en cuadro m - [agric] checkrow(ing)
— f — montoncillo(s) m - [agric] hill dropping
— f — montones m - [agric] hill dropping
Siemens-Martin m - [metal-prod] open hearth
— — básico - [metal-prod] basic open hearth
siempre adv - . . .; all the way; at all times every time • absolutely
— adv creciente - always, o ever, growing, o increasing
— adv mayor(es) - ever, higher, o larger
— adv menor(es) - ever, lower, o smaller
— adv que - if and when; provided; insofar as; whenever; as long as • providing
— adv que sea posible - whenever, o where(ever) possible
— adv — resulte - whenever
— adv — — factible - whenever feasible
— adv — — práctico,ca - whenever practical
— adv y cuando - if and when
siendo adv - being • where
sierra f alternativa - jigsaw
— f — con hoja f intercambiable - [herram] jig-, o saber-, saw
— f basculante - [herram] drop saw
— f bayoneta f - [herram] bayonet, o jig-, saw
— f caladora - [herram] keyhole, o jig-, saw
— f cilíndrica - [herram] barrel, o drum. saw
— f circular - [herram] buzz, o bench, o cir- cular, saw
— f — oscilante - [herram] drunken (drum) saw
— f con banda f - [herram] band saw
— f con bastidor m - [herram] buck saw

sierra f con bastidor m deslizante - [herram]
traveling, frame, o buck, saw
— f — calibre,brador - [herram] gage saw
— f — cinta f - [herram] band saw
— f — diamante(s) m - [herram] diamond, saw, o
blade
— f — — m para corte m en seco - [herram] dry
cutting diamond, blade, o saw
— f con graduador - [herram] gage saw
— f — vaivén - [herram] jig saw
— f continua - [herram] belt, o band, saw
— f — carborundo m - [herram] carborundum saw
— f — — m impulsada con motor - [herram] motor driven carborundum saw
— f deslizante - [metal-lam] traveling saw
— f — en caliente - [metal-lam] traveling hot
saw
— f — para corte m en caliente - [metal-lam]
traveling hot saw
— f en caliente - [metal-lam] hot saw
— f — — con bastidor m - [metal-lam] frame
hot saw
— f — — — m deslizante - [metal-lam]
traveling frame hot saw
— f — frío - [metal-lam] cold saw
— f — — deslizante - [metal-lam] traveling, o
sliding, cold saw
— f escopleadora - [herram] plunge saw
— f impulsada con motor m - [herram] motor
driven saw
— f lapidaria - [herram] lapidary saw
— f levadiza - [herram] drop saw
— f manual - [herram] manual, o hand (held) saw
— f mecánica - [herram] power (driven) saw
— f — alternativa - [herram] power(ed) saber
saw
— f — con vaivén m - [herram] power(ed), o
power driven saber saw
— f mecanizada - [herram] power(ed) saw
— f motorizada - [herram] motor (driven) saw
— f oscilante - [herram] drunken saw
— f para baldosa(s) f - [herram] tile saw
— f — contornear - [herram] counter, o jig,
saw • sweep saw
— f — cortar con veta f - [herram] ripsaw
— f — corte m en caliente—[metal-lam] hot saw
— f — — m en caliente con bastidor m -
[herram] frame hot saw; hot frame saw
— f — — m — — — m deslizante - [metal-
-lam] traveling frame hot saw
— f — — m — frío - [metal-lam] cold saw
— f — — m — — deslizante - [metal-lam]
sliding cold saw
— f — hender v - [herram] ripsaw
— f — hormigón m - [herram] concrete saw
— f — mampostería f - [herram] masonry saw
— f — marquetería f - [herram] keyhole saw;
hacksaw
— f — metal(es) m - [herram] hacksaw; steel
cutting saw
— f — hender n - [herram] ripsaw
— f — teja(s) f - [herram] tile saw
— f — tronzar v - [herram] cutoff saw
— f — trozar v - [herram] crosscut saw
— f perforadora - [herram] hole saw
— f por fricción f - [herram] friction saw
— f — — f regulada - [herram] controlled
frction saw
— f — — f — con computador m - [ind] computer controlled friction saw
— f rápida "Quickie" - [herram] Quickie saw
— f semiesférica - [herram] hemisphere,ric saw
— f sin fin - [herram] band saw • belt saw
— f tipo inglete - [herram] miter type saw
— f trozadora - [herram] crosscut, o cut-off, o
buck, saw
— f universal - [herram] frame saw
sifón m - [tub] . . . . [metal-prod] skimmer •
[petról] gooseneck • [sanit] trap
— m cloacal - [sanit] (barrel) sewer siphon
— m — con conducto m triple - [constr] triple
barrel sewer siphon

sifón m cloacal para impulsión f - [constr]
force sewer, o sewer force, siphon
— m — triple - [constr] triple sewer siphon
— m con cuello m de cisne - [petról] gooseneck
(siphon)
— m — chorro m - [sanit-inodoros) siphon jet
— m — conducto m triple - [constr] triple barrel siphon
— m — impulsión f - [constr] force siphon
— m invertido - [tub] inverted siphon
— m llenado - [tub] filled siphon
— m lleno - [tub] full siphon
— m para dosificación f - [sanit] dosing siphon
— m principal - [constr] main siphon
— m para ruta f - [metal-prod] skimmer
— m triple - [tub] triple siphon
— m vacío - [tub] empty siphon
sigilo m - . . . • confidentiality
sigla(s) f - [legal] initials • acronym
— f para consultor m - consultant(s) acronym
siglo m catorce - véase siglo m décimocuarto
— cuarto - fourth century
— m de cien - one hundreds; second century
— m — cuatrocientos - four hundreds; fifth
century
— m — doscientos m - two hundreds; third century
— m — mil - ten hundreds; eleventh century
— m — — cien - eleven hundreds; twelfth
century
— m — — doscientos - twelve hundreds; thirteenth century
— m — novecientos - nineteen hundreds;
twentieth century
— m — ochocientos - eighteen hundreds;
nineteenth century
— m — — quinientos - fifteen hundreds; sixteenth century
— m — — seiscientos - sixteen hundreds; seventeenth century
— m — — setecientos - seventeen hundreds;
eighteenth century
— m — — trescientos - thirteen hundreds;
fourteenth century
— m — novecientos - nine hundreds; tenth century
— m — ochocientos - eight hundreds; ninth
century
— m — quinientos - five hundreds; sixth century
— m — seiscientos - six hundreds; seventh
century
— m — setecientos - seven hundreds; eighth
century
— m — trescientos - three hundreds; fourth
century
— m décimo - tenth century; nine hundreds
— m décimocuarto - fourteenth century
— m décimonono - véase siglo m décimonoveno
— m décimonoveno - nineteenth century
— m décimoprimero - véase siglo m undécimo
— m décimoquinto m - fifteenth century; four-
teen hundreds
— m décimosegundo - twelfth century; eleven
hundreds
— m décimoséptimo - seventeenth century; sixteen hundreds
— m décimosexto - sixteenth century; fifteen
hundreds
— m décimotercero - thirteenth century; twelve
hundreds
— m diecinueve - véase siglo décimono(ve)no
— m dieciocho - véase siglo m décimooctavo
— n dieciocho - véase siglo m décimooctavo
— m dieciseis - véase siglo décimosexto
— m diecisiete - véase siglo m décimoséptimo
— m doce - véase siglo m décimosegundo
— m duodécimo - véase siglo m décimosegundo
— m no(ve)no - ninth century; eight hundreds
— m octavo - eighth century; seven hundreds
— m once - véase siglo m undécimo
— m primero - first century; one hundreds

siglo m quince - véase siglo m décimoquinto
— m segundo - second century; one hundreds
— m séptimo - seventh dentury; six hundreds
— m sexto - sixth century; five hundreds
— m tercero - third century; two hundreds
— m trece - véase siglo m décimotercero
— m undécimo - eleventh century; ten hundreds
— m veinte - véase siglo m vigésimo
— m vigésimo - twentieth century; nineteen hundreds
— m I - véase siglo m, primero, o uno
— m II - véase siglo m, segundo, o dos
— m III - véase siglo m, tercero, o tres
— m IV - véase siglo m, cuarto, o cuatro
— m V - véase siglo, quinto, o cinco
— m VI - véase siglo, sexto, o seis
— m VII - véase siglo, séptimo, o siete
— v VIII- véase siglo m, octavo, u ocho
— m IX - véase siglo m, noveno, o nueve
— m X - véase siglo, décimo, o diez
— m XI - véase siglo m, undécimo, u once
— m XII - véase siglo, décimosegundo, o doce
— m - XIII - véase siglo, décimotercero, o trece
— m XIV - véase siglo, décimocuarto, o catorce
— m XV - véase siglo, décimoquinto, o quince
— m XVI - véase siglo m, décimosexto, o dieciseis
— m XVII - véase siglo, décimoséptimo, o diecisiete
— XVIII - véase siglo, décimoctavo, o dieciocho
— m XIX - véase siglo, décimono(ve)no, o diecinueve
— m XX - véase siglo, vigésimo, o veinte
signado,da a - signed • signalled
signar s - to sign • to signal
signar v - . . . • to signal
signatura f - . . .; signing • [public] byline
significación f de configuración f - [electr-oper] pattern, significance, o meaning
— f de escala f - scale significance
— f normal - normal, significance, o meaning
significado - . . .; significance • translated (in)to
— m amplio - broad meaning
— m — ubicación f - location significance
— m normal - normal meaning
— n relativo - relative meaning
— m similar - similar meaning
significado,da a - meant • translated (in)to
significante - . . .; meaningful
significar v - . . .; to, imply, o denote • to translate (in)to
— v normalmente - to mean normally
significativo,va - . . .; impressive; meaningfiul • relevant; remarkable
signo m diacrítico - [gram] diacritical mark
significativamente a mayor - significantly greater
significativo,va a - . . .; considerable
signo m - . . . • [comput] character
sigue v en página f - continues on page . . .
— v en peso m - [adv] (que le) next heaviest
— v de cerca - [fam] hot on @ heel(s)
siguiente adv - . . . • [public] below
silbato m para alarma f—[segurid] alarm whistle
silbido m - . . .; whistling; hiss(ing) sound
silenciado,da a—silenced • [electrón] squelched
silenciador m - . . . • [acúst] sound damper • [electrón] squelch(er); damper
— m aislado - [comb.int] insulated muffler
— f aprobado - [comb.int] approved muffler
— f controlado - [electrón] controlled, muffler, o squelch(er)
— m — con tono m - [electrón] tone controlled squelch(er)
— m corriente - [comb.int] standard muffler
— m estándar - [comb.int] standard muffler
— f guardachispas - [comb.int] spark arresting muffler
— m — aprobado - [comb.int] approved spark arresting muffler
— m común - [comb.int] common muffler

silenciador m guardachispas aprobado para uso m marítimo - [comb.int] approved marine spark arresting muffler
— m para automóvil(es) m - [autom] automobile muffler
— m — escape m - [autom] (exhaust), muffler, o silencer
— m — — m para motor m - [comb.int] engine (exhaust) muffler
— m — motor m - [comb.int] engine, o exhaust, muffler, o arrester
— m regulado - [electrón] controlled squelch
— m — con tono m continuo - [electrón] continuous tone controlled squelch
— m según norma f - [comb.int] standard muffler
— m simultáneo—[electrón] simultaneous squelch
— m — continuo - [electrón] simultaneous continuous squelch
— m — — m regulado - [electrón] controlled simultaneous continuous squelch
— m — — — con tono m continuo - [electrón] simultaneous continuous tone controlled squelch
silenciamiento* m - [electrón] squelching
silenciar v - to silence • [electrón] to squelch
silencioso,sa - . . . • [válv] non slam
silicato m de calcio m - [quím] calcium silicate
— m — hierro m - [quím] iron silicate
— m — manganeso - [quím] manganese silicate
— m — sodio m - [quím] sodium silicate • water glass
sílice m alto - [metal] high silica
— m bajo - [metal] low silica
silícico,ca a - [geol] silicic
silicificado,da a - [geol] silicified
silicio m alto - [metal] high silicon
— m bajo—[metal] low silicon
— m elevado - [metal] high silicon
silício-manganeso m - [metal] silico-manganese
— m mediano - [metal] medium silicon
— m reducido - [metal] low silicon
silícioso,sa a - véase silícico,ca
silicium m - véase silicio
silicona f - [quím] silicone
silicono m - [quím] silicona
silicoso,sa a - [quím] siliceous
silicificación f - [geol] silicification
silo m - . . .; (storage) bin; bunker
— m anular - [miner] annular bunker
— m de metal - [ind] metal bin
— m — — m para, cereal(es) m, o grano(s) m - [agric] metal grain bin
— m de tolva(s) f - [metal-prod] high line
— m íntegramente metálico - [metal-fabr] all metal bin
— m para agregado m - [constr] aggregate bin
— m — almacenamiento m - [ind] stock, o storage, bin
— m — carbón m - [ind] coal, bin, o bunker
— m — cemento m - [constr] cement silo
— m para clasificación f - [ind] classification bin
— m — f de carbón m - [ind] coal classification bin
— m — depósito m - [ind] storage, o stock, bin
— m — grano(s) m - [agric] grain bin
— m — — m íntegramente metálico - [metal-fabr] all metal grain bin
— m — — m metálico m - [metal-fabr] metal grain bin
— m — homogeneización f - [ind] homogenizing bin
— m — mezcla f - [ind] mixing, o blending, bin
— m — planta f - [ind] plant bin
— m — — f para sinterización f - [metal-prod] sintering plant bin
silueta f - . . . • outline • skyline
silla f con rueda(s) f - [mec] wheelchair
— f cromada - [domést] chrome chair
— f G E O (para riel m) - [f.c.] G E O plate
— f para cocina f - [Domést] kitchen chair
— f — comedor m - [domést] dining room chair
— f — escritorio m - [domést] desk chair

silla f para patio m - [domést] deck, o lawn,
chair
— f paralela - [mec] parallel seat; prefiérase
asiento m paralelo
— f plegadiza - [domést] folding chair
— f poltrona - [domést] easy chair
— f recta - [domést] straight (back) chair
silleta f - [domést] chair • [mec] saddle block
• [tub] saddle • [f.c.] tie plate
— f acronitrilo-butadieno-estirénico - [tub]
acronitrile-butadiene-styrene saddle
— f de acronitrilo-butadieno-estireno m - [tub]
acronitrile-butadiene-styrene saddle
— f — Armco m - [tub] Armco saddle
— f para asiento m - [mec] saddle branch
— f — — m estándar - [tub] standard saddle
branch
— f — riel(es) m - [f.c.] rail stool
— f — soporte m - [constr] support saddle
— f — — m para tubería f - [constr] pipe support saddle
— f — tubería f - [tub] pipe saddle
— f — — f entramada - [tub] truss pipe saddle
— m — tubo(s) m - [tub] pipe saddle
— f prefabricada - [tub] prefabricated saddle
— f — de acronitrilo-butadieno-estireno—[tub]
prefabricated acrinitrile-butadiene-styrene
saddle
— f — Armco - [tub] Armco prefabricated, o
prefabricated Armco, saddle
— f según norma f - [tub] standard pipe saddle
— f — — f entramada - [tub] standard
truss pipe saddle
sillón m acolchado - [domést] (over)stuffed
chair
— m almohadón - [domést] bean bag
— m con respaldo m - [domést] back chair
— m — — m con ala(s) f - [domést] wing back
chair
— m para escritorio m - [domést] desk chair
— m reclinante - [domést] chaise longue
— m rellenado - [domést] (over)stuffed chair
simbolizado,da a - symbolized
símbolo m - . . .; emblem
— m básico - basic symbol
— m — para soldadura f—[sold] basic weld(ing)
symbol
— m complementario - supplementary symbol
— m de A W S - [sold] American Welding Society,
o A W S, welding symbol
— m — — — — para soldadura f - [sold] American Welding Society welding symbol
— m — Asociación f Estadounidense para Normas
f - American Standards Association symbol
— m — confiabilidad f - dependability symbol
— m — contorno m - [sold] contour symbol
— m — ranura f - [sold] groove symbol
— m — — f en U - [sold] U groove symbol
— m — — f — V - [sold] V groove symbol
— m — — f para soldadura f - [sold] groove
weld symbol
— m — Sociedad f Estadounidense para Soldadura
f - [sold] American Welding Society symbol
— m eléctrico - [electr] electric(al) symbol
— m — A N S I - [electr] véase Símbolo m Eléctrico según Instituto m Nacional Estadounidense para Normas f
— m eléctrico de Asociación f Estadounidense
para Normas f - [electr] American Standards
Association electrical symbol
— m — según Instituto m Nacional (Estadounidense) para Normas f - [electr] American National Standards Institute, o A N S I, electrical symbol
— m estándar - standard symbol
— m europeo - European symbol
— v — para velocidad f - [autom-neumát] European speed symbol
— m — año m - year symbol
— m para circuito m - [electrón] circuit symbol
— m — seguridad f - [segurid] safety symbol
— m — soldadura f - [sold] weld(ing) symbol
— m — — f de contorno m - [sold] weld around
symbol

símbolo m para soldadura f en todo contorno m -
[sold] weld-all-around symbol; all around
weld(ing) symbol
— m — — f doble - [sold] double weld(ing)
symbol
— m — — f en ángulo m (interior) - [sold]
double fillet weld(ing) symbol
— m — — f en ángulo m (interior) - [sold]
fillet weld(ing) symbol
— m — — f obra f - [sold] field weld(ing)
symbol
— m — — f ranura f - [sold] groove weld-
-(ing) symbol
— m — — f — f con chaflán m - [sold]
bevel groove weld(ing) symbol
— m — — f — f — m doble - [sold]
double bevel groove weld(ing) symbol
— m — — f — v en U - [sold] U groove
weld(ing) symbol
— m — — f — f — doble - [sold] double
U groove weld(ing) symbol
— m — — f — f — sencilla - [sold]
single U groove welding symbol
— m — — f — f — V - [sold] V groove
weld(ing) symbol
— m — — f — f — doble - [sold] double
V groove weld(ing) symbol
— m — — f — f — sencilla - [sold]
single V groove weld(ing) symbol
— m — — f intermitente - [sold] intermittent
weld(ing) symbol
— m — — f en tresbolillo m - [sold] staggered intermitent weld(ing) symbol
— m — — f — — — m en ángulo m (interior) -
[sold] staggered intermittent fillet weld-
-(ing) symbol
— m — terminación f - [sold] finish(ing) symbol
— m — tierra f - [electr-instal] (to) ground
symbol
— m — velocidad f - speed symbol
— m — reemplazado - replaced symbol
— m retenido - retained symbol
— m según Instituto m Nacional (Estadounidense)
para Normas f - American National Standards
Istitute symbol
— m típico - typical symbol
— m — para soldadura f - [sold] typical weld-
-(ing) symbol
— m verificado - checked symbol
simbología* f - véase símbolos m
simetría f ideal - ideal symmetry
simétrico,ca a - . . .; equivalent • [mec]
push-pull
— a con respecto a centro m - [metal-lam] symmetric to @ centerline
— a efectivo,va - [electr] symmetrical effective
similar a - . . .; comparable; like
— a a caucho m - rubberlike
— a aprobado - approved equal
similaridad* f - véase similitud f
— f importante - important similarity
simple a - . . . • [fig] green
simplificación f - . . .; simplifying; uncom-
plicating,tion • simplicity
— f de alineación f - [mec] alignment simplification,fying
— f de conservación f - [ind] maintenance simplification,fying
— f — contabilización f - [contab] accounting
simplification,fying
— f — instalación f - installation simplification,fying
— f — procedimiento(s) m contable(s) - [cont]
accounting procedure(s) simplification,fying
— f — trabajo m - work simplification,fying
— f excesiva - excess(ive) simplification,fying
• oversimplification,fyin
simplificado,da a - simplified; uncomplicated
— a excesivamente - oversimplified
simplificar v - . . .; to uncomplicate
— v alineación f - [mec] to simplify @ alignment

simplificar v armado m - to simplify @ assembly,bling
— v conexión f - to simplify @ connection
— f — f para alimentadora f - [sold] to simplify @ feeder connection
— f — f —— f para alambre m - [sold] to simplify @ wire feeder connection
— v conservación f - [ind] to simplify @ maintenance
— v contabilización f - [contab] to simplify @ accounting
— v excesivamente - to oversimplify
— v instalación f - to simplify @ installation
— v procedimiento m - to simplify @ procedure
— v —— contable - [contab] to simplify @ accounting procedure
— v reencendido m - [sold] to simplify @ restriking
simposio m - symposium
simulación f - . . . . • dissembling
— f de circuito m - [comput] circuit sumulation
— f — ensayo m - test simulating,tion
— f — par motor - [mec] torque simulating,tion
— f — proceso m—[ind] process simulating,tion
— f — prueba f - test simulating,tion
simulado,da a - simulated • dissembled • dummy • dead • [arquit] blind
simulador m - . . .; simulator
simular v ajuste - to simulate @ tightening
— v — m de tuerca f - [mec] to simulate @ nut tightening
— v — m —— f hasta par m motor de . . . . [mec] to simulate @ nut torque,quing to . . .
— v circuito m - [electrón] to simulate @ circuit
— v ensayo m - to simulate @ test(ing)
— v par m motor - [mec] to simulate @ torque
— v proceso m - to simulate @ process
— v prueba f - to simulate @ test
simultaneidad f - . . .; synchronism
simultáneamente adv - . . .; at @ same time • [sold] on @ fly
simultáneo,nea - . . .; at @ same time • [sold] on @ fly
sin adv - . . . . • no(ne) • minus
— adv abarrotamiento m - uncluttered
— adv abastecer - [int.comb] dry; without fuel
— adv abocinar v - [mec] flareless; unflared
— adv abovedar v - [vial] uncrowned; without crown(ing)
— adv abrir v - unopened
— adv acabar - unfinished
— adv accesorio(s) m - without accessory,ries
— adv aceitar v - [mec] unoiled; not oiled; dry
— adv aceite m - [mec] oilless
— adv acepillar v - [mec] without shaving
— adv aclarar v - unclear(ed)
— adv acojinar v - [mec] unsprung
— adv achaflanar v - [mec] unbeveled • square
— adv aderezar v - [culin] without dressing
— adv aditamento(s) m - [quím] straight
— adv admitir alternativa(s) f - wihtout alternate,tive
— adv afectar - unaffected • unassigned
— adv afiliación f - unaffiliated
— adv — con (ningún) partido m político—[pol] non-partisan (basis)
— adv — política - [pol] non-partisan basis
— adv aforo m - [fisc] unclassified
— adv agujero(s) m - hole-free
— adv aire - without air; air-free
— adv alcanzar v hasta pie m - [mec] clear of @ toe
— adv — v pie m de diente m - [mec] clear of @ tooth toe
— adv aleación f - without alloy(s); unalloyed
— adv ambaje(s) - straightforward; flatly; directly; without beating about @ bush
— adv amortiguar v - [mec] undamp(en)ed
— adv analizar v - unanalyzed
— adv anclar v - unanchored
— adv antecedente(s) n - record setting
— adv apagar v - [miner] unslaked

sin adv aplastar v - uncrushed
— adv aplicar - unapplied • unappropriated; unallocated
— adv aportación f - [sold] without @ inflow
— adv — f de material m - [sold] without @ material inflow
— adv apoyo m - unsupported
— adv apresuramiento m - at leisure
— adv aprobación f - not, o non, approved; unapproved • non-code
— adv aprobar - unapproved • [ind] non-code
— adv apropiar - unappropriated
— adv apuntalar v - [constr] unstrutted
— adv armadura f - [electr-cond] unarmored; without @ armor • [constr] unreinforced
— adv armar - unassembled; knocked down
— adv arrancar v - [ind] unstarted; not started • [agric] not uprooted
— adv arrugar v - wrinkle free
— adv asignar v - unassigned
— adv autorización v - unauthorized
— adv avanzar v - [sold] travel off
— adv aviso m - without, notice, o warning
— adv — m previo - without prior, notice, o warning, o advice
— adv ayuda f - unaided; without @ aid
— adv beneficiar - [miner] unbeneficiated; without beneficiation
— m beneficio m - without, benefit, o aid
— adv — m de soldadura f - without benefit of @ weld(ing)
— adv bobina f - coilless • lacking @ coil
— adv bocamanga f - [vest] cuffless
— adv bruñir v - unburnished
— adv burbuja(s) f - bubble free
— adv cabeza f - headless • [clavos] countersunk head
— adv cable m - without, o no, cable
— adv caída f libre - without @ free fall; non-freefall
— adv calce m - [mec] without @ shim; shimless
— adv calcinar - [combut] uncalcined
— adv calefacción f - unheated
— adv calentar v - unheated
— adv calmar - [metal-prod] unkilled
— adv cámara f - [autom-neumát] tubeless
— adv cambo(s) m - unchanged
— adv carácter m limitativo - [legal] without limitation
— adv característica(s) f - without @ characteristc(s)
— adv carga f - [mec] unloaded; loadless; without @ load; under no load • [electr-mot] no load; at idle
— adv — de soldadura f - [sold] without @ welding load
— adv cargar v - uncharged; without @ charge
— adv cargo - [contab] position free • at no, expense, o charge
— adv — m a cliente m - at no charge, o not chargeable, to @ customer
— adv — m consumidor m - not chargeable, o at no charge, to @ consumer
— adv — m adicional - [com] at no extra charge
— adv — m alguno - at no charge; not chargeable; at no, cost, o expense
— adv carrete(s) m - [transp] without reel(s)
— adv carril(es) m - without rail(s); trackless
— adv centrar v - not centered; off center
— adv cepillar v - [mec] unshaved; without shaving
— adv circulación f - without circulation
— adv — f de corriente f - [electr-oper] de--energized
— adv clasificar - unclassified • [geol] bank run
— adv cohesión f - cohesionless
— adv colar - [metal-prod] untapped; uncast; without, tapping, o casting
— adv combadura f - without @ camber
— adv — f previa - without precamber(ing)
— adv combustible m - [combust] fuelless; out of fuel

sin adv compendiar v - unabridged
— adv compensación f - uncompensated; without compensation • for free
— adv complicación(es) f - uncomplicated
— adv compromiso m - without obligation; charge free
— adv concentrar v - unconcentrated
— adv condensación f - noncondensed
— adv condensador m - without @ condensor
— adv conectar v - unconnected; not connected • [electr-oper] not turned on
— adv — v eléctricamente - [electr-oper] electrically, disconnected, o independent
— adv conexión f con tierra f - [electr-instal] ungrounded
— adv confinar - unconfined • unenclosed
— adv conservar v - unmaintained
— adv contactador m - [electr-instal] without @ contactor
— adv contar v - not counting; exclusive of
— adv — v con - without; not counting
— adv contener(se) v - without restraint • [autom] flat out
— adv corregir v - uncorrected
— adv corriente - [electr-oper] electrically cold; dead; de-energized
— adv cortar v - uncut; without cutting
— adv costo m - without cost
— adv costura f - seamless
— adv cribar - [constr] unscreened; bank run
— adv cruceta f - [mec] without @ spider
— adv cucharón m - [constr] bucketless
— adv — m para pala f - [constr-equip] without @ bucket shovel
— adv cuidado m - without care • perfunctorily
— adv curvar v - uncurved
— adv chaflán n - [mec] square edge; without @ bevel
— adv chicoteo m - [sold] without whipping
— adv decapar v - [metal-trat] unpickled
— adv defecto(s) m - without @ defect(s); flawless
— adv deformar(se) v - without deflection
— adv dejar(se) v vencer - not to be outdone
— adv denunciar - unreported
— adv derecho(s) - without @ right(s)
— adv — m a reembolso m - [legal] without @ (rights to) reimbursement
— adv desalentar v - undaunted
— adv desbarbar v - [metal-trat] not deburred
— adv desbastar v - [metal-lam] unbloomed; rough
— adv descanso m - [labor] around @ clock
— adv despuntar v - [mec] unclipped • [metal-lam] uncropped
— adv detener(se) v - [mec] without stopping
— adv devanado m - [electr-equip] coilless
— adv devengar v - [contab] unearned; unaccrued
— adv dificultad f - trouble-free • [fig] to be @ snap
— adv diluir - undiluted • [sold] all weld
— adv dirección f - undirected; unguided
— adv disputar v - undisputed; uncontested
— adv distribuir - [fin] undistributed; undivided
— adv disyuntor - [electr-equip] without @ breaker
— adv dividir v - undivided; without dividing
— adv documentar - undocumented
— edificar v - unbuilt
— adv efecto m - without effect; null and void
— v electrizar v - [sold] cold
— adv embalar v - unpacked
— adv — v en fábrica f (de origen) - [transp] mill unpacked
— adv embargo - nevertheless; however; but
— adv emitir v - [fin] unissued
— adv empaquetadura f - [mec] without @ packing
— adv encerrar v - unenclosed; unconfined
— adv enterrar v - unburied
— adv escala - not to scale
— adv escarfar - [metal-lam] unscarfed
— adv escuadrar - unsquared

sin adv esfuerzo(s) m - effortlessly
— adv encender v - [combust] unlit
— adv especificar v - unspecified
— adv estabilizador(es) m - [grúas] without @ outrigger(s)
— adv estabilizar v - unstabilized
— adv estañar v - [metal-trat] untinned
— adv estar v en rotación f - not rotating
— adv estopa f - [mec] without @ packing
— adv estrella(s) f - [metal-trat] unspangled
— adv estructura(s) f - [constr] without structure(s)
— adv excavación f - [hidr] ditchless
— adv excepción f - without @ exception
— adv expandir v - [metal-trat] non-expanded
— adv experiencia f - inexperienced
— adv — suficiente - insufficiently experienced
— adv experimentar v problema(s) m - to experience no problem(s)
— adv facturar v - [com] uninvoiced
— adv fermentar v - [quim] unfermented
— adv fijar - unfixed • unrestrained
— adv filo m - [mec] blunted
— adv fin m - [mec] worm (gear); (drive) screw • [herram] auger | a - endless; helical
— adv — de lucro m - [fin] not for profit
— adv — m para descarga f - [agric] unloading auger
— adv — m — mezcladora f - [ind] mixing screw
— adv — y engranaje m - [mec] worm and gear
— adv firma f - without @ signature
— adv fiscalización f - without control; uncontrolled
— adv fiscalizar v - uncontrolled
— adv formar v - [mec] not formed • blank
— adv — pella(s) f - unpelletized
— adv forro m - [vest] unlined • [grúas] unlagged; without @ lagging
— adv franquicia f - without exception
— adv fuego m - [cald] unfired
— adv operar v - inoperative; stopped
— adv fundamento m - unfounded
— adv fusible(s) m - [electr-instal] unfused; without @ fuse(s)
— adv galvanizar v - [metal-trat] ungalvanized • [cabl] bright
— adv gancho(s) m - [mec] hookless; without @ hook(s)
— adv garantía f - [com] unguaranteed; unwarrated • unsecured
— adv gasto(s) m - expense free
— adv giro - turnless
— adv golpeteo m - [mec] knock-free
— adv grieta(s) f - crackless; without @ crack
— adv guardabarro(s) m - [autom] without @ fender; fenderless; open wheel
— adv guiar v - unguided
— adv gunitar v - [metal-trat] without guniting
— adv habitante(s) m - [pol] uninhabited
— adv hacer v contacto m - not closing
— adv hilos m - [electrón] wireless
— adv holgura f - (just) snug
— adv hueco(s) m - void free
— adv ilustrar v - unillustrated; not illustrated
— adv impacto(s) m - [electrón] non-impact
— adv impedimento m - unrestricted
— adv impedir v - unimpeded
— adv importancia f - unimportant; irrelevant
— adv imputar v - unallocated; unassigned
— adv incluir v - without, o not, including; not included
— adv inconveniente(s) m - trouble free; without @ hitch • well • uneventful
— adv indemnización f - [legal] without indemnity
— adv inflar - uninflated
— adv inhibición(es) f - uninhibited
— adv intermediario(s) m - without intermediaries
— adv interrupción f - uninterrupted(ly); con-

tinuously; interruption free; without interruption
— <u>adv</u> **interruptor** <u>m</u> - [electr-equip] without @
@ breaker
— <u>adv</u> — <u>m</u> **automático** - [electr-instal] without
@ contactor
— <u>adv</u> — <u>m</u> — **optativo** - [sold] without @, contactor option, o @ optional contactor
— <u>adv</u> **intersticio(s)** <u>m</u> - [metal-prod] without
interstices; interstitial free
— <u>adv</u> **investigar** <u>v</u> - uninvestigated
— <u>adv</u> **laminar** <u>v</u> - [metal-lam] to dummy through
— <u>adv</u> **limitación** <u>f</u> - unlimited • [autom] flat
out • without limitation
— <u>adv</u> — <u>f</u> **de espesor(es)** <u>m</u> - [metal-lam] unlimited thickness(es)
— <u>adv</u> — <u>f</u> — **fuerza** <u>f</u> - [ind] unlimited power
— <u>adv</u> **limitar(se)** <u>v</u> - [autom] flat out
— <u>adv</u> **limpiar** <u>v</u> - unclean(ed)
— <u>adv</u> **lograr(se)** <u>v</u> - unattained
— <u>adj</u> **lubricación** <u>f</u> - [mec] without lubrication
— <u>adv</u> **lubricar** <u>v</u> - unlubricated
— <u>adv</u> **lugar a duda(s)** - undoubtedly; clearly
— <u>adv</u> **lustrar** <u>v</u> - unpolished • unburnished
— <u>adv</u> **llegar** <u>v</u> **a fallar** - without failure
— <u>adv</u> **llenar** <u>v</u> - unfilled; without filling
— <u>adv</u> **masilla** <u>f</u> - [constr] without putty
— <u>adv</u> **material(es)** <u>n</u> - without material(s)
— <u>adv</u> **mazarota** <u>f</u> - [metal-prod] without, @ hot
top, o capping
— <u>adv</u> **mejorar** - unimproved
— <u>adv</u> **mengua** <u>f</u> - without sacrifice
— <u>adv</u> **menoscabo** <u>m</u> - consistent with
— <u>adv</u> **modificación** <u>f</u> - without @ change(s) •
[autom] stock
— <u>adv</u> **modificar** <u>v</u> - without changin • [autom]
showroom
— <u>adv</u> — **en salón** <u>m</u> **para venta(s)** <u>f</u> - [autom]
showroom stock
— <u>adv</u> **molestar** - undisturbed
— <u>adv</u> **montaje** <u>m</u> - [trefil] without @ casing
— <u>adv</u> **montar** <u>v</u> - [mec] unassembled • loose
— <u>adv</u> **movimiento** <u>m</u> **de efectivo** <u>m</u> - [fin] not
involving cash
— <u>adv</u> **muelle(s)** <u>m</u> - [mec] springless
— <u>adv</u> **muro(s)** <u>m</u> - [constr] without wall(s)
— <u>adv</u> — <u>m</u> **de cabecera** <u>f</u> - [constr] without,
o no, headwall
— <u>adv</u> **nivelar** <u>v</u> - [constr] unleveled
— <u>adv</u> **nombre** <u>m</u> - nameless
— <u>adv</u> **normalizar** <u>v</u> - unstandarized
— <u>adv</u> **novedad** <u>f</u> - [ind] operating normally
— <u>adv</u> **núcleo** <u>m</u> - coreless
— <u>adv</u> **observación** <u>f</u> - [transp] clear
— <u>adv</u> **obstrucción** <u>f</u> - unobstructed
— <u>adv</u> **odorizar** <u>v</u> - unodorized
— <u>adv</u> **ondulación(es)** <u>f</u> - [sold] ripple free
— <u>adv</u> **opción** <u>f</u> - without @ option
— <u>adv</u> **orientar** <u>v</u> - nonoriented
— <u>adv</u> **otro(s) comentario(s)** <u>m</u> - without further
ado
— <u>adv</u> **pago** <u>m</u> - without (any) payment
— <u>adv</u> **palanca(s)** <u>f</u> - [mec] leverless
— <u>adv</u> **papel** <u>m</u> - without paper • paper free
— <u>adv</u> **paralelo** <u>m</u> - unparalleled
— <u>adv</u> **parangón** <u>m</u> - unmatched
— <u>adv</u> **parar máquina** <u>f</u> - [sold] on-the-fly
— <u>adv</u> **pared(es)** <u>f</u> - without @ wall(s)
— <u>adv</u> — **posterior** - [constr] without @ rear
wall
— <u>adv</u> **paridad** <u>f</u>—[comput] without, o no, parity
— <u>adv</u> **pegar(se)** <u>v</u> - without sticking
— <u>adv</u> **pena(lidad)** <u>f</u> - without @ penalty
— <u>adv</u> **perforar** <u>v</u> - unperforated • unpunched
— <u>adv</u> **perjuicio** <u>m</u> - despite; not withstanding
— <u>adv</u> — <u>m</u> **de ello** <u>m</u> - despite @ foregoing
— <u>adv</u> **perno(s)** <u>m</u>—[mec] without bolts; boltless
— <u>adv</u> **pestaña(s)** <u>f</u> - [f.c.-ruedas] flangeless
— <u>adv</u> **picaporte** <u>m</u> - unlatched
— <u>adv</u> **pintar** <u>v</u> - unpainted
— <u>adv</u> **poder(se)** <u>v</u> **reparar** - beyond repair
— <u>adv</u> **polvo** <u>m</u> - dustless
— <u>adv</u> **porosidad** <u>f</u> - porosity free

sin <u>adv</u> **precedente(s)** - unprecedented; first-
-ever; record setting
— <u>adv</u> **premura** <u>f</u> - at leisure
— <u>adv</u> **preocupación** <u>f</u> - without, care, o worry
— <u>adv</u> **preocupar(se)** <u>v</u> - without, care, o worry
— <u>adv</u> **presión** <u>f</u> - without pressure; pressurless
— <u>adv</u> — <u>f</u> **alta** - without high pressure
— <u>adv</u> — <u>f</u> — **en tragante** <u>m</u> - [metal-prod]
without high pressure in @ throat
— <u>adv</u> — <u>f</u> **en tragante** <u>m</u> - [metal-prod] without
pressure in @ throat
— <u>adv</u> **problema** <u>m</u> - trouble-free; without pro-
blem(s)
— <u>adv</u> — <u>m</u> **mayor** - without major problem(s) •
quite well
— <u>adv</u> — **para, conservación** <u>f</u>, **o entretenimien-**
to <u>m</u>, **o mantenimiento** <u>m</u> - trouble free; without maintenance problems
— <u>adv</u> **protección** <u>f</u> - unprotected • [electr-
-cond] unshielded
— <u>adv</u> **púa(s)** <u>f</u> - [alambre] barbless
— <u>adv</u> **pulir** <u>v</u> - unpolished; unburnished
— <u>adv</u> **pulsación(es)** <u>f</u> - nonpulsating
— <u>adv</u> **pulverizar** - uncrushed; unpowdered
— <u>adv</u> **punta** <u>f</u> - blunted
— <u>adv</u> **punzonar** <u>v</u> - unpunched
— <u>adv</u> **purificar** <u>v</u> - unpurified
— <u>adv</u> **quemar** <u>v</u> - [combust] unburned
— <u>adv</u> **ranurar** <u>v</u> - grooveless
— <u>adv</u> **rasguño(s)** <u>m</u> - unscratched
— <u>adv</u> **razón** <u>f</u> - without reason
— <u>adv</u> — <u>f</u> **aparente** - without, o for no apparent, reason
— <u>adv</u> **rebabado** <u>m</u> - [metal-fabr] without debur-
ring; not deburred
— <u>adv</u> **rebabar** <u>v</u> - [metal-fabr] without debur-
rig; not deburred
— <u>adv</u> **recalentamiento** <u>m</u> - without reheating •
without overheating
— <u>adv</u> **recalentar** <u>m</u> - without reheating • without overheating
— <u>adv</u> **recargo** <u>m</u> - [labor] straight line
— <u>adv</u> — <u>m</u> **en precio** <u>m</u> - at no extra charge
— <u>adv</u> **recargue** <u>m</u> - [sold] without build-up
— <u>adv</u> **reclamar** <u>v</u> - unclaimed
— <u>adv</u> **recocer** <u>v</u> - [metal-trat] unannealed;
green
— <u>adv</u> — <u>v</u> **magnético,ca** - [metal-trat] unannealed magnetic
— <u>adv</u> **recocido** <u>m</u> - [metal-trat] unannealed
— <u>adv</u> — **después de corte** <u>m</u> - [metal-trat] unannealed after shearing
— <u>adv</u> — **magnético,ca** - [metal-trat] unannealed
magnetic
— <u>adv</u> **rectificar** <u>v</u> - [metal-trat] rough cored
— <u>adv</u> — <u>v</u> **para diámetro(s)** <u>m</u> **según norma** <u>f</u> -
[alambre] rough cored to @ standard size
— <u>adv</u> **recubrimiento** <u>m</u> - [,etal-trat] uncoated;
without coating • [tub] bare condition •
[grúas] unlagged; without lagging
— <u>adv</u> — <u>m</u> **alguno** - uncoated • [alambre] bare
condition
— <u>adv</u> **recubrir** <u>v</u> - uncoated • uncovered
— <u>adv</u> **reembolso** <u>m</u> - [legal] without reimbursement; unreimbursed
— <u>adv</u> **regalía** <u>f</u> - without @ royalty
— <u>adv</u> **regulación** <u>f</u> - uncontrolled
— <u>adv</u> **remover** <u>v</u> - unremoved • undisturbed
— <u>adv</u> **rotar** <u>v</u> - not rotating
— <u>adv</u> **rescuadrar** <u>v</u> - not resquared
— <u>adv</u> **resorte(s)** <u>m</u> - springless
— <u>adv</u> **responsabilidad** <u>f</u> - unresponsible
— <u>adv</u> — <u>f</u> **mayor** - without @ further responsibility
— <u>adv</u> **restricción(es)** <u>f</u> - unrestricted
— <u>adv</u> **restringir** <u>v</u> - unrestrained
— <u>adv</u> **retención** <u>f</u> - [com] clean on board
— <u>adv</u> **retorno** <u>m</u> - [electr-oper] no, o without,
return
— <u>adv</u> — **de voltaje** <u>m</u> - [electr-oper] no, o
without, @ voltage return
— <u>adv</u> **revestimiento** <u>m</u> - unlined
— <u>adv</u> **riel(es)** <u>m</u> - trackless

sin adv roscar v - [tub] threadless
— adv rozar - not rubbing • clear(ed)
— adv rugosidad(es) - wrinkle free
— adv sacrificar v - without sacrifice,cing
— adv — v excesivamente - without excessive sacrifice,cing
— adv sacrificio m - without @ sacrifice,cing
— adv — excesivo - without @ excessive sacrifice,cing
— adv salida - no, o dead, end • [fig] blind
— adv salpicadura(s) f - without spatter
— adv — f excesiva(s) - [sold] without @ excessive spatter(ing)
— adv saltar,to - without skip(ping)
— adv saturar v - unsaturated
— adv seleccionar - without selecting • [miner] run of @ mine
— adv separación f - [mec] without, o no, gap
— adv servodirección f - [autom] without power steering
— adv sobrecalentamiento m—without overheating
— adv socavación f - [sold] without undercut(ting)
— adv soldadura f - [sold] weldless
— adv soldar v - [sold] unwelded
— adv soportar v - unsupported
— adv soporte(s) m - unsupported
— adv sostén m - unsupported
— adv subsidio(s) m - unsubsidized; selfsupporting
— adv sujeción f - unclamped • unfastened
— adv tapa f - uncovered • open top
— adv taponar v - not plugged • uncovered
— adv tener(se) v en cuenta - regardless
— adv — — f profundidad f - regardless of @ depth
— adv terminación,nar - without finishing; unfinished • crude • [mec] blank
— adv tizón(es) m - [constr] stretcher bond
— adv tocar v - untouched • cleared • [hormig] undisturbed
— adv tocar extremo m - without touching, o clear of, @ end
— adv — v — m inferior - without touching, o clear of, @ toe
— adv — — m — de diente m - [mec] without touching, o clear of, @ tooth toe
— adv — pie m - [mec] without touching, o clear of, @, foot, o toe
— adv — pie m de diente m - [mec] without touching, o clear of, @ tooth toe
— adv — trabajo m - [sold] without touching, o off of, @, work, o toe
— adv ton ni son - without rhyme or reason
— adv trámite m alguno - with no procedure whatsoever
— adv — m judicial o extrajudicial alguno - [legal] without any judicial or extrajudicial procedure whatsoever
— adv transferir v - untransferred
— adv tratamiento m - untreated • [sanit] raw
— adv — m térmico - [metal-trat] not heat treated
— adv tratar v térmicamente - not heat treated • without heat treatment
— adv triturar v - uncrushed
— adv usar - unused • [ind] inoperative
— adv utilizar v - unused
— adv valor m - without, o no, value
— adv — a par f - [fin] no par value
— adv — m comercial - [com] without any commercial value
— adv ventaja f - without @ advantage
— adv — r apreciable - not by much
— adv ventilación f - unventilated
— adv vidrio(s) m - unglazed
— adv vuelta f - [vest] cuffless
sinad m - [electrón] signal to noise ratio; sinad
sinclonorio m - [geol] synclinorium
sincronización f - . . . .; timing
— f adelantada - [comb.int] early timing
— f atrasada - [comb.int] late timing

sincronización f atrasada - [comb.int] late timing
— f automática - [electrón] automatic timing
— f con precisión f - [electrón] precision timing
— f de cronómetro m - [comput] timer synchronization
— f — — m externo - [comput] external timer synchronization
— f — encabezadora f - [trefil] header timing
— f — encendido m - [comput] ignition timing
— f — trefilador m - [trefil] wire drawer timing
— f — tono m - [comput] tone timing
— f — válvula f - [válv] valve timing
— f demorada - [comb.int] late timing
— f incorrecta - [comb.int] incorrect timing
— f nueva - [comb.int] retiming
— f — f de encendido m - [comb.int] ignition retiming
— f retrasada - [mec] late, o delayed, timing
— f temprana - [comb.int] early timing
sincronizado,da a - synchronized • [comb.int] timed
— a automáticamente - [electrón] automatically timed
— a con precisión f - [electrón] precision timed
— a nuevamente - [comb.int] retimed
sincronizador m - . . . . • [sold] timer • [electrón] timer
— m automático - [instrum] automatic timer
— m con retardo m - [sold] delay(ed) timer
— m — — m regulado - [sold] adjustable delay timer
— m eléctrico - [instrum] electric timer
— m fallado - [instrum] faulty timer
— m interdigital - [comput] interdigit(al) timer
— m no desbaratable - [instrum] undefeatable timer
— m para cortar corriente f - [sold] (current) shut-off timer
— m — corte m - [sold] shut-off timer
— m — — m para agua m - [hidr] water shut-off timer
— m — — m retardado m - [sold] shut-off timer
— m — — m — para flujo m - [sold] shut-off and afterflow timer
— m — — m — — m para agua m - [sold] water shut-off and afterflow timer
— m — — m — — m — gas m - [sold] gas shut-off and afterflow timer
— m — — m — — m — m y agua m—[sold] gas and water shut-off and afterflow timer
— m — detención f - [instrum] shutdown timer
— m — — f automática - [instrum] automatic shutdown timer
— m — detener motor m - [electr] (motor) shut-off timer
— m — — m para alimentación f - [sold] feed motor shut-off timer
— m — — m — — f para alambre m - [sold] wire feed(ing) motor shut-off timer
— m — flujo m retardado - afterflow timer
— m — — m retardado - afterflow timer
— m — — m — para agua m - [instrum] water afterflow timer
— m — Hetzios m - [electrón] cycle(s) timer
— m — — m para soldadura f - [sold] welding cycle(s) timer
— m — motor m - [electr-mot] motor timer - [comb.int] engine timer
— m — — m para alimentación f - [sold] feed-(ing) motor timer
— m — — m — f para alambre m - [sold] wire feed(ing) motor timer
— m — puntada(s) f - [sold] stitch timer
— m — punto(s) m - [sold] spot timer
— m — purga(s) f - [instrum] purge,ging timer
— m — f fallado - [combust] faulty purge-ging timer

sincronizador m **para soldadura** f - [sold] welding timer
— m — — f **por punto(s)** m - [sold] spot timer
— m — **purga(s)** f - [combust] purge,ging timer
— m — — f **reemplazado** - [combust] replaced purge,ging timer
— m **purgado** - [combust] purged timer
— m **reemplazado** - [instrum] replaced timer
sincronizador,ra a - synchronizing • timing
sincronizar v - . . . • [comb.int] to time
— v **automáticamente** - [electrón] to time automatically
— v **con precisión** f - [electrón] to precision time
— v **encabezadora** f - [trefil] to time @ header
— v **encendido** - [comb.int] to time @ ignition
— v **operación** f - [comput] to time @ operation
— v **tono** m - [comput] to time @ tone
— v **trefiladora** f - [treil] to time @ wire drawer
— v **válvula** f - [combust] to time @ valve
— v **velocidad** f - to synchronize @ speed
sindicado,da a - syndicated • [penal] accused
sindical a - [labor] labor (union)
sindicato m - [legal] syndicate • [labor] union
— m **para diamante(s)** m - diamond(s) syndicate
— m **profesional** - professional syndicate • guild
síndico m - . . . • stockholders, representative, o delegate • auditor • [G.B.] registrar
— m **titular** - [legal] titular, stockholders, representative, o delegate; titular syndic
— m **suplente** - [legal] acting, o substitute, stockholders, representativem o delegate; acting, o substitute. stockholders, representative, o delegate; acting syndic
sinestrórsum adv - left
sinfín m - . . . • [mec] auger
— m **para elevación** f - [naút] worm hoist
— m — — f **de botalón** m - [náut] worm boom hoist
— m — — f — **aguilón** m - [grúas] worm boost hoist
— m — **gato** m - [mec] jack screw
— m — **grano** m - [agric-equip] grain auger
— m — **mecanismo** m **para dirección** f - [autom] steering worm
— m — **motor** m - [mec] motor worm
singular a - . . .; unique; unusual; exceptional
singularidad f - . . .; uniqueness
singularmente adv - . . .; particularly
— adv **corrosivo,va** - particularly corrosive
siniestro m - [seguros] occurence; accident; disaster; loss; calamity; catastrophe • damage
— m **causado** - [seguros] caused loss
— m **impagado** - [seguros] unpaid loss
— m **por pagar** v - [seguros] unpaid loss
— m **y gasto(s)** m **incurrido(s)** - [seguros] incurred loss and expenses
siniestro m - . . . | a - left (handed) • [instrum] counterclockwise
sínter m - [miner] sinter; véase también **conglomerado** m
— m **con basicidad** f **mayor** - [metal-prod] higher basicity sinter
— m **frágil** - [miner] brittle sinter
— m **friable** - [miner] brittle sinter
— m **normal** - [miner] normal sinter
— m **para horno** m - [metal-prod] furnace sinter
— m — — m **alto** - [metal-prod] blast furnace sinter
— m **quebradizo** m - [miner] brittle sinter
— m **recirculado** - [metal-prod] recirculated sinter
— m **sin cribar** - [metal-prod] unscreened sinter
sinterización f - [miner] sintering • sintering plant
— f **de aglomeración** f - [metal-prod] agglomeration sintering
— f **en caliente** - [miner] hot sintering
sinterizado* m - [miner] véase **conglomerado** m
sinterizado,da a - [miner] sintered
sinterizar v - [miner] to sinter
sinterizar v **aglomeración** f - [miner] to sinter @ agglomeration
síntesis f - . . . • summary
— f **de amoníaco** n - [quím] ammonia synthesis
— m — **diamante(s)** m - [quím] diamond synthesis
— m — **proyecto** m - project synthesis
— f — — m **técnico** - technical project synthesis
— f **técnica** - technical synthesis
sintético,ca a - . . .; synthetic • man made
sintetizado,da a - synthesized • summarized
sintetizar v - . . . • to summarize
síntoma m **de envenenamiento** m - [medic] poisoning symptom
— m **evidente** - [medic] evident symptom
— m **primero** - first symptom
sintonización f **ajustada** - [electrón] fine tuning
— f **ajustada con reóstato** m - [electrón] rheostat fine tuning
— f **exacta** - [electrón] fine tuning
— f — **con reóstato** m - [electrón] rheostat fine tuning
— f **precisa** - [electrón] fine tuning
— f — **con reóstato** m - [electrón] rheostat fine tuning
sinuosidad f - . . . • ess(es) • meander(ing)
— f **leve** - slight meandering
— f — **en cauce** m - [hidr] slight channel meandering
sinuoso,sa a - . . .; meandering; wandering; winding • circuitous
sinusoidal a - [electrón] . . .; sine
siquiera adv - . . . • just • only
sirena f **para alarma** f - [segurid] alarm siren
sírva(n)se interj - please
sisa f - . . . • shortage
sísal m - [textil] sisal
sismógrafo m **por reflexión** f - [instrum] reflection seismograph
sistema m - . . . • approach; practice • type • policy • [comput] software • [electr-instal] véase **red** f
— m **a tierra** f - [electr-instal] ground(ing) system
— m **accionado** - activated system
— m **activado** - activated system
— m **actual** - present system
— m **actualizado** - updated system
— m **adecuado** - adequate system
— m — **para catalogación** f - adequate cataloguing system
— m **administrativo** - administrative, o management, system
— m — **(según) Allen** - [admin] Allen management system
— m — **integrado** - [admin] integrated management system
— m — — **creado** - [admin] developed integrated management system
— m — **lógico** - [admin] logical management system
— m — — **creado** - [admin] developed logical management system
— m **aéreo** - [aeron] aerial system
— m **aeroespacial** - [aeron] aerospace system
— m **alimentador** - [mec] feeding system
— m — **con tornillo** m **sin fin** - [mec] worm gear feeding system
— m — m **semiautomático** - [ind] semi-automatic feeding system
— m — — **para alambre** m - [sold] semi-automatic wire feeding system
— m **Allen** - [admin] Allen system
— m — **para administración** f - [admin] Allen management system
— m **Americano** - véase **sistema estadounidense** • U S A system
— m **analógico** - analogical system
— m — **para control** m - [ind] analogical control system
— m **antieconómico** - uneconomical system
— m **antigolpe(s)** - [mec] antiknock system
— m — **con ariete** m - [tub] antiknock ram

sistema m anunciador - annunciator system
— n aprobado - approved system
— m — para aislación f - [electr-instal] approved insulation system
— m — para protección f - [segurid] approved protection system
— m — — — f con aislación f - [segurid] approved insulation protection system
— m — — — f — — f doble - [segurid] approved double insulation protection system
— m apropiado - appropriate, o proper, system
— m atascado - jammed system
— m autoiniciado - [comput] self-initiated system
— m — para, condición f, o estado m - [comput] self-initiated status system
— m — — posición f - [comput] self initiated status system
— m automático - automatic system
— m — para carga f - [metal-prod] automatic charging system
— m — — para engrase m - [mec] automatic greasing system
— m — — entrada f - automatic intake system
— m — — — f para fuel oil m - [metal-prod] automatic fuel oil intake system
— m — — lubricación f - [mec] automatic lubricating,tion system
— m — — muestreo m - [ind] automatic samplin system
— m — — — m para planta f - [ind] plant automatic sampling system
— m — — — — f para sínter(ización) - [metal-prod] sinter(ing) plant automatic sampling system
— m — — nivelación f - [ind] automatic leveling system
— m — — regulación f - [ind] automatic control(ling) system
— m — — soldadura f - [sold] automatic welding system
— m automatizado - [ind] automated system
— m — para regulación f - [ind] automated control(ling) system
— m — — — f para espesor m - [metal-lam] automated gage control(ling) system
— m auxiliar - auxiliary system
— m — para seguridad f - [segurid] safety aid, o auxiliary safety, system
— m bajo nivel m de mar m - [petról] subsea system
— m bajo tierra f - [petról] subsurface system
— m básico - basic system
— m binario - [electrón] binary system
— m bueno - good system
— m cambiable - changeable system
— m Camtrol - Camtrol system
— m canadiense - Canadian system
— m — para perforación f - [petról] Canadian pole (drilling) system
— m Cardox - [ind] Cardox system
— m — contra incendio(s) m - [ind] Cardox fire protection system
— m carretero - [vial] highway system
— m central - central system
— m — para lubricación f - [ind] central lubricating,tion system
— m centralizado - centralized system
— m — automático - [mec] automatic centralized system
— m — — — m para lubricación f - [mec] centralized lubricating,tion system
— m — para limpieza f - [electr] central(ized cleaning system
— m — — — f por vacío m - [electr] centralized vacuum cleaner system
— m — para lubricación f - [ind] central(ized) grease,sing system
— m cloacal - [sanit] sanitary sewer system
— m cohesivo - cohesive system
— m colector - [ind] collecting,tor system
— m — festoneado - [grúas] festooned collector system

sistema m colocado - [ind] installed system
— m combinado - combined system
— m — de suelo m y acero m - [constr] composite soil and steel system
— m completado - completed system
— m completamente (de)bajo nivel m de mar - [petról] total subsea system
— m completo - complete system
— m complicado - complicated, o sophisticated, system
— m — para control m - [comput] sophisticated control system
— m — — — m con microprocesador(es) m - [comput] sophisticated microprocessor control system
— m comprobado - checked system • proven system
— m común - common, o routine, system
— m con base f para vapor m - [metal-prod] steam based system
— m — calidad f - [ind] quality system
— m — — f alta - [ind] high-grade system
— m — banda circunferencial - [autom-neumát] belt(ed) system
— m — — f — con cuerda(s) f de rayón m - [autom-neumát] rayon cord belt system
— m — — f — de fibra f de vidrio m - [autom-neumát] fiberglass belt system
— m — — f — — — — f plegada(s) - [autom-neumát] folded fiberglass belt system
— m — — f — plegada(s) - [autom-neumát] folded belt system
— m — — f circunferencial f plegada con cuerda(s) f - [autom-neumát] cord folded belt system
— m — — — — f de rayón m - [autom-neumát] rayon cord folded belt system
— m — base f refrigerada - [metal-prod] cooled base system
— m — batería(s) - [electr-prod] battery,ries system
— m — bomba(s) f - [bombas] pump system
— m — — f libre(s) - [petról] free pump system
— m — cabeza f - [comb.int] véase sistema m con culata f
— m — — f sólida - [comb.int] véase sistema m con culata f sólida
— m — cable(s) m - [ind] cable(d) system
— m — — m flojo(s) - [metal-prod] slack cable system
— m — cambio(s) m diferencial(es) - [fin] differential exchange system
— m — cebador m - [int.comb] throttle system
— m — . . . conductor(es) m - [electr-instal] . . .-wire system
— m — conexión f - [electr-instal] connected system
— m — — f con tierra f - [electr-instal] grounded,ding system
— m — — f — — f negativa - [electr-instal] negative ground(ing) system
— m — — f — — f positiva - [electr-instal] positive ground(ing) system
— m — — f negativa - [electr-instal] negative connecting system
— m — — f — con tierra f - [electr-instal] negative ground(ing) system
— m — — f positiva - [electr-instal] positive connecting system
— m — — f — con tierra f - [electr-instal] negative ground(ing) system
— m — correa(s) f - [ind] belted system
— m — — f para mando m - véase correaje m, o conjunto m de correas f, para mando m
— m — corriente f - [electr-oper] energized system
— m — — f alterna - [electr-instal] alternating current system
— m — — f continua - [electr-instal] direct current system
— m — corte m transversal - [coque] crosswise cutting system
— m — costo m alto - [comput] high cost system

sistema f con costo m bajo - [comput] low cost
system
— m — culata f - [petról] head, system, o arrangement
— — f sólida f - [petról] solid head, arrangement, o system
— m — decantación f - [hidr] settling system
— m — densidad f - [comput] density system
— m — f alta - [comput] high density system
— m — f baja - [comput] low density system
— m — disco(s) m - [mec] disk system
— m — múltiple(s) m - [mec] multiple
disk(s) system
— m m — para transmisión f - [mec] multiple disk(s) drive,ving system
— m — m para transmisión f - [mec] disk
drive,ving system
— m — dos producto(s) m químico(s) - [quím]
two-chemical(s) system
— m — empaquetadura f - [mec] packing system
— — f regulable m - [petról] adjustable
packing, system, o arrangement
— m — engranaje(s) m - [mec] gear(s) system
— m — m cicloidal(es) - [mec] cycloid(al)
gear system
— m — equipo(s) m especializado(s) - [ind]
task force approach
— m — equipo(s) m - [ind] equipment system
— — m vertical(es) - [metal-prod] vertical equipment system
— m — estanqueidad f para aceite m - [inf]
oil seal (system)
— m — fibra(s) f de vidrio m - [autom-neumát]
fiberglass system
— — f — m plegada(s)—[autom-neumát]
folded fiberglass system
— m — ficha(s) f para accionar y cerradura f -
[mec] card-key lock system
— m — flexibilidad f - flexibility system
— m — flotador m - [comb.int] float system
— m — freno(a) m - [mec] brake,king system
— m — m con disco(s) m - [mec] disk brake
system
— m — m múltiple(s) - [mec] multiple disk
brake system
— m — m gradual(es) - [grúas] stepless
brake system
— m — m — para grúa f - [grúas] stepless
crane brake system; crane stepless brake $ys-
tem
— m — gráfico(s) m - graph(ics) system
— m — lanza(s) f - [metal-prod] lance system
— — f para oxígeno m - [metal-prod] oxygen lance system
— m — líquido(s) m - liquid(s) system
— — m penetrante(s) m - [liquid] pene-
trant(s) system
— m — llave(s) f - [constr] keying
— m — multifrecuencia(s) f - [electrón] multifrequency system
— m — f con tono m - [electrón] tone multifrequency system
— m — — — m doble - [electrón] dual
tone multifrequency system
— m — f — — m — para despacho m -
[electrón] dual tone multifrequency dis-
patch(ing) system
— m — . . . paso(s) m - . . .-step approach
— m — patín m - [electr-instal] shoe, system,
o arrangement
— m — m colector - [electr-instal] collector shoe, system, o arrangement
— m — pila f - [constr] pile system
— m — f agotada - [coque] finished pile
system
— m — plantilla(s) f - [mec] die(s) system
— — f modular(es) - [mec] modular die(s)
system
— m — polímero(s) m - [quím] polymer(s) system
— m — precipitador m - [ind] precipitator system
— m — m por vía f húmeda - [ind] wet precipitator system

sistema m con presión f - pressure system
— m — f para lubricación f - [ind] pressure lubricating,tion system
— m con pulsador(es) m - [electrón] push button
system
— m — — m para intercomunicación - [electrón]
push button (operated) intercommunication
system
— m — punta(s) f coladora(s) f - [hidr] well
point system
— m — raspador m - [metal-lam] friction piece
system (pack opener)
— m — regulador(es) m - [electrón] control
system
— m — m con transistor(es) m - [electrón]
solid state, o transistorized, control system
— m — resina f - resin system
— m — f poliestérica - [autom-neumát] polyester resin system
— m — riostra(s) f - [constr] bracing system
— m — f cruzada(s) f - [constr] waler, o
cross bracing system
— m — f — (mayormente) horizontales -
[constr] waler system
— m — solvente - [metalprod] solvent system
— m — tablestacado m - [constr-pil] sheetpiling system
— m — m anclado - [constr-pil] anchored
sheetpiling system
— m — tapón m - [metal-prod] stopper system
— m — m de varilla f - [metal-prod] rod
stopper system
— m — tres conductores m - [electr-instal]
three-phase system
— m — vapor m - [ind] steam system
— m — — m e hidrocarburo(s) m - [metal=prod]
steam-hydrocarbon system
— m — m — — m para reforma(ción) f -
[metal-prod] steam-hydrocarbon(s) reforming
system
— m conservado - [ind] maintained system
— m continuo - [mec] continuous system
— m — para arriostramiento m - [constr] continuous bracing system
— m — — m horizontal - [constr] horizontal
continuous bracig system
— m contra incendio(s) m - [segurid] fire-
fighting, o (fire) protection, system
— m convencional - conventional system •
[coque] multi-spot system
— m — de patín(es) m colector(es) - [electr-
-instal] conventional collector shoe, system, o arrangement
— m corriente - conventional, o standard, system
— m — para aspiración f - [ambient] current, o
average, exhaust system
— m — de líquido(s) m penetrante(s) - [metal]
commercial, o current, penetrant(s) system
— m — — murete(s) m y sumidero(s) m - [hidr]
conventional dike-and-catch-basin system
— m — por aspiración f - [ambient] current, o
average, exhaust system
— m Cowper - [ind] Cowper system
— m — para calefacción f - Cowper heating system
— m creado - created, o developed, system
— m crítico - [comput] critical system
— m — para control m - [comput] critical con-
trol(ling) system
— m D T M F para despacho m - [electrón] dual
tone multi-frequency dispatching system
— m — abanico m - fan system • [electr-instal]
radial system
— m de acelerador m—[comb.int] throttle system
— m — acumulador(es) m - [electr-prod] battery,ries system
— m — agregado(s) m - [quím] filler(s) system
— m — alcantarillado(s) m - [hidr] sew(er)age
system
— m — alcantarilla(s) f - [hidr] sew(er)age
system

sistema m̲ de altavoces m̲ ▸ [electrón] loud
speaker system
— m̲ — — m̲ para conferencia(s) f̲ - [electrón]
p̲ublic address system
— m̲ — baranda(s) f̲ - [constr] rail(ing) system
— m̲ — — f̲ confiable(s) - [constr] reliable
rail(ing)̲ system
— m̲ — — f̲ para puente(s) m̲ - [constr] re-
l̲iable bridge rail(ing) syst̲em
— m̲ — — f̲ para puente(s) m̲ - [constr] bridge
rail(ing)̲ system
— m̲ — bobina(s) f̲ - [electr-equip] coil system
— m̲ — bomba(s) f̲ - [bombas] pump system
— m̲ — — f̲ para descascarillado m̲ - [metal-lam]
descaling pump system
— m̲ — buza(s) f̲ con corredera f̲ - [metal-prod]
s̲lide,ding gate system (cucharas)
— m̲ — canal(es) m̲ - [hidr] canal system
— m̲ — — m̲ para barcaza(s) f̲ - [hidr] barge
canal system
— m̲ — caja(s) f̲ - [petról] cage system
— m̲ — carretera(s) f̲ - [vial] highway system
— m̲ — cascada f̲ - [electr] cascade system
— m̲ — cloaca(s) f̲ - [sanit] sew(er)age system;
s̲anitary sewer(s̲) system
— m̲ — conductores m̲ - [electr-instal] cable
s̲ystem
— m̲ — conducto(s) m̲ - [electr-distrib] conduit
s̲ystem
— m̲ — costo(s) m̲ - [contab] cost(s) system
— m̲ — — m̲ según norma f̲ - standard cost(s)
s̲ystem
— m̲ — desagües m̲ - sew(er)age, o̲ drainage,
s̲vstem
— m̲ — — m̲ pluviales - [hidr] storm, sewer, o̲
drainage, system
— m̲ — engranaje(s) m̲ - [mec] gearing; gear(s)
s̲ystem
— m̲ — estímulo(s) m̲—[ind] incentive(s) system
— m̲ — incentivo(s) m̲ - [ind] incentive(s)
s̲ystem
— m̲ — instalación(es) f̲ - [ind] facility,ties
s̲ystem
— m̲ — — f̲ en miniplanta f̲ - [metal-prod] miniplant facility,ties syst̲em
— m̲ — libertad f̲ - system of freedom
— m̲ — cambio(s) m̲ para eje m̲ - [autom-mec]
axle shift syst̲em
— m̲ — — m̲ — m̲ con . . . velocidad(es) f̲ -
[autom-mec] . . .-speed axle shift system
— m̲ — cloaca(s) f̲ - [sanit] sewer(s) system
— m̲ — cojinete(s) m̲ - [mec] bearing(s) system
— m̲ — m̲ para turboalimentador m̲—[comb.int]
turbocharger bearing(s) system
— m̲ — concesión(es) f̲ - [com] concession(s)
s̲ystem
— m̲ — — f̲ para exploración f̲ - [miner] explo-
r̲ation concession system
— m̲ — — f̲ — explotación f̲ - [miner] development concession(s) system
— m̲ — defensa(s) f̲ - [vial] guardrails system
— m̲ — — f̲ lateral(es) - [vial] guardrail(s)
s̲ystem
— m̲ — — f̲ fuera de puente(s) m̲ - [constr]
off-deck guardrail(s) system
— m̲ — — f̲ — semirrígida(s) - [constr] semirigid guardrail(s) system
— m̲ — dispositivo m̲ para cambio(s) m̲ - [autom-
-mec] shift unit syst̲em
— m̲ — empalme(s) m̲ - [autom-mec] linkage sys- n
t̲em
— m̲ — — m̲ para freno m̲ - [autom-mec] brake
l̲inkage (system)
— m̲ — empresa f̲ libre - [econ] free enterprise
s̲ystem
— m̲ — encofrado m̲ - [constr] form system
— m̲ — — m̲ de acero m̲ - [constr] steel form
s̲ystem
— m̲ — — m̲ — — m̲ galvanizado - [const] gal-
v̲anized steel form system
— m̲ — — m̲ para puente m̲ - [constr] galva-
n̲ized steel bridge form system
— m̲ — — m̲ — — m̲ para puente(s) - [constr]

steel bridge form system
sistema m̲ de encofrado m̲ de acero m̲ para puentes
— m̲ - [constr] steel bridge form system
— m̲ — — m̲ para puente(s) m̲ - [constr] bridge
form system
— m̲ — engranaje(s) m̲ - [mec] gearing; gear
s̲ystem
— m̲ — esfuerzo(s) m̲ - [mec] stress(es) system
— m̲ — — m̲ con contacto m̲ - [mec] contact
stress(es) system
— m̲ — filtro m̲ - [mec] filter system
— m̲ — — m̲ con tela(s) f̲ - [ind] baghouse
(filter)̲ system
— m̲ — freno(s) m̲ - [mec] braking system
— m̲ — indicación(es) f̲ - [instrum] reading(s)
s̲ystem
— m̲ — matriz,ces f̲ - [mec] die(s) system
— m̲ — — f̲ cambiable(s) - [mec] changeable die
s̲ystem
— m̲ — matriz,ces f̲ - [mec] die(s) system
— m̲ — — f̲ dura(s) - [mec] hard dies system
— m̲ — — f̲ — cambiable(s) - [mec] changeable
hard dies system
— m̲ — — f̲ — intercambiable(s) - [mec] interchangeable hard die(s) system
— m̲ — — f̲ intercambiable(s) - [mec] interchangeable die(s) system
— m̲ — — f̲ modular(es) - [mec] modular dies
s̲ystem
— m̲ — — f̲ para corte m̲ de muesca(s) f̲ - [mec]
notching die(s) system
— m̲ — — f̲ — perforación f̲ - [mec] punching
die(s) system
— m̲ — — f̲ — — f̲ y corte m̲ (de muescas f̲) -
[mec] punching and notching die(s) system
— m̲ — medida(s) f̲ - [metric] measurement(s)
s̲ystem
— m̲ — — f̲ inglesa(s) - [metric] English mea-
surement(s) system
— m̲ — — f̲ métrica(s) - [metric] metric mea-
surement(s) system ▪ [autom-neumát] metric
sizing system
— m̲ — microprocesador(es) m̲ - [comput] micro-
processor(s) system
— m̲ — molde(s) m̲ - [constr] form(s) system
— m̲ — — m̲ de acero m̲ - [constr] steel forms
s̲ystem
— m̲ — — m̲ — — m̲ galvanizado - [constr]
galvanized steel form(s) system
— m̲ — — m̲ — — m̲ para puente(s) m̲ -
[constr] galvanized steel bridge form(s) system
— m̲ — — m̲ — — m̲ para puente(s) m̲ - [constr]
steel bridge form(s) system
— m̲ — — m̲ para puente(s) m̲ - [constr] bridge
form(s) system
— m̲ — murete m̲ - [constr] dike system
— m̲ — — m̲ y sumidero m̲ - [hidr] dike and
catch basin system
— m̲ — navegación f̲ - [nav] navigation system
— m̲ — — f̲ interior - [nav] seaway system
— m̲ — — f̲ — por Río m̲ San Lorenzo - [nav]
St. Lawrence seaway
— m̲ — oleoducto m̲ - [petról] pipeline (system)
— m̲ — pista(s) f̲ para maniobra(s) f̲ y aparcamiento m̲ - [aeron] taxiway-apron system
— m̲ — — f̲ para aparcamiento m̲ - [aeron]
apron system
— m̲ — — f̲ — maniobra(s) f̲ - [aeron] taxiway
s̲ystem
— m̲ — polea(s) f̲ - [mec] pulley(s) system
— m̲ — — f̲ para vagoneta f̲ - [metal-prod] car
pulley system
— m̲ — — f̲ — — f̲ para montacargas - [metal-
-prod] skip car pulley system
— m̲ — — f̲ — — f̲ — — f̲ para horno m̲ alto -
[metal-prod] blast furnace skip car pulley
system
— m̲ — premio(s) m̲ - [labor] bonus, o̲ premium,
system
— m̲ — — m̲ para todo personal m̲ - [labor] general bonus system
— m̲ de prima(s) f̲ - [labor] bonus system

sistema m de proveedor m para garantía f de calidad f̲ - [ind] supplier('s) quality guaranty
  system
— m̲ — quemador(es) m̲ - [combust] burner(s)
  system
— m̲ — — m para máxima f̲ - [metal-prod] maximum flow burner system
— m̲ — — m̲ — mínima f̲ - [metal-prod] minimum
  flow burner system
— m̲ — — m — flujo m̲ máximo - [metal-prod]
  maximum flow burner system
— m̲ — — — m̲ mínimo - [metal-prod] minimum flow burner system
— m̲ — radar - [electrón] radar system
— m̲ — regulador(es) m - control(ling) system
— m̲ — — m̲ con transistor(es) m̲ - [electrón]
  solid state control system
— m̲ — resina(s) f̲ - resin system
— m̲ — — f̲ con calidad f̲ alta - high grade
  resin(s) system
— m̲ — rociador(es) m̲ - [hidr] sprinkler system
— m̲ — rodamientos m̲ - [cojin] bearings system
— m̲ — — m̲ para turboalimentador m̲ - [comb.-
  -int] turbocharger bearing system
— m̲ — sello(s) m seal(s) system
— m̲ — suelo m - [constr] soil system
— m̲ — — m̲ y acero m̲ - [constr] soil and steel
  system
— m̲ — sujetador(es) m̲ - [mec] fastener(s) system
— m̲ — tela(s) - [autom-neumát] ply system
— m̲ — — f̲ circunferenciales - [autom-neumát]
  belt ply system
— m̲ — — f̲ — de cristal m̲ hilado - [autom-
  -neumát] fiberglass (ply) belt system
— m̲ — — f̲ — — fibra f̲ de vidrio m̲ - [autom-
  -neumát] fiberglass (ply) belt system
— f̲ — — f̲ — de vibrolana f̲ - [autom-neumát]
  fiberglass (ply) belt system
— m̲ — tipo m̲ eléctrico - [autom-mec] electric
  (type) system
— m̲ — — m̲ — para cambio(s) m̲ - [autom-mec]
  electric (type) shift(ing) system
— m̲ — — m̲ neumático - [autom-mec] air (type)
  system
— m̲ — — m̲ — para cambio(s) m̲ - [autom-mec]
  air (type) shift(ing) system
— m̲ — transmisor m̲ - [electrón] transmitter
  system
— m̲ — — m̲ y respondedor m̲ - [electrón] transponder system
— m̲ — transpondedor m̲ - [electrón] transponder
  system
— m̲ — transportador(es) m̲—[ind] conveying,yor
  system
— m̲ — — m̲ con correa(s) f̲ - [ind] (belt) conveying,yor system
— m̲ — tubería(s) f̲ - [tub] tubing, o tubes,
  system
— m̲ — válvula(s) f̲ - [válv] valve(s) system
— m̲ — varilla(s) f̲ - [mec] rod(s) system
— m̲ — — f̲ para conexión f̲ con tierra f̲ -
  [electr-inst] grounding rod system
— m depurado - clean(ed), o cleansed, system
— m depurador - [ind] scrubber system
— m̲ — con Venturi - [ind] Venturi scrubber
  syssytem
— m — provisto con Venturi - [ind] Venturi
  (provided) scrubber system
— m desarrollado - developed system
— m̲ detector - [ind] detecting,tor system
— m̲ — para temperatura(s) f̲ - [instrum] temperature detecting,tor system
— m directo m - direct system
— m̲ — de prima(s) m̲ - [labor] direct, premium,
  o bonus, system
— m diseñado - designed, o engineered, system
— m̲ — hidráulico m̲ - [petról] designed, o engineered, hydraulic system
— m̲ — — bajo tierra f̲ - [petról] subsurface,
  designed, o engineered, hydraulic system
— m̲ — — — f̲ Oilmaster - [petról] Oilmaster subsurface, designed, o engineered, hy-

draulic system
sistema m̲ disolvente - [quím] dissolving, o
  solvent, system
— m distinto - different system
— m̲ doble - double, o dual, o duplex, system
— m̲ — para refrigeración f̲ - [ind] double
  cooling system
— m̲ — — — f̲ para rodillo(s) m̲ - [metal-lam]
  double roll cooling system
— m económico - economical, o low cost, system
— m̲ Edison - [electr-instal] Edison system
— m̲ — con ... conductor(es) m̲ - [electr-
  -instal] . . .-wire Edison system
— m̲ — ...fásico - [electr-instal] Edison
  . . .-phase system
— m̲ — . . .fásico con . . . conductor(es) m̲ -
  [electr] . . .-wire Edison . . .-phase system
— m eléctrico - [electr] electric(al) system
— m̲ — para cambio(s) - [mec] electric
  shift(ing) system
— m̲ — — dispositivo m̲ para cambio(s) m̲ -
  [autom-mec] shift(ing) unit electrical system
— m̲ — — distribución f̲ - [electr-distrib]
  electric distributing,tion system
— m̲ — — supervisión f̲ - [electr-oper] electric supervision system
— m̲ — — — f̲ de distribución f̲—[electr-oper]
  electric distribution supervision system
— m en desarrollo m̲ - developing system
— m̲ — . . . etapa(s) f̲ - . . . stage system
— m̲ — . . . — para reducción f̲ - [mec] . . .
  stage reduction system
— m̲ — pulgada(s) f̲ - [metric] inch system
— n encerrado - enclosed system
— m̲ — en tambor m̲—[mec] drum enclosed system
— m̲ enfriado - cooled (off) system
— m̲ estable - stable system
— m enteramente manual - completely manual system
— m̲ especial - special system
— m̲ — para lubricación f̲ - special oil(ing),
  plan, o system
— m̲ estadounidense - American system
— m̲ — para perforación f̲ - [petról] American
  drilling system
— m̲ estándar - standard system
— m̲ — para perforación f̲ - [petról] standard
  (drilling) system
— m̲ estatal - [pol] state system
— m̲ — de carretera(s) f̲ - [vial] state highway system
— m estático - [comput] static system
— m̲ — para regulación f̲ - [comput] static control system
— m̲ estrangulador - [comb.int] choking system
— m̲ estructural para retención f̲ - [constr]
  structural, retaining, o retention, o contention, o containing, o containment, system
— m̲ — sostén m̲ - [constr] structural sup-
  port(ing) system
— m̲ existente - existing system
— m̲ — empleado - existing system used
— m̲ — para aire m̲ comprimido - [ind] existing
  compressed air system
— m̲ — — alarma m̲ - [comput] existing
  alarm system
— m̲ — — condición f̲ - [electrón] existing
  status system
— m̲ — — estado m̲ - [electrón] existing status
  system
— m̲ — — posición f̲ - [electrón] existing
  status system
— m̲ — usado - existing method used
— m̲ extrusor* - extruder,ding system
— m̲ extintor - [segurid] extinguishing system
— m̲ Farval - [metal-prod] Farval system
— m̲ — para engrase m̲ - [metal-prod] Farval,
  grease,sing, o lubricating,tion, system
— m̲ . . .fásico - [electr-instal] . . .-phase
  system
— m̲ ferroviario - [f.c.] railway system
— m̲ fijo - [ind] stationary system

sistema m **fijo para depuración** f - [ind] stationary cleansing system
— m — — — f **para gas** m - [ind] stationary
gas cleaning system
— m **financiero** - [fin] financial system
— m **fiscalizado** - controlled system
— m — **para acero** m - [metal-prod] controlled
steel, system, o practice
— m — **para producción** f **de acero** m - [metal-
-prod] controlled steel practice
— m **flector** m - [metal-lam] spindle
— m — **para rodillo** m—[metal-lam] roll spindle
— m — — — m **para trabajo** m - [metal-lam]
work roll spindle
— m **flexible** - flexible system
— m **Flocon** - [col.cont] Flocon system
— m **fluvial** - [nav] water, o river, transportation system
— m **general** - general system
— m — **para ventilación** f - [ind] general ventilating,tion system
— m **gráfico** - [electrón] graphic(s) system
— m — **con computador(es)** m - [electrón] computer graphic(s) system
— m **guiador** - [ind] guidance,ding system
— m — **según norma** f - [sold] standard guidance
system
— m **guiador normal** - [sold] standard guidance
system
— m **hidráulico** - hydraulic ststem • [petról]
fluid end
— m — **auxiliar** - auxiliary hydraulic system
— m — **bajo tierra** f - [petról] subsurface hydraulic system
— m — — — f **Oilmaster** - [petról] Oilmaster
subsurface hydraulic system
— m — — — f — **proyectado** - [petról] engineered Oilmaster subsurface hydraulic system
— m — — — f **proyectado** - [petról] engineered
subsurface hydraulic system
— m — **Oilmaster** - [petról] Oilmaster hydraulic
system
— m — **para accionamiento** m - [mec] hydraulic,
drive,ving, o power system
— m - — **entrada** f - [metal-prod] feed end hydraulic system
— m — **impulsión** f - [mec] hydraulic drive,-
ving system
— m — **nivelación** f - [constr] hydraulic
leveling system
— m — **pozo** m - [petról] well hydraulic system
— m — — m **único** - [petról] single well hydraulic system
— m — **presión** f - [metal-prod] pressure hydraulic, o hydraulic pressure, system
— m — — f **alta** - [metal-prod] high pressure hydraulic system
— m — **sección** f - [ind] section hydraulic
system
— m — — — f **para entrada** f - [metal-lam]
feed(ing) end hydraulic system
— m — **sistema** n - [ind] system hydraulic
system
— m — — — m **flector** - [ind] spindle hydraulic system
— m — — — m — **para rodillo** m - [metal-lam]
roll spindle hydraulic system
— m — — — m — — — m **para trabajo** m -
[metal-lam] work roll spindle hydraulic system
— m — **proyectado** - [petról] engineered hydraulic system
— m **hidrostático** - hydrostatic system
— horizontal - horizontal system
— m — **para arriostramiento** m - [constr] horizontal bracing system
— m — — — m **continuo,nuado** - [constr] continuous horizontal bracing system
— m **inadecuado** - inadequate system
— m **independiente** - independent system
— m **independiente para agua** m **para emergencia** f
- [hidr] emergency water independent system

sistema m **indicador** - [ind] indicating system
— m — **para carga** f - [grúas] load indicating
system
— m **indirecto** - indirect system
— m **informático*** - [comput] informatics system
— m **inglés** - English, o British, system
— m **iniciado** - initiated, o started, system
— m — **por operador** m - operator initiated system
— m **inoperante** - inoperative, o nonoperating,
system
— m **ininterrumpido** - uninterrupted system
— m — **para suministro** m (**de energía** f) -
[electr-distrib] uninterrupted (power) supply
system
— m **inseguro** - [segurid] unsafe system
— m **instalado** - installed system
— m **integrado** - [ind] integrated system
— m — **para administración** f - [admin] integrated management system
— m — — — f **creado** - [admin] developed integrated management system
— m — **lubricación** f - [ind] integrated lubricating,tion system
— m — **totalmente** - [ind] fully integrated system
— m **integral** - integrated system
— m **íntegro** - entire system
— m — **de viga(s)** f - [constr] entire, beams, o
rails, system
— m **intercambiable** - (inter)changeable system
— m **Interestatal** - [vial] Interstate System
— m — **de Carreteras** f - [vial] Interstate
Highway System
— m **internacional** - international system
— m — **de Unidad(es)** f - International Units
System
— m **interno** - internal system
— m — **para fiscalización** f - [admin] internal control
system
— m — — — f **de calidad** f - internal quality
control system
— m **introducido** - introduced system
— m **inyector** - injecting,tion system
— m **lateral** - lateral system
— m — **para refrigeración** f - [metal-prod] additional cooling system
— m **Ledex** - [electrón] Ledex system
— m — **para control** m - [electrón] Ledex control system
— m **lento** - slow system
— m **limitador** - limiter,ting system
— m — **para ángulo** m - [grúas] angle limiter
system
— m — — — m **alto** - [grúas] high angle limiter system
— m — — — m — **para aguilón** m - [grúas] high
boom angle limiter system
— m — — — m **bajo de aguilón** m - [grúas] low
boom angle limiter system
— m — — — m **para aguilón** m - [grúas] boom
angle limiter system
— m **libre** - free system
— m **limpiado** - cleaned system
— m **limpio** - clean system
— m **lógico** - logical system
— m — **para administración** f - [admin] logical
management system
— m — — — f **creado** - [admin] developed logical management system
— m **luminoso** m - luminous system
— m **llenado** - (re)filled system
— m **Magna-Die** - [mec] Magna-Die system
— m — — **de Whistler** - [mec] Whistler Magna-
-Die system
— m **malo** - bad, o poor, system
— m **mantenido** - [ind] maintained system
— m **manual** - manual system
— m — **instalado** - installed manual system
— m — **para deshojado** m - [metal-lam] manual
opening system
— m **mecánico** - [mec] mechanical system
— m — **auxiliar** - [mec] auxiliary mechanical

sistema m mecánico de levas f - [mec] mechanical cam(s) system
— m **mejor** - better system • best system
— m **métrico** - [metric] metric system
— m **mixto** - mixed system • double system
— m — **para refrigeración** f - [ind] mixed cooling system • double cooling system

— m **modernizado** - modern(ized) system
— m **moderno** - modern system
— m **modular** - modular system • [mec] die system
— m — **con plantilla(s)** f - [mec] modular die system
— m — **Magna-Die** - [mec] Magna-Die (modular) system
— m — — — — **de whistler** - [mec] Whistler Magna-Die (modular) ststem
— m **monofásico** - [electr-instal] single phase system
— m — **con . . . conductor(es)** m - [electr-instal] . . .-lead(s) single phase system
— m **Morgoil** - [metal-lam] Morgoil system
— m — **para lubricación** f - [metal-lam] Morgoil lubricating,tion system
— m **movible** - mobile, o movable, system
— m — **proyectado** - [electrón] designed mobile system
— m **multifrecuencia** - [electrón] multifrequency system
— m — **con tono** m - [electrón] multifrequency tone, o tone multifrequency
— m — — m **doble** - [electrón] dual tone multifrequencey system
— m — — m — **para despacho** m - [electrón] dual tone multifrequency dispatch(ing) system
— m **nacional** - national system
— m — **de carretera(s)** f - [vial] national highway system
— m — — — í **interestat al(es)** - [vial] national interstate highway(s) system
— m — — — — f — **y para Defensa** f - [vial] National System of Interstate and Defense Highways
— m — **Estadounidense de Carreteras** f **Interestatales y para Defensa** f - [vial] National System of Interstate and Defense Highways
— m **natural** - [ind] natural system
— m — **para alimentación** f - natural feed(ing) system
— m **neumático** - [ind] pneumatic system
— m — **para cambio(s)** m - [mec] air shift(ing) system
— m **nuevo** - new system • [ind] new process
— m — **para alarma** f - [electrón] new alarm, o alarm new, system
— m — —, **condición** f, o **estado** m, o **posición** f - [electrón] new status system
— m **numérico** - numerical system
— m **ofertado** - offered system
— m **ofrecido** m - offered system
— m **Oilmaster** - [petról] Oilmaster system
— m — **para bombeo** m - [petról] Oilmaster pumping system
— m — — — m **hidráulico** - [petról] Oilmaster hydraulic, o hydraulic Oilmaster, pumping system
— m — — — m **neumático** - [petról] Oilmaster pneumatic pumping system
— m — — — m **submarino** - [petról] Oilmaster subsurface pumping system
— m — **Unidraulic** - [petról] Oilmaster Unidraulic system
— m — **Unidraulic para bombeo** m - [petról] Oilmaster Unidraulic pumping system
— m **operado** - [ind] operated, o driven, system
— m **operante** - operating,tive system
— v **optativo** - [ind] optional system
— m — **para fundente** m - [sold] optional flux system
— m **para acabado** m - [pint] finishing system
— m — — m **con pulverización** f - [pint] spray finishing system
— m — **acceso** m - [comput] access system
— m — — m **discado** - [comput] dial up (access)

system
sistema m para acceso m **mediante disco,cado** m - [comput] dial up (access) system
— m — **accionamiento** m - [mec] drive,ving system • power system
— m — — m **hidráulico** - [mec] hydraulic power system
— m — **aceite** m - [lubric] oil(ing) system
— m — **con presión** f - [comb.int] pressure oil(ing) system
— m — — m **para regulación** f - regulating oil(ing) system
— m — **acondicionamiento** m - [air.cond] conditioning system
— m — — m **de aire** m - [air.cond] air conditioning system
— m — **acopio** m - [hidr] catchment system
— m — — m **para agua** m - [hidr] water datchnent system
— m — — m — — f **de Estado** - [hidr] State Water (Catchment) Project
— m — **acoplamiento** m - [mec] coupling system
— m — **adherencia** f - adhesion system
— m — **administración** f - [admin] management system
— m — **aislación** f - [electr-instal] insulation system
— m — **ajuste(s)** m - adjustment system
— m — — m **de precio(s)** m - [com] price escalation system
— m — **alarma** f - [segurid] alarm, o warning, system
— m — — f **autoiniciado** - [comput] self-initiated alarm system
— m — — f **contra carga(s)** f **excesiva(s)** - [grúas] overload warning system
— m — — f **y regulación** f - alarm and control system
— m — **alimentación** f - [ind] feed(ing) system
— m — — f **automática** - [sold] automatic feeding equipment
— m — — f — **para fundente** m - [sold] automatic flux feeding, equipment, o system
— m — — f **continua** - [sold] continuous feed(ing) system
— m — — f — **para fundente** m - [sold] continuous flux feed(ing) system
— m — — f **para alambre** m - [sold] wire feed(ing) system
— m — — f — — m **con velocidad** f **constante** - [sold] constant speed wire feed(ing) system
— m — — f — — m — — f **variable** - [sold] variable speed wire feed(ing) system
— m — — f **para cadena** f - [mec] chain feed(ing) system
— m — — f — — f **para sinterización** f - [metal-prod] sintering chain feed(ing) system
— m — — f — **fundente** - [sold] flux feed(ing) system
— m — — f **fluidificada** - [sold] fluidized feed(ing) system
— m — — f — **para fundente** m - [sold] fluidized flux feed(ing) system
— m — — f **natural** - [comb.int] natural feed(ing) system
— m — **almacenamiento** m - storage,ring system
— m — — m **en acumulador(es)** m - [electr-prod] battery storage system
— m — — m **en batería(s)** f - [electr-prod] battery storage system
— m — **anclaje** m - [constr] anchoring system
— m — — m **en suelo** m - [constr] soil anchoring system
— m — **apagamiento** m - [combust] quenching system
— m — **apertura** f - [mec] opening system
— m — — f **y cierre** m - [mec] opening and closing system
— m — — f — — m **de chapín** m - [metal-prod] bleeder opening and closing system
— m — **apoyo** m - support(ing) system
— m — **aprovechamiento** m - utilization system
— m — — m **para gas** m - [combust] gas utilization system

sistema m para archivo,var - [com] filing system
— m — — m de dato(s) m - data filing system
— m — — m — información f - information filing system • [electrón] data system
— m — arrancador m - starter system
— m — — m eléctrico - [comb.int] electric
starter system
— m — — m con . . . voltio(s) m - [comb.-
-int] . . .-volt electric starter,ting system
— m — — m según norma f - [comb.int] standard electric starter,ting system
— m — arranque m - [comb.int] starter,ting, o
start-up, system
— m — — m con cuerda f - [comb.int] rope
starting system
— m — — m para temperatura(s) f baja(s) -
[comb.int] low temperature start-up system
— m — arrastrar v - [f.c.] haul(ing) system
— m — — v vagón(es) m - [f.c.] car haul(ing)
system
— m — arrastre m - [cabl] haulage system
— m — — m de vagón(es) m - [f.c.] car
haul(ing) system
— m — arriostramiento m - [constr] bracing
system
— m — — m continuo,nuado - [constr] continuous bracing system
— m — — m horizontal - [constr] horizontal
bracing system
— m — — m — continuo,nuado - [constr] horizontal continuous bracing system
— m — — m para edificio(s) m - [constr]
building bracing system
— m — asfalto m - [nucl] asphalt system
— m — auxilio m - [segurid] aid system
— m — — m para seguridad f - [segurid] safety
aid system
— m — bloqueo m - [mec] blocking system
— m — — m para entrada f - [tub] intake, o
incoming, blocking system
— m — — m de salida f - [tub] outlet, o outgoing, blocking system
— m — bobina(s) f - [instrum] coil(s) system
— m — — f Helmholz - [instrum] Hemholtz coil
system
— m — bombeo m - [bombas] pumping system •
[petról] lift(ing) system
— m — — m artificial - [petról] artificial
lift(ing) system
— m — — económico - [petról] economical
artificial lift(ing) system
— m — — m bajo tierra f - [petról] subsurface
pumping system
— m — — m hidráulico - [petról] hydraulic
pumping system
— m — — m para pozo m único - [petról]
single well hydraulic pumping system
— m — — m neumático - [petról] pneumatic
pumping system
— m — — m para pozo m único - [petról] single
well pumping system
— m — — m subterráneo - [petról] subsurface
pumping system
— m — cabeza f - [comb.int] véase sistema m
para culata f
— m — — f para cilindro m - [comb.int] cylinder head system
— m — — f — pozo m (de)bajo de nivel m de
mar m - [petrol] subsea wellhead system
— m — — f — — m submarino - [petról] subsea
wellhead system
— m — calefacción f - [constr] heating system
— m — — f con resistencia f eléctrica -
[constr] electrical resistance heating system
— m — calentamiento m - [ind] heating system
— m — — m para baño m - [ind] bath heating
system
— m — — m — — m para decapado m - [metal-
-trat] pickling bath heating system
— m — — n por combustión f sumergida - [ind]
submerged combustion heating system
— m — — m con inyección f - [ind] injection
heating system

sistema m para calentamiento m con inyección f de
vapor m - [ind] steam injection heating system
— m — — — f directa - [ind] direct injection heating system
— m — — — f — de vapor m - [ind] direct steam injection heating system
— m — calidad f - quality system
— m — calificación f - rating system
— m — cambio(s) m - [mec] shift(ing) system •
[fin] exchange system
— m — — m de rodillo(s) m - [metal-lam] roll
change,ging system
— m — — m neumático - [autom-mec] air shift-
-(ing) system
— m — — m para eje(s) m - [autom-mec] axle
shift(ing) system
— m — — m — — m con . . . velocidad(es) f -
[autom-mec] . . .-speed axle shift(ing) system
— m — — m rápido m de rodillo(s) m - [metal-
-lam] quick roll change,ging system
— m — captación f - [ambient] collection system
— m — — f de contaminante(s) m - [ambient]
pollutant collection system
— m — carbonato m de calcio m - [quím]calcium
carbonate system
— m — carga f - loading system • [electr-oper]
charging system
— m — — f de camión(es) m - [transp] truck
loading system
— m — — f — carbón m - [miner] coal loading
system
— m — — f — coque m - [coque] coke charging
system
— m — — f para horno m - [ind] furnace charging system
— m — — f — — m alto - [metal-prod] blast
furnace charging system
— m — — f —, bolsas f, o sacos m - [transp]
bag, o sack, loading system
— m — — f para tren m - [f.c.] train loading
system
— m — — f — — m enterizo - [f.c.] unit train
loading system
— m — catalogación f - cataloguing system
— m — centrado* m - centering system
— m — — m para desbaste(s) m - [metal-lam]
bloom centering system
— m — cierre m - closing, o seal(ing), o locking, system, o device
— m — — m para grasa f - [mec] grease seal-
-(ing) system
— m — — m — — f para tragante m - [metal-
-prod] throat grease seal(ing) system
— m — clasificación f - classification, o
classifying, system • grading system
— m — — f para viscosidad(es) f - [lubric]
viscosity classification system
— f — — f — — f según Organización f Internacional para Normas f - [lubric] International Standards Organization viscosity classification system
— m — cobertura f - covering system
— m — codificación f - coding system
— m — colección f - collecting,tion system
— m — — f de polvo(s) m - [metal-prod] dust
collection system
— m — combustible - [combust] fuel system
— m — — m purgado - [comb.int] bled fuel system
— m — — m regulado con flotador m - [comb.int]
float controlled fuel system
— m — — m taponado - [comb.int] clogged fuel
system
— m — combustión f - [combust] combustion system
— m — — f sumergida - [ind] submerged combustion system
— m — comparación f - comparison system
— m — comprobación f - checking system
— m — — f de rodillo(s) m - [metal-lam] rolls
checking system
— m — cómputo(s) m - scoring system
— m — concesión(es) f - [com] concession(s)
system

sistema m para condición f - [electrón] status
  system
— m —— f autoiniciado - [comput] self-initiated status system
— m — condensado m - [ind] condensate system
— m — conducción f - [admin] management system
— m — conexión f - [electr-instal] connecting
  system
— m —— f con tierra f - [electr-instal]
  ground(ing) system
— m —— f —— f con varilla(s) f de cobre m
  - [electr-instal] copper rod grounding system
— m — conservación f—[ind] maintenance system
— m — contacto m - contact system
— m — control m - control system
— m —— m analógico - [ind] analogical control system
— m —— m con microprocesador(es) m—[comput]
  microprocessor control system
— m —— m crítico - [comput] critical control
  system
— m —— m para calidad f - [ind] quality control system
— m —— m —— f para sínter m—[metal-prod]
  sinter quality control system
— m —— m — combustión f - [combust] combustion control system
— m — corte m - [mec] cutting system
— m — costo(s) m - [contab] cost(s) system
— m —— m estándar - [contab] standard
  cost(s) system
— m —— m según norma f - [Contab] standard
  cost(s) system
— m — cubierta f - [puentes] deck system
— m — culata f - [int.comb] head system
— m —— f para cilindro m - [petról] cylinder
  head, system, o arrangement
— m — curación f - [mec] cure,ring system
— m — defensa - [puentes] guard, o rail,
  system
— m —— f para puente(s) m - [constr] bridge,
  guard, o rail, system
— m — demolición f - [constr] wrecking system
— m — depuración f - cleaning, o purification,
  system • [ind] scrubbing system
— f —— f para aire m - air cleaning system
— m —— f — gas m - [ind] gas, cleaning, o
  scrubbing, system
— m —— f —— m para horno m - [ind] furnace gas scrubbing system
— m —— f —— m —— m alto - [metal-prod]
  blast furnace gas scrubbing system
— m — descarga f - [ind] unloading, o dumping, system • discharge,ging system
— m —— f automática - [ind] automatic, discharge,ging, o unloading, system
— m —— f para barco(s) m - [transp] ship, o
  vessel, o boat, unloading system
— m —— f — condensado m - [ind] condensate
  dump(ing) system
— m —— f —— m para depurador m - [ind]
  scrubber condensate dump(ing) system
— m — descarga f - [ind] unloading system
— m —— f mecánica - [ind] mechanical unloading system
— m — descascarillado m—[metal-lam] descaling
  system
— m —— m hidráulico - [metal-lam] hydraulic
  descaling system
— m — desconexión f - [electr-oper] tripping
  system • [mec] disconnecting system • [f.c.]
  uncoupling system
— m —— f selectiva - [electr-oper] selective
  tripping system • [mec] selective disconnecting system • [f.c.] selective uncoupling system
— m — desespumado m - [metal-prod] dross, o
  foam, removel, system, o bucket
— m — deshojamiento m - [metal-lam] opening
  system
— m — despacho m - [electrón] dispatching system • [ind] shipping system
— m — determinación f - determination system

sistema m para determinación f de humedad f -
  moisture determination system
— m —— f —— f en mezcla f - blend moisture determination method
— m — dirección f - [admin] management system
  • [autom-mec] steering system
— m —— f con potencia f hidrostática -
  [autom-mec] hydrostatic power steering system
— m — disecación f - drying system
— m — elaboración f - [ind] processing system
— m —— f para gas m - [petról] gas processing system
— m —— f —— m en yacimiento m - [petról]
  gas field processing system
— m —— f para petróleo m - [petról] oil processing system
— m —— f —— m en yacimiento m - [petról]
  oil field processing system
— m —— f —— m y gas m - [petról] oil and
  gas processing system
— m —— f —— m en yacimiento m -
  [petról] oil and gas field processing system
— m —— f en yacimiento m - [petról] field
  processing system
— m — elevación f - [mec] hoisting system •
  [petról] lift(ing) system
— m —— f artificial - [petról] artifical
  lift(ing) system
— m —— f con gato(s) m - [mec] jack lifting
  system
— m —— f — presión f - [grúas] pressure
  hoist(ing) system
— m —— f —— f alta - [grúas] high pressure hoist(ing) system
— f —— f —— f elevada - [grúas] high
  pressure hoist(ing) system
— m — eliminación f - [ind] dump(ing), o abatement system
— m —— f para ceniza(s) f - [ind] ash disposal system
— m —— f — condensado m - [ind] condensate
  dump(ing) system
— m —— f —— m para depurador m - [ind]
  scrubber condensate dump(ing) system
— m — embarque m - [transp] shipping system
— m —— m de saco(s) m - [ind] bag shipping
  system
— m — embolsamiento m - [ind] bagging system
— m — embolso m - [ind] bagging system
— m — emergencia f - [ind] emergency system •
  back-up system
— m —— f con motor m con combustión f interna - [ind] internal combustion engine emergency system
— m — encendido m - [combust] ignition system
  • lighting system
— m —— m para chispa f - [comb.int] spark
  ignition system
— m — enclavamiento m - [ind] interlocking, o
  lock and block, system
— m — enfriamiento m - [ind] cooling system
— m —— m atascado - [comb.int] clogged cooling system
— m —— m con aceite m - [ind] oil cooling
  system
— m —— m — torre(s) f para refrigeración f
  - [ind] cooling tower (cooling) system
— m —— n enjuagado - [comb.int] flushed
  cooling system
— m —— m limpiado - [ind] cleaned cooling
  system
— m —— m para, banda f, o cinta f, o chapa
  f, o fleje m - [metal-lam] strip cooling system
— m —— m — horno m - [ind] furnace cooling
  system
— m —— m — rodillo(s) m - [metal-lam] roll
  cooling system
— m —— m —— m para trabajo m - [metal-
  lam] work roll(er) cooling system
— m —— m —— n y lubricación f para, banda f, o cinta f, o chapa f, o fleje m -
  [metal-lam] roll cooling and strip lubrica-

ting system
sistema m **para enfriamiento** m **purgado** - [comb.--int] drained, o flushed, cooling system
— m — **engrase** m - [lubric] greasing system
— m — m **para distribuidor** m - [mec] distributor lubrication system
— m — **enganche** m - [mec] coupling system
— m — **engrase** m - [mec] greasing system; [ind] lubricating,tion system
— m — m **automático** - [mec] automatic greasing system
— m — m **centralizado** - [ind] central(ized) grease,sing system
— m — m, **colocado**, o **instalado** - [ined] installed grease,sing system
— m — m **para caja** f - [ind] case lubricating,tion system
— m — m — f **con engranaje(s)** m - [mec] gear case lubricating,tion system
— m — m — f — **piñón(es)** m - [mec] pinion stand lubricating,tion system
— m — **enrase** m - [constr] strike-off method
— m — **ensacado** m - [ind] bagging system
— m — **ensacamiento** m - [ind] bagging system
— m — **entrada** f - [ind] intake system
— m — m **para fuel oil** m - [ind] fuel oil intake system
— m — f — **laminador** m - [metal-lam] rolling mill intake system
— m — **equilibrio** m - [metal-lam] (roll) balancing system
— m — m **para rodillo** m - [metal-lam] roll balancing system
— m — **escape** m - [comb.int] exhaust system
— m — m **aislado con asbesto** m - [comb.int] asbestos wrapped exhaust system
— m — **especificación(es)** f - specification(s), o specifying, system
— m — f **para cadena(s)** f - [cadenas] chains specification,fying system
— m — **estabilización** f - stabilizing,zation system
— m — f, **química**, o **con producto(s)** m **químico(s)** - [quím] chemical stabilizing,zation system
— m — f **para suelo** m - [suelos] soil stabilization system
— m — f — m **con . . . producto(s)** m **químico(s)** - [suelos] . . .-chemical(s) soil stabilization system
— m — f **química** - [quím] chemical stabilization system
— m — f — **para suelo** m - [suelos] chemical soil, o soil chemical, stabilization system
— m — **estado** m - [comput] status system
— m — m **autoiniciado** - [comput] self-initiating,ted status system
— m — **estimación(es)** f—[fin] budgeting system
— m — f **para planificación** f - [fin] planning budgeting system
— m — f — f **para programa(s)** m - [fin] program planning budgeting system
— m — **estrangulación** f - [comb.int] choke,king system
— m — **estrangulador** m—[comb.int] choke system
— m — m **conservada** - [petról] maintained choke,king system
— m — m **mantenido** - [petró] maintained choke,king system
— m — m — **perforación** t - [petról] drilling choke system
— m — m — f, **conservado**, o **mantenido** - [petról] maintained drilling choke system
— m — **evacuación** f - [ind] removal system
— m — f **para escamilla(s)** f - [metal-lam] scale removal system
— m — f — f **para laminador** m **para palanquilla(s)** f - [metal-lam] (billet) mill scale removal system
— m — **evaporación** f—[nucl] evaporation system
— m — f **y fijación** f - [nucl] evaporation--fixing system
— m — f — f **en asfalto** m - [nucl] as-

phalt evaporation-fixing system
— m — **excitación** f - [electr-instal] exciting, tation system
— m — **extensión** f - [constr] extending, o spreading, system
— m — **extracción** f - [combust] exhaust, o extraction, system • removal system
— m — f **de ceniza(s)** - [combust] ash removal system
— m — f — **costra(s)** f - [metal-prod] crust removal system
— m — f — f **de horno** m - [metal-prod] furnace crust removal system
— m — f — f — m **alto** - [metal-prod] blast furnace crust removal system
— m — f **para humo(s)** m - [combust] fume(s) exhaust system
— m — f **en tajo** m **abierto** - [miner] open pit extraction system
— m — **fijación** f - fastening system
— m — f **en cemento** m - [constr] cement, o solidification, system
— m — **filtración** f - [hidr] filtering system
— m — **fiscalización** f - [ind] monitoring, o checking, system
— m — **de calidad** f - [ind] quality control system
— m — f **de granulometría** f - [miner] size monitor(ing) system
— m — f — m **de partícula(s)** f - [miner] particle size monitor(ing) system
— m — f — **partícula(s)** f - [miner] particle monitor(ing) system
— m — f — **tamaño** m - [ind] size monitoring system
— m — f — m **de partícula(s)** f - [ind] particle(s) size monitor(ing) system
— m — **frenado** m - [mec] braking system
— m — m **para automóvil** m - [autom-mec] car braking system
— m — m — m **para carrera(s)** f - [deport] race,cing car braking system
— m — **freno** m - [mec] brake,king system
— m — m **colocado** - [mec] installed brake system **instalado** - [mec] installed braking
— m — m **instalado** - installed braking system
— m — m **y engrase** m - [mec] brake and grease system
— m — **fundición** f - [metal-prod] melt(ing) system
— m — f **eléctrica** - [metal-prod] electric melting system
— m — f — **proyectado** - [metal-prod] designed electric(al) melting system
— m — **fundente** m - [sold] flux system
— m — **garantía** f - guaranty system
— m — f **para calidad** f - [ind] quality guaranty system
— m — **giro** m - [grúas] rotating system
— m — **halar** - [cabl] hauling, o pulling, system; haulage system
— m — m — v **vagón(es)** m - [f.c.] car haul(ing) system
— m — **hojalata** f - [metal-lam] tin strip system
— m — **homogeneización** f - homogenizing system
— m — **humo(s)** m - [combust] exhaust, o flue, system
— m — **identificación** f identifying,fication system
— m — f **para electrodo(s)** m - [sold] electrode(s) identification system
— m — **impulsión** f - [mec] drive,ving system • drive
— m — **inclinación** f - [ind] tilting system
— m — f **para cañón** m - [mec] gun tilting system
— m — **incorporación** f - [nucl] incorporation system
— m — f **para asfalto** m - [nucl] asphalt incorporation system
— m **para indicación** f - [ind] indicating system

sistema m para indicación f para carga f -
[grúas] load indicating system
— m — — f de radio m = [grúas] radius indicating system
— m — indicar v carga f - [grúas] load indicating system
— m — — radio m - [grúas] radius indicating
system
— m — inducción f—[comb.int] induction system
— m — información f - informing,mation, o re-
port(ing), system
— m — inhibición f - inhibiting system
— m — — f para corrosión f - [metal] corrosion, o rust, inhibiting system
— m — interacción f - interaction system
— m — — f entre suelo m y tierra f - [constr]
soil-structure interaction system
— m — — estructura f y suelo m -
[constr] soil-structure interaction system
— m — — f — suelo m y estructura f—[constr]
soil-structure interaction system
— m — intercambio m - interchange, o exchange,
system
— m — — m de ión(es) m - [nucl] ion(s), exchange, o interchange, system
— m — intercomunicación f - [electrón] intercommunication, o intercom, system
— m — interconexión f - interconnecting,tion,
o interlocking, system
— m — — f para caldera(s) f - [cald] boiler
interlocking system
— m — inversión f - [ind] inversion system •
[fin] investment system
— m — inyección f—[ind] injecting,tion system
— m — — f directa - [ind] direct injection
system
— m — — f para combustible m - [comb.int]
fuel injecting,tion system
— m — jalar vagón(es) m - [f.c.] car haul(ing)
system
— m — jubilación(es) f - [labor] pension plan
— m — lámina(s) f - [metal-lam] sheet system
— m — limpieza f - cleaning system
— m — — f para bóveda(s) f - [ind] arch
cleaning system
— m — — f para recuperador m -
[combust] recuperator arch cleaning system
— m — — f — — m túnel m - [combust] tunnel arch cleaning system
— m — — f lingote m - [metal-prod] ingot
cleaning system
— m — — f lingotera f - [metal-prod] ingot
mold cleaning system
— m — línea f - [mec] line system
— m — — f para lámina(s) f - [mec] sheet line
system
— m — — f — plancha(s) f - [metal-lam] sheet
line system
— f — lubricación f - [ind] lubricating,tion
system
— f — — f con aceite m - oil lubricating,tion
system
— m — — f con presión f - [mec] pressure lubricating,tion system
— m — — f forzada - [ind] forced lubricating,tion system
— m — — f hidrostática - [metal-lam] hydrostatic lubricating,tion system
— m — — f interior - [comb.int] inner lubricating,tion system
— m — — f mediante rociadura f - [petról]
spray lubricating,tion system
— m — — f — en (toda) circunferencia f
[petról] (full) circle spray lubricating,tion
sysem
— m — — f para caja f - [ind] case, o stand,
lubricating,tion system
— m — — f — — f para engranaje(s) m - [mec]
gear(s) stand lubricating,tion system
— m — — f — — f piñón(es) m - [mec] pinion stand lubricating,tion system
— m — — f — todo círculo m - [petról] full
circle spray lubricating,tion system

sistema m para lubricación f por alimentación f
forzada - [mec] forced feed lubrication system
— m — — f — aspersión f - spray lubrication
system
— m para manipuleo m - [ind] handling system
— m — — m de material(es) m - [ind] mate-
rial(s) handling system
— m — mantenimiento m - [ind] maintenance
system
— m — — m preventivo - [ind] preventive
maintenance system
— m — marcha f lenta - [comb.int] idle system
— m — — f sin carga f - [comb.int] idle system
— m — medición f - [ind] measuring,rement
system
— m — — f y regulación f - [ind] measurement
and control system
— m — mezcla f - [metal-prod] blending system
• [mec] mixing system
— m — — f de carbón(es) m - [metal-prod]
coal(s) blending system
— m — muestreo m - [ind] sampling system
— m — — m automático - [ind] automatic sampling system
— m — — m en planta f para sínter(ización) - [metal-prod] sinter(ing) plant automatic sampling system
— m — — m en planta f para sínter(ización) -
[metal-prod] sinter(ing) plant sampling system
— m — molienda f - [cuerp.moled] grinding
system
— m — — f para cemento m - [cement] cement
grinding system
— m — — f — materia(s) f prima(s)—[cement]
raw material(s) grinding system
— m — montaje m - [mec] mounting system
— m — — m para eje m - [autom-mec] axle
mounting system
— m — — m — freno m - [mec] brake mounting
system
— m — multifrecuencia(s) f - multifrequency
system
— m — — f con tono m doble - [electrón] dual
tone multifrequency system
— m — nivelación f - [constr] leveling system
— m — numeración f - numbering system
— m — — f de, A W S, o American Welding Society [sold] A W S, o American Welding Society, numbering system
— m — — de Sociedad f Estadounidense para
Soldadura f - [sold] American Welding Society numbering system
— m — odorización f - odorizing system
— m — operación f - operating system
— m — — f con lanza f para oxígeno m -
[metal-prod] oxygen lance operating system
— m — perforación f—[petról] drilling system
— m — — f conservado - [petról] maintained
drilling system
— m — — f mantenido - [petról] maintained
drilling system
— m — — f por percusión f - [petról] percussion drilling system
— m — — f según norma f - [petról] standard
(drilling) system
— m — pesaje m - [ind] weighing system
— m — piso m - [constr] floor system
— m — plancha(s) f - [mec] sheet system
— m — planeamiento m—[admin] planning system
— m — planificación f - [admin] planning system
— m — plástico(s) m - [plat] plastic(s) sys-
tem(s)
— m — polea(s) f - [mec] pulley(s) system
— m — — f para montacarga(s) m—[metal-prod]
skip pulley system
— m — — f — — m para horno m - [ind] furnace skip pulley system
— m — — f — — m — — m alto—[metal-prod]
blast furnace skip pulley system

sistema m̲ para polea(s) para vagoneta(s)—metal-
prod]⁻car pulley system
— f̲ — f̲ — — f̲ para montacargas m̲—[metal-
-prod] skip car pulley system
— m̲ — f̲ — f̲ — — m̲ para horno m̲ -
[metal-prod] furnace skip car pulley system
— m̲ — f̲ — f̲ — — m̲ — — m̲ alto -
[metal-prod] blast furnace skip car pulley
system
— m̲ — posición f̲ - [comput] status system
— m̲ — f̲ autoiniciado - [comput] self-initiated status system
— m̲ — pozo m̲ - [mec] well system
— m̲ — — m̲ único - single well system
— m̲ — mantenimiento m̲ - [ind] maintenance system
— m̲ — precalentamiento m̲ - [ind] preheating
system
— m̲ — — m̲ para rodillo(s) m̲ - [metal-lam]
roll preheating system
— m̲ — procesamiento m̲ - processing system
— m̲ — — m̲ para dato(s) m̲ - [comput] data
processing system
— m̲ — producción f̲ - [ind] production system •
[petról] lift(ing) system
— m̲ — f̲ artificial - [petról] artificial
lift(ing) system
— m̲ — — f̲ económica - [petról] economical
artificial lift(ing) system
— m̲ — — f̲ de acero m̲ - [metal-prod] steel,
production system, o practice
— m̲ — programación f̲ - [admin] programming, o
scheduling, system
— m̲ — — f̲ para estimación(es) f̲ - [admin]
budgeting, programming, o planning, o scheduling, system
— m̲ — — f̲ — planificación f̲ - [admin] planning, programming, o scheduling, system
— m̲ — — — f̲ para estimación(es) f̲ -
[admin] budgeting planning, programming, o
scheduling, system
— m̲ — propulsión f̲ - [mec] drive,ving system
— m̲ — — f̲ eléctrica - [mec] electric(al)
drive,ving system
— m̲ — purificación f̲ - purification system
— m̲ — — f̲ para agua m̲ [hidr] water purification system
— m̲ — — f̲ — gas m̲ - [combust] gas scrubbing
system
— m̲ — — f̲ — — m̲ para horno m̲ - [metal-prod]
furnace gas scrubbing system
— m̲ — — f̲ — — m̲ alto - [metal-prod]
blast furnace gas scrubbing system
— m̲ — recirculación f̲ - recirculating,tion
system
— m̲ — — f̲ para aceite m̲ - [metal-lam] oil recirculating,tion system
— m̲ — — f̲ — — m̲ para laminación f̲ - [metal-
-lam] rolling oil recirculating,tion system
— m̲ — recogida f̲ - collection, o gathering,
system
— m̲ — recolección f̲ - [ind] collecting, o
gathering, system
— m̲ — reconstrucción f̲ - [constr] rebuilding,
o reconstructing,tion, system • [metal-prod]
relining system
— m̲ — recuperación f̲ - recovery,ring system
— m̲ — — f̲ para derrame m̲ - [ind] spillage recovery,ring system
— m̲ — reducción f̲—[mec] reducing,ction system
— m̲ — — f̲ con . . . etapa(s) f̲ - [mec] . . .
stage reducing,ction system
— m̲ — — f̲ encerrado - [mec] enclosed reducing,ction system
— m̲ — — f̲ encerrado totalmente - [mec] completely, o totally, enclosed reducing,ction
system
— m̲ — — f̲ y regulación f̲ - [metal-lam] pressure reducing,ction and control(ling) system
— m̲ — — f̲ para energía f̲ - [electr-oper] energy reducing,ction system
— m̲ — — f̲ directa - [metal-prod] direct reducing,ction, system, o design

sistema m̲ para reducción f̲ en consumo m̲ - [ind]
use reduction system
— m̲ — — f̲ — — m̲ de energía f̲ - [ind] energy
use reduction system
— m̲ — referencia f̲ - reference system
— m̲ — reforma(ción) f̲ - [metal-prod]
reform(ing) system
— m̲ — f̲ con base f̲ de vapor m̲ - [metal-
-prod] steam based reform(ing) system
— m̲ — — f̲ con vapor m̲ - [metal-prod] steam
(operated) reform(ing) system
— m̲ — — — m̲ e hidrocarburos(s) m̲—[metal-
-prod] steam-hydrocarbon(s) reforming system
— m̲ — — f̲ operada con vapor m̲ - [miner] steam
operated reform(ing) system
— m̲ — refrigeración f̲ - [ind] refrigeration, o
cooling, system
— m̲ — — f̲ enjuagado - [ind] flushed cooling
system
— m̲ — — f̲ inoperante - [ind] nonoperating
refrigeration system
— m̲ — — f̲ limpio - [comb.int] clean cooling
system
— m̲ — — f̲ limpiado - [comb.int] cleaned cooling system
— m̲ — — f̲ operante - [ind] operating,tive refrigeration system
— m̲ — — f̲ para rodillo(s) m̲ - [metal-lam]
roll cooling system
— m̲ — — f̲ purgado - [comb.int] flushed cooling system
— m̲ — regeneración f̲ - regenerating,tion system
— m̲ — — f̲ de arcilla f̲ - [petról] clay, regenerating, o revivifying, system
— m̲ — registración f̲ - [instrum] recording, o
registering,tration, system
— m̲ — registro m̲—[instrum] record(ing) system
— m̲ — regulación f̲ - control(ling) system
— m̲ — — f̲ analógica - [comput] analogical
control(ling) system
— m̲ — — f̲ automática - [ind] automatic con-
trol(ling) system
— m̲ — — f̲ — para espesor m̲ - [metal-lam] automatic gage control(ling) system
— m̲ — — f̲ dañado - [ind] damaged control system
— m̲ — — f̲ de demanda f̲ - demand control(ling)
system
— m̲ — — f̲ para demanda f̲ para usina f̲ - [ind]
power plant demand control(ling) system
— m̲ — — f̲ de espesor m̲ - [metal-lam] gage
control(ling) system
— m̲ — — f̲ digital - [comput] digital con-
trol(ling) system
— m̲ — — f̲ electrónica - [electrón] electronic
control(ling) system
— m̲ — — f̲ en sala f̲ - [ind] room control
system
— m̲ — — f̲ — — f̲ para regulación f̲ - [ind]
control room control(ling) system
— f̲ — — f̲ estática - [comput] static con-
trol(ling) system
— m̲ — — f̲ hidráulica - [hidr] hydraulic con-
trol(ling) system
— m̲ — — f̲ para agua m̲ - [hidr] water con-
trol(ling) system
— m̲ — — f̲ — caldera(s) f̲ - [cald] boiler(s)
control(ling) system
— m̲ — — f̲ — combustión f̲ - [combust] combustion control(ling) system
— m̲ — — f̲ — contaminación f̲ - [ind] pollution control(ling) system
— m̲ — — f̲ — — f̲ para agua m̲ - [hidr] water
pollution control(ling) system
— m̲ — — f̲ — — agua m̲ - [hidr] water
pollution control(ling) system
— m̲ — — f̲ para fluido(s) m̲ - [ind] fluid(s)
control(ling) system
— m̲ — — f̲ — velocidad f̲ - [mec] speed con-
trol(ling) system
— m̲ — — f̲ — voltaje m̲ - [electr-instal]
voltage control(ling) system

sistema m para regulación f y alimentación f para alambre m - [sold] wire control and feeding
system
— m — regulación f - [comb.int] throttling system
— m — reserva f - [ind] backup system
— m — respaldo m - support(ing) system
— m — respiración f breather,thing system
— m — retención f - retention, o withholding,
system • [hidr] reservoir system
— m — f de camisa f - [petról] liner retention system
— m — riego m - [ind] spray(ing) system •
[agric] watering system
— m — — m para montacarga(s) m - [metal-prod]
skip spray(ing) system
— m — — m — skip m - [metal-prod] skip spray
system
— m — rociadura f - [ind] spray(ing) system
— m — — f con presión f - [hidr] pressure
spray(ing) system
— m — rodaje m - [ind] dolly train
— m — salpicadura f - [ind] splash(ing) system
— m — salpique m - [ind] splash(ing) system
— m — sangría f - [ind] bleeding system
— m — seguimiento m - followup, o following,
system
— m — seguridad f - [segurid] safety system
— m — sellado,dura - sealing system
— m — señalización f - [ind] signalling system
— m — — f luminosa - [ind] luminous signalling
system
— m — sobrepaso* m - [petról] override,ding
system
— m — — m para estrangulador m - [petról]
choke overrid,ding system
— m — soldadura f - [sold] welding system
— m — — f automática - [sold] automatic welding system
— m — — f — con regulador(es) m con transistores m - [sold] solid state control automatic welding system
— m — — f — regulador(es) m con transistores
m - [sold] solid state control(led weld(ing) ·
system
— m — soplado m - [ind] blowing system
— m — — m para aire m - [ind] air blowing system
— m — soporte m - support(ing) system
— m — sostén m - [mec] support(ing) system
— m — sujeción f - [mec] fastening, o holding,
system
— m — suministro m - [ind] supply(ing) system
— m —
          m para agua m - [hidr] water sup-
ply(ing) system
— m — — m — — m para tobera f - [metal-prod]
tuyere water supply(ing) system
— m — — m — — m — — f para Venturi m -
[metal-prod] Venturi tuyere water supply system
— m — — m para energía f - [electr-prod] power
water supply(ing) system
— m — supervisión f - [ind] supervision system
— m — — f para distribución f - [electr-prod]
distribution supervision system
— m — supresión f - supression, o abatement,
system
— m — suspensión f - [mec] suspension system
— m — — f flexible - [mec] flexible suspension
system
— m — — f para eje m - [autom-mec] axle suspension system
— m — — f — vehículo m - [autom-mec] vehicle
suspension system
— m — terminación f - [pint] finishing system
— m — — f con pulverización f - [pint] spray
finishing system
— m — termoformación f - [plást] thermoforming
system
— m — tiro m - [mec] pulling system • [transp]
hauling system • [combust| draft system
— m — toma(s) f - [electr-instal] collector
system

sistema f para toma(s) f para corriente f -
[electr-instal] (current) collector system
— m — — f — — f para grúa(s) f - [grúas]
crane (current) collector system
— m — trabajo m - work system
— m — transición f - [mec] transition system
— m — transmisión f - [mec] transmission, o
drive,ving, system
— m — transporte m - [transp] transportation
system • transportation mode
— m — — m aéreo - [ind] aerial conveying
system
— m — — m de materia(s) f prima(s) - [ind]
raw material(s) transportation system
— m — tubo(s) m - [mec] pipe(s) system
— m — vapor n - [vapor] steam, system, o installation
— m — venteo m - [ind] venting system
— m — ventilación f - [ind] ventilating,tion
system
— m — — f con tiro m ascendente - [ind]
up-draft ventilating,tion system
— m — — f — — m ascendente - [ind] down
draft ventilating,tion system
— m — — f general - [ind] general ventilating,tion system
— m — verificación f - checking system
— m — — f para calidad f - [ind] quality
control system
— m — vigilancia t - monitoring system
— m — viscosidad f - [lubric] viscosity system
— m — — f para lubricante(s) m - [lubric] lu-
bricant(s) viscosity system
— m — — f — — m para fluido(s) m - [lubric]
fluid(s) lubricant(s) viscosity system
— m — — f — — — m industrial(es) m -
[lubric] industrial fluid(s) lubricant(s)
viscosity system
— m patentado - [ind] patented system •
[comput] proprietary software
— m penetrado - penetrated system
— m peor - worse system • worst system
— m pérmico - [geol] Permian system
— m perno-tuerca - [mec] bolt-nut system
— m planetario - [astron] planetary system
— m — con . . . etapa(s) f - [mec] . . . stage
planetary system
— m — encerrado - [mec] enclosed planetary
system
— m — — en tambor m - [mec= drum enclosed
planetary system
— m — — totalmente - [mec] totally, o completely, enclosed planetary system
— m — para reducción f - [mec] planetary reducing,duction system
— m — — f, con, o en, . . . etapa(s) f
[mec] . . . stage planetary reducing,ction
system
— m — — — f encerrado - [ind] enclosed
planetary reduction system
— m — — — f —, con, o en, . . . etapa(s) f
- [mec] enclosed . . . stage planetary reducing,ction system
— m — — — f encerrado totalmente - [mec]
totally, o completely, enclosed planetary
reducing,ction system
— m — — f, con, o en, . . . etapa(s) f encerrado en tambor m - [mec] . . . stage drum
enclosed planetary reducing,ction system
— m — — f, —, o — . . . — f — totalmente en tambor m - [mec] . . .-stage, totally, o completely, enclosed planetary reducing,ction system
— m por percusión f - [petról] percussion system
— m por aspiración f - [ambient] exhaust system
— m positivo m - [mec] positive system
— m — m para acoplamiento m - [mec] positive
coupling system
— m precalentador m - preheating system
— m primario - [electr-instal] primary system
— m — para distribución f - [electr-instal]

primary (distribution) system
sistema m principoal - principal, o main, system
— m — para combustible m - [comb.int] main fuel system
— m propuesto - proposed, o (pr)offered, system
— m protector - protective system
— m proyectado - projected, o designed, o engineered, system
— m purgado - purged system
— m químico - [quím] chemical system
— m Raky - [petról] Raky system
— m rápido - rapid, o fast, o quick, system
— m rascador - scraper,ping system
— m recomendado - recommended system
— m refractario - [metal-prod] refractory system
— m — para recuperación f - [metal-prod] (refractory) checker system
— m registrado - registered, o recorded, system
  • [electrón] proprietary software
— m regulado - controlled, o set up, system
— m — con flotador m - [comb.int] float controlled system
— m — — m para combustible m - [comb.int] float controlled fuel system
— m — para viento m - [metal-prod] air blast control system
— m respirador - breather,thing system
— m retrointegrado - retrointegrated system
— m rígido - rigid system
— m rociador - [mec] spray(ing) system
— m — para camisa f - [mec] liner spray(ing) system
— m — — émbolo m - [mec] piston spray(ing) system
— m — — m principal - [mec] main piston spray(ing) system
— m — — m y camisa f - [mec] piston and liner spray(ing) system
— m — — — f accionado con motor m eléctrico - [mec] electric motor driven piston-liner spray(ing) system
— m — — — — m — montado en parte f posterior - [mec] rear mounted electric motor driven piston-liner spray(ing) system
— m — — m — f — — m — — — parte f superior - [mec] top mounted electric motor driven piston-liner spray(ing) system
— m rotativo - rotary system
— m secundario - secondary system
— m — para recuperación f - [petról] secondary recovery system
— m según norma f - standard system
— m seguro - [segurid] safe system
— m — para regulación f - [ind] safe control system
— m — — — f automática - [ind] safe automatic control(ling) system
— m semiautomático - semi-automatic system
— m — para alimentación f de alambre m - [sold] semi-automatic wire feed(ing) system
— m semirrígido - semirigid system
— m sencillo - simple system
— m — para regulación f - [ind] simple control(ling) system
— m — — — f automática - [ind] simple automatic control(ling) system
— m séptico - [sanit] septic system
— m silenciador m - [electrón] squelching system
— m — m controlado - [electrón] controlled squelching system
— m — — con tono m - [electrón] tone controlled squelch(ing) system
— m — regulado - [electrón] controlled squelching system
— m — — con tono m - [electrón] tone controlled squelch(ing) system
— m — — — — m continuo - [electrón] continuous (tone) controlled squelch(ing) system
— m — simultáneo - [electrón] simultaneous squelch(ing) system
— m — — continuo m - [electrón] continuous simultaneous squelch(ing) system

sistema m silenciador simultáneo continuo regulado con tono m continuo - [electrón] simultaneous continuos tone controlled squelch(ing) system
— m sin gas m - [combust] off-gas system
— m single spot - [coque] single spot system
— m singular - [ind] unique system
— m — para reforma(ción) f - [metal-prod] unique reforming system
— m — — — f con vapor m e hidrocarburos m - [metal-prod] unique steam-hydrocarbon reforming system
— m sintónico - [electrón] syntonic system
— m sonoro - [electrón] sound system
— m — electrónico - [electrón] electronic sound system
— m — para automóvil m - [electrón] automobile('s), o car('s) sound system
— m soplador - blowing system
— m submarino - submarine system • [petról] subsea system
— m subterráneo - underground system • [petról] subsurface system
— m suelo-estructura - [constr] soil-structure system
— m suplementario - make-up system
— m — para provisión f - make-up supply system
— m — — de aire m - make-up, air supply, o supply air, system
— m suspendido - [petról] suspended,nsion system
— m — en fondo m de mar m - [petról] mudline suspension system
— m T E C O - [cerám] T E C O system
— m taponado - clogged system
— m telefónico - [telecom] telephone system
— m telegráfico - [telecom] telegraph system
— m termoformador—[plást] thermoforming system
— m tetrafilar - [electr] four wire system
— m — para distribución f - [electr-instal] four wire distribution system
— m — — f de energía f - [electr-instal] four wire power distribution system
— m total - total, o entire, system
— m — de viga(s) f - [constr] entire rail system
— m totalmente automático - [ind] full(y) automatic system
— m — — para alimentación f - [sold] full(y) automatic feed(ing) system
— m — — — — f de alambre m - [sold] full(y) automatic wire feed(ing) system
— m — automatizado - fully automated system
— m trabador - [mec] locking system
— m Trabon - [metal-prod] Trabon system
— m — para cierre m - [metal-prod] Trabon seal(ing) system
— m — — — para grasa f - [metal-prod] Trabon grease seal(ing) system
— m — — — — f para tragante m - [metal-prod] Trabon throat grease seal(ing) system
— m — — engrase m - [metal-prod] Trabon lubricating,tion system
— m transmisor - [electrón] transmission system
— m — y respondedor - [electrón] transponder system
— m — movible - [electrón] mobile transponder system
— m — móvil - [electrón] mobile transponder, o transponder mobile, system
— m — — proyectado - [electrón] designed mobile transponder system
— m — proyectado - [electrón] designed transponder system
— m transportador - [ind] conveyor system
— m — aéreo - [ind] aerial conveyor system
— m trifásico - [electr-instal] three-phase system
— m troncal - [hidr] trunk system
— m Unidraulic - [petról] Unidraulic system
— m — para bombeo m - [petról] Unidraulic pumping system
— m vertical - vertical system

sistema m vial - [vial] highway system
— m viejo - old system
sistemática f - systematics • véase también
plan m; programa f
sitiado,da a - [milit] besieged
sitio n - room; place • area • position • location; site; scene • [pol] locality • [milit]
siege; besieging • [constr] property; lot
— m abierto - open, space, o area
— m ácido - [suelos] acid(ic) site
— m acordado - agreed location
— m áspero - rough, o rugged, spot
— m atendido - attended site
— m — regularmente - regularly attended site
— m bajo - [topogr] low site
— m bien ventilado - well ventilated, location,
o area, o site
— m cenagoso - swampy, location, o site
— m cómodo - comfortable, site, o place
— m con atención f - [electrón] attended site
— m — frecuente - [electrón] frequently attended site
— m — f infrecuente - [electrón] infrequently attended site
— m — f regular - regularly attended site
— m — contenido m moderado - moderate, o
slight, content location
— m — m — de polvo m - [ind] moderate, o
slight, dust content location
— m — corriente f (fuerte de aire m) - [ind]
drafty location
— m — polvo m - dusty, o dirty, location
— m — tierra f - dirty location
— m convenido - agreed location
— m conveniente - convenient location
— m crítico - critical location
— m cualquiera - any location; anywhere
— m dado - given, location, o site
— m de accidente - [segurid] accident site
— m — empleo m - [labor] employment, place, o
site, o location
— m — fuga f - leakage site
— m delicioso - beautiful, site, o place
— m demarcado - established, site, o place
— m despejado - clear, o uncluttered, site, o
space
— m determinado - determined, o specific, location, o site
— m distante - distant, o remote, site, o place
— m (en) particular - particular site
— m equivocado - wrong, site, o place
— m específico - specific, site, o location
— m — remoto - particular, o remote, location,
o site
— m estrecho - cramped area; tight quarter(s)
— m evaluado - [constr] evaluated site
— m — para conducto m - [constr] evaluated
conduit site
— m exacto - exact, location, o site
— m excepcionalmente sucio - particularly dirty
location
— m excesivamente polvoriento - excessively
dirty location
— m fresco - cool, site, o place
— m geológico - [geol] geological, location, o
site
— m histórico - [hist] historical site
— m — nacional - [hist] national historical
site
— m húmedo - damp, o wet, o humid, place, o location, o site
— m ideal - ideal, site, o location
— m individual - individual, site, o location
— m — para aparcamiento m - [constr] parking
space
— m interrogado - [comput] interrogate(d) space
— m limpio - clean, o clear, space, o site, o
place
— m mejor - best site • better site
— m normal - normal, site, o place, o location
— m — de empleo m - [labor] normal employment
place
— m para alcantarilla f - [hidr] culvert, site,

o location
sitio m para almacenamiento m - storage,ring,
area, o space
— m — m permanente - permanent storage,ring
place
— m — amperímetro m - [electr-instrum] ammeter,
space, o location
— m — m y voltímetro m - [instrum] ammeter
and voltmeter location
— m — aparcamiento m - [constr] parking (area,
o space)
— m — m acuático - [hidr] aquatic parking,
area, o place, o space, o lot
— m — m para automóvil(es) m - [constr]
automobile, o car, parking area
— m — apilamiento m - [ind] piling area
— m — caja f - [ind] case, o box, location
— m — f para mando m - [ind] control box
location
— m — f — telemando m - remote control
box location
— m — camping m - [deport] camping site
— m — canal m - [hidr] channel, o canal, location, o site
— m — carrera f - [deport] race,cing site
— m — cauce m - [hidr] channel location
— m — conducto m - [constr] conduit site
— m — construcción f - [constr] construction
site
— m — descanso m - rest(ing) area; place to
relax
— m — desecho(s) m - [ind] waste site
— m — m radiactivo(s) - [nucl] radioactive
waste site
— m — desenvolver(se) - [ind] working, space,
o area
— m para descarga f - unloading, o dumping,
area, o space, o location
— m — disposición - disposal area; disposal
site
— m — establecimiento m - location site
— m — ensayo(s) m - [ind] testing, location, o
site, o area
— m — fabricación f - [ind] manufacturing,
site, o area
— m — espectador(es) m - [deport] viewing,
point, o area; spectator area
— m — estacionamiento m - [constr] parking
(area, o space)
— m — explotación f - [miner] pit location
— m — feria f - [pol] fairgrounds
— m — inspección f inspection, site, o area
— m — instalación f - [ind] installation site
• [constr] job site
— m — f para tubería f - [constr] pipe,ping
installation site
— m — f — f corrugada - [constr] corrugated pipe installation site
— m — interrogación f - [comput] interrogate
site
— m — mezcla(do) - [constr] mixing site
— m — mina f - [miner] mine site
— m — montaje m - [ind] mounting site
— m — m para caja f - [ind] box, o case,
mounting, site, o location
— m — m — f para mando m - [ind] con-
trol(ling) box mounting location
— m — m — f — telemando m - [ind] remote control(ling) box mounting location
— m — obra f - [ind] work, o job, site • construction , o project, o installation, site;
site • yard
— m — pesca f - [deport] fishing site
— m — pilote m - [constr-pil] pile location
— m — planta f - [ind] plant, site, o location
— m — precipitación f - precipitation site
— m — proyección f - [constr] design, site, o
point
— m — recreación f - [deport] recreation area
— m — f y descanso m - [deport] recreation
type area
— m — recreo m - playground • resort
— m — m en montaña f - [deport] mountain
resort area

resort area
sitio m para repetición f - repeating area
— m — residencia f - residence, o home, site
— m — residuo(s) m - [ind] waste(s) site
— m — — m radiactivo(s) m - [nucl] radioactive
waste site
— m — reunión(es) m - meeting, o gathering,
place
— m — soldadura f - [sold] welding, place, o
site • weld, location, o placement
— m — trabajo m - [ind] work(ing), o job, site,
o area, o station
— m — vaciado m - [constr] dumping, o pouring,
area, o place
— m — vuelco m - dump(ing), site, o location
— m particular - specific site
— m — remoto - specific remote, area, o place
— m permanente - permanent, location, o place
— m polvoriento - dusty, o dirty, location
— m popular - popular place
— m — para reunión(es) f - popular, meeting, o
gathering, place
— m propuesto - proposed, site, o place
— m — para alcantarilla f - [constr] proposed
culvert site
— m — para mina f - [miner] proposed mine site
— m remoto - remote, site, o location, o place;
remote
— m repetidor - [comput] repeater site
— m seco - dry, site, o place, o location
— m — apropiado - suitable, o appropriate, dry
place
— m — y limpio - clean dry place
— m sobre cubierta f - [náut] deck space
— m solitario - lonely, site, o place
— m sucio - dirty, place, o location
— m suficiente - sufficient, space, o room
— m templado - warm, place, o location
— m tierra f adentro - [petról] on land site
— m tierra f afuera - [petról] off shore site
— m típico - typical, site, o location
— m valioso - valuable, site, o space
— m — sobre cubierta f - [náut] valuable deck
space
— m ventilado - ventilated area
sitios m distantes múltiples - [comput] multiple
remote site(s)
— m donde obtener probeta(s) f (para ensayos m)
- [sold] test specimen(s) location
— m múltiples - multiple sites
— m remotos múltiples - multiple remote sites
situación f - . . .; positioning; locating; placing • delivery; delivering; véase también
ubicación
— f actual - present, condition, o position
— f arbitraria - arbitrary situation
— f buena - good condition • good placement
— f codiciable - enviable condition
— f confundente - confusing situation
— f confusa - confusing situation
— f controvertida - [legal] misunderstanding;
difference
— f costosa - costly situation
— f de alambre m - [cabl] wire position(ing)
— f — apuro m tight, o tough, situation
— f — carga f - [ind] charge location
— f — crecimiento - growth condition
— f — — estable - stable growth condition
— f — inconvertibilidad - [fin] inconvertibi-
lity, condition, o situation
— f — — f de moneda f - [fin] currency incon-
vertibility situation
— f — retiro m - [labor] retirement] adv -
(en) retired
— f — termocupla f - [metal-prod] thermocouple
location
— f — — f para crisol m - [metal-prod] hearth
thermocouple location
— f descri(p)ta - described situation
— f económica - [econ] economic, situation, o
position
— f — actual - [econ] current, o present, economic situation

situación f estratégica - strategic location
— f estudiada - studied situation
— f evaluada - evaluated, o valued, situation
— f financiera - [fin] financial position
— f — consolidada - [fin] consolidated financial position
— f futura - future, position, o situation
— f ideal - ideal situation
— f insegura - unsafe, situation, o condition
— f mejor - better condition • best condition
— f para venta f - [com] selling situation
— f particular - particular, o peculiar, situation, o position
— f patrimonial - [fin] equity position
— f regresiva - regressive situation
— f verosímil - [fin] operating situation
situado,da a - situated; located; placed • delivered
— a bien - well placed
— a convenientemente - located conveniently
— a estratégicamente - located strategically
situador,ra a - locator • locating
situar v - . . .; to, situate, o position • to
deliver
— v bien - to place well
— convenientemente - to locate conveniently
— f estratégicamente - to locate strategically
skip m - [metal-prod] skip; véase también carro
m; vagoneta f; cubeta f (para skip m)
slurry m - véase pasta f aguada
snout m - [metal-trat] véase conducto m para
bajada f
sobordo m para carga f - [transp] freight, o cargo, manifest
— m — — separado - [transp] separate, freight,
o cargo, manifest
sobra a - . . .; spare
sobrante m - . . .; excess; remainder • cut off
• odd quantity
— m de chatarra f - [metalprod] scrap surplus
sobrar v - . . .; to spare • to remain
sobre adv banda f - [culin] over @ band
— adv bobina f - [electr-install] (up)on @
coil
— adv cabeza f - [sold] overhead
— adv camino m de tierra f - [transp] on @ dirt
road • off-road
— adv camión m - [transp] on (board) @ truck
— adv canal m - [hidr] over @, canal, o channel
— adv carrete m - [mec] on @ reel | adv - reel
packed
— carretera f - on @ road
— m cerrado - closed envelope
— m con instrucción(es) f - instruction envelope
— m — — f para instalación f installation
(instructions) envelope
— adv derecha f - on @ right (side)
— m aparte - separate envelope
— m diferente - different envelope
— m dirigido - addressed envelope
— adv eje m - on @ axis
— adv — m neutro - about a neutral axis
— adv escala f completa - over @ (full) range
— adv — f — — de temperatura(s) f - over @
(full) temperature range
— adv — de temperatura(s) f - over @ tempera-
ture(s) range
— adv ficha f - carded
— adv — f de cartón m - [com] carded
— adv izquierda f - on @ left (side)
— m lacrado - [com] sealed envelope
— adv lago - [hidr] lakeside
— adv larguero m - [constr] over, o on, @
stringer
— adv marcha f - [mec] while, operating, o in
operation • [Chi.] (while) in motion • [ind]
with @ hot furnace
— adv mismo,ma - there(up)on
— adv movimiento m - while in motion
— adv muelle m - [transp] dockside; on @ dock
— adv neumático(s) m - [autom] rubber tired;
tire mounted | adv - wheel type

sobre adv nivel m de mar - above @ sea level
— adv — m — suelo m - above @ ground (level)
— adv oruga f - [constr] track, o crawler, type
— adv parte f inferior - on @ bottom
— adv — superior - on @ top
— adv pasada(s) f de raíz f - [sold] over, o on, @ root pass(es)
— adv pasta f - [metal-prod] over @ paste
— adv patín(es) m - [ind] on @ skid(s)
— adv pavimento m - over, o on, @ pavement
— adv — m mojado - [autom] in @ wet; on @ wet vement
— adv — m seco - [autom-neumát] in @ dry; on @ dry pavement
— adv pedestal m - on @ pedestal • free standing
— adv pucho m - on @ spot
— adv punta f - on @ end
— adv — f delantera - [autom] on @ nose
— adv punto m - over @ spot
— adv rasante f - above @ grade
— adv rodaje m - [sold] carriage mounted
— m rotulado - [com] return address envelope • marked envelope
— adv rueda(s) f - on wheel(s)
— adv seco - [coque] dry base
— m sellado - [com] sealed envelope
— m separado - [com] separate envelope
— adv suelo m - on, o above, @, ground, o floor
— adv superficie f - above @ surface
— adv — f de terreno m - above @ ground level
— adv tierra f - above ground; on @ land • [petról] land
— adv título m - over @ title
— adv todo - above all • especially
— adv torno m - [petról] on @ cathead
— adv trabajo m - [mec] over @, work, o job
— adv vagón m (ferroviario) - [transp] on (board) @ (railway) car
sobrealimentación f - overfeeding • [sold] (electrode) overrun; overrunning
— f de alambre m - [sold] wire overrun(ning)
— f — — m a finalizar(se) v soldadura f - [sold] weld end wire overrun(ning)
— f — electrodo m - [sold] electrode overrun
— f — motor m para alimentación f - [sold] feed(ing) motor overrun(ning)
— f — m — — f para alambre m - [sold] wire feeding motor overrun(ning)
sobrecabeza adv - [sold] overhead
sobrecalefacción f - overheating
sobrecalefaccionado,da a - overheated
sobrecalefaccionar v - to overheat
— v localmente - to overheat locally
sobrecalentado,da a - overheated • superheated
— a localmente - overheated locally
sobrecalentador m - superheater; véase también recalentador m
— m para vapor m - [cald] steam (de)superheater
sobrecalentamiento m - overheating • desuperheating • [sold] high heat
— m de asfalto m - [constr] asphalt overheating
— m — barra(s) f - [metal-trat] bar(s) overheating
— m — bisagra f - [mec] hinge overheating
— m — motor m - electr-mot] motor overheating • [comb.int] engine overheating
— m — rectificador m - [electr-oper] rectifier overheating
— m — soldadora f - [sold] welder overheating
— m — transformador m - [electr-oper] transformer overheating
— m — válvula f - [comb.int] valve overheating
— m — vapor m - [cald] steam, overheating, o (de)superheating
— m local - local overheating
sobrecalentar v - to superheat; to overheat; to desuperheat
— v asfalto m - [constr] to overheat @ asphalt
— v bisagra f - to overheat @ hinge
— v localmente - to overheat locally
— v motor m - [electr-mot] to overheat @ motor • [comb.int] to overheat @ engine
— v válvula f - [comb.int] to overheat @ valve

sobrecalentar v vapor m - [cald] to (de)superheat @ steam
sobrecama f de mimbre m - [domést] wicker bedspread
sobrecapa f - [suelos] overburden
sobrecarga f - . . .; overloading • [electr-oper] overcharge,ging; surcharge,ging load • [comb.-int] overcharge,ging • [constr] surcharge
— f continuada - consistent overloading
— f de arena f cargada - [Constr] dredged sand surcharge
— f — construcción f - [constr] construction surcharge,ging
— f de nivel m - [constr] level surcharge,ging
— f durante construcción f - [constr] construction surcharge,ging
— f frecuente - frequent overload(ing)
— f magnética - [electr-oper] magnetic overload
— f infinita - infinite surcharge,ging
— f leve - (s)light surcharge,ging
— f limitada - limited surcharge,ging
— f nivelada - level(ed) surcharge,ging
— f para compresor m - [ind] compressor overload(ing)
— f — — m sin conectar v - [ind] open compressor overload(ing)
— f — conductor m - [electr-oper] cable, o lead, overload(ing)
— f — grúa f - [grúas] crane overload(ing)
— f — motor m - [comb.int] engine overloading • [electr-mot] motor overload(ing)
— f — rectificador m - [electr-oper] rectifier overload(ing)
— f — secador m - [ind] drier overload(ing)
— f pareja - level surcharge,ging
— f permanente - [constr] permanent surcharge
— f posible - possible overload(ing)
— f repentina - [electr-oper] surge
— f súbita - sudden overload(ing)
— f, sospechada, o supuesta - [mec] suspected overload(ing)
— f térmica - [electr-oper] thermal overload
— f termomangética - [electr] thermal-magnetic overload(ing)
— f termostática - [electr-oper] thermostatic overload(ing)
sobrecargado,da a - overloaded; overburdned; surcharged • [constr] weighted
sobrecargar v - to surcharge • to overload • to overcharge • to overburden
— v motor m - [comb.int] to overload @ engine • [electr-mot] to overload @ motor
— v secador m - [ind] to overload @ drier
sobrecebado,da a - [comb.int] overprimed
sobrecorriente f - [electr-oper] overcurrent • [sold] overload
— f a tierra f - [electr-oper] ground overcurrent
— f — f monofásica - [electr-oper] single phase ground overcurrent
— f de fase f a fase f - [electr-oper] phase to phase overcurrent
— f — f — tierra f - [electr-oper] phase to ground overcurrent
— f en elemento m - [electr-oper] element overload(ing)
— f — m térmico - [sold] thermal element overload(ing)
— f monofásica - [electr-oper] single phase overcurrent
sobrecristal m - [instrum] cover(ing) lens
— m claro - [óptica] clear cover glass
— m — sencillo m - plain clear cover glass
sobrediámetro m - excess diameter | oversize
sobredimensionado,da a - oversized
sobredimensionar v - to oversize
sobredoblado,da a - overbent
sobredoblamiento m - overbending
— m de varilla f - [mec] rod overbending
sobredoblar v - to overbend
— v varilla f - [mec] to overbend @ rod
sobreelevación f - [electr-oper] surge
— v de temperatura f - temperature overelevation

sobreelevación f de temperatura f en devanado m -
[elecr-oper] winding temperature overelevation
— f — f — m para estator m—[electr-mot]
stator winding temperature overelevation
— f — f — m — rotor m - [electr-equip]
rotor winding temperature overelevation
— f máxima - [electr-oper] maximum overelevation
— f — promedio - [electr-mot] temperature maximum average overelevation
— f — — para temperatura f para devanado m -
[electr-mot] winding temperature maximum
average overelevation
sobreelevar v - to overelevate
— v temperatura f—to overelevate @ temperature
— v — f en devanado m - [electr-oper] to overelevate @ winding temperature
sobreengrasado,da a - [mec] overgreased; grease
overpacked
sobreengrasar v - [mec] to overgrease; to grease
overpack
sobreengrase m - [lubric] overgreasing
sobreentender v - to imply
— v garantía f—to imply @, guaranty, o warranty
sobreentendido,da a - implied
sobreentendimiento m - implying; implication
— m de garantía f - guaranty, o warranty, implying, o implication
sobreesfuerzo m - overstress(ing)
sobreespesor m - excess thickness • [sold] over-
weld(ing); reinforcement,cing
— m — cordón m - [sold] bead excess thickness
sobreestadía f - [transp] . . ; demurrage
sobreestante m - [labor] supervisor
sobreexcavación f - [constr] over excavating,tion
— f de horadación f - [constr] bore over excavating,tion
sobreexcavado,da a - [constr] overexcavated
sobreexcavar v - [constr] to overexcavate
— v horadación f - [constr] to overexcavate @
bore
sobrehenchido,da a - overfilled
spbrehenchimiento m - overfilling
sobrehenchir v - to overfill
sobreimpuesto,ta a - superimposed
sobreinclinación f - [ventil] overpitching
sobreinclinado,da a - [ventil] overpitched
sobreinclinar v - [ventil] to overpitch
sobrellenado m - [metal-lam] overfill(ing)
sobremedida f - oversize
sobremonta f - [sold] overlap(ping); overhang
— f apropiada - [mec] proper, o appropriate,
overlap(ping)
sobreoxigenación f - [metal-prod] oxygen enriching,chment
sobreoxigenado,da a - [metal-prod] oxygen enriched
sobreoxigenar v - [metal-prod] to oxygen enrich
sobrepasada* f - véase sobrepaso m
sobrepasado,da a - exceeded; surpassed
sobrepasar v - . . .; to surpass; to be over • to
surpass • to, overlap, o overhang • to run off
— v cuota f - to surpass @ quota
— v curva f - [deport] to overshoot @, curve, o
turn
— v nivel m - to overfill
— v presión f - [mec] to, surpass, o override, o
exceed, @ pressure
— v recorrido m - [metal-prod] to exceed @ run
— v rosca f - [mec] to clear @ thread
— v válvula f - [mec] to override @ valve
— v vida f - to, surpass, o be larger than, life
sobrepaso m - surpassing
— m automático - [mec] automatic override,ding
— m de cuota f - quota surpassing
— m — curva f - [deport] curve overshooting
— m — presión f - [mec] pressure override,ding
— m — f en tubería f para entubación f -
[petról] casing pressure override,ding
— m — f — f — — f activado - [petról]
activated casing pressure override,ding
— m — válvula f - [petról] valve override,ding
— m en tubería f - [petról] pipe,ping override
— m — f para entubación f - [petról] casing

override,ding
sobrepaso m para estrangulador m - [petról] choke
override,ding
sobreponer v - . . . . • to superimpose • to overcome
— v cordón m - [sold] to weld over
— v parcialmente - to partially superimpose
— v — escala f - to partially overlap @ range
— v — límite(s) m - to partially overlap @
range
sobreponerse v - to out-qualify; to override; to
outtough • to, overpower, o overcome • to
handle, o win • [deport] to, nose out, o dominate
— v a lance m - [deport] to handle @ experience
— v — resistencia f contra rodadura f (cuesta
abajo) - [autom-neumát] to overcome @ rolling
resistance
— v a rugosidad f - to overcome @ roughness
sobreposición f - . . . . • [deport] dominating;
domination; overcoming; nosing out
sobreprecio m - . . .; surcharge; overprice •
price increase
sobrepresión f - overpressure; excess pressure
— f en cambiador n - [mec] changer overpressure
— f — compartimento m en cambiador m - [mec]
changer compartment overpressure
— f — gabinete m - cabinet overpressure
— f — — m para cambiador n - [electr-oper]
tap changer cabinet overpressure
— f — horno m - [ind] furnace overpressure
— f liberada - vented overpressure
sobreprotección f - overprotecting,tion
sobreprotector,ra a - overprotecting,tive
sobreproteger v - to overprotect
sobreprotegido,da a - overprotected
sobreprovisión f - oversupply(ing)
sobrepuesto,ta a - superimposed • piled on •
overpowered; overcome • handled • [deport]
dominated; nosed out
— a parcialmente - partially superimposed
sobresalido,da a - sticking out; projected;
protruding,ded
— a desde relleno m - [constr] projected from
@ fill
sobresaliente a - outstanding; excellent; remarkable • jutting; overhead; projected,ting;
protruding; offset
— a desde terraplén m - [constr] projected,ting
beyond @ fill
— a — relleno m - [constr] projected,ting
from @ fill
sobresalir v - . . .; to stick out
— v apropiadamente - to project properly
— v borde m - [mec] to protrude @ lip
— v de oscilador m - [electr-instal] to project
from @ oscillator
— v — suelo m - [suelos] to pierce @ soil
— v desde relleno m - [constr] to project from
a fill
— v — terraplén m - [constr] to project beyond
@ fill
— v más - [sold] to project, more, o farther
sobresalirse v - to protrude
sobresalto m - . . .; thrill
sobresoplado,plo m - [metal-prod] afterblow
sobrestadía f - [transp] demurrage
sobrestante n - [labor] . . .; foreman
sobretamaño m - oversize
sobretasa f - [fisc] surtax • extra charge
— f por exceso m de peso m - [transp] heavy
lift charge
sobretensión f - [electr-oper] . . .; (voltage)
surge • [constr] overstress
sobretiempo m - [labor] overtime; véase también
tiempo m suplementario; hora(s) f, extra, o
extraordinaria(s)
sobrevelocidad f - overspeed
— f garantizada - guaranteed overspeed
— f máxima - maximum overspeed
— f — garantizada - maximum guaranteed overspeed
sobrevenir v - . . .; to befall; to occur

sobrevivido,da a - survived; lived through
sobrevivir v - . . .; to, live through, o outlast
sobrevoltaje m - [electr-oper] overvoltage •
surge
— m inicial - [electr-oper] initial overvoltage
• [sold] hot starting circuit
— m momentáneo - [electr-oper] surge
— m repentino - [electr-oper] surge
sobreyacente a - overlying
socalzar v - véase también recalzar v
socavación f - undermining • [sold] undercut;
undercutting; underwash(ing) • [hidr] washout; scouring
— f de alcantarilla f - [hidr] culvert undermining
— f — calzada f - [vial] roadway washout
— f — moral f - morale lowering
— f — terraplén m - [hidr] embankment undermining
— f — tubería f - [constr] pipe undermining
— f en salida f - [constr] outlet undermining
— f evitada - [constr] avoided, o prevented,
undermining
— f exterior - [sold] external undercut(ting)
— f interior - [sold] internal undercut(ting)
— f interna - [sold] internal undercut(ting)
— f leve - [sold] (s)light undercut(ting);
wagon track(s)
— v posible - [hidr] possible, undercut(ting),
o washout
— f prevenida a - [constr] prevented undermining
— f resistida - [hidr] resisted undermining
socavado,da a - undermined; undercut
socavadura f - undermining; undercutting
socavar v - . . . • [sold] to undercut
— v alcantarilla f - [hidr] to undermine @ culvert
— v terraplén m - [hidr] to undermine @ embankment
— v tierra f - [constr] to dig (@ soil)
— v tubería f - [constr] to undermine @ pipe
socavón m - . . . . • [miner] gallery; pit •
[constr] heading • advance digging • [miner]
adit; drift
social a - . . . . • [legal] corporate
sociedad f accidental - [legal] joint venture;
combine
— f — constructora - [constr] joint venture
contractor
— f — justificada - justified joint venture
— f afiliada - [legal] affiliated concern
— f andina f - [legal] Andean corporation
— f anónima - [legal] . . .; corporation; nameless company; limited liability, corporation,
o company
— f — con capital m fijo - [legal] fixed capital corporation
— f — — — m variable - [legal] variable capital corporatioon
— f — — participación f restringida - [legal]
closed corporation
— f anterior - [legal] former corporation
— f aparte - [legal] separate corporation
— f asociada - [legal] associated corporation
— f colectiva - [legal] general partnership
— f comanditaria f - [legal] . . . . • limited
partnership
— f — simple - [legal] general partnership
— f con capital m mixto - [legal] mixed capital, corporation, o company
— f — — m variable - [legal] variable capi-
pital, corporation, o company
— f — responsabilidad f limitada - [legal] limited liability, corporation, o company
— f controlante - [legal] controlling corporation
— f de actuario(s) m - [legal] actuarial corporation; Society of Actuaries
— f de capital m e industria f - [legal] capital and industry, corporation, o partnership;

limited partnership
sociedad f de granjeros m retirados - [agric]
Green Thumbs Incorporated; retired farmers
guild
— f — ingenieros m - engineers,ring, society,
o institute
— f — — m Automotores m - [autom] Society of
Automotive Engineers
— f — — m para Alumbrado m de Asociación f de
Compañías f para Alumbrado Edison - Association for Edison Illuminating Companies Illuminating Engineering Society
— f — — m para Industria f Automotriz - Society of Automotive Engineers
— f — responsabilidad f limitada - [legal] limited liability, company, o corporation
— f en comandita - [legal] commandite partnership; partnershipo with @ silent partner;
silent partner, company, o corporation
— f — — por acciones f - [legal] joint stock,
corporation, o company
— f — participación f - [legal] joint venture
— f — — f en país m - [legal] domestic joint
venture (company, o corporation)
— f — — f nacional - [legal] domestic joint
venture (company, o corporation)
— f estadounidense - [legal] American, corporation, o company, o concern
— f — de Fabricantes de Engranajes m - American Gear Manufacturers Association
— f — — Ingenieros m Civiles m - American Society of Civil Engineers; A S C E
— f — — — m para Alumbrado m - Illuminating
Engineering Society
— f — — — m — Calefacción f y Ventilación f
- American Society of Heating and Ventilation
Engineers
— f — — — m — Industria f Automotriz - Society of Automotive Engineers
— f — — — m Mecánicos - American Society of
Mechanical Engineers
— f — — — m Profesionales m - National Society of Professional Engineers
— f — — Traductores m (Públicos) - [legal]
American Translators Association
— f — para Ensayos m y Materiales m - [ing]
American Society for Testing and Materials
— f — — Soldadura f - [sold] American Welding Society; A W S
— f extranjera - [legal] foreign, corporation, o
company, o concern
— f financiera - [fin] financial, corporation, o
company, o concern
— f histórica - [hist] historical society •
conservation society
— f importante - [legal] large, corporation, o
company, o concern
— f limitada - [legal] limited, corporation, o
company, o concern
— f liquidada - [legal] liquidated, corporation,
o company, o concern
— f matriz - [legal] parent corporation
— f mejicana - [legal] Mexican corporation
— f mercantil - [legal] commercial corporation
— f — con capital m variable - [legal] variable capital (commercial), corporation, o company, o concern
— f mixta - [legal] mixed (capital), corporation, o company, o concern • binational corporation
— f multinacional - [legal] multinational, corporation, o company, o concern
— f — andina - [legal] Andean multinational,
corporation, o company, o concern
— f nacional - [legal] domestic, corporation, o
company, o concern
— f por acciones f - [legal] stock, company, o
corporation; corporation
— f privada - [legal] privately held corporation
— f pública f - [legal] public(ly held) corporation
— f respectova - respective corporation

sociedad f separada - [legal] separate, corporation, o company, o concern
— f subsidiaria - [legal] subsidiary, corporation, o company, o concern; subsidiary
— f — en ultramar - [legal] overseas subsidiary, corporation, o company, o concern
— f — extranjera - [legal] foreign subsidiary,
(corporation, o company, o concern)
— f técnica - technical, association, o corporation, o company, o concern
societario,ria a - [legal] corporate
socio m - [legal] . . • [com] associate; member
— m accionista - [legal] stockholding partner
— m comanditario - [legal] general partner •
silent partner
— m con responsabilidad f financiera - [legal]
financially liable partner
— m constituyente - [legal] incorporating partner; incorporator
— m de club - club member
— m fundador - [legal] founding partner; incorporator
— m gerente - [com] managing partner
— m plenario - [legal] full partner
— m principal - [legal] senior, o main, partner
socio,cia a - [legal] associate
socorrer v - . . .; to rescue
socorro m - . . .; rescue; aid
— m de emergencia f - [segurid] emergency rescue
soda f - [quím] . . .; óxido m de sodio m
soda f cáustica - [quím] caustic soda; sodium hy-
droxide
sodio m bicrómico - [quím] sidium bichromate
sofá m - [domést] . . .; couch; chaise longue
— m acolchado - [domést] overstuffed couch
sofisticación f - . . . • véase complicación f
sofisticado,da a - véase complicado,da
sofisticar v - véase complicar f
sofocación f - . . .; stifling
sofocado,da a - suffocated; stifled
sofrenado,da a - . . . . • [autom] backed off
sofrenar v - . . .; to back off
soga f - . . .; line
— f de abacá m - [cabl] manila rope
— f — cáñamo m - [cabl] manila rope
— f — fibra f - [cabl] fiber rope
— f — — f vegetal—[cabl] vegetable fiber rope
— f — manila f - [cabl] manila rope
— f destrenzada - [cabl] unbraided rope • un-
ravelled rope
— f para remolcar,lque - [cabl] tow rope
soja f - [botán] . . . • [metal-lam] streak; scab
• sliver • spill
sol m alto - [meteorol] high sun
— m ardiente—[meteorol] blazing, o burning, sun
— m bajo - [meteorol] low sun
sola adv discreción f - sole discretion
— adv firma f - [fin] (a) unguaranteed
— adv presentación f - [legal] simple presentation; presentation alone
— adv propiedad f - sole ownership | a wholly
owned
solamente adv - . . . • simply
— adv una manera - only one way
solapa f - . . . • [sold] lap joint
— f igual - [mec] same (size) lap
solapadamente adv - . . .; underhandedly
solapado,da a - . . . . • [mec] (hammer) lapped
solapante a - overlapping
solapar v - . . .; to overhang; to (hammer) lap
— v circunferencialmente - to overlap circumferentially
solape m - [mec] overlap
solapo m - lap • [mec] hammer lap(ping) • [sold]
lap joint
— m a través de plancha f - [sold] through lap
— m de viga f - [vial] guardrail lap
— m en ángulo m (interior) - [sold] fillet lap
— m — borde f de plancha f - [sold] edge lap
— f — pasada f única - [sold] single pass lap
— m — — con avance m rápido - [sold] high
speed single pass lap
solar m - . . .; plot

solazado,da a - solaced; relaxed
solazarse v - to relax
soldabilidad f - [sold] weldability
— f excelente - [sold] excellent weldability
soldable a - [sold] weldable; welding quality
— a entre sí - [sold] weldable to each other
soldado,da a - welded • soldered • [mec] fabricated
— a a ancho m - [sold] welded across
— a a ángulo m - [sold] welded to @ angle
— a a armadura f - [sold] welded to @ frame
— a a barra f - [sold] welded to @ bar
— a a bastidor m - [sold] welded to @ frame
— a a bomba f - [mec] welded to @ pump
— a a caja f - [sold] welded to @, body, o case
— a a defensa f - [mec] welded to @ guard
— a a hierro m - [sold] welded to @ iron
— a — — m angular = [sold] welded to @ angle
(iron)
— a a larguero m - [constr] welded to @ stringer
— a a sesgo - [sold] scarf welded
— a a motor m - [electr-mot] welded to @ motor •
[comb.int] welded to @ engine
— a a tope - [sold] butt welded; welded end to
end
— a — — continuamente - [sold] continuously
butt welded
— a con amperaje m bajo - [sold] low current
welded
— a — aportación f (de material m) - [sold] material inflow welded
— a — bronce m - [sold] brazed
— a — — m con antorcha f con dos electrodos m
- [sold] arc torch brazed
— a — electrodo m - [sold] electrode welded
— a — — m con corriente f negativa - [sold]
electrode negative welded
— a — — m — — f positiva - [sold] electrode
positive welded
— a — — m con corriente f negativa - [sold]
electrode negative welded
— a — — m — — f positiva - [sold] electrode
positive welded
— a — estaño m - [sold] soldered
— a — — m con antorcha f - [sold] arc torch
soldered
— a — — m — — f con dos electrodos m -
[sold] arc torch soldered
— a — éxito m - [sold] welded successfully
— a — gas m - [sold] brazed
— a — latón m - [sold] brazed
— a — llama f - [sold] flame brazed
— a — — f de gas m - [sold] gas brazed
— a — plomo m - [sold] (lead) soldered
— a — proceso m con arco m sumergido - [sold]
submerged arc process welded
— a — régimen m pleno - [sold] full rate output
welded
— a — solapo m - [sold] lap welded
— a — soldadora f - [sold] welder welded
— a continua(da)mente - [sold] welded, steadily,
o continuously
— a de derecha f a izquierda f - [sold] welded
from @ right to @ left
— a — izquierda f a derecha f - [sold] welded
from @ left to @ right
— a económicamente - [sold] welded economically
— a eléctricamente - [sold] welded electrically
— a en argón m (gaseoso) - [sold] argon (gas)
welded
— a — atmósfera f regulada - [sold] controlled
atmosphere welded
— a — — f de argón m (gaseoso) - [sold]
argon (gas) controlled atmosphere welded
— a — — f — — gas m de argón m - [sold] argon gas controlled atmosphere welded
— a — espiral - [sold] spiral welded
— a — exterior m - [sold] outside welded
— a — forma f plana f - [sold] flat welded
— a — — f plana con punto(s) m - [sold] flat
tack(ed) (welded)
— a — frío m - [sold] cold welded
— a — gas m - [sold] gas welded

soldado,da en gas m de argón m - [sold] argon
  gas welded
soldado,do a — horno m - [sold] furnace welded
 — a — obra f - [sold] field welded; welded
  during @ erection
 — a — posición f (fija) - [sold] position
  welded
 — a — recinto m (reducido) - [sold] welded in
  @ confined space
 — a — sitio m - [sold] in place welded
 — a — — m en obra f - [sold] in place welded
  during @ erection
 — a — taller m - [sold] shop welded
 — a — torno - [sold] girth welded; welded,
  around , o round about
 — a entre sí - [sold] welded together
 — a firmemente - [sold] firmly, o securely,
  welded
 — a fuera de posición f - [sold] out of position welded
 — a helicoidalmente - [sold] spiral, o helically, welded
 — a integral - [sold] welded in(to)
 — a íntegramente - [sold] all welded
 — a manualmente - [sold] welded manually
 — a mediante reacción f química - [quím] chemically welded; chemical reaction welded
 — a para formar trozo(s) m (con largo m) de
  . . . - [sold] welded into . . . length(s)
 — a por ambos lados m - [sold] both sides weld=
  ed • conductos] double jointing
 — a — arco m - [sold] arc welded
 — a — — m a tope - [sold] electrically, o
  arc, butt welded
 — a — — m sumergido - [sold] submerged arc
  welded
 — a — fusión f (eléctrica) - [sold] (electric)
  fusion welded
 — a — punto(s) m - [sold] spot welded
 — a — recubrimiento m - [sold] véase soldado,da a con solapo m
 — a — resistencia f - [sold] (electric) resistance welded
 — a — — f — (y) reducido,da por estiramiento (en) caliente - [tub] electric resistance
  welded hot-stretched-reduced
 — a sin aportación f (de material m) - [sold]
  without material inflow welded
 — a sobre - [sold] welded on
 — a totalmente - [sold] totally, o all, welded
soldador m - [sold] . . .; welder; weldor • fa-
  bricator; operator • burner • builder
 — m aficionado - [sold] amateur, o nonprofessional, weldor
 — m bien calificado - [sold] well trained, operator, o weldor
 — m calificado - [sold] qualified, o trained,
  weldor, o operator
 — m competente - [sold] competent, o qualified,
  weldor
 — m con electrodo(s) m - [sold] electrode(s)
  weldor
 — m — — m con alma m fundente - [sold] Innershield weldor
 — m — Innershield m - [sold] Innershield,
  weldor, o operator
 — m — soplete m - [sold] blowtorch weldor •
  blowtorch
 — m concienzudo - [sold] conscientious weldor
 — m de tubería(s) f - [sold] pipe weldor •
  pipeliner
 — m en obra f - [sold] field weldor
 — m hábil - [sold] competent weldor
 — m individual - [sold] individual weldor
 — m Innershield - [sold] Innershield, weldor, o
  operator
 — m manual - [sold] manual weldor; stick operator
 — m no profesional - [sold] nonprofessional
  weldor
 — m novato - [sold] welding beginner
 — m para reconstrucción f - [sold] rebuilder;
  rebuilding weldor

soldador m por arco m - [sold] arc weldor
 — m — destajo m - [sold] job weldor
 — m principiante - [sold] beginner,ning weldor
 — m — no profesional - [sold] nonprofessional
  beginning weldor
 — m profesional principiante - [sold] professional beginning, o beginning professional,
  weldor
 — m reconstructor - [sold] rebuilder, o rebuilding, weldor
soldadora f - [sold] welder; arc welding machine; machine • blowtorch
 — f a prueba de roedor(es) - [sold] vermin
  resistant welder
 — f — tope m—[sold] end-to-end welder
 — f — — m con destello(s) m - [sold] flash
  welder
 — f — m — chispa(s) f - [sold] flash
  welder
 — f accionada con electricidad f - [sold]
  electrically driven welder
 — f — eléctricamente - [sold] electrically
  driven, welder, o machine
 — f adaptable - [sold] versatile welder
 — f anticuada - [sold] old, o obsolete, welder
 — f antigua - [sold] old(er), o early, o obsolete, welder
 — f apilada—[sold] stacked, welder, o machine
 — f automática - [sold] (fully) automatic,
  welder, o machine
 — f — completa - [sold] complete automatic
  welder
 — f — con electrodo m único - [sold] single,
  electrode, o wire, automatic welder
 — f — (de) Lincoln - [sold] Lincoln automatic (welder)
 — f — para corriente f alterna - [sold] automatic alternating current (arc) welder •
  fully automatic alternating current welder
 — f — — — f con arco m sumergido—[sold]
  submerged arc alternating current automatic
  welder
 — f — por arco m sumergido - [sold] (fully)
  automatic submerged arc, welder, o welding
  head
 — f — fija - [sold] stationary automatic
  welder
 — f — Lincolnweld - [sold] Lincolnweld automatic welder
 — f — — para arco m sumergido - [sold] Lincolnweld submerged arc automatic welder
 — a — para arco m sumergido - [sold] submerged arc automatic welder
 — f — formación f de sección(es) doble(s)
  - sold] double ending automatic welder
 — f — — f en planta f de sección(es) f
  doble(s) - [sold-tub] stationary double
  ending automatic welder
 — f — por gas m - [sold] automatic gas,
  welder, o welding machine
 — f buena - [sold] fine, o good, welder
 — f — para conexión f con red f general -
  [sold] fine utility welder
 — f colocada - [sold] placed welder
 — f combinada - [sold] combined, welder, o
  machine
 — f — para voltaje(s) variable o constante -
  [sold] combined,nation variable-constant
  voltage, welder, o machine
 — f compacta - [sold] compact, welder, o machine
 — f con alimentación f automática - [sold] mechanized, o squirt, welder
 — f — — f — de electrodo m - [sold] (semi-)
  automatic, o squirt, welder, o machine, o
  feeder
 — f — — f — — — m para aplicación(es) f
  múltiple(s) - [sold] multi-purpose squirt
  welder
 — f — — f — — — m para cualquier procedimiento m - [sold] multi-purpose Squirt
  welder
 — f — arco m gemelo - [sold] Twin-Arc welder

soldadora f con arco m gemelo con electrodo(s) m grueso(s) - [sold] large wire Twin-Arc welder
— f — — m protegido (Shield-Arc) - [sold] Shield-Arc (welder)
— f — avance m manual - [sold] hand travel, welder, o unit
— f — — m — mecanizado - [sold] mechanized hand travel, welder, o unit
— f — capacidad f grande - [sold] heavy duty welder
— f — — f — con motor m con combustión f interna - [sold] heavy duty internal combustion engine welder
— f — — f — — m con gasolina f - [sold] heavy duty gasoline engine welder
— f — — f — — diésel - [sold] heavy duty diesel engine welder
— f — mediana - [sold] medium duty welder
— f — correa f - [sold] belted, o belt driven, welder
— f — costo m alto - [sold] high cost welder
— f — — m bajo - [sold] low cost welder
— f — — m bajo con motor m con combustión f interna - [sold] low cost engine driven welder
— f — destello(s) m - [sold] flash welder
— f — devanado m especial - [Sold] special(ly) wound welder
— f — — m — con transformador m - [sold] special(ly) wound transformer welder
— f — dispositivo m para frecuencia f alta - [sold] high frequency unit welder
— f con electrodo m - [sold] electrode welder
— f — — m con alma m fundente - [sold] Inner-shield, welder, o machine
— m — — m — — m — y alimentación f automática (de electrodo m) - [sold] innershield squirt welder
— f — impulsión f - [sold] driven welder
— f — — f con correa f - [sold] belted, o belt driven, welder
— f — — f — motor m - [sold] engine driven welder • motor driven welder
— f — — f — m con combustión f interna - [sold] engine driven welder
— f — motogenerador m - [sold] motor-generator welder
— m — — n para corriente f alternada - [sold] alternating current motor generator welder
— f — — m — — f continua - [sold] direct current motor-generator welder
— f — motor m con combustión f interna—[sold] (internal combustion) engine driven welder
— f — — m — — f enfriado con agua m - [sold] water cooled engine driven welder
— f — — m con gasolina f - [sold] gasoline (engine) driven welder
— f — — m diesel - [sold] diesel engine (driven) welder
— f — — m eléctrico - [sold] (electric) motor driven (arc) welder • motor-generator (arc) welder
— f — — m — Shield-Arc (de Lincoln) - [sold] (Lincoln) Shield-Arc motor-generator welder
— f — peso m reducido - [sold] light-weight welder
— f — potencia f reducida - [sold] light duty, o utility, welder
— f — — f — para trabajo(s) m general(es) - [sold] (light-weight) utility welder
— f — transformador m - [sold] transformer welder • welder transformer
— f — — m para corriente f alterna - [sold] alternating current transformer welder
— f — — m — voltaje(s) m bajo(s) - [sold] low voltage transformer welder
— f — — m — — m — en circuito m abierto - [sold] open circuit low voltage transformer welder
— f conectada - [sold] connected, o on, welder
— f — con tierra f - [sold] grounded welder
— f — — en paralelo - [sold] parallel connected welder
— f confiable - [sold] dependable welder

soldadora f construida - [sold] built welder
— f corriente - [sold] standard welder
— f de gasolina f - [sold] gasoline (engine) welder
— f — f con fuente f de fuerza f motriz - [sold] (gasoline) engine welder and power unit
— f — f generadora f de fuerza f motriz - [sold] (gasoline) engine driven power generator and welder
— f — f — — f en corriente alterna - [sold] (gasoline) engine driven alternating current power generator and welder
— f — medio - [sold] middle welder
— f — modelo m anterior - [sold] older welder
— f — — m anticuado - [sold] older welder
— f — — m viejo - [sold] old(er) model welder
— f — tamaño m medio,diano - [sold] medium size(d) welder
— f — tipo m con correa f - [sold] belted, o belt driven, welder
— f — — m industrial - [sold] industrial type welder
— f — — m motogenerador con amperaje(s) muy bajo(s) - [sold] aircraft motor generator type welder
— f — — transformador-rectificador - [sold] transformer-rectifier type welder
— f desconectada - [sold] disconnected, o turned off, o tripped, welder
— f — de línea f - [sold] line tripped off welder
— f — — — f por sí sóla - [sold] tripped off @ line welder
— f — por sí sóla - [sold] tripped off welder
— f — — — — de línea f - [sold] tripped off @ line welder
— f destruida - [sold] destroyed, o burned up, welder
— f detenida - [sold] turned, o shut, off, o stopped, welder
— f diseñada - [sold] designed welder
— f disponible - [sold] available welder
— f doble - [sold] double welder
— f económica - [sold] economic(al) welder
— f — con gas m de tungsteno - [sold] economical tungsten gas welder
— f — — m inerte - [sold] economical inert gas welder
— f — — m — de tungsteno - [sold] economical tungsten inert gas welder
— f eficiente - [sold] efficient, o fine, welder
— f eléctrica - [sold] electric welder
— f — fija - [sold] stationary electric welder
— f embarcada - [sold] shipped welder
— f en atmósfera f regulada - [sold] controlled atmosphere welder
— f en circuito m abierto - [Sold] open circuit welder
— f en funcionamiento - [sold] welder on
— f — marcha - [sold] on, o running, welder
— f — paralelo - [sold] parallel(ed) welder
— f enfriada - [sold] cooled welder
— f — con motor m con combustión f interna - [sold] cooled engine driven welder
— f estándar - [sold] standard welder
— f exterior - [tub] outside (diameter) welder
— f — para tubería f - [tub] outside (diameter) welder
— f ...fásica - [sold] . . .-phase welder
— f fija - [sold] stationary welder
— f — con motor m de gasolina f - [sold] stationary (gasoline) engine welder
— f — — — m diésel - [sold] stationary diesel engine welder
— m — impulsada por motor m de gasolina f - stationary gasoline engine driven welder
— f — — — — m diésel - [sold] stationary diesel engine, driven, o powered, welder
— f — — — m eléctrico - [sold] stationary electri motor, driven, o powered, welder
— f fundida - [sold] burned-up welder

soldadora f generadora - [sold] welder generator
— f — confiable - [sold] dependable welder generator
— f — Lincolnwelder - [sold] Lincolnwelder (welder-generator)
— f — para transporte m fácil - [sold] easily movable welder-generator
— f — recia - [sold] rugged welder generator
— f — Weldanpower - [sold] Weldanpower welder generator
— t grande - [sold] large welder
— f Idealarc - [sold] Idealarc (welder)
— f — para ambas corrientes f - [sold] Idealarc alternating/direct current welder
— f — para corriente f alterna - [sold] Idealarc alternating current welder
— f — — — f continua - [sold] Idealarc direct current welder
— f impulsada - [sold] driven welder
— f — con correa f - [sold] belted, o belt driven, welder
— f — motor m con combustión f interna - [sold] engine driven welder
— f — — — m con enfriamiento m con agua m - [sold] water cooled engine driven welder
— f — — m con gasolina f - [sold] gasoline engine, driven, o powered, welder
— f — — m diesel - [sold] diesel engine (driven) welder
— f — — m eléctrico - [sold] electric motor, driven, o powered, welder
— f industrial - [sold] industrial welder
— f — con capacidad f grande - [sold] heavy duty industrial welder
— f — para corriente f alterna - [sold] industrial alternating current welder
— f — — — f continua - [sold] industrial direct current wilder
— f — recia - [sold] rugged industrial welder; industrial welding workhorse
— f inferior - [sold] lower, o bottom, welder
— f instalada - [sold] installed welder
— f interior - [sold] inside diameter welder
— f L N 4 de combinación f - [sold] combination LN-4 welder
— f — — — f para soldadura f - [sold] LN-4 combination welder
— f — de combinación f para soldadura f - [sold] LN-4 combination (welder)
— f — — — f — — — f por arco m sumergido - [sold] LN-4 submerged arc combination welder
— f — — — — — f — — m — con electrodo m con alma m fundente Innershield - [sold] [sold] Submerged-Arc and Innershield combination LN-4 welder
— f — para soldadura f con electrodo m con alma m fundente Innershield - [sold] Innershield LN-4
— f larga - [sold] long welder
— f limpia - [sold] clean welder
— f limpiada - [sold] cleaned welder
— f — con sopladura f de aire m - [sold] blown out welder
— f Lincoln - [sold] Lincoln welder
— f liviana - [sold] light(weight) welder
— f más antigua - [sold] older, o earlier, welder • oldest, o earliest, welder
— f más popular - [sold] more popular welder • most popular welder
— f — — corriente f alterna - [sold] most popular alternating current welder
— f — — — — f continua - [sold] most popular direct current welder
— f — reciente - [sold] later, o more recent, welder • latest, o most recent, welder
— f mayor - [sold] larger, o largest, welder
— f mecanizada - [sold] mechanized welder
— f menor - [sold] smaller, o smallest, welder
— f monofásica - [sold] single phase welder
— f motogeneradora - [sold] motor-generator welder
— f montada - [sold] mounted welder
— f movible - [sold] portable welder

soldadora f muy móvil - [sold] easily movable welder
— f — portátil - [sold] highly movable welder
— f — transportable - [sold] highly portable welder
— f no arranca v - [sold] welder does not start
— f nueva - [sold] new welder
— f obsoleta - [sold] obsolete welder
— f operada - [sold] operated welder
— f para ambas corrientes f - [sold] alternating current/direct current welder
— f — transformadora-rectificadora - [sold] alternating current/direct current, o AC/DC, transformer rectifier
— f — . . . amperio(s) m - [sold] . . . ampere(s) welder
— f — . . . — en corriente f alterna - [sold] . . . ampere(s) alternating current welder
— f — . . . — — — f continua - [sold] . . . ampere(s) direct current welder
— f — . . . — para ambas corrientes f - [sold] . . . amperes, alternating current/ direct current, o AC/DC, welder
— f — aplicación(es) f múltiple(s) [sold] multi-purpose welder
— f — aviación f - [sold] aircraft welder
— f — avión(es) m - [sold] aircraft welder
— f — banda(s) f - [metal-trat] strip (seam) welder
— f — corriente f alterna - (sold) alternating current welder
— f — — f con motor m con combustión f interna - [sold] alternating current engine driven welder
— f — — — m con explosión f - [sold] alternating current engine driven welder
— f — — f — — — m diesel - [sold] diesel engine alternating current welder
— f — — f — — — m eléctrico - [sold] alternating current motor driven welder
— f — — — impulsada con motor m diesel - [sold] diesel engine driven alternating current welder
— t — — f — tipo transformador - [sold] alternating current transformer
— f — — f continua - [sold] direct current welder
— f — — f con motor m para combustión f interna - [sold] direct current engine driven welder
— f — — f — — — m — explosión f - [sold] direct current engine driven welder
— f — — — m eléctrico - [sold] direct current motor driven welder
— f — — f — — — m diesel - [sold] diesel engine driven direct current welder
— f — — f impulsada con motor m diesel - [sold] diesel engine driven direct current welder
— f — — f — — — m para corriente f alterna - [sold] alternating current motor driven direct current welder
— f — — f — tipo m transformador - [sold] direct current transformer
— f — — f ...fásica - [sold] . . .-phase welder
— f — — f trifásica - [sold] three-phase welder
— f — costura(s) t - [sold] seam welder
— f — — f en, banda f, o cinta f, o chapa f, o fleje m - [metal-fabr] strip seam welder
— f — cualquier procedimiento m - [sold] multi-process welder
— f — chapa(s) f - [sold] sheet welder • [metal-trat] strip seam welder
— f — decapado m - [metal-trat] pickling welder
— f — dos voltajes m - [sold] dual voltage welder
— f — electrodo(s) m—[sold] electrodes welder
— f — — m manual(es) - [sold] stick electrode(s) welder

soldadora f para fleje(s) m - [metal-trat] strip (seam) welder
— f — formación f en planta f de sección(es) f doble(s) - [sold] stationary double ending welder
— f — frecuencia f alta - [sold] high frequency welder
— f — gas m de tungsteno m - [sold] tungsten gas welder
— f — m inerte - [sold] inert gas welder
— f — montaje m - [sold] assembly welder
— f — remiendo(s) m - [sold] patch(ing) welder
— f — rendimiento m máximo - [sold] maximum yield welder
— f — servicio m - [sold] service welder
— f — m pesado - [sold] heavy duty welder
— f — soldadura f - [sold] welding welder
— f — — f mecanizada f - [sold] mechanized welding welder
— f — transporte m fácil - [sold] easily movable welder
— f — tubos m - [sold] pipe welder
— f — m para caldera(s) f - [cald] (boiler) flue welder
— f — un (sólo) voltaje m - [sold] single voltage welder
— f — voltaje m alto - [sold] high voltage welder
— f — m bajo - [sold] low voltage welder
— f — m — en circuito m abierto - [sold] low open cicuit voltage welder
— f — m constante - [sold] constant, o fixed, voltage welder
— f — m en circuito m abierto - [sold] open circuit voltage welder
— f — — m de entrada f alto - [sold] high input (voltage) welder
— f — m — bajo - [sold] low input (voltage) welder
— f — m variable - [sold] adjustable voltage welder
— v — m único - [sold] single voltage welder
— f — . . . voltio(s) m - [sold] . . . volt(s) welder
— f parada - [sold] stopped, o (turned) off, welder
— f paralela - [sold] parallel welder
— f — con motor m para combustión f interna - [sold] parallel engine driven welder
— f pequeña - [sold] small welder
— f — para trabajos m generales - [sold] utility welder
— f popular - [sold] popular welder
— f — para servicio m - [sold] popular service welder
— f — — — m liviano - [sold] popular light duty welder
— f — — — m pesado - [sold] popular heavy duty welder
— f — portátil—[sold] popular portable welder
— f — — para servicio m liviano - [sold] popular light duty portable welder
— f — — — m pesado - [sold] popular heavy duty portable welder
— f por arco - [sold] arc, welder, o welding machine
— f — m con motor m para gasolina f—[sold] gasoline engine driven arc welder
— f — m — — m diesel - [sold] diesel engine driven arc welder
— f — m — — m eléctrico - [sold] electrical (motor) driven arc welder; motor-generator arc welder
— f — (de) Lincoln - [sold] Lincoln arc welder
— f — — m eléctrico - [sold] electric arc welder
— f — — m ferroviaria - [sold] railroad(er) arc welder
— f — — m industrial - [sold] industrial arc welder
— f — — m — recia - [sold] (rugged) indus-

trial arc welder (workhorse)
soldadora f por arco m industrial recia para electrodo(s) m manual(es) - [sold] industrial manual electrode arc welder (workhorse)
— f — — para corriente f alterna - [sold] alternating current arc welder
— f — — m — f continua - [sold] direct current arc welder
— f — — m para ferrocarril(es) m - [sold] railroad(er) arc welder
— f — — m protegido (Shield-arc) - [sold] Shield-arc welder
— f — — m — (de) Lincoln - [sold] Lincoln Shield-arc welder
— f — — m sumergido - [sold] submerged arc welder
— f — — — con alimentación f automática de electrodo m - [sold] submerged arc squirt welder
— f — — m — en tándem - [sold] tandem submerged arc welder
— f — — m recia f - [sold] rugged arc welder
— f — — m — para electrodo(s) m manual(es) - [sold] rugged, o workhorse, manual electrode arc welder
— f por chispa(s) f - [sold] flash welder
— f — gas m - [sold] gas welder
— f — — m inerte - [sold] inert gas welder
— f — — m de tungsteno m - [sold] tungsten inert gas welder
— f — inducción f - [sold] induction welder
— f — punto(s) - [sold] tack welder; stitcher
— f — rueda f - [sold] wheel, o seam, welder
— f portátil - [sold] portable welder
— f — antigua - [sold] old portable welder
— f — más antigua - [sold] older portable welder • oldest portable welder
— f — popular - [sold] more popular portable welder; most popular portable welder
— f — — para corriente f alterna - [sold] most popular alternating current portable welder
— f — — — — f continua - [sold] most popular direct current portable welder
— f — para . . . amperio(s) m - [sold] . . . amperes portable welder
— f — — corriente f alterna - [sold] alternating current portable welder
— f — — — f más popular - [sold] most popular portable alternating current welder
— f — — — f — — para servicio m pesado - [sold] most popular portable heavy duty alternaing current welder
— f — — — f — para servicio m pesado - [sold] heavy duty alternating current portable welder
— f — — — f continua - [sold] direct current portable welder
— f — — — — para servicio m pesado - [sold] heavy duty direct current portable welder
— f — — — — f — más popular f - [sold] most popular portable direct current welder
— f — — — f — — — para servicio m pesado - [sold] most popular portable heavy duty direct current welder
— f portátil - [sold] portable welder
— f — popular—[sold] popular portable welder
— f — — para corriente f continua - [sold] popular portable direct current welder
— f — — — f — para servicio m pesadp - [sold] popular portable heavy duty direct current welder
— f — — — servicio m pesado - [sold] heavy duty portable welder
— f precisa - [sold] precise, o fine, welder
— f primera - [sold] early, o first, welder
— f proyectada - [sold] designed welder
— f puesta en marcha f - [sold] started welder
— f — — paralelo - [sold] paralleled welder
— f que no arranca - [sold] welder that will not start
— f quemada - [sold] burned (out) welder

soldadora f recalentada - [sold] overheated, o hot, welder
— f recia - [sold] rugged welder; workhorse welder; welding workhorse
— f — para electrodo(s) m manuales - [sold] rugged, o workhorse, manual electrode welder
— f reciente - [sold] recent, o late, welder
— f rectificadora - [sold] rectifier,fying welder
— f ribeteadora - [sold] seam welding stitcher
— S A E con arco m protegido - [sold] Shield-Arc S A E welder
— f según norma f - [sold] standard welder
— f semi o totalmente automática - [sold] fully and semiautomatic welder
— f —— —— para corriente f continua - [sold] direct current semiautomatic and fully automatic welder
— f semiautomática - semi-automatic, o mechanized squirt, welder
— f — para corriente f alterna - [sold] alternating current semiautomatic welder
— f — — — f continua - [sold] direct current semiautomatic welder
— f — Squirt welder - [sold] semiautomatic, o mechanized Squirt welder
— f — con alimentación f automática de electrodo - [sold] semi-automatic squirt wire feeder (welder)
— f — Squirt con alimentación f automática de electrodo - [sold] semi-automatic Squirt wire feeder (welder)
— f Shield-Arc - [sold] Shield-Arc welder
— f —— —— con arco m protegido—[sold] Shield-Arc welder
— f —— —— de Lincoln - [aols] Lincoln Shield-Arc welder
— f sin arrancar v - [sold] unstarted, o not started, welder
— f sin carga f - [sold] idle welder
— f sólida - [sold] rugged welder
— f soplada - [sold] blown (out) welder
— Squirt - [sold] Squirt welder
— f — con alimentación f automática de electrodo m - [sold] Squirt wire feeder
— f — — — —— —— para aplicaciones múltiples - [sold] multi-purpose Squirt welder
— f —— —— f —— —— m — procedimiento(s) m múltiples - [sold] multiprocess Squirt welder
— f — para cualquier procedimiento m - [sold] multi-process Squirt welder
— f —— procedimientos m múltiples - [sold] multiprocess Squirt welder
— f ——Welder - [sold] (mechanized) Squirt welder
— f superior - [sold] superior welder • upper, o top, welder
— f totalmente automática - [sold] full(y) automatic welder
— f —— —— para corriente f alterna - [sold] fully automatic alternating current welder
— f —— —— —— f continua - [sold] fully automatic direct current welder
— f tractora - [sold] tractor (welder)
— f transformadora - [sold] welder transformer:
— f —— corriente f alterna - [sold] alternating current transformer
— f — — — — f continua - [sold] direct current transformer
— f — para voltaje m alto en circuito m abierto - [sold] high open circuit voltage transformer welder
— f —— —— —— m bajo en circuito abierto m - [sold] low open circuit voltage transformer welder welder
— f —— —— m —— —— m — para corriente f alterna - [sold] low open circuit voltage alternating current transformer welder
— f —— grande - [sold] large transformer welder
— f —— corriente f alterna - [sold] alternating current transformer welder
— f —— —— f continua - [sold] direct current transformer welder

soldadora f transformadora pequeña - [sold] small transformer (welder)
— f —— con voltaje m bajo - [sold] small voltage transformer welder
— f —— —— en circuito m abierto - [sold] small low open circuit voltage transformer welder
— f —— —— —— m —— —— m para corriente f alterna - [sold] small low open circuit voltage alternating current transformer welder
— f — rectificadora - [sold] transformer rectifier welder
— f transportable - [sold] portable welder
— f transportada - [sold] transported welder
— f trifásica - [sold] three phase welder
— f Twinarc - [sold] Twinarc welder
— f — con arco(s) m gemelo(s) - [sold] Twinarc welder
— f —— —— m — con electrodos m gruesos - [sold] large wire Twinarc welder
— f ubicada - [sold] located, o placed, welder
— f única - [sold] single welder
— f universal - [sold] universal welder
— f — con alimentación f automática (de electrodo m) - [sold] universal Squirt welder
— f versátil - [sold] versatile welder
— f vertical - [sold] vertical welder
— f — en ángulo m (interior) - [sold] vertical fillet weld
— f —— —— m (—) con una (sóla) pasada f - [sold] one, o single, pass (only) vertical fillet weld
— f vieja - [sold] old, o antique, welder
— f Weldanpower - [sold] Weldanpower (welder)
— f — de Lincoln - [sold] Lincoln Weldanpower (welder)
soldadoras f varias - [sold] miscellaneous welders
soldados,das a entre sí - [sold] welded together
soldadura f - [sold] weld(ing); joining; fabricating,tion; securing • joint; junction • deposit • solder • electrode; rod
— f a ancho m - [sold] (a)cross welding
— f — ángulo m - [sold] welding to @ angle
— f — armadura f - [sold] welding to @ frame
— f — barra f - [sold] welding to @ bar
— f — bastidor m - [sold] welding to @ frame
— f — bomba f - [mec] welding to @ pump
— f — caja f - [sold] welding to @ case
— f — defensa f - [sold] welding to @ guard
— f — hierro m - [sold] welding to @ iron
— f — m angular - [sold] welding to @ angle (iron)
— f — inversa - [sold] reverse welding; véase soldadura f a revés m
— f — larguero m - [constr] welding to @ stringer
— f — motor m - [mec] welding to @, motor, o engine
— f — nivel m - [sold] véase soldadura f a ras
— f — ras m - [sold] flush weld(ing)
— f — sesgo m—[sold] scarf, weldinng, o joint
— f — tope m - [sold] butt. o end-to-end, weld(ing) • jump joint
— f —— m circunferencial - [sold] circumferential butt weld(ing)
— f —— m con borde m angular - [sold] square edge butt weld(ing)
— f —— m —— chaflán m - [sold] beveled butt end weld(ing)
— f —— m —— m en V - [sold] V-butt (beveled), weld, o joint
— f — destello(s) m - [sold] flash butt, o butt flash, weld(ing)
— f —— m — dos pasadas f - [sold] double, o dual, pass butt weld
— f —— m — frecuencia f alta - [sold] high frequency butt weld(ing)
— f —— —— junta f con borde(s) m biselado(s) - [sold] bevel(ed) joint butt weld(ing)
— f —— —— separación f (considerable) - [sold] gapped butt weld(ing)

soldadura f a tope m continua - [sold] continuous butt weld(ing)
— f — — m — con penetración f total - [sold]
continuous full penetration butt weld
— f — — m de chapa f - [sold] sheet butt weld
— f — — m — f de metal m - [sold] sheet
metal butt weld(ing)
— f — — m — f metálica - [sold] sheet
metal butt weld(ing)
— f — — m — junta f - [sold] joint butt weld
— f — — m — pasada(s) f múltiple(s) - [sold]
multiple pass butt weld
— f — — m — f en ranura f profunda -
[sold] multiple pass deep groove butt weld
— f — — m — f única con avance rápido -
[sold] high speed single pass butt weld
— f — — m — penetración f total—[sold] full
penetration butt weld
— f — — m — plancha(s) f (de acero m) -
[sold] (steel) sheet metal butt weld
— f — — m — resistencia f comparable—[sold]
equal strength butt(s) (welds)
— f — — m doble - [sold] double butt weld
— f — — en ángulo m (interior) - [sold] angle
butt weld
— f — — m en posición f plana - [sold] flat
position butt weld
— f — — m — ranura f - [sold] groove butt
weld
— f — — m — — f profunda - [sold] deep
groove butt weld
— f — — m — con chaflán m - [sold]
bevel(ed) deep groove butt weld
— f — — m — f — — m pequeño - [sold]
small bevel(ed) deep groove butt weld
— f — — m — con pasada(s) f múlti-
ple(s) - [sold] multiple pass deep groove
butt weld
— f — — m — f — — f única - [sold]
single pass deep groove butt weld
— v — — m en V - [sold] V-butt, weld, o joint
— v — — m horizontal - [sold] horizontal butt
(weld)
— f — — m — en ranura f - [sold] groove horizontal butt weld
— f — — m — — f profunda - [sold] deep
groove horizontal butt weld
— f — — m plana - [sold] flat butt weld
— f — — m — en chapa(s) f - [sold] flat butt
weld in sheet metal
— f — — m — — ranura f profunda - [sold]
deep groove flat butt (weld)
— f — — m por arco m sumergido en chapa(s) f
de acero m - [sold] sheet metal submerged arc
butt weld(ing)
— f — — m — chispa(s) f - [sold] flash butt
weld(ing)
— f — — m — resistencia f (eléctrica) [sold]
resistance butt weld(ing)
— f — — m sin chaflán m - [sold] square edge
butt, weld, o joint
— f — — m sobrecabeza - [sold] overhead butt
weld(ing)
— f — — m vertical—[sold] vertical butt weld
— f — — m — ascendente - [sold] vertical-up
butt weld(ing)
— f — — m — descendente - [sold] vertical
down butt weld(ing)
— f — través (de plancha f) - [sold] through
weld(ing)
— f, acabado m e inspección f - [tub] welding,
finishing and inspecting,tion
— f acelerada - [sold] speed(y), o accelerated,
weld(ing), o soldering
— f aceptable - [sold] acceptable weld(ing)
— f acetilénica - [sold] acetylene weld(ing)
— f aireacetilénica - [sold] air-acetylene
weld(ing)
— f alimentada - [sold] fed solder
— f — en junta f - [sold] solder fed into @
joint
— f alineada - [sold] aligned weld
— f aliviada - [sold] relieved weld

soldadura f aliviada de tensión f - [sold]
stress relieved weld(ing)
— f — — — f residual - [sold] residual
stress relievewd weld
— f alternada - [sold] staggered weld(ing)
— f — en ángulo m (interior) - [sold] staggered fillet weld(ing)
— f aluminotérmica - [sold] thermit weld(ing)
— f — con presión f - [sold] thermit pressure
weld(ing)
— f allanada - [sold] rolled weld
— f amarilla - [sold] véase soldadura f con
bronce
— f ancha - [sold] wide weld(ing)
— f — a tope m - [sold] wide butt weld(ing)
— f — — — m con chaflán m - [sold] wide
beveled butt welding
— f — — — m en V - [sold] wide
V-butt weld(ing)
— f — — — m (en posición f) plana - [sold]
wide flat position butt weld(ing)
— f — — — m en V - [sold] wide V-butt weld
— f — en ángulo m (interior) - [sold] wide
fillet weld(ing)
— f — interior m de ángulo m - [sold] wide
fillet weld(ing)
— f ortogonal - [sold] wide fillet weld(ing)
— f — Spread-Arc - [sold] soldadura Spread-Arc
— f angosta - [sold] narrow weld(ing)
— f angular - [sold] angle weld(ing); square
edge weld(ing)
— f — a tope m - [sold] square edge butt weld
— f aplicada - [sold] applied weld(ing)
— f aportada - [sold] deposited weld • weld deposit • fed solder
— f — en junt(ur)a f - [sold] solder, feeding,
o fed, into @ joint
— f autógena - [sold] autogenous weld(ing);
gas, torch, o fusion, weld(ing) • weld(ing)
— f — de costura(s) f - [sold] gas seam welding
— f — — — f por punto(s) m - [sold] gas seam
spot weld(ing)
— f — por punto(s) m - [sold] gas spot weld-
-(ing)`
— f automática - [sold] automatic weld(ing);
machine weld(ing)
— f — con alambre m - [sold] wire (electrode)
automatic weld(ing)
— f — — — m con alma m fundente - [sold]
automatic Innershield weld(ing)
— f — con arco m en tándem - [sold] tandem arc
automatic weld(ing)
— f — — — m gemelo - [sold] twin arc automatic weld(ing)
— f — — — m único - [sold] single arc automatic weld(ing); automatic single arc welding
— f — — regulador(es) m con transistor(es) m
- [sold] solid state control(s) automatic
weld(ing)
— f — gemela - [sold] twin automatic weld(ing)
— f — Innershield - [sold] Innershield welding
— f — — con alambre m con alma m fundente --
[sold] automatic Innershield weld(ing)
— f — — — m sumergido - [sold] automatic
submerged arc welding
— f — — — m — con corriente f alterna -
[sold] alternating current automatic submerged arc welding
— f — — — m — — f continua - [sold] direct current automatic submerged arc welding
— f — por puntos m - [sold] progressive spot
welding
— f automática y semiautomática - [sold] automatic and semiautomatic welding
— f autoprotegida - [sold] self-shielded welding
— f — con alimentación f automática - [sold]
semiautomatic self-shielded welding
— f — — f — de electrodo m - [sold] semi-
automatic self-shielded welding
— f — — f — — m con alma m fundente -
[sold] semiautomatic self-shielded flux-cored
(arc) welding

soldadura f autoprotegida con electrodo m con
   núcleo fundente - [sold] self-shielded flux-
   -cored welding
— f —— — m con núcleo m fundente - [sold]
   self-shielded weld(ing) • [Innershield self-
   -shielded welding
— f —— — m con alma f fundente - [sold] Innershield self-shielded (flux-cored) welding
— f —— — alma m fundente - [sold] self-shielded flux-cored (arc) welding
— f —— — núcleo m fundente - [sold] self-
   -shielded flux-cored (arc) welding
— f basta - [sold] rough, o coarse, welding
— f blanda - [sold] solder(ing); brazing
— f — acelerada - [sold] sped soldering
— f — con antorcha f con dos electrodos m -
   [sold] arc torch soldering
— f buena - [sold] good, o sound, welding •
   good, o sound, soldering
— f — por arco m - [sold] good arc weld(ing)
— f —— —— m sumergido - [sold] submerged arc weld(ing)
— f caliente - [sold] hot weld(ing)
— f cilindrada - [sold] véase soldadura f laminada
— f circunferencial - [sold] circumferential, o
   girth, weld(ing); roundabout weld(ing)
— f con . . . pasada(s) f - [sold] . . .
   pass(es) girth weld
— f — en ranura f - [sold] roundabout groove
   weld(ing)
— f — exterior - [tub] outside diameter girth
   weld(ing)
— f — en ángulo m interior - [sold] roundabout
   fillet weld(ing)
— f — con solapo m - [sold] roundabout lap
   weld(ing)
— f — primera - [sold] first girth weld
— f — segunda - [sold] second girth weld
— f — tercera - [sold] third girth weld
— f combinada - [sold] combined weld
— f completa - [sold] complete, o entire, weld
— f completada - [sold] complete(d), o finished, weld • final weld
— f compuesta - [sold] composite weld(ing)
— f común - [sold] common weld
— f con acero m con plata f - [sold] silver-
   -steel weld(ing)
— f — alambre m - [sold] wire weld(ing)
— f — aleación f - [sold] alloy weld(ing)
— f —— f con latón m - [sold] brass alloy
   weld(ing)
— f —— f — aluminio m - [sold] aluminum
   alloy weld(ing)
— f —— f — cobre m - [sold] copper alloy
   welding
— f —— f — plata f - [sold] silver alloy
   weld(ing)
— f —— f — plomo m - [sold] soldering
— f — alimentación f automática (de alambre m)
   - [sold] semi-automatic Squirt welding
— f —— f —— m para relleno m rápido -
   [sold] Fill-Freeze Squirt welding
— f — alivio m - [sold] relieved weld(ing)
— f —— m de tensión(es) g - [sold] stress
   relieved weld(ing)
— f — aluminio m - [sold] aluminum weld(ing)
— f — amperaje(s) m alto(s) - [sold] high, amperage, o current, weld(ing)
— f —— m bajo(s) - [sold] low, amperage, o
   current, weld(ing)
— f —— m distinto - [sold] different amperage weld(ing)
— f —— m elevado - [sold] high, amperage, o
   current, weld(ing)
— f — aportación f - [sold] inflow weld(ing)
— f —— f de material - [sold] material inflow weld(ing)
— f —— f rápida - [sold] high deposition
   weld(ing)
— f — arco m abierto - [sold] open arc welding
— f —— m — con alimentación f automática de
   electrodo m - [sold] open arc squirt welding

soldadura f con arco m autoprotegido con núeldo
   fundente - [sold] self shielded flux-cored
   (arc) welding
— f —— m en tándem - [sold] tandem arc
   weld(ing)
— f —— m —— totalmente automática—[sold]
   fully automatic tandem arc welding
— f —— m entre metal(es) m protegida—[sold]
   shielded metal arc weld(ing)
— f —— m gemelo - [sold] twin arc welding
— f — arco(s) m sumergido(s) m - [sold] twin
   arc submerged arc weld(ing)
— f —— metálico - [sold] metal(lic) arc
   weld(ing)
— f —— múltiple(s) - [sold] multiple arc(s)
   welding
— f —— m protegido - [sold] shielded arc
   welding
— f —— m — entre metal(es) m - [sold]
   shielded metal arc weld(ing)
— f —— m sumergido - [sold] submerged arc
   weld(ing)
— f —— m con alimentación f automática de
   alambre m - [sold] semi-automatic Squirt
   submerged arc weld(ing)
— f —— m —— f —— electrodo m -
   [sold] submerged arc squirt welding
— f —— m con intensidad f alta - [sold]
   high current submerged arc weld(ing)
— f —— m semiautomática - [sold] semi-
   -automatic submerged arc Squirt welding
— f —— m único - [sold] single arc weld(ing)
— f —— m —— totalmente automática - [sold]
   (fully) automatic single arc welding
— f — argón m - [sold] argon (arc) welding
— f — arrastre m - [sold] drag weld(ing)
— f —— m ascendente - [sold] up-drag
   weld(ing)
— f —— m descendente - [sold] down drag
   weld(ing)
— f — aspecto m bueno - [sold] good looking
   weld
— f —— m excelente—[sold] good looking weld
— f — bisel m - véase soldadura f con chaflán
— f — borde m angular - [sold] square edge
   weld
— f — bronce m—[sold] bronze welding; brazing
— f —— m con antorcha f con dos electrodos m
   - [sold] arc torch welding
— f — calidad f - [sold] quality weld(ing)
— f —— f alta - [sold] high quality welding
— f —— f — con arco m sumergido - [sold]
   high quality sumerged arc welding
— f —— f — producida - [sold] produced high
   quality weld(ing)
— f —— f básica - [sold] basic quality
   weld(ment)
— f —— f básica 18/8 - [sold] basic 18/8
   grade weld(ment)
— f —— f con 18% de cromo y 8% de níquel m - [sold] basic 18/8 grade weld(ment)
— f —— f buena - [sold] good, o high, quality welding)
— f —— f constante - [sold] constant, o
   consistent, quality weld(ing)
— f —— f destacada—[sold] high quality weld
— f —— f en cualquier posición f - [sold]
   all-position, quality weld, o weld quality
— f —— f más alta - [sold] highest quality
   weld • higher quality weld
— f —— f máxima - [sold] maximum, o top,
   quality weld
— f —— f para inspección f con rayos-X -
   [sold] X-ray (quality) weld
— f —— f para producción f - [sold] quality
   production weld(ing)
— f —— f —— f con mando m con botonera f
   - [sold] quality pushbutton production weld
— f —— f —— f —— m —— f con calidad
   constante - [sold] consistent quality pushbutton production welding
— f —— f por arco m sumergido - [sold] high
   quality submerged arc weld(ing)

soldadura f con calidad f por arco m sumergido
con corriente f continua—[sold] high quality
direct current submerged arc weld(ing)
— f con calidad f según código m - [sold] code
quality weld(ing)
— f — — f sin problema(s) m - [sold] trouble-
-free quality weld(ing)
— f — — f superior - [sold] top (notch) quality weld
— f — cordón m - [sold] fillet weld
— f — — con sección f triangular - [sold]
triangular fillet weld
— f — corriente f alterna - [sold] alternating
current weld(ing); A C weld(ing)
— f — — f continua - [sold] direct current, o
D C, welding
— f — costura f - [sold] seam weld(ing)
— f — — f traslapada - [sold] lap(ped) seam
weld(ing)
— f — costo m reducido - [sold] low cost weld-
-(ing)
— f — — m — de plancha(s) f - [sold] low
cost plate weld(ing)
— f — cromo-níquel m - [sold] chromium-nickel
weld(ing)
— f — cubrejunta f - [sold] reinforced, o
strap, weld(ing)
— f — chaflán f - [sold] beveled weld(ing)
— f — — m doble - [sold] double bevel, o V
butt, weld(ing)
— f con destello(s) m - [sold] flash weld(ing)
— f — electrodo m - [sold] electrode weld(ing)
— f — — m automática - [sold] automatic electrode weld(ing)
— f — — m autoprotegido - [sold] Innershield
self-shielded weld(ing)
— f — — m — con alma m fundente - [sold] Innershield self-shielded flux cored weld(ing)
— f — — m con hidrógeno m bajo - [sold] low
hydrogen electrode weld(ing)
— m — m con corriente f negativa - [sold]
electrode negative weld(ing)
— f — — m — — f positiva - [sold] electrode
positive weld(ing)
— m — — m con diámetro m reducido - [sold]
small wire electrode weld(ing)
— f — — m — manganeso m - [sold] manganese
electrode weld(ing)
— f — — m núcleo m fundente - [sold] flux-
-cored weld(ing)
— f — — m de acero m - [sold] steel electrode
weld(ing)
— f — — m — — m inoxidable - [sold] stainless steel electrode weld(ing)
— f — — m — alambre m - [sold] wire electrode weld(ing)
— f — — m — — m con diámetro m reducido -
[sold] small wire electrode weld(ing)
— f — — m manual - [sold] manual, o stick,
electrode weld(ing)
— f — — m corriente - [sold] normal, stick,
o manual, electrode weld(ing)
— f — — m múltiple - [sold] multi-wire electrode weld(ing)
— f — — m muy sobresaliente - [sold] long
stickout, o Linc-Fill, electrode weld(ing)
— f — — m poco sobresaliente - [sold] short
stickout (electrode) weld(ing)
— f — — m sobresaliente - [sold] Linc-Fill
weld(ing)
— f — — m único - [sold] single electrode
weld(ing)
— f — — — con velocidad f de avance m rápida - [sold] fast travel speed single electrode weld(ing)
— f — estaño m - [sold] (soft) solder(ing)
— f — — m con aleación f de plata f - véase
soldadura f con plata f
— f — — m con antorcha f con dos electrodos m
- [sold] arc torch solder(ing)
— f — estelita f - [sold] stellite weld(ing)
— f — filete m - véase soldadura f en ángulo m
(interior)

soldadura f con filete(s) m alternado(s) a tresbolillo - [sold] véase soldadura f a tresbolillo en ángulo m interior
— f — m en cadena f - [véase] soldadura f
en ángulo m en cadena f intermitente
— f — frecuencia f alta - [sold] high frequency weld(ing)
— f — fusión f - [sold] fusion weld(ing)
— f — — f total - [sold] full fusion weld
— f — gas m - [sold] gas weld(ing)
— f — — m de agua m - [sold] water gas welding
— f — — m de costura(s) f - [sold] seam gas
weld(ing)
— f — — — — f con punto(s) m - [sold]
gas seam spot weld(ing)
— f — — m inerte - [sold] inwert gas welding
• helium, o argon, arc welding; Heli-Arc
weld(ing)
— f — — — m — de tungsteno - [sold] tungsten
inert gas weld(ing)
— f — — m por punto(s) m - [Sold] gas spot
welding
— f — generador m con potencia f constante -
[sold] constant power source welding
— f con hidrógeno m - [sold] hydrogen weld(ing)
— f — — m bajo - [sold] low hydrogen welding
— f — impacto m - [sold] percussion weld(ing)
— f — sobreintensidad f - [sold] overvoltage
welding
— f — latón m - [sold] brazing; bronze welding
• véase también soldadura f con bronce m
— f — mando m con botonera f - [sold] pushbutton welding
— f — — f para producción f - [sold] bush-
button production welding
— f — martillo m - [sold] hammer weld(ing)
— f — movimiento m amplio de tejido m - [sold]
wide weave,ving weld(ing)
— f — — m corto de retroceso m - [sold]
back-step weld(ing); backstepping
— f — — m de traslación f y alimentación f
de electrodo m automático(s) - [sold] automatic welding
— f — núcleo m ácido - [sold] acid core welding
— f — ondulación(es) f - [sold] ripple(d)
weld(ing)
— f — . . . pasada(s) f - [sold] . . . pass
weld(ing)
— f — — f con avance m rápido - [sold] high
speed pass weld
— f — — f múltiple(s) - [sold] multiple pass
weld(ing)
— f — — f — en posición f cualquiera—[sold]
multiple pass all position weld(ing)
— f — — f — — f — con alimentación f
automática de electrodo m - [sold] multiple
pass all position semiautomatic weld(ing)
— f — — f única - [sold] single pass welding
— f — — f — en posición f cualquiera—[sold]
single pass all position weld(ing)
— f — — — — — í automática de
electrodo m - [sold] single pass all position
semiautomatic weld(ing)
— f — — f en plancha(s) f - [sold] plate
single pass weld(ing)
— f — penetración f escasa - [Sold] low penetrating,tion weld
— f — pistola f - [sold] gun weld(ing)
— f — — f para alimentación f automática de
electrodo m - [sold] automatic gun weld(ing)
— f — plancha(s) f en posición f de V - [sold]
trough weld(ing)
— f — potencia f constante - [sold] constant
power weld(ing)
— f — presión f - [sold] pressure weld(ing)
— f — — f con termita f - [sold] pressure
thermit weld(ing)
— f — proceso m con arco m sumergido - [sold]
submerged arc process weld(ing)
— f — prolongación f normal (de electrodo m) -
[sold] normal stickout weld(ing)

soldadura f con reborde m - [sold] bead weld
— f — recalcadura f - [sold] upset weld(ing)
— f — régimen m pleno - [sold] full rate output weld(ing)
— f — regulador(es) m - [sold] control(led) weld(ing)
— f —— m con transistor(es) m - [sold] solid state control(led) welding
— f — resistencia f a tensión f - [sold] tensile strength weld
— f — f — f de . . . - [sold] . . . psi tensile strength weld
f con retroceso m - [sold] back step welding
— f — régimen m pleno - [sold] fully rated output weld(ing)
— f — soldadora f - [sold] welder welding
— f — soplete m - [sold] torch weld(ing)
— f —— m con hidrógeno m - [sold] hydrogen (torch) weld(ing)
— f —— m —— m atómico - [sold] atomic hydrogen weld(ing)
— f — tejido - [sold] weave (weld)
— f —— m angosto - [sold] small weave (weld)
— f —— m con caja f - [sold] box weave weld
— f —— m en caja f - [sold] box weave weld
— f —— m triangular - [sold] triangular weave weld
— f — tenor m muy bajo de silicio n - [sold] extra low silicon weld
— f — termita - [sold] thermit weld(ing)
— f — transferencia f de metal m por medio m de corte(s) m en circuito m - [sold] short circuiting type transfer weld(ing)
— f — transformador m - [sold] transformer welding
f con traslapo m - [sold] lap weld(ing)
— f — tungsteno m - [sold] tungsten weld(ing)
— f —— m y gas m inerte - [sold] tungsten inert gas welding
— f — m —— m — con corriente f alterna - [sold] alternating current tungsten inert gas welding
— f — m —— m —— f continua—[sold] direct current tungsten inert gas welding
— f — m —— m —— frecuencia f alta [sold] high frequency tungsten inert gas weld(ing)
— f — velocidad f alta - [sold] high speed weld(ing)
— f —— f — por arco m sumergido - [sold] high speed submerged arc weld(ing)
— f — f de avance m alta - [sold] high travel speed welding
— f — f grande - [sold] high speed welding
— f — f grande para avance m - [sold] high speed weld(ing)
— f — una (sóla) pasada f - [sold] one pass weld(ing)
— f — voltaje m constante - [sold] constant voltage weld(ing)
— f —— m variable - [sold] variable voltage weld(ing)
— f cóncava - [sold] concave weld(ing)
— f — en ángulo m interior - [sold] concave fillet weld(ing)
— f— gruesa - [sold] large concave weld
— f — pequeña - [sold] small concave weld
— f considerablemente mejor - [sold] considerably better weld(ing)
— f contaminada - [sold] contaminated weld
— f continua - [sold] continuous, o line, (filled) weld(ing) • all-day, o steady, welding • seam (welding)
— f — en ángulo m interior - [sold] continuous fillet weld(ing)
— f — larga - [sold] long continuous weld(ing)
— f continuada - [sold] sustained weld(ing)
— f — con velocidad f alta - [sold] sustained high speed weld(ing)
--- f contrahorizontal* - [sold] véase soldadura f sobrecabeza
— f convencional - [sold] conventional welding
— f convexa - [sold] convex weld(ing)

soldadura f convexa en ángulo interior - [sold] convex fillet weld
— f — gruesa - [sold] large convex weld
— f — pequeña - [sold] small convex weld
— f correspondiente f - corresponding weld • chain weld(ing)
— f — en ángulo m interior - [sold] chain, fillet, o intermittent, weld
— f corriente - [sold] conventional, o common, o routine, weld(ing)
— f — en ángulo m interior - [sold] conventional fillet (weld)
— f corta - [sold] short, weld(ing), o work
— f — gruesa - [sold] short heavy weld
— f cortada - [sold] cut (off) weld
— f crítica - [sold] critical welding,dment
— f cronometrada - [sold] timed weld(ing)
— f curva - [sold] curved weld(ing)
— f de acero m - [sold] steel weld(ing)
— f —— m aleado - [sold] alloy steel welding
— f —— m cadmiado - [sold] cadmium plated) steel weld(ing)
— f —— m común - [sold] common steel weld(ing)
— f —— m con aleación f - [sold] alloy steel weld(ing)
— f —— m —— f —— f baja - [sold] low alloy steel fabricating,tion, o welding
— f —— m — carbono m - [sold] carbon steel weld(ing)
— f —— m —— m alto - [sold] high carbon steel weld(ing)
— f —— m —— m bajo - [sold] low carbon steel weld(ing)
— f —— m con elemento(s) m de aleación f - [sold] alloy steel welding
— f —— m con plata f - [sold] silver-steel weld(ing)|
— f —— m con resistencia f alta—[sold] high strength steel weld(ing)
— f —— m —— f alta a tensión f - [sold] high strength steel weld(ing)
— f —— m con resistencia f alta a tensión f templado,- high strength tempered steel weld(ing)
— f —— m de tipo m ferrítico - [sold] ferritic type steel welding,dment
— f —— m dulce - [sold] mild steel weld(ing)
— f —— m emplomado - [sold] lead plated steel weld(ing)
— f —— m enchapado - [sold] plated steel weld(ing)
— f —— m — con plomo m - [sold] lead plated steel weld(ing)
— f —— m enfriado - [sold] cooled steel weld(ing)
— f —— m — por inmersión f—[sold] quenched steel weld(ing)
— f —— m —— f con resistencia f alta (a tensión) f - [sold] high strength quenched steel weld(ing)
— f —— m —— f y templado - [sold] high strength quenched and tempered steel welding
— f —— m —— f — con resistencia f alta a tensión f - [sold] high strength quenched and tempered steel weld(ing)
— f —— m ferrítico - [sold] ferritic steel welding,dment
— f —— m galvanizado - [sold] galvanized steel weld(ing)
— f —— m inoxidable - [sold] stainless steel welding,dment, o fabricating
— f —— m — de tipo m ferrítico - [sold] ferritic type stainless steel welding,dment
— f —— m — delgado - [sold] thin stainless steel weld(ing)
— f —— m — ferrítico - [sold] ferritic stainless steel welding,dment
— f —— m intemperizado - [sold] weathering steel weld(ing)
— f —— m martensítico - [sold] martensitic steel welding,dment
— f —— m pintado - [sold] painted steel

welding
soldadura f de acero m plaqueado - [sold] clad
steel weld(ing)
— f — — m **plástico** - [sold] plastic steel
welding (rod)
— f — — m **problemático** - [sold] problem(atic)
steel weld(ing)
— — — m **recubierto con cadmio m** - [sold]
cadmium plated steel welding
— f — — m **sucio** - [sold] dirty steel welding
— f — — m **templado** - [sold] tempered steel
weld(ing)
— f — — m — **con resistencia f alta** - [sold]
high strength tempered steel weld(ing)
— f — — m — **a tensión f** - [sold]
high tensile strength tempered steel welding
— f — **alambre m** - [sold] wire, welding, o soldering
— f — **aleación f** - [sold] alloy weld(ing)
— f — — f **para temperatura(s) f baja(s)** -
[sold] low temperature alloy welding
— f — **aluminio m** - [sold] aluminum weld(ing)
— f — — f **con aleación f** - [sold] aluminum
alloy weld(ing)
— f — — m **en plancha(s) f** - [sold] aluminum
plate welding
— f — **aportación f** - [sold] deposit(ion) welding
— f — — f **rápida** - [sold] fast-fill, o high
deposit(ion), weld(ing)
— f — — f **con alimentación f automática**
(de electrodo m) - [sold] high deposition
Squirt weld(ing)
— f — — f **Fast-Fill** - [sold] Fast-Fill
high deposition weld(ing)
— f — — f **Fast-Fill con alimentación f automática de electrodo m** - [sold] Fast-Fill
high deposition Squirt weld(ing)
— f — — f **y solidificación f rápidas** - [sold]
fill-freeze weld(ing)
— f —, **anillo m, o aro m** - [sold] ring welding
— f — **arista(s) f** - [sold] edge weld(ing)
— f — **avance m rápido** - [sold] high speed weld
— f — — m **en chapa(s) f** - [sold] high
speed sheet weld(ing)
— f — — m **de acero m** - [sold] high
speed sheet metal weld(ing)
— f — — m — **en pasada f única** - [sold] high
speed single pass weld(ing)
— f — **banda f** - [sold] strip welding
— f — **bastidor m** - [sold] frame weld(ing)
— f — — m **para automóvil m** - [autom] automobile, o car, frame weld(ing)
— f — — m — **camión m** - [autom] truck frame
weld(ing)
— f — **borde m** - [sold] edge weld(ing)
— f — — m **de solapo m** - [sold] lap edge weld
— f — **borne m** - [mec] stud weld(ing)
— f — **botón m** - [sold] bead weld(ing)
— f — **bronce m** - [sold] bronze weld(ing)
— f — **cable m** - [cabl] rope weld(ing)
— f — — m **de alambre m** - [cabl] wire rope
weld(ing)
— f — **cadena f** - [sold] chain weld(ing)
— f — **campana f** - [sold] bell weld(ing)
— f — — f **y espiga f** - [sold] bell and spigot
weld(ing)
— f — **canto m** - [sold] edge weld(ing)
— f — — m **de chapa(s) f** - [sold] sheet edge(s)
weld(ing)
— f — — m — — f **de acero m** - [sold] sheet
metal edge weld(ing)
— f — — m **sobrecabeza** - [sold] overhead edge
weld(ing)
— f — **caños, ñerías** - [sold] pipe weld(ing)
— f — — **de cobre m** - [sold] copper pipe,ping,
soldering, o welding
— f — **cobre m** - [sold] copper, welding, o soldering
— f — — m **con aleación f** - [sold] copper alloy weld(ing)
— f — — m — **antorcha f** - [sold] copper torch
weld(ing)

**soldadura f de cobre m con antorcha f con dos e-
lectrodo(s) m** - [sold] arc torch copper
weld(ing)
— f — **collar(es) m** - [petról] collar welding
— f — **conducto m** - [sold] conduit weld(ing)
— f — **conductor m** - [electr-instal] wire, o
lead, soldering
— f — **conjunto m** - [sold] assembly weld(ing)
— f — — m **de columna f** - [sold] column assembly weld(ing)
— f — — f **de caja f** - [sold] box
column assembly weld(ing)
— f — **coraza f** - [sold] shell weld(ing)
— f — — f **para horno m** - [ind] furnace shell
weld(ing)
— f — **cordón m** - [sold] seam weld(ing) • bead
weld(ing)
— f — — m **para cierre m** - [sold] cover pass
weld(ing)
— f — — m **final** - [sold] cover pass weld(ing)
— f — — m **inicial** - [sold] initial pass
weld(ing)
— f — **chapa f** - [sold] strip weld(ing) • sheet
weld(ing)
— f — — f **de acero m** - [sold] sheet, steel, o
metal, weld(ing)
— f — — f **metálica** - [sold] sheet metal,
welding, o soldering
— f — **chaveta f** - [mec] key weld(ing)
— f — — f **para cilindro m** - [mec] barrel key
weld(ing)
— f — **depósito m** - [sold] tank weld(ing)
— f — **derecha f a izquierda f** - [sold] right
to left weld(ing)
— f — **envase m** - [sold] container, welding, o
soldering
— f — **espárrago m** - [sold] stud weld(ing)
— f — **éxito** - [sold] successful weld(ing)
— f — **extremo m** - [sold] end weld(ing)
— f — — m **de cable m** - [sold] rope, o cable,
end weld(ing)
— f — — m **de alambre m** - [cabl] wire
rope end weld(ing)
— f — **fleje m** - [metal-lam] strip weld(ing)
— f — **forja f** - [sold] forge weld(ing)
— f — **fundición f** - [sold] cast iron weld(ing)
— f — — f **blanca** - [sold] white cast iron
weld(ing)
— f — **hierro m** - [sold] iron weld(ing)
— f — — m **cobrizo** - [sold] copper iron welding
— f — — m **fundido blanco** - [sold] white cast
iron weld(ing)
— f — — m — **caliente** - [sold] hot cast iron
weld(ing)
— f — — m — **labrable** - [sold] machinable
cast iron weld(ing) • machinable cast iron
electrode
— f — — m **gris** - [sold] grey cast iron
weld(ing)
— f — **izquierda f a derecha f** - [sold] left to
right weld(ing)
— f — **junta f** - [sold] joint weld(ing)
— f — **latonero m** - véase **soldadura f blanda**
— f — **magnesio m** - [sold] magnesium weld(ing)
— f — **mano f** - [sold] véase **soldadura f manual**
— f — **medida(s) f máxima(s) en pasada f única**
- [sold] maximum single pass weld(ing)
— f — — f — — f — **en ángulo m interior**
- [sold] maximum single pass fillet weld
— f — **metal(es) m** - [sold] metal(s) weld(ing)
— f — — m **no ferroso(s)** - [sold] nonferrous
metal(s) weld(ing)
— f — — m — **en horno m** - [sold] furnace
brazing
— f — — m **para edad f espacial** - [sold]
space,cial age metal(s) weld(ing)
— f — — m **vertical** - [sold] vertical metal
welding
— f — **mismo tipo m** - [sold] same type, joint,
o weld(ing)
— f — **orej(et)a f** - [sold] tab weld(ing)
— f — **pasada f final** - [sold] final, o cover,
pass weld(ing)

soldadura f de pasada(s) f múltiple(s) - [sold]
multiple pass, o multipass, weld(ing)
— f —— f en ángulo m (interior) - [sold]
multiple pass fillet weld
— f —— f —— plancha(s) f—(sold) multiple
pass welding on plate(s)
— f —— f —— f en caliente - [sold] hot
multiple pass plate(s) welding
— f —— f —— f —— posición f cualquiera
- [sold] all position(s) plate(s) multiple
pass(es) weld(ing)
— f —— f ←— ranura(s) f profunda(s) - [sold]
deep groove multipass weld(ing)
— f —— f para cierre m - [sold] cover pass
weld(ing)
— f —— f —— m en junta f -[sold] joint
cover pass weld(ing)
— f —— f —— —— f en tubería f -
pipe joint cover pass weld(ing)
— f —— f para relleno m - [sold] fill pass
weld(ing)
— f —— f paralela(s) - [sold] parallel
passes weld(ing)
— f —— f con movimiento m reducido de tejido m - [sold] split weave weld(ing)
— f —— f primera - [sold] first pass weld
— f —— f única - [sold] single pass weld
— f —— f única con costo m reducido - [sold]
economical, o low cost, single pass weld(ing)
— f —— f —— en posición f cualquiera - [sold]
all position single pass weld(ing)
— f —— f —— en plancha(s) f - [sold] single
pass plate welding
— f —— f —— —— f gruesa(s) - [sold] heavy
plate single pass weld(ing)
— f —— f —— ranura f - [sold] groove
single pass weld(ing)
— f —— f —— —— f profunda - [sold] deep
groove single pass weld(ing)
— f —— pedazo m de hierro m - [sold] iron piece
welding
— f —— penetración f - [sold] penetration weld
— f —— f completa - [sold] full penetration
weld
— f —— f plena—[Sold] full penetration weld
— f —— f profunda - [sold] deep penetration,
weld, o process
— f —— f —— en ángulo m (interior) - [sold]
deep penetration fillet (weld)
— f —— f total—[sold] full penetration weld
— f —— pieza f - [sold] part weld(ing)
— f —— f de cobre m - [sold] copper part,
welding, o soldering
— f —— f —— hierro m - [sold] iron part,
welding, o soldering
— f —— f —— m fundido - [sold] cast iron,
part, o piece, welding
— f —— f —— m — caliente - [sold] hot
cast iron, part, o piece, weld(ing)
— f —— f —— m — frío - [sold] cold cast
iron, part, o piece, welding
— f —— f en taller m - [sold] part, o piece,
shop welding
— f —— f estañada - [sold] tinned part soldering
— f —— f galvanizada - [sold] galvanized
part soldering
— f —— f pequeña - [sold] small part welding
— f —— pieza f - [sold] assembly welding
— f —— placa(s) f - [sold] plate(s) weld(ing)
— f —— plancha(s) f - [sold] plate(s) weld(ing)
— f —— —— f con pasada(s) f múltiple(s)—[sold]
multiple pass(es) plate(s) weld(ing)
— f —— f de acero m - [sold] steel plate(s)
welding
— f —— f —— m dulce - [sold] mild steel
plate(s) welding
— f —— f — aluminio m - [sold] aluminum
plate(s) welding
— f —— f de espesor m ilimitado - [sold] unlimited thickness plate(s) welding
— f —— f delgada(s) - [sold] thin plate(es)
weld(ing)

soldadura f de plancha(s) f gruesa(s) - [sold]
thick, o heavy, plate(s) welding
— f —— f liviana(s) - [sold] light plate(s)
welding
— f —— f con velocidad f alta - [sold]
high speed light weight welding
— f —— f más gruesa(s) - [sold] thicker, o
heavier, plate(s) welding
— f —— f plana(s) - [sold] flat plate(s)
welding
— f — plata f - [sold] silver brazing
— f — punto(s) m—[sold] tack welding; tacking
— f — rendimiento m máximo - [sold] maximum
duty weld(ing)
— f — resalto m - [sold] (resistance) project-
tion welding • upset welding
— f —— m por punto(s) m - [sold] projection
spot welding
— f — resistencia f - [sold] véase soldadura f
resistente
— f —— f plena - [sold] full, strength, o
resistance, weld
— f —— f total - [sold] full strength weld
— f — retroceso m - [sold] back-step weld(ing)
• backstepping
— f —— m salteado - [sold] skip back-step
weld(ing)
— f — revés - [sold] backhand(ed) weld(ing)
— f —— m pesado - [sold] heavy duty welding
— f — solapa,pe,po - [sold] lap (joint) weld;
lap
— f —— a través de plancha(s) f - [sold]
through @ lap weld
— f —— m con rebajo m - [sold] offset lap
weld
— f —— m con pasada f única - [sold] single
pass lap weld
— f —— m —— f múltiple(s) - [sold] multiple pass lap weld
— f —— m en ángulo m (interior) - [sold]
fillet lap weld
— f —— m — borde m - [sold] edge lap weld
— f —— m —— m de chapa(s) f - [sold]
(sheet) edge lap weld
— f —— m —— f de acero m - [sold] steel
sheet lap weld
— f —— m en pasada f única con avance m rápido - [sold] high speed single pass lap weld
— f — solidificación f - [sold] freeze welding
— f —— f rápida - [sold] fast freeze welding
— f —— f rápida con alimentación f automática de electrodo m - [sold] Fast-Freeze
Squirt welding
— f —— f en posición f cualquiera con
alimentación f automática de electrodo m -
[sold] Fast-Freeze all position Squirt
welding
— f — soplete m - [sold] torch soldering
— f —— m con bronce m - [sold] torch brazing
— f —— m — metal(es) m no ferroso(s) -
[sold] torch brazing
— f — tamaño m cabal - [sold] full size weld
— f —— demasiado reducido - [sold] too small,
o undersized, weld
— f —— m excesivo - [sold] oversize weld
— f —— m exigüo - [sold] undersized weld
— f — tambor m - [sold] drum weld(ing)
— f — tapón m - [sold] plug weld
— f — techo m - [sold] overhead weld(ing);
véase también soldadura f sobrecabeza
— f — terminal - [electrón] lug soldering
— f — termita f - [sold] thermit welding
— f —— f con presión f - [sold] thermit
pressure, o pressure thermit, welding
— f — tipo m diferente - [sold] different type
weld
— f —— m distinto - [sold] different type
weld
— f —— m diverso - [sold] different type
weld
— f — tubería(s) f - [tub] pipe,ping welding
— f —— f con corriente f alterna - [sold]

alternating current pipe welding
soldadura <u>f</u> de tubería <u>f</u> de acero <u>m</u> - [sold]
steel pipe welding
— <u>f</u> —— <u>f</u> — **cobre <u>m</u>** - [sold] copper, pipe, <u>o</u>
piping, solder(ing)
— <u>f</u> —— <u>f</u> **para conducción <u>f</u>** - [sold] pipeline
welding
— <u>f</u> —— <u>f</u> —— <u>f</u> **en obra <u>f</u>** - [tub] cross
country (pipeline) welding
— <u>f</u> —— <u>f</u> —— <u>f</u> **en taller <u>m</u>**—[tub] in-plant
pipe welding
— <u>f</u> —— <u>f</u> **con pared <u>f</u> delgada** - [sold] thin
wall tubing weld(ing)
— <u>f</u> —— <u>f</u> —— <u>f</u> **gruesa** - [sold] thick, <u>o</u>
heavy, wall tube weld(ing)
— <u>f</u> —— <u>f</u> —— <u>f</u> — **de acero <u>m</u>**—[sold] steel
heavy wall tube,bing weld(ing)
— <u>f</u> —— <u>f</u> — **resistencia <u>f</u> alta** - [sold] high
strength pipe weld(ing)
— <u>f</u> —— <u>f</u> —— <u>f</u> — **a tensión <u>f</u>**—[sold] high
tensile pipe welding
— <u>f</u> —— <u>f</u> **en planta**(s) <u>f</u> **industrial**(es) -
[sold] in-plant pipe welding
— <u>f</u> —— <u>f</u> — **taller <u>m</u>** - [sold] in-plant pipe
welding • shop pipe welding
— <u>f</u> —— <u>f</u> **para presión**(es) <u>f</u> **alta**(s)—[sold]
high pressure pipe(line) weld(ing)
— <u>f</u> —— <u>f</u> **para temperatura**(s) <u>f</u> **alta**(s) -
[sold] high temperature pipe,ping weld(ing)
— <u>f</u> —— <u>f</u> —— <u>f</u> **y presión**(es) **alta**(s) -
[sold] high temperature high pressure
pipe,ping weld(ing)
— <u>f</u> — **tubo**(s) <u>m</u> - [sold] pipe(s) weld(ing)
— <u>f</u> —— <u>m</u> **para caldera**(s) <u>f</u> - [cald] flue(s)
welding
— <u>f</u> — **tuerca**(s) <u>f</u> - [sold] nut(s) welding
— <u>f</u> — **una (sóla) pasada** - [sold] one pass weld
— <u>f</u> — **varias pasadas <u>f</u>** - [sold] split weld
— <u>f</u> —— — **con movimiento <u>m</u> reducido de teji-**
do <u>m</u> - [sold] split weave weld
— <u>f</u> — **velocidad <u>f</u> alta** - [sold] high speed
weld(ing)
— <u>f</u> —— <u>f</u> — **en chapa**(s) <u>f</u> **de acero <u>m</u>**—[sold]
high speed sheet metal weld(ing)
— <u>f</u> —— <u>f</u> **uniforme** - [sold] uniform speed
weld(ing)
— <u>f</u> — **viga <u>f</u>** - [sold] beam weld(ing)
— <u>f</u> —— <u>f</u> **creciente** - [sold] tapered beam
welding
— <u>f</u> —— <u>f</u> **decreciente** - [sold] tapered beam
welding
— <u>f</u> **defectuosa** - [sold] defective, <u>o</u> unsound,
weld(ing)
— <u>f</u> **defectuosa** - [sold] defective weld(ing)
— <u>f</u> — **cortada** - [sold] cut off defective weld
— <u>f</u> — **reparada**—[sold] repaired defective weld
— <u>f</u> **deficiente** - [sold] defective weld(ing)
— <u>f</u> **delgada** - [sold] thin, <u>o</u> shallow, weld
— <u>f</u> **demasiado buena** - [sold] too good weld
— <u>f</u> — **reducida** - [sold] undersized weld(ing)
— <u>f</u> **densa** - [sold] dense weld(ing)
— <u>f</u> — **con calidad <u>f</u> para inspección <u>f</u> (con**
rayo(s) **X** - [sold] dense X-ray quality weld
— <u>f</u> **descendente** - [sold] downhill weld(ing)
— <u>f</u> — **con pasada <u>f</u> única** - [sold] single pass
downhill (weld)
— <u>m</u> —— **relleno <u>m</u> rápido** - [sold] fast-fill
downhill, weld, <u>o</u> joint
— <u>f</u> —— **solapo <u>m</u>**—[sold] downhill lap (weld)
— <u>f</u> —— **en ángulo <u>m</u> (interior)** - [sold] downhill
fillet weld
— <u>f</u> —— — <u>m</u> (—) **con pasada <u>f</u> única** - [sold]
single pass downhill fillet (weld)
— <u>f</u> —— **tubería**(s) <u>f</u> - [sold] pipe downhill
weld(ing)
— <u>f</u> — **vertical** - [sold] vertical downhill weld
— <u>f</u> —— — **con pasada <u>f</u> única** - [sold] single
pass vertical (down) weld
— <u>f</u> **desde abajo** - [sold] overhead weld(ing)
— <u>f</u> **desviada** - [sold] roundabout weld(ing)
— <u>f</u> **detenida** - [sold] stopped weld(ing)
— <u>f</u> **difícil** - [sold] difficult, <u>o</u> hard, welding
 • tough weld(ing)

soldadura <u>f</u> difícil para producción <u>f</u> - [sold]
difficult production weld(ing)
— <u>f</u> —— **reparación <u>f</u>** - [sold] difficult
maintenance weld(ing)
— <u>f</u> **desde un lado** - [sold] weld(ing) from one
side
— <u>f</u> **directa** - [sold] direct, <u>o</u> forehand, weld
— <u>f</u> **doble** - [sold] double weld(ing)
— <u>f</u> —— **en ángulo <u>m</u> interior** - [sold] double
fillet weld(ing)
— <u>f</u> **dúctil** - [sold] ductile weld
— <u>f</u> —— **con cromo-níquel**—[sold] ductile chromium-nickel weld(ing)
— <u>f</u> **dura** - [sold] brazing
— <u>f</u> **económica** - [sold] economical, <u>o</u> low cost,
weld(ing)
— <u>f</u> — **con pasada <u>f</u> única** - [sold] economical
single pass weld(ing)
— <u>f</u> — **en posición <u>f</u> horizontal** - [sold] low
cost horizontal position weld(ing)
— <u>f</u> —— — <u>f</u> **plana** - [Sold] low cost flat
position weld(ing)
— <u>f</u> —— — <u>f</u> **vertical** - [sold] low cost vertical position weld(ing)
— <u>f</u> **efectuada** - [sold] performed, <u>o</u> done, weld
— <u>fejecutada</u> **automáticamente** - [sold] performed automatic weld(ing)
— <u>f</u> — **debidamente** - [sold] properly performed
weld
— <u>f</u> — **íntegramente desde un (sólo) lado <u>m</u>** -
[sold] 100% weld from one side
— <u>f</u> **eléctrica** - [sold] electric, <u>o</u> arc, weld-
-(ing); **véase también soldadura <u>f</u> por arco <u>m</u>**
— <u>f</u> — **con antimonio <u>m</u>** - [sold] antimony
weld(ing)
— <u>f</u> — **de hierro <u>m</u> fundido** - [sold] cast iron
arac weld(ing)
— <u>f</u> — **por arco <u>m</u>** - [sold] (electric) arc
weld(ing)
— <u>f</u> — **por costura <u>f</u>** - [sold] seam weld(ing)
— <u>f</u> —— **chispa <u>f</u>** - [sold] electric flash
weld(ing)
— <u>f</u> —— **fusión <u>f</u>** - [sold] electric fusion
weld(ing)
— <u>f</u> —— **punto**(s) <u>m</u> - [sold] (electric) spot
weld(ing)
— <u>f</u> —— **resistencia <u>f</u>** - [sold] (electric)
resistance weld(ing)
— <u>f</u> **electrónica** - [sold] electronic weld(ing)
— <u>f</u> **empleada** - [sold] used weld(ing)
— <u>f</u> — **más comúnmente** - [sold] most widely
used weld(ing)
— <u>f</u> **en aleación <u>f</u>** - [sold] alloy weld(ing)
— <u>f</u> — **aluminio <u>m</u>** - [sold] aluminum weld(ing)
— <u>f</u> — **ángulo <u>m</u>** - [sold] angle, <u>u</u> fillet, weld
— <u>f</u> —— <u>m</u> **exterior** - [sold] corner (joint)
weld(ing)
— <u>f</u> —— <u>m</u> **interior** - [sold] fillet (joint)
weld(ing); fillet
— <u>f</u> —— <u>m</u> **con cadena <u>f</u>** - [sold] chain fillet
weld(ing)
— <u>f</u> —— — <u>m</u> —— <u>f</u> **intermitente** - [sold] intermittent chain fillet weld(ing)
— <u>f</u> —— — <u>m</u> — **pasada <u>f</u> única** - [sold] single
pass fillet weld
— <u>f</u> —— — <u>m</u> —— <u>f</u> **múltiple**(s) - [sold] multiple pass(es) fillet weld
— <u>f</u> —— <u>m</u> **en forma <u>f</u> de canaleta <u>f</u>** - [sold]
trough weld
— <u>f</u> —— — <u>m</u> — **posición <u>f</u> plana** - [sold]
trough weld
— <u>f</u> —— <u>m</u> **exterior** - [sold] corner weld(ing)
— <u>f</u> —— — <u>m</u> — **con pasada <u>f</u> única** - [sold]
single pass corner weld
— <u>f</u> —— — <u>m</u> **con relleno <u>m</u> completo** - [sold]
full size corner weld
— <u>f</u> —— — <u>m</u> —— **tamaño <u>m</u> cabal** - [sold] full
size corner weld
— <u>f</u> —— — <u>m</u> **en posición <u>f</u> horizontal** -
[sold] horizontal position corner weld(ing)
— <u>f</u> —— — <u>m</u> —— — <u>f</u> **plana** - [sold] level, <u>o</u>
flat, position corner weld
— <u>f</u> —— — <u>m</u> **interior <u>m</u>** - [sold] fillet weld(ing)

soldadura f en ángulo m interior con ranura f -
[sold] groove fillet weld
— f — — m — con superficie(s) f para fusión
igual(es) - [sold] equal legged fillet weld
— f — — m — en chapa f de acero m - [sold]
sheet metal fillet (weld)
— f — — m — — f metálica(s) - [sold]
sheet metal corner weld(ing)
— f — — m — fuerza f comparable - [sold]
equal strength fillet (weld)
— f — — m — pasada f única - [sold] single
pass fillet weld
— f — — m — — f — con avance m rápido -
[sold] high speed single pass fillet (weld)
— f — — m — — f en posición f horizontal - [sold] single pass horizontal position fillet (weld)
— f — — m — — f en posición f plana -
sold] single pass flat position fillet (weld)
— f — — m — pasada(s) f múltiple(s) -
[sold] multiple pass fillet (weld)
— f — — m — de pieza(s) f en posición f para
soldadura f plana - [sold] positioned fillet
(weld)
— f — — ¬ — — f puesta(s) en posición f
para soldadura f plana - [sold] positioned
fillet (weld)
— f — — m de resistencia f comparable -
[sold] equal strength fillet (weld)
— f — — m — en posición f de canaleta f -
[sold] trough position fillet (weld)
— f — — m — — f plana - [sold] flat po-
fillet (weld)
— f — — m intermitente—[sold] intermittent
fillet (weld)
— f — — m salteada - [sold] intermittent fillet (weld)
— f en argón m - [sold] argon weld(ing)
— f — — m gaseoso—[sold] argon gas weld(ing)
— f — atmósfera f regulada—[sold] controlled
atmosphere weld(ing)
— f — — f — con argón m gaseoso - [sold] argon gas controlled atmosphere weld(ing)
— f — — f — gas m de argón m - [sold] argon
gas controlled atmosphere weld(ing)
— f — calidad f con manganeso m - [sold] manganese grade welding,dment
— f — canaleta f - [sold] trough weld(ing)
— f — capa(s) f múltiple(s) - [sold] multi-
layer weld(ing)
— f — coraza f - [ind] shell weld(ing)
— f — — f para horno m - [sold] furnace shell
welding,dment
— f — chaflán m - [sold] groove wled(ing)
— f — descubierto m - [sold] open weld(ing)
— f — dirección ↑ ascendente - [sold] upwards
weld(ing) • vertical up weld(ing)
— f — — f descendente - [sold] down(wards), o
downhill, weld(ing)
— f — ebullición f - [sold] boiling, weld(ing),
o solder
— f — ejecución f - [sold] weld(ing) in, process, o progress, o being performed
— f — escala f grande - [sold] high production
weld(ing) • heavy duty weld(ing)
— f — — f industrial - [sold] production,
weld(ing), o work
— f — exterior m - [sold] outside weld(ing)
— f — frío - [sold] cold soldering
— f — gas m - [sold] gas weld(ing)
— f — — m de argón m - [sold] argon gas
weld(ing)
— f — — m inerte - [sold] inert gas weld(ing)
— f — general - [sold] general (purpose)
weld(ing)
— f — — de aleación(es) f - [sold] alloy general purpose weld(ing)
— f — — — f con resistencia f alta -
[sold] alloy general purpose high strength
weld(ing)
— f — — f — — f — — f a tensión f -
[sold] high tensile strength general purpose
weld(ing)

soldadura f en general de plancha(s) f - [sold]
general purpose plate weld(ing)
— f — grieta(s) f - [sold] groove weld(ing)
— f — — f profunda(s) - [sold] deep groove
weld(ing)
— f — horno m - [sold] furnace weld(ing)
— f — — m de metal(es) m - [sold] furnace
metal(s) welding • metal(s) furnace weld(ing)
— f — — m no ferroso(s) - [sold] furnace brazing • non ferrous metals furnace
brazing
— f — intemperie f - [sold] out-of-doors
weld(ing)
— f — interior m - [sold] inside weld(ing)
— f — — m de ángulo m - [sold] fillet joint
weld(ing
— f — muesca f - [sold] véase soldadura f en
ranura f
— f — obra f - [sold] field, o job (site),
weld(ing)
— f — — f con alivio m de tensión(es) f -
[sold] stress relieve field weld(ing)
— f — — f con penetración f total - [sold]
full penetration field weld(ing)
— f — posición f - [sold] position weld(ing)
— f — — f plana f - [sold] flat (position)
weld(ing)
— f — — f — con punto(s) m - [sold] flat
tacking
— f — plancha(s) f - [sold] plate weld(ing)
— f — — f con pasada f única - [sold] single
pass plate weld(ing)
— f — — f inclinada(s) f - [sold] inclined, o
slanting, plate(s) welding
— f — planta f - [sold] in, plant, o shop,
weld(ing)
— f — — f industrial - [sold] industrial, o
in-plant, weld(ing)
— f — — f para fabricación f - [sold] fabrication plant, o in-plant, welding
— f — posición f de canaleta f - [sold]
trough position weld(ing)
— f — — f apropiada - [sold] appropriate position weld(ing)
— f — — f cualquiera - [sold] all-position,
weld(ing), o operating,tion
— f — — f — con alimentación f automática
de electrodo m - [sold] all position, semi-
-automatic (arc), o Squirt, weld(ing)
— f — — f horizontal - [sold] horizontal position weld(ing)
— f — — f — en ángulo m exterior - [sold]
horizontal position corner weld(ing)
— f — posición f inclinada - [sold] inclined,
o slanting, position weld(ing)
— f — — f otra que plana f - [sold] out-of-
-position, weld(ing), o work
— f — — f plana - flat, o downhand, o level,
position weld(ing)
— f — — f — sobre ángulo m exterior—[sold]
flat position corner weld(ing)
— f — — f vertical - [sold] vertical position
weld(ing)
— f — ranura f - [sold] groove, o slot, weld-
-(ing)
— f — — f achaflanada - [sold] bevel(ed)
groove weld(ing)
— f — — f angosta - [sold] narrow groove
weld(ing)
— f — — f con chaflán m - [sold] beveled
narrow joint weld
— f — — f — — m en V - [sold] narrow V
joint (groove) weld(ing)
— f — — f — y profunda - [sold] deep narrow
groove weld(ing)
— f — — f biselada - [sold] bevel(ed) groove
weld(ing)
— f — — f con chaflán m - [sold] bevel(ed)
groove weld(ing)
— f — — f — — m doble - [sold] double beveled groove weld(ing)
— f — — f con chaflán m - [sold] beveled
groove weld(ing)

soldadura f en ranura f con chaflán m̲ en V -
  (sold) V̄-joint weld(ing)
— f — — f — ángulo(s) m̲ recto(s) - [sold]
  square groove weld(ing)
— f — — f en U - [sold] U groove weld(ing)
— f̄ — — f̄ — — doble - [sold] double U
  groove weld(ing)
— f — — f — — sencilla - [sold] single U
  groove weld(ing)
— f — — f — V - [sold] V-groove weld(ing)
— f̄ — — f̄ — — doble - [sold] double V
  groove weld(ing)
— f — — f — — sencilla - [sold] single V
  groove weld(ing)
— f — — f profunda - [sold] deep grove weld-
  -(ing) • narrow groove weld(ing)
— f — — — con pasada(s) f múltiple(s) -
  [sold] multiple pass, o multipass, deep
  groove weld(ing)
— f — — f — sobre pasada(s) f en raíz f -
  [sold] deep groove weld over @ root pass(es)
— f recinto m̲ - [sold] enclosed weld(ing)
— f̄ — — m̲ con corriente(s) f (de aire m̲) -
  [sold] drafty location weld(ing)
— f — — m̲ reducido - [sold] confined space
  weld(ing)
— f — serie f - [sold] production weld(ing) •
  production application
— f — — f a tope m̲ - [sold] production butt
  weld(ing)
— f — — con pasada f única - [sold] single
  pass production weld(ing)
— f — — — f — con avance m̲ rápido -
  [sold] high speed single pass production
  weld(ing)
— f — — f en ángulo m̲ interior con pasada f
  única - [sold] single pass production fillet
  weld(ing)
— f — — — f — — m̲ — — — f — con avance m̲
  rápido - [sold] high speed single pass production fillet weld(ing)
— f — — f según norma f - [sold] standard
  production weld(ing)
— f — sitio n con corriente(s) f (de aire m̲) -
  [sold] drafty location weld(ing)
— f — sitio m̲ - [sold] in-place weld(ing)
— f̄ — taller m̲ - [sold] shop, o in-plant,
  weld(ing)
— f — — m̲ con alivio m̲ de tensión(es) m̲ -
  [sold] stress relieved shop weld(ing)
— f — — m̲ con penetración f total - [sold]
  full penetration shop weld(ing)
— f — — m̲ para reparación f - [sold] shop repair weld(ing)
— f — terreno m̲ - [sold] on-site weld(ing)
— f̄ — toda posición f - [sold] all-position
  weld(ing)
— f — — contorno m̲ - [sold] all-around weld
— f̄ — — torno m̲ - [sold] (all) around weld(ing)
— f̄ — — tresbolillo - [sold] staggered weld(ing)
— f̄ — — — m̲ en ángulo m̲ (interior) - [sold]
  staggered fillet weld(ing)
— f — tubería(s) f (para conducción f) -
  [sold] pipeline weld(ing)
— f en vaivén m̲ - [sold] weave,ving weld(ing)
— f̄ — zig-zag - [sold] weave,ving weld(ing)
— f̄ — zurco m̲ - [sold] groove weld(ing)
— f̄ enfriada - [sold] cooled, o chilled, weld
— f̄ entre sí - [sold] welding together
— f̄ esquinada - [sold] corner weld(ing)
— f̄ — de chapa(s) f metálica(s) - [sold] sheet
  metal corner weld(ing)
— f estanca(dora) - [sold] véase soldadura f
  para obturación f
— f excesiva - [sold] overweld(ing); exces-
  s(ive) weld(ing)
— f excesivamente penetrante - [sold] excessively penetrating weld(ing)
— f exigente - [sold] demanding, o fussy, weld
— f̄ extendida - [sold] long, weld, o seam
— f̄ exterior - [sold] external, o outside diameter, o girth, weld(ing)
— f fácil - [sold] easy weld(ing); easy opera-

ting,tion (welding)
soldadura f Fast-Freeze para solidificación f
  rápida (con alimentación f automática de e-
  lectrodo m̲ - [sold] Fast-Freeze Squirt
  weld(ing)
— f — — con solidificación f rápida (en cualquier posición f [con alimentación f automá-
  ricaz de electrodo m̲) - [sold] Fast-Freeze
  all position Squirt weld(ing)
— f, exitosa, o feliz - [sold] successful weld
— f̄ firme - [sold] secure weld(ing)
— f̄ fluida - [sold] fluid weld
— f̄ fresable - [sold] machinable weld
— f̄ fría - [sold] cold weld(ing)
— f̄ fuera de posición f - [sold] out-of-position, weld(ing), o work
— f — — — f con calidad f para inspección f -
  [sold] out-of-position inspection quality
  weld(ing)
— f — — — f — — f — — f con rayos-X m̲ -
  [sold] out-of-position X-ray quality weld(ing)
— f̄ fuerte - [sold] strong weld(ing) • brazing
— f̄ — con gas m̲ - [sold] gas brazing
— f̄ — — llama f - [sold] (flame) brazing
— f̄ — — — f de gas m̲ - [sold] gas (flame)
  brazing
— f — — temple m̲ - [sold] (tempered) brazing
— f̄ — por inducción f - [sold] (induction)
  brazing
— f gemela - [sold] twin (arc) weld(ing)
— f̄ — por arco m̲ sumergido - [sold] twin submerged arc weld(ing)
— f general - [sold] general weld(ing)
— f̄ — para fabricación f - [sold] general fabrication weld(ing)
— f — para montaje m̲ - [sold] general assembly
  weld(ing)
— f — — producción f - [sold] general fabrication weld(ing)
— f — — reparación f - [sold] general repair
  weld(ing)
— f gobernada f - [sold] governed weld
— f̄ — por código m̲ - [sold] code, covered, o
  established, weld
— f grande - [sold] large weld
— f̄ — con una (sóla) pasada f - [sold] one, o
  single, pass large weld; large, one, o single, pass weld(ment)
— f — en ángulo m̲ interior - [sold] large fillet weld(ment)
— f — levemente cóncava - [sold] large slightly concave weld
— f — — convexa f - [sold] large slightly
  convex weld
— f gruesa - [sold] thick, o heavy, weld
— f̄ — corta - [sold] short heavy weld
— f̄ hacia ángulo m̲—[sold] toward @ corner weld
— f̄ — conexión f con tierra a - [sold] toward
  @ ground(ing) weld
— f hecha automáticamente - [sold] automatically, o machine made, weld
— f hervida - [sold] boiled solder
— f̄ hidrógeno-atómica - [sold] véase soldadura
  f por hidrógeno m̲ atómico
— f̄ horizontal - [sold] horizontal weld(ing)
— f̄ — a tope m̲ - [sold] horizontal butt weld
— f̄ — — — m̲ en plancha(s) f vertical(es).-
  [sold] three o'clock position weld
— f — con solapo m̲ - [sold] horizontal lap
  (weld)
— f — — — m̲ con pasada f única - [sold] single pass horizontal lap weld
— f — en ángulo m̲ (interior) - [sold] horizontal fillet (weld)
— f — — — m̲ (—) con pasada f única f—[sold]
  single pass horizontal fillet (weld)
— f — — — m̲ — — f múltiple(s) - [sold]
  multiple pass(es) horizontal fillet (weld)
— f — — interior m̲ de ángulo m̲ - [sold] horizontal fillet (weld)
— f — ortogonal - [sold] horizontal fillet
  (weld)
— f inclinada - [sold] inclined, o slanting,

weld

soldadura f inclinada ascendente - [sold] uphill weld(ing)
— f — descendente - [sold] downhill weld(ing)
— f industrial - [sold] industrial welding • welding fabricating,tion
— f inferior - [sold] inferior weld(ing) • bottom weld(ing)
— f iniciada - [sold] started welding • started weld(ing)
— f ininterrumpida - [sold] uninterrupted, o sustained, weld(ing)
— f   pareja - [sold] uninterrupted even weld--(ing) • steady weld(ing) output
— f inmovilizada - [sold] restrained weld(ing)
— f Innershield - [sold] Innershield weld(ing)
— f — con electrodo m autoprotegido con alma m fundente - [sold] Innershield self-shielded flux cored weld(ing)
— f — — — m con alma m fundente - [sold] Innershield weld(ing)
— f — — con alma m fundente con arco m abierto - [sold] Innershield open arc welding
— f — — — m — — m — con voltaje m constante - [sold] Innershield constant voltage welding
— f inoxidable - [sold] stainless weld(ing)
— f insegura - [sold] unsound weld(ing)
— f íntegra - [sold] integral, o entire, weld
— f integral - [sold] welding in
— f interior - [sold] internal, o inside, weld--(ing • inside diameter, o I D, weld(ing)
— f — circunferencial - [sold] inside, circumferential, o girth, weld(ing)
— f intermitente—[sold] intermittent weld(ing)
— f — alternada - [sold] staggered intermittent weld(ing)
— f — — en ángulo m (interior) - [sold] staggered intermittent fillet weld(ing)
— f — a tresbolillo - [sold] staggered intermittent weld(ing)
— f — — — m en ángulo m (interior) - [sold] staggered intermittent fillet weld(ing)
— f — correspondiente - [sold] intermittent corresponding, o corresponding intermittent, weld(ing)
— f — — en ángulo m (interior) - [sold] intermittent chain fillet weld(ing)
— f interrumpida - [sold] intermittent welding • stopped weld(ing)
— f irregular - [sold] irregular weld(ing) • roundabout weld(ing)
— f labrable - [sold] machinable weld(ing)
— f lenta - [sold] slow weld(ing)
— f levemente descendente - [sold] slightly downhill weld(ing)
— f libre de grieta(s) f - [sold] crack-free weld(ing)
— f — — poros(idad) - [sold] porosity free weld(ing) • non-porous weld
— f limpia - [sold] clean weld(ing)
— f limpiada - [sold] cleaned weld(ing)
— f Linc-Fill - [sold] Linc-Fill, o long stickout, weld(ing)
— f — — con electrodo m muy sobresaliente - [sold] Linc-Fill long stickout weld(ing)
— f lisa - [sold] smooth weld(ing)
— f liviana - [sold] light weld(ing)
— f — para producción f - [sold] light, fabricating,tion, o production, weld(ing)
— f longitudinal - [sold] longitudinal, o linear, weld(ing)
— f — íntegra - [sold] entire longitudinal weld(ing)
— f manual - [sold] manual, o hand, o stick, weld(ing)
— f — con alimentación f automática de, alambre m, o electrodo m - [sold] manual semiazutomatic, Squirt weld(ing)
— f — con amperaje m alto - [sold] high, amperage, o current, manual weld(ing)
— f — —, — m, o —, elevado - [sold] high, amperage, o current, manual weld(ing)

soldadura f manual corriente - [sold] conventional, manual, o hand, weld(ing)
— f maquinable - [sold] machinable weld(ing)
— f más común - [sold] most common weld(ing)
— f — fluida - [sold] more fluid weld(ing)
— f — grande - [sold] larger weld(ing) • larger bead
— f — lenta - [sold] slower weld(ing)
— f — rápida - [sold] faster weld(ing)
— f — resistente - [sold] stronger weld(ing)
— f — — a abrasión f - [sold] most, o more, abrasion resisting,tant weld
— f máxima - [sold] maximum weld(ing)
— f — en pasada f única - [sold] maximum single pass weld(ing)
— f — — — f única en ángulo m (interior) - [sold] maximum single pass fillet (weld)
— f mayor m - [sold] larger, weld, o bead • largest, weld, o bead
— f mecanizada - [sold] mechanized weld(ing)
— f — con alimentación f automática de, alambre, o electrodo m - [sold] mechanized semiautomatic Squirt weld(ing)
— f — con corriente f alterna - [sold] alternating current mechanized weld(ing)
— f — — — f — con voltaje m constante - [sold] alternating current constant voltage mechanized weld(ing)
— f — — — f continua - [sold] direct current mechanized weld(ing)
— f — — — f — con voltaje m constante - [sold] direct current constant voltage mechanized weld(ing)
— f mejor - [sold] best weld(ing) • better weld(ing)
— f mejorada - [sold] improved weld(ing)
— f menos fluida - [sold] less fluid weld(ing)
— f metálica - [sold] metal(lic) weld(ing)
— f — vertical - [sold] vertical metal(lic) weld(ing)
— f múltiple - [sold] multiple weld(ing)
— f — con punto(s) m - [sold] multiple spot weld(ing)
— f muy restringida - [sold] highly restrained weld(ing)
— f no continua - [sold] non-continuous, o skip (step), weld(ing)
— f no fresable - [sold] non-machinable weld
— f — labrable - [sold] non-machinable weld
— f — maquinable - [sold] non-machinable weld
— f — sólida - [sold] unsound weld(ing)
— f — trabajable - [sold] non-machinable weld
— f normal - [sold] normal weld(ing)
— f — con electrodo(s) m manual(es) - [sold] normal, manual, o hand, o stick, electrode(s) weld(ing)
— f nueva - [sold] new weld(ing) • reweld(ing)
— f ocasional - [sold] part time welding
— f — para producción f - [sold] part time production welding
— f — — reparación f - [sold] part time repair weld(ing)
— f ondulada - [sold] véase soldadura f con ondulación(es)
— f óptima - [sold] quality weld(ing)
— f — con electrodo(s) m manual(es) - [sold] stick electrode quality weld(ing)
— f orotogonal - [sold] fillet (joint) weld; fillet
— f oxiacetilénica - [sold] (oxy)acetyline weld(ing)
— f — automática - [sold] automatic oxiacetylene weld(ing)
— f para aplicación f general - [sold] general purpose weld(ing)
— f — — f — en obra f - [sold] general purpose field weld(ing)
— f — avance m rápido - [sold] fast follow weld(ing)
— f — cada día m - [sold] each, o every, day's weld(ing)
— f — cierre m - [sold] cover weld(ing) • seal weld(ing)

soldadura f para cierre m en posición f vertical ascendente - [sold] vertical up cover welding
— f — — m — f — descendente - [sold] vertical down cover weld(ing)
— f — — vertical - [sold] vertical cover weld(ing)
— f — — m — ascendente - [sold] vertical up cover weld(ing)
— f — — m — descendente - [sold] vertical down cover weld(ing)
— f — conservación f - [sold] maintenance weld(ing)
— f — — f de calidad f - [sold] quality maintenance weld(ing)
— f — — f óptima - [sold] high quality maintenance weld(ing)
— f — costura f - [sold] seam weld(ing)
— f — — f por punto(s) m - [sold] seam spot weld(ing)
— f — empalme m - splice weld(ing) • welded splice
— f — endurecimiento m - [sold] hardening weld(ing)
— f — — m de superficie f - [sold] hardsurfacing weld(ing)
— f — ensayo(s) m - [sold] test(ing) weld • practice, o mock-up, weld(ing)
— f — — m en producción f - [sold] production test weld(ing)
— f — erección f - [sold] erection welding
— f — — f en obra f - [sold] field erection weld(ing)
— f — — f — — f con electrodo(s) m (Innershield) con alma m fundente - [sold] Innershield erection weld(ing)
— f — fabricación f - [sold] fabricating,tion weld(ing)
— f — — f con pasada(s) f múltiple(s)—[sold] multiple pass fabricating,tion (welding)
— f — — f en taller m - [sold] shop fabricating,tion weld(ing)
— f — fijación f—[sold] holding, o fastening, weld
— f para fin(es) m general(es) - [sold] general purpose weld(ing)
— f — formar trozo(s) m con largo(r) de . . . - [sold] welding into . . . length(s)
— f — granja(s) f - [sold] farm(ing) weld(ing)
— f — mantenimiento m - [sold] maintenance weld(ing)
— f — montaje m - [sold] assembly weld(ing) • véase también soldadura f en obra f
— f — — m corta—[sold] short assembly welding
— f — — m en posición f sobrecabeza - [sold] overhead fabricating
— f — obturación f - [sold] seal(ing) weld
— f — producción f - production weld(ing) • fabrication (welding)
— f — — f con calidad f óptima - [sold] high quality production weld(ing)
— f — — m pasada f única - [sold] single pass production weld(ing)
— f — — f eficiente - [sold] efficient production weld(ing)
— f — — f en escala f, grande, o industrial - [sold] high production welding
— f — — f mediana - [sold] medium duty production (welding)
— f — — f en posición f cualquiera - [sold] all position fabrication weld(ing)
— f — — f — — f — para fin(es) m general(es) - [sold] general purpose all position fabrication weld(ing)
— f — — f industrial - [sold] production weld(ing)
— f — — f para fin(es) m general(es) - [sold] - all purpose fabrication weld(ing)
— f — — f granja f - [sold] farm fabrication weld(ing)
— f — — f uso(s) m general(es) - [sold] all, o general, purpose fabrication weld(ing)
— f — prueba(s) f - [sold] test(ing) weld(ing)
— f — raíz f - [sold] root weld(ing)

soldadura f para ranura(s) f - [sold] groove weld(ing)
— f — — f en J - [sold] J groove weld(ing)
— f — — f — U - [sold] U groove weld(ing)
— f — — f — V - [sold] V groove weld(ing)
— f — — f recta - véase soldadura f para ranura f con ángulo(s) m recto(s)
— f — recipiente(s) n - [sold] container weld(ing)
— f — relleno m - [sold] fill bead
— f — — m continuo - [sold] continuous filler weld(ing)
— f — — m rápido - [sold] Fast-Fill weld(ing)
— f — remiendo m - [sold] patch weld(ing)
— f — reparación f - [sold] repair weld(ing)
— f — — f en posición f cualquiera - [sold] all position repair weld(ing)
— f — — f — — f — para fin(es) m general(es) - [sold] general purpose all position repair weld(ing)
— f — — f para fin(es) m general(es) - [sold] general purpose repair weld(ing)
— f — — f — granja(s) f - [sold] farm repair weld(ing)
— f — repetición f - [sold] repetitive, o repeat, weld(ing)
— f — resistencia f - [sold] véase soldadura f resistente
— f — — f, alta, o plena—[sold] full strength weld
— f —, sellado m, o sellar - véase] soldadura f para obturación f
— f — sello m - [sold] seal(ing) bead
— f — toda posición f - [sold] all position weld(ing)
— f — uso m (muy) común - [sold] widely used weld
— f — velocidad f alta - [sold] high speed weld(ing)
— f — — f sostenida - [sold] sustained high speed weld(ing)
— f paralela - [sold] parallel, o side-by-side, weld(ing)
— f pareja - [sold] smooth, o even, weld(ing) • consistent weld(ing) • steady weld(ing)
— f penetrante - [sold] penetrating weld(ing)
— f — en ángulo m interior - [sold] penetration fillet (weld)
— f pequeña f - [sold] small weld
— f — para avance m rápido - [sold] small high speed weld(ing)
— f — — pasada f única - [sold] small single pass weld(ing)
— f — en ángulo m (interior) - [sold] small fillet weld(ing)
— f — levemente cóncava - [sold] small slightly concave weld(ing)
— f — — convexa - [sold] small slightly convex weld(ing)
— f perimétrica - [sold] girth weld(ing) • round about weld(ing)
— f pesada - [sold] heavy weld(ing)
— f — en escala f grande - [sold] heavy production weld(ing)
— f — — — f industrial - [sold] heavy production weld(ing)
— f — — ranura f - [sold] heavy groove weld
— f — — — f profunda - [sold] heavy deep groove weld
— f — para fabricación f - [sold] heavy production weld(ing)
— f — — producción f - [sold] heavy production weld(ing)
— f — — trabajo m en serie f - [sold] heavy production weld(ing)
— f plana - [sold] flat weld(ing); downhand position weld(ing)
— f — a tope m - [sold] flat, o downhand position, weld(ing)
— f — — m con chaflán m - [sold] flat, o downhand, groove(d) butt weld(ing)
— f — — m en ranura f profunda - [sold] flat, o downhand, deep groove butt weld(ing)

soldadura f plana con pasada(s) f múltiple(s) -
[sold] multiple pass, o multipass, downhand
weld(ing)
— f — — **solapo m** - [sold] flat lap weld(ing)
— f — — **con pasada f única** - [sold] single pass flat lap weld(ing)
— f — **en ángulo m** - [sold] flat fillet (weld)
— f — — — **m con pasada(s) f múltiple(s)** -
[sold] multiple pass(es) flat fillet (weld)
— f — — — f **interior** - [sold] flat fillet
(weld)
— f — — — m — **con pasada f única** - [sold]
single pass flat fillet (weld)
— f — — **interior m de ángulo m** - [sold] flat
fillet (weld)
— f — **en ranura f** - [sold] flat deep groove,
weld, o joint
— f — — — f **profunda** - [sold] flat deep
groove, weld, o joint
— f — **grande** - [sold] large flat weld
— f — — — **en ángulo m** - [sold] large flat fillet (weld)
— f — — — — m **interior** - [sold] large flat
fillet weld(ing)
— f — **horizontal** - [sold] horizontal flat weld
— f — **ortogonal** - [sold] flat fillet (weld)
— f — **sobre ángulo m exterior** - [sold] flat, o
level, position corner weld
— f — **y ascendente** - [sold] downhand and uphill weld(ing)
— f — — **descendente** - [sold] downhand and
downhill weld(ing)
— f **pobre** - [sold] poor weld(ing)
— f — **por arco m sumergido** - [sold] poor submerged arc weld(ing)
— f **poco profunda** - [sold] shallow weld(ing)
— f **por arco m** - [sold] arc, o electric, weld,
o welding
— f — — — m **abierto** - [sold] open arc weld(ing)
— f — — — m — **autoprotegido m** - [sold]
shielded open arc welding
— f — — m — **con alimentación f automática**
(de electrodo n) - [sold] open arc Squirt
weld(ing)
— f — — — m **automática** - [sold] automatic arc
weld(ing)
— f — — — m **autoprotegida** - [sold] self-shielded arc weld(ing)
— f — — m — **con alimentación f automática de
electrodo m** - [sold] semi-automatic self-
-shielded (arc) weld(ing)
— f — m — — — f — — — f **con alma m
fundente** - [sold] semi-automatic self-shielded flux-cored (arc) weld(ing)
— f — m — — — f — — — m — **núcleo m
fundente** - [sold] semi-automatic selfshielded
flux-cored (arc) weld(ing)
— f — — — m **con electrodo m autoprotegido y
(con) alma m fundente** - [sold] Innershield
self-shielded flux cored arc weld(ing)
— f — — — m — — m **con alma m fundente** - [sold]
Innershield arc welding
— m — — — — m — — — **núcleo m fundente** - [sold]
self-shielded flux-cored arc weld(ing)
— f — — — m **carbónico** - [sold] véase **soldadura
por arco m entre carbones m**
— f — — m **con aleación f con cobre m** - [sold]
copper alloy weld(ing)
— f — — — m — **latón m** - [sold] arc brazing
— f — — — m **con pistola f** - [sold] gun arc
weld(ing)
— f — — — m **de hierro m fundido** - [sold] cast
iron arc weld(ing)
— f — — m — — — **labrable** - [sold] machinable cast iron arc weld(ing)
— f — — — m **de plasma f** - [sold] plasma arc
weld(ing)
— f — — — m **eléctrico** - [sold] electric arc
weld(ing)
— f — — m — **con corriente f alterna** - [sold]
alternating current weld(ing)
— f — — m — — — f **continua** - [sold] direct
current weld(ing)

soldadura f por arco m en descubierto - [sold]
open arc weld(ing)
— f — —(s) **en tándem** - [sold] tandem arc
weld(ing)
— f — — m **entre carbón(es) m** - [sold] carbon
arc weld(ing)
— f — — — m — — m **bajo gas m** - [sold] (under)
gas carbon arc weld(ing)
— f — — — — — — m **protector m** [sold]
protective, o inert, gas carbon arc weld(ing)
— f — — m — **electrodo(s) m de carbono m** -
[sold] carbon arc weld(ing)
— f — — — m **entre metal(es) m** - [sold] metal
arc weld(ing)
— f — — — m — — m **bajo gas m protector** -
[sold] protective, o inert, gas metal arc
weld(ing)
— f — — — m **metálico** - [sold] metal(lic) arc
weld(ing)
— f — — — m — **con gas m** - [sold] gas metal(lic)
arc weld(ing)
— f — — — m **múltiple** - [sold] multiple arc
weld(ing)
— f — — m **protegido** - [sold] shielded arc
weld(ing)
— f — — m **sumergido** - [sold] submerged arc
weld(ing)
— f — — m — **con alimentación f automática
de electrodo m** - [sold] submerged arc Squirt
weld(ing)
— f — — m — — **corriente f alterna** - [sold]
alernating current submerged arc weld(ing)
— f — — — — f **continua** - [sold] direct
current submerged arc weld(ing)
— f — — m — **de acero m** - [sold] steel submerged arc weld(ing)
— f — — — — m **dulce** - [sold] mild
steel submerged arc weld(ing)
— f — — m — — **calidad f alta** - [sold] high
quality submerged arc weld(ing)
— f — — — m — **plancha(s) f** - [sold] plate(s)
submerged arc weld(ing)
— f — — — — f **pesada(s)** - [sold] heavy
plate submerged arc weld(ing)
— f — — — — f **plana(s)** - [sold] flat
plate(s) submerged arc weld(ing)
— f — — — m **de chapa(s) f** - [sold] sheet
metal(s) submerged arc weld(ing)
— f — — — m — — f **de acero m** - [sold] steel
sheet metal submerged arc weld(ing)
— f — — — m **totalmente automática** - [sold] fully
automatic arc weld(ing)
— f — **capilaridad f** - [sold] flow weld(ing)
— f — **chispa(s) f** - [sold] flash weld(ing)
— f — **electroescoria* f** - [sold] electroslag*
weld(ing)
— f — **fusión f** - [sold] fusion weld(ing)
— f — — — f **con gas m** - [sold] gas fusion
weld(ing)
— f — — — f **con termita f** - [sold] fusion thermit, o thermit fusion, weld(ing)
— f — — f **eléctrica** - [sold] electric fusion
weld(ing)
— f — **gas m** - [sold] gas weld(ing)
— f — **haz m electrónico** - [sold] electronic beam
weld(ing)
— f — **láser** - [sold] laser beam weld(ing)
— f — **hidrógeno m** - [sold] hydrogen weld(ing)
— f — — m **atómico** - [sold] atomic hydrogen
weld(ing)
— f — **inmersión f** - [sold] immersion weld(ing)
— f — — f **en fundente m no ferroso** - [sold]
flux weld(ing)
— f — **percusión f** - [sold] percussion weld(ing)
— f — **presión f** - [sold] pressure weld(ing)
— f — **proyección f** - [sold] projection weld(ing)
— f — **punto(s) m** - [sold] tack, o spot, weld-
-(ing); tacking; stitching; stitch weld(ing) •
intermittent weld(ing)
— f — — m **automática** - véase **soldadura f automática por punto(s) f**
— f — — m **con precalentamiento m** - [sold] tack
weld(ing) with preheat(ing)

soldadura f por punto(s) m para costura f traslapada - [sold] lap(ped) seam spot weld(ing)
— f — — n para (volver v a) terminar en continua - [sold] back-step weld(ing)
— f — — m por resistencia f - [sold] resistance spot weld(ing)
— f — — m por resistencia f de costura f -
[sold] seam resistance spot weld(ing)
— f — — — f — — f traslapada - [sold]
lap(ped) seam resistance spot weld(ing)
— f — — m proyectado(s) m - [sold] véase
soldadura f de resalto m por punto(s) m -
— f — — m sin precalentamiento m—[sold] tack
weld(ing) without preheat(ing)
— f — resistencia f - [sold] resistance
weld(ing)
— f — — f eléctrica - [sold] electric resistance weld(ing)
— f — retroceso m - [sold] back-step weld(ing)
— f — — m salteado - [sold] skip back-step
weld(ing)
— f — rueda f - [sold] (wheel) seam weld(ing)
— f — sumersión f - [sold] dip brazing
— f — — f — fundente m - [sold] dip brazing
— f — — f — — m no ferroso - [sold] dip
brazing
— f — transferencia f - [sold] transfer
weld(ing)
— f — — f de metal m - [sold] metal transfer
welding
— f — — f — — m mediante corte(s) m en circuito m - [sold] short circuiting metal
transfer
— f — — f — — m — — m — — m de gas m -
[sold] short circuiting metal transfer in gas
welding
— f primera - [sold] first weld(ing)
— f problemática - [sold] problem weld(ing)
— f — de acero m - [sold] problem(atic) steel
weld(ing)
— f producida - [sold] produced weld(ing)
— f progresiva - [sold] progressive weld(ing)
— f — por punto(s) m - [sold] progressive spot
weld(ing)
— f protegida - [sold] protected, o shielded,
weld(ing)
— f — contra corrosión f - [sold] corrosion
protected weld(ing)
— f provisional - [sold] temporary weld(ing) •
tack weld(ing); tacking
— f provisoria - [sold] tack weld(ing); tacking
— f rápida - [sold] quick, o fast, o speedy,
weld(ing)
— f — de plancha(s) f - [sold] high speed
plate(s) weld(ing)
— f — — — f liviana(s) - [sold] high speed,
o speedy, o quick, light plate(s) weld(ing)
— f reallanada - [sold] rerolled weld(ing)
— f realmente automática - [sold] truly automatic weld(ing)
— f recilindrada - [sold] rerolled weld(ing)
— f recta - [sold] straight weld(ing)
— f — exigente - [sold] demanding, o fussy,
weld(ing), o seam
— f rechazada - [sold] weld(ing) reject(ing)
— f reforzada - [sold] reinforced weld(ing)
— f regulada - [sold] controlled weld(ing)
— f — eléctricamente - [sold] electrically
controlled weld(ing)
— f reiniciada - [sold] restarted weld(ing)
— f relaminada - [sold] rerolled weld(ing)
— f relativamente delgada - [sold] relatively
thin weld(ment)
— f reparada - [sold] repaired weld(ing)
— f repetida - [sold] repeated, o repetitive,
weld(ing)
— f — corta - [sold] short, repeated, o repetitive, weld(ing), o work
— f — económica - [sold] low cost repetitive
weld(ing)
— f requerida - [sold] required weld(ing)
— f residencial - [sold] residential, o home,
weld(ing)

soldadura f residencial para producción f -
[sold] residential, o home fabrication,
weld(ng)
— f — reparación f - [sold] residential, o
home, repair weld(ing)
— f resistente - [sold] strong, o full
strength, weld(ing)
— f — contra abrasión f - [sold] abrasion resistant weld(ing)
— f — agrietamiento m - [sold] crack resisting,tant weld(ing)
— f — tensión f - [sold] strong, o tensile
resisting,tant, weld(ing)
— f restringida - [sold] restrained weld(ing)
— f revirada - [sold] lap(ped) weld(ing)
— f rota - [sold] broken weld(ing)
— f saliente - [sold] protruding weld
— f — en interior - [sold] inside weld protrusion
— f salteada - [sold] skip (step), o intermittent, weld(ing)
— f sana - [sold] sound weld(ing)
— f según código m para recipiente(s) f para
presión f - [sold] pressure vessel code work
— f segunda - [sold] second weld(ing)
— f semiautomática - [sold] semi-automatic
weld(ing)
— f — con pistola f - [sold] semi-automatic
gun weld(ing)
— f — manual - [sold] semi-automatic, o
Squirt, weld(ing)
— f — con arco m sumergido - [sold] (semi-
-automatic) Squirt weld(ing)
— f — mecanizada - [sold] mechanized semi-automatic weld(ing)
— f — Squirt - [sold] (semi-automatic)
Squirt weld(ing)
— f simple - [sold] simple weld
— f sin aportación f (de material m) - [sold]
weld(ing) without @ material inflow
— f — chaflán m - [sold] square edge weld(ing)
— f — defecto(s) m - [sold] flawless weld(ing)
— f — espesor m suficiente - [sold] shallow
weld(ing)
— f — fusión f - [sold] brazing
— f — guiar - [sold] unguided weld(ing)
— f — porosidad f - [sold] nonporous, o
pore,rosity free, weld(ing)
— f — presión f - [sold] nonpressure, o pressure free, weld(ing)
— f — problema(s) m - [sold] trouble free
weld(ing)
— f sobre - [sold] weld(ing) on (to)
— f — acero m - [sold] steel, weld(ing), o
weldment
— f — ángulo m - [sold] corner weld(ing)
— f — — m de chapa(s) f - [sold] sheet corner
weld(ing)
— f — — m — — f, de metal m, o metálica -
[sold] sheet metal corner weld(ing)
— f — banco m - [sold] bench weld(ing)
— f — modelo m - [sold] make-up weld(ing)
— f — plancha(s) f - [sold] plate weld(ing)
— f — — f inclinada(s) - [sold] weld(ing) on
@ inclined plate(s)
— f sobrecabeza - [sold] overhead weld(ing)
— f — con pasada(s) f múltiple(s) - [sold]
multiple pass(es), o multipass, overhead
weld(ing)
— f — de plancha(s) f - [sold] overhead plate
weld(ing)
— f — — — f con pasada(s) f múltiple(s) -
[sold] multiple pass(es), o multipass, overhead weld(ing)
— f — — — f única - [sold] single pass overhead weld(ing)
— f — difícil - [sold] tough, o difficult,
overhead weld(ing)
— f — en ángulo m - [sold] overhead fillet
weld(ing)
— f — — — m interior - [sold] overhead fillet (weld)
— f — — — m exterior - [sold] overhead cor-

ner weld(ing)

soldadura f sobrecabeza en ángulo m interior -
[sold] overhead fillet weld
— f sobrecabeza horizontal - [sold] horizontal
overhead weld
— f solapada - [sold] welded lap • lap weld
— f sometida a tratamiento m - [sold] treated
weld
— f sostenida - [sold] sustained weld
— f Squirt - [sold] Squirt weld(ing)
— f — con alimentación f automática de electrodo - [sold] Squirt weld(ing)
— f — — — f — — m para relleno m rápido - [sold] Fill-Freeze Squirt weld(ing)
— f — en posición f cualquiera - [sold] all
position Squirt weld(ing)
— f suave - [sold] smooth, o soft, performance
— f superior - [sold] upper, o top, weld(ing);
superior, o top notch, weld(ing)
— f tercera - [sold] third weld
— f, terminación f, e inspección f - [sold]
welding, finishing, and inspecting
— f terminada - [sold] finished, weld, o work
— f típica - [sold] typical weld(ing)
— f totalmente automática - [sold] fully automatic weld(ing)
— f — — con arcos m en tándem - [sold] tandem arc fully automatic weld(ing)
— f — — — — m gemelo(s) - [sold] twin arc
fully automatic weld(ing)
— f — — — — m único - [sold] single arc
fully automatic weld(ing)
— f — — por arco m sumergido - [sold] submerged arc full(y) automatic weld(ing)
— f inmovilizada - [sold] fully, o highly,
restrained weld(ing)
— f — sólida - [sold] entirely sound weld
— f trabajable - [sold] machinable weld(ing)
— f transversal - [sold] transverse, o crosswise, weld(ing)
— f traslapada - [sold] overlapping weld(ing)-
— f Twin-Arc con arco(s) m gemelo(s) m—[sold]
Twin-Arc weld(ing)
— f — — — — m — con electrodo(s) m gruesos - [sold] large wire Twin-Arc weld(ing)
— f — — — m — — m delgado(s) -
[sold] small wire Twin-Arc welding
— f última - [sold] last weld(ing)
— f uniforme - [sold] uniform weld(ing)
— f — para avance m rápido - [sold] uniform
high speed weld(ing)
— f vertical - [sold] vertical weld(ing)
— f — a tope m - [sold] vertical butt weld
— f — — — — m con chaflán m - [sold] vertical
straight end butt weld
— f — — m — — m en V - [sold] vertical
V-butt weld(ing)
— f — — — — m — — — m — — para pasada f
primera - [sold] first pass vertical V-butt
weld(ing)
— f — — — — m — ranura con una sóla pasada f
- [sold] one pass (only) vertical V-butt
weld(ing)
— f — — — m — m en V - [sold] vertical V-butt
weld(ing)
— f — — — m — — para pasada f primera -
[sold] first pass vertical V-butt weld(ing)
— f — ascendente - [sold] vertical-up weld-
-(ing)
— f — — a tope m - [sold] vertical-up butt
weld(ing)
— f — — con . . . pasada(s) f - [sold] . . .
pass vertical-up weld(ing)
— f — — de chapa(s) f metálica(s) - [sold]
vertical-up sheet metal weld(ing)
— f — — — plancha(s) f - [sold] vertical-up
plate weld(ing)
— f — — — — ángulo m (interior) - [sold] vertical-up fillet weld(ing)
— f — — en tubería(s) f - [sold] vertical-up
pipe weld(ing)
— f — — triangular - [sold] vertical-up triangular weave

soldadura f vertical con pasada(s) f múlti-
ple(s) - [sold] multiple pass, o multipass,
vertical weld(ing)
— f — — — f única - [sold] single pass vertical weld(ing)
— f — de chapa(s) f metálica(s) - [sold] vertical sheet metal weld(ing)
— f — — plancha(s) f - [sold] vertical plate,
o plate vertical, weld(ing)
— f — — — f con pasada(s) f múltiple(s) -
[sold] multiple pass(es), o multipass, vertical plate weld(ing)
— f — tubería(s) f - [sold] vertical pipe
weld(ing)
— f — descendente - [sold] vertical down
weld(ing); welding vertical down
— f — — con . . . pasada(s) f - [sold] . . .-
-pass vertical down weld(ing)
— f — — de chapa(s) f metálica(s) - [sold]
vertical-down sheet metal weld(ing)
— f — — — plancha(s) f - [sold] vertical
plate(s) down weld(ing)
— f — — en ángulo m interior - [sold] vertical down fillet (weld)
— f — — — — m — — f única - [sold]
single pass vertical down fillet (weld)
— f — — — — m — — f múltiple(s) - [sold]
multiple pass, o multipass, vertical down
fillet (weld)
— f — — tubería(s) f - [sold] vertical
down pipe weld(ing)
— f — triangular - [sold] vertical down
triangular weld(ing)
— f — en ángulo m en pasada f primera - [sold]
first pass vertical fillet weld(ing)
— f — en ángulo m exterior - [sold] vertical
corner weld(ing)
— f — — — m interior - [sold] vertical fillet weld(ing)
— f — — — — m — en pasada f primera - [sold]
first pass vertical fillet weld(ing)
— f — — interior m de ángulo m —[sold] vertical fillet weld(ing)
— f — — — — m — — m en pasada f primera -
[sold] first pass vertical fillet weld(ing)
— f — — plancha(s) f con pasada f única -
[sold] single pass vertical plate weld(ing)
— f — ortogonal - [sold] vertical fillet weld
— f — para cierre m - [sold] vertical cover
weld(ing)
— f visible - [sold] visible weld(ing)
soldar v - [sold] . . .; to fuse; to weld together • to secure • to fabricate
— v a ángulo m - [sold] to weld to @ angle
— v — bastidor m - [sold] to weld to @ frame
— v — bomba f - [sold] to weld to @ pump
— v — caja f - [sold] to weld to @, case, o
box, o body
— v — defensa f - [sold] to weld to @ guard
— v — hierro m - [sold] to weld to @ iron
— v — — — m angular - [sold] to weld to @ angle
(iron)
— v — larguero m—[sold] to weld to @ stringer
— v — motor m - [comb.int] to weld to @ engine
• [electr-mot] to weld to @ motor
— v — sesgo m - [sold] to scarf weld
— v — solapa,pe,po - [sold] to lap weld
— v — tope m - [sold] to butt weld • to weld
end-to-end
— v — — continuamente - [sold] to butt weld
continuously
— v acercándo(se) - [sold] to weld towards
— v — a ángulo m - [sold] to weld
toward(s) @ corner
— v — a conexión f con tierra f - [sold] to
weld toward(s) @ ground
— v alambre m - [sold] to, weld, o solder, @
wire
— v aleación f - [sold] to weld @ alloy
— v — f con aluminio m - [sold] to weld @ aluminum alloy
— v — f — cobre m - [sold] to weld a copper

alloy
soldar v alejándose - [sold] to weld away (from)
— v — de ángulo m - [sold] to weld away from @ corner
— v — — conexión f con tierra f - [sold] to weld away from @ ground
— v aluminio m - [sold] to weld @ aluminum
— v — n con antorcha f con dos electrodos m - [sold] to weld @ aluminum with @ arc torch
— v armadura f - [weld] to weld @ frame
— v aro m - [sold] to weld @ ring
— v barra f - [sold] to weld @ bar
— v bastidor m - [sold] to weld @ frame
— v — m para automóvil m - [autom-mec] to weld @, automobile, o car, frame
— v — m para camión m - [autom-mec] to weld @ truck frame
— v bien - [sold] to weld, well, o successfully
— v borne m - [mec] to weld @ stud
— v cadena f - [sold] to weld @ chain
— v cobre m - [sold] to, weld, o solder, copper
— v — m con antorcha f - [sold] to torch weld
— v — m — — f con dos electrodos m - [sold] to weld copper with @ arc torch
— v colector m - [electr] to weld @ collector
— v con aleación f - [sold] to braze
— v — — f con cobre m - [sold] to braze
— v — — f — plata f - [sold] to braze
— v — — f con plomo m - [sold] to solder
— v — amperaje m alto - [sold] to weld with @ high current
— v — m bajo - [sold] to weld with @ low current
— v — aportación f - [sold] to weld with @ inflow
— v — — f de material m - [sold] to material inflow weld
— v — bronce m - [sold] to braze
— v — con antorcha f - [sold] to, braze with @ torch, o torch braze
— v — — m — f con dos electrodos m - [sold] to braze with @ arc torch
— v — electrodo m - [sold] to electrode weld
— v — — m con corriente f negativa - [sold] to weld with @ negative current • to electrode negative weld
— v — — m — — f positiva - [sold] to weld with @ electrode positive; to electrode positive weld
— v — estaño m - [sold] to solder
— v — — m con antorcha f - [sold] to solder with @ torch
— v — — — f con dos electrodos m - [sold] to solder with @ arc torch
— v — exceso m - [sold] to overweld
— v — éxito m - [sold] to weld successfully
— v — gas m - [sold] to gas, weld, o braze
— v — — costura f - [sold] to seam gas weld
— v — — m con costura f con punto(s) m - [sold] to seam gas spot weld
— v — — m con punto(s) m - [sold] to gas spot weld
— v — inclinación v ascendente - [sold] to weld uphill
— v — — f descendente - [sold] to weld downhill
— v — latón m — [sold] to braze
— v — llama f - [sold] to flame weld
— v — — m de gas m - [sold] to gas flame braze
— v — pasada(s) f múltiple(s) - [sold] to, multiple pass, o multipass, weld
— v — plomo m - [sold] to solder
— v — procedimiento m - [to weld with @ procedure
— v — proceso m - [sold] to process weld
— v — m con arco m sumergido - [sold] to submerged arc process weld
— v — punto(s) m - [sold] to, stitch, o tack, weld
— v — régimen r pleno - [sold] to weld at @ full rate output
— v — soldadora f - [sold] to welder weld

soldar v con velocidad f alta - [sold] to, weld at @ high speed, o high speed weld
— v — f para avance m - [sold] to weld at @ high travel speed(s)
— v conductor m - [electr-instal] to, weld, o solder, @ wire
— v continua(da)mente - [sold] to weld, continuously, o steadily
— v contorno m - [sold] to weld around
— v cordón m - [sold] to weld @ bead
— f — m para obturación f - [sold] to weld @ seal bead
— v — m — — f en trabajo(s) m plano(s) - [sold] to weld @ seal bead on @ flat work
— v — m para sello m - [sold] to weld @ seal bead
— v — m — — m — trabajo(s) m plano(s) - [sold] to weld @ seal bead on @ flat work
— v correctamente - to weld, correctly, o O K
— v costura f - [sold] to weld @ seam
— v — f con punto(s) m - [sold] to spot weld @ seam; to seam spot weld
— v — f — m por resistencia f - [sold] to resistance spot weld @ seam
— v traslapada - [sold] to weld @ lap seam
— v chapa f - [sold] to weld @ sheet
— v — f a petaca f - [metal-prod] to weld @ (metal) sheet to @ cooling plate
— v — f para coraza f - [sold] to weld @ shell plate
— v — f — petaca f - [metal-prod] to weld @ (metal) sheet to @ cooling plate
— v chaveta f - [sold] to weld @ key
— v — f para cilindro m - [mec] to weld @ barrel key
— v de nuevo - [sold] to reweld
— f — derecha f - [sold] to weld from @ right
— v — — f a izquierda f - [sold] to weld from @ right to @ left
— f — izquierda f - [sold] to weld from @ left
— v — — f a derecha f - [sold] to weld from @ left to @ right
— v dentro - [sold] to weld, into, o within
— v — de plancha f - [sold] to weld into @ plate
— v depósito m - [sold] to weld @ tank
— v desde - [sold] to weld (away) from
— v — ángulo m - [sold] to weld from @ corner
— v — conexión f - to weld (away) from @ connection
— v — — f con tierra f - [sold] to weld (away) from @ ground(ing) (connection)
— v — y no hacia - [sold] to weld away from and not toward(s)
— v — — — f conexión f con tierra f - to weld away from (and not towards) @ ground (connection)
— v económicamente - [sold] to weld economically
— v eléctricamente - [sold] to weld electrically
— v electrodo m - [sold] to weld @ electrode
— v electrónicamente - [electrón] to weld electronically
— v en ángulo m (interior) - [sold] to fillet
— v — argón m - [sold] to argon weld
— v — — m gaseoso - [sold] to argon gas weld
— v — atmósfera f - [sold] to weld in @ open
— v — — f regulada - [sold] to weld in @ controlled atmosphere; to controlled atmoshere weld
— v — — f — con argón m gaseoso - [sold] to argon controlled atmosphere weld
— v — — f — gas m de argón m - [sold] to argon gas controlled atmosphere weld
— v — dirección f inclinada - [sold] to weld on @ slant
— f — — f inclinada ascendente - [sold] to weld up hill
— v — — f — descendente - [sold] to weld down hill
— v — — f vertical descendente - [sold] to weld vertical down(wards)
— v — exterior - [sold] to weld outside
— v — forma f plana - [sold] to weld flat

soldar v en forma f plana con punto(s) m   — 1496 —

soldar v en forma f plana con punto(s) m -
[sold] to tack flat
— v — f — sin combadura f - [sold] to tack
flat without camber(ing)
— f — f — — — f previa - [sold] to tack
flat without precamber(ing)
— v — frío - [sold] to weld cold; to cold weld
— v — gas m - [sold] to gas weld
— v — m de argón m - [sold] to argon gas
weld
— v — horno m - [sold] to furnace weld
— v — interior m - [sold] to weld inside
— v — obra f - [sold] to, field, o job site,
weld; to weld during @ erection
— v — tubería t - [sold] to field weld @
pipe,ping
— v — recinto n - [sold] to weld in @ enclosure
— v — m reducido - [sold] to weld in @ confined, area, o space, o place
— v — sitio m - [sold] to weld in @ place
— v — en obra f - [sold] to weld in @
place during @ erection
— v — taller m - [sold] to shop weld
— v — torno - [sold] to weld around
— v entre sí - [sold] to weld together
— v espárrago m - [sold] to weld @ stud
— v — m roscado - [sold] to weld @ stud
— v — m en (su) sitio m - [sold] to weld @
stud in (a) place
— v firmemente - [sold] to weld, securely, o
solidly
— v fuera de posición f - [sold] to weld out of
position
— v fuga f - [sold] to stop @ leak by welding
— f fundación f - [sold] to weld @ cast iron
— v — f gris - [sold] to weld @ grey cast iron
— v grieta f - [sold] to weld @ crack
— v hierro m - [sold] to weld @ iron
— v — m fundido - [sold] to weld @ cast iron
— v interiormente - [sold] to inside weld
— v — tobera f - [sold] tuyere inner fusing
— v junta f - [sold] to weld @ joint
— v manualmente - [sold] to weld manually
— v muy de prisa - [sold] to weld, hurriedly, o
too fast
— v nuevamente - [sold] to, reweld, o weld
again
— v para formar trozo(s) m con largo(r) m de -
[sold] to weld into ... length(s)
— v pieza f - [sold] to, weld, o solder, @ part
— v f de cobre n - [sold] to, weld, o solder, @
copper part
— v — f en taller m - [sold] to shop weld @
part
— v — f estañada - [sold] to, weld, o solder,
@ tinned part
— v — f galvanizada - [sold] to solder @ galvanized part
— v plancha f - [sold] to weld @ plate
— v — f con punto(s) m - [sold] to tack @
plate(s)
— v — f — — m de soldadura f - [sold] to
tack @ plate
— v — f — — m (— — f) en ambos extremos m
- to tack @ plate at both ends
— v — f — — m sin combadura f previa—[sold]
to tack @ plate without precamber(ing)
— v — f delgada - [sold] to weld @ thin plate
— v — f en forma f plana - [sold] to weld @
flat plate
— v — f — — f sin combadura f (previa) -
[sold] to tack @ plate without (pre)cambering
— v plancha f - [sold] to weld @ plate
— v — f gruesa - [sold] to weld @ heavy plate
— v — f metálica - [sold] to weld @ metal
plate
— v por arco m - [sold] to (arc) weld
— v — — m sumergido - [sold] to submerged arc
weld
— v — punto(s) m—[sold] to, tack, o spot weld
— v — — m costura f - [sold] to tack weld @
seam

soldar v por punto(s) m costura f traslapada -
[sold] to spot weld @ seam
— f — — m por resistencia f - [sold] to resistance weld
— v — m — — f costura f - [sold] to resistance spot weld @ seam
— v — m — — f f traslapada - [sold] to
resistance spot weld @ lap seam
— v provisoriamente - [sold] to weld temporarily • to tack (weld)
— v recipiente m - [sold] to, weld, o solder, @
container
— v en - [sold] to weld, on, o in
— v — ancho(r) m - [sold] to weld across
— v tambor m - [sold] to weld @ drum
— v terminal - [electrón] to, weld, o solder, @
lug
— v todo contorno m - [sold] to weld all around
— v tope m - [sold] to weld @ stop
— v tubería f - [sold] to weld @ pipe,ping
— v tuerca f - [sold] to weld @ nut
— v tubo - to, weld, o solder, @ pipe
— v — m de cobre m - [sold] to solder @ copper
pipe
— v verticalmente - [sold] to weld vertically
— v — en sentido m ascendente - [sold] to weld
vertically up
— v — — — m descendente - [sold] to weld
vertically down
soldarse v - [sold] to weld itself; to fuse
soldeo* m - welding; véase soldadura f
solenoide m desarmado - [electr-equip] taken
apart solenoid
— m eléctrico - [electr-equip] electrical solenoid
— m — para línea f para combustible m - [comb.-
-int] fuel line electrical, o electrical fuel
line, solenoid
— m limpiado - [electrón] cleaned solenoid
— m limpio - [electrón] clean solenoid
— m para adaptador m - [electrón] adapter solenoid
— m — — m para velocímetro m - [autom-mec]
speedometer adapter solenoid
— m — arrancador m - [comb.int] (self)starter
solenoid
— m — corriente f alterna—[electr-equip] alternating current solenoid
— m — — f continua - [electr-equip] direct
current solenoid
— m — corte m de combustible m - [comb.int]
fuel shut-down solenoid
— m — — m — — m para bomba f - [comb.int]
pump fuel shut-down solenoid
— m — depósito m - [comb.int] tank solenoid
— m — — m para combustible m - [comb.int]
fuel tank solenoid
— m — detención f - [comb.int] shut-down solenoid
— m — línea f - [comb.int] line solenoid
— m — — f para combustible m - [comb.int]
fuel line solenoid
— m — llama f - [combust] flame solenoid
— m — — f piloto - [combust] pilot flame solenoid
— m — regulación f - [mec] control(ling) solenoid
— m — — f para altura f para cabezal m -
[agric-equip] header height control solenoid
— m — — — f espigadora f - [agric-
equip] header height control solenoid
— m — regulador m - [mec] control(ling) solenoid
— m — — m para marcha f sin carga f - [comb.-
-int] idler solenoid
— m — válvula f - [ind] valve solenoid
— m — — f igualadora - [metal-prod] equalizer
valve solenoid
— m quitado - [electr-instal] removed solenoid
— m reemplazado - [electrón] replaced solenoid
— m removido - [electr-instal] removed solenoid
— m sacado - [electr-instal] removed solenoid
soler v llamar - to (be) know(n) as

solera f - . . . • sill; floor; hearth; bottom;
foreplate • [mec] bolster • sole • [constr]
wall plate
— f básica - [metal-prod] basic bottom
— f de fondo m - [hidr] bed(ding)
— f — fosa f - [metal-lam] pit bottom
— f — — f plana - [metal-lam] flat pit bottom
— f — razonablemente plana - [metal-lam] reasonably flat pit bottom
— f — horno m - [combust] furnace, hearth, o
bottom
— f — — m de fosa f - [metal-lam] (soaking)
pit bottom
— f — — m — — f razonablemente plana -
[metal-lam] reasonably flat soaking pit bottom
— f — — m razonablemente plana - [metal-lam]
reasonably flat furnace bottom
— f — puerta f - [metal-prod] foreplate •
[constr] door sill
— f — ventana f - [arquit] window sill
— f inclinada - [combust] slanting hearth
— f en frente m - [petról] nose sill
— f inclinada - [metal-prod] sloping hearth
— f movible - [metal-prod] movable hearth
— f — de horno m - [metal-prod] furnace
movable hearth
— f móvil - [combust] movable hearth
— f para apoyo m - [petról] mud sill
— f — — m para torre f para perforación f -
[petról] mud sill
— f — larguero m - [petról] reel tail sill
— f — motor m - [petról] engine (end) sill
— f — piso m - [grúas] floor sill
— f — — m para torre f - [petról] derrick
floor sill
— f — — m — — f para perforación f -
[petról] derrick floor sill
— f plana - [combust] flat hearth
— f — para horno m de fosa f - [metal-lam]
flat soaking pit bottom
— f sobre tierra f - [mec] mud sill
— f — — f para motor m - [petról] engine mud
sill
solerse v - most people
solicitación f - . . . • [metal] stress(ing);
strain(ing)
— f absorbida - [metal] absorbed stress
— f crítica - [constr] critical stress
— f de carga f - [mec] load stress
— f — corte m - [mec] shear stress
— f de relleno m - [constr] backfill, stress, o
force
— f — suelo m—[constr] soil, strain, o stress
— f — tubería f - [constr] pipe, strain, o
stress
— f distribuida - [constr] distributed, strain,
o force, o stress
— f — uniformemente - [constr] uniformly distributed, force(s), o stress(es)
— f moderada - [mec] moderate stress
— f por tracción f - [mec] traction stress
— f vibración f - [mec] vibrational stress
solicitada f - [legal] paid advertisement
solicitado,da a - solicited; requested • [mec]
under stress • [metal] stressed
— a a compresión f - in compression stress
— a a tracción f - in tension
— a altamente - [mec] highly stressed
— a por esfuerzo m - [mec] in, o under, stress
— a — — m compresivo - [mec] in compression
stress
— a — — m neutral - [mec] in neutral stress
— a tracción f - [mec] in, o under, stress
solicitar v - . . . • to, request, o requisition
• [mec] to stress
— v crédito m - [fin] to request @ credit
— v información f - to request @ information
— v repuesto m - [ind] to, request, o requisition, @ part
— v — f para repuesto m - [ind] to, request, o
requisition, @ spare part
— v propuesta(s) v - [legal] to, request, o

open for, o call for, bid(s)
solicitar v prueba(s) f - to request @ test(s)
— v repuesto m - [ind] to request @ spare part
— v tolerancia f - to request @ tolerance
solicitud f - . . . • [pers] application • [com]
requisition
— f de clasificación f - [pol] classification
request
— f — — f industrial - [pol] industrial classification request
— f concesión f - concession request • [min]
claim
— f — crédito m - [fin] credit request
— f — — m en cuenta f abierta - [fin] open account credit request
— f — equipo m - [ind] equipment requisition
— f — información f - information request(ing)
— f — garantía f - [com] guaranty request
— f — — f para reembolso m - [fin] reimbursement guaranty request
— f — modificación f - [legal] amendment request
— f — patente f - [legal] patent application
— f — — f de invención f - [legal] patent application
— f — propuesta f - [com] inquiry
— f — prueba f - test request
— f — reembolso m - [fin] reimbursement request
— f — registro m de patente f - [legal] patent,
application, o request
— f — repuesto(s) m - [ind] spare(s) part(s),
requisition, o request
— f — — m para equipo m - [ind] equipment
(spare) part(s) requisition
— f para crédito m - [fin] credit application
— f — — m especial - [fin] special credit application
— f — — m — para comprador m - [fin] buyers'
special credit application
— f — — m para comprador m - [fin] buyers
credit application
— f — información f - information request
— m — límite m crediticio - [fin] credit limit
application
— f — — m — especial - [fin] special credit
limit(ation) application
— f — — m — — para comprador m - [fin]
special buyer credit limit application
— f — — m — para comprador m - [fin] buyer
credit limit application
— f — patente f - [legal] patent application
— f — — f de invención f - [legal] patent application
— f — pieza f - [ind] part request
— f — — f para repuesto m - [mec] spare part
request
— f — repuesto m - [ind] spare request
— f — trabajo m - [ind] work, request, o requisition
— f recibida - received application
solidaridad f - . . . • team, work, o spirit;
esprit de corps
solidario,ria a - . . . • [mec] built in; integral
solidez f - . . .; ruggedness • [metal] soundness
— f brindada - provided, o afforded, strength
— f comercial - [com] commercial, o business,
strength, o soundness
— f de material(es) m - [ind] material(s)
soundness
— f — soldadura f - [sold] weld(ing) soundness
— f estructural - structural soundness
— f interna - internal soundness
— f maximizada - maximized ruggedness
— f provista - provided, o afforded, strength
solidificación f - . . . • [sold] freeze,zing
— f completa - [metal-prod] complete, o total,
solidification
— f — cemento m - [constr] cement solidification
— f — ión(es) m - [nucl] ion solidification
— f de soldadura f - [sold] weld freezing

solidificación f **de soldadura** f **a tope** m -
[sold] butt weld freezing
— f —— f **en ángulo** m **(interior)** - [sold]
fillet weld freezing
— f —— f **en ranura** f - [sold] groove weld
freezing
— f **eventual** - eventual solidification
— f **lenta** - [sold] flow freezing
— f **más lenta** - [sold] slower freezing
— f — **rápida** - [sold] faster freezing
— f **rápida** f - [sold] fast, o quick, freezing;
fast freeze
solidificado,da a - solidified • [sold] frozen
solidificar v - . . . • [sold] to freeze
— v **cráter** m - [sold] to, solidify, o freeze,
@ puddle
— v **ion(es)** m - [nucl] to solidify @ ion(s)
— v **metal** m - [sold] to, solidify, o freeze, @
metal
— v — m **fundido** - [sold] to, solidify, o
freeze, @, puddle, o molten metal
— v **rápidamente** - [sold] to freeze rapidly
sólido m **bombeado** - [bombas] pumped, o handled,
solid(s)
— m **disuelto** - [hidr] dissolved solid(s)
— m **en agua** m **depurada** - [hidr[treated water
solid(s)
— m **fluido** - fluid solid(s)
sólido,da a - . . . • rugged • sound • tight •
substantial • story • strong
— a **desde punto** m **de vista estructural** •
structurally sound
sólido m **bombeados** - [bombas] pumped, o handled, solids
— m **disueltos** - [hidr] dissolved solids
— m **en gas** m - [combust] gas solids
— m —— m **para horno** m - [combust] furnace
gas solids
— m —— — m **alto** - [combust] blast furnace gas solids
— m — **suspensión** f - [sanit] suspended solids
— m —— f **en gas** m - [combust] gas suspended
solids
— m —— f —— m **para horno** m - [combust]
furnace gas suspended solids
— m —— f —— — m **alto** - [metal-prod]
blast furnace gas suspended solids
— m **manipulados** - [ind] handled solids
— m **suspendidos** m - [hidr] suspended solids
— m —— **en gas** m - [combust] gas suspended
solids
— m —— — m **para horno** m - [combust] furnace
gas suspended solids
— m —— —— m —— m **alto** - [metal-prod]
blast furnace gas suspended solids
soliviado,da a - . . .; boosted; (up)lifted
soliviar v - . . .; to boost; to (up)lift
solivio m - . . .; boosting; (up)lifting
sólo,la adv - . . .; simple,ply • single
— adv **asegurador** - [seguros] sole insurer
soltado,da a - loose(ne)d; released; unlatched;
disengaged; slacked off; let out • broken
free • broken from
— a **automáticamente** - released automatically
— a **fácilmente** - released easily
— a **manualmente** - hand released; released manually
— a **suficiente(mente)** - [mec] loose(ne)d, far
enough, o sufficiently
soltador m - [mec] . . . • release
— m **con resorte** m - [mec] spring release
— m **para bloque** m - [mec] block release
— m —— m **sujetador** - [mec] grip box release
— m —— **caja** f - [mec] box, o case, release
— m —— — f **sujetadora** - [trefil] grip box release
— m —— **sujetador** m - [mec] grip release
soltar v - . . .; to break free; to slack off;
to bust loose; to disengage; to release; to
unlatch; to loosen; to let out • [mec] to
trip
— v **abrazadera** f - [mec] to, release, o open, @
clamp

soltar v **abrazadera** f **para brida** f - [mec] to
release @ clamp flange
— v — f — **pestaña** f - [mec] to release @
clamp flange
— v **alambre** m - [mec] to slack off @ wire
— v — m **para estrangulador** m - [comb.int] to
slack off @ choke wire
— v **automáticamente** - to release automatically
— v **botón** m - to release @ button
— v — m **para rerregulación** f - [mec] to release @ reset(ting) button
— v **conductor** m - [electr-instal] to release @
lead
— v — m **a electrodo** m - [sold] to, release, o
loosen, @ electrode lead
— v — m **a tierra** f - [sold] to, release, o
loosen, @ ground lead
— v **correa** f - [mec] to loosen @ belt
— v **embrague** m - [mec] to, release, o let out,
@ clutch
— v **fácilmente** - to release easily
— v **freno** m - [mec] to release @ brake
— v **gancho** m - [mec] to, release, o unhook, o
unlatch, @ hook
— v **manguera** f - [mec] to loosen @ hose
— v **gatillo** m - [mec] to release @ trigger
— v — m **para pistola** f - [sold] to release @
gun trigger
— v **manija** f - [mec] to release @ handle
— v **manualmente** - [mec] to release, manually, o
with @ hand; to hand release
— v **mordaza** f - [mec] to release @ clamp
— v **palanca** f - [mec] to release @ lever
— v **pedal** m - [mec] to release @ pedal
— v **perno** m - [mec] to loosen @ bolt
— v — m **para sujeción** f - [mec] to loosen @
hold down bolt
— v **presión** f - [mec] to release @ pressure
— v — f **de aire** m - [neumát] to release @ air
pressure
— v **suficiente(mente)** - [mec] to, release, o
loosen, sufficiently, o far enough
— v **sujetador** m - [mec] to, loosen, o release,
@, fastener, o clamp
— v **tensión** f - [mec] to release @ tension
— v — f **de resorte** m - [mec] to, release, o
loosen, @ spring tension
— v **válvula** f - [mec] to release @ valve
soltura f - loosening • [gram] fluency
— f **de tierra** f - [constr] earth, o ground,
loosening
solubilidad f **alta** - high solubility
— f — **para hidrógeno** m - [quím] high hydrogen
solubility
— f **de cobre** m - [miner] copper solubility
— f **para hidrógeno** m - [quím] hydrogen solubility
solución f - . . .; solving • handling • answer
• remedy • payoff • [ind] . . .; liquid • véase también **disolución** f
— f **a plazo** m **corto** - short term solution
— f — — m **largo** - long term solution
— f **ad hoc** - ad hoc solution • véase también
solución f **específica**
— f **adaptable** - versatile, o adaptable, solution, o answer
— f **alternativa** - alternate,tive solution
— f **buena** - good solution
— f **completa** - complete solution
— f **comprensiva** - comprehensive, o overall, solution
— f **considerada** - considered solution
— f **consistente** - solid solution
— f **de cierre** m - [mec] closure solution
— f — **ecuación** f - [mat] equation solving
— f — **enfriamiento** m - cooling solution
— f — **problema** m - problem, solving, o solution • [ind] trouble shooting
— f **diferente** - different solution
— f **distinta** - different solution
— f **para presentación** f - [sold] fit-up handling
— f —— f **pobre** - [sold] poor fit-up handling

solución f peor - worse solution - worst solution
— f positiva - positive, o solid, solution
— f práctica - practical solution
— f reparativa - [ind] repair(-type) solution
— f prevista - foreseen, o considered, solution
— f propuesta - proposed solution
— f provisoria - temporary, o provisional, solution
— f sensata - sensible, solution, o answer
— f sólida - solid solution
— f temporaria - temporary solution
— f única - single solution
solucionado,da a - solved • handled • right
— a a medias - half solved; half right
solucionador m - solver
— m para problema m - problem solver
solucionador,ra a - solving
solucionar v - . . .; to deal with; to handle
— v ecuación f - [mat] to solve @ equation
— v presentación f - [sold] to, handle, o
solve, @ fit-up
— v — f pobre - [sold] to, solve, o handle, @
poor fit-up
— v problema m - to solve @ problem
— v — m de erosión f - to solve @ erosion problem
solvencia f fiscal - [fisc] fiscal solvency
— f municipal - [fisc] municipal solvency
— f tributaria - [fisc] tax solvency
solvente a - . . . . • [fin] self-supporting
sombra f alargada - elongated, o linear, shadow
— f en penetrámetro m - penetrameter shadow
— f obscura - dark shadow
— f redondeada - rounded shadow
sombreado m - shadowing • [imprent] (cross-)
hatching
sombreado,da a - shaded • [dib] cross-hatched
sombrear v - . . . • [dib] to cross-hatch
sombrerete m - [mec] . . .; hood; cap section •
[comb.int] spark arrester
— m cónico - [mec] conical hood; cowl
— m contra polvo m - [ambient] dust hood
— m para cierre m - [tub] cap screw
sombrilla f para patio m - [comést] patio umbrella
someramente adv - . . .; casually
somero,ra a - . . .; quick
someter v - . . . . • to present • to subject •
to undergo
— v a acción f centrífuga - to centrifuge; to
spin
— v — aprobación f - to submit for @ approval
— v — ataque m - to submit to @ attack
— v — m con ácido(s) m - [metal] to etch
— v — — m según procedimiento(s) m
según norma f - [metal-trat] to etch by
standard procedure(s)
— v a carga f - [mec] to subject to @ load(s)
— v — — f hasta fatiga f - [sold] to subject
to @ fatigue load(ing)
— v — compresión f - to subject to compression
— v — consulta f - to subject to inquiry
— v — previa - to subject to (prior) inquiry
— v — esfuerzo m - [mec] to stress
— v — juicio m político - [pol] to impeach
— v — presión f - to subject to @ pressure
— v — tratamiento m - [ind] to subject to @
treatment
— v — — m soldadura f—[sold] to treat @ weld
— v — — m térmico - [sold] to, heat treat, o
subject to @ heat treatment
— v — información f - to submit @ information
— v informe m - to submit @ report
— v macromuestra f a ataque m con ácido(s) m -
to etch @ macrospecimen
— v para aprobación f—to submit for @ approval
sometido,da a - submitted • subjected
— a acción f centrífuga - centrifuged
— a a juicio m político - [pol] impeached
— a — tratamiento m térmico - [metal-trat]
submitted to heat treatment; heat treated

sometido,da a tratamiento m térmico antes de
soldar(se) - [sold] heat treated before
welding
— a para aprobación f - submitted for approval
sometimiento m - . . .; submitting
— m a presión f - subjecting,tion to pressure;
pressurization
— m — tratamiento m - [metal-prod| submitting
to @ treatment
— m — — m térmico - [metal-trat] submitting
to @ heat treating,tment
— m de información f - information submitting
— m — soldadura f a tratamiento m térmico -
[sold] heat treating,tment
— m para aprobación f - submitting for approval
son n - . . . | adv - [fin] to wit
— m de bocina f - [electrón] horn, sound(ing),
o blow(ing)
— m ni ton m - rhyme (n)or reason
sonajear v - to rattle
sonar v alarma f - [segurid] to sound @ alarm
— v alerta f - [electrón] to sound @ alert
— v — f audible - [slectrón] to sound @ audible alert
— v bocina f - [electrón] to, sound, o blow, @
horn
— v — f en vehículo m - [transp] to sound @
vehicle horn
— v señal f - to sound @, signal, o alarm
— v tono m para alerta f - [electrón] to sound
@ alert tone
— v — — f audible - [electrón] to sound
@ audible alert tone
sonda f - . . . . • [metal-prod] test, o stock,
rod; stockline, indicator, o recorder; rod •
[electr] probe • [medic] . . .; probe •
[int.comb] dip stick • [petról] drill
— f para aceite m - [mec] (oil) dip stick
— f cilíndrica - [instrum] rod gage
— f con neutrón(es) m - [instrum] neutron
probe
— f detectora - [instrum] sensing probe
— f este - [metal-prod] east stock rod
— f insertada - [instrum] inserted probe
— f manual .- [metal-prod] manual stock rod
— f oeste.- [metal-prod] west stock rod
— f para horno m - [metal-prod] furnace, stock,
o test(ing), rod
— f — horno m alto - [metal-prod] blast furnace (stock) rod
— f para uso m en obra f - [instrum] field
probe
— f — — m — — f insertada - [instrum] inserted field probe
— f — — m — — removida - [instrum] removed
field probe
— f — — m — — f retirada - [instrum] withdrawn field probe
— f removida - [instrum] removed probe
— f retirada - [instrum] withdrawn probe
— f rota - [metal-prod] broken stock rod
— f ultrasónica - [electrón] ultrasonic probe
— f útil - [metal-prod] stock rod in operation
sondaje m - véase sondeo m
sondeado,da a - sounded • [petról] drilled
sondear v - . . . to drill
— v con munición(es) f - [miner] shot drilled
— v — peridogón(es) m - [miner] to shot drill
sondeo m - . . . . • probe,bing; test, o trial,
bore,ring; bore hole • [geol] boring test •
[petról] drilling
— m con munición(es) m - [miner] shot drilling
— m — perdigón(es) m - [miner] shot drilling
— m para prueba f - [petról] trial bore (hole)
— m — testigo(s) m - [miner] core drilling
sonido m - . . . . • [electrón] beep
— m de bocina f - [electrón] horn, sound(ing),
o blowing
— m — — f en vehículo m - [electrón] vehicle
horn sound(ing)
— m débil - weak sound
— m fuerte - loud sound
— m hueco - hollow sound

sonido m percibido - heard, o percieved, sound
— m sentido - heard sound
— m sibilante - hissing, o whistling, sound, o noise
— m, silbador, o silbante - hissing, o whistling, sound, o noise
— m zumbante - buzzing, sound, o noise
— m débil - weak buzzing, sound, o noise
— m fuerte - loud buzzing, sound, o noise
sonreido,da a - smiled
sonriente a - smiling
sonsacar v - . . . .; to pull out; to elicit
sonso,sa* a - véase zonzo,za a • green
sopapa f - [válv] foot valve
soplado m - . . . ; blowing • [ind] (air) blast; wind blow(ing) • [tub] blowdown • [autom-neumát] blow(ing) out • véase también soplido m; sopladura f
— m acortado - [ind] reduced (air) blast
— m alternativo - [ind] alternate blow(ing)
— m con aire m - [ind] blowing out with air
— m — — m con presión f alta - [neumát] blowing out with high pressure air
— m — — m — — f baja - [neumát] blowing out with low pressure air
— m — — m seco - [neumát] blowing out with dry air
— m — arena f - [mec] sand blast(ing)
— m — munición(es) f - [mec] shot blast(ing)
— m — nitrógeno m - [ind] nitrogen blow(ing)
— m — vapor m - [ind] steam blow(ing)
— m cortado - [ind[cut (off) blast
— m de condensador m - [aire-acond] condenser blowing (out)
— m — oxígeno m - [metal-prod] oxygen blast
— m — regenerador(es) m [metal-prod] checker blowing (out)
— m directo - [ind] direct, blowing, o blast
— m — para cobre m - [metal-prod] direct copper, o copper direct, blast, o blowing
— m en seco - [ind] dry blowing; blowing dry
— m — — con aire m comprimido - [ind] blowing dry with compressed air
— m enriquecido - [ind] enriched blast
— m — con oxígeno m - [ind] oxygen enriched blast
— m hasta secar(se) - [ind] blowing dry
— m — — con aire m - [ind] blowing dry with air
— m — — — — m comprimido - [ind] blowing dry with compressed air
— m inferior - [combust] bottom blast
— m normal - [ind] normal blast
— m periódico - [ind] periodical blowing (out)
— m reducido - [ind] reduced (air) blast
— m superior - [combust] top blast
— m total - [ind] total blast
soplado,da a - blown • [ind] blown orf • [autom-neumát] blown out
— a con aire m - blown out with @ air
— a — — m comprimido - [ind] blown (out) with compressed air • [hormig] blown clean
— a — — m con presión f alta - blown out with high pressure air
— a — — m — — f baja - [ind] blown out with low pressure air
— a en seco - blown dry
— a — — con aire m - [ind] blown dry with air
— a — — — — m comprimido - [ind] blown dry with compressed air
— a hasta secar(se) v - [ind] blown dry
— a — — v con aire m comprimido - [ind] blown dry with compressed air
— a periódicamente - [ind] blown (out) periodically
soplador m - [mec] . . .; booster • [metal-prod] bustle; tuyere • véase también soplante m
— m a distancia f - [combust] distance, o remote, blower
— m auxiliar - [ind] auxiliary blower
— m centrífugo—[ind] centrifugal (type) blower
— m con fuerza f motriz - [mec] power blower
— m mecánico—[ind] mechanical, o power, blower

soplador m para aire m - [ind] air, o wind, blower
— m — — m para combustión f - [combust] combustion air blower
— m — calentador m - [ind] heater blower
— m — circulación f - circulation blower
— m — combustión f - [combust] combustion blower
— m — — f puesto en marcha - [combust] started combustion blower
— m — desplazamiento m positivo - [coque] Root type blower
— m — emparrillado m - [ind] grill blower
— m — fragua f - [metal] forge blower
— m — recirculación f - [mec] recirculation blower
— m — horno m - [metal-prod] furnace blower
— m — viento m - [ind] wind blower
— m principal - [mec] main blower
— m regulado - [mec] set blower
sopladura f - [metal-prod] blow, o gas, hole; blowhole; blister • [sold] gas pocket
— f alternativa - [metal-prod] alternative blow(ing)
— f con aire m - [ind] air blow(ing)
— f — — m comprimido - [ind] compressed air blow(ing)
— f — arena f - [ind] sand blast(ing)
— f de aire m - [neumát] air blow(ing)
— f — filtro m - [ind] filter blowing
— f — piquera f - [metal-prod] tap hole blowing
— f por abajo - [combust] bottom blast
— f — arriba - [combust] top blast
soplante m - [metal-prod] (air) blower; booster • [electr] booster | a - blower; blowing
— m axial - [metal-prod] axial (type) blower
— f para recirculación f - recirculating,tion blower
— m — reserva f - [ind] stand-by blower
soplar v - . . . . • to blow (off, o out) • [metal-prod] to blow; to leak; to blast
— v a través de - to blow, out, o through
— v aire m - to blow @ air
— v con aire m - to blow (out) with air
— v — — m comprimido - to blow with compressed air; to compressed air blow • to blow, out, o clean
— v — — m con presión f alta - to blow (out) with high pressure air
— v — — m — — f baja - to blow (out) with low pressure air
— v — — m soldadora f - [sold] to blow out @ welder
— v — manguera f - to blow with @ hose
— v — — f para aire m - [ind] to blow (out) with @ air hose
— v — — f neumática - [ind] to blow (out) with @ air hose
— v condensador m - [aire.acond] to blow (out) @ condenser
— v en falso - [metal-prod] to blast through @ taphole (during @ cast)
— v — seco m - to blow dry
— v — — m con aire m - [ind] to blow dry with air
— v — — — — m comprimido - [ind] to blow dry with @ compressed air
— v escorial m - [metal-prod] to, blow, o leak (a slag notch)
— v filtro m - [mec] to blow @ filter
— v fuga f - [metal-prod] to blow (through) @ break
— v gas m - [metal-prod] to blow gas
— v — m colector m de polvo m - [metal-prod] to blow gas a dustcatcher
— v hasta secar(se) v - to blow dry
— v — — v con aire m - to blow dry with air
— v — — v — — m comprimido - to blow dry with @ compressed air
— v hollín m - [combust] to blow @ soot
— v interior m - [mec] to blow out
— v — m de soldadora f - [sold] to blow out @

welder

soplar v̲ **junta** f̲ - [metal-prod] to blow through
@ joint
— v̲ **periódicamente** - to blow (out) periodically
— v̲ **petaca** F̲ _ [metal-prod] to, leak, o̲ blow,
(@ cooling plate)
— v̲ **por tubo(s)** m̲ **petaca** f̲ - [metal-prod] to
blow through @ cooling plate pipe
— v̲ **por junta** f̲ - [metal-prod] to blow through
@ joint
— v̲ **regnerador(es)** m̲ - [metal-prod] to blow
(through) @ checker(s)
— v̲ **regulador(es)** m̲ - to blow out @ control(s)
— v̲ — m̲ **con corriente** f̲ **de aire** m̲ - [ind] to
blow out @ control(s)
— v̲ — — — **manguera** f̲ - [ind] to blow out @
control(s) with @ hose
— v̲ — — — f̲ **para aire** m̲ - [ind] to blow out
@ control(s) with @ air hose
— v̲ — — — f̲ **neumática** - [ind] to blow out @
control(s) with @ air hose
— v̲ **soldadora** f̲ - [sold] to blow out @ welder
— v̲ — f̲ **con manguera** f̲ - [sold] to blow out @
welder with @ hose
— v̲ — — f̲ **neumática** - [sold] to blow out @
welder with @ air hose
— v̲ — f̲ — — f̲ **para aire** m̲ - [sold] to blow
out @ welder with @ air hose
soplete m̲ - . . . . • [sold] (flame) torch
— m̲ **acetilénico**—[sold] acetylene, o̲ gas, torch
— m̲ **cortador** - [sold] cutting torch
— m̲ **montado** - [sold] mounted cutting torch
— m̲ — — **sobre carrito,rillo** m̲ - [sold] carriage mounted cutting torch
— m̲ **manual** - [herram] manual, o̲ hand, torch
— m̲ **montado** - [sold] mounted torch
— m̲ **neumático** - [neumát] air gun
— m̲ **oxiacetilénico** - [sold] oxyacetylene torch
— m̲ **para precalentamiento** - [sold] oxyacetylene preheating, o̲ preheating oxyacetylene,
torch
— m̲ **para acetileno** m̲ - [sold] acetylene torch
— m̲ — **aire** m̲ - [herram] air gun
— m̲ — — m̲ **comprimido** - [sold] compresseed air
gun
— m̲ — **corte** m̲ - [sold] cutting torch
— m̲ — — m̲ **acetilénico** - [sold] acetylene cutting torch
— m̲ — **gas** m̲ - [sold] gas torch
— m̲ — — m̲ **inerte** - [sold] inert gas torch
— m̲ — — m̲ **natural** - [ind] natural gas torch
— m̲ — — m̲ **y acetileno** m̲ - [sold] natura;
gas and acetylene torch
— m̲ — **oxiacetileno** m̲ - [sold] oxyacetylene
torch
— m̲ — **precalentamiento** m̲ - [sold] preheating
torch
— m̲ — **pintar** - [pint] paint gun
— m̲ **por hidrógeno** m̲ - [sold] hydrogen torch
— m̲ — — m̲ **atómico** - [sold] atomic hydrogen
torch
— m̲ **precalentador** - [sold] preheating torch
— m̲ **T E C** - [sold] T E C torch
— m̲ **T E C para gas** m̲ - [sold] T E C gas torch
— m̲ — — — m̲ **inerte** - [sold] T E C inert gas
torch
sopleteado* m̲ - [metal-trat] scarfing • [sold]
burn(ing) (off)
— m̲ **con gunita*** f̲ - [metal-prod] guniting
— m̲ — **refractario(s)** - [metal-prod] guniting
— m̲ — **extremo(s)** m̲ - [sold] end burn(ing)-off
— m̲ **de rodillo(s)** m̲ - [metal-lam] véase **granallado** m̲ **de rodillo(s)** m̲
— m̲ **para remiendo(s)** m̲—[sold] repair burn-off
— m̲ — — m̲ **grande(s)** - [sold] heavy repair
burn-off
sopleteado,da* a̲ - [metal-trat] scarfed
sopleteador* m̲ - [sold] scarfing, o̲ burn(ing)
off, machine
sopletear* v̲ - [metal-trat] to, scarf, o̲ burn
off
sopleteo* m̲ - [metal-lam] véase **granallado** m̲
sopletero* m̲ - [metal-prod] burner operator;

véase también **operador** m̲ **para soplante** m̲ •
[combust] véase **operador** m̲ **para quemador(es)** •
[sold] oxyacetylene torch operator • [metal-
prod] véase **cortador** m̲
sopletero m̲ **auxiliar** - [metal-lam] assistant
bruner operator; burner operator helper
— m̲ **para oxicorte** m̲ - [sold] oxyacetylene cutter
— m̲ — — m̲ **automático** - [sold] automatic oxyacetylene cutter
— m̲ — **parque** m̲ **intermedio** - [metal-prod] intermediate yard burner
— m̲ — **playa** f̲ - [metal-prod] yard burner
— m̲ — — f̲ **intermedia** - [metal-prod] intermediate yard burner
soplo m̲ **de aire** m̲ - [ind] air, blast, o̲ blow(Ing)
— m̲ — **arco** m̲ - [sold] arc blow(ing)
— m̲ — — m̲ **hacia atrás** - [sold] backward(s) arc
blow(ing)
— m̲ **enérgico** - [ind] strong, blast, o̲ blow(ing)
— m̲ **magnético** - [sold] magnetic blow(ing)
— m̲ — **de arco** m̲ - [sold] arc (magnetic) blow
— m̲ — — — m̲ **hacia atrás** - [sold] backward(s)
arc blow(ing)
— m̲ — **severo** - [sold] severe (magnetic) blow
— m̲ — — **de arco** m̲ - [sold] severe arc (mag-
metic) blow(ing)
— m̲ — — — **arco** m̲ - [sold] severe (magnetic)
arc blow(ing)
soportado,da a̲ - supported • shouldered; held;
sustained • accomodated
— a̲ **adecuadamente** - supported adequately
— a̲ **con seguridad** f̲ - supported, o̲ withstood,
safely
soportal m̲ - . . . . • [metal-prod] mantle
soportar v̲ - . . . ; to shoulder; to sustain • to
withstand • to hold • to accomodate
— v̲ **adecuadamente** - to support adequately
— v̲ **cable** m̲ - [cabl] to support @ (wire) rope
— v̲ — m̲ **de alambre** m̲ - [cabl] to support @
wire rope
— v̲ **cabo** m̲ - [cabl] to support @ rope
— v̲ **carga** f̲ - [mec] to, support, o̲ bear, o̲ withstand, o̲ carry, o̲ accept, o̲ sustain, @ load
— v̲ — f̲ **de tránsito** m̲ - [vial] to, support, o̲
bear, o̲ withstand, @ traffic load
— v̲ — f̲ **viva** - [constr] to, support, o̲ carry,
@ live load
— v̲ **con seguridad** f̲ - to withstand safely
— v̲ **choque** m̲ - to withstand @ shock
— v̲ — m̲ **térmico** - to withstand @ thermal shock
— v̲ **estructura** f̲ - [constr] to support @ structure
— v̲ **impacto** m̲ - [mec] to support @ impact
— m̲ **muro** m̲ - [constr] to support @ wall
— v̲ **nivel** - to carry @ level
— v̲ — m̲ **elevado** - [constr] to carry @ high
level
— v̲ — — **de solicitación** f̲ - [constr] to carry @ high stress level
— v̲ **pared** f̲ - [constr] to support @ wall
— v̲ **pavimento** m̲ - [constr] to support @ pavement
— v̲ **presión** f̲ - [mec] to hold @ pressure
— v̲ **tratamiento** m̲ - to witstand @ treatment
— v̲ — m̲ **rudo** - to withstand @ rough treatment
— v̲ **tubería** f̲ - [constr] to support @ pipe
— v̲ **vía** f̲ **(férrea)** - [constr] to support @
(railway) track
soporte m̲ - . . . . • stand; leg; stay; brace,cing;
arm; rack; skid; stall; mount(ing); retainer;
hanger; bracket; slip; mounting; bolster; carriage; carrier; jig; cradle,ling • [comb.intl]
(choke) bracket • [constr] hanger • standard;
pedestal; holder; neck; stand; bent • sustaining
— m̲ **a caja** f̲ - [mec] bracket to @ frame
— m̲ **abatible** - [constr] break-away support
— m̲ — **para alumbrado** m̲ - [vial] break-away
luminaire support
— m̲ — — **letrero** m̲ - [vial] break-away sign
support
— m̲ **abisagrado** - [constr] hinged support
— m̲ **adecuado** - [constr] adequate support

soporte m̲ adicional - additional support
— m̲ ajustado - [mec] adjusted support
— m̲ angular - [mec] angle,gular support
— m̲ anterior - [mec] previous support • front,
o̲ forward, support
— m̲ asegurado en pared f̲ - [mec] bracket
— m̲ cambiado - [mec] changed support
— m̲ central - [mec] center,tral, support, o̲
brace,cing
— m̲ — para cadena f̲ - [mec] center,tral chain
support
— m̲ cerrado - [mec] closed, support, o̲ mounting
— m̲ — para rollo m̲—[mec] closed reel mounting
— m̲ — — m̲ de alambre m̲ - [mec] closed wire
reel mounting
— m̲ colgado - [mec] suspended bracket
— m̲ colgante - [mec] suspended bracket • [tub]
hanging support; hanger
— m̲ — con muelle m̲ - [tib] spring hanger
— m̲ — — — m̲ para tubería f̲ - [tub] spring
pipe hanger
— m̲ — para tubería f̲ - [tub] pipe hanger
— m̲ — según norma f̲ - [tub] standard pipe
hanger
— m̲ colocado - [mec] installed holder
— m̲ combinado - [mec] compound, support, o̲ rest
— m̲ — para cojinete n̲,- [mec] combined bearing, o̲ bearing combined, support, o̲ rest
— m̲ compuesto m̲ - [mec] compound, support, o̲
rest
— m̲ — para cojinete n̲ - [mec] compound bearing, o̲ bearing compound, support, o̲ rest
— m̲ — — carro,rillo m̲ - [tornos] compound
slide rest
— m̲ — — deslizador m̲ - [tornos] compound
slide rest
— m̲ con cuello m̲ de cisne m̲ - [mec] gooseneck
support
— m̲ con seguridad f̲ - safe withstanding
— m̲ continuado - continuous support
— m̲ contra deformación f̲ - [constr] deflection
support
— m̲ corriente - [tub] standard, o̲ pipe, support
— m̲ — para carrete m̲ - [sold] standard reel,
support, o̲ mounting
— m̲ — — m̲ para alambre m̲ - [sold] standard
wire reel mounting
— m̲ creciente - increasing support
— m̲ de acero m̲ - [constr] steel support • steel
plate cradle
— m̲ — — m̲ fundido - [mec] cast steel gib
— m̲ — — m̲ galvanizado - [[mec] galvanized
steel support
— m̲ — — — por inmersión f̲ (en caliente) -
[mec] (hot)-dip galvanized steel support
— m̲ — — m̲ intemperizado - [constr] weathering
steel support
— m̲ — — m̲ para alumbrado m̲ - [constr] steel
light support
— m̲ — — m̲ para conducto m̲ - [constr] conduit
steel support
— m̲ — aluminio m̲ - [mec] aluminum support
— m̲ — carga f̲ - load, withstanding, o̲ sustaining, o̲ acceptance
— m̲ — — f̲ viva - [constr] live load, sup-
port(ing), o̲ carrying
— m̲ — catgoría f̲ - [mec] quality support
— m̲ — — f̲ . . . . • [constr] . . . class, o̲
class . . ., support, o̲ bedding
— m̲ — caucho m̲ - [mec] rubber support
— m̲ — clase f̲ . . . - [constr] class . . ., o̲
. . . class, support, o̲ bedding
— m̲ — choque m̲ - [mec] shock withstanding
— m̲ — — m̲ térmico - [mec] thermal shock withstanding
— m̲ — hormigón m̲ - [constr] concrete cradle
— m̲ — — m̲ armado - [constr] reinforced concrete, support, o̲ cradle
— m̲ — — m̲ simple - [constr] plain concrete,
cradle , o̲ support
— m̲ — madera f̲ - [constr] timber, support, o̲
cradle
— m̲ — micarta f̲—[electr-inst] micarta support

soporte m de suelo m̲ - [constr] soil support
— m̲ delantero - [mec] front, o̲ forward, support
— m̲ derecho - [mec] right bracket
— m̲ — para rectificador m̲ - [electr-instal]
rectifier right bracket
— m̲ desigual - unequal, o̲ uneven, support
— m̲ deslizable - [tub] sliding support
— m̲ — para tubería f̲ - [tub] sliding pipe,ping
support
— m̲ desparejo m̲ - uneven support
— m̲ elástico - elastic support
— m̲ en extremo m̲ - [mec] end support
— m̲ — — m̲ activo - [mec] live end support
— m̲ encerrado - [mec] enclosed, support, o̲
mounting
— m̲ — para montaje m̲ - [mec] enclosed support
mounting
— m̲ — — m̲ para carrete m̲ - [sold] enclosed
reel, mounting, o̲ support
— m̲ — — m̲ — — m̲ para alambre m̲ - [sold]
enclosed wire reel, support, o̲ mounting
— m̲ — rollo m̲—[mec] enclosed reel mounting
— m̲ — — m̲ de alambre m̲ - [mec] enclosed
wire reel mounting
— m̲ engoznado - [constr] hinged support
— m̲ estructural - [constr] structural support;
buckstay
— m̲ — construido - [constr] constructed, o̲
built, structural support
— m̲ — para indicador(es) m̲ - [constr-vial]
sign structural support
— m̲ — — m̲ para carretera(s) f̲ - [constr-
-vial] highway sign structural support
— m̲ — proyectado - [constr-vial] designed
structural support
— m̲ exterior - [mec] outside supposrt
— m̲ — para carrete m̲ - [mec] outside reel support
— m̲ — — molinete m̲ - [mec] outside reel support
— m̲ flexible - [mec] flexible support
— m̲ — de fundente m̲ - [sold] flexible flux
support
— m̲ frontal - [mec] front support
— m̲ — para motor m̲ - [electr-mot] front motor,
o̲ motor front, support • [comb.int] front engine, o̲ engine front, support
— m̲ galvanizado - [mec] galvanized support
— m̲ — por inmersión f̲ (en caliente) - [constr]
(hot-)dip galvanized support
— m̲ giratorio - swivel, o̲ rotating, bracket •
pinwheel
— m̲ igual - equal, o̲ even, support
— m̲ inferior - [mec] bottom, o̲ lower, support,
o̲ bracket
— m̲ — para montaje m̲ - [mec] lower mounting
bracket
— m̲ — — muelle m̲ - [mec] lower spring support
— m̲ ininterrumpido - [constr] continuous bearing
— m̲ instalado - [mec] installed, support, o̲
holder, o̲ bracket
— m̲ intermedio - [mec] intermediate support
— m̲ izquierdo - [mec] left, bracket, o̲ support
— m̲ — para rectificador m̲ - [electr-instal]
rectifier left bracket
— m̲ lateral - [mec] side support • side stand
— m̲ — apropiado - [mec] proper side support
— m̲ — efectivo - effective side support
— m̲ liviano - [mec] lightweight mounting
— m̲ — para carrete m̲ - [sold] lightweight reel
mount(ing)
— m̲ — — montaje m̲ para carrete - [sold] lightweight wire reel mounting
— m̲ — — — — m̲ — — — m̲ para alambre m̲ - [sold]
lightweight wire reel mounting
— m̲ — — rollo m̲ - [mec] lightweight roll
mounting
— m̲ — — — m̲ de alambre - [sold] lightweight
wire roll mounting
— m̲ mediante cordón m̲ mensajero - [electr-inst]
messenger support
— m̲ monolítico - [constr] monolithic, support, o̲
cradle

soporte m monolítico de hormigón m - [constr] concrete monolithic cradle
— m — — — m armado - [constr] reinforced concrete monolithic cradle
— m — — — m simple - [constr] plain concrete monolithic cradle
— m — reforzado - [constr] reinforced monolithic cradle
— m — simple—[constr] plain monolithic cradle
— m para accionamiento m - [mec] drive support
— m — — m con varilla f - [mec] rod drive support
— m — m — — f para tracción f - [mec] pullrod drive support
— m — acumulador m—[comb.int] battery bracket
— m — altavoz n - [electrón] loudspeaker support
— m — — m portátil - [electrón] mobile loudspeaker support
— m — alternador m - [comb.int] alternator bracket
— m — alumbrado m - [vial] luminaire support
— m — apoyo m - [mec] supporting rest; support
— m — árbol m - [mec] shaft, o arbor, support, o bracket; (spindle) carrier
— m — — m para desplazador m - [mec] shifter shaft support
— n — — m — mecanismo m para cambio(s) m - [mec] shifter shaft support
— m — artefacto m - [electr-instal] appliance support
— m — — m para alumbrado m - [electr-instal] luminaire, o lighting appliance, support
— m — m — iluminación f - [electr-instal] luminaire, o lighting appliance, support
— m — asiento m - [mec] seat support
— m — — m para punzón m - [mec] punch seat support
— m — balancín m - [mec] center iron; Samson post
— m — banda f - [mec] strap bracket
— m — barra f - [electr-instal] bus support
— m — barra f apalancadora - [mec] pry bar support
— m — — f colectora - [electr-instal] bus bar bracing; bus support
— m — — f — horizontal - [electr-install] horizontal bus bar brace,cing
— m — — f — vertical - [electr-instal] vertical bus bar brace,cing
— m — — f horizontal - [electr-inst] horizontal bar brace,cing
— m — — f vertical - [electr-instal] vertical bar brace,cing
— m — bisagra f - [mec] hinge support
— m — bobina f - [comb.int] coil bracket
— m — — f para encendido m - [comb.int] ignition coil bracket
— m — bomba f - [mec] pump support
— m — — f para inyección f - [comb.int] injection pump support
— m — — — f para combustible m - [comb-int] fuel injection pump support
— m — — f rociadora - [mec] spray(ing) pump support
— m — — f — para émbolo m - [mec] piston spray(ing) pump support
— m — — f — — — m y camisa f - [mec] piston-liner spray(ing) pump support
— m — botellón m - [combust] bottle support
— m — — m para gas m - [combust] gas bottle support
— m — bóveda f - [constr] arch support
— m — brazo m - [mec] arm support
— m — — m para polea f - [mec] pulley arm support
— m — — m — — f tensora - [mec] idler arm support
— m — cable m - [cabl] (wire) rope, o cable, support(ing)
— m — — m de alambre m - [cabl] wire rope support(ing)

soporte m para cable m para elevación f—[grúas] hoist line support
— m — cabo m - [cabl] rope support(ing)
— m — cadena f - [mec] chain support(ing)
— m — — f central—[mec] center chain support
— m — canal m - [mec] trough brace
— m — — m para aceite m - [mec] oil trough brace
— m — m — — m crudo - [mec] crude oil trough brace,cing
— m — m — — m para lubricación f - [mec] lubricating oil trough brace,cing
— m — canaleta f - [mec] chute support(ing)
— m — — f para lubricación f - [mec] feeding chute support(ing)
— m — carrete m - [sold] reel mount(ing)
— m — — m para alambre m - [sold] wire reel support(ing)
— m — — m — electrodo m - [sold] wire reel mount(ing)
— n — carro m - [tornos] slide rest
— m — cinta f - [mec] strip support(ing) • [metal-prod-pigcaster] chain fork
— m — cojinete m - [mec] bearing support; chock
— m — — m para árbol m - [mec] main shaft bearing support
— m — — m — — m principal - [mec] (main) shaft bearing support
— m — — m — — bancada f - [mec] main bearing support
— m — — m — — contraárbol m - [mec] jackshaft bearing support
— m — — m — — contraeje m - [mec] jackshaft bearing support
— m — — m para costado m - [mec] side bearing support
— m — — m — — m hacia mesa f rotatoria - [petról] rotary side bearing support
— m — — m — — — m perforador m - [mec] driller('s) side bearing support
— m — — m — — — m mesa f rotatoria - [petról] rotary side bearing support
— m — — m — — m hacia perforador m - [petról] driller('s) side bearing support
— m — — m — rodillo(s) m - [metal-lam] roll chock
— m — colador m - [mec] strainer support
— m — — m para aceite m - [mec] oil strainer support
— m — colector m - [mec] sump support
— m — — m vertical - [electr-instal] vertical bus brace,cing
— m — conducto m - [constr] conduit support
— m — cono m - [cojin] cone stand
— m — contraárbol m - [mec] jackshaft support
— m — contraeje m - [mec] jackshaft support
— m — copa f - [cojin] cup, stand, o support
— m — — f con pestaña f - [cojin] flanged cup (cup) stand
— m — cordón m - [cabl] strand support
— m — cruz f - [mec] cross support
— m — — f para colador m - [mec] strainer cross support
— m — — f — — m para descarga f - [mec] discharge strainer cross support
— m — cubierta f—[mec] roof, o cover, support
— m — cuchilla f - [mec] blade support
— m — cursor m - [trefil] slide holder
— m — choque m - [mec] shock withstanding
— m — defensa f - [mec] guard support
— m — depósito m - [mec] tank support
— m — — m para fundente m - [sold] flux tank support
— m — deslizador m - [torno] slide rest
— m — — m para torno m - [torno] (lathe) slide rest
— m — desplazador m - [mec] shifter support
— m — dispositivo m - [mec] device support
— m — — m para cambio(s) m - [mec] shifter (device) support
— m — distribuidor m - [mec] manifold, o distributor, support

soporte m̲ **para distribuidor** m̲ **para tubería** f̲ -
 [mec] pipe manifold support
— m̲ — **eje** m̲ - [mec] shaft, o̲ arbor, support
— m̲ — **eje** m̲ - [mec] shaft, support, o̲ bracket;
 (spindle) carrier
— m̲ — **elevador** m̲ - [grúas] hoist support
— m̲ — **encofrado** m̲ - [constr] form support
— m̲ — **enfriador** m̲ - [metal-prod] cooler holder
— m̲ — m̲ **para tobera** f̲ - [metal-prod] tuyere
 cooler holder
— m̲ — **engranaje** m̲ - [mec] gear support
-- m̲ — **espejo** m̲ - [domést] mirror support
— m̲ — **estrangulador** m̲ - [comb.int] choke
 bracket
— m̲ — **estructura** f̲ - [constr] structure sup-
 port(ing) • side support(ing)
— m̲ — **farol** m̲ - [constr] lamp bracket
— m̲ — **fiador** m̲ - [mec] retainer, bracket, o̲
 holder
— m̲ — m̲ **para cierre** m̲ - [mec] seal retainer
 holder
— m̲ — m̲ — **sello** m̲ - [mec] seal retainer
 holder
— m̲ — **fijación** f̲ - [mec] bracket mounting
— m̲ — **freno** m̲ - [mec] brake support
— m̲ — m̲ **para malacate** m̲ - [petról] hoisting
 machine brake support
— m̲ — m̲ — m̲ **para cuchara** f̲ - [petról]
 back brake support
— m̲ — m̲ **para torno** m̲ **para cuchara** f̲ -
 [petról] back brake support
— m̲ — **fundente** m̲ - [sold] flux support
— m̲ — **generador** m̲ - [electr-equip] generator
 bracket
— m̲ — m̲ **para soldadura** f̲ - [sold] welder
 generator bracket
— m̲ — **gorrón** m̲ - [mec] pivot support
— m̲ — m̲ **principal**—[mec] main pivot support
— m̲ — **grifo** m̲ - [mec] cock support
— m̲ — m̲ **para regulación** f̲ - [mec] control
 cock support
— m̲ — **guía** f̲ - [mec] guide, support, o̲ holder
— m̲ — f̲ **para válvula** f̲ - [válv] valve guide
 holder
— m̲ — **guiadera** f̲ - [mec] guide, support, o̲
 holder
— m̲ — f̲ **para válvula** f̲ - [válv] valve guide
 holder
— m̲ — **hilera** f̲ - [trefil] die holder
— m̲ — **indicador** m̲ - [constr-vial] sign support
— m̲ — m̲ **clavado en tierra** f̲ - [constr]
 ground-mounted sign support
— m̲ — m̲ **grande** - [constr-vial] large sign
 support
— n̲ — m̲ **pequeño** - [constr-vial] small sign
 support
— m̲ — m̲ **para carretera(s)** f̲ - [vial] highway sign support
— m̲ — **lámpara** f̲ - [ilumin] lamp bracket;
 lampholder
— m̲ — **letrero** m̲ - [vial] sign support
— m̲ — m̲ **clavado en tierra** f̲ - [constr]
 ground-mounted sign support
— m̲ — m̲ **de acero** m̲ - [vial] steel sign support
— m̲ — m̲ **grande** - [constr-vial] large sign
 support
— m̲ — m̲ **pequeño** - [vial] small sign support
— m̲ — **leva** f̲ - [mec] cam support
— m̲ — **madrastra** f̲ - [metal-prod] mantle (support)
— m̲ — **malacate** m̲ - [petról] wheel post
— m̲ — m̲ **para herramienta(s)** f̲ - [petról]
 bull wheel post
— m̲ — m̲ **para tubería** f̲ - [petról] calf
 wheel post
— m̲ — **mandril** m̲ - [mec] puppet
— m̲ — **matriz** f̲ - [mec] die, holder, o̲ retainer
— m̲ — m̲ **modular**—[mec] modular die retainer
— m̲ — **ménsula** f̲ - [mec] bracket support
— m̲ — f̲ **para ventilador** m̲ - [mec] fan
 bracket support
— m̲ — **molde** m̲ - [constr] form support

soporte m̲ **para montaje** m̲—[mec] mounting bracket
— m̲ — m̲ **para acumulador** m̲ - [comb.int] battery mounting bracket
— m̲ — m̲ — **alternador** m̲ - [electr-prod]
 alternator mounting bracket
— m̲ — m̲ — **carrete** m̲ - [sold] wire reel
 mounting bracket
— m̲ — m̲ — **filtro** m̲ - [mec] filter mounting bracket
— m̲ — m̲ — m̲ **para aire** m̲ - [comb.int]
 air filter mounting bracket
— m̲ — m̲ — **instrumento(s)** m̲ - [sold] instrument mounting bracket
— m̲ — m̲ — **relé** m̲ - [electr-instal] relay
 mounting bracket
— m̲ — m̲ **para tablero** m̲ - [electr-instal]
 panel mounting bracket
— m̲ — m̲ — m̲ **para instrumento(s)** m̲ -
 [electr-instal] selector panel mounting
 bracket
— m̲ — m̲ **para tapa** f̲ - [sold] cover mounting
 clip
— m̲ — m̲ — f̲ **para carrete** m̲ - [sold]
 reel cover mounting clip
— m̲ — m̲ — **transformador** m̲ - [electr-
 instal] transformer mounting bracket
— m̲ — m̲ **soldado** - [sold] welded mounting
 bracket
— m̲ — m̲ — **a base** f̲ - [sold] foot mounted
 bracket welded to @ base
— m̲ — **motor** m̲ - [electr-mot] motor support •
 [comb.int] engine support
— m̲ — **muelle** m̲ - [mec] spring support
— m̲ — **múltiple** m̲ - [mec] manifold support
— m̲ — **muñonera** f̲ - [mec] trunnion bracket
— m̲ — **muro** m̲ - [constr] wall support
— m̲ — **palanca** f̲ - lever support
— m̲ — f̲ **desplazadora** - [mec] shifter shaft
 support
— m̲ — **palanquilla(s)** f̲ - [metal-lam] billet
 skid
— m̲ — **pared** f̲ - [constr] wall support
— m̲ — **pavimento** m̲ - [constr] pavement support
— m̲ — **pilote(s)** m̲ - [constr-pil] pile,ling
 support
— m̲ — m̲ **de tubería** f̲ - [pil] pipe piling
 support
— m̲ — m̲ — f̲ **ahusado** - [constr-pil]
 tapered pipe piling support
— m̲ — **plancha(s)** f̲ - [constr] plate(s) cradle
— m̲ — f̲ **de acero** m̲ - [constr] steel plates
 cradle
— m̲ — **polea** f̲ - [mec] pulley support
— m̲ — f̲ **tensora** - [mec] idler support
— m̲ — **portaescobilla(s)** f̲ - [electr-mot]
 brushholder bracket
— m̲ — **poste** m̲ - [electr-instal] pole support
— m̲ — m̲ **de hormigón** m̲ - [electr-instal]
 concrete pole support
— m̲ — **puente(s)** m̲ - [constr] bridge support
— m̲ — m̲ **para indicador(es)** m̲ - [constr]
 sign bridge support
— m̲ — **punzón** m̲ - [mec] punch, support, o̲ retainer
— m̲ — m̲ **modular** - [mec] modular punch retainer
— m̲ — **radiador** m̲ - [comb.int] radiator pad
— m̲ — **recipiente** m̲ - [constr] container support
— m̲ — m̲ **para fluido(s)** m̲ - [constr] fluid
 container support
— m̲ — **rectificador** m̲ - [electr-instal] rectifier bracket
— m̲ — **regenerador** m̲ - [metal-prod] regenerator
 support
— m̲ — m̲ **para canal** m̲ - [metal-prod] duct
 regenerator support
— m̲ — m̲ **para solera** f̲—[metal-prod]
 hearth duct regenerator bracket
— m̲ — **regulador** m̲ - [mec] throttle support
— m̲ — **rejilla** f̲ - [mec] grid support
— m̲ — **retén** m̲ - retainer holder
— m̲ — m̲ **para cierre** m̲ - [mec] seal retainer

holder
soporte m **para retén** m **para sello** m - [mec] seal
retainer holder
— m — **riel** m - [constr] track support • [f.c.]
rail stool
— m — **rociadora** f - [mec] spray(er) support
— m — **rodillo** m - [mec] roll, support, o stand
— m — — m **estabilizador** - [mec] stabilizing
roll support
— m — **rotor** m - [mec] rotor support
— m — **salida** f - [constr] outlet support
— m — f **en voladizo** - [constr] cantilever
outlet support
— m — **suplemento** m - [mec] insert holder
— m — — m **para guía(dera)** f - [válv] guide
insert holder
— m — m — — f **para válvula** f - [válv]
valve guide insert holder
— m — **tambor** m - [sold] drum, support, o
mounting
— m — — m **para desenrollamiento** m **rápido** -
[mec] Speed Feed drum mounting
— m — — m — **de alambre** m - [sold]
Speed Feed (wire) drum mounting
— m — **taza** f - [cojin] cup stand
— m — **tobera** f - [metal-prod] tuyere holder
— m — **tope** m - [mec] top support
— m — **torno** m - [herram] lathe support •
[petról] wheel post
— m — **travesaño** m - [constr] beam, support, o
hanger
— m — — m **para tablero** m - [constr] floor
beam, support, o hanger
— m — **tubería** f - [tub] pipe support
— m — — f **elevada** - [tub] serial line support
— m — **tubo(s)** m - [tub] pipe support
— m — **tuerca** f - [mec] nut support
— m — — f **para ajuste** m - [mec] adjustment
nut support
— m — — f — — m **para banda** f (**para freno** m)
- (brake) band adjustment nut support
— m — **varilla** f - [mec] rod support
— m — — f **para aceite** m - [comb.int] oil rod
support
— m — — f **para medición** f - [comb.int] measuring rod support
— m — — — f **para aceite** m - [comb.int]
oil (measuring) rod support
— m — **vástago** m - [mec] rod support
— m — — m **para regulador** m - [mec] throttle
rod support
— m — **ventilador** m - [mec] fan bracket
— m — **vía** f - [f.c.] track support • [grúas]
runway support
— m — — f **para rodamiento** m - [grúas] runway
support
— m — **viga(s)** f - [constr] beam, support, o
hanger
— m — — f **transversal(es)** - [constr] cross, o
floor, beam(s), support, o hanger
— m — — f **vigueta(s)** f - [constr] floor beam(s)
hanger
— m — **voladizo** m - [constr] cantilever support
— m — **zapata(s)** f - [mec] shoe support
— m **parejo** m - [mec] even support
— m **posterior** - [mec] back, o rear, support •
[tornos] (lathe) back rest
— m — **en torno** m - [tornos] lathe back rest
— m **principal** - [mec] main, support, o pedestal
— m — **para gorrón** m - [mec] main pivot support
— m **propio** - [constr] own, o self, support(ing)
— m **que cede a flexión** f - [constr] hinged support
— m **quitado** - [mec] removed, support, o holder
— m **ranurado** - [mec] slotted, support, o shank
— m **reforzado** - [constr] reinforced, support, o
cradle
— m **removido** - [mec] removed, support, o holder
— m **rígido** - [constr] rigid support
— m — **para letrero** m - [vial] rigid sign support
— m **sacado** - [mec] removed, support, o holder
— m **satisfactorio** - satisfactory support

soporte m **seguro** - [mec] safe support
— m — **para artefacto** m (**para iluminación** f)
- [electr-instal] safe luminaire support
— m **simple** - [constr] plain, support, o cradle
— m **sin reforzar** - [constr] unreinforced cradle
— m **superior** - [mec] top, o upper, support, o
bracket
— m — **para montaje** m - [mec] top, o upper.
mounting bracket
— m — **para muelle** m - [mec] upper spring, o
spring upper, support
— m **suspendido** - [mec] suspende, o hanging,
support, o bracket
— m **terminal** - [constr] terminal, bracket, o
hanger
— m — **para travesaño(s)** m - [constr] end beam
hanger(s)
— m — — m **para tablero** m - [constr] end
floor beam hanger
— m — — **viga(s)** f (**transversal(es)** - [constr]
end floor beam hanger
— m — — **vigueta(s)** f - [constr] end floor
beam hanger(s)
— m **total** - [constr] full bearing
— m — **ininterrumpido** - [constr] full continuous bearing
— m **trasero** - [mec] rear support
— m — **para cubierta** f - [mec] rear roof support
— m — — **motor** m - [electr-mot] rear motor, o
motor rear, support • [comb.int] rear engine, o engine rear, support
— m **único** - [constr] single support
— m — **de acero** m - [constr] single steel support
— m **uniforme** - uniform support
— m **vertical** - [constr] vertical support •
[mec] pedestal bearing
— m — **intermedio** - [constr] intermediate
vertical support
sordina f - [acúst] . . .; sound damper
— f **para escape** m - [comb.int] exhaust silencer
sorghum m **vulgaris sudanensis**—[botán] Sudangrass
sorgo m **cafre** - [botán] Kafir (corn, o maize)
sorprendentemente adv - surprisingly; amazingly
— adv **alto,ta** - surprisingly high
— adv **bajo,ja** - surprisingly low
— adv **bien** - surprisingly well
sorprender v - . . .; to catch by surprise
sorpresa f **deparada** - provided surprise
sorpresivamente adv - surprisingly
sorpresivo,va a - surprising
sorteado,da a - drawn
sorteo m **especial** - special drawing
— m **para posición** f - [deport] spot drawing
— m — f **para largada** f - [deport] starting
spot drawing
sosegado,da a - . . .; gentle
sosiego m - . . . • assurance
— m **de motorista** m - [transp] motorist, o driver, assurance
soslayado,da a - bypassed; ignored; missed
soslayar v - . . . • to by-pass; to dodge, o shy
away; to miss; to ignore
— v **problema** m - [legal] to by-pass @ problem
— v — m **legal** - [legal] to by-pass @ legal
problem
soslayo m - . . . • bypassing; ignoring; missing
sospecha f - . . .; suspicioning; suspecting
sospechado,da a - suspected; suspicioned
sospechador m - suspecter; suspicioner
sospechar v - . . .; to suspicion • to have @
idea • to imagine
sostén m - . . . • prop • bracket; cradle •
shank; shoulder; hanger • bearing; backing •
bent
— m **adecuado** - adequate support(ing)
— m **adicional** - additional, o extra, support
— m **cedente** - [constr] yielding support
— m **central** - [mec] central, support, o brace
— m **con pilote(s)** m - [constr-pil] piling sup-
port(ing)
— m — — m **de tubería** f - [constr-pil] pipe

piling support
sostén m con pilotes m tubulares - [constr-pil]
pipe piling support
— m — **puntal(es)** m - [constr] shoring support
— m **de campana** f - [mec] bell, o hood, support
— m — **carga** f - load, support(ing), o bearing
— m — — f **muerta** - [constr] dead load bearing
• truss dead load
— m — **conducto** m - [mec] conduit support(ing)
• hood support(ing)
— m — — m **movible** - [electr-equip] movable
contact support(ing)
— m — **fundente** m - [sold] flux supporting
— m — **hormigón** m - [constr] concrete cradle •
concrete support(ing)
— m —, **hierba** f, o **pasto** m - [botán] grass
support(ing)
— m — **peso** m - [mec] weight support(ing)
— m — **presión** f - [constr] pressure supporting
— m — — f **de suelo** m - [constr] soil pressure
support(ing)
— m — **suelo** m - [constr] ground, o soil, sup-
port(ing)
— m **desigual** - [mec] uneven support(ing)
— m **desparejo** - [mec] uneven support(ing)
— m **estructural** - [constr] structural support
— m **igual** - [mec] even, o equal, support(ing)
— m **ininterrumpido** - [mec] continuous, o uninterrupted, support(ing)
— m **mejor** - [mec] better support(ing)
— m **para banda** f - [mec] strip support(ing)
— m — — f **para rodamiento** m - [autom-neumát]
tread support(ing)
— m — **cable** m - [cabl] cable, o rope, support
— m — — m **de alambre** m - [cabl] wire rope
support(ing)
— m — **cabo** m - [cabl] rope support(ing)
— m — **cojinete** m - [mec] bearing bracket
— m — **conducto** m - [electr-instal] conduit
support(ing)
— m — **contacto** m - [electr-equip] contact sup-
port(ing)
— m — **engranaje** m - [mec] gear support(ing)
— m — **equipo** m - [mec] equipment support(ing)
— m — — m **eléctrico** - [electr-instal] electrical equipment support(ing)
— m — **estructura** f - [constr] structure sup-
port(ing)
— m — **fundación** f - [constr] foundation sup-
port(ing)
— m — — f **para estructura** f - [constr] structure foundation support(ing)
— m — **grifo** m - [mec] cock support(ing)
— m — — m **para regulación** f - [mec] control
cock support(ing)
— m — **muro** m - [constr] wall support(ing)
— m — **pared** f - [constr] wall support(ing)
— m — **parte** f **cilíndrica** - [tub] barrel sup-
port(ing)
— m — **pavimento** m - [vial] pavement support
— m — **tubería** f - [constr] pipe support(ing)
— m — — f **para entubación** f - [petról] casing, support, o suspender
— m — **ventilador** m - [mec] fan bracket
— m — **vía** f - [f.c.] track support(ing)
— m — — f **férrea** - [[f.c.] railway track sup-
port(ing)
— m **parejo** - even support(ing)
— m **peor** - worse support • worst wupport
— m **por presión** f - [constr] pressure support-
-(ing)
— m — — f **de suelo** m - [constr] soil pressure
support(ing)
— m **proporcionado** - [mec] provided support(ing)
— m **provisorio**—[constr] temporary support(ing)
— m **provisto** - provided support
— m **relativamente cedente** - [constr] relatively
yielding support
— m **satisfactorio** - satisfactory support(ing)
— m **sin reforzar** - [constr] unreinforced, sup-
port(ing), o cradle
— m **sobre pilar(es)** m - [constr] bent sup-
port(ing)

sostén m **suficiente** - enough, o sufficient, sup-
port(ing)
— m **uniforme** - uniform support(ing)
— m — **para parte** f **cilíndrica (de tubería** f) -
[tub] (pipe) barrel uniform support(ing)
— m — — **tubería.bo** - [tub] uniform pipe sup-
port(ing)
sostenedor m - [mec] support
— m **de cabriola** f - [mec] purlin support(ing)
sostener v - . . .; to, bear, o uphold, o hold
up, o shoulder, o hold • to claim • to position
— v **adecuadamente** - to support adequately
— v **antorcha** f - [sold] to hold @ torch
— v — **con dos electrodos** m (de carbono m)
[sold] to hold @ arc torch
— v **cable** m - [cabl] to support @ (wire) rope
— v — m **de alambre** m - [cabl] to support @
wire rope
— v **cabo** m - [cabl] to support @ rope
— v **campana** f - [mec] to support @ bell • to
support @ hood
— v **carga** f - [mec] to, support, o withstand, @
@ load
— v — f **de tránsito** m - [vial] to, support, o
withstand, @ traffic load
— v **con mano** f **derecha** - to hold with @ right
hand
— v — — f **izquierda**—to hold with @ left hand
— v — — **presión** f [constr] to pressure support
— v — — f **de suelo** m - [constr] to soil pressure support
— v — **puntal(es)** m - [constr] to shoring support
— v — **seguridad** f - to support securely
— v **conducto** m - [tub] to support @ duct
— v **conjunto** m - [mec] to support @ assembly
— v — m **de portadiferencial** m - [autom-mec] to
support @ differential carrier assembly
— v **contacto** m - [electr-equip] to maintain @
contact • to support @ contact
— v — m **movible** - [electr-equip] to support @
movable contact
— v **cuba** f - [cabl] to hold @ bucket
— v **cuchara** f - [cabl] to hold @ bucket
— v **cucharón** m - [cabl] to hold @ bucket
— v **distribuidor** m - [mec] to support @ divider
— v — m **para fuerza** f - [autom-mec] to support
@ power divider
— v **electrodo** m - [sold] to hold @ electrode
— v **firmemente** - to hold securely
— v **en un punto** m - to hold at @ point
— v **encima** - to hold over
— v **injerto** m - [mec] to hold @ insert
— v **muro** m - [constr] to support @ wall
— v **parcialmente** - to carry part(ial)ly
— v **pared** f - [constr] to support @ wall
— v **paridad** f - to maintain @ parity
— v **pasto** m - to support @ grass
— v **pavimento** m - [constr] to support @ pavement
— v **peso** m - [mec] to support @ weight
— v **sobre pilar(es)** m - [constr] to bent support
— v **presión** f - [mec] to hold @ pressure
— v **rígidamente** - to support rigidly
— v **suelo** m - [constr] to support @ ground
— v **suplemento** m - [mec] to hold @, supplement,
o insert
— v **tubería** f - [constr] to support @ pipe • to
cushion @ pipe
— v **tubo** m - [tub] to hold @ pipe
— v **uniformemente** - to support uniformly
— v **vía** f - [f.c.] to support @ track
— v — f **férrea** - [constr] to support @ (railway) track
sostenerse v - to, hold, o stand
sostenidamente adv - sustainedly
sostenido,da a - . . .; held; shouldered; held
up; upheld
— a **adecuadamente** - supported adequately
— a **con puntal(es)** m - [constr] shoring supported

sostenido,da a en descubierto • [electr-instal]
  open supported
— a en un punto m - held at @ point
— a firmemente - held securely
— a parcialmente - carried part(ial)ly
— m con presión f - [constr] pressure supported
— a — f de suelo m - [constr] soil pressure
  supported
— a por rueda f - [mec] wheel supported
— a sobre pilar(es) m - [constr] bent supported
— a uniformemente - uniformly supported
sostenimiento m - . . .; holding (up); upholding
— m de campana f - [mec] bell support(ing) •
  hood supporting
— m — carga f - [constr] load, carrying, o
  support(ing)
— m — — f de tránsito m - [vial] traffic
  load, supporting, o withstanding
— m — conducto m - [mec] duct support(ing)
— m — conjunto m - [mec] assembly support(ing)
— m — — m de portadiferencial m - [autom-mec]
  differential carrier assembly support(ing)
— m — contacto m movible - [electr-equip] movable contact support(ing)
— m — distribuidor m - [mec] divider sup-
  port(ing)
— m — — m para fuerza f - [autom-mec] power
  divider support(ing)
— m — injerto m - [mec] insert holding
— m — paridad f - parity maintaining,tenance
— m — pasto m - grass support(ing)
— m — pavimento m - [constr] pavement sup-
  port(ing)
— m — peso m - [mec] weight support(ing)
— m — presión f - [mec] pressure, holding, o
  maintaining
— m — suplemento m - [mec] insert holding
— m — tubería f - [constr] pipe supporting •
  pipe cushioning
— m — tubo m - [mec] pipe, supporting, o holding
— m — vía f (férrea) - [f.c.] {railway) track
  support(ing)
— m en un punto m - holding at @ point
— m firme - secure holding
— m parcial - partial carrying
sotabanco m - [constr] . . .; skewback
sótano m - [constr] . . .; underground • [ind]
  pit
— m para aceite m - [ind] oil, pit, o cellar
— m anegado - [hidr] flooded, cellar, o basement
— m — engrase m - [ind] oil, cellar, o pit
— m — — m para cizalla f - [ind] shear oil
  cellar
— m — — m — tijera f—[ind] shear oil cellar
— m — hojalata f - [metal-trat] tinning mill,
  basement , o cellar
— m — lubricación f - [ind] lubrication, pit,
  o cellar
— m — tren m - [metal-lam] mill basement
— m — — m para hojalata f - [metal-prod] tinning mill basement
sotavento m - [náut] lee side; leeward
soterrado,da a - buried • underground
— a directamente - buried directly
soterramiento m directo - direct burial
sotomuración f - véase recalzo m
sotomurar v - véase recalzar v
spiegel m - [metal] véase arrabio m con contenido, alto, o elevado, de manganeso m
squirtmobile m - [sold] Squirtmobile
— m para avance m mecanizado - [sold] Squirtmobile
standard m - véase norma f; estándar m | a - vé-
  ase según norma f
standardización f - véase estandarización f
stock* m - stock(s); véase también existencia(s)
  • inventario m
— m en salón m para venta(s) f - [com] showroom
  stock
stoke m - [lubric] stoke
su adv - . . . • thereof

su adv comprensión f - their, o it, o his, o
  her, understanding; understanding, them, o
  it, o him, o her
suave a - . . . . • mild • downy
— a para tacto m - smooth, o soft, to @ touch
suavemente adv - smoothly; gently; softly
suavidad f - softness
— f de arco m - [sold] arc softness
suavizado,da a - softened; smoothed out
suavizador m para agua m - [hidr] water softener
— m — alimentación f - [hidr]
  feed(er) water softener
— m — m — f para caldera f - [cald]
  boiler feed water softener
suavizamiento m - softening; smoothing out
— f — superficie f - [vial] surface smoothing
  out
— m — — f de carretera f - [vial] road surface smoothing out
suavizar v - [mec] . . .; to, smooth out, o
  limber • to loosen
— v chapín m - [metal-prod] to loosen @ bleeder
— v — m para lavador m - [metal-prod] to
  loosen @ (gas) washer bleeder
— v mariposa f - [metal-prod] to lloosen @ butterfly (valve)
— v superficie f - to smooth out @ surface
— v — f de carretera f - [vial] to smooth out
  @ road surface
— v válvula f - [metal-prod] to loosen @ valve
— v — f para chimenea f - [metal-prod] to
  loosen @ stack valve
— v — f reguladora - [metal-prod] to loosen @,
  control(ling), o regulating, valve
sub- adv - [legal] assistant; under-
—-base - [mec] sub-base
—-prensa f - [mec] sub-press
—-prensado - [mec] sub-press(ing)
— adv procurador m general - [pol] assistant
  attorney general
subácueo,a - underwater; véase también sumergido,da
subadaptador m - [mec] subadapter
subalterno m - [labor] subordinate; underling
subalterno,ña a - subordinate; secondary
subangular a - subangular
subárea m - subarea
subarrendado,da a - subleased
subarrendamiento m - . . .; subleasing
subarriendo m - . . .; subleasing
subbase f - [constr] subbase
— f saturada - [hidr] saturated subbase
subcapataz m - [labor] assistant foreman
— m para turno m - [labor] turn assistant foreman
— m — — m para colada f—[metal-prod] casting
  assistant turn foreman
— m — — m — — f continua - [metal-prod]
  [col.cont] continuous casting assistant turn
  foreman
— m — — m para convertidor(es) m - [metal-
  -prod] converter(s) assistant turn foreman
subcategoría f - [ind] subcategory
— f de trabajo m - [labor] work subcategory
— f — — m administrativo - [admin] management
  work subcategory
— f — — m humano - [admin] human work subcategory
— f definitiva - [admin] definit(iv)e subcategory
— f determinada - determined subcategory
subclase f - subclass
subcomisión f administrativa - administrative
  subcommittee
— f técnica - [ind] technical subcommittee
subcomité* m - véase subcomisión f
subconjunto m - [ind] sub-assembly • [ind] unit
  assembly
— m de cigüeñal m y engranaje m - [mec] crankshaft and gear subassembly
— m individual - [ind] individual, subassembly,
  o component
— m para cuerpo m - [ind] body sub-assembly

subconjunto m para émbolo m - [comb.int] piston sub-assembly
— m — regulación f - [mec] control(ling), sub--assembly. o unit assembly
subcontratación f - [legal] subcontracting
— f de pieza f - [ind] part subcontracting
— f — servicio(s) m - [labor] service(s) sub-contracting
subcontratar v pieza f - [ind] to subcontract @ part
subcontratista m experimentado - [constr] expe-rienced subcontractor
— m extranjero - foreign subcontractor
— m local - local subcontractor
— m nacional - domestic subcontractor
subcontrato m por servicio(s) m - [legal] ser-vice(s) subcontract
subcuenta f - [constab] subaccount
— f de ganancia(s) f y pérdida(s) - [contab] profit and loss subaccount
subdelegación f - [pol] (sub-)district office
subdesarrollado,da a - . . .; véase también des-arrollado,da incompletamente; con desarrollo m incompleto
subdesarrollar v - . . .; véase también desarro-llar incompletamente
subdesarrollo m - . . . • véase también desarro-llo m incompleto
subdirección f - [admin] subdirector's, o assis-tant director's, office • [pol] department
— f general - [admin] general assistant direc-tor('s) office
— f para laminación f - [metal-lam] rolling de-partment
— f — mantenimiento m - [ind] maintenance de-partment
— f — — m y energía f - [ind] maintenance and power department
— f — producttión f - [ind] production depart-ment
— m — — f de acero m - [metal-prod] steel producing,ction department; open hearth shop
— f — — f — hierro m - [metal-prod] iron production, o blast furnace, department
— m — programación f - [ind] programming, o scheduling, department
— m — — f y control m - [ind] scheduling and control(ling) department
— f técnica - [ind] technical department
subdirector m artístico - [public] assistant, o associate, arc director
— m delegado - [legal] executive vice president
— m general - [admin] general assistant di-rector
— m para obra(s) f pública(s) - [pol] public works assistant director
— m — — f — en condado m - [pol] county pub-lic works assistant director
— m principal - [ind] assistant general super-intendent
subdividido,da a - subdivided; broken down • niche • [admin] divisionalized
— a adicionalmente - further, o additionally, subdivided
— a en división(es) f - [admin] divisionalized
subdividir v - . . . • [admin] to divisionalize
subdividir v - . . . • [metal-lam] to split
— v adicionalmente - to subdivide further
— v cuadro m orgánico - [admin] to division-alize @ organization(al) structure
— v en división(es) f—[admin] to divisionalize
— v estructura f orgánica - [admin] to subdi-vide,visionalize @ organization(al) structure
— v — f para organización f - [admin] to sub-divide,visionalize @ organization(al) struc-ture
— v proporcionalmente - to subdivide propor-tionally • [ambient] to piece proportionally
subdivisión f - . . . • [admin] divisionaliza-tion • [metal-lam] slitting
— f adicional - additional, o further, subdivi-sion
— f de cuadro m orgánico - [admin] organization

structure divisionalizing
subdivisión f en división(es) f - [admin] divi-sionalization,zing
— f de estructura f - [admin] structure divi-sion(alizing)
— f — — f orgánica - [admin] organization(al) structure, subdividing, o divisionalizing
— f — para organización f - [admin] or-ganization(al) structure, subdividing, o di-visionalizing
— f de función f - [admin] function, subdivi-ding, o divisionalizing
— f — — f administrativa - [admin] management function, subdividing, o subdivision
subdivisora f - [metal-lam] subdivider; slitter
subdrén n - [hidr] subdrain
— m de tubería f - [hidr] pipe subdrain
— m — — f perforada - [hidr] perforated pipe subdrain
subdrenaje m - [hidr] subdrainage,ning
— m adecuado—[hidr] adequate subdrainage,ning
— m apropiado - [hidr] appropriate, o proper, subdrainage,ning
— m debajo de, pavimento m, o zona f pavimen-tada - [hidr] pavement subdrain(age,ning)
— m — — — m para campo m para aviación f - [hidr] airfield pavement subdrain(age,ning)
— m estabilizado - [constr] stabilized sub-drainage,ning
— m exigido - [hidr] required subdrainage,ning
— m ferroviario - [f.c.] railway subdrainage
— m normal - [hidr] normal subdrainage,ning
— m para campo m para aviación f - [hidr] air-field subdrain(age)
— m — carretera f - [vial] highway subdrainage
— m — ferrocarril(es) m - [f.c.] railway sub-drainage
— m pobre - [hidr] poor subdrainage
subdrenar v - [hidr] to subdrain
subensamble,laje m - sub-assembly
subestación f eléctrica - [electr-distrib] elec-trical substation
— f existente - [electr-distrib] existing sub-station
— f para planta f - [electr-distrib] plant substation
— f — servicio m - [electr-distrib] service substation
— f — — m público - [electr-distrib] utility substation
— f — taller m - [electr-distrib] shop sub-station
— f — — m para conservación f - [electr--distrib] maintenance shop substation
— f — — m — mantenimiento m - [electr--distrib] maintenance shop substation
— f — transformador(es) m - [electr-distrib] transformer(s) substation
— f — unidad f de centro m para distribución f - [electr-distrib] load center (unit) sub-station
— f principal - [electr-distrib] main sub-station
— f — para planta f - [electr-distrib] plant main substation
— f proyectada - [electr-distrib] designed sub-station
— f rectificadora - [electr-distrib] recti-fier,fying substation
— f — para laminador m - [metal-lam] mill rec-tifier,fying substation
— f — — — m para palanquilla(s) f - [metal--lam] billet mill rectifier,fying substation
— f reubicada - [electr-distrib] relocated sub-station
— f ubicada - [electr-distrib] located substa-tion
subestimación f—underestimation; understatement
subestimado,da a - undersestimated • understated
subestimar v - to underestimate; to understate
subestrato m - [feol] substratum,ta
subestructura f - [constr] substructure; véase también infraestructura f

subestructura f de acero m - [constr] steel sub, o under,structure
— f —— m para puente(s) m - [constr] bridge steel undertructure
— f para puente m - [constr] bridge understructure
— f —— m de acero m - [constr] steel bridge understructure
— f — torre f (para perforación f) - [petról] derrick substructure
subfluvial a - subfluvial; underriver
subgerencia f - [admin] assistant management
— f para planta f - [ind] plant assistant manager(ship)
— f —— f termoeléctrica - [electr-prod] thermoelectric plant, manager(ship), o division, o department
subgerente n - [admin] assistant manager; submanager • [ind] assistant (area) superintendent
— m para establecimiento m - [ind] assistant plant manager • [com] assistant store manager
— m — movimiento m - [ind] assistant traffic manager
— m — servicio(s) m - [ind] service(s) submanager
— m —, m general(es) - [admin] (general) service(s) manager
— m —, tráfico m, o tránsito m - assistant traffic manager
subgrupo m - [ind] subgroup; subassembly
subhorizontal* a - subhorizontal
subida f - . . .; say up; ascending,nsion; upswing • uptake • [topogr] acclivity
— f de émbolo m - [mec] piston, ascent, o stairway
— f — gancho m - [grúas] hook raising
— f — gas m - [metal-prod] uptake
— f — metal - [metal-prod] metal rising
— f —— m a(l) enfriar(se) v - [metal-prod] véase dispersión f
— f — motón m - [grúas] block raising
— f — ramal m - [ambient] branch riser
— f y bajada f - rise,sing and lowering • ups and downs
— f —— f de aguilón m - [grúas] booming
subido,da a - raised
subinspector m - . . .; assistant inspector • [pol] assistant commissioner
subir v - . . .; to raise; to (pull) up; to ascend
— v a tope m - to rise to @ top
— v —— m temperatura f en tragante m - [metal-prod] to raise @ shaft temperature to @ maximum • excessive shaft temperature increase
— v carga f - to raise @, load, o burden
— v con manivela f - [mec] to crank up
— v cuesta f - to, climb, o go up, @ hill
— v émbolo eyector - [mec] to raise 2 ejector piston • [bombas] to release
— v gancho m - [grúas] to raise @ hook
— v hasta - to, climb, o go up, to
— v motón m - [grúas] to raise @ block
— v o bajar v - [mec] to raise o lower
— v o bajar aguilón m - [grás] to boom; to raise or lower @ boom
— v pistón m - [grúas] to raise @ piston
— v — m eyector - [grúas] to raise @ ejector piston
— v temperatura f - to raise @ temperature
— v — f en tragante m - [metal-prod] furnace throat temperature
— v —— m de horno m - [metal-prod] to increase @ furnace throat temperature
— v — f hasta tope m - [metal-prod] to raise @ temperature to @ maximum
— v tubería f - [tub] to, raise, o pull up; @ pipe
— v — f de núcleo m hincador - [constr-pil] to pull up @ drive,ving core pipe

subir v tubería f sobre núcleo m hincador - [constr-pil] to pull up @ pipe over @ drive,ving core
— v y bajar v - to raise and (to) lower
— v —— cabeza f - to raise and lower @ head
subir(se) v escoria f - [metal-prod] slag rise
— v carga f - [metal-prod] to raise @ burden
subítem m - [legal] subitem
subjefe m - [labor] . . .; assistant, head, o supervisor, o superintendent
— m para agrupación f - [ind] assistant group, supervisor, o head; group assistant head
— m — departamento m - [ind] department assistant superintendent
— m — equipo m - [labor] assistant group, head, o supervisor
— m — mantenimiento m - [ind] maintenance assistant, head, o supervisor
— m —— m asignado - [ind] assigned maintenance assistant, head, o supervisor
— m — sección f - [ind] assistant section, head, o supervisor
— m —— f para nave f para colada f - [metal-prod] pouring bay section assistant, head, o supervisor
— m — turno m - [ind] assistant turn, head, o supervisor
— m —— colada f - [metal-prod] pouring bay assistant, head, o supervisor
— m —— m —— f continua - [metal-prod] continuous casting assistant turn supervisor
— m —— m para convertidor(es) m - [metal-prod] converter assistant turn supervisor
— m —— m para soplado m - [metal-prod] blast assistant turn supervisor
sublevación f - [pol] . . . • [hidr] uplift(ing)
submarino m nuclear - [milit] nuclear submarine
submarino,na a - submarine; undersea • [petról] subsea
submersión f - submersion • immersion
submontaje m - [mec] subassembly
— m soldado - [sold] subweldment; welded subassembly
submurar v - véase recalzar
subordinado,da a - . . .; ancillary
subpedido m - [com] suborder
— m cursado - issued suborder
subpresión f - subpressure • [geol] flotation • [hidr] uplift(ing)
subproducto m combustible - [coque] combustible, o fuel, by-product
— m de coque m - [coque] coke by-product
— m — petróleo m - [petról] oil by-product
— m — proceso m - [ind] process by-product
— m —— m siderúrgico - [metal-prod] steel process by-product
— m — trigo m - [agric] wheat by-product
subproveedor m - subcontractor • véase también subcontratista m
subrasante f - [constr] subgrade top • [hidr] subgrade
— f blanda - [constr] soft subgrade
— f conservada - [constr] maintained subgrade
— f debajo de calzada f - [constr] under pavement subgrade
— f —— pavimento m - [constr] underpavement subgrade
— f débil - [constr] weak subgrade
— f debilitada - [constr] weakened subgrade
— f estable - [constr] stable subgrade
— f mantenida - [constr] maintained subgrade
— f mejor - [constr] better subgrade
— f saturada - [hidr] saturated subgrade
subrayado m - underscoring • stressing
subrayado,da a - underscored • stressed
subrayar v - . . . • to stress
subregión f - [geogr] subregion
subregional a - [geogr] subregional
subsanación f - exculpating • correction,ting; remedy,ding
subsanado,da a - exculpated; corrected; remedied
subsanar v - . . .; to remedy
— v defecto m - to, remedy, o correct, @ fault

subsanar v falla f - to correct @ defect
subscribir v - to subscribe • to (under)sign
— v contrato m - [legal] to sign @ contract
— v convenio m - [legal] to sign @ agreement
— v en mi presencia f - [legal] to subscribe,
in my presencew, o before me
— v obligación f - [fin] to sign @ bond
subscripción f - subscribing; subscription; signature; signing
— f de acción(es) f - [legal] share(s), o
stock, subscribing, o subscription
— f — f privada(s) f - [legal] private
shares, subscribing, o subscription
— f — f pública(s) - [legal] public shares,
subscribing, o subscription
— f — contrato m - [legal] contract signing
— f — convenio m - [legal] agreement, signing,
o signature, o subscribing, o subscription
— f — obligación f - [fin] bond, signing, o
signature; commitment, o obligation, signing;
debenture, subscription, o signing
— f en mi presencia f - [legal] signing before
me; subscribing, o signing, in my presence
— f no integrada - [legal] unpaid subscription
— f privada - [fin] private subscription
— f pública - [fin] public subscription
subscrito,ta a - [legal] subscribed
— a en mi presencia f - [legal] subscribed. in
my presence, o before me
— a y jurado,da a - [legal] subscribed, o
signed, and sworn to
subsecretario m - undersecretary; subsecretary;
assistant secretary
— m de estado m - [pol] assistant secretary of
state
— m — recursos m naturales - [pol] expendable
resources undersecretary
— m — transportes m - [pol] transportation undersecretary
subscripto,ta a - subscribed • undersigned
subsecuente a - véase subsiguiente a
subsidiado,da a - subsidized
subsidiaria f - subsidiary; véase también filial
— f consolidada - [fin] consolidated subsidiary
— f de propiedad f exclusiva - [legal] wholly
owned subsidiary
— f en operación f - operating subsidiary
— f en propiedad f absoluta - [legal] wholly
owned subsidiary
— f extranjera - [legal] foreign subsidiary
— f — consolidada - [fin] consolidated foreign
subsidiary
— f manufacturera - [find] manufacturing subsidiary
— f para comercialización f - [com] marketing,
subsidiary, o arm
subsidiario,ria a - subsidiary • auxiliary
subsidio m - . . .; subvention • allocation •
[com] bonus • [fin] grant; funding
— m a exportación f - [fin] export bonus
— m especial - special, subsidy, o subvention
— m apreciable - [pol] substantial allocation
— m considerable - [pol] substantial allocation
— m eventual - eventual, subsidy, o subvention
— m gubernamental - [fin] government allocation
— m — apreciable - [pol] substantial government allocation
— m — considerable - [pol] substantial government allocation
— m para desarrollo m - [pol] development aid
— m — desempleo m - [labor] unemployment
benefit
— m — planificación f - [pol] planning grant
— m sobre exportación f - [com] export bonus
— m — — f de mineral(es) m - [miner] ore export bonus
subsistencia f - . . .; subsisting
subsistido,da a - subsisted
subsistir v - . . . • to stand up
subsolar a - . . . ; véase también bajo tierra f
substancia f aglomerante - agglomerating substance
— f combustible - [combust] combustible sub-

substancia f combustible en estado m líquido -
[quím] liquid state, combustible, o fuel,
substance
— f compatible - compatible substance
— f convencional - conventional substance
— f descri(p)ta - described substance
— f en interior - inside substance
— f — estado m gaseoso - gaseous state substance
— f — — m líquido - liquid state substance
— f — — m sólido - solid state substance
— f espesante - thickening substance
— f extraña - foreign, o strange, substance
— f gaseosa - gaseous substance
— f gelificante* - jellying substance
— f líquida - liquid substance
— f nociva - noxious, o harmful, substance
— f penetrante - penetrating substance
— f perjudicial - harmful substance
— f que contiene carbono m - [quím] carbon containing substance
— f — reacciona - reacting,tive substance
— f química - [quím] chemical (substance)
— f silicónica - silicone containing substance
— f — aglomerante - silicone agglomerating
substance
— f sólida - solid substance
— f tenso-activa - [quím] tenso-active substance
— f — convencional - [quím] conventional
tenso-active substance
— f volátil - volatile (substance)
— f — peligrosa - dangerous volatile (substance)
substancial a - . . . . • véase también considerable
substancialmente adv - . . . . • véase también
considerablemente adv
substitución f - . . .; substituting • replacing
• replacement
— f conveniente - convenient substituting,tion
— f de alcantarilla f - [constr] culvert replacement,cing
— f componente m - [quím] component substituting,tion
— f — cristal(es) m - crystal(s) substituting,tion
— f — — m para tablero m - [electrón] board
crystal substituting,tion
— f — — m — — m para manipulador m -
[electrón] keyer board crystal substituting,tion
— f — — m — — m — — m para tono(s) m -
[electrón] tone keyer board crystal substituting,tion
— f — equipo m - [ind] equipment substituting,tion
— f — defectuoso - [ind] defective equipment substituting,tion
— f — importación(es) f - [econ] import(s)
substituting,tion
— f — manipulador m - [electrón] keyer substituting,tion
— f — — m para tono(s) m - [electrón] tone
keyer substituting,tion
— f — material(es) m - [ind] material(s), substituting,tion, o replacement,cing
— f — — prerreducido(s) m - [miner] prereduced material(s), substituting,tion, o replacement,cing
— f — mineral m - [miner] ore, substituting,tion, o replacement,cing
— f — neumático(s) m - [autom-neumát] tire(s),
substituting,tion, o replacement,cing
— f — pieza(s) f - [ind] part(s), substituting,tion, o replacement,cing
— f — — f para equipo m - [ind] equipment
part(s) substituting,tion, o replacement,cing
— f — retención f - [fin] withholding substituting,tion
— f — servicio(s) m - [ind] service(s) substituting,tion

substitución f̲ de servicio(s) m̲ técnico(s) -
[ind] technical service(s) substituting,tion
— f̲ — tablero m̲ - [electrón] board substituting,tion
— f̲ — — m̲ para manipulador m̲ - [electrón]
keyer board substituting,tion
— f̲ — — m̲ — — m̲ para tono(s) m̲ - [electrón]
tone keyer board substituting,tion
— f̲ efectuada - performed, o̲ made, substitution
— f̲ hecha - made, o̲ performed substitution
— f̲ menor - minor substitution
— f̲ para alcantarilla f̲ - [constr] culvert,
substituting,tion, o̲ replacement,cing
— f̲ — importación(es) f̲ - [com] import(ation)
substituting,tion
— f̲ rápida - rapid, o̲ quick, substituting,tion
substituido,da a̲ - substituted; replaced
— a̲ convenientemente - substituted conveniently
— a̲ rápidamente - substituted quickly
substituidor m̲ para componente(s) n̲ - [quím]
component(s) substituting,tion
— m̲ — equipo m̲ - [ind] equipment substituting
— m̲ — pieza f̲ - [ind] part substituting,tion
— m̲ — — f̲ para equipo m̲ - [ind] equipment
part substituting,tion
— m̲ — retención f̲ - [fin] withholding substituting,tion
substituir v̲ - . . . • [deport] to drive for
— v̲ componente m̲ - [quím] to substitute @ component
— v̲ convenientemente - to substitute conveniently
— v̲ convertidor m̲ - [metal-prod] to replace @
converter
— v̲ cristal - to substitute @ crystal
— v̲ — m̲ para tablero m̲ - [electrón] to substitute @ board crystal
— v̲ — m̲ — — m̲ para manipulador m̲—[electrón]
to substitute @ keyer board crystal
— v̲ — m̲ — — m̲ — — m̲ para tono(s) m̲ -
[electrón] to substitute @ tone keyer board
crustal
— v̲ equipo m̲ - [ind] to, substitute, o̲ replace,
@ equipment
— v̲ — m̲ defectuoso - [ind] to substitute @ defective equipment
— v̲ importación(es) f̲ - to substitute @ imports
— v̲ manipulador m̲ - [electrón] to substitute @
keyer
— v̲ — m̲ para tono(s) m̲ - [electrón] to substitute @ tone keyer
— v̲ neumático m̲ - [autom-neumát] to substitute
@ tire
— v̲ pieza f̲ - [ind] to substitute @ part
— v̲ — f̲ para equipo m̲ - [ind] to substitute @
equipment part
— v̲ rápidamente - to substitute quickly
— v̲ retención f̲ - [fin] to substitute @ withholding
— v̲ tablero m̲ — [electrón] to substitute @ board
— v̲ — m̲ para manipulador   [electrón] to,
substitute, o̲ replace, @ keyer board
— v̲ — m̲ — — m̲ para tono m̲ - [electrón] to,
substitute, o̲ replace, @ tone keyer board
substitutivo,va a̲ - . . . .; substitutive; alternate,tive
substituto m̲ - . . . .; replacement; alternate;
second choice
— m̲ para aguarrás m̲ - [petról] turpentine substitute
— m̲ — fluorita f̲ - [miner] fluorite substitute
— m̲ — importación f̲ [com] import(s) substitution
— m̲ — trementina f̲ - [petról] turpentine substitute
substracción f̲ - . . . • tearing away
substraer v̲ - . . . . • to tear away
substraerse v̲ - to tear away
substraido,da a̲ - subtracted • torn away
subsuelo m̲ - [suelos] . . . .; undersoil; subsurface • [constr] basement; cellar
— m̲ determinado - [constr] specific subsoil
— m̲ estable - [suelos] stable subsoil

subsuelo m̲ explorado - [constr] explored, subsurface, o̲ subsoil
— m̲ inestable - [suelos] unstable soil
subsuperficial a̲ - sub-surface,ficial; véase tam
bién submarino,na; suberráneo,nea; subál-
veo,vea; bajo tierra
subsuperficie f̲ - sub-surface
subterráneo m̲ [f.c.] underground
subterráneo,nea a̲ - subterranean; subsurface;
underground
subtesorero m̲ - under, o̲ assistant, treasurer
subtítulo m̲ - [public] . . .; subhead(ing)
— m̲ destacado - [public] display subhead(ing)
— m̲ ornamental - [public] display subhead(ing)
— m̲ para adorno m̲ - [public] display subheading
subunidad f̲ - subunit
— f̲ productiva - [ind] production,tive subunit
suburbio m̲ residencial - [pol] residential, o̲
bedroom, suburb
subusina f̲ - [electr-distrib] substation
subutilización f̲ - under use
subvención f̲ - . . . . • support
— f̲ especial - special subvention
— f̲ eventual - eventual subvention
subvenido,da a̲ - subventioned; supported
subvenir v̲ - . . .; to support
subyacer v̲ - to underlie
subzona f̲ - subzone; subarea
succión f̲ - . . .; sucking • draft
— f̲ ascendente - [comb.int] up-draft
— f̲ buena - good suction
— f̲ de bomba f̲ - [petról] pump suction
— f̲ — — f̲ para aceite m̲ - [bombas] oil pump
suction
— f̲ — — f̲ — — m̲ liviano - [bombas] light
oil pump suction
— f̲ — — f̲ — fuel oil - [bombas] fuel oil
pump suction
— f̲ doble - double suction
— f̲ inducida - [fís] induced, o̲ charged, suction
— f̲ natural - [fís] natural suction
— f̲ promedio - average suction
— f̲ simple - simple, o̲ single, suction
succionado,da a̲ - véase chupado,da; aspirado,da
succionador m̲ - véase chupador m̲
succionador,ra a̲ - véase chupador,ra
succionar v̲ - véase chupar; aspirar
suceder v̲ - . . .; to, happen, o̲ occur, o̲ happen • to befall • to work out
sucedido m̲ - . . . . | a̲ - happened; occurred •
occurring • worked out
sucesión f̲ - . . . . • sequence
— f̲ ordenada - sequence
sucesivo,va a̲ - . . . . • sequential
suceso m̲ - . . . . • occurrence
— m̲ imprevisto - unexpected, o̲ unforeseen, happening, o̲ event
— m̲ inesperado - unexpected, occurrence, o̲
event, o̲ happening
sucesor m̲ - successor
— m̲ legal - [legal] legal successor
— m̲ societario m̲ - [legal] corporate successor
suciedad f̲ acumulada - accumulated dirt
— f̲ — purgada - drained accumulated dirt
— f̲ arrastrada - drawn, o̲ entrained, dirt
— f̲ depositada - deposited, dirt, o̲ filth
— f̲ — en conducto(s) m̲ - [ambient] deposited,
dirt, o̲ dust, o̲ filth
— f̲ en agua m̲ - [hidr] water impurity,ties
— f̲ — aire m̲ - air, o̲ entrained, dirt
— f̲ — — m̲ para entrada f̲ - [comb.int] intake
air, dirt, o̲ dust
— f̲ — — m̲ — alimentación f̲ - [ind] intake
air, dirt, o̲ dust
— f̲ — calle f̲ - [hidr] street, dirt, o̲ wash
— f̲ excesiva - excessive dirt
— f̲ introducida - drawn, dirt, o̲ dust
— f̲ negra - black dirt(iness)
— f̲ purgada - withdrawn, o̲ drained, dirt
— f̲ quitada - removed dirt
— f̲ removida - removed dirt
— f̲ sacada - removed dirt

suciedad f verificada - checked dirt
sucio* m - véase suciedad f
sucio,cia a - . . . . • black
sucursal f - . . .; branch office
— f de correo m - [comunic] branch post office
— f — entidad f extranjera - [com] foreign entity branch
sudado,da a - . . .; sweated • oozed
sudamérica f - [geogr] South America
sudcentral a - south center,tral
sudoccidental a - [geogr] Southwest(ern)
sudoccidente - [geogr] Southwest
sudoeste - [geogr] Southwest
sudor m - . . . . • oozing
sudoriental a - [geogr] Southeast(ern)
sudoriente m - [geogr] Southeast
suela f - . . . . • [mec] base plate
— f de horno m - [metal-prod] furnace bottom
— f refrigerada - [ind] cooled base plate
sueldo m devengado - [labor] accrued salary
— m industrial - [labor] industrial salary
— m horario - [labor] hourly, salary, o rate
— m mínimo - [labor] minimum, salary, o wage
— m pagado - [labor] paid salary
— m total - [labor] total salary
— m vital - [labor] living wage
— m — mínimo - [labor] minimum living wage
sueldos m para personal m - [labor] personnel, salary,ries, o wage(s)
— m — m en casa f matriz - [labor] head office, o headquarters, salary,ries, o payroll
— m y salarios m - [labor] payroll; salary,ies and wage(s)
suelo m - [suelos] . . .; terrain • bottom • [constr] ground space
— m ácido m - [suelos] acid soil • [tub] sour service
— m adyacente - adjacent ground
— m agresivo - [suelos] aggressive soil
— m agrícola - [agric] agricultural soil
— m — corriente - [suelos] average agricultural soil
— m altamente ácido - [suelos] highly acid soil
— m ambiente - [suelos] soil environment
— m anegado - [suelos] waterlogged gound
— m apropiado - [suelos] appropriate soil
— m apto - [suelos] suitable, o apt, soil
— m — para fundación(es) f - [constr] (suitable, o apt) foundation(al) soil
— m arcilloso - [suelos] clayey, o claylike, soil
— m asentado - [constr] subsided, o settled, soil
— m — para base f - [constr] subsided, o settled, foundation(al) soil
— m — — fundación f - [constr] subsided, o settled, foundation(al) soil
— m asentante - [constr] subsiding soil
— m — para base f - [constr] subsiding foundation(al) soil
— m — — fundación f - [constr] subsiding foundation(al) soil
— m bien compactado - [constr] well compacted soil
— m blando - [suelos] soft soil • soft spot
— m — horadado - [constr] tunneled soft ground
— m bueno - [suelos] good soil
— m cambiante - [suelos] changing soil • [constr] drifting soil
— m para, base f, o fundación f - [constr] shifting foundation soil
— u circundante - [suelos] surrounding soil
— m cohesivo - [suelos] cohesive soil
— m compactado - [constr] compacted soil
— m compresible - [geol] compressible soil
— m con grava f - [suelos] gravelly soil
— m contenido - [constr] contained soil • retained soil
— m corriente - [suelos] current, o average, soil
— m corrosivo - [geol] corrosive soil
— m dado - [suelos] given soil
— m de arena f - [suelos] sand soil

suelo m de base f - [constr] base soil
— m — canto(s) m rodado(s) - [suelos] gravel soil
— m — cimentación f - [suelos] foundation soil
— m — formación f suelta - [suelos] loose foundation soil; loose soil formation
— m — fundación f - [constr] foundation soil
— m — lugar m - [suelos] soil at @, site, o location
— m — océano m - [marit] ocean floor
— m — subrasante f - [constr] subgrade soil
— m — zona f - [suelos] area soil
— m debajo de calzada f - [constr] under @ pavement, o sub-pavement, soil
— m — — carga f - [suelos] under @ load soil
— m — — f elevada - [constr] soil under @ high load
— m — — pavimento m - [constr] under @ pavement, o sub-pavement, soil
— m descripto - [suelos] described soil
— m desigual - [constr] uneven ground
— m desparejo m - [suelos] uneven ground
— m diferente - [suelos] different soil
— m duro - [geol] hard soil
— m empleado - [suelos] soil used
— m — en construcción f - [constr] constructional soil
— m — — ensayo(s) m - [constr] test soil
— m en condición f natural - [suelos] undisturbed soil
— m — lugar m - [suelos] soil at @, site, o location
— m — relleno m - [constr] (back)fill soil
— m — sitio m - [suelos] soil at @, site, o location
— m encontrado - [suelos] found, o encountered, soil
— m ensayado - [suelos[tested soil
— m erosionado - [suelos] eroded soil
— m — fácilmente - [suelos] easily eroded soil
— m estabilizado - [suelos] stabilized soil
— m estable - [geol] stable, soil, o ground
— m estándard - [suelos] standard soil
— m evaluado - [suelos] evaluated soil
— m expansivo - [suelos] expansive soil
— m expuesto - [constr] exposed ground
— m fino - [suelos] fine soil
— m firme - [suelos] firm soil
— m fluido - [suelos] fluid soil
— m geológico - [geol] geological soil
— m granular - [suelos] granular soil
— m — fino - [suelos] fine, granular, o grained, soil
— m — grueso - [suelos] coarse, granular, o grained, soil
— m permeable - [suelos] pervious granular soil
— m grueso - [suelos] coarse soil
— m guijoso - [suelos] gravelly soil
— m — compactado - [suelos] compacted gravelly soil
— m — compacto—[suelos] compact gravelly soil
— m húmedo m - [suelos] humid, o damp, o wet, soil, o ground
— m — pesado - [suelos] heavy wet soil • weighed wet soil
— m impermeable - [suelos] impervious soil
— m — pesado - [suelos] heavy impervious soil
— m inestable - [suelos] unstable soil
— m — para cimiento(s) m - [constr] unstable foundation soil
— m — — fundación f - [constr] unstable foundation soil
— m levemente permeable - [suelos] slightly pervious soil
— m mantenido - [constr] maintained, o held, ground
— m moderadamente permeable - [suelos] moderately pervious soil
— m mojado m - [suelos] wet, soil, o ground
— m — para cimiento(s) m - [constr] wet foundation soil
— m movedizo - [suelos] shifting soil

suelo m movido ~ [suelos] shifted soil
— m nacional - [pol] national, o native, soil
— m natural - [suelos] natural, o undisturbed, soil, o ground • native soil
— m — en sitio m - [suelos] location, o site, natural soil
— m no cohesivo - [suelos] non-cohesive soil
— m orgánico - [geol] organic soil
— m para base f - [constr] foundation soil
— m — cimentación f - [constr] foundation soil
— m — f reemplazado - [constr] replaced foundation soil
— m — cimiento(s) m - [constr] foundation soil
— m — ensayo(s) m - [constr] test(ing) soil
— m — fundación f - [constr] foundation soil
— m — relleno m - [constr] (back)fill soil
— m — — m lateral - [constr] side-fill soil
— m permeable - [geol] pervious soil
— m — granular - [suelos] granular, o grained, pervious soil
— m — — grueso - [suelos] coarse, granular, o grained, pervious soil
— m — grueso - [suelos] coarse pervious soil
— m pesado - [suelos] heavy soil • [constr] weighed soil
— m plástico - [suelos] plastic soil
— m pobre - [suelos] poor soil
— m — para cimentación f - [constr] poor foundation soil
— m predominantemente arcilloso - [suelos] predominantly clayey soil
— m que se hincha - [suelos] heaving soil
— m — — con helada(s) f - [suelos] frost heaving soil
— m quitado - [suelos] removed soil
— m reemplazado - [suelos] replaced soil
— m removido - [suelos] removed soil
— m refractario - [refract] refractory soil
— m residual - [suelos] residual soil
— m retentivo - [suelos] retentive soil
— m — de humedad f - [suelos] moisture retentive soil
— m resistente - [suelos] resisting,tant soil
— m — a erosión f - [suelos] erosion resistant soil
— m retenido - [constr] retained soil
— m rocoso - [geol] rocky, soil, o ground
— m — compactado - [suelos] compacted rocky soil
— m — compacto - [suelos] compact rocky soil
— m sacado - [constr] removed soil
— m saturado - [suelos] saturated soil
— m seco - [suelos] dry soil
— m — pesado - [suelos] heavy dry soil • weighed dry soil
— m sin cohesión f - [suelos] cohesionless soil
— m — remover - [geol] undisturbed soil
— m soportado - [constr] supported, soil, o ground
— m sostenido - [constr] supported ground
— m standard - [suelos] standard soil
— m subyacente - [constr] underlying soil
— m suelto - [constr] loose(ned) soil
— m superficial - [constr] surface soil
— m tocado - [constr] touched, soil, o ground
— m uniforme - [constr] uniform soil
— m usado - [suelos] soil used • used soil
— m virgen - [geol] undisturbed, o virgin, soil, o earth
suelta f - . . .; release,sing • disengaging; letting out; unlatching • breaking free
— f automática - automatic release,sing
— f de botón m - [mec] button release,sing
— f — m para regulación f - [mec] control button release,sing
— f — — m para rerregulación f - [mec] reset(ting) button release,sing
— f — carga f - [mec] load release,sing
— f — conductor m - [sold] lead, loosening, o release,sing
— f — — m a electrodo m - [sold] electrode lead, loos(en)ing, o release,sing
suelta f de embrague m - [mec] clutch, letting out, o release,sing
— f — freno m - [mec] brake release,sing
— f — gancho m - [mec] hook, release,sing, o unlatching , o unlocking
— f — lengüeta f - [mec] tongue release,sing
— f — manguera f - [mec] hose, loosening, o release,sing
— f — manija f - [mec] handle release,sing
— f — sujetador m - [mec] fastener, o clamp, release,sing
— f — mordaza f - [mec] clamp release,sing
— f — palanca f - [mec] lever release,sing
— f — pedal m - [mec] pedal release,sing
— f — perno m - [mec] bolt loosening
— f — — m para sujeción f - [mec] hold-down bolt, loosening, o release,sing
— f — presión f - pressure release,sing
— f — — f de aire m - [neumát] air pressure release,sing
— f — sujetador m - [mec] fastener, o grip, release,sing
— f — tensión f - [mec] tension release,sing
— f — — f de resorte m - [mec] spring tension release,sing
— f — válvula f - [mec] valve release,sing
— f fácil - [mec] easy release,sing
— f hidráulica - [mec] hydraulic release,sing
— f manual - [mec] hand, o manual, release,sing
— f rápida - quick, loosening, o release,sing
— f suficiente - [mec] sufficient release,sing
suelto m - [miner] loose (ore)
— m cargado - [miner] loaded loose (ore)
suelto,ta a - . . . • lax; slack • released • separate(ly)
sueño m acariciado - (long) standing dream
— m temerario m - wild dream
suerte f - . . . • [fam] lady luck
— f agria - bitter, o sour luck
— f amarga - bitter luck
— f buena - good luck
— f — en carrera f - [deport] good racing luck
— f en carrera(s) f - [deport] racing luck
— f mala - bad, o poor, o sour, luck
— f — en carrera f - [deport] poor racing luck
— f parecida - similar luck • similar fate
— f similar - similar luck • similar fate
suéter m - [vest] sweater
— m liviano - [vest] light sweater
— m mediano - [vest] medium (weight sweater
— m para tenis m - [deport] tennis sweater
— m pesado - [vest] heavy sweater
suficiencia f - . . . • plenty
suficiente adv - . . . • enough; ample, plenty; plenteous • high enough • far enough
suficientemente adv - . . . • plenty • ample • far enough • high enough
— adv alto,ta - high enough
— adv ancho,cha - wide enough; sufficiently wide
— adv apretado,da - tight enough
— adv bueno,na - good enough
— adv caliente - hot, o warm. enough
— adv exacto,ta - accurate enough
— adv flexible - flexible enough; sufficiently flexible
— adv frío,ria - cold enough
— adv fuerte - strong enough; sufficiently strong
— adv grande - large, o big, enough
— adv largo,ga - long enough; amply long
— adv lento,ta - slow enough; sufficiently slow
— adv liviano,na - [aceites] thin enough
— m malo,la - bad enough
— adv pequeño,ña - small enough
— adv potente - powerful, o strong, enough
— m resistente - resistant, o strong, enough
sufijo m - [gram] suffix
— m para número m - number suffix
— m — — m de serie f - serial number suffix

sufragado,da a - [pol] voted; cast @ vote
sufragante m - [pol] voter
sufragar v - [fin] to, pay, o cover • [pol] to (cast @) vote
— v gasto(s) m - [fin] to meet, o cover, o pay, @, cost(s), o expense(s)
sufragio m - [pol] vote,ting
sufridera f - [mec] die • [domést] vase opening
sufrido,da a - suffered • endured
sufrimiento m - . . . . • endurance,ring
— m de daño m - damage, suffering, o sustaining
sufrir v - . . . • to take place
— v consecuencia f - to suffer @ consequence
— v daño m - to, suffer, o sustain, @ damage
— v deformación f - to, sustain, o undergo, @ deformation • [metal] to undergo @ yielding
— v — f plástica - [metal] to undergo @ plastic, deformation, o yielding
— v — f (más) severa - [metal] to undergo @ (most) severe(st) deformation
— v pena f - to suffer @ penalty
— v penalidad f - to suffer @ penalty
sugerencia f - . . .; proposal; tip; pointer; hint
— f de, empleado(s) m, o personal m - employee suggestion
— f de valor m - value(able), suggestion, o proposal
— f específica - specific suggestion
— f para aplicación f - application suggestion
— f — instalación f - installation suggestion
— f — proyección f - design suggestion
— f — solución f - solution suggesting,tion
— f — vestimenta - [vest] dress suggestion
— f — f apropiada - [vest] (proper) dress (code) suggestion
— f sobre neumático(s) m - [autom-neumát] tire, suggestion, o tip
— f — seguridad f—[segurid] safety suggestion
— f técnica—[ind] technical, suggestion, o tip
sugerente a - . . . . • tentative
sugerentemente adv - suggestingly; tentatively
sugerido,da - suggested; proposed
sugerir v - . . .; to propose
— v, ancho(r), o anchura, para llanta f - [autom-neumát] to suggest @ rim width
— v aplicación f - to suggest @ application
— v proyección f to suggest @ design(ing)
— v respuesta f - to suggest @, answer, o response
— v solución f - to suggest @ solution
— v valor m - to, suggest, o propose, @ value
sugestión f - . . . . • [fam] tip
— f para regulación f - control(ling) hint
— f — f de distorsión f - [sold] distortion control hint
sugestionado,da a - suggested
sugestionar v - . . .; to suggest
sui generis a - . . .; unusual
suiche* m - véase conmutador m; interruptor m
sujeción f - . . . . • holding (down); clamping; locking; latching; attaching,chment; grip; gripping; anchoring; fastening • snapping; bite,ting • maintaining • conforming
sujeción f - . . .; subjecting
— f a - clamping to • subjecting,tion to
— f — banco m - [mec] clamping to @ table
— f — cadena f - [mec] attaching to @ chain
— f — caja f - [mec] attaching, o fastening, to @, case,sing, o body, o housing
— f — f para eje m - [autom-mec] fastening to @ axle housing
— f — trabajo m - [mec] clamping to @ work
— f ajustada - [mec[tight, o adjusted, grip
— f automática - [mec] automatic grip(ping)
— f con arandela f - [mec] washer fastening
— f — f para seguridad f - [mec] lockwasher fastening; fastening with @ lockwasher
— f — firmeza f - [mec] tight grip(ping)
— f — f de tornillo m para banco m - [mec] viselike grip(ping)
— f — perno(s) m - [mec] fastening with bolts
— f — resorte m - [mec] snapping; spring fast-

ening
sujeción f con tuerca(s) f - [mec] fastening with @ nut(s)
— f con tornillo m - [mec] fastening with @ screw
— f — — m con tornillo m con casquete m - [mec] fastening with @ cap screw
— f de acero m - [mec] steel, fastening, o securing
— f de alambre m - wire, fastening, o gripping
— f — m trefilado - [trefil] drawn wire grip(ping)
— f — árbol m - [mec] shaft, holding, o gripping
— f — banda f - [mec] band attaching • [metal-lam] strip fastening
— f — — f para acoplamiento m - [mec] connecting band, attaching, o fastening
— f — cable m (de alambre m) - [cabl] wire rope, clamping, o fastening
— f de cadena f - [mec] chain, attaching, o fastening
— f — camisa f - [mec] liner clamping
— f — convertidor m - [metal-prod] converter fastening
— f — cremallera f - [mec] rack, fastening, o clamping
— f — cubierta f - [mec] cover fastening
— f — dispositivo m - [mec] device fastening • [electr-instal] fixture fastening
— f — horquilla f - [mec] yoke holding
— f — — f para árbol m - [autom-mec] shaft yoke, fastening, o holding
— f — — f — — m para aportación f (de fuerza f) - [autom-mec] input shaft yoke holding
— f — línea f - line holding
— f — madera f - [mec] wood fastening
— f — mandíbula f - [mec] jaw, holding, o clamping, o gripping
— f — material m - [mec] material, fastening, o holding, o gripping, o fastening
— f — oscilador m - [mec] rocker clamping
— f — portadiferencial m - [autom-mec] differential carrier fastening
— f — relé m - [electrón] relay latching
— f — tipo triple - [mec] triple type fastening
— f — tubería f - [tub] pipe holding
— f — uña f - [mec] latching
— f — — f para cañón m - [metal-prod] mud gun latch(ing)
— f — varilla f - [trefil] rod grip(ping)
— f — volante m - [autom-mec] steering (wheel) lock
— f directa - [mec] direct, attaching,chment, o holding
— f en mandíbula(s) f - [mec] jaw grip(ping)
— f — con presión f de tornillo m para banco m - [mec] viselike jaw grip
— f — sitio m - in place holding
— f firme - firm, o positive, o tight, o secure, hold(ing), o tightening, o clamping, o anchoring, o grip(ping)
— f fuerte - [mec] tight, holding, o grip(ping)
— f magnética - magnetic, holding, o latching
— f momentánea - [mec] momentary, holding, o latching, o grip(ping)
— f perfecta - perfect, fastening, o holding
— f — de convertidor m - [metal-prod] converter, o perfect converter, fastening, o holding
— f permanente - [mec] permanent. fastening, o holding
— f positiva - positive, fastening, o hold(ing)
— f segura - secure, fastening, o holding, o bite,ting
— f transversal - crosswise grip(ping)
— f — a través de orificio m—[mec] grip(ping) across @ port
— f triple - [mec] triple fastening
sujeta m cable(s) - [electr-instal] cable, o wire, retainer, o holder
sujetado,da a - fastened; held; latched; locked;

clamped • snapped • maintained
sujetado,da a a - subjected to
— a **a banco** m - [mec] clamped, o fastened, to @, bench, o table
— a — **cable** m - [sold] fastened to @ lead
— a — — **conductor** - [sold] taped, o fastened, to @ conductor cable
— a — **cadena** f - [mec] fastened, o attached, to @ chain
— a — **caja** f - [mec] fastened to @, housing, o body, o case, o box
— a — f **para eje** m - [autom-mec] fastened to @ axle housing
— a — **trabajo** m - [mec] clamped, o fastened, to @ work
— a **con arandela** f - [mec] fastened with @ washer
— a — — f **para seguridad** f - [mec] fastened with @ lockwasher
— a **con perno(s)** m - [mec] fastened with @ bolt(s); bolted down; bolt-held
— a — **resorte** m - [mec] spring held • snapped
— a — **seguridad** f - fastened securely
— a — **tornillo(s)** m - [mec] screw fastened
— a — — m **con casquete** m - [mec] fastened with @ cap screw(s)
— a — **tuerca(s)** - [mec] fastened with @ nut(s); nut held
— a **directamente** - [mec] attached directly
— a **en sitio** m - [mec] held in @ place
— a **firmemente** - [mec] fastened, o held, o clamped, o achored, o gripped, tightly, o firmly, o securely, o positively • tightened
— a **fuertemente** - [mec] gripped tightly
— a **magnéticamente** - [mec] held, o fastened, o latched, magnetically
— a **permanentemente** - fastened permanently
— a **transversalmente** - fastened across
— a — **a través de orificio** m - [mec] gripped across @ port
sujetador m - [mec] . . . • fastener; brace; clip; clamp; holder; holddown; griplet; locking slip • stitch • cam - [sold] jig
— m **abierto** - [mec] open, grip, o fastener
— m **accionado** - [mec] operated clamp
— m — **con cilindro** m - [mec] cylinder operated clamp
— m — — — m **neumático** - [mec] air cylinder operated clamp
— m **aflojado** - [mec] loosened, o released, fastener, o clamp
— m **ajustable** - [mec] adjustable clamp
— m — **con tornillo** m - [mec] screw adjusable, clamp, o cam
— m **ajustado** - [mec] adjusted, clamp, o grip
— m **ancho** - [mec] wide, clamp, o fastener, o cam
— m **apretado** - [mec] tightened, fastener, o clamp
— m **automático** - [mec] automatic, grip, o clamp
— m — **para alambre** m - [trefil] automatic wire grip
— m **central** - [mec] center,tral, fastener, o clamp
— m **colocado** - [mec] placed clamp
— m **con ajuste** m - [mec] adjustable, clamp, o cam
— m — — m **con tornillo** m - [mec] screw adjusted, clamp, o cam
— m — **cabeza** f - [mec] headed fastener
— m — **estría(s)** f **(circulares)** - [mec] serrated cam
— m — **frente** m **ancho** - [mec] wide face(d) cam
— m — **gancho** m - [mec] hook fastener
— m — — m **y ojo** m - [mec] hook and eye fastener
— m — **resorte** m - [mec] spring (loaded) cam
— m — — m **integral** - [mec] spring-loaded cam
— m — **tuerca** f - [tub] hook and eye, bolt, o fastener; hook bolt
— m **convexo** - [mec] convex cam
— m **cuneiforme** - [mec] wedge, fastener, o grip
— m — **para avance** m **gradual** - [trefil] inching

wedge grip
sujetador m **de acero** m - [mec] steel fastener
— m — — m **con carbono** m **alto** - [mec] high carbon steel fastener
— m — — m — — m **bajo** - [mec] low carbon steel fastener
— m — — m — — m — **roscado exteriormente** - [mec] low carbon steel externally threaded fastener
— m — — m — — m — — **interiormente** - [mec] low carbon steel internally threaded fastener
— m — — m **roscado exteriormente** - [mec] externally threaded steel fastener
— m — — m — **interiormente** - [mec] internally threaded steel fastener
— m — — m **según norma** f - [mec] standard steel, o steel standard, fastener
— m **alambre** m - [mec] wire, clamp, o clip, o fastener
— m **aluminio** m - [mec] aluminum clamp
— m **bronce** - [mec] bronze fastener
— m **corchete** m - [tub] hook and eye, bolt, o fastener
— m **fibra** f - [mec] fiber retainer
— m — f **para devanado** - [electr-equip] coil fiber retainer
— m — f — m **protector** - [electr-equip] release coil fiber retainer
— m — — f — m — **para tensión** f **nula** - [electr-equip] no voltage release coil fiber retainer
— m **metal** - [mec] metal fastener
— m — — m **liso** - [mec] smooth metal cam
— m **nilón** m - [mec] nylon clamp
— m **resorte** m - [mec] spring, clip, o holder
— m **tipo** m **con sillete** m - [cabl] chair type clip
— m — — m — — m **para cable(s)** m - [cabl] chair type wire rope clip
— m — — m **triple** - [mec] triple type fastener
— m **dentado** - [mec] serrated, clip, o cam
— m **elevado** - [mec] raised clamp
— m **empaquetado** - [mec] packed, grip, o fastener
— m **en enchufe** m - [electr-equip] clamp, in, o at, @ plug
— m **engranado** - [mec] engaged cam
— m **epicéntrico** - [mec] over-center latch
— m **especial** - [mec] special(ty) fastener
— m **estándard** - [mec] standard fastener
— m — **para velocidad** f **reducida** - [mec] standard low, speed, o velocity, fastener
— m **flojo** - [mec] loose fastener
— m **hidráulico** - [mec] hydraulic, fastener, o holddown
— m **inferior** - [mec] lower, o bottom, clamp, o fastener
— m — **para cable** m - [electr-instal] lower, o bottom, cable clamp
— m **inspeccionado** - [mec] inspected, clamp, o cam, o fastener
— m **interior** - [sold] internal clamp
— m — **para alineación** f - [sold] internal line-up clamp
— m — — **alivio** m **de tensión** f - [mec] internal strain relief clamp
— m **interno** - [tub] internal clamp
— m — **con accionador** m **eléctrico** - [tub-instal] electrically powered internal clamp
— m **laminado** - [f.c.] rolled clip
— m — **para riel** m - [f.c.] rolled rail clip
— m **liso** - [mec] smooth cam
— m **mantenido** - [mec] held cam
— m **momentáneo** - [mec] momentary latch
— m **movido** - [mec] moved cam
— m **para acoplamiento** m - [mec] connect(ing) fastener
— m — — f **rápido** - [mec] quick-coonect(ing) fastener
— m — — m **y desacoplamiento** m **rápido(s)** - [mec] quick-connect and (quick-) release fastener
— m — **acumulador** m - [comb.int] battery, hold-

er, o̲ bracket
sujetador m̲ **para ajuste** m̲ ‒ [mec] adjusting,ment
fastener
— m̲ — — m̲ **rápido** - [mec] speed fastener
— m̲ — **alabe** m̲ - cam holder
— m̲ — **alambre** m̲ - [trefil] wire grip(ping)
— m̲ — **alineación** f̲ - [tub] line-up clamp
— m̲ — **alivio** m̲ - [mec] relief clamp
— m̲ — — m̲ **para tensión** f̲ - [mec] strain, relief, o̲ relieving, clamp
— m̲ — m̲ — — m̲ **de tracción** f̲ - [mec]
strain, relief, o̲ relieving, clamp
— m̲ — — — — f̲ **ajustado** - [mec] tightened
strain, relief, o̲ relieving, clamp
— m̲ — m̲ — — f̲ **apretado** - [mec] tightened
strain, relief, o̲ relieving, clamp
— m̲ — — — **tracción** f̲ - [mec] strain, relief, o̲ relieving, clamp
— m̲ — **anillo** m̲ - [mec] ring fastener
— m̲ — **aro** m̲ - [mec] ring fastener
— m̲ — **avance** m̲ - advance,cing grip
— m̲ — — m̲ **gradual** - [mec] inch(ing) grip
— m̲ — — m̲ **abierto**—[trefil] open inch(ing)
grip
— m̲ — **balancín** m̲ - [mec] rocker clamp
— m̲ — **banda** f̲ - [metal-lam] strip holder
— m̲ — — f̲ **para entrada** f̲ - [metal-lam] feeding end strip holder
— m̲ — — f̲ — **extremo** m̲ **para entrada** f̲ -
[metal-lam] feeding end strip holder
— m̲ — — f̲ — — m̲ **para salida** f̲ - [metal-lam]
delivery end holder
— m̲ — — f̲ — **salida** f̲ - [metal-lam] delivery
(end) strip holder
— m̲ — **batería** f̲ - [mec] battery, clamp, o̲
bracket
— m̲ — **bobina** f̲ - [metal-lam] coil holder
— m̲ — — f̲ **para entrada** f̲ - [metal-lam] feeding end coil holder
— m̲ — — f̲ — **extremo** m̲ **para entrada** f̲ -
[metal-lam] feeding end coil holder
— m̲ — **boquilla** f̲ - [comb.int] jet holder •
[sold] nozzle clamp
— m̲ — — f̲ **principal** - [comb.int] main jet
holder
— m̲ — **buje** m̲ - [mec] bushing, holder, o̲ clamp
— m̲ — — m̲ **para cursor** m̲ - [mec] slide bushing
clamp
— m̲ — — — — m̲ **para hilera** f̲ - [mec] die
slide bushing clamp
— m̲ — **cable** m̲ - [cabl] cable, o̲ wire rope,
holder, o̲ clip • [grúas] line holder •
[electr-instal] cable clamp
— m̲ — — m̲ **muerto** - [petról] slack line holder
— m̲ — — m̲ **para cabrestante** - [petról] catline
grip
— m̲ — — m̲ **para maniobra(s)** f̲ - [petról] catline grip
— m̲ — — m̲ — **torno** m̲ - [petról] catline grip
— m̲ — **camisa** f̲ - [mec] liner clamp
— m̲ — — f̲ **colocado** - [mec] installed liner
clamp
— m̲ — — f̲ **elevado** - [mec] raised liner clamp
— m̲ — — f̲ **instalado** - [mec] installed liner
clamp
— m̲ — — f̲, **quitado**, o̲ **removido**, o̲ **sacado** -
[mec] removed liner clamp
— m̲ — — f̲ **verificado** - [mec] checked liner
clamp
— m̲ — — f̲ — **nuevamente** - [mec] rechecked
liner clamp
— m̲ — — f̲ **y vástago** m̲ - [mec] liner and rod
clamp
— m̲ — — f̲ **y vástago** m̲ **verificado** - [mec]
checked liner and rod clamp
— m̲ — — f̲ — — m̲ — **nuevamente** - [mec] rechecked liner and rod clamp
— m̲ — **caño,ñería** - [tub] culvert pipe fastener
— m̲ — **cielo** m̲ **raso** - [constr] ceiling clip
— m̲ — **cinta** f̲ - [mec] lining, fastener, o̲
keeper
— m̲ — — f̲ **para freno** m̲ - [mec] brake lining,
fastener, o̲ keeper

sujetador m̲ **para cojinete** m̲—[mec] bearing clamp
— m̲ — — m̲ **para biela** f̲ - [mec] pitman bearing clamp
— m̲ — **conducto** m̲ - [mec] conduit clamp
— m̲ — **conductor** m̲ - [electr-instal] cable, o̲
lead, clamp, o̲ clip
— m̲ — — m̲ **para entrada** f̲ - [sold] input lead
clamp
— m̲ — — m̲ — **salida** f̲ - [sold] output lead
clamp
— m̲ — **conectar** v̲ - [electr-instal] connecting,tion clamp
— m̲ — — v̲ **cable** m̲ - [electr-instal] cable
connecting,tion clamp
— m̲ — — v̲ **conductor** m̲ - [electr-instal] cable
connecting clamp
— m̲ — **conexión** f̲ - [electr-instal] connecting,tion clampo
— m̲ — — f̲ **de cable** m̲ - [electr-instal] cable
connecting,tion clamp
— m̲ — — f̲ — **conductor** m̲ - [electr-instal]
cable connecting,tion clamp
— m̲ — **conjunto** m̲ - [mec] assembly holder •
[comput] (reset) group latch
— m̲ — **corona** f̲ - [mec] ring fastener
— m̲ — — f̲ **dentada** - [mec] ring gear fastener
— m̲ — **cremallera** f̲ - [mec] rack clamp
— m̲ — **cursor** m̲ - [mec] slide clamp
— m̲ — — m̲ **para hilera** f̲ - [mec] die slide
clamp
— m̲ — **chapa** f̲ - [metal-lam] strip holder
— m̲ — **desacoplamiento** m̲ (**rápido**) - [mec]
(quick-)release fastener
— m̲ — **disco** m̲ - [mec] disk, clamp, o̲ clip
— m̲ — **engranaje** m̲ - [mec] gear fastener
— m̲ — **fleje** m̲ - [metal-lam] strip holder •
[transp] signode, clip, u̲ clamp, o̲ sealer
— m̲ — — m̲ **para entrada** f̲ - [metal-lam] feeding end strip holder
— m̲ — — m̲ **en extremo** m̲ **para entrada** f̲ -
[metal-lam] feeding end strip holder
— m̲ — — m̲ — — m̲ **para salida** f̲ - [metal-lam]
delivery end strip holder
— m̲ — **guardaojal** m̲ - [mec] grommet clip
— m̲ — — m̲ **para conductor** m̲ - [electr-instal]
wire, o̲ lead, grommet clip
— m̲ — — m̲ — — m̲ **para carga** f̲ - [electr-
instal] charge,ging wire grommet clip
— m̲ — **lápiz** m̲ - pencil, holder, o̲ clip
— m̲ — **manubrio** m̲ - crank holder bracket
— m̲ — **montaje** m̲ - [mec] mounting clip
— m̲ — **mordaza** f̲ - [mec] clamp, cam, o̲ clip
— m̲ — **perno** m̲ - [mec] bolt, o̲ locking, clip
— m̲ — **plancha** f̲ - [mec] plate, holder, o̲
holddown
— m̲ — **punzón** m̲ - [mec] punch, retainer, o̲
holder
— m̲ — **reajuste** m̲ - [comput] reset(ting) latch
— m̲ — **resorte** m̲—[mec] spring, holder, o̲ clamp
— m̲ — **riel** m̲ - [f.c.] rail clip
— m̲ — **rodillo** m̲ - [mec] roll clamp
— m̲ — — m̲ **para trabajo** m̲ - [metal-lam] work
roll clamp
— m̲ — **seguridad** f̲ (**para tubería** f̲) - [petról]
liner catcher
— m̲ — **soporte** m̲ - [mec] support, clip, o̲ clamp
— m̲ — — m̲ **para defensa** f̲ - [mec] guard support, clip, o̲ clamp
— m̲ — **soldadura** f̲ - [sold] welding clamp •
welding jig
— m̲ — **soldar** v̲ - [sold] welding jig
— m̲ — — v̲ **collar(es)** m̲ - [petról] collar(s)
welding jig
— m̲ — **surtidor** m̲ - [comb.int] jet holder
— m̲ — **tapa** f̲ - [mec] cover, o̲ lid, holder, o̲
clamp
— m̲ — — f̲ **para cursor** m̲ - [mec] slide cover
clamp
— m̲ — — f̲ — — m̲ **para hilera** f̲ - [trefil]
die slide cover clamp
— m̲ — — f̲ — **hilera** f̲ - [trefil] die cover
— m̲ — **tornillo** m̲ - [mec] screw, holder, o̲ clip
— m̲ — — m̲ **para acoplamiento** m̲ - [mec] cou-

pling screw clip
sujetador <u>m</u> **para tubería** <u>f</u> - [mec] tube, <u>o</u> pipe, clamp
— <u>m</u> — <u>m</u> **para alimentación** <u>f</u> - [mec] feed-(ing) tube clamp
— <u>m</u> — **tubo(s)** <u>m</u> - [tub] pipe,ping clamp • [petról] pipe hanger
— <u>m</u> — — f **colgante** - [petról] liner hanger
— <u>m</u> — — <u>f</u> **para bombeo** <u>m</u> - [petról] tubing catcher
— <u>m</u> — — f **flexible** - [mec] loom clip
— <u>m</u> — — <u>m</u> **para alcantarilla(s)** <u>f</u> - [tub] culvert pipe fastener
— <u>m</u> — **velocidad** <u>f</u> - speed fastener
— <u>f</u> — — f **alta** - [mec] high velocity fastener
— <u>m</u> — — <u>f</u> **baja** - [mec] low velocity fastener
— <u>m</u> — — <u>f</u> **estándar(d)** - [mec] standard velocity fastener
— f — — f **reducida** - [mec] low velocity fastener
— <u>m</u> **para zuncho(s)** <u>m</u> - [mec] Signode, clip, <u>o</u> sealer
— m **perimétrico** - [sold] girth clamp
— <u>m</u> — **para soldadura** <u>f</u> - [sold] girth welding clamp
— m **porta bobina(s)** <u>f</u> - [electr-equip] coil support bracket
— m **quitado** - [mec] removed fastener
— <u>m</u> **recubierto** <u>m</u> - [mec] coated clamp
— <u>m</u> **reemplazado** - [mec] replaced cam
— <u>m</u> **removido** - [mec] removed fastener
— <u>m</u> **roscado** - [mec] threaded fastener
— <u>m</u> — **de acero** <u>m</u> - [mec] threaded steel fastener
— <u>n</u> — — — <u>m</u> **con carbono** <u>m</u> **alto**—[metal-fabr] high carbon steel threaded fastener
— <u>m</u> — — <u>m</u> — — **bajo** - [metal-fabr] low carbon steel threaded fastener
— <u>m</u> — — <u>m</u> **según norma** <u>f</u> - [metal-fabr] threaded steel standard fastener
— <u>m</u> — **exteriormente** - [metal-fabr] externally threaded fastener
— <u>m</u> — **interiormente** - [metal-fabr] internally threaded fastener
— <u>m</u> **sacado** - [mec] removed fastener
— <u>m</u> **según norma** <u>f</u> - [mec] standard fastener
— <u>m</u> — — <u>f</u> **de acero** <u>m</u>—**standard** steel fastener
— <u>m</u> — — <u>f</u> — — <u>m</u> **con carbono** <u>m</u>—[metal-fabr] carbon steel standard fastener
— <u>m</u> — — <u>t</u> — — <u>m</u> — <u>m</u> **alto** - [metal-fabr] high carbon steel standard fastener
— <u>m</u> — <u>f</u> — — <u>m</u> — — <u>m</u> **bajo** - [metal-fabr] low carbon steel standard fastener
— <u>m</u> — <u>f</u> — — <u>m</u> — — — **roscado exteriormente** - [metal-fabr] low carbon steel externally threaded standard fastener
— <u>m</u> — <u>f</u> — <u>m</u> — — — **interiormente** - [metal-fabr] low carbon steel internally threaded standard fastener
— <u>m</u> — — <u>f</u> **roscado** - [metal-fabr] standard threaded fastener
— <u>m</u> — — <u>f</u> — **exteriormente** - [metal-fabr] externally threaded standard fastener
— <u>m</u> — — <u>f</u> — **interiormente** - [metal-fabr] internally threaded standard fastener
— <u>m</u> **serrado** - [metal-fabr] serrated cam
— <u>m</u> **soltado** - [mec] released clamp
— <u>m</u> **suelto** - [mec] loose fastener
— <u>m</u> **superior** - [mec] upper, <u>o</u> top, clamp
— <u>m</u> **para cable** <u>m</u> - [electr-instal] upper, <u>o</u> top, cable clamp
— <u>m</u> **tipo horquilla** <u>f</u> - [mec] forked latch
— <u>m</u> **trabado** - [mec] locked cam
— <u>m</u> **triple** - triple fastener
— <u>m</u> **verificado** - [mec] checked, fastener, <u>o</u> clamp
— <u>m</u> — **nuevamente** - [mec] rechecked clamp
sujetadores <u>m</u> **y herramientas** <u>f</u> - [herram] fasteners and tools
sujetador,ra <u>a</u> - holder,ding; fastener,ning; clamp; clip; dog; support
sujetar <u>v</u> - . . . .; to grip; to attach; to anchor • to secure; to hold; to bite; to fix; to attach • to tighten • to clip; to clamp; to

lock • to latch • to position • to snap; to hold down • [cabl] to seize • to subject
sujetar <u>v</u> **a** - to subject to
— <u>v</u> — **banco** <u>m</u> - [mec] to clamp to @, bench, <u>o</u> table
— <u>v</u> — **cadena** <u>f</u> - [mec] to attach to @ chain
— <u>v</u> — **caja** <u>f</u> - [mec] to fasten to @, box, <u>o</u> case,sing, <u>o</u> body, <u>o</u> housing
— <u>v</u> — — <u>f</u> **para eje** <u>m</u> - [autom-mec] to fasten to @ axle housing
— <u>v</u> — **trabajo** <u>m</u> - [mec] to clamp to @ work
— <u>v</u> — **variación(es)** <u>f</u> - to subject to variation(s)
— <u>v</u> — — <u>f</u> **específica(s)** - to subject to @ specific variation(s)
— <u>v</u> — — <u>f</u> — **permitida(s)** - to subject to @ specific permissible variation(s)
— <u>v</u> — — <u>f</u> **permitida(s)** - to subject to @ permissible variation(s)
— <u>v</u> **acero** <u>m</u> - [mec] to fasten @ steel
— <u>v</u> **alambre** <u>n</u> - [mec] to grip @ wire
— <u>v</u> — **trefilado** - [trefil] to grip @ drawn wire
— <u>v</u> **árbol** <u>m</u> - [mec] to hold @ shaft
— <u>v</u> — <u>m</u> **para aportación** <u>f</u> (**de fuerza** <u>f</u>) - [autom-mec] to hold @ input shaft
— <u>v</u> **banda** <u>f</u> - [mec] to hold @, band, <u>o</u> strip • to attach @, band, <u>o</u> strip
— <u>v</u> — <u>f</u> **para acoplamiento** <u>m</u> - [mec] to attach @ connecting band
— <u>v</u> **cable** <u>m</u> - [cabl] to to, fasten, <u>o</u> clamp, @, cable, <u>o</u> wire rope
— <u>v</u> — <u>m</u> **de alambre** <u>m</u> - [cabl] to, fasten, <u>o</u> clamp, @ wire rope
— <u>v</u> **cadena** <u>f</u> - [mec] to to, fasten, <u>o</u> attach, @ chain
— <u>v</u> **camisa** <u>f</u> - [mec] to to, attach, <u>o</u> clamp, @ liner
— <u>v</u> **con arandela** <u>f</u> - [mec] to fasten with @ washer
— <u>v</u> — — <u>f</u> **para seguridad** <u>f</u> - [mec] to fasten with @ lockwasher
— <u>v</u> **perno** <u>m</u> - [mec] to fasten with @ bolt
— <u>v</u> **resorte** <u>m</u> - [mec] to snap
— <u>v</u> **seguridad** <u>f</u> - to fasten scurely
— <u>v</u> **tornillo** <u>m</u> - [mec] to fasten with @ screw; to screw down
— <u>v</u> — <u>m</u> **con casquete** - [mec] to fasten with @ cap screw
— <u>v</u> **tuerca** <u>f</u> - [mec] to fasten with @ nut
— <u>v</u> **convertidor** <u>m</u> - [metal-prod] to fasten @ converter
— <u>v</u> **cremallera** <u>f</u> - [mec] to clamp @ rack
— <u>v</u> **cubierta** <u>f</u> - [mec] to fasten @ cover
— <u>v</u> **directamente** - to, fasten, <u>o</u> grip, <u>o</u> attach, directly
— <u>v</u> **dispositivo** <u>m</u> - [mec] to fasten @, fixture, <u>o</u> device
— <u>v</u> **elemento** <u>m</u> - [mec] to fasten @ element
— <u>v</u> — <u>m</u> **para escoria** <u>m</u> - [metal-prod] to fasten @ slag notch element
— <u>v</u> **en sitio** <u>m</u> - to hold in @ place
— <u>v</u> **firmemente** - [mec] to, hold, <u>o</u> clamp, <u>o</u> grip, <u>o</u> anchor, securely, <u>o</u> firmly, <u>o</u> tightly, <u>o</u> positively • to tighten securely
— <u>v</u> **fuertemente** - [mec] to grip tightly
— <u>v</u> **horquilla** <u>f</u> - to hold @ yoke
— <u>v</u> — <u>f</u> **para árbol** <u>m</u> - [mec] to hold @ shaft yoke
— <u>v</u> — <u>f</u> — — <u>m</u> **para aportación** <u>f</u> (**de fuerza** <u>f</u>) - [autom-mec] to hold @ input shaft yoke
— <u>v</u> **madera** <u>f</u> - [mec] to fasten @ wood
— <u>v</u> **magnéticamente** - [mec] to latch magnetically
— <u>v</u> **mandíbula** <u>f</u>—[mec] to, hold, <u>o</u> clamp, @ jaw
— <u>v</u> **material** <u>m</u> - [mec] to fasten @ material
— <u>v</u> **perfectamente** - [mec] to, fasten, <u>o</u> attach, perfectly
— <u>v</u> — **convertidor** <u>m</u> - [metal-prod] to, fasten, <u>o</u> attach, perfectly @ converter
— <u>v</u> **permanentemente** - [mec] to fasten permanently
— <u>v</u> **portadiferencial** <u>m</u> - [autom-mec] to fasten @ differential carrier

sujetar v relé m - [electrón] to, fasten, o latch, @ relay
— v transversalmente - [mec] to grip arcoss
— v — a través de orificio m - [mec] to grip across @ port
— v tubería f - [tub] to hold @, tubing, o piping
— v tubo m - [tub] to hold @, tube, o pipe
— v varilla f - [trefil] to grip @ rod
sujeto,ta a - subject(ed) • conformed
— a ajuste m - subject to @ adjustment
— a — aprobación f—subject to @ approval
— a — cambio(s) - subject to @ change(s)
— a — consulta f - subject to inquiry
— a — f previa - subject to prior inquiry
— a — fallar v - subject to @ failure
— a — deslizamiento,zar - [topogr] subject to @ landslide(s)
— a — filtración f - subject to @ seepage
— a — juicio m - subject to @ judgement • liable to @ permit
— a — litigación,gio - [legal] litigable
— a — multa f - subject to @ fine
— a — presencia f de aceite(s) m - [ind] subject to @ oil, presence, o condition(s)
— a — reajuste - [com] subject to @ escalation • escalatable
— a — revisión f - revisable; subject to @ revision • escalatable
— a — tolerancia f - [ind] subject to @ tolerance
— a — f en fabricación f - [ind] subject to @ manufacturing tolerance(s)
— a — variación(es) f - subject to @ variation(s)
— a — — f específica(s) - subject to specific variations
— a — — f — permitida(s) - [ind] subject to @ specific allowable specification(s)
— a — — f permitida(s) - subject to, allowable, o permissible, variation(s)
sulfatación f - [quím] [quím] sulfatizing
sulfatizante a - [miner] sulfatizing
sulfato m alumínico - [quím] aluminum sulfate
— m — potásico - [quím] aluminum pottasium sulfat e
— m amónico - [quím] ammonic sulfate
— n cúprico - [quím] cupric sulfate
— m de aluminio m - [quím] aluminum sulfate
— m — amonio m - [quím] ammonium sulfate
— m — bario m - [quím] barium sulfate
— m — — m con gravedad f específica de 4,3 a 4,6 - [quím] baryte
— m — cinc - [quím] zinc sulfate
— m — cobre m - [quím] copper sulfate
— n — hierro m - [quím] iron sulfate
— m — magnesio - [quím] magnesium sulfate
— m — manganeso m - [quím] manganese sulfate
— m — potasio m - [quím] potassium sulfate
— m — sodio m - [quím] sodium sulfate
— m férrico - [quím] ferric sulfate
— m ferroso - [quím] ferrous sulfate
— m magnésico - [quím] magnesic sulfate
— m láurico - [quím] lauric sulfate
— m — de sodio m—[quím] sodium lauric sulfate
— m manganésico - [quím] manganesic sulfate
— m potásico - [quím] potassic,sium sulfate
sulfito m de amoníaco m—[quím] ammonium sulfite
sulfonítrico,ca a - [quím] sulfonitric
sulfurado,da a - [quím] sulfurized; sulfurated; sulfureted
sulfúrico,ca a - [quím] sulfuric
sulfuro m - [quím] sulfide
— m de amoníaco m - [quím] ammonium sulfide
— m — hidrógeno m - [quím] hydrogen sulfide
— m — hierro m - [quím] iron sulfide
— m hidrogenado - [quím] hydrogen(ated) sulfide
— m liviano - [quím] light sulfide
— f pesado - [quím] heavy sulfide
sulfuroso,sa a - [metal] sulfur bearing; sulfurous
suma f - . . . . • combination

suma f asegurada - [seguros] insured value
— f de capacidad f - [ind] output sum
— f — — f nominal - [ind] rated output sum
— f — desviación(es) f - deviation(s) sum
— f — peso(s) m - combined, o total, weight(s)
— f — producción f - [ind] output sum
— f — — f nominal - [ind] rated output sum
— f — rendimiento m - yield, o output, sum
— f — — m nominal - [ind] rated output sum
— f global - lump sum • [com] fixed price
— f otorgada - granted, o given, sum, o amount
— f — como anticipo m - amount given as @ advance
— f retenida - withheld amount
— f total - total, o lump, sum
sumamente adv - highly • remarkably
— adv corrosivo,va a - severely corrosive
— adv crítico,ca a - very critical
— adv cuidadoso,sa - very, o extra, careful
— adv exigente - extraordinarily demanding
— adv impermeable - very, o highly, impermeable
— adv irregular - very, o highly, irregular
— adv liso,sa a - very, o extra, smooth
— adv sencillo,lla a - extremely simple
— adv sensible a - extra, o very, sensitive
— adv tenaz - [metal] very, o exceptionally, tough
— adv variable - very, o highly, variable
sumar v - to add • to amount to
— v salida f - [electrón] to sum @ output
sumarse v a - to join (to)
sumergente a - submerging,rsing
sumergido,da a - . . .; submerged • subaqueous; underwater • dipped • plunged
— a en aceite m - oil, immersed, o dipped; dipped in oil
— a — agua m - water dipped; dipped in water
— a — asfalto m - asphalt dipped
— a — — m (en) caliente - [constr] hot asphalt dipped
— a — barniz m - [constr] varnish dipped
— a — — m aislador - insulating varnish dipped
— a — bonderita f - [constr] bonderite (solution) dipped
— a — caliente - hot dipped
— a — — en asfalto m - [constr] hot asphalt, dipped, o immersed
— a parcialmente - partially submerged
— a — en caliente - partially hot dipped
— repetidamente - dipped repeatedly
— a totalmente - totally, o completely, dipped, o submerged, o immersed
— a — en fundente m - [sold] completely flux submerged
sumergir v - . . .; to immerse; to dip
— v alcantarilla f - [constr] to submerge @ culvert
— v bomba f - to submerge,rse @ pump
— v corona f - [mec] to submerge,rse @ crown
— v — f de alcantarilla f - [hidr] to submerge,rse @ culvert crown
— v electrodo m - to,submerge, ó immerse, @ electrode
— v en aceite m - [mec] to, dip in oil, o oil dip
— v — agua m - to dip in water; to water dip
— v — asfalto m - [constr] to submerge in asphalt
— v — — m en caliente - [constr] to hot asphalt dip
— v — barniz m - [constr] to varnish dip
— v — — m aislador - [constr] to dip in insulating varnish
— v — caliente - to hot dip
— v — disolución f - [quím] to dip in @ solution
— v — — f de bonderita f - [quím] to dip in @ bonderite solution
— v — solución f - to dip in @ solution
— v — — f de bonderita f - [quím] to dip in @ bonderite solution
— v parcialmente - to dip partially

sumergir v repetidamente - to dip repeatedly
— v transformador m - [electr-equip] to dip @ transformer
sumersión f - . . .; dip(ping); plunge,ging
— f de alcantarilla f - [constr] culvert submerging,rsing,rsion
— f — corona f - [constr] crown submerging,rsing,rsion
— f — — f de alcantarilla f - [hidr] culvert crown, sumberging,rsing,rsion
— f — electrodo m - [sold] electrode, immersing,sion, o submerging,rsing,rsion
— m — transformador m - [electr-equip] transformer dipping
— f en aceite m - oil dipping; dipping in oil
— f — agua m - water dipping; dipping in water
— f — asfalto m - [constr] asphalt dipping
— f — en caliente - [constr] asphalt hot, o hot asphalt, dipping
— f — barniz m - [constr] varnish dip(ping)
— f — — m aislador - [constr] insulating varnish dipping
— f — caliente - hot dip(ping)
— f repetida - repeated dip(ping)
sumidero m - [hidr] . . .; catch basin; drain, pit, o pocket, o sump • [sanit] sump (pit); sink (hole); floor, o sump, drain • [petról] sump • [sanit] cesspit; cesspool; [constr] receptacle; drop inlet • inlet • [comb.int] well • [hidr] inlet
— m colector - [hidr] collector basin; drainage island
— m con reja(s) f - [hidr] open grill
— m en seco - dry sump
— m estándar - [hidr] standard catchbasin
— m húmedo - [hidr] wet sump
— m hundido - [hidr] subsurface collector
— m interceptor - [hidr] intercepting sump
— m para aceite m - [comb.int] oil sump
— m — alimentación f - [mec] feed(ing) sump
— m — captación f - [hidr] catchbasin • drainage island
— m — condensado(s) m - [ind] condensate(s) sump
— m — desagüe m - [constr] drainage, sump, o receptacle
— m — — m pluvial - [constr] storm drainage, sump, o receptacle
— m — — m de alquitrán m - [ind] tar drainage sump
— m — torre f - [petról] tower sump
— m — — f para enfriamiento m cooling tower sump
— m — tubería f - [hidr] pipe catchbasin
— m — — f de acero m - [hidr] steel pipe catchbasin
— m — — f — — m corrugado - [hidr] corrugated steel pipe catchbasin
— m seco - dry sump
— m típico - [hidr] typical catchbasin
— m — de tubería f - [hidr] typical pipe catchbasin
— m — — — f de acero m - [hidr] typical steel pipe catchbasin
— m — — — f — — m corrugado - [hidr| typical corrugated steel pipe catchbasin
— m tubular - [constr] pipe, catchbasin, o drop inlet
— m — corrugado - [constr] corrugated pipe, catchbasin, o drop inlet
— m — — de acero m - [constr] corrugated steel pipe, catchbasin, o drop inlet
— m — de acero m - [constr] steel pipe, catchbasin, o drop inlet
suministrado,da a - supplied • [com] delivered
suministrador m - . . .; supplier
— m extranjero - foreign supplier
— m nacional - national, o domestic, supplier
suministrar v - . . . • to deliver
suministro m - . . .; provision • delivery
— m amplio - ample supply
— m — de energía f - [electr-prod] ample power supply(ing)

suministro m adecuado - appropriate, o adequate, supply
— m de agua m - [hidr] water supply
— m — m para tobera f - [metal-prod] tuyere water supply
— m — m — f de Venturi - [metal-prod] Venturi tuyere water supply
— m — arena f - [ind] sand feeding
— m — f para mezcladora f - [constr] mixer sand feeding
— m — coque m - [metal-prod] coke supply
— m — electricidad f - [electr-prod] electrical,city supply
— m — electrodo(s) m - [sold] electrode(s) supply
— m — — m restringido - [sold] tight(ened) electrode supply
— m — energía f - [electr-prod] power supply
— m — — f amplio - [electr-prod] ample power supply
— m — equipo m - [ind] equipment supply • equipment delivery
— m — — m eléctrico - [electr-instal] electrical equipment, supply, o provision
— m — — m mecánico - [ind] mechanical equipment, supply,ling, o provision,viding
— m — gas m - [combust] gas supply(ing)
— m — hidrógeno m - [ind] hydrogen supply(ing)
— m — importación f - [ind] imported supply
— m — material(es) m - material(s) supply(ing)
— m — mineral m - [miner] ore supply(ing) • ore shipment
— m — — m de hierro m - [minet] iron ore supply(ing)
— m — — m — — m desde exterior m - [miner] foreign iron ore supply(ing)
— m — pellet(s) m - [metal-prod] pellet(s) supply(ing)
— m — pieza(s) f - [ind] part(s) supply(ing)
— m — — f para repuesto m - [ind] spare parts supply(ing)
— m — regulador(es) m [equip] regulator(s) supply(ing)
— m — — m para voltaje m - [electr-equip] voltage regulator(s) supply(ing)
— m — respuesto(s) m - [ind] spare(s) (parts) supply(ing)
— m — servicio(s) m - [ind] service(s) supply(ing)
— m debido - [ind] proper supply(ing)
— m — de aire m - [ind] proper air supply(ing)
— m desactivado - deactivated supply(ing) • [electr-distrib] disconnected supply(ing)
— m desconectado - [electr-distrib] disconnected supply(ing)
— m desde exterior m - foreign supply(ing)
— m extranjero m - foreign supply(ing)
— m financiado - [fin] financed supply(ing)
— m igual - equal supply(ing)
— m independiente - independent supply(ing) • [electr-distrib] isolated supply(ing)
— m ininterrumpido - [electr-distrib] uninterrupted supply(ing)
— m — de energía f - [electr-distrib] uninterrupted power supply(ing)
— m interrumpido - interrupted supply(ing)
— m licitado - [legal] tendered (supply)
— m local - [com] local supply(ing)
— m insuficiente - [com] short supply
— m nacional - national, o domestic, supply(ing)
— m normal - normal supply • [electr-distrib] normal service
— m — de energía f - [electr-distrib] normal electric(al) supply(ing)
— m optativo - optional supply
— m — para filtro m - [electrón] optional filter supply(ing)
— m para equipo m - [ind] equipment supply(ing)
— m — — m perforador - [petról] rig supply(ing)
— m — filtro m - [electrón] filter supply(ing)

suministro m para filtro m optativo - [electrón] optional filter supply
— m — manipulador m - [electrón] keyer supply
— m — mantenimiento m - [ind] maintenance supply(ing)
— m — fabricante m - [ind] fabricator('s) supply(ing)
— m posterior—after supply • [sold] after flow
— m — de gas m - [sold] (gas) after flow
— m — — — m inerte - [sold] inert gas after-flow
— m regulado - controlled supply(ing)
— m restringido - tight(ened) supply
— m similar - similar supply(ing)
— m total - total supply(ing)
sumisión f - submission; submitting
summum m - [fig] pinnacle
sumo,ma a - . . .; extreme • utmost
suntuario,ria a - . . .; sumptuous,uary; luxurious
super a - super; ultra; extra
Super-Span a - [constr] Super-Span
— — — con porte m grande - [constr] Super-Span
— — m proyectado - [constr] designed Super--Span
superabrasivo m - [ind] superabrasive
superabrasivo,va a - [ind] superabrasive
superabundancia f - . . .; overabundance
superación f - exceeding; excelling • bridging • overcoming • [hidr] overtopping • [deport] sweep(ing)• surpassing
— f de crisis f - crisis overcoming
— f — expectativa - expectation surpassing
— f — pendiente f - [topogr] grade overcoming
— f — — fuerte - steep grade overcoming
— f — terraplén m - [hidr] embankment overtopping
superado,da - overcome • exceeded • excelled • bridged • [deport] swept
superagitador m - [miner] super agitator
superar v . . . • to exceed • to bridge • to cope with • to offset • [deport] to beat; to sweep • to overcome
— v crisis f - to overcome @ crisis
— v expectativa - to surpass @ expectation
— v luz f - [constr] to, exceed, o arch, @ span
— v pendiente f - to overcome @ grade
— v — f fuerte - to overcome @ steep grade
— v record* - to, surpass, o exceed, @ record
— v resistencia f - to overcome @ resistance
— v — f contra rodadura f - [autom-neumát] to overcome @ rolling resistance
superastro* m - [teatro] superstar
superautofundente m - [sold] super-self-fluxing
superávit m capitalizado - [fin] capitalized surplus
— m en operación(es) f - [contab] operating surplus
— m — revaluación f—[fin] revaluation surplus
supercarretera f - [vial] . . .; express highway • [G.B.] motorway; main arterial road • dual carriageway
— f alternativa f - [vial] expressway bypass
— f — con acceso m limitado - [vial] expressway bypass; limited access bypass highway
— f con nivel m alto - [vial] skyway
— f con peaje m - [vial] turnpike
— f — — m de Pensilvania - [vial] Pennssylvania Turnpike
— f interestatal - [vial] Interstate Highway
— f para circunvalación f - [vial] loop expressway; outerbelt freeway
— f — velocidad f, alta, o elevada, o grande - [vial] high speed, expressway, o superhighway
— f urbana - [vial] expressway • freeway
superestrella f - [teatro] superstar
superextrarresistente adv - extra extra strong
superficial a - . . .; surface • upper, o top, surface • véase también de, o en, o sobre, superficie f
superficie f - . . . • face [constr] floor space • top surface; square footage • [sold] (weld) face; outer face • [agric] acreage

superficie f a nivel - level suface
— f acanalada - [mec] grooved surface
— f achaflanada - [sold] beveled surface
— f ahusada - [mec] tapered surface
— f — para rodadura f - [mec] tapered tread
— f aislada - insulated surface
— f alta - high surface
— f anormal - abnormal surface
— f antideslizante - [constr] non-slip surface
— f — corrugada - [constr] non-slip corrugated surface
— f — para tránsito m - [constr] shoegripping walking surface
— f apareada - [mec] mating surface
— f aplanada - flattened surface
— f — de cable m - [cabl] flattened, cable, o rope, surface
— f — mecanizada - [mec] machined flat (surface)
— f apropiada - appropriate, o proper, surface
— f — de cordón m - [sold] proper bead surface
— f — para rodadura f - [autom-neumát] proper, o appropriate, tread
— f arada - [agric] plowed surface
— f asentada - settled surface
— f asfáltica - [constr] asphalt surface
— f áspera - rough surface • serrated surface
— f baja - low surface
— f bajo cielo m abierto - outdoor surface
— f bituminosa - [constr] bituminous, o black-top, surface; blacktop
— f — triple - [constr] triple bituminous surface
— f blanca - white surface
— f — grisácea - white-gray surface
— f brillante - brilliant, o glossy, surface
— f bruñida - burnished surface
— f buena - good surface
— f caliente - [segurid] hot surface
— f cargada - loaded surface • [constr] load bearaing area
— f casi plana - [sold] nearly flat surface
— f cerámica - ceramic surface
— f cilíndrica - [geom] cylindrical surface
— f coincidente - coinciding, o mating, surface
— f — de acero m - [mec] steel mating surface
— f — — caja f para eje m - [autom-mec] axle housing mating surface
— f de acero m para base f - [metal] base steel surfacew
— f — agua m - [hidr] water surface
— f — — m en entrada f - [hidr] water surface at @ inlet
— f — aleación f - [sold] alloy surface
— f — aluminio m - [mec] aluminum surface
— f — asiento m - [petról] seating surface
— f — — m con estelita f - [válv] stellite(d) seating surface
— f — banda f - [mec] band surface • [metal--lam] strip surface
— f — — f para rodamiento m - [autom-neumát] tread face
— f — barra f - [metal-lam] bar surface
— f — bloque m - [constr] block, surface, o top
— f — bobina f - [mec] coil surface
— f — — f para medición f - [instrum] measuring coil surface
— f — cable m - [cabl] cable, o wire rope, surface
— f — — m de alambre m - [cabl] wire rope surface
— f — calzada f - [vial] road(way), top, o surface, o surface top
— f — — f cambiada - [vial] changed road surface
— f — camino m - [vial] road surface • [deport] course surface
— f — camisa f—[petról] liner, face, o surface
— f — canal m - [metal-prod] heating surface
— f — carretera f - [vial] road, o highway, surface
— f — — f alisada - [vial] smoothed (out) road surface

superficie f de carretera f alisada - [vial]
smoothed (out) road surface
— f — — f suavizada - [vial] smoothed (out)
road surface
— f — cerámica - ceramic surface
— f — cilindro m - [comb.int] cylinder surface
— f — cinta f - [metal-lam] strip surface
— f — cobre m - [metal] copper surface
— f — — m preestañada - [metal-trat] pre-
tinned copper surface
— f — cojinete m - [cojín] bearing surface
— f — — m para rotación f - [grúas] swing
bearing surface
— f — colector m - [electr-mot] commutator
surface
— f — conducto m - [constr] conduit surface
— f — contrapiso m - [constr] subsurface surface
— f — — m de hormigón m - [constr] concrete
subfloor surface
— f — cordón m - [sold] bead, face, o surface
— f — — m casi plana - [sold] nearly flat
bead surface
— f — — m cóncavo - [sold] concave bead surface
— f — — m convexo - [sold] convex bead surface
— f — — m plana - [sold] flat bead surface
— f — crisol m - [metal-prod] hearth surface
— f — cubierta f - cover surface
— f — — f para caja f - [mec] housing cover
surface
— f — — f para engranaje(s) m - [mec] gear
cover surface
— f — cuenca f - [hidr] watershed area
— f — — f colectora - [hidr] watershed (collecting) area
— f — — f hidrográfica - [hidr] watershed
area
— f — cuerpo m - [anatom] body surface • body
area
— f — — m de mandíbula f - [mec] jaw body
surface
— f — chapa f - [metal-lam] strip surface
— f — — f corrugada - [metal-lam] corrugated
sheet surface
— f — — f — de acero m - [metal-lam] corrugated steel sheet surface
— f — — f metálica - [metal-lam] sheet metal
surface
— f — — f — corrugada - [metal] corrugated
sheet metal surface
— f — diente m - tooth surface
— f — — m en rueda f dentada - [mec] sprocket
tooth surface
— f — — — f — endurecida - [mec] hardened sprocket tooth surface
— f — — m endurecida - [mec] hardened tooth
surface
— t — disipador m - [electr] sink surface
— f — — m de aluminio m - [electr- aluminum
sink surface
— f — — m para calor m - [electr] heat sink
surface
— f — — m — — m de aluminio m - [electr]
aluminum heat sink surface
— f — electrodo m - [electr-equip] electrode
surface
— f — embalse m - [hidr] pool area
— f — émbolo m - [mec] piston surface • piston
area
— f —, empaquetadura f, o guarnición f - [mec]
packing, o gasket, surface
— f — epoxia f - epoxy surface
— f — fiador m - [mec] retainer surface
— f — — m para cierre m - [mec] seal(ing) retainer surface
— f — fleje m - [metal-lam] strip surface
— f — franja f - [mec] strip surface
— f — — f divisoria - [vial] median strip
surface
— f — fundación f - [ind] foundation surface
— f — fusión f - [metal-prod] fusion, o melting, area• [sold] leg
superficie f de fusión f de cordón m - [sold]
weld leg
— f — — f desigual - [sold] unequal leg
— f — — f igual - [sold] equal leg
— f — — f mayor - [sold] larger leg
— f — — f menor - [sold] smaller leg
— f — guarnición f - [mec] gasket surface
— f — hormigón m - [constr] concrete surface
— f — — m expuesta - [constr] exposed concrete surface
— f — — m — permanentemente - [constr] permanently exposed concrete surface
— f — huella f - [autom-neumát] footprint
(area)
— f — intercambiador m - [ind] interchanger, o
exchanger, surface
— f — — m para calor m - [ind] heat, interchanger, o exchanger, surface
— f — — m — — m para condensador m - [ind]
condenser heat, interchanger, o exchanger,
surface
— f — junta f - [sold] joint surface • [cald]
gasket surface
— f — — f para conexión f - [cald] nozzle
gasket surface
— f — mandíbula f - [mec] jaw surface
— f — — f, giratoria, o rotativa - [mec]
swiveling jaw surface
— f — material - material surface
— f — metal m - metal surface
— f — — m de base f - [sold] base metal surface
— f — — m — paleta f - [mec] blade metal
surface
— f — muestra f - sample surface
— f — paleta f - [mec] blade surface
— f — — f ajustable - [mec] adjustable blade
surface
— f — — f fija - [mec] stationary blade surface
— f — pared f - [constr] wall surface
— f — — f exigida - [constr] required wall,
surface, o area
— f — — f exterior - [constr] outer wall surface
— f — — f interior - [constr] inner wall surface
— f — — f sin recubrimiento m - [constr] uncoated wall, surface, o area
— f — pasarela f - [constr] walkway surface
— f — pavimento m - [constr] pavement (top)
surface
— f — — m de pista f - [constr] runway pavement surface
— f — — m flexible - [constr] flexible pavement (top) surface
— f — pieza f - [mec] part surface
— f — piso m - [constr] floor surface • floor
space
— f — — m para carga f - [metal-prod] charging floor top
— f — — m — operación(es) - [ind] operating
floor, o floor operating, surface
— f — pista f - [deport] track surface •
[aeron] runway surface
— f — placa f - plate area • plate surface
— f — plancha f - [mec] plate area • [sold]
plate surface
— f — planchón m - [metal-lam] slab surface
— f — polea f - [mec] sheave surface
— f — producto m - [ind] product surface
— f — — m semiterminado - [metal-lam] semifinished product surface
— f — radiador m - [comb.int] radiator surface
— f — — m exterior - [comb.int] external radiator surface
— f — — m (en) interior m - [comb.int] internal radiator surface
— f — raíz f - [sold] root face
— f — recipiente m - vessel surface
— f — reactor m - [nucl] reactor surface
— f — —-evaporador m - [nucl] reactor-evapo-

rator, o mixer-evaporator, surface
superficie f de recuperador m - [metal-prod]
checker surface
— f — retén m - [mec] retainer surface
— f — — m para cierre m - [mec] seal retainer
surface
— f — — n — grava s - [constr] gravel stop
surface
— f — roca f - rock surface
— f — rodillo(s) m - [mec] roll(s) surface
— f — rosca f - [mec] thread surface
— f — — f de tapa f - [mec] cover thread surface
— f — — f — — f para válvula f - [válv]
valve cover thread surface
— f — rotor m - [electr-mot] rotor surface
— f — rotura f - [suelos] break(ing) surface
— f — serpentín m - [tub] coil surface
— f — soldadura f - [sold] weld, surface, o
(outer) face, o top • groove face
— f — suelo m - ground, surface, o line •
[suelos] soil upper surface
— f — — m congelada - [meteorol] frozen
ground surface
— f — sujetador m - [mec] cam surface
— f — tablestaca f - [constr] sheeting surface
— f — tablestacado n - [constr] sheeting surface
— f — taladro m - [mec] bore surface
— f — — m en cojinete m - [mec] bearing bore
surface
— f — — m — rodamiento m - [cojin] bearing
bore surface
— f — — m para cojinete m - [mec] bearing
bore surface
— f — — m — rodamiento m - [mec] bearing
bore surface
— f — tambor m - [mec] drum surface
— f — techo m - [constr] roof surface
— f — terreno m - land area
— f — tubería f - [tub] tube, o pipe, surface
• [ambient] duct surface • [constr] conduit
surface
— f — viga f - [constr] beam surface
— f — — f para empuje m - [constr] thrust
beam surface
— f — zapata f - [mec] shoe surface • [constr-
-equip] tread plate area
— f decapada - [metal-trat] pickled surface
— f defectiva - [mec] defective surface
— f deficiente—deficient, o defective, surface
— f demasiado alta - too high surface
— f — baja - too low surface
— f densa f - [constr-hormig] dense surface
— f deseada - desired surface
— f desigual - uneven surface
— f — de camino m - [vial] uneven road surface
— f — — carretera f - [vial] uneven. highway,
o road, surface
— f desnuda - bare surface
— f — de alambre m - [alambre] wire bare, o
bare wire, surface
— f despareja - uneven surface
— f deteriorada - deteriorated surface
— f determinada - specified surface
— f disponible - available surface
— f — para carga f - [transp] loading area
— f distorsionada - [sold] distorted surface
— f diversa - different surface
— f drenada - [hidr] drained area • drainage
area
— f dura - hard surface
— f — para banda f para rodamiento m - [autom-
-neumát] hard surface tread
— f efectiva - effective surface
— f en condición f buena - good condition surface
— f — — f mala - bad condition surface
— f — contacto m - contact(ing), o mating,
surface
— f — tensión f - [sold] stressed, o in tension, surface, o outer face
— f encespedada - [constr] grassed,sy surface

superficie f endurecida - hard(ened) (sur)face
— f — f de diente m - [mec] tooth hardened, o
hardened tooth, surface
— f — — — m de rueda f (dentada) - [mec]
sprocket tooth hardened surface
— f — para prolongar vida f - [metal-trat] hard
surface for @ long life
— f — estría(s) f - [mec] land
— f entre ranura(s) f - [comb.int] (ring) land
— f — — f de émbolo m - [comb.int] piston ring
land
— f equipotencial - equipotential surface
— f esmerilada - [mec] ground surface
— f — hasta quedar lisa - [mec] smooth ground
surface; surface ground smooth
— f — lisa f - [mec] smooth ground surface
— f especificada - [mec] specified surface
— f estimada - estimated, o projected, surface,
o area
— f — para artefacto m - [electr-ilumin] luminaire projected area
— f estriada - [mec] grooved surface
— f examinada - examined surface
— f exigida - required area
— f expuesta - exposed surface
— f — de hormigón m - [constr] exposed concrete
surface
— f — dos veces f - double, o twice. exposed
surface
— f — permanentemente - permanently exposed
surface
— f — — a vista f - [constr] permanently exposed to view surface
— f — recubierta f - [mec] coated exposed surface
— f exterior - external, o exterior, o outisde,
o outer, surface - [sold] outer face
— f — de cubierta f - [mec] cover outer surface
— f — — f para engranaje(s) m - [mec] gear
cover outer surface
— f — — cuerpo m de mandíbula f - [mec] jaw
body outside surface
— f — — pared f - [constr] wall outer surface
— f — — soldadura f - [sold] weld outer face
— f — — tambor m - [mec] drum outside surface
— f — — tubería f - [tub] tubing, o piping,
outer surface; pipe external surface -
[a.acond] duct, exterior, o outside, surface
— f — limpiada con vapor m - steam cleaned, external, o outside, surface
— f — lisa - smooth, exterior, o outside, surface
— f — lubricada - lubricated outside surface
— f externa - outside surface
— f — de polea f - [mec] sheave outside surface
— f fija - [constr] stationary surface
— f filtrante - [hidr] filtering surface
— f final - final, o finished, surface
— f — de hormigón m - [constr] finished concrete surface
— f — propuesta f - [constr] proposed finished
surface
— f — — de hormigón m - [constr] proposed finished concrete surface
— f firme - firm surface
— f frontal - [constr] face square footage
— f — total - [constr] total face square footage
— f galvanizada - [metal-trat] galvanized surface
— f gibosa - [sold] humpy, surface, o start
— f glaseada - [constr] glazed surface
— f golpeable - [mec] striking surface
— f grisácea - gray surface
— f — blanca - gray(ish)-white surface
— f — — mate - matte, o dull, gray(ish)-white
surface
— f horizontal - horizontal surface
— f — de viga f - [constr] beam. top, o horizontal, surface
— f — — f para empuje m - [constr] thrust
beam, top, o horizontal, surface
— f hormigonada - [constr] concrete(d) area

superficie f húmeda - moist, o wet. surface
— f — de plancha f - [sold] plate wet, o wet plate, surface
— f igual - equal surface • equal area
— f imprimada - [pint] primed surface
— f inferior - lower, o bottom, surface, o edge
— f — de fiador m - [mec] retainer bottom surface
— f — — — para cierre m - [mec] seal retainer, bottom, o lower, surface
— f — — reten m para cierre m - [mec] sealing retainer, bottom, o lower, surface
— f — — — m — sello m - [mec] seal(ing) retainer, bottom, o lower, surface
— f inspeccionada - [ind] inspected surface
— f interior - inside, o inner, o interior, o internal, surface - [petról] pack-off surface
— f — de cuerpo m - [mec] body inside surface
— f — — — m de mandíbula f - [mec] jaw body inside surface
— f — reactor-evaporador m - [nucl] reactor-evaporator, o mixer-evaporator, inside surface
— f — de tubería f - [tub] internal, pipe, o tube, o, tube, o pipe, internal, surface; pipe, o tube, inner, o inside, surface
— f — lisa f - [tub] smooth, inside, o interior, surface
— f — lisa - [tub] smooth interior surface
— f — — de revestimiento m - [tub] lining smooth, inside, o interior, surface
— f — — — m con asbesto m - [tub] asbestos lining smooth , inside, o interior, surface
— f — lubricada - [tub] lubricated, inside, o interior, surface
— f irregular - irregular, o rough, surface
— f labrable - [mec] machinable surface
— f labrada - [mec] machined surface
— f — final - [mec] final machined surface
— f laminada - [metal-lam] rolled surface
— f laqueada - [pint] laquer(ed), surface, o finish
— f libre m - free surface
— f — de agua m - [hidr] water free surface
— f licuada - [metal-trat] liquated surface
— f limpia - clean surface
— f limpiada - cleaned surface
— f — con vapor m - steam cleaned surface
— f lisa - smooth, o level, surface, o face
— f — entre estría(s) f - [comb.int] piston's land
— f lisa y limpia - [mec] smooth clean surface
— f lubricada - lubricated surface
— f lustrada - polished, o burnished, surface
— f más externa - outermost surface
— f — fría - cooler surface • coolest surface
— f mate - matte, o dull, surface
— f máxima - maximum, surface, o area
— f — de contacto m - [mec] maximum bearing area
— f — — apoyo m - [grúas] maximum bed space
— f mayor - greater surface • greatest surface • [agric] more acreage
— f — para portar carga(s) f - [mec] larger load bearing surface • largest load bearing surface
— f mecanizada - [mec] machined surface
— f — de cubierta f - [mec] cover machined surface
— f — — — f para engranaje m - [mec] gear cover machined surface
— f — plana - [mec] machined flat (surface)
— f metálica - metal(lic) surface
— f mínima - minimum, surface, o area
— f — de contacto m - [mec] minimum contact area
— f — para apoyo m - [grúas] minimum bed space
— f mojada - wet surface
— f — de plancha f - [sold] plate wet surface
— f mordedora - [mec] gripping surface
— f — de mandíbula f - [mec] jaw gripping sur-

superficie f mordedora de mandíbula f giratoria - [mec] swiveling jaw gripping surface
— f muy endurecida - [sold] badly hardened surface
— f — — por trabajo m - [sold] (work) badly hardened surface
— f muy distorsionada - [sold] badly distorted surface
— f nivelada - level(ed) surface
— f no conductora - non-conducting surface
— f expuesta - unexposed surface
— f original - original surface
— f para apoyo m - [mec] bearing, surface, o area, o face
— f — m para rotación f - [grúas] swing bearing surface
— f — artefacto m - [electr-ilumin] luminaire area
— f — asiento m - [cojin] bore
— f — calentamiento m - heating surface
— f — m en emparrillado - [metal-prod] checker heating surface
— f — — m — escaqueado m - [metal-prod] checker heating surface
— f — — m — recuperador m - [metal-prod] checker heating surface
— f — — m — reticulado m - [metal-prod] checker heating surface
— f — — m para planchón(es) m - [metal-lam] slab heating surface
— f — cierre m - [petról] sealing surface
— f — combustión f - [combust] combustion surface
— f — compactación f - [herram] compacting face
— f — comprobar resbalamiento m - [deport] skidpad
— f — conexión f - connecting,tion surface
— f — — f con tierra f - [comb.int] ground(ing) surface
— f — contacto m - [electr-equip] contact surface; interface • [mec] bearing, o contact, area, o surface, o face • [autom-neumát] tread • [mec] working surface
— f — — m de sujetador m - [mec] cam working surface
— f — — m en apoyo m [mec] pad contact area
— f — — m — banda f para rodamiento m - [autom-neumát] tread patch
— f — — m — conmutador m - [electr-equip] switch contact surface
— f — — m en escobilla f - [electr-mot] (motor brush) bearing surface
— f — — m para zapata f - [constr-equip] (tread) plate contact area
— f — — m para soporte m - [mec] support contact surface
— f — — m — — m, anterior, o delantero - [mec] forward support contact surface
— f — corte m - [metal-lam] cropping surface
— f — desgaste m - [mec] wear(ing) surface
— f — — m de diente m - [mec] tooth wear(ing) surface
— f — — m — m de rueda f dentada - [mec] sprocket tooth wear(ing) surface
— f — m — m — f — endurecida - [mec] haradened sprocket tooth wear(ing) surface
— f — — m lento - [autom-neumát] long wearing surface
— f — drenaje m - [hidr] drainage area
— m — escurrimiento m - [domést] drain(ing) board
— f — montaje m - [mec] mounting surface
— f — — m para rueda f - [mec] wheel mounting surface
— f — — m para cubierta f - [mec] cover mounting surface
— f — — m — f para distribuidor m (para fuerza f) - [autom-mec] (power) divider cover mounting surface
— f — portar (carga f) - [ind] (load) bearing surface
— f — refrigeración f - cooling surface

superficie f para rodadura f - [mec-ruedas]
tread (surface)
— f —— f ahusada - [mec] tapered tread
— f —— f para rueda f—[mec-ruedas] wheel
tread
— f —— f —— f para riel(es) m - [mec-
-ruedas] track wheel tread
— f —— f desgastada—[mec-ruedas] worn tread
— f —— f plana - [mec-ruedas] flat tread
— f —— f radial - [mec=ruedas] radial tread
— f — rodamiento m - [aeron] runway surface •
[sold] riding surface • [mec] bearing surface
• wearing surface • [cojin] rolling surface
— f —— m cilíndrica - [mec-ruedas] straight
tread
— f —— m de cojinete m - [cojin] bearing
rolling surface
— f — m — m con bola(s) f - [cojin]
roller bearing rolling surface
— f —— m — riel m - [f.c.] rail top
— f — rozamiento m - [mec] bearing surface
— f — selladura f - [petról] sealing surface
— f — soldar(se) v - [sold] surface to be
welded
— f — subdrenaje m - [hidr] subdrainage area
— f — sujeción f - [mec] gripping surface
— f — soporte m - [mec] support(ing) surface
— f —— m anterior - [mec] forward support-
-(ing) surface
— f — sustentación f - [constr] support(ing),
o bearing, surface, o area
— f — sustentación f - [mec] bearing area
— f — transferencia f - transfer surface
— f —— f para calor m - [termol] heat transfer surface
— f paralela - parallel surface
— f pareja - even surface
— f pasivada* - [metal-trat] passivated surface
• scaled surface
— f pavimentada - [constr] paved surface
— f — bajo cielo m abierto - [constr] paved
outdoor surface
— f — de pista f - [constr] runway paved surface
— f pequeña f - small surface
— f perfectamente lisa - perfectly smooth surface
— f perturbada—perturbed, o disturbed, surface
— f pintada - [pint] painted surface
— f plana - level, o flat, surface
— f — de cable m - [cabl] (wire) rope flat-
-(tened) surface; flat(tened) rope surface
— f — mecanizada - [mec] machined flat (area)
— f plástica - [constr] plastic surface
— f — nivelada - [constr] level(ed) plastic
surface
— f pobre - poor surface
— f portante - [mec] bearing surface
— f pulida - [mec] polished surface
— f radial - [mec] radial surface
— f — para rodadura f - [mec] radius tread
— f recubierta - [mec] coated surface
— f reducida - reduced surface • small surface
— f, resbaladiza, o resbalosa—slippery surface
— f resistente - strong, o tough, surface
— f rocosa - rocky surface
— f rota - broken, surface, o face
— f — de letrero m - [com] broken sign face
— f rugosa - rough, o knurled, surface
— f satinada - [metal-lam] satin surface
— f saturada - saturated surface
— f seca - dry surface
— f — imprimada - [pint] primed dry, o dry
primed, surface
— f secada - dried surface
— f semibrillante - semigloss(y) surface
— f semimate - semimate, o dull, surface
— f sin bruñir - unburnished, o unpolished,
surface
— f — lustrar - unpolished surface
— f — obstrucción(es) f - unobstructed (land)
view
— f — protección f - unprotected surface

superficie f sin pulir - unpolished, o unburnished, surface
— f — tensión f - [sold] not-in-stress,-ten-
sion, surface
— f — usar - unused, o not-in-use, surface
— f sólida - solid surface
— f subrasante - [constr] sub-grade surface
— f sumamente lisa - extra smooth surface
— f superimpuesta - superimposed surface
— f superior - upper surface; top land
— f — de diente m - [mec] tooth top land
— f — fiador m - [mec] retainer top surface
— f —— m para cierre m - [mec] seal retainer top surface
— f —— pavimento m - [constr] pavement top
surface
— f —— m flexible - [constr] flexible
pavement top surface
— f — retén m - [mec] retainer top surface
— f —— m para cierre m - [mec] seal retainer top surface
— f — de suelo m - [constr] soil top surface
— f —— viga f - [constr] beam top surface
— f —— f para empuje m - [constr] thrust
beam top surface
— f supuesta - presumed surface
— f — de roca f - [geol] presumed rock surface
— f — tránsito m - walking surface
— f techada - [constr] roof(ed) surface
— f — impermeable - [constr] watertight roof
surface
— f tenaz - tought surface
— f — de banda f para rodamiento m - [autom-
-neumát] tough, surface tread, o tread surface
— f terminada - finished surface
— f — de hormigón m - [constr] finished concrete surface
— f tratada - treated surface
— f triple - triple surface
— f única - single surface
— f uniforme - [constr] uniform surface
— f — (y) lisa - [constr] uniform (and)
smooth surface
— f útil - useful surface
— f — de bobina f - coil useful surface
— f —— f para medición f [instrum]
measuring coil useful surface
— f vertical - vertical surface
— f viciada - [ind] fouled surface
— f — de intercambiador m - [ind] fouled, exchanger, o interchanger, surface
— f —— m para calor m - [ind] fouled
heat, exchanger, o interchanger, surface
— f —— m —— m para condensador m -
[ind] fouled condenser gear, exchanger, o interchanger, surface
— f vidriosa - [constr] glassy, o glazed, surface
— f Z - Z surface
superficies f diversas - various surfaces
— f varias - various, o sundry, surfaces
supergás m - [petról] low pressure gas
supergrupo* m - supergroup*
superimpuesto,ta a - superimposed
superintendente m - . . . .; véase también jefe m;
manager
— m auxiliar - [admin] assistant manager
— m para laminador,ra - [metal-lam] mill superintendent
— m —— para banda(s) f - [metal-lam] strip
mill superintendent
— m —— —— f en caliente - [metal-lam] hot
strip mill superintendent
— m —— —— f en frío - [metal-lam] cold
strip mill superintendent
— m — montaje m - [ind] erection superintendent
superior a - . . . .; extra high • top notch • in
excess
— a a ~ over; above; in excess of
— a — nominal - above rated
— a — normal - higher than normal

superior a todos menos . . . .—second only to ...
— a — — menos uno,na - second highest
— a derecho,cha - upper, o top, right (hand)
— a inmediato - [labpr] immediate, o first
line, supervisor
— a izquierdo,da - upper, o top, left (hand)
— a para accionamiento m - [mec] high drive
— a — impulsión f - [mec] high drive
superioridad f - . . . • [admin] administration;
management; top, administration, o nanagement
superponer v - to superimpose • to overwrite •
véasse también sobreponer v
superposición f . . .; superimposing • overwrit-
ing • [sold] overlap; overhang • [transp]
stacking • [constr] lap(ping)
— f parcial - partial overlapping
— f — de escala f - [sold] partial range overlapping
superpotencia f económica - [econ] economic superpower
— f industrial - [ind] industrial superpower
superpuesto,ta a - superimposed • overwritten
— a parcialmente - partially overlapped
superresistente a - extra strong
superunidad* f - superunit*
supervisado,da a - supervised: overseen;
[admin] managed • monitored
supervisar v - . . .; to oversee • [admin] to
manage; to monitor
— v artículo m - [ind] to supervise @ item
— v instalación f - to supervise @ installation
— v — f de bomba f - [bombas] to supervise @
pumpo installation
— v montaje m - to supervise @ erection
— v operación f - to supervise @ operation
— v puesta f en marcha—tp supervise @ start-up
— v — — f de bomba f - [bombas] to supervise @ pump start-up
— v técnicamente - to supervise technically
supervisibilidad f - supervisibility
supervisible a - supervisible
supervisión f - . . .; control(ling): overseeing
• monitoring
— v apropiada - appropriate, o proper, supervising,sion
— f automática - automatic, supervising,sion, o
monitoring
— f cuidadosa - careful, o close, supervision
— f de artículo m - [ind] item supervising,sion
— f — distribución f - [electr-distrib] distribution supervising,sion
— f — — f eléctrica - [electr-distrib] electric distribution supervising,sion
— f — instalación f - [ind] installation supervising,sion
— f — — f de bomba f - [bombas] pump installation supervising,sion
— f — jerarquía f superior - [admin] high(er)
level supervising,sion
— f — modificación(es) f - [ind] change(s)
supervising,sion
— f — montaje m - [ind] erection supervision
— f — — m de línea f - [ind] line erection
supervising,sion
— f — — m — — f de cizalla(s) f - [metal-
-lam] shear(ing) line erection, control-
-(ling), o supervising,sion
— f — operación f - [ind] operation, supervising,sion, o overseeing,sight, o control
— f — personal - [pers] personnel, supervising,sion, o overseeing,sight
— f — planta f - [ind] plant supervising,sion
— f — producción f - [ind] production, o operating,tion, supervising,tion, o oversight
— f — proyecto m - [ind] project supervision
— f — puesta f en marcha - [ind] start-up supervising,sion
— f — — f — — f de bomba f - [bombas] pump
start-up supervising,sion
— f — seguridad f - [segurid] safety, supervising,sion, o monitoring
— f departamental - [ind] departmental supervising,sion

supervisión f en departamento m - [admin] departmental supervising,sion
— f — obra f - (project) site supervising,sion
— f metalúrgica - [metal-prod] metallurgical
supervision
— f para construcción f - [constr] construction supervising,sion
— f — montaje m - [mec] erection supervision
— f — — m para sistema m - system erection
supervising,sion
— f — personal m - [labor] personnel supervising,sion
— f — — m para control m para calidad f -
[ind] quality control personnel supervision
— f — — m — fiscalización f para calidad f -
[ind] quality control personnel supervision
— f — — m — verificación f para calidad f -
[ind] quality control personnel supervision
— f — planta f - [ind] plant supervising,sion
— f — puesta f en marcha f - [ind] start-up
supervising,sion
— f — — — de línea f - [metal-lam]
line start-up supervising,sion
— f — — f — — — — f de cizalla(s) f -
[metal-lam] shear line start-up supervision
— f — torre f - [petról] rig supervisimg,sion
— f — — f para perforaci•on f - [petról]
(drilling) rig supervising,sion
— f propia - self, o own, supervising,sion
— f superior - [labor] higher supervising,sion
— f técnica - [ind] technical supervising,sion
— f — cuidadosa - [ind] close technical supervising,sion
— f — para modificación(es) f - [ind] change
technical supervising,sion
— f — — proyecto m - [ind] project technical
supervising,sion
— f — — — m especial - [ind] special project
technical supervising,sion
supervisor m - . . . . • [ind] operator • véase
también jefe m; encargado m
— m capacitado - [labor] qualified, o trained,
supervisor
— f capaz - [ind] capable, o knowledgable, supervisor
— m concienzudo - [labor] conscientious supervisor
— m conocedor - knowing, o knowledgable, supervisor
— m — jerarquía f - [labor] higher supervisor
— m — — f superior - [labor] oversight, o
general, supervisor
— m directo - [labor] direct supervisor • first
line supervisor
— m electricista - [electr] supervising electrician
— m experimentado - [labor] experienced supervisor
— m — en fundición f - [metal-prod] experienced melt(ing) shop supervisor
— m general - [labor] general supervisor
— m inmediato - [labor] immediate, o subordinate, supervisor
— m médico - [med] medical supervisor
— m para armado m - [ind] erection supervisor
— m — — m para fabricante m - [ind] manufacturer's erection supervisor
— m — — m — — m de equipo m - [ind] equipment manufacturer's erection supervisor
— m — conservación f - [ind] maintenance supervisor
— m — crédito(s) m - [fin] credit(s) supervisor
— m — cuenta(s) f - [contab] account(s) supervisor
— m — diseño(s) m - [ind] design(s) supervisor
— m — división f - [ind] division supervisor
— m — — f para instalación(es) f - [ind] in-
stallation(s) division supervisor
— m — fiscalización f - [ind] control(ling)
supervisor
— m — — f para construcción(es) f - [ind]
construction control(ling) supervisor

supervisor m **para fiscalización** f **para reconstrucción** f - reconstruction control supervisor • [metal-prod] relining control supervisor
— m — **fundición** m - [metal-prod] foundry, o melt(ing) shop, supervisor
— m — **grúa(s)** f - [ind] crane(s) supervisor
— m — **instalación** f - [ind] installation supervisor
— m — — f **de tubería(s)** f [constr] pipe,ping (installation), supervisor, o foreman
— m — **línea** f - [ind] line supervisor
— m — — f **de cizalla(s)** f - [metal-lam] shear-(ing) line supervisor
— m — **mantenimiento** m - [ind] maintenance supervisor
— m — — m **eléctrico** - [ind] electrical maintenance supervisor
— m — — m **mecánico** - [ind] mechanical maintenance supervisor
— m — — m **preventivo** - [ind] preventive maintenance supervisor
— m — **montaje** m - [ind] erection supervisor
— m — — m **para fabricante** m - [ind] manufacturer's erection supervisor
— m — — m — — m **de equipo** m - [ind] equipment manufacturer's erection supervisor
— m — **obra** f - [ind] project supervisor
— m — **producción** f - [ind] production, o operating, supervisor
— m — — m **fabricado** - [ind] fabricated product supervisor
— m — **puesta** f **en marcha** - [ind] start-up supervisor
— m — — f — — **de línea** f - [ind] line start-up supervisor
— m — — — — — f **de cizalla(s)** f - [metal-lam] shear(s) line start-up supervisor
— m — **relación(es)** f - [labor] relation(s) supervisor
— m — — f **laboral(es)** - [labor] labor relation(s) supervisor
— m — **reparación(es)** f - [ind] repair(s) supervisor
— m — — f **electrica(s)** - [electr] electric repair(s) supervisor
— m — — f **electromecánica(s)** - [ind] electro--mechanical repair(s) supervisor
— m — **soldadura(s)** f—[sold] welding supervisor
— m — **turno** m - [ind] turn, supervisor, o foreman
— m — — m **para producción** f - [ind] production turn, supervisor, o foreman • operating,tion turn foreman
— m **que asigna tarea(s)** f - [labor] job assigning supervisor
supervisor,ra a - supervisory,vising
supervisorio,ria a - véase **supervisor**,ra a
supervivencia f - . . . . • living through
suplementación f - supplementing,tation; extending,nsion • segmenting,tation • [metal--prod] bottom build(ing) up
— f **de acoplamiento** m - [mec] coupling, suplementing,tation
— f **para borde** m **reforzado** - [mec] reinforced edge supplementing,tation
— f — **chatarra** f - [metal-prod] scrap supplementing,tation
suplementado,da a - supplemented • extended; increased; augmented
suplementador,ra a - supplementing; extending; augmenting
suplementar v - to, supplement, o extend, o augment, o increase • [mec] to shim
— v **acoplamiento** m - [mec] to supplement @ coupling
— v **alcantarilla** f - [constr] to supplement @ culvert
— v **borde** m **reforzado** - [mec] to supplement @ reinforced edge
— v **chatarra** f - [metal-prod] to supplement @ scrap
suplementario,ria - . . . .; supplemental; additional; extra • make-up
suplemento m - . . . . • [mec] insert; extension;

filler; shim; space • liner • [tub] tail piece
suplemento m **colocado** - [mec] installed, supplement, o insert, o shim
— m **de** . . . **metro(s)** m - [mec] . . . meter(s), insert, o supplement (úsense feet)
— m — . . . — **para aguilón** m - [grúas] . . . feet boom insert
— m — . . . — — m **para pescante** m - [grúas] . . . foot jib boom insert
— m — . . . — **para pescante** m - [grúas] . . . meters jib insert • (úsense feet)
— m — **poliuretano** m - [válv] polyurethane, supplement, o insert
— m — — m **reemplazable** - [mec] replaceable polyurethane, insert, o supplement
— m **desgastado** - [mec] worn, supplement, o insert
— m **desplazable** - [grúas] swing away extension
— m — **para aguilón** m - [grúas] swing away boom extension
— m **frangible** - [mec] frangible insert
— m **friable** - brittle insert
— m **instalado** - [mec] installed insert
— m **manual** - [mec] manual insert
— m — **para aguilón** m - [grúas] boom manual insert
— m — **accesorio** m - [tub] fitting tail piece
— m **para aguilón** m - [grúas] boom insert
— m — — m **para pescante** m - [grúas] jib boom insert
— m — — m **para torre** f - [Grúas] tower boom insert
— m — **asiento** m - [mec] seat insert
— m — — m **para válvula** f - [válv] valve seat insert
— m — — m — — f **para escape** m - [comb.int] exhaust valve seat insert
— m — **barra** f - [herram] bar insert
— m — — f **para ajuste** m - [herram] torque bar insert
— m — — f — **torsión** f - [herram] torque bar insert
— m — **boquilla** f - [sold] nozzle insert
— m — **brazo** m - [mec] arm insert
— m — — m **tirador** - [mec] draw(ing) bar insert
— m — **busa** f - [metal-prod] blowpipe supplement
— m — **cable** m - [cabl] cable, o wire rope, insert, o supplement
— m — **carrete** m - [mec] reel extension
— m — — m **para alambre** m - [sold] wire reel extension
— m — **codo** m - [metal-prod] gooseneck supplement
— m — **fijación** f - [mec] tap-lok insert
— m — **guía** f - [sold] guide insert
— m — — f **para entrada** f - [sold] incoming guide insert
— m — — f **para salida** f - [sold] outgoing guide insert
— m — **guiadera** f - [sold] guide tube insert
— m — — m **para entrada** f - [mec] incoming guide insert
— m — — m — **salida** f - [sold] outgoing guide insert
— m — — f — **válvula** f - [mec] valve guide insert
— m — **llave** f - [herram] wrench, extension, o liner
— m — — f **de cuadrado** - [herram] tool wrench liner
— m — **mesa** f - [domést] table leaf
— m — **pescante** m - [grúas] jib insert
— m — — m **para torre** f - [grúas] tower jib insert
— m — **soporte** m - [mec] holder insert
— m — — m **para fiador** m - [mec] retainer holder insert
— m — — m **para guia(dera)** f - [mec] guide holder insert
— m — — m — — f **para válvula** f - [válv] valve guide holder insert

suplemento m para soporte m para retén m - [mec]
  retainer holder insert
— m — torre f - [mec] tower insert
— m — tubo m - [tub] tube insert
— m — m guiador - [sold] guide,ding tube
  insert
— m — válvula f - [válv] valve insert
— m — f, colocado, o instalado - [válv] installed valve insert
— m — — f, quitado, o removido, o sacado -
  [válv] removed valve insert
— m primaveral—[public] spring(time) supplement
— m quitado - [mec] removed insert
— m reemplazable - [mec] replaceable insert
— m, removido, o sacado - [mec] removed insert
— m sostenido - [mec] held (up) insert
suplencia f filling in • [labor] back-up
suplente m - alternate; substitute • [labor]
  stand-in; replacement • spell | a - deputy;
  acting
supletorio,ria a - suppletory
suplicado,da a - entreated; begged
suplido,da a - supplied; furnished • given • included • serviced • filled in
— a ampliamente - amply, o properly, supplied
— a con - supplied with
— a normalmente - supplied normally
— a por comprador m—purchaser, o user, supplied
— a — consumidor m - user supplied
— a — usuario m - user supplied
— a previo acuerdo m—supplied after @ agreement
— a separadamente - supplied separately
— a únicamente - supplied only
— a — previo acuerdo m - supplied only upon a-
  greement
suplir v - . . .; to meet • to give • to fill in •
  to include • to service
— v agua m - [hidr] to supply water
— v . . . amperio(s) m - [electr-distrib] to
  supply . . . ampere(s)
— v . . . — m de energía f - [electr-distrib]
  to supply . . . ampere(s) power
— v . . . — m en forma f continuada - [electr
  -distrib] to supply . . . power continuously
— v apropiadamente - to supply, appropriately, o
  properly
— v asesoramiento m - to supply @ advice
— v conocimiento(s) m - [ind] to supply @, knowledge, o know-how
— v electrodo m - [sold] to supply @ electrode
— v en forma f continua(da) - to supply continuously
— v energía f - [electr-distrib] to supply power
— v — f de . . . voltio(s) m - [electr-distrib]
  to supply . . . volt power
— v equipo m - [ind] to supply @ equipment
— v exigencia f - to, supply, o meet, @, need, o
  requirement
— v lubricación f.- [mec] to provide lubrication
— v necesidad f - to supply @ need
— v normalmente - to supply normally
— v pericia f - to provide @ expertise
— v previo acuerdo m - to supply on agreement
— v producto m—to, supply, o provide, @ product
— v separadamente - to, supply, o provide, separately • to service separately
— v servicio m—to, supply, o provide, @ service
— v supervisión f - [labor] to, supply, o provide, @ supervision
— v tope m - [mec] to, supply, o provide, @ stop
— v únicamente - to, supply, o provide, only
— v — previo acuerdo - to, supply, o provide,
  only on agreement
— v voltaje m - [electr-distrib] to, supply, o
  provide, @ voltage
— v — en salida f - [electr-distrib] to, supply, o provide, @ output voltage
suponedor,ra - . . . . • guessestimator
suponer v - . . . . • to presume • to hypothesize;
  to guess(es)timate
— v condición f - to, presume, o assume, @ condition
— v quemado,da a - [ind] to presume burned (out)

suposición f -. . . .; presuming,mption; assuming,sumption • guessestimate,mating,mation
  • hypothesizing
— f de condición f - condition assuming
suprimido,da a - supressed; omitted; eliminated
  • abatement
— f de interruptor m - [electr-instal] breaker,
  elimination, o omission
— f — reóstato m - [electr-instal] rheostat,
  elimination, o omission
supresor m de onda(s) f - [electrón] wave suppressor • surge suppressor
suprimido,da a - suppressed • [legal] waived
suprimir v - . . .; to omit; to eliminate • to
  cut out • [legal] to waive
— v reóstato m - [electr-instal] to, omit, o e-
  liminate, @ rheostat
— v separación f - to eliminate @ separation •
  [sold] to eliminate @ gap
supuesto m - . . .; supposed case
supuesto,ta a - supposed; assumed; presumed; assumed • guess(es)timated • hypothesized
supuración f - [med] . . .; suppurating •
  drainage,ning
supurado,da a - [med] suppurated; drained
supurante a - [med] . . .; draining
supurar v - [mec] . . .; to drain
sur a norte a - [vial] northbound
— m franco - due south
Suramérica f - véase América f del Sur; Sudamérica
surcado,da a - [agric] furrowed
surco m - . . . . • row • groove
surgencia f - surging
— f por inyección f de gas m - [petról] gas
  lift(ing)
surgido,da a - surged; arisen
surgimiento m - surging; arising
surgir v - . . .; to surge
surtido m - . . . .; set; selection; line; range
— m amplio - wide,o broad, range, o line, o
  selection
— m — de producto(s) - [ind] product(s), wide,
  o broad, range, o assortment
— m — — m de alambre m - [alambre] wire
  poroducts wide, range, o assortment
— m — — — m laminado(s) - [metal-lam]
  rolled product(s), wide, o broad, range
— m — — m plano(s) - [metal-lam] flat
  product(s), wide, o broad, range
— m — — — m — laminado(s) - [metal-lam]
  flat rolled product(s) wide, o broad, range
— m completo - complete, o full, o broad,
  range, o set. o assortment, o line
— m — de matriz,ces f - [mec] complete die(s),
  assortment, o facilties
— m — — acabado(s) m - finish(es) range
— m — — accesorio(s) m - accessory,ries, range,
  o assortment
— m — — camisa(s) f - [petról] liner(s) range •
  liner size range
— m — diámetro(s) n - [ind] size(s), o dia-
  meter(s), assortment, o range • assorted,
  sizes, o diameter(s)
— m — matriz,ces - [mec] die(s) assort,emt
— m — opción(es) f - option(s), range, o line
— m — producto(s) m - [ind] product(s) range
— m — — m de alambre m - [alambre] wire pro-
  duct(s) range
— m — — m laminado(s) - [metal-lam] rolled
  product(s) range
— m — — m plano(s) - [metal-lam] flat pro-
  duct(s) range
— m — — — m — laminado(s) m - [metal-lam] flat
  rolled product(s) range
— m — repuesto(s) m - [ind] spare part(s) assortment
— m — tipo(s) m - type(s) assortment • assorted type(s)
surtido,da a - assorted • [ind] serviced
— a separadamente - serviced separately
surtidor m - . . . . • [mec] . . .; atomizer,
  sprayer; spout ? injector • [comb.int] jet;

nozzle • [tub] standpipe
surtidor m **atorado** - [comb.int] clogged jet
— m **auxiliar** - [comb.int] accelerating well
— m — **para carburador** m - [comb.int] accelerating well
— m **de aire** m - [comb.int] air, jet, o nozzle
— m **economizador** - [mec] economizer jet
— m **para aire** m - [comb.int] air nozzle
— m — **embolsar** v - [mec] sacking spout
— m — **gasolina** f - [comb.int] gasoline jet • gasoline pump
— m **piloto** - pilot jet
— m **principal** - [mec] main jet
— m — **para aire** m - [comb.int] main air jet
surtimiento m - . . . • [com] servicing
surtir v - . . . • [mec] to service • to stock
— v **separadamente** - to supply separately • to service separately
sus adv **efectos** - its, ends, o purposes
susceptibilidad f - . . . • [metal] sensitivity
— f **a agrietamiento** m—[sold] crack sensitivity
susceptible a - . . .: apt
— m **a agrietamiento** m - crack sensitive
— adv **a ajuste** m - [mec] adjustable
— adv — **fallar** v - failure susceptible
— adv — **regulación** f - regulatable; adjustable
suscitación f - . . . • arising • arousing
suscitado,da a - aroused; arisen; arising
suscitante a - arising • arousing
suscitar v - . . . • to arise • to arouse
susodicho,cha a - . . . • above • said
suspender v - . . . • [labor] to lay off • [legal] to recess
— v **alambre** m - to suspend @ wire
— v **artefacto** m - [electr-instal] to hang @ fixture
— v **cable** m—[cabl] to suspend @, rope, o cable
— v — m **de alambre** m - [cabl] to suspend @ wire rope
— v **cielo** m **raso** - [constr] to, hang, o suspend, @ ceiling
— v **cordón** m - [cabl] to, hang, o suspend, @ strand
— v — m **de alambre** m - [cabl] to suspend @ wire strand
— v **energía** f - [electr-distrib] to, suspend, o cut off, @ power
— v — f **eléctrica** - [electr-distrib] to, suspend, o cut off, @ electric power
— v **inclinador** m - [mec] to suspend @ tilter
— v **muestra** f - to, suspend, o hang, @ sample
— v **objeto** m **de arte** m - [domést] to hang @ work of art • to hang @ novelty
— v **operación** f - to, suspend, o stop, @ operation
— v **volcador** m - to suspend @ tilter
suspendido,da a - suspended; hung • supported • overhead • [legal] recessed
— a **de fondo** m **de mar** m - [petról] mudline suspension
— a **en aire** m - air suspended
suspensión f - . . . • support • mounting • shackle bar • lifting; raising • [autom] ride • [legal] recess(ing) • day(s) off
— f **acuosa** - water suspension
— f — **coloidal** - colloidal water suspension
— f **ajustable** - [autom] adjustable ride
— f **bajada** - [autom] lowered suspension
— f **cambiada** - [autom-mec] changed suspension
— f **coloidal** - [hidr] colloidal suspension
— f **con gancho** m - [mec] hook suspending,nsion
— f **cómoda** - [autom-mec] comfortable ride
— f **con resorte(s)** m - [mec] spring loading
— f **con rodaje** m - [constr] trolley
— f — m **para portón** m - [constr] gate trolley
— f — **viga(s)** f - [autom] beam suspension
— f — — f H - [autom] I-beam suspension
— f — — f — **gemela(s)** f - [autom-mec] twin I-beam suspension
— f **dañada** - damaged suspension
— f **de alambre** m - [mec] wire suspending,nsion
— f — **artefacto** m - [domést] fixture sus-

suspensión f **de bogie** m—[mec] bogie suspension
— f — — m **sólido** - [mec] solid bogie suspension
— f — **cable** m - [cabl] cable, o rope, suspending
— f — — a **de alambre** m - [cabl] wire rope suspending,nsion
— f — **caja** f - [mec] case suspending,nsion
— f — — f **para transmisión** f - [mec] transmission case suspending,nsion
— f — **campana** f - [metal-prod] bell, hanging, o suspending,nsion, (system)
— f — **cardán** m - [autom-mec] véase **suspensión** f **transmisión** f
— f — **carga** f - [grúas] overhead lifting - [metal-prod] scaffolding
— f — **cielo** m **raso** - [constr] ceiling, hanging, o suspending,nsion
— f — **columpio** m - [deport] swing, hanging, o suspension
— f — **cordón** m - [cabl] strand suspending
— f — m **de alambre** m - [cabl] wire strand suspending
— m — **despacho** m - [transp] (shipment) holding
— f — **dispositivo** - device, o fixture, suspension, o hanging
— f — **eje** m - [autom-mec] axle suspension
— f — **energía** f - [electr-distrib] power suspending,sion
— f — — f **eléctrica** - [electr-distrib] electric power suspending,nsion
— f — **hamaca** f - [domést] hammock suspension
— f — **inclinador** m - [mec] tilter suspending
— f — **línea** f - [electr-instal] line suspending,nsion
— f — — f **telefónica** - [telecom] telephone line suspending,nsion
— f — **muestra** f - sample, hanging, o suspending,nsion
— f — **objeto** m **de arte** - [domést] work of art hanging • novelty hanging
— f — **operación(es)** f - [ind] operation(s) suspending,nsion
— f — **peso(s)** m - [grúas] overhead lifting
— f — **vehículo** m - [autom] vehicle('s) suspending,nsion
— f — — m **rebotada** - [autom] bounced vehicle('s) suspending,nsion
— f — **volcador** m - [mec] dumper, o tilter, suspending,nsion
— f **delantera** - [autom-mec] front suspension
— f — **derecha** - [autom-mec] right front suspension
— f — **independiente** - [autom] independent front suspension
— f — **izquierda** f - [autom-mec] left front suspension
— f **desde extremo(s)** m - [electr-instal] end suspending,nsion
— f — — m **sin corriente** f - [electr-instal] dead end(s) suspending,nsion
— f **elevada** - [autom-mec] raised suspension
— f **en líquido** m - liquid suspending,nsion
— f — . . . **parte(s)** f - [mec] . . . part(s) suspending,nsion
— f — . . . **punto(s)** m - [mec] . . . point(s) suspending,nsion
— f — **punto** m **intermedio** - [mec] mid point suspension
— f — **tres punto(s)** m - [mec] three point suspending,nsion
— f **independiente** - [autom] independent suspension
— f — **de rueda(s)** f - [autom-mec] wheel(s) independent, o independente wheel(s), suspension
— f **levantada** - [autom] raised suspension
— f **líquida** - liquid suspending,nsion
— f **mala** - [autom-mec] poor, o off, suspension
— f **mejor** - [autom-mem] better suspension
— f **mejorada** - [autom-mec] improved, o refined, suspension
— f **modificada** - [autom] modified suspension

suspensión f̱ para aguilón m̱ - [grúas] boom suspending,ñsion
— f̱ — servicio m̱ pesado - [autom-mec] heavy
duty suspension
— f̱ — transmisión f̱ - [autom-mec] gimbal
— f̱ — vehículo m̱ - [autom-mec] vehicle suspending,ñsion • suspension system
— f̱ permitida - [autom-mec] swapped suspension
— f̱ proyectada—[autom-mec] designed suspension
— f̱ refrozada—[autom-mec] stiffened suspension
— f̱ retráctil m̱ - [autom-mec] retractable suspension
— f̱ — para peso m̱ pesado - [autom-mec] heavy
duty retractable suspension
— f̱ trasera - [autom-mec] rear suspension
— f̱ — derecha f̱ - [autom-mec] right rear suspension
— f̱ — independiente - [autom-mec] independent
rear suspension
— f̱ — izquierda - [autom-mec] left rear suspension
suspensor m̱ - [mec] hanger
— m̱ para tubería f̱ - [tub] pipe,ping, o̱
tube,bing, suspender, o̱ hanger
— m̱ — f̱ entubación f̱ - [petról] casing,
suspender, o̱ hanger
sustancia f̱ activa—active, o̱ forcing, substance
— f̱ — extraña - active foreign substance
— f̱ antiespumante - antifoaming, substance, o̱
agent
— f̱ carbónica - [quím] carbonic substance
— f̱ cáustica - [ind] caustic substance
— f̱ extraña - extraneous, o̱ foreign, substance
— f̱ lixiviante - [miner] lixiviating substance
— f̱ radiactiva - [nucl] radioactive substance
— f̱ recarburante - recarburizing substance
sustancialmente adv - véase substancialmente adv
sustentación f̱ - [mec] . . .; bearing
— f̱ de carga f̱ - [mec] load, bearing, o̱ supporting, o̱ withstanding
— f̱ en obra f̱ - [constr] field support(ing)
— f̱ exigida - [constr] required, bearing, o̱
support(ing)
sustentado,da a̱ - [mec] supported
— a̱ en emulsión f̱ - emulsion supported
sustentamiento m̱—[constr] support(ing); bearing
— m̱ en emulsión f̱ - emulsion support(ing)
— m̱ exigido - [constr] required, support(ing),
o̱ bearing
sustentante a̱ - . . .; supporting
sustentar v̱ en emulsión f̱ - to emulsion support
sustento m̱ - . . . .; bearing
sustraído,da a̱ - subtracted
suyo(s) adv - . . . . • thereof
suyo,ya a̱ - its; their(s)
— adv propio,pia - his, o̱ her, o̱ its, own
symposium m̱ - véase simposio m̱

T

T/A a̱ - [autom-neumát] T/A (marca de fábrica f̱
de BFGoodrich)
T de latón m̱ - [tub] brass tee
— macho - [tub] male tee
— m̱ — y hembra - [tub] street tee
— m̱ — — de latón m̱ - [tub] brass street T
— m̱ para tubería f̱ - [tub] tee pipe
— m̱ — — f̱ abierto - [tub] open tee pipe
tabicado m̱ - [constr] bricking, o̱ walling, up
— m̱ de madrastra f̱ - [metal-prod] bustle pipe
bricking up
— m̱ — tobera f̱ - [metal-prod] tuyere bricking
up
tabicado,da a̱ - [constr] bricked up
— a̱ con ladrillo(s) m̱ - [constr] bricked up
tabicar v̱ - [Constr] . . .; to brick up
— v̱ con ladrillo(s) m̱ - [constr] to brick up

tabicar v̱ escorial m̱ - [metal-prod] to brick up
a̱ slag notch
— v̱ — con ladrillo(s) m̱ - [metal=prod] to
brick up @ slag notch
tabique m̱ - [constr] partition; curtain wall •
middle wall; bulkhead • [metal-prod] baffle •
[electr-instal] barrier
— m̱ aislador - [electr-instal] barrier
— m̱ alrededor de tubería f̱ - [constr] partition around @ pipe
— m̱ colgado - [constr] curtain wall
— ṉ con forma f̱ de media luna - [mec] crescent
(shaped) partition
— f̱ — parte f̱ superior lisa - [constr] flush
top partition
— f̱ deflector - [mec] baffle, o̱ deflecting,
plate
— m̱ en forma f̱ de media luna - [constr] crescent partition
— m̱ intermedio - [constr] middle wall
— m̱ montado - [constr] erected. o̱ mounted,
partition
— m̱ — sobre piso m̱ - [constr] floor mounted
partition
— m̱ oara canal m̱ - [constr] channel baffle
— m̱ — m̱ para humo(s) m̱ - [metal-prod] flue
baffle
— m̱ — entrada f̱ para hombre(s) m̱ - [metal-
-prod] man hole baffle
— m̱ — escorial m̱ - [metal-prod] slag notch,
baffle, o̱ closing wall
— m̱ — excusado m̱ - [snit] toilet partition
— m̱ — — m̱ con parte f̱ superior lisa f̱ -
[constr] flush top toilet partition
— m̱ — — m̱ montado sobre piso m̱ - [constr]
floor mounted toilet partition
— m̱ — registro m̱ - [constr] manhole baffle
— m̱ — soporte m̱ - [constr] bearing partition
— m̱ soportante - [constr] bearing partition
— m̱ vertical - [mec] vertical baffle
tabla f̱ - [carp] board; lumber; sheet - schedule
• chart • [petról] cant • [public] table
— f̱ a pie - [public] table below
— f̱ ajustadora - [mec] adjuster table
— f̱ astillero - [petról] finger board
— f̱ basada sobre experiencia f̱ - experience
(based) table
— f̱ básica - basic table
— f̱ — para estímulo m̱ - [labor] basic incentive table
— f̱ — — incentivo,vación f̱ - [labor] basic
incentive table
— f̱ Camtrol - Camtrol, table, o̱ chart
— f̱ comparativa - comparative table
— f̱ — de diámetros m̱ de alambre m̱ - [alambre]
wire size(s) comparative table
— f̱ con cantidad(es) f̱ - quantity,ties table
— f̱ — especificación(es) f̱ - specification(s),
chart, o̱ table
— f̱ — — f̱ para ancho,chura f̱ - [mec] width(s)
specification(s), chart, o̱ table
— f̱ — — f̱ — — para llanta f̱ - [autom-neum]
rim width(s) specification, table, o̱ chart
— f̱ — fecha(s) - [cronol] date(s) table
— f̱ — — f̱ para montaje m̱ - erection date(s)
table
— f̱ de altura(s) f̱ - [constr] height(s) table
— f̱ — — f̱ para relleno m̱ - [constr] fill
height(s) table
— f̱ — — — terraplén m̱ - [constr] fill
height(s) table
— f̱ — calibre(s) f̱ —[metal-lam] gage(s) table
— f̱ — — f̱ basada en experiencia f̱ - [metal-
-lam] experience gage table
— f̱ — — m̱ empírica - [mec] empirical gage(s)
table
— f̱ — — m̱ según norma - [metal-lam] standard
gage(s) table
— f̱ — carga(s) f̱ - [mec] load, chart, o̱ table
— f̱ — — f̱ muerta(s) - [constr] dead load(s)
table
— f̱ — — f̱ para grúa f̱ - [grúas] crane load(s)
table

tabla f de carga(s) f para grúa f hidráulica -
[grúas] hydraulic crane(s) load(s) table
— f — — f — — f — sobre pedestal m -
[grúas] hydraulic pedestal crane loads chart
— f — — — f sobre pedestal m - [grúas]
pedestal crane load(s) chart
— f — — f viva(s) - [mec] live load(s) table
— f — cobertura(s) f - [constr] cover table
— f — — f mínima(s) - [constr] minimum co-
ver(s) table
— f — conexión(es) f - [electr-instal] con-
nection(s) table
— f — — f para dos voltajes m - [electr-
-instal] dual voltage connection(s) table
— f — — f — voltaje m único - [electr-
-instal] single voltage connection(s) table
— f — curvatura(s) f - curvature(s), table, o
chart
— f — diámetro(s) m - diameter(s), table, o
chart
— f — — m admisible(s) - allowable diame-
ter(s), table, o chart
— f — — m mínimo(s) - minimum diameter(s),
table, o chart
— f — — m admisible(s) - minimum allowable diameter(s), table, o chart
— f — dimensión(es) f - dimension(s), o
size(s), table, o chart
— f — especificación(es) f - specification(s)
table, o chart
— f — — f para motor m - [mec] motor speci-
fication(s), table, o chart
— f — — f — par m motor—[mec] torque spec-
ification(s), table, o chart
— f — espesor(es) m - thickness(es), o
gage(s), table, o chart
— f — — m empírica - empirical gage, table, o
chart
— f — — m para pared(es) f—[constr] wall(s)
thickness gage,s • [tub] schedule
— f — fricción(es) f - [mec] friction(s),
table, o chart
— f — — f contra pared f - [hidr] wall friction, table, o chart
— f — presión(es) f - [mec] pressure(s),
table, o chart
— f — — f para ensayo(s) m - [mec] test(ing)
pressure(s), table, o chart
— f — — — m hidrostático(s) m - [mec]
hydrostatig test pressure(s) table, o chart
— f — gasto(s) m—[hidr] flow, table, o chart
— f — — m para regulación f - [hidr] control
flow, table, o chart
— f — — m — f para entrada f - [hidr]
inlet control flow, table, o chart
— f — — f — salida f - [hidr] outflow, o outlet, control flow chart
— f — intensidad(es) - [hidr] intensity chart
— f — — f de precipitación f (pluvial) -
[hidr] rainfall intensity, table, o chart
— f — — f — — f — en localidad f - [hidr]
local rainfall intensity, table, o chart
— f — intercambio(s) n - interchange(s),
table, o chart
— f — madera f - [maderas] wood(en) board
— f — — f para entibación f - [constr] wood
spiling board
— f — materias f - [public] content(s) table;
index
— f — — f cambiada - changed contents table
— f — — f modificada - [public] changed con-
tetn(s) table
— f — medida(s) f - measurement(s) table
— f — — f para neumático(s) m - [autom-neum]
tire size(s) table
— f — — f — — m para substitución f -
[autom-neumát] tire size(s) substitution,
table, o chart
— f — micarta f - [electr-instal] micarta
board
— f — pandeo(s) m - [constr] buckling, table,
o chart
— f — par(es) m motor(es) - [mec] torque(s),

table, o chart
tabla f de perfil(es) m estructural(es)—[metal-
-lam] structural shapes, table, o chart
— f — peso(s) m - weight(s), table, o chart •
[mec] load(s), table, o chart
— f — — m promedio(s) m - average weight(s),
table, o chart
— f — precipitación f pluvial - [hidr] rainfall (intensity), table, o chart
— f — — f — en localidad f - [hidr] local
rainfall (intensity), table, o chart
— f — presión(es) f - [mec] pressure(s),
table, o chart
— f — — f para neumático(s) m - [autom-
-neumát] tire pressure(s) (usage), table, o
chart
— f — — f para prueba(s) f - test(ing) pres-
sure(s), table, o chart
— f — — f — — f hidrostática(s) f - [hidr]
hydrostatic test(ing) pressure, table, o
chart
— f — procedimiento(s) m - procedure(s),
table, o charat
— f — — m para substitución f - alternate
procedure(s), table, o chart
— f — punto(s) m para ajuste m - [mec] adjustment point(s), table, o chart
— f — rendimiento(s) m - performance, table, o
chart
— f — resistencia(s) f - resistance(s), table,
o chart
— f — — f contra corrosión f - [metal] corrosion resistance, table, o chart
— f — salario(s) m - [labor] wage(s), table, o
chart
— f — temperatura(s) f - temperature(s),
table, o chart
— f — tolerancia(s) f - tolerance(s), table, o
chart
— f — — f dimensional(es) - dimensional tol-
erance(s), table, o chart
— f — — f para cojinete(s) m - [cojin] bear-
ing(s) tolerance(s), table, o chart
— f — — f — rodamiento(s) m - [cojin] bear-
ing(s) tolerance(s), table, o chart
— f — valor(es) - value(s), table, o chart
— f — velocidad(es) - speed(s), o velocity,
table, o chart
— f — — f para cauce m - [hidr] channel velocity, table, o chart
— f empírica - empirical, table, o chart
— f — de calibre(s) m - [metal empirical
gage, table, o chart
— f experimental - experience, o experimental,
table, o chart
— f — de calibre(s) m - experimental, o experience, gage, table, o chart
— f freática f - véase capa f freática
— f general - general, table, o chart
— f hawaiana - [deport] surfboard; surf board
— f interpretada - interpreted table
— f machihembrada f - [madera] matchboard
— f matemática - [mat] mathmatic(al) table
— f métrica - [matem] metric table
— f — europea f - [autom-neumát] European metric, table, o chart, o guide
— f para ajuste m - [mec] adjustment, o fitment, chart
— f — altura f - height(s), table, o chart
— f — — f de cobertura f - [constr] cover, o
fill, height(s) table
— f — cálculo m - [matem] calculation table
— f — — m para carga f - [metal-prod] charge
calculation table
— f — carga(s) f - [mec] load(s) table
— f — cilindro m - [metal-lam] (roll) barrel
— f — combinación f de neumático(s) m—[autom-
-neumát] tire combination chart
— f — combustión f - [combust] combustion,
table, o chart
— f — conversión(es) f - converion(s) table
— f — correspondencia f - [mec] fitment chart
— f — desgaste m - [mec] wear, table, o chart

tabla f para detección f de falla(s) f - [ind]
 troubleshooting chart
— f — entibación f - [constr] spiling board
— f — entibar v - [constr] spiling board
— f — estado m - [electrón] status table
— f — guía f - guide, table, o section
— f — — f para substitución f - substitution
 guide, table, o section
— f — — — f de neumático(s) m—[autom-
 -neumát] tire substitution guide, table, o
 section
— f — inflación f - [autom-neumát] inflation
 table
— f — — f de neumático(s) m - [autom-neumát]
 tire inflation guide
— f — — f — m de acuerdo con carga f -
 [autom-neumát] tire/load inflation table
— f — — f según carga f - [autom-neumát]
 load inflation table
— f — inglete(s),tear - [mec] miter board
— f — intercambio(s) m - interchange,ging
 chart
— m — juego(s) m - [mec] play chart
— f — — m longitudinal - [mec] longitudinal,
 o end, play chart
— m — — de árbol m - [autom-mec] shaft
 end play chart
— f — m — — — m para aportación f [de
 fuerza f) - [autom-mec] input shaft end play
 chart
— f — — m — — eje m - [autom-med] shaft
 end play chart
— f — — m — — m para aportación f (de
 fuerza f) - [autom-mec] input shaft end play
 chart
— f — lubricación f - [ind] lubrication chart
— f — mantenimiento m - [ind] maintenance
 chart
— f — momento m - [mec] momentum chart
— f — — m torsional - [mec] torque chart
— f — par m motor - [mec] torque chart
— f — pared f - wall chart • [constr] wall
 board
— f — piso m - [constr] floor board • [autom]
 footboard
— f — planchar v - [domést] ironing board
— f — reducción(es) f - reduction table
— f — reemplazo m - [mec] replacement chart
— f — — m de neumático(s) m - [autom-neumát]
 tire replacement chart
— f — revestimiento m - [constr] poling board
— f — servicio m - [ind] service chart
— f — tablazón m - [constr] sheathing board
— f — terraplén(es) m - [constr] overfill
 table
— f — zócalo m - [constr] baseboard
— f que sigue - following table; table below
— f rígida - [maderas] rigid board
— f segura - safe board • safe table
— f — para carga f - safe loader,ding table
— f siguiente - [public] following table
— f solapada - [constr] weatherboard
— f traducida - translated table
tablado m - [constr] plank flooring • platform
 • [puentes] deck(ing)
— m de hormigón m - [constr] concrete deck
— m — — m pretensado - [constr] prestressed
 concrete deck
— m — — m prevaciado - [constr] precast concrete deck
— m — madera f - [constr] wood decking
tablazón m - [constr] sheathing; board form
— f de madera f - [constr] wood board form
— f — encofrado(s) m - [constr] form board
— m — — m para superficie(s) f - [constr]
 face form lumber
— m para frente(s) m - [constr] breast board
tableado m - [metal-lam] shattering
— m de rodillo(s) m - [metal-lam] roll shattering
— m — — m para apoyo m - [metal-lam] back-up
 roll shattering
— m — — m — — m para laminador m para tem-

ple m - [metal-lam] temper mill back-up roll
 shatter(ing)
tableado m de rodillo(s) m para apoyo m para
 tren m para temple m - [metal-lam] temper
 mill back-up roll shatter(ing)
tableado,da a - [metal-lam] shattered
tablear v - [metal-lam] to shatter
tablero m - . . . . • transp] pallet • [electr]
 switchgear; switchboard • [constr] deck(ing);
 floor(ing) • [autom] dashboard; instrument
 panel • [electr-instal] pawl • [electrón]
 véase tablilla f
— m auxiliar - [electr-instal] sub-panel
— m blindado - [electr-instal] metal clad , o
 armored, switchgear
— m para voltaje m alto - [electr-equip]
 high voltage switchgear
— m — — m bajo - [electr-equip] low voltage switchgear
— m central - [mec] center deck
— completo - [electr-instal] complete, o full,
 panel, o switchboard
— m con borne(s) m - [electr-instal] switchboard • terminal, block, o board
— m — — m para alimentación f - [electr-inst]
 input terminal block
— m — — m — regulación f - [electr-instal]
 switchboard terminal block
— m — — m — salida f - [electr-instal] output (stud) panel
— m — — m para selector m - [electr-instal]
 selector stud panel
— m — — m — — m para voltaje(s) m—[electr-
 -instal] voltage selector stud panel
— m — — m con tamaño m inferior a normal -
 [electr-instal] undersized terminal block
— m — — m — m en miniatura f - [electr-instal]
 miniature terminal block
— m — — m para circuito(s) m - [electr-inst]
 circuit terminal board
— m — — m — — m para regulación f—[electr-
 -instal] control circuit terminal block
— m — — m reforzado - [electr-equip] heavy
 duty terminal block
— m — circuito m - [electrón] circuit board
— m — — m estampado m -[electr-instal]
 printed circuit board
— m — — m — conectado - [electr-instal]
 connected printed circuit board
— m — — m — desconectado - [electr-instal]
 disconnected printed circuit board
— m — — m — para circuito m - [sold] circuit
 printed circuit board
— m — — m — — m para encendido m -
 [electr-instal] firing circuit printed circuit board
— m — — m — — m — protección f para
 voltaje m momentáneo m - [electr-instal]
 transient voltage protection circuit printed
 circuit board
— m — — m — — m para regulación f -
 [sold] control circuit printed circuit board
— m — — m — para regulación f - [electr-
 -instal] printed circuig control board
— m — — m — — — m para marcha f sin carga
 f - [electr-instal] idler printed circuit
 board
— m — — m — reemplazado - [electrón] replaced printed circuit board
— m — — m — conectado - [electr-instal]
 connected printed circuit board
— m — — m — — m sin conectar v - [electr-instal]
 - disconnected, o unconnected, printed circuit board
— m — — m — m impreso - [electron] printed circuit board
— m — — m — — conectado - [electr-instal] connected printed circuit board
— m — — m — desconectado - [electr-instal]
 disconnected printed circuit board
— m — — m — para filtro m pasabanda -
 [electrón] bandpass filter printed circuit
 board

tablero m con circuito m impreso para filtro m
pasabanda optativo - [electrón] optional
bandpass filter printed circuit board
— m — — m — — regulador m—[electr-instal]
printed circuit control board
— m — — — m para regulador m para marcha f
sin carga f - [electr-instal] idler printed
circuit control board
— m — — — m — reemplazado - [electrón] replaced printed circuit board
— m — — m — conectado - [electr-instal]
connected printed circuit board
m — — m —, desconectado, o sin conectar -
[electr-instal] disconnected, o not connected, printed circuit board
— m — fusible(s) m - [electr-equip] fuse(d)
panel
— m — — m en consola f - [electr-instal]
console fuse(d) panel
— m — — m para distribución f - [electr-dis
tribution fuse(d) board
— m con indicador m (sonoro) - [ind] anunciator panel
— m — — m para alarma f - [ind] anunciator
alarm panel
— m con luz,ces f - [instrum] light(ed) panel
— m — — f condición f - [instrum] lighted
status panel
— m — — f — estado m - [instrum] lighted
status panel
— m — medidor(es) m - meter(ed) panel
— m — regulador(es) m - [electr-instal] con-
trol(ling), board, o panel
— m — — m para cráter m - [sold] crater control, board, o panel
— m — terminal(es) m - [electr-instal] terminal strip
— m — transistor(es) m - [sold] solid state,
o transistorized, panel
— m — — m para determinación f de falla(s) f
- [sold] solid state troubleshooting panel
— m conveniente - [autom-instrum] convenient
panel
— m — para regulación f - [autom-instrum]
convenient control panel
— m dañado - [electr-instal] damaged panel
— m de tipo m con dispositivo(s) m en lado m
de atrás - [electr-instal] dead front type
switchgear
— m — — m consola - [electr-instal] console
type panel
— m — madera f - [constr] wood deck(ing)
— m deprimido - [instrum] recessed panel
— m eléctrico - [electr-instal] electrical
panel
— m — para trefilador m - [trefil] (wire)
drawer electrical panel
— m en cabina f - [electr-instal] cab console
— m — consola f - [electr-instal] console
panel
— m enchufable - [electrón] plug-in, circuit
bord, o panel
— m — con circuito m impreso - [electrón]
plug-in printed circuit board
— m existente - existing, panel, o console
— m frontal - [instrum] front panel
— m — para regulación f - [sold] front control panel
— m — deprimido - [ind] recessed front panel
— m — hundido - [ind] recessed front panel
— m hundido - [mec] recessed panel
— m indicador - [instrum] indicator panel
— m — para estado m - [ind] status, panel, o
table
— m liso - [constr] flush, o smooth, panel
— m optativo - [ind] optional, board, o panel
— m — ajuste m - [ind] adjusting,tment, panel, o board
— m — — m preciso - [ind] vernier board
— m — alimentación f - [electr-instal] input
panel - [ind] feed,der,ding panel
— m — amplificación f - [electr-instrum] amplifier,fying panel

tablero m para amplificación f magnética -
[electr- magnetic amplifier panel
— m — — f para dispositivo m para avance
m - [sold] travel magnetic amplifier panel
— m — amplificador m—[electr] amplifier panel
— m — — m magnético - [electr] magnetic amplifier panel
— m — — m para avance m - [sold] travel
magnetic amplifier panel
— m — — m para avance m - [sold] travel amplifier panel
— m — anuncio(s) m - bulletin board
— m — artesa f - véase esparavel m
— m — avance m - [sold] travel panel
— m — aviso(s) m - bulletin board
— m — bomba f - [metal-prod] pump (instrument)
panel
— m — borne(s) m - [electr-instal] terminal,
block, o panel
— m — campo m - [electr] field panel
— m — — m — motor m - [electr] motor field
panel
— m — — m — — m sincrónico - [electr] synchronous motor field panel
— m — — m sincrónico - [electr] synchronous
field panel
— m — cargador m - [metal-prod] skip hoist
panel
— m — carro m - [metal-prod] car panel
— m — — m báscula - [metal-prod] scale car
panel
— m — cierre m - [hidr] stop log(s)
— m — circuito(s) m - [electr] circuit board
— m — — m para carga f - [metal-prod] charging circuit panel
— m — comando m - command, o control, panel
— m — conexión(es) f - [electr-instal] connection panel
— m — — f para entrada f - [electr-instal]
input connection(s) panel
— m — control m - [instrum] control, o instrument, panel
— m — corriente f alterna - [electr-equip]
alternating current panel
— m — — f continua - [electr-equip] direct
current panel
— m — determinación f de falla(s) f - [sold]
trouble shooting, o fault finder, panel
— m — dibujo m - [dib] drafting, o drawing,
board; véase también mesa f para dibujo m
— m — diseño m - [dib] drawing board
— m — distribución f - [electr-instal] distribution, panel, o board
— m — encofrado(s) m - [constr] form board
— m — — m para superficie(s) f - [constr]
face form, panel, o board
— m — energía f - [electr-instal] power
switchboard
— m — entrada f - [electr-instal] input panel
— m — estado m - [electrón] status table
— m — estrangulador m - [mec] choke console
— m — estufa f - [metal-prod] stove (control)
panel
— m — excitación f - [electr-instal] exciter
panel
— m — filtro m - [electrón] filter, panel, o
board
— m — — m pasabanda - [electrón] bandpass
filter, board, o panel
— m — — m — optativo m - [electrón] optional bandpass filter, board, o panel
— m — fiscalización f—[instrum] control panel
— m — — f de tiempo(s) m - [instrum] timer
panel
— m — indicador,cación f - sign, panel, o
background
— m — instrumento(s) m - [instrum] instrument,
board, o panel
— m — — m completo - [mec] full instrument
panel
— m — — m iluminado - [autom-electr] lighted
instrument panel
— m — — m para automóvil m - [autom-electr]
automotive instrument panel

tablero m para interconexión f—[electr-instal]
interconnecting,tion panel
— m — interruptor m - [comb.int] switch panel
— m — llave f - [comb.int] switch panel
— m — f para chispa f - [comb.int] ignition switch panel
— m — f — encendido m - [comb.int] ignition switch panel
— m — f — ignición f - [comb.int] ignition switch panel
— m — mando m - [electr-instal] (operator)
control, panel, o board, o bank; bench board
• switchgear
— m — m para motor n - [electr-mot] motor
control panel • [comb.int] engine control
panel
— m — maniobra(s) f - [electr-instal] switchgear unit
— m — manipulador m - [electrón] keyer board
— m — m para tono(s) m - [electrón] tone
keyer board
— m — — m substituido - [electrón]
substituted tone keyer board
— m — máquina f - [electr-instal] machine
panel
— m — medidor(es) m - [sold] meter panel
— m — montacarga(s) f - [metal-prod] skip
panel
— m — montaje m - [instrum] mounting panel
— m — m interior - [sold] internal mounting panel
— m — motor m - [electr-mot] motor panel -
[comb.int] engine panel
— m — m sincrónico - [electr-mot] synchronous motor panel
— m — planta f - [ind] mill console
— m — precipitador m - precipitator, panel, o
board
— m — protección f - [mec] protection,tive
panel
— m — puente(s) m - [constr] bridge deck(ing)
— m — reconexión f - [electr-instal] reconnecting,tion panel
— m — f para voltaje m - [electr-instal]
voltage reconnect(ing),tion) panel
— m — f — entrada f - [electr-instal] input reconnecting,tion panel
— m — f rápida - [electr-instal] quick
reconnect(ing,tion) panel
— m — regulación f - [electr-instal] control,
panel, o (switch)board; bench board; instrument panel; controller; console
— m — f hundido - [electr-instal] recessed
control panel
— m — f maestro - [electr-instal] master
control panel
— m — f — para instalación f soldadora -
[sold] master fixture control panel
— m — f observado - [electr-oper] observed
control panel
— m — f para amplificador m magnético -
[sold] magnetic amplifier control panel
— m — f arrancador,nque m - [comb.int]
starter control panel
— m — f — bomba f - [bombas] pump control
panel
— m — f — cráter m - [sold] crater control, panel, o board
— m — f — dispositivo m - [sold] device
control, panel, o board
— m — f — m para fijación f - [sold]
fixture console
— m — f — f — m sujeción f - [sold]
fixture control panel
— m — f — horno m - [combust] furnace
control panel
— m — f — m eléctrico - [metal-prod]
arc furnace control panel
— m — f para instalación f soldadora -
[sold] fixture, console, o control panel
— m — f para laminador m - [metal-lam]
mill control panel
— m — f — m para tubo(s) m - [metal-

-lam] tube mill control panel
tablero m para regulación f para motor m -
[comb.int] engine control panel • [electr-
-mot] motor control panel
— m — f para proceso m - [ind] process control panel
— m — f puesta f en marcha - [ind]
start(ing), start-up, control, panel, o board
— m — f — tren m - [metal-lam] mill control panel
— m — f — válvula f - [mec] valve control
panel
— m — f — f para agua m - [hidr] water
valve control panel
— m — f — velocidad f - [sold] speed control panel
— m — f por operador m - [electr-oper]
operator('s) control panel
— m — sala f para cañón m - [metal-prod] mud,
o clay, gun control room panel
— m — selección f - [electr-instal] selector
panel
— m — f de banda(s) f de voltaje m - [sold]
voltage range selector panel
— m — f — escala(s) f de voltaje m -
[sold] voltage range selector panel
— m — f — límite(s) de voltaje m - [sold]
voltage range selector panel
— m — f — voltaje(s) m - [sold] voltage
selector panel
— m — skip* m - [metal-prod] skip (hoist)
panel
— m — suministro m de energía f - [electrón]
power supply, panel, o board
— m — f típico - [instrum] typical control
panel
— m — f tipo escritorio m - [instrum] desk
type control panel
— m — trefilador m - [trefil] (wire) drawer
(electrical) panel
— m — unidad f - [electr-instal] unit panel
— m — voltaje(s) m - [electr-instal] voltages,
panel, o switchgear
— m — m alto - [electr-instal] high voltage, panel, o switchgear
— m — m bajo - [electr-instal] low voltage,
panel, o switchgear
— m — m variable - [electr-instal] variable
voltage, panel, o board, o switchgear
— m parcial - [electr-instal] partial, o sub-
-panel
— m principal - [electr-instal] main panel
— m — para regulación f - [ind] main control
panel
— m — — f para proceso m - [ind] main process, o process main, control panel
— m substituido - [electrón] substituted, panel, o board
— m típico - typical panel
— m tipo armario - [ind] cabinet, panel, o
board
— m — consola - [electr-instal] console type
panel
— m — escritorio - [instrum] desk type panel
— m único - [ind] single panel
tablestaca f - [constr-pil] . . .; sheet.piling;
pile • [vial] sheeting • [hidr] sheeting
— f Armco - [constr-pil] Armco, sheeting, o
sheet pile,ling
— f — de acero m - [metal-fabr] Armco steel
sheeting
— f con, brida f, o pestana f - [hidr] flange
type sheeting
— f con peso m reducido - [constr] lightweight
sheeting
— f — traba(s) f - [metal-fabr] interlocking
sheeting
— f corrugada—[metal-fabr] corrugated sheeting
— f — de acero m - [constr] corrugated steel,
sheeting, o piling
— f — hincada - [constr]pil] driven corrugated
sheet (pile)
— f — liviana - [constr-pil] lightweight cor-

rugated sheet(ing)
tablestaca f corta - [constr] short sheet pile
— **f dañada** - [constr] damaged sheet(ing)
— **f de acero m** - [hidr] steel sheet(ing)
— **f — — m corrugado** - [hidr] corrugated steel sheet(ing)
— **f — — m — liviana** - [constr] lightweight corrugated steel sheet(ing)
— **f — — f hincada** - [constr] driven steel sheet(ing)
— **f embridada** - [constr] flanged sheet(ing)
— **f engatillada** - [constr] interlocking sheet- -(ing) panel
— **f hincada** - [constr] driven, sheet(ing), o panel
— **f laminada** - [metal-lam] rolled sheet(ing)
— **f — de acero m** - [metal-lam] rolled steel sheet(ing)
— **f — en caliente** - [metal-lam] hot rolled, sheet(ing), o piling
— **f larga** - [constr] long sheet pile,ling
— **f liviana** - [constr] lightweight sheeting; light sheet pile
— **f — de acero m** - [constr] lightweight steel piling
— **f movida** - [constr] moved, panel, o sheeting
— **f pesada** - [constr] heavy, sheet pile, o piling
— **f — laminada** - [constr] heavy rolled piling
— **f — — en caliente** - [constr] heavy hot rolled piling
tablestacado m - [constr] sheet piling (struc- ture; piling • lining • shoring • [hidr] sheeting • [náut] bulkhead sheeting
— **m ajustado** - [constr] adjusted, o tight sheeting
— **m anclado**—[constr-pil] anchored sheetpiling
— **m colocado** - [constr-pil] placed, o located, sheetpiling
— **m con peso m reducido** - [constr-pil] light- weigh sheeting
— **n — traba f** - [constr] interlocking piling
— **m — — f de acero m** - [constr] interlocking sheet steel piling
— **m corrugado** - [constr] corrugated, piling, o sheeting
— **m — de acero m** - [constr] corrugated steel, sheeting, m o piling
— **m — hincado** - [constr-pil] driven corruga- ted, sheeting, o piling
— **m — liviano** - [constr] lightweight corruga- ted, sheeting, o piling
— **m dañado** - [constr] damaged, sheeting, o piling
— **m de acero m** - [constr] sheet steel piling; steel sheeting • steel sheet piling
— **m — — m corrugado** - [constr] corrugated (sheet) steel piling
— **m — — corrugado hincado** - [constr-pil] driven corrugted steel, sheeting, o piling
— **m — — m — liviano** - [constr] lightweight corrugated steel, sheeting, o piling
— **m — — — m hincado** - [constr] driven steel, sheeting, o piling
— **m engatillado** - [constr] interlock(ing), sheeting, o piling
— **m — de acero m** - [constr] interlocking steel, sheeting, o piling
— **m entrelazado** - [constr] interlocking, sheeting, o piling
— **m — de acero m** - [constr] interlocking steel, sheeting, o piling
— **m hincado** - [constr] driven, sheeting, o piling
— **m laminado** - [metal-lam] rolled sheeting
— **m — en caliente** - [metal-lam] hot rolled, sheeting, o piling
— **m liviano** - [metal-lam] lightweigh, sheet- ing, o piling
— **m — de acero m** - [constr] lightweight steel, sheeting, o piling
— **m movido**—[constr] moved, sheeting, o piling
— **m para zanja(s) f** - [constr] trench sheeting

tablestacado m pesado - [constr] heavy piling
— **m — laminado** - [metal-lam] heavy rolled, o rolled heavy, piling
— **m — — en caliente** - [metal-lam] heavy hot rolled piling
— **m — — — frío m** - [metal-lam] heavy cold rolled, o cold rolled heavy, piling, o sheet- ing
— **m trabado** - [metal-lam] interlocking piling
— **m — de acero m** - [metal-lam] interlocking sheet steel piling
tablestacar v - [constr] to sheet
tableta f - [electrón] véase **tablero m**
tablilla f - slat • [electrón] véase **tablero m** • [constr-persianas] blade
— **f con circuito m** - [electrón] circuit board
— **m — m estampado** - [electrón] printed circuit board
— **m — — m para, arranque m, o arrancar v** - [electrón] start-up printed circuit board
— **f — — m — para circuito m** - [electrón] circuit printed circuit board
— **f — — m — — m para energía f**—[electr- -instal] power (circuit) printed circuit board
— **f — — m — — m —, aportación f, o en- trada f** - [electrón] power circuit printed circuit board
— **f — — — m para regulación f** - electrón] control circuit printed circuit board
— **f — — m — para cráter m** - [sold] crater printed circuit board
— **f — — m — — encendido f** - [sold] start- -up printed circuit board
— **f — — m — regulación f** - [sold] con- trol printed circuit, o printed circuit con- trol, board
— **f — — m — — f de cráter m** - [sold] crater control printed circuit board
— **f — — m — para regular cráter m** - [sold] crater control printed circuit board
— **f — — — — v encendido m** - [sold] start(ing) control printed circuit board
— **f — — m — voltaje m** - [sold] voltage printed circuit board
— **f — — m — — m constante** - [sold] constant voltage printed circuit board
— **m — — — — m variable** - [sold] varia- ble voltage printed circuit board
— **f — — m impreso m** - [electrón] printed cir- cuit board
— **f — — m — arranque,ncar** - [electrón] start(ing) printed circuit board
— **f — — m — para cráter m** - [sold] crater printed circuit board
— **m — — m — — encendido m** - [sold] start- int printed circuit board
— **f — — m — energía f** - [electr-instal] power printed circuit board
— **f — — — circuito m para energía f** - [electr-instal] power circuit printed circuit board
— **f — — m — — — m para, aportación f, o entrada f** -[electrón] power, o input, cir- cuit printed circuit board
— **f — — m — para regulación f** - [electrón] control circuit printed circuit board
— **f — — m — — regulación f**—[electr-instal] printed circuit control board
— **f — — m — — f para cráter m** - [sold] crater fill(ing) printed circuit board
— **m — — — regular v cráter m** - [sold] crater control printed circuit board
— **f — — m — — v encendido m** - [sold] start(ing) control printed circuit board
— **f — — m — voltaje m constante** - [sold] constant voltage printed circuit board
— **f — — m — variable** - [sold] variable voltage printed circuit board
— **f — — m para voltaje m constante** - [electr] constant voltage circuit board

tablilla f con circuito m para voltaje m variable - [electr] variable voltage circuit board
— f — regulador(es) m para cráter m - [sold]
crater control board
— f de acero m - [mec] steel slat
— f — — m engatillada - [mec] interlocking
steel slat
— f — — — a prueba de intemperie f—[carp]
weatherproof interlocking steel slat
— f — — m galvanizado [metal-trat] galvanized
steel slat
— f — — m — engatillada - [metal-lam] interlocking galvanized steel slat
— f — — — a prueba de intemperie f -
[mec] weatherproof interlocking galvanized
steel slat
— f engatillada de acero m - [mec] interlocking
steel slat
— f galvanizada - [metal-trat] galvanized slat
— f — engatillada - [metal-fabr] interlocking
galvanized, o galvanized interlocking, slat
— f para circuito m - [electrón] circuit card
— f — energía f - [electrón] power board
— f — — f con circuito m estampado - [electr-
-instal] printed circuit power board
— f — — f — — m impreso - [electr-instal]
printed circuit power board
— f — persiana f - [mec] louver blade
— f — recordatorio(s) m - [ind] poster
board
— f — regulación f - [ind] control board
— f — — f con circuito m estampado - [electr-
-instal] printed circuit control board
— m — — f — — m — para energía f—[electr-
-instal] power printed circuit control board
— f — — — m — — potencia f - [electr-
-instal] power printed circuit control board
— f — — f — — m impreso - [electr-instal]
printed circuit control board
— f — — f — — m para energía f—[electr-
-instal] power printed circuit control board
— f — — f para cráter m - [sold] crater control board
— m — — f soldadora f - [sold] welder
control, board, o panel
— f por medio - alternate slat
tablón m de madera f - [constr] wood, o lumber,
plank
— m — — f desgastado - [constr] worn wood
plank
— m — anuncio(s) - bulletin board
— m para cierre m - [constr] stop, log, o plank
tabulación f - tabulation • framework • record
— f efectuada en obra f - on-@-job record
tabulado,da a - tabulated • recorded
tabular v - to tabulate; to record | a - . . .
taco m - . . . • [mec] lug; plug; pad; block •
[autom-neumát] lug • [metal-prod] stem
— m con ranura f - grooved block
— m — — f en V - [sold] V-block
— m — calendario m - [imprent] calendar pad
— f fornido m - [autom-neumát] beefy lug
— m — para tracción f - [autom-neumát] beefy
traction lug
— m fuerte—[autom-neumát] strong, o beefy, lug
— m — para tracción f - [autom-neumát] strong,
o beefy, traction lug
— m inferior - [mec] lower, o bottom, pad
— m para fijación f - [mec] lower, o bottom,
fastening pad
— m para borde m - [autom-neumát] shoulder lug
— m — de banda f para rodamiento m -
[autom-neumát] shoulder lug
— m — fijación f - [mec] fastening, o holding,
o stop(ping), block, o pad
— m — muñón m - [mec] trunnion block
— m — nivelación f - [mec] leveling block
— m — sujeción f - [mec] fastening, o holding,
block
— m — tracción f - [autom-neumát] traction lug
— m superior - [mec] top, o upper, pad, o block
— m — para fijación f [mec] top, o upper,
fastening, o holding, pad, o block

taco m tope - [mec] top block
— m transversal - [metal-trat] transverse stem
— m y pasador m - [mec] block and pin
— m — — m para muñón m - [mec] trunnion block
and pin
tacómetro m para fin(es) m múltiple(s) -
[instrum] multi-function tachometer
táctica f de vender y huir - [com] sell and run
tactic(s)
tacha f - . . .; spot
— f negruzca - [metal] blackish spot
tachado,da - censured • erased • deleted
tachar v - . . . • to erase • to delete
tacho m - . . .; pot
— m para escoria f - [metal-prod] slag, o cinder, pot; véase también cuba f, o pote m,
para escoria f
— m —, volteo m, o vuelco m - [ind] tilting, o
dumping, pot
— m — — m frontal - [ind] front tilting pot
tachuelado m - [mec] tacking
tachuelado,da a - [mec] tacked
tachuelar v - [mec] to tack
tafilete m - [vest] sweat, o hat, o head, band
— m — máscara f (protectora) - [segurid]
shield head band
— f — — f para soldador(es) m - [segurid]
weld(ing) shield head band
— m perforado - [segurid] perforated head band
taja v tubo(s) m - [petról] casing, ripper, o
splitter
tajada f de torta f - [culin] cake slice • [fam]
pie piece
tajadera f - [herram] . . .; chisel; howel
— f en caliente - [herram] hot chisel
tajadora f - . . . • [herram] butt howel •
[agric-equip] chopper; shredder
tajar v - . . .; to notch
tajea f - [hidr] . . .; cross drain
tajo m - . . . • [labor] job; work • area •
[miner] cut; pit
— m abierto m - [miner] open pit
— m en mina f - [miner] mine cut
— m limpiado - [miner] cleaned cut
— m — para arranque m - [miner] cleaned ripping cut
— m limpio m - [miner] clean cut
— m para arranque,ncadura - [miner] ripping,
cut, o pit
— m — — m limpio,pia - [miner] clean ripping
cut
— m — carga f - [miner] loading, pit, o cut
— m — — f limpiado - [miner] cleaned loading,
cut, o pit
— m — — f limpio - [miner] clean loading cut
— m reparado - [miner] repaired cut
tal(es) adv - . . .; like that • such
— adv como - just as • that is • such as
— adv — acabado,da de fundir - [metal-prod]
rough cored
— adv — construido,da a - as built
— adv — está - as is
— adv — queda al soldar(se) - [sold] as welded
— adv — — luego de soldar(se) - [sold] as
welded
— adv — sale(n) - as produced; right out of
— adv — se expide(n) - [ind] as shipped
— adv — soldado,da - [sold] as welded
— adv — recibido,da - as received
— adv cual - as is • as presented
— adv — queda depositado,de - [sold] as welded
— adv efecto - that, o this, end, o purpose
— adv vez - maybe; perhaps
tala f - [maderas] . . .; logging
— f de bosque(s) m - [mader] (forest) logging
taladrado m - [mec] drilling; boring • [tub]
tapping • [comb.int] véase diámetro m • véase
también taladro m
— m apareado - [mec] match(ed) drilling
— a automático - [mec] automatic, drilling, o
boring
— m de agujero m - [mec] hole, drilling, o
boring

taladrado m de barreno m - [mec] hole drilling
— m — cilindro m - [comb.int] cylinder boring
— m — cubo m - [mec-fuedas] hub boring
— m — plantilla f - [mec] templet boring
— m fácil - [mec] easy, boring, o drilling
— m y refrentado m - [mec] boring and facing
taladrado,da a - [mec] drilled • bored
— a apareadamente - [mec] match drilled
— a con torno m - [mec] lathe, bored, o drilled
— a en basto - [mec] rough bored
taladradora f - [mec] boring mill; drilling machine; drill (press); piercing mill; boring machine
— f automática - [mec] automatic drill(ing) machine • [metal-prod] bott mechanism; taphole drill
— f con columna f - [mec] drill press
— f de precisión f - [herram] jig borer
— f horizontal - [mec] horizontal drill(ing machine)
— f múltiple - [herram] multiple drill
— f para cilindro(s) m - [comb.int] cylinder, boring machine, o drill
— f — corcho(s) m - [herram] cork borer
— f — piquera f [metal-prod] iron notch drill
— f — plantilla(s) f - [mec] jig borer
— f — riel(es) m - [herram] rail drill
— f radial - [herram] radial drill
— f y refrentadora f - [mec] boring, o drilling, and facing mill
taladrar v - [mec] ... • to pink
— v agujero m - [mec] to, drill, o bore, @ hole
— v apareadamente - [mec] to match drill
— v barreno m - [mec] to drill @ hole
— v carbón m - [miner] to bore (through @) coal
— v cuba f - [metal-prod] to, drill, o perforate, @ furnace stack
— v plantilla f - [mec] to bore @ templet
— v roca f - [constr] to bore (through @) rock
— v y volar v - [explos] to drill and shoot
taladro m - [herram] ... .; boring tool; bit • [metal-prod] (drilled) hole • [geol] boring • [petról] drill bit; drilling rig • [mec] barrel
— m ahusado - [mec] tapered bore
— m anular - [herram] annular, o ring, borer
— m automático - [herram] automatic drill
— m combinado - [hydr] combined drill
— m con percusión f y rotación f - [hidr] combined percussion and rotation drill
— m con aire m comprimido - [miner] compressed air (driven) (rock) drill
— m — m para roca f - [miner] compressed air (driven) (rock) drill
— m — diamante(s) m - [petról] diamond drill
— m — punta f con diamante(s) m - [petról] diamond (tipped) drill
— m cónico - [mec] tapered bore
— m de armadura f - [electr-mot] armature bore
— m — cilindro m - [comb.int] cylinder bore
— m — ... pulgada(s) f - [mec] ... .-inch bore
— m desgastado - [mec] worn bore
— m eléctrico - [herram] electric drill
— m en acelerador m - [comb.int] throttle bore
— m — cojinete m - [mec] bearing bore
— m — cuba f - [metal-prod] bosh, bore, o hole
— m — mesa f - [mec] table bore
— m — f rotativa - [petról] (rotary) table bore
— m — f — inspeccionado - [petról] inspected (rotary) table bore
— m — portadiferencial m - [autom-mec] (differential) carrier bore
— m — prensaestopa(s) m - [mec] packing box bore
— m — rodamiento m - [cojin] bearing bore
— m expansible - [herram] expanding auger
— m giratorio - [petról] rotary drill
— m montado - [herram] mounted auger
— m — sobre (auto)camión m - [herram] truck mounted auger

taladro m montado sobre camión m automóvil - [herram] truck-mounted auger
— m múltiple - [herram] multiple drill
— m neumático - [herram] pneumatic, o air, drill
— m para árbol m - [mec] shaft bore
— m — m para aportación f de fuerza f - [mec] input shaft bore
— m — m — f — en cubierta f - [mec] cover input shaft bore
— m — m — f — f para distribuidor m para fuerza f - [autom-mec] power divider cover input shaft bore
— m — cojinete m - [mec] bearing bore
— m — horadación f - [herram] boring auger
— m — mano f - [herram] hand, drill, o auger
— m — pecho m - [herram] breast, borer, o auger
— m — piquera f - [metal-prod] taphole drill
— m — regulador m - [comb.int] throttle bore
— m — roca f - [herram] rock drill
— m — rodamiento m - [mec] bearing bore
— m — rueda f - [mec-ruedas] wheel bore • wheel boring mill
— m — tierra f - [constr] soil auger
— m — — f montado sobre (auto)camión m - [constr] truck mounted earth auger
— m — velocidad f alta - [herram] high speed drill
— m por percusión f - [hidr] percussion drill
— m radial - [hidr] radial, o rotation, drill
— m sobre oruga(s) f - [constr] crawler drill
— m tubular - [petról] core drill
talante n - mood
talco m - ... .; talcum
— m con granulometría f fina - [miner] fine grain talc(um)
— m — f muy fina - [miner] very fine grain talc(um)
— m en polvo m - [miner] powdered talc(um); talcum powder
talentos m diversos - diverse talents
tales adv - such • those
talón m - .... - [imprent] stub • [com] record book stub • [electr-cond] lug • [autom-neum] rim; bead • [mec] lug • eking
— m de cordón - [sold] bead, rim, o heel
— m — diente m - [mec] tooth, heel, o rim
— m — neumático m - [autom-neumát] tire bead
— m — reja f - [agric-equip] share heel
— m — f para arado m - [agric-equip] plow share heel
— m — soldadura f - [sold] weld heel
— m desplazado - [autom-neumát] displaced, o pushed off, bead
— m embutido - [autom-neumát] countersunk lug
— m — cubierta f - [autom-neumát] tire rim
— m soldado - [electr-instal] soldered lug
— m terminal - [electr-instal] terminal lug
— m — con presión f - [Electr-instal] pressured terminal lug
— m — — f de tipo m aprobado - [electr-instal] pressure type terminal lug
— m — — f — m recomendado - [electr-instal] approved pressure type terminal lug
talonario m - [imprent] stub (book)
— m — cheque(s) m - [fin] check (stub) book
talud m - .... [topogr] (side) slope; embankment; sidehill; apron; bank slope; talus; bank; scarp; declivity; inclination; ascent
— m amplio - [constr] broad, o long, slope
— m aproximado - approximate slope
— m con ángulo m recto - [constr] square cut slope
— m — m recto para extremo m de alcantarilla f - [constr] square cut culvert end slope
— m declive m - [topogr] sloping hillside
— m — m pronunciado - [topogr] steep(ly) sloping hillside
— m —, hierba f, o cesped m, o pasto m - [constr] grassed (pasture) backslope
— m contenido - [hidr] controlled slope • [constr] retained slope

talud m cortado - [constr] cut slope
— m de cauce m - [hidr] channel slope
— m — — m protegido - [hidr] protected channel slope
— m — corte m - [topogr] back slope
— m — hormigón m - [constr] concrete apron
— m — presa f - [hidr] dam slope
— m — relleno m - [constr] fill slope
— m — ribera f - [hidr] bank slope
— m — roca f - [hidr] rock slope • [constr]
rock slope protection
— m — suelo m - [constr] earth slope
— m — terraplén m - [constr] embankment slope
— m — — m de tierra f - [constr] earthwork
fill slope
— m detrítico - [geol] talus
— m empedrado - [constr] riprap slope
— m encachado - [hidr] riprap slope
— m erosionado - [topogr] eroded slope
— m especificado - [constr] specified slope
— m — para terraplén m - [constr] specified
embankment slope
— m estabilizado - [constr] stabilized slope
— m estable - [topogr] stable slope
— m establecido - [constr] established, grade,
o slope
— m exterior - [constr] backslope; outer,tside
slope
— m — con, césped, o hierba f, o pasto m -
[constr] grassed (pasture) backslope
— m inclinado - [constr] slanting slope
— m inestable - [constr] unstable slope
— m ingleteado - [constr] mitered slope
— m — en extremo m—[constr] mitered end slope
— m — — — m de alcantarilla f - [constr]
mitered culvert end slope
— m lateral - [constr] side slope
— f — de terraplén m - [constr] embankment
side slope
— m natural -[topogr] natural ground • [constr]
natural slope
— m plano - [topogr] flat slope
— m protegido - [hidr] protected slope
— m río abajo - [hidr] downstream slope
— m — — de presa f - [hidr] dam downstream
slope
— m — arriba - [hidr] upstream slope
— m — — de presa f—[hidr] dam upstream slope
— m seguro - [constr] safe slope
tallado m - [mec] cutting; shaping • engraving
— m de engranaje(s) m - [mec] gear cutting
talladora f - cutter, o engraving, machine
— f para engranaje(s) m - [mec] gear, cutter, o
cutting machine
taller m - . . .; shop • [ind] plant • mill; facility
— m automovilístico - [autom-mec] automotive
shop
— m — para rendimiento m alto - [autom-mec]
automotive performance shop
— m con accesorio(s) m para tubería f - [tub]
pipefitting shop
— m con prensa(s) f - [public] press shop
— m convencional - conventional, o traditional,
shop
— m corriente - [ind] traditional shop
— m de aficionado - amateur, o home, work shop
— m — calderería f - [cald] boiler shop
— m — subcontratista - [ind] subcontractor's
shop
— m — proveedor m - supplier('s) shop
— m eléctrico - [ind] electrical (repair) shop
— m — para mantenimiento m - [ind] electrical
maintenance shop
— m en granja f - [mec] farm shop
— m — obra f - [mec] field shop
— m grande - [mec] large shop
— m — para soldadura(s) f - [sold] large welding shop
— L D - [metal-prod] basic oxygen, o Linz Donawitz, o L D, shop
— m mecánico—[ind] mechanical, o machine, shop
— m — en granja f - [mec] farm (machine) shop

taller m mecánico para aficionado m - amateur, o
home work, shop
— m movible - [ind] mobile shop
— m para neumático(s) m - [autom-neumát] mobile tire shop
— m naval - [náut] boat, o naval, yard
— m óptimo - [ind] first class shop
— m para acabado m - [ind] finishing shop
— m — — m de plancha(s) f - [metal-lam]
plate finishing (shop)
— m — amoladura f - [metal-lam] grinding shop
— m — — f de rodillo(s) m - [metal-lam] roll
grinding shop
— m — ánodos m - [ind] anode(s), shop, o
plant
— m — armado m - [ind] assembly,bling shop
— m — automóvil(es) m - [autom-mec] automobile
shop; garage
— m — carpintería f - [carp] carpenter,try, o
woodworking, shop
— m — clasificación f - [ind] (as)sorting
house • [f.c.] classification, o switch(ing),
yard
— m — clavos m - [ind] nail shop
— m — colada f - [metal-prod] cast house
— m — conservación f - [ind] maintenance,
shop, o facility
— f — — f eléctrica - [ind] electrical maintenance shop
— m — — f mecánica - [ind] mechanical maintenance shop
— m — — f ubicado estratégicamente - [ind]
strategically located maintenance facility
— m — convertidor(es) m - [metal-prod] converter shop
— m — corte - [ind] cutting shop
— m — — m acetilénico - [sold] oxyacetylene
cutting shop
— m — cribado m - [ind] screeing, shop, o
building
— m — eclisa(s) f - [metal-lam] splice bar(s),
plant, o shop
— m — engrase m - [ind] grease,sing shop;
oil(ing) house
— m — entretenimiento m - [ind] maintenance
shop
— m — — m eléctrico - [ind] electrical maintenance shop
— m — — m mecánico - [ind] mechanical maintenance shop
— m — estiramiento m de alambre m - [trefil]
wiredrawing shop
— m — estirar v alambre m - [Trefil] wiredrawing shop
— m — fabricación f - [ind] manufacturing, o
fabricating,tion, plant, o shop
— m — — f en escala f grande - [sold] large
scale, o production, shop
— m — — f — — f industrial - [sold] production shop
— m — — f en serie - [sold] production shop
— m — forja(dura) f - [ind] forge,ging shop •
blacksmith shop
— m — — f y plancha(s) f - [metal-prod] plate
and forge,ging shop
— m — fundición f - [metal-prod] foundry (shop)
• melting shop; cast house
— m — granallado m - [metal-trat] shot blasting shopl
— m — — m de rodillo(s) m - [metal-prod]
roll(s) shot blasting shop
— m — herramienta(s) f - [ind] tool shop; véase tambien pañol m para herramienta(s) f
— m — hojalatería f - [ind] tin shop
— m — inspección f - [ind] inspection, o control, shop, o house
— m — — f para palanquilla(s) f - [metal-lam]
billet(s) control house
— m — — f y transferencia f (de palanquilla) -
[metal-lam] billet control and transfer house
— m — laminación f - [metal-lam] (rolling) mill
— m — laminado m - [metal-lam] rolling shop;
rolled product(s) mill

taller m para locomotoras f - [ind] roundhouse;
locomotive shop
— m —— f y equipo m móvil - [ind] locomotive and mobile equipment shop
— m — mantenimiento m—[ind] maintenance shop
— m —— m eléctrico - [ind] electrical maintenance shop
— m —— m general - [ind] general maintenance shop
— m —— m mecánico - [ind] mechanical maintenance shop
— m —— m para locomotora(s) f - [f.c.] locomotive maintenance shop
— m — máquina(s) f - [mec] machine shop
— m — mecanización f - [mec] machine shop
— m — modelo(s) m - [ind] pattern shop
— m —— m y carpintería f - [metal-prod]
pattern and carpenter shop
— m — moldeo m - [metal-prod] molding shop •
cast house
— m — montaje m - [ind] assembly,bling shop •
rigger('s) shop
— m —— m y tubería(s) f - [ind] rigger('s)
and pipe(s) shop
— m — neumático(s) m - [autom-neumát] tire
shop
— m — oxicorte m - [sold] oxyacetylene cutting shop
— m — placa(s) f - [metal-lam] plate shop
— m —— f para asiento m y eclisa(s) f -
[metal-lam] tie plate and splice bar plant
— m — plancha(s) f - [metal-prod] plate shop
— m — posicionador(es) m [sold] jib shop
— m — prensado m - [metal-prod] press shop
— m — prensadura f - [mec] press shop
— m — reparación f - [ind] press, o dressing,
shop • repair shop
— m — preparación f - [ind] preparation shop
— m —— f para lingoteras f - [metal-prod]
mold preparation, building, o shop
— m — producción f - [ind] preparation, shop,
o floor, o facility,ties
— m — f en escala f industrial - [sold] production shop
— m —— f en serie - [ind] production shop
— m — producto(s) m plano(s) - [metal-prod]
flat product(s) shop
— m — rebabar,bado - [metal-lam] fettling
shop
— m — rectificación f - [metal-lam] grinding shop
— m —— f para chapa(s) f [metal-lam] (strip
mill) grinding shop
— m —— f —— f en caliente - [metal-lam]
hot strip mill grinding shop
— m —— f para rodillo(s) m - [metal-lam]
roll grinding shop
— m — reducción f - [metal-lam] reducing, o
reduction, mill
— m — reparación f - [mec] repair shop
— m —— f para automóvil(es) m - [autom-mec]
automobile repair shop
— m —— f cuchara(s) f - [metal-prod]
ladle repair, house, o shop
— n —— f para material(es) m - [ind] mate-
rial(s) repair shop
— m —— f —— m ferroviario(s) - [f.c.]
railway (materials) repair shop
— m —— f — vagón(es) m - [f.c.] car repair
shop
— m —— f —— m ferroviario(s) - [f.c.]
(railway) car repair shop
— m —— f — vagoneta(s) f - [ind] buggy
repair shop
— m —— f —— f y vagones m ferroviarios -
[ind] car and buggy repair shop
— m —— f ferroviaria - [f.c.] railway repair shop; roundhouse
— m —— f — edificio(s) m - [ind] building
repair shop
— — n —— reparaciones f eléctricas - [ind]
electrical repair shop
— m —— f eléctricas - [ind] mechanical re-

pair shop
taller m para reparación f para locomotora(s) f
- [f.c.] locomotive repair shop
— m — rodillo(s) m - [metal-lam] roll shop
— m —— m para laminación f - [metal-lam]
strip mill roll shop
— m —— m —— f en caliente - [metal-lam]
hot strip mill roll shop
— m —— m —— f — frío - [metal-lam] cold
(rolling) mill roll shop
— m — servicio m - [ind] (field) service shop
— m —— m autorizado - [ind] authorized service shop
— m —— m para locomotora(s) f - [f.c.] locomotive service, shop, o station
— m — sillete(s) m y eclisa(s) f - [metal-lam]
tie plate and splice bar plant
— m — soldadura f - [sold] weld(ing) shop
— m —— f a destajo m - [sold] job (weld)
shop
— m —— f grande—[sold] large weld(ing) shop
— m —— f y forja(dura) f - [metal-trat] tem-
per(ing) y forge,ging, shop
— m — temple m - [metal-trat] temper(ing), o
hardening, shop
— f — terminación f - [metal-lam] finishing
shop
— m —— f para plancha(s) f - [metal-lam]
plate finishing (shop)
— m — torneado m - [mec] turning shop
— m —— m para rodillo(s) m - [metal-lam]
roll turning shop
— m — trabajo(s) m - [ind] work shop
— m —— m a destajo - [ind] job shop
— m —— m en escala f grande - [sold] production shop
— m —— m —— f industrial - [ind] production shop
— m —— m en serie - [ind] production shop
— m — transferencia f - [ind] transfer house
— m —— f para palanquilla(s) f - [metal-lam]
billet transfer house
— m — trefilería f - [metal-alambre] (wire)
drawing shop
— m — tubería f - [tub] pipe,ping shop
— m pequeño - [ind] small shop
— m — para soldadura(s) f - [sold] small welding shop
— m principal - [ind] main. o headquarters,
shop, o building
— m Siemens - [metal-prod] open hearth (shop)
— m — Martin - [metal-prod] open hearth shop
tallo m - [bot] . . .; sprig
tamaño m aceptable - acceptable size
— m aconsejado - [ind] recommended size
— m — para cable m - [electr-cond] recommended
cable size
— m —— m para salida f - [electr-cond]
recommended output cable (size)
— m adecuado - adequate, o suitable, size
— m adicional - additional size
— m amplio - ample size
— m anterior - former, o older, size
— m apropiado - appropriate, o proper, o sufficient, o suitable, size
— m — de cono m - [sold] appropriate cone size
— m —— m para fundente m - [sold] appropriate flux cone size
— m — de cordón m - [sold] appropriate, o proper, bead size
— m aproximado - approximate size
— m — para alcantarilla f - [tub] approximate
culvert size
— m cabal - full size • prototype size
— m — para tubería f - [tub] full (line) size
— m comercial - commercial, o trade, size
— m compacto - compact size
— m comparable - comparable size
— m comparado - compared size
— m considerable - considerable size
— m correcto - correct size
— m — para neumático(s) m - [autom-neumát]
correct tire size

tamaño m correspondiente - corresponding size
— m corriente - current, o standard, size
— m de abertura f - [mec] opening size
— m — agarratubos m - [grúas] pipe grab size
— m — agujero m - [mec] hole size
— m — alambre n - [cabl] wire size
— m — bola(s) f - [cuerp.moled] ball(s) size
— m — boquilla f - [mec] nozzle size
— m — — f empleado - [mec] nozzle size used
— m — — f para cuchara f - [metal-prod] ladle nozzle size
— m — — f usado - [mec] nozzle size used
— n — bóveda f - [constr] arch size
— m — — f de plancha(s) f (estructurales) - [constr](structural) plate arch size
— m — — f — — f de acero m - [constr] steel structural plate arch size
— m — brazo m - arm size • [mec] fork size
— m — brida f - [mec] flange size
— m — briqueta f - [miner] briquette size
— m — cabeza f - head size
— m — cable m - [electr-cond] cable size
— m — cadena f - [cadenas] chain size
— m — caja f - [med] case size
— m — camisa f - [mec] liner size
— m — carcasa f - [electr-mot] frame size
— m — cojinete m - [cojin] bearing size
— m — conductor m - [electr-cond] wire size
— m — — m de cobre m - [electr-cond] copper wire size
— m — conjunto m de calce m - [mec] shim pack size
— m — — m — laminilla f - [mec] shim pack size
— m — cordón m - [sold] bead size • [tub] bead extent • bead size
— m — — m en ángulo m (interior) - [sold] fillet size
— m — — — m (—) medido con plantilla f - fillet gage size
— m — — regulado - [sold] controlled, o regulated, bead size
— m — corrugación(es) f - [mec] corrugation size
— m — chumacera f - [mec] bearing size
— m — defecto m - defect size
— m — edificio m - [constr] building size
— m — émbolo m buzo - [petról] plunger size
— m — engranaje m - [mec] gear size
— m — estrella f - [astron] star size • [metal-trat] spangle size
— m — estructura f - [constr] structure size
— m — expandimiento m - [metal-fabr] expansion size
— m — fino(s) m - [miner] fine(s) size
— m — — m de sinterización f - [metal-prod] sinter(ing) fine(s) size
— m — garganta f - [sold] throat size
— m — gota f - drop size
— m — gotita f - [sold] droplet size
— m — grano(s) m - [metal-prod] grain size; véase también granulometría f
— m — horno m - [ind] furnace size
— m — — m alto - [metal-prod] blast furnace size
— m — horquilla f - [mec] fork size
— m — imán m - magnet size
— m — ladrillo m - [cerám] brick size
— m — letrero m - [constr] sign size
— m — lingote m - [metal-prod] ingot size
— m — lingotera f - [metal-prod] ingot mold size
— m — llanta f - [autom-neumát] rim size
— m — malla f - [mec] mesh, o screen size
— m — mango,guito m - [mec] sleeve size
— m — manguito m para barra f taponadora - [metal-prod] stopper rod sleeve size
— m — máquina f - [mec] machine size
— m — material(es) m - material(s) size
— m — — m filtrante(s) - filter(ing) material(s) size
— m — — m para subdrenaje m - [hidr] subdrainage filtering material(s) size

tamaño m de mineral(es) m - [miner] ore, o mineral, size
— n — molde m - [metal-prod] mold size
— m — muestra f - sample size
— m — neumático m - [autom-neumát] tire size
— m — — m inaceptable - [autom-neumát] unacceptable tire size
— m — — m para reemplazo m - [autom-neumát] substitute, o replacement, tire size
— m — orificio m - [mec] opening, o hole, size
— m — palanca f - [mec] lever size
— m — partícula(s) f - particle size
— m — f de mineral m - [miner] ore particle size
— m — pedido m - order size
— m — perno m - [mec] bolt size
— m — — m para anclaje m - [constr] anchor bolt size
— m — pieza f - [mec] component, o part, size
— m — plancha f - [metal-lam] plate size
— m — planta f - [ind] plant size
— m — poste m - [constr] post size
— m — presa f - [hidr] dam size
— m — producción f - [ind] production size
— m — — f normal - [ind] normal production size
— m — quemador m - [combust] burner size
— m — remache m - [mec] rivet size
— m — — m circunferencial - [mec] circumferential rivet size
— m — revestimiento m - [petról] liner size
— m — riel m - [metal-lam] rail size
— m — rueda f - [mec-ruedas] wheel size
— m — separador m - [mec] spacer size
— m — soldadura f - [sold] weld, o joint, size
— m — — f horizontal - [sold] horizontal weld size
— m — — f — en ángulo m interior - [sold] horizontal fillet (weld) size
— m — sujetador m - [mec] fastener size
— m — tubería f - [tub] pipe size • [ambient] duct size • piping size
— m — — f abovedada - [constr] pipe-arch size
— m — — f circular - [tub] round pipe size
— m — tubo m - [tub] pipe size
— m determinado - determined, o specified, o given, size
— m disminuido - diminished, o reduced, size
— m distinto - different size • alternate size
— m diverso - different size
— m efectivo - effective size
— m — m de garganta f - [sold] effective throat size
— m equivalente - equivalent size
— m especial - special size
— m especificado - specified size
— m estándar - standard size
— m excepcional - exceptional, o unusual, size
— m excesivo - exceptional size; oversize
— m exigido - required size
— m exiguo - too small size; undersize
— m extraordinario - oversize
— m final - [ind] production, o final, size
— m grande - large size
— m igual - equal, o same, size
— m inaceptable - unacceptable size
— m indicado - specified size
— m inferior - undersize
— m — a normal - below normal size
— m inmediatamente mayor - next larger size
— m — menor - next smaller size
— m insuficiente - undersize
— m intermedio - intermediate, o medium, size
— m máximo - maximum size
— m — admisible - maximum acceptable size
— m — para horquilla f - [mec] fork maximum size
— m mayor - larger size • largest size
— m mediano - medium size
— m medio - average size
— m menor - smaller size • smallest size
— m métrico - [metric] metric size
— m mínimo - minimum, o smallest, size

tamaño m mínimo admisible - minimum acceptable
size
— m natural - natural size - full size - full
scale
— m necesario - needed, o required, size
— m nominal - nominal size - [cadenas] trade
size
— m normal - normal size - standard size
— m — de planta f - [ind] normal plant, o
plant normal, size
— m oficio - [papel] legal size
— m normal - normal, o standard, size
— m — para neumático m - [autom-neumát] tire
standard, o standard tire, size
— m nuevo - new size
— m optativo - optional size
— m óptimo - optimum, o ideal, size
— m — para partícula f - particle optimum, o
optimum particle, size
— m original - original size
— m para flotación f - [autom-neumát] flotation size
— m — horquilla f - [mec] fork size
— m — neumático m - [autom-neumát] tire size
— m — — m estándar - [autom-neumát] standard
tire size
— m — producción f - [ind] production size
— m — serie f - [ind] series size
— m — — f . . . - [autom-neumát] . . . series
size
— m pequeño - small size
— m popular - popular size
— m promedio - average size
— m — de partícula f de mineral - [miner] ore
particle average size
— m real - true, o actual, o real, size
— m recomendado - recommended size
— m — para rueda f - [mec-ruedas] recommended
wheel, o wheel recommended, size
— m reducido—reduced, o compact, o small, size
— m — de cordón m - [sold] bead, controlled, o
regulated, size • [cabl] strand controlled
size
— m representativo - representative size
— m — de bóveda f - [constr] arch representative size
— m — — f de plancha(s) f - [tub] arch representative size
— m — — f — — f estructural(es) - [tub]
structural plate arch representative size
— m — — — f — — f — de acero m - [constr]
steel strucural, o structural steel, plate
representative size
— m según norma f - [mec] standard size
— m siguiente - next, o following, size
— m similar - similar size
— m suficiente - sufficient size; size enough
— m — para producción f—[ind] production size
— m sugerido - suggested size
— m — para rueda f - [mec-riedas] wheel suggested, o suggested wheel, size
— m tal - such size
— m único - single size • one size
— m único de partícula(s) f - single particle
size
— m uniforme - uniform size
— m — de partícula(s) f - uniform particle, o
particle uniform, size
— m útil - useful size
— m variable - variable size
— m — para bloque m - [autom-neumát] block
variable, o variable block, size
tamaños m diversos - different, o sundry, sizes
• several sizes
— m varios - various, o several, sizes
tambaleado,da a - staggered • wobbled • reeled
tambaleante a - staggering • wobbling • reeling
tambalear v - • to wobble
tambaleo m - . . . . • wobble,ling
— m de rueda f - [mec] wheel wobble,ling
tambo m - [ganad] . . .; véase también vaquería
tambor m - [mec] . . . • reel • [transp] drum;
barrel; container; can • [mec] arbor

tambor m acanalado - [cabl] grooved drum
— m — en hélice f—[cabl] spiral grooved drum
— m accionado - [mec] driven drum
— m ahusado - [mec] tapered drum
— m anterior - [mec] front drum • previous drum
— m calentado - [mec] heated drum
— m cerrado - [mec] closed drum
— m cilíndrico - [mec] cylindrical, barrel, o
drum
— m con brida(s) f - [mec] flanged drum
— m — fondo m - [mec] drum with @ head
— m — forro m - [grúas] lagged drum
— m — huella(s) f - [cabl] corrugated drum
— m — recubrimiento m -"[grúas] lagged drum
— m — superficie f acanalada - grooved surface
drum
— m — — f lisa f - [mec] smooth, surface, o
faced, drum
— m — — f lisa para elevación f - [grúas]
smooth, surface, o faced, hoisting drum
— m cónico - [cabl] conical, o tapered, drum
— m — para desenrollamiento m - [metal-lam]
cone type pay-off reel
— m corrugado - [cabl] corrugated drum
— m cortado - [mec] cut drum
— m dañado - [mec] damaged drum
— m de acero m - [mec] steel drum
— m — — m rotado - [mec] rolled steel drum
— m — caldera f - [cald] boiler drum
— m — tipo m constrictor - [mec] constricting
type drum
— m delantero - [mec] front drum • [constr-
-equip] front roller
— m desgastado - [mec] worn drum
— m doble - [mec] double drum
— m elevado - [mec] raised, o hoisted, drum
— m embridado - [mec] flanged drum
— m en rotación f - [mec] rotating, o turning,
drum
— m — tándem - [mec] tandem drum
— m — — m sin recubrimiento m - [mec] unlagged tandem drum
— m engranado - [mec] engaged drum
— m — radialmente - [mec] radially engaged
drum
— m especificado - [mec] specified drum
— m específico - [mec] specific drum
— m forrado - [grúas] lagged drum
— m frontal - [mec] front drum
— m girado - [mec] turned drum
— m giratorio - [mec] revolving drum • [metal-
-alambre] block
— m — para estiradora f - [metal-alambre] machine block
— m — — — f para alambre m - [metal-alambre]
wiredrawing machine block
— m — — trefiladora f [para alambre m] -
[metal-alambre] wiredrawing machine block
— m — — estiramiento m [de alambre m] - wiredrawing block
— m— Hyster m - [cabl] Hyster winch
— m inclinado - [mec] slanting, o tilted, drum
— m inferior - [mec] low(er), o bottom, drum
— m liso - [mec] smooth drum
— m — para elevación f - [grúas] smooth hoisting drum
— m llenado - filled drum
— m lleno - full drum
— m manipulado - [mec] handled drum
— m marinero - [petról] cathead
— m mezclador - [ind] mixing drum
— m para abobinado m - [mec] reeling drum
— m — accionamiento m - [mec] drive,ving drum
— m — — m solidario - [mec] solidary
drive,ving drum
— m — aglomeración f [en frío] - [miner]
(cold) balling drum
— m — alambre n - [mec] wire drum
— m — — m "Speed Feed" - [sold] Speed Feed
drum
— m — alimentación f - [mec] feeding drum
— m — arena f - [ind] sand (storage) drum
— m — — f seca—[ind] dry sand (storage) drum

tambor m para arrollamiento m - [cabl] winding
drum
— m — balance m - [mec] balancing drum
— m — bobina(s) f - [mec] coil drum
— m — cable m - [grúas] cable, o (wire)
rope, drum • hoist(ing), o reel, drum
— m — — m de alambre m—[cabl] wire rope drum
— m — — m — cola f - [grúas] tag line drum
— m — — m — fijo - [grúas] fast line drum
— m — — m para aparejo m - [cabl] hoisting
drum
— m — — m — cuchara f - [petról] sand line,
spool, o reel
— m — — m — elevación f - [grúas] hoist(ing)
drum
— m — — m — entubación f—[petról] calf reel
— m — — m — gancho(s) m - [grúas] whip drum
— m — — m — m verificado - [grúas]
checked whip drum
— m — — m — herramienta(s) f - [petról] bull
reel
— m — — m para maniobra(s) f - [petról] calf
reel
— m — — m — motón m - [grúas] main, o hoist,
drum
— m — — m — — m verificado—[grúas] checked
main drum
— m — — m — perforación f - [petról] coring
reel
— m — — m — sacanúcleo(s) m - [petról] coring reel
— m — — m — sacasondas m - [petról] coring
reel drum
— m — — m — sacatestigos m - [petról] coring
reel (drum)
— m — — m — tubería f - [petról] calf reel
— m — — m — — f para entubación - [petról]
casing reel
— m — cabo m - [cabl] rope drum
— m — cabrestante m - [mec] winch drum
— m — cambiador m - [mec] shifter drum
— m — — m para embrague m - [mec] clutch
shifter drum
— m — cargador m—[metal-prod] skip hoist drum
— m — carrete m - [mec] reel drum
— m — — m para núcleo m - [petról] core reel
drum
— m — cilindro m - [petról] cylinder barrel
— m — cuchareo m - [petról] sand reel
— m — desabobinamiento m - [metal-lam] uncoiling reel
— m — desenrollamiento m - [metal-lam] decoiling, o pay-off, reel; uncoiling reel
— m — — m rápido - [metal-alambre] speed feed
drum
— m — — m — de alambre m - [sold] Speed Feed
drum
— m — elevación f - [mec] hoist(ing) drum
— m — — f para aguilón m - [grúas] boom
hoist(ing) drum
— m — elevador m - [grúas] hoist drum
— m — — m con cable m - [grúas] line hoist
drum
— m — — m — — m fijo - [grúas] fast line
hoist drum
— m — — m para aguilón m - [grúas] boom
hoist(ing) drum
— m — — m — cable m a motón m - [grúas] main
hoist (line) drum
— m — — m — montacarga(s) m - [metal-prod]
skip hoist drum
— m — — m principal - [grúas] main hoist drum
— m — embrague m - [mec] clutch drum
— m — empalme m - [constr-pil] splice,cing can
— m — enrollamiento m - [mec] reeling drum
— m — — m para cable m - [grúas] hoist drum
— m — para entrada f - [mec] intake reel
— m — entrega f - [mec] pay-off reel
— m — — f de tipo m con cono(s) m - [mec]
cone type pay-off reel
— m — equilibrio m - [mec] balancing drum
— m — freno m - [mec] brake, drum, o wheel
— m — herramienta(s) f - [petról] bull wheel

tambor m para hormigonera f - [constr] concrete
mixer drum
— m — imán m - [grúas] magnet reel
— m — izar v - [cabl] hoist(ing) reel
— m — v ranurado - [cabl] grooved hoisting
drum
— m — malacate m - [grúas] winch drum
[petról] drawworks drum
— m — — m auxiliar - [grúas] auxiliary winch
drum
— m — — m para herramienta(s) f - [petról]
bull wheel, spool, o drum
— m — — m para maniobra(s) f - [petról] calf
wheel, spool, o drum
— m — — m para tubería f - [petról] calf
wheel, spool, o drum
— m — — m principal - [grúas] main winch drum
— m — mando m - [cabl] power takeoff drum
— m — — m solidario - [mec] solidary drive
drum
— m — máquina f para skip m - [metal-prod]
skip hoist drum
— m — mezcla f - [mec] mixing drum
— m — montacargas m - [mec] hoist drum
— m — peletización f—[miner] pelletizing drum
— m — potencia f - [cabl] power takeoff drum
— m — propulsión f -]cabl] power takeoff drum
— m — reprocesamiento m - [ind] reprocessing,
o recycling, drum
— m — residuo(s) m - [vial] litter barrel
— m — salida f - [mec] delivery reel
— m — tensión f - [mec] tension drum
— m — toma f para potencia f - [cabl] power
takeoff drum
— m — torno m - [herram] winch drum -
[petról] cathead drum
— m — tracción f - [mec] capstan; véase tam-
-bien cabrestante m
— m posterior - [mec] rear drum
— m primero - [mec] first drum
— m principal - [mec] main drum
— m ranurado - [mec] grooved drum
— m — — elevación f - [mec] grooved hoisting drum
— m recibido - [ind] received drum
— m recubierto - [grúas] lagged drum
— m reprocesador - [ind] reprocessing, o recycling, drum
— m resellado - [mec] resealable drum
— m resistente - [transp] strong, o solid. drum
— m rodado - [mec] rolled drum
— m rotado - [mec] rotated drum
— m roto - [mec] broken drum
— m segundo - [mec] second drum
— m sencillo - [mec] single drum
— m sin abrir v - unopened drum
— m — fondo m - [mec] headless drum
— m sin forro m - [grúas] unlagged drum
— m — ranura(s) f - [mec] smooth faced drum;
ungrooved, o grooveless, drum
— m — recubrimiento m - [grúas] unlagged drum
— m soldado - [sold] welded drum
— m sólido, o sturdy, drum
— m Speed Feed - [sold] Speed Feed drum
— m superior - [mec] high(er) drum
— m — para cadena f - [metal-prod] tell-tale
driving drum
— m — — cinta f - [metal-prod] tell-tale
driving drum
— m — para sonda f - [metal-prod] tell-tale
driving drum
— m tercero - [mec] third drum
— m torneado - [mec] turned drum
— m trasero - [mec] rear drum • [constr-equip]
rear roller
— m único - [mec] single drum
— m vaciado - [mec] emptied drum
— m vacío - [mec] empty drum • bare drum
— m ventilado - [mec] ventilated drum
— m verificado - [mec] checked drum
— m volcado - dumped drum • tilted drum
tamborar* v - véase tumbar v
tamboreado,da a - drummed

tamborear v - to drum
tamboreo m - drumming
tamiz m con malla f - [mec] mesh sieve
— m para acemite m - [culin] graham flour, sieve, o screen, o bolter
— m para criba f - [herram] screen
— m — — f para sinterización f—[coque] sintering screen
— f — — f — — f en caliente - [coque] hot sintering screen
— m — desagüe m - [hidr] dewatering screen
— m — referencia f—[miner] basic mesh (size)
tamizado,da a - screened
tan adv bien como (se) podría—as best it could
— adv fuerte - as strong
— adv importante - as important; all important
— adv poco(s) - as little • as few
— adv rápidamente - as rapidly
— adv — como posible - as, rapidly, o fast, as possible
— adv sólo,la - solely
tanda f - [ind] . . .; charge • run • burn • rank
— f de producción f - [ind] production run
tándem m - . . . | adv - tándem
tanería* f - [ind] véase curtiembre f
tangencialmente adv - tangentially; véase también en forma f tangencial
tangente a de ángulo m - [matem] angle tangent
tangible a - . . . • [contab] corporeal
tanque m - . . . . • reservoir; véase también depósito m • cuba f
— m almacenador - véase tanque m para depósito
— m australiano - [constr] bottomless, o Australian, water tank
— m cilíndrico - [ind] cylindrical tank
— m — horizontal - [ind] horizontal cylindrical, o cylindrical horizontal, tank
— m — vertical - vertical cylindrical, o cylindrical vertical, tank
— m compresor m - [neumát] compressor tank
— m — para aire m - [ind] air receiver
— m de aluminio m - aluminum tank
—* m de concreto* m - véase depósito m de hormigón m
— m inferior - [hidr] lower tank
— m decapante - [metal-trat] pickling tank
— m digestor - [sanit] digester,tion tank
— m elevado - [hidr] elevated, o high level, tank
— m — para depósito m - [hidr] elevated, o highj level, storage tank
— m — — m para agua m - [hidr] elevated, o high level, water storage tank
— m empernado - [ind] bolted tank
— m — cilíndrico - [petról] cylindrical bolted tank
— m — — m horizontal - [ind] horizontal cylindrical, o cylindrical horizontal, bolted tank
— m — vertical - [ind] vertical bolted tank
— m medidor - [ind] measuring, o meter, tank
— m — para ácido m - [ind] acid, measuring, o meter(ing), tank
— m para aceite m - [ind] oil tank
— m — — m hidráulico - [ind] hydraulic oil tank
— m — — m crómico - [metal-trat] chromic acid tank
— m — agua m - [hidr] water tank
— m — — desmineralizada - [cald] demineralized water tank
— m — aire m - [neumát] air tank
— m — — m comprimido - [neumát] compressed air tank
— m — alimentación f - [ind] feeder,ding tank
— m — — f de emulsión f - [nucl] emulsion feeding tank
— m — — f — — f asfáltica - [nucl] asphalt emulsion feed(ing) tank
— m — — f de vacío m - [neumát] vacuum feed(ing) tank
— m — almacenamiento m - [ind] storage tank

tanque m para almacenamiento m para agua m - [hidr] water storage tank
— m — — m — — desmineralizada - [cald] demineralized water storage tank
— m — apagado m - [ind] quenching tank
— m — — m para cal f - lime quenching tank
— m — baño m - [ind] bath, vat, o tank
— m — — ácido - [metal-trat] acid bath, vat, o tank
— m — bencina f - [comb.int] gasoline tank
— m — captación f - [hidr] reservoir
— m — clasificación f - classification tank
— m — — f de desecho(s) m - [sanit] waste classification tank
— m — — f — residuo(s) m - [sanit] waste classification tank
— m — combustible m - [comb.int] fuel, o gasoline, tank
— m — — m con capacidad f para . . . - [comb.-int] . . . capacity fuel tank
— m — combustión f - [combust] combustion, o oil, tank
— m — condensado m - [tub] condensate tank
— m — decapar v - [metal-trat] pickling tank
— m — depósito m - storage tank
— m — — m esférico - [ind] spherical storage tank
— m — — m para agua m - [hidr] water storage tank
— m — destilación f - [ind] distilling tank
— m — disolución f - [metal-trat] solution, vat, o tank
— m — — f para hojalata f - [metal-trat] tinning solution, vat, o tank
— m — drenaje m - [cald] flash, o drainage. tank
— m — — m continuo = [cald] continuous, flash, o drainage, tank
— m — esencia f - [comb.int] gasoline tank
— m — expansión f - [tub] flash tank
— m — — f para condensado m - [tub] condensate flash tank
— m — guerra f - [milit] (battle) tank
— m — inmersión f - [ind] immersion, o dipping, tank
— m — lavado m - [ind] washing, o rinsing, tank
— m — nafta f - [comb.int] gasoline tank
— m — purga f - [hidr] flash tank
— m — — f para drenaje m - [cald] blowdown flash tank
— m — — f — — m continuo - [cald] blowdown flash tank
— m — — m — — m continuo m - [cald] continuous blowdown flash tank
— m — purgación f - véase tanque m para purga
— m — petróleo m - [petról] oil tank
— m — radiador m - [Comb.int] radiator tank
— m — recogida f - collecting,tion tank
— m — — f para, desecho(s) m, o residuo(s) m - waste collecting,tion tank
— m — remojo m - soaking tank • [metal-lam] bosh tank
— m — suministro m - [ind] supply(ing) tank
— m — tratamiento m - conditioning tank
— m — — m químico - [sanit] chemical conditioning tank
— m — vacío m - [neumát] vacuum tank
— m receptor - [ind] receiver (tank)
— m — para aire m - [ind] air receiver,ving (tank)
— m secundario - [ind] secondary tank
— m séptico - [sanit] septic tank
— m — doméstico - [sanit] domestic septic tank
— m — industrial m - [sanit] industrial septic tank
— m superior - [hidr] upper, o top, tank
— m — para aceite m - [ind] upper, o top, oil, tank, o vessel
— m vertical - [ind] vertical tank
tántalo m - . . . • [metal] tantalum
tanteado,da a - tried; tested • felt • probed
tantear v - . . .; to probe • to feel

tanteo m - . . . . feel and touch • tentative test(ing) • probe • trial (and error) • [deport] score; strike
— m combinado - [deport] combined score
tanto m - [deport] strike; score
— m alzado - lump sum
— adv calor m - so much heat
— adv como — as much as • as well as
— adv — (sea) posible - as much as possible
— m consular - [com] consular copy
— n en contra - [deport] strike against
— m en tanto adv - from time to time
— adv es así - in fact
— m por ciento - per cent; percentage
— m — — de utilidad f - [com] per cent (of), earning(s), o profit
tantos adv - as many
tañido m como (de) plata f - silver sound
tapa f - . . . . • top • [mec] shield • [metal--prod] véase mazarota f
— f a prueba f de goteo m - [mec] drip proof cover
— f abisagrada - [mec] hinged cover
— f ajustada - [mec] tightened cover • adjusted cover
— f apretada - [mec] tightened cover
— f armada - [mec] made up, o assembled, cover
— f ciega - [electr-instal] blind, o blank, cover
— f — de acero m - [electr-instal] blank steel cover
— f — — m galvanizado - [electr-instal] galvanized blank steel cover
— f circular - [mec] round cover
— f colocada—[mec] placed, o installed, cover
— f con rejilla f - [mec] screen, o grill, cover
— f con resorte m - [mec] spring cover
— f — m para válvula f - [válv] valve spring cover
— f contra explosión(es) f - [mec] explosion, cover, o door
— f — lluvia f - [mec] rain, cover, o cap
— f de acero m - [mec] steel cover
— f — chapa f - [mec] strip, cover, o cap
— f — fundición f - [mec] cast cap
— f — grafito m - [mec] graphite, cover, o cap • graphite nozzle
— f hierro m - [mec] iron cover
— f — m fundido - [mec] cast iron cover
— f — metal m - [mec] metal cover
— f deformada - [mec] warped cover
— f delantera - [mec] front cover
— f desarmada - [mec] disassembled cover
— f empernada f - [válv] bolted, o bolt-on, o bolt-down, cover
— f — inferior - [mec] lower, o bottom, bolt--on cover
— f — superior - [mec] top, o upper. bolt-on cover
— f en soldadora f sin arrancador m - [sold] starterless welder cover
— f enclavada—[electr-oper] interlocked cover
— f escorial - [metal-prod] bott
— f exterior - [mec] outer, cover, o cap
— f — contra polvo m - [mec] outer dust cap
— f — para consola f - [electr-equip] console outer,tside cover
— f frontal - [mec] front cover (plate)
— f — para caja f - [mec] case front cover
— f — — — f para engranaje(s) m - [mec] gear case front cover
— f grande - [mec] large cover
— f —, quitada, o removida, o sacada - [mec] removed large cover
— f guardapolvo(s) - [electr-mot] dust cap
— m — para extremo m - [electr-mot] end dust cap
— m — — m exterior - [electr-mot] outer end dust cap
— f — — — m — de motor m - [electr-mot] motor outer end dust cap
— f — — — m interior - [electr-mot] inner

tapa f guardapolvo(s) para extremo m interior para colector m - [electr-mot] commutator inner end dust cap
— guía - [metal-prod] (stock rod) guide cover
— f guiadera - [mec] guide cover
— f hexagonal - [mec] hexagon(al) (head) cap
— f inclinada - [constr] slanting, o sloping, top
— f inferior - [mec] bottom, o lower, cover
— f — para generador m - [electr-prod] generator bottom cover
— f — — turbina f - [turb] turbine bottom cover
— f inspeccionada - [mec] inspected cover
— f instalada - [mec] installed cover
— f intercambiable - [mec] interchangeable, top, o cover
— f interior - [mec] inside, o inner, cap
— f lateral - [mec] side cover • [metal-prod] side (manhole) cover • [autom] wheel skirt
— f — para cúpula f [metal-prod] dome side manhole cover
— f — — — f para estufa f - [metal-prod] stove dome side manhole cover
— f — — estufa f - [metal-prod] stove side (manhole) cover
— f levantada - [mec] lifted, o raised, cover
— f limpia - clean cover
— f limpiada - cleaned cover
— f metálica - [mec] metal cover
— f optativa - [mec] optional cover
— f — para carrete m - [sold] optional reel cover
— f — — m para alambre m - [sold] optional wire reel, o wire reel optional, cover
— f para abertura f - [mec] opening cover
— f — — f para regulación f - control, o timer, o timing, hole cover
— f — — f — dispositivo m para cambio(s) m - [mec] shift(ing) unit opening cover
— f — — f — magneto m - [mec] magneto, hole, o opening, cover
— f — acceso m - [mec] access cover
— f — — m a cruceta f - [mec] crosshead access cover
— f — — m — — f inferior - [mec] bottom crosshead access cover
— f — — m — — f superior - [mec] top crosshead access cover
— f — — m inferior - [mec] bottom access cover
— f — — m superior - [mec] top access cover
— f — acoplamiento m - [mec] coupling cover
— f — acumulador m - [electr-acumul] battery, top, o cover
— f — agujero m - [mec] hole cover
— f — — m para, henchir, o llenar - [comb.-int] filling hole cover
— t — — m — — m de depósito m - [comb.int] tank filling hole cover
— f — — — m — — m para combustible m - [comb.int] fuel tank filling hole cover
— f — amortiguador m - [mec] damp(en)er cover
— f — — m para línea f - [mec] line damp-(en)er cover
— f — — m — — f para aspiración f - [mec] suction line damp(en)er cover
— f — armadura f [mec] frame cover • [electr.-mot] armature, top, o cover
— f — — f principal - [mec] main frame cover
— f — arrancador m - [mec] starter cover
— f — balancín m - [electr-mot] rocker cover
— f — biela f - [comb.int] connecting rod cap
— f — bloque m - [mec] block cover
— f — — m para contacto m - [electr-equip] contact block cover
— f — boca f - [mec] hole, o opening, cover
— f — — f para carga f - [metal-prod] coke hole cover
— f — — f — inspección f - [ind] manhole cover
— f — — f — registro m - [ind] manhole cover(ing) (plate)

tapa f para boca f para registro m lateral -
[metal-prod] side manhole cover
— f — — f — — m para estufa f - [metal-
-prod] stove side manhole cover
— f — — — m inferior - [metal-prod]
lower side manhole cover
— f — — — m superior - [metal-prod] upper side manhole cover
— f — — — m lateral - [metal-prod]
side manhole cover
— f — — — — para estufa f—[metal-
-prod] upper stove side manhole cover
— f — bomba f - [bombas] pump cover
— f — — f para aceite m - [bombas] oil pump
cover
— f — — f — agua n - [hidr] water pump cover
— f — caja f - [mec] case, o box, o housing, o
cabinet, top, o cover
— f — — f alta - [mec] high case cover
— f — — f auxiliar - [mec] auxiliary cabinet,
cover, o top
— f — — f — para regulación f - [mec]
auxiliary control cabinet cover
— f — — f para cadena f - [mec] chain case top
— f — — f — cambio(s) m - [mec] shift housing
cover
— f — — f — carrete m - [mec] reel housing,
top, o cover
— f — — f — — m para alambre m - [sold] wire
reel housing, top, o cover
— f — — f cojinete m - [mec] bearing box
cover
— f — — f — — m no expansible - [mec] non-
-expanding bearing box cover
— f — — f — conmutador m - [electr-instal]
switch housing cover
— f — — f — engranaje(s) m - [mec] gear box
cover
— f — — f — gobierno m - [electr-instal] control, o switch, box, cover
— f — — f — interruptor m - [electr-instal]
switch, housing, o box, cover
— f — — f — interruptor m automático -
[electr-equip] contactor box cover
— f — — f — manómetro m - [ind] gage housing
cover
— f — — f — — m para presión f - [ind] pressure gage housing cover
— f — — f — regulación f - [ind] control box,
cover, o top
— f — — f — telemando n - [sold] remote control box cover
— f — — f — torno m - [petról] cathead
housing cover
— f — — f — trampa f - [mec] trap housing,
cover, o top
— f — — f — transmisión f - [mec] transmission case cover
— f — — f sujetadora - [mec] grip(ping), box,
o case, top, o cover
— f — cámara f - [mec] chamber cover
— f — — f para válvula f - [válv] valve chamber cover
— f — — f para vástago m - [mec] rod chamber
cover
— f — — f subterránea - [electr-instal] manhole, o underground chamber, cover
— f — canal m para espesador m - [metal-prod]
slurry flume cover
— f — carcasa f - [mec] housing, cover, o plate
— f — carburador m—[comb.int] carburetor cover
— f — carrete m - [sold] reel cover
— f — — m optativo—[sold] optional reel cover
— f — — m — para alambre m - [sold] optional
wire reel cover
— f — — m para alambre m - [sold] wire reel
cover
— f — cartela f - [mec] bracket cover
— f — — f para rodillo m - [mec] roll bracket
cover
— f — — f — — m para alimentación f - [mec]
feed(ing) roll(er) bracket cover
— f — célula f - cell, cap, o cover

tapa f para célula f de acumulador m - [electr]
battery cell cap
— f — cilindro m - [comb.int] cylinder head
— f — cojinete m - [mec] bearing, cover, o cap
— f — — m para biela f - [mec] connecting rod
cap
— f — — m transportador - [mec] carrier
bearing cap
— f — colector m - [electr-mot] commutator
cover - [hidr] sump cover
— f — — m, colocada, o instalada - [mec] installed sump cover
— f — — m quitada - [mec] removed sump cover
— f — — m removible - [mec] removable sump
cover
— f — — m, removida, o sacada - [mec] removed sump cover
— f — conjunto m - [mec] assembly cover
— f — — m de rodillo m - [sold] roll assembly
cover
— f — — m — m impulsor - [sold] drive,
o driving, roll assembly cover
— f — — m óptico - [metal-prod] peepsight
cover
— f — conmutador m - [electr-instal] switch
cover • [electr-mot] commutator cover
— f — consola f—[electr-instal] console cover
— f — cuerpo m - [válv] body, o shell, cover
— f — — m de válvula f - [válv] valve shell
cover
— f — — m — — f para viento m (caliente) -
[metal-prod] (hot) blast valve body cover
— f — cursor m - [mec] slide cover
— f — — m para hilera f - [trefil] die slide
cover
— f — chumacera f - [mec] bearing, cover, o
cap
— f — depósito m - [autom] tank, top, o cover
— f — — m para combustible m - [comb.int]
fuel, o gas(oline) tank, cover, o cap
— f — — m para fundente m - [sold] flux
tank cover
— f — desventeo* m - [mec] vent(ing) cap
— f — dispositivo - [mec] device, o unit,
cover
— f — distribuidor m - [comb.int] distributor
cap
— f — embrague m - [mec] clutch cover
— f — engranaje(s) m - [mec] gear(s) cover
— f — — m a bloque m - [comb.int] gear cover
to block
— f — — m a espárrago m roscado - [comb.int]
gear cover to @ stud
— f — — m a pieza f para relleno m - [comb.-
-int] gear cover to filler block
— f — — m a plato m - [comb.int] gear cover
to end plate
— f — equipo m - [mec] equipment cover
— f — — m para interrupción f automática -
[sold] contactor kit top
— f — escotilla f - [nav] hatch(way) cover
— f — — f para acceso m - [nav] access hatch-
way cover
— f — — f de hormigón m - [nav] concrete
hatch(way) cover
— f — — f removible - [constr] removable
hatch(way) cover
— f — — f — de hormigón m - [constr] removable concrete hatch(way) cover
— f — estufa f - [metal-prod] stove cover
— f —, excitador m, o excitatriz f - [electr-
-mot] exciter cover
— f — extremo m - [mec] end, cover, o cap
— f — generador m - [electr-prod] generator
cover
— f — gollete m - [radiador] filler cap
— f — — m para henchimiento m - [comb.int]
filler (neck) cap
— f — — m — — m con aceite m - [comb.int]
oil filler cap
— f — — m radiador m - [comb.int] radiator filler cap
— f — hilera f - [trefil] die cover

tapa f para horno m - [ind] furnace cover
— f — — n de fosa f - [metal-lam] (soaking) pit cover
— f — — m para igualación f de temperatura f [metal-lam] soaking pit cover
— f — m para recocido m - [metal-trat] annealing (furnace) cover
— f — impulsor m - [mec] drive(r) cover
— f — — f para piñón m - [mec] pinion drive cover
— f — — m — — m para arrancador m - [comb.-iny] starter pinion drive cover
— f — inspección f - [mec] inspection cover • [constr] construction cover; manhole
— f — — f para válvula f - [comb.int] valve inspection cover
— f — interruptor m - [electr-instal] breaker, o switch, cover, o plate
— f — libro m - [public] book cover
— f — limpieza f - [mec] clean out cover
— f — lingotera f - [metal-prod] (ingot) mold cap
— f — lluvia f - rain cap
— f — manómetro m - [mec] gage cover
— f — — m para presión f - [mec] pressure gage cover
— f — mensula f - [mec] bracket cover
— f — — f para rodillo m - [mec] roll bracket cover
— f — — f — — m para alimentación f - [mec] feed roll backet cover
— f — mesa f - [mec] table cover
— f — — f, quitada, o removida, o sacada - removed table cover
— f — — f rotatoria - [petról] rotary table cover
— f — mirilla f - [metal-prod] peep sight, o tuyere, cap
— f — molde m - [metal-prod] mold cap
— f — motor m - [electr-mot] motor cover • [comb.int] engine, cover, o hood • [mec] machine cover
— f — — m para ventilador m - [mec] fan motor cover
— f — orificio m - [mec] hole cover
— f — — m en bloque m - [comb.int] block hole cover
— f — — m — — m para bomba f - [bombas] pump block hole cover
— f — — m — — m — — f para combustible m - [comb.int] fuel pump block hole cover
— f — — m — cruceta f - [mec] crosshead hole cover
— f — — m para acceso m - [mec] access hole cover
— f — — m para inspección f - [mec] inspection hole cover
— f — — m — llave f - [mec] wrench hole cover - keyhole cover
— f — — m para mano f - [mec] hand hole cover
— f — — m — — f en cruceta f - [mec] crosshead hand hole cover
— f — piñón m - [mec] pinion cover
— f — — m para arrancador m - [comb.int] starter pinion cover
— f — platino m - [comb.int] point cover
— f — — m para motor m - [comb.int] breaker point(s) cover
— f — portafusible(s) m - [electr-equip] fuse holder, cap, o cover
— f — presión f - [comb.int] pressure cap
— f — — f para radiador m - [comb.int] radiator pressure cap
— f — quemador m - [combust] burner cover
— f — — m para estufa f - [metal-prod] stove burner cover
— f — radiador m - [comb.int] radiator cap
— f — reactor m - reactor cover
— f — recipiente m - [mec] container cap
— m — registro m - [mec[(man)hole cover plate
— f — — m manual - [mec] hand (operated) hole cover
— f — — m — en cruceta f - [mec] crosshead hand (operated) hole cover

tapa f para registro m para inspección f - [mec] inspecion hole cover
— f — relé m - [electr-instal] relay cover
— f — reloj m - [instrum] clock cover
— f — resorte m - [mec] spring cover
— f — retención f - [mec] retaining cover
— f — — f para tubería f para entubación f - [petról] casing bridge plug
— f — rodillo m - [mec] roll(er) cover
— f — — para formación f - [metal-lam] forming roll cover
— f — — m — impulsión f - [sold] drive,ving roll cover
— f — rueda f - [mec] wheel cover(ing) • [autom-mec[hubcap
— f — seguridad f - [segurid] safety cover
— f — — f para depósito m - [ind] safety tank cover
— f — — f — — m para combustible m - [comb-int] safety fuel tank cover
— f — soldadora f - [sold] welder, cover, o top; machine top
— f — soporte m - [mec] bracket cover
— f — sujeción f - [mec] holding cover - face plate
— f — tambor m - [mec] drum, cover, o head
— f — trampa f - [válv] trap cap
— f — transmisión f - [mec] transmission (case) cover
— f — — f colocada - [mec] installed transmission cover
— f — — f instalada - [mec] installed transmission cover
— f — — f, quitada, o removida, o sacada - [mec] removed transmission cover
— f — tubo m - [tub] pipe, cap, o lid
— f — — m para engrase m - [mec] grease pipe cap
— f — — m para escape m - [comb.int] exhaust pipe cap • rain cap
— f — — m — — m para lluvia f - [comb.int] exhaust rain, o rain exhaust, pipe
— f — — m para subida f de agua m - [tub] water ascending,nsion pipe lid
— f — turbina f - [turb] turbine cover
— f — unidad f - [mec] unit cover
— f — — f para cambio(s) m - [mec] shift(ing) unit cover
— f — válvula f - [mec] valve cover
— f — — f aflojada - [válv] loose(ened) valve cover
— f — — f ajustada - [válv] tightened valve cover
— f — — f apretada - [válv] tightened valve cover
— f — — f colocada - [válv] installed valve cover
— f — — f con resorte m - [válv] valve spring, o spring valve, cover
— f — — f inspeccionada - [válv] inspected valve cover
— f — — f instalada - [valv] installed valve cover
— f — — f limpia - [válv] clean valve cover
— f — — f limpiada - [válv] cleaned valve cover
— f — — f operada - [válv] operated valve cover
— f — — f para cambio m rápido - [válv] fast, o quick, change valve cover
— f — — f — succión f - [petról] suction valve cover
— f — — f — trabajo m - [válv] working valve cover
— f — — f — tubería f para descarga f - [tub discharge pipe valve cover
— f — — f para viento m caliente - [metal-prod] hot blast valve cover
— f — — f — — m — para estufa f - [metal-prod] stove hot blast valve cover
— f — — f quitada — [válv] removed valve cover
— f — — f recubierta - [válv] coated valve cover
— f — — f, removida, o sacada - [válv] re-

moved valve cover

tapa f para vasija f - container, o can, cover
— f — **vástago m** - [mec] rod cover
— f — **ventilador m** - [mec] fan cover
— f **pequeña f** - [mec] small cover
— f **posterior** - [mec] rear cover • [public] back cover
— f — **removible** - [mec] removable rear cover
— f — **superior** - [mec] upper rear cover
— f **prefabricada** - [mec] prefabricated cover • precast cover
— f **preparada** [mec] prepared, o made up, cover
— f **principal** - [mec] main cover
— f — **instalada** - [mec] installed main cover
— f —, **quitada, o removida, o sacada** - [mec] removed main cover
— f **protectora** - [mec] protective, cover, o cap • [sold] top shield • [segurid] (top) guard
— f — **para brida f** - [mec] flange protective cover
— f — — **carrete m** - [mec] reel protective cover • cover to protect @, reel, o spool
— f — — — m **pequeño m** - [mec] small, spool, o reel, protective cover
— f **quitada** [mec] removed, cover, o cap, o top
— f **recubierta** - [mec] coated cover
— f **redonda** - [mec] round, cover, o cap, o top
— f — **de hierro m** - [mec] round iron cover
— f — — — m **fundido** - [constr] round cast iron, o cast iron round, cover
— f — **pequeña** - [mec] small round cover
— f **reforzada** - [mec] reinforced cover
— f — **prefabricada** - [mec] prefabricated, o precast, reinforced, cover
— f **removible** - [mec] removable cover
— f — **de acero m** - [mec] removable steel cover
— f — — **hormigón m** - [constr] removable concrete cover
— f — **para colector m** - [mec] removable sump cover
— f — — **dispositivo m** - [mec] device removable cover
— f — — m — **cambio(s) m** - [autom-mec] shift(ing) unit (opening) removable cover
— f — — **horquilla f** - [autom-mec] shift(ing) fork (opening) removable cover
— f **removida** - [mec] removed, cover, o cap
— f **retenedora** - [mec] retaining, cover, o cap
— f — **para rodillo m** - [mec] roller retaining, cap, o cover
— f — — **rueda f** - [mec] wheel retaining cap
— f — — — f **loca** - [mec] idler retaining cap
— f — — — f — **con rodillo(s) m** - [mec] roller idler retaining cap
— f **roscada** - [tub] threaded, o screw(ed), cap
— f **sacada** - [mec] removed, cover, o cap, o top
— f **seguidora** - [cojin] follower cup • cup follower
— f **selladora** - [mec] cover(ing), seal, o cap
— f **superior** - [mec] top, cover, o shield
— f — **lateral** - [mec] upper side cover
— f — — **para estufa f** - [metal-prod] stove side upper manhole cover
— f — **para cuerpo m** - [válv] body upper cover
— f — — — m **de válvula f** - [metal-prod] valve upper body cover
— f — — — m — — f **para viento m caliente** - [metal-prod] hot blast valve upper body cover
— f — — **generador m** - [electr-prod] generator top cover
— f — — **turbina f** - [turb] turbine top cover
— f **terminal** - [mec] end cover • [tub] end closure section
— f **trasera** - [mec] rear cover
tapas f y fondos m intercambiables - interchangeable tops and bottoms
tapabarro(s) m - [autom] véase **guardabarros m; guardafangos m**
tapada f - [constr] (back)fill • [constr] cover
— f **impermeable** - [constr] sealed top
tapadera f - . . . . • [mec] cover(ing); door
tapado m - . . . . • véase también **taponamiento m** - [metal-prod] capping

tapado m de lingote(s) m - [metal-prod] ingot capping
— m **mecánico** - [metal-prod] mechanical capping
— m **químico** - [metal-prod] chemical capping
tapado,da a - covered; capped • closed • [metal-prod] plugged; capped
— a **con cuidado** - covered carefully
— a **mecánicamente** - [metal-prod] mechanically capped
tapador m - conver(ing) [metal-prod] bott (mechanism)
— m **automático para escoria m** - [metal-prod] automatic slag notch plugger
— m **para escoria m** - [metal-prod] bott; slag notch, cover, o plugger
tapadora f con disco(s) m - [agric-equip] disk coverer
tapadura f con cuidado - careful covering
— f **cuidadosa** - careful covering
— f **de acero m** - [metal-prod] steel capping
— f — — m **agitado** - [metal-prod] rimming steel capping
— f — — m **efervescente** - [metal-prod] rimming steel capping
— f — **vasija f** - container, o can, covering
**tapajunta(s) - . . . . • [constr] flashing; joint, cover(ing), o strip, o filler
— f **de membrana f** - [constr] flashing membrane
— f — — **especial** - [constr] special membrane flashing
— f — — f — **reforzada** - [constr] special reinforced membrane flashing
— f — **plomo m** - [constr] lead flashing
— f **especial** - [constr] special flashing
— f — **reforzada** - [constr] special reinforced flashing
— f **metálica** - [constr] metal(lic) flashing
— f **para cielo m raso** - [constr] ceiling, flashing, o joint cover(ing)
— f — **ventilador m** - [constr] ventilator, o fan, flashing
— m **reforzado** - [constr] reinforced flashing
— m — **de membrana f** - [constr] reinforced membrane flashing
tapamiento m, con cuidado, o cuidadoso - [constr] careful covering
— m **de reactor m** - reactor covering
tapaporos m - [pint] sealer
— m **imprimador** - [pint] primer,ming sealer
tapar v - . . . .; to close • [metal-prod] to, plug, o pack, o grout
— v **acero m** - [metal-prod] to cap @ steel
— v — m **agitado** - [metal-prod] to cap @ rimming steel
— v — m **efervescente** - [metal-prod] to cap @ rimming steel
— v **boca f** - to cover @ mouth • [metal-prod] to plug @ notch
— v **cañón m** - [metal-prod] to plug @, clay, o mud gun
— v **con arcilla f** - [metal-prod] to clay plug
— v — **cañón m** - [metal-prod] to plug with @ (mud) gun
— v **con cuidado** - to cover carefully
— v — **mano f** - to cover with @ hand • [metal-prod] to plug, manually, o by hand
— v **cuidadosamente** - to cover carefully• [metal-prod] to plug carefully
— v **de nuevo** - to cover again - [metal-prod] to, replug; o plug again
— v — — **boca f** - [metal-prod] to (re)plug @ notch
— v **escorial m** - [metal-prod] to plug @ slag notch
— v **mal** - [metal-prod] to plug poorly
— v —, **escorial m** - [metal-prod] to plug poorly @ slag notch
— v — **piquera f** - [metal-prod] to plug poorly @ iron notch
— v **piquera f** - [metal-prod] to plug @ iron notch
— v **por sí sólo,la** - [metal-prod] to plug itself
— v **reactor m** - to cover @ reactor

tapar v sin salir hierro m - [metal-prod] to plug (@ furnace) without casting
— v — sangrar v - [metal-prod] to plug (@ tap-hole) without casting
— v vasija f - to cover @, container, o can
taparse v - to become clogged
tapete m - . . . . • [domést] matting
tapicería f - . . .; upholstering
tapiz m - [domést] . . .; wall piece
tapizado m - [domést] upholstering
tapizar v - . . . . • to upholster
tapón - . . . . • [metal-prod] (stopper rod) head; stopper; bott • cap • stopple • [tub] plug button • [mec] plug; insert • [petról] packer
— m abierto - [mec] open, plug, o stopper
— m ahuecado - [mec] cored plug
— m — estándar - [mec] standard cored plug
— m — sgún norma f - [mec] standard cored plug
— m ahusado - [mec] tapered plug
— m — de hormigón m - [constr-pil] tapered concrete plug
— m — — m precolado - [constr=pil] tapered precast concrete plug
— m — precolado - [constr-pil] tapered precast plug
— m autocentrado - [mec] self-centered plug
— m autocentrante - [mec] self-centering plug
— m cegador m - obturating plug • [petról] bridge plug
— m — para entubación f - [petról] casing bridge plug
— m cerrado - [mec] closed plug
— m ciego - [mec] blind plug • [petról] bull plug
— m colocado - [mec] installed plug
— m con cabeza f - [mec] head plug
— m — — f ahuecada - [mec] cored head plug
— m — — f cuadrada - [mec] square headed plug
— m — — f — ahuecada - [mec] cored square head(ed) plug
— m — con respiradero m - [mec] square head(ed) vented plug
— m — — f embebida - [mec] socket head plug
— m — — f — estándar - [mec] standard countersunk plug
— m — — f —para tubería f - [tub] standard countersunk pipe plug
— m — — f — estándar - [mec] standard countersunk plug
— m — — f — para tubería f - [tub] standard counersunk pipe plug
— m — — f para tubería f - [tub] countersunk pipe plug
— m — — f hendida - [mec] slotted, o split, head plug
— m — — f hexagonal - [mec] hexagon(al) head plug
— m — — f — para tubería f - [mec] hexagon(al) head pipe plug
— m — — f hueca - [mec] socket head plug
— m — — f ranurada - [mec] slotted head plug
— m — — f sólida - [mec] solid head plug
— m — — f — cuadrada - [mec] solid square head plug
— m corriente - [mec] standard plug
— m — con cabeza f -[mec] standard head plug
— m — — f cuadrada - [mec] square head(ed) standard plug
— m — — f — ahuecada - [mec] cored square head(ed) standard plug
— m — de acero m - [mec] standard steel plug
— m — con cabeza f cuadrada - [mec] square head(ed) standard steel plug
— m — — f — f sólida - [mec] solid head standard steel plug
— m — — — cuadrada - [mec] solid square head(ed) standard steel plug
— m cuadrado - [mec] square plug
— m — estándar - [mec] standard square plug
— m — según norma f - [mec] standard square plug
— m de acero m - [mec] steel plug

tapón m de acero m con cabeza f = [mec] head(ed) steel plug
— f — — m — — f almenada - [mec] slotted head steel plug
— m — — m — — f cuadrada - [mec] square head steel plug
— m — — m — — f embebida - [mec] socket head steel plug
— m — — m — — f encastillada—[mec] slotted head steel plug
— m — — m — — f hueca - [mec] socket head steel plug
— m — — m — — f ranurada - [mec] slotted head steel plug
— m — — m hexagonal - [mec] hexagon(al) steel plug
— m — — m para cuchara f - [metal-prod] steel ladle stopper plug
— m — — m — tubería f - [tub] steel pipe,ping plug
— m — — m ranurado - [mec] slotted steel plug
— m — arcilla f - [metal-prod] clay plug
— m — barra f - [tub] bar plug
— f — f taponadora - [metal-prod] stopper rod head
— m — — f ahuecado - [metal-prod] cored bar plug
— m — — f macizo - [tub] solid bar plug
— m — bronce m - [tub] bronze plug
— m — hielo m - [hidr] ice jam
— m — corcho m - [mec] cork stopper
— m — grafito m - [metal-prod] graphite stopper
— m — hormigón m - [constr-pil] concrete plug
— m — — m con forma f ahusada - [constr-pil] cork shaped, o tapered, concrete plug
— m — — m — f de cono m truncado—[constr-pil] cork shaped concrete plug
— m — — m encajado - [constr=pil] embedded concrete plug
— m — — m — en pilote m - [constr-pil] concrete plug embedded in @ pile
— m — — m precolado - [constr-pil] precast concrete plug
— m — — m pretensado - [constr-pil] prestressed concrete plug
— m — — m — circunferencialmente - [constr-pil] circumferentially prestressed concrete plug
— m — latón m - [tub] brass plug
— m — madera f - [mec] wood(en) plug
— m — (material) plástico m - [mec] plastic plug
— m — varilla f - [metal-prod] rod stopper
— m debajo de caja f - [mec] housing bottom plug
— m embutido - [mec] countersunk plug
— m — de acero m—[mec] countersunk steel plug
— m — estándar - [mec] standard countersunk plug
— m — hexagonal - [mec] hexagonal countersunk plug
— m — para tubería f - [tub] countersunk pipe plug
— m — — — f de acero m - [tub] countersunk steel pipe plug
— m — según norma f - [mec] standard countersunk plug
— m encajado - [constr=pil] embedded plug
— m — en pilote m - [constr-pil] pile embedded plug
— m estándar - [mec] standard plug
— m — con cabeza f ahuecada - [mec] standard cored head plug
— m — — f cuadrada - [mec] standard square head(ed) plug
— m — — f — ahuecada - [mec] standard cored square head plug; cored square head(ed) standard plug
— m — — — f sólida - [mec] solid head standard plug
— n — — f — cuadrada - [mec] solid square head(ed) standard plug
— m — de acero m - [mec] standard steel plug
— m — — m con cabeza f cuadrada - [mec]

square head(ed) standard steel plug

tapón m **estándar de acero** m **con cabeza** f **sólida**
- [mec] solid head(ed) standard steel plug
— m — — m — **cuadrado** - [mec] solid square head(ed) standard steel plug
— m **expansible** - [mec] expansion plug
— m **flotante** - [mec] floating plug
— m **fusible** m - [electr-equip] fuse,sible plug
— m **galvanizado** - [mec] galvanized plug
— m — **con cabeza** f - [mec] head(ed) galvanized plug
— f — — — f **cuadrada** - [mec] square head(ed) galvanized plug
— m — — — f — **ahuecada** - [mec] cored square head(ed) galvanized plug
— m — **corriente** - [mec] standard galvanized plug
— m — **con cabeza** f **cuadrada** - [mec] square head(ed) standard galvanized plug
— m — — — f — **ahuecada** - [mec] cored square head(ed) standard galvanized plug
— m **giratorio** - [válv] swivel plug
— m **Hansen** - [mec] Hansen plug
— m **hembra** - [tub] (female) cap
— m **hexagonal** - [mec] hexagon(al) plug
— m — **de acero** m - [mec] hexagonal steel plug
— m — — m **para tubo,bería** f - [tub] hexagon(al) steel pipe,ping plug
— m **embutido** - [mec] hexagon(al) countersunk plug
— m — — **de acero** m - [mec] hexagon(al) countersunk steel plug
— m — — — m **para tubo,bería** - [mec] hexagon(al) countersunk steel pipe,ping plug
— m — **para tubo,bería** f - [mec] hexagon(al) countersunk pipe,ping plug
— m — **para tubo,bería** f - [mec] hexagon(al) pipe,ping plug
— m — — — f **de acero** m - [mec] hexagon(al) steel pipe plug
— m **horadado** - [tub] drilled plug; insert
— m **de nilón** m - [sold] nylon insert
— m **para guiadera** f - [sold] guide tube insert
— m — — — f **para entrada** f - [sold] incoming guide (tube) insert
— m — — — f — **salida** f - [sold] outgoing guide (insert, o tube)
— m — — **tubo** m **guiador** - [sold] guide,ding tube insert
— m — — — m — **para entrada** f - [sold] incoming guide,ding tube insert
— m — — — m — **salida** f - [sold] outgoing guide,ding tube insert
— m **inferior** - [mec] bottom, o lower, plug
— m **inoxidable** - [tub] stainless plug
— m — **para cabeza** f - [mec] head pipe plug
— m — — — f **hexagonal** - [mec] hexagon(al) head pipe plug
— m — — — f **para tubo,bería** - [mec] stainless hexagon(al) head pipe plug
— m **instalado** - [mec] installed plug
— m **lubricado** - [tub] lubricated plug
— m **macho** m - [tub] (male) plug
— m — **con cabeza** f **embutida** - [tub] countersunk plug
— m — — — f **cuadrada** - [tub] square head(ed) (male) plug
— m — — — f — **ahuecada** - [tub] cored square head(ed) plug
— m — — — f — **sólida** - [tub] solid square head(ed) plug
— m — — — f **hexagonal** - [tub] hexagon(al) head(ed) plug
— m — — — f **hueca** - [tub] countersunk plug
— m — — — f **redonda** - [tub] round headed plug
— m **magnético** - [mec] magnetic plug
— m — **para henchimiento** m - [mec] magnetic fill(ing) plug
— m — **purga** f - [mec] magnetic drain plug
— m **mirilla** - [mec] (peep)sight plug
— m **para aceite** m - [mec] oil plug • cap screw
— m — **agua** m - [mec] water plug
— m — — — m **para fondo** m - [mec] bottom water

plug

tapón m **para alivio** m - [mec] relief plug
— m — — m **para grasa** f - [mec] grease relief plug
— m — **aportación** f - [mec] filler plug
— f — — f **para aceite** m - [mec] oil filler plug
— m — **caja** f - [bombas] housing plug
— m — — f **principal**—[mec] main housing plug
— m — **caño,ñería** - [tub] pipe,ping plug; plug
— m — **cebadura** f - [mec] priming plug
— m — **cementación** f - cementing plug
— m **cierre** m - closing,sure plug
— m — — m **ahusado** - [constr] tapered closure plug
— m — — m — **de hormigón** m - [constr-pil] tapered concrete closure plug
— m — — m **de hormigón** m - [constr-pil] concrete closure plug
— m — **cobertura** f - [tub] cover(ing) plug
— m — **colector** m - [mec] sump plug
— m — — m **para aceite** m - [comb.int] oil pan plug
— m — — m **para aspiración** f - [mec] suction manifold plug
— m — **contacto** m - [electr-oper] contact plug • attachment plug
— m — **cubierta** f - [mec] cover plug
— m — **cuchara** f - [metal-prod] (ladle) stopper
— m — **depósito** m - [comb.int] tank cap
— m — — m **para combustible** m - [comb.int] fuel tank cap
— m — **dilatación** f - [mec] expansion plug
— m — — f **inferior** - [mec] lower, o bottom, expansion plug
— m — — f **superior** - [mec] upper expansion plug
— m — **drenaje** m - [mec] drain plug • [combust.-int] crankcase drain(er)
— m — — m **instalado** - [comb.int] installed drain plug
— m
— m — — m **para bandeja** f - [comb.int] pan drain plug
— m — — — f **para aceite** m - [comb.int] oil pan drain plug
— m — — m **para colector** m **para aceite** m - [comb.int] oil pan drain plug
— m — — m — **respiradero** m - [mec] vent, o breather, drain plug
— m — — m, **quitado, o removido** - [mec] removed drain plug
— m — — m **repuesto** - [mec] replaced drain plug
— m — **sacado** - [mec] removed drain plug
— m — **ensayo(s)** m - [mec] test(ing) plug
— m — **escoria** f - [metal-prod] cinder bott
— m — **escorial** m - [metal-prod] slag, notch, o bott, o stopper
— m — **estrangulador** m - [mec] choke plug
— m — **expansión** f - [mec] expansion plug
— m — — f **en cubierta** f - [mec] cover expansion plug
— f — — f — — f **para distribuidor** m **(para fuerza** f) - [mec] power divider cover expansion plug
— m — — f — — f — — m (— — f) **verificado** - [autom-mec] checked power divider cover expansion plug
— m — — f **inferior** - [mec] lower, o bottom, expansion plug
— m — — f **superior** - [mec] top, o upper, expansion plug
— f — — f **verificado** - [mec] checked expansion plug
— m — **extremo** m - [mec] end plug
— m — — m **de caja** f - [mec] housing end plug
— m — — m — — f **para árbol** m - [mec] shaft housing end plug
— m — **exterior de caja** f **para árbol** m - [mec] shaft housing outer end plug
— m — — m **de colector** m - [mec] sump end plug
— m — **filtro** m - [mec] filter plug

tapón m para filtro, colocado, o instalado -
[mec] installed filter plug
— m — — m para aceite m - [autom-equip] oil
filter plug
— m — — m, quitado, o removido, o sacado -
[mec] removed filter plug
— m — fondo m - [mec] bottom plug
— m — — m de caja f - [mec] housing bottom
plug
— m — — m — — f principal - [mec] main
housing bottom plug
— m — freno m - [mec] brake plug
— m — — m, colocado, o instalado - [mec] installed brake plug
— m — — m, quitado, o removido, o sacado -
[mec] removed brake plug
— m — grasa f - [mec] - grease plug
— m — llenador m - [autom-mec] filler plug
— m — — m magnético - [mec] magnetic filler
— m — muelle m - [mec] spring plug
— m — — m para anillo m - [mec] ring spring
plug
— m — — m — — m portarodillo(s) m - [mec]
roller cage spring plug
— m — muestra f - [mec] sample, o test, plug
— m — múltiple m - [mec] manifold plug
— m — — m para tubería f - [mec] pipe manifold plug
— m — nivel m - [comb.int] level plug
— m — — m para aceite m—[mec] oil level plug
— m — — m — — m, colocado, o instalado -
[mec] installed oil level plug
— m — — m — — m en caja f - [mec] case oil
level plug
— m — — m — — m — — f, colocado, o instalado - [mec] installed case oil level plug
— m — — m — — m en caja f para cadena f -
[mec] chain case oil level plug
— m — — m — — m — — f, colocado,
o instalado • [mec] installed chain case oil
level plug
— m — — m — — m — — f, quitado, o
removido, o sacado - [mec] removed chain case
oil level plug
— m — — m — — m — — f, colocado, o instalado - [mec] installed case oil level plug
— m — — m — — m — — f, quitado, o removido, o sacado - [mec] removed case oil level
plug
— m — — m — — m, colocado, o instalado -
[mec] installed oil level plug
— m — — m — — m, quitado, o removido, o sacado - [mec] removed oil level plug
— m — observación f - [mec] peep sight; sight
plug
— m — obturación f - [constr-pil] closure plug
— m — — f ahusado - [constr-pil] tapered closure plug
— m — — f ahusado de hormigón m - [constr-
-pil] tapered concrete closure plug
— m — — f — — m precolado - [constr-pil]
tapered precast concrete closure plug
— m — — f ahusado precolado - [constr-pil]
tapered precast closure plug
— m — — f de hormigón m - [constr-pil] concrete closure plug
— m — — f — — m precolado - [constr-pil]
precast concrete closure plug
— m — — f precolado - [constr-pil] precast
closure plug
— m — oído m - [segurid] ear plug
— m — orificio m - [mec] port, o opening, plug
— m — — m para abastecimiento n - [mec] filler plug
— m — — m — — m de aceite m - [mec] oil
filler (hole) plug
— m — orificio m - [mec] filler plug
— m — — m para aceite m - [comb.int] (oil)
filler plug
— m — — m para drenar v - [mec] draining plug
— m — — m — — v aceite m - [comb.int] (oil)
drain(ing) plug
— m — — m — lubricación f - [mec] lubrication plug

tapón m para orificio m para lubricación f, colocado, o instalado - [mec] installed lubrication hole plug
— m — — m para nivel m de aceite m - [comb-
-int] oil level plug
— m — petaca f - [metal-prod] cooling plate
plug
— m — piquera f (para escoria f) - [metal-
-prod] slag notch stopper; cinder bott
— m — poste m - post plug
— m — — m magnético - [mec] magnetic post
plug
— m — prueba(s) f - test(ing) plug
— m — purga f - [mec] drain plug
— m — — f abierto - [mec] open drain plug
— m — — f cerrado - [mec] closed drain plug
— m — — f inspeccionado - [comb.int] inspected drain plug
— m — — f limpiado - [mec] cleaned drain
plug
— m — — f limpio m - [mec] clean drain plug
— m — — f para colector m - [mec] manifold
drain plug
— m — — f — — m para aspiración f - [mec]
suction manifold drain plug
— m — — f — — m — — f instalado - [mec]
installed suction manifold drain plug
— m — — f — — m — — f, quitado, o removido, o sacado - [mec] removed suction manifold drain plug
— m — — f — mesa f - [mec] table drain plug
— m — — f — — f rotativa - [petról] rotary
table drain plug
— m — — f para placa f - [metal-prod] plate
drain plug
— m — — f — — f para refrigeración f -
[metal-prod] cooling plate drain plug
— m — — f — respiradero m - [ind] breather,
drain plug, o vent (plug)
— m — — f — purgador m - [ind] drain plug
— m — — m magnético - [mec] magnetic drain
plug
— m — rebalse m - [mec] overflow plug
— m — respiradero m - [mec] vent(ing), o
breather, plug
— m — retención f - [mec] retaining plug
— m — — f para tubería f para ademe m -
[petról] casing bridge plug
— m — sangría f - [ind] bleed(ing) plug
— m — seguridad f - [mec] safety plug
— m — soporte m - [mec] support(ing) plug
— m — — m para grifo m - [mec] cock support-
-(ing) plug
— m — — m — — m para regulación f - [mec]
control(ling) cock support(ing) plug
— m — sostén m - [mec] support(ing) plug
— m — — m para grifo m - [mec] cock support-
-(ing) plug
— m — — m — — f para regulación f - [mec]
control(ling) cock support(ing) plug
— m — tapa f - [válv] cover plug
— m — — f para válvula f - [válv] valve cover plug
— m — — f — — v sellado dentro de cilindro
m - [válv] valve cover plug sealed in @ cylinder
— m — tensión f - [petról] bridge plug
— m — tobera f - [metal-prod] tuyere plug
— m — tubería f - [tub] pipe,ping plug
— m — — f de acero m - [tub] steel pipe,ping
plug
— m — — f embutido - [ind] countersunk pipe
plug
— m — — f hexagonal - [tub] hexagon(al) pipe
plug
— m — — f para cobertura f - [tub] cover(ing)
pipe plug
— m — — f para entubación f - [petról] casing
plug
— m — tubo m - [tub] pipe plug • pipe cap
— m — — m colocado—[tub] installed pipe plug
— m — — m instalado - [mec] installed pipe
plug
— m — — m para colector m - [mec] sump pipe

plug
tapón m para tubo m para cubierta f - [mec]
cover(ing) pipe plug
— m — — m **para descarga** f - [tub] outlet
tube, cap, o plug
— m — — — m **para extremo** m—[mec] end pipe plug
— m — — — — m **para colector** m - [mec]
sump end pipe plug
— m — — m — **salida** f - [tub] outlet, tube, o
pipe, cap, o plug
— m — — m, **quitado, o removido, o quitado** -
[mec] removed pipe plug
— m — **vaciar,ciamiento** m - [mec] drain plug
— m — — **batea** f - [comb.int] pan drain plug
— m — — — v — f **para aceite** m - [comb.int] oil
pan drain plug
— m — — v **colector** m - [comb.int] pan drain
plug
— m — — v **colector** m **para aceite** m - [comb.-
-int] oil pan drain plug
— m — **válvula** f - [mec] valve plug
— m — f **para alivio** m—[mec] relief valve plug
— m — — f — m **para presión** f - [mec]
pressure relief valve plug
— m — — f — m — f **para aceite** m -
[mec] oil pressure relief valve plug
— m — **verificación** f - [mec] check(ing) plug
— m **precolado** - [constr-pil] precast plug
— m **pretensado** - [constr-pil] prestressed plug
— m — **circunferencialmente** - [constr-pil] circumferentially prestressed plug
— m **principal** - [mec] main plug
— m **purgador** m - [mec] drain(ing) plug
— m **quitado** m - [mec] removed plug
— m **ranurado** - [mec] slotted plug
— m **reemplazado** - [mec] replaced plug
— m **removible** - [mec] removable plug
— m **removido** - [mec] removed plug
— m **repuesto** m - [mec] replaced plug
— m **roscado** - [mec] screw plug
— m **roto** - [mec] broken plug
— m **sacado** - [mec] removed plug
— m **según norma** f - [mec] standard plug
— m — — f **con cabeza** f **ahuecada** - [mec] standard cored head plug
— m — — f — f **cuadrada** - [mec] standard
square head plug
— m — — f — — f **ahuecada** f - [mec]
standard cored square head plug
— m **semicónico** - [válv] semi-cone plug
— m **similar** - [mec] similar plug
— m **sólido** - [mec] solid plug
— m — **cuadrado** - [mec] solid square plug
— m **superior** - [mec] superior plug • upper plug
— m y **cadena** f **para radiador** m - [autom-mec]
radiator cap and chain
taponado,da a - plugged • clogged
— a **con salpicadura(s)** f - [sold] spatter
clogged
taponador m - [mec] stopper • [metal-prod] gun
— m **por aire** m **comprimido** - compressed air
driven stopper
— m — — m — **para descarga** f **para escoria** f -
[metal-prod] air operated cinder notch stopper
— m — — m — **operación** f **para escoriero** m
- [metal-prod] air operated cinder notch
stoppoer
— m — — m — — **orificio** m **para escoria** f -
[metal-prod] air operataed slag notch stooper
— m — — m — **piquera** f (**para escoria** f) -
[metal-prod] air operated slag notch stopper
taponador,ra a - stopping; plugging • stopper
taponamiento m - . . . • [mec] plugging; stopping; clogging
— m **de agujero** m - [mec] hole, plugging, o
closing
— m — — m **para colada** f - [metal-prod] tap
hole, plugging, o closing
— m — **conducto** - conduit, stopping, o clogging
— m — **pico** m - [mec] nozzle plugging
— m — — m **rociador** m - [hidr] spray nozzle
plugging

taponamiento m **de piquera** f - [metal-prod] tap
hole, plugging, o closing
— m — **reactor** m - [nucl] reactor, o mixer,
covering
— m — **red** f - [ind] system clogging
— m — — f **para combustible** m - [comb.int]
fuel system clogging
— m — **sistema** m - [ind] system clogging
— m **hermético** - [ind] (tight) sealing
taponar v - . . . • [ind] to plug • to clog
— v **agujero** m - [mec] to plug @ hole • [sold]
to repair @ hole
— v **con arcilla** f - [ind] to plug with clay
— v — **mano** f - [metal-prod] to plug by hand
— v **fuga** f - [metal-prod] to plug @ leak
— v **herméticamente** - to seal (hermetically)
— v **red** f - [ind] to clog @ system
— v — f **para combustible** m - [comb.int] to
clog @ fuel system
— v **sistema** m - [ind] to clog @ system
— v **tobera** f - [metal-prod] to clog @ tuyere
taquete m - [extrusion de electrodos] slug
taqueteadora f - [extrusión f de electrodos]
slug press
taquidactilógrafía f - [com] stenotyping
taquidactilógrafo,fa - [com] stenotypist;
shorthand typist; stenographer
taquimecanógrafo,fa - [com] shorthand typist
taquímetro m - [instrum] . . .; véase también
velocímetro m
tardanza f **creciente** - [labor] increasing late-
ness
tarde f - [chronol] . . .; evening | a - . . .;
tardy
— f **apacible** - [meteorol] calm evening
— adv **en carrera** f - [deport] late in @ race
— f **fresca** - [meteorol] cool evening
tardecita f - [chronol] . . .; evening
tarea f - . . .; assignment; job • duty
— f **aministrativa** - [admin] management, work, o
task, o job
— f **asignada** - [labor] assigned, task, o work,
o job
— f **auxiliar** - [labor] auxiliary task
— f **conductiva** - [admin] leadership, o management, work, o task
— f **conexa** - [labor] related, duty, o task
— f **de índole** f **técnica** - [labor] technical,
task, o work, o job
— f **determinada** - [labor] specific task
— f **directiva** - [admin] management work
— f **ejecutada** - performed, task, o work, o job
— f — **frecuentemente** - frequently performed
task
— f **especializada** - specialized, work, o task
— f **específica** - specific task
— f **fácil** - easy task
— f **liviana** - light, o easy. task
— f **necesaria** - necessary, task, o work, o job
— f **para administración** f - [admin] management
work
— f — **conducción** f [smin] lead(ing) work •
[autom] driving chore
— f — **conservación** f - [ind] maintenance task
— f — — f **acelerada** - faster maintenance work
— f — — f **ejecutada** - [ind] performed maintenance, task, o work
— f — — f — **frecuentemente** - [ind] frequently performed maintenance, work, o task
— f — **constatación** f - [admin] controlling,
task, o work
— f — **diseño** m - [dib] design(ing) task
— f — **fiscalización** f - [admin] controlling
work
— f — **organización** f - [admin] organizing work
— f — **planificación** f - [admin] planning,
work, o job
— f — **proyección** f - design(ing) task
— f — **sección** f - [labor] section, o unit,
task, o work
— f — **verificación** f - [admin] controlling,
work, o task, o job
— f **pendiente** - pending, job, o task

tarea f̲ **programada** - [labor] scheduled, task, o̲
job
— f̲ **sencilla** f̲ - [ind] simple, task, o̲ job
— f̲ **similar** - [ind] similar, task, o̲ job
— f̲ **técnica** - technical, task, o̲ job, o̲ work
tarifa f̲—. . .; rate • fee • schedule • [fisc]
tariff
— f̲ **de avalúo(s)** m̲ - appraisal, rete, o̲ fee
— f̲ — **derecho(s)** m̲ - fee(s) schedule
— f̲ **diaria** - daily rate
— f̲ **establecida** - established rate
— f̲ **fija** - fixed rate
— f̲ — **diaria**—daily fixed, o̲ fixed daily, rate
— f̲ **para avalúo(s)** m̲ - valuation rate
— f̲ — **habitación** f̲—room rate
— f̲ — — f̲ **en hotel** m̲ - hotel room rate
— f̲ — — f̲ **individual** - single room rate
— f̲ — **hotel** m̲ - hotel rate
— f̲ — **pieza** f̲ - room rate
— f̲ — — f̲ **en hotel** m̲ - hotel room rate
— f̲ — — f̲ **individual** - single room rate
— f̲ — **servicio(s)** n̲ - service(s) rate
— f̲ — — m̲ **técnico(s)** - [ind] technical ser-
vice(s) rate
— f̲ **rebajada** - reduced, o̲ lowered, rate
— f̲ **vigente** - [transp] prevailing rate
tarima f̲ - . . .; platform; stand; stage •
[constr] plank flooring • [transp] pallet
— f̲ **alubias** f̲ - [agric] bean platform
— f̲ — **cereal(es)** m̲ - [agric] grain platform
— f̲ **espectador(es)** m̲ - [deport] spectators'
stage; bleachers
— f̲ — **fríjol(es)** m̲ - [agric] bean platform
tarjeta f̲ - . . .; tag
— f̲ **comercial** - [com] business card
— f̲ **con circuito** m̲, **estampado**, o̲ **impreso** -
[electrón] printed circuit card
— f̲ — **corredera** f̲ - [matem] slide card
— f̲ **metálica** - metal tag
— f̲ — **cálculo** m̲ - [matem] slide card
— f̲ — — m̲ **para bolsillo** m̲ - [metam] pocket
slide card
— f̲ — **control** m̲ - [com] control card
— f̲ — **crédito** m̲ - [fin] credit card
— f̲ — **identidad** f̲ - [pol] identity card
— f̲ — **identificación** f̲ - identification, card,
o̲ label, o̲ tag; tag
— f̲ — — f̲ **para bobina** f̲ - [metal-lam] coil
identification tag
— f̲ — **registro** m̲ - record card
— f̲ — **ubicación** f̲ - location, o̲ lift, tag
tarraja f̲ - [mec] **véase terraja** f̲
tarro m̲ - . . .; can; tin
— m̲ **para basura** f̲ - [sanit] trash, can, o̲ tin
tartán m̲ - [textil] . . .; plaid
tarugo m̲ - . . .; peg; pin • knob; plug button
— f̲ **de madera** f̲ - [mec] wood(en), peg, o̲ plug
— f̲ — — f̲ **creosotado** - [f.c.] creosoted tie
plug
— m̲ — — f̲ **para durmiente** m̲ - [f.c.] creosoted, tie, o̲ sleeper, plug
— m̲ — f̲ — — **traviesa(s)** f̲ - [f.c.] creosoted, tie, o̲ sleeper, plug
— m̲ — **plástico** m̲ - [mec] plastic plug
— m̲ **metálico** m̲ - [mec] metal plug
— m̲ **para pared** f̲ - [constr] metal wall plug
— m̲ — **chavetero** m̲ - [mec] keyway plug
— m̲ — — m̲ **para polea** f̲ - [mec] pulley keyway
plug
— m̲ — — m̲ — — f̲ **para impulsión** f̲ - [mec]
drive pulley keyway plug
— m̲ — m̲ — — f̲ — — f̲ **para ventilador** m̲ -
[mec] fan drive pulley keyway plug
— m̲ — **desplazador** m̲ - [mec] shift(er) plug
— m̲ — **durmiente** m̲ - [f.c.] tie plug
— m̲ — **pared** f̲ - [constr] wall plug
— m̲ — — f̲ **metálica** - [constr] metal wall plug
— m̲ — **traviesa** f̲ - [f.c.] tie plug
— m̲, **tapagujero(s)**, o̲ **tapahueco(s)** - [mec] hole
filling plug button
tasa f̲ - . . .; rating • fee; charge • [ind]
output • [fisc] . . .; tax; import • taking •
[fin] receipts

tasa f̲ **aceptable** - acceptable,ted rate
— f̲ — **para infiltración** f̲ - [hidr] accepted
infiltration rate
— f̲ **aceptada** - accepted rate
— f̲ — **en general** - commonly, o̲ generally, accepted rate
— f̲ **alta** - high rate
— f̲ **anual** - annual, o̲ yearly, rate
— f̲ — **para consumo** m̲—annual consumption rate
— f̲ — — **crecimiento** m̲ - annual growth rate
— f̲ — — m̲ **para consumo** m̲ - annual consumption growth rate
— f̲ — **interés** m̲ - [fin] annual, o̲ yearly,
interest rate
— f̲ — **uniforme** - [matem] uniform, annual, o̲
yearly, rate, o̲ index
— f̲ **de abrasión** f̲ - [mec] abrasion rate
— f̲ — **corrosion** f̲ - corrosion rate
— f̲ — — f̲ **y abrasión** f̲ - [ind] corrosion-
-abrasion rate
— f̲ — **crecimiento** m̲ - growth, rate, o̲ index
— f̲ — — m̲ **anual** - annual, o̲ yearly, growth
rate
— f̲ — — m̲ **de consumo** m̲ - consumption growth,
rate, o̲ index
— f̲ — **descarga** f̲ - [ind] discharge,ging rate
— f̲ — **desgaste** m̲ - [mec] wear rate
— f̲ — — m̲ **predicha** - [cuerp.moled] predicted wear rate
— m̲ — **escurrimiento** m̲ - [hidr] runoff rate
— f̲ — **frecuencia** f̲ - [ind] frequency rate
— f̲ — **gasto** m̲ - [hidr] flow rate
— f̲ — **gravedad** f̲ - [segur] severity rate
— f̲ — **impuesto(s)** m̲ - [fisc] tax rate
— f̲ — — m̲ **en exterior** m̲ - [fisc] foreign tax
rate
— f̲ — **incremento** m̲ - increase rate
— f̲ — **interés** m̲ - [fin] tax rate
— f̲ — **pérdida** f̲ - loss rate
— f̲ — — f̲ **de metal** m̲ - [quím] metal loss
rate
— f̲ — — f̲ **de metal** m̲ **observada** - [quím]
observed metal loss rate
— f̲ — — f̲ **observada** - observed loss rate
— f̲ — **precipitación** f̲ - [meteorol] rainfall
rate
— f̲ — **reacción** f̲ - reaction rate
— f̲ — **reducción** f̲ - metal-lam] reduction rate
— f̲ — **salida** f̲ - [mec] discharge rate
— f̲ **económica** - economic rate
— f̲ — **para retorno** m̲ - [econ] economic return
rate
— f̲ **elevada** - high rate
— f̲ **establecida** - established rate
— f̲ — **para impuesto** m̲ - [fisc] statutory, o̲
established, tax rate
— f̲ — — m̲ **nacional** - [fisc] statutory
federal tax rate
— f̲ — — m̲ **sobre, réditos** m̲, o̲ **renta** f̲ -
[fisc] statutory income tax rate
— f̲ **fijada** - [fin] set, o̲ established, rate •
fixed rate
— f̲ **impositiva**—[fisc] tax, assessment, o̲ rate
— f̲ — **máxima** - [fisc] maximum tax rate
— f̲ — **mínima**—[fisc] minimum tax rate
— f̲ **individual** - individual rate
— f̲ **inferior** - lower, o̲ inferior, rate
— f̲ **intercambiaria** - [fin] interchange rate
— f̲ — **en Londres** - [fin] London interchange
rate
— f̲ **interna** - internal rate
— f̲ — **para retorno** m̲ - [fin] internal return
rate
— f̲ — — — m̲ **económico** - [econ] economic return internal rate
— f̲ **legal** - legal rate
— f̲ — **para impuesto** m̲ - [fisc] legal tax rate
— f̲ — — — m̲ **nacional** - [fisc] federal income tax rate
— f̲ — — — m̲ **sobre, rédito(s)** m̲, o̲ **renta**
f̲ - [fisc] statutory federal income tax rate
— f̲ — — — m̲ **sobre, rédito(s)** m̲, o̲ **renta** f̲ -
[fisc] statutory income tax rate

tasa f más alta - higher rate
— f — elevada - higher rate
— f máxima - maximum, o peak, rate
— f — de quemador m para combustión f - [combust] burner maximum firing rate
— f — para combustión f - [combust] maximum firing rate
— f — — escurrimiento m - [hidr] peak runoff rate
— f — — impuesto m - [fisc] maximum tax rate
— m mayor - higher rate • highest rate
— f media - average rate
— f — prevista - average, foreseen, o forecast, rate
— f mínima - minimum rate
— f — de quemador m para combustión f - [combuts] burner minimum firing rate
— f — para combustión f - [combust] minimum firing rate
— f — — impuesto m - [fisc] minimum tax rate
— f nominal - [fin] nominal rate • [electr-prod] rated output
— f observada - observed rate
— f — de pérdida f - observed loss rate
— f — — — f de metal m - [quím] observed metal loss rate
— f para combustión f - [combust] firing rate
— f — infiltración f - [hidr] infiltration rate
— f — legalización f - [fisc] legalization fee
— f — quemador m - [combust] burner rate
— f — — m para combustión f - [combust] burner firing rate
— f — retorno m - [fin] return(s) rate
— f — utilización f - use, o utilization, rate
— f por derecho(s) m - fee(s), o due(s), rate
— f portuario(s) m - [transp] port dues fee
— f — servicio(s) m - service(s) fee
— f — — m aduanero(s) - [fisc] custom(s) service(s) fee
— f — — — m sanitario(s) - [sanit] sewer fee(s)
— f promedio - average rate • average fee
— f — de pérdida f - average loss rate
— f — — — — f de metal m - [quím] average metal loss rate
— f — — — — f — — m observada - [quím] observed average metal loss rate
— f — observada - [quím] observed average rate
— f — — de pérdida f de metal m - [quím] observed average metal loss rate
— f retributiva - [mec] pay, o retribution rate
— f — por servicio(s) f - [fisc] services retribution rate
— f tributaria - [fisc] tax rate
— f uniforme - uniform rate
tasación f - valuation; assessment; appraisal • [fisc] tax(ing); taxation
— f impositiva - [pol] tax valuation
tasado,da a - appraised • [fisc] taxed
taxonómica f - taxonomics
taza f - . . . • [comb.int] cup; well; bowl • [cabl] basket • [petról] basket
— f aceitadora - [mec] oil(ing) cup
— f con rejilla f - [mec] screen(ed) cup
— f de cerámica f - [sold] ceramic cup
— f — material m compuesto - [válv] composition cup
— f guardapolvo(s) - [mec] dust, cap, o cup
— f inferior - [mec] lower, o bottom, cup
— f para base f - base, o seating, cup
— f para carburador m - [int.comb] carburetor, cup, o bowl, o well
— f — casquillo m - [cabl] socket basket
— f — cojinete m - [cojin] bearing cup
— f — eje m - [mec] axle, o shaft, cup
— f — — m para acelerador m - [comb.int] throttle shaft cup
— f — enchufe m - [petról] socket bowl
— f — éter m - [comb.int] ether cup
— f — rueda f - [autom-mec] hubcap
— f — sedimento(s) m - [comb.int] sediment cup
— f — — m en línea f para combustible m -

[comb.int] gas line sediment bowl
taza f para semilla(s) f - [agric] seed bowl
— f — válvula f - [petról] valve pot
— f — f para cambio m rápido - [petról] fast change valve pot
— f semillera f - [agric-equip] seed cup
— f sin sujeción f - [mec] unclamped cup
— f superior - [mec] top, o upper, cup, o bowl
— f y aza f - [comést] cup and handle • [petról] basket and bail
— f — — f en una (sóla) pieza f - [cabl] integral basket and bail
— f — — integrales - [cabl] integral basket and bail
tazón m - . . . [mec] pot bowl • [sanit] bowl; basin
— m de porcelana f - [sanit] china bowl
— m — — f vítrea - [sanit] vitreous china bowl
— m para cámara f - [comb.int] chamber bowl
— m — — f para flotador m - [comb.int] float chamber bowl
— m — combustible m - [comb.int] fuel bowl
— m — cuña(s) f - [petról] slip bowl
— m — — recoger sedimento(s) m - [comb.int] sediment collecting,tion bowl
— m — sedimento(s) m - [comb.int] sediment bowl
— m pequeño - [sanit] small bowl
— m vítreo - [cerám] vitreous bowl
te m - . . . . • [tub] T(ee)
— m corriente - [mec] standard T(ee)
— m embridado - [mec] flanged T(ee)
— m estándar - [mec] standard T(ee)
— m galvanizado - [mec] galvanized T(ee)
— m — corriente - [mec] standard galvanized T(ee)
— m — estándar - [mec] standard galvanized T(ee)
— imperial - [mec] imperial T(ee)
— m para línea f para retorno m - [tub] return line T(ee)
— m para reducción f - [tub] reducing T(ee)
— m — tubería f - [mec] tube T(ee)
— m reductor - [tub] reducing T(ee)
— m Stockham - [tub] Stockham T(ee)
tecla f - . . . . • [electrón] key; keyset
— f deprimida - [electrón] depressed key(set)
teclado m - . . . . • [electrón] keyboard; keyset; keypad
— m con retorno m - [comput] feedback keypad
— m — — m positivamente sensible a tacto m - [comput] positive tactile feedback keypad
— m — — m sensible a tacto m - [comput] tactile feedback keypad
— m configurado - [electrón] configured keypad
— m deprimido - [electrón] depressed keypad
— m — manipulador m - [electrón] keyer board
— m — — — m para tono m - [electrón] tone keyer board
— m sensible a tacto m - [electrón] tactile keypad
técnica f - . . .; practice • [mec] design
— f a tope m - [sold] butt technique
— f — — horizontal - [sold] horizontal butt technique
— f — — vertical - [sold] vertical butt technique
— f administrativa - [admin] management, technique, o skill
— f aportada - supplied, o provided, o contributed, technique
— f apropiada - appropriate, o proper, technique
— f — para establecer amperaje m - [sold] proper current setting technique
— f — para fijar v amperaje m - [sold] proper current setting technique
— f — para línea(s) f para conducción f - [sold] proper pipeline technique
— f — — soldadura f - [sold] proper welding technique

técnica f̲ apropiada para soldadura f̲ para líneas
  para conducción f̲ - [sold] appropriate, o̲
  proper, pipeline (welding) technique
— f̲ — — tubería(s) f̲ para conducción f̲ -
  [sold] proper pipeline tehcnique
— f̲ avanzada - [ind] advanced, o̲ latest, technique
— f̲ — para fabricación f̲ - [ind] advanced, o̲
  latest, manufacturing technique
— f̲ básica - [ind] basic technique
— f̲ — para operación f̲ - [ind] basic operating
  technique
— f̲ — — soldadura f̲ - [sold] basic welding
  technique
— f̲ — — venta(s) f̲ - [com] basic sale(s)
  technique
— f̲ cierta - certain, o̲ sure, t echnique
— f̲ comprobada - proved,ven technique
— f̲ con adición(es) - [ind] addition(s) technique
— f̲ — — f̲ de aceite m̲ - [metal-prod] oil ad-
  dition(s) technique
— f̲ — arrastre m̲ - [sold] drag, o̲ contact,
  technique
— f̲ — — m̲ descendente—[sold] down drag(ging)
  technique
— f̲ — — n̲ ascendente - [sold] up drag(ging)
  technique
— f̲ — — m̲ horizontal - [sold] horizontal
  drag(ging) technique
— f̲ — — m̲ vertical - [sold] vertical dragging
  technique
— f̲ — — m̲ — ascendente - [sold] vertical-up
  drag(ging) technique
— f̲ — — m̲ — descendente - [sold] vertical
  down drag(ging) technique
— f̲ — cordón m̲ recto - [sold] stringer (bead)
  technique
— f̲ — — m̲ — con arco m̲ corto - [sold] short
  arc stringer bead technique
— f̲ — partícula(s) f̲ magnética(s) - [metal-
  -prod] magnetic particle(s) technique
— f̲ — pasada f̲ única - [sold] single pass
  technique
— f̲ contribuida - contributed technique
— f̲ convencional - conventional, o̲ standard,
  technique
— f̲ corriente - current, o̲ standard, technique
— f̲ creada - [ind] created technique
— f̲ de progresión f̲ - [ind] progression technique
— f̲ — — f̲ directa - [sold] straight, o̲
  stringer bead, technique
— f̲ — — f̲ — con movimiento m̲ ligero de tejido m̲ - [sold] slight weave stringer bead
  technique|
— f̲ deficiente - [ind] faulty technique
— f̲ destructiva - destructive technique
— f̲ efectiva - effective technique
— f̲ — para estabilización f̲ - [ind] effective
  stabilization technique
— f̲ — — — f̲ de suelo m̲ - [suelos] effective
  soil stabilization technique
— f̲ empleada - technique used
— f̲ — para excavación f̲ - [miner] digging
  technique
— f̲ escalonada - [sold] stepping technique
— f̲ — rápida - [sold] rapid, o̲ speedy, stepping technique
— f̲ especial - special technique
— f̲ — para protección f̲ - [mec] special
  shielding technique
— f̲ específica - specific technique
— f̲ especificada - [ind] specified technique
— f̲ establecida - established technique
— f̲ extranjera - [ind] foreign technique
— f̲ general - [ind] general technique
— f̲ individual - individual technique
— f̲ — para operario m̲ - [sold] operator('s)
  individual technique
— f̲ — para soldadura f̲ - [sold] individual
  welding technique
— f̲ — — — f̲ para operador m̲ - [sold] opera-

tor('s) individual welding technique
técnica f̲ internacional - [ind] international
  technique
— f̲ manual - [sold] manual, o̲ hand, technique,
  o̲ method
— f̲ más avanzada - [ind] latest technique
— f̲ — reciente - [ind] latest technique
— f̲ minera f̲ - [miner] mining technique
— f̲ — para concentración f̲ - [miner] mining
  concentration technique
— f̲ — — transporte m̲ - [miner] mining transportation technique
— f̲ nacional - [ind] national technique
— f̲ nueva - new technique
— f̲ operativa - [ind] operational technique
— f̲ — para combustión f̲ - [combust] combustion operational technique
— f̲ para amoladura f̲—[ind] grinding technique
— f̲ — — m̲ de rodillo(s) m̲ - [mec] roll grinding technique
— f̲ — antifalseo* m̲ -[Comput] antifalsing
  technique
— f̲ — antorcha f̲ - [sold] torch technique
— f̲ — — f̲ con dos electrodos m̲ (de carbono m̲)
  - [sold] arc torch technique
— f̲ — aplicación f̲ - [ind] application technique
— f̲ — circulación f̲ - [vial] traffic technique
— f̲ — — f̲ vial - [vial] highway traffic
  technique
— f̲ — colada f̲ - [metal-prod] casting technique
— f̲ — — f̲ continua - [metal-prod] continuous
  casting technique
— f̲ — colocación f̲ - [ind] placement,cing
  technique
— f̲ — — f̲ coordinada - [ind] coordinated
  placement,cing technique
— f̲ — — f̲ muy coordinada - [ind] highly coordinated placement,cing technique
— f̲ — conducción f̲ - [autom] driving technique
— f̲ — combustión f̲ - [combust] combustion
  technique
— f̲ — comercialización f̲ - [com] marketing, o̲
  merchandising, technique
— f̲ — — f̲ en salón m̲ para venta(s) f̲ - [com]
  in-store merchandising technique
— f̲ — concentración f̲ - [miner] concentration
  technique
— f̲ — conducción f̲ - [autom] driving technique
— f̲ — — f̲ para rendimiento m̲ alto - [autom]
  performance driving technique
— f̲ — conformación f̲ - [metal-fabr] forming, o̲
  shaping, technique
— f̲ — — f̲ automática - [metal-fabr] automatic
  forming, o̲ shaping, technique
— f̲ — — f̲ en caliente - [metal-fabr] hot,
  forming, o̲ shaping, technique
— f̲ — — f̲ — frío - [metal-trat] cold forming
  technique
— f̲ — — f̲ — frío y en caliente - [metal-
  fabr] hot and cold forming technique
— f̲ — conservación f̲ - [ind] maintenance technique
— f̲ — construcción f̲ - [constr] building
  technique
— f̲ — — f̲ de grúas f̲ - [grúas] crane building, technique, o̲ technology
— f̲ — control* m̲ para calidad f̲ - [ind] quality control technique
— f̲ — cronometraje m̲ - [labor] time study
  technique
— f̲ — chicoteo m̲ - [sold] whipping technique
— f̲ — — m̲ corto - [sold] short whipping
  technique
— f̲ — — m̲ en soldadura(s) f̲ sobrecabeza -
  [sold] overhead whip(ping) technique
— f̲ — — m̲ — — f̲ en ángulo m̲ interior -
  [sold] fillet weld(s) whipping technique;
  whipping technique for fillet weld(ing)
— f̲ — — m̲ rápido - [sold] quick whip(ping)
  technique
— f̲ — desoxidación f̲ - deoxidation practice

técnica f para desulfuración f - [metal-prod] desulfurization technique
— f —— f en cuchara f - [metal-prod] ladle desulfurization technique
— f — distribución f - [ind] distribution technique
— f —— f de combustible(s) m - [combust] fuel(s) distribution technique
— f —— f — servicio(s) m - [ind] utili-ty,ties distribution tehcnique
— f — electrodo m - [sold] electrode technique
— f — m muy sobresaliente - [sold] long stick-out (electrode) technique
— f — encendido m - [sold] starting tcchnique
— f — n con raspado m - [sold] scratch(ing) starting technique
— f — ensayo(s) m - testing technique
— f —— m no destructivo(s) - [ind] non-des-tructive testing technique
— f — estabilización f - stabilization, tech-nique, o skill
— f —— f de suelo m - [suelos] soil stabili-zation, technique, o skill
— f — establecimiento m - setting, technique, o skill
— f —— m de amperaje m - [sold] current set-ting technique
— f — evaluación f - [admin] evaluation tech-nique
— f —— f y revista f de programa - [admin] program evolution and review technique; PERT
— f — excavación f - digging, o excavating,-cavation, technique
— f — exhibición f - [com] exhibiting,tion, o merchandising, technique
— f — fabricación f - [ind] manufacturing, o fabrication, technique
— f — fijación f - [sold] setting technique
— f —— f para amperaje m - [sold] current setting technique
— f — formación f - [ind] forming technique
— f —— f de pella(s) f - [miner] pelletizing technique
— f — granallado m - [mec] shot blasting tech-nique
— f —— m de rodillo(s) m - [mec] roll shot blasting technique
— f — gunitado* m - [metal-prod] guniting technique
— f — hincadura f - [constr-pil] driving tech-nique
— f —— f con núcleo m - [constr-pil] core driving technique
— f — horno m - [metal-prod] furnace, tech-nique, o practice
— f — inspección f—[ind] inspection technique
— f —— f con anilina f - [sold] dye inspec-tion technique
— f — f —— f penetrante - [sold] dye penetrant , o penetrating dye, inspection technique
— f —— f para calidad f - [ind] quality con-trol technique
— f — instalación f - installation, o instal-ling, technique
— f —— f para conducción f - [sold] pipeline technique
— f —— f para montaje m - [ind] assembly line technique
— f — manejo m - handling, o operating, tech-nique
— f —— m de material(es) m - [ind] materials handling technique
— f — manipulación f—[ind] handling technique
— f —— f de material(es) m - [ind] materials handling technique
— f — mantenimiento m - [ind] maintenance technique
— f — material m - [ind] material technique
— f — m revelada - [ind] revealed material technique
— f — mercadeo* m - [com] marketing, o mer-chandising, technique

técnica f para negociación(es) f - [labor] bar-gaining technique
— f — operación f - operating,tion, o opera-tor, technique
— f —— f con electrodo m - [sold] electrode operating technique
— f —— f —— m Jetweld - [sold] Jetweld (electrode) operating technique
— f —— f — Jetweld - [sold] Jetweld opera-ting technique
— f — organización f - [ind] industrial en-gineering · organizing,zation technique
— f — peletización f - [miner] pelletizing technique
— f — penetrante(s) m - [metal] penetrant(s) technique
— f — presentación f - [com] mechandising, o display, technique
— f — predicción f - forecasting technique
— f — prerreducción f - [metal-prod] prere-duction technique
— f — previsión f - forecasting technique
— f — producción f - production tehcnique
— f —— f en general - [ind] general produc-tion technique
— f — protección - protection, o shielding technique
— f — proyección f - design(ing) technique
— f — f revelada - revealed design technique
— f — raspado m - [sold] scratch technique
— f — rectificación f - [comb.int] grinding technique
— f —— f de rodillo(s) m - [metal-lam] roll grinding technique
— f — rectificación f - [miner] recovery, technique, o technology
— f —— f de baritina f - [miner] barytine recovery, technology, o technique
— f — regulación f - [sold] current setting technique
— f —— f para amperaje m - [sold] current, o amperage, setting technique
— f — relleno,namiento - [constr] fill(ing) technique
— f — reparación f - repair(ing) technique
— f — retroceso m - [sold] backstep(ping) technique
— f — revista f - review(ing) technique
— f —— f de evaluación f - [admin] evalua-tion review(ing) technique
— f —— f —— f para programa m - [admin] program evaluation review(ing) technique
— f — seguridad f - [segur] safety technique
— f —— f para distribución f - [ind] dis-tribution safety technique
— f —— f —— f de combustible(s) m - [combust] fuel(s) distribution safety tech-nique
— f —— f —— f — servicio(s) m - [ind] utility,ties distribution safety tehcnique
— f — sinterización f - [miner] sintering technique
— f — soldadura f - [sold] welding, o joining, technique
— f —— f a tope - [sold] butt weld(ing) technique
— f —— f —— horizontal - [sold] horizon-tal butt weld(ing) technique
— f —— f con bronce - [sold] brazing tech-nique
— f —— f — latón m - [sold] brazing tech-nique
— f —— f —— m con antorcha f (con dos e-lectrodos m (de carbono m) - [sold] arc torch brazing technique
— f —— f para aluminio m - [sold] aluminum welding technique
— f —— f — línea(s) f para conducción f - [sold] pipeline (welding) technique
— f —— f — tuberías f para conducción f - [sold] pipeline (welding) technique
— f —— f fuerte - [sold] brazing technique
— f —— f — con antorcha f con dos electro-

dos (de carbono m) - [sold] arc torch brazing technique
técnica f para soldadura f para operario m - [sold] operator's welding technique
— f — tubería(s) f - [sold] pipe weld-ing technique
— f — f en ángulo m (interior) - [sold] fillet weld(ing) technique
— f — f salteada - [sold] skip weld(ing) technique
— f — f sobrecabeza - [sold] overhead weld--(ing) technique
— f — f — horizontal - [sold] horizontal overhead weld(ing) technique
— f — f vertical - [sold] vertical welding technique
— f — f — ascendente - [sold] vertical up (welding) technique
— f — f — — a tope - [sold] vertical up butt weld(ing) technique
— f — f — en ángulo m (interior) - [sold] fillet (weld) vertical up technique
— f — f — — para tubería(s) f - [sold] vertical up pipe (welding) technique
— f — f — descendente - [sold] vertical down weld(ing) technique
— f — f — — para tubería(s) f - [sold] vertical down pipe welding technique
— f — f — para tubería(s) f - [sold] vertical pipe weld(ing) technique
— f — suministro(s) m - [ind] supply(ing) technique
— f — tejido m - [sold] weave,ving technique
— f — — m angosto - [sold] narrow, o small, weave technique
— f — — m completo - [sold] full weave,ving technique
— f — — m — angosto - [sold] narrow full, o full narrow, weave(ing) technique
— f — — m pequeño m - [sold] small weave,ving technique
— f — — m — y rápido - [sold] small quick weave,ving technique
— f — — m rápido m - [sold] quick weave,ving technique
— f — — m triangular - [sold] triangular weave,ving technique
— f — tiempo m - time technique
— f — torneado m - [mec] turning technique
— f — — m de rodillo(s) m - [mec] roll turn-ing technique
— f — trabajo n - [ind] work technique
— f — — m establecida - [ind] established, work, o operating, technique
— f — transporte m - [transport] trans-porting,tation technique
— f — tratamiento m - [ind] treating,tment technique
— f — — m térmico - [metal-trat] heat treat-ing,tment technique
— f — tubería f - [tub] pipe,ping, o tube,bing technique
— f — — f para conducción f - [sold] pipeline technique
— f — vaciado m - [metal-prod] casting tech-nique • [constr] pouring technique
— f — venta(s) f - [com] sale(s) technique
— f particular - particular, o specific, tech-nique
— f perfecta - perfect technique
— f personal - [ind] personal technique
— f pobre - [ind] poor technique
— f por medio de partículas f magnéticas - [metal] magnetic particle(s) technique
— f precisa - precise, o certain, technique
— f preferida - preferred, o favored, technique
— f radiográfica - [electrón] radiographic technique
— f reciente - [ind] recent, o late, technique
— f recta - [sold] straight technique
— f repetida - [ind] repeated technique
— f revelada - [ind] revealed technique
— f según norma f - [ind] standard technique

técnica f según norma f para ensayo(s) m - [ind] standard testing technique
— f siguiente - following technique
— f singular - unique technique
— f — para vaciado m - [constr] unique, pour-ing, o casting, technique
— f sobrecabeza - [sold] overhead technique
— f — horizontal - [sold] horizontal overhead technique
— f térmica - [termol] thermal technique
— f — aplicada - [termol] applied thermal technique
— f — — como mantenimiento m - [termol] ap-plied thermal technique maintenance
— f — — — m preventivo - [termol] ap-plied thermal technique preventive mainte-nance
— f triangular - [sold] triangular technique
— f última - [ind] latest technique
— f ultrasónica - [electrón] ultrasonic,sound technique
— f — aplicada - [electrón] applied ultr-sonic,sound technique
— f — — como mantenimiento m - [electrón] ap-plied ultrasonic,sound maintenance technique
— f — — — m preventivo - [electrón] ultrasonic,sound technique preventive main-tenance
— f uniforme - uniform technique
— f — para fabricación f - [ind] uniform fab-rication technique
— f usual - [ind] usual technique
— f — para soldadura f - [sold] usual, o general, welding technique
— f vertical - vertical technique
— f — ascendente - [sold] vertical-up tech-nique
— m — con tejido m triangular - [sold] tri-angular weave vertical-up technique
— f — descendente - [sold] vertical-down tech-nique
— f vial - [vial] highway technique
técnicas f - ins and outs
— f varias - [ind] various techniques
técnico m - technician • expert; specialist • scientist • [com] product specialist
— m asesor - consulting, o design, specialist
— m calificado - qualified, o trained, techni-cian
— m capacitado - qualified, o trained, techni-cian; skilled technician
— m electrónico - [electrón] electronics tech-nician
— m — calificado - [electrón] qualified, o trained, electronics technician
— m especializado - specialized, o skilled, technician
— m extranjero - [inf] foreign technician
— m en fábrica f - [ind] factory technician
— m especialista - [ind] specialized technician • technical specialist
— m — altamente - [ind] highly specialized technician
— m experimentado - [ind] experienced techni-cian
— m muy especializado - [ind] highly specia-lized technician
— m nacional - national technician
— m para estabilización f - [suelos] stabili-zation, technician, o expert
— m — — f de suelo(s) m - [suelos] soil sta-bilization. expert, o technician
— m — fabricación f - [ind] operating, o fa-bricating,tion, o manufacturing, technician
— m — mantenimiento m - [ind] maintenance technician
— m — operación f - [ind] operating,tion tech-nician
— f — organización f - [ind] organization technician • industrial engineer
— m — prospección f - [miner] prospecting technician
— m — rayos-X - [electrón] X-ray technician

técnico m para seguridad f - [segurid] safety, expert, o technician
— m — — f para automóvil(es) m - [segurid] automotive safety expert
— m — siderurgia f - [metal-prod] steelmaking technician; knowledgeable steelman
— m — sistema(s) m - [ind] systems technician
— m — — m para descascarillado m - [metal-trat] descaling system(s) technician
— m electrónico - [electrón] electronic(s) technician
— m especialista - specialized technician
— n — en sistema(s) m para descascascarillado - [metal-trat] descaling system(s) specialized technician
— m metalúrgico,gista - [metal-prod] metallurgical technician
— m minero - [miner] mining technician
— m montador - [ind] erection technician
— m para montaje m - [ind] erection technician
— m — f para línea f - [ind] line erection technician
— m — — — f de cizallas f - [metal-lam] shear line erection technician
— m — puesta f a punto m - [ind] tune-up technician
— m — — f — — m para línea f - [metal-lam] line tune-up technician
— m — — — — — f de cizalla(s) f - [metal-lam] shear line tune-up technician
— m — — f en marcha f - [ind] start-up technician
— m práctico - [ind] practical technician
— m radiográfico - [electrón] radiographic,phy technicina
— m supervisor - [ind] supervising technician
tecnicolor m - [photogr] . . .; full color
tecnificación* f - [ind] upgrading
tecnificado,da* a - [ind] upgraded
tecnificar* v - [ind] to upgrade
tecnología f actual - [ind] current technology
— f actualizada - [ind] updated, o state-of-@-art, technology
— f adecuada - [ind] adequate technology
— f alta - [ind] high, o advanced, technology
— f aplicada - [ind] applied technology
— f avanzada - [ind] advanced, o high, technology
— f automotriz - [autom] automotive technology
— f automovilística - [autom] automotive technology
— f aplicada - [ind] applied technology
— f avanzada - [ind] advanced technology
— f — para neumático(s) m - [autom-neumát] advanced tire technology
— f — — — m radial(es) - [autom-neumát] advanced radial (tire) technology
— f con temperatura(s) f y presión(es) f alta(s) - [ind] high temperature/high pressure technology
— f contratada - contracted technology
— f definitiva - [ind] final technology
— f — en escala f industrial - [ind] industrial scale final technology
— f desarrollada - [ind] developed technology
— f corriente - [ind] current, o present, technology
— f exigida - [ind] required technology
— f extranjera - [ind] foreign technology
— f generada - [ind] generated technology
— f importada - [ind] imported technology
— f incorporada - [ind] incorporated technology
— f industrial - [ind] industrial technology
— f metalúrgica - [metal] metallurgical technology
— f minera - [miner] mining technology
— f nacional - [ind] domestic, o national, technology
— f nueva - [ind] new technology
— f — aplicada - [ind] applied new technology
— f para compuesto(s) m - [quím] compound(s) technology
— f — hélice(s) - [mec] screw technology

tecnología f para horno m alto - [metal-prod] blast furnace technology
— f — laminación f - [metal-lam] rolling technology
— f — lingote(s) - [metal-prod] ingot technology
— f — — m de acero m - [metal-prod] steel ingot(s) technology
— f — — m — — m efervescente - [metal-prod] rimming steel ingot(s) technology
— f — — m grande(s) m - [metal-prod] large ingot(s) technology
— f — — m de acero m - [metal-prod] large steel ingot technology
— f — — m — — m efervescente - [metal-prod] large rimming steel ingot technology
— f — neumático(s) m - [autom-neumát] tire technology
— f — — m para carrera(s) - [autom-neumát] race,cing tire technology
— f — — m radial(es) - [autom-neumát] radial (tire) technology
— f — proceso m - [ind] process technology
— f — — m para sinterización f - [metal-prod] sintering process technology
— f — proyección f - design(ing) technology
— f procesamiento m - [ind] processing technology
— f — rayo(s) m laser - [electrón] laser technology
— f — recirculación f - [ind] recycling technique • recirculation technology
— f — — f para asfalto m - [constr] asphalt recycling technique
— f — recuperación f - [miner] recovery technology
— f — — f para cobre m - [miner] copper recovery technology
— f — rendimiento m - [ind] performance technology
— f — reprocesamiento m - [constr] recycling technology • reprocessing technology
— f — — m para asfalto m - [constr] asphalt recycling technology
— f — tornillo m - [mec] screw technology
— f pertinente - [ind] related technology
— f producida - [ind] produced technology
— f progresista - [ind] progresive technology • advance(d) technology
— f propia - [ind] own, o private, technology
— f siderúrgica - steel making technology
— f transferida - [legal] tranferred technology
— f vieja - [ind] old technology
— f vinculada con minería f - [miner] mine related technology
tectónica f - tectonics
tectónico,ca - tectonic
techado m - [constr] . . .; roofing
— m en rollo(s) m - [constr] roll roofing
— m exterior - [constr] outside, o exterior, roofing
— m para cabina f - [mec] cab(in) roof
— m — túnel m - [constr] tunnel roof
— m preanunciado - [constr] preannounced roof
— m prearmado - [constr] preassembled roofing • roll roofing
— m preparado - [constr] roll roofing
techado,da a - [constr] roofed
techo m - [constr] roof; ceiling • canopy
— m abovedado - [constr] arched roof
— m — para túnel m - [constr] tunnel arch(ed) roof
— m armado - [constr] built-up roofing - assembled roofing
— m con cuatro aguas f - [constr] hip roof
— m — — vertientes f - [constr] hip roof
— m — dos aguas f - [constr] gable roof
— m cónico - [constr] cone,nical roof
— m de acero m - [constr] steel roof(ing)
— m — bóveda - [constr] arched roof
— m — cabina f - cab roof
— m — cámara f - [constr] chamber roof
— m — — f subterránea f - [ind] manhole roof

techo m flotante - [constr] floating roof
— m impregnado - [constr] (impreganted) roofing
— m — en rollo(s) m - [constr] impregnated roofing roll
— m limpiado - [miner] cleaned, cover, o roof
— m limpio - [miner] clean cover
— m metálico - [constr] (metal) deck
— m — terminado - [constr] finished (metal) deck
— m no conductor - [constr] nonconducting roof
— m para capa f - [miner] seam cover
— m — f limpiado - [miner] cleaned seam cover
— m — f limpio - [miner] clean seam cover
— m — estanque m - [petról] tank, roof, o cover
— m — intercalación f - [miner] intercalation cover
— m — f limpiado - [miner] cleaned intercalation cover
— m — f limpio - [miner] clean intercalation cover
— m — torre f - [constr] tower roof • [petról] derrick roof
— m — f para perforación f - [petról] derrick roof
— m — túnel - [constr] tunnel roof
— m plano - [constr] flat roof
— m — de acero m - [constr] flat steel roof
techumbre m - [constr] . . .; roofing
teflón m - [plást] teflon
teja f colonial - [constr] mission tile
— f — canal - [constr] channel tile
— f — media caña - [constr] channel tile
— f española f - [constr] Mexican tile
— f para cumbre - [constr] crest tile
— f — revestimiento m - [ceram] facing tile
tejadillo m - [constr] protruding roof; eaves
— m en parte f inferior de cuba f—[metal-prod] roof around @ shell over @ cast house
tejado m - [constr] . . .; roofing • tiling
— m con declive m único - [constr] single slope roof
— m — dos aguas - [constr] peaked, o gable, roof
— m — pendiente f única - [constr] single slope roof
— m inclinado - [constr] sloped roof
tejamaní(1) m - [constr] . . .; wood shake
tejerse v - [sold] to weave (itself)
tejido m - . . .; lacing • [constr] screen • [metal-fabr] fabric • [constr] texture
tejer v con arco m—[sold] to weave (with) @ arc
tejido m - . . . • lacing; fabric • texture
— m ascendente - [sold] up weave,ving
— m — triangular - [sold] triangular up weave
— m circular - [sold] circular weave,ving
— m completo - [sold] full weave,ving
— m con unión(es) m tipo doble (re)torcido - [alambre] double wound type fastening netting
— m contra insecto(s) m—[constr] insect screen
— m de alambre m - [alambre] wire net(ting) • wire mesh • [constr] fence; wire fabric • screen cloth
— m — m con unión(es) m tipo doble (re)torcido - [alambre] double wound type fastening netting
— m — m galvanizado - [alambre] galvanized fence,cing
— m — m soldado - [alambre] welded wire, fence,cing, o fabric
— m — arco m - [sold] straight whip(ping)
— m — en caja f - [sold] box weave,ving
— m — vaivén m—[sold] side-to-side weave,ving
— m — zigzagueo—[sold] véase tejido m en caja
— m — simple—[sold] straight weave,ving
— m epitelial - [botan] epithelial tissue
— m — gramíneo m - [botán] gramineous epithelial tissue
— m hacia costado(s) m - [sold] side-to-side weave,ving
— m — lado(s) - [sold] side-to-side weave,ving
— m lateral - [sold] sideways weave,ving

tejido m leve - [sold] slight weave,ving
— m — en cráter m - [sold] slight crater weave,ving
— m mosquitero - [constr] mosquito netting
— m para filtro m - [mec] filter screen
— m recto - [sold] straight weave,ving
— m simple - [sold] simple weave,ving
— f — de vaivén - [sold] simple side-to-side weave,ving
— m triangular - [sold] triangular weave,ving
— m — ascendente - [sold] triangular vertical-up weave,ving
— m vertica; - [sold] vertical weave,ving
— f — ascendente - [sold] vertical-up weave
— f — descendente - [sold] vertical-down weave,ving
— m — triangular - [sold] triangular vertical weave,ving
tejido,da a - weaved • woven
— a en caja - [sold] box, weave, o woven
tejuelo m - [mec] . . . • blank
tela f - . . . • [alambre] wire mesh; screen mat • [papel] web • [autom-neumát] ply,lies
— f a bies - [autom-neumát] bias ply
— f — con perfil m bajo - [autom-neumát] low profile bias ply
— f circunferencial f - [autom-neumát] belt; belting
— f — de cristal m hilado - [autom-neumát] (folded) fiberglass belt
— f — virulana f - [autom-neumát] (folded) fiberglass belt
— f — plegada - [autom-neumát] folded belt
— f — alambre m - [alambre] wire, fence,cing, o cloth; netting
— m — m galvanizada - [alambre] galvanized netting
— f — — hexagonal - [alambre] galvanized hexagon(al) (wire) netting
— f — — m hexagonal - [alambre] hexagon(al) netting
— f de araña f - [entomol] spider web
— f — esmeril m - [mec] emery cloth
— f — fibra f - [textil] fiber cloth
— f — — f de vidrio m - [cerám] fiberglass cloth
— f — nilón m - [autom-neumát] nylon ply
— f — poliestero m - [autom-neumát] polyester ply
— f — rayón m - [autom-neumát] rayon ply
— f — — m para banda f (circunferencial) - [autom-neumát] rayon belt ply
— f — — m — carcasa f - [autom-neumát] carcass rayon ply
— f — — m — — f con dos telas f - [autom-neumát] dual carcass rayon ply
— f — — m para carcasa f doble - [autom-neumát] dual carcass rayon ply
— f debajo de banda f para rodamiento m - [autom-neumát] tread ply,lies
— f diagonal - [autom-neumát] bias ply
— f doble - [autom-neumát] dual, o double, ply
— f — de rayón m - [autom-neumát] dual rayon ply
— f elástica - [autom-neumát] resilient ply
— f — — m para banda f circunferencial - [autom-neumát] resilient rayon belt ply
— f — para banda f circunferencial - [autom-neumát] resilient belt ply
— f — resistente - [autom-neumát] rugged resilient ply
— f — — para banda f circunferencial - [autom-neumát] rugged resilient belt ply
— f en banda f - [autom-neumát] belt ply
— f — — f circunferencial - [autom-neumát] belt ply
— f en pared f lateral - [autom-neumát] sidewall ply,lies
— f exterior - [autom-neumát] cap, o outside, ply
— f — circunferencial - [autom-neumát] circumferential cap ply
— f — — de nilón m - [autom-neumát] circumfe-

ferential nylon cap ply

tela f **exterior de nilón** m - [autom-neumát] nylon cap ply
— f **gruesa** f - [textil] thick, o heavy, cloth • [autom-neumát] blanket
— f — **de amianto** m - asbestos blanket
— f **metálica** - [alambre] wire, screen, o cloth, o mesh • [metal-fabr] hardware cloth
— f **para banda** f - [mec] belt fabric
— f — — f **circunferencial** - [autom-neumát] belt ply
— f — **carcasa** f - [autom-neumát] carcass ply
— f — **correa** f - [mec] belt fabric
— f — **criba** f - [alambre] mesh cloth
— f — **neumático(s)** m - [autom-neumát] tire fabric
— f — **refuerzo** m - reinforcement fabric
— f **radial** - [autom-neumát] radial ply
— f — **para carcasa** f - [autom-neumát] radial carcass ply
— f **recia** f - [autom-neumát] rugged ply
— f — **de rayón** m - [autom-neumát] rugged rayon ply
— f — — — m **para banda** f **circunferencial** - [autom-neumát] rugged rayon belt ply
— f — — f — f — **para rodamiento** m - [autom-neumát] rugged rayon belt ply
— f **resistente** — [autom-neumát] rugged ply
— f — **de rayón** m - [autom-neumát] rugged rayon ply
— f — — — m — — f **circunferencial** - [autom-neumát] rugged rayon belt ply
— f — **elástica** f - [autom-neumát] rugged resilient ply
— f — — **de rayón** m - [autom-neumát] rugged resilient rayon belt ply
— f — — — m — — f **circunferencial** - [autom-neumát] rugged resilient rayon belt ply
— f — **para banda** f **para rodamiento** m - [autom-neumát] rugged belt ply
— f **saturada** - [constr] saturated, cloth, o ply
— f — **con alquitrán** m - [constr] tar saturated, cloth, o ply
— f **superior** - [autom-neumát] overhead ply
— f — **entera** -autom-neumát] full overhead ply
— f — — **de nilón** m - [autom-neumát] full overhad nylon ply
— f — **total** - [autom-neumát] full ply overhead
— f — — **de nilón** m - [autom-neumát] full ply nylon overehad
— f **única** - [autom-neumát] monoply; single ply
telecontrol m - [electrón] . .; véase también **telemando** m
telediscado m - [telecom] direct, o long distance, dialing
teleférico m - [miner] aerial tramway; cable railway
teleférico,ca a - [mech] funicular
telefónico,ca a - [telecom] . . .; telephone
teléfono m **con tono** m **estándar** - [electrón] standard tone telephone
— m **contestado** - [telecom] answered (tele)phone
— m **estándar** - [telecom] standard telephone
— m **sonoro** - [telecom] sound (powered) telephone
telegobernado,da a - [telecom] remote controlled
telegobierno m - [electrón] remote control
— m **optativo** - [electrón] optional remote control
telegrafiar v - [telecom] . . .; to wire
telégrafo m - [telecom] . .: telegraphy
— m **automático** - [telecom] machine telegraphy
— m **magnético** - [telecom] magnetic telegraph(y)
telegrama m - [telecom] . . .; wire
— m **colacionado** - [com] collated telegram
teleimpresión f - [electrón] teleprinting
teleimpreso,sa a - [electrón] teleprinted
teleimpresor m - [electrón] teleprinter
— m **convencional** - [electrón] conventional teleprinter
— m **estándar** - [electrón] standard teleprinter
teleimprimir v - [electrón] to teleprint
telemandado,da a - [electrón] remotely controlled

telemandar v - [mec] to remotely control
telemando m - [electrón] remote control | a - [electrón] remote (controlled)
— m **con transistor(es)** m - [sold] solid state remote controlled
— m **conectado** - [electrón] connected remote control
— m **instalado** - [electrón] installed remote control
— m **optativo** - [ind] optional remote control
— m — **conectado** - [sold] connected optional remote control
— m — **instalado** - [sold] installed optional remote control
— m **para amperaje** m - [sold] remote current control
— m — **contactador** m - [electr-instal] contactor remote, operation, o control
— m — **estabilizador** m - [grúas] (crane) stabilizer remote control
— m — — m **corriente** - [grúas] standard stabilizer remote control
— m — — m **según norma** f - [grúas] standard stabilizer remote control
— m — **interruptor** m - [electr-instal] breaker remote control
— m — — m **automático** - [electr-instal] contactor remote, control, o operation
— m — **polaridad** f - [sold] remote polarity control
— m **transistorizado*** - [sold] solid state remote control
— m — **vehículo** m **(transportador)** - [grúas] carrier remote control
telemensaje m - [telecom] telex (message)
teleproceso* m - [ind] remote process
telerregulación f - [electrón] remote control
telerregulador m - [electrón] remote control • [sold] Lincontrol
— m **con transistor(es)** m - [electrón] solid state remote control
— m — — m **en campo** m - [sold] solid state remote field control
— m **optativo** - [sold] optional remote control
— m — **para amperaje** m - [sold] optional remote current control
— m — **campo** m - [sold] remote field control
— m — **con transistor(es)** m - [sold] solid state remote field control
telescópico,ca a - . . .; telescoping; véase también **retráctil**
televisión f - [electrón] . . .; video
televisor m - [electrón] television (set)
— m **con circuito** m **cerrado** - [electrón] closed circuit television (set)
— f — **pantalla** f **grande** - [electrón] wide screen television
— m **en color(es)** m - [electrón] color television
— m **portátil** - [electrón] portable television
telex m **de intención** f - [legal] intent telex
telón m **de acero** m - iron curtain
— m **de fondo** m - [teatro] backdrop
— m **grande** - [electrón] wide screen
tema m - . . . . • topic • [pint] scheme
— m **aparte** - separate, o another, subject, o discussion
— m **central** - main, o central, subject • highlight
— m **complejo** - complex subject
— m **conjunto** - complete subject
— m **de lógica** f - logic theme
— m **general** - general, theme, o subject
— m **lateral** - side, o lateral, subject, o theme
— m **lógico** - logic(al) theme
— m **nuevo** - new, subject, o theme • [pint] new scheme
— m **para estudio** m - study, subject, o theme
— m — **instrucción** f - instruction, o training, subject
— m **principal** - main theme
— m **prometedor** - promising subject
— m **siguiente** - following subject

tema m sobre seguridad f - [segurid] safety topic
temario m para instrucción f - [pers] training
agenda
— m — plan m - plan agenda
— m — — m para entrenamiento m - [pers] training plan agenda
— m — — m — instrucción f - [pers] training
plan agenda
temario m - . . .; general, theme, o subject
temerario,ria a - . . .; wild
temido,da a - feared
temor m reverencial - reverential fear
— m reverente - awe
témpano m - [hidr] . . .; iceberg
— m de hielo m - [hidr] iceberg
temper m - [metal] véase temple m; templadura f
• revenido m
temperabilidad f - [metal-trat] quenchability
temperación f - [metal-trat] temperability
temperado,da a - tempered; seasoned
temperatura f a colar v - [metal-prod] temperature when pouring
— f — v en lingotera f - [metal-prod] temperature when pouring in(to) @ mold
— f a entrada f - [ind] incoming temperature
— f — — f para aceite m - [ind] oil incoming
temperature
— f — — f — m para trabajo m - [ind] working oil incoming temperature
— f a principio m - [ind] starting temperature
— f — que se emplea - [metal-prod] service temperature
— f — salida f - outgoing temperature
— f — — f para aceite m - [ind] oil outgoing
temperature
— f — — f — — m para trabajo m - [ind]
working oil outgoing temperature
— f admisible - allowable temperature
— f alcanzada - attained, o reached, temperature
— f alta - [ind] high, temperature, o heat
— f — de aceite m hidráulico - [ind] high hydraulic oil temperature
— f — — m para convertidor m para torsión f
- [mec] high torque converter oil temperature;
torque converter oil high temperature
— f — — agua m - high water temperature
— f — — aire m - high air temperature
— f — — precalentamiento m - [sold] high pre-
heat(ing) temperature
— f — — viento m - [metal-prod] high temperature blast
— f — — — m soplado - [metal-prod] blast high
temperature
— f — en cojinete m - [cojin] high bearing
temperature
— f — — laboratorio m - high laboratory temperature
— f — — rodamiento m - [cojin] high bearing
temperature
— f — para descarga f - [ind] high discharge
temperature
— f — — fusión f - [metal-prod] high melting
point
— f ambiente - atmospheric, o outside, o ambient, o environmental, o room, temperature
— f — baja - low, ambient, o room, temperature
— f — común - ordinary, room, o ambient, temperature
— f — . . . grado(s) m - [ind] . . . de-
gree(s) ambient temperature
— f — elevada - high, ambient, o room, temperature
— f — inferior - [meteorol] lower, ambient, o
outside, temperature
— f — media - average ambient temperature
— f — reducida - low(er), ambient, o room, temperature
— f — superior - [meteorol] higher ambient
temperature
— f anual - [meteorol] annual, o yearly. temperature
— f — máxima - [meteorol] maximum annual temperature

temperatura f anual media - [meteorol] average
annual temperature
— f — mínima - [meterol] minimum annual temperature
— f apropiada - appropriate, o proper, o right,
temperature
— f — para laminación f - [metal-lam] proper
rolling temperature
— f atmosférica - atmosphere,ric temperature •
outside temperature
— f aumentada - increased temperature
— f baja - low temperature
— f — de aceite m - [mec] low oil temperature
— f — — aire m - low air temperature
— f — — horno m - [metal-prod] low furnace
temperature
— f — para descarga f - [ind] low discharge
temperature
— f bajada - lowered temperature
— f bajo cero m - [meteorol] temperature below
freezing; freezing, temperature, o weather
— f — presión f - pressured temperature
— f cálida - [meteorol] warm, temperature, o
weather
— f calurosa - [meteorol] hot, weather, o temperature
— f caliente - hot temperature
— f cambiada - changed temperature
— f comprobada - checked, o proven, temperature
— f — de coraza f - [metal-prod] checked shell
temperature
— f — — crisol m - [metal-prod] checked
hearth temperature
— f constante - constant, o sustained, o
continuous, temperature
— f — de cobre m - [electr-cond] continuous
copper temperature
— f continua - continuous temperature
— f correcta - correct, o right, temperature
— f — estimada - estimated correct temperature
— f — para laminación f - [metal-lam] proper
rolling temperature
— f — — precalentamiento m - [sold] correct
preheat(ing) temperature
— f — — — m estimada - [sold] estimated correct preheat(ing) temperature
— f crítica - [metal-prod] critical temperature
— f cualquiera - any temperature
— f de aceite m - oil temperature
— f — — m en entrada f - oil incoming temperature
— f — — m — salida f - oil outgoing temperature
— f — — m comprobada - [comb.int] checked oil
temperature
— f — — m — convertidor m - [mec] converter
checked oil temperature
— f — — m — — m para torsión f - [mec]
torque converter oil temperature
— f — — m en transmisión f - [mec] transmission oil temperature
— f — — m hidráulico - [comb.int] hydraulic
oil temperature
— f — — m en caja f para engranaje(s) m -
[mec] gear, box, o case, oil temperature
— f — — m hidráulico - [mec] hydraulic oil
temperature
— f — — m para convertidor m - [mec] converter oil temperature
— f — — — m para torsión f alta - [mec]
high torque converter oil temperature
— f — — m verificada - [comb.int] checked oil
temperature
— f — ácido m - acid temperature
— f — — m en entrada - inflowing acid temperature
— f — — m — salida f - outflowing acid temperature
— f — agua m - water temperature
— f — — m en motor m - [comb.int] engine
water temperature
— f — — m para enfriamiento m - cooling water
temperature

temperatura f de agua m para refrigeración f -
[metal-prod] cooling water temperature
— f — — m depurada - [sanit] treated water
temperature
— f — — m industrial - [hidr] industrial water temperature
— f — — m potable - [hidr] drinking water
temperature
— f — — m refrigerante - [ind] cooling water
temperature
— f — aire m - air temperature
— f — — m de descarga f - [combust] exhaust
air temperature
— f — — m ambiente - [meteorol] ambient, o
outside, air temperature
— f — — m en entrada f - [ind] inlet air temperature
— f — — m — salida f - [ind] outlet air temperature
— f — — m para refrigeración f - [ind] cooling air temperature
— f — arrabio m - [metal-prod] pig iron temperature
— f — artesa f - [metal-col.cond] tundish
temperature
— f — aspiración f - suction, o intake, temperature
— f — baño m - [ind] bath temperature
— f — — m decapante - [metal-trat] pickling
bath temperature
— f — — m fundido - [metal-prod] molten
bath temperature
— f — — m líquido - [metal-prod] molten bath
temperature
— f — barra f - [metal-lam] bar temperature
— f — bobina f - [metal-lam] coil temperature
— f — boca f - [ind] intake temperature
— f — — f para aspiración f - [ind] air, inlet, o intake, temperature
— f — boquilla f - [metal-prod] nozzle temperature
— f — brea f - [constr] pitch temperature
— f — caja f - [ind] case temperature
— f — — f para cambio(s) m - [mec] gear box
temperature
— f — — f para velocidad(es) f - [mec] gear
box (oil) temperature
— f — calentador m - [ambient] heater temperature
— f — cámara f - [metal-trat] chamber temperature
— f — — f para combustión f - [combust] combustion chamber temperature
— f — — m — canal m — humo(s) m - [combust]
flue temperature
— f — caucho m - [plást] rubber temperature
— f — ceniza f - ash temperature
— f — ciclo m para recocido m - [metal-trat]
annealing cycle temperature
— f — colada f - [metal-prod] casting, o pouring, temperature
— f — — f de lingotera f - [metal-prod] teeming temperature; temperature when pouring into
@ mold
— f — conducto m - [combust] flue temperature
— f — coraza f - [metal-prod] shell temperature
— f — — f comprobada - [metal-prod] checked
shell temperature
— f — crisol m - [metal-prod] hearth temperature
— f — — m comprobada - [metal-prod] checked
hearth temperature
— f — culata f - [comb.int] head temperature
— f — — f de cilindro m - [comb.int] cylinder
head temperature
— f — chimenea f - [combust] stack, o flue,
temperature
— f — destape m - [metal-prod] casting temperature
— f — diseño m - [cald] design temperature
— f — — m de recipiente m - [cald] vessel design temperature

temperatura f de disolución f - solution temperature
— f — electrodo m - [sold] electrode temperature
— f — entrada f - intake, o incoming, temperature
— f — escoria f - [metal-prod] slag temperature
— f — esfuerzo m - [metal] stress(ing) temperature
— f — etapa f - [ind] stage temperature
— f — fase f - [metal-prod] stage, o phase,
temperature
— f — fleje m - [metal-lam] strip temperature
— f — fuel oil m - [combust] fuel oil temperature
— f — gas m - [combut] gas temperature
— f — — m de escape m - [comb.int] exhaust
temperature
— f — — m para reformación f - [metal-prod]
reforming gas temperature
— f — hasta . . . . • temperature up to . . .
— f — fondo m - [ind] bottom temperature
— f — — m de horno m - [metal-prod] furnace
bottom temperature
— f — — m — — m alto - [metal-prod] blast
furnace bottom temperature
— f — gas - [combust] gas temperature
— f — — m horno m - [combust] furnace gas
temperature
— f — — m — — m alto - [combust] blast furnace gas temperature
— f — — m — tragante m - [metal-prod] furnace
throat gas temperature
— f — — m residual - [combust] residual gas
temperature
— f — humo(s) m - [combust] fume(s) temperature
— f — lecho m - [miner] bed temperature
— f — lingote m - [metal-prod] ingot temperature • (ingot) mold temperature
— f — — f para colada f - [metal-prod] mold
pouring temperature
— f — líquido m - liquid temperature
— f — — m refrigerante - [comb.int] coolant
temperature
— f — local m - [ind] room temperature
— f — llama f - [combust] flame temperature
— f — marcha f - [ind] operating temperature
— f — — f para horno m - [metal-prod] furnace
operating temperature
— f — marmita f - [ind] kettle temperature
— f — material m - material temperature
— f — metal m - [ind] metal temperature
— f — — m de base f - [sold] base metal tem=
perature
— f — — m en cojinete m - [mec] bearing metal
temperature
— f — motor m - [comb.int] engine temperature •
[electr-mot] motor temperature
— f — — m aumentada - [comb.int] increased engine temperature • [electr-mot] increased
motor temperature
— f — — m reducida - [comb.int] reduced, o
decreased, engine temperature • [electr-mot]
reduced, o decreased, motor temperature
— f — pared f - wall temperature
— f — — f de cuba f - [metal-prod] bosh wall
temperature
— f — — f — etalaje m - [metal-prod] bosh
wall temperature
— f — piel m - skin temperature
— f — — f de lingote m - [metal-prod] ingot
skin temperature
— f — pieza f - [sold] part temperature
— f — planchón m - [metal-lam] slab temperature
— f — precalentador m - [metal-trat] preheater
temperature
— f — salida f - exit temperature
— f — — f — — m para tocho(s) m - [metal-
lam] blooming mill delivery temperature
— f — — f para gas m - [metal-prod] gas outgoing temperature
— f — — f — — m para reforma(ción) f -
[metal-prod] reforming gas outgoing tempera-

ture
temperatura f **de salida** f **para gas** m **para refor-ma(ción)** f̄ - [metal-prod] reforming gas exhaust temperature
— f —— f **para planchón** m - [metal-lam] slab exit temperature
— f — **sistema** m - system temperature
— f̄ — **tambor** m̄ - [mec] drum temperature
— m̄ — **termopar** m - [instrum] thermocouple temperature
— f — **tubería** f - [tub] pipe temperature
— f̄ — **vidrio** m̄ - [cerám] glass temperature
— f̄ — **unos -20°** ·sub-zero temperature
— f̄ — **vapor** m - [cald] steam temperature
— f̄ —— m **saturado** - [cald] saturated steam temperature
— f —— m **supercalentado** - [cald] superheated steam temperature
— f —— m **vivo**—[cald] live steam temperature
— f̄ — **viento** m—[metal-prod] blast temperature
— f̄ —— m **soplado** - [metal-prod] blown blast temperature
— f **demasiado baja** f - [metal-prod] too low temperature
— f — **elevada** - [metal-prod] too high temperature
— f **deseada** - desired, o sought, temperature
— f̄ — **para operación** f - [combust] desired operating temperature
— f **desequilibrada** - unbalanced temperature
— f̄ **disminuida** - decreased temperature
— f̄ **disponible** - available temperature
— f̄ **elevada** - high, o elevated, temperature
— f̄ — **de aceite** m - [mec] high oil temperature
— f̄ —— m **hidráulico** - [mec] hydraulic oil temperature
— f —— m **para convertidor** m - [mec] converter oil high temperature
— f —— m **para torsión** f - [mec] torque converter oil high temperature
— f — **agua** m - high water temperature
— f̄ — **horno** m - [combust] high furnace, o furnace high, temperature
— f —— m **de fosa** - [metal-lam] high soaking pit, o soaking pit high, temperature
— f — **para servicio** m - high service temperature
— f — **rápidamente** - quickly raised temperature
— f̄ **en aumento** - increasing temperature • [sold] creeping heat
— f — **boca** f **para aspiración** f - intake opening temperature
— f — **bóveda** f - [metal-prod] roof temperature
— f̄ —— f **de horno** m - [metal-prod] furnace roof temperature
— f —— m **fijo** - [metal-prod] non-tilting furnace roof temperature
— f —— f —— m — **Siemens-Martin** - [metal-prod] non-tilting open hearth furnace roof temperature
— f — **canal** m - [hidr] channel temperature • [metal-prod] runner temperature
— f — **chimenea** f - [combust] stack temperature
— f̄ — **devanado** m - [electr-mot] winding temperature
— f —— m **para estator** m - [electr-mot] stator winding temperature
— f —— m — **rotor** m - [electr-mot] rotor winding temperature
— f —— m **sobreelevada** - [electr-mot] over-elevated temperature winding
— f — **exceso** m - temperature, above, o over
— f̄ — **fondo** m - bottom temperature
— f̄ — **fosa** f - [metal-prod] pit temperature
— f̄ —— f **a(l) cargar** v - [metal-prod] pit charging temperature
— f — **grado(s)** m - degree(s) temperature
— f̄ —— m **Celsius** - [termol] Celsius degrees temperature
— f —— m **centígrado(s)** m - [termol] Celsius degrees temperature
— f —— m **Fahrenheit** - [termol] Farenheit - [termol] Fahrenheit degrees temperature

temperatura f **en horno** m - [combust] furnace temperature
— f —— m **alto** - [metal-prod] blast furnace temperature
— f —— m **correcta** - [combust] correct furnace temperature
— f —— m **con fosa** f - [metal-prod] soaking pit temperature
— f —— m —— f **elevada** - [metal-prod] annealing furnace temperature
— f —— m —— f — **rápidamente** - [metal-lam] quickly raised soaking pit temperature
— f —— m **elevada** - [combust] raised pit temperature
— f —— m — **rápidamente** = [metal-lam] quickly raised furnace temperature
— f —— m **para recocido** m - [metal-trat] annealing furnace temperature
— f —— m **reducida** - [metal-prod] lowered furnace temperature
— f — **intemperie** - outside temperature
— f̄ — **laboratorio** m - [ind] laboratory temperature
— f̄ — **parte** f **media de escala** f - [ind] mid-range temperature
— f — **punto** m **de cristalización** f - [quím] crystallization point temperature
— f — **regenerador** m - [combust] regenerator temperature
— f̄ — **sitio** m - in-place temperature
— f̄ — **superficie** f - surface temperature
— f̄ —— f **exterior** f - outside surface temperature
— f —— f — **de cubierta** f - [ind] outside cover surface temperature
— f —— f **interior** - inside surface temperature
— f —— f — **de cubierta** f - inside cover surface temperature
— f — **tope** m - [ind] top temperature
— f̄ — **tragante** - [metal-prod] (furnace) throat temperature
— f —— m **de horno** m - [metal-prod] furnace throat temperature
— f — **transmisión** f - [mec] transmission temperature
— f **entre pasada(s)** f - [sold] interpass temperature
— f —— f **sucesiva(s)** - [sold] interpass temperature
— f **establecida** - [combust] set temperature
— f̄ **estimada** - estimated temperature
— f̄ **exacta** - exact temperature
— f̄ **excesiva** - excess(ive), o high, temperature
— f **exterior** - outside temperature
— f̄ **extraordinaria** - unusual temperature
— f̄ **extrema** - extreme temperature
— f̄ **extremadamente alta** - extremely high temperature
— f̄ — **baja** - extremely low temperature
— f̄ **final** - final temperature • [metal-prod] finishing temperature
— f — **para laminación** f - [metal-lam] final rolling, o rolling end, temperature
— f **final para soplado** m - [metal-prod] blow end temperature
— f **fresca** - [meteorol] cool, weather, o temperature
— f **fría** f - [meteorol] cold, weather, o temperature
— f **gélida** - [meteorol] freezing, o sub-zero, temperature
— f **glacial** - glacial temperature
— f̄ **habitual** - habitual temperature
— f̄ — **de aire** m - [ind] habitual air temperature
— f —— m **en descarga** f - [combust] exhaust air habitual temperature
— f **ideal** - [termol] ideal temperature
— f̄ — **en fosa** f - [metal-prod] ideal pit temperature
— f **igual** - equal temperature

temperatura f igualada - equalized temperature
— f impropia - improper temperature
— f — de ceniza f - [combust| imprper ash temperature
— f — para fusión f - [combust] improper fusion temperature
— f — — — f de ceniza f - [combust] improper ash fusion temperature
— inapropiada - inappropriate temperature
— f — para ceniza f - [combust] inappropriate ash temperature
— f —
sion temperature
— f — — — f de ceniza f - [combus] inappropriate fusion temperature
— f incrementada - [combut] increased, o incremented, temperature
— f inferior - lower, o colder, temperature
— f — a . . .o. - temperature below . . .
— f — — 18° bajo cero n - [meterol] sub-zero temperature
— f inicial - initial, o starting, temperature
— f — de piel m - [metal-prod] initial skin temperature
— f — — — f de lingote m - [metal-lam] initial ingot skin
— f inicial en centro m - [metal-lam] initial center,tral temperature
— f inicial — fosa f [metal-prod] initial pit temperature
— f interior - [ind] inside, o room, temperature
— f irregular - irregular temperature
— f más alta - higher temperature • highest temperature
— f   apacible - [termol] more placid, temperature
— f — baja - lower, o colder, temperature • lowest, o coldest, temperature
— f más fría - colder temperature • coldest temperature
— f máxima - maximum temperature
— f — admisible - maximum, allowable, o permissible, temperature
— f — alcanzada - maximum temperature reached
— f — ideal - [termol] ideal maximum temperature
— f — mantenida - maximum maintained temperature
— f — para agua m - maximum water temperature
— f — — — m depurada - [hidr] treated water maximum temperature
— f — — — m industrial - [hidr] industrial water maximum temperature
— f — — — m para enfriamiento m - [hidr] maximum cooling water temperature
— f — — — m potable - [hidr] drinking water maximum temperature
— f — — aire m - maximum air temperature
— f — — — m comprimido - [ind| maximum compressed air temperature
— f — — cobre m - [mrysl] maximum copper temperature
— f — — llama f - [combust] maximum flame temperature
— f — para metal m - [ind] metal maximum temperature
— f máxima — — m en cojinete m - [mec] bearing metal maximum temperature
— f máxima para operación f - [inc] maximum operating temperature
— f — — — f segura - [ind] maximum safe operating,tion level
— f — precalentador m - [metal-trat] maximum preheater temperature
— f — permisible - maximum allowable temperature
— f — — para llama f - [combust] maximum allowable flame temperature
— f máxima segura - maximum safe temperature
— f mayor - higher, o greater, temperature
— f media - average, o mean, temperature ÷ medium temperature

temperatura f media de agua m - [hidr] water average, o average water, temperature
— f — — — m potable - [hidr] average drinking water, o drinking water average, tempera= ture
— f — — aire m - average air, o air average, temperature
— f — lingote(s) m - [metal-prod] average ingot, o ingot average, temperature
— f — — trabajo m - [ind] average, o normal, operating temperature
— f menor de pieza f entre pasada(s) f - [sold] interpass temperature
— f menor - lower tempoerature • lowest temperature
— f — entre pasada(s) f - [sp;d] interpass temperature
— f mínima - minimum temperature
— f — admisible - minimum allowable temperature
— f — alcanzada - minimum, attained, o reached, temperature
— f — de agua m - minimum water temperature
— f — — — m para enfriamiento m - [ind] cooling water minimum, o cooling water minimum, temperature
— f — mantenica - maintained minimum temperature
— f mínima para agua m depurada - [sanit] treated water minimum temperature
— f — — — m industrial - [hidr] industrial water minimum temperature
— f — — — m potable - [hidr] drinking water minimum, o minimum drinking water, temperature
— f — — aire m - minimum air, o air minimum, temperature
— f — — precalentador m - [metal-trat] minimum preheater.
— f — operación f - minimum operating temperature
— f — — segura - minimum safe temperature
— f — — parâ operación f - minimum safe operating temperatura
— f moderada - [termol] moderate temperature • [meteorol] moderate weather
— f muy alta - very high temperature • too high temperature
— f — baja - [metal-prod] too low temperature; cryogènics
— f nocturna - [meterol] night(time) temperature
— f normal - normal temperature
— f — para operación f - [me] normal operating temperature
— f — trabajo m - [ind] normal operating temperature
— f óptima - optimum temperature
— f para acabado m - [ind] end temperature
— m — para banda f caliente - [metal-lam] hot strip end temperature
— f — — — para desbaste m - [metal-lam] end roughing, o roughing end, mill, temperature
— m — acondicionador m (para aire m) [ambient] air condiioner temperature
— f para afino m - [metal-prod] refining temperature
— f — aspiración f - exhaust(ing), o suction, temperature
— f — bobinado m - [metal-lam] coiling temperature
— f — calebtamiento m - [metal-prod] warming, o heating, temperature
— f — m previo - [metal-sold] preheating, o heat(ing) temperature
— f — carga f - [combust] charge,ging temperature
— f para colada f - [metal-prod] pouring temperature
— f — — f en lingotera f - [metal-prod] ingot mold pouring temperature
— f — combustión f - [combust] combustion temperature

temperatura f para condensación f - [metal-trat] condensing temperature
— f — cristalización f - [quím] crystalizing temperature
— f — descomposición f - [sold] break down temperature
— f — f de recubrimiento m - [sold] covering break down temperature
— f — descarga f - discharging, o unloading, temperature
— f — ebullición f - boiling temperature
— f — forja(dura) f - [metal-prod] forging temperature
— f — f óptima - [metal-prod] optimum forge,ging temperature
— f — f optimizada - [metal-prod] optimized forge,ging temperature
— f — fusión f - [metal-prod] melting point • melting temperature
— f — f para ceniza f - [combust] ash. fusion, o melting, temperature
— f — f de metal - [sold] metal, fusion, o melting, temperature
— f — f — m para aportación f - [sold] filler metal melting temperature
— f — gobierno m - control temperature
— f — laminación f - [metal-lam] rolling temperature
— f — mando m - command, o control, temperature
— f — normalización f - [metal-trat] normalization,lizing temperature
— f — operación f - operating temperature
— f — f para cojinete m = [cojín] bearing operating temperature
— f — f máxima - maximum operating temperature
— f — f máxima f - maximum operating temperature
— f — f mínima - minimum operating temperature
— f — f normal - normal operating temperature
— f — f para horno m - [metal-prod] furnace operating temperature
— f — f segura - [ind] safe opearting temperature
— f — f — máxima - [ind] maximum safe operating temperature
— f — f — mínima - minimum safe operating temperature-
— f — precalentamiento m - [metal-trat] preheat(ing) temperature
— f — f correcta - [sold] correct preheat(ing) temperature
— f — f — m correcta estimada - [sold] estimated correct preheat(ing) temperature
— f — proyección f - [ind] design temperature
— f — recocido m - [metal-trat] annealing temperature
— f — recristalización f - [metal-prod] recrystallization temperature
— f — recubrimiento m - [sold] covering temperature
— f — reducción f - [metal-prod] reduction temperature
— f — refinación f - [metal-prod] refining temperature
— f — reforma(ción) f - [metal-prod] reforming temperature
— f — refrigeración f - [comb.int] coolant temperature
— f — m para motor m - [comb.int] engine coolant temperature
— f — regulación f - control(ling) temperature
— f — solicitación f - [mec] stressing temperature
— f — soplado m - [metal-prod] blast, o blowing, temperature
— f — succión f - suction temperature
— f — suministro m - supply(ing) temperature
— f — temple,pladura f - [sold] drawing temperature
temperatura f para tensión f - stress(ing temperature

rature
temperatura f para terminación f para banda f - [metal-lam] strip, finishing, o end, temperature
— f — f — laminación f - [metal-lam] finishing rolling temperature
— f — f — cinta f - [metal-lam] strip, finishing, o end, temperature
— f — f — chapa f - [metal-lam] strip. finishing, o end, temperature
— f — f para fleje m - [metal-lam] strip, finishing, o end, temperature
— f — trabajo m - [ind] work(ing), o operating, temperature • service temperature
— f — f m admisible - allowable working temperature • allowable working temperature
— f — transformación f - [sold] transformation temperature
— f — vertido m - [electr-instal] pouring temperature
— f patrón m - standard tempoerature
— f peligrosa - dangerous temperature
— f prefijada - preset, o preestablished, temperature
— f perfecta - perfect temperature
— f precisa - precise, o accurate, o exact. temperature
— f — para bobinado m - [metal-lam] precise, o exact, rolling temperature
— f real - true temperature
— f reducida - low(er), o reduced, o decreased, temperature
— f — de aceite m - [mec] low(ered) oil temperature
— f — en horno m - [metal-prod] low furnace temperature
— f regulada - controlled, o set, temperature
— f — para pieza f entre pasada(s) f - [sold] controlled interpass temperature
— f rigurosa - extreme temperature
— f segura - safe temoerature
— f — para operación f - safe operating temperature
— f sensibilizadora - [metal] sensitizing temperatura
— f sobreelevada - overelevated temperature
— f suficiente - sufficient temperature • [sold] sufficient heat
— f superior - higher, o hotter, temperature
— f superior a . . . . - temperature above . . .
— f típica - typical temperature
— f tomada - [instrum] taken temperature
— f tope - maximum temperature
— f — para trabajo m - maximum operating temperature
— f — segura - maximum safe temperature
— f — — para trabajo n - maximum safe operating temperature
— f uniforme - uniuform temperature
— f — de lingote(s) m - [metal-prod] uniform ingot temperature
— f — para bobinado m - [metal-lam] uniform coiling temperature
— f variable - variable, o varying, temperature
terminado,da a - finished • ended; completed
— a en licor m - [metal-trat] liquor finished
— a — planta f - [metal-lam] mill, o plant, finished
— a felizmente - finished, o ended, happily
terminador n para cable - [electr-cond] pothead; end bell
terminador,ra a - finisher
terminadora f - [herram] finisher; finishing machine
— f autopropulsora - [constr] selfpropelled (asphalt) finisher
— f — para asfalto m - [constr] selfpropelled asphalt finisher
— f para safalto m - [constr] asphalt finisher
— f — m autropropulsada - [constr] self-propelled asphalt finisher
— f para pavimento(s) m asfáltico(s) - [constr] asphalt finisher

terminadora f vibratoria - [herram] vibrator
 finisher
terminal m - . . . . • [electr-cond] (battery)
 connector; lug | a - . . .; final
— m aéreo - [electr-instal] aerial terminal
 (rod)
— m afianzado - [electr-instal] clamped terminal
— m aislado - [electr-instal] insulated terminal
— m — para carga f - [electr-instal] insulated
 load terminal
— m con aro m - [electr-instal] ring terminal
— m — costo m reducido - [electr cond] inexpensive, o low cost, terminal, o lug
— m — tornillo(s) m - [electr-instal]
 screw(ed) terminal
— m de latón m - [electr-instal] brass terminal
— m eléctrico - [electr-instal] electrical terminal
— m en armario m - [instrum] cabinet terminal
— m — interruptor m - [electr-instal] switch
 lug
— m — — para línea f - [electr-instal] line
 switch lug
— m — losa f - [electr-instal] apron terminal
— m — — f trasera - [electr-instal] rear
 apron terminal
— m — panel m - [electr-instal] panel lug
— m — — m posterior - [electr-instal] rear
 panel ground(ing) lug
— m — — m para conexión f con tierra f -
 [electr-instal] rear panel ground(ing) lug
— m entorchado - [electr-instal] (wire) wrapped
 terminal
— m hundido - [electr-instal] recessed terminal
— m identificado - [electr-i stal] identified
 terminal
— m individual - [electr-instal] individual
 terminal
— m instalado - [electr-instal] installed terminal • attached terminal
— m interactivo - [electrón] interactive terminal
— m laminado - [electr-instal] laminated terminal
— m laminar - [electr-instal] laminar,nated
 terminal
— m marítimo - [transp] maritime terminal
— m montado - [electr-instal] mounted, o attached, terminal
— m neutral - [electr-instal] neutral terminal
— m para acumulador m - [comb.int] battery connector
— n — cable m - [electr-instal] cable lug
— m — — m a electrodo m - [sold] electrode
 cable lug
— m — — m a tierra f - [sold] ground cable
 lug
— m — — m eléctrico - [electr-cond] electric
 cable lug
— m — — m para soldadura f - [sold] welding
 cable lug
— m — — m — — f para conexión f rápida -
 [sold] quick connecting,tion welding cable
 lug
— m — carga f - [transp] load(ing) terminal
— m — compresión f - [electr-instal] pressure
 stud
— m — conectador m - [electr-instal] connector
 terminal
— m — conexión f - [electr-instal] tterminal
 stud
— m — — f con tierra f - [electr-instal]
 ground(ing) lug
— m — — f — — f en panel m posterior -
 [electr-instal] rear panel ground(ing) lug
— m — — f rápida - [electr-instal] quick connecting, lug, o terminal
— m — conmutador m - [electr-equip] switch
 terminal
— m — corriente f para salida f - [electr-
 -instal] output terminal

terminal m para elemento m para telemando m -
 [sold] control end terminal
— m — — m portátil para telemando m - [sold]
 control pod terminal
— f — entrada f - [electr-instal] intake terminal
— m — exportación f - [transp] export(ation)
 terminal
— m — — f para mineral m - [miner] ore ex-
 port(ation) terminal
— m — motor m - [electr-mot] motor terminal
— m — pasajero(s) m - [transp] passenger terminal
— m — salida f - [electrón] output lug
— m — — f hundido - [electr-equip] recessed
 output. terminal, o lug
— m — — para . . . tono(s) m - [electrón]
 . . . tone output lug
— soldado m - [electr-instal] soldered, terminal, o lug
— m unificado - [transp] union terminal
terminar v - . . .; to finish; to end (up); to
 complete; to draw to @ close
— v artículo m - [ind] to finish @ article
— v asiento m - [mec] to finish @ seat
— v bien(es) m - to, finish, o end, @ goods
— v cambio m - [mwx] ro, xomplwrw, o end, @,
 repolacement, o change
— v carga f - [ind] top, end, o finish, @
 charage,ging, o load(ing)
— v carrera f - [deport] to come away from @
 race
— v con licor m - [metal-trat] to liquor-
 -finish
— v contrato m - to complete @ contract
— v de soldar v - [Sold] to complete a weld
— v ejercicio m - [fin] to end @ (fiscal) year
— v empleo m - [labor] to terminate
— v — m de energía f auxiliar - [sold] to stop
 @ auxiliary power use
— v en dos lugares m primeros - [deport] to finish one-two
— v — — puestos m primero(s) - [deport] to
 finish one-two
— v — lugar primero - to finish first
— v — posición(es) f primera(s) - [deport]
 to finish in @ money
— v equipo m - [ind] to finish @ esquipment
— v felizmente - to end happily
— v junta f - [sold] to finish @ joint
— v línea f - [comput] to, finish, o terminate,
 @ line
— v material m - to finish @ material
— v mediante esmerilado m - [Sold] to finish by
 grinding
— v montaje m - [ind] to complete @ erection
— v muro m - [constr] to finish @ wall
— v parada f - [ind] to end @ shutdown
— v pared f - [Constr] to finish @ wall
— v placé - [deport] to finish in @ money
— v segundo - [deport] to finish second
— v soldadura f - [sold] to, finish, o stop, @
 weld(ing)
— v trabajo m - to complete @, work, o project,
 o job
tempestad f con recurrencia f probable cada . .
 . . años m - [mryrotol] . . .-year storm
— f probable cada . . . años m - [meteorol]
 . . .-year storm
templabilidad* f - [metal-trat] hardenability
— f alta - [metal-trat] high hardenability
— f baja - [metal-trat] low hardenability
templable* a - [metal-trat] hardenable
templado m - [metal-trat] tempering; (air) hardening • quenching; véase también temple m;
 templadura f
templado,da a - [sold] hardened • [metal-trat]
 tempered → [geogr] temperate
— a con llama(s) f - [metal-trat] flame hardened
— a con revenido m subsiguiente - [metal-trat]
 hardened with subsequent drawing
— a con temperatura f ambiente -[metal-trat]

room temperature tempering
templado,da a en aceite m - [metal-trat] oil, tempering, o quenching
— **a — aire** m - [metal-trat] air hardened
— **a en caliente** - [metal-trat] hot quenched
— **a totalmente** - [metal-trat] full tempered
— **a — y bonderizado,da** - [metal-trat] full tempered bonderized
— **a — — galvanizado,da** - [metal-trat] full tempered galvanized
templadura f - . . . • [metal-sold] drawing
— **f de acero** m - [metal-prod] steel tempering
— **f — m efervescente** - [metal-trat] rimming steel tempering
— **f — barreno,na** - [petról] bit hardening
— **f — bobina** f - [metal-trat] coil tempering
— **f — — f de acero** m - [metal-trat] steel coil tempering
— **f — — f — n efervescente** m - [metal-trat] rimming steel coil tempering
— **f — — f laminada** - [metal-trat] rolled steel tempering
— **f — — f — en caliente** - [metal-trat] hot rolled coil tempering
— **f — hojalata** f - [metal-lam] tin strip tempering
— **f — trépano** m - [petról] bit tempering
— **f directa** - [metal-trat] direct quenching
— **f en aceite** m - [metal-trat] oil, tempering, o hardening
— **f en agua** m - [metal-trat] water hardening
— **f húmeda** - [metal-trat] wet tempering
— **f — en temples** m - [metal-trat] aqua tempering
— **f mecánica** - véase **temple** m **mecánico**
templar v - [metal-trat] . . .; to draw • to (air) harden
— **v acero** m - [metal-trat] to, temper, o harden, @ steel
— **v con llama** f - [metal-trat] to flame harden
— **v material** - [metal-trat] to harden @ material
— **v por enfriamiento** m - [metal-trat] to harden by quenching
— **v rueda** f - [mec]ruedas] to, temper, o harden, @ wheel
temple n - ⌊metal-prod] temper(ing); hardening; quench hardening • [metal-trat] skin pass
— **m aumentado** - [metal-trat] increased tempering
— **m blando** - [metal-trat] soft temper(ing)
— **f con llama(s)** f - [metal-trat] flame hardening
— **m con revenido** m **posterior** - [metal-trat] hardening with subsequent drawing
— **m — — m subsiguiente** - [metal-trat] quenching with subsequent hardening
— **m de acero** m - ⌊metal-trat] sttel hardening
— **m — — m con llama(s)** f - [metal-trat] steel flame hardening
— **m — barreno,na** - [petról] bit hardening
— **m — bobina(s)** f - [metal-trat] steel hardening; coil tempering
— **m — chapa(s)** f - [metal-trat] sheet tempering
— **m — hojalata** f - [metal-trat] tin plate tempering
— **m — lámina(s)** f - [metal-trat] sheet tempering
— **n — material(es)** m [metal-trat] material(s) hardening
— **m — rodillo(s)** m - [metal-trat] roll(s) temper(ing)
— **m — rueda(s)** f - [mec-ruedas] wheel, temper(ing), o hardening
— **m — superficie** f - [mec-ruedas] surface hardening
— **m — — f para rodadura** f - [mec-ruedas] tread surface hardeninf,dness
— **m — trépano(s)** m - [petról] bit hardening
— **m doble** - [metal-trat] double temper(ing)
— **m duro** - [metal-trat] hard temper(ing)
— **m — pleno** = [metal-trat] full hard temper(ing)
— **m en aire** m - [metal-trat] air hardening

temple m **en caliente** - [metal-trat] hot, tempering, o quenching
— **m especial** - [metal-trat] special tempering
— **m incrementado** - [metal-trat] increased hardness
— **m mecánico** - [metal-trat] mechanical tempering
— **m por inducción** f - [metal-trat] induction, tempering, o hardening
— **m profundo** - [metal-trat] deep hardening
— **m rápido** - [metal-trat] quick, o quench, hardening
— **m secundario** - [metal-trat] secondary hardening
— **m simple** - [metal-trat] single tempering
— **m suave** - [metal-trat] soft temper(ing)
— **m y revenido** m - [metal-trat] quenching and drawing
— **m superficial** - [metal-trat] casehardening
templillo m - [metal-prod] (slag notch) intermediate (tuyere) cooler
— **m para escorial** m - [metal-prod] slag notch; monkey
— **m — válvula** f **para viento** m **caliente** - [metal-prod] hot blast valve seat ring
temporada f - season - adv - (de) seasonal; part time
— **f actual** - present, o current, season
— **f de lluvia(s)** f - [meteorol] rainy season
— **f lluviosa** - [meteorol] rainy season
— **f para carrera(s)** - [deport] racing season
— **f — — f sobre carretera** f - [deport] road racing season
— **f — turismo** m - [deport] tourist season
— **f pasada** - [cronol] past season
— **f presente** - present, o current, o this, season
— **f Pro Rally** - [deport] Pro Rally season
— **f próxima** - [cronol] next, o coming, season
— **f siguiente** - [cronol] next season
— **f turística** - tourist season
— **f última** - last season
temporal m - . . . | a - part time
— **a total** - [segur] temporary total
temporario,ria a - . . .; provisional; interim
temporización f - . . .; idling
temporizado,da a - idled
temporizar v - . . .; to idle
temprano a en carrera f - [deport] early on (in a race)
— **a en día** m - early in @ day
tenacidad f - . . . • [metal] tensile strength
— **f a entalle** m - [metal] notch toughness
— **f a rotura** f - fracture toughness
— **f contra fractura(s)** f - [metal] fracture toughness
— **f contra rotura** f - [metal] fracture toughness
— **f excepcional** - exceptional toughness
— **f — contra fractura(s)** f - [metal] exceptional fracture toughness
— **f — — rotura** f - [metal] exceptional fracture toughness
— **f máxima** - [mec] maximum toughness
tenaz a - tenacious; tough • dogged • hard
tenaza(s) f - [herram] . . .; tong(s); pincers
— **f abierta(s)** - [mec] open(ed) tong(s)
— **f cerrada(s)** - [mec] closed tong(s)
— **f desconectadora(s)** - [petról] breakout tongs
— **f expansible(s)** - [comb.int] expander tool
— **f para accesorio** m - [herram] fitting tongs
— **f — cadena** f - [herram] boll weevil; tongs
— **f — calentador** m - [herram] heat tongs
— **f — contrafuerza** f - [petról] backup tongs
— **f — desconexión,nectar** - [petról] breakout tongs
— **f — desenroscado,dura** - [petról] breakout tongs
— **f — desenroscamiento,cadura** - [petról] breakout tongs
— **f — deslingoteado*** m - [metal-prod] tripping tongs

tenaza(s) f para desmoldeo m - [metal-prod]
tripping tongs ingot stripper
— f — forja f - [metal-prod] gad tongs
— f — enroscadura f - [petról] make-up tongs
— f — enroscar - [petról] back up tongs
— f — tubería f - [mec] tubing tongs
— f — — f para bombeo m - [petról] pipe fitting, o tubing, tongs
— f — f — — entubación f - [tub] pipe,
fitting, o casing, tongs
— f — tubo(s) m - [herram] pipe tongs
— f — — m y accesorio(s) m - [herram] pipe
and fitting tongs
tendencia f a - prone(ness) to
— f a abrir(se) - spreading tendency
— f — agrietamiento m - cracking tendency
— f — agrietar(se) v - [metal] cracking tendency
— f — arrugamiento m - wrinkling tendency
— f — combustión f - combustion tendency
— f — — f espontánea f - [combuts] pyrophoric,
o spontaneous combustion, tendency
— f — correr(se) v - sliding, o slipping, tendency
— f — chocar electrodo m contra trabajo m -
[sold] stubbing tendency
— f — distorsión f - [sold] distortion,ting
tendency
— f — pegar(se) v - [sold] sticking (tendency)
— f — picaduras f superficiales - [sold] fur-
face hole tendency
— f — porosidad f - [sold] porosity tendency
— f — — f en superficie f - [sold] surface,
holes, o porosity, tendency
— f — variar - (sold] creeping tendency
— f actual - current, tendency, o trend
— f ascendente - upward(s) tendency, o trend
— f cambiante - change,ging, tendency, o trend
— f de costos,tas - [ind] cost(s) trend
— f — productividad f - [ind] productivity.
tendency, o trend
— f — — f internacional - [ind] international
productivity. trend, o tendency
— f — relación f - relationship, tendency, o
trend
— f descendente - downward(s), tendency, o trend
— f económica - [com] economic, tendency, o
trend
— f futura - future, tendency, o trend
— f mayor - greater, tendency, o trend
— f — a distorsión f - [sold] greater distortion, tendency, o trend
— f pirofórica - pyrophoric, o spontaneous combustion, tendency, o trend
— f prevista - foreseen, o projected. tendency,
o trend
— f — para productividad f - [ind] projected,
o foreseen, productivity, tendency, o trend
— f — — — f internacional - [ind] projected,
o foreseen, international productivy, trend, o
tendency
— f reducida - reduced, tendency, o trend
— f variante - changing, tendency, o trend
tender v - . . . .; to trend; to tend • [constr] to
lay - [electr-instal] to route
— v a calentar(se) - [electr-oper] to run hot
— v — correr(se) v - to, tend, o trend, to,
run, o slip, o slide
— v — — delante - [sold] to tend to run ahead
— v — corrobrar - to tend to corroborate
— v — evitar - to discourage
— v — picar v - [sold] to tend to pit
— v — — v superficie f - [sold] to tend to
pit @ surface
— v a perforar v - [sold] to tend to burn thru
— v a producir v - to tend to produce
— v — variar v - [sold] to creep
—, v alinear y, conectar, o empalmar v - [mec]
to lay, line and join
— v tubería f - [tub] to lay @ pipe
ténder m - [f.c.] tender; coal car
tendido m - tending • [constr] laying; wiring
— m de cable m - [cabl] cable laying

tendido,da a - tended • horizontal • [constr]
laid
—, alineación f y empalme - [tub] lay, line
and join • aligning and joining
—, alineado,da y empalmado,da - [tub] laid,
lined and joined
— m con ojal(es) m - dead end suspension
— m de conductor(es) m - [electr-instal] wiring
— m — tubería f - [tub] pipe laying
— m — — f para conducción f - [tub] pipeline
laying
tendiente a - tending; with @ tendency to
— a hacia objetivo m - goal seeking
tendencialmente* adv - tendentially*
tenedor m de acción(es) f - [legal] shareholder
— m — libro(s) m - [contab] bookkeeper
— m para cinta f - [metal-prod] chain fork (pig
casting machine)
— m — — f para máquina f para colar v -
[metal-prod] pig casting machine strand holder
— m — máquina f para colar v - [metal-prod]
pig casting machine strand holder
teneduría f de libros m - [contab] bookkeeping
tenencia f - [fin] holding
tener v - . . . . • to be equipped with • to assume • [legal] to entitle
— v a cargo m - [labor] to manage • to be in
charge of
— v a mano f - to, keep, o have, on hand
— v acción(es) f - [fin] to hold, stock, o
shares
— v actuación f destacada - to figure heavily
— v almacenado,da - to have, stored, o in stock
— v ángulo m - to have @ angle; to be angled
— v . . . año(s) m - to be . . . years old
— v apariencia f buena - to look good
— v — f impresionante - to look terrific
— v aplicación f - to, apply, o be applicable
— v aspecto m aplanado - [cadenas] to lie flat
— v billete m - [fin] to have @ bill • [transp]
to, hold, o have, @ ticket
— v capacidad f - to be able to hold • to be
capable
— v — f nominal - to be rated
— v — para - to (be able) to handle
— v compromiso m - to be committed; to be tied
up
— v conocimiento(s) m - to be knowledgeable
— v contacto m con - to be in contact with; to
bear
— v — — m con colector m - [electr-prod] to
bear on @ commutator
— v cortocircuito m - to be shorted
— v cuidado m - to be careful; to take care
— v — m especial - to take particular care
— v disponible - to, have, o keep, available, o
on hand
— v efecto - to, affect, o have effect • to
inure
— v efecto m adverso - to have @ adverse effect
— v en cartera f - [fin] to hold;to have in hand
— v — cuenta f - to take into account; to bear
in mind; to consider; to figure in; to make
allowance
— v — — f efecto m - to, consider, o affect,
o influence
— v — — f influencia f - to consider @ influence
— v — seguridad f - with safety in mind
— v — existencia f - [com] to (have in) stock
— v — mano f - to have in hand
— v — mano f derecha - to hold in @ right hand
— v — f izquierda - to hold in @ left hand
— v — mente f - to have in mind
— v enganchado picaporte m - [constr] latched
— v escoria f - [metal-prod] to, have, o accu-
mulate, @ slag
— v existencia(s) f - to have in stock
— v éxito m - to be successful
— v experiencia - to have experience
— v factor m para utilización f - [sold] to,
rate, o be rated

tener v **falla(s)**f - to be at fault
— v **hierro** m - [metal-prod] to be iron plugged
— v — m **busa** f - [metal-prod] iron plugged blowpipe
— v **lugar** - to, happen, o occur • to hold • to take
— v **medida(s)** f (necesarias) - to size
— v **montado,da** - to hold
— v **movimiento** m - [mec] to be powered
— v — m **independiente** - [mec] to move under it's own power
— v **mucha tracción** f - [autom-neumát] to have much pull
— v **ocasión** f - to, have, o get, @ chance
— v **pasada tuerca** f - [mec] overripe nut
— v **personal** - [pers] to be staffed
— v — m **escaso** - [labor] to be short handed
— v **por** - (to be) deemed
— v **por cierto,ta** - to expect
— v — **costumbre** f - to make (it) @ practice
— v — **fin** - to intend
— v **presente** - to, keep in mind, o remember
— v **producto** m - to have @ product
— v **profundidad** f **suficiente** - to be deep enough
— v· — f — **de carga** f **fría** - [metal-prod] to have sufficient depth of chilled charge
— v **que** - to have to
— v — **reparar** v - to have to repair
— v — **tapar** v - [metal-prod] to have, o to be forced to, plug
— v **que ver** - to involve
— v **rendimiento** m **mayor** - to outperform
— v **resistencia** f - to be strong
— v — f **mayor a tensión** f **de tubería** f - [tub] to "beat @ pipe"
— v — f **suficiente** - to be strong enough
— v **responsabilidad** f - be responsible; to bear, o to have, @ responsibility
— v **rotación** f - [mec] to have rotation
— v — f **direccional** - [mec] to have directional rotation
— v **salinidad** f - to have sodic chloride
— v — f **en soplante** m - [metal-prod] sodic chloride in @ turbo-blower condensing pipe
— v **seguro(s)** m - [seguros] to carry insurance
— v **sensación** f - to feel
— v — f **de normal(es)** - to feel normal
— v **suerte** f - to be lucky • to fare well
— v **tendencia** f - to have @ tendency
— v **tensión** f - to have tension • to be stressed
— v — f **apropiada** - [mec] to have @ proper tension
— v — f — **un resorte** m - [mec] to have @ proper spring tension
— v **todo menos que** - to have everything but
— v **tolerancia** f - (to be) subject to @ tolerance
— v **trabajando** - [mec] to have in operation
— v **una sóla fase** f - [electr-equip] (to be) single phased
— v **valor** m - to be of value • to rate
— v — m **nominal** - to be rated
— v **viscosidad** f **baja** - to have low viscosity • [sold] to wet
tenería f - [ind] véase **curtiembre** f
tenido,da a **disponible** - had available; kept on hand
— a **a mano** - kept on hand
— a **en cuenta** f - taken into account; figured in; considered
teniendo adv - having
— adv **en cuenta** - figuring in
teniente n - [mil] lieutenant
— m **coronel de estado** m **mayor** - [milit] staff lieutenant colonel
— m **navío** m - [milit] lieutenant commander
tenis m - [deport] tennis
— m **medioeval** - [deport] medieval, tennis, o court
tenor m - . . . . • [legal] content
— m **alto** - [quím] high (content)

tenor m **alto de azufre** m - [metal] high sulfur content
— m — **carbono** m - [metal] high carbon content
— m — **volátil(es)** m - [metal] high volatility; high volatile(s) content
— m **bajo** - [metal] low (content)
— m — **de aleación** f - [metal] low alloyed
— m — **carbono** m - [metal] low carbon content)
— m — **hidrógeno** m - [,etal] low hydrogen (content)
— m **de aluminio** m - [metal] aluminum content
— m — m **antes de corrosión** f - [metal] aluminum content before corrosion
— m — m **después de corrosión** f - [metal] aluminum content after corrosion
— m — **hierro** m - [miner] iron content
— m — m **en agua** m - [hidr] water iron content
— m — m — m **depurada** - [hidr] treated water iron conent
— m — m — m **industrial** - [hidr] industrial water iron content
— m — m — m **potable** - [hidr] drinking water iron content
— m — m **máximo** - [metal] maximum iron content
— n — m **en agua** m **industrial** - [hidr] industrial water maximum iron content
— m — m **potable** - [hidr] drinking water maximum iron content
— m — m **medio** - [metal] average iron content
— m — m **en agua** m - [hidr] water average iron content
— m — m — m **industrial** - [hidr] industrial water average iron content
— m — m — m **potable** - [hidr] drinking water average iron content
— m — m **mínimo** - [metal] minimum iron content
— m — m **en agua** m - [miner] water minimum iron content
— m — m — m **industrial** - [hidr] industrial water minimum iron content
— m — m — m **potable** - [hidr] drinking water minimum iron content
— m **óxido** m - [metal] oxide content
— m — m **de hierro** m - [metal] iron oxide content
— m **oxígeno** m - [quím] oxygen content
— m — m **en baño** m - [metal-prod] bath oxygen content
— m **después de corrosión** f - [metal] content after corrosion
— m **mediano** - [quím] medium content
— m — **de carbono** m - [quím] medium carbon (content)
— m **medio** - [quím] medium content
— m — **de hierro** m - [quím] average iron content
— m — — m **en agua** m - [hidr] average water iron content
— m — — m **depurada** - [hidr] average treated water iron content
— m **muy alto** - [metal] very, o extra, high, content
— m — **de carbono** m - [quím] very, o extra, high carbon content
— m — **bajo** - [quím] extra low (content)
— m — **de carbono** m - [sold] extra low carbon
— m — — **silicio** m [metal] extra low silicon (content)
tenorio m - flirtl ladies' man
tensado,da a - [mec] tensioned; stressed
tensar v - . . . .; to tension; to stress
— v **cadena** f - [mec] to, tension, o tighten, o stretch, @ chain
— v — f **transportadora** - [mec] to, tension, o adjust, o tighten, @ conveyor chain
tensil a - [mec] véase a **tensión** f

tensiómetro m - [instrum] tensiometer
tensión f - . . . • [metal] stress • [electr] véase **voltaje** m • [metal] stress • [mec] take up • [papel] M.D.; machine tension resistance
— f **admisible** - [suelos] allowable stress • design stress
— f — **prevista** - limiting design stress
— f — **unitaria** - design unit stress
— f — — **prevista** - limiting design unit stress
— f **aflojada** - [mec] released, tension, o stress
— f **alta** - high tension • [metal] high stress • [electr] véase **voltaje** m **alto**
— f **apropiada** - appropriate, o proper, o correct, stress
— f — **para cadena** f - [mec] proper chain, tension, o stress
— f — — **correa** f - [mec] proper belt tension
— f — — **muelle** m - [mec] proper spring tension
— f — — **resorte** m - [mec] proper spring tension
— f **asignada** - [electr] véase **voltaje** m **asignado**;rated voltage
— f **aumentada** - [electr] véase **voltaje** m **aumentado**;increased voltage
— f **baja** - [mec] low tension • [electr] véase **voltaje** m **bajo**;low voltage • under voltage
— f — **en línea** f - [electr-distrib] low line volage; low volts
— f **bloqueadora** - blocking tension
— f **comprobada** - checked, o proven, tension
— f **con resorte** m - [mec] spring loaded
— f **conservada** - [mec] retained tension
— f **constante** - [electr] constant potential
— f **continua** - [mec] continuous, o steady, tension
— f **correcta** - [mec] correct, o proper, tension
— f **de banda** f - [metal-lam] strip tension
— f — **cable** m - [cabl] cable tension
— f — — m **de acero** m - [cabl] steel cable tension
— f — — m — — m **a rotura** f - [cabl] steel rope breaking strength
— f — **cadena** f - [cadenas] chain tension
— f — — f **afectada** - [mec] affected chain tension
— f — — f **ajustada** - [mec] adjusted chain tension
— f — — f **comprobada** - [cadenas] checked chain tension
— f — — f **verificada** - [mec] checked chain tension
— f — **correa** f - [mec] belt tension
— f — **carril** m - [constr-equip] rail tension
— f — **colada** f - [metal-prod] casting strain
— f — **contracción** f - [metal] contraction stress
— f — **correa** f - [mec] belt tension
— f — — f **para compresor** m - [ind] compressor belt tension
— f — **corte** m - [mec] shear(ing) stress
— f — **deformación** f - [mec] flow stress
— f — **enfriamiento** m - [metal] cooling stress
— m — **enrollado,llamiento** m - [metal-lam] coil(ing) stress
— f — **entrada** f - [mec] entry tension • [electr] voltage input
— f — **expansión** f - [metal] expansion stress
— f — **fluencia** f - [metal] yield stress
— f — **forja(dura)** f - [metal] forging strain
— f — **freno** m - [mec] brake,king tension
— f — — m **para eje** m - [mec] spindle brake tension
— f — **grilla** f - [electr] grid voltage
— f — **impulso** n **pleno** - [electr] full wave impulse
— f — **muelle** m - [mec] spring tension
— f — **rejilla** f - [electr] grid voltage
— f — **resorte** f - [mec] spring tension
— f — — m **soltada** - [mec] released spring tension

tensión f **de rotura** f - [mec] tension, o tensile, o (ultimate) breaking, o rupture, stress
— f — — f **para cable** m - [cabl] rope breaking strength
— f — — f — — m **de acero** m - [cabl] steel rope breaking strength
— f — — f — — m — **alambre** m **de acero** m - [cabl] steel rope breaking strength
— f — **ruptura** f - [metal] ultimate stress
— f — **salida** f - [metal-lam] delivery tension
— f — **salto** m - [electr] flashover voltage
— f — **suelo** m - [suelos] soil strain
— f — **tierra** f - [electr] ground tension
— f — **utilización** f - [electr] utilization voltage
— f **debida a presión** f - [mec] pressure stress
— f — — — f **de suelo** m - [suelos] soil pressure stress
— f **disruptiva** - [electr] flashover voltage
— f **disruptora** - [electr] flashover voltage
— f — **en seco** - [electr] dry flashover voltage
— f **distribuida** - [mec] distributed stress
— f **dorsal** - back tension
— f **en cara** f **de sello** m - [mec] seal face tension
— f — **límite** m **de fluencia** f - [metal] (normal) yield strength
— f — **línea** f - [electr-distrib] line, volts, o voltage
— f — **superficie** f - surface tension
— f **entre caja(s)** f - [metal-lam] tension between stands
— f — **fases** f - [electr] interpass voltage
— f **excesiva** - excessive tension • [constr] overstress • [mec] overtightening
— f — **para correa** f - [mec] belt overtightening
— f **generada** - generated tension
— f **interior** - [mec] internal stress
— f — **alta** - [metal] high internal stress
— f **interna** - [metal] (internal) stress
— f — **baja** - [mec] low internal stress
— f — **mayor** - [metal] higher internal stress
— f — **mayor** - [metal] greater, o increased, internal stress
— f — **menor** - [metal] reduced internal stress
— f **máxima** - maximum, tension, o stress
— f — **en límite** m **de fluencia** f - [metal] maximum yield strength
— f — **para rotura** f - [metal] (ultimate) (tensile breaking) strength
— f — **para tendido** m - [electr-instal] maximum pulling stress
— f — **ejercida sobre conductor** m - [electr-instal] maximum conductor tension
— f **mayor** - [mec] more tension
— f **mecánica** - [mec] mechanical tension
— f — **de banda** f - [metal-lam] strip mechanical tension
— f **mediana** - [mec] medium tension • [electr] medium voltage
— f **menor** - less tension • reduced stress
— f **mínima** - [constr] minimum stress
— f — **en límite** m **de fluencia** f - [metal] minimum yield strength
— f **nominal** - [electr] nominal, o rated, voltage; voltage rating
— f — **en límite** m **de fluencia** f - [metal] normal yield strength
— f **nula** - [electr] no voltage
— f **para servicio** m - [electr] service voltage
— f — — m **entre conductor(es)** m - [electr] between conductor(s) service voltage
— f — **tendido** m - [electr-instal] tensile, o pulling, stress
— f — **trabajo** m - [suelos] work(ing) stress • [electr] service, o working, voltage, o tension
— f **plena** - [electr] full voltage
— f **por adherencia** f - [hormig] bond stress
— f **por unidad** f **de superficie** f **de sección** f **transversal** - [cabl] unit stress

tensión f predeterminada - predetermined tension
— f previa - pre-tension
— f prevista - expected stress • limiting stress
— f primaria - [electr] primary voltage
— f — asignada—[electr] rated primary voltage
— f — nominal - [electr] rated primary voltage
— f reducida - [electr] reduced voltage
— f repentina - sudden tension • [electr-oper] sudden voltage
— f residual - residual, tension, o stress
— f — aliviada - [sold] relieved residual stress
— f — en superficie f - [metal] residual surface tension
— f residual superficial - [metal] residual surface tension
— f retenida—[mec] retained, tension, o stress
— f secundaria f - [electr] secondary voltage
— f — asignada - [electr] rated secondary voltage
— f — nominal - [electr] rated secondary voltage
— f soltada - [mec] released tension
— f súbita - sudden tension • [electr] sudden voltage
— f superficial - [mec] surface tension
— f — bloqueada - [quím] blocked surface tension
— f — bloqueadora - [mec] blocking surface tension
— f — — de burbuja(s) f - [explos] bubble blocking surface tension
— f — — — — f de aire m - [explos] air bubble blocking surface tension
— f trasera - [mec] back tension
— f unitaria - unit, tension, o stress
— f — admisible - limiting unit stress
— f — prevista - design unit stress
— f verificada - [mec] checked, o proven, tension, o stress
— f vertical - [mec] vertical, stress, o strain
— f — de suelo m - [suelos] vertical soil, o soil vertical, stress, o strain
tensionado,da* - véase solicitado,da
tensor m - [mec] . . .; draw bar; tie rod; bracket; brace; guy; tension, member, o device • tightener • [metal-trat-galvanización] guard
— m con rosca f - [mec] turnbuckle
— m — tornillo m - turnbuckle (rod)
— m hidráulico - [mec] hydraulic tension device
— m para carril m - [mec] rail hydraulic tension device (oruga)
— f para carril m - [mec] rail hydraulic tension device (oruga)
— m para banda f - [mec] belt tightener
— m — cable m - [cabl] wire rope turnbuckle
— m — cadena f - [mec] chain tightener
— m — codo m - [metal-prod] (tuyere) stock tension member
— m — — m para tobera f - [metal-prod] tuyere stock tension member
— m — correa f - [mec] belt tightener • belt idler
— m — ensamblar v - [tub] section fitter
— m — — v sección(es) f - [tub] sections pulling together draw bar
— m — entrada f - [mec] entry tension device
— m — placa f - [mec] plate holding device
— m — — f para asiento m para busa f—[metal-prod] blowpipe seat plate holding device
— m — portavientos m - [metal-prod] bustle pipe tension member
— m — poste n - [constr] post brace
— m — — m — — m para maniobra(s) f - [petról] wheel post brace
— m — — m — — m — tubería f - [petról| wheel post brace
— m — — m — torno m (para maniobra(s) f) - [petról] wheel post brace
— m — — m — malacate m para tubería f - [petról] wheel post brace

tensor m para soporte m para tambor m para herramientas f—[petról] bull wheel post brace
— m — — m — — m — maniobra(s) - [petról] calf wheel post brace
— m — — m para maniobra(s) f - [petról] wheel post brace
— m — — m — — m — tubería f - [petról] wheel post brace
— m — tobera f - [metal-prod] tuyere bracket • tuyere bridle
tentado,da a - tried • probed
tentar v - . . .; to test • to try @ hand
— v fortuna f - [juego] to test @ luck
— v suerte f - to test @ luck
tentativa f - . . .; tentative; try; attempt
tentativamente adv - tentatively
tentativo,va a - . . . . • in principle
tenue a - tenuous; thin; fine
teodolito m - [instrum] . . .; transit
teorema m de Bernoulli - [hidr] Bernouilli('s) theorem
teoría f - . . . . • speculation
— f de carga(s) f cerrada(s) - [constr] external load theory
— f — — f sobre conducto m cerrado - [constr] closed conduit external load theory
— f — circuito m - [electrón] circuit theory
— f — contorno m - [autom-neumát] contour theory
— f — — m neutral - [autom-neumát] neutral contour theory
— f — Coulomb - [suelos] Coulomb theory
— f — elasticidad f - elastic,ity theory
— f — estabilidad f - stability theory
— f — — f elástica - elastic stability theory
— f — — f — de Timoshenko - Timoshenko's elastic stability theory
— f — Rankine - [suelos] Rankine theory
— f — Rowe - [constr-pil] Rowe theory
— f hidrostática - hydrostatic theory
— f nueva - new theory
— f para cola(s) f - queuing theory
— f — juegos m - game, theory, o method • Montecarlo theory
— f plástica - plastic theory
— f vieja - old theory
tercelete m - [arquit] tierceron
tercero m - third • other(s) • [ind] third helper • [legal] third oarty
— m para convertidor(es) m - [metal-prod] converter third helper
— m — horno m - [metal-prod] (furnace) third helper
— m — — m basculante - [metal-prod] tilting furnace third helper
— m — — m con fosa f - [metal-lam] soaking pit third helper
— m — — m para mezcla(dura) f - [metal-prod] mixing furnace third helper
— m — — m fijo - [combust] stationary furnace third helper
— m — — m mezclador - [metal-prod] mixing furnace third helper
— m — máquina f - [ind] machine third helper
tercio m - [ind] third shift
terco,ca a - . . . .; willful
térmica f - [electr-prod] steam power plant
— f central - [electr-prod] power house
térmicamente adv - thermally
térmico,ca a - . . . . • heat(ing)
terminación f - . . . .; finish(ing); drawing to @ close • [constr] topping off
— f adicional - additional finish(ing)
— f — para extremo m - [constr] additional end finish(ing)
— f áspera - rough finish(ing)
— f atractiva - [tub] attractive (end) finish
— f autocoloreada - [metal-trat] self colored finish
— f autocoloreante* - [metal-trat] self colored,loring finish
— f brillante - bright, o glossy, finish(ing)
— f — de superficie f - glossy surface finish

terminación f brunida - [metal-prod] burnished,
o buff, finish
— f comercial - [metal-lam] commercial finish .
merchant finish
— f — brillante - [metal-prod] bright commercial finish
— f — semibrillante - [metal-prod] semiglossy
commercial, o commercial semiglossy, finish
— f con esmalte m - [pint] enamel(ed) finish
— f — m de vinilo m - [pint] vinyl enamel
finish
— f — m vinílico - [pint] vinyl enamel finish
— f — nitrógeno m - [metal-trat] nitrogen fin-
ish(ing)
— f — pulverización f - [pint] spray fin-
ish(ing)
— f — superficie f pulida - [metal-lam] polish surface finish
— f de artículo m - [ind] article finish(ing)
— f — asiento m - [mec] seat finish(ing)
— f — aspecto m bueno - [constr] neat end fin-
ish(ing)
— f — banda f - [metal-lam] strip finish(ing)
— f — — f gruesa - [metal-lam] thick strip
finish(ing)
— f — barra f - [metal-lam] bar finish(ing)
— f — bien(es) m - [ind] good(s) finish(ing)
— f — bobina(s) f - [metal-trat] coil fin-
ish(ing)
— f — — f laminada(s) - [metal-trat] rolled
coil finish(ing)
— f — — f — en caliente - [metal-trat] hot
rolled coil finish(ing)
— f — camisa f - [mec] liner finish(ing)
— f — — f para árbol m - [mec] shaft liner
finish(ing)
— f — carga f - [ind] charge, end(ing), o
finish(ing), o completion • [transp] load-
-(ing), end(ing), o finish(ing), o completion
— f — cartón m - [papel] cardboard finishing
— f — ciclo m - cycle completion
— f — — m para operación f - [ind] operating
cycle completion
— f — cinta f - [metal-lam] strip finish(ing)
— f — — f gruesa - [metal-lam] thick strip
finish(ing)
— f — construcción f - [constr] construction,
o building, completion
— f — contrato m - [legal] contract completing,tion • agreement termination
— f — convenio m - [legal] agreement termination
— f — chapa f - [metal-lam] strip finish(ing)
— f — — f en frío - [metal-lam] cold strip
finish(ing)
— f — — f gruesa - [metal-lam] thick strip
finish(ing)
— f — — f en caliente - [metal-lam] hot,
sheet, o strip, finish(ing)
— f — deshornamiento m - [coque] push(ing)
completion
— f — empleo m - employment termination
— f — — m de energía f auxiliar - [sold]
auxiliary power use ending
— f — equipo m - [ind] equipment finish(ing)
— f — extremo(s) m - [constr] end finish(ing)
— f — — m de alcantarilla f - [hidr] culvert
end treatment
— f — fleje m - [metal-lam] strip finish(ing)
— f — — m en caliente - [metal-lam] hot
strip finish(ing)
— f — — m grueso - [metal-lam] thick strip
finish(ing)
— f — fundición f - [metal-prod] heat completing,tion
— f — hojalata f - [metal-trat] tin (strip)
finish(ing)
— f — huelga f - [labor] strike termination
— f — — f de estibador(es) m - [labor] long-
shoremen's strip finish(ing)
— f — junta f - [mec] joint end(ing)
— f — — f calafateada - [constr] caulked

joint finish(ing)
terminación f de laminación f - [metal-lam]
mill finish[ing[
— f — — f en frío - [metal-lam] cold rolling
finish(ing)
— f — línea f - [comput] line termination
— f — material m - material finish(ing)
— f — mecanización f - [mec] machining finishing
— f — montaje m - [ind] erection completing,tion
— f — muro m - [constr] wall finishing
— f — obra f - [constr] project completion;
job end(ing)
— f — operación(es) f - operation(s) completing,tion
— f — — f restante(s) - [ind] remaining, o
other, operation(s), completing,tion
— f — papel m - [papel] paper finishing •
role ending
— f — parada f - [ind] shutdown end(ing)
— f — pared f - [constr] wall finishing
— f — parte f - part completing,tion
— f — — f de trabajo m - work, o project,
part completing,tion
— f — partida f - [ind] run end
— f — perfil(es) m - [metal-lam] structural
finishing
— f — — m estructural(es) - [metal-lam]
structural(s) finishing
— f — plancha(s) f - [metal-lam] plate(s)
finishing
— f — riel(es) m - [metal-lam] rail finishing
— f — rotación f - [grúas] rotation completing,tion
— f — soldadura f - [sold] weld finishing
— f — superficie f - surface finish(ing)
— f — — f brillante - [metal-lam] glossy
surface finish(ing)
— f — — f comercial - [metal-lam] commercial,
o merchant, surface finish(ing)
— f — — f — semibrillante - [metal-lam]
commercial, o merchant, semiglossy surface
finish(ing)
— f — tanda f - [ind] run, end, o finish(ing)
— f — tarea f - job end
— f — temple m - [metal-lam] temper finishing
— f — — m blando - [metal-trat] soft temper
finish(ing)
— f — — m suave - [metal-trat] soft temper
finish(ing)
— f — trabajo m - work finish(ing)
— f — vertedero m - [constr] spillway (end)
finish(ing)
— f eficiente - [tub] efficient end finish(ing)
— f en caliente - [metal-lam] hot finish(ing)
— f — — de banda f - [metal-lam] hot strip
finish(ing)
— f — — — cinta f - [metal-lam] hot strip
finish(ing)
— f — — — fleje m - [metal-lam] hot strip
finish(ing)
— f — frío - [metal-lam] cold finish(ing)
— f — planta f - [metal-lam] mill finish(ing)
— f enchapada - [mec] plated finish(ing)
— f especial - [ind] special finish(ing)
— f — para extremo m - [constr] special end
finish(ing)
— f estándar - [ind] standard finish(ing)
— f exterior - outside, o exterior, finishing
— f feliz - happy end(ing)
— f final - final, end(ing), o finish(ing)
— f galvanizada - [metal-trat] galvanized
finish
— f — en caliente - [metal-trat] hot galvanized finish(ing)
— f granallada - [metal-trat] peened, o Pangborn, finish
— f — en caliente - [metal-trat] hot peened
finish(ing)
— f laqueada - [pint] lacquer finish
— f leve - slight finish(ing)

terminación f lisa - [mec] smooth finish(ing)
— f manual - [ind] hand finish(ing)
— f — posterior a embutición f - [metal-fabr] after-drawing hand finishing
— f — — estampado m - [metal-fabr] after-drawing hand finishing
— f mate - [metal-lam] matte finish(ing)
— f mejor - better finish(ing)
— f mínima - minimum finish(ing)
— f nominal - nominal finish(ing)
— f Pangborn - [metal-lam] Pangborn finish(ing)
— f para alcantarilla f - [tub] culvert finish
— f — cloaca f - [tub] sewer (end) finish
— f — descarga f - [tub] outfall end finish
— f — — f para cloaca f - [tub] sewer outfall end finish(ing)
— f — extremo m - [constr] end treatment
— f — — m para entrada f - [constr] inlet end treatment
— f — — m — salida f - [constr] outlet end treatment
— f por abrasión f - [mec] abrasion finish(ing)
— f posterior a embutición f - [metal-fabr] after-drawing finish(ing)
— f — estampado m - [metal-fabr] after-drawing finishing
— f pulida f - [metal-lam] polished finish(ing)
— f reciente - recent completion
— f rugosa - [metal-lam] textured, o Pangborn, finish • rough finish(ing)
— f — áspera - [metal-lam] rough textured finish
— f semimate - [metal-lam] semimatte finish
— f silenciosa - [comput] quiet termination
— f sin ruido(s) m - [comput] quiet, o noiseless, termination, o ending
— f variada - varied finish(ing)
terminado* m - véase terminación f; acabado m • finish(ing)
terminado,da a - finished; completed; ended • drawn to @ close • after
— a en frío - [metal-lam] cold finished
terminal m para elemento m (para telemando m) - [sold] (control) pod terminal
— m — — m portátil (para telemando m) - [sold] control pod terminal
— m — entrada f - intake terminal
— m — exportación f de mineral(es) m - [miner] ore exportation terminal
— m — motor m - [electr-mot] motor terminal
— m — pasajero(s) m - [transp] passenger terminal
— m — salida f - [electrón] output, terminal, o lug
— m — — f hundido - [electron] recessed output terminal
— m — — f para . . . tono(s) m - [electrón] . . . tone output, terminal, o lug
— m soldado - [electr-instal] solder(ed) terminal, o lug
— m unificado - [transp] union terminal
terminar v - . . .; to end up; to draw to @ close • [mec] to abut
— v artículo m - to finish @ article
— v asiento m - [mec] to finish @ seat
— v bien(es) m - [ind[to finish @ good(s)
— v cambio m - [mec] to complete @ replacement
— v carga f - [ind] to, end, o finish, @ charge • [transp] to, end, o finish, @ load(ing)
— v carrera f - [educ] to complete @ career • [mec] to complete @ stroke • [deport] to, come away, o withdraw, from @ race
— v con licor m - [metal-trat] to liquor-finish
— v contrato m - [legal] to complete @ contract
— v de soldar v - [sold] to complete, @ weld, o welding
— v ejercicio m - [fin] to end @ (fiscal) year
— v empleo m - [labor] to terminate
— v — m de energía f auxiliar - [ind] to stop @ auxiliary power use
— v en dos lugares m primeros - [deport] to finish one-two
— v — — puestos m primeros - [deport] to finish one-two

terminar v en lugar m primero - to finish first
— v — posiciones f primeras - [deport] to finish in @ money
— v equipo m - [ind] to finish @ equipment
— v felizmente - to end happily
— v junta f - [sold] to finish @ joint
— v línea f - [comput] to terminate @ line
— v material m - to finish @ material
— v mediante esmerilado - [sold] to finish by grinding
— v montaje m - [ind] to complete @ erection
— v muro m - [constr] to finish @ wall
— v parada f - [ind] to end @ shutdown
— v pared f - [constr] to finish @ wall
— v placé v - [deport] to finish in @ money
— v segundo - [deport] to finish second
— v soldadura f - [sold] to, finish, o stop, @ weld(ing) • stopping
— v trabajo m - to complete @, work, o job, o project
término m - . . . . • time, length, o period • . . .; expression
— m administrativo - [admin] management term
— m breve - short, term, o time length
— m claro - [gram] clear term
— m con significado m amplio - [gram] loosely applied term
— m de compensación f - [labor] compensation term
— m — carrera f - [deport] race end
— m — contrato m - [legal] contract term
— m — — m para garantía f - [legal] guaranty contract term
— m — cronometraje m - [labor] time study end
— m — ecuación f - [matem] equation term
— m — — f de corriente f - [hidr] flow equation term
— m — experiencia f propia - own experience term
— m — fórmula f - [matem] formula term
— m — garantía f - warranty, o guaranty, term
— m — orden - order term
— m — pago m - véase condición f para pago m
— m — pedido m - order term
— f — pérdida f de energía f - energy loss term
— m — póliza f - [seguros] policy term
— m — servicio(s) m - [legal] service(s), . term, o period
— m — sigilo m - [legal] confidentiality term
— m definido - defined term
— m — contradictoriamente - inconsistently defined term
— m — inconsecuentemente - inconsistently defined term
— m desusado - [filol] unusual, o out of use, term, o word
— m específico - specific term
— m fijado - established, time, o period
— m fijo - fixed, o set, time
— m final - final term
— m — para referencia f - final reference term
— m general - general term | adv - (en) by rule of thumb; basically
— m — para crédito m - [fin] general credit, o credit general, term
— m internacional - international term
— m más breve - shorter, test time (period)
— m medio - average
— m menor - shorter, test, term, o time period
— m municipal - [pol] municipal district
— m necesario - necessary term
— m para ciencia f administrativa - [admin] management term
— m — — f para administración f - [admin] management term
— m — crédito m - [fin] credit term
— m — motor(es) m - [electr-mot] motor term
— m — — m para ventilador(es) m - [ventil] fan motor term
— m — paleta f - [ventil] blade term
— m — referencia f - reference term
— m — — f interpretado - interpreted refer-

ence term
término m químico - [quím] chemical, o expression, o term
— **m raro** - [filol] rare, o unusual, term
— **m reordenado** - rearranged term
— **m sencillo** - simple term
— **m simple** - simple term
— **m técnico** - technical, o engineering, term
— **m — para automóvil(es) m** - [auton] automobile, technical, o engineering, term
— **m — neumático(s) n** - [autom-neumát] tire, engineering, o technical, term
términos m y condiciónes f - [ind] terms and conditions
terminología f aceptada - accepted terminology
— **f aclarada** - clarified terminology
— **f administrativa** - [admin] management terminology
— **f lógica** - logical terminology
— **f para cadena(s) f** - [cadenas] chain terminology
— **f — mando n** - [mec] control terminology
— **f — regulador n** - [mec] control terminology
termita f - [sold] thermit(e)
— **f para forja f** - [sold] forging thermit(e)
— **f simple** - [sold] plain thermit(e)
termo m - thermos; vacuum bottle
— **m para metal m caliente** - [metal-prod] hot metal car
termocupla f - thermocouple
— **f de cromel constantan m** - constantan chromel thermocouple
— **f inferior** - bottom, o lower, thermocouple
— **f — para cuba f** - [metal-prod] bosh level, lower, o bottom, thermocouple
— **f — etalaje m** - [metal-prod] bosh level, lower, o bottom, thermocouple
— **f instalada** - installed thermocouple
— **f para baño m** - [metal-prod] bath thermocouple
— **f — base f** - [instrum] base thermocouple
— **f — carcasa f** - [instrum] frame thermocouple
— **f — cojinete m** - [instrum] bearing thermocouple
— **f — conducto m** - [combust] conduit thermocouple
— **f — crisol m** - [metal-prod] hearth thermocouple
— **f — chumacera f** - [instrum] bearing thermocouple
— **f — fondo m** - bottom thermocouple
— **f — — m de crisol m** - [metal-prod] hearth bottom thermocouple
— **f — m — horno m** - [metal-prod] furnace bottom thermocouple
— **f — horno m** - [metal-prod] furnace thermocouple
— **f — inmersión f** - [instrum] immersion thermocouple
— **f — inversión f** - inversion thermocouple
— **f — motor m** - [electr-mot] motor thermocouple
— **f — — m eléctrico** - [electr-mot] electric motor thermocouple
— **f platino-platino** - [instrum] platinum-platinum thermocouple
— **f — para inmersión f** - [instrum] platinum-platinum immersion thermocouple
— **f superior** - [instrum] top thermocouple
termoeléctrico,ca a - thermoelectric
termoestabilización f - [plást] thermosetting
— **f de caucho m** - [plást] rubber thermosetting
termoestabilizado,da a - [plást] thermoset
termoestabilizar v - [plást] to thermoset
— **v caucho m** - [plást] to thermoset @ rubber
termofluencia f - [metal] creep(ing)
termoformación f - [plást] thermoforming
— **f de pieza f** - [plást] part thermoforming
termoformado,da a - [plást] thermoformed
termoformador m - [plást] thermoformer
— **m continuo** - [plást] continuous thermoformer
termoformar v - [plást] to thermoform

termoformar v pieza f - [plást] to thermoform @ part
termología f - thermology
termomagnético,ca - thermomagnetic; thermal-magnetic
termómetro m - [instrum] . . .; temperature, gage, o indicator
— **m apropiado** - [instrum] appropriate, o suitable, thermometer
— **m — legible** - [instrum] suitable legible thermometer
— **m con ampolleta f** - [instrum] bulb thermometer
— **m — — f seca** - [instrum] dry bulb thermometer
— **m — — f húmeda** - [instrum] wet bulb thermometer
— **m — — f — y (ampolleta f) seca** - [instrum] wet and dry bulb thermometer
— **m — cuadrante m** - [instrum] dial thermometer
— **m con contacto(s) m** - [instrum] contact(s) thermometer
— **m — — m para alarma f** - [instrum] alarm contact(s) thermometer
— **m — esfera f** - [instrum] dial thermometer
— **m de tipo m con cuadrante m** - [ind] dial type thermometer
— **m — — m con esfera f** - [ind] dial type thermometer
— **m indicador** - [instrum] indicating thermometer
— **m — para refrigerante m** - [comb.int] coolant indicating thermometer
— **m — — — m para motor m** - [comb.int] engine coolant indicating thermometer
— **m — — temperatura f** - [instrum] temperature indicating thermometer
— **m — — — f para aceite m** - [instrum] oil temperature indicating thermometer
— **m — — — f — — m en caja f para cambios m** - [instrum] gear box oil temperature indicating thermometer
— **m — — — f — — m — — f — engranajes m** - [instrum] gear box oil temperature indicating thermometer
— **m — — — f — — m — — f — velocidades f** - [instrum] gear box oil temperature indicating thermometer
— **m — — — f — — m — convertidor m** - [instrum] converter oil temperature indicating thermometer
— **m — — — f — — m — — m para torsión f** - [instrum] torque converter oil temperature indicating thermometer
— **m — — — f para refrigerante m** - [instrum] coolant temperature indicating thermometer
— **m — — — f — — m en motor m** - [comb.int] engine coolant temperature indicating thermometer
— **m legible** - [instrum] legible thermometer
— **m para aceite m** - [instrum] oil thermometer • temperature gage
— **m — — m en convertidor m** - [instrum] converter oil gage temperature
— **m — — m — — m para torsión f** - [instrum] torque converter oil thermometer
— **m — graduación f máxima y mínima** - [instrum] véase **termómetro m para máxima y mínima**
— **m — líquido(s) m** - [instrum] liquid(s) thermometer
— **m — — m refrigerante** - [comb.int] coolant thermometer
— **m — máxima f** - [instrum] maximum thermometer
— **m — — f y mínima f** - [instrum] maximum-minimum thermometer
— **m — mínima f** - [instrum] minimum thermometer
— **m — refrigerante m** - [comb.int] coolant thermometer
— **m — — m para motor n** - [comb.int] engine coolant thermometer

termómetro m para temperatura f - [instrum] temperature thermometer
— m — f para aceite m - [instrum] oil temperature thermometer
— m — — m para convertidor m - [instrum] converter oil temperature thermometer
— m — — f — — m — — m para torsión f - [instrum] torque converter oil temperature thermometer
— m — — f — agua m - [comb.int] water temperature, thermometer, o gage
— m — — f en caja f para cambio(s) n - [instrum] gear box oil temperature thermometer
— m — — f — — f para velocidad(es) f - [instrum] gear box oil temperature thermometer
— m — — f — — f — engranaje(s) m - [instrum] gear box oil temperature thermometer
— m — — f para líquido(s) n - [instrum] liquid temperature thermometer
— m — — f — — m refrigerante - [comb.int] coolant temperature thermometer
— m — — f — refrigerante m - [comb.int] coolant temperature thermometer
— m — — f — — m en motor m - [comb.int] engine coolanbt temperature thermometer
termopar* m - [instrum] thermocouple; véase también termocupla f
— m instalado m - [instrum] installed thermocouple
termoplástico,ca a - thermoplastic
termopozo m - thermowell
termorecuperador m - (ind) heat exchanger
termosensible a - temperature responsive
termostáticamente adv - thermostatically
termostato m - véase termóstato m
termóstato m - . . . . • [electr] heater link
— m abierto - [instrum] open(ed) thermostat
— m ajustado - [instrum] adjusted thermostat
— m bajado - [instrum] lower(ed) thermostat
— m bimetálico - [instrum] bi-metallic thermostat
— m — sucio - [instrum] dirty bi-metallic thermostat
— m con acción f ultrarrápida - [sold] snap-action thermostat
— m con gas m - [instrum] gas thermostat
— m desconectado - [instrum] tripped thermostat
— m eléctrico - [instrum] electric thermostat
— m elevado - [instrum] raised thermostat
— m integral - [instrum] built-in thermostat
— m Klixon - [instrum] Klixon thermostat
— m operado en base f a corriente f - [instrum] current operated thermostat
— m — — f — temperatura f - [instrum] temperature operated thermostat
— m para acondicionador m para aire m - [ambient] air conditioner thermostat
— m — bobina f—[electr-equip] coil thermostat
— m — — f primaria - [instrum] primary coil thermostat
— m — — f secundaria - [electr-instal] secondary coil thermostat
— m — obturación f - [instrum] blanking thermostat
— m — — f con derivación f - [instrum] by-pass blanking thremostat
— m rectificación,cador - [instrum] rectifier,fying thermostat
— m — reemplazo m - [instrum] replacement thermostat
— m sobrecarga(s) f - [electrón] overload thermostat
— m — — f en reóstato m - [electr-instal] rectifier overload thermostat
— m primario - [instrum] primary thermostat
— m — para bobina f - [sold] primary coil thermostat
— m — — f primaria - [sold] primary coil thermostat
— m — — f secundaria f-[sold] secondary coil thermostat

termóstato m protector - [instrum] protective thermostat
— m — integral - [instrum] built-in protective thermostat
— m secundario - [instrum] secondary thermostat
— m — para bobina f - [sold] secondary coil thermostat
— m sucio - [instrum] dirty thermostat
termotratado,da* a - véase tratado,da térmicamente
termotratamiento m - véase tratamiento m térmico
termotratar* v - véase tratar térmicamente
terna n - circuit • [legal] slate (of three)
terne m - [metal-trat] terne; tin and lead, o lead and tin, alloy
terracear v - [constr] to terrace; véase terraplenar; abancalar
terraja f - [herram] threading machine • head; die, head, o holder; diestock • [tornos] stock • screw plate, o stock • tap(s) and die(s) • breach
— f con filete m doble - [herram] screw auger
— f mecánica - [herram] threading machine
— f para roscar f madera f - [herram] devil
— f — tubo(s) m - [tub] pipe threader
— f radial - [herram] extensible, radial template, o modelling board
— f torsa f - [herram] screw auger
— f y cojinete m - [mec] stock and die
— f y dado m - [med] stock and die
terraplén m - [constr] . . .; bank; fill; bank, o earth, fill • offset; offset; mound • [petról] earth wall
— m ancho - [constr] wide embankment
— m angosto - [constr] narrow embankment
— m bajo - [constr] low, o shallow, fill, o embankment
— m blando m - [constr] soft fill
— m carretero - [constr] highway, o roadway, embankment, o fill
— m circular - [miner] circular fill
— m — plano m - [miner] circular flat fill
— m — práctico - [constr] practical round fill
— m circundante - [constr] abutting embankment
— m con altura f considerable - [constr] high fill
— m — — f mayor - [constr] higher fill
— m — — f mediana - [constr] moderately high, o medium, fill, o embankment
— m conservado - [constr] maintained fill
— m consolidado - [constr] consolidated fill
— m contenido - [constr] retained fill
— m dañado - [constr] damaged, o harmed, fill
— m de altura f especificada - [constr] given, o specified, height fill
— m — — f reducida - [constr] shallow fill
— m — — f relativamente reducida - [constr] comparatively, o relatively, shallow fill
— m de escoria f - [constr] slag embankment
— m — suelo m - [constr] soil fill
— m — — m inestable - [constr] unstable soil fill
— m — tierra f - [constr] earth, o soil, embankment, o fill; earthwork
— m — — f compactada - [constr] compacted earth, embankment, o fill
— m elevado - [constr] elevated, o built up, embankment, o fill
— m escurridizo m - [constr] slippery fill
— m especificado - [constr] specified, embankment, o fill
— m estable - [constr] stable embankment
— m ferroviario - [f.c.] railroad, embankment, o fill
— m más alto - [constr] higher fill
— m inestable - [constr] unstable fill
— m para acceso m - [constr] approach, fill, o embankment
— m — — m para puente m - [constr] bridge approach, embankment, o fill
— m — calzada f - [constr] roadway embankment
— m — camino m - [constr] road embankment

terraplén m para ferrocarril m - [f.c.] railway,
  embankment, o bed
— m — paso m superior - [constr] overpass fill
— m — puente m - [constr] bridge embankment
— m — seguridad f - [petról] safety (earth)
  wall
— m — vía f - [f.c.] track, o railway, bed
— m pavimentado - [constr] paved, o surfaced,
  fill
— m perdido - [constr] lost embankment
— m plano - [constr] flat fill
— m protegido - [constr] protected embankment
— m que se está asentando - [constr] settling
  embankment
— m — extiende - [constr] spreading, embankment, o fill
— m simulado - [constr] simulated embankment
— m sin consolidar v - [constr] unconsolidated,
  o soft, fill
— m — pavimentar v - [constr] unpaved, o un-
  surfaced, fill
— m sobre tubería f - [constr] pipe overfill
— m socavado - [hidr] undermined embankment
— m sólido - [constr] solid earth fill
— m vial - [constr] highway, o roadway, fill, o
  embankment
terraplenado m - [constr] backfilling • [vial]
  approach work
— m existente - [constr] existing fill
— m para estructura f - [constr] structural
  backfilling
terraplenar v - véase también terracear; abancalar
— v estructura f - [constr] to backfill @
  structure
terreno m - . . .; real estate; property; lot •
  terrain
— m accidentado - [geol] rough, o rugged, terrain
— m adyacente - adjacent, land, o property
— m agradable - agreeable terrain
— m agrícola - [agric] farmland
— m aguas abajo - [hidr] downstream land
— m — arriba - [hidr] upstream land
— m aluvional - [geol] bench land
— m arcilloso - [geol] clayey, terrain, o soil
— m áspero - [topogr] rough terrain
— m castigador - [vial] punishing terrain
— m circundante - surrounding terrain
— m corrosivo - [hidr] corrosive soil
— m — altamente - [hidr] highly corrosive soil
— m cuidado - attractive, o well kept, grounds
— m cultivado - [agric] cultivated, land, o soil
— m colindante - adjacent acreage
— n congelado - frozen terrain
— m de acarreo m - [hidr] drift
— m — castillo m - castle ground(s)
— m — colegio m - [educ] college grounds;
  campus
— m desigual - uneven ground; rough terrain
— m deportivo - [deport] playground
— m desigual - [constr] unlevelled land
— m desnivelado - [constr] unlevelled land
— m desparejo - [constr] uneven ground
— m despoblado - open, o unsettled, country
— m — ondulado - open rolling country
— m escabroso m - [topogr] rough, o rugged,
  terrain
— m firme - [suelos] firm soil
— m húmedo - wet, o humid, soil, o ground
— m inclinado - [topogr] sloping terrain
— m — ligeramente - slightly, o gently, rol-
  ;ing, land
— m libre - free, o undeveloped, land
— m llano - [topogr] flat, area, o land
— m maderable - [geogr] timberland
— m marginal - [topogr] marginal land
— m montañoso - [geogr] mountainous land
— m — accidentado - [topogr] rugged mountain
  terrain
— m muy accidentado - [topogr] very rugged terrain
— m nacional - [pol] national soil

terreno m natural - natural, terrain, o soil
— m no nivelado - [constr] unlevelled terrain
— m ondulado - [topogr] rolling, terrain, o
  land, o country; hilly country
— m original - [constr] original soil
— m pantanoso - [suelos] marshy land • swamp
— m para camping m - [deport] camping, grounds,
  o site
— m para feria(s) f - fairgrounds
— m — recubrimiento m - [hidr] cover
— m petrolífero m - [petról] oil land
— m — comprobado - [petról] proven oil land
— m plano - [topogr] level, terrain, o country
  • flat land
— m posiblemente petrolífero - [petról] probable, o prospective, oil land
— f quebrado - [topogr] hilly, o broken, terrain, o land
— m recuperado - [topogr] reclaimed land
— m residual - [suelos] residual soil
— m revaluado - [fin] revalued land
— m seguro - [segurid] safe ground
— m sin nivelar - [constr] unlevelled terrain
— m submarginal - submarginal land
— m traicionero - treacherous terrain
— m valioso - valuable land
— m variado - [topogr] varied terrain
— m vecino - adjacent land
terreno,na a - earthly • terrain
terrenos m - [educ] campus
— m, cercos y desvío m ferroviario - [f.c.]
  land, fences and railway spur
terrero m - [miner] . . .; tailing(s) soil bank
terrestre a - . . . • on land • [transp] domestic; inland
terrible a - . . .; awesome • terrific
territorio m establecido - established territory
— m — cuidadosamente - [admin] carefully established territory
— m nacional - [pol] national, territory, o
  boundary,ries
terromontero m -,[topogr] . . .; butte
terrón m - . . . • [suelos] clump
— m aglomerado - [coque] lump
— m — arcilla f - [suelos] clay lump
— m — hierro m - [miner] iron lump
— m — óxido m - [miner] oxide lump
— m — m de hierro m - [miner] iron oxide
  lump
— m reducido - [metal-prod] reduced lump
terror m - . . . • awe
terrorífico,ca a - terrifying; frightening
tesis f doctoral - [educ] master's, o doctor's,
  (degree) thesis
tesla f - [electr] tesla
tesoro m público - [pol] public treasury; ex-
  chequer
testarudo m - . . .; willful
testero m - . . . • [mec] front; end, piece, o
  wall • [miner] back stope
testigo m - . . . . • [mec] test plug; sample •
  [petról] (drill) core; core sample
— m cilíndrico - [petról] coring sample
— m — sacado - [petról] taken core sample
— m — tomado - [petról] removed core sample
— m de accidente m - [segurid] accident witness
— m — coquización f - [coque] button coke
— m — ensayo - test sample
— m — m de coquización f - [coque] coke
  button
— m — oídas - hearsay, o auricular, witness
— m — perforación f - [petról] drill core
— m — vista - eyewitness
— m directo - direct witness
— m doble - [petról] double core
— m indirecto - indirect witness
— m instrumental - [legal] attesting witness
— m nombrado - [legal] named witness
— m, nombrado, o nominado - [legal] named witness
— m ocular - eyewitness
testimonio m - . . . . • [legal] . . .; copy;
  (official) transcript

tesura f contra bamboleo m⸺[mec] rail stiffness
— f — rodadura f - [mec] roll stiffness
tetóna m - . . . • [mec] lug
tetracentrado,da a - [arquit] four centered
tetracloruro m de carbono m - [quím] carbon tetrachloride
— m — titanio m—[quím] titanium tetrachloride
tetraedrita f - [miner] tetrahedrite
tetraedro m - tetrahedron
tetrafilar a - [electr-cond] four wire(d)
tetralobulado,da a - [arquit] four, cusped, o foiled
tetralobular a—[arquit] four, cusped, o foiled
texto m - . . . • wording • script
— m actualizado - [public] latest, o updated revised, text
— m castellano - [gram] Spanish; o Castilian, text
— m de alumno m - [educ] student's textbppl; workbook
— m español m - [gram] Spanish text
— m excelente - [public] excellent text(book)
— m inglés - [gram] English text(book)
— m propuesto m - [gram] proposed wording
textura f afanítica - [geol] aphanitic texture
— f arenosa - [geol] gritty texture
— f — fina - [constr] fine gritty texture
— f corriente - common texture
— f cremosa - creamy texture
— f de azulejo(s) m - [cerám] tile texture
— f — suelo m - [suelos] soil texture
— f extremadamente lisa - extra-smooth texture
— f granoblástica - [feol] granoblastic texture
— f liviana - light texture
— f original - original structure
— f pesada - heavy texture
— f piroclástica - [geol] pyroclastic texture
— f suave - soft texture
— f porfiri(ti)ca - [miner] porphyritic texture
texturación f - texturing
texturado,da a - textured
texturador,ra a - texturing
texturar v - to texture
tez f pálida - pale complexion
ticket m - véase boleta f; billete m
Thomas a - [metal-prod] basic Bessemer
thyrector m - [electr] Thyrector
tibio,bia a - . . .; warm
tiempo m - . . .; time, lapse, o length • space • [comb.int] cycle
— m adecuado - adequate time
— m — permitido - allowed adequate time
— m administrado - [admin] managed time
— m adverso - [meteorol] adverse, o undesirble, weather
— m ahorrado - saved time
— m ajustable - adjustable time
— m ajustado - adjusted time
— m amenazante - [meteorol] menacing, o threatening, weather
— m antes de bloqueo m - [metal-prod] before blocking time; time before blocking
— m — — colada f - [metal-prod] time before tapping
— m — — sangría f - [metal-prod] time before tapping
— m apropiado - appropriate, o adequate, o proper, time
— m — para curación f - [autom-neumát] appropriate curing time
— m aproximado - approximate time
— m asignado - assigned time
— m asoleado - [meteorol] sunny weather
— m breve - brief, o short, time
— m bruto - gross time • brutal weather
— m — para coquización f - [coque] gross coking time
— m bueno - [meteorol] good weather
— m cálido - [meteorol] warm, o hot, weather
— m caliente - [meteorol] hot weather
— m caluroso - [meteorol] hot, o warm, weather
— m claro - [meteorol] clear weather

tiempo m con flujo m máximo - [metal-prod] maximum flow time
— m — — m mínimo - [metal-prod] minimum flow time
— m concedido - [labor] allowed time
— m considerable - quite a long time
— m corriente - normal time
— m corto - short time • quick time
— m cronometrado - [labor] stop watch time • measured time
— m cronométrico - [labor] stop watch time
— m dado - given time
— m de asistencia f - [labor] attendance time
— m — baja - [segur] time off; down time
— m — compra f - purchase, time, o date
— m — demora f - [ind] delay time
— m — — f para carga f - [ind] charging time delay
— m — — f para descarga f - [ind] discharging, o drawing, time delay
— m — detención f - [ind] down time
— m — — f para prensa f - [ind] press down time
— m — — f — — f eliminado - [mec] eliminated press down time
— m — — f — torre f para perforación f - [petról] rig down time
— m — — f — — f minimizado - [petról] minimized rig down time
— m — duración f - [ind] duration time; véase también vida f
— m — encendido m - [int.comb] véase distribución f de encendido m
— m — entrega f - delivery, time, o period
— m — — f total - total delivery period
— m — espera f - waiting, period, o time; frlsy • [metal-prod] track time • [ind] lost time
— m — estadía f - [transp] demurrage; lay time
— m — exposición f - [fotogr] exposure time
— m — — f exigido - [fotogr] required exposure time
— m — inactividad f - [ind] downtime; down time; inactive,vity time
— m — — f para equipo m - [ind] equipment downtime
— m — — f — prensa f - [ind] press down time
— m — — f reducido - [ind] reduced down time
— m — — f total - [ind] total inactive,vity time
— m — interrupción f - [electr-oper] interrupting,tion time
— m — parada f - [ind] down time
— m — — f para equipo m - [ind] equipment down time
— m — — f — reenhebrado m - [ind] re-threading equipment down time
— m — — f programado - [ind] scheduled downtime
— m — perforación f - [petról] rig time
— m — permanencia f - [ind] holding time
— m — — f en horno m - [metal-prod] furnace holding, o pit, time
— m — predominio m - easy time(s)
— m — reposo m - [metal-prod] véase tiempo a en vía f
— m — retención f - [ind] holding, o retention, time; detention period
— m — retroceso m - [mec] back up time
— m — — m para émbolo m - [comb.int] piston back up, stroke, o time
— m — . . . segundo(s) m - . . . second(s) time
— m — . . . — m para entrada f - [electrón] . . . second(s) input time
— m — servicio m - [ind] service, time, o period, o length • [labor] seniority
— m — — m continuado - [labor] continuous service time
— m — — m determinado - fixed service period
— m — viaje m - [metal-prod] véase tiempo m

en vía f̲
tiempo m̲ dedicado - dedicated, o̲ devoted, time
— m̲ deficitario - [labor] deficit time
— m̲ desastroso - [meteorol] disastrous weather
— n̲ desde colada f̲ hasta desmoldeo m̲ - [metal-pro] time in @ mold
— m̲ — fin m̲ de reposo m̲ hasta desmoldeo m̲ - [metal-prod] track time
— m̲ desfavorable - [meteorol] unfavorable, o̲ poor, weather
— m̲ desperdiciado - [labor] wasted time
— m̲ destinado a trabajo m̲ administrativo - [admin] time devoted to, management, o̲ administrative, work
— m̲ — — — m̲ técnico - [admin] time devoted to technical work; technical work time
— m̲ diáfano - [meteorol] clear weather
— m̲ disponible - available time
— m̲ empleado - time, used, o̲ spent
— m̲ — para descarga f̲ - [transp] unloading time
— m̲ en aire m̲ - [metal-prod] time after teeming
— m̲ — fosa f̲ - [metal-lam] pit time
— m̲ — horas f̲ - [labor] time in hours
— m̲ — horno m̲ - [combust] furnace time
— m̲ — — m̲ de fosa f̲ - [metal-lam] soaking pit time
— m̲ — — m̲ — — f̲ vacío - [metal-lam] empty soaking pit time
— m̲ — — m̲ vacío - [combust] empty furnace time
— m̲ — lingotera f̲ - [metal-prod] mold time
— m̲ — minuto(s) m̲ - [labor] time in minute(s)
— m̲ — postrimerías f̲ de otoño m̲ - [meteorol] late fall weather
— m̲ — prueba(s) f̲ para calificación f̲ — [deport] qualifying time
— m̲ — reposo m̲ - [metal-prod] holding time • track time
— m̲ — — m̲ para acero m̲ - [metal-prod] steel, track, o̲ holding, time
— m̲ — segundo(s) m̲ - time in second(s)
— m̲ — tiempo adv - (de) from time to time
— m̲ — tránsito m̲ - [metal-prod] track time
— m̲ — — recomendado - [metal-prod] recommended track time
— m̲ — vía f̲ - [metal-prod] track time
— m̲ — — f̲ corto — [metal-prod] short track time
— m̲ — — f̲ distinto - [metal-prod] varying, o̲ different, track time
— m̲ — — f̲ equivalente - [metal-prod] equivalent track time
— m̲ — — f̲ óptimo - [ind] optimum track time
— m̲ — — f̲ recomendado - [metal-prod] recommended track time
— m̲ — —/tiempo m̲ en fosa f̲ - [metal-prod] track time/pit time
— m̲ — viaje m̲ - véase tiempo m̲ en vía f̲
— m̲ entre colada(s) f̲ - [metal-prod] tapping to tapping time; time between tappings
— m̲ entre dígito(s) m̲ - [electrón] interdigital time
— m̲ establecido - established, o̲ allotted, time
— m̲ estándar - [labor] standard time
— m̲ — para operación f̲ - [ind] standard operating time
— m̲ — — reconstrucción f̲ - [constr] standard rebuilding time • [metal-prod] standard relining time
— m̲ estimado - estimated, o̲ anticipated, time
— m̲ — para capacitación f̲ - estimated training time
— m̲ excelente - [deport] excellent, o̲ quick, time • [meteorol] excellent weather
— m̲ exigido - [labor] required time
— m̲ — normal - [labor] standard required time
— m̲ extremadamente frío - [meteorol] extremely cold weather
— m̲ — húmedo - [meteorol] extremely, humid, o̲ damp, weather
— m̲ final - end(ing) time
— m̲ fiscalizado - [labor] controlled time
— m̲ fresco - [meteorol] fresh, o̲ cool, weather
— m̲ frío m̲ - [meteorol] cold weather
— m̲ gastado - time spent • time wasted

tiempo m̲ hasta perforación f̲ - time to @ perforation • [petról] time till @ drilling
— m̲ — — f̲ de fondo m̲ - [constr] time to @ invert perforation
— m̲ — — f̲ primera - [constr] time to @ first perforation
— m̲ ido - time gone • long ago • old(en) days
— m̲ improductivo - [ind] nonproductive, o̲ down, time • idle time
— m̲ — mayor - longer, o̲ more, down time
— m̲ — menor - less(er), o̲ shorter, down time
— m̲ imputable - [labor] chargeable time
— m̲ inactivo - [ind] down time; downtime
— m̲ — menor - [ind] shorter, o̲ less, downtime
— m̲ — reducido - [ind] reduced down time
— m̲ inclemente - [meteorol] inclement weather
— m̲ indeseable - [meteorol] undesirable weather
— m̲ íntegro - whole, o̲ entire, time
— m̲ interdigital - [electrón] interdigit(al) time
— m̲ inusitado - unheardof time • [meteorol] unfavorable weather
— m̲ liberado - liberated time • [telecom] given up, o̲ freed, (air) time
— m̲ libre - free time • idle time
— m̲ lluvioso - [meteorol] rainy weather
— m̲ malgastado - [labor] wasted time
— m̲ malo - bad time • [meteorol] bad weather
— máquina f̲ - [labor] machine-time
— m̲ máximo - maximum, time, o̲ lapse
— m̲ — en fosa f̲ - [metal-prod] maximum pit time
— m̲ — — vía f̲ - [metal-prod] maximum track time
— m̲ — para calentamiento m̲ - [ind] maximum heating time
— m̲ — — — m̲ de lingote(s) m̲ - [metal-prod] maximum ingot heating time
— m̲ — para cierre m̲ - maximum closing time
— m̲ — — deslingotado m̲ - [metal-prod] maximum stripping time
— m̲ — reposo m̲ - [metal-prod] maximum track time
— m̲ mayor - more, o̲ longer, time • slower time
— m̲ — para calentamiento m̲ - longer heating, time, o̲ period
— m̲ — — soldadura f̲ - [sold] longer, o̲ increased, arc, o̲ weld(ing), time
— m̲ medio - average time
— m̲ — en fosa f̲ - [metal-lam] averge pit time
— m̲ — para labrado m̲ - [mec] average machining time
— m̲ — — maquinado* m̲ - [mec] average machining time
— m̲ — — mecanización f̲ - [mec] average machining time
— m̲ — — reconstrucción f̲ - [sold] average rebuilding time • [metal-prod] avrage relining time
— m̲ — — relleno , - [sold] average rebuilding time
— m̲ mejor - best time • [deport] top time
— m̲ menor - shorter, o̲ quicker, time • fastest, o̲ shortest, o̲ quickest, o̲ least, time • less time
— m̲ — para calentamiento m̲ - [ind] shorter heating, time, o̲ period
— m̲ mínimo - minimum, o̲ least, time, o̲ lapse
— m̲ — en vía f̲ - [metal-prod] minimum, o̲ shortest, track time
— m̲ — para calentamiento m̲ - [ind] minimum, o̲ shortest, heating time
— m̲ — — — m̲ de lingote(s) m̲ - [metal-prod] minimum, o̲ shortest, ingot heating time
— m̲ — — cierre m̲ - minimum closing time
— m̲ — — deslingotado m̲ - [metal-prod] minimum stripping time
— m̲ — — proyección f̲ - minimum design(ing) time
— m̲ — — reposo m̲ - [metal-prod] minimum track time
— m̲ — — terminación f̲ - minimum finishing

time
tiempo m **miserable** - [meteorol] miserable weather
— m **moderado** - [meteorol] moderate, weather, o temperature
— m **moderno** - [cronol] modern time
— m **muerto** - [labor] down, o inactive, o nonproductive, time • shut down
— m **muy frío** - [meteorol] very cold weather
— m **necesario** - necessary, o needed, time
— m **neto** - net time
— m — **para coquización** f - [coque] net coking time
— m — — **montaje** m - [ind] net erection time
— m **no imputable** - [labor] nonchargeable time, o hour(s)
— m — **productivo** - [labor] nonproductive time
— m **normal** - [cronol] standard time • [meteorol] normal weather
— m **exigido** - [labor] standard required time
— m **ocupado** - time spent; spent time
— m **oficial** - [deport] official time
— m — **para etapa** f - [deport] qualifying lap time
— m — — **vuelta** f - [deport] qualifying lap time
— m **ominoso** - [meteorol] ominous weather
— m **operativo** - [ind] operating,tion time
— m **oportuno** - opportune, o proper, time
— m **largo** m - long time
— m — **en vía** f - [metal-prod] long track time
— m **para adición** f **de metal** m **caliente** - [metal-prod] hot metal addition time
— m — **acondicionamiento** m - conditioning time
— m — **actividad(es)** f - [labor] ativity,ties time
— m **para afino** m - [metal-prod] refining time
— m — **apertura** f - opening time • [electr-oper] interrupting,tion time
— m — **arco** m - [sold] arc time
— m — — m **apagado** - [sold] arc-off time
— m — — m **encendido** - [sold] arc-on time
— m — **armado** m - [mec] assembly,bling time
— m — — m **para plancha(s)** f - [mec] plate(s) assembly,bling time
— m — **armadura** f - [mec] assembly,bling time
— m — **armar** v - [mec] assembly,bling time
— m — **calentamiento** m - [ind] heating, time, o period
— m — — m **de lingote(s)** m - [metal-lam] ingot heating time
— m — **calificación** f - [deport] qualifying time
— m — **capacitación** f - training time
— m — **carga** f - [ind] charging time • [transp] loading time
— m — — f **de sólido(s)** m - [ind] solid(s) charging time
— m — — f — **líquido(s)** m - [metal-prod] liquids charge,ging time
— m — **cierre** m closing time
— m — **colada** f - [metal-prod] tapping, o pouring, time • [ind] bleeding time
— m — **comienzo** m - starting, o beginning, time
— m — **concentración** f - [hidr] concentration time
— m — **conducción** f - [vial] driving time
— m — **construcción** f - construction, o building, time • lead time
— m — **conversión** f - [electrón] conversion,ting time
— m — — f **de onda(s)** f **digital(es) en sinusoidal(es)** - [electrón] digital to sine wave conversion,rting timing
— m — **coquización** f - [coque] coking time
— m — **cronometraje** m - [labor] time study time
— m — **curación** f - [quím] cure,ring time
— m — **descanso** m - [labor] rest(ing), time, o allowance
— m — — m **permitido** - [labor] rest allowance
— m — **descarga** f - [transp] unloading time • [ind] discharge,ging time • [metal-prod] drawing time
— m — **deslingotado** m - [metalprod] st

time
tiempo m **para desmoldeo** m - [metal-prod] stripping time
— m — **detención** f - [ind] down time
— m — **ejecución** f - [ind] performance time • [admin] activity time; schedule
— m — **embarque** m - [transp] shipment,pping time
— m — **empapamiento** m - [metal-trat] hold(ing) time • [metal-prod] soaking time
— m — **endurecimiento** m - hardening time
— m — **ensamblaje,le** m - [ind] assembly,bling, time, o period
— m — — m **final** - [ind] final assembly,bling time
— m — **ensayo(s)** - [ind] test(ing) time • [ind] practice time
— m — **entrega** f - [transp] delivery, date, o time, o period
— m — **erección** f - [mec] erection, o assembly,bling, time
— m — — f **de plancha(s)** f - [mec] plate(s) assembly,bling time
— m — **etapa** f - [deport] lap time
— m — — f **para clasificación** f - [deport] qualifying lap time
— m — — f **para ensayo(s)** m - [deport] practice lap time
— m — **fermentación** f - [quím] fermenting,tation time
— m — **flotación** f - [miner] flotation time
— m — **fraguado** m - [constr] setting, o hardening, time
— m — **fusión** f - [metal] melting time
— m — **hincadura** f - [constr=pil] driving time
— m — **igualación** f - [metal-trat] soaking, time, o period
— m — — f **en ciclo** m **experimental** - [metal-prod] experimental cycle soaking time
— m — **inicio,ciación** - beginning time
— m — **instalación** f - installation time
— m — **inversión** f - [mec] inversion time • [fin] investment time
— m — **labrado** m - [mec] machining time
— m — **laminación** f - [metal-lam] rolling, o mill(ing), time
— m — **largada** f - [deport] starting time
— m — **limpieza** f - cleaning time
— m — **mantenimiento** m - [ind] maintenance time • holding time
— m — **máquina** f - [ind] machine time
— m — — f **fiscalizado** - [labor] controlled machine time
— m — **mecanización** f - [mec] machining time
— m — **medición** f - measuring time
— m — **mezcla,clado,cladura** - [constr] mixing time
— m — **montaje** m - [mec] erection, o assembly, time
— m — **navegación** f - [nab] navigation time
— m — **operación** f - operating,tion time
— m — — f **para servomotor** m - [electr-oper] servomotor operating,tion time
— m — **pesada** f - [ind] weighing time
— m — **prensa(s)** f - [ind] press time
— m — **preparación** f - preparation time
— m — — f **para cucharón(es)** m - [metal-prod] ladle preparation time
— m — **proyección** f - design(ing) time
— m — **puesta** f **en, funcionamiento** m, o **marcha** f - [ind] start-up time
— m — **reacción** f - reaction, o response, time
— m — — f **a caracterización** f - [admin] identification response time
— m — — f **de combinación** f **neumático-vehículo** - [autom-neumát] tire-vehicle system tire response
— m — — f **para motorista** m - [autom] motorist('s) response time
— m — **realización** f - performance, time, o period

tiempo m para recocido m - [metal-trat] annealing time
— m — reconstrucción f - [constr] reconstruction, o rebuilding, time • [metal-prod] relining time
— m — recuperación f - recovery time • [cald]
recuperation time
— m — reducción f - [metal-prod] reduction, o
reducing, time
— m — relleno m - [sold] rebuilding time
— m — reparación f - [ind] repair time; down
time
— m — reposo m - rest(ing) time
— m — sangría f - [metal-prod] tapping time •
[combust] bleeding time
— m — secado,camiento m - drying time
— m — — m recomendado - recommended drying
time
— m — sellado,llamiento m - [metal-prod]
capping time
— m — sincronización f - [labor] synchronization, o timing, time
— m — sinterización f - [miner] sintering,ri-
zation, time
— m — — f de material(es) m - [miner] mate-
rial(s) sintering,terization, time
— m — soldadura f - [sold] welding, o arc,
time
— m — soplado,dura - [metal-prod] blast, o
blowing, time
— m — tapado,dura - [metal-prod] covering time
— m — — de lingotera(s) f - [metal-prod] ingot mold covering time
— m — semiterminación f - [ind] finishing time
— m — trabajo m - [ind] work(ing), o operating,tion, time
— m — — m para ingeniería f - [ind] engineering time
— m — traslado m - [ind] transfer time
— m — vaciado m - [metal-prod] pouring time
— m — validez f - validity time
— m — venta f - sale, o selling, time
— m — vigencia f - validity, time, o period
— m — vuelta f - return time • [deport] lap
time
— m — — f para clasificación f - [deport] lap
qualifying time
— m — — f — ensayo m - [deport] lap practice time
— m pasado - passed time • spent time
— m perdido - lost time; time lost • [ind]
down time
— m — en procesamiento m - [labor] process, o
processing, allowance
— m — por (causa f de) lluvia f - [constr]
rain, shutdown, o down time
— m — valioso - costly down time
— m perfecto m - [meteorol] perfect weather
— m permitido - allowed time
— m pésimo - [meteorol] miserable, o dismal,
weather
— m preestablecido - preestablished, o preset,
time • adjusted time
— m prefijado - preset time
— m previsto - foreseen time
— m programado - scheduled time
— m prolongado - extended time (period)
— m promedio - average time
— m — para adición f de arrabio m (líquido) -
[metal-prod] hot metal addition average time
— m — — — f — metal m caliente - [metal-
-prod] hot metal addition average time
— m — para bloqueo m - [combust] average
blocking, o blocking average, time
— m — — colada f - [metal-prod] average,
bleeding, o tapping, time
— m — — fase f - [ind] phase average time
— m — — mecanización f - [mec] average machining time
— m — reconstrucción f - [sold] average rebuilding time • [metal-prod] average relining
time
— m — — relleno m - [sold] average rebuilding

tiempo m promedio para reparación(es) f - [ind]
average repair(ing) time
— m — — sangría f - [ind] average bleeding
time
— m rápido - quick, o short, time
— m — generado - generated quick time
— m real - real, o actual, o true, time
— m — para igualación f - [metal-prod] actual
soaking time
— m — — laminación f - [metal-lam] actual
rolling time
— m — para soldadura f - [sold] actual arc
time
— m — — trabajo m - [ind] real work(ing) time
— m recomendado - recommended time
— m record* - [deport] record, o fastest, time
— m registrado - [labor] recorded time
— m regulado - adjusted, o controlled, time
— m relativamente breve - relatively, brief, o
short, time
— m — corto - relatively short time
— requerido - required time
— m seco - [meteorol] dry weather
— m según norma f - [ind] standard time
— m — — f para operación f - [ind] standard
operating time
— m — — f — reconstrucción f - [constr]
standard, reconstruction, o rebuilding, time
• [metal-prod] standard relining time
— m sin perforar v - [petról] handling time
— m sobrado - time to spare
— m sobre vía f - [metal-prod] track time
— m suficiente - sufficient, o adequate, time
| adv long enough
— m — permitido - allowed, sufficient, o adequate, time
— m sugerido - suggested time
— m suplementario - supplementary time •
[labor] overtime
— m total - total time
— m — de apertura f - [electr-oper] total, o
complete, interrupting,tion time
— m — — arco m - [sold] total arc time
— m — — inactividad f - [ind] total inactive,tivity time
— m — interrupción f - [ind] total, o
complete, interrupting,tion time
— m — en fosa f - [metal-prod] total pit time
— m — — construcción f - [constr] total,
building, o construction, time • [metal-
-prod] total relining time
— m — entrega f - total delivery period
— m — — soldadura f - [sold] total, welding,
o arc, time
— m transcurrido - transpired, o elapsed, time
— m unitario - [ind] unit time
— m útil - useful, o productive, time
— m variable - variable,rying time
— m verificado - [labor] checked, o controlled,
time
tiempos m - [ind] timing•• [int.comb] cycle(s)
tienda f - [com] . . .; boutique
— f con artículo(s) m para automovilistas m -
[autom] auto(mobile) center
— f — — m — turista(s) m - [com] tourist, o
specialty, shop
— f — — m profesional(es) - [deport] pro shop
— f — comestible(s) m - [com] grocery (store)
— f — ramos m varios - [com] department store
— f para campaña f - . . . . • [constr] canvass
shelter
tiento m - [cueros] rawhide
tierra f - . . . . • dirt; dust • [electr-instal]
ground(ing); ground connection
— f absoluta - absolute earth
— f acumulada - accumulated dirt
— f adentro - [geogr] inland | a - on @ land |
adv - [petról] land
— f aluvional - [hidr] alluvium
— f apisonada - [constr] tamped earth
— f apretada - [constr] packed earth
— f apropiada - [suelos] suitable ground

tierra f arcillosa - [geol] clayey, o claylike, soil, o soil, o earth
— f arenosa - [geol] sandy soil
— f — mediana - [geol] average sandy soil
— f — pendiente - [topogr] steep sandy soil
— f — plana - [topogr] flat sandy soil
— f armada - [constr] reinforced soil
— f blanda - [suelos] soft, soil, o earth, o ground
— f circundante - surrounding, soil, o dirt
— f colectada - [constr] collected dirt
— f compacta - [suelos] campacted, o heavy, dirt, o soil
— f — mediana - [suelos] average, compacted, o heavy, soil
— f — plana - [suelos] flat heavy soil
— f — pendiente - [topogr] steep heavy soil
— f compactada - [constr] compacted, earth, o soil
— f — acabadamente - [constr] thoroughly, compacted, o tamped, soil, o earth
— f bien - [constr] throroughly, compacted, o tamped, soil, o earth
— f completamente seca - [constr] perfectly dry earth
— f común - [suelos] common, earth, o soil
— f con bosque(s) m nacional(es) - [maderas] national forest land
— f — calidad f buena - [constr] high grade, o good quality, soil, o earth
— f — — f pobre - [constr] low grade, o poor quality, soil
— f — — f — para relleno,nar - [constr] low grade, o poor quality, backfill(ing) soil
— m — — — m para tubería(s) f - [constr] low grade pipe backfill(ing) soil
— f con cemento m - [constr] soil cement
— f confinada—[constr] confined, earth, o soil
— f contigua - [constr] abutting soil
— f corriente - [constr] ordinary earth
— f — nivelada - [constr] graded ordinary, earth, o soil
— f — — cuidadosamente - [constr] carfully, o smoothly, graded ordinary, earth, o soil
— f cultivada - [agric] cultivated land
— f de aluvión m - [hidr] alluvium
— f — batán(ero) - [petról] fuller's earth
— f — camino m - [vial] road dirt
— f — Fuller - [petról] Fuller's earth
— f debajo de rasante f - [constr] below ground level earth
— f dentro de terraplén m - [constr] earth within @ embankment
— f desnuda - bare earth
— f eléctrica - [sold] electrical ground
— f en terraplén m - [constr] embankment earth
— f encespedada - [constr] sodded earth
— f endurecida - [vial] hard earth • hardpan
— f excavada - [constr] excavated earth
— f excesiva - [constr] excessive dirt
— f extraída - [constr] drilled out, o removed, earth, o ground
— f firme - [geogr] terra firme; mainland
— f forestada - [maderas] forest(ed) land
— f hullera - [miner] coal, land, o country
— f húmeda - moist, o humid, land, o soil, o earth
— f infusoria* - [suelos] infusorial earth
— f laterítica - [geol] laterite (soil)
— f levemente húmeda - [constr] slightly moist earth
— f llana - [topogr] flat land; flatland
— f movediza f - [suelos] shifting soil • quicksand
— f movida - [constr] moved earth
— f nivelada - [constr] levelled, o graded, earth
— f — cuidadosamente - [constr] smoothly graded earth
— f ondulada - [topogr] rolling land
— f — cultivada - [agric] rolling cultivated land
— f ordinaria - ordinary soil • ordinary dirt

tierra f ordinaria removida - [constr] removed ordinary dirt
— f pantanosa - [topogr] marshy land; marshland
— f para batanero m - [suelos] fuller's earth
— f — moldeo m - [geol] molding clay
— f — pastoreo m - [agric] pasture, o range, land
— f para relleno m - [constr] backfill(ing), earth, o soil; earth fill;
— f pobre - [geol] poor, earth, o soil • [sold] poor ground(ing)
— f quitada - [constr] removed dirt
— f rara - [miner] rare earth • rare metal
— f refractaria - [refract] refractory, earth, o clay
— m registrado - [cronol] clocked time
— f relativamente húmeda - relatively, o fairly, moist, earth, o ground
— f removida - [constr] (re)moved, earth, o ground, o soil, o dirt
— f sacada - [constr] removed dirt
— f seca - dry, earth, o dirt
— f sobre piso m - [domest] floor dirt
— f suelta f - [constr] loose(ned), earth, o dirt, o soil, o ground
— f turbosa - [suelos] muck
— f útil - useful, o usable, land, o dirt
— f valiosa - valuable, land, o soil, o dirt
— f vegetal - [suelos] topsoil; loam
— f zarandeada - [constr] screened earth
tieso,sa a - stiff
tiesura f - stiffness
— f máxima - maximum stiffness
— f mínima - minimum stiffness
— f para manipuleo m - handling stiffness
— f verificada - checked stiffness
tijera f - .... • snipper(s) • rein(s) • [petról] jar
— f ascendente - [metal-lam] upcut shear
— f circular—[mec] circle,cular shear; slitter
— f — para hojalata f - [metal-trat] tin plate slitter
— f con disco(s) m - [metal-lam] disc shear
— f — guillotina f - [mec] guillotine, o gate, shear • down-and-upcut shear • flying shear
— f — — f para barra(s) f - [metal-lam] flying bar shear
— f — — f — palanquilla(s) f y barras f - [metal-lam] flying billet and bar shear
— f — mandíbula f - [mec] alligator shear
— f conformadora f - [mec] forming shear
— f desbastadora - [metal-lam] blooming shear
— f descabezadora - [metal-lam] cropping shear
— f descendente - [mec] downcut shear
— f — y ascendente - [metal-lam] down and upcut(ting) shear
— f despuntadora - [metal-lam] cropping shear
— f destrabadora - [mec] bumper shear
— f eléctrica - [herram] electric shear
— f — para mano f - [herram] electric hand shear
— f en caliente - [metal-lam] hot shear
— f escuadradora - [mec] (re)squaring shear
— f final - [metal-lam] final shear
— f grande - [herram] (large) shears
— f guillotina - [metal-lam] guillotine shear
— f hidráulica - [mec] hydraulic shear
— f lateral - [metal-lam] side cut shear
— f longitudinal - [metal-lam] slitter shear
— f — para fleje m - [metal-lam] slitter shear
— f múltiple - [metal-lam] gang shear
— f para barra(s) f - [metal-lam] bar shear
— f — bordes m - [metal-lam] edger; side, trimmer, o slitter
— m — canto(s) m - [metal-lam] side trimmer
— f — corte m - [mec] cutoff shear; snippers
— m — — m ascendente - [mec] upcut shear
— f — — m hacia arriba - [mec] upcut shear
— f — — m transversal - [metal-lam] cross cut shear
— f — chatarra f - [metal-lam] scrap shear • salvage shear

tijera f para desbaste(s) m - [metal-lam] bloom(ing) (mill) shears • cropping shears
— f — escuadrar,drado - [mec] resquaring shear
— f — fleje(s) n - [metal-lam] band shear
— f — formas f - [metal-lam] shape,ping shear • [mec] circle shear
— f — f irregular(es) - [metal-lam] sketch shear
— m — lado para entrada f - [metal-lam] feeding end shear
— f — lingote(s) m - [metal-lam] ingot shear
— f — mano - [herram] hand shear; snipper(s)
— f — palanquilla f - [metal-lam] billet shear
— f — perforación f - [petról] drilling jar
— f — pesca f - [petról] fishing jar
— f — recortadura f - [metal-lam] trimming shear
— f — f lateral - [metal-lam] side trimming shear
— f — v borde(s) m - [metal-lam] side trimming shear
— f — rechazos m - [metal-lam] salvage shear
— f — recuperación f - [metal-prod] salvage shear
— f — redondo(s) m - [herram] circle shear
— f — tocho(s) m - [metal-lam] bloom(ing) shear
— f — recorte,tado m lateral - [metal-lam] side trimming shear
— f rebordeadora - [metal-lam] side trimmer
— f recortadora f - [metal-lam] trimming shear
— f — para extremo(s) m - [metal-lam] véase despuntadora f
— f reforzada - [metal-lam] heavy shears
— f rotativa - [metal-lam] rotary trimmer
— f — para borde(s) m - [metal-lam] rotary side trimmer
— f seccionadora f - [metal-lam] slitting shear
— f — y escuadradora - [metal-lam] slitting and squaring shear
— f transversal - [metal-lam] cross cut shear
— f voladiza - véase tijera f volante
— f volante - [metal-lam] flying shear
— f — para barra(s) f - [metal-lam] flying bar shear
— f — palanquilla f - [metal-lam] flying billet shear
— f — f y barra(s) f - [metal-lam] flying billet and bar shear
tijerero* m - véase cortador m; operador m para tijera f
tildado* m - checking
tildado,da a - checked
tildar v - . . . . • to check
— v ítem m - to check @ item
tilde* m de ítem m - item check(ing)
tiller m - véase guardín m
timbre m - [electr-equip] bell; ringer
— m auxiliar - [electr-instal] auxiliary ringer
— m conectado - [electr-instal] connected, o turned on, bell, o ringer
— m de ley f - [fisc] documentary (tax) stamp
— m desconectado - [electr-instal] disconnected, o turned off, bell, o ringer
— m fiscal - [fisc] documentary, o revenue, stamp
— m tocado - sounded, alarm, o ringer
timón m para arado m - [agric-equip] plow beam
timorato,ta a - . . .; faint of heart
timpa f - [metal-prod] . . .; tymp
— f de cubilote m - [metal-prod] recess
tímpano m - . . . . • [constr] gable
tingladillo m - [constr] weatherboard • siding
— m imitación f ladrillo m - [constr] pressed brick siding
tinglado m - [constr] . . .; lean-to
— m abierto - [constr] open shelter
— m adosado - [constr- lean to
tiniebla(s) f - . . .; blackness
tinta f penetrante - [ind] penetrating dye
tinte m - . . . . • shade
— m metálico - [pint] metallic, paint, o tint
tintorería f . . .; dying • dry cleaning establishment

tíovivo m vertical - [deport] ferris wheel
típicamente adv - typically
típico,ca a - typical
tipificar v - to typify
tipo m - . . .; kind; specie; group; order • grade • standard • [fin] rate • [electr-mot] frame • sizing
— m a prueba f de goteo m - non-leaking type
— m abatible - [constr] breakaway type • breakaway design
— m abierto - open type
— m acodado - [válv] angle type
— m activamente penetrante - [sold] forceful digging type
— m actual - late, o current, type, o style
— m ajustable - adjustable type
— m almeja - [mec] clam type
— m alto - high type
— m — montable - high mountable type
— m angular - angle, type, o design
— m anterior - previous, o former, type • early, type, o style
— m antisalpicadura(s) f - non-spatter type
— m antiguo - old, o early, type
— m aprobado - approved type
— m apropiado - appropriate, o proper, o suitable, type
— m — de banda f - [mec] suitable band type
— m arancelario - [fisc] tariff, type, o rate
— m armado - [mec] erected type
— m ascendente - [válv] lift(ing) type
— m asequible - available, o obtainable, type
— m atomizador - [mec] atomizer,zing type
— m atornillable - [mec] screw-on type
— m autotrabador - [mec] self-locking type
— m axial - axial type
— m babero - bib, style, o type
— m bajo - low type
— m — montable - low mountable type
— m basculante - [mec] tilting type
— m básico - basic type
— m bayoneta - [mec] bayonet type
— m bituminoso - bituminous type
— m — doble - double bituminous type
— m — triple - triple bituminous type
— m blindado - armored type
— m — con ventilación f - ventilated armored type
— m — sin ventilación f - [electr-mot] non-ventilated, armored, o enclosed, type
— m — totalmente - [electr-mot] totally, enclosed, o armored, type
— m — — sin ventilación f - [electr-mot] totally, enclosed, o armored, non-ventilated type
— m — — con enfriamiento m - [electr-mot] totally enclosed cooled type
— m — — — — m con ventilador m - [electr-mot] totally enclosed fan cooled type
— m — y enfriado con ventilador m - [electr-mot] enclosed fan cooled type
— m botonera - [electr-mot] push button type
— m buffet - [culin] buffet (type, o style)
— m caballete adv - [mec] A-frame
— m cajón adv - bin-type
— m cartucho adv - cartridge type
— m cáscara f de naranja adv - orange peel type
— m celosía adv - louver type
— m celular adv - [mec] bin type
— m centrífugo adv - [mec] centrifugal, o flyball, type
— m cilindro adv - [tub] barrel type
— m cerrado - closed type
— m cohesivo - cohesive type
— m colmena - beehive type
— m combinado - combined type
— m — para voltaje m variable o constante - [sold] combination variable voltage/constant voltage
— m comercial - commercial type • [fin] commercial, o business, rate
— m compuerta - [hidr] sluice (gate) type
— m común - common type • [public] lightface (type)

tipo m̲ común cortado con alambre m̲ - [cerám]
plain wire cut type
— m̲ con - design
— m̲ — acceso m̲ - [ind] access type
— m̲ — — m̲ interior - [grúas] walk-in, type,
o̲ design
— m̲ — — m̲ limitado - [vial] Interstate, o̲ limited access, variety, o̲ type
— m̲ — arco m̲ y tangente m̲ - arc-and-tangent
type
— m̲ — banda f̲ - [mec] band type
— m̲ — brida(s) f̲ - [mec] flange(d) type
— m̲ — carbono m̲ - [metal-prod] carbon type
— m̲ — — m̲ alto—[metal-prod] high carbon type
— m̲ — — m̲ bajo m̲—[metal-prod] low carbon type
— m̲ — — m̲ muy alto - [metal-prod] extra high
carbon type
— m̲ — — m̲ — bajo - [metal-prod] extra low
carbon type
— m̲ — correa f̲ - [mec] belt(ed) type
— m̲ —, cuadrante m̲, o̲ esfera f̲ - [instrum]
dial type
— m̲ — — m̲ calibrado - [instrum] calibrated
dial type
— m̲ — — m̲ no calibrado - [mec] uncalibrated
dial type
— m̲ — cuello m̲ - [mec] neck type
— m̲ — — m̲ para soldar v̲ - [tub] weld(ing)
neck type
— m̲ — culata f̲ - [comb.int] head type
— m̲ — — f̲ en T - [comb.int] T-head type
— m̲ — chaflán m̲ - [mec] bevel type
— m̲ — — m̲ helicoidal - [mec] helical, o̲ spiral, bevel type
— m̲ — destello(s) m̲ - flashing type
— m̲ — disco(s) m̲ - [mec] disk type
— m̲ — — m̲ basculante - [válv] tilting disk
type
— m̲ — dispositivo(s) m̲ dorsal(es) m̲ -
[electr-equip] dead front type
— m̲ — — m̲ frontal(es) - [electr-equip] dead
rear type
— m̲ — dos zapatas f̲ - [mec] two shoe type
— m̲ — elemento(s) m̲ descartable(s) - [ind]
throwaway element type
— m̲ — engranaje(s) m̲ - [mec] gear(ed), type, o̲
design
— m̲ — espiga f̲ - [sold] jack type
— m̲ — — m̲ y espiga f̲ - [sold] jack and plug
type
— m̲ — filo m̲ continuo - [mec] ridge type
— m̲ — gancho m̲ - [f.c.] hook type
— m̲ — — central - [f.c.] central hook type
— m̲ — grieta(s) f̲ - [mec] crack(ed) type
— n̲ — indicador(es) m̲ - [instrum] indicator
— m̲ — — m̲ con cuadrante m̲ - [instrum] dial
indicator type
— m̲ — — m̲ — esfera f̲ - [instrum] dial indicator type
— m̲ — inducido m̲ - [electr-mot] armature type
— m̲ — límite m̲ - [electr-equip] limit(ing)
type
— m̲ — — m̲ para corriente f̲ - [electr-equip]
current limit(ing) type
— m̲ — líquido m̲ - [segurid] liquid type
— m̲ — — m̲ vaporizante - [segurid] vaporizing
liquid type
— m̲ — motor m̲ - [ind] motor (equipped) type
— m̲ — — m̲ blindado - [ind] enclosed motor
type
— m̲ — muesca(s) f̲ - [mec] notched type
— m̲ — palanca f̲ - [mec] lever (equipped) type
— m̲ — paragolpe(s) m̲ - [f.c.] buffer type
— m̲ — — m̲ y gancho m̲ central - [f.c.] buffer-
-central hook type
— m̲ — parte f̲ superior lisa - flush type top
— m̲ — pasador m̲ - [mec] pin (equipped) type
— m̲ — pasta f̲ aguada - slurry type
— m̲ — . . . pieza(s) f̲ - . . . piece type
— m̲ — . . . piñón(es) m̲ - [mec] pinion design
— m̲ — — m̲ y engranaje(s) m̲ lateral(es) -
[mec] . . . pinion and side gear design
— m̲ — presión f̲ - [mec] pressure type

tipo m̲ con refuerzo m̲ - [mec] reinforced type
— m̲ — — m̲ angular - [mec] gusseted type
— m̲ — regulación f̲ - controlled type
— m̲ — — f̲ electrónica - [electrón] electronically controlled type
— m̲ — — f̲ graduada,dable - [mec] step control type
— m̲ — resalto(s) m̲ - [mec] notched type •
[metal-fabr] offset, design, o̲ type
— m̲ — resorte m̲ - [mec] spring type
— m̲ — rodillo(s) m̲ - [mec] roller type
— m̲ — tablero m̲ - [instrum] panel type
— m̲ — — m̲ para regulación f̲ - [instrum] control panel type
— m̲ — talón m̲ - [mec] lug type
— m̲ — — m̲ embutido - [mec] countersunk lug
type
— m̲ — tapón m̲ - [tub] plug type
— m̲ — — m̲ semicónico - [válv] semi-cone plug
type
— m̲ — torcido m̲ - [alambre] wound type
— m̲ — — m̲ triple m̲ - [alambre[triple wound
type
— m̲ — torsión f̲ - [mec] twist lock type
— m̲ — trinquete m̲ - [mec] ratchet type
— m̲ — trole m̲ - [mec] trolley type
— m̲ — varilla(s) f̲ - [mec] rod type
— m̲ — — f̲ y oreja(s) f̲ = [mec] rod and lug
type
— m̲ — viga f̲ - [constr] beam type
— m̲ — — f̲ fuerte - [constr] strong beam type
— m̲ — — f̲ — sobre poste m̲ débil - [constr]
strong-beam weak-post system
— m̲ — — f̲ para puente m̲ - [grúas] bridge
girder type
— m̲ — — f̲ — — m̲ para grúa f̲ - [grúas]
bridge girder type
— m̲ — viscosidad f̲ - viscosity type
— m̲ — — f̲ alta - high viscosity type
— m̲ — — f̲ baja - low viscosity type
— m̲ — voltaje m̲ - [electr] voltage type; type
with voltage
— m̲ — — m̲ variable - [electr] variable voltage type
— m̲ cónico - cone type
— m̲ constrictor - [mec] constricting,tor type
— m̲ constructivo - constructing,tion type
— m̲ contra salpicadura(s) f̲ - non-spatter type
— m̲ convencional - conventional type
— m̲ corredizo - [mec] draw-out, o̲ sliding,
type
— m̲ corriente - current, o̲ regular, type •
[public] lightface type • regular grade
— m̲ continuo - continuous type
— m̲ cortado - cut type
— m̲ — con alambre m̲ - [cerám] wire cut (type)
— m̲ corto - short type
— m̲ cualquiera - any type
— m̲ de acabado - finish, type, o̲ class
— m̲ — accidente m̲ - [segurid] accident type
— m̲ — accionamiento m̲ - [ind] drive type
— m̲ — aceite m̲ - [lubric] oil type
— m̲ — — m̲ para motor m̲ - [comb.int] engine
oil type • [electr-mot] motor oil type
— m̲ — acequia f̲ - [constr] ditch, o̲ channel,
type
— m̲ — acero m̲ - [metal-prod] steel type
— m̲ — — m̲ con carbono m̲ - [metal-prod] carbon
steel type
— m̲ — — — m̲ con cuello m̲ para soldar v̲
- [tub] forged carbon steel neck weld(ing)
type
— m̲ — — m̲ — — m̲ soldado - [tub] carbon
steel weld(ing) type
— m̲ — — — — m̲ a tope m̲ - [tub] carbon
steel but weld(ing) type
— m̲ — — m̲ producido - [metal-prod] produced
steel type; steel type produced
— m̲ — acoplamiento - [mec] coupling type
— m̲ — adorno m̲ - ornament type • [public] display type
— m̲ — aguja f̲ - needle type

tipo m de aislación f - [electr-cond] insulation type
— m — aislación f de papel m - [electr-cond] paper insulation type
— m — — f para devanado m - [electr-instal] winding insulation type
— m — — m — — m para estator m - [electr--instal] stator winding insulation type
— m — — f — — m — rotor m - [electr--mot] rotor winding insulation type
— m — alambre m - [sold] wire type
— m — alcantarilla f - [vial] culvert type
— m — alimentación f - [mec] feed(ing) type
— m — ambiente m - environment type
— m — anillo m - [me] ring type
— m — — m lubricante - [mec] dip ring type
— m — — m octagonal - [mec] octagonal ring type
— m — — m ovalado - [mec] oval ring type
— m — aplicación f - application type
— m — apoyo m - [mec] support type
— m — — m para zapata f - [electr-oper] shoe support type
— m — armadura f - [electr-equip] armor type
— m — arrecife m - [nav] reef type
— m — arrendamiento m - rental, o lease, type
— m — artefacto m - [electr-instal] appliance type • [ilumin] luminaire type
— m — — m para iluminación f - [electr-ilum] luminaire type
— m — asiento m - [constr] bedding class
— m — aspiración f - [mec] suction type
— m — atomizador m - atomizer type
— m — banda f - [mec] band type • belt type
— m — — f conect(ad)ora - [mec] connecting band type
— m — — f para acoplamiento m - coupling, o connecting, band type
— m — basurero m - [sanit] landfill type
— m — blindaje m - sheath(ing) type
— m — bloqueo m - [metal-prod] blocking type
— m — — m empleado - [metal-prod] blocking type used
— m — bola f - [cuerp.moled] ball, type, o style
— m — — f moledora - [cuerp.moled] grinding ball type
— m — bomba f - [bombas] pump type
— m — — f para aceite m - [bombas] oil pump type
— m — — f — — m lubricante - [bombas] lubricating oil pump type
— m — boquilla f - [mec] nozzle type
— m — — f, empleado, o usado - [mec] nozzle type used
— m — bóveda f - [arquit] arch type
— m — bulto m - [transp] bundle, o parcel, o package, type, o kind
— m — cable m - [cabl] cable type
— m — — m de alambre m - [cabl] wire rope, type, o consruction
— m — — m de alambre m - [cabl] wire rope type
— m — cadena f - [cadenas] chain type
— m — caja f - [metal-trat] box, o case, type
— m — cajón m - [transp] box type
— m — calzada f - [vial] roadway type
— m — calle f - [vial] street type
— m — cambiador m - [mec] changer type
— m — — m para toma(s) f - [electr-instal] tap changer, type, o design
— m — cambio m - [fin] exchange rate
— m — — m comercial - [fin] commercial exchange rate
— m — — m para dólar m - [fin] dollar exchange rate
— m — — m financiero - [fin] financial exchange rate
— m — — m libre - [fin] free exchange rate
— m — — m vigente - [fin] current, o prevailing, exchange rate
— m — campana f - [mec] bell type
— m — canal m - [hidr] channel type

tipo m de carcasa f - [electr-mot] frame type
— m — carga f = [ind] charge,ging type • [constr] load(ing) type • [hidr] load(ing) type • [transp] load type
— m — — f explosiva - [explos] pow(d)er load type
— m — — f utilizado m - charge type used
— m — carrera f - [deport] race type - racing form
— m — carretel m - [mec] spool, o reel, type
— m — carretera f - [vial] road, o thoroughfare, type
— m — cauce m - [hidr] channel type
— m — celosía f - [constr] lattice type
— m — cinta f - ribbon, o strip, type
— m — — f de papel m - [papel] paper ribbon type
— m — — f — — m metalizado - [electr-cond] metallized paper ribbon type
— f — circulación f - circulation type
— m — — f forzada - forced circulation type
— m — cloaca f - [sanit] sewer type
— m — cojinete m - [cojin] bearing type
— m — — m con empuje m - [mec] thrust bearing type
— m — — m — — m axial - [mec] axial thrust bearing type
— m — — m para piñón m - [mec] pinion bearing, type, o design
— m — — m Timken - [cojin] Timken bearing type
— m — colocación f directa - [constr] set-on type
— m — combustible m - [combust] fuel type
— m — combustión f - [combust] combustion type
— m — compañía f - [legal] type of company
— m — componente m - [ind] component type
— m — condición f - condition type
— m — conducto m - [electr-instal] conduit type
— m — conectador m - [electr-instal] connector type
— m — conexión f - (joint) connection type
— m — — f transversal - [ambient] transverse (joint) connection type
— m — conjunto m - [mec] assembly type
— m — — m de engranaje(s) m - [autom-mec] gearing type
— m — consideracion f - consideration type
— m — construcción f - [constr] construction type
— m — — f básica - basic construction type
— m — — f diferente - [autom-neumát] different basic construction type
— m — — f diferente - [autom-neumát] different consruction (type)
— m — — f para neumático(s) m - [autom-neumát] tire construction (type)
— m — contrato m - [legal] contract type
— m — cordón m - [cabl] strand pattern
— m — correa f - [mec] belt type
— m — corriente f - [electr-prod] current type
— m — corrugación f - [mec] corrugation type
— m — costura f - [mec] seam type
— m — — f estándar - [mec] standard seam type
— m — — f para tubería f - [tub] standard pipe seam type
— m — — f para tubería f - [tub] pipe seam, type, o construction
— m — crédito - [fin] credit, type, o form
— m — cuadrante m - [instrum] dial type
— m — cubo m - [mec-ruedas[hub type
— m — — m para rueda f - [mec-ruedas] wheel hub type
— m — cuenca f - [hidr] watershed type
— m — cuerno m - [mec] horn type
— m — cultivo m - [agric] vegetation type
— m — chatarra f - [metal-prod] scrap type
— m — — f, empleado, o usado - [metal-prod] scrap type used
— m — chumacera f - [mec] bearing type
— m — — f con empuje m - [cojin] thrust

bearing type
tipo m de chumacera f con empuje m axial =
[cojín] axial thrust bearing type
— m — defecto m - defect type
— m — — m en soldadura f - [sold] weld defect type
— m — diamante m - [miner] diamond type
— m — — m alterado - [miner] altered diamond
tipo
— m — — m mejorado - [miner] improved diamond
type
— m — distribuidor m - [comb.int] distributor
type
— m — ecuación f - [matem] equation type
— m — eje m - [mec] axle type
— m — electrodo m - [sold] electrode type •
rod type
— m — elemento m - element type
— m — embalaje - [transp] packing type
— m — embrague m - [mec] clutch type
— m — empaquetado - [ind] pack(ag)ing, type, o
class
— m — empaquetadura f - [mec] packing type
— m — empresa f - [legal] concern, o enterprise, type
— m — engranaje(s) m - [mec] gearing (type)
— m — equipo m - equipment type
— m — escoria f - [sold] slag type
— m — esfera f - [metal-prod] ball style
— m — esfuerzo m - [mec] stress type
— m — eslinga f - [cabl] sling type
— m — — f empleado - [cabl] sling type used
— m — — f usado - [cabl] sling type used
— m — estructura f - [constr] structure type
— m — excavación f - [constr] digging type
— m — explosivo m - [explos] explosive type
— m — extremo m - [tub] finish type; end finish • [constr] end finish type
— m — — m estándar - [mec] standard end type
— m — fabricación f - [ind] fabrication type
— m — falla f - [ind] failure, type, o mode
— m — freno m - [mec] brake type
— m — fuente f para energía f - [ind] power,
type, o system
— m — fuerza t motriz exigido - [electr-pród]
required power system
— m — — f — requerido - [electr-prod] required power system
— m — función f - function type
— m — fundente m - [sold] flux type
— m — fusible - [electr-instal] fuse type
— m — garganta f - [grúas] tread type
— m — gasolina - [petról] gasoline, type o
grade
— m — — f sin plomo m - [petról] unleaded
type gasoline
— m — grieta f - crack type
— m — grúa f - [grúas] crane, type, o series
— m — guía f - [metal-lam] guide type
— m — — f lateral - [metal-lam] side guide
type
— m — guillotina f - [válv] guillotine type
— m — hincadura f - [constr] driving type
— m — hoja f - [mec] blade type
— m — — f con V invertida - [herram] inverted
V blade type
— m — horno m - [ind] furnace type
— m — identificación f - identification type
— m — impacto m - [mec] impact type
— m — inclusión f - [metal] inclusion type
— m — información f - information type • coverage type
— m — inspección f - inspection type
— m — instalación f - [ind] installation type
• system type
— m — interés m - [fin] interest rate
— m — jaula f - [ind] cage type
— m — — f de ardilla f - [electr-mot]
squirrel cage type
— m — jornal m - [labor] pay rate
— m — junta f - [sold] joint type
— m — larguero m - [constr] stringer type

tipo m de lecho m - [constr] bed(ding) type
— m — letra f - [public] face (type)
— m — — f modificado - [public] modified
type face
— m — leva f - [mec] cam type
— m — liga f - [metal-prod] bond formulation
— m — lingote m - [metal-prod] ingot type
— m — — m colado - [metal-prod] ingot type
poured
— m — — m vaciado - [metal-prod] ingot type
poured
— m — lingotera f - [metal-prod] ingot mold
type; mold type
— m — listón m - [mec] flight type
— m — lubricación f - [mec] lubrication, o
oiling, type
— m — lubricante - [lubr] lubricant type
— m — mando m - [mec] drive type • control
type
— m — — f para bomba f - [bombas] pump control type
— m — — m — — f para aceite m - [bombas]
oil pump conrol type
— m — — m — — f — — m lubricante -
[bombas] lubricating oil pump control type
— m — — m — — f principal para aceite m
lubricante - [bombas] main lubricating oil
pump control type
— m — mantenimiento m - [metal-prod] holding
type • [ind] maintenance type
— m — manubrio m - [mec] crank type
— m — máquina f - [mec] machine type
— m — maquinaria f— [mec] machinery type
— m — — f minera - [miner] mining machinery
type
— m — máscara f - [segurid] mask type
— m — masilla f - [constr] putty type
— m — material m - material type
— m — — m empleado - [ind] material type used
— m — — m para revestimiento m - [constr] lining material type
— m — — m usado - [ind] material type used
— m — metal m - [metal] metal type
— m — — m de base f - [sold] base metal type
— m — mezcla f - blend(ing) type • [constr]
mortar type
— m — — f empleado—[constr] mortar type used
— m — — f usado - [constr] mortar type used
— m — mineral(es) m - [miner] ore type
— m — modelo m - model type
— m — molde m - [metal-prod] mold type
— m — mordaza f - [mec] clamp, type, o style
— m — mortero m - [constr] mortar type
— m — — m empleado—[constr] mortar type used
— m — motor m - [electr-mot] motor type •
[comb.int] engine type
— m — muesca f - [mec] notch type
— m — muro m - [constr] wall type
— m — neumático(s) m - [autom-neumát] tire
type
— m — — m para carretera f - [autom-neumát]
highway, o road, type tire
— m — núcleo m - [electr-mot] core type
— m — — m movible - [electr-mot] movable core
type
— m — obra f - [constr] work, o project, type
— m — obturador m stopper type
— m — — m renovable - [válv] renewable plug
type
— m — operación f - operation type
— m — organización f - [legal] organization
type
— m — oruga f - [constr-equip] crawler type
— m — paleta f - [turb] blade type
— m — paragolpe(s) m - [f.c.] buffer type
— m — pared f - [constr] wall type
— m — pavimento - [constr] pavement type
— m — permeabilidad f - permeability type
— m — — f de suelo m - [hidr] soil permeability type
— m — permeámetro m - [instrum] permeameter
type

tipo m de pestillo m - [mec] latch type
— m — m pasante - [constr] passing latch type
— m — pieza f - [mec] part(s) type
— m — f de acero m - [mec] steel part type
— m — m para extremo m - [constr] steel end finish type
— m — f para extremo m - [constr] end finish (type)
— m — pilote m - [constr-pil] pile type
— m — m disponible - [constr-pil] available pile type
— m — m tubular - [constr=pil] pipe pile type
— m — m — disponible - [constr-pil] available pipe pile type
— m — pistón - [válv] piston type
— m — placa f - [metal-lam] (ingot) mold stool type
— m — f para lingotera f - [metal-prod] z (ingot) mold stool type
— m — plancha f - [metal-lam] plate type
— m — f para revestimiento m - [constr] liner plate type
— m — plantilla f - [sold] gage type
— m — plataforma f - [constr] platform type | a - platform mounted,ting
— m — política - [admin] policy type
— m — poste m - post, o pole, type
— m — potencia f - power, type, o sistem
— m — f exigida - [electr-prod] required power system
— m — f requerida - [electr-prod] required power system
— m — presión f - pressure type
— m — producto m - product type
— m — m explosivo - [explos] explosive product type
— m — propaganda - [com] advertising form
— m — protección f - protection type
— m — proyecto m - [constr] project, o work, type
— m — pulsador m - [electr-instal] push button type
— m — quemador m - [combust] burner type
— m — ramal m - [cabl] strand, type, o pattern
— m — recepтion - reception type • [com] ac-ceptance type
— m — reclasificación f—reclassification type
— m — recubrimiento m - coating, type, o class
— m — m para placa f - [metal-trat] steel, lining, o cover(ing), type; plate covering type
— m — refractario(s) m - [cerám] refrac-tory,ries type
— m — refrigerante n - refrigerant, ọ coolant, type
— m — reglamentación f - regulation, o ruling, type
— m — regulación f - regulation type • opera-ting,tion variable
— m — relé m - [electr-equip] relay type
— m — m para sobrecorriente f - [electr--equip] overcurrent relay induction type
— m — m — f monofásica - [Electr-equip] single phase overcurrent relay induction type
— m — relleno m - [constr] fill type • [sold] fill type
— m — reóstato m—[electr-equip] rheostat type
— m — resorte m - [mec] spring type
— m — revestimiento m - [constr] surfacing type • [mec] liner,ning type
— m — m para cauce m - [hidr] channel sur-facing type
— m — m — lingotera f - [metal-prod] (in0 got) mold lining type
— m — m — placa f - [metal-prod] (ingot) mold steel lining type
— m — rociadura f - spray(ing) type
— m — rodamiento m - [cojin] bearing type
— m — Timken—[cojin| Timken bearing type
— m — rodillo m - [mec] roll(er) type
— m — rueda f - [mec] wheel type

tipo m de salida f—[electr-instal] outlet type
— m — f para alcantarilla f - [constr] cul-vert outlet type
— m — f regulada - regulated, o controlled, output, o outlet, type
— m — salpicadura f - [comb.int] splash type
— m — sello m - [mec] seal type
— m — servicio m - service type
— m — sistema m - system type
— m — m propuesto - proposed system type
— m — sociedad f - [legal] corporation type
— m — soldadura f - [sold] weld(ment).o joint, type
— m — solidificación f - [sold] freeze,zing type
— m — soporte m - [constr] support type
— m — sostén m - [constr] support type
— m — succión f [mec] suction type
— m — suelo m - [suelos] soil type • [hidr] terrain type
— m — superficie f - surface type
— m — f para rodadura f - [mec-ruedas] tread type
— m — suspensión f - [mec] suspension type
— m — tapón m - [tub] plug type • [metal-prod] stopper type
— m — m de grafito m - [metal-prod] graph-ite stopper type
— m — tarea f - [admin] task, o work, type
— m — tejido m - [sold] weave pattern
— m — terminación f - finish type • end finish type
— m — terminal m - [electr-cond] terminal type
— m — terreno m - [suelos] soil type • [hidr] terrain type
— m — tolerancia f - tolerance type
— m — trabajo m - [labor] work type
— m — m a destajo m - [labor] piece type work • piece rate
— m — m producido - [ind] work produced type
— m — transmisión f - [mec] transmission type
— m — traslapo m - [mec] lap(ping) type
— m — m substitutivo - [mec] alternate lap type
— m — trefilador m - [trefil] wire drawer type
— m — trinquete m - [mec] ratchet type
— m — trole m - [electr-equip] trolley type
— m — tubería f - [tub] pipe,ping, type, o design • [mec] line type
— m — f disponible - [constr-pil] available pipe,ping type
— m — f propuesto - [tub] proposed pipe type
— m — tubo m - [tub] tube, o pipe, type
— m — unidad f - unit type
— m — uso m - use type • exposure level
— m — vagón m - [f.c.] car type
— m — varilla f - [mec] rod type
— m — vegetación f - [botán] vegetation type • ground cover type
— m — vehículo m - [autom] vehicle type
— m — venta f - [com] sale type
— m — viga f - [constr] beam, o girder, type
— m — f profunda - [constr] deep beam type
— m — voltaje m - [electr] voltage type
— m — m constante - [electr-prod] constant voltage type
— m — zapata f - [mec] shoe type
— m — zapato m - [vest] shoe type • [mec] shoe type
— m — m para pilote m - [constr-pil] pile shoe type
— m decauville - [f.c.] dinkey type
— m deflector - [metal-prod] bull nose type
— m descartable - throwaway type | a - expend-able type
— m diferente - different, type, o kind
— m — de suelo m - [suelos] different, o vary-ing, soil type
— m discontinuo - discontinuous type • [ind] batch type
— m disponible - available type

tipo m distinto - different, type, o kind |
  pl - different, o various, types
— m diverso - different type
— m doble - double, type, o kind, o style •
  duplex type
— m — torcido - [alambre] double wound type
— m dúplice* - duplex type
— m Elliot - Elliot type
— m embridado - [mec] flange(d) type
— m embutido - [constr] mortise(d) type
— m empernado - [mec] bolted type
— m empleado - type used
— m en baño m de aceite m - [mec] oil bath type
— m — estilo helvético - [public] Helvetica
  face type
— m encajable - [tub] nestable, type, o design
— m encerado - wax coated, o waxed, type
— m enchufable - [mec] bayonet type • [electr-
  -instal] plug-in (draw-out) type
— m enganchador - hooking, o latching, type
— m engatillado - [mec] interlocking type
— m enrejado - [mec] lattice type
— m enroscable - [mec] screw-on type
— m entramado - [mec] truss type • truss shaped
— m epoxídico* - [pint] epoxy type
— m esclusa - [hidr] sluice type
— m escritorio - desk type
— m escudete - escutcheon type
— m especial - special type
— m específico - specific, o particular, type,
  o kind • specific series
— m — de suelo m - [suelos] specific, o particular, soil type
— m estabilizado mO [sold] stabilized type
— m estacionario - stationary type
— m estándar - standard type
— m estrella - star type
— m —-triángulo - [electr-equip] star-delta,
  type, o design
— m excéntrico - eccentric type | a - [mec]
  cam acting
— m exclusivo - exclusive type
— m explosivo - explosive type
— m favorable - favorable type • [fin] favorable rate
— m — de cambio m - [fin] favorable exchange
  rate
— m ferrítico - ferritic type • ferrite form
— m fijo - [mec] stationary type • [válv] non-
  -rising type • [fin] fixed rate
— m — para cambio m - [fin] fixed exchange
  rate
— m financiero - [fin] financial rate
— m flauta - flute type
— m fusible - [mec] fuse type
— m general - general type
— m giratorio—[mec] turning, o turntable, type
— m helvético - [public] Helvetica face type
— m hermético a aceite - oil-tight type
— m hongo - mushroom type
— m impulsor - [mec] driver,ving type
— m inclinable - [mec] tilting type
— m industrial - [ind] industrial type
— m inglete - miter type
— m inoxidable - [metal] stainless type
— m insertable - stab-type
— m interior - inside, o indoor, o inner, type
— m jaula - cage type
— m largo - long type
— m Libor - [fin] Libor rate
— m libre - free type
— m magnético - [electr] magnetic type
— m manubrio m - [mec] crank-type
— m marcado - marked type
— m más favorable - [fin] prime, o most favorable, rate
— m — reciente - latest, o most recent, type
— m material - material type
— m mecánico - mechanical type
— m medio - average type • [fin] average rate
— m — de interés m - [fin] average interest
  rate

tipo m medio embutido - [mec] half mortise(d)
  (template) type
— m mejorado - improved type
— m — de diamante m - [miner] improved diamond type
— m menos prominente - [public] light(er)
  typeface
— m monopila - [metal-trat] single stack, type,
  o design
— m montable - mountable type
— m motogenerador - [sold] motor generator
  type
— m motor generador - motor generator type
— m negrilla,rita - [public] bold face type
— m no abatible - non-breakaway type
— m no ascendente - [valv] non-rising type
— m — bruñido - unburnished type
— m — drenante - non-draining type
— m — con viscosidad f alta - [electr-
  -instal] high viscosity non-leaking type
— m — elevable - [válv] non-rising type
— m — productivo - nonproductive type
— m — tragador - [mec] non-locking, type, o
  style
— m normal - normal type • [fin] standard rate
— m nuevo - new, type, o style
— m obtenible - available type
— m obturador - [mec] plug(ging) type
— m — refrentable - [válv] regrindable plug
  type
— m original - original, o early, type
— m ornamental - [public] display type
— m Paintgrip m - [metal-trat] Paintgrip type
— m — con calidad f comercial - [metal-trat]
  commercial quality Paintgrip type
— m para ambas corrientes f - combination
  alternating (current)/direct current type
— m — aspiración f - suction type
— m — — f doble - double suction type
— m — — f simple - single suction type
— m — cambio m fijo - [fin] fixed exchange
  rate
— m — — m variable,ado - [fin] variable exchange rate
— m — carga(s) f pesada(s) - [constr] bearing
  type
— m — construcción f - [constr] construction
  type
— m — desconexión f - [electr-equip] disconnecting type
— m — — f rápida - [electr-equip] quick dis-
  connect(ing) type
— m — distribuidor m - [comb.int] distributor
  type
— m — encendido m - [comb.int] ignition type
— m — enchufe m - [electr-equip] socket type
— m — excavación f - [constr-equipo] excavating,tion, o digging, type
— m — flujo m - flow type
— m — — m mixto - mixed flow type
— m — impacto(s) m - [mec] impact type
— m — impulsión f - [mec] driver,ving type
— m — intemperie f - outdoor, o open air, type
— m — interconexión f - [vial] interconnecting
  type
— m — f — supercarretera(s) f - [vial]
  superhighway, o interstate highway, interchange type
— m — laminación f - [metal-lam] mill, o rolling, type
— m — plato(s) m - [herram] face type
— m — reajuste m - [ind] reset(ting) type
— m — — m manual - [ind] manual resetting
  type
— m — rendimiento m - [ind] performance type
— m — — m alto - [ind] (high) performance
  type
— m — señalización f - [comput] signalling
  type
— m — succión f - suction type
— m — — f doble - double suction type
— m — — f simple - single suction type

tipo m para seguimiento m - [sold] follow(ing)
type
— m — señalamiento,lización f - [comput] signalling type
— m — servicio(s) m - [ind] service type
— m — — m pesado - [ind] heavy duty type
— m — soldar v - [sold] weld(ing) type
— m — taller m - [ind] shop type
— m — tonelaje(s), alto(s), o elevado(s) -
[ind] high, o large, tonnage, type, o style
— m — — m reducido(s) - [ind] small, o low,
tonnage, type, o style
— m — tubería f—[tub] pipe,ping, o line, type
— m — voltaje - [electr-equip] voltage type
— m — — m constante - [electr-equip] constant
voltage type
— n — — m variable - [electr-equip] variable
voltage type
— m particular - specific, type, o kind
— m — de suelo m - [suelos] specific soil type
— m pechera - bib-style
— m permanente - permanent type
— m planetario - [mec] planetary type
— m popular - popular, type, o style
— m por gravedad f - gravity type
— m — inducción f - [electr] induction type
— m — salpicadura(s) f—[ind] splash(ing) type
— m prensaestopa(s) - [mec] stuffing box type
— m primero - first type
— m primitivo - primitive, o early, type
— m principal - principal, o main, type
— m productivo - productive type
— m promedio - [fin] average rate
— m — de interés m - [fin] average interest
rate
— m propuesto - proposed type • proposed rate
— m — de tubería f - [tub] proposed pipe type
— m pulsación-eco - [electrón] pulse-echo type
— m puntiforme - [public] point type
— m rebalse - [hidr] overflow type
— m reciente - recent, o late, type
— m recomendado - recommended type
— m rectificador - [electr-equip] rectifier
type
— m recto - straight type • [sanit] stall type
— m recubierto - [ind] coated type
— m — con aluminio m - [comput] aluminum
coated type
— m recuperable - recoverable, o reusable, type
— m reforzado - reinforced, o heavy duty, type
— m refrigerante - cooling type
— m regenerativo - regenerative type
— m regulado - controlled type
— m — eléctricamente - electrically controlled
type
— m remachado - [mec] riveted type
— m réplica - replica type
— m reversible - reversible,sing type
— m rígido - rigid type
— m — con brida(s) f - [mec] rigid flange(d)
type
— m rosa f - [constr] staggered pattern
— m saliente - protruding type
— m seco - dry type
— m segundo - second type
— m sencillo m - simple style • single type
— m señalador - [comput] signalling type
— m severo - severe type
— m sin abocinar v - flareless type
— m — cámara f - [autom-neumát] tubeless,
type, o construction
— m — plomo m - [petról] unleaded type
— m — ventilación f - unventilated type
— m soldado - [sold] welded type
— m — a tope m - [sold] butt welded type
— m substitutivo - alternate type
— m superior - superior type
— m — para estampado,par - [metal-prod] high
drawing, quality, o type
— m tambor - [mec] drum type
— m temprano - early type
— m térmico - thermal type
— m torcido - [electr-equip] wound type

tipo m torpedo - torpedo type
— m torrecilla f - [torno] turret type
— m totalmente embutido - totally, o full(y),
mortise(d) type
— m trabador - [mec] latching, o locking, type
— m — para tonelaje(s) m elevado(s) m - [mec]
large tonnage locking, type, o style
— m — — — m reducido(s) - [mec] small tonnage locking, type, o style
— m tractor - [ind] tractor type
— m transformador m - [electr-prod] transformer type
— m —rectificador - [sold] transformer-rectifier (type)
— m triple - triple type
— m tubular - tube,bular type
— m túnel - tunnel type
— m único - single, type, o style
— m unitario - [ind] unit type
— m usado - type used; used type
— m utilizado - type used; used type
— m vagón m - [f.c.] car type
— m vertical - vertical type
— m — con dos líneas f - [metal-col.cont] two-
strand vertical type
— m viejo - old, type, o style
— m vigente - [fin] current, o prevailing, rate
— m vinculado - [admin] related type
— m — de tarea(s) f administrativa(s) - [adm]
management work related, type, o kind
tira f - . . .; strip; band • [electr-cond]
strip • [sold] strip; strap • [metal-lam]
[mec] slit; véase también banda f; cinta f;
chapa f; fleje m • [mec] slip • list
— f circunferencial - [tub] banding strip
— f con borne(s) m - [electr-instal] terminal
strip
— f — — m accesible - [sold] handy, o accessible, terminal strip
— m — — de tipo m reparable - [electr-
-equip] repairable type terminal strip
— f — — m debajo m de arrancador m - [sold]
terminal panel below @ starter
— f — — m para elemento m para telemando m -
[sold] control pad terminal strip
— f — — m pequeña - [electr-instal] small
terminal strip
— f — — m reparable - [electr-instal] repairable terminal strip
— contra fricción f - [autom-neumát] rub strip
— f contra resplandor m de sol m en parte f
superior de parabrisa m - [autom] windshield
sun screen
— f cortada - [metal-lam] cut strip • cut
length
— f de acero m - [metal-lam] steel strap •
[electr-cond] steel strip
— f — — m para respaldo m - [sold] steel
back-up, bar, o strip
— f — asfalto m - [constr] asphalt, strip, o
ribbon
— f — — m fundido - [constr] steeped asphalt
ribbon
— f — barra(s) f - [petról] (rod) stand
— f — barrera(s) f - [electr-instal] rod(s)
strip
— f —barrera(sentrada f - [electr-instal]
input barrier rod strip
— f — — f — salida f - [electr-instal] output barrier rod strip
— f — caucho m - [mec] rupper strip
— f — cobre n - [mec] copper strip
— f — — f desnudo - [electr-instal] bare copper strip
— f — — m para respaldo m - [sold] copper
back-up strip
— f — hierro m - [mec] iron strip
— f — — m magnético - [metal-lam] magnetic
iron strip
— f — material m - [mec] naterial strip
— f — — m sellante - [tub] seam-sealant tape
— f — metal - [metal-lam] metal strip

tira f de metal m arrollada - [electr-cond] wound metal band
— f —— m — en espiral - [electr-cond] spiral(ly) wound metal band
— f —— m —— ininterrumpidamente - [electr-cond] wound continuous metal band
— f —— m —— en espiral - [electr-cond] spirally wound continuous metal band
— f — neopreno m—[electr-cond] neoprene strip
— f — nilón m - [textil] nylon strip
— f —— m con ángulo m cero - [autom-neumát] zero-degree nylon strip
— f — plomo m - [mec] lead tape • [electr-instal] lead strip
— f — tubo(s) m - [petról] (pipe) stand
— f delgada - [mec] thin strip
— f — de metal m - [mec] thin metal strip
— f desnuda - [metal-trat] bare strip
— f gruesa de metal m - [metal-lam] thick metal strip
— f ininterrumpida - [metal-lam] continuous, o uninterrupted, metal, strip, o band
— f — de metal m - [electr] continuous, o uninterrupted, metal band
— f interior - [mec] internal, strip, o strap
— f — de cobre m - [mec] internal copper, strap, o strip
— f lateral - [mec] side, strap, o strip • [puentes] end dam
— f — soldada - [puentes] welded end dam
— f metálica - [electr-cond] metal(lic) strip
— f —— a borne m - [electr-instal] metal strip to @ terminal
— f —— relé m - [electr-instal] metal strip to @ relay
— f —— m para sobrecarga f - [electr-instal] metal strip to @ overload relay
— f para borne(s) m - [electr-instal] terminal, strip, o panel
— f — cierre m - [constr] closure strip
— f — conexión f - [mec] connecting,tion, strip, o strap
— f — costura f - [mec-correas] lacing
— f —— f para banda f - [mec] belt lacing
— f —— f — correa f - [mec] belt lacing
— f — estancamiento - [constr] dam
— f — protección f - [mec] grommet strip
— f — respaldo m - [sold] back-up, strip, o bar
— f —— m de retazo(s) m - [sold] scrap back-up strip
— f —— m —— m de acero m - [sold] steel scrap back-up strip
— f — sujetar cable(s) m - [electr-instal] wire retainer,taining strap
— f pequeña - [mec] small strip
— f selladora - [mec] sealing, strip, o band
— f — de material m plástico - [plást] plastic (material) sealing, strip, o band
— f terminal - terminal strip
tiracable(s) m - [cabl] cable puller
tirada f - [constr] lift • [public] issue; circulation; edition; printing
— f — conductor m - [electr-instal] cable pulling
— f de tablazón(es) m - [constr] form lift
tirado,da a - thrown (away) • pulled; lugged; drawn; towed
— a excesivamente - pulled excessively
— a hacia abajo - pulled down
— a — adelante - pulled forward(s)
— a — afuera - pulled out
— a — arriba - pulle, o drawn, o lifted, up
— a — atrás - pulled back(wards)
— a — costado m - pulled, sideways, o toward @ side
— a — operador m - [ind] pulled towards @ operator
— a hasta apartar v - pulled, apart, o away from
— a — sacar - pulled, off, o away, from
— a — separar - pulled apart
— a — tensar - pulled, taut, o tight

tirado,da a rápidamente - pulled, rapidly, o quickly
tirador m - . . . . • [mec] puller • pulling knob • [trefil] drawer • [vest] suspenders
— m nilón - [mec] nylon puller • [vest] pl - nylon suspenders
— m alambre m - [mec] wire, drawer, o puller
— m cable(s) m - [cabl] cable puller
— m guía f - [mec] guide puller
— m guiadera f - [mec] guide puller
tirafondo m - [mec] . . .; lag screw
tiraje m - [combust] véase tiro m
tiramiento m - . . . . • pulling; drawing • véase también tiro m
— m de buje m - [mec] bushing pulling
— m — cable m - [mec] cable pulling
— m — conductor m - [electr-cond] cable pulling
— m — fiador m - [mec] retainer pull(ing)
— m —— m para cierre m - [mec] seal retainer pulling
— m —— m para cierre m - [mec] closing, o seal(ing), retainer pull(ing)
— m hacia operador m - [ind] pulling toward(s) @ operator
— m hasta apartar - pulling, apart, o away from
— m — separar v - pulling apart
tirante m - . . . . • [constr] stay; strut; brace' tie beam; truss; chord; tie rod; girder; truss bar; tension rod; buckstay • [metal-trat] mancer rod | a - taut
tirante m inferior - [constr] lower chord
— m — recto - [constr] straight lower chord
— m para columna f - [constr] column base tie
— m — encofrado m - [constr] form tie (rod)
— m — petaca f - [metal-prod] cooling plate tie rod
— m — retención f - [metal-prod] hold down, screw, o tie rod
— m — suspensión f - [metal-prod] hanger
— m inferior - [constr] lower chord
— m — curvo - [constr] curved lower chord • fish belly
— m superior - [constr] upper chord
— m — curvo - [constr] curved upper chord
— m — recto m - [constr] straight upper chord
tirantez f - . . . .; strain
tirar v - . . . .; to lug
— v alambre m - [mec] to, draw, o pull, @ wire
— v asidero m - [mec] to pull @ handle
— v buje m - [mec] to pull @ bushing
— v cable m - [mec] to pull @ cable • [cabl] to pull @ line
— v cadena f - [mec] to pull @ chain
— v — f hasta que esté tensar - [mec] to pull @ chain, tight, o taut
— v conjunto m - [mec] to pull @ assembly
— v —— m de árbol n - [autom-mec] to pull @ shaft assembly
— v —— m —— m para salida f (de fuerza f) - [auto-mec] to pull @ output shaft assembly
— m conductor m - [electr-cond] to pull @ cable
— v directamente - [mec] to pull, straight, o directly
— v — hacia adelante - to pull straight ahead
— v —— atrás - to pull straight back
— v estrangulador m - [comb.int] to pull @ choke
— v excesivamente - to pull excessively
— v hacia abajo - to pull down(wards)
— v — adelante - to pull forward(s)
— v — afuera - to pull out • to pull to @ out position
— v — arriba - to, pull, o draw, o lift, up, o upwards
— v — atrás - to pull back(wards)
— v —— columna f - [mec] to pull @ column back(wards)
— v —— f para dirección f - [autom] to pull back @ steering column
— v — costado m - [mec] to pull sideways
— v — operador m - to pull towards @ operator
— v hasta apartar v - to pull apart

tirar v hasta sacar - to pull, off, o out (of)
— v — separar - to pull apart
— v — tensar v - [mec] to pull, tight, o taut
— v línea f - [cabl] to pull @ line
— v objeto m - [mec] to pull @ object • to throw (out) @ object
— v palanca f—[mec] to pull @, lever, o handle
— v — f hacia adelante - [mec] to pull @, lever, o handle, forward(s)
— v — f — atrás - [mec] to pull @, lever, o handle, backward(s)
— v rápidamente - to pull, quickly, o rapidly
— v regulador m - [mec] to pull @, regulator, o control, o throttle
— v — m hacia afuera - [mec] to pull out @, regulator, o control, o throttle
— v retén m - [mec] to pull @ retainer
— v — n para cierre n - [mec] to pull @, seal, o closing, retainer
— v — m — sello m - [mec] to pull @ seal retainer
tírese v imper - pull
— v para arrancar v imper - pull to start
— v — marchar v imper - pull to run
— v imper — parar - pull to stop
tirilla f - . . . . • lacing
tiristor* m - [electrón] tiristor
tiristorizado,da* a - [electrón] tiristorized
tiro m - . . .; pull(ing) • [combust] draft • [miner] air draft • [mec] . . .l draw • [transp] tow(ing) • [metal-fabr] length (of pipe, rail, etc)
— m ascendente - [combust] up draft
— m cuadruple - [petról] fourble
— m de alambre - [mec] wire pulling • [trefil] wire drawing
— m — asidero m - [mec] handle pulling
— m — buje m - [mec] bushing pull(ing)
— m — conjunto m - [mec] assembly pulling
— m — — m de árbol m - [mec] shaft assembly pulling
— m — — m — m para salida f (de fuerza f) - [autom-mec] output shaft assembly pulling
— n — émbolo m - [mec] plunger pull(ing)
— m — fiador m - [mec] retainer pulling
— m — — m para cierre m - [mec] seal retainer pull(ing)
— m — hilera f - [trefil] die pull(ing)
— m — horno m - [metal-prod] furnace draft
— m — matriz f - [trefil] die pull(ing)
— m — palanca f - [mec] lever pull(ing)
— m — retén m - [mec] retainer pull(ing)
— m — — m para cierre m - [mec] seal retainer pull(ing)
— m — ventilador m - [ventil] fan draft
— m descendente - [combust] down draft • [hidr] drawdown
— m en chimenea f - [combust] stack draft
— m — línea f - [cabl] line pull(ing)
— m forzado - [combust] forced draft
— m hacia abajo - down pull(ing) • [combust] down draft
— m — adelante - forward pull(ing) • [combust] forward draft
— m — afuera - out pull(ing) • [combust] outward draft
— m — arriba - up(ward) pull(ing) • [combust] up(ward) draft
— m — atrás - back pull(ing)• [combust] back--(ward) draft
— m — — de columna f - [mec] column backwards pull(ing)
— m — — f para dirección f - [mec] steering column back(wards) pulling
— m — costado m - sideways,wise pull(ing)
— m hasta sacar - pull(ing), off, o out, (of)
— m horizontal - [combust] horizontal draft
— m inducido - [combust] induced, o forced, draft
— m lateral - [mec[side(ways,wise) pull(ing)
— m liviano - [combust] light draft
— m mayor - [mec] greater pull(ing) • [combust] greater draft

tiro m natural - [combust] natural draft
— m normal - [combust] normal draft
— m para achique m - [mec] bail pull(ing)
— m pesado - [trefil] heavy draft
— m promedio - average pull(ing)
— m rápido - rapid, o fast, pulling • [milit] quick shooting
— m sobre varilla f - [trefil] rod draft
tirón m - . . . .; jerk(ing); lug(ging)
— m firme - firm pull(ing)
— m parejo - steady pull(ing)
— m primero - [mec] first pull(ing)
— m segundo - second pull(ing)
tironeado,da a - jerked; lugged
tironamiento* m - jerk(ing), lug(ging)
tironear v - to, jerk, o lug
titánio-calcio m - [miner] titanium-calcium
titración f - [metal-lam] titration
titular m - . . .; owner • [pol] incumbent • [public] headline; display heading • [legal] holder • owner • [pol] manager • [fin] drawee
— m de cartera f - [pol] department secretary
— m — derecho m de propiedad f intelectual - [legal] copyright holder
— m — dirección f - [pol] Authority manager
— m — — f de desagües m - [pol] Sewer Authority Manager
— m — patente f - [legal] patent holder
— m — propiedad f intelectual - [legal] copyright holder
— m ornamental - [public] display headline
— m para adorno m - [pib;oc] display heading
título m - . . . . • designation • [educ] degree • [fin] certificate; bond • security • stock(s)
— m académico - [educ] academic degree
— m a costo m amortizado - [fin] bond at amortized cost; amortized cost bond
— m amortizado - [fin] amortized bond • depreciataed bond
— m autoamortizante - [fin] self, liquidating, o amortizing, o liquidating, o supporting, bond • revenue bond
— m — por anticipación f de ingreso(s) m - [fin] self-liquidating revenue bond
— m colectivo - [legal] collective certificate
— m con escena f pintoresca - [fotogr] scene title
— m — plazo corto - [fin] short term bond
— m — — m largo - [fin] long term bond
— m de clase f primera - [fin] first class bond
— m — corredor m - [deport] driver title
— m — — m mejor absoluto - [deport] overall driver title
— m — documento m - [public] document title
— m — estado - [fin] government bond
— m — gobierno m - [fin] government bond
— m — principio m - principle title
— m — propiedad f - [legal] property title; deed
— m — renta f pública - [fin revenue bond
— m escénico - [public] scene,nic title
— m excepcional - [public] exceptional title
— m extraviado - [legal] lost, o misplaced, certificate, o title
— m ilustrativo - illustrative purpose; example
— m individual - [legal] individual certificate
— m limitativo - limiting purpose(s)
— m meramente ilustrativo - illustrative purpose(s) (only
— m nacional m - [fin] national, o government, security, o title - [educ] national, degree, o title
— m negociable - [fin] negotiable certificate
— m no negociable - [fin] nonnegotiable certificate
— n nuevo - new title • [educ] new degree
— m orientativo - orientating purpose
— m ornamental - [public] display. head(ing), o headline
— m para adorno m - [public] display, headline, o heading
— m perdido m - [legal] lost certificate

título m perdido - [legal] lost certificate
— m para categoría f - [deport] class title
— m por anticipación f de ingresos m - [fin]
revenue bond
— m — f — m para servicio(s) m público(s)
- [fin] utility revenue bond
— m privado - [fin] private bond
— m — de clase f primera - [fin] first class
private bond
— m profesional - [educ] professional degree
— m público - [fin] (government) bond
— m universitario - [educ] university degree •
college degree • bachelor's degree
tizón m - . . . . • [constr] header
toalla f de papel - [papel] paper towell
toba f - [geol] . . .; tuff
— f Breccia - [geol] Breccia, toba, o tuff
— f Brecha - [geol] véase toba f, o tuff f,
Breccia
tobera f - nozzle • [metal-prod] tuyere • bustle
• [tub] (pipe) nipple
— f atascada - [metal-prod] plugged (up) tuyere
— f, busa f y codo m - [metal-prod] tuyere,
blowpipe, and gooseneck
— f colectora f—[metal-prod] collecting tuyere
— f con morro m - [metal-prod] deflector type
tuyere • nose tuyere
— f — m inclinado - [metal-prod] deflector
type tuyere
— f — pico m flauta - [metal-prod] flute type
tuyere
— f — perfil m especial - [metal-prod] special
profile tuyere
— f de tipo m deflector - [metal-prod] deflector type tuyere
— f — m flauta - [metal-prod] flute type
tuyere
— m — m normal - [metal-prod] normal (type)
tuyere
— f — m recto.- [metal-prod] straight
(type) tuyere
— f deflectora - [metal-prod] deflector (type),
o angle nose, o bull nose, tuyere
— f en ángulo m - [metal-prod] angled tuyere
— f inclinada - [metal-prod] angle nose, o
slanting, tuyere
— f intermedia - [metal-prod] intermediate tuyere • intermediate (slag notch) cooler
— f no quenada - [metal-prod] non burned-out
tuyere
— f normal - [metal-prod] normal (type) tuyere
— f para aire m - [ind] air nozzle
— f — escorial m - [metal-prod] slag notch intermediate cooler
— f — horno m - [ind] furnace tuyere
— f — m para coque m - [coque] coke oven
tuyere
— f — inyección f - [metal-prod] injection
tuyere
— f — f de aire m - [metal-prod] air injection tuyere
— f — f — gas m - [metal-prod] gas injection tuyere
— f — trefilería f - [trefil] (wire) drawing
die
— f — venturi m - [metal-prod] Venturi tuyere
— f principal - [comb.int] main tuyere
— f quemada - [metal-prod] burned (out) tuyere
— f recta - [metal-prod] straight tuyere
— f reducida - [metal-prod] curtailed, o reduced, o cut back, tuyere
— f soldada - [metal-prod] inner fused tuyere
— f — con hierro m - [metal-prod] iron plugged
tuyere
— f — interiormente - [metal-prod] inner fused
plugged tuyere
— f tabicada - [metal-prod] bricked up tuyere
— f tapada - [metal-prod] plugged tuyere
— f, toberón m y toberín m (para escorial m) -
[metal-prod] (slag notch) nozzle, intermediate cooler, and cooler
— f — m y busa f - [metal-prod] tuyere, cooler
and blowpipe

tobera f Venturi - [comb.int] (carburetor) Venturi
— f y toberín m para escorial - [metal-prod]
slag notch nozzle and cooler
toberín m - [metal-prod] (slag noth) nozzle •
tuyere stock • intermediate cooler
— m para escorial m - [metal-prod] slag notch
nozzle
— m — m quemado - [metal-prod] burned (out)
slag notch nozzle
— m no quemado - [metal-prod] non burned out
slag notch (nozzle
— m quemado - [metal-prod] burned (out) slag
notch
toberón m - [metal-prod] (tuyere) cooler
— m no quemado m - [metal-prod] non burned out
cooler
— m para escorial m - [metal-prod] slag notch
cooler
— m quemado [metal-prod] burned (out) cooler
— m, tobera f y busa f - [metal-prod] cooling
plate, tuyere, and blowpipe
— m, — f, y toberín m - [metal-prod] slag
notch cooler and nozzle
— m, — f, y toberón m para escorial m -
[metal-prod] slag notch cooler, holder and
nozzle
tobín m - [metal-prod] notch
— m para escoria f - [metal-prod] slag notch
tobogán m - [deport] . . .; slide
tocadisco(s) m - [electrón] . . .; turntable;
pick-up
— m doble - [electrón] dual turntable
tocado,da a - . . . . • probed
— a simultáneamente - touched simultanously
tocador m - [domést] . . .; dresser | a -
toilet(te); toiletry,ries
— m para cinta f magnética - [domést] tape deck
— a a penas - touched slightly
tocando a ligeramente - touching lightly
— a sólo ligeramente - touching only lightly
tocar v - . . .; to contact • to butt together •
to probe • to be up to - [mús] to play
— v bocina f - [segurid] to, sound, o play, o
operate, o blow, @ horn
— v carbono m - [sold] to short @ carbon(s)
— v con alambre m - [electr-oper] to probe
with @ wire
— m — conductor m - [electr-oper] probe with @
wire
— v — electrodo m - [sold] to touch with @
electrode
— v — mordaza f - [electr-oper] to touch with
@ clamp
— v cuerda f sensible - [fig] to ring @ bell
— v metal m - [sold] to touch @ metal
— f música f - [mús] to play @ music
— v simultáneamente - to touch simultaneously •
[mús] to play simultanously
— v suelo m - to touch @ ground
— v timbre m - to sound @ ringer
— v trabajo m - [sold] to touch @ work • [mec]
to touch @ job
— v — con electrodo m - [sold] to touch @
electrode to @, job, o work
tocarse v - to butt together
— v ligeramente - [sold] to touch slightly
tocón m metálico - [mec] metal, stub, o stump
— m clasificado - [metal-lam] graded bloom
— m con calidad f para laminación f - [metal-
-lama] (re)rolling quality bloom
— m degradado - [metal-prod] downgraded bloom
— m elaborado - [metal-lam] rolled bloom
— m identificado - [metal-lam] identified bloom
— m laminado - [metal-lam] rolled bloom
— m nuevo - [metal-lam] new, billet, o bloom
— m plano- [metal-lam] flat bloom
— m primero - [metal-lam] first bloom
— m redondo - [metal-lam] round bloom
— m sin analizar v - [metal-lam] unanalyzed
bloom
— m último - [metal-lam] last bloom
tochos m y desbastes m planos - [metal-lam]

blooms and slabs
tochos m y planchones m - [metal-lam] blooms and
slabs
toda f altura f - full height
— f — f **útil** — full, useful height, o clearance
— f **escala** f - entire range
— f **dirección(es)** f - all direction(s)
— f **estación(es)** - [meteorol] all season(s)
— f **hora** - around @ clock
— f **ley(es)** - [legal] all @ law(s)
— f **marcha** f - full speed; full throttle
— f **noche** f - all @ night
— f **ocasión** f - every occasion
— f **opotunidad** f - every opportunity
— f **posición** f - [sold] all positions; all set-
tings
— f **precaución** f - every precaution; every
effort
— f — f **posible** - every possible, precaution,
o effort
— f **zona** f - every zone; entire area
todas f partes - everywhere • throughout
— f **pasadas** f - [sold] all @ passes
— f **regla(s)** f - all @ rule(s)
— f **reglamentación(es)** - all @ ruling(s)
todo m color m - full color • all @ color
— m **concepto** m - every concept • every kind
— m **contorno** m - all around
— m **contrario** m - everything but
— m **diámetro(s)** - all diameter(s)
— m **equipo** m - [ind] all @ equipment
— m **esfuerzo** m - all, o every, effort
— m — m **posible** - every possible effort
— m **material** - [ind] all @ material
— m **menos** - everything but
— m **menos que fácil** - anything but easy
— m **metal m aportado** - [sold] all @ weld metal
— m **mundo** m - all @ world; whole
world; everyone; everybody • worldwide
— m **país** f - whole country
— a **pavimentado** - all, o fully, paved
— m **personal** m - [labor] all @ personne;;
everybody; everyone
— m **que puede fallar fallará** - [fig] Murphy's
law
— m **riesgo** - all, o every, risk(s)
— m **sentido** m - every sense; all around
— m **tiempo** m - all @ time; whole time
— m **uso** - every use • all purpose • general
purpose(s)
— m **volumen** m - [public] (thruoughout) @ full
volume • the whole volume
— m — m **de material n** - throughout @ material
volume
todos,das - everyone; everybody
— m **modelo(s)** m - all models
— m **pasos** - all @ steps
— m — **para soldadura** f - [sold] complete welding sequence; all weld(ing) steps
— m **restante(s)** - everyone else
toldo m - . . . • shelter • [autom] top
tolerado,da a - tolerated • allowed,wable
tolerancia f - . . .; variance; plus or minus •
allowable; limit • relaxation; clearance
— f **aceptable** - acceptable, o correct, tolerance, o allowance
— f — **en juego m** - [mec] play allowance
— f — — m **longitudinal** - [mec] acceptable,
longitudinal, o end play, tolerance
— f **aceptada** - accepted tolerance
— f — **corrientemente** - [mec] currently, o
presently, accepted tolerance
— f — **para diámetro** - [mec] accepted diameter
tolerance
— f **admisible** - admissible tolerance • accepted
variance
— f **admitida** - accepted tolerance
— f **ajustada** - adjusted tolerance • close tolerance
— f **amplia** - [ind] ample, o wide, tolerance
— f **apropiada** - proper, o correct, tolerance
— f **calculada** - calculated tolerance
— f **considerable** - [constr] large clearance

tolerancia f aceptable para túnel - [constr] acceptable tunnel clearance
— f **correcta** - [mec] correct tolerance
— f **corriente** - ordinary, o current, toler=
ance, o allowance
— f — **aceptada** - currently accepted tolerance
— f — — **para diámetro m** - [metal-fabr] currently accepted diameter tolerance
— f — **en especificación(es)** f - ordinary
specification(s) tolerance(s)
— f **determinada** - [ind] specified tolerance
— f — **para construcción** f - specified construction tolerance
— f — — **servicio m** - specified service tolerance
— f **diametral** - diameter tolerance
— f **dimensional** - dimensional, o size, tolerance
— F — **en altura** f - [mec] dimensional height
tolerance
— m — — **ancho,chura** f - [mec] dimensional
width tolerance
— f — **en espesor m** - [metal-lam] dimensional
gage, o gage dimensional, tolerance
— f — — **largo(r) m** - dimensional length
tolerance
— f **en alargamiento m** - elongation tolerance
— f — **alma m** - [metal-lam] web tolerance
— f — — **de perfil m** - [metal-prod] shape
web tolerance
— f — — m — — m **soldado** - [metal-lam] welded shape web tolerance
— f — **análisis m** - analysis tolerance
— f — — m **para comprobación** f - check(ing)
analysis tolerance
— f — **ancho,chura** - [metal-lam] width tolerance
— f — — m **de banda** f - [metal-lam] strip
width tolerance
— f — — — **bobina** f - [metal-lam] coil width
tolerance
— f — — m — f **con borde m de laminación**
f - [metal-lam] rolled ege coil width tolerance
— f — — m — f **de chapa** f - [metal-lam]
strip mill width tolerance
— f — — m — f — f **sin escuadrar v** -
[metal-lam] unsquared coil strip width tolerance
— f — — m — f — f — — v **con borde m**
cortado - [metal-lam] trimmed edge unsquared
strip coil width tolerance
— f — — m — — **de borde m de laminación f**
- [meta-lam] rolled edge sheet width tolerance
— f — — m — **cinta** f - [metal-lam] strip
width tolerance
— f — — m — **chapa** f - [metal-lam] strip
width tolerance
— f — — m — f **en bobina** f - [metal-lam]
coil strip width tolerance
— f — — m — f — **hojas** f - [metal-lam]
strip sheet width tolerance
— f — — m — f — f **escuadrada(s)** -
[metal-lam] squared strip sheet width tolerance
— f — — m — **hoja f de chapa f sin escuadrar**
v - [metal-lam] unsquared sheet strip width
tolerance
— f — — m — f — f — — v **con borde m**
cortado - [metal-lam] trimmed edge unsquared
strip sheet width tolerance
— f — — m — **fleje m** - [metal-lam] strip
width tolerance
— f — — m — f **para bobinas** f - [metal-
lam] coil strip width tolerance
— f — — m — — f **con borde m de la-**
minación f - [metal-lam] rolled edge coil
strip width tolerance
— f — — m — m — **chapas** f - [metal-lam]
sheet strip width tolerance
— f — — m — m — f **con borde m de la-**
minación f - [metal-lam] rolled edge sheet

strip width tolerance

tolerancia f en ancho m de hoja f de banda f - [metal-lam] strip sheet width tolerance

— f — m — f **cinta f** - [metal-lam] strip sheet width tolerance

— f — m — f **chapa f** - [metal-lam] strip sheet width tolerance

— f — m — **patín m** - [metal-lam] shape base width tolerance

— f — m — f **fleje m** - [metal-lam] strip sheet base width tolerance

— f — m — **patín m** - [metal-lam] base width tolerance

— f — m — m **para perfil m** - [metal-lam] shape base width tolerance

— f — m — m — m **soldado** - [metal-lam] welded shape base width tolerance

— f — **plancha f** - [metal-lam] plate width tolerance

— f — **anchura f** - width tolerance

— f — f **nominal** - nominal width tolerance

— f — **banda f** - [metal-lam] strip tolerance

— f — **bobina f** - [metal-lam] coil tolerance

— f — f **de chapa f** - [metal-lam] strip coil tolerance

— f — f — f **sin escuadrar v** - [metal-lam] unsquared strip coil tolerance

— f — **borde m** - [metal-lam] edge tolerance

— f — **fleje m** - [metal-lam] coil strip width tolerance— f — **borde m de chapa f** - [metal-lam] strip edge tolerance

— f — m **en bobina f** - [metal-lam] coil strip edge tolerance

— f — **cinta f** - [metal-lam] strip tolerance

— f — **combadura f** - [mec] camber tolerance

— f — f **de alma m** - [metal-lam] web camber tolerance

— f — f — m **de perfil m** - [metal-lam] shape web camber tolerance

— f — f — **patín m** - [metal-lam] base camber tolerance

— f — f — m **para perfil m** - [metal-lam] shape base camber tolerance

— f — **concavidad f** - [metal-lam] concavity, tolerance, o allowance

— f — **convexidad f** - convexity, tolerance, o allowance

— f — **curvatura f** - curvature, tolerance, o allowance

— f — f **transversal** - [metal-lam] transverse curvature, tolerance, o allowance

— f — **chapa f** - [metal-lam] strip, o sheet, tolerance

— f — f **en hojas f** - [metal-lam] strip sheet tolerance

— f — f — f **escuadrada** - [metal-lam] squared strip sheet tolerance

— f — **defecto m** - low side

— f — **descentrado m** - off center tolerance

— f — **descentramiento m de alma m** - [metal-lam] web, off-center, o out of center, tolerance

— f — m — m **perfil m** - [metal-lam] [metal-lam] shape web off-center tolerance

— f — **diámetro m** - diameter tolerance

— f — m **exterior** - [mec] outside diameter tolerance

— f — m **de taza f** - [cojin] cup outside diameter tolerance

— f — m **interior** - [mec] inside diameter tolerance

— f — m **de cono m** - [cojin] cone bore tolerance

— f — **dimensión f** - dimension tolerance

— f — m **de ancho m** - width dimension tolerance

— f — f — m **de patín f** - [metal-lam] base width dimension tolerance

— f — f — m — m **para perfil m** - [metal-lam] shape base width dimension tolerance

— f — f — m — m — m **soldado** - [metal-lam] welded shape base width dimension tolerance

tolerancia f en dimensión f de ancho m de patín m - [metal-lam] base width dimension tolerance

— f — f — m **para perfil m** - [metal-lam] shape base width dimension rolerance

— f — f — **peralte m** - [metal-lam] rise dimension tolerance

— f — f — m **de perfil m** - [metal-lam] shape rise dimension tolerance

— f — f — m — m **soldado** - [metal-lam] welded shape rise dimension tolerance

— f — f — **perfil m** - [metal-prod] shape dimension tolerance

— f — f — **sección f** - [metal-lam] section dimension tolerance

— f — f — f **transversal** - [metal-lam] cross section dimension tolerance

— f — f — m **de perfil m** - [metal-lam] shape cross section dimension tolerance

— f — f — f — m **soldado** - [metal-lam] welded shape cross section dimension tolerance

— f — **especificación f** - specification tolerance

— f — **espesor m** - [mec] thickness, o gage, tolerance

— f — m **de bobina f** - [metal-lam] coil gage tolerance

— f — m — f **de chapa f** - [metal-lam] strip coil gage tolerance

— f — m — f — f **sin escuadrar** - [metal-lam] unsquared coil, strip, o sheet, coil gage tolerance

— f — m — f — f **con borde m cortado** - trimmed edge unsquared strip coil gage tolerance

— f — m **chapa f de bobina f** - [metal-lam] coil strip gage tolerance

— f — m **chapa f** - [metal-lam] strip gage tolerance

— f — m — f **en hoja(s) f** - [metal-lam] strip sheet gage tolerance

— f — m — f **escuadrada** - [metal-lam] squared strip sheet gage tolerance

— f — m — **fleje m** - [metal-lam] strip gage tolerance

— f — m — **hoja f** - [metal-lam] sheet gage tolerance

— f — m — f **de chapa f** - [metal-lam] strip sheet gage tolerance

— f — m — f — f **sin escuadrar v** - [metal-lam] unsquared strip sheet gage tolerance

— f — m — f — f — v **con borde m cortado** - [metal-lam] trimmed edge unsquared strip sheet gage tolerance

— f — m — **plancha f** - [metal-lam] plate thickness tolerance

— f — m — f **de perfil m** - [metal-lam] shape plate thickness tolerance

— f — m — f — m **soldado** - [metal-lam] welded shape plate thickness tolerance

— f — m **nominal** - [mec] nominal thickness tolerance

— f — **exceso m** - [mec] high side (tolerance)

— f — **falta f de redondez f** - [tub] out-of-round tolerance

— f — **flecha f** - [metal-lam] camber tolerance

— f — **fleje m** - [metal-lam] strip tolerance

— f — **fuera de escuadra f** - [metal-lam] out-of-square tolerance

— f — f **de patín m** - [metal-lam] base out-of-square tolerance

— f — **hoja f** - [metal-lam] sheet tolerance

— f — f **de banda f** - [metal-lam] strip sheet tolerance

— f — f **cinta f** - [metal-lam] strip sheet tolerance

— f — f **de chapa f** - [metal-lam] strip sheet tolerane

— f — f **fleje m** - [metal-lam] strip sheet tolerance

— f — **juego m longitudinal** - [mec] end play

tolerance

tolerancia f **en largo(r)** m - length tolerance

— f —— m **de banda** f - [metal-lam] strip length tolerance

— f —— m —— **bobina** f - [metal-lam] coil length tolerance

— f —— m —— f **de chapa** f - [metal-prod] strip coil length tolerance

— f —— m —— f —— f **sin escuadrar** - [metal-lam] unsquaared coil strip length tolerance

— f —— m —— f —— f —— f **con borde** m **cortado** - [metal-lam] trimmed edge unsquared coil strip length tolerance

— f —— m —— **cinta** f - [metal-lam] strip length tolerance

— f —— m —— **chapa** f - [metal-lam] strip length tolerance

— f —— m —— f **en bobina** f - [metal-lam] coil strip length tolerance

— f —— m —— **fleje** m - [metal-lam] strip length tolerance

— f —— m —— **hoja** f - [metal-lam] sheet length tolerance

— f —— m —— **de chapa** f - [metal-lam] strip sheet length tolerance

— t —— m —— f —— f **sin escuadrar** v - [metal-lam] unsquared strip sheet length tolerance

— f —— m —— f —— f —— v **con borde** m **cortado** - [metal-lam] trimmed edge unsquared strip sheet length tolerance

— f —— m —— **perfil** m - [metal-lam] shape length tolerance

— f —— m —— m **soldado** - [metal-lam] welded shape length tolerance

— **f** — **largura** f - [metal-lam] véase **tolerancia en largo(r)** m

— f — **longitud** f - véase **tolerancia** f **en largo(r)**

— f —— f **de cubo** m - [f.c.-ruedas] hub length tolerance

— f —— f **de perfil** m - [metal-lam] shape length tolerance

— f —— f —— m **soldado** - [metal-lam] welded shape length tolerance

— f — **más** - [mec] plus tolerance; amount over

— f —— **en ancho** m - [metal-lam] plus width tolerance

— f —— **flecha** f - [metal-lam] plus camber tolerance

— f —— **largo(r)** m - [metal-lam] plus length tolerance

— f —— **peso** m - plus weight tolerance

— f —— **o (en menos)** - [mec] plus or minus (tolerance)

— f — **material(es)** m - [ind] material(s) tolerance

— f — **medida(s)** f - meassurement(s), o size, tolerance(s)

— f —— f **para palanquilla** f - [metal-lam] billet measurement(s) tolerance(s)

— f —— f —— f **por colada** f **continua** - [col.cont] continuous cast billet(s) measurement(s) tolerance

— f — **menos** - minus tolerance; amount under

— f —— **en ancho(r)** m - [metal-lam] minus width tolerance

— f —— **flecha** f - [metal-lam] minus camber tolerance

— f —— **largo(r)** m - [metal-lam] minus length tolerance

— f — **ovalamiento** m - [metal-lam] out-of-round tolerance

— f — **ovalidad** f—[tub] out-of-round tolerance

— f — **paralelismo** m - parallelism tolerance

— f —— m **de ala(s)** f - [metal-lam] flange(s) parallelism tolerance

— f —— m — **perfil(es)** m - [metal-lam] shape parallelism tolerance

— f —— m **de perfil** m **soldado** - [metal-lam] welded shape parallelism tolerance

tolerancia f **en patín** m - [metal-lam] base tolerance

— f —— m **de perfil** m - [metal-lam] shape base tolerance

— f —— m —— m **soldado** - [metal-lam] welded shape base tolerance

— f — **peralte** m - vertical elongation, o rise, tolerance

— f —— m **para perfil** m - [metal-lam] shape rise tolerance

— f —— m —— m **soldado** - [metal-lam] welded shape rise tolerance

— f — **perfil** m - [metal-lam] shape tolerance

— f —— m **soldado** - [,eta;-lam] welded shape tolerance

— g — **peso** m - weight tolerance

— f —— m **de lote** m - lot weight tolerance

— f — **plancha** f - [metal-lam] weight tolerance

— f —— f **de perfil** m - [metal-lam] shape plate tolerance

— f —— f —— m **soldado** - welded shape plate tolerance

— f — **rectitud** - straightness tolerance

— f —— f **de perfil** m - [metal-lam] shape straightness tolerance

— f —— f —— m **soldado** - [metal-lam] welded shape straightness tolerance

— f — **sección** f - [metal-lam] section tolerance

— f —— f **transversal** - [metal-lam] cross section tolerance

— f — **tamaño** m - size tolerance

— f — **varilla(s)** f - [metal-lam] rod tolerance

— f **especial** - special tolerance

— f **ajustada** - adjusted, o special, close tolerance

— f **más ajustada** - special closer tolerance

— f **especificada** - [ind] specified tolerance

— f **para construcción** f - specified construction tolerance

— f — **servicio** m - soecified service tolerance

— f **estándar** - standard tolerance

— f **estrecha** - close tolerance

— f **exacta** - [mec] exact, o fine, tolerance

— f **exigente** - exacting tolerance

— f **exterior** - outside tolerance

— f **interior** - inside tolerance

— f **más exacta** - more exact tolerance

— f — **reducida** - closer tolerance

— f **máxima** - maximum tolerance

— f — **en corrosión** f - maximum corrosion, tolerance, o allowance

— f —— **flecha** f - camber maximum tolerance

— f —— f **de bobina** f - [metal-lam] coil camber maximum tolerance

— f —— f —— f **de chapa** f - [metal-lam] strip coil camber maximum tolerance

— f **mínima** f - minimum tolerance

— f — **en flecha** f - [metal-lam] camber minimumn tolerance

— f —— f **de hoja** f - [metal-lam] sheet camber minimium tolerance

— f **nominal** - nominal, tolerance, o allowance

— f — **para calza** f - [mec] nominal shim, tolerance, o allowance

— f **normal** - standard, tolerance, o variation

— f **para ajuste** m - [mec] adjustment tolerance • clearance

— f — **amoladura** f - [mec] grinding tolerance

— f —— f **para rodillo(s)** m - [metal-lam] roll grinding tolerance

— f — **ancho,chura** - [mec] width tolerance

— f —— m **total** - [cojin] bearing total width tolerance

— f —— m —— **rodamiento** m - [cojin] bearing total width tolerance

— f — **borde** m - [metal-lam] edge tolerance

— f —— m **de bobina** f - [metal-lam] coil edge tolerance

— f —— m —— f **de chapa** f - [metal-lam] strip coil edge tolerance

— f —— m **de hoja** f - [metal-lam] sheet edge

tolerance
tolerancia f para borde m de hoja f de chapa f - [metal-lam] strip sheet edge tolerance
— f — calza f - [mec] shim, tolerance, o allowance
— f — cojinete(s) m—[cojin] bearing tolerance
— f — combadura f - [metal-lam] camber tolerance
— f — construcción f - construction tolerance
— f — corrosión f - corrosion, tolerance, o allowance
— f — f para recipiente m - [cald] vessel corrosion, tolerance, o allowance
— f — f — m con presión f - [cald] pessure vessel corrosion, tolerance, o allowance
— f — diámetro m - diameter tolerance
— f — m exterior - [mec] outside diameter tolerance
— f — m para pestaña f - [cojin] flange outside diameter tolerance
— f — — f de taza f - [cojin] cup flange outside diameter tolerance
— f — fabricación - [ind] fabricating,tion, o manufacturing, tolerance
— f — fabricante m - [ind] manufacturer's tolerance
— f — m de cronómetro(s) m - [mec] watchmaker('s) tolerance
— f — flecha f - [mec] camber tolerance
— f — f de bobina f - [metal-lam] coil camber tolerance
— f — f — f de chapa f - [metal-lam] strip coil camber tolerance
— f — de hoja f - [metal-lam] sheet camber tolerance
— f — f — f de chapa f - [metal-lam] strip sheet camber tolerance
— f — granallado - [metal-trat] sheet, peening, o (shot) blasting, tolerance
— f — m de rodillo(s) m - [metal-lam] roll shot blasting tolerance
— f — inserción f - [constr] threading, clearance, o tolerance
— f — laminación f - [metal-lam] mill, o rolling, tolerance
— f — material(es) m - [ind] material(s) tolerance
— f — patín m - [metal-lam] flange tolerance
— f — rectificación f - [metal-trat] grinding tolerance
— f — f de rodillo(s) m - [metal-lam] roll grinding tolerance
— f — relojero(s) - [mec] watchmaker('s) tolerance
— f — residuo(s) m - [ind] residues tolerance
— f — rodillo m - [mec] roll tolerance
— f — m para sostén m - [metal-lam] supporting roll tolerance
— f — m detrás de banda f - [mec] band roller, clearance, o tolerance
— f — servicio m - service tolerance
— f — soporte m - [cojin] support tolerance
— f — m — cono m - [cojin] core stand tolerance
— f — m — taza f - [cojin] cup stand tolerance
— f — torneadura f - [metal-lam] turning tolerance
— f — f de rodillo(s) m - [metal-lam] roll turning tolerance
— f — torsión f - [mec] twist(ing) tolerance
— f — i de tono m - [electrón] tone twist(ing) tolerance
— f — tubo(s) m - [tub] pipe tolerance
— f — m de acero m - [tub] steel tube tolerance
— f — m — m soldado - [tub] welded steel tube tolerance
— f — túnel m - [constr] tunnel tolerance
— f — viga f - [constr] beam tolerance
— f — f compensadora - [mec] equalizing beam, clearance, o tolerance

tolerancia f precisa - [mec] fine, o exacting, tolerance
— f recomendada - recommended tolerance
— f reducida - [mec] close tolerance
— f restringida - restricted toleranca • close tolerance
— f según Sociedad f Estadounidense para Ensayos m y Materiales m - [ind] American Society for Testing and Materials tolerance
— f solicitada - requested tolerance
— f transversal - transverse tolerance
— f verificada - [mec] checked, o proved,ven tolerance
— f vertical - vertical tolerance
tolerar v inconducta f - to tolerate @ misconduct
— v residuo(s) m - [ind] to tolerate @ residual,due
tolerarse v - to be, tolerable, o permissible
tolueno m - . . .; toluol
tolva f - . . .; bin; bunker • pan • [ind] storage bunker • [metal-prod] (scale car) pocket
— f adecuada - [ind] adequate bin
— f alimentadora f - [ind] feeding hopper
— f — para alambre m - [ind] wire feeding hopper
— f — — arena f - [ind] sand feeding hopper
— f anular - [miner] annular bunker
— f baja - [mec] shallow, o low, bin
— f bajo vía f - [ind] undertrack bin
— f basculante - [ind] dump(ing) hopper
— f cargadora - [ind] loading, pocket, o bin
— f — portátil - [ind] portable loading hopoper
— f combinada - [mec] combined hopper
— f — para maíz m y arvejas f - [agric] combined, corn, o maize, and pea hopper
— f — — y quisantes m - [agric] combined, corn, o maize, and pea hopper
— f cerrada - [mec] closed bin
— f con caída f - [ind] receiving hopper
— f — frente m cerrado - [constr] closed-face bin
— f con manga,guera f - [ind] tremie
— f — tubería f - [ind] tremie
— f de hormigón m - [ind] concrete bin
— f — metal m - [ind] metal bin
— f — m para, cereal(es) m, o grano(s) m - [agric] metal grain bin
— f — plancha(s) f - [ind] plate(s) bin
— f — f pesada(s) f - [ind] heavy plate(s) bin
— f — f — corrugada(s) - [ind] heavy corrugated plate(s) bin
— f — f — curva(s) - [ind] heavy curved corrugated plate(s) bin
— f — f — f curva(s) de acero m - [ind] heavy curved corrugated steel plate(s) bin
— f — f — — de acero m para almacenamiento m - [ind] heavy curved corrugated steel plate(s) storage bin
— f elevada - [ind] high bin • bulk bin
— f doble - [ind] double hopper
— f — para depósito m - [ind] double (bell) feeder hopper
— f — para granel m - [mec] overhead bulk bin
— f giratoria - [ind] revolving, o rotary, bin, o hopper
— f inferior - [ind] lower, o bottom, hopper
— f medidora - [constr] batcher; batching, bin, o hopper; measuring, hopper, o bin
— f metálica - [ind] metal, bin, o hopper
— f mezcladora - [ind] mixing, bin, o hopper
— f moledora - [mec] grinding, bin, o pan
— f movible - [ind] movable hopper
— f para adición(es) f - [ind] addition(s), bin, o hopper
— f — f para cuchara f - [metal-prod] ladle addition(s) hopper
— f — aditivo(s) m - [ind] additive(s), bin, o hopper
— f — m para convertidor m - [metal-prod]

tolva f para agregado(s) m — 1594 —

converter additive(s) bin
tolva f para agregado(s) m - [constr] aggre-
gate(s) bin
— f — alambre m - [sold] wire hopper
— f — alimentación f - [ind] feeding, hopper,
o bin
— f —— f para arena f - [constr] sand feeding hopper
— f —— f —— f para mezcladora f -
[constr] mixer sand feeding hopper
— f —— f para campana f - [metal-prod] bell
feeder hopper
— f —— f —— f doble - [metal-prod] double
bell feeder hopper
— f — alimentación f - [ind] feeding hopper
— f para almacenamiento m - [ind] storage, o
stock. bin, o hopper, o bunker
— f — m para agregado(s) m (para mezcla f)
- [constr] aggregate(s) storage bin
b— f para arena f - [constr] sand hopper
— f —— f para mezcladora f - [constr] mixer
sand hopper
— f — campana - [metal-prod] bell hopper
— f —— f doble - [metal-prod] double bell
(feeder) hopper
— f —— f grande - [metal-prod] large bell
hopper
— f —— f pequeña - [metal-prod] small bell
hopper
— f — cárbón m - [coke] coal hopper • oven
bin
— f —— m — horno m para coque m - [coque]
oven coal bin
— f — carga f - [metal-prod] receiving, o
batch, hopper • [ind] loading gate
— f — carro m báscula - [coque] scale car
pocket
— f — cemento m - [ind] cement, bin, o hopper
— f — cereal(es) m - [agric] grain bin
— f — cinta f - [ind] conveyor hopper
— f — clasificación f - [ind] classification,
o classifying, o grading, bin
— f —— f de cárbón m - [metal-prod] coal,
classification, o classifyin, o grading, bin
— f — convertidor(es) m - [metal-prod] converter bin
— f — coque m - [metal-prod] coke bin
— f — depósito m - [ind] storage, o receiving,
o stopck, bin
— f — derrame m—[ind] spillage, hopper, o bin
— f —— m para carbón m - [metal-prod] spillage coal hopper
— f — descarga f por arriba - [ind] top dump
hopper
— f — entrega f - [ind] delivery bin
— f —— f de mineral - [metal-prod] ore (delivery) bin
— f — escoria f - [metal-prod] slag bin
— f — finos m de coque m - [metal-prod] coke,
breeze, o fines, bin
— f — fundente m - [sold] flux, hopper, o bin
— f — granel m - [ind] bulk bin
— f — horno m - [metal-prod] furnace hopper
— f —— m alto - [metal-prod] blast furnace,
hopper, o bin
— f — máquina f - [ind] machine hopper
— f — mezcla f - [ind] mixer bin
— f — mezcladora f - [ind] mixer hopper
— f — mineral m - [metal-prod] ore bin
— f — molino m - [ind] crusher bin
— f — pella(s) f - [metal-prod] pellet(s),
fin, o hopper
— f — recepción f - [ind] receiving bin
— f — segadora f - [agric] reaoer bin
— f —— f segadora - [agric] combine bin
— f — semilla f - [agric] seed hopper
— f — separadora f - separator hopper
— f — sínter m - [metal-prod] sinter, bin, o
hopper
— f — sistema m - [ind] system, bin, o hopper
— f —— m para carga f - [ind] charging system bin • loading system bin
— f — m —— f para horno m -[metal-prod]

furnace charging system, bin, o hopper
tolva f para sistema m para carga f de horno m
alto - [metal-prod] blast furnace charging
system bin
— f — suministro m - [ind] supply(ing) hopper
— f — m para arena f - [constr] sand feeding hopper
— f — m —— f para mexcladora f -
[constr] mixer sand feeding hopper
— f — tragante m - [metal-prod] (furnace) top
hopper • furnace throat hopper • small bell
hopper
— f — m para horno m - [metal-prod] furnace, top, o throat, hopper
— f — m —— m alto - [metal-prod] blast
furnace top hopper
— f — vertido m - [ind] (pouring) hopper
— f — vía f - [f.c.] track hopper
— f — pesadora - [metal-prod] weigh(ing) hopper
— f — para coque m - [metal-prod] coke
weighing hopper
— f portátil - [mec] portable hopper
— f profunda - [ind] deep, bin, o hopper
— f receptora - [metal-prod] receiving hopper
— f — para campana f - [metal-prod] bell receiving hopper
— f —— línea f elevada (de silos m) - [metal
-prod] high bin line receiving hopper
— f — de tragante m - [metal-prod] upper
throat, o furnace throat (receiving) hopper
— f —— coque m - [metal-prod] coke receiving hopper
— f rotatoria - [mec] rotating hopper
— f sobre piso m - floor hopper • floor drain
— f subterránea - underground, hopper, o bin
— f superior - [ind] top, o upper, hopper
— f suplementaria - [mec] supplemental hopper
— f volcadora - [ind] dump(ing), hopper, o bin
tolvilla f - [ind] hopper; chute
— f metálica - [ind] metal, hopper, o chute
toma f - [hidr] inlet; intake (connection) •
hydrant • [fin] taking out; drawing • assuming; assumption • outlet; (off)take • [mec]
grip(ping); hold(ing) • take • [electr-inst]
tap
— f automática - [ind] automatic, taking, o
sampling
— f buena de viraje(s) m - [autom] good cornering
— f cambiada - [electr-instal] changed tap
— f con capacidad f plena - [electr-transform]
full capacity tap
— f con cuidado - careful, taking, o holding
— f con poca profundidad f - [ind] shallow bin
— f cuidadosa - careful, taking, o holding
— f de acción f - [admin] action taking
— f —— f efectiva - effective action taking
— f —— f eficiente—efficient action taking
— f — cargo m - [admin] taking, over, o charge
— f — conocimiento m - cognizance taking
— f — curva(s) f - [autom] cornering
— f —— f con pavimento m mojado - [autom]
wet cornering
— f —— f —— m muy mojado - [autom] very
wet cornering
— f —— f —— m seco - [autom] dry cornering
— f —— f excelente - [autom] excellent cornering
— f —— f mejorada - [autom] improved cornering
— f —— f por neumático(s) m - [autom] tire
cornering
— f — dato(s) m - data taking
— f — decisión(es) - decision making
— f — información f - information taking
— f — junta(s) f - [constr] pointing
— f —— f con enlechado - [constr] grouting
in
— f — lectura(s) f - [instrum] reading taking
— f —— f de presión f - [instrum] pressure
reading (taking)
— f — masa f - [sold] ground connection

toma f de medida(s) - measurement(s) taking -
action, o measure, taking
— f —— f para plazo m corto - short term
measure taking
— f —— f —— m largo - long term measure
taking
— f — muestra(s) f - [ind] sample taking; sampling
— f —— f de atmósfera f - [ambient] atmosphere sampling
— f —— f —— f en horno m - [combust]
furnace atmosphere, sample taking, o sampling
— f —— f de gas m - [combust] gas sampling
— f —— f —— m en tragante m - [combust]
throat gas, sample taking, o sampling
— f —— f espectrográfica(s) - [instrum]
spectrographic sampling
— f — obligación(es) f - [fin] bond(s), assuming, o assumption
— f — paso(s) m - step(s) taking
— f — presión(es) f - pressure taking—[metal-
-prod] pressure intake (connection)
— f —— f de agua m - water pressure taking
— f —— f —— m en entrada f - [metal-prod]
water intake pressure measuring (connection)
— f —— f —— m —— salida f - [metal-prod]
water outlet pressure measuring (connection)
— f — préstamo(s) m - [fin] load taking • borrowing
— f — probeta(s) f - sampling
— f — raíz f - [agric] root taking • [sold]
weld root; root (pass) bead
— f — riesgo m - [segurid] chance, o risk,
taking
— f — temperatura f - [instrum] temperature
taking
— f — testigo(s) - [ind] sample taking
— f —— m cilíndrico(s) - [petról] core sample taking
— f — tiempo(s) m - time taking
— f — tierra f - [electr-instal] ground connecting,tion; grounding
— f —— f para pararrayos m - [electr-instal]
lightning, conductor, o rod, ground cable
— f — vapor m - steam intake • throttle
— f — Venturi m - [metal-prod] Venturi intake
— f — viraje(s) m - [autom] cornering
— f — voltaje m - [electr-oper] voltage taking
— f — conjunto - taking together
— f — cuenta f - taking into, account, o consideration; considering
— f — intemperie f - outside intake
— f — préstamo m - [finl borrowing
— f —— de chatarra f marginal - [metal-prod]
marginal scrap borrowing
— f para acetileno m - [ind] acetylene intake
— f — agua m - [hidr] water, intake, o inlet,
o connection
— f —— m de río m—[hidr] river water intake
— f —— m para servicio m - [hidr] service
water intake
— f — aire m - air, inlet, o intake, o connection, o port • prefiérase admisión f de aire
— f —— m desde exterior m - [ambient] outside air intake
— f —— m exterior - [ambient] outside air
intake
— f —— m para volante m - [comb.int] flywheel air intake
— f — bomba f - [bombas] pump intake
— f — casa f de bomba(s) f - [bombas] pump
house intake
— f — corriente f - [electr-instal] collector;
tap • véase también tomacorriente m
— f — derivación f - [electr-distrib] tap
— f — energía f - [electr-instal] power source
— f — ensayo(s) m - sampling (for tests)
— f — fuerza f - [electr-oper] power take-off
— f — gas m - [metal-prod] gas intake
— f —— m con presión f alta - [metal-prod]
high pressure gas intake
— f —— m para Venturi m - [metal-prod] Venturi gas intake

toma f para oxígeno m - [ind] oxygen, intake, o
connection— f para potencia f - [mec] power
take-off
— f para provisión f de agua m - [hidr] water
supply intake
— f preliminar - preliminary take,king
— f regulada - [electr-oper] set tap
tomacorriente m - [electr-instal] . . .; receptacle; outlet; tap • contact shoe; current
collector • conductor, o contact, rail
— m apropiado - [electr-instal] appropriate, o
matching, receptacle
— m con conexión f con tierra f - [sold]
grounded receptacle
— m con . . . polo(s) - [electr-instal] . . .-
-pronged receptacle
— m — puesta f a tierra f - [electr-instal]
grounding type receptacle
— m — tres polos m - [electr-instal] three-
-prong(ed) receptacle
— m —— m (uno) para conexión f con tierra f - [electr-instal] three-pronged
— m — puntos - [electr-instal] three-
-pronged, o three-wire, receptacle
— m conectado - [electr-instal] connected receptacle
— m — con tierra f - [electr-instal] grounded
receptacle
— m — por comprador m - [electr-instal] customer connected receptacle
— m conveniente - [electr-instal] convenient,
receptacle, o outlet
— m — sin fusible(s) - [electr-instal] unfused convenience outlet
— m dañado - [electr-instal] damaged receptacle
— m de tipo m para torsión f - [electr-instal]
twist-lock receptacle
— m doble - [electr-instal] double, o duplex,
receptacle
— m — conectado con tierra f - [electr-instal]
duplex grounded receptacle
— m —— f para corriente f alterna -
[electr-instal] alternating current duplex
grounded receptacle
— m —— f —— f continua - [electr-
-instal] direct current duplex grounded receptacle
— m — para corriente f alterna - [electr-
-instal] alternating current duplex receptacle
— m —— f continua - [electr-instal] direct current duplex receptacle
— f — para servicio m - [electr-instal] duplex service receptacle
— m —— m conectado con tierra f -
[electr-insal] duplex grounded service receptacle
— m —— m —— f para corriente f alterna - [electr-instal] alternating current
duplex grounded service receptacle
— m —— m —— f —— f continua -
[electr-instal] direct current duplex
grounded service receptacle
— m —— m para corriente f alterna -
[electr-instal] alternating current duplex
service receptacle
— m —— m —— f alterna - [electr-inst]
alternating current duplex service receptacle
— m —— m —— f continua - [electr-
-instal] direct current duplex service receptacle
— m en, lugar, o sitio m, conveniente -
[electr-instal] convenient receptacle
— m estándar - [electr-instal] standard receptacle
— m inferior - [electr-instal] lower receptacle
— m montado - [electr-instal] mounted receptacle
— m no reversible - [electr-instal] non-rever-
sible, receptacle, o outlet
— m para carro m balanza - [metal-prod] scale
car contact shoe
— m — . . . clavija(s) f - [electr-instal]

. . .-prong(ed) receptacle
tomacorriente m **para . . . clavija(s)** f **para e-
nergía** f - [electr-instal] . . .-prong(ed), o
. . .-wire, power outlet
— m — **conexión** f **con tierra** f -
[electr-instal] grounding type receptacle
— m — **corriente** f **alterna** - [electr-instal]
alternating current receptacle
— m — — f — **conectado con tierra** f -
[electr-instal] alternating current grounded
receptacle
— m — — f — — — — f **para servicio** m -
[electr-instal] alternating current grounded
service receptacle
— m — — f — **para servicio** m - [electr-inst]
alternating current service receptacle
— m — — f **auxiliar** - [electr-instal] auxiliary power receptacle
— m — — f **continua** - [electr-instal] direct
current, receptacle, o outlet
— m — **encendido** m - [sold] starting receptacle
— m — — m **para arco** m - [sold] arc start(ing)
receptacle
— m — **energía** f - [electr-instal] power, outlet, o plug, o receptacle
— m — — f **auxiliar** - [electr-instal] auxiliary power, receptacle, o outlet
— m — — f **en corriente** f **alterna** - [electr-
-instal] alternating current power receptacle
— m — — f — — f **continua** - [electr-instal]
direct current power receptacle
— m — **frecuencia** f **alta** - [electr-instal] high
frequency receptacle
— m — **fuerza** f **auxiliar** - [electr-instal] auxiliary power receptacle
— f — — f **motriz** - [electr-instal] (utility)
power, receptacle, o outlet, o plug
— m — — f — **en corriente** f **alterna**—[electr-
-instal] alternating current power receptacle
— m — — f — — f **continua** - [electr-
-instal] direct current power receptacle
— m — **pared** f—[electr-instal] wall receptacle
— m — **potencia** f **auxiliar** - [electr-instal]
auxiliary power receptacle
— m — **regulación** f **de artefacto** m - [electr-
-inst] convenience control (electrical) receptacle
— m — **servicio** m - [electr-instal] service receptacle
— m — — m **con conexiaón** f **con tierra** f -
[electr-instal] grounded service receptacle
— m — — m **conexión** f **con tierra** f - [electr-
-instal] grounded service receptacle
— m — — m **conectado con tierra** f - [electr-
-inst] grounded service receptacle
— m — — f — — f **para corrien-
te** f **alterna** - [electr-instal] alternating
current grounded service receptacle
— m — — m — — — f — — f **continua** -
[electr] direct current grounded service receptacle
— m — **soldadura** f - [sold] welding (circuit)
outlet
— m — **tablero** m - [electrón] board receptacle
— m — — m **con circuito** m -
[electrón] circuit board receptacle
— m — — m —, **estampado**, o
impreso - [electrón] printed circuit board
receptacle
— m — — f **para mando** m - [electr-instal] control panel receptacle
— m — — m — **regulación** f - [sold] control
panel receptacle
— m — **tablilla** f **para circuito** m, **estampado**, o
impreso - [electrón] printed circuit board
receptacle
— m — **telemando** m - [electrón] remote control
receptacle
— m — — m **y encendido** m **de arco** m - [sold]
remote control and arc start receptacle
— m — **telerregulador** m **Lincolntrol** - [sold]
Lincontrol receptacle
— m **para uso** m **general** - [electr-instal] conve-

nience outlet
tomacorriente m **para . . . voltio(s)** m -
[electr-instal] . . . volt(s) receptacle
— m **polarizado** - [electr-instal] polarized receptacle
— m **práctico** - [electr-instal] convenience receptacle
— m — **sin fusible** m - [electr-instal] unfused
convenience outlet
— m **recalentado** - [electr-oper] overheated
receptacle
— m **reemplazado** - [electr-instal] replaced receptacle
— m **remoto** m - [electr-instal] remote outlet
— m **sacado** - [electr-instal] removed, o
pulled, receptacle
— m **sin fusible** - [electr-instal] unfused, o
non fused, outlet, o receptacle
— m **superior** - [electr-instal] upper, o top,
outlet, o receptacle
— m **tripolar** - [electr-instal] thrre-prong(ed)
receptacle
tomado,da a - taken • gripped; held • [fin]
drawn; taken out
— a **cargo** m - taken, charge, o over
— a **con cuidado** m - held carefully
— a **con escoria** f - [metal-prod] slag blocked;
obsructed, o plugged, o filled, o blocked,
with slag
— a **cuidadosamente** - taken, o held, carefully,
o with care
— a **de escoria** f - [metal-prod] taken from @
slag • véase también **tomado,da con escoria** f
— a **en conjunto** m - taken (all) together
— a — **consideración** f - taken into account
— a — **cuenta** - taken into account; considered
— a — **préstamo** m - borrowed
tomador m - taker
— m **de tiempo(s)** m - [labor] time study man
tomar v - . . .; to grip; to hold • to catch •
to take over • to establish • to assume •
[educ] to, give, o administer, @ test
— v **a azar** - to, take, o pick, at random
— v **acción** f - to take action
— v — **efectiva** - [admin] to take effective,
action, o measure(s)
— v — f **eficaz** - [admin] to take effective,
action, o measure(s)
— v — **eficiente** - [admin] to take, effective,
o efficient, action, o measure(s)
— v — f **regularmente** - to act regularly
— v — **sobre** - **to act upon**
— v **alambre** m - to, take hold of, o pick up, @
wire
— v **cargo** m - to take, charge, o over
— v **carta(s)** f - to, pursue, o follow up • to
become involved; to intervene
— v **como de quien viene** - [fam] to, consider,
o take into account, @ source
— v **con cuidado** - to, take, o hold, carefully
— v **conformación** f - to take @, form, o shape
— v — **de cola** f **de pescado** m - to (take @)
form @ fish tail
— v **conocimiento** m - to take, knowledge, o cognizance; to become informed
— v **contacto** m - to (form, o make, @) contact
— v **cuidado** - to take care
— v — m **especial** - to take special care
— v **cuidadosamente** - to, take, o hold, care-
fully|
— v **curva** f - [autom] to take @ curve; to (go
around @) corner
— v **dato(s)** - to take, information, o data
— v **decisión** f - to make @ decision
— v **desprevenido,da** - to take by surprise
— v **dirección** f - [autom] to take over @, wheel,
o driver's seat
— v **en conjunto** m - to take (all) together
— v — **consideración** f - to consider; to take
into, account, o consideration
— v **en préstamo** m - to borrow
— v **examen** m - [educ] to, give, o administer,
@, test, o examination

tomar v̲ **forma** f̲ **de cola** f̲ **de pescado** m̲ - to take
@ form of @ fishtail
— v̲ **fotografía** f̲ - [fotogr] to photograph, o̲
take @, photograph, o̲ picture
— v̲ **fuertemente** - to take, o̲ hold, tightly
— v̲ **información** f̲ - to take, @ information, o̲
data
— v̲ **juntas** f̲ - [constr] to point
— v̲ — f̲ **con enlechado** m̲ - [constr] to seal @
joint
— v̲ **lectura** f̲ - [instrum] to take @ reading
— v̲ — f̲ **de presión** f̲ - [instrum] to take @
pressure reading
— v̲ **lugar** m̲ - to take @ place of; to replace
— v̲ **medida(s)** f̲ - to, measure, o̲ take a mea-
surement(s) • [admin] to take @ action • to
make @ arrangement(s)
— v̲ — **para plazo** m̲ **corto** - to take @ short
term measure(s)
— v̲ — f̲ — — m̲ **largo** - [admin] to take @
long term measure(s)
— v̲ **minuto** m̲ - to take @, minute, o̲ moment
— v̲ **momento** m̲ - to take @, moment, o̲ minute
— v̲ **mucho espacio** m̲ - to take (up) much, space,
o̲ room
— v̲ **muestra(s)** f̲ - [ind] to, sample, o̲ take @
sample(s)
— v̲ **nota** f̲ - to, record, o̲ make note of
— v̲ **obligación** f̲ - to assume @ obligation •
[fin] to assume @ bond
— v̲ **por separado** - to take separately
— v̲ **precaución** f̲ - [seguridj] to take @ precaution
— v̲ **puntería** f̲ - to aim
— v̲ **poco espacio** m̲ - to take (up) little, room,
o̲ space
— v̲ **por tangente** f̲ - to digress
— v̲ **razón** f̲ - to make note of
— v̲ **riesgo** m̲ - [seguridj] to take @, risk, o̲
chance
— v̲ **señal** f̲ - to take @ signal
— v̲ **temperatura** f̲ - [instrum] to take @ temperature
— v̲ **testigo** - [ind] to (take @) sample
— v̲ — m̲ **cilíndrico** - [petról] to take @ core
sample
— v̲ **volante** m̲ - [autom] to take (over) @, driving, o̲ driver('s) seat, o̲ wheel
tomarse v̲ **mortero** m̲ - [constr] to point @ mortar
tomo m̲ - . . .: book
— m̲ **separado** - [jmprent] separate volume
tonel m̲ - [transp] . . .; hogshead
— m̲ **para tabaco** m̲ - [transp] tobacco hogshead
tonelada f̲ **acabada** - [ind] finished ton
— f̲ **anual** - [ind] yearly ton
— f̲ — **de producción** f̲ - [ind] yearly production ton
— f̲ — — **producto** m̲ - [ind] yearly product ton
— f̲ **bruta** - gross ton
— f̲ **cumplida** - full, o̲ sinished, ton
— f̲ **de concentrado** m̲ - [miner] concentrate(s)
ton(s)
— f̲ — **dos mil libras** f̲ - [ind] short ton(s)
[ind] (short) ton(s) per mile
— f̲ — **lingote(s)** m̲ - [metal-prod] ingot ton(s)
— f̲ — — m̲ **de acero** m̲ - [metal-prod] steel ingot ton(s)
— f̲ — **producción** f̲ - [ind] production ton
— f̲ — **producto** - [ind] product ton
— f̲ — **2000 libras** f̲ - [metric] short ton
— f̲ — **2240 libras** f̲ - [metric] long ton
— f̲ **exportada** f̲ - [com] exported ton
— f̲ **medida** - [miner] measured ton
— f̲ **métrica bruto** - gross metric ton
— f̲ — **producida** - [ind] metric ton, produced,
o̲ of production
— f̲ — **por día** m̲ - [metric] metric ton per day
— f̲ — **por kilómetro** m̲ - [trans] metric ton(s)
per kilometer
— f̲ **neta** - [ind] net ton
— f̲ — **de arrabio** m̲ - [metal-prod] net ton(s)
of pig iron
— f̲ **original** - [ind] original ton
— f̲ — **de capacidad** f̲ - [ind] original ton ca-

pacity
tonelada f̲ **por año** m̲ - ton(s) per year
— f̲ — **día** m̲ - [metric- ton(s) per day
— f̲ — **kilómetro** m̲ - ton(s) per kilometer
— f̲ **producida** - [ind] ton produced
— f̲ **terminada** - [inf] finished ton
— m̲ **anual** - [ind] annual, o̲ yearly, tonnage
— m̲ — **de producción** f̲ - [ind] annual, o̲ yearly production tonnage
— m̲ — — **producto(s)** - annual, o̲ yearly, product ton(nage)
— m̲ — **cascarón** m̲ - [metal-prod] scab tonnage
— m̲ — **lingote(s)** m̲ - [metal-prod] ingot, ton,
o̲ tonnage
— m̲ — **producción** f̲ - [ind] production tonnage
— m̲ **elevado** - large tonnage
— m̲ **mayor** - greater, o̲ increased, tonnage
— m̲ **menor** - smaller, o̲ decreased, tonnage
— m̲ **nuevo** - new tonnage
— m̲ **para feriado(s)** m̲ - [ind] holiday tonnage
— m̲ — — m̲ **laboral** - [ind] labor holiday tonnage
— m̲ **producido** - [ind] production tonnage
— m̲ **reducido** - reduced, o̲ small, tonnage
tonelada/hora f̲ - [ind] ton/hour
— f̲ **efectiva** - [labor] effective ton/hour
tono m̲ **alto** - high tone
— m̲ **audible** - [electrón] audible tone
— m̲ — **en salida** f̲ - [comput] sidetone output
— m̲ **bajo** - low tone
— m̲ **codificado** - [comput] coded tone
— m̲ **comprobado** - [electrón] proven, o̲ checked,
tone
— m̲ **continuo** - [electrón] continuous tone
— m̲ **controlado** - [electrón] controlled tone
— m̲ — **con cristal** m̲ - [comput] cystal controlled tone
— m̲ — **para sistema** m̲ **silenciador** - [electrón]
controlled squech(ing) system tone
— m̲ **corto** - [electrón] short tone
— m̲ — **audible** - [electrón] short audible tone
— m̲ **cronometrado** - [comput] timed tone
— m̲ **(de) decodificador** m̲ - [comput] decoder tone
— m̲ — **prueba** f̲ **a bordo** - [comput] on board test
tone
— m̲ — — m̲ — — m̲ **de . . . hercios** m̲ -
[comput] on board . . . Hertz test tone
— m̲ — **salida** f̲ - [electrón] output tone
— m̲ — — f̲ **audible** - [comput] side tone
— m̲ — **sistema** m̲ **silenciador** - [electrón]
squelch(ing) system tone
— m̲ — — **regulado a por tono** m̲ **continuo** -
[electrón] continuous tone controlled squelch
system
— m̲ — **teléfono** m̲ - [electrón] telephone tone
— m̲ — **zumbador** m̲ - [electrón] buzzer tone
— m̲ **económico** - [econ] economic tone
— m̲ **emitido** - [electrón] tone out
— m̲ **en sistema** m̲ - [electrón] system tone
— m̲ — — m̲ **para silenciamiento***m̲ - [electrón]
squelching system tone
— m̲ — — ⅲ **silenciador** - [electrón] squelching
system tone
— m̲ — — — **controlado con tono** - [electrón]
tone controlled squelch system tone
— m̲ — — m̲ **silenciador regulado** - [electrón]
conrolled squelch system tone
— m̲ — — m̲ — — m̲ **continuo** - [electrón]
continuous tone controlled squelch system tone
— m̲ **equilibrado** - [electrón] balanced tone
— m̲ **escogido** - [electrón] selected tone
— m̲ **estándar** - [comput] standard system
— m̲ — **doble** - [electrón] standard dual tone
— m̲ **externo para prueba(s)** f̲ - [comput] external
test tone
— m̲ **lateral** - [comput] sidetone
— m̲ **inferior a . . .** [electrón] tone below . . .
— m̲ **recibido** - [comput] tone in
— m̲ — **en decodificador** m̲ - [comput] decoder
tone in
— m̲ **manipulado** - [electrón] keyed tone
— m̲ — **para cambio** m̲ **de audiofrecuencia** f̲ -
[electrón] audio frequency shift keyed tone
— m̲ — — **corrimiento*** m̲ **de audiofrecuencia** f̲ -
[electrón] audio frequency shift keyed tone

tono m manipulado para desplazamiento m de audiofrecuencia f - [electrón] audio frequency
shift keyed, o A F S K, tone
— m marca-espacio - [electrón] mark-space tone
— m opuesto m - [electrón] opposite tone
— m oscilado - [electrón] oscillated tone
— m para alerta f - [electrón] alert tone
— m — — f audible - [electrón] audible alert
tone
— m — — f audible sonado - [electrón] sounded audible alert tone
— m — decodificador m - [comput] decoder tone
— m — entrada f para decodificador m -
[comput] decoder tone in
— m — espacio(s) m - [electrón] space tone
— m — marca f - [electrón] mark tone
— m — marca-espacio m - [electrón] mark-space
tone
— m — — m para salida f - [electrón] output
mark-space tone
— m — prueba f a bordo - [comput] on-board
test tone
— m — prueba f - [comput] test tone
— m — — f para . . . hercio(s) m - [comput]
. . . Hertz test tone
— m — — f integral - [comput] integral test
tone
— m — — f — para . . . hercios m - [comput]
integal, o on-board, . . . Hertz test tone
— m paralelo - [comput] parallel, o side, tone
— m primero - [electrón] first tone
— m segundo - [electrón] second tone
— m simultáneo - [electrón] simultaneous tone
— m — continuo - [electrón] simultaneous continuous tone
— m — en sistema m silenciador m - [electrón]
simultaneous squelch system tone
— m — — — m — regulado por tono m continuo
- [electrón] simultaneous continuous tone
controlled squelch system
— m sincronizado - [comput] timed tone
— m — precisamente - [comput] precision timed
tone
— m sintetizado - [comput] synthetized tone
— m — controlado digitalmente - [electrón]
digitally controlled synthetized tone
— m — digitalmente - [comput] digitally controoled synthetized tone
— m — — controlado con cristal m - [electrón]
cristal-controlled digitally synthetized tone
— m superior - superior, o higher, tone
— m — a - [electrón] tone above . . .
— m verificado - [electrón] checked tone
tonto,ta a - . . . . • [fam] green
topado,da a - collided • encountered • hit
topadora f - [constr-equip] bulldozer; dozer;
pushdozer • blade
— f con oruga f - [constr] caterpillar mounted
(bull) dozer
— f con rueda(s) f - [constr-equip] wheeled, o
wheel mounted, (bull)dozer
— f para bodega f - [nav] hold bulldozer
topamiento m - colliding; hitting
topar v - . . .; to encounter; to hit
— v con - to, collide, o butt into
— v con discontinuidad(es) f to encounter @
discontinuity,ties
— v con problema m - to, run into, o meet, o
encounter, o experience, o have, @ problem
tope m - . . . . • stop • bumper • dog • abutment • arrester • butt • backstop • buffer •
pad • detent; gab; catch; lug; peg; toe •
threshold • [cojin] banking shoulder
— a tope m - [constr] bumper to bumper
— m abierto - open top
— m ajustado - [mec] adjusted stop
— m central - [constr] center stop
— m cerrado - [sold] tight butt
— m con chaflán m - [sold] V-butt
— m con forma f de cuello m - [mec] neck shape
stop
— m con ranura f profunda - [sold] deep groove
butt

tope m de acero m - [constr] steel backstop
— m — para armadura f - [constr] steel
framing backstop
— m — bóveda f - [constr] arch top
— m — cilindro m - [mec] cylinder top
— m — cimiento m - [constr] foundation top
— m — columna f - [mec] column top
— m — conducto m - [combust] conduit top
— m — — m para humo(s) m — [combust] flue top
— m — — m — — m para calentamiento m -
[combust] heating-up flue, top, o end
— m — escala,lera f - [constr] ladder top
— m — fundamento m—[constr] foundation top
— m — horno m - [combust] furnace, top, o
butt
— m — — m para reducción f - [metal-prod]
reduction furnace top
— m — horquilla f - [mec] fork top
— m — madero m pesado m - [constr] heavy
timber backstop
— m — panel m - [mec] panel top
— m — pilote m - [constr-pil] pile top
— m — — m de acero m - [constr-pil] steel
pile top
— m — — m tubular - [constr-pil] pipe pile
top
— m — — m — de acero m - [constr-pil] steel
pipe pile top
— m — portadiferencial m - [autom-mec] differencial carrier top
— m — presa f - [hidr] dam top
— m — recipiente m - container, o can, top
— m — riel m - [f.c.] rail top
— m — tapón m - [mec] plug top
— m — — m de hormigón m - [constr-pil] concrete plug top
— m — — m impulsor m - [mec] drive,ving type
top • drive,ving type pad
— m — trampa f - [mec] trap top
— m — transbordador m - [grúas] transfer
crane stop
— m — tubo,bería f - [constr] pipe,ping top
— m — — abovedado,da - [constr] pipe-arch
top
— m — vástago m - [mec] plunger, o rod, stop
— m — ventilador m - [mec] fan, o ventilator,
top
— m elástico m - [mec] elastic stop
— m en ajustador m - [mec] adjuster lug
— m — cárter - [comb.int] crankcase stop
— m — ranura f - [sold] groove butt
— m — — f profunda - [sold] deep groove butt
— m — — f — con chaflán m - [sold] bevel
deep groove butt
— m — — f — — f pequeño - [sold] small
bevel deep groove butt
— m — V - [sold] V-butt
— m final - [mec] end stop
— m — para apiladora f - [metal-lam] piler end
stop
— m giratorio - [mec] swivel pad
— m horizontal - [mec] horizontal stop • [sold]
horizontal butt
— m — en ranura f profunda f - [sold] deep
groove horizontal butt
— m limitador m - [mec] (travel) limit (butt)
— m lubricado m - [mec] lubricated stop •
[f.c.] lubricated buffer
— m máximo m - [fin] maximum, limit, o ceiling
— m — para ajuste m alzado - maximum escalation limit
— m — para reajuste m - maximum escalation
limit
— m movible - [mec] movable stop
— m para acelerador m - [comb.int] throttle
stop
— m — aguilón m - [grúas] boom stop
— m — — m lubricado - [grúas] lubricated
boom stop
— m — — m para torre f - [grúas] tower boom
stop
— m — ajustador m - [mec] adjuster stop
— m — — m para cabeza f - [mec] head ad-

juster stop
tope m **para ajustador** m **vertical** - [mec] vertical adjuster stop
— m — **ajuste** m **para posición** f - [mec½ position stop adjustment
— m — — m **alzado** - escalation limit
— m — **ala** m - [constr] flange top
— m — **alma** m - [constr] web top
— m — **anclaje** m - [mec] anchor(ing) top
— m — **apiladora** f - [metal-lam] piler stop
— m — **apoyo** m - [mec] support(ing) stop
— m — **árbol** m - [mec] shaft buffer
— m — **bastidor** m - [mec] frame top
— m — **bobina** f - [metal-lam] coil stop
— m — **brazo** m - [mec] arm stop
— m — — m **para palanca** f - [mec] lever arm stop
— m — — m **protector** m **para presión** f **nula** [electr- equip] no voltage release arm stop
— m — **cama** f - [mec] cradle stop
— m — **canaleta** f - [mec] trough stop
— m — **cerrojo** m - [mac] latch stop
— m — **cilindro** m - [mec] cylinder top
— m — **compuerta** f - [mec] (sluice) gate stop
— m — — f **para emergencia** f - [metal-prod] emeregency gate stop
— m — f — — f **para tolva** f - [metal-prod] bin emergency gate stop
— m — f — — — f **para coque** m - [metal-prod] coke bin emergency gate stop
— m — **contacto** m - [mec] contact bumper
— m — — m **en tuerca** f - [mec] nut contact bumper
— m — **contención** f - [mec] backstop
— m — — f **lubricado** - [mec] lubricated backstop
— m — — f **para aguilón** m - [grúas] boom backstop
— m — — — m **lubricado** - [grúas] lubricated boom backstop
— m — — f **reforzada** - [mec] heavy duty backstop
— m — — f **tubular** - [mec] tubular backstop
— m — — f — **reforzado** - [mec] heavy duty tubular backstop
— m — **chapa(s)** f - [sold] sheet stop
— m — **desplazador** m - [mec] shifter stop
— m — **dirección** f - [autom] steering stop
— m — **émbolo** m - [mec] piston stop
— m — **empuje** m - [mec] tappet
— m — **escuadrar** v - [mec] squaring stop
— m — **grúa** f - [grúas] crane, stop, o bumper
— m — **mandil** m - [constr] apron stop
— m — — f **frontal** - [constr] (front) apron stop
— m — **manguito** m - [mec] sleeve stop
— m — — m **para cilindro** m - [mec] cylinder sleeve stop
— m — **ménsula** f - [mec] bracket stop
— m — — f **para desplazador** m - [mec] shifter bracket stop
— m — **muesca** f - [mec] notch stop
— m — — f **para regulación** f - [mec] control notch stop
— m — **palanca** f - [mec] lever stop
— m — — f **para mando** m - [comb.int] control lever stop
— m — — f — **regulación** f - [mec] control lever stop
— m — **parachoque(s)** m - [autom] bumper guard
— m — **paragolpe(s)** m - [autom] bumper guard
— m — **pasada** f **única** - [sold] single pass butt
— m — — **con avance** m **rápido** - [sold] high speed single pass butt
— m — **pedal** m - [mec] pedal stop
— m — — m **para embrague** m - [mec] clutch pedal stop
— m — **perno** m - [mec] bolt stop
— m — **pistón** m - [mec½ piston stop
— m — **portón** m - [constr] gate stop
— m — **posición** f - [mec] position stop
— m — **posicionador*** m - [mec] positioner stop
— m — **reajuste** m - [com] escalation limit

tope m **para regulación** f - [mec] control stop
— m — — f **para velocidad** f - [mec] speed control stop
— m — f — — **para traslación** f = gound speed control stop
— m — **regulador** m - [mec] control stop
— m — — m **para palanca** f - [mec] lever control stop
— m — — m **vertical** - [mec] vertical adjuster stop
— m — — m **para cabeza** f - [sold] vertical head adjuster stop
— m — **resorte** m - [mec] spring, stop, o block
— m — **respaldo** m - [constr] backstop
— m — **riel** m - [mec] rail stop
— m — **tensión** f - [mec] tension, stop, o tab
— m — **torre** f - [mec] tower stop
— m — — f **para perforación** f - [petról] mast head
— m — **tubería** f - [tub] pipe,ping top
— m — — f **para agua** m - [petról] wash pipe top
— m — **tubo** m - [tub] pipe top
— m — — m **indicador** - [instrum] sight glass stop
— m — **vagón(es)** m - [f.c.] car bumper
— m — **viraje(s)** m - [autom] steering stop
— m **plano** - [sold] flat butt
— m — **en ranura** f - [sold] groove flat butt
— m — — — f **profunda** - [sold] deep groove flat butt
— m **plateado** - silver top • silver stop
— m **provisto** - [mec] provided stop
— m **retráctil** - [mec] disappearing stop
— m — **para grúa** f - [grúas] crane disappearing stop
— m — **para transbordador** m - [grúas] transfer crane disappearing stop
— m **riel(es)** m - [mec] rail stop
— m **sin diente(s)** - [mec] toothless pad
— m **soldado** - [sold] welded stop
— m **sólido** - [constr] substantial backdrop • solid stop
— m **suplido** - [mec] supplied, o provided, stop
— m **terminal** - [med] terminal stop • end bumper
— m **vertical** - [mec] vertical stop
— m **y botador** m - [mec] stop and kick(er)
— m — **disparador** m - [mec] stop and kick(er)
— m — **lanzador** m - [mec] stop and kick(er)
toque m - . . . . • probing
— m **de bocina** f - [electrón] horn, sound(ing), o blowing
— m — **metal** m - [sold] metal touch(ing)
— m — **música** f - [mús] music playing
— m — **suelo** m - ground touching
— m — **timbre** m - ringer, o alarm, sounding
— m **final** - final, o finishing, touch
— m **personal** - personal touch
— m **simultáneo** - simultaneous touch(ing)
torca f - [topogr] sink hole
torcedura f - . . . .; twisting; camber(ing); torsion • kink(ing) • [med] sprain(ing)
— f **de brazo** m - arm bending
— f — **cable** m - [cabl] rope kink(ing)
— f — — m **de alambre** m - [cabl] wire rope kink(ing)
— f — **pasador** m - [mec] pin distortion
— f — — m **para aguilón** m - [grúas] boom pin distortion
— f — **pieza** f - [mec] part, o member, distortion
— f — **roldana** f - [cabl] roller distortion
— f — — f **para cable** m - [cabl] wire rope roller distortion
— f **hacia costado** m - [autom-neumát] rolling over
— f — — m **sobre pared** f **de neumático** m - [autom-neumát] rolling over
torcer v - . . . . • [autom] to swerve
— v **árbol** m - [mec] to, bend o deflect, @ shaft
— v **brazo** m - to, bend, o twist, @ arm
— v **contacto** m - [electr-instal] to twist @

contact
torcer v eslabón m - [cadenas] to, bend, o
  twist, @ link
— v — m **conectador** - [cadenas] to, bend, o
  twist, @ connecting link
— v **gaza** f - [mec] to twist @ loop
— v **hacia costado** m - [autom-neumát] to roll
  over
— v — — m **sobre pared** f (**de neumático** m) -
  [autom-neumát] to roll over
— v **lazo** m - [mec] to twist @ loop
— v **pasador** m - [grúas] to, bend, o distort, @
  pin
— v — m **para aguilón** m - [grúas] to distort @
  boom pin
— v **pieza** f - [mec] to, twist, o distort, @,
  part, o member
— v **pista** f - [cojin] to cock @ cup
— v **roldana** f - [mec] to distort @ roller
— v — f **para cable** m - [cabl] to distort @
  wire rope roller
— v **sobre sí mismo,ma** - to wind, o twist,
  upon, o around, itself
torcido m - . . . . • [cabl] lay (direction)
torcido,da a - bent; wound; skew(ed); warped;
  cocked • deflected • [cabl] kinked • [med]
  sprained • [autom] swerved
— a **alrededor de sí mismo,ma** - wound, twisted,
  around itself
— m **triple** - [cabl] triple, winding, o wound
torcimiento m - . . . .; cocking • deflecting,tion
  • [autom] swerving • véase también **torcedura**
— m — **árbol** m - [mec] shaft deflecting,tion
— m — **brazo** m - arm, bending, o twisting
— m — **contacto** m - [electr-equip] contact
  twisting
— m **hacia costado** m - [autom-neumát] rolling
  over
— m — — m **sobre pared** f**de neumático**
  m - [autom-neumát] rolling over
— m — **eslabón** m - [cadenas] link, twisting, o
  bending
— m — — m **conectador** - [cadenas] connecting
  link, bending, o twisting
— m **de gaza** f - [cabl] loop twist(ing)
— m — **lazo** m - [cabl] loop twist(ing)
— m — **pista** f - [cojin] cup cocking
toriado,da a - thoriated
tórico,ca a - toric
toricónico,ca a - [mec] dished • [geom] tori-
  conical
torisférico,ca a - [geom] torispherical
tormenta f **de invierno** m - [meteorol] winter
  storm
— f **frecuente** - [meteorol] frequent storm
— f **intensa** - [meteorol] intense storm
— f **invernal** - [meteorol] winter storm
— f **más frecuente** - [meteorol] more frequent
  storm
— f — **intensa** - [meteorol] high intensity
  storm
— f **menos frecuente** - [meteorol] less frequent
  storm
— f — **intensa** - [meteorol] less intense storm
— f — **cual se ha ideado,da** - [hidr] design
  storm
— f **siguiente** - [meteorol] next storm • following storm
— f **última** - [meteorol] last storm
tormento m - . . . . • bother(ing)
tornado,da a - become
— a **accesible** - made accessible
— a **práctico,ca** - made practical
— a **obscuro,ra** - turned dark
tornador m **de borde(s)** m - [mec] turn roll
tornamesa* f - [mec] véase **mesa** f **giratoria**
tornamiento m **de borde(s)** m - [mec] edge turning • turn rolling
tornando adv - turning; becoming
— adv **accesible** - making accessible
tornapunta f - [mec] . . .; strut
— f **para compresión** f - [constr] construction
  strut

tornapunta f **para malacate** m - [mec] capstan
  brace
— f — — m **para herramienta(s)** f - [petról]
  bull wheel brace
— f — — m — **maniobra(s)** f - [petról] calf
  wheel brace
— f — — m — **tubería** f - [petról] calf wheel
  brace
— f/— **poste** m - [mec] jack post brace
— m — — m **maestro** - [petról] Samson post
  brace
— f — — m — **malacate** n (**para herramientas**)
  f - m - [petról] bull wheel post brace
— f — — m — — n — **maniobra(s)** f—[petról]
  wheel post brace
— f — — m — — m **para tubería** f - [petról]
  calf wheel post brace
— f — — m **para rueda** f (**motriz**) - [petról]
  jack post brace
— f — — m — **tambor** m **para herramienta(s)** f
  - [petról] bull wheel post brace
— f — — m — **torno** m **para maniobra(s)** f -
  [petról] calf bull wheel brace
— f — — — — m — **tubería** f - [petról]
  calf wheel post brace
— f — **soporte** m - [petról] support brace
— f — — m **para balancín** m - [petról] Samson
  post brace
— f — — m **para torno** m (**para maniobras** f) -
  [petról] wheel post brace
— f — — m — — m **para tubería** f [petról]
  wheel post brace
— f — **tambor** m - [petról] wheel brace
— m — — m **para herramienta(s)** f = [petról]
  bull wheel brace
— f — — m **para maniobra(s)** f - [petról] calf
  wheel brace
— f — — m — **tubería** f - [petról] calf wheel
  brace
tornar v **accesible** - to make accessible
— m **borde** m - [mec] to turn roll
— v **obscuro** - to darken; to become dark
— v **práctico,ca** - to make practical
— v **rojo** - to, redden, o turn red
tornarse v - to become
— v **aparente** - to become apparent
— v **difícil** - to become difficult
torneado m - [mec] turning; machining • véase
  también **torneadura** f
torneado,da a - [mec] turned; machine cut •
  ground
torneador m - [mec] turn roll
— m **para borde(s)** m - [mec] edge turn roll
torneadura f - [mec] grinding
— f **de rodillo(s)** m - [metal-lam] roll, turning,
  o machining
— f — **rodillo** m **Bridle** - [metal-lam] bridle
  roll turning
— f — — m **loco** - [metal-lam] idle roll turning
— f — — m **para apoyo** m - [metal-lam] back-up
  roll turning
— f — — m **para trabajo** m - [metal-lam] work
  roll turning
tornear v - [mec] . . .; to machine, to lathe;
  to grind
— v **borde** m - [mec] to edge turnroll
— v **colector** m - [electr-mot] to turn @ collector
— v **rodillo** m - [mec] to grind @ roll
— v **rueda** f - [mec] to turn @ wheel
— v **tambor** m - [mec] to turn @ drum
torneo m - [mec] . . . • turnaround; turning
— m **anterior** - [deport] previous tournament
— m **de golf** - [deport] golf tournament
— m — **tracción** f - [autom] traction tournament
— m — **neumático(s)** m - [autom-neumát] tire
  tournament
— m **primero** •• [deport] first tournament; opening round
— m — **de temporada** - [Deport] season('s) opening round
tornería f - [metal-lam] lathe shop

tornero m - [mec] lathe operator; turner • [metal-lam] (roll) grinder
— m para rodillos m - [metal-lam] roll turner
tornillería f - [mec] bolts and nuts; nuts and bolts; bolts
tornillo m - [mec] . . . • bolt
— m aflojado - [mec] loosened, o backed off, screw
— m ajustado - [mec] adjusted, o tightened, screw
— — en forma f pareja - [mec] evenly, adjusted, o tightened, screw
— m alimentador m - [mec] feeding screw • tempering screw
— m Allen - [mec] Allen screw—
— m amortiguador - [mec] buffer screw
— m apretado - [mec] tightened screw
— — en forma f pareja - [mec] evenly tightened screw
— m atornillado m - [mec] screwed (in) screw
— m — hacia abajo - [mec] turned down screw
— m autorroscante - [mec] self-tapping screw
— m — con cabeza f plana - [mec] self-tapping pan (headed) screw
— m — para montaje m - [mec] mounting self-tapping screw
— m — — — m de deflector m - [mec] baffle mounting self-tapping screw
— m — — — — respaldo m - [mec] back mounting self-tapping screw
— m — — — — soporte m - [mec] bracket mounting self-tapping screw
— m — — — m — transformador m - [electr-instal] transformer mounting self-tapping screw
— m — — — m para amplificador m - [electrón] amplifier mounting self-tapping screw
— m — — — m — — m magnético - [electrón] magnetic amplifier mounting self-tapping screw
— m — — — m sobre respaldo m - [mec] back mounting self-tapping screw
— m — y arandela f - [mec] self-tapping screw and washer
— m cadmiado - [mec] cadmium plated screw
— m — con cabeza f plana - [mec] cadmium plated flat head screw
— m — — — f — para metal(es) m - [mec] cadmium plated flat head machine screw
— m — con casquete m - [mec] cadmium plated cap screw
— m — — — m con cabeza f hexagonal - [mec] cadnium plated hexagonal head cap screw
— m — para metal(es) m - [mec] cadmium plated machine screw
— n con aleta(s) f - [mec] butterfly screw thumb screw
— m — arandela f - [mec] washered screw
— m — f con saliente(s) m exterior(es) - [mec] Sems screw
— m — — prearmado - [mec] screw with @ permanent washer; Sems screw
— m — f premontada - [mec] preassembled washer, o Sems, screw
— m — bola f - [mec] ball screw
— m — cabeza f - [mec] machine screw • cap screw
— f — — f ancha f - [mec] broad head screw
— m — — f avellanada - [mec] countersunk headed screw
— m — — f cilíndrica - [mec] cheese, o fillister, head screw
— m — — f ranurada - [mec] fillister head screw
— f — — f — — convexa - [mec] convex fillister head screw
— m — — f — — plana - [mec] flat top fillister head screw
— m — — f — — para metal(es) m - [mec] fillister head machine screw
— f — — f con concavidad f hexagonal - [mec] Allen (head) screw
— m — — f con talón m - [mec] plow screw

tornillo m con cabeza f cuadrada - [mec] square head(ed) screw
— m — — f chata - [mec] flat, o cheese, head(ed), screw
— m — — f — ranurada - [mec] slotted cheese head(ed) screw
— m — — f como gota f de sebo m - [mec] round head(ed) screw
— m — — f embutida - [mec] flat head(ed), o countersunk, screw
— m — — f — para metal(es) m - [mec] flat head(ed) countersunk machine screw
— f — — f embutida - [mec] countersunk, o flush head(ed), screw
— m — — f fresada - [mec] countersunk, o machined head, screw
— m — — f hueca - [mec] socket screw
— m — — f — hexagonal - [mec] hexagonal socket head(ed) screw; Sems cap screw
— m — — f — para polo m a caja f - [electr-instal] pole to frame hexagonal head screw
— f — — f — soporte m para caja f - [mec] bracket to frame hexagonal head screw
— m — — f — — tapa f - cover screw hexagonal head screw
— m — — f — — — f para engranaje(s) m - [mec] gear cover hexagonal head screw
— m — r — — f para relleno m - [comb.-int] filler block hexagonal head screw
— m — — f hueca - [mec] socket (head) screw
— m — — f ovalada - [mec] oval head(ed) screw
— m — — f — para metal(es) m - [mec] oval head(ed) machine screw
— m — — f Phillips - [mec] Phillips head screw
— m — — f — con arandela f prearmada—[mec] Sems Phillips head screw
— m — — f plana - [mec] binding head screw
— m — — f plana - [mec] flat head(ed) screw • pan (headed) screw
— m — — f — biselada - [mec] pan head(ed) screw
— m — — f — para madera f - [mec] flat head wood screw
— m — — f — ranurada - [mec] flat fillister, o cheese, head screw
— m — — f rasa - [mec] flush head(ed) screw
— m — — f redonda - [mec] round headed screw
— m — — f — con arandela f prearmada - [mec] Sems round head(ed) screw
— m — — f — para montaje m - [mec] mouning Sems round head(ed) screw
— m — — f — — f — — — m de núcleo m laminar - [electr-equip] lamination mounting Sems round head(ed) screw
— m — — f — hueca - [mec] button head socket screw
— m — — f — para madera f - [mec] round head(ed) wood screw
— m — — f — para metal(es) m - [mec] round head(ed) machine screw
— m — — f segmental - [mec] truss head screw
— m — — f taladrada - [mec] drilled head screw
— m con casquete m - [mec] cap screw • socket (head) screw
— m — — m aflojado - [mec] loosened cap screw
— m — — m ajustado - [mec] tightened cap screw
— m — — m apretado - [mec] tightened cap screw
— m — — m — hasta par m motor de . . . - [mec] cap screw torqued to . . .
— m — — m con carbono m bajo - [mec] low carbon cap screw
— m — — m con cabeza f cilíndrica ranurada - [mec] fillister head cap screw
— m — — m — — f con hueco m hexagonal - [mec] hexagonal socket head cap screw
— m — — n con cabeza f hexagonal - [mec] hexagon head cap screw
— m — — m — — f — horadada - [mec] drilled

hexagonal head cap screw
tornillo m con casquete m con cabeza f hexago-
nal tratado térmicamente - [mec] heat treated hexagon head cap screw
— m — — m — — f **hueca** - [mec] socket head
cap screw
— m — — m — — f — **ranurada** - [mec] fluted
socket head cap screw
— m — — m — — f **redonda** - [mec] round head
cap screw • button head cap screw
— m — — m — — t **con arandela t prearma-**
da - [mec] Sams round head cap screw
— m — — m — — f **taladrada** - [mec] drilled
head cap screw
— m — — m **con carbono m bajo** - [mec] low
carbon cap screw
— m — — m **con concavidad f (hexagonal)** -
[mec] Allen head cap screw
— m — — m **corto** - [mec] short cap screw
— m — — m **instalado** - [mec] installed cap
screw
— m — — m **largo** - [mec] long cap screw
— m — — m **para caja f** - [mec] body, o case, o
housing, cap screw
— m — — m — — f **para cojinete m** - [mec]
bearing cage cap screw
— m — — m — — f — **diferencial m** - [mec]
[mec] differential case cap screw
— m — — m **para cobertura f** - [mec] cover cap
screw
— m — — m — **cojinete m** - [cojin] bearing
cap screw
— m — — m — **cubierta f** -[mec] cover cap
screw
— m — — m — — f **para caja f** - [mec] body
cover cap screw
— m — — m — — f — **cojinete m** - [mec]
bearing cover cap screw
— m — — m — — f — — **bomba f** - [bombas]
pump cover cap screw
— m — — m — — f — — f **para aceite m** -
[bombas] oil pump cover cap screw
— m — — m — — f — — m — **árbol m para en-**
trada f - [autom-mec] input shaft bearing
cover cap screw
— m — — m — — f — **diferencial m** -
[autom-mec] housing cover cap screw
— m — — m — — f — **cuerpo m** - [mec] body
cover cap screw
— m — — m — — f — **distribuidor m para**
fuerza f - [autom-mec] power divider cover
cap screw
— m — — m — **cuerpo m** - [mec] body cap screw
— m — — m — — f **removido** - [autom-mec] removed cover cap screw
— m — **diferencial m** - [mec] differential cap screw
— m — — m — **jaula f** - [mec] cage cap screw
— m — — m — — f **para cojinete m** - [cojin]
bearing cage cap screw
— m — — m — **motor m** - [mec] motor cap screw
— m — — m **para pista f** - [cojin] bearing race
cap screw
— m — — m — **portadiferencial m** - [autom-mec]
differential carrier cap screw
— m — — m — **rascador m** - [mec] scraper cap
screw
— m — — m — **raspador m** - [mec] scraper cap
screw
— m — — m — — m **para aceite m** - [mec] oil
scraper cap screw
— m — — m — **tapa f** - [mec] cover cap screw
— m — — m — — f **para cojinete m** - [mec]
bearing cover cap screw
— m — — m — **traba f** - [mec] lock cap screw
— m — — m — — f **para ajustador m** - [mec]
adjuster,ting lock cap screw
— m — — m — — f — — m **para cojinete m** -
[mec] bearing adjuster lock cap screw
— m — — m — **transformador m** - [mec] carrier
cap screw

tornillo m con casquete m para transportador m
diferencial - [mec] differential carrier
cap screw
— m — — m **quitado** - [mec] removed cap screw
— m — — m **removido** - [mec] removed cap screw
— m — — m **sacado** - [mec] removed cap screw
— m — **casquillo m** - [mec] socket screw
— m — — m **hexagonal** - [mec] hexagonal
socket screw
— m — **collar m** - [mec] collar screw
— m — — m **para barra f** - [mec] bar collar
screw
— m — — m — — f **retenedora** - [mec] retaining bar collar screw
— m — **concavidad f** - [mec] socket screw
— m — — f **hexagonal** - [mec] Allen (head)
screw
— m — **enrosque m propio** - [mec] váse tornillo
m autoroscante
— m **con filo m angular** - [mec] cup point screw
— m — — m **triangular** - [mec] angular screw
— m — **garra f (con cabeza f hendida)** - [mec]
prong screw
— m — **mariposa f** - [mec] wing, o butterfly,
screw
— m — **muletilla f** - [mec] tommy screw
— m — **núcleo m prolongado** - [mec] dog point
screw
— m — **ojo m** - [mec] eyebolt
— m — **oreja(s) f** - [mec] thumb screw
— m — **paso m fino** - [mec] fine thread screw
— m — — m **izquierdo** - [mec] left handed
thread screw
— m — . . . **paso(s) m por pulgada f** - [mec]
. . . (threads) per inch screw
— m — — m **sencillo** - [mec] single thread
screw
— m — — m **triangular** - [mec] (tri)angular
thread(ed) screw
— m — **punta ahuecada** - [mec] cup point screw
— m — — f — **para presión f** - [mec] cup point
set screw
— m — **resalto m cuadrado** - [mec] square
thread screw
— m — — m **fino** - [mec] fine thread screw
— m — — m **macho** - [mec] male thread screw
— m — — m **matriz** - [mec] female thread screw
— m — — m **redondeado** - [mec] round(ed) thread
screw
— m — — m **triangular** - [mec] triangular, o
V, thread screw
— m — — m **trapezoidal** - [mec] butress thread
screw
— m — — m **triangular** - [mec] angular thread
screw
— m — **rosca f** - [mec] threaded screw
— m — — f **con paso m fino** - [mec] fine pitch
screw
— m — — f **hacia derecha f** - [mec] right hand
screw
— m — — f — **izquierda f** - [mec] left hand
pitch screw
— m — — f **hembra** - [mec] female thread screw
— m — — f **macho** - [mec] male thread screw
— m — — f **redonda,deada** - [mec] round(ed)
thread screw
— m — — f **sencilla** - [mec] single thread
screw
— m — **tratamiento m térmico** - [mec] heat
treated screw
— m — **tuerca f** - [mec] bolt
— m **corto** - [mec] short screw
— m **con casquete m** - [mec] short cap screw
— m **de acero m** - [mec] steel screw
— m — m **con cabeza f redonda** - [mec] round
head(ed) steel screw
— m — **acero m inoxidable** - [mec] stainless
steel screw
— m — — m — **con cabeza f redonda** - [mec]
round head(ed) stainles steel machine screw

tornillo m de acero m inoxidable con casquete m - [mec] stainless steel cap screw
— m — m — m con cabeza f embutida - [mec] stainless steel socket head cap screw
— m — m — m — f hexagonal - [mec] stainless steel hexagonal socket head cap screw
— m — m para madera f - [mec] steel wood screw
— m — aluminio m - [mec] aluminum screw
— m — m con cabeza f hexagonal - [mec] aluminum hexagon head screw
— m — Arquímedes m - [hidr] Archimides screw • screw conveyor
— m — cobre m - [mec] copper screw
— m — latón m - [mec] brass screw
— m — m cabeza f hexagonal - [mec] brass hexagon head screw
— m — m con cabeza f redonda - brass round head, o round head brass, screw
— m — m — casquete m - [mec] brass cap screw
— m — m — m con cabeza f hexagonal - [mec] brass hexagon head cap screw
— m — con casquete m con cabeza f plana - [mec] brass flat head cap screw
— m — Palmer - [instrum] micrometric caliper
— m — polo m a caja f - [electr-instal] pole to frame screw
— m desatornillado - [mec] unscrewed, o backed out, screw
— m descendente - [mec] véase tornillo m para ajuste m
— m destornillado - [mec] unscrewed, o backed out, screw
— m diferencial - [mec] differential screw
— m embebido - [mec] countersunk (head) screw
— m empotrable - [constr] embedable screw
— m — en base f de hormigón m - [constr] concrete base embedable screw
— m — hormigón m - [constr] concrete embedable screw
— m empotrado - [mec] embedded screw
— m en bomba f - [bombas] pump screw
— m espaciador - [mec] spacing, o spreading, screw
— m esparcidor - [mec] spreading screw
— m estándar - [mec] standard screw
— m exterior - [mec] outside screw
— m para montaje m - [mec] outer mounting screw
— m flojo - [mec] loose screw
— m gemelo - [mec] twin screw
— m girado - [mec] turned screw
— m hecho girar v - [mec] turned screw
— m helicoidal - [mec] helical, o auger, screw; suger
— m hendido - [mec] prong, o split, screw
— m horadado - [mec] drilled screw
— m hueco - [mec] hollow screw
— m — para caja f - [mec] box hollow screw
— m — m — f sujetadora - [mec] grip box hollow screw
— m impulsor - [mec] drive,ving screw
— m inferior - [mec] lower, o bottom, screw
— m inoxidable - [mec] stainless (steel) screw
— m — con casquete m - [mec] stainless cap screw
— m — m con cabeza f hexagonal - [mec] stainless hexagonal socket head cap screw
— m — para ajuste m manual - [mec] stainless thumb screw
— m — para montaje m - [mec] inner, o inside, mounting screw
— m largo - [mec] long screw
— m — con casquete m - [mec] long cap screw
— m limitador m - [combust] stop screw
— m — para acelerador m - [comb.int] throttle stop screw
— m — estrangulador m - [comb.int] throttle stop screw
— m — regulación f - [comb.int] throttle stop screw

tornillo m limitador para regulador m - [comb.-int] throttle stop screw
— m lógico - [electrón] logic(al) screw
— m positivo/negativo - [electrón] logic(al) positive/negative screw
— m mal ajustado - [mec] misdajusted screw
— m mariposa - [mec] wing screw • thumb screw
— m micrométrico - [mec] micrometric screw • [insrum] micrometric caliper
— m obturador - [mec] plug(ging) screw
— m para aceite m - [mec] oil screw
— m — acoplamiento m - [mec] coupling screw
— m — adaptador m - [mec] adapter screw
— m — ajuste m - [mec] adjusting,tment,ter screw; set screw • [metal-prod] screwdown
— m — m con cabeza f cuadrada - [mec] square head, set, o adjuster, screw
— m — m — f con punta f hueca - [mec] square head cup point set screw
— m — m — f embutida f - [mec] socket head set screw
— m — m — f — hexagonal - [mec] hexagonal socket head set screw
— m — m — f — con filo m anular - [mec] hexagonal socket head cup point set screw
— m — m — f — punta f ahuecada - [mec] hexagonal socket (head) cup point set screw
— m — m — f con hueco m hexagonal - [mec] hexagon(al) socket set screw
— m — m — f cuadrada - [mec] square head set screw
— m — m — f hexagonal - [mec] hexagon(al) head set screw
— m — m — f hueca - [mec] socket set screw
— m — m — casquillo m hexagonal - [mec] Allen socket set screw
— m — m — filo m anular - [mec] cup point set screw
— m — m motor m - [mec] motor operated screwdown
— m — m — núcleo m prolongado - [mec] dog point set screw
— m — m — punta f ahuecada - [mec] cup point set screw
— m — m de cuña f - [mec] wedge adjustment screw
— m — m derecho - [lam] right screwdown
— m — m hexagonal - [mec] hexagonal set screw
— m — m — con casquete m - [mec] hexagonal cap set screw
— m — m — m y punta f ahuecada - [mec] hexagonal (head) socket cup point set screw
— m — m hueco - [mec] hollow set screw
— m — m — con filo m angular - [mec] cup point hollow setscrew
— m — m izquierdo - [metal-lam] left screwdown
— m — m manual - [mec] thumb screw
— m — m — abierto - [mec] opened thumbscrew
— m — m — aflojado - [mec] loosened, o opened, thumb screw
— m — m — ajustado - [mec] tightened thumbscrew
— m — m — apretado - [mec] tightened thumbscrew
— m — m para caja f - [mec] case setscrew
— m — m — camisa f - [mec] liner setscrew
— m — m — carrete m - [sold] reel adjusing screw
— m — m — m para alambre m - [mec] wire reel adjusting (set)screw
— m — m — cilindro m - [petról] cylinder set screw
— m — m para cuña f - [mec] wedge adjusting screw
— m — m para chumacera f - [mec] pillow block adjusting screw

tornillo m para ajuste m para empaquetadura f -
[petról] packing adjuster screw
— m — — m — — f para camisa f - [petról|
liner packing set screw
— m — — m para freno m - [mec] brake adjusting screw
— m — — m — — m para carrete m - [sold]
reel brake adjusting screw
— n — — m — — m — — m para alambre m -
[solkd] wire reel brake adjusting screw
— m — — m para inyector m de baja f - [mec]
idle jet adjusting screw
— m — — m — — m para marcha f en vacío -
[mec] idle jet adjusting screw
— m — — m para laminador m - [metal-lam]
mill screwdown
— m — — m — — m terminador - [metal-lam]
finishing mill screwdown
— m — — m para palanca f - [mec] throttle
lever set screw
— m — — m — — f — — m para velocidad f
alta en vacío m - [comb.int] high idle speed
throttle lever set screw
— m — — m — — f — — m — — f baja en
vacío - [comb.int] low idle speed throttle
lever set screw
— m — — m para pasador m - [mec] pin adjusting screw
— m — — m — — m excéntrico - [mec] eccentric pin adjusting screw
— m — — m — programador m - [metal-prod]
screwdown porogrammer
— m — — m — — regulador m - [mec] governor
speed screw
— m — — m — — resorte m - [mec] spring adjustment screw
— m — — m — — m para regulador n - [mec]
governor spring adjustment screw
— m — — m — — m — — m malajustado - [mec]
misajusted governor spring adjustment screw
— m — — m — cuña f - [mec] wedge adjusting
screw
— m — — m para tapa f - [mec] cover adjusting screw
— m — — m para velocidad f - [mec] speed (adjusting,tment) screw
— m — — m — — f regulado - [mec] adjusted
speed screw
— m — alambre m - [mec] wire screw
— m — — m para mando m - [mec] control wire
screw
— m — alzaválvula(s) m - [comb.int] valve tappet screw
— m — amplitud f - [electrón] amplitude screw
— m — — f girado - [electrón] rotated amplitude screw
— m — apretadura f - [mec] tightening screw
— m — — f manual - [mec] thumb screw
— m — arado m - [mec] plow screw
— m — árbol m - [mec] shaft screw
— m — — m para leva(s) f - [mec] camshaft
screw
— m — — m para rodillo m loco - [mec] idle
roll shaft screw
— m — armadura f - [mec] armature screw
— m — armar v - assembly,bling screw
— m — v paleta f - [ventil] blade assembling screw
— m — avance m - [mec] lead screw • temper
screw
— m — bajada f - [mec] screw down
— m — banco m - [mec] vise • check screw
— m — m con base f giratoria - [mec]
swivel base vise
— m — barra f - [mec] bar screw
— m — — f retenedora - [mec] retaining bar
screw
— m — bastidor m - [mec] frame screw
— m para bloqueo m - [mec] (b)locking screw
— m — bomba f - [bombas] pump screw
— m — — f para purga f - [bombas] pump
bleed(ing) screw

tornillo m para bomba f para sangría f -
[bombas] pump bleed(ing) screw
— m — botador m - [mec] tappet screw
— m — — m para alzaválvula(s) m - [comb.-
-int] valve tappet screw
— m — cable m - [mec] cable screw
— m — — m para estrangulador m - [comb.int]
choke cable screw
— m — caja f - [mec] case, o box, o housing,
screw
— m — — f diferencial - [mec] differential
case screw
— m — — f para cojinete m - [mec] bearinq
cage (cap) screw
— m — — f — diferencial m - [mec] differential case (cap) screw
— m — — f — embrague m - [mec] clutch housing screw
— m — — f — pistola f - [sold] gun housing
screw
— m — — f sujetadora - [trefil] grip (box)
screw
— m — campana f - [ventil] canopy screw
— m — carruaje m - [mec] carriage bolt
— m — casquete m - [mec] cap screw
— m — — m para cojinete m - [mec] bearing
cap screw
— m — — m — extremo m — [mec] end cover screw
— m — — m — — m para excitador m - [electr-
-prod] exciter end cover screw
— m — cierre m - [mec] closure, o locking,
screw; seal bolt • [metal-prod] seal(ing)
bolt
— m — — m ajustado - [mec] tight(ened) locking screw
— m — — m apretado - [mec] tightened locking screw
— m — — m para gas m - [metal-prod] gas seal
bolt
— m — — m — tapa f - [mec] cover lock(ing)
screw
— m — — m — — f ajustado - [mec] tight-
-(ened) cover locking screw
— m — — m — — f apretado - [mec] tight-
-(ened) cover locking screw
— m — — m — — f para válvula f - [mec]
valve cover locking screw
— m — — m — — f — — f ajustado - [mec]
tight(ened) valve cover locking screw
— m — — m — — f — — f apretado - [mec]
tight(ened) valve cover locking screw
— m — — m — — f — — f verificado - [mec]
checked valve cover locking screw
— m — — m — — f verificado - [mec] checked
cover locking screw
— m — — m verificado - [mec] checked locking
screw
— m — cinta f - [frenos] lining screw
— m — — f para freno m - [mec] brake lining
screw
— m — clavar v - [mec] drive,ving screw
— m — — v con cabeza f redonda - [mec] round
head drive,ving screw
— m — coche m - [mec] carriage bolt
— m — collar m - [mec] collar screw
— m — conectar v - [electr-instal] connecting
screw
— m — — sujetador m - [electr-instal] clamp
connecting screw
— m — conexión f - [mec] connecting,tion screw
— m — — f con tierra f — [mec] grounding screw
— m — — f de conductor(es) m - [electr-insta]
lead connecting screw
— m — conmutador m - [electr-equip] switch
screw
— m — cremallera f - [mec] rack screw
— m — — f para fijación f - [mec] clamping
rack screw
— m — — f — sujeción f - [mec] clamping rack
screw
— m — cubierta f - [mec] cover screw
— m — — f para caja f — [mec] case cover screw

tornillo m para cubierta f para cojinete m -
[mec] bearing cover screw
— m — — f — — m para árbol m - [mec] shaft
bearing cover screw
— m — — f — — m — — m para entrada f -
[autom-mec] input shaft bearing cover screw
— m — — f para caja f - [mec] body cover
screw
— m — — f — engranaje m - [comb.int] gear
cover screw
— m — — f — inspección f - [mec] inspection
cover screw
— m — — f — portadiferencial m - [autom-mec]
differentiala carrier cap screw
— m — — f — válvula f - [petról] valve cover
screw
— m — cubresoporte m - [mec] bracket cover
screw
— m — cuña f - [mec] wedge screw
— m — cúpula f - [ventil] canopy screw
— m — chapa f - [mec] sheet (metal) screw
— m — f metálica(s) f - [mec] sheet metal
screw
— m — deflector m - [mec] baffle screw
— m — desplazador m - [mec] shifter screw
— m — — m para embrague m - [mec] clutch
shifter screw
— m — diferencial m - [mec] differential (cap)
screw
— m — disco m - [mec] disk screw
— m — — m a cubo m - [mec] disk to hub screw
— m — distribuidor m - [metal-prod] distributor screw
— m — embrague m - [mec] clutch screw
— m — embutir v - [mec] countersunk headed
screw
— m — estator m - [electr-mot] stator screw
— m — estrangulador m - [comb.int] choke screw
— m — excéntrico m - [mec] eccentric screw
— m — fijación f - [mec] lock(ing) screw; setscrew; stay bolt
— m — — f ajustado - [mec] adjusted, o tightened, setscrew
— m — — f — hacia adelante - [mec] setscrew
adjusted forward(s)
— m — — f — — atrás - [mec]
screw adjusted backwards
— m — — f — camisa f - [mec] liner locking
screw
— m — — f para conexión f - [mec] connection
locking screw
— m — — f — — f eléctrica - [electr-instal]
electrical connection locking screw
— m — freno m - [mec] brake screw
— m — — m para carrete m - [mec] reel
brake screw
— m — — m — — m para alambre m - [sold]
wire reel brake screw
— m — gato m - [mec] jack screw
— m — globo m - [electr-instal] globe screw
— m — grillete m - [mec] clevis clamp screw
— m — guardaventilador m - [electr-mot] fan
guard screw
— m — guia(dera) f - [mec] guide tube screw
— m — f para entrada f - [mec] ingoing guide
tube screw
— m — horquilla f - [mec] clevis (clamp) screw
— m — impulsión f - [mec] drive,ving screw
— m — — f hacia derecha f - [mec] right hand
drive,ving screw
— m — — f — izquierda f - [mec] left hand
drive,ving screw
— m — — f para accionador m - [mec] actuator
drive,ving screw
— m — — f — — m para cuadrante m - [instr]
actuaor to quadrant drive,ving screw
— m — interruptor m - [electr-equip] switch, o
breaker, screw
— m — jaula f - [mec] cage screw
— m — — f para cojinete m - [mec] bearing
cage screw
— m — limitación f - [combust] limiting, o
stop(ping) screw

tornillo m para limitación f para carga f - [mec]
load limit(ing) screw
— m — madera f - [mec] wood screw
— m — mandíbula f - [mec] jaw screw
— m — — f de cobre m - [mec] copper jaw screw
— m — máquina f - [mec] machine bolt • véase
también tornillo m para metal m
— m — marcha f sin carga f - [comb.int] idle
screw
— m — ménsula f - [mec] bracket screw
— m — — f — — m de platino(s) m - [comb.int]
point(s) bracket mounting screw
— m — metal m - [mec] machine screw
— m — — m con canto m con cabeza f redpmdeada
- [mec] pan head machine screw
— m — — m — — f cilíndrica ranurada - [mec]
flat top fillister head machine screw
— m — — f sujetadora - [mec] binding
head machine screw
— m — montaje m - [mec] mounting screw
— m — — m aflojado - [mec] loosened mounting
screw
— m — — m ajustado - [mec] tightened mounting screw
— m — — m apretado - [mec] tightened mounting
screw
— m — — m exterior - [mec] outer, o outside,
mounting screw
— m — — m interior - [mec] inner, o inside,
mounting screw
— m — — m para adaptador m - [mec] adapter
mounting screw
— m — — m — arrancador m - [comb.int]
starter mounting screw
— m — — m con arrollamiento m automático - [comb.int] recoil starter mounting
screw
— m — — m para bloque m - [mec] block mounting screw
— m — — m — — m para enclavamiento m -
[electr-instal] interlock(ing) block mounting
screw
— m — — m — caja f - [mec] housing mounting
screw
— m — — m — casquete m - [mec] cover mounting screw
— m — — m — — m para extremo m - [mec] end
cover mounting screw
— m — — m — — m de excitador m -
[electr-instal] exciter end cover mounting
screw
— m — — m — condensador m - [comb.int] condenser mounting screw
— m — — m — cubierta f - [mec] cover, o
shroud, mounting screw
— m — — m — — f para engranaje(s) m - [mec]
gear cover mounting screw
— m — — m — cubresoporte m - [mec] bracket
cover mounting screw
— m — — m — deflector m - [mec] baffle
mounting screw
— m — — m — depurador m para aire m -
[comb.int] air cleaner mounting screw
— m — — m — equipo m - [mec] kit mounting
screw
— m — — m — — m para excitador m - [electr-
-prod] exciter end mounting screw
— m — — m — horquilla f - [mec] yoke mounting screw
— m — — m — interruptor m - [comb-int] breaker, o switch, mounting screw
— m — — m — — m para detención f - [comb.-
int] stop(ping) switch mounting screw
— m — — m — motor m - [electr-mot] motor
mounting screw • [comb.int] engine mounting
screw
— m — — m — — m para arranque m - [electr-
-mot] starting motor mounting screw
— m — — m — núcleo m laminar - [electr-mot]
lamination mounting screw
— m — — m — placa f - [mec] plate mounting
screw
— m — — m — — f para escobilla f - [electr-

-equip] brush plate mounting screw
tornillo m **para montaje** m **para polea** f - [mec] pulley mounting screw
— m —— f **para arranque** m - [mec] start-up pulley mounting screw
— f —— m **para rejilla** f - [mec] screen mounting screw
— m —— f **para volante** m - [comb.- int] flywheel screen mounting screw
— m —— m **respaldo** m - [mec] back mount- ing screw
— m —— m **tapa** f - [mec] cover mounting screw
— m —— f **para engranaje(s)** m - [comb.int] gear cover mounting screw
— m —— f **para platino(s)** m - [comb- int] point(s) cover mounting screw
— m —— m **sobre respaldo** m - [mec] back mounting screw
— m — **mordaza** f - [mec] clamp bar
— m — **motor** m - [electr-mot] motor screw
— m — **muelle** m **accionador para anillo** m **porta- rrodillos** m - [mec] roller cage actuating spring screw
— m — **nivelación** f - [mec] leveling screw
— m — **núcleo** m - [electr-mot] core screw
— m —— m **para polo** m - [electr-mot] pole core screw
— m — **operación** f - [tornillo para banco] operating bar
— m —— f **de mordaza** f - [mec] clamp operat- ing bar
— m —— f **sin carga** f - [comb.int] idle (operating) screw
— m — **oscilador** - [mec] oscillator screw
— m —— m **a cubo** m—[mec] rocker to hub screw
— m — **paleta** f - [ventil] blade screw
— m — **pasador** m - [mec] pin screw
— m —— m **para excéntrico** - [mec] eccentric pin screw
— m — **paso** m - [mec] passage screw
— m —— m **para aceite** m - [mec] oil passage screw
— m — **pieza** f - [mec] part screw
— m —— f **para relleno** m - [mec] filler block screw
— m — **placa** f - [mec] plate screw
— m —— f **giratoria** - [mec] swivel screw
— m — **oscilante** - [mec] swivel screw
— m — **plato** m - [comb.int] end plate screw
— m — **portaescobilla(s)** m **a caja** f [electr-mot] brush holder to frame screw
— m — **posicionamiento*** m - [mec] positioning screw
— m —— m **para rodillo** m - [metal-lam] roll positioning screw
— m — **presión** f - [mec] clamping, o lock, screw; setscrew
— m —— f **para muelle** m - [mec] spring pres- sure screw
— m —— f —— m **para rodillo** m **loco**—[sold] idle roll spring pressure screw
— m — **puesta** † **a tierra** f - [electr-instal] grounding screw
— m — **purga** f - [comb.int] bleeding screw
— m —— f **aflojado** - [mec] loosened bleeding screw
— m — **rascador** m - [mec] scraper screw
— m — **reactor** m—[electr-instal] reactor screw
— m —— m **para regulación** f - [electr-instal] control reactor screw
— m —— f **para amperaje** m - [electr- insal] ampere(s) control reactor screw
— m — **regulación** f - [mec] adjustment, o con- trol, screw
— m —— f **para carburador** m - [comb.int] car- buretor adjusting screw
—— m —— f —— m **ajustado** - [comb.int] ad- justed carburetor adjusting screw
— m —— f — **velocidad** f - [mec] speed (ad- jusment) screw
— m — **regulador** m - [mec] misadjusted governor screw

tornillo m **para regulador** m **ajustado** - [mec] adjusted governor screw
— m — **resorte** m - [med] spring screw
— m —— m **ajustado** - [mec] adjusted spring screw
— m —— m **mal ajustado** - [mec] misadjusted spring screw
— m —— m **para presión** f - [mec] pressure spring screw
— m —— m — **regulador** m - [mec] governor spring screw
— m —— m —— m **mal ajustado** - [mec] mis- adjusted governor spring screw
— m — **respaldo** m - [mec] back screw
— m — **retención** f - [mec] retaining, o hold- ing, screw
— m —— f **aflojado** - [mec] loosened retain- ing screw
— m —— f **ajustado** - [mec] tightened, retain- ing, o holding, screw
— m —— f **apretado** - [mec] tightened, re- taining, o holding, screw
— m — **sangría** f - [comb.int] bleed(ing) screw
— m —— f **aflojado** - [mec] loosened bleeding screw
— m — **seguridad** f - [mec] safety screw • lock- screw
— m —— f **para ajuste** m - [mec] safety set- screw
— m — **separación** f **y refuerzo** m - [mec] stay bolt
— m —— f **para placa** f - [mec] plate lock- screw
— m —— f —— f **para sostén** m - [mec] plate support(ing) lockscrew
— m —— f —— f —— m **para contacto** m - [comb.int] contact support plate lockscrew
— m —— f **para tapón** m - [mec] plug lock screw
— m — **sombrerete** m - [mec] cap screw
— m — **soporte** m - [mec] bracket screw
— m —— m **a caja** f - [mec] bracket to frame screw
— m — **sujeción** f [mec] clamping, o holding, o hold down screw; setscrew
— m —— f **ajustado** - [mec] adjusted setscrew; tightened holding screw
— m —— f **apretado** - [mec] tightened hold- ing screw
— m —— f **para conexión** f **eléctrica** - elec- trical connection clamping screw
— m —— f — **guiadera** f **para entrada** f [mec] tube clamping screw
— m —— f —— f **para entrada** f - [mec] in- coming guide tube clamping screw
— m —— f —— m **pistola** f - [sold] gun tube clamping screw
— m — **tablero** m - [sold] panel screw
— m —— m **para mando** m - [sold] control panel screw
— m — **tapa** f - [mec] cover, o lid, screw
— m —— f **para cojinete** m - [cojin] bearing, cap, o cover, screw
— m —— f — **engranaje** m - [comb.int] gear cover screw
— m —— f —— m **a bloque** m - [comb.int] gear cover to block screw
— m —— f —— m **a pieza** f **para relleno** m - [comb.int] gear cover to filler block screw
— m —— f —— m — **plato** m - [comb.int] gear cover to end plate screw
— m —— f — **válvula** f - [mec] valve cover screw
— m —— f —— f **para succión** f - [petról] suction valve cover screw
— m — **tapón** m - [mec] plug screw
— m —— m **para agujero** m **para inyector** m - [carburador] nozzle hole plug screw
— m —— m —— m — **pulverizador** m - [comb- int] nozzle hole plug screw
— m —— m —— m — **surtidor** m - [comb.- int] nozzle hole plug screw
— m — **tensión** f - [mec] tension screw

tornillo m para tensión f para freno m - [mec] brake tension screw
— m — — m para eje m - [mec] spindle brake tension screw
— m — f — rodillo m loco - [mec] idle roll tension screw
— m — terminal m - [electr-instal] terminal screw
— m — tope m - [mec] stop, screw, o pin
— m — — m para acelerador m - [comb.int] throttle stop screw
— m — — m — conmutador m - [electr-instal] switch stop screw
— m — — m — — m para selección f - [electr-instal] selector switch stop screw
— m — — — palanca f—[mec] lever stop screw
— m — — m — f para mando m - [comb.int] control lever stop switch
— m — tornillo m para banco m - [herram] vise screw
— m — traba f - [mec] lockscrew
— m — trabar v - [mec] lock(ing) screw
— m — — v alambre m - [mec] wire lock screw
— m — — v — m para mando m - [mec] control wire lock screw
— m — tracción f - [mec] traction screw • cap screw
— m — — f con cabeza f hexagonal - [mec] hexagon(al) head cap screw
— m — transportador m - [mec] carrier screw
— m — — m diferencial - [mec] differential carrier (cap) screw
— m — unir v - [mec] joining screw
— m — — v parte(s) f - [mec] part(s) joining screw • cap screw
— m — válvula f - [válv] valve screw
— m — — f para aceleración f - [comb.int] throttle valve screw
— m — — f para estrangulación f - [comb.int] choke valve screw
— m — — f — — f y aceleración f - [comb.int] choke and throttle valve screw
— m — vástago m - [mec] rod screw
— m — — m para desplazador m - [mec] shifter rod screw
— m — velocidad f - [mec] speed screw
— m — — f alta - [mec] high speed screw
— m — — f — sin carga f - [mec] high idle speed screw
— m — — f — — — ajustado - [sold] adjusted high idle speed screw
— m — f baja - [mec] low speed screw
— m — — f — sin carga f - [sold] low idle speed screw
— m — — f — — — f ajustado—[sold] adjusted low idle speed screw
— m — f sin carga f - [mec] idle speed screw
— m — — f — — f ajustado - [mec] adjusted idle speed screw
— m — — f — — f mal ajustado - [mec] misadjusted idle speed screw
— m — volante m - [mec] flywheel screw
— m — — m a árbol m para leva(s) f - [mec] flywheel to crankshaft screw
— m piloto m - [mec] pilot screw
— m — para acelerador m - [int.comb] throttle pilot screw
— m principal - [mec] lead, o main, screw • [mec·tornos] lead screw
— m provisto - [mec] provided screw
— m regulador - [mec] temper screw
— m — para paso m - [mec] passage screw
— m — — — m para aceite m - [mec] oil passage screw
— m remachado - [mec] rivited screw
— m — para anclaje m - [mec] (riveted) anchor bolt
— m removido - [mec] removed bolt
— m roscador m - [mec] thread cutting screw
— m — para montaje m - [electr-instal] mounting thread cutting bolt
— m — — — m de transformador m - [electr-instal] transformer mounting thread cutting

tornillo m roscador para montaje m de transformador m - [electr-instal] transformer mounting thread cutting screw
— m — para soporte m - [mec] bracket mounting thread cutting screw
— m — — soporte m - [mec] bracket thread cutting screw
— m sacado - [mec] removed screw
— m seguro - [mec] secure, o safe, screw
— m — para campana f - [mec] secure canopy screw
— m Sems - [mec] Sems screw; screw with @ preassembled washer
— m separador m - [mec] spacing, o spreading, screw
— m set - [mec] set screw
— m sin cabeza f - [mec] headless, o plug, screw
— m sin fin - [mec] worm (gear); helical, o drive, screw • auger
— m — engranado - [mec] engaged auger
— m — — m para avance m - [mec] screw feed
— m — — m — descarga f - [agric] unloading auger
— m — — para remover v polvo m - [metal-prod] pug mill
— m — — — v — m de base f de extractor m (de polvo m de horno m alto) - [metal-prod] pug mill
— m — pintar v - [mec] unpainted screw
— m — — v para bastidor m - [mec] unpainted frame screw
— m — — v — tablero m - [sold] unpainted panel screw
— m — — v — — m para mando m - [sold] unpainted control panel screw
— m — tuerca f - [mec] (nutless) screw
— m superior - [mec] top screw
— m trabador - [mec] lock(ing) screw
— m — para cremallera f - [mec] rack locking screw
— m — — — f para fijación f - [mec] clamping rack locking screw
— m transportador - [mec] screw conveyor; conveyor screw
— m tratado m (térmicamente) - [mec] heat treated screw
— m tuerto - [mec] véase tornillo m zurdo
— m verificado - [mec] checked screw
— m y arandela f - [mec] screw and washer
— m — — f de seguridad f - [mec] screw and lockwasher
— m zurdo - [mec] left handed screw
torniquete m - . . . . • [mec] swivel
torno m - . . . . • [herram] vise • [metal-lam] (roll) grinder • [petról] cathead • bull block
— m automático - [mec] automatic lathe • [petról] automatic cathead
— m colocado - [petról] installed cathead
— m con plato m - [herram] chuck, o face plate, lathe
— m con punta(s) f - [herram] center lathe
— m con tambor m - [mec] drum winch
— m con torrecilla f - [herram] turret lathe
— m copiador - [herram] contour lathe
— m corriente - [mec] common, o engine, lathe
— m de tipo m corriente - [mec] regular type lathe
— m elevador - [metal-prod] hoist
— m — manual - [mec] (hand) winch
— m enroscador - [petról] make-up cathead
— m hidráulico - [mec] hydraulic winch
— m horizontal - [mec] engine lathe
— m Hyster - [cabl] Hyster winch
— m instalado - [mec] installed winch • [petról] installed cathead
— m lubricado - [mec] lubricated winch • [petról] lubricated cathead
— m manual - [mec] hand hoist
— m — para elevación f - [mec] hand operated hoist
— m mecánico - [mec] engine lathe

torno m mecánico de tipo m corriente - [mec] regular type engine lathe
— m **para armado** m - [petról] make-up cathead
— m — **banco** m - [mec] bench lathe • vise
— m — **bobinado** m **de armadura(s)** f - [electr] armaure machine
— m — m — **inducido(s)** m - [electr-mot] armature machine
— m — **cañónes** m - [mec] gun(barrel) lathe
— m — **desconexión** f - [petról] breakout cathead
— m — **desenroscadura** f - [petról] breakout cathead
— m — **desenroscamiento** m - [petról] breakout cathead
— m — **devanado** m - [electr] armature machine
— m — **elevación** f - [constr] hoisting winch • jack roll
— m — **enroscadura** f - [petról] make-up cathead
— m — **enroscamiento** m - [petról] make-up cathead
— m — **fabricación** f **(en serie)** - [herram] manufacturing lathe
— m — **inducido(s)** m—[electr] armature machine
— m — **inversión** f - [valv] butterfly valve
— m — **lingote(s)** m - [metal-prod] ingot lathe
— m — **mano** f - [mec] windlass; winch
— m — **mesa** f - [mec] bench lathe
— m — **producción** f - [herram] manufacturing lathe
— m — **refrentar** v - [herram] facing lathe
— m — **reserva** - [petról] make-up cathead
— m — **rodillo(s)** m - [metal-lam] roll (turning) lathe
— m — **roscar** v - [herram] engine lathe
— m **paralelo** m - [mec] engine lathe
— m **revólver** - [herram] turret lathe
— m **vertical** - [mec] (vertical) bull lathe
torón m -[quim] . . . • [cabl] véase **cordón** m; **ramal** m
torpedeado,da a - [milit] torpedoed • shot; fired
torpedeamiento m **de pozo** m - [petról] well shooting
torpedear v - [milit] . . . . • [explos] to shoot
— v **pozo** m - [petról] to shoot @ well
torpedeo m - . . . • [petról] shooting
— m **de pozo** m - [petról] well shooting
torpedo m - . . . . • [metal-prod] hot metal car • [autom] cowl
— m **para limadora** f - [mec] shaper ram
torque* m - véase **torsión** f
torre f - . . . . • [petról] rig; derrick
— f **aislada** - [metal-prod] isolated tower
— f **catalizadora** f - catalytic tower
— f **con extensión** f - [petról] telescoping derrick
— f — f **enchufada** - [petról] telescoping derrick
— f — **sección(es)** f **enchufada(s)** - [petról] telescoping derrick
— f **costa afuera** - [petról] off-shore rig
— f **de acero** m - [constr] steel tower
— f — m **corrugado** - [constr] corrugated steel tower
— f — m **estructural** - [constr] structural steel tower
— m — **Glover** - [petról] Glover's tower
— f — **Gay Lussac** - [fís] Gay Lussac tower
— f **destiladora** - [petról] distillation,ling tower
— f **elevada** - elevated, o high, tower
— f — **extremo** m **final de línea** f - [electr-instal] dead end tower
— f **enfriadora** - [inc] cooling tower
— f **exenta** - [metal-prod] isolated tower
— f **fraccionadora** - [petról] fractioning tower
— f **húmeda** f - [ind] wet tower
— f **lavadora** - [coque] separating,tor tower
— f — **para aceite** m - [coque] oil(s) separator tower
— f **movida** - [constr] moved tower
— f **para absorción** f - absorption tower

torre f **para absorción** f **para amoníaco** m - [sub.-prod] ammonia absorbing,rption column
— f — **alumbrado** m - [electr-instal] lighting tower
— m — **anclaje** m - anchor(ing) tower
— m — **apagamiento** m - [coque] quenching tower
— f — **burbujeo** m - [petról] bubble,ling tower
— f — **cabeza** f **para línea** f - [electr-instal] dead end tower
— f — **carbón** m - [coque] coal tower • oven bin
— f — m **vegetal** - [ind] charcoal tower
— f — m **activado** - [ind] activated charcoal tower
— f — **carga** f - [ind] loading tower
— f — **compuerta** f - [constr] gate, tower, o wall
— m — **contacto** m - contact tower
— m — **depósito** m - [ind] tank tower
— m — m **elevado** - [hidr] elevated tank tower
— f — **empalme** m - [electr-instal] junction tower
— f — **enfriamiento** m - [ind] cooling tower
— f — **enlace** m - [electr-instal] junction tower
— f — **fraccionamiento** m - [petról] fractioning tower
— f — **hacer granalla** f - [metal-prod] shot tower
— f — v **munición(es)** f - [metal-prod] shot tower
— f — **lavado** m - washing tower • [coque] separator tower
— m — m **para aceite** m - [coque] oil separator tower
— m — m — m **liviano** - [coque] light oil(s) separator tower
— f — **muestreo(s)** m - [ind] sampling tower
— f — **perforación** f - [petról] (drilling) rig; drawworks; derrick
— f — f **actualizada** - [petról] advanced design derrick
— f — f **comprobada** - [petról] proven derrick • tested derrick
— f — **en yacimiento** m **petrolífero** - [petról] oil field derrick
— f — f **específica** - [petról] particular, o specific, rig , o derrick
— f — **precalentamiento** m - [ind] preheating tower
— f — **refrigeración** f - [ind] cooling tower
— f — **separación** f - [coque] separating,tion tower
— f — **teledifusión** f - [telecom] television tower
— f — **toma** f - [hidr] shaft intake
— f — **transferencia** f - [ind] tramsfer tower
— f — **transmisión** f - [telecomun] transmission,mitting tower
— f — **unión** f - [electr-instal] junction, o joining, tower
— f **plegadiza** f - [petról] telescoping derrick
— f **seleccionada** - chosen tower
— f **terminal** - [electr-instal] dead end tower
— f **transmisora** - [telecomun] transmission,mitting tower
— f **trasladada** - (re)moved tower
torrecilla f - [tornos] turret
— f **con ocho estación(es)** f - [tornos] turret; (tool) carriage
torrente f **desbordante** - [hidr] overflowing torrent
torrero m - . . . . • [petról] derrick man
torsión f - [mec] . . .; bending • torque
— f **alta** - [mec] high torque,quing
— f **aplicada** - [mec] applied torque,quing
— f **baja** - [mec] low torque,quing
— f **cruzada** - [mec] cross torquing
— f **de muñeca** f - wrist, twist(ing), o turning
— f — **neumático** m - [autom-neumát] tire rollover

torsión f de perno m - [mec] bolt torquing
— f — — m de por sí—[mec] bolt self torquing
  • mere torquing
— f — tono m - [comput] tone twist(ing)
— f elevada - [mec] high torque,quing
— f — tuerca f - [mec] nut torque,quing
— f excesiva - [mec] excessive, twist(ing), o
  torque(quing)
— f leve - [mec] slight twist(ing)
— f — de muñeca f - [mec] slight wrist motion
— f máxima - [mec] maximum, twist(ing), o
  tarque,quing
— f mínima - [mec] minimum, twist(ing), o
  toque,quing
— f muy leve de muñeca f - wrist very slight,
  motion, o twist(ing), o torque,quing
— f necesaria - [mec] needed, o necessary,
  torque,quing • required torque,quing
— f por impacto(s) m - [mec] torsion impact
— f requerida - [mec] required torque,quing
— f — para apertura f - [mec] required opening
  torque,quing
torsionalmente adv - [mec] torsionally
— f mezclada - [culin] mixed cake • marble(d)
  cake
— f para filtro m - [mec] filter cake
tortilla f de escoria f - [metal-prod] (slag)
  (pan)cake
tortuoso,sa a - . . . . ; twisting,ty; circuitous;
  meandering
tosca f - . . . . • [geol] hardpan
— f arcillosa f - [geol] clayey hardpan; hardpan clay
— f — azul - [geol] hardpan blue clay
tosco,ca a - rough • coarse
tosquilla f - [geol] agglomeration • fine
  grained, hardpan, o compact subsoil
tostación - [miner] roasting
— f clorurante - [miner] chloridizing roasting
— f de concentrado(s) m— [miner] concentrates
  roasting
— f sulfatizante - [miner] sulfatizing roasting
tostado m - . . . ; roasting • tan(ning)
tostado,da a - toasted • roasted • brown(ed);
  tan(ned)
— a de sol m - sun baked
tostador,ra a - toasting; roasting
— f para wáfeles m - [culin] waffle, grill, o
  iron
tostadura f - . . . • tan(ning)
total m a favor - [com] total credit
— m aportado - total, input, o invested
— m de acero m - [metal-prod] total steel
— m — — m elaborado - [metal-prod] total
  steel produced
— m — exportación(es) f - [com] exports total
— m — inducción f - [electr] induction total
— m — — f magnética - [electr] magnetic induction total
— m — ingreso(s) m - [com] total income
— m — — m vario(s) - [contab] total sundry
  income
— m — producto m - [ind] total product
— m — — m terminado - [ind] total finished
  product
— m — punto(s) m - point(s) total
— m — venta(s) f - [com] total sale(s)
— m — — f hasta fecha f - [com] total sales
  to date
— m — — f para mes m - [com] total sales for
  @ month
— m discriminado - broken down total
— m disponible - available total
— m máximo - maximum total
— m mundial - world total
— m para país m - [mec] country total
— m por tonelada f - [ind] total per ton
— m — — f producida - [ind] total per ton
  produced
— m prerregulado* - preset total
— m — alcanzado - reached preset total
— m producido - total produced
totalidad f de garganta f - [sold] full throat
  section

totalidad f de garganta f de cordón m - [sold]
  full throat section
— f — — f — — m ligeramente convexa f -
  [sold] slightly convex bead full throat section
— f — pieza f - [mec] entire, o full part
— f — — f de hierro m - [metal] entire, o
  full, iron part
— f — — f — — m colado - [metal-prod] entire, o full, cast iron part
— f — — f — — m fundido - [metal-prod]
  entire, o full, cast iron part
— f — servicio(s) m - total service(s)
— f — suministro(s) m - [ind] total supplies
totalización f - totalizing,zation
totalizado,da a - totalized
totalizador m - totalizer; totalizing device
totalizador,ra a - totalizing
totalizante a - totalizing
totalizar v - . . . ; to total(ize)
totalmente adv - . . . ; integrally: entirely;
  throughout; absolutely; all @ way; distinct-
  ly; solidly; one hundred per cent
— adv a derecha f - totally, o extremely, right
— adv a izquierda f - totally, o extremely,
  left
— adv a obscuras - in total darkness
— adv abierto,ta - fully open
— adv aceptable - totally, o fully, acceptable
— adv aislado,da - fully insulated
— adv ajustable - fully adjustable
— adv automático,ca - fully automatic
— adv automatizado - fully automated
— adv bajo techo m - [constr] fully, roofed,
  o under @ roof
— adv blindado,da - fully armored
— adv calcinado,da - [miner] dead burned
— adv cansado,da - bone tired
— adv cerrado,da - completely, o fully, (en)
  closed
— adv colado,da - [metal-prod] integrally cast
— adv de acero m - [metal] all steel
— adv — aluminio m - [metal] all aluminum
— adv — chatarra f - [metal] all scrap
— adv encablado,da - [electr-instal] comletely
  wired
— adv endurecido,da - [metal-prod] (fully)
  hard tempered
— adv — mediante temple m - [metal-prod]
  full(y) hard tempered
— adv engrasado,da - [mec] fully greased
— adv extendido,da - fully extended
— adv flojo,ja - [mec] fully, loose, o released
— adv flotante - [mec] totally floating
— adv fundido,da - [petról] integrally cast
— adv galvanizado,da - [metal] fully galvanized
— adv inmovilizado,da - fully restrained •
  highly restrained
— adv inseguro,ra - totally unsafe
— adv insertado - fully inserted; inserted all
  @ way
— adv integrado,da - [metal-prod] fully integrated
— adv libre - totally, o fully, free
— adv metálico,ca - all metal(lic)
— adv no destructivo - completely nondestructive
— adv nuevo,va - totally, o completely, o all,
  o brand, new
— adv portante - fully conveying
— adv regulable - fully adjustable
— adv retraído,da - fully withdrawn • [mec]
  fully collapsed
— adv revestido,da - fully, o 100%, lined
— adv satisfactorio,ria - totally, o completely satisfactory • most enjoyable
— adv seguro,ra - absolutely, o fully, safe
— adv soldado,da - [sold] totally, o fully,
  welded
— adv tragado,da - [mec] completely swallowed
— adv universal - fully, o totally, universal
toza f - [petról] cant
tozar v - to insist

traba f - [mec] . . .; lock; locking; binder;
binding; freeze; freezing • drag • ⌐constr⌐
key; brace,cing • [electr] interlocking •
[electr-mot] rotor locking
— f **activada** - [mec] engaged lock
— f **ajustadora** - [mec] adjuster,ting lock
— f — **para cojinete** n - [mec] bearing adjuster,ting lock
— f **con** . . . **posición(es)** f - [grúas] . . .
position lock
— f — . . . — f **contra rotación** f - [grúas]
. . . position swing lock
— f **contra rotación** f - [grúas] swing lock
— f **de alambre** m - [mec] wire lock(ing)
— f — **armadura** f - [electr-mot] armature,
binding, o freezing
— f — **árbol** m - [mec] shaft binding
— f — **cojinete** m - [mec] bearing freezing
— f — **cuerda** f - [cabl] rope lock(ing)
— f — **engranaje** m - [mec] gear binding
— f — **rueda** f - [mec] wheel locking
— f — **vapor** m - vapor lock(ing)
— f **derecha** - [mec] right lock
— f **desconectada** - [mec] released lock
— f **desengranada** - [mec] disengaged lock
— f **en batalla** f - [deport] battle locking
— f — **posición** f **cerrada** - [mec] closed locking
— f **engranada** - [mec] engaged lock
— f **fijada** - [mec] set lock
— f **hidráulica** - [mec] hydraulic lock(ing)
— f — **para cabina** f - [grúas] hydraulic cab
lock
— f **impulsora** - [mec] drive,ving lock
— f **insertable** - [mec] drive lock
— f **izquierda** - [mec] left lock
— f **momentánea** - [mec] momentary latch(ing)
— f **para aguilón** m - [grúas] boom lock
— f — — m **fijada** - [grúas] set boom lock
— f — **ajustador** m - [mec] adjuster lock
— f — m **para cojinete** m - [mec] bearing adjuster lock
— f — **ajuste** m - [mec] adjuster,ting lock
— f — m **para cojinete** m - [mec] bearing adjuster,ting lock
— f — **alambre** m - [mec] wire lock
— f — **ariete** m - [mec] ram lock
— f — **cabello** m - [vest] bobby pin
— f — **cabina** f - [grúas] cab lock
— f — **cojinete** m - [mec] bearing lock
— f — **contratuerca** f - [mec] jam nut lock(ing)
— f — **desplazador** n - [mec] shifter lock
— f — **fiador** m - [mec] retainer lock
— f — m **para resorte** m - [mec] spring retainer lock
— f — m — — m **para válvula** f - [mec]
valve spring retainer lock
— f — **freno** m - [mec] brake lock
— f — m **conectado** - [mec] connected brake
lever
— f — m **desconectado** - [mec] released brake
lever
— f — m **engranada**—[mec] engaged brake lock
— f — m **para estacionamiento** m - [autom]
parking brake lock
— f — m — — m **engranada** - [mec] engaged
parking brake lock
— f — m **sacada** - [mec] released brake lock
— f — **gato** m - [mec] jack lock
— f — **leva** f - [mec] cam lock
— f — **mesa** f - [petról] table lock
— f — f **rotatoria** - [petról] table lock
— f — **palanca** f - [mec] lever lock
— f — f **para desplazador** m - [mec] shifter
lever lock
— f — **perno** m - [mec] bolt lock • locking clip
— f — **puerta** f - [mec] door, lock, o latch
— f — **resorte** m - [mec] spring lock
— f — — m **para válvula** f - [válv] valve
spring lock
— f — **retén** m - [mec] retainer lock
— f — **sujetador** m - [mec] cam lock
— f — **trinquete** m - [mec] pawl lock(ing)

traba f **para válvula** f - [válv] valve lock
— f **positiva** - [mec] positive lock
— f — **contra rotación** f - [grúas] positive
swing lock
— f — — **sujetador** m - [mec] positive cam
lock
— f **sacada** - [mec] released lock
— f **verificada** - [mec] checked lock
trabado,da a - [mec] frozen; locked; bound;
seized; jammed • [constr] shoved
— a **en lugar** m - [mec] locked in @ place; in @
place locked
— a — **movimiento** m - locked in, operation, o
movement
— a — **posición** f - [mec] locked in position
— a — — f **abierta** - [mec] open position
locked
— a — — f **cerrada** - [mec] closed position
locked
— a **entre sí** - [mec] interlocked
— a **totalmente** - [mec] totally locked •
[constr] full shored
trabador m - [mec] lock
— m **de alambre** m - [mec] wire lock; lockwire
— m **para serrucho** m - [herram] saw set
— m — **sierra** f - [herram] saw set
trabador,ra a - [electr-oper] (inter)locking
trabadura f - [mec] freeze; freezing; lock(ing);
binding; seizing,zure
— f **de acelerador** m - [comb.int] throttle
sticking
— f — **aguja** f - [instrum] pointer sticking
— f — — **para amperaje** m - [sold] current, o
amperage, pointer, binding, o sticking
— f — **árbol** m - [mec] shaft binding
— f — **armadura** f - [electr-mot] armature,
binding, o freezing
— f — **cadena** f - [mec] chain binding
— f — **cojinete** m - [mec] bearing freeze,zing
— f — **contratuerca** f - [mec] jam nut locking
— f — **émbolo** m - [mec] pinion jamming
— f — **engranaje** m - [mec] gear binding
— f — **freno** m - [mec] brake locking
— f — **gato** m - [mec] jack locking
— f — **palanca** f - [mec] lever locking
— f — — f **para carrete** m - [agric-equip]
spool lever locking
— f — f — **freno** m - [mec] brake
locking
— f — **rotor** m - [electr-mot] rotor locking
— f — **rueda** f - [mec] wheel locking
— f — **tuerca** f - [mec] nut locking
— f — **posición** f **abierta** - [mec] open locking
— f — — f **cerrada** - [mec] closed (position)
locking
— f **entre sí** - [mec] interlocking
— f — **de hebra(s)** f - [plást] fiber
strands interlocking
— f — — — f **de fibra(s)** f **de vidrio** m -
[plást] fiberglass strand(s) interlocking
trabajable a - [mec] machinable; véase también
labrable; elaborable; mecanizable*
trabajado,da a - worked • [ind] operated • [mec]
machined
— a **bajo** - [labor] reported,ting to
— a — **órdenes de** - [labor] worked under; reported to
— a **con torno** m - [mec] lathed; vease también
torneado,da
— a **costa** f **adentro** - [petról] operated on
shore
— a — f **afuera** - [petról] operated offshore
— a **eficazmente** - [labor] worked effectively
— a **eficientemente** - [labor] worked efficiently
— a **en forma** f **eficiente** - [labor] worked effectively
— a **en caliente** - [metal-lam] worked hot
— a **frío** - [metal] worked cold
— a **junto a** - worked alongside
— a **junto(s)** - [labor] worked together
— a **óptimamente** - [labor] worked optimally
— a **sólo** - worked along
— a **satisfactoriamente** - worked, o operated,

satisfactorily\
trabajador m **calificado** - [labor] craftsman
— m **inexperto** - [labor] unskilled workman
trabajar v - . . .; to operate; to habdle •
to, machine, o form, o fabricate • [constr]
to tool • [hogmit] to, puddle, o work •
[ind] to perform
— v **a ciegas** f - to work, blindly, o in @ dark
— v — **deshoras** f - to work overtime
— v — **reglamento** m - [labor] to work, by @
book, o to @ rule(s); to go slow
— v **a unísono** - to work in unison
— v **afiebradamente** - tp work feverishly
— v **alrededor de barra(s)** f - [constr-hormig]
to work around @ reinforcement(s)
— v **apropiadamente** - [labor] to, work, o operate, properly
— v **arduamente** - to work (very) hard
— v **bajo** - [labor] to work under
— v — **carga** f - [ind] to work under @ load
— v — — f **con presión** f - [hidr] to, work, o
flow, under @ pressure head
— f — **órdenes** f - [labor] to work under; to
report to
— v **cadena** f - [mec] to work @ chain
— v **carrera** f - [deport] to work @ race
— v **como equipo** m - [labor] to work as @ team
— v **con** - to work with • to handle
— v — **empeño** m - [fig] to work @ heart out
— v — **máquina** f - [ind] to machine
— v — **presupuesto** m - [contab] to work with @
budget
— v — **seguridad** f - [labor] to work safely
— v **confiablemente** - [ind] to work reliably
— v **continuamente** - to work, continuously, o
steadily
— v **correctamente** - to work correctly
— v — **desde principio** m - to start right
— v **costa** f **adentro** - [ind] to work on shore •
[petról] to operate on shore
— v — f **afuera** - [petról] to work offshore
— v **debidamente** - to, work, o operate, right
— v **día** m **entero** - to, work, o operate, all day
— v **duramente** - to work hard
— v **efectivamente** - to work effectively
— v **eficazmente** - to work efficiently
— v **eficientemente** - to work efficiently
— v **en contraposición** f - to work, against, o
at crosspurpose(s)
— v — **escorial** m - [metal-prod] to work, in, o
on @ slag notch
— v — **forma** f **constante** - to work, steadily, o
constantly
— v — — f **continua** - to work continuously
— v — — f **eficiente** - [labor] to work, efficiently, o effectively
— v — **interior** - [ind] to work on @ inside
— v — — m **de soldadora** f - [sold] to work inside @ welder
— v **en oposición** f - to work against (each
other)
— v — **piquera** f - [metal-prod] to work on @
iron notch
— v — **ruta** f—[metal-prod] to work on @ runner
— v **febrilmente** - to work feverishly • to rush
— v **fuera de horario** m - [labor] to work overtime • to work outside @ regular hours
— v **hacia** - to work towards
— v **hasta** - to work till • to work into
— v — **ángulo** m - [hormig] to work into @ angle
— v — **rincón** m - to work into @ corner
— v **ininterrumpidamente** - to work ceaselessly •
to work around @ clock
— v **junto(s)** - to work together • to work
alongside
— v **libremente** - to work freely
— v **material** m - [ind] to work @ material
— v **óptimamente** - to work optimally
— v **productivamente** - to work productively
— v **regularmente** - to work regularly
— v **reserva(s)** f - [miner] to work @ reserve(s)
— v **satisfactoriamente** - to, work, o operate,
satisfactorily

trabajar v **sin descanso** m - to work tirelessly
— v — **presión** f - to work without pressure
— v — — f **alta** - to work without high pressure • [metal-prod] to operate without high
pressure
— v — **seguridad** f - [segurid] to work unsafely
— v **sólo** - to work alone
— v **unido(s)** - [labor] to work together
trabajo m - . . . . • need • working • application • job • paper • operating,tion • [educ]
project • [sold] work; structure • [constr]
tooling • phase
— m **a desgano** m - [labor] slow down
— m — **destaje,jo** m - [labor] piece work
— m — **efectuar(se)** v - work to be, done, o
performed, o accomplished
— m — **generador** m - [sold] work to generator
— m — **reglamento** m - [labor] work by @, book,
o rule; (book) slowdown; slow, o delayed,
motion work
— m — **velocidad** f **alta** - [sold] high speed
work
— m **acelerado** - accelerated, o sped up, work
— m **administrativo** - [admin] management (work)
— m — **de sección** f - [admin] unit management
work
— m — **descompuesto** - [admin] broken down management work
— m — **desdoblado** - [admin] broken down management work
— m — **efectuado** - [admin] performed management work
— m — **ejecutado** - [admin] performed management work
— m — **para conducción** f - [admin] leading
management work
— m — — **control** - [Admin] control(ling) management work
— m — — **hoy** - [admin] today's management work
— m — — **planificación** f - [admin] planning
management, o management planning, work
— m — — **organización** f - [admin] organization management, o management organization,
work
— m — — **verificación** f - [admin] control-
-(ling) management, o management control-
-(ling) work
— m — **realzado** - [admin] emphasized management
work
— m — **realizado** - [admin] performed management
work
— m **ahorrado** - [labor] work saved
— m **apropiado** - appropriate work • proper operation
— m **arduo** - arduous, o hard, work
— m **ascendente** - [sold] uphill work
— m **asignado** - [labor] assigned, o allocated, o
given, work; job assignment
— m — **lógicamente** - [admin] work logically,
assigned, o allocated
— m **aumentado** - increased work
— m **autorizado** - [admin] authorized work
— m **bajo** - work(ing) under
— m — **carga** f - work under @ load
— m — — f **con presión** f - [hidr] flow under @
pressure head
— m — **órden(es)** f - work(ing) under; reporting
to
— m — **techo** m - [constr] inside, o interior
work; work under @ roof
— m **básico** - basic, work, o job
— m — **de construcción** f - [constr] basic construction, work, o job
— m **codificado** - [sold] code work
— m **comercial** - [com] commercial work
— m — **en general** - [sold] general commercial
work
— m **complementario** - [labor] complementary work
— m **completado** - completed, o finished, work
— m — **en oficina** f **matriz** - [admin] home office labor completion
— m **completo** - complete, work, o job

trabajo m con acero m estructural - [constr]
structural steel work
— m — amortiguador(es) m - [autom-med] shock
(absorber) work
— m — arco m - [sold] arc work
— m — — m sumergido - [sold] submerged arc
work
— m — armazón(es) m - [constr] frame work
— m — calidad f - [ind] quality work
— m — — f alta - [ind] high quality work
— m — — f baja - low quality work
— m — — f pobre - [ind] poor quality work
— m — compuesto(s) m - [quím] compound(s) work
— m — — m experimental(es) - [quím] experimental compound(s) work
— m — conducto(s) m - [constr] duct work
— m — — m tubular(es) - [tub] pipe duct work
— m — corriente f alterna - [sold] alternating
current, work, o operation
— m — — f continua - [sold] direct current,
opoeration, o work
— m — chapa(s) f - [metal-fabr] sheet work
— m — — f metálica(s) - [metal-fabr] sheet
metal work(ing)
— m — elemento(s) m de acero m estructural -
[constr] structural steel work
— m — encendido m difícil - [sold] tough
starting job
— m — espectrómetro m - [instrum| spectrometer
work
— m — esqueleto m - [constr] structural frame
work
— m — freno(s) m - [mec] brake work
— m — fricción f - [mec] friction work
— m — — f de diente m - [mec] tooth friction
work
— m — — f — — m de engranaje m - [mec] gear
tooth friction work
— m — hormigón m - [constr] concrete work
— m — máquina(s) - [admin] machine work
— m — material - [ind] material work(ing)
— m — personal - [labor] personnel work
— m — suspensión f - [autom] suspension work
— m conductivo - [admin] management, o leadership, work
— m conectado con tierra f - [sold] grounded
work
— m conocido - known, work, o job
— m continuo - [sold] continuous, work, o duty
— m contratado - [constr] contracted work
— m — en curso m - [constr] contracted work in
progress
— m correlacionado - correlated work
— m corriente - current work • [sold] common,
work, o job
— m — para soldadura f - [sold] common welding
job; common welding work
— m — — — f de acero m - [sold] common
steel welding, job, o work
— m — — f — — m inoxidable - [sold] common stainless (steel) welding, job, o work
— m corto - [sold] short, work, o job
— m costa f adentro - [petról] on shore, work,
= o operation
— m — † afuera - [petról] offshore, work, o
operation
— m erectivo - [constr] erection,ive work
— m crítico - critical work
— m cuidadoso - careful work • good job
— m — para apisonamiento m - [constr] careful,
o good, tamping, work, o job
— m — para relleno m - [constr] good, o careful, backfilling job
— m cumplido - accomplished, o finished, job
— m dado - given, work, o job
— m de burro m en ruta f - [metal-prod] hard
runner cleaning work
— m — calidad f - [ind] quality work
— m — — f según código m - [sold] quality
code, o code quality, work
— m — cantera f - [miner] quarry, work, o job
— m — carpintería f - [carp] woodworking (job)
— m — cimentación f - [constr] foundation work

trabajo m de componedor m - compounder('s) work
— m — cuadrilla f - [;abor] crew('s). work, o
activity
— m — — f en obra f - [constr] field, o
site, crew work
— m — chapistería f - [sold] sheet metal work •
[autom-mec] auto(mobile) body, work, o job
— m — departamento m - [ind] department('s),
work, o operation
— m — — m para amoladura f - [metal-lam]
grinding department, work, o job
— m — — — f de rodillo(s) m - [metal-
-lam] roll grinding department operation(s)
— m — — m — granallado m - [metal-lam] shot
blasting department operation(s)
— m — — m — — m de rodillo(s) m - [metal-
-lam] roll shot blasting department operation
— m — — m — rectificación f - [metal-lam]
grinding department operation
— m — — — f de rodillo(s) m - [metal-
-lam] roll grinding department operation
— m — — m — torneado m - [metal-lam] turning
deparment operation
— m — — — m de rodillo(s) m - [metal-
-lam] roll turning department operation
— m — — m — — rodillo(s) m - [metal-
-lam] roll turning department operation
— m — ingeniería f - [ind] engineering work
— m — — f para construcción f - [constr] construction engineering (work)
— m — — f — planta f - [ind] plant engineering work
— m — — f — — f industrial - [ind] indus=
trial plant engineering work
— m — laboratorio m - laboratory work
— m — mampostería f - [constr] masonry (work)
— m — minería f - [miner] mining (work)
— m — — f subterránea - [miner] underground
mining (work)
— m — operario m - [labor] operator('s) work
— m — personal - [labor] personnel, o force,
work
— m — prensa f - [ind] press work • [metal-
-fabr] drawing
— m — proyectista m - designer('s) work
— m — radiografía f - [electrón] radiography
— m — repetición f - [ind] repeat work
— m — substitución f - substitute, o replacement, work, o job
— m — temporada f - seasonal, p part time, work
— m — zapa f - pioneer work
— m defectuoso - [ind] defective work
— m deficiente - defficient, o defective, o inadequate, work
— m — corregido - corrected defective work
— m — no corregido - uncorrected defective,
work, o job
— m — sin corregir v - uncorrected deficient
work
— m — subsanado - corrected defective work
— m delegado - [labor] delegated, work, o job
— m delicado - delicate, o critical, work
— m demorado - delayed work
— m diario - daily work
— m diferente - different, work, o job
— m difícil - hard, o tough, work, o job
— m directivo - [admin] management, o directive, work
— m dispuesto - [admin] arranged work
— m distribuido - distributed, o divided, work
— m dividido - divided, work, o task, o job
— m duro - hard work
— m efectivo - effective work • actual work
— m efectuado - work, performed, o done
— m — actualmente - work currently performed
— m — corrientemente - work currently performed
— m — por administrador m - [admin] manager('s)
work performed
— m eficaz - effective work(ing)
— m eficiente - efficiente work(ing)
— m ejecutado - work performed
— m — bien - workmanlike job
— m — en hora(s) f extraordinaria(s) - [ind]

overtime performed workk
trabajo m **eléctrico** - [electr] electrical work
— m **empírico** - empirical work
— m **experimental** - experimental work
— m **en acero** m - [metal-fabr] steelwork
— m — — m **con cromo** m **níquel** - [sold] chromium-nickel steel, work, o structure
— m **en conjunto** m - team work(ing)
— m — **curso** - [contab] work in, process, o progress
— m — **ejecución** f - work, in progress, o being, done, o performed
— m — — f **evaluado** - [admin] assesed work in progress
— m — — f **regulado** - regulated, o controlled, work in progress
— m — — f **valorado** - [admin] assesed work in progress
— m — **equipo** m - team work(ing)
— m — **escala (grande)** - [sold] production, work, o welding
— m — — f **industrial** - [ind] volume work • [sold] production, welding, o work
— m — **estructura(s)** f **metálica(s)** - [constr] steel work; steelwork
— m — **fosa** f - [autom-mec] pit work
— m — **frío** m - [metal-lam] cold work(ing)
— m — **intemperie** f - exterior, o outisde, work • open air work
— m — **interior** - [mec] inside work
— m — — m **de soldadora** f - [sold] work inside @ welder
— m — **obra** f - field, work, o job, o operation, o labor, o service • project site work • outside work • portable operation
— n — — f **por empresa** f **para servicio(s)** m **público(s)** - portable service operation
— m — **paralelo** - [ind] parallel, work, o operation
— m — **partida(s)** f **intermitente(s)** - [ind] intermittent run work•] batch work
— m — **piquera** f - [metal-prod] iron notch work
— m — **posición** f **ascendente** - [sold] uphill poisition(ed), operation, o work
— m — — f **cualquiera** • [sold] any, o all, position, operation, o work
— m — — f **descendente** - [sold] downhill positioned, operation, o work
— m — — f **inclinada** - [sold] inclined, o slant(ing) position, operation, o work
— m — — f — **ascendente** - [sold] uphill (inclined) position(ed), operation, o work
— m — — f — **descendente** - [sold] downhill (inclined) position(ed), operation, o work
— m — — f **nivelada** - [sold] level(ed) position(ed), operation, o work
— m — — f **otra que plana** - [sold] out-of-position, operation, o work
— m — **prensa** f - [mec] press work • work, o job, being printed
— m — **ruta** f - [vial] route work • [meta;--prod] runner work
— m — **sección** f - [ind] section work • unit time
— m — **serie** f - [ind] production work • [sold] production weld(ing)
— m — **taller** - [ind] shop work • inside work
— m — — m **mecánico** m - [autom-mec] garage work
— m — **tubería(s)** f [tub] pipe,ping work
— m — — f **para conducción** f - [sold] pipeline work
— m — **yacimiento** m - [petról] field, operation, o work
— m **especial** - special work • special project
— m — **de ingeniería** f - special engineering work
— m — — — f **emprendida** - special engineering work undertaken
— m — — **investigación** f - special investigating work
— m — — — — f **emprendido** - special investigating work undertaken

trabajo m **especializado** - specialized sork
— m **específico** - specific, o particular, work, o job
— m **estandarizado*** - standardized work
— m **evaluado** - evaluated, work, o job
— m **exigente** - exacting, o demanding, work
— m **experimental** - [ind] experimental work
— m — **con compuesto(s)** m - [quím] experimental compound work
— m **expuesto** - [constr] exposed work
— m **exterior** - [constr] outside work
— m — **expuesto** - [constr] exposed, exterior, o outside, work
— m **fácil** - easy, work, o job, o operation
— m **fascinante** - fascinating work
— m **final** - final work
— m — **para perforación** f - [petról] tailing in
— m **finalizado** - finished, work, o job
— m **físico** - [admin] physical work
— m **forestal** - [Smader] logging; forestry
— m **fuera de posición** f - [sold] out-of-position work
— m **futuro** - [ind] future, o succeding, work, o job
— m **garantizado** - guaranteed, work, o job
— m **general** - general work
— m — **en taller** m - [sold] general (purpose) shop work
— m — **para conservación** f - [ind] general servicing • general maintenance
— m **grande** - large, o big, job
— m — **para reparación** f - large repair job
— m **gratuito** - [labor] free, work, o labor
— m **hecho** - work, o job, done
— m **humano** - [admin] human, work, o labor
— m **identificado** - identified work
— m **importante** - important work
— m **imprevisto** - unforseen, o emergency, work, o job
— m **inadecuado** - inadequate, work, o job
— m **inapropiado** - inappropriate work
— m **inaceptable** - unacceptable work
— m **incluido** - included work
— m **indirecto** - [labor] indirect, work, o labor • overhead
— m **individual** - individual, work, o labor
— m **infrecuente** - infrequent, o occasional, job
— m — **para producción** f - [ind] infrequent fabrication job
— m — **reparación** f - [sold] occasional repair job
— m **inicial** - initial work • pioneer work
— m **inmediato** - immediate work(ing)
— m **intensivo** - intensive work • hard work
— m **inseguro** - [segurid] unsafe work • working unsafely
— m **interior** - [constr] inside, o interior, work, o job
— m — **expuesto** - [constr] exposed, inside, o interior, work
— m **intermitente** - [ind] intermittent work
— m **junto(s)** - work(ing) together • work(ing) alongside
— m **levemente ascendente** - [sold] slightly uphill work
— m **libre** - [labor] free, work, o labor
— m **limitado** - limited work
— m **liviano** - light work • light duty
— m **lógico** - logical work
— m **llevado a cabo** - work performed
— m **manual** - [admin] manual, work, o labor • handwork
— m **máximo** - maximum work
— m **mayor** - major work • largest job
— m — **para conservación** f - [ind] large, o major, maintenance, work, o operation
— m — — **entretenimiento** m - [inc] major maintenance operation
— m — **mantenimiento** m - [ind] major, o large, maintenance, work, o operation
— m — — **reparación** f - major repair, work, o job
— m **mecánico** - mechanical, o machine, work(ing)

trabajo m mediano ~ medium, work, o duty
— m mejor - best work • better work
— m mejorado - improved, work, o job
— m mensual - monthly work
— m mental - [admin] mental work
— m mínimo - minimum work
— m necesario - mecessary, o required, work
— m nivelado - [sold] level(ed) work
— m no administrativo - [admin] non-adminis-
trative, o non-management. work
— m — autorizado - unauthorized work
— m — corregido - uncorrected work
— m — incluido - non-included work
— m nocturno - night(time) work
— m normalizado - standardized work
— m nuevo - new work • new job
— ocasional - occasional, o part time, work, o
job
— m óptimo - optimum work
— m ordenado - [admin] arranged, o prdered,
work • orderly work
— m organizado - organized work
— m — lógicamente - [admin] logically organized work
— m para acceso m - [constr] approach work
— m — administración f - [admin] administration, o management, work
— m — adquisición(es) f - procurement, work,
o activity,ties
— m — amoladura f - [metal-lam] grinding activity
— m — — m de rodillo(s) m - [metal-lam]
roll(s) grinding activity
— m — apilamiento m - [ind] piling, work, o
duty
— m — apisonamiento m - [constr] tamping job
— m — aplicación f - [educ] application project
— m — atención f - [ind] servicing job
— m — carrera f - [deport] race work
— m — conducción f - [admin] leading,dership
work • [autom] driving work
— m — conformación f - [metal] forming work
— m — conservación f - conservation work •
servicing • [ind] maintenance, o repair, work
— m — — f en interior m de soldadora f -
[sold] welder inside maintenance work
— m — — f exigido - [ind] required, maintenance, o service
— m — — f de rutina - [ind] routine servicing
— m — — f en general - [ind] general maintenance (work, o operation)
— m — — f de parte f hidráulica - [petról]
fluid end maintenance
— m — constatación f - [admin] control(ling)
work
— m — construcción f - [constr] construction,
work, o job • construction requirement
— m — desarrollo m - development work
— m — determinación f de falla(s) f - [ind]
trouble shooting work
— m — elevación f - [grúas] lifting, o hoisting, work, o job
— m — emergencia f - [ind] emergency, work, o
job
— m — empalme,madura - [electr-instal] splicing, work, o job
— m — endurecimiento m - hardening work
— m — — m de superficie(s) f - [sold] hardsurfacing job
— m — entrada - [constr] approach work
— m — entretenimiento m - [ind] maintenance,
o repair, work
— m — — m en general - [ind] geneal maintenance, work, o operation
— m — evaluación f - evaluation (work)
— m — excavación f - [constr] excavation
(work) • earth moving,vement
— m — exploración f - [miner] exploration,
work, o job(s)
— m — explotación f - [ind] operation (work)
— m — fabricación f - [sold] fabrication work
— m — formación f - [metal-fabr] forming work

trabajo m para fiscalización f - [admin] controlling work
— m — hincadura f - [constr-pil] driving,
work, o phase
— m — — m de pilote(s) m - [constr-pil] pile
driving, work, o phase
— m — ingeniería f - [ind] engineering (work)
— m — — f efectuado - [ind] performed engineering (work)
— m — inserción f (con gatos m) - [constr]
jacking job
— m — instalación f - installing,lation work
— m — investigación f - investigation, o research, o deveopment, work
— m — — f emprendido - investigating,tive
work undertaken
— m — laminación f - [metal-lam] rolling, o
mill, work
— m — limpieza f - cleaning work
— m — — f para pozo m - [petról] dlean-out
work
— m — montaje m - [ind] erection work
— m — — m electromecánico - [electr-equip]
electromechanical   erection work
— m — mantenimiento m - [ind] maintenance work
• repair work
— m — — m en general - [ind] general maintenance (operation)
— m — manutención f - [ind] maintenance, o repair, work
— m — organización f - [admin] organizing;
organizing,zation work
— m — perfeccionamiento m - development work
— m — — m de neumático(s) m - [autom-neumát]
tire development work
— m — perforación f - [petról] drilling, operation, o work
— m — — f eficiente - [petról] efficient drilling, operation, o work
— m — f suave - [petról] smooth drilling,
operation, o work
— m — planificación f - [admin] planning work
— m — producción f - [sold] fabrication work
— m — f en serie - [sold] production work
— f — en taller m - [sold] factory production work
— m — proyección f - [ind] design(ing) work
— m — recargue m - [sold] build-up job
— m — — m con soldadura f - [sold] (welding)
build-up job
— m — — m — — f con níquel m y manganeso m
- [sold] nickel-manganese build-up job
— m — — m — f — aleación f baja - [sold]
low-alloy build-up job
— m — reconstrucción f - [constr] remodeling
work • reconstruction, work, o job
— m — rectificación f - [metal-lam] grinding,
work, o activity
— m — recubrimiento m duro - [sold] hardsurfacing job
— m — recuperación f - [ind] recuperation, o
recovery, o salvage, work, o operation
— m — reemplazo m - [ind] replacement job
— m — — m para conservación f - [constr]
maintenance replacement job
— m — — m — puente m - [constr] bridge replacement job
— m — relleno m - [constr] backfilling, work,
o operation, o job; fill work • backfill
— m — — m cumplido - [constr] good backfilling job • completed backfilling job
— m — remoción f - [constr] removal,ving, job,
o operation
— m — reparación f - [ind] repair, work, o job
— m — — f en planta f - [ind] plaint repair,
work, o job
— m — restauración f - restoration work
— m — sección f - [ind] unit work
— m — servicio m - [ind] service,cing job
— m — soldadura f - [sold] welding, job, o
operation
— m — — f automática - [sold] automatic
welding, work, o job

trabajo m.para soldadura f corriente - [sold] standard welding job
— m — — f de acero m - [sold| steel welding job
— m — — f — — m inoxidable - [sold] stainless (steel) welding job
— m — — f para fabricación f - [sold] fabrication welding job
— m — — f — instalación f - [sold] installation welding job
— m — — f — producción f - [sold] fabrication welding job
— m — — f — reparación f - [sold] repair welding job
— m — — f manual - [sold] manual welding job
— m — supervisión f - [ind] supervisory,sion, work, o job
— m — terraplenamiento m - [constr] backfilling (work, o job)
— m — túnel m - [constr] tunnel job
— m — — m pequeño - [constr] small tunnel job
— m — urbanización f - [pol] (urban) development (job)
— m — venta f - [com] sales, work, o job
— m — verificación f - [admin] control(ling), work, o job
— m — vía(s) f - [f.c.] track work
— m parcial - [com] partial, work, o job
— m pendiente - pending, work, o job
— m perfecto - perfect, job, o work
— m pequeño - small, job, o work
— m — para horadación f - [constr] small tunneling job
— m — — reparación f - small repair job
— m pesado - heavy, work, o job • tough job | adv - heavy duty
— m por administración f—company performed job
— m por costa f y costas f - cost plus work job
— m portuario - [transp] port, o dock, work
— m producido - [ind] work performed
— m productivo - [ind] productive work
— m profesional - professional, work, o job
— m programado - [ind] scheduled work
— m propio - own work
— m puesto a tierra f - [sold] grounded work
— m push-pull - [mec] push-pull work
— m rápido - quick, o fast, work
— m real - actual, work, o operation(s)
— m — en yacimiento m - [petról] actual field, work, o operation(s)
— m realizado - work, performed, o accomplished
— m — realmente - [labor] work, truly, o actually, performed
— m — verdaderamente - [labor] work truly performed
— m recíproco - [mec] interlock
— m rectificado - corrected, o rectified, work
— m reducido - reduced, o diminished, work
— m regular - regular work • [petról] assessment work
— m relacionado - [admin] related work
— m repetidor - repetitive work
— m — estandarizado - [admin] standardized repetetive work
— m repetitivo - [ind] repetitive work
— m retrasado - delayed, work, o job
— m rudo - rough, work, o job
— m satisfactorio - satisfactory, work(ing), o operating, o job
— m según esoecificación(es) f [ind] specification work
— m — — f de código m - [sold] code work
— m — — f — para recipientes cor presión f - [sold] pressure vessel code work
— m seguro - [segurid] safe, work, o job
— m severo - tough, o severe, work, o service
— m similar - similar work
— m sin corregir - uncorrected work
— m sísmico - [petról] seismic work
— m — por sistema m de reflexión f - [petról] reflection seismic work
— m sobre cabeza f - [sold] overhead, work, o job

trabajo m solo -working alone
— m subsiguiente - subsequent, o succeding, job
— m suplementario - extra, o additional, work
— m técnico - technical work
— m técnico de sección f - [admin] unit technical work
— m — — vendedor m - [com] salesman('s) technical work
— m efectuado - performed technical work
— m ejecutado - performed technical work
— m temporal - temporary, o part time, job
— m terminado - finished, o completed, work, o job
— m típico - [ind] typical, work, o job
— m total - total, work, o job
— m unido - [labor] worki(ing), together, o united(ly)
— m vertical - [sold] vertical, work, o job
trabajos m relacionados - related jobs
— m — entre sí - [admin] interrelated work
— m repetidos - repeated, o repetitive, work
— m — corto(s) - [sold] short repetitive work
— m varios - sundry, o miscellaneous, work, o jobs

trabajosamente adv - slowly; laboringly
trabajoso,sa a - cumbersome • time consuming
trabamiento m - [mec] freezing; locking; seizing • binding
— m de acelerador m - [int.comb] throttle sticking
— m — armadura f - [electr-mot] armature freezing
— m — cojinete - [mec] bearing freezing
— m — cuerda f - rope lock(ing)
— m — engranaje m - [mec] gear binding
— m — rotor m - [electr-mot] rotor locking
— m — trinquete m - [mec] pawl locking
— m — tuerca f - [mec] nut locking
— m — en posición f abierta - [mec] open locking
— m — — f cerrada - [mec] closed locking
— m evitado - [mec] avoided, o prevented, seizure
— m momentáneo - [mec] momentary locking
— m prevenido - [mec] avoided seizure
trabar v - [mec] . . .; to freeze; to jam • to gall
— v acelerador m - [comb.int] throttle sticking
— v aguja v - [instrum] to, lock, o bind, @, pointer, o needle
— v — para amperaje m - [sold] to, lock, o bind, @, current, o amperes, pointer, o needle
— v árbol m - [mec] to bind @ shaft
— v ariete m - [mec] to, bind, o lock, @ ram
— v armadura f - [electr-mot] to, bind, o freeze, @ armature
— v cadena f - [mec] to bind @ chain
— v cojinete m - [cojin] to, seize, o freeze, o bind, @ bearing
— v contratuerca f - [mec] to lock @ jam nut
— v — f contra horquilla f - [mec] to lock @ jam nut against @ clevis
— v cuerda f - to lock @ rope
— v dispositivo m - [mec] to lock @ device
— v — m para marcha f sin carga f - [sold] o lock @ idler
— v émbolo m - [mec] to jam @ piston
— v en batalla f - [deport] to lock in @ battle
— v — posición f abierta - [mec] to lock (in @) open position
— v — — cerrada - [mec] to lock (in @) closed (position)
— v engranaje m - [mec] to bind @ gear
— v entre sí - to interlock
— v — — hebra(s) f - to interlock @ strand(s)
— v — — — f de fibra f - to interlock @ fiber strand(s)
— v — — — f — — f de vidrio m - [plást] to interlock @ fiberglass strand(s)
— v freno m - [mec] to lock @ brake
— v gato m - [mec] to lock @ jack
— v leva f - [mec] to lock @ cam
— v palanca f - [mec] to lock @ lever
— v — f para freno m - [mec] to lock @ brake lever

trabar v palanca f para carrete m—[agric-equip]
  to lock @, spool, o reel, lever
— v rotor m - [electr-mot] to lock @ rotor
— v sujetador m - [mec] to lock @ cam
— v tromquete m - [mec] to lock @ pawl
— v tuerca f - [mec] to lock @ nut
trabazón f - [mec] . . . • binding; • galling
— f — espiga f - herringbone bond
— f de rotor m - [electr-mot] rotor locking
trabilla f - [vest] bobby pin
trabe m - [constr] . . .; girder
tracción f - . . . • pulling; towing • [electr-
  -insal] pull(ing); tow(ing) • [electr-install
  pull • [metal] pull; tension • [autom] drive •
  speed(s) range
— f adicional - additional, o extra, traction
— f animal - [transp] animal traction | adv •
  horse and buggy
— f aumentada - increased traction • [autom-
  -neumát] added traction
— biaxil - [metal] biaxial pull
— f buena - [transp] good traction
— f con oruga(s) f - [constr-equip] crawler,
  tread, o traction
— f con pavimento m mojado - [autom-neumát]
  wet traction
— f — — seco - [autom-neumát] dry traction
— f de banda f para rodamientom—[autom-neumát]
  tread traction
— f — grúa f - [constr] crane traction
— f — superficie f para rodadura f - [autom-
  -neumát] tread traction
— f — viento m - [constr] wind traction
— f delantera - [autom] front, drive, o traction • front-wheel, drive, o traction
— f directa - [mec] direct, traction, o tension
— f disminuida - diminished, o decreased, traction
— f doble - double traction • [autom-mec] four-
  -wheel drive
— f en cuatro ruedas f - [autom] four wheel
  drive; FWD; 4WD
— f — punta f - tip traction
— f — . . . rueda(s) f - [autom] . . .-wheel
  drive
— f — rueda(s) f delantera(s) - [autom] front
  wheel drive
— f — — f trasera(s) - [autom] rear wheel
  drive
— f — seco - [autom-neumát] dry traction
— f — — excelente—[autom] great dry traction
— v — viraje(s) m - [autom] cornering traction
— f especificada - [mec] specified traction
— f excelente - [autom] great, o excellent,
  traction
— f — sobre pavimento m húmedo - [autom] excellent wet traction
— f — — — n mojado - [autom] great, o excellent, wet traction
— f — — — m seco - [autom] excellent dry
  traction
— f excesiva - excessive, traction, o pull(ing)
— f extraordinaria - [autom-neumát] extraordinary, o outstanding, o terrific, traction
— m — sobre pavimento m mojado—[autom-neumát]
  terrific wet traction
— f — — — m seco - [autom-neumát] terrific
  dry traction
— f frontal - [autom] front-wheel drive
— f fuera de carretera f - [autom-neumát] off-
  -@-road traction
— f horizontal - [mec] horizontal traction
— f lateral - [mec] side(ways) traction
— f máxima - maximum traction
— f mayor - major traction • increased traction
— f mejor - better, o improved, traction
— f mejorada - [mec] improved traction
— f menor - less, o decreased, traction
— f mínima - minimum traction
— f perdida - [autom] lost, o broken, traction
— f posterior - [autom] rear (wheel), drive, o
  traction
— f requerida - required, traction, o pull

tracción f rota - [autom] lost, o broken, traction
— f segura - [mec] safe teaction • [autom-
  -neumát] rugged traction
— f sobre cuatro ruedas f - [autom] four-
  -wheel drive
— f — dos ruedas f - [autom] two-wheel drive
— f — nieve f - [autom-neumát] snow traction
— f — pavimento m mojado - [autom-neumát]
  wet traction
— f — — m seco - [autom-neumát] dry traction
— f — rueda(s) f - [autom] . . .-wheel,
  drive, o traction
— f sobresaliente - outstanding traction
— f sólida - [autom-neumát] rugged traction
— f total - [grúas] full, o total, traction
— f — de grúa f - [grúas] full crane traction
— f — oruga f - [constr-equip] full crawler drive
— f — de viento m - [constr] full wind traction
— f trasera - [autom] rear, traction, o drive
tractil* a - véase a tracción f; para tracción f
tractor m automático - [sold] full(y) automatic
  tractor
— m caminero - [vial] road tractor
— m con ariete m - [ind] ram tractor • véase
  también carro m para transferencia f de bobinas f
— m — carriles m - [constr] caterpillar, o
  crawler, tractor
— m — combustión f interna - [agric-equip]
  internal combustion tractor
— m — horquilla f - [ind] forklift truck
— m — oruga(s) f - [ind] track type, o crawler, tractor; caterpillar tractor
— m — remolque m - [autom] tractor trailer
— m — ruedas f - [agric-equip] wheel tractor
— m — topadora f - [constr-equip] (bull)dozer
  tractor
— m — — m con oruga(s) f - [constr-equip]
  crawler type tractor
— m grúa - [grúas] tractor crane
— m Lincolnweld - [sold] Lincolnweld tractor
— m para gasolina f - [agric-equip] gasoline
  tractor
— m — huerta(s) f - [agric-equip] garden
  tractor
— m — nafta f—[agric-equip] gasoline tractor
— m sobre carril(es) m - [mec] track type
  tractor • [f.c.] railway tractor
— m tipo m oruga f - [mec] crawler (type)
  tractor
— m y semirremolque m - [transp] tractor and
  semi-trailer (unit); eighteen wheeler
tractor,ra a - tractive
tradicional a - . . .; conventional
traducción f - . . . • interpretation
— f anexa - attached translation
— f cuidadosa - careful translation
— f de instrucción(es) f - [ind] instruction(s)
  translation
— f tabla f - table translation
— f fiel - [legal] faithful, o true, translation
— f libre - free translation
traducido,da a - translated • interpreted
traducir v - . . . • to interpret
— v instrucción(es) f - [ind] to translate @
  instruction(s)
— v tabla - to translate @ table
traer v consigo - to bring along
tráfago m - . . .; fast stepping action
tráfico m - . . . • [transp] véase tránsito m
— m aéreo - [aeron] air traffic
— m — interrumpido - [aeron] interrupted air
  traffic
— m aumentado - [transp] increased traffic
— m aumentante - [transp] increasing traffic
— m de avión(es) m - [aeron] air traffic
— m — empresa f - [fisc] company turnover
— m — flota f - [electrón] fleet traffic
— m — frente m - [transp] oncoming traffic

tráfico m̱ de frente evitado - [vial] avoided on-coming traffic
— m̱ — rutina f - [electrón] routine traffic
— m̱ — tránsito - [transp] through, o̱ transit, traffic
— m̱ durante hora f de mayor afluencia - [vial] rush hour traffic
— m̱ en aumento - [transp] increasing traffic
— m̱ hidráulico - [hidr] hydraulic traffic
— m̱ iniciado - initiated, o̱ begún, traffic • [electrón] initiated, traffic, o̱ communication
— m̱ intenso - intense, o̱ lots of, traffic
— m̱ verbal - [electrón] voice traffic
— m̱ — de flota f - [electrón] fleet voice traffic
— m̱ — rutina - [electrón] routine voice traffic
— m̱ — f de flota f - [electrón] fleet routine voice traffic
— m̱ — iniciado - [electrón] initiated voice, traffic, o̱ communication
— m̱ — minimizado - [electrón] minimized voice traffic
— m̱ vocal - [electrón] voice traffic; véase también tráfico m̱ verbal
tragado,da a̱ - swallowed
tragaluz m̱ - [constr] . . . • fan light; transom
tragante m̱ - . . . • [metal-prod] . . .; stack; shaft; furnace, top, o̱ throat • hopper; top; upper stack
— m̱ con presión - [metal-prod] pressurized top
— m̱ de horno m̱ - [metal-prod] furnace, throat, o̱ top
— m̱ — m̱ alto - [metal-prod] blast furnace, throat, o̱ top, o̱ ahaft, o̱ stack
tragarse f petaca f - [metal-prod] cooling plate swallowing
traicionero,ra a̱ - . . .; tricky
traída f - . . . • bringing
— f consigo - bringing along
trailer* m̱ - [transp] trailer; véase acoplado m̱
trailla f Carryall - [constr-equip] Carryall scraper
— f para arrancadura f - [miner] ripper scraper
— f — arrastre m̱ - [constr] drag scraper
— f volcadora de arrastre m̱ - [constr] roll over scraper
traje m̱ confeccionado - [vest] off-@-rack suit
— m̱ para baño m̱ - [vest] bathing suit
— m̱ — mecánico - [vest] (machinest's) overalls
— m̱ — deporte(s) m̱ - [deport] sports suit
— m̱ — — m̱ invernal(es) - [vest] wintersports, o̱ snowmobile, suit
tramitación f - . . . • handling; negotiation; negituatubg; [fisc] formality,ties; red tape
— f de propuesta f - proposal handling
— f judicial - [legal] court formalities
tramitado,da a̱ - negotiated • handled • trans-acted
tramitar v - to handle (@ formality,ties) • to process • to transact; to negotiate, to ob-tain; to take steps
— v propuesta f - to handle @ proposal
trámite m̱ - . . . • [fisc] formality,ties; proc-cess; procedure; step • red tape | adv - (en) pending; requested
— m̱ aduanero - [fisc] custom(s) transaction(s)
— m̱ exigido - required, steps, o̱ formalities
— m̱ extrajudicial - [legal] extrajudicial, pro-cedure(s), o̱ formality,ties
— m̱ judicial - [legal] judicial, procedure(s), o̱ formality,ties
— m̱ jurídico - [legal] judicial, o̱ legal, pro-cedure(s), o̱ formality,ties, o̱ stop(s), o̱ pro-cess
— m̱ para importación f - [com] importation for-mality,ties
— m̱ pertinente m̱ - related, o̱ pertinent, steps
— m̱ respectivo - respective step(s)
tramo m̱ - . . . • length; section; segment; run; span • [tub] pipe stand • [hidr] reach • [vial] stretch • [deport] lap; leg
— m̱ adecuado - adequate, section, o̱ length

tramo m̱ admisible - [constr] allowable span
— m̱ aéreo m̱ - [aeron] aerial, section, o̱ leg • [electr-instal] aerial, o̱ overhead, run
— m̱ bajo tierra f - [electr-instal] under-ground run
— m̱ con grava f - [vial] gravel(led) section
— m̱ — hielo m̱ - [vial] icy, section, o̱ spot
— m̱ con nivel m̱, alto, o̱ elevado - [vial] high level span
— m̱ con viga(s) f - [constr-puentes] beam span
— m̱ — f para acceso m̱ - [constr-puentes] beam approach span
— m̱ conectado - [mec] connected, span, o̱ length
— m̱ continuo - continuous, o̱ endless, span, o̱ section, o̱ length
— m̱ continuo de cordón m̱ - [cabl] continuous strand length • [vial] continuous curb, sec-tion, o̱ length
— m̱ — — metal m̱ - [mec] continuous metal length
— m̱ — único - [metal-alambre] single conti-nuous length
— m̱ conveniente - [tub] convenient length
— m̱ corto - short, span, o̱ section, o̱ length • short distance
— m̱ cuádruple - [petról] fourble
— m̱ curvo - curved section
— m̱ de acero m̱ - [constr] steel span
— m̱ — cable m̱ (de alambre m̱) - [cabl] wire rope, span, o̱ section
— m̱ — — m̱ para tranvía m̱ - [cabl] track strand length
— m̱ — cabo m̱ - [cabl] rope, span, o̱ length
— m̱ — cadena f - [mec] chain length
— m̱ — f conectado - [mec] connected chain length
— m̱ — canaleta f - [mec] trough, section, o̱ length; gutter length
— m̱ — carrera f - [deport] race leg
— m̱ — cordón m̱ - [cabl] strand length
— m̱ — m̱ continuo - [Cabl] continuous strand lenth
— m̱ — — m̱ — — cable m̱ (de alambre m̱) - [cabl] wire rope continuous strand length
— m̱ — conducto m̱ - [electr-instal] conduit, o̱ duct, run
— m̱ — — m̱ celular - [electr-instal] duct bank section
— m̱ — — m̱ continuo - [cabl] continuous strand length
— m̱ — hormigón m̱ - [constr] concrete span
— m̱ — madera f - [constr] wood span
— m̱ — metal m̱ continuo - [mec] continuous metal length
— m̱ — pared f - [cpmstr] wall section
— m̱ — puente m̱ - [constr] bridge, span, o̱ section
— m̱ — — m̱ con vigas f - [constr] truss bridge span
— m̱ — — m̱ — f maestra - [constr] beam bridge span
— m̱ — — — viga(s) f armada(s) - [constr] truss bridge span
— m̱ — — m̱ — — f armada(s) de acero m̱ - [consr] steel truss bridge span
— m̱ — — — f maestra - [constr] girder span bridge
— m̱ — — — f — de hormigón m̱—[constr] concrete girder span bridge
— m̱ — sección f semirrígida - [constr] semi-rigid section length
— m̱ — tres tubos m̱ en pie - [petról] pipe stand
— m̱ — tubería f - [tub] pipe,ping, stand, o̱ length, o̱ section • [ambient] duct run
— m̱ — — f galvanizada - [tub] galvanized pipe length
— m̱ — — f perforada - [tub] perforated pipe section
— m̱ — — f — de acero m̱ - [tub] perforated steel pipe section
— m̱ — tubo m̱ - [tub] pipe length

tramo m de vía f - [f.c.] track section
— m — f **recta** - [f.c.] straight track section
— m — **viga** f - [constr] beam section
— m — — f **armada** - [constr] truss span
— m — — f — **de acero** m - [constr] steel truss span
— m — f — m **para acceso** m - [constr] steel beam approach span
— m — — f **maestra** - [constr] girder span
— m — — f — **con alma** m **llena** - [constr] web plate girder span
— m **debajo de río** m - [constr] river crossing
— m **embutido en hormigón** m - [constr] concrete embedded run
— m **en pie** m **de tubería** f - [petról] pipe stand
— m — — n — f **para perforación** f - [petról] drill pipe stand
— m **entre postes** m - [electr-instal] section between poles
— m — **soportes** m - [mec] span between bents
— m **estándar** - [constr] standard span
— m **exterior** - exterior span • [electr-instal] outdoor run
— m **final** - [deport] homebound leg
— m **húmedo** - [constr] wet section
— m **interior** - interior, o inside, span
— m **irregular** - irregular, section, o span
— m **largo** - long, length, o distance
— m — **de tubería** f - [constr] long pipe line
— m — **y liviano** - [tub] long light(weight), length, o section
— m **levadizo** - [puentes] lift span; draw span
— m — **de viga** f **armada** - [puent] truss, lift, o draw, span
— m — **vertical** - [puent] vertical lift span
— m — — **de viga** f **armada** - [puent] truss vertical lift span
— m — — f — **de acero** m - [puent] steel truss vertical lift span
— m **mayor** - [tub] longer length
— m **menor** - [tub] shorter length
— m **para acoplamiento** m - [mec] coupling section
— m — **anclaje** m - [constr] anchor(ing) span
— m — **conexión** f - [constr] transition section
— m — **empalme** m - [f.c.] connecting section
— m — **ensayo(s)** m - [tub] test(ing) section
— m — **escalera** f - [constr] stair(s) flight
— m — **transición** f - transition section
— m **paralelo** - [electr-cond] parallel length
— m **pavimentado** - [vial] paved section
— m **razonable** - reasonable, length, o section
— m **recto** - [tub] straight, section, o length
— m — **de tubería** f - [tub] straight pipe length
— m — **entre bridas** f - [mec] straight section between flanges
— m **reforzado** - [mec] reinforced span
— m **rígido** - [constr] rigid, o unyielding section
— m **sin fin** - [cabl] endless length
— m **subterráneo** m - [constr] underground length
— m — **de conducto** m **celular** - [electr-instal] underground duct bank
— m **suspendido** - [constr] suspended,nsion span
— m **tan corto como posible** - short as possible length
— m **último** - last, o final, section
— m **útil** - [tub] pay length
— m — **de tubería** f - [tub] pipe pay length
— m **vertical** - vertical span
trampa f - . . . • [metal-prod] dust catcher; véase también **colector** m **para polvo** f - [f.c.] derail switch • [constr] door
— f **cebada** - [mec] primed trap
— f **con balde** m - [válv] bucket trap
— f — — m **invertido** - [válv] inverted bucket trap
— f — **cubo** m - [válv] bucket trap
— f — — m **invertido** - [válv] inverted bucket trap
— f **estratigráfica** - [geol] stratigraphic trap

trampa f **fácil** - easy trap
— f **limpia** - [mec] clean trap
— f **limpiada** - [mec] cleaned trap
— f **manual** - [f.c.] hand derail
— f **para agua** m - [hidr] water trap
— f — — m **y polvo** m - [sold] water and dust trap
— f — **condensación** f - [cald] condensate trap
— f — **cucharón** m - [constr-equip] ladle door
— f — **drenaje** m - [sanit] drain(age) trap
— f — — m **inoperante** - [ind] nonoperating drain(age) trap
— f — **gas** m - [ind] gas trap
— f — **grasa** f - [sanit] grease trap
— f — **grava** f - [hidr] gravel trap
— f — **líquido(s)** m - liquid(s) trap
— f — **piedra(s)** f - [constr] stone(s) trap
— f — **polvo** m - dust trap
— f — **tierra** f - dirt trap
— f — **vapor** m - [tub] steam trap
— f **reemplazada** - [mec] replaced trap
— f **removida** - [mec] removed trap
— f **tipo arrecife** m - [mec] reef type trap
tranquil m - . . .] adv - (por) rampant; uneven spring
tranquilidad f - . . .; security
— f **de motorista** m - [transp] motorist('s) assurance
tranquilo,la a - . . . • restful
transacción f **bancaria** - [fin] bank transaction • banking
— f — **desde automóvil** m - [fin] drive-in banking
— f **comercial** - [com] commercial, o business, transaction
— f **comprometida** - committed transaction
— f **con gobierno** - government, transaction, o business
— f — **industria** f - [ind] industrial transaction
— f — **sector** m **privado** - [com] private sector, transaction, o business
— f — m **público** - [com] public sector, transaction, o business
— f **en comercio** m - [com] commercial transaction
— f — **moneda** f **extranjera** - [fin] foreign currency transaction
— f **entre empresas** f **afiliadas** - [com] intercompany transaction
— f **exterior** - [com] foreign transaction
— f **formal** - [com] formal, o no-nonsene, transaction
— f **gubernamental** - [com] government transaction
— f **privada** - [com] private, transaction, o business
— f **pública** - [com] public, transaction, o business
— f **significativa** - [com] significant transaction
— f — **entre empresas** f - [com] significant intercompany transaction
— f — — — f **afiliada(s)** - [com] significant affiliated companies transaction
transado,da a - negotiated; compromised
transalpino,na a - . . .; beyond, o north of, @ Alps
transamazónico,ca - transamazon(ic)
transbordado,da a - [transp] transhipped
transbordador m - . . . • traverse • [mav] ferry
— m **de calibre** m **de cizalla** f - [mec] shear gage traverse
— m — — m — **cuchilla** f - [mec] blade gage traverse
— m — — m — **tijera** f - [mec] shear gage traverse
transbordo m **no permitido** - [transp] no transshipment allowed
— m **permitido** - [transp] transshipment allowed
— m **prohibido** - [transp] transshipment not allowed
transcanadiense a - transcanada,dian

transcripción f - . . .; transcript
transcurrido,da <u>a</u> - elapsed • ago • transpired
transcurrir <u>v</u> - . . .; to transpire • to go
— <u>v</u> como reloj <u>m</u> - to go like @ clock(work)
transcurso <u>m</u> - . . . • elapsing
— <u>m</u> de laminación <u>f</u> - [metal-lam] rolling operation
transductor <u>m</u> - [petról] transducer
— <u>m</u> electrohidráulico - [electr-
-equip] electrohydraulic transducer
— <u>m</u> para fin <u>m</u> de carrera <u>f</u> - [electr-equip]
stroke end transducer
— <u>m</u> — posición <u>f</u> - [electr-equip] position
transducer
— <u>m</u> — <u>f</u> y fin de carrera <u>f</u> - [electr-
-equipo] position stroke end transducer
— <u>m</u> — término <u>m</u> de carrera <u>f</u> - [electr-equip]
stroke end transducer
transferencia f - . . .; move; moving; conveying • [mec] shedding • [fin] translation •
[contab] <u>véase</u> <u>también</u> comprobante <u>m</u> para
transferencia; asiento <u>m</u> para diario <u>m</u> •
[public] offset
— f a exterior <u>m</u> - [fin] transfer out of @
country
— f bancaria - [fin] bank transfer
— f buena - good transfer
— f — de calor <u>f</u> - good heat transfer
— f cablegráfica - [fin] cable transfer
— f de acción(es) <u>f</u> - [fin] stock, <u>o</u> shares.
transfer
— f — aceite <u>m</u> - oil trasnfer
— f — activo <u>m</u> - [contab] asset(s) transfer
— f — arrabio <u>m</u> - [metal-prod] pig iron
transfer
— f — bulto(s) <u>m</u> - [transp] bag(gage} <u>o</u> parcel, transfer
— f — calor <u>m</u> - [ind] heat transfer
— f — <u>m</u> de superficie <u>f</u> - [metal-prod]
surface heat transfer; transfer of surface
heat
— f — desecho(s) <u>m</u> - [nucl] waste(s) transfer
— f — — <u>m</u> con radiactividad <u>f</u> alta - [nucl]
high level (radioactive) waste(s) transfer
— f — divisa(s) <u>f</u> - [fin] (foreign) exchange, transfer, <u>o</u> movement
— f — empuje <u>m</u> - [mec] thrust transfer
— f — energía <u>f</u> - energy transfer
— f — — <u>f</u> para hincadura <u>f</u> - [constr-pil]
driving energy transfer
— f — — <u>f</u> efectiva - [constr-pil] effective
energy transfer
— f — — <u>f</u> para hincadura <u>f</u> - [constr-pil]
effective driving energy transfer
— f — fabricación <u>f</u> - [ind] manufacture,ring
transfer
— f — fondo(s) <u>m</u> - [fin] fund(s) transfer
— f — grano(s) <u>m</u> - [agric] grain transfer
— f — — <u>m</u> a granel - [agric] bulk grain
transfer
— f — información <u>f</u> - [electrón] information
transfer(ring)
— f — interés <u>m</u> - [legal] interest assignment
• [fin] interest transfer
— f — líquido(s) <u>m</u> - liquid(s) transfer(ring)
— f — mercadería(s) <u>f</u> - [contab] transfer of,
merchandise, <u>o</u> goods
— f — metal - [sold] metal transfer
— f — — <u>m</u> a través de arco <u>m</u> - [sold] metal
transfer through @ arc
— f — — <u>m</u> con corte(s) <u>m</u> en circuito <u>m</u> -
[sold] short circuiting (type) metal trans-
fer
— f — — <u>m</u> mediante corte(s) <u>m</u> en circuito <u>m</u>
en gas <u>m</u> - [sold] gas short circuiting metal
transfer
— f — orden f para compra - [coml] purchase
order transfer(ring)
— f — palanquilla <u>f</u> - [metal-lam] billet(s)
transfer(ring)
— f — peso <u>m</u> - [mec] weight transfer(ring)
— f — petróleo <u>m</u> - [petról] oil transfer
— f — planchón(es) <u>m</u> - [metal-lam] slab(s)
transfer

transferencia <u>f</u> de presión <u>f</u> - pressure trans-
fer(ence)
— f — — f distribuida - distributed pressure transfer(ence)
— f — producto <u>m</u> - [ind] product transfer
— f — — <u>m</u> laminado - [metal-lam] rolled prod-
uct(s) transfer
— f — propiedad <u>f</u> - [legal] property, <u>o</u> ownership, transfer
— f — recurso(s) - [econ] resource(s) transfer
— f — residuo(s) <u>m</u> - waste(s) transfer
— f — — <u>m</u> con radiactividad <u>f</u> alta - [nucl]
high level waste(s) transfer
— f — — — f baja - [nucl] low level
waste(s) transfer
— f — — <u>m</u> — f elevada - [nucl] high
level (radioactivity) waste(s) transfer
— f — riel(es) <u>m</u> - [metal-lam] rail(s) transfer
— f — tecnología <u>f</u> - technology transfer
— f — tensión(es) <u>f</u> - stress(es) transfer
— f — — f por adherencia <u>f</u> - [hormig] stress
transfer by bond(ing)
— f — tubería <u>f</u> - [tub] pipe,ping transfer
— f — — f entre cinta(s) f transportadora(s)
para entrada <u>f</u> y salida <u>f</u> - [tub] pipe,ping
transfer between @ in-conveyor and @ out-
-conveyor
— f — utilidad(es) <u>f</u> - [fin] profit(s) transfer
— f — — f a exterior <u>m</u> - [fin] profit(s)
transfer out of @ country
— f directa - direct, <u>o</u> straight, transfer
— f inicial - initial transfer
— f máxima - maximum transfer
— f — de energía <u>f</u> - energy maximum transfer
— f — — f para hincadura <u>f</u> - [constr-pil]
driving energy maximum tansfer
— f — — f efectiva - [constr-pil] effective
energy maximum transfer
— f — — — f para hincadura <u>f</u> - [constr-
-pil] effective driving energy maximum transfer
— f mediante corte(s) <u>m</u> en circuito <u>m</u> - [sold]
short circuitng transfer
— f parcial - partial transfer(ring)
— f permitida - [electrón] permitted transfer
— f pobre - poor transfer(ring)
— f — de calor <u>m</u> - [termol] poor heat transfer
— f provista - [electrón] provided transfer
— f restringida - [fin] restricted transfer
— f — de acción(es) <u>f</u> - [fin] restricted,
stock, <u>o</u> share(s), transfer
— f — — propiedad <u>f</u> - [legal] restricted
ownership transfer
— f tecnológica - technological transfer
— f térmica - [metal-prod] heat transfer
— f — eficiente - [metal-prod] efficient heat
transfer
— f total - total transfer(ring)
— f — a exterior <u>m</u> - [fin] total transfer out
of @ country
— f y promoción <u>f</u> - [labor] transfer and promotion
transferibilidad <u>f</u> - [legal] transferability;
assignability
transferido,da <u>a</u> - transferred • moved; conveyed • [legal] assigned
— a exterior <u>m</u> - [fin] transferred out of @
country
— a parcialmente - transferred partially
— a totalmente - transferred totally
transferidor <u>m</u> para tubería <u>f</u> - [tub] pipe
transfer
transferidor <u>m</u> - transferer; assigner
— <u>m</u> propuesto - [legal] proposed, transferer,
<u>o</u> assigner
transferidora <u>f</u> - [mec] transfer
transferir <u>v</u> - . . .; to move; to carry •
[legal] to assign • [mec] to shed
— <u>v</u> a exterior <u>m</u> - [fin] to transfer out of @
country

transferir v a formación f de suelo m -
[constr] to transfer to @ soil formation
— v **acción(es)** f - [fin] to transfer @, stock,
o shares
— v **aceite** m - to transfer @ oil
— v **activo(s)** m - to transfer @ asset(s)
— v **calor** m - to transfer @ heat
— v **carga** f - to transfer @ load • [electr]
to transfer @ charge
— v — f **a formación** f **de suelo** m - [constr] to
transfer @ load to @ soil formation
— v **desecho(s)** m - [nucl] to transfer @ wastes
— v — m **con radiactividad** f - [nucl] to transfer @ radioactive waste(s)
— v — m — f **alta** - [nucl] to transfer @
high level waste(s)
— t — m — — **baja** - [nucl] to transfer @ low
level waste(s)
— v — m — f **elevada** - [nucl] to transfer @
high level waste(s)
— v — m — **f, intermedia, o mediana**—[nucl]
to transfer @ intermediate level waste(s)
— v — m — — f **reducida** - [nucl] to transfer
@ low level waste(s)
— v **divisa(s)** f - [fin] to transfer @ exchange
— v **empuje** m - [constr] to transfer @ thrust
— f **fabricación** f - [ind] to transfer @, fabricating,tion, o manufacturing
— v **fondo(s)** m - [fin] to transfer @ fund(s)
— v **información** f - [comput] to transfer @ information
— v **interés(es)** m - [legal] to, transfer, o assign, @ interest
— v **líquido(s)** m - to transfer @ liquid(s)
— v **orden** f **para compra** f - [legal] to transfer
@ purchase order
— v **parcialmente** - to transfer partially
— v **peso** m - to transfer @ weight
— v **presión** f - to transfer @ pressure
— v **propiedad** f - [legal] to transfer @ property
— v **recurso(s)** v - [econ] to transfer @ re-
source(s)
— v **residuo(s)** m - [nucl] to transfer @ wastes
— v — m **con radiactividad** f **alta** - [nucl] to
transfer @ hig h level waste(s)
— v — m — f **baja** - [nucl] to transfer @
low level waste(s)
— v — m — f **elevada** - [nucl]
to transfer @ high level waste(s)
— v — m — **f, intermedia, o mediana**—[nucl]
to transfer @ intermediate level waste(s)
— v — — — f **reducida** - [nucl] to transfer @
low level waste(s)
— v **tecnología** f - [legal] to transfer @ technology
— v **tensión** f - [hormig] to transfer @ stress
— v — f **por adherencia** f - [hormig] to transfer @ stress by bond
— v **totalmente** - to transfer totally
transformación f **de corriente** f - [electr-oper]
current transformation
— f — **energía** f - [electr-distrib] power
transformation
— f — **entrada** f - [electrón] input transformation
— f — **ganancia(s)** f - [econ] profit transformation
— f — **materia(s)** f **prima(s)** - [ind] raw mate-
rial(s) transformation,ming
— f — **motoestibadora** f - [mec] fork lift truck
transformation,ming
— f — **rayos-X en rayos** m **luminosos**—[electrón]
X-rays conversion into light rays
— f — **reserva(s)** f - [fin] reserve(s) transformation
— f — **salida** f - [electrón] output transformation
— f — **suministro** m **de energía** f - [electrón]
power supply transformation,ming
— f — **tubería** f - [tub] tubing transformation
— f — **utilidad(es)** f—profit(s) transformation
— f **en capital** m - [fin] transformation into

capital
transformado,da a - transformed
— a **en capital** m - [fin] transformed into capital
transformador m **activado** - [electr-oper] activated, o energized, transformer
— m **aéreo** - [electr-equip] platform transformer
— m **aislado** - [electr-equip] insulated transformer
— m — **por medio de líquido** m - [electr-equip]
liquid insulated transformer
— m **armado** - [electr-equip] assembled transformer
— m **auxiliar** - [electr-equip] auxiliary transformer
— m **combinado** - [electr-equip] combined transformer
— m — **para ambas corrientes** f - [sold] combination, o combined, alternating/direct current transformer
— m — **para corriente** f **alternada o continua** -
[electr-equip] combined,nation alternating/
direct current transformer
— m **con aceite** m - [electr-equip] oil transformer
— m — **base** f **de líquido** - [electr-equip]
liquid filled transformer
— m — **circulación** f **de corriente** f - [electr-
-oper] energized transformer
— m — **devanado** m - [electr-equip] winding
transformer
— m — m **doble** - [electr-equip] two-winding transformer
— m — m **especial** - [electr-equip] especially wound transformer
— m — **gatillo** m - [sold] trigger transformer
— m — **núcleo** m **de aire** m - [electr-equip] air
core transformer
— m — m — m **conectado** - [electr-equip]
connected air core transformer
— m — m — m **desconectado** - [electr-
-equip] disconnected air core transformer
— m — m — **hierro** m - [electr-equip] iron
core transformer
— m — m **movible** - [electr-equip] movable
core transformer
— m **constante** - [electr-equip] constant transformer
— m **corriente** - [electr-equip] standard transformer
— m **dañado** - [electr-equip] damaged transformer
— m **de tipo** m **con núcleo** m **movible** - [electr-
-equip] movable core type transformer
— m — m **para puerta(s)** f **afuera** - [electr-
-equip] outdoor type transformer
— m — m **seco** - [electr-equip] dry type
transformer
— m — — **sellado** - sealed type transformer
— m — — **con nitrógeno** m - [electr-
-equip] nitrogen sealed type transformer
— n — — m **sumergido en aceite** m - [electr-
-equip] oil immersed type transformer
— m **deficiente** - [electr-equip] defective
transformer
— m — **para regulación** f - [electr-equip] defective control transformer
— m — — m **con devanado** m **doble** - [electr-
-equip] two-winding type transformer
— m **desactivado** - [electr-equip] de-energized
transformer
— m **desconectado** - [electr-oper] disconnected
transformer
— m **directo** - [electr-equip] straight transformer
— m — **para corriente** f **alterna** - [electr-
-equip] straight alternating current transformer
— m **eficiente** - [electr-oper] efficient transformer
— m — **para energía** f - [electr-equip] efficient power transformer
— m — — — f **trifásica** - [electr-equip] ef-

ficient three phase power transformer
transformador m **eficiente para, fuerza** f **motriz.**
o **potencia** f - [electr--equip] efficient
power transformer
— m **eléctrico** - [electr-equipo] electric transformer
— m **en serie** - [electr-equip] series transformer
— m **enfriado con agua** m - [electr-equip] water
cooled transformer
— m **ensayado**—[electr-equip] tested transformer
— m **grande** - [electr-equip] large transformer
— m **Idealarc** - [sold] Idealarc (transformer)
— m **instalado** - [electr-equip] installed transformer
— m **mayor** - [electr-equip] larger transformer
— m **mediano** - [electr-equip] medium transformer
— m **medidor** - [electr-equip] metering transformer
— m **monofásico** - [electr-equip] single phase
transformer
— m — **para regulación** f - [electr-equip] single phase (controlling) transformer
— m **multifásico** - [electr-equip] multiphase
transformer
— m **para aislación** f - [electr-equip] isolation
transformer
— m — — f **para entrada** f - [electrón] input
isolation transformer
— m — — f — **salida** f - [electrón] output
isolation transformer
— m — **alimentación** f - [electr-equip] feeding
transformer
— m — **amperaje** m - [electr-prod] amperage
transformer
— m — **audio** m - [electrón] speaker
— m — **bobinadora** f - [metal-lam] coil transformer
— m — **cambio** m **para fase(s)** f - [electr-equip]
phase shifting transformer
— m — **corriente** f - [electr-equip] current
transformer
— m — — f **alterna** - [electr-equip] alternating current transformer
— m — — f — **para servicio** m **pesado** - [sold]
heavy duty alternating current transformer
— m — — f — — **soldadora** f - [sold] welder
alternating current transformer
— m — — f — — **taller(es)** m **pequeños**—[sold]
alternating current small shop transformer
— m — — f **continua** - [electr-equip] direct
current transformer
— m — — f **para acción** f **rápida** - [electr-
-equip] quick action current transformer
— n — — f — — f **ultrarápida** -
[electr-equip] snap-action current transformer
— m — — f — **soldadora** f - [sold] welder current transformer
— m — **distribución** f - [electr-equip] distribution transformer
— m — **encendido** m - [comb.int] ignition transformer
— m — **energía** f - [electr-equip] power transformer
— m — — f **monofásica** = [electr-equip] single
phase power transformer
— m — — f **trifásica** - [electr-equipo] three, o
triple, phase power transformer
— m — — f **entrada** f - [electr-equip] input transformer
— m — **filtraje** m - [electr-equip] filtering
transformer
— m — **frecuencia** f, **alta**, o **elevada** - [electr-
-equip] high frequency transformer
— m — **gatillo** m - [sold] trigger transformer
— m — — m **para voltaje** m **bajo** - [sold] low
voltage trigger transformer
— m — **horno** m - [electr-equip]
furnace transformer
— m — — m **eléctrico** - [metal-prod] arc furnace transformer
— m **para instrumento(s)** m - [electr-equip] in-

strument transformer
— m — **intensidad** f - [electr-equip] intensity transformer • amperage, o current, transformer
— m — . . . **kilovoltamperio(s)** m - [electr-
-distrib] . . . kilovoltampere transformer
— m — **línea** f - [electr-equip] line transformer
— m — **mando** m - [electr-equip] control transformer
— m — — m **para interruptor** m - [electr-
-instal] switchgear, o breaker, transformer
— m — **materia(s)** f **prima(s)** - [ind] raw ma-
erial(s) transformer
— m — **pararrayos** m - [electr-instal] lightning (rod) arrester
— m — **potencia** f - [electr-equip] power
transformer
— m — — f **aislado** - [electr-equip] insulated
power transformer
— m — — f — **por medio** m **de líquido** m -
[electr-equip] liquid insulated power transformer
— m — — f **para puertas** f **adentro** - [electr-
-equip] indoor(s) power transformer
— m — — f — — f — **con aislación** f -
[electr-equip] insulated indoor power transformer
— m — — f — — f — — — f **por medio de**
líquido m - [electr-equip] liquid insulated
indoor power transformer
— m — — f **monofásico** - [electr-equip] single
phase power transformer
— m — — f **trifásico** - [electr-equip] three
phase power transformer
— m — **potencial** m - [electr-equip] potential
transformer
— m — — m **constante** - [electr-equip] constant potential transformer
— m — **protección** f - [electr-equip] protecting,tive transformer • relaying transformer
— m — **puertas** f **adentro** - [electr-equip] indoor transformer
— m — — f — **con aislación** f - [electr-
-equip] indoor insulated transformer
— m — — f — — — f **por medio de líquido** m
- [electr-equip] indoor liquid insulated
transformer
— m — — f **afuera** - [electr-equip] outdoor
transformer
— m — **regulación** f - [electr-equip] control
transformer
— m — — f **para carga** f - [metal-prod] charging control transformer
— m — **retransmisión** f - [electr-equip] retransmitting, o relaying, transformer
— m — **salida** f - [electrón] output transformer
— m — **servicio** m **pesado** - [sold] heavy duty
transformer
— m — **sobrecorriente** f - [electr-equip] overcurrent transformer
— m — — f **a neutro** - [electr-equip] overcurrent transformer to neutral
— m — — f — **para devanado** — **secundario**
- [electr-equip] overcurrent transformer to
secondary neutral winding
— m — **suministro** m - [electr-equip] supply
transformer
— m — — m **para energía** f - [Electr-equipp
power supply transformer
— m — **tablero** m - [electr-instal] panel
transformer
— m — — m **para mando** m - [electr-instal]
switchgear transformer
— m — **taller** m - [electr-equip] shop transformer
— m — — m **pequeño** - [electr-equip] small
shopl transformer
— m — **tensión** f - [electr-equip] tension, o
potential, o voltage, transformer
— m — — f **alta** - [electr-equip] high voltage
transformer

transformador m para tensión f alta con frecuen
  cuencia f, alta, o elevada - [electr-equip]
  high frequency high voltage transformer
— m — termostato m - [electr-equip] thermostat
  transformer
— m — voltaje m - [electr-equip] voltage
  transformer
— m — m alto - [electr-equip] high voltage
  transformer
— m pequeño m - [electr-equip] small transformer
— m perdido - [electr-oper] knocked out transformer
— m piloto - [electr-equip] pilot transformer
— m portátil - [electr-equip] portable transformer
— m principal - [electr-equip] main transformer
— m — para energía f - [electr-equip] main
  power transformer
— m reactivado - [electr-oper] re-energized
  transformer
— m rectificador - [sold] transformer rectifier
  (welder)
— m — combinado - [sold] combined,nation
  transformer rectifier
— m — — para ambas corrientes f - [electr-
  -equip] combined,nation alternating/direct
  current transformer rectifier
— m — — corriente f alterna o continua -
  [sold] combined,nation alternating/direct
  current transformer rectifier
— m — para corriente f continua - [electr-
  -equip] direct current transformer rectifier
— m — — voltaje m constante - [sold] constant
  voltage transformer rectifier
— m reemplazado - [electr-equip] replaced
  transformer
— m seco - [electr-equip] dry transformer
— m sellado - [electr-equip] sealed transformer
— m — con nitrógeno m - [electr-equip] nitrogen sealed transformer
— m sin circulación f (de corriente f) -
  [electr-equip] de-energized transformer
— m soldador - [sold] welding transformer
— m — monofásico - [electr-equip] single phase
  welding transformer
— m — — para corriente f alterna - [sold]
  single phase alternating current welding
  transformer
— m subterráneo - [electr-equip] underground
  transformer
— m sumergido - [electr-equip] immersed. o submerged, transformer
— m — en aceite m - [electr-equip] oil immersed transformer
— m — — líquido m - [electr-equip] liquid immersed transformer
— m temporario - [electr-equip] temporary
  transformer
— m térmico - [electr-equip] thermal, o thermo,
  transformer
— m trifásico - [electr-equip] three-phase
  transformer
— m probado - [electr-oper] tested transformer
transformar v - . . .; to convert • [geol] to
  metamorphose
— v corriente f - [electr-oper] to transform @
  current
— v en capital m - [fin] to transform into capital
— v energía f - [electr-oper] to transform @
  power
— v entrada f - [electrón] to transform @ input
— v materia(s) f prima(s) - [ind] to transform
  @ raw material(s)
— v motoestibadora f - [constr-equip] to
  transform @ fork lift (truck)
— v rayos-X - [electrón] to transform @ X-rays
— v — — en rayo(s) m luminoso(s) - [electrón]
  to, transform, o convert, X-rays into light
  rays
— v salida f - [electrón] to transform @ output
— v suministro m - [electr-prod] to transform @

supply
— v suministro m de energía f - [electr-
  -distrib] to transform @ power supply
— v tubería f - [tub] to transform @, pipe, o
  piping, o tube,bing
transgredido,da a - transgressed • departed
transgredir v - . . . • to depart
transgresión f - . . .; transgressing • departing,ture
transición f - . . . • [cald] knuckle
— f de corte m a terraplén m - [constr] change
  from @ cut to @ fill
— f efectiva - effective transition
— f — en rigidez f - [constr] effective stiffness transition
— f efectuada - made transition
— f en rigidez f - rigidity, o stiffness,
  transition
— f entre corte m y terraplén m - [constr] cut
  to fill transition
— f entre sección(es) f con corte m y con terraplén m - [constr] cut to fill transition
— f — sistema(s) m - between system(s) transition
— f gradual - gradual transition
— f hecha - made transition
— f para ensanche m - widening transition
— f — — m de calzada f - [vial] roadway
  widening transition
— f estrechamiento - narrowing transition
— f — — m de calzada f - [vial] roadway
  narrowing transition
— f pareja f - even o smooth, transition
— f proyectada - [hidr] designed transition
— f sencilla - simple transition
— f simple - simple transition
— f suave - smooth transition
— f — en rigidez f - smooth rigidity transition
— f típica - typical transition
transigido,da a - compromised
transistor m conmutado - [electrón] switched
  transistor
transistorizado,da* a - [electrón] véase con
  transistor(es) m
transitado,da a - [vial] travelled; busy
transitar v etapa f - [deport] to score @ stage
— v sendero m - [vial] to beat @ path
tránsito m - . . .; travel(ling) • traffic •
  [transp] traffic flow
— m a prever(se) v - [vial] foreseeable traffic; traffic expectation
— m aéreo - [aeron] air traffic
— m — interrumpido - [aeron] interrupted air
  traffic
— m — en aeropuerto m - [aeron] airport air
  traffic
— m atendido - [transp] handled traffic
— m aumentado - [transp] increased traffic
— m aumentante - [transp] increasing traffic
— m automotor m - [vial] auto(mobile) traffic
— m conducido - [transp] carried traffic
— m congestionado - [vial] congested traffic •
  lapped traffic
— m convergente - [vial] converging, o mixing,
  o blending, traffic
— m de automotor(es) m - [vial] automotive
  traffic
— m — avión(es) m - [aeron] air traffic
— m — — en aeropuerto m - [aeron] airport
  air traffic
— m — equipo m móvil - [ind] mobile equipment
  traffic
— m — etapa f - [deport] stage, covering, o
  scoring
— m — ganado m - [ganad] livestock, o cattle,
  movement,ving
— m — mercadería f - [transp] transit, o
  movement of, merchandise, o goods
— m — peatón(es) m - [vial] pedestrian traffic
— m — vehículo(s) m - [vial] vehicle,cular,
  movement, o traffic
— m — sendero m - path beating

tránsito m en aeropuerto m - [aeron] airport
    traffic
— m en aumento - [transp] increasing traffic
— m — . . . sentido(s) m - [vial] . . . way
    traffic
— m — una (sóla) dirección f - [vial] one way
    traffic
— m escaso - [transp] scarce traffic; low traf-
    fic volume
— m ferroviario - [f.c.] railway traffic
— m intensificado - [vial] intensified, o in-
    creased, traffic
— m intenso - [vial] intense, o heavy, traffic
— m interferido - [transp] interferred traffic
— m interrumpido - [transp] interrupted, o dis-
    rupted, traffic
— m libre - [vial] free transit
— m liviano - [transp] light traffic
— m local - [vial] local traffic
— m máximo - [transp] maximum traffic
— m mínimo - [transp] minimum traffic
— m para fin m de semana f - [transp] weekend
    traffic
— m pasante - [transp] passing traffic
— m pedestre - [vial] pedestrian, o foot, traf-
    fic
— m peligroso - [transp] dangerous, o hazard-
    ous, traffic
— m perturbado - [transp] disturbed traffic
— m pesado - [vial] heavy, o intense, traffic
— m por agua m - [transp] water, travel, o
    transportation
— m — camino m - [transp] road travel
— m rápido - [vial] rapid transit • fast moving
    traffic
— m reencausado - [transp] rerouted traffic
— m regulado - [vial] controlled traffic
— m seguro - [transp] safe, travel, o movement
— m separado - [transp] separate(d) traffic
— m vehicular - [Transp] vehicular traffic
— m vial - [transp] highway traffic
transitorio, ria a - . . .; transient; temporary
transmisión f - . . .; transmitting • [autom-
    mec] driveshaft; propeller,ling shaft; drive-
    line • [electrón] sending • [mec] flow(ing);
    shafting; line shaft • véase también acciona-
    miento m - [fin] transfer
— f a interior m - transmitting,mission into
— f a vehículo m - [transp] transmission to @
    vehicle
— f audible - [electrón] audio, o audible, out-
    put
— f — de salida f - [electrón] transmitted
    audio output
— f automática - [autom-mec] automatic trans-
    mission; power drive • [telecom] automatic
    sending
— f — de ráfaga - [electrón] automatic burst
    transmission
— f — — — f de impulso(s) m - [electrón] au-
    tomatic (impulse) burst transmission
— f — de retorno m - [telecom] sending back
    automatically
— f auxiliar - [mec] auxiliary transmission
— f — para . . . velocidades f - [mec] . . .
    speed(s) auxiliary transmission
— f buena - good transmission,mittal
— f — de calor m - good heat transmission
— f cambiada - [autom] changed transmission
— f colmada - [mec] (over)filled transmission
— f — con aceite m - [mec] oil (over)filled
    transmission
— f colocada - [mec] installed transmission
— f — en posición f neutral - [mec] transmis-
    sion, put, o placed, in @ neutral (position)
— f con cable m - [mec] rope drive
— f — cadena f - [mec] chain, transmission, o
    drive,ving • sprocket drive,ving
— f — — i' y rueda(s) f dentada(s) - [mec]
    sprocket and chain transmission
— f — correa f - [mec] belt, transmission, o
    drive
— f — disco(s) m - [mec] disk drive

transmisión f con discos m múltiples - [mec]
    multiple disk drive
— f — engranaje(s) m - [mec] gear drive,ving
— f — fluido m - [petról] fluid drive
— f — palanca f - [autom] manual transmission
— f — proceso m nuevo - [mec] new process
    transmission
— f — sin fin - [mec] worm drive
— f — sistema m nuevo - [mec] new process
    transmission
— f conectada - [autom-mec] connected transmis-
    sion
— f corrediza - [mec] sliding transmission
— f de calor m - [termol] heat transmittal
— f — carga f - [constr] load transmitting
— f — código m - [electrón] code, transmittal,
    o transmission, o sending
— f — corriente - [electr-distrib] current
    transmission
— f — dígito m - [electrón] digit transmis-
    sion
— f — — m primero - [electrón] first digit
    transmission
— f — energía f - [electr-distrib] power
    transmission
— m — estado m - [electrón] status, trans-
    mission,mitting, o sending
— f — falla f - [electr-oper] fault trans-
    mission,mitting
— f — fuerza f - [mec] force transmission •
    power flow • drive
— f — impacto(s) m - [mec] impact(s) trans-
    mission,mitting
— m — momento m torsional - [mec] torque
    transmission,mitting
— f — par motor m - [mec] torque trans-
    mission,mitting
— f — nivel m - [instrum] level trans-
    mission,mitting
— f — potencia f - [mec] power flow
— f — propiedad f - [legal] property transfer
— f — ráfaga f - [electrón] burst trans-
    mission,mitting
— f — retorno m - [electrón] transmission
    back • [telecom] sending back
— f — rutina f - [telecom] routine trans-
    mission,mitting
— f — señal(es) f - [instrum] signal trans-
    mission,mitting
— f — vibración(es) f - [constr] vibration)s)
    transmission,mitting
— f — — f infinita(s) - [mec] (positive)
    infinite vibration transmission
— f demorada - [electrón] delayed trans-
    mission,mitting
— f desconectada - [autom-mec] disconnected,
    transmission, o driveline
— f detenida - [mec] stationary, ó stopped,
    transmission
— f drenada - [mec] drained tranmission
— f eléctrica f - [electr-oper] electric
    transmission
— f — de señal(es) f - [electr-oper] elec-
    tric(al) signal transmission
— f en ángulo m - [mec] angle transmission
— f — m con engranaje(s) m - [mec] angle
    drive gearing
— f — caja f para transferencia f - [autom]
    transfer box transmission
— f — cuatro ruedas f - [autom-mec] four wheel
    drive
— f — . . . rueda(s) f - [autom-mec] . . .-
    wheel drive
— f engranada - [mec] engaged transmission
— f entre eje(s) m - [mec] inter-axle driveline
— f — — m conectada - [autom-mec] connected
    inter-axle driveline
— f — — m desconectada - [autom-mec] discon-
    nected inter-axle driveline
— f especial - [autom-mec] special transmission
— f — para . . . velocidad(es) f - [autom-mec]
    special . . .-speed transmission
— f estacionaria - stationary transmission

transmisión f facsimilar - [electrón] facsimile, o fax, transmission
— f hidráulica - [mec] hydraulic, transmission, o drivel fluid (transmission) drive
— f — para elevación f - [grúas] hoist hydaru- lic, o hidraulic hoist(ing), transmission
— f — para elevación f - [grúas] hoist(ing) hydraulic transmission
— f hidrostática - [mec] hydrostatic transmis- sion
— f Hotchkiss - [autom] Hotchkiss, drive, o transmission
— f improvisada - [autom] improvised, o make- shift, transmission
— f independiente - [mec] independent drive
— f — para bomba f - [petról] independent pump drive
— f instalada - [mec] installed transmission
— f intermedia—[mec] intermediate transmission
— f — para movimiento m - [mec] jack
— f inversora - [mec] reversing transmission
— f longitudinal - [electr-transm] longitudinal transmission
— f — de corriente f - [electr-equip] longi- tudinal corrent transmission
— f manual - [autom-mec] manual transmission
— f mecanizada - [autom-mec] power shift trans- mission
— f — para cambios m para velocidad f - [mec] power shift transmission
— f — — m — f con . . . escala(s) f - [mec] . . . range power shift transmission
— f mediante cadena f y rueda(s) f dentada(s) - [mec] chain and procket transmission
— f neumática - [mec] pneumatic transmission
— f — de señal(es) f - [instrum] pneumatic signal(s) transmission
— f para aparejo m para maniobras f - [petról] drawworks drive
— f — bomba f - [petról] pump drive
— f — cambio m de marcha - [mec] shift trans- mission
— f — — m — f mecanizado - [mec] power shift transmission
— m — m — velocidad(es) f - [mec] shift transmission
— f — — m — f mecanizado - [mec] power shift transmission
— f — contramarcha f - [mec] reverse,sing transmission
— f — cuadro m para maniobra(s) f - [petról] drawworks drive
— f — elevación f - [grúas] hoist transmission
— f — malacate m - [petról] drawworks trans- mission
— f — — m con cadena f y ruedas f dentadas - [petról] drawworks chain and aprocket trans- mission
— f — transferencia f - [autom-mec] transfer transmission
— f — . . . velocidad(es) hacia adelante - [autom-mec] . . . forward speeds transmission
— f — velocidad(es) f múltiple(s) - [autom- -mec] multiple speed(s) transmission
— f perdida - [autom-mec] lost transmission
— f pobre [termol] poor transmission
— f — de calor m - [termol] poor heat trans- mission,mittal
— v por correa(s) en V - [mec] V-belt transmis- sion
— v — engranaje(s) m - [mec] gear drive
— f — manivela f - [mec] crank, gear, o drive
— f — tornillo m sin fin - [mec] worm drive
— f — tubería f - [tub] pipe transmission
— f — — f para conducción f - [tub] pipeline transmission
— f — teléfono m - [telecom] phoning (in)
— f prematura - premature transmission
— f — evitada - prevented. o avoided, prema- ture transmission
— f principal - [autom-mec] main, transmission, o driveline

transmisión f principal para . . . velocidades f - [autom-mec] . . . speed main transmission
— f — conectada - [autom-mec] connected main, transmission, o driveline
— f — desconectada - [autom-mec] disconnected, o disengaged, main, transmission, o driveline
— f provisional - [mec] provisional, o tempo- rary, o interim, transmission, o driveline
— f puesta en posición f neutral - [mec] trans- mission (placed) in @ neutral (position)
— f rebelde - [autom-mec] balky transmission
— f reemplazada - [mec] replaced transmission
— f selectiva - [mec] selective transmission
— f temporaria - [mec] temporary, o provisio- nal, o interim, transmission, o driveline
— f tipo Hotchkiss - [autom-mec] Hotchkiss (type), transmission, o drive(line)
— f variable - [mec] variable, transmission, o drive(line)
— f verbal - [telecom] vocal transmission
— f vocal - [telecom] voice transmission
transmisor m ciego - [comput] blind transmitter
— m de momento m de torsión f - [mec] torque, transmitter, o converter
— m — torsión f - [mec] torque converter
— m específico - [electrón] specific transmit- ter
— m estación f - [comput] field transmitter
— m nivel m - [instrum] level transmitter
— m para presión f - [instrum] pressure trans- mitter
— m — f diferencial - [instrum] differen- tial pressure transmitter
— m para volumen m - volume transmitter
— m remoto - [electrón] remote transmitter
— m respondedor - [electrón] transponder
— m — instalado - [electrón] installed trans- ponder
— m — fácilmente - [electrón] easily in- stalled transponder
— m — movible - [electrón] mobile transponder
— m — en vehículo m (automotor) - [electrón] vehicle mobile transponder
— m — rudo m - [electrón] rugged transponder
— m Selsyn - [electrón] Selsyn transmitter
— m — con husillo m para ajuste m - [electrón] screw-down Selsyn transmitter
— m sincrónico - [electrón] synchronous trans- mitter; Selsyn
— m sincrónico - [electrón] synchronous trans- mitter; Selsyn
— m — Selsyn - [electrón] Selsyn; selsyn
transmisor,ra a - [mec] carrier
transmitido,da a - transmitted • [mec] flowed • [telecom] sent
— a a interior m - transmitted into
— a a vehículo m - [electrón] transmitted to @ vehicle
— a automáticamente - [electrón] transmitted, o sent, automatically
— a — en retorno - [telecom] sent back auto- matically
— a en retorno - [telecom] transmitted, ó sent back
— a por teléfono m - [telecom] phoned in
— a prematuramente - transmitted prematurely
transmitir v - to transmit • to relay • [tele- com] to send
— v a vehículo m - [electrón] to transmit to @ vehicle
— v automáticamente - [electrón] to, transmit, o send, automatically
— v — ráfaga f - [electrón] to automatically transmit @ burst
— f — — f de impulso(s) m - [electrón] to automatically transmit @ impulse(s) burst
— v — en retorno - [electrón] to, transmit, o send, back, automatically
— v carga f - [constr] to, transfer, o trans- mit, @ load
— v código m - [electrón] to, transmit, o send, @ code
— v — m para reconocimiento m - [electrón] to

transmitir v dígito - [electrón] to, transfer, o transmit, @ digit
— v — m primero - [electrón] to, transfer, o transmit, @ first digit
— v en retorno - [electrón] to send back
— v energía f - [electr-distrib] to transmit @ power
— v esfuerzo m - [mec] to transmit @, force, o effort
— v — m compresivo - [mec] to transmit @ compressive force
— v — m — total - to transmit @ full compressive force
— v — m total - to transmit @, full, o total, force
— v estado m - [electrón] to, transmit, o send, @ status
— v falla f - [electr-oper] to transmit @ fault
— v fuerza f - [mec] to transmit @ force
— v impacto m - [mec] to transmit @ impact
— v llamada f - [electrón] to transmit @, call, o ring(ing)
— v nivel m - [instrum] to transmit @ level
— v par m motor - [mec] to transmit @ torque
— v — m torsional—[mec] to transmit @ torque
— v por teléfono m - [telecom] to phone in
— v prematuramente - to transmit prematurely
— v presión f - [mec] to transmit @ pressure
— v — f diferencial - [mec] to transmit @ differential pressure
— v propiedad f - [legal] to transfer @ property
— v ráfaga f - [electrón] to, transmit, o send, @ burst
— v retorno m -[electrón] to transmit back
transparencia f - transparency • [dib] reproducible
transparente a - transparent • [dib] reproducible
transpirado,da a - transpired • perspired
transpondedor m - [electrón] transponder
— m decodificador - [electrón] transponder decoder; decoding transponder
— m instalado fácilmente - [electrón] easily installed transponder
— m movible - [electrón] movable, o mobile, transponder
— m programado - [electrón] programmed transponder
— m rudo - [electrón] rugged, o solid, transponder
transpondencia* f - [electrón] transpondency; transponding
— f de reconocimiento m - [electrón] acknowledgment transponding
transponder* v - [electr] to transpond; véase también transmisor m respondedor
— v reconocimiento m - [electrón] to transpond @ acknowledgment
transpondido,da a - [electrón] transponded
transponer v - . . . .; to move through; to exit
— v calzada f - [vial] to span @ highway
— v canal m - [hidr] to cross @ channel
— v curso m de agua m - [hidr] to cross @ (water) stream
— v depresión f - to cross @ depression
— v laguna f - [hidr] to cross @, lagoon, o lake, o pond
— v zanjón m - [hidr] to, brdige, o cross, @ ravine
transportable adv - . . . .; portable
transportación f - . . . .; transporting; conveying; véase también transporte m
transportado,da a - [transp] transported; conveyed; hauled • shipped
— a a emplazamiento m - [transp] transported to @, location, o site
— a — — m para obra f - [constr] transported to @ project site
— a dentro de - [transp] transported (with)in
— a — de sitio de obra f - [transp] trans-

ported within @, project, o work, site
transportado,da a económicamente - [transp] transported, o hauled, economically
— a fácilmente - [transp] transported, o shipped, easily
— a hasta destino m - [transp] transported tto @ destination
— a — muelle m - [ind] conveyed to @ wharf
— a — puerto m - [transp] transported to @ port
— a — sitio m de obra f - [transp] transported to @, project, o work, o job, site
— a longitudinalmente - transported longitudinally
— a por agua m - [transp] transported by water
— a — camión f - [transp] transported by truck
— a — ferrocarril m - [transp] transported by rail(way)
— a — tierra f - [transp] (over)land transported
— a sobre distancia f corta - [transp] short (distance) hauled
— a — patín(es) m - [transp] conveyed on @, skid(s), o pallet(s)
transportador m - [transp] . . .; carryall • conveyor • [quím] carrier • [constr] earth mover • [dib] protractor
— m aéreo - [ind] aerial conveyor
— m — con cable m - [transp] (aerial) cableway
— m circular - [ind] circular conveyor
— m con aguilón m - [mec] boom conveyor
— m — banda f - [ind] belt conveyor
— m — f articulada - [ind] apron conveyor
— f — f sobre cama f de rodillo(s) m - [ind] apron conveyor
— m bandeja(s) f - [mec] pallet conveyor
— m cable m - [ind] cable, o rope, transfer, o conveyor; cableway • [f.c.] funicular, o cable, railway
— m cadena f - [mec] chain conveyor
— m cajas f - [ind] pallet, o apron, conveyor
— m canal m - [ind] chute conveyor
— m cangilón(es) m - [ind] bucket conveyor
— m cinta f - [mec] belt conveyor
— m correa f - [ind] belt conveyor
— m escala f - [dib] ladder
— m lanzadera f - [mec] shuttle conveyor
— m mandril m - [ind] apron conveyor
— m pluma f - [ind] boom conveyor
— m rodillo(s) m - [ind] gravity conveyor • roller conveyor
— m tornillo m - [mec] screw conveyor
— m — m helicoidal - [ind] helical screw conveyor
— m — m sin fin - [mec] (helical) screw conveyor
— m — vertedero m - [ind] chute conveyor
— m elevado - [ind] aerial conveyor
— m enfriador - [ind] cooling conveyor
— m funicular - [f.c.] funicular, o cable, railway
— m mezclador - [mec] mixing conveyor
— m monocarril - [mec] monorail conveyor
— m monorriel - [ind] monorail conveyor
— m móvil - [ind] shuttle conveyor
— m para almacén(aje) m - [ind] storage conveyor
— m — bobinas f - [metal-lam] coil, conveyor, o carrier
— m — cascarilla f - [metal-lam] scale conveyor
— m — coque· m - [coque] coke conveyor
— m — cristalización f - [metal-trat] spangle conveyor
— m — depósito m - [ind] storage conveyor
— m — — m para bobinas f - [metal-lam] coil storage conveyor
— m — — m — f en bruto - [metal-lam] rough coil storage, transfer, o conveyor

transportador <u>m</u> para inspección <u>f</u> - [metal-lam]
inspection table
— <u>m</u> — lámina(s) <u>f</u> - [metal-lam] sheet carrier
— <u>m</u> — perfil(es) <u>m</u> estructural(es) <u>m</u> - [metal-
-lam] structural transfer
— <u>m</u> — planchas <u>f</u> - [metal-lam] plate, transfer, <u>o</u> conveyor
— <u>m</u> — <u>f</u> de tipo <u>m</u> con cadena <u>f</u> y gatillo <u>m</u>
[ind] chain and dog type plate transfer
— <u>m</u> — planchones <u>m</u> - [metal-lam] slab transfer
— <u>m</u> — recuperador <u>m</u> para carbón <u>m</u> - [ind] coal
recovery conveyor
— <u>m</u> — remoción <u>f</u> - [metal-lam] runout conveyor
train
— <u>m</u> — retorno <u>m</u> - [mec] return conveyor
— <u>m</u> — rieles <u>m</u> - [metal-lam] rail transfer
— <u>m</u> — salida <u>f</u> - [ind] runout conveyor (train)
— <u>m</u> — <u>f</u> a nivel <u>m</u> de piso <u>m</u> - [ind] runout
floor conveyor
— <u>m</u> — secado,camiento <u>m</u> - [ind] drier conveyor
— <u>m</u> — tocho(s) <u>m</u> - [metal-lam] bloom transfer
— <u>m</u> — transferencia <u>f</u> - [mec] transfer, <u>o</u> by-
-pass, conveyor
— <u>m</u> por arrastre <u>m</u> - [ind] drag conveyor
— <u>m</u> — gravedad <u>f</u> - [ind] gravity conveyor
— <u>m</u> — <u>f</u> con rodillo(s) <u>m</u> - [ind] gravity
transfer table
— <u>m</u> portátil - [mec] shuttle conveyor
— <u>m</u> público - [transp] public carrier
— <u>m</u> reversible - [ind] shuttle conveyor
— <u>m</u> secador - [ind] drier conveyor
— <u>m</u> telesférico - [f.c.] funicular, <u>o</u> cable,
railway
— <u>m</u> vibratorio - [mec] vibratory conveyor
transportar <u>v</u> - . . . . to, haul, <u>o</u> move • to
transfer • to ship
— <u>v</u> agua <u>m</u> tratada - [hidr] to transport @
treated water
— <u>v</u> automóvil <u>m</u> - [transp] to transport @ automobile
— <u>v</u> cajón <u>m</u> - [mec] to transport @ box
— <u>v</u> camión <u>m</u> - [transp] to transport @ truck
— <u>v</u> cañería(s) <u>f</u> - [tub] to convey @ pipe,ping
— <u>v</u> — <u>f</u> sobre patín(es) <u>m</u> - [transp] to convey @ pipe,ping on @ skid(s)
— <u>v</u> caño(s) <u>m</u> - [tub] to convey @ pipe(s)
— <u>v</u> carbón <u>m</u> - [miner] to, transport, <u>o</u> haul, @
coal
— <u>v</u> combustible <u>m</u> - [combust] to, transport, <u>o</u>
carry, @ fuel
— <u>v</u> — <u>m</u> diesel - [petról] to, transport, <u>o</u>
carry, @ diesel fuel
— <u>v</u> cordón <u>m</u> - [cabl] to transport @ strand
— <u>v</u> dentro de - [transp] to transport within
— <u>v</u> — — sitio <u>m</u> de obra <u>f</u> - [transp] to
transport within @ work site
— <u>v</u> económicamente - to, handle, <u>o</u> transport,
economically
— <u>v</u> equipo <u>m</u> - [transp] to transport @ equipment
— <u>v</u> fácilmente - to transport easily • [transp]
to ship easily
— <u>v</u> hasta destino <u>m</u> - [transp] to transport to
@ destination
— <u>v</u> — puerto <u>m</u> - [transp] to transport to @
port
— <u>v</u> — sitio <u>m</u> [transp] to transport to @ site
— <u>v</u> — — <u>m</u> de obra <u>f</u> - [transp] to transport
to @, worksite, <u>o</u> jobsite
— <u>v</u> implemento - [transp] to transport @ implement
— <u>v</u> — <u>m</u> agrícola - [transp] to transport @
farm implement
— <u>v</u> longitudinalmente - to transport longitudinally
— <u>v</u> máquina <u>f</u> - [transp] to transport @ machine
— <u>v</u> — pesada - [transp] to transport @ heavy
machine
— <u>v</u> material(es) <u>m</u> - [ind] to, transport, <u>o</u>
carry, @ material(s)
— <u>v</u> — <u>m</u> suelto(s) - [transp] to, transport, <u>o</u>
carry, <u>o</u> haul, @, loose, <u>o</u> bulk, material(s)
— <u>v</u> personal - [labor] to transport @ personnel

transportar <u>v</u> por agua <u>m</u> - [transp] to transport by water
— <u>v</u> — ferrocarril - [transp] to transport by
rail(way)
— <u>v</u> — camión <u>m</u> - [transp] to transport by
truck
— <u>v</u> sobre patín(es) <u>m</u> - [transp] to, transport, <u>o</u> convey, on @ skid(s)
— <u>v</u> soldadora <u>f</u> - [sold] to, transport, <u>o</u>
haul, @ welder
— <u>v</u> sólido(s) <u>m</u> - [transp] to transport @
solid(s)
— <u>v</u> — <u>m</u> suspendido(s) - [hidr] to transport
@ suspended solid(s)
— <u>v</u> tubería <u>f</u> - [tub] to, transport, <u>o</u> convey, @ pipe,ping
— <u>v</u> — <u>f</u> sobre patín(es) <u>m</u> - [transp] to,
transport, <u>o</u> convey, @ pipe,ping on @
skid(s)
— <u>v</u> tubo(s) <u>m</u> - [tub] to, transport, <u>o</u> convey,
@ pipe(s)
transporte <u>m</u> - . . .; transporting; conveying;
hauling • [transp] shipment; shipping •
transportation means • [contab] brought, <u>o</u>
carried, forward; balance
— <u>m</u> a aeropuerto <u>m</u> - [transp] airport delivery
— <u>m</u> a emplazamiento <u>m</u> - [constr] transportation to @ site
— <u>m</u> — — para obra <u>f</u> - [constr] transportation to @, project, <u>o</u> work, site
— <u>m</u> — granel - [transp] bulk transportation
— <u>m</u> — obra <u>f</u> - [transp] transportation to @
site; field handling
— <u>m</u> a planta <u>f</u> - [transp] transportation to @
plant
— <u>m</u> acelerado - [transp] sped up transportation • [f.c.] rapid transit
— <u>m</u> aéreo - [transp] air, <u>o</u> aerial, trans-
port(ation), <u>o</u> travel • grúas] overhead
lifting
— <u>m</u> — internacional - [aeron] international
air transportation
— <u>m</u> — nacional - [aeron] domestic, <u>o</u> internal, air transportation
— <u>m</u> aumentado - [transp] increased transportation
— <u>m</u> automotor - [transp] auto(motive) transportation
— <u>m</u> con correa <u>f</u> - [ind] belt, <u>o</u> conveyor,
transportation
— <u>m</u> contratado - [transp] contracted transportation
— <u>m</u> de agua <u>m</u> tratada - [hidr] treated water
transportation
— <u>m</u> — arrabio <u>m</u> - [metal-prod] pig iron
transportation
— <u>m</u> — automóvil(es) <u>m</u> - [transp] automobile
transportation
— <u>m</u> — bien(es) <u>m</u> - [transp] good(s) transportation
— <u>m</u> de bordo <u>m</u> alto - [transp] deep water
transportation
— <u>m</u> — cable <u>m</u> - [cabl] cable, <u>o</u> rope, transportation
— <u>m</u> — cajón(es) <u>m</u> - [mec] box(es), conveying,
<u>o</u> carrying, <u>o</u> transporting,tation
— <u>m</u> — camión(es) <u>m</u> - [transp] truck transportation
— <u>m</u> — cañería(s) <u>f</u> - [tub] pipe,ping, conveying, <u>o</u> transport(ation)
— <u>m</u> — — <u>f</u> sobre patín(es) <u>m</u> - [tub] pipe
conveying on skids
— <u>m</u> — carbón <u>m</u> - [miner] coal, transporting,
<u>o</u> transportation, <u>o</u> hauling • coal shipping
— <u>m</u> — carga <u>f</u> - [transp] load, <u>o</u> freight,
transporting,tation, <u>o</u> carrying
— <u>m</u> — — <u>f</u> varia - [transp] multiple load
transportation
— <u>m</u> — combustible <u>m</u> - [transp] fuel transporting,tation, <u>o</u> carrying, <u>o</u> hauling
— <u>m</u> — — <u>m</u> diesel - [transp] diesel fuel,
transporting,tation, <u>o</u> carrying
— <u>m</u> — container(s) <u>m</u> - [transp] container(s)

transporting,tation, o drayage
transporte m **de cordón** m - [cabl] strand transporting,tation
— m — **chatarra** f - [metal-prod] scrap transporting,tation
— m — **desecho(s)** m - [nucl] waste(s) transporting,tation
— m — m **radiactivo(s)** - [nucl] radioactive
waste transporting,tation
— m — **equipo** m - [transp] equipment transporting,tation
— m — **implemento(s)** m - [transp] implement(s)
transporting,tation
— m — m **agrícola(s)** - [transp] farm imple-
ment(s) transporting,tation
— m — **lingote(s)** m - [metal-lam] ingot transporting,tation
— m — m **en acería** f - [metal-lam] steel
plant ingot transporting,tation
— m — **líquido(s)** m - [transp] liquid(s) transporting,tation
— m — **madera(s)** f - [transp] timber, o lumber,
transporting,tation
— m — **máquina** f - [ind] machine transportation
— m — f **pesada** - [ind] heavy machine transporting,tation
— m — **masas** f - [transp] mass transportation
— m — **materia(s)** f **prima(s)** - [ind] raw mate-
rial(s) transporting,tation
— m — **material(es)** m - [ind] material(s),
transporting,tation, o carrying
— m — — m **suelto(s)** m - [miner] loose, o
bulk, materials transporting,tation
— m — **mercadería(s)** f - [transp] merchandise
transporting,tation
— m — **mineral(es)** m - [miner] ore transporting,tation
— m — — m **de hierro** m - [miner] iron ore
transporting,tation
— m — **petróleo** m - [petról] petroleum transporting,tation
— m — **residuo(s)** m - [nucl] waste(s) transporting,tation
— m — — m **radiactivo(s)** - [nucl] radioactive
waste(s) transporting,tation
— m — **soldadora** f - [sold] welder transporting,tation
— m — **tubería** f - [transp] pipe,ping, transporting,tation, o conveying
— m — — f **sobre patín(es)** m - [tub] pipe,ping
conveyance,ying, o transporting,tation, on @
skid(s)
— m **dentro de** - [transp] transporting,tation
within @
— m — — **sitio** m - [transp] transporting,ta-
tion within @ worksite
— m **desviado** - [transp] diverted transportation
— m **económico** - economical, handling, o transporting,tation
— m **efectuado** - [transp] transportation performed
— m **en cantera** f - [miner] quarry, transporting,tation, o haulage,ling
— m — **clase** f **primera** - [transp] first class
transporting,tation
— m — **exterior** - [transp] foreign transporting,tation, o travel(ing)
— m — **interior** - [transp] domestic transporting,tation
— m — **masa** f - [transp] mass transportation
— m — **país** m — [transp] domestic transportation
— m — **taller** m - [ind] in plant, o on site,
transporting,tation
— m **especial** - [transp] special transportation
— m **externo** - [transp] outside transportation
— m **fácil** - easy, transportation, o shipment •
portability • easy shipping,pment
— m **fuera de carretera** f - [transp] off-highway
transporting,tation
— m **hasta destino** m - [transp] transportation
to @ destination
— m **hasta obra** f - [transp] transportation to @,
project, o job, site

transporte m **hasta puerto** m - [transp]
transporting,tation to @ port
— m **hasta sitio** m - [stransp] transporting
to @, worksite, o jobsite, o project
— m **hidráulico** - [hidr] hydraulic transporting,tation; sluicing
— m **interno** - [ind] internal transportation
— m — **de personal** m - [labor] personnel
internal transporting,tation
— m **interrumpido** - [transp] interrupted transporting,tation
— m **local** - [transp] local transportation
— m **longitudinal** - [transp] longitudinal
transporting,tation
— m **marítimo** - [transp] marine, o maritime, o
deep water, transporting,tation, o shipping
• overseas shipping,pment
— m **no urbano** - [Transp] nonurban transporting,tation
— m **normal** - [transp] normal transporting,tation
— m **para material(es)** m - [ind] material(s)
transporting,tation
— m — **mineral** m **de hierro** m - [transp] ore,
transporting,tation, o boat
— m **para personal** m - [labor] personnel
transporting,tation
— m **pesado** - [transp] heavy transport(ation)
— m — **por camión** m - [transp] heavy duty,
trucking, o transporting,tation
— m **por agua** m - [transp] water transporting,tation
— m — **camión** m - [transp] truck transporting,tation
— m — **ferrocarril** m - [transp] rail(way)
transporting,tation
— m — **mar** m - [transp] sea transport(ation)
— m — — m **alta** - [transp] deep sea, o overseas, o ocean, transporting,tation, o
shipping,pment
— m — **terreno(s)** m **accidentado(s)** - [transp]
rough terrain, transporting,tation, o hauling
— m — **tonelada** f - [transp] transportation
per ton
— m **público** - [transp] public. transit, o
transport(ing,tation)
— m **sobre patín(es)** m - [Transp] conveying on
@ skid(s)
— m **suspendido** - [transp] suspended transporting,tation • [grúas] overhead lifting
— m **terrestre** - [transp] land, o inland, o
overland, transporting,tation
— m **urbano** - [transp] urban transportation
transportista m - [transp] transporter; carrier
— m **aéreo** - [aeron] airline carrier
transposición f - . . . . • [vial] going, o passing, over; spanning
— f **de calzada** f - [vial] roadway spanning
transvasado,da a - repacked; transferred
transvasador m - repackager
— m **para arrabio** m - [metal-prod] pig iron
transfer operator
transvasamiento m - véase **transvase** m
— m **de ácido(s)** m - acid(s) transfer
transvasar v - . . . .; to repackage
— v **líquido(s)** m - to transfer @ liquid(s)
transvase m - transfer • repackaging
— m **de líquido(s)** m - liquid(s) transfer
transversal a - . . . .; crosswise; sideways •
[deport] cross country
tranvía m - [transp] . . . . • [G.B.] tram
— m **aéreo** - cableway; aerial tramway
trapa f - [miner] trap rock
trapiche m - . . . . • [miner] grinding mill
— m **con correa** f - [agric] belt (powered)
cane mill
— m **con sangre** f **para caña** f **de azucar** m -
[agric] animal power cane mill
— m **hidráulico para caña** f **de azucar** m -
[agric] water power, o hydraulic, cane mill
— m **para caña** f **de azucar** m - [agric] (sugar)
cane mill

trapiche m para caña f de azucar m y sorgo m -
  [aric] (sugar) cane and sorghum mill
— m — sorgo m - [agric] sorghum mill
trapo m empapado - soaked rag
— m con agua m - water soaked rag
trasbordo m - [transp] transshipment
trascendencia f - trascendency,ding
— f legal - [legal] legal transcendence,dency
trascendente a - trascending,dent ; far reaching
trascender v a práctica f - to carry over into
  practice
trascendido,da a - . . .; trascended
trasdos m - . . .; outside face
— m apuntado - [arquit] pointed extrados
— m en punta f - [arquit] pointed extrados
— m escarzado - [arquit] stilted extrados
— m peraltado - [arquit] stilted extrados
— m realzado - [arquit] stilted extrados
trasegamiento m - transfer
— m de ácido(s) m - acid(s) transfer
trasegar v - . . .; to transfer
trasero,ra a - . . .; rear end • on @ back
traslación f - . . .; transfer; movement,ving;
  travel(ling) • [grúas] traverse speed { adv -
  (de) translational
— f de grúa f - [grúas] crane, travel(ling), o
  movement,ving
— f de mesa f - [mec] table travel(ing)
— f — f corrediza f - [mec] traveling table
  travel
— f — f deslizante - [mec] traveling table
  travel
— f — f movible - [mec] traveling table
  travel
— f lenta - [mec] slow, translation, o travel
— f longitudinal - [mec] longitudinal, o
  lengthwise, translation, o travel
— f transversal - [mec] crosswise, translation,
  o travel
trasladado,da a - moved; conveyed • relocated •
  [legal] shifted • [filol] translated
— a exterior - moved, outside, o outdoors;
  moved out of @ country
— a — intemperie - [mec] moved outdoors
— a fácilmente - moved easily
trasladar v - . . .; to, travel, o convey • to
  pass on • to relocate • [grúas] . . .; to
  transport • [legal] to shift • [filol] to
  translate
— v cañería f - [tub] to convey @ pipe
— v caño m - [tub] to convey @ pipe
— v a exterior - [ind] to move outdoors •
  [transp]| to move out of @ country
— v — intemperie f - to move outdoors
— v fácilmente - to move easily
— v torre f - [ind] to move @ tower
— v tubería f - [tub] to convey @, pipe, o tube
— v tubo m - [tub] to convey @, pipe, o tube
traslado m - . . .; moving; conveying; move; mobility; relocating,tion • travel • [legal]
  shift(ing) • [deport] transit • [filol]
  translation
— m de cañería f - [tub] pipe conveying
— m — equipo m - [ind] equipment transfer
— m — mercadería f - [ind] merchandise, o
  good(s) transfer
— m — personal - [labor] personnel transfer
— m — producto m - [ind] product transfer(ing)
— m — rodillo(s) m - [metal-lam] roll transfer
— m — torre f - [ind] tower move,ving
— m — tubería f - [tub] pipe, conveying, o
  transfer
— m eficiente - [transp] efficient transfer
— m — de equipo m - [ind] efficient equipment
  transfer
— m — mercadería f - [ind] efficient, merchandise, o goods, transfer
— m — personal - [labor] efficient personnel transfer
— m fácil - easy, transfer, o moving
— m — en obra f - [ind] on site mobility
— m horizontal - [ind] horizontal translation
— m longitudinal - [ind] longthwise translation

traslado m seguro - [transp] safe movement
— m — de material(es) m - [segurid] safe ma-
  terial(s) movement
— m — — personal - [segurid] safe, person-
  -nel, o human, movement
traslapado,da a - [mec] lapped; lap joint(ed)
traslapadora f - [mec] lapper
— f transversal - [culin] cross grain lapper
traslapante a - overlapping
traslapar v - [mec] . . .; to lap • [mec] to
  scarf
— v blindaje m - [electr-cond] to overlap @
  shield(ing)
— v costura f - [mec] to (over)lap @ seam
— v extremo(s) m - [mec] to overlap @ end(s)
— v — m de tubería f - [tub] to overlap @
  pipe end(s)
— v junta f - [mec] to lap @ joint
— v — longitudinal - [mec] to lap @ longitudinal joint
— v pieza f - [mec] to (over)lap @ part
traslaparse v sobre sí mismo,ma - [mec] to lap
  (over) itself
traslapo m - [mec] . . .; (over)lap; overhang •
  [mec] scarfing
— m bueno - [sold] good overlap(ping)
— m — en soldadura f - [sold] good overlap
  weld
— m — — f corcunferencial - [sold] good
  overlap on @ roundabout
— m cementado - cemented overlap(ping)
— m de blindaje m - [electr-cond] shielding
  overlap(ping)
— m — costura f - [mec] seam (over)lap(ping)
— m — extremo m - [mec] end overlap(ping)
— m — — m de tubería f - [tub] pipe end
  overlap(ping)
— m — junta f - [mec] joint (over)lap(ping)
— m — — f longitudinal - [mec] longitudinal,
  o lengthwise, joint (over)lap(ping)
— m — plancha(s) f - [constr] plate(s), o
  sheet(s), (over)lap(ping)
— m — punta f [mec] end (over)lap(ping)
— m en soldadura f - [sold] weld overlapping
— m — — f circunferencial - [sold] overlap
  on @ roundabout
— m exterior - [mec] outer lap(ping)
— m interior - [mec] inner, o inside, (over)-
  lap(ping)
— m lateral - [mec] side (over)lap(ping)
— m longitudinal - [mec] longitudinal, o
  lengthwise (over)lap(ping)
— m remachado - [mec] riveted (over)lap(ping)
— m substitutivo - [mec] alternate lap(ping)
traspasado,da a - conveyed • [contab] carried.
  o brought, forward
traspasar v - . . .; to convey • [contab] to,
  carry, o bring, forward
traspaso m - . . .; conveying • cross-over •
  [contab] carrying, o bringing, forward •
  [legal] transfer
— m de acción(es) m - [legal] stock, o
  share(s) transfer
— m — calor m - [metal-prod] heat transfer
— m — interés m - [fin] interest transfer
traspie n - . . . • tripping
traspilado,da a - transfered from one pile to
  another
traspilamiento m - [ind] transfer (from one
  pile to another); véase también traspile m
traspilar v - to transfer (from one pile to
  another)
traspile n - transfering from one pile to another
— m de arena f - [constr] sand, handling, o
  moving
— m — dolomita f [miner] dolomite, handling,
  o moving
— m — mineral m de hierro m - [miner] iron
  ore, handling, o moving
trasplantado,da a - transplanted
trasplantadora f - [agric] transplanter
trasplantar v - [agric] to transplant

trasponedor m - véase **transponedor** m
trasponer v - to transpose; to go, over, o across • to, cross, o drive, over • [deport]
to, cover, o come off • [arquit] to bridge
— y **muro** - to go over @ wall
— v **paso** m - [vial] to go over @ pass
— v **primero llegada** f - [Deport] to take @ flag
traspuesto,ta - transposed • crossed; gone over
traspuntín m - [autom] . . . . • folding seat
trastabillar v - . . .; to stumble
trastornado,da a - . . .; disturbed
trastorno m **meteorológico** - [meteorol] metereo-
logical disturbance; nature('s) freak
— m **circulatorio** - [mec] cir-
culatory disorder
— m **digestivo** - [mec] digestive disorder
— m **respiratorio** - [mec] respiratory disorder
trastrocable adv - véase **reversible**
trasvase m - pouring; decanting
— m **de arrabio** m - [metal-prod] pig iron transfer
tratado m **sobre hidráulica** - [hidr] hydraulics
treatise
tratado,da a - treated • discussed • tried
— m **bajo presión** f - [maderas] pressure treated
— a **con boro** - [metal-trat] boron treated
— a **con calor** m - véase **tratado,da térmicamente**
• **con tratamiento** m **térmico**
— a **con creosota** f - [maderas] creosote treated
— a — f **bajo presión** f - [maderas] creosote
pressure treated
— a **con gas** m - gas treated
— a **estéticamente** - aesthetically treated
— a **para protección** f - [metal] protection
treated
— a — f **contra corrosión** f - [metal-trat]
corrosion protection treated
— a **por fluidificación** f - [metal-trat] fluidizing treated
— a **rudamente** - treated, o handled, roughly, o
rudely
— a **térmicamente** - [metal-trat] heat treated
tratador m - . . . . • [petról] treater
tratamiento m - . . . .; **treating; processing;
handling** • [miner] beneficiation
— m **anticipado** - pretreatment
— m — **de líquido(s)** m - liquid(s) pretreating
— m — — — m **cloacal(es)** - [sanit] effluents
pretreatment
— m **bajo presión** f - [maderas] pressure treatment
— m **bituminoso** - bituminous treatment
— m — **de superficie** f - surface bituminous
treatment
— m — **de tipo** m **triple** - [vial] triple bituminous type treatment
— m — **triple** - [vial] triple bituminous treatment
— m — — m **para superficie** f - [vial] triple
bituminous surface treatment
— m **con aceite** m - oil treating,tment
— m — **bombardeo** m **de perdigón(es)** m - [metal-
-trat] cloudburst treatment
— m — **boro** m - [metal-trat] boron treatment
— m — **creosota** f - [maderas] creosote treatment
— m — — f **bajo presión** f - [maderas] creosote
pressure treating,tment
— m — **gas** m - gas treating,tment • [med] gas-
sing
— m — **intercambio** m **de ión(es)** m - [nucl] ion
exchange treating,tment
— m — **temperatura** f **alta** - [metal-trat] high
temperature treating,tment
— m — — f **baja** - [metal-trat] low temperature
treatment
— m — **vacío** m - [metal-prod] vacuuum treatment
— m **continuo** m - continuous treating,tment
— m **correspondiente** - corresponding treatment
— m **de líquido(s)** m - [ind] liquid(s), treating,ment, o handling
— m — **recocido** m - [metal-trat] annealing
treating,tment

tratamiento m **de tipo** m **bituminoso** - bituminous
type treating,tment
— m — **wolmanizado** - [maderas] wolmanized
treating,tment
— m **especial** - special treating,tment
— m **estabilizador** - [metal-prod] stabilizing
treatment
— m **estético** - aesthetical treating,tment
— m **fraccionado** - split treatment
— m — **para reducción** f **de dureza** f - [hidr]
hardness reduction split treatment
— m **general** - general treating,tment
— m **ininterrumpido** - [ind] continuous, o uninterrupted, treating,tment
— m **intermedio,diario** - intermediate
treating,tment
— m **interrumpido** - interrupted treating,tment
— m **malo** - bad, treatment, o handling
— m **médico** - [mec] medical treating,tment
— m **metalúrgico** - [metal-prod] metallurgical
treating,tment
— m **para aceite** m - oil treating,tment
— m — **acero** m - [metal-trat] steel
treating,tment
— m — **con boro** m - [metal-trat] boron
steel treating,tment
— m — **agua** m - [hidr] water treating,tment
— m — — **circulante** - [hidr] circulating
water treating.tment
— m — — m **de pozo** m - [hidr] well water
treating,tment
— m — — m **para industria** f - [hidr] industrial water treating,tment
— m — m — — f **y otros usos** m - [hidr]
industry and other uses water treating,ment
— m — **berma** f - [constr] shoulder treatment
— m — **camino** m - [vial] road treatment
— m — **combustible** m - [combut] fuel
treating,tment
— m — — m **irradiado** - [nucl] radiated fuel
treating,tment
— m — **decapado** m - [metal-trat] pickling
treatment
— m — **descarga** f - [sint] effluent treatment
— m — **desecho(s)** m - [nucl] waste(s)
treating,tment
— m — — m **con radiactividad** f - [nucl]
radioactive waste treating,tment
— m — m — — f **alta** - [nucl] high level
waste treatment
— m — m — — f **baja** - [mucl] low level
waste treating,tment
— m — m — — f **elevada** - [nucl] high
level waste treating,tment
— m — m — — f **mediana** - [nucl] intermediate level waste treating,tment
— m — — m **radiactivo(s)** - [nucl] radioactive
waste(s) treating,tment
— m — **líquido(s)** - [nucl] liquid
radioactive waste(s) treating,tment
— m — **escoria** f - [metal-prod] slag
treating,tment
— m — **estabilización** f - [metal-prod] satibi-
lizing,zation treating,tment
— m — **extremo(s)** m - end(s) treating,tment
— m — **gas** m - [coque] gas, handling, o
treating,tment
— m — **gasolina** f - [petról] gasoline, handling, o treating,tment
— m — **hierro** m - iron, treating,tment, o
handling
— m — — m **prerreducido** - [metal-prod] prereduced iron, handling, o treating,tment
— m — **líquido(s)** m - liquid(s), handling, o
treating,tment
— m — — m **cloacal(es)** - [sanit] sewage,
handling, o treating,tment
— m — **madera** f - [mader] wood, handling, o
treating,tment
— m — **mineral(es)** m - [miner] ore, handling,
o treating,tment
— m — — m **de hierro** m - [miner] iron ore,
handling, o treating,tment

tratamiento m para pasivación - [metal-trat]
  scaling, o pasivating treating,tment
— m — protección f - protection treating,tment
— m — — f contra corrosión f - [metal-trat]
  corrosion protection treating,tment
— m — recocido m - [metal-trat] annealing
  treating,tment
— m — residuo(s) m - [nucl] waste(s) treatment
— m — — m con radiactividad f alta - [nucl]
  high level waste(s) treating,tment
— m — — m — f baja - [nucl] low level
  waste(s) treating,tment
— m — — m — f elevada - [nucl] high level
  waste(s) treating,tment
— m — — f, intermedia, o mediana f -
  intermediate (level) waste(s) treating,tment
— m — — m radiactivo(s) - [nucl] radioactive
  weste(s) treating,tment
— m — — — líquido(s) - [nucl] liquid radioactive waste(s) treating,tment
— m — sensibilización f - [ind] sensitizing
  treating,tment
— m — superficie f - surface treating,tment
— m pobre - poor, treating,tment, o handling
— m por fluidificación f - [metal-trat] fluidizing, treating,tment
— m — nitruración f - [metal-trat] nitriding
  treating,tment
— m — pérdida f dieléctrica - [metal-trat]
  dielectric loss treating,tment
— m posterior - later treating,tment
— m preservativo - [maderas] preservative,tion
  treating,tment
— m primario - primary treating,tment
— m protector - protective treating,tment
— m químico - [quím] chemical treating,tment
— m rudo - rude, o rough, treating,tment
— m — resistido - resisted, o withstood, rough
  treating,tment
— m — soportado - withstood rough treatment
— m secundario - [sanit] secondary treatment
— m similar - [naderas] similar treating,tment
— m tal - such treating,tment
— m terciario - [Sanit] tertiary treating,tment
— m térmico - [metal-trat] heat treating,tment
— m — con disolución f - [metal-trat] solution
  heat treating,tment
— m — especializado - [metal-trat] specialized
  heat treating,tment
— m — integral - [f.c.-ruedas] integral heat
  treating,tment
— m — luego de soldado,da - [metal-trat] heat
  treating,tment after welding
— m — normalizador—[metal-trat] normalizing
  heat treating,tment
— m — original - [metal-trat] original heat
  treating,tment
— m — para acero m - [metal-prod] steel heat
  treating,tment
— m — — — m con aleación f - [metal-prod]
  alloy steel heat treating,tment
— m — — alivio m - [metal-trat] relief heat
  treating,tment
— m — — — m de tensión(es) f - [metal-trat]
  stress relief heat treatment
— m — — cuerpo m - [mec] body heat treatment
— m — — material(es) m - [metal-trat] mate-
  rial(s) heat treating,tment
— m — — pieza f - [mec] part heat treatment
— m — — recocido m - [metal-trat] annealing
  heat treating,tment
— m — — tornillo m - [mec] screw heat treating,tment
— m — — tubería f - [tub] pipe heat treating,tment
— m — — varilla f - [metal-trat] rod heat
  treating,tment
— m — por inducción f - [metal-trat] induction
  heat treating,tment
— m — — — m de material(es) m - [metal-trat]
  material(s) induction heat treating,tment
— m — — pérdida f dieléctrica - [metal-trat]

dielectric loss heat treating,tment
tratamiento m térmico preferencial - [metal-
  -trat] preferential heat treating,tment
— m — regulado - [metal-trat] controlled
  heat treating,tment
— m — variado - [metal-trat] varied heat
  treating,tment
— m triple - [metal-trat] triple treatment
— m — de superficie f - [vial] triple surface, treating,tment
— m ulterior - later treating,tment
— m vial - [vial] road treating,tment
tratar v - . . .; to try • to, negotiate, o
  deal
— v aceite m - to treat @ oil
— v acero m - [metal] to treat @ steel
— v — m con boro m - [metal-trat] to boron
  treat @ steel
— v agua m - [hidr] to treat @ water
— v — de pozo m - [hidr] to treat @ well
  water
— v bajo presión f - [maderas] to pressure
  treat; to treat under pressure
— v anticipadamente - to pretreat
— v — líquido m - [sanit] to pretreat @
  liquid
— v — — m cloacal - [sanit] to pretreat @
  effluent
— v camino m - [vial] to treat @ road
— v combustible m - [combust] to treat @ fuel
— v — m irradiado - [nucl] to pretreat @
  (ir)radiated fuel
— v con - to treat with - to deal with
— v — agua m - [hidr] to treat with @ water
— v — — m fuerte - [quím] to etch
— v — boro m - [metal-trat] to treat with
  boron
— v — creosota f - [maderas] to (treat with)
  creosote
— v — — f bajo presión f - [maderas] to creosote pressure treat
— v con gas m - to gas treat
— v chatarra f - [metal-prod] to treat @ scrap
— v de operar v - to try to operate
— v desecho(s) m - [nucl] to t reat @ waste(s)
— directamente v - to treat directly • to deal
  directly
— v económicamente - [ind] to treat economically
— v ecuánimemente - to deal fairly
— v eluído m - [sanit] to treat @ effluent
— v escoria f - [metal-prod] to treat @ slag
— v estéticamente - to treat aesthetically
— v líquido m - to treat @ liquid
— v — m cloacal - [sanit] to treat @ sewage
— v para protección v - to treat for @ protection
— v — — f contra corrosión f - [metal-trat]
  to treat for corrosion protection
— v por fluidificación f - [metal-trat] to,
  fluidizing treat, o treat by fluidifica-
  tion
— v residuo(s) m - [nucl] to treat @ waste(s)
— v rudamente - to, treat, o handle, rudely, o
  roughly
— v superficie f - [metal-trat] to treat @
  surface
— v térmicamente - [metal-trat] to heat treat
— v — acero m - [metal-prod] to heat treat @
  steel
— v — — m con aleación f - [metal-trat] to
  heat treat @ alloys steel
— m — cuerpo m - [mec] to heat treat @ body
— v — mediante recalentamiento m - [sold] to
  reheat treat
— v — pieza f - [mec] to heat treat @ part
— v — rodillo m - [mec] to heat treat @ screw
— v — tubería f - [tub] to heat treat @ pipe
— v — varilla f - [metal-trat] to heat treat
  @ rod
tratativa f - negotiating,tion; bargaining
trato m - . . . • handling

trato m ecuánime - fair deal(ing)
— m en obra f - field handling
— m especial - special treatment
— m justo - square deal
— m rudo - rough, handling, o treatment
— m — en obra f—[constr] rough field handling
traumatismo m articular - [mec] articular trauma
— m muscular - [mec] muscular traume
traversa f - [constr] . . • bolster
través adv - . . .; across
— adv de - throughout
— adv — arco m - [sold] through @ arc
— adv — . . . fase(s) f - [electr-instal]
across . . . phase(s)
travesaño m - . . .; girt; cross, member, o bar,
o tie • [puertas] rail • [escaleras] rung
— m central - [mec] center cross member
— m delantero - [mec] front cross member
— m — para bastidor m - [autom-mec] frame
front cross member
— m en cruz - [mec] cross member
— m — X - [mec] cross member
— m metálico - [constr] metal rail • [mec]
metal rail
— m metálico - [constr] metal, rail, o cross
piece
— m — hueco - [constr] hollow metal rail
— m para bastidor m - [autom-mec] frame cross
member
— m — calibrador m - [mec] gage travel
— m — — m graduado - [mec] shear gage travel
— m — cierre m - [mec] yoke bar
— m — chapín m - [metal-prod| bleeder cross-
piece
— m — — m central - [metal-prod] central
bleeder crosspiece
— m — tablero m - [constr] floor beam
— m — torre f - [petról] derrick girt
— m — f para perforación f - [petról] derrick girt
— m portapoleas - [petról] crown winch
— m trasero - [mec] back cross member
— m — para bastidor m - [autom-mec] frame rear
cross member
travesero m - [mec] véase travesaño m
— m — polea f - [petról] véase corona f para
torre f
traviesa f - . . . • cross
— f de acero m - [f.c.] steel tie
— f — madera f - [f.c.] wood, o lumber, tie
— f — mina f - [miner] mine tie
trayectoria f - . . . • path • distance
— f de agua m - [hidr] water trajectory
— f eléctrica - [electr-oper] electrical path
— f — con metal m - [electr-oper] metallic
electrical path
— f horizontal - horizontal, trajectory, o path
trazo m - . . . • trace • outline - [topogr]
survey line
— f de arcilla f - clay, trace, o line
— f y nivel m - [constr] line and grade
trazabilidad f - traceability
— f de artículo m - article, o item, traceability
trazado m - . . .; outline; layout • plat •
routing; plot(ting); survey; marking • marking off • route - [constr] alignment • route
• [f.c.] roadbed • design • path
— m de curva f - curve plotting
— m — linea f - line drawing
— m — — f perpendicular - [dib] perpendicular
line drawing
— m — placa(s) f - [metal-fabr] plate layout
— m — plancha(s) f - [metal-fabr] plate(s)
layout
— f — — f de carretera(s) f - [vial] roadway
design
— m ferroviario - [f.c.] railway (re) alignment
— n nuevo - relocation; realignment • rerouting
— m — para camino m - [vial] road relocation
— m — — — m vecinal - [vial] township road
relocation
— m — para carretera f - [constr] highway re-

location
trazado m para instalación f - installation
layout
trazado,da a - . . .; drawn; outlined; plotted
trazador m - . . . • [herram] tracer; plotting
machine • [constr] layout man
— m — placa(s) f - [metal-fabr] plate layout
man
— m — plancha(s) f - [metal-fabr] plate layout man
trazamiento* m - véase trazado m
trazar v - . . . • to outline; to draw; to plot
— v curva f - [dib] to plot @ curve
— v línea f - [dib] to draw @ line
— v — f perpendicular - [dib] to draw @ perpendicular line
— v pieza f - [dib] to trace @ part
trazo m fino - [dib] light line
— m grueso - [dib] thick, o heavy, line
— m lleno - [dib] full, o solid, line
— m único - [dib] single line
trébol m - . . . • [metal-lam] pod • spangle
— m en rodillo m - [metal-lam] roll pod
— m para unión f - [metal-lam] wobbler
— m pie m de pájaro m - birdsfoot trefoil
— m sueco - [botán] alsike clover
treceavo,va a - thirteenth
trecho m - . . .; stretch; lap • span
— m culebreante m - [vial] twisty; snaky
— m para velocidad f elevada - [deport] high
speed, stage, o section
— m resbaladizo - [vial] slippery section
trefilación* f - véase trefilería f
trefilado* m - véase trefilería f
trefilado,da a - [trefil] drawn
— a con aceite m - [trefil] oil drawn
— a con precisión f - [trefil] sized accurately
— a en aceite m - [trefil] oil drawn
— a en frío - [alambre] cold drawn
— a normalmente - [trefil] drawn normally
trefilador m - [alambre] (wire) drawer
— m avanzado - [trefil] advanced drawer
— m — gradualmente - [trefil] jogged drawer
— m básico - [trefil] basic drawer
— m con carrera f corta f - [trefil] short
stroke (wire) drawer
— m — — f larga - [trefil] long stroke
(wire) drawer
— m en continuo - [trefil] continuous, drawer,
o drawing machine
— m grande - [tefil] large (wire) drawer
— m mayor - [trefil] larger (wire) drawer
— m menor - [trefil] smaller (wire) drawer
— m para barra(s) f - [trefil] bar. drawer
— m — reducción f doble - [trefil] double
draft (wire) drawer
— m — — f alta - [trefil] high speed drawing machine
— m pequeño - [trefil] small (wire) drawer
— m sincronizado - [Trefil] timed wire drawer
— m típico - [trefil] typical (wire) drawer
trefiladora f - [trefil] wiredrawing machine;
wire mill
trefilar v - [trefil] to draw
— v barra f - [trefil] to draw @ bar
— v con aceite m - [trefil] to oil draw; to
draw with oil
— v — precisión f - [trefil] to size accurately
— v en aceite m - [trefil] to oil draw
— v — frío m - [trefil] to cold draw
— v — — varilla f - [trefil] to cold, draw,
o roll, @ rod
— v material m - [trefil] to draw @ material
— v — m para labrar v - [trefil] to draw @
stock
— v normalmente - [trefil] to draw normally
— v varilla f - [trefil] to draw @ rod
trefilería f - [trefil] (wire) drawing, (factory, o mill)
— v con aceite m - [trefil] oil drawing
— f de alambre m - [trevil] wire drawing

trefilería f de alambre m para electrodo(s) m -
[trefil] electrode wire drawing
— f — **alambrón** m - [trefil] wire rod drawing
— f — **barra(s)** f - [trefil] bar drawing
— f — **material** n para labrar v - [trefil]
stock drawing
— f — m — **trefilar** v - [tefil] stock drawing
— f — m — plano - [Trefil] flat stock
drawing
— f — **varilla(s)** f - [trefil] rod(s), drawing,
o draft
— f en aceite m - [trefil] oil drawing
— f — **continuo** - [trefil] continuous drawing
— f — **frío** m - [metal-lam] cold drawing
— f — de varilla(s) f - [trefil] wire (rod)
cold drawing
— f normal - [trefil] normal drawing
— f precisa - [trefil] accurate sizing
tremendo,da a - . . . ; horrendous
tren m - . . . • [f.c.] . . . ; bogie • [autom-
mec] assembly; end • [metal-lam] mill •
[electrón] burst
— m a entrar v - [f.c.] inbound train
— m — salir v - [f.c.] outbound train
— m **acabador** m - [metal-lam] finishing mill
— m **aarenero** - [constr] sand train
— m **articulado** - [f.c.] bogie
— m **blooming** - [metal-lam] blooming mill
— m —slabbing - [metal-lam] blooming-slabbing
mill
— m **canteador** - [metal-lam] edging mill
— m **carbonero** m - [f.c.] coal train
— m — **lanzadera** f - [f.c.] coal shuttle train
— m **cargado** - [f.c.] loaded train
— m — con carbón m - [f.c.] coal loaded train
— m **comercial** - [metal-lam] merchant mill
— m **completo** - [f.c.] unit train
— m **con** . . . caja(s) f - [metal-lam] . . .
stand(s) mill
— m — . . . — para laminación f - [metal-lam]
. . . stand (rolling) mill
— m — . . . f — en caliente - [metal-
lam] . . . stand hot mill
— m — . . . f — f — frío - [metal-lam]
. . . stand cold mill
— m **continuo** - [metal-lam] continuous mill
— m — en tándem - [metal-lam] continuous tandem mill
— m — para alambre m - [metal-lam] continuous
wire (rod) mill
— m — — alambrón m - [metal-lam] continuous
wire rod mill
— m — — barra(s) f - [metal-lam] continuous
rod mill • continuous bar mill
— m — — f para estirar v - [metal-lam]
continuous wire rod mill
— m — — fleje(s) m - [metal-lam] continuous
strip mill
— m — — m ancho(s) - [metal-lam] continuous wide strip mill
— m — — laminación f - [metal-lam] continuous
rolling mill
— m — — f — frío - [metal-lam] continuous
cold strip mill
— m — — laminación f - [metal-lam] continuous
(rolling, o strip,) mill
— m — — — f en caliente - [metal-lam] continuous hot (strip, o rolling,) mill
— m — — f — frío - [metal-lam] continuous
cold, (strip, o rolling,) mill
— m — para palanquilla f - [metal-lam] continuous billet mill
— **cuarto** - [metal-lam] four-high, stand, o mill
— m — para bobina(s) f - [metal-lam] four-high
coil mill
— m — — chapa(s) f - [metal-lam] four-high
sheet mill
— m — — temple m - [metal-lam] four-high temper mill
— m — — — m de chapa(s) f - [metal-lam] four
high sheet temper mill
— m de cajas f en líneas f paralelas - [metal-

-lam] cross country mill
tren m de caja(s) f en líneas f paralelas para
barras f - [metal-lam] cross country bar
mill
— m — — f — para palanquilla f - [metal-
-lam] cross country billet mill
— m — **engranaje(s)** m - [mec] gear train;
gearing
— m — **impulso(s)** m - [electrón] pulse train;
burst
— m — — m para dato(s) m - [electrón] data
pulse train
— m — — m — — m dividido por novecientos -
[electrón] divide by nine-hundred data pulse
train
— m de lanzadera f - [f.c.] shuttle train
— m — ruedas f - [f.c.] véase **bogie** m
— m **delantero** m - [autom-mec] front end
— m **desbastador** - [metal-lam] roughing, o
blooming, o slabbing, o cogging, mill
— m — con . . . cajas f - [metal-lam] . . .
stand roughing mill
— m — — palanquilla(s) f - [metal-lam] billet
roughing mill
— m — — planchón(es) m - [metal-lam] slab
roughing mill
— m — — m y palanquilla f - [metal-lam]
slab and billet roughing mill
— m — — tocho(s) m - [metal-lam] blooming, o
bloom roughing, mill
— m — **secundario** - [metal-lam] secondary
blooming. o breakdown, mill
— m **desbocado** - [f.c.] runaway train
— **descascarillador** - [metal-lam] descaling mill
— m **discontinuo** - [metal-lam] cross country
mill
— m **dúo** - [metal-lam] two-high mill
— m — para temple m - [metal-lam] two-high
temper mill
— m **eléctrico** - [f.c.] electric train
— m en caliente m - [metal-lam] hot mill
— m — — para banda(s) f - [metal-lam] hot
strip mill
— m — — cinta f - [metal-lam] hot strip
mill
— m — — chapa(s) f - [metal-lam] hot strip
mill
— m — frío - [metal-lam] cold mill
— m — — m para banda(s) f - [metal-lam] cold
strip mill
— m — — m para cinta f - [metal-lam] cold
strip mill
— m — — — fleje m - [metal-lam] cold
strip mill
— m — tándem - [metal-lam] tandem mill
— m **enterizo** - [f.c.] unit train
— m — cargado con carbón m - [f.c.] coal loaded unit train
— m **entrado** m - [f.c.] inbound train
— m **especial** - [f.c.] special train
— m **estructural** - [metal-lam] structural mill
— m **fermachine** - [metal-lam] véase **tren** m para, barra(s) f, o varilla(s) f (de acero m)
— m **inversor** [ind] véase **tren** m reversible
— m **laminador** - [metal-lam] rolling mill
— m — cuarto - [metal-lam] four high mill
— m — en caliente - [metal-lam] hot mill
— m — — para barra(s) f - [metal-lam] hot
bar mill
— m — — m para banda(s) f - [metal-lam]
hot strip mill
— m — — — cinta f - [metal-lam] hot
strip mill
— m — — m para chapa(s) f - [metal-lam]
hot strip mill
— m — — — fleje m - [metal-lam] hot
strip mill
— m — en frío - [metal-lam] cold rolling mill
— m — — para banda(s) f - [metal-lam]
cold strip mill
— m — — — cinta f - [metal-lam] cold
strip mill
— m — — — — chapa f - [metal-lam] cold

strip mill
tren m **laminador en frío** m **para fleje** m -
[metal-lam] cold stip mill
— m —— **chapa(s)** f - [metal-lam] sheet rolling
mill
— m —— **para pasada** f **para endurecimiento** m **su-**
perficial - [metal-trat] temper pass mill
— m —— **perfil(es)** m **liviano(s)** m - [metal-
-lam] light section mill
— m —— **planchón(es)** m - [metal-lam] slabbing
mill
— m —— **temple** m **en frío** - [metal-trat] cold
roll temper mill
— m ——— m **con . . . caja(s)** f - [metal-
trat] . . . stand temper mill
— **lanzadera** - [f.c.] shuttle train
— m **para acabado** m - [metal-lam] finishing mill
— m ——— m **para banda** f - [metal-lam] strip
finishing mill
— m ——— m — **chapa** f - [metal-lam] strip finishing mill
— m ——— m — **fleje** m - [metal-lam] strip finishing mill
— m — **alambrón** m - [metal-lam] wire rod mill
— m — **aterrizaje** m - [aeron] landing gear;
undercarriage
— m — **banda(s)** f - [metal-lam] strip mill
— m —— f **ancha** - [metal-lam] wide strip mill
— m —— f **con . . . caja(s)** f - [metal-lam]
. . . stand strip mill
— m —— f **en caliente** - [metal-lam] hot strip
mill
— m —— f — **frío** - [metal-lam] cold strip
mill
— f —— f **gruesa** - [metal-lam] thick strip
mill
— m —— f **reversible(s)** m - [metal-lam] reversible strip mill
— m — **barra(s)** f - [metal-lam] bar, o rod,
mill
— m —— f **de acero** m - [metal-lam] steel rod
mill
— m —— f ——— m **con carbono** m - [metal-lam]
carbon steel rod mill
— m —— f **para estirar** v - [metal-lam] wire
rod mill
— m — **carga(s)** f - [f.c.] freight, o goods,
train
— m — **cinta** f - [metal-lam] strip mill
— m —— f **ancha** - [metal-lam] wide strip mill
— m —— f **con . . . caja(s)** f - [metal-lam]
. . . stand strip mill
— m —— f **gruesa** - [metal-lam] thick strip
mill
— f — **chapa** f - [metal-lam] strip mill • sheet
mill
— m —— f **ancha** - [metal-lam] wide strip mill
— m —— f **con . . . caja(s)** f - [metal-lam]
. . . stand strip mill
— m —— f **en caliente** - [metal-lam] hot strip
mill
— m —— f **gruesa** - [metal-lam] thick strip
mill
— m — **descascarillado** m - [metal-lam] descaling mill
— m — **descascarillar** v - [metal-lam] scale
breaker
— m — **endurecimiento** m - [metal-lam] hardening
mill
— m ——— m **de superficie** f - [metal-lam] skin
pass mill
— m — **fleje** m - [metal-lam] strip mill
— m ——— ʒ **ancho** - [metal-lam] wide strip mill
— m —— f **con . . . caja(s)** f - [metal-lam]
. . . stand strip mill
— m —— m **grueso** - [metal-lam] thick strip
mill
— m — **impulsión** f - [mec] drive,ving train
— m — **laminación** f - [metal-lam] rolling,
mill, o train
— m —— f **continua** - [metal-lam] continuous
rolling, mill, o line
— m ——— f **cuarto** - [metal-lam] four-high

tren m **para varillas** f **de acero** m **con carbono** m

stand
tren m **para laminación** f **de chapa(s)** f - [metal=
-lma] strip mill
— m ——— f ——— f **en caliente** m - [metal-lam]
hot strip mill
— m ——— f ——— f **frío** - [metal-lam] cold
strip mill
— m ——— f **en caliente** - [metal-lam] hot
(strip) mill
— m ——— f — **frio** m - [metal-lam] cold
(strip) rolling mill
— m ——— f **en frío con . . . cajas** f - [metal-
lam] . . . stand cold rolling mill
— m ——— f **para perfil(es)** m - [metal-lam]
section mill
— m ——— f ——— m **liviano(s)** - [metal-lam]
light section mill
— m ——— f ——— m **mediano(s)** - [metal-lam]
medium section mill
— m ——— f ——— m **pesados** m - [metal-lam]
heavy section mill
— m ——— f — **planchón(es)** m - [metal-lam]
slabbing mill
— m ——— f — **plancha(s)** f - [metal-lam] plate
mill
— m ——— f ——— f **liviana(s)** - [metal-lam]
light plate(s) mill
— m ——— f ——— f **mediana(s)** - [metal-lam]
medium plate(s) mill
— m ——— f ——— f **pesada(s)** f - [metal-lam]
medium plate(s) mill
— m ——— f — **palanquilla(s)** f - [metal-lam]
billet mill
— m ——— f **semicontinua** - [metal-lam] semicontinuous (rolling) mill
— m — **mercancías** f - [f.c.] freight, o goods,
train
— m — **propulsión** f - [ind] drive,ving train
— m — **redondo(s)** m [metal-lam] round(s) mill •
rod mill
— m — **reducción** f - [metal-lam] reducing,ction
mill
— m ——— f **en caliente** - [metal-lam] hot reducing,ction mill
— m ——— f — **frío** m - [metal-lam] cold reducing,ction mill
— m — **riel(es)** m - [metal-lam] rail mill
— m ——— m **y perfiles** m (**estructurales**) -
[metal-lam] rail and structural mill
— m — **temple** m - [metal-lam] temper mill •
skin pass mill
— m ——— m **con . . . caja(s)** f - [metal-lam]
. . . stand skin pass mill
— m ——— m **cuarto** - [metal-lam] four-high
temper mill
— m ——— m **para chapas** f - [metal-lam] sheet
temper mill
— m ——— m **para terminación** f - [metal-lam]
finishing temper mill
— m ——— m **en caliente** - [metal-lam]
hot roll(ing) temper mill
— m — **terminación** f - [metal-lam] finishing
mill
— m ——— f **para banda** f - [metal-lam] strip
finishing mill
— f ——— f — **chapa** f - [metal-lam] strip
finishing mill
— m ——— f — **fleje** m - [metal-lam] strip
finishing mill
— m ——— f — **perfil(es)** m (**estructurales**) -
[metal-lam] structural finishing mill
— m — **tocho(s)** m - [metal-lam] blooming mill
— m ——— m **y desbaste(s)** m **plano(s)** m -
[metal-lam] blooming and slabbing mill
— m ——— m **y palanquilla** f - [metal-lam]
slab(bing) and billet mill
— m ——— m **y planchón(es)** m - [metal-lam]
slabbing and blooming mill
— m — **varilla(s)** f - [metal-lam] rod mill
— m ——— f **de acero** m - [metal-lam] steel rod
mill
— m ——— f ——— m **con carbono** m - [metal-lam]
carbon steel rod mill

tren m posterior - [autom-mec] rear assembly
— m preparador - [metal-lam] intermediate mill
— m reductor - [metal-lam] reducing,ction mill
— m reversible - [metal-lam] reversible,sing mill
— f — para velocidad f alta - [metal-lam] high speed reversible,sing mill
— m rodante m - [autom-mec] rolling gear • [f.c.] rolling stock • [mec] undercarriage • [grúas] running gear
— m — optativo - [mec] optional undercarriage
— m semicontinuo - [metal-lam] semicontinuous mill
— m — para banda(s) f - [metal-lam] semicontinuous strip mill
— m — banda(s) f - [metal-lam] semicontinuous strip mill
— m — — f en caliente - [metal-lam] semicontinuous hot strip mill
— m — — perfil(es) m - [metal-lam] semicontinuous shape(s) mill
— m — — m comercial(es) f - [metal-lam] semicontinuous merchant mill
— m Sendzimir - [metal-lam] Sendizimir mill
— m slabbing - [metal-lam] slabbing mill
— m Steckel - [metal-lam] Steckel mill
— m tandem - [metal-lam] tandem mill
— m — con . . . caja(s) f - [metal-lam] . . . stand tandem mill
— m — — dos cajas f para laminación f de hojalata f - [metal-lam] two-stand tandem tin plate mill
— m — — — — f — — f de plancha(s) f - [metal-lam] two stand tandem plate mill
— m — — — — f — — f en frío - [metal-lam] two-stand tandem cold mill
— m — — — — f — — f — — de hojalata f - [metal-lam] two-stand tandem tin plate cold mill
— m — — — — f — — f — — plancha(s) f - [metal-lam] two-stand tandem plate cold mill
— m — — continuo - [metal-lam] continuous tandem mill
— m — — con dos cajas f para laminación f en frío de hojalata f doble reducida - [metal-lam] two-stamd tandem cold mill for thin tin plate
— m — — — — f — — f — — m de hojalata f fina - [metal-lam] two-stand tandem cold mill for thin tin plate
— m — en frío para banda(s) f - [metal-lam] tandem cold strip mill
— m — — — — chapas f - [metal-lam] tandem cold strip mill
— m — — — — fleje m - [metal-lam] tandem cold strip mill
— m — — — en frío - [metal-lam] tandem cold rolling mill
— m temper - [metal-lam] temper mill; véase también tren m para temple,pladura
— m — cuarto - [metal-lam] four-high temper mill
— m — para laminación f - [metal-lam] temper rolling mill
— m — — — f en frío - [metal-lam] cold temper rolling mill
— m temperador - [metal-lam] tempering mill; véase también tren m para temple m
— m temple - [metal-lam] temper mill; skin pass
— m terminador - [metal-lam] finishing mill
— m — con . . . cajas f - [metal-lam] . . .- stand finishing mill
— m transversal—[metal-lam] cross country mill
— m trasero - [autom-mec] rear, assembly, o end
— m trío - [metal-lam] three-high stand
trenza f de vidrio m - [plást] glass braid
— f — — m impregnada - [electr-cond] impregnated glass braid
trenzado m - . . . - [cabl] lay twist • construction • stranding; lay(ing); weave,ving
— m Albert - [cabl] Albert('s), o Lang, lay

trenzado m alternado - - [cabl] alternate lay
— m — alrededor de alma m de fibra f - [cabl] alternate lay(ing) around @ fiber core
— m cerrado - [cabl] locked coil
— m cruzado - [cabl] cross(ed) lay
— m de alambre m - [cabl] wire lay(ing) • [tub] wire braid
— m — — m de acero m - [cabl] steel wire braid
— m — — m con resistencia f alta a tensión f - [alambre] high tensile steel wire braid
— m de cable m - [cabl] cable, o rope, fabricating,tion, o constructing,tion, o lay(ing)
— m — — m alrededor de alma m - [cabl] laying @ cable around @ core
— m — — — f de fibra f - [cabl] laying @ cable around @ fiber core
— m — — m de alambre m - [cabl] wire rope construction
— m — cabo m - [Cabl] rope construction
— m — cordón m - [cabl] strand construction
— n — par(es) m - [telecom-cond] pair(s) braiding
— f — capa(s) - [cabl] layer laying
— m — — f coaxial(es) - [cabl] concentric layer laying
— m — — f concéntrica(s) - [cabl] concentric layer(s) laying
— m en resorte m - [cabl] spring construction
— m grueso - [cabl] coarse lay
— m hacia derecha f - [cabl] right (hand) lay
— m — izquierda f - [cabl] left (hand) lay
— m helicoidal - [cabl] helical, braiding, o lay
— m inverso - [cabl] reverse lay
— m invertido - [cabl] reverse lay
— m lang - [cabl] lang lay
— m — hacia derecha f - [cabl] right lang lay
— m — — izquierda f - [cabl] left lang lay
— m normal - [cabl] regular, o normal, lay
— m regular - [cabl] regular lay
— m — hacia derecha f - [cabl] regular right lay
— m — — izquierda f - [cabl] regular left lay
— m Seale - [cabl] Seale('s) lay
— m sencillo - [cabl] single braid(ed)
— m — de alambre m - [alambre] single wire braid(ed)
— m — — — m de acero m - [alambre] single steel wire braid(ed)
— m — — — — muy resistente a tensión f - [alambre] single high tensile steel wire braid(ed)
— — Warrington - [cabl] Warrington construction
trenzado,da a - braided • [cabl] constructed • laced • [autom] tagged
— a alternadamente - [cabl] laid alternately
— a — alrededor de alma m de fibra f - [cabl] laid alternately around @ fiber cord
— m de cobre m - stranded copper
— m en cable m de alambre m - [cabl] laid into @ (wire) rope
— a — cabo(s) m - [cabl] laid into @ rope(s)
— a — capa(s) f - [cabl] layer laid
— a — — f coaxial(es) - [Cabl] concentric layer laid
— a — — f concéntrica(s) f - [cabl] concentric layer laid
— a — cordón(es) m - [cabl] laid into strands
— a — dirección f opuesta - [cabl] laid in @ opposite direction
— a — hélice f - [cabl] helically braided; laid in @ helix
— a — misma dirección f - [cabl] laid in @ same direction
— a — mismo sentido m - [Cabl] laid in @ same direction
— a — resorte m - [Cabl] spring lay
— a — sentido m opuesto - [cabl] laid in @ opposite direction
— a hacia derecha f - [cabl] right (hand) lay
— m — izquierda f - [cabl] left hand laid
— a helicoidalmente - [cabl] helically braided

trenzar v - . . .; to lace • [cabl] to, construct, o lay, o weave • [autom] to tag
— v alambre m - [cabl] to lay @ wire
— m — en cordón(es) m - [cabl] to ley @ wire
in(to) a strand
— v alternadamente - [cabl] to lay alternately
— v — alrededor de alma m de fibra f - [cabl]
to lay alternately around @ fiber core
— v cable m - to, construct, o build, o braid,
o fabricate, @, cable, o rope
— m cordón m - [cabl] to lay @ strand
— v cordones m en cable m - [cabl] to lay @
strands into @ wire rope
— v — m en cabo(s) m - [cabl] to lay @ strand
into @ wire rope
— v en cable(s) f - [cabl] to lay into @ (wire)
rope(s)
— v — capa(s) f - [cabl] to weave into layers
— v — — f coaxiales - [cabl] to lay into coaxial layers
— v — — f concéntrica(s) - [cabl] to lay in
concentric layers
— v — cordón(es) m - [cabl] to lay into
strand(s)
— v par m - [telecom-cond] to braid @ pair
trepa f - . . . . • climb(ing)
— f hasta nube(s) f - [deport] climb (in)to @
cloud(s)
— f sobre - climbing on (to)
trepado,da - climbed • [imprent] perforated
— a sobre - climbed on (to)
— m de línea f - [constr] lineman's climber(s)
trepanación f - trepanning
trépano m - [petról] drill(ing bit)
— m — cola f de pescado m - [petról] fish tail
bit
— m — arrastre m - [petról] drag bit
— m — cono m - [petról] cone rock bit
— m — fricción f - [petról] drag bit
— m para perforación f - [petról] drilling bit
— m — — f con cable m - [petról] cable drilling bit
— m — roca f - [constr] rock cutter
— m rotatorio - [petról] rotary drill
— m — con punta f con diamante(s) m - [petról]
diamond bit rotary drill
— v cuesta f - to climb @ hill
— v montaña f - to climb @ mountain
— v sobre - to climb on (to)
trepidar v - . . . . • [fig] to be leery
tres adv arteria(s) v - [electr-instal] three
conductor
— cables para alimentación f - [electr-instal]
three wire input power
— cajas f - [metal-lam] three stand
— cantos m - three edges
— centros m - three centers | adv - [arquit]
three centered
— m — peraltado(s) - [arquit] raised three
centered
— cilindros m - [bombas] three cylinder(s);
triplex
— m conductos m - [tub] three barrel(s)
— conductor(es) - [electr-cond] three conductor
— fosos m - [coque] three hole(s)
— niveles m - three levels; trilevel
— pasos m - three step
— piezas f - three, piece, o part
— posición(es) f - three position
— punto(s) m - three point
— ruedas - three wheel(ed)
— velocidad(es) - [mec] three speed
— vías - three way • [f.c.] three track •
[vial] three lane(s)
triac m - [electr-equip] triac
triángulo m de conexión(es) f - [electr-instal]
connection trinagle
— m — selector m - [electr-instal] selector
triangle
— m instalado - [electr-instal] installed triangle
— m isósceles - [geom] isosceles trinagle
— m para conexión(es) f - [electr-instal] con-

nection(s) triangle
triángulo m para conexión f para selector m para
voltaje m - [sold] voltage selector connection triangle
— m — — f para tablero m - [sold] panel connection triangle
— m — — — m para selección f de voltaje
m - [sold] voltage selector panel connection
triangle
— m — selección f de voltaje m - [electr-inst]
voltage selector triangle
— m recto - [geom] right triangle
— m isósceles - [geom] isosceles right
triangle
triaxial a - triaxial
tribu m - . . . . • (indian) nation
tribuna f - . . . . • [deport] (grand)stand • soap-
box
— f — observación f - [constr] observation deck
— f principal - [deport] main grandstand
tribunal m competente - [legal] competent court
— m de alzada f - [legal] appeal(s), o appel-
late, court
— m federal - [legal] federal court
— m nacional - [legal] national, o federal,
court
— m para arbitraje m - [legal] arbitration court
— m — condado m - [legal] county court
— m — registro m - [legal] court of record
tributario m - [hidr] . . .; branch
tributario,ria a - . . . . • [fisc] taxation
tributo m sobre constitución f - [fisc] incorporation tax
— m — — f de sociedad f - [fisc] corporate
organization(al) tax
— m — organización f - [fisc] organization tax
tricéntrico,centrado - [arquit] three centered
— a peraltado,da - [arquit] raised three centered
triciclo m motorizado - [deport] all terrain
cycle, A T C
tricloroetilénico,ca a - [quím] trichloroethylene
tricloroetileno m - [quím] trichloroethylene
triennial a - . . .; three-year
trifásico,ca a conectado,da en triángulo m -
[electr-instal] delta three phase
— a de tipo m de jaula f de ardilla f - [electr-
equip] three-phase squirrel cage type
— a en triángulo m - [electr-instal] delta three
-phase
trilingüismo n - trilingualism
trilobulado,da a - [arquit] three, foiled, o
lobed
trilla f - [agric] . . .; threshing
trillado,da a - [agric] thrashed; threshed
trilladora f - [agric-equip] thrasher; thresher;
thrashing, o threshing, machine, o rig
— f para arroz n - [agric-equip] rice thresher
— f — hacina(s) f - [agric-equip] shuck thresher
— f — parva(s) f - [agric-equip] stack thrasher
trillar v - [agric] . . .; to thresh
trilobular a - véase trilobulado,da
trimestralmente adv - . . .; on @ quarterly basis
trimestre m anticipado - quarter(ly) in advance
— m calendario - [cronol] calendar quarter
trincado,da a - [transp] lashed
trinchera f abierta - [constr] open, trench, o
ditch, o cutting
— f con agua m [constr] wet cut
— f para cimentación f - [constr] footing trench
— f — — f con hormigón m - [constr] cement(ed)
footing trench
— f — — f en roca f - [constr] rock footing
trench
— f — — f excavada en roca f - [constr] excavated rock footing trench
— f — desagüe,guadero m - [hidr] drain trench
— f — — en franja f medianera - [vial] median
drain(age) trench
trincheradora f - [constr-equip] véase (ex)cava-
dora para, trincheras f, o zanja(s) f
trineo m - . . . . • [transp] skid
— m mecanizado - [deport] snowmobile

trinquete <u>m</u> - [mec] ratchet; tripper; latch;
  stay; monkey; release; lock pawl; finger •
  [constr] keeper
— <u>m</u> **automático** - [mec] automatic pawl
— <u>m</u> — **para traba** <u>f</u> - [mec] automatic lock pawl
— <u>m</u> **derecho** - [mec] right pawl
— <u>m</u> **desengranado** - [mec] disengaged pawl
— <u>m</u> **engranado** - [mec] engaged pawl
— <u>m</u> — **automáticamente** - [mec] automatically
  engaged pawl
— <u>m</u> **fijado** - [mec] set pawl
— <u>m</u> **izquierdo** - [mec] left pawl
— <u>m</u> **para accionamiento** <u>m</u>—[mec] drive,ving pawl
  • drive,ving ratchet
— <u>m</u> — **agarre** <u>m</u> - [mec] holding, pawl, <u>o</u> dog
— <u>m</u> — **aguilón** <u>m</u> - [grúas] boom pawl
— <u>m</u> — — <u>m</u> **fijado** - [grúas] set boom pawl
— <u>m</u> — — <u>m</u> **verificado** - [grúas] checked boom
  pawl
— <u>m</u> — **arranque** <u>m</u> - [comb.int] starting, jaw, <u>o</u>
  pawl
— <u>m</u> — — <u>m</u> **para cigüeñal** <u>m</u> - [comb.int] crank
  starting, jaw, <u>o</u> pawl
— <u>m</u> — **elevador** <u>m</u> - [mec] hoist pawl
— <u>m</u> — — <u>m</u> **para aguilón** <u>m</u> - [grúas] boom hoist
  pawl
— <u>m</u> — — <u>m</u> — <u>m</u>, **comprobado**, <u>o</u> **verificado** -
  [grúas] checked boom hoist pawl
— <u>m</u> — **freno** <u>m</u> - [mec] brake pawl
— <u>m</u> — **impulsión** <u>f</u> - [mec] drive,ving ratchet
— <u>m</u> — <u>f</u> **insertado** - [mec] inserted
  drive,ving ratchet
— <u>m</u> — **mesa** <u>f</u> - [mec] table pawl
— <u>m</u> — **operación** <u>f</u> - [mec] operating pawl
— <u>m</u> — — <u>f</u> **automática** - [mec] automatically
  operated pawl
— <u>m</u> — **palanca** <u>f</u> - [mec] lever pawl
— <u>m</u> — — <u>f</u> **para freno** <u>m</u> - [mec] brake lever
  pawl
— <u>m</u> — **retén** <u>m</u> - [mec] lock(ing) pawl
— <u>m</u> — — <u>m</u> **para mesa** <u>f</u> **(rotatoria)** - [petról]
  table lock(ing) pawl
— <u>m</u> — **retención** <u>f</u> **para mesa** <u>f</u> **(rotatoria)** -
  [petról] table lock(ing) pawl
— <u>m</u> — **seguridad** <u>f</u> - [mec] safety pawl
— <u>m</u> — **sujeción** <u>f</u> - [mec] dog; locking pawl
— <u>m</u> — **traba** <u>f</u> - [mec] (ratchet) lock(ing) pawl
— <u>m</u> — — <u>f</u> **con operación** <u>f</u> **independiente** -
  [mec] independently operated ratchet lock(ing)
  pawl
— <u>m</u> — — <u>f</u> **accionado** - [mec] operated, <u>o</u>
  flipped, lock(ing) pawl
— <u>m</u> — — <u>f</u> **derecho**—[mec] right lock(ing) pawl
— <u>m</u> — — <u>f</u> **desengranado** - [mec] disengaged
  lock(ing) pawl
— <u>m</u> — — <u>f</u> **engranado** - [mec] engaged lock(ing)
  pawl
— <u>m</u> — — <u>f</u> **engrasado** - [mec] greased lock(ing)
  pawl
— <u>m</u> — — <u>f</u> **fijado** - [mec] set lock(ing) pawl
— <u>m</u> — — <u>f</u> **hacia derecha** <u>f</u> - [mec] right lock-
  (ing) pawl
— <u>n</u> — — <u>f</u> — **izquierda** <u>f</u> - [mec] left lock-
  (ing) pawl
— <u>m</u> — — <u>f</u> **para elevador** <u>m</u> - [grúas] hoist
  lock(ing) pawl
— <u>m</u> — — <u>f</u> — — <u>m</u> **para aguilón** <u>m</u> - [grúas]
  boom hoist lock(ing) pawl
— <u>m</u> — — <u>f</u> — — <u>m</u> — —, **comprobado**, <u>o</u> **verificado** - [grúas] checked boom hoist lock(ing)
  pawl
— <u>m</u> — — <u>f</u> **independiente** - [mec] independent
  ratchet lock(ing) pawl
— <u>m</u> — — <u>f</u> **para mesa** <u>f</u> **rotatoria** - [petról]
  table lock(ing) pawl
— <u>m</u> — — <u>f</u> **para operación** <u>f</u> - [mec] operating
  lock(ing) pawl
— <u>m</u> — — <u>f</u> — — <u>f</u> **automática** - [mec] automa-
  cally operated lock(ing) pawl
— <u>m</u> — — <u>f</u> **verificado**—[mec] checked lock(ing)
  pawl
— <u>m</u> — **trabadura** <u>f</u> - [mec] lock(ing) pawl
— <u>m</u> **quitado** - [mec] removed, pawl, <u>o</u> ratchet

trinquete <u>m</u> **removido** - [mec] removed, pawl, <u>o</u>
  ratchet
— <u>m</u> **sacado** - [mec] removed ratchet
— <u>m</u> **trabado** - [mec] locked pawl
— <u>m</u> **y contracuña** <u>f</u> - [mec] tripper and gib assembly; tripping device
— <u>m</u> — **fiador** <u>m</u> - [mec] ratchet and pawl
— <u>m</u> — **retén** <u>m</u> - [mec] ratchet and pawl
trío <u>m</u> - . . . | <u>a</u> - [metal-lam] three-high
triple <u>a</u> - . . .; threefold • [mec] triplex •
  three-pronged
— <u>m</u> **de superficie** <u>f</u> - triple surface
triplicación <u>f</u> - . . .; tripling
triplicado <u>a</u> - triplicate
triplicado,da <u>a</u> - . . . • tripled; threefold
triplice* <u>a</u> - véase **triplex**
triplex <u>a</u> - triplex
tripulación <u>f</u> - . . . . • manning
tripulado,da <u>a</u> - manned
tripular <u>v</u> - to man
triscado,da <u>a</u> - [herram-sierra] set
triscadura <u>f</u> - [herram-sierra] tooth setting
trisódico,ca <u>a</u> - [quím] trisodium
trituración <u>f</u> - . . .; crushing; grinding;
  breaking; chewing up
— <u>f</u> **de arcilla** <u>f</u> - [cerám] clay crushing
— <u>f</u> — **carbón** <u>m</u> - [miner] coal crushing
— <u>f</u> — **mineral** - [miner] ore crushing
— <u>f</u> — — <u>m</u> **de hierro** <u>m</u> - [miner] iron ore
  crushing
— <u>f</u> — **roca** <u>f</u> - [miner] rock crushing
— <u>f</u> **primaria** - [miner] primary crushing
— <u>f</u> **secundaria** - [miner] secondary crushing
— <u>f</u> **y cribado** <u>m</u> - [miner] crushing and screening
— <u>f</u> — — <u>m</u> **de mineral** <u>m</u> - [miner] ore crushing and screening
triturado,da <u>a</u> - crushed • chewed up
trituradora <u>f</u> - [miner] . . .; breaker; grinder;
  lump breaker
— <u>f</u> **con anillo(s)** <u>m</u> - [miner] ring crusher
— <u>f</u> — — <u>m</u> **y rodillo(s)** <u>m</u> - [miner] rings and
  roll(s) crusher
— <u>f</u> — **bolas** <u>f</u> [miner] ball, mill, <u>o</u> crusher
— <u>f</u> — **cono(s)** <u>m</u> - [miner] cone crusher
— <u>f</u> — **dos tambores** <u>m</u> - [miner] two-, <u>o</u> double,
  roll crusher
— <u>f</u> — **efecto** <u>m</u> **doble** - [miner] jaw crusher
— <u>f</u> — **mandíbulas** <u>f</u> - [miner] jaw crusher
— <u>m</u> — **martillo(s)** <u>m</u> - [miner] hammer mill
— <u>f</u> — **rodillo(s)** <u>m</u> - [miner] roll crusher
— <u>f</u> — **tambor(es)** <u>m</u> - [miner] roll crisher
— <u>f</u> **cónica** - [miner] cone,nical crusher
— <u>f</u> **fija** - [miner] stationary crusher
— <u>f</u> **giratoria** <u>f</u> - [miner] rotary,tating crusher
— <u>f</u> — **Kennedy** - [constr] Kennedy rotary
  crusher
— <u>f</u> **girogravilladora** - [constr] (rotary) fine
  gravel crusher|
— <u>f</u> **para arcilla** <u>f</u> - [cement] clay crusher
— <u>f</u> — **grano(s)** <u>m</u> - [agric] grain crusher •
  feed grinder
— <u>f</u> — **mineral** - [miner] ore crusher
— <u>f</u> — **reducción** <u>f</u> - [miner] reducing,ction
  crusher
— <u>f</u> — **remolque** <u>m</u> - [constr] crusher trailer
  unit
— <u>f</u> **por compresión** <u>f</u> - [mec] compression
  crusher
— <u>f</u> **portátil** - [constr] portable crusher
— <u>f</u> **primaria** - [coque] primary crusher
— <u>f</u> **rotatoria** - [constr] rotary crusher
— <u>f</u> **secundaria** - [coque] secondary crusher
— <u>f</u> **semiportátil** - [miner] semiportable crusher
— <u>f</u> **terciaria** - [miner] tertiary crusher
— <u>f</u> **tipo R** - [miner] R-type crusher
triturar <u>v</u> - . . . • to grind; to chew up
— <u>v</u> **arcilla** <u>f</u> - [cerám] to crush @ clay
— <u>v</u> **roca** <u>f</u> - [constr] to crush @ rock(s)
triunfado,da <u>a</u> - triumphed; won
— <u>a</u> **cómodamente** - won comfortably
triunfador <u>m</u> - [deport] winner
triunfador,ra <u>a</u> - winner,ning

triunfante a - . . .; triumphing • [deport] red
hot • winning
triunfar v - [deport] . . .; to notch @ win
— v cómodamente - to win comfortably
— v en carrera f - [deport] to win @ race
— v — f para campeonato m - [deport] to win
@ championship race
— v — categoría f - [deport] to win @ class
— v — forma f absoluta - [deport] to win overall
— v sobre - [deport] to, win over, o beat
triunfo m - . . . • [deport] win(ning); achievement • first
— m absoluto - [deport] first overall (finish);
overall, win, o victory
— m admirable - admirable, o remarkable, win
— m aplastante - [deport] smashing victory
— m asegurado - [deport] assured, o secured,
triumph, o win
— m bien merecido - [deport] well deserved, o
hard earned, win
— m cómodo - comfortable, win, o victory
— m consecutivo - [deport] consecutive, win, o
victory; straight victory; winning streak
— m desértico - [deport] desert, win, o triumph
— m deslumbrante - eye-opening win
— m en carrera f - [deport] race, win(ning), o
victory
— m — clase f única - [deport] single class win
— m — etapa f - [deport] lap, o stage, win
— m — lucha f ardua - [deport] hard fought
victory
— m — primer lugar - [deport] first class win
— m esperado - [deport] expected win
— m fácil - [deport] easy, win, o victory
— m familiar - [deport] family sweep
— m fuera de carretera f - [autom] off road
victory
— m impresionante - [deport] impressive victory
• good performance
— m limitado - [deport] limited success
— m logrado - [deport] scored, o reeled in, win
— m mayor - [deport] major victory; best finish
— m mejor - best finish
— m para año m - [deport] year('s) victory
— m — categoría f - [deport] class, victory, o
win, o title
— m — f Gran Turismo - [deport] Gran Turismo class victory
— m seguido - [deport] straight win
— m — para categoría f - [deport] straight
class win
— m seguro - [deport] secure win
trivalencia f - [quím] trivalance
trocar v nombre m - to change @ name; to rename
troceador,ra a - [mec] cross cutting
troceadora f - cross, cutter, o cutting, shear
trocear v - [mec] to cross cut
troceo m - [mec] cross cutting
trocha f - . . . • [petról] right-of-way • tread
• lane • [f.c.] gage; track width
— f ancha - [f.c.] wide, o broad, gage, o track
[autom] wide tread
— f angosta - [f.c.] narrow gage - [vial] narrow t read
— f de via f - [f.c.] track gage
— f — 1,435 m - [f.c.] standard gage
— f media - [f.c.] standard gage
— f normal - [f.c.] normal, o standard, gage
— f — europea - [f.c.] standard gage
— f para peatón(es) m - [vial] pedestrian walkway
— f trasera - [autom] rear track
trofeo m - . . .; crown; cup
— m para campeón m - [deport] championship,
award, o trophy
trole m común - [mec] plain trolley
— m Yale - [mec] Yale trolley
trolera* f - véase carro m
trompa f - [zool] snout
tronada f con chabusco m - [meteorol] thundershower
— f retumbante - [meteorol] rolling thunder
— f — con chabusco m - [meteorol] rolling
thundershower

tronadura f - véase voladura f
tronco m - [maderas] log; trunk • [botán] trunk
• [geom] frustum
tronco-cónico - [geom] truncated cone
— m de cono m - [geom] truncated cone
— m — tubería f (para empalme m) - [tub] stub
tronchable adv - break-away
tronchado,da a - sheared off
tronchable a - breakaway
tronchado,da a - severed; sheared (off)
tronchadura f - sever(ing); shear(ing)
— f de línea f para combustible(s) m - [comb.-
-int] fuel line severance, ring
— f — eje m - [mec] shaft shearing
tronchar v - to, sever, o shear
— v eje m - [mec] to shear @ shaft
— v línea f para combustible m - [comb.int]
to sever @ fuel line
tronchar v - . . .; to shear (off)
— v serie f - to break @ streak
tronera f - . . .; port; opening
— f para depósito m - [hidr] tank vent
tronzado,da a - sheared • broken
tronzadura f - shear(ing); break(ing)
— f de pasador m - [mec] pin shearing
— f — m para seguridad f - [mec] safety
shear pin break(ing)
tronzar v - . . .; to shear • [mader] to cross
cut
— v pasador m - [mec] to shear @ pin
— v — m para seguridad f - [mec] to break @
safety shear pin
tropas f auxiliares - [milit] auxiliary troops
— f para desembarco m - [milit] marines
tropezado,da a - stumbled • encountered
tropezar v - . . . • to encounter
— v busa f en tubo m - [metal-prod] to strike @
blow pipe against @ tube
— v con - to encounter
— v — obstáculo m - to meet @ obstacle
tropiezo m - . . .; stumbling; tripping • en-
countering • [fig] hitch
troquel m - . . . • [mec] (die) insert
— m abierto - [mec] open die
— m acanalado - [metal-fabr] grooved die
— m — de acero m - [herram] grooved steel die
— m cerrado - [mec] closed die
— m encabezador - [mec] heading insert
— m fijo - [mec] fixed, o stationary, die
— m Magna Die - [mec] Magna Die (die)
— m — de Whistler - [mec] Whistler Magna
Die (die)
— m movible - [mec] movable die
— m para cierre n - [mec] closing die
— m — embutir v forma f - [mec] form drawing
die
— m — estampar v forma f - [mec] form drawing
die
— m — extrusión f - [mec] extruding, o extrusion, die
— m — m para electrodo(s) m - [sold] ele-
trode extrusion die
— m — forja(dura) f - [mec] forging die
— m — formar v pestaña(s) f - [mec] flange
bending die
— m — freno(s) m - [mec] brake die
— m — marcación f - [mec] marking die
— m — f de huella(s) f - [mec] groove marking die
— m — moldeo m - [mec] molding die
— m — m por extrusión f - [mec] extruding
(molding) die
— m — punzonado,zamiento m - [mec] punching
die
— m — m de agujero(s) m - [mec] hole punching die
— m — recalcado m - [mec] swaging die
— m — recalcadora f - [mec] heading tool, die,
o insert
— m — reducción f - [mec] reducing die • die
taper
— m — respaldo m - [mec] back(ing) die
— m — m ahusado m - [mec] taper back(ing)
die

troquel m para trefilería f - [alambre] wire-drawing die
— m **preciso** - [mec] accurate die
— m **progresivo** - [metal-fabr] progressive die
— m **proyectado** - [mec] designed die
— m **semicilíndrico** - [mec] half die
— m **sólido** - [mec] solid, o single, die
— m **único** - [metal-fabr] single die
troquelado m - [metal-fabr] stampind
— m **extra profundo** - [metal-fabr] extra deep drawing
— m **profundo** - [metal-fabr] deep drawing
troquelar v - [metal-fabr] to, stamp, o swage
trozado m de lingote m - [metal-prod] ingot slicing
trozar v - . . . . • [metal-prod] to slice
trozador m - [mec] slicer
— m **tallo(s) m** - [agric] stalk cutter
trozo m - . . . . • bit; scrap; stub; length; section - [electr-instal] length (of cable) • [metal-prod] slice
— m **continuo** - continuous, section, o length
— m — **de cordón m** - [cabl] continuous. strand, o length, o section
— m — — **ramal** - [cabl] continuous strand section
— m **cortado** - [metal-lam] cut length
— m **corto** - [mec] short, length, o stub
— m — **de conductor m** [electr-cond] short stub
— m **de alambre m** - [alambre] wire length • [trefil] wire blank
— m — **por labrar(se) v** - [trefil] blank
— m — **árbol m** - [maderas] véase **rollizo m** • [mec] shaft, piece, o length
— m — — m **para transmisión f** - [mec] line-shaft, section, o piece
— m — **asfalto m** - [constr] asphalt chunk
— m — **barra f** - [mec] bar, o rod, piece, o section
— m — **cable m** - [electr-cond] cable length • [cabl] rope length
— m — — m **de alambre m** - [cabl] wire rope length
— m — **carbón** - [miner] coal, piece, o lump, o chunk
— m — **cordón m** - [cabl] strand, length, o section
— m — — m **continuo** - [cabl] continuous strand length
— m — — m — **de cable m (de alambre m)** - [cabl] wire rope continuous strand length
— m — **hormigón m** - [constr] concrete chunk
— m — **ladrillo m** - piece of brick(work)
— m — — m **y gunita f** - [metal-prod] brickwork and gunite piece
— m — **línea f** - [dib] line section
— m — — f **perpendicular** - [dib] perpendicular line section
— m — **lingote m** - [metal-prod] ingot slice
— m — **pilote m** - [constr-pil] pile section
— m — — m **disponible** - [constr-pil] available pile section
— m — — m **tubular** - [constr-pil] pipe pile section
— m — — m — **disponible** - [constr-pil] available pipe pile section
— m — **plancha f** - [metal-lam] plate section
— m — **ramal m** - [cabl] strand section
— m — **tramo m** - [tub] section length
— m — — m **recto** - [tub] straight section length
— m — **tubería f** [tub] pipe, section, o length
— m — — f **corrugada** - [tub] corrugated pipe, length, o stub
— m — — f **disponible** - [tub] available pipe section
— m — — f **pesada** [tub] heavy tubular section
— m — — f **pesada hincado con núcleo m** - [constr-pil] core driven heavy tube section
— m — **tubo m que se extiende debajo de tubería f para bombeo m** - [petról] anchor
— m **disponible** - available section

trozo m disponible de pilote m - [constr=pil] available pile section
— m — — — m **tubular** - [constr-pil] available pipe pile section
— m — **tubería f** - [tub] available pipe section
— m **final** - end section • [cabl] whip line
— m — **de cable m** - [cabl] whip line
— m **igual** - [mec] equal length
— m — **de cable m** - [cabl] equal cable length
— m — — m **para soldadura f** - [sold] equal length welding cable
— m **para rueda f** - [f.c.-ruedas] wheel block
— m **por labrar demasiado corto** - [metal-fabr] (too) short blank
— m **tubular** - [tub] tube,bular section
— m **único** - single piece
truco m - [fam] stunt
truchado,da a - véase **atruchado,da**
trueque m - changeover
truncado,da a - . . . . • cut (too) short
truncamiento m - . . . .; cutting (too) short
truncar v - to lop off; to cut (too) short
tubería f - [tub] . . .; tube; pipe • line; conduit • hose • pipe line • [petról] casing • [ambient] duct • [electr-instal] duct • [tub] barrel • [constr] shell
— f **abovedada** - [constr] pipe arch
— f **abovedada armada en planta f** - shop-fabricated pipe-arch
— f — **con corrugación(es) f interior(es) cubierta(s)** - [metal-fabr] Smooth-Flo pipe arch
— f — **con muro m para cabecera f** - [constr] pipe-arch with @ headwall
— f — **recubrimiento m bituminoso** - [tub] bituminous coated pipe arch
— f — **corrugada de acero m con cobre m galvanizado** - [tub] galvanized copper bearing steel corrugated pipe arch
— f — — **de acero m estándar** - [tub] standard corrugated steel pipe arch
— f — **con interior m liso (Smooth-Flo)** • [tub] Smooth-Flo pipe arch
— f — — **encajable** - [tub] corrugated nestable pipe arch
— f — — **galvanizada** - [metal-fabr] galvanized corrugated pipe arch
— f — **cubierta con capa(s) f** - [constr] layer covered pipe-arch
— f — **de acero m corrugado** - [metal-fabr] corrugated steel pipe arch
— f — — **helicoidalmente** - [metal-fabr] helically corrugated steel pipe arch
— f — — **acero m recubierta** - [metal-fabr] coated steel pipe arch
— f — — m **revestida** - [tub] paved steel pipe arch
— f — **plancha(s) f estructural(es)** - [tub] structural plate pipe arch
— f — — f — **corrugadas de acero m** - [tub] steel structural plate pipe arch
— f — — f **múltiples** - [metal-fabr] Multi-Plate pipe arch
— f — **encajable** - [tub] nestable pipe-arch
— f — — **galvanizada** - [metal-fabr] galvanized corrugated nestable pipe-arch
— f — — **de acero m con cobre m galvanizada** - [metal-fabr] galvanized copper bearing steel corrugated nestable pipe arch
— f — — **de acero m con cobre m de tipo m con muesca(s) f** - [metal-fabr] copper bearing notched type nestable pipe arch
— f — — — m — m **galvanizada** - [metal-fabr] galvanized copper bearing steel nestable pipe arch
— f — — — m **con cobre m** - [tub] copper bearing steel pipe arch
— m — — — — m **puro** - [tub] ingot iron nestable pipe arch
— f — — — — m **galvanizado** - [metal-fabr] galvanized steel nestable pipe arch

tubería f abovedada encajable de piezas f intercambiables - [metal-fabr] interchangeable
  nestable pipe arch
— f —— galvanizada - [metal-fabr] galvanized
  nestable pipe arch
— f ——— de tipo m con muesca(s) - [metal-
  -fabr] galvanized notched type nestable pipe
  arch
— f ——— m intercambiable - [metal-fabr] interchangeable nestable pipe-arch
— f ——— de tipo m con muesca(s) f -
  [metal-fabr] interchangeable notch-type nestable pipe arch
— f — fallante - [constr] failing pipe-arch
— f — galvanizada - [metal-fabr] galvanized
  pipe-arch
— f — gemela - [metal-fabr] twin pipe arch
— f — f de plancha(s) f estructural(es) -
  [constr] twin structural plate pipe arch
— f — grande - [constr] large pipe-arch
— f — intercambiable - [constr] interchangeable pipe-arch
— f — de hierro m puro - [constr] ingot
  iron interchangeable pipe arch
— f — Multi-Plate - [constr] Multi-Plate pipe-
  -arch
— f abovedada pequeña—[constr] small pipe-arch
— f — peraltada - [constr] vertically elongated pipe-arch
— f — que fluye llena - [hidr] full flowing
  pipe-arch
— f — recubierta - [constr] coated pipe arch
— f — rellenada—[Constr] backfilled pipe arch
— f — remachada - [tub] riveted pipe arch
— f — de acero m - [tub] steel riveted pipe
  arch
— f ———— m corrugado - [tub] corrugated
  steel riveted pipe arch
— f — prefabricada - [metal-fabr] prefabricated riveted pipe-arch
— f ——— de acero m - [metal-fabr] prefabricated riveted steel pipe-arch
— f — revestida - [tub] paved pipe arch
— f — sin muro m para cabecera f - [constr]
  pipe-arch without @ headwall
— f — Smooth-Flo - [metal-fabr] Smooth-Flo
  -arch
— f — unida - [constr] joined pipe-arch(es)
— f — acanalada - [metal-fabr] corrugated pipe
— f — encajable - [metal-fabr] nestable corrugated pipe
— f — intercambiable - [metal-fabr] interchangeable nestable corrugated pipe
— f acoplada - [tub] coupled pipe
— f acronitrilo-butadieno-estirénica - [tub]
  acronitirle-butadiene-styrene pipe
— f——— con pared f sólida - [tub] solid
  wall acronitrile-butadiene-styrene pipe
— f adyacente - [tub] adjacent, o adjoining,
  pipe
— f aérea - [tub] aerial, pipe, o tube • pipe
  bridge
— f aferente - [tub] intake,king line
— f ahusada - [tub] tapered pipe
— f aislada - [tub] isolated piping • [metal-
  fabr] insulated pipe,ping
— f para alimentación f - [ind] insulated
  feed(ing) tube
— f aisladora - [mec] insulating tube,bing
— f — para agua m—[hidr] insulated water pipe
  • water string
— f aislante - [electr-instal] conduit
— f anular - [tub] circle pipe
— f — inferior - [tub] lower circle pipe
— f — superior - [tub] upper circle pipe
— f aplastada - [tub] collapsed pipe • flattened pipe
— f aplicada - [tub] applied tube,bing
— f apropiada - [tub] suitable pipe,ping
— f — para conformación f - [tub] suitable
  forming pipe
— f ——— embridar f - [tub] suitable flanging
  pipe

tubería f apropiada para operación(es) f de
  conformación f - [tub] forming operation(s)
  suitable pipe
— f —— plegado m - [tub] bending suitable
  pipe
— f —— plegadura f - [tub] bending suitable pipe
— f —— serpentina(s) f - [tub] coiling
  suitable pipe
— f —— soldadura f - [tub] weld(ing) suitable pipe
— f ——— f a tope m - [tub] butt welding
  suitable pipe
— f —— soldar v - [sold] weld(ing) suitable
  pipe
— f armada - [tub] reinforced pipe • fabricated
  pipe • [metal-fabr] assembled pipe
— f — con diámetro m pequeño - [tub] small
  diameter reinforced pipe
— f ——— m reducido - [tub] small diameter
  reinforced pipe
— f — de acero m - [tub] fabricated steel pipe
— f ——— m corrugado - [metal-fabr] fabricated corrugated steel pipe
— f — en fábrica f - [metal-fabr] factory,
  made, o fabricated, pipe
— f —— planta f - [metal-fabr] shop fabricataed, o factory made, pipe
— f — en obra f - [tub] field assembled pipe
— f —— taller n - [tub] shop fabricated pipe
— f Armco - [metal-fabr] Armco pipe
— f — de acero m - [metal-fabr] Armco steel
  pipe
— f ——— m para agua m - [metal-fabr] Armco
  steel water pipe
— m — para agua n - [tub] Armco water pipe
— f ——— planta(s) f industrial(es) f - [ind]
  Armco industrial plant pipe,ping
— f — soldada - [metal-fabr] Armco welded pipe
— f —— de acero m -[metal-fabr] Armco welded
  steel pipe
— f ——— m para agua n - [metal-fabr] Armco
  welded steel water pipe
— f — Smooth-Flo - [metal-fabr] Armco Smooth-
  -Flo pipe
— m ——— con liga f de asbesto m (Asbestos-
  -Bonded) - [metal-fabr] Armco Asbestos-Bonded
  Smooth-Flo pipe
— f arrollada f - [metal-fabr] coiled tubing
— f — de acero m - [tub] coiled steel tubing
— f ——— m revestida - [metal-fabr] coated
  coiled steel tubing
— f ——— m con (material) plástico m -
  [metal-fabr] plastic (material) coated coiled
  steel tubing
— f — revestida - [metal-fabr] coated coiled
  tubing con (material) plástico m - [metal-
  -fabr] plastic coated coiled tubing
— f Asbestos-Bonded - [metal-fabr] Asbestos-
  -Bonded tubing
— f —— Smooth-Flo - [metal-fabr] Asbestos-
  -Bonded Smooth-Flo tubing
— f ———— m con corrugaciones
  interiores cubiertas - [tub] Asbestos-Bonded
  Smooth-Flo pipe
— f ——— f —— m e interior liso
  - [metal-fabr] Asbestos-Bonded Smooth-Flo
  pipe
— f ———para cloaca(s) f - [tub] Asbestos-Bonded Smooth-Flo sewer pipe
— f ascendente - [tub] rising, pipe, o unit
— f aspiradora - [tub] suction pipe
— f atacada - [quím] attacked pipe
— f atascada - [tub] plugged, o clogged, pipe
— f — con lodo(s) m - [tub] mud plugged pipe
— f austenítica - [tub] austenitic pipe
— f — sin costura f - [tub] seamless austenitic pipe
— f autosoportante - [tub] self-supporting,
  pipe, o line
— f — de acero m - [tub] self-supporting
  steel pipe
— f bajo tierra f [constr] underground pipe ►

buried, o sub-surface, pipe
tubería f bañada con asfalto m - [tub] asphalt
dipped pipe
— f bien aislada - [tub] well insulated pipe
— f Bundy - [tub] Bundy tubing
— f — con diámetro m reducido - [tub] small
diameter Bundy tubing
— f calentada - [tub] heated pipe
— f calorifugada* - [tub] véase tubería f, aislada, o con aislación f
— f calorizada*- calorized* piping
— f camisa - [tub] casing pipe; pipe casing
— f capilar - [tub] capillary tube,bing
— f captadora - [petról] gathering (line)
— f ciega - [tub] casing
— f circular - [metal-fabr] circular, o (full)
round, pipe • [metal-prod] bustle pipe
— f con cuerda f equivalente - [tub] equivalent span round pipe
— f — luz f equivalente - [tub] equivalent
span round pipe
— f conformada en frío - [tub] cold formed
round tubing
— f — de acero m - [tub] round steel pipe
— f — lámina f de acero m - [metal-fabr]
steel strip round pipe
— f — — f — — m soldada - [tub] welded
steel strip round pipe
— f — plancha(s) f estructural(es) - [tub]
structural plate circular pipe
— f equivalente - [constr] equivalent round
pipe
— f para agua m - water circle, o circular
water, pipe
— f para horno m alto - [metal-prod] blast
furnace bustle pipe
— f circular - véase también tubería f redonda
— f cloacal - [sanit] sewer, pipe, o line
— f — aérea - [sanit] aerial sewer line
— f — con infiltración f reducida - [sanit]
low infiltration sewer line
— f — elevada - [sanit] aerial sewer line
— f — para impulsión f - [sanit] sewer force
main
— f — para urbanización f - [constr] development sewer line
— f colectora f - [tub] collector, pipe, o line
• [tub] pipe header • [petról] gathering line
— f — de acero m - [tub] steel collector pipe
— f — — m corrugado - [tub] corrugated
steel collector, line, o pipe
— f — maestra - [sanit] trunk sewer
— f colmada - [tub] full pipe • running full
— f colocada - [constr] placed, pipe, o line
— f — en posición f - [tub] positioned pipe
— f — — f (apropiada) para soldadura f por
puntos m - [tub] pipe positioned for @ tack
weld(ing)
— f combada - [tub] cambered, o bent, pipe
— f comercial - [tub] commercial (pipe) size
— f completa - [tub] complete pipe,ping
— f compuesta - [petról] compound pipe,ping;
multi string
— f común - [tub] common, o standard, pipe
— f — para abastecimiento m - [hidr] common
supply, pipe, o line
— f — — provisión f - [mec] common supply,
pipe, o line
— f — — suministro m - [mec] common supply,
pipe, o line
— f con aleación f - [tub] alloy, tubing, o
piping
— f — — f alta - [tub] high alloy, tubing, o
pipe
— f — — f baja - [tub] low alloy, tubing, o
pipe
— f — — f — para resistencia f alta (a tensión f) - [tub] high-strength low-alloy, tub-
ign, o pipe
— f — ángulo m - [tub] angle pipe
— f — — m sesgado - [tub] skewed angle pipe
— f — calidad f especial - [petról] specialty
grade pipe

tubería f con calidad f especial para entubación f - specialty grade casing (type)
— f — — f — — — f para industria f petrolera - [petról] specialty grade oil country casing pipe
— f — — f nuclear - [tub] nuclear quality
.. tubing
— f — — f para plancha(s) f - [tub] plate
grade pipe
— f — — f superior - [tub] superior quality
pipe
— f — — f — en superficie f - [tub] superior
surface quality pipe
— f — campana f - [tub] bell pipe
— f — — f y espiga f - [tub] bell and spigot
pipe
— f — capa f - [tub] coated pipe
— f — — f aplicada en fábrica - [tub] mill
coated pipe
— f — — f bitumástica - [tub] bitumastic
coated pipe
— f — — f final - [tub] seal coated pipe
— f — — f negra - [tub] black coated pipe
— f — — f — aplicada en fábrica f - [tub]
black mill coated pipe
— f — — f para sello m - [tub] seal coated
pipe
— f — — f protectora - [tub] (protection)
coated pipe
— f — — f — aplicada en fábrica f - [tub]
mill (protection) coated pipe
— f — característica(s) f superior(es) -
[tub] su;erior quality pipe
— f — i — en superficie f - [tub] superior surface quality pipe
— f — conformación f - [tub] shaped pipe
— f — — f cuadrada - [tub] square shaped,
tube,bing, o pipe,ping
— f — — f elíptica - [tub] elliptically
shaped, pipe, o structure
— f — — f redonda - [tub] round shaped,
pipe, o structure
— f — cordón m [interior]—[tub] flash-in
tubing
— f — corrugación(es) f - [tub] corrugated
pipe
— f — — f helicoidal(es) - [tub] helically
corrugated pipe
— f — — f — para alcantarilla(s) f - [tub]
helically corrugated, o Hel-Cor, culvert
pipe
— f — — f — desagüe(s) m pluvial(es) -
[tub] Hel-Cor storm sewer pipe
— f — — f — cubiertas [Smooth-Flo] - [tub]
Smooth-Flo, o smooth-lined, pipe
— f — costura f - [tub] welded, tubing, o pipe
— f — — f en espiral m - [tub] spiral seam
pipe
— f — — f engargolada - [tub] lock-seam pipe
— f — — f helicoidal - [tub] spiral welded
pipe
— f — — f — (Hel-Cor) con corrugación(es) f
interiores cubiertas (Smooth-Flo)—[tub] Hel-
Cor Smooth-Flo pipe
— F — — f recta - [tub] straight seam pipe
— f — — f soldada f - [tub] welded seam pipe
— f con carbono m - [tub] carbon (steel) pipe
— f — — y aleación f - [tub] carbon (steel)
and alloy pipe
— f — — m — — f para (uso m con temperatu-
ra(s) alta(s) - [tub] high temperature(s)
(service) carbon and alloy pipe
— f — — m — — f para temperatura(s) f ba-
ja(s) - [tub] low temperature carbon (steel)
and alloy pipe
— f — conexión(es) f embridada(s) - - [tub]
flanged connection tubing
— f — — f mediante brida(s) f - [tub] flanged
connection tubing
— f — configuración f especial - [tub] spe-
cial(ly) shape(d) tubing
— f — cuerda f equivalente - [constr] equivalent span pipe

tubería f con diámetro m dado - [tub] given size
pipe
— f — — m estándar - [tub] standard size pipe
— f — — m grande - [tub] large diameter pipe
— f — — m mínimo - [tub] minimum diameter
pipe
— f — — m nominal - [tub] nominal diameter
pipe
— m — — m pequeño - [tub] small diameter pipe
— f — — m reducido—[tub] small diameter pipe
— f — — m — para subdrenaje m - [tub] small
diameter subdrainage pipe
— f — dos hileras f de remaches m - [tub]
double riveted pipe
— f — espesor m de pared f corriente - [tub]
ordinary, o common, wall thickness pipe
— f — espiga f - [tub] spigot pipe
— f — — f y campana f - [tub] spigot and bell
pipe
— f — — f — — con guarnición f de caucho m
[tub] Stab-Joint pipe
— f — — f — — f — — m para agua m -
[tub] Stab-Joint waer pipe
— f — extremo m abridado - [tub] flanged end
pipe
— f — — m con boquilla f para soldar v—[tub]
socket welding end pipe
— f — extremo(s) m liso(s) - [tub] plain end
pipe
— f — — m para soldadura f - [tub] weld(ing)
end pipe
— f — — m — — f a tope - [tub] butt weld-
-(ing) end pipe
— f — fondo m pavimentado - [tub] paved invert
pipe
— f — forma f de herradura f - [tub] horseshoe
(shaped) pipe
— f — imprimación f - [tub] primed pipe
— f — — f exterior - [tub] outside primed
pipe
— f — — f — — únicamente - [tub] outside primed
only pipe
— f — junta f de espiga f y campana - [tub]
bell and spigot pipe
— f — — f — — f — — f con guarnición f de
caucho m - [tub] stab-joint pipe
— f — — f — — f — campana f de acero m -
[tub] steel stab-joint pipe
— f — junta f lisa - [tub] flush joint pipe
— f — — f tipo Dresser - [tub] dresser
coupled pipe
— f — lado m plano - [tub] flat side tubing
— f — liga f de asbesto m - [tub] asbestos
bonded pipe
— f — — f — — m con corrugación(es) f interiores cubiertas - [tub] Asbestos-Bonded
Smooth-Flo pipe
— f — — f — — m e interior n liso - [tub]
Asbestos-Bonded Smooth-Flo pipe
— f — — f — — para cloaca(s) f - [tub] Asbestos-Bonded Smooth-Flo sewer pipe
— f — — f — — m y fondo m encachado - [tub]
Asbestos-Bonded Paved-Invert pipe
— f — luz f equivalente - [constr] equivalent
span pipe
— m — muro m para cabecera f - [constr] pipe
with @ headwall
— f — pared f delgada - [tub] thin, o light,
wall, pipe, o tubing
— f — — f guesa - [tub] thick wall pipe,ping
— f — — f — para conducción f - [tub] thick,
o heavy, wall line pipe
— f — — f lisa - [tub] smooth wall(ed) pipe
— f — — f liviana - [tub] light wall, pipe, o
tube,bing
— f — pavimento m - [tub] paved pipe
— f — — m bituminoso - [tub] bituminous paved
pipe
— f — peso m reducido - [tub] lightweight pipe
— f — presión f - [tub] pressurized pipe,ping
— f — — f entubada - [tub] encased pressurized pipe

tubería f con presión f baja • [tub] low pressure, line, o pipe
— f — recubrimiento m asfáltico - [tub] asphalt coated pipe
— f — — m bituminoso - [metal-fabr] bituminous coated pipe
— f — — m bitumástico - [tub] bitumastic
coated pipe
— f — resistencia f alta - [tub] high, resistance, o strength, pipe, o tubing
— f — — f a tensión f - [tub]
high tensile line pipe
— f — — f — — f para conducción f -
[tub] high tensile line pipe
— f — — f con aleación f baja - [tub] high
strength low alloy, o HSLA, pipe, o tubing
— f — — f — conducción f - [tub] high
tensile line pipe
— f — — f excepcional - [tub] extra high
strength, piping, o material
— f — — f a tensión f - [tub] extra high
tensile strength, material, o pipe,ping
— v — revestimiento m - [tub] lined pipe
— f — — m asfáltico - [tub] asphalt lined
pipe
— f — — m centrifugado - [tub] spun lined
pipe
— f — — m interior - [tub] (inner) lined
pipe
— f — — m — total - [tub] fully lined pipe
— f — rosca f - [tub] threaded pipe
— f — — f ahusada - [tub] taper tapped pipe
— f — — f cónica - [tub] taper tapped pipe
— f — — f y manguito - [tib] thread and
coupling (provided) pipe
— f con sección f especial - [tub] special
shape tubing
— f — tamaño m cabal - [tub] full scale pipe
— f — unión f lisa - [tub] flush joint pipe
— f conductora para calor m - [tub] heat carrying line
— f — — — m o frío m - [tub] heat or cold
carrying line
— f — — frío m - [tub] cold carrying line
— f — — inyección f - [petról] mud conveying
line
— f conectada - [tub] connected tubing
— f conformada - [tub] shaped pipe
— f congelada - frozen, o line
— f contigua - [tub] contiguous, o adjoining,
o adjacent, pipe,ping, o line
— f convencional - [tub] conventional pipe,ping
— f corriente - [tub] standard pipe,ping
— f — armada en fábrica f - [tub] standard
factory-made pipe
— f — de hormigón m - [constr] conventional
concrete pipe
— f — hecha en fábrica f - [tub] factory,
made, o assembled, standard pipe
— f — metálica - [tub] conventional metal pipe
— f — sin costura f - [tub] standard seamless
pipe
— f corroída - [tub] corroded pipe
— f corrugada - [tub] corrugated pipe • corrugated conduit
— f — anular(mente) - [tub] annular(ly) corrugated pipe
— f — armada en fábrica - [tub] fabricated
corrugated pipe
— f — bajo tierra f - [constr] buried, o underground, corrugated pipe
— f — con corrugación(es) f interior(es) cu-
bierta(s) con asfalto m - [tub] asphalt
coated Smooth-Flo corrugated pipe
— f — con diámetro m estándar - [tub] standard size corrugated pipe
— f — — diámetro m mayor - [tub] larger corrugated pipe
— f — — muesca(s) f - [tub] notch type corrugated pipe
— f — — revestimiento m centrifugado - [tub]
spun-lined corrugated pipe

tubería f corrugada con revestimiento m interior total - [constr] fully-lined corrugated pipe
— f — — m total - [constr] fully lined corrugated pipe
— f — corriente - [tub] standard corrugated pipe
— f — de acero m - [tub] corrugated steel pipe • corrugated steel tubing
— f — — m con cobre m - [tub] copper bearing steel corrugated pipe
— f — — m — — m galvanizado - [tub] galvanized copper bearing steel corrugated pipe
— f — m armada - [tub] assembled, o fabricated, steel pipe
— f — m en fábrica f - [tub] fabricated steel pipe
— f — m austenítico - [tub] austenitic steel, tubing, o pipe
— f — — soldada - [tub] welded austenitic steel pipe
— f — m bajo tierra f - [constr] buried, o underground, steel pipe
— f — — m con aleación f - [tub] alloy(ed) steel, tubing, o pipe
— f — m con aleación f baja - [tub] low alloy steel, pipe, o tubing
— f — m — f — con resistencia f alta (a tensión f) - [tub] high strength low alloy steel, pipe, o tubing
— f — m — f sin costura f - [tub] seamless alloy steel tubing
— f de acero n - [tub] steel pipe
— f — — con calidad . . . • [tub] . . . grade steel pipe
— f — m — — . . . imprimida con minio m - [tub] red primed . . . grade steel pipe
— f — m — — imprimida en fábrica - [tub] mill primed . . . grade steel pipe
— f — — m — — soldada - [tub] . . . grade steel welded pipe
— f — m — f . . . — en espiral - [tub] . . . grade steel spiral welded pipe
— f — m con carbono m—[tub] carbon (steel) pipe
— f — m — — m para alambique m - [tub] carbon steel still tube
— f — m — — m sin costura f - [tub] seamless carbon steel pipe
— f — m — — m soldada a tope m - [tub] butt welded carbon steel pipe
— f — m — m y aleación f - [tub] carbon and alloy steel pipe
— f — m — f para temperatura(s) f baja(s) - [tub] low temperature carbon and alloy steel pipe
— f — m — f para uso(s) con tempertura(s) f alta(s) - [tub] high temperature service carbon and alloy steel pipe
— f — m — m — f — temperatura(s) f baja(s) - [tub] low temperature service carbon and alloy steel pipe
— f — m — m — f — m — f elevada(s) - [tub] high temperature service carbon and alloy steel pipe
— f — m con cobre m - [tub] copper bearing steel pipe
— f — — m galvanizado - [tub] galvanized, copper bearing, o Zinc-Grip, steel pipe
— f — — — m — corrugado - [tub] copper bearing steel galvanized corrugated pipe
— f — — m — — galvanizado corrugada helicoidalmente - [tub] copper bearing steel Zinc-Grip Hel-Cor pipe
— f — m — f interiores cubiertas (Smooth-Flo) recubierta con asbesto m - [tub] asphalt-coated, Smooth-Flo, o smooth-lined, steel pipe
— f — — m — f perforada - [tub] helically corrugated perforated steel pipe
— f — — m con costura f - [tub] welded steel pipe

tubería f de acero m con costura f por resistencia f eléctrica - ['tub] electric resistance welded steel pipe
— f — — m con cromo-molibdeno - [tub] chromium-molybdenum (steel) pipe
— f — — m junta f de espiga f y campana f - [tub] bell and socket joint steel pipe
— f — — m — f — f y — f con guarnición f de caucho m - [tub] stab-joint steel pipe
— f — m — — f tipo m Dresser - [tub] Dresser-coupled steel pipe
— f — — m pared(es) f de . . . • [tub] . . . thick wall steel, pipe, o tube
— f — — m — f gruesa - [tub] heavy wall steel, pipe, o tubing
— f — — m — pared(es) f lisa(s) - [constr] smooth wall(ed) steel pipe
— f — m con proporción f alta de silicio m - [tub] high silicon (steel) pipe
— f — m — protección f católica - [tub] cathodically protected steel pipe
— f — — m resistencia f alta - [tub] high strength steel, pipe, o tubing
— f — — m con revestimiento m - [tub] lined steel pipe
— f — — m — — m centrifugado - [tub] spun lined steel pipe
— f — — m de mortero m de cemento m - [tub] cement mortar lined steel pipe
— f — — m — m interior - [constr] lined steel pipe
— f — — — — m — total - [constr] fully lined steel pipe
— f — — m corrugado - [tub] corrugated steel pipe
— f — — m — anularmente - [tub] annularly corrugated steel pipe
— f — — m bajo tierra f - [constr] buried, o underground, corrugated steel pipe
— f — — m con diámetro m reducido - [tub] small diameter corrugated steel pipe
— f — — m — diámetro m reducido revestido con asfalto m - [Tub] small diameter asphalt-coated corrugated steel pipe
— f — — m con fondo m pavimentado - [tub] Paved-Invert corrugated steel pipe
— f — — — m — galvanización f simple - [tub] plain galvanized corrugated steel pipe
— f — — m — con interior m liso - [tub] Smooth-Flo, o smooth lined, corrugated steel pipe
— f — — m — — — m liso recubierta con asbesto m - [tub] asphalt-coated, smooth lined. o Smooth-Flo, corrugated steel pipe
— f — — — m — junta f lisa - [tub] Smooth-Flo corrugated steel pipe
— f — — m — — — m — Smooth-Flo recubierta con asbesto m - [tub] asphalt-coated Smooth-Flo corrugated steel pipe
— m — — m — — recubrimiento m bituminoso - [tub] bituminous coated corrugated steel pipe
— f — — m — en sección(es) f larga(s) - [tub] long length corrugat ed steel pipe
— f — — m corrugada en hélice f - [tub] helically corrugated steel pipe
— f — — m — encachada - [constr] Paved-Invert corrugated steel pipe
— f — — m encajable - [tub] corrugated steel nestable pipe
— f — — m — — intercambiable - [tub] corrugated steel interchangeable nestable pipe
— f — — m — — — con muesca(s) f - [tub] corrugated steel interchangeable notch type nestable pipe
— f — — m corrugada enterrada - [constr] buried corrugated steel pipe
— f — — m — estándar - [constr] standard corrugated steel pipe
— f — — m intercambiable - [tub] [tub] interchangeable nestable steel pipe
— f — — m enterrada - buried steel pipe

tubería f de acero m especial - [tub] special-
-(ty) steel pipe
— f — m soldada en espiral m - [tub]
special(ty) steel spiral welded pipe
— f — m estructural - [tub] structural
steel tube,bing
— f — m extra reforzada - [tub] extra
strong steel pipe
— f — m fundido gris - [metal-fabr] gray
cast steel pipe
— f — m galvanizado - [tub] galvanized
steel, pipe, o tube, o duct, o conduit
— f — m con corrugación(es) f helicoidales - [tub] galvanized helically corrugated
steel pipe
— f — m liso con costura f helicoidal
engatillada - [tub] Smooth-Lok pipe
— f — m simple - [tub] plain
galvanized steel pipe
— f — m Hel-Cor - [tub] Hel-Cor steel pipe
— f — con corrugación(es) f helicoidales -
[tub] Hel-Cor corrugated steel pipe
— f — m imprimada - [tub] primed steel pipe
— f — m con minio m - [tub] red primed
steel pipe
— f — m inoxidable - [tub] stainless steel,
pipe,ping, o tube,bing, o line
— f — m austenítico - [tub] austenitic
stainless steel pipe
— f — m soldada - [tub] welded austenitic stainless steel pipe
— f — m con calidad f para alcantarillas
f - [constr] culvert grade stainless steel
pipe
— f — m para alcantarilla f - [constr]
culvert grade stainless steel pipe
— f — m soldada - [tub] welded stainless
steel, pipe, o tubing
— f — m inspeccionada - [tub] inspected
steel pipe
— f — m instalada - [tub] installed steel
pipe
— f — m laminado - [tub] rolled steel pipe
— f — m en caliente - [tub] hot rolled
steel pipe
— f — m con filete m interior - [tub]
flash-in hot rolled steel pipe
— f — m con filete m interior redondeado sin rebabar v - [tub] not deburred
flash-in aluminized steel square tubing
— f — m m con filete m interior
sin rebabar - [tub] not deburred flash-in
welded hot rolled steel tubing
— f — m m con filete m redondeado
- [tub] round(ed) flash-in welded hot rolled
steel tubing
— f — m soldada - [tub] welded hot
rolled steel tubing
— f — en frío - [tub] cold rolled steel
tubing
— f — m soldado con filete inte-
rior sin rebabar - [tub] not deburred flash-
in welded cold rolled steel tubing
— f — m m redondeado -
[tub] round(ed) flash-in welded cold rolled
steel tubing
— f — m m sin reba-
bar - [tub] not deburred round(ed) flash-in
welded cold rolled steel tubing
— f — en frío con filete m interior -
[tub] flash-in cold rolled steel tubing
— f — m negro - [tub] black steel pipe
— f — m descartable - [tub] expendable
black steel pipe
— f — m reforzado - [tub] heavy duty
black steel pipe
— f — m reforzado descartable - [tub]
expendable heavy duty black steel pipe
— f — m para agua m—[tub] steel water pipe
— f — m m con revesti-
miento m de mortero m de cemento m - [tub]
[tub] cement mortar lined steel water pipe

tubería f de acero m para agua m congelada—[tub]
frozen steel water pipe
— f — m — descongelada - [tub]
thawed frozem steel water pipe
— f — m con espiga f y campana f -
[tib] bell and spigot steel water pipe
— f — f y campana f con
guarnición f de caucho m - [tub] Stab-Joint
steel water pipe
— f — m — cloaca f - [tub] steel sewer
pipe
— f — m — conducción f de agua m - [tub]
steel water line
— f — m — desagüe m - [sanit] steel drainage pipe
— f — m — m pluvial(es) - [sanit] steel
storm drainage line
— f — m — entubación f - [electr-instal]
steel pipe casing
— f — m — drenaje m - [tub] steel drainage
pipe
— f — m — pilote(s) m - [tub] steel pile
pipe
— f — m para plancha(s) f estructurales -
[tub] structural plate steel pipe
— f — m presión f - [tub] steel pressure
pipe
— f — m — subdrenaje m - [tub] steel subdrainage pipe; steel pipe subdrain
— f — m para uso m con presión f - [tub]
steel pressure pipe
— f — m perforada - [tub] perforated steel
pipe
— f — m — con corrugación(es) f helicoi-
dal(es) - [constr] helically corrugated perforated steel pipe
— f — m — para subdrenaje m - [tub] perforated steel (pipe) subdrain
— f — m proyectada - [tub] designed steel
pipe
— f — m recia - [tub] rugged steel pipe
— f — m reforzada - [tub] reinforced, o extra strong, steel pipe
— f — m rellenada a con hormigón m - [tub]
concrete filled steel pipe
— f — m revestida con asfalto m - [tub] asphalt coated steel pipe
— f — cobre m - [tub] copper coated
steel, pipe, o tubing
— f — m — mortero m de cemento m -
[tub] cement mortar lined steel pipe
— f — m — m con espiga f y
campana con guarnición f de caucho m - [tub]
cement mortar lined Stab-Joint steel pipe
— f — m con junta f de
espiga f y campana f con guarnición f de caucho Stab-Joint - [tub] cement mortar lined
Stab-Joint steel pipe
— f — m sin costura f - [tub] seamless
steel pipe
— f — m soldada - [tub| welded steel, pipe,
o tubing
— f — m con filete m interior - [tub]
flash-in welded steel, pipe, o tubing
— f — m — redondeado - [tub]
round(ed) flash-in welded steel tubing
— f — m — m sin rebabar - [tub]
not deburred flash-in welded steel tube,bing
— f — m — eléctricamente - [tub] elec-
troc(ally) welded steel tubing
— f — m — en espiral - [tub] spiral welded
steel, pipe, o tubing
— f — m — para pilote(s) m - [tub]
spiral welded steel pipe pile,ling
— f — m — helicoidalmente - [tub] spiral
welded steel pipe
— f — m — con arco m - [tub] arc welded, o
electric fusion welded, steel pipe
— f — m — revestida con mortero m - [tub]
mortar lined welded steel pipe
— f — acronitrilo-butadieno-estireno m - [tub]
acronitrile-butadiene-styrene pipe

tubería f de aleación f - [tub] alloy, tubing, o pipe
— f —— f alta - [tub] high alloy, pipe, o tubing
— f —— f baja - [tub] low alloy, pipe, o tubing
— f — aluminio m - [tub] aluminum, pipe, o tubing
— f — arcilla f - [tub] clay, pipe, o tubing • [constr] clay tile
— f —— f vitrificada - [constr] vitrified, clay, o tile, pipe
— f — barro m - [tub] clay (drain) pipe
— f — calidad f . . . . • [tub] . . . grade pipe
— f —— f diversa - [tub] various, o different, grade pipe
— f — carbono m alto • [tub] high carbon pipe
— f — cemento m amiantado - [tub] asbestos-cement pipe
— f —— m soldada - [tub] welded cement pipe
— f — cloruro m de polivinilo m - [tub] polyvinyl chloride, tube, o pipe,ping
— f — cobre m - [tub] copper, tube,bing, o pipe, o line
— f —— m acampanada - [tub] flared copper, pipe, o tubing
— f —— m recocido - [tub] annealed copper, pipe, o tubing
— f —— m soldada - [tub] soldered copper, pipe,ping, o tubing
— f — conformación f circular - [tub] round shape, pipe, o tubing, o conduit
— f —— f para aplicación(es) f mecánica(s) - [tub] round shape(d), conduit, o mechancal, pipe, o tubing
— f —— f especial - [tub] special shape(d), pipe, o tubing, o conduit
— f —— f para aplicación(es) f mecánica(s) - [tub] special shape(d), mechanical tubing, o conduit
— f — chapa f de acero m - [ambient] steel sheet, pipe, o tubing
— f —— f metálica - [ambient] (metal strip) ductwork
— f — fundición f liviana - [tub] light weight (service) cast iron soil pipe
— f — tipo m entramado - [tub] truss (shaped) pipe
— f debajo de pavimento m - [tub] under pavement, pipe, o conduit, o main
— f —— m rígido - [tub] pipe, o conduit, o main, under @ rigid pavement
— m —— terraplén m - [constr] embankment conduit
— f deformada - [tub] deformed pipe
— f delgada - [tub] thin, pipe,ping, o tubing • [constr-pil] thin shell
— f — para pilote(s) m - [constr-pil] thin pile shell
— f desacoplada - [tub] uncoupled pipe
— f desarmable - [tub] disassemblable pipe • nestable pipe • portable pipe
— f descargada - [transp] unloaded pipe • [ind] discharged, o empty, pipe, o line
— f descartable - [tub] expendable pipe
— f — de acero m - [tub] expendable steel pipe
— f —— m negro - [tub] expendable black steel pipoe
— f —— m — reforzado - [tub] expendable heavy duty black steel pipe
— f desconectada - [tub] disconnected, pipe, o tubing
— f desgastada - [tub] worn (out), pipe, o tube
— f desocupada - [tub] empty,tied pipe, o tube
— f dilatada - [tub] expanded pipe
— f diversa - [tub] diverse pipe
— f doble - [tub] double pipe
— f — extra fuerte - [tub] double extra strong pipe
— f domiciliaria f - [tub] domestic piping
— f — para calefacción f - [constr] domestic heating, pipe,ping, o line

tubería f eferente - [tub] exit, pipe, o line
— f elaborada - [tub] fabricated pipe,ping
— f — con calidad f para plancha(s) f - [tub] plate grade fabricated pipe
— f — en fábrica - [tub] factory fabricated pipe
— f — planta f - [tub] plant, o shop, fabricated, pipe, o conduit
— f — en taller m—[tub] shop fabricated pipe
— f elevada - [tub] hoisted pipe • overhead pipe,ping; aerial line
— f — para agua m - [tub] overhead water pipe
— f —— m de pozo m - [ind] overhead well water pipe,ping
— f —— m — servicio m - [tub] overhead service water pipe,ping
— f elíptica - [tub] elliptical, o ellipsed, pipe,ping
— f — convencional - [tub] conventional elliptic(al),psed pipe,ping
— f — ensanchada horizontalmente - [constr] horizontally ellipsed pipe
— f —— verticalmente - [tub| verticallu ellipsed pipe
— f — horizontal - [tub] horizontally ellipsed pipes
— f — horizontalmente - [tub] horizontally ellipsed pipe
— f — vertical - [tub] vertical ellipsed pipe
— f — convencional - [tub] conventional vertically ellipsed pipe
— f — verticalmente - [tub] vertical(ly) ellipsed,ptical pipe
— f embridada - [tub] flanged, pipe, o duct
— f embutida - [tub] embedded pipe
— f empujada - [constr] pushed, o driven, pipe, o conduit
— f en largo(s) m diverso(s) - [tub] random, o varied, length(s) pipe
— m —— m diverso(s) desde fábrica f - [tub] random mill length pipe
— f — rotación f - [mec] turning, o rotating, pipe
— f — sección(es) f - [tub] sectional pipe
— f —— f corta(s) - [tub] short section(al) pipe
— f —— m para desagüe(s) m [hidr] sectional drain pipe
— f — tramo(s) m corto(s) - [tub] sectional short pipe,ping
— f — zanja f - [tubb trench, pipe,ping, o conduit
— f encachada - [tub] (invert) paved pipe
— f encajable - [tub] nestable pipe
— f — con cobre m - [tub] copper bearing nestable pipe,ping
— f —— m de tipo m con muesca(s) f - [tub] copper bearing notched type nestable pipe,ping
— f ——— m galvanizado m - [[metal-fabr] copper bearing galvanized pipe
— m ——— m de tipo m con muesca(s) f - [metal-fabr] copper bearing galvanized notched type nestable pipe
— f — corrugada - [metal-fabr] corrugated nestable pipe
— f —— de acero m - [metal-fabr] steel corrugated nestable pipe
— f ——— m con cobre m - [metal-fabr] copper bearing steel corrugated nestable pipe
— f ———— m —— galvanizada - [metal-fabr] galvanized copper bearing steel corrugated nestable pipe
— f ———— m puro - [metal-fabr] iron corrugated nestable pipe
— f ———— m puro - [metal-fabr] ingot iron corrugated nestable pipe
— f —— galvanizada - [metal-fabr] galvanized corrugated nestable pipe
— f —— intercambiable - [metal-fabr] corrugated interchangeable nestable pipe

tubería f encajable corrugada intercambiable (de tipo) con muescas f - [metal-fabr] corrugated interchangeable notch type nestable pipe

— f ——— (—) —— f de acero m con cobre ⋈ galvanizado - [metal-fabr] galvanized copper bearing steel corrugated notch type interchangeable nestable pipe

— f — de acero m - [metal-fabr] steel nestable pipe; nestable steel pipe

— f ——— m con cobre m—[metal--fabr] copper bearing steel nestable pipe

— f ——— m —— m galvanizado - [metal--fabr] galvanized copper bearing nestable steel pipe

— f ——— m —— m de tipo m con muescas f - [metal-fabr] copper bearing galvanized notched type nestable steel pipe

— f ——— m —— m intercambiable (de tipo m) con muesca(s) f - [metal-fabr] copper bearing steel interchangeable notch type nestable pipe

— f ——— m —— m de tipo m con muesca(s) f - [metal-fabr] copper bearing steel notched type nestable pipe

— f ——— m acanalado - [metal-fabr] nestable corrugated steel pipe

— f ——— m corrugado - [metal-fabr] nestable corrugated steel pipe; corrugated nestable steel pipe

— f ——— n galvanizado - [metal-fabr] galvanized steel nestable pipe

— f ——— m ondulado - [metal-fabr] nestable corrugated, o corrugated nestable, steel pipe

— f —— hierro m - [tub] iron nestable pipe

— f —— m puro - [tub] ingot iron nestable pipe

— f — de pieza(s) f intercambiable(s)—[metal--fabr] interchangeable nestable pipe

— f —— tipo m con muesca(s) f - [metal-fabr] notched type nestable pipe

— f ——— resalto(s) m - [tub] notched type nestable pipe

— f — galvanizada - [metal-fabr] galvanized nestable pipe

— f ——— de tipo m con muesca(s) f - [metal-fabr] galvanized notched type nestable pipe

— f — (e) intercambiable - interchangeable nestable, o nestable interchangeable, pipe

— f — f (—) — acanalado - [metal-fabr] nestable interchangeable corrugated pipe

— f —— corrugada - [metal-fabr] nestable interchangeable corrugated pipe

— f — (—) — de acero m acanalado - [metal--fabr] nestable interchangeable corrugated steel pipe

— f — (—) ——— m corrugado - [metal-fabr] nestable interchaneable corrugated steel pipe

— f — (—) ——— m ondulado - [metal-fabr] nestable interchangeable corrugated steel pipe

— f ——— de tipo m con muesca(s) f - [tub] interchangeable notch-type nestable pipe

— f —— ondulado - [metal-fabr] nestable interchangeable corrugated pipe

— f encogible—[tub] shrinkable, pipe, o tubing

— f engarbolada - [metal-fabr] stove pipe

— f ensayada - [tub] tested pipe,ping

— f enterrada - [constr] buried pipe,ping

— f — con profundidad f escasa - [constr] shallow buried pipe,ping

— f ——— f grande - [constr] deep(ly) buried pipe,ping

— f entramada - [tub] truss pipe

— f — de acronitrilo-butadieno-estireno m - [tub] acronitirle-butadiene-styrene truss pipe

— F — Truss-Pipe - [tub] truss pipe

— f entubada - [constr] encased, pipe, o line

— f equivalente - [tub] equivalent pipe

— f escariada - [tub] reamed pipe

— f — y mandrilada - [tub] reamed and drifted pipe

tubería f escogida - [tub] selected, o chosen, pipe

— f esencialmente horizontal - [ambient] essentially horizontal duct

— f — vertical - [ambient] essentially vertical duct

— f especial - [tub] special(ty) pipe

— f específica - [tub] specific, o given, pipe

— f especificada - [tub] specified pipe

— f estándar - [tub] standard pipe

— f — de hierro m - [tub] standard iron pipe

— f ——— m fundido - [tub] standard cast iron pipe

— f estañada - [tub] tinned, pipe, o tubing

— f estriada - [tub] grooved pipe

— f — para agua m - [petról] grooved wash pipe

— f estructural - [tub] structural, o construction, pipe, o tubing

— f — de acero m - steel structural, o structural steel, tubing

— f ——— m con aleación f baja - [tub] low alloy steel structural, pipe, o tubing

— f ——— m —— f — y resistencia f alta (a tensión f) - [tub] high-strength low-alloy steel structural tubing

— f ——— f ——— f —— f —— soldada - [tub] welded high strength low alloy structural tubing

— f ——— m —— f ——— f —— tensión f —— f conformada en caliente - [tub] hot formed seamless low-alloy high strength structural steel, pipe, o tubing

— f ——— f ——— f —— f conformada en caliente - [tub] hot formed welded high strength low-alloy structural tubing

— f — con carbono m - [tub] carbon steel structural tubing

— f ——— m —— m sin costura f - [tub] seamless carbon steel structural tubing

— f ——— m —— m —— f conformada en caliente m - [tub] hot formed seamless carbon steel structural tubing

— f ——— m ——— f conformada en caliente - [tub] hot formed seamless steel structural tubing

— f — conformada en caliente - [tub] hot formed structural tubing

— f ——— soldada - [tub] welded steel structural tubing

— f ——— m — conformada en caliente m - [tub] hot formed welded steel structural tubing

— f — con pared f delgada - [tub] light wall structural, pipe, o tubing

— f — corrugada - [constr] corrugated structural pipe,ping

— f — liviana - [tub] light structural pipe

— f — pesada - [tub] heavy structural tubing

— f — redonda - [tub] round structural tubing

— f —— sin costura f - [tub] seamless round structural tubing

— f ——— f con carbono m - [tub] round seamless carbon steel structural tubing

— f ——— f ——— m conformada en frío - [tub] cold formed round seamless carbon steel structural tubing

— f — sin costura f - [tub] seamless structural tube,bing

— f ——— f con aleación f baja - [tub] low alloy seamless structural tubing

— f ——— f —— f — y resistencia f alta (a tensión f) - [tub] seamless high strength low alloy structural tubing

— f ——— f configurada - shaped seamless structural tubing

— f ——— f conformada en caliente - [tub] hot formed seamless structural tubing

— f ——— f de acero m - [tub] steel structural tubing

— f — soldada - [tub] welded structural tubing

— f ——— conformada en caliente - [tub] hot

formed welded structural tubing

tubería f estructural soldada de acero m con carbono m conformada en frío - [tub] cold formed welded carbon steel structural tubing

— f **examinada** - [tub] examined, pipe, o tubing

— f **exigida** - [tub] required, pipe, o tubing

— f **existente** - [tub] existing, pipe, o tubing, o line

— f — **para aire** m - [ind] existing air line

— f **expandida** - [tub] expanded, pipe, o tubing

— f — **en frío** - [tub] cold expanded, pipe, o line

— f **extendida** - [constr] long, o extended, line

— f **exterior** - [tub] outdoor, o exterior, line, o pipe,ping · conduit; casing · [constr] liner pipe; (tunnel) encasement

— f **para tubería f para desagüe** m - [constr] sewer encasement

— f — — f — — m **pluvial** - [constr] storm sewer encasement

— f **extrafuerte** - [tub] extra strong pipe,ping

— f — **de latón** m **rojo** - [tub] extra strong red brass pipe

— f — — — m — **sin costura** f - [tub] extra strong seamless red brass pipe

— f — — m **sin costura** f - [tub] extra strong seamless brass pipe

— f — **sin costura** f - [tub] extra strong seamless pipe

— f **extraída** - [petról] pulled up pipe

— f **extrapesada** - [tub] extra strong pipe

— f — **sin costura** f - [tub] extra strong seamless pipe

— f **extrarresistente para protección** f **contra tumbo(s)** m - [tub] roll over protective structures, o ROPS, tube, o structure

— f **extrasuave** - [tub] dead soft tubing

— f **extruida** - [tub] extruded pipe

— f — **de resina** f **acronitrilo-butadieno-estirénica** - [tub] extruded acronitile-butadiene--styrene pipe

— f **fabricada** - [tub] fabricated pipe

— f — **en planta** f - [tub] shop, o plant, fabricated pipe

— f — — **taller** m - [tub] shop fabricated pipe

— f **fallada** - [tub] failed pipe

— f — **revestida** - [constr] lined failed pipe

— f **fallante** - [constr] failing pipe

— f **flexible** - [tub] flexible, pipe, o tubing

— f — **de acero** m - [tub] flexible steel pipe

— f **fundida** - [tub] cast pipe

— f — **centrifugada** - [tub] centrifugally cast pipe

— f — **para temperatura(s)** f **alta(s)** - [tub] high temperature cast pipe

— f **galvanizada** - [tub] galvanized, pipe, o duct · Zinc-Grip pipe

— f — **con corrugación(es)** f **helicoidal(es)** - helically corrugated galvanized pipe

— f — **con extremo(s)** m **liso(s)** - [tub] galvanized plain end pipe

— f — **corriente** - [tub] standard galvanized pipe

— f — **corrugada helicoidalmente** - [tub] galvanized helically corrugated pipe

— f — — **de hierro** m **puro** - [tub] ingot iron galvanized corrugated pipe

— f — — **helicoidalmente** - [tub] galvanized Zinc-Grip Hel-Cor pipe

— f — **de acero** m - [tub] galvanized steel pipe

— f — — — m **con corrugación(es)** f **helicoidal(es)** - [tub] galvanized helically corrugated steel pipe

— f — — — m **para entubación** f - [tub] galvanized steel pipe casing

— f — — **cubierta** f **para vástago** m - [compuertas] galvanized pipe stem cover

— f — — **hierro** m **puro** - [tub] ingot iron galvanized pipe

— f — **de plancha(s)** f **múltiples (Multi-Plate)** - [tub] galvanized Multi-Plate pipe

— f — **desarmable** - [tub] portable, o disassemblable, galvanized pipe

tubería f galvanizada estándar - [tub] standard galvanized pipe

— f — **para agua** m - [tub] galvanized water pipe

— f — **para conducción** f **de aire** m - [tub] air conduction galvanized pipe

— f — — **drenaje** - [tub] galvanized drainage, pipe, o tubing

— f — — **pozo(s)** m **para agua** m - [tub] galvanized water well casing

— f — **revestimiento** m **de pozo(s)** m - [tub] galvanized well casing

— f — **perforada** - [tub] galvanized, o Zinc--Grip, perforated pipe

— f — **roscada y con manguito(s)** m - [tub] galvanized thread and coupling pipe

— f **sencilla** - [tub] plain galvanized pipe

— f **soldada** - [tub] welded galvanized pipe

— f — — **para tubos** m **para revestimiento** m - [tub] welded galvanized casing

— f — **Zinc-Grip** - [tub] Zinc-Grip pipe

— f **gemela** - [tub] twin pipe,ping

— f **común** - [tub] twin common pipe

— f — **de plancha(s)** f **estructural(es)** - [tub] twin structural plate pipe(s)

— f — — f **múltiple(s)** - [tub] twin sectional plate pipe,ping

— f **giratoria** - [petról] spinning line

— f **golpeadora** - [tub] véase tubería f para hincar

— f **grande** - [tub] large pipe

— f — **para descarga** f - [hidr] large outfall line

— f **guía** - [petról] surface casing

— f **hecha en fábrica** - [tub] factory made pipe

— f **Hel-Cor** - [tub] Hel-Cor pipe

— f — — **para alcantarilla(s)** f - [metal-fabr] Hel-Cor culvert pipe

— f **helicoidal** - [tub] helical pipe

— f **hermética** - [ambient] airtight duct

— f **hexagonal** - [tub] hexagonal pipe

— f **hidráulica** - [tub] hydraulic, o water, pipe

— f **hidrostática** - hydrostatic line

— f **hincada** - [tub] driven, pipe, o shell

— f — **con mandril** m - [constr] mandrel driven shell

— f **hincadora** - [petról] drive,ving pipe

— f **horizontal** - [tub] horizontal, pipe, o line

— f **hueca** - [tub] hollow, tube, o pipe,ping

— f — **de acero** m - [tub] hollow steel pipe

— f **igualadora** - [tub] equalizing pipe · [metal-prod] mixer main

— f **imprimada** - [tub] primed pipe

— f — **con minio** m - [tub] red primed pipe

— f **impulsada** - [tub] pushed pipe

— f **industrial** - [tub] industrial pipe,ping

— f **inoxidable** - [tub] stainless pipe,ping

— f — **austenítica** - [tub] austenitic stainless pipe

— f **insertada** - [tub] jacked, o threaded, pipe

— f — **con gato(s)** m - [constr] jacked pipe

— f — — — m **hidráulico(s)** - [constr] hydraulically jacked pipe

— f **inspeccionada** - [tub] inspected pipe,ping

— f **instalada** - [tub] installed pipe,ping

— f — — **hoyo** m **pretaladrado** - [constr] drop--in pile shell

— f **íntegra** - [tub] entire pipe

— f **intemperizada** - [tub] weathering tube · Earth Tube

— f **intercambiable** - [tub] interchangeable pipe

— f **interior** - [tub] interior, o inside, piping

— f — **y exterior** - [tub] line and conduit

— f **laminada** - [metal-lam] rolled pipe

— f — **en caliente** - [metal-lam] hot rolled pipe

— f **larga** - [tub] long, pipe, o line

— f — **para descarga** f - [tub] long discharge line

— f **limpia** - [tub] clean pipe

— f **limpiada** - [tub] cleaned pipe

— f **liriforme** - [tub] lyre shaped pipe

— f **lisa** - [tub] smooth pipe

tubería f lisa corrugada anularmente - [constr] smooth annular(ly) corrugated pipe
— f —— helicoidalmente - [constr] smooth helical(ly) corrugated pipe
— f — de hormigón m armado - [constr] reinforced smooth concrete pipe
— f — de plástico - [tub] smooth plastic pipe
— f liviana - [tub] light, pipe, o tubing
— f — de acero m - [metal-fabr] light(weight) steel pipe
— f longitudinal - [tub] lengthwise pipe
— f maestra - [tub] main (pipe, o tube,bing)
— f mayor - [tub] larger pipe
— f mecánica - [tub] mechanical, pipe, o tubing
— f — sin costura f - seamless mechanical pipe
— f menor - [tub] smaller pipe
— f metálica - [metal-fabr] metal pipe
— f — aislada - [tub] insulated metal pipe
— f — corrugada - [tub] corrugated metal pipe
— f — — corriente - [tub] standard corrugated metal pipe
— f — — para descarga f - [constr] corrugated steel outfall pipe
— f — delgada—[constr-pil] (thin) metal shell
— f — para pilote(s) m - [constr-pil] thin metal pile shell
— f — Electrunite - [electr-instal] Electrunite electrical metallic tubing
— f — larga - [tub] long metal pipe,ping
— f — longitudinal - [tub] lengthwise metal pipe,ping
— f — para alcantarilla(s) f - [metal-fabr] metal culvert pipe,ping
— f — — instalación(es) f - [constr] metallic installation tubing
— f — — — f eléctrica(s) - [electr-instal] electrical installation metallic tubing
— f monolítica - [tub] monolithic, tube, o pipe,ping
— f — de hormigón m - [tub] monolithic concrete pipe,ping
— f Multi-Plate de plancha(s) f múltiple(s) - [metal-fabr] Multi-Plate pipe,ping
— f múltiple - [tub] multiple, pipe,ping, o line
— f — instalada—[tub] installed multiple line
— f negra - [tub] black, pipe, o tubing
— f para agua m - [tub] water, pipe, o main
— f — para lavador m - [metal-prod] washer water main (water) pipe
— f — — m con presión f - [tub] pressurized descaling water, pipe, o tubing
— m — — m congelada - [tub] frozen water pipe
— f — — m para Venturi - [metal-prod] scrubber water, pipe, o tubing
— f — — m helada - [tub] frozen water pipe
— f — — m para incendio(s) m - [segurid] fire main
— f — aire m - [ind] air main • [metal-prod] (air) blast main • air main • ventilation duct • air line
— f — — m caliente - [metal-prod] hot (air) blast main
— f — — m para soplante m - [metal-prod] blast main
— f — alambique m - [petról] still tube
— f — alimentación f - [ind] feed(ing) pipe
— f — — f calentada - [tub] heated feed(ing) pipe
— f — — f para agua m - [ind] water feed(ing) pipe
— f — — f agua m para precipitador m - [metal-prod] precipitator water feed pipe
— f — — f de aire m—[ind] air feed(ing) line
— f — — f larga - [ind] long feed(ing) tube
— f — alivio m - [ambient] vent(ing) duct
— f — — m autolimpiador - [ambient] self-cleaning vent duct
— f — — m con inglete m único - [ambient] single miter self-cleaning vent duct
— f — análisis m - [ind] analysis, pipe, o tube
— f — — m especial - [metal-prod] special analysis pipe

tubería f para aspiración f - [tub] suction, o intake, pipe
— f — avenamiento m - [hidr] drainage pipe
— f — bomba f - [tub] pump piping
— f — — f para servicio m - [tub] service pump piping
— f — — — m para planta f - [tub] mill, o plant, service pump piping
— f — bombeo m - [petról] tubing
— f — calentamiento m - [tub] heating pipe
— f — — m para serpentín m - [tub] heating, pipe coil, o coil pipe
— f — camisa f - [tub] casing, o liner, pipe; pipe, casing, o sleeve
— f — — f para barreno m - [herram] auger casing
— f — combustible m - [comb.int] fuel line
— f — conducción f - [tub] piping • [hidr] transmission line • [petról] pipeline; line pipe
— f — — f para artesa f a espesador m - [metal-prod] pipe from @ trough to @ thickenner
— f — — gas m - [combust] gas piping
— f — — f con resistencia f alta - [tub] high test line pipe
— f — — f para líquido m - [tub] liquid carrier pipe
— m — — f — m cloacal - [sanit] sewage carrier pipe
— f — — f — petróleo m - [petról] petroleum pipeline
— f — — f para distancia f larga - [petról] cross county pipeline
— f — — f soldada - [tub] welded, pipeline,o (steel) conduit
— f — — f — de acero m galvanizado - [tub] welded galvanized steel conduit
— f — — f de acero m galvanizado - [tub] galvanized steel conduit
— f — cubierta f - [tub] pipe cover
— f — — f para vástago m - [compuertas] pipe stem cover
— f — — f — m ascendente - [compuertas] rising pipe stem cover
— f — derivación f - [ambient] branch duct
— f — descarga f - [hidr] outfall • [metal-prod] discharge, o exhaust, pipe,ping
— f — — f a océano m - [hidr] ocean outfall
— f — — f para agua m - [hidr] water discharge pipe
— f — — f — m para Venturi m - [metal-prod] Venturi water discharge pipe
— f — — f — bomba f - [hidr] pump discharge pipe,ping
— f — — f — f para alimentación f - [tub] feed(ing) pump discharge pipe,ping
— f — — f — presión f alta - [metal-prod] high pressure discharge pipe,ping
— f — — f — f para separador m - [metal-prod] separator exhaust pipe,ping
— f — — f — — m para Venturi m - [metal-prod] scrubber separator exhaust pipe,ping
— f — — f directamente a océano m - [hidr] (direct) ocean outfall
— f — — f normal - [ind] normal discharge pipe,ping
— f — — f — para separador m - [metal-prod] separator normal discharge pipe,ping
— f — — f — — Venturi m - [metal-prod] Venturi (separator) normal discharge pipe
— f — destilación f - [tib] distillery tube
— f — — f para refinería f - [tub] refinery distillery, tube, o pipe,ping
— f — distribución f - [tub] distribution, line, o system
— f — — f para vapor m - [cald] steam distribution pipe,ping
— f — — f m para planta f - [cald] plant steam distribution pipe,ping
— f — drenaje m - [hidr] drainage pipe,ping
— f — — para calentador m - [cald] heat drainage pipe,ping

tubería f para drenaje m para calentador m para
  para presión f alta—hi pressure heater drain
— f — eje m - [tub] axle, o shaft, pipe
— f —— m horizontal - [tub] horizontal,
  axle, o shaft, pipe,ping
— f —— m vertical - [tub] vertical, axle, o
  shaft, pipe,ping
— f — elaboración f - [ind] process(ing) pipe
— f — elevación f - [tub] rising pipe,ping •
  riser unit
— f — enfriador m - [tub] cooler,ling tube
— f —— m primario - [emtal-prod] primary
  cooler,ling tube
— f — engrase m - [mec] greasing, o lubricating, pipe
— f — entrada f - [tub] inflow, o inlet, o intake, o supply(ing), pipe, o line
— f —— f para agua m - [tub] water, inlet, o
  feed(ing) pipe
— f — f —— m para tobera f para escorial
  n - [metal-prod]| slag notch intermediate
  cooler water inlet pipe
— f —— f —— m — toberín m para escorial
  m - [metal-prod] slag notch cooling water inlet pipe
— f —— f — petaca f - [metal-prod] cooling
  plate intake pipe
— f — entubación f - [tub] pipe casing
— f —— f de acero m - [electr-instal] steel
  (pipe) casing
— f —— f —— m soldada - [tub] welded
  steel pipe casing
— f —— f —— m — en espiral - [tub] spiral welded steel pipe casing
— f —— f con rosca f - [tub] threaded casing
— f —— f — f redonda - [petról] round
  thread casing
— f —— f subterránea - [tub] underground
  casing
— f — equipo m - [constr] equipment piping
— f — escape m - [comb.int] exhaust pipe,ping
— f — estufa f - [metal-prod] stove pipe,ping
— f — extremo m - [tub] end pipe,ping
— f —— m abridado - [tub] flanged end pipe
— f —— m — normal - [tub] standard flanged
  end pipe
— f — flujo m - [petról] flow, pipe, o line
— f — fuel oil m - [comb.int] fuel oil line
— f — gas m - [tub] gas pipe,ping • [coque]
  gas, line, o main
— f —— m para estufa f - [metal-prod] stove
  gas main
— f —— m — horno m alto - [metal-prod]
  blast furnace gas main
— f —— m lavado - [ind] clean gas main
— f —— m limpio m - [metal-prod] clean gas,
  pipe, o main
— f —— m semilimpio—[metal-prod] semi-clean
  gas, pipe, o main
— f — grasa f - [metal-prod] grease tube,bing
— f —— f para cierre m para distribuidor m -
— f —— f igualación f - [tub] equalizer tube,bing
— f —— f para nivel m de lago m - [hidr]
  lake water level equalizing tube
— f — impulsión f - [bombas] force,cing main •
  discharge,qing pipe • [hidr]
 pressure pipe,ping
— f —— f de hormigón m - [hidr] concrete,
  force, o pressure, pipe,ping, o main
— f — indicador m para nivel m en Venturi m -
  [metal-prod] Venturi level indicator tube
— f — inyección f - [ind] injection pipe
— f — f para vapor m - [metal-prod] steam
  injection pipe
— f —— f — viento m - [metal-prod] blast
  injection pipe
— f — irrigación f - [hidr] irrigation pipe
— f — lado m este - [constr] east (side) pipe
— f —— m norte [constr] north (side) pipe
— f —— m oeste - [constr] west (side) pipe
— f —— m sud - [constr] south (side) pipe
— f — lavador m - [metal-prod] washer pipe
— f — lechada f - [constr] grout(ing) pipe

tubería f para llenado m - [tub] filling pipe
— f para mezcla f - [metal-prod] mixing, pipe,
  o main
— f — oxígeno m - [ind] oxygen, pipe, o line
— f — perforación f—[petról] drilling pipe •
  drive pipe
— f — pozo m - [petról] well tubing
— f — presión f - [metal-fabr] pressure,
  tube,bing, o pipe,ping
— f —— f alta - [tub] high pressure, tube,
  o pipe,ping
— f — producción f - [petról] (production),
  liner, o tubing
— f — purga(r) - purge,ging pipe; bleeder line
— f — ramal m - [ambient.] branch duct
— f — rebose m - [tub] overflow pipe
— f —— m en separador m - [metal-prod] separator overflow pipe
— f — refrigeración f - [ind] cooling water
  main
— f —— f para cuba f - [metal-prod] bosh
  cooling system pipe,ping
— f —— f — etalaje m - [metal-prod] bosh
  cooling system pipe,ping
— f —— f — placa f - [metal-prod] plate
  cooling water main
— f —— f —— f para crisol m—[metal-prod]
  hearth stave(s) cooling water main
— f —— f — tobera f - [metal-prod] tuyere
  cooling, pipe, o main
— f — regulación f - [ind] governor, o con-
  trol(ling), main, o pipe,ping
— f —— f — agua m limpia - [metal-prod]
  level control clean water, main, o pipe
— f — resistencia f alta a tensión f - [metal]
  high (tensile) resistance pipe
— f —— f mayor a tensión f - [tub] higher
  (tensile) strength pipe
— f para retorno m - [tub] return line •
  [sanit] drain return line
— f —— m —— m de calentador m - [Cald]
  heater drip return piping
— f — revestimiento m - [tub] lining pipe •
  water well casing • casing (pipe)
— f —— m de pozo(s) m - [petról] casing
— f —— m galvanizada - [tub] galvanized
  casing
— f —— m para pozo(s) m - [tub] well casing
— f —— m —— m para agua m - [tub] water
  well casing
— f —— m soldada - [tub] welded well casing
— f —— m superior - [petról] surface casing
— f — salida f - [comb.int] exhaust pipe,ping;
  outlet pipe
— f —— f para agua m - [tub] water outlet
  pipe
— f —— f — petaca f - [metal-prod] cooling
  plate outlet pipe
— f — secado m - [metal-prod] drying pipe
— f — seguridad f - [petról] safety casing •
  surface casing
— f — separador m - [tub] separator pipe
— f — servicio m - [ind] service pipe,ping
— f —— m para planta f - [tub] plant, o
  mill, service, pipe,ping
— f — sistema m - [tub] system pipe
— f —— m Morgoil - [ind] Morgoil system pipe
— f —— m para enfriamiento m - [ind] cooling
  system pipe,ping
— f —— m —— m de rodillo(s) m - [metal-
  lam] roll cooling system pipe,ping
— f —— m — rodillos m para trabajo m -
  [metal-lam] work(ing) roll(s) cooling system
  pipe,ping
— f — soplante m - [metal-prod] blast main
— f — subdrenaje m - [hidr] subdrain(age) pipe
  • pipe subdrain
— f —— m con perforación(es) hacia abajo -
  [tub] perforations down subdrainage pipe
— f — subida f - [tub] ascending, o up, pipe
— f —— f para agua m - [tub] water, ascending, o up, pipe
— f — succión f - suction, o draft, pipe, o

tube
tubería f **para sumidero** m - [sanit] sump drain
pipe
— f — **suministro** m—[tub] supply, pipe, o line
— f — — m **para aire** m - [combust] air supply
pipe, o line
— f — — m **para gas** m - [combust] gas supply,
pipe, o line
— f — **surgencia** f - [petról] surge,ging tubing
— f — **temperatura(s)** f **alta(s)** - [tub] high
temperature pipe,ping
— f — — f **baja(s)**—[tub] low temperature pipe
— f — — f **elevada** - [petról] high temperature
pipe,ping
— f — — f **y presión** f **altas** - [tub] high temperature high pressure pipe,ping
— f — **terraplén** m—[constr] embankment conduit
— f — **tobera** f - [metal-prod] tuyere (cooling)
pipe
— f — **transmisión** f - [tub] transmission
tube, o pipe, o line
— f — — f **hidrostática** - [mec] hydrostatic
transmission line
— f — — **para gas** m - [tub] gas transmission,
pipe,ping, o line
— f — **transporte** m - [petról] transportation,
pipe,ping, o line
— f — — m **para petróleo** m - [petról] petroleum transportation, pipe,ping, o line
— f — — m — — m **líquido** - [petról] liquid
petroleum transportation, pipe,ping, o line
— f — — m **con temperatura(s)** f **alta(s)** -
[tub] high temperature service, pipe, o line
— f — — m — — f **baja(s)** - [tub] low temperature service pipe
— f — — m — — f **elevada(s)** f - [tub] high
temperature(s) service pipe
— f — — m **general** - [tub] general service tubing
— f — — m **mecánico** - [tub] mechanical tubing
— f — — m **para conducción** f - [tub] transmission service pipe
— f — — m — — f **bajo presión(es)** f **eleva-
da(s)** - [tub] high pressure transmission service pipe
— f — — m — — f — **presión(es)** f **alta(s)** -
[tub] high pressure transmission service pipe
— f — **vacío** m - [ind] vacuum line
— f — **vapor** m - [tub] steam, line, o pipe
— f — — m **con presión** f **alta** - [cald] high
pressure steam line
— f — — m — — f **baja** - [cald] low pressure
steam, pipe,ping, o line
— f — — m **debajo de campana** f - [metal-prod]
steam pipe below @ bell
— f — — m — — f **grande** - [metal-prod]
steam pipe below @ big bell
— f — — **para desaireador** m - [cald] deareator
steam pipe
— f — — m **para distribución** f - [cald] distribution steam, o steam distribution, pipe
— f — — m — — f **en planta** f - [cald] plant
distribution steam pipe
— f — — m **para extracción** f - [cald] extraction steam pipe
— f — — m — — f **con presión** f **alta** - [cald]
high pressure extraction steam pipe
— f — — m — — f — — f **baja** - [cald] low
pressure extraction steam pipe
— f — **ventilación** m - [ambient] ventilation, o
vent, pipe,ping, o duct(work)
— f — — f **por aspiración** f - [ambient] exhaust (ventilation) ductwork
— f — **Venturi** - [metal-prod] Venturi pipe
— f — **viento** m - wind pipe • [metal-prod]
bustle pipe; blast main
— f — — m **frío** - [metal-prod] cold, blast, o
air, main, o duct
— f — **viento** m **para estufa** f - [metal-prod]
(cold) blast main to @ stove
— f — — m **caliente** - [metal-prod] hot blast
duct
— f — — m **para soplante** m - [metal-prod]

blast main
tubería f **paralela** - [tub] parallel, pipe, o
main • multiple pipe line
— f **patentada** - [tub] patented pipe
— f — **con fondo** m **pavimentado** - [tub] patented Paved-Invert pipe
— f — **Paved-Invert con fondo** m **encachado** -
[tub] patented Paved-Invert pipe
— f — — — **con fondo** m **pavimentado** - [tub]
patented Paved-Invert pipe
— f **pavimentada** - [tub] paved pipe
— f **pequeña** - [tub] small pipe
— f **peraltada** - [constr] vertically elongated
pipe
— f **perforada** - [tub] perforated pipe
— f — **de acero** m - [tub] perforated steel
pipe
— f — — — m **con cobre** m - [metal-fabr] copper bearing steel perforated pipe
— f — — — m — — m **galvanizado** - [tub]
copper bearing steel Zinc-Grip perforated
pipe
— f — — — m — — m **corrugada helicoidalmente** - [tub] copper bearing steel Zinc-
Grip Hel-Cor perforated pipe
— f — **Hel-Cor con corrugación helicoidal** -
[metal-fabr] perforated Hel-Cor pipe
— f — **para subdrenaje** m - [hidr] perforated,
underdrain, o subdrainage pipe
— f **periférica** - [hidr] peripheral pipe
— f **pesada** - [tub] heavy pipe
— f **plástica** - [tub] plastic, pipe, o tubing
— f **por gravedad** f - [hidr] gravity line
— f — — **en descubierto** m - [hidr] exposed
gravity line
— f — — f — — **para agua** m - [hidr] exposed
gravity water line
— f — — f **expuesta** - [hidr] exposed gravity
line
— f — — f — **para agua** m - [hidr] exposed
gravity water line
— f — — f **para agua** m - [hidr] gravity
water line
— f **portacables** - [electr-instal] conduit
— f **portante** - [ambient] conveying duct
— f **prearmada** - [constr] pre-assembled pipe
— f — **de plancha(s)** f **estructural(es)** -
[constr] pre-assembled structural plate pipe
— f **precalentada** - [tub] preheated pipe
— f **pretensada** - [tub] prestressed pipe
— f **prevaciada** - [tub] precast pipe,ping
— f — **de hormigón** f - [tub] precast concrete
pipe,ping
— f **principal** - [tub] main (pipe) • [ambient]
main, duct, o run
— f — **de acero** m - [tub] steel main
— f — — — m **corrugado** - [tub] corrugated
steel main (pipe)
— f — — — m **parra cloaca** f - [sanit]
corrugated steel sewer main
— f — **para agua** m - [hidr] water main
— f — — **aire** m - [ind] air main
— f — — — m **caliente** - [metal-prod] hot
blast main
— f — — **conducción** f - [tub] large transmission line
— f — — **gas** n - [combust] gas main
— f — — **red** f - [tub] system main pipe
— f — — — f **de tubería(s)** f - [ambient]
duct system main run
— f — — **soplado** m - [metal-prod] blast main
— m — — m **para horno** m **alto** - [metal-prod]
blast main
— f — — **viento** m - [metal-prod] blast main
— f — **para viento** m **frío** - [metal-prod] cold
air, blast, o main
— f **producida** - [tub] fabricated pipe
— f — **en planta** f - [tub] shop-fabricated pipe
— f — — **taller** m - [tub] shop-fabricated pipe
— f **protegida** - [tub] protected, pipe, o tube
— f — **contra corrosión** f - [tub] corrosion
protected pipe
— f **proviniente de captador** m **de polvo** m -

[ambient] duct from @ dust collector
tubería f proyectada - [constr] designed barrel
— **f purgada** - [mec] discharged line
— **f que fluye llena** - [hidr] full flowing pipe
— **f quitada** - [tub] removed tubing
— **f ranurada** - [tub] grooved, o slotted, pipe
— **f recalcada** - [tub] upset tubing
— **f — exteriormente** - [tub] external(ly) upset tubing
— **f recia** - [tub] rugged pipe
— **f — de acero m** - [tub] rugged steel pipe
— **f recocida** - [tub] annealed pipe
— **f recolectora** - [petról] gathering (line)
— **f recta** - [tub] straight, pipe, o tube, o duct
— **f rectangular** - [tub] rectangular(ly shaped), pipe,ping, o tube,bing • [ambient] rectangular duct
— **f — con lado(s) m plano(s)** - [tub] flat side rectangular tube,bing x
— **f — conformada en frio** - [tub] cold formed rectangular, pipe, o tube,bing
— **f recubierta** - [tub] coated pipe,ping
— **f — con asfalto m** - [tub] asphalt coated pipe
— **f — con plástico m** - [tub] plastic coated, pipe,ping, o tube,bing
— **f — totalmente** - [tub] full(y) coated pipe
— **f redonda** - [tub] round, tube,bing, o pipe • [ambient] round duct
— **f — conformada en frío** - [tub] cold formed round tubing
— **f — de acero m** - [tub] steel round tubing
— **f — m aluminizado** - [tub] aluminized steel round tubing
— **f — — m soldado** - [tub] welded aluminized steel round tubing
— **f — — m con filete m sin rebabar** - [tub] not deburred flash-in welded aluminized steel round tubing
— **f — — m eléctricamente** - [tub] electric(ally) welded aluminized steel round tubing
— **f — — — con filete m interior** - [tub] flash-in electric(ally) welded aluminized steel round tubing
— **f — — m — — m sin rebabar** - [tub] not deburred flash in electric(ally) welded aluminized steel round tubing
— **f — — m laminada** - [tub] rolled steel round tubing
— **f — — — m en caliente** - [metal-fabr] hot rolled steel round tubing
— **f — — — m — frío** - [tub] cold rolled steel round tubing
— **f — — — m — — soldado** - [tub] welded cold rolled steel round tubing
— **f — — m — — con filete interior** - [tub] flash-in welded cold rolled steel round tubing
— **f — — — — — — m — sin rebabar** - [tub] not deburred flash-in welded cold rolled steel round tubing
— **f — — m soldada eléctricamente** - [tub] electric(ally) welded steel round tubing
— **f — — m soldado con filete m interior** - [tub] round flash-in welded steel tubing
— **f — o abovedada** - [tub] round, o pipe arch, tubing
— **f — — — encajable** - [metal-fabr] nestable pipe and pipe arch
— **f — — intercambiable** - [metal-fabr] interchangeable nuestable pipe and pipe arch
— **f — que fluya llena** - [hidr] full flowing round pipe
— **f — remachada** - [metal-fabr] riveted round pipe
— **f — soldada** - [metal-fabr] welded round, pipe, o tubing
— **f — — con filete m interior** - [metal-fabr] flash-in welded round tubing
— **f reducida** - [metal-fabr] reduced tube,bing]
— **f — por estiramiento m** - [tub] stretched reduced pipe
— **f — — — m en caliente** [tub] hot stretched reduced pipe

tubería f reducida por estiramiento m en continuo - continuous(ly) stretch reduced pipe
— **f — — — m — — para industria f petrolera** - [tub] continuous(ly) stretch(ed) reduced oil country pipe
— **f reemplazada** - [tub] replaced, pipe,ping, o tube,bing, o line
— **f reforzada** - [tub] reinforced, tube,bing, o pipe,ping
— **f — con diámetro m grande** - [tub] large diameter reinforced pipe
— **f — de hierro m negro** - [tub] heavy duty black steel pipe
— **f rellenada** - [constr] backfilled pipe
— **f — con hormigón m** - [tub] concrete filled pipe
— **f remachada** - [tub] riveted pipe • stove pipe
— **f — de acero m** - [tub] riveted steel pipe
— **f — — m corrugado** - [tub] riveted corrugated steel pipe
— **f — — m — con galvanización f** - [tub] galvanized corrugated steel pipe
— **f — — — m corrugado — — f simple** - [metal-fabr] plain galvanized corrugated steel riveted pipe
— **f — — — m — prefabricada** - [metal-fabr] prefabricated riveted corrugated steel pipe
— **f — por punto(s) m** - [tub] spot welded pipe
— **f — prefabricada** - [metal-fabr] prefabricated riveted pipe
— **f — de acero m corrugado** - [metal-fabr] prefabricated riveted corrugated steel pipe
— **f removida** - [tub] removed, tubing, o pipe
— **f revestida** - [tub] lined pipe • coated, o paved, pipe
— **f — con asfalto m** - [metal-fabr] asphalt lined pipe
— **f — cemento m** - [tub] cement lined pipe
— **f — — (material) plástico m** - [tub] plastic coated, tube,bing, o pipe,ping
— **f — mortero m de cemento m** - [tub] cement mortar lined pipe
— **f — con (material) plástico m** - [tub] plastic coated, pipe,ping, o tube,bing
— **f — interiormente** - [tub] internally lined pipe
— **f — totalmente** - [tub] full(y), lined, o paved, pipe
— **f — y pavimentada** - [tub=fabr] coated and paved pipe
— **f rígida** - [tub] rigid pipe
— **f — sin armar** - [tub] non-reinforced rigid pipe • unassembled rigid pipe
— **f — sin reforzar** - [tub] unreinforced rigid pipe
— **f rodeada** - [constr] surrounded pipe
— **f rotante** - [mec] rotating pipe
— **f sacada** - [tub] removed tubing
— **f sanitaria** - [sanit] sanitary tubing
— **f seca** - dry, pipe, o tubing
— **f secada** - [tub] dried, pipe, o tubing
— **f secundaria** - [ambient] secondary tubing; submain
— **f según medida f** - [tub=fabr] tailor-made pipe,ping
— **f — norma** - [tub] standard pipe,ping
— **f serpentín** - [tub] pipe coil • coil pipe
— **f — para calentamiento m** - [tub] heating pipe coil
— **f sin aislar** - [ambient] uninsulated duct
— **f — armar v** - [tub] unassembled pipe • unreinforced pipe
— **f — costura f** - [tub] seamless, tube, o pipe
— **f — — f con aleación f baja** - [tub] seamless low alloy tubing
— **f — — f — — f baja y resistencia f alta (a tensión f)** - [tub] seamless high strength low alloy tubing
— **f — — f — espesor m grande de pared f** - [tub] thick wall seamless pipe
— **f — — f con pared f gruesa** - [tub] thick wall seamless pipe

tubería f sin costura f de latón m rojo - [tub]
seamless red brass pipe
— f — f extrarreforzado - [tub] thick wall
seamless pipe • schedule . . . pipe
— f — f de acero m austenítico - [tub]
seamless austenitic steel pipe
— f — — m con aleación f - [tub]
seamless alloy steel, pipe, o tubing
— f — f — m carbono m - [tub] seamless carbon steel pipe
— f — f — m estirado - [tub]
drawn seamless steel, pipe, o tubing
— f — — m en caliente—[metal-lam]
hot drawn seamless steel tubing
— f — — m — frío - [metal-lam]
cold drawn seamless steel tubing
— f — — m inoxidable - [tub] seamless
stainless steel tubing
— f — — m inoxidable austenítico -
seamless austenitic stainless steel pipe
— f — f — m para temperatura(s) f al-
ta(s) - [tub] seamless high temperature
steel pipe
— f — f — m — f baja(s) - [tub]
seamless low temperature steel pipe
— f — f para hincar v - [tub] seamless
drive,ving pipe
— f — f — uso(s) m con temperatura(s) f
alta(s) - [tub] seamless high temperature
service pipe
— f — f — m — baja(s) - [tub] low
temperature seamless service pipe
— f — f — — f elevada(s) - [tub]
seamless high temperature service pipe
— f sin enterrar v - [constr] unburied pipe
— f — expandir v - [tub] unexpanded pipe
— f — muro m para cabecera f - [constr] pipe
without @ headwall
— f — perforar v - [tub] unperforated pipe,ping
— f — presión f [ind] pressureless pipe,ping
— f — recubrir v - [tub] uncoated pipe,ping
— f — reforzar v - [tub] unreinforced pipe
— f Smooth-Flo - [tub] Smooth-Flo pipe,ping
— f — con liga f de asbesto m (Asbestos
Bonded) - [tub] Asbestos-Bonded Smooth-Flo
pipe,ping
— f — para cloaca(s) f - [metal-fabr]
Smooth-Flo sewer pipe
— f —Lock - [metal-fabr] Smooth-Lock pipe
— f socavada - [constr] undermined pipe
— f soldada - [sold] welded, pipe, o tube,bing
— f — con aleación f - [sold] alloy welded
tube,bing, o pipe
— f — — f alta - [tub] high alloy steel,
tube,bing, o pipe,ping
— f — — f baja - [tub] low alloy welded,
tube,bing, o pipe,ping
— f — — f baja con tratamiento m térmico -
[tub] heat treated alloy welded tube,bing
— f — calidad f . . . - [tub] . . ., grade,
o quality, welded pipe
— f — capa f aplicada en fábrica - [tub]
(black) mill coated welded pipe
— f — — f porotectora - [tub] coated welded pipe
— f — — f aplicada en fábrica - [tub]
(black) mill coated welded pipe
— f — costura f - seam welded pipe
— f — — f en espiral m - [tub] spiral seam
welded pipe
— f — — f recta - [tub] straight seam
welded pipe
— f — — filete m interior - [tub] flash-in
welded tubing
— f — — m — redondeado - [tub] rounded
flash-in welded tubing
— f — solapo m - [tub] lap welded pipe
— f — corriente - [tub] standard welded pipe
— f — de acero m - [tub] welded steel, tubing,
o pipe
— f — — m austenítico - [tub] austenitic
steel welded tube,bing
— f — de acero m - [tub] welded steel tubing

tubería f soldada de acero m con aleación f -
[tub] welded alloy steel, pipe, o tube,bing
— f — — m — f baja (a tensión f) -
[tub] high-strength low alloy welded steel,
pipe, o tubing
— f — — — carbono m - [tub] carbon
steel welded, o welded carbon steel, pipe
— f — — m — filete m interior - [tub]
flash-in welded steel tubing
— f — — m — junta f con espiga f y campana f - [tub] welded bell and spigto steel
pipe
— f — — m — f — f — f con
guarnición f de caucho m - [tub] welded
Stab-Joint steel pipe
— f — — — m — f tipo Dresser - [tub]
welded Dresser-coupled steel pipe
— f — — m corrugado - [tub] welded corrugated steel pipe
— f — — m inoxidable - [tub] stainless
steel welded, o welded stainless steel,
pipe, o tubing
— f — — — austenítico - [tub] welded
austenitic stainless steel, pipe, o tubing
— f — — m laminada en caliente - [tub]
hot rolled steel welded tubing
— f — — m — frío - [tub] cold rolled
steel welded, pipe, o tubing
— f — — m — frío con filete m interior - [tub] flash-in welded cold rolled
steel tubing
— f — — m para agua m - [tub] welded
steel water pipe
— f — — m revestida - [tub] lined welded
steel, pipe, o tubing
— f — — — con mortero m - [tub] mortar
lined welded steel, pipe, o tubing
— f — — — m de cemento m - [tub]
cement mortar lined welded steel pipe
— f — — — m — f con junta f
de espiga f y campana f con guarnición f de
caucho m - [tub] cement mortar lined Stab-
Joint welded steel pipe
— f — eléctricamente - [tub] electric(ally)
welded tubing
— f — en espiral - [tub] spiral welded pipe
— f — — con capa f aplicada en fábrica f
- [tub] (black) mill coated spiral welded
pipe
— f — — — — f protectora aplicada en
fábrica - [tub] (black) mill coated spiral
welded pipe
— f — — horno m - [tub] furnace welded pipe
— f — galvanizada - [tub] galvanized welded,
o welded galvanized, pipe, o tubing
— f — — corriente - [tub] standard welded
galvanized, pipe, o tubing
— f — — para revestimiento m de pozo(s) m -
[tub] welded galvanized well casing
— f — — — m — — m para agua m - [tub]
welded galvanized water well casing
— f — laminada f - [tub] rolled welded tubing
— f — — en caliente - [tub] hot rolled welded tubing
— f — — — frío - [tub] cold rolled welded
tubing
— f — helicoidalmente - [tub] spiral, o heli-
cal(ly) welded, pipe
— f — para pilote(s) m - [tub] Hel-Cor
pile shell
— f — para agua m - [tub] welded water pipe
— f — — pilote(s) m - [tub] welded pile
pipe,ping
— f — revestimiento m para pozo(s) m -
— f — — m — m para agua m - [tub]
welded water well, o water well welded, casing
— f — por fusión f - [tub] fusion welded pipe
— f — — — f eléctrica - [sold] electric
fusion welded pipe
— f — — f para conducción f - [tub] fusion
welded line pipe
— f — — puntos m - [tub] spot welded pipe

tubería f soldada por puntos m por resistencia f
— - [tub] resistance spot-welded pipe,ping
— f — por resistencia f (eléctrica) - [tub] (electric) resistance welded pipe
— f — — para uso(s) m con temperatura(s) f baja(s) - [tub] electric resistance welded low temperature service pipe
— f — — f — reducia por estiramiento m en caliente - [tub] electric resistance welded hot-stretch-reduced tube,bing
— f — Smooth-Flo con interior m liso - [tub] welded seam Smooth-Flo pipe
— f soportada - [constr] supported pipe
— f sostenida - [constr] supported pipe • cushioned pipe
— f Stab-Joint con espiga f y campana f - [tub] bell and spigot joint pipe
— f — — — — f — — con guarnición f de caucho m - Stab-Joint pipe
— f subacuática - [tub] underwater, o subaqueous, pipe, o tube, o line
— f — para agua m - [tub] underwater, o subcqueous, water, pipe, o line
— f subterránea f - [tub] underground, o subsurface, pipe, o line
— f — para agua m - [tub] underground water, pipe, o line
— f — — cloaca f - [tub] underground sewer (line)
— f sucia - [tub] dirty pipe
— f — extraída - [petról] pulled up dirty pipe
— f sud - [constr] south, pipe, o line
— f sumergida - [tub] submerged, o underwater. pipe, o tube, o line
— f superior - [tub] upper, pipe, o line
— f suspendida - [constr] suspended pipe,ping; pipe bridge
— f tapada - covered pipe • plugged pipe
— f terminada - [tub] final, o finished, pipe
— f totalmente portante - [ambient] fully conveying duct
— f transformada - [tub] transformed tubing
— f transportada - [tub] transported, o conveyed, pipe, o tube,bing
— f — sobre patín(es) m - [transp] pipe, o tubing, transported, o conveyed, on @ skid(s)
— f trasladada - [tub] transfered, o conveyed, pipe
— f tratada térmicamente - [metal-trat] heat treated, pipe, o tube,bing
— f triple - [constr] tiple pipe
— f troncal - [tub] trunk pipeline
— f tubular - [tub] tubular pipe(line)
— f única - [constr] single pipe,ping
— f unida - [tub] united, o joined, pipe(s)
— f uniforme - [tub] uniform pipe,ping
— f vacía - [tub] empty pipe,ping
— f vacía - [tub] empty pipe
— f vaciada - [tub] emptied pipe
— f vertical - [tub] vertical, pipe, o unit • (pipe) riser • [ambient] vertical duct(work) • [petról] stand pipe
— f — corrugada - [tub] vertical, o riser, corrugated pipe
— f — — de acero m - [tub] corrugated, vertical, o riser, steel pipe
— f — metálica - [constr] metal, vertical, o riser, pipe
— f — — corrugada - [constr] corrugated metal, vertical, o riser, pipe
— f — para alimentación f - [petról] stand pipe • vertical feed(ing) pipe
— f — típica - [constr] typical (pipe) riser
— f vitrificada - [tub] vitrified, tube, o pipe • clay pipe,ping
— f y tubería f abovedada - [tub] pipe and pipe arch
— f — — f — encajable - [metal-fabr] nestable pipe and pipe arch
— f — — f — — de acero m - [metal-fabr] nestable steel pipe and pipe arch
— f — — f — — — m con cobre m - [metal-fabr] copper bearing steel pipe and pipe arch

tubero m - [tub] pipefitter; pipe man
tubo m - . . . . • round pipe • cylinder • lamp, tube, o chimney • [mec] spout • [comb.int] véase cámara f
— m a atomizador m - [Comb.int] tube to @ atomizer
— m — — m reemplazado - [tub] replaced tube to @ atomizer
— m — válvula f - [tub] tube to @ valve
— m — — f reemplazado - [tub] replaced tube to @ valve
— m abocinado - [mec] flared, tube, o pipe
— m abovedado - [metal-fabr] pipe arch
— m — con liga f de asbesto m - [metal-fabr] asbestos-bonded pipe-arch
— m — con fondo m pavimentado - [metal-fabr] paved-invert pipe-arch
— m — continuo - [metal-fabr] continuous pipe-arch
— m — encajable - [metal-fabr] nestable pipe-arch
— m — corrugado - [metal-fabr] corrugated steel nestable pipe-arch
— f — engrampado - [metal-fabr] stitch joined pipe-arch (tube)
— m — flexible - [metal-fabr] flexible pipe-arch
— m — galvanizado - [metal-fabr] galvanized pipe-arch
— m — — corriente - [metal-fabr] plain galvanized pipe-arch
— m — Hel-Cor - [metal-fabr] Hel-Cor pipe arch
— m — — plancha(s) f múltiple(s) m - [metal-fabr] Multi-Plate pipe-arch
— m — Multi-Plate - [metal-fabr] Multi-Plate pipe-arch
— m — remachado - [metal-fabr] riveted pipe-arch
— m abridado - [hidr] thimble
— m acanalado - [metal-fabr] corrugated pipe
— m — encajable - [metal-fabr] nestable corrugated, pipe, o tube
— m — — intercambiable - [metal-fabr] nestable interchangeable corrugated pipe
— m — helicoidalmente - [metal-fabr] Hel-Cor, o helically corrugated, pipe
— m aceitador - [mec] oiling, o oiler, pipe
— m — inferior - [mec] lower oiler pipe
— m — superior - [mec] upper oiler pipe
— m aislador m - [tub] insulating tube
— m ascendente - [tub] ascending, o rising, pipe
— m celantado - [metal-fabr] heated tube
— m calorifugado - véase tubo m calorizado
— m calorizado - [tub] calorized* pipe
— m cargado - [tub] loaded pipe
— m central - [mec] central, tube, o pipe
— m — de papel m endurecido - [electr-equip] hard paper central tube
— m circular - [tub] circular, o round, pipe
— m colador - [tub] screen, o straining, pipe, o liner
— m colector - [mec] collecting,tor pipe • head receiver • header
— m completo - [tub] complete pipe
— m con aleación f - [tub] alloy, pipe, o tube
— m — campana f y enchufe m - [tub] bell and spigot pipe
— m — campana f y espiga f - [tub] bell and spigot pipe
— f — corrugación(es) f - [tub] corrugated pipe
— f — — f helicoidal(es) - [metal-fabr] Hel-Cor, o helically corrugated, pipe
— m — — f — revestida interiormente - [metal-fabr] Hel-Cor, o helically corrugated, lined pipe
— f — — f interiores cubiertas - [metal-fabr] Smooth-Flo pipe
— m con costura f - [tub] welded pipe
— m — — f engargolada - [metal-fabr] lock seam tube,bing

tubo <u>m</u> con costura <u>f</u> engatillada <u>f</u> - [metal-
-fa<u>b</u>r] lock seam tubing
— <u>f</u> con diámetro <u>m</u> mayor - [tub] larger diameter pipe
— <u>m</u> — — <u>m</u> reducido—[tub] small diameter pipe
— <u>m</u> — — extremo <u>m</u> recalcado - [petról] upset end
<u>p</u>ipe
— <u>m</u> — — <u>m</u> exteriormente ᵧ [petról] exter-
na<u>l</u>(ly) upset end pipe
— <u>m</u> — fondo <u>m</u> pavimentado - [metal-fabr] paved
<u>i</u>nvert pipe
— <u>m</u> — grasa <u>f</u> - [lubr] grease tube
— <u>m</u> — liga <u>f</u> de asbesto <u>m</u> - [metal-fabr] asbestos bonded pipe
— <u>m</u> — ondulación <u>f</u> helicoidal - [hidr] Hel-Cor
<u>p</u>ipe
— <u>m</u> — pared <u>f</u> delgada - [tub] thin wall, tube,
<u>o</u> pipe
— <u>m</u> — — <u>f</u> doble - [tub] double wall pipe
— <u>m</u> — — <u>f</u> fina - [metal-fabr] thin wall pipe
— <u>f</u> — — <u>f</u> gruesa - [tub] thick wall(ed) pipe
— <u>f</u> — presión <u>f</u> - [tub] pressurized pipe
— <u>m</u> — recubrimiento <u>m</u> - [tub] coated pipe
— <u>m</u> — — <u>m</u> bituminoso—[metal-fabr] bituminous
coated pipe
— <u>m</u> — resistencia <u>f</u> alta a tensión <u>f</u> - [metal-
-fabr] high (tensile) strength pipe
— <u>m</u> — — <u>f</u> corriente - [electrón] standard resistance tube,bing
— <u>m</u> — rosca <u>f</u> - [metal-fabr] threaded, <u>o</u>
tapped, pipe
— <u>m</u> — — <u>f</u> cónica - [tub] tapered tap pipe
— <u>m</u> — sección <u>f</u> circular - [metal-fabr] (full)
round pipe
— <u>m</u> — — <u>f</u> circular para drenaje <u>m</u> - [metal-
-fabr] full-round drainage pipe
— <u>m</u> — titanio <u>m</u> - [metal-fabr] titanium tube
— <u>f</u> — — <u>m</u> para hervidor <u>m</u> - [nucl] boiler
titanium tube
— <u>m</u> conductor - [sold] (conductor) tube, <u>o</u> pipe
— <u>m</u> — para boquilla <u>f</u> - [sold] nozzle conductor tube
— <u>m</u> conectado - [mec] connected tube
— <u>m</u> corriente - [electrón] standard tube
— <u>m</u> corrugado - [metal-fabr] corrugated pipe
— <u>m</u> — encajable - [metal-fabr] nestable
corrugated pipe
— <u>m</u> — — intercambiable - [metal-fabr] nestable interchangeable corrugated pipe
— <u>m</u> — helicoidalmente - [metal-fabr] Hel-cor,
<u>o</u> helically corrugated, pipe
— <u>m</u> — — con corrugación(es) <u>f</u> interior(es)
cubierta(s) - [metal-fabr] Hel-Cor Smooth-Flo
pipe
— <u>m</u> curvado - [mec] curved, <u>o</u> U-, tube
— <u>m</u> — para intercambiador <u>m</u> (para calor <u>m</u>) -
[metal-fabr] heat exchanger U-tube
— <u>m</u> curvo - [tub] U-tube
— <u>m</u> de acero <u>m</u> - [mec] steel, tube, <u>o</u> pipe •
[constr-poil] steel shell
— <u>m</u> — — acanalado - [tub] corrugated steel
tube, <u>o</u> pipe
— <u>f</u> — — <u>m</u> para revestimiento <u>m</u> interior -
[tub] corrugated steel liner (pipe)
— <u>m</u> — — <u>m</u> austenítico - [tub] austenitic
steel tube
— <u>f</u> — — — común - [tub] plain steel pipe
— <u>m</u> — — — con carbono <u>m</u> - [tub] plain carbon steel pipe
— <u>f</u> — — <u>m</u> con carbono <u>m</u> - [tub] carbon steel,
tube, <u>o</u> pipe
— <u>n</u> — — <u>m</u> — —-molibdeno - [tub] carbon-mo-
libdenum steel tube
— <u>m</u> — — <u>m</u> para caldera(s) <u>f</u> - [tub]
carbon steel boiler tube
— <u>m</u> — — <u>m</u> costura <u>f</u> - [tub] welded steel
pipe
— <u>m</u> — — <u>m</u> con extremo <u>m</u> cerrado—[constr-pil]
closed end steel pipe
— <u>m</u> — — <u>m</u> corriente - [tub] plain steel pipe
— <u>m</u> — — — <u>m</u> con carbono <u>m</u> - [tub] plain carbon steel pipe
— <u>m</u> — — <u>m</u> corrugado - [tub] corrugated steel,

tube, <u>o</u> pipe
tubo <u>m</u> de acero <u>m</u> corrugado con liga <u>f</u> de asbesto <u>m</u> - [metal-fabr] asbestos bonded corrugated steel pipe
— <u>m</u> — — <u>m</u> — con pared(es) <u>f</u> delgada(s) -
[metal-fabr] thin wall(ed) corrugated steel
pipe
— <u>m</u> — — — (y) galvanizado - [metal-fabr]
galvanized corrugated steel pipe
— <u>m</u> — — <u>m</u> — helicoidalmente - [metal-fabr]
helically corrugated steel, pipe, <u>o</u>
tube
— <u>m</u> — — <u>m</u> — para revestimiento <u>m</u> (interior)
- [metal-fabr] corrugated steel liner pipe
— <u>m</u> — — <u>m</u> en espiral para semilla(s) <u>f</u> -
[agric-equip] steel ribbon grain tube
— <u>m</u> — — <u>m</u> encarrujado - [metal-fabr] corrugated steel pipe
— <u>m</u> — — <u>m</u> — para revestimiento <u>m</u> (interior)
- [metal-fabr] corrugated steel liner (pipe)
— <u>m</u> — — <u>m</u> hueco - [tub] hollow steel tube
— <u>m</u> — — <u>m</u> — obturado - [tub] capped, <u>o</u> clogged,
hollow steel tube
— <u>m</u> — — <u>m</u> inoxidable - [tub] stainless steel,
tube, <u>o</u> pipe
— <u>m</u> — — <u>m</u> — obturado - [tub] capped, <u>o</u>
clogged, stainless steel, tube, <u>o</u> pipe
— <u>m</u> — — <u>m</u> — austenítico - [tub] austenitic
stainless steel, tube, <u>o</u> pipe
— <u>m</u> — — <u>m</u> obturado - [tub] capped, <u>o</u> clogged,
(hollow) steel, tube
— <u>m</u> — — <u>m</u> ondulado - [metal-fabr] corrugated
steel, tube, <u>o</u> pipe
— <u>m</u> — — <u>m</u> para revestimiento <u>m</u> (interior)
- [metal-fabr] corrugated steel liner (pipe)
— <u>m</u> — — <u>m</u> para filtro <u>m</u> - [sold] steel filter
tube
— <u>m</u> — arcilla <u>f</u> - [tub] clay, <u>o</u> soil, pipe
— <u>m</u> — barro <u>m</u> - [tub] tile
— <u>m</u> — — <u>m</u> cocido - [tub] soil, <u>o</u> terracota,
pipe
— <u>m</u> — caucho <u>m</u> - [plást] rubber tube
— <u>m</u> — cobre <u>m</u> - [mec] copper, tube, <u>o</u> pipe
— <u>m</u> — — <u>m</u> soldado - [tub] soldered copper,
tube, <u>o</u> pipe
— <u>m</u> — fundición <u>f</u> - [tub] cast iron pipe
— <u>m</u> — hierro <u>m</u> fundido - [tub] cast iron pipe
— <u>f</u> — hormigón <u>m</u> - [constr] concrete pipe
— <u>m</u> — — <u>m</u> con pared <u>f</u> fina - [constr] thin
walled concrete pipe
— <u>m</u> — — — — <u>f</u> gruesa - [constr] thick
walled concrete pipe
— <u>m</u> — medio punto <u>m</u> - [tub] half circle pipe
— <u>m</u> — metal - [tub] metal tube,bing
— <u>m</u> — — <u>m</u> para agua <u>m</u>—[tub] metal water pipe
— <u>m</u> — papel <u>m</u> - [papel] paper tube
— <u>m</u> — — — <u>m</u> endurecido - [papel] hard(ened)
paper tube
— <u>m</u> — Pitot - [petról] Pitot tube
— <u>m</u> — plancha(s) <u>f</u> - [constr] plate(s) pipe
— <u>m</u> — — ᵣ estructural(es) <u>m</u> - [metal-fabr]
structural plate(s) pipe
— <u>f</u> — — <u>f</u> múltiple(s) - [metal-fabr] Multi-
-Plate (pipe)
— <u>m</u> — titanio - <u>m</u> - [mec] titanium tube
— <u>m</u> — — <u>m</u> para hervidor <u>m</u> - [cald] titanium
boiler tube
— <u>m</u> delgado - [tub] thin, tube, <u>o</u> pipe
— <u>m</u> descendente - [metal-prod] downcomer
— <u>m</u> desconectado - [mec] disconnected tube
— <u>m</u> difusor <u>m</u> - [hidr] diffuser,sing tube
— <u>m</u> — movible - [hidr] swing diffuser
— <u>m</u> distribuidor - [mec] distributing tube •
manifold
— <u>m</u> electrónico - [electrón] electron tube •
vacuum tube
— <u>m</u> — para recepción <u>f</u> - [electrón] receiving
(type) electron tube
— <u>m</u> elevador <u>m</u> - [agric-equip] blower tube
— <u>m</u> emparvador - [agric-equip] stacker pipe
— <u>m</u> en espiral - [tub] ribbon, <u>o</u> gooseneck,
tube
— <u>f</u> — — <u>m</u> de acero <u>m</u> - steel ribbon tube

tubo m̲ en espiral para semilla(s) f̲ - [agric-
-equip] ribbon grain tube
— m̲ — extremo m̲ - [mec] end tube
— m̲ — — m̲ de pistola f̲ - [sold] gun end tube
— m̲ — — m̲ posterior - [sold] back end tube
— m̲ — — m̲ — de pistola f̲ - [sold] gun back
end tube
— n — — m̲ — — — f̲ soldadora - [sold] welding gun back end tube
— m̲ — lingote m̲ - [metal-prod] véase rechupe m̲
en lingote m̲
— m̲ — movimiento m̲ - [tub] moving pipe
— m̲ — S - [tub] (pipe) offset
— m̲ en sección(es) f̲ - [metal-fabr] sectional
pipe
— m̲ — — f̲ corta(s) - [metal-fabr] short sectional pipe
— m̲ encachado - [metal-fabr] paved invert pipe
— m̲ encajable - [metal-fabr] nestable pipe
— m̲ — con brida(s) f̲ - flanged (type) nestable
pipe
— m̲ — de acero m̲ - [metal-fabr] nestable steel
pipe
— m̲ — — — m̲ acanalado - [metal-fabr] nestable corrugated steel pipe
— m̲ — — — m̲ corrugado - [metal-fabr] nestable corrugated steel pipe
— m̲ — — — m̲ ondulado - [metal-fabr] nestable
corrugated steel pipe
— m̲ — intercambiable - [metal-fabr] interchangeable nestable pipe
— m̲ — — acanalado - [metal-fabr] nestable interchangeable corrugated pipe
— m̲ — — corrugado - [metal-fabr] nestable interchangeable corrugated pipe
— f̲ — — de acero m̲ ondulado,- [metal-fabr]
nestable interchangeable corrugated steel pipe
— m̲ — — encarrujado - [metal-fabr] nestable
interchangeable corrugated pipe
— m̲ — — ondulado - [metal-fabr] nestable interchangeable corrugated pipe
— m̲ encojido - [tub] shrunk tube
— m̲ — térmicamente - [tub] heat shrunk tube
— m̲ endurecido - [tub] hardened tube
— m̲ — de papel m̲ - [tub] hardened paper tube
— m̲ ensayado - [tub] tested tube
— m̲ entramado - [tub] truss, tube, o pipe
— m̲ estabilizado - [tub] stabilized, tube, o
pipe
— m̲ estriado - [tub] rifled pipe
— m̲ filtro - [tub] filter pipe • screen liner
— m̲ flexible - [tub] flexible pipe • loom
— m̲ — de acero m̲ - [tub] flexible steel pipe
— m̲ — — — m̲ acanalado - [tub] flexible corrugated steel pipe
— m̲ — — — m̲ corrugado - [tub] flexible corrugated steel pipe
— m̲ — — — m̲ encarrujado - [tub] flexible
corrugated steel pipe
— m̲ — — — m̲ ondulado - [tub] flexible corrugated steel pipe
— m̲ fotoeléctrico - [electrón] photoelectric
tube; phototube
— m̲ frangible - [mec] frangible tube
— m̲ fuerte - [tub] strong, tube, o pipe
— m̲ galvanizado - [metal-fabr] galvanized,
tube, o pipe
— m̲ — remachado - [metal-fabr] riveted galvanized, tube, o pipe
— m̲ gemelo - [constr] twin, tube, o pipe
— m̲ guía(dor) - [mec] guide,ding, tube, o pipe
— m̲ — para entrada - [mec] incoming guide tube
— m̲ — — salida f̲ - [mec] outgoing guide tube
— m̲ — para alambre m̲ - [sold] wire guide tube
— m̲ — para cabeza f̲ - [trefil] head guide,ding
tube
— m̲ — — — f̲ para extrusión f̲ - [trefil] extrusion head guide,ding tube
— m̲ — — entrada f̲ - [mec] incoming guide tube
— m̲ hacinador - [agric-equip] stacking pipe
— m̲ Hel-Cor - [metal-fabr] Hel-Cor pipe
— m̲ — revestido (interiormente) - [metal-
-fabr] Hel-Cor lined pipe

tubo m Hel-Cor Smooth-Flo - [metal-fabr] Hel-
-Cor Smooth-Flo pipe
— m̲ hincado - [constr-pil] driven pipe
— m̲ — en suelo m̲ - [constr-pil] ground driven
pipe
— m̲ indicador - [tub] indicator, pipe, o tube
— m̲ — para nivel m̲ - [metal-prod] level indicator tube
— m̲ inmovilizado - [tub] immobilized, o stabilized, tube, o pipe
— m̲ íntegro - [tub] entire pipe
— m̲ interior - [metal-prod] inner pipe
— m̲ inyector m̲ - [tub] jet pipe - [metal-prod]
grouting pipe
— m̲ limpiado - [tub] cleaned pipe
— m̲ — con aire m̲ - [tub] blown, o cleaned,
out tube
— m̲ — — aspiración f̲ - [tub] vacuum cleaned
pipe
— m̲ — — vacío m̲ - [tub] vacuum cleaned pipe
— m̲ limpio - [tub] clean, tube, o pipe
— m̲ lleno - [tub] full, o filled, tube
— m̲ — con fundente - [sold] flux filled pipe
— m̲ mandrilado - [tub] expanded, tube, o pipe
— m̲ montante - [tub] upright pipe
— m̲ Multi-Placa—[metal-fabr] Multi-Plate pipe
— m̲ Multi-Plate - [tub-fabr] Multi-Plate pipe
— m̲ — — de plancha(s) f̲ múltiple(s) - [metal-
-fabr] Multi-Plate pipe
— m̲ neumático - [neumát] pneumatic tube
— m̲ no protegido - [tub] unprotected tube
— m̲ obturado - [tub] capped, o clogged, tube
— m̲ ondulado - [metal-fabr] corrugated pipe
— m̲ — encajable - [metal-fabr] nestable corrugated pipe
— m̲ — (y) intercambiable - [metal-fabr]
nestable interchangeable corrugated pipe
— m̲ — helicoidalmente - [metal-fabr] Hel-Cor,
o helically corrugated, pipe
— m̲ original - [metal-fabr] original, o starting, tube
— m̲ oscilador - [electrón] oscillating,tor tube
— m̲ para aceite m̲ - [mec] oil, tube, o pipe
— m̲ — — m̲ y tapa f̲ - [mec] oil pipe and cover
— m̲ — admisión f̲ - [comb.int] intake, o inlet,
tube, o pipe
— m̲ — — f̲ para filtro m̲ - [comb.int] filter
inlet tube
— m̲ — — f̲ — — m̲ para aire m̲ - [comb.int]
air filter inlet tube
— m̲ — agua m̲ - [tub] water, tube, o pipe
— m̲ — — m̲ fría - [Tub] cold water, tube, o
pipe
— m̲ — — m̲ para nivel m̲ - [instrum] gage
water column
— m̲ — aire m̲ - air tube
— m̲ — — m̲ de caucho m̲ - [mec] rubber air tube
— m̲ — — m̲ — — m̲ vulcanizado - [ind] vul
canized rubber air tube
— m̲ — aislación f̲ - [tub] insulation tube
— m̲ — alambique m̲ - [tub] still tube
— m̲ — — m̲ para refinería f̲ - [petról] refinery still tube
— m̲ — alimentación f̲ - [ind] feed(er, o feeding, tube, o pipe
— m̲ — — f̲ para ácido m̲ - [ind] acid feed(er),
tube, o pipe
— m̲ — — f̲ — — m̲ sulfúrico - [ind] sulfuric
acid feed(er), tube, o pipe
— m̲ — — f̲ — anhidrido m̲ sulfuroso - [ind]
hydrogen sulfide feed(ing), tube, o pipe
— m̲ — — f̲ para combustible m̲ - [comb.int]
fuel feed(ing), pipe, o line, o tube
— m̲ — anhidrido m̲ sulfúrico - [ind] hydrogen
sulfide pipe
— m̲ — árbol m̲ para transmisión f̲ - [autom-mec]
(shaft) torque tube
— m̲ — aspiración f̲ -[ind] suction, tube, o
pipe • [comb.int] vacuum line
— m̲ — — f̲ para aire m̲ - [comb.int] air intake, tube, o pipe
— m̲ — avenamiento m̲ - [hidr] drainage pipe
— m̲ — bajada f̲ - [metal-prod] downcomer

tubo m para bajada f para agua m - [constr] downpipe; downspout
— m —— f para gas m - [metal-prod] (gas) downcomer
— m — boquilla f - [sold] nozzle tube
— m — burbuja(s) f - [ind] bubble(r) pipe • [petról] cracker pipe
— m — burbujeo m - [petról] cracker pipe
— m — caldera f - [cald] boiler tube (• boiler (flue) pipe
— m — calefacción f - heater,ting tube
— m —— f para agua m - water heater tube
— m —— f —— m para alimentación f - [cald] feed(ing) water heater tube
— m — calentamiento m - [cald] heater tube
— m — camisa f - [tub] shell tube
— m —— f de acero m corrugado - [metal-fabr] corrugated steel shell
— m —— f liso - [Constr] smooth pipe casing
— m —— f para pilote m - [metal-fabr] pile shell
— m —— f perdida - [tub] shell
— m —— f — corrugado - [metal-fabr] corrugated shell
— m —— f —— de acero m - [metal-fabr] corrugated steel shell
— m — cimentación f - [constr] foundation pipe
— m — colada f - [metal-prod] casting tube
— m —— f sumergido - [metal-prod] submerged casting tube
— m — condensador m - [ind] condenser tube
— m — conducción f para gas m - gas pipe
— m — conexión f - [tub] connecting,tion tube
— m — contrapeso m - [metal-prod] counterweight tube
— m — derrame m - [hidr] overflow pipe
— m — desagüe m - [hidr] discharge, tube, o pipe, o spout
— m —— m para madrastra f - [metal-prod] mantle discharge pipe
— m — descarga f - [tub] leg; spout • [hidr] discharge line
— m —— f para suciedad f - [tub] dirt leg
— m — desmontaje m - [mec] disassembly tube
— m — distribución f - [ind] distribution tube • manifold
— m — dragado m—[metal-fabr] dredge,ging tube
— m — drenado,naje - [hidr] drainage pipe
— m — eje m - [mec] shaft, o axle,
— m —— m trasero—[autom-mec] rear axle tube
— m — enfriamiento m - [metal-prod] chill
— m —— m con agua m - [sold] water cooling tube
— m —— entre parte f móvil y fija - [metal-prod] furnace and port end chill (tube)
— m — engrase m - grease,sing, tube, o pipe
— m —— m para tragante m - [metal-prod] furnace throat, grease,sing, o lubricating, tube
— m — entrada f [metal-prod] lead in tube • [comb.int] inlet, o intake, tube; véase también tubo m para admisión f
— m —— f para aceite m - [comb.int] oil inlet tube; inlet oil line
— m —— f — filtro m - [ind] filter inlet tube
— m —— f para filtro m para aire m - [comb.-int] air filter inlet tube
— m —— f — toberín m - [metal-prod] nozzle intake, tube, o pipe
— m —— f —— m para escorial—[metal-prod] slag notch nozzle intake, tube, o pipe
— m — escape m - [Comb.int] exhaust pipe • [autom] tail pipe
— m —— m para caldera f - [cald] boiler exhaust pipe
— m —— m —— f para recuperación f—[cald] recuperating boiler exhaust pipe
— m —— m —— f —— f para calor m - [cald] heat recuperation boiler exhaust pipe
— m — expulsión f - [cald] blowoff
— m — filtro m - [mec] filter tube
— m —— m para aire m - [comb.int] air filter tube

tubería f para fundente m - [sold] flux tube
— m — gas m - [combust] gas tube • [petról] gas bottle
— m — gasoducto m - [metal-prod] gas pipe • [petról] gas line pipe
— m — guía f - [mec] guide tube
— m —— f para entrada f - [mec] incoming guide tube
— m —— f — salida f - [mec] outgoing guide tube
— m —— f desde tolva f para alambre m hasta cabeza f para extrusión f - [extrusión de electrodos] wire hopper to extruding head guide tube
— m — henchimiento m—[comb.int] filler tube
— m —— m para aceite m - [comb.int] oil filler tube
— m — hervidor m - [nucl] boiler tube
— m — indicador m para Venturi m - [metal-prod] Venturi indicating,tor tube
— m — inmersión f - [petról] immersion, o dip, pipe
— m — intercambiador m [tub] interchanger, o exchanger, tube
— m —— m para calor m - [tub] heat exchanger tube
— m — lubricación f - [mec] lubrication, tube, o line
— m — llenado m - véase tubo m para henchimiento m
— m — nivel m - [instrum] level tube; gage, column, o glass
— m — oleoducto m - [metal-prod] oil tube • [petról] pipeline tube,bing
— m —— m y gasoducto m - [petról] oil and gas line pipe
— m — perforación f - [petról] drillng, o drive,ving, pipe
— m —— f bajado - [petról] lowered drilling pipe
— m —— f levantado - [petról] raised drilling pipe
— m — pilote m - [metal-fabr] pile shell
— m —— m compuesto - [metal-fabr] composite pipe shell
— m —— m de acero m - [metal-fabr] steel pipe shell
— m —— m para cimentación f [metal-fabr] foundation pipe
— m — piquera f - [metal-prod] iron notch pipe
— m — presión f - [tub] pressure pipe
— m — protección f - protection tube
— m — purga f - [hidr] discharge, line, o tube • [metal-prod] bleeder, o blow-off, tube
— m — rayo(s) m catódico(s) - [electrón] cathode ray tube
— m —— m — para televisión f - [electrón] video, o television, cathode ray tube
— m — rebose,samiento m - [hidr] overflow tube
— m — recipiente(s) m para presión f - [tub] pressure vessel tube
— m — red f - [tub] system pipe,ping
— m —— f para agua m - [tub] water system pipe
— m — refinería f - [ind] refinery tube
— m — refrigeración f - [ind] cooling tube
— m —— f exterior - [ind] external cooling pipe
— m —— f para coraza f - [metal-prod] external shell cooling pipe
— m — resistencia f - resistance tube
— m — respiración f - breather,thing tube • riser
— m — revestimiento m - [constr] casing
— m —— m para pozo m - [constr] well casing
— m — riego m - sprinkler, o spray, pipe
— m —— m superior - [metal-prod] top, o upper, spray pipe
— m —— m para cuba f - [metal-prod] top, o upper, furnace shell spray pipe
— m — salida f - [tub] outlet (tube)
— m —— f para aceite m - outlet oil tube
— m — sangría f - [tub] bleeder, tube, o pipe

tubo m para semilla(s) f   - 1656 -

tubo m para semilla(s) f - [agric-equip] grain tube
— m — soplado m - [metal-prod] blast pipe
— m — soporte m - [metal-lam] skid pipe
— m — subida f - [tub] ascending pipe • [constr] riser (unit)
— m — — f para gas m - [metal-prod] (gas) uptake
— m — succión f - [mec] suction pipe
— m — sujeción f - [mec] clamping tube
— m — tobera f - [metal-prod] tuyere pipe
— f — toberín m - [metal-prod] (slag) notch tube
— m — toma f - [metal-prod] uptake (tube)
— m — — f para gas m - [metal-prod] gas uptake; gas intake pipe
— m — transición f - transition tube
— m — vacío m—[comb.int] vacuum, line, o tube
— m — ventilación f - [tub] vent (tube)
— m — Venturi - [metal-prod] Venturi tube
— m pequeño - [tub] small, tube, o pipe
— m pesado - [constr] heavy, tube, o pipe
— m — de hormigón m - [constr] heavy concrete, tube, o pipe
— m perforado - [tub] perforated, tube, o pipe
— m portatestigo(s) - [petról] core barrel
— m prevaciado (de hormigón m) - [constr] precast (concrete) pipe
— m principal - [tub] main, tube, o pipe
— m — para ventilación f - [tub] main vent, tube, o pipe
— m protector - [tub] protection tube
— m — para termopar m - [combust] thermocouple protection tube
— m quitado - [tub] removed tube
— m radiante - [electrón] radiant tube
— m rayado - [tub] riffled pipe
— m recolector - [agric-equip] wiper tube
— m rectangular - [tub] rectangular tube
— m recubierto - [tub] covered, o coated, tube
— m — con material m bituminoso - [metal-fabr] bituminous coated, tube, o pipe
— m redondo - [tub] round tube
— m reemplazado - [tub] replaced tube
— m refrigerante - [ind] cooling tube
— m remachado - [metal-fabr] riveted pipe
— m — de acero m - [metal-fabr] riveted steel pipe
— m — — m acanalado - [metal-fabr] riveted corrugated steel pipe
— m — — m corrugado - [metal-fabr] riveted corrugated steel pipe
— m — — m encarrujado - [metal-fabr] riveted corrugated steel pipe
— m — — m ondulado - [metal-fabr] riveted corrugated steel pipe
— m removido - [tub] removed tube
— m resorte - [tub] spring, tube, o pipe
— m respirador - [tub] breather,thing tube
— m revestido - [tub-fabr] lined, tube, o pipe
— m — con liga f de asbesto m - [metal-fabr] asbestos-bonded, tube, o pipe
— m rígido - [tub] rigid, tube, o pipe
— m rociado—[metal-fabr] sprayed, tube, o pipe
— m roto - [tub] broken, tube, o pipe
— m sacado - [tub] removed. tube, o pipe
— m sacanúcleo(s) m - [petról] core barrel
— m — doble - [petról] double core barrel
— m secado - [tub] dried, tube, o pipe
— m seco - [tub] dry, tube, o pipe
— m según medida f - [metal-fabr] tailor made, o mode to size, tube, o pipe
— m semicircular - [tub] semicircle, o half circle, tube, o pipe
— m separador - [mec] separator, o spacer. tube
— m sin costura f - [metal-fabr] seamless, tube, o pipe
— f — — f de acero m - [metal-prod] steel seamless, o seamless steel, tube, o pipe
— m — — f — m con carbono m—[metal-prod] carbon steel seamless, tube, o pipe
— m — — f de acero m aleado - [metal-prod] alloy steel seamless, tube, o pipe

tubo m sin costura f de acero m común - [metal-prod] common steel seamless pipe
— f — — f de diámetro grande - [metal-fabr] large diameter seamless, tube, o pipe
— m — — f — m pequeño - [metal-fabr] small diameter seamless, tube, o pipe
— m — — f — hierro m común - [metal-prod] common iron seamless, tube, o pipe
— m — — f recto - [metal-fabr] straight seamless pipe
— m sin protección f - [tub] unprotected pipe
— m — soldadura f - [tub] weldless tube
— m Smooth-Flo - [tub]fabr] Smooth-Flo, tube, o pipe, o line
— m — de acero m corrugado - [metal-fabr] Smooth-Flo corrugated steel pipe
— m soldado - [metal-fabr] welded pipe
— m — de acero m - [metal-fabr] welded steel, tube, o pipe
— m — para agua m - [tub] welded water pipe
— m soplador - [agric-equip] blower pipe
— m sostenido - [mec] held pipe
— m sujetado - [tub] held pipe
— m terminado - [tub] finished, tube, o pipe
— m transportador - [ind] conveyor tube
— m unido - [tub] joined, tube, o pipe
— m uniforme - [tub] uniform, tube, o pipe
— m vertedero - [tub] tube, o pipe, o spillway
— m — corrugado - [hidr] corrugated pipe spillway
— m — — de acero m - [hidr] steel corrugated pipe spillway
— m — — — m corrugado - [hidr] corrugated steel pipe spillway
— m vertical - [tub] vertical, tube, o pipe • [hidr] standpipe
— m — grande - [tub] large vertical pipe
— m — pequeño - [tub] small vertical pipe
— m — supletorio - [hidr] standpipe
— m viga - [metal-lam] tube-rail
tubos m y accesorios m - [metal-prod] pipes and fittings
tubular a - . . .; cylindrical
— a acodado - [tub] elbow tube
— a triple - [tub] triple tubular
tuerca f abocinada - [mec] flared nut
— f acampanada - [mec] flared nut
— f acanalada - [mec] slotted nut
— f afianzada - [mec] locked nut
— f aflojada - [mec] loose(ned) nut
— f ajustada - [mec] adjusted, o tight(ened) nut
— f — en mas - [mec] overtightened nut
— f — — menos - [mec] undertightened nut
— f — excesivamente - [mec] overtightened nut
— f — hasta momento m torsional de . . . - [mec] nut torqued to . . .
— f — — par m motor - [mec] torqued nut
— a ajustadora - [mec] adjuster,ting nut
— f — para empaquetadura f - [petról] packing adjuster,ting nut
— f almenada - [mec] castle, o slotted, nut • castellated nut
— f ancha - [mec] wide, o full height, nut
— f angosta - [mec] narrow, o half height, nut
— f apretada - [mec] tightened, o torqued, nut
— f — ajustadamente - [mec] snugly tightened nut
— f — en más - [mec] overtightened nut
— f — — menos - [mec] undertightened nut
— f — excesivamente - [mec] overtightened nut
— f — hasta momento m torsional de . . . • [mec] not torqued to . . .
— f — — par m motor correcto - [mec] nut tightened to @ correct torque
— f — — — m — de - [mec] nut torqued to . .
— f apropiada - [mec] appropriate, o suitable, nut
— f atascada - [mec] jammed nut
— f autotrabadora - [mec] self locking nut
— f — ancha - [mec] wide, o full height, self-locking nut
— f — instalada - [mec] installed self-locking nut

tuerca f autotrabadora para piñón m - [mec]
   self-locking pinion nut
— f autotrabante - [mec] self-locking nut
— f — para piñón m - [mec] self-locking pinion
   nut
— f cadmiada - [mec] cadmium plated nut
— f castillo n - [mec] véase tuerca f ranurada,
   o almenada
— f cementada (en caja f) - [mec] casehardened
   nut
— f ciega - [mec] cap nut • acorn nut
— f colocada - [mec] installed nut
— f con aleta(s) f - [mec] wing(ed) nut
— f — — f para biela f - [mec] pitman knockoff nut
— f — base f - [mec] flanged nut
— f — brida f - [mec] flanged nut
— f — horquilla f - [mec] toggle clamp nut
— f — mariposa f - [mec] spinner, o wing, nut
— f — — f para palanca f - [mec] lever wing
   nut
— f — — f — — f para, gobierno, o regulación f - [comb.int] control lever wing nut
— f — ojo m - [mec] eye nut
— f — pestaña f - [mec] (flanged) weld nut
— f — — f soldada - [mec] welded flanged, o
   flanged welded, nut
— f — — f soldada para conexión f de triángulo m - [mec] triangle connection welded
   flange nut
— f — — — — triángulo m [mec]
   triangle connection welded flange nut
— f — presión f - [mec] pressure, o jam, nut
— f — ranura(s) f - [mec] castellated nut
— f — reborde m - [mec] flange(d) nut
— f — resistencia f alta - [mec] high strength
   nut
— f — resorte m - [mec] spring nut
— f — rosca f - [mec] thread(ed) nut
— f — — f fina - [mec] fine thread(ed) nut
— f — — f gruesa - [mec] coarse thread(ed) nut
— f — hacia derecha f - [mec] right hand
   (threaded) nut
— m — — f — izquierda f - [mec] left hand
   (threaded) nut
— f — saliente(s) m - [mec] horn nut
— f — tratamiento m térmico - [mec] heat
   treated nut
— f corriente - [mec] standard nut
— f cuadrada - [mec] square nut
— f — corriente - [mec] standard, o regular,
   square nut
— f — de acero m - [mec] steel square nut
— f — para tornillo(s) m para me-
   tal(es) m - [mec] square machine screw nut
— f — semiterminada - [mec] semi-finished
   square nut
— f de acero m - [mec] steel nut
— f — — m con carbono m - [mec] carbon steel
   nut
— f — — m — — m con resistencia f alta -
   [mec] high strength carbon steel nut
— f — — — m — — f para junta(s)
   estructural(es) f - [constr] high strength
   carbon steel structural joint(s) nut
— f — — m — — m para junta(s) f estructu-
   ral(es) - [constr] carbon steel structural
   joint nut
— f — — m con tratamiento m térmico - [mec]
   heat treated steel nut
— f — — m inoxidable - [mec] stainless steel
   nut
— f — aluminio m - [mec] aluminum nut
— f — — m con cabeza f hexagonal - [mec]
   aluminum hexagon(al) head nut
— f bronce m - [mec] bronze nut
— f — cadmio m - [mec] cadmium nut
— f — — m plaqueado - [mec] plated cadmium nut
— f — latón m - [mec] brass nut
— f — — almenada - [mec] slotted brass nut
— f — (material) plástico - [mec] plastic nut
— f — tipo m para mamparo m - [mec] bulkhead
   nut

tuerca f de tipo m para mamparo m para cable(s)
   m - [mec] cable bulkhead nut
— f — — m — — m — — m a regulador m -
   [mec] throttle cable bulkhead nut
— f — — — tabique m - [mec] bulkhead nut
— f — — m — — m — cable m - [comb.int]
   cable bulkhead nut
— f — tornillo m - [mec] screw box
— f delgada - [mec] thin nut
— f — almenada - [mec] thin slotted nut
— f desajustada - [mec] loose(ned) nut
— f doble - [mec] double nut
— f electrogalvanizada - [mec] electrogalva-
   nized nut
— f encastillada - [mec] slotted nut
— f — de latón m - [mec] brass slotted nut
— f — delgada - [mec] thin slotted nut
— f — hexagonal - [mec] hex(agohal) slotted
   nut
— f — — de latón m - [mec] brass hex(agonal)
   slotted nut
— f enchapada - [mec] plated nut
— f endurecida (superficialmente) - [mec]
   véaase tuerca f cementada
— f ensanchada - [mec] flared nut
— f entallada - [mec] castle,tellated nut
— f especial - [mec] special nut
— f estampada - [mec] pressed nut
— f — en caliente - [mec] hot pressed nut
— f estriada - [mec] knurled nut
— f expansible - [mec] expanding, o expansion,
   nut
— f — de material m plástico - [mec] plastic
   expansion nut
— f exterior - [mec] outer, o outside, nut
— f fabricada - [mec] manufactured, o made, nut
— f fiadora - [mec] jam, o lock, nut
— f Flexloc - [mec] Flexloc nut
— f floja - [mec] loose(ned) nut
— f — para cojinete m - [cojin] loose bearing
   nut
— f galvanizada - [mec] galvanized nut
— f — con resistencia f alta - [mec] galvanized high strength nut
— f — para defensas f laterales para caminos
   m - [vial] (roadway) guardrail galvanized nut
— f girada - [mec] turned nut
— f gruesa - [mec] thick nut
— f hendida - [mec] split nut
— f herrumbrada - [mec] rusty nut
— f — colocada - [mec] installed rusty nut
— f —, quitada, o removida, o sacada - [mec]
   removed rusty nut
— f hexagonal - [mec] hex(agonal) nut
— f — almenada - [mec] hexagonal slotted nut
— f — — de latón m - [mec] brass hex(agonal)
   slotted nut
— f — — pesada - [mec] heavy hex(agonal)
   slotted nut
— f — cadmiada - [mec] cadmium plated hex-
   -(agonal) nut
— f —, común, o corriente - [mec] regular, o
   common, o standard, hex(agonal) nut
— f — — según norma f - [mec] standard regular hex(agonal) nut
— f — de acero m - [mec] hex(agonal) steel nut
— f — — — m inoxidable - [mec] stainless
   steel hex(agonal) nut
— f — — aluminio m - [mec] aluminum hex-
   -(agonal) nut
— f — — latón m - [mec] hex(agonal) brass nut
— f — enchapada - [mec] plated hex(agonal) nut
— f — galvanizada - [mec] galvanized hex-
   -(agonal) nut
— f — gruesa - [mec] thick hex(agonal) nut
— f — para cuerpo m - [bombas] body hex(agonal)
   nut
— f — — — m para bomba f - [bombas] pump body
   hex(agonal) nut
— f — — — — f para aceite m - [comb.-
   int] oil pump body hex(agonal) nut
— f — — — m — — f — m para cojinete m -
   bearing oil pump body hex(agonal) nut

tuerca f hexagonal para engranaje m - [mec] gear hex(agonal) nut
— f — — — m para árbol m - [mec] shaft gear hex(agonal) nut
— f — — — m — — m para leva(s) f - [comb-int] camshaft gear hexagonal nut
— f — — múltiple m - [comb.int] manifold hex-(agonal) nut
— f — para placa f - [comb.int] plate hex-(agonal) nut
— f — — — f para empuje m—[comb.int] thrust plate hex(agonal) nut
— f — — — f — — m para árbol m—[comb.int] shaft thrust plate (hex)agonal nut
— f — — — f — — — m con leva(s) f - [comb.int] camshaft thrust plate hex(agonal) nut
— f — — espárrago m (roscado) m - [mec] stud hex(agonal) nut
— f — — montaje m—[mec] mounting hex(agonal) nut
— f — — — m para soporte m - [mec] bracket mounting hex(agonal) nut
— f — — — m — resistencia f - [electr-inst] resistor mounting hex(agonal) nut
— f — — sello m - [mec] seal hex(agonal) nut
— f — — soporte m - [mec] bracket hex(agonal) nut
— f — — tapa f - [mec] cover hex(agonal) nut
— f — — — f para engranaje m - [comb.int] gear cover hex(agonal) nut
— f — — — f — — — m para espárrago m (roscado) - [comb.int] stud gear cover hex(agonal) nut
— f — — — f — cámara f - [válv] chamber cover hexagonal nut
— f — — — f — — f para válvula f - [válv] valve chamber cover hexagonal nut
— f — — tornillo m para metal(es) m - [mec] hex(agon) machine screw nut
— f — pesada - [mec] heavy hex(agonal) nut
— f — semiterminada - [mec] heavy hex(agon-al) semifinished nut
— f — punzonada—[mec] punched hex(agonal) nut
— f — — en frío - [mec] cold punched hex-(agonal) nut
— f — reforzada - [mec] heavy hex(agonal) nut
— f — semiterminada - [mec] semi-finished hex(agonal) nut
— f — — f punzonada - [mec] punched semi-finished hex(agonal) nut
— f — — en frío - [mec] cold punched semifinished hex(agonal) nut
— f impulsora - [mec] drive,ving nut
— f instalada - [mec] installed nut
— f interior - [mec] inside, o inner, nut
— f invertida - [mec] inverted, o reversed, nut
— f mariposa - [mec] wing, o butterfly, o spin-ner, nut • thumb screw
— f — inoxidable - [mec] stainless wing nut
— f — para palanca f - [mec] lever wing nut
— f — — f para mando m - [comb.int] con-trol lever wing nut
— f moleteada - [mec] knurled nut
— f movida - [mec] moved nut
— f necesaria - necessary, o needed, nut
— f negra - [mec] black nut
— f — para tubería f - [mec] pipe black nut
— f — — f de plancha(s) f múltiple(s) (Multi-Plate) - [tub] black Multi-Plate nut
— f ovalada - [mec] oval, o acorn, nut
— f para abrazadera f—[mec] clip, o clamp, nut
— f — acción f rápida - [mec] quick acting nut
— f — ajuste m - [mec] adjuster,ting,tment nut
— f — — m ajustada - adjusted, o tightened, nut
— m — — m apretada - [mec] tightened adjust-ing nut
— f — — m para banda f - [mec] band adjusting nut
— f — — m — — f para freno m - [mec] brake band adjusting nut
— f — — — m — cojinete m - [cojin] bearing ad-juster,ting,tment nut

tuerca f para ajuste m para cojinete m para cubo m - [mec] hub bearing adjuster,ting nut
— f — — m — collarín m [para prensaesto-pas m) - [petról] gland adjusting nut
— f — — m — empaquetadura f - [petról] pack-ing adjuster nut
— f — — m — perno m - [mec] bolt jam nut
— f — — m — — m para anclaje m - [mec] an-chor bolt jam nut
— f — — m para resorte m - [mec] spring ad-juster,ting,tment nut
— f — — m — tapa f - [válv] cover adjusting nut
— f — — m — — f para válvula f - [válv] valve cover adjuster,ting,tment nut
— f — — m — — f aflojada - [válv] loosened valve cover adjuster,ting,tment nut
— f — — m — — f aflojada - [válv] loose-ned valve cover adjuster,ting,tment nut
— f — — m — — f ajustada - [mec] tightened valve cover adjuster,ting,tment nut
— f — — m — — f apretada - [mec] tightened valve cover adjuster,ting,tment nut
— f — — m — — f floja - [válv] loose valve cover adjuster,ting,tment nut
— f — — m — — v suelta - [válv] loose valve cover adjuster,ting,tment nut
— f — — m rápido - [mec] quick acting, o speed, nut; spring clip; tinnerman's nut
— f — anclaje m - [mec] anchor(ing) nut
— f — árbol m - [mec] shaft nut
— f — — m para cambio(s) m - [autom-mec] gear shaft nut
— f — — m para entrada f (para fuerza f) - [autom-mec] input shaft nut
— f — — m — — f — — instalada - [autom-mec] installed input shaft nut
— f — — m — — f removida - [autom-mec] removed input shaft nut
— f — — m piñón m - [mec] pinion shaft nut
— f — — m — salida f (de fuerza f) - [autom-mec] output shaft nut
— f — bastidor m - [mec] frame nut
— f — biela f - [comb.int] connecting rod, o push rod, o pitman, nut
— f — boquilla f - [mec] nozzle nut
— f — borne m - [mec] stud nut
— f — — m para salida f - [electr-instal] output stud nut
— f — brida f - [mec] flange nut
— f — — f para tubería f - [tub] tube,bing flange nut
— f — — f hidráulica - [tub] hydraulic tubing flange nut
— f — buje m - [mec] bushing nut
— f — — m piloto - [mec] pilot bushing nut
— f — — m para camisa f - [mec] liner pilot bushing nut
— f — caballete m - [grúas] gantry nut
— f — cable m - [mec] cable nut
— f — — m para regulación f - [comb.int] throttle cable nut
— f — — m — — f ajustada - [comb.int] ad-justed throttle cable nut
— f — — m para regulador m - [comb.int] throttle cable nut
— f — camisa f - [mec] liner nut
— f — — f para árbol m—[mec] shaft liner nut
— f — cigüeñal m - [mec] crankshaft nut
— f — codo m - [mec] elbow nut
— f — — m para salida f de agua m - [comb.-int] water outlet elbow nut
— f — — m — — f — — m de cilindro(s) m a culata f - [comb.int] cylinder water outlet elbow to head nut
— f — cojinete m - [mec] bearing nut
— f — — m flojo - [mec] loose bearing nut
— f — — m para cubo m - [mec] hub bearing nut
— f — — m para rotación f - [grúas] swing bearing nut
— f — conductor(es) - [electr-instal] wire nut
— f — conjunto m - [mec] assembly nut

tuerca f para conjunto m para árbol n - [mec] shaft assembly nut
— f — corona f (dentada) - [autom-mec] ring gear nut
— f — contención f - [mec] restraining, o holding, nut
— f — cubierta f - [mec] cover nut
— f — — f para caja f - [autom-mec] housing cover nut
— f — f — — f para eje m - [autom-mec] axle housing cover nut
— f — cubo m - [mec] bearing nut
— f — cuchilla f - [mec] blade nut
— f — cuerpo m - [mec] body nut
— f — — m de bomba f - [bombas] pump body nut
— f — — — f para aceite m - [comb.int] oil pump body nut
— f — m — f — — m para cojinete m - [comb.int] oil pump body bearing nut
— f — culata f - [comb.int] head nut
— f — f para cilindro m - [comb.int] cylinder head nut
— f — defensa(s) f laterales (para caminos m) - [vial] (road) guardrail nut
— f — diferencial m - [mec] differential nut
— f — — m entre eje(s) m - [autom-mec] inter--axle differential nut
— f — dispositivo m - [mec] device nut
— m — m para cambio(s) m - [auto-mec] shifting unit nut
— m — eje m - [mec] axle nut • shaft nut
— f — — m delantero - [autom-mec] forward axle nut
— f — empuje m - [mec] push(ing), o thrust, nut
— f — engranaje m - [mec] gear nut
— f — — m aflojada - [mec] loosened gear nut
— f — — m impulsor - [autom-mec] drive,ving gear nut
— f — — m para bomba f - [bombas] drive, o driving, gear nut
— f — — m para árbol m - [mec] shaft gear nut
— f — — — m para leva(s) f - [comb.int] camshaft gear nut
— f — — m — impulsión f - [mec] drive,ving gear nut
— f — — m — — f para bomba f - [bombas] pump drive,ving gear nut
— f — — m — — f para bomba f para lubricante m - [bombas] lubricant pump drive,ving gear nut
— f — — m para reducción f - [mec] gear reducer nut
— f — — m — — f para rotación f - [grúas] swing gear reducer nut
— f para espárrago m - [mec] stud nut
— f — — m roscado - [mec] (threaded) stud nut
— f — espiga f - [mec] dowel nut
— f — — f roscada - [mec] dowel screw nut
— f — — f — para tapa f - [mec] cover dowel screw nut
— f — — f — — — f para engranaje m - [comb-int] gear cover dowel screw nut
— f — expansión f - [mec] expansion nut
— f para extremo m - [mec] kep nut
— f — grillete m - [mec] shackle nut
— f — horquilla f - [mec] clevis nut
— f — — f — entrada f (de fuerza f) - [mec] input yoke nut
— f — husillo m - [mec] spindle nut
— f — — m para émbolo m - [mec] piston screw nut
— f — — m — — m para cañón m - [metal-prod] mud gun piston screw nut
— f — huso m - [mec] spindle nut
— f — inyector m - [mec] injector nut
— f — junta f - [mec] splice, o joint, nut
— f — — f para manguera f - [mec] hose joint nut
— f — línea f - [mec] line nut
— f — f para inyección f - [comb.int] injection line nut
— f — llanta f - [autom-mec] rim nut

tuerca f para llanta f para rueda f dentada - [mec] sprocket rim nut
— f — mampara f - [nav] bulkhead nut
— f — — f para cable(s) m - [comb.int] cable(s) bulkhead nut
— f — — f — — m para regulación f - [comb-int] throttle cable bulkhead nut
— f — manguera f - [mec] hose nut
— f — mecanismo m - [mec] mechanism nut
— f — — m para cambio(s) m [autom-mec] shift(ing), device, o unit, nut
— f — montaje m - [mec] mounting nut
— f — — m para carburador m - [comb.int] carburetor mounting nut
— f — — m — depósito m - [comb.int] tank mounting nut
— f — — m — — m sobre ménsula f - [comb.-int] tank to bracket mounting nut
— f — — m — fiador m - [mec] retainer mounting nut
— f — — m para interruptor m - [electr-inst] switch mounting nut
— f — — m — — m magnético - [electr-inst] magnetic switch mounting nut
— f — — m para ménsula f - [comb.int] bracket mounting nut
— f — — m — — f sobre culata f - [comb.int] bracket to head mounting nut
— f — — m — silenciador m - [comb.int] muffler mounting nut
— f — — m — soporte m - [mec] bracket mounting nut
— f — — m para volante m - [mec] flywheel mounting nut
— f — múltiple m a bloque m - [comb.int] manifold to block nut
— f — operación,rar - [mec] operating nut
— f — operar v válvula f - [válv] valve operating nut
— f — orificio m - [mec] hole nut
— f — — m en bloque m - [comb.int] block hole nut
— f — — m — — m para bomba f - [comb.int] pump block hole nut
— f — — m — m — — f para combustible m - [comb.int] fuel pump block hole nut
— f — palanca f - [mec] lever nut
— f — — f para regulador m - [comb.int] governor lever nut
— f — pasador m - [mec] stud nut
— f — — m central - [mec] center pin nut
— f — perno m - [mec] bolt nut • stud nut
— f — — m común - [mec] standard bolt nut
— f — — m — con gancho m y ojo m - [mec] standard hook and eye bolt nut
— f — — m con cabeza f ranurada - [mec] stove bolt nut
— f — — m galvanizado - [mec] galvanized bolt nut
— f — — m para poste(s) m - [mec] galvanized post bolt nut
— f — — m para anclaje m - [mec] anchor bolt nut
— f — — m para biela f - [comb.int] connecting rod, o pitman, bolt nut
— f — — m para volante m - [mec] flywheel bolt nut
— f — piloto m - [mec] pilot nut
— f — — m para camisa f - [mec] liner pilot nut
— f — piñón m - [mec] pinion nut
— f — — m apretado - [mec] tightened pinion nut
— f — — m — hasta momento m torsional indicado - [mec] torqued pinion nut
— f — — m — — par m motor indicado - [mec] pinion nut torqued to @ specification
— f — — m impulsor - [mec] drive,ving pinion nut
— f — — m para eje m - [mec] axle pinion nut
— f — — m — m delantero - [autom-mec] forward axle pinion nut
— f — plancha f - [mec] plate nut

tuerca f para plancha f estructural - [const] structural plate nut
— f — plato m - [mec] plate nut
— f — — m fınal - [comb.int] end plate nut
— f — portadiferencial m - [autom-mec] differential carrier nut
— f — poste m - [mec] post nut
— f — prensaestopa(s) m - [mec] packing nut
— f — prisionero m - [mec] stud nut
— f — — m roscado - [mec] (threaded) stud nut
— f — puerta f - [constr] door nut
— f — — f para quemador m - [metal-prod] burner door nut
— f — — m — — m para estufa f - [ind] stove burner door nut
— f — regulación f - [mec] control(ling) nut • adjusting nut
— f — — f para resorte m - [mec] spring adjusting,tment nut
— f — rematar,te - [mec] kep nut
— f — retén(er) - [mec] retaining nut; véase también contratuerca f
— f — retención f - [mec] retaining, o check, o holding, o hold down, nut
— f — — f atornillada - [mec] screwed (on), retaining, o holding, nut
— f — — f corrida - [mec] slipped, retaining, o holding, nut
— f — — f deslizada - [mec] slipped, retaining, o holding, nut
— f — rodillo m - [mec] roll nut
— f — — m para alimentación f - [mec] feeding roll nut
— f — rollo m - [sold] coil nut
— f — rueda f - [mec] wheel nut
— f — —, comprobada, o verificada - [autom-mec] checked wheel nut
— f — seguridad f - [mec] jam, o lock, o check, nut
— f — — f doble - [mec] two-way, o double, lock nut
— f — — estándar - [mec] standard locknut
— f — — f flexible - [mec] flexlock nut
— f — — f hexagonal - [mec] hex(agonal) lock nut
— f — — f — galvanizada - [mec] galvanized hex(agonal) lock nut
— f — — f — — para tubería f - [mec] pipe galvanized hex(agonal) lock nut
— f — — f para cojinete m - [mec] bearing lock nut
— f — — f — para tubería f - [mec] pipe hex(agonal) lock nut
— f — — f para borne m - [electr-equip] terminal lock nut
— f — — f — culata f - [comb.int] head lock nut
— f — — f — — f para cilindro m - [comb.-int] cylinder head lock nut
— f — — f — engranaje m — [mec] gear lock nut
— f — — f — tablero m (con circuito m) - [electr-instal] circuit board lock nut
— f — — f — terminal m - [electr-instal] terminal lock nut
— f — — f — tornillo m - [mec] screw lock nut
— f — — f — — m impulsor - [mec] drive,ving screw lock nut
— f — — f — — m sin fin - [mec] drive,ving screw lock nut
— f — sello,lladura - [mec] seal(ing) nut
— f — — hexagonal [mec] hex(agonal) seal nut
— ı — soporte m — [mec] support(inq), o bracket, nut
— f — sujeción f - [mec] hold down, o holding, nut
— m — — f aflojada - [mec] loose(ned) hold down nut
— f — — f ajustada - [mec] tight(ened) hold down nut
— f — — f apretada - [mec] tight(ened) hold down nut

tuerca f para suspensión f - [autom-mec| suspensíon nut
— f — — f delantera - [autom-mec] front suspension nut
— f — para vástago m para campana f - [metal-prod] bell rod suspension nut
— f — — f trasera - [autom-mec] rear suspension nut
— f — tabique m - [mec] bulkhead nut
— f — — m para cable m - [comb.int] cable bulkhead nut
— f — — — m para regulación f - [comb.int] throttle cable bulkhead nut
— f — tapa f - [mec] cover nut
— f — — f para cámara f - [válv] chamber cover nut
— f — — f — — f para válvula f - [Válv] valve chamber cover nut
— f — — f — engranaje m - [comb.int] gear cover nut
— f — — f — — m a espárrago m roscado - [comb.int] threaded stud gear cover nut
— f — — f — — m a plato m - [comb.int] gear cover to end plate nut
— f — — f — válvula f - [válv] valve cover nut
— f — tope m - [mec] stop nut • keep nut
— f — — m elástico - [mec] elastic stop nut
— f — tornillo m - [mec] screw nut
— f — — m limitador - [mec] stop screw nut
— f — — m para ajuste m - [mec] adjusting screw nut
— f — — m — — m para tapa f - [válv] cover screw nut
— f — — m — — m — — f para válvula f - [válv] valve cover adjusting screw nut
— f — — m para banco m - [herram] vise nut
— f — — m — tope m - [mec] stop screw nut
— f — — m para palanca f - [mec] lever stop screw nut
— f — traba(r) - [mec] lock nut; véase también tuerca f trabadora; contratuerca f
— f — tubería f [tub] pipe nut
— f — — f con plancha(s) f múltiple(s) - [tub] Multi-Plate pipe nut
— f — unión f - [mec] coupling, union, o nut
— f — — f para tubo(s) m - [tub] tube nut
— f — varilla f - [mec] rod nut
— f — — f para empuje m - [mec] push rod nut
— f — vástago m - [petról] rod nut
— f — — m para émbolo m - [mec] piston rod nut
— f — ventilador m - [ind] blower nut
— f — volante m - [mec] flywheel nut
— f — — m a árbol m para leva(s) f - [mec] flywheel to crankshaft nut
— f partida - [mec] split nut
— f pasada - [mec] overripe nut
— f pesada - [mec] heavy nut
— f — almenada - [mec] heavy slotted nut
— f — semiterminada - [mec] heavy semifinished nut
— f pestañada - [mec] flanged nut • weld nut
— f punzonada - [mec] punched nut
— f — en frío - [mec] cold punched nut
— f quitada - [mec] removed nut
— f ranurada - [mec] slotted, o castle,tellated nut
— f reapretada - [mec] retightened nut
— f removida - [mec] removed nut
— f reforzada - [mec] reinforced, o heavy, nut
— f resistente - [mec] resisting nut
— f — contra vibración(es) f - [mec] vibration resisting nut
— f restante - [mec] remaining nut
— f retenida - [mec] retained nut
— f sacada - [mec] removed nut
— f semiterminada - [mec] semifinished nut
— f — con cabeza f hexagonal - [mec] semifinished hex(agonal) head nut
— f sin ajustar v - [mec] untightened nut
— f — apretar v - [mec] untightened nut
— f — pintar v - [mec] unpainted nut

tuerca f sin pintar para bastidor m - [mec] unpainted frame nut
— f soldada - [sold] welded nut
— f̄ sujetadora - [mec] holding nut • knockoff
nut
— f — con aleta(s) f - [mec] knockoff wing nut
— f̄ — para biela f - [mec] pitman knockoff nut
— f̄ tapón m - [mec] flanged nut
— f̄ terminada - [mec] finished nut
— f̄ trabada - [mec] locked nut
— f̄ trabadora - [mec] jam nut; lock nut
— f̄ universal - [mec] universal nut
— f̄ vuelta a ajustar - [mec] retightened nut
— f̄ — —, comprobar, o verificar - [mec] rechecked nut
— f y contratuerca f - [mec] double locknut
tufa v - véase toba f̄ - [geol] tufa; tuff
tufo m - [geol] tuff; tufa
tumbado,da a - dumped • canted • [autom] flipped
tumbador m - [metal-lam] manipulator; flipper
— m para bobina(s) f - [metal-lam] coil upender
— m — inspección f - [metal-lam] inspection
turn-up
— m — — f de placa(s) f - [metal-lam] plate
inspection turn-up
— m — — f — plancha(s) f - [metal-lam] plate
inspection turn-up
— m — recipiente(s) m - [mec] container dump
tumbadora f - [metal-lam] upender • tilter
tumbamiento m - [metal-lam] upending
tumbar v - . . . . • [mec] to turn up; to upend •
[metal-lam] to flip
tumbarse v - to cant • [autom] to roll over
— v sobre costado m - [autom] to roll onto @
side
tumbo m - . . . . • [autom] rollover; overturning;
canting
tuna f - [botán] . . .; cactus pear
— f de cacto,tus m - [botán] cactus pear
túnel m - . . . . [constr] encasement • [vial]
underpass • [miner] shaft
— m ancho - [constr] wide, o broad, tunnel
— m angosto - [constr] narrow tunnel
— m aplastado - [constr] flattened tunnel
— m carretero - [contr] highway tunnel
— m circular - [constr] circular tunnel
— m completado - [constr] completed tunnel
— m con ancho(r) m excepcional - [constr] extra
wide tunnel
— m con conducto m insertado con gato(s) m -
[constr] jacked-in-place tunnel
— m — inserción f con gato(s) m - [constr]
jacked-in-place tunnel
— m de acero m - [constr] steel tunnel
— m — — m corrugado - [constr] corrugated
steel tunnel
— m — — m para servicio m interior - [ind]
corrugated steel service tunnel
— m — — m estructural - [constr] structural
steel tunnel
— m — hormigón m - [constr] concrete tunnel
— m — plancha(s) f - [constr] plate(s) tunnel
— f — corrugada(s) - [constr] corrugated
plate(s) tunnel
— m — — f para revestimiento m - [constr]
corrugated line plate(s) tunnel
— m — — f de acero m—[constr] steel plate(s)
tunnel
— m — — f — — m estructural - [constr]
structural steel plate(s) tunnel
— m — — f — estructural para recuperación
f - [constr] structural steel plate(s) reclaim tunnel
— m — — f — — m para recuperación f -
[constr] steel plate(s) reclaim tunnel
— m — — f estructural(es) - [constr] structural plate(s) tunnel
— m — — f — de acero m - [constr] structural
plate(s) steel tunnel • [vial] structural
plate(s) steel underpass
— m — — f — para recuperación f - [constr]
structural plate(s) reclaim tunnel
— m — — f múltiple(s) m - [constr] Multi-

-Plate tunnel
túnel m de plancha(s) f para revestimiento m -
[constr] liner plate tunnel
— m — — f múltiple(s) Multi-Plate - [constr]
Multi-Plate tunnel
— m — — f — — para peatón(es) m -
[constr] Multi-Plate pedestrian tunnel
— m — tubería f - [constr] pipe tunnel
— m — — f de acero m - [constr] steel pipe
tunnel
— m — — f — — m corrugado - [constr] corrugated steel pipe tunnel
— m — — f — — m — cinta f transportadora
- [constr] corrugated steel pipe convoyer
tunnel - [constr] corrugated steel pipe conveyor tunnel
— m — — f — — m - de planchas f gruesas -
[constr] heavy gage corrugated steel pipe
tunnel
— m — — f — — m — — f — para cinta f
transportadora -[constr] heavy gage corrugated steel pipe conveyor tunnel
— m — — f de planchas f estructurales -
[constr] structural plate pipe tunnel
— m deformado - [constr] deformed, o deflected,
tunnel
— m encubado - [constr] liner plate tunnel
— m estrecho - [constr] narrow tunnel
— m estructural - [constr] structural (plate)
tunnel
— m — agigantado - [constr] giant structural
tunnel
— m — gigantesco* - [constr] véase túnel estructural agigantado
— m — para recuperación f - [constr] structural
reclaim tunnel
— m excavado - [constr] excavated tunnel
— m expuesto - [constr] exposed tunnel
— m ferroviario - [f.c.] railroad,lway, o
train, tunnel
— m gigantesco* - [constr] véase túnel agigantado; giant tunnel
— m — de acero m - [constr] giant steel tunnel
— m — — — m estructural - [constr] giant
structural steel tunnel
— m — — plancha(s) f de acero m - [constr]
giant steel plate tunnel
— m — — f — — m estructural - [constr]
giant structural steel plate tunnel
— m — — f — — m — para recuperación f -
[constr] giant structural steel plate reclaim
tunnel
— m — — f — — m para recuperación f -
[constr] giant steel plate reclaim tunnel
— m — — f estructurales - [constr] giant
structural plate(s) tunnel
— m — — f estructurales - [constr] giant
structural plate(s) tunnel
— m — — f — para recuperación f - [constr]
giant structural plate steel reclaim tunnel
— m — para recuperación f - [constr] giant reclaim tunnel
— m horadado - [constr] pierced, o excavated, o
dug, tunnel
— m importante - [constr] important tunnel
— m insertado (con gatos m) - [constr] jacked-in-
-place tunnel
— m muy ancho - [constr] extra-wide tunnel
— m no rígido - [constr] non-rigid tunnel
— m para aire m fresco - [miner] fresh air tunnel
— m — barco(s) m - [constr] boat tunnel
— m — carga(s) f - [transp] freight tunnel •
[miner] reclaim(ing) tunnel
— m — — f de carbón m - [transp] coal loading
tunnel
— m — cinta f transportadora - [constr] conveyor, tunnel, o housing
— m — — f — cubierta - [ind] enclosed conveyor, tunnel, o housing
— m — cloaca f - [sanit] sewer tunnel
— m — derivación f - [constr] by-pass tunnel
— m — desviación f - by-pass tunnel

túnel m para embarcación(es) f - [constr] boat tunnel
— m — entubación f - [constr] casing tunnel
— m — — f de cloaca(s) f - [sanit] sewer casing tunnel
— m — evacuación f - [miner] escapeway
— m — peatón(es) m - pedestrian tunnel
— m — recuperación f - [constr] reclaim(ing), o recovery, tunnel
— m — — f para agregado m - [constr] aggregate reclaim(ing) tunnel
— m — red f - [ind] system tunnel
— m — — f para servicio(s) m público(s) - [constr] utility tunnel
— m — servicio(s) m - [constr] service tunnel
— m — — m interior(es) - [constr] service tunnel
— m — tren(es) m - [f.c.] train tunnel
— m — tubería f - [constr] pipe,ping tunnel
— m — — f para vapor m - [constr] steam (pipe) tunnel
— m — vapor m - [constr] steam tunnel
— m subterráneo - [constr] underground tunnel
— m temporario - [constr] temporary tunnel
— m — para desviación f - [constr] temporary by-pass tunne;
tunelar v - [constr] to tunnel • to bore
tungsteno m con carburo m - [metal-prod] carbide tungsten
— m fundido - [metal] cast tungsten
— m toriado - [sold] thoriated tungsten
— m y gas m inerte - [sold] tungsten (and) inert gas
tupí m - [herram] molding machine
tupidez f - [botán] cover
— f promedia - [botán] average cover
tupir v - . . .; to clog
turbera f - [suelos] . . .; peat bed
turbidez* f - véase turbieza f: turbiedad f • turbidity
turbiedad f - turbidity
— f de agua m - [hidr] water turbidity
— f — — m en salida f - [hidr] outflowing water turbidity
— f — — m potable - [hidr] drinking water turbidity
— f máxima - maximum turbidity
— f — de agua m - [hidr] water maximum, o maximum water, turbidity
— f — de agua m industrial - [hidr] maximum industrial water turbidity
— f — — — m potable - [hidr] maximum drinking water turbidity
— f media - [hidr] average turbidity
— f — de agua m - [hidr] water average. o average water, turbidity
— f — — — m industrial - [hidr] average industrial water turbidity
— f — — — m potable - [hidr] drinking water average turbidity
— f mínima - [hidr] minimum turbidity
— f — — agua m - [hidr] minimum water, o water minimum, turbidity
— f — — — m industrial - [hidr] minimum industrial water turbidity
— f — — — m potable - [hidr] drinking water minimum turbidity
turbieza f - turbidity
— f de agua m - [hidr] water turbidity
— f — — m industrial - [hidr] industrial water turbidity
— f — — m potable - [hidr] drinking water turbidity
— f máxima - [hidr] maximum turbidity
— f — de agua m - [hidr] water maximum,o maximum water turbidity
— f — de agua m industrial - [hidr] industrial water maximum turbidity
— f — — — m potable - [hidr] drinking water maximum turbidity
— f media - [hidr] average turbidity
— f — de agua m - [hidr] water average, o average water, turbidity

turbieza f media de agua m industrial - [hidr] industriual water average turbidity
— f — — — m potable - [hidr] drinking water average turbidity
— f mínima - [hidr] minimum turbidity
— f — de agua m - [hidr] minimum water, o water minimum, turbidity
— f — — — m industrial - [hidr] industrial water minimum turbidity
— f — — — m potable - [hidr] drinking water minimum turbidity
— f — — — m potable - [hidr] minimum drinking water turbidity
turbina f con reacción f a impulsión f - [turb] impulse reaction turbine
— f motriz - [electr-prod] driving turbine
— f para accionamiento m - [hidr] driving turbine
— f — condensación f - [hidr] condensation, o condensing, turbine
— f — gas m - [electr-prod] gas turbine
— f — impulsión f - [hidr] driving, o impulse, turbine
— f — vapor m - [cald] steam turbine
— f — — m de mercurio m - mercury vapor turbine
— f por reacción f - [turb] reaction turbine
turbio,bia a - . . .; unclear
turboalimentación f - [comb.int] turbocharging
turboalimentado,da a - [comb.int] turbocharged
turboalimentador m - [comb.int] turbocharger
— m enfriado - [comb.int] cooled turbocharger
— m mantenido - [comb.int] maintained turbocharger
— m para regulación f - [comb.int] controlling charger
— m — — f electrónica - [comb.int] electronic control turbocharger
— m — — f de mezcla - [comb.int] electronic mixture control turbocharger
turboalimentar v - [comb.int] to turbocharge
turbocompresor m - [turb] turbocompressor
turboconvertidor m - [turb] turboconverter
turbogenerador m - [turb] turbogenerator; turbine driven generator
— m hidráulico - [turv] hydraulic turbogenerator
— m — auxiliar - [electr-prod] auxiliary hydraulic turbogenerator
turbosobrealimentado,da a - [comb.int] turbo--charged
turbosoplador m - [metal-prod] turbo blower
turbosoplador,ra a - [metal-prod] turboblowing
turbosoplante m - [metal-prod] turbo blower
— m — reserva - [metal-prod] stand-by turbo-blower
— m para horno m alto - [metal-prod] blast furnace turbo blower
turbulencia f de gas m - gas turbulence
turgita f - [geol] turgite
turismo m - . . .; sightsseing
turista m - . . .; sightseer
turnado,da a - alternated
— a en volante m - [autom] alternated behind @ wheel
turnarse v - [autom] to alternate; to swap
— v en volante m - [autom] to alternate behind @ wheel
turno m - . . . • [labor] shift; tour
— m corto - [labor] short, turn, o shift
— m de noche f - night, o graveyard, shift
— m diurno m - [labor] dary shift • day crew
— m largo - [labor] long, turn, o shift
— m nocturno - [labor] night, shift, o turn • night crew
— m para horno m - [ind] furnace, turn, o shift
— m — — m para recocido m - [metal-trat] annealing furnace turn
— m — laminación f - [metal-lam] rolling turn
— m — laminador m - [,etal-lam] mill turn
— m — — m en caliente - [metal-lam] hot strip turn
— m — — m terminador - [metal-lam] finishing mill turn

turno m para soplado m ≈ [metal-prod] blast turn
— m — tren m - [metal-lam] mill turn
— m — — m terminador - [metal-lam] finishing
  mill turn
— m programado - [ind] scheduled turn
— m — para laminador m - [metal-lam] scheduled
  mill turn
— m — — m terminador - [metal-lam] scheduled finishing mill turn
— m — tren m terminador - [metal-lam]
  scheduled finishing mill turn
. . . turno(s) m por día - [labor] . . . turn(s)
  per day
tusa f - [agric] cob

U

U I T P - véase Unión f Internacional de (los)
  Transportes m Púlicos
U P U f - véase Unión f Postal Universal
U I T f - véase Unión f Internacional de Tele-
  comunicación(es) f
ubicación f - . . . . • placing,cement; positioning; locating • storing • domicile
— f accesible - accessible, location, o area
— f ácida - [suelos] acidic site
— f actual - [ind] present site
— f — de planta f - [ind] present plant site
— f alejada - remote location
— f — para planta f - [ind] remote plant loca-
  tion|
— f alternativa - [ind] alternate location
— f — para planta f - [ind] alternate plant
  location
— f apartada - [ind] remote location
— f — para planta f - [ind] remote plant location
— f apropiada - proper, o appropriate, o suitable, location
— f apropiada de cilindro m - [mec] proper cylinder location
— f — asegurada - insured proper location
— f — para cojinete m - [mec] proper bearing
  location
— f — — excavación f - [constr] proper, excavation, o trench, location
— f buena - good location • good placement
— f central - central location
— f céntrica - central location
— f conveniente - convenient location
— f correcta - correct, location, o positioning
— f corriente - current, o usual, location
— f de agujero m - [mec] hole location
— f — alcantarilla f - [constr] culvert location
— f — amperímetro m - [instrum] ammeter location
— f — — m y voltímetro m - [instrum] ammeter-
  -voltmeter location
— f — arandela - [mec] washer location
— f — artefacto m - [electr-instal] luminaire
  location • fixture location
— f — automóvil m - [autom] car locating,tion
— f — barreno m - [constr] hole location
— f — boquilla f - [sold] nozzle location
— f — botonera f - [electr-instal] pushbutton
  location
— f — — f individual - [electr-instal] individual pushbutton location
— f — cilindro m - [mec] cylinder location
— f — cojinete m - [mec] bearing location
— f — — m (de)bajo de carga f - [mec] bearing
  location under @ load
— f — comercio m - [com] store, o business,
  location
— f — componente m - [mec] component location

ubicación f de conexión f con tierra f - [sold]
  ground(ing) location
— f — cordón m - [sold] bead location
— f — diagrama - diagram location
— f — elemento m - element location
— f — — m para símbolo m - symbol element
  location
— f — — m — — m para soldadura f - [sold]
  welding symbol element location
— f — engranaje m - [mec] gear, location, o
  positioning
— f — equipo m - [ind] equipment location
— f — estructura f - structure location
— f — espiga f - [mec] dowel (pin) location
— f — estación f - [ind] station locating,tion
— f — estructura f - [constr] structure, locating,tion, o situating,tion
— f — excavación f - [constr] excavation, o
  trench, location
— f — explotación f - [miner] pit location
— f — grieta f - [mec] crack location
— f — horno m - [ind] furnace location
— f — — m alto - [metal-prod] blast furnace
  location
— f — motor m - [electr-instal] motor location • [comb.int] engine location
— f — — m individual - [electr-instal] individual motor location • [comb.int] individual engine location
— f — número m - number location
— f — — m de serie f - [ind] serial number
  location
— f — obra f - [constr] project location
— f — orificio m - [mec] hole location
— f — — m en cubierta f - [mec] cover hole
  location
— f — — m — f delantera - [mec] front
  cover hole location
— f — — m — f posterior - [mec] rear
  cover hole location
— f — — m para válvula f - [autom-mec] valve
  hole location
— f — pasador m - [mec] dowel pin location
— f — — m en cubierta f - [mec] cover dowel
  pin location
— f — — m — — f para distribuidor m - [mec]
  divider cover dowel pin location
— f — — m — — f — — m para fuerza f -
  [mec] power divider cover dowel pin location
— f — pieza f - [mec] part location
— f — — f componente - [mec] component (part)
  location
— f — — f constitutiva - [mec] component
  (part) location
— f — pilote m - [constr-pil] pile location
— f — planta f - [ind] plant, site, o location
— f — población f - [pol] townsite; town location
— f — rueda f - [mec] wheel location
— f — — f cambiada - [mec] changed wheel location
— f — — f delantera - [mec] front wheel location
— f — — f directriz - [agric-equip] guide
  wheel location
— f — — f guía - [agric-equip] guide wheel
  location
— f — — f trasera - [mec] rear wheel location
— f — símbolo m - symbol location
— f — — m para soldadura f - [sold] welding
  symbol location
— f — soldadura f - [sold] weld location •
  solder(ing) location
— f — — f en sitio m apropiado - [sold] good
  weld placement
— f — — f — — m preciso - [sold] precise
  weld placement
— f — subestación f - [electr-distrib] substation locating,tion
— f — tablero m - [electrón] board location
— f — — m con circuito m - [electrón] circuit
  board location
— f — tablilla f - [electrón] board location

ubicación f de tablilla f̱ con circuito m̱ -
[electrón] circuit board location
— f̱ —— —— m̱ estampado - [electrón]
printed circuit board location
— f̱ —— f̱ —— m̱ impreso - [electrón] printed circuit bord location
— f̱ — termocupla f̱ - [metal-prod] thermocouple
location
— f̱ —— f̱ para crisol m̱ - [metal-prod] hearth
thermocouple location
— f̱ definitiva - [mec] final, o̱ permanent, location
— f̱ desacostumbrada - unusual location
— f̱ desusada - unusual situation
— f̱ distante - remote location
— f̱ distinta - different location
— f̱ en diagrama - diagram location
— f̱ — ladera f̱ - hillside location
— f̱ —— f̱ con pendiente f̱ fuerte - [constr]
steep hillside location
— f̱ — región f̱ alejada - [ind] remote location
— f̱ —— f̱ — de planta f̱ - [ind] remote plant
location
— f̱ esquemática - schematic location
— f̱ estándar - standard location
— f̱ — para orificio m̱ - hole standard location
— f̱ —— m̱ para válvula f̱ - [autom-neumát]
standard valve hole location
— f̱ estratégica - strategic(al) location
— f̱ final - final location
— f̱ geográfica - geographic, location, o̱ situation
— f̱ geológica - [geol] geologic(al), location,
o̱ situation, o̱ site
— f̱ inapropiada - inappropriate, o̱ poor, location
— f̱ inconveniente - inconvenient location
— f̱ independiente - independent, location, o̱
situation, o̱ position(ing)
— f̱ indicada - indicated, o̱ given, position, o̱
location
— f̱ individual - individual location
— f̱ mejor - better, o̱ best, site, o̱ location
— f̱ mostrada - shown, location, o̱ site
— f̱ no estándar - non-standard location
— f̱ —— para orificio m̱ - [mec] non standard
hole location
— f̱ —— —— m̱ para válvula f̱ - [autom-neum]
non-standard valve hole location
— f̱ para alcantarilla f̱ - culvert location
— f̱ — canal m̱ - [hidr] channel location
— f̱ — cauce m̱ - [hidr] channel location
— f̱ — cojinete m̱ - [mec] bearing, location, o̱
position(ing)
— f̱ — conducto m̱ - [constr] conduit location
— f̱ — planta f̱ - [ind] plant location
— f̱ permanente - permanent, o̱ final, location
— f̱ polvorienta - dusty location
— f̱ por estación(es) f̱ - location by station(s)
— f̱ precisa - precise, o̱ exact, location
— f̱ —— de soldadura f̱ - [sold] good weld placement; precise weld location
— f̱ presunta - presumed location
— f̱ provisoria - temporary location
— f̱ remota - remote location
— f̱ —— de planta f̱ - [ind] remote plant location
— f̱ segura - [segurid] safe location
— f̱ separada - separate, location, o̱ placement
— f̱ singular - exceptional location
— f̱ sobre ladera f̱ - hillside location
— f̱ —— portadiferencial - [autom-mec] location
on @ differential carrier
— f̱ substitutiva - alternate location
— f̱ — prevista - foreseen alternate location
— f̱ — provista - provided alternate location
ubicado,da a̱ - located; positioned; situated •
stored • domiciled • stradling • working
— a̱ antes - up-stream
— a̱ apropiadamente - properly, o̱ well, located,
o̱ situated, o̱ positioned, o̱ placed
— a̱ convenientemente - located conveniently
— a̱ correctamente - correctly, positioned, o̱
located
ubicado,da a̱ dentro de puerta f̱ delantera—[mec]
located inside of @ front door
— a̱ —— —— f̱ — de caja f̱ - [mec] located
inside @ cabinet front door
— a̱ después - [hidr] downstream
— a̱ en eje m̱ - stradling @ centerline
— a̱ —— —— m̱ vertical - stradling @ vertical
centerline
— a̱ estratégicamente - located strategically
— a̱ sobre portadiferencial - [autom-mec] located on @ differential carrier
ubicar v - . . . • to localize • to store • to,
position, o̱ set • to domicile
— f̱ apropiadamente - to, locate, o̱ position,
appropriately, o̱ properly
— v̱ arandela - [mec] to, place, o̱ locate, @
washer
— v̱ automóvil m̱ - to, locate, o̱ position, @,
automobile, o̱ car
— v̱ bien - to, locate, o̱ place, well
— v̱ cojinete m̱ - [mec] to position @ bearing
— v̱ conducto m̱ - [constr] to locate @ conduit
— v̱ convenientemente - to locate conveniently
— v̱ correctamente - to position correctly
— v̱ engranaje m̱ - [mec] to position @ gear
— v̱ equipo m̱ - [ind] to locate @ equipment
— v̱ espiga f̱ - [mec] to locate @ dowel pin
— v̱ estación f̱ - to locate @ station
— v̱ estratégicamente - to locate strategically
— v̱ estructura f̱ - to, situate, o̱ locate, @
structure
— v̱ fuga f̱ - to locate @ leak
— v̱ planta f̱ - [ind] to locate @ plant
— v̱ sobre portadiferencial - [autom-mec] to
locate on @ differential carrier
— v̱ soldadura f̱ - [sold] to, locate, o̱ place,
@, weld, o̱ solder(ing)
— v̱ subestación f̱ - [electr-distrib] to locate
@ substation
uescentral a̱ - west center,tral
ulterior a̱ - . . .; later
último m̱ - . . .; latter
último,ma a̱ - . . .; terminal • rear • [ind]
late model
— a̱ de varios - latter
ultraalto,ta a̱ - ultrahigh
ultraelevado,da a̱ - ultrahigh
ultraliviano,na a̱ - ultra light
ultramar m̱ - overseas
ultramarino,na a̱ - overseas; across @ sea(s);
across @ ocean
ultramoderno,na a̱ - . . .; state-of-@-art
ultrapesado,da a̱ - ultraheavy; very heavy
ultrarrápido,da a̱ - extra fast
ultrasensible a̱ - extra sensitive
ultrasónicamente adv - [electrón] ultrasonically
ultrasónico,ca a̱ - ultrasonic; ultrasound
ultrasonido m̱ - [electrón] ultrasound
ultrasonoscópico,ca a̱ - ultrasonoscopic
umbral m̱ - [constr] . . .; (door) sill
— m̱ de bronce m̱ - [constr] bronze threshold
— m̱ detectado - [electrón] detected threshold
un a̱ centenar - one hundred
— a̱ conductor m̱ - [electr-instal] single conductor
— a̱ lado m̱ - one side
— a̱ — solamente - only one side
— a̱ mismo sentido m̱ - one (same), o̱ single,
direction
— a̱ plano solamente - only one plane
— a̱ sentido - one, sense, o̱ direction
— a̱ sólo conductor m̱ - single conductor
— a̱ tercio - one third
una adv caja f̱ - [metal-lam] one stand
— adv dirección f̱ - one direction • [electr]
single throw
— adv misma dirección f̱ - one (same) direction
— adv pieza f̱ - single, o̱ one, piece. o̱ part
— f̱ — por vez - single unit
— adv por una - one at @ time
— adv punta f̱ - single, tip, o̱ prong
— adv sóla pieza f̱ - single piece

una adv sóla vez - once, o one time, only
— adv sóla vía f - [vial] one lane • one way • [fin] single copy
— adv vez - once; one time
— adv — f cada año m - once, @, o each, year; yearly • véase anualmente
— adv — f por dia m - once, @, o each, day; daily
— adv — f mes m - once each month; monthly
— adv — f quincena - once each fortnight; fortnightly
— adv — f semana f - once each week; weekly
— adv — f turno m - [labor] once each, turn, o shift
— adv vía - [vial] one lane • [f.c.] one track • [fin] one copy
— f vida f - one life • [f.c.-ruedas] one wear
— adv y otra f vez f - again and again; time and again
únicamente - . . . . • alone • all around
— adv con cromo m - [metal-prod] straight chromium
único,ca adv - . . .; alone • single
— adv en clase f - one of @ kind
— a — tipo m - unique
— a repetido,da - single repeat
unidad f - . . . . • article; item • [ind] equipment; unit • [mec] package • kit
— f abaco - [cement] abacus unit
— f activada - activated unit
— f adquirible - available, o obtainable, unit
— f — en plaza f - commercially available unit
— f alimentadora f - [ind] feed(er) unit
— f — para alambre m - [sold] wire, drive,ving, o feed(ing), unit, o box, o equipment
— f análoga - analogous, o similar, unit
— f — para repuesto m—[ind] similar spare unit
— f antigua - old, o ancient, unit
— f apiladora - [ind] piling unit
— f apoyada - supported unit
— f — sobre fondo m - [petról] bottom supported unit
— f automática - [int] automatic unit
— f — completa—[ind] complete automatic system
— f — para alimentación f de alambre m - [sold] complete(ly) automatic wire feed(ing) system
— f autónoma - [mec] autonomous, o self-supporting, unit
— f autoportante - [constr] self-supporting unit
— f — continua - [constr] standard self-supporting unit
— f autosoportante - [electr] free standing unit
— f bajo tierra f - underground unit
— f — f para bombeo m - [petról] subsurface, pumping, o production, unit
— f — f — producción f - [petról] subsurface production unit
— f base - [ind] base unit
— f básica - [ind] basic unit
— f bombeadora - [bombas] pump(ing) unit
— f calibradora - [culin] sizer unit
— f cambiadora - [mec] changer,ging unit
— f — para toma(s) m - [electr-equip] tap changer,ging unit
— f cementadora - [nucl] cementing unit
— f cizalladora - [ind] shearing unit
— f combinada - [ind] combined,nation unit
— f — de soldadora f para corriente f alterna y generador m - [sold] combination alternating current welder and power unit
— f — — — f continua y generador m - [sold] combination direct current welder and power unit
— f — para soldadura f o fuerza f motriz - [sold] combination welder and power unit
— f comercial - [com] commercial, o business, unit
— f — corriente—[com] standard commercial unit
— f — — para rayos m gamma - [electrón] commercial gamma (ray) unit
— f — para rayos m gamma - [electrón] commercial gamma (ray) unit

unidad f comercial para rayos-X - [electrón] commercial X-ray unit
— f compacta - [ind] compact unit
— f completa - [ind] complete unit
— f — armada - [ind] complete assembled unit
— f — de distribuidor m para fuerza f - [autom-mec] complete assembled power divider unit
— f independiente - [ind] complete independent, o self contained, unit
— f — para arranque m,o arrancadura f - [comb.int] complete starting, unit, o equipment, o arrangement
— f — bombeo m - [bombas] complete pumping unit
— f — — m para agua m - [bombas] water complete pumping unit
— f — — m — m para alimentación f - [bombas] feed water complete pumping unit
— f compresora - [mec] compressor,sing unit
— f — para aire m - [neumát] air compressing unit
— f con ángulo m amplio - [ind] wide angle unit; broad angle unit
— f — — m — para pulverización f - [ind] wide angle pulverizing unit
— f — — m grande - [ind] wide angle unit
— f — costo m alto - [ind] high cost unit
— f — — m bajo - [ind] low cost unit
— f — — m — con motor m con, combustión f interna, o explosión f - [ind] low cost, engine, o internal combustion, driven unit
— f con circuito m cerrado - [ind] closed circuit unit
— f — depósito m - [mec] tank unit
— f — . . . eje(s) m - [autom-mec] . . . axle unit
— f — eje m horizontal - [mec] horizontal, axis, o axle, unit
— f — — m para carrete m horizontal - [trefil] horizonal axis reel unit
— f — enfriamiento m con agua m - [ind] water cooled unit
— f — motor m - [electr-mot] motor driven unit • [comb.int] engine driven unit
— f — número m de serie f - [ind] serial(ly) numbered unit
— f — pantalla f - [electrón] display(ing) unit
— f — f para escritorio m - [electrón] desktop display unit
— f — f — m para informe m único - [electrón] single report desktop display unit
— f — f para informe m único - [electrón] single report display unit
— f — rodillo(s) m - [metal-lam] roll unit
— f — — m para agarre m - [metal-lam] pinch roll unit
— f — — m — arrastre m - [metal-lam] pinch roll unit
— f — m (de) tipo n (con) carrete m para tracción f - [mec] spool roller traction unit
— f — tambor m - [mec] drum unit
— f — — m sencillo - [mec] single drum unit
— f — — m único - [mec] single drum unit
— f — . . . tambor(es) m - [mec] . . .-drum unit
— f — tubo(s) m - [mec] tube unit
— f — — m para rayos m catódicos - [electrón] cathode ray tube, o CRT, unit, o device
— f conectada - [mec] connected unit • [electr-instal] wired unit
— f con red f - [electr-instal] system connected unit
— f — — f de torre f para enfriamiento m - [ind] unit piped into @ cooling tower system
— f conmutadora - [electr-instal] switching, unit, o device
— f — en estado m sólido - [electr-instal] solid state, o transistorized, switching unit
— f continua - [metal-lam] continuous unit
— f — para redondo(s) m - [metal-lam] rounds continuous unit

unidad f continua para redondo(s) m pequeño(s) -
[metal-lam] small rounds continuous mill
— f corriente - [ind] commercial, o standard,
unit
— f — para rayo(s) m gamma - [electrón] commercial, o standard, gamma (ray) unit
— f cronometrada - [ind] timed unit
— f de albañilería f - [constr] masonry unit
— f — — f de hormigón m - [constr] concrete
masonry unit
— f — — f — m hueco - [constr] hollow
concrete masonry unit
— f — — f — m — para portar carga(s) f -
[constr] hollow load bearing concrete masonry
unit
— f — — f — m — soportar carga(s) f -
[constr] load bearing concrete masonry unit
— f — area m de sección f - section area unit
— f — arrancador m - [electr] starter unit
— f — — m combinado—[electr] combined,nation
starter unit
— f — base f - base unit
— f — cabezal m - [agric-equip] header unit
— f — calor m [fís] heat unit
— f — cambiador m - [autom] shifter,ting unit
— f — — m eléctrico - [autom-mec] electric
shifting unit
— f — cemento m - [constr] cement unit
— f — cobalto m . . . - [electrón] cobalt
. . . unit
— f — columna(s) f - [constr] column(s) unit
— f — diamante(s) - [quím] diamond(s) unit
— f — eje m - [mec] axle, o shaft, unit
— f — — m delantero - [mec] front axle unit
— f — — m trasero - [mec] rear axle unit
— f — energía f - [electr] power unit
— f — engranaje(s) m - [mec] gear(s) unit
— f — — m para reducción f - [mec] gear(s)
reduction unit; reduction gear(s) unit
— f — equipo m - [ind] equipment unit
— f — — m enfriada - [ind] cooled equipment
unit
— f — — m — con agua m - [ind] water cooled
(equipment) unit
— f — — m sumergido - [ind] immersed equipment
(unit)
— f — — m — en aceite m - [ind] oil immersed
equipment (unit)
— f — espigadora f - [agric-equip] header unit
— f — filamento m - [electr-ilumin] filament
unit
— f — hierro m - [miner] iron unit
— f — intensidad f - intensity unit
— f — — f de campo m - field intensity unit
— f — — f — m magnético -
[electr] magnetic field intensity unit
— f — — f magnética - [electr] magnetic intensity unit
— f — manguera f - [mec] hose unit
— f — manipulador m - [electrón] keyer unit
— f — — m para tono(s) m - [electrón] tone
keyer unit
— f — maquinaria f - [ind] machinery, unit, o
piece
— f — potencia f - [mec] power unit
— f — prioridad(es) f - priority,ties unit
— f — reserva - [ind] reserve, o spare, o
stand-by, unit
— f — rodillo(s) m - [mec] roll(s) unit
— f — — m para arrastre m - [mec] pinch roll
unit
— f — — m — tracción f - [metal-lam] pinch
roll unit
— f — superficie f - surface unit
— f — tiempo m - [cronol] time unit
— f — tipo m original - early type unit
— f — — m primitivo - early type unit
— f — tonelada f - [ind] ton unit
— f — — f de hierro m - [metal] iron ton unit
— f — trabajo m - [labor] work unit
— f — transformador m - [electr-equip] transformer unit
— f — — m para horno m - [electr-equip] furnace transformer unit

unidad f de transformador m para horno m eléctrico - [electr-equip] electric, o arc, furnace transformer unit
— f — válvula f - [válv] valve unit
— f — — f principal - [mec] main valve unit
— f desconectada - [electr-oper] disconnected,
o turned off, unit
— f deschaladora - [agric-equip] husking unit
— f desenrolladora - [metal-lam] pay-off unit
— f — con eje m horizontal - [trefil] horizontal axle (reel) pay-off unit
— f — de tipo m sobre caballete m - [trefil]
A-frame type pay-off unit
— f — con eje m para carrete m vertical -
[trefil] horizontal axis reel pay-off unit
— f — para carrete m - [trefil] reel pay-off
unit
— f — sobre caballete m - [trefil] A-frame
pay-off unit
— f despinochadora - [agric-equip] husking unit
— f económica - [econ] economic(al) unit
— f eficiente - [ind] efficient unit
— f — para economizar energía f - [ind] energy
efficient unit
— f — reducir costo(s) m - [ind] cost efficient unit
— f eléctrica - [electr-equip] electric(al)
unit
— f — para cambio(s) m - [autom-mec] electric
shift(ing) unit
— f elevada - [agric-equip] raised unit
— f embarcada - [transp] shipped, o loaded,
unit
— f embobinadora - [mec] coiling unit
— f en base a pila(s) f - [ind] battery powered
unit
— f en punto m remoto - [electrón] remote site
unit
— f — sitio m remoto - [electrón] remote site
unit
— f — vehículo m - [electrón] mobile unit
— f enchufable - [electrón] plug-in unit
— f enchufada - [electrón] plugged in unit
— f enderezadora - [metal-lam] straightening
unit
— f — con rodillo(s) m para, agarre, o arrastre - [metal-lam] straightener-pinch roll
unit
— f — — m — arrastre m - [metal-lam]
straightener-pinch roll unit
— f enfriadora - [ind] cooling unit
— f — para motor m - [comb.int] engine cooling unit
— f — con agua m - [ind] water cooled unit
— f — — aire m - [ind] air cooled unit
— f escarpadora f - [metal-lam] scarfing unit
— f esclava - [electrón] slave unit
— f — adicional - [electrón] additional slave
unit
— f — — con tubo m para rayos m catódicos -
[electrón] additional slave cathode ray tube
unit
— f — con tubo m para rayos m catódicos -
[electrón] slave cathode ray tube (unit)
— f estampadora - [mec] stamping unit
— f examinada - [ind] examined unit
— f excitadora - [metal-prod] exciter unit
— f — para criba f - [metal-prod] (sinter)
screen exciter unit
— f — — f para sínter m - [metal-prod]
sinter screen exciter unit
— f excitatriz - [electrón] exciter,ting, unit,
o machine
— f — — — f para sinterización f en caliente
- [miner] hot sinter(ing) screen exciter,ting
unit
— f — con . . . voltio(s) m - [electrón]
. . . volt(s) exciter machine
— f fabril - [ind] manufacturing unit
— f fija - [ind] stationary unit
— f — corriente - stationary commercial unit
— f — — para rayos m gamma - [electrón] stationary commercial gamma ray unit

unidad f fija obtenible -- [ind] available commercial unit
— f — — en plaza f -- [ind] stationary commercial unit; locally obtainable stationary unit
— f — para rayo(s) m gamma - [electrón] stationary gamma ray unit
— f — — m gamma obtenible en plaza f - [electrón] stationary commercial gamma ray unit; locally obtainable stationary gamma ray unit
— f filtradora - [ind] filtering unit
— f —, lubricadora y reguladora - [ind] filter-luricator-regulator unit
— f flotante - [petról] floating unit
— f forjadora - [metal-prod] forging unit
— f funcional - [ind[operating unit
— f generadora - [electr-prod] power unit
— f — para fuerza f motriz - [electr-prod] power unit
— f — — f — con soldadora f - [sold] power unit and welder
— f — — — f para corriente f alterna - [sold] power unit with alternating current welder
— f — — — f — continua - [sold] power unit with direct current welder
— f giratoria - [mec] rotating unit
— f Hi-Freq - [sold] Hi(gh) Freq(uency unit
— f hidráulica - [petról] hydraulic unit
— f — en producción f - [petról] hydraulic production unit
— f Idealarc - [sold] Idealarc, machine, o unit
— f impulsora - [sold] drive,ving unit
— f — para alambre m - [sold] wire drive,ving unit
— f — — electrodo m - [sold] wire drive,ving unit
— f — variable - [mec] variable speed drive,ving unit
— í independiente - [mec] independent unit • module
— f Innershield - [sold] Innershield unit
— f — para alimentación f - [sold] Innershield feeding, unit, o system
— f — — — f de alambre - [sold] Innershield wire feeding system
— f — — — — m con alma m fundente - [sold] Innershield wire feeding system
— f íntegra - whole unit
— f intercambiable - [ind] interchangeable unit
— f limpiadora - cleaning unit
— f Lincoln - [sold] Lincoln unit
— f — para alimentación f de alambre m—[sold] Lincoln wire feeding equipment
— f motogeneradora - [electr-prod] motor generator,ting unit
— f motriz - [electr-prod] power unit
— f móvil,vible - [ind] mobile unit
— f — de cobalto m . . . - [instrum] mobile cobalt . . . unit
— f — para reparación(es) f - [ind] mobile, o portable, repair, unit, o rig
— f normal - [ind] normal, o standard, unit
— f obtenible en plaza f - [ind] market available unit • commercial unit
— f Oilmaster - [petról] Oilmaster unit
— f — para producción f - [petról] Oilmaster production unit
— f — para producción f subterránea - [petról] Oilmaster subsurface production unit
— f operativa - [ind] operating,tional unit
— f para aire m acondicionado - [ambient] air conditioning unit
— f — alimentación f - [ind] feeding unit
— f — — f para alambre m - [sold] wire feeding unit
— f — — — m con alma m fundente - [sold] Innershield wire feeding unit
— f — amortiguación f - [ind] damping unit
— f — aplanamiento m - [metal-lam] leveling unit; leveler
— f — arrancadura f - [comb.int] starting unit
— f — arrastre m - [constr] drag(ging) unit

unidad f para avance m - [sold] travel unit
— f — — m mecanizado - [sold] mechanized travel(ing) unit
— f — bombeo m - [bombas] pumping unit
— f — — m para agua m - [bombas] water pumping unit
— f — — — m para alimentación f - [bombas] feed water pumping unit
— f — calentamiento m - [ind] heating unit
— f — — m en planta(s) f siderúrgica(s) - [metal-prod] steel plant(s) heating unit(s)
— f — cambio(s) m - [autom-mec] shifting unit
— f — — m neumática - [autom-mec] air shift unit
— f — — m con aire m para émbolo m - [mec] piston air shift unit
— f — — m eléctrica - [autom-mec] electric shift(ing) unit
— f — — m neumática - [mec] air shift unit
— f — — m para eje(s) m - [autom-mec] axle shift unit
— f — — m para pistón m - [mec] piston air shift unit
— m — carrete m - [mec] reel unit
— f — — m para alambre m - [sold] wire reel unit
— f — circuito m - [electr] circuit unit
— f — — m para motor m - [electr] motor circuit unit
— f — cementación f - [nucl] cementing, o cement incorporation, unit
— f — centrado*m - [metal-lam] centering unit
— f — — m para debobinador* m - [metal-lam] decoiler centering unit
— f — centro m para distribución f - [electr] load center(ing) unit
— f — cocción f - [culin] cooking, o baking, unit
— f — — f de galletita(s) f - [culin] cookie baking unit
— f — conservación f - [ind] maintenance, o service, kit
— f — corriente f alterna - [sold] alternating current, unit, o machine
— f — — f continua - [sold] direct current, unit, o machine
— f — depósito m - [mec] tank unit
— f — — m para crema f - [culin] cream tank unit
— f — — m para relleno m - [culin] cream tank unit
— f — encendido m - [ind] starting, o ignition, unit, o kit
— f — — m para rectificador m - [electr] rectifier starting unit
— m — enderezamiento m - [mec] straightening unit
— f — — m de extremo m - [mec] end straightening, o back bending, unit
— t — energía f - [electr-instal] energy, o power, unit
— f — — f almacenada - [electr-instal] stored energy unit
— m — enfriamiento m - [ambient] cooling unit
— f — ensayo(s) m - [ind] testing unit
— f — — m con tierra f - [instrum] ground-(ing) tester
— f — escritorio m - desk(top) unit
— f — estañadura f - [metal-trat] tinning unit
— f — — f por inmersión f - [metal-trat] dip tinning unit
— m — — m — — f con paso m (di)recto - [metal-trat] straight away hot dip tinning unit
— f — flotación f - [emtal-trat] floating unit
— f — formación f - [ind] forming, o shaping, unit
— f — — f de galletitas f - [culin] cookie, forming, o shaping, unit
— f — frecuencia f alta - [electrón] hi(gh) freq(uency) unit
— f — fuerza f motriz f - [sold] power unit
— f — — f motriz con soldadora f - [sold]

power unit and welder
unidad f para fuerza f motriz con soldadora f
  para corriente f alterna - [sold] power unit
  with @ alternating current welder
— f —— f —— f —— f continua—[sold]
  power unit with @ direct current welder
— f — giro m - [grúas] rotating,tion unit •
  revolving unit
— f —— m con motor m hidráulico
  - [grúas] hydraulic motor powered revolving
  unit
— f — impulsión f - [mec] drive,ving unit
— f —— f con motor m—[mec] motor drive,ving
  unit
— f — incorporación f—incorporating,tion unit
— f —— f en cemento m - [nucl] cement incorporating,tion unit
— f — informe m - [electrón] report(ing) unit
— f —— m único - [electrón] single re-
  port(ing) unit
— f — inmersión f - [metal-trat] dipping unit
— f —— f en ácido m - [metal-trat] acid
  dip(ping) unit
— f — inspección f—[ind] inspecting,tion unit
— f —— f con partícula(s) f magnética(s) -
  [electrón] magnetic particle inspection unit
— f — intercambio m—interchange(ability) unit
— f — lavado m - [ind] wash(ing), o scrub-
  -(bing) unit
— f — limpieza f - cleaning unit
— f —— f con afrecho m - [metal-trat] bran
  cleaning unit
— f —— f — salvado m - [metal-trat] bran
  cleaning unit
— f — medición f - measuring unit
— f — medida f - measuring,rement unit •
  [labor] unit measurement
— f —— f para trabajo m - [labor] work measurement unit
— f — pedido(s) m - [com] order unit
— f — plancheado m - [metal-trat] plating unit
— f —— m electrolítico - [metal-trat] electrolytic plating uni
— f — procesamiento m - [ind] processing unit
— f — producción f - production unit • [ind]
  operating unit
— f —— f producida - [labor] production unit
  produced
— f —— f subterránea - [petról] subsurface
  production unit
— f — prolongación f - [sold] extension unit
— f — pulverización f - [mec] pulverizing unit
— f — rayo(s) m gamma - [electrón] gamma (ray)
  unit
— f —— m — obtenible en plaza f—[electrón]
  commercial gamma (ray) unit
— f — rayo(s)-X - [electrón] X-ray, unit, o
  machine
— f — recuperación f - recovery,ring unit
— f — reducción f - [ind] reducing,ction unit
— f —— f con engranaje(s) m - [mec] gear
  reducing,ction unit
— f —— f —— m helicoidal(es) - [mec] worm
  gear reducing,ction unit
— f —— f —— m sin fin - [mec] worm gear
  reducing,ction unit
— f —— f de sólido(s) m - [ind] solid(s) reducing,ction unit
— f —— f — velocidad f - [mec] speed reducing,ction unit
— f —— f simple—[mec] single reducing,ction
  unit
— f — refrigeración f - [ind] cooling unit
— f —— f con agua m - [ind] water cooling
  unit
— f —— f para agua m - [hidr] water cooling
  unit
— f — regulación f - [ind] control(ling) unit
— f —— f para caudal m - [hidr] flow con-
  trol(ling) unit
— f —— f para centrado* m - [metal-lam]
  centering control(ling) unit
— f —— f —— m de chapa f - [metal-lam]

strip centering control(ling) unit
unidad f para regulación f para combustión f -
  [combust] combustion control(ling) unit
— f —— f — potencia f - [ind] power con-
  trol(ling) unit
— f —— f tipo m túnel - [ind] tunnel type
  control(ling) unit
— f — reparación(es) f - [mec] repair(ing),
  unit, o rig
— f — reproducción f - [electrón] display unit
— f —— f para alarma f - [electrón] alarm
  display(ing) unit
— f —— f para estado m - [electrón] status
  display(ing) unit
— f —— f —— m conectada - [electrón]
  connected status display(ing) unit
— f —— f —— m de alarma f - [electrón]
  alarm status display(ing) unit
— f — rotación f - [grúas] rotating,tion unit
— f — soldadura f - [sold] welding unit
— f — f para producción f - [sold] production welding, unit, o equipment
— f — terminación f - [ind] finishing unit
— f — tracción f - [mec] traction,tor unit
— f —— f lateral - [mec] side, o lateral,
  traction,tor unit
— f — transferencia f - [mec] transfer unit
— f —— f para regulación f - control(ling)
  transfer unit
— f — tratamiento m - [metal-trat] treatment
  unit
— f —— m térmico - [metal-trat] heat treatment unit
— f — velocidad f - [mec] speed (control) unit
— f —— f variable - [ind] variable speed
  (control) unit
— f — ventilación f - ventilating,tion unit
— f — vigilancia f [mec] care unit
— f —— f intensiva - [ind] intensive care
  unit
— f — viscosidad f - [lubric] viscosity unit
— f patinadora f - [mec] skid(ding) unit
— f perforadora - [mec] perforating unit •
  [petról] drilling unit
— f por peso m - [ind] weight unit
— f —— m para venta f de lámina(s) f -
  [metal-prod] base box
— f portátil - [ind] portable, unit, o equipment
— f — corriente - current portable unit • portable commercial unit
— f — en base f de pila(s) f - [ind] battery
  powered portable unit
— f — corriente - [electrón] portable commercial unit
— f —— para rayos m gamma - [electrón] portable commercial gamma ray unit
— f — obtenible en plaza f - (available) portable commercial unit
— f —— inspección f - [ind] portable inspection unit
— f —— f de partícula(s) f magnética(s) -
  [electrón] portable magnetic particle inspection unit
— f — para rayos m gamma - [electrón] portable
  gamma ray unit
— f primera - [ind] first unit
— f procesadora - [ind] processing unit
— f producida - [ind] unit produced
— f productiva - [ind] production,tive unit
— f — de fuerza f motriz - [electr-prod] power
  (producing) unit
— f —— f — con soldadora f - [sold]
  power unit and welder
— f —— f — para soldadora f - [sold]
  power unit and welder
— f —— f —— f para corriente f alterna - [sold] power unit with @ alternating
  current welder
— f —— f —— f —— f continua -
  [sold] power unit with @ direct current
  welder

unidad f que requiere solo,lamente enchufarse -
[electr] simple plug-in unit
— f quitada - removed unit
— f radiográfica - [electrón] radiographic unit
— f — portátil - [electrón] portable radiographic unit
— f respectiva - respective unit
— f refrigeradora - [criog] cooling unit • refrigerating unit
— f — para agua m - [criog] water cooling unit
— f — — — m destilada - [criog] distilled
water cooling unit
— f — — — m — en circuito m cerrado -
[hidr] closed circuit distilled water cooling unit
— f reductora - [mec] reducing,ction unit
— f para velocidad f - [mec] speed reducing
unit
— f reguladora f - [mec] control(ling) unit
— f — para potencia f - [mec] power control
unit
— f removida - [ind] removed unit
— f retrocedida - [autom-mec] reversed unit
— f recia f - [ind] rugged unit
— f rotadora - [cuerp.moled] rotator,ting unit
— f —/alimentadora - [cuerp.moled] rotator/-
feeder, unit, o carriage
— f rotatoria - [mec] rotating unit
— f sacada - [mec] removed unit
— f semiintegrada - [ind] semi-integrated unit
— f sencilla - [ind] simple unit
— f sencilla para incorporación f - [nucl]
simple incorporation unit
— f — — — f en cemento m - [nucl] simple
cement incorporation unit
— f sensora - [electrón] sensing,sor unit
— f — de falla f a tierra f - [electr] ground
sensing,sor, unit, o equipment
— f separada - separate(d) unit • [mec] module
— f similar - similar unit
— f sobre caballete m - [grúas] A-frame unit
— f soldadora - [sold] welding unit
— f térmica - [termol] thermal, o heat, unit
— f térmica británica - [termol] B T U; British
thermal unit
— f — inglesa - [termol] B T U; British
thermal unit
— f tipo abaco m - [cement] abacus (type) unit
— f tractora - [mec] tractor unit
— f transportadora - [mec] transporting, o conveying, unit
— f turbogeneradora - [electr-prod] turbogen-
erating unit
— f variable - variable unit
— f vertical - vertical unit
— f — partida - [constr] split vertical unit
unidad-hora f - unit-hour
unido,da a - . . . . [mec] spliced • fused •
linked (together)
— a acero m con acero m - [sold] steel to steel
— a ajustadamente - [mec] joined tightly
— a con soldadura f - [sold] welded together
— a económicamente - united economically
— a entre sí - joined, o fitted, together
— a fácilmente - joined easily
— a firmemente - [mec] joined tightly
— a mecánicamente - fastened mechanically
— a mediante fusión f - [sold] fuse joined;
fused together
— a — presión f - [mec] snapped (together)
— a metal m con metal m - [mec] joined metal-
-to-metal
— a metal(es) entre sí - [sold] joined metal-
-to-metal
— a perpendicularmente - joined perpendicularly
Unidraulic a - Unidraulic
unificación f - . . . . • unification; unitization
• [f.c.] (gage) standardizing,zation
— f de coeficiente(s) - coefficient unification
— f de trocha(s) f - [f.c.] gage(s) standardizing,zation
— f ferroviaria—[f.c.] railway standardization
— f total - total, o complete, unification, o

unitizing
unificado,da a - unified; unitized; united
— a totalmente - unified, o unitized, o united,
completely
unificar v - . . .; to unitize • to unite
— v coeficiente(s) m - to unify @ coefficients
— v totalmente - to, unify, o unitize, completely
unifilar a - unifilar • [dib] single line
uniformación f - . . . . • conforming • evening
— f de método(s) m - method(s), unifying, o
conforming
— f — — m contable(s) - [contab] accounting
method(s), conforming, o unifying
uniformado,da a - made uniform; conformed •
evened • equalized
uniformar v - . . . . • to unitize • to conform •
to even • to equalize • to match
— v métodos m - to make @ method(s) uniform
— v — m contable(s) - [contab] to, conform, o
make uniform, @ accounting method(s)
uniform m - . . . | a - uniform; even; regular
• standard • steady • matched
uniformemente adv - . . .; evenly
uniformidad f de inducción f - [electrón] induction uniformity
— f — — f magnética - [electrón] magnetic
inducion uniformity
— f de metalización f - [metal-prod] metallization uniformity
— f — norma(s) f - standardization
— f en operación f - [ind] operation uniformity
— f — ductilidad f - ductility uniformity
— f — espsor m de pared f - [constr] wall
thickness unifomity
— f operación f de horno m - [metal-prod]
furnace operation uniformity
— f en redondez f - roundness uniformity
uniformización f - uniforming,mizing; equalizing • matching
unilateralidad f - unilaterality
unilateralmente adv - unilaterally
unión f - . . . • [mec] coupler,ling; tie;
splice; collar; connection,tor; junction;
joint • fusing; linking; joining • [tub]
nipple • [sold] weld
— f a tope m - [sold] butt joint
— f — — m con chaflán m - [sold] butt joint
— f — m en V - [sold] V-butt weld
— f abierta - [mec] open(ed) joint
— f — en tubería f - [tub] pipe open joint
— f — — — f rígida - [tub] rigid pipe open
joint
— f aceptable - [tub] acceptable, joint, o
coupling
— f acodillada - [mec] toggle joint
— f adecuada - [mec] adequate. joint, o coupling
— f — especial - [mec] adequate special coupling
— f ajustada - [mec] tight, joint, o coupling
— f articulada - [mec] articulated, o knuckle,
o link, joint, o coupling
— f atornillada - [mec] screwed, coupling, o
connecion
— f circular - [sold] circular, o circumferential, o girth, coupling, o joint
— f comun - [tub] common, o standard, coupling
— f con brida(s) f - [tub] flange(d), joint, o
coupling
— f — — f Kewane - [tub] Kewane('s) flange(d)
joint
— f — campana f - [tub] bell coupling
— f — — f doble - [tub] double bell coupling
— f — — f y espiga f - [tub] bell and apigot
coupling
— f — capa bitumástica - [tub] bitumastic
coated,ting coupling
— f — conexión f - [tub] connection coupling
— f — — f positiva - [tub] positive connection coupling
— f — extremo(s) m rebajado(s) - [tub] offset
end joint

unión f con fuga(s) f - [mec] leaking,ky joint
— f — guarnición f - [tub] gasketed joint
— f — — f anular—[tub] O-ring gasketed joint
— f — — f anular de caucho m - [tub] O-ring (type) rubber gasketed joint
— f — f de caucho m - [tub] rubber gasketed joint
— f — junta f—[tub] joint, union, o coupling
— f — con caja f y espiga - [mec] tongue and groove joint, union, o coupling
— f — — f machihembrada - [mec] tongue and groove joint (union, o coupling)
— f — liga f química - [mec] solvent welded joint
— f — recubrimiento m bitumástico - [tub] bitumastic coated coupling
— f — taladro m de . . . - [tub] . . . bore coupling
— f — válvula f - [hidr] valve joint
— f — — f para regular v circulación f - [hidr] circulation (controlling) joint
— f cónica - [tub] cone,nic(al) joint
— f — para herramienta(s) f - [petról] tool joint
— f — — f para perforación f - [herram] tool joint
— f — tubería f - [tub] pipe,ping conic(ai) joint
— f — — f para perforación f - [petról] tool joint rotary system
— f — sólida - [petról] solid, conical, o tool, joint
— f — — para herramienta f - [petról] conical tool joint
— f — — f para cable m - [petról] cable tool joint
— f corriente - [tub] standard, joint, o coupling
— f — de hierro m - [mec] standard iron coupling
— f — — m galvanizado - [mec] standard galvanized iron coupling
— f de acero m - [tub] steel coupling
— f — m común - [tub] standard steel coupling
— f — — m con(tra) acero m - [sold] steel to steel, coupling, o joining
— f — — m corriente - [mec] standard steel coupling
— f — — m según norma f - [tub] standard steel coupling
— f — acoplamiento m tipo m Erickson—[electr-equip] Erickson coupling, union, o joint
— f — ala(s) f - [mec] flange splice,cing
— f — borde(s) f - [mec] edge(s), o side(s), joining
— f — brida(s) f - [tub] flange, joining, o coupling
— f — — f hidráulica(s) - [tub] hydraulic flange(s), joining, o coupling
— f — bronce m—[tub] bronze, joint, o fitting
— f — — m con pestaña f - [tub] bronze flanged, joint, o fitting
— f — cable(s) m - [cabl] cable(s) joining • hook and eye joint
— f — cloruro m de polivinilo - [tub] polyvinyl chloride coupling
— f — cubierta f con cabeza f - [mec] cover/-head, joint, o coupling
— f — charnela - [mec] hinge, o knuckle, joint
— f — dos mitades f - [mec] two-piece coupling
— f — — — f empernadas - [mec] two-piece bolted coupling
— f — — piezas f - [mec] two-piece coupling
— f — — sección(es) para formar una sóla pieza f [tub] double jointing
— f — fondo m - [tub] bottom coupling
— f — hierro m - [tub] iron joint
— f — — m galvanizado - [tub] galvanized iron joint
— f — — m maleable - [tub] malleable iron joint
— f — junta f cónica—[tub] ground joint union

unión f de latón m—[mec] brass, joint, o union
— f — metal(es) m - [sold] metal(s) joining
— f — — m con metal m - [sold] metal to metal to metal joining
— f — — m en escala f comercial - [sold] commerciazl metal(s) joining
— f — — m entre sí - [sold] metal to metal joining
— f — rótula f - [mec] articulated joint
— f — sección(es) f - [mec] section joining
— f — sindicato(s) m - [ind] professional guild
— f — — m profesional(es) m - [ind] professional (syndicate), guild, o union
— f — solapa f - [sold] lap joint
— f — tipo m torcido - [alambre] wound type, fastening, o joint
— f — tope en V - [sold] V-butt joint
— f — tubería f - [constr] pipe, joint, o joining
— f — — f rígida - [tub] rigid pipe,ping, joint, o joining
— f — tubo(s) m - [tub] tube(s), joint, o coupling
— f — válvula f - [válv] valve, joint, o joining
— f — — f para viento m - [metal-prod] blast valve, joint, o joining
— f — — f — — m caliente - [metal-prod] hot blast valve, joint, o joining
— f — — f — — m de estufa f - [metal-prod] stove hot blast valve, joint, o joining
— f — — f — — m — — — f con coraza f - [metal-prod] stove hot blast valve to shell connection
— f — vástago m - [mec] rod, joint, o joining
— f — — m para regulador m—[mec] throttle rod coupling
— f deteriorada - [tub] deteriorated, joint, o joining
— f defectuosa - [tub] defective, joint, o joining
— f doble - [mec] double, joint, o joining • [herram] tool joint
— f — para barra f para sondeo m - [petról] tool joint
— f económica -[mec] economic(al), o low cost, union, o joint, o joining
— f efectiva - [mec] effective, union, o joint, o joining
— f efectuada - [mec] made, joint, o joining
— f — en obra f - [tub] field, joint, o joining
— f elástica - [mec] elastic, o flexible, joint, o joining, o coupling
— f embridada - [tub] flanged, union, o joint, o joining
— f empernada - [mec] bolted, joint, o joining, o coupling
— f — de dos mitades f - [mec] two-piece bolted coupling
— f — rápida - [mec] quick bolted, joint, o joining
— f — bisel - [sold] véase soldadura f a sesgo - [autom-carroc] véase borde m afinado
— m — extremo n — [mec] end, joint, o joining
— f — fabricación f - [metal-fabr] manufacturing, joint, o joining
— f en obra f - [sold] field, connecting,tion, o splice,cing, o joint, o joining
— f — T - [mec] T-. joint, o joining
— f — taller m - [sold] shop, weld(ing), o connecting,tion, o splice,cing
— f esférica - [mec] ball joint coupling
— f especial - [mec] special, joint, o coupling
— f — adecuada - [mec] adequate special coupling
— f estándar - [mec] standard coupling
— f — de acero m - [mec] steel standard coupling
— f — — — m galvanizado - [mec] standard galvanized steel coupling

unión f estructural - [constr] structural joint
— f expansible - [mec] expansion, joint, o
  splice
— f fácil - easy joining
— f firme - [mec] tight, joint, o joining
— f flexible—[mec] flexible, joint, o coupling
— f fuera de eje m - [tub] offset, joint, o
  coupling
— f galvanizada - [mec] galvanized, joint, o
  coupling
— f — corriente - [mec] standard galvanized,
  union, o coupling
— f — estándar - [mec] standard galvanized
  coupling
— f — hembra - [mec] female galvanized, union,
  o coupling
— f — — f para tubería f - [mec] pipe galvanized female, union, o coupling
— f giratoria - [mec] swing, joint, o coupling
— f — acampanada - [mec] flared swivel joint
— f — hidráulica - [hidr] water swivel
— f Hansen - [mec] Hansen, joint, o coupling
— f — para tapón m - [mec] Hansen plug coupling
— f hecha en obra f - [tub] field joint
— f hembra - [mec] female, joint, o union
— f — doble - [tub] double female, union, o
  joint, o coupling
— f — — con asiento m plano - flat seat
  double female, union, o joint, o coupling
— f — — — — m — de hierro m (galvanizado)
  maleable - [mec] flat seat end double female
  malleable iron, coupling, o joint
— f — Stockham - [mec] Stockham female, union,
  o joint, o coupling
— f hermética - [mec] watertight joint
— f integral - [tub] integral coupling
— f intermedia - [mec] intermediate, joint, o
  coupling
— f Internacional para Comunicación(es) f -
  [pol] International Communications Union
— f — — Telecomunicación(es) f - [pol] International Telecommunications Union; I T U
— f — — Transportes m Públicos - [transp] International Public Transportation Union
— f invertida - [tub] inverted connector
— f Kewane - [tub] Kewane('s), union, o coupling
— f lisa - [tub] smooth, joint, o coupling
— f machihembrada - [mec] tongue and groove,
  joint, o coupling
— f mala - [tub] bad, o poor, joint, o coupling
— f mecánica - [mec] mechanical, union, o coupling
— f — integral - [tub] integral mechanical,
  joint, o coupling
— f mediante fusión f [sold] fuse(d) joining
— f — presión f - [mec] snapping
— f mixta - Kewane's, union, o coupling
— f muy restringida - [sold] highly restricted,
  joint, o joining, o coupling
— f — rígida - [sold] highly restrained joint
— f oscilante - [mec] swing(ing) joint
— f para ajuste m - [tub] adjusting coupling
— m — — m automático - [tub] automatic adjusting coupling
— f barreno,na - [petról tool joint
— f — bifurcación f - [tub] wye fitting
— f — cambio m rápido - [mec] quick change,
  union, o coupling
— f — carburador m - [comb.int] carburetor,
  joint, o coupling
— f — circulación f - [hidr] circulation joint
— f — construcción f - [tub] construction coupling
— f — crédito - [fin] credit, union, o consor-
  tium
— f — conductor(es) m - [electr-cond] conduc-
  tor(s) joint • [labor] driver's union
— m — herramienta(s) f - tool joint • [petról]
  full hole tool joint
— f — f para cable m - [petról] cable tool
  joint

unión f para reemplazo m rápido - [mec] quick
  change, union, o coupling
— f — regulador m - [mec] regulator, o
  throttle, joint, o coupling
— f — reparación f - [mec] repair coupling
— f — tapón m - [mec] plug coupling
— f — tubería f - [tub] tube,bing, o pipe, o
  piping, union, o joint, o coupling
— f — f desconectada - [tub] disconnected
  pipe, union, o coupling
— f — — f para grúa f - [grúas] crane pipe,
  union, o coupling
— f — vástago m - [mec] rod coupling
— f — m para válvula f - [válv] valve rod,
  coupling, o substitute
— f — ventosa f - [válv] air valve coupling
— f pareja f - [tub] smooth, o even, joint
— f Perkins - [petról] Perkins joint
— f pobre - [tub] poor, joint, o coupling
— f positiva - [mec] positive joint
— f postal - [telecom] postal union
— f — Universal - [pol] Universal Postal Union
  • U P U
— f rápida - [mec] quick, joint, o joining
— f rebajada - [mec] offset joint • reduced
  joint
— f remachada - [mec] riveted joint
— f resistente - [mec] strong joint
— f rígida - [sold] rigid, o restrained, joint
— f roscada - [tub] threaded, joint, o coupling
— f rugosa - [mec] rough joint
— f según norma f - [tub] standard, joint, o
  coupling
— f sencilla f - [mec] simple joint
— f simple - [tub] simple joint
— f sin roscar v - [tub] unthreaded coupling
— f soldada - [sold] welded joint
— f — con disolvente m - solvent welded joint
— f — en obra f - [constr] field welded, connection, o joint
— f Stockham - [mec] Stockham, joint, o union
— f suave - [mec] smooth, o soft, joint
— f T - [mec] T-joint
— f típica - [mec] typical joint
— f tipo Dresser - [tub] Dresser coupling
— f — doble torcida - [alambre] double wound
  type, joint, o fastening
— f — triple torcida - [alambre] triple wound
  fastening
— f torcida - [alambre] wound, o twisted,
  fastening
— f — triple -- [alambre] triple wound fastening

unipersonal a - . . .; one-man
unipolar a - . . . . • single conductor
— a con tres cuchillas f - [electr-instal]
  single-pole triple-throw
— a — — vías f - [electr-instal] single-
  -pole triple-throw
— a — — direcciones f - [electr-instal]
  single-pole triple throw
— a — dos cuchillas f - [electr-instal] single-pole double throw
— a — — vías f - [electr-instal] single-pole
  double-throw
— a — — dirección(es) f - [electr-instal]
  single-pole double-throw
— a — una (sóla) cuchilla f - [electr-instal]
  single-pole single-throw
— a — — (—) vía f - [electr-instal] single-
  -pole single-throw
— a — — (—) dirección f - [electr-instal]
  single-pole single-throw
unir v - . . . . • to chain; to link • to splice •
  to fuse • to tie • to, link, o hold, together
— v acero m con acero m - [sold] to join steel
  to steel
— v ajustadamente - [mec] to join tightly
— v bajo presión f - to join under pressure
— v borde m - to join @, edge, o side
— v bóveda f - [tub] to join @ arch

unir v con perno(s) m - [mec] to bolt together)
— v — soldadura f - [sold] to weld (together)
— v económicamente - to unite economically
— v fácilmente - to join easily
— v firmemente - [mec] to join, firmly, o
tightly
— v mecánicamente - [constr] to, join, o
fasten, mechanically
— v mediante fusión f - [sold] to fuse together
— v — perno(s) m - [mec] to bolt (together)
— v — presión f - [mec] to snap (together)
— v — soldadura f - [sold] to weld together
— v metal(es) m - [sold] to join @ metal(s)
— v — m con metal(es) - [sold] to join metal
to metal
— v — m entre sí - [sold] to join metal to
metal
— v plancha(s) f con soldadura f - [sold] to
weld @ plate(s) together
— v sección(es) f - [mec] to join @
section(s)
— v tubería f - [tub] to join @ pipe
— v — f abovedada(s) f - [tub] to join @
pipe-arch(es)
— v tubo m - [tub] to join @ pipe
unitario,ría a - . . .; unit
universal a . . .; versatile; véase también para
fines m múltiples
universidad f - [educ] . . . • college
— f estatal - [educ] state, university, o college • government, university, o college
— f nacional - [educ] national university
— f privada - [educ] private university
uno adv por uno - one by one; one at @ time
— adv — medio - alternate • every other
— m para otro m - for each other
— adv por sistema - one per system
— adv contra otro adv - against each other
— adv detrás de otro adv - [deport] nose-to-tail
— adv cuántos - several
untado,da a - coated • [lubric] oiled
— a liberalmente - coated, o oiled, liberally
— a — con aceite - [mec] oiled liberally
— a — con grasa f - [mec] greased liberally
untador,ra a - [sold] buttering
untadora f - [culin] spreader
— f para crema f - [culin] cream spreader
— f — — f para oblea(s) f - [culin] wafer
cream spreader
untadura f - . . . • [sold] buttering
— f con acero m inoxidable - [sold] stainless
steel buttering
untamiento m liberal - liberal coating
— m — con grasa f - [lubric] liberal grease
coating
untar v - . . . • [sold] to butter
— f liberalmente v - [sold] to coat liberally
— v — con grasa f - [mec] to coat liberally
with grease
untura f - . . . • [sold] buttering
uña f - . . . • [mec] finger; lug; paw; fang •
catch • [constr-equip] ripper
— f accionada con resorte m - [mec] spring controlled finger
— f de pie m - [anat] toenail
— f metálica - [mec] metal(lic) finger
— f para avance m gradual—[mec] inching finger
— f — basculación f - [mec] tilting lub
— f — cañón m - [metal-prod] mud-gun catch
— f — cursor m - [mec] slide,ding finger
— f — — m con mordaza f - [trefil] grip slide
finger
— f — — m para avance m gradual -
[trefil] inching grip slide finger
— f — — m para sujeción f - [trefil] grip
slide finger
— f — elevación f - [mec] lift(ing) finger
— f — empuje m - [mec] pusher,shing finger
— f para escarificadora f - [constr-equip] sca-
rifier ripper
— f — manipulador m - [metal-lam] manipulator
finger
— f — — m para entrada f - [metal-lam] entry

manipulator finger
uña f para manipulador m para salida f -
[metal-lam] delivery manipulator finger
— f — posicionador m - [mec] positioner, o
placer, finger
— f — mordaza f - [trefil] grip finger
— f — — f para avance m - [trefil] advance,
o foward movement, grip finger
— f — — f — m gradual - [trefil] inching grip finger
— f — sujetador m para avance m - [trefil]
advance, o forward movement, grip finger
— f — — m — m gradual - [trefil] inching
grip finger
— f — tablilla f - [carp] slat end
— f — tumbador m - [metal-lam] manipulator
finger
— f — volteador m - [metal-lam] manipulator
finger
— f — — m para entrada f - [metal-lam] entry
manipulator finger
— f — — m — salida f - [metal-lam] delivery
manipulator finger
— f transportadora - [mec] transporting, o
traveling, finger
uptake* m - [metal-prod] véase conducto m ascendente
uranio m enriquecido - [nucl] enriched uranium
urbanización f - [constr] . . . • development;
housing, o community, o urban, o real estate,
development; subdivision • developed real estate • housing project
— f comercial - [constr] commercial development
— f con financiamiento m privado - [constr]
privately financed development
— f — financiación f privada - [constr] privately financed development
— f — instalación(es) f recreativa(s) - [dep]
recreational development
— f industrial—[constr] industrial development
— f — recreo m - [constr] recreational development
— f prevista f - [pol] anticipated development
— f residencial - [constr] residential, o housing, development
— f ribereña - [constr] waterfront development
urbanizador m - [constr] developer
— m de inmueble(s) m ribereño(s) - [constr]
waterfront developer
urbanizar v - . . . • [constr] to develop (@
real estate)
urbe f - [pol] . . .; major city
urgir v - to urge • to press
urna f - . . .; slotted box
— f biclave - [pol] double lock urn
usable a - usable
usado,da a - . . . . • applied • operated •
second hand
— a acostumbradamente - used commonly
— a anteriormente—used, previously, o formerly
— a como fuente f de energía f - [electr- used,
for power, o as @ power source
— a comunmente - used commonly
— a con - used with
— a — éxito - used successfully
— a — seguridad f - [segurid] used safely
— a constantemente - used constantly
— a corrientemente—used, currently, o commonly
— a independientemente - used independently
— a de nuevo - used again; reused
— a en forma f continua(da) - used continuously
— a extendidamente - used extensively
— a extensamente - used extensively
— a inapropiadamente - used improperly
— a indebidamente - misused
— a ininterrumpidamente- used, uninterruptedly,
o continuously
— a mayormente - used mainly
— a mucho - used extensively
— a normalmente - used normally
— a nuevamente - used again; reused
— a ocasionalmente - used occasionally
— a sobre llanta f - [autom-neumát] used on @

rim

usado,da a **ventajosamente** - used advantageously
— a **obligatoriamente** - used mandatorily
— a **para** - used for
— a — **energía** f - [electr] used for power
— a — **soldadura** f - [sold] used for welding
— a **posteriormente** - used, later, o afterwards
— a **repetidamente** - used repeatedly; reused
— a **sobre riel(es)** m - [med] used on @ rail(s)
— a **típicamente** - used typically
— a **universalmente** - used universally
usador,ra a - user
— a **de recurso(s)** m - [econ] resource(s) user
usar v - . . . ; to consume; to utilize; to employ; to apply; to make use of; to operate • to incorporate • [electr-oper] to draw
— v **acelerador** m - [comb.int] to use @ throttle
— v **acostumbradamente** - to use, normally, o commonly
— v **anteojos** m - [segurid] to wear @ glass(es)
— v — **para seguridad** f - [segurid] to wear @ safety glasses
— v **anteriormente** - to use previously
— v **bomba** f - [bombas] to use @ pump
— v **cabeza** f - to use @ head
— v — f **para inyección** f - [petról] to use @ swivel
— v **cable** m a **tierra** f - [sold] to use @ ground
— v — — — f **con ramal(es)** m - [sold] to split @ ground
— v **cadena** f - [cadenas] to use @ chain
— v **cal** f - [constr] to use @ lime
— v **calibre,brador** - [mec] to use @ gage
— v — m **para ranura(s)** f - [mec] to use @ groove gage
— v **como fuente** f **para energía** f—[electr-oper] to use for power
— v **comúnmente** - to use commonly
— v **con** - to, use, o be used, with
— v — **candado** m - [segurid] to use with @ padlock
— v — **éxito** - to use successfully
— v — **seguridad** f - [segurid] to use safely
— v **constantemente** - to use constantly
— v **corrientemente** - to use, currently, o commonly, o usually
— v **crédito** m - [fin] to use @ credit
— v **criterio** m (**sano**) - to exercise judgment
— v **de nuevo** - to reuse
— v **derecho(s)** m - to (make) use (of) @ right
— v — m **de patente** f (**de invención** f) - [legal] to (make) use (of) @ patent right(s)
— v — m **literario(s)** - [legal] to (make) use (of) @ copyright
— v **electrodo** m - [sold] to use @ electrode
— v — m **con alma** m **fundente** - [sold] to use @ Innershield (electrode)
— v **en forma continua(da)** - to use continuously
— v **energía** f - [electr-oper] to use @ power
— v **engranaje** m - [mec] to use @ gear(ing)
— v **explosivo(s)** m - [explos] to use @ explosive(s)
— v **extendidamente** - to use extensively
— v **extensamente** - to use extensively
— v **filtro** m - [ind] to use @ filter
— v **fluido** m - to use @ fluid
— v — m **en neumático** m - [agric-equip] to use @ fluid in @ tire
— v **freno** m - [mec] to, use, o apply, @ brake
— v **fuerza** f - to use, force, o strength
— v **gafas** f - [segurid] to use @ goggles
— v — f **para seguridad** f - [segurid] to use @ safety, goggles, o glasses
— v **guante(s)** m - [segurid] to wear @ glove(s)
— v **implemento** m - [agric] to use @ implement
— v **impuesto(s)** m - [fisc] to use @ tax(es)
— v **inapropiadamente** - to use improperly
— v **indebidamente**—to, use improperly, o misuse
— v **independientemente** - to use independently
— v **ininterrumpidamente** - to use, uninterruptedly, o continuously
— v **Innershield** m - [sold] to use @ Innershield (electrode)

usar v **letra(s)** f **de imprenta** f - to print
— v **lingotera(s)** f - [metal-prod] to use @ ingot mold(s)
— v **mandril** m - [mec] to use @ drift
— v — m **de latón** m - [mec] to use a brass drift
— v **manguito** m - [mec] to use @ sleeve
— v — m **apropiado** - [mec] to use @ appropriate sleeve
— v **marca** f - [legal] to use @ (trade) mark
— v — f **de fabricación** f - [legal] to use @ trade mark
— v — f **industrial** - [legal] to use @ trade, mark, o name
— v — f **registrada** - [legal] to use @ trade mark
— v **mayormente** - to use mainly
— v **mayúscula(s)** f - [gram] to capitalize
— v **mazo** m - [mec] to use @ mallet
— v **medidor** m - [instrum] to use @ meter
— v **misma pieza** f - [mec] to use @ same part
— v **mucho** - to use, much, o extensively
— v **nombre** m **comercial** - [legal] to use @ trade name
— v **nomograma** m - to use @ nomograph,gram
— v **normalmente** - to use normally
— v **nuevamente** - to use again; to reuse
— v **obligatoriamente** - to use mandatorily
— v **ocasionalmente** - to use occasionally
— v **para** - to use for
— v — **energía** f - [electr] to use for power
— f — **soldadura** f - [sold] to use for welding
— v **patente** f (**de invención** f) - to use @ patent
— v **pieza** f - [mec] to use @ part
— v — f **de otra máquina** f - [mec] to cannibalize @ machine
— v — f **otro vehículo** m - [autom] to cannibalize @ vehicle
— v — f **nueva** - [mec] to use @ new part
— v — f **para reparar** v **otra máquina** f - [mec] to cannibalize
— v — f — — v **otro vehículo** m - [autom] to cannibalize @ vehicle
— v — f **usada** - [ind] to use @ (re)used part
— v **posteriormente** - to use later
— v **prensa** f - [mec] to use @ press
— v **producto** m - to use @ product
— v **prolongación** f - [mec] to use @ extension
— v **propiedad** f **intelectual** - [legal] to use @ copyright
— v **recurso(s)** m - [econ] to use @ resource(s)
— v — m **humano(s)** - [admin] to use @ human resource(s)
— v **refuerzo** m - [mec] to use @, reinforcement, o brace
— v **repetidamente** - to use repeatedly
— v **repuesto** m - (ind) to use @ spare part
— v **respirador** m - [segurid] to use @ respirator
— v **servicio(s)** m - to use @ service(s)
— v — m **contratado(s)** - [ind] to use @ contracted service(s)
— v **seso(s)** m - to use @, brain, o head
— v **sobre llanta** f - [autom-neumát] to use on @ rim
— v **sobre riel(es)** m - [mec-ruedas] to use on @ rail(s)
— m **suelo** v - [suelos] to use @ soil • [ind] to use @, floor, o ground
— v **típicamente** - to use typically
— v **universalmente** - to use universally
— v **vehículo** - to use @ vehicle
— v **ventajosamente** - to use advantageously
usina f - [ind] véase **planta** f - [electr-prod] power, plant, o house
— f **eléctrica** - [electr-prod] (electric) power, plant, o house
— f **hidroeléctrica** - [electr-prod] hydraulic power plant
— f **industrial** - [electr-prod] industrial (power) plant

usina f industrial en planta f siderúrgica -
[electr-prod] steel plant industrial power
plant
— f **termoeléctrica** - [electr-prod] thermoelectric, o steam, power plant
uso m - . . .; using; utilization •
application • consumption • service • operation • purpose • operating
— n **acostumbrado** - customary, o accustomed, o
customary, o common, o normal, use
— m **adicional** - additional use
— m **amplio** - ample, o broad, use
— m **anterior** - previous use
— m **aumentado** - increased use
— m **callejero** - [autom-neumát] street use
— m **cesante** - loss of use
— m **comercial** - commercial use
— m **— en general** - general commercial use
— m **como fuente f para energía f**—[electr-prod]
use as @ power source
— m **común** - common use
— m **con** - use with
— m **— éxito m** - successful use
— m **— pozo(s) m para agua m** - [hidr] water
well, use, o service
— m **— presión f** - [ind] pressure service
— m **— f alta** - [ind] high pressure service
— m **— f baja** - [ind] low pressure service
— m **— f elevada** - [ind] high pressure service
— m **— f reducida** - [ind] low pressure service • reduced pressure service
— m **— temperatura f alta** - [ind] high temperature service
— m **— f baja** - low temperature service
— m **— f elevada** - [ind] high temperature
service
— m **— f reducida** - low temperature service
• reduced temperature service
— m **— contrapeso m** - [mec] use with @ counterweight
— m **— seguridad f** - [segurid] safe, use, o
utilization
— m **constante** - constant use
— m **continuado** - continuous, use, o usage
— m **continuo** - continuous use
— m **contra muelle m** - [petról] dockside use
— m **corriente** - current, o common, o ordinary,
o routine, use, o utilization
— m **costa adentro m** - [petról] on shore use
— m **— afuera** - [petról] offshore use
— m **de cliente m** - customer('s) use
— m **— comprador m** - customer('s) use
— m **— consumidor** - consumer('s) use
— m **— chapa f** - [metal-fabr] strip use
— m **— equipo m** - [ind] equipment, use, o utilization
— m **— — m fabril** - [ind] manufacturing equipment, use, o utilization
— m **— para amoladura f** - [mec] grinding
equipment, use, o utilization
— m **— — m — f de rodillo(s) m** - [metal-
-lam] roll grinding equipment, use, o utilization
— m **— m para granallado m** - [mec] shot
blasting equipment, use, o utilization
— m **— — m de rodillo m**
- [metal-lam] roll shot blasting equipment,
use, o utilization
— m **— — m para rectificación f** - [mec] grinding equipment, use, o utilization
— m **— — m — f de rodillo(s) m** - [metal-
-lam] roll grinding equipment, use, o utilization
— m **— — m — torneado m** - [mec] turning
equipment, use, o utilization
— m **— — m de rodillo m** - [metal-
-lam] roll turning equipment utilization
— m **— fondo(s) m** - [fin] fund(s) use
— m **— instalación(es) f** - [ind] equipment,
use, o utilization
— m **— mano f de obra f** - [ind] manpower, use,
o utilization

uso m **de mano f de obra f para amoladura f** -
[metal-lam] grinding manpower, use, o utilization
— m **— — f — — f — — f de rodillo(s)** -
[metal-lam] roll grinding manpower, use, o
utilization
— m **— — f — — f — granallado m** - [metal-
-lam] shot blasting manpower utilization
— m **— — f — — f de rodillos m** -
[metal-lam] roll shot blasting manpower,
use, o utilization
— m **— — f — — f — rectificación f** [metal-
-lam] grinding manpower, use, o utilization
— m **— — f — — f — f de rodillo(s) m** -
[metal-lam] roll grinding manpower, use, o
utilization
— m **— . — f — — f — torneado m** - [mec]
turning manpower, use, o utilization
— m **— — — — — m de rodillo(s) m** -
[metal-lam] roll turning manpower, use, o
utilization
— m **— marca f (de fábrica f)** - trade mark use
— m **— material(es) m** - [ind] material(s), o
supply, lies, use
— m **— mineral m** - [miner] ore use
— m **— — m, cribado, o zarandeado** - [miner]
screened ore use
— m **— mordaza f** - [mec] clamp use
— m **— patente f** - [legal] use of @ patent;
patent use
— m **— personal m** - [labor] personnel use
— m **— pieza f** - [ind] part use
— m **— — f para (reparar) otra máquina f** -
[mec] cannibalization, zing
— m **— — f — (—) — otro vehículo m** -
[autom-mec] vehicle cannibalizing, zation
— m **— prerreducción f** - [miner] prereduction,
use, o practice
— m **— recurso(s) m** - [fin] resource(s) use
— m **— regulación f** - [ind] use of control
— m **— — f de demanda f** - [ind] demand control
use
— m **— sínter m** - [metal-prod] sinter use
— m **— — m autofundente** - [metal-prod] self-
-fluxing sinter use
— m **— vehículo m** - [autom] vehicle use
— m **debido** - proper use
— m **— de personal m** - [labor] proper personnel
use
— m **definido** - defined, o established, use
— m **definitivo** - end, o final, use
— m **deportivo** - [deport] recreational use
— m **durante todo año m** - year-round use
— f **efectivo** - effective use
— m **— de recurso(s) m** - [admin] effective resources use
— m **— — m humanos** - [admin] effective
human resource(s) use
— m **eficaz** - effective, use, o utilization
— m **eficiente** - efficient, o effective, use, o
utilization
— m **— de producto** - efficient, o effective,
product, use, o utilization
— m **— servicio** - efficient, o effective,
service, use, o utilization
— m **eléctrico** - [electr-oper] electrical use
— m **carretera f** - [vial] highway use
— m **en caso m de emergencia f** - emergency use
— m **— cuchara f** - [metal-prod] ladle use
— m **— eje m** - [mec] axle use
— m **— — m para dirección f** - [autom-mec]
steering axle use
— m **— forma continua** - continuous use
— m **— horno m** - [combust] furnace use
— m **— obra f** - field, o job site, use
— m **— planta f** - [ind] (in-)plant use
— m **— taller m** - [ind] shop use
— m **específico** - specific use • specific need
— m **estructural** - [constr] structural use
— m **exacto** - exact, use, o usage
— m **excesivo** - excessive use
— m **exclusivo** - exclusive use
— m **extendido** - extended, nsive use

uso m **extenso** - extended,nsive use
— m **extenso** - extended,nsive use
— m **externo** - external use
— m **fácil** - easy use | adv - (de) easy to use
— m — **en instalación(es)** - easy facility,ties use
— m **final** - final, o end, use
— m — **indicado** - indicated final use
— m — **propuesto** - [ind] final intended use
— m **fuera de carretera** f - [vial] off-road use
— m **general** - general, use, o purpose - all, o general, duty, o purpose, use
— m — **en taller** m - [sold] general, o all, purpose, shop use
— m — **para servicio(s)** m **público(s)** - general utility use
— m **habitual** - habitual use • current use
— m **inadecuado** - inadequate, o improper, use
— m **inapropiado** - inappropriate, o improper, use
— m **indebido** - improper, o inappropriate, use; misuse
— m **independiente** - independent use
— m **industrial** - [ind] industrial use
— m — **vario(s)** - sundry industrial use(s)
— m **inicial** - initial, use, o usage
— m **ininterrumpido** - uninterrupted, o continuous, use
— m **intensivo** - intensive use • heavy use
— m — **de oxígeno** m—[ind] intensive oxygen use
— m — — — m **industrial** - [ind] intensive industrial oxygen use
— m **interno** - internal use
— m **legal** - [legal] legal use
— m **local** - local use
— m **normal** - normal use
— m **máximo** - maximum use
— m **mayormente** - use mainly
— m **metalúrgico**—[metal-prod] metallurgical use
— m **múltiple** - multiple use; wide application
— m **normal** - normal use
— m **nuevo** - new use
— m — **evitado** - prevented re-use
— m **ocasional** - occasional use
— m **otro que para túnel** m - [constr] non-tunnel use
— m **para** - use for
— m — **conducción** f - [tub] transmission, use, o service
— m — — f **bajo presión(es)** f **alta(s)** - [tub] high pressure transmission service
— m — — f — — f **elevada(s)** - [tub] high pressure transmission service
— m — **conducto(s)** m - [tub] conduit(s) use
— m — **energía** f - [electr-distrib] power use
— m — **fin(es)** m **general(es)** - general, purpose, o utility, use
— m — **muro(s)** m - [constr] wall use
— m — **neumático** m - [autom-neumát] tire use
— m — **pared(es)** - wall use
— m — **servicio** m **pesado** - heavy duty use
— m — **soldadura** f - [sold] welding use
— m **posterior** - later use
— m **potable** - [hidr] potable use
— m **preciso** - precise, o exact, o specific, use
— m **principal** - principal, o main, use
— m **promedio** - average use
— m **propietario** - [legal] proprietary use
— m **propuesto,-** proposed, o intended, use
— m **racional** - rational use
— m **radiotelefónico** - [comput] radio(telephone) use
— m **real** - actual, o true, use, o practice
— m **recreativo** - [deport] recreational use
— m **relativo** - relative use
— m — **para equipo** m - [ind] relative equipment utilization
— m **repetido** - repeat(ed) use; re-use; re-using
— m **restringido** - restricted use
— m **rudo** - rough, o rugged, use
— m — **fuera de carretera** f - [autom-neumát] rough off-road use
— m **segundo** - second, use, o hand
— m **sobre llanta** f—[autom-neumát] use on @ rim

uso m **sobre muelle** m - [transp] dock(side) use
— m — **riel(es)** m [mec-ruedas] use on @ rail
— m **sobrecabeza** - overhead use
— m **temporario** - temporary use
— m **telefónico** - [electrón] telephone use
— m **típico** - typical use
— m **único** - single, use, o application
— m **uniforme** - uniform use • pl uniform customs
— m **universal** - universal use
— m **ventajoso** - advantageous use
usos m **varios** - sundry, o miscellaneous, use(s)
— m **y prácticas** f **uniformes** - uniform customs and practices
usual a - . . .; common; general
usuario m - [ind] consumer • [ind] field
— m **común** - average consumer
— m **corriente** - average, o common, consumer
— m **de agua** m - [hidr] water, user, o customer
— m — **cadena(s)** f - [cadenas] chain, user, o customer
— m — **herramienta(s)** f - [ind] tool user
— m — — f **superabrasiva(s)** - [herram] superabrasive tool(s), user, o customer
— m — **neumático(s)** m - [autom-neumát] tire, consumer, o customer
— m — **refractario(s)** m - [refract] refractory,ries, user, o customer
— m — **sistema** m - system user
— m **final** - final, o end, user
usuario,ria a - . . . • user; consumer; patron; customer
usufructar* v - véase **usufructuar**
usufructuar v - to usufruct
ut a supra - [legal] . . .; above (written)
útil a - . . .; helpful; usable; serviceable
— a **particularmente** - particularly useful
útiles m - fixtures
— m **para escritorio** m - office, fixtures, o supplies • desk items
— m — **oficina** f - office supplies
utilaje* m - véase **útiles** m - [herram] tools; devices; outfit; equipment
utilidad f - . . .; usability; serviceability • earnings; returns; profits; profitability • value • performance
utilidades f - profit(s)
— f **acumuladas** - [fin] accumulated profits
— f — **no distribuida(s)** - [fin] undistributed accumulated profits
— f **adicionales** - [fin] additional profits
— f — **con recuperación** f - [ind] additional recovery profits
— f — — — f **de vapor** m - [petról] vapor recovery additional profits
— f **altas** - [econ] high profits
— f **antes de impuesto** m **a rédito(s)** m - [fin] profit, o income, before income tax(es)
— f — — — m **sobre renta** f - [econ] profit, o income, before @ income tax(es)
— f **baja(s)** - [com] low, profit, o income
— f **bruta(s)** - [fin] gross profit(s)
— f — **acumulada(s)** - [fin] accumulated gross profit(s)
— f **buena(s)** - [fin] good, profit(s), o return
— f **comprobada(s)** - [fin] proven profit(s)
— f **con recuperación** f **(de vapores** m**)** • [ind] [vapor] recovery profit(s)
— f **contabilizada(s)** - [contab] recognized profits
— f **contable(s)** - [contab] book profit(s)
— f **creada(s)** f - created, o earned, profit(s)
— f **de inversión** f - [contab] investment(s) profit
— f — — f **extranjera** - [fin] foreign investment(s) profit(s)
— f — — f — **directa** - [fin] profit(s) from @ direct foreign invesment(s)
— f **sobre operación(es)** f - [fin] operating profit(s)
— f **de sociedad(es)** f **anónima(s)** - [contab] corporate profit(s)
— f **después de impuesto** m **sobre rédito(s)** m - [fin] profit, o income, after @ income taxes

utilidad(es) f después de impuesto m sobre renta f - [fin] profit, o income, after @ income
taxes
— f devengada(s) - [fin] accrued profit(s)
— f diferida(s) - [fin] deferred, income, o
profit(s)
— f elevada(s) - [com] high, o large, profit(s)
— f en operación(es) f - operating profit(s)
— f — contrato m - [legal] contract profit(s)
— f — — m contabilizada(s) - [contab] recognized contract profit(s)
— f estimada(s) - estimated profit(s)
— f excesiva(s) - [contab] excess profit(s)
— f final(es) - final profit(s); serviceability
— f grande(s) - [com] large profit(s)
— f gravable(s) - [fisc] taxable profit(s)
— f incluida(s) - included profit(s)
— f líquida(s) - [contab] net, profit(s), o
earning(s)
— f mayor(es) - greater, o better, profit(s)
— f mejor(es) - better earning(s)
— f menor(es)—smaller, profit(s), o earning(s)
— f neta(s) - [fin] net, income, o profit, o
earning(s)
— f — comprobada(s) - [fin] proven net profit
— f — final(es) - [fin] final net profit(s)
— f — no realizada(s) - [contab] unrealized
net, gain, o profit(s)
— f — — — sobre valor(es) para especulación
f - [fin] net unrealized gain, o profit(s);
(equity) securities
— f neta realizada - [fin] net realized profit
— f — sobre venta f de valor(es) m - [fin]
securities sale net, gain, o profit
— f no distribuida(s) f [fin] undistributed, o
undivided, profit(s), o earning(s)
— f — realizada - [fin] unrealized, profit,
o gain
— f — — sobre cambio(s) m - [fin] unrealized
exchange, gain, o profit
— f — remitida(s) - [fin] unremitted, pro-
fit(s), o earning(s)
— f — transferida(s) - [fin] untransferred
profit(s)
— f obtenida(s) - [fin] obtained profit(s)
— f operativa - operating profit
— f pendiente(s) de aplicación f - [fin] unallocated, o undistributed, profit(s)
— f pequeña(s) - [fin] small profit(s)
— f percibida(s) - [fin] profit(s) received
— f por aplicar - [contab] unallocated, o unappropriated, o undistributed, profit(s)
— f por conversión f de divisas f - [fin] currency translation profit(s)
— f — — f de divisa(s) en exterior m - [fin]
foreign currency translation profit(s)
— f — — f — — f extranjera(s) - f - [fin]
foreign currency translation profit(s)
— f — venta f de título(s) m público(s) -
[fin] public bond(s) sale profit(s)
— f práctica - practical usefulness
— f prolongada - prolonged, o extended, usefulness, o, profit(s)
— v proviniente de inversión(es) f en exterio-
m - [fin] foreign investment(s) profit(s)
— f — f — — f extranjera(s) directa(s) -
profits from direct foreign investment(s)
— f realizada(s) - realized, o converted, o accrued, profit(s), o earning(s), o gain(s)
— f reducida(s) - [fin] small, o low, pro-
fit(s) • reduced profit(s)
— f reconocida(s) - [contab] recognized profits
— f reinvertida(s) - [fin] reinvested profit(s)
— f restaurada(s) - restored profit(s)
— f retenida(s) - [fin] retained, o withheld,
profit(s)
— f — disminuida(s) - [contab] decreased retained, profit(s), o earning(s)
— f — reducida(s) - [contab] reduced, o decreased, retained, profit(s), o earning(s)
— f sin aplicar - [fin] unallocated, o undistributed, profit(s), o earning(s)
— f — distribuir - [fin] undistributed, o undivided, profits, o earning(s)

utilidad(es) f sin transferir - [fin] untransferred profit(s)
— f sobre activo m - [fin] return on asset(s)
— f — — m neto - [fin] return on @ net as-
set(s); R O N A
— f — operación(es) f • [contab] operation(s),
return, o profits
— f — — f ordinaria(s) - [contab] common oper-
ation(s), return, o profit(s)
— f — venta(s) f - sales, gain, o profit(s)
— f — — de activo m - [contab] asset(s)
sales profit
— f — — f — — m fijo - [contab] profit on
fixed assets sales
— f — — f — material(es) m - [com] materials
sales profit(s)
— f — — f — valor(es) m - [fin] securities
sale, gain, o profit(s)
— f total(es) - [ind] complete serviceability •
total profit(s)
— f transferible(s) - [fin] transferable profits
— f — a exterior m - [fin] profit(s) transferable out of @ country
— f transferida(s) - [fin] transferred profit(s)
— f — a exterior m - [fin] profit(s) transferred out of @ country
utilizable a - usable; useful; available
utilización f - utilizatioh; use • tapping • following
— f absoluta - absolute utilization
— f — de equipo m - [ind] absolute equipment
utilization
— f — — instalación(es) f - [ind] absolute
equipment utilizaion
— f como fuente f para energía f - [electr-prod]
use as power (source)
— f constante - constant, utilization, o use
— f de agua m - [hidr] water, utilization, o use
— f — — m de río m - [hidr] river water, utilization, o use
— f — aire m - air, utilization, o use
— f — aluminio m - [metal] aluminum, utilization, o use
— f — boro m - [metal-prod] boron, utilization,
o use
— f — combustible - [combust] fuel, utilization, o use
— f — computador m - [comput] computer use
— f — container(s) m - [transp] containerization; container, utilization, o use
— f — crédito m - [fin] credit use
— f — chapa(s) f - [metal-fabr] strip use
— f — equipo m - [ind] equipment, utilization,
o use
— f — espacio m - space, utilization, o use
— m — hierro m - [metal] iron, utilization, o
use
— f — — m esponja - [metal] sponge iron utilization, o use
— f — estimulador m - [ind] stimulator use
— f — impuesto(s) m - [fin] tax(es) use
— f — lingotera f - [metal-prod] (ingot) mold
use
— f — instalación(es) - [ind] equipment, o
facility,ties, use, o utilization
— f — mano f de obra f - [labor] manpower, use,
o utilization
— f — mineral m - [metal-prod] ore use
— f — — m prerreducido = [metal-prod] prereduced ore use
— f — moneda f - [fin] currency use
— f — ordenadora - [comput] computer use
— f — oxígeno m - [ind] oxygen use
— m — — m biológico - [sanit] biological oxygen, use, o demand
— m — — m — exigido - [sanit] biological oxygen demand use
— f — recurso(s) m - [fin] resource(s), use, o
utilization
— f — — m de préstamo m - [fin] loan resources
utilization
— f — servicio(s) m - [ind] service(s),use, o
utilization

utilización f de terreno(s) m - land use
— f — torneado m - [mec] turning utilization
— f definida - defined, use, o utilization
— f eficiente - efficient, utilization, o use
— f — de producto m - effective product, use,
o utilization
— f — — recurso(s) m - effective resource(s),
use, o utilization
— m — — — m humano(s) - [admin] effective
human resource(s), use, o utilization
— f — servicio(s) m - effective service(s),
use, o utilization
— f en cuchara f - [metal-prod] ladle use
— f — horno m - [metal-prod] furnace use
— f exclusiva - exclusive, use, o utilization
— f fácil - easy, use, o operation
— f inmediata - immediate, use, o usability
— f máxima - maximum, utilization, o use
— f obligatoria - mandatory, use, o utilization
— f óptima - optimum, use, o utilization
— f para - use for
— f — energía f - [electr-oper] use for power
— f — soldadura f - [sold] use for welding
— f provechosa - beneficial use
— f relativa - relative, use, o utilization
— f — de equipo m - [ind] relative equipment,
use, o utilization
— f — instalación(es) f - [ind] relative,
equipment, o facilities, use, o utilization
— f — operario m - [labor] relative operator, use, o utilization
— f — servicio(s) m - [labor] relative ser-
vice(s), use, o utilization
utilizado,da a - utilized; used • tapped • followed
— a como fuente f para energía f—[electr-oper]
used as @ power (source)
— a constantemente - used constantly
— a obligatoriamente - used mandatorily
— a para - used for
— a — soldadura f - [sold] used for welding
utilizador,ra a - utilizer; user
utilizante a - utilizing
utilizar v - . . .; to use; to employ • to tap •
to follow
— v como fuente f para energía f—[electr-oper]
to use as @ power (source)
— v constantemente - to, utilize, o use, constantly
— v crédito m - [fin] to use @ credit
— v electrodo(s) m - to use @ electrode(s)
— v — m con alma f fundente - [sold] to use @
Innershield (electrode)
— v espacio m - to, utilize, o use, @ space
— v fácilmente - to, use, o operate, easily
— v impuesto(s) m - [fisc] to use @ tax(es)
--- v Innershield m - [sold] to use @ Innershield
(electrode)
— v lingotera f - [metal-prod] to use @ ingot
mold
— v moneda f - [fin] to use @ currency
— v obligatoriamente - to use mandatorily
— v para - to, utilize, o use, for
— v — energía v - [electr-oper] to, use, o
utilize, for power
— v soldadura f - [sold] to use for welding
— v recurso(s) m - [econ] to, use, o utilize, @
resource(s)
— v — m humano(s) m - [admin] to, use, o utilize, @ human resource(s)
utiliaje* m - [ind] tools; equipment; outfit

V

V invertida - inverted V
V para velocidad f - [autom-neumát] V-speed
vacación(es) f - . . .; holiday(s)

vacación f breve - brief, o short, vacation
— f concedida—[labor] granted vacation (time)
— f corta - short vacation • busman('s) holiday
— f de fin m de semana f - weekend vacation
— f impaga(da) - [labor] unpaid vacation
— f paga(da) - [labor] paid vacation
vaciada f - [metal-prod] . . .; casting; tapping; teeming
vaciadero m - [ind] . . .; dump; spoil, area, o
bankj • [miner] tipple; spoilbank
— m para desperdicio(s) m - [miner] waste dump
— m — escoria f - [metal-prod] slag dump
— m — residuo(s) m - [miner] waste(s) dump
— m — vagoneta f - [miner] tipple
— m — f para mina f - [miner] mine tipple
vaciado m - . . . . • emptying • pouring; casting;
filling • placing • deposit
— m de acero m - [metal-prod½ steel pouring
— m — — m en lingotera f - [metal-prod]
steel pouring into @ (ingot) mold(s)
— m — arrabio m (líquido) - [metal-prod] molten pig iron charge,ging
— m — artesa f -[col.cont] tundish emptying
— m — asentamiento m - [mec] settled scale;
véase también cascarilla f asentada
— m — cimiento(s) m - [constr] foundation
pour(ing)
— m — — de hormigón m - [constr] concrete
foundation pouring
— m — collar m - [constr] collar casting
— m — — m de hormigón m - [constr] concrete
collar casting
— m — estribo(s) f - [constr] abutment pouring
— m — hormigón m - [constr] concrete pouring
— m — lingote(s) m - [metal-prod] ingot pouring
— m — tambor m - drum emptying
— m — metal m - [metal-prod] metal pouring
— m — — m caliente - [metal-prod] hor metal
pouring
— m — muro m - [constr] wall pour(ing)
— m — — m de hormigón m - [constr] concrete
wall pouring
— m — — m para retención f - [constr] retaining wall pour(ing)
— m — piso m - [constr] floor pour(ing)
— m — viga f - [constr] beam pouring
— m — — f de hormigón m - [constr] concrete
beam pouring
— m — — f — — m para empuje m - [constr]
concrete thrust beam pouring
— m — — f para empuje m - [constr] thrust
beam, pouring, o casting
— m monolítico - [constr] monolithic pouring
— m singular - [constr] unique pouring
vaciado,da a - emptied • drained • [constr]
poured • filled • pumped out
— a alrededor de extremo m - [constr] poured,
o cast, around @ end
— a — — — m achaflanado - [constr] poured,
o cast, around a beveled end
— a con cuidado - emptied carefully • [constr]
poured carefully
— a en obra f - [constr] cast in, place, o situ
— a in situ - [constr] cast in place
— a periódicamente - emptied, o pumped out,
periodically • [constr] poured periodically
vaciamiento m - . . . . . • pumping out • [hidr]
draining; drainage
— m — aceite m - [comb.int] oil draining
— m — — m en cráter m - [comb.int] crankcase
oil draining
— m — cárter m - [comb.int] crankcase draining
— m — filtro m - [mec] filter draining
— m — — m para aceite m - [comb.int] oil filter draining
— m — tubería f - [ind] pipe emptying
— m — viga f - [constr] beam, pouring, o casting
— m — — f para empuje n - [constr] thrust
beam, pouring, o casting
— m — zona f - [ind] area pumping out

vaciamiento m de zona f para asentamiento m -
[comb.int] settling area pumping out
— m **periódico** - periodic pumping out
vaciar v - . . . • to dump • to dig out • to
drain • to pump out • [constr] to, fill, o
place, o cast • to pink
— v **aceite** m - [comb.int] to drain @ oil
— v — m **en cárter** m - [comb.int] to drain @
crankcase oil
— v **alrededor de extremo** m - [constr] to cast
around @ end
— v — — — m **achaflanado** - [constr] to cast
around @ beveled end
— v **cárter** m - [comb.int] to drain @ crankcase
— v **cimiento** m - [constr] to pour @ foundation
— v — m **de hormigón** m - [constr] to pour @
concrete foundation
— v **con cuidado** - to empty carefully
— v **cuchara** f - [metal-prod] to empty @ ladle
— v **cuidadosamente** - to empty carefully
— v **filtro** m - [mec] to drain @ filter
— v — m **para aceite** m - [comb.int] to drain @
oil filter
— v **hormigón** m - [constr] to pour @ concrete
— v **horno** m - [metal-prod] to dig out @ furnace
— v **lingote** m - [metal-prod] to pour @ ingot
— v **muro** m - [Constr] to pour @ wall
— v — m **de hormigón** m - [constr] to pour @
concrete wall
— v — m **para retención** f - [constr] to pour @
retaining wall
— v **periódicamente** - to pump out periodically
— v **piso** m - [constr] to pour @ floor
— v **tambor** m - to empty @ drum
— v **tubería** f - [ind] to empty @ pipe,ping
— v **tubo** m - [ind] to empty @ pipe
— v **viga** f - [constr] to, pour, o cast, @ beam
— v — f **de hormigón** m - [constr] to pour @
concrete beam
— v — f — m **para empuje** m - [constr] to
pour @ concrete thrust beam
— v — f **para empuje** m - [constr] to, cast, o
pour, @ thrust beam
— v **zona** f - to empty @ area
— v — f **para asentamiento** m - [comb.int] to,
empty, o pump out, @ settling area
vacilación f - . . .; vacillating,tion; fluctuating,tion • wavering
— f **marcada** - [sold] distinct hesitation
vacilado,da a—vacillated; hesitated; wavered;
fluctuated; flickered
vacilante a - . . .; hesitant • flickering
— m **para delegar** v - hesitant to delegate
vacilar v - . . .; to vacillate • to flicker; to
flutter • to be, reluctant, o shy
— v **encendiéndo(se) y apagándo(se)** - [sold] to
flutter on and off
vacío m - . . . - vacuum
— m **actual** - [neumát] present vacuum
— m **equivalente a columna** f **de mercurio** m de
. . . . mercury vacuum
— m **parcial** - partial vacuum
vacuómetro m - [instrum] véase **manómetro** m de
vacío m
vadear v - . . . • to, wade, o ford
vademécum m - . . .; data book; notebook; manual
• [ind] practice manual
— m **de ingeniería** f - [ind] engineering practice manual
— m — **producto(s)** m **para construcción** f -
[constr] construction products handbook
— m — — m — **drenaje** m - [hidr] drainage products handbook
— m — **soldadura** f - [sold] welding handbook
— m — — f **metálica** - [sold] metallic welding
handbook
— — — — f — **para Capacitación** f **de Montado-**
res de Tubería f bajo Reglas f de A S M E -
[sold] Pipefitter Welder's Review of Metallic
Welding for Qualification under A S M E Rules
— m **para construcción** f - [constr] construction
handbook
— m — — f **vial** - [vial] highway construction

handbook
vademécum m **para plancha(s)** f **en sección(es)** f -
[estruct] sectional plate(s) handbook
— m — — **producto(s)** m - [ind] product(s) handbook
— m — — m **de acero** m - [metal-prod] steel
product(s) handbook
— m — — m — m **para drenaje** m - [constr]
drainage steel product(s) handbook
— m — m — — m — **construcción** f **vial** -
[vial] highway construction steel products
handbook
— m **técnico** - [ing] technical handbook
— m — — m — — f **vial** - [vial] highway
construction product(s) handbook
— m — — m **para drenaje** m - [constr] drainage
products handbook
— m **sobre motor(es)** f - [electr-mot] motor(s)
data book • [comb.int] engine(s) data book
vado m **pavimentado** - [vial] paved ford
vagabundeado,da - rovered; wandered
vagabundear v - . . . • to wander
vagabundeo m - . . .; wandering
vagabundo,da a - . . . . • [electr] stray; vaga-
bond
vagancia f - . . .; wandering; meandering
vagar v - . . . . • to meander
vagón m - [f.c.] (railway) car • piggy back car
— m **abierto** - [f.c.] gondola (car)
— m **accionado eléctricamente** - [ind] electrically driven car
— m **balanza** - [ind] scale car
— m **Barney** - [cabl] Barney car
— m **basculante** - [f.c.] dump, o tilting, car
— m **batea** - [f.c.] gondola (car)
— m **cargado** - [f.c.] loaded car
— m **cerrado** - [f.c.] box car
— m **cocodrilo** - [miner] deep well car
— m **completo** - [f.c.] carload lot
— m **con borde(s)** m **bajo(s)** - [f.c.] gondola car
— m **con carga** f **plena** - [f.c.] carload lot
— m **con fondo** m **tipo tolva** - [f.c.] hopper bottom car
— m **convencional** - [f.c.] conventional car •
[coque] conventional, o multi-spot, car
— m — **para apagamiento** m - [coque] conventional, o multi-spot, quench(ing) car
— m **corriente para apagamiento** m - [coque]
standard quench(ing) car
— m **cubierto** - [f.c.] box car
— m **decauville** - [f.c.] narrow gage car
— m **descarrilado** - [f.c.] derailed car
— m **distribuidor** - [metal-prod] larry car
— m **entero** - [f.c.] carload, o whole car, lot
— m **especial** - [f.c.] special car
— m **estanque** - [f.c.] tank car
— m **ferroviario** - [f.c.] railway, o railroad,
car
— m — **cargado** - [f.c.] loaded freight car
— m **lingotero** - [metal-prod] (ingot) mold car
— m **mezclador** - [metal-prod] mixer, o mixing,
car
— m **para apagamiento** m - [vowur] quenching car
— m — — m **detenido** - [coque] stopped
quench(ing) car
— m — — m **para deshornamiento** m - [coque]
single-spot quench(ing) car
— m — **arrabio** m - [metal-prod] (pig) iron car
— m — — m **caliente** - [metal-prod] hot metal
car
— m — — m **fundido** m - [metal-prod] hot metal
car
— m — — m **líquido** - [metal-prod] hot metal car
— m — **bandeja(s)** f **con chatarra** f - [metal-
prod] scrap box, buggy, o car
— m — f — — f **accionado eléctricamente** -
[metal-prod] electrically driven scrap box
car
— f — f **con carga** f - [metal-lam] charging
buggy
— m — **carbón** m - [f.c.] coal car
— m — **carbonilla** f - [metal-prod] cinder car
— f — **carga** f - [f.c.] freight car • [metal-
prod] ingot buggy

vagón m para carga f para carbón m - [metal-
-prod] coal charging car
— m — — f muy pesada - [f.c.] heavy duty
freight car
— m — ceniza(s) f - [combust] cinder car
— m — coque m - [f.c.] coke car
— m — cuchara(s) - [metal-prod] ladle car
— m — — f para escoria f - [metal-prod] slag
ladle car
— m — — f para metal m caliente - [metal-
-prod] hot metal ladle car
— m — — f para metal, fundido, o líquido -
[metal-prod] hot metal ladle car
— m — enfriamiento m - [coke] quenching car
— m — escoria f - [metal-prod] slag car
— f — ferrocarril m - [f.c.] railway car
— m — lado m para carga f - [metal-prod]
charging side car
— m — — m deshornamiento m - [coque] discharge side car
— m — lingote(s) m - [metal-prod] ingot, car,
o buggy
— m — lingotera(s) f - [metal-prod] ingot mold
car; ingot, car, o buggy
— m — mercancía(s) f - [f.c.] freight car
— m — metal, caliente, o fundido—[metal-prod]
hot metal car
— m — — m líquido m - [metal-prod] hot metal
car
— m — pote(s) m - [metal-prod] pot car
— m — — m para escoria f - [metal-prod] slag
pot car
— m — semirremólque(s) m carretero(s) - [f.c.]
piggyback car
— m — subterráneo m - [f.c.] subway car
— m — transferencia f - [f.c.] transfer car
— m — — f para caldero(s) m - [metal-prod]
(ladle) transfer car
— m — — f lingote(s) m - [metal-prod] (ingot) transfer car
— m — transporte m - [f.c.] transportation car
— m — — f para cuchara(s) f - [metal-prod]
ladel, cara, o buggy
— m — — m — bobina(s) f - [metal-prod] coil
transfer car
— m — — m — carbón m - [miner] coal car
— m — — m — semirremolque(s) m carreteros -
[f.c.] piggyback car
— m — trocha f ancha - [f.c.] brpad gage car
— f — — f angosta - [f.c.] narrow gage car
— m — — f media - [f.c.] medium, o standard,
gage car
— m pequeño - [f.c.] small car
— m plataforma f - [f.c.] flat(bed) car
— m — corriente - [f.c.] standard flat car
— m playo m - [f.c.] flat car
— m portaconos - [metal-prod] thimble car
— m portalingoteras - [metal-prod] (ingot) mold
car
— m postal - [f.c.] mail car
— m relativamente pequeño - [f.c.] relatively
small car
— m sin freno m - [f.c.] brakeless car
— m single spot para apagamiento m - [coque]
single-spot quench(ing) car
— m tanque m - [f.c.] tank car
-— m termo - [metal-prod] thermos car
— m tolva - [f.c.] hopper car • bottom dump car
— m — con descarga f inferior - [f.c.] bottom
dump hopper car
— m — — — f por arriba - [f.c.] top dump
hopper car
— m torpedo - [f.c.] torpedo (type) car
— m transbordador - [metal-prod] transfer
(scale) car
— m transportador - [metal-prod] transfer car
— m volcador - [f.c.] dump car
vagoneta f - [ind] . . .; hand truck; car; carriage • [metal-lam] buggy (car) • [metal-
-prod] skip (car) • [cabl] carrier, trolley •
[f.c.] narrow gage car
vagoneta f autovolcadora—[ind] self dumping car
• larry car

vagoneta f autovolcadora para escoria f - [metal-
-prod] self dumping slag car
— f Barney - [cabl] Barney car
— f lingotera - [metal-prod] ingot, transfer
carriage, o mold car
— f minera - [miner] mine car
— f montacarga(s) - [metal-prod] skip car
— f oeste - [metal-prod] west skip car
— f — de montacarga(s) f - [metal-prod] west
skip car (hopper)
— f para caja(s) f para carga f - [metal-prod]
charging box, car, o buggy
— f — cantera f - [miner] quarry, car, o tram
— f — carga f - [metal-prod] charging box car
• skip (car)
— f — — f para carbón m - [metal-prod] coal
charging car
— f — cargar carbón m - [metal-prod] coal
charging (skip) car
— f — descarga f inferior - [metal-prod] bottom dumpo skip car
— f — — f superior - [metal-prod] top dump
skip (car)
— f — despuntes m - [metal-lam] crop bucket
— f — elevación f - [mec] lifting car
— m — elevador m - [agric] elevator wagon
— f — escoria f - [metal-prod] slag car
— f — finos m - [metal-prod] coke breeze car
— f — — m de coque m - [metal-prod] coke
breeze skip
— f — horno m - [metal-prod] furnace car
— f — limpieza f - [metal-prod] cleaning, o
service, skip
— f — lingotera(s) f - [metal-prod] ingot mold
car
— f — lingote(s) m - [metal-prod] ingot, car,
o chariot, o transfer carriage
— f — mina f - [miner] mine car
— f — montacarga(s) m - [metal-prod] skip car
— m — — m para horno m - [metal-prod] furnace
skip:car
— f — — — m — — m alto - [metal-prod] blast
furnace skip car
— f — punta(s) f - [metal-lam] crop bucket
— f — recorte(s) m - [metal-lam] crop bucket
— f — sínter m - [metal
— f — transportador m—[mec] conveyor carriage
— f — — m aéreo - [ind] aerial conveyor carriage
— f — traslado m - [ind] transfer car
— f — — m de rodillo(s) m - [metal-lam] roll
transfer car
— f pequeña - [ind] small car
— f playa f - [constr] flat car(t)
— f portacuchara(s) f - [metal-prod] ladle car
— f portalingotera(s) - [metal-prod] ingot mold
car
— f relativamente pequeña - [ind] relatively
small car
— f volcadora - [metal-prod] dump skip car
vaho m ácido - [ind] acid fume
— m corrosivo - [ind] corrosive, mist, o vapor
— m de aceite m - [mec] oil mist
— m — hidrocarburo(s) m - [segurid] hydrocar-
bon(s) vapor
— m — — m clorinado(s) - [segurid] chlorinated hydrocarbon(s) vapor
— m disolvente - [ind] solvent vapor
— m inflamable - [combust] flammable vapor
— m inhalado - [segurid] inhaled vapor
vaina f - . . .; sheathing; casing | a - (con)
sheathed; jacketed
— f adicional - additional, sheath, o jacket
— f aislante - [electr-cond] insulating sheath
— f — de neoprene - [electr-cond] neoprene
insulating sheath(ing)
— f anticorrosiva - [electr-cond] anticorrosive
sheath(ing)
— f — de (material) plástico - [electr-cond]
anticorrosive plastic (material) sheath(ing)
— f de cloruro m—[electr-cond] chloride sheath
— f — — m de polivinilo m - [electr-cond] polyvinyl (chloride), sheath(ing), o jacket

vaina f de material m plástico - [electr-cond]
  plastic (material) sheathing
— f — neoprene m - [electr-cond] neoprene,
  jacket, o sheath(ing)
— f — plomo m - [electr-cond] lead, sheath, o
  casing
— f electrostática - [electr-cond] electrostatic sheathing
— f exterior - [electr-cond] outer sheath(ing)
— f — de cloruro m de polivinilo m - [electr-
  -cond] polyvinyl chloride outer sheathing
— f — de ndeoprene m - [electr-cond] outer
  neoprene, o neoprene outer, sheath(ing)
— f — — — trenza f de vidrio m impregnada -
  [electr-cond] impregnated glass braid outer
  sheath(ing)
— f — para cable m - [electr-cond] cable outer
  sheath(ing)
— f — — — m para regulación f—[electr-cond]
  control cable outside sheath(ing)
— f fibrosa flexible - [electr-cond] loom
— f flexible - [electr-cond] loom
— f — de fibra f - [electr-cond] loom
— f no inflamable - [electr-cond] non-flammable
  sheath(ing)
— f para cable m - [electr-cond] cable sheath
— f — conductor m - [electr-cond] conductor
  sheath(ing)
— f — prolongación f - extension sheath
— f protectora—[electr-cond] protective sheath
— f — de neoprene m - [electr] [electr-cond]
  neoprene protective sheath
— f resistente a aceite m - [electr-cond] oil
  resisting sheath(ing)
vaivén m de herramienta f - [mec] tool swing
— m formando caja f - [sold] box weaving
vajilla f - [domést] tableware; silverware •
  dishes
— f de acero m - [domést] steel ware
— f — — m inoxidable - [domést] stainless
  steel ware
— f — (material) plástico m - [domést] plastic
  (material), dishes, o tableware
— f para cocina f - [domést] kitchenware
— f — excursión(es) f - [deport] picnic items
— f — mesa f - [domést] tableware
— f usada - [domést] used . . .ware
— f — para cocina f—[domést] used kitchenware
vale m - [contab] I O U; I owe you
— adv decir - to wit • that is
valente adv - [quím] valent
valer v - . . . . • to earn
— v para - to rate; to be worth; to be good for
valerse v - to (make) use (of) • to, make, o
  rely on • to feature • to capitalize • to
  take advantage of; to avail ...self
— v de oportunidad f - to rise to @ occasion
— v — ventaja f - to capitalize; to take advantage
validación f - . . . ; authentication
validado,da a - validated • authenticated
validar v - . . . . • to authenticate
validez f - effectiveness
— f asegurada - assured validity
— f de acuerdo m - [legal] agreement validity •
  agreement term
— f — carta f de crédito m - [fin] letter of
  credit validity
— f — crédito m - [fin] credit validity
— f — garantía f - [legal] warranty validity
— f — oferta f - [com] proposal, o bid, o offer, validity
— f — patente f - [legal] patent validity
— f — póliza f - [seguros] policy, validity, o
  term
— f — propuesta f - [com] proposal validity
— f dudosa - doubtful validity • [seguros]
  judgment case
— f para cotización f - quotation validity
— f — oferta f - [com] offer validity
válido,da a - . . .; effective • available •
  earned; capitalized • authentic • made
— a hasta - valid (un)til(1)

válido,da para - valid for • rated as
valimiento m - . . . . • capitalizing,zation •
  making
valor m a determinar(se) v - value to be deter-
  minar(se)
— m — estimar(se) - value to be estimated
— m — par f - par value
— m — plazo m corto - [fin] short term security
— m — — m largo - [fin] long term security
— m absoluto - absolute (value)
— m acordado - agreed value
— m actualizado - updated value
— m adjudicado - [com] awarded value
— m admisible - allowable, o accepted, value
— m — para proyección f - allowable design
  value
— m — — — f de junta f empernada - [mec]
  bolted joint allowable design value
— m agregado - [econ] added value
— m — mayor - greater added value
— m — menor - smaller added value
— m ajustado - adjusted value
— m amortizado - amortized value
— m anormal - abnormal value
— m añadido - added value
— m aprobado - approved value
— m — para contrato m - approved contract
  value
— m apropiado - appropriate value
— m aproximado - approximate value
— m — en mercado m - approximate market value
— m asegurable - [seguros] insurable value
— m asegurado - [seguros] insured value
— m aumentado - increased value • added value •
  value added
— m autorizado - authorized value
— m bruto - gross value
— m — según factura f - [contab] gross invoice value
— m bursátil - [fin] (stock) market value •
  stock price
— m — en fecha f - [fin] market value on @
  date
— m — — — f de balance m - [fin] statement
  date, o related, market value
— m calorífico - [combust] heat, o calor(if)ic,
  value
— m científico - scientific value
— m clarificado - clarified value
— m comercial - [com] commercial, o business,
  value
— m compatible - compatible value
— m comprobado - proven, o checked, value
— m contable - [contab] book value
— m — por acción f - [contab] book value per
  share
— m contra corte m - [mec] shear value
— m correcto - correct value
— m correspondiente - corresponding value
— m (de) costo m, seguro m y flete - [transp]
  cost, insurance and freight value; C I F
  value
— m cotizado - [com] quoted value
— m dado - given value • above value
— m de adjudicación f - [com] award value
— m — adquisición f - purchase value
— m — aislación f - [electr-instal] insulation value
— m — caída f - drop, o fall, value
— m — — f de tensión f - [electr-oper] voltage drop value
— m — — f — voltaje m - [electr-oper] voltage drop value
— m — cambio - change value • [fin] exchange
  value
— m — campo m magnético - [electr-oper] magnetic field value
— m — capital m - [fin] capital value
— m — — m exportable - [fin] exportable
  capital value
— m — carga f - load value • measured load
— m — — f necesaria - [mec] required measured
  load

load.

valor m de carta f de crédito m - [fin] credit letter value

— m — **coeficiente** - coefficient value

— m — — m **para dilatación f** - expansion factor value

— m — — m **para presión f** - pressure coefficient value

— m — — m **para rugosidad f** - roughness coefficient value

— m — **compactación f** - compaction value

— m — — f **de suelo m** - [constr] soil (compaction) value

— m — **compra f** - purchase value

— m — **contrato m** - contract value

— m — — m **modificado** - [legal] changed contract value

— m — **corona f** - [metal-lam] crown value

— m — **corriente f** [electr-oper] current value

— m — **costo m** - cost value

— m — **crédito m** - credit value

— m — **cresta f** - [electr-oper] crest value

— m — **densidad f** - [mec] density value

— m — — f **de suelo m** - [suelos] soil density value

— m — **diseño m** - design value

— m — **dureza f** - [coque] hardness value

— m — **emisión f** - [fin] issue value

— m — **ensayo m** - test(ing) value

— m — **esfuerzo m** - [constr] stress value

— m — **estabilidad f** - stability value

— m — **exportación f** - [fin] export value

— m — — f **por cobrar v** - [fin] export recievable

— m — **factura f** - [com] invoice value

— m — **flete m** - [transp] freight value

— n — — m **marítimo** - [transp] ocean freight value

— m — **humedad f en ceniza f** - [combust] ash moisture value

— m — **impermeabilidad f** - [constr] imperviousness value

— m — — f **relativa** - [constr] relative imperviousness value

— m — **índice m** - [labor] index value

— m — **inducción f** - [electr-oper] induction value

— m — **interrupción f** [com] interruption value

— m — — f **comercial** - [com] commercial, o business, inerruption value

— m — — f **mercantil** - [com] commercial, o business, interruption value

— m — — f **en negocio(s) m** - [com] business, o commercial, interruption value

— m — **resistencia f** - [mec] resistance value

— m — — f **de aislación f** - [electr-instal] insulation resistance value

— m — **inversión f** - [fin] investment value • invested value

— m — **momento m torsional** - [mec] torque value

— m — **obra f** - [constr] project, value, o cost

— m — **par m motor** - [mec] torque value

— m — **porcentaje m** - percentage value

— m — **precipitación f** - [hidr] rainfall rate

— f — — f **horaria** - [hidr] (hourly) rainfall rate

— m — **primera f (clase)** - [fin] forst class security

— m — **producción f** - production value

— m — — f **de baritina f** - [miner] barytine production value

— m — — f — **cobre m** - [miner] copper production value

— m — — f **metálica f** - [miner] metal(lic), production, o output, value

— m — **propiedad f** - property value

— m — **propuesta f** - proposal('s) value

— m — **recuperación f** - [hidr] recovery value

— m — **relación f** - ratio value

— m — **rendimiento m** - yield value

— m — **resistencia f** - [mec] resistance, o strength, value

— m — — f **a impacto(s) m** - [metal] impact value

valor m de resistencia f eléctrica - [suelos] (soil) resistivity value

— f — — f — **de agua m** [quím] water resistivity value

— m — — f — **suelo m** - [quím] soil resistivity value

— m — — f **efectiva** - [quím] (effective) resistivity value

— m — — f — **agua m** - [quím] water resistivity value

— m — — f — **de suelo m** - [quím] soil resistivity value

— m — — f **final** - ultimate strength value

— m — — f **máxima** - ultimate strength value

— m — **sal(es) f fijada(s)** - [nucl] fixed tailing(s) value

— m — **saldo m no utilizado** - [fin] unused balance value

— m — **sección f** - [metal-fabr] area value

— m — — f **transversal** - [metal-fabr] end area value

— m — **seguro m** - [seguros] insurance value

— m — **terreno(s) m** - land value

— m — **trabajo m** - work, o job, value

— m — **velocidad f** - speed value

— m — — f **de viento m** - [meteorol] wind speed value

— m — **venta f** - [contab] sale(s) value

— m — **voltaje m** - [electr-oper] voltage value

— m — — m **regulado** - [electr-oper] controlled voltage value

— m **decorativo** - decorative value

— m **determinado** - determined, o established, value; chosen value

— m **diferente** - different,ring, o various, value

— m — **para densidad f** - [suelos] differing soil value

— m **disponible** - avilable value

— m **distinto** - different value

— m **diverso** - different value

— m **dudoso** - dubious, o doubtful, value

— m E - [constr] E value

— m **educativo,cacional** - educational value

— m **en mercado m** - [com] market value

— m — — m **aproximado** - [fin] approximate market value

— m — — m **en fecha f** - [fin] today's market value

— m — — m — — f **de balance m** - [fin] related market value

— m — **plaza f** - [com] market value

— m — **tabla f** - table value

— m **especificado** - specified value

— m **establecido** - established value

— m **estimado** - estimated value

— m **exacto** - exact, o precise, value

— m — **para voltaje m** - [electr-oper] exact, o correct, voltage value

— m — — m **regulado** - [electr-oper] correct, o exact, controlled voltage value

— m **exento m** - [fin] exempt security

— m — **de impuesto m** - [fisc] tax exempt security

— m — — m **sobre, rédito(s) m, o renta f** - [fisc] income tax exempt security

— m **exigible** - [contab] account payable • account receivable • demand account

— m **exigido** - required, value, o amount

— m **extinguido** - [fin] amortized value

— m **extremo** - extreme value

— m — **de hidrógeno m** - [quím] extreme hydrogen value

— m — — — m **sulfurado** - [coque] extreme sulfurized hydrogen value

— m — **hora f** - hour extreme value

— m — **humedad f** - [combust] moisture extreme value

— m — — f **en ceniza f** - [combust] ash moisture extreme value

— m F O B **en puerto m** - [transp] free on board, o F O B, port value

— m **facturado** - [com] invoiced value

valor m fijo - fixed, o set, value
— m final - final, o ultimate, value
— m garantizado - guaranteed value
— m grande - great value | a - (con) invaluable
— m identificado - identified value
— m — específicamente - specifically identified value
— m industrial - [ind] industrial value
— m inferior - lower,west value
— m inmobiliario - [contab] fixed, intangible,
o incorporeal, asset
— m inmovilizado - immobilized asset • [fin]
fixed asset
— m intangible - [contab] intangible (asset)
— m interpretativo - interpretative value
— m — de imagen f - image interpretative value
— m invertido - inverted value • [fin] invested
value
— m investigado - investigated, o researched,
value
— m libre a bordo—[transp] free on board, o
F O B, value
— m — — — en, fábrica f, o planta f - [ind]
free on board, mill, o plant, value
— m — — costado m de barco m - [com] free
along side, o F A S, value
— m límite - limit value • threshold value
— m — inferior m - lower,est limit value
— m — máximo - maximum limit value
— m — — para espesor m [metal-lam] maximum
gage limit value
— m — mínimo - minimum limit value
— m — — para espesor m - [metal-lam] minimum
gage limit value
— m — para espesor m - [metal-lam] gage limit
value
— m — superior - higher,est, o upper, limit
value
— m líquido - net value
— m — invertido - [fin] net invested amount •
equity
— m magnético - [metal] magnetic value
— m matemático - [matem] mathmatical quantity
— m máximo - maximum, o ultimate, value • peak
— m — de punta - [electr-oper] maximum peak
value
— m — para proyección f - [dib] maximum, o
ultimate, design value
— m — — — f de junta f - [mec] joint, maximum, o ultimate, design value
— m — — — f — — f empernada - [mec] bolted
joint, maximum, o ultimate, design value
— m — recomendado - maximum recommended, o recommended maximum, value
— m mayor - greater,est, o higher,est, value
— m mecánico - [mec] mechanical value
— m — alto - [metal-lam] high mechanical value
— m — bajo - [metal-lam] low mechanical value
— m — garantizado - [metal] guaranteed, o warranted, mechanical value
— m medio - average, o mean, value
— m — de sal(es) f - [nucl] tailing(s) average
value
— m — — — f fijada(s) - [nucl] fixed tail-
ing(s) average value
— m — mensual - average monthly value
— m menor - smaller,est, o lower,est, value
— m metalúrgico - [metal] metallurgical value
— m — en extremo m - [metal-prod] end metallurgical value
— m — — — m de salida f - [metal-prod] output end metallurgical value
— m mínimo - minimum value
— m — de resistencia f eléctrica (efectiva) -
minimum resistivity value
— m — — f — (—) para agua m - [hidr]
minimum water resistivity value
— m — — — f — (—) — suelo m - [suelos]
minimum soil, o soil minimum, resistivity
value
— m mobiliario m - [fin] stock(s) and bond(s) •
unregistered security
— m monetario - monetary value • dollar value

valor m monetario de economía f - [fin] economy,
o savings, dollar value
— m nacional - [fin] national security
— m negociable - [fin] negotiable security
— m — con plazo m corto - [fin] short term
marketable security
— m — — — m largo - [fin] long term marketable security
— m neto - [fin] net value • net worth
— m — de, compañía f, o empresa f - [fin] corporate, o company, net worth
— m — — factura f - [contab] net invoice, o
invoide net, value
— m — — sociedad f - [fin] corporate net worth
— m — real - [fin] true net worth
— m — según registro(s) m contable(s) -
[contab] net book value
— m — total - [contab] total net worth
— m nominal - [fin] nominal value • [electr]
rated current
— m — de corriente f - [electr] rated current
— m — — f en transformador m - [electr]
rated transformer current
— m — grantizado - guaranteed nominal value
— m — según chapa f para identificación f -
nameplate rating
— m — según norma(s) f estadounidense(s) -
[ind] American rating
— m normal - normal, value, o rating
— m — de índice m - normal index value
— m — — rendimiento m - normal yield value
— n — en corriente f alterna - [sold] normal
alternating current value
— m — — — f continua - [sold] normal direct
current value
— m numérico - numerical value
— m obtenido - obtained value
— m óptimo - optimum value
— m — de índice m - optimum index value
— m — — rendimiento m - [labor] optimum yield
value
— m originado - originated value
— m original - original value
— m p-H - [hidr] p-H value
— m par - par value
— m para aduana f - [fisc] costums value
— m — — f únicamente - [fisc] value for cus-
ton purpose(s) only
— m — apretadura f - [mec] tightening value
— m — — f para momento m torsional - [mec]
torque tightening value
— m — — f — — para sujetador m - [mec]
fastener torque tightening value
— m — — f — par m motor - [mec] torque tightening value
— m — — — m — para sujetador m - [mec]
fastener torque tightening value
— m — cambio m - change value • [fin] exchange
value
— m — capacidad f - capacity value
— m — carga f - load value • load speed
— m — coeficiente m - coefficient value
— m — densidad f - density value
— m — — f de suelo m - [suelos] soil density
value
— m — ensayo(s) m - test value
— m — estadística f - statistical (purposes)
value
— m — fin(es) m aduanero(s) - [fisc] customs
purpose value
— m — — m — únicamente - [fisc] value for
custons purpose(s) only
— m — flujo m - [metal] flow, o flux, value
— m — — m para inducción f - [metal] induction
flux value
— m — fuerza f - [mec] force value
— m — — f para tracción f - [hidr] tractive
(force) value
— m — guía f - guideline
— m — límite m - limit value
— m — — m inferior - low(est) limit value
— m — — m menor - low(est) limit value
— m — — m superior - high(est) limit value

valor m para oferta f - offer, o bid, value
— m — proyección f - design value
— m — f para junta f - [mec] joint design value
— m — f — f empernada - [mec] bolted joint design value
— m — rendimiento m - yield value
— m — m para oferta f - [fin] offer yield value
— m — resistencia f - [mec] strength value
— m — f a tracción f - [metal] tensile strength value
— m — rugosidad f - [mec] roughness value
— m parcial - partial value
— m de carta f de crédito m - [fin] partial credit letter value
— m percibido - perceived, o received, value
— m por cobrar - [fin] receivable
— m exportación f por cobrar - [fin] export receivable
— m — unidad f - unit value; value per unit
— m porcentual - percentage value
— m prefijado - pre-established value
— m presumido - design, o presumed, value
— m previsto - foreseen value • projected value • design value
— m primitivo - [fin] original value
— m privado - [fin] private security
— m probable - probable value
— m promedio - average, value, o figure(s)
— m — para consumo m de coque m - [combust] average coke, requirement, o consumption
— m proporcional - proportional value
— m propuesto - proposed value
— m prorrateado - prorated value
— m publicado - published value
— m punto - [labor] value point
— m puntual - short term value
— m que no puede cargar(se) a consumidor m - value not chargeable to @ consumer
— m "E" - [metal-prod] "E" value
— m Ra - Ra value
— m razonable - reasonable, o fair, value
— m — para propiedad f - property fair value
— m realizable m—[contab] convertible security
— m registrado - registered value
— m real - true, value, o worth • actual value
— m — de activo m - [fin] true asset value
— m — capital - [fin] true capital value
— m — — m exportable - [fin] exportable capital(s) true value
— m — en mercado m - true, o actual, market value
— m recomendado - recommended value
— m — para momento m torsional - [mec] recommended torque value
— m — par m motor - [mec] recommended torque value
— m relativo - [fin] related,tive value
— m — en mercado m—[fin] related market value
— m revisado - revised, o checked, value
— m — aprobado - revised approved value
— m — — para contrato m - contract revised approved value
— m — para contrato m - contract revised value
— m según libro(s) m - [contab] book value
— m — — contable(s) m - [contab] book value
— m según Manning - [hidr] Manning('s) value
— m — registro(s) m contable(s) m - [contab] book value
— m siguiente m - following value
— m simétrico - [electr] symmetrical value
— m sin embalar - [transp] unpacked value
— m sugerido - proposed, o suggested, value
— m superior - higher,est value
— m tabulado - tabular,lated value
— m tecnológico - technological value
— m típico - typical value
— m total - total value
— m — C I F - [com] total C I F value
— m — de costo m - [com] total cost value
— m — — — m, seguro m, y flete m - [com] total, cost, insurance and freight, o C I F, value

valor m total de inversión f - [fin] total investment value
— m — — seguro(s) m - [seguros] total insurance value
— m — en puerto m - [transp] total port value
— m — estimado - total estimated cost
— m — libre a bordo - [transp] total, free on board, o F O B, value
— m — — — en puerto m - [transp] total, free on board, o F O B, at port value
— m unitario - unit value
— m valioso - valuable value
— m — para cambio m - [fin] valuable exchange value
valores m varios m - [fin] miscellaneous, o sundry, securities
valoración f - . . .; valuation; rating; evaluating • assessing,sment • taking into account
— f de cantidad f de trabajo m - [labor] work quantity rating
— f — elemento(s) m - individual item(s) valuing
— f — esfuerzo(s) m - effort(s) assessment
— f — — m logrado(s) - result(s) secured assessment
— f — resultado(s) m - result(s) assessment
— f — — m obtenido(s) - obtained, o attained, result(s) assessment
— f — trabajo m - [labor] work, rating, o assessment
— f — — m en ejecución f - [labor] work in progress assessment
— f por elemento(s) m - individual item(s) valuation
valorado,da a - valued • assessed • taken into account
— a por elemento(s) m - valued by individual item(s)
valorar v - . . .; to, rate, o evaluate, o assess; to assign @ value; to take into account
— v económicamente - to, establish, o assign, o set. @ value
— v por elemento(s) m - to value by individual item(s)
— v resultado(s) m - to assess @ result(s)
— v — m obtenido(s) - [admin] to assess @ result(s), obtained, o secured
— v trabajo m - to assess @, work, o job
— v — m en ejecución f - [admin] to assess @ work in progress
valores m - [fin] securities
valorización f - valuating,tion • pricing
— f de inventario m - [com] inventory, pricing, o valuating,tion
valorizado,da a - valued
valorizar v ÷ to value; véase también valorar v
valuación f - . . .; evaluation; appraisal; assessing,sment
— f — bien(es) f raíz,ces - real estate appraisal
— f — inmueble(s) m - real estate appraisal
— f — inventario m - [contab] inventory, valuation, o pricing
— f de moneda f - [fin] currency valuation
— f — resultado(s) m - result(s) assessment
— f — — m obtenido(s) m - attained, o obtained, result(s) assessment
valuado,da a - valued; assessed
valuar v - . . . • to assess
— v inventario m - [contab] to value @ inventory
— v resultado(s) m - [admin] to assess @ result(s)
— v — m obtenido(s) - to assess @, attained, o secured, result(s)
válvula f - [válv] . . . • [metal-prod] release valve
— f a prueba f de falla(s) f - [válv] foolproof valve
— f abierta f - [válv] open(ed), o turned on, valve
— f accesoria - [válv] accessory valve
— f — para regulación f - [válv] accessory control valve

válvula f̲ accionada - activated valve
— f̲ acodada - [válv] angle valve • elbow valve
— f̲ — para retención f̲ - [válv] elbow check valve
— f̲ activada - [válv] turned on valve
— f̲ — con solenoide m̲ - [válv] solenoid, activated, o̲ turned on, valve
— f̲ actuada - [válv] actuated valve
— f̲ aflojada - [válv] loosened valve
— f̲ ajustable - [válv] adjustable valve
— f̲ — para regulación f̲ - [válv] adjustable control valve
— f̲ — — — f̲ de bajada f̲ - [válv] adjustable lowering control valve
— f̲ — — — elevación f̲ - [válv] adjustable raising control valve
— f̲ ajustada - [válv] adjusted valve • tightened valve
— f̲ — en fábrica f̲ - [válv] factory set valve
— f̲ alabeada - [válv] warped valve
— f̲ amortiguadora - [válv] dampener,ning, o̲ cushion, valve
— f̲ angular v - [válv] angle,gular valve
— f̲ — Stockham - [válv] Stockham angle valve
— f̲ anteojo m̲ - [metal-prod] goggle valve
— f̲ — para colector m̲ - [metal-prod] dust catcher goggle valve
— f̲ — — — m̲ primario - [metal-prod] primary dust catcher goggle valve
— f̲ — — — m̲ secundario - [metal-prod] secondary dust catcher goggle valve
— f̲ — — precipitador m̲ - [metal-prod] precipitator goggle valve
— f̲ — — — primario m̲ - [metal-prod] primary dust catcher goggle valve
— f̲ — — secundario m̲ - [metal-prod] secondary dust catcher goggle valve
— f̲ anuladora - [válv] disabling valve
— f̲ — para sobrepaso* m̲ - [válv] override disabling valve
— f̲ apretada - [válv] tightened valve
— f̲ armada - [válv] assembled valve
— f̲ Askania - [metal-prod] Askania valve
— f̲ — normal - [metal-prod] normal Askania valve
— f̲ — para emergencia f̲ - [metal-prod] emergency Askania valve
— f̲ — f̲ — — f̲ para Venturi m̲ - [metal-prod] Venturi emergency Askania valve
— f̲ — — — Venturi - [metal-prod] Venturi Askania valve
— f̲ autocomandada - [válv] self-driven valve
— f̲ automática - [válv] automatic valve
— f̲ — para desaeración f̲ - [válv] automatic deaeration valve
— f̲ — para fundente m̲ - [sold] automatic flux valve
— f̲ — — retención f̲ - [válv] automatic check valve
— f̲ — — — f̲ para estabilizador m̲ - [grúas] outrigger automatic check valve
— f̲ automotriz - [comb.int] automotive valve
— f̲ auxiliar - [válv] auxiliary valve
— f̲ — con . . . circuito(s) m̲ - . . .-circuit auxiliary valve
— f̲ — — — motor m̲ - [válv] motor auxiliary, o̲ auxiliary motor, valve
— f̲ — — — m̲ hidráulico - [hidr] hydraulic motor auxiliary valve
— f̲ bisagra f̲ - [valv] swing valve
— f̲ — para retención f̲ - [válv] swing check valve
— f̲ bola f̲ - [válv] globe valve
— f̲ cerrada - [válv] closed, o̲ turned off, valve
— f̲ colocada - [válv] installed valve
— f̲ combinada - [valv] combined,nation valve
— f̲ — para drenaje m̲ y filtración f̲ - [válv] combined,nation drain and filter valve
— f̲ compensada - [válv] compensated valve
— f̲ — para presión f̲ - [válv] compensated pressure, o̲ pressure compensated, valve
— f̲ compensadora - [válv] equalizing valve
— f̲ compuerta f̲ - [válv] goggle valve • [tub] sluice valve • [metal-prod] gate valve

válvula f̲ compuerta para descaarga f̲ - [metal--prod] exhaust gate valve
— f̲ — — — f̲ para Venturi m̲ - [metal-prod] Venturi, o̲ scrubber, exhaust gate valve
— f̲ con abertura f̲ - [válv] port valve
— f̲ — — f̲ graduable - [válv] adjustable port valve
— f̲ — aguja f̲ - [válv] needle valve
— f̲ — — f̲ para combustible m̲ - [comb.int] fuel needle valve
— f̲ — — f̲ — entrada f̲ para combustible m̲ - [comb.int] (fuel) needle valve
— f̲ — — — — m̲ en carburador m̲ - [comb.int] float needle valve
— f̲ — aleta(s) f̲ - [válv] feather valve
— f̲ — anillo m̲ - [válv] ring valve
— f̲ — — m̲ de teflón m̲ - [válv] teflon ring valve
— f̲ — anteojo m̲ - [válv] goggle valve
— f̲ — — m̲ para colector m̲ - [metal-prod] dust catcher goggle valve
— f̲ — — m̲ — m̲ primario - [metal-prod] primary dust catcher goggle valve
— f̲ — — m̲ — m̲ secundario - [metal-prod] secondary dust catcher goggle valve
— f̲ — — m̲ — gas m̲ - [metal-prod] gas goggle valve
— f̲ — — m̲ — — m̲ para estufa f̲ - [metal--prod] stove gas goggle valve
— f̲ — asiento m̲ - [valv] seated valve
— f̲ — — m̲ de caucho m̲ [válv] rubber seated valve
— f̲ — — m̲ — — m̲ para cierre m̲ - [válv] shut off rubber seated valve
— f̲ — — m̲ — — m̲ — — m̲ perfecto m̲ - [válv] bubble tight shut off rubber seated valve
— f̲ — — m̲ doble - [válv] double seat(ed) valve
— f̲ — — m̲ renovable - [válv] renewable seat valve
— f̲ — bola f̲ - [tub] globe valve
— f̲ — bonete m̲ - [válv] bonnet valve
— f̲ — — m̲ para sello m̲ - [válv] seal bonnet valve
— f̲ — — m̲ — — m̲ para presión f̲ - [válv] pressure seal bonnet valve
— f̲ — capuchón m̲ - [telecom-cond] cap valve
— f̲ — . . . carrete(s) m̲ - [válv] . . . spool valve
— f̲ — . . . — m̲ para segadora f̲ trilladora f̲ - [agric] combine . . . spool valve
— f̲ — cierre m̲ total - [válv] ram gate
— f̲ — compuerta f̲ - [válv] gate valve
— f̲ — con asiento m̲ - [válv] seat(ed) gate valve
— f̲ — — — m̲ paralelo - [válv] parallel seat(ed) gate valve
— f̲ — — f̲ con disco m̲ - [válv] disk gate valve
— f̲ — — — m̲ doble - [válv] double disk gate valve
— f̲ — — f̲ — — m̲ — y asiento m̲ paralelo - [válv] double disk parallel seat gate valve
— f̲ — cono m̲ - [válv] plug, o̲ cone, valve
— f̲ — — m̲ para Venturi m̲ - [metal-prod] Venturi cone valve
— f̲ — cuatro vías f̲ - [válv] four-way valve
— f̲ — cuña f̲ - [válv] wedge valve
— f̲ — — f̲ de disco m̲ - [válv] wedge disk, o̲ disk wedge, valve
— f̲ — charnela f̲ - [válv] swing valve • articulated valve
— f̲ — dardo m̲ - [válv] dart valve
— f̲ — desacoplamiento m̲ - [válv] release valve
— f̲ — — m̲ rápido - [válv] quick release valve
— f̲ — desconexión f̲ - [válv] release valve
— f̲ — — f̲ rápida - [válv] quick release valve
— f̲ — diafragma m̲ - [válv] diaphragm valve
— f̲ — disco m̲ - [válv] disk valve
— f̲ — — m̲ basculante - [válv] tilting disk valve
— f̲ — — m̲ de cuña f̲ - [válv] wedge disk valve
— f̲ — — — m̲ con cuña f̲ de bronce m̲ - [válv]

bronze wedge disk valve

válvula f **con disco** m **con cuña** f **de hierro** m -
[válv] iron wedge disk valve
— f —— m **de tipo** m **de aguja** f - [válv] needle type disk valve
— f —— — m **con jaula** f - [válv] cage
type disk valve
— f —— m **doble** - [válv] double disk valve
— f —— m **tipo jaula** f - [válv] cage type
disk valve
— f — **dos electrodos** m - [electrón] véase
diodo m
— f — **engranaje(s)** m - [válv] gear valve
— f —— m **para mando** m - [válv] gear operated
valve
— f — **escape** m - [válv] leaky,king valve
— f — **esclusa** f - [válv] gate valve
— f —— f **maestra** - [válv] master gate valve
— f —— f **principal** - [válv] master gate valve
— f — **espiga** f - [válv] needle valve
— f — **expansión** f - [válv] expansion valve
— f —— f **térmica** - [válv] thermal expansion
valve
— f — **fuga** f - [válv] leaky,king valve
— f — **globo** m - [válv] globe valve
— f — **injerto** m - [válv] insert valve
— f —— m **de poliuretano** - [válv] polyurethane insert valve
— f — **mando** m - [válv] operating,tion valve
— f —— m **con solenoide** m - [válv] solenoid
operating,tion valve
— f —— m **mecánico** - [válv] mechanically operated valve
— f — **membrana** f - [válv] diaphragm valve
— f — **momento** m **torsional alto** - [válv] high
torque valve
— f —— m — **bajo** - [valv] low torque valve
— f — **movimiento** m **vertical** - [válv] tappet, o
poppet, o lift(ing) valve
— f — **obturación** f - [válv] plug(ging) valve
— f —— f **con disco** m - [válv] disk plug
valve
— f —— f —— m **de bronce** m - [válv] bronze
plug disk valve
— f — **obturador** m - [válv] poppet valve
— f —— m **flotante** - [válv] floating poppet
valve
— m —— m **para respiración** f - [válv]
floating poppet breather valve
— f — **orificio** m - [válv] orifice valve
— f —— m **variable** - [válv] variable orifice
valve
— f — **palanca** f - [válv] lever valve
— f —— f **para seguridad** f - [cald] lever
safety valve
— f — **par** m **motor, alto, o elevado** - [válv]
high torque valve
— f —— m — **bajo** - [valv] low torque valve
— f — **pedal** m - [válv] pedal, o foot, valve
— f — **pérdida** f - [válv] leaky,king valve
— f — **platillo** m - [válv] tappet valve
— f — **plato** m - [válv] disk valve
— f — **reajuste** m - [válv] reset valve
— f —— m **manual** - [válv] manual reset valve
— f —— m — **para cierre** m - [válv] manual
reset shut-off valve
— f — **resorte** m - [válv] spring (loaded) valve
— f — **suplemento** m - [valv] insert valve
— m —— m **de poliuretano** m - [válv] polyurethane insert valve
— f — **trampa** f - [válv] trap valve
— f —— f **con balde** m - [válv] bucket trap
valve
— f —— f —— m **invertido** - [válv] inverted
bucket trap valve
— f — **tres electrodos** m - [electrón] véase
triodo m
— f — **vías** f - [válv] three-way valve
— f —— — f **para cilindro** m **neumático** -
[planta para extrusión de electrodos] - pneumatic cylkinder three-way valve
— f — **vástago** m - [válv] stem valve

válvula f **con disco** m **asendente** - [válv] rising
stem valve
— f — m **hueco** - [válv] hollow stem valv
— f — f **no levadizo** - [válv] non-rising
stem valve
— f . . . **vía(s)** f - [válv] . . .-way valve
— f **cónica** - [válv] miter valve • [petról] dart
valve
— f **contra explosión** f - [válv] explosion valve
— f — **polvo** m - [válv] dust valve
— f **corredera,diza** - [válv] sleeve, o slide,
valve • piston valve
— f **charnela** - [valv] swing (type) valve
— f — **para retención** f - [válv] swing (type)
check valve
— f **checadora*** - [válv] check valve
— f **chapaleta** - [válv] flap valve
— f **check** - [válv] véase **válvula para escape** m
— f **de acero** m - [válv] steel valve
— f — m **soldado** - [válv] fabricated, o
welded, steel valve
— f — m **para dilatación** f - [válv] fabricated, o welded, expansion valve
— f — **hierro** m **fundido** - [válv] cast iron
valve
— f — m — **con disco** m - [válv] cast iron
disk valve
— f — m —— m **doble** - [Válv½ double disk
cast iron valve
— f **estelita** f - [válv] stellite valve
— f — **tipo** m **abierto** - [válv] open type valve
— f — m **con mando** m **mecánico** - [válv]
mechanically operated open type valve
— f — m **acodado** - [Válv] angle type valve
— f — m **con bonete** m - [válv] bonnet type
valve
— f —— m —— m **para sello** m **para presión** f
- [válv] pressure seal bonnet type valve
— f — m **con obturador** m - [válv] poppet
type valve
— f — m —— m **flotante** - [válv] floating
poppet type valve
— f — m — **charnela** f - [válv] swing type
valve
— f — m — **disco** m - [válv] disk type valve
— f — m —— m **basculante** - [válv] tilting
disk type valve
— f — m — **émbolo** m - [válv] piston type
valve
— f — m — **pistón** m - [válv] piston type
valve
— f — m **obturador** - [válv] poppet type
valve
— f — m **para cámara** f - [autom-neumát] inner tube type valve
— f — m —— f **para automóvil** - [autom-
-neumát] automobile inner tube type valve
— f — **vaivén** m - [válv] shuttle valve
— f **deformada** - [válv] deformed, o warped,
valve
— f **desactivada** - [válv] turned off, o disconnected, valve
— f **deslizante** - [válv] sliding valve
— f **desventadora** - [válv] vent)ing) valve
— f **distribuidora** - [válv] distributing,tion
valve
— f — **para aire** m - [combust] air distributing,tion valve
— f —— **gas** m - [combust] gas distributing,tion valve
— f **eléctrica** - [válv] electric valve
— f — **para fundente** m - [sold] electric flux
valve
— f — **para tolva** f - [ind] hopper electric
valve
— f —— — f **para fundente** m - [sold] flux
hopper electric valve
— f **electrónica** - [electrón] electronic valve;
vacuum tube
— f — **con dos electrodos** m - [electrón] véase
diodo m
— f —— **tres electrodos** m - [véase] triodo m

válvula f electrónica de tipo m para recepción f
- [electrón] receiving type electron(ic) tube
— f en ángulo m - [válv] angle valve
— f — codo m - [válv] angle valve
— f — culata f - [comb.int] (in) head valve •
overhead valve
— f — fondo m - [válv] bottom valve
— f — — m de colector m - [metal-prod] dust
catcher bottom volve
— f — m — — m para polvillo m - [metal-
-prod] dust catcher bottom valve
— f — m lavador m - [metal-prod] gas,
washer, o scrubber, bottom valve
— f — posición f abierta - [válv] open position valve
— f encerrada - [válv] enclosed, o built-in,
valve
— f — para mezcla f - [hidr-ducha] built-in
mixing valve
— f — — f para ducha f - [hidr] built-in
shower mixing valve
— f esclusa - [válv] sluice, o gate, valve •
[hidr] sluice gate
— f — con cuña f - [válv] wedge gate valve
— f — — — f de disco m - [válv] wedge disk
gate valve
— f — — disco m - [válv] disk gate valve
— f — — — m de cuña f - [válv] wedge disk
gate valve
— f — — — m — — f de bronce - [válv]
bronze wedge disk gate valve
— f — — m — — f — hierro m - [válv]
iron wedge disk gate valve
— f — — — m doble - [válv] double disk gate
valve
— f — — — m de hierro m (fundido) - [válv]
(cast) iron disk gate valve
— f — de hierro m (fundido) m - [válv] (cast)
iron gate valve
— f — — m (—) — con aleación f baja -
[válv] low alloy (cast) iron gate valve
— f — — — m — con disco m doble - [válv]
double disk cast iron gate valve
— f — — — — m de hierro m fundido -
cast iron double disk gate valve
— f — hidráulica - [válv] hydraulic sluice
valve
— f — para horno m alto - [metal-prod] blast
furnace sluice gate valve
— f — tipo m con guillotina f - [válv]
shear type gate valve
— f — — — m — de hierro m fundido con aleación f baja - [válv] low alloy cast iron
shear type gate valve
— f esférica - [válv] globe valve
— f — para descarga f - [válv] ball discharge
valve
— f — — — f con disco m - [válv] disk globe
plug discharge valve
— f esférica para retención f - [mec] ball
check(ing) valve
— f esmerilada - [válv] ground valve
— f estranguladora - [válv] throttle valve
— f estrictora - [válv] pinching valve
— f excitadora - [válv] activating valve
— f exigida - [válv] required valve
— f — para regulación f - [válv] required con-
trol(ling) valve
— f extrarreforzada - [válv] heavy duty valve
— f filtro - [válv] filter valve
— f — para henchimiento m - [válv] filter
filling valve
— f floja - [válv] loose(ned) valve
— f gaseosa - [electrón] gas(eous) tube
— f girada - [válv] turned, o rotated, valve
— f globo - [válv] globe valve
— f hidráulica - [válv] hydraulic valve
— f — para cierre m - [válv] hydraulic shut-
-off valve
— f horizontal - [válv] horizontal valve
— f hueca - [válv] hollow valve
— f igualadora - [valv] equalizing valve -
mixer,xing valve

válvula f igualadora para estufa f - [metal-
-prod] stove mixer valve
— f indicadora f - válv] indicating valve
— f inferior - [válv] lower valve
— f — colocada - [válv] installed lower valve
— f — para colector m - [metal-prod] dust
catcher, lower, o discharge, valve
— f — — — m para polvo m - [metal-prod]
dust catcher, lower, o discharge, valve
— f — — — m primario - [metal-prod] primary
dust catcher, lower, o discharge, valve
— f — — — m secundario - [metal-prod] secondary dust catcher, lower, o discharge,
valve
— f — — polvillo m - [metal-prod] lower dust
(catching) valve
— f — — — m en colector m - [metal-prod]
dust catcher lower (dust) valve
— f — purga f - [metal-prod] lower discharge,ging valve
— f — — — f para colector m (para polvo m) -
[metal-prod] dust catcher lower dust valve
— f — — polvo m - [metal-prod] lower, dust,
o discharge, valve
— f — — — m para botellón m - [metal-prod]
dust catcher lower, dust, o discharge, valve
— f — — salida f - [ind] lower discharge
valve
— f — — — f para colector m (para polvillo
m) - [metal-prod] dust catcher lower discharge valve
— f — quitada - [válv] removed lower valve
— f — removida - [válv] removed lower valve
— f — sacada - [válv] removed lower valve
— f inoperante - [válv] non-operating valve
— f inspeccionada - [válv] inspected valve
— f instalada - [válv] installed, o mounted,
valve
— f — en cabina f - [válv] cab mounted valve
— f íntegramente de bronce m - [válv] all
bronze valve
— f — — hierro m - [válv] all iron valve
— f intercomunicante - [válv] intercommuni-
cating,tion valve • [petról] compounding
valve
— f interruptora f - [válv] shut-off valve
— f — para seguridad f - [válv] safety shut-
-off valve
— f introducida - [válv] driven valve
— f inversora - [válv] reversing valve
— f lenteja f - [válv] disk valve
— f limitadora - [válv] limiting, o trim(ming),
valve
— f — para gas m - [válv] gas, limiting, o
trim(ming), valve
— f — — — m ácido - [petról] sour gas
trim(ming) valve
— f limpia f - [válv] clean valve
— f limpiada - [válv] cleaned valve
— f magnética - [válv] magnetic valve
— f — para mando m - [válv] magnetic (command)
valve
— f mandada - [válv] activated valve
— f manual - [válv] hand, o manual(ly), (operated) valve
— f — para aire m - [válv] hand air valve
— f — — combustible m - [combust] manual fuel
valve
— f — — — m para quemador m - [combust]
manual burner fuel valve
— f — — fundente m - [sold] manual flux valve
— f — — gas m - [combust] manual gas valve
— f — — suministro m - [combust] manual supply valve
— f — — — m de gas m (combustible) - [comb]
(fuel) gas supply, manual, o hand, valve
— f — principal - [válv] main, manual, o hand,
valve
— f mariposa - [válv] butterfly valve
— f — montada - [válv] mounted butterfly valve
— f mariposa normal - [válv] normal butterfly
valve
— f — — para Venturi - [metal-prod] Venturi

normal butterfly valve
válvula f **mariposa para agua** m - [válv] water
butterfly valve
— f — — **askania** f - [metal-prod] Askania butterfly valve
— f — — **guarda** f - [mec] guard butterfly
valve
— f — — — f **para turbina** f - [válv] turbine
guard butterfly valve
— r — — f **para turbina** f - [válv] turbine
guard butterfly valve
— f — **para lavadora** f - [metal-prod] scrubber,
o washer, butterfly (valve)
— f — — — **nivel** m - [válv] level butterfly valve
— f — — **toma** f — [válv] intake butterfly valve
— f — — — f **montada** - [valv] mounted butterfly intake valve
— f — **reguladora** - [válv] control(ling) butterfly valve
— f — — **para temperatura** f - [metal-prod]
temperature, control(ling), o regulating,
butterfly valve
— f — — — f **para viento** m - [metal-prod]
blast temperature, controlling, o regulating,
butterly valve
— f **para admisión** f - [válv] intake, o inlet, o
admission, valve
— f **mecánica** - [válv] mechanical valve
— f — **para alivio** m - [válv] mechanical relief
valve
— f — — — m **para presión** f - [Válv] mechanical pressure relief valve
— f **medidora** - [válv] measuring, o metering,
valve
— f **mezcladora** - [válv] mixing, o blending,
valve
— f **moduladora** - [electrón] modulating,tory
valve
— f **montada** - [válv] mounted valve • assembled
valve
— f **motorizada** - [válv] motorized valve
— f — **para apertura** f - [valv] motorized opening valve
— f — **para cierre** m - [válv] motorized closing
valve
— f **movible** - [válv] movable,ving valve
— f **móvil** - [válv] moving, o traveling, valve
— f **mudada** - [válv] moved valve
— f **necesaria** - [válv] needed, o necessary, o
required, valve
— f — **para regulación** f - [válv] necessary, o
required, control valve
— f **neumática** - [válv] air, o pneumatic, valve
— f — **para cambio(s)** m - [autom-mec] air
shifter,ting valve
— f — — **comando** m - [neumat] pneumatic control valve
— f — — **regulación** f - [válv] pneumatic control valve
— f — — — f **para gas** n - [combust] pneumatic
gas control (air) valve
— f — — **torno** - [mec] cathead air valve
— f **normal** - [válv] normal, o standard, valve
— f — **para Venturi** m - [metal-prod] Venturi
normal, o normal Venturi, valve
— f **obturadora** - [válv] poppet (type), o plug-
(ging), valve
— f — **de acero** m - [válv] steel plug(ging)
valve
— f — — m **fundido** - [válv] cast steel
plug(ging) valve
— f — **lubricada** - [válv] lubricated plug(ging)
valve
— f — **para respiración** f - [válv] poppet
(type) breather valve
— f **ocular** - [válv] goggle valve
— f **operada** - [válv] operated valve • switched
valve
— f — **con llave** f - [válv] wrench operated
valve
— f **optativa** - [válv] optional valve
— f **osciladora** - [electrón] oscillating,tor,
tube, o valve

válvula f **para abastecimiento** m - [válv] supply
valve
— f — — m **para aire** m - [válv] air supply
valve
— f — — m — — m **para equipo** m **perforador** -
[petról] rig air supply valve
— f — — m — **equipo** m **perforador** - [petról]
rig supply valve
— f — **acción** f **rápida** - [válv] quick operating
valve • quick action valve
— f — **aceleración** f - [comb.int] throttle
valve
— f — **acelerador** m - [comb.int] throttle
valve • butterfly valve
— f — **activación** f - [válv] activating valve
— f — — f **para estrangulador** m - [comb.int]
choke operating valve
— f — — f **y desactivación** f - [válv] on-off
valve
— f — — f — — f **para abastecimiento** m -
[válv] supply on-off valve
— f — — f — — f — — m **para aire** m -
[válv] air supply on-off valve
— f — — f — — f — — m — — m **para equipo** m **perforador** - [petról] rig air supply
on-off valve
— f — — f — **equipo** m **perforador** - [petróll]
[petról] rig on-off valve
— f — — f — — f — **provisión** f - [mec]
supply on-off valve
— f — — f — — f — — f **para aire** m - [mec]
air supply on-off valve
— f — — f — — f — **suministro** m - [mec]
supply on-off valve
— f — — f — — f — — m **para aire** m -
[mec] air supply on-off valve
— f — — f — — f — — m — — m **para equipo** m **perforador** - [petról] rig air supply
on-off valve
— f — **achicador** m - [petról] bailer valve
— f — **admisión** f - [valv] intake valve •
[comb.int] feed(ing) valve
— f — — f **para carburador** m - [comb.int½]
carburetor, feed, o intake, valve
— f — — f **pegada** - [comb.int] stuck intake
valve
— f — **agua** m - [válv] water valve
— f — — m **para alimentación** f - [ind] feed
water valve
— f — — m **para septum** m - [metal-prod] septum water valve
— f — — m — **Venturi** - [metal-prod] Venturi
water valve
— f — **aire** m - [válv] air valve
— f — — m **caliente** - hot air valve • [metal-
-prod] hot blast valve
— f — — m **comprimido** - [ind] compressed air
valve
— f — — m — **desde exterior** m - [ind] external compressed air valve
— f — — m — **exterior** - [ind] external compressed air valve
— f — — m **con mando** m **con solenoide** m -
[válv] solenoid operated air valve
— f — — m **frío** - [ind] cold air valve •
[metal-prod] cold blast valve
— f — — m **para torno** m - [petról] cathead
air valve
— f — — m **reparada** - [válv] repaire air valve
— f — **aislamiento** m - [válv] cut-off valve
— f — — m **para enfriador** m - [válv] cooler
cut-off valve
— f — — m — **filtro** m - [válv] filter cut-
-off valve
— f — **alimentación** f - [válv] feed(ing) valve
— f — — f **para carburador** m - [comb.int]
carburetor feed(ing) valve
— f — — f — **combustible** m - [comb.int] fuel
feed(ing) valve
— f — **alivio** m - [válv] relief valve • [tub]
air release valve
— f — — m **para explosión(es)** f - [metal-prod]

explosion door

válvula f para alivio m mecánica - [válv] mechanical relief valve

— f —— m **para aire** m - [tub] air release valve

— f —— m — **presión** f - [mec] pressure relief valve

— f —— m —— f **para aceite** m - [mec] oil pressure relief valve

— f —— m — **sobrepresión** f - [válv] surge, o overpressure, relief valve

— f ——m—— f **para émbolo** m - [mec] piston, surge, o overpressure, (relief) valve

— f —— m — f —— m **deslizante** - [Válv] sliding piston, surge, o overpressure, (relief) valve

— f — **amortiguación** f - [cutom-mec] shock absorbing valve

— f — **amortiguador** m - [autom] shock absorbing valve

— f — **aprovisionamiento** m - [válv] supply(ing) valve

— f —— m **para quemador** m - [combust] burner supply(ing) valve

— f — **asiento** m - [válv] seat valve

— f —— m **chato** - [válv] flat seat valve

— f —— m **plano** - [válv] flat seat valve

— f — **aspiración** f - [válv] suction valve

— f — **bajada** f - [válv] lowering valve

— f — **balanceo** m - [metal-prod] snort valve • hot blast valve • check valve

— f — **bloqueo** m - [válv] blocking valve

— f —— m **para entrada** f - [válv] incoming, o intake, blocking valve

— f —— m **esférica** - [válv] spherical blocking valve

— f — **caída** f - [valv] drop valve

— f —— f **para línea** f - [metal-prod] main drop valve

— f —— f —— f **para viento** m—[metal-prod] blast main drop valve

— f —— f —— f —— m **por mezclar** - [metal-prod] mixer main drop valve

— f —— f — **tubería** f - [metal-prod] main drop valve

— f —— f —— f **para viento** m **frío**—[metal-prod] cold blast main drop valve

— f —— f **y mezcla** f - [metal-prod] hot blast mixer valve

— f — **caja** f **para bomba** f - [bombas| fang

— f — **calentador** m - [válv] heater valve

— f — **cámara** f - [autom-neumát] inner tube valve

— f — **cambiador** m - [válv] shifter, o changer, valve

— f —— m **neumático** - [mec] air shifter valve

— f — **cambio(s)** m - [autom-mec] shifter valve

— f —— m **neumático** - [autom-mec] air shifter valve

— f — **carburador** m - [comb.int] carburetor valve

— f — **carrete** m - [válv] reel, o spool, valve

— f —— m **para núcleo** m - [petról] core reel, o reel clutch, valve

— f — **cierre** m - [mec] shut-off valve

— f —— m **con reajuste** m **manual** - [válv] manual reset(ting) shut-off valve

— f —— m **rápido** - [válv] quick closing valve

— f —— m **vertical** - [válv] lift valve

— f — **colector** m - [válv] drain valve

— f —— m **para lavador** m - [ind] scrubber, o washer, drain valve

— f — **combustible** m - [comb.int] fuel valve

— f —— m **para quemador** m - [combust] burner fuel valve

— f — **compartimiento** m - [Válv] compartment valve

— f —— m **para cambiador** m - [válv] changer compartment valve

— f — **compresión** f - [válv] compression valve

— f —— f **para ducha** f - [válv] compression shower valve

— f — **comprobación** f - [válv] test(ing) valve

válvula f para comprobación f para, red f, o sistema m, para sobrepaso(s) m - [petról] override system test(ing) valve

— f —— f **sobrepaso(s)** m - [petról] override test(ing) valve

— f — **compuerta** f - [válv] gate valve

— f —— f **montada** - [válv] mounted gate valve

— f —— f **— en obra** f - [mec] field mounted gate valve

— f — **conexión** f **y desconexión** f - [válv] on-off valve

— f — **contrapresión** f - [válv] back, o counter, pressure valve

— f — **contrasoplado** m - [válv] counter blast valve

— f — **corte** m - [válv] cut-, o shut-,off valve

— f —— m **mediante bloqueo** m - [combust] glocking shut-off valve

— f —— m **para abastecimiento** m - [válv] supply shut-off valve

— f —— m —— m **de gas** m - [xombust] gas supply shut -off valve

— r —— m —— m **combustible** - [combust] fuel gas supply shut-off valve

— f —— m **para combustible** m - [combust] fuel shut-off valve

— f —— m — **gas** m - [combust] gas shut-off valve

— f —— m —— m **combustible** - [combust] fuel gas shut-off valve

— f —— m —— m **para hovno** m **alto** - [metal-prod] blast furnace gas shut-off valve

— f —— m —— m — **por presión** f **baja o alta** - [metal-prod] high or low pressure blast furnace gas shut-off valve

— f —— m **para seguridad** f - [válv] safety shut-off valve

— f —— m **por sobrepresión** f - overpressure shut-off valve

— f — **cuba** f - [metal-prod] stack valve

— f — **chimenea** f - [metal-prod] stack valve

— f — **decompresión** f - [metal-prod] decompression valve

— f — **derivación** f - [metal-prod] bleeder valve • by-pass valve

— f — **desagüe** m - discharge valve

— f —— f **para emergencia** f - emergency discharge valve

— f —— m —— f **para lavador** m - [metal-prod] scrubber emergency (water) discharge valve

— f —— m —— f — **Venturi** m - [metal-prod] Venturi scrubber emergency (water) discharge valve

— f — **desahogo** m - [válv] relief valve

— f — **descarga** f—discharge, o exhaust, o relief valve • blow-off, o unloading, valve • véase también **válvula f para purga** f - [tub] blowoff valve• [metal-prod] (stove) chimney valve • [sanit-inodoro] flush valve

— f —— f — **agua** m **en Venturi** - [metal-prod] scrubber water, discharge, o exhaust, valve

— f — **bomba** f - [bombas] pump release valve

— f —— f —— f **para inyección** f - [petról] mud pump release valve

— f —— f —— f — **lodo** m - [petról] mud pump release valve

— f —— f — **colector** m **para polvo** m - [metal-prod] dust catcher discharge valve

— f —— f —— m **primario** - [metal-prod] primary (dust catcher) discharge valve

— f —— f —— m **secundario** - [metal-prod] secondary (dust catcher) discharge valve

— f —— f — **compresor** m **para gas** f—[metal-prod] gas booster discharge valve

— f —— f — **corte** m - [metal-prod] check, o snort, discharge valve

— f —— f — **emergencia** f - [metal-prod] emergency discharge valve

válvula f para descarga f para freno m - [mec]
brake discharge valve
— f — — f — polvo m - [metal-prod] dust exhaust valve
— f — — f — presión f - [mec] pressure relief valve
— f — — f — Venturi - [metal-prod] scrubber,
discharge, o exhaust, valve
— f — — f rápida - [válv] quick exhaust valve
— f — — f refretntada con estelita f - [comb.
-int] stellite faced exhaust valve
— f — desconexión f rápida - [válv] quick
release valve
— f — desviación f - [válv] by-pass valve
— f — desvío m - [válv] valve to @ by-pass
— f — — m para colector m - [válv] drain detour valve
— f — — — m para lavador m - [metal-
-prod] scrubber, o washer, drain detour
valve
— f — dilatación f - [válv] expansion valve
— f — — f térmica - [válv] thermal expansion
valve
— f — distribución f - [válv] distribution, o
manifold, valve • slide valve • manifold
— f — — f para gas m - [combust] gas distribution valve
— f — — f — presión f - [válv] pressure distribution valve
— f — — f — — f para ventilador m -
[combust] fan pressure distribution valve
— f — — f — — — m para aire m -
[combust] air fan pressure distribution valve
— f — — f — — — m — gas m -
[combust] gas fan pressure distribution valve
— f — distribuidor m - [válv] manifold valve
— f — dos posiciones f - [válv] on-off valve
— f — drenaje m - [válv] drain(age) valve
— f — — m y filtración f - [válv] drain(age)
and filter(ing) valve
— f — émbolo m - [comb.int] piston valve
— f — embrague m - [mec] clutch valve
— f — — m para carrete m - [petról] reel
clutch valve
— f — — m — — m para núcleo m - [petról]
core reel clutch valve
— f — empuje m - [válv] push(ing) valve
— f — emergencia f - [válv] emergency valve
— f — engrase m - [mec] grease,sing valve
— f — entrada f - intake, o inlet, valve
— f — — f para agua m - [válv] water, intake,
o inlet, valve
— m — — — m para Venturi m - [metal-
-prod] scrubber water,intake, o inlet, valve
— f — — f — aire m - [válv] air, intake, o
inlet, valve
— f — — f — — m comprimido - [mec] compressed air, intake, o inlet, valve
— f — — f — — m — (desde) exterior m -
[mec] external compressed air, intake, o inlet, valve
— f — — f — — m en carburador m - [comb.-
-int] float needle valve
— f — — f — tapón m - [tub] plug, intake, o
inlet, valve
— f — — f — — m lubricado - [tub] lubricated
plug, intake, o inlet, valve
— f — equipo m - [mec] equipment valve
— f — — m perforador - [petról] rig valve
— f — escape m - [válv] exhaust, o blowoff, o
release, o relief, valve • véase también válvula f para purga f - [metal-prod] snort
valve
— f — — m para aire m - [válv] air, cock, o
release valve
— f — escape m libre - [comb.int] cut out
— f — — m para presión f - [mec] pressure
relief valve
— f — — m — viento m - [metal-prod] (furnace)
check valve
— f — — m pegada - [comb.int] stuck exhaust
valve
— f — esclusa f - [válv] gate valve

válvula f para esclusa f con disco m - [válv]
disk gate valve
— f — — f — — m de hierro m - [válv] iron
disk gate valve
— f — estrangulación f - [comb.int] throttle,
o choke,king, valve
— f — estrangulador m - [válv] choke valve •
throttle,ling valve •   [comb.int] butterfly valve
— f — — m operada - [mec] switched choke
valve
— f — evacuación f - [válv] exhaust valve
— f — expansión f - [válv] expansion valve
— f — — f térmica - [metal-prod] thermal expansion valve
— f — explosión f - [metal-prod] safety valve
— f — extremo m - [válv] end valve
— f — — m hidráulico - [petról] fluid (end)
valve
— f — filtro m - [válv] filter valve
— f — — f inferior - [Válv] lower, o bottom,
filter valve
— f — flotador m - [válv] float valve
— f — fondo m - [válv] lower, o bottom, valve
— f — — m para achicador m - [petról] bailer
valve
— f — freno m - [válv] brake valve
— f — — m auxiliar - [mec] auxiliary brake
valve
— f — — m para mesa f - [petról] table brake
valve
— f — — m — — f rotatoria - [petról] rotary (table) brake valve
— f — — m por inercia f - [mec] inertia
brake valve
— f — fundente m - [sold] flux valve
— f — — m en pistola f - [sold] gun flux
valve
— f — gas m - [válv] gas valve • [electrón]
gas(eous), tube, o valve
— f — — m ácido - [petról] sour gas valve
— f — — m para celantamiento m - [combust]
heating gas valve
— f — — m para coquería f - [combust| coke
plant gas valve
— f — — m — estufa f - [metal-prod] stove
gas valve
— f — — m rico - [metal-prod] rich gas valve
— f — — — m para mechero m - [metal-prod]
rich gas burner valve
— f — control m - [válv] control valve
— f — — m para dirección f - [válv] steering
control valve
— f — igualación f - [válv] equalizing valve
— f — inversión f - [metal-prod] reversing
valve
— f — — f para aire m - [válv] air reversing
valve
— m — lavador m - [metal-prod] scrubber, o
washer, valve
— m — — m para gas m - [metal-prod] gas,
scrubber, o washer, valve
— f — liberación f - [válv] release valve
— f — — f rápida - [Válv] quick release valve
— f — línea f - [tub] line valve
— f — — f para aspiración f - [bombas] suction line valve
— f — — f — — f para bomba f - [bombas]
pump suction line valve
— f — — f — — f — — f para inyección f -
[bombas] charge pump suction line valve
— f — — f — — f para bomba f inyectora -
[bombas] charge pump section line valve
— f — — m combustible m - [comb.int] fuel
line valve
— f — — f para derivación f - [válv] by-pass
line valve
— f — — f — — m para horno m - [metal-
-prod] oven, o furnace, gas line valve
— f — — f — — m para coque m -
[metal-prod] coke oven gas line valve
— f — llenado m - [mec] filling valve
— f — mando m - [Válv] control(ling) valve

válvula f para manómetro m - [válv] gage valve
— f —— m para presión f - [válv] pressure
gage valve
— m —— f para evaporador m - [válv]
evaporator pressure gage valve
— f — medición f - [válv| measuring, o meter-
ing, valve
— f —— f precisa f - [válv] fine measuring
valve
— f — mezcla(dura) f - [válv] mixer,xing valve
— f —— f para ducha f - [metal-prod] shower
mixer,xing valve
— f —— f —— estufa v - [metal-prod] stove
mixing valve
— f — motor m - [comb.int] engine valve •
[electr] motor valve
— f —— m hidráulico - [válv] hydraulic motor
valve
— f — neumático m - [autom-neumát] tire valve
— f — nivel - m - [válv] level valve
— f —— m de agua f—[válv] water level valve
— f — obturación f - [válv] closure, o plugging, valve
— f — operación f—[válv] operating,tion valve
— f —— f para estrangulador n - [válv] choke
operating,tion valve
— f — paso m - [válv] shut-off valve
— f —— m para calentador m - [válv] heater
shut-off valve
— f —— m —— m para cabina f - [autom] cab
heater shut-off valve
— f —— m para línea f - [comb.int] line shut-
-off valve
— f —— m —— f para combustible m - [comb.
-int] fuel line shut-off valve
— f — pie n - [válv] foot valve
— f — pistón n - [válv] véase válvula f para
émbolo
— f — polvo m - [válv] dust valve
— f —— m para colector m - [metal-prod] dust
catcher valve
— f —— m —— m primario - [metal-prod]
primary dust catcher (dust) valve
— f —— m —— m secundario - [metal-prod]
secondary dust catcher (dust) valve
— f — presión f - [válv] pressure valve
— f —— f activadora - [válv] activating
pressure valve
— f —— f excitadora - [valv] exciter,ting
pressure valve
— f — provisión f - [válv] supply(ing) valve
— f —— f para aire m - [válv] air supply
valve
— f —— f —— m para equipo m - [Válv]
equipment air supply valve
— f —— f —— m perforador -
[petrol] rig air supply valve
— f —— f —— m para trefilador m—[trefil]
wire drawer,wing air supply valve
— f —— f — equipo m perforador - [petról]
rig supply valve
— f —— f — gas m - [válv] gas supply valve
— f — purga f - [válv] blow-down, o drain, o
release, o blow-off, valve • [metal-prod] release, o decompression, valve • discharge
valve
— f —— f para aire m - [válv] air cock
— f —— f — bomba f - [bombas] pump release
valve
— f —— f —— f para inyección f - [petról]
mud pump release valve
— f —— f —— f — lodo m - [petról] mud
pump release valve
— f —— f para caldera f - [cald] boiler
bleeder valve
— f —— f para cono m - [metal-prod] cone
(emergency) discharge valve
— f —— f —— m para Venturi m - [metal-
-prod] Venturi cone (emergency) discharge
valve
— f —— f — freno n - [válv] brake discharge
valve
— f —— f — lavador m - [metal-prod] scrub-

ber, o washer, discharge,ging valve
válvula f para purga f para lavador m - [válv]
scrubber, o washer, discharge valve
— f — quemador m - [combust] burner valve
— f —— m para estufa f - [metal-prod] stove
burner valve
— f — rebose m - [metal-prod] overflow valve
— f —— f de colector m para lavador m -
[metal-prod] dustcatcher, scrubber, o washer,
overflow valve
— f —— m final - [válv] final overflow valve
— f —— m para colector m para lavador m -
[válv] scrubber, o washer, dustcatcher overflow valve
— f — red f - [válv] system valve
— f —— f para aceite m - [lubric] oil system
valve
— f —— f —— m para lubricación f -
[lubric] control(ling) oil system valve
— f —— f — agua m - [hidr] water system
valve
— f —— f —— m para enfriamiento m -
[hidr] cooling water system valve
— f —— f — regulación f- [válv] control-
-(ling) system valve
— f —— f —— f para aceite m - [lubric]
oil control(ling) system valve
— f — reducción f - [válv] reducing,cction
valve
— f —— f para presión f - [válv] pressure
reducing,cction valve
— f — regulación f - [válv] regulating,tion,
o control(ling), valve¢ choke valve
— f —— f automática - [válv] automatic con-
trol(ling) valve
— f —— f —— f para caudal m - [válv] automatic flow control valve
— f —— f —— f —— m de agua m - [sold] automatic water flow control valve
— f —— f exigida - [válv] required control
valve
— f —— f girada - [válv] turned control
valve
— f —— f para agua n - [hidr] water regulating valve
— f —— f — aire m - [válv] air controlling
valve
— f —— f — altura f [mec] height control valve
— f —— f —— f para cabezal m - [agric-
-equip] header height control valve
— f —— f —— f — espigadora f - [agric-
-equip] header height control valve
— f —— f para antepozo m - [petról] cellar
control valve
— f —— f — bajada f - [mec] lowering control valve
— f —— f — caída f - [mec] drop control
valve
— f —— f —— f para cabezal m - [agric-
-equip] header drop control valve
— f —— f —— f — espigadora f - [agric-
-equip] header drop control valve
— f —— f para carrete m - [grúas] spool
control valve
— f —— f — caudal - [válv] flow control
valve
— f —— f —— m para agua m - [hidr] water
flow control valve
— f —— f —— m variable - [Válv] variable
flow control valve
— f —— f — descarga f - [válv] discharge
control valve
— f —— f —— f para separador m - [metal-
-prod] separator, discharge, o exhaust, control valve
— f —— f —— f —— m para Venturi -
[metal-prod] scrubber, o washer, separator,
discharge, o exhaust, control valve
— f —— f para dirección f - [autom-mec]
steering control valve
— f —— f — flujo m - [válv] flow control
valve

válvula f para regulación f de freno m - [válv]
brake control(ling) valve
— f — — f — gas m - [combust] gas control
valve
— f — — f — hoja f - [mec] blade control
valve
— f — — f — f para topadora f
- [constr] dozer blade control valve
— f — — f — mezcla f - [mec] ratiotrol
— f — — f — presión f - [válv] pressure
control valve
— f — — f — purga f- [válv] blowdown control
valve
— f — — f — quemador m - [combust] burner
control valve
— f — — f — sótano m - [petról] cellar control valve
— f — — f — suministro m - [válv] supply
control valve
— f — — f — — m de gas m - [combust] gas
supply control valve
— f — — f — temperatura f - [ind] temperature control valve
— f — — f — — f viento m caliente -
[metal-prod] hot blast temperature control
valve
— f — — f — velocidad f - [mec] speed control valve
— f — — f — viento m caliente - [metal-prod]
hot blast control valve
— f — — f — para estrangulador m -
[mec] choke speed control valve
— f — — f requerida - [Válv] required control
valve
— f — regular v - [válv] regulating valve
— f — — v circulación f - [petról] circulating head
— f — repuesto m - [válv] spare (part) valve
— f — — m para aire m - [neumát] spare air
valve
— f — resoplado m - [metal-prod] snort valve
— f — respiración f - [válv] breather,thing
valve
— f — — f con obturador m - [válv] poppet
breather,thing valve
— f — — f de tipo m con obturador m - [válv]
poppet (type) breather,thing valve
— f — — f — m flotante - [válv]
floating poppet type breather,thing valve
— f — respiradero m - [constr] vent valve
— f — retención f - [mec] check valve • [válv]
non-return, o drop, valve
— f — — f automática - [válv] automatic check
valve
— f — — f para amortiguador m - [válv] deamp-
ener check valve
— f — — f — bomba f - [bombas] pump check
valve
— f — — f — caudal m - [válv] flow check
valve
— f — — f — — m excesivo - [válv] excess
flow check valve
— f — — f — cierre m vertical - [válv]
lift(ing) check valve
— f — — f — estabilizador m - [grúas] outrigger check valve
— f — — f— furnia* f igualadora f - [metal-
-prod] equalizing main check valve
— f — — f — línea f igualadora—[metal-prod]
equalizing main check valve
— f — — f — tubería f igualadora - [metal-
-prod] equalizing, o mixer, main check valve
— f — — f — vapor m - [válv] steam non-
-return valve
— f — salida f - [válv] exhaust, o outlet,
valve
— f — — f para aire m - [válv] air, outlet,
o exhaust, o release, valve
— f — — f — m comprimido - [válv] compressed air outlet valve
— f — — f — polvillo m - [metal-prod] dust
removal valve
— f — — f — — m en colector m - [metal-
-prod] dust catcher dust removal valve

válvula f para sangría f - [válv] bleeder valve
— f — seccionamiento m - [válv] isolating
valve
— f — seguridad f - [válv] safety, o blow-off,
valve
— f — — f con disparo m - [válv] pop safety
valve
— f — — f para aire m - [válv] air safety
valve
— f — — f — alivio m - [válv] safety relief,
o relief safety, valve
— f — — f para cierre m - [válv] safety shut-
-off valve
— f — — f — m con reajuste m manual -
[válv] manual reset shut-off (safety) valve
— f — — t — m — — m — para gas m -
[combust] gas manual reset shut-off safety
valve
— f — — f — corte m para bloqueo m para gas
m combustible - [combust] fuel gas blocking
safety shut-off valve
— f — — f — corte m - [válv] safety shut-off
valve
— f — — f — m mediante bloqueo m - [válv]
blocking safety shut-off valve
— f — — f — m para combustible m - [válv]
fuel safety shut-off valve
— f — — f — m — gas m - [combust] gas
safety shut-off valve
— f — — f para gas m—[válv] gas safety valve
— f — — f — contrapeso m - [mec] dead
weight safety valve
— f — servicio m - [válv] service valve
— f — m para derivación f - [válv] by-pass
service valve
— f — — m — refrigeración f - [ind] refrigeration service valve
— f — sobrecarga f - [válv] overload valve
— f — sobrepaso m - [petról] override valve
— f — sobrepresión f - [electr-equip] overpressure valve
— f — con contacto(s) m para alarma f -
[válv] overpressure valve with @ alarm con-
tact(s)
— f — — f en cámara f para (inter)cambiador
m - [metal-prod] exchanger chamber overpressure valve
— f — — f — cambiador m - [mec] changer
overpressure valve
— f — — f — compartimento m para cambiador m
- [mec] changer compartment overpressure valve
— f — succión f - [válv] suction valve
— f — suministro m - [válv] supply valve
— f — — m para aire m - [válv] air supply
valve
— f — — m — — m para equipo m perforador -
[petról] rig air supply valve
— f — — m — gas m - [válv] gas supply valve
— f — — m — — m combustible - [combust]
fuel gas supply valve
— f — — f — reajuste n - [válv] reset safety
valve
— f — — f — — m manual - [combust] manual
reset safety valve
— f — — f principal—[válv] main safety valve
— f — — f — para corte m para gas m - [válv]
main gas safety shut-off valve
— f — — f — reajuste m manual - [combust]
main manual reset safety valve
— f — sistema n - [válv] system valve
— f — — m para aceite m - [lubric] oil system
valve
— f — — m — — m para regulación f - [válv]
control(ling) oil system valve
— f — — m — — f para aceite m - [lubric]
oil control(ling) system valve
— f — — m para sobrepaso m - [petról] override system valve
— f — tolva f - [válv] hopper, o bin, valve
— f — — f para fundente m - [sold] flux hopper valve
— f — torre f - [petról] cathead valve
— f — tragante m - [metal-prod] throat valve
— f — tubería f - [válv] pipe valve

válvula f̲ para tubería f̲ para descarga f̲—[válv] discharge pipe valve
— f̲ — **vapor** m̲ - [valv] steam valve
— f̲ — — f̲ **para tubería** f̲ - [tub] main steam valve
— f̲ — — m̲ — — f̲ **general** - [metal-prod] main steam valve
— f̲ — — f̲ — **para estufa(s)** f̲ - [metal-prod] stoves main steam valve
— f̲ — — m̲ — **tragante** m̲ - [metal-prod] throat steam valve
— f̲ — **venteo** m̲ - [válv] venting valve
— f̲ — — m̲ — **gas** m̲ - [combust] gas vent(ing) valve
— f̲ — — m̲ — — m̲ **combustible** - [combust] fuel gas vent(ing) valve
— f̲ — **Venturi** m̲ - [metal-prod] Venturi valve
— f̲ — **viento** m̲ **caliente** - [metal-prod] hot blast valve
— f̲ — — m̲ — **de estufa** f̲ - [metal-prod] stove hot blast valve
— f̲ — — m̲ **frío** - [metal-prod] cold blast valve
— f̲ — — m̲ — **para estufa** f̲ - [metal-prod] stove cold blast valve
— f̲ **pegada** - [válv] stuck valve
— f̲ **pegadiza** - [comb.int] sticky valve
— f̲ **picada** - [comb.int] pitted valve
— f̲ **piloto** - [válv] pilot valve
— f̲ — **instalada** - [válv] installed pilot valve
— f̲ — **solenoide** - [válv] solenoid pilot valve
— f̲ **plana** f̲ - [válv] flat valve
— f̲ **por disparo** m̲ - [válv| pop valve
— f̲ **precisa** - [válv] fine, o precise, valve
— f̲ — **para medición** f̲ - [válv] fine metering valve
— f̲ **primaria** - [válv] primary valve
— f̲ — **para polvo** m̲ - [válv] primary dust valve
— f̲ — — **colector** m̲ **(para polvo** m̲**)** - [metal--prod] primary (dust) catcher dust valve
— f̲ **principal** - [válv] main valve
— f̲ — **para cierre** m̲ **para gas** m̲ - [combust] main gas latch valve
— f̲ — — **corte** m̲ - [válv] main cut off valve
— f̲ — — — m̲ **para abastecimiento** m̲ - [combust] main gas supply cut-off valve
— f̲ — — m̲ — — m̲ **de gas** m̲ **(combustible)** - [combust] main fuel gas supply shut-off valve
— f̲ — — **gás** m̲ - [combust] main gas valve
— f̲ — — **regulación** f̲-[válv] main control valve
— f̲ — **sobrecarga** f̲ - [válv] main overload valve
— f̲ — — **suministro** m̲ - [ind] main supply valve
— f̲ — — — m̲ **para gás** m̲ - [combust] main gas supply valve
— f̲ — — — m̲ — — m̲ **combustible** - [combust] main fuel gas supply valve
— f̲ **provista con resorte** m̲ - [mec] spring loaded valve
— f̲ **proyectada** - [válv] designed valve
— f̲ **purgada** - [válv] drained valve
— f̲ **purgadora** - [válv] draining valve
— f̲ **que funciona** - [válv] functioning valve
— f̲ **quemada** - [comb.int] burned valve
— f̲ **quitada** - [válv] removed valve
— f̲ **recubierta** - [válv] coated valve
— f̲ **reductora** - [válv] reducing valve
— f̲ — **para presión** f̲ - [metal-prod] snort valve pressure reducing valve
— f̲ **refrentada** - [válv] faced valve
— f̲ — **con estelita** f̲ - [comb.int] stellite faced valve
— f̲ **regulable** - [válv] adjustable valve
— f̲ — **con mano** f̲ - [válv] hand adjustable valve
— f̲ **regulada** - [válv] controlled, o adjusted, valve
— f̲ — **para estrangulación** f̲ - [válv] controlled choke,king valve
— f̲ **reguladora** - [válv] regulating, o control, valve
— f̲ — **ajustada** - [válv] adjusted control valve
— f̲ — — **en fábrica** f̲ - [válv] factory set, regulating, o control, valve

válvula f̲ **reguladora para agua** m̲ - [válv] water regulating valve
— f̲ — — — — — m̲ **ajustada** - [Válv] adjusted, o set, water regulating valve
— f̲ — — — — m̲ — **en fábrica** f̲ - [Válv] factory, adjusted, o set, water regulating valve
— f̲ — — **aire** m̲ - [válv] air control valve
— f̲ — — **combustible** m̲ - [combust] fuel control valve
— f̲ — — **flujo** m̲ - [válv] flow control valve
— f̲ — — **nivel** m̲ - [válv] height control valve - level control(ling) valve
— f̲ — — **presión** f̲ = [válv] pressure control valve
— f̲ — — **temperatura** f̲ - [valv] temperature control valve
— f̲ — — — f̲ **para viento** m̲ - [metal-prod] blast temperature control valve
— f̲ — — **viento** m̲ - [metal-prod] blast control valve
— f̲ — **reductora** - [válv] reducing control valve
— f̲ — — **presión** f̲ - pressure reducing control valve
— f̲ **remota** - [válv] remote valve
— f̲ — **con . . . circuito(s)** m̲ - [valv] . . . circuit remote valve
— f̲ **removida** - [válv] removed valve
— f̲ **reparada** - [válv] repaired valve
— f̲ **requerida** - [válv] required valve
— f̲ — **para regulación** f̲ - [válv] required control valve • valve required for @ control
— f̲ **respiradora** - [válv] breather,thing valve
— f̲ **reversible** - [válv] reversible,sing valve
— f̲ — **para aire** m̲ - [válv] air reversing, o reversing air, valve
— f̲ **roscada** - [válv] threaded valve
— f̲ — **hacia adentro** - [válv] threaded in valve
— f̲ — — **afuera** - [válv] threaded out valve
— f̲ **sacada** - [válv] removed valve
— f̲ **sangrada** - [válv] bled valve
— f̲ **sangradora** - [válv] bleeder valve
— f̲ **secundaria para colector** m̲ **para polvo** m̲ - [metal-prod] secondary dust catcher (dust) valve
— f̲ **selectora** - [válv] selector valve
— f̲ — **girada** - [válv] switched, o turned, selector valve
— f̲ — **operada** - [válv] switched, o operated, selector valve
— f̲ — — **para estrangulador** m̲ - [válv] choke selector valve
— f̲ — — — m̲ **operada** - [válv] switched choke selector valve
— f̲ **séptum** - [metal-prod] septum valve
— f̲ **sifón** - [válv] siphon valve
— f̲ **silenciosa** - [válv] silent, o noisless, o non-slam, valve
— f̲ **sincronizada** - [comb.int] synchronized, o timed, valve
— f̲ **snort** - [metal-prod] snort vakve
— f̲ **sobrecalentada** - [válv] overheated valve
— f̲ **sobrepasada** - [válv] overriden valve
— f̲ **solenoide** - [válv] solenoid valve
— f̲ — **para (llama** f̲**) piloto** - [válv] pilot gas solenoid valve
— f̲ — — **granuladora** f̲ - [metal-prod] granulator solenoid valve
— f̲ — — **línea** f̲ - [comb.int] line solenoid valve
— f̲ — — — f̲ **para combustible** - [comb.int] fuel line solenoid valve
— f̲ — — **luz** f̲ - [válv] light solenoid valve
— f̲ — — — f̲ **piloto** - [combust] pilot (light) solenoid valve
— f̲ — — **llama** f̲ - [combust] flame solenoid valve
— f̲ — — — f̲ **piloto** - [combust] pilot (flame) solenoid valve
— f̲ — — **piloto** m̲ - [combust] pilot solenoid valve
— f̲ — — — **seguridad** f̲ - [combust] solenoid safety valve

válvula f solenoide para seguridad f para venteo
  m de gas m - [combust] venting gas solenoid
  safety valve
— f — — venteo m - [combust] venting solenoid valve
— f — — — m de gas m - [combust] gas venting
  solenoid valve
— f soltada - [válv] released valve
— f Stockham - [válv] Stockham valve
— f suelta - [valv] loose valve
— f superior - [válv] upper, o higher, valve
— f — colocada - [válv] installed upper valve
— f — con filtro m - [válv] upper filtering
  valve
— f — — — m para henchimiento m - [válv] upper filter filling valve
— f — instalada - [válv] installed upper valve
— f — para henchimiento m - [válv] upper filling valve
— f — quitada - [válv] removed upper valve
— f — removida - [válv] removed upper valve
— f — sacada - [válv] removed upper valve
— f tapón - [válv] plug valve
— f — de acero m - [válv] steel plug valve
— f — — — m fundido - [válv] cast steel plug
  valve
— f — operada con llave f - [válv] wrench
  operated plug valve
— f térmica - [válv] thermal valve
— f tipo charnela para retención f - [válv]
  swing type check valve
— f — diafragma para regulación f - [válv]
  diaphragm control valve
— f ubicada en cabina f - [válv] cab-mounted
  valve
— f vacía - [válv] empty valve
— f variable - [válv] variable valve
— f vertical - [válv] vertical valve • poppet
  valve
— f viajera - [válv] traveling valve
valla f contra escoria f - [metal-prod] slag
  shield
— f de acero m - [constr] steel railing
— f — — m para puente(s) m - [constr] steel
  bridge railing
— f — nuemático(s) m - [deport] tire wall
— f divisoria - [vial] median barrier
— f — con dos frentes f - [constr-vial] double
  -faced median barrier
— f — de viga f en caja f - [constr] box beam
  median barrier
— f flexible - [constr] flexible barrier
— f imaginaria - imaginary, barrier, o fence
— f lateral - [vial] guard rail
— f — para defensa f - [vial] guard rail
— f medianera - [vial] median guardrail
— f para barrera(s) f - [constr] barrier beam
— f — contención f - [vial] guardrail
— f — puente m - [constr] bridge railing
— f rígida - [constr] rigid barrier
— f semirrígida - [constr] semirigid barrier
— f tubular - [constr] tubular railing
— f — de acero m - [constr] tubular steel
  railing
— f — — m para puente(s) m - [constr] tubular steel bridge railing
vallado,da a - enclosed
vallar v - . . .; to enclose
valle m angosto - [topogr] narrow valley
— f de arroyo m - [topogr] creek valley
— m — — m rural - [topogr] rural creek valley
— m corrugación f - [mec] corrugation, valley, o low point
— m estrecho - [topogr] narrow valley
— m fértil - [topogr] fertile valley
— m fluvial - [topogr] river valley
— m — angosto - [topogr] narrow river valley
— m playo - [topogr] shallow valley
vanguardia f - . . .; lead; forefront; leader
— f en fabricación f - [ind] manufacturing
  leadership
— f — ingeniería f - [ind] engineering leadership

vanguardia f en maquinaria f - [ind] machinery
  leadership
— f — — f para (material) m plástico -
  [plást] plastic(s) machinery leadership
— f — plástico(s) m - [plást] plastic(s)
  leadership
— f — servicio m - [ind] service,cing lead-
  erhip
— f — sistema m - [ind] system's leadership
— f — — m para (materiales) plásticos m -
  [plást] plastics systems leadership
vano m - [constr] bent; bay • opening; span
— m adyacente - [constr] adjacent bent
— m arqueado - [archit] arch
— m final - [constr] end bay
— m para puerta f - [constr] doorway
— m transversal - [constr] transverse bent
— m — adyacente - [constr] adjacent transverse bent
vapor m - . . . . • [nav] . . .; ship; vessel •
  [segurid] vapor; fume(s)
— m calentado - [cald] heated steam
— m circulado - [cald] circulated steam
— m clorinado - [quím] chlorinated vapor
— m combustible - [combust] fuel vapor
— m — de propano m - [combust] propane, fuel
  vapor, o vapor fuel
— m con presión f - [cald] pressurized steam
— m — — f alta - [Vald] high pressure steam
— m — — f baja - [cald] low pressure steam
— m corrosivo - [cald] corrosive steam
— m cortado - [vapor] cut, o turned, off steam
— m de agua m - [cald] (water), vapor, o steam
— m — calor m residual - [cald] waste steam
  heat
— m — desareador m - [sold] deareator steam
— m — escape m - [cald] exhaust steam
— m — expansión f - [cald] flash steam
— m — extracción f - [cald] extraction steam
— m — — f bajo presión f - [cald] pressure
  extraction steam
— m — — f con presión f alta - [cald] high
  pressure extraction steam
— f — — f — — f baja - [cald] low pressure
  extraction steam
— m — hidrocarburo(s) m - [quím] hydrocarbon
  vapor
— m — — m clorinado(s) - [quím] chlorinated
  hydrocarbon vapor
— m — mercurio m - [electr] mercury vapor
— m — precipitador m - precipitator steam
— m — propano m - [combust] propane vapor
— m — — m purgado - [combust] purged propane
  vapor
— m — yodoformo m - [quím] iodoform vapor
— m descontaminado - [cald] decontaminated
  steam
— m despedido - [cald] given off steam •
  [comb.int] sprayed out steam
— m disolvente - [cald] solvent vapor
— m — depósito m - [cald] tank steam
— m — — m para almacenamiento m - [petról]
  stock tank vapor
— m excedente - [cald] excess steam
— m manejado - [cald] handled steam
— m para calefacción f - [cald] heating steam
— m — distribución f - [cald] distribution
  steam
— m — — f en planta f - [cald] plant distribution steam
— m posible - [cald] possible steam • [petról]
  possible vapor(s)
— m — en depósito m - [petról] stock tank
  vapor
— m — — — m para almacenamiento m -
  [petról] possible stock tank vapor(s)
— m purgado - [quím] purged vapor
— m recalentado - [cald] reheated steam
— m residual - [cald] residual steam
— m saturado - [cald] saturated steam
— m seco - [cald] dry steam
— m sobrecalentado m - [cald] superheated, o
  overheated, steam

vapor m sobrecalentado con presión f alta -
[cald] ñigh pressure superheated steam
— m — — — f baja - [cald] low pressure superheated steam
— m vivo - [cald] live steam
vaporización f - [cald] . . .; given off steam •
flashing • [pint] véase pulverización f
— f alta - [comb.int] high vaporization
— f baja - [comb.int] low vaporization
— f de gasolina f - [comb.int] gasoline vaporization
— f de agua m - [cald] water vaporization
— f — — m en caldera f - [cald] boiler water
vaporization
— f instantánea f - [cald] instant(aneous) vaporization; flashing
vaporizado,da a - vaporized
vaporizador m para propano m - [combust] propane
vaporizer
vaporizante a - vaporizing
vaporizar v - . . . • [cald] to, generate, o
give off, steam; to steam
— v gasolina - [comb.int] to vaporize @ gasoline
vaquería f - [ganad] . . .; dairy
vara f extrusionada - [metal-fabr] extruded rod
— f para aforar v - [hidr] water gage
— f — aforo m - [hidr] water gage
— f — agrimensor m - [topogr] range pole
— f — medición f - [metric] measuring, rod, o
stick • [comb.iny] measuring stick
— f — remolque m - [autom] tow bar
— f y lanza f combinada - [transp] combined pole
and thill
varadero m - . . . . • [mader] skidway
varado,da a - [nav] aground; grounded
variabilidad f - . . . • variance; tolerance
— f en granulometría f - [miner] particle (size)
variability
— f real - actual, variability, o variance
variable f - . . . | a - . . .; varying; changing
• random
— f ajustable - fit variable
— f en altura f - depth variable
— f — calidad f - [ind] quality variable
— f — — f de producción f - [ind] manufacturing quality variable
— f — combustión f - [combust] combustion variable
— f — costa(s) - [contab] cost(s) variable
— f — proceso m - [ind] process variable
— m — procedimiento m—[ind] procedure variable
— f — producción f - [ind] production variable
• manufacture,ring variable
— f — soldadura f - [sold] welding variable
— f — trabajo m - [ind] work, o operation, variable
— f externa - external variable
— f — definida - [comput] defined external
variable
— f flexible - flexible variable
— f identificable - identifiable variable
— f — regulada - controlled identifiable variable
— f manipulada - manipulated variable
— f para operación f - [ind] operating variable
— f — proceso m - process variable
— f — — m para soldadura f - [sold] welding
process variable
— f reflejada - reflected variable
— f regulada - controlled variable
— f — mecánicamente - [mec] mechanically controlled variable
— f restante - remaining, o other, variable
— f sin restricción f - unrestricted ariable |
a - infinitely variable
— f suavemente - smoothly variable
variación f - . . .; range,ging • departure •
variance • deviation • [fig] swing(ing) •
differing
— f admisible - allowable variation
— v — en temperatura f - allowable temperature
variation
— f — — — f de bobina(s) f - [metal-lam]

coil temperature allowable variation
variación f amplia - wide variation • wide la-
titude
— f en tolerancia f - wide tolerance variation
— f de alambre m - [alambre] wire variation
— f con aplicación f - variation, o varying,
with @ application
— f constante - constant, variation, o change
— f contractual - contract variation
— f debida a operario m - operator variation
— f desde - variation, o range, from
— f detectable - detectable, o noticeable,
variation
— f dimensional - dimensional variation
— f eléctrica - [electr-oper] electrical, variation, o variance
— f en adición f - addition varying,riation
— f — ajuste m - adjustment. variation, o
varying • setting, change,ging
— f — alargamiento m - [metal] strain, o
elongation, variation, o range
— f — altura f - height varying,riation •
[hidr] head
— f — amperaje m - [electr-oper] amperage variation • [sold] current variation
— f — amplitud f - amplitude varying,riation
— f — análisis m - [quím] analysis variation
— f — — f de cuchara f - [metal-prod] ladle
analysis variation
— f — — m para comprobación f - check(ing)
analysis variation
— f acnho,chura - width varying,riation
— f — ángulo m - angle variation
— f — arrastre m - drag variation
— f — — m de alambre n - [sold] wire drag
variation
— f — capacidad f - capacity variation
— f — característica(s) f - characteristic(s)
variation
— f — — f de arco m - [sold] arc character-
istic(s) variation
— f — caudal m - [hidr] flow variation
— f — consumo m - [ind] consumption variation
— f — — m de material(es) m - [ind] mate-
rial(s) consumption variation
— f — contragolpe m - [mec] backlash, variation, o change
— f — corriente f - [electr] current variation
— f — — f primaria - [electr-distrib] primary current variation
— f — costo(s) m - cost(s) variation
— f — densidad f - density variation
— f — — f de carga f - [metal-prod] charge
density variation
— f — diámetro m - diameter, variation,riating
— f — dimensión f - [mec] dimension variation
— f — diseño m - design variation
— f — eficiencia - [ind] efficiency variation
— f — electrodo m—[sold] electrode variation
— f — — m de alambre m - [sold] wire (electrode) variation
— f — elevación f - height variation • [hidr]
head
— f — entrehierro m - [electr-mot] air gap,
variation, o change,ging
— f — esfuerzo m - effort variation • [metal]
stress, variation, o change
— f — espesor m - thickness variation
— f — estría f - spline variation
— f — — f en engranaje m - [mec] gear spline
variation
— f — — f — — m lateral - [mec] side gear
spline variation
— f — — f para diferencial m - [mec] differential side gear spline variation
— f — exigencia f - requirement variation
— f — exigencia f - requirement variation •
[com] inventory variation
— f — fabricación f - [ind] fabrication, o
manufacturing, variation
— f — — f de pieza f - [ind] part manufac-

turing variation
variación f **en fabricación** f **de pieza** f **indivi-**
dual - [ind] individual part manufacturing
variation
— f **factor** m - factor variable
— f — m **para operación** f - [ind] operating
(factor) variable
— t — **fase** f - phase variation
— f — **flujo** m - flow variation
— f — m **magnético** - [electr] magnetic flow
variation
— f — **frecuencia** f - frequency, vaying,ria-
tion, o change
—   — **fuerza** f - force, o strength, o effort,
varying,riation
— f — **para compactación** f compaction,tive
effort varying,riation
— f — **gasto** m - [contab] expense varying •
[hidr] flow varying,ration
— f — **generador** m - [electr-prod] generator
variation
— f — **humedad** f - humidity, o moisture,
varying,riation
— f — **inclinación** f - slant varying,ration •
[ventil] pitch varying,riation
— f — **inducción** f - [electr-mang] induction
varying,riation
— f — — f **magnética** - [electr-magnet] magnetic induction variation
— f — **inventario** m - inventory variation
— f — **juego** m - play varying,riation
— f — — m **longitudinal** - [mec] end play
varying,riation
— f — **largo(r)** m - length variation
— f — **material** m - material varying,riation
— f — **neumático(s)** m - [autom-neumát] tire
varying,riation
— f — **nivel** m - level varying,riation
— f — — m **de combustible** m - [comb.int] fuel
level varying,riation
— f — **operación** f - operating,tion variable
— f — **orden** f - [com] order varying,riation
— f — **parámetro(s)** - [ind] parameter(s)
varying,riation
— f — **paso** m - [mec] pitch varying,riation
— f — **peso** m - weight varying,riation
— f — — m **efectivo** - effective weight
varying,riation
— f — — m — **lineal** - linear effective weight
varying,riation
— f — **pieza** f - [ind] part varying,riation
— f — — f **individual** - [ind] individual part
varying,variation
— f — **porcentaje** m—percentage varying,riation
— f — **precio** m - price varying,riation
— f — **presión** f - pressure varying,riation •
pressure variance
— f — **procedimiento** m - [ind] procedure
varying,riation
— f — — m **para soldadura** f - [sold] welding
procedure varying,riation
— f — **productividad** f - [ind] productivity
varying,riation
— f — **profundidad** f - depth varying,riation
— f — **proporción** f—proportion varying,riation
• [metal-prod] content varying,riation
— f — — f **de fósforo** m - [quím] phosphorus
content varyin,riation
— f — — f — — m **en arrabio** m - [metal-prod]
hot metal, o pig iron, phosphorus content
varying,riation
— f — — f — — m — **metal** m **caliente** -
[metal-prod] hot metal phosphorus content
varying,riation
— f — **proyección** f - design varying,riation
— f — **ranura** f - [mec] groove, o spline,
varying,riation
— f — — f **en engranaje** m - [mec] gear spline
varying,riation
— f — — f — — m **lateral** - [mec] side gear
spline varying,riation
— f — **recorrido** m - course varying,riation
— f — — m **de cauce** m - [hidr] channel course,
varying,riation, o change,ging

variación f **en regulación** f - setting,
varying,riation, o change,ging
— f — **rendimiento** m - [ind] yield variation
• ratio variation
— f — **rueda** f - [autom] wheel variation
— f — **temperatura** f - temperature variation
— f — — f **de bobina** f - [metal-lam] coil
temperature varying,riation
— f — **tendencia** f - tendency, o trend,
varying,riation, o change,ging
— f — **tensión** f - tensión varying,riation •
[electr-prod] voltage spread
— f — **terminación** f - [ind] finish variation
— f — **terreno** m - [topogr] terrain variation
— f — **tiempo** m - time variation
— f — **tipo** m - type varying,riation
— f — — m **para cambio** m - [fin] exchange
rate variation
— f — **tolerancia** f - [mec] tolerance variance
— f — **tratamiento** m - treatment variation
— f — — m **térmico** - [metal-trat] heat
treatment varying,riation
— f — **tubería** f - [tub] tube, o pipe, variation
— f — **tuerca** f - [mec] nut variation
— f — — f **para piñón** m - [mec] pinion nut
variation
— f — — f — — m **impulsor**—[mec] drive,ving
pinion nut variation
— f — **velocidad** f - speed varying,riation
— f — — f **de eje** m - [mec] axle speed
varying,riation
— f — **voltaje** m - [electr-prod] voltage
varying,riation
— f — — m **para entrada** f - [sold] line
voltage variation
— f — — m **en circuito** m **abierto** - [sold]
open circuit voltage variation
— f — — m — — f **para entrada** f - [electr]
input line voltage variation
— f — **volumen** m - [ind] volume variation
— f **entre neumático(s)** m **y rueda(s)** f -
[autom-neumát] tire-wheel variation
— f **estándar** - standard variation
— f — **admisible** - standard permissible
variation
— f — **permitida** - standard, permissible, o
allowable, variation
— f **gradual** - gradual varying,riation
— f **grande** - great, o large, variation
— f **infinita** - infinite variation
— f **isobárica** - [geogr] isobaric variation
— f **leve** - slight variation
— f **máxima** - maximum variation
— f — **admisible** - maximum allowable variation
— f — — **en temperatura** f - maximum allowable
temperature variation
— f — — — — f **para bobina(s)** f - [metal-
-lam] coil temperature maximum allowable
variation
— f — **en nivel(es)** m - maximum level variation
— m — — **temperatura** f - maximum temperature
variation
— f — — — f **de bobina(s)** f - [metal-lam]
coil temperature maximum variation
— f — **entre nivel(es)** m - maximum level
variation
— f — **permisible** - maximum allowable variation
— f — **permitida** - maximum allowable variation
— f **mínima** - minimum variation
— f — **admisible** - minimum allowable variation
— f — — **en temperatura** f - minimum allowable
temperature variation
— f — **en nivel** m - minimum level variation
— f — — **temperatura** f - minimum temperature
variation
— f **normal** - normal, o standard, variation
— f — **admisible** - normal, o standard, permissible variation
— f — **permitida** - standard permissible variation
— f **permisible** - permissible, o allowable, va
riation
— f **permitida** - permitted, o allowed, variation

variación f plena - full variation
— f positiva - positive variation
— f — infinita - infinite positive, o positive infinite, variation
— f razonable - reasonable variation
— f repetida - repeated variation
— f — en esfuerzo(s) m - [constr] repeated stress(es) variation
— f sin precedentes m - unprecedented, variation, o alteration
— f substancial - substantial variation
— f técnica - technical variation
— f total - total, o full, variation
variado,da a - . . . . • varied; differed • assorted; ranging • [fig] swung
— f con aplicación f—varied with @ application
— a constantemente - varied, o changed, constantly
— a desde - varied. o ranged, from
— a en algo - varied somewhat
— a gradualmente - varied gradually
variador m - variator
— m hidráulico - hydraulic variator
— m — completo - complete hydraulic variator
— m — — para velocidad f - complete hydraulic speed variator
— m para velocidad f - hydraulic speed, o speed hydraulic, variator
— m para frecuencia(s) f - frequency, variator, o changer
— m para velocidad f - speed variator
variante f - variant; alternate,tive; variable • deviation • [f.c.] relocation | a - varied; ranging
— a desde - varying, o ranging, from
— f dimensional - dimensional variant
variar v - . . . .; to alternate; to range; to run; to differ • [fig] to swing
— v adición f - [metal] to vary @ addition
— v ajuste m - to, vary, o change, @, adjustment, o setting
— v altura f - to vary @ height
— v amperaje m - [electr] to vary @ amperage • [sold] to vary @ current
— v — m para soldadura f - [sold] to very @ welding, amperage, o current
— v amplitud f - to vary @ amplitude
— v ancho(r) m - to vary @ width
— v ángulo m - to vary @ angle
— v con aplicación f - to vary with @ application
— v característica(s) f - to vary @ characteristic(s)
— v — f de arco m - [sold] to vary @ arc characteristic(s)
— v caudal m - [hidr] to vary @ flow
— v con espesor m - to vary with @ thickness
— v constantemente - to, very, o change, constantly
— v contragolpe m - [mec] to, vary, o change. @ backlash
— f corriente f - [electr-oper] to, very, o change, @ current
— v desde - to, vary, o range, from
— v diámetro m - to vary @ diameter
— v efecto m - to vary @ effect
— v — f de construcción f - [mec] to vary @ pinch(ing) effect
— v en amplitud f - to vary in amplitude
— v — largo(r) m - to vary in @ length
— v — tamaño m - to range in @ size
— v entrehierro m - [electr-mot] to, vary, o change, @ (air) gap
— v frecuencia f - to, vary, o change, @ frequency
— v fuerza f - to vary @ strength
— v — f de compactación f - to vary @ compaction effort
— v gasto m - to vary @ expense • [hidr] to vary @ flow
— v gradualmente - to vary gradually
— v grandemente - to vary, greatly, o widely
— v humedad f - to vary @, humidity, o moisture

variar v inclinación f - [mec] to vary @ inclination • [ventil] to vary @ pitch
— v juego m - [mec] to vary @ play
— v — m longitudinal - [mec] to vary @ end play
— v largo(r) m - to vary @ length
— v nivel m - to vary @ level
— v — m de combustible - [comb.int] to vary @ fuel level
— v peso m - to vary @ weight
— v precio m - [com] to vary @ price
— v presión f - to vary, o change, @ pressure
— v profundidad f - to vary @ depth
— v recorrido m - to vary @ course
— v — m de cauce m - [hidr] to, vary, o change, @ channel
— v regulación f - to, vary, o change, @, adjustment, o setting
— v tendencia f - to, vary, o change, @ tendency, o trend
— f terminación f - [ind] to vary @ finish
— v terreno m - [topogr] to vary @ terrain
— v tipo m - to, vary, o change, @ type
— v — m para cambio m - [fin] to vary @ exchange rate
— v tratamiento m - [ind] to vary @ treatment
— v — m térmico - [metal-trat] to vary @ heat treatment
— v voltaje m - [electr-oper] to vary @ voltage
— v — m en circuito m abierto - [sold] to vary @ open circuit voltage
varias adv variable(s) - several variables
— adv veces f - several times
variedad f amplia - ample, o wide, o large, range, o variety
— f — de aplicación(es) f - wide application, range, o variety
— f — — mordedura(s) f - [mec] wide grip range
— f — — vehículo(s) m - [autom] wide vehicle range
— f — acoplamiento(s) m - [med] coupling variety
— f — aplicación(es) f - application(s) variety
— f — capacidad(es) - capacity,ties range
— f — — f para inyección f - [plást] shot capacity range
— f — diámetro(s) m - diameter(s), o size(s), variety
— f — equipo(s) m - equipment variety
— f — estilo(s) m - style(s) variety
— f — material(es) m - [ind] material(s) variety
— f — medida(s) f - measurement(s), o size(s), variety
— f — modelo(s) m - model(s), variety, o choice
— f — opción(es) f - option(s) choice
— f — razón(es) f - number of reason(s)
— f — tamaño(s) m - size(s), variety, o range, o change, o choice
— f — vehículo(s) m - [autom] vehicle(s) range
— f — velocidad(es) f - speed(s) range
— f estrecha - narrow variety
— f — de aplicaciones f - narrow applications variety
— f grande - large, o wide, variety • [fig] multitude
— f identificable - identifiable variety
— f infinita - infinite variety
— f más amplia—broadest, o widest, o greatest, variety
varilla f - . . . . • [mec] shaft • [metal-prod] lance • [sold] rod; stick
— f adelgazada - [trefil] prepointed rod
— f alimentada - [trefil] fed rod
— f alineada - [mec] aligned rod
— f avanzada - [trefil] advanced rod
— f calentada - [sold] heated rod
— f baja en hidrógeno m - [sold] low hydrogen rod

varilla f caliente - [sold] hot rod
— f cambiada - [sold] changed rod
— f circunferencial - [mec] circumferential rod
— f con carbono m - [sold] carbon rod
— f — — m mediano - [sold] medium carbon rod
— f — — m para soldadura f - [sold] medium carbon welding rod
— f — diámetro m grande - [metal-lam] large, diameter, o size, rod
— f — — m mayor - [metal-lam] larger, size, o diameter, rod
— f — — m menor - [metal-lam] smaller, diameter, o size, rod
— f — — m reducido - [metal-lam] small, size, o diameter, rod
— f con gancho m - [mec] operating rod
— f continua - [metal-lam] continuous rod
— f — de acero m - [metal-lam] continuous steel rod
— f corrugada - [metal-lam] corrugated rod
— f de acero m - [metal-lam] steel rod
— f — — m aleado - [mec] alloy steel rod
— f — — m con aleación f - [metal-lam] alloy steel rod
— f — — m — — f esfereoidizada* - [metal--lam] spheroidized* alloy steel rod
— f — — m con carbono m - [metal-lam] carbon steel rod
— f — — m con proporción f mayor de carbono m - [metal-lam] higher carbon steel rod
— f — — m dulce - [netak-lam] mild steel rod
— f — — m con carbono m - [metal-prod] mild carbon steel rod
— f — — m sin carbono m - [metal-lam] mild carbonless steel rod
— f — — m esferoidizada* - [metal-lam] spheroidized* steel rod
— f — — m inoxidable - [metal-lam] stainless steel rod
— f — — m para soldadura f - [sold] stainless steel welding rod
— f — — m roscada - [metal-fabr] threaded steel rod
— f — — m y plata f - [metal-lam] silver--steel rod
— f — — m — — f con recubrimiento m—[sold] coated silver-(and) steel rod
— f — — m — — f — — m para soldadura f - [sold] coataed silver-(and) steel welding rod
— f — bronce m - [metal-lam] bronze rod
— f — — m para soldadura f - [sold] bronze welding rod
— f — — — f autógena - [sold] bronze autogenous welding rod
— f — carbono m - [sold] carbon, rod, o stick
— f — — — m para regulador m - [comb.int] throttle coated carbon rod
— f — carburo m - [mec] carbide rod
— f — cobre m - [metal-lam] copper rod
— f — — m para conexión f con tierra f - [electr-instal] copper grounding rod
— f — hierro m con carbono m - [metal-lam] carbon steel rod
— f — — m magnético - [metal-prod] magnetic rod
— f decapada - [metal-trat] pickled rod
— f deformada - [hormig½] deformed, rod, o bar
— f — para hormigón m - [constr] deformed reinforcing, rod, o bar
— f derecha f - [mec] straight rod • right rod
— f descascarillada - [metal-trat] descaled rod
— f — mecánicamente - [metal-trat] mechanically descaled rod
— f despojadora f - [agric-equip] stripper plate
— f deflectora - [mec] deflector rod
— f doblada - [mec] bent rod
— f — excesivamente - [mec] overbent rod
— f encalada - [metal-trat] limed rod
— f encorvada - [trefil] buckled, o bent, rod
— f enderezada - [mec] staightened rod
— f enfriada - [metal-lam] cooled rod
— f enterrada - [constr] buried rod
— f — para conexión f con tierra f - [electr-instal] buried ground rod

varilla f esferoidizada* - [metal-lam] spherodized rod
— f estándar - [constr] standard rod
— f — para soldadura f - [sold] standard welding rod
— f — — f con latón m - [sold] standard brazing rod
— f — — — f — — m con llama f con gas m - [sold] standard gas brazing rod
— f — — — f fuerte - [sold] standard brazing rod
— f — — — f — con llama f (con gas m) - [sold] standard gas (flame) brazing rod
— f fosfatada - [metal-trat] phoscoated rod
— f fundente - [sold] welding rod
— f indicadora - [comb.int] dipstick
— f — para nivel m - [mec] gage rod; dipstick
— f — — m de aceite m - [comb.int] oil, gage rod, o dipstick
— f interconectada - [mec] interconnected rod
— f — para conexión f con tierra f - [electr-instal] interconnected grounding rod
— f laminada - [metal-lam] rolled rod
— f — en frío m - [metal-lam] cold rolled rod
— f limpia - [trefil] clean rod
— f — para regulador m - [comb.int] clean throttle shaft
— f — — m para carburante m - [comb.int] clean (fuel) throttle shaft
— f limpiada - [trefil] cleaned rod
— f limpiadora f - [mec] cleaning rod
— f longitudinal - [hormig] longitudinal rod
— f — deformada - [constr] deformed longitudinal rod
— f medidora - measuring rod • [comb.int] dip stick; bayonet gage; gage rod
— f — para aceite m - [comb.int] dip stick; bayonet gage
— f — — m en cárter m - [comb.int] crankcase, dip stick, o bayonet gage
— f — — m en extremo m hacia motor m - [petrol] power end dip stick
— f — — m en motor m - [comb.int] engine oil dip stick
— f — — m en transmisión f - [mec] transmission dip stick
— f — — m verificada - [comb.int] checked dip stick
— f — cárter m - [comb.int] crankcase bayonet
— f — colector m - [mec] sump dipstick
— f — extremo m - [mec] end dipstick
— f — — m hacia motor m - [petról] power end dipstick
— f — mesa f rotativa - [petról] rotary table dipstick
— f — motor m - [electr-mot] motor, bayonet gage, o dipstick • [comb.int] engine, bayonet gage, o dipstick
— f — orificio m para aceite m - [comb.-int] oil filler dipstick
— f — — nivel m para aceite m - [comb.int] oil level dipstick
— f — tapón m para gollete m - [comb.int] filler plug (attached), gage, o dipstick
— f — transmisión f - [mec] transmission dipstick
— f no ascendente - [válv] nonrising stem
— f para aceite m - [comb.int] dipstick
— f — acelerador m - [comb.int] throttle, o accelerator, o gas, rod
— m alambre m - [metal-prod] wire rod
— f — allanar v - [mec] leveling rod
— f — anclaje m - [constr] anchor(ing) rod
— f — aportación f - [sold] filler rod
— f — — aplicada - [sold] applied filler rod
— f — armadura f - [constr] reinforcing bar
— f — arrastre m - [mec] pull(ing) rod
— f — biela f - [comb.int] connecting rod
— f — biselar - [sold] chamfer(ing) rod
— f — bomba f - [petról] sucker rod
— f — cambio m de marcha f - [mec] shift(ing) rod • reach rod

varilla f̲ para conexión f̲ - [mec] connecting rod
´link
— f̲ —— f̲ **con tierra** f̲ - [electr-instal]
grounding rod
— f̲ —— f̲ **de carburador** m̲ - [comb.int] carburetor link
— f̲ — **derivación** f̲ - [electr-instal] tap box
— f̲ — **descarga** f̲ - [mec] dumper rod
— f̲ — **detención** f̲ - [mec] stop(ping) rod
— f̲ —— f̲ **e inversión** f̲ - [mec] stop(ping)
and reverse,sing rod
— f̲ — **dirección** f̲ - [mec] drag link
— f̲ — **émbolo** m̲ - [bombas] piston rod
— f̲ — **empuje** n̲ - [mec] push(ing) rod; tappet
— f̲ —— m̲ **para horquilla** f̲ - [autom-mec]
fork push(ing) rod
— f̲ —— m̲ —— f̲ **para cambio(s)** m̲ - [autom-
-mec] shift fork push(ing) rod
— f̲ —— m̲ — **válvula** f̲ - [válv] valve push
rod
— f̲ — **estrangulador** m̲ - [comb.int] choke rod
— f̲ — **freno** m̲ - [mec] brake rod
— f̲ — **gobierno** m̲ - [mec] control(ling) rod
— f̲ — **inversión** f̲ - [mec] reverse,sing rod
— f̲ — **levantaválvulas** m̲ - [mec] valve lifter
rod • tappet rod
— f̲ — **llama(s)** f̲ - [combust] flame rod
— f̲ — **mando** m̲ - [mec] control rod • operating
rod
— f̲ —— m̲ **para freno** m̲ - [mec] brake, driving, o̲ activating, rod
— f̲ — **medición** f̲ **para aceite** m̲ - [comb.int]
oil gage rod
— f̲ — **medición** f̲ **para nivel** m̲ - [comb.int]
oil gage rod
— f̲ — **molienda** f̲ - [curp.moled] grinding rod
— f̲ — **oxígeno** - [metal-prod] oxygen lance
— f̲ — **piquera** f̲ - [metal-prod] tap(ping),
bar, o̲ rod
— f̲ — **recargo** m̲ - [sold] filler rod
— f̲ — **refuerzo** m̲ - [constr] reinforcing, rod,
o̲ bar • hog rod
— f̲ — **regulación** f̲ - [mec] control(ling) rod
— f̲ — **regulador** m̲ - [comb.int] throttle, rod,
o̲ shaft
— f̲ —— m̲ **para carburante** m̲ - [comb.int]
(fuel) throttle shaft
— f̲ — **relleno,namiento** m̲ - [sold] filler rod
— f̲ — **soldadura** f̲ - [sold] soldering, o̲ welding, rod
— f̲ —— f̲ **eléctrica** - [sold] welding rod
— f̲ —— f̲ **con latón** m̲ - [sold] brazing rod
— f̲ —— f̲ **de acero** m̲ **inoxidable** - [sold]
stainless steel welding rod
— f̲ —— f̲ **fuerte** - [sold] brazing rod
— f̲ —— f̲ — **calentada** - [sold] heated brazing rod
— f̲ —— f̲ — **con latón** m̲ - [sold] brazing
rod
— f̲ —— f̲ —— m̲ **con llama** f̲ **de gas** m̲ -
[sold] gas brazing rod
— f̲ —— f̲ **para acero** m̲ - [sold] steel welding rod
— f̲ —— f̲ —— m̲ **inoxidable** - [sold]
stainless steel welding rod
— f̲ —— f̲ **para aportación** f̲ **rápida** - [sold]
Fast Fill welding rod
— f̲ —— f̲ —— f̲ **para acero** m̲ - [sold]
Fast Fill steel welding rod
— f̲ —— f̲ —— f̲ —— m̲ **con manganeso** m̲
- [sold] Fast Fill manganese (steel) welding
rod
— f̲ —— f̲ — **rellenar** v̲ - [sold] Fast Fill
welding rod
— f̲ — **soldar** v̲ - [sold] welding rod; electrode
— f̲ — **sondeo** m̲ - [metal-prod] stock, o̲ feeling, rod
— f̲ —— m̲ **para horno** m̲ **alto** - [metal-prod]
blast furnace stock rod
— f̲ **para succión** f̲ - [petról] sucker rod
— f̲ — **suspensión** f̲ - suspension rod
— f̲ — **tensión** f̲ - [mec] tension rod
— f̲ — **tornillo** m̲ - [herram] vise rod

varilla f̲ para tornillo m̲ para banco m̲ -
[herram] vise rod
— f̲ — **torsión** f̲ - [mec] torque rod
— f̲ — **tracción** f̲ - [mec] pull(ing), o̲ traction, rod
— f̲ — **trituración** f̲ - [cuerp.moled] grinding
rod
— f̲ — **válvula** f̲ - [válv] valve rod
— f̲ — **verificación** f̲ - [mec] checking rod
— f̲ —— f̲ **para nivel** m̲ - [mec] level checking dipstick
— f̲ —— f̲ —— m̲ **de aceite** m̲ - [mec]
oil level check(ing) dipstick
— f̲ **pequeña** - [metal-lam] small rod
— f̲ **recomendada** - [sold] recommended rod
— f̲ — **para aportación** f̲ - [sold] recommended filler rod
— f̲ **recta** - [mec] straight rod
— f̲ **redonda** - [metal-lam] round rod
— f̲ **reemplazada** - [sold] changed rod • replaced rod
— f̲ **repuesta** - replaced shaft
— f̲ **resorte** - [mec] spring shaft
— f̲ **roscada** - [mec] threaded rod
— f̲ **sin recubrimiento** m̲ - [metal-lam] uncoated
rod
— f̲ **sobredoblada*** - [mec] overbent rod
— f̲ **sonda (para aceite** m̲**)** - [mec] dip stick
— f̲ **sujetada** - [trefil] held. o̲ gripped, rod
— f̲ **tensora** - [mec] tie rod • truss rod
— f̲ **tipo** m̲ **bayoneta** f̲ - [mec] bayonet gage;
dip stick
— f̲ —— m̲ —— f̲ **para aceite** m̲ - [comb.int] (oil)
dip stick • bayonet gage
— f̲ **tratada** - [metal-lam] treated rod
— f̲ — **térmicamente** - [metal-trat] heat
treated rod
— f̲ **trefilada** - [trefil] drawn rod
— f̲ — **en frío** - [trefil] cold drawn rod
— f̲ **trituradora** - [cuerp.moled] grinding rod
varillaje m̲ - [mec] . . . . • rib(s) • linkage
— f̲ **para regulación** f̲ - [mec] governor linkage
varios,rias adv - various; several; few;
miscellaneous • other • [contab] sundry
— adv **años** m̲ - several years
— adv **golpes** m̲ - [mec] several strokes
— adv **grados** m̲ - [termol] several degrees
— adv —— m̲ **bajo cero** - [meteorol] low teens
— adv **minutos** - several, o̲ few, minutes
varmeter* m̲ - varmeter
vasar m̲ - [culin] . . .; mantle piece
vaselina f̲ **de petróleo** m̲ - [petról] petroleum
jelly|
vasija f̲ - [culin] . . .; bowl; container; can
— f̲ **con tapa** f̲ - [culin] covered, dish, o̲ pan
— f̲ **limpiada** f̲ - [culin] cleaned vessel
— f̲ **limpia** f̲ - [domést] clean, vessel, o̲ container
— f̲ **tapada** - [domést] covered, vessel, o̲ container
vaso m̲ **comunicante** - communicating vessel
— m̲ **de papel** m̲ - [papel] paper cup
— m̲ **para medir** v̲ - measuring, cup, o̲ vessel
vástago m̲ - . . . . • finger; spindle; tong •
[mec] stud • [petról] stem; reach • [bomba]
rod • [herram] tang • [cerraduras] shank •
[hidr-compuertas] stem - [metal-prod]
stopper rod • [remaches] stem
— m̲ **ajustable** - [mec] adjustable stem
— m̲ **ajustable** m̲ - [mec] adjustable stem
— m̲ — **para prolongación** f̲ - [mec] adjustable
extension stem
— m̲ **ascendente** - [válv] rising stem
— m̲ **asentado** - [válv] seated, o̲ tapped, stem
— m̲ — **con golpe(s)** m̲ **ligero(s)** - [válv]
tapped down stem
— m̲ **Blue Chrome** - [petról] Blue Chrome (piston) rod
— m̲ **Blue Stripe** - [petról] Blue Stripe (piston) rod
— m̲ **colocado** - [mec] installed rod • [válv]
installed stem
— m̲ — **en posición** f̲ - [mec] positioned rod

vástago m compactable - [válv] compactible stem
— m completo - [mec] complete rod • [válv] complete stem
— m con abrazadera f - [mec] rod clamp
— m — extremo m con horquilla f - [mec] rod with clevis end
— m corredizo - [válv] rising stem
— m Cromo m Azul - [petról] Blue Chrome (piston) rod
— m cuadrado m - [petról] square, Kelly, o stem
— m de acero m - [mec] steel stem • [remaches] linear-grip stem
— m — — m dulce - [mec] mild steel stem
— m — — m inoxidable - [mec] stainless steel stem
— m — bronce m - [mec] bronze stem
— m — grapa f - [clavos] staple shank
— m — tipo m fijo - [válv] non-rising type stem
— m elevable - [valv] rising stem
— m embridado - [mec] flanged rod
— m enfriado - [mec] cooled rod
— m esmerilado - [comb.int] ground stem
— m — para válvula f - [comb.int] ground valve stem
— m estriado - [mec-remaches] grooved stem
— m fijo - [válv] non-rising stem
— m guiado - [mec] guided stem
— m guiador - [mec] guide,ding stem
— m — para disco m para válvula f - [válv] valve disk guide,ding stem
— m hexagonal - [mec] hexagonal stem • [petról] hexagonal Kelly
— f — para perforación f - [petról] hexagonal Kelly
— m hueco - [mec] hollow stem
— m inferior - [mec] low(er) rod
— m — para embrague m - [mec] clutch low rod
— m — — m para conexión f - [mec] tie clutch low rod
— m — — m — unión f - [mec] tie clutch low rod
— m inoxidable - [mec] stainless, stem, o rod
— m inspeccionado - [mec] inspected rod
— m instalado - [mec] installed rod • [válv] installed stem
— m intermedio - [petról] intermediate rod
— m — colocado - [mec] installed intermediate rod
— m — conectado - [mec] connected intermediate rod
— m — instalado - [mec] installed intermediate rod
— m — lubricado - [petról] lubricated intermediate rod
— m — para bombeo m - [petról] intermediate pumping rod
— m — quitado - [mec] removed intermediate rod
— m — removido - [mec] removed intermediate rod
— m — sacado - [mec] removed intermediate rod
— m levantador - [mec] lifter,ting rod
— m limpiado m - [mec] clean, rod, o shaft
— m — para regulador m - [comb.int] cleaned control shaft
— m — — — m para carburante m - [comb.int] cleaned fuel throttle shaft
— m limpio m - [comb.int] clean shaft
— m — para regulador m - [comb.int] clean control shaft
— m — — — m para carburante m - [comb.int] clean fuel throttle shaft
— m — regulado - [mec] controlled clean shaft
— m liso - [mec] smooth stem
— m lubricado - [mec] lubricated, rod, o stem
— m mecanizado - [mec] machined, rod, o stem
— m no ascendente - [Válv½] non-rising stem
— m — levadizo - [hidr] non-rising stem
— m — — m para regulador m - [mec] governor control rod
— m para apoyo m - [mec] pad, rod, o tong
— m — — m para reemplazo m - [mec] replacement pad, rod, o tonh
— m — arrancador n - [mec] puller stem

vástago m para barreno,na - [petról] drilling, auger, o stem
— m — bomba f - [bombas] pump rod
— m — bombeo m - [petról] pumping, o sucker, rod • polished rod
— m — caja f - [ventil] housing stem
— m — f para conmutador m - [ventil] switch housing stem
— m — cambiador m - [mec] shifter rod
— m — m para embrague m - [mec] clutch shifter rod
— m — m — — m inferior - [mec] low(er) clutch shifter rod
— m — m — — m superior = [mec] high, o upper, clutch shifter rod
— m — campana f - [metal-prod] bell rod
— m — — f grande - [metal-prod] large bell rod
— m — — f pequeña - [metal-prod] small bell rod
— m — carrete m - [petról] reel reach
— m — — m para cuchareo m - [petról] sand reel reach
— m — cilindro m - [mec] (drive) cylinder rod
— m — — m para accionamiento m - [mec] drive cylinder rod
— m — — m — — m para hoja f - [constr] blade drive cylinder rod
— m — — m — — f para topadora f - [constr] dozer blade drive cylinder rod
— m — compuerta f - [hidr] gate, stem, o rod
— m — conexión f - [mec] tie rod
— m — f para carburador m - [comb.int] carburetor link
— m — crisol m - [sold] solder pot post
— m — — m para aleación f - [sold] alloy solder pot post
— m — cuña f - [mec] wedge rod
— m — desplazador m - [mec] shifter rod
— m — — m para embrague m - [mec] clutch shifter rod
— m — — m — — m inferior - [mec] low(er) clutch shifter rod
— m — — m — — m superior - [mec] high, o upper, clutch shifter rod
— m — disco m - [mec] disk stem
— m — — m para válvula f - [válv] valve disk stem
— m — émbolo m - [mec] plunger, rod, o shaft • [bombas] piston rod
— m — — m Blue Chrome - [petról] Blue Chrome piston rod
— m — — m Blue Stripe - [petról] Blue Stripe piston rod
— m — — m conectado - [mec] connected piston rod
— m — — m Cromo Azul - [petról] Blue Chrome piston rod
— m — — m en dos piezas f - [mec] two-piece piston rod
— m — — m para solenoide m - [comb.int] solenoid plunger, shaft, o rod
— m — — m Raya f Azul - [petról] Blue Stripe piston rod
— m — embrague m - [mec] clutch rod
— m — — m para conexión f - [mec] tie clutch rod
— m — — m — unión f - [mec] tie clutch rod
— m — empuje m - [mec] push(ing) rod
— f — — m para válvula f - [válv] valve push(ing) rod
— n — — — f neumática - [válv] air valve push(ing) rod
— f — — m — f para aire m - [válv] air valve push(ing) rod
— m — freno m - [mec] brake rod
— m — guiadera f - [mec] guide,ding stem
— m — — f para disco m - [mec] disk guide,ding stem
— m — horno m - [combust] furnace rod
— m — — m alto - [metal-prod] blast furnace stock rod
— m — horquilla f - [mec] yoke rod

vástago m para horquilla f para desplazador m -
[mec] shifter yoke rod
— n — levantaválvula(s) m - [mec] valve lifter
rod • [comb.int] tappet stem
— m — llama f - [combust] flame rod
— m — — f de quemador m - [combust] burner
flame rod
— m — malacate m - [petról] reel reach
— m — — m para cuchara f - [petról] sand reel
reach
— m — perforación f - [petról] drill(ing), o
auger, stem, o grief; Kelly
— m — — f cuadrado - [petról] square Kelly
— m — perilla f - [mec] knob shaft
— m — prolongación f - [mec] extension stem
— m — regulación f - [mec] control rod
— m — — f para carburador m - [comb.int] carburetor control rod
— m — — f para carburador m - [comb.int] governor to carburetor control rod
— m — regulador m - [mec] throttle, o control,
o governor, rod
— f — m para carburante - [comb.int] carbu-
rant, o fuel, throttle shaft
— m — — m para marcha f sin carga f - [comb.-
-int] idler control rod
— m — sujeción f - [mec] holding, o anchor,
rod • [metal-prod] (slag notch) clamp
— m — tambor m - [mec] drum rod • [petról] reel
reach
— m — — m para cuchareo m - [petról] sand reel
reach
— m — tapón m - [metal-prod] stopper rod
— m — tiro,ramiento m - [mec] pull(ing) rod
— m — transmisión f - [petról] véase vástago m
para perforación f
— m — válvula f - [válv] valve, stem, o shaft,
o rod
— m — — f ajustable - [válv] adjustable valve,
stem, o shaft
— m — — f colocado - válv] installed valve
stem
— m — — f guiado - [válv] guided valve stem
— m — — f inferior - [válv] lower valve stem
— m — — f — colocado - [válv] installed lower
valve stem
— m — — f — instalado - [valv] installed
lower valve stem
— m — — f — quitado - [válv] removed lower
valve stem
— m — — f — removido - [válv] removed lower
valve stem
— m — — f — sacado - [válv] removed lower
valve stem
— m — — f inmovilizado (en sitio m) - [válv]
locked(-in-place) valve stem
— m — — f instalado - [válv] installed valve
stem
— f — — f mariposa - [válv] butterfly valve,
shaft, o stem
— m — — f neumática - [válv] air valve, stem.
o rod
— m — — f para aire m - [válv] air valve stem
— m — — f — empuje m - [válv] push valve,
stem, o rod
— m — — f — fondo m - [válv] bottom valve,
stem, o rod
— m — — f — — m de lavador m - [metal-prod]
(gas), scrubber, o washer, bottom valve, rod.
o stem
— m — — f para viento m caliente—[metal-prod]
hot blast valve, stem, o rod
— m — — f — para estufa f - [metal-
-prod] stove hot blast valve, stem, o rod
— m — — f quitado - [válv] removed valve stem
— m — — f regulable - [válv] adjustable valve,
stem, o rod
— m — — f removido - [válv] removed valve,
stem. o rod
— m — — f sacado - [válv] removed valve, stem,
o rod
— m — — f superior - [válv] upper valve, stem,
o rod

vástago m para válvula f superior colocado -
[válv] installed upper valve, stem, o rod
— m — — f — instalado - [válv] installed
upper valve, stem, o rod
— m — — f — quitado - [válv] removed upper
valve, stem, o rod
— m — — f — removido - [válv] removed upper
valve, stem, o rod
— m — — f — sacado - [válv] removed upper
valve, stem, o rod
— m — succión f - [petról] sucker rod
— m — unión f - [mec] tie rod
— m piloto - [válv] pilot, stem, o rod
— m pulido - [metal-lam] polished rod
— m quitado - [mec] removed rod • [válv] removed stem
— m Raya f Azul - [petról] Blue Stripe (piston) rod
— m regulable - [mec] adjustable stem
— f — para prolongación f - [mec] adjustable
extension stem
— m removido - [mec] removed rod • [válv] removed stem
— m roto m - [mec] broken rod • [válv] broken
stem
— m sacado - [mec] removed rod • [válv] removed stem
— m saliente - [hidr] rising stem
— m Schroeder - [válv] Schroeder core
— m sesgado - [mec] skew(ed) stem
— m — para compuerta f - [hidr] skew(ed) gate
stem
— m superior - [mec] upper, o high, rod
— m — para embrague m - [mec] cluth, high, o
upper, rod
— m — — m para conexión f - [mec] tie
clutch, upper, o high, rod
— m — — m — unión f - [mec] tie clutch,
upper, o high, rod
— m totalmente regulable - [hidr] fully adjustable stem
— m verificado - [mec] checked, stem, o rod
— m — nuevamente - [mec] rechecked rod
— m y grillete m - [mec] rod and clevis
— m y horquilla f - [mec] rod and clevis
— m — — f para válvula f - [válv] valve rod
and clevis
vataje* m - [electr] véase vatiaje m
vatiaje m alto - [electr-oper] high wattage
— m bajo - [electr-oper] low wattage
— m de planta f para alumbrado m - [electr-
-prod] light plant wattage
— m elevado - [electr-oper] high wattage
— m exigido - [electr-oper] required wattage
— m — para planta f para alumbrado m -
[electr-oper] required light plant wattage
— m exigido - [electr-oper] required wattage
— m reducido - [electr-oper] reduced, o low,
wattage
vaticinado,da - forecast; projected
vaticinar v - . . . ; to forecast; to project
vaticinio m - . . . ; projecting,tion
vatímetro m indicador - [instrum] indicating
wattmeter
vatiohorámetro m - [instrum] véase contador m
para vatiohora(s) f
veces f - times
. . . — f diámetro m de barra f - [constr] . . .
times @ bar diameter
vecinal a - . . .; neighborhood
vecino m de - [pol] living in
vector m de curvatura f - gradient curvature
— m — gravedad f - (gravity) gradient
vegetación f - [botán] . . .; ground cover
— f acuática - [botán] aquatic vegetation •
[sanit] aquatic weed nuisance
— f en cauce m - [hidr] channel vegetation
— f existente - [botán] existing, vegetation,
o ground cover
— f resistente a erosión f - [constr] erosion
resisting,tant vegetation
— f selvática - [botán] jungle growth

vegetación f tropical - [botán] tropical vegetation; jungle growth
vehicular a - vehicular
vehículo m - . . . .; carriage
— m acabado de salir m de fábrica - [autom]
factory-fresh vehicle
— m accidentado - [autom] wreck
— m afectado - [autom] involved vehicle
— m agotado - [autom] spent vehicle
— m alineado - [autom] aligned vehicle
— m ambulancia - [segurid] ambulance
— m automotor - [autom] motor vehicle • automotive vehicle
— m campeón - [deport] champion vehicle
— m — en carrera f entre pilones m - [autom]
autocross champion vehicle
— m con carrocería f tubular - [autom] buggy
— m — cuatro ruedas f - [autom] four wheeler
— m — — f motrices - [autom] four-wheel-
-drive vehicle
— m — impulsión f en cuatro ruedas f - [autom]
four-wheel-drive vehicle
— m — . . . ruedas f - [autom] . . .-wheel
vehicle
— m — tracción f en cuatro ruedas f - [autom]
four-wheel-drive vehicle
— m — — — f frontal - [autom] front wheel drive
vehicle
— m — — — f en rueda(s) f delantera(s) -
[autom] front wheel drive vehicle
— m — — — f delanteras - [autom] front
wheel drive vehicle
— m — — — f trasera - [autom] rear-wheel drive
vehicle
— m conducido - [autom] driven, o handled, vehicle • [vial] moved vehicle
— m contra incendio(s) m - [segurid] fire,
vehicle, truck, o wagon
— m chocante - [transp] colliding, o crash-
ing, vehicle
— m dado - [transp] given vehicle
— m dañado - [transp] damaged vehicle
— m de cliente m - [Transp] customer('s) vehicle
— m — fabricación f nacional - [Autom] domestic (manufactured) vehicle
— m de serie f - [autom] stock vehicle
— m — — f con tracción f en . . . ruedas f -
[autom] stock . . .-wheel-drive vehicle
— m — stock* m - [autom] stock vehicle
— m — — m con tracción f en . . . rueda(s) f -
[autom] . . .-wheel-drive stock vehicle
— m deportivo - [autom] sport, o recreational,
vehicle
— m — y para carga(s) f liviana(s) - [autom]
sport-utility vehicle
— m — — — servicio m general - [autom]
sport(s)-utility vehicle
— m determinado - [transp] certain, o given, o
specific, vehicle
— m en franja f paralela - [transp] adjacent
traffic
— m equipado con freno(s) sin disco(s) m -
[autom] non-disk brake equipped vehicle
— m espacial - [astronáut] space vehicle
— m especializado - specialized, o ad-hoc, vehicle
— m específico - [autom] specific, o particular, vehicle
— m estabilizado - [autom] stabilized vehicle
— m estacionado - [transp] parked, o standing,
vehicle
— m estadounidense - [autom] United States, o
American*, vehicle • domestic vehicle
— m excluido - [transp] excluded vehicle
— m exótico - [autom] exotic vehicle
— m extranjero - [autom] foreign vehicle
— m incapacitado - [autom] disabled vehicle
— m industrial - [transp] industrial vehicle
— m — especializado - [ind] specialized industrial vehicle
— m instrumentado—[autom] instrumented vehicle
— m — totalmente - [autom] fully instrumented
vehicle

vehículo m levantado - [autom] raised vehicle
— m liviano - [autom] light vehicle
— m — con neumáticos m desproporcionados -
[autom-deport] dune buggy
— m llamado - [electrón] called vehicle • signal vehicle
— m manejado - [autom] handled, o driven, vehicle
— m maniobrado - [autom] maneuvered vehicle
— m más raudo - [autom] faster,test vehicle
— m motorizado - [autom] motorized vehicle
— m normal - [autom] normal vehicle
— m operado - [autom] operated, o driven, vehicle
— m para agricultura f - [agric-equip] agricultural vehicle
— m — carga(s) f liviana(s) - [autom] utility vehicle
— m — carrera(s) f - [autom] race,cing vehicle
— m — — f entre pilones m - [autom] autocross vehicle
— m — — f fuera de carretera(s) f - [autom]
off-road race,cing vehicle
— m — — f sobre carretera f - [autom] road
racing vehicle
— m — conservación f - [autom] maintenance
vehicle
— m — construcción f - [constr] construction
vehicle
— m — fines m generales - [autom] utility
vehicle
— m — industria f - [autom] industrial vehicle
— m — mantenimiento m - [autom] maintenance
vehicle
— m — nieve f - [deport] snowmobile
— m — recreo m - [autom] recreational vehicle;
R V • sports vehicle
— m — rendimiento m alto - [autom] high performance vehicle
— m — servicio m general - [autom] utility
vehicle
— m — todo terreno m - [autom] all terrain
vehicle
— n — transporte m - [transp] transportation
vehicle
— m — — m de pasajeros m - [autom] passenger
transportation vehicle
— m — — m público - [transp] public transit
vehicle
— m — turismo m - [autom] recreational vehicle • camper
— m pesado - [autom] heavy (duty) vehicle
— m potente - [autom] powerful vehicle
— m preparado para carrera(s) f - [deport]
race prepared vehicle
— m que siguen - [transp] following traffic
— m raudo - [transp] fast vehicle
— f redirigido - [transp] redirected vehicle
— m reencausado - [Transp] redirected vehicle
— m remolcador - [transp] towing vehicle
— m sin carrocería f - [deport] buggy
— m — gobierno m - [transp] uncontrolled, o
errant, vehicle
— m — — m redirigido - [transp] redirected
errant vehicle
— m — piezas f empleadas para (reparar) otros
vehículos m - [autom] cannibalized vehicle
— m stock - [autom] stock vehicle
— m transmisor - [electrón] transmitting vehicle • [medic] carrier
— m transportador - [transp] transporting vehicle; transporter • [grúas] carrier
— m — con motor m diesel - [Transp] disel
engine carrier
— m — optativo - [grúas] optional, o alternate, carrier
— m usado - [transp] vehicle used • used vehicle
veinticuatro horas f - . . .; twice around @
clock
vejado,da a - . . . • plagued
vejamen m - . . . .; plague,guing
vejar v = . . .; to plague

velador m   - 1702 -

velada - . . . • evening • evening out
velador m - . . . • [domést] night stand • table
  lamp • night light
velamen m pleno - [nav] full sail • square rigging
velar v - . . . • [labor] to supervise
velo m de arcilla f - [geol] clay veil
velocidad f - . . .; (rate of) speed • [mec]
  ratio • [autom] gear - véase también rapidez
— f acelerada - increased speed
— f — para marcha f en vacío m - [comb.int]
  high idle speed
— f — — — f sin carga f - [comb.int] high
  idle speed
— f — sin carga f - high idle speed
— f admisible - allowable, o permissible,
  speed, o velocity
— f ajustada - [mec] adjusted speed
— f alcanzable - [mec] attainable speed
— f alta - high, o full, velocity, o speed •
  [autom] high gear
— f en vacío m [ind] high idle speed
— f — para accionamiento m [mec] high drive
  speed
— f — — — m para motor m - [mec] high drum
  drive,ving speed
— f — para alambre m - [sold] fast wire speed
— f — — alimentación f para alambre m -
  [sold] fast wire speed feed(ing)
— f — — avance m - [sold] high welding speed
— f — — cable m - [grúas] high line, o line
  high, speed
— f — — gancho m - [grúas] high hook speed
— f — — marcha f en vacío m - [comb.int] high
  idle speed
— f — — — f sin carga f - [comb.int] high
  idle speed
— f — operación f - high operating speed
— f — — transmisión f - [mec] transmission
  high, o high transmission, speed
— f — sin carga f - [mec] high idle (no load)
  speed
— f aminorada - slowing down • reduced speed
— f apropiada - appropriate, o proper, speed •
  [mec] high ratio
— f — para tambor m -[mec] appropriate, o proper, drum speed
— f — motor m - [electr-mot] appropriate, o
  proper, motor speed • [comb.int] appropriate,
  o proper, engine speed
— f — seleccionada - appropriate, o proper,
  selected speed
— f aumentada - increased, speed, o velocity
— f — automáticamente - automatically increased speed
— f — para motor n - [electr-mot] increased
  motor speed • [comb.int] increased engine
  speed
— f — para movimiento m de estrangulador m -
  [mec] increased choke,king movement speed
— f autolimpiadora - self-cleaning speed
— f baja - [mec] (s)low, speed, o velocity, o
  rate • idling • [autom] low gear
— f — para cable m - [grúas] low line, o line
  low, speed
— f — — gancho m - [grúas] low hook speed
— f — — operación f - low operating speed
— f — — transmisión f - [mec] transmission
  low, o low transmission, speed
— f — sin carga f - [comb.int] low (idle)
  (no load) speed
— f — — — f para motor m - [electr-mot] low
  (idle) (no load) motor speed • [comb.int] low
  (idle) (no load) engine speed
— f cambiada - changed speed • [mec] shifted,
  (gear, o speed), o tranmission gear
— f centrífuga - [mec] centrifugal, o outward,
  speed
— f centrípeta - centripetal, o inward, speed
— f con carga f - [mec] loaded speed
— f — — f completa - full load speed
— f — — f plena - vull load speed
— f — referencia f a tierra f - ground speed

velocidad f con referencia a tierra f deseada -
  desired ground speed
— f — — f — — f verificada - [mec] checked
  ground speed
— f constante - constant speed
— f — para alambre m - [sold] constant wire
  speed
— f — — alimentación f - [sold] constant
  feed(ing) speed
— f — — — f para alambre m - [cold] constant wire feed(ing) speed
— f — — avance m - [sold] constant travel
  speed
— f — — motor m - [electr-mot] constant motor
  speed • [comb.int] constant engine speed
— f continua - continuous speed
— f — determinada - established continuous
  speed
— f — máxima - maximum continuous speed
— f — mínima - minimum continuous speed
— f contrastada - contrasted speed
— f controlada - controlled speed
— f correcta - correct speed
— f — para avance m - correct travel speed
— f — lograda - obtained correct speed
— f crítica - critical speed
— f — para enfriamiento m - critical cooling
  speed
— f correcta - correct speed
— f — para soldadora f - [sold] correct
  welder speed
— f — soldadura f - [sold] correct
  welding speed
— f — — unidad f - [sold] unit('s) correct
  speed
— f — — — r para procesamiento m - [ind]
  processing unit correct speed
— f — provista - [mec] provided correct speed
— f correspondida - [mec] matched speed
— f creciente - increasing speed
— f crítica - critical speed • [hidr] critical
  velocity
— f — primaria - primary critical speed
— f cuarta f - [mec] fourth speed • [autom]
  fourth gear
— f de acoplado m - [transp] trailer speed
— f — agua m - [hidr] water, speed, o velocity
— f — — m reducida - [hidr] reduced water,
  speed, o velocity
— f — aire m - [neumát] air speed
— f — — m comburente - [combust] burning air
  speed
— f — árbol m - [mec] shaft speed
— f — — m con carga f plena - [mec] full load
  shaft speed
— f — — m sin carga f - [mec] no load shaft
  speed
— f — arco m - [sold] arc speed
— f — — m apropiada - [sold] appropriate arc
  speed
— f — — m en centímetros m por hora f -
  [sold] arc speed in centimeters per hour
— f — — m — pulgadas f por hora f - [sold]
  arc speed in inches per hour
— f — — m mayor - [sold] faster arc speed
— f — — m menor - [sold] slower arc speed
— f — — m por minuto m - [sold] arc speed
  per minute
— f — arrastre m - [vial] dragging speed
— f — aumento m - speed increase
— f — — m de temperatura - temperature rise,
  speed, o rate
— f — carrera f - [deport] race,cing speed
— f — carrete m - [mec] reel, o spool, speed
— f — — m para segadora f (trilladora) -
  [agric] combine reel speed
— f — carrillo m - [grúas] trolley speed
— f — carro m - [transp] wagon speed •
  [sold] carriage speed
— f — — m para montacargas m - [metal-prod]
  skip car speed
— f — cilindro m - [mec] cylinder speed
— f — cinta f transportadora f - [mec] con-

veyor (belt) speed
velocidad f **de colada** f - [metal-prod] pouring,
o casting, speed
— f — **combustión** f - [combust] combustion, o
burning, speed
— f —— f **de aire** m - [combust] air combustion speed
— f — **corriente** f - [hidr] current, o flow,
velocity
— f —— f **en entrada** f - [hidr] approach velocity
— f —— f **en conducto** m **tubular** - [hidr]
barrel flow velocity
— f — **corrosión** f - [metal] corrosion speed
— f — **curso** m **de agua** m - [hidr] stream velocity
— f —— m —— m **natural** - [hidr] natural
stream velocity
— f — **chapa** f - [metal-lam] strip speed
— f — **cangrejo** m - snail('s) pace
— f — **cubeta** f **de skip** - [metal-prod] skip car
speed
— f — **deposición** f - [sold] véase **velocidad**
f **de aportación** f
— f — **derrubio** m - [hidr] scour(ing) rate
— f — **desagote,tamiento** - [hidr] dewatering
velocity
— f — **desbaste,tadura** - [metal-lam] slabbing
speed
— f — **descarga** f - [hidr] flow velocity
— f —— f **segura** - [hidr] safe exit velocity
— f — **descascarillado** m - [metal-trat] scaling
rate
— f — **desplazamiento** m - [mec] travel(ing), o
line, speed
— f — **difusión** f - [metal-prod] diffusion rate
— f —— f **de carbono** m - [metal-prod] carbon
diffusion rate
— f — **eje** m - [mec] axle, o shaft, speed
— f —— m **con carga** f **plena** - [mec] full load
shaft speed
— f —— m **sin carga** f - [mec] no load shaft
speed
— f — **elevador** m - [mec] elevator speed
— f —— m **para vagón(es)** m - [agric] wagon
elevator speed • (f.c.) (freight) car elevator speed
— f —— m **regulada** - [mec] controlled elevator speed
— f — **émbolo** m - [comb.int] piston speed
— f — **enfriamiento** m - cooling, rate, o speed
— f — **entrada** f - [mec] incoming, o feeding,
speed • input • [hidr] inlet velocity
— f —— f **aumentada** - [hidr] increased inlet
velocity
— f —— f **computada** - [hidr] computed inlet
velocity
— f —— f **reducida** - [hidr] reduced inlet velocity
— f — **envejecimiento** m - [metal-prod] aging
speed
— f — **equipo** m - (ind) equipment speed
— f — **escarpado** m - [metal-lam] scarfing speed
— f — **estrangulador** m - [mec] choke,king speed
— f — **evaporación** f - evaporation speed
— f — **extensión** f - [mec] extension speed •
extending speed
— f — **fleje** m - [metal-lam] strip speed
— f — **flujo** m - flow, rate, o speed
— f — **función** f - function speed
— f — **funcionamiento** m - [mec] operating,tion
speed
— f —— m **lento** - [ind] slow operating speed
— f — **fusión** f - [metal-prod] fusion, o melting, speed • [sold] melt-off rate
— f —— f **de alambre** m - [sold] wire melt-off
rate
— f — **gancho** m - [grúas] hook speed
— f — **giro** m - [fin] turnover (frequency •
[grúas] swing speed
— f — **grúa** f—[grúas] crane, velocity, o speed
— f — **husillo** m - [herram] spindle speed
— f — **impacto** m - [mec] impact velocity

velocidad f **de laminación** f - [metal-lam] rolling speed
— f — **laminador** m - [metal-lam] rolling speed
• mill speed
— f — **licuación** f - [metal-prod] melting, o
fusion, speed
— f — **línea** f - [ind] line speed
— f —— f **para laminación** f - [metal-lam]
rolling line speed
— f — **lingoteado** m - [metal-lam] slabbing, o
blooming, speed
— f — **manguito** m - [mec] sleeve, o bushing,
o spindle, speed
— f — **marcha** f - [mec] operating,tion speed
— f —— f **alta** - [mec] high operating speed
— f —— f — **sin carga** f - [mec] high idle
operating speed
— f —— f **baja** - [mec] low operating speed
— f —— f — **sin carga** f - [mec] low idle
operating speed
— f —— f **con carga** f **plena** f - [mec] full
load operating speed
— f —— f **en vacío** m - [comb.int] idle speed
— f —— f **sin carga** f—[comb.int] idle speed
— f — **material(es)** m - [ind] material(s)
speed
— f — **medidor** m - [instrum] meter speed
— f — **mesa** f - [mec] table speed
— f — **montacarga(s)** m - [metal-prod] skip
(car) speed
— f — **motor** m - [electr-mot] motor speed •
[comb.int] engine speed • revolutions per
minute
— f —— m **aumentada** - [comb.int] increased
engine speed
— f —— m — **automáticamente** - [comb.int]
automatically increased engine speed
— f —— m **disminuida** - [electr-mot] reduced
motor speed • [comb.int] reduced engine
speed
— f —— m — **automáticamente** - [electr-mot]
automatically decreased motor speed •
[comb.int] automatically decreased engine
speed
— f —— m **hidráulico** - [mec] hydraulic motor
speed
— f —— m **para laminador** m - [metal-lam] mill
motor speed
— f —— m **sin carga** f - [comb.int] no load
engine speed
— f — **movimiento** m - movement speed
— f —— n **de estrangulador** m - [mec] choke
movement speed
— f — **operación** f - [mec] operating,tion,
speed, o rate • funcion(ing), speed, o rate
— f —— f **alta** - [mec] high operating speed
— f —— f — **sin carga** f - [mec] high idle
operating speed
— f —— f **baja** - [mec] low operating speed
— f —— f — **sin carga** f - [mec] low idle
operating speed
— f —— f **con carga** f **plena** - [mec] full
load operating speed
— f —— f **constante** - [mec] constant operating speed
— f —— f **de soldadora** f - [Wold] welder
operating speed
— f —— f **lenta** - [mec] slow operating speed
— f —— f **sin carga** f - [mec] no load, o
idle, operating speed
— f — **oscilación** f - [grúas] swing speed
— f — **par** m **motor** - [mec] torque speed
— f — **pistón*** m - [comb.int] piston speed
— f — **prensadura** f - [mec] pressing speed
— f — **provisión** f - [ind] supply(ing) speed
— f —— f **de aire** m - [combust] air supply,
velocity, o speed
— f — **puente** m - [grúas] bridge speed
— f —— m **de grúa** f - [grúas] crane bridge
speed
— f — **ráfaga** f - [meteorol] gust speed
— f — **reacción** f - reaction, speed, o rate
— f — **régimen** - speed rate

velocidad f de respuesta f - [electrón] response speed
— f — retracción f - [mec] retraction speed
— f — retroceso m - backwards, o reverse, speed
— f — — m de electrodo m (desde trabajo m) - [sold] away from @ work inching speed
— f — rotación f -[mec] rotation speed - [grúas] swing speed
— f — — f de grúa f - [grúas] crane swing speed
— f — salida f - [mec] delivery, o outgoing, speed; exit speed
— f — — f aumentada - [hidr] increased outlet velocity
— f — — f computada - [hidr] computed outlet velocity
— f — — f de planchón m - [metal-lam] slab exit speed
— f — — f reducida - [hidr] reduced outlet velocity
— f — segadora f - [agric-equip] reaper speed
— f — — f trilladora - [agric-equip] combine speed
— f — sistema m - system, velocity, o speed
— f — soldador m - [sold] weldor's speed
— f — soldadora f - [sold] welder speed
— f — soldadura f - [sold] welding, speed, o rate
— f — — f (blanda) - [sold] soldering speed
— f — subida f - climb(ing) speed
— f — — f de temperatura f - [ind] tempera- ture rise, rate, o speed
— f — tambor m - [mec] drum speed
— f — — m seleccionada - [mec] selected drum speed
— f — trabajo m - [mec] operating, speed, o rate
— f — tracción f - [mec] traction speed
— f — transición f - transition speed
— f — transmisión f - [mec] transmission speed
— f — transportador m - [ind] conveyor speed
— f — traslación f - travel(ing) speed • [autom] ground speed • translational speed
— f — — f deseada - desired ground speed
— f — — f precisa - [autom] precise, o accu- rate, ground speed
— f — — f — mantenida - [autom] maintained accurate ground speed
— f — — f verificada - [mec] checked ground speed
— f — traslado m - [mec] translational speed
— f — tren m - [f.c.] train speed • [metal- -lam] mill, o rolling, speed
— f — turbina f - [turb] turbine speed
— f — unidad f - [mec] unit('s) speed
— f — — f para procesamiento m - [ind] pro- cessing unit('s) speed
— f — vaciado m - [metal-prod] pouring speed
— f — vagoneta f - [ind] car speed
— f — — f para montacarga(s) f - [metal-prod] skip car speed
— f — variación f - variation speed
— f — — f de campo m - [electr-oper] field variation speed
— f — viento m - [meteorol] wind, speed, o ve- locity
— f — — m para proyección f - [meteorol] de- sign wind speed
— f decreciente - decreasing, velocity, o speed • dropping speed
— f demasiado reducida - [mec] too low speed
— f — de tambor m - [mec] too low drum speed
— f deseada - desired speed
— f — regulada - controlled desired speed
— f — — para estrangulador m - [mec] set desired choke speed
— f destructiva - destructive speed
— f determinada - determined, o established, speed
— f disminuida - reduced, o dropped, speed
— f — automáticamente - automatically, de-

creased. o reduced, speed
velocidad f disminuida de movimiento m de es- trangulador m - [mec] decreased choke move- ment speed
— f distinta - different speed
— f elevada - high, velocity, o speed
— f — sin carga f - [mec] high idle speed
— f en camino m - [transp] road speed
— f — carretera f - [transp] highway speed
— f — cauce m - [hidr] channel velocity
— f — instante - instant velocity
— f — — m de impacto m - [segurid] impact instant velocity
— f — pie(s) m - velocity in feet
— f — recta f - [deport] straightway speed
— f — red f - system velocity
— f — salida f - outgoing, o output, speed
— f — tacómetro m - [instrum] tachometer speed
— f — tubería f - [tub] pipe velocity • [ambient] duct velocity
— f — vacío m - [sold] idle speed
— f — alta - [comb.int] high idle speed
— f — — ajustada - [comb.int] adjusted high idle speed
— f — — para motor m - [electr-mot] high no load idle motor speed • [comb.int] high idle no load engine speed
— f — — recomendada - [mec] recommended high idle speed
— f — — sin carga f - high no load speed
— f — — — f para motor m - [electr- -mot] high idle no load motor speed • [comb.- -int] high idle no load engine speed
— f — baja - [comb.int] low idle speed
— f — — ajustada - [comb.int] adjusted low idle speed
— f — — recomendada - [comb.int] recom- mended low idle speed
— f — — sin carga f - [comb.int] no load low idle engine speed
— f — — — f para motor m - [comb.int] low idle no load engine speed
— f — de motor m - [comb.int] engine idle speed
— f — — máxima f - [comb.int] high idle no load speed
— f — — para motor m - [comb.int] high idle no load engine speed
— f específica - specific speed
— f — máxima - specific maximum speed
— f — mínima - minimum specific speed
— f estándar - [mec] standard speed
— f — para motor m - [electr-mot] standard motor speed • [comb.int] standard engine speed
— f excesiva - excessive, speed, o velocity • overspeed
— f — de avance m - [sold] excessive travel speed
— f — — cable m - [cabl] excessive, cable, o rope, speed
— f — — operación f - excessive, o too high, operating speed
— f — — tambor m - [mec] excessive drum speed
— f — en camino m - [vial] excessive road speed
— f — — carretera f - [vial] excessive road speed
— f exigida - required speed
— f exigua - too slow speed
— f — de operación f - too low operating speed
— f — — tambor m - too (s)low drum speed
— f extraordinaria - extraordinary speed
— f — de aportación f - [sold] exceptionally high deposit rate
— f fija - [mec] fixed speed
— f — predeterminada - [mec] predetermined fixed speed
— f final - final speed • output speed
— f fiscalizada - [mec] monitored speed
— f hacia adelante - [autom] forward speed
— f — adentro - inward speed
— f — afuera - outward speed
— f — atrás - reverse, o backward, speed

velocidad f hacia exterior m - outward speed
— f — interior m - inward speed
— f ideal - ideal speed
— f — para tambor m - [mec] ideal drum speed
— f ilimitada - unlimited speed
— f incrementada - increased speed
— f — en movimiento m - [mec] increased movement speed
— f — — — m de estrangulador m - [mec] increased choke movement speed
— f inicial - initial speed; input
— f intermedia - intermediate speed
— f inutilizable,- [mec] unused speed
— f lenta - slow, speed, o rate • [comb.int] idle speed • low speed
— f — para avance m - [sold] slow travel speed
— f — — — cable m - [grúas] slow line speed
— f — — — f en vacío - [comb.int] (s)low idle speed
— f — — — f sin carga f - [comb.int] (s)low idle speed
— f — motor m - [electr-mot] slow motor speed • [comb.int] slow engine speed
— f — sin carga f - [mec] (low) idle speed
— f limitada - [mec] limited speed
— f — para alimentación f - [mec] limited feed(ing) speed
— f — — — f para alambre m - [sold] wire feed(ing) limited speed
— f lineal - linear speed
— f — para barra f - [cuerp.moled] bar linear speed
— f — — émbolo m - [int.comb] piston linear speed
— f — — pistón* m - [comb.int] véase velocidad f linear para émbolo m
— f lograda - attained, o obtained, speed
— f mantenida - maintained, speed, o velocity
— f máxima - [mec] maximum, speed, o velocity, o revolutions per minute • full, o top, speed • [transp] speed limit
— f — aceptable - maximum allowable speed
— f — admisible - maximum allowable speed
— f — alcanzable - [transp] top, o maximum, attainable speed
— f — con carga f plena - [mec] full load maximum speed
— f — continua f - maximum continuous speed
— f — — determinada - established maximum continuous speed
— f — para aportación f - [sold] maximum deposit rate
— f — árbol m - [mec] maximum shaft speed; maximum axle speed
— f — — — m con carga f plena - [mec] full load maximum shaft speed
— f — — — m sin carga f - [mec] no load maximum shaft speed
— f — — banda f - [metal-lam] maximum strip speed
— f — — eje m - [mec] maximum, axle, o shaft, speed
— f — — — m sin carga f - [mec] maximum no load shaft speed; no load maximum shaft speed
— f — — fleje m - [metal-lam] maximum strip speed
— f — — gancho m - [grúas] maximum hook speed
— f — — grúa f - [grúas] maximum crane speed
— f — — línea f - [metal-lam] maximum line speed
— f — — marcha f en vacío - [ind] high idle speed
— f — — — f sin carga f - [ind] high idle speed
— f — — soldadura f - [sold] maximum welding speed
— f — — torsión f [mec] maximum torque,quing speed
— f — — tubería f - [ambient] maximum duct velocity • [tub] maximum pipe,ping velocity
— f — permisible - maximum allowable speed
— f — — para motor m - [electr-mot] maximum allowable motor speed • [comb.int] maximum

allowable engine speed
velocidad f máxima sin carga f - [mec] maximum no load speed
— f mayor - higher, o faster, speed, o velocity
— f — para avance m - [sold] faster (travel) speed
— f — — deposición f - [sold] higher,ghest deposit rate
— f — que - speed, o velocity over
— f media - average, o mean, speed, o velocity • [autom] middle gear
— f — con carga f plena - [mec] full load average speed
— f — de árbol m - [mec] average shaft, o shaft average, speed
— f — — — m con carga f plena f - [mec] full load average shaft speed
— f — — — m sin carga f - [mec] no load average, o average no load, shaft speed
— f — corriente f - [hidr] flow, average, o mean, speed, o velocity
— f — — — m en conducto m tubular - [hidr] barrel flow, average, o mean, speed, o velocity
— f — para eje - [mec] average shaft speed
— f — — — m con carga f plena - [mec] full load average shaft speed
— f — — — m sin carga f - [mec] no load average, o average no load, shaft speed
— f menor - (s)lower, speed, o velocity
— f — que - speed, o velocity, under
— f — para avance m - [sold] slower travel speed
— f mínima - minimum, speed, o velocity • low speed • lowest speed
— f — con carga f plena - [mec] full load minimum speed
— f — continua - continuous minimum, speed, o velocity
— f — — determinada - established minimum continuous speed
— f — en vacío - [mec] low idle speed
— f — — — sin carga f - [mec] slow idle no load speed
— f — para árbol m - [mec] minimum shaft speed
— f — — — m con carga f plena - [mec] full load minimum shaft speed
— f — — — m sin carga f - [mec] no load minimum, o minimum no load, shaft speed
— f — — banda f - [metal-lam] minimum strip speed
— f — — eje m - [mec] minimum shaft speed
— f — — — m con carga f plena - [mec] full load minimum shaft speed
— f — — — m sin carga f - [mec] minimum no load, o no load minimum, shaft speed
— f — — fleje m - [metal-lam] minimum strip speed
— f — — gancho m - [grúas] minimum hook speed
— f — — grúa f - [grúas] minimum crane speed
— f — — operación f - [mec] minimum operating speed
— f — — tubería f - [tub] minimum pipe speed • [ambient] minimum duct velocity
— f — sin carga f - [comb.int] minimum, idle, o no load, speed
— f moderada - moderate speed
— f múltiple - multiple speed
— f nominal - [mec] nominal, o rated, speed
— f — europea - [autom-neumát] European speed rating
— f — según norma(s) f estadounidense(s) - [autom-neumát] American speed rating
— f — — f europeas - [autom-neumát] European speed rating
— f — V - [autom-neumát] V-speed rating
— f normal - normal, o standard, speed
— f observada - observed, o monitored, speed
— f operativa - [ind] operating,tion speed
— f para accionamiento m - [mec] drive,ving speed
— f — — m para cabezal m para torno m -

[petról] catshaft drive,ving speed
velocidad f̲ para accionamiento m̲ para mesa f̲ rotatoria - [petról] rotary drive,ving speed
— f̲ —— m̲ **para rotación f̲** - [mec] rotary drive,ving speed
— f̲ — **alimentación f̲** - [mec] feed(ing) speed
— f̲ — — f̲ **con engranaje(s) m̲** - [mec] geared feed(ing) rate
— f̲ — — f̲ **para alambre m̲**,- [sold] wire feed(ing) speed
— f̲ — — f̲ — — m̲ **para soldadura f̲** - [sold] welding wire feed(ing) speed
— f̲ — — f̲ **para electrodo m̲** - [sold] wire feed(ing) speed
— f̲ — — f̲ — — m̲ **de alambre m̲** [sold] wire electrode, o̲ welding wire, feed(ing) speed
— f̲ — — f̲ **rápida** - [mec] fast feeding speed
— f̲ — **alimentadora f̲ para alambre m̲** - [sold] wire feed(ing) speed
— f̲ — **aportación f̲** - [sold] deposition speed
— f̲ **para avance m̲** - [mec] advance,cing speed • [sold] forward, o̲ operating, o̲ travel, speed • inching speed • travel, o̲ welding, rate • [autom] forward speed
— f̲ — — m̲ **adecuada** - [sold] adequate travel speed
— f̲ — — m̲ **alta** - [sold] high welding speed; high travel speed
— f̲ — — f̲ **aumentada** - [sold] increased, weld-ing, o̲ travel, speed
— f̲ — — f̲ **considerable** - [sold] fast travel speed
— f̲ — — m̲ **para arco m̲** - [sold] arc (travel) speed
— f̲ — — m̲ — — m̲ **excesiva** - [sold] excessive arc (travel) speed
— f̲ — — n̲ — **barra f̲** - [mec] bar advancing speed
— f̲ — — m̲ — **electrodo m̲** - [sold] electrode advancing speed
— f̲ — — m̲ — **material** - [ind] material(s) forward speed
— f̲ — — m̲ — **soldadura f̲ con alimentación f̲ automática para electrodo m̲** - [sold] Squirt welding travel speed
— f̲ — — m̲ — — f̲ **Squirt con alimentación f̲ automática para electrodo m̲** - [sold] Squirt welding travel speed
— f̲ — — m̲ — **tractor m̲** - [sold] tractor trav-el speed
— f̲ — — m̲ **demasiado lenta** - [sold] excessive-ly (s)low (travel) speed
— f̲ — — m̲ **elevada** - [sold] fast travel speed
— f̲ — — m̲ **excesiva** - [sold] excessive(ly fast) travel speed
— f̲ — — m̲ **excesivamente rápida** - [sold] ex-cessively fast travel speed
— f̲ — — m̲ **exigua** - [sold] too slow travel speed
— f̲ — — m̲ **gradual** - [mec] inch(ing) (travel) speed
— f̲ — — m̲ — **ajustada** - [sold] adjusted inching speed
— f̲ — — m̲ — **alta** - [sold] high inch(ing) speed
— f̲ — — m̲ — **baja** - [sold] slow inch(ing) speed
— f̲ — — m̲ — **para electrodo m̲** - [sold] elec-trode inching speed
— f̲ — — m̲ — **lenta f̲** - [sold] slow (rate) inch(ing) speed
— f̲ — — m̲ — **reducida** - [sold] slow inch(ing) speed
— f̲ — — m̲ **incrementada** - [sold] increased travel speed
— f̲ — — m̲ **lenta** - [sold] slow travel speed
— f̲ — — m̲ **mayor** - [sold] fast(er), o̲ high(er) travel speed
— f̲ — — m̲ **limitada** - [sold] limited travel speed
— f̲ — — m̲ **menor** - [sold] slower travel speed
— f̲ — — m̲ **pareja** - [sold] even travel speed
— f̲ — — m̲ **prefijada** - [sold] preset travel

speed
velocidad f̲ para avence m̲ rápida - [sold] fast trevel speed
— f̲ — — m̲ **reducida** - [sold] slow travel speed • reduced travel speed
— f̲ — **bajada f̲** - [mec] down, o̲ drop(ping) speed
— f̲ — **banda f̲** - [metal-lam] strip speed
— f̲ — — f̲ **transportadora** - [ind] conveyor (belt) speed
— f̲ — **barra f̲** - [cuerp.moled] bar speed
— f̲ — **(a)bobinadora f̲** - [metal-lam] coiler speed
— f̲ — **bomba f̲** - [bombas] pump speed
— f̲ — **cabezal m̲ para torno m̲** - [petról] cat-shaft speed
— f̲ — **cable m̲** - [cabl] cable, o̲ line, o̲ rope, speed
— f̲ — — m̲ **a gancho m̲** - [grúas] whip line speed
— f̲ — — m̲ — **motón m̲** - [grúas] hoist line speed
— f̲ — — m̲ — — m̲ **principal** - [grúas] main hoist line speed
— f̲ — — m̲ — **elevador m̲** - [grúas] hoist line speed
— f̲ — — m̲ — — m̲ **principal** - [grúas] main hoist line speed
— f̲ — **cadena f̲** - [mec] chain speed
— f̲ — **caída f̲** - [mec] drop(ping) speed
— f̲ — — f̲ **para motón m̲** - [grúas] block drop(ping) speed
— f̲ — **caja f̲** - [metal-lam] stand speed
— f̲ — **calentamiento m̲** - [termol] heating speed
— f̲ — **carga f̲** - [electr-oper] charge,ging rate [mec] load(ing) velocity • [hidr] velocity head
— f̲ — **cauce m̲** - [hidr] channel velocity
— f̲ — — m̲ **para agua m̲** - [hidr] streambed velocity
— m̲ — — m̲ — **para curso m̲ para agua m̲** - [hidr] streambed velocity
— f̲ — **cinta f̲ transportadora** - [ind] conveyor (belt) speed
— f̲ — **conducción f̲** - [autom] driving speed
— m̲ — **corte m̲** - [mec] cutting speed
— m̲ — **derretimiento m̲** - [metal-lam] melting speed
— m̲ — **desplazamiento** - travel(ling) speed
— f̲ — **elevación f̲** - [grúas] hoist(ing) speed • lifting speed
— m̲ — **embalamiento m̲** - run-up speed
— f̲ — **enhebrado m̲** - threading speed
— m̲ — **ensayo m̲** - test(ing) speed
— f̲ — — m̲ **para tracción f̲** - traction test speed
— f̲ — **entrada f̲** - [hidr] inlet velocity
— f̲ — **fresado,dura** - [mec] machining speed
— f̲ — — **lento,ta** - [mec] slow machining speed
— f̲ — **giro m̲** - [mec] revolving, o̲ rotating, speed • [grúas] swing(ing) speed
— f̲ — **implemento m̲** - [agric] implement speed
— f̲ — **labrado m̲** - [mec] machining speed
— m̲ — — m̲ **lento** - [mec] slow machining speed
— m̲ — — m̲ **rápido** - [mec] fast machining speed
— f̲ — **lecho m̲ para curso m̲ para agua m̲** - [hidr] streambed velocity
— f̲ — **línea f̲** - [cabl] line speed
— f̲ — **marcha f̲** - [ind] operating,tional speed
— f̲ — — f̲ **para motor m̲** - [electr-mot] motor operating speed • [comb.int] engine operating speed
— f̲ — **mecanización f̲** - [mec] machining speed
— f̲ — **operación f̲** - operating,tion speed
— f̲ — **recuperación f̲** - recovery rate
— f̲ — — f̲ **de capital m̲** - [fin] capital re-covery rate
— f̲ — **remolque m̲** - [transp] towing speed
— f̲ — — m̲ **para acoplado m̲** - [transp] trailer towing speed
— f̲ — **rendimiento m̲** - [ind] yield(ing), speed, o̲ velocity
— f̲ — **resolución f̲** - solution, o̲ solving, speed
— m̲ — **retorno m̲** - [mec] return(ing) speed
— f̲ — **retroceso m̲** - [mec] back(ing) up speed

velocidad f para salida f - [hidr] outlet, velocity, o speed
— f — señalización f - [comput] signalling speed
— f — separación f - [mec] separation speed
— f — — f para mandíbula f - jaw separation speed
— f — tracción f - traction speed • backhaul speed
— f — unidad f - [mec] unit('s) speed
— f perdida - lost speed • [autom] lost gear
— f permisible - allowable speed
— f — máxima - [comb.int] maximum allowable speed
— f — para motor - [electr-mot] allowable motor speed • [comb.int] allowable engine speed
— f plena - full speed
— f — para soldadura f - [sold] full welding speed
— f — sin carga f - [mec] full idle speed
— f por diferencia f en altura f - [hidr] velocity head; head velocity
— f — — f en elevación f - [hidr] velocity head; head velocity
— f posible - possible speed
— f — con electrodo(s) m Jetweld (con polvo m de hierro m) [sold] Jetweld speed
— f práctica - practical speed
— f práctica para proceso m - [metal-lam] practical processing speed
— f precisa - precise, o exact, o accurate, speed
— f predeterminada - [mec] predetermined speed
— f preestablecida - pre-established, o preset, speed
— f prefijada - preset speed
— f prevista - design, speed, o velocity
— f primera - [mec] first speed • [autom] low, o first, gear
— f promedia - average, speed, o velocity
— f proporcional - proportional speed
— f — correcta - correct proportional speed
— f provista - provided, o supplied, speed
— f que no permite decantación f (de limo m) - [hidr] non-silting velocity
— f quinta - [autom] fifth gear
— f rápida - fast speed • full speed
— f — para avance m - [sold] fast travel speed
— f razonable - reasonable speed
— f real - true, o actual, speed
— f — para tambor m - [mec] actual drum speed
— f recomendada—recommended, speed, o velocity
— f — para motor m - [electr-mot] recommended motor speed • [comb.int] recommended engine
— f — — m sin carga f - [comb.int] recommended engine idle speed
— f — — tubería f - [tub] recommended, tube, o pipe, velocity • [ambient] recommended duct velocity
— f reducida - reduced, o slow, speed, o velocity • dropped speed • [fotogr] slow motion
— f — para alimentación f - [mec] reduced feed(ing) speed
— f — — — f para alambre m - [sold] reduced wire feed(ing) speed
— f — — motor m - [electr-mot] reduced, o decreased, motor speed • [comb.int] reduced, o decreased, engine speed • low engine speed
— f — — sin carga f - [mec] low idle speed
— f — — — movimiento m - reduced, o decreased, movement speed
— f — — — m de estrangulador m - [mec] decreased choke movement speed
— f registrada - [instrum] recorded speed
— f regulada - controlled, o adjusted, o governed, o set, speed
— f — para elevador m - [ind] controlled elevator speed
— f — — — m para vagón(es) m - [agric-equip] controlled wagon elevator speed
— f — — estrangulador m - [mec] controlled, o set, choke speed

velocidad f regulada para operación f - governed operating,tion speed
— f relativa - relative, o proportional, speed
— f — correcta - correct, relative, o proportional, speed
— f relativamente alta - relatively high speed
— f — baja - relatively low speed
— f según norma f - [mec] standard speed
— f segunda - [mec] second speed • [autom-mec] second gear
— f segura - safe, velocity, o speed
— f — para cauce m para agua m - [hidr] safe streambed velocity
— f — descarga f - [hidr] safe exit velocity
— f seleccionada - [mec] selected speed
— f — apropiada - [mec] selected proper speed
— f sin carga f - [mec] no load speed • [sold] idle speed
— f sincrónica - synchronic,nous speed
— f — para motor m - [electr-mot] synchronous motor speed • [comb.int] synchronous engine speed
— f sobre pista f caminera - [autom] road course speed
— f suficiente - [mec] sufficient, o high enough, speed
— f tercera - [mec] third speed • [autom] third gear
— f transversal - [mec] transverse speed
— f uniforme - uniform speed
— f útil - [autom-mec] usable, o useful, gear
— f utilizable - [autom-mec] usable gear
— f V - [autom-neumát] V-speed
— f valiosa - [deport] valuable, o precious, speed
— f variable - variable speed
— f — para motor m - [electr-mot] motor variable speed • [comb.int] engine variable speed
— f — — m hidráulico - [mec] hydraulic motor variable speed
— f — — alimentación f - [sold] variable feed speed
— f variada - varied speed
— f verificada - [mec] checked speed
velocidades f altas - [autom-mec] high range
— f bajas - [autom-mec] low range
velocímetro m - [instrum] . . .; speed, recorder, o indicator
— m accionado con transmisión f - [autom] transmission drive(n) speedometer
— m calibrado - [instrum] calibrated speedometer
velocipédico,ca a - [transp] velocipede
velódromo m - [deport] . . .; (bi)cycle, speedway, o (race) track
— m peraltado - [deport] banked (bi)cycle track
velómetro m - [instrum] velometer
veloz a - . . . .; speedy
vena f - [miner] streak; thin seam; vein
vencedor m moral - moral victor
— m en categoría f - [deport] class winner
vencer v - . . . . • [legal] to expire • [mec] to spread
— v — lugar primero - to, mature, o expire, first
— v primero - [legal] to expire first
— v uno sobre otro - to win over each other
vencerse v - [mec] to fail • [legal] to expire
— v mandíbula v - [mec] to spread @ jaw
vencido,da a - vanquished; conquered; subdued • [mec] failed • sagging • obsolete • [fin] matured; expired • outdated• [mec] spread • [com] outdated
vencimiento m - [com] . . .; due date; deadline; maturity (date) • [mec] spread(ing)
— m de contrato m - [legal] contract, termination, o expiration, o maturity
— m — convenio m - agreement termination
— m — mandíbula f - [mec] jaw spreading
— m — obligación f - [fin] obligation maturity
— m — — f con plazo corto - [fin] short term obligation maturity
— m — — f — m largo - [fin] long term obligation maturity

vencimiento m de período - period termination
— m — — m para garantía f - [seguros] warranty period, expiration, o termination
— m — plazo m - [fin] period, maturity, o termination
— m en base f - [mec] base spread(ing)
— m simple - [fin] simple maturity
vendado,da a - [medic] bandaged
vendedor m - [com] . . .; sales person
— m comercial - [com] commercial salesperson
— m de bienes f raíces - [com] realtor
— m — neumático(s) m - [autom-neumát] tire, salesman, o vendor; tireman
— f — — m profesional - [autom-neumát] tire professional salesperson
vendedor,ra m/f - . . . .; vendor; sales person
vender v - . . . . • to betray
— v a par f - [fin] to sell at par
— v acción t - [fin] to sell @, share, o stock
— v automóvil m - [autom] to sell @, automobile, o car
— v compañía f - [legal] to sell @ company
— v con compromiso m para retroventa - [com] to sell with @ recourse
— v conversión f - [autom] to sell @ conversion
— v derecho(s) m - to sell @ right(s)
— v — m literario(s) - [legal] to sell @ copyright
— v empresa f - [legal] to sell @ concern
— v en conjunto m - to sell in @ block; to package together
— m — — m de dos - to sell as @ pair
— v inversión f - [fin] to sell @ investment
— v marca f - [legal] to sell @ name
— v — f industrial - [legal] to sell @ trade mark
— v — f registrada - [legal] to sell @ trade mark
— v neumático(s) m - [autom-neumát] to sell @ tire(s)
— v nombre m comercial - [legal] to sell @ trade name
— v patente f - [legal] to sell @ patent
— v — f de invención f - [legal] to sell @ patent
— v producto m - [com] to sell @ product
— v propiedad f - [com] to sell @ property
— v — f intelectual - [legal] to sell a copyright
— v separadamente - to sell separately
— v sociedad f - [legal] to sell @ corporation
vendibilidad f salability; merchantability
vendido,da a - sold
— a con compromiso f para retroventa f - [com] sold with @ recourse
— a en conjunto m - sold as @ block; packaged together
— a en juego m de dos - [com] sold in pair(s)
— a separadamente - sold separately
veneciana f - [domést] jalousie • louver
venenoso,sa a - . . . .; toxic
venia f - . . . . • [legal] go-ahead; approval
venida f de arrabio m - [metal-prod] pig iron appearance
— f — — m en piquera f - [metal-prod] pig iron appearance at @ taphole
venido,da a - come • [metal-prod] appeared
venidero,ra a - . . . .; upcoming
venir v - . . . . • to enter • to appear
— v a tobera f - [metal-prod] to, enter, o appear in, @ tuyere
— v ante - [legal] to, come, o appear, before
— v antes - to come before
— v como anillo m a dedo m - to fit as @ (finger) ring • [fam] to not hurt
— v colada - [metal-prod] pig iron appearance
— v coque m - [coque] to appear (@ coke)
— v — n a piquera f - [metal-prod] to appear (coke at @ iron notch)
— m — m menudo - [metal-prod] breexe, o fines, appearance
— v — m — a piquera f - [metal-prod] coke, breeze, o fines, appearance, at @ tap hole
— v de perilla(s) v - to, come, o be, handy

venir v de perilla(s) - to, be, o come in, handy
— v desde atrás - [deport] to come up from behind
— v en medida(s) f - to come in @ size(s)
— v escoria - [metal-prod] slag appearamce
— v — a tobera f - [metal-prod] slag appearance at @ taphole
— v hierro m - [metal-prod] iron appearance
— v — a portafientos m - [metal-prod] iton appearance in @ blowpipe
— v — en caída f - [metal-prod] iron appearamce (when @ burden falls)
— v — m y escoria f a tobera f - [metal-prod] iron and slag appearance at @ tuyere
— v nuevamente sangría f - [metal-prod] (iron, o slag) reappearance when tapping
— v sangría f - [metal-prod] taphole breakthrough
— v violenta (sangría f) - [metal-prod] violent (casting) breakthrough
venirse v - to come (down, o back)
venita f - [miner] (small) vein
— f de cuarzo m - [geol] quartz vein
— f — pirita - [geol] pyrite vein
venoso,sa a - [medic] blue, vein, o skin
venta f - [com] . . . . • account • business • betrayal,ying
— f a cliente m - [com] customer sale
— f — contado - [com] cash sale
— f — detalle m - [com] retail, sale, o transaction, o business
— f — distribuidor m - [com] retailer sale
— f — exterior m - [com] foreign sale; sale abroad
— f — fecha f - [com] sale(s) to date
— f — futuro - [com] future(s) sale
— f — ganadero(s) m - [ganad] cattlemen's sale
— f — gobierno m - [com] government sale
— f — industria(s) f - [com] sale to @ industry; industrial sale
— f — menudeo - [com] retail sale
— f — por menor - [com] retail sale
— f — revendedor(es) m - sale to retailer(s) • commercial sale(s)
— f aceptada - [com] accepted sale
— f acumulada - [com] accumulated sale
— f — hasta fecha f - [com] accumulated sale to date
— f adicional - [com] additional, o add-on, sale
— f anual - [com] annual sale
— f aumentada - [com] increased sale
— f bruta - [com] gross sale
— f cerrada - [com] closed sale
— f con cambio m - [com] exchange sale
— f — compromiso para retroventa f - [com] recourse sale
— f — importación f - import, o indent, sale
— f — — f directa - [com] indent sale
— f — — f por comprador m - [com] indent sale
— f — permuta f - [com] exchange sale
— f de acción(es) f - [fin] share(s), o stock, sale
— f — f en mercado m abierto - [fin] open, o public, market share(s) sale
— f — activo(s) - [contab] assets sale
— f — m fijo - [contab] fixed assets sale
— f — automóvil(es) - [autom] car sale(s)
— f — m nuevo(s) - [autom] new car sale(s)
— f — m usado(s) - [autom] used car(s), business, o sale(s)
— f — cambio(s) m - [fin] exchange sale
— f — conversión f - conversion, sale, o selling
— f — derecho(s) m - [legal] right(s) sale
— f — — m literario(s) m - [legal] copyright sale
— f — desecho(s) m - [ind] salvage, o scrap, sale(s)
— f — divisa(s) f - [fin] exchange sale
— f — equipo(s) m - [com] equipment sale
— f — — m industrial(es) - [com] industrial equipment sale(s)
— f — — m original - [com] original equipment, sale(s), o business

venta f de estaño m • [miner] tin sale
— f — inversión f - [fin] investment sale
— f — marca f para fabricación f - [legal]
trade mark sale
— f — — f industrial - [legal] trade mark
sale
— f — — f registrada—[legal] trade mark sale
— f — material(es) m - [com] materials sale
— f — — m de desecho m - [ind] salvage sale
— f — mes m - [com] month('s) sale
— f — mineral m - [miner] ore sale
— f — — m puesto(s) en mercado m - [miner]
ore sales on @ market
— f — neumático(s) - [autom-neumát] tire sale
• tire, business, o trade
— f — — m a detalle - [autom-neumát] retail
tire, sale, o business
— f — — m radial(es) - [autom-neumát] radial
(tires) sale
— f — nombre m comercial - [legal] trade name
sale
— f — participación f - [legal] participation,
o share, sale; partnership sale
— f — patente f - [legal] patent sale
— f — — f de invención f—[legal] patent sale
— f — producto m - [com] product sale
— f — — m no plano(s) - [metal-lam] non flat
product(s) sale
— f — — m plano(s) - [metal-lam] flat pro-
duct(s) sale
— f — título(s) m - [fin] bond(s) sale
— f — — m nacional(es) - [fin] public, o government, bond(s) sale
— f — propiedad f -,[legal] property sale
— f — — f intelectual—[legal] copyright sale
— f — valor(es) m - [fin] security,ties sale
— f directa - [com] direct sale • retail sale
— f — a cliente(s)—[com] direct customer sale
— f en aumento - [com] growing, o increasing,
sale, o business
— f — conjunto - [com] package sale
— f — equipo(s) m - [com] equipment sale
— f — — m original - [com] original equipment
sale
— f — exterior - [com] overseas, o foreign,
sale
— f — extranjero m - [com] overseas, o foreign, sale
— f — juego(s) - [com] sale in sets
— f — — m de dos - [com] sale in pairs
— f — mercado m - [com] market sale
— f — — m abierto - [fin] open, o public,
market sale
— f entre empresa(s) f - [legal] intercompany
sale
— f — — f afiliada(s) - [legal] (associated)
intercompany sale
— f especial - [com] (special) sale
— f facturada - [com] invoiced sale
— f finalizada - [com] finalized sale
— f incrementada - [com] increased sale
— f internacional - [com] international sale
— f neta - [com] net sale
— f no comercial - [com] noncommercial sale
— f nueva - [com] new, sale, o business
— f pagada - [com] paid sale
— f — por anticipado - [com] prepaid sale
— f para exportación f - [com] export sale
— f perdida - [com] lost sale
— f por comisión f - [com] commission sale
— f por (parte de) distribuidor m - [com] dis-
tributor('s), sale, o marketing
— f separado - [com] separate sale
— f previa - [com] prior sale
— f separada - [com] separate sale
— f total - [com] total sale
— f — hasta fecha f - [com] total sale to date
— f — para mes m - [com] total sale for month
ventaja f - . . . • lead • head start; handicap
• plus • benefit; inducement; feature •
[deport] heading; gap; outstanding; edge
— f adicional - additional advantage; added
plus • spinoff benefit

ventaja f apreciable - appreciable, o large, advantage, o lead
— f aumentada - increased advantage
— f brindada - [com] offered advantage • [fig]
paid dividend
— f clara - clear advantage
— f comparativa - comparative advantage
— f competitiva - competitive advantage
— f comprobada - proven, o (time) tested, advantage
— f durante año(s) m - time tested advantage
— f considerable - significant, o substantial,
advantage, o lead
— f de ahorro m en espacio m - space saving advantage
— f liberación f - exemption advantage
— f proceso m - [ind] process advantage
— f — — m Innershield - [sold] Innershield
process advantage
— f derivada - spinoff, advantage, o benefit
— f económica - cost, o economic, advantage
— f mayor - greater, o major, cost advantage
— f en construcción f - [constr] construction
advantage
— f costo m - cost advantage
— f — — m de existencia(s) f - [ind] inventory cost advantage
— f — — m de flete m - [Transp] freight cost
advantage
— f — gasto(s) m general(es) - [ind] overhead
cost(s) advantage
— f práctica f - functional advantage
— f rapidez f - speed advantage
— f trabajo m - [labor] work, o job, gain
— f uso m - functional advantage
— f velocidad f - speed advantage
— f — — f de soldadura f - [sold] welding
speed advantage
— f escasa - scant advantage • [deport] thin, o
scant, lead
— f específica - specific advantage
— f fiscal - [fisc] fiscal advantage
— f funcional - functional, o operating, advantage
— f hidráulica - [hidr] hydraulic advantage
— f importante - important advantage
— f inicial - initial advantage; head start
— f lograda - attained advantage • [deport]
built lead
— f mantenida - maintained advantage
— f marcada - distinct advantage
— f mayor - major, o greater, advantage
— f mecánica - [mec] mechanical advantage
— f mínima - minimum advantage • [deport] thin
lead
— f natural - natural advantage
— f normal - normal advantage
— f ofrecida - offered advantage
— f para corredor m - [deport] runner('s), o
racer('s), advantage
— f — descuento m - [com] discount advantage
— f — labrado m - [mec] machining advantage
— f — mecanización f—[mec] machining advantage
— f — operación f - operating,tion advantage
— f particular - particular, o specific, o
special, advantage
— f plena - full advantage
— f principal - principal, o main, advantage
— f recreativa - [deport] recreational advantage
— f resultante - resulting advantage • spinoff
benefit
— f técnica - [ind] technical advantage
ventajosamente adv - . . .; profitably
ventana f con secciones f - [constr] section
window
— f — — f abisagrada(s) - [constr] casement
section window
— f — — f intermedia(s) - [constr] intermediate section window
— f — — f — abisagrada(s) f - [constr] intermediate casement section window
— f de aluminio m - [constr] aluminum window
— f — guillotina f - [constr] sash window

ventana f de guillotina f con dos secciones f -
[constr] double hung window sash
— f en ampuesa f para apoyo m -[metal-lam]
back-up screwdown window
— f —— f rodillo m para apoyo m - [metal-
-lam] back-up roll screwdown window
— f en coraza f - [metal-prod] shell window
— f — envolvente f - [metal-prod] shell opening; jacket, o casing, window, o opening
— f limpia f - [constr] clean window
— f limpiada - [Constr] cleaned window
— f metálica - [constr] metal window
— f para operación f - [ind] peep hole; vision
window; peep sight
— f — otear - [ind] peep sight
— f saliente - [constr] protruding, o projecting, window
— f sucia - [constr] dirty window
ventanal m en tubería f igualadora - [metal-
-prod] mixer main window
ventanero m - . . . • window peeper
ventanilla f - window • port
— f astillada - [autom] shattered window
— f automática - [autom] power window
— f auxiliar - [electrón] auxiliary port
— f mecanizada - [autom] power window
— f para automovilista m - [fin] drive-in
window
— f — entrada f - [electrón] input port
— f — inspección f - [ind] inspection port
— f — salida f - [electrón] output port
venteado,da a -[mader] split
ventear v - . . . • to vent • [mader] to split
venteo m - . . . • venting • [mader] splitting
ventilación f - . . .; ventilating; venting •
vent; air, port, o vent
— f a atmósfera f - venting to @ outside
— f adecuada - adequate, o proper, ventilation
— f apropiada - appropriate ventilation
— f — impedida - hindered appropriate ventilation
— f buena - good ventilation
— f con tiro m descendente - [ind] down draft
ventilation
— f extractiva - extractive, o exhaust, ventilation
— f extractora - extractive, o exhaust, ventilation
— f forzada - (fan) forced ventilation
— f — para extracción f - forced exhaust ventilation
— f — para salida f - forced exhaust ventilation
— f general - general ventilating,tion
— f impedida - hindered ventilation
— f industrial - [ind] industrial ventilation
— f natural - natural ventilation
— f para criba f - [coque] screen exhaust
— f —— f para coque m - [coque] coke screen
exhaust
— f — extracción f - exhaust, o extraction,
ventilation
— f — planta f - [ind] plant, o mill, ventilation
— f —— f para acero m - [metal-prod] steel,
plant, o mill, ventilation
— f — sitio m - site, o area, ventilation
— f — tambor m - [ind] drum ventilating,tion
— f — válvula f - [valv] valve vent(ilation)
— f —— f para alivio m - [cald] relief valve
vent(ilation)
— f —— f — seguridad f - [válv] safety
valve vent(ilation)
— f — zona f - [ind] area ventilation
— f por aspiración f - [ambient] exhaust ventilation
— f principal - [ind] main vent(ilation)
ventilado,da a - ventilated; vented
— a adecuadamente - adequately ventilated
— a apropiadamente - appropriately, o adequately, ventilated
— a bien - ventilated well; well ventilated
— a inadecuadamente - inadequately ventilated

ventilado,da malamente - poorly ventilated
ventilador m - . . .; (blower) fan • vent
— m abisagrado -[constr] open-out ventilator
— m armado - [mec] assembled fan
— m aspirador - [mec] suction fan • exhaust fan
— m — inferior - [mec] lower suction fan
— m — para cabina f - [grúas] cab suction fan
— m —— f inferior - [grúas] lower cab
suction fan
— m —— f superior - [grúas] upper cab
suction fan
— m — superior - [ventil] upper suction fan
— m auxiliar - [mec] auxiliary fan
— m — para casilla f para mando m para cargador m - [metal-prod] hoist house auxiliary fan
— f blindado - [mec] enclosed fan
— m centrífugo - [mec] centrifugal fan
— m comprobado - [mec] tested fan
— m con aire m - [ind] air fan
— m . . . aletas f - [mec] . . . blade fan
— m . . . aspas f - [mec] . . . blade fan
— m — fuerza f motriz - [mec] power blower
— m — motor m - [electr-equip] motor fan
— m —— m hermético - [electr-equip] sealed
motor fan
— m . . . paletas f - [mec] . . . blade fan
— n — pedestal - [mec-ventil] standing fan
— m — regulación f - [mec-ventil] controlled
fan
— m —— f con relé(s) m [electr-equip] relay controlled fan
— m dañado - [mec] damaged fan
— m de tipo m con filo m continuo - [constr]
(continuous) ridge type ventilator
— m —— m estacionario - [constr] stationary
type ventilator
— f —— m redondo - [constr] round type ventilator
— m desarmado - [mec-ventil] unassembled fan
— m detenido - [mec-ventil] non operating, o
off, fan
— m doble - [mec-ventil] double fan
— m empujado - [mec=ventil] pushed fan
— m en funcionamiento - [mec-ventil] operating
fan
— m — marcha - [mec=ventil] operating, o on,
fan
— m — operación f - [mec-ventil] operating fan
— m ensayado - [mec=ventil] tested fan
— m espantainsectos m - [ind] bug blower
— m extractor - [ventil] exhaust fan
— m — para criba f - [mec] screen exhaust fan
— m —— f para coque m - [coque] coke
screen exhaust fan
— m fuera de alcance m - [constr] out of reach
ventilator
— m hecho operar - [mec] run, o operated, fan
— m helicoidal - [mec] helical blower
— m henchido - [mec-ventil] filled fan
— m impulsado - [mec-ventil] driven fan
— m inactivo - [mec-ventil] nonoperating fan
— m individual - [mec-ventil] individual, fan,
o ventilator
— m — para techo m - [constr] individual roof
ventilator
— m —— m por gravedad f - [constr] individual gravity roof ventilator
— m inferior - [mec] low(er) fan
— m instalado - [ventil] installed fan
— m — inapropiadamente - [ventil] improperly
installed fan
— m inclinado - [ventil] tilted, o tipped, fan
— m inoperante - [ventil] nonoperating, o lost,
ventilator, o fan
— m lubricado - [ventil] lubricated, ventilator, o fan
— m llenado - [ventil] filled, ventilator, o
fan
— m mecánico - [ventil] power blower
— m montado - [ventil] mounted, ventilator, o
fan
— m operado - [ventil] operated, ventilator, o
fan

ventilador m operado en frío - [ventil] cold
  operated fan
— m desenganchado - [Ventil] unhooked fan
— m — acondicionador m para aire m - [ambient]
  air conditioner, blower, o fan
— m — acoplamiento m—[ventil] coupling blower
— m — cabina f - [autom] cab fan
— m — caja f—[ventil] case, ventilator, o fan
— m — caldera f - [cald] furnace blower
— m — — f para calefacción f - [cald] heating
  furnace blower
— m — calefaccionador m - [ventil] heater
  blower
— m — calefactor m - [combust] furnace, o
  space heater, blower
— m — calentador m - heater, blower, o fan
— m — — m para cabina f - [autom] cab heater
  fan
— m — — m y desempañador m - [autom-electr]
  heater and defroster fan
— m — cárter m - [comb.int] crankcase, fan, o
  ventilator
— m — casilla f para mando m - [metal-prod]
  house air fan
— m — — f — m para cargador m - [metal-
  -prod] hoist house (air) fan
— m — circulación f - circulation, blower, o
  fan
— m — — f — aire m [ambient] air circulation
  blower
— m — — f — — fresco - [ind] fresh air circulation, blower, o fan
— m — combustión f - [combust] combustion,
  fan, o blower
— m — compresión f—[ventil] pressurizer (fan)
— m — — f operado - [ventil] operated pressurizer (fan)
— m — condensador m - [ind] condenser fan
— m — criba f - [coque] screen fan
— m — — f para coque m - [coque] coke screen
  fan
— m — cumbrera f - [constr] ridge ventilator
— m — chimenea f - [combust] stack blower
— m — desempañador m - [autom-electr] defog-
  ger, o defroster, fan
— m — — m para cabina f - [autom] cab defroster fan
— m — enfriamiento m - cooling fan
— m — estufa f - [metal-prod] stove fan
— m — — f regeneradora - [metal-prod] (regenerating) stove burner fan
— m — extracción f - exhaust fan
— m — gas m - [combust] gas, fan, o blower
— m — generador m - [sold] generator fan
— m — — m para soldadora f - [sold] welder
  generator fan
— m — horno m alto - [metal-prod] blast furnace, blower, o fan
— m — impulsión f - [mec] ventilating fan
— m — motor m - [electr-mot] motor fan •
  [comb.int] engine fan
— m — — m lubricado - [electr-mot] lubricated
  motor fan • [comb.int] lubricated engine fan
— m — radiador m - [comb.int] radiator fan
— m — recirculación f—[combust] recirculating
  fan
— m — regulación f - [ventil] regulating fan
— m — — f automática - [ventil] automatic
  controlling fan
— m — — f de temperatura f - [ventil] auto-
  automatic temperature controlling fan
— f — f — — f por medio m de relés m -
  [ventil] automatic temperature relay controlling fan
— m — renovación f - [ventil] renewing fan
— m — — f de aire m - [ventil] air renewal
  fan • make-up air tower
— m — retorno m - [combust] return fan
— m — soldadora f - [sold] welder fan
— m — soplador m - [ventil] blower fan
— m — — m para aire m - [ventil] air blower
  fan
— m — succión f - [ventil] suction fan

ventilador m para techo m - [constr] roof fan •
  ceiling fan
— m — — por gravedad f - [constr] gravity
  roof, o roof gravity, ventilator
— m — — temperatura f alta - [ventil] high temperature fan
— m — — f para elemento m calentador -
  [ind] heating element high temperature fan
— m — tiro m forzado - (forced) draft fan
— m — — m inducido - [ind] induced draft fan
— m — torpedo m - [autom] cowl ventilator
— m — — f enfriada por aire m - [ind] air
  cooled unit fan
— m — velocidad f variable - [mec] variable
  speed fan
— m perdido - [ventil] lost fan
— m por aspiración f - [ventil] sucker fan
— m — — f para cabina f - [autom] cab sucker
  fan
— m — — f — — f inferior - (grúas) lower
  cab sucker fan
— m — — f — — f superior - [grúas] upper
  cab sucker fan
— m — gravedad f - [constr] gravity ventilator
— m principal - [ventil] main fan
— m — para casilla f para mando m - [metal-
  -prod] house main fan
— m — — f — — m para cargador m—[metal-
  -prod] hoist house main fan
— m — — combustión f - [comb] main combustion
  fan
— m recirculante - [ind] recirculating fan
— m — para soplador m (para aire m)—[combust]
  air blower recirculating fan
— m redondo - [ventil] round, fan, o ventilator
— m — — m estacionario - [constr] round
  type stationary, fan, o ventilator
— m — estacionario - [ventil] round stationary,
  ventilator, o fan
— m reemplazado - [ventil] replaced fan
— m regulado - [ventil] set blower
— m — con relés m - [ventil] relay controlled
  fan
— m roto - [ventil] broken fan
— m sucio - [ventil] dirty fan
— m superior - [ventil] upper fan
— m tipo soplador - [ventil] blower type fan
— m — — con . . . paleta(s) f - [ventil]
  blower type . . . blade fan
— m variable - [ventil] variable fan
— m verificado - [mec] checked, o tested, fan
— m y acoplamiento m - [ventil] coupling blower
ventilar v - . . .; to vent
— v adecuadamente - [ventil] to ventilate adequately
— v apropiadamente - [ventil] to ventilate
  appropriately
— v bien - [ventil] to ventilate well
— v inadecuadamente - [ventil] to ventilate inadequately
— v sitio m - [ventil] to ventilate @ area
— v tambor m - [ventil] to ventilate @ drum
— v zona f - [ventil] to ventilate @ area
ventisca f - [meteorol] . . .; driven,ving snow
ventosa f - . . . . • [válv] air valve
— f con acción f doble - [válv] double action
  air valve
Venturi m - [metal-prod] (Venturi) scrubber
— m primario - [carburad] primary Venturi
ver v - . . .; to review
— v desde - to see from
— v — abajo - to view from below
— f — arriba - to view from above
— v — atrás - to view from behind
— v ilustración f - to see @ illustration
— v oscilación f - [electrón] to see @ oscillation
— v que - to see that • to make sure
— retrospectivamente - to see retrospectively •
  to flash back
— v tabla f - to see @, table, o chart
vera f - . . .; [constr] bank
— f de camino m - [constr] roadside

vera f̱ de carretera f̱ - [vial] highway roadside
— f̱ — zanja f̱ - [constr] ditch bank
verano m - . . .; summertime
— m antártico - [geogr] antarctic summer
— m ártico - [geogr] arctic summer
— m̱ pleno - [meterol] mid-summer
veraz a - . . .; true
verbalmente a̱ - verbally | [adv] by telling
verbeneado,da a - abounded
verbigracia adv - . . .; that is; i. e.
verbo m auxiliar - [gram] auxiliary verb
verdaderamente adv eficiente - truly efficient
verde a claro - light, o pale, green
— a̱ grisáceo - grayish green
— m̱ obscuro - dark green
vereda f - [constr] . . .; walk(way); sidewalk •
   trail • véase acera f̱
— f̱ para peatón(es) m̱ - [constr] pedestrian,
   sidewalk, o walkway
vergüenza f - . . .; embarrassment
verificación f - . . .; verifying; looking for;
   check(ing) out; determination; checking;
   examination; review' adjusting,tment; proofing; check(ing); control(ling) • inspection;
   tally(ing) • research(ing)
— f a azar m̱ - [ind] spot check(ing)
— f̱ — intervalo(s) m̱ - interval(s) check(ing)
— f̱ — m regular(es) - regular interval(s)
   check(ing)
— f̱ anterior - previous check(ing)
— f̱ — a puesta f̱ en marcha f̱ - [mec] pre-
   -start(ing) check-up
— f̱ — uso m̱ - [mec] pre-starting check-up
— f̱ antes de entrega f̱ - pre-delivery check-up
— f̱ — remachar v̱ - [mec] pre-riverting
   check-up
— f̱ colacionada - [comput] report back verification
— f̱ — de comando m̱ - [comput] reported back
   command verification
— f̱ — — mandato m̱ - [comput] reported back
   command verification
— f̱ — de orden f̱ - [comput] reported back command verificaion
— f̱ comparada - compared check(ing)
— f̱ completa f - [ind] complete check(ing)
— f̱ con amperímetro m̱ - [instrum] ammeter
   check(ing)
— f̱ — diagrama m̱ - check(ing) against @ diagram
— f̱ — empresa f̱ para energía f̱ - [electr-prod]
   check(ing) with @ power company
— f̱ — lámpara f - [electr-oper] check(ing)
   with @ lamp; lamp check(ing)
— f̱ — timbre m̱ - [electr-oper] ringer checking
— f̱ — voltímetro m̱ - [instrum] voltmeter
   check(ing)
— f̱ conforme - O. K. check(ing)
— f̱ constante - constant check(ing)
— f̱ — de calidad f̱ - constant quality checking
— f̱ contra placa f̱ para identificación f̱ -
   [ind] check(ing) against @ nameplate
— f̱ cuidadosa - careful, check(ing), o control
— f̱ de accesorio m̱ - [mec] fitting check(ing)
— f̱ — — m̱ para línea f̱ - [ind] line fitting
   check(ing)
— f̱ — accionamiento m̱ - [mec] drive check(ing)
— f̱ — — m̱ para bomba f̱ - [bombas] pump drive
   check(ing)
— f̱ — m̱ — — triple - [petról] triple pump
   drive check(ing)
— f̱ — — m̱ para motor m̱ - [electr-mot] motor
   drive check(ing) • [comb.int] engine drive
   check(ing)
— f̱ — aceite m̱ - [mec] oil check(ing)
— f̱ — — m̱ en bomba f̱ - [bombas] pump oil
   check(ing)
— f̱ — — m̱ — — f̱ acon impulsión f̱ con cadena
   f̱ - chain driven pump oil check(ing)
— f̱ — — m̱ en caja f̱ - [mec] case oil check
— f̱ — — m̱ — — f̱ para cadena f̱ -
   [mec] chain case oil check(ing)

verificación f̱ de aceite en cárter m̱ - [comb.-
   -int] crankcase oil check(ing)
— f̱ — — m̱ — conjunto m̱ - [mec] assembly oil
   check(ing)
— f̱ — — — m̱ en árbol m̱ para aguilon m -
   [grúas] boom shaft assembly oil check(ing)
— f̱ — m̱ — m̱ — — m̱ — cable m̱ a gancho
   - [grúas] whip shaft assembly oil check(ing)
— f̱ — m̱ — — m̱ — — m̱ — motón m̱ -
   [grúas] main shaft assembly oil check(ing)
— f̱ — — m̱ lubricador m̱ - [petról] lubricator oil check(ing)
— f̱ — — m̱ — m̱ para línea f̱ - [petról]
   line lubricator oil check(ing)
— f̱ — — — — — f̱ neumática - [petról]
   air line lubricator oil check(ing)
— f̱ — — m̱ motor m̱ - [electr-mot] motor oil
   check(ing) • [comb.int] engine oil check(ing)
— f̱ — — m̱ hidráulico - [petról] hydraulic oil
   check(ing)
— f̱ — aceleración f̱ - [autom] acceleration monitoring
— f̱ — — f̱ lateral - [autom] lateral acceleration monitoring
— f̱ — acoplamiento m̱—[mec] coupling check(ing)
— f̱ — — m̱ para accionamiento m̱ - [mec] drive
   coupling check(ing)
— f̱ — — m̱ para accionamiento m̱ - [mec] drive
   coupling check(ing)
— f̱ — — m̱ — m̱ para motor m̱ - [electr-mot]
   motor drive coupling check(ing) • [comb.int]
   engine drive coupling check(ing)
— f̱ — actuación f̱ - [admin] performance assessment
— f̱ — acumulador m̱ - [electr] battery check
— f̱ — administración f̱ - [admin] management,
   check(ing), o control(ling)
— f̱ — agua m̱ - [hidr] water check(ing)
— f̱ — aguilón m̱ - [grúas] boom check(ing)
— f̱ — aire m̱ - air check(ing)
— f̱ — aislación f̱ - [electr-instal] insulation
   check(ing)
— f̱ — — m̱ para toro m̱ - [mec] torus insulation
   check(ing)
— f̱ — ajuste m̱ - adjustment, o setting, check-
   -(ing) • check(ing) for @ fit
— f̱ — — m̱ apropiado - [mec] proper tightness
   check(ing)
— f̱ — — m̱ — freno m̱ - [mec] brake adjustment
   check(ing)
— f̱ — — m̱ — sujetador m - [mec] fastener, o
   clamp, tightness check(ing)
— f̱ — — m̱ — tornillo m̱ - [mec] screw tightness check(ing)
— f̱ — — m̱ — — m̱ para cierre m̱ - [mec] locking screw tightness check(ing)
— f̱ — — m̱ — — m̱ — cierre m̱ para tapa f̱ -
   [mec] cover lock(ing) screw tightness check-
   -(ing)
— f̱ — — m̱ — — m̱ — — m̱ — — f̱ para válvula f̱ - [mec] valve cover locking screw tightness check(ing)
— f̱ — — m̱ debido - [mec] proper adjustment
   check(ing)
— f̱ — — m̱ lateral - [cojin] lateral, adjustment, o setting, check(ing)
— f̱ — — m̱ para cojinete m̱ - [cojin] bearing (lateral), adjustment, o setting, check-
   -(ing)
— f̱ — — m̱ — — rodamiento m̱ - [cojin] bearing (lateral), adjustment, o setting, check-
   -(ing)
— f̱ — — m̱ manual - [mec] hand tightness
   check(ing)
— f̱ — alargamiento m̱ - [mec] elongation check-
   -(ing)
— f̱ — — m̱ de pieza f̱ - [mec] part elongation
   check(ing)
— f̱ — alguna posible conexión f̱ con tierra f̱ -
   [electr-instal] check(ing) for @ ground
— f̱ — alineación f̱ - [autom] alignment check-
   -(ing)

verificación f de alineación f angular - [mec] angular alignment check(ing)
— f — — f desplazada - [mec] offset, o displaced, alignment check(ing)
— f — amortiguador m - [mec] damp(en)er check(ing)
— f — — m para pulsación f - [mec] pulsation damp(en)er check(ing)
— f — — — f para línea f - [mec] line pulsation damp(en)er check(ing)
— f — — m — — f — — f para aspiración f - [mec] suction line pulsation damp(en)er check(ing)
— f — amperaje - [electr-oper] amperage check(ing) • [sold] current check(ing)
— f — m en salida f - [sold] current output check(ing)
— f — — m recomendado - [electr-oper] recommended amperage check(ing) • [sold] recommended current check(ing)
— f — apoyo m - [mec] bearing, o support, check(ing)
— f — apretadura f—[mec] tightness check(ing)
— f — — f de sujetador m - [mec] clamp tightness check(ing)
— f — — tornillo m - [mec] screw tightness check(ing)
— f — — f — — m para cierre m - [mec] locking screw tightness check(ing)
— f — — m — — — m para tapa f - [mec] cover locking screw tightness check(ing)
— f — — f — — — m — — f para válvula f - [mec] valve cover locking screw tightness check(ing)
— f — aprovisionamiento m - [ind] supply check(ing)
— f — m de combustible m - [comb.int] fuel supply check(ing)
— f — arandela f - [mec] washer check(ing)
— f — — f de cuero m - [mec] leather washer check(ing)
— f — — m para aguilón m - [grúas] boom shaft check(ing)
— f — — m para cable m - [grúas] cable shaft check(ing)
— f — — — m para gancho m - [grúas] whip shaft check(ing)
— f — — m — — m — motón m - [grúas] main shaft check(ing)
— f — armadura f - [electr-mot] armature check(ing)
— f — arrollamiento m - coiling, o spooling, check(ing)
— f — m apropiado - [mec] proper, o appropriate, coiling, o spooling, check(ing)
— f — asidero m - [mec] handle check(ing)
— f — — — m para regulación f - [mec] control handle check(ing)
— f — — m — — f para rotación f - [grúas] swing control handle check(ing)
— f — — m — — m para cable m a gancho m - [grúas] crane hoist control handle check(ing)
— f — — m — — m — m a motón m - [grúas] main hoist control handle check(ing)
— f — — m — — m — elevador m - [grúas] hoist(ing) control handle check(ing)
— f — — m — — m para cable m a gancho m - [grúas] whip hoist control handle check(ing)
— f — — m — — m para regulador m - grúas] hoist control handle check(ing)
— f — — m — — m — m principal - [grúas] main hoist control handle check(ing)
— f — — m — — — f para aguilón m - [grúas] boom hoist control handle check(ing)
— f — — m — — m para rotación f - [grúas] swing control handle check(ing)
— f — base f - [mec] base check(ing)
— f — bomba f - [bombas] pump check(ing)
— f — — m con impulsión f - [bombas] drive(n) pump check(ing)
— f — — f — — f con cadena f - [bombas] chain drive(n) pump check(ing)

verificación f de bomba f hidráulica - [bombas] hydraulic pump check(ing)
— f — — f triple - [bombas] triple pump check(ing)
— f — boquilla f - [mec] nozzle check(ing)
— f — — f para rociadura f - [mec] spray nozzle check(ing)
— f — caballete m - [mec] sawhorse check(ing) • [grúas] gantry check(ing)
— f — cable m - [cabl] cable check(ing)
— f — — m a gancho m - [grúas] whip (hoist) check(ing)
— f — — m — motón m - [grúas] main (hoist) check(ing)
— f — cadena f - [mec] chain check(ing)
— f — — f para accionamiento m - [mec] drive,ving chain check(ing)
— f — — m para bomba f - [bombas] pump drive,ving chain check(ing)
— f — caja f - [mec] case check(ing)
— f — — f para cadena f - [mec] chain case check(ing)
— f — calidad f - [ind] quality, control(ling), o check(ing)
— f — — f durante fabricación f - [ind] quality control during, fabrication, o manufacturing
— f — camisa f - [mec] liner check(ing)
— f — — f y vástago m - [mec] liner and rod check(ing)
— f — campo m - [electr-instal] field checking
— f — capa f - [metal-trat] coating check(ing)
— f — — f galzanizada - [metal-trat] galvanized coating check(ing)
— f — — f — en alambre m - [alambre] wire galvanized coating check(ing)
— f — capacitancia f - [electr-oper] capacitancy check(ing)
— f — carburación f - [comb.int] carburation check(ing)
— f — carburador m - [comb.int] carburetor check(ing)
— f — carga f - [transp] load check(ing) • [metal-prod] charge check(ing)
— f — — f completa - [electr-oper] full charge check(ing)
— f — — f plena - [electr-oper] full charge check(ing)
— f — caudal m - flow check(ing)
— f — — m de aire m - air flow check(ing)
— f — centro m para control m - [cpmput] control center, verification, o check(ing)
— f — cilindro m - [mec] cylinder check(ing)
— f — — m para éter m - [comb.int] ether cylinder check(ing)
— f — cinta f - [metal-lam] strip check(ing) • [mec-frenos] (brake) lining check(ing)
— f — circuito m - [electr-instal] circuit check(ing)
— f — — m con voltímetro m - [electr-instal] voltmeter circuit testing
— f — — m — lamparilla f - [electr-instal] lamp circuit test(ing)
— f — — — timbre m - [electr-instal] ringer circuit, testing, o check(ing)
— f — — m de campo m - [electr-mot] field circuit check(ing)
— f — — m — — m de excitador m - [electr-mot] exciter field circuit check(ing)
— f — — m en excitador m - [electr-mot] exciter circuit check(ing)
— f — cojinete m - [cojin] bearing check(ing)
— f — — m para motor m - [electr-mot] motor bearing check(ing)
— f — — m para rotación f - [grúas] swing bearing check(ing)
— f — colador m - [mec] strainer check(ing)
— f — compatibilidad f - compatibility checking
— f — compensación f - compensation check(ing)
— f — composición f - composition, check(ing), o control
— f — — f química - [quím] chemical composition, check(ing), o control

verificación f de condensador m - [ind] condenser check(ing)
— f —— m **incrustado** - [ind] scaled condenser
check(ing)
— f —— m **sucio** - [ind] dirty condenser
check(ing)
— f — **condición** f - condition check(ing)
— f — **conductor** m - [electr-instal] lead, o
conducor, check(ing)
— f — **conexión** f - [electr-instal] connection,
o wiring, check(ing)
— f —— f **para acumulador** m - [electr-instal]
battery connection check(ing)
— f —— f — **entrada** f - inlet connection
check(ing)
— f —— f — **salida** f - outlet connection
check(ing)
— f —— f **pobre** - poor connection check(ing)
— f — **conjunto** m - [mec] assembly check(ing)
— f —— m **de árbol** m - [mec] shaft assembly
check(ing)
— f —— m **para aguilón** n - [grúas]
boom shaft assembly check(ing)
— f —— m — **cable** m - [grúas] cable
shaft assembly check(ing)
— f —— m —— m **a gancho** m -
[grúas] whip shaft assembly check(ing)
— f — m —— m **a motón** m - [grúas]
main shaft assembly check(ing)
— f —— m — **diodo** m - [electr-instal] diode
assembly check(ing)
— f — m — **eje** m - [autom-mec] axle assembly check(ing)
— f — **conmutador** m - [electr-oper] switch
check(ing)
— f —— m **regulador** - [electr-oper] control-
-(ling) switch check(ing)
— f — **contacto** m - [mec] contact check(ing)
— f —— m **de diente** m - [mec] tooth contact
check(ing)
— f — **contaminante** - contaminant check(ing)
— f —— m **en refrigerante** m - [ind] coolant,
o refrigerant, contaminant check(ing)
— f — **continuidad** f - continuity check(ing)
— f — **contragolpe** m—[mec] backlash check(ing)
— f — **cordón** m - [electr-oper] cord check(ing)
— f — m **a línea** f - [electr-oper] line cord
check(ing)
— f — **corriente** f - [electr-oper] current
check(ing)
— f — **corriente** f **en campo** m - [electr-oper]
field current check(ing)
— f — **cumplimiento** m - compliance check(ing)
— f —— m **con norma** f - [ind] standard(s)
compliance check(ing)
— f —— m **para operación** f - [ind]
operating standard(s) compliance check(ing)
— f — **costura** f - [mec] seam check(ing)
— f —— f **empernada** - [mec] bolted seam
check(ing)
— f — **chapa** f - [metal-lam] strip check(ing)
— f —— f **para identificación** f - [ind] nameplate check(ing)
— f — **chatarra** f - [metal-prod] scrap
check(ing)
— f — **chaveta** f - [mec] cotter pin check(ing)
— f —— f **desplazada** - [mec] displaced, o
dislocated, cotter pin check(ing)
— f —— f **faltante** - [mec] missing cotter pin
check(ing)
— f — **depósito** m - [ind] tank check(ing)
— f —— m **para aceite** m - [mec] oil tank
check(ing)
— f —— m — **fluido** m - [ind] fluid tank
check(ing)
— f —— m — m **hidráulico** - [petról] hydraulic fluid tank check(ing)
— f — **depurador** m - [ind] purifier, o cleaner,
check(ing)
— f —— m **para aire** m - [mec] air cleaner
check(ing)
— f —— m —— m **para motor** m - [comb.int]
engine air cleaner check(ing)

verificación f de desbaste(s) m - [metal-lam]
bloom, control(ling), o check(ing)
— f — **desempeño** m - performance, verification,
o verifying, o check(ing)
— f — m **de departamento** m - department(al)
performance check(ing)
— f — **desgaste** m - [mec] wear check(ing)
— f — m — **cinta** f - [mec-frenos] lining
wear check(ing)
— f — m — **disco** m - [mec-embrague] plate
wear check(ing)
— f — m —— m **para embrague** m - [mec]
clutch plate wear check(ing)
— f — m — **plato** m - [mec-embrague] plate
wear check(ing)
— f — m —— m **para embrague** m - [mec]
clutch plate wear check(ing)
— f — **devanado** m - [electr-mot] winding
check(ing)
— f —— m **para campo** m - [electr-mot] field
winding check(ing)
— f — **diagrama** m - [ind] diagram check(ing)
— f — m **de conexión(es)** f - [electr-instal]
wiring diagram check(ing)
— f — **diámetro** m - [mec] diameter, check(ing),
o verification,fying
— f — **diente** m - [mec] tooth check(ing)
— f — m **de cojinete** m - [mec] bearing tooth
check(ing)
— f — m —— m **para rotación** f - [grúas]
swing bearing tooth check(ing)
— f — **diodo** m - [electr] diode check(ing)
— f — **dispositivo** m - [ind] device check(ing)
— f — m **eléctrico** - [electr-equip] electrical device check(ing)
— f — **distribuidor** m - [mec] distributor, o
manifold, check(ing)
— f —— m **rociador** - [mec] spray manifold
check(ing)
— f — **efectividad** f - efectiveness check(ing)
— f — f **de entrevista** - [labor] contact,
quality, o effectiveness, check(ing)
— f — **eficiencia** f - efficiency, review, o
check(ing)
— f — **elevador** m - [grúas] hoist check(ing)
— f — m **para aguilón** m - [grúas] boom hoist
check(ing)
— f —— m — **cable** m - [grúas] cable hoist
check(ing)
— f —— m —— m **a gancho** m - [grúas] whip
hoist check(ing)
— f —— m **principal** - [grúas] main hoist
check(ing)
— f — **embarque** m - [transp] shipment, o shipping, check(ing)
— f — **encaminamiento** m - [mec] routing, o
tracking, check(ing)
— f — **engranaje** m - [mec] gear check(ing)
— f — m **para reducción** f - [mec] reducing
gear, o gear reducer, check(ing)
— f — m — f **para rotación** f - [grúas]
swing gear reducer check(ing)
— f — m — **rotación** f - [grúas] swing gear
check(ing)
— f — **entrehierro** m - [mec] air gap check(ing)
— f — **equipo** m - [ind] equipment check(ing)
— f — **estado** m - [comput] status. check(ing),
o update,ting
— f — **estructura** f - [const] structure
check(ing)
— f — **excitador** m - [electr-mot] exciter
check(ing)
— f — **exigencia** f - requirement check(ing)
— f — **fabricante** m - manufacturer's, control,
o check(ing)
— f — **filtro** m - [mec] filter check(ing)
— f — **fluido** m - fluid check(ing)
— f — m **en accionamiento** m - [ind] drive
fluid check(ing)
— f — m — m **para bomba** f - [bombas] pump
drive,ving fluid check(ing)
— f — m — m —— f **triple** m - [petról]
triple pump drive,ving fluid check(ing)

verificación f de fluido m hidráulico -
[petról] hydraulic fluid check(ing)
— f — frecuencia f - [electrón] frequency
check(ing)
— f — f de tono m - [electrón] tone frequency check(ing)
— f — f — — m para salida f - [electrón]
tone output frequency check(ing)
— f — freno m - [mec] brake check(ing)
— f — — m para cable m - [grúas] cable brake
check(ing)
— f — — m — — m a gancho m - [grúas] whip
brake check(ing)
— f — — m — — m — motón f - [grúas]
main brake check(ing)
— f — — m — — m a aguilón m - [Grúas] boom
brake check(ing)
— f — — m — — m principal - [grúas] main
brake check(ing)
— f — — m — — m secundario - [grúas] whip
brake check(ing)
— m — — m — tambor m - [grúas] drum brake
check(ing)
— f — — m — — m para cable m - [grúas]
cable drum brake check(ing)
— m — — m — — m — — m a gancho m—[grúas]
whip drum brake check(ing)
— f — — m principal - [mec] main brake
check(ing)
— f — funcionamiento m - operation check(ing)
— f — — m de bocina f - [segurid] horn('s)
operation check(ing)
— f — fusible m—[electr-oper] fuse check(ing)
— f — — m quemado - [electr-oper] blown fuse
check(ing)
— f — — m saltado - [electr-oper] blown fuse
check(ing)
— f — galvanización f - [metal-trat] galvanization testing
— f — — f en alambre m - [alambre] wire galvanizing testing
— f — gancho m - [grúas] hook check(ing)
— f — garrucha f - [mec] sheave check(ing)
— f — hallazgo m - finding check(ing)
— f — holgura f - [mec] clearance check(ing)
— f — humedad f - moisture check(ing)
— f — — f en recubrimiento m - [sold] coating
moisture check(ing)
— f — — m para electrodo m - [sold]
electrode coating moisture check(ing)
— f — — f — m — — m para soldadura f -
[sold] welding electrode coating moisture
check(ing)
— f — indicación f de carga f excesiva - [ind]
overload sign(s) check(ing)
— f — — f — desgaste m - [mec] wear sign(s)
check(ing)
— f — — f sobrecarga f - [ind] overload
sign(s) check(ing)
— f — indicador m - indicator check(ing)
— f — m de posición f - position indicator
check(ing)
— f — información f - reserach(ing) • [comput]
information, verificación, o check(ing)
— f — informe m - report, check(ing), o review
— f — sobre eficiencia f - efficiency report, o review, o check(ing)
— f — — m sobre metalurgia f - [metal] metallurgy report, review, o check(ing)
— f — — m — producción f - [ind] production
report, review, o check(ing)
— f — inventario(s) m—[ind] inventory control
— f — irregularidad(es) f - irregularity,ties,
control, o check(ing)
— f — — f interna(s) f - internal irregularity,ties, control, o check(ing)
— f — ítem m - item check(ing)
— f — — juego m - [mec] play check(ing)
— f — — m longitudinal - [mec] end play
check(ing)
— f — — m para árbol m - [autom-mec]
shaft end play check(ing)

verificación f de montante m para caballete m
verificación f de juego m longitudinal para
árbol m para aportación f de fuerza f -
[autom-mec] input shaft end play check(ing)
— f — laminación f - [metal-lam] rolling,
control, o check(ing)
— f — línea f - line check(ing)
— f — — f a carburador m - [comb.int] carburetor line check(ing)
— f — líquido m - [ind] liquid check(ing)
— f — — m refrigerante - [comb.int] coolant,
o cooling liquid, check(ing)
— f — — m — para camisa f - [petról] liner
coolant check(ing)
— f — — m — — émbolo m - [petról] piston
coolant check(ing)
— f — — m — — m para camisa f - [ind]
liner piston coolant check(ing)
— f — — m — motor m - [comb.int] engine
coolant check(ing)
— f — limpiador m - [ind] cleaner, o scrubber, check(ing)
— f — — m para aire m - air, cleaner, o
scrubber, check(ing)
— f — — f desconectada - [electr-instal]
open line check(ing)
— f — — f para aire m - [comb.int] air line
check(ing)
— f — lista f - list check(ing)
— f — lubricación f - [mec] lubrication
check(ing)
— f — — f para acoplamiento m - [mec] coupling lubrication check(ing)
— f — — m para accionamiento m -
[mec] drive,ving coupling lubrication check-
-(ing)
— f — — f — m — — m para motor m—[mec]
motor drive,ving coupling lubrication
check(ing)
— f — lubricador m para línea f - [ind] line
lubricator check(ing)
— f — — m — — f neumática - [petról] air
line lubricant check(ing)
— f — lubricante m - [lubric] lubricant
check(ing)
— f — — m para acoplamiento m - [mec] coupling lubricant check(ing)
— f — — f — m — — m para accionamiento m para
motor m - [mec] motor drive,ving coupling
lubricant check(ing)
— f — luz f - [mec] clearance check(ing) •
[ilumin] light check(ing)
— f — mandato m - [comput] command check(ing)
— f — — m mediante retorno m de informe m -
[comput] report back command verification
— f — manómetro m - [instrum] gage check(ing)
— f — — m para motor m - [comb.int] engine
gage check(ing)
— f — — m — operación f apropiada -
[instrum] proper operation gage check(ing)
— f — — m — presión f en manómetro m -
[instrum] evaporator pressure gage check(ing)
— f — marca f - [ind] mark(ing) check(ing)
— f — marcación f - [ind] marking check(ing)
— f — material(es) m - material(s) control
— f — — m para reparación f - [ind] repair
material(s), control, o check(ing)
— f — medidor m - [instrum] meter check(ing)
— f — — m para combustión f - [combust] combustion meter check(ing)
— f — mella(dura)s - [mec] nick check(ing)
— f — metalurgia f - [metal] metallurgy, verifying, o check(ing)
— f — método(s) m - [ind] method(s) check(ing)
— f — — m para ensayo(s) m - test(ing)
method(s) check(ing)
— f — mirilla f - [ind] peep sight check(ing)
— f — — f para compresor m - [ind] compressor (peep) sight (glass) check(ing)
— f — — f — evaporador m - [ind] evaporator
peep) sight (glass) check(ing)
— f — montante m - [grúas] leg check(ing)
— f — — m para caballete m - [grúas] gantry
leg check(ing)

verificación f de motón m - [grúas] block
check(ing)
— f —— m **viajero** - [Grúas] traveling block
check(ing)
— f — **motor** m - [electr-mot] motor check(ing) •
[comb.int] engine check(ing)
— f — **nivel** m - level check(ing)
— f —— m **de aceite** m - [comb.int] oil level
checking
— f —— m —— m **en bomba** f - [bombas] pump
oil level check(ing)
— f —— m —— m —— f **con impulsión** f **con
cadena** f - [bombas] chain driven pump oil
level check(ing)
— f —— m —— m **en caja** f - [mec] case oil
level check(ing)
— f —— m —— m —— f **para cadena** f—[mec]
chain case oil level check(ing)
— f —— m —— m —— **cárter** m - [comb.int]
crankcase oil level check(ing)
— f —— m —— m —— **conjunto** m - [mec] assembly oil level check(ing)
— f —— m —— m —— m **de árbol** m **para
aguilón** m - [grúas] boom shaft assembly oil
level check(ing)
— f —— m —— m —— m —— m **para cable** m
a gancho m - [Grúas] whip shaft assembly oil
level check(ing)
— f —— m —— m —— m —— m —— m a
motón m - [grúas] main shaft assembly oil
level check(ing)
— f —— m —— m —— **lubricador** m **para línea** f
neumática - [grúas] air line lubricator oil
level check(ing)
— f —— m —— m **en motor** m - [comb.int] engine oil level check(ing)
— f —— m —— m **hidráulico** - [mec] hydraulic
oil level check(ing)
— f —— m —— **agua** m - water level check(ing)
— f —— m —— m **acumulador** m - [electr]
battery water level check(ing)
— f —— m — **fluido** m - [mec] fluid level
check(ing)
— f —— m —— m **en accionamiento** m **para bomba** f - [petról] pump drive,ving fluid level
check(ing)
— f —— m —— m **hidráulico** - [petról]
hydraulic fluid level check(ing)
— f —— m —— **líquido** m **refrigerante**—[comb.int]
coolant (liquid) level check(ing)
— f —— m —— m **para motor** m - [comb.int]
engine coolant level check(ing)
— f —— m **de refrigerante** m - [comb.int] coolant level check(ing)
— f —— m —— m **en motor** m - [comb.int]
engine coolant level check(ing)
— f —— m **en depósito** m—tank level check(ing)
— f —— m —— m **hidráulico** m - [petról]
hydraulic oil tank level check(ing)
— f —— m **hidráulico** - [mec] hydraulic level
check(ing)
— f — **núcleo** m - [int.comb] core check(ing)
— f —— m **de radiador** m - [int.comb] radiator
core check(ing)
— f —— m **para tono** m **de salida** f - [electrón]
tone output level check(ing)
— f — **obstrucción** f - obstruction check(ing)
— f —— f **para aire** m - [mec] air obstruction
check(ing)
— f — **ohmímetro** m - [instrum] ohmmeter checking
— f — **operación** f - operation check(ing)
— f —— f **de bocina** f - [segurid] horn('s)
operation check(ing)
— f —— f **debida** - [mec] proper operation
check(ing)
— f — **orden** f - [comput] command check(ing)
— f —— f **mediante retorno** m **de informe** m -
[compu] report back command check(ing)
— f — **página** f - page check(ing)
— f — **par** m **motor** - [mec] torque check(ing)
— f —— m — **para rotación** f - [mec] rolling
torque check(ing)
— f — **parámetro(s)** m - parameter(s) check(ing)

verificación f de pasador m - [mec] pin, o
cotter, check(ing)
— f —— m **faltante** - [mec] missing, pin, o
cotter, check(ing)
— f —— m **para aguilón** m - [grúas] boom pin
check(ing)|
— f —— m **para caballete** m - [grúas] gantry
pin check(ing)
— f —— m —— **montante** m **para caballete** m -
[grúas] gantry leg pin check(ing)
— f — **patrón** m - [mec] pattern check(ing)
— f —— m **para contacto** m - [mec] contact
pattern check(ing)
— f —— m —— m **de diente** m - [mec] tooth
contact pattern check(ing)
— f — **pérdida** f - loss check(ing) • [petrúl]
by-pass check(ing)
— f —— f **de aceite** m - [mec] oil leak
check(ing)
— f — **perforación** f - [metal-fabr] punching
proofing
— f — **perno** m **suelto** - [mec] loose bolt
check(ing)
— f — **peso** m - [mec] weight, check(ing), o
control
— f —— m **de lingote** m - [metal-prod] ingot
weight, control, o check(ing)
— f — **pieza** f - [mec] part check(ing)
— f —— f **conexa** - [mec] related part
check(ing)
— f —— f **para ajuste** m - [mec] fitting
check(ing)
— f — **planilla** f **para ruta** f - [transp] log
check(ing)
— f — **polaridad** f - [electr-oper] polarity
check(ing)
— f — **polea** f - [mec] pulley, o sheave,
check(ing)
— f — **posibilidad** f **de acumulador** m **descargado** - [autom-electr] check(ing) for @ dead
battery
— f —— f **de alambre** m **pelado** - [electr-
-oper] check(ing) for @ bare wire
— f —— f — **circuito** m **sobrecargado** -
[electr-oper] check(ing) for @ overloaded
circuit
— v —— **conductor** m **pelado** - [electr-oper]
check(ing) for @ bare wire
— f —— f — **conexión** f **pobre** - [electr-
-instal] check(ing) for @ poor connection
— f —— f — **cortocircuito** m - [electr-oper]
check(ing) for @ short circuit
— f —— f — **fuga** f - [comb.int] check(ing)
for @ leak
— f —— f —— f **de combustible** m - [comb.-
-int] check(ing) for @ fuel loss
— f —— f **de voltaje** m **alto** - [electr-oper]
check(ing) for @ high voltage
— f —— f —— m **bajo** - [electr-oper]
check(ing) for @ low voltage
— f — **posición** f - position check(ing)
— f —— f **de válvula** f - [válv] valve position check(ing)
— f — **precarga** f - [mec] preload check(ing)
— f —— f **para cojinete** m - [mec] bearing
preload check(ing)
— f — **presión** f **de aire** m - air pressure
check(ing)
— f — **presencia** f **de energía** f **en dispositivo**
m **para cambio(s)** m - [autom-mec] power at
@ shift(ing) unit check(ing)
— f — **presión** f - pressure check(ing) •
[autom-neumát] inflation check(ing)
— f —— f **de aceite** m - [comb.int] oil pressure check(ing)
— f —— f —— m **en depósito** m - [comb.int]
reservoir oil pressure check(ing)
— f —— f — **freno** m - [mec] brake pressure
check(ing)
— f —— f —— m **para pie** m - [mec] foot
brake pressure check(ing)
— f —— f —— m —— m **para línea** f **principal** - [mec] main line foot brake pressure

check(ing)

verificación f de presión f de freno m para pie m sobre cable m a motón m - [mec] main line foot brake pressure check(ing)

— f —— f — m — tambor m para cable m a motón m - [grúas] main drum brake pressure check(ing)

— f —— f en evaporador m - [ind] evaporator pressure check(ing)

— f —— f hidráulica - hydraulic pressure check(ing)

— f —— f — apropiada - [mec] appropriate, o proper, hydraulic presure check(ing)

— f —— f — correcta - [mec] correct hydraulic pressure check(ing)

— f — para inyección f - [bombas] charge pressure, verification, o check(ing)

— f — proyección f - design verification,fying

— f — para regulación f - control(ling) pressure, verification, o check(ing)

— f — producción f - [ind] production review • [electr-prod] output check(ing)

— f — productor m - manufacturer('s) control

— f — progreso m - progress check(ing)

— f — radiador m - [comb.int] radiator check-(ing)

— f —— m exterior - [comb.int] external radiator check(ing)

— f — ranura f - [mec] groove check(ing)

— f —— f en garrucha f - [mec] sheave groove check(ing)

— f —— f — polea f - [mec] sheave groove check(ing)

— f —— f — roldana f - [mec] sheave groove check(ing)

— f — razón f - reason check(ing)

— f — recorrido m - [mec] travel check(ing)

— f —— m de resorte m - [mec] spring travel check(ing)

— f — rectitud mediante luz f - [constr] lamping

— f — recubrimiento m - [metal-trat] coating, check(ing), o testing

— f —— m galvanizado - [metal-trat] galvanized coating, check(ing), o testing

— f —— m en alambre n - [alambre] wire galvanized coating, check(ing), o testing

— f — red eléctrica - [electr-oper] electrical system check(ing)

— f — para aceite m - [mec] oil system check(ing)

— f —— f — enfriamiento m - [comb.int] cooling system check(ing)

— f —— f — refrigeración f - [comb.int] cooling system check(ing)

— f — refrigerante m - [comb.int] coolant, o refrigerant, check(ing)

— f —— m para camisa f - [petról] liner coolant check(ing)

— f —— m émbolo m - [petról] piston coolant check(ing)

— f —— m —— m para camisa f - [petról] liner piston coolant check(ing)

— f —— m — motor m - [comb.int] engine coolant check(ing)

— f — registros m—[contab] records check(ing)

— f — contables - [contab] accounting records check(ing)

— f — regulación f - setting check(ing)

— f —— f para elevación f - [grúas] hoist-(ing) control check(ing)

— f —— f — elevador m para cable m a gancho m - [grúas] whip hoist control check(ing)

— f — regulador m - [mec] control check(ing)

— f —— m para cable m - [grúas] whip hoist control check(ing)

— f —— m —— m a gancho m - [grúas] whip hoist control check(ing)

— f —— m —— m — motón m - [grúas] main hoist control check(ing)

— f —— m para combustión f - [combust] combustion control check(ing)

— f —— m — elevación - [grúas] hoist control check(ing)

verificación f de regulador m para elevación de aguilón m - [grúas] boom hoist control check(ing)

— f —— — f de cable m a gancho m - [grúas] whip hoist control check(ing)

— f —— m — elevador m principal - [grúas] main hoist control check(ing)

— f —— m — marcha f sin carga f - [mec] idler check(ing)

— f —— m para voltaje m - [electrón] voltage regulator check(ing)

— f — resistencia f - [electr-oper] resistance check(ing)

— f — respiradero m - breather check(ing)

— f — resultado(s) m - result(s), o finding, checking

— f — revoluciones f por minuto - [mec] revolutions per minute, o r.p.m., check(ing)

— f — rigidez f - [mec] rigidity, o stiffness, check(ing)

— f —— f para manipulación f - [mec] handling stiffness check(ing)

— f — roldana f - [mec] sheave check(ing)

— f — rotación f - [mec] rotation check(ing) • [grúas] swing check(ing)

— f —— f correcta - [mec] correct rotation check(ing) • [grúas] correct swing check(ing)

— f —— f de compresor m - [mec] compressor rotation check(ing)

— f — rueda f - [mec] wheel check(ing)

— f —— f dentada—[mec] sprocket check(ing)

— f —— f — para cadena f - [mec] chain sprocket check(ing)

— f — ruido m - noise, o sound, check(ing)

— f — rutina f - routine check(ing)

— f — seguridad f - safety check(ing)

— f — selección f - selection check(ing)

— f — sello m - [mec] seal check(ing)

— f — sentido m de rotación f - [mec] rotation check(ing)

— f —— m — f de compresor m - [mec] compressor rotation check(ing)

— f — separador m—[mec] separator check(ing)

— f —— m para agua m - [mec] water separator check(ing)

— f — símbolo m - symbol check(ing)

— f — suciedad f - dirt(iness) check(ing)

— f — sujetador m - [mec] clamp, o fastener, check(ing)

— f —— m para camisa f - [mec] liner clamp check(ing)

— f —— m —— f y vástago m - [mec] liner and rod clamp check(ing)

— f —— m de agua m - water supply check(ing)

— f —— m para condensador m - [ind] condenser water supply check(ing)

— f —— m de combustible m - [comb.int] fuel supply check(ing)

— f — tambor m - [mec] drum check(ing)

— f —— m para cable m - [grúas] (cable) drum check(ing)

— f —— m —— m a gancho m - [grúas] cable drum check(ing)

— f —— m —— m a gancho m - [grúas] whip drum check(ing)

— f —— m —— m — motón m - [grúas] main drum check(ing)

— f — tapón m - [mec] plug check(ing)

— f —— m cónico - [mec] expansion plug check(ing)

— f —— m — en cubierta f para distribuidor m para fuerza f - [autom-mec] power divider cover expansion ring check(ing)

— f —— m para expansión f - [mec] expansion plug check(ing)

— f —— m — en cubierta f para distribuidor m para fuerza f - [autom-mec] power divider cover expansion plug check(ing)

— f — temperatura f - [instrum] temperature check(ing)

— f —— f de aceite m - [comb.int] oil temperature check(ing)

verificación f de temperatura f de coraza f -
[metal-prod] shell temperature check(ing)
— f — tensión f - [mec] tension check(ing)
— f — — f de cadena f - [mec] chain tension
check(ing)
— f — tiesura f - stiffness check(ing)
— f — tolerancia f [mec] tolerance check(ing) •
clearance check(ing)
— f — tono m - [electrón] tone check(ing)
— f — tornillo m - [mec] screw check(ing)
— f — — m para cierre m - [mec] locking screw
check(ing)
— f — — m — — m para tapa f - [mec] cover
locking screw check(ing)
— f — — m — — m — — m para válvula f -
[mec] valve cover locking screw check(ing)
— f — toro m - torus check(ing)
— f — traba f - [mec] lock check(ing)
— f — — f para elevador m - [grúas] boom hoist
lock check(ing)
— f — trabajo m - work, control, o check(ing)
— f — trinquete m - [mec] pawl check(ing)
— f — — m para aguilón m - [grúas] boom pawl
check(ing)
— f — — m elevador m - [grúas] hoist pawl
check(ing)
— f — — m — — m para aguilón m - [grúas]
boom hoist pawl check(ing)
— f — — m — traba f - [mec] lock pawl check-
-(ing)
— f — — m — — f para aguilón n - [grúas]
boom lock pawl check(ing)
— f — — m — — f para elevador m para aguilón
m - [grúas] boom hoist lock pawl check(ing)
— f — tubo m - [tub] tube, o pipe, check(ing)
— f — tuerca f - [mec] nut check(ing)
— f — — f para rueda f - [autom] wheel nut
check(ing)
— f — varilla f - [mec] rod check(ing)
— f — — f medidora (para aceite m)—[comb.int]
dipstick check(ing)
— f — — f — para colector m - [mec] sump dipstick check(ing)
— f — — f — — transmisión f - [mec] transmission dipstick check(ing)
— f — vástago m - [mec] rod check(ing)
— f — velocidad f - [mec] speed check(ing)
— f — — f con referencia f a tierra f - [mec]
ground speed check(ing)
— f — — f de traslación f - [mec] ground speed
check(ing)
— f — ventilador m - [mec] fan check(ing)
— f — vista f - visual check(ing)
— f — — f de garrucha f - [mec] sheave visual
check(ing)
— f — — f — polea f - [mec] sheave visual
check(ing)
— f — — f — roldana f - [mec] sheave visual
check(ing)
— f — ranura f - [mec] groove visual check-
-(ing)
— f — — f — — f en garrucha f - [mec] sheave
groove visual check(ing)
— f — — f — — polea f - [mec] sheave
groove visual check(ing)
— f — — f — — f — roldana f - [mec] sheave
groove visual check(ing)
— f — voltaje m - [electr] voltage check(ing)
— f — — m alto - [electr] high voltage check-
-(ing)
— f — — m bajo - [electr] low voltage check-
-(ing)
— f — — m con placa f para identificación f -
[electr-oper] voltage check(ing) against @
nameplate
— f — — m en salida f - [electr-oper] output
voltage check(ing)
— f — — m — — f de excitador m - [electr-
-prod] exciter output voltage check(ing)
— f — — m — corriente alterna - [electr] alternating current voltage check(ing)
— f — — m — — f continua - [electr- direct
current voltage check(ing)

verificación f de zanja f - [constr] ditch
check(ing)
— f diaria - dialy, check(ing), o review
— f — de equipo m - [ind] draily equipment,
check(ing), o inspecting,tion
— f — — informe(s) m - daily report(s), review, o check(ing)
— f — — m metalúrgico - [metal-prod] daily
metallurgical report, review, o check(ing)
— f — — m sobre eficiencia f - [ind] daily
efficiency report, check(ing), o review
— f directa - direct, o first hand, check(ing)
— f efectuada - [ind] made, o performed, check
— f en busca f de - check(ing) for
— f — — f — agrietamiento m - [mec] check-
-(ing) for @ crack(ing)
— f — — f — alargamiento m - [mec] checking
for @ elongation
— f — — f — alineación f pobre - check(ing)
for @ misalignment
— f — — f — conexión f floja - [electr]
checking for @ loose connection
— f — — f — contaminación f - [ind] check-
-(ing) for @ contamination
— f — — f — corrosión - [metal] check(ing)
for @ corrosion
— f — — f — chaveta f - [mec] check(ing) for
@ cotter pin
— f — — f — — f desplazada - [mec] check-
-(ing) for @, displaced, o dislocated, cotter pin
— f — — f — — f faltante - [mec] check(ing)
for @ missing cotter pin
— f — — f — deformación(es) f - [mec] check-
-(ing) for deformations,mities
— f — — r — desgaste m - [mec] check(ing) for
wear
— f — — f — — m en cojinete m - [mec]
check(ing) for @ bearing wear
— f — — f — deterioro,ración f - check(ing)
for deterioration
— f — — f — flojedad f - check(ing) for
looseness
— f — — f — grieta(s) f - [mec] check(ing)
for @ crack(s)
— f — — f — — f en armadura f - [mec]
check(ing) for @ body crack(s)
— f — — f de pasador(es) m flojo(s) - [mec]
check(ing) for @ loose pin(s)
— f — — f — suciedad f - check(ing) for dirt
— f — caliente m - hot check(ing)
— f — libro(s) m - book check(ing)
— f — — m de pieza(s) f - [ind] part(s) book
check(ing)
— f — playa f - [ind] yard check(ing)
— f específica - [ind] specific check(ing)
— f estadística - [ind] statistical check(ing)
— f — de calidad f - [ind] statistical quality,
check(ing), o control
— f moderna - [ind] modern statistical, control, o check(ing)
— f general - general, check(ing), o control
— f importante - important. check(ing), o control, o inspecting,tion
— f ininterrumpida - uninterrupted, o constant,
check(ing), o control
— f — de calidad f - [ind] constant quality,
check(ing), o control
— f manual - manual, check(ing), o control
— f mediante ensayo(s) m - [ind] proof test(ing)
— f — informe(s) m recuperado(s) - [comput]
report back verification
— f — inspección f - [admin] inspection con-
trol(ling)
— f metalográfica - [metal] metallographic
check(ing)
— f metalúrgica - [metal] metallurgical control
— f moderna - modern control(ling)
— f no destructiva - nondestructive, control, o
check(ing)
— f — — integral - [ind] integral nondestructive, check(ing), o control
— f normal - [contab] normal check(ing)

verificación f nueva - recheck(ing); double check(ing)
— f — **de aceite** m - [lubric] oil recheck(ing)
— f — — **camisa** f - [mec] liner recheck(ing)
— f — — f **y vástago** m - [mec] liner and rod rechecking
— f — — **contragolpe** m - [mec] backlash re-checking
— f — — **sujetador** m—[mec] clamp recheck(ing)
— f — — — m **para camisa** f - [mec] liner clamp recheck(ing)
— f — — — m **para camisa** f **y vástago** m - [mec] liner and rod clamp(s) recheck(ing)
— f — — **vástago** m - [mec] rod recheck(ing)
— f — **proyección** f - design, checking, o veri-fication
— f **parcial** - [ind] random test(ing)
— f **periódica** - periodical check(ing)
— f — **de caja** f - [com] periodic cash check
— f **por administración** f - [admin] management, check(ing), o control(ling)
— f — **concesionarjo** m - [com] dealer('s) check(ing)
— f — **representante** m - [com] dealer('s) check
— f **posible** - possible check(ing)
— f **positiva** - positive verification.
— f **positiva** - positive verification
— f — **para proyección** f - positive design ve-rification
— f **posterior** - later verification; follow-up
— f — a **intrevista** f - [labor] interview fol-low-up
— f **primera** - first, verification, o inspection
— f **química** - [quím] chemical control
— f **reciente** - recent check(ing)
— f **recuperada** f - [comput] reported back veri-fication
— f — **de mandato** m - [comput] reported back command verification
— f — — **orden** f - [comput] reported back com-mand verification
— f **regular** - regular check(ing)
— f **retornada** - [comput] reported back verifi-cation
— f **satisfactoria** - satisfactory verification
— f **según norma** f - standard check(ing)
— f **semanal** - weekly, o once @ week, check(ing)
— f **sorpresiva** - [adm] surprise check(ing)
— f **última** - last, o latest, check(ing)
— f **visual** - visual check(ing) - [comput] visual verification
— f — **de garrucha** f - [mec] visual sheave check(ing)
— f — — **polea** f - [mec] sheave visual check
— f — — **ranura** f - [mec] visual groove check
— f — — **roldana** - [mec] visual sheave check
— f — **final** - [ind] final visual inspection
verificado,da a - verified; inspected; checked (out); tested; proven; examined; looked for; reviwed; determined; confirmed • researched
— a a **intervalo(s)** m **regular(es)** - checked at regular interval(s)
— a **antes de entrega** f—checked before delivery
— a **con** - checked with
— a — **lamparilla** f - [electr] lamp checked
— a — **voltímetro** m - [electr-instrum] volt-meter checked
— a **conforme** - checked O.K.
— a **contra** - checked against
— a **cuidadosamente** - checked carefully
— a **de vista** - checked visually
— a **diariamente** - checked daily
— a **en busca** f **de** - checked for
— a — **caliente** - hot checked
— a **específicamente** - checked specifically
— a **manualmente** - checked manually
— a **mediante ensayo** m - proof tested
— a — **inspección** f - inspection tested
— a **nuevamente** - rechecked; double checked
— a **periódicamente** - checked periodically
— a **por excepción** f - checked by exception
— a **positivamente** - verified positively
— a **recientemente** - recently checked

verificado,da regularmente - checked regularly
— a **semanalmente** - checked, weekly, o once @ week
— a **sólo por excepción** f - checked only by ex-ception
— a **visualmente** - checked visually • [comput] verified visually
verificador m - [ind] controller
— m **de desbastes** m - [metal-lam] bloom con-troller
— f — **lingote(s)** m - [metal-prod] ingot con-troller
— m — **producción** f - [ind] production con-troller
— m — **taller** m - [ind] shop controller
verificador,ra a - [ind] checker; inspector
verificar v - . . . .; to inspect; to investigate; to check against; to control; to ascertain; to evaluate; to review; to look for; to ex-amine • to tally • to determine • to research
— v a **intervalos** m **regulares** - to check at @ regular interval(s)
— v **accesorio** m - [mec] to check @ fitting
— v **accionamiento** m - [mec] to check @ drive
— v **aceite** m - [mec] to check @ oil
— v — m **hidráulico** - [mec] to check @ hydrau-lic oil
— v **aceleración** f - to check @ acceleration
— v **acoplamiento** m - [mec] to check @ coupling
— v **actuación** f - [ind] to check @ performance
— v **acumulador** m - [electr] to check @ battery
— v **agua** m - [hidr] to check @ water
— v **aguilón** m - [grúas] to check @ boom
— v **aire** m - to check @ air
— v **ajuste** m - to check @, adjustment, o set-ting
— v — m **de freno** m - [mec] to check @ brake adjustment
— v — m — **sujetador** m - [mec] to check @ clamp tightness
— v — m — **tornillo** m - [mec] to check @ screw tightness
— v — m — — m **para cierre** m - [mec] to check @ locking screw tightness
— v — **debido** - [mec] to check (for) @ proper adjustment
— v — m **lateral** - [mec] to adjust @ lateral adjustment
— v — m **para cojinete** m - [cojin] to check @ bearing, adjustment, o setting
— v — m **manual** - [mec] to check @ hand tight-ness
— v **alargamiento** m - to check @ elongation
— v **alineación** f - [autom] to check @ alignment
— v — f **angular** - [mec] to adjust @ angular alignment
— v — **desplazada** - [mec] to check @ offset alignment
— v **amortiguador** m - to check @ damp(en)er
— v **amperaje** m - [electr-oper] to check @ am-perage • [sold] to check @ current
— v **antes de entrega** f - [com] to check before @ delivery
— v — — **remachar(se)** v - [mec] to check be-fore riveting
— v **apretadura** f - [mec] to check @ tightness
— v — **de tornillo** m - [mec] to check @ screw tightness
— v **aprovisionamiento** m - to check @ supply
— v — m **de combustible** m - [comb.int] to check @ fuel supply
— v **arandela** f - [mec] to check @ washer
— v **árbol** m - [mec] to check @ shaft
— v **armadura** f - [electr-mot] to check @ arma-ture
— v **arrollamiento** m - [mec] to check @ spooling
— v **asidero** m - [mec] to check @ handle
— v — **para regulador** m - [grúas] to check @ control hanbdle
— v **base** f - [mec] to check @ base
— f **bomba** f - [bombas] to check @ pump
— v — f **con impulsión** f **con cadena** f—[bombas] to check @ chain driven pump

verificar v̲ bomba f̲ hidráulica - [bombas] to check @̲ hydrauic pump
— v̲ boquilla f̲ - [mec] to check @ nozzle
— v̲ — f̲ para rociadura f̲ - [mec] to check @ spray nozzle
— f̲ caballete m̲ - [grúas] to check @ gantry
— v̲ cable m̲ - [cabl] to check @ cable
— v̲ — m̲ a gancho m̲ - [grúas] to check @ whip
— v̲ — m̲ a motón m̲ - [grúas] to check @ main (hoist)
— v̲ cadena f̲ - [mec] to check @ chain
— v̲ — f̲ para accionamiento m̲ - [mec] to check @ drive,ving chain
— v̲ — f̲ —— m̲ para bomba f̲ - [bombas] to check @ pump drive,ving chain
— v̲ caja f̲ - [mec] to check @ case
— v̲ — f̲ para cadena f̲ - [mec] to check @ chain case
— v̲ calidad f̲ - [ind] to, check, o̲ control, @ quality
— v̲ camisa f̲ - [mec] to check @ liner
— v̲ — f̲ y̲ vástago m̲ - [mec] to check @ liner and rod
— v̲ campo m̲ - [electr-instal] to check @ field
— v̲ capacitancia f̲ - to check @ capacitancy
— v̲ carburador m̲ - [comb.int] to check @ carburetor
— v̲ carga f̲ - [mec] to check @ load • [electr-oper] to check @ charge
— v̲ — completa - [mec] to check (for) @ full load • [electr] to check (for) @ full charge
— v̲ — f̲ plena f̲ - [electr] to check (for) @ full charge
— v̲ carrera f̲ - [mec] to check @ travel - [deport] to check @ race
— v̲ caudal m̲ - [hidr] to check @ flow
— v̲ — m̲ de aire m̲ - [ind] to check @ air flow
— v̲ ciclo m̲ - to check @ cycle
— v̲ cilindro m̲ - [mec] to check @ cylinder
— v̲ — m̲ para éter m̲ - [comb.int] to check @ ether cylinder
— v̲ cinta f̲ - [metal-lam] to check @ strip • [mec-frenos] to check @ lining
— v̲ circuito m̲ - to check @ circuit
— v̲ — m̲ con lamparilla f̲ - [electr-instal] to lamp check @ circuit
— v̲ — m̲ — timbre m̲ - [electr-instal] to ringer check @ circuit
— v̲ — m̲ — voltímetro m̲ - [electr-instal] to voltmeter check @ circuit
— v̲ — m̲ de campo m̲ - [electr-mot] to check @ field circuit
— v̲ — m̲ en excitador m̲ - [electr-mot] to check @ exciter field circuit
— v̲ cojinete m̲ - [mec] to check @ bearing
— v̲ — m̲ para motor m̲ - [electr-mot] to check @ motor bearing • [comb.int] to check @ engine bearing
— v̲ — m̲ — rotación f̲ - [grúas] to check @ swing bearing
— v̲ colador m̲ - [mec] to check @ strainer
— v̲ combustión f̲ - [combust] to check @ combustion
— v̲ compatibilidad f̲ - to check @ compatibility
— v̲ compensación f̲ - to check @ compensation
— v̲ con - to check with
— v̲ — amperímetro m̲ - [electr-oper] to check with @ ammeter; to ammeter check
— v̲ — diagrama m̲ - to check against @ diagram
— v̲ — empresa f̲ para energía f̲ - [electr-prod] to check with @ power company
— v̲ — timbre m̲ - [electr-instal] to check with @ ringer
— v̲ condensador m̲ - [ind] to check @ condenser
— v̲ condición f̲ - to check @ condition
— v̲ conductor m̲ - [electr-instal] to check @ lead
— v̲ conexión f̲ - [electr-instal] to check, @ connection, o̲ @ wiring
— v̲ — f̲ con diagrama m̲ - [electr-instal] to check @ wiring against @ diagram
— v̲ — f̲ entre fuente f̲ para energía f̲ y̲ enchufe m̲ - [electr-oper] to check @ power supply and plug connections

verificar v̲ conexión f̲ para acumulador m̲ - [electr] to check @ battery connection
— v̲ — f̲ para entrada f̲ - to check @ inlet connection
— v̲ — f̲ — salida f̲ - to check @ outlet connection
— v̲ — f̲ pobre - [electr-instal] to check @ poor connection
— v̲ — f̲ posible - [electr] t check for @ possible connection
— v̲ conforme - to check O. K.
— v̲ conjunto m̲ - [mec] to check @ assembly
— v̲ — m̲ de árbol m̲ - [mec] to check @ shaft assembly
— v̲ — m̲ —— m̲ para aguilón m̲ - [grúas] to check @ boom shaft assembly
— v̲ — m̲ --- cable m̲—[electr-install to check @ shaft assembly
— v̲ — m̲ — m̲ — m̲ a gancho m̲ - [grúas] to check @ whip shaft assembly
— v̲ — m̲ — m̲ — motón m̲ - [grúas] to check @ main shaft assembly
— v̲ — m̲ — diodo m̲ - [electr] to check @ diode assembly
— v̲ — m̲ para eje m̲ - [autom-mec] to check @ axle assembly
— v̲ conmutador m̲ - [electr-oper] to check @ switch
— v̲ — m̲ regulador m̲ - [electr-oper] to check @ control switch
— v̲ constantemente - to check constantly
— v̲ contacto m̲ - [mec] to check @ contact
— v̲ — m̲ de diente m̲ - [mec] to check @ tooth, contact, o̲ pattern
— v̲ contaminante m̲ - [ind] to check @ contaminant
— v̲ — m̲ en refrigerante m̲ - [ind] to check @, refrigerant, o̲ coolant, contaminant
— v̲ continuidad f̲ - [electr-oper] to check @ continuity
— v̲ contra - to, check, o̲ verify, against
— v̲ — placa f̲ para identificación f̲ - to check against @ nameplate
— v̲ contragolpe m̲ - [mec] to check @ backlash
— v̲ cordón m̲ - [electr-instal] to check @ cord
— v̲ — m̲ a línea f̲ - [electr-oper] tp check @ line cord
— v̲ corriente f̲ - [electr-oper] to check @ current • [hidr] to check @ stream
— f̲ — f̲ en campo m̲ - [electr-instal] to check @ field current
— f̲ — f̲ — m̲ con amperímetro m̲ - [electr] to ammeter check @ field current
— v̲ costo(s) m̲ - to control @ cost(s)
— v̲ costura f̲ - [mec] to check @ seam
— v̲ — f̲ empernada - [mec] to check @ bolted seam
— v̲ cuenta(s) f̲ - [contab] to audit (@ account)
— v̲ cuidadosamente - to check carefully; to double check
— v̲ cumplimiento m̲ - to police • [admin] to assess @, compliance, o̲ performance
— v̲ chapa f̲ - [metal-lam] to check @ strip
— v̲ — f̲ para identificación f̲ - [ind] to check @ nameplate
— v̲ de vista - to check visually
— v̲ —— f̲ polea f̲ - [mec] to visually check @ sheave
— v̲ — — f̲ garrucha f̲ - [mec] to visually check @ sheave
— f̲ —— f̲ ranura f̲ - [mec] to visually check @ groove
— f̲ —— f̲ — f̲ en garrucha f̲ - [mec] to visually check @ sheave groove
— f̲ —— f̲ — f̲ — polea f̲ - [mec] to visually check @ sheave groove
— f̲ —— f̲ — f̲ — roldana f̲ - [mec] to visually check @ sheave groove
— v̲ — — f̲ roldana f̲ - [mec] to visually check @ sheave
— f̲ depósito m̲ - to check @ tank
— f̲ — m̲ para aceite m̲ - to check @ oil tank

verificar v depósito m para aceite m hidráulico
m - to check @ hydraulic oil tank
— f depurador m - to check @ purifier
— v — m para aire m - to check @ air cleaner
— v — m para motor m - [comb.int] to
check @ engine air cleaner
— v desempeño m - to, verify, o assess, o review, @ performance
— v desgaste n - [mec] to check @ wear
— v — m de cinta f - [mec] to check @ lining
wear • [metal-lam] to check @ strip wear
— v — m disco m - [mec] to check @ plate
wear
— v m — plato para embrague m - [mec] to
check @ clutch plate wear
— v — m — m agarre m - [autom-mec]
clutch plate wear
— v devanado m—[electr-mot] to check @ winding
— v — m para campo m - [electr-mot] to check @
field winding
— v diagrama m - to check @ diagram
— v — m de conexión(es) f - [electr-instal]
to check @ wiring diagram
— v diámetro m - [mec] to, verify, o check, @
diameter
— v diariamente - to check daily
— v diente m - [mec] to check @ tooth
— v — m de cojinete m - [mec] to check @ bearing tooth
— v — m — m para rotación f - [grúas] to
check @ swing bearing tooth
— v diodo m - [electr-instal] to check @ diode
— v dispositivo m - [ind] to check @ device
— v — m eléctrico - [electr-oper] to check @
electrical device
— v rociador m - [mec] to check @ manifold
— v — m rociador - [mec] to check @ spray(ing)
manifold
— v elevador m - [grúas] to check @ hoist
— v — m para aguilón m - [grúas] to check @
boom hoist
— v — m — cable m - [grúas] to check @ cable
hoist
— v — — m a gancho m - [grúas] to check @
whip hoist
— v embarque m - [transp] to check @ shipment
— v en busca f de - to check for
— v — caliente - [ind] to hot check
— v — libro m - to check in @ book
— f — m de pieza(s) f - [ind] to check in @
part(s) book
— v — obra f—[constr] to verify at @ job site
— v — sitio m - [ind] to verify at @ site
— v — — m de obra f - [constr] to verify at @
job site
— v encaminamiento m—[mec] to check @ tracking
— v engranaje m - [mec] to check @ gear
— v — m para reducción f - [mec] to check @
gear reducer,ction
— v — m — f para rotación f - [grúas] to
check @ swing gear reducer,ction
— v — v — rotación f - [grúas] to check @
swing gear
— v existencia f - to check @ existence
— v — f de energía f - [ind] to check @ availability of @ power
— v enladrillado m - [constr] to check @ brickwork
— v entrehierro m - [mec] to check @ (air) gap
— v específicamente - [ind] to check specifically
— v estado m - to check @, status, o condition
— v estructura f—[constr] to check @ structure
— v excitador m - [electr] to check @ exciter
— v exigencia f - to check @ requirement
— v filtro m - [mec] to check @ filter
— v finalmente - to final-check
— v fluido m - [ind] to check @ fluid
— v — m en accionamiento m - [mec] to check @
drive,ving fluid
— v — m — m para bomba f - [bombas] to
check @ pump drive,ving fluid
— v frecuencia f - to check @ frequency

verificar f frecuencia f de tono m - [electrón]
to check @ tone frequency
— v — f — — m de salida f - [electrón] to
check @ tone output frequency
— v freno m - [mec] to check @ brake
— v — m — cable m - [grúas] to check @ cable
brake
— v — — — m a gancho m - [grúas] to check
@ whip brake
— v — m — — motón m - [grúas] to check
@ main brake
— v — m — m para aguilón m - [grúas] to
check @ boom cable brake
— v — m — m principal - [grúas] to check
@ main brake
— v — m — m secundario m - [grúas] to
check @ whip brake
— m — m — tambor m - [grúas] to check @ drum
brake
— v — m — m para cable m - [grúas] to
check @ cable drum brake
— v — m — m a gancho m - [grúas] to
check @ whip drum brake
— f — m — — m a motón m - [grúas] to
check @ main (cable) drum brake
— v — m principal - [mec] to check @ main
brake
— v funcionamiento m - to check @ operation
— v — m de bocina f - [segurid] to check @
horn's operation
— v fusible m - [electr-oper] to check @ fuse
— v — m quemado - [electr-oper] to check @
blown fuse
— v — m saltado - [electr-oper] to check @
blown fuse
— v gancho m - [grúas] to check @ hook
— v garrucha f - [mec] to check @ sheave
— v hallazgo m - to check @ find(ing)
— v holgura f - [mec] to check @ clearance
— v indicación f - to check @ sign
— v — de carga f excesiva - to check @ overload sign
— v — — desgaste m - [mec] to check @ wear
sign
— v — f — sobrecarga f - [transp] to check @
overload sign
— v indicador m de posición f - [instrum] to
check @ position, indicator, o sign
— v información f - to, verify, o research, o
check, @ information
— v instrumento m - to check @ instrument
— v irregularidad f - to check @ irregularity
— v — f interna - to check @ internal irregularity
— v item m - to check @ item
— v juego m - [mec] to check @ play
— v — m longitudinal - [mec] to check @ end
play
— v — m — de árbol m - [mec] to check @
shaft end play
— v — m — — — m para aportación f (de
fuerza f) - [autom-mec] to check @ input
shaft end play
— v limpiador m para aire m - [ind] to check @
air cleaner
— v línea f - to check @ line
— v — f a carburador m - [comb.int] to check
@ caraburetor line
— v — f conectada - [electr-instal] to check
@, connected, o closed, line
— v — f desconectada - [electr-instal] to
check @, open, o disconnected, line
— v — f para aire m - [comb.int] to check @
air line
— v líquido m - to check @ liquid
— v — m refrigerante - [comb.int] to check @
coolant
— v — m — para camisa f - [petról] to check
@ liner coolant
— v — m — — émbolo m - [petról] to check @
piston coolant
— v — m — — — m para camisa f - [petról]
to check @ liner piston coolant

verificar v lista f - to check @ list
— v lubricacaión f - to check @ lubrication
— v — f de acoplamiento m - [mec] to check @ coupling lubrication
— v — f — m para accionamiento n - [mec] to check @ drive coupling lubricarion
— f — f — — m — m para motor m - [mec] to check @ motor drive coupling lubrication
— v lubricador m para línea f neumática - [petról] to check @ air line lubricator
— v lubricante m—[lubr] to check @ lubricant
— v — m para acoplamiento m - [mec] to check a coupling lubricant
— v — m — — m para accionamiento m - [mec] to check @ drive,ving coupling lubricant
— v — m — — m para motor m - [mec] to check @ motor drive coupling lubricant
— v luz f - [mec] to check @ clearance
— f mampostería f - [constr] to check @ brick-work
— v manómetro m - [mec] to check @ gage
— v — m para motor m - [comb.int] to check @ engine gage
— v — — presión f - [mec] to check @ pressure gage
— v — m — — f en evaporador m - to check @ evaporator pressure gage
— f manualmente - to check manually
— v marca(ción) f - [ind] to check @ mark(ing)
— v mediante ensayo m - to proof test
— v — inspección f - [admin] to inspection, check, o control
— v medidor m - [mec] to check @ meter
— v — m para combustión f - [combust] to check @ combustion meter
— f mella f - [mec] to check @, nick, o dent
— v metalurgia f - [metal] to, verify, o check, @ metallurgy
— v método m - to check @ method
— v — m para ensayo m - to check @ test method
— v mirilla f - [ind] to check @ (peep) sight (glass)
— v — f en evaporador m - [ind] to check @ evaporator (peep) sight (glass)
— v montante m - to check @ upright • [grúas] to check @ leg
— v — m para caballete m - [grúas] to check @ gantry leg
— v motón m - [grúas] to check @ block
— v — m viajero - [grúas] to check @ traveling block
— v motor m - [electr-mot] to check @ motor • [comb.int] to check @ engine
— v muestra f - to check @ sample
— v nivel m - to check @ level
— v — m de aceite m - [comb.int] to check @ oil level
— v — m — — m en bomba f - [bombas] to check @ pump oil level
— v — m — — — f con impulsión f con cadena f - [bombas] to check @ chain drive(n) pump oil level
— v — m — — — caja f - [mec] to check @ case oil level
— v — m — — m — — f para cadena f - [mec] to check @ chain case oil level
— v — m — — — cárter m - [comb.int] to check @ crankcase oil level
— v — m — — — conjunto m - [mec] to check @ assembly oil level
— v — m — — m de árbol m - [grúas] to check @ shaft assembly oil level
— v — m — — m — — m para aguilón m - [grúas] to check @ boom shaft assembly oil level
— v — m — — m — m — — m para cable m a motón m - [grúas] to check @ main shaft assembly oil level
— v — m — — m — lubricador m - [grúas] to check @ lubricator oil level
— v — m — — m — — m para línea f neumática - [grúas] to check @ air line lubricator oil level

verificar v nivel m de aceite m en motor m - [comb.int] to check @ engine oil level
— v — m — m hidráulico - [mec] to check @ hydraulic oil level
— v — m — agua m - [hidr] to check @ water level
— v — m — — m en acumulador m - [electr] to check @ battery water level
— v — m — fluido m - [bombas] to check @ fluid level
— v — m — — m en accionamiento m para bomba f - [bombas] to check @ pump drive fluid level
— v — m — — m hidráulico - [petról] to check @ hydraulic fluid level
— v — m — líquido m - [mec] to check @ liquid level
— v — m — — m refrigerante - [comb.int] to check @ coolant level
— v — m — — — en motor m - [comb.int] to check @ engine coolant level
— v — m — refrigerante m - [comb.int] to check @ coolant level
— v — m — — m en motor m - [comb.int] to check @ engine coolant level
— f — m en depósito m - to check @ tank level
— v — m — — m para aceite m - [comb.int] to check @ oil tank level
— v — m — — m hidráulico m - [mec] to check @ hydraulic oil tank level
— v — m — — m — fluido m hidráulico - [petról] to check @ hydraulic fluid tank level
— v — m — tono m - [electrón] to check @ tone level
— v — m — — m de salida f - [electrón] to check @ tone ouput level
— v — m hidráulico - [mec] to check @ hydraulic level
— v núcleo m - [mec] to check @ core
— v — m de radiador m - [comb.int] to check @ radiator core
— v nuevamente - to, check again, o double check, o re-check
— v — aceite m - [mec] to recheck @ oil
— v — ajuste m - [mec] to recheck @ tightness
— v — vástago m - [mec] to recheck @ rod
— v obstrucción f - to check @ obstruction
— v ohmímetro m - [instrum] to check @ ohmmeter
— v operación f - [ind] to check @ operation
— v — f de bobina f - [segurid] to check @ horn('s) operation
— v para determinar v - to check, for, o to determine
— v — f debida - [mec] to check for @ proper operation
— v página f - to check @ page
— v par motor m - [mec] to check @ torque
— v — — m para rotación f - [mec] to check @ rolling torque
— v para ajuste m - to check for @ fit
— v — comprobar v - to check to prove
— v — v arrollamiento m apropiado - [mec] to check for @ proper spooling
— v parámetro(s) m - to check @ parameter(s)
— v pasador m - [mec] to check @ pin, o cotter
— v — m faltante - [mec] to check @ missing, pin, o cotter
— v — m para aguilón m - [grúas] to check @ boom pin
— m — m — caballete m - [grúas] to check @ gantry pin
— f — m para montante m - [mec] to check @ upright pin
— v — m — — m para caballete m - [grúas] to check @ gantry leg pin
— v patrón m - to check @ pattern
— v — m para contacto m - [mec] to check @ contact pattern
— v — m — — m para diente m - [mec] to check @ tooth contact pattern
— v pérdida f - to check (for) loss
— v — f de aceite m - to check for @ oil leak

verificar <u>v</u> **periódicamente** - to check periodically
— <u>v</u> **perno** <u>m</u> - [mec] to check @ bolt
— <u>v</u> — <u>m</u> **suelto** - [mec] to check @ loose bolt
— <u>v</u> **peso** <u>m</u> - [mec] to check @ weight
— <u>v</u> **pieza** <u>f</u> - [mec] to check @ part
— <u>v</u> — <u>f</u> **conexa** - [mec] to check @ related part
— <u>v</u> — <u>f</u> **para ajuste** <u>m</u> - [mec] to check @, adjusting part, o fitting
— <u>v</u> **planilla** <u>f</u> - to check @, list, o log
— <u>v</u> **polaridad** <u>f</u> - [electr] to check @ polarity
— <u>v</u> **polea** <u>f</u>—[mec] to check @, pulley, o sheave
— <u>v</u> **por excepción** <u>f</u> - to, check, o control, by exception
— <u>v</u> **posibilidad** <u>v</u> - to check @ possibility
— <u>v</u> — **de acumulador** <u>m</u> **descargado** - [autom-electr] to check for @ dead battery
— <u>v</u> — <u>f</u> — **alambre** <u>m</u> **pelado** - [electr-instal] to check for @ bare wire
— <u>v</u> — <u>f</u> — **circuito** <u>m</u> **sobrecargado** - [electr-oper] to check for @ overloaded circuit
— <u>v</u> — <u>f</u> — **condensador** <u>m</u> **con vicio** <u>m</u> - [ind] to check for @ fouled condenser
— <u>v</u> — <u>f</u> — — <u>m</u> **incrustado** - [ind] to check for @ scaled condenser
— <u>f</u> — <u>f</u> — **conductor** <u>m</u> **pelado** - [electr-oper] to check for @ bare, lead, o wire
— <u>v</u> — <u>f</u> — **conexión** <u>f</u> **pobre** - [electr-instal] to check for @ poor connection
— <u>v</u> — <u>f</u> — **contaminación** <u>f</u> - [ind] to check for @ contaminant
— <u>v</u> — <u>f</u> — **cortocircuito** <u>m</u> - [electr-oper] to check for @ short circuit
— <u>v</u> — <u>f</u> — **fuga(s)** <u>f</u> - to check for @ leak(s)
— <u>v</u> — <u>f</u> — — <u>f</u> **de combustible** <u>m</u> - [comb.int] to check for @ fuel leak(s)
— <u>v</u> — <u>f</u> — **voltaje** <u>m</u>, **alto**, **o elevado** - [electr-oper] to check for @ high voltage
— <u>v</u> — <u>f</u> — <u>m</u> **bajo** - [electr-oper] to check for @ low voltage
— <u>v</u> **posible obstrucción** <u>f</u> **de aire** <u>m</u> - [mec] to check for @ air obstruction
— <u>f</u> **posición** <u>f</u> - to check @ position
— <u>v</u> — <u>f</u> **de válvula** <u>f</u> - [Válv] to check @ valve position
— <u>v</u> **positivamente** - to verify positively
— <u>f</u> **precarga** <u>f</u> - to check @ preload
— <u>v</u> — <u>f</u> **para cojinete** <u>m</u> - [mec] to check @ bearing preload
— <u>v</u> **presión** <u>f</u> - to check @ pressure • [autom-neumát] to check @ inflation
— <u>v</u> — <u>f</u> **de aceite** <u>m</u> - [comb.int] to check @ oil pressure
— <u>v</u> — <u>f</u> — — <u>m</u> **en depósito** <u>m</u> - [comb.int] to check @ reservoir oil pressure
— <u>v</u> — <u>f</u> — **aire** <u>m</u> - to check @ air pressure
— <u>v</u> — <u>f</u> — **bomba** <u>f</u> - [bombas] to, verify, o check, @ pump pressure
— <u>v</u> — <u>f</u> **en evaporador** <u>m</u> - [ind] to check @ evaporator pressure
— <u>v</u> — <u>f</u> — **freno** <u>m</u> - [mec] to check @ brake pressure
— <u>f</u> — <u>f</u> — — <u>m</u> **para pie** <u>m</u> - [mec] to check @ foot brake pressure
— <u>v</u> — <u>f</u> — — <u>m</u> — — <u>m</u> **sobre cable** <u>m</u> **a motón** <u>n</u> - to check @ main line foot brake pressure
— <u>f</u> — — — <u>m</u> — — <u>m</u> — **línea** <u>f</u> **principal** - [mec] to check @ main line foot brake pressure
— <u>v</u> — — — <u>m</u> **sobre tambor** <u>m</u> **para cable** <u>π</u> **a gancho** <u>m</u> - [grúas] to check @ whip drum brake pressure
— <u>v</u> — <u>f</u> — <u>m</u> — — <u>m</u> **para cable** <u>m</u> **a motón** <u>m</u> - [grúas] to check @ main drum brake pressure
— <u>v</u> — <u>f</u> **hidráulica** - to check @ hydraulic pressure
— <u>v</u> — <u>f</u> — **apropiada** - [mec] to check @ appropriae hydraulic pressure
— <u>v</u> — <u>f</u> — **correcta** - [mec] to check @ correct hydraulic pressure
— <u>v</u> — <u>f</u> **para inyección** <u>f</u> - [bombas] to, check, o verify, @ charge pressure
— <u>v</u> — <u>f</u> **para regulación** <u>f</u> - to, verify, o

check, @ control(ling) pressure
verificar <u>v</u> **producción** <u>f</u> - [ind] to check @ production • [electr-prod] to check @ output
— <u>v</u> **programa** <u>m</u> - [ind] to, check, o review, @, program, o schedule
— <u>v</u> **progreso** <u>m</u> - to check @ progress
— <u>f</u> **proyección** <u>f</u> - to verify, o check, @ design
— <u>v</u> **quemador** <u>m</u> - [combust] to check @ burner
— <u>v</u> **radiador** <u>m</u> - [comb.int] to check @ radiator
— <u>v</u> — <u>m</u> **exterior** - [comb.int] to check @ external radiator
— <u>v</u> **ranura** <u>f</u> - [mec] to check @ groove
— <u>v</u> — <u>f</u> **en garrucha** <u>f</u> - [mec] to check @ sheave groove
— <u>v</u> — <u>f</u> — **polea** <u>v</u> - [mec] to check @ sheave groove
— <u>f</u> — <u>f</u> — **roldana** <u>f</u> - [mec] to check @ sheave groove
— <u>v</u> **razón** <u>f</u> - to check (for) @ reason
— <u>v</u> **recorrido** <u>m</u> - [mec] to check @ travel
— <u>v</u> — <u>m</u> **de resorte** <u>m</u> - [mec] to check @ spring travel
— <u>v</u> **rectificador** <u>m</u> - [electr-equip] to check @ rectifier
— <u>f</u> — <u>m</u> **para puente** <u>m</u> - [electr-p½er] to check @ bridge rectifier
— <u>v</u> **red** <u>f</u> - [electr-instal] to check @ system
— <u>v</u> — <u>f</u> **eléctrica** - [electr-instal] to check @ electrical system
— <u>v</u> — <u>f</u> **para aceite** <u>m</u> - [mec] to check @ oil system
— <u>v</u> — <u>f</u> — **enfriamiento** <u>m</u> - [comb.int] to check @ cooling system
— <u>v</u> — <u>f</u> — **refrigeración** <u>f</u> - [comb.int] to check @ cooling sysem
— <u>v</u> **refrigerante** <u>m</u> - [ind] to check @ coolant
— <u>v</u> — <u>m</u> **para camisa** <u>f</u> - [petról] to check @ liner coolant
— <u>v</u> — <u>m</u> **para émbolo** <u>m</u> - [petról] to check @ piston coolant
— <u>v</u> — <u>m</u> — — <u>m</u> **para camisa** <u>f</u> - [petról] to check @ liner piston coolant
— <u>v</u> **regulación** <u>f</u> - [ind] to check @ setting
— <u>v</u> — <u>f</u> **para cable** <u>m</u> - [grúas] to check @, hoist, o cable, control(ling)
— <u>f</u> — <u>f</u> — — <u>m</u> **a gancho** <u>m</u> - [grúas] to check @ whip hoist control(ling)
— <u>v</u> — <u>f</u> **para elevación** <u>f</u> - [grúas] to check @ hoist(ing) control(ling)
— <u>v</u> **regulador** <u>m</u> - [mec] to check @ control(ler)
— <u>v</u> — <u>m</u> **para cable** <u>m</u> - [grúas] to check @ hoist(ing) control(ler)
— <u>v</u> — <u>m</u> — — <u>m</u> **a gancho** <u>m</u> - [grúas] to check @ whip hoist(ing) control(ler)
— <u>v</u> — <u>m</u> **para combustión** <u>f</u> - [combust] to check @ combustion control(ler)
— <u>v</u> — <u>m</u> — **elevación** <u>f</u> - [grúas] to check @ hoisting control(ler)
— <u>v</u> — <u>m</u> — <u>f</u> **para aguilón** <u>m</u> - [grúas] to check @ boom hoist(ing) control(ler)
— <u>v</u> — <u>m</u> — **elevador** <u>m</u> - [grúas] to check @ hoist control(ler
— <u>v</u> — <u>m</u> — <u>m</u> **principal** - [grúas] to check @ main hoist(ing) control(ler)
— <u>v</u> — <u>m</u> — **marcha** <u>f</u> **sin carga** <u>f</u> - [mec] to check @ idler
— <u>v</u> — <u>m</u> — **voltaje** <u>m</u> - [electrón] to check @ voltage control(ler)
— <u>v</u> **regularmente** - to check regularly
— <u>v</u> **resistencia** <u>f</u> - [electr-oper] to check @ resistance
— <u>v</u> **respiradero** <u>m</u> - to check @ breather
— <u>v</u> **resultado** <u>m</u> - to check, @ result(s), o find(ing)
— <u>v</u> **revolución(es)** <u>f</u> - [mec] to check @ revolution(s)
— <u>v</u> — <u>m</u> **por minuto** <u>n</u> - [ind] to check @ revolution(s) per minute
— <u>v</u> **rigidez** <u>f</u> - to check @ stiffness
— <u>v</u> — <u>f</u> **para manipulación** <u>f</u> - [mec] to check @ handling stiffness
— <u>v</u> **roldana** <u>f</u> - [mec] to check @ sheave
— <u>v</u> **rotación** <u>f</u> - [mec] to check @ rotation •

[guías] to check @ swing
verificar v **rotación** f **correcta** - to check @
correct rotation
— v — f **de compresor** n - [mec] to check @ compressor rotation
— v **rueda** f - [mec] to check @ wheel
— f — f **dentada** - [mec] to check @ sprocket
— f — f — **para cadena** f - [mec] to check @
chain sprocket
— v **ruido** m - to check @ noise
— v **seguridad** - [segurid] to check @ safety
— v **selección** v - to check @ selection
— v **sello** m - [mec] to check @ seal • to look
for @ seal
— v **semanalmente** - to check, weekly, o once
each week
— v **sentido** m **de rotación** f to check for @ rotation
— v — m — — f **correcto** - [mec] to check @
correct rotation
— v — m — — f **de compresor** - [mec] to check
@ compressor rotation
— v **separador** n - [mec] to check @ separator
— v — m **para agua** m - [mec] to check @ water
separator
— v **símbolo** m - to check @ symbol
— v **sólo por excepción** f - to control (only) by
exception
— v **suciedad** f - to check @, dirt, o filth
— v **sujetador** m - [mec] to check @ clamp
— v — m **para camisa** f - [mec] to check @ liner
clamp
— v — m — — f **para vástago** m - [mec] to
check @ liner rod clamp
— v **suministro** m - to check @ supply
— v — **de agua** m - to check @ water supply
— v — — — m **para condensador** m - [ind] to
check @ condenser water supply
— v — m **de combustible** m - [comb.int] to check
@ fuel supply
— v **tambor** m - [mec] to check @ drum
— v — m **para cable** m - [grúas] to check @
cable drum
— v — m — — m **a gancho** m - [grúas] to check
@ whip drum
— v — m — — m **a motón** m - [grúas] to check @
main drum
— v **tapón** n - [mec] to check @ plug
— v — m **cónico** - [mec] to check @ expansion
plug
— v — m — **en cubierta** f - [autom-mec] to
check @ cover expansion, ring, o plug
— v — m — — m — — f **para distribuidor** m -
[autom-mec] to check @ divider cover ring
— v — m — — — f **para distribuidor** m **para**
fuerza f - [autom-mec] to check @ power divider cover expansion, ring, o plug
— v — m **para expansión** f - [mec] to check @
expansion, ring, o plug
— , **temperatura** f - to check @ temperature
— v — f **de aceite** m - [comb.int] to check @
oil temperature
— v **tensión** f - [mec] to check @ tension •
[electr-oper] to check @ voltage
— v — f **en cadena** f - [mec] to check @ chain
tension
— v **tiesura** f - [mec] to check @ stiffness
— v **tolerancia** f - to check @, tolerance, o
variation • [mec] to check @ clearance
— v **tono** m - [electrón] to check @ tone
— v **tornillo** m - [mec] to check @ screw
— v — m **para cierre** m - [mec] to check @ locking screw
— v — m — — m **para tapa** f - [mec] to check @
cover locking screw
— v — m — — m — — f **para válvula** f - to
check @ valve cover locking screw
— v **traba** f - [mec] to check @, clamp, o lock
— v — f **para elevador** m - [grúas] to check @
hoist lock
— v — m — — m **para aguilón** m - [grúas] to
check @ boom hoist lock
— v **trinquete** m - [mec] to check @ pawl

verificar v **trinquete** m **para aguilón** m -
[grúas] to check @ boom pawl
— v — m — **elevador** m - [grúas] to check @
hoisting pawl
— v — m — — m **para aguilón** m - [grúas] to
check @ boom hoist pawl
— v — m **para traba** f - [mec] to check @ lock
pawl
— v — m — — f **para aguilón** m - [grúas] to
check @ boom lock pawl
— v — m — — f — **elevador** m **para aguilón** m
- [grúas] to check @ boom hoist lock pawl
— f **tubería** v - [tub] to check @ pipe,ping
— v **tuerca** f - [mec] to check @ nut
— f — f **para rueda** f - [autom] to check @
wheel nut
— v **ubicación** f - [mec] to check @ location
— v — f **para arandela** f - [mec] to check @
washer location
— v **varilla** f - [mec] to check @ rod
— v — f **medidora (para aceite** m) - [comb.int]
to check @ (oil) dipstick
— v — f — **colector** m - [mec] to check @
sump dipstick
— v — f — — **transmisión** f - [mec] to check
@ transmission dipstick
— v **vástago** m - [mec] to check @ rod
— v **velocidad** f - [mec] to check @ speed
— v — f **con referencia** f **a tierra** f - [mec]
to check @ ground speed
— v — f **de traslación** f - [mec] to check @
ground speed
— v **ventilador** m - [mec] to check @ fan
— v **visualmente**—to, check, o verify, visually
— v — **garrucha** f - [mec] to visually check @
sheave
— v — **polea** f - [mec] to visually check @
sheave
— v — **ranura** f - [mec] to visually check @
groove
— v — — f **en garrucha** f - [mec] to visually
check @ sheave groove
— v — — f — **polea** f - [mec] to visually
check @ sheave groove
— v — — f — **roldana** f - [mec] to visually
check @ sheave groove
— v — **roldana** f - [mec] to visually check @
sheave
— v **voltaje** m—[electr-oper] to check @ voltage
— v — m **alto** - [electr-oper] to check @ high
voltage
— v — m **bajo** - [electr-oper] to check @ low
voltage
— v — m **contra placa** f **para identificación** f
- [sold] to check @ voltage against @ nameplate
— v — m **de salida** f - [electr-prod] to check @
output voltage
— v — m — — f **en excitador** m - [electr-prod]
to check @ exciter output voltage
— v — m **en corriente** f **alterna** - [electrón] to
check @ alternating current voltage -
— v **zanja** f - [constr] to check @ ditch
verosímil a - realistic; likely
versación f - versing
versar v - to verse • to deal with
— v **sobre** - to deal with
versatilidad f **de proyección** f - design versatility
— f **mayor** - greater versatility
— f — **colocación** f **de accesorio(s)** m - versatile fixturing
verse v **apremiado** - to have all one can handle
— v **desplazado,da** - [eport] to be headed
— v **detenido,da** - [deport] to be stopped
— v **eliminado,da** - to fall by @ wayside
— v **embestido,da** - [autom] to be tagged
— v **obligado,da** - to have to; to be obliged to
versión f - . . . . • type • [public] edition
— f **condensada** - condensed version
vertedera f - [constr] moldboard
vertedero m - [ind] . . . .; spout; outfall; apron
• dump; chute • [metal-prod] pouring nozzle

[metal-prod] cone bottom • pouring box ••
  runner • [agric-equip] moldboard
vertedero ɯ **abierto** - [hidr] open spillway
— m **auxiliar** - [hidr] auxiliary spillway
— m **basculante** - [metal-prod] tilting chute
— m **con sumidero** <u>m</u> - [hidr] drop inlet spillway
— m —— m **tubular** - [hidr] pipe drop inlet
  <u>spillway</u>
— m —— m — **de metal** <u>m</u> - [hidr] metal pipe
  <u>drop inlet spillway</u>
— m —— m —— — m **corrugado** - [hidr] cor-
  <u>rugated metal pipe drop inlet spillway</u>
— m — **zanja** <u>f</u> - [hidr] ditch spillway
— <u>m</u> **conformado** - [miner] shaped dump(ing) area
— <u>m</u> **corrugado** - [hidr] corrugated, spillway, <u>o</u>
  chute
— m **de acero** <u>m</u> - [ind] steel, spillway, <u>o</u> chute
— <u>m</u> —— m **corrugado** - [constr] corrugated
  <u>steel, spillway, o chute</u>
— m —— m **en plancha(s)** <u>f</u> - [hidr] plate
  <u>steel, o steel plate, spillway</u>
— <u>m</u> — **hormigón** m - [hidr] concrete spillway
— <u>m</u> —— **plancha(s)** <u>f</u> - [hidr] plate spillway
— <u>m</u> —— f **de acero** <u>m</u> - [hidr] steel plate(s)
  <u>spillway</u>
— v —— f **estructural(es)** - [hidr] structural
  <u>plate steel spillway</u>
— m —— f —— — **curva(s)** - [hidr] curved
  <u>structural steel plate(s) spillway</u>
— m —— f — **de acero** <u>m</u> - [hidr] structural
  <u>plate steel spillway</u>
— <u>n</u> — **roca** f - [hidr] rock spillway
— <u>n</u> — **tubería** <u>f</u> - [hidr] pipe,ping spillway
— <u>m</u> —— f **corrugada** - [constr] cprrugated
  <u>pipe,ping spillway</u>
— <u>m</u> —— f **de acero** <u>m</u> - [hidr] steel pipe,ping
  <u>spillway</u>
— m —— f —— m **corrugado** - [hidr] corru-
  <u>gated steel pipe,ping spillway</u>
— m — **tubo(s)** <u>m</u> [hidr] pipe spillway
— <u>m</u> **descubierto** - [hidr] open spillway
— <u>m</u> **en descubierto** - [hidr] open spillway
— <u>v</u> — **V** - [hidr] V, trough, <u>o</u> spillway
— <u>m</u> **girado** - [mec] swing spout
— <u>m</u> **inclinado** - [hidr] chute
— m **lateral** - [hidr] lateral, o side, spillway
— <u>m</u> **macizo** - [agric-equip] solid moldboard
— <u>m</u> **para arrabio** ɯ - [metal-prod] (pig) iron, <u>o</u>
  <u>hot metal, runner</u>
— m — **banqueta** <u>f</u> - [vial] shoulder spillway
— <u>m</u> — **basura(s)** <u>f</u> - [sanit] garbage dump
— <u>m</u> — **berma** <u>f</u> - [vial] shoulder spillway
— <u>m</u> —— f **para supercarretera** <u>f</u> - [vial] ex-
  <u>pressway shoulder spillway</u>
— m — **cuchara** <u>f</u> - [metal-prod] ladle trough
— <u>m</u> — **descarga** <u>f</u> - [mec] discharge chute
— <u>m</u> —— f **para lingoteadora** <u>f</u> - [metal-prod]
  <u>pig casting machine discharge chute</u>
— m —— f **para máquina** <u>f</u> **para colar** <u>v</u> -
  [metal-prod] pig casting machine discharge
  chute
— m — **emergencia** <u>f</u> - [hidr] emergency spillway
— <u>m</u> — **entrega** <u>f</u> - [mec] delivery chute
— <u>m</u> —— f **para rompedora** <u>f</u> - [miner] breaker
  <u>delivery chute</u>
— m — **escoria** <u>f</u> - [metal-prod] slag chute
— <u>m</u> — **forraje** <u>m</u> - [agric-equip] forage spout
— <u>m</u> — **franja** <u>f</u> **medianera** - [vial] median
  <u>spillway</u>
— m —— f — **para supercarretera** <u>f</u> - [vial]
  <u>expressway median spillway</u>
— m — **lago** <u>m</u> - [hidr] lake spillway
— <u>m</u> — **lingoteadora** <u>f</u> - [metal-prod] pig cast-
  <u>ing machine chute</u>
— m — **máquina** <u>f</u> **para colar** - [metal-prod] pig
  <u>casting machine chute</u>
— m — **rompedora** <u>f</u> - [miner] breaker chute
— <u>m</u> — **segadora** <u>f</u> **(para forraje** <u>m</u>**)** - [agric-
  <u>-equip] (forage) harvester spout</u>
— m — **tajadora** <u>f</u> - [agric-equip] chopper spout
— <u>m</u> —— f **y canalón** <u>m</u> - [constr] pipe and
  <u>flume spillway</u>
— <u>m</u> **principal** - [constr] main spillway

vertedero <u>m</u> **proyectado** - [Constr] designed chute
— <u>m</u> **triple** - [hidr] triple (line) spillway
— <u>m</u> **tubular** - [constr] tubular, <u>o</u> pipe, spillway
— <u>m</u> — **con caída** <u>f</u> - [constr] drop pipe spillway
— <u>f</u> — **de metal** <u>m</u> - [tub] metal pipe spillway
— <u>m</u> — ⌐ **corrugado** - [constr] corrugated
  <u>metal pipe spillway</u>
— <u>m</u> —— m — **para drenaje** <u>m</u> - [constr]
  <u>corrugated metal drain pipe spillway</u>
— <u>m</u> — **metálico** - [constr] metal pipe spillway
— <u>m</u> —— **para drenaje** <u>m</u> - [constr] metal drain
  <u>pipe spillway</u>
— <u>m</u> — **para drenaje** <u>m</u> - [constr] drain pipe
  <u>spillway</u>
— m **medidor** - [hidr] weir
verter <u>v</u> - . . . • [sold] to fall away • [filol]
  to translate
— <u>v</u> **agua** <u>m</u> - to, spill, <u>o</u> shed, @ water
— <u>v</u> **cauce** <u>m</u> - [constr] to pour @ channel
— <u>v</u> **dentro de molde** <u>m</u> - [metal-prod] yo, teem, <u>o</u>
  <u>pour into @ mold</u>
— <u>v</u> — **caliente** - [metal] to pour hot
— <u>v</u> — **molde** <u>m</u> - [metal-prod] to, teem, <u>o</u> pour
  <u>into @ mold</u>
— v **hormigón** <u>m</u> - [constr] to pour @ concrete
— <u>v</u> **líquido** <u>m</u> - to pour @ liquid
— <u>v</u> **metal** <u>m</u> - [metal-prod] to pour @ metal
vertical <u>a</u> - . . . • upright • [tub] drop inlet
— <u>a</u> **tope** <u>a</u> - [sold] vertical butt
— <u>v</u> — **con chaflán** <u>a</u> - [sold] vertical butt
  <u>beveled</u>
— <u>a</u> —— m —— m **en V** - [sold] vertical
  <u>V-butt beveled</u>
— <u>a</u> —— m —— m —— **de pasada** <u>f</u> **primera** -
  [sold] first pass vertical V-butt
— <u>a</u> —— m **con ranura** <u>f</u> - [sold] vertical
  <u>V-pass butt</u>
— <u>a</u> —— m —— f **con pasada** <u>f</u> **única** - [sold]
  <u>single pass vertical V-pass butt</u>
— <u>a</u> **achatado** - [sold] vertical flat
— <u>a</u> **ascendente** - [sold] vertical-up
— <u>a</u> — **con tejido** <u>m</u> **triangular** - [sold] trian-
  <u>gular weave vertical up fillet</u>
— <u>a</u> — **triangular** - [sold] triangular vertical
  up; vertical up triangular
— <u>a</u> **con ángulo** <u>m</u> **interior** - [sold] single pass
  <u>vertical fillet</u>
— <u>a</u> —— m — **con una (sóla) pasada** - [sold]
  <u>single pass vertical fillet</u>
— <u>a</u> **con dos líneas** <u>f</u> - [metal-prod] two-strand
  <u>vertical</u>
— <u>a</u> **de pasada** <u>f</u> **primera** - [sold] first pass ver-
  <u>tical fillet</u>
— <u>a</u> — **piquera** <u>f</u> - [metal-prod] vertical over @
  <u>iton notch</u>
— <u>a</u> — **tobera** <u>f</u> - [metal-prod] vertical over @
  <u>tuyere</u>
— <u>a</u> — **tope** <u>m</u> **en V** - [sold] vertical V-butt
— <u>a</u> —— m —— ī **de pasada** <u>f</u> **primera** - [sold]
  <u>first pass vertical V-butt</u>
— — <u>a</u> — **tubo(s)** <u>m</u> - [comb.int] vertical tube
— — <u>a</u> —— m **achatado(s)** - [comb.int] vertical
  <u>flat tube</u>
— <u>a</u> **descendente** - [sold] vertical down
— <u>a</u> — **en ángulo** <u>m</u> **interior** - [sold] vertical
  <u>down fillet</u>
— <u>a</u> —— m — **con pasada** <u>f</u> **única** - [sold]
  <u>single pass vertical down fillet</u>
— <u>a</u> — **triangular** - [sold] vertical down triangular
verticalidad <u>f</u> **de tablestacado** <u>m</u> - [constr-pil]
  sheetpiling verticality
vértice <u>f</u> - . . . • corner • [arquit] vortex
— <u>f</u> **aguda** - sharp corner
— <u>f</u> **de cabeza** <u>f</u> **de riel** <u>m</u> - [metal-lam] rail
  <u>head corner</u>
— — <u>f</u> — **cordón** <u>m</u> - [sold] head heel
— <u>f</u> — **junta** <u>f</u> - [sold] corner of @ joint
— <u>f</u> — **plancha** - [sold] weld corner
— <u>ī</u> — **soldadura** <u>f</u> - [sold] wele heel
vertido <u>m</u> - effluent; residual liquid • dumping
vertido,da <u>a</u> - poured (out) • tilted • dumped •
  [sold] fallen away

vertido a en caliente   - 1726 -

vertido,da en caliente - [constr] poured hot
— a en frío - [constr] poured cold
— a en molde m - [metal-prod] poured into @
  mold
vertiente f - . . . • [topogr] bowl shaped area;
  watershed • [constr] eave(s). • [hidr] spring
— f lateral - [topogr] side slope
— f para tejado m - [constr] eave(s)
vertiginoso,sa a - . . . . • [fiql] torrid
vertimiento m - . . . • pouring out • dumping •
  spillage
— m de agua m - water, shedding, o spilling, o
  pouring (out)
— m — cauce m - [constr] channel pouring (out)
— m — líquido m - liquid, dumping, o pouring
— m — metal m - [metal-prod] metal pour(ing)
— m dentro de molde m - [metal-prod] pouring
  into @ mold
vesánico,ca a - . . .; demented
vestíbulo m de hotel m - [constr] hotel lobby
vestido m para diario - [vest] daytime dress
— m — ambiente m - outside weather clothing
— m — azufre - [miner] sulfur trace
— m — carbono m - [quím] carbon trace
— m — fósforo m - [miner] phosphorus trace
— n — humo m - [combust] smoke trace
— m menudo - minute trace
vestimenta f - . . .; clothing; attire • dress-
  -(ing) (code)
— f apropiada - [vest] appropriate dress(ing) •
  dress(ing) code
vestuario m - . . . . • [ind] dressing, o locker,
  room • change house
veta f de arena f--[geol] sand, streak, o stripe
— f — carbón m - [miner] coal seam
— f — madera f - [mader] wood, vein, o grain
veteado,da - [miner] banded • [mader] grained
veterano,na a - . . . .; old time; elder(ly)
— a en carrera(s) f - [deport] racing veteran
— a — — f de regularidad f - [deport] veteran
  rallyist
vetusto,ta a - . . .; veteran • obselete • antique • rickety
vez,ces f cada año - time(s) @ year
— f anterior - last, o previous, time
— f primera - first time
— f próxima - next time
— f segunda - second time
— f tras vez f—over and over, o time and, again
vía f - . . . • [vial] lane • runway • [transp]
  vía; by way of • route • [hidr] opening •
  [legal] copy
— v administrativa - [admin] by administration;
  administrative procedure
— f ancha - [f.c.] wide, o broad, gage
— f angosta - [f.c.] narrow gage (track)
— f arterial - [vial] arterial highway
— f auxiliar - [vial] auxiliary lane
— f — para tránsito m - [vial auxiliary traffic
  lane
— f bajo cielo m abierto - [f.c.] outdoor(s)
  track
— f cambiada - [vial] changed (traffic) lane •
  [f.c.] changed track
— f central - [f.c.] central (running) track
— f combinada para transporte m - [transp] common roadway • taffic corridor
— f con cremallera f - [f,c,] rack railway
— f — dirección f este - [vial] eastbound lane
  • [f.c.] eastbound track
— f — f norte - [vial northbound lane •
  [vial] northbound lane • [f.c.] northbound
  track
— f — f oeste - [vial] westbound lane •
  [f.c.] westbound track
— f — f sud - [vial] southbound lane •
  [f.c.] southbound track
— f de agua m - [nav] waterway • [hidr] water
  flow
— f — comparación f - way of comparison
— f — desarrollo m - [econ] developing
— f — rodillo(s) m - [mec] roller track
— f — m para plataforma f giratoria - [mec]

turntable roller, path, o track
vía f decauville - [f.c.] narrow gage track
— f despareja - [f.c.] uneven, o rough, track
— f directa - direct way • [f.c] through, o direct, track
— f doble - [f.c.] double, line, o track
— f elevada - [f.c.] raised track
— f en sección(es) f - [f.c.] panel track
— f — f prearmadas - [f.c.] panel track
— f este - [f.c.] east track
— f estrecha - [f.c.] narrow (gage) track
— f existente - [f.c.] existing track(work)
— f exterior - [f.c.] outside, o outdoor,
  track
— f férrea - [f.c.] railway; railroad (track)
— f — con trocha f ancha - [f.c.] wide, o
  broad, gage railroad track
— f — soportada - [constr] supported (railway)
  track
— f sostenida - [constr] supported track
— f ferroviaria - [f.c.] railroad, o railway,
  track, o line
— f — para arrabio m caliente - [metal-prod]
  hot metal railroad track
— f — — trene(s) m con arrabio m caliente -
  [metal-prod] hot metal railroad track
— f — peligrosa - [f.c.] dangerous railway
  track
— f fluvial - [hidr] waterway | a - [hidr] by
  river
— f hacia este m - [f.c.] eastbound track
— f — norte m - [f.c.] northbound track
— f — oeste m - [f.c.] westbound track
— f — sur m - [f.c.] southbound track
— f húmeda f - [ind] wet, method, o process •
  [f.c.] wet track
— f — de rutina f - [quím] routine wet method
— f interior - [f.c.] inside track
— f jardín - [vial] parkway
— f legalizda - [legal] legalized copy
— f levantada - [constr] raised track
— f marítima - sea
— f muerta - [f.c.] spur; stub, o dead, end
  (track)
— f múltiple - [f.c.] multiple track(s)
— f navegable - [hidr] navigable waterway
— f norte - [f.c.] north track
— f oeste - [f.c.] west track
— f oral - by way of mouth
— f para acero m - [f.c.] access, o lead, track
— f — apagado,gamiento m - [coque] quenching (car) track
— f — arrabio n - [metal-prod] pig iron track
— f — — m líquido - [metal-prod] hot metal
  track
— f — carrito m para lingote(s) m - [metal-
  -pro] ingot chariot track
— f — circulación f - [vial] (traffic) lane
— f — — f cambiada - [vial] changed traffic
  lane
— f — clasificación f - [f.c.] classification
  track
— f — conexión f - [vial] connecting lane
— f — drenaje m - [hidr] drainage opening
— f — cuchara f - [metal-prod] ladle (car)
  track
— d — deshornadora f - [coque] pusher track
— f — distribuidor m - [metal-prod] larry
  (car) track
— f — enlace m - [f.c.] crossover
— f — expedición f - [f.c.] departing track
— f — formación f - [f.c.] make-up, o forwarding, track
— f — — f de tren(es) m - [f.c.] make-up, o
  forwarding, track
— f — fosa f - [deport] pit, track, o lane •
  [f.c.] (soaking) pit track
— f — grúa f - [grúa] crane runway
— f — maniobras f - [f.c.] switch(ing), o
  drill, track
— f — máquina f sacapuertas - [metal-prod]
  door machine track
— f — permanencia f - [ind] holding track

vía f para reposo m - [metal-prod] holding track
— f — rodillo(s) m - [mec] roller path
— f — retardación f - [metal-prod] retarding
track
— f — — f eléctrica - [f.c.] electric retarding track
— f — rodadura f - [mec] runway
— f — rodamiento m - [grúas] runway
— f — — m ancha - [grúas] wide (span) runway
— f — — m para grúa f - [grúas] crane runway
— f — salida f - [f.c.] outgoing track
— f — substitución f - [vial] detour
— f — — f para velocidad(ds) alta(s) - [vial]
high speed detour
— f — tránsito m - [vial] traffic lane
— f — — m con acceso m limitado - [vial] limited access, lane, o facility
— f — — m (más) veloz - [vial] fast lane
— f — — vagón(es) m - [f.c.] car track
— f — — m para apagamiento m - [coque] quench-
-(ing) car track
— f — — m portacuchara(s) - [metal-prod] ladle
car track
— f — — m distribuidor - [metal-prod] larry
(car) track
— f — vagoneta(s) f - [f.c.] narrow gage, railway, o track
— f — — f para lingote(s) m - [metal-prod] ingot chariot track
— f — — f para montacarga(s) m - [metal-prod]
skip car track
— f paralela - [f.c.] parallel track
— f peligrosa - [f.c.] dangerous track
— f permanente - [f.c.] right of way
— f principal - [vial] arterial highway - [f.c.]
mainline track
— f pública - [vial] highway; public road(way)
— f química - [quím] chemical process
— f recta - [f.c.] straight track
— f — paralela - [f.c.] straight parallel track
— f seca - [f.c.] dry track • [ind] dry process
— f sostenida - [f.c.] supported track
— f suburbana f - [vial] suburban highway
— f sur - [f.c.] south(ern) track
— f terrestre - [transp] land
— f transitada - [vial] traveled (high)way
— f única - [vial] single lane • [f.c.] single
track
— f urbana - [vial] urban highway
— f variable - [f.c.] varying track
viabilidad f de construcción f - [constr] construction feasibility
— f — proyecto m - project feasibility
— f — — m para explotación f - development
project feasibility
— f — regulación f para demanda f - [electr-
-prod] demand control feasibility
— f económica - [com] economic feasibility
— f estudiada - studied feasibility
viaducto m - [constr] . . .; causeway; structure
• [f.c.] . . .; (track) trestle
— m de caballete(s) m - [f.c.] trestle
— m — madera f - [f.c.] timber trestle
— m —, terraplén, o tierra f - [vial] earth-
-fill viaduct
— m ferroviario - [f.c.] railway, viaduct, o
trestle
— m — sobre caballete(s) m - [f.c.] railroad
trestle
— m — de madera f - [f.c.] timber railway,
trestle, o bridge
— m — reemplazado - [v.c.] replaced railway
trestle
— m reemplazado - [constr] replaced, viaduct, o
trestle
viajado,da a - traveled • ridden
— a por barcaza f - [transp] barge traveled
viajante m de comercio m - [com] traveling, o
outside, salesman
viajar v - . . . . • to ride
— v por - to travel over • to ramble
— v — agua m - [transp] to travel by water
— v — barcaza f - [transp] to travel by barge

viajar v sobre - to, travel, o ride, by, o on
viaje m - . . .; ride,ding • round trip • [autom]
ride • [metal-prod] track time
— f a exterior m - [transp] foreign travel
— m aéreo - [transp] air travel
— m consecutivo - [transp] consecutive trip
— m costoso - [Transp] costly trip
— m de cucharón m - [constr] dipper trip •
[metal-prod] ladle trip
— m — ida (solamente) - [transp] single (way)
trip
— m — personal - [labor] personnel travel
— m en automóvil m - [autom] drive; car trip
— m — exterior - [transp] foreign travel
— m — helicóptero m - [aeron] helicopter ride
— m — interior m - [transp] domestic travel
— m innecesario - unnecessary, travel, o trip
— m local - [transp] local trip
— m — país m - [transp] domestic travel
— m no urbano - [transp] non-urban trip
— m por barcaza f - [transp] barge travel(ing)
— m — placer m - [transp] pleasure, ride, o
trip
— m redondo - [transp] round trip • [electrón]
round trip
— m — medido - [comput] measured round trip
— m seguro - [transp] safe travel • safe, ride,
o trip
— m sencillo - [transp] single, o one way,
travel, o trip
— m sobre - [transp] travel, by, o on; ride,ing
on
— m urbano - [transp] urban travel
viajero m que arriba - [transp] arriving, traveler, o passenger
— m — parte - [transp] departing. traveler, o
passenger
vías f - [f.c.] trackage
— f múltiples - multiple ways | a - multi-lane
— f y obras f - [f.c.] tracks and structures •
maintenance
viático m - . . .; per diem
— m correspondiente - respective per diem
— m diario - per diem (allowance)
vibración f - . . .; vibrating • [mec] chat-
ter(ing); jarring • whip(ping) - [metal]
fretting
— f de cigüeñal m - [comb.int] crankshaft vibration,ting
— f — encofrado m - [constr] form vibration
— f — frecuencia f alta - [electr] high frequency vibration
— f — hormigón m - [constr] concrete vibrating
— f — malacate m - [petról] drawworks vibrating
— f — máquina f - [mec] machine vibrating,tion
— f — plataforma f - [constr] platform vibrating,tion
— f — — f de acero m - [mec] steel platform
vibrating,tion
— f — tubería f - [mec] tube,bing vibration
— f — tubo m - [mec] tube vibrating,tion
— f — — m para conexión f - [tub] connecting, o
connection, tube vibrating,tion
— f — ventilador m - [ventil] fan vibration
— f evitada - avoided, o prevented, vibration
— f excesiva - excessive vibrating.tion
— f — de ventilador m - [ventil] excessive fan
vibrating,tion
— f interna - internal vinration
— f máxima - [mec] maximum vibration • maximum
whipping
— f mecánica - mechanical vibration
— f mínima - [mec] minimum vibration • minimum
whipping
— f reflejada - [mec] reflected vibration
— f terminal - [mec] terminal vibration
vibrado,da - vibrated • [mec] whipped • [metal]
fretted
— a excesivamente - vibrated excessively
vibrador m con frecuencia f alta - [constr] high
frequency vibrator
— m para reserva - [constr] standby vibrator
— m — vagón(es) m - [f.c.] car shaker

vibrador m̲ interno - [constr] internal vibrator
vibrador,ra a̲ - [constr] vibrator,ting
vibrar v̲ - . . . • [mec] to whip • [metal] to
    fret
— v̲ excesivamente .- to vibrate excessively
— v̲ hormigón m̲ - [constr] to vibrate @ concrete
— v̲ malacate m̲ - [petról] to vibrate @ drawworks
vibratorio,ria a̲ - . . . . • vibrational
vibro* m̲ - [mec] véase vibrador m̲
vibroalimentadora f̲ - [metal-prod] vibrating
    feeder
vibrocorrído,da a̲ - [metal-prod] fretted
vibrocorrer* v̲ - [metal] to fret
vibrocorrosión* f̲ - [metal] fretting (corrosion)
vice versa adv - véase viceversa
vicegobernación f̲ - [pol] liutenant governor's,
    office, o̲ dignity
vicepresidente m̲ adjunto - [pol] assistant vice-
    -president
— m̲ encargado de venta(s) f̲ - [com] marketing
    vice president
— m̲ para, comercialización f̲, o̲ mercadeo* m̲ -
    [legal] marketing vice president
— m̲ — desarrollo m̲ - [legal] development vice-
    president
— m̲ — grupo m̲ - [legal] group vice president
— m̲ — ingeniería f̲ - [legal] engineering vice-
    president
— m̲ — investigación(es) m̲ - [legal] research
    vice president
— m̲ — nercadeo* m̲ - [legal] marketing vice pre-
    sident
— m̲ regional - [legal] regional vice president
viceversa adv - vice versa
viciado,da a̲ - . . . .; fouled • flawed
viciar v̲ - . . .; to foul • to flaw
vicio m̲ - . . . . • flaw
victoria f̲ aplastante - crushing, o̲ smashing,
    victory, o̲ win
— f̲ consecutiva - [deport] consecutive victory
— f̲ en clase f̲ producción f̲ - [autom] production
    class, victory, o̲ win
— f̲ fácil - easy, victory, o̲ win
— f̲ para categoría f̲ - [deport] class win
— f̲ fuera de carretera f̲ - [autom] off-road vic-
    tory
— f̲ seguida - [deport] consecutive win
— f̲ única - single, o̲ lone, victory
vida f̲ - . . . . • career • [f.c.-ruedas] wear •
    [ind] véase vida f̲ útil
— f̲ acortada - [biol] shortened life
— f̲ — de acumulador m̲ - [electr] shortened
    battery life
— f̲ adicional - additional, o̲ added. life
— f̲ aproximada - approximate life
— f̲ — de troquel m̲ - [mec] approximate die life
— f̲ aumentada - increased, o̲ lengthened, life
— f̲ breve - short life
— f̲ buena - good life
— f̲ corta - short life
— f̲ — de acumulador m̲ - [electr] short battery
    life
— f̲ — — banda f̲ para rodamiento m̲ - [autom-
    -neumát] (tread) short road life
— f̲ — camión m̲ - [autom] short truck life
— f̲ de acumulador m̲ - [electr] battery life
— f̲ — — m̲ acortada - [electr] shortened bat-
    tery life
— f̲ — banda f̲ para rodamiento m̲ - [autom-neumát]
    -neumát] tread life
— f̲ — cable m̲ - [cabl] cable life
— f̲ — cadena f̲ - [cadenas] chain life
— f̲ — calzada f̲ - [constr] pavement life
— f̲ — camión m̲ - [autom] truck life
— f̲ — correa f̲ - [mec] belt life
— f̲ — cucharón m̲ - [metal-prod] ladle life
— f̲ — herramienta f̲ - [herram] tool life
— f̲ — hilera f̲ - [trefil] die life
— f̲ — horno m̲ - [combust] furnace life
— f̲ — juego m̲ - [mec] set life
— f̲ — — m̲ de engranaje(s) m̲ - [mec] gear set
    life
— f̲ — lingotera f̲ - [metal-prod] (ingot) mold
    life
vida f̲ de mordaza f̲ - [mec] jaw, o̲ clamp, life
— f̲ — motor m̲ - [comb.int] engine life -
    [electr-mot] motor life
— f̲ — — m̲ acortada - [comb.int] shortened
    engine('s) life • [electr-mot] shortened mo-
    tor('s) life
— f̲ — neumático m̲ - [autom-neumát] tire life
— f̲ — pieza f̲ - [mec] part life
— f̲ — — f̲ para horno m̲ - [combust] furnace
    part life
— f̲ — producto m̲ - [ind] product life
— f̲ — quemador m̲ - [combust] burner life
— f̲ — refractario(s) m̲ - [combust] refrac-
    tory,ries life
— f̲ — respuesto m̲ - [mec] spare (part) life
— f̲ — — m̲ de estelita - [ind] Stellite part
    life
— f̲ — — m̲ — — f̲ reconstruido -[mec] rebuilt
    Stellite part life
— f̲ — rodillo m̲ - [metal-lam] roll life
— f̲ — — f̲ para conformación f̲ - [mec] forming
    roll life
— f̲ — troquel m̲ - [mec] die life
— f̲ descansada - restful, o̲ casual, life
— f̲ diaria - daily life
— f̲ efectiva - actual life
— f̲ — para sujeción f̲ - [mec] gripping life
— f̲ en aire m̲ libre - outdoor, o̲ open air, liv-
    ing, o̲ life
— f̲ esperada - life expectancy
— f̲ estimada - estimated life • esteemed life
— f̲ — para anillo m̲ - [mec] estimated ring
    life
— f̲ excelente - excellent life
— f̲ — de banda f̲ para rodamiento m̲ - [autom-
    -neumát] excellent tread life
— f̲ extendida - extended, o̲ prolonged, o̲ maxi-
    mized, o̲ stretched, life
— f̲ indefinida - indefinite life
— f̲ laboral - [labor] working life
— f̲ — diaria - [labor] daily working life
— f̲ larga - long life
— f̲ — para acumulador m̲ - [electr] long bat-
    tery life
— f̲ — — bando f̲ para rodamiento m̲ - [autom-
    -neumát] long tread life
— f̲ — — calzada f̲ - [vial] pavement long life
— f̲ — — motor m̲ - [electr-mot] long motor
    life • [combust.int] long engine life
— f̲ — soldadora f̲ - [sold] long welder life
— f̲ — sin inconveniente(s) m̲ - [ind] long
    trouble free life
— f̲ — — — m̲ para soldadora f̲ - [sold] long
    trouble free welder life
— f̲ — — problema(s) m̲ - [ind] long, trouble,
    o̲ problem, free life
— f̲ — — — m̲ para soldadora f̲ - [sold] long
    trouble free welder life
— f̲ libre - free life
— f̲ — de preocupación(es) f̲ - worry free life
— f̲ más corta - shorter,test life
— f̲ — larga - longer,gest life
— f̲ máxima - maximum life
— f̲ — para cable m̲ - [cabl] maximum cable life
— f̲ mayor - longer,gest life
— f̲ media - average life
— f̲ múltiple - multiple life • [f.c.-ruedas]
    multiple wear
— f̲ normal - normal life
— f̲ — para cable m̲ - [cabl] normal, rope, o̲
    cable, life
— f̲ — — — m̲ de alambre m̲ - [cabl] normal
    wire rope life
— f̲ — — cadena f̲-[cadenas] chain normal life
— f̲ óptima - optimum life
— f̲ — para eje m̲ - [mec] maximum axle life
— f̲ para anillo m̲ - [mec] ring life
— f̲ — contrato m̲ - [legal] contract life
— f̲ — operación f̲ - [ind] service life
— f̲ — servicio m̲ - [ind] service life
— f̲ prevista - life expectancy • design life
— f̲ — exigida - [constr] required design life

vida f prevista máxima - maximum life expectancy
— f — mínima - minimum life expectancy
— f — para estructura f - [constr] structure
life expectancy
— f privada - private life
— f probable - probable life; life expectancy
— f prolongada - prolonged, o extended, o long,
o increased, o stretched, life
— f — para banda f para rodamiento m - [autom-
-neumát] long tread wear
— f — motor - [comb.int] long, o extended,
engine life • [electr-mot] long, o extended,
motor life
— f — pieza f - [mec] prolonged, o extended, part life
— f proyectada - design life
— f recreativa - recreational, life, o living
— f — en aire m libre - [deport] outdoor recreational living
— f ribereña - waterfront living
— f sin problema(s) m - problem, o trouble,
free, life, o living
— f — m para conservación f trouble, o
maintenance, tree living
— f — m — entretenimiento m - trouble-free
life
— f — m — mantenimiento m - trouble free
life
— f — m — soldadora f - [sold] trouble
free welder life
— f suburbana - suburban, life, o living
— f útil - useful, o service, life • working, o
industrial, life; life expectancy
— f — acortada - [mec] shortened (useful) life
— f — adecuada - adequate, useful, o service,
life
— f — adicional - [labor] added useful life
— f — anticipada - expected, useful, o service, life
— f — apropiada - proper, o adequate. useful,
o service, life
— f — para sello m - [mec] appropriate, o
proper, seal life
— f — aproximada - approximate (useful) life
— f — para troquel m - [mec] approximate
die life
— f — asegurada - assured useful life
— f — aumentada - [mec] increased useful life
— f — — para motor m - [comb.int] increased
engine life • [electr-mot] increased motor
life
— f — breve - [ind] short useful life
— f — — para camión m - [autom] truck short
(useful) life
— f — contra fatiga f - [metal-ruedas] useful
fatigue life
— f — — f de superficie f para radadura f -
[mec-ruedas] (rolling) contact fatigue life
— f — — — f de rueda(s) f
para riel(es) m - [mec-ruedas] track wheel
rolling contact fatigue life
— f — corta f - [ind] short useful life
— f — para acumulador m - [electr] short
battery life
— f — — banda f para rodamiento m—[autom-
-neumát] short tread life
— f — esperada - expected useful life
— f — estimada - [ind] estimated, useful, o
service, life
— f — para calcantarilla f - [constr] estimated culvert life
— f — — anillo m - [mec] estimated ring
life
— f — excelente - excellent (useful) life
— f — — para banda f para rodamiento m -
[autom-neumát] excellent tread life
— f — exigida - required, useful, o service,
life
— f — extendida - [ind] extended, o long, o
maximized, o lengthened, (service) life
— f — para motor m - [comb.int] increased
engine life • [electr-mot] increased motor
life

vida f útil indefinida - indefinite, o unde=
fined, useful, o service life
— f — larga - [labor] long, o extended, service, o useful, life
— f — — f de acumulador m - [electr] long
battery (useful) life
— f — de banda f para rodamiento m—[autom-
-neumát] long, o extended, tread live
— f — para calzada f - [constr] long pavement (useful) life
— f — — soldadura f - [sold] long, welder,
o solderer, useful, o service, life
— f — — camión m - [autom] long truck
(useful) life
— f — sin problema(s) m - [mec] long trouble free (useful) life
— f — — — — — para soldadora f - [sold]
long trouble free welder life
— f — — — — — para conservación f - long
trouble free, life, o operation
— f — más corta - shorter (useful) life
— f — — larga - longer (useful) life
— f — — prolongada - [ind] longer, useful, o
service, life • greater life
— f — máxima - maximum, useful, o service, o
expectancy, life
— f — para rueda f - [mec-ruedas] maximum
wheel life
— f — mayor - longer,gest (useful) life
— f — mínima - minimum, useful life, o life
expectancy
— f — normal - normal (useful) life
— f — para cable m - [cabl] normal (useful)
rope life
— f — — m de alambre m - [cabl] normal
wire rope (useful) life
— f — óptima - [ind] optimum, useful, o service, life
— f — para eje m - [mec] optimum useful
axle life
— f — para acumulador m - [electr] battery
(useful) life
— f — — alcantarilla f - [vial] culvert,
useful, o service, life
— f — — anillo m - [mec] ring (useful) life
— f — — banda f para rodamiento m - [autom-
-neumát] tread (useful) life
— f — — cadena f - [cadenas] chain, useful, o
service, life
— f — — calzada f - [vial] pavement (useful)
life
— f — — camión f - [autom] truck (service)
life
— f — — carrera(s) f - [deport] race life
— f — — correa f - [mec] belt (useful) life
— f — — cucharón m - [metal-prod] ladle (useful) life
— f — — eje m - [mec] axle (useful) life
— f — — equipo m - [ind] equipment (useful)
life
— f — — estructura f - [constr] structure,
(useful) life(time)
— f — — — f para drenaje m - [hidr] drainage
structure (useful) life(time)
— f — — fatiga f - [ind] useful fatigue life
— f — — — f para superficie f para rodadura
f (para rueda f) - [mec-ruedas] wheel rolling
contact fatigue life
— f — — herramienta(s) f - [herram] tool(s)
(useful) life
— f — — hilera f - [trefil] die (useful) life
— f — — horno m - [combust] furnace (useful)
life
— f — — juego m - [mec] set (useful) life
— f — — — m de engranajes m - [mec] gear
set (useful) life
— f — — lanza f - [combust] lance (useful)
life
— f — — matriz,ces f - [mec] die(s) (useful)
life
— f — — mordaza f - [mec] clamp (useful) life
— f — — motor m - [comb.int] engine (useful)
life • [electr-mot] motor (useful) life

vida f útil de motor m acortada - [comb.int] shortened engine's (useful) life • [electr--mot] shortened motor's (useful) life
— f — — **neumático m**—[autom-neum] tire's (useful) life
— f — — **pieza f** - [mec] part's, useful, o service, life
— f — — — f **para horno m** - [combust] furnace part's (useful) life
— f — — **punzón m**—[mec] punch part's useful life
— f — — **quemador m** - [combust] burner's (useful) life
— f — — **m para encendido m** - [combust] ignition burner's (useful) life
— f — — **refractario(s) m** - [refract] refractory,ries (useful) life
— f — — **m para quemador m** - [refract] burner refractories (useful) life
— f — — **m** — **m para encendido m** - [refract] ignition burner's refractories (useful) life
— f — — — **m** — **recubrimiento m para quemador m para encendido m** - [refract] ignition burner covering refractories useful life
— f — — **rodillo m** - [metal-lam] roll (useful) life
— f — — **m para conformación f** - [metal--lam] forming roll (useful) life
— f — — **rueda f** - [mec-ruedas] wheel (useful) life
— f — — — f **para riel(es) m** - [mec-ruedas] track wheel (useful) life
— f — — **sello m** - [mec] seal (useful) life
— f — — **troquel m** - [mec] die (useful) life
— f — — **tubería f** - [tub] pipe,ping (useful) life
— f — — **tubo m** - [tub] pipe (useful) life
— f — — **turboalimentador m** - [comb.int] turbocharger's, (useful, o service), life
— f — **probable** - [ind] probable, o expected, (useful) life
— f — **prolongada** - [ind] prolonged, o extended, o lengthened, (useful, o service), life
— f — — **para motor m** - [electr-mot] extended motor (useful) life
— f — — — **pieza f** - [mec] extended part's (useful) life
— f — **restante** - [ind] remaining (useful) life • life expectancy
— f — **satisfactoria** - [ind] satisfactory (useful) life
— f — **sin problema(s) m** - [ind] trouble free (useful) life
— f — — — **m para soldadora f** - [sold] trouble free welder (useful) life
videograbación* f - [electrón] (video)taping
videograbado,da* a - [electrón] (video)taped
videograbar* v - [electrón] to (video)tape
vidriera f - [constr] . . .; sash
— f **abisagrada** - [constr] casement
— f — **verticalmente** - [constr] casement
— f **comercial** - [constr] commercial sash
— f — **de acero m** - [constr] commercial steel sash
— f — — — m **de tipo m fijo** - [constr] steel commercial fixed type sash
— f — — **m** — — **m saliente** - [constr] steel commercial protruding type sash
— f — — **tipo m fijo** - [constr] commercial fixed type sash
— f — — — **m saliente** - [constr] commercial protruding type sash
— f — **tipo m fijo** - [constr] fixed type sash
— f **fija** - [constr] fixed sash
— f **movible** - [constr] movable, o operating, sash
— f **que puede abrir(se)** - [constr] operating sash
— f **saliente** - [constr] protruding sash
vidrio m claro - [plást] clear glass
— m **coloreado** - [plást] tinted glass
— m **de calidad f** - [plást] quality glass

vidrio m filtrante - [cerám] filtering glass
— m **fundido** - [cerám] molten glass
— m **impregnado** - [electr-cond] impregnated glass
— m — **de calidad f** - [electr-cond] impregnated quality glass
— m — **trenzado** - [electr-cond] braided impregnated glass
— m — — **de calidad f** - [electr-cond] braided impregnated quality glass
— m **inastillable** - [plást] shatterproof glass
— m **para seguridad f** - [cerám] safety glass
— m — — f **inastillable** - [plást] shatterproof safety glass
— m **Pyrex** - [plást] Pyrex glass
— m **reforzado** - [plást] reinforced glass
— m — **con malla f** - [plást] mesh reinforced glass
— m — — — f **de alambre m** - [plást] wire mexh reinforced glass
— m **soluble** - [cerám] solluble glass • [quím] water glass; sodium silicate
— m **volcánico** - [geol] volcanic glass
vidrioso,sa a - . . . .; glazed
viene v de página f . . . - [public] continued from page . . .
viento m - . . . . • air • [ind] (air) blast
— m **atravesado** - [aeron] cross-wind
— m **caliente** - hot wind • [metal-prod] hot, air, o blast
— m **de costado m** - [meteorol] side wind
— m **enriquecido** - [metal-prod] enriched blast
— m — **con oxígeno m** - [metal-prod] oxygen enriched blast
— m **entumecedor** - [meteorol] numbing wind
— m **flojo** - [metal-prod] slack, wind, o blast
— m **frío** - [meteorol] cold wind
— m **fuerte** - [meteorol] strong, o high, wind
— m **lateral** - [meteorol] side wind
— m **liviano** - [meteorol] (s)light wind • [metal-prod] slack, wind, o blast
— m **moderado** - [meteorol] moderate wind • [metal-prod] moderate blast
— m **normal** - [meteorol] normal wind
— m **pleno** - [meteorol] full wind
— m **predominante** - [meteorol] prevailing, o predominant, wind
— m **prevaleciente** - [meteorol] prevailing wind
— m **prevalente*** - [meteorol] prevailing wind
— m **sobreoxigenado** - [metal-prod] oxygen enriched blast
— m **soplado** - [metal-prod] blast
vientre m - . . . . • [mec] underside • camber • [metal-prod] mantle; stack; belly • [metal--lam] loop (vibración)
— m **de horno m** - [metal-prod] furnace, belly, o mantle
viga f - [constr] . . .; stringer • [pil] cap • [mrysl-lsm] rail • [constr] rail
— f **acanalada** - [metal-lam] channel (beam)
— f — **conformada** - [metal-lam] formed, o shaped, channel (beam)
— f — — **recta** - [metal-lam] straight channel (beam)
— f — — **recia** - [metal-lam] rugged, formed, o shaped, channel (beam)
— f — **corriente** - [metal-lam] standard channel (beam)
— f — **estándar** - [metal-lam] standard channel (beam)
— f — **extra reforzada** - [metal-lam] extra reinforced channel (beam)
— f — **para bóveda f** - [metal-lam] arch channel (beam)
— f — — — f **asentada** - [constr] anchored arch channel (beam)
— f — **recia**—[metal-lam] rugged channel (beam)
— f — **reforzada** - [metal-lam] reinforced channel (beam)
— f **accionada** - [constr] driven beam
— — — **hidráulicamente** - [constr] hydraulically driven beam
— f **ajustable** - [constr] adjustable beam

viga f ancha f - [metal-lam] wide beam
— f anular - [constr] ring beam
— f — de acero m - [metal-lam] steel ring beam
— f angular con alas f desiguales - [metal-lam] unequal flange angle beam
— f — — m — f iguales - [metal-lam] equal flange angle beam
— f arbotante m - [grúas] outrigger beam
— f armada - [constr] truss, o built up, beam; truss panel
— f — de acero m - [constr] steel truss (beam)
— f arriostrada - [constr] stayed girder
— f atirantada - [constr] truss beam
— f calada - [estruct] lattice, beam, o girder
— f clavada - [constr] encased beam
— f compensada - [mec] equalized beam
— f compensadora - [mec] equalizer,zing beam
— f — para freno m - [mec] brake euqlizer,zing beam
— f compuesta - [constr] built up girder; composite beam; girder
— f compuesta - [constr] built-up girder
— f con ala m ancha - [metal-lam] wide flange, o broad flanged, beam
— f con alas f - [metal-lam] flange beam
— f — — desigual(es) - [metal-lam] unequal flange beam
— f — — f iguales - [metal-lam] equal flange beam
— f — — f inclinadas - [metal-lam] slanted flange(s) beam
— f — — f — laminada en caliente - [metal-lam] hot rolled slanted flange beam
— f — alma m abierta - [constr] lattice girder
— f — — m doble - [constr] box girder
— f — — m llena - [metal-lam] solid, o plate, web, beam, o girder
— f — conformación f en W - [constr] W, beam, o rail
— f — — f — — montada en voladizo m - [constr] blocked out, o cantilevered, W beam
— f conectadora - [constr] connecting channel
— f continua - [constr] continuous beam
— f — con alma m llena - [constr] continuous plate girder
— f curva(da)—[constr] curved, beam, o element
— f de acero m - [constr] steel, beam, o rail, o girder
— f — — m con conformación f en W - [metal-lam] steel W-beam
— f — — m embutida - [constr] encased steel beam
— f — — m en U f - [metal-lam] steel channel
— f — — m — — con alas f desiguales - [metal-lam] unbalanced steel channel
— f — — m — — laminada en caliente - [metal-lam] hot rolled steel channel
— f — — m enclavada - [constr] encased steel beam
— f — — m — — hormigón m - [constr] concrete encased steel beam
— f — — m instalada - [constr] installed concrete beam
— f — — m para armadura f - [constr] steel truss
— f — celosía f - [constr] lattice, beam, o girder
— f — hormigón n - [constr] concrete beam
— f — — m instalada - [constr] installed concrete beam
— f — — m para empuje m - [constr] concrete thrust beam
— f — — m pretensada* - [constr] prestressed concrete, beam, o girder
— f — — m vaciado - [constr] pouted concrete beam
— f — patín m ancho - [metal-lam] wide base beam
— f — — m — laminada en caliente m - [metal-lam] hot rolled wide base beam
— f — — m — y recto - [metal-lam] wide and straight base beam
— f — — m — — laminada en caliente -

[metal-lam] hot rolled wide (and) straight base beam
viga f de patín m recto - [metal-lam] straight base beam
— f — — m — laminada en caliente m - [metal-lam] hot rolled straight base beam
— f débil - [constr] weak beam
— f decreciente - [constr] tapered beam
— f delantera f - [mec] front(al) beam
— f distribuidora - [mec] spreader beam
— f divisoria - [vial] median rail
— f — para viga f en caja f - [constr] box beam median rail
— f doble T - [metal-lam] I beam
— f elevada - [constr] overhead, beam, o girder • lifted beam
— f embutida - [constr] encased beam
— f — en hormigón m - [constr] concrete encased beam
— f en caja f - [constr] box beam • box section
— f — — f de acero m - [constr] steel box beam
— f — — f — hormigón m - [constr] concrete box girder
— f — — f — — m pretensado* - [constr] prestressed concrete box girder
— f — cajón f - [constr] box girder
— f — forma f de caja f - [constr] box beam
— f — H - [constr-pil] H pile
— f — hilera f - [constr] soldier beam
— f — T - [metal-lam] T, beam, o section
— f — U - [constr] channel (beam) • structural, o steel, o ship, channel
— f — — con alas f desiguales - [metal-lam] unbalanced, u unequal sides, channel
— f — — instalada - [constr] installed, U beam, o channel
— f — — laminada en caliente - [metal-lam] hot rolled channel
— f — — ranurada - [constr] grooved channel
— f — — sin equilibrar - [metal-lam] unbalanced channel
— f — Z - [metal-lam] Z beam
— f enclavada - [constr] encased beam
— f — en hormigón m - [constr] concrete encased beam
— f entre pilares m - [constr] beam, o rail, between, piers, o pillars
— f estampada - [constr] formed channel • stamped channel
— f estructural - [metal-lam] structural, beam, o channel
— f — con conformación f desusada - [metal-lam] odd shaped structural beam
— f — — — f insólita - [metal-lam] odd shaped structural beam
— f — de acero m - [metal-lam] structural steel, beam, o shape
— f — — m en U - [metal-lam] structural steel channel
— f — — m — laminada en caliente - [metal-lam] hot rolled structural steel beam
— f — en U - [metal-lam] structural channel
— f — grande - [metal-lam] large structural beam
— f — laminada - [metal-lam] rolled structural beam
— f — — en caliente - [metal-lam] hot rolled structural, beam. o shape
— f — — frío - [metal-lam] cold rolled structural, beam, o shape
— f — pequeña - [metal-lam] small structural, beam, o shape
— f flexible - [metal-lam] flexible beam
— f fuerte - [metal-lam] strong beam
— f — sobre poste m débil - [metal-lam] strong beam on @ weak post
— f gemela - [constr] twin beam
— g — de hormigón f - [constr] twin concrete beam
— f grande - [constr] large, o deep, beam
— f H - [metal-lam] H beam • [constr-pil] H pile

viga f para patín m delantero - [mec] front skid
beam
— f — — m, posterior, o trasero - [mec] rear
skid beam
— f — piso m—[constr] floor beam • floor sill
— f — — m para torre f - [petról] derrick
(floor) sill
— f — — m — — f para perforación f -
[petról] derrick (floor) sill
— f — puente m—[constr] bridge (floor) girder
— f — referencia f - reference beam
— f — riel(es) m - [f.c.] track girder •
[grúas] runway girder
— f — m para grúa f - crane runway girder
— f — soporte m - [mec] support, beam, o rail
— f — — m para colector m - [mec] sump
support rail
— f patentada - [constr] patented beam
— f pesada - [constr] heavy beam
— f principal—[constr] main, o principal, beam
— f — para armadura f - [constr] truss main, o
main truss, beam
— f — — piso m - [constr] main floor beam
— f — — retén m - [petról] (stop) sill
— f profunda - [metal-lam- deep beam
— f recta - [metal-lam] straight beam
— f rebajada - [constr] pony, beam, o bent
— f reforzada - [metal-lam] reinforced beam •
haunched beam
— f refrigerada - [metal-lam] (front wall)
backstay
— f resistente - [constr] strong, o resistant,
beam
— f reticulada - [constr] lattice, beam, o girder
— f Salmer - [metal-prod] (door frame) bolster
beam
— f secundaria - [constr] secondary beam
— f sobrecabeza f - [constr] overhead beam
— f soldada - [metal-lam] welded beam
— f T - [constr] T beam; T section
— f — doble - [metal-lam] H beam
— f — — estándar—[metal-lam] standard I beam
— f — para acensor(es) m - [metal-lam] elevator, T, o tee
— f seccionada - [constr] split tee (beam)
— f transversal - [constr] cross, beam, o rail •
floor beam • (piling) cap
— f — para base f - [constr] base cross rail
— f trasera - [constr] rear beam
— f triangular - [constr] triangular beam •
concrete beam
— f — de hormigón m - [constr] triangular concrete beam
— f tubular - [constr] tubular, beam, o rail
— f U - [constr] channel, beam, o rail
— f — para deslizamiento m - [constr] channel
rub(bing) rail
— f vaciada - [constr] poured, o cast, beam
— f voladiza f - [constr] cantilever (beam)
— f W estándar - [constr] standard W beam
vigencia f - . . . . • duration; life; term • application; enforcement
— f de acuerdo m - [legal] agreement, term, o
duration, o validity
— f — contrato m - [legal] contract, duration,
o validity
— f — crédito m - [fin] credit validity
— f — póliza f—[seguros] policy validity
vigente a - . . .; valid; prevailing; existing;
current; applying • unexpired • with @ unexpired term
vigía f . . . . • [ind] groundman; watchman
vigilado,da a - watched; checked; monitores; observed - taken care of
— a continuamente - monitored continuously
— a remotamente - monitored remotely
vigilancia f - . . .; monitoring; check(ing);
suoervising;observation • taking care of •
[pol] security • [ind] police
— f ambiental - [ambient] environmental, monitoring, o surveilance
— f continua - continuous monitoring

vigilancia f de calidad f - quality check(ing)
— f — — f de línea f - [electrón] line quality, control(ing), o monitoring
— f — instrumentación f - [instrum] instrumentation monitoring
— f — línea f - [electrón] line monitoring
— f — seguridad f - [segurid] safety, monitoring, o observation
— f — señal(es) f - signal(s) monitoring
— f intensiva - [mec] intensive care
— f planeada - [segurid] planned monitoring
— f — para seguridad f - [segurid] planned
safety monitoring
— f remota - remote monitoring
vigilante m - . . . . • [ind] groundman; véase
también inspector m | a - . . . .
— m de carga f - [metal-prod] feeder bin
stocker
vigilar v - . . .; to, supervise, o observe, o
monitor • to take care of
— v calidad f - to, observe, o monitor, quality
— v — f de línea f - [electrón] to monitor @
line quality
— v clavija f - [electrón] to monitor @ pin
— v continuamente - to monitor continuously
— v instrumentación f - [ind] to monitor @ instrumentation
— v línea f - [electrón] to monitor @ line
— v remotamente - to monitor remotely
— v señal f - to monitor @ signal
vigor m - . . .; force
— m pleno - full force
vigorizante a - . . . . • crisp
vigorosamente adv - vigorously • crisply
vigoroso,sa a - vigorous • crisp
vigueta f - [constr] . . .; rafter; purlin •
[metal-lam] rail • [mec] skid
— f continua - [mec] continuous small beam;
beam type purlin
— f de acero m - [constr] steel, joist, o purlin
— f — celosía f - [constr] open web joist
— f — — f de acero m - [constr] steel open
web, o open web steel, joist
— f para asiento m - [comb.int] engine pony
sill
— f — depósito m - [ind] tank, rail, o joist
— f — — m para combustible m - [combust] fuel
tank, rail, o joist
— f — motor m - [comb.int] engine pony sill -
[electr-mot] motor pony sill
— f — prensa f enderezadora - [mec] gag press
skid
— f — — f para alinear - [mec] gag press skid
— f transversal -- [constr] crossbeam; fore-
beam; riser
villa f - [pol] . . .; borough; township; village • [resid] second, o country, home
vinculación f - . . . • connection; relation-
(ship) • [com] association • contact • [hidr]
pipeline
— f directa - direct relation(ship)
— f establecida - established relation(ship)
— f personal - personal relation(ship)
— f sólida - solid, o sound, personal rela-
tion(ship)
— f sólida - sound relation(ship)
vinculado,da a - affiliated; associated; connected; related
— a con - associated, o related, o connected,
to, o with
— a con defensa f - [milit] defense related
— a — minería f - [miner] mining related
— a — neumático(s) m - [autom-neumát] tire related
— a — personal m - [labor] personnel related;
man-related
— a — trabajo m - work-, o job-, related • on
@ job
— a — vehículo(s) m - [autom] vehicle, o car,
related
— a directamente - related directly
vincular v - . . .; to, associate, o relate • to

connect
vincular v directamente - to relate directly
vínculo m - . . . .; relation; connection • [legal] instrument • [com] contact • [mat] bracket
vinílico,ca a - [plást] vinyl(ic)
violación f- . . . . • infringement,ging
— f — patente f - [legal] patent, trespassing, o violating,tion, o infringement,ging
violado,da a - violated • [legal] infringed
violar v patente f - [legal] to infringe @ patent
violencia f de explosión f - [explos] explosion force
virado,da a - veered; turned • [autom] swerved
— a hacia derecha f - [autom] turned to(ward) @ right
— a — izquierda f - [autom] turned to(ward) @ left
— a lentamente - [autom] turned slowly
viraje m - . . . . • [autom] swerving • cornering
— m hacia derecha f - [autom] turn(ing) toward @ right
— n — izquierda f - [autom] turning toward @ left
— m amplio - [autom] wide turn(ing)
— m — efectuado - [autom] made wide turn
— m brusco - [autom] sharp, swerving, o cornering
— m bueno - [autom] good, turn(ing), o cornering
— m cerrado - [autom] tight cornering
— m con pavimento m mojado - [autom] wet cornering
— m — — m seco - [autom] dry cornering
— m corto - [autom] short turn(ing)
— m efectuado - [autom] made turn
— m excelente - [autom] excellent cornering
— m hacia adentro - [deport] turn(ing)-in
— m — afuera - [deport] turn(ing)-out
— m — derecha f - [autom] right(-hand) turning
— m — izquierda f - [autom] left(-hand) turning
— m inicial - [autom] initial turn(ing)
— m lento - [autom] slow turn(ing)
— m máximo - [autom] full turn(ing)
— m — hacia derecha f - [autom] full right turn
— m — — izquierda f - [autom] full left turn
— m mejorado - [autom] improved, turn(ing), o cornering
virando adv - [mec] turn(ing) • idling
virar v - to turn • [autom] to, turn, o swerve • [fotogr] to develop
— v hacia derecha f - to turn to(ward) @ right
— v — izquierda f - to turn to(ward) @ left
— v lentamente - to, turn, o swerve, slowly
virgen f - . . . | a - . . . . • undisturbed; untapped
virola f - [mec] . . . .; ferrule; hoop; slid sleeve; ring • burr • [metal-prod] curved plate; hoop • [cald] shell • [constr] jointer runner
— f de tornillo m - [mec] screw ferrule
virotillo m - [mec] stay bolt
virtualmente adv - . . . . • essentially
viruela f - [med] . . . . • [metal-prod] pock(ing); pitt(ing) • jet scale
virulana f - [plást] fiberglass
viruta f - [mec] . . . .; turning; dirt; chipping • [metal-fabr] cutting; bossing • [mec] screw
— f de acero m - [metal-fabr] steel wool (ball)
— f — fundición f - [metal-prod] cast iron shaving
— f — hierro m - [mec] iron, turning, o bossing
— f — torneadura f - [metal-fabr] machine turning
virutamiento m - [mec] chip forming
— m en labrado m con rodillo(s) m - [metal-fabr] roll maching chip forming
— m — mecanización f con rodillo(s) m - [metal-fabr] roll machining chip forming
visación f - [pol] visa(ing)
— f consular - [pol] consular visa
— f de documento m - document visa(ing)
— f temporaria - [pol] temporary visa
visado m - [pol] véase visación f
visado,da a - [pol] visaed

visado,da a por cónsul(ado) m - [pol] visaed by @ consul(ate)
visar v - [poll] . . .; to visa
viscómetro* m - véase viscosímetro m
viscosidad f - [lubric] . . . .; body; weight
— f absoluta f - [combust] absolute viscosity
— f de combustible - [combust] fuel absolute viscosity
— f — — fuel oil m - [combust] fuel oil absolute viscosity
— f afectada - [lubric] affected viscosity
— f alta - [lubric] high viscosity
— f baja - [lubric] low viscosity
— f cinemática - [lubric] kinematic viscosity
— f correcta - [lubric] correct viscosity
— f — para aceite m - [lubric] correct oil viscosity
— f de aceite m - [lubric] oil, viscosity, o weight
— f — — m para base f - [lubric] base oil viscosity
— f — combustible m - [combust] fuel viscosity
— f — escoria f - [metal-prod] slag viscosity
— f — fuel oil m - [combust] fuel oil viscosity
— f — lubricante m - [lubric] lubricant viscosity
— f — — m para eje(s) m - [lubric] axle lubrication viscosity
— f — — f para transmisión f - [lubric] transmission lubricant viscosity
— f — papilla f - [explos] pap viscosity
— f específica - [lubric] specific viscosity
— f — de aceite m - [lubric] specific oil viscosity
— f I S O - [lubric] I S O viscosity
— f múltiple - [lubric] multiple viscosity
— f reducida - [lubric] reduced viscosity • lowered viscosity
— f Saybolt - [petról] Saybolt viscosity
— f — —Furol - [lubric] Saybolt-Furol viscosity
— f — universal - [petról] Saybolt universal viscosity
— f según Organización f Internacional para Normas - [lubric] International Standards Organization. o I S O, viscosity
— f — Sociedad f de Ingenieros m Automotores - [lubric] Society of Automotive Engineers, o S A E, viscosity
— f única - [lubric] single viscosity
— f universal - [petról] universal viscosity
viscosímetro m - [instrum] visco(si)meter
— m Furol - [instrum] Furol visco(si)meter
— m — de Saybolt - [instrum] Saybolt Furol visco(si)meter
— m Redwood - [instrum] Redwood visco(si)meter
— m universal - [instrum] universal viscometer
viscoso,sa a - [lubric] . . .; viscose • tough
visera f - . . . . • [autom] sun visor
— f antireflectante - [autom] antiglare visor
visibilidad f - . . . . • exposure
— f alta - high visibility
— f baja - low visibility
— f buena - good visibility • [meteorol] clear condition(s)
— f — para conducción f - [vial] good driving condition(s)
— f de establecimiento m - [com] business exposure
— f desmejorada - [vial] impaired visibility
— f excelente - excellent visibility
— f hacia adelante - front visibility
— f — arriba - overhead visibility
— v — atrás - rear visibility
— f horizontal - horizontal visibility
— f — excelente - excellent horizontal visibility
— f mala - bad, o poor, visibility
— f mayor - greater visibility
— f mejor - better visibility • [vial] clear condition(s)
— f — para conducción f - [vial] clearer driving condition(s)

visibilidad f mejorada - improved visibility
— f menor - lesser, o reduced, visibility
— f para conducción f—[vial] driving visibility
— f — operador m - operator('s) visibility
— f peor - worse,st visibility
— f pobre - poor visibility
— f reducida - reduced, o low, visibility
— f suficiente - sufficient visibility
— f total - total, o full, vision,sibility
— f vertical - vertical visibility
visible a - . . . • apparent • exposed • visual
visión f - . . .; eyesight • overview
— f anticipada - preview
— f buena - good eyesight
— f general - general overview
— f interdisciplinaria - interdisciplinary view
— f mala - bad, o poor, eyesight
— f pobre - poor eyesight
visita f a obra f - [ind] field visit
— f de emergencia f - emergency, visit, o call
— f — rutina - routine, visit, o call
— f en frío - [admin] cold call
— f esporádica - sporadic, visit, o call
— f facilitada - facilitated visit
— f final - final, visit, o call
— f inicial - initial, visit, o call
— f mensual - monthly, visit, o call
— f para inspección f—inspection, visit, o call
— f — — f final - final inspection visit
— f periódica - periodic, visit, o call
— f permitida - permitted, o allowed, visit
— f preliminar - preliminary, visit, o call
— f programada - scheduled, visit, o call
— f regular - regular, visit, o call
— f rutinaria - rutinary, visit, o call
— f subsiguiente - subsequent, visit, o call
visitado,da a - visited; called (en)
visitante m - . . . • patron
— m pernoctador - overnight, visitor, o caller
visitar v planta f - [ind] to visit @ plant
— v — de cliente m - to visit @ customer's plant
vista f - . . . • scene • eyes(ight) • review • [fotogr] ; shot
— f a vuelo m de pájaro m - bird's eye view
— f aérea - aerial, o air, view
— f — desde cerca - close-up air view
— f ampliada - [fotogr] enlarged, o exploded, o close-up, view
— f — de portadiferencial m - [autom-mec] differential carrier exploded view
— f anticipada - preview • overview
— f de bomba f - [bombas] pump view
— f — cámara f - [mec] chamber view
— f — — f para vástago m - [mec] rod chamber view
— f — campo m petrolífero - [petról] oil field, view, o scene
— f — cilindro m - [mec] cylinder view
— f — conjunto m - general assembly (view)
— f — costado m - side view
— f — distribución f - plan view; layout
— f — yacimiento m petrolífero - [petról] oil field, view, o scene
— f desde cerca - [fotogr] close-up view
— f — — de cabeza f para inyección f—[petról] swivel close-up (view)
— f — — — conjunto m—[ind] assembly close-up
— f — — — cordón m - [sold] bead close-up
— f — detrás - rear view; view from behind
— f — extremo m - end view
— f — frente m - front(al) view
— f despiezada - exploded view
— f desplegada - exploded view
— f detallada - detailed, o exploded, view
— f dorsal - rear view; view from @ rear
— f en escala f reducida - small size view
— f expandida - expanded, o exploded, view
— f exterior - outside, o exterior, view
— f frontal - front view; view from @ front
— f general - overall view; overview
— f —, abarcante, o comprensiva - comprehensive overview

vista f inferior - bottom view; view from below
— f interior - inside, o internal, view
— f invernal - winter(time), view, o scene
— f lateral - side view • [dib] elevation
— f marina - seascape
— f parcial - partial, o sectional, view
— f — de cimiento(s) m [constr] fundation sectional view
— f — fundamento m - [constr] foundation sectional view
— f perpendicular - [dib] perpendicular view
— f perspectiva - perspective view
— f pintoresca - picturesque, o scenic, view
— f posterior - rear view • later view
— f — de malacate m - [petról] drawworks, rear, o back, view
— f previa - preview
— f protegida - [segurid] protected eyes
— f radiográfica - [med] X-ray
— f — de soldadura f - [electrón] weld X-ray
— f recortada - [dib] cutaway view
— f retrospectiva - retrospect; flash back
— f seccional - sectional, o partial, view
— f simple - simple view
— f superior - top view; view from above
— f — de cilindro m - [mec] cylinder top view
— f — ménsula f - [mec] bracket top view
— f — — relé m—[electr-equip] relay top view
— f típica - typical view
— f — de ménsula f—[mec] bracket typical view
— f — — f superior - [mec] top bracket typical view
— f tomada desde cerca - [fotogr] close-up view
— f transversal - transverse, o crosswise, view
visto adv - [legal] whereas; in view of; considering
— adv y considerando - [legal] whereas
visto,ta a - seen • reviewed
— a de frente m - front view
— a desde - seen, o looking, from
— a — abajo - bottom view; view from below
— a — arriba - top view; looking down; view from above
— a — atrás - rear view; looking from behind
— a — costado m - side view; view from @ side
— a — detrás - rear view; view from behind
— a retrospectivamente - viewed retrospectively; flashed back
vistoso,sa a - . . .; ornate
visual f - [topogr] survey line • [óptica] line of vision; visual line
visualización f - visualization,zing • [comput] display(ing)
— f con tubo m para rayos-X m - [comput] cathode ray tube, o C R T, display(ing)
visualizador m - [comput] display
— m para condición f - [comput] status display
— m — escritorio m - [comput] desktop display
— m — — m para condición f - [comput] desktop status display
— m — — m estado m - [comput] desktop atatus display
— m — estado m - [comput] status display
visualmente adv - visually
vitalicio,cia a - . . . • lifetime; for life
vitola f - [instrum] (ring) gage
— f para sierra(s) f - [herram] saw gage
vítor m - cheer • [pl] whoops and hollers
vitoreado,da a - cheered
vítreo,rea a - . . . • glazed
vitrificado,da a - vitrified • glazed
vitrificar v - . . . • to become vitreous
vitrina f - [com] display case
vivaz a - . . .] [botán] perennial
viveza f - . . .; wit • intelligence
vívidamente adv - vividly • starkly
vívido,da a - . . .; colorful
vivienda f - [constr] . . .; home • housing
— f rural - [constr] country, house, o home
vivo,va a - . . . • [mec] sharp; keen • [miner] unslaked
vocabulario m común - [gram] common vocabulary
— m corriente - current, o common, vocabulary

vocación f - (a)vocation; calling • occupation
vocal m - . . . . • [legal] . . .; member; director
— m de junta f - [legal] board member
voladizo m - [constr] cantilever
voladizo,za a - [constr] offset; overhanging;
projecting; cantilever(ed) • véase volante a
— m de conducto m - [constr] conduit overhang
volado,da a - flown • blasted, o blown, off
volador,ra a - vease volante
voladura f de roca f - [constr] rock blasting
— f nuclear - [constr] nuclear blast(ing)
volante m - [mec] flywheel; regulator • handwheel
• wheel• [autom] steering wheel | a - flying
— m ajustado - [autom] adjusted steering wheel
— m girado - [mec] turned flywheel • [autom]
turned steering wheel
— m inclinable - [autom] tilting steering wheel
— m instalado - [mec] installed flywheel •
[autom] installed steering wheel
— m mantenido - [autom] held steering wheel
— m manual - [mec] handwheel
— m — sobre pedestal m - [hidr] pedestal mounted hand wheel
— m movido - [mec] moved flywheel • [autom]
moved steering wheel
— m para árbol m - [mec] shaft flywheel
— m — m con leva(s) f - [mec] crankshaft
flywheel
— m — cadena(s) f - [mec] chain (fly)wheel
— m — dirección f - [autom] steering wheel
— m — mano f - [mec] handwheel
— m — separador m - [mec] separator flywheel
— m — m modificado - [mec] changed separator
flywheel
— m — telemando m - [mec] remote control, o
telegraph, (fly)wheel
— m retráctil—[autom] telescopic steering wheel
— m rotado - [mec] rotated flywheel
— m rotante - [comb.int] rotating flywheel
— m vuelto - [autom] turned steering wheel
volar v - . . . . • [explos] to, blast, o set off,
o blow, up, o off
— v roca(s) f - [miner] to blast @ rock(s)
volátil m - [coque] volatile | a - volatile
— m captado - [coque] entrained volatile
volatilidad f alta - [coque] high volatility
— f baja - [coque] low volatility
— f reducida—[coque] low, o reduced, volatility
volcadero m - [ind] dump (site); véase también
vaciadero m; basurero m • [miner] tipple
— m para escoria(s) f - [metal-prod] slag dump
— m municipal - [sanit] city dump
— m para vagoneta(s) f - [miner] (dump car)
tipple
— m — f para mina f - [miner] mine tipple
volcado,da a - dumped; emptied
— a desde altura f (grande) - high dumped
— a en masa f - [constr] mass dumped
volcador m - [ind] dumper; tilter • dumping, o
tilting, mechanism • turn up • [ind] skip -
[constr] dump truck dumper
— m hidráulico - hydraulic dump(er)
volcador m para bobina(s) f - [metal-lam] coil
tilter
— m — cuchara(s) f - [metal-prod] ladle tilter
— m — inspección f - [metal-lam] inspection
turn-up
— m — f para chapa(s) f [metal-lam] sheet
inspection turn-up
— m — f — placas f - [metal-lam] plate inspection turn-up
— m — máquina f para colar - [metal-prod] pig-
-casting (machine) dumper
— m — plancha(s) f - [metal-lam] plate turnover
— m — tambor(es) m - [mec] drum tilter
— m — vagón(es) m - [f.c.] (car) dump(st)er
— m suspendido - [ind] suspended tilter
volcador,ra a - tilting • dumping
volcán m de lodo m - [geol] mud volcano
volcar v - . . . . • to dump
— v desde altura f (grande) - [ind] to high dump
— v líquido - to, dump, o pour, @ liquid
— v tambor m - to, dump, o tilt, @ drum

voleo m - [agric] broadcast(ing)
Volkswagen m con carrocería f tubular - [autom]
Baja Bug
— m con neumático(s) m desproporcionado(s) -
[autom] Baja Bug; dune buggy
— m sin carrocería f - [autom] Baja Bug
volquete m - . . . . • [autom] dump truck
— m trasero - [autom] rear dump truck
voltaje m a través de ventilador m - [electr]
across @ fan voltage
— m — usar(se) - [electr-oper] utilization
voltage
— m ajustado - [electr-oper] adjusted voltage
• [sold] fine voltage
— m alto - [electr-oper] high voltage
— m — en arco m - [sold] high arc voltage
— m — — circuito m - [electr-oper] high
circuit voltage
— m — — m abierto - [electr-oper] high
open circuit voltage
— m — — línea f - [electr-oper] high line
voltage
— m — para encencido m (de arco m) - [sold]
arc striking high (level) voltage
— m — verificado - [electr-oper] checked
high voltage
— m aplicado - [electr-oper] applied voltage
— m apropiado - [electr-oper] appropriate, o
proper, o correct, voltage • utilization
voltage
— m — en línea f - [electr-oper] proper line
voltage
— m — para horno m - [metal-prod] furnace,
appropriate, o utilization, voltage
— m aproximado - [sold] approximate, o coarse,
voltage
— m asignado - [electr-oper] rated voltage
— m — para arco m - [sold] rated volts arc
— m aumentado - [electr-oper] increased voltage
— m bajo - [electr-oper] low voltage
— m — en circuito m - [electr-oper] low circuit voltage
— m — — — m abierto - [electr-oper] low
open circuit voltage
— m — en línea f - [electr-oper] low line
voltage
— m — para soldadura f - [sold] low welding
voltage
— m — verificado - [electr-oper] checked low,
o low checked, voltage
— m bloqueado - [electr-oper] blocked voltage
— m bloqueador - [electr-oper] blocking voltage
— m comprobado - [electr-oper] checked voltage
— m compuesto - [electrón] composite voltage
— m con carga f nominal - [sold] voltage
at @ rated level
— m conocido - [electr-oper] known voltage
— m constante - [sold] constant, o steady,
voltage, o potential • fixed voltage
— m — con regulación f en arco m - [sold] arc
controlled constant voltage
— m — — velocidad f constante - [sold] constant speed voltage
— m — — f — para alimentación f de electrodo m - [sold] constant wire feed speed
constant voltage
— m — con corriente f alterna - [sold] alternating current constant voltage
— m — — f constante - [sold] constant
voltage direct current
— m — — f continua - [sold] direct current constant voltage
— m — o variable - [sold] constant/variable
voltage
— m — regulado electrónicamente - [sold]
electronically controlled constant, voltage, o
potential
— m — satisfactorio - [sold] satisfactory
constant voltage
— m — y variable - [sold] constant and variable voltage

voltaje **m** contrastado - [electr-oper] contrasted
voltage
— **m** — **con placa** f **para identificación** f -
[electr-instal] voltage checked against @
nameplate
— **m convertido** - [electr-oper] converted voltage
— **m correcto** - [electr-oper] correct voltage
— **m** — **en arco** m - [sold] correct arc voltage
— **m de corriente** f **aportada** - [electr-oper] input power voltage
— **m** — **energía** f **aportada** - [electr-oper] input
power voltage
— **m** — — f **recibida** - [electr-oper] input power
voltage
— **m** — **entrada** f - [electr-oper] input voltage
— **m** — — f **compensado** - [electr-oper] compensated input voltage
— f — — f **débil** - [electr-oper] low, o weak,
input voltage
— **m** — — f **desde fuente** f **de energía** f - [sold]
power source input voltage
— **m** — — f **máximo** - [electr-oper] maximum input
voltage
— **m** — — f **mínimo** - [electr-oper] minimum input
voltage
— **m** — — f **para motor** m - [electr-oper] motor,
input, o supply, voltage
— **m** — — f **para soldadora** f - [sold] welder input voltage
— **m** — — f **medio** - [electr-oper] medium voltage
input
— **m** — — f **reducido** - [electr-oper] low input
voltage
— **m** — **salida** f - [electr-oper] output
volts,tage • secondary voltage
— **m** — — f **en excitador** m - [electr-prod] exciter output voltage
— **m** — — **f máximo** - [electr-oper] maximum output voltage
— **m** — — f **mínimo** - [electr-oper] minimum output voltage
— **m** — **techo** m - [electr-oper] ceiling voltage
— **m** — — m **para excitación** f - [electr-oper]
exciting ceiling voltage
— **m** — **tierra** f - [electr-oper] ground voltage
— **m** — **voltímetro** m - [electr-oper] (volt)meter
voltage
— **m desconocido** - [electr-oper] unknown voltage
— **m deseado** - [electr-oper] desired voltage
— **m desequilibrado** - [electr-oper] unbalanced
voltage
— **m disminuido** - [electr-oper] lowered, o de-
ceased, voltage
— **m disponible** - [electr-oper] available voltage
— **m** — **para fuerza** f **motriz** - [electr-oper] -
power voltage
— **m disruptor** - [electr-oper] flash-over voltage
— **m** — **en seco** - [electr-oper] dry flash-over
voltage
— **m doble**—[electr-oper] dual, o double, voltage
— **m elevado** - [electr-oper] high voltage
— **m** — **en circuito** m - [electr-oper] high circuit voltage
— **m** — — — m **abierto** - [electr-oper] high open
circuit voltage
— **m** — **en línea** f - [electr-oper] high line
voltage
— **m empleado** - [electr-oper] voltage used
— **m en arco** m - [sold] arc (stream) voltage
— **m** — **bobina** f - [instrum] coil voltage
— **m** — — f **para medición** f - [instrum] measuring coil voltage
— **m** — **borne** m - [electr-oper] terminal voltage
— **m** — — m **en generador** m - [electr-prod] generator terminal(s) voltage
— **m** — **botón** m **para arranque** m - [sold] starter
push button voltage
— **m** — **campo** m - [electr-oper] field voltage
— **m** — **circuito** m **abierto** - [electr] open circuit voltage
— **m** — — m — **de fuente** f **de energía** f -
[elect] power source open circuit voltage
— **m** — — m **para ventilador** m - [sold] voltage
across @ fan

voltaje **m** en corriente - [electr-oper] current
voltage
— **m** — — f **alterna** - [electr-oper] alternating current voltage
— **m** — — f **continua** - [electr-oper] direct
current voltage
— **m** — **excitador** m - [electr-oper] exciter
voltage
— **m** — — m **con potencia** f **constante** - [electr-
-oper] constant potential exciter voltage
— f — **fuente** f **para energía** f - [sold] power
source voltage
— **m** — **generador** m - [electr-prod] generator
voltage
— **m** — **grilla** f - [electr-oper] grill voltage
— **m** — **línea** f - [electr-distrib] line voltage
— **m** — — f **para entrada** f - [sold] input line
voltage
— **m** — — f **para alimentación** f - [electr-
-oper] power line voltage
— **m** — **planta** f - [electr-oper] plant voltage
— **m** — **red** f - [electr-oper] system voltage
— **m** — **rejilla** f - [electr] grid voltage
— **m** — **tomacorriente** m - [electr-oper] receptacle voltage
— **m entre fase(s)** f - [electr-oper] interphase
voltage
— **m equilibrado** - [electr-oper] balanced voltage
— **m equivocado** - [electr-oper] wrong voltage
— **m establecido** - [electr-oper] established, o
rated, voltage
— **m excesivo** - [electr-oper] excessive voltage;
overvoltage; high voltage
— **m excesivo en entrada** f - [electr-oper] excessive, o high, input voltage
— **m** — — — f **a motor** m - [electr-oper] excessive, o high, motor input voltage
— **m exigido** - [electr-oper] required voltage
— **m exiguo** - [electr-oper] (too) low voltage;
under-voltage
— **m** — **entrada** f - [electr-oper] low input
voltage
— **m final** - [electr-oper] final, o finishing,
voltage
— **m flotante** m—[electr-oper] floating voltage
— **m** — **para cargador** m - [electr-oper] charger
floating voltage
— **m** — — — m **para acumulador** m - [electr-
-oper] battery charger floating voltage
— **m** — — — m — **batería** f - [electr-oper]
battery charger floating voltage
— **m graduable**—[electr-oper] adjustable voltage
— **m idéntico** - [electr-oper] identical voltage
— **m igual** - [electr-oper] equal, o same. voltage
— **m impropio** - [electr-oper] improper voltage
— **m inapropiado** - [electr-oper] wrong voltage
— **m indicado** - indicated, o listed, voltage
— **m** — **en placa** f **para identificación** f -
[electr-oper] nameplate voltage
— **m inicial** - [elecrr-oper] initial voltage •
[sold] starting voltage
— **m listado** - [electr-oper] listed voltage
— **m máximo** - [electr-oper] maximum voltage
— **m** — **en circuito** m **abierto** - [electr-oper]
maximum open circuit voltage
— **m mayor** - [electr-oper] higher voltage
— **m medio,diano** - [electr-oper] medium voltage
— **m** — **en circuito** m **abierto** - [electr-oper]
medium open circuit voltage
— **m medido** - [electr-oper] measured voltage
— **m menor** - [electr-oper] lower voltage
— **m** — **en circuito** m **abierto** - [electr-oper]
lower open circuit voltage
— **m mínimo** - [electr-oper] minimum voltage
— **m** — **en entrada** f - [electr-oper] minimum input voltage
— **m** — — **salida** f - [electr-oper] minimum output voltage
— **m** — **en circuito** m **abierto** - [electr-oper]
minimum open circuit voltage
— **m momentáneo** - [electr-oper] transient, o
surge, o momentary, voltage

voltaje m nominal - [electr-oper] nominal, o rated, o listed, voltage, o volt(s)
— m normal - [electr-oper] normal voltage
— m operativo - [electr-oper] operating voltage
— m optativo - [e;ectr-oper] optional voltage
— m para alimentación f - [electr-oper] feeding, o power line, voltage
— m — alumbrado m - [electr-oper] lighting voltage
— m — consumo m - [electr=oper] working voltage
— m — encendido m - [sold] starting voltage
— m — — m en circuito m abierto—[sp;d] starting open circuit voltage
— m — ensayo(s) - [electr-oper] test voltage
— m — entrada f - [electr-oper] input voltage
— m — horno m - [combust] furnace voltage
— m — inducción f - [electr] induction voltage
— m — motor m - [electr-mot] motor voltage
— m — — m para corriente f alterna - [electr-mot] alternating current motor voltage
— m — — m — — f continua - [electr-mot] direct current motor voltage
— m — operación f - [sold] operating voltage
— m — regulación f—[electr-oper] control(ling) voltage
— m — relé - [electr] relay voltage
— m — servicio m - [electr-oper] service voltage
— m — — m entre conductor(es) m—[electr-oper] between conductor(s) service voltage
— m — soldadora f - [sold] welder voltage
— m — soldadura f - [sold] welding voltage
— m — — f aproximado - [sold] approximate welding voltage
— m — suministro m - [electr-oper] supply voltage
— m — trabajo m - [electr-oper] working, o operating, o service, voltage
— m — utilización f - utilization voltage
— m permisible—[electr-oper] allowable voltage
— m pleno - [electr-oper] full voltage
— m preciso - [electr] precise, o fine, voltage
— m — para soldadura f - [sold] exact welding voltage
— m prefijado - [sold] pre-set voltage
— m presente - [electr-oper] present voltage
— m primario - [electr-primary voltage
— m — asignado - [electr-oper] rated primary voltage
— m — establecido - [electr-oper] established primary voltagae
— m — nominal - [elecrr] nominal, o rated, primary voltage
— m — para planta f - [electr-oper] plant primary voltage
— m que indica medidor m - [electr-oper] meter voltage
— m reactivo - [electr-oper] reactive voltage
— m real - [electr-oper] actual voltaqe
— m — para arco m - [sold] actual arc voltage
— m — soldadura f—[sold] actual welding voltage
— m recibido - [electr-oper] input voltage
— m reducido - [electr] reduced, o low(er), o decreased, o cut, voltage
— m — en circuito m - [electr-oper] low circuit voltage
— m — — — m abierto - [sold] low open circuit voltage
— m — en línea f - [electr-oper] low line voltage
— m regulado - [electr-oper] regulated, o controlled, o adjusted, o set, voltage
— m — para arco m - [sold] controlled arc voltage
— m — — en corriente f alterna - [electrón] regulated alternating current voltage
— m — — — f continua - [electrón] regulated direct current voltage
— m requerido m - [electr-oper] required voltage
— m satisfactorio—[electr] satisfactory voltage
— m secundario m - [electr] secondary voltage
— m según chapa f para identificación f—[elect]

identification plate (liste) voltage
voltaje m sin carga f - no load voltage
— m sostenido - [electr-oper] steady voltage
— m superior - [electr-oper] higher voltage
— m terminal - [electr-oper] terminal voltage
— m total - [electr-oper] total voltage
— m — en circuito m abierto - [electr-oper] total open circuit voltage
— m traslapante m — [electr-oper] overlapping voltage
— m único - [electr-oper] single voltage
— m variable - [electr-oper] variable voltage
— m — con factor m de utilización f de . . . - [sold] . . . duty cycle variable voltage
— m — con velocidad f constante - [sold] constant speed variable voltage
— m — en corriente f alterna - [electr-oper] alternating current variable voltage
— m — — — f continua - [electr-oper] direct current variable voltage
— m variado - [electr-oper] varied voltage
— m verificado - [electr-oper] checked voltage
— m — con chapa f para identificación f - [electr] voltage checked against @ nameplate
volteada f - [metal-lam] turning
— f de desbaste m - [metal-lam] roll turning
volteado,da a - [mec] turned; tumbled; flipped
volteador m - [metal-lam] turner; flipper • kick off • [metal-lam] manipulator • upender; downender • dumper
— m en entrada f - [metal=lam] approach turner; entry manipulator
— m — salida f - [metal-lam] delivery turner; delivery manipulator
— m para bobinas f - [metal-lam] coil upender
— m — desbastes m - [metal-lam] bloom turner
— m — vagón(es) - [f.c.] car dumper
volteador,ra a -turner,ning; tilter,ting
volteadora f - [mec] tilter; turnover
— f en enderezadora f - [mec] straightener turner
— f — entrada f - [metal-lam] approach turner
— f — salida f - [metal-lam] delivery turner
— f lateral - [metal-lam] side tilter
volteo m de desbastes m - [metal-lam] bloom turning
— m evitado - avoided overturning
— m frontal - [metal-lam] front tilting
— m resistido - [mec] resisted overturning
voltiamperímetro m - [instrum] voltammeter
voltímetro-amperímetro m - [instrum] voltammeter
— m analógico - [instrum] analogue voltammeter
— m — según norma f - [instrum] standard analogue voltmeter
— m con tubo m electrónico - [electrón] vacuum tube voltmeter
— m — válvula f electrónica - [electrón] calibrated vacuum tube voltmeter
— f para alimentadora f - feeder voltmeter
— m electrónico—[instrum] electronic voltmeter
— m indicador - [instrum] indicating voltmeter
— m magnético - [instrum] magnetic voltmeter
— m miniatura - [instrum] miniature voltmeter
— m optativo - [instrum] optional voltmeter
— m para carga f y descarga f - [instrum] zero center voltmeter
— m — corriente f alterna - [instrum] alternating current voltmeter
— m — — f continua - [instrum] direct current voltmeter
— m según norma f—[instrum] standard voltmeter
voltio-amperio m reactivo - [instrum] reactive volt-ampere
voltios m en entrada f - [electr] input volts
— m — circuito m abierto - [sold] open circuit volts,tage
— m — corriente f continua - [electr-oper] alternating current volts
— m — — f continua - [electr-oper] direct current volts
— m nominales - [electr-oper] nominal volts
— m — en corrientef alterna - [electr-oper] alternating current nominal volts

voltios m nominales en corriente f continua -
[electr-oper] direct current nominal volts
volumen m a régimen m de soplado m [metal-prod]
blast/rate volume
— m — — m normal de soplado m - [metal-prod]
normal blast/rate volume
— m actual - current, o actual, volume
— m almacenado - stored volume
— m alto - high volume
— m — de producción f - [ind] high production
volume
— m aspirado - volume, inhaled, o taken in
— m aumentado de escoria f - [metal-prod] increased slag volume
— m — — tránsito m - increased traffic volume
— m bajo - low volume
— m — de producción f - low production volume
— m de acero m - [metal-prod] steel volume
— m — admisión f - intake volume; volume in
— m — aire m - air volume • [ind] blast volume
— m — soplado - [ind] blowing, o blast, rate
— m — aportación f - [sold] weld deposit volume
— m — chatarra f - [metal-prod] scrap volume
— m — descarga f - discharge volume
— m — desplazamiento m - displacement volume
— m — entrada f - [ind] volume in
— m — — f y salida f - in-and-out volume
— m — escoria f - [metal-prod] slag volume
— m — escurrimiento m - [hidr] run off volume
— m — gas m - [combust] gas volume
— m — inspección f - inspection volume
— m — producción f - [ind] production volume
— m — tránsito m - [vial] traffic volume
— m — viento m - [metal-prod] blast volume
— m elevado - high volume
— m enlechado - [constr] grouted volume
— m específico - specific volume
— m físico - [fís] physical volume
— m útil - useful volume
volúmico,ca a - véase volumétrico,ca
voluntad f - . . . • willingness
voluntario m - . . .; volunteer; willing
voluta f - [arquit] . . .; spiral; helix
— f centrífuga - [geom] centrifugal spiral
volvedor v - . . .; véase también volcador m
— m para riel(es) - [metal-lam] rail turner
volvedor,ra a - turning - [zool] homing
volvedora f para lingote(s) m - [metal-lam] ingot
turner
— m — planchón(es) m - [metal-lam] slab turner
volver v - . . . • to return • to render • to,
come, o head, o switch, back
— v a apretar v - [mec] to retighten
— v — comprobar - to recheck
— v — conectar - to reconnect • to rewire
— v — fija(s) f uno - to return to page one •
to wipe @ slate clean
— v aceptable - to, qualify, o make acceptable
— v de por sí - [mec] to spring back
— v hacia - to turn towards
— v oído m sordo - to turn @ deaf ear
— v pasivo,va - to passivate
volverse v - to become; get
— v malo,la - to turn bad • to turn sour
— m plástico,ca a - to run plastic
— v — a con humedad f - to turn plastic with @
moisture
— v problemático,ca a - to become problematic
— v responsable - to become liable
— v rojo,ja - to turn red (hot)
votado,da a - voted
voto m afirmativo - affirmative vote
— m calificado - [legal] qualified vote
— m de calidad f - [legal] qualifying vote •
deciding, o tie-breaking, vote
— m de confianza - confidence vote
— m desfavorable - [legal] unfavorable vote;
vote against
— m para desempate m - deciding, o tie-breaking,
vote
— m positivo - positive vote
voz f acoplada - [comput] coupled voice
— f común - [fig] hearsay

vuelcaplanchas m - [metal-lam] plate turner
vuelco m - . . .; dump(ing) • [mec] turnover
— m de carga f - [transp] load dumping
— m — ceniza(s) f - [ind] ash dumping
— m — líquido m - liquid dumping
— m — plancha(s) - [metal-lam] plate turnover
— m — tambor m - [mec] drum dumping • drum
tilting
— m desde altura f (grande) - high dumping
— m en masa - [constr] mass dumping
— m repetido - [mec] repetitive dumping
vuelo m - . . . • [agric] broadcast
— m corto - [aeron] short (range) flight
— m de pájaro m - [ornit] bird's flight | a -
[fotogr] bird's eye view
— m — retorno m - [aeron] return flight
— m largo - [aeron] long (range) flight
vuelta f - . . . • turn(ing); twist(ing) • coming, o heading, back • becoming • bend; loop;
lap • [mec] resetting • [ind] wrap(ping) •
[deport-boxeo] round • [vest] cuff • [public]
(sheet) back (side)
— f a ajustar v - [mec] retightening
— f — — v de perno m - [mec] bolt retightening
— f — — v — tuerca f - [mec] nut retightening
— f — armar v - [mec] reassembly,ling
— f — centrar v - [mec] recentering
— f — colocar v - repositioning
— f — comprobar - recheck(ing)
— f — conectar v - [electr-instal] rewiring
— f — (con)formar v - [mec] re-forming
— f — insertar v - reinserting,tion
— f — montar v - [mec] re-mounting
— f — negociar v - renegotiating,tion
— f — posición f neutral - [mec] return(ing)
to @ neutral (position); snapping back
— f — rojo m - [metal] turn(ing) red
— f — verificar v - recheck(ing)
— f automática - [mec] automatic return
— f cerrada - [transp] short, o sharp, turn •
[vial] hairpin turn
— f completa - [mec] complete turn
— f corta - short, o sharp, turn
— f de foja f - [public] sheet back (side)
— f — muñeca f - wrist, turn, o twist
— f — por sí - [mec] spring(ing), o snap(ping)
back • return(ing) on its own
— f — pista f - [deport] race, o track, lap
— f — siglo m - [cronol] turn of @ century
— f — volante m - [autom] steering wheel turn
— f — — m para dirección f - [autom] steering
wheel turn(ing)
— f descontada - [deport] clocked off lap
— f desde posición f para marcha f atrás — [mec]
return(ing) from @ reverse (position)
— f final - [Cabl] (extreme) end wrap •
[deport] final, o last, turn, o lap
— f — en tambor m - [cabl] drum end wrap
— f inicial - [deport-boxeo] opening round
— f invertida - [cabl] reverse bend
— f lenta - [deport] slow lap
— f semisuperpuesta - [electr-cond] half lap
winding
— f tras vuelta f - time and again • [deport]
lap after lap
— f U - U turn
vuelto,ta a - returned • [mec] reset • headed
back
— a automáticamente - returned automatically
— a de por sí - returned, o snapped, o sprung,
back, by itself, o automatically
— a hacia - turned toward(s)
— a inactivo,va - inactivated; passivated
— a pasivo,va - passivated
— a plástico,ca - turned, o become, plastic
— f — con humedad f - turned plastic with @
moisture|
— a rojo,ja - [metal] turned red (hot)
vulcanizado,da a - [plást] vulcanized
vulgarizado,da a - vulgarized
vulnerado,da a - vulnerated

W

wáfel - [culin] waffle
Wánkel a - véase rotativo,va
wátio/kilogramo m - [electr] watt/kilogram
wolframio m - [metal] véase volframio m
wolmanizado,da a - [mader] wolmanized
wustita f - [quím] wustite

X

xenolita f - [geol] xenolite
— f de tufa f - [beol] tuff xenolite
xenolítico,ca a - xenolithic
xilol m - [quím] xylol

Y

Y f - Y
ya adv - already
— adv existente - already existing
— adv instalado,da - (already) installed
— adv no - no longer
— adv programado,da - in-programmed
— adv que - because; whereas • since
— adv - sea - either; whether
— adv solidificado,da - new, o already, solid
yaciente a - lying
yacimiento n - [miner] . . .; mineral deposit;
range
— m adjudicado - [miner] awarded, o assigned,
field, o deposit
— m amplio - [miner] extensive deposit
— m argentífero - [miner] silver deposit
— m carbonífero - [miner] coal deposit
— m de cinc m - [miner] zinc deposit
— m — cobre m - [miner] copper deposit
— m — gas m - [petról] gas field
— m — hierro m - [miner] iron deposit
— m — ley f alta - [miner] high grade deposit
— m — — f baja - [miner] low grade deposit
— m — lignito m - [miner] soft coal stratum
— m — mena f - [miner] ore deposit
— m — mineral m - [miner] ore, bed, o range
— m — — rocoso - [miner] rocky ore bed
— m — petróleo m - [miner] ore, field, o deposit
— m — plata f - [miner] silver deposit
— m — plomo m - [miner] lead deposit
— m — sal f - [miner] salt deposit
— m detrítico - [miner] dedtrital,tus field
— m explorado - [miner] explored field
— m explotado - [miner] developed deposit
— m extenso - [miner] extensive deposit
— m ferrífero - [miner] iron (bearing) deposit
— m mineral - [miner] mineral, o ore, deposit,
o range
— m minero - [miner] ore deposit
— m mioceno - [geol] miocene deposit
— m petrolero,lífero - [petról] oil field
— m radioactivo - [nucl] radioactive deposit
ye f bridada - [tub] flanged lateral
yerba f paraguaya - [botán] Paraguayan tea
yeso m terciario - [geol] tertiary gypsum
yoduro m de potasio - [quím] potassium iodide
yugo m - . . . . • [grúas] spider

yugo m para horquilla f - [mec] fork yoke
— m — — f para cambio(s) m - [autom-mec]
shift fork yoke
yunque m de diamante m - [metal-trat] diamond
anvil
— m para afiladura f - [petról] dressing anvil
— m — — para barreno(s) m - [petról] bit
dressing anvil
— m — herrero m - [herram] smith's anvil
— m — hincadura f - [constr-pil] driving anvil
— m tipo puente - [mec] bridge (type) anvil
yunta f - . . .; team
— f de caballos m - horse team
yute m impregnado - [electr-cond] impregnated
jute

Z

zacatón m - [botán] sacaton
zafado,da a - slipped; broken loose
zafadura f - slipping,page • [cabl] jump(ing)
zafarse v - to slip, out, o away
zaguero,ra a - . . . . • trailing
zamarra f - [metal-prod] bloom; slab
zampa f - [constr-pil] bearing pile
zanca f - [mec] brace
zancajera f - . . . • [autom] running board
zangoloteado,da a - dangled
zanja f - . . .; channel; pit; gully
— f a costado m de carretera f - [vial] roadside ditch
— f abierta f - [constr] open(ed), ditch, o
trench, o channel
— f ataludada - [constr] angled ditch
— f atascada - [hidr] clogged ditch
— f cavada - [constr] dug, ditch, o trench
— f — en declive - [constr] angled ditch
— f colectora - [hidr] catch basin • drainage
island
— f con cubierta f impermeable - [constr]
sealed top trench
— f — fondo m plano - flat bottom(ed) ditch
— f — profundidad f reducida - shallow ditch
— f cubierta - [constr] covered, ditch, o trench
— f desagotada - [constr] dewatered trench
— f en superficie f - [hidr] surface ditch
— f encachada - [hidr] rock trench • French
drain
— f encespedada - [hidr] turf covered ditch
— f entubada - [hidr] (en)closed ditch
— f excavada - [constr] excavated, o dug, ditch
[hidr] gouged ditch
— f para aproximación f approach trench
— f — desagüe - [constr] dewatering trench
— f — subdrenaje m - [hidr] subdrainage ditch
— f — tubería f - [constr] pipe trench
— f revestida - [constr] sheeted trench
— f transversal - [hidr] cross ditch
zanjadora f - [constr] ditcher; trencher;
trench hoe
zanjón m - [topogr] . . .; gully; ravine
zapata f - . . . . • [frenos] pad; shoe • [constr]
(track) pad • shoe; boot • [constr] grouser •
footing; tread plate • [herram] punch holder
• [transp] skid
— f arrastrada - [frenos] dragged pad
— f con garra f - [constr] grouser shoe
— f con recubrimiento m - [mec] covered shoe
— f de acero m - [constr] steel shoe
— f desgastada - [frenos] worn shoe
— f forjada - [mec] forged shoe
— f fresadora - [mec] milling shoe
— f metálica - [mec] metallic, shoe, o pad
— f motriz - [mec] power shoe
— f para cojinete m - [cojin] bearing shoe
— f — desarraigadora f - [constr] rooter shoe
— f — freno(s) m [mec] brake shoe
— f — — m por fricción f - [mec] friction

brake shoe

zapata f para freno m por inercia f - [mec] inertia brake shoe
— **f — fricción f** - [mec] friction, pad, o shoe
— **f — guía f** - [petról] guide shoe
— **f —— f para tubería f** - [petról] pipe guide shoe
— **f — f —— f para entubación f** - [petról] (casing) guide shoe
— **f — oruga f** - [constr] crawler, shoe, o tread • track shoe
— **f —— f para pala f mecánica** - [constr] shovel track pad
— **f — pierna f** - [constr] shank shoe
— **f — tubería f para entubación f** - [petról] casing shoe
— **f — rodillo m** - [mec] roller shoe
— **f plana** - [mec] flat shoe
— **f reguladora f para profundidad**—[agric-equip] gage shoe
— **f rotatoria para cortar tubo(s) m** - [mec] milling shoe
— **f sembradora** - [agric-equip] planter shoe
zapato m de cliente m - [vest] customer's shoe
— **m de cuero n** - [vest] leather shoe
— **m desgastado** - [vest] worn shoe
— **m tapón** - [petról] casing plug
zaranda f - . . . • [agric-zarandón] sleeve
— **f con extremo m vibratorio** - [miner] vibrating end screw
— **f — fondo m doble**—[mec] double bottom screen
— **f — doble** - [mec] double screen
— **f — para inyección f** - [petról] mud screen
— **f —— lodo(s) m** - [petról] mud screen
— **f — maleza(s) f** - [agric] weed screen
— **f — sínter m** - [metal-prod] sinter screen
— **f — trigo m** - [agric-equip] wheat screen
— **f — yuyo(s) m** - [agric] weed screen
— **f — zarandón** - [agric-equip] shaffer sleeve
— **f rotatoria** - [mec] rotary, tating screen
— **f — para maleza(s) f** - [agric-equip] rotary weed screen
— **f según Sociedad f Estadounidense para Ensayos m y materiales m** - [mec] American Society for Testing and Materials, screen, o sieve
— **f rotatoria** - [mec] vibrating screen • shale shaker • [petról] vibrating (mud) screen
— **f — para inyección f** - [petról] (vibrating) mud screen
zarandeado,da a - screened • [constr] graded
zarandear v - . . . • to screen • to grade
zarandeo m - screening • grading
— **m de mineral m** - [miner] ore screening
— **M — pellet(s) m** - [metal] pellet(s) screening
— **m en húmedo** - [miner] wet screening
zarandón m - [agric-equip] shaffer
zarzo m - . . . • [agric] hayrack; straw rack
zeolita f - [geol] zeolite
zeolítico,ca a - [geol] zeolitic
zeolitización f - [geol] zeolitization
zigzagueo m - zigzagging • [sold] (slight) whipping
— **m simple m** - [sold] straight whip(ping)
zinc m - [metal] véase cinc
zócalo m - [constr] . . .; baseboard; socle; base • footer,ting • toe guard • [segurid] apron • [autom] splashguard; splash shield
— **m en estribo m** - [autom] running board splashguard
zocateado,da a - [botán] overripe(ned)
zocatear v - [botán] to overripen
zona f - . . .; region • section • vicinity • range • [metal-prod] quadrant
— **f a drenar(se) v** - [hidr] drainage area
— **f — rojo** - [metal-prod] (red) hot spot
— **f — soldar(se) v** - [sold] area to be welded
— **m — subdrenarse v** - area to be subdrained
— **f abierta** - open area
— **f abollada** - [mec] dented area
— **f accesible** - accessible area
— **f acorralada** - impunded area
— **f adyacente** - adjacent, area, o zone
— **f aislada** - insulated area • isolated area

zona f ajardinada - landscaped area
— **f alta** - high, o upper, area, o zone
— **f alterada** - [Geol] altered zone
— **f amolada** - [mec] ground area
— **f amplia** - ample, u wide, zone
— **f andina** • [geogr] Andean area
— **f anegadiza** - [geol] marshy area; flood plain
— **f apropiada** - appropriate, o suitable, area
— **f — para trabajo m** - suitable work(ing) area
— **f arbolada** - wooded area
zona f árida - [geogr] arid zone
— **f áspera** - rough, o rugged, area
— **f atacada (con ácido m)** • [metal] atched area
— **f bajo techo m** - covered, o roofed, area
— **f calentada** — heated area
— **f caliente** - hot area • [sold] hot spot
— **f céntrica** - [pol] downtown (area)
— **f comercial** - commercial, o business, area • shopping area
— **f — céntrica** - [com] downtown business area
— **f — vecinal** - [com] neighborhood business area
— **f con calefacción f** - heated area
— **f — casas f independientes** - [constr] single family area
— **f — declive m** - [topogr] sloping area
— **f --- filtración f** - [suelos] seepage area
— **f --- mejora(s) f** - improved area
— **f — precipitación f abundante** - [hidr] heavy rainfall area
— **f — presión f** - [bombas] pressure zone
— **f — succión f** - [bombas] suction zone
— **f — tránsito intenso** - [pol] high traffic area
— **f — m liviano** -[pol] light traffic area
— **f contaminada** - [segurid] contaminated area
— **f corroída** - [metal] corroded area
— **f crítica** - critical, area, o zone, o range
— **f — de fondo m** - [tub] critical invert area
— **f — soldadura f** - critical weld area
— **f cubierta** - covered, o roofed, area
— **f de borde m** - [metal-lam] edge, area, o zone
— **f — brida f** - [tub] flange area
— **f — cobertizo(s) y conservación f** - [aeron] hangar and service area
— **f — concentración f alta de azufre m** - [metal] high sulfur concentration area
— **f — conducto(s) m ascendente(s)** - [metal-prod] uptake, area, o zone
— **f — coquería f** - [coque] coke oven area
— **f — cráter m** - [sold] crater area
— **f — cuba f** - [metal-prod] bosh area
— **f — densidad f crítica** - critizal density, area, o zone
— **f — falda f** - [topogr] hillside area
— **f — fondo m** - [tub] invert area
— **f — heladas f** - [geogr] frost zone
— **f — horno m** - [metal-prod] furnace area
— **f — influencia f térmica** - [sold] heat affected zone
— **f — interacción f** - interaction zone
— **f — interés m** - interest, zone, o area
— **f — interfaz** - [electr] interface area
— **f — interferencia f** - interference area
— **f — ladera f** - [topogr] hillside zone
— **f — medio oeste** - [geogr] midwestern area
— **f — mercado m** - [com] market area
— **f — metal** - metal, area, o zone
— **f —— m de aportación f** [sold] weld ,eta; zone
— **f —— base f** - [sold] base metal zone
— **f --- mina f** - [miner] mine,ning area
— **f — muelle m** - [mec] spring area • [pl] [transp] dock area
— **f — pandeo m** - [mec] buckling, area, o zone
— **f — peligro m** - danger zone
— **f — plataforma f** - [f.c.] rail dock area
— **f --- quemador m** - [ind] burner zone
— **f — recolección f** - [hidr] contributing area
— **f — seguridad f** - [segurid] safety zone
— **f explotada** - [miner] developed area
— **f franca** - [fisc] free zone
— **f gunitada** - [metal-prod] gunited area

zona f hormigonada - [constr] concrete area
— f impermeable - [hidr] impervious area
— f industrial - [ind] industrial area
— f íntegra - entire area
— f interior - inner zone • [deport] infield
— f mal ventilada - poorly ventilated area
— f minera bajo cielo m abierto - [miner] open pit mining area
— f orogénica - [geol] orogenic, zone, o area
— f para acumulación f - [hidr] ponding area
— f — acumulador(es) m - [ind] battery zone
— f — almacenamiento m - [ind] storage area
— f — aparcamiento m - [constr] parking area
— f — apilamiento m - [constr] stock area
— f — apoyo m - [mec] bearing area
— f — asentamiento m - [hidr] settling area
— f — aterrizar,zaje - [aeron] landing area
— f — barrilete m - [coque] collector zone
— f — cabotaje m - [nav] coastal shipping area
— f — calcinación f - calcining area
— f — camping m - [deport] camping area
— f — cimiento(s) m - [constr] foundation area
— f — circulación f intensa - [transp] heavy traffic area
— f — combustible(s) m - [ind] fuel area
— f — m para avión(es) m - [aeron] aircraft fuel storage area
— f — f en culata f de cilindro m - [comb.-int] cylinder head combustion area
— f — comercialización f - [com] marketing area
— f — contacto m - [mec] contact area
— f — corte m - [ind] cutting, o shearing, area
— f — decantación f - [hidr] settling area
— f — decapado m - [metal-trat] pickling area
— f — decarburación f - decarbonization area
— f — desbastadora f—[metal-lam] roughing area
— f — descarte m - [ind] spoil, area, o bank
— f — desembarco - [milit] landing area
— f — deslingoteado m - [metal] stripping area
— f — desmoldeo m - [metal-prod] stripping area
— f — disipador m - [electr] sink(ing) area
— f — embalaje m - [ind] packaging area
— f — empaquetado m - [ind] packaging area
— f — encendido m - [sold] striking area
— f — enfriamiento m - [ind] cooling area
— f — ensayo(s) m - [ind] testing area
— f — entrega f de equipaje(s) m - [transp] baggage claim area
— f — escarpado m - [metal-prod] scarfing area
— f — escurrimiento m - [hidr] seepage area
— f — esquí(aje) m - [deport] ski(ing) area
— f — estacionamiento m - [constr] parking area
— f — fabricación f - [ind] manufacturing area
— f — fusión f - [sold] fusion, o melting, area
— f — grúa f - [grúas] crane area
— m — horno m fundidor - [ceram] melter area
— f — igualación f - [metal-prod] soaking area
— f — industria f liviana - light industry area
— f — laminación f - [metal-lam] rolling area
— f — lingote(s) m - [metal-prod] ingot area
— f — marcación f - [ind] marking area
— f — mecanización f por abrasión f - [mec] abrasion machining area
— f — mercadeo m - [com] marketing area
— f — mezcla(dura) f - [ind] blending area
— f — montaje m - [constr] staging area
— f — práctica f - [deport] practice green
— f — producción f - [ind] production area
— f — reacción f - [mec] reaction area
— f — recargue m - [sold] buildup area
— f — recreo m - [deport] playground (area)
— f — refinamiento m - [sold] refining area
— f — reparación(es) f - [ind] repair area
— f — servicio m - [ind] service area
— f — m eléctrico - [electr-instal] electrical service area
— f — soldadura f - [sold] weld(ing) area
— f — f circunferencial - [sold] girth welding area
— f — m perimétrica - [sold] girth welding area
— f — suministro m - [ind] supply area
— f — m eléctrico - [electr-prod] electrical

supply area
zona f para trabajo m - [ind] work(ing) area
— f para transformación f transformation area
— f — tránsito m - [vial] traffic area
— f — transmisión f - [mec] transmission area
— f — venta(s) f - [com] sales, o marketing, o trading, area
— f — vestuario(s) m - [labor] dressing rooms area
— f — vuelco m - disposal area; dump site
— f peligrosa - [segurid] dangerous, o hazardous, area
— f permeable - [hidr] pervious, area, o zone
— f petrolífera - [petról] oil, zone, o country
— f plana - [topogr] flat area
— f poblada - [pol] populated area
— f recreativa - [deport] recreational area
— f recubierta - covered, o coated, area
— f reducida - small, o reduced, area
— f reemplazable - replaceable area
— f relevada - surveyed area
— f residencial - [constr] residential area
— f — con casa(s) f independiente(s)—[constr] single family residential area
— f restringida f - [pol] restricted area
— f rural - [pol] rural area
— f segura f - [ind] safe area
— f semiárida - semi-arid area
— f sin mejoras,rar - unimproved area
— f soldada - [sold] welded area
— f sombreada - [instrum] crosshatched area
— f terminal - [aeron] terminal area
— f total - total, o entire, area
— f transitada - [vial] traveled, area, o way
— f urbana f - [pol] urban area
— f — céntrica - [pol] downtown area
— f vecinal - [pol] neighborhood area
— f ventilada - [constr] ventilated area
zonal a - zonal
zonificado,da a - zoned
zonificar v - to zone
zorra f - [ind] . . .; hand truck
— f con cabrestante m - [constr] winch truck
— f elevadora - [ind] lift(ing) truck
— f para tubo(s) m - [tub] pipe carriage
zubia f dañada - [hidr] damaged flume
— f de hormigón m - [hidr] concrete flume
— f — m prevaciado - [hidr] precast concrete flume
— f para irrigación f - [hidr] irrigation flume
zulaque m - mastic
— m aplicado en frío - cold applied mastic
zumbado,da a - hummed; buzzed
zumbador,ra a - . . .; hummer
— m audible - [electrón] audible buzzer
— m continuo - [electr-equip] continuous buzzer
— m intermitente - [electrón] intermittent buzzer
— m momentáneo - [electrón] momentary buzzer
zumbido m para alerta(r) - [segurid] alert(ing) buzz(er)
— m pulsante - [electrón] pulsating buzzer • pulsating humm
zunchado m - [ind] strapping; banding
— m automático - [ind] automatic banding
zunchado,da a - [ind] strapped; banded
— a fuertemente - [ind] banded tightly
— a únicamente - [ind] banded only
zunchador m - [ind] bander
zunchadora f - [ind] bander
zunchar v - [ind] to, strap, o band
zuncho m - [ind] . . .; tie band; strap; ring; banding strip; girdle • Signode band
— m de cinc n - [electr-cond] zinc band
— m longitudinal - [ind] longitudinal band
— m metálico - [ind] metal, strap, o band
— m para tensión f - [mec] drawband
— m — f de acero m - [mec] steel drawband
— m para tonel(es) m - [ind] hogshead hoop
— m — m para tabaco m - [metal-lam] tobacco hogshead hoop
— m tonelería f - [ind] cooperage hoop
— m perimetral - [ind] perimetrical metal band